Nanotribology and Nanomechanics
2nd edition

Bharat Bhushan
Editor

Nanotribology and Nanomechanics

An Introduction

2nd edition

 Springer

Professor Bharat Bhushan
Ohio Eminent Scholar and The Howard D. Winbigler Professor
Director, Nanoprobe Laboratory for Bio- & Nanotechnology and Biomimetics (NLB^2)
201 W. 19th Avenue
Ohio State University
Columbus, Ohio 43210-1142
USA
bhushan.2@osu.edu

ISBN 978-3-540-77607-9   e-ISBN 978-3-540-77608-6

DOI 10.1007/978-3-540-77608-6

Library of Congress Control Number: 2008921404

© 2008 Springer-Verlag Berlin Heidelberg

This work is subject to copyright. All rights are reserved, whether the whole or part of the material is concerned, specifically the rights of translation, reprinting, reuse of illustrations, recitation, broadcasting, reproduction on microfilm or in any other way, and storage in data banks. Duplication of this publication or parts thereof is permitted only under the provisions of the German Copyright Law of September 9, 1965, in its current version, and permission for use must always be obtained from Springer. Violations are liable to prosecution under the German Copyright Law.

The use of general descriptive names, registered names, trademarks, etc. in this publication does not imply, even in the absence of a specific statement, that such names are exempt from the relevant protective laws and regulations and therefore free for general use.

Typesetting and production: le-tex publishing services oHG, Leipzig, Germany
Cover design: WMXDesign, Heidelberg, Germany

Printed on acid-free paper

9 8 7 6 5 4 3 2 1

springer.com

Foreword

The invention of the scanning tunneling microscope in 1981 has led to an explosion of a family of instruments called scanning probe microscopes (SPMs). One of the most popular instruments in this family is the atomic force microscope (AFM), which was introduced to the scientific community in 1986. The application of SPMs has penetrated numerous science and engineering fields. Proliferation of SPMs in science and technology labs is similar to optical microscopes fifty years ago. SPMs have even made it into some high school science labs. Evolution of nanotechnology has accelerated the use of SPMs and vice versa. The scientific and industrial applications include quality control in the semiconductor industry and related research, molecular biology and chemistry, medical studies, materials science, and the field of information storage systems.

AFMs were developed initially for imaging with atomic or near atomic resolution. After their invention, they were modified for tribological studies. AFMs are now intensively used in this field and have lead to the development of the field of nanotribology. Researchers can image single lubricant molecules and their agglomeration and measure surface topography, adhesion, friction, wear, lubricant film thickness, and mechanical properties all on a micrometer to nanometer scale. SPMs are also used for nanofabrication and nanomachining. Beyond as an analytical instrument, SPMs are being developed as industrial tools such as for data storage.

With the advent of more powerful computers, atomic-scale simulations have been conducted of tribological phenomena. Simulations have been able to predict the observed phenomena. Development of the field of nanotribology and nanomechanics has attracted numerous physicists and chemists. I am very excited that SPMs have had such an immense impact on the field of tribology.

I congratulate Professor Bharat Bhushan in helping to develop this field of nanotribology and nanomechanics. Professor Bhushan has harnessed his own knowledge and experience, gained in several industries and universities, and has assembled a large number of internationally recognized authors. The authors come from both academia and industry.

Professor Bharat Bhushan's comprehensive book is intended to serve both as a textbook for university courses as well as a reference for researchers. It is a timely addition to the literature on nanotribology and nanomechanics, which I anticipate will stimulate further interest in this important new field. I expect that it will be well received by the international scientific community.

IBM Research Division
Rueschlikon, Switzerland
Nobel Laureate Physics, 1986

Prof. Dr. Gerd Binnig

Preface

Tribology is the science and technology of interacting surfaces in relative motion and of related subjects and practices. The nature and consequences of the interactions that take place at the moving interface control its friction, wear and lubrication behavior. Understanding the nature of these interactions and solving the technological problems associated with the interfacial phenomena constitute the essence of tribology. The importance of friction and wear control cannot be overemphasized for economic reasons and long-term reliability.

The recent emergence and proliferation of proximal probes, in particular tip-based microscopies and the surface force apparatus and of computational techniques for simulating tip-surface interactions and interfacial properties, has allowed systematic investigations of interfacial problems with high resolution as well as ways and means for modifying and manipulating nanostructures. These advances provide the impetus for research aimed at developing a fundamental understanding of the nature and consequences of the interactions between materials on the atomic scale, and they guide the rational design of material for technological applications. In short, they have led to the appearance of the new field of nanotribology and nanomechanics.

The field of tribology is truly interdisciplinary. Until 1980s, it had been dominated by mechanical and chemical engineers who conduct macro tests to predict friction and wear lives in machine components and devise new lubricants to minimize friction and wear. Development of the field of nanotribology has attracted many more physicists, chemists, and material scientists who have significantly contributed to the fundamental understanding of friction and wear processes and lubrication on an atomic scale. Thus, tribology and mechanics are now studied by both engineers and scientists. The nanotribology and nanomechanics fields are growing rapidly and it has become fashionable to call oneself a "tribologist." The tip-based microscopies have also been used for materials characterization as well as for measurement of mechanical and electrical properties all on nanoscale. Since 1991, international conferences and courses have been organized on this new field of nanotribology, nanomechanics, and nanomaterials characterization.

There are also new applications which require detailed understanding of the tribological and mechanics processes on macro- to nanoscales. Since early 1980s, tribology of magnetic storage systems (rigid disk drives, flexible disk drives, and tape drives) has become one of the important parts of tribology. Microelectromechanical Systems (MEMS)/Nanoelectromechanical Systems (NEMS) and biodevices, all part of nanotechnology, have appeared in the marketplace in the 1990s which present new tribological challenges. Another emerging area of importance is biomimetics. It involves taking ideas from nature and implementing them in an application. Examples include Lotus effect and gecko adhesion. Tribology of processing systems such as copiers, printers, scanners, and cameras is important although it has not received much attention. Along with the new industrial applications, there has been development of new materials, coatings and treatments such as synthetic diamond, diamondlike carbon films, self assembled monolayers, and chemically grafted films, to name a few with nanoscale thicknesses.

It is clear that the general field of tribology has grown rapidly in the last thirty years. Conventional tribology is well established but nanotribology and nanomechanics are evolving rapidly and have taken the center stage. New materials are finding use. Furthermore, new industrial applications continue to evolve with their unique challenges.

Very few tribology handbooks exist and these are dated. They have focused on conventional tribology, traditional materials and matured industrial applications. No mechanics handbook exists. Nanotribology, nanomechanics, and nanomaterial characterization are becoming important in many nanotechnology applications. A primer to nanotribology, nanomechanics, and nanomaterial characterization is needed. The purpose of this book is to present the principles of nanotribology and nanomechanics and applications to various applications. The appeal of the subject book is expected to be broad.

The chapters in the book have been written by internationally recognized experts in the field, from academia, national research labs and industry, and from all over the world. The book integrates the knowledge of the field from mechanics and materials science points of view. In each chapter, we start with macroconcepts leading to microconcepts. We assume that the reader is not expert in the field of nanotribology and nanomechanics, but has some knowledge of macrotribology/mechanics. It covers various measurement techniques and their applications, and theoretical modeling of interfaces. Organization of the book is straightforward. The first part of the book covers fundamental experimental and theoretical studies. The latter part covers applications.

The book is intended for three types of readers: graduate students of nanotribology/nanomechanics/nanotechnology, research workers who are active or intend to become active in this field, and practicing engineers who have encountered a tribology and mechanics problem and hope to solve it as expeditiously as possible. The book should serve as an excellent text for one or two semester graduate courses in scanning probe microscopy/applied scanning probe methods, nanotribology/nanomechanics/nanotechnology in mechanical engineering, materials science, or applied physics.

I would like to thank the authors for their excellent contributions in a timely manner. And I wish to thank my wife, Sudha, my son, Ankur, and my daughter, Noopur, who have been very forbearing during the preparation of this book.

Powell, Ohio *Bharat Bhushan*
Oct. 16, 2007

Contents

**1 Introduction –
Measurement Techniques and Applications**
Bharat Bhushan 1
1.1 Definition and History of Tribology 1
1.2 Industrial Significance of Tribology 3
1.3 Origins and Significance of Micro/Nanotribology 4
1.4 Measurement Techniques 6
    1.4.1 Scanning Probe Microscopy 6
    1.4.2 Surface Force Apparatus (SFA) 13
    1.4.3 Vibration Isolation 23
1.5 Magnetic Storage Devices and MEMS/NEMS 23
    1.5.1 Magnetic Storage Devices 23
    1.5.2 MEMS/NEMS 27
1.6 Role of Micro/Nanotribology and Micro/Nanomechanics
    in Magnetic Storage Devices and MEMS/NEMS 30
1.7 Organization of the Book 31
References 31

Part I Scanning Probe Microscopy

**2 Scanning Probe Microscopy – Principle of Operation,
Instrumentation, and Probes**
Bharat Bhushan, Othmar Marti 37
2.1 Introduction 37
2.2 Scanning Tunneling Microscope 39
    2.2.1 The STM Design of Binnig et al. 41
    2.2.2 Commercial STMs 42
    2.2.3 STM Probe Construction 45
2.3 Atomic Force Microscope 46
    2.3.1 The AFM Design of Binnig et al. 50

		2.3.2	Commercial AFMs	50
		2.3.3	AFM Probe Construction	59
		2.3.4	Friction Measurement Methods	66
		2.3.5	Normal Force and Friction Force Calibrations of Cantilever Beams	71
	2.4		AFM Instrumentation and Analyses	74
		2.4.1	The Mechanics of Cantilevers	74
		2.4.2	Instrumentation and Analyses of Detection Systems for Cantilever Deflections	79
		2.4.3	Combinations for 3-D Force Measurements	93
		2.4.4	Scanning and Control Systems	94
References				103

3 Probes in Scanning Microscopies
Jason H. Hafner 111
3.1 Introduction 112
3.2 Atomic Force Microscopy 113
    3.2.1 Principles of Operation 113
    3.2.2 Standard Probe Tips 114
    3.2.3 Probe Tip Performance 115
    3.2.4 Oxide-Sharpened Tips 118
    3.2.5 FIB tips 118
    3.2.6 EBD tips 119
    3.2.7 Carbon Nanotube Tips 120
3.3 Scanning Tunneling Microscopy 129
    3.3.1 Mechanically Cut STM Tips 130
    3.3.2 Electrochemically Etched STM Tips 131
References 131

4 Noncontact Atomic Force Microscopy and Related Topics
Franz J. Giessibl, Yasuhiro Sugawara, Seizo Morita, Hirotaka Hosoi,
Kazuhisa Sueoka, Koichi Mukasa, Akira Sasahara, Hiroshi Onishi 135
4.1 Introduction 135
4.2 Atomic Force Microscopy (AFM) 136
    4.2.1 Imaging Signal in AFM 136
    4.2.2 Experimental Measurement and Noise 137
    4.2.3 Static AFM Operating Mode 138
    4.2.4 Dynamic AFM Operating Mode 139
    4.2.5 The Four Additional Challenges Faced by AFM .. 140
    4.2.6 Frequency-Modulation AFM (FM-AFM) 142
    4.2.7 Relation Between Frequency Shift and Forces ... 143
    4.2.8 Noise in Frequency Modulation AFM: Generic Calculation ... 145
    4.2.9 Conclusion 146
4.3 Applications to Semiconductors 146
    4.3.1 Si(111)-(7×7) Surface 147

		4.3.2	Si(100)-(2×1) and Si(100)-(2×1):H Monohydride Surfaces	149
		4.3.3	Metal Deposited Si Surface	151
	4.4	Applications to Insulators		155
		4.4.1	Alkali Halides, Fluorides and Metal Oxides	156
		4.4.2	Atomically Resolved Imaging of a NiO(001) Surface	162
	4.5	Applications to Molecules		165
		4.5.1	Why Molecules and Which Molecules?	166
		4.5.2	Mechanism of Molecular Imaging	166
		4.5.3	Perspectives	170
	References			171

5 Low-Temperature Scanning Probe Microscopy
Markus Morgenstern, Alexander Schwarz, Udo D. Schwarz 179

	5.1	Introduction		180
	5.2	Microscope Operation at Low Temperatures		181
		5.2.1	Drift	181
		5.2.2	Noise	181
		5.2.3	Stability	182
		5.2.4	Piezo Relaxation and Hysteresis	182
	5.3	Instrumentation		182
		5.3.1	A Simple Design for a Variable-Temperature STM	183
		5.3.2	A Low Temperature SFM Based on a Bath Cryostat	185
	5.4	Scanning Tunneling Microscopy and Spectroscopy		187
		5.4.1	Atomic Manipulation	188
		5.4.2	Imaging Atomic Motion	189
		5.4.3	Detecting Light from Single Atoms and Molecules	191
		5.4.4	High-Resolution Spectroscopy	191
		5.4.5	Imaging Electronic Wave Functions	199
		5.4.6	Imaging Spin Polarization: Nanomagnetism	205
	5.5	Scanning Force Microscopy and Spectroscopy		206
		5.5.1	Atomic-Scale Imaging	209
		5.5.2	Force Spectroscopy	214
		5.5.3	Atomic Manipulation	218
		5.5.4	Electrostatic Force Microscopy	218
		5.5.5	Magnetic Force Microscopy	220
	References			225

6 Dynamic Modes of Atomic Force Microscopy
A. Schirmeisen, B. Anczykowski, Harald Fuchs 235

	6.1	Motivation: Measurement of a Single Atomic Bond		236
	6.2	Harmonic Oscillator: A Model System for Dynamic AFM		242
	6.3	Dynamic AFM Operational Modes		245
		6.3.1	Amplitude-Modulation/Tapping-Mode AFM	246
		6.3.2	Self-Excitation Modes	255
	6.4	Q-Control		262

6.5	Dissipation Processes Measured with Dynamic AFM	268
6.6	Conclusion	273
References		274

7 Molecular Recognition Force Microscopy: From Simple Bonds to Complex Energy Landscapes
Peter Hinterdorfer, Ziv Reich 279

7.1	Introduction	279
7.2	Ligand Tip Chemistry	280
7.3	Immobilization of Receptors onto Probe Surfaces	283
7.4	Single-Molecule Recognition Force Detection	285
7.5	Principles of Molecular Recognition Force Spectroscopy	288
7.6	Recognition Force Spectroscopy: From Isolated Molecules to Biological Membranes	291
	7.6.1 Forces, Energies, and Kinetic Rates	291
	7.6.2 Complex Bonds and Energy Landscapes	294
	7.6.3 Live Cells and Membranes	299
7.7	Recognition Imaging	300
7.8	Concluding Remarks	303
References		303

Part II Nanotribology and Nanomechanics: Fundamental Studies

8 Nanotribology, Nanomechanics and Materials Characterization
Bharat Bhushan 311

8.1	Introduction	311
8.2	Description of AFM/FFM and Various Measurement Techniques	314
	8.2.1 Surface Roughness and Friction Force Measurements	314
	8.2.2 Adhesion Measurements	320
	8.2.3 Scratching, Wear and Fabrication/Machining	321
	8.2.4 Surface Potential Measurements	321
	8.2.5 In Situ Characterization of Local Deformation Studies	322
	8.2.6 Nanoindentation Measurements	323
	8.2.7 Localized Surface Elasticity and Viscoelasticity Mapping	324
	8.2.8 Boundary Lubrication Measurements	330
8.3	Surface Imaging, Friction and Adhesion	330
	8.3.1 Atomic-Scale Imaging and Friction	330
	8.3.2 Microscale Friction	336
	8.3.3 Directionality Effect on Microfriction	341
	8.3.4 Surface-Roughness-Independent Microscale Friction	346
	8.3.5 Velocity Dependence on Micro/Nanoscale Friction	351
	8.3.6 Nanoscale Friction and Wear Mapping	354
	8.3.7 Adhesion and Friction in a Wet Environment	356

		8.3.8	Separation Distance Dependence of Meniscus and van der Waals Forces	363

       8.3.9   Scale Dependence in Friction 364
8.4 Wear, Scratching, Local Deformation, and Fabrication/Machining ... 372
       8.4.1   Nanoscale Wear 372
       8.4.2   Microscale Scratching 373
       8.4.3   Microscale Wear 374
       8.4.4   In Situ Characterization of Local Deformation 381
       8.4.5   Nanofabrication/Nanomachining 385
8.5 Indentation 385
       8.5.1   Picoindentation 386
       8.5.2   Nanoscale Indentation 387
       8.5.3   Localized Surface Elasticity and Viscoelasticity Mapping 391
8.6 Boundary Lubrication 392
       8.6.1   Perfluoropolyether Lubricants 392
       8.6.2   Self-Assembled Monolayers 401
       8.6.3   Liquid Film Thickness Measurements 406
8.7 Closure 408
References 410

9 Surface Forces and Nanorheology of Molecularly Thin Films
Marina Ruths, Jacob N. Israelachvili 417
9.1 Introduction: Types of Surface Forces 417
9.2 Methods Used to Study Surface Forces 420
       9.2.1   Force Laws 420
       9.2.2   Adhesion Forces 422
       9.2.3   The SFA and AFM 422
       9.2.4   Some Other Force-Measuring Techniques 425
9.3 Normal Forces Between Dry (Unlubricated) Surfaces 427
       9.3.1   Van der Waals Forces in Vacuum and Inert Vapors........... 427
       9.3.2   Charge-Exchange Interactions 430
       9.3.3   Sintering and Cold Welding 431
9.4 Normal Forces Between Surfaces in Liquids 432
       9.4.1   Van der Waals Forces in Liquids....................... 432
       9.4.2   Electrostatic and Ion Correlation Forces 434
       9.4.3   Solvation and Structural Forces 437
       9.4.4   Hydration and Hydrophobic Forces 440
       9.4.5   Polymer-Mediated Forces 444
       9.4.6   Thermal Fluctuation Forces........................... 447
9.5 Adhesion and Capillary Forces 448
       9.5.1   Capillary Forces 448
       9.5.2   Adhesion Mechanics 449
       9.5.3   Effects of Surface Structure, Roughness, and Lattice Mismatch 451
       9.5.4   Nonequilibrium and Rate-Dependent Interactions:
                  Adhesion Hysteresis................................ 453

9.6 Introduction: Different Modes of Friction and the Limits
of Continuum Models 455
9.7 Relationship Between Adhesion and Friction Between Dry
(Unlubricated and Solid Boundary Lubricated) Surfaces 458
    9.7.1 Amontons' Law and Deviations from It Due to Adhesion:
The Cobblestone Model 458
    9.7.2 Adhesion Force and Load Contribution to Interfacial Friction . 460
    9.7.3 Examples of Experimentally Observed Friction of Dry Surfaces 466
    9.7.4 Transition from Interfacial to Normal Friction with Wear 474
9.8 Liquid Lubricated Surfaces 475
    9.8.1 Viscous Forces and Friction of Thick Films: Continuum Regime 475
    9.8.2 Friction of Intermediate Thickness Films 477
    9.8.3 Boundary Lubrication of Molecularly Thin Films:
Nanorheology 479
9.9 Effects of Nanoscale Texture on Friction 493
    9.9.1 Role of the Shape of Confined Molecules 493
    9.9.2 Effects of Surface Structure 495
References 497

10 Interfacial Forces and Spectroscopic Study of Confined Fluids
Y. Elaine Zhu, Ashis Mukhopadhyay, Steve Granick 517
10.1 Introduction 517
10.2 Hydrodynamic Force of Fluids Flowing in Micro- to Nanofluidics:
A Question About No-Slip Boundary Condition 518
    10.2.1 How to Quantify the Amount of Slip 519
    10.2.2 The Mechanisms that Control Slip in Low-Viscosity Fluids ... 520
    10.2.3 Experimental 522
    10.2.4 Slip Can Be Modulated by Dissolved Gas 525
    10.2.5 Slip Past Wetted Surfaces 527
    10.2.6 The Purposeful Generation of Slip 527
    10.2.7 Outlook 528
10.3 Hydrophobic Interaction and Water at a Hydrophobicity Interface 528
    10.3.1 Experimental 529
    10.3.2 Hydrophobic Interaction 530
    10.3.3 Hydrophobicity at a Janus Interface 532
10.4 Ultrafast Spectroscopic Study of Confined Fluids:
Combining Ultra-Fast Spectroscopy with Force Apparatus 537
    10.4.1 Challenges 539
    10.4.2 Principles of FCS Measurement 540
    10.4.3 Experimental Set-up 541
10.5 Contrasting Friction with Diffusion in Molecularly Thin Films 543
10.6 Diffusion of Confined Molecules During Shear 548
10.7 Summary 550
References 551

11 Friction and Wear on the Atomic Scale
Enrico Gnecco, Roland Bennewitz, Oliver Pfeiffer, Anisoara Socoliuc,
Ernst Meyer 557
11.1 Friction Force Microscopy in Ultrahigh Vacuum 557
    11.1.1 Friction Force Microscopy 558
    11.1.2 Force Calibration 559
    11.1.3 The Ultrahigh Vacuum Environment 562
    11.1.4 A Typical Microscope Operated in UHV 563
11.2 The Tomlinson Model... 565
    11.2.1 One-Dimensional Tomlinson Model 565
    11.2.2 Two-Dimensional Tomlinson Model 566
    11.2.3 Friction Between Atomically Flat Surfaces 568
11.3 Friction Experiments on the Atomic Scale 569
    11.3.1 Anisotropy of Friction 572
11.4 Thermal Effects on Atomic Friction 574
    11.4.1 The Tomlinson Model at Finite Temperature 575
    11.4.2 Velocity Dependence of Friction 579
    11.4.3 Temperature Dependence of Friction 582
11.5 Geometry Effects in Nanocontacts 582
    11.5.1 Continuum Mechanics of Single Asperities 583
    11.5.2 Dependence of Friction on Load........................ 584
    11.5.3 Estimation of the Contact Area 584
11.6 Wear on the Atomic Scale 588
    11.6.1 Abrasive Wear on the Atomic Scale 588
    11.6.2 Contribution of Wear to Friction........................ 590
11.7 Molecular Dynamics Simulations of Atomic Friction and Wear 591
    11.7.1 Molecular Dynamics Simulations of Friction Processes 592
    11.7.2 Molecular Dynamics Simulations of Abrasive Wear 594
11.8 Energy Dissipation in Noncontact Atomic Force Microscopy 595
11.9 Conclusion 600
References 600

12 Nanomechanical Properties of Solid Surfaces and Thin Films
Adrian B. Mann 607
12.1 Introduction 607
12.2 Instrumentation 608
    12.2.1 AFM and Scanning Probe Microscopy 608
    12.2.2 Nanoindentation 609
    12.2.3 Adaptations of Nanoindentation 612
    12.2.4 Complimentary Techniques............................ 613
    12.2.5 Bulge Tests 614
    12.2.6 Acoustic Methods.................................... 614
    12.2.7 Imaging Methods 617
12.3 Data Analysis 617
    12.3.1 Elastic Contacts 618

12.3.2　Indentation of Ideal Plastic Materials 619
　　　12.3.3　Adhesive Contacts 620
　　　12.3.4　Indenter Geometry 621
　　　12.3.5　Analyzing Load/Displacement Curves 622
　　　12.3.6　Modifications to the Analysis 626
　　　12.3.7　Alternative Methods of Analysis 629
　　　12.3.8　Measuring Contact Stiffness 630
　　　12.3.9　Measuring Viscoelasticity 631
　12.4　Modes of Deformation 632
　　　12.4.1　Defect Nucleation 632
　　　12.4.2　Variations with Depth 634
　　　12.4.3　Anisotropic Materials 635
　　　12.4.4　Fracture and Delamination 635
　　　12.4.5　Phase Transformations 637
　12.5　Thin Films and Multilayers 639
　　　12.5.1　Thin Films 640
　　　12.5.2　Multilayers 644
　12.6　Developing Areas 646
References 647

13　Computer Simulations of Nanometer-Scale Indentation and Friction
Susan B. Sinnott, Seong-Jun Heo, Donald W. Brenner, Judith A. Harrison .. 655
　13.1　Introduction 655
　13.2　Computational Details 657
　　　13.2.1　Energies and Forces 657
　　　13.2.2　Important Approximations 660
　13.3　Indentation 664
　　　13.3.1　Surfaces 665
　　　13.3.2　Thin Films 680
　13.4　Friction and Lubrication 688
　　　13.4.1　Bare Surfaces 689
　　　13.4.2　Decorated Surfaces 698
　　　13.4.3　Thin Films 702
　13.5　Conclusions 727
References 727

14　Mechanical Properties of Nanostructures
Bharat Bhushan 741
　14.1　Introduction 742
　14.2　Experimental Techniques for Measurement of Mechanical Properties
　　　of Nanostructures 744
　　　14.2.1　Indentation and Scratch Tests Using Micro/Nanoindenters 744
　　　14.2.2　Bending Tests of Nanostructures Using an AFM 745
　　　14.2.3　Bending Tests of Micro/Nanostructures Using a Nanoindenter . 751
　14.3　Experimental Results and Discussion 752

14.3.1 Indentation and Scratch Tests of Various Ceramic and Metals Using a Micro/Nanoindenter 752
14.3.2 Bending Tests of Ceramic Nanobeams Using an AFM 758
14.3.3 Bending Tests of Metallic Microbeams Using a Nanoindenter . 764
14.3.4 Indentation and Scratch Tests of Polymeric Microbeams Using a Nanoindenter 766
14.3.5 Bending Tests of Polymeric Microbeams Using a Nanoindenter 770
14.4 Finite Element Analysis of Nanostructures with Roughness and Scratches................................... 773
14.4.1 Stress Distribution in a Smooth Nanobeam 776
14.4.2 Effect of Roughness in the Longitudinal Direction 776
14.4.3 Effect of Roughness in the Transverse Direction and Scratches 777
14.4.4 Effect on Stresses and Displacements for Materials Which are Elastic, Elastic-Plastic or Elastic-Perfectly Plastic .. 782
14.5 Closure 783
References 785

15 Scale Effect in Mechanical Properties and Tribology
Bharat Bhushan, Michael Nosonovsky 791
15.1 Nomenclature 791
15.2 Introduction 793
15.3 Scale Effect in Mechanical Properties......................... 796
  15.3.1 Yield Strength and Hardness 797
  15.3.2 Shear Strength at the Interface 800
15.4 Scale Effect in Surface Roughness and Contact Parameters 804
  15.4.1 Scale Dependence of Roughness and Contact Parameters 804
  15.4.2 Dependence of Contact Parameters on Load 807
15.5 Scale Effect in Friction 809
  15.5.1 Adhesional Friction 810
  15.5.2 Two-Body Deformation 813
  15.5.3 Three-Body Deformation Friction 814
  15.5.4 Ratchet Mechanism 817
  15.5.5 Meniscus Analysis 819
  15.5.6 Total Value of Coefficient of Friction and Transition from Elastic to Plastic Regime 820
  15.5.7 Comparison with the Experimental Data 822
15.6 Scale Effect in Wear 828
15.7 Scale Effect in Interface Temperature 829
15.8 Closure 831
15.A Statistics of Particle Size Distribution...................... 832
  15.A.1 Statistical Models of Particle Size Distribution 832
  15.A.2 Typical Particle Size Distribution Data 835
References 837

Part III Molecularly-Thick Films for Lubrication

16 Nanotribology of Ultrathin and Hard Amorphous Carbon Films
Bharat Bhushan 843
16.1 Introduction 843
16.2 Description of Common Deposition Techniques 848
    16.2.1 Filtered Cathodic Arc Deposition 849
    16.2.2 Ion Beam Deposition 851
    16.2.3 Electron Cyclotron Resonance Chemical Vapor Deposition ... 852
    16.2.4 Sputtering Deposition 853
    16.2.5 Plasma-Enhanced Chemical Vapor Deposition 853
16.3 Chemical and Physical Coating Characterization 854
    16.3.1 EELS and Raman Spectroscopy 855
    16.3.2 Hydrogen Concentrations 859
    16.3.3 Physical Properties 860
    16.3.4 Summary 862
16.4 Micromechanical and Tribological Coating Characterization 862
    16.4.1 Micromechanical Characterization 862
    16.4.2 Microscratch and Microwear Studies 875
    16.4.3 Macroscale Tribological Characterization 885
    16.4.4 Coating Continuity Analysis 891
16.5 Closure 893
References 894

17 Self-Assembled Monolayers (SAMs) for Controlling Adhesion, Friction, and Wear
Bharat Bhushan 901
17.1 Introduction 901
17.2 A Brief Organic Chemistry Primer 906
    17.2.1 Electronegativity/Polarity 906
    17.2.2 Classification and Structures of Organic Compounds 908
    17.2.3 Polar and Nonpolar Groups 913
17.3 Self-Assembled Monolayers: Substrates, Spacer Chains; and End Groups in the Molecular Chains 913
17.4 Tribological Properties of SAMs 918
    17.4.1 Measurement Techniques 922
    17.4.2 Hexadecane Thiol and Biphenyl Thiol SAMs on Au(111) ... 923
    17.4.3 Alkylsilane and Perfluoroalkylsilane SAMs on Si(100) and Alkylphosphonate SAMS on Al 934
    17.4.4 Degradation and Environmental Studies 947
17.5 Closure 951
References 953

18 Nanoscale Boundary Lubrication Studies
Bharat Bhushan, Huiwen Liu 959
18.1 Introduction 959
18.2 Lubricants Details 960
18.3 Nanodeformation, Molecular Conformation, and Lubricant Spreading 963
18.4 Boundary Lubrication Studies 966
    18.4.1 Friction and Adhesion 967
    18.4.2 Rest Time Effect 972
    18.4.3 Velocity Effect 976
    18.4.4 Relative Humidity and Temperature Effect 979
    18.4.5 Tip Radius Effect 983
    18.4.6 Wear Study 985
18.5 Closure 988
References 989

Part IV Biomimetics

19 Lotus Effect: Roughness-Induced Superhydrophobic Surfaces
Bharat Bhushan, Michael Nosonovsky and Yong Chae Jung 995
19.1 Introduction 995
19.2 Modeling of Contact Angle for a Liquid in Contact
     with a Rough Surface 1001
    19.2.1 Contact Angle Definition 1001
    19.2.2 Heterogeneous Interfaces and the Wenzel
          and Cassie–Baxter Equations 1002
    19.2.3 Contact Angle Hysteresis 1010
    19.2.4 The Cassie–Wenzel Wetting Regime Transition 1013
19.3 Lotus-Effect and Water-Repellent Surfaces in Nature 1017
    19.3.1 Water-Repellent Plants 1017
    19.3.2 Characterization of Hydrophobic
          and Hydrophilic Leaf Surfaces 1019
19.4 Wetting of Micro- and Nanopatterned Surfaces 1031
    19.4.1 Experimental Techniques 1031
    19.4.2 Micro- and Nanopatterned Polymers 1033
    19.4.3 Micropatterned Si Surfaces 1037
    19.4.4 Self-Cleaning 1056
19.5 Role of Hierarchical Roughness for Superhydrophobicity 1056
19.6 How to Make a Superhydrophobic Surface 1057
    19.6.1 Roughening to Create One-level Structure 1058
    19.6.2 Coating to Create One-level Hydrophobic Structures 1063
    19.6.3 Methods to Create Two-Level (Hierarchical)
          Superhydrophobic Structures 1064
19.7 Closure 1065
References 1065

20 Gecko Feet: Natural Hairy Attachment Systems for Smart Adhesion – Mechanism, Modeling and Development of Bio-Inspired Materials
Bharat Bhushan 1073
- 20.1 Introduction 1073
- 20.2 Hairy Attachment Systems 1074
- 20.3 Tokay Gecko 1077
  - 20.3.1 Construction of Tokay Gecko 1077
  - 20.3.2 Adaptation to Rough Surfaces 1078
  - 20.3.3 Peeling 1081
  - 20.3.4 Self Cleaning 1083
- 20.4 Attachment Mechanisms 1084
  - 20.4.1 Van der Waals Forces 1085
  - 20.4.2 Capillary Forces 1086
- 20.5 Experimental Adhesion Test Techniques and Data 1087
  - 20.5.1 Adhesion under Ambient Conditions 1088
  - 20.5.2 Effects of Temperature 1090
  - 20.5.3 Effects of Humidity 1090
  - 20.5.4 Effects of Hydrophobicity 1092
- 20.6 Adhesion Modeling 1092
  - 20.6.1 Spring Model 1092
  - 20.6.2 Single Spring Contact Analysis 1094
  - 20.6.3 The Multi-Level Hierarchical Spring Analysis 1095
  - 20.6.4 Adhesion Results of the Multi-level Hierarchical Spring Model 1099
  - 20.6.5 Capillary Effects 1103
  - 20.6.6 Adhesion Results that Account for Capillary Effects .. 1104
- 20.7 Modeling of Biomimetic Fibrillar Structures 1106
  - 20.7.1 Fiber Model 1107
  - 20.7.2 Single Fiber Contact Analysis 1108
  - 20.7.3 Constraints 1108
  - 20.7.4 Numerical simulation 1112
  - 20.7.5 Results and discussion 1114
- 20.8 Fabrication of Biomimetric Gecko Skin 1118
  - 20.8.1 Single Level Hierarchical Structures 1119
  - 20.8.2 Multi-Level Hierarchical Structures 1123
- 20.9 Closure 1125
- 20.A Typical Rough Surfaces 1127
- References 1130

Part V Applications

21 Micro/Nanotribology and Micro/Nanomechanics of Magnetic Storage Devices
Bharat Bhushan 1137
21.1 Introduction 1138
    21.1.1 Magnetic Storage Devices 1138
    21.1.2 Micro/Nanotribology and Micro/Nanomechanics and Their Applications 1142
21.2 Experimental 1142
    21.2.1 Description of AFM/FFM 1142
    21.2.2 Test Specimens 1146
21.3 Surface Roughness 1146
21.4 Friction and Adhesion 1151
    21.4.1 Magnetic Head Materials 1151
    21.4.2 Magnetic Media 1153
21.5 Scratching and Wear 1156
    21.5.1 Nanoscale Wear 1156
    21.5.2 Microscale Scratching 1157
    21.5.3 Microscale Wear 1167
21.6 Indentation 1182
    21.6.1 Picoscale Indentation 1182
    21.6.2 Nanoscale Indentation 1183
    21.6.3 Localized Surface Elasticity 1185
21.7 Lubrication 1187
    21.7.1 Boundary Lubrication Studies 1189
21.8 Closure 1192
References 1194

22 Nanotribology and Materials Characterization of MEMS/NEMS and BioMEMS/BioNEMS Materials and Devices
Bharat Bhushan 1199
22.1 Introduction 1199
    22.1.1 Introduction to MEMS 1202
    22.1.2 Introduction to NEMS 1204
    22.1.3 BioMEMS/BioNEMS 1204
    22.1.4 Tribological Issues in MEMS/NEMS and BioMEMS/BioNEMS 1205
22.2 Tribological Studies of Silicon and Related Materials 1224
    22.2.1 Virgin and Treated/Coated Silicon Samples 1225
    22.2.2 Tribological Properties of Polysilicon Films and SiC Film 1231
22.3 Lubrication Studies for MEMS/NEMS 1235
    22.3.1 Perfluoropolyether Lubricants 1235
    22.3.2 Self-Assembled Monolayers (SAMs) 1240

  22.3.3 Hard Diamond-Like Carbon (DLC) Coatings 1245
22.4 Tribological Studies of Biological Molecules
  on Silicon-Based Surfaces and of Coated Polymer Surfaces 1245
  22.4.1 Adhesion, Friction, and Wear of Biomolecules
    on Si-Based Surfaces 1245
  22.4.2 Adhesion of Coated Polymer Surfaces..................... 1251
22.5 Nanopatterned Surfaces 1253
  22.5.1 Analytical Model and Roughness Optimization 1253
  22.5.2 Experimental Validation 1256
22.6 Component-Level Studies 1261
  22.6.1 Surface Roughness Studies of Micromotor Components 1261
  22.6.2 Adhesion Measurements of Microstructures 1263
  22.6.3 Microtriboapparatus for Adhesion, Friction,
    and Wear of Microcomponents 1264
  22.6.4 Static Friction Force (Stiction) Measurements in MEMS 1269
  22.6.5 Mechanisms Associated with Observed Stiction Phenomena
    in Digital Micromirror Devices (DMD)
    and Nanomechanical Characterization 1273
22.7 Conclusion 1278
22.A Appendix Micro/Nanofabrication Methods 1279
  22.A.1 Top-Down Methods 1279
  22.A.2 Bottom-Up Fabrication (Nanochemistry) 1284
References 1285

23 Mechanical Properties of Micromachined Structures
Harold Kahn 1297
23.1 Measuring Mechanical Properties of Films on Substrates 1297
  23.1.1 Residual Stress Measurements 1298
  23.1.2 Mechanical Measurements Using Nanoindentation 1299
23.2 Micromachined Structures for Measuring Mechanical Properties 1299
  23.2.1 Passive Structures 1299
  23.2.2 Active Structures 1304
23.3 Measurements of Mechanical Properties 1314
  23.3.1 Mechanical Properties of Polysilicon 1314
  23.3.2 Mechanical Properties of Other Materials 1318
References 1320

24 Structural, Nanomechanical, and Nanotribological Characterization of Human Hair Using Atomic Force Microscopy and Nanoindentation
Bharat Bhushan, Carmen LaTorre 1325
24.1 Introduction 1325
24.2 Human Hair, Skin, and Hair Care Products 1330
  24.2.1 Human Hair and Skin 1330
  24.2.2 Hair Care: Cleaning and Conditioning Treatments,
    and Damaging Processes 1339

24.3 Experimental 1345
    24.3.1 Experimental Procedure 1348
    24.3.2 Hair and Skin Samples 1361
24.4 Structural Characterization Using an AFM 1363
    24.4.1 Structure of Hair Cross-section & Longitudinal Section 1363
    24.4.2 Structure of Various Cuticle Layers 1367
    24.4.3 Summary 1373
24.5 Nanomechanical Characterization Using Nanoindentation,
     Nanoscratch, and AFM 1375
    24.5.1 Hardness, Young's Modulus, and Creep 1375
    24.5.2 Scratch Resistance 1387
    24.5.3 In-Situ Tensile Deformation Studies
           on Human Hair Using AFM 1395
    24.5.4 Summary 1401
24.6 Multi-Scale Tribological Characterization 1402
    24.6.1 Macroscale Tribological Characterization 1402
    24.6.2 Nanotribological Characterization Using an AFM 1410
    24.6.3 Scale Effects 1439
    24.6.4 Summary 1449
24.7 Conditioner Thickness Distribution and Binding Interactions
     on Hair Surface 1451
    24.7.1 Conditioner Thickness and Adhesive Force Mapping 1452
    24.7.2 Effective Young's Modulus Mapping 1459
    24.7.3 Binding Interactions Between Conditioner and Hair Surface .. 1462
    24.7.4 Summary 1464
24.8 Surface Potential Studies of Human Hair Using Kelvin Probe
     Microscopy 1465
    24.8.1 Effect of Physical Wear and Rubbing with Latex
           on Surface Potential 1465
    24.8.2 Effect of External Voltage and Humidity on Surface Potential . 1471
    24.8.3 Summary 1476
24.9 Closure 1476
24.A Shampoo and Conditioner Treatment Procedure 1479
    24.A.1 Shampoo Treatments 1479
    24.A.2 Conditioner Treatments 1479
24.B Conditioner Thickness Approximation 1480
References 1481

The Editor 1487

Index 1489

List of Contributors

Dr. Boris Anczykowski
nanoAnalytics GmbH
Gievenbecker Weg 11
48149 Muenster
Germany
anczykowski@nanoanalytics.com

Prof. Roland Bennewitz
McGill University
Ernest Rutherford Physics Building
3600 rue University
Montréal, QC, H3A 2T8
Canada
roland.bennewitz@mcgill.ca

Prof. Bharat Bhushan
Ohio Eminent Scholar and
The Howard D. Winbigler Professor
Director, Nanoprobe Laboratory
for Bio- & Nanotechnology
and Biomimetics (NLB^2)
201 W. 19th Avenue
Ohio State University
Columbus, Ohio 43210-1142
USA
bhushan.2@osu.edu

Prof. Donald W. Brenner
North Carolina State University
Department of Materials Science
and Engineering
Raleigh, North Carolina 27695-7909
USA
brenner@ncsu.edu

Prof. Harald Fuchs
University of Muenster
Institute of Physics
Wilhem-Klemm-Str. 10
48149 Muenster
Germany
fuchsh@uni-muenster.de

Prof. Dr. Franz J. Giessibl
University of Regensburg
Institute of Experimental
and Applied Physics
Physics Building, Room PHY 1.1.22
Universitätsstr. 31
93053 Regensburg
Germany
*franz.giessibl@
 physik.uni-regensburg.de*

Dr. Enrico Gnecco
University of Basel
Institute of Physics
Klingelbergstr. 82
4056 Basel
Switzerland
enrico.gnecco@unibas.ch

Prof. Steve Granick
University of Illinois
at Urbana-Champaign
Department of Materials Science
and Engineering
104 S. Goodwin Ave
Urbana, IL 61801
USA
sgranick@uiuc.edu

Prof. Jason H. Hafner
Rice University
Physics & Astronomy – MS61
PO Box 1892
Houston, TX 77251-1892
USA
hafner@rice.edu

Prof. Judith A. Harrison
U.S. Naval Academy
Chemistry Department, MS 9B
572 Holloway Road
Annapolis, Maryland 21402
USA
jah@usna.edu

Mr. Seong-Jun Heo
3400 Stevenson Blvd. J18
Fremont, CA 94538
USA
heogyver@ufl.edu

Prof. Peter Hinterdorfer
University of Linz
Institute for Biophysics
Altenbergerstr. 69
4040 Linz
Austria
peter.hinterdorfer@jku.at

Dr. Hirotaka Hosoi
Hokkaido University
Creative Research Initiative Sousei
Kita 21, Nishi 10, Kita-ku, Sapporo
Japan
hosoi@cris.hokudai.ac.jp

Prof. Jacob N. Israelachvili
University of California
Department of Chemical Engineering
and Materials
Santa Barbara, CA 93106
USA
jacob@engineering.ucsb.edu

Yong Chae Jung
The Ohio State University
Nanotribology Laboratory
for Information Storage
and MEMS/NEMS
W390 Scott Laboratory
201 W. 19th Avenue
Columbus, OH 43210-1142
USA
jung.181@osu.edu

Dr. Harold Kahn
Case Western Reserve University
Department of Materials Science
and Engineering
10900 Euclid Ave.
Cleveland, OH 44106-7204
USA
hxk29@po.cwru.edu

Carmen LaTorre
Owens Corning
Insulating Systems Business
2790 Columbus Road
Route 16 (Bldg 20-1)
Granville, OH 43023
USA
carmen.latorre@owenscorning.com

Dr. Huiwen Liu
Seagate Technology
Mailstop NRW 130
7801 Computer Avenue
Bloomington, MN 55435
USA
huiwen.liu@seagate.com

Prof. Adrian B. Mann
Rutgers University
Department of Ceramic
& Materials Engineering
and Biomedical Engineering
607 Taylor Road
Piscataway, NJ 08854
USA
abmann@rci.rutgers.edu

Prof. Othmar Marti
University of Ulm
Department of Experimental Physics
Albert-Einstein-Allee 11
89069 Ulm
Germany
othmar.marti@uni-ulm.de

Prof. Ernst Meyer
University of Basel
Institute of Physics
Klingelbergstr. 82
4056 Basel
Switzerland
ernst.meyer@unibas.ch

Prof. Markus Morgenstern
RWTH Aachen University
II. Institute of Physics B
52056 Aachen
Germany
mmorgens@physik.rwth-aachen.de

Prof. Seizo Morita
Osaka University
Department of Electronic Engineering
Graduate School of Engineering
Yamada-Oka 2-1, Suita 565-0871
Japan
smorita@ele.eng.osaka-u.ac.jp

Prof. Koichi Mukasa
Hokkaido University
Nanoelectronics Laboratory
Graduate School of Engineering
Nishi-8, Kita-13, Kita-ku, Sapporo
060-8628
Japan
mukasa@nano.eng.hokudai.ac.jp

Prof. Ashis Mukhopadhyay
Wayne State University
Deptartment of Physics
Detroit, MI
USA
ashis@physics.wayne.edu

Dr. Michael Nosonovsky
Stevens Institute of Technology
Department of Mechanical Engineering
Castle Point on Hudson
Hoboken, NJ 07030
USA
michael.nosonovsky@stevens.edu

Prof. Hiroshi Onishi
Kobe University
Department of Chemistry
Rokko-dai, Nada-ku, Kobe 657-8501
Japan
oni@kobe-u.ac.jp

Dr. Oliver Pfeiffer
University of Basel
Institute of Physics
Klingelbergstr. 82
4056 Basel
Switzerland
oliver.pfeiffer@unibas.ch

Prof. Marina Ruths
University of Massachusetts Lowell
Department of Chemistry
1 University Avenue
Lowell, MA 01854
USA
marina_ruths@uml.edu

Akira Sasahara
Kanagawa Academy of Science
and Technology
Surface Chemistry Laboratory
KSP East 404, 3-2-1 Sakado
Takatsu-ku, Kawasaki-shi
Kanagawa 213-0012
Japan
sasahara@kobe-u.ac.jp

Dr. Andre Schirmeisen
University of Muenster
Institute of Physics
Wilhelm-Klemm-Str. 10
48149 Muenster
Germany
schira@uni-muenster.de

Dr. Alexander Schwarz
University of Hamburg
Institute of Applied Physics
Jungiusstr. 11
20355 Hamburg
Germany
aschwarz@physnet.uni-hamburg.de

Prof. Udo D. Schwarz
Yale University
Department of Mechanical Engineering
P.O. Box 208284
15 Prospect Street, Rm. 213
New Haven, CT 06520-8284
USA
udo.schwarz@yale.edu

Prof. Susan B. Sinnott
University of Florida
Deptartment of Materials Science
and Engineering
154 Rhines Hall
P.O. Box 116400
Gainesville, Florida, 32611-6400
USA
sinnott@mse.ufl.edu

Dr. Anisoara Socoliuc
University of Basel
Institute of Physics
Klingelbergstr. 82
4056 Basel
Switzerland
a.socoliuc@unibas.ch

Prof. Yasuhiro Sugawara
Osaka University
Department of Applied Physics
Graduate School of Engineering
Yamada-Oka 2-1, Suita 565-0871
Japan
sugawara@ap.eng.osaka-u.ac.jp

Prof. Y. Elaine Zhu
Deptartment of Chemical
& Biomolecular Engineering
182 Fitzpatrick Hall
University of Notre Dame
Notre Dame, IN 46556
USA
yzhu3@nd.edu

List of Abbreviations

2-DEG	two-dimensional electron gas
AC	alternating current
AFAM	atomic force acoustic microscopy
AFM	atomic force microscopy
AM	amplitude modulation
BFP	biomembrane force probe
bioMEMS	biomedical microelectromechanical systems
BSA	bovine serum albumium
BW	bonded washed
CD	compact disc
CD	critical dimension
CDW	charge density wave
CE	capillary electrophoresis
CG	controlled geometry
CMOS	complementary metal oxide semiconductor
CSM	continuous stiffness measurement
CVD	chemical vapor deposition
DC	direct current
DFM	dynamic force microscopy
DFS	dynamic force spectroscopy
DI	deionized
DLC	diamond like carbon
DLP	digital light processing
DLVO	Derjaguin–Landau–Verwey–Overbeek
DMD	digital micromirror device
DMT	Derjaguin–Muller–Toporov
DNA	deoxyribonucleic acid
DOS	density of state
DSP	digital signal processor
EAM	embedded atom method

EBD	electron beam deposition
ECR-CVD	electron cyclotron resonance chemical vapor depostion
EELS	electron energy loss spectroscopy
EFM	electrostatic force microscopy
EHD	electrohydrodynamic
FC	flip-chip
FCA	filtered cathodic arc
FCP	force calibration plot
FEM	finite element modeling
FESP	force modulation etched Si probe
FET	field effect transistor
FFM	friction force microscope
FIB	focused ion beam
FKT	Frenkel–Kontorova–Tomlinson
FM	frequency modulation
FM-AFM	force modulation mode atomic force microscopy
FM-SFM	force modulation mode scanning force microscopy
FS	force spectroscopy
GMR	giant magnetoresistance
HDT	hexadecanethiol
HF	hydrofluoric acid
HOP	highly oriented pyrolytic
HOPG	highly oriented pyrolytic graphite
HtBDC	hexa-tert-butyl-decacyclene
HTCS	high-temperature superconductivity
IBD	ion beam deoposition
IC	integrated circuit
ICAM	interellular adhesion molecule
ISE	indentation size effect
JKR	Johnson–Kendall–Roberts
KPFM	Kelvin probe force microscopy
LB	Langmuir–Blodgett
LDOS	local density of state
LFA	leukocyte function-associated antigen
LFM	lateral force microscopy
LJ	Lennard–Jones
LN	liquid-nitrogen
LPCVD	low-pressure chemical vapor deposition
LVDT	linear variable differential transformer
MAP	manifold absolute pressure
MD	molecular dynamics
ME	metal-evaporated
MEMS	microelectromechanical systems
MFM	magnetic force microscopy

MP	metal particle
MRFM	magnetic resonance force microscopy
MRFM	molecular recognition force microscopy
MRI	magnetic resonance imaging
MWNT	multiwall nanotube
NC-AFM	noncontact AFM
NEMS	nanoelectromechanical systems
NP	silicon nitride probe
NSOM	near-field scanning optical microscopy
NTA	nitrilotriacetate–hexahistidine 6
OTS	octadecyltrichlorosilane
PBC	periodic boundary condition
PDMS	polydimethylsiloxane
PDP	2-pyridyldithiopropionyl
PECVD	plasma enhanced chemical vapor deposition
PEG	poly(ethylene glycol)
PES	photoemission spectroscopy
PET	poly(ethylene terephthalate)
PFPE	perfluoropolyether
PMMA	polymethylmethacrylate
PS	polystyrene
PSD	position-sensitive detector
PSGL	P-selectin glycoprotein ligand
PTFE	polytetrafluoroethylene
PZT	lead zirconium titanate
QCM	quartz crystal microbalance
RF	radio-frequency
RH	relative humidity
RICM	reflection interference contrast
RNA	ribonucleic acid
RPM	revolutions per minute
SAM	scanning acoustic microscopy
SAM	slef-assembled monolayer
SCPM	scanning chemical potential microscopy
SEcM	scanning electrochemical microscopy
SEFM	scanning electrostaitc force microscopy
SEM	scanning electron microscopy
SFA	surface force apparatus
SFAM	scanning force aoustic microscopy
SFD	shear flow detachment
SFM	scanning force microscopy
SICM	scanning ion conductance microscopy
SKPM	scanning Kelvin probe microscopy
SMM	scanning magnetic microscopy

SNOM	scanning near field optical microscopy
SPM	scanning probe microscopy
SThM	scanning thermal microscopy
STM	scanning tunneling microscopy
SWCNT	single-wall carbon nanotube
SWNT	single-wall nanotube
TEM	transmission electron microscopy
TESP	tapping-mode etched silicon probe
TIRM	total internal reflection microscopy
TTF	tetrathiafulvalene
UHV	ultrahigh vacuum
vdW	van der Waals

1

Introduction – Measurement Techniques and Applications

Bharat Bhushan

Summary. In this introductory chapter, the definition and history of tribology and their industrial significance and origins and significance of an emerging field of micro/nanotribology are described. Next, various measurement techniques used in micro/nanotribological and micro/nanomechanical studies are described. The interest in micro/nanotribology field grew from magnetic storage devices and latter the applicability to emerging field micro/nanoelectromechanical systems (MEMS/NEMS) became clear. A few examples of magnetic storage devices and MEMS/NEMS are presented where micro/nanotribological and micro/nanomechanical tools and techniques are essential for interfacial studies. Finally, reasons why micro/nanotribological and micro/nanomechanical studies are important in magnetic storage devices and MEMS/NEMS are presented. In the last section, organization of the book is presented.

1.1 Definition and History of Tribology

The word tribology was first reported in a landmark report by *Jost* [1]. The word is derived from the Greek word tribos meaning rubbing, so the literal translation would be "the science of rubbing". Its popular English language equivalent is friction and wear or lubrication science, alternatively used. The latter term is hardly all-inclusive. Dictionaries define tribology as the science and technology of interacting surfaces in relative motion and of related subjects and practices. Tribology is the art of applying operational analysis to problems of great economic significance, namely, reliability, maintenance, and wear of technical equipment, ranging from spacecraft to household appliances. Surface interactions in a tribological interface are highly complex, and their understanding requires knowledge of various disciplines including physics, chemistry, applied mathematics, solid mechanics, fluid mechanics, thermodynamics, heat transfer, materials science, rheology, lubrication, machine design, performance and reliability.

It is only the name tribology that is relatively new, because interest in the constituent parts of tribology is older than recorded history [2]. It is known that drills made during the Paleolithic period for drilling holes or producing fire were fitted

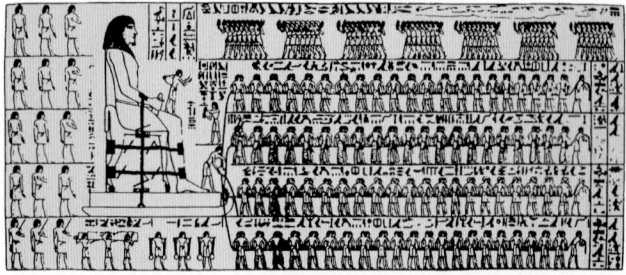

Fig. 1.1. Egyptians using lubricant to aid movement of colossus, El-Bersheh, circa 1800 BC

with bearings made from antlers or bones, and potters' wheels or stones for grinding cereals, etc., clearly had a requirement for some form of bearings [3]. A ball thrust bearing dated about AD 40 was found in Lake Nimi near Rome.

Records show the use of wheels from 3500 BC, which illustrates our ancestors' concern with reducing friction in translationary motion. The transportation of large stone building blocks and monuments required the know-how of frictional devices and lubricants, such as water-lubricated sleds. Figure 1.1 illustrates the use of a sledge to transport a heavy statue by egyptians circa 1880 BC [4]. In this transportation, 172 slaves are being used to drag a large statue weighing about 600 kN along a wooden track. One man, standing on the sledge supporting the statue, is seen pouring a liquid (most likely water) into the path of motion; perhaps he was one of the earliest lubrication engineers. (*Dowson* [2] has estimated that each man exerted a pull of about 800 N. On this basis, the total effort, which must at least equal the friction force, becomes 172×800 N. Thus, the coefficient of friction is about 0.23.) A tomb in Egypt that was dated several thousand years BC provides the evidence of use of lubricants. A chariot in this tomb still contained some of the original animal-fat lubricant in its wheel bearings.

During and after the glory of the Roman empire, military engineers rose to prominence by devising both war machinery and methods of fortification, using tribological principles. It was the renaissance engineer-artist Leonardo da Vinci (1452–1519), celebrated in his days for his genius in military construction as well as for his painting and sculpture, who first postulated a scientific approach to friction. Da Vinci deduced the rules governing the motion of a rectangular block sliding over a flat surface. He introduced for the first time, the concept of coefficient of friction as the ratio of the friction force to normal load. His work had no historical influence, however, because his notebooks remained unpublished for hundreds of years. In 1699, the French physicist Guillaume Amontons rediscovered the rules of friction after he studied dry sliding between two flat surfaces [5]. First, the friction force that resists sliding at an interface is directly proportional to the normal load. Second, the amount of friction force does not depend on the apparent area of contact. These observations were verified by French physicist Charles-Augustin Coulomb (better known for his work on electrostatics [6]). He added a third rule that the friction force is independent of velocity once motion starts. He also made a clear distinction between static friction and kinetic friction.

Many other developments occurred during the 1500s, particularly in the use of improved bearing materials. In 1684, Robert Hooke suggested the combination of steel shafts and bell-metal bushes as preferable to wood shod with iron for wheel bearings. Further developments were associated with the growth of industrialization in the latter part of the eighteenth century. Early developments in the petroleum industry started in Scotland, Canada, and the United States in the 1850s [2–7].

Though essential laws of viscous flow were postulated by Sir Isaac Newton in 1668; scientific understanding of lubricated bearing operations did not occur until the end of the nineteenth century. Indeed, the beginning of our understanding of the principle of hydrodynamic lubrication was made possible by the experimental studies of *Tower* [8] and the theoretical interpretations of *Reynolds* [9] and related work by *Petroff* [10]. Since then developments in hydrodynamic bearing theory and practice were extremely rapid in meeting the demand for reliable bearings in new machinery.

Wear is a much younger subject than friction and bearing development, and it was initiated on a largely empirical basis. Scientific studies of wear developed little until the mid-twentieth century. *Holm* made one of the earliest substantial contributions to the study of wear [11].

The industrial revolution (1750–1850 A.D.) is recognized as a period of rapid and impressive development of the machinery of production. The use of steam power and the subsequent development of the railways in the 1830s led to promotion of manufacturing skills. Since the beginning of the twentieth century, from enormous industrial growth leading to demand for better tribology, knowledge in all areas of tribology has expanded tremendously [11–17].

1.2 Industrial Significance of Tribology

Tribology is crucial to modern machinery which uses sliding and rolling surfaces. Examples of productive friction are brakes, clutches, driving wheels on trains and automobiles, bolts, and nuts. Examples of productive wear are writing with a pencil, machining, polishing, and shaving. Examples of unproductive friction and wear are internal combustion and aircraft engines, gears, cams, bearings, and seals.

According to some estimates, losses resulting from ignorance of tribology amount in the United States to about 4% of its gross national product (or about $ 200 billion dollars per year in 1966), and approximately one-third of the world's energy resources in present use appear as friction in one form or another. Thus, the importance of friction reduction and wear control cannot be overemphasized for economic reasons and long-term reliability. According to *Jost* [1, 18], savings of about 1% of gross national product of an industrialized nation can be realized by research and better tribological practices. According to recent studies, expected savings are expected to be on the order of 50 times the research costs. The savings are both substantial and significant, and these savings can be obtained without the deployment of large capital investment.

The purpose of research in tribology is understandably the minimization and elimination of losses resulting from friction and wear at all levels of technology where the rubbing of surfaces is involved. Research in tribology leads to greater plant efficiency, better performance, fewer breakdowns, and significant savings.

Tribology is not only important to industry, it also affects day-to-day life. For example, writing is a tribological process. Writing is accomplished by a controlled transfer of lead (pencil) or ink (pen) to the paper. During writing with a pencil there should be good adhesion between the lead and paper so that a small quantity of lead transfers to the paper and the lead should have adequate toughness/hardness so that it does not fracture/break. Objective during shaving is to remove hair from the body as efficiently as possible with minimum discomfort to the skin. Shaving cream is used as a lubricant to minimize friction between a razor and the skin. Friction is helpful during walking and driving. Without adequate friction, we would slip and a car would skid! Tribology is also important in sports. For example, a low friction between the skis and the ice is desirable during skiing.

1.3 Origins and Significance of Micro/Nanotribology

At most interfaces of technological relevance, contact occurs at numerous asperities. Consequently, the importance of investigating a single asperity contact in studies of the fundamental tribological and mechanical properties of surfaces has been long recognized. The recent emergence and proliferation of proximal probes, in particular tip-based microscopies (e. g., the scanning tunneling microscope and the atomic force microscope) and of computational techniques for simulating tip-surface interactions and interfacial properties, has allowed systematic investigations of interfacial problems with high resolution as well as ways and means for modifying and manipulating nanoscale structures. These advances have led to the development of the new field of microtribology, nanotribology, molecular tribology, or atomic-scale tribology [15, 16, 19–22]. This field is concerned with experimental and theoretical investigations of processes ranging from atomic and molecular scales to microscales, occurring during adhesion, friction, wear, and thin-film lubrication at sliding surfaces.

The differences between the conventional or macrotribology and micro/nanotribology are contrasted in Fig. 1.2. In macrotribology, tests are conducted on components with relatively large mass under heavily loaded conditions. In these tests, wear is inevitable and the bulk properties of mating components dominate the tribological performance. In micro/nanotribology, measurements are made on components, at least one of the mating components, with relatively small mass under lightly loaded conditions. In this situation, negligible wear occurs and the surface properties dominate the tribological performance.

The micro/nanotribological studies are needed to develop fundamental understanding of interfacial phenomena on a small scale and to study interfacial phe-

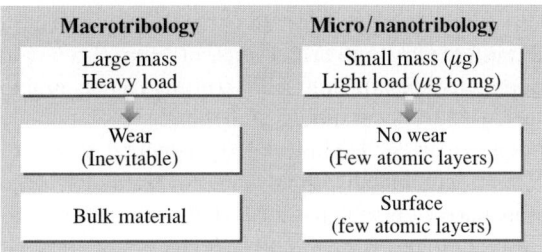

Fig. 1.2. Comparisons between macrotribology and micro/nanotribology

nomena involving ultrathin films (as low as 1–2 nm) and in micro/nanostructures, both used in magnetic storage systems, micro/nanoelectromechanical systems (MEMS/NEMS) and other industrial applications. The components used in micro- and nanostructures are very light (on the order of few micrograms) and operate under very light loads (smaller than one microgram to a few milligrams). As a result, friction and wear (on a nanoscale) of lightly loaded micro/nanocomponents are highly dependent on the surface interactions (few atomic layers). These structures are generally lubricated with molecularly thin films. Micro/nanotribological techniques are ideal to study the friction and wear processes of ultrathin films and micro/nanostructures. Although micro/nanotribological studies are critical to study ultrathin films and micro/nanostructures, these studies are also valuable in fundamental understanding of interfacial phenomena in macrostructures to provide a bridge between science and engineering.

The probe-based microscopes (scanning tunneling microscope, the atomic force and friction force microscopes) and the surface force apparatus are widely used for micro/nanotribological studies [16, 19–23]. To give a historical perspective of the field, the scanning tunneling microscope (STM) developed by *Binnig* and *Rohrer* and their colleagues in 1981 at the IBM Zurich Research Laboratory, Forschungslabor, is the first instrument capable of directly obtaining three-dimensional (3-D) images of solid surfaces with atomic resolution [24]. STMs can only be used to study surfaces which are electrically conductive to some degree. Based on their design of STM, in 1985, *Binnig* et al. [25, 26] developed an atomic force microscope (AFM) to measure ultrasmall forces (less than 1 µN) present between the AFM tip surface and the sample surface. AFMs can be used for measurement of all engineering surfaces which may be either electrically conducting or insulating. AFM has become a popular surface profiler for topographic measurements on micro- to nanoscale. AFMs modified to measure both normal and friction forces, generally called friction force microscopes (FFMs) or lateral force microscopes (LFMs), are used to measure friction on micro- and nanoscales. AFMs are also used for studies of adhesion, scratching, wear, lubrication, surface temperatures, and for measurements of elastic/plastic mechanical properties (such as indentation hardness and modulus of elasticity [14, 16, 19, 21].

Surface force apparatuses (SFAs), first developed in 1969, are used to study both static and dynamic properties of the molecularly thin liquid films sandwiched between two molecularly smooth surfaces [16, 20–22, 27]. However, the liquid under study has to be confined between molecularly-smooth optically-transparent or sometimes opaque surfaces with radii of curvature on the order of 1 mm (leading to poorer lateral resolution as compared to AFMs). Only AFMs/FFMs can be used to study engineering surfaces in the dry and wet conditions with atomic resolution.

Meanwhile, significant progress in understanding the fundamental nature of bonding and interactions in materials, combined with advances in computer-based modeling and simulation methods, have allowed theoretical studies of complex interfacial phenomena with high resolution in space and time [16, 20–22]. Such simulations provide insights into atomic-scale energetics, structure, dynamics, thermodynamics, transport and rheological aspects of tribological processes. Furthermore, these theoretical approaches guide the interpretation of experimental data and the design of new experiments, and enable the prediction of new phenomena based on atomistic principles.

1.4 Measurement Techniques

1.4.1 Scanning Probe Microscopy

Family of instruments based on STMs and AFMs, called Scanning Probe Microscopes (SPMs), have been developed for various applications of scientific and industrial interest. These include – STM, AFM, FFM (or LFM), scanning electrostatic force microscopy (SEFM), scanning force acoustic microscopy (SFAM) (or atomic force acoustic microscopy, AFAM), scanning magnetic microscopy (SMM) (or magnetic force microscopy, MFM), scanning near field optical microscopy (SNOM), scanning thermal microscopy (SThM) scanning electrochemical microscopy (SEcM), scanning Kelvin Probe microscopy (SKPM), scanning chemical potential microscopy (SCPM), scanning ion conductance microscopy (SICM), and scanning capacitance microscopy (SCM). Family of instruments which measure forces (e. g. AFM, FFM, SEFM, SFAM, and SSM) are also referred to as scanning force microscopies (SFM). Although these instruments offer atomic resolution and are ideal for basic research, yet these are used for cutting edge industrial applications which do not require atomic resolution.

STMs, AFMs and their modifications can be used at extreme magnifications ranging from $10^3 \times$ to $10^9 \times$ in x-, y-, and z-directions for imaging macro to atomic dimensions with high-resolution information and for spectroscopy. These instruments can be used in any environment such as ambient air, various gases, liquid, vacuum, low temperatures, and high temperatures. Imaging in liquid allows the study of live biological samples and it also eliminates water capillary forces present in ambient air present at the tip–sample interface. Low temperature imaging is useful for the study of biological and organic materials and the study of low-temperature phenomena such as superconductivity or charge-density waves. Low-temperature operation is also advantageous for high-sensitivity force mapping due to the reduction in thermal vibration. These instruments also have been used to image liquids such as liquid crystals and lubricant molecules on graphite surfaces. While the pure imaging capabilities of SPM techniques dominated the application of these methods at their early development stages, the physics and chemistry of probe–sample interactions and the quantitative analyses of tribological, electronic, magnetic, biological, and chemical surfaces are commonly carried out. Nanoscale science and technology are strongly driven by SPMs which allow investigation and manipulation of surfaces down to the atomic scale. With growing understanding of the underlying interaction mechanisms, SPMs have found applications in many fields outside basic research fields. In addition, various derivatives of all these methods have been developed for special applications, some of them targeting far beyond microscopy.

A detailed overview of scanning probe microscopy – principle of operation, instrumentation, and probes is presented in a later chapter (also see [16, 20–23]). Here, a brief description of commercial STMs and AFMs follows.

Commercial STMs

There are a number of commercial STMs available on the market. Digital Instruments, Inc. located in Santa Barbara, CA introduced the first commercial STM, the Nanoscope I, in 1987. In a recent Nanoscope IV STM for operation in ambient air, the sample is held in position while a piezoelectric crystal in the form of a cylindrical tube (referred to as PZT tube scanner) scans the sharp metallic probe over the surface in a raster pattern while sensing and outputting the tunneling current to the control station, Fig. 1.3. The digital signal processor (DSP) calculates the desired separation of the tip from the sample by sensing the tunneling current flowing between the sample and the tip. The bias voltage applied between the sample and the tip encourages the tunneling current to flow. The DSP completes the digital feedback loop by outputting the desired voltage to the piezoelectric tube. The STM operates in both the "constant height" and "constant current" modes depending on a parameter selection in the control panel. In the constant current mode, the feedback gains are set high, the tunneling tip closely tracks the sample surface, and the variation in the tip height required to maintain constant tunneling current is measured by the change in the voltage applied to the piezo tube. In the constant height mode, the feedback gains are set low, the tip remains at a nearly constant height as it sweeps over the sample surface, and the tunneling current is imaged.

Physically, the Nanoscope STM consists of three main parts: the head which houses the piezoelectric tube scanner for three dimensional motion of the tip and the preamplifier circuit (FET input amplifier) mounted on top of the head for the tunneling current, the base on which the sample is mounted, and the base support, which supports the base and head [16, 21]. The base accommodates samples up to 10 mm by 20 mm and 10 mm in thickness. Scan sizes available for the STM are $0.7\,\mu\text{m} \times 0.7\,\mu\text{m}$ (for atomic resolution), $12\,\mu\text{m} \times 12\,\mu\text{m}$, $75\,\mu\text{m} \times 75\,\mu\text{m}$ and $125\,\mu\text{m} \times 125\,\mu\text{m}$.

The scanning head controls the three dimensional motion of tip. The removable head consists of a piezo tube scanner, about 12.7 mm in diameter, mounted into an invar shell used to minimize vertical thermal drifts because of good thermal match between the piezo tube and the Invar. The piezo tube has separate electrodes for

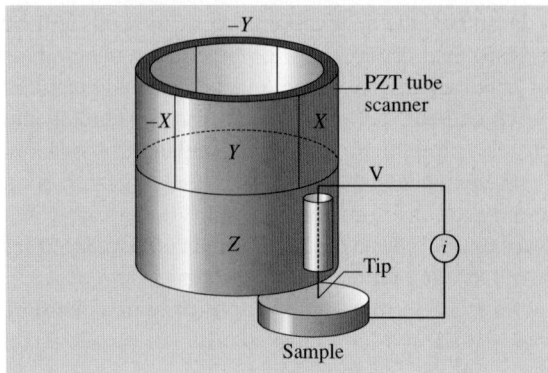

Fig. 1.3. Principle of operation of a commercial STM, a sharp tip attached to a piezoelectric tube scanner is scanned on a sample (from [16])

X, Y and Z which are driven by separate drive circuits. The electrode configuration (Fig. 1.3) provides x and y motions which are perpendicular to each other, minimizes horizontal and vertical coupling, and provides good sensitivity. The vertical motion of the tube is controlled by the Z electrode which is driven by the feedback loop. The x and y scanning motions are each controlled by two electrodes which are driven by voltages of same magnitudes, but opposite signs. These electrodes are called $-Y$, $-X$, $+Y$, and $+X$. Applying complimentary voltages allows a short, stiff tube to provide a good scan range without large voltages. The motion of the tip due to external vibrations is proportional to the square of the ratio of vibration frequency to the resonant frequency of the tube. Therefore, to minimize the tip vibrations, the resonant frequencies of the tube are high about 60 kHz in the vertical direction and about 40 kHz in the horizontal direction. The tip holder is a stainless steel tube with a 300 μm inner diameter for 250 μm diameter tips, mounted in ceramic in order to keep the mass on the end of the tube low. The tip is mounted either on the front edge of the tube (to keep mounting mass low and resonant frequency high) (Fig. 1.3) or the center of the tube for large range scanners, namely 75 and 125 μm (to preserve the symmetry of the scanning.) This commercial STM accepts any tip with a 250 μm diameter shaft. The piezotube requires X–Y calibration which is carried out by imaging an appropriate calibration standard. Cleaved graphite is used for the small-scan length head while two dimensional grids (a gold plated ruling) can be used for longer range heads.

The Invar base holds the sample in position, supports the head, and provides coarse x–y motion for the sample. A spring-steel sample clip with two thumb screws holds the sample in place. An x–y translation stage built into the base allows the sample to be repositioned under the tip. Three precision screws arranged in a triangular pattern support the head and provide coarse and fine adjustment of the tip height. The base support consists of the base support ring and the motor housing. The stepper motor enclosed in the motor housing allows the tip to be engaged and withdrawn from the surface automatically.

Samples to be imaged with STM must be conductive enough to allow a few nanoamperes of current to flow from the bias voltage source to the area to be scanned. In many cases, nonconductive samples can be coated with a thin layer of a conductive material to facilitate imaging. The bias voltage and the tunneling current depend on the sample. Usually they are set at a standard value for engagement and fine tuned to enhance the quality of the image. The scan size depends on the sample and the features of interest. Maximum scan rate of 122 Hz can be used. The maximum scan rate is usually related to the scan size. Scan rate above 10 Hz is used for small scans (typically 60 Hz for atomic-scale imaging with a 0.7 μm scanner). The scan rate should be lowered for large scans, especially if the sample surfaces are rough or contain large steps. Moving the tip quickly along the sample surface at high scan rates with large scan sizes will usually lead to a tip crash. Essentially, the scan rate should be inversely proportional to the scan size (typically 2–4 Hz for 1 μm, 0.5–1 Hz for 12 μm, and 0.2 Hz for 125 μm scan sizes). Scan rate in length/time, is equal to scan length divided by the scan rate in Hz. For example, for 10 μm × 10 μm scan size scanned at 0.5 Hz, the scan rate is 10 μm/s. The 256 × 256 data formats are most commonly used. The lateral resolution at larger scans is approximately equal to scan length divided by 256.

Commercial AFM

A review of early designs of AFMs is presented by *Bhushan* [21]. There are a number of commercial AFMs available on the market. Major manufacturers of AFMs for use in ambient environment are: Digital Instruments Inc., a subsidiary of Veeco Instruments, Inc., Santa Barbara, California; Topometrix Corp., a subsidiary of Veeco Instruments, Inc., Santa Clara, California; and other subsidiaries of Veeco Instruments Inc., Woodbury, New York; Molecular Imaging Corp., Phoenix, Arizona; Quesant Instrument Corp., Agoura Hills, California; Nanoscience Instruments Inc., Phoenix, Arizona; Seiko Instruments, Japan; and Olympus, Japan. AFM/STMs for use in UHV environment are primarily manufactured by Omicron Vakuumphysik GMBH, Taunusstein, Germany.

We describe here two commercial AFMs – small sample and large sample AFMs – for operation in the contact mode, produced by Digital Instruments, Inc., Santa Barbara, CA, with scanning lengths ranging from about 0.7 μm (for atomic resolution) to about 125 μm [28–31]. The original design of these AFMs comes from *Meyer* and *Amer* [32]. Basically the AFM scans the sample in a raster pattern while outputting the cantilever deflection error signal to the control station. The cantilever deflection (or the force) is measured using laser deflection technique, Fig. 1.4. The DSP in the workstation controls the z-position of the piezo based on the cantilever deflection error signal. The AFM operates in both the "constant height" and "constant force" modes. The DSP always adjusts the height of the sample under the tip based on the cantilever deflection error signal, but if the feedback gains are low the piezo remains at a nearly "constant height" and the cantilever deflection data is collected. With the high gains, the piezo height changes to keep the cantilever deflection nearly constant (therefore the force is constant) and the change in piezo height is collected by the system.

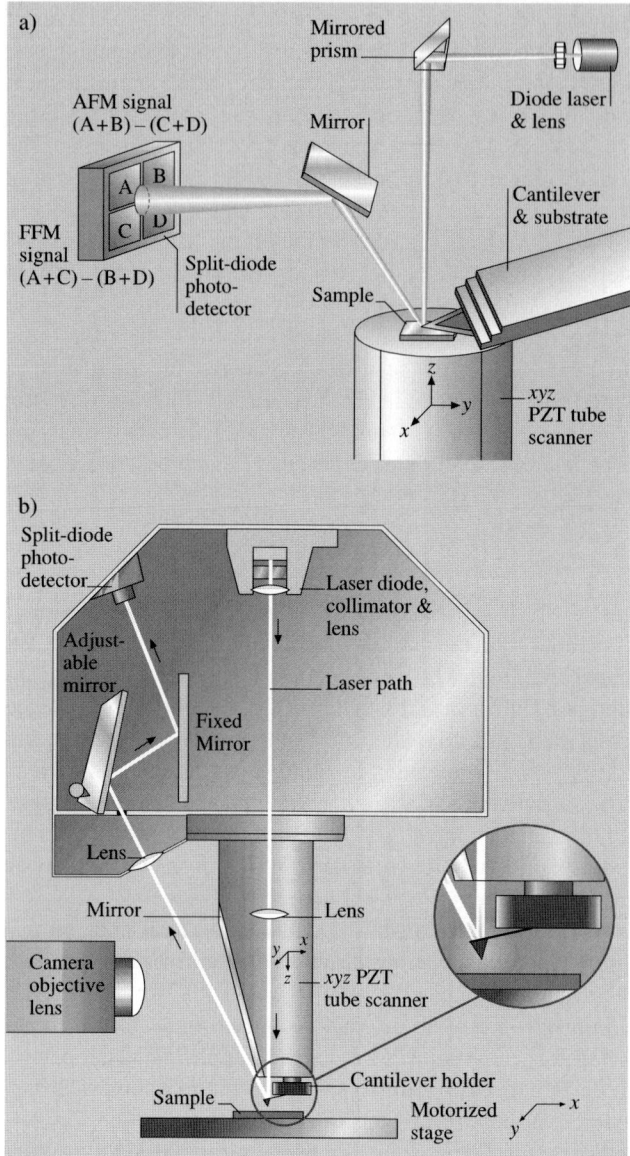

Fig. 1.4. Principles of operation of (**a**) a commercial small sample AFM/FFM, and (**b**) a large sample AFM/FFM (from [16])

To further describe the principle of operation of the commercial small sample AFM shown in Fig. 1.4a, the sample, generally no larger than 10 mm × 10 mm, is mounted on a PZT tube scanner which consists of separate electrodes to scan precisely the sample in the x-y plane in a raster pattern and to move the sample in the vertical (z) direction. A sharp tip at the free end of a flexible cantilever is brought in contact with the sample. Features on the sample surface cause the cantilever to deflect in the vertical and lateral directions as the sample moves under the tip. A laser beam from a diode laser (5 mW max peak output at 670 nm) is directed by a prism onto the back of a cantilever near its free end, tilted downward at about 10° with respect to the horizontal plane. The reflected beam from the vertex of the cantilever is directed through a mirror onto a quad photodetector (split photodetector with four quadrants, commonly called position-sensitive detector or PSD, produced by Silicon Detector Corp., Camarillo, California). The differential signal from the top and bottom photodiodes provides the AFM signal which is a sensitive measure of the cantilever vertical deflection. Topographic features of the sample cause the tip to deflect in the vertical direction as the sample is scanned under the tip. This tip deflection will change the direction of the reflected laser beam, changing the intensity difference between the top and bottom sets of photodetectors (AFM signal). In the AFM operating mode called the height mode, for topographic imaging or for any other operation in which the applied normal force is to be kept a constant, a feedback circuit is used to modulate the voltage applied to the PZT scanner to adjust the height of the PZT, so that the cantilever vertical deflection (given by the intensity difference between the top and bottom detector) will remain constant during scanning. The PZT height variation is thus a direct measure of the surface roughness of the sample.

In a large sample AFM, both force sensors using optical deflection method and scanning unit are mounted on the microscope head, Fig. 1.4b. Because of vibrations added by cantilever movement, lateral resolution of this design is somewhat poorer than the design in Fig. 1.4a in which the sample is scanned instead of cantilever beam. The advantage of the large sample AFM is that large samples can be measured readily.

Most AFMs can be used for topography measurements in the so-called tapping mode (intermittent contact mode), also referred to as dynamic force microscopy. In the tapping mode, during scanning over the surface, the cantilever/tip assembly is sinusoidally vibrated by a piezo mounted above it, and the oscillating tip slightly taps the surface at the resonant frequency of the cantilever (70–400 Hz) with a constant (20–100 nm) oscillating amplitude introduced in the vertical direction with a feedback loop keeping the average normal force constant, Fig. 1.5. The oscillating amplitude is kept large enough so that the tip does not get stuck to the sample because of adhesive attractions. The tapping mode is used in topography measurements to minimize effects of friction and other lateral forces and/or to measure topography of soft surfaces.

Topographic measurements are made at any scanning angle. At a first instance, scanning angle may not appear to be an important parameter. However, the friction force between the tip and the sample will affect the topographic measurements in a parallel scan (scanning along the long axis of the cantilever). Therefore a perpendicular scan may be more desirable. Generally, one picks a scanning angle which

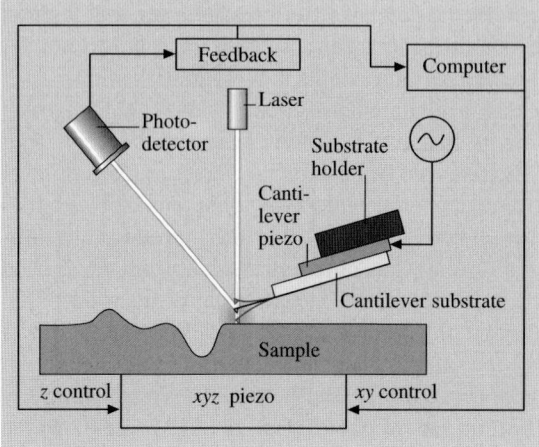

Fig. 1.5. Schematic of tapping mode used for surface roughness measurement (from [16])

gives the same topographic data in both directions; this angle may be slightly different than that for the perpendicular scan.

For measurement of friction force being applied at the tip surface during sliding, left hand and right hand sets of quadrants of the photodetector are used. In the so-called friction mode, the sample is scanned back and forth in a direction orthogonal to the long axis of the cantilever beam. A friction force between the sample and the tip will produce a twisting of the cantilever. As a result, the laser beam will be reflected out of the plane defined by the incident beam and the beam reflected vertically from an untwisted cantilever. This produces an intensity difference of the laser beam received in the left hand and right hand sets of quadrants of the photodetector. The intensity difference between the two sets of detectors (FFM signal) is directly related to the degree of twisting and hence to the magnitude of the friction force. This method provides three-dimensional maps of friction force. One problem associated with this method is that any misalignment between the laser beam and the photodetector axis would introduce error in the measurement. However, by following the procedures developed by *Ruan* and *Bhushan* [30], in which the average FFM signal for the sample scanned in two opposite directions is subtracted from the friction profiles of each of the two scans, the misalignment effect is eliminated. By following the friction force calibration procedures developed by *Ruan* and *Bhushan* [30], voltages corresponding to friction forces can be converted to force unites. The coefficient of friction is obtained from the slope of friction force data measured as a function of normal loads typically ranging from 10 to 150 nN. This approach eliminates any contributions due to the adhesive forces [33]. For calculation of the coefficient of friction based on a single point measurement, friction force should be divided by the sum of applied normal load and intrinsic adhesive force. Furthermore, it should be pointed out that for a single asperity contact, the coefficient of friction is not independent of load.

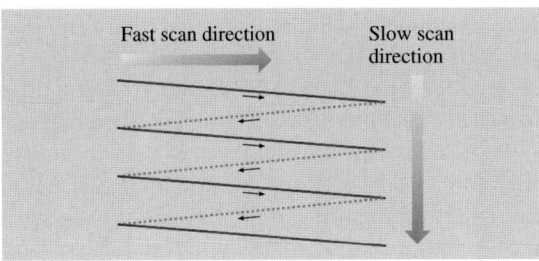

Fig. 1.6. Schematic of triangular pattern trajectory of the AFM tip as the sample is scanned in two dimensions. During imaging, data are recorded only during scans along the solid scan lines (from [16])

The tip is scanned in such a way that its trajectory on the sample forms a triangular pattern, Fig. 1.6. Scanning speeds in the fast and slow scan directions depend on the scan area and scan frequency. Scan sizes ranging from less than 1 nm × 1 nm to 125 μm × 125 μm and scan rates from less than 0.5 to 122 Hz typically can be used. Higher scan rates are used for smaller scan lengths. For example, scan rates in the fast and slow scan directions for an area of 10 μm × 10 μm scanned at 0.5 Hz are 10 μm/s and 20 nm/s, respectively.

1.4.2 Surface Force Apparatus (SFA)

Surface Force Apparatuses (SFAs) are used to study both static and dynamic properties of the molecularly-thin liquid films sandwiched between two molecularly smooth surfaces. The SFAs were originally developed by *Tabor* and *Winterton* [27] and later by *Israelachvili* and *Tabor* [34] to measure van der Waals forces between two mica surfaces as a function of separation in air or vacuum. *Israelachvili* and *Adams* [35] developed a more advanced apparatus to measure normal forces between two surfaces immersed in a liquid so thin that their thickness approaches the dimensions of the liquid molecules themselves. A similar apparatus was also developed by *Klein* [36]. The SFAs, originally used in studies of adhesive and static interfacial forces were first modified by *Chan* and *Horn* [37] and later by *Israelachvili* et al. [38] and *Klein* et al. [39] to measure the dynamic shear (sliding) response of liquids confined between molecularly smooth optically-transparent mica surfaces. Optically transparent surfaces are required because the surface separation is measured using an optical interference technique. *Van Alsten* and *Granick* [40] and *Peachey* et al. [41] developed a new friction attachment which allow for the two surfaces to be sheared past each other at varying sliding speeds or oscillating frequencies while simultaneously measuring both the friction force and normal force between them. *Israelachvili* [42] and *Luengo* et al. [43] also presented modified SFA designs for dynamic measurements including friction at oscillating frequencies. Because the mica surfaces are molecularly smooth, the actual area of contact is well defined and measurable, and asperity deformation do not complicate the analysis. During sliding experiments, the area of parallel surfaces is very large compared to the thickness of the sheared film and this provides an ideal condition for studying shear behavior because it permits one to study molecularly-thin liquid films whose thickness is well defined to the resolution of an angstrom. Molecularly thin liquid films cease to behave as a structural continuum with properties different from that of the bulk material [40, 44–47].

Tonck et al. [48] and *Georges* et al. [49] developed a SFA used to measure the static and dynamic forces (in the normal direction) between a smooth fused borosilicate glass against a smooth and flat silicon wafer. They used capacitance technique to measure surface separation; therefore, use of optically-transparent surfaces was not required. Among others, metallic surfaces can be used at the interface. *Georges* et al. [50] modified the original SFA so that a sphere can be moved towards and away from a plane and can be sheared at constant separation from the plane, for interfacial friction studies.

For a detailed review of various types of SFAs, see *Israelachvili* [42, 51], *Horn* [52], and *Homola* [53]. SFAs based on their design are commercially available from SurForce Corporation, Santa Barbara, California.

Israelachvili's and Granick's Design

Following review is primarily based on the papers by *Israelachvili* [42] and *Homola* [53]. Israelachvili et al.'s design later followed by Granick et al. for oscillating shear studies, is most commonly used by researchers around the world.

Classical SFA The classical apparatus developed for measuring equilibrium or static intersurface forces in liquids and vapors by *Israelachvili* and *Adams* [35], consists of a small, air-tight stainless steel chamber in which two molecularly smooth curved mica surfaces can be translated towards or away from each other, see Fig. 1.7. The distance between the two surfaces can also be independently controlled to

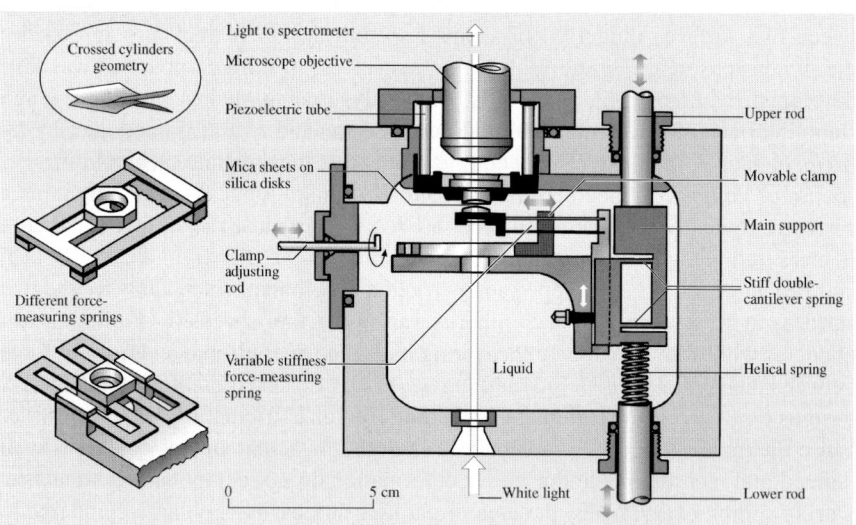

Fig. 1.7. Schematic of the surface force apparatus that employs the cross cylinder geometry [35, 42]

within ±0.1 nm and the force sensitivity is about 10 nN. The technique utilizes two molecularly smooth mica sheets, each about 2 μm thick, coated with a semi reflecting 50–60 nm layer of pure silver, glued to rigid cylindrical silica disks of radius about 10 mm (silvered side down) mounted facing each other with their axes mutually at right angles (crossed cylinder position), which is geometrically equivalent to a sphere contacting a flat surface. The adhesive glue which is used to affix the mica to the support is sufficiently compliant, so the mica will flatten under the action of adhesive forces or applied load to produce a contact zone in which the surfaces are locally parallel and planar. Outside of this contact zone the separation between surfaces increases and the liquid, which is effectively in a bulk state, makes a negligible contribution to the overall response. The lower surface is supported on a cantilever spring which is used to push the two surfaces together with a known load. When the surfaces are forced into contact, they flatten elastically so that the contact zone is circular for duration of the static or sliding interactions. The surface separation is measured using optical interference fringes of equal chromatic order (FECO) which enables the area of molecular contact and the surface separation to be measured to within 0.1 nm. For measurements, white light is passed vertically up through the two mica surfaces and the emerging beam is then focused onto the slit of a grating spectrometer. From the positions and shapes of the colored FECO fringes in the spectrogram, the distance between the two surfaces and the exact shape of the two surfaces can be measured (as illustrated in Fig. 1.8), as can the refractive index of the liquid (or material) between them. In particular, this allows for reasonably accurate determinations of the quantity of material deposited or adsorbed on the surfaces and the area of contact between two molecularly smooth surfaces. Any changes may be readily observed in both static and sliding conditions in real time (applicable to the design shown in Fig. 1.8) by monitoring the changing shapes of these fringes.

The distance between the two surfaces is controlled by use of a three-stage mechanism of increasing sensitivity: coarse control (upper rod) allows positioning of within about 1 μm, the medium control (lower rod, which depresses the helical

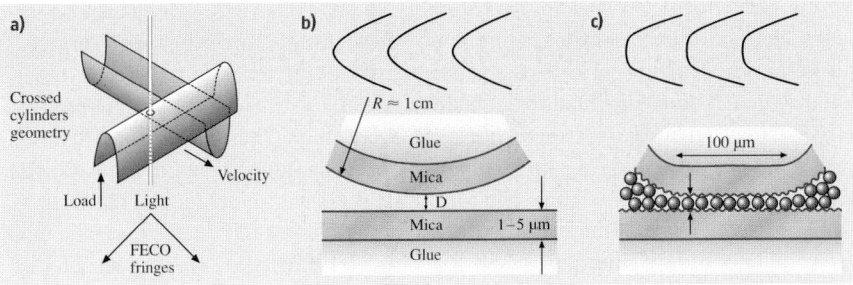

Fig. 1.8. (a) Cross cylinder configuration of mica sheet, showing formation of contact area. Schematic of the fringes of equal chromatic order (FECO) observed when two mica surfaces are (b) separated by distance D and (c) are flattened with a monolayer of liquid between them [54]

spring and which in turn, bends the much stiffer double-cantilever spring by 1/1000 of this amount) allows positioning to about 1 nm, and the piezoelectric crystal tube – which expands or controls vertically by about 0.6 nm/V applied axially across the cylindrical wall – is used for final positioning to 0.1 nm.

The normal force is measured by expanding or contracting the piezoelectric crystal by a known amount and then measuring optically how much the two surfaces have actually moved; any difference in the two values when multiplied by the stiffness of the force measuring spring gives the force difference between the initial and final positions. In this way both repulsive and attractive forces can be measured with a sensitivity of about 10 nN. The force measuring springs can be either single-cantilever or double-cantilever fixed-stiffness springs (as shown in Fig. 1.7), or the spring stiffness can be varied during an experiment (by up to a factor of 1000) by shifting the position of the dovetailed clamp using the adjusting rod. Other spring attachments, two of which are shown at the top of the figure, can replace the variable stiffness spring attachment (top right: nontilting nonshearing spring of fixed stiffness). Each of these springs are interchangeable and can be attached to the main support, allowing for greater versatility in measuring strong or weak and attractive or repulsive forces. Once the force F as a function of distance D is known for the two surfaces of radius R, the force between any other curved surfaces simply scales by R. Furthermore, the adhesion energy (or surface or interfacial free energy) E per unit area between two flat surfaces is simply related to F by the so-called Derjaguin approximation [51] $E = F/2\pi R$. We note that SFA is one of the few techniques available for directly measuring equilibrium force-laws (i.e., force versus distance at constant chemical potential of the surrounding solvent medium) [42]. The SFA allows for both weak or strong and attractive or repulsive forces.

Mostly the molecularly smooth surface of mica is used in these measurements [55], however, silica [56] and sapphire [57] have also been used. It is also possible to deposit or coat each mica surface with metal films [58, 59], carbon and metal oxides [60], adsorbed polymer layers [61], surfactant monolayers and bilayers [51, 58, 62, 63]. The range of liquids and vapors that can be used is almost endless.

Sliding Attachments for Tribological Studies So far we have described a measurement technique which allows measurements of the normal forces between surfaces, that is, those occurring when two surfaces approach or separate from each other. However, in tribological situations, it is the transverse or shear forces that are of primary interest when two surfaces slide past each other. There are essentially two approaches used in studying the shear response of confined liquid films. In the first approach (constant velocity friction or steady-shear attachment), the friction is measured when one of the surfaces is traversed at a constant speed over a distance of several hundreds of microns [38, 39, 45, 54, 60, 64, 65]. The second approach (oscillatory shear attachment) relies on the measurement of viscous dissipation and elasticity of confined liquids by using periodic sinusoidal oscillations over a range of amplitudes and frequencies [40, 41, 44, 66, 67].

For the constant velocity friction (steady-shear) experiments, the surface force apparatus was outfitted with a lateral sliding mechanism [38, 42, 46, 54, 64, 65] allow-

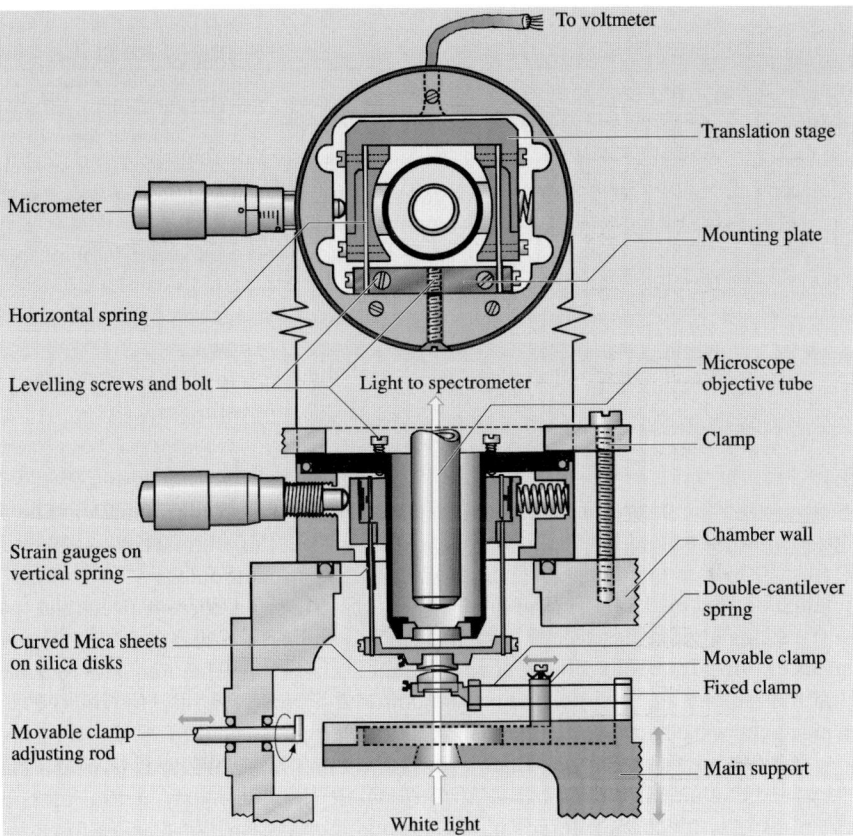

Fig. 1.9. Schematic of shear force apparatus. Lateral motion is initiated by a variable speed motor-driven micrometer screw that presses against the translation stage which is connected through two horizontal double-cantilever strip springs to the rigid mounting plate [38, 42]

ing measurements of both normal and shearing forces (Fig. 1.9). The piezoelectric crystal tube mount supporting the upper silica disk of the basic apparatus shown in Fig. 1.7, is replaced. Lateral motion is initiated by a variable speed motor-driven micrometer screw that presses against the translation stage, which is connected via two horizontal double-cantilever strip springs to the rigid mounting plate. The translation stage also supports two vertical double-cantilever springs (Fig. 1.10) that at their lower end are connected to a steel plate supporting the upper silica disk. One of the vertical springs acts as a frictional force detector by having four resistance strain gages attached to it, forming the four arms of a Wheatstone bridge and electrically connected to a chart recorder. Thus, by rotating the micrometer, the translation stage deflects, causing the upper surface to move horizontally and linearly at a steady rate. If the upper mica surface experiences a transverse frictional or viscous shearing force, this will cause the vertical springs to deflect, and this deflection can be

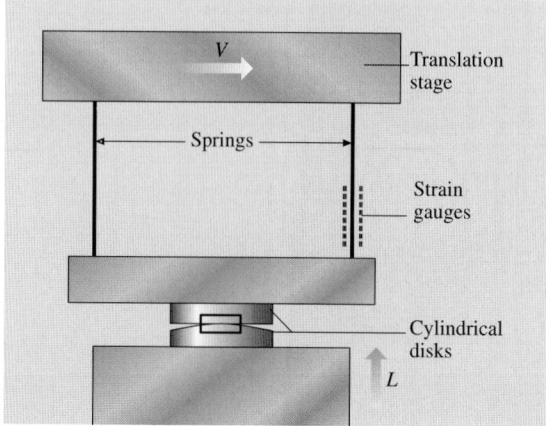

Fig. 1.10. Schematic of the sliding attachment. The translation stage also supports two vertical double-cantilever springs, which at their lower end are connected to a steel plate supporting the upper silica disk [46]

measured by the strain gages. The main support, force-measuring double-cantilever spring, movable clamp, white light, etc., are all parts of the original basic apparatus (Fig. 1.7), whose functions are to control the surface separation, vary the externally applied normal load, and measure the separation and normal force between the two surfaces, as already described. Note that during sliding, the distance between the surfaces, their true molecular contact area, their elastic deformation, and their lateral motion can all be simultaneously monitored by recording the moving FECO fringe pattern using a video camera and recording it on a tape [46].

The two surfaces can be sheared past each other at sliding speeds which can be varied continuously from 0.1 to 20 µm/s while simultaneously measuring both the transverse (frictional) force and the normal (compressive or tensile) force between them. The lateral distances traversed are on the order of a several hundreds of micrometers which correspond to several diameters of the contact zone.

With an oscillatory shear attachment, developed by *Granick* et al., viscous dissipation and elasticity and dynamic viscosity of confined liquids by applying periodic sinusoidal oscillations of one surface with respect to the other can be studied [40, 41, 44, 66, 67]. This attachment allows for the two surfaces to be sheared past each other at varying sliding speeds or oscillating frequencies while simultaneously measuring both the transverse (friction or shear) force and the normal load between them. The externally applied load can be varied continuously, and both positive and negative loads can be applied. Finally the distance between the surfaces, their true molecular contact area, their elastic (or viscoelastic) deformation and their lateral motion can all be simultaneously by recording the moving interference fringe pattern using a video camera-recorder system.

To produce shear while maintaining constant film thickness or constant separation of the surfaces, the top mica surface is suspended from the upper portion of the apparatus by two piezoelectric bimorphs. A schematic description of the surface force apparatus with the installed shearing device is shown in Fig. 1.11 [40, 41, 44, 66, 67]. *Israelachvili* [42] and *Luengo* et al. [43] have also

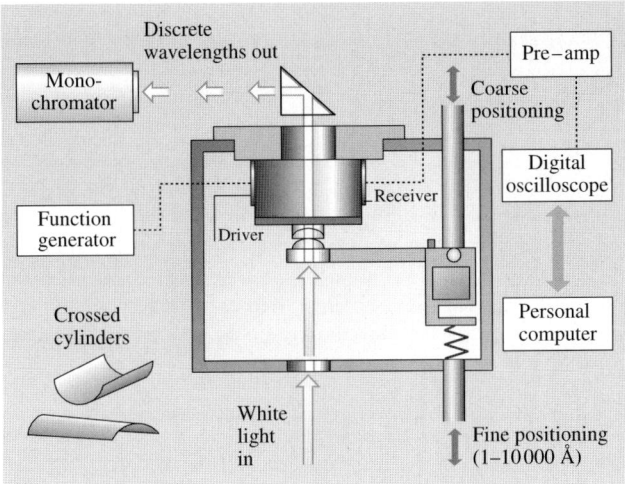

Fig. 1.11. Schematic of the oscillatory shearing apparatus [40]

presented similar designs. The lower mica surface, as in the steady-shear sliding attachment, is stationary and sits at the tip of a double cantilever spring attached at the other end to a stiff support. The externally applied load can be varied continuously by displacing the lower surface vertically. An AC voltage difference applied by a signal generator (driver) across one of the bimorphs tends to bend it in oscillatory fashion while the frictional force resists that motion. Any resistance to sliding induces an output voltage across the other bimorphs (receiver) which can be easily measured by a digital oscilloscope. The sensitivity in measuring force is on the order of a few μN and the amplitudes of measured lateral displacement can range from a few nm to 10 μm. The design is flexible and allows to induce time-varying stresses with different characteristic wave shapes simply by changing the wave form of the input electrical signal. For example, when measuring the apparent viscosity, a sine wave input is convenient to apply. Figure 1.12a shows an example of the raw data, obtained with a hexadecane film at a moderate pressure, when a sine wave was applied to one of the bimorphs [66]. By comparing the calibration curve with the response curve, which was attenuated in amplitude and lagged in phase, an apparent dynamic viscosity can be inferred. On the other hand, a triangular waveform is more suitable when studying the yield stress behavior of solid-like films as of Fig. 1.12b. The triangular waveform, showing a linear increase and decrease of the applied force with time, is proportional to the driving force acting on the upper surface. The response waveform, which represents a resistance of the interface to shear, remains very small indicating that the surfaces are in a stationary contact with respect to each other until the applied stress reaches a yield point. At the yield point the slope of the response curve increases dramatically, indicating the onset of sliding.

Homola [53] compared the two approaches – steady shear attachment and oscillating shear attachment. In experiments conducted by Israelachvili and his co-

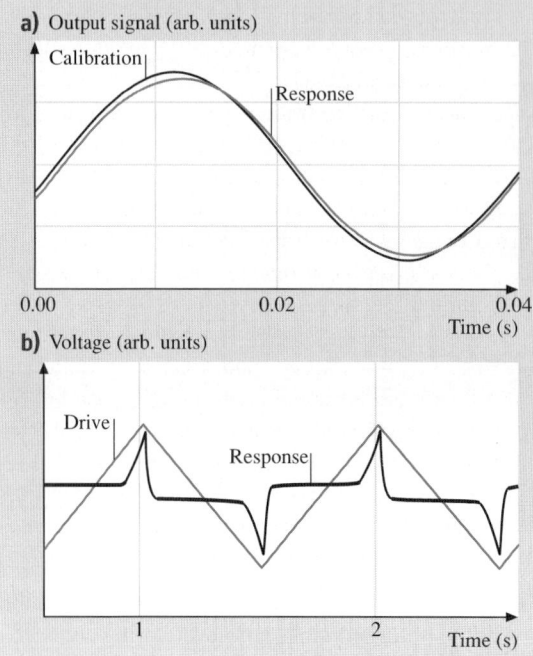

Fig. 1.12. (a) Two output signals induced by an applied sine wave (not shown) are displaced. The "calibration" waveform is obtained with the mica sheets completely separated. The response waveform is obtained with a thin liquid film between the sheets, which causes it to lag the calibration waveform, (b) the oscilloscope trace of the drive and response voltages used to determine critical shear stress. The drive waveform shows voltage proportional to induced stress on the sheared film and the response waveform shows voltage proportional to resulting velocity. Spikes in the response curve correspond to the stick–slip event [66]

workers, the steady-shear attachment was employed to focus on the dynamic frictional behavior of the film after a sufficiently high shear stress was applied to exceed the yield stress and to produce sliding at a constant velocity. In these measurements, the film was subjected to a constant shearing force for a time sufficiently long to allow them to reach a dynamic equilibrium, i.e., the molecules, within the film, had enough time to order and align with respect to the surface, both normally and tangentially. Under these conditions, dynamic friction was observed to be "quantized" according to the number of liquid layers between the solid surfaces and independent of the shear rate [38]. Clearly, in this approach, the molecular ordering is optimized by a steady shear which imposes a preferred orientation on the molecules in the direction of shear.

The above mode of sliding is particularly important when the sheared film is made of a long chain lubricant molecules requiring a significantly long sliding time to order and align and even a longer time to relax (disorder) when sliding stops. This suggests that a steady-state friction is realized only when the duration of sliding exceeds the time required for an ensemble of the molecules to fully order in a specific shear field. It also suggests, that static friction should depend critically on the sliding time and the extend of the shear induced ordering [53].

In contrast, the oscillatory shear method, which utilizes periodic sinusoidal oscillations over a range of amplitudes and frequencies, addresses a response of the system to rapidly varying strain rates and directions of sliding. Under these conditions, the molecules, especially those exhibiting a solid-like behavior, cannot respond sufficiently fast to stress and are unable to order fully during duration of a single pass, i.e., their dynamic and static behavior reflects and oscillatory shear induced ordering which might or might not represent an equilibrium dynamic state. Thus, the response of the sheared film will depend critically on the conditions of shearing, i.e., the strain, the pressure, and the sliding conditions (amplitude and frequency of oscillations) which in turn will determine a degree of molecular ordering. This may explain the fact that the layer structure and "quantization" of the dynamic and static friction was not observed in these experiments in contrast to results obtained when velocity was kept constant. Intuitively, this behavior is expected considering that the shear-ordering tendency of the system is frequently disturbed by a shearing force of varying magnitude and direction. Nonetheless, the technique is capable of providing an invaluable insight into the shear behavior of molecularly thin films subjected to non-linear stresses as it is frequently encountered in practical applications. This is especially true under conditions of boundary lubrication where interacting surface asperities will be subjected to periodic stresses of varying magnitudes and frequencies [53].

Georges et al.'s Design

The SFA developed by *Tonck* et al. [48] and *Georges* et al. [49] to measure static and dynamic forces in the normal direction, between surfaces in close proximity, is shown in Fig. 1.13. In their apparatus, a drop of liquid is introduced between a macroscopic spherical body and a plane. The sphere is moved towards and away from a plane using the expansion and the vibration of a piezoelectric crystal. Piezoelectric crystal is vibrated at low amplitude around an average separation for dynamic measurements to provide dynamic function of the interface. The plane specimen is supported by a double-cantilever spring. Capacitance sensor C_1 measures the elastic deformation of the cantilever and thus the force transmitted through the liquid to the plane. Second capacitance sensor C_2 is designed to measure the relative displacement between the supports of the two solids. The reference displacement signal is the sum of two signals: first, a ramp provides a constant normal speed from 50 to 0.01 nm/s, and, second, the piezoelectric crystal is designed to provide a small sinusoidal motion, in order to determine the dynamic behavior of sphere-plane interactions. A third capacitance sensor C measures the electrical capacitance between the sphere and the plane. In all cases, the capacitance is determined by incorporating the signal of an oscillator in the inductive–capacitance (L–C) resonant input stage of an oscillator to give a signal-dependent frequency in the range of 5–12 MHz. The resulting fluctuations in oscillation frequency are detected using a low noise fre-

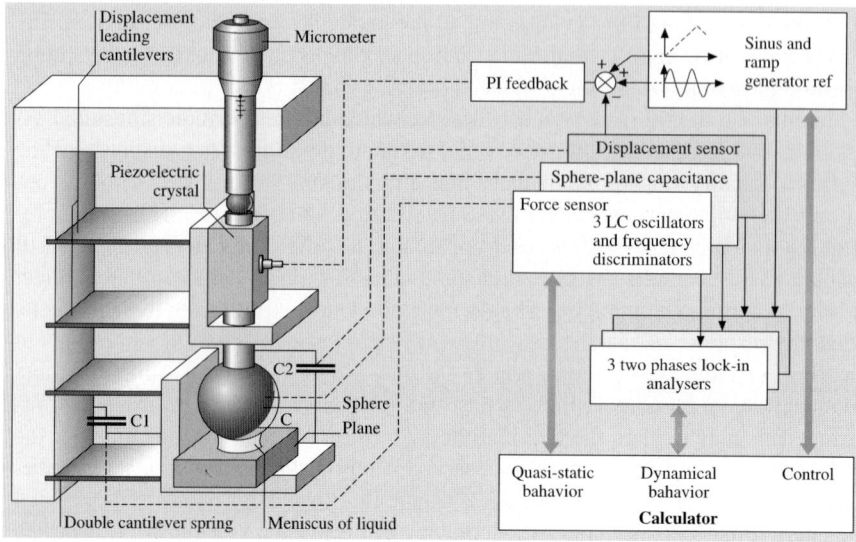

Fig. 1.13. Schematic of the surface force apparatus that employs a sphere–plane arrangement [49]

quency discriminator. Simultaneous measurements of sphere-plane displacement, surface force, and the damping of the interface allows an analysis of all regimes of the interface [49]. *Loubet* et al. [68] used SFA in the crossed-cylinder geometry using two freshly-cleaved mica sheets similar to the manner used by Israelachvili and coworkers.

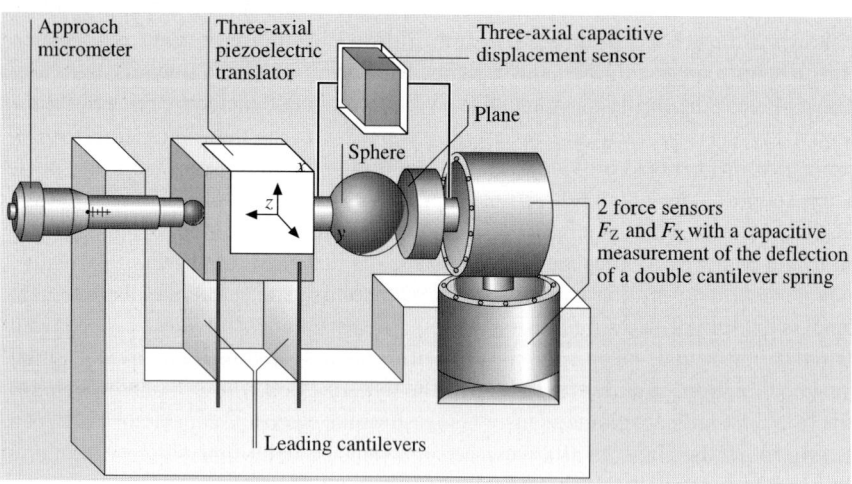

Fig. 1.14. Schematic of shear force apparatus [50]

Georges et al. [50] modified their original SFA to measure friction forces. In this apparatus, in addition to having sphere move normal to the plane, sphere can be sheared at constant separation from the plane. The shear force apparatus is shown in Fig. 1.14. Three piezoelectric elements controlled by three capacitance sensors permit accurate motion control and force measurement along three orthogonal axes with displacement sensitivity of 10^{-3} nm and force sensitivity of 10^{-8} N. Adhesion and normal deformation experiments are conducted in the normal approach (z-axis). Friction experiments are conducted by introducing displacement in the X-direction at a constant normal force. In one of the experiment, *Georges* et al. [50] used 2.95 mm diameter sphere made of cobalt-coated fused borosilicate glass and a silicon wafer for the plane.

1.4.3 Vibration Isolation

STM, AFM and SFA should be isolated from sources of vibration in the acoustic and sub-acoustic frequencies especially for atomic-scale measurements. Vibration isolation is generally provided by placing the instrument on a vibration isolation air table. For further isolation, the instrument should be placed on a pad of soft silicone rubber. A cheaper alternative consists of a large mass of 100 N or more, suspended from elastic "bungee" cords. The mass should stretch the cords at least 0.3 m, but not so much that the cords reach their elastic limit. The instrument should be placed on the large mass. The system, including the microscope, should have a natural frequency of about 1 Hz or less both vertically and horizontally. Test this by gently pushing on the mass and measure the rate at which its swings or bounces.

1.5 Magnetic Storage Devices and MEMS/NEMS

1.5.1 Magnetic Storage Devices

Magnetic storage devices used for storage and retrieval are tape, flexible (floppy) disk and rigid disk drives. These devices are used for audio, video and data storage applications. Magnetic storage industry is some $ 60 billion a year industry with $ 20 billion for audio and video recording (almost all tape drives/media) and $ 40 billion for data storage. In the data storage industry, magnetic rigid disk drives/media, tape drives/media, flexible disk drives/media, and optical disk drive/media account for about $ 25 B, $ 6 B, $ 3 B, and $ 6 B, respectively. Magnetic recording and playback involves the relative motion between a magnetic medium (tape or disk) against a read-write magnetic head. Heads are designed so that they develop a (load-carrying) hydrodynamic air film under steady operating conditions to minimize head–medium contact. However, physical contact between the medium and head occurs during starts and stops, referred to as contact-start-stops (CSS) technology [13, 14, 69]. In the modern magnetic storage devices, the flying heights (head-to-medium separation) are on the order of 5 to 20 nm and roughnesses of head and medium surfaces are on the order of 1–2 nm RMS. The need for ever-increasing recording densities requires that surfaces be as smooth as possible and the flying heights be as low as possible. High stiction (static friction) and wear are the limiting technology to future of this industry. Head load/unload (L/UL) technology has recently been used as an alternative to CSS technology in rigid disk drives that eliminates stiction and wear failure mode associated with CSS. Several contact or near contact recording devices are at various stages of development. High stiction and wear are the major impediments to the commercialization of the contact recording.

Magnetic media fall into two categories: particulate media, where magnetic particles (γ-Fe_2O_3, Co-γFe_2O_3, CrO_2, Fe or metal (MP), or barium ferrite) are dispersed in a polymeric matrix and coated onto a polymeric substrate for flexible media (tape and flexible disks); thin-film media, where continuous films of magnetic materials are deposited by vacuum deposition techniques onto a polymer substrate for flexible media or onto a rigid substrate (typically aluminium and more recently glass or glass ceramic) for rigid disks. The most commonly used thin magnetic films for tapes are evaporated Co–Ni (82–18 at.%) or Co–O dual layer. Typical magnetic films for rigid disks are metal films of cobalt-based alloys (such as sputtered Co–Pt–Ni, Co–Ni, Co–Pt–Cr, Co–Cr and Co–NiCr). For high recording densities, trends have been to use thin-film media. Magnetic heads used to date are either conventional thin-film inductive, magnetoresistive (MR) and giant MR (GMR) heads. The air-bearing surfaces (ABS) of tape heads are generally cylindrical in shape. For dual-sided flexible-disk heads, two heads are either spherically contoured and slightly offset (to reduce normal pressure) or are flat and loaded against each other. The rigid-disk heads are supported by a leaf spring (flexure) suspension. The ABS of heads are almost made of Mn–Zn ferrite, Ni–Zn ferrite, Al_2O_3–TiC and calcium titanate. The ABS of some conventional heads are made of plasma sprayed coatings of hard materials such as Al_2O_3–TiO_2 and ZrO_2 [13, 14, 69].

Figure 1.15 shows the schematic illustrating the tape path with details of tape guides in a data-processing linear tape drive (IBM LTO Gen1) which uses a rectangular tape cartridge. Figure 1.16a shows the sectional views of particulate and thin-film magnetic tapes. Almost exclusively, the base film is made of semicrystalline

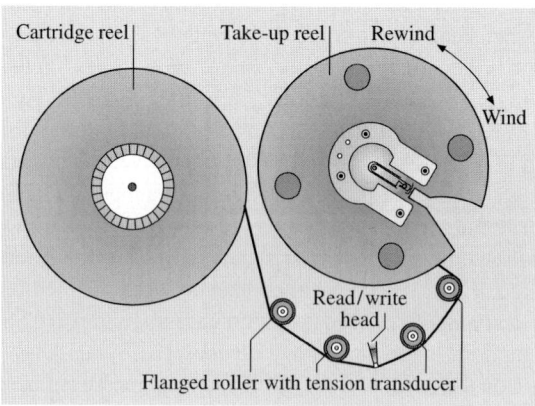

Fig. 1.15. Schematic of tape path in an IBM Linear Tape Open (LTO) tape drive

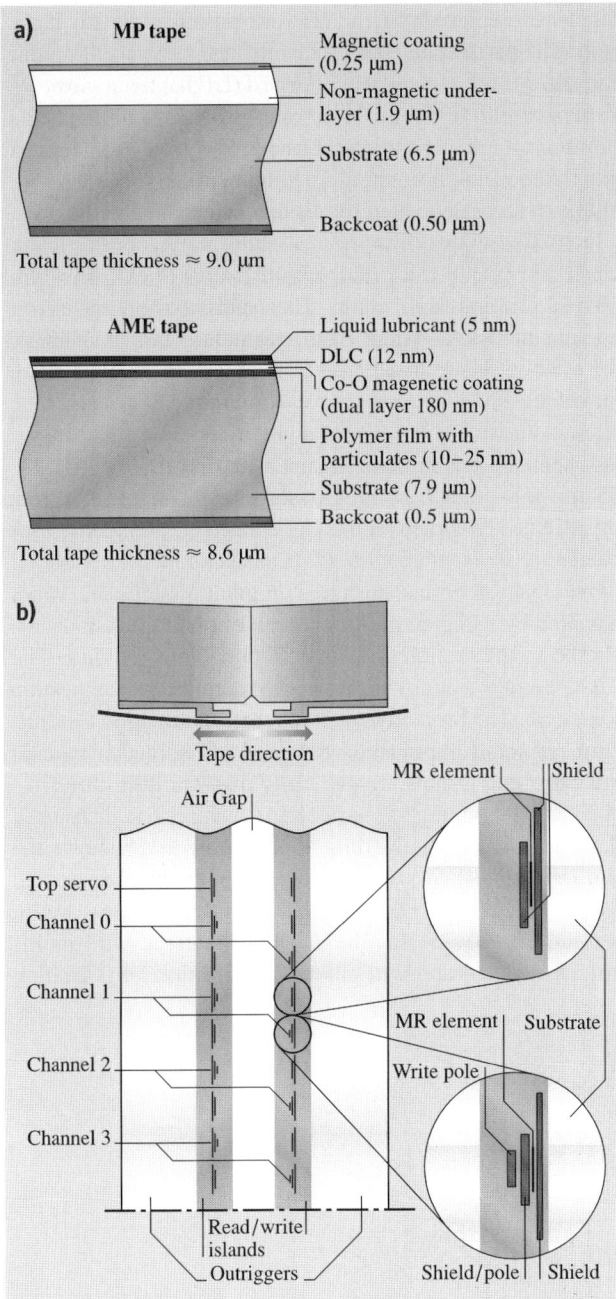

Fig. 1.16. (a) Sectional views of particulate and thin-film magnetic tapes, and (b) schematic of a magnetic thin-film read/write head for an IBM LTO Gen 1 tape drive

biaxially-oriented poly (ethylene terephthalate) (or PET) or poly (ethylene 2,6 naphthalate) (or PEN) or Aramid. The particulate coating formulation consists of binder (typically polyester polyurethane), submicron accicular shaped magnetic particles (about 50 nm long with an aspect ratio of about 5), submicron head cleaning agents (typically alumina) and lubricant (typically fatty acid ester). For protection against wear and corrosion and low friction/stiction, the thin-film tape is first coated with a diamondlike carbon (DLC) overcoat deposited by plasma enhanced chemical vapor deposition, topically lubricated with primarily a perfluoropolyether lubricant. Figure 1.16b shows the schematic of an 8-track (along with 2 servo tracks) thin-film read-write head with MR read and inductive write. The head steps up and down to provide 384 total data tracks across the width of the tape. The ABS is made of Al_2O_3–TiC. A tape tension of about 1 N over a 12.7 mm wide tape (normal pressure ≈ 14 kPa) is used during use. The RMS roughnesses of ABS of the heads and tape surfaces typically are 1–1.5 nm and 5–8 nm, respectively.

Figure 1.17 shows the schematic of a data processing rigid disk drive with 21.6, 27.4, 48, 63.5, 75, and 95 mm form factor. Nonremovable stack of multiple disks mounted on a ball bearing or hydrodynamic spindle, are rotated by an electric motor at constant angular speed ranging from about 5000 to in excess of 15,000 RPM, dependent upon the disk size. Head slider-suspension assembly (allowing one slider for each disk surface) is actuated by a stepper motor or a voice coil motor using a rotary actuator. Figure 1.18a shows the sectional views of a thin-film rigid disk. The substrate for rigid disks is generally a non heat-treatable aluminium–magnesium alloy 5086, glass or glass ceramic. The protective overcoat commonly used for thin-film disks is sputtered DLC, topically lubricated with perfluoropolyether type of lubricants. Lubricants with polar-end groups are generally used for thin-film disks in order to provide partial chemical bonding to the overcoat surface. The disks used for CSS technology are laser textured in the landing zone. Figure 1.18b shows the schematic of two thin-film head picosliders with a step at the leading edge, and GMR read and inductive write. "Pico" refers to the small sizes of 1.25 mm × 1 mm. These sliders use Al_2O_3–TiC (70–30 wt%) as the substrate material with multi-layered thin-film head structure coated and with about 3.5 nm thick DLC coating

Fig. 1.17. Schematic of a data-processing magnetic rigid disk drive

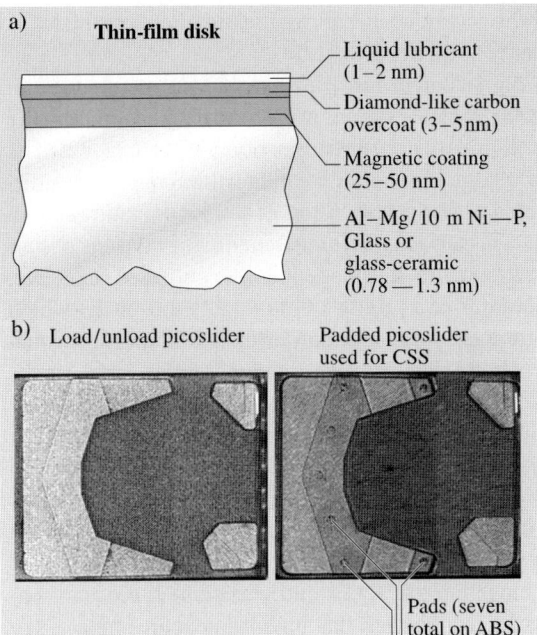

Fig. 1.18. (a) Sectional views of a thin-film magnetic rigid disk, and (b) schematic of two picosliders – load/unload picoslider and padded picoslider used for CSS

to prevent the thin film structure from electrostatic discharge. The seven pads on the padded slider are made of DLC and are about 40 μm in diameter and 50 nm in height. A normal load of about 3 g is applied during use.

1.5.2 MEMS/NEMS

The advances in silicon photolithographic process technology led to the development of MEMS in the mid-1980s [16]. More recently, lithographic and nonlithographic processes have been developed to process nonsilicon (plastics or ceramics) materials. MEMS for mechanical applications include acceleration, pressure, flow, and gas sensors, linear and rotary actuators, and other microstructures of microcomponents such as electric motors, gear trains, gas turbine engines, nozzles, fluid pumps, fluid valves, switches, grippers, and tweezers. MEMS for chemical applications include chemical sensors and various analytical instruments. Microoptoelectromechanical systems (or MOEMS) include micromirror arrays and fiber optic connectors. Radio frequency MEMS or RF-MEMS include inductors, capacitors, and antennas. High-aspect ratio MEMS (HARMEMS) have also been introduced. BioMEMS include biofluidic chips (microfluidic chips or bioflips or simple biochips) for chemical and biochemical analyses (biosensors in medical diagnostics, e.g., DNA, RNA, proteins, cells, blood pressure and assays, and toxin identification), and implantable drug delivery. Killer applications include capacitive-type silicon accelerometers for automotive sensory applications and digital micromirror devices for projection displays. Any component requiring relative motions needs to be optimized for stiction and wear [16, 20, 22, 69, 70].

Figure 1.19 also shows two digital micromirror device (DMD) pixels used in digital light processing (DLP) technology for digital projection displays in portable and home theater projectors as well as table top and projection TVs [16, 71, 72]. The entire array (chip set) consists of a large number of rotatable aluminium micromirrors (digital light switches) which are fabricated on top of a CMOS static random access memory integrated circuit. The surface micromachined array consists of half of a million to more than two million of these independently controlled reflective, micromirrors (mirror size on the order of 14 μm × 14 μm and 15 μm pitch) which flip backward and forward at a frequency of on the order of 5000 times a second. For the binary operation, micromirror/yoke structure mounted on torsional hinges is rotated ±10° (with respect to the horizontal plane) as a result of electrostatic attraction between the micromirror structure and the underlying memory cell, and is limited by a mechanical stop. Contact between cantilevered spring tips at the end of the yoke (four present on each yoke) with the underlying stationary landing sites is required for true digital (binary) operation. Stiction and wear during a contact between aluminium alloy spring tips and landing sites, hinge memory (metal creep at high operating temperatures), hinge fatigue, shock and vibration failure, and sensitivity to particles in the chip package and operating environment are some of the important issues affecting the reliable operation of a micromirror device. Perfluorodecanoic acid (PFDA) self-assembled monolayers are used on the tip and landing sites to reduce stiction and wear. The spring tip is used in order to use the spring stored energy to pop up the tip during pull-off. A lifetime estimate of over one hundred thousand operating hours with no degradation in image quality is the norm.

NEMS are produced by nanomachining in a typical top-down approach (from large to small) and bottom-up approach (from small to large) largely relying on nanochemistry [16]. The top-down approach relies on fabrication methods including advanced integrated-circuit (IC) lithographic methods – electron-beam lithography, and STM writing by removing material atom by atom. The bottom-up approach

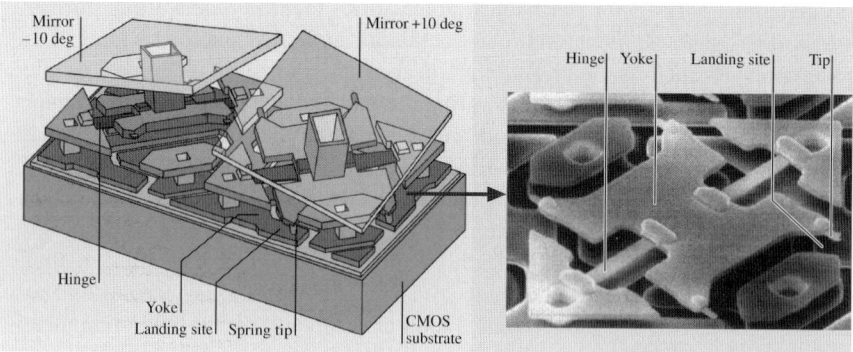

Fig. 1.19. Digital micromirror devices for projection displays (from [16])

includes chemical synthesis, the spontaneous "self-assembly" of molecular clusters (molecular self-assembly) from simple reagents in solution, or biological molecules (e.g., DNA) as building blocks to produce three dimensional nanostructures, quantum dots (nanocrystals) of arbitrary diameter (about $10-10^5$ atoms), molecular beam epitaxy (MBE) and organometallic vapor phase epitaxy (OMVPE) to create specialized crystals one atomic or molecular layer at a time, and manipulation of individual atoms by an atomic force microscope or atom optics. The self-assembly must be encoded, that is, one must be able to precisely assemble one object next to another to form a designed pattern. A variety of nonequilibrium plasma chemistry techniques are also used to produce layered nanocomposites, nanotubes, and nanoparticles. NEMS field, in addition to fabrication of nanosystems, has provided impetus to development of experimental and computation tools.

Examples of NEMS include nanocomponents, nanodevices, nanosystems, and nanomaterials such as microcantilever with integrated sharp nanotips for STM and AFM, AFM array (Millipede) for data storage, AFM tips for nanolithography, dip-pen nanolithography for printing molecules, biological (DNA) motors, molecular gears, molecularly-thick films (e.g., in giant magnetioresistive or GMR heads and magnetic media), nanoparticles (e.g., nanomagnetic particles in magnetic media), nanowires, carbon nanotubes, quantum wires (QWRs), quantum boxes (QBs), and quantum transistors [16]. BIONEMS include nanobiosensors – microarray of silicon nanowires, roughly few nm in size, to selectively bind and detect even a single biological molecule such as DNA or protein by using nanoelectronics to detect the slight electrical charge caused by such binding, or a microarray of carbon nanotubes to electrically detect glucose, implantable drug-delivery devices – e.g., micro/nanoparticles with drug molecules encapsulated in functionized shells for a site-specific targeting applications, and a silicon capsule with a nanoporous membrane filled with drugs for long term delivery, nanodevices for sequencing single molecules of DNA in the Human Genome Project, cellular growth using carbon nanotubes for spinal cord repair, nanotubes for nanostructured materials for various applications such as spinal fusion devices, organ growth, and growth of artificial tissues using nanofibers.

Figure 1.20 shows AFM based nanoscale data storage system for ultrahigh density magnetic recording which experiences tribological problems [73]. The system uses arrays of several thousand silicon microcantilevers ("Millipede") for thermomechanical recording and playback on an about 40-nm thick polymer (PMMA) medium with a harder Si substrate. The cantilevers are integrated with integrated tip heaters with tips of nanoscale dimensions. Thermomechanical recording is a combination of applying a local force to the polymer layer and softening it by local heating. The tip heated to about 400 °C is brought in contact with the polymer for recording. Imaging and reading are done using the heater cantilever, originally used for recording, as a thermal readback sensor by exploiting its temperature-dependent resistance. The principle of thermal sensing is based on the fact that the thermal conductivity between the heater and the storage substrate changes according to the spacing between them. When the spacing between the heater and sample is reduced

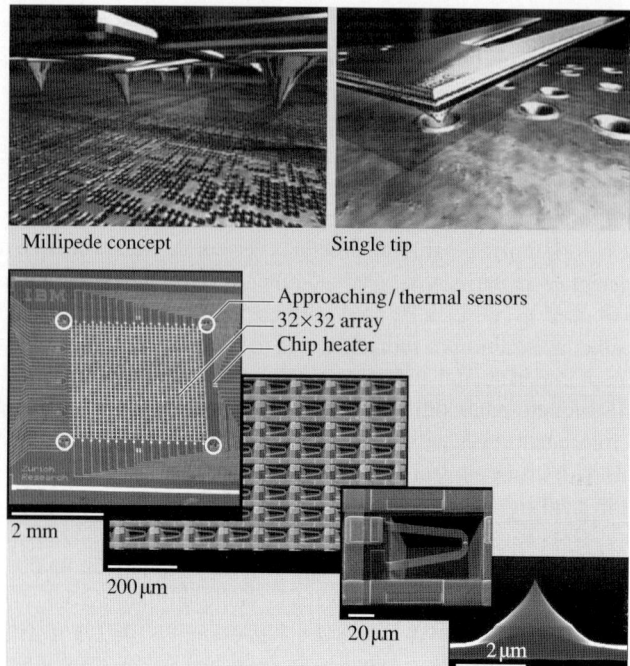

Fig. 1.20. AFM based nanoscale data storage system with 32×32 tip array – that experiences a tribological problem (from [16])

as the tip moves into a bit, the heater's temperature and hence its resistance will decrease. Thus, changes in temperature of the continuously heated resistor are monitored while the cantilever is scanned over data bits, providing a means of detecting the bits. Erasing for subsequent rewriting is carried out by thermal reflow of the storage field by heating the medium to 150 °C for a few seconds. The smoothness of the reflown medium allows multiple rewriting of the same storage field. Bit sizes ranging between 10 and 50 nm have been achieved by using a 32 × 32 (1024) array write/read chip (3 mm × 3 mm). It has been reported that tip wear occurs by the contact between tip and Si substrate during writing. Tip wear is considered a major concern for the device reliability.

1.6 Role of Micro/Nanotribology and Micro/Nanomechanics in Magnetic Storage Devices and MEMS/NEMS

The magnetic storage devices and MEMS/NEMS are the two examples where micro/nanotribological and micro/nanomechanical tools and techniques are essential for studies of micro/nano scale phenomena. Magnetic storage components continue to shrink in physical dimensions. Thicknesses of hard solid coating and liquid lubricant coatings on the magnetic disk surface continue to decrease. Number of contact recording devices are at various stages of development. Surface roughnesses of the storage components continue to decrease and are expected to approach to about 0.5 nm RMS or lower. Interface studies of components with ultra-thin coatings can be ideally performed using micro/nanotribological and micro/nanomechanical tools and techniques.

In the case of MEMS/NEMS, the friction and wear problems of ultrasmall moving components generally made of single-crystal silicon, polysilicon films or polymers need to be addressed for high performance, long life, and reliability. Molecularly-thin films of solid and/or liquids are used for low friction and wear in many applications. Again, interfacial phenomena in MEMS/NEMS can be ideally studied using micro/nanotribological and micro/nanomechanical tools and techniques.

1.7 Organization of the Book

The introductory book integrates knowledge of nanotribology and nanomechanics. The book starts with the definition of tribology, history of tribology and micro/nanotribology, its industrial significance, various measurement techniques employed, followed by various industrial applications. The remaining book is divided into four parts. The first part introduces scanning probe microscopy. The second part provides an overview of nanotechnology and nanomechanics. The third part provides an overview of molecularly-thick films for lubrication. And the last part focuses on nanotribology and nanomechanics studies conducted for various industrial applications.

References

1. P. Jost. *Lubrication (Tribology) – A Report on the Present Position and Industry's Needs*. Dept. of Education and Science (H.M. Stationary Office), 1966.
2. D. Dowson. *History of Tribology*. Inst. Mech. Engineers, 2nd edition, 1998.
3. C. S. C. Davidson. Bearings since the stone age. *Eng.*, 183:2–5, 1957.
4. A. G. Layard. *Discoveries in the Ruins of Nineveh and Babylon*, volume I, II. Murray, 1853.
5. G. Amontons. De la resistance causee dans les machines. *Mem. Acad. R. A*, 257–282, 1699.
6. C. A. Coulomb. Theorie des machines simples, en ayant regard an frottement de leurs parties et a la roideur des cordages. *Mem. Math. Phys. X, Paris*, 161–342, 1785.
7. W. F. Parish. Three thousand years of progress in the development of machinery and lubricants for the hand crafts. *Mill Factory*, 16, 17, 1935.
8. B. Tower. Report on friction experiments. *Proc. Inst. Mech. Eng.*, 632, 1884.
9. O. O. Reynolds. On the theory of lubrication and its applications to mr. beauchamp tower's experiments. *Philos. Trans. R. Soc. (London)*, 117:157–234, 1886.
10. N. P. Petroff. Friction in machines and the effects of the lubricant. *Eng. J.*, 1883.
11. R. Holm. *Electrical Contacts*. Springer, 1946.

12. F. P. Bowden and D. Tabor. *The Friction and Lubrication of Solids*, volume 1, 2. Clarendon, 1950, 1964.
13. B. Bhushan. *Tribology and Mechanics of Magnetic Storage Devices*. Springer, 2nd edition, 1996.
14. B. Bhushan. *Mechanics and Reliability of Flexible Magnetic Media*. Springer, 2nd edition, 2000.
15. B. Bhushan. *Introduction to Tribology*. Wiley, 2002.
16. B. Bhushan (Ed.). *Springer Handbook of Nanotechnology*. Springer, 2004.
17. B. Bhushan and B. K. Gupta. *Handbook of Tribology: Materials, Coatings, and Surface Treatments*. McGraw-Hill, 1991.
18. P. Jost. *Economic Impact of Tribology*, volume 423 of *NBS Spec. Pub*. Proc. Mechanical Failures Prevention Group, 1976.
19. B. Bhushan, J. N. Israelachvili, and U. Landman. Nanotribology: Friction, wear and lubrication at the atomic scale. *Nature*, 374:607–616, 1995.
20. B. Bhushan. *Micro/Nanotribology and its Applications*. NATO ASI Series E: Applied Sciences **330**. Kluwer Academic, 1997.
21. B. Bhushan. *Handbook of Micro/Nanotribology*. CRC, 2nd edition, 1999.
22. B. Bhushan. *Fundamentals of Tribology and Bridging the Gap Between the Macro- and Micro/Nanoscales*, volume 10 of *NATO Science Series II: Mathematics, Physics, and Chemistry*. Kluwer Academic, 2001.
23. B. Bhushan, H. Fuchs, and S. Hosaka. *Applied Scanning Probe Methods*. Springer, 2004.
24. G. Binnig, H. Rohrer, Ch. Gerber, and E. Weibel. Surface studies by scanning tunnelling microscopy. *Phys. Rev. Lett.*, 49:57–61, 1982.
25. G. Binnig, C. F. Quate, and Ch. Gerber. Atomic force microscope. *Phys. Rev. Lett.*, 56:930–933, 1986.
26. G. Binnig, Ch. Gerber, E. Stoll, T. R. Albrecht, and C. F. Quate. Atomic resolution with atomic force microscope. *Europhys. Lett.*, 3:1281–1286, 1987.
27. D. Tabor and R. H. S. Winterton. The direct measurement of normal and retarded van der waals forces. *Proc. R. Soc. Lond. A*, 312:435–450, 1969.
28. S. Alexander, L. Hellemans, O. Marti, J. Schneir, V. Elings, and P. K. Hansma. An atomic-resolution atomic-force microscope implemented using an optical lever. *J. Appl. Phys.*, 65:164–167, 1989.
29. B. Bhushan and J. Ruan. Atomic-scale friction measurements using friction force microscopy: Part ii – application to magnetic media. *ASME J. Tribol.*, 116:389–396, 1994.
30. J. Ruan and B. Bhushan. Atomic-scale friction measurements using friction force microscopy: Part i – general principles and new measurement techniques. *ASME J. Tribol.*, 116:378–388, 1994.
31. J. Ruan and B. Bhushan. Atomic-scale and microscale friction of graphite and diamond using friction force microscopy. *J. Appl. Phys.*, 76:5022–5035, 1994.
32. G. Meyer and N. M. Amer. Novel optical approach to atomic force microscopy. *Appl. Phys. Lett.*, 53:1045–1047, 1988.
33. B. Bhushan, V. N. Koinkar, and J. Ruan. Microtribology of magnetic media. *Proc. Inst. Mech. Engineers. Part J: J. Eng. Tribol.*, 208:17–29, 1994.
34. J. N. Israelachvili and D. Tabor. The measurement of van der waals dispersion forces in the range of 1.5 to 130 nm. *Proc. R. Soc. Lond. A*, 331:19–38, 1972.
35. J. N. Israelachvili and G. E. Adams. Measurement of friction between two mica surfaces in aqueous electrolyte solutions in the range 0–100 nm. *Chem. Soc. J., Faraday Trans. I*, 74:975–1001, 1978.
36. J. Klein. Forces between mica surfaces bearing layers of adsorbed polystyrene in cyclohexane. *Nature*, 288:248–250, 1980.

37. D. Y. C. Chan and R. G. Horn. The drainage of thin liquid films between solid surfaces. *J. Chem. Phys.*, 83:5311–5324, 1985.
38. J. N. Israelachvili, P. M. McGuiggan, and A. M. Homola. Dynamic properties of molecularly thin liquid films. *Science*, 240:189–190, 1988.
39. J. Klein, D. Perahia, and S. Warburg. Forces between polymer-bearing surfaces undergoing shear. *Nature*, 352:143–145, 1991.
40. J. van Alsten and S. Granick. Molecular tribology of ultrathin liquid films. *Phys. Rev. Lett.*, 61:2570–2573, 1988.
41. J. Peachey, J. van Alsten, and S. Granick. Design of an apparatus to measure the shear response of ultrathin liquid films. *Rev. Sci. Instrum.*, 62:463–473, 1991.
42. J. N. Israelachvili. Techniques for direct measurements of forces between surfaces in liquids at the atomic scale. *Chemtracts – Anal. Phys. Chem.*, 1:1–12, 1989.
43. G. Luengo, F. J. Schmitt, R. Hill, and J. N. Israelachvili. Thin film bulk rheology and tribology of confined polymer melts: Contrasts with build properties. *Macromol.*, 30:2482–2494, 1997.
44. J. van Alsten and S. Granick. Shear rheology in a confined geometry – polysiloxane melts. *Macromol.*, 23:4856–4862, 1990.
45. A. M. Homola, J. N. Israelachvili, M. L. Gee, and P. M. McGuiggan. Measurement of and relation between the adhesion and friction of two surfaces separated by thin liquid and polymer films. *ASME J. Tribol.*, 111:675–682, 1989.
46. M. L. Gee, P. M. McGuiggan, J. N. Israelachvili, and A. M. Homola. Liquid to solid-like transitions of molecularly thin films under shear. *J. Chem. Phys.*, 93:1895–1906, 1990.
47. S. Granick. Motions and relaxations of confined liquids. *Science*, 253:1374–1379, 1991.
48. A. Tonck, J. M. Georges, and J. L. Loubet. Measurements of intermolecular forces and the rheology of dodecane between alumina surfaces. *J. Colloid Interface Sci.*, 126:1540–1563, 1988.
49. J. M. Georges, S. Millot, J. L. Loubet, and A. Tonck. Drainage of thin liquid films between relatively smooth surfaces. *J. Chem. Phys.*, 98:7345–7360, 1993.
50. J. M. Georges, A. Tonck, and D. Mazuyer. Interfacial friction of wetted monolayers. *Wear*, 175:59–62, 1994.
51. J. N. Israelachvili. *Intermolecular and Surface Forces*. Academic, 2nd edition, 1992.
52. R. G. Horn. Surface forces and their action in ceramic materials. *Am. Ceram. Soc.*, 73:1117–1135, 1990.
53. A. M. Homola. *Interfacial friction of molecularly thin liquid films*, pages 271–298. World Scientific, 1993.
54. A. M. Homola, J. N. Israelachvili, P. M. McGuiggan, and M. L. Gee. Fundamental experimental studies in tribology: The transition from interfacial friction of undamaged molecularly smooth surfaces. *Wear*, 136:65–83, 1990.
55. R. M. Pashley. Hydration forces between solid surfaces in aqueous electrolyte solutions. *J. Colloid Interface Sci.*, 80:153–162, 1981.
56. R. G. Horn, D. T. Smith, and W. Haller. Surface forces and viscosity of water measured between silica sheets. *Chem. Phys. Lett.*, 162:404–408, 1989.
57. R. G. Horn and J. N. Israelachvili. Molecular organization and viscosity of a thin film of molten polymer between two surfaces as probed by force measurements. *Macromol.*, 21:2836–2841, 1988.
58. H. K. Christenson. Adhesion between surfaces in unsaturated vapors – a reexamination of the influence of meniscus curvature and surface forces. *J. Colloid Interface Sci.*, 121:170–178, 1988.

59. C. P. Smith, M. Maeda, L. Atanasoska, and H. S. White. Ultrathin platinum films on mica and measurement of forces at the platinum/water interface. *J. Phys. Chem.*, 95:199–205, 1988.
60. S. J. Hirz, A. M. Homola, G. Hadzioannou, and S. W. Frank. Effect of substrate on shearing properties of ultrathin polymer films. *Langmuir*, 8:328–333, 1992.
61. S. S. Patel and M. Tirrell. Measurement of forces between surfaces in polymer fluids. *Annu. Rev. Phys. Chem.*, 40:597–635, 1989.
62. J. N. Israelachvili. Solvation forces and liquid structure – as probed by direct force measurements. *Accounts Chem. Res.*, 20:415–421, 1987.
63. J. N. Israelachvili and P. M. McGuiggan. Forces between surface in liquids. *Science*, 241:795–800, 1988.
64. A. M. Homola. Measurement of and relation between the adhesion and friction of two surfaces separated by thin liquid and polymer films. *ASME J. Tribol.*, 111:675–682, 1989.
65. A. M. Homola, H. V. Nguyen, and G. Hadzioannou. Influence of monomer architecture on the shear properties of molecularly thin polymer melts. *J. Chem. Phys.*, 94:2346–2351, 1991.
66. J. van Alsten and S. Granick. Tribology studied using atomically smooth surfaces. *Tribol. Trans.*, 33:436–446, 1990.
67. W. W. Hu, G. A. Carson, and S. Granick. Relaxation time of confined liquids under shear. *Phys. Rev. Lett.*, 66:2758–2761, 1991.
68. J. L. Loubet, M. Bauer, A. Tonck, S. Bec, and B. Gauthier-Manuel. *Nanoindentation with a surface force apparatus*, pages 429–447. Kluwer Academic, 1993.
69. B. Bhushan. *Macro- and microtribology of magnetic storage devices*, volume 2, pages 1413–1513. CRC, 2001.
70. B. Bhushan. *Tribology Issues and Opportunities in MEMS*. Kluwer Academic, 1998.
71. L. J. Hornbeck and W. E. Nelson. *Bistable deformable mirror devices*, volume 8 of *OSA Technical Digest Series*, pages 107–110. OSA, 1988.
72. L. J. Hornbeck. A digital light processing™ update – status and future applications. *Proc. SPIE*, 3634:158–170, 1999.
73. P. Vettinger, J. Brugger, M. Despont, U. Dreschier, U. Duerig, and W. Haeberie. Ultra-high density, high data-rate nems based afm data storage systems. *Microelectron. Eng.*, 46:11–27, 1999.

Part I

Scanning Probe Microscopy

2

Scanning Probe Microscopy – Principle of Operation, Instrumentation, and Probes

Bharat Bhushan and Othmar Marti

Summary. Since the introduction of the STM in 1981 and the AFM in 1985, many variations of probe-based microscopies, referred to as SPMs, have been developed. While the pure imaging capabilities of SPM techniques initially dominated applications of these methods, the physics of probe–sample interactions and quantitative analyses of tribological, electronic, magnetic, biological, and chemical surfaces using SPMs have become of increasing interest in recent years. SPMs are often associated with nanoscale science and technology, since they allow investigation and manipulation of surfaces down to the atomic scale. As our understanding of the underlying interaction mechanisms has grown, SPMs have increasingly found application in many fields beyond basic research fields. In addition, various derivatives of all these methods have been developed for special applications, some of them intended for areas other than microscopy.

This chapter presents an overview of STM and AFM and various probes (tips) used in these instruments, followed by details on AFM instrumentation and analyses.

2.1 Introduction

The scanning tunneling microscope (STM), developed by *Dr. Gerd Binnig* and his colleagues in 1981 at the IBM Zurich Research Laboratory in Rueschlikon (Switzerland), was the first instrument capable of directly obtaining three-dimensional (3-D) images of solid surfaces with atomic resolution [1]. *Binnig* and *Rohrer* received a Nobel Prize in Physics in 1986 for their discovery. STMs can only be used to study surfaces which are electrically conductive to some degree. Based on their design of the STM, in 1985, *Binnig* et al. developed an atomic force microscope (AFM) to measure ultrasmall forces (less than 1 μN) between the AFM tip surface and the sample surface [2] (also see [3]). AFMs can be used to measure any engineering surface, whether it is electrically conductive or insulating. The AFM has become a popular surface profiler for topographic and normal force measurements on the micro- to nanoscale [4]. AFMs modified in order to measure both normal and lateral forces are called lateral force microscopes (LFMs) or friction force microscopes (FFMs) [5–11]. FFMs have been further modified to measure lateral forces in two orthogonal directions [12–16]. A number of researchers have modified and improved the original AFM and FFM designs, and have used these improved systems to measure the adhesion and friction of solid and liquid surfaces on micro- and nanoscales [4, 17–30]. AFMs have been used to study scratching and wear, and to measure elastic/plastic mechanical properties (such as indentation hardness and the modulus of elasticity) [4, 10, 11, 21, 23, 26–29, 31–36]. AFMs have been used to manipulate individual atoms of xenon [37], molecules [38], silicon surfaces [39] and polymer surfaces [40]. STMs have been used to create nanofeatures via localized heating or by inducing chemical reactions under the STM tip [41–43] and through nanomachining [44]. AFMs have also been used for nanofabrication [4, 10, 45–47] and nanomachining [48].

STMs and AFMs are used at extreme magnifications ranging from 10^3 to 10^9 in the x-, y- and z-directions in order to image macro to atomic dimensions with high resolution and for spectroscopy. These instruments can be used in any environment, such as ambient air [2,49], various gases [17], liquids [50–52], vacuum [1,53], at low temperatures (lower than about 100 K) [54–58] and at high temperatures [59, 60]. Imaging in liquid allows the study of live biological samples and it also eliminates the capillary forces that are present at the tip–sample interface when imaging aqueous samples in ambient air. Low-temperature (liquid helium temperatures) imaging is useful when studying biological and organic materials and low-temperature phenomena such as superconductivity or charge-density waves. Low-temperature operation is also advantageous for high-sensitivity force mapping due to the reduced thermal vibration. They also have been used to image liquids such as liquid crystals and lubricant molecules on graphite surfaces [61–64]. While applications of SPM techniques initially focused on their pure imaging capabilities, research into the physics and chemistry of probe–sample interactions and SPM-based quantitative analyses of tribological, electronic, magnetic, biological, and chemical surfaces have become increasingly popular in recent years. Nanoscale science and technology is often tied to the use of SPMs since they allow investigation and manipulation of surfaces down to the atomic scale. As our understanding of the underlying interaction mechanisms has grown, SPMs and their derivatives have found applications in many fields beyond basic research fields and microscopy.

Families of instruments based on STMs and AFMs, called scanning probe microscopes (SPMs), have been developed for various applications of scientific and industrial interest. These include STM, AFM, FFM (or LFM), scanning electrostatic force microscopy (SEFM) [65, 66], scanning force acoustic microscopy (SFAM) (or atomic force acoustic microscopy (AFAM)) [21, 22, 36, 67–69], scanning magnetic microscopy (SMM) (or magnetic force microscopy (MFM)) [70–73], scanning near-field optical microscopy (SNOM) [74–77], scanning thermal microscopy (SThM) [78–80], scanning electrochemical microscopy (SEcM) [81], scanning Kelvin probe microscopy (SKPM) [82–86], scanning chemical potential microscopy (SCPM) [79], scanning ion conductance microscopy (SICM) [87, 88] and scanning capacitance microscopy (SCM) [82, 89–91]. When the technique is used to measure forces (as in AFM, FFM, SEFM, SFAM and SMM) it is also referred to as scanning force microscopy (SFM). Although these instruments offer atomic resolution and

Table 2.1. Comparison of various conventional microscopes with SPMs

	Optical	SEM/TEM	Confocal	SPM
Magnification	10^3	10^7	10^4	10^9
Instrument Price (U.S. $)	$10k	$250k	$30k	$100k
Technology Age	200 yrs	40 yrs	20 yrs	20 yrs
Applications	Ubiquitous	Science and technology	New and unfolding	Cutting edge
Market 1993	$800 M	$400 M	$80 M	$100 M
Growth Rate	10%	10%	30%	70%

are ideal for basic research, they are also used for cutting-edge industrial applications which do not require atomic resolution. The commercial production of SPMs started with the STM in 1987 and the AFM in 1989 by Digital Instruments Inc (Santa Barbara, CA, USA). For comparisons of SPMs with other microscopes, see Table 2.1 (Veeco Instruments, Inc., Santa Barbara, CA, USA). Numbers of these instruments are equally divided between the U.S., Japan and Europe, with the following split between industry/university and government laboratories: 50/50, 70/30, and 30/70, respectively. It is clear that research and industrial applications of SPMs are expanding rapidly.

2.2 Scanning Tunneling Microscope

The principle of electron tunneling was first proposed by *Giaever* [93]. He envisioned that if a potential difference is applied to two metals separated by a thin insulating film, a current will flow because of the ability of electrons to penetrate a potential barrier. To be able to measure a tunneling current, the two metals must be spaced no more than 10 nm apart. *Binnig* et al. [1] introduced vacuum tunneling combined with lateral scanning. The vacuum provides the ideal barrier for tunneling. The lateral scanning allows one to image surfaces with exquisite resolution – laterally to less than 1 nm and vertically to less than 0.1 nm – sufficient to define the position of single atoms. The very high vertical resolution of the STM is obtained because the tunnel current varies exponentially with the distance between the two electrodes; that is, the metal tip and the scanned surface. Typically, the tunneling current decreases by a factor of 2 as the separation is increased by 0.2 nm. Very high lateral resolution depends upon sharp tips. *Binnig* et al. overcame two key obstacles by damping external vibrations and moving the tunneling probe in close proximity to the sample. Their instrument is called the scanning tunneling microscope (STM). Today's STMs can be used in ambient environments for atomic-scale imaging of surfaces. Excellent reviews on this subject have been presented by *Hansma* and *Tersoff* [92], *Sarid* and *Elings* [94], *Durig* et al. [95]; *Frommer* [96], *Güntherodt* and

Wiesendanger [97], *Wiesendanger* and *Güntherodt* [98], *Bonnell* [99], *Marti* and *Amrein* [100], *Stroscio* and *Kaiser* [101], and *Güntherodt* et al. [102].

The principle of the STM is straightforward. A sharp metal tip (one electrode of the tunnel junction) is brought close enough (0.3–1 nm) to the surface to be investigated (the second electrode) to make the tunneling current measurable at a convenient operating voltage (10 mV–1 V). The tunneling current in this case varies from 0.2 to 10 nA. The tip is scanned over the surface at a distance of 0.3–1 nm, while the tunneling current between it and the surface is measured. The STM can be operated in either the constant current mode or the constant height mode, Fig. 2.1. The left-hand column of Fig. 2.1 shows the basic constant current mode of operation. A feedback network changes the height of the tip z to keep the current constant. The displacement of the tip, given by the voltage applied to the piezoelectric drive, then yields a topographic map of the surface. Alternatively, in the constant height mode, a metal tip can be scanned across a surface at nearly constant height and constant voltage while the current is monitored, as shown in the right-hand column of Fig. 2.1. In this case, the feedback network responds just rapidly enough to keep the average current constant. The current mode is generally used for atomic-scale images; this mode is not practical for rough surfaces. A three-dimensional picture $[z(x,y)]$ of a surface consists of multiple scans $[z(x)]$ displayed laterally to each other in the y direction. It should be noted that if different atomic species are present in a sample, the different atomic species within a sample may produce different tunneling currents for a given bias voltage. Thus the height data may not be a direct representation of the topography of the surface of the sample.

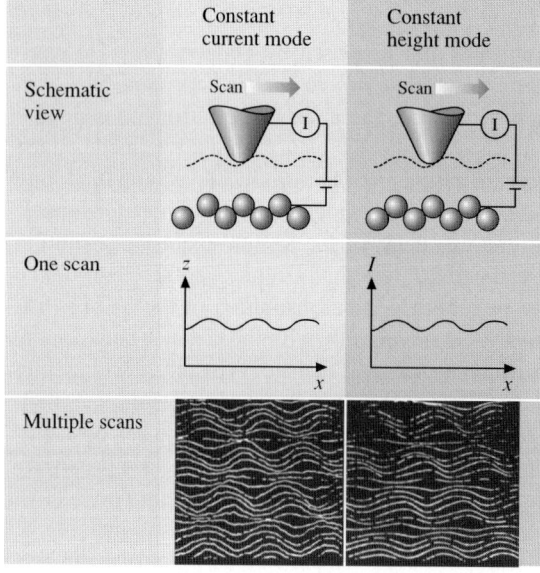

Fig. 2.1. An STM can be operated in either the constant-current or the constant-height mode. The images are of graphite in air [92]

2.2.1 The STM Design of Binnig et al.

Figure 2.2 shows a schematic of an AFM designed by Binnig and Rohrer and intended for operation in ultrahigh vacuum [1, 103]. The metal tip was fixed to rectangular piezodrives P_x, P_y, and P_z made out of commercial piezoceramic material for scanning. The sample is mounted via either superconducting magnetic levitation or a two-stage spring system to achieve a stable gap width of about 0.02 nm. The tunnel current J_T is a sensitive function of the gap width d where $J_T \propto V_T \exp(-A\phi^{1/2}d)$. Here V_T is the bias voltage, ϕ is the average barrier height (work function) and the constant $A = 1.025 \text{ eV}^{-1/2}\text{Å}^{-1}$. With a work function of a few eV, J_T changes by an order of magnitude for an angstrom change in d. If the current is kept constant to within, for example, 2%, then the gap d remains constant to within 1 pm. For operation in the constant current mode, the control unit CU applies a voltage V_z to the piezo P_z such that J_T remains constant when scanning the tip with P_y and P_x over the surface. At a constant work function ϕ, $V_z(V_x, V_y)$ yields the roughness of the surface $z(x, y)$ directly, as illustrated by a surface step at A. Smearing the step, δ (lateral resolution) is on the order of $(R)^{1/2}$, where R is the radius of the curvature of the tip. Thus, a lateral resolution of about 2 nm requires tip radii on the order of 10 nm. A 1-mm-diameter solid rod ground at one end at roughly 90° yields overall tip radii of only a few hundred nanometers, the presence of rather sharp microtips on the relatively dull end yields a lateral resolution of about 2 nm. In situ sharpening of the tips, achieved by gently touching the surface, brings the resolution down to the 1-nm range; by applying high fields (on the order of 10^8 V/cm) for, say, half an hour, resolutions considerably below 1 nm can be reached. Most experiments have been performed with tungsten wires either ground or etched to a typical radius of 0.1–10 μm. In some cases, in situ processing of the tips has been performed to further reduce tip radii.

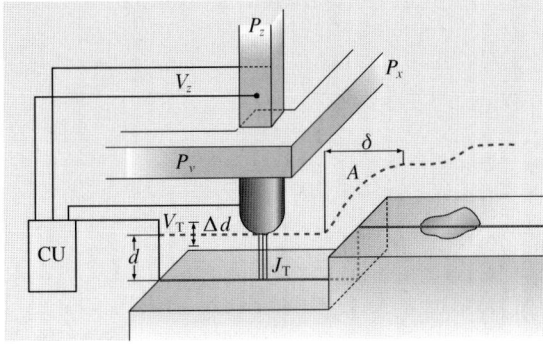

Fig. 2.2. Principle of operation of the STM, from *Binnig* and *Rohrer* [103]

2.2.2 Commercial STMs

There are a number of commercial STMs available on the market. Digital Instruments, Inc. introduced the first commercial STM, the Nanoscope I, in 1987. In the recent Nanoscope IV STM, intended for operation in ambient air, the sample is held in position while a piezoelectric crystal in the form of a cylindrical tube (referred to as a PZT tube scanner) scans the sharp metallic probe over the surface in a raster pattern while sensing and relaying the tunneling current to the control station (Fig. 2.3). The digital signal processor (DSP) calculates the tip–sample separation required by sensing the tunneling current flowing between the sample and the tip. The bias voltage applied between the sample and the tip encourages the tunneling current to flow. The DSP completes the digital feedback loop by relaying the desired voltage to the piezoelectric tube. The STM can operate in either the "constant height" or the "constant current" mode, and this can be selected using the control panel. In the constant current mode, the feedback gains are set high, the tunneling tip closely tracks the sample surface, and the variation in the tip height required to maintain constant tunneling current is measured by the change in the voltage applied to the piezo tube. In the constant height mode, the feedback gains are set low, the tip remains at a nearly constant height as it sweeps over the sample surface, and the tunneling current is imaged.

Physically, the Nanoscope STM consists of three main parts: the head, which houses the piezoelectric tube scanner which provides three-dimensional tip motion and the preamplifier circuit for the tunneling current (FET input amplifier) mounted on the top of the head; the base on which the sample is mounted; and the base support, which supports the base and head [4]. The base accommodates samples which are up to 10 mm by 20 mm and 10 mm thick. Scan sizes available for the STM are 0.7 µm (for atomic resolution), 12 µm, 75 µm and 125 µm square.

The scanning head controls the three-dimensional motion of the tip. The removable head consists of a piezo tube scanner, about 12.7 mm in diameter, mounted into an Invar shell, which minimizes vertical thermal drift because of the good thermal match between the piezo tube and the Invar. The piezo tube has separate electrodes

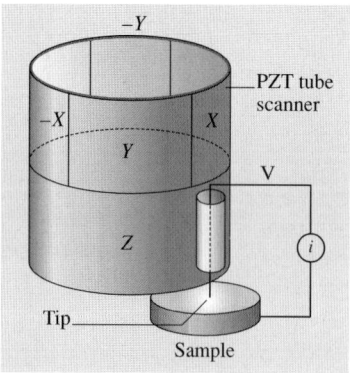

Fig. 2.3. Principle of operation of a commercial STM. A sharp tip attached to a piezoelectric tube scanner is scanned on a sample

for x, y and z motion, which are driven by separate drive circuits. The electrode configuration (Fig. 2.3) provides x and y motions which are perpendicular to each other, it minimizes horizontal and vertical coupling, and it provides good sensitivity. The vertical motion of the tube is controlled by the Z electrode, which is driven by the feedback loop. The x and y scanning motions are each controlled by two electrodes which are driven by voltages of the same magnitude but opposite signs. These electrodes are called $-y$, $-x$, $+y$, and $+x$. Applying complimentary voltages allows a short, stiff tube to provide a good scan range without the need for a large voltage. The motion of the tip that arises due to external vibrations is proportional to the square of the ratio of vibration frequency to the resonant frequency of the tube. Therefore, to minimize the tip vibrations, the resonant frequencies of the tube are high: about 60 kHz in the vertical direction and about 40 kHz in the horizontal direction. The tip holder is a stainless steel tube with an inner diameter of 300 µm when 250 µm-diameter tips are used, which is mounted in ceramic in order to minimize the mass at the end of the tube. The tip is mounted either on the front edge of the tube (to keep the mounting mass low and the resonant frequency high) (Fig. 2.3) or the center of the tube for large-range scanners, namely 75 and 125 µm (to preserve the symmetry of the scanning). This commercial STM accepts any tip with a 250 µm-diameter shaft. The piezotube requires x–y calibration, which is carried out by imaging an appropriate calibration standard. Cleaved graphite is used for heads with small scan lengths while two-dimensional grids (a gold-plated rule) can be used for long-range heads.

The Invar base holds the sample in position, supports the head, and provides coarse x–y motion for the sample. A sprung-steel sample clip with two thumb screws holds the sample in place. An x–y translation stage built into the base allows the sample to be repositioned under the tip. Three precision screws arranged in a triangular pattern support the head and provide coarse and fine adjustment of the tip height. The base support consists of the base support ring and the motor housing. The stepper motor enclosed in the motor housing allows the tip to be engaged and withdrawn from the surface automatically.

Samples to be imaged with the STM must be conductive enough to allow a few nanoamperes of current to flow from the bias voltage source to the area to be scanned. In many cases, nonconductive samples can be coated with a thin layer of a conductive material to facilitate imaging. The bias voltage and the tunneling current depend on the sample. Usually they are set to a standard value for engagement and fine tuned to enhance the quality of the image. The scan size depends on the sample and the features of interest. A maximum scan rate of 122 Hz can be used. The maximum scan rate is usually related to the scan size. Scan rates above 10 Hz are used for small scans (typically 60 Hz for atomic-scale imaging with a 0.7 µm scanner). The scan rate should be lowered for large scans, especially if the sample surfaces are rough or contain large steps. Moving the tip quickly along the sample surface at high scan rates with large scan sizes will usually lead to a tip crash. Essentially, the scan rate should be inversely proportional to the scan size (typically 2–4 Hz for a scan size of 1 µm, 0.5–1 Hz for 12 µm, and 0.2 Hz for 125 µm). The

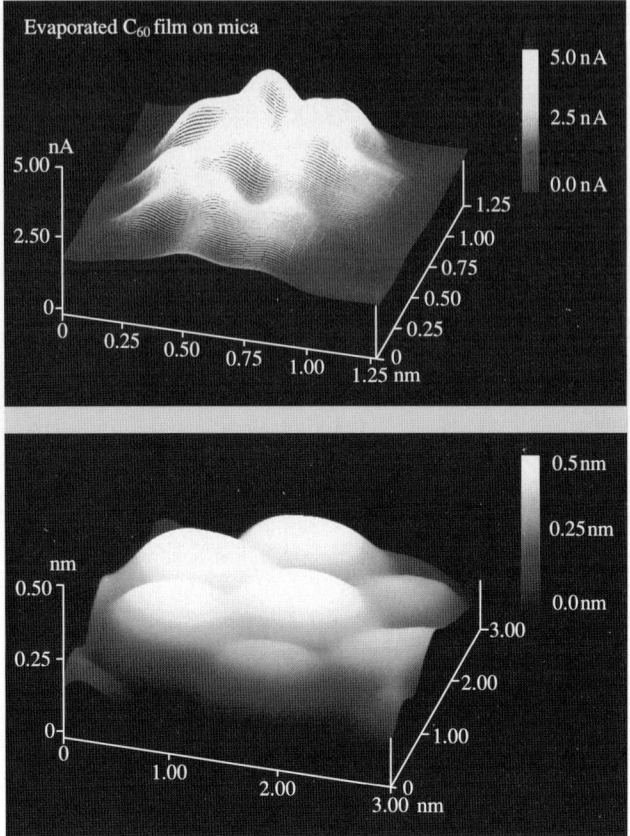

Fig. 2.4. STM images of evaporated C_{60} film on gold-coated freshly cleaved mica obtained using a mechanically sheared Pt–Ir (80-20) tip in constant height mode [104]

scan rate (in length/time) is equal to the scan length divided by the scan rate in Hz. For example, for a scan size of 10 μm × 10 μm scanned at 0.5 Hz, the scan rate is 10 μm/s. 256 × 256 data formats are the most common. The lateral resolution at larger scans is approximately equal to scan length divided by 256.

Figure 2.4 shows sample STM images of an evaporated C_{60} film on gold-coated freshly-cleaved mica taken at room temperature and ambient pressure [104]. Images were obtained with atomic resolution at two scan sizes. Next we describe some STM designs which are available for special applications.

Electrochemical STM

The electrochemical STM is used to perform and monitor the electrochemical reactions inside the STM. It includes a microscope base with an integral potentiostat, a short head with a 0.7 μm scan range and a differential preamp as well as the software required to operate the potentiostat and display the result of the electrochemical reaction.

Standalone STM

Standalone STMs are available to scan large samples. In this case, the STM rests directly on the sample. It is available from Digital Instruments in scan ranges of 12 and 75 μm. It is similar to the standard STM design except the sample base has been eliminated.

2.2.3 STM Probe Construction

The STM probe has a cantilever integrated with a sharp metal tip with a low aspect ratio (tip length/tip shank) to minimize flexural vibrations. Ideally, the tip should be atomically sharp, but in practice most tip preparation methods produce a tip with a rather ragged profile that consists of several asperities where the one closest to the surface is responsible for tunneling. STM cantilevers with sharp tips are typically fabricated from metal wires (the metal can be tungsten (W), platinum-iridium (Pt-Ir), or gold (Au)) and are sharpened by grinding, cutting with a wire cutter or razor blade, field emission/evaporation, ion milling, fracture, or electrochemical polishing/etching [105, 106]. The two most commonly used tips are made from either Pt-Ir (80/20) alloy or tungsten wire. Iridium is used to provide stiffness. The Pt-Ir tips are generally formed mechanically and are readily available. The tungsten tips are etched from tungsten wire by an electrochemical process, for example by using 1 M KOH solution with a platinum electrode in a electrochemical cell at about 30 V. In general, Pt-Ir tips provide better atomic resolution than tungsten tips, probably due to the lower reactivity of Pt. However, tungsten tips are more uniformly shaped and may perform better on samples with steeply sloped features. The tungsten wire diameter used for the cantilever is typically 250 μm, with the radius of curvature ranging from 20 to 100 nm and a cone angle ranging from 10 to 60° (Fig. 2.5). The wire can be bent in an L shape, if so required, for use in the instrument. For calculations of the normal spring constant and the natural frequency of round cantilevers, see *Sarid* and *Elings* [94].

High aspect ratio, controlled geometry (CG) Pt-Ir probes are commercially available to image deep trenches (Fig. 2.6). These probes are electrochemically etched from Pt-Ir (80/20) wire and are polished to a specific shape which is consistent from tip to tip. The probes have a full cone angle of approximately 15°, and a tip radius of less than 50 nm. To image very deep trenches (> 0.25 μm) and nanofeatures, focused ion beam (FIB)-milled CG probes with extremely sharp tips

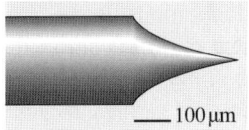

Fig. 2.5. Schematic of a typical tungsten cantilever with a sharp tip produced by electrochemical etching

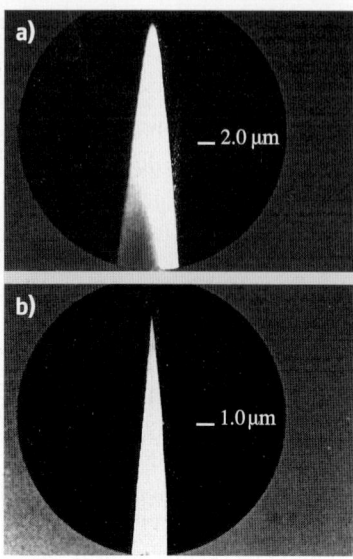

Fig. 2.6. Schematics of (**a**) CG Pt–Ir probe, and (**b**) CG Pt–Ir FIB milled probe

(radii < 5 nm) are used. The Pt/Ir probes are coated with a nonconducting film (not shown in the figure) for electrochemistry. These probes are available from Materials Analytical Services (Raleigh, NC, USA).

Pt alloy and W tips are very sharp and give high resolution, but are fragile and sometimes break when contacting a surface. Diamond tips have been used by *Kaneko* and *Oguchi* [107]. TDiamond tips made conductive by boron ion implantation were found to be chip-resistant.

2.3 Atomic Force Microscope

Like the STM, the AFM relies on a scanning technique to produce very high resolution, 3-D images of sample surfaces. The AFM measures ultrasmall forces (less than 1 nN) present between the AFM tip surface and a sample surface. These small forces are measured by measuring the motion of a very flexible cantilever beam with an ultrasmall mass. While STMs require the surface being measured be electrically conductive, AFMs are capable of investigating the surfaces of both conductors and insulators on an atomic scale if suitable techniques for measuring the cantilever motion are used. During the operation of a high-resolution AFM, the sample is generally scanned instead of the tip (unlike for STM) because the AFM measures the relative displacement between the cantilever surface and the reference surface and any cantilever movement from scanning would add unwanted vibrations. However, for measurements of large samples, AFMs are available where the tip is scanned and the sample is stationary. As long as the AFM is operated in the so-called contact mode, little if any vibration is introduced.

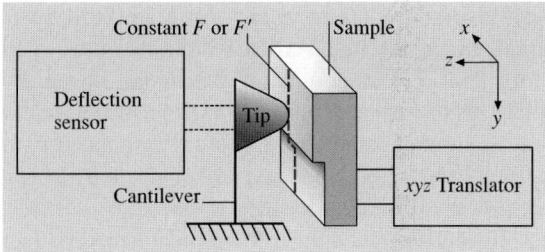

Fig. 2.7. Principle of operation of the AFM. Sample mounted on a piezoelectric scanner is scanned against a short tip and the cantilever deflection is usually measured using a laser deflection technique. The force (in contact mode) or the force gradient (in noncontact mode) is measured during scanning

The AFM combines the principles of the STM and the stylus profiler (Fig. 2.7). In an AFM, the force between the sample and tip is used (rather than the tunneling current) to sense the proximity of the tip to the sample. The AFM can be used either in the static or the dynamic mode. In the static mode, also referred to as the repulsive or contact mode [2], a sharp tip at the end of the cantilever is brought into contact with the surface of the sample. During initial contact, the atoms at the end of the tip experience a very weak repulsive force due to electronic orbital overlap with the atoms in the surface of the sample. The force acting on the tip causes the cantilever to deflect, which is measured by tunneling, capacitive, or optical detectors. The deflection can be measured to within 0.02 nm, so a force as low as 0.2 nN (corresponding to a normal pressure of ~ 200 MPa for a Si_3N_4 tip with a radius of about 50 nm against single-crystal silicon) can be detected for typical cantilever spring constant of 10 N/m. (To put these number in perspective, individual atoms and human hair are typically a fraction of a nanometer and about 75 µm in diameter, respectively, and a drop of water and an eyelash have masses of about 10 µN and 100 nN, respectively.) In the dynamic mode of operation, also referred to as attractive force imaging or noncontact imaging mode, the tip is brought into close proximity to (within a few nanometers of), but not in contact with, the sample. The cantilever is deliberately vibrated in either amplitude modulation (AM) mode [65] or frequency modulation (FM) mode [65, 94, 108, 109]. Very weak van der Waals attractive forces are present at the tip–sample interface. Although the normal pressure exerted at the interface is zero in this technique (in order to avoid any surface deformation), it is slow and difficult to use, and is rarely used outside of research environments. The surface topography is measured by laterally scanning the sample under the tip while simultaneously measuring the separation-dependent force or force gradient (derivative) between the tip and the surface (Fig. 2.7). In the contact (static) mode, the interaction force between tip and sample is measured by monitoring the cantilever deflection. In the noncontact (or dynamic) mode, the force gradient is obtained by vibrating the cantilever and measuring the shift in the resonant frequency of the cantilever. To obtain topographic information, the interaction force is either recorded directly, or used as a control parameter for a feedback circuit that maintains the force or force derivative at a constant value. Using an AFM operated in the contact mode, topographic images with a vertical resolution of less than 0.1 nm (as low as 0.01 nm) and a lateral resolution of about 0.2 nm have been obtained [3,50,110–114]. Forces of 10 nN to 1 pN are measurable with a displacement sensitivity of 0.01 nm. These forces are comparable to the forces associated with chemical bonding, for example 0.1 µN for an ionic bond and 10 pN for a hydrogen bond [2]. For further reading, see [94–96, 100, 102, 115–119].

Lateral forces applied at the tip during scanning in the contact mode affect roughness measurements [120]. To minimize the effects of friction and other lateral forces on topography measurements in the contact mode, and to measure the topographies of soft surfaces, AFMs can be operated in the so-called tapping or force modulation mode [32, 121].

The STM is ideal for atomic-scale imaging. To obtain atomic resolution with the AFM, the spring constant of the cantilever should be weaker than the equivalent spring between atoms. For example, the vibration frequencies ω of atoms bound in a molecule or in a crystalline solid are typically 10^{13} Hz or higher. Combining this with an atomic mass m of approximately 10^{-25} kg gives an interatomic spring constant k, given by $\omega^2 m$, of around 10 N/m [115]. (For comparison, the spring constant of a piece of household aluminium foil that is 4 mm long and 1 mm wide is about 1 N/m.) Therefore, a cantilever beam with a spring constant of about 1 N/m or lower is desirable. Tips must be as sharp as possible, and tip radii of 5 to 50 nm are commonly available.

Atomic resolution cannot be achieved with these tips at normal loads in the nN range. Atomic structures at these loads have been obtained from lattice imaging or by imaging the crystal's periodicity. Reported data show either perfectly ordered periodic atomic structures or defects on a larger lateral scale, but no well-defined, laterally resolved atomic-scale defects like those seen in images routinely obtained with a STM. Interatomic forces with one or several atoms in contact are 20–40 or 50–100 pN, respectively. Thus, atomic resolution with an AFM is only possible with a sharp tip on a flexible cantilever at a net repulsive force of 100 pN or lower [122]. Upon increasing the force from 10 pN, *Ohnesorge* and *Binnig* [122] observed that monoatomic steplines were slowly wiped away and a perfectly ordered structure was left. This observation explains why mostly defect-free atomic resolution has been observed with AFM. Note that for atomic-resolution measurements, the cantilever should not be so soft as to avoid jumps. Further note that performing measurements in the noncontact imaging mode may be desirable for imaging with atomic resolution.

The key component in an AFM is the sensor used to measure the force on the tip due to its interaction with the sample. A cantilever (with a sharp tip) with an extremely low spring constant is required for high vertical and lateral resolutions at small forces (0.1 nN or lower), but a high resonant frequency is desirable (about 10 to 100 kHz) at the same time in order to minimize the sensitivity to building vibrations, which occur at around 100 Hz. This requires a spring with an extremely low vertical spring constant (typically 0.05 to 1 N/m) as well as a low mass (on the order of 1 ng). Today, the most advanced AFM cantilevers are microfabricated from silicon or silicon nitride using photolithographic techniques. Typical lateral dimensions are on the order of 100 µm, with thicknesses on the order of 1 µm. The force on the tip due to its interaction with the sample is sensed by detecting the deflection of the compliant lever with a known spring constant. This cantilever deflection (displacement smaller than 0.1 nm) has been measured by detecting a tunneling current similar to that used in the STM in the pioneering work of *Binnig* et al. [2] and later used by *Giessibl* et al. [56], by capacitance detection [123, 124], piezoresistive detection [125, 126], and by four optical techniques, namely (1) optical interferometry [5, 6, 127, 128] using optical fibers [57, 129] (2) optical polarization detection [72, 130], (3) laser diode feedback [131] and (4) optical (laser) beam deflection [7, 8, 53, 111, 112]. Schematics of the four more commonly used detection systems are shown in Fig. 2.8. The tunneling method originally used by *Binnig* et al. [2] in the first version of the AFM uses a second tip to monitor the deflection of the cantilever with its force sensing tip. Tunneling is rather sensitive to contaminants and the interaction between the tunneling tip and the rear side of the cantilever can become comparable to the interaction between the tip and sample. Tunneling is rarely used and is mentioned mainly for historical reasons. *Giessibl* et al. [56] have used it for a low-temperature AFM/STM design. In contrast to tunneling, other deflection sensors are placed far from the cantilever, at distances of microns to tens of millimeters. The optical techniques are believed to be more sensitive, reliable and easily implemented detection methods than the others [94, 118]. The optical beam deflection method has the largest working distance, is insensitive to distance changes and is capable of measuring angular changes (friction forces); therefore, it is the most commonly used in commercial SPMs.

Almost all SPMs use piezo translators to scan the sample, or alternatively to scan the tip. An electric field applied across a piezoelectric material causes a change in the crystal structure, with expansion in some directions and contraction in others. A net change in volume also occurs [132]. The first STM used a piezo tripod for scanning [1]. The piezo tripod is one way to generate three-dimensional movement of a tip attached at its center. However, the tripod needs to be fairly large (∼ 50 mm)

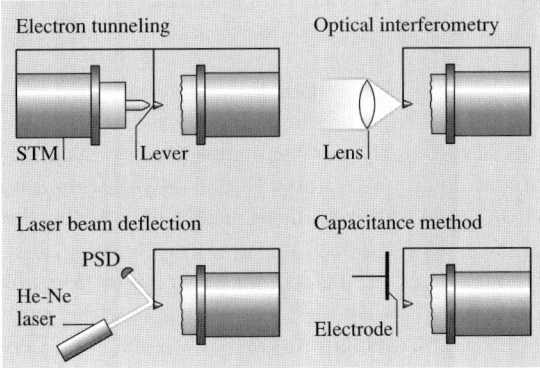

Fig. 2.8. Schematics of the four detection systems to measure cantilever deflection. In each set-up, the sample mounted on piezoelectric body is shown on the right, the cantilever in the middle, and the corresponding deflection sensor on the left [118]

to get a suitable range. Its size and asymmetric shape makes it susceptible to thermal drift. Tube scanners are widely used in AFMs [133]. These provide ample scanning range with a small size. Electronic control systems for AFMs are based on either analog or digital feedback. Digital feedback circuits are better suited for ultralow noise operation.

Images from the AFMs need to be processed. An ideal AFM is a noise-free device that images a sample with perfect tips of known shape and has a perfectly linear scanning piezo. In reality, scanning devices are affected by distortions and these distortions must be corrected for. The distortions can be linear and nonlinear. Linear distortions mainly result from imperfections in the machining of the piezo translators, causing cross-talk between the Z-piezo to the x- and y-piezos, and vice versa. Nonlinear distortions mainly result from the presence of a hysteresis loop in piezoelectric ceramics. They may also occur if the scan frequency approaches the upper frequency limit of the x- and y-drive amplifiers or the upper frequency limit of the feedback loop (z-component). In addition, electronic noise may be present in the system. The noise is removed by digital filtering in real space [134] or in the spatial frequency domain (Fourier space) [135].

Processed data consists of many tens of thousand of points per plane (or data set). The outputs from the first STM and AFM images were recorded on an x-y chart recorder, with the z-value plotted against the tip position in the fast scan direction. Chart recorders have slow responses, so computers are used to display the data these days. The data are displayed as wire mesh displays or grayscale displays (with at least 64 shades of gray).

2.3.1 The AFM Design of Binnig et al.

In the first AFM design developed by *Binnig* et al. [2], AFM images were obtained by measuring the force exerted on a sharp tip created by its proximity to the surface of a sample mounted on a 3-D piezoelectric scanner. The tunneling current between the STM tip and the backside of the cantilever beam to which the tip was attached was measured to obtain the normal force. This force was kept at a constant level with a feedback mechanism. The STM tip was also mounted on a piezoelectric element to maintain the tunneling current at a constant level.

2.3.2 Commercial AFMs

A review of early designs of AFMs has been presented by *Bhushan* [4]. There are a number of commercial AFMs available on the market. Major manufacturers of AFMs for use in ambient environments are: Digital Instruments Inc., Topometrix Corp. and other subsidiaries of Veeco Instruments, Inc., Molecular Imaging Corp. (Phoenix, AZ, USA), Quesant Instrument Corp. (Agoura Hills, CA, USA), Nanoscience Instruments Inc. (Phoenix, AZ, USA), Seiko Instruments (Chiba, Japan); and Olympus (Tokyo, Japan). AFM/STMs for use in UHV environments are manufactured by Omicron Vakuumphysik GMBH (Taunusstein, Germany).

We describe here two commercial AFMs – small-sample and large-sample AFMs – for operation in the contact mode, produced by Digital Instruments, Inc., with scanning lengths ranging from about 0.7 μm (for atomic resolution) to about 125 μm [9,111,114,136]. The original design of these AFMs comes from *Meyer* and *Amer* [53]. Basically, the AFM scans the sample in a raster pattern while outputting the cantilever deflection error signal to the control station. The cantilever deflection (or the force) is measured using a laser deflection technique (Fig. 2.9). The DSP in the workstation controls the z position of the piezo based on the cantilever deflection error signal. The AFM operates in both "constant height" and "constant force" modes. The DSP always adjusts the distance between the sample and the tip according to the cantilever deflection error signal, but if the feedback gains are low the piezo remains at an almost "constant height" and the cantilever deflection data is collected. With high gains, the piezo height changes to keep the cantilever deflection nearly constant (so the force is constant), and the change in piezo height is collected by the system.

In the operation of a commercial small-sample AFM (as shown in Fig. 2.9a), the sample (which is generally no larger than 10 mm × 10 mm) is mounted on a PZT tube scanner, which consists of separate electrodes used to precisely scan the sample in the x–y plane in a raster pattern and to move the sample in the vertical (z) direction. A sharp tip at the free end of a flexible cantilever is brought into contact with the sample. Features on the sample surface cause the cantilever to deflect in the vertical and lateral directions as the sample moves under the tip. A laser beam from a diode laser (5 mW max peak output at 670 nm) is directed by a prism onto the back of a cantilever near its free end, tilted downward at about 10° with respect to the horizontal plane. The reflected beam from the vertex of the cantilever is directed through a mirror onto a quad photodetector (split photodetector with four quadrants) (commonly called a position-sensitive detector or PSD, produced by Silicon Detector Corp., Camarillo, CA, USA). The difference in signal between the top and bottom photodiodes provides the AFM signal, which is a sensitive measure of the cantilever vertical deflection. The topographic features of the sample cause the tip to deflect in the vertical direction as the sample is scanned under the tip. This tip deflection will change the direction of the reflected laser beam, changing the intensity difference between the top and bottom sets of photodetectors (AFM signal). In a mode of operation called the height mode, used for topographic imaging or for any other operation in which the normal forceapplied is to be kept constant, a feedback circuit is used to modulate the voltage applied to the PZT scanner in order to adjust the height of the PZT, so that the cantilever vertical deflection (given by the intensity difference between the top and bottom detector) will remain constant during scanning. The PZT height variation is thus a direct measure of the surface roughness of the sample.

In a large-sample AFM, force sensors based on optical deflection methods or scanning units are mounted on the microscope head (Fig. 2.9b). Because of the unwanted vibrations caused by cantilever movement, the lateral resolution of this design is somewhat poorer than the design in Fig. 2.9a in which the sample is scanned

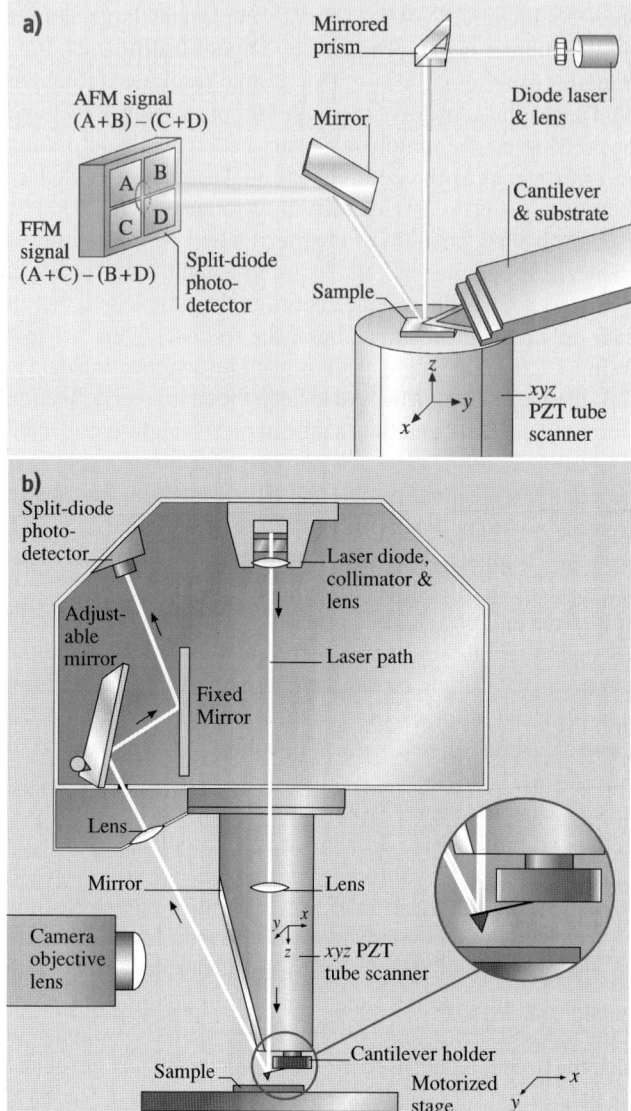

Fig. 2.9. Principles of operation of (**a**) a commercial small-sample AFM/FFM, and (**b**) a large-sample AFM/FFM

instead of the cantilever beam. The advantage of the large-sample AFM is that large samples can be easily measured.

Most AFMs can be used for topography measurements in the so-called tapping mode (intermittent contact mode), in what is also referred to as dynamic force microscopy. In the tapping mode, during the surface scan, the cantilever/tip assembly

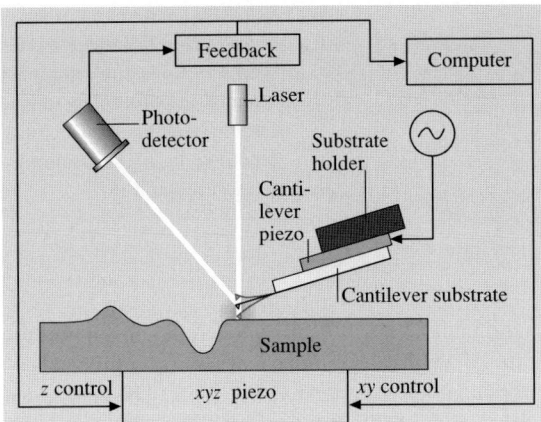

Fig. 2.10. Schematic of tapping mode used for surface roughness measurements

is sinusoidally vibrated by a piezo mounted above it, and the oscillating tip slightly taps the surface at the resonant frequency of the cantilever (70–400 kHz) with a constant (20–100 nm) amplitude of vertical oscillation, and a feedback loop keeps the average normal force constant (Fig. 2.10). The oscillating amplitude is kept large enough that the tip does not get stuck to the sample due to adhesive attraction. The tapping mode is used in topography measurements to minimize the effects of friction and other lateral forces to measure the topography of soft surfaces.

Topographic measurements can be made at any scanning angle. At first glance, the scanning angle may not appear to be an important parameter. However, the friction force between the tip and the sample will affect the topographic measurements in a parallel scan (scanning along the long axis of the cantilever). This means that a perpendicular scan may be more desirable. Generally, one picks a scanning angle which gives the same topographic data in both directions; this angle may be slightly different to that for the perpendicular scan.

The left-hand and right-hand quadrants of the photodetector are used to measure the friction force applied at the tip surface during sliding. In the so-called friction mode, the sample is scanned back and forth in a direction orthogonal to the long axis of the cantilever beam. Friction force between the sample and the tip will twist the cantilever. As a result, the laser beam will be deflected out of the plane defined by the incident beam and the beam is reflected vertically from an untwisted cantilever. This produces a difference in laser beam intensity between the beams received by the left-hand and right-hand sets of quadrants of the photodetector. The intensity difference between the two sets of detectors (FFM signal) is directly related to the degree of twisting and hence to the magnitude of the friction force. This method provides three-dimensional maps of the friction force. One problem associated with this method is that any misalignment between the laser beam and the photodetector axis introduces errors into the measurement. However, by following the procedures developed by *Ruan* and *Bhushan* [136], in which the average FFM signal for the sample scanned in two opposite directions is subtracted from the friction profiles of

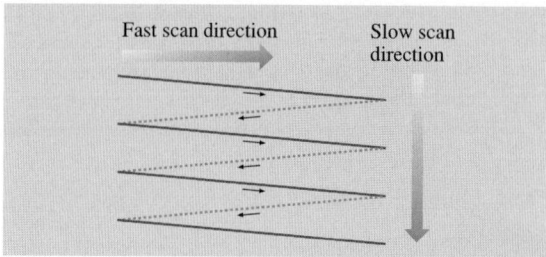

Fig. 2.11. Schematic of triangular pattern trajectory of the AFM tip as the sample is scanned in two dimensions. During imaging, data are only recorded during scans along the solid scan lines

each of the two scans, the misalignment effect can be eliminated. By following the friction force calibration procedures developed by *Ruan* and *Bhushan* [136], voltages corresponding to friction forces can be converted to force units. The coefficient of friction is obtained from the slope of the friction force data measured as a function of the normal load, which typically ranges from 10 to 150 nN. This approach eliminates any contributions from adhesive forces [10]. To calculate the coefficient of friction based on a single point measurement, the friction force should be divided by the sum of the normal load applied and the intrinsic adhesive force. Furthermore, it should be pointed out that the coefficient of friction is not independent of load for single-asperity contact,. This is discussed in more detail later.

The tip is scanned in such a way that its trajectory on the sample forms a triangular pattern (Fig. 2.11). Scanning speeds in the fast and slow scan directions depend on the scan area and scan frequency. Scan sizes ranging from less than 1 nm × 1 nm to 125 µm × 125 µm and scan rates of less than 0.5 to 122 Hz are typically used. Higher scan rates are used for smaller scan lengths. For example, the scan rates in the fast and slow scan directions for an area of 10 µm × 10 µm scanned at 0.5 Hz are 10 µm/s and 20 nm/s, respectively.

We now describe the construction of a small-sample AFM in more detail. It consists of three main parts: the optical head which senses the cantilever deflection; a PZT tube scanner which controls the scanning motion of the sample mounted on one of its ends; and the base, which supports the scanner and head and includes circuits for the deflection signal (Fig. 2.12a). The AFM connects directly to a control system. The optical head consists of a laser diode stage, a photodiode stage preamp board, the cantilever mount and its holding arm, and the deflected beam reflecting mirror, which reflects the deflected beam toward the photodiode (Fig. 2.12b). The laser diode stage is a tilt stage used to adjust the position of the laser beam relative to the cantilever. It consists of the laser diode, collimator, focusing lens, baseplate, and the x and y laser diode positioners. The positioners are used to place the laser spot on the end of the cantilever. The photodiode stage is an adjustable stage used to position the photodiode elements relative to the reflected laser beam. It consists of the split photodiode, the base plate, and the photodiode positioners. The deflected beam reflecting mirror is mounted on the upper left in the interior of the head. The cantilever mount is a metal (for operation in air) or glass (for operation in water) block which holds the cantilever firmly at the proper angle (Fig. 2.12d). Next, the tube scanner consists of an Invar cylinder holding a single tube made of piezoelectric crystal

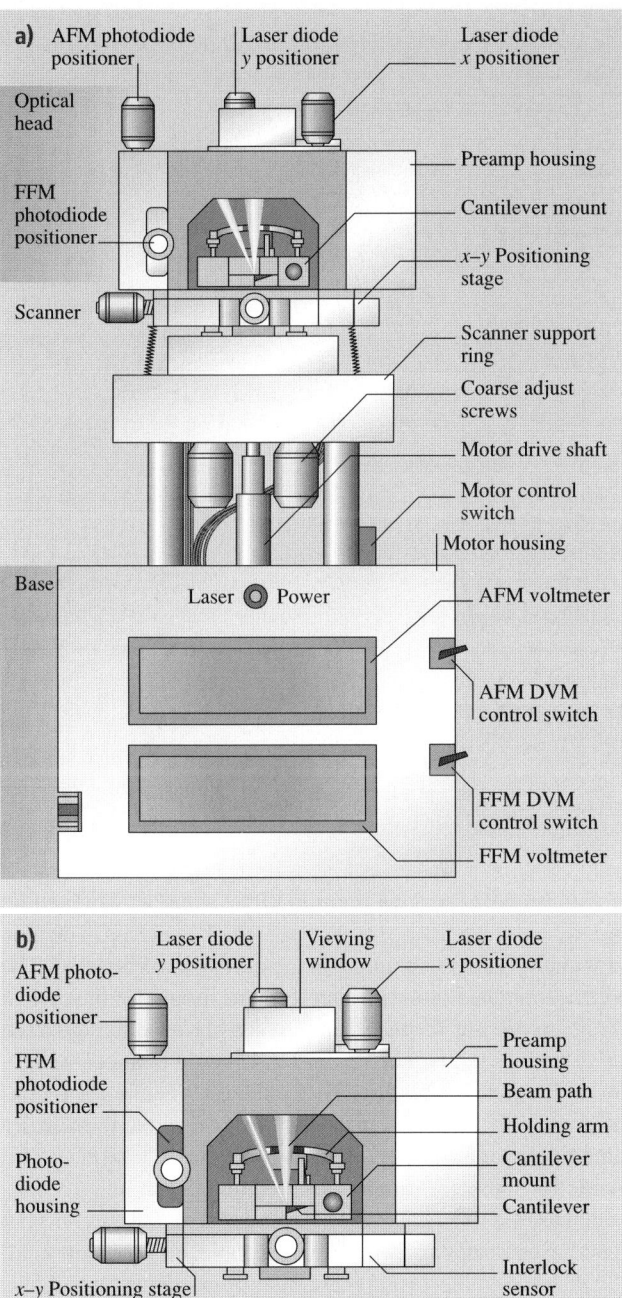

Fig. 2.12a,b. Schematics of a commercial AFM/FFM made by Digital Instruments Inc. (**a**) Front view, (**b**) optical head

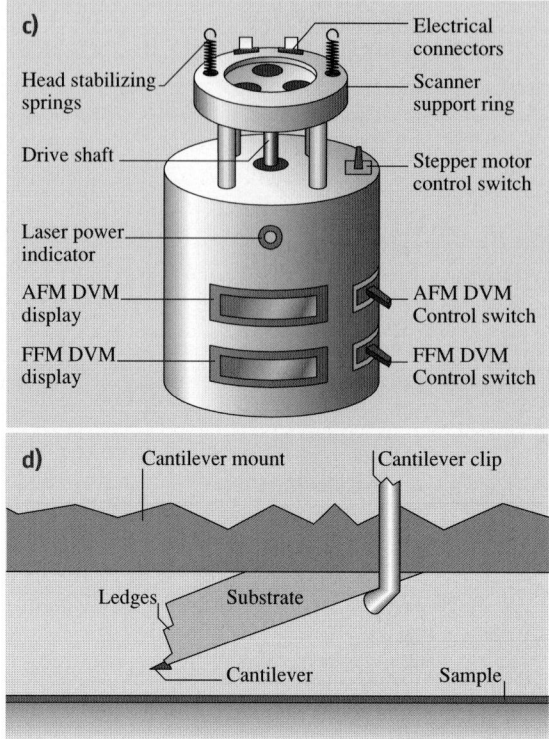

Fig. 2.12c,d. Schematics of a commercial AFM/FFM made by Digital Instruments Inc. (**c**) base, and (**d**) cantilever substrate mounted on cantilever mount (not to scale)

which imparts the necessary three-dimensional motion to the sample. Mounted on top of the tube is a magnetic cap on which the steel sample puck is placed. The tube is rigidly held at one end with the sample mounted on the other end of the tube. The scanner also contains three fine-pitched screws which form the mount for the optical head. The optical head rests on the tips of the screws, which are used to adjust the position of the head relative to the sample. The scanner fits into the scanner support ring mounted on the base of the microscope (Fig. 2.12c). The stepper motor is controlled manually with the switch on the upper surface of the base and automatically by the computer during the tip-engage and tip-withdraw processes.

The scan sizes available for these instruments are 0.7 μm, 12 μm and 125 μm. The scan rate must be decreased as the scan size is increased. A maximum scan rate of 122 Hz can be used. Scan rates of about 60 Hz should be used for small scan lengths (0.7 μm). Scan rates of 0.5 to 2.5 Hz should be used for large scans on samples with tall features. High scan rates help reduce drift, but they can only be used on flat samples with small scan sizes. The scan rate or the scanning speed (length/time) in the fast scan direction is equal to twice the scan length multiplied by the scan rate in Hz, and in the slow direction it is equal to the scan length multiplied by the scan rate in Hz divided by number of data points in the transverse direction. For example, for a scan size of 10 μm × 10 μm scanned at 0.5 Hz, the scan rates in the fast and slow scan directions are 10 μm/s and 20 nm/s, respectively. Normally 256 × 256 data points are taken for each image. The lateral resolution at larger scans is approximately equal to the scan length divided by 256. The piezo tube requires x–y calibration, which is carried out by imaging an appropriate calibration standard. Cleaved graphite is used for small scan heads, while two-dimensional grids (a gold-plated rule) can be used for long-range heads.

Examples of AFM images of freshly cleaved highly oriented pyrolytic (HOP) graphite and mica surfaces are shown in Fig. 2.13 [50, 110, 114]. Images with near-atomic resolution are obtained.

The force calibration mode is used to study interactions between the cantilever and the sample surface. In the force calibration mode, the x and y voltages applied to the piezo tube are held at zero and a sawtooth voltage is applied to the z electrode of the piezo tube, Fig. 2.14a. At the start of the force measurement the cantilever is in its rest position. By changing the applied voltage, the sample can be moved up and down relative to the stationary cantilever tip. As the piezo moves the sample up and down, the cantilever deflection signal from the photodiode is monitored. The force–distance curve, a plot of the cantilever tip deflection signal as a function of

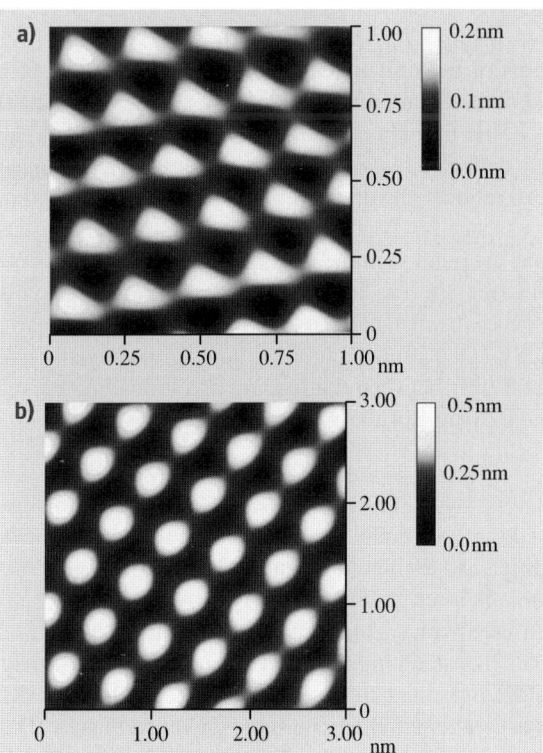

Fig. 2.13. Typical AFM images of freshly-cleaved (**a**) highly oriented pyrolytic graphite and (**b**) mica surfaces taken using a square pyramidal Si_3N_4 tip

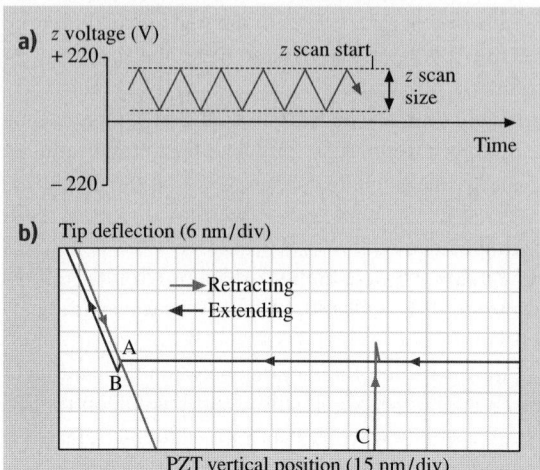

Fig. 2.14. (a) Force calibration Z waveform, and (b) a typical force–distance curve for a tip in contact with a sample. Contact occurs at point B; tip breaks free of adhesive forces at point C as the sample moves away from the tip

the voltage applied to the piezo tube, is obtained. Figure 2.14b shows the typical features of a force–distance curve. The arrowheads indicate the direction of piezo travel. As the piezo extends, it approaches the tip, which is in mid-air at this point and hence shows no deflection. This is indicated by the flat portion of the curve. As the tip approaches the sample to within a few nanometers (point A), an attractive force kicks in between the atoms of the tip surface and the atoms of the surface of the sample. The tip is pulled towards the sample and contact occurs at point B on the graph. From this point on, the tip is in contact with the surface, and as the piezo extends further, the tip gets deflected further. This is represented by the sloped portion of the curve. As the piezo retracts, the tip moves beyond the zero deflection (flat) line due to attractive forces (van der Waals forces and long-range meniscus forces), into the adhesive regime. At point C in the graph, the tip snaps free of the adhesive forces, and is again in free air. The horizontal distance between points B and C along the retrace line gives the distance moved by the tip in the adhesive regime. Multiplying this distance by the stiffness of the cantilever gives the adhesive force. Incidentally, the horizontal shift between the loading and unloading curves results from the hysteresis in the PZT tube [4].

Multimode Capabilities

The multimode AFM can be used for topography measurements in the contact mode and tapping mode, described earlier, and for measurements of lateral (friction) force, electric force gradients and magnetic force gradients.

The multimode AFM, when used with a grounded conducting tip, can be used to measure electric field gradients by oscillating the tip near its resonant frequency. When the lever encounters a force gradient from the electric field, the effective spring constant of the cantilever is altered, changing its resonant frequency. Depending on which side of the resonance curve is chosen, the oscillation amplitude of the cantilever increases or decreases due to the shift in the resonant frequency. By recording the amplitude of the cantilever, an image revealing the strength of the electric field gradient is obtained.

In the magnetic force microscope (MFM), used with a magnetically coated tip, static cantilever deflection is detected when a magnetic field exerts a force on the tip, and MFM images of magnetic materials can be obtained. MFM sensitivity can be enhanced by oscillating the cantilever near its resonant frequency. When the tip encounters a magnetic force gradient, the effective spring constant (and hence the resonant frequency) is shifted. By driving the cantilever above or below the resonant frequency, the oscillation amplitude varies as the resonance shifts. An image of the magnetic field gradient is obtained by recording the oscillation amplitude as the tip is scanned over the sample.

Topographic information is separated from the electric field gradient and magnetic field images using the so-called lift mode. In lift mode, measurements are taken in two passes over each scan line. In the first pass, topographical information is recorded in the standard tapping mode, where the oscillating cantilever lightly taps the surface. In the second pass, the tip is lifted to a user-selected separation (typically 20–200 nm) between the tip and local surface topography. By using stored topographical data instead of standard feedback, the tip–sample separation can be kept constant. In this way, the cantilever amplitude can be used to measure electric field force gradients or relatively weak but long-range magnetic forces without being influenced by topographic features. Two passes are made for every scan line, producing separate topographic and magnetic force images.

Electrochemical AFM

This option allows one to perform electrochemical reactions on the AFM. The technique involves a potentiostat, a fluid cell with a transparent cantilever holder and electrodes, and the software required to operate the potentiostat and display the results of the electrochemical reaction.

2.3.3 AFM Probe Construction

Various probes (cantilevers and tips) are used for AFM studies. The cantilever stylus used in the AFM should meet the following criteria: (1) low normal spring constant (stiffness); (2) high resonant frequency; (3) high cantilever quality factor Q; (4) high lateral spring constant (stiffness); (5) short cantilever length; (6) incorporation of components (such as mirror) for deflection sensing, and; (7) a sharp protruding tip [137]. In order to register a measurable deflection with small forces, the cantilever must flex with a relatively low force (on the order of few nN), requiring vertical spring constants of 10^{-2} to 10^2 N/m for atomic resolution in the contact profiling mode. The data rate or imaging rate in the AFM is limited by the mechanical resonant frequency of the cantilever. To achieve a large imaging bandwidth, the AFM cantilever should have a resonant frequency of more than about 10 kHz (30–100 kHz is preferable), which makes the cantilever the least sensitive part of the system. Fast imaging rates are not just a matter of convenience, since the effects of thermal drifts are more pronounced with slow scanning speeds. The combined requirements of a low spring constant and a high resonant frequency are met by reducing the mass of the cantilever. The quality factor $Q \, (= \omega_R/(c/m)$, where ω_R is the resonant frequency of the damped oscillator, c is the damping constant and m is the mass of the oscillator) should have a high value for some applications. For example, resonance curve detection is a sensitive modulation technique for measuring small force gradients in noncontact imaging. Increasing the Q increases the sensitivity of the measurements. Mechanical Q values of 100–1000 are typical. In contact modes, the Q value is of less importance. A high lateral cantilever spring constant is desirable in order to reduce the effect of lateral forces in the AFM, as frictional forces can cause appreciable lateral bending of the cantilever. Lateral bending results in erroneous topography measurements. For friction measurements, cantilevers with reduced lateral rigidity are preferred. A sharp protruding tip must be present at the end of the cantilever to provide a well-defined interaction with the sample over a small area. The tip radius should be much smaller than the radii of the corrugations in the sample in order for these to be measured accurately. The lateral spring constant depends critically on the tip length. Additionally, the tip should be centered at the free end.

In the past, cantilevers have been cut by hand from thin metal foils or formed from fine wires. Tips for these cantilevers were prepared by attaching diamond fragments to the ends of the cantilevers by hand, or in the case of wire cantilevers, electrochemically etching the wire to a sharp point. Several cantilever geometries for wire cantilevers have been used. The simplest geometry is the L-shaped cantilever, which is usually made by bending a wire at a 90° angle. Other geometries include single-V and double-V geometries, with a sharp tip attached at the apex of the V, and double-X configuration with a sharp tip attached at the intersection [31, 138]. These cantilevers can be constructed with high vertical spring constants. For example, a double-cross cantilever with an effective spring constant of 250 N/m was used by *Burnham* and *Colton* [31]. The small size and low mass needed in the AFM make hand fabrication of the cantilever a difficult process with poor reproducibility. Conventional microfabrication techniques are ideal for constructing planar thin-film structures which have submicron lateral dimensions. The triangular (V-shaped) cantilevers have improved (higher) lateral spring constants in comparison to rectangular cantilevers. In terms of spring constants, the triangular cantilevers are approximately equivalent to two rectangular cantilevers placed in parallel [137]. Although the macroscopic radius of a photolithographically patterned corner is seldom much less than about 50 nm, microscopic asperities on the etched surface provide tips with near-atomic dimensions.

Cantilevers have been used from a whole range of materials. Cantilevers made of Si_3N_4, Si, and diamond are the most commonl. The Young's modulus and the density are the material parameters that determine the resonant frequency, aside from the geometry. Table 2.2 shows the relevant properties and the speed of sound, indicative of the resonant frequency for a given shape. Hardness is an important

Table 2.2. Relevant properties of materials used for cantilevers

Property	Young's Modulus (E) (GPa)	Density (ρg) (kg/m^3)	Microhardness (GPa)	Speed of sound ($\sqrt{E/\rho}$) (m/s)
Diamond	900–1050	3515	78.4–102	17,000
Si$_3$N$_4$	310	3180	19.6	9900
Si	130–188	2330	9–10	8200
W	350	19,310	3.2	4250
Ir	530	–	≈ 3	5300

indicator of the durability of the cantilever, and is also listed in the table. Materials used for STM cantilevers are also included.

Silicon nitride cantilevers are less expensive than those made of other materials. They are very rugged and well suited to imaging in almost all environments. They are especially compatible with organic and biological materials. Microfabricated triangular silicon nitride beams with integrated square pyramidal tips made using plasma-enhanced chemical vapor deposition (PECVD) are the most common [137]. Four cantilevers, marketed by Digital Instruments, with different sizes and spring constants located on cantilever substrate made of boron silicate glass (Pyrex), are shown in Figs. 2.15a and 2.16. The two pairs of cantilevers on each substrate measure about 115 and 193 µm from the substrate to the apex of the triangular cantilever, with base widths of 122 and 205 µm, respectively. The cantilever legs, which are of the same thickness (0.6 µm) in all the cantilevers, are available in wide and narrow forms. Only one cantilever is selected and used from each substrate. The calculated spring constants and measured natural frequencies for each of the configurations are listed in Table 2.3. The most commonly used cantilever beam is the 115 µm-long, wide-legged cantilever (vertical spring constant = 0.58 N/m). Cantilevers with smaller spring constants should be used on softer samples. The pyramidal tip is highly symmetric, and the end has a radius of about 20–50 nm. The side walls of the tip have a slope of 35 deg and the lengths of the edges of the tip at the cantilever base are about 4 µm.

An alternative to silicon nitride cantilevers with integrated tips are microfabricated single-crystal silicon cantilevers with integrated tips. Si tips are sharper than Si$_3$N$_4$ tips because they are formed directly by anisotropic etching of single-crystal

Table 2.3. Measured vertical spring constants and natural frequencies of triangular (V-shaped) cantilevers made of PECVD Si$_3$N$_4$ (data provided by Digital Instruments, Inc.)

dimension	Spring constant (k_z) (N/m)	Natural frequency (ω_0) (kHz)
115-µm long, narrow leg	0.38	40
115-µm long, wide leg	0.58	40
193-µm long, narrow leg	0.06	13–22
193-µm long, wide leg	0.12	13–22

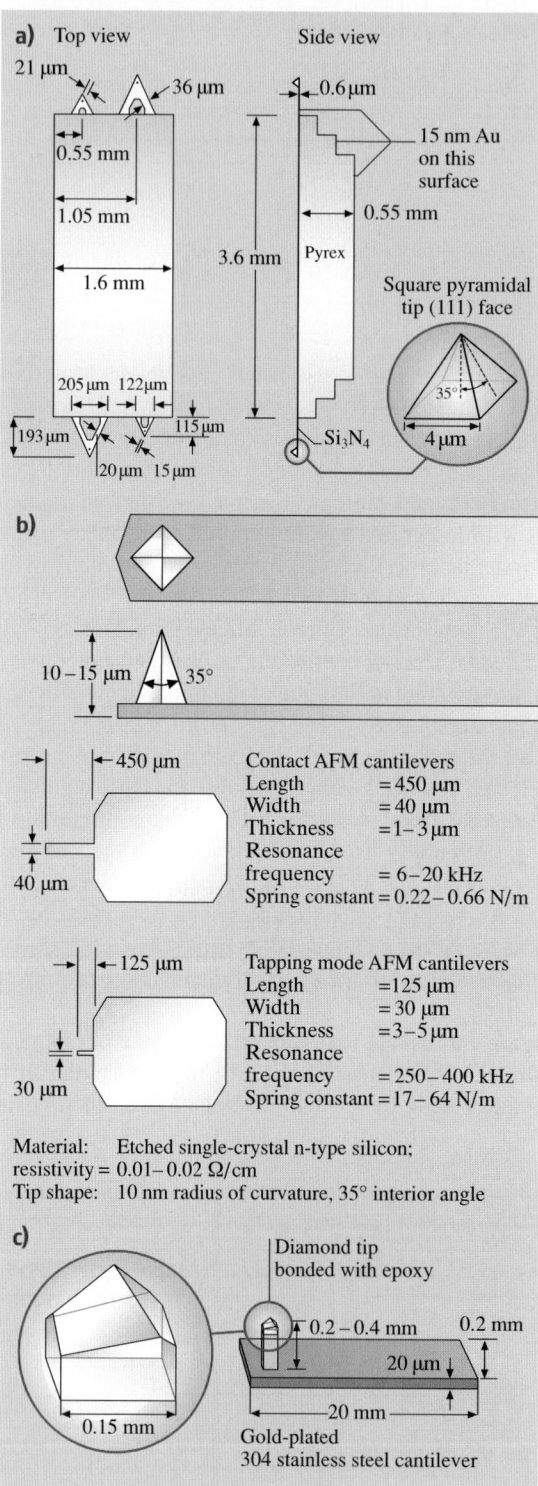

Fig. 2.15. Schematics of **(a)** triangular cantilever beam with square-pyramidal tips made of PECVD Si_3N_4, **(b)** rectangular cantilever beams with square-pyramidal tips made of etched single-crystal silicon, and **(c)** rectangular cantilever stainless steel beam with three-sided pyramidal natural diamond tip

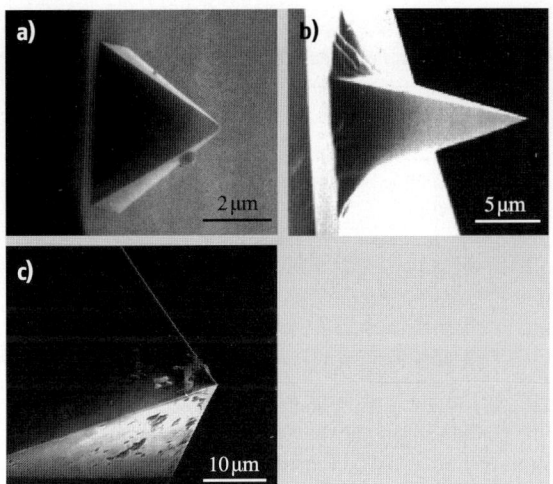

Fig. 2.16. SEM micrographs of a square-pyramidal PECVD Si_3N_4 tip (**a**), a square-pyramidal etched single-crystal silicon tip (**b**), and a three-sided pyramidal natural diamond tip (**c**)

Si, rather than through the use of an etch pit as a mask for the deposited material [139]. Etched single-crystal *n*-type silicon rectangular cantilevers with square pyramidal tips of radii <10 nm for contact and tapping mode (tapping-mode etched silicon probe or TESP) AFMs are commercially available from Digital Instruments and Nanosensors GmbH, Aidlingen, Germany, Figs. 2.15b and 2.16. Spring constants and resonant frequencies are also presented in the Fig. 2.15b.

Commercial triangular Si_3N_4 cantilevers have a typical width:thickness ratio of 10 to 30, which results in spring constants that are 100 to 1000 times stiffer in the lateral direction than in the normal direction. Therefore, these cantilevers are not well suited for torsion. For friction measurements, the torsional spring constant should be minimized in order to be sensitive to the lateral force. Rather long cantilevers with small thicknesses and large tip lengths are most suitable. Rectangular beams have smaller torsional spring constants than the triangular (V-shaped) cantilevers. Table 2.4 lists the spring constants (with the full length of the beam used) in three directions for typical rectangular beams. We note that the lateral and torsional spring constants are about two orders of magnitude larger than the normal spring constants. A cantilever beam required for the tapping mode is quite stiff and may not be sensitive enough for friction measurements. *Meyer* et al. [140] used a specially designed rectangular silicon cantilever with length = 200 μm, width = 21 μm, thickness = 0.4 μm, tip length = 12.5 μm and shear modulus = 50 GPa, giving a normal spring constant of 0.007 N/m and a torsional spring constant of 0.72 N/m, which gives a lateral force sensitivity of 10 pN and an angle of resolution of 10^{-7} rad. Using this particular geometry, the sensitivity to lateral forces can be improved by about a factor of 100 compared with commercial V-shaped Si_3N_4 or the rectangular Si or Si_3N_4 cantilevers used by *Meyer* and *Amer* [8], with torsional spring constants of ~ 100 N/m. *Ruan* and *Bhushan* [136] and *Bhushan* and *Ruan* [9] used 115 μm-long, wide-legged V-shaped cantilevers made of Si_3N_4 for friction measurements.

Table 2.4. Vertical (k_z), lateral (k_y), and torsional (k_{yT}) spring constants of rectangular cantilevers made of Si (IBM) and PECVD Si_3N_4 (source: Veeco Instruments, Inc.)

Dimensions/stiffness	Si cantilever	Si_3N_4 cantilever
Length (L) (μm)	100	100
Width (b) (μm)	10	20
Thickness (h) (μm)	1	0.6
Tip length (ℓ) (μm)	5	3
k_z (N/m)	0.4	0.15
k_y (N/m)	40	175
k_{yT} (N/m)	120	116
ω_0 (kHz)	~ 90	~ 65

Note: $k_z = Ebh^3/4L^3$, $k_y = Eb^3h/4\ell^3$, $k_{yT} = Gbh^3/3L\ell^2$, and $\omega_0 = [k_z/(m_c + 0.24bhL\rho)]^{1/2}$, where E is Young's modulus, G is the modulus of rigidity [$= E/2(1+\nu)$], ν is Poisson's ratio], ρ is the mass density of the cantilever, and m_c is the concentrated mass of the tip (~ 4 ng) [94]. For Si, $E = 130$ GPa, $\rho g = 2300$ kg/m^3, and $\nu = 0.3$. For Si_3N_4, $E = 150$ GPa, $\rho g = 3100$ kg/m^3, and $\nu = 0.3$

For scratching, wear and indentation studies, single-crystal natural diamond tips ground to the shape of a three-sided pyramid with an apex angle of either 60° or 80° and a point sharpened to a radius of about 100 nm are commonly used [4, 10] (Figs. 2.15c and 2.16). The tips are bonded with conductive epoxy to a gold-plated 304 stainless steel spring sheet (length = 20 mm, width = 0.2 mm, thickness = 20 to 60 μm) which acts as a cantilever. The free length of the spring is varied in order to change the beam stiffness. The normal spring constant of the beam ranges from about 5 to 600 N/m for a 20 μm-thick beam. The tips are produced by R-DEC Co., Tsukuba, Japan.

High aspect ratio tips are used to image within trenches. Examples of two probes used are shown in Fig. 2.17. These high aspect ratio tip (HART) probes are produced from conventional Si_3N_4 pyramidal probes. Through a combination of focused ion beam (FIB) and high-resolution scanning electron microscopy (SEM) techniques, a thin filament is grown at the apex of the pyramid. The probe filament is approximately 1 μm long and 0.1 μm in diameter. It tapers to an extremely sharp point (with a radius that is better than the resolutions of most SEMs). The long thin shape and sharp radius make it ideal for imaging within "vias" of microstructures and trenches (> 0.25 μm). This is, however, unsuitable for imaging structures at the atomic level, since probe flexing can create image artefacts. A FIB-milled probe is used for atomic-scale imaging, which is relatively stiff yet allows for closely spaced topography. These probes start out as conventional Si_3N_4 pyramidal probes, but the pyramid is FIB-milled until a small cone shape is formed which has a high aspect ratio and is 0.2–0.3 μm in length. The milled probes permit nanostructure resolution without sacrificing rigidity. These types of probes are manufactured by various manufacturers including Materials Analytical Services.

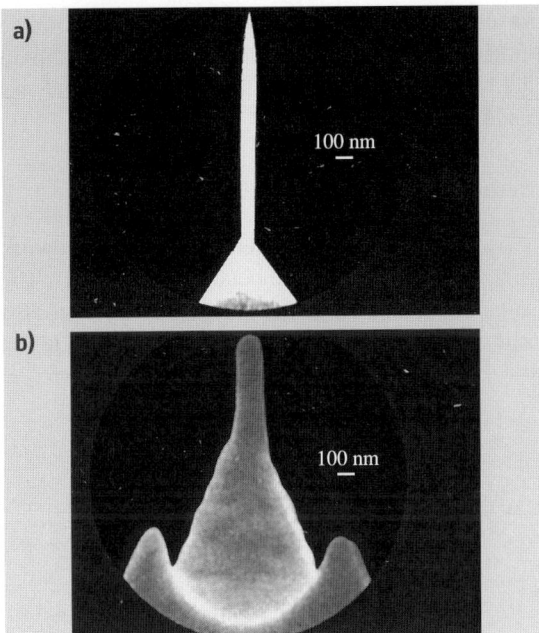

Fig. 2.17. Schematics of (**a**) HART Si$_3$N$_4$ probe, and (**b**) an FIB-milled Si$_3$N$_4$ probe

Carbon nanotube tips with small diameters and high aspect ratios are used for high-resolution imaging of surfaces and of deep trenches, in the tapping mode or the noncontact mode. Single-wall carbon nanotubes (SWNTs) are microscopic graphitic cylinders that are 0.7 to 3 nm in diameter and up to many microns in length. Larger structures called multiwall carbon nanotubes (MWNTs) consist of nested, concentrically arranged SWNTs and have diameters of 3 to 50 nm. MWNT carbon nanotube AFM tips are produced by manual assembly [141], chemical vapor deposition (CVD) synthesis, and a hybrid fabrication process [142]. Figure 2.18 shows a TEM micrograph of a carbon nanotube tip, ProbeMax, commercially produced by mechanical assembly by Piezomax Technologies, Inc. (Middleton, WI, USA). To fabricate these tips, MWNTs are produced using a carbon arc and they are physically attached to the single-crystal silicon, square-pyramidal tips in the SEM, using a manipulator and the SEM stage to independently control the nanotubes and

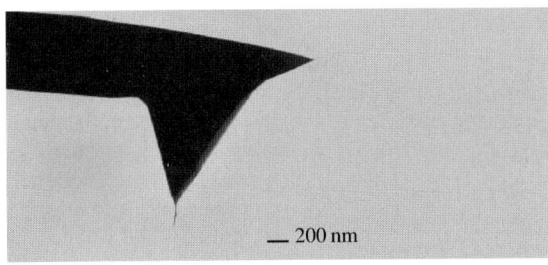

Fig. 2.18. SEM micrograph of a multiwall carbon nanotube (MWNT) tip physically attached to a single-crystal silicon, square-pyramidal tip (courtesy of Piezomax Technologies, Inc.)

the tip. When the nanotube is first attached to the tip, it is usually too long to image with. It is shortened by placing it in an AFM and applying voltage between the tip and the sample. Nanotube tips are also commercially produced by CVD synthesis by NanoDevices (Santa Barbara, CA, USA).

2.3.4 Friction Measurement Methods

The two methods for performing friction measurements that are based on the work by *Ruan* and *Bhushan* [136] are now described in more detail (also see [8]). The scanning angle is defined as the angle relative to the y-axis in Fig. 2.19a. This is also the long axis of the cantilever. The zero-degree scanning angle corresponds to the sample scan in the y-direction, and the 90-degree scanning angle corresponds to the

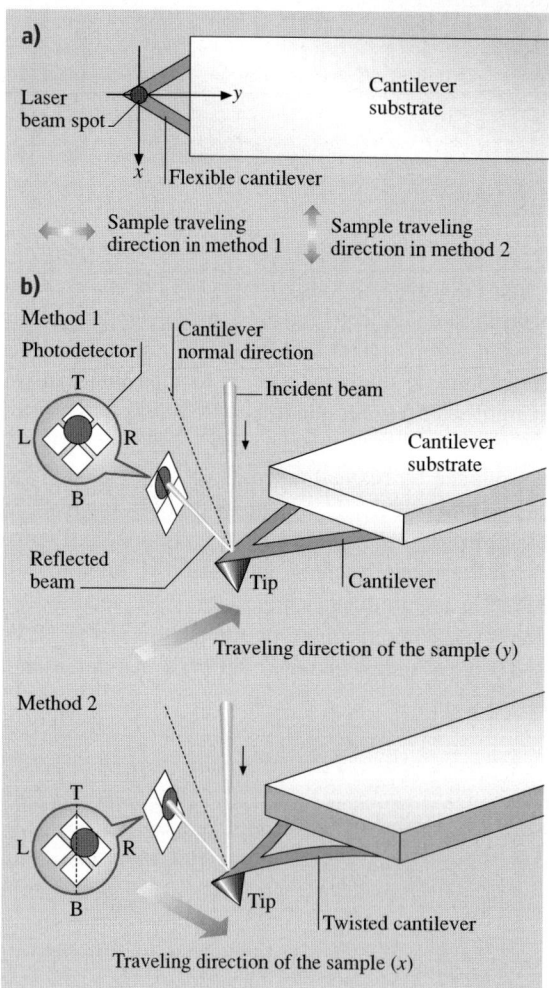

Fig. 2.19. (a) Schematic defining the x- and y-directions relative to the cantilever, and showing the direction of sample travel in two different measurement methods discussed in the text. (b) Schematic of deformation of the tip and cantilever shown as a result of sliding in the x- and y-directions. A twist is introduced to the cantilever if the scanning is performed in the x-direction ((b), *lower part*) [136]

sample scan perpendicular to this axis in the xy-plane (along x-axis). If both the y- and $-y$-directions are scanned, we call this a "parallel scan". Similarly, a "perpendicular scan" means that both the x- and $-x$-directions are scanned. The direction of sample travel for each of these two methods is illustrated in Fig. 2.19b.

Using method 1 ("height" mode with parallel scans) in addition to topographic imaging, it is also possible to measure friction force when the sample scanning direction is parallel to the y-direction (parallel scan). If there was no friction force between the tip and the moving sample, the topographic feature would be the only factor that would cause the cantilever to be deflected vertically. However, friction force does exist on all surfaces that are in contact where one of the surfaces is moving relative to the other. The friction force between the sample and the tip will also cause the cantilever to be deflected. We assume that the normal force between the sample and the tip is W_0 when the sample is stationary (W_0 is typically 10 nN to 200 nN), and the friction force between the sample and the tip is W_f as the sample is scanned by the tip. The direction of the friction force (W_f) is reversed as the scanning direction of the sample is reversed from the positive (y) to the negative ($-y$) direction ($W_{f(y)} = -W_{f(-y)}$).

When the vertical cantilever deflection is set at a constant level, it is the total force (normal force and friction force) applied to the cantilever that keeps the cantilever deflection at this level. Since the friction force is directed in the opposite direction to the direction of travel of the sample, the normal force will have to be adjusted accordingly when the sample reverses its traveling direction, so that the total deflection of the cantilever will remain the same. We can calculate the difference in the normal force between the two directions of travel for a given friction force W_f. First, since the deflection is constant, the total moment applied to the cantilever is constant. If we take the reference point to be the point where the cantilever joins the cantilever holder (substrate), point P in Fig. 2.20, we have the following

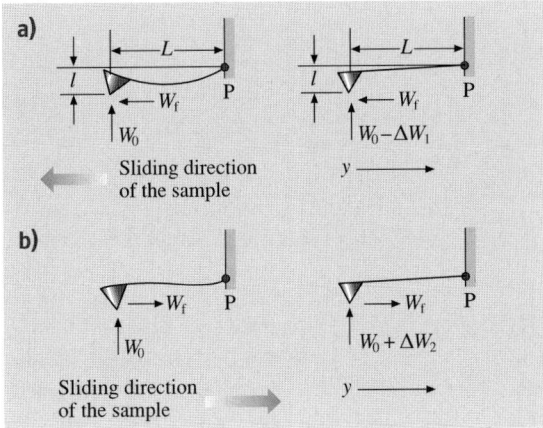

Fig. 2.20. (a) Schematic showing an additional bending of the cantilever due to friction force when the sample is scanned in the y- or $-y$-directions (*left*). (b) This effect can be canceled out by adjusting the piezo height using a feedback circuit (*right*) [136]

relationship:

$$(W_0 - \Delta W_1)L + W_f\ell = (W_0 + \Delta W_2)L - W_f\ell \tag{2.1}$$

or

$$(\Delta W_1 + \Delta W_2)L = 2W_f\ell. \tag{2.2}$$

Thus

$$W_f = (\Delta W_1 + \Delta W_2)L/(2\ell), \tag{2.3}$$

where ΔW_1 and ΔW_2 are the absolute values of the changes in normal force when the sample is traveling in the $-y$- and y-directions, respectively, as shown in Fig. 2.20; L is the length of the cantilever; ℓ is the vertical distance between the end of the tip and point P. The coefficient of friction (μ) between the tip and the sample is then given as

$$\mu = \frac{W_f}{W_0} = \left(\frac{(\Delta W_1 + \Delta W_2)}{W_0}\right)\left(\frac{L}{2\ell}\right). \tag{2.4}$$

There are adhesive and interatomic attractive forces between the cantilever tip and the sample at all times. The adhesive force can be due to water from the capillary condensation and other contaminants present at the surface, which form meniscus bridges [4, 143, 144] and the interatomic attractive force includes van der Waals attractions [18]. If these forces (and the effect of indentation too, which is usually small for rigid samples) can be neglected, the normal force W_0 is then equal to the initial cantilever deflection H_0 multiplied by the spring constant of the cantilever. $(\Delta W_1 + \Delta W_2)$ can be derived by multiplying the same spring constant by the change in height of the piezo tube between the two traveling directions (y- and $-y$-directions) of the sample. This height difference is denoted as $(\Delta H_1 + \Delta H_2)$, shown schematically in Fig. 2.21. Thus, (2.4) can be rewritten as

$$\mu = \frac{W_f}{W_0} = \left(\frac{(\Delta H_1 + \Delta H_2)}{H_0}\right)\left(\frac{L}{2\ell}\right). \tag{2.5}$$

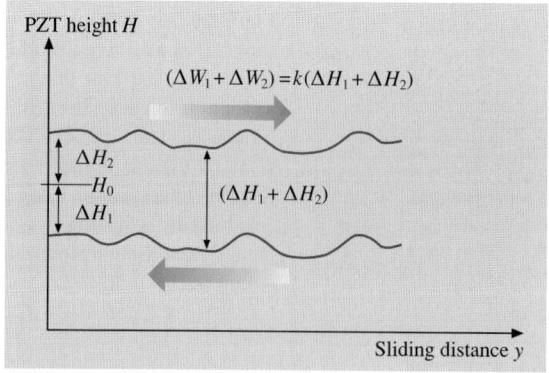

Fig. 2.21. Schematic illustration of the height difference for the piezoelectric tube scanner as the sample is scanned in the y- and $-y$-directions

Since the vertical position of the piezo tube is affected by the topographic profile of the sample surface in addition to the friction force being applied at the tip, this difference must be found point-by-point at the same location on the sample surface, as shown in Fig. 2.21. Subtraction of point-by-point measurements may introduce errors, particularly for rough samples. We will come back to this point later. In addition, precise measurements of L and ℓ (which should include the cantilever angle) are also required.

If the adhesive force between the tip and the sample is large enough that it cannot be neglected, it should be included in the calculation. However, determinations of this force can involve large uncertainties, which is introduced into (2.5). An alternative approach is to make the measurements at different normal loads and to use $\Delta(H_0)$ and $\Delta(\Delta H_1 + \Delta H_2)$ in (2.5). Another comment on (2.5) is that, since only the ratio between $(\Delta H_1 + \Delta H_2)$ and H_0 enters this equation, the vertical position of the piezo tube H_0 and the difference in position $(\Delta H_1 + \Delta H_2)$ can be in volts as long as the vertical travel of the piezo tube and the voltage applied to have a linear relationship. However, if there is a large nonlinearity between the piezo tube traveling distance and the applied voltage, this nonlinearity must be included in the calculation.

It should also be pointed out that (2.4) and (2.5) are derived under the assumption that the friction force W_f is the same for the two scanning directions of the sample. This is an approximation, since the normal force is slightly different for the two scans and the friction may be direction-dependent. However, this difference is much smaller than W_0 itself. We can ignore the second-order correction.

Method 2 ("aux" mode with perpendicular scan) of measuring friction was suggested by *Meyer* and *Amer* [8]. The sample is scanned perpendicular to the long axis of the cantilever beam (along the x- or $-x$-direction in Fig. 2.19a) and the outputs from the two horizontal quadrants of the photodiode detector are measured. In this arrangement, as the sample moves under the tip, the friction force will cause the cantilever to twist. Therefore, the light intensity between the left and right (L and R in Fig. 2.19b, right) detectors will be different. The differential signal between the left and right detectors is denoted the FFM signal $[(L-R)/(L+R)]$. This signal can be related to the degree of twisting, and hence to the magnitude of friction force. Again, because possible errors in measurements of the normal force due to the presence of adhesive force at the tip–sample interface, the slope of the friction data (FFM signal vs. normal load) needs to be measured for an accurate value of the coefficient of friction.

While friction force contributes to the FFM signal, friction force may not be the only contributing factor in commercial FFM instruments (for example, NanoScope IV). One can see this if we simply engange the cantilever tip with the sample. The left and right detectors can be balanced beforehand by adjusting the positions of the detectors so that the intensity difference between these two detectors is zero (FFM signal is zero). Once the tip is engaged with the sample, this signal is no longer zero, even if the sample is not moving in the xy-plane with no friction force applied. This would be a detrimental effect. It has to be understood and eliminated from the data acquisition before any quantitative measurement of friction force is made.

One of the reasons for this observation is as follows. The detectors may not have been properly aligned with respect to the laser beam. To be precise, the vertical axis of the detector assembly (the line joining T–B in Fig. 2.22) is not in the plane defined by the incident laser beam and the beam reflected from the untwisted cantilever (we call this plane the "beam plane"). When the cantilever vertical deflection changes due to a change in the normal force applied (without the sample being scanned in the xy-plane), the laser beam will be reflected up and down and form a projected trajectory on the detector. (Note that this trajectory is in the defined beam plane.) If this trajectory is not coincident with the vertical axis of the detector, the laser beam will not evenly bisect the left and right quadrants of the detectors, even under the condition of no torsional motion of the cantilever, see Fig. 2.22. Thus, when the laser beam is reflected up and down due a change in the normal force, the intensity difference between the left and right detectors will also change. In other words, the FFM signal will change as the normal force applied to the tip is changed, even if the tip is not experiencing any friction force. This (FFM) signal is unrelated to friction force or to the actual twisting of the cantilever. We will call this part of the FFM signal "FFM_F", and the part which is truly related to friction force "FFM_T".

The FFM_F signal can be eliminated. One way of doing this is as follows. First the sample is scanned in both the x- and the $-x$-directions and the FFM signals for scans in each direction are recorded. Since the friction force reverses its direction of

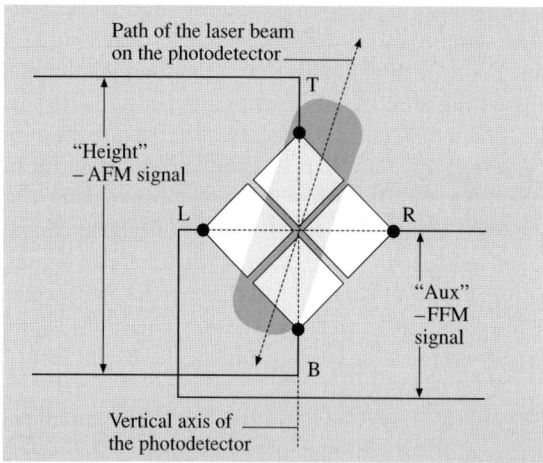

Fig. 2.22. The trajectory of the laser beam on the photodetectors as the cantilever is vertically deflected (with no torsional motion) with respect to the laser beam for a misaligned photodetector. For a change of normal force (vertical deflection of the cantilever), the laser beam is projected to a different position on the detector. Due to a misalignment, the projected trajectory of the laser beam on the detector is not parallel with the detector vertical axis (the line T–B) [136]

action when the scanning direction is reversed from the x- to the $-x$-direction, the FFM_T signal will change signs as the scanning direction of the sample is reversed ($FFM_T(x) = -FFM_T(-x)$). Hence the FFM_T signal will be canceled out if we take the sum of the FFM signals for the two scans. The average value of the two scans will be related to FFM_F due to the misalignment,

$$FFM(x) + FFM(-x) = 2FFM_F. \qquad (2.6)$$

This value can therefore be subtracted from the original FFM signals of each of these two scans to obtain the true FFM signal (FFM_T). Or, alternately, by taking the difference of the two FFM signals, one gets the FFM_T value directly:

$$\begin{aligned} FFM(x) - FFM(-x) &= FFM_T(x) - FFM_T(-x) \\ &= 2FFM_T(x). \end{aligned} \qquad (2.7)$$

Ruan and *Bhushan* [136] have shown that the error signal (FFM_F) can be very large compared to the friction signal FFM_T, so correction is required.

Now we compare the two methods. The method of using the "height" mode and parallel scanning (method 1) is very simple to use. Technically, this method can provide 3-D friction profiles and the corresponding topographic profiles. However, there are some problems with this method. Under most circumstances, the piezo scanner displays hysteresis when the traveling direction of the sample is reversed. Therefore, the measured surface topographic profiles will be shifted relative to each other along the y-axis for the two opposite (y and $-y$) scans. This would make it difficult to measure the local difference in height of the piezo tube for the two scans. However, the average difference in height between the two scans and hence the average friction can still be measured. The measurement of average friction can serve as an internal means of friction force calibration. Method 2 is a more desirable approach. The subtraction of the FFM_F signal from FFM for the two scans does not introduce any error into local friction force data. An ideal approach when using this method would be to add the average values of the two profiles in order to get the error component (FFM_F) and then subtract this component from either profile to get true friction profiles in either directions. By performing measurements at various loads, we can get the average value of the coefficient of friction which then can be used to convert the friction profile to the coefficient of friction profile. Thus, any directionality and local variations in friction can be easily measured. In this method, since topography data are not affected by friction, accurate topography data can be measured simultaneously with friction data and a better localized relationship between the two can be established.

2.3.5 Normal Force and Friction Force Calibrations of Cantilever Beams

Based on *Ruan* and *Bhushan* [136], we now discuss normal force and friction force calibrations. In order to calculate the absolute values of normal and friction forces in Newtons using the measured AFM and FFM_T voltage signals, it is necessary to first have an accurate value of the spring constant of the cantilever (k_c). The spring constant can be calculated using the geometry and the physical properties of the cantilever material [8, 94, 137]. However, the properties of the PECVD Si_3N_4 (used to fabricate cantilevers) can be different from those of the bulk material. For example, using ultrasonics, we found the Young's modulus of the cantilever beam to be about 238 ± 18 GPa, which is less than that of bulk Si_3N_4 (310 GPa). Furthermore, the thickness of the beam is nonuniform and difficult to measure precisely. Since the stiffness of a beam goes as the cube of thickness, minor errors in precise measurements of thickness can introduce substantial stiffness errors. Thus one should measure the spring constant of the cantilever experimentally. *Cleveland* et al. [145] measured normal spring constants by measuring resonant frequencies of beams.

For normal spring constant measurement, *Ruan* and *Bhushan* [136] used a stainless steel spring sheet of known stiffness (width = 1.35 mm, thickness = 15 μm, free hanging length = 5.2 mm). One end of the spring was attached to the sample holder and the other end was made to contact with the cantilever tip during the measurement, see Fig. 2.23. They measured the piezo travel for a given cantilever deflection. For a rigid sample (such as diamond), the piezo travel Z_t (measured from the point where the tip touches the sample) should equal the cantilever deflection. To maintain the cantilever deflection at the same level using a flexible spring sheet, the new piezo travel $Z_{t'}$ would need to be different from Z_t. The difference between $Z_{t'}$ and Z_t corresponds to the deflection of the spring sheet. If the spring constant of the spring sheet is k_s, the spring constant of the cantilever k_c can be calculated by

$$(Z_{t'} - Z_t)k_s = Z_t k_c$$

or

$$k_c = k_s(Z_{t'} - Z_t)/Z_t. \qquad (2.8)$$

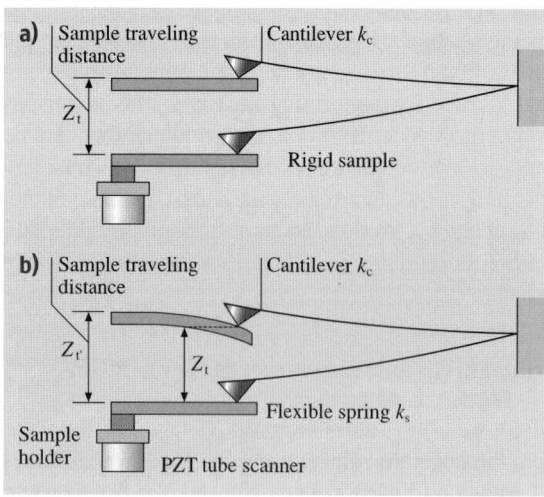

Fig. 2.23. Illustration showing the deflection of the cantilever as it is pushed by (**a**) a rigid sample, (**b**) a flexible spring sheet [136]

The spring constant of the spring sheet (k_s) used in this study is calculated to be 1.54 N/m. For the wide-legged cantilever used in our study (length = 115 μm, base width = 122 μm, leg width = 21 μm and thickness = 0.6 μm), k_c was measured to be 0.40 N/m instead of the 0.58 N/m reported by its manufacturer – Digital Instruments Inc. To relate the photodiode detector output to the cantilever deflection in nanometers, they used the same rigid sample to push against the AFM tip. Since the cantilever vertical deflection equals the sample traveling distance measured from the point where the tip touches the sample for a rigid sample, the photodiode output observed as the tip is pushed by the sample can be converted directly to the cantilever deflection. For these measurements, they found the conversion factor to be 20 nm/V.

The normal force applied to the tip can be calculated by multiplying the cantilever vertical deflection by the cantilever spring constant for samples that have very small adhesion with the tip. If the adhesive force between the sample and the tip is large, it should be included in the normal force calculation. This is particularly important in atomic-scale force measurements, because the typical normal force that is measured in this region is in the range of a few hundreds of nN to a few mN. The adhesive force could be comparable to the applied force.

The conversion of friction signal (from FFM_T) to friction force is not as straightforward. For example, one can calculate the degree of twisting for a given friction force using the geometry and the physical properties of the cantilever [53, 144]. One would need information about the detector such as its quantum efficiency, laser power, gain and so on in order to be able convert the signal into the degree of twisting. Generally speaking, this procedure can not be accomplished without having some detailed information about the instrument. This information is not usually provided by the manufacturer. Even if this information is readily available, errors may still occur when using this approach because there will always be variations as a result of the instrumental set-up. For example, it has been noticed that the measured FFM_T signal varies for the same sample when different AFM microscopes from the same manufacturer are used. This means that one can not calibrate the instrument experimentally using this calculation. *O'Shea* et al. [144] did perform a calibration procedure in which the torsional signal was measured as the sample was displaced a known distance laterally while ensuring that the tip did not slide over the surface. However, it is difficult to verify that tip sliding does not occur.

A new method of calibration is therefore required. There is a simpler, more direct way of doing this. The first method described above (method 1) of measuring friction can provide an absolute value of the coefficient of friction directly. It can therefore be used as an internal calibration technique for data obtained using method 2. Or, for a polished sample, which introduces the least error into friction measurements taken using method 1, method 1 can be used to calibrate the friction force for method 2. Then this calibration can be used for measurements taken using method 2. In method 1, the length of the cantilever required can be measured using an optical microscope; the length of the tip can be measured using a scanning electron microscope. The relative angle between the cantilever and the horizontal sample surface can be measured directly. This enables the coefficient of friction to be measured with few unknown parameters. The friction force can then be calculated by multiplying the coefficient of friction by the normal load. The FFM_T signal obtained using method 2 is then converted into the friction force. For their instrument, they found the conversion to be 8.6 nN/V.

2.4 AFM Instrumentation and Analyses

The performance of AFMs and the quality of AFM images greatly depend on the instrument available and the probes (cantilever and tips) in use. This section describes the mechanics of cantilevers, instrumentation and analysis of force detection systems for cantilever deflections, and scanning and control systems.

2.4.1 The Mechanics of Cantilevers

Stiffness and Resonances of Lumped Mass Systems

All of the building blocks of an AFM, including the body of the microscope itself and the force-measuring cantilevers, are mechanical resonators. These resonances can be excited either by the surroundings or by the rapid movement of the tip or the sample. To avoid problems due to building- or air-induced oscillations, it is of paramount importance to optimize the design of the AFM for high resonant frequencies. This usually means decreasing the size of the microscope [146]. By using cube-like or sphere-like structures for the microscope, one can considerably increase the lowest eigenfrequency. The fundamental natural frequency ω_0 of any spring is given by

$$\omega_0 = \frac{1}{2\pi}\sqrt{\frac{k}{m_{\text{eff}}}}, \qquad (2.9)$$

where k is the spring constant (stiffness) in the normal direction and m_{eff} is the effective mass. The spring constant k of a cantilever beam with uniform cross-section (Fig. 2.24) is given by [147]

$$k = \frac{3EI}{L^3}, \qquad (2.10)$$

where E is the Young's modulus of the material, L is the length of the beam and I is the moment of inertia of the cross-section. For a rectangular cross-section with a width b (perpendicular to the deflection) and a height h one obtains the following expression for I

$$I = \frac{bh^3}{12}. \qquad (2.11)$$

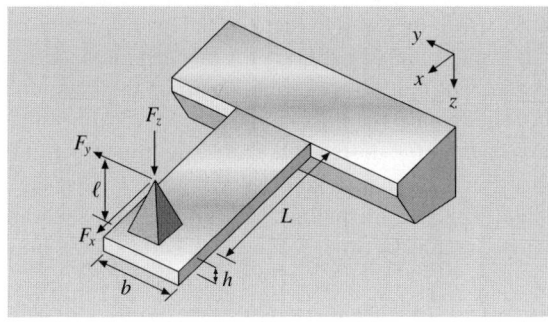

Fig. 2.24. A typical AFM cantilever with length L, width b, and height h. The height of the tip is ℓ. The material is characterized by the Young's modulus E, the shear modulus G and the mass density ρ. Normal (F_z), axial (F_x) and lateral (F_y) forces exist at the end of the tip

Combining (2.9), (2.10) and (2.11), we get an expression for ω_0

$$\omega_0 = \sqrt{\frac{Ebh^3}{4L^3 m_{\text{eff}}}}. \tag{2.12}$$

The effective mass can be calculated using Raleigh's method. The general formula using Raleigh's method for the kinetic energy T of a bar is

$$T = \frac{1}{2} \int_0^L \frac{m}{L} \left(\frac{\partial z(x)}{\partial t}\right)^2 dx. \tag{2.13}$$

For the case of a uniform beam with a constant cross-section and length L, one obtains for the deflection $z(x) = z_{\max}[1-(3x/2L)+(x^3/2L^3)]$. Inserting $z_{\max}$ into (2.13) and solving the integral gives

$$T = \frac{1}{2} \int_0^L \frac{m}{L} \left[\frac{\partial z_{\max}(x)}{\partial t}\left(1 - \frac{3x}{2L}\right) + \left(\frac{x^3}{L^3}\right)\right]^2 dx$$
$$= \frac{1}{2} m_{\text{eff}}(z_{\max} t)^2,$$

which gives

$$m_{\text{eff}} = \frac{9}{20} m. \tag{2.14}$$

Substituting (2.14) into (2.12) and noting that $m = \rho Lbh$, where ρ is the mass density, one obtains the following expression:

$$\omega_0 = \left(\frac{\sqrt{5}}{3}\right) \sqrt{\frac{E}{\rho}} \frac{h}{L^2}. \tag{2.15}$$

It is evident from (2.15) that one way to increase the natural frequency is to choose a material with a high ratio E/ρ; see Table 2.2 for typical values of $\sqrt{E/\rho}$

for various commonly used materials. Another way to increase the lowest eigenfrequency is also evident in (2.15). By optimizing the ratio h/L^2, one can increase the resonant frequency. However, it does not help to make the length of the structure smaller than the width or height. Their roles will just be interchanged. Hence the optimum structure is a cube. This leads to the design rule that long, thin structures like sheet metal should be avoided. For a given resonant frequency, the quality factor Q should be as low as possible. This means that an inelastic medium such as rubber should be in contact with the structure in order to convert kinetic energy into heat.

Stiffness and Resonances of Cantilevers

Cantilevers are mechanical devices specially shaped to measure tiny forces. The analysis given in the previous section is applicable. However, to better understand the intricacies of force detection systems, we will discuss the example of a cantilever beam with uniform cross-section, Fig. 2.24. The bending of a beam due to a normal load on the beam is governed by the Euler equation [147]

$$M = EI(x)\frac{d^2z}{dx^2}, \tag{2.16}$$

where M is the bending moment acting on the beam cross-section. $I(x)$ is the moment of inertia of the cross-section with respect to the neutral axis, defined by

$$I(x) = \int_z \int_y z^2 \, dy \, dz. \tag{2.17}$$

For a normal force F_z acting at the tip,

$$M(x) = (L-x)F_z \tag{2.18}$$

since the moment must vanish at the endpoint of the cantilever. Integrating (2.16) for a normal force F_z acting at the tip and observing that EI is a constant for beams with a uniform cross-section, one gets

$$z(x) = \frac{L^3}{6EI}\left(\frac{x}{L}\right)^2\left(3 - \frac{x}{L}\right)F_z. \tag{2.19}$$

The slope of the beam is

$$z'(x) = \frac{Lx}{2EI}\left(2 - \frac{x}{L}\right)F_z. \tag{2.20}$$

From (2.19) and (2.20), at the end of the cantilever (for $x = L$), for a rectangular beam, and by using an expression for I in (2.11), one gets,

$$z(L) = \frac{4}{Eb}\left(\frac{L}{h}\right)^3 F_z, \tag{2.21}$$

$$z'(L) = \frac{3}{2}\left(\frac{z}{L}\right). \tag{2.22}$$

Now, the stiffness in the normal (z) direction, k_z, is

$$k_z = \frac{F_z}{z(L)} = \frac{Eb}{4}\left(\frac{h}{L}\right)^3. \tag{2.23}$$

and the change in angular orientation of the end of cantilever beam is

$$\Delta\alpha = \frac{3}{2}\frac{z}{L} = \frac{6}{Ebh}\left(\frac{L}{h}\right)^2 F_z. \tag{2.24}$$

Now we ask what will, to a first-order approximation, happen if we apply a lateral force F_y to the end of the tip (Fig. 2.24). The cantilever will bend sideways and it will twist. The stiffness in the lateral (y) direction, k_y, can be calculated with (2.23) by exchanging b and h

$$k_y = \frac{Eh}{4}\left(\frac{b}{L}\right)^3. \tag{2.25}$$

Therefore, the bending stiffness in the lateral direction is larger than the stiffness for bending in the normal direction by $(b/h)^2$. The twisting or torsion on the other hand is more complicated to handle. For a wide, thin cantilever ($b \gg h$) we obtain torsional stiffness along y-axis, k_{yT}

$$k_{yT} = \frac{Gbh^3}{3L\ell^2}, \tag{2.26}$$

where G is the modulus of rigidity ($= E/2(1+\nu)$; ν is Poisson's ratio). The ratio of the torsional stiffness to the lateral bending stiffness is

$$\frac{k_{yT}}{k_y} = \frac{1}{2}\left(\frac{\ell b}{hL}\right)^2, \tag{2.27}$$

where we assume $\nu = 0.333$. We see that thin, wide cantilevers with long tips favor torsion while cantilevers with square cross-sections and short tips favor bending. Finally, we calculate the ratio between the torsional stiffness and the normal bending stiffness,

$$\frac{k_{yT}}{k_z} = 2\left(\frac{L}{\ell}\right)^2. \tag{2.28}$$

Equations (2.26) to (2.28) hold in the case where the cantilever tip is exactly in the middle axis of the cantilever. Triangular cantilevers and cantilevers with tips which are not on the middle axis can be dealt with by finite element methods.

The third possible deflection mode is the one from the force on the end of the tip along the cantilever axis, F_x (Fig. 2.24). The bending moment at the free end of the cantilever is equal to $F_x\ell$. This leads to the following modification of (2.18) for forces F_z and F_x

$$M(x) = (L-x)F_z + F_x\ell. \tag{2.29}$$

Integration of (2.16) now leads to

$$z(x) = \frac{1}{2EI}\left[Lx^2\left(1 - \frac{x}{3L}\right)F_z + \ell x^2 F_x\right] \quad (2.30)$$

and

$$z'(x) = \frac{1}{EI}\left[\frac{Lx}{2}\left(2 - \frac{x}{L}\right)F_z + \ell x F_x\right]. \quad (2.31)$$

Evaluating (2.30) and (2.31) at the end of the cantilever, we get the deflection and the tilt

$$z(L) = \frac{L^2}{EI}\left(\frac{L}{3}F_z - \frac{\ell}{2}F_x\right),$$

$$z'(L) = \frac{L}{EI}\left(\frac{L}{2}F_z + \ell F_x\right). \quad (2.32)$$

From these equations, one gets

$$F_z = \frac{12EI}{L^3}\left[z(L) - \frac{Lz'(L)}{2}\right],$$

$$F_x = \frac{2EI}{\ell L^2}[2Lz'(L) - 3z(L)]. \quad (2.33)$$

A second class of interesting properties of cantilevers is their resonance behavior. For cantilever beams, one can calculate the resonant frequencies [147, 148]

$$\omega_n^{\text{free}} = \frac{\lambda_n^2}{2\sqrt{3}}\frac{h}{L^2}\sqrt{\frac{E}{\rho}} \quad (2.34)$$

with $\lambda_0 = (0.596864\ldots)\pi$, $\lambda_1 = (1.494175\ldots)\pi$, $\lambda_n \to (n + 1/2)\pi$. The subscript n represents the order of the frequency, such as the fundamental, the second mode, and the nth mode.

A similar equation to (2.34) holds for cantilevers in rigid contact with the surface. Since there is an additional restriction on the movement of the cantilever, namely the location of its endpoint, the resonant frequency increases. Only the terms of λ_n change to [148]

$$\lambda'_0 = (1.2498763\ldots)\pi, \quad \lambda'_1 = (2.2499997\ldots)\pi,$$

$$\lambda'_n \to (n + 1/4)\pi. \quad (2.35)$$

The ratio of the fundamental resonant frequency during contact to the fundamental resonant frequency when not in contact is 4.3851.

For the torsional mode we can calculate the resonant frequencies as

$$\omega_0^{\text{tors}} = 2\pi\frac{h}{Lb}\sqrt{\frac{G}{\rho}}. \quad (2.36)$$

For cantilevers in rigid contact with the surface, we obtain the following expression for the fundamental resonant frequency: [148]

$$\omega_0^{\text{tors,contact}} = \frac{\omega_0^{\text{tors}}}{\sqrt{1+3(2L/b)^2}}. \tag{2.37}$$

The amplitude of the thermally induced vibration can be calculated from the resonant frequency using

$$\Delta z_{\text{therm}} = \sqrt{\frac{k_B T}{k}}, \tag{2.38}$$

where k_B is Boltzmann's constant and T is the absolute temperature. Since AFM cantilevers are resonant structures, sometimes with rather high Q values, the thermal noise is not as evenly distributed as (2.38) suggests. The spectral noise density below the peak of the response curve is [148]

$$z_0 = \sqrt{\frac{4k_B T}{k\omega_0 Q}} \quad (\text{in m}/\sqrt{\text{Hz}}), \tag{2.39}$$

where Q is the quality factor of the cantilever, described earlier.

2.4.2 Instrumentation and Analyses of Detection Systems for Cantilever Deflections

A summary of selected detection systems was provided in Fig. 2.8. Here we discuss the pros and cons of various systems in detail.

Optical Interferometer Detection Systems

Soon after the first papers on the AFM [2] appeared, which used a tunneling sensor, an instrument based on an interferometer was published [149]. The sensitivity of the interferometer depends on the wavelength of the light employed in the apparatus. Figure 2.25 shows the principle of such an interferometeric design. The light incident from the left is focused by a lens onto the cantilever. The reflected light is collimated by the same lens and interferes with the light reflected at the flat. To separate the reflected light from the incident light, a $\lambda/4$ plate converts the linearly polarized incident light into circularly polarized light. The reflected light is made linearly polarized again by the $\lambda/4$-plate, but with a polarization orthogonal to that of the incident light. The polarizing beam splitter then deflects the reflected light to the photodiode.

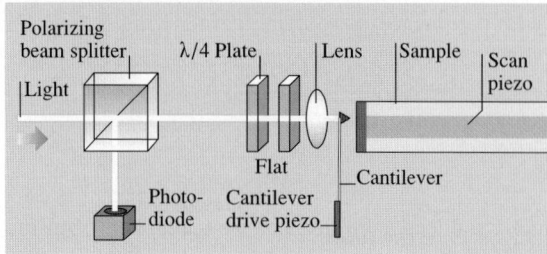

Fig. 2.25. Principle of an interferometric AFM. The light from the laser light source is polarized by the polarizing beam splitter and focused onto the back of the cantilever. The light passes twice through a quarter-wave plate and is hence orthogonally polarized to the incident light. The second arm of the interferometer is formed by the flat. The interference pattern is modulated by the oscillating cantilever

Homodyne Interferometer. To improve the signal-to-noise ratio of the interferometer, the cantilever is driven by a piezo near its resonant frequency. The amplitude Δz of the cantilever as a function of driving frequency Ω is

$$\Delta z(\Omega) = \Delta z_0 \frac{\Omega_0^2}{\sqrt{\left(\Omega^2 - \Omega_0^2\right)^2 + \frac{\Omega^2 \Omega_0^2}{Q^2}}}, \tag{2.40}$$

where Δz_0 is the constant drive amplitude and Ω_0 the resonant frequency of the cantilever. The resonant frequency of the cantilever is given by the effective potential

$$\Omega_0 = \sqrt{\left(k + \frac{\partial^2 U}{\partial z^2}\right)\frac{1}{m_{\text{eff}}}}, \tag{2.41}$$

where U is the interaction potential between the tip and the sample. Equation (2.41) shows that an attractive potential decreases Ω_0. The change in Ω_0 in turn results in a change in Δz (2.40). The movement of the cantilever changes the path difference in the interferometer. The light reflected from the cantilever with amplitude $A_{\ell,0}$ and the reference light with amplitude $A_{r,0}$ interfere on the detector. The detected intensity $I(t) = [A_\ell(t) + A_r(t)]^2$ consists of two constant terms and a fluctuating term

$$2A_\ell(t)A_r(t) = A_{\ell,0}A_{r,0} \sin\left[\omega t + \frac{4\pi\delta}{\lambda} + \frac{4\pi\Delta z}{\lambda}\sin(\Omega t)\right]\sin(\omega t). \tag{2.42}$$

Here ω is the frequency of the light, λ is the wavelength of the light, δ is the path difference in the interferometer, and Δz is the instantaneous amplitude of the cantilever, given according to (2.40) and (2.41) as a function of Ω, k, and U. The time average of (2.42) then becomes

$$\langle 2A_\ell(t)A_r(t)\rangle_T \propto \cos\left[\frac{4\pi\delta}{\lambda} + \frac{4\pi\Delta z}{\lambda}\sin(\Omega t)\right] \approx \cos\left(\frac{4\pi\delta}{\lambda}\right) - \sin\left[\frac{4\pi\Delta z}{\lambda}\sin(\Omega t)\right]$$

$$\approx \cos\left(\frac{4\pi\delta}{\lambda}\right) - \frac{4\pi\Delta z}{\lambda}\sin(\Omega t) \,. \tag{2.43}$$

Here all small quantities have been omitted and functions with small arguments have been linearized. The amplitude of Δz can be recovered with a lock-in technique. However, (2.43) shows that the measured amplitude is also a function of the path difference δ in the interferometer. Hence, this path difference δ must be very stable. The best sensitivity is obtained when $\sin(4\delta/\lambda) \approx 0$.

Heterodyne Interferometer. This influence is not present in the heterodyne detection scheme shown in Fig. 2.26. Light incident from the left with a frequency ω is split into a reference path (upper path in Fig. 2.26) and a measurement path. Light in the measurement path is shifted in frequency to $\omega_1 = \omega + \Delta\omega$ and focused onto the cantilever. The cantilever oscillates at the frequency Ω, as in the homodyne detection scheme. The reflected light $A_\ell(t)$ is collimated by the same lens and interferes on the photodiode with the reference light $A_r(t)$. The fluctuating term of the intensity is given by

$$2A_\ell(t)A_r(t) = A_{\ell,0}A_{r,0}\sin\left[(\omega + \Delta\omega)t + \frac{4\pi\delta}{\lambda} + \frac{4\pi\Delta z}{\lambda}\sin(\Omega t)\right]\sin(\omega t)\,, \tag{2.44}$$

where the variables are defined as in (2.42). Setting the path difference $\sin(4\pi\delta/\lambda) \approx 0$ and taking the time average, omitting small quantities and linearizing functions with small arguments, we get

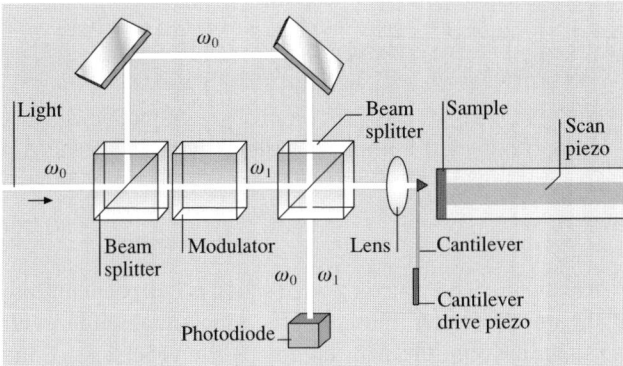

Fig. 2.26. Principle of a heterodyne interferometric AFM. Light with frequency ω_0 is split into a reference path (upper path) and a measurement path. The light in the measurement path is frequency shifted to ω_1 by an acousto-optical modulator (or an electro-optical modulator). The light reflected from the oscillating cantilever interferes with the reference beam on the detector

$$\langle 2A_\ell(t)A_r(t)\rangle_T \propto \cos\left[\Delta\omega t + \frac{4\pi\delta}{\lambda} + \frac{4\pi\Delta z}{\lambda}\sin(\Omega t)\right]$$

$$= \cos\left(\Delta\omega t + \frac{4\pi\delta}{\lambda}\right)\cos\left[\frac{4\pi\Delta z}{\lambda}\sin(\Omega t)\right] - \sin\left(\Delta\omega t + \frac{4\pi\delta}{\lambda}\right)\sin\left[\frac{4\pi\Delta z}{\lambda}\sin(\Omega t)\right]$$

$$\approx \cos\left(\frac{4\pi\delta}{\lambda}\right) - \sin\left[\frac{4\pi\Delta z}{\lambda}\sin(\Omega t)\right]$$

$$\approx \cos\left(\Delta\omega t + \frac{4\pi\delta}{\lambda}\right)\left[1 - \frac{8\pi^2\Delta z^2}{\lambda^2}\sin(\Omega t)\right] - \frac{4\pi\Delta z}{\lambda}\sin\left(\Delta\omega t + \frac{4\pi\delta}{\lambda}\right)\sin(\Omega t)$$

$$= \cos\left(\Delta\omega t + \frac{4\pi\delta}{\lambda}\right) - \frac{8\pi^2\Delta z^2}{\lambda^2}\cos\left(\Delta\omega t + \frac{4\pi\delta}{\lambda}\right)\sin(\Omega t)$$

$$- \frac{4\pi\Delta z}{\lambda}\sin\left(\Delta\omega t + \frac{4\pi\delta}{\lambda}\right)\sin(\Omega t)$$

$$= \cos\left(\Delta\omega t + \frac{4\pi\delta}{\lambda}\right) - \frac{4\pi^2\Delta z^2}{\lambda^2}\cos\left(\Delta\omega t + \frac{4\pi\delta}{\lambda}\right)$$

$$+ \frac{4\pi^2\Delta z^2}{\lambda^2}\cos\left(\Delta\omega t + \frac{4\pi\delta}{\lambda}\right)\cos(2\Omega t) - \frac{4\pi\Delta z}{\lambda}\sin\left(\Delta\omega t + \frac{4\pi\delta}{\lambda}\right)\sin(\Omega t)$$

$$= \cos\left(\Delta\omega t + \frac{4\pi\delta}{\lambda}\right)\left(1 - \frac{4\pi^2\Delta z^2}{\lambda^2}\right)$$

$$+ \frac{2\pi^2\Delta z^2}{\lambda^2}\left\{\cos\left[(\Delta\omega + 2\Omega)t + \frac{4\pi\delta}{\lambda}\right] + \cos\left[(\Delta\omega - 2\Omega)t + \frac{4\pi\delta}{\lambda}\right]\right\}$$

$$+ \frac{2\pi\Delta z}{\lambda}\left\{\cos\left[(\Delta\omega + \Omega)t + \frac{4\pi\delta}{\lambda}\right] + \cos\left[(\Delta\omega - \Omega)t + \frac{4\pi\delta}{\lambda}\right]\right\}. \quad (2.45)$$

Multiplying electronically the components oscillating at $\Delta\omega$ and $\Delta\omega + \Omega$ and rejecting any product except the one oscillating at Ω we obtain

$$A = \frac{2\Delta z}{\lambda}\left(1 - \frac{4\pi^2\Delta z^2}{\lambda^2}\right)\cos\left[(\Delta\omega + 2\Omega)t + \frac{4\pi\delta}{\lambda}\right]\cos\left(\Delta\omega t + \frac{4\pi\delta}{\lambda}\right)$$

$$= \frac{\Delta z}{\lambda}\left(1 - \frac{4\pi^2\Delta z^2}{\lambda^2}\right)\left\{\cos\left[(2\Delta\omega + \Omega)t + \frac{8\pi\delta}{\lambda}\right] + \cos(\Omega t)\right\}$$

$$\approx \frac{\pi\Delta z}{\lambda}\cos(\Omega t). \quad (2.46)$$

Unlike in the homodyne detection scheme, the recovered signal is independent from the path difference δ of the interferometer. Furthermore, a lock-in amplifier with the reference set $\sin(\Delta\omega t)$ can measure the path difference δ independent of the cantilever oscillation. If necessary, a feedback circuit can keep $\delta = 0$.

Fiber-Optical Interferometer. The fiber-optical interferometer [129] is one of the simplest interferometers to build and use. Its principle is sketched in Fig. 2.27. The light of a laser is fed into an optical fiber. Laser diodes with integrated fiber pigtails are convenient light sources. The light is split in a fiber-optic beam splitter into two

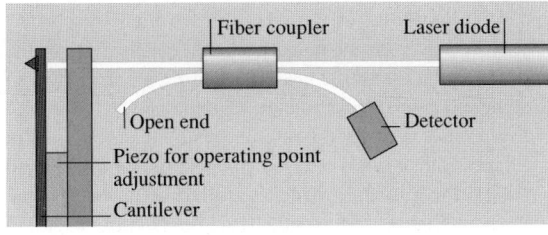

Fig. 2.27. A typical set-up for a fiber-optic interferometer readout

fibers. One fiber is terminated by index-matching oil to avoid any reflections back into the fiber. The end of the other fiber is brought close to the cantilever in the AFM. The emerging light is partially reflected back into the fiber by the cantilever. Most of the light, however, is lost. This is not a big problem since only 4% of the light is reflected at the end of the fiber, at the glass–air interface. The two reflected light waves interfere with each other. The product is guided back into the fiber coupler and again split into two parts. One half is analyzed by the photodiode. The other half is fed back into the laser. Communications grade laser diodes are sufficiently resistant to feedback to be operated in this environment. They have, however, a bad coherence length, which in this case does not matter, since the optical path difference is in any case no larger than 5 μm. Again the end of the fiber has to be positioned on a piezo drive to set the distance between the fiber and the cantilever to $\lambda(n + 1/4)$.

Nomarski-Interferometer. Another way to minimize the optical path difference is to use the Nomarski interferometer [130]. Figure 2.28 shows a schematic of the microscope. The light from a laser is focused on the cantilever by lens. A birefringent crystal (for instance calcite) between the cantilever and the lens, which has its optical axis 45° off the polarization direction of the light, splits the light beam into two paths, offset by a distance given by the length of the crystal. Birefringent crystals have varying indices of refraction. In calcite, one crystal axis has a lower index than

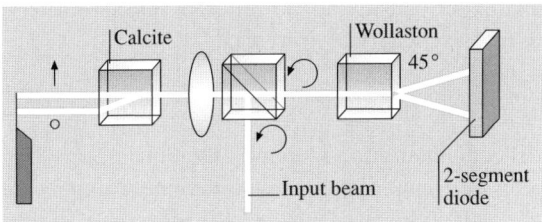

Fig. 2.28. Principle of Nomarski AFM. The circularly polarized input beam is deflected to the left by a nonpolarizing beam splitter. The light is focused onto a cantilever. The calcite crystal between the lens and the cantilever splits the circular polarized light into two spatially separated beams with orthogonal polarizations. The two light beams reflected from the lever are superimposed by the calcite crystal and collected by the lens. The resulting beam is again circularly polarized. A Wollaston prism produces two interfering beams with a $\pi/2$ phase shift between them. The minimal path difference accounts for the excellent stability of this microscope

Table 2.5. Noise in interferometers. F is the finesse of the cavity in the homodyne interferometer, P_i the incident power, P_d is the power on the detector, η is the sensitivity of the photodetector and RIN is the relative intensity noise of the laser. P_R and P_S are the power in the reference and sample beam in the heterodyne interferometer. P is the power in the Nomarski interferometer, $\delta\theta$ is the phase difference between the reference and the probe beam in the Nomarski interferometer. B is the bandwidth, e is the electron charge, λ is the wavelength of the laser, k the cantilever stiffness, ω_0 is the resonant frequency of the cantilever, Q is the quality factor of the cantilever, T is the temperature, and δi is the variation in current i

	Homodyne interferometer, fiber-optic interferometer	Heterodyne interferometer	Nomarski interferometer
Laser noise $\langle \delta i^2 \rangle_L$	$\frac{1}{4}\eta^2 F^2 P_i^2$ RIN	$\eta^2 \left(P_R^2 + P_S^2\right)$ RIN	$\frac{1}{16}\eta^2 P^2 \delta\theta$
Thermal noise $\langle \delta i^2 \rangle_T$	$\frac{16\pi^2}{\lambda^2}\eta^2 F^2 P_i^2 \frac{4k_B T B Q}{\omega_0 k}$	$\frac{4\pi^2}{\lambda^2}\eta^2 P_d^2 \frac{4k_B T B Q}{\omega_0 k}$	$\frac{\pi^2}{\lambda^2}\eta^2 P^2 \frac{4k_B T B Q}{\omega_0 k}$
Shot noise $\langle \delta i^2 \rangle_S$	$4 e \eta P_d B$	$2 e \eta (P_R + P_S) B$	$\frac{1}{2} e \eta P B$

the other two. This means that certain light rays will propagate at different speeds through the crystal than others. By choosing the correct polarization, one can select the ordinary ray or the extraordinary ray or one can get any mixture of the two rays. A detailed description of birefringence can be found in textbooks (e. g., [150]). A calcite crystal deflects the extraordinary ray at an angle of 6° within the crystal. Any separation can be set by choosing a suitable length for the calcite crystal.

The focus of one light ray is positioned near the free end of the cantilever while the other is placed close to the clamped end. Both arms of the interferometer pass through the same space, except for the distance between the calcite crystal and the lever. The closer the calcite crystal is placed to the lever, the less influence disturbances like air currents have.

Sarid [116] has given values for the sensitivities of different interferometeric detection systems. Table 2.5 presents a summary of his results.

Optical Lever

The most common cantilever deflection detection system is the optical lever [53, 111]. This method, depicted in Fig. 2.29, employs the same technique as light beam deflection galvanometers. A fairly well collimated light beam is reflected off a mirror and projected to a receiving target. Any change in the angular position of the mirror will change the position where the light ray hits the target. Galvanometers use optical path lengths of several meters and scales projected onto the target wall are also used to monitor changes in position.

In an AFM using the optical lever method, a photodiode segmented into two (or four) closely spaced devices detects the orientation of the end of the cantilever. Initially, the light ray is set to hit the photodiodes in the middle of the two subdiodes. Any deflection of the cantilever will cause an imbalance of the number of photons

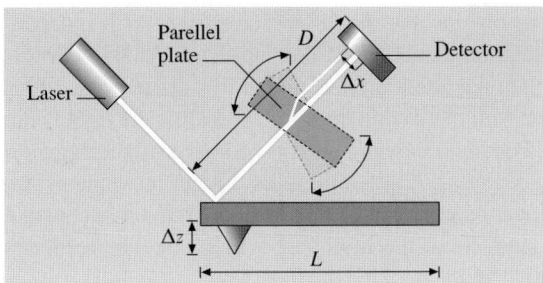

Fig. 2.29. Set-up for an optical lever detection microscope

reaching the two halves. Hence the electrical currents in the photodiodes will be unbalanced too. The difference signal is further amplified and is the input signal to the feedback loop. Unlike the interferometeric AFMs, where a modulation technique is often necessary to get a sufficient signal-to-noise ratio, most AFMs employing the optical lever method are operated in a static mode. AFMs based on the optical lever method are universally used. It is the simplest method for constructing an optical readout and it can be confined in volumes that are smaller than 5 cm in side length.

The optical lever detection system is a simple yet elegant way to detect normal and lateral force signals simultaneously [7,8,53,111]. It has the additional advantage that it is a remote detection system.

Implementations. Light from a laser diode or from a super luminescent diode is focused on the end of the cantilever. The reflected light is directed onto a quadrant diode that measures the direction of the light beam. A Gaussian light beam far from its waist is characterized by an opening angle β. The deflection of the light beam by the cantilever surface tilted by an angle α is 2α. The intensity on the detector then shifts to the side by the product of 2α and the separation between the detector and the cantilever. The readout electronics calculates the difference in the photocurrents. The photocurrents, in turn, are proportional to the intensity incident on the diode.

The output signal is hence proportional to the change in intensity on the segments

$$I_{\text{sig}} \propto 4\frac{\alpha}{\beta} I_{\text{tot}}. \tag{2.47}$$

For the sake of simplicity, we assume that the light beam is of uniform intensity with its cross-section increasing in proportion to the distance between the cantilever and the quadrant detector. The movement of the center of the light beam is then given by

$$\Delta x_{\text{Det}} = \Delta z \frac{D}{L}. \tag{2.48}$$

The photocurrent generated in a photodiode is proportional to the number of incoming photons hitting it. If the light beam contains a total number of N_0 photons, then the change in difference current becomes

$$\Delta(I_R - I_L) = \Delta I = \text{const } \Delta z \, D \, N_0. \tag{2.49}$$

Combining (2.48) and (2.49), one obtains that the difference current ΔI is independent of the separation of the quadrant detector and the cantilever. This relation is true if the light spot is smaller than the quadrant detector. If it is greater, the difference current ΔI becomes smaller with increasing distance. In reality, the light beam has a Gaussian intensity profile. For small movements Δx (compared to the diameter of the light spot at the quadrant detector), (2.49) still holds. Larger movements Δx, however, will introduce a nonlinear response. If the AFM is operated in a constant force mode, only small movements Δx of the light spot will occur. The feedback loop will cancel out all other movements.

The scanning of a sample with an AFM can twist the microfabricated cantilevers because of lateral forces [5, 7, 8] and affect the images [120]. When the tip is subjected to lateral forces, it will twist the cantilever and the light beam reflected from the end of the cantilever will be deflected perpendicular to the ordinary deflection direction. For many investigations this influence of lateral forces is unwanted. The design of the triangular cantilevers stems from the desire to minimize the torsion effects. However, lateral forces open up a new dimension in force measurements. They allow, for instance, two materials to be distinguished because of their different friction coefficients, or adhesion energies to be determined. To measure lateral forces, the original optical lever AFM must be modified. The only modification compared with Fig. 2.29 is the use of a quadrant detector photodiode instead of a two-segment photodiode and the necessary readout electronics, see Fig. 2.9a. The electronics calculates the following signals:

$$U_{\text{Normal Force}} = \alpha \left[\left(I_{\text{Upper Left}} + I_{\text{Upper Right}} \right) - \left(I_{\text{Lower Left}} + I_{\text{Lower Right}} \right) \right],$$
$$U_{\text{Lateral Force}} = \beta \left[\left(I_{\text{Upper Left}} + I_{\text{Lower Left}} \right) - \left(I_{\text{Upper Right}} + I_{\text{Lower Right}} \right) \right]. \quad (2.50)$$

The calculation of the lateral force as a function of the deflection angle does not have a simple solution for cross-sections other than circles. An approximate formula for the angle of twist for rectangular beams is [151]

$$\theta = \frac{M_t L}{\beta G b^3 h}, \quad (2.51)$$

where $M_t = F_y \ell$ is the external twisting moment due to lateral force, F_y, and β, a constant determined by the value of h/b. For the equation to hold, h has to be larger than b.

Inserting the values for a typical microfabricated cantilever with integrated tips

$b = 6 \times 10^{-7}$ m,
$h = 10^{-5}$ m,
$L = 10^{-4}$ m,
$\ell = 3.3 \times 10^{-6}$ m,
$G = 5 \times 10^{10}$ Pa,
$\beta = 0.333$ (2.52)

into (2.51) we obtain the relation

$$F_y = 1.1 \times 10^{-4} \,\text{N} \times \theta. \tag{2.53}$$

Typical lateral forces are of the order of 10^{-10} N.

Sensitivity. The sensitivity of this set-up has been calculated in various papers [116, 148, 152]. Assuming a Gaussian beam, the resulting output signal as a function of the deflection angle is dispersion-like. Equation (2.47) shows that the sensitivity can be increased by increasing the intensity of the light beam I_{tot} or by decreasing the divergence of the laser beam. The upper bound of the intensity of the light I_{tot} is given by saturation effects on the photodiode. If we decrease the divergence of a laser beam we automatically increase the beam waist. If the beam waist becomes larger than the width of the cantilever we start to get diffraction. Diffraction sets a lower bound on the divergence angle. Hence one can calculate the optimal beam waist w_{opt} and the optimal divergence angle β [148, 152]

$$w_{opt} \approx 0.36 b,$$
$$\theta_{opt} \approx 0.89 \frac{\lambda}{b}. \tag{2.54}$$

The optimal sensitivity of the optical lever then becomes

$$\varepsilon\,[\text{mW/rad}] = 1.8 \frac{b}{\lambda} I_{tot}\,[\text{mW}]. \tag{2.55}$$

The angular sensitivity of the optical lever can be measured by introducing a parallel plate into the beam. Tilting the parallel plate results in a displacement of the beam, mimicking an angular deflection.

Additional noise sources can be considered. Of little importance is the quantum mechanical uncertainty of the position [148, 152], which is, for typical cantilevers at room temperature

$$\Delta z = \sqrt{\frac{\hbar}{2m\omega_0}} = 0.05\,\text{fm}, \tag{2.56}$$

where $\hbar$ is the Planck constant ($= 6.626 \times 10^{-34}$ J s). At very low temperatures and for high-frequency cantilevers this could become the dominant noise source. A second noise source is the shot noise of the light. The shot noise is related to the particle number. We can calculate the number of photons incident on the detector using

$$n = \frac{I\tau}{\hbar\omega} = \frac{I\lambda}{2\pi B\hbar c} = 1.8 \times 10^9 \frac{I[\text{W}]}{B[\text{Hz}]}, \tag{2.57}$$

where I is the intensity of the light, τ the measurement time, $B = 1/\tau$ the bandwidth, and c the speed of light. The shot noise is proportional to the square root of the number of particles. Equating the shot noise signal with the signal resulting from the deflection of the cantilever one obtains

$$\Delta z_{\text{shot}} = 68 \frac{L}{w} \sqrt{\frac{B[\text{kHz}]}{I[\text{mW}]}} \, [\text{fm}] \,, \tag{2.58}$$

where w is the diameter of the focal spot. Typical AFM set-ups have a shot noise of 2 pm. The thermal noise can be calculated from the equipartition principle. The amplitude at the resonant frequency is

$$\Delta z_{\text{therm}} = 129 \sqrt{\frac{B}{k[\text{N/m}] \omega_0 Q}} \, [\text{pm}] \,. \tag{2.59}$$

A typical value is 16 pm. Upon touching the surface, the cantilever increases its resonant frequency by a factor of 4.39. This results in a new thermal noise amplitude of 3.2 pm for the cantilever in contact with the sample.

Piezoresistive Detection

Implementation. A piezoresistive cantilever is an alternative detection system which is not as widely used as the optical detection schemes [125,126,132]. This cantilever is based on the fact that the resistivities of certain materials, in particular Si, change with the applied stress. Figure 2.30 shows a typical implementation of a piezoresistive cantilever. Four resistances are integrated on the chip, forming a Wheatstone bridge. Two of the resistors are in unstrained parts of the cantilever, and the other two measure the bending at the point of the maximal deflection. For instance, when an AC voltage is applied between terminals a and c, one can measure the detuning of the bridge between terminals b and d. With such a connection the output signal only varies due to bending, not due to changes in the ambient temperature and thus the coefficient of the piezoresistance.

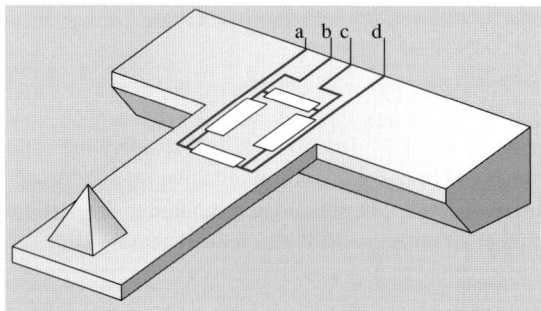

Fig. 2.30. A typical set-up for a piezoresistive readout

Sensitivity. The resistance change is [126]

$$\frac{\Delta R}{R_0} = \Pi \delta, \tag{2.60}$$

where Π is the tensor element of the piezo-resistive coefficients, δ the mechanical stress tensor element and R_0 the equilibrium resistance. For a single resistor, they separate the mechanical stress and the tensor element into longitudinal and transverse components:

$$\frac{\Delta R}{R_0} = \Pi_t \delta_t + \Pi_l \delta_l. \tag{2.61}$$

The maximum values of the stress components are $\Pi_t = -64.0 \times 10^{-11}\,\mathrm{m^2/N}$ and $\Pi_l = -71.4 \times 10^{-11}\,\mathrm{m^2/N}$ for a resistor oriented along the (110) direction in silicon [126]. In the resistor arrangement of Fig. 2.30, two of the resistors are subject to the longitudinal piezo-resistive effect and two of them are subject to the transversal piezo-resistive effect. The sensitivity of that set-up is about four times that of a single resistor, with the advantage that temperature effects cancel to first order. The resistance change is then calculated as

$$\frac{\Delta R}{R_0} = \Pi \frac{3Eh}{2L^2} \Delta z = \Pi \frac{6L}{bh^2} F_z, \tag{2.62}$$

where $\Pi = 67.7 \times 10^{-11}\,\mathrm{m^2/N}$ is the averaged piezo-resistive coefficient. Plugging in typical values for the dimensions (Fig. 2.24) ($L = 100\,\mu\mathrm{m}$, $b = 10\,\mu\mathrm{m}$, $h = 1\,\mu\mathrm{m}$), one obtains

$$\frac{\Delta R}{R_0} = \frac{4 \times 10^{-5}}{\mathrm{nN}} F_z. \tag{2.63}$$

The sensitivity can be tailored by optimizing the dimensions of the cantilever.

Capacitance Detection

The capacitance of an arrangement of conductors depends on the geometry. Generally speaking, the capacitance increases for decreasing separations. Two parallel plates form a simple capacitor (see Fig. 2.31, upper left), with capacitance

$$C = \frac{\varepsilon \varepsilon_0 A}{x}, \tag{2.64}$$

where A is the area of the plates, assumed equal, and x is the separation. Alternatively one can consider a sphere versus an infinite plane (see Fig. 2.31, lower left). Here the capacitance is [116]

$$C = 4\pi \varepsilon_0 R \sum_{n=2}^{\infty} \frac{\sinh(\alpha)}{\sinh(n\alpha)} \tag{2.65}$$

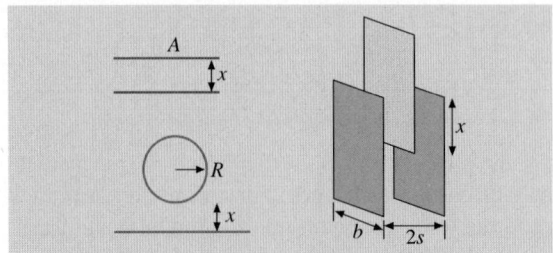

Fig. 2.31. Three possible arrangements of a capacitive readout. The *upper left* diagram shows a cross-section through a parallel plate capacitor. The *lower left* diagram shows the geometry of a sphere versus a plane. The *right-hand* diagram shows the linear (but more complicated) capacitive readout

where R is the radius of the sphere, and α is defined by

$$\alpha = \ln\left(1 + \frac{z}{R} + \sqrt{\frac{z^2}{R^2} + 2\frac{z}{R}}\right). \tag{2.66}$$

One has to bear in mind that the capacitance of a parallel plate capacitor is a non-linear function of the separation. One can circumvent this problem using a voltage divider. Figure 2.32a shows a low-pass filter. The output voltage is given by

$$U_{\text{out}} = U_\approx \frac{\frac{1}{j\omega C}}{R + \frac{1}{j\omega C}} = U_\approx \frac{1}{j\omega CR + 1}$$
$$\cong \frac{U_\approx}{j\omega CR}. \tag{2.67}$$

Here C is given by (2.64), ω is the excitation frequency and j is the imaginary unit. The approximate relation at the end is true when $\omega CR \gg 1$. This is equivalent to the statement that C is fed by a current source, since R must be large in this set-up. Plugging (2.64) into (2.67) and neglecting the phase information, one obtains

$$U_{\text{out}} = \frac{U_\approx x}{\omega R \varepsilon \varepsilon_0 A}, \tag{2.68}$$

which is linear in the displacement x.

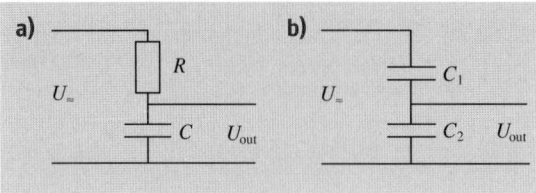

Fig. 2.32. Measuring the capacitance. (**a**) Low pass filter, (**b**) capacitive divider. C (*left*) and C_2 (*right*) are the capacitances under test

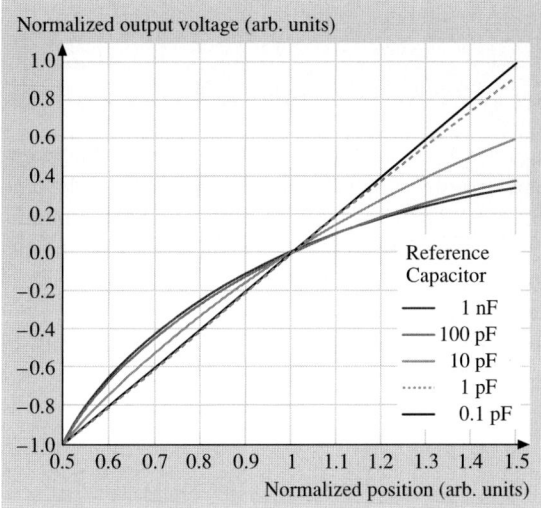

Fig. 2.33. Linearity of the capacitance readout as a function of the reference capacitor

Figure 2.32b shows a capacitive divider. Again the output voltage U_{out} is given by

$$U_{\text{out}} = U_{\approx} \frac{C_1}{C_2 + C_1} = U_{\approx} \frac{C_1}{\frac{\varepsilon \varepsilon_0 A}{x} + C_1}. \tag{2.69}$$

If there is a stray capacitance C_s then (2.69) is modified as

$$U_{\text{out}} = U_{\approx} \frac{C_1}{\frac{\varepsilon \varepsilon_0 A}{x} + C_s + C_1}. \tag{2.70}$$

Provided $C_s + C_1 \ll C_2$, one has a system which is linear in x. The driving voltage $U_{\approx}$ must be large (more than 100 V) to gave an output voltage in the range of 1 V. The linearity of the readout depends on the capacitance C_1 (Fig. 2.33).

Another idea is to keep the distance constant and to change the relative overlap of the plates (see Fig. 2.31, right side). The capacitance of the moving center plate versus the stationary outer plates becomes

$$C = C_s + 2\frac{\varepsilon \varepsilon_0 b x}{s}, \tag{2.71}$$

where the variables are defined in Fig. 2.31. The stray capacitance comprises all effects, including the capacitance of the fringe fields. When the length x is comparable to the width b of the plates, one can safely assume that the stray capacitance is constant and independent of x. The main disadvantage of this set-up is that it is not as easily incorporated into a microfabricated device as the others.

Sensitivity. The capacitance itself is not a measure of the sensitivity, but its derivative is indicative of the signals one can expect. Using the situation described in

Fig. 2.31 (upper left) and in (2.64), one obtains for the parallel plate capacitor

$$\frac{dC}{dx} = -\frac{\varepsilon\varepsilon_0 A}{x^2}. \tag{2.72}$$

Assuming a plate area A of 20 μm by 40 μm and a separation of 1 μm, one obtains a capacitance of 31 fF (neglecting stray capacitance and the capacitance of the connection leads) and a dC/dx of 3.1×10^{-8} F/m = 31 fF/μm. Hence it is of paramount importance to maximize the area between the two contacts and to minimize the distance x. The latter however is far from being trivial. One has to go to the limits of microfabrication to achieve a decent sensitivity.

If the capacitance is measured by the circuit shown in Fig. 2.32, one obtains for the sensitivity

$$\frac{dU_{\text{out}}}{U_\approx} = \frac{dx}{\omega R \varepsilon \varepsilon_0 A}. \tag{2.73}$$

Using the same value for A as above, setting the reference frequency to 100 kHz, and selecting $R = 1$ GΩ, we get the relative change in the output voltage U_{out} as

$$\frac{dU_{\text{out}}}{U_\approx} = \frac{22.5 \times 10^{-6}}{\text{Å}} \times dx. \tag{2.74}$$

A driving voltage of 45 V then translates to a sensitivity of 1 mV/Å. A problem in this set-up is the stray capacitances. They are in parallel to the original capacitance and decrease the sensitivity considerably.

Alternatively, one could build an oscillator with this capacitance and measure the frequency. RC-oscillators typically have an oscillation frequency of

$$f_{\text{res}} \propto \frac{1}{RC} = \frac{x}{R\varepsilon\varepsilon_0 A}. \tag{2.75}$$

Again the resistance R must be of the order of 1 GΩ when stray capacitances C_s are neglected. However C_s is of the order of 1 pF. Therefore one gets $R = 10$ MΩ. Using these values, the sensitivity becomes

$$df_{\text{res}} = \frac{C\,dx}{R(C+C_s)^2 x} \approx \frac{0.1\,\text{Hz}}{\text{Å}} dx. \tag{2.76}$$

The bad thing is that the stray capacitances have made the signal nonlinear again. The linearized set-up in Fig. 2.31 has a sensitivity of

$$\frac{dC}{dx} = 2\frac{\varepsilon\varepsilon_0 b}{s}. \tag{2.77}$$

Substituting typical values ($b = 10$ μm, $s = 1$ μm), one gets $dC/dx = 1.8 \times 10^{-10}$ F/m. It is noteworthy that the sensitivity remains constant for scaled devices.

Implementations. Capacitance readout can be achieved in different ways [123, 124]. All include an alternating current or voltage with frequencies in the 100 kHz to 100 MHz range. One possibility is to build a tuned circuit with the capacitance of the cantilever determining the frequency. The resonance frequency of a high-quality Q tuned circuit is

$$\omega_0 = (LC)^{-1/2}. \tag{2.78}$$

where L is the inductance of the circuit. The capacitance C includes not only the sensor capacitance but also the capacitance of the leads. The precision of a frequency measurement is mainly determined by the ratio of L and C

$$Q = \left(\frac{L}{C}\right)^{1/2} \frac{1}{R}. \tag{2.79}$$

Here R symbolizes the losses in the circuit. The higher the quality, the more precise the frequency measurement. For instance, a frequency of 100 MHz and a capacitance of 1 pF gives an inductance of 250 µH. The quality then becomes 2.5×10^8. This value is an upper limit, since losses are usually too high.

Using a value of $dC/dx = 31$ fF/µm, one gets $\Delta C/\text{Å} = 3.1$ aF/Å. With a capacitance of 1 pF, one gets

$$\frac{\Delta \omega}{\omega} = \frac{1}{2} \frac{\Delta C}{C},$$

$$\Delta \omega = 100 \,\text{MHz} \times \frac{1}{2} \frac{3.1 \,\text{aF}}{1 \,\text{pF}} = 155 \,\text{Hz}. \tag{2.80}$$

This is the frequency shift for a deflection of 1 Å. The calculation shows that this is a measurable quantity. The quality also indicates that there is no physical reason why this scheme should not work.

2.4.3 Combinations for 3-D Force Measurements

Three-dimensional force measurements are essential if one wants to know all of the details of the interaction between the tip and the cantilever. The straightforward attempt to measure three forces is complicated, since force sensors such as interferometers or capacitive sensors need a minimal detection volume, which is often too large. The second problem is that the force-sensing tip has to be held in some way. This implies that one of the three Cartesian axes is stiffer than the others.

However, by combining different sensors it is possible to achieve this goal. Straight cantilevers are employed for these measurements, because they can be handled analytically. The key observation is that the optical lever method does not determine the position of the end of the cantilever. It measures the orientation. In the previous sections, one has always made use of the fact that, for a force along one of the orthogonal symmetry directions at the end of the cantilever (normal force, lateral force, force along the cantilever beam axis), there is a one-to-one correspondence of the tilt angle and the deflection. The problem is that the force along the cantilever beam axis and the normal force create a deflection in the same direction. Hence, what is called the normal force component is actually a mixture of two forces. The deflection of the cantilever is the third quantity, which is not considered in most of the AFMs. A fiber-optic interferometer in parallel with the optical lever measures the deflection. Three measured quantities then allow the separation of the three orthonormal force directions, as is evident from (2.27) and (2.33) [12–16].

Alternatively, one can put the fast scanning direction along the axis of the cantilever. Forward and backward scans then exert opposite forces F_x. If the piezo movement is linearized, both force components in AFM based on optical lever detection can be determined. In this case, the normal force is simply the average of the forces in the forward and backward direction. The force, F_x, is the difference in the forces measured in the forward and backward directions.

2.4.4 Scanning and Control Systems

Almost all SPMs use piezo translators to scan the tip or the sample. Even the first STM [1, 103] and some of its predecessors [153, 154] used them. Other materials or set-ups for nanopositioning have been proposed, but they have not been successful [155, 156].

Piezo Tubes

A popular solution is tube scanners (Fig. 2.34). They are now widely used in SPMs due to their simplicity and their small size [133, 157]. The outer electrode is segmented into four equal sectors of 90 degrees. Opposite sectors are driven by signals of the same magnitude, but opposite sign. This gives, through bending, two-dimensional movement on (approximately) a sphere. The inner electrode is normally driven by the z signal. It is possible, however, to use only the outer electrodes for scanning and for the z-movement. The main drawback of applying the z-signal to the outer electrodes is that the applied voltage is the sum of both the x- or y-movements

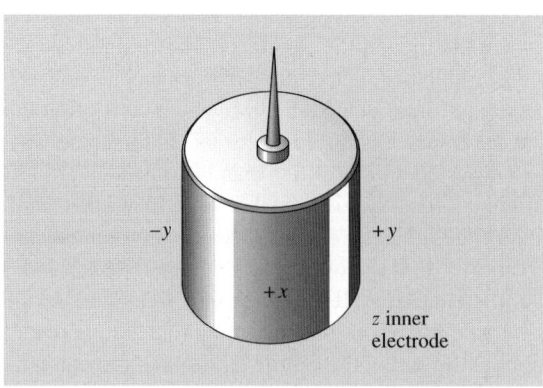

Fig. 2.34. Schematic drawing of a piezoelectric tube scanner. The piezo ceramic is molded into a tube form. The outer electrode is separated into four segments and connected to the scanning voltage. The z-voltage is applied to the inner electrode

and the z-movement. Hence a larger scan size effectively reduces the available range for the z-control.

Piezo Effect

An electric field applied across a piezoelectric material causes a change in the crystal structure, with expansion in some directions and contraction in others. Also, a net volume change occurs [132]. Many SPMs use the transverse piezo electric effect, where the applied electric field E is perpendicular to the expansion/contraction direction.

$$\Delta L = L(\mathbf{E} \cdot \mathbf{n})d_{31} = L\frac{V}{t}d_{31}, \qquad (2.81)$$

where d_{31} is the transverse piezoelectric constant, V is the applied voltage, t is the thickness of the piezo slab or the distance between the electrodes where the voltage is applied, L is the free length of the piezo slab, and $\mathbf{n}$ is the direction of polarization. Piezo translators based on the transverse piezoelectric effect have a wide range of sensitivities, limited mainly by mechanical stability and breakdown voltage.

Scan Range

The the scanning range of a piezotube is difficult to calculate [157–159]. The bending of the tube depends on the electric fields and the nonuniform strain induced. A finite element calculation where the piezo tube was divided into 218 identical elements was used [158] to calculate the deflection. On each node, the mechanical stress, the stiffness, the strain and the piezoelectric stress were calculated when a voltage was applied on one electrode. The results were found to be linear on the first iteration and higher order corrections were very small even for large electrode voltages. It was found that, to first order, the x- and z-movement of the tube could be reasonably well approximated by assuming that the piezo tube is a segment of a torus. Using this model, one obtains

$$dx = (V_+ - V_-)|d_{31}|\frac{L^2}{2td}, \qquad (2.82)$$

$$dz = (V_+ + V_- - 2V_z)|d_{31}|\frac{L}{2t}, \qquad (2.83)$$

where $|d_{31}|$ is the coefficient of the transversal piezoelectric effect, L is the tube's free length, t is the tube's wall thickness, d is the tube's diameter, V_+ is the voltage on the positive outer electrode, while V_- is the voltage of the opposite quadrant negative electrode and V_z is the voltage of the inner electrode.

The cantilever or sample mounted on the piezotube has an additional lateral movement because the point of measurement is not in the endplane of the piezotube. The additional lateral displacement of the end of the tip is $\ell \sin\varphi \approx \ell\varphi$, where ℓ is the tip length and φ is the deflection angle of the end surface. Assuming that the sample or cantilever is always perpendicular to the end of the walls of the tube, and calculating with the torus model, one gets for the angle

$$\varphi = \frac{L}{R} = \frac{2dx}{L}, \quad (2.84)$$

where R is the radius of curvature of the piezo tube. Using the result of (2.84), one obtains for the additional x-movement

$$dx_{add} = \ell\varphi = \frac{2dx\ell}{L}$$
$$= (V_+ - V_-)|d_{31}|\frac{\ell L}{td} \quad (2.85)$$

and for the additional z-movement due to the x-movement

$$dz_{add} = \ell - \ell\cos\varphi = \frac{\ell\varphi^2}{2} = \frac{2\ell(dx)^2}{L^2}$$
$$= (V_+ - V_-)^2|d_{31}|^2\frac{\ell L^2}{2t^2 d^2}. \quad (2.86)$$

Carr [158] assumed for his finite element calculations that the top of the tube was completely free to move and, as a consequence, the top surface was distorted, leading to a deflection angle that was about half that of the geometrical model. Depending on the attachment of the sample or the cantilever, this distortion may be smaller, leading to a deflection angle in-between that of the geometrical model and the one from the finite element calculation.

Nonlinearities and Creep

Piezo materials with a high conversion ratio (a large d_{31} or small electrode separations with large scanning ranges) are hampered by substantial hysteresis resulting in a deviation from linearity by more than 10%. The sensitivity of the piezo ceramic material (mechanical displacement divided by driving voltage) decreases with reduced scanning range, whereas the hysteresis is reduced. Careful selection of the material used for the piezo scanners, the design of the scanners, and of the operating conditions is necessary to obtain optimum performance.

Passive Linearization: Calculation. The analysis of images affected by piezo non-linearities [160–163] shows that the dominant term is

$$x = AV + BV^2, \quad (2.87)$$

where x is the excursion of the piezo, V is the applied voltage and A and B are two coefficients describing the sensitivity of the material. Equation (2.87) holds for scanning from $V = 0$ to large V. For the reverse direction, the equation becomes

$$x = \tilde{A}V - \tilde{B}(V - V_{max})^2, \quad (2.88)$$

where $\tilde{A}$ and $\tilde{B}$ are the coefficients for the back scan and V_{max} is the applied voltage at the turning point. Both equations demonstrate that the true x-travel is small at the beginning of the scan and becomes larger towards the end. Therefore, images are stretched at the beginning and compressed at the end.

Similar equations hold for the slow scan direction. The coefficients, however, are different. The combined action causes a greatly distorted image. This distortion can be calculated. The data acquisition systems record the signal as a function of V. However the data is measured as a function of x. Therefore we have to distribute the x-values evenly across the image. This can be done by inverting an approximation of (2.87). First we write

$$x = AV\left(1 - \frac{B}{A}V\right). \tag{2.89}$$

For $B \ll A$ we can approximate

$$V = \frac{x}{A}. \tag{2.90}$$

We now substitute (2.90) into the nonlinear term of (2.89). This gives

$$x = AV\left(1 + \frac{Bx}{A^2}\right),$$
$$V = \frac{x}{A}\frac{1}{(1 + Bx/A^2)} \approx \frac{x}{A}\left(1 - \frac{Bx}{A^2}\right). \tag{2.91}$$

Hence an equation of the type

$$x_{\text{true}} = x(\alpha - \beta x/x_{\text{max}}) \quad \text{with } 1 = \alpha - \beta \tag{2.92}$$

takes out the distortion of an image. α and β are dependent on the scan range, the scan speed and on the scan history, and have to be determined with exactly the same settings as for the measurement. x_{max} is the maximal scanning range. The condition for α and β guarantees that the image is transformed onto itself.

Similar equations to the empirical one shown above (2.92) can be derived by analyzing the movements of domain walls in piezo ceramics.

Passive Linearization: Measuring the Position. An alternative strategy is to measure the positions of the piezo translators. Several possibilities exist.

1. The interferometers described above can be used to measure the elongation of the piezo elongation. The fiber-optic interferometer is especially easy to implement. The coherence length of the laser only limits the measurement range. However, the signal is of a periodic nature. Hence direct use of the signal in a feedback circuit for the position is not possible. However, as a measurement tool and, especially, as a calibration tool, the interferometer is without competition. The wavelength of the light, for instance that in a HeNe laser, is so well defined that the precision of the other components determines the error of the calibration or measurement.

2. The movement of the light spot on the quadrant detector can be used to measure the position of a piezo [164]. The output current changes by $0.5\,\mathrm{A/cm} \times P(\mathrm{W})/R(\mathrm{cm})$. Typical values ($P = 1\,\mathrm{mW}$, $R = 0.001\,\mathrm{cm}$) give $0.5\,\mathrm{A/cm}$. The noise limit is typically $0.15\,\mathrm{nm} \times \sqrt{\Delta f(\mathrm{Hz})/H(\mathrm{W/cm^2})}$. Again this means that the laser beam above would have a 0.1 nm noise limitation for a bandwidth of 21 Hz. The advantage of this method is that, in principle, one can linearize two axes with only one detector.
3. A knife-edge blocking part of a light beam incident on a photodiode can be used to measure the position of the piezo. This technique, commonly used in optical shear force detection [75, 165], has a sensitivity of better than 0.1 nm.
4. The capacitive detection [166, 167] of the cantilever deflection can be applied to the measurement of the piezo elongation. Equations (2.64) to (2.79) apply to the problem. This technique is used in some commercial instruments. The difficulties lie in the avoidance of fringe effects at the borders of the two plates. While conceptually simple, one needs the latest technology in surface preparation to get a decent linearity. The electronic circuits used for the readout are often proprietary.
5. Linear variable differential transformers (LVDT) are a convenient way to measure positions down to 1 nm. They can be used together with a solid state joint set-up, as often used for large scan range stages. Unlike capacitive detection, there are few difficulties in implementation. The sensors and the detection circuits LVDTs are available commercially.
6. A popular measurement technique is the use of strain gauges. They are especially sensitive when mounted on a solid state joint where the curvature is maximal. The resolution depends mainly on the induced curvature. A precision of 1 nm is attainable. The signals are low – a Wheatstone bridge is needed for the readout.

Active Linearization. Active linearization is done with feedback systems. Sensors need to be monotonic. Hence all of the systems described above, with the exception of the interferometers, are suitable. The most common solutions include the strain gauge approach, capacitance measurement or the LVDT, which are all electronic solutions. Optical detection systems have the disadvantage that the intensity enters into the calibration.

Alternative Scanning Systems

The first STMs were based on piezo tripods [1]. The piezo tripod (Fig. 2.35) is an intuitive way to generate the three-dimensional movement of a tip attached to its center. However, to get a suitable stability and scanning range, the tripod needs to be fairly large (about 50 mm). Some instruments use piezo stacks instead of monolithic piezoactuators. They are arranged in a tripod. Piezo stacks are thin layers of piezoactive materials glued together to form a device with up to 200 µm of actuation range. Preloading with a suitable metal casing reduces the nonlinearity.

If one tries to construct a homebuilt scanning system, the use of linearized scanning tables is recommended. They are built around solid state joints and actuated

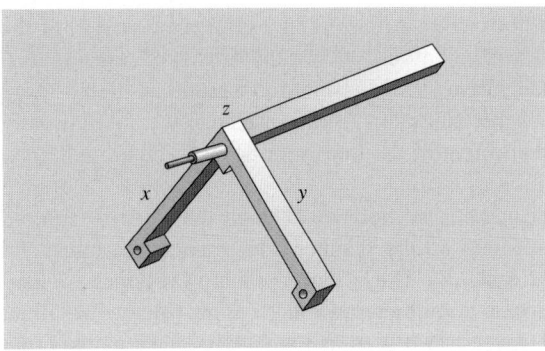

Fig. 2.35. An alternative type of piezo scanner: the tripod

by piezo stacks. The joints guarantee that the movement is parallel with little deviation from the predefined scanning plane. Due to the construction it is easy to add measurement devices such as capacitive sensors, LVDTs or strain gauges, which are essential for a closed loop linearization. Two-dimensional tables can be bought from several manufacturers. They have linearities of better than 0.1% and a noise level of 10^{-4} to 10^{-5} for the maximal scanning range.

Control Systems

Basics. The electronics and software play an important role in the optimal performance of an SPM. Control electronics and software are supplied with commercial SPMs. Electronic control systems can use either analog or digital feedback. While digital feedback offers greater flexibility and ease of configuration, analog feedback circuits might be better suited for ultralow noise operation. We will describe here the basic set-ups for AFMs.

Figure 2.36 shows a block schematic of a typical AFM feedback loop. The signal from the force transducer is fed into the feedback loop, which consists mainly of a subtraction stage to get an error signal and an integrator. The gain of the integrator (high gain corresponds to short integration times) is set as high as possible without

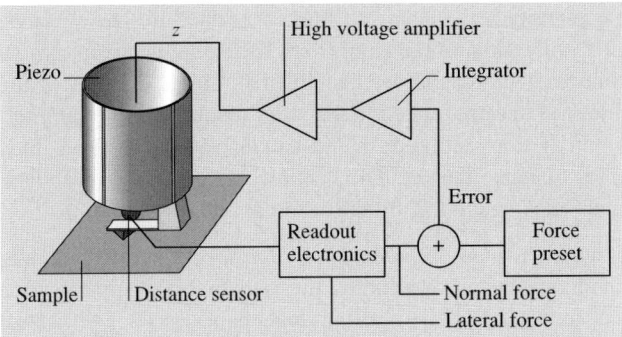

Fig. 2.36. Block schematic of the feedback control loop of an AFM

generating more than 1% overshoot. High gain minimizes the error margin of the current and forces the tip to follow the contours of constant density of states as well as possible. This operating mode is known as constant force mode. A high-voltage amplifier amplifies the outputs of the integrator. As AFMs using piezotubes usually require ±150 V at the output, the output of the integrator needs to be amplified by a high-voltage amplifier.

In order to scan the sample, additional voltages at high tension are required to drive the piezo. For example, with a tube scanner, four scanning voltages are required, namely $+V_x$, $-V_x$, $+V_y$ and $-V_y$. The x- and y-scanning voltages are generated in a scan generator (analog or computer-controlled). Both voltages are input to the two respective power amplifiers. Two inverting amplifiers generate the input voltages for the other two power amplifiers. The topography of the sample surface is determined by recording the input voltage to the high-voltage amplifier for the z-channel as a function of x and y (constant force mode).

Another operating mode is the variable force mode. The gain in the feedback loop is lowered and the scanning speed increased such that the force on the cantilever is no longer constant. Here the force is recorded as a function of x and y.

Force Spectroscopy. Four modes of spectroscopic imaging are in common use with force microscopes: measuring lateral forces, $\partial F/\partial z$, $\partial F/\partial x$ spatially resolved, and measuring force versus distance curves. Lateral forces can be measured by detecting the deflection of a cantilever in a direction orthogonal to the normal direction. The optical lever deflection method does this most easily. Lateral force measurements give indications of adhesion forces between the tip and the sample.

$\partial F/\partial z$ measurements probe the local elasticity of the sample surface. In many cases the measured quantity originates from a volume of a few cubic nanometers. The $\partial F/\partial z$ or local stiffness signal is proportional to Young's modulus, as far as one can define this quantity. Local stiffness is measured by vibrating the cantilever by a small amount in the z-direction. The expected signal for very stiff samples is zero: for very soft samples one also gets, independent of the stiffness, a constant signal. This signal is again zero for the optical lever deflection and equal to the driving amplitude for interferometric measurements. The best sensitivity is obtained when the compliance of the cantilever matches the stiffness of the sample.

A third spectroscopic quantity is the lateral stiffness. It is measured by applying a small modulation in the x-direction on the cantilever. The signal is again optimal when the lateral compliance of the cantilever matches the lateral stiffness of the sample. The lateral stiffness is, in turn, related to the shear modulus of the sample.

Detailed information on the interaction of the tip and the sample can be gained by measuring force versus distance curves. The cantilevers need to have enough compliance to avoid instabilities due to the attractive forces on the sample.

Using the Control Electronics as a Two-Dimensional Measurement Tool. Usually the control electronics of an AFM is used to control the x- and y-piezo signals while several data acquisition channels record the position-dependent signals. The control electronics can be used in another way: they can be viewed as a two-dimensional function generator. What is normally the x- and y-signal can be used to control two

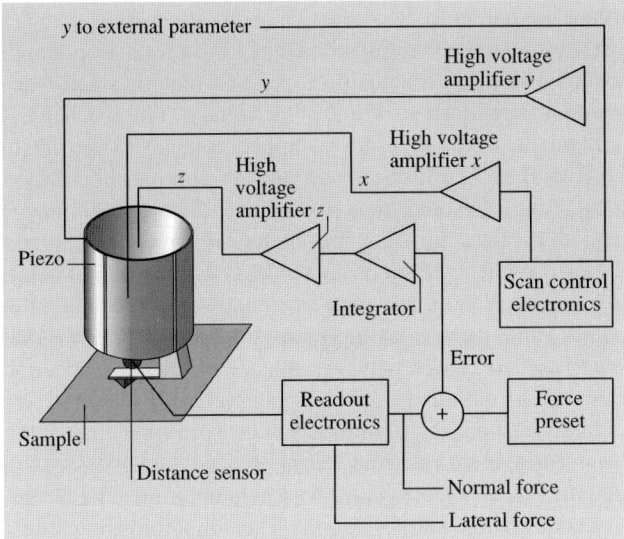

Fig. 2.37. Wiring of an AFM to measure friction force curves along a line

independent variables of an experiment. The control logic of the AFM then ensures that the available parameter space is systematically probed at equally spaced points. An example is friction force curves measured along a line across a step on graphite.

Figure 2.37 shows the connections. The z-piezo is connected as usual, like the x-piezo. However, the y-output is used to command the desired input parameter. The offset of the y-channel determines the position of the tip on the sample surface, together with the x-channel.

Some Imaging Processing Methods

The visualization and interpretation of images from AFMs is intimately connected to the processing of these images. An ideal AFM is a noise-free device that images a sample with perfect tips of known shape and has perfect linear scanning piezos. In reality, AFMs are not that ideal. The scanning device in an AFM is affected by distortions. The distortions are both linear and nonlinear. Linear distortions mainly result from imperfections in the machining of the piezotranslators causing crosstalk from the z-piezo to the x- and y-piezos, and vice versa. Among the linear distortions, there are two kinds which are very important. First, scanning piezos invariably have different sensitivities along the different scan axes due to variations in the piezo material and uneven electrode areas. Second, the same reasons might cause the scanning axes to be non-orthogonal. Furthermore, the plane in which the piezoscanner moves for constant height z is hardly ever coincident with the sample plane. Hence, a linear ramp is added to the sample data. This ramp is especially bothersome when the height z is displayed as an intensity map.

The nonlinear distortions are harder to deal with. They can affect AFM data for a variety of reasons. First, piezoelectric ceramics do have a hysteresis loop, much like ferromagnetic materials. The deviations of piezoceramic materials from linearity increase with increasing amplitude of the driving voltage. The mechanical position for one voltage depends on the previously applied voltages to the piezo. Hence, to get the best positional accuracy, one should always approach a point on the sample from the same direction. Another type of nonlinear distortion of images occurs when the scan frequency approaches the upper frequency limits of the x- and y-drive amplifiers or the upper frequency limit of the feedback loop (z-component). This distortion, due to the feedback loop, can only be minimized by reducing the scan frequency. On the other hand, there is a simple way to reduce distortions due to the x- and y-piezo drive amplifiers. To keep the system as simple as possible, one normally uses a triangular waveform to drive the scanning piezos. However, triangular waves contain frequency components as multiples of the scan frequency. If the cut-off frequencies of the x- and y-drive electronics or of the feedback loop are too close to the scanning frequency (two or three times the scanning frequency), the triangular drive voltage is rounded off at the turning points. This rounding error causes, first, a distortion of the scan linearity and, second, through phase lags, the projection of part of the backward scan onto the forward scan. This type of distortion can be minimized by carefully selecting the scanning frequency and by using driving voltages for the x- and y-piezos with waveforms like trapezoidal waves, which are closer to a sine wave. The values measured for x, y or z piezos are affected by noise. The origin of this noise can be either electronic, disturbances, or a property of the sample surface due to adsorbates. In addition to this incoherent noise, interference with main and other equipment nearby might be present. Depending on the type of noise, one can filter it in real space or in Fourier space. The most important part of image processing is to visualize the measured data. Typical AFM data sets can consist of many thousands to over a million points per plane. There may be more than one image plane present. The AFM data represents a topography in various data spaces.

Most commercial data acquisition systems implicitly use some kind of data processing. Since the original data is commonly subject to slopes on the surface, most programs use some kind of slope correction. The least disturbing way is to subtract a plane $z(x,y) = Ax + By + C$ from the data. The coefficients are determined by fitting $z(x,y)$ to the data. Another operation is to subtract a second-order function such as $z(x,y) = Ax^2 + By^2 + Cxy + Dx + Ey + F$. Again, the parameters are determined with a fit. This function is appropriate for almost planar data, where the nonlinearity of the piezos caused the distortion.

In the image processing software from Digital Instruments, up to three operations are performed on the raw data. First, a zero-order flatten is applied. The flatten operation is used to eliminate image bow in the slow scan direction (caused by a physical bow in the instrument itself), slope in the slow scan direction, and bands in the image (caused by differences in the scan height from one scan line to the next). The flattening operation takes each scan line and subtracts the average value of the height along each scan line from each point in that scan line. This brings each scan line to the same height. Next, a first-order plane fit is applied in the fast scan direction. The plane-fit operation is used to eliminate bow and slope in the fast scan direction. The plane fit operation calculates a best fit plane for the image and subtracts it from the image. This plane has a constant non-zero slope in the fast scan direction. In some cases a higher order polynomial "plane" may be required. Depending upon the quality of the raw data, the flattening operation and/or the plane fit operation may not be required at all.

References

1. G. Binnig, H. Rohrer, Ch. Gerber, E. Weibel: Surface studies by scanning tunneling microscopy, Phys. Rev. Lett. **49**, 57–61 (1982)
2. G. Binnig, C.F. Quate, Ch. Gerber: Atomic force microscope, Phys. Rev. Lett. **56**, 930–933 (1986)
3. G. Binnig, Ch. Gerber, E. Stoll, T.R. Albrecht, C.F. Quate: Atomic resolution with atomic force microscope, Europhys. Lett. **3**, 1281–1286 (1987)
4. B. Bhushan: *Handbook of Micro/Nanotribology*, 2nd edn. (CRC, Boca Raton 1999)
5. C.M. Mate, G.M. McClelland, R. Erlandsson, S. Chiang: Atomic-scale friction of a tungsten tip on a graphite surface, Phys. Rev. Lett. **59**, 1942–1945 (1987)
6. R. Erlandsson, G.M. McClelland, C.M. Mate, S. Chiang: Atomic force microscopy using optical interferometry, J. Vacuum Sci. Technol. A **6**, 266–270 (1988)
7. O. Marti, J. Colchero, J. Mlynek: Combined scanning force and friction microscopy of mica, Nanotechnology **1**, 141–144 (1990)
8. G. Meyer, N.M. Amer: Simultaneous measurement of lateral and normal forces with an optical-beam-deflection atomic force microscope, Appl. Phys. Lett. **57**, 2089–2091 (1990)
9. B. Bhushan, J. Ruan: Atomic-scale friction measurements using friction force microscopy: Part II – Application to magnetic media, ASME J. Tribol. **116**, 389–396 (1994)
10. B. Bhushan, V.N. Koinkar, J. Ruan: Microtribology of magnetic media, Proc. Inst. Mech. Eng., Part J: J. Eng. Tribol. **208**, 17–29 (1994)
11. B. Bhushan, J.N. Israelachvili, U. Landman: Nanotribology: Friction, wear, and lubrication at the atomic scale, Nature **374**, 607–616 (1995)
12. S. Fujisawa, M. Ohta, T. Konishi, Y. Sugawara, S. Morita: Difference between the forces measured by an optical lever deflection and by an optical interferometer in an atomic force microscope, Rev. Sci. Instrum. **65**, 644–647 (1994)
13. S. Fujisawa, E. Kishi, Y. Sugawara, S. Morita: Fluctuation in 2-dimensional stick-slip phenomenon observed with 2-dimensional frictional force microscope, Jpn. J. Appl. Phys. **33**, 3752–3755 (1994)
14. S. Grafstrom, J. Ackermann, T. Hagen, R. Neumann, O. Probst: Analysis of lateral force effects on the topography in scanning force microscopy, J. Vacuum Sci. Technol. B **12**, 1559–1564 (1994)
15. R.M. Overney, H. Takano, M. Fujihira, W. Paulus, H. Ringsdorf: Anisotropy in friction and molecular stick-slip motion, Phys. Rev. Lett. **72**, 3546–3549 (1994)
16. R.J. Warmack, X.Y. Zheng, T. Thundat, D.P. Allison: Friction effects in the deflection of atomic force microscope cantilevers, Rev. Sci. Instrum. **65**, 394–399 (1994)

17. N.A. Burnham, D.D. Domiguez, R.L. Mowery, R.J. Colton: Probing the surface forces of monolayer films with an atomic force microscope, Phys. Rev. Lett. **64**, 1931–1934 (1990)
18. N.A. Burham, R.J. Colton, H.M. Pollock: Interpretation issues in force microscopy, J. Vacuum Sci. Technol. A **9**, 2548–2556 (1991)
19. C.D. Frisbie, L.F. Rozsnyai, A. Noy, M.S. Wrighton, C.M. Lieber: Functional group imaging by chemical force microscopy, Science **265**, 2071–2074 (1994)
20. V.N. Koinkar, B. Bhushan: Microtribological studies of unlubricated and lubricated surfaces using atomic force/friction force microscopy, J. Vacuum Sci. Technol. A **14**, 2378–2391 (1996)
21. V. Scherer, B. Bhushan, U. Rabe, W. Arnold: Local elasticity and lubrication measurements using atomic force and friction force microscopy at ultrasonic frequencies, IEEE Trans. Magn. **33**, 4077–4079 (1997)
22. V. Scherer, W. Arnold, B. Bhushan: Lateral force microscopy using acoustic friction force microscopy, Surf. Interf. Anal. **27**, 578–587 (1999)
23. B. Bhushan, S. Sundararajan: Micro/Nanoscale friction and wear mechanisms of thin films using atomic force and friction force microscopy, Acta Mater. **46**, 3793–3804 (1998)
24. U. Krotil, T. Stifter, H. Waschipky, K. Weishaupt, S. Hild, O. Marti: Pulse force mode: A new method for the investigation of surface properties, Surf. Interf. Anal. **27**, 336–340 (1999)
25. B. Bhushan, C. Dandavate: Thin-film friction and adhesion studies using atomic force microscopy, J. Appl. Phys. **87**, 1201–1210 (2000)
26. B. Bhushan: *Micro/Nanotribology and its Applications* (Kluwer, Dordrecht 1997)
27. B. Bhushan: *Principles and Applications of Tribology* (Wiley, New York 1999)
28. B. Bhushan: *Modern Tribology Handbook Vol. 1: Principles of Tribology* (CRC, Boca Raton 2001)
29. B. Bhushan: *Introduction to Tribology* (Wiley, New York 2002)
30. M. Reinstaedtler, U. Rabe, V. Scherer, U. Hartmann, A. Goldade, B. Bhushan, W. Arnold: On the nanoscale measurement of friction using atomic force microscope cantilever torsional resonances, Appl. Phys. Lett. **82**, 2604–2606 (2003)
31. N.A. Burnham, R.J. Colton: Measuring the nanomechanical properties and surface forces of materials using an atomic force microscope, J. Vacuum Sci. Technol. A **7**, 2906–2913 (1989)
32. P. Maivald, H.J. Butt, S.A.C. Gould, C.B. Prater, B. Drake, J.A. Gurley, V.B. Elings, P.K. Hansma: Using force modulation to image surface elasticities with the atomic force microscope, Nanotechnology **2**, 103–106 (1991)
33. B. Bhushan, A.V. Kulkarni, W. Bonin, J.T. Wyrobek: Nano/Picoindentation measurements using capacitive transducer in atomic force microscopy, Philos. Mag. A **74**, 1117–1128 (1996)
34. B. Bhushan, V.N. Koinkar: Nanoindentation hardness measurements using atomic force microscopy, Appl. Phys. Lett. **75**, 5741–5746 (1994)
35. D. DeVecchio, B. Bhushan: Localized surface elasticity measurements using an atomic force microscope, Rev. Sci. Instrum. **68**, 4498–4505 (1997)
36. S. Amelio, A.V. Goldade, U. Rabe, V. Scherer, B. Bhushan, W. Arnold: Measurements of mechanical properties of ultra-thin diamond-like carbon coatings using atomic force acoustic microscopy, Thin Solid Films **392**, 75–84 (2001)
37. D.M. Eigler, E.K. Schweizer: Positioning single atoms with a scanning tunnelling microscope, Nature **344**, 524–528 (1990)

38. A.L. Weisenhorn, J.E. MacDougall, J.A.C. Gould, S.D. Cox, W.S. Wise, J. Massie, P. Maivald, V.B. Elings, G.D. Stucky, P.K. Hansma: Imaging and manipulating of molecules on a zeolite surface with an atomic force microscope, Science **247**, 1330–1333 (1990)
39. I.W. Lyo, Ph. Avouris: Field-induced nanometer-to-atomic-scale manipulation of silicon surfaces with the STM, Science **253**, 173–176 (1991)
40. O.M. Leung, M.C. Goh: Orientation ordering of polymers by atomic force microscope tip-surface interactions, Science **225**, 64–66 (1992)
41. D.W. Abraham, H.J. Mamin, E. Ganz, J. Clark: Surface modification with the scanning tunneling microscope, IBM J. Res. Dev. **30**, 492–499 (1986)
42. R.M. Silver, E.E. Ehrichs, A.L. de Lozanne: Direct writing of submicron metallic features with a scanning tunnelling microscope, Appl. Phys. Lett. **51**, 247–249 (1987)
43. A. Kobayashi, F. Grey, R.S. Williams, M. Ano: Formation of nanometer-scale grooves in silicon with a scanning tunneling microscope, Science **259**, 1724–1726 (1993)
44. B. Parkinson: Layer-by-layer nanometer scale etching of two-dimensional substrates using the scanning tunneling microscopy, J. Am. Chem. Soc. **112**, 7498–7502 (1990)
45. A. Majumdar, P.I. Oden, J.P. Carrejo, L.A. Nagahara, J.J. Graham, J. Alexander: Nanometer-scale lithography using the atomic force microscope, Appl. Phys. Lett. **61**, 2293–2295 (1992)
46. B. Bhushan: Micro/Nanotribology and its applications to magnetic storage devices and MEMS, Tribol. Int. **28**, 85–96 (1995)
47. L. Tsau, D. Wang, K.L. Wang: Nanometer scale patterning of silicon(100) surface by an atomic force microscope operating in air, Appl. Phys. Lett. **64**, 2133–2135 (1994)
48. E. Delawski, B.A. Parkinson: Layer-by-layer etching of two-dimensional metal chalcogenides with the atomic force microscope, J. Am. Chem. Soc. **114**, 1661–1667 (1992)
49. B. Bhushan, G.S. Blackman: Atomic force microscopy of magnetic rigid disks and sliders and its applications to tribology, ASME J. Tribol. **113**, 452–458 (1991)
50. O. Marti, B. Drake, P.K. Hansma: Atomic force microscopy of liquid-covered surfaces: atomic resolution images, Appl. Phys. Lett. **51**, 484–486 (1987)
51. B. Drake, C.B. Prater, A.L. Weisenhorn, S.A.C. Gould, T.R. Albrecht, C.F. Quate, D.S. Cannell, H.G. Hansma, P.K. Hansma: Imaging crystals, polymers and processes in water with the atomic force microscope, Science **243**, 1586–1589 (1989)
52. M. Binggeli, R. Christoph, H.E. Hintermann, J. Colchero, O. Marti: Friction force measurements on potential controlled graphite in an electrolytic environment, Nanotechnology **4**, 59–63 (1993)
53. G. Meyer, N.M. Amer: Novel optical approach to atomic force microscopy, Appl. Phys. Lett. **53**, 1045–1047 (1988)
54. J.H. Coombs, J.B. Pethica: Properties of vacuum tunneling currents: Anomalous barrier heights, IBM J. Res. Dev. **30**, 455–459 (1986)
55. M.D. Kirk, T. Albrecht, C.F. Quate: Low-temperature atomic force microscopy, Rev. Sci. Instrum. **59**, 833–835 (1988)
56. F.J. Giessibl, Ch. Gerber, G. Binnig: A low-temperature atomic force/scanning tunneling microscope for ultrahigh vacuum, J. Vacuum Sci. Technol. B **9**, 984–988 (1991)
57. T.R. Albrecht, P. Grutter, D. Rugar, D.P.E. Smith: Low temperature force microscope with all-fiber interferometer, Ultramicroscopy **42–44**, 1638–1646 (1992)
58. H.J. Hug, A. Moser, Th. Jung, O. Fritz, A. Wadas, I. Parashikor, H.J. Güntherodt: Low temperature magnetic force microscopy, Rev. Sci. Instrum. **64**, 2920–2925 (1993)
59. C. Basire, D.A. Ivanov: Evolution of the lamellar structure during crystallization of a semicrystalline-amorphous polymer blend: Time-resolved hot-stage SPM study, Phys. Rev. Lett. **85**, 5587–5590 (2000)

60. H. Liu, B. Bhushan: Investigation of nanotribological properties of self-assembled monolayers with alkyl and biphenyl spacer chains, Ultramicroscopy **91**, 185–202 (2002)
61. J. Foster, J. Frommer: Imaging of liquid crystal using a tunneling microscope, Nature **333**, 542–547 (1988)
62. D. Smith, H. Horber, C. Gerber, G. Binnig: Smectic liquid crystal monolayers on graphite observed by scanning tunneling microscopy, Science **245**, 43–45 (1989)
63. D. Smith, J. Horber, G. Binnig, H. Nejoh: Structure, registry and imaging mechanism of alkylcyanobiphenyl molecules by tunnelling microscopy, Nature **344**, 641–644 (1990)
64. Y. Andoh, S. Oguchi, R. Kaneko, T. Miyamoto: Evaluation of very thin lubricant films, J. Phys. D **25**, A71–A75 (1992)
65. Y. Martin, C. C. Williams, H. K. Wickramasinghe: Atomic force microscope-force mapping and profiling on a sub 100-A scale, J. Appl. Phys. **61**, 4723–4729 (1987)
66. J. E. Stern, B. D. Terris, H. J. Mamin, D. Rugar: Deposition and imaging of localized charge on insulator surfaces using a force microscope, Appl. Phys. Lett. **53**, 2717–2719 (1988)
67. K. Yamanaka, H. Ogisco, O. Kolosov: Ultrasonic force microscopy for nanometer resolution subsurface imaging, Appl. Phys. Lett. **64**, 178–180 (1994)
68. K. Yamanaka, E. Tomita: Lateral force modulation atomic force microscope for selective imaging of friction forces, Jpn. J. Appl. Phys. **34**, 2879–2882 (1995)
69. U. Rabe, K. Janser, W. Arnold: Vibrations of free and surface-coupled atomic force microscope: Theory and experiment, Rev. Sci. Instrum. **67**, 3281–3293 (1996)
70. Y. Martin, H. K. Wickramasinghe: Magnetic imaging by force microscopy with 1000 Å resolution, Appl. Phys. Lett. **50**, 1455–1457 (1987)
71. D. Rugar, H. J. Mamin, P. Guethner, S. E. Lambert, J. E. Stern, I. McFadyen, T. Yogi: Magnetic force microscopy – General principles and application to longitudinal recording media, J. Appl. Phys. **63**, 1169–1183 (1990)
72. C. Schoenenberger, S. F. Alvarado: Understanding magnetic force microscopy, Z. Phys. B **80**, 373–383 (1990)
73. U. Hartmann: Magnetic force microscopy, Annu. Rev. Mater. Sci. **29**, 53–87 (1999)
74. D. W. Pohl, W. Denk, M. Lanz: Optical stethoscopy-image recording with resolution lambda/20, Appl. Phys. Lett. **44**, 651–653 (1984)
75. E. Betzig, J. K. Troutman, T. D. Harris, J. S. Weiner, R. L. Kostelak: Breaking the diffraction barrier – optical microscopy on a nanometric scale, Science **251**, 1468–1470 (1991)
76. E. Betzig, P. L. Finn, J. S. Weiner: Combined shear force and near-field scanning optical microscopy, Appl. Phys. Lett. **60**, 2484 (1992)
77. P. F. Barbara, D. M. Adams, D. B. O'Connor: Characterization of organic thin film materials with near-field scanning optical microscopy (NSOM), Annu. Rev. Mater. Sci. **29**, 433–469 (1999)
78. C. C. Williams, H. K. Wickramasinghe: Scanning thermal profiler, Appl. Phys. Lett. **49**, 1587–1589 (1986)
79. C. C. Williams, H. K. Wickramasinghe: Microscopy of chemical-potential variations on an atomic scale, Nature **344**, 317–319 (1990)
80. A. Majumdar: Scanning thermal microscopy, Annu. Rev. Mater. Sci. **29**, 505–585 (1999)
81. O. E. Husser, D. H. Craston, A. J. Bard: Scanning electrochemical microscopy – high resolution deposition and etching of materials, J. Electrochem. Soc. **136**, 3222–3229 (1989)

82. Y. Martin, D. W. Abraham, H. K. Wickramasinghe: High-resolution capacitance measurement and potentiometry by force microscopy, Appl. Phys. Lett. **52**, 1103–1105 (1988)
83. M. Nonnenmacher, M. P. O'Boyle, H. K. Wickramasinghe: Kelvin probe force microscopy, Appl. Phys. Lett. **58**, 2921–2923 (1991)
84. J. M. R. Weaver, D. W. Abraham: High resolution atomic force microscopy potentiometry, J. Vacuum Sci. Technol. B **9**, 1559–1561 (1991)
85. D. DeVecchio, B. Bhushan: Use of a nanoscale Kelvin probe for detecting wear precursors, Rev. Sci. Instrum. **69**, 3618–3624 (1998)
86. B. Bhushan, A. V. Goldade: Measurements and analysis of surface potential change during wear of single-crystal silicon (100) at ultralow loads using Kelvin probe microscopy, Appl. Surf. Sci. **157**, 373–381 (2000)
87. P. K. Hansma, B. Drake, O. Marti, S. A. C. Gould, C. B. Prater: The scanning ion-conductance microscope, Science **243**, 641–643 (1989)
88. C. B. Prater, P. K. Hansma, M. Tortonese, C. F. Quate: Improved scanning ion-conductance microscope using microfabricated probes, Rev. Sci. Instrum. **62**, 2634–2638 (1991)
89. J. Matey, J. Blanc: Scanning capacitance microscopy, J. Appl. Phys. **57**, 1437–1444 (1985)
90. C. C. Williams: Two-dimensional dopant profiling by scanning capacitance microscopy, Annu. Rev. Mater. Sci. **29**, 471–504 (1999)
91. D. T. Lee, J. P. Pelz, B. Bhushan: Instrumentation for direct, low frequency scanning capacitance microscopy, and analysis of position dependent stray capacitance, Rev. Sci. Instrum. **73**, 3523–3533 (2002)
92. P. K. Hansma, J. Tersoff: Scanning tunneling microscopy, J. Appl. Phys. **61**, R1–R23 (1987)
93. I. Giaever: Energy gap in superconductors measured by electron tunneling, Phys. Rev. Lett. **5**, 147–148 (1960)
94. D. Sarid, V. Elings: Review of scanning force microscopy, J. Vacuum Sci. Technol. B **9**, 431–437 (1991)
95. U. Durig, O. Zuger, A. Stalder: Interaction force detection in scanning probe microscopy: Methods and applications, J. Appl. Phys. **72**, 1778–1797 (1992)
96. J. Frommer: Scanning tunneling microscopy and atomic force microscopy in organic chemistry, Angew. Chem. Int. Ed. **31**, 1298–1328 (1992)
97. H. J. Güntherodt, R. Wiesendanger (eds): *Scanning Tunneling Microscopy I: General Principles and Applications to Clean and Adsorbate-Covered Surfaces* (Springer, Berlin, Heidelberg 1992)
98. R. Wiesendanger, H. J. Güntherodt (eds): *Scanning Tunneling Microscopy, II: Further Applications and Related Scanning Techniques* (Springer, Berlin, Heidelberg 1992)
99. D. A. Bonnell (ed): *Scanning Tunneling Microscopy and Spectroscopy – Theory, Techniques, and Applications* (VCH, New York 1993)
100. O. Marti, M. Amrein (eds): *STM and SFM in Biology* (Academic, San Diego 1993)
101. J. A. Stroscio, W. J. Kaiser (eds): *Scanning Tunneling Microscopy* (Academic, Boston 1993)
102. H. J. Güntherodt, D. Anselmetti, E. Meyer (eds): *Forces in Scanning Probe Methods* (Kluwer, Dordrecht 1995)
103. G. Binnig, H. Rohrer: Scanning tunnelling microscopy, Surf. Sci. **126**, 236–244 (1983)
104. B. Bhushan, J. Ruan, B. K. Gupta: A scanning tunnelling microscopy study of fullerene films, J. Phys. D **26**, 1319–1322 (1993)

105. R. L. Nicolaides, W. E. Yong, W. F. Packard, H. A. Zhou: Scanning tunneling microscope tip structures, J. Vacuum Sci. Technol. A **6**, 445–447 (1988)
106. J. P. Ibe, P. P. Bey, S. L. Brandon, R. A. Brizzolara, N. A. Burnham, D. P. DiLella, K. P. Lee, C. R. K. Marrian, R. J. Colton: On the electrochemical etching of tips for scanning tunneling microscopy, J. Vacuum Sci. Technol. A **8**, 3570–3575 (1990)
107. R. Kaneko, S. Oguchi: Ion-implanted diamond tip for a scanning tunneling microscope, Jpn. J. Appl. Phys. **28**, 1854–1855 (1990)
108. F. J. Giessibl: Atomic resolution of the silicon(111)–(7×7) surface by atomic force microscopy, Science **267**, 68–71 (1995)
109. B. Anczykowski, D. Krueger, K. L. Babcock, H. Fuchs: Basic properties of dynamic force spectroscopy with the scanning force microscope in experiment and simulation, Ultramicroscopy **66**, 251–259 (1996)
110. T. R. Albrecht and C. F. Quate: Atomic resolution imaging of a nonconductor by atomic force microscopy, J. Appl. Phys. **62**, 2599–2602 (1987)
111. S. Alexander, L. Hellemans, O. Marti, J. Schneir, V. Elings, P. K. Hansma: An atomic-resolution atomic-force microscope implemented using an optical lever, J. Appl. Phys. **65**, 164–167 (1989)
112. G. Meyer, N. M. Amer: Optical-beam-deflection atomic force microscopy: The NaCl(001) surface, Appl. Phys. Lett. **56**, 2100–2101 (1990)
113. A. L. Weisenhorn, M. Egger, F. Ohnesorge, S. A. C. Gould, S. P. Heyn, H. G. Hansma, R. L. Sinsheimer, H. E. Gaub, P. K. Hansma: Molecular resolution images of Langmuir–Blodgett films and DNA by atomic force microscopy, Langmuir **7**, 8–12 (1991)
114. J. Ruan, B. Bhushan: Atomic-scale and microscale friction of graphite and diamond using friction force microscopy, J. Appl. Phys. **76**, 5022–5035 (1994)
115. D. Rugar, P. K. Hansma: Atomic force microscopy, Phys. Today **43**, 23–30 (1990)
116. D. Sarid: *Scanning Force Microscopy* (Oxford Univ. Press, Oxford 1991)
117. G. Binnig: Force microscopy, Ultramicroscopy **42–44**, 7–15 (1992)
118. E. Meyer: Atomic force microscopy, Surf. Sci. **41**, 3–49 (1992)
119. H. K. Wickramasinghe: Progress in scanning probe microscopy, Acta Mater. **48**, 347–358 (2000)
120. A. J. den Boef: The influence of lateral forces in scanning force microscopy, Rev. Sci. Instrum. **62**, 88–92 (1991)
121. M. Radmacher, R. W. Tillman, M. Fritz, H. E. Gaub: From molecules to cells: Imaging soft samples with the atomic force microscope, Science **257**, 1900–1905 (1992)
122. F. Ohnesorge, G. Binnig: True atomic resolution by atomic force microscopy through repulsive and attractive forces, Science **260**, 1451–1456 (1993)
123. G. Neubauer, S. R. Coben, G. M. McClelland, D. Horne, C. M. Mate: Force microscopy with a bidirectional capacitance sensor, Rev. Sci. Instrum. **61**, 2296–2308 (1990)
124. T. Goddenhenrich, H. Lemke, U. Hartmann, C. Heiden: Force microscope with capacitive displacement detection, J. Vacuum Sci. Technol. A **8**, 383–387 (1990)
125. U. Stahl, C. W. Yuan, A. L. Delozanne, M. Tortonese: Atomic force microscope using piezoresistive cantilevers and combined with a scanning electron microscope, Appl. Phys. Lett. **65**, 2878–2880 (1994)
126. R. Kassing, E. Oesterschulze: Sensors for scanning probe microscopy. In: *Micro/Nanotribology and Its Applications*, ed. by B. Bhushan (Kluwer, Dordrecht 1997) pp. 35–54
127. C. M. Mate: Atomic-force-microscope study of polymer lubricants on silicon surfaces, Phys. Rev. Lett. **68**, 3323–3326 (1992)
128. S. P. Jarvis, A. Oral, T. P. Weihs, J. B. Pethica: A novel force microscope and point contact probe, Rev. Sci. Instrum. **64**, 3515–3520 (1993)

129. D. Rugar, H.J. Mamin, P. Guethner: Improved fiber-optical interferometer for atomic force microscopy, Appl. Phys. Lett. **55**, 2588–2590 (1989)
130. C. Schoenenberger, S.F. Alvarado: A differential interferometer for force microscopy, Rev. Sci. Instrum. **60**, 3131–3135 (1989)
131. D. Sarid, D. Iams, V. Weissenberger, L.S. Bell: Compact scanning-force microscope using laser diode, Opt. Lett. **13**, 1057–1059 (1988)
132. N.W. Ashcroft, N.D. Mermin: *Solid State Physics* (Holt Reinhart and Winston, New York 1976)
133. G. Binnig, D.P.E. Smith: Single-tube three-dimensional scanner for scanning tunneling microscopy, Rev. Sci. Instrum. **57**, 1688 (1986)
134. S.I. Park, C.F. Quate: Digital filtering of STM images, J. Appl. Phys. **62**, 312 (1987)
135. J.W. Cooley, J.W. Tukey: An algorithm for machine calculation of complex Fourier series, Math. Comput. **19**, 297 (1965)
136. J. Ruan, B. Bhushan: Atomic-scale friction measurements using friction force microscopy: Part I – General principles and new measurement techniques, ASME J. Tribol. **116**, 378–388 (1994)
137. T.R. Albrecht, S. Akamine, T.E. Carver, C.F. Quate: Microfabrication of cantilever styli for the atomic force microscope, J. Vacuum Sci. Technol. A **8**, 3386–3396 (1990)
138. O. Marti, S. Gould, P.K. Hansma: Control electronics for atomic force microscopy, Rev. Sci. Instrum. **59**, 836–839 (1988)
139. O. Wolter, T. Bayer, J. Greschner: Micromachined silicon sensors for scanning force microscopy, J. Vacuum Sci. Technol. B **9**, 1353–1357 (1991)
140. E. Meyer, R. Overney, R. Luthi, D. Brodbeck: Friction force microscopy of mixed Langmuir–Blodgett films, Thin Solid Films **220**, 132–137 (1992)
141. H.J. Dai, J.H. Hafner, A.G. Rinzler, D.T. Colbert, R.E. Smalley: Nanotubes as nanoprobes in scanning probe microscopy, Nature **384**, 147–150 (1996)
142. J.H. Hafner, C.L. Cheung, A.T. Woolley, C.M. Lieber: Structural and functional imaging with carbon nanotube AFM probes, Prog. Biophys. Mol. Biol. **77**, 73–110 (2001)
143. G.S. Blackman, C.M. Mate, M.R. Philpott: Interaction forces of a sharp tungsten tip with molecular films on silicon surface, Phys. Rev. Lett. **65**, 2270–2273 (1990)
144. S.J. O'Shea, M.E. Welland, T. Rayment: Atomic force microscope study of boundary layer lubrication, Appl. Phys. Lett. **61**, 2240–2242 (1992)
145. J.P. Cleveland, S. Manne, D. Bocek, P.K. Hansma: A nondestructive method for determining the spring constant of cantilevers for scanning force microscopy, Rev. Sci. Instrum. **64**, 403–405 (1993)
146. D.W. Pohl: Some design criteria in STM, IBM J. Res. Dev. **30**, 417 (1986)
147. W.T. Thomson, M.D. Dahleh: *Theory of Vibration with Applications*, 5th edn. (Prentice Hall, Upper Saddle River 1998)
148. J. Colchero: Reibungskraftmikroskopie. Ph.D. Thesis (University of Konstanz, Konstanz 1993)
149. G.M. McClelland, R. Erlandsson, S. Chiang: Atomic force microscopy: General principles and a new implementation. In: *Review of Progress in Quantitative Nondestructive Evaluation*, Vol. 6B, ed. by D.O. Thompson, D.E. Chimenti (Plenum, New York 1987) pp. 1307–1314
150. Y.R. Shen: *The Principles of Nonlinear Optics* (Wiley, New York 1984)
151. T. Baumeister, S.L. Marks: *Standard Handbook for Mechanical Engineers*, 7th edn. (McGraw-Hill, New York 1967)
152. J. Colchero, O. Marti, H. Bielefeldt, J. Mlynek: Scanning force and friction microscopy, Phys. Stat. Sol. **131**, 73–75 (1991)

153. R. Young, J. Ward, F. Scire: Observation of metal-vacuum-metal tunneling, field emission, and the transition region, Phys. Rev. Lett. **27**, 922 (1971)
154. R. Young, J. Ward, F. Scire: The topographiner: An instrument for measuring surface microtopography, Rev. Sci. Instrum. **43**, 999 (1972)
155. C. Gerber, O. Marti: Magnetostrictive positioner, IBM Tech. Discl. Bull. **27**, 6373 (1985)
156. R. Garcìa Cantù, M. A. Huerta Garnica: Long-scan imaging by STM, J. Vacuum Sci. Technol. A **8**, 354 (1990)
157. C. J. Chen: In situ testing and calibration of tube piezoelectric scanners, Ultramicroscopy **42–44**, 1653–1658 (1992)
158. R. G. Carr: Finite element analysis of PZT tube scanner motion for scanning tunnelling microscopy, J. Microsc. **152**, 379–385 (1988)
159. C. J. Chen: Electromechanical deflections of piezoelectric tubes with quartered electrodes, Appl. Phys. Lett. **60**, 132 (1992)
160. N. Libioulle, A. Ronda, M. Taborelli, J. M. Gilles: Deformations and nonlinearity in scanning tunneling microscope images, J. Vacuum Sci. Technol. B **9**, 655–658 (1991)
161. E. P. Stoll: Restoration of STM images distorted by time-dependent piezo driver aftereffects, Ultramicroscopy **42–44**, 1585–1589 (1991)
162. R. Durselen, U. Grunewald, W. Preuss: Calibration and applications of a high precision piezo scanner for nanometrology, Scanning **17**, 91–96 (1995)
163. J. Fu: In situ testing and calibrating of Z-piezo of an atomic force microscope, Rev. Sci. Instrum. **66**, 3785–3788 (1995)
164. R. C. Barrett, C. F. Quate: Optical scan-correction system applied to atomic force microscopy, Rev. Sci. Instrum. **62**, 1393 (1991)
165. R. Toledo-Crow, P. C. Yang, Y. Chen, M. Vaez-Iravani: Near-field differential scanning optical microscope with atomic force regulation, Appl. Phys. Lett. **60**, 2957–2959 (1992)
166. J. E. Griffith, G. L. Miller, C. A. Green: A scanning tunneling microscope with a capacitance-based position monitor, J. Vacuum Sci. Technol. B **8**, 2023–2027 (1990)
167. A. E. Holman, C. D. Laman, P. M. L. O. Scholte, W. C. Heerens, F. Tuinstra: A calibrated scanning tunneling microscope equipped with capacitive sensors, Rev. Sci. Instrum. **67**, 2274–2280 (1996)

3

Probes in Scanning Microscopies

Jason H. Hafner

Summary. Scanning probe microscopy (SPM) provides nanometer-scale mapping of numerous sample properties in essentially any environment. This unique combination of high resolution and broad applicability has lead to the application of SPM to many areas of science and technology, especially those interested in the structure and properties of materials at the nanometer scale. SPM images are generated through measurements of a tip-sample interaction. A well-characterized tip is the key element to data interpretation and is typically the limiting factor.

Commercially available atomic force microscopy (AFM) tips, integrated with force sensing cantilevers, are microfabricated from silicon and silicon nitride by lithographic and anisotropic etching techniques. The performance of these tips can be characterized by imaging nanometer-scale standards of known dimension, and the resolution is found to roughly correspond to the tip radius of curvature, the tip aspect ratio, and the sample height. Although silicon and silicon nitride tips have a somewhat large radius of curvature, low aspect ratio, and limited lifetime due to wear, the widespread use of AFM today is due in large part to the broad availability of these tips. In some special cases, small asperities on the tip can provide resolution much higher than the tip radius of curvature for low-Z samples such as crystal surfaces and ordered protein arrays.

Several strategies have been developed to improve AFM tip performance. Oxide sharpening improves tip sharpness and enhances tip asperities. For high-aspect-ratio samples such as integrated circuits, silicon AFM tips can be modified by focused ion beam (FIB) milling. FIB tips reach 3 degree cone angles over lengths of several microns and can be fabricated at arbitrary angles. Other high resolution and high-aspect-ratio tips are produced by electron beam deposition (EBD) in which a carbon spike is deposited onto the tip apex from the background gases in an electron microscope. Finally, carbon nanotubes have been employed as AFM tips. Their nanometer-scale diameter, long length, high stiffness, and elastic buckling properties make carbon nanotubes possibly the ultimate tip material for AFM. Nanotubes can be manually attached to silicon or silicon nitride AFM tips or "grown" onto tips by chemical vapor deposition (CVD), which should soon make them widely available. In scanning tunneling microscopy (STM), the electron tunneling signal decays exponentially with tip-sample separation, so that in principle only the last few atoms contribute to the signal. STM tips are, therefore, not as sensitive to the nanoscale tip geometry and can be made by simple mechanical cutting or electrochemical etching of metal wires. In choosing tip materials, one prefers hard, stiff metals that will not oxidize or corrode in the imaging environment.

3.1 Introduction

In scanning probe microscopy (SPM), an image is created by raster scanning a sharp probe tip over a sample and measuring some highly localized tip-sample interaction as a function of position. SPMs are based on several interactions, the major types including scanning tunneling microscopy (STM), which measures an electronic tunneling current; atomic force microscopy (AFM), which measures force interactions; and near-field scanning optical microscopy (NSOM), which measures local optical properties by exploiting near-field effects (Fig. 3.1). These methods allow the characterization of many properties (structural, mechanical, electronic, optical) on essentially any material (metals, semiconductors, insulators, biomolecules) and in essentially any environment (vacuum, liquid, or ambient air conditions). The unique combination of nanoscale resolution, previously the domain of electron microscopy, *and broad applicability* has led to the proliferation of SPM into virtually all areas of nanometer-scale science and technology.

Several enabling technologies have been developed for SPM, or borrowed from other techniques. Piezoelectric tube scanners allow accurate, sub-angstrom positioning of the tip or sample in three dimensions. Optical deflection systems and microfabricated cantilevers can detect forces in AFM down to the picoNewton range. Sensitive electronics can measure STM currents less than 1 picoamp. High transmission fiber optics and sensitive photodetectors can manipulate and detect small optical signals of NSOM. Environmental control has been developed to allow SPM imaging in UHV, cryogenic temperatures, at elevated temperatures, and in fluids. Vibration and drift have been controlled such that a probe tip can be held over a single molecule for hours of observation. Microfabrication techniques have been developed for the mass production of probe tips, making SPMs commercially available and allowing the development of many new SPM modes and combinations with other characterization methods. However, of all this SPM development over the past 20 years, what has received the least attention is perhaps the most important aspect: the probe tip.

Interactions measured in SPMs occur at the tip-sample interface, which can range in size from a single atom to tens of nanometers. The size, shape, surface chemistry, electronic and mechanical properties of the tip apex will directly influence the data signal and the interpretation of the image. Clearly, the better charac-

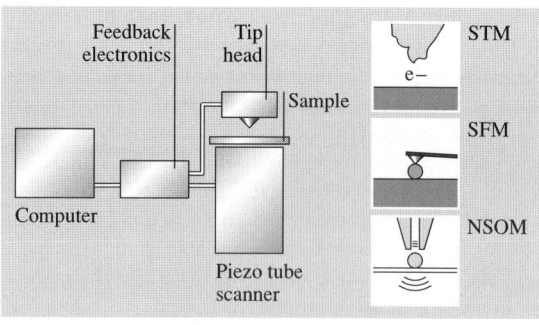

Fig. 3.1. A schematic of the components of a scanning probe microscope and the three types of signals observed: STM senses electron tunneling currents, AFM measures forces, and NSOM measures near-field optical properties via a sub-wavelength aperture

terized the tip the more useful the image information. In this chapter, the fabrication and performance of AFM and STM probes will be described.

3.2 Atomic Force Microscopy

AFM is the most widely used form of SPM, since it requires neither an electrically conductive sample, as in STM, nor an optically transparent sample or substrate, as in most NSOMs. Basic AFM modes measure the topography of a sample with the only requirement being that the sample is deposited on a flat surface and rigid enough to withstand imaging. Since AFM can measure a variety of forces, including van der Waals forces, electrostatic forces, magnetic forces, adhesion forces and friction forces, specialized modes of AFM can characterize the electrical, mechanical, and chemical properties of a sample in addition to its topography.

3.2.1 Principles of Operation

In AFM, a probe tip is integrated with a microfabricated force-sensing cantilever. A variety of silicon and silicon nitride cantilevers are commercially available with micron-scale dimensions, spring constants ranging from 0.01 to 100 N/m, and resonant frequencies ranging from 5 kHz to over 300 kHz. The cantilever deflection is detected by optical beam deflection, as illustrated in Fig. 3.2. A laser beam bounces off the back of the cantilever and is centered on a split photodiode. Cantilever deflections are proportional to the difference signal $V_A - V_B$. Sub-angstrom deflections can be detected and, therefore, forces down to tens of picoNewtons can be measured. A more recently developed method of cantilever deflection measurement is through a piezoelectric layer on the cantilever that registers a voltage upon deflection [1].

A piezoelectric scanner rasters the sample under the tip while the forces are measured through deflections of the cantilever. To achieve more controlled imaging conditions, a feedback loop monitors the tip-sample force and adjusts the sample Z-position to hold the force constant. The topographic image of the sample is then

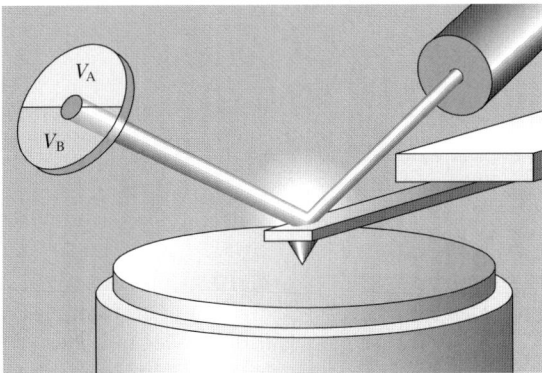

Fig. 3.2. An illustration of the optical beam deflection system that detects cantilever motion in the AFM. The voltage signal $V_A - V_B$ is proportional to the deflection

taken from the sample Z-position data. The mode described is called contact mode, in which the tip is deflected by the sample due to repulsive forces, or "contact". It is generally only used for flat samples that can withstand lateral forces during scanning. To minimize lateral forces and sample damage, two AC modes have been developed. In these, the cantilever is driven into AC oscillation near its resonant frequency (tens to hundreds of kHz) with amplitudes of 5 to tens of s. When the tip approaches the sample, the oscillation is damped, and the reduced amplitude is the feedback signal, rather than the DC deflection. Again, topography is taken from the varying Z-position of the sample required to keep the tip oscillation amplitude constant. The two AC modes differ only in the nature of the interaction. In intermittent contact mode, also called tapping mode, the tip contacts the sample on each cycle, so the amplitude is reduced by ionic repulsion as in contact mode. In noncontact mode, long-range van der Waals forces reduce the amplitude by effectively shifting the spring constant experienced by the tip and changing its resonant frequency.

3.2.2 Standard Probe Tips

In early AFM work, cantilevers were made by hand from thin metal foils or small metal wires. Tips were created by gluing diamond fragments to the foil cantilevers or electrochemically etching the wires to a sharp point. Since these methods were labor intensive and not highly reproducible, they were not amenable to large-scale production. To address this problem, and the need for smaller cantilevers with higher resonant frequencies, batch fabrication techniques were developed (see Fig. 3.3). Building on existing methods to batch fabricate Si_3N_4 cantilevers, *Albrecht* et al. [2] etched an array of small square openings in an SiO_2 mask layer over a (100) silicon surface. The exposed square (100) regions were etched with KOH, an anisotropic etchant that terminates at the (111) planes, thus creating pyramidal etch pits in the silicon surface. The etch pit mask was then removed and another was applied to define the cantilever shapes with the pyramidal etch pits at the end. The Si wafer was then coated with a low stress Si_3N_4 layer by LPCVD. The Si_3N_4 fills the etch pit, using it as a mold to create a pyramidal tip. The silicon was later removed by etching to free the cantilevers and tips. Further steps resulting in the attachment of

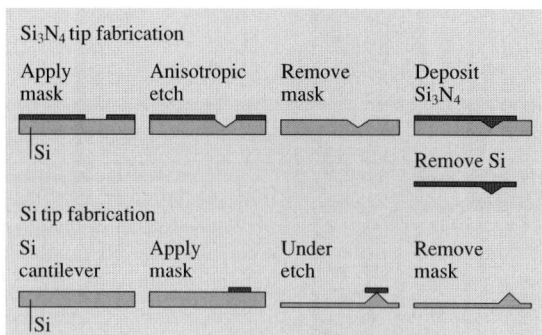

Fig. 3.3. A schematic overview of the fabrication of Si and Si_3N_4 tip fabrication as described in the text

the cantilever to a macroscopic piece of glass are not described here. The resulting pyramidal tips were highly symmetric and had a tip radius of less than 30 nm, as determined by scanning electron microscopy (SEM). This procedure has likely not changed significantly, since commercially available Si_3N_4 tips are still specified to have a curvature radius of 30 nm.

Wolter et al. [3] developed methods to batch fabricate single-crystal Si cantilevers with integrated tips. Microfabricated Si cantilevers were first prepared using previously described methods, and a small mask was formed at the end of the cantilever. The Si around the mask was etched by KOH, so that the mask was under cut. This resulted in a pyramidal silicon tip under the mask, which was then removed. Again, this partial description of the full procedure only describes tip fabrication. With some refinements the silicon tips were made in high yield with curvature radii of less than 10 nm. Si tips are sharper than Si_3N_4 tips, because they are directly formed by the anisotropic etch in single-crystal Si, rather than using an etch pit as a mask for deposited material. Commercially available silicon probes are made by similar refined techniques and provide a curvature typical radius of < 10 nm.

3.2.3 Probe Tip Performance

In atomic force microscopy the question of resolution can be a rather complicated issue. As an initial approximation, resolution is often considered strictly in geometrical terms that assume rigid tip-sample contact. The topographical image of a feature is broadened or narrowed by the size of the probe tip, so the resolution is approximately the width of the tip. Therefore, the resolution of AFM with standard commercially available tips is on the order of 5 to 10 nm. *Bustamante* and *Keller* [4] carried the geometrical model further by drawing an analogy to resolution in optical systems. Consider two sharp spikes separated by a distance d to be point objects imaged by AFM (see Fig. 3.4). Assume the tip has a parabolic shape with an end radius R. The tip-broadened image of these spikes will appear as inverted parabolas. There will be a small depression between the images of depth Δz. The two spikes are considered "resolved" if Δz is larger than the instrumental noise in the z direction. Defined in this manner, the resolution d, the minimum separation at which the spikes are resolved, is

$$d = 2\sqrt{2R(\Delta z)}, \tag{3.1}$$

where one must enter a minimal detectable depression for the instrument (Δz) to determine the resolution. So for a silicon tip with radius 5 nm and a minimum detectable Δz of 0.5 nm, the resolution is about 4.5 nm. However, the above model assumes the spikes are of equal height. *Bustamante* and *Keller* [4] went on to point out that if the height of the spikes is not equal, the resolution will be affected. Assuming a height difference of Δh, the resolution becomes:

$$d = \sqrt{2R}\left(\sqrt{\Delta z} + \sqrt{\Delta z + \Delta h}\right). \tag{3.2}$$

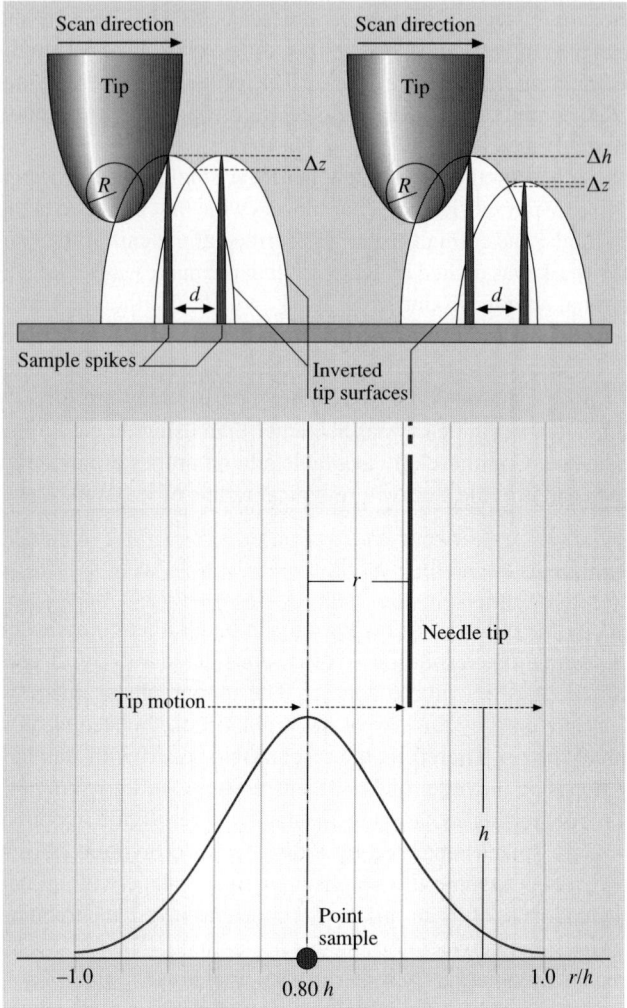

Fig. 3.4. The factors that determine AFM imaging resolution in contact mode (*top*) and non-contact mode (*bottom*), adapted from [4]

For a pair of spikes with a 2 nm height difference, the resolution drops to 7.2 nm for a 5 nm tip and 0.5 nm minimum detectable Δz. While geometrical considerations are a good starting point for defining resolution, they ignore factors such as the possible compression and deformation of the tip and sample. *Vesenka* et al. [5] confirmed a similar geometrical resolution model by imaging monodisperse gold nanoparticles with tips characterized by transmission electron microscopy (TEM).

Noncontact AFM contrast is generated by long-range interactions such as van der Waals forces, so resolution will not simply be determined by geometry because the tip and sample are not in rigid contact. *Bustamante* and *Keller* [4] have derived an expression for the resolution in noncontact AFM for an idealized, infinitely thin "line" tip and a point particle as the sample (Fig. 3.4). Noncontact AFM is sensitive to the gradient of long-range forces, so the van der Waals force gradient was calculated as a function of position for the tip at height h above the surface. If the resolution d is defined as the full width at half maximum of this curve, the resolution is:

$$d = 0.8h. \tag{3.3}$$

This shows that even for an ideal geometry, the resolution is fundamentally limited in noncontact mode by the tip-sample separation. Under UHV conditions, the tip-sample separation can be made very small, so atomic resolution is possible on flat, crystalline surfaces. Under ambient conditions, however, the separation must be larger to keep the tip from being trapped in the ambient water layer on the surface. This larger separation can lead to a point where further improvements in tip sharpness do not improve resolution. It has been found that imaging 5 nm gold nanoparticles in noncontact mode with carbon nanotube tips of 2 nm diameter leads to particle widths of 12 nm, larger than the 7 nm width one would expect assuming rigid contact [8]. However, in tapping mode operation, the geometrical definition of resolution is relevant, since the tip and sample come into rigid contact. When imaging 5 nm gold particles with 2 nm carbon nanotube tips in tapping mode, the expected 7 nm particle width is obtained [9].

The above descriptions of AFM resolution cannot explain the sub-nanometer resolution achieved on crystal surfaces [10] and ordered arrays of biomolecules [11] in contact mode with commercially available probe tips. Such tips have nominal radii of curvature ranging from 5 nm to 30 nm, an order of magnitude larger than the resolution achieved. A detailed model to explain the high resolution on ordered membrane proteins has been put forth by [6]. In this model, the larger part of the silicon nitride tip apex balances the tip-sample interaction through electrostatic forces, while a very small tip asperity interacts with the sample to provide contrast (see Fig. 3.5). This model is supported by measurements at varying salt concentrations to vary the electrostatic interaction strength and the observation of defects in the ordered samples. However, the existence of such asperities has never been confirmed by independent electron microscopy images of the tip. Another model, considered

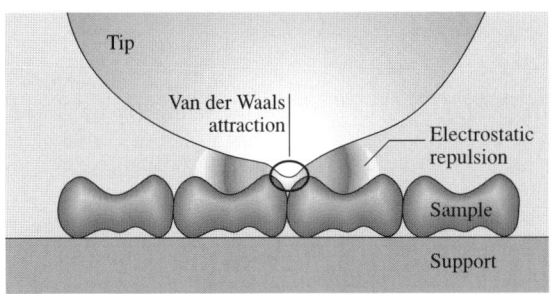

Fig. 3.5. A tip model to explain the high resolution obtained on ordered samples in contact mode, from [6]

especially applicable to atomic resolution on crystal surfaces, assumes the tip is in contact with a region of the sample much larger than the resolution observed, and that force components matching the periodicity of the sample are transmitted to the tip, resulting in an "averaged" image of the periodic lattice. Regardless of the mechanism, the structures determined are accurate and make this a highly valuable method for membrane proteins. However, this level of resolution should not be expected for most biological systems.

3.2.4 Oxide-Sharpened Tips

Both Si and Si_3N_4 tips with increased aspect ratio and reduced tip radius can be fabricated through oxide sharpening of the tip. If a pyramidal or cone-shaped silicon tip is thermally oxidized to SiO_2 at low temperature ($< 1050\,°C$), Si-SiO_2 stress formation reduces the oxidation rate at regions of high curvature. The result is a sharper, higher-aspect-ratio cone of silicon at the high curvature tip apex inside the outer pyramidal layer of SiO_2 (see Fig. 3.6). Etching the SiO_2 layer with HF then leaves tips with aspect ratios up to 10:1 and radii down to 1 nm [7], although 5–10 nm is the nominal specification for most commercially available tips. This oxide sharpening technique can also be applied to Si_3N_4 tips by oxidizing the silicon etch pits that are used as molds. As with tip fabrication, oxide sharpening is not quite as effective for Si_3N_4. Si_3N_4 tips were reported to have an 11 nm radius of curvature [12], while commercially available oxide-sharpened Si_3N_4 tips have a nominal radius of < 20 nm.

3.2.5 FIB tips

A common AFM application in integrated circuit manufacture and MEMs is to image structures with very steep sidewalls such as trenches. To accurately image these features, one must consider the micron-scale tip structure, rather than the nanometer-scale structure of the tip apex. Since tip fabrication processes rely on anisotropic etchants, the cone half-angles of pyramidal tips are approximately 20 degrees. Images of deep trenches taken with such tips display slanted sidewalls and may not reach the bottom of the trench due to the tip broadening effects. To image such samples more faithfully, high-aspect-ratio tips are fabricated by focused ion beam (FIB)

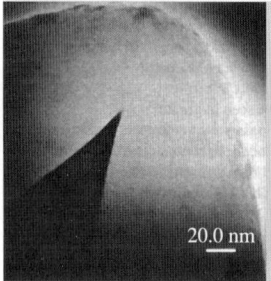
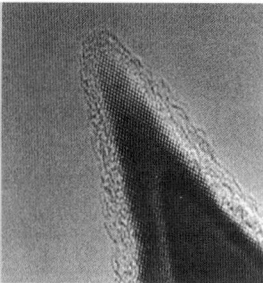

Fig. 3.6. Oxide sharpening of silicon tips. The *left image* shows a sharpened core of silicon in an outer layer of SiO_2. The *right image* is a higher magnification view of such a tip after the SiO_2 is removed. Adapted from [7]

machining a Si tip to produce a sharp spike at the tip apex. Commercially available FIB tips have half cone angles of < 3 degrees over lengths of several microns, yielding aspect ratios of approximately 10:1. The radius of curvature at the tip end is similar to that of the tip before the FIB machining. Another consideration for high-aspect-ratio tips is the tip tilt. To ensure that the pyramidal tip is the lowest part of the tip-cantilever assembly, most AFM designs tilt the cantilever about 15 degrees from parallel. Therefore, even an ideal "line tip" will not give an accurate image of high steep sidewalls, but will produce an image that depends on the scan angle. Due to the versatility of the FIB machining, tips are available with the spikes at an angle to compensate for this effect.

3.2.6 EBD tips

Another method of producing high-aspect-ratio tips for AFM is called electron beam deposition (EBD). First developed for STM tips [13, 14], EBD tips were introduced for AFM by focusing an SEM onto the apex of a pyramidal tip arranged so that it pointed along the electron beam axis (see Fig. 3.7). Carbon material was deposited by the dissociation of background gases in the SEM vacuum chamber. *Schiffmann* [15] systematically studied the following parameters and how they affected EBD tip geometry:

Deposition time :	0.5 to 8 min
Beam current :	3–300 pA
Beam energy :	1–30 keV
Working distance :	8–48 mm .

EBD tips were cylindrical with end radii of 20–40 nm, lengths of 1 to 5 µm, and diameters of 100 to 200 nm. Like FIB tips, EBD tips were found to achieve improved imaging of steep features. By controlling the position of the focused beam, the tip geometry can be further controlled. Tips were fabricated with lengths over 5 µm and aspect ratios greater than 100:1, yet these were too fragile to use as a tip in AFM [13].

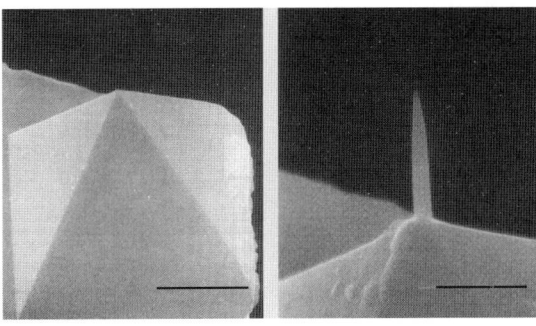

Fig. 3.7. A pyramidal tip before (*left*, 2-µm-scale bar) and after (*right*, 1-µm-scale bar) electron beam deposition, adapted from [13]

3.2.7 Carbon Nanotube Tips

Carbon nanotubes are microscopic graphitic cylinders that are nanometers in diameter, yet many microns in length. Single-walled carbon nanotubes (SWNT) consist of single sp^2 hybridized carbon sheets rolled into seamless tubes and have diameters ranging from 0.7 to 3 nm.

Carbon Nanotube Structure

Larger structures called multiwalled carbon nanotubes (MWNT) consist of nested, concentrically arranged SWNT and have diameters ranging from 3 to 50 nm. Figure 3.8 shows a model of nanotube structure, as well as TEM images of a SWNT and a MWNT. The small diameter and high aspect ratio of carbon nanotubes suggests their application as high resolution, high-aspect-ratio AFM probes.

Carbon Nanotube Mechanical Properties

Carbon nanotubes possess exceptional mechanical properties that impact their use as probes. Their lateral stiffness can be approximated from that of a solid elastic rod:

$$k_{\text{lat}} = \frac{3\pi Y r^4}{4l^3}, \tag{3.4}$$

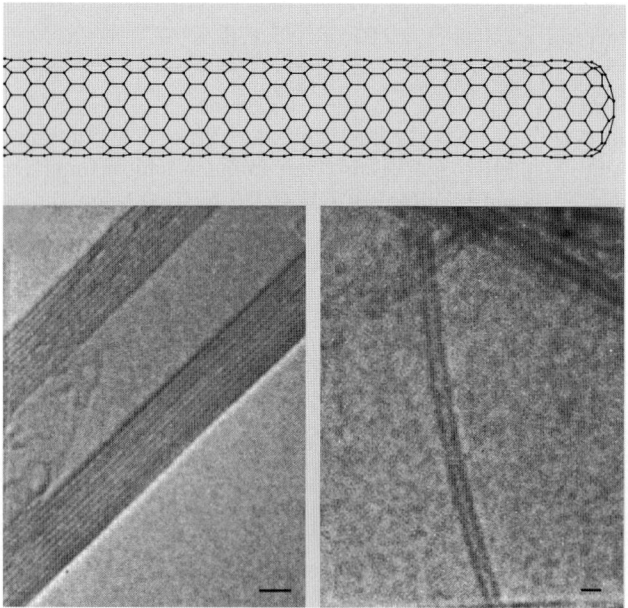

Fig. 3.8. The structure of carbon nanotubes, including TEM images of a MWNT (*left*) and a SWNT (*right*), from [16]

where the spring constant k_{lat} represents the restoring force per unit lateral displacement, r is the radius, l is the length, and Y is the Young's modulus (also called the elastic modulus) of the material. For the small diameters and extreme aspect ratios of carbon nanotube tips, the thermal vibrations of the probe tip at room temperature can become sufficient to degrade image resolution. These thermal vibrations can be approximated by equating $\frac{1}{2}k_B T$ of thermal energy to the energy of an oscillating nanotube:

$$\frac{1}{2}k_B T = \frac{1}{2}k_{lat}a^2, \tag{3.5}$$

where k_B is Boltzmann's constant, T is the temperature, and a is the vibration amplitude. Substituting for k_{lat} from (3.4) yields:

$$a = \sqrt{\frac{4k_B T l^3}{3\pi Y r^4}}. \tag{3.6}$$

The strong dependence on radius and length reveals that one must carefully control the tip geometry at this size scale. Equation (3.6) implies that the stiffer the material, i.e., the higher its Young's modulus, the smaller the thermal vibrations and the longer and thinner a tip can be. The Young's moduli of carbon nanotubes have been determined by measurements of the thermal vibration amplitude by TEM [18, 19] and by directly measuring the forces required to deflect a pinned carbon nanotube in an AFM [20]. These experiments revealed that the Young's modulus of carbon nanotubes is 1–2 TPa, in agreement with theoretical predictions [21]. This makes carbon nanotubes the stiffest known material and, therefore, the best for fabricating thin, high-aspect-ratio tips. A more detailed and accurate derivation of the thermal vibration amplitudes was derived for the Young's modulus measurements [18, 19].

Carbon nanotubes elastically buckle under large loads, rather than fracture or plastically deform like most materials. Nanotubes were first observed in the buckled state by transmission electron microscopy [17], as shown in Fig. 3.9. The first experimental evidence that nanotube buckling is elastic came from the application of nanotubes as probe tips [22], described in detail below. A more direct experimental observation of elastic buckling was obtained by deflecting nanotubes pinned to a low friction surface with an AFM tip [20]. Both reports found that the buckling force could be approximated with the macroscopic Euler buckling formula for an elastic column:

$$F_{Euler} = \frac{\pi^3 Y r^4}{4l^2}. \tag{3.7}$$

The buckling force puts another constraint on the tip length: If the nanotube is too long the buckling force will be too low for stable imaging. The elastic buckling property of carbon nanotubes has significant implications for their use as AFM probes. If a large force is applied to the tip inadvertently, or if the tip encounters a large step in sample height, the nanotube can buckle to the side, then snap back without degraded imaging resolution when the force is removed, making these tips highly robust. No other tip material displays this buckling characteristic.

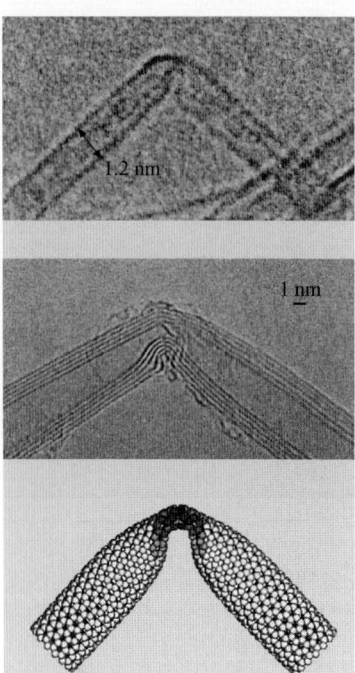

Fig. 3.9. TEM images and a model of a buckled nanotube, adapted from [17]

Manually Assembled Nanotube Probes

The first carbon nanotube AFM probes [22] were fabricated by techniques developed for assembling single-nanotube field emission tips [23]. This process, illustrated in Fig. 3.10, used purified MWNT material synthesized by the carbon arc procedure. The raw material, which must contain at least a few percent of long nanotubes (> 10 μm) by weight, purified by oxidation to approximately 1% of its original mass. A torn edge of the purified material was attached to a micromanipulator by carbon tape and viewed under a high power optical microscope. Individual nanotubes and nanotube bundles were visible as filaments under dark field illumination. A commercially available AFM tip was attached to another micromanipulator opposing the nanotube material. Glue was applied to the tip apex from high vacuum carbon tape supporting the nanotube material. Nanotubes were then manually attached to the tip apex by micromanipulation. As assembled, MWNT tips were often too long for imaging due to thermal vibrations and low buckling forces described in Sect. 3.2.7. Nanotubes tips were shortened by applying 10 V pulses to the tip while it was near a sputtered niobium surface. This process etched ∼ 100 nm lengths of nanotube per pulse.

The manually assembled MWNT tips demonstrated several important nanotube tip properties [22]. First, the high aspect ratio of the MWNT tips allowed the accurate imaging of trenches in silicon with steep sidewalls, similar to FIB and EBD tips. Second, elastic buckling was observed indirectly through force curves (see

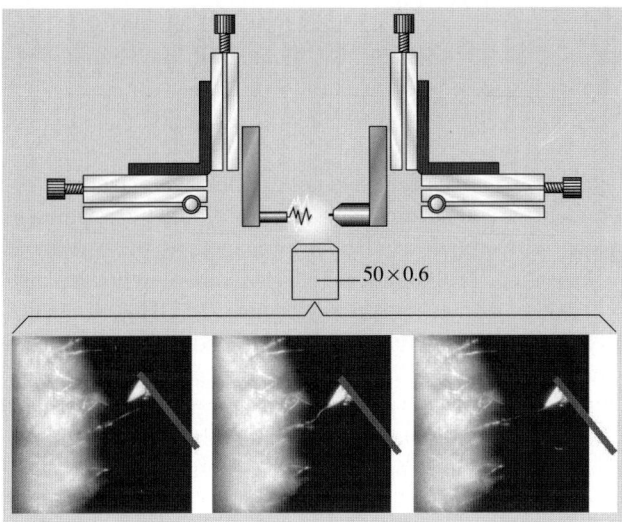

Fig. 3.10. A schematic drawing of the setup for manual assembly of carbon nanotube tips (*top*) and optical microscopy images of the assembly process (the cantilever was drawn in for clarity)

Fig. 3.11). Note that as the tip taps the sample, the amplitude drops to zero and a DC deflection is observed, because the nanotube is unbuckled and is essentially rigid. As the tip moves closer, the force on the nanotube eventually exceeds the buckling force. The nanotube buckles, allowing the vibration amplitude to partially recover,

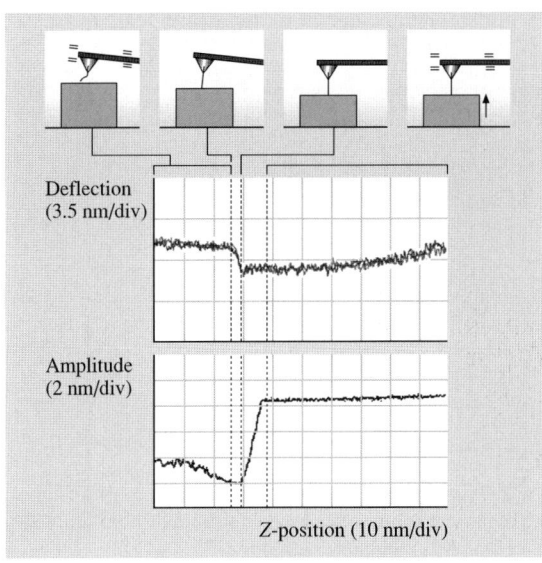

Fig. 3.11. Nanotube tip buckling. *Top* diagrams correspond to labeled regions of the force curves. As the nanotube tip buckles, the deflection remains constant and the amplitude increases, from [16]

and the deflection remains constant. Numeric tip trajectory simulations could only reproduce these force curves if elastic buckling was included in the nanotube response. Finally, the nanotube tips were highly robust. Even after "tip crashes" or hundreds of controlled buckling cycles, the tip retained its resolution and high aspect ratio.

Manual assembly of carbon nanotube probe tips is straightforward, but has several limitations. It is labor intensive and not amenable to mass production. Although MWNT tips have been made commercially available by this method, they are about ten times more expensive than silicon probes. The manual assembly method has also been carried out in an SEM, rather than an optical microscope [24]. This eliminates the need for pulse-etching, since short nanotubes can be attached to the tip, and the "glue" can be applied by EBD. But this is still not the key to mass production, since nanotube tips are made individually. MWNT tips provided a modest improvement in resolution on biological samples, but typical MWNT radii are similar to that of silicon tips, so they cannot provide the ultimate resolution possible with a SWNT tip. SWNT bundles can be attached to silicon probes by manual assembly. Pulse etching at times produces very high resolution tips that likely result from exposing a small number of nanotubes from the bundle, but this is not reproducible [25]. Even if a sample could be prepared that consisted of individual SWNT for manual assembly, such nanotubes would not be easily visible by optical microscopy or SEM.

CVD Nanotube Probe Synthesis

The problems of manual assembly of nanotube probes discussed above can largely be solved by directly growing nanotubes onto AFM tips by metal-catalyzed chemical vapor deposition (CVD). The key features of the nanotube CVD process are illustrated in Fig. 3.12. Nanometer-scale metal catalyst particles are heated in a gas

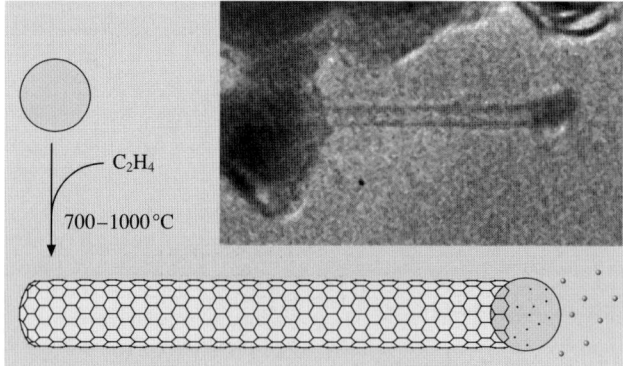

Fig. 3.12. CVD nanotube synthesis. Ethylene reacts with a nanometer-scale iron catalyst particle at high temperature to form a carbon nanotube. The *inset* in the upper right is a TEM image showing a catalyst particle at the end of a nanotube, from [16]

mixture containing hydrocarbon or CO. The gas molecules dissociate on the metal surface, and carbon is adsorbed into the catalyst particle. When this carbon precipitates, it nucleates a nanotube of similar diameter to the catalyst particle. Therefore, CVD allows control over nanotube size and structure, including the production of SWNTs [26] with radii as low as 3.5 Angstrom [27].

Several key issues must be addressed to grow nanotube AFM tips by CVD: (1) the alignment of the nanotubes at the tip, (2) the number of nanotubes that grow at the tip, and (3) the length of the nanotube tip. *Li* et al. [28] found that nanotubes grow perpendicular to a porous surface containing embedded catalyst. This approach was exploited to fabricate nanotube tips by CVD [29] with the proper alignment, as illustrated in Fig. 3.13. A flattened area of approximately 1–5 μm² was created on Si tips by scanning in contact mode at high load (1 μN) on a hard, synthetic diamond surface. The tip was then anodized in HF to create 100 nm-diameter pores in this flat surface [30]. It is important to only anodize the last 20–40 μm of the cantilever, which includes the tip, so that the rest of the cantilever is still reflective for use in the AFM. This was achieved by anodizing the tip in a small drop of HF under the view of an optical microscope. Next, iron was electrochemically deposited into the pores to form catalyst particles [31]. Tips prepared in this way were heated in low concentrations of ethylene at 800 °C, which is known to favor the growth of thin nanotubes [26]. When imaged by SEM, nanotubes were found

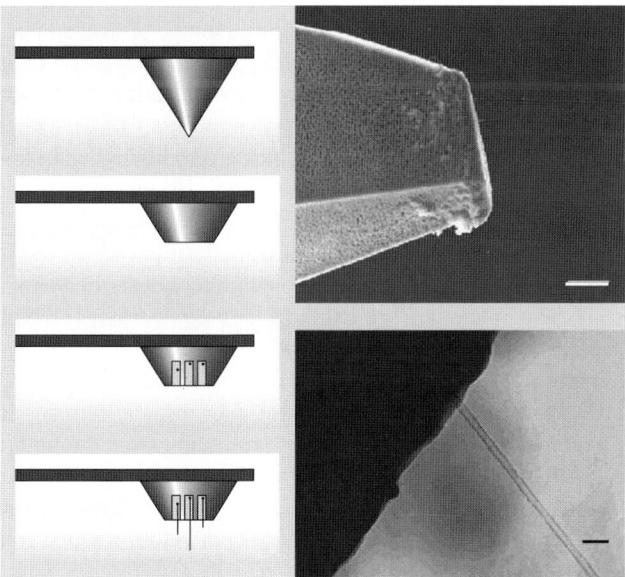

Fig. 3.13. Pore-growth CVD nanotube tip fabrication. The *left panel*, from top to bottom, shows the steps described in the text. The *upper right* is an SEM image of such a tip with a small nanotube protruding from the pores (scale bar is 1 μm). The *lower right* is a TEM of a nanotube protruding from the pores (scale bar is 20 nm), from [16]

to grow perpendicular to the surface from the pores as desired (Fig. 3.13). TEM revealed that the nanotubes were thin, individual, multiwalled nanotubes with typical radii ranging from 3–5 nm. If nanotubes did not grow in an acceptable orientation, the carbon could be removed by oxidation, and then CVD repeated to grow new nanotube tips.

These "pore-growth" CVD nanotube tips were typically several microns in length – too long for imaging – and were pulse-etched to a usable length of < 500 nm. The tips exhibited elastic buckling behavior and were very robust in imaging. In addition, the thin, individual nanotube tips enabled improved resolution [29] on isolated proteins. The pore-growth method demonstrated the potential of CVD to simplify the fabrication of nanotube tips, although there were still limitations. In particular, the porous layer was difficult to prepare and rather fragile.

An alternative approach for CVD fabrication of nanotube tips involves direct growth of SWNTs on the surface of a pyramidal AFM tip [32, 33]. In this "surface-growth" approach, an alumina/iron/molybdenum-powdered catalyst known to produce SWNT [26] was dispersed in ethanol at 1 mg/mL. Silicon tips were dipped in this solution and allowed to dry, leaving a sparse layer of ~ 100 nm catalyst clusters on the tip. When CVD conditions were applied, single-walled nanotubes grew along the silicon tip surface. At a pyramid edge, nanotubes can either bend to align with the edge, or protrude from the surface. If the energy required to bend the tube and follow the edge is less than the attractive nanotube-surface energy, then the nanotube will follow the pyramid edge to the apex. Therefore, nanotubes were effectively steered toward the tip apex by the pyramid edges. At the apex, the nanotube protruded from the tip, since the energetic cost of bending around the sharp silicon tip was too high. The high aspect ratio at the oxide-sharpened silicon tip apex was critical for good nanotube alignment. A schematic of this approach is shown in Fig. 3.14. Evidence for this model came from SEM investigations that show that a very high yield of tips contains nanotubes only at the apex, with very few protruding elsewhere from the pyramid. TEM analysis demonstrated that the tips typically consist of small SWNT bundles that are formed by nanotubes coming together from different edges of the pyramid to join at the apex, supporting the surface growth model described above (Fig. 3.14). The "surface growth" nanotube tips exhibit a high aspect ratio and high resolution imaging, as well as elastic buckling.

The surface growth method has been expanded to include wafer-scale production of nanotube tips with yields of over 90% [34], yet one obstacle remains to the mass production of nanotube probe tips. Nanotubes protruding from the tip are several microns long, and since they are so thin, they must be etched to less than 100 nm. While the pulse-etching step is fairly reproducible, it must be carried out on nanotube tips in a serial fashion, so surface growth does not yet represent a true method of batch nanotube tip fabrication.

Hybrid Nanotube Tip Fabrication: Pick-up Tips

Another method of creating nanotube tips is something of a hybrid between assembly and CVD. The motivation was to create AFM probes that have an *individual*

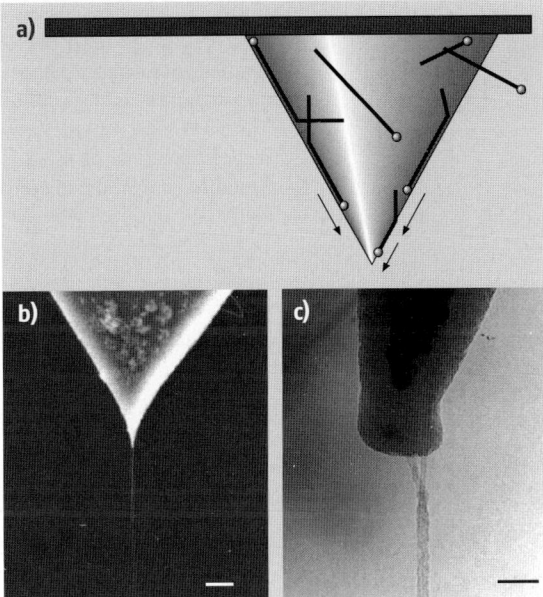

Fig. 3.14a–c. Surface-growth nanotube tip fabrication. (a) Schematic represents the surface growth process in which nanotubes growing on the pyramidal tip are guided to the tip apex. (b),(c) Images show (b) SEM (200-nm-scale bar) and (c) TEM (20-nm-scale bar) images of a surface growth tip, from [16]

SWNT at the tip to achieve the ultimate imaging resolution. In order to synthesize isolated SWNT, they must be nucleated at sites separated farther than their typical length. The alumina-supported catalyst contains a high density of catalyst particles per 100 nm cluster, so nanotube bundles cannot be avoided. To fabricate completely isolated nanotubes, isolated catalyst particles were formed by dipping a silicon wafer in an isopropyl alcohol solution of $Fe(NO_3)_3$. This effectively left a submonolayer of iron on the wafer, so that when it was heated in a CVD furnace, the iron became mobile and aggregated to form isolated iron particles. During CVD conditions, these particles nucleated and grew SWNTs. By controlling the reaction time, the SWNT lengths were kept shorter than their typical separation, so that the nanotubes never had a chance to form bundles. AFM analysis of these samples revealed 1–3 nm-diameter SWNT and un-nucleated particles on the surface (Fig. 3.15). However, there were tall objects that were difficult to image at a density of about 1 per 50 µm^2. SEM analysis at an oblique angle demonstrated that these were SWNTs that had grown perpendicular to the surface (Fig. 3.15).

In the "pick-up tip" method, these isolated SWNT substrates were imaged by AFM with silicon tips in air [9]. When the tip encountered a vertical SWNT, the oscillation amplitude was damped, so the AFM pulled the sample away from the tip. This pulled the SWNT into contact with the tip along its length, so that it became attached to the tip. This assembly process happened automatically when imaging in tapping mode – no special tip manipulation was required. When imaging a wafer with the density shown in Fig. 3.15, one nanotube was attached per 8 µm × 8 µm scan at 512 × 512 and 2 Hz, so a nanotube tip could be made in about 5 min. Since the as-formed SWNT tip continued to image, there was usually no evidence in the

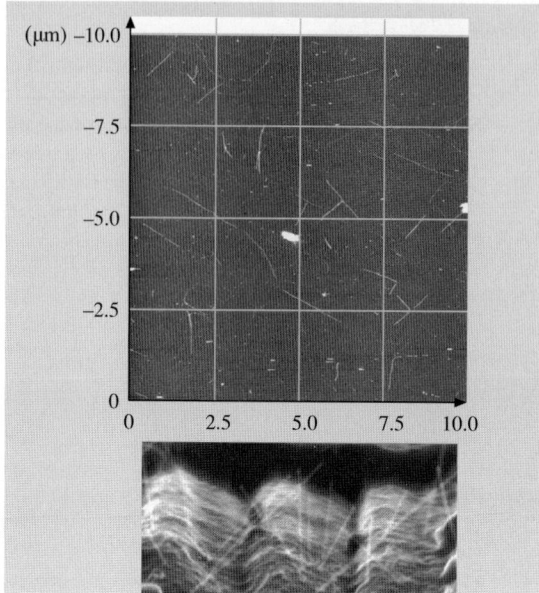

Fig. 3.15. Atomic force microscopy image (*top*) of a wafer with isolated nanotubes synthesized by CVD. The SEM view provides evidence that some of these nanotubes are arranged vertically

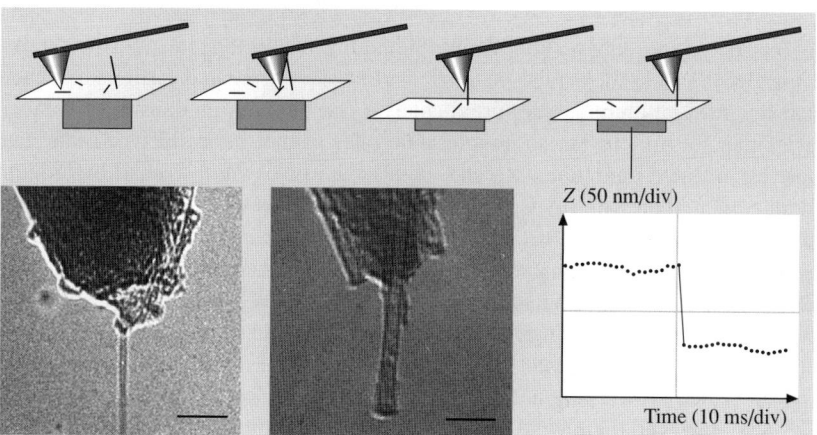

Fig. 3.16. Pick-up tip assembly of nanotube probes. The *top* illustrates the nanotube pick-up process that occurs while imaging vertical nanotubes in an AFM, including a trace of the Z-position during a pick-up event. The *lower left* TEM images show single nanotubes (diameters 0.9 nm and 2.8 nm) on an AFM tip fabricated by this method, adapted from [9]

topographic image that a nanotube had been attached. However, the pick-up event was identified when the Z-voltage suddenly stepped to larger tip-sample separation due to the effective increase in tip length.

Individual SWNT tips must be quite short, typically less than 50 nm in length, for reasons outlined above. Pulse-etching, which removes 50–100 nm of nanotube

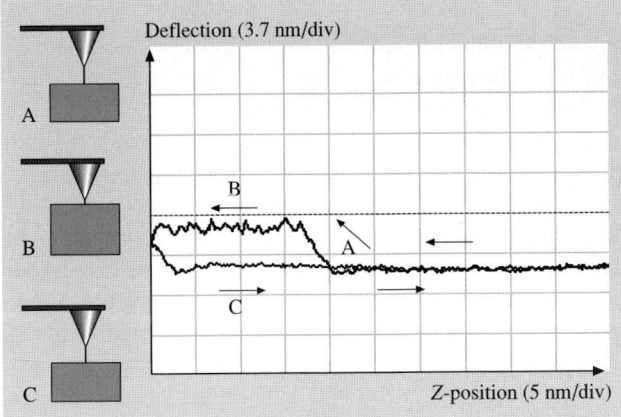

Fig. 3.17. The process by which nanotube tips can be shortened in AFM force curves. The hysteresis in the deflection trace (*bottom*) reveals that ~ 17 nm were removed

length at a time, lacked the necessary precision for shortening pick-up tips, so the tips were shortened through force curves. A pick-up tip force curve is shown in Fig. 3.17. When the tip first interacted with the sample, the amplitude decreased to zero, and further approach generated a small deflection that ultimately saturated. However, this saturation was not due to buckling. Note that the amplitude did not recover, as in the buckling curves. This leveling was due to the nanotube sliding on the pyramidal tip, which was confirmed by the hysteresis in the amplitude and deflection curves. If the force curve was repeated, the tip showed no deflection or amplitude drop until further down, because the tip was essentially shorter.

Pick-up SWNT tips achieve the highest resolution of all nanotube tips, since they are always individuals rather than bundles. In tapping mode, they produce images of 5 nm gold particles that have a *full width* of ~ 7 nm, the expected geometrical resolution for a 2 nm cylindrical probe [9]. Although the pick-up method is serial in nature, it may still be the key to the mass production of nanotube tips. Note that the original nanotube tip length can be measured electronically from the size of the Z-piezo step. The shortening can be electronically controlled through the hysteresis in the force curves. Therefore, the entire procedure (including tip exchange) can be automated by computer.

3.3 Scanning Tunneling Microscopy

Scanning tunneling microscopy (STM) was the original scanning probe microscopy and generally produces the highest resolution images, routinely achieving atomic resolution on flat, conductive surfaces. In STM, the probe tip consists of a sharpened metal wire that is held 0.3 to 1 nm from the sample. A potential difference of 0.1 V to 1 V between the tip and sample leads to tunneling currents on the order of

0.1 to 1 nA. As in AFM, a piezo-scanner rasters the sample under the tip, and the Z-position is adjusted to hold the tunneling current constant. The Z-position data represents the "topography", or in this case the surface, of constant electron density. As with other SPMs, the tip properties and performance greatly depend on the experiment being carried out. Although it is nearly impossible to prepare a tip with a known atomic structure, a number of factors are known to affect tip performance, and several preparation methods have been developed that produce good tips.

The nature of the sample being investigated and the scanning environment will affect the choice of the tip material and how the tip is fabricated. Factors to consider are mechanical properties – a hard material that will resist damage during tip-sample contact is desired. Chemical properties should also be considered – formation of oxides, or other insulating contaminants will affect tip performance. Tungsten is a common tip material because it is very hard and will resist damage, but its use is limited to ultrahigh vacuum (UHV) conditions, since it readily oxidizes. For imaging under ambient conditions an inert tip material such as platinum or gold is preferred. Platinum is typically alloyed with iridium to increase its stiffness.

3.3.1 Mechanically Cut STM Tips

STM tips can be fabricated by simple mechanical procedures such as grinding or cutting metal wires. Such tips are not formed with highly reproducible shapes and have a large opening angle and a large radius of curvature in the range of 0.1 to 1 μm (see Fig. 3.18a). They are not useful for imaging samples with surface roughness above a few nanometers. However, on atomically flat samples, mechanically cut tips can achieve atomic resolution due to the nature of the tunneling signal, which drops exponentially with tip-sample separation. Since mechanically cut tips contain many small asperities on the larger tip structure, atomic resolution is easily achieved as long as one atom of the tip is just a few angstroms lower than all of the others.

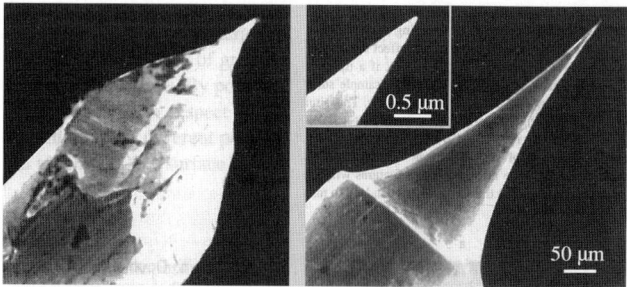

Fig. 3.18. A mechanically cut STM tip (*left*) and an electrochemically etched STMtip (*right*), from [35]

3.3.2 Electrochemically Etched STM Tips

For samples with more than a few nanometers of surface roughness, the tip structure in the nanometer-size range becomes an issue. Electrochemical etching can provide tips with reproducible and desirable shapes and sizes (Fig. 3.18), although the exact atomic structure of the tip apex is still not well controlled. The parameters of electrochemical etching depend greatly on the tip material and the desired tip shape. The following is an entirely general description. A fine metal wire (0.1–1 mm diameter) of the tip material is immersed in an appropriate electrochemical etchant solution. A voltage bias of 1–10 V is applied between the tip and a counterelectrode such that the tip is etched. Due to the enhanced etch rate at the electrolyte-air interface, a neck is formed in the wire. This neck is eventually etched thin enough so that it cannot support the weight of the part of the wire suspended in the solution, and it breaks to form a sharp tip. The widely varying parameters and methods will be not be covered in detail here, but many recipes are found in the literature for common tip materials [36–39].

References

1. R. Linnemann, T. Gotszalk, I. W. Rangelow, P. Dumania, E. Oesterschulze: Atomic force microscopy and lateral force microscopy using piezoresistive cantilevers, J. Vac. Sci. Technol. B **14**(2), 856–860 (1996)
2. T. R. Albrecht, S. Akamine, T. E. Carver, C. F. Quate: Microfabrication of cantilever styli for the atomic force microscope, J. Vac. Sci. Technol. A **8**(4), 3386–3396 (1990)
3. O. Wolter, T. Bayer, J. Greschner: Micromachined silicon sensors for scanning force microscopy, J. Vac. Sci. Technol. B **9**(2), 1353–1357 (1991)
4. C. Bustamante, D. Keller: Scanning force microscopy in biology, Phys. Today **48**(12), 32–38 (1995)
5. J. Vesenka, S. Manne, R. Giberson, T. Marsh, E. Henderson: Colloidal gold particles as an incompressible atomic force microscope imaging standard for assessing the compressibility of biomolecules, Biophys. J. **65**, 992–997 (1993)
6. D. J. Muller, D. Fotiadis, S. Scheuring, S. A. Muller, A. Engel: Electrostatically balanced subnanometer imaging of biological specimens by atomic force microscope, Biophys. J. **76**(2), 1101–1111 (1999)
7. R. B. Marcus, T. S. Ravi, T. Gmitter, K. Chin, D. Liu, W. J. Orvis, D. R. Ciarlo, C. E. Hunt, J. Trujillo: Formation of silicon tips with < 1 nm radius, Appl. Phys. Lett. **56**(3), 236–238 (1990)
8. J. H. Hafner, C. L. Cheung, C. M. Lieber: unpublished results (2001)
9. J. H. Hafner, C. L. Cheung, T. H. Oosterkamp, C. M. Lieber: High-yield assembly of individual single-walled carbon nanotube tips for scanning probe microscopies, J. Phys. Chem. B **105**(4), 743–746 (2001)
10. F. Ohnesorge, G. Binnig: True atomic resolution by atomic force microscopy through repulsive and attractive forces, Science **260**, 1451–1456 (1993)
11. D. J. Muller, D. Fotiadis, A. Engel: Mapping flexible protein domains at subnanometer resolution with the atomic force microscope, FEBS Lett. **430**(1–2 Special Issue SI), 105–111 (1998)
12. S. Akamine, R. C. Barrett, C. F. Quate: Improved atomic force microscope images using microcantilevers with sharp tips, Appl. Phys. Lett. **57**(3), 316–318 (1990)

13. D.J. Keller, C. Chih-Chung: Imaging steep, high structures by scanning force microscopy with electron beam deposited tips, Surf. Sci. **268**, 333–339 (1992)
14. T. Ichihashi, S. Matsui: In situ observation on electron beam induced chemical vapor deposition by transmission electron microscopy, J. Vac. Sci. Technol. B **6**(6), 1869–1872 (1988)
15. K.I. Schiffmann: Investigation of fabrication parameters for the electron-beam-induced deposition of contamination tips used in atomic force microscopy, Nanotechnology **4**, 163–169 (1993)
16. J.H. Hafner, C.L. Cheung, A.T. Woolley, C.M. Lieber: Structural and functional imaging with carbon nanotube AFM probes, Prog. Biophys. Mol. Biol. **77**(1), 73–110 (2001)
17. S. Iijima, C. Brabec, A. Maiti, J. Bernholc: Structural flexibility of carbon nanotubes, J. Chem. Phys. **104**(5), 2089–2092 (1996)
18. M.M.J. Treacy, T.W. Ebbesen, J.M. Gibson: Exceptionally high Young's modulus observed for individual carbon nanotubes, Nature **381**, 678–680 (1996)
19. A. Krishnan, E. Dujardin, T.W. Ebbesen, P.N. Yianilos, M.M.J. Treacy: Young's modulus of single-walled nanotubes, Phys. Rev. B **58**(20), 14013–14019 (1998)
20. E.W. Wong, P.E. Sheehan, C.M. Lieber: Nanobeam mechanics – elasticity, strength, and toughness of nanorods and nanotubes, Science **277**(5334), 1971–1975 (1997)
21. J.P. Lu: Elastic properties of carbon nanotubes and nanoropes, Phys. Rev. Lett. **79**(7), 1297–1300 (1997)
22. H.J. Dai, J.H. Hafner, A.G. Rinzler, D.T. Colbert, R.E. Smalley: Nanotubes as nanoprobes in scanning probe microscopy, Nature **384**(6605), 147–150 (1996)
23. A.G. Rinzler, Y.H. Hafner, P. Nikolaev, L. Lou, S.G. Kim, D. Tomanek, D.T. Colbert, R.E. Smalley: Unraveling nanotubes: Field emission from atomic wire, Science **269**, 1550 (1995)
24. H. Nishijima, S. Kamo, S. Akita, Y. Nakayama, K.I. Hohmura, S.H. Yoshimura, K. Takeyasu: Carbon-nanotube tips for scanning probe microscopy: Preparation by a controlled process and observation of deoxyribonucleic acid, Appl. Phys. Lett. **74**(26), 4061–4063 (1999)
25. S.S. Wong, A.T. Woolley, T.W. Odom, J.L. Huang, P. Kim, D.V. Vezenov, C.M. Lieber: Single-walled carbon nanotube probes for high-resolution nanostructure imaging, Appl. Phys. Lett. **73**(23), 3465–3467 (1998)
26. J.H. Hafner, M.J. Bronikowski, B.R. Azamian, P. Nikolaev, A.G. Rinzler, D.T. Colbert, K.A. Smith, R.E. Smalley: Catalytic growth of single-wall carbon nanotubes from metal particles, Chem. Phys. Lett. **296**(1–2), 195–202 (1998)
27. P. Nikolaev, M.J. Bronikowski, R.K. Bradley, F. Rohmund, D.T. Colbert, K.A. Smith, R.E. Smalley: Gas-phase catalytic growth of single-walled carbon nanotubes from carbon monoxide, Chem. Phys. Lett. **313**(1–2), 91–97 (1999)
28. W.Z. Li, S.S. Xie, L.X. Qian, B.H. Chang, B.S. Zou, W.Y. Zhou, R.A. Zhao, G. Wang: Large-scale synthesis of aligned carbon nanotubes, Science **274**(5293), 1701–1703 (1996)
29. J.H. Hafner, C.L. Cheung, C.M. Lieber: Growth of nanotubes for probe microscopy tips, Nature **398**(6730), 761–762 (1999)
30. V. Lehmann: The physics of macroporous silicon formation, Thin Solid Films **255**, 1–4 (1995)
31. F. Ronkel, J.W. Schultze, R. Arensfischer: Electrical contact to porous silicon by electrodeposition of iron, Thin Solid Films **276**(1–2), 40–43 (1996)
32. J.H. Hafner, C.L. Cheung, C.M. Lieber: Direct growth of single-walled carbon nanotube scanning probe microscopy tips, J. Am. Chem. Soc. **121**(41), 9750–9751 (1999)

33. E.B. Cooper, S.R. Manalis, H. Fang, H. Dai, K. Matsumoto, S.C. Minne, T. Hunt, C.F. Quate: Terabit-per-square-inch data storage with the atomic force microscope, Appl. Phys. Lett. **75**(22), 3566–3568 (1999)
34. E. Yenilmez, Q. Wang, R.J. Chen, D. Wang, H. Dai: Wafer scale production of carbon nanotube scanning probe tips for atomic force microscopy, Appl. Phys. Lett. **80**(12), 2225–2227 (2002)
35. A. Stemmer, A. Hefti, U. Aebi, A. Engel: Scanning tunneling and transmission electron microscopy on identical areas of biological specimens, Ultramicroscopy **30**(3), 263 (1989)
36. R. Nicolaides, L. Yong, W.E. Packard, W.F. Zhou, H.A. Blackstead, K.K. Chin, J.D. Dow, J.K. Furdyna, M.H. Wei, R.C. Jaklevic, W.J. Kaiser, A.R. Pelton, M.V. Zeller, J.J. Bellina: Scanning tunneling microscope tip structures, J. Vac. Sci. Technol. A **6**(2), 445–447 (1988)
37. J.P. Ibe, P.P. Bey, S.L. Brandow, R.A. Brizzolara, N.A. Burnham, D.P. DiLella, K.P. Lee, C.R.K. Marrian, R.J. Colton: On the electrochemical etching of tips for scanning tunneling microscopy, J. Vac. Sci. Technol. A **8**, 3570–3575 (1990)
38. L. Libioulle, Y. Houbion, J.-M. Gilles: Very sharp platinum tips for scanning tunneling microscopy, Rev. Sci. Instrum. **66**(1), 97–100 (1995)
39. A.J. Nam, A. Teren, T.A. Lusby, A.J. Melmed: Benign making of sharp tips for STM and FIM: Pt, Ir, Au, Pd, and Rh, J. Vac. Sci. Technol. B **13**(4), 1556–1559 (1995)

4

Noncontact Atomic Force Microscopy and Related Topics

Franz J. Giessibl, Yasuhiro Sugawara, Seizo Morita, Hirotaka Hosoi,
Kazuhisa Sueoka, Koichi Mukasa, Akira Sasahara, and Hiroshi Onishi

Summary. Scanning probe microscopy (SPM) methods such as scanning tunneling microscopy (STM) and noncontact atomic force microscopy (NC-AFM) are the basic technologies for nanotechnology and also for future bottom-up processes. In Sect. 4.2, the principles of AFM such as its operating modes and the NC-AFM frequency-modulation method are fully explained. Then, in Sect. 4.3, applications of NC-AFM to semiconductors, which make clear its potential in terms of spatial resolution and function, are introduced. Next, in Sect. 4.4, applications of NC-AFM to insulators such as alkali halides, fluorides and transition-metal oxides are introduced. Lastly, in Sect. 4.5, applications of NC-AFM to molecules such as carboxylate (RCOO$^-$) with R=H, CH_3, $C(CH_3)_3$ and CF_3 are introduced. Thus, NC-AFM can observe atoms and molecules on various kinds of surfaces such as semiconductors, insulators and metal oxides with atomic or molecular resolution. These sections are essential to understand the state of the art and future possibilities for NC-AFM, which is the second generation of atom/molecule technology.

4.1 Introduction

The scanning tunneling microscope (STM) is an atomic tool based on an electric method that measures the tunneling current between a conductive tip and a conductive surface. It can electrically observe individual atoms/molecules. It can characterize or analyze the electronic nature around surface atoms/molecules. In addition, it can manipulate individual atoms/molecules. Hence, the STM is the first generation of atom/molecule technology. On the other hand, the atomic force microscopy (AFM) is a unique atomic tool based on a mechanical method that can even deal with insulator surfaces. Since the invention of noncontact AFM (NC-AFM) in 1995, the NC-AFM and NC-AFM-based methods have rapidly developed into powerful surface tools on the atomic/molecular scales, because NC-AFM has the following characteristics: (1) it has true atomic resolution, (2) it can measure atomic force (so-called atomic force spectroscopy), (3) it can observe even insulators, and (4) it can measure mechanical responses such as elastic deformation. Thus, NC-AFM is the second generation of atom/molecule technology. Scanning probe microscopy

136   Franz J. Giessibl et al.

(SPM) such as STM and NC-AFM is the basic technology for nanotechnology and also for future bottom-up processes.

In Sect. 4.2, the principles of NC-AFM will be fully introduced. Then, in Sect. 4.3, applications to semiconductors will be presented. Next, in Sect. 4.4, applications to insulators will be described. And, in Sect. 4.5, applications to molecules will be introduced. These sections are essential to understanding the state of the art and future possibilities for NC-AFM.

4.2 Atomic Force Microscopy (AFM)

The atomic force microscope (AFM), invented by *Binnig* [1] and introduced in 1986 by *Binnig*, *Quate* and *Gerber* [2] is an offspring of the scanning tunneling microscope (STM) [3]. The STM is covered in several books and review articles, e. g. [4–9]. Early in the development of STM it became evident that relatively strong forces act between a tip in close proximity to a sample. It was found that these forces could be put to good use in the atomic force microscope (AFM). Detailed information about the noncontact AFM can be found in [10–12].

4.2.1 Imaging Signal in AFM

Figure 4.1 shows a sharp tip close to a sample. The potential energy between the tip and the sample V_{ts} causes a z component of the tip–sample force $F_{ts} = -\partial V_{ts}/\partial z$. Depending on the mode of operation, the AFM uses F_{ts}, or some entity derived from F_{ts}, as the imaging signal.

Unlike the tunneling current, which has a very strong distance dependence, F_{ts} has long- and short-range contributions. We can classify the contributions by their range and strength. In vacuum, there are van-der-Waals, electrostatic and magnetic forces with a long range (up to 100 nm) and short-range chemical forces (fractions of nm).

The van-der-Waals interaction is caused by fluctuations in the electric dipole moment of atoms and their mutual polarization. For a spherical tip with radius R

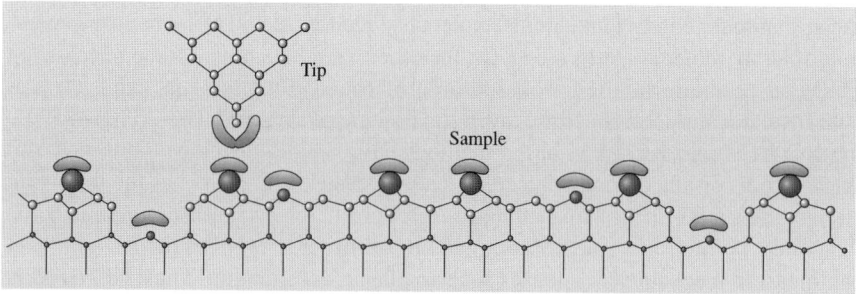

Fig. 4.1. Schematic view of an AFM tip close to a sample

next to a flat surface (z is the distance between the plane connecting the centers of the surface atoms and the center of the closest tip atom) the van-der-Waals potential is given by [13]:

$$V_{\text{vdW}} = -\frac{A_H}{6z}. \tag{4.1}$$

The Hamaker constant A_H depends on the type of materials (atomic polarizability and density) of the tip and sample and is of the order of 1 eV for most solids [13].

When the tip and sample are both conductive and have an electrostatic potential difference $U \neq 0$, electrostatic forces are important. For a spherical tip with radius R, the force is given by [14]:

$$F_{\text{electrostatic}} = -\frac{\pi\varepsilon_0 R U^2}{z}. \tag{4.2}$$

Chemical forces are more complicated. Empirical model potentials for chemical bonds are the Morse potential (see e. g. [13]).

$$V_{\text{Morse}} = -E_{\text{bond}}\left(2e^{-\kappa(z-\sigma)} - e^{-2\kappa(z-\sigma)}\right) \tag{4.3}$$

and the Lennard–Jones potential [13]:

$$V_{\text{Lennard–Jones}} = -E_{\text{bond}}\left(2\frac{\sigma^6}{z^6} - \frac{\sigma^{12}}{z^{12}}\right) \tag{4.4}$$

These potentials describe a chemical bond with bonding energy E_{bond} and equilibrium distance σ. The Morse potential has an additional parameter: a decay length κ.

4.2.2 Experimental Measurement and Noise

Forces between the tip and sample are typically measured by recording the deflection of a cantilever beam that has a tip mounted on its end (see Fig. 4.2). Today's microfabricated silicon cantilevers were first created in the group of *Quate* [15–17] and at IBM [18].

The cantilever is characterized by its spring constant k, eigenfrequency f_0 and quality factor Q.

For a rectangular cantilever with dimensions w, t and L (see Fig. 4.2), the spring constant k is given by [6]:

$$k = \frac{E_Y w t^3}{4L^3} \tag{4.5}$$

where E_Y is the Young's modulus. The eigenfrequency f_0 is given by [6]:

$$f_0 = 0.162 \frac{t}{L^2}\sqrt{\frac{E}{\rho}} \tag{4.6}$$

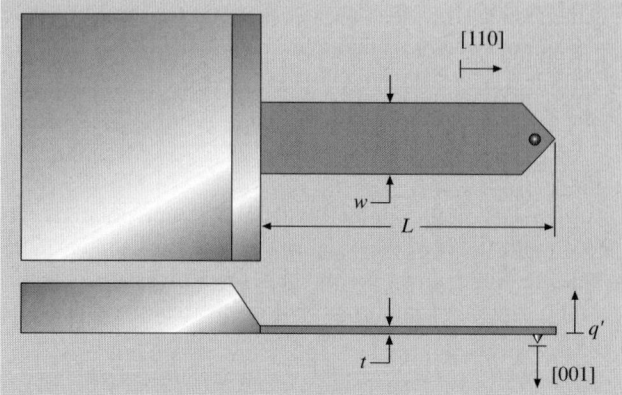

Fig. 4.2. Top view and side view of a microfabricated silicon cantilever (schematic)

where ρ is the mass density of the cantilever material. The Q-factor depends on the damping mechanisms present in the cantilever. For micromachined cantilevers operated in air, Q is typically a few hundred, while Q can reach hundreds of thousands in vacuum.

In the first AFM, the deflection of the cantilever was measured with an STM; the back side of the cantilever was metalized, and a tunneling tip was brought close to it to measure the deflection [2]. Today's designs use optical (interferometer, beam-bounce) or electrical methods (piezoresistive, piezoelectric) to measure the cantilever deflection. A discussion of the various techniques can be found in [19], descriptions of piezoresistive detection schemes are found in [17, 20] and piezoelectric methods are explained in [21–24].

The quality of the cantilever deflection measurement can be expressed in a schematic plot of the deflection noise density versus frequency as in Fig. 4.3.

The noise density has a $1/f$ dependence for low frequency and merges into a constant noise density (white noise) above the $1/f$ corner frequency.

4.2.3 Static AFM Operating Mode

In the static mode of operation, the force translates into a deflection $q' = F_{ts}/k$ of the cantilever, yielding images as maps of $z(x, y, F_{ts} = \text{const.})$. The noise level of the force measurement is then given by the cantilever's spring constant k times the noise level of the deflection measurement. In this respect, a small value for k increases force sensitivity. On the other hand, instabilities are more likely to occur with soft cantilevers (see Sect. 4.2.1). Because the deflection of the cantilever should be significantly larger than the deformation of the tip and sample, the cantilever should be much softer than the bonds between the bulk atoms in the tip and sample. Interatomic force constants in solids are in the range 10–100 N/m; in biological samples, they can be as small as 0.1 N/m. Thus, typical values for k in the static mode are 0.01–5 N/m.

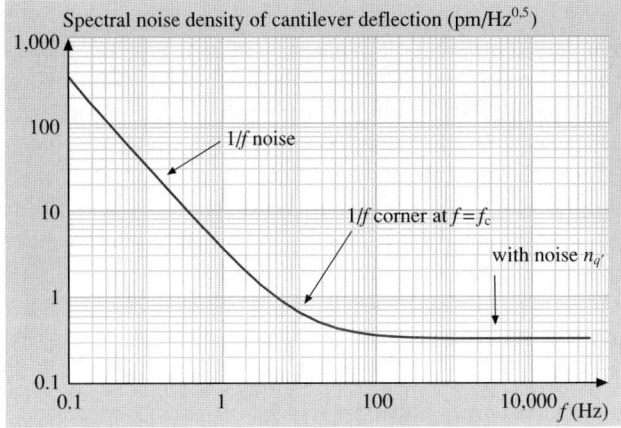

Fig. 4.3. Schematic view of $1/f$ noise apparent in force detectors. Static AFMs operate in a frequency range from 0.01 Hz to a few hundred Hz, while dynamic AFMs operate at frequencies around 10 kHz to a few hundred kHz. The noise of the cantilever deflection sensor is characterized by the $1/f$ corner frequency f_c and the constant deflection noise density $n_{q'}$ for the frequency range where white noise dominates

Even though it has been demonstrated that atomic resolution is possible with static AFM, the method can only be applied in certain cases. The detrimental effects of $1/f$-noise can be limited by working at low temperatures [25], where the coefficients of thermal expansion are very small or by building the AFM using a material with a low thermal-expansion coefficient [26]. The long-range attractive forces have to be canceled by immersing the tip and sample in a liquid [26] or by partly compensating the attractive force by pulling at the cantilever after jump-to-contact has occurred [27]. *Jarvis* et al. have canceled the long-range attractive force with an electromagnetic force applied to the cantilever [28]. Even with these restrictions, static AFM does not produce atomic resolution on reactive surfaces like silicon, as the chemical bonding of the AFM tip and sample poses an unsurmountable problem [29, 30].

4.2.4 Dynamic AFM Operating Mode

In the dynamic operation modes, the cantilever is deliberately vibrated. There are two basic methods of dynamic operation: amplitude-modulation (AM) and frequency-modulation (FM) operation. In AM-AFM [31], the actuator is driven by a fixed amplitude A_{drive} at a fixed frequency f_{drive} where f_{drive} is close to f_0. When the tip approaches the sample, elastic and inelastic interactions cause a change in both the amplitude and the phase (relative to the driving signal) of the cantilever. These changes are used as the feedback signal. While the AM mode was initially used in a noncontant mode, it was later implemented very successfully at a closer distance range in ambient conditions involving repulsive tip–sample interactions.

The change in amplitude in AM mode does not occur instantaneously with a change in the tip–sample interaction, but on a timescale of $\tau_{AM} \approx 2Q/f_0$ and the AM mode is slow with high-Q cantilevers. However, the use of high Q-factors reduces noise. *Albrecht* et al. found a way to combine the benefits of high Q and high speed by introducing the frequency-modulation (FM) mode [32], where the change in the eigenfrequency settles on a timescale of $\tau_{FM} \approx 1/f_0$.

Using the FM mode, the resolution was improved dramatically and finally atomic resolution [33, 34] was obtained by reducing the tip–sample distance and working in vacuum. For atomic studies in vacuum, the FM mode (see Sect. 4.2.6) is now the preferred AFM technique. However, atomic resolution in vacuum can also be obtained with the AM mode, as demonstrated by *Erlandsson* et al. [35].

4.2.5 The Four Additional Challenges Faced by AFM

Some of the inherent AFM challenges are apparent by comparing the tunneling current and tip–sample force as a function of distance (Fig. 4.4).

The tunneling current is a monotonic function of the tip–sample distance and has a very sharp distance dependence. In contrast, the tip–sample force has long- and short-range components and is not monotonic.

Jump-to-Contact and Other Instabilities

If the tip is mounted on a soft cantilever, the initially attractive tip–sample forces can cause a sudden jump-to-contact when approaching the tip to the sample. This

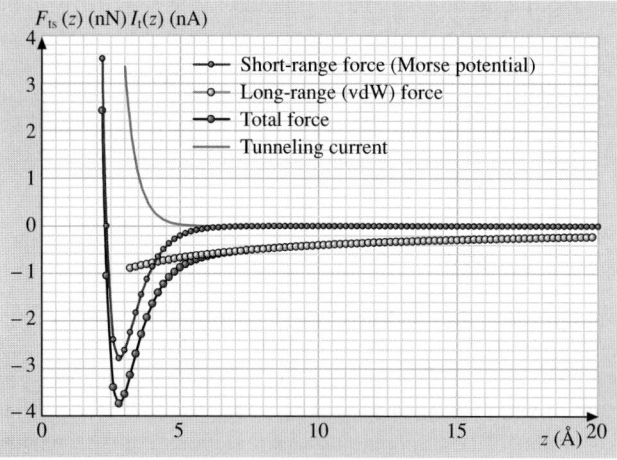

Fig. 4.4. Plot of the tunneling current I_t and force F_{ts} (typical values) as a function of the distance z between the front atom and surface atom layer

instability occurs in the quasistatic mode if [36, 37]

$$k < \max\left(-\frac{\partial^2 V_{ts}}{\partial z^2}\right) = k_{ts}^{max} \,. \tag{4.7}$$

Jump-to-contact can be avoided even for soft cantilevers by oscillating at a large enough amplitude A [38]:

$$kA > \max(-F_{ts}) \,. \tag{4.8}$$

If hysteresis occurs in the $F_{ts}(z)$-relation, energy ΔE_{ts} needs to be supplied to the cantilever for each oscillation cycle. If this energy loss is large compared to the intrinsic energy loss of the cantilever, amplitude control can become difficult. An additional approximate criterion for k and A is then

$$\frac{kA^2}{2} \geq \frac{\Delta E_{ts} Q}{2\pi} \,. \tag{4.9}$$

Contribution of Long-Range Forces

The force between the tip and sample is composed of many contributions: electrostatic, magnetic, van-der-Waals and chemical forces in vacuum. All of these force types except for the chemical forces have strong long-range components which conceal the atomic force components. For imaging by AFM with atomic resolution, it is desirable to filter out the long-range force contributions and only measure the force components which vary on the atomic scale. While there is no way to discriminate between long- and short-range forces in static AFM, it is possible to enhance the short-range contributions in dynamic AFM by proper choice of the oscillation amplitude A of the cantilever.

Noise in the Imaging Signal

Measuring the cantilever deflection is subject to noise, especially at low frequencies ($1/f$ noise). In static AFM, this noise is particularly problematic because of the approximate $1/f$ dependence. In dynamic AFM, the low-frequency noise is easily discriminated when using a bandpass filter with a center frequency around f_0.

Non-monotonic Imaging Signal

The tip–sample force is not monotonic. In general, the force is attractive for large distances and, upon decreasing the distance between tip and sample, the force turns repulsive (see Fig. 4.4). Stable feedback is only possible on a monotonic subbranch of the force curve.

Frequency-modulation AFM helps to overcome challenges. The non-monotonic imaging signal in AFM is a remaining complication for FM-AFM.

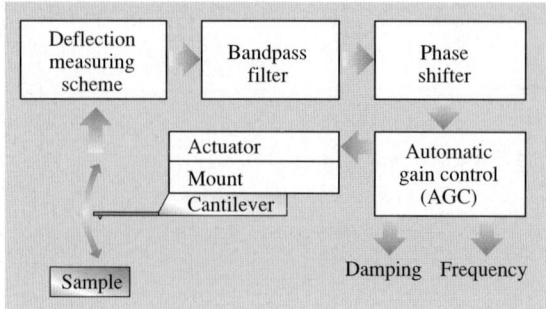

Fig. 4.5. Block diagram of a frequency-modulation force sensor

4.2.6 Frequency-Modulation AFM (FM-AFM)

In FM-AFM, a cantilever with eigenfrequency f_0 and spring constant k is subject to controlled positive feedback such that it oscillates with a constant amplitude A [32], as shown in Fig. 4.5.

Experimental Set-Up

The deflection signal is phase-shifted, routed through an automatic gain control circuit and fed back to the actuator. The frequency f is a function of f_0, its quality factor Q, and the phase shift ϕ between the mechanical excitation generated at the actuator and the deflection of the cantilever. If $\phi = \pi/2$, the loop oscillates at $f = f_0$. Three physical observables can be recorded: (1) a change in the resonance frequency Δf, (2) the control signal of the automatic gain control unit as a measure of the tip–sample energy dissipation, and (3) an average tunneling current (for conducting cantilevers and tips).

Applications

FM-AFM was introduced by *Albrecht* and coworkers in magnetic force microscopy [32]. The noise level and imaging speed was enhanced significantly compared to amplitude-modulation techniques. Achieving atomic resolution on the Si(111)-(7×7) surface has been an important step in the development of the STM [39] and, in 1994, this surface was imaged by AFM with true atomic resolution for the first time [33] (see Fig. 4.6).

The initial parameters which provided true atomic resolution (see caption of Fig. 4.6) were found empirically. Surprisingly, the amplitude necessary to obtain good results was very large compared to atomic dimensions. It turned out later that the amplitudes had to be so large to fulfill the stability criteria listed in Sect. 4.2.5. Cantilevers with $k \approx 2000$ N/m can be operated with amplitudes in the Å range [24].

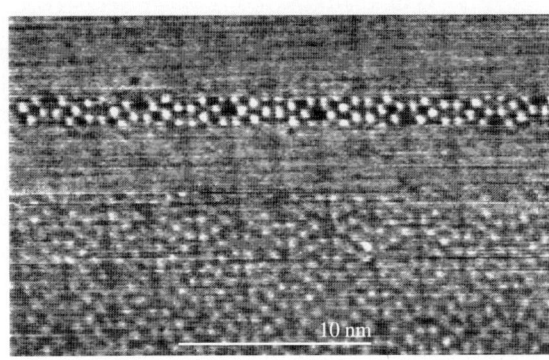

Fig. 4.6. First AFM image of the Si(111)-(7×7) surface. Parameters: $k = 17$ Nm, $f_0 = 114$ kHz, $Q = 28{,}000$, $A = 34$ nm, $\Delta f = -70$ Hz, $V_t = 0$ V

4.2.7 Relation Between Frequency Shift and Forces

The cantilever (spring constant k, effective mass m^*) is a macroscopic object and its motion can be described by classical mechanics. Figure 4.7 shows the deflection $q'(t)$ of the tip of the cantilever: it oscillates with an amplitude A at a distance $q(t)$ from a sample.

Generic Calculation

The Hamiltonian of the cantilever is:

$$H = \frac{p^2}{2m^*} + \frac{kq'^2}{2} + V_{ts}(q) \tag{4.10}$$

where $p = m^* \, dq'/dt$. The unperturbed motion is given by:

$$q'(t) = A\cos(2\pi f_0 t) \tag{4.11}$$

and the frequency is:

$$f_0 = \frac{1}{2\pi}\sqrt{\frac{k}{m^*}} \, . \tag{4.12}$$

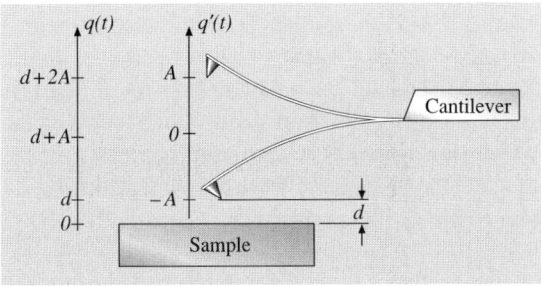

Fig. 4.7. Schematic view of an oscillating cantilever and definition of geometric terms

If the force gradient $k_{ts} = -\partial F_{ts}/\partial z = \partial^2 V_{ts}/\partial z^2$ is constant during the oscillation cycle, the calculation of the frequency shift is trivial:

$$\Delta f = \frac{f_0}{2k} k_{ts}. \tag{4.13}$$

However, in classic FM-AFM k_{ts} varies over orders of magnitude during one oscillation cycle and a perturbation approach, as shown below, has to be employed for the calculation of the frequency shift.

Hamilton–Jacobi Method

The first derivation of the frequency shift in FM-AFM was achieved in 1997 [38] using canonical perturbation theory [40]. The result of this calculation is:

$$\Delta f = -\frac{f_0}{kA^2} \langle F_{ts} q' \rangle$$

$$= -\frac{f_0}{kA^2} \int_0^{1/f_0} F_{ts}(d + A + q'(t)) q'(t) \, dt. \tag{4.14}$$

The applicability of first-order perturbation theory is justified because, in FM-AFM, E is typically in the range of several keV, while V_{ts} is of the order of a few eV. *Dürig* [41] found a generalized algorithm that even allows one to reconstruct the tip–sample potential if not only the frequency shift, but the higher harmonics of the cantilever oscillation are known.

A Descriptive Expression for Frequency Shifts as a Function of the Tip–Sample Forces

With integration by parts, the complicated expression (4.14) is transformed into a very simple expression that resembles (4.13) [42].

$$\Delta f = \frac{f_0}{2k} \int_{-A}^{A} k_{ts}(z - q') \frac{\sqrt{A^2 - q'^2}}{\frac{\pi}{2} kA^2} \, dq'. \tag{4.15}$$

This expression is closely related to (4.13): the constant k_{ts} is replaced by a weighted average, where the weight function $w(q', A)$ is a semicircle with radius A divided by the area of the semicircle $\pi A^2/2$ (see Fig. 4.8). For $A \to 0$, $w(q', A)$ is a representation of Dirac's delta function and the trivial zero-amplitude result of (4.13) is immediately recovered. The frequency shift results from a convolution between the tip–sample force gradient and weight function. This convolution can easily be reversed with a linear transformation and the tip–sample force can be recovered from the curve of frequency shift versus distance [42].

The dependence of the frequency shift on amplitude confirms an empirical conjecture: small amplitudes increase the sensitivity to short-range forces. Adjusting the

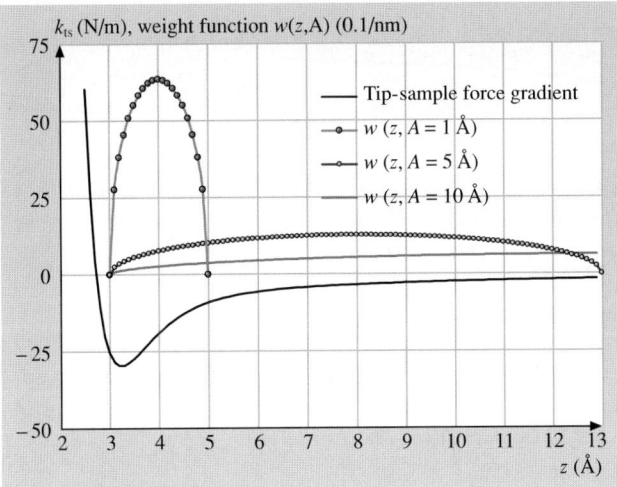

Fig. 4.8. The tip–sample force gradient k_{ts} and weight function for the calculation of the frequency shift

amplitude in FM-AFM is comparable to tuning an optical spectrometer to a passing wavelength. When short-range interactions are to be probed, the amplitude should be in the range of the short-range forces. While using amplitudes in the Å range has been elusive with conventional cantilevers because of the instability problems described in Sect. 4.2.5, cantilevers with a stiffness of the order of 1000 N/m like those introduced in [23] are well suited for small-amplitude operation.

4.2.8 Noise in Frequency Modulation AFM: Generic Calculation

The vertical noise in FM-AFM is given by the ratio between the noise in the imaging signal and the slope of the imaging signal with respect to z:

$$\delta z = \frac{\delta \Delta f}{\left|\frac{\partial \Delta f}{\partial z}\right|} . \tag{4.16}$$

Figure 4.9 shows a typical curve of frequency shift versus distance. Because the distance between the tip and sample is measured indirectly through the frequency shift, it is clearly evident from Fig. 4.9 that the noise in the frequency measurement $\delta \Delta f$ translates into vertical noise δz and is given by the ratio between $\delta \Delta f$ and the slope of the frequency shift curve $\Delta f(z)$) (4.16). Low vertical noise is obtained for a low-noise frequency measurement and a steep slope of the frequency-shift curve.

The frequency noise $\delta \Delta f$ is typically inversely proportional to the cantilever amplitude A [32, 43]. The derivative of the frequency shift with distance is constant for $A \ll \lambda$ where λ is the range of the tip–sample interaction and proportional to

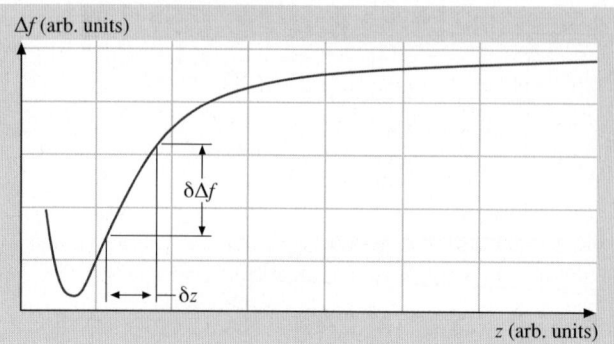

Fig. 4.9. Plot of the frequency shift Δf as a function of the tip–sample distance z. The noise in the tip–sample distance measurement is given by the noise of the frequency measurement $\delta\Delta f$ divided by the slope of the frequency shift curve

$A^{-1.5}$ for $A \gg \lambda$ [38]. Thus, minimal noise occurs if [44]:

$$A_{\text{optimal}} \approx \lambda \qquad (4.17)$$

for chemical forces, $\lambda \approx 1$ Å. However, for stability reasons, (Sect. 4.2.5) extremely stiff cantilevers are needed for small-amplitude operation. The excellent noise performance of the stiff cantilever and the small-amplitude technique has been verified experimentally [24].

4.2.9 Conclusion

Dynamic force microscopy, and in particular frequency-modulation atomic force microscopy has matured into a viable technique that allows true atomic resolution of conducting and insulating surfaces and spectroscopic measurements on individual atoms [10,45]. Even true atomic resolution in lateral force microscopy is now possible [46]. Challenges remain in the chemical composition and structural arrangement of the AFM tip.

4.3 Applications to Semiconductors

For the first time, corner holes and adatoms on the Si(111)-(7×7) surface have been observed in very local areas by *Giessible* using pure noncontact AFM in ultrahigh vacuum (UHV) [33]. This was the breakthrough of true atomic-resolution imaging on a well-defined clean surface using the noncontact AFM. Since then, Si(111)-(7×7) [34, 35, 45, 47], InP(110) [48] and Si(100)-2×1 [34] surfaces have been successively resolved with true atomic resolution. Furthermore, thermally induced motion of atoms or atomic-scale point defects on a InP(110) surface have been observed at room temperature [48]. In this section we will describe typical results of atomically resolved noncontact AFM imaging of semiconductor surfaces.

4.3.1 Si(111)-(7×7) Surface

Figure 4.10 shows the atomic-resolution images of the Si(111)-(7×7) surface [49]. Here, Fig. 4.10a (type I) was obtained using the Si tip without dangling, which is covered with an inert oxide layer. Figure 4.10b (type II) was obtained using the Si tip with a dangling bond, on which the Si atoms were deposited due the mechanical soft contact between the tip and the Si surface. The variable frequency shift mode was used. We can see not only adatoms and corner holes but also missing adatoms described by the dimer–adatom–stacking (DAS) fault model. We can see that the image contrast in Fig. 4.10b is clearly stronger than that in Fig. 4.10a.

Interestingly, by using the Si tip with a dangling bond, we observed contrast between inequivalent halves and between inequivalent adatoms of the 7×7 unit cell. Namely, as shown in Fig. 4.11a, the faulted halves (surrounded with a solid line) are brighter than the unfaulted halves (surrounded with a broken line). Here, the positions of the faulted and unfaulted halves were determined from the step

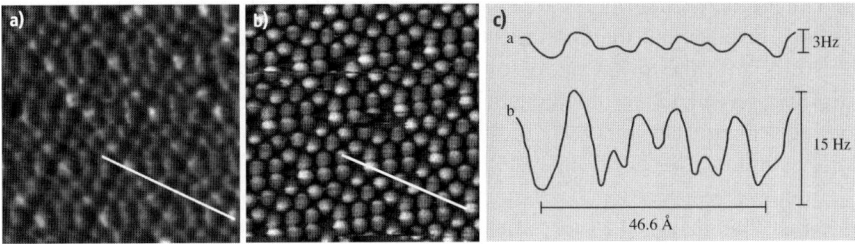

Fig. 4.10. Noncontact-mode AFM images of a Si(111)-(7×7) reconstructed surface obtained using the Si tips (**a**) without and (**b**) with a dangling bond. The scan area is 99 Å × 99 Å. (**c**) The cross-sectional profiles along the long diagonal of the 7×7 unit cell indicated by the *white lines* in (**a**) and (**b**)

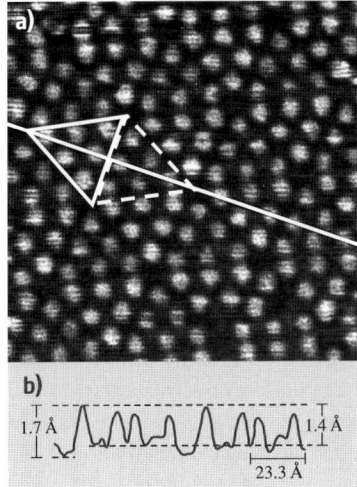

Fig. 4.11. (**a**) Noncontact mode AFM image with contrast of inequivalent adatoms and (**b**) a cross-sectional profile indicated by the *white line*. The halves of the 7×7 unit cell surrounded by the *solid line* and *broken line* correspond to the faulted and unfaulted halves, respectively. The scan area is 89 Å × 89 Å

direction. From the cross-sectional profile along the long diagonal of the 7×7 unit cell in Fig. 4.11b, the heights of the corner adatoms are slightly higher than those of the adjacent center adatoms in the faulted and unfaulted halves of the unit cell. The measured corrugation are in the following decreasing order: Co – F > Ce – F > Co – U > Ce – U, where Co – F and Ce – F indicate the corner and center adatoms in faulted halves, and Co – U and Ce – U indicate the corner and center adatoms in unfaulted halves, respectively. Averaging over several units, the corrugation height differences are estimated to be 0.25 Å, 0.15 Å and 0.05 Å for Co – F, Ce – F and Co – U, respectively, with respect to to Ce – U. This tendency, that the heights of the corner adatoms are higher than those of the center adatoms, is consistent with the experimental results using a silicon tip [47], although they could not determine the faulted and unfaulted halves of the unit cell in the measured AFM images. However, this tendency is completely contrary to the experimental results using a tungsten tip [35]. This difference may originate from the difference between the tip materials, which seems to affect the interaction between the tip and the reactive sample surface. Another possibility is that the tip is in contact with the surface during the small fraction of the oscillating cycle in their experiments [35].

We consider that the contrast between inequivalent adatoms is not caused by tip artifacts for the following reasons: (1) each adatom, corner hole and defect was clearly observed, (2) the apparent heights of the adatoms are the same whether they are located adjacent to defects or not, and (3) the same contrast in several images for the different tips has been observed.

It should be noted that the corrugation amplitude of adatoms ≈ 1.4 Å in Fig. 4.11b is higher than that of 0.8–1.0 Å obtained with the STM, although the depth of the corner holes obtained with noncontact AFM is almost the same as that observed with STM. Moreover, in noncontact-mode AFM images, the corrugation amplitude of adatoms was frequently larger than the depth of the corner holes. The origin of such large corrugation of adatoms may be due to the effect of the chemical interaction, but is not yet clear.

The atom positions, surface energies, dynamic properties and chemical reactivities on the Si(111)-(7×7) reconstructed surface have been extensively investigated theoretically and experimentally. From these investigations, the possible origins of the contrast between inequivalent adatoms in AFM images are the followings: the true atomic heights that correspond to the adatom core positions, the stiffness (spring constant) of interatomic bonding with the adatoms corresponding to the frequencies of the surface mode, the charge on the adatom, and the chemical reactivity of the adatoms. Table 4.1 summarizes the decreasing orders of the inequivalent adatoms for individual property. From Table 4.1, we can see that the calculated adatom heights and the stiffness of interatomic bonding cannot explain the AFM data, while the amount of charge of adatom and the chemical reactivity of adatoms can explain the our data. The contrast due to the amount of charge of adatom means that the AFM image is originated from the difference of the vdW or electrostatic physical interactions between the tip and the valence electrons at the adatoms. The contrast due to the chemical reactivity of adatoms means that the AFM image is originated

Table 4.1. Comparison between the adatom heights observed in an AFM image and the variety of properties for inequivalent adatoms

	Decreasing order	**Agreement**
AFM image	Co-F > Ce-F > Co-U > Ce-U	–
Calculated height	Co-F > Co-U > Ce-F > Ce-U	×
Stiffness of inter-atomic bonding	Ce-U > Co-U > Ce-F > Co-F	×
Amount of charge of adatom	Co-F > Ce-F > Co-U > Ce-U	○
Calculated chemical reactivity	Faulted > Unfaulted	○
Experimental chemical reactivity	Co-F > Ce-F > Co-U > Ce-U	○

from the difference of covalent bonding chemical interaction between the atoms at the tip apex and dangling bond of adatoms. Thus, we can see there are two possible interactions which explain the strong contrast between inequivalent adatoms of 7×7 unit cell observed using the Si tip with dangling bond.

The weak-contrast image in Fig. 4.10a is due to vdW and/or electrostatic force interactions. On the other hand, the strong-contrast images in Figs. 4.10b and 4.11a are due to a covalent bonding formation between the AFM tip with Si atoms and Si adatoms. These results indicate the capability of the noncontact-mode AFM to image the variation in chemical reactivity of Si adatoms. In the future, by controlling an atomic species at the tip apex, the study of chemical reactivity on an atomic scale will be possible using noncontact AFM.

4.3.2 Si(100)-(2×1) and Si(100)-(2×1):H Monohydride Surfaces

In order to investigate the imaging mechanism of the noncontact AFM, a comparative study between a reactive surface and an insensitive surface using the same tip is very useful. Si(100)-(2×1):H monohydride surface is a Si(100)-(2×1) reconstructed surface that is terminated by a hydrogen atom. It does not reconstruct as metal is deposited on the semiconductor surface. The surface structure hardly changes. Thus, the Si(100)-(2×1):H monohydride surface is one of most useful surface for a model system to investigate the imaging mechanism, experimentally and theoretically. Furthermore, whether the interaction between a very small atom such as hydrogen and a tip apex is observable with noncontact AFM is interested. Here, we show noncontact AFM images measured on a Si(100)-(2×1) reconstructed surface with a dangling bond and on a Si(100)-(2×1):H monohydride surface on which the dangling bond is terminated by a hydrogen atom [50].

Figure 4.12a shows the atomic-resolution image of the Si(100)-(2×1) reconstructed surface. Pairs of bright spots arranged in rows with a 2×1 symmetry were observed with clear contrast. Missing pairs of bright spots were also observed, as indicated by arrows. Furthermore, the pairs of bright spots are shown by the white dashed arc and appear to be the stabilize-buckled asymmetric dimer structure. Furthermore, the distance between the pairs of bright spots is 3.2 ± 0.1 Å.

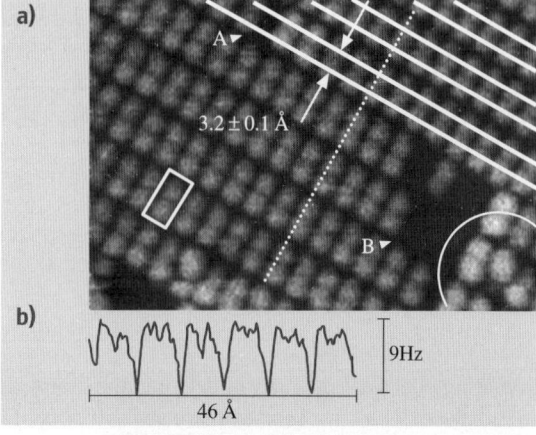

Fig. 4.12. (a) Noncontact AFM image of a Si(001)-(2×1) reconstructed surface. The scan area was 69×46 Å. One 2×1 unit cell is outlined with a *box*. *White rows* are superimposed to show the bright spots arrangement. The distance between the bright spots on the dimer row is 3.2 ± 0.1 Å. On the *white arc*, the alternative bright spots are shown. (b) Cross-sectional profile indicated by the *white dotted line*

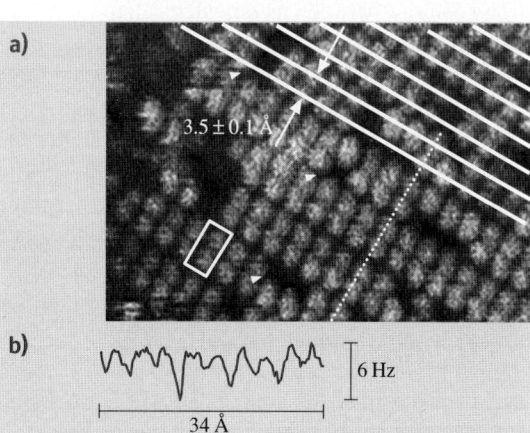

Fig. 4.13. (a) Noncontact AFM image of Si(001)-(2×1):H surface. The scan area was 69×46 Å. One 2×1 unit cell is outlined with a *box*. *White rows* are superimposed to show the bright spots arrangement. The distance between the bright spots on the dimer row is 3.5 ± 0.1 Å. (b) Cross-sectional profile indicated by the *white dotted line*

Figure 4.13a shows the atomic-resolution image of the Si(100)-(2×1):H monohydride surface. Pairs of bright spots arranged in rows were observed. Missing paired bright spots as well as those paired in rows and single bright spots were observed, as indicated by arrows. Furthermore, the distance between paired bright spots is 3.5 ± 0.1 Å. This distance of 3.5 ± 0.1 Å is 0.2 Å larger than that of the Si(100)-(2×1) reconstructed surface. Namely, it is found that the distance between bright spots increases in size due to the hydrogen termination.

The bright spots in Fig. 4.12 do not merely image the silicon-atom site, because the distance between the bright spots forming the dimer structure of Fig. 4.12a, 3.2 ± 0.1 Å, is lager than the distance between silicon atoms of every dimer structure model. (The maximum is the distance between the upper silicones in an asymmetric dimer structure 2.9 Å.) This seems to be due to the contribution to the imaging of the chemical bonding interaction between the dangling bond from the apex of the silicon tip and the dangling bond on the Si(100)-(2×1) reconstructed surface. Namely, the chemical bonding interaction operates strongly, with strong direction dependence, between the dangling bond pointing out of the silicon dimer structure on the Si(100)-(2×1) reconstructed surface and the dangling bond pointing out of the apex of the silicon tip; a dimer structure is obtained with a larger separation than between silicones on the surface.

The bright spots in Fig. 4.13 seem to be located at hydrogen atom sites on the Si(100)-(2×1):H monohydride surface, because the distance between the bright spots forming the dimer structure (3.5 ± 0.1 Å) approximately agrees with the distance between the hydrogens, i.e., 3.52 Å. Thus, the noncontact AFM atomically resolved the individual hydrogen atoms on the topmost layer. On this surface, the dangling bond is terminated by a hydrogen atom, and the hydrogen atom on the topmost layer does not have chemical reactivity. Therefore, the interaction between the hydrogen atom on the topmost layer and the apex of the silicon tip does not contribute to the chemical bonding interaction with strong direction dependence as on the silicon surface, and the bright spots in the noncontact AFM image correspond to the hydrogen atom sites on the topmost layer.

4.3.3 Metal Deposited Si Surface

In this section, we will introduce the comparative study of force interactions between a Si tip and a metal-deposited Si surface, and between a metal adsorbed Si tip and a metal-deposited Si surface [51, 52]. As for the metal-deposited Si surface, Si(111)-($\sqrt{3}\times\sqrt{3}$)-Ag (hereafter referred to as $\sqrt{3}$-Ag) surface was used.

For the $\sqrt{3}$-Ag surface, the honeycomb-chained trimer (HCT) model has been accepted as the appropriate model. As shown in Fig. 4.5, this structure contains a Si trimer in the second layer, 0.75 Å below the Ag trimer in the topmost layer. The topmost Ag atoms and lower Si atoms form covalent bonds. The interatomic distances between the nearest-neighbor Ag atoms forming the Ag trimer and between the lower Si atoms forming the Si trimer are 3.43 Å and 2.31 Å, respectively. The apexes of the Si trimers and Ag trimers face the [11$\bar{2}$] direction and the direction tilted a little to the [$\bar{1}\bar{1}$2] direction, respectively.

In Fig. 4.15, we show the noncontact AFM images measured using a normal Si tip at a frequency shift of (a) −37 Hz, (b) −43 Hz and (c) −51 Hz, respectively. These frequency shifts correspond to tip–sample distances of about 0–3 Å. We defined the zero position of the tip–sample distance, i.e., the contact point, as the point at which the vibration amplitude began to decrease. The rhombus indicates the $\sqrt{3}\times\sqrt{3}$ unit cell. When the tip approached the surface, the contrast of the noncontact AFM images become strong and the pattern changed remarkably. That is, by approaching the tip toward the sample surface, the hexagonal pattern, the trefoil-like pattern composed of three dark lines, and the triangle pattern can be observed sequentially. In Fig. 4.15a, the distance between the bright spots is 3.9 ± 0.2 Å. In Fig. 4.15c, the distance between the bright spots is 3.0 ± 0.2 Å, and the direction of the apex of all the triangles composed of three bright spots is [11$\bar{2}$].

In Fig. 4.16, we show the noncontact AFM images measured by using Ag-absorbed tip at a frequency shift of (a) −4.4 Hz, (b) −6.9 Hz and (c) −9.4 Hz, respectively. The tip–sample distances Z are roughly estimated to be $Z = 1.9$ Å,

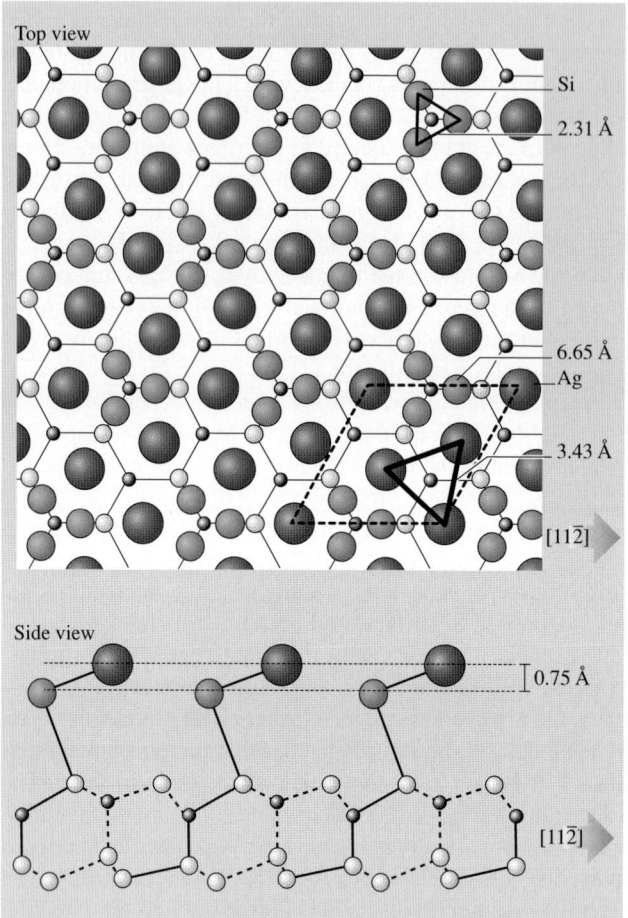

Fig. 4.14. HCT model for the structure of the Si(111)-($\sqrt{3} \times \sqrt{3}$)–Ag surface. *Black closed circle, gray closed circle, open circle,* and *closed circle with horizontal line* indicate Ag atom at the topmost layer, Si atom at the second layer, Si atom at the third layer, and Si atom at the fourth layer, respectively. The *rhombus* indicates the $\sqrt{3} \times \sqrt{3}$ unit cell. The *thick, large, solid triangle* indicates an Ag trimer. The *thin, small, solid triangle* indicates a Si trimer

0.6 Å and ≈ 0 Å (in the noncontact region), respectively. When the tip approached the surface, the pattern of the noncontact AFM images did not change, although the contrast become clearer. A triangle pattern can be observed. The distance between the bright spots is 3.5 ± 0.2 Å. The direction of the apex of all the triangles composed of three bright spots is tilted a little from the [$\overline{11}2$] direction.

Thus, noncontact AFM images measured on Si(111)-($\sqrt{3} \times \sqrt{3}$)-Ag surface showed two types of distance dependence in the image patterns depending on the atom species on the apex of the tip.

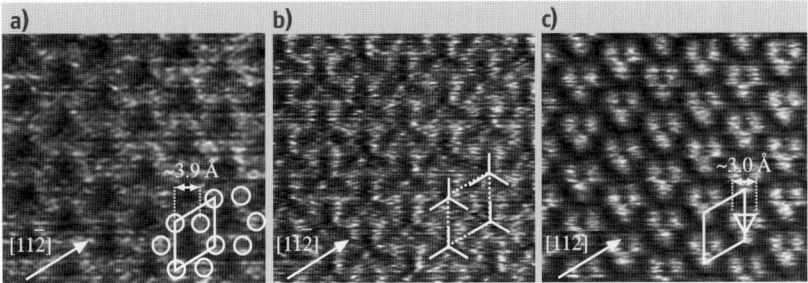

Fig. 4.15. Noncontact AFM images obtained at frequency shifts of (**a**) −37 Hz, (**b**) −43 Hz, and (**c**) −51 Hz on a Si(111)-($\sqrt{3} \times \sqrt{3}$)–Ag surface. This distance dependence was obtained with a Si tip. The scan area is 38 Å × 34 Å. A *rhombus* indicates the $\sqrt{3} \times \sqrt{3}$ unit cell

By using the normal Si tip with a dangling bond, in Fig. 4.15a, the measured distance between the bright spot of 3.9 ± 0.2 Å agrees with the distance of 3.84 Å between the centers of the Ag trimers in the HCT model within the experimental error. Furthermore, the hexagonal pattern composed of six bright spots also agrees with the honeycomb structure of the Ag trimer in HCT model. So the most appropriate site corresponding to the bright spots in Fig. 4.15a is the site of the center of Ag trimers. In Fig. 4.15c, the measured distance of 3.0 ± 0.2 Å between the bright spots forming the triangle pattern agrees with neither the distance between the Si trimer of 2.31 Å nor the distance between the Ag trimer of 3.43 Å in the HCT model, while the direction of the apex of the triangles composed of three bright spots agrees with the [11$\bar{2}$] direction of the apex of the Si trimer in the HCT model. So the most appropriate site corresponding to the bright spots in Fig. 4.15c is the intermediate site between the Si atoms and Ag atoms. On the other hand, by using the Ag-adsorbed tip, the measured distance between the bright spots of 3.5 ± 0.2 Å in Fig. 4.16 agrees with the distance of 3.43 Å between the nearest-neighbor Ag atoms forming the Ag trimer in the topmost layer in the HCT model within the experimental error. Furthermore, the direction of the apex of the triangles composed of three bright spots also

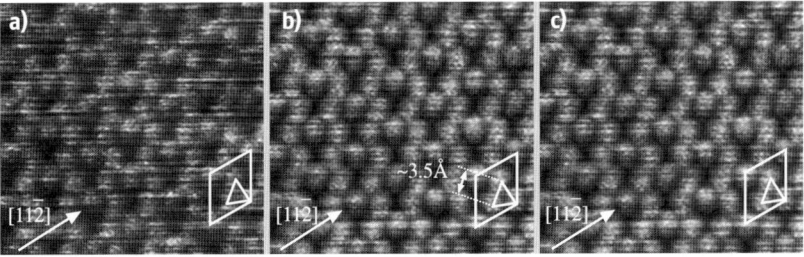

Fig. 4.16. Noncontact AFM images obtained at frequency shifts of (**a**) −4.4 Hz, (**b**) −6.9 Hz, and (**c**) −9.4 Hz on a Si(111)-($\sqrt{3} \times \sqrt{3}$)-Ag surface. This distance dependence was obtained with the Ag-adsorbed tip. The scan area is 38 Å × 34 Å

agrees with the direction of the apex of the Ag trimer, i.e., tilted [$\overline{11}2$], in the HCT model. So, the most appropriate site corresponding to the bright spots in Fig. 4.16 is the site of individual Ag atoms forming the Ag trimer in the topmost layer.

It should be noted that, by using the noncontact AFM with a Ag-adsorbed tip, for the first time, the individual Ag atom on the $\sqrt{3}$-Ag surface could be resolved in real space, although by using the noncontact AFM with an Si tip, it could not be resolved. So far, the $\sqrt{3}$-Ag surface has been observed by a scanning tunneling microscope (STM) with atomic resolution. However, the STM can also measure the local charge density of states near the Fermi level on the surface. From first-principle calculations, it was proven that unoccupied surface states are densely distributed around the center of the Ag trimer. As a result, bright contrast is obtained at the center of the Ag trimer with the STM.

Finally, we consider the origin of the atomic-resolution imaging of the individual Ag atoms on the $\sqrt{3}$-Ag surface. Here, we discuss the difference between the force interactions when using the Si tip and the Ag-adsorbed tip. As shown in Fig. 4.17a, when using the Si tip, there is a dangling bond pointing out of the topmost Si atom on the apex of the Si tip. As a result, the force interaction is dominated by physical bonding interactions, such as the Coulomb force, far from the surface and by chemical bonding interaction very close to the surface. Namely, if a reactive Si tip with a dangling bond approaches a surface, at distances far from the surface the Coulomb force acts between the electron localized on the dangling bond pointing out of the topmost Si atom on the apex of the tip, and the positive charge distributed around the center of the Ag trimer. At distances very close to the surface, the chemical bonding interaction will occur due to the onset of orbital hybridization between the dangling bond pointing out of the topmost Si atom on the apex of the Si tip and

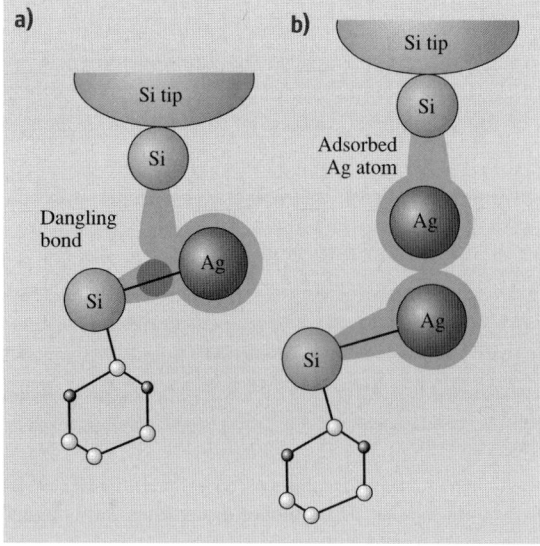

Fig. 4.17. Schematic illustration of (**a**) the Si atom with dangling bond and (**b**) the Ag-adsorbed tip above the Si–Ag covalent bond on a Si(111)-($\sqrt{3} \times \sqrt{3}$)-Ag surface

a Si–Ag covalent bond on the surface. Hence, the individual Ag atoms will not be resolved and the image pattern will change depending on the tip–sample distance. On the other hand, as shown in Fig. 4.17b, by using the Ag-adsorbed tip, the dangling bond localized out of topmost Si atom on the apex of the Si tip is terminated by the adsorbed Ag atom. As a result, even at very close tip–sample distances, the force interaction is dominated by physical bonding interactions such as the vdW force. Namely, if the Ag-adsorbed tip approaches the surface, the vdW force acts between the Ag atom on the apex of the tip and the Ag or Si atom on the surface. Ag atoms in the topmost layer of the $\sqrt{3}$-Ag surface are located higher than the Si atoms in the lower layer. Hence, the individual Ag atoms (or their nearly true topography) will be resolved, and the image pattern will not change even at very small tip–sample distances. It should be emphasized that there is a possibility to identify or recognize atomic species on a sample surface using noncontact AFM if we can control the atomic species at the tip apex.

4.4 Applications to Insulators

Insulators such as alkali halides, fluorides, and metal oxides are key materials in many applications, including optics, microelectronics, catalysis, and so on. Surface properties are important in these technologies, but they are usually poorly understood. This is due to their low conductivity, which makes it difficult to investigate them using electron- and ion-based measurement techniques such as low-energy electron diffraction, ion-scattering spectroscopy, and scanning tunneling microscopy (STM). Surface imaging by noncontact atomic force microscopy (NC-AFM) does not require a sample surface with a high conductivity because NC-AFM detects a force between the tip on the cantilever and the surface of the sample. Since the first report of atomically resolved NC-AFM on a Si(111)-(7×7) surface [33], several groups have succeeded in obtaining "true" atomic resolution images of insulators, including defects, and it has been shown that NC-AFM is a powerful new tool for atomic-scale surface investigation of insulators.

In this section we will describe typical results of atomically resolved NC-AFM imaging of insulators such as alkali halides, fluorides and metal oxides. For the alkali halides and fluorides, we will focus on contrast formation, which is the most important issue for interpreting atomically resolved images of binary compounds on the basis of experimental and theoretical results. For the metal oxides, typical examples of atomically resolved imaging will be exhibited and the difference between the STM and NC-AFM images will be demonstrated. Also, theoretical studies on the interaction between realistic Si tips and representative oxide surfaces will be shown. Finally, we will describe an antiferromagnetic NiO(001) surface imaged with a ferromagnetic tip to explore the possibility of detecting short-range magnetic interactions using the NC-AFM.

4.4.1 Alkali Halides, Fluorides and Metal Oxides

The surfaces of alkali halides were the first insulating materials to be imaged by NC-AFM with "true" atomic resolution [53]. To date, there have been reports on atomically resolved images of (001) cleaved surfaces for single-crystal NaF, RbBr, LiF, KI, NaCl, [54], KBr [55] and thin films of NaCl(001) on Cu(111) [56]. In this section we describe the contrast formation of alkali halides surfaces on the basis of experimental and theoretical results.

Alkali Halides

In experiments on alkali halides, the symmetry of the observed topographic images indicates that the protrusions exhibit only one type of ions, either the positive or negatively charged ions. This leads to the conclusion that the atomic contrast is dominantly caused by electrostatic interactions between a charged atom at the apex of the tip and the surface ions, i.e. long-range forces between the macroscopic tip and the sample, such as the van der Waals force, are modulated by an alternating short-range electrostatic interaction with the surface ions. Theoretical work employing the atomistic simulation technique has revealed the mechanism for contrast formation on an ionic surface [57]. A significant part of the contrast is due to the displacement of ions in the force field, not only enhancing the atomic corrugations, but also contributing to the electrostatic potential by forming dipoles at the surface. The experimentally observed atomic corrugation height is determined by the interplay of the long- and short-range forces. In the case of NaCl, it has been experimentally demonstrated that a blunter tip produces a lager corrugation when the tip–sample distance is shorter [54]. This result shows that the increased long-range forces induced by a blunter tip allow for more stable imaging closer to the surface. The stronger electrostatic short-range interaction and lager ion displacement produce a more pronounced atomic corrugation. At steps and kinks on an NaCl thin film on Cu(111), the corrugation amplitude of atoms with low coordination number has been observed to increase by a factor of up to two more than that of atomically flat terraces [56]. The low coordination number of the ions results in an enhancement of the electrostatic potential over the site and an increase in the displacement induced by the interaction with the tip.

Theoretical study predicts that the image contrast depends on the chemical species at the apex of the tip. *Bennewitz* et al. [56] have performed the calculations using an MgO tip terminated by oxygen and an Mg ion. The magnitude of the atomic contrast for the Mg-terminated tip shows a slight increase in comparison with an oxygen-terminated tip. The atomic contrast with the oxygen-terminated tip is dominated by the attractive electrostatic interaction between the oxygen on the tip apex and the Na ion, but the Mg-terminated tip attractively interacts with the Cl ion. In other words, these results demonstrated that the species of the ion imaged as the bright protrusions depends on the polarity of the tip apex.

These theoretical results emphasized the importance of the atomic species at the tip apex for the alkali halide (001) surface, while it is not straightforward to define the nature of the tip apex experimentally because of the high symmetry of the surface structure. However, there are a few experiments exploring the possibilities to determine the polarity of the tip apex. *Bennewitz* et al. [58] studied imaging of surfaces of a mixed alkali halide crystal, which was designed to observe the chemically inhomogeneous surface. The mixed crystal is composed of 60% KCl and 40% KBr, with the Cl and Br ions interfused randomly in the crystal. The image of the cleaved $KCl_{0.6}Br_{0.4}(001)$ surface indicates that only one type of ion is imaged as protrusions, as if it were a pure alkali halide crystal. However, the amplitude of the atomic corrugation varies strongly between the positions of the ions imaged as depressions. This variation in the corrugations corresponds to the constituents of the crystal, i.e. the Cl and Br ions, and it is concluded that the tip apex is negatively charged. Moreover, the deep depressions can be assigned to Br ions by comparing the number with the relative density of anions. The difference between Cl and Br anions with different masses is enhanced in the damping signal measured simultaneously with the topographic image [59]. The damping is recorded as an increase in the excitation amplitude necessary to maintain the oscillation amplitude of the cantilever in the constant-amplitude mode [56]. Although the dissipation phenomena on an atomic scale are a subject under discussion, any dissipative interaction must generally induce energy losses in the cantilever oscillation [60,61]. The measurement of energy dissipation has the potential to enable chemical discrimination on an atomic scale. Recently, a new procedure for species recognition on a alkali halide surface was proposed [62]. This method is based on a comparison between theoretical results and the site-specific measurement of frequency versus distance. The differences in the force curves measured at the typical sites, such as protrusion, depression, and their bridge position, are compared to the corresponding differences obtained from atomistic simulation. The polarity of the tip apex can be determined, leading to the identification of the surface species. This method is applicable to highly symmetric surfaces and is useful for determining the sign of the tip polarity.

Fluorides

Fluorides are important materials for the progress of an atomic-scale-resolution NC-AFM imaging of insulators. There are reports in the literature of surface images for single-crystal BaF_2, SrF_2 [63], CaF_2 [64–66] and a CaF bilayer on Si(111) [67]. Surfaces of fluorite-type crystals are prepared by cleaving along the (111) planes. Their structure is more complex than the structure of alkali halides, which have a rock-salt structure. The complexity is of great interest for atomic-resolution imaging using NC-AFM and also for theoretical predictions of the interpretation of the atomic-scale contrast information.

The first atomically resolved images of a $CaF_2(111)$ surface were obtained in topographic mode [65], and the surface ions mostly appear as spherical caps. *Barth* et al. [68] have found that the $CaF_2(111)$ surface images obtained by using the constant-height mode, in which the frequency shift is recorded with a very low loop gain, can be categorized into two contrast patterns. In the first of these the ions appear as triangles and in the second they have the appearance of circles, similar to the contrast obtained in a topographic image. Theoretical studies demonstrated that these two different contrast patterns could be explained as a result of imaging with tips of different polarity [68–70]. When imaging with a positively charged (cation-terminated) tip, the triangular pattern appears. In this case, the contrast is dominated by the strong short-range electrostatic attraction between the positive tip and the negative F ions. The cross section along the [121] direction of the triangular image shows two maxima: one is a larger peak over the F(I) ions located in the topmost layer and the other is a smaller peak at the position of the F(III) ions in the third layer. The minima appear at the position of the Ca ions in the second layer. When imaging with a negatively charged (anion-terminated) tip, the spherical image contrast appears and the main periodicity is created by the Ca ions between the topmost and the third F ion layers. In the cross section along the [121] direction, the large maxima correspond to the Ca sites because of the strong attraction of the negative tip and the minima appear at the sites of maximum repulsion over the F(I) ions. At a position between two F ions, there are smaller maxima. This reflects the weaker repulsion over the F(III) ion sites compared to the protruding F(I) ion sites and a slower decay in the contrast on one side of the Ca ions.

The triangular pattern obtained with a positively charged tip appears at relatively large tip–sample distance, as shown in Fig. 4.18a. The cross section along the [121] direction, experiment results and theoretical studies both demonstrate the large-peak and small-shoulder characteristic for the triangular pattern image (Fig. 4.18d). When the tip approaches the surface more closely, the triangular pattern of the experimental images is more vivid (Fig. 4.18b), as predicted in the theoretical works. As the tip approaches, the amplitude of the shoulder increases until it is equal to that of the main peak, and this feature gives rise to the honeycomb pattern image, as shown in Fig. 4.18c. Moreover, theoretical results predict that the image contrast changes again when the tip apex is in close proximity to surface. Recently, *Giessibl* et al. [71] achieved atomic imaging in the repulsive region and proved experimentally the predicted change of the image contrast. As described here, there is good correspondence in the distance dependency of the image obtained by experimental and theoretical investigations.

From detailed theoretical analysis of the electrostatic potential [72], it was suggested that the change in displacement of the ions due to the proximity of the tip plays an important role in the formation of the image contrast. Such a drastic change in image contrast, depending on both the polarity of the terminated tip atom and on the tip–sample distance, is inherent to the fluoride (111) surface, and this image-contrast feature cannot be seen on the (001) surface of alkali halides with a simple crystal structure.

The results of careful experiments show another feature: that the cross sections taken along the three equivalent [121] directions do not yield identical results [68]. It is thought that this can be attributed to the asymmetry of the nanocluster at the tip apex, which leads to different interactions in the equivalent directions. A better understanding of the asymmetric image contrast may require more complicated modeling of the tip structure. In fact, it should be mentioned that perfect tips on an

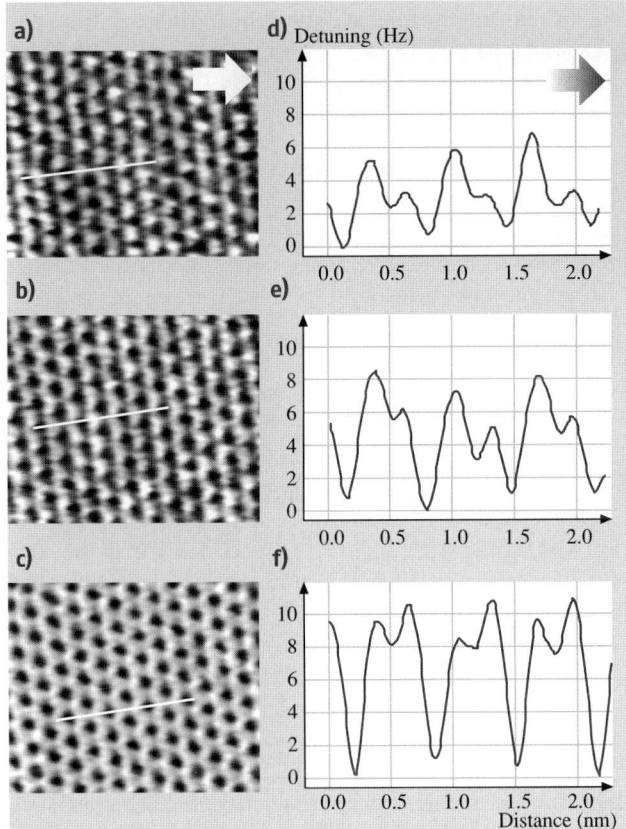

Fig. 4.18. (a)–(c) CaF$_2$(111) surface images obtained by using the constant-height mode. From (a) to (c) the frequency shift was lowered. The *white lines* represent the positions of the cross section. (d)–(f) The cross section extracted from the Fourier-filtered images of (a)–(c). The *white* and *black arrows* represent the scanning direction. The images and the cross sections are from [68]

atomic scale can occasionally be obtained. These tips do yield identical results in forward and backward scanning, and cross sections in the three equivalent directions taken with this tip are almost identical [74].

The fluoride (111) surface is an excellent standard surface for calibrating tips on an atomic scale. The polarity of the tip-terminated atom can be determined from the image contrast pattern (spherical or triangular pattern). The irregularities in the tip structure can be detected, since the surface structure is highly symmetric. Therefore, once such a tip has been prepared, it can be used as a calibrated tip for imaging unknown surfaces.

The polarity and shape of the tip apex play an important role in interpreting NC-AFM images of alkali halide and fluorides surfaces. It is expected that the achievement of good correlation between experimental and theoretical studies will help to advance surface imaging of insulators by NC-AFM.

Metal Oxides

Most of the metal oxides that have attracted strong interest for their technological importance are insulating. Therefore, in the case of atomically resolved imaging of metal oxide surfaces by STM, efforts to increase the conductivity of the sample are needed, such as, the introduction of anions or cations defects, doping with other atoms and surface observations during heating of the sample. However, in principle, NC-AFM provides the possibility of observing nonconductive metal oxides without these efforts. In cases where the conductivity of the metal oxides is high enough for a tunneling current to flow, it should be noted that most surface images obtained by NC-AFM and STM are not identical.

Since the first report of atomically resolved images on a $TiO_2(110)$ surface with oxygen point defects [75], they have also been reported on rutile $TiO_2(100)$ [76–78], anatase $TiO_2(001)$ thin film on $SrTiO_3(100)$ [79] and on $LaAO_3(001)$ [80], $SnO_2(110)$ [81], $NiO(001)$ [82,83], $SrTiO_3(100)$ [84], $CeO_2(111)$ [85] and $MoO_3(010)$ [86] surfaces. Also, *Barth* et al. have succeeded in obtaining atomically resolved NC-AFM images of a clean $\alpha-Al_2O_3(0001)$ surface [73] and of a UHV cleaved $MgO(001)$ [87] surface, which are impossible to investigate using STM. In this section we describe typical results of the imaging of metal oxides by NC-AFM.

The $\alpha-Al_2O_3(0001)$ surface exists in several ordered phases that can reversibly be transformed into each other by thermal treatments and oxygen exposure. It is known that the high-temperature phase has a large ($\sqrt{31} \times \sqrt{31})R \pm 9°$ unit cell. However, the details of the atomic structure of this surface have not been revealed, and two models have been proposed. *Barth* et al. [73] have directly observed this reconstructed $\alpha-Al_2O_3(0001)$ surface by NC-AFM. They confirmed that the dominant contrast of the low-magnification image corresponds to a rhombic grid representing a unit cell of ($\sqrt{31} \times \sqrt{31})R + 9°$, as shown in Fig. 4.19a. Also, more details of the atomic structures were determined from the higher-magnification image (Fig. 4.19b), which was taken at a reduced tip–sample distance. In this atomically resolved image, it was revealed that each side of the rhombus is intersected by ten atomic rows, and that a hexagonal arrangement of atoms exists in the center of the rhombi (Fig. 4.19c). This feature agrees with the proposed surface structure that predicts order in the center of the hexagonal surface domains and disorder at the domain boundaries. Their result is an excellent demonstration of the capabilities of the NC-AFM for the atomic-scale surface investigation of insulators.

The atomic structure of the $SrTiO_3(100)$-($\sqrt{5} \times \sqrt{5})R26.6°$ surface, as well as that of $Al_2O_3(0001)$ can be determined on the basis of the results of NC-AFM imaging [84]. $SrTiO_3$ is one of the perovskite oxides, and its (100) surface exhibits the many different kinds of reconstructed structures. In the case of the ($\sqrt{5} \times \sqrt{5})R26.6°$ reconstruction, the oxygen vacancy–Ti^{3+}–oxygen model (where the terminated surface is TiO_2 and the observed spots are related to oxygen vacancies) was proposed from the results of STM imaging. As shown in Fig. 4.20, *Kubo* et al. [84] have

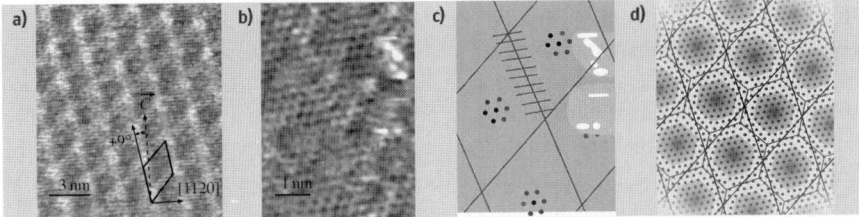

Fig. 4.19. (a) Image of the high-temperature, reconstructed clean $\alpha-Al_2O_3$ surface obtained by using the constant-height mode. The *rhombus* represents the unit cell of the $(\sqrt{31} \times \sqrt{31})R + 9°$ reconstructed surface. (b) Higher-magnification image of (a). Imaging was performed at a reduced tip–sample distance. (c) Schematic representation of the indicating regions of hexagonal order in the center of reconstructed rhombi. (d) Superposition of the hexagonal domain with reconstruction rhombi found by NC-AFM imaging. Atoms in the *gray shaded regions* are well ordered. The images and the schematic representations are from [73]

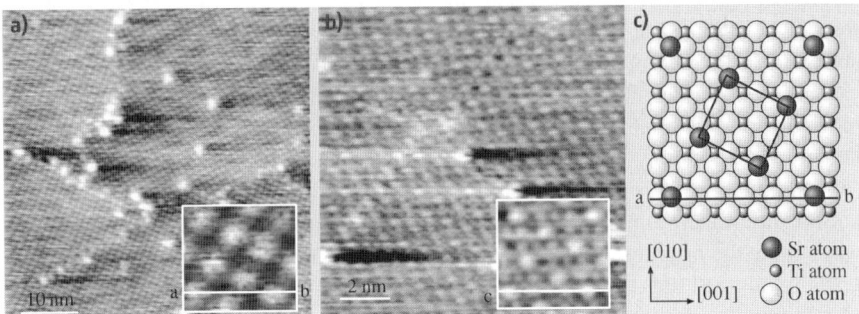

Fig. 4.20. (a) STM and (b) NC-AFM images of a $SrTiO_3(100)$ surface. (c) A proposed model of the $SrTiO_3(100)$-$(\sqrt{5} \times \sqrt{5})R26.6°$ surface reconstruction. The images and the schematic representations are from [84]

performed measurements using both STM and NC-AFM, and have found that the size of the bright spots as observed by NC-AFM is always smaller than that for STM measurement, and that the dark spots, which are not observed by STM, are arranged along the [001] and [010] directions in the NC-AFM image. A theoretical simulation of the NC-AFM image using first-principles calculations shows that the bright and dark spots correspond to Sr and oxygen atoms, respectively. It has been proposed that the structural model of the reconstructed surface consists of an ordered Sr adatom located at the oxygen fourfold site on the TiO_2-terminated layer (Fig. 4.20c).

Because STM images are related to the spatial distribution of the wave functions near the Fermi level, atoms without a local density of states near the Fermi level are generally invisible even on conductive materials. On the other hand, the NC-AFM image reflects the strength of the tip–sample interaction force originating from chemical, electrostatic and other interactions. Therefore, even STM and

NC-AFM images obtained using an identical tip and sample may not be identical generally. The simultaneous imaging of a metal oxide surface enables the investigation of a more detailed surface structure. The images of a $TiO_2(110)$ surface simultaneously obtained with STM and NC-AFM [78] are a typical example. The STM image shows that the dangling-bond states at the tip apex overlap with the dangling bonds of the $3d$ states protruding from the Ti atom, while the NC-AFM primarily imaged the uppermost oxygen atom.

Recently, calculations of the interaction of a Si tip with metal oxides surfaces, such as $Al_2O_3(0001)$, $TiO_2(110)$, and $MgO(001)$, were reported [88, 89]. Previous simulations of AFM imaging of alkali halides and fluorides assume that the tip would be oxides or contaminated and hence have been performed with a model of ionic oxide tips. In the case of imaging a metal oxide surface, pure Si tips are appropriate for a more realistic tip model because the tip is sputtered for cleaning in many experiments. The results of ab initio calculations for a Si tip with a dangling bond demonstrate that the balance between polarization of the tip and covalent bonding between the tip and the surface should determine the tip–surface force. The interaction force can be related to the nature of the surface electronic structure. For wide-gap insulators with a large valence-band offset that prevents significant electron-density transfer between the tip and the sample, the force is dominated by polarization of the tip. When the gap is narrow, the charge transfer increase and covalent bonding dominates the tip–sample interaction. The forces over anions (oxygen ions) in the surface are larger than over cations (metal ions), as they play a more significant role in charge transfer. This implies that a pure Si tip would always show the brightest contrast over the highest anions in the surface. In addition, *Foster* et al. [88] suggested the method of using applied voltage, which controls the charge transfer, during an AFM measurement to define the nature of tip apex.

The collaboration between experimental and theoretical studies has made great progress in interpreting the imaging mechanism for binary insulators surface and reveals that a well-defined tip with atomic resolution is preferable for imaging a surface. As described previously, a method for the evaluation of the nature of the tip has been developed. However, the most desirable solution would be the development of suitable techniques for well-defined tip preparation and a few attempts at controlled production of Si tips have been reported [24, 90, 91].

4.4.2 Atomically Resolved Imaging of a NiO(001) Surface

The transition metal oxides, such as NiO, CoO, and FeO, feature the simultaneous existence of an energy gap and unpaired electrons, which gives rise to a variety of magnetic property. Such magnetic insulators are widely used for the exchange biasing for magnetic and spintronic devices. NC-AFM enables direct surface imaging of magnetic insulators on an atomic scale. The forces detected by NC-AFM originate from several kinds of interaction between the surface and the tip, including magnetic interactions in some cases. Theoretical studies predict that short-range magnetic interactions such as the exchange interaction should enable the NC-AFM to image magnetic moments on an atomic scale. In this section, we will describe imaging of the antiferromagnetic NiO(001) surface using a ferromagnetic tip. Also, theoretical studies of the exchange force interaction between a magnetic tip and a sample will be described.

Theoretical Studies of the Exchange Force

In the system of a magnetic tip and sample, the interaction detected by NC-AFM includes the short-range magnetic interaction in addition to the long-range magnetic dipole interaction. The energy of the short-range interaction depends on the electron spin states of the atoms on the apex of the tip and the sample surface, and the energy difference between spin alignments (parallel or anti-parallel) is referred to as the exchange interaction energy. Therefore, the short-range magnetic interaction leads to the atomic-scale magnetic contrast, depending on the local energy difference between spin alignments.

In the past, extensive theoretical studies on the short-range magnetic interaction between a ferromagnetic tip and a ferromagnetic sample have been performed by a simple calculation [92], a tight-binding approximation [93] and first-principles calculations [94]. In the calculations performed by *Nakamura* et al. [94], three-atomic-layer Fe(001) films are used as a model for the tip and sample. The exchange force is defined as the difference between the forces in each spin configuration of the tip and sample (parallel and anti-parallel). The result of this calculation demonstrates that the amplitude of the exchange force is measurable for AFM (about 0.1 nN). Also, they forecasted that the discrimination of the exchange force would enable direct imaging of the magnetic moments on an atomic scale. Foster et al. [95] have theoretically investigated the interaction between a spin-polarized H atom and a Ni atom on a NiO(001) surface. They demonstrated that the difference in magnitude in the exchange interaction between opposite-spin Ni ions in a NiO surface could be sufficient to be measured in a low-temperature NC-AFM experiment. Recently, first-principles calculation of the interaction of a ferromagnetic Fe tip with an NiO surface has demonstrated that it should be feasible to measure the difference in exchange force between opposite-spin Ni ions [96].

Atomically Resolved Imaging Using Non-coated and Fe-coated Si Tips

The detection of the exchange interaction is a challenging task for NC-AFM applications. An antiferromagnetic insulator NiO single crystal that has regularly aligned atom sites with alternating electron spin states is one of the best candidates to prove the feasibility of detecting the exchange force for the following reason. NiO has an antiferromagnetic AF_2 structure as the most stable below the Néel temperature of 525 K. This well-defined magnetic structure, in which Ni atoms on the (001) surface are arranged in a checkerboard pattern, leads to the simple interpretation of an image containing the atomic-scale contrast originating in the exchange force. In addition, a clean surface can easily be prepared by cleaving.

Figure 4.21a shows an atomically resolved image of a NiO(001) surface with a ferromagnetic Fe-coated tip [97]. The bright protrusions correspond to atoms spaced about 0.42 nm apart, consistent with the expected periodic arrangement of

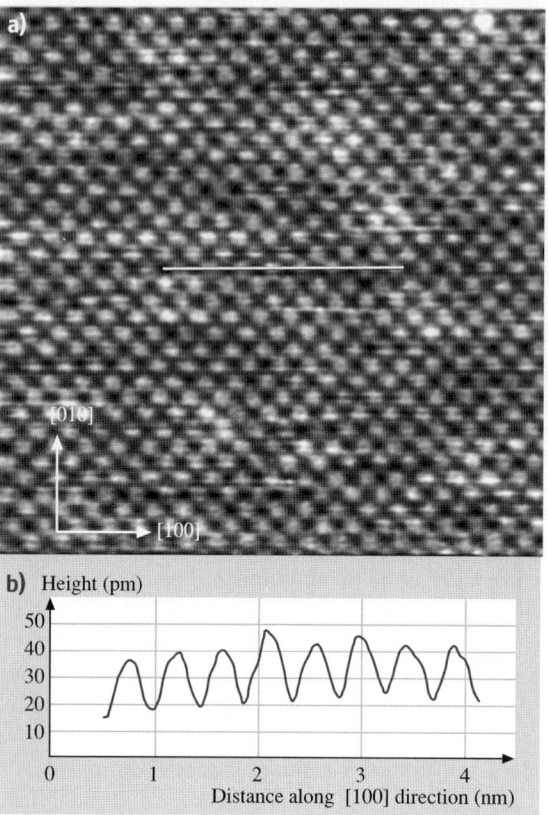

Fig. 4.21. (a) Atomically resolved image obtained with an Fe-coated tip. (b) Shows the cross sections of the middle part in . Their corrugations are about 30 pm

the NiO(001) surface. The corrugation amplitude is typically 30 pm, which is comparable to the value previously reported [82, 83, 98–100], as shown in Fig. 4.21b. The atomic-resolution image (Fig. 4.21b), in which there is one maximum and one minimum within the unit cell, resembles that of the alkali halide (001) surface. The symmetry of the image reveals that only one type of atom appears to be at the maximum. From this image, it seems difficult to distinguish which of the atoms are observed as protrusions. The theoretical works indicate that a metal tip interacts strongly with the oxygen atoms on the MgO(001) surface [95]. From this result, it is presumed that the bright protrusions correspond to the oxygen atoms. However, it is still questionable which of the atoms are visible with a Fe-coated tip.

If the short-range magnetic interaction is included in the atomic image, the corrugation amplitude of the atoms should depend on the direction of the spin over the atom site. From the results of first-principles calculations [94], the contribution of the short-range magnetic interaction to the measured corrugation amplitude is expected to be about a few percent of the total interaction. Discrimination of such small perturbations is therefore needed. In order to reduce the noise, the corrugation amplitude was added on the basis of the periodicity of the NC-AFM image. In addition, the topographical asymmetry, which is the index characterizing the difference in atomic corrugation amplitude, has been defined [101]. The result shows that the value of the topographical asymmetry calculated from the image obtained with an Fe-coated Si tip depends on the direction of summing of the corrugation amplitude, and that the dependency corresponds to the antiferromagnetic spin ordering of the NiO(001) surface [101, 102]. Therefore, this result implies that the dependency of the topographical asymmetry originates in the short-range magnetic interaction. However, in some cases the topographic asymmetry with uncoated Si tips has a finite value [103]. The possibility that the asymmetry includes the influence of the structure of tip apex and of the relative orientation between the surface and tip cannot be excluded. In addition, it is suggested that the absence of unambiguous exchange contrast is due to the fact that surface ion instabilities occur at tip–sample distances that are small enough for a magnetic interact [100]. Another possibility is that the magnetic properties of the tips are not yet fully controlled because the topographic asymmetries obtained by Fe- and Ni-coated tips show no significant difference [103]. In any cases, a careful comparison is needed to evaluate the exchange interaction included in an atomic image.

From the aforementioned theoretical works, it is presumed that a metallic tip has the capability to image an oxygen atom as a bright protrusion. Recently, the magnetic properties of the NiO(001) surface were investigated by first-principles electronic-structure calculations [104]. It was shown that the surface oxygen has finite spin magnetic moment, which originates from symmetry breaking. We must take into account the possibility that a metal atom at the ferromagnetic tip apex may interact with a Ni atom on the second layer through a magnetic interaction mediated by the electrons in an oxygen atom on the surface.

The measurements presented here demonstrate the feasibility of imaging magnetic structures on an atomic scale by NC-AFM. In order to realize explicit detection of exchange force, further experiments and a theoretical study are required. In particular, the development of a tip with well-defined atomic structure and magnetic properties is essential for *exchange force microscopy.*

4.5 Applications to Molecules

In the future, it is expected that electronic, chemical, and medical devices will be downsized to the nanometer scale. To achieve this, visualizing and assembling individual molecular components is of fundamental importance. Topographic imaging of nonconductive materials, which is beyond the range of scanning tunneling microscopes, is a challenge for atomic force microscopy (AFM). Nanometer-sized domains of surfactants terminated with different functional groups have been identified by lateral force microscopy (LFM) [107] and by chemical force microscopy (CFM) [108] as extensions of AFM. At a higher resolution, a periodic array of molecules, Langmuir–Blodgett films [109] for example, was recognized by AFM. However, it remains difficult to visualize an isolated molecule, molecule vacancy, or the boundary of different periodic domains, with a microscope with the tip in contact.

Fig. 4.22. The constant frequency-shift topography of domain boundaries on a C_{60} multilayered film deposited on a Si(111) surface based on [105]. Image size: 35×35 nm^2

4.5.1 Why Molecules and Which Molecules?

Access to individual molecules has not been a trivial task even for noncontact atomic force microscopy (NC-AFM). The force pulling the tip into the surface is less sensitive to the gap width (r), especially when chemically stable molecules cover the surface. The attractive potential between two stable molecules is shallow and exhibits r^{-6} decay [13].

High-resolution topography of formate (HCOO$^-$) [110] was first reported in 1997 as a molecular adsorbate. The number of imaged molecules is now increasing because of the technological importance of molecular interfaces. To date, the following studies on molecular topography have been published: C_{60} [105, 111], DNAs [106, 112], adenine and thymine [113], alkanethiols [113, 114], a perylene derivative (PTCDA) [115], a metal porphyrin (Cu–TBPP) [116], glycine sulfate [117], polypropylene [118], vinylidene fluoride [119], and a series of carboxylates (RCOO$^-$) [120–126]. Two of these are presented in Figs. 4.22 and 4.23 to demonstrate the current stage of achievement. The proceedings of the annual NC-AFM conference represent a convenient opportunity for us to update the list of molecules imaged.

4.5.2 Mechanism of Molecular Imaging

A systematic study of carboxylates (RCOO$^-$) with R = H, CH$_3$, C(CH$_3$)$_3$, C $\equiv$ CH, and CF$_3$ revealed that the van der Waals force is responsible for the molecule-dependent microscope topography despite its long-range (r^{-6}) nature. Carboxylates adsorbed on the (110) surface of rutile TiO$_2$ have been extensively studied as a prototype for organic materials interfaced with an inorganic metal oxide [128]. A carboxylic acid molecule (RCOOH) dissociates on this surface to a carboxylate (RCOO$^-$) and a proton (H$^+$) at room temperature, as illustrated in Fig. 4.24. The pair of negatively charged oxygen atoms in the RCOO$^-$ coordinate two positively charged Ti atoms on the surface. The adsorbed carboxylates create a long-range ordered monolayer. The lateral distances of the adsorbates in the ordered monolayer

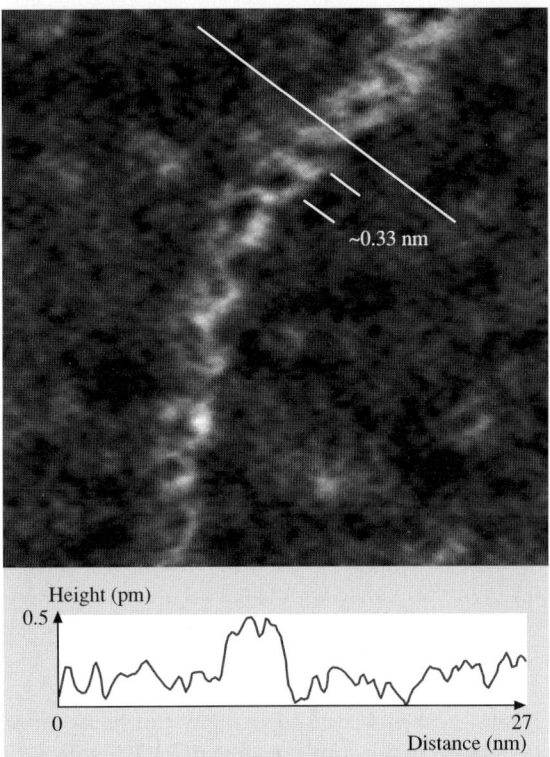

Fig. 4.23. The constant frequency-shift topography of a DNA helix on a mica surface based on [106]. Image size: 43×43 nm^2. The image revealed features with a spacing of 3.3 nm, consistent with the helix turn of B-DNA

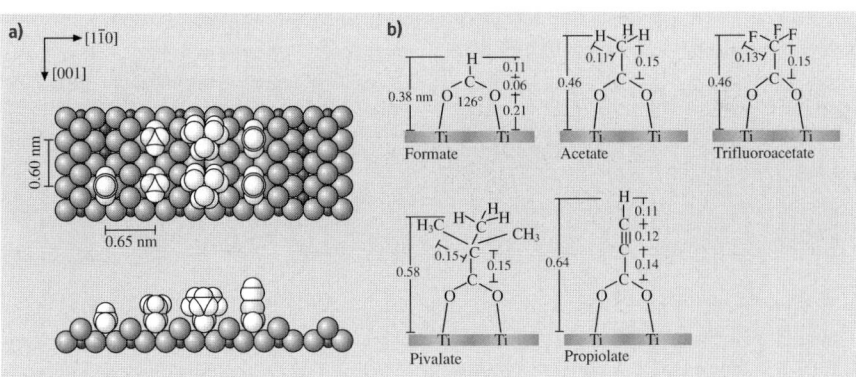

Fig. 4.24. The carboxylates and TiO$_2$ substrate. (**a**) Top and side view of the ball model. *Small shaded* and *large shaded balls* represent Ti and O atoms in the substrate. Protons yielded in the dissociation reaction are not shown. (**b**) Atomic geometry of formate, acetate, pivalate, propiolate, and trifluoroacetate adsorbed on the TiO$_2$(110) surface. The O–Ti distance and O–C–O angle of the formate were determined in the quantitative analysis using photoelectron diffraction [127]

are regulated at 0.65 and 0.59 nm along the [110] and [001] directions. By scanning a mixed monolayer containing different carboxylates, the microscope topography of the terminal groups can be quantitatively compared while minimizing tip-dependent artifacts.

Figure 4.25 presents the observed constant frequency-shift topography of four carboxylates terminated by different alkyl groups. On the formate-covered surface of panel (a), individual formates (R=H) were resolved as protrusions of uniform brightness. The dark holes represent unoccupied surface sites. The cross section in the lower panel shows that the accuracy of the height measurement was 0.01 nm or better. Brighter particles appeared in the image when the formate monolayer was exposed to acetic acid (CH_3COOH) as shown in panel (b). Some formates were exchanged with acetates (R=CH_3) impinging from the gas phase [129]. Because the number of brighter spots increased with exposure time to acetic acid, the brighter particle was assigned to the acetate [121]. Twenty-nine acetates and 188 formates were identified in the topography. An isolated acetate and its surrounding formates exhibited an image height difference of 0.06 nm. Pivalate is terminated by bulky R=$(CH_3)_3$. Nine bright pivalates were surrounded by formates of ordinary brightness in the image of panel (c) [123]. The image height difference of an isolated pivalate over the formates was 0.11 nm. Propiolate with C≡CH is a needle-like adsorbate of single-atom diameter. That molecule exhibited in panel (d) a microscope topography 0.20 nm higher than that of the formate [125].

The image topography of formate, acetate, pivalate, and propiolate followed the order of the size of the alkyl groups. Their physical topography can be assumed based on the C–C and C–H bond lengths in the corresponding RCOOH molecules in the gas phase [130], and is illustrated in Fig. 4.24. The top hydrogen atom of

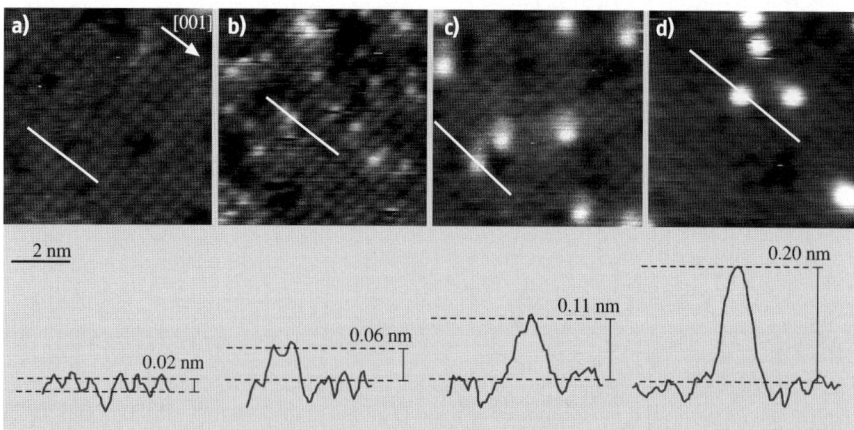

Fig. 4.25. The constant frequency-shift topography of carboxylate monolayers prepared on the $TiO_2(110)$ surface based on [121, 123, 125]. Image size: 10×10 nm^2. (**a**) Pure formate monolayer; (**b**) formate–acetate mixed layer; (**c**) formate–pivalate mixed layer; (**d**) formate–propiolate mixed layer. Cross sections determined on the *lines* are shown in the lower panel

the formate is located 0.38 nm above the surface plane containing the Ti atom pair, while three equivalent hydrogen atoms of the acetate are more elevated at 0.46 nm. The uppermost H atoms in the pivalate are raised by 0.58 nm relative to the Ti plane. The H atom terminating the triple-bonded carbon chain in the propiolate is at 0.64 nm. Figure 4.26 summarizes the observed image heights relative to the formate, as a function of the physical height of the topmost H atoms given in the model. The straight line fitted the four observations [122]. When the horizontal axis was scaled with other properties (molecular weight, the number of atoms in a molecule, or the number of electrons in valence states), the correlation became poor.

On the other hand, if the tip apex traced the contour of a molecule composed of hard-sphere atoms, the image topography would reproduce the physical topography in a one-to-one ratio, as shown by the broken line in Fig. 4.26. However, the slope of the fitted line was 0.7. A slope of less than unity is interpreted as the long-range nature of the tip–molecule force. The observable frequency shift reflects the sum of the forces between the tip apex and individual molecules. When the tip passes above a tall molecule embedded in short molecules, it is pulled up to compensate for the increased force originating from the tall molecule. Forces between the lifted tip and the short molecules are reduced due to the increased tip–surface distance. Feedback regulation pushes down the probe to restore the lost forces.

This picture predicts that microscope topography is sensitive to the lateral distribution of the molecules, and that was in fact the case. Two-dimensionally clustered acetates exhibited enhanced image height over an isolated acetate [121]. The tip–molecule force therefore remained nonzero at distances over the lateral separation of the carboxylates on this surface (0.59–0.65 nm). Chemical bond interactions cannot be important across such a wide tip–molecule gap, whereas atom-scale images of Si(111)(7×7) are interpreted with the fractional formation of tip–surface chemical bonds [24,45,49]. Instead, the attractive component of the van der Waals force is probable responsible for the observed molecule-dependent topography. The absence of the tip–surface chemical bond is reasonable on the carboxylate-covered surface terminated with stable C–H bonds.

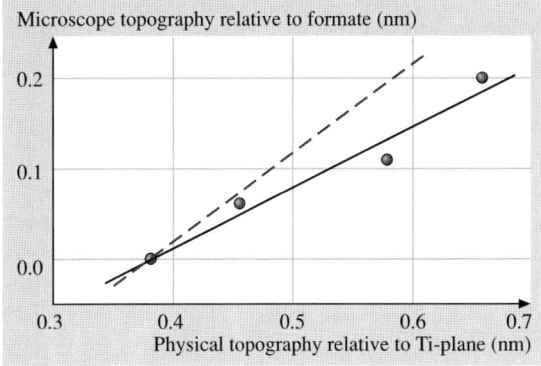

Fig. 4.26. The constant frequency-shift topography of the alkyl-substituted carboxylates as a function of their physical topography given in the model of Fig. 4.3 based on [123]

The attractive component of the van der Waals force contains electrostatic terms caused by permanent-dipole/permanent-dipole coupling, permanent-dipole/induced-dipole coupling, and induced-dipole/induced-dipole coupling (dispersion force). The four carboxylates examined are equivalent in terms of their permanent electric dipole, because the alkyl groups are non-polar. The image contrast of one carboxylate relative to another is thus ascribed to the dispersion force and/or the force created by the coupling between the permanent dipole on the tip and the induced dipole on the molecule. If we further assume that the Si tip used exhibits the smallest permanent dipole, the dispersion force remains dominant to create the NC-AFM topography dependent on the non-polar groups of atoms. A numerical simulation based on this assumption [125] successfully reproduced the propiolate topography of Fig. 4.25d. A calculation that does not include quantum chemical treatment is expected to work, unless the tip approaches the surface too closely, or the molecule possesses a dangling bond.

In addition to the contribution of the dispersion force, the permanent dipole moment of molecules may perturb the microscope topography through electrostatic coupling with the tip. Its possible role was demonstrated by imaging a fluorine-substituted acetate. The strongly polarized C–F bonds were expected to perturb the electrostatic field over the molecule. The constant frequency-shift topography of acetate (R=CH_3) and trifluoroacetate (R=CF_3) was indeed sensitive to the fluorine substitution. The acetate was observed to be 0.05 nm higher than the trifluoroacetate [122], although the F atoms in the trifluoroacetate as well as the H atoms in the acetate were lifted by 0.46 nm from the surface plane, as illustrated in Fig. 4.24.

4.5.3 Perspectives

The experimental results summarized in this section prove the feasibility of using NC-AFM to identify individual molecules. A systematic study on the constant frequency-shift topography of carboxylates with R=CH_3, $C(CH_3)_3$, $C \equiv CH$, and CF_3 has revealed the mechanism behind the high-resolution imaging of the chemically stable molecules. The dispersion force is primarily responsible for the molecule-dependent topography. The permanent dipole moment of the imaged molecule, if it exists, perturbs the topography through the electrostatic coupling with the tip. A tiny calculation containing empirical force fields works when simulating the microscope topography.

These results make us optimistic about analyzing physical and chemical properties of nanoscale supramolecular assemblies constructed on a solid surface. If the accuracy of topographic measurement is developed by one more order of magnitude, which is not an unrealistic target, it may be possible to identify structural isomers, chiral isomers, and conformational isomers of a molecule. Kelvin probe force microscopy (KPFM), an extension of NC-AFM, provides a nanoscale analysis of molecular electronic properties [118, 119]. Force spectroscopy with chemically modified tips seems promising for the detection of a selected chemical force. Operation in a liquid atmosphere [131] is required for the observation of biochemical materials in their natural environment.

References

1. G. Binnig: Atomic force microscope, method for imaging surfaces with atomic resolution, US Patent 4,724,318 (1986)
2. G. Binnig, C.F. Quate, C. Gerber: Atomic force microscope, Phys. Rev. Lett. **56**, 930–933 (1986)
3. G. Binnig, H. Rohrer, C. Gerber, E. Weibel: Surface studies by scanning tunneling microscopy, Phys. Rev. Lett. **49**, 57–61 (1982)
4. G. Binnig, H. Rohrer: The scanning tunneling microscope, Sci. Am. **253**, 50–56 (1985)
5. G. Binnig, H. Rohrer: In touch with atoms, Rev. Mod. Phys. **71**, S320–S330 (1999)
6. C.J. Chen: *Introduction to Scanning Tunneling Microscopy* (Oxford Univ. Press, Oxford 1993)
7. H.-J. Güntherodt, R. Wiesendanger (Eds.): *Scanning Tunneling Microscopy I–III* (Springer, Berlin, Heidelberg 1991)
8. J.A. Stroscio, W.J. Kaiser (Eds.): *Scanning Tunneling Microscopy* (Academic, Boston 1993)
9. R. Wiesendanger: *Scanning Probe Microscopy and Spectroscopy: Methods and Applications* (Cambridge Univ. Press, Cambridge 1994)
10. S. Morita, R. Wiesendanger, E. Meyer (Eds.): *Noncontact Atomic Force Microscopy* (Springer, Berlin, Heidelberg 2002)
11. R. Garcia, R. Perez: Dynamic atomic force microscopy methods, Surf. Sci. Rep. **47**, 197–301 (2002)
12. F.J. Giessibl: Advances in atomic force microscopy, Rev. Mod. Phys. **75**, 949–983 (2003)
13. J. Israelachvili: *Intermolecular and Surface Forces*, 2nd edn. (Academic, London 1991)
14. L. Olsson, N. Lin, V. Yakimov, R. Erlandsson: A method for in situ characterization of tip shape in AC-mode atomic force microscopy using electrostatic interaction, J. Appl. Phys. **84**, 4060–4064 (1998)
15. S. Akamine, R.C. Barrett, C.F. Quate: Improved atomic force microscopy images using cantilevers with sharp tips, Appl. Phys. Lett. **57**, 316–318 (1990)
16. T.R. Albrecht, S. Akamine, T.E. Carver, C.F. Quate: Microfabrication of cantilever styli for the atomic force microscope, J. Vac. Sci. Technol. A **8**, 3386–3396 (1990)
17. M. Tortonese, R.C. Barrett, C. Quate: Atomic resolution with an atomic force microscope using piezoresistive detection, Appl. Phys. Lett. **62**, 834–836 (1993)
18. O. Wolter, T. Bayer, J. Greschner: Micromachined silicon sensors for scanning force microscopy, J. Vac. Sci. Technol. **9**, 1353–1357 (1991)
19. D. Sarid: *Scanning Force Microscopy*, 2nd edn. (Oxford Univ. Press, New York 1994)
20. F.J. Giessibl, B.M. Trafas: Piezoresistive cantilevers utilized for scanning tunneling and scanning force microscope in ultrahigh vacuum, Rev. Sci. Instrum. **65**, 1923–1929 (1994)
21. P. Güthner, U.C. Fischer, K. Dransfeld: Scanning near-field acoustic microscopy, Appl. Phys. B **48**, 89–92 (1989)
22. K. Karrai, R.D. Grober: Piezoelectric tip–sample distance control for near field optical microscopes, Appl. Phys. Lett. **66**, 1842–1844 (1995)
23. F.J. Giessibl: High-speed force sensor for force microscopy and profilometry utilizing a quartz tuning fork, Appl. Phys. Lett. **73**, 3956–3958 (1998)
24. F.J. Giessibl, S. Hembacher, H. Bielefeldt, J. Mannhart: Subatomic features on the silicon (111)-(7×7) surface observed by atomic force microscopy, Science **289**, 422–425 (2000)

25. F. Giessibl, C. Gerber, G. Binnig: A low-temperature atomic force/scanning tunneling microscope for ultrahigh vacuum, J. Vac. Sci. Technol. B **9**, 984–988 (1991)
26. F. Ohnesorge, G. Binnig: True atomic resolution by atomic force microscopy through repulsive and attractive forces, Science **260**, 1451–1456 (1993)
27. F.J. Giessibl, G. Binnig: True atomic resolution on KBr with a low-temperature atomic force microscope in ultrahigh vacuum, Ultramicroscopy **42-44**, 281–286 (1992)
28. S.P. Jarvis, H. Yamada, H. Tokumoto, J.B. Pethica: Direct mechanical measurement of interatomic potentials, Nature **384**, 247–249 (1996)
29. L. Howald, R. Lüthi, E. Meyer, P. Guthner, H.-J. Güntherodt: Scanning force microscopy on the Si(111)7×7 surface reconstruction, Z. Phys. B **93**, 267–268 (1994)
30. L. Howald, R. Lüthi, E. Meyer, H.-J. Güntherodt: Atomic-force microscopy on the Si(111)7×7 surface, Phys. Rev. B **51**, 5484–5487 (1995)
31. Y. Martin, C.C. Williams, H.K. Wickramasinghe: Atomic force microscope—force mapping and profiling on a sub 100 Å scale, J. Appl. Phys. **61**, 4723–4729 (1987)
32. T.R. Albrecht, P. Grutter, H.K. Horne, D. Rugar: Frequency modulation detection using high-Q cantilevers for enhanced force microscope sensitivity, J. Appl. Phys. **69**, 668–673 (1991)
33. F.J. Giessibl: Atomic resolution of the silicon (111)-(7×7) surface by atomic force microscopy, Science **267**, 68–71 (1995)
34. S. Kitamura, M. Iwatsuki: Observation of silicon surfaces using ultrahigh-vacuum noncontact atomic force microscopy, Jpn. J. Appl. Phys. **35**, 668–L671 (1995)
35. R. Erlandsson, L. Olsson, P. Martensson: Inequivalent atoms and imaging mechanisms in AC-mode atomic-force microscopy of Si(111)7×7, Phys. Rev. B **54**, R8309–R8312 (1996)
36. N. Burnham, R.J. Colton: Measuring the nanomechanical and surface forces of materials using an atomic force microscope, J. Vac. Sci. Technol. A **7**, 2906–2913 (1989)
37. D. Tabor, R.H.S. Winterton: Direct measurement of normal and related van der Waals forces, Proc. R. Soc. London A **312**, 435 (1969)
38. F.J. Giessibl: Forces and frequency shifts in atomic resolution dynamic force microscopy, Phys. Rev. B **56**, 16011–16015 (1997)
39. G. Binnig, H. Rohrer, C. Gerber, E. Weibel: 7×7 reconstruction on Si(111) resolved in real space, Phys. Rev. Lett. **50**, 120–123 (1983)
40. H. Goldstein: *Classical Mechanics* (Addison Wesley, Reading 1980)
41. U. Dürig: Interaction sensing in dynamic force microscopy, New J. Phys. **2**, 5.1–5.12 (2000)
42. F.J. Giessibl: A direct method to calculate tip–sample forces from frequency shifts in frequency-modulation atomic force microscopy, Appl. Phys. Lett. **78**, 123–125 (2001)
43. U. Dürig, H.P. Steinauer, N. Blanc: Dynamic force microscopy by means of the phase-controlled oscillator method, J. Appl. Phys. **82**, 3641–3651 (1997)
44. F.J. Giessibl, H. Bielefeldt, S. Hembacher, J. Mannhart: Calculation of the optimal imaging parameters for frequency modulation atomic force microscopy, Appl. Surf. Sci. **140**, 352–357 (1999)
45. M.A. Lantz, H.J. Hug, R. Hoffmann, P.J.A. van Schendel, P. Kappenberger, S. Martin, A. Baratoff, H.-J. Güntherodt: Quantitative measurement of short-range chemical bonding forces, Science **291**, 2580–2583 (2001)
46. F.J. Giessibl, M. Herz, J. Mannhart: Friction traced to the single atom, Proc. Nat. Acad. Sci. USA **99**, 12006–12010 (2002)
47. N. Nakagiri, M. Suzuki, K. Oguchi, H. Sugimura: Site discrimination of adatoms in Si(111)-7×7 by noncontact atomic force microscopy, Surf. Sci. Lett. **373**, L329–L332 (1997)

48. Y. Sugawara, M. Ohta, H. Ueyama, S. Morita: Defect motion on an InP(110) surface observed with noncontact atomic force microscopy, Science **270**, 1646–1648 (1995)
49. T. Uchihashi, Y. Sugawara, T. Tsukamoto, M. Ohta, S. Morita: Role of a covalent bonding interaction in noncontact-mode atomic-force microscopy on Si(111)7 × 7, Phys. Rev. B **56**, 9834–9840 (1997)
50. K. Yokoyama, T. Ochi, A. Yoshimoto, Y. Sugawara, S. Morita: Atomic resolution imaging on Si(100)2 × 1 and Si(100)2 × 1-H surfaces using a non-contact atomic force microscope TS11, Jpn. J. Appl. Phys. **39**, L113–L115 (2000)
51. Y. Sugawara, T. Minobe, S. Orisaka, T. Uchihashi, T. Tsukamoto, S. Morita: Noncontact AFM images measured on Si(111) $\sqrt{3} \times \sqrt{3}$-Ag and Ag(111) surfaces, Surf. Interface Anal. **27**, 456–461 (1999)
52. K. Yokoyama, T. Ochi, Y. Sugawara, S. Morita: Atomically resolved Ag imaging on Si(111) $\sqrt{3} \times \sqrt{3}$-Ag surface with noncontact atomic force microscope, Phys. Rev. Lett. **83**, 5023–5026 (1999)
53. M. Bammerlin, R. Lüthi, E. Meyer, A. Baratoff, J. Lü, M. Guggisberg, Ch. Gerber, L. Howald, H.-J. Güntherodt: True atomic resolution on the surface of an insulator via ultrahigh vacuum dynamic force microscopy, Probe Microscopy **1**, 3–7 (1997)
54. M. Bammerlin, R. Lüthi, E. Meyer, A. Baratoff, J. Lü, M. Guggisberg, C. Loppacher, Ch. Gerber, H.-J. Güntherodt: Dynamic SFM with true atomic resolution on alkali halide surfaces, Appl. Phys. A **66**, S293–S294 (1998)
55. R. Hoffmann, M. A. Lantz, H. J. Hug, P. J. A. van Schendel, P. Kappenberger, S. Martin, A. Baratoff, H.-J. Güntherodt: Atomic resolution imaging and force versus distance measurements on KBr(001) using low temperature scanning force microscopy, Appl. Surf. Sci. **188**, 238–244 (2002)
56. R. Bennewitz, A. S. Foster, L. N. Kantotovich, M. Bammerlin, Ch. Loppacher, S. Schär, M. Guggisberg, E. Meyer, A. L. Shluger: Atomically resolved edges and kinks of NaCl islands on Cu(111): Experiment and theory, Phys. Rev. B **62**, 2074–2084 (2000)
57. A. I. Livshits, A. L. Shluger, A. L. Rohl, A. S. Foster: Model of noncontact scanning force microscopy on ionic surfaces, Phys. Rev. **59**, 2436–2448 (1999)
58. R. Bennewitz, O. Pfeiffer, S. Schär, V. Barwich, E. Meyer, L. N. Kantorovich: Atomic corrugation in nc-AFM of alkali halides, Appl. Surf. Sci. **188**, 232–237 (2002)
59. R. Bennewitz, S. Schär, E. Gnecco, O. Pfeiffer, M. Bammerlin, E. Meyer: Atomic structure of alkali halide surfaces, Appl. Phys. A **78**, 837–841 (2004)
60. M. Gauthier, L. Kantrovich, M. Tsukada: Theory of energy dissipation into surface viblationsed. In: *Noncontact Atomic Force Microscopy*, ed. by S. Morita, R. Wiesendanger, E. Meyer (Springer, Berlin, Heidelberg 2002) pp. 371–394
61. H. J. Hug, A. Baratoff: Measurement of dissipation induced by tip–sample interactions. In: *Noncontact Atomic Force Microscopy*, ed. by S. Morita, R. Wiesendanger, E. Meyer (Springer, Berlin, Heidelberg 2002) pp. 395–431
62. R. Hoffmann, L. N. Kantorovich, A. Baratoff, H. J. Hug, H.-J. Güntherodt: Sublattice identification in scanning force microscopy on alkali halide surfaces, Phys. Rev. B **92**, 146103/1–4 (2004)
63. C. Barth, M. Reichling: Resolving ions and vacancies at step edges on insulating surfaces, Surf. Sci. **470**, L99–L103 (2000)
64. R. Bennewitz, M. Reichling, E. Matthias: Force microscopy of cleaved and electron-irradiated $CaF_2(111)$ surfaces in ultra-high vacuum, Surf. Sci. **387**, 69–77 (1997)
65. M. Reichling, C. Barth: Scanning force imaging of atomic size defects on the $CaF_2(111)$ surface, Phys. Rev. Lett. **83**, 768–771 (1999)
66. M. Reichling, M. Huisinga, S. Gogoll, C. Barth: Degradation of the $CaF_2(111)$ surface by air exposure, Surf. Sci. **439**, 181–190 (1999)

67. A. Klust, T. Ohta, A. A. Bostwick, Q. Yu, F. S. Ohuchi, M. A. Olmstead: Atomically resolved imaging of a CaF bilayer on Si(111): Subsurface atoms and the image contrast in scanning force microscopy, Phys. Rev. B **69**, 035405/1–5 (2004)
68. C. Barth, A. S. Foster, M. Reichling, A. L. Shluger: Contrast formation in atomic resolution scanning force microscopy of $CaF_2(111)$: experiment and theory, J. Phys. Condens. Matter **13**, 2061–2079 (2001)
69. A. S. Foster, C. Barth, A. L. Shulger, M. Reichling: Unambiguous interpretation of atomically resolved force microscopy images of an insulator, Phys. Rev. Lett. **86**, 2373–2376 (2001)
70. A. S. Foster, A. L. Rohl, A. L. Shluger: Imaging problems on insulators: What can be learnt from NC-AFM modeling on CaF_2?, Appl. Phys. A **72**, S31–S34 (2001)
71. F. J. Giessibl, M. Reichling: Investigating atomic details of the $CaF_2(111)$ surface with a qPlus sensor, Nanotechnology **16**, S118–S124 (2005)
72. A. S. Foster, C. Barth, A. L. Shluger, R. M. Nieminen, M. Reichling: Role of tip structure and surface relaxation in atomic resolution dynamic force microscopy: $CaF_2(111)$ as a reference surface, Phys. Rev. B **66**, 235417/1–10 (2002)
73. C. Barth, M. Reichling: Imaging the atomic arrangements on the high-temperature reconstructed α-Al_2O_3 surface, Nature **414**, 54–57 (2001)
74. M. Reichling, C. Barth: Atomically resolution imaging on Fluorides. In: *Noncontact Atomic Force Microscopy*, ed. by S. Morita, R. Wiesendanger, E. Meyer (Springer, Berlin, Heidelberg 2002) pp. 109–123
75. K. Fukui, H. Ohnishi, Y. Iwasawa: Atom-resolved image of the $TiO_2(110)$ surface by noncontact atomic force microscopy, Phys. Rev. Lett. **79**, 4202–4205 (1997)
76. H. Raza, C. L. Pang, S. A. Haycock, G. Thornton: Non-contact atomic force microscopy imaging of $TiO_2(100)$ surfaces, Appl. Surf. Sci. **140**, 271–275 (1999)
77. C. L. Pang, H. Raza, S. A. Haycock, G. Thornton: Imaging reconstructed $TiO_2(100)$ surfaces with non-contact atomic force microscopy, Appl. Surf. Sci. **157**, 223–238 (2000)
78. M. Ashino, T. Uchihashi, K. Yokoyama, Y. Sugawara, S. Morita, M. Ishikawa: STM and atomic-resolution noncontact AFM of an oxygen-deficient $TiO_2(110)$ surface, Phys. Rev. B **61**, 13955–13959 (2000)
79. R. E. Tanner, A. Sasahara, Y. Liang, E. I. Altmann, H. Onishi: Formic acid adsorption on anatase $TiO_2(001)$-(1×4) thin films studied by NC-AFM and STM, J. Phys. Chem. B **106**, 8211–8222 (2002)
80. A. Sasahara, T. C. Droubay, S. A. Chambers, H. Uetsuka, H. Onishi: Topography of anatase TiO_2 film synthesized on $LaAlO_3(001)$, Nanotechnology **16**, S18–S21 (2005)
81. C. L. Pang, S. A. Haycock, H. Raza, P. J. Møller, G. Thornton: Structures of the 4×1 and 1×2 reconstructions of $SnO_2(110)$, Phys. Rev. B **62**, R7775–R7778 (2000)
82. H. Hosoi, K. Sueoka, K. Hayakawa, K. Mukasa: Atomic resolved imaging of cleaved NiO(100) surfaces by NC-AFM, Appl. Surf. Sci. **157**, 218–221 (2000)
83. W. Allers, S. Langkat, R. Wiesendanger: Dynamic low-temperature scanning force microscopy on nickel oxide (001), Appl. Phys. A **72**, S27–S30 (2001)
84. T. Kubo, H. Nozoye: Surface Structure of SrTiO3(100)-($\sqrt{5} \times \sqrt{5}$)–$R26.6°$, Phys. Rev. Lett. **86**, 1801–1804 (2001)
85. K. Fukui, Y. Namai, Y. Iwasawa: Imaging of surface oxygen atoms and their defect structures on $CeO_2(111)$ by noncontact atomic force microscopy, Appl. Surf. Sci. **188**, 252–256 (2002)
86. S. Suzuki, Y. Ohminami, T. Tsutsumi, M. M. Shoaib, M. Ichikawa, K. Asakura: The first observation of an atomic scale noncontact AFM image of $MoO_3(010)$, Chem. Lett. **32**, 1098–1099 (2003)

87. C. Barth, C. R. Henry: Atomic resolution imaging of the (001) surface of UHV cleaved MgO by dynamic scanning force microscopy, Phys. Rev. Lett. **91**, 196102/1–4 (2003)
88. A. S. Foster, A. Y. Gal, J. M. Airaksinen, O. H. Pakarinen, Y. J. Lee, J. D. Gale, A. L. Shluger, R. M. Nieminen: Towards chemical identification in atomic-resolution noncontact AFM imaging with silicon tips, Phys. Rev. B **68**, 195420/1–8 (2003)
89. A. S. Foster, A. Y. Gal, J. D. Gale, Y. J. Lee, R. M. Nieminen, A. L. Shluger: Interaction of silicon dangling bonds with insulating surfaces, Phys. Rev. Lett. **92**, 036101/1–4 (2004)
90. T. Eguchi, Y. Hasegawa: High resolution atomic force microscopic imaging of the Si(111)-(7×7) surface: Contribution of short-range force to the images, Phys. Rev. Lett. **89**, 266105/1–4 (2002)
91. T. Arai, M. Tomitori: A Si nanopillar grown on a Si tip by atomic force microscopy in ultrahigh vacuum for a high-quality scanning probe, Appl. Phys. Lett. **86**, 073110/1–3 (2005)
92. K. Mukasa, H. Hasegawa, Y. Tazuke, K. Sueoka, M. Sasaki, K. Hayakawa: Exchange interaction between magnetic moments of ferromagnetic sample and tip: Possibility of atomic-resolution images of exchange interactions using exchange force microscopy, Jpn. J. Appl. Phys. **33**, 2692–2695 (1994)
93. H. Ness, F. Gautier: Theoretical study of the interaction between a magnetic nanotip and a magnetic surface, Phys. Rev. B **52**, 7352–7362 (1995)
94. K. Nakamura, H. Hasegawa, T. Oguchi, K. Sueoka, K. Hayakawa, K. Mukasa: First-principles calculation of the exchange interaction and the exchange force between magnetic Fe films, Phys. Rev. B **56**, 3218–3221 (1997)
95. A. S. Foster, A. L. Shluger: Spin-contrast in non-contact SFM on oxide surfaces: Theoretical modeling of NiO(001) surface, Surf. Sci. **490**, 211–219 (2001)
96. T. Oguchi, H. Momida: Electronic structure and magnetism of antiferromagnetic oxide surface—First-principles calculations, J. Surf. Sci. Soc. Jpn. **26**, 138–143 (2005)
97. H. Hosoi, M. Kimura, K. Sueoka, K. Hayakawa, K. Mukasa: Non-contact atomic force microscopy of an antiferromagnetic NiO(100) surface using a ferromagnetic tip, Appl. Phys. A **72**, S23–S26 (2001)
98. H. Hölscher, S. M. Langkat, A. Schwarz, R. Wiesendanger: Measurement of three-dimensional force fields with atomic resolution using dynamic force spectroscopy, Appl. Phys. Lett. **81**, 4428–4430 (2002)
99. S. M. Langkat, H. Hölscher, A. Schwarz, R. Wiesendanger: Determination of site specific interaction forces between an iron coated tip and the NiO(001) surface by force field spectroscopy, Surf. Sci. **527**, 12–20 (2003)
100. R. Hoffmann, M. A. Lantz, H. J. Hug, P. J. A. van Schendel, P. Kappenberger, S. Martin, A. Baratoff, H.-J. Güntherodt: Atomic resolution imaging and frequency versus distance measurement on NiO(001) using low-temperature scanning force microscopy, Phys. Rev. B **67**, 085402/1–6 (2003)
101. H. Hosoi, K. Sueoka, K. Hayakawa, K. Mukasa: Atomically resolved imaging of a NiO(001) surface. In: *Noncontact Atomic Force Microscopy*, ed. by S. Morita, R. Wiesendanger, E. Meyer (Springer, Berlin, Heidelberg 2002) pp. 125–134
102. K. Sueoka, A. Subagyo, H. Hosoi, K. Mukasa: Magnetic imaging with scanning force microscopy, Nanotechnology **15**, S691–S698 (2004)
103. H. Hosoi, K. Sueoka, K. Mukasa: Investigations on the topographic asymmetry of non-contact atomic force microscopy images of NiO(001) surface observed with a ferromagnetic tip, Nanotechnology **15**, 505–509 (2004)
104. H. Momida, T. Oguchi: First-principles studies of antiferromagnetic MnO and NiO surfaces, J. Phys. Soc. Jpn. **72**, 588–593 (2003)

105. K. Kobayashi, H. Yamada, T. Horiuchi, K. Matsushige: Structures and electrical properties of fullerene thin films on Si(111)-7×7 surface investigated by noncontact atomic force microscopy, Jpn. J. Appl. Phys. **39**, 3821–3829 (2000)
106. T. Uchihashi, M. Tanigawa, M. Ashino, Y. Sugawara, K. Yokoyama, S. Morita, M. Ishikawa: Identification of B-form DNA in an ultrahigh vacuum by noncontact-mode atomic force microscopy, Langmuir **16**, 1349–1353 (2000)
107. R.M. Overney, E. Meyer, J. Frommer, D. Brodbeck, R. Lüthi, L. Howald, H.-J. Güntherodt, M. Fujihira, H. Takano, Y. Gotoh: Friction measurements on phase-separated thin films with amodified atomic force microscope, Nature **359**, 133–135 (1992)
108. D. Frisbie, L.F. Rozsnyai, A. Noy, M.S. Wrighton, C.M. Lieber: Functional group imaging by chemical force microscopy, Science **265**, 2071–2074 (1994)
109. E. Meyer, L. Howald, R.M. Overney, H. Heinzelmann, J. Frommer, H.-J. Guntherodt, T. Wagner, H. Schier, S. Roth: Molecular-resolution images of Langmuir–Blodgett films using atomic force microscopy, Nature **349**, 398–400 (1992)
110. K. Fukui, H. Onishi, Y. Iwasawa: Imaging of individual formate ions adsorbed on $TiO_2(110)$ surface by non-contact atomic force microscopy, Chem. Phys. Lett. **280**, 296–301 (1997)
111. K. Kobayashi, H. Yamada, T. Horiuchi, K. Matsushige: Investigations of C_{60} molecules deposited on Si(111) by noncontact atomic force microscopy, Appl. Surf. Sci. **140**, 281–286 (1999)
112. Y. Maeda, T. Matsumoto, T. Kawai: Observation of single- and double-strand DNA using non-contact atomic force microscopy, Appl. Surf. Sci. **140**, 400–405 (1999)
113. T. Uchihashi, T. Ishida, M. Komiyama, M. Ashino, Y. Sugawara, W. Mizutani, K. Yokoyama, S. Morita, H. Tokumoto, M. Ishikawa: High-resolution imaging of organic monolayers using noncontact AFM, Appl. Surf. Sci **157**, 244–250 (2000)
114. T. Fukuma, K. Kobayashi, T. Horiuchi, H. Yamada, K. Matsushige: Alkanethiol self-assembled monolayers on Au(111) surfaces investigated by non-contact AFM, Appl. Phys. A **72**, S109–S112 (2001)
115. B. Gotsmann, C. Schmidt, C. Seidel, H. Fuchs: Molecular resolution of an organic monolayer by dynamic AFM, Euro. Phys. J. B **4**, 267–268 (1998)
116. Ch. Loppacher, M. Bammerlin, M. Guggisberg, E. Meyer, H.-J. Güntherodt, R. Lüthi, R. Schlittler, J.K. Gimzewski: Forces with submolecular resolution between the probing tip and Cu-TBPP molecules on Cu(100) observed with a combined AFM/STM, Appl. Phys. A **72**, S105–S108 (2001)
117. L.M. Eng, M. Bammerlin, Ch. Loppacher, M. Guggisberg, R. Bennewitz, R. Lüthi, E. Meyer, H.-J. Güntherodt: Surface morphology, chemical contrast, and ferroelectric domains in TGS bulk single crystals differentiated with UHV non-contact force microscopy, Appl. Surf. Sci. **140**, 253–258 (1999)
118. S. Kitamura, K. Suzuki, M. Iwatsuki: High resolution imaging of contact potential difference using a novel ultrahigh vacuum non-contact atomic force microscope technique, Appl. Surf. Sci. **140**, 265–270 (1999)
119. H. Yamada, T. Fukuma, K. Umeda, K. Kobayashi, K. Matsushige: Local structures and electrical properties of organic molecular films investigated by non-contact atomic force microscopy, Appl. Surf. Sci **188**, 391–398 (2000)
120. K. Fukui, Y. Iwasawa: Fluctuation of acetate ions in the (2×1)-acetate overlayer on $TiO_2(110)$-(1×1) observed by noncontact atomic force microscopy, Surf. Sci. **464**, L719–L726 (2000)
121. A. Sasahara, H. Uetsuka, H. Onishi: Singlemolecule analysis by non-contact atomic force microscopy, J. Phys. Chem. B **105**, 1–4 (2001)

122. A. Sasahara, H. Uetsuka, H. Onishi: NC-AFM topography of HCOO and CH_3COO molecules co-adsorbed on $TiO_2(110)$, Appl. Phys. A **72**, S101–S103 (2001)
123. A. Sasahara, H. Uetsuka, H. Onishi: Image topography of alkyl-substituted carboxylates observed by noncontact atomic force microscopy, Surf. Sci. **481**, L437–L442 (2001)
124. A. Sasahara, H. Uetsuka, H. Onishi: Noncontact atomic force microscope topography dependent on permanent dipole of individual molecules, Phys. Rev. B **64**, 121406(R) (2001)
125. A. Sasahara, H. Uetsuka, T. Ishibashi, H. Onishi: A needle-like organic molecule imaged by noncontact atomic force microscopy, Appl. Surf. Sci. **188**, 265–271 (2002)
126. H. Onishi, A. Sasahara, H. Uetsuka, T. Ishibashi: Molecule-dependent topography determined by noncontact atomic force microscopy: Carboxylates on $TiO_2(110)$, Appl. Surf. Sci. **188**, 257–264 (2002)
127. S. Thevuthasan, G.S. Herman, Y.J. Kim, S.A. Chambers, C.H.F. Peden, Z. Wang, R.X. Ynzunza, E.D. Tober, J. Morais, C.S. Fadley: The structure of formate on $TiO_2(110)$ by scanned-energy and scanned-angle photoelectron diffraction, Surf. Sci. **401**, 261–268 (1998)
128. H. Onishi: Carboxylates adsorbed on $TiO_2(110)$. In: *Chemistry of Nano-molecular Systems*, ed. by T. Nakamura (Springer, Berlin, Heidelberg 2002) pp. 75–89
129. H. Uetsuka, A. Sasahara, A. Yamakata, H. Onishi: Microscopic identification of a bimolecular reaction intermediate, J. Phys. Chem. B **106**, 11549–11552 (2002)
130. D.R. Lide: *Handbook of Chemistry and Physics*, 81st edn. (CRC, Boca Raton 2000)
131. K. Kobayashi, H. Yamada, K. Matsushige: Dynamic force microscopy using FM detection in various environments, Appl. Surf. Sci. **188**, 430–434 (2002)

5

Low-Temperature Scanning Probe Microscopy

Markus Morgenstern, Alexander Schwarz, and Udo D. Schwarz

Summary. This chapter is dedicated to scanning probe microscopy (SPM) operated at cryogenic temperatures, where the more fundamental aspects of phenomena important in the field of nanotechnology can be investigated with high sensitivity under well defined conditions. In general, scanning probe techniques allow the measurement of physical properties down to the nanometer scale. Some techniques, such as the scanning tunneling microscope and the scanning force microscope even go down to the atomic scale. Various properties are accessible. Most importantly, one can image the arrangement of atoms on conducting surfaces by scanning tunneling microscopy and on insulating substrates by scanning force microscopy. But the arrangement of electrons (scanning tunneling spectroscopy), the force interaction between different atoms (scanning force spectroscopy), magnetic domains (magnetic force microscopy), the local capacitance (scanning capacitance microscopy), the local temperature (scanning thermo microscopy), and local light-induced excitations (scanning near-field microscopy) can also be measured with high spatial resolution. In addition, some techniques even allow the manipulation of atomic configurations.

Probably the most important advantage of the low-temperature operation of scanning probe techniques is that they lead to a significantly better signal-to-noise ratio than measuring at room temperature. This is why many researchers work below 100 K. However, there are also physical reasons to use low-temperature equipment. For example, the manipulation of atoms or scanning tunneling spectroscopy with high energy resolution can only be realized at low temperatures. Moreover, some physical effects such as superconductivity or the Kondo effect are restricted to low temperatures. Here, we describe the design criteria of low-temperature scanning probe equipment and summarize some of the most spectacular results achieved since the invention of the method about 20 years ago. We first focus on the scanning tunneling microscope, giving examples of atomic manipulation and the analysis of electronic properties in different material arrangements. Afterwards, we describe results obtained by scanning force microscopy, showing atomic-scale imaging on insulators, as well as force spectroscopy analysis. Finally, the magnetic force microscope, which images domain patterns in ferromagnets and vortex patterns in superconductors, is discussed. Although this list is far from complete, we feel that it gives an adequate impression of the fascinating possibilities of low-temperature scanning probe instruments.

In this chapter low temperatures are defined as lower than about 100 K and are normally achieved by cooling with liquid nitrogen or liquid helium. Applications in which SPMs are operated close to 0 °C are not covered in this chapter.

5.1 Introduction

More than two decades ago, the first design of an experimental setup was presented where a sharp tip was systematically scanned over a sample surface in order to obtain local information on the tip-sample interaction down to the atomic scale. This original instrument used the tunneling current between a conducting tip and a conducting sample as a feedback signal and was named *scanning tunneling microscope* accordingly [1]. Soon after this historic breakthrough, it became widely recognized that virtually any type of tip-sample interaction can be used to obtain local information on the sample by applying the same general principle, provided that the selected interaction was reasonably short-ranged. Thus, a whole variety of new methods has been introduced, which are denoted collectively as *scanning probe methods*. An overview is given by *Wiesendanger* [2].

The various methods, especially the above mentioned scanning tunneling microscopy (STM) and scanning force microscopy (SFM) – which is often further classified into subdisciplines such as the topography-reflecting atomic force microscopy (AFM), the magnetic force microscopy (MFM), or the electrostatic force microscopy (EFM) – have been established as standard methods for surface characterization on the nanometer scale. The reason is that they feature extremely high resolution (often down to the atomic scale for STM and AFM), despite a principally simple, compact, and comparatively inexpensive design.

A side effect of the simple working principle and the compact design of many scanning probe microscopes (SPMs) is that they can be adapted to different environments such as air, all kinds of gaseous atmospheres, liquids, or vacuum with reasonable effort. Another advantage is their ability to work within a wide temperature range. A microscope operation at higher temperatures is chosen to study surface diffusion, surface reactivity, surface reconstructions that only manifest at elevated temperatures, high-temperature phase transitions, or to simulate conditions as they occur, e.g., in engines, catalytic converters, or reactors. Ultimately, the upper limit for the operation of an SPM is determined by the stability of the sample, but thermal drift, which limits the ability to move the tip in a controlled manner over the sample, as well as the depolarization temperature of the piezoelectric positioning elements might further restrict successful measurements.

On the other hand, low-temperature (LT) application of SPMs is much more widespread than operation at high temperatures. Essentially five reasons make researchers adapt their experimental setups to low-temperature compatibility. These are: (1) the reduced thermal drift, (2) lower noise levels, (3) enhanced stability of tip and sample, (4) the reduction in piezo hysteresis/creep, and (5) probably the most obvious, the fact that many physical effects are restricted to low temperature. Reasons (1) to (4) only apply unconditionally if the whole microscope body is kept at low temperature (typically in or attached to a bath cryostat, see Sect. 5.3). Setups in which only the sample is cooled may show considerably less favorable operating characteristics. As a result of (1) to (4), ultrahigh resolution and long-term stability can be achieved on a level that significantly exceeds what can be accomplished at room temperature even under the most favorable circumstances. Typical examples for (5) are superconductivity [3] and the Kondo effect [4].

5.2 Microscope Operation at Low Temperatures

Nevertheless, before we devote ourselves to a short overview of experimental LT-SPM work, we will take a closer look at the specifics of microscope operation at low temperatures, including a discussion of the corresponding instrumentation.

5.2.1 Drift

Thermal drift originates from thermally activated movements of the individual atoms, which are reflected by the thermal expansion coefficient. At room temperature, typical values for solids are on the order of $(1-50) \times 10^{-6}\,\mathrm{K}^{-1}$. If the temperature could be kept precisely constant, any thermal drift would vanish, regardless of the absolute temperature of the system. The close coupling of the microscope to a large temperature bath that maintains a constant temperature ensures a significant reduction in thermal drift and allows for distortion-free long-term measurements. Microscopes that are efficiently attached to sufficiently large bath cryostats, therefore, show a one- to two-order-of-magnitude increase in thermal stability compared with non-stabilized set-ups operated at room temperature.

A second effect also helps suppress thermally induced drift of the probing tip relative to a specific location on the sample surface. The thermal expansion coefficients at liquid-helium temperatures are two or more orders of magnitude smaller than at room temperature. Consequently, the thermal drift during low-temperature operation decreases accordingly.

For some specific scanning probe methods, there may be additional ways in which a change in temperature can affect the quality of the data. In *frequency-modulation SFM* (FM-SFM), for example, the measurement principle relies on the accurate determination of the eigenfrequency of the cantilever, which is determined by its spring constant and its effective mass. However, the spring constant changes with temperature due to both thermal expansion (i.e., the resulting change in the cantilever dimensions) and the variation of the Young's modulus with temperature. Assuming drift rates of about 2 mK/min, as is typical for room-temperature measurements, this effect might have a significant influence on the obtained data.

5.2.2 Noise

The theoretically achievable resolution in SPM often increases with decreasing temperature due to a decrease in thermally induced noise. An example is the thermal noise in SFM, which is proportional to the square root of the temperature [5, 6]. Lowering the temperature from $T = 300\,\mathrm{K}$ to $T = 10\,\mathrm{K}$ thus results in a reduction of the thermal frequency noise by more than a factor of five. Graphite, e.g., has been imaged with atomic resolution only at low temperatures due to its extremely low corrugation, which was below the room-temperature noise level [7, 8].

Another, even more striking, example is the spectroscopic resolution in *scanning tunneling spectroscopy* (STS). This depends linearly on the temperature [2] and is consequently reduced even more at LT than the thermal noise in AFM. This provides the opportunity to study structures or physical effects not accessible at room temperature such as spin and Landau levels in semiconductors [9].

Finally, it might be worth mentioning that the enhanced stiffness of most materials at low temperatures (increased Young's modulus) leads to a reduced coupling to external noise. Even though this effect is considered small [6], it should not be ignored.

5.2.3 Stability

There are two major stability issues that considerably improve at low temperature. First, low temperatures close to the temperature of liquid helium inhibit most of the thermally activated diffusion processes. As a consequence, the sample surfaces show a significantly increased long-term stability, since defect motion or adatom diffusion is massively suppressed. Most strikingly, even single xenon atoms deposited on suitable substrates can be successfully imaged [10, 11], or even manipulated [12]. In the same way, the low temperatures also stabilize the atomic configuration at the tip end by preventing sudden jumps of the most loosely bound, foremost tip atom(s). Secondly, the large cryostat that usually surrounds the microscope acts as an effective cryo-pump. Thus samples can be kept clean for several weeks, which is a multiple of the corresponding time at room temperature (about 3–4 h).

5.2.4 Piezo Relaxation and Hysteresis

The last important benefit from low-temperature operation of SPMs is that artifacts from the response of the piezoelectric scanners are substantially reduced. After applying a voltage ramp to one electrode of a piezoelectric scanner, its immediate initial deflection, l_0, is followed by a much slower relaxation, Δl, with a logarithmic time dependence. This effect, known as piezo relaxation or *creep*, diminishes substantially at low temperatures, typically by a factor of ten or more. As a consequence, piezo nonlinearities and piezo hysteresis decrease accordingly. Additional information is given by *Hug* et al. [13].

5.3 Instrumentation

The two main design criteria for all vacuum-based scanning probe microscope systems are: (1) to provide an efficient decoupling of the microscope from the vacuum system and other sources of external vibrations, and (2) to avoid most internal noise sources through the high mechanical rigidity of the microscope body itself. In vacuum systems designed for low-temperature applications, a significant degree of complexity is added, since, on the one hand, close thermal contact of the SPM and cryogen is necessary to ensure the (approximately) drift-free conditions described above, while, on the other hand, good vibration isolation (both from the outside world, as well as from the boiling or flowing cryogen) has to be maintained.

Plenty of microscope designs have been presented in the last 10–15 years, predominantly in the field of STM. Due to the variety of the different approaches, we will, somewhat arbitrarily, give two examples at different levels of complexity that might serve as illustrative model designs.

5.3.1 A Simple Design for a Variable-Temperature STM

A simple design for a variable-temperature STM system is presented in Fig. 5.1 (similar systems are also offered by Omicron (Germany) or Jeol (Japan)). It should give an impression of what the minimum requirements are, if samples are to be investigated successfully at low temperatures. It features a single ultrahigh vacuum (UHV) chamber that houses the microscope in its center. The general idea to keep the set-up simple is that only the sample is cooled, by means of a flow cryostat that ends in the small liquid-nitrogen (LN) reservoir. This reservoir is connected to the sample holder with copper braids. The role of the copper braids is to attach the LN reservoir thermally to the sample located on the sample holder in an effective manner, while vibrations due to the flow of the cryogen should be blocked as much as possible. In this way, a sample temperature of about 100 K is reached. Alternatively, with liquid-helium operation, a base temperature of below 30 K can be achieved, while a heater that is integrated into the sample stage enables high-temperature operation up to 1000 K.

A typical experiment would run as follows. First, the sample is brought into the system by placing it in the so-called *load-lock*. This small part of the chamber can be separated from the rest of the system by a valve, so that the main part of the system can remain under vacuum at all times (i.e., even if the load-lock is opened to introduce the sample). After vacuum is reestablished, the sample is transferred to the main chamber using the transfer arm. A linear-motion feedthrough enables the storage of sample holders or, alternatively, specialized holders that carry replacement tips for the STM. Extending the transfer arm further, the sample can be placed on the sample stage and subsequently cooled down to the desired temperature. The scan head, which carries the STM tip, is then lowered with the scan-head manipulator onto the sample holder (see Fig. 5.2). The special design of the scan head (see [14] for details) allows not only a flexible positioning of the tip on any desired location on the sample surface, but also compensates to a certain degree for the thermal drift that inevitably occurs in such a design due to temperature gradients.

In fact, thermal drift is often much more prominent in LT-SPM designs, where only the sample is cooled, than in room-temperature designs. Therefore, to benefit fully from the high stability conditions described in the introduction, it is mandatory

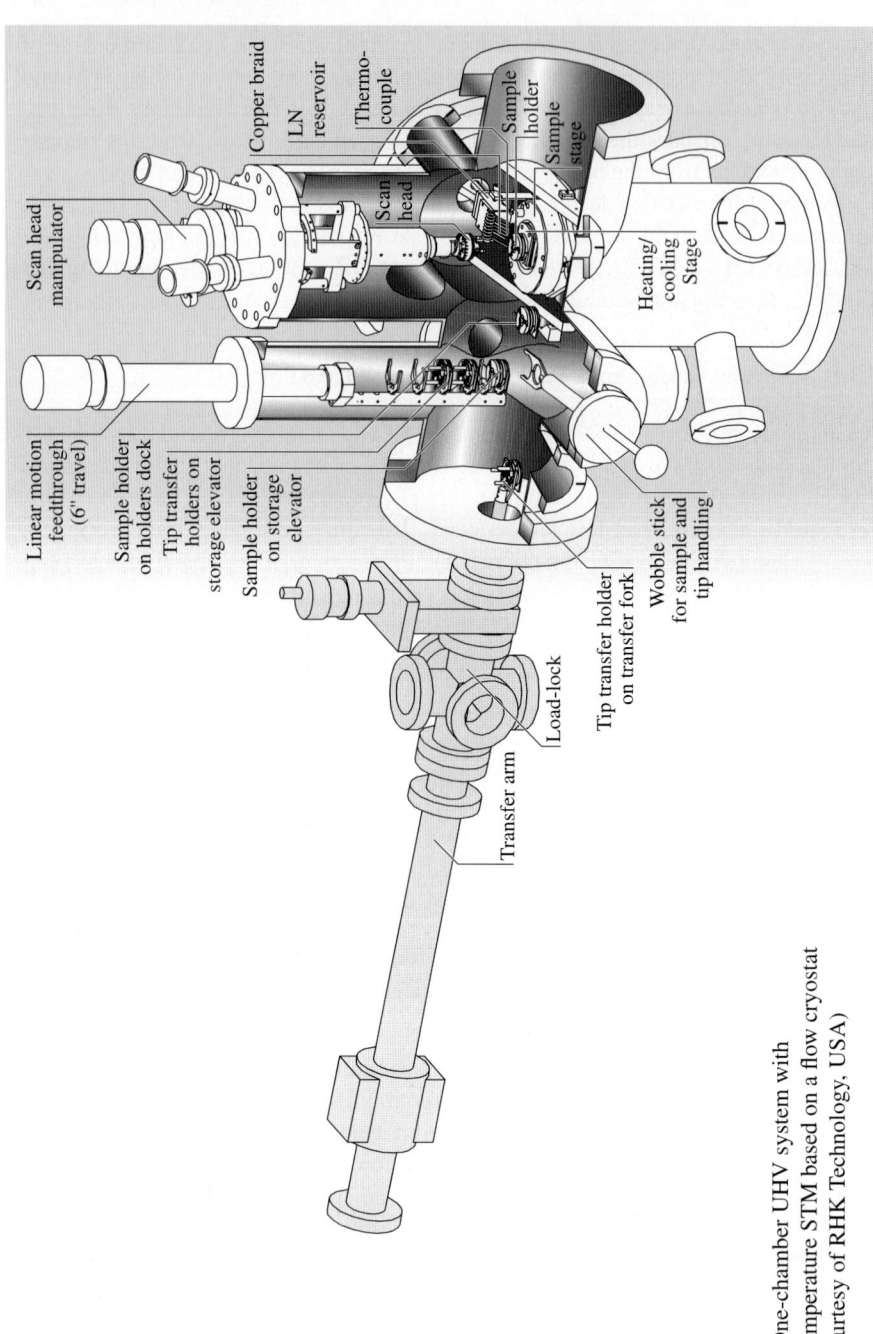

Fig. 5.1. One-chamber UHV system with variable-temperature STM based on a flow cryostat design (courtesy of RHK Technology, USA)

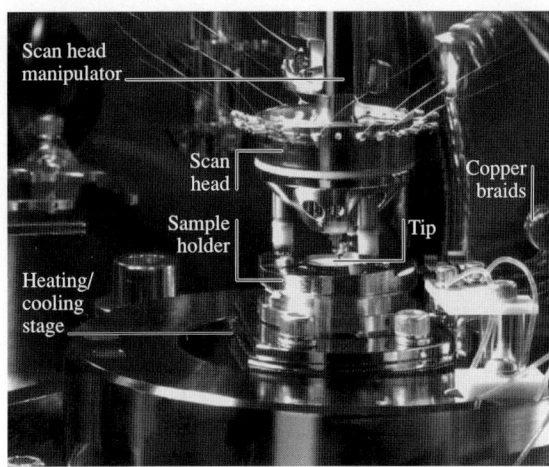

Fig. 5.2. Photograph of the STM located inside the system sketched in Fig. 5.1. After the scan head has been lowered onto the sample holder, it is fully decoupled from the scan head manipulator and can be moved laterally using the three piezo legs on which it stands (courtesy of RHK Technology, USA)

to keep the whole microscope at the exact same temperature. This is mostly realized by using bath cryostats, which add a certain degree of complexity.

5.3.2 A Low Temperature SFM Based on a Bath Cryostat

As an example of an LT-SPM set-up based on a bath cryostat, let us take a closer look at the LT-SFM system sketched in Fig. 5.3, which has been used to acquire the images on graphite, xenon, NiO, and InAs presented in Sect. 5.5. The force microscope is built into a UHV system that comprises three vacuum chambers: one for cantilever and sample preparation, which also serves as a transfer chamber, one for analysis purposes, and a main chamber that houses the microscope. A specially designed vertical transfer mechanism based on a double chain allows the lowering of the microscope into a UHV-compatible bath cryostat attached underneath the main chamber. To damp the system, it is mounted on a table carried by pneumatic damping legs, which, in turn, stand on a separate foundation to decouple it from building vibrations. The cryostat and dewar are separated from the rest of the UHV system by a bellow. In addition, the dewar is surrounded by sand for acoustic isolation.

In this design, tip and sample are exchanged at room temperature in the main chamber. After the transfer into the cryostat, the SFM can be cooled by either liquid nitrogen or liquid helium, reaching temperatures down to 10 K. An all-fiber interferometer as the detection mechanism for the cantilever deflection ensures high resolution, while simultaneously allowing the construction of a comparatively small, rigid, and symmetric microscope.

Figure 5.4 highlights the layout of the SFM body itself. Along with the careful choice of materials, the symmetric design eliminates most of the problems with drift inside the microscope encountered when cooling or warming it up. The microscope body has an overall cylindrical shape with a height of 13 cm and a diameter of

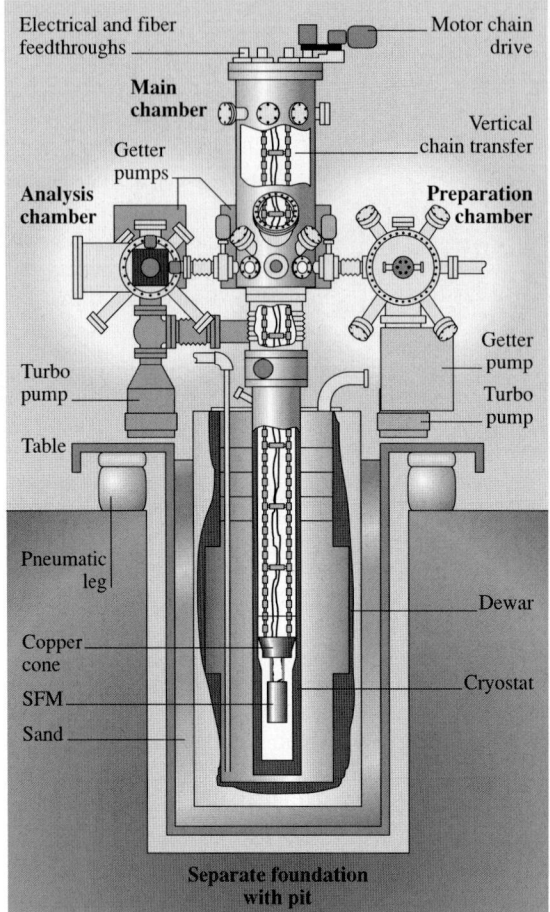

Fig. 5.3. Three-chamber UHV and bath cryostat system for scanning force microscopy, front view

6 cm and exact mirror symmetry along the cantilever axis. The main body is made of a single block of macor, a machinable glass ceramic, which ensures a rigid and stable design. For most of the metallic parts titanium was used, which has a temperature coefficient similar to macor. The controlled but stable accomplishment of movements, such as coarse approach and lateral positioning in other microscope designs, is a difficult task at low temperatures. The present design uses a special type of piezo motor that moves a sapphire prism (see the *fiber approach* and the *sample approach* labels in Fig. 5.3); it is described in detail in [15]. More information regarding this design is given in [16].

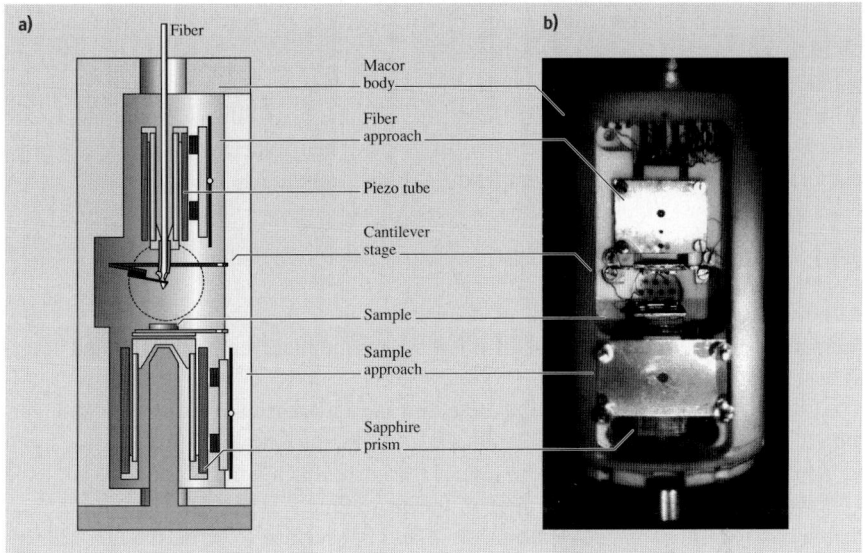

Fig. 5.4. The scanning force microscope incorporated into the system presented in Fig. 5.3. (**a**) Section along plane of symmetry. (**b**) Photo from the front

5.4 Scanning Tunneling Microscopy and Spectroscopy

In this section, we review some of the most important results achieved by LT-STM. After summarizing the results, placing emphasis on the necessity for LT equipment, we turn to the details of the different experiments and the physical meaning of the results obtained.

As described in Sect. 5.2, the LT equipment has basically three advantages for scanning tunneling microscopy (STM) and spectroscopy (STS). First, the instruments are much more stable with respect to thermal drift and coupling to external noise, allowing the establishment of new functionalities of the instrument. In particular, the LT-STM has been used to move atoms on a surface [12], cut molecules into pieces [17], reform bonds [18], and, consequently, establish new structures on the nanometer scale. Also, the detection of light resulting from tunneling into a particular atom [19] and the visualization of thermally induced atomic movements [20] partly require LT instrumentation.

Second, the spectroscopic resolution in STS depends linearly on temperature and is, therefore, considerably reduced at LT. This provides the opportunity to study physical effects unaccessible at room temperature. Obvious examples are the resolution of spin and Landau levels in semiconductors [9], or the investigation of lifetime broadening effects of particular electronic states on the nanometer scale [21]. More spectacularly, electronic wave functions have been imaged for the first time in real space using an LT-STM [22], and vibrational levels giving rise to additional inelastic tunneling have been detected [23] and localized within particular molecules [24].

Third, and most obviously, many physical effects, in particular, effects guided by electronic correlations, are restricted to low temperature. Typical examples are superconductivity [3], the Kondo effect [4], and many of the electron phases found in semiconductors [25]. Here, LT-STM provides the possibility to study electronic effects on a local scale, and intensive work has been done in this field, the most elaborate with respect to high-temperature superconductivity [26].

5.4.1 Atomic Manipulation

Although manipulation of surfaces on the atomic scale can be achieved at room temperature [27], only the use of LT-STM allows the placement of individual atoms at desired atomic positions [28].

The usual technique to manipulate atoms is to increase the current above a certain atom, which reduces the tip–atom distance, then to move the tip with the atom to a desired position, and finally to reduce the current again in order to decouple the atom and tip. The first demonstration of this technique was performed by *Eigler* and *Schweizer* [12], who used Xe atoms on a Ni(110) surface to write the three letters "IBM" (their employer) on the atomic scale (Fig. 5.5a). Nowadays, many laboratories are able to move different kinds of atoms and molecules on different surfaces with high precision. An example featuring CO molecules on Cu(110) is shown in Fig. 5.5b–g. Basic modes of controlled motion, pushing, pulling, and sliding of the molecules, have been established that depend on the tunneling current, i.e., the distance and the particular molecule–substrate combination [29]. It is believed that the electric field between the tip and molecule is the strongest force moving the molecules, but other mechanisms such as electromigration caused by the high current density [28] or modifications of the surface potential due to the presence of the tip [30] have been put forth as important for some of the manipulation modes.

Meanwhile, other types of manipulation on the atomic scale have been developed. Some of them require inelastic tunneling into vibrational or rotational modes of the molecules or atoms. They lead to controlled desorption [31], diffusion [32], pick-up of molecules by the tip [18], or rotation of individual entities [33, 34]. Also, dissociation of molecules by voltage pulses [17], conformational changes induced by dramatic change of the tip–molecule distance [35], and association of pieces into larger molecules by reducing their lateral distance [18] have been shown. Figure 5.5h–m shows the production of biphenyl from two iodobenzene molecules [36]. The iodine is abstracted by voltage pulses (Fig. 5.5i,j), then the iodine is moved to the terrace by the pulling mode (Fig. 5.5k,l), and finally the two phenyl parts are slid along the step edge until they are close enough to react (Fig. 5.5m). The chemical identification of the components is not deduced straightforwardly from STM images and partly requires detailed calculations of their apparent shape.

Low temperatures are not always required in these experiments, but they increase reproducibility because of the higher stability of the instrument, as discussed in Sect. 5.2. Moreover, rotation or diffusion of entities could be excited at higher temperatures, making the intentionally produced configurations unstable.

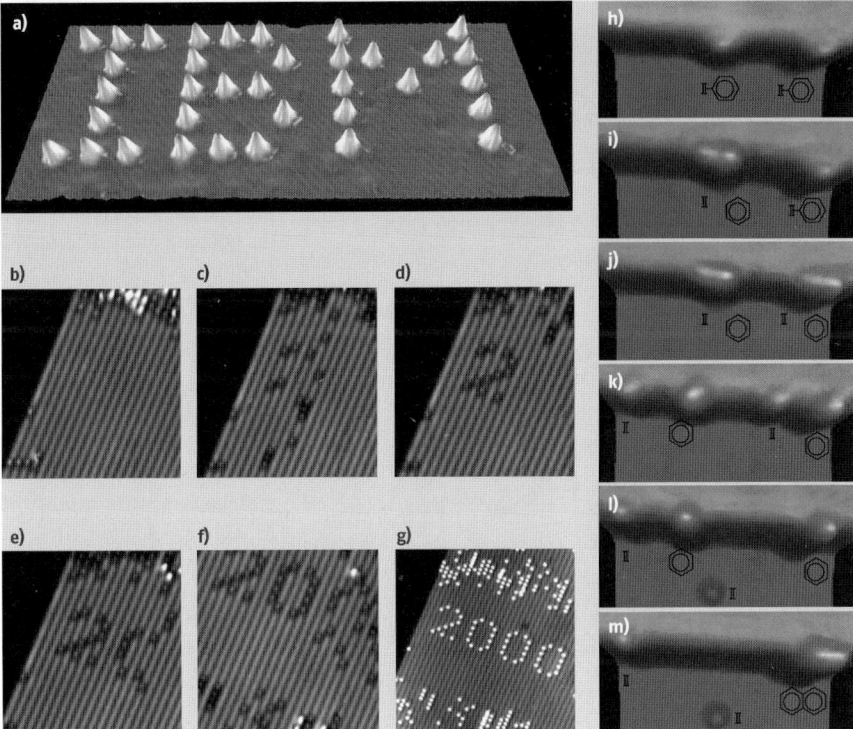

Fig. 5.5. (a) STM image of single Xe atoms positioned on a Ni(110) surface in order to realize the letters IBM on the atomic scale (courtesy of Eigler, IBM); (**b**)–(**f**) STM images recorded after different positioning processes of CO molecules on a Cu(110) surface; (**g**) final artwork greeting the new millennium on the atomic scale ((**b**)–(**g**) courtesy of Meyer, Berlin). (**h**)–(**m**) Synthesis of biphenyl from two iodobenzene molecules on Cu(111): First, iodine is abstracted from both molecules (**i**),(**j**), then the iodine between the two phenyl groups is removed from the step (**k**), and finally one of the phenyls is slid along the Cu-step (**l**) until it reacts with the other phenyl (**m**); the line drawings symbolize the actual status of the molecules ((**h**)–(**m**) courtesy of Hla and Rieder, Ohio)

5.4.2 Imaging Atomic Motion

Since individual manipulation processes last seconds to minutes, they probably cannot be used to manufacture large and repetitive structures. A possibility to construct such structures is self-assembled growth. This partly relies on the temperature dependence of different diffusion processes on the surface. A detailed knowledge of the diffusion parameters is required, which can be deduced from sequences of STM images measured at temperatures close to the onset of the process of interest [37]. Since many diffusion processes have their onset at LT, LT are partly required [20]. Consecutive images of so-called hexa-tert-butyl-decacyclene (HtBDC) molecules on Cu(110) recorded at $T = 194$ K are shown in Fig. 5.6a–c [38]. As indicated by the

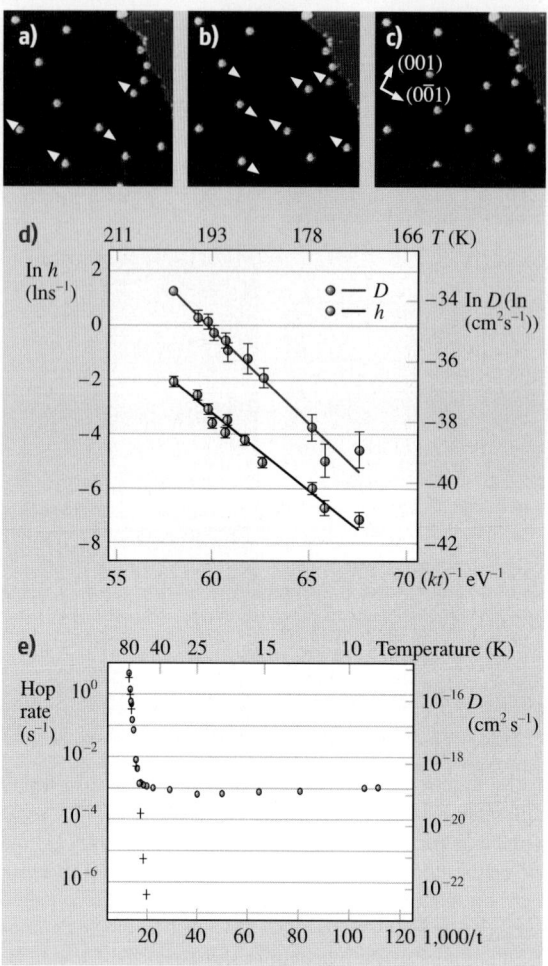

Fig. 5.6. (a)–(c) Consecutive STM images of hexa-tert-butyl decacyclene molecules on Cu(110) imaged at $T = 194$ K; *arrows* indicate the direction of motion of the molecules between two images. (d) Arrhenius plot of the hopping rate h determined from images like (a)–(c) as a function of inverse temperature (*grey symbols*); the *brown symbols* show the corresponding diffusion constant D; lines are fit results revealing an energy barrier of 570 meV for molecular diffusion ((a)–(d) courtesy of M. Schuhnack and F. Besenbacher, Aarhus). (e) Arrhenius plot for D (*crosses*) and H (*circles*) on Cu(001). The constant hopping rate of H below 65 K indicates a non-thermal diffusion process, probably tunneling (courtesy of Ho, Irvine)

arrows, the position of the molecules changes with time, implying diffusion. Diffusion parameters are obtained from Arrhenius plots of the determined hopping rate h, as shown in Fig. 5.6d. Of course, one must make sure that the diffusion process is not influenced by the presence of the tip, since it is known from manipulation experiments that the presence of the tip can move a molecule. However, particularly at low tunneling voltages, these conditions can be fulfilled.

Besides the determination of diffusion parameters, studies of the diffusion of individual molecules showed the importance of mutual interactions in diffusion, which can lead to concerted motion of several molecules [20], or, very interestingly, the influence of quantum tunneling [39]. The latter is deduced from the Arrhenius plot of hopping rates of H and D on Cu(001), as shown in Fig. 5.6e. The hopping rate of H levels off at about 65 K, while the hopping rate of the heavier D atom goes down to nearly zero, as expected from thermally induced hopping.

Other diffusion processes such as the movement of surface vacancies [40] or of bulk interstitials close to the surface [41], and the Brownian motion of vacancy islands [42] have also been displayed.

5.4.3 Detecting Light from Single Atoms and Molecules

It had already been realized in 1988 that STM experiments are accompanied by light emission [43]. The fact that molecular resolution in the light intensity was achieved at LT (Fig. 5.7a,b) [19] raised the hope of performing quasi-optical experiments on the molecular scale. Meanwhile, it is clear that the basic emission process observed on metals is the decay of a local plasmon induced in the area around the tip by inelastic tunneling processes [44, 45]. Thus, the molecular resolution is basically a change in the plasmon environment, largely given by the increased height of the tip with respect to the surface above the molecule [46]. However, the electron can, in principle, also decay via single-particle excitations. Indeed, signatures of single-particle levels are observed. Figure 5.7c shows light spectra measured at different tunneling voltages V above a nearly complete Na monolayer on Cu(111) [47]. The plasmon-mode peak energy (arrow) is found, as usual, to be proportional to V, but an additional peak that does not move with V appears at 1.6 eV (p). Plotting the light intensity as a function of photon energy and V (Fig. 5.7d) clearly shows that this additional peak is fixed in photon energy and corresponds to the separation of quantum-well levels of the Na ($E_n - E_m$).

Light has also been detected from semiconductors [48], including heterostructures [49]. There, the light is mostly caused by single-particle relaxation of the injected electrons, allowing a very local source of electron injection.

5.4.4 High-Resolution Spectroscopy

One of the most important modes of LT-STM is STS. It detects the differential conductivity dI/dV as a function of the applied voltage V and the position (x,y). The dI/dV signal is basically proportional to the local density of states (LDOS) of the sample, the sum over squared single-particle wave functions Ψ_i [2]

$$\frac{dI}{dV}(V,x,y) \propto \text{LDOS}(E,x,y) = \sum_{\Delta E} |\Psi_i(E,x,y)|^2 , \qquad (5.1)$$

where ΔE is the energy resolution of the experiment. In simple terms, each state corresponds to a tunneling channel if it is located between the Fermi levels (E_F) of the tip and the sample. Thus, all states located in this energy interval contribute to I, while $dI/dV(V)$ detects only the states at the energy E corresponding to V. The local intensity of each channel depends further on the LDOS of the state at the corresponding surface position and its decay length into vacuum. For s-like tip states, *Tersoff* and *Hamann* have shown that it is simply proportional to the LDOS at the position of the tip [50]. Therefore, as long as the decay length is spatially constant, one measures the LDOS at the surface (5.1). Note that the contributing

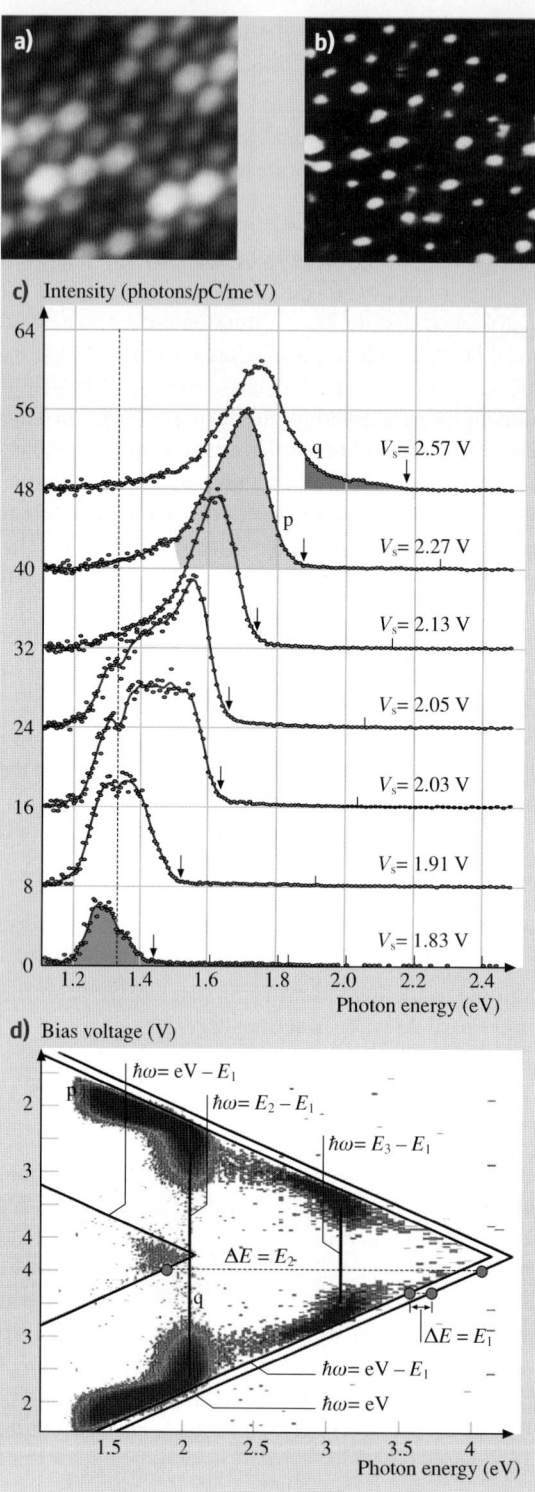

Fig. 5.7. (a) STM image of C_{60} molecules on Au(110) imaged at $T = 50$ K. (b) STM-induced photon intensity map of the same area; all photons from 1.5 eV to 2.8 eV contribute to the image, tunneling voltage $V = -2.8$ V ((a),(b) courtesy of Berndt, Kiel). (c) STM-induced photon spectrum measured on 0.6 monolayer of Na on Cu(111) at different tunneling voltages as indicated. Besides the shifting plasmon mode marked by an *arrow*, an energetically constant part named p is recognizable. (d) Greyscale map of photon intensity as a function of tunnelling voltage and photon energy measured on 2.0 monolayer Na on Cu(111). The energetically constant photons are identified with intersubband transitions of the Na quantum well, as marked by $E_n - E_m$ ((c),(d) courtesy of Hoffmann, Hamburg)

states are not only surface states, but also bulk states. However, surface states usually dominate if present. *Chen* has shown that higher orbital tip states lead to the so-called derivation rule [51]: p_z-type tip states detect $d(LDOS)/dz$, d_{z^2}-states detect $d^2(LDOS)/dz^2$, and so on. As long as the decay into vacuum is exponential and spatially constant, this leads only to an additional factor in dI/dV. Thus, it is still the LDOS that is measured (5.1). The requirement of a spatially constant decay is usually fulfilled on larger length scales, but not on the atomic scale [51]. There, states located close to the atoms show a stronger decay into vacuum than the less localized states in the interstitial region. This effect can lead to corrugations that are larger than the real LDOS corrugations [52].

The voltage dependence of dI/dV is sensitive to a changing decay length with V, which increases with V. Additionally, $dI/dV(V)$-curves might be influenced by possible structures in the DOS of the tip, which also contributes to the number of tunneling channels [53]. However, these structures can usually be identified, and only tips free of characteristic DOS structures are used for quantitative experiments.

Importantly, the energy resolution ΔE is largely determined by temperature. It is defined as the smallest energy distance of two δ-peaks in the LDOS that can still be resolved as two individual peaks in $dI/dV(V)$-curves and is $\Delta E = 3.3\,kT$ [2]. The temperature dependence is nicely demonstrated in Fig. 5.8, where the tunneling gap of the superconductor Nb is measured at different temperatures [54]. The peaks at the rim of the gap get wider at temperatures well below the critical temperature of the superconductor ($T_c = 9.2$ K).

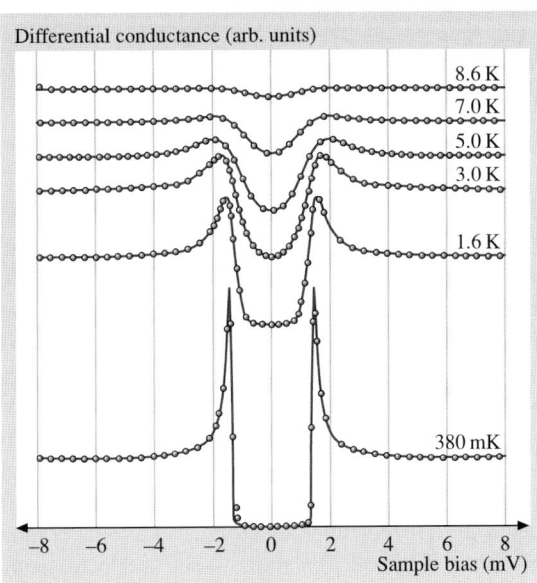

Fig. 5.8. Differential conductivity curve $dI/dV(V)$ measured on a Au surface by a Nb tip (*circles*). Different temperatures are indicated; the lines are fits according to the superconducting gap of Nb folded with the temperature-broadened Fermi distribution of the Au (courtesy of Pan, Houston)

Lifetime Broadening

Besides ΔE, intrinsic properties of the sample lead to a broadening of spectroscopic features. Basically, the finite lifetime of the electron or hole in the corresponding state broadens its energetic width. Any kind of interaction such as electron–electron interaction can be responsible. Lifetime broadening has usually been measured by photoemission spectroscopy (PES, but it turned out that lifetimes of surface states on noble metal surfaces determined by STS (Fig. 5.9a,b) are up to a factor of three larger than those measured by PES [55]. The reason is probably that defects broaden the PESpectrum. Defects are unavoidable in a spatially integrating technique such as PES thus, STS has the advantage of choosing a particularly clean area for lifetime measurements. The STS results can be successfully compared to theory, highlighting the dominating influence of intraband transitions for the surface-state lifetime on Au(111) and Cu(111), at least close to the onset of the surface band [21].

With respect to band electrons, the analysis of the width of the band onset in $dI/dV(V)$-curves has the disadvantage of being restricted to the onset energy. Another method circumvents this problem by measuring the decay of standing electron waves scattered from a step edge as a function of energy [56]. Figure 5.9c,d shows the resulting oscillating dI/dV-signal measured for two different energies. To deduce the coherence length L_Φ, which is inversely proportional to the lifetime τ_Φ, one has to consider that the finite ΔE in the experiment also leads to a decay of the standing wave away from the step edge. The dotted fit line using $L_\Phi = \infty$ indicates this effect and, more importantly, shows a discrepancy from the measured curve. Only including a finite coherence length of 6.2 nm results in good agreement, which, in turn, determines L_Φ and thus τ_Φ, as displayed in Fig. 5.9c. The found $1/E^2$-dependence of τ_Φ points to a dominating influence of electron–electron interactions at higher energies in the surface band.

Landau and Spin Levels

Moreover, the increased energy resolution at LT allows the resolution of electronic states that are not resolvable at room temperature (RT). For example, Landau and spin quantization appearing in a magnetic field B have been probed on InAs(110) [9, 57]. The corresponding quantization energies are given by $E_\text{Landau} = \hbar eB/m_\text{eff}$ and $E_\text{spin} = g\mu B$. Thus InAs is a good choice, since it exhibits a low effective mass $m_\text{eff}/m_\text{e} = 0.023$ and a high g-factor of 14 in the bulk conduction band. The values in metals are $m_\text{eff}/m_\text{e} \approx 1$ and $g \approx 2$, resulting in energy splittings of only 1.25 meV and 1.2 meV at $B = 10$ T. This is obviously lower than the typical lifetime broadenings discussed in the previous section and also close to $\Delta E = 1.1$ meV achievable at $T = 4$ K.

Fortunately, the electron density in doped semiconductors is much lower, and thus the lifetime increases significantly. Figure 5.10a shows a set of spectroscopy curves obtained on InAs(110) in different magnetic fields [9]. Above E_F, oscillations with increasing intensity and energy distance are observed. They show the separation expected from Landau quantization. In turn, they can be used to deduce

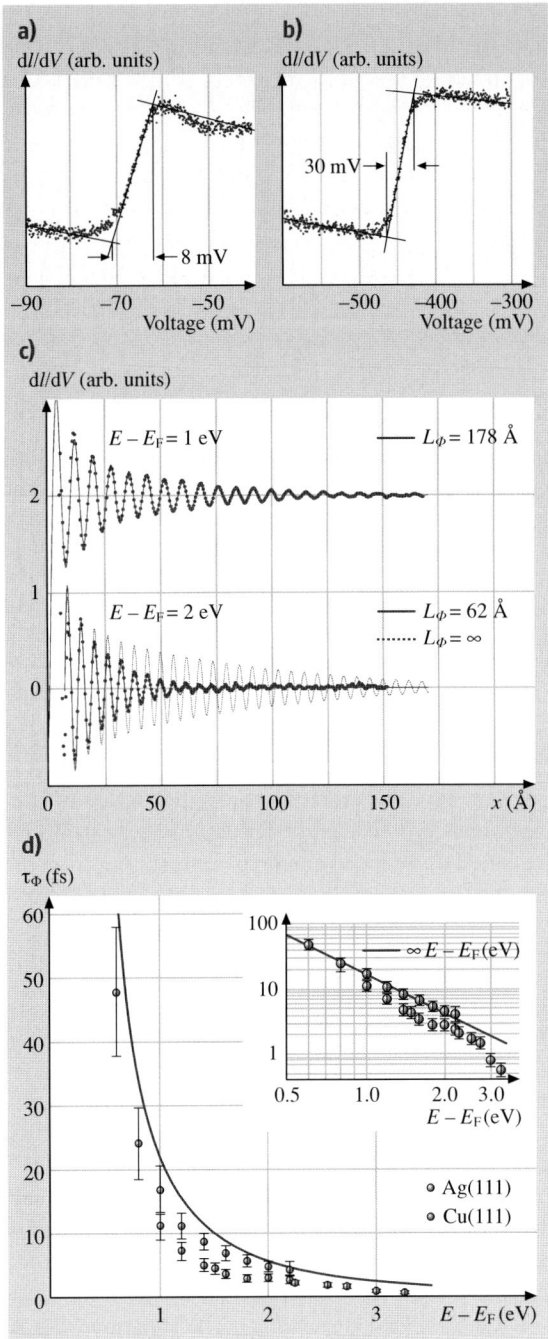

Fig. 5.9. (a),(b) spatially averaged $dI/dV(V)$-curves of Ag(111) and Cu(111); both surfaces exhibit a surface state with parabolic dispersion starting at -65 meV and -430 meV, respectively. The lines are drawn to determine the energetic width of the onset of these surface bands ((**a**),(**b**) courtesy of Berndt, Kiel); (**c**) dI/dV-intensity as a function of position away from a step edge of Cu(111) measured at the voltages $(E - E_F)$, as indicated (*points*); the lines are fits assuming standing electron waves with a phase coherence length L_Φ as marked; (**d**) resulting phase coherence time as a function of energy for Ag(111) and Cu(111). *Inset* shows the same data on a double logarithmic scale evidencing the E^{-2}-dependence (*line*) ((**c**),(**d**) courtesy of Brune, Lausanne)

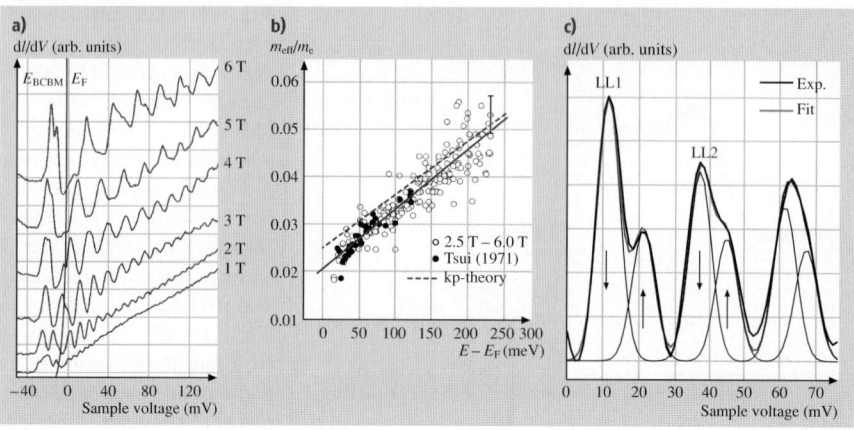

Fig. 5.10. (a) dI/dV-curves of n-InAs(110) at different magnetic fields as indicated; E_{BCBM} marks the bulk conduction band minimum; oscillations above E_{BCBM} are caused by Landau quantization; the double peaks at $B = 6$ T are caused by spin quantization. (b) Effective-mass data deduced from the distance of adjacent Landau peaks ΔE according to $\Delta E = heB/m_{\text{eff}}$ (*open symbols*); filled symbols are data from planar tunnel junctions (Tsui), the *solid line* is a mean-sqare fit of the data and the *dashed line* is the expected effective mass of InAs according to kp-theory. (c) Magnification of a dI/dV-curve at $B = 6$ T exhibiting spin splitting; the Gaussian curves marked by *arrows* are the fitted spin levels

m_{eff} from the peak separation (Fig. 5.10b). An increase of m_{eff} with increasing E has been found, as expected from theory. Also, at high fields spin quantization is observed (Fig. 5.10c). It is larger than expected from the bare g-factor due to contributions from exchange enhancement [58]. A detailed discussion of the peaks revealed that they belong to the so-called tip-induced quantum dot resulting from the work function difference between the tip and sample.

Vibrational Levels

As discussed with respect to light emission in STM, inelastic tunneling processes contribute to the tunneling current. The coupling of electronic states to vibrational levels is one source of inelastic tunneling [23]. It provides additional channels contributing to $dI/dV(V)$ with final states at energies different from V. The final energy is simply shifted by the energy of the vibrational level. If only discrete vibrational energy levels couple to a smooth electronic DOS, one expects a peak in d^2I/dV^2 at the vibrational energy. This situation appears for molecules on noble-metal surfaces. As usual, the isotope effect on the vibrational energy can be used to verify the vibrational origin of the peak. First indications of vibrational levels have been found for H_2O and D_2O on TiO_2 [59], and completely convincing work has been performed for C_2H_2 and C_2D_2 on Cu(001) [23] (Fig. 5.11a). The technique can be used to identify individual molecules on the surface by their characteristic vibrational levels. In particular, surface reactions, as described in Fig. 5.5h–m, can be directly verified. Moreover, the orientation of complexes with respect to the surface

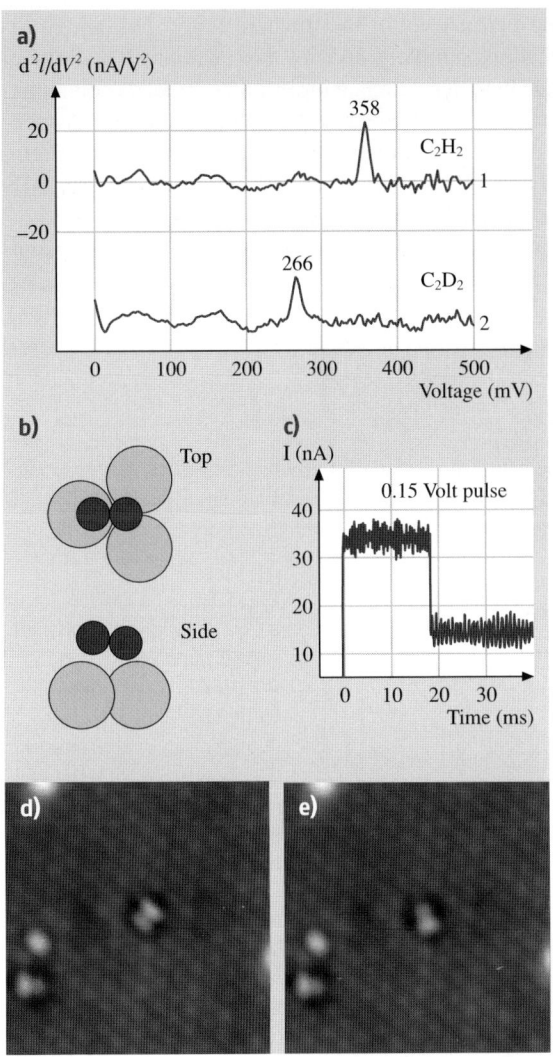

Fig. 5.11. (a) d^2I/dV^2-curves taken above a C_2H_2 and a C_2D_2 molecule on Cu(100); the peaks correspond to the C–H, respectively, C–D stretch-mode energy of the molecule. (b) Sketch of O_2 molecule on Pt(111). (c) Tunnelling current above an O_2 molecule on Pt(111) during a voltage pulse of 0.15 V; the jump in current indicates rotation of the molecule. (d), (e) STM image of an O_2 molecule on Pt(111) ($V = 0.05$ V) prior and after rotation induced by a voltage pulse to 0.15 V ((a)–(e) courtesy of Ho, Irvine)

can be determined to a certain extent, since the vibrational excitation depends on the position of the tunneling current within the molecule. Finally, the excitation of certain molecular levels can induce such corresponding motions as hopping [32], rotation [34] (Fig. 5.11b–e), or desorption [31], leading to additional possibilities for manipulation on the atomic scale.

Kondo Resonance

A rather intricate interaction effect is the Kondo effect. It results from a second-order scattering process between itinerate states and a localized state [60]. The two states exchange some degree of freedom back and forth, leading to a divergence of the scattering probability at the Fermi level of the itinerate states. Due to the divergence, the effect strongly modifies sample properties. For example, it leads to an unexpected increase in resistance with decreasing temperature for metals containing magnetic impurities [4]. Here, the exchanged degree of freedom is the spin. A spectroscopic signature of the Kondo effect is a narrow peak in the DOS at the Fermi level, disappearing above a characteristic temperature (the Kondo temperature). STS provides the opportunity to study this effect on the local scale [61, 62].

Figure 5.12a–d shows an example of Co clusters deposited on a carbon nanotube [63]. While a small dip at the Fermi level, which is probably caused by curva-

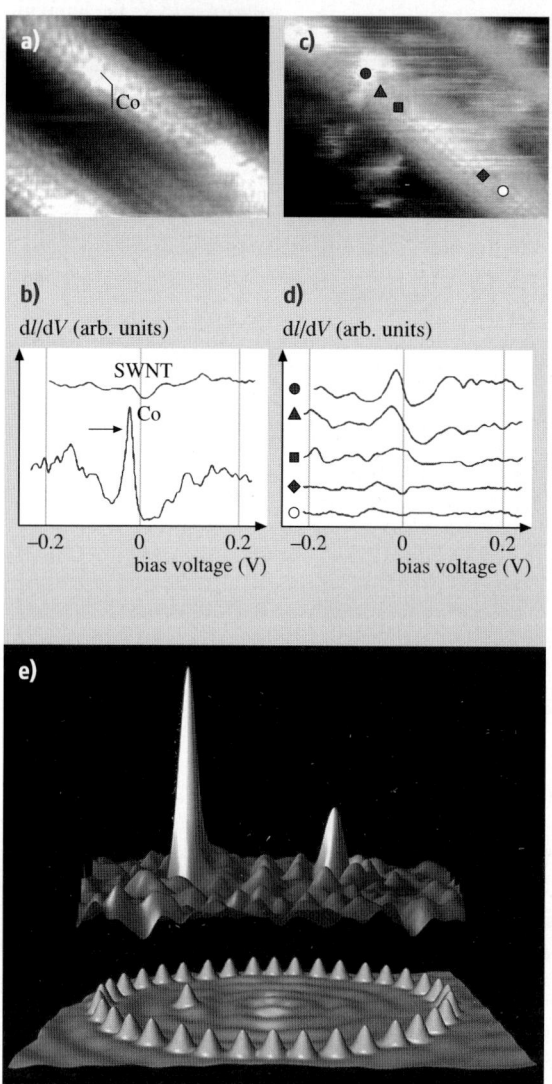

Fig. 5.12. (a) STM image of a Co cluster on a single-wall carbon nanotube (SWNT). (b) dI/dV-curves taken directly above the Co cluster (Co) and far away from the Co cluster (SWNT); the *arrow* marks the Kondo peak. (c) STM image of another Co cluster on a SWNT with *symbols* marking the positions where the dI/dV-curves displayed in (d) are taken. (d) dI/dV-curves taken at the positions marked in (c) ((a)–(d) courtesy of Lieber, Cambridge). (e) *Lower part:* STM image of a quantum corral of elliptic shape made from Co atoms on Cu(111); one Co atom is placed in one of the foci of the ellipse. *Upper part:* map of the strength of the Kondo signal in the corral; note that there is also a Kondo signal in the focus, which is not covered by a Co atom ((e) courtesy of Eigler, Almaden)

ture influences on the π-orbitals, is observed without Co (Fig. 5.12b) [64], a strong peak is found around a Co cluster deposited on top of the tube (Fig. 5.12a, arrow). The peak is slightly shifted with respect to $V = 0$ mV due to the so-called Fano resonance [65], which results from interference of the tunneling processes into the localized Co-level and the itinerant nanotube levels. The resonance disappears within several nanometers of the cluster, as shown in Fig. 5.12d.

The Kondo effect has also been detected for different magnetic atoms deposited on noble-metal surfaces [61, 62]. There, it disappears at about 1 nm from the magnetic impurity, and the effect of the Fano resonance is more pronounced, contributing to dips in $dI/dV(V)$-curves instead of peaks.

A fascinating experiment has been performed by *Manoharan* et al. [66], who used manipulation to form an elliptic cage for the surface states of Cu(111) (Fig. 5.12e, bottom). This cage was constructed to have a quantized level at E_F. Then, a cobalt atom was placed in one focus of the elliptic cage, producing a Kondo resonance. Surprisingly, the same resonance reappeared in the opposite focus, but not away from the focus (Fig. 5.12e, top). This shows amazingly that complex local effects such as the Kondo resonance can be guided to remote points.

Orbital scattering as a source of the Kondo resonance has also been found around a defect on Cr(001) [67]. Here, it is believed that itinerate sp-levels scatter at a localized d-level to produce the Kondo peak.

5.4.5 Imaging Electronic Wave Functions

Since STS measures the sum of squared wave functions (5.1), it is an obvious task to measure the local appearance of the most simple wave functions in solids, namely, Bloch waves.

Bloch Waves

The atomically periodic part of the Bloch wave is always measured if atomic resolution is achieved (inset of Fig. 5.14a). However, the long-range wavy part requires the presence of scatterers. The electron wave impinges on the scatterer and is reflected, leading to self-interference. In other words, the phase of the Bloch wave becomes fixed by the scatterer.

Such self-interference patterns were first found on Graphite(0001) [68] and later on noble-metal surfaces, where adsorbates or step edges scatter the surface states (Fig. 5.13a) [22]. Fourier transforms of the real-space images reveal the k-space distribution of the corresponding states [69], which may include additional contributions besides the surface state [70]. Using particular geometries as the so-called quantum corrals to form a cage for the electron wave, the scattering state can be rather complex (Fig. 5.13b). Anyway, it can usually be reproduced by simple calculations involving single-particle states [71].

Bloch waves in semiconductors scattered at charged dopants (Fig. 5.13c,d) [72], Bloch states confined in semiconductor quantum dots (Fig. 5.13e–g) [73], and Bloch waves confined in short-cut carbon nanotubes (Fig. 5.13h,i) [74, 75] have been visualized.

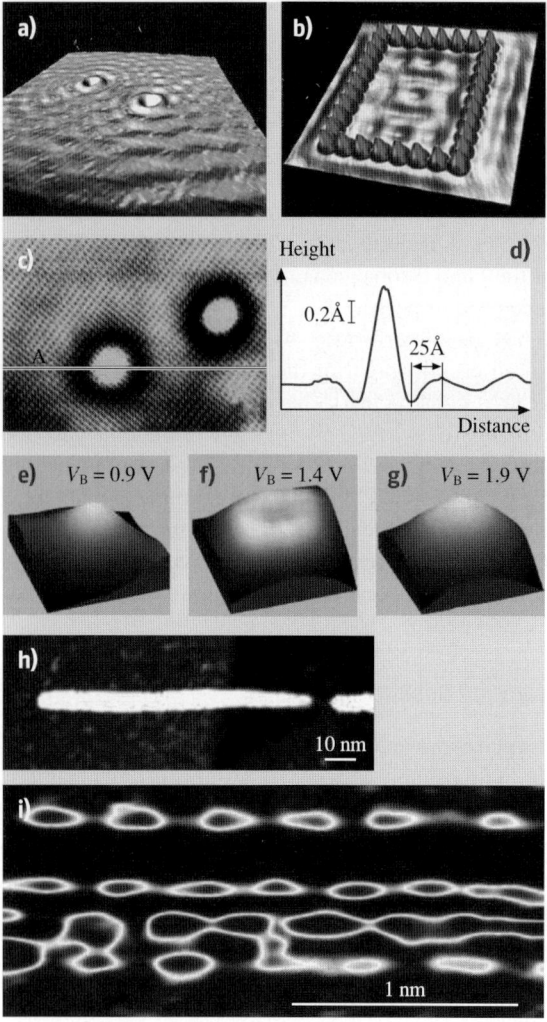

Fig. 5.13. (a) Low-voltage STM image of Cu(111), including two defect atoms; the waves are electronic Bloch waves scattered at the defects. (b) Low-voltage STM image of a rectangular quantum corral made from single atoms on Cu(111); the pattern inside the corral is the confined state of the corral close to E_F; ((a),(b) courtesy of Eigler, Almaden). (c) STM image of GaAs(110) around a Si donor, $V = -2.5$ V; the line scan along A shown in (d) exhibits an additional oscillation around the donor caused by a standing Bloch wave; the grid like pattern corresponds to the atomic corrugation of the Bloch wave ((c),(d) courtesy of van Kempen, Nijmegen). (e)–(g) STM images of an InAs/ZnSe-core/shell-nanocluster at different V. The image is measured in the so-called constant-height mode, i.e., the images display the tunneling current at constant height above the surface; the hill in (e) corresponds to the s-state of the cluster, the ring in (f) to the degenerate p_x- and p_y-state and the hill in (g) to the p_z-state ((e)–(g) courtesy of Millo, Jerusalem). (h) STM-image of a short-cut carbon nanotube. (i) Colour plot of the dI/dV intensity inside the short-cut nanotube as a function of position and tunneling voltage; four wavy patterns of different wavelength are visible in the voltage range from -0.1 to 0.15 V ((h),(i) courtesy of Dekker, Delft)

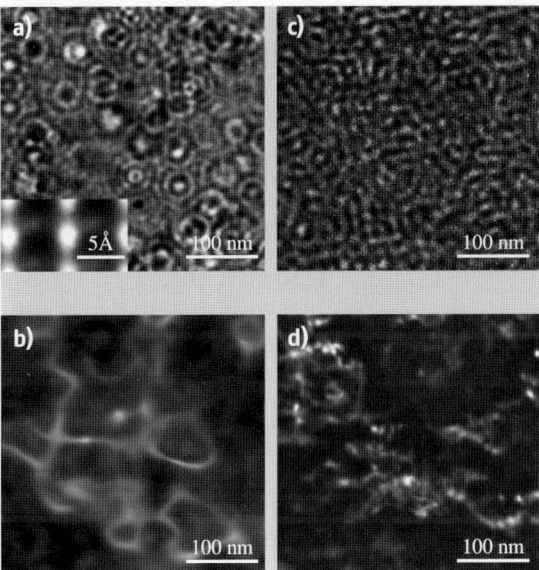

Fig. 5.14. (a) dI/dV-image of InAs(110) at $V = 50\,\text{mV}$, $B = 0\,\text{T}$; circular wave patterns corresponding to standing Bloch waves around each S donor are visible; inset shows a magnification revealing the atomically periodic part of the Bloch wave. (b) Same as (a), but at $B = 6\,\text{T}$; the stripe structures are drift states. (c) dI/dV-image of a 2-D electron system on InAs(110) induced by the deposition of Fe, $B = 0\,\text{T}$. (d) Same as (c) but at $B = 6\,\text{T}$; note that the contrast in (a) is increased by a factor of ten with respect to (b)–(d)

Drift States

More-complex wave functions result from interactions. A nice playground to study such interactions is doped semiconductors. The reduced electron density with respect to metals increases the importance of electron interactions with potential disorder and other electrons. Applying a magnetic field quenches the kinetic energy, further enhancing the importance of interactions. A dramatic effect can be observed on InAs(110), where three-dimensional (3-D) bulk states are displayed. While the usual scattering states around individual dopants are observed at $B = 0\,\text{T}$ (Fig. 5.14a) [76], stripe structures are found in high magnetic field (Fig. 5.14b) [77]. They run along equipotential lines of the disorder potential. This can be understood by recalling that the electron tries to move in a cyclotron orbit, which is accelerated and decelerated in electrostatic potential, leading to a drift motion along an equipotential line [78].

The same effect has been found in two-dimensional (2-D) electron systems (2-DES) of the same substrate, where the scattering states at $B = 0\,\text{T}$ are, however, found to be more complex (Fig. 5.14c) [79]. The reason is the tendency of a 2-DES to exhibit closed scattering paths [80]. Consequently, the self-interference does not result from scattering at individual scatterers, but from complicated self-interference paths involving many scatterers. However, drift states are also observed in the 2-DES at high magnetic fields (Fig. 5.14d) [81].

Charge Density Waves

Another interaction modifying the LDOS is the electron–phonon interaction. Phonons scatter electrons between different Fermi points. If the wave vectors connecting Fermi points exhibit a preferential orientation, a so-called Peierls instability oc-

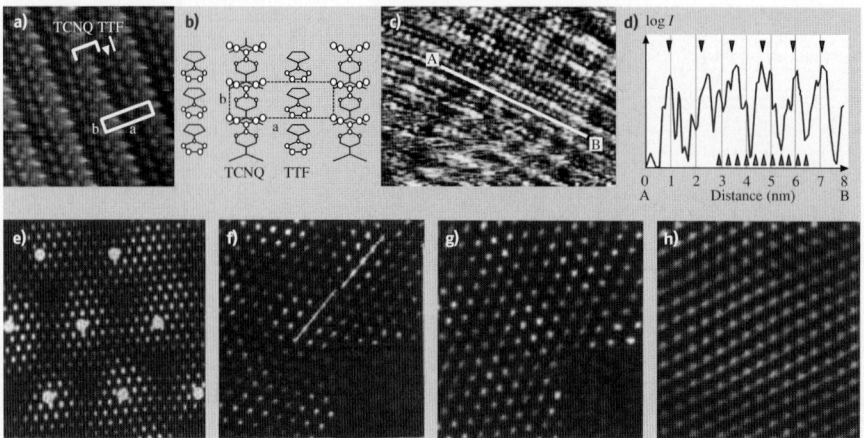

Fig. 5.15. (a) STM image of the ab-plane of the organic quasi-1-D conductor TTF-TCNQ, $T = 300$ K; while the TCNQ chains are conducting, the TTF chains are insulating. (b) Stick-and-ball model of the ab-plane of TTF-TCNQ. (c) STM image taken at $T = 61$ K, the additional modulation due to the Peierls transition is visible in the profile along line A shown in (d); the *brown triangles* mark the atomic periodicity and the *black triangles* the expected CDW periodicity ((**a**)–(**d**) courtesy of Kageshima, Kanagawa). (**e**)–(**h**) Low-voltage STM images of the two-dimensional CDW-system 1 T-TaS$_2$ at $T = 242$ K (**e**), 298 K (**f**), 349 K (**g**), 357 K (**h**). A long-range, hexagonal modulation is visible besides the atomic spots; its periodicity is highlighted by *large white dots* in (**e**); the additional modulation obviously weakens with increasing T, but is still apparent in (**f**) and (**g**), as evidenced in the lower magnification images in the insets ((**e**)–(**h**) courtesy of Lieber, Cambridge)

curs [82]. The corresponding phonon energy goes to zero, the atoms are slightly displaced with the periodicity of the corresponding wave vector, and a charge density wave (CDW) with the same periodicity appears. Essentially, the CDW increases the overlap of the electronic states with the phonon by phase-fixing with respect to the atomic lattice. The Peierls transition naturally occurs in one-dimensional (1-D) systems, where only two Fermi points are present, and, hence, preferential orientation is pathological. It can also occur in 2-D systems if large areas of the Fermi line run in parallel.

STS studies of CDWs are numerous (e. g., [83, 84]). Examples of a 1-D CDW on a quasi-1-D bulk material and of a 2-D CDW are shown in Fig. 5.15a–d and Fig. 5.15e, respectively [85, 86]. In contrast to usual scattering states, where LDOS corrugations are only found close to the scatterer, the corrugations of CDWs are continuous across the surface. Heating the substrate toward the transition temperature leads to a melting of the CDW lattice, as shown in Fig. 5.15f–h.

CDWs have also been found on monolayers of adsorbates such as a monolayer of Pb on Ge(111) [87]. These authors performed a nice temperature-dependent study revealing that the CDW is nucleated by scattering states around defects, as one might expect [88]. 1-D systems have also been prepared on surfaces showing Peierls transitions [89, 90]. Finally, the energy gap occurring at the transition has been studied by measuring $dI/dV(V)$-curves [91].

Superconductors

An intriguing effect resulting from electron–phonon interaction is superconductivity. Here, the attractive part of the electron–phonon interaction leads to the coupling of electronic states with opposite wave vector and mostly opposite spin [92]. Since the resulting Cooper pairs are bosons, they can condense at LT, forming a coherent many-particle phase, which can carry current without resistance. Interestingly, defect scattering does not influence the condensate if the coupling along the Fermi surface is homogeneous (s-wave superconductor). The reason is that the symmetry of the scattering of the two components of a Cooper pair effectively leads to a scattering from one Cooper pair state to another without affecting the condensate. This is different if the scatterer is magnetic, since the different spin components of the pair are scattered differently, leading to an effective pair breaking, which is visible as a single-particle excitation within the superconducting gap. On a local scale, this effect was first demonstrated by putting Mn, Gd, and Ag atoms on a Nb(110) surface [93]. While the nonmagnetic Ag does not modify the gap shown in Fig. 5.16a, it is modified in an asymmetric fashion close to Mn or Gd adsorbates, as shown in Fig. 5.16b. The asymmetry of the additional intensity is caused by the breaking of the particle–hole symmetry due to the exchange interaction between the localized Mn state and the itinerate Nb states.

Another important local effect is caused by the relatively large coherence length of the condensate. At a material interface, the condensate wave function cannot stop abruptly, but overlaps into the surrounding material (proximity effect). Consequently, a superconducting gap can be measured in areas of non-superconducting material. Several studies have shown this effect on the local scale using metals and doped semiconductors as surrounding materials [94, 95].

While the classical type-I superconductors are ideal diamagnets, the so-called type-II superconductors can contain magnetic flux. The flux forms vortices, each containing one flux quantum. These vortices are accompanied by the disappearance of the superconducting gap and, therefore, can be probed by STS [96]. LDOS maps measured inside the gap lead to bright features in the area of the vortex core. Importantly, the length scale of these features is different from the length scale of the magnetic flux due to the difference between London's penetration depth and the coherence length. Thus, STS probes another property of the vortex than the usual magnetic imaging techniques (see Sect. 5.5.4). Surprisingly, first measurements of the vortices on $NbSe_2$ revealed vortices shaped as a sixfold star [97] (Fig. 5.16c). With increasing voltage inside the gap, the orientation of the star rotates by $30°$ (Fig. 5.16d,e). The shape of these stars could finally be reproduced by theory, assuming an anisotropic pairing of electrons in the superconductor (Fig. 5.16f–h) [98]. Additionally, bound states inside the vortex core, which result from confinement by the surrounding superconducting material, are found [97]. Further experiments investigated the arrangement of the vortex lattice, including transitions between

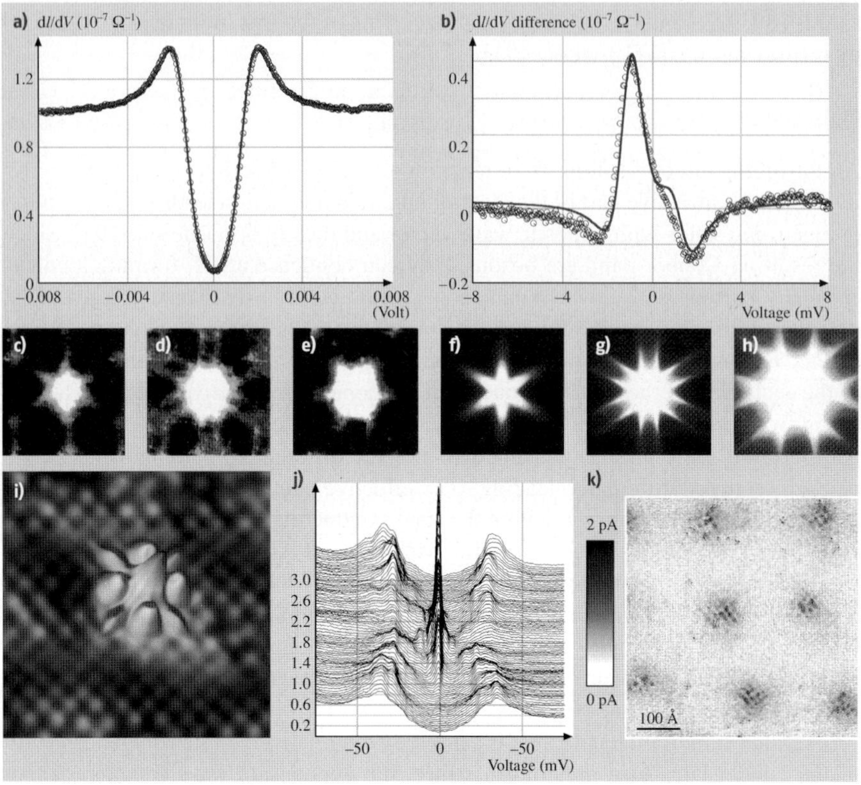

Fig. 5.16. (a) dI/dV-curve of Nb(110) at $T = 3.8$ K (*symbols*) in comparison with a BCS fit of the superconducting gap of Nb (*line*). (b) Difference between the dI/dV-curve taken directly above a Mn-atom on Nb(110) and the dI/dV-curve taken above the clean Nb(110) (*symbols*) in comparison with a fit using the Bogulubov-de Gennes equations (*line*) ((**a**),(**b**) courtesy of Eigler, Almaden). (**c**)–(**e**) dI/dV-images of a vortex core in the type-II superconductor 2H-NbSe$_2$ at 0 mV (**c**), 0.24 mV (**d**), and 0.48 mV (**e**) ((**c**)–(**e**) courtesy of H.F. Hess). (**f**)–(**h**) Corresponding calculated LDOS images within the Eilenberger framework ((**f**)–(**h**) courtesy of Machida, Okayama). (**i**) Overlap of an STM image at $V = -100$ mV (background 2-D image) and a dI/dV-image at $V = 0$ mV (overlapped 3-D image) of optimally doped Bi$_2$Sr$_2$CaCu$_2$O$_{8+\delta}$ containing 0.6% Zn impurities. The STM image shows the atomic structure of the cleavage plane, while the dI/dV-image shows a bound state within the superconducting gap, which is located around a single Zn impurity. The fourfold symmetry of the bound state reflects the d-like symmetry of the superconducting pairing function; (**j**) dI/dV-curves taken at different positions across the Zn impurity; the bound state close to 0 mV is visible close to the Zn atom; (**k**) LDOS in the vortex core of slightly overdoped Bi$_2$Sr$_2$CaCu$_2$O$_{8+\delta}$, $B = 5$ T; the dI/dV-image taken at $B = 5$ T is integrated over $V = 1-12$ mV, and the corresponding dI/dV-image at $B = 0$ T is subtracted to highlight the LDOS induced by the magnetic field. The checkerboard pattern within the seven vortex cores exhibits a periodicity, which is fourfold with respect to the atomic lattice shown in (**i**) and is thus assumed to be a CDW ((**i**)–(**k**) courtesy of S. Davis, Cornell and S. Uchida, Tokyo)

hexagonal and quadratic lattices [99], the influence of pinning centers [100], and the vortex motion induced by current [101].

The understanding of high-temperature superconductivity (HTCS) is still an important topic. An almost accepted property of HTCS is its d-wave pairing symmetry. In contrast to s-wave superconductors, scattering can lead to pair breaking, since the Cooper-pair density vanishes in certain directions. Indeed, scattering states (bound states in the gap) around nonmagnetic Zn impurities have been observed in $Bi_2Sr_2CaCu_2O_{8+\delta}$ (BSCCO) (Fig. 5.16i,j) [26]. They reveal a d-like symmetry, but not the one expected from simple Cooper-pair scattering. Other effects such as magnetic polarization in the environment probably have to be taken into account [102]. Moreover, it has been found that magnetic Ni impurities exhibit a weaker scattering structure than Zn impurities [103]. Thus, BSCCO shows exactly the opposite behavior to that of Nb discussed above (Fig. 5.16a,b). An interesting topic is the importance of inhomogeneities in HTCS materials. Evidence for inhomogeneities has indeed been found in underdoped materials, where puddles of the superconducting phase are shown to be embedded in non-superconducting areas [104].

Of course, vortices have also been investigated in HTCS materials [105]. Bound states are found, but at energies that are in disagreement with simple models, assuming a Bardeen–Cooper–Schrieffer (BCS)-like d-wave superconductor [106, 107]. Theory predicts, instead, that the bound states are magnetic-field-induced spin density waves, stressing the competition between antiferromagnetic order and superconductivity in HTCS materials [108]. Since the spin density wave is accompanied by a charge density wave of half wavelength, it can be probed by STS [109]. Indeed, a checkerboard pattern of the right periodicity has been found in and around vortex cores in BSCCO (Fig. 5.16k). It exceeds the width of an individual vortex core, implying that the superconducting coherence length is different from the antiferromagnetic one.

Complex Systems (Manganites)

Complex phase diagrams are not restricted to HTCS materials (cuprates). They exist with similar complexity for other doped oxides such as manganites. Only a few studies of these materials have been performed by STS, mainly showing the inhomogeneous evolution of metallic and insulating phases [110, 111]. Similarities to the granular case of an underdoped HTCS material are obvious. Since inhomogeneities seem to be crucial in many of these materials, a local method such as STS might continue to be important for the understanding of their complex properties.

5.4.6 Imaging Spin Polarization: Nanomagnetism

Conventional STS couples to the LDOS, i.e., the charge distribution of the electronic states. Since electrons also have spin, it is desirable to also probe the spin distribution of the states. This can be achieved if the tunneling tip is covered by a ferromagnetic material [112]. The coating acts as a spin filter or, more precisely, the tunneling current depends on the relative angle α_{ij} between the spins of the tip

Fig. 5.17. (**a**)–(**d**) Spin-polarized STM images of 1.65 monolayer of Fe deposited on a stepped W(110) surface measured at different B-fields, as indicated. Double-layer and monolayer Fe stripes are formed on the W substrate; only the double-layer stripes exhibit magnetic contrast with an out-of-plane sensitive tip, as used here. *White* and *grey areas* correspond to different domains. Note that more white areas appear with increasing field. (**e**) STM image of an antiferromagnetic Mn monolayer on W(110). (**f**) Spin-polarized STM-image of the same surface (in-plane tip). The insets in (**e**) and (**f**) show the calculated STM and spin-polarized STM images, respectively, and the stick-and-ball models symbolize the atomic and the magnetic unit cell ((**a**)–(**f**) courtesy of M. Bode, Hamburg). (**g**) Spin-polarized STM image of a 6-nm-high Fe island on W(110) (in-plane tip). Four different areas are identified as four different domains with domain orientations, as indicated by the *arrows*. (**h**) Spin-polarized STM image of the central area of an island; the size of the area is indicated by the rectangle in (**g**); the measurement is performed with an out-of-plane sensitive tip showing that the magnetization turns out-of-plane in the center of the island

and the sample according to $\cos(\alpha_{ij})$. In ferromagnets, the spins mostly have one preferential orientation along the so-called easy axis, i.e., a particular tip is not sensitive to spin orientations of the sample that are perpendicular to the spin orientation of the tip. Different tips have to be prepared to detect different spin orientations of the sample. Moreover, the magnetic stray field of the tip can perturb the spin orientation of the sample. To avoid this, a technique using antiferromagnetic Cr as a tip coating material has been developed [113]. This avoids stray fields, but still provides a preferential spin orientation of the few atoms at the tip apex that dominate the tunneling current. Depending on the thickness of the Cr coating, spin orientations perpendicular or parallel to the sample surface are prepared.

So far, the described technique has been used to image the evolution of magnetic domains with increasing B-field (Fig. 5.17a–d) [114], the antiferromagnetic order of a Mn monolayer on W(110) (Fig. 5.17e,f) [115], and the out-of-plane orientation predicted for a magnetic vortex core as it exists in the center of a Fe island exhibiting four domains in the flux closure configuration (Fig. 5.17g,h) [116].

Besides the obvious strong impact on nanomagnetism, the technique might also be used to investigate other electronic phases such as the proposed spin density wave around a HTCS vortex core.

5.5 Scanning Force Microscopy and Spectroscopy

The examples discussed in the previous section show the wide variety of physical questions that have been tackled with the help of LT-STM. Here, we turn to the other prominent scanning probe method that is applied at low temperatures, namely, SFM, which gives complementary information on sample properties on the atomic scale.

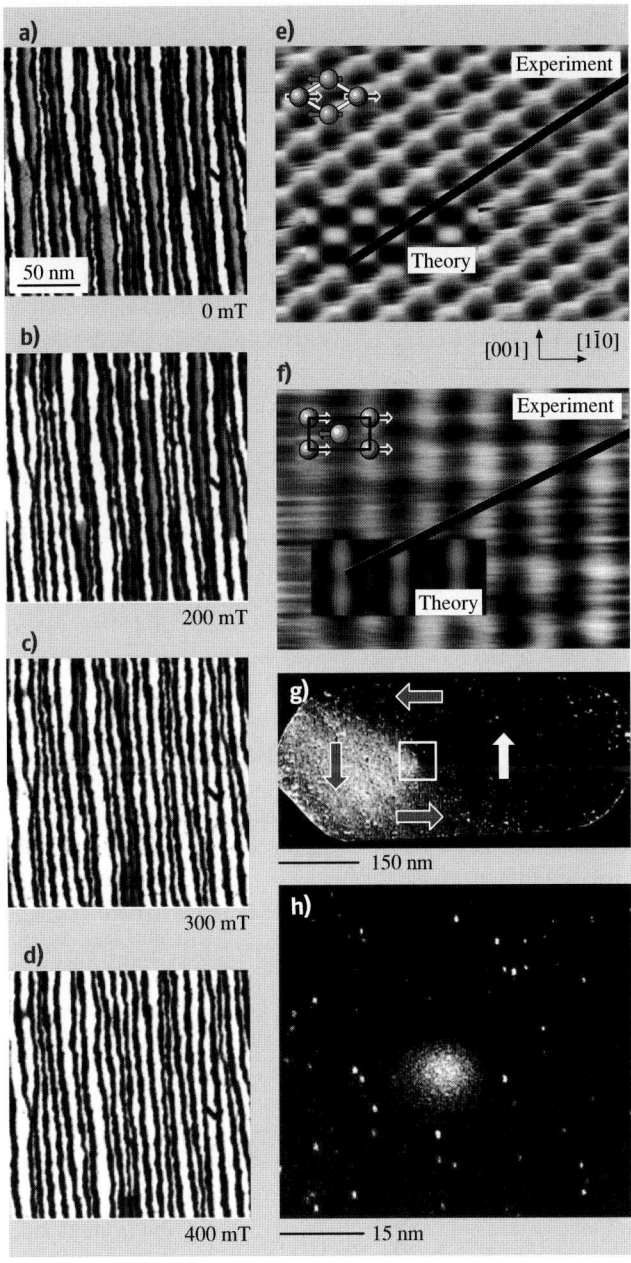

The ability to detect *forces* sensitively with spatial resolution down to the atomic scale is of great interest, since force is one of the most fundamental quantities in physics. Mechanical force probes usually consist of a cantilever with a tip at its free end that is brought close to the sample surface. The cantilever can be mounted parallel or perpendicular to the surface (general aspects of force probe designs are described in Chap. 3). Basically, two methods exist to detect forces with cantilever-based probes: the *static* and the *dynamic* mode (see Chap. 2). They can be used to generate a laterally resolved image (*microscopy* mode) or determine its distance dependence (*spectroscopy* mode). One can argue about the terminology, since spectroscopy is usually related to energies and not to distance dependencies. Nevertheless, we will use it throughout the text, because it avoids lengthy paraphrases and is established in this sense throughout the literature.

In the static mode, a force that acts on the tip bends the cantilever. By measuring its deflection Δz the tip–sample force F_{ts} can be directly calculated with Hooke's law: $F_{ts} = c_z \cdot \Delta z$, where c_z denotes the spring constant of the cantilever. In the various dynamic modes, the cantilever is oscillated with amplitude A at or near its eigenfrequency f_0, but in some applications also off-resonance. At ambient pressures or in liquids, amplitude modulation (AM-SFM) is used to detect amplitude changes or the phase shift between the driving force and cantilever oscillation. In vacuum, the frequency shift Δf of the cantilever due to a tip–sample interaction is measured by the frequency-modulation technique (FM-SFM). The nomenclature is not standardized. Terms like tapping mode or intermittent contact mode are used instead of AM-SFM, and NC-AFM (noncontact atomic force microscopy) or DFM (dynamic force microscopy) instead of FM-SFM or FM-AFM. However, all these modes are *dynamic*, i.e., they involve an oscillating cantilever and can be used in the noncontact, as well as in the contact, regime. Therefore, we believe that the best and most consistent way is to distinguish them by their different detection schemes. Converting the measured quantity (amplitude, phase, or frequency shift) into a physically meaningful quantity, e.g., the tip–sample interaction force F_{ts} or the force gradient $\partial F_{ts}/\partial z$, is not always straightforward and requires an analysis of the equation of motion of the oscillating tip (see Chaps. 4 and 6).

Whatever method is used, the resolution of a cantilever-based force detection is fundamentally limited by its intrinsic *thermomechanical* noise. If the cantilever is in thermal equilibrium at a temperature T, the equipartition theorem predicts a thermally induced *root mean square* (rms) motion of the cantilever in the z direction of $z_{rms} = (k_B T/c_{eff})^{1/2}$, where k_B is the Boltzmann constant and $c_{eff} = c_z + \partial F_{ts}/\partial z$. Note that usually $dF_{ts}/dz \gg c_z$ in the contact mode and $dF_{ts}/dz < c_z$ in the noncontact mode. Evidently, this fundamentally limits the force resolution in the static mode, particularly if operated in the noncontact mode. Of course, the same is true for the different dynamic modes, because the thermal energy $k_B T$ excites the eigenfrequency f_0 of the cantilever. Thermal noise is *white* noise, i.e., its spectral density is flat. However, if the cantilever transfer function is taken into account, one can see that the thermal energy mainly excites f_0. This explains the term *thermo* in thermomechanical noise, but what is the *mechanical* part?

A more detailed analysis reveals that the thermally induced cantilever motion is given by

$$z_{\text{rms}} = \sqrt{\frac{2k_B T B}{\pi c_z f_0 Q}},\qquad(5.2)$$

where B is the measurement bandwidth and Q is the quality factor of the cantilever. Analogous expressions can be obtained for all quantities measured in dynamic modes, because the deflection noise translates, e.g., into frequency noise [5]. Note that f_0 and c_z are correlated with each other via $2\pi f_0 = (c_z/m_{\text{eff}})^{1/2}$, where the effective mass m_{eff} depends on the geometry, density, and elasticity of the material. The Q-factor of the cantilever is related to the external damping of the cantilever motion in a medium and on the intrinsic damping within the material. This is the *mechanical* part of the fundamental cantilever noise.

It is possible to operate a low-temperature force microscope directly immersed in the cryogen [117, 118] or in the cooling gas [119], whereby the cooling is simple and very effective. However, it is evident from (5.2) that the smallest fundamental noise is achievable in vacuum, where the Q-factors are more than 100 times larger than in air, and at low temperatures.

The best force resolution up to now, which is better than 1×10^{-18} N/Hz$^{1/2}$, has been achieved by *Mamin* et al. [120] in vacuum at a temperature below 300 mK. Due to the reduced thermal noise and the lower thermal drift, which results in a higher stability of the tip–sample gap and a better signal-to-noise ratio, the highest resolution is possible at low temperatures in ultrahigh vacuum with FM-SFM. A vertical rms noise below 2 pm [121, 122] and a force resolution below 1 aN [120] have been reported.

Besides the reduced noise, the application of force detection at low temperatures is motivated by the increased stability and the possibility to observe phenomena that appear below a certain critical temperature T_c, as outlined on page 181. The experiments, which have been performed at low temperatures until now, were motivated by at least one of these reasons and can be roughly divided into four groups: (i) atomic-scale imaging, (ii) force spectroscopy, (iii) investigation of quantum phenomena by measuring electrostatic forces, and (iv) utilizing magnetic probes to study ferromagnets, superconductors, and single spins. In the following, we describe some exemplary results.

5.5.1 Atomic-Scale Imaging

In a simplified picture, the dimensions of the tip end and its distance to the surface limit the lateral resolution of force microscopy, since it is a near-field technique. Consequently, atomic resolution requires a stable single atom at the tip apex that has to be brought within a distance of some tenths of a nanometer to an atomically flat surface. The latter condition can only be fulfilled in the dynamic mode, where the additional restoring force $c_z A$ at the lower turnaround point prevents the jump-to-contact. As described in Chap. 4, by preventing the so-called jump-to-contact *true* atomic resolution is nowadays routinely obtained in vacuum by FM-AFM. The nature of the short-range tip–sample interaction during imaging with atomic resolution has been studied experimentally as well as theoretically. Si(111)-(7×7) was the first surface on which true atomic resolution was achieved [123], and several studies have been performed at low temperatures on this well-known material [124–126]. First-principles simulations performed on semiconductors with a silicon tip revealed that *chemical* interactions, i.e., a significant charge redistribution between the dangling bonds of the tip and sample, dominate the atomic-scale contrast [127–129]. On V–III semiconductors, it was found that only one atomic species, the group V atoms, is imaged as protrusions with a silicon tip [128, 129]. Furthermore, these simulations revealed that the sample, as well as the tip atoms are noticeably displaced from their equilibrium position due to the interaction forces. At low temperatures, both aspects could be observed with silicon tips on indium arsenide [121, 130]. On weakly interacting surfaces the short-range interatomic van der Waals force has been believed responsible for the atomic-scale contrast [131–133].

Chemical Sensitivity of Force Microscopy

The (110) surface of the III–V semiconductor indium arsenide exhibits both atomic species in the top layer (see Fig. 5.18a). Therefore, this sample is well suited to study the chemical sensitivity of force microscopy [121]. In Fig. 5.18b, the usually observed atomic-scale contrast on InAs(110) is displayed. As predicted, the arsenic atoms, which are shifted by 80 pm above the indium layer due to the (1 × 1) relaxation, are imaged as protrusions. While this general appearance was similar for most tips, two other distinctively different contrasts were also observed: a second protrusion (c) and a sharp depression (d). The arrangement of these two features corresponds well to the zigzag configuration of the indium and arsenic atoms along the [1$\bar{1}$0]-direction. A sound explanation would be as follows: the contrast usually obtained with one feature per surface unit cell corresponds to a silicon-terminated tip, as predicted by simulations. A different atomic species at the tip apex, however,

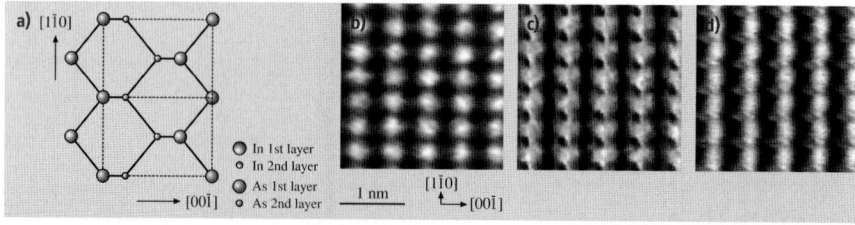

Fig. 5.18. The structure of InAs(110) as seen from above (**a**) and three FM-AFM images of this surface obtained with different tips at 14 K (**b**)–(**d**). In (**b**), only the arsenic atoms are imaged as protrusions, as predicted for a silicon tip. The two features in (**c**) and (**d**) corresponds to the zigzag arrangement of the indium and arsenic atoms. Since force microscopy is sensitive to short-range chemical forces, the appearance of the indium atoms can be associated with a chemically different tip apex

can result in a very different charge redistribution. Since the atomic-scale contrast is due to a chemical interaction, the two other contrasts would then correspond to a tip that has been accidentally contaminated with sample material (an arsenic or indium-terminated tip apex). Nevertheless, this explanation has not yet been verified by simulations for this material.

Tip-Induced Atomic Relaxation

Schwarz et al. [121] were able to visualize directly the predicted tip-induced relaxation during atomic-scale imaging near a point defect. Figure 5.19 shows two FM-AFM images of the same point defect recorded with different constant frequency shifts on InAs(110), i.e., the tip was closer to the surface in (b) compared to (a). The arsenic atoms are imaged as protrusions with the silicon tip used. From the symmetry of the defect, an indium-site defect can be inferred, since the distance-dependent contrast is consistent with what is expected for an indium vacancy. This expectation is based on calculations performed for the similar III–V semiconductor GaP(110), where the two surface gallium atoms around a P-vacancy were found to

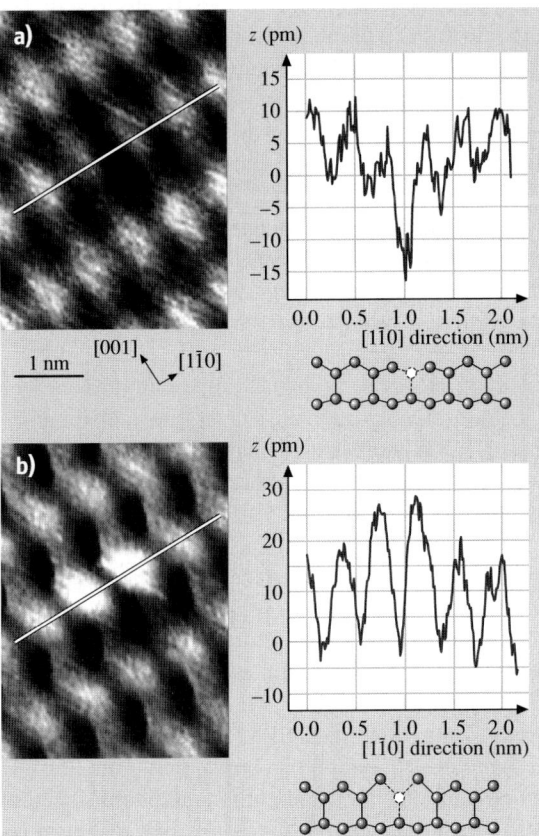

Fig. 5.19. Two FM-AFM images of the identical indium-site point defect (presumably an indium vacancy) recorded at 14 K. If the tip is relatively far away, the theoretically predicted inward relaxation of two arsenic atoms adjacent to an indium vacancy is visible (**a**). At a closer tip–sample distance (**b**), the two arsenic atoms are pulled farther toward the tip compared to the other arsenic atoms, since they have only two instead of three bonds

relax downward [134]. This corresponds to the situation in Fig. 5.19a, where the tip is relatively far away and an inward relaxation of the two arsenic atoms is observed. The considerably larger attractive force in Fig. 5.19b, however, pulls the two arsenic atoms toward the tip. All other arsenic atoms are also pulled, but they are less displaced, because they have three bonds to the bulk, while the two arsenic atoms in the neighborhood of an indium vacancy have only two bonds. This direct experimental proof of the presence of tip-induced relaxations is also relevant for STM measurements, because the tip–sample distances are similar during atomic-resolution imaging. Moreover, the result demonstrates that FM-AFM can probe elastic properties on an atomic level.

Imaging of Weakly Interacting van der Waals Surfaces

For weakly interacting van der Waals surfaces, much smaller atomic corrugation amplitudes are expected compared to strongly interacting surfaces of semiconductors. A typical example is graphite, a layered material, where the carbon atoms are covalently bonded and arranged in a honeycomb structure within the (0001) plane. Individual graphene layers stick together by van der Waals forces. Due to the *ABA* stacking, three distinctive sites exist on the (0001) surface: carbon atoms with (*A*-type) and without (*B*-type) neighbor in the next graphite layer and the *hollow site* (H-site) in the hexagon center. In static contact force microscopy as well as in STM the contrast exhibits usually a trigonal symmetry with a periodicity of 246 pm, where *A*- and *B*-site carbon atoms could not be distinguished. However, in high-resolution FM-AFM images acquired at low temperatures, a large maximum and two different minima have been resolved, as demonstrated by the profiles along the three equivalent [1-100] directions in Fig. 5.20a. A simulation using the Lennard–Jones potential, given by the short-range interatomic van der Waals force, reproduced these three features very well (dotted line). Therefore, the large maximum could be assigned to the H-site, while the two different minima represent *A*- and *B*-type carbon atoms [132].

Compared to graphite, the carbon atoms in a single-walled carbon nanotube (SWNT), which consists of a single rolled up graphene layer, are indistinguishable. For the first time *Ashino* et al. [133] successfully imaged the curved surface of a SWNT with atomic resolution. Note that for geometric reasons, atomic resolution is only achieved on the top (see Fig. 5.20b). Indeed, as shown in Fig. 5.20b, all profiles between two hollow sites across two neighboring carbon atoms are symmetric [135]. Particularly, curve 1 and 2 exhibit two minima of equal depth, as predicted by theory (cf., dotted line). The assumption used in the simulation (dotted lines in the profiles of Fig. 5.20) that interatomic van der Waals forces are responsible for the atomic-scale contrast has been supported by a quantitative evaluation of force spectroscopy data obtained on SWNTs [133].

Interestingly, the image contrast on graphite and SWNTs is inverted with respect to the arrangement of the atoms, i.e., the minima correspond to the position of the carbon atoms. This can be related to the small carbon–carbon distance of only 142 pm, which is in fact the smallest interatomic distance that has been re-

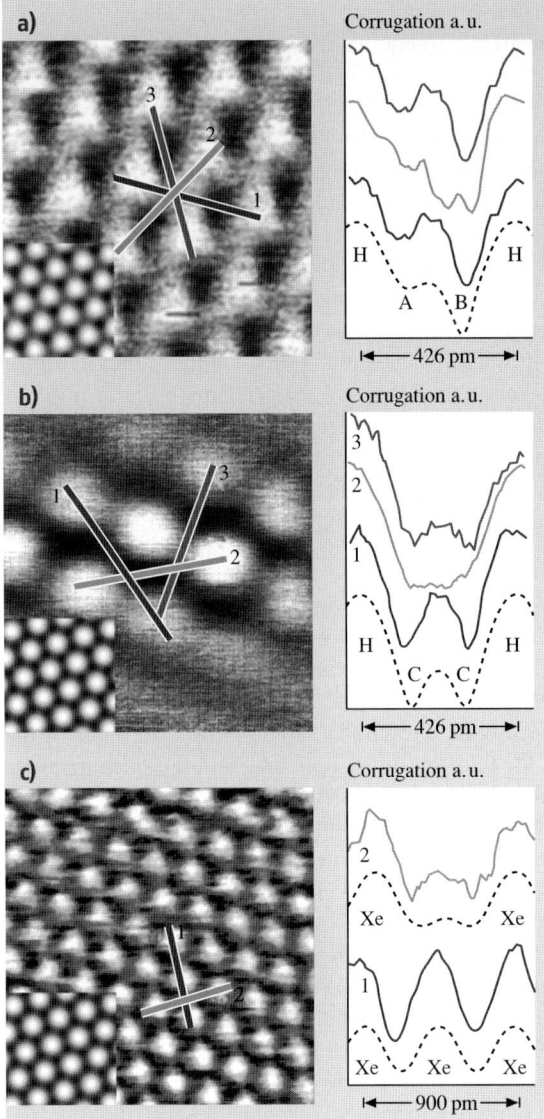

Fig. 5.20. FM-AFM images of graphite(0001) (**a**) a single-walled carbon nanotube (SWNT) (**b**) and Xe(111) (**c**) recorded at 22 K. On the right side, line sections taken from the experimental data (*solid lines*) are compared to simulations (*dotted lines*). A- and B-type carbon atoms, as well as the hollow site (H-site) on graphite can be distinguished, but are imaged with inverted contrast, i.e., the carbon sites are displayed as minima. Such an inversion does not occur on Xe(111)

solved with FM-AFM so far. The van der Waals radius of the front tip atom, (e.g., 210 pm for silicon) has a radius that is significantly larger than the intercarbon distance. Therefore, next-nearest-neighbor interactions become important and result in a contrast inversion [135].

While experiments on graphite and SWNTs basically take advantage of the increased stability and signal-to-noise ratio at low temperatures, solid xenon (melting temperature $T_m = 161$ K) can only be observed at sufficient low temperatures [8]. In addition, xenon is a pure van der Waals crystal and, since it is an insulator, FM-AFM is the only real-space method available today that allows the study of solid xenon on the atomic scale.

Allers et al. [8] adsorbed a well-ordered xenon film on cold graphite(0001) ($T < 55$ K) and studied it subsequently at 22 K by FM-AFM (see Fig. 5.20c). The sixfold symmetry and the distance between the protrusions corresponds well with the nearest-neighbor distance in the close-packed (111) plane of bulk xenon, which crystallizes in a face-centered cubic structure. A comparison between experiment and simulation confirmed that the protrusions correspond to the position of the xenon atoms [132]. However, the simulated corrugation amplitudes do not fit as well as for graphite (see sections in Fig. 5.20c). A possible reason is that tip-induced relaxations, which were not considered in the simulations, are more important for this pure van der Waals crystal xenon than they are for graphite, because in-plane graphite exhibits strong covalent bonds. Nevertheless, the results demonstrated for the first time that a weakly bonded van der Waals crystal could be imaged nondestructively on the atomic scale. Note that on Xe(111) no contrast inversion exists, presumably because the separation between Xe sites is about 450 pm, i.e., twice as large as the van der Waals radius of a silicon atom at the tip end.

Atomic Resolution Using Small Oscillation Amplitudes

All the examples above described used spring constants and amplitudes on the order of 40 N/m and 10 nm, respectively, to obtain atomic resolution. However, *Giessibl* et al. [137] pointed out that the optimal amplitude should be on the order of the characteristic decay length λ of the relevant tip–sample interaction. For short-range interactions, which are responsible for the atomic-scale contrast, λ is on the order of 0.1 nm. On the other hand, stable imaging without a jump-to-contact is only possible as long as the restoring force $c_z A$ at the lower turnaround point of each cycle is larger than the maximal attractive tip–sample force. Therefore, reducing the desired amplitude by a factor of 100 requires a 100 times larger spring constant. Indeed, *Hembacher* et al. [136] could demonstrate atomic resolution with small amplitudes (about 0.25 nm) and large spring constants (about 1800 N/m) utilizing a qPlus sensor [138]. Figure 5.21 shows a constant-height image of graphite recorded at 4.9 K within the repulsive regime. Note that compared to Fig. 5.20a,b the contrast is inverted, i.e., the carbon atoms appear as maxima. This is expected, because the imaging interaction is switched from attractive to repulsive regime [131, 135].

5.5.2 Force Spectroscopy

A wealth of information about the nature of the tip–sample interaction can be obtained by measuring its distance dependence. This is usually done by recording the measured quantity (deflection, frequency shift, amplitude change, phase shift) and applying an appropriate voltage ramp to the z-electrode of the scanner piezo, while the z-feedback is switched off. According to (5.2), low temperatures and high Q-factors (vacuum) considerably increase the force resolution. In the static mode,

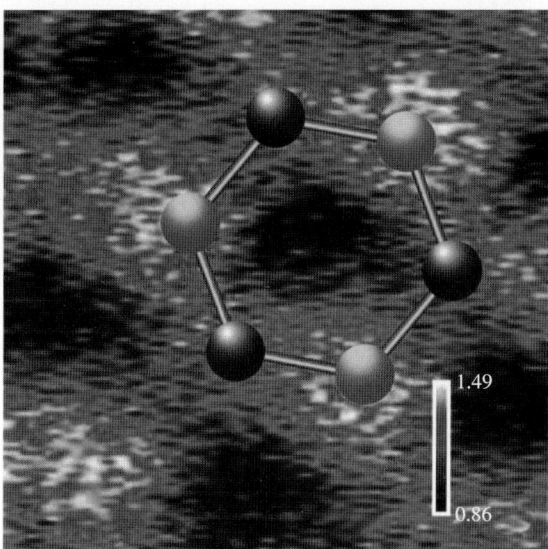

Fig. 5.21. Constant-height FM-AFM image of graphite(0001) recorded at 4.9 K using a small amplitude ($A = 0.25$ nm) and a large spring constant ($c_z = 1800$ N/m). As in Fig. 5.20a, A- and B-site carbon atoms can be distinguished. However, they appear as maxima, because imaging has been performed in the repulsive regime (courtesy of F. J. Giessibl; cf. [136])

long-range forces and contact forces can be examined. Force measurements at small tip–sample distances are inhibited by the *jump-to-contact* phenomenon: If the force gradient $\partial F_{ts}/\partial z$ becomes larger than the spring constant c_z, the cantilever cannot resist the attractive tip–sample forces and the tip snaps onto the surface. Sufficiently large spring constants prevent this effect, but reduce the force resolution. In the dynamic modes, the jump-to-contact can be avoided due to the additional restoring force ($c_z A$) at the lower turnaround point. The highest sensitivity can be achieved in vacuum by using the FM technique, i.e., by recording $\Delta f(z)$-curves. An alternative FM spectroscopy method, the recording of $\Delta f(A)$-curves, has been suggested by *Hölscher* et al. [139]. Note that, if the amplitude is much larger than the characteristic decay length of the tip–sample force, the frequency shift cannot simply be converted into force gradients by using $\partial F_{ts}/\partial z = 2c_z \cdot \Delta f / f_0$ [140]. Several methods have been published to convert $\Delta f(z)$ data into the tip–sample potential $V_{ts}(z)$ and tip–sample force $F_{ts}(z)$ (see, e.g., [141–144]).

Measurement of Interatomic Forces at Specific Atomic Sites

FM force spectroscopy has been successfully used to measure and determine quantitatively the short-range chemical force between the foremost tip atom and specific surface atoms [109, 145, 146]. Figure 5.22 displays an example for the quantitative determination of the short-range force. Figure 5.22a shows two $\Delta f(z)$-curves measured with a silicon tip above a corner hole and above an adatom. Their position is indicated by arrows in the inset, which displays the atomically resolved Si(111)-(7×7) surface. The two curves differ from each other only for small tip0-sample distances, because the long-range forces do not contribute to the atomic-scale contrast. The low, thermally induced lateral drift and the high stability at low

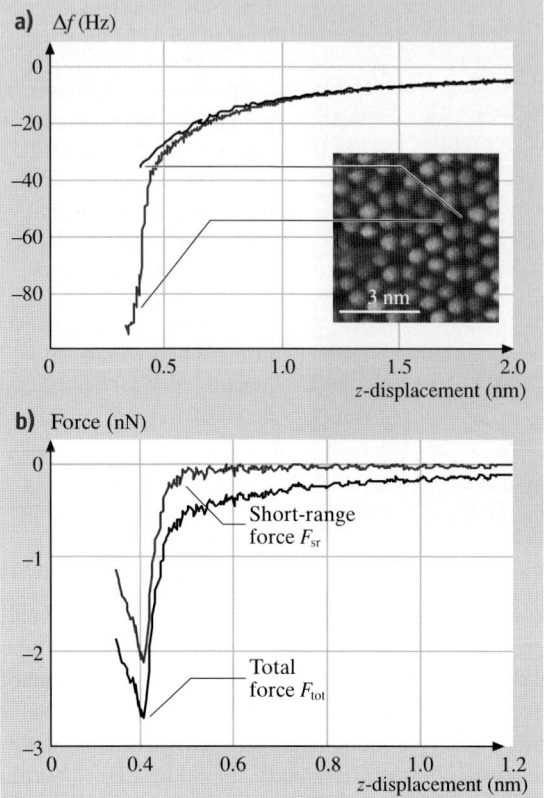

Fig. 5.22. FM force spectroscopy on specific atomic sites at 7.2 K. In (**a**), an FM-SFM image of the Si(111)-(7×7) surface is displayed together with two $\Delta f(z)$-curves, which have been recorded at the positions indicated by the *arrows*, i.e., above the corner hole (*brown*) and above an adatom (*black*). In (**b**), the total force above an adatom (*brown line*) has been recovered from the $\Delta f(z)$-curve. After subtraction of the long-range part, the short-range force can be determined (*black line*) (courtesy of H. J. Hug; cf. [145])

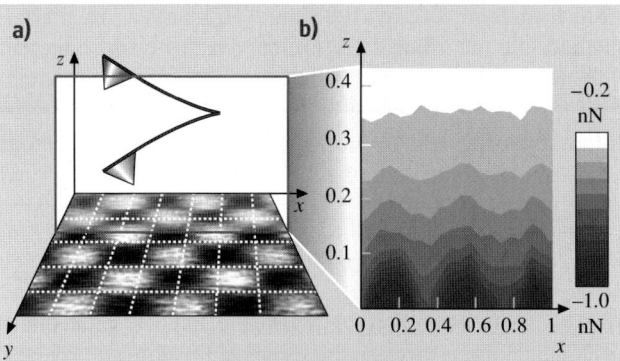

Fig. 5.23. Principle of the 3-D force field spectroscopy method (**a**) and a 2-D cut through the 3-D force field $F_{ts}(x,y,z)$ recorded at 14 K (**b**). At all 32 × 32 image points of the 1 nm × 1 nm scan area on NiO(001), a $\Delta f(z)$-curve has been recorded. The $\Delta f(x,y,z)$ data set obtained is then converted into the 3-D tip–sample force field $F_{ts}(x,y,z)$. The *shaded slice* $F_{ts}(x, y = \text{const}, z)$ in (**a**) corresponds to a cut along the [100]-direction and demonstrates that atomic resolution has been obtained, because the distance between the protrusions corresponds well to the lattice constant of nickel oxide

temperatures were required to precisely address the two specific sites. To extract the short-range force, the long-range van der Waals and/or electrostatic forces can be subtracted from the total force. The grey curve in Fig. 5.22b has been reconstructed from the $\Delta f(z)$-curve recorded above an adatom and represents the total force. After removing the long-range contribution from the data, the much steeper black line is obtained, which corresponds to the short-range force between the adatom and the atom at the tip apex. The measured maximum attractive force (-2.1 nN) agrees well with first-principles calculations (-2.25 nN).

Three-Dimensional Force Field Spectroscopy

Further progress with the FM technique has been made by *Hölscher* et al. [147]. They acquired a complete 3-D force field on NiO(001) with atomic resolution (*3-D force field spectroscopy*). In Fig. 5.23a, the atomically resolved FM-AFM image of NiO(001) is shown together with the coordinate system used and the tip to illustrate the measurement principle. NiO(001) crystallizes in the rock-salt structure. The distance between the protrusions corresponds to the lattice constant of 417 pm, i.e., only one type of atom (most likely the oxygen) is imaged as a protrusion. In an area of 1 nm × 1 nm, 32 × 32 individual $\Delta f(z)$-curves have been recorded at every (x,y) image point and converted into $F_{ts}(z)$-curves. The $\Delta f(x,y,z)$ data set is thereby converted into the 3-D force field $F_{ts}(x,y,z)$. Figure 5.23b, where a specific x–z-plane is displayed, demonstrates that atomic resolution is achieved. It represents a 2-D cut $F_{ts}(x,y=\text{const},z)$ along the [100]-direction (corresponding to the shaded slice marked in Fig. 5.23a). Since a large number of curves have been recorded, *Langkat* et al. [146] could evaluate the whole data set by standard statistical means to extract the long- and short-range forces. A possible future application of 3-D force field spectroscopy could be to map the short-range forces of complex molecules with functionalized tips in order to resolve locally their chemical reactivity. A first step in this direction has been accomplished on SWNTs. Its structural unit, a hexagonal carbon ring, is common to all aromatic molecules. Like the constant frequency-shift image of an SWNT shown in Fig. 5.20b the force map shows clear differences between hollow sites and carbon sites [133]. Analyzing site-specific individual force curves extracted from the 3-D data revealed a maximum attractive force of about -0.106 nN above H-sites and about -0.075 nN above carbon sites. Since the attraction is one order of magnitude weaker than on Si(111)-(7×7) (cf., Fig. 5.22b), it has been inferred that the short-range interatomic van der Waals force and not a chemical force is responsible for atomic-scale contrast formation on such nonreactive surfaces.

Noncontact Friction

Another approach to achieve small tip–sample distances in combination with high force sensitivity is to use soft springs in a perpendicular configuration. The much higher cantilever stiffness along the cantilever axis prevents the jump-to-contact, but the lateral resolution is limited by the magnitude of the oscillation amplitude. However, with such a set-up at low temperatures, *Stipe* et al. [148] measured the

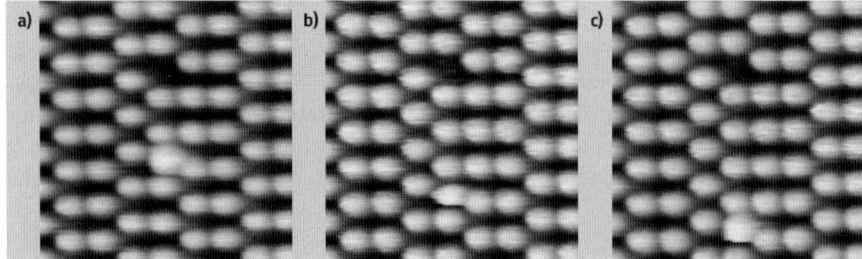

Fig. 5.24. Consecutively recorded FM-AFM images showing the tip-induced manipulation of a Ge adatom on Ge(111)-c(2 × 8) at 80 K. Scanning was performed from bottom to top (courtesy of N. Oyabu; cf. [150])

distance dependence of the very small force due to noncontact friction between the tip and sample in vacuum. The effect was attributed to electric charges, which are moved parallel to the surface by the oscillating tip. Since the topography was not recorded in situ, the influence of contaminants or surface steps remained unknown.

5.5.3 Atomic Manipulation

Nowadays, atomic-scale manipulation is routinely performed using an STM tip (see Sect. 5.4.1). In most of these experiments an adsorbate is dragged with the tip due to an attractive force between the foremost tip apex atoms and the adsorbate. By adjusting a large or a small tip–surface distance via the tunneling resistance, it is possible to switch between imaging and manipulation. Recently, it has been demonstrated that controlled manipulation of individual atoms is also possible in the dynamic mode of atomic force microscopy, i.e., FM-AFM. Vertical manipulation was demonstrated by pressing the tip in a controlled manner into the Si(111)-(7 × 7) surface [149]. The strong repulsion leads to the removal of the selected silicon atom. The process could be traced by recording the frequency shift and the damping signal during the approach. For lateral manipulation a *rubbing* technique has been utilized [150], where the slow scan axis is halted above a selected atom, while the tip–surface distance is gradually reduced until the selected atom hops to a new stable position. Figure 5.24 shows a Ge adatom on Ge(111)-c(2 × 8) that was moved during scanning in two steps from its original position (a) to its final position (c). In fact, manipulation by FM-AFM is reproducible and fast enough to write nanostructures in a bottom-up process with single atoms [151].

5.5.4 Electrostatic Force Microscopy

Electrostatic forces are readily detectable by a force microscope, because the tip and sample can be regarded as two electrodes of a capacitor. If they are electrically connected via their back sides and have different work functions, electrons will flow between the tip and sample until their Fermi levels are equalized. As a result, an electric field and, consequently, an attractive electrostatic force exists between them at zero bias. This *contact potential difference* can be balanced by applying an appropriate bias voltage. It has been demonstrated that individual doping atoms in semiconducting materials can be detected by electrostatic interactions due to the local variation of the surface potential around them [152, 153].

Detection of Edge Channels in the Quantum Hall Regime

At low temperatures, electrostatic force microscopy has been used to measure the electrostatic potential in the quantum Hall regime of a *two-dimensional electron gas* (2-DEG) buried in epitaxially grown GaAs/AlGaAs heterostructures [154–157]. In the 2-DEG, electrons can move freely in the x–y-plane, but they cannot move in z-direction. Electrical transport properties of a 2-DEG are very different compared to normal metallic conduction. Particularly, the Hall resistance $R_H = h/ne^2$ (where h represents Planck's constant, e is the electron charge, and $n = 1, 2, \ldots$) is quantized in the quantum Hall regime, i.e., at sufficiently low temperatures ($T < 4$ K) and high magnetic fields (up to 20 T). Under these conditions, theoretical calculations predict the existence of *edge channels* in a Hall bar. A Hall bar is a strip conductor that is contacted in a specific way to allow longitudinal and transversal transport measurements in a perpendicular magnetic field. The current is not evenly distributed over the cross section of the bar, but passes mainly along rather thin paths close to the edges. This prediction has been verified by measuring profiles of the electrostatic potential across a Hall bar in different perpendicular external magnetic fields [154–156].

Figure 5.25a shows the experimental set-up used to observe these edge channels on top of a Hall bar with a force microscope. The tip is positioned above the surface

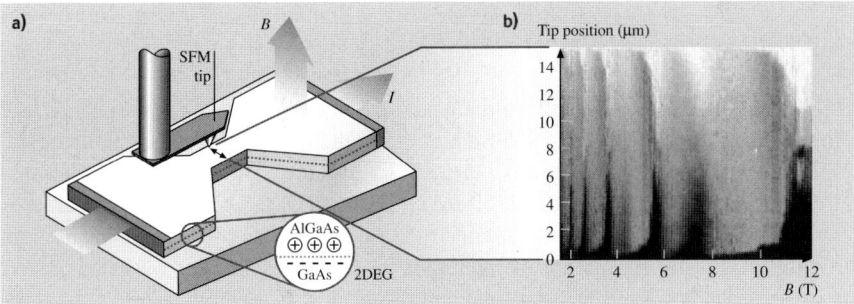

Fig. 5.25. Configuration of the Hall bar within a low temperature ($T < 1$ K) force microscope (**a**) and profiles (y-axis) at different magnetic field (x-axis) of the electrostatic potential across a 14-µm-wide Hall bar in the quantum Hall regime (**b**). The external magnetic field is oriented perpendicular to the 2-DEG, which is buried below the surface. *Bright* and *dark regions* reflect the characteristic changes of the electrostatic potential across the Hall bar at different magnetic fields and can be explained by the existence of the theoretically predicted edge channels (courtesy of E. Ahlswede; cf. [156])

of a Hall bar under which the 2-DEG is buried. The direction of the magnetic field is oriented perpendicular to the 2-DEG. Note that, although the 2-DEG is located several tens of nanometers below the surface, its influence on the electrostatic surface potential can be detected. In Fig. 5.25b, the results of scans perpendicular to the Hall bar are plotted against the magnitude of the external magnetic field. The value of the electrostatic potential is grey-coded in arbitrary units. In certain field ranges, the potential changes linearly across the Hall bar, while in other field ranges the potential drop is confined to the edges of the Hall bar. The predicted edge channels can explain this behavior. The periodicity of the phenomenon is related to the filling factor v, i.e., the number of Landau levels that are filled with electrons (see also Sect. 5.4.4). Its value depends on $1/B$ and is proportional to the electron concentration n_e in the 2-DEG ($v = n_e h/eB$, where h represents Planck's constant and e the electron charge).

5.5.5 Magnetic Force Microscopy

To detect magnetostatic tip–sample interactions with magnetic force microscopy (MFM), a ferromagnetic probe has to be used. Such probes are readily prepared by evaporating a thin magnetic layer, e.g., 10 nm iron, onto the tip. Due to the in-plane shape anisotropy of thin films, the magnetization of such tips lies predominantly along the tip axis, i.e., perpendicular to the surface. Since magnetostatic interactions are long-range, they can be separated from the topography by scanning in a certain constant height (typically around 20 nm) above the surface, where the z-component of the sample stray field is probed (see Fig. 5.26a). Therefore, MFM is always operated in noncontact mode. The signal from the cantilever is directly recorded while the z-feedback is switched off. MFM can be operated in the static mode or in the dynamic modes (AM-MFM at ambient pressures and FM-MFM in vacuum). A lateral resolution below 50 nm can be routinely obtained.

Observation of Domain Patterns

MFM is widely used to visualize domain patterns of ferromagnetic materials. At low temperatures, *Moloni* et al. [158] observed the domain structure of magnetite below its Verwey transition temperature ($T_V = 122$ K), but most of the work concentrated on thin films of $La_{1-x}Ca_xMnO_3$ [159–161]. Below T_V, the conductivity decreases by two orders of magnitude and a small structural distortion is observed. The domain structure of this mixed-valence manganite is of great interest, because its resistivity strongly depends on the external magnetic field, i.e., it exhibits a large colossal-magnetoresistive effect. To investigate the field dependence of the domain patterns under ambient conditions, electromagnets have to be used. They can cause severe thermal drift problems due to Joule heating of the coils by large currents. Fields on the order of 100 mT can be achieved. In contrast, much larger fields (more than 10 T) can be rather easily produced by implementing a superconducting magnet in low-temperature set-ups. With such a design, *Liebmann* et al. [161] recorded the

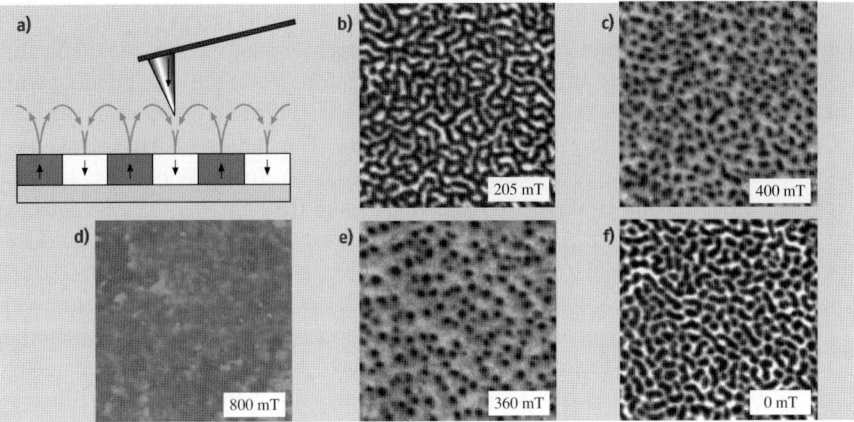

Fig. 5.26. Principle of MFM operation (**a**) and field-dependent domain structure of a ferromagnetic thin film (**b**)–(**f**) recorded at 5.2 K with FM-MFM. All images were recorded on the same 4 μm × 4 μm scan area. The $La_{0.7}Ca_{0.3}MnO_3/LaAlO_3$ system exhibits a substrate-induced out-of-plane anisotropy. *Bright* and *dark* areas are visible and correspond to attractive and repulsive magnetostatic interactions, respectively. The series shows how the domain pattern evolves along the major hysteresis loop from, i.e., zero field to saturation at 600 mT and back to zero field

domain structure along the major hysteresis loop of $La_{0.7}Ca_{0.3}MnO_3$ epitaxially grown on $LaAlO_3$ (see Fig. 5.26b–f). The film geometry (the thickness is 100 nm) favors an in-plane magnetization, but the lattice mismatch with the substrate induces an out-of-plane anisotropy. Thereby, an irregular pattern of strip domains appears at zero field. If the external magnetic field is increased, the domains with antiparallel orientation shrink and finally disappear in saturation (see Fig. 5.26b,c). The residual contrast in saturation (d) reflects topographic features. If the field is decreased after saturation (see Fig. 5.26e,f), cylindrical domains first nucleate and then start to grow. At zero field, the maze-type domain pattern has evolved again. Such data sets can be used to analyze domain nucleation and the domain growth mode. Moreover, due to the negligible drift, domain structure and surface morphology can be directly compared, because every MFM can be used as a regular topography-imaging force microscope.

Detection of Individual Vortices in Superconductors

Numerous low-temperature MFM experiments have been performed on superconductors [162–169]. Some basic features of superconductors have been mentioned already in Sect. 5.4.5. The main difference of STM/STS compared to MFM is its high sensitivity to the electronic properties of the surface. Therefore, careful sample preparation is a prerequisite. This is not so important for MFM experiments, since the tip is scanned at a certain distance above the surface.

Superconductors can be divided into two classes with respect to their behavior in an external magnetic field. For type-I superconductors, any magnetic flux is entirely excluded below their critical temperature T_c (Meissner effect), while for type-II superconductors, cylindrical inclusions (*vortices*) of normal material exist in a superconducting matrix (*vortex* state). The radius of the vortex *core*, where the Cooper-pair density decreases to zero, is on the order of the coherence length ξ. Since the superconducting gap vanishes in the core, they can be detected by STS (see Sect. 5.4.5). Additionally, each vortex contains one magnetic quantum flux $\Phi = h/2e$ (where h represents Planck's constant and e the electron charge). Circular supercurrents around the core screen the magnetic field associated with a vortex; their radius is given by the London penetration depth λ of the material. This magnetic field of the vortices can be detected by MFM. Investigations have been performed on the two most popular copper oxide high-T_c superconductors, $YBa_2Cu_3O_7$ [162, 163, 165] and $Bi_2Sr_2CaCu_2O_8$ [163, 169], on the only elemental conventional type-II superconductor Nb [166, 167], and on the layered compound crystal $NbSe_2$ [164, 166].

Most often, vortices have been generated by cooling the sample from the normal state to below T_c in an external magnetic field. After such a *field-cooling* procedure, the most energetically favorable vortex arrangement is a regular triangular Abrikosov lattice. *Volodin* et al. [164] were able to observe such an Abrikosov lattice on $NbSe_2$. The intervortex distance d is related to the external field during B cool down via $d = (4/3)^{1/4}(\Phi/B)^{1/2}$. Another way to introduce vortices into a type-II superconductor is vortex penetration from the edge by applying a magnetic field at temperatures below T_c. According to the Bean model, a vortex density gradient exists under such conditions within the superconducting material. *Pi* et al. [169] slowly increased the external magnetic field until the vortex front approaching from the edge reached the scanning area.

If the vortex configuration is dominated by the *pinning* of vortices at randomly distributed structural defects, no Abrikosov lattice emerges. The influence of pinning centers can be studied easily by MFM, because every MFM can be used to scan the topography in its AFM mode. This has been done for natural growth defects by *Moser* et al. [165] on $YBa_2Cu_3O_7$ and for $YBa_2Cu_3O_7$ and niobium thin films, respectively, by *Volodin* et al. [168]. *Roseman* et al. [170] investigated the formation of vortices in the presence of an artificial structure on niobium films, while *Pi* et al. [169] produced columnar defects by heavy-ion bombardment in a $Bi_2Sr_2CaCu_2O_8$ single crystal to study the strong pinning at these defects.

Figure 5.27 demonstrates that MFM is sensitive to the polarity of vortices. In Fig. 5.27a, six vortices have been produced in a niobium film by field cooling in $+ 0.5$ mT. The external magnetic field and tip magnetization are parallel, and, therefore, the tip–vortex interaction is attractive (bright contrast). To remove the vortices, the niobium was heated above T_c (≈ 9 K). Thereafter, vortices of opposite polarity were produced by field-cooling in -0.5 mT, which appear dark in Fig. 5.27b. The vortices are probably bound to strong pinning sites, because the vortex positions are identical in both images of Fig. 5.27. By imaging the vortices at different scanning

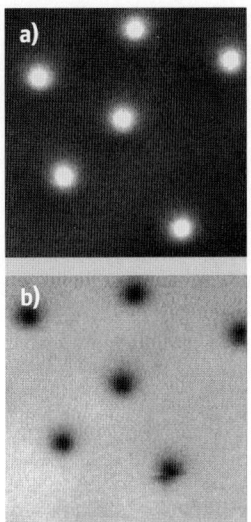

Fig. 5.27. Two 5 μm × 5 μm FM-MFM images of vortices in a niobium thin film after field-cooling at 0.5 mT (**a**) and −0.5 mT (**b**), respectively. Since the external magnetic field was parallel in (**a**) and antiparallel in (**b**) with respect to the tip magnetization, the vortices exhibit reversed contrast. Strong pinning dominates the position of the vortices, since they appear at identical locations in (**a**) and (**b**) and are not arranged in a regular Abrikosov lattice (courtesy of P. Grütter; cf. [167])

heights, *Roseman* et al. [167] tried to extract values for the London penetration depth from the scan height dependence of their profiles. While good qualitative agreement with theoretical predictions has been found, the absolute values do not agree with published literature values. The disagreement was attributed to the convolution between the tip and vortex stray fields. Better values might be obtained with calibrated tips.

Single Spin Detection

So far, only collective magnetic phenomena like ferromagnetic domains have been observed via magnetostatic tip–sample interactions detected by MFM. However, magnetic ordering exists due to the magnetic exchange interaction between the electron spins of neighboring atoms in a solid. The most energetically favorable situation can be either ferromagnetic (parallel orientation) or antiferromagnetic (antiparallel orientation) ordering. It has been predicted that the magnetic exchange force between an individual spin of a magnetically ordered sample and the spin of the foremost atom of a magnetic tip can be detected at sufficiently small tip–sample distances [172, 173].

The experimental realization, however, is very difficult, because the magnetic exchange force is about a factor of ten weaker and of even shorter range than the chemical interactions that are responsible for the atomic-scale contrast. FM-AFM experiments with a ferromagnetic tip have been performed on the antiferromagnetic NiO(001) surface at room temperature [174] and with a considerable better signal-to-noise ratio at low temperatures [122]. Although it was possible to achieve atomic resolution, a periodic contrast that could be attributed to the antiferromagnetically ordered spins of the nickel atoms could not be observed.

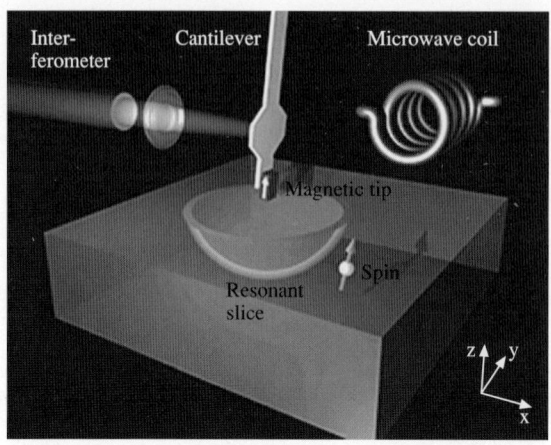

Fig. 5.28. MRFM set-up. The cantilever with the magnetic tip oscillates parallel to the surface. Only electron spins within a hemispherical slice, where the stray field of the tip plus the external field matches the condition for magnetic resonance, can contribute to the MRFM signal due to cyclic spin inversion (courtesy of D. Rugar; cf. [171])

Even more ambitious is the proposed detection of individual nuclear spins by magnetic resonance force microscopy (MRFM using a magnetic tip [175,176]. Conventionally, nuclear spins are investigated by nuclear magnetic resonance (NMR), a spectroscopic technique to obtain microscopic chemical and physical information about molecules. An important application of NMR for medical diagnostics of the inside of humans is magnetic resonance imaging (MRI). This tomographic imaging technique uses the NMR signal from thin slices through the body to reconstruct its three-dimensional structure. Currently, at least 10^{12} nuclear spins must be present in a given volume to obtain a significant MRI signal. The ultimate goal of MRFM is to combine aspects of force microscopy with MRI to achieve true 3-D imaging with atomic resolution and elemental selectivity.

The experimental set-up is sketched in Fig. 5.28. An oscillating cantilever with a magnetic tip at its end points toward the surface. The spherical resonant slice within the sample represents those points where the stray field from the tip and the external field match the condition for magnetic resonance. The cyclic spin flip causes a slight shift of the cantilever frequency due to the magnetic force exerted by the spin on the tip. Since the forces are extremely small, very low temperatures are required.

To date, no individual nuclear spins have been detected by MRFM. However, the design of ultrasensitive cantilevers has made considerable progress, and the detection of forces below 1×10^{-18} N has been achieved [120]. Therefore, it has become possible to perform nuclear magnetic resonance [177], and ferromagnetic resonance [178] experiments of spin ensembles with micrometer resolution. Moreover, in SiO_2 the magnetic moment of a single electron, which is three orders of magnitude larger than the nuclear magnetic moment, could be detected [171] using the set-up shown in Fig. 5.28 at 1.6 K. This major breakthrough demonstrates the capability of force microscopy to detect single spins.

References

1. G. Binnig, H. Rohrer, Ch. Gerber, E. Weibel: Surface studies by scanning tunneling microscopy, Phys. Rev. Lett. **49**, 57–61 (1982)
2. R. Wiesendanger: *Scanning Probe Microscopy and Spectroscopy* (Cambridge Univ. Press, Cambridge 1994)
3. M. Tinkham: *Introduction to Superconductivity* (McGraw–Hill, New York 1996)
4. J. Kondo: Theory of dilute magnetic alloys, Solid State Phys. **23**, 183–281 (1969)
5. T.R. Albrecht, P. Grütter, H.K. Horne, D. Rugar: Frequency modulation detection using high-Q cantilevers for enhanced force microscope sensitivity, J. Appl. Phys. **69**, 668–673 (1991)
6. F.J. Giessibl, H. Bielefeld, S. Hembacher, J. Mannhart: Calculation of the optimal imaging parameters for frequency modulation atomic force microscopy, Appl. Surf. Sci. **140**, 352–357 (1999)
7. W. Allers, A. Schwarz, U.D. Schwarz, R. Wiesendanger: Dynamic scanning force microscopy at low temperatures on a van der Waals surface: graphite(0001), Appl. Surf. Sci. **140**, 247–252 (1999)
8. W. Allers, A. Schwarz, U.D. Schwarz, R. Wiesendanger: Dynamic scanning force microscopy at low temperatures on a noble-gas crystal: atomic resolution on the xenon(111) surface, Europhys. Lett. **48**, 276–279 (1999)
9. M. Morgenstern, D. Haude, V. Gudmundsson, C. Wittneven, R. Dombrowski, R. Wiesendanger: Origin of Landau oscillations observed in scanning tunneling spectroscopy on n-InAs(110), Phys. Rev. B **62**, 7257–7263 (2000)
10. D.M. Eigler, P.S. Weiss, E.K. Schweizer, N.D. Lang: Imaging Xe with a low-temperature scanning tunneling microscope, Phys. Rev. Lett. **66**, 1189–1192 (1991)
11. P.S. Weiss, D.M. Eigler: Site dependence of the apparent shape of a molecule in scanning tunneling micoscope images: Benzene on Pt{111}, Phys. Rev. Lett. **71**, 3139–3142 (1992)
12. D.M. Eigler, E.K. Schweizer: Positioning single atoms with a scanning tunneling microscope, Nature **344**, 524–526 (1990)
13. H. Hug, B. Stiefel, P.J.A. van Schendel, A. Moser, S. Martin, H.-J. Güntherodt: A low temperature ultrahigh vacuum scanning force microscope, Rev. Sci. Instrum. **70**, 3627–3640 (1999)
14. S. Behler, M.K. Rose, D.F. Ogletree, F. Salmeron: Method to characterize the vibrational response of a beetle type scanning tunneling microscope, Rev. Sci. Instrum. **68**, 124–128 (1997)
15. C. Wittneven, R. Dombrowski, S.H. Pan, R. Wiesendanger: A low-temperature ultrahigh-vacuum scanning tunneling microscope with rotatable magnetic field, Rev. Sci. Instrum. **68**, 3806–3810 (1997)
16. W. Allers, A. Schwarz, U.D. Schwarz, R. Wiesendanger: A scanning force microscope with atomic resolution in ultrahigh vacuum and at low temperatures, Rev. Sci. Instrum. **69**, 221–225 (1998)
17. G. Dujardin, R.E. Walkup, Ph. Avouris: Dissociation of individual molecules with electrons from the tip of a scanning tunneling microscope, Science **255**, 1232–1235 (1992)
18. H.J. Lee, W. Ho: Single-bond formation and characterization with a scanning tunneling microscope, Science **286**, 1719–1722 (1999)
19. R. Berndt, R. Gaisch, J.K. Gimzewski, B. Reihl, R.R. Schlittler, W.D. Schneider, M. Tschudy: Photon emission at molecular resolution induced by a scanning tunneling microscope, Science **262**, 1425–1427 (1993)

20. B. G. Briner, M. Doering, H. P. Rust, A. M. Bradshaw: Microscopic diffusion enhanced by adsorbate interaction, Science **278**, 257–260 (1997)
21. J. Kliewer, R. Berndt, E. V. Chulkov, V. M. Silkin, P. M. Echenique, S. Crampin: Dimensionality effects in the lifetime of surface states, Science **288**, 1399–1401 (2000)
22. M. F. Crommie, C. P. Lutz, D. M. Eigler: Imaging standing waves in a two-dimensional electron gas, Nature **363**, 524–527 (1993)
23. B. C. Stipe, M. A. Rezaei, W. Ho: Single-molecule vibrational spectroscopy and microscopy, Science **280**, 1732–1735 (1998)
24. H. J. Lee, W. Ho: Structural determination by single-molecule vibrational spectroscopy and microscopy: Contrast between copper and iron carbonyls, Phys. Rev. B **61**, R16347–R16350 (2000)
25. C. W. J. Beenakker, H. van Houten: Quantum transport in semiconductor nanostructures, Solid State Phys. **44**, 1–228 (1991)
26. S. H. Pan, E. W. Hudson, K. M. Lang, H. Eisaki, S. Uchida, J. C. Davis: Imaging the effects of individual zinc impurity atoms on superconductivity in $Bi_2Sr_2CaCu_2O_{8+\delta}$, Nature **403**, 746–750 (2000)
27. R. S. Becker, J. A. Golovchenko, B. S. Swartzentruber: Atomic-scale surface modifications using a tunneling microscope, Nature **325**, 419–42 (1987)
28. J. A. Stroscio, D. M. Eigler: Atomic and molecular manipulation with the scanning tunneling microscope, Science **254**, 1319–1326 (1991)
29. L. Bartels, G. Meyer, K. H. Rieder: Basic steps of lateral manipulation of single atoms and diatomic clusters with a scanning tunneling microscope, Phys. Rev. Lett. **79**, 697–700 (1997)
30. J. J. Schulz, R. Koch, K. H. Rieder: New mechanism for single atom manipulation, Phys. Rev. Lett. **84**, 4597–4600 (2000)
31. T. C. Shen, C. Wang, G. C. Abeln, J. R. Tucker, J. W. Lyding, Ph. Avouris, R. E. Walkup: Atomic-scale desorption through electronic and vibrational excitation mechanisms, Science **268**, 1590–1592 (1995)
32. T. Komeda, Y. Kim, M. Kawai, B. N. J. Persson, H. Ueba: Lateral hopping of molecules induced by excitations of internal vibration mode, Science **295**, 2055–2058 (2002)
33. Y. W. Mo: Reversible rotation of antimony dimers on the silicon(001) surface with a scanning tunneling microscope, Science **261**, 886–888 (1993)
34. B. C. Stipe, M. A. Rezaei, W. Ho: Inducing and viewing the rotational motion of a single molecule, Science **279**, 1907–1909 (1998)
35. F. Moresco, G. Meyer, K. H. Rieder, H. Tang, A. Gourdon, C. Joachim: Conformational changes of single molecules by scanning tunneling microscopy manipulation: a route to molecular switching, Phys. Rev. Lett. **86**, 672–675 (2001)
36. S. W. Hla, L. Bartels, G. Meyer, K. H. Rieder: Inducing all steps of a chemical reaction with the scanning tunneling microscope tip: Towards single molecule engineering, Phys. Rev. Lett. **85**, 2777–2780 (2000)
37. E. Ganz, S. K. Theiss, I. S. Hwang, J. Golovchenko: Direct measurement of diffusion by hot tunneling microscopy: Activations energy, anisotropy, and long jumps, Phys. Rev. Lett. **68**, 1567–1570 (1992)
38. M. Schuhnack, T. R. Linderoth, F. Rosei, E. Laegsgaard, I. Stensgaard, F. Besenbacher: Long jumps in the surface diffusion of large molecules, Phys. Rev. Lett. **88**, 156102, 1–4 (2002)
39. L. J. Lauhon, W. Ho: Direct observation of the quantum tunneling of single hydrogen atoms with a scanning tunneling microscope, Phys. Rev. Lett. **85**, 4566–4569 (2000)

40. N. Kitamura, M. Lagally, M. B. Webb: Real-time observation of vacancy diffusion on Si(001)-(2 × 1) by scanning tunneling microscopy, Phys. Rev. Lett. **71**, 2082–2085 (1993)
41. M. Morgenstern, T. Michely, G. Comsa: Onset of interstitial diffusion determined by scanning tunneling microscopy, Phys. Rev. Lett. **79**, 1305–1308 (1997)
42. K. Morgenstern, G. Rosenfeld, B. Poelsema, G. Comsa: Brownian motion of vacancy islands on Ag(111), Phys. Rev. Lett. **74**, 2058–2061 (1995)
43. B. Reihl, J. H. Coombs, J. K. Gimzewski: Local inverse photoemission with the scanning tunneling microscope, Surf. Sci. **211–212**, 156–164 (1989)
44. R. Berndt, J. K. Gimzewski, P. Johansson: Inelastic tunneling excitation of tip-induced plasmon modes on noble-metal surfaces, Phys. Rev. Lett. **67**, 3796–3799 (1991)
45. P. Johansson, R. Monreal, P. Apell: Theory for light emission from a scanning tunneling microscope, Phys. Rev. B **42**, 9210–9213 (1990)
46. J. Aizpurua, G. Hoffmann, S. P. Apell, R. Berndt: Electromagnetic coupling on an atomic scale, Phys. Rev. Lett. **89**, 156803, 1–4 (2002)
47. G. Hoffmann, J. Kliewer, R. Berndt: Luminescence from metallic quantum wells in a scanning tunneling microscope, Phys. Rev. Lett. **78**, 176803, 1–4 (2001)
48. A. Downes, M. E. Welland: Photon emission from Si(111)-(7×7) induced by scanning tunneling microscopy: atomic scale and material contrast, Phys. Rev. Lett. **81**, 1857–1860 (1998)
49. M. Kemerink, K. Sauthoff, P. M. Koenraad, J. W. Geritsen, H. van Kempen, J. H. Wolter: Optical detection of ballistic electrons injected by a scanning–tunneling microscope, Phys. Rev. Lett. **86**, 2404–2407 (2001)
50. J. Tersoff, D. R. Hamann: Theory and application for the scanning tunneling microscope, Phys. Rev. Lett. **50**, 1998–2001 (1983)
51. C. J. Chen: *Introduction to Scanning Tunneling Microscopy* (Oxford Univ. Press, Oxford 1993)
52. J. Winterlin, J. Wiechers, H. Brune, T. Gritsch, H. Hofer, R. J. Behm: Atomic-resolution imaging of close-packed metal surfaces by scanning tunneling microscopy, Phys. Rev. Lett. **62**, 59–62 (1989)
53. A. L. Vazquez de Parga, O. S. Hernan, R. Miranda, A. Levy Yeyati, N. Mingo, A. Martin-Rodero, F. Flores: Electron resonances in sharp tips and their role in tunneling spectroscopy, Phys. Rev. Lett. **80**, 357–360 (1998)
54. S. H. Pan, E. W. Hudson, J. C. Davis: Vacuum tunneling of superconducting quasiparticles from atomically sharp scanning tunneling microscope tips, Appl. Phys. Lett. **73**, 2992–2994 (1998)
55. J. T. Li, W. D. Schneider, R. Berndt, O. R. Bryant, S. Crampin: Surface-state lifetime measured by scanning tunneling spectroscopy, Phys. Rev. Lett. **81**, 4464–4467 (1998)
56. L. Bürgi, O. Jeandupeux, H. Brune, K. Kern: Probing hot-electron dynamics with a cold scanning tunneling microscope, Phys. Rev. Lett. **82**, 4516–4519 (1999)
57. J. W. G. Wildoer, C. J. P. M. Harmans, H. van Kempen: Observation of Landau levels at the InAs(110) surface by scanning tunneling spectroscopy, Phys. Rev. B **55**, R16013–R16016 (1997)
58. M. Morgenstern, V. Gudmundsson, C. Wittneven, R. Dombrowski, R. Wiesendanger: Nonlocality of the exchange interaction probed by scanning tunneling spectroscopy, Phys. Rev. B **63**, 201301(R), 1–4 (2001)
59. M. V. Grishin, F. I. Dalidchik, S. A. Kovalevskii, N. N. Kolchenko, B. R. Shub: Isotope effect in the vibrational spectra of water measured in experiments with a scanning tunneling microscope, JETP Lett. **66**, 37–40 (1997)

60. A. Hewson: *From the Kondo Effect to Heavy Fermions* (Cambridge Univ. Press, Cambridge 1993)
61. V. Madhavan, W. Chen, T. Jamneala, M. F. Crommie, N. S. Wingreen: Tunneling into a single magnetic atom: Spectroscopic evidence of the Kondo resonance, Science **280**, 567–569 (1998)
62. J. Li, W. D. Schneider, R. Berndt, B. Delley: Kondo scattering observed at a single magnetic impurity, Phys. Rev. Lett. **80**, 2893–2896 (1998)
63. T. W. Odom, J. L. Huang, C. L. Cheung, C. M. Lieber: Magnetic clusters on single-walled carbon nanotubes: the Kondo effect in a one-dimensional host, Science **290**, 1549–1552 (2000)
64. M. Ouyang, J. L. Huang, C. L. Cheung, C. M. Lieber: Energy gaps in metallic single-walled carbon nanotubes, Science **292**, 702–705 (2001)
65. U. Fano: Effects of configuration interaction on intensities and phase shifts, Phys. Rev. **124**, 1866–1878 (1961)
66. H. C. Manoharan, C. P. Lutz, D. M. Eigler: Quantum mirages formed by coherent projection of electronic structure, Nature **403**, 512–515 (2000)
67. O. Y. Kolesnychenko, R. de Kort, M. I. Katsnelson, A. I. Lichtenstein, H. van Kempen: Real-space observation of an orbital Kondo resonance on the Cr(001) surface, Nature **415**, 507–509 (2002)
68. H. A. Mizes, J. S. Foster: Long-range electronic perturbations caused by defects using scanning tunneling microscopy, Science **244**, 559–562 (1989)
69. P. T. Sprunger, L. Petersen, E. W. Plummer, E. Laegsgaard, F. Besenbacher: Giant Friedel oscillations on beryllium(0001) surface, Science **275**, 1764–1767 (1997)
70. P. Hofmann, B. G. Briner, M. Doering, H. P. Rust, E. W. Plummer, A. M. Bradshaw: Anisotropic two-dimensional Friedel oscillations, Phys. Rev. Lett. **79**, 265–268 (1997)
71. E. J. Heller, M. F. Crommie, C. P. Lutz, D. M. Eigler: Scattering and adsorption of surface electron waves in quantum corrals, Nature **369**, 464–466 (1994)
72. M. C. M. M. van der Wielen, A. J. A. van Roij, H. van Kempen: Direct observation of Friedel oscillations around incorporated Si_{Ga} dopants in GaAs by low-temperature scanning tunneling microscopy, Phys. Rev. Lett. **76**, 1075–1078 (1996)
73. O. Millo, D. Katz, Y. W. Cao, U. Banin: Imaging and spectroscopy of artificial-atom states in core/shell nanocrystal quantum dots, Phys. Rev. Lett. **86**, 5751–5754 (2001)
74. L. C. Venema, J. W. G. Wildoer, J. W. Janssen, S. J. Tans, L. J. T. Tuinstra, L. P. Kouwenhoven, C. Dekker: Imaging electron wave functions of quantized energy levels in carbon nanotubes, Nature **283**, 52–55 (1999)
75. S. G. Lemay, J. W. Jannsen, M. van den Hout, M. Mooij, M. J. Bronikowski, P. A. Willis, R. E. Smalley, L. P. Kouwenhoven, C. Dekker: Two-dimensional imaging of electronic wavefunctions in carbon nanotubes, Nature **412**, 617–620 (2001)
76. C. Wittneven, R. Dombrowski, M. Morgenstern, R. Wiesendanger: Scattering states of ionized dopants probed by low temperature scanning tunneling spectroscopy, Phys. Rev. Lett. **81**, 5616–5619 (1998)
77. D. Haude, M. Morgenstern, I. Meinel, R. Wiesendanger: Local density of states of a three-dimensional conductor in the extreme quantum limit, Phys. Rev. Lett. **86**, 1582–1585 (2001)
78. R. Joynt, R. E. Prange: Conditions for the quantum Hall effect, Phys. Rev. B **29**, 3303–3317 (1984)
79. M. Morgenstern, J. Klijn, C. Meyer, M. Getzlaff, R. Adelung, R. A. Römer, K. Rossnagel, L. Kipp, M. Skibowski, R. Wiesendanger: Direct comparison between potential landscape and local density of states in a disordered two-dimensional electron system, Phys. Rev. Lett. **89**, 136806, 1–4 (2002)

80. E. Abrahams, P.W. Anderson, D.C. Licciardello, T.V. Ramakrishnan: Scaling theory of localization: absence of quantum diffusion in two dimensions, Phys. Rev. Lett. **42**, 673–676 (1979)
81. M. Morgenstern, J. Klijn, R. Wiesendanger: Real space observation of drift states in a two-dimensional electron system at high magnetic fields, Phys. Rev. Lett. **90**, 056804, 1–4 (2003)
82. R.E. Peierls: *Quantum Theory of Solids* (Clarendon, Oxford 1955)
83. C.G. Slough, W.W. McNairy, R.V. Coleman, B. Drake, P.K. Hansma: Charge-density waves studied with the use of a scanning tunneling microscope, Phys. Rev. B **34**, 994–1005 (1986)
84. X.L. Wu, C.M. Lieber: Hexagonal domain-like charge-density wave of TaS_2 determined by scanning tunneling microscopy, Science **243**, 1703–1705 (1989)
85. T. Nishiguchi, M. Kageshima, N. Ara-Kato, A. Kawazu: Behaviour of charge density waves in a one-dimensional organic conductor visualized by scanning tunneling microscopy, Phys. Rev. Lett. **81**, 3187–3190 (1998)
86. X.L. Wu, C.M. Lieber: Direct observation of growth and melting of the hexagonal-domain charge-density-wave phase in $1T$-TaS_2 by scanning tunneling microscopy, Phys. Rev. Lett. **64**, 1150–1153 (1990)
87. J.M. Carpinelli, H.H. Weitering, E.W. Plummer, R. Stumpf: Direct observation of a surface charge density wave, Nature **381**, 398–400 (1996)
88. H.H. Weitering, J.M. Carpinelli, A.V. Melechenko, J. Zhang, M. Bartkowiak, E.W. Plummer: Defect-mediated condensation of a charge density wave, Science **285**, 2107–2110 (1999)
89. H.W. Yeom, S. Takeda, E. Rotenberg, I. Matsuda, K. Horikoshi, J. Schäfer, C.M. Lee, S.D. Kevan, T. Ohta, T. Nagao, S. Hasegawa: Instability and charge density wave of metallic quantum chains on a silicon surface, Phys. Rev. Lett. **82**, 4898–4901 (1999)
90. K. Swamy, A. Menzel, R. Beer, E. Bertel: Charge-density waves in self-assembled halogen-bridged metal chains, Phys. Rev. Lett. **86**, 1299–1302 (2001)
91. J.J. Kim, W. Yamaguchi, T. Hasegawa, K. Kitazawa: Observation of Mott localization gap using low temperature scanning tunneling spectroscopy in commensurate $1T$-$TaSe_2$, Phys. Rev. Lett. **73**, 2103–2106 (1994)
92. J. Bardeen, L.N. Cooper, J.R. Schrieffer: Theory of superconductivity, Phys. Rev. **108**, 1175–1204 (1957)
93. A. Yazdani, B.A. Jones, C.P. Lutz, M.F. Crommie, D.M. Eigler: Probing the local effects of magnetic impurities on superconductivity, Science **275**, 1767–1770 (1997)
94. S.H. Tessmer, M.B. Tarlie, D.J. van Harlingen, D.L. Maslov, P.M. Goldbart: Probing the superconducting proximity effect in $NbSe_2$ by scanning tunneling micrsoscopy, Phys. Rev. Lett **77**, 924–927 (1996)
95. K. Inoue, H. Takayanagi: Local tunneling spectroscopy of Nb/InAs/Nb superconducting proximity system with a scanning tunneling microscope, Phys. Rev. B **43**, 6214–6215 (1991)
96. H.F. Hess, R.B. Robinson, R.C. Dynes, J.M. Valles, J.V. Waszczak: Scanning-tunneling-microscope observation of the Abrikosov flux lattice and the density of states near and inside a fluxoid, Phys. Rev. Lett. **62**, 214–217 (1989)
97. H.F. Hess, R.B. Robinson, J.V. Waszczak: Vortex-core structure observed with a scanning tunneling microscope, Phys. Rev. Lett. **64**, 2711–2714 (1990)
98. N. Hayashi, M. Ichioka, K. Machida: Star-shaped local density of states around vortices in a type-II superconductor, Phys. Rev. Lett. **77**, 4074–4077 (1996)
99. H. Sakata, M. Oosawa, K. Matsuba, N. Nishida: Imaging of vortex lattice transition in YNi_2B_2C by scanning tunneling spectroscopy, Phys. Rev. Lett. **84**, 1583–1586 (2000)

100. S. Behler, S.H. Pan, P. Jess, A. Baratoff, H.-J. Güntherodt, F. Levy, G. Wirth, J. Wiesner: Vortex pinning in ion-irradiated $NbSe_2$ studied by scanning tunneling microscopy, Phys. Rev. Lett. **72**, 1750–1753 (1994)
101. R. Berthe, U. Hartmann, C. Heiden: Influence of a transport current on the Abrikosov flux lattice observed with a low-temperature scanning tunneling microscope, Ultramicroscopy **42-44**, 696–698 (1992)
102. A. Polkovnikov, S. Sachdev, M. Vojta: Impurity in a d-wave superconductor: Kondo effect and STM spectra, Phys. Rev. Lett. **86**, 296–299 (2001)
103. E.W. Hudson, K.M. Lang, V. Madhavan, S.H. Pan, S. Uchida, J.C. Davis: Interplay of magnetism and high-T_c superconductivity at individual Ni impurity atoms in $Bi_2Sr_2CaCu_2O_{8+\delta}$, Nature **411**, 920–924 (2001)
104. K.M. Lang, V. Madhavan, J.E. Hoffman, E.W. Hudson, H. Eisaki, S. Uchida, J.C. Davis: Imaging the granular structure of high-T_c superconductivity in underdoped $Bi_2Sr_2CaCu_2O_{8+\delta}$, Nature **415**, 412–416 (2002)
105. I. Maggio-Aprile, C. Renner, E. Erb, E. Walker, Ø. Fischer: Direct vortex lattice imaging and tunneling spectroscopy of flux lines on $YBa_2Cu_3O_{7-\delta}$, Phys. Rev. Lett. **75**, 2754–2757 (1995)
106. C. Renner, B. Revaz, K. Kadowaki, I. Maggio-Aprile, Ø. Fischer: Observation of the low temperature pseudogap in the vortex cores of $Bi_2Sr_2CaCu_2O_{8+\delta}$, Phys. Rev. Lett. **80**, 3606–3609 (1998)
107. S.H. Pan, E.W. Hudson, A.K. Gupta, K.W. Ng, H. Eisaki, S. Uchida, J.C. Davis: STM studies of the electronic structure of vortex cores in $Bi_2Sr_2CaCu_2O_{8+\delta}$, Phys. Rev. Lett. **85**, 1536–1539 (2000)
108. D.P. Arovas, A.J. Berlinsky, C. Kallin, S.C. Zhang: Superconducting vortex with antiferromagnetic core, Phys. Rev. Lett. **79**, 2871–2874 (1997)
109. J.E. Hoffmann, E.W. Hudson, K.M. Lang, V. Madhavan, H. Eisaki, S. Uchida, J.C. Davis: A four unit cell periodic pattern of quasi-particle states surrounding vortex cores in $Bi_2Sr_2CaCu_2O_{8+\delta}$, Science **295**, 466–469 (2002)
110. M. Fäth, S. Freisem, A.A. Menovsky, Y. Tomioka, J. Aarts, J.A. Mydosh: Spatially inhomogeneous metal–insulator transition in doped manganites, Science **285**, 1540–1542 (1999)
111. C. Renner, G. Aeppli, B.G. Kim, Y.A. Soh, S.W. Cheong: Atomic-scale images of charge ordering in a mixed-valence manganite, Nature **416**, 518–521 (2000)
112. M. Bode, M. Getzlaff, R. Wiesendanger: Spin-polarized vacuum tunneling into the exchange-split surface state of Gd(0001), Phys. Rev. Lett. **81**, 4256–4259 (1998)
113. A. Kubetzka, M. Bode, O. Pietzsch, R. Wiesendanger: Spin-polarized scanning tunneling microscopy with antiferromagnetic probe tips, Phys. Rev. Lett. **88**, 057201, 1–4 (2002)
114. O. Pietzsch, A. Kubetzka, M. Bode, R. Wiesendanger: Observation of magnetic hysteresis at the nanometer scale by spin-polarized scanning tunneling spectroscopy, Science **292**, 2053–2056 (2001)
115. S. Heinze, M. Bode, A. Kubetzka, O. Pietzsch, X. Xie, S. Blügel, R. Wiesendanger: Real-space imaging of two-dimensional antiferromagnetism on the atomic scale, Science **288**, 1805–1808 (2000)
116. A. Wachowiak, J. Wiebe, M. Bode, O. Pietzsch, M. Morgenstern, R. Wiesendanger: Internal spin-structure of magnetic vortex cores observed by spin-polarized scanning tunneling microscopy, Science **298**, 577–580 (2002)
117. M.D. Kirk, T.R. Albrecht, C.F. Quate: Low-temperature atomic force microscopy, Rev. Sci. Instrum. **59**, 833–835 (1988)

118. D. Pelekhov, J. Becker, J.G. Nunes: Atomic force microscope for operation in high magnetic fields at milliKelvin temperatures, Rev. Sci. Instrum. **70**, 114–120 (1999)
119. J. Mou, Y. Jie, Z. Shao: An optical detection low temperature atomic force microscope at ambient pressure for biological research, Rev. Sci. Instrum. **64**, 1483–1488 (1993)
120. H.J. Mamin, D. Rugar: Sub-attoNewton force detection at milliKelvin temperatures, Appl. Phys. Lett. **79**, 3358–3360 (2001)
121. A. Schwarz, W. Allers, U.D. Schwarz, R. Wiesendanger: Dynamic mode scanning force microscopy of n-InAs(110)-(1×1) at low temperatures, Phys. Rev. B **61**, 2837–2845 (2000)
122. W. Allers, S. Langkat, R. Wiesendanger: Dynamic low-temperature scanning force microscopy on nickel oxide(001), Appl. Phys. A **72**, S27–S30 (2001)
123. F.J. Giessibl: Atomic resolution of the silicon(111)-(7×7) surface by atomic force microscopy, Science **267**, 68–71 (1995)
124. M.A. Lantz, H.J. Hug, P.J.A. van Schendel, R. Hoffmann, S. Martin, A. Baratoff, A. Abdurixit, H.-J. Güntherodt: Low temperature scanning force microscopy of the Si(111)-(7×7) surface, Phys. Rev. Lett. **84**, 2642–2465 (2000)
125. K. Suzuki, H. Iwatsuki, S. Kitamura, C.B. Mooney: Development of low temperature ultrahigh vacuum force microscope/scanning tunneling microscope, Jpn. J. Appl. Phys. **39**, 3750–3752 (2000)
126. N. Suehira, Y. Sugawara, S. Morita: Artifact and fact of Si(111)-(7×7) surface images observed with a low temperature noncontact atomic force microscope (LT-NC-AFM), Jpn. J. Appl. Phys. **40**, 292–294 (2001)
127. R. Peréz, M.C. Payne, I. Štich, K. Terakura: Role of covalent tip–surface interactions in noncontact atomic force microscopy on reactive surfaces, Phys. Rev. Lett. **78**, 678–681 (1997)
128. S.H. Ke, T. Uda, R. Pérez, I. Štich, K. Terakura: First principles investigation of tip–surface interaction on GaAs(110): Implication for atomic force and tunneling microscopies, Phys. Rev. B **60**, 11631–11638 (1999)
129. J. Tobik, I. Štich, R. Peréz, K. Terakura: Simulation of tip–surface interactions in atomic force microscopy of an InP(110) surface with a Si tip, Phys. Rev. B **60**, 11639–11644 (1999)
130. A. Schwarz, W. Allers, U.D. Schwarz, R. Wiesendanger: Simultaneous imaging of the In and As sublattice on InAs(110)-(1×1) with dynamic scanning force microscopy, Appl. Surf. Sci. **140**, 293–297 (1999)
131. H. Hölscher, W. Allers, U.D. Schwarz, A. Schwarz, R. Wiesendanger: Interpretation of 'true atomic resolution' images of graphite (0001) in noncontact atomic force microscopy, Phys. Rev. B **62**, 6967–6970 (2000)
132. H. Hölscher, W. Allers, U.D. Schwarz, A. Schwarz, R. Wiesendanger: Simulation of NC-AFM images of xenon(111), Appl. Phys. A **72**, S35–S38 (2001)
133. M. Ashino, A. Schwarz, T. Behnke, R. Wiesendanger: Atomic-resolution dynamic force microscopy and spectroscopy of a single-walled carbon nanotube: characterization of interatomic van der Waals forces, Phys. Rev. Lett. **93**, 136101, 1–4 (2004)
134. G. Schwarz, A. Kley, J. Neugebauer, M. Scheffler: Electronic and structural properties of vacancies on and below the GaP(110) surface, Phys. Rev. B **58**, 1392–1499 (1998)
135. M. Ashino, A. Schwarz, H. Hölscher, U.D. Schwarz, R. Wiesendanger: Interpretation of the atomic scale contrast obtained on graphite and single-walled carbon nanotubes in the dynamic mode of atomic force microscopy, Nanotechnology **16**, 134–137 (2005)
136. S. Hembacher, F.J. Giessibl, J. Mannhart, C.F. Quate: Local spectroscopy and atomic imaging of tunneling current, forces, and dissipation on graphite, Phys. Rev. Lett. **94**, 056101, 1–4 (2005)

137. F.J. Giessibl, H. Bielefeldt, S. Hembacher, J. Mannhart: Calculation of the optimal imaging parameters for frequency modulation atomic force microscopy, Appl. Surf. Sci. **140**, 352–357 (1999)
138. F.J. Giessibl: High-speed force sensor for force microscopy and profilometry utilizing a quartz tuning fork, Appl. Phys. Lett. **73**, 3956–3958 (1998)
139. H. Hölscher, W. Allers, U.D. Schwarz, A. Schwarz, R. Wiesendanger: Determination of tip–sample interaction potentials by dynamic force spectroscopy, Phys. Rev. Lett. **83**, 4780–4783 (1999)
140. H. Hölscher, U.D. Schwarz, R. Wiesendanger: Calculation of the frequency shift in dynamic force microscopy, Appl. Surf. Sci. **140**, 344–351 (1999)
141. B. Gotsman, B. Anczykowski, C. Seidel, H. Fuchs: Determination of tip–sample interaction forces from measured dynamic force spectroscopy curves, Appl. Surf. Sci. **140**, 314–319 (1999)
142. U. Dürig: Extracting interaction forces and complementary observables in dynamic probe microscopy, Appl. Phys. Lett. **76**, 1203–1205 (2000)
143. F.J. Giessibl: A direct method to calculate tip–sample forces from frequency shifts in frequency-modulation atomic force microscopy, Appl. Phys. Lett. **78**, 123–125 (2001)
144. J.E. Sader, S.P. Jarvis: Accurate formulas for interaction force and energy in frequency modulation force spectroscopy, Appl. Phys. Lett. **84**, 1801–1803 (2004)
145. M.A. Lantz, H.J. Hug, R. Hoffmann, P.J.A. van Schendel, P. Kappenberger, S. Martin, A. Baratoff, H.-J. Güntherodt: Quantitative measurement of short-range chemical bonding forces, Science **291**, 2580–2583 (2001)
146. S.M. Langkat, H. Hölscher, A. Schwarz, R. Wiesendanger: Determination of site specific forces between an iron coated tip and the NiO(001) surface by force field spectroscopy, Surf. Sci. (2002)in press
147. H. Hölscher, S.M. Langkat, A. Schwarz, R. Wiesendanger: Measurement of three-dimensional force fields with atomic resolution using dynamic force spectroscopy, Appl. Phys. Lett. (2002)in press
148. B.C. Stipe, H.J. Mamin, T.D. Stowe, T.W. Kenny, D. Rugar: Noncontact friction and force fluctuations between closely spaced bodies, Phys. Rev. Lett. **87** (2001)
149. N. Oyabu, O. Custance, I. Yi, Y. Sugawara, S. Morita: Mechanical vertical manipulation of selected single atoms by soft nanoindentation using near contact atomic force microscopy, Phys. Rev. Lett. **90**, 176102, 1–4 (2004)
150. N. Oyabu, Y. Sugimoto, M. Abe, O. Custance, S. Morita: Lateral manipulation of single atoms at semiconductor surfaces using atomic force microscopy, Nanotechnology **16**, 112–117 (2005)
151. Y. Sugimoto, M. Abe, S. Hirayama, N. Oyabu, O. Custance, S. Morita: Atom inlays performed at room temperature using atomic force microscopy, Nature Mater. **4**, 156–160 (2005)
152. C. Sommerhalter, T.W. Matthes, T. Glatzel, A. Jäger-Waldau, M.C. Lux-Steiner: High-sensitivity quantitative Kelvin probe microscopy by noncontact ultra-high-vacuum atomic force microscopy, Appl. Phys. Lett. **75**, 286–288 (1999)
153. A. Schwarz, W. Allers, U.D. Schwarz, R. Wiesendanger: Dynamic mode scanning force microscopy of n-InAs(110)-(1×1) at low temperatures, Phys. Rev. B **62**, 13617–13622 (2000)
154. K.L. McCormick, M.T. Woodside, M. Huang, M. Wu, P.L. McEuen, C. Duruoz, J.S. Harris: Scanned potential microscopy of edge and bulk currents in the quantum Hall regime, Phys. Rev. B **59**, 4656–4657 (1999)

155. P. Weitz, E. Ahlswede, J. Weis, K. v. Klitzing, K. Eberl: Hall-potential investigations under quantum Hall conditions using scanning force microscopy, Physica E **6**, 247–250 (2000)
156. E. Ahlswede, P. Weitz, J. Weis, K. v. Klitzing, K. Eberl: Hall potential profiles in the quantum Hall regime measured by a scanning force microscope, Physica B **298**, 562–566 (2001)
157. M.T. Woodside, C. Vale, P.L. McEuen, C. Kadow, K.D. Maranowski, A.C. Gossard: Imaging interedge-state scattering centers in the quantum Hall regime, Phys. Rev. B **64** (2001)041310-1–041310-4
158. K. Moloni, B.M. Moskowitz, E.D. Dahlberg: Domain structures in single crystal magnetite below the Verwey transition as observed with a low-temperature magnetic force microscope, Geophys. Res. Lett. **23**, 2851–2854 (1996)
159. Q. Lu, C.C. Chen, A. de Lozanne: Observation of magnetic domain behavior in colossal magnetoresistive materials with a magnetic force microscope, Science **276**, 2006–2008 (1997)
160. G. Xiao, J.H. Ross, A. Parasiris, K.D.D. Rathnayaka, D.G. Naugle: Low-temperature MFM studies of CMR manganites, Physica C **341–348**, 769–770 (2000)
161. M. Liebmann, U. Kaiser, A. Schwarz, R. Wiesendanger, U.H. Pi, T.W. Noh, Z.G. Khim, D.W. Kim: Domain nucleation and growth of $La_{07}Ca_{0.3}MnO_{3-\delta}/LaAlO_3$ films studied by low temperature MFM, J. Appl. Phys. **93**, 8319–8321 (2003)
162. A. Moser, H.J. Hug, I. Parashikov, B. Stiefel, O. Fritz, H. Thomas, A. Baratoff, H.J. Güntherodt, P. Chaudhari: Observation of single vortices condensed into a vortex-glass phase by magnetic force microscopy, Phys. Rev. Lett. **74**, 1847–1850 (1995)
163. C.W. Yuan, Z. Zheng, A.L. de Lozanne, M. Tortonese, D.A. Rudman, J.N. Eckstein: Vortex images in thin films of $YBa_2Cu_3O_{7-x}$ and $Bi_2Sr_2Ca_1Cu_2O_{8-x}$ obtained by low-temperature magnetic force microscopy, J. Vac. Sci. Technol. B **14**, 1210–1213 (1996)
164. A. Volodin, K. Temst, C. van Haesendonck, Y. Bruynseraede: Observation of the Abrikosov vortex lattice in $NbSe_2$ with magnetic force microscopy, Appl. Phys. Lett. **73**, 1134–1136 (1998)
165. A. Moser, H.J. Hug, B. Stiefel, H.J. Güntherodt: Low temperature magnetic force microscopy on $YBa_2Cu_3O_{7-\delta}$ thin films, J. Magn. Magn. Mater. **190**, 114–123 (1998)
166. A. Volodin, K. Temst, C. van Haesendonck, Y. Bruynseraede: Imaging of vortices in conventional superconductors by magnetic force microscopy images, Physica C **332**, 156–159 (2000)
167. M. Roseman, P. Grütter: Estimating the magnetic penetration depth using constant-height magnetic force microscopy images of vortices, New J. Phys. **3**, 24.1–24.8 (2001)
168. A. Volodin, K. Temst, C. van Haesendonck, Y. Bruynseraede, M.I. Montero, I.K. Schuller: Magnetic force microscopy of vortices in thin niobium films: Correlation between the vortex distribution and the thickness-dependent film morphology, Europhys. Lett. **58**, 582–588 (2002)
169. U.H. Pi, T.W. Noh, Z.G. Khim, U. Kaiser, M. Liebmann, A. Schwarz, R. Wiesendanger: Vortex dynamics in $Bi_2Sr_2CaCu_2O_8$ single crystal with low density columnar defects studied by magnetic force microscopy, J. Low Temp. Phys. **131**, 993–1002 (2003)
170. M. Roseman, P. Grütter, A. Badia, V. Metlushko: Flux lattice imaging of a patterned niobium thin film, J. Appl. Phys. **89**, 6787–6789 (2001)
171. D. Rugar, R. Budakian, H.J. Mamin, B.W. Chui: Single spin detection by magnetic resonance force microscopy, Nature **430**, 329–332 (2004)

172. K. Nakamura, H. Hasegawa, T. Oguchi, K. Sueoka, K. Hayakawa, K. Mukasa: First-principles calculation of the exchange interaction and the exchange force between magnetic Fe films, Phys. Rev. B **56**, 3218–3221 (1997)
173. A. S. Foster, A. L. Shluger: Spin-contrast in non-contact AFM on oxide surfaces: Theoretical modeling of NiO(001) surface, Surf. Sci. **490**, 211–219 (2001)
174. H. Hoisoi, M. Kimura, K. Hayakawa, K. Sueoka, K. Mukasa: Non-contact atomic force microscopy of an antiferromagnetic NiO(100) surface using a ferromagnetic tip, Appl. Phys. A **72**, S23–S26 (2001)
175. J. A. Sidles, J. L. Garbini, G. P. Drobny: The theory of oscillator-coupled magnetic resonance with potential applications to molecular imaging, Rev. Sci. Instrum. **63**, 3881–3899 (1992)
176. J. A. Sidles, J. L. Garbini, K. J. Bruland, D. Rugar, O. Züger, S. Hoen, C. S. Yannoni: Magnetic resonance force microscopy, Rev. Mod. Phys. **67**, 249–265 (1995)
177. D. Rugar, O. Züger, S. Hoen, C. S. Yannoni, H. M. Vieth, R. D. Kendrick: Force detection of nuclear magnetic resonance, Science **264**, 1560–1563 (1994)
178. Z. Zhang, P. C. Hammel, P. E. Wigen: Observation of ferromagnetic resonance in a microscopic sample using magnetic resonance force microscopy, Appl. Phys. Lett. **68**, 2005–2007 (1996)

6

Dynamic Modes of Atomic Force Microscopy

A. Schirmeisen, B. Anczykowski, and Harald Fuchs

Summary. This chapter presents an introduction to the concept of the dynamic operational modes of the atomic force microscope (AFM). While the static (or contact-mode) AFM is a widespread technique to obtain nanometer-resolution images on a wide variety of surfaces, true atomic-resolution imaging is routinely observed only in the dynamic mode. We will explain the jump-to-contact phenomenon encountered in static AFM and present the dynamic operational mode as a solution to avoid this effect. The dynamic force microscope is modeled as a harmonic oscillator to gain a basic understanding of the underlying physics in this mode.

Under closer inspection dynamic AFM comprises a whole family of operational modes. A systematic overview of the different modes typically found in force microscopy is presented, and special attention is paid to the distinct features of each mode. Two modes of operation dominate the application of dynamic AFM. First, the amplitude-modulation mode (also called the tapping mode) is shown to exhibit an instability, which separates the purely attractive force interaction regime from the attractive–repulsive regime. Second, the self-excitation mode is derived and its experimental realization is outlined. While the tapping mode is primarily used for imaging in air and liquid, the self-excitation mode is typically used under ultrahigh vacuum (UHV) conditions for atomic-resolution imaging. In particular, we explain the influence of different forces on spectroscopy curves obtained in dynamic force microscopy. A quantitative link between the experimental spectroscopy curves and the interaction forces is established.

Force microscopy in air suffers from the small quality factors of the force sensor (i.e., the cantilever beam), which are shown to limit the resolution. Also, the aforementioned instability in the amplitude-modulation mode often hinders imaging of soft and fragile samples. A combination of the amplitude modulation with the self-excitation mode is shown to increase the quality, or Q-factor, and extend the regime of stable operation. This so-called Q-control module allows one to increase as well as decrease the Q-factor. Apart from the advantages of dynamic force microscopy as a nondestructive, high-resolution imaging method, it can also be used to obtain information about energy-dissipation phenomena at the nanometer scale. This measurement channel can provide crucial information on electric and magnetic surface properties. Even atomic-resolution imaging has been obtained in the dissipation mode. Therefore, in the last section, the quantitative relation between the experimental measurement channels and the dissipated power is derived.

6.1 Motivation: Measurement of a Single Atomic Bond

The direct measurement of the force interaction between two distinct molecules has been a challenge for scientists for many years now. The fundamental forces responsible for the solid state of matter can be directly investigated ultimately between defined single molecules. However, it has not been until very recently that the chemical forces could be quantitatively measured for a single atomic bond [1]. How can we reliably measure forces that may be as small as one billionth of 1 N? How can we identify one single pair of atoms as the source of the force interaction?

The same mechanical principle that is used to measure the gravitational force exerted by your body weight (e. g., with the scale in your bathroom) can be employed to measure the forces between single atoms. A spring with a defined elasticity is compressed by an arbitrary force (e.g., your weight). The compression Δz of the spring (with spring constant k) is a direct measure of the force F exerted, which in the regime of elastic deformation obeys Hooke's law:

$$F = k \cdot \Delta z . \tag{6.1}$$

The only difference with regard to your bathroom scale is the sensitivity of the measurement. Typically springs with a stiffness of 0.1–10 N/m are used, which will be deflected by 0.1–100 nm upon application of an interatomic force of some nN. Experimentally, a laser-deflection technique is used to measure the movement of the spring. The spring is a bendable cantilever microfabricated from silicon wafers. If a sufficiently sharp tip, usually directly attached to the cantilever, approaches within some nanometers of a surface, we can measure the interacting forces through changes in the deflected laser beam. This is a static measurement; hence it is called *static AFM*. Alternatively, the cantilever can be excited to vibrate at its resonant frequency. Under the influence of tip–sample forces the resonant frequency (and consequently also the amplitude and phase) of the cantilever will change and serve as measurement parameters. This is called *dynamic AFM*. Due to the multitude of possible operational modes, expressions such as noncontact mode, intermittent-contact mode, tapping-mode, frequency-modulation (FM)-mode, amplitude modulation (AM)-mode, self-excitation, constant-excitation, or constant-amplitude-mode AFM are found in the literature, which will be systematically categorized in the following paragraphs.

In fact, the first AFMs were operated in the dynamic mode. In 1986, *Binnig, Quate* and *Gerber* presented the concept of the atomic force microscope [2]. The deflection of the cantilever with the tip was measured with sub-angstrom precision by an additional scanning tunneling microscope (STM). While the cantilever was externally oscillated close to its resonant frequency, the amplitude and phase of the oscillation were measured. If the tip approaches the surface, the oscillation parameters, amplitude and phase, are influenced by the tip–surface interaction, and can, therefore, be used as feedback channels. Typically, a certain set point for the amplitude is defined, and the feedback loop will adjust the tip–sample distance such that the amplitude remains constant. The controller parameter is recorded as a function of the lateral position of the tip with respect to the sample, and the scanned image essentially represents the surface topography.

What then is the difference between the static and the dynamic mode of operation for the AFM? The static-deflection AFM directly gives the interaction force between tip and sample using (6.1). In the dynamic mode, we find that the resonant frequency, amplitude, and phase of the oscillation change as a consequence of the interaction forces (and also dissipative processes, as discussed in the last section).

In order to get a basic understanding of the underlying physics, it is instructive to consider a very simplified case. Assume that the vibration amplitude is small compared to the range of force interaction. Since van der Waals forces range over typical distances of 10 nm, the vibration amplitude should be less than 1 nm. Furthermore, we require that the force gradient $\partial F_{ts}/\partial z$ does not vary significantly over one oscillation cycle. We can view the AFM set-up as a coupling of two springs (see Fig. 6.1). Whereas the cantilever is represented by a spring with spring constant k, the force interaction between tip and surface can be modeled by a second spring. The derivative of the force with respect to the tip–sample distance is the force gradient and represents the spring constant k_{ts} of the interaction spring. This spring constant k_{ts} is constant only with respect to one oscillation cycle, but varies with the average tip–sample distance as the probe approaches the sample. The two springs are effectively coupled in parallel, since the sample and tip supports are rigidly connected for a given value of z_0. Therefore, we can write for the total spring constant of the AFM system:

$$k_{\text{total}} = k + k_{ts} = k - \frac{\partial F_{ts}}{\partial z}. \tag{6.2}$$

From the simple harmonic oscillator (neglecting any damping effects) we find that the resonant frequency ω of the system is shifted by $\Delta\omega$ from the free resonant frequency ω_0 due to the force interaction:

$$\omega^2 = (\omega_0 + \Delta\omega)^2 = k_{\text{total}}/m^* = \left(k - \frac{\partial F_{ts}}{\partial z}\right)/m^*. \tag{6.3}$$

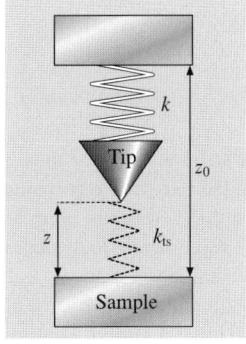

Fig. 6.1. Model of the AFM tip while experiencing tip–sample forces. The tip is attached to a cantilever with spring constant k, and the force interaction is modeled by a spring with a stiffness equal to the force gradient. Note that the force interaction spring is not constant, but depends on the tip–sample distance z

Here m^* represents the effective mass of the cantilever. A detailed analysis of how m^* is related to the geometry and total mass of the cantilever can be found in the literature [3]. In the approximation that $\Delta\omega$ is much smaller than ω_0, we can write:

$$\frac{\Delta\omega}{\omega_0} \cong -\frac{1}{2k} \cdot \frac{\partial F_{ts}}{\partial z}. \tag{6.4}$$

Therefore, we find that the frequency shift of the cantilever resonance is proportional to the force gradient of the tip–sample interaction.

Although this consideration is based on a very simplified model, it shows qualitatively that in dynamic force microscopy we will find that the oscillation frequency depends on the force gradient, while static force microscopy measures the force itself. In principle, we can calculate the force curve from the force gradient and vice versa (neglecting a constant offset). It seems, therefore, that the two methods are equivalent, and our choice will depend on whether we can measure the beam deflection or the frequency shift with better precision at the cost of technical effort.

However, we have neglected one important issue for the operation of the AFM so far: the mechanical stability of the measurement. In static AFM, the tip approaches the surface slowly. The force between the tip and the surface will always be counteracted by the restoring force of the cantilever. In Fig. 6.2, you can see a typical force–distance curve. As the tip approaches the sample, the negative attractive forces, rep-

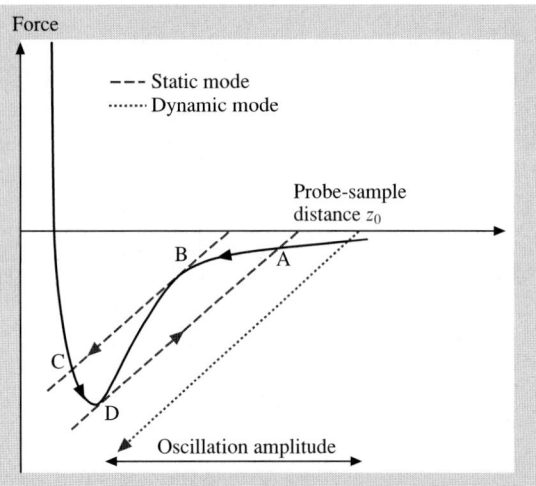

Fig. 6.2. Force–distance curve of a typical tip–sample interaction. In static-mode AFM the tip would follow the force curve until point B is reached. If the slope of the force curve becomes larger than the spring constant of the cantilever (*dashed line*) the tip will suddenly jump to position C. Upon retraction a different path will be followed along D and A again. In dynamic AFM the cantilever oscillates with a preset amplitude. Although the equilibrium position of the oscillation is far from the surface, the tip will experience the maximum attractive force at point D during some parts of the oscillation cycle. However, the total force is always pointing away from the surface, therefore avoiding an instability

resenting van der Waals or chemical interaction forces, increase until a maximum is reached. This turnaround point is due to the onset of repulsive forces caused by Coulomb repulsion, which will start to dominate upon further approach. The spring constant of the cantilever is represented by the slope of the straight line. The position of the z-transducer (typically a piezo element), which moves the probe, is at the intersection of the line with the horizontal axis. The position of the tip, shifted from the probe's base due to the lever bending, can be found at the intersection of the cantilever line with the force curve. Hence, the total force is zero, i.e., the cantilever is in its equilibrium position (note that the spring constant line here shows attractive forces, although in reality the forces are repulsive, i.e., pulling the tip back from the surface). As soon as the position A in Fig. 6.2 is reached, we find two possible intersection points, and upon further approach there are even three force equilibrium points. However, between points A and B the tip is at a local energy minimum and, therefore, will still follow the force curve. But at point B, when the adhesion force upon further approach would become larger than the spring restoring force, the tip will suddenly jump to point C. We can then probe the predominantly repulsive force interaction by further reducing the tip–sample distance. When retracting the tip, we will pass point C, because the tip is still in a local energy minimum. Only at position D will the tip suddenly jump to point A again, since the restoring force now exceeds the adhesion. From Fig. 6.2 we can see that the sudden instability will happen at exactly the point where the slope of the adhesion force exceeds the slope of the spring constant. Therefore, if the negative force gradient of the tip–sample interaction at any point exceeds the spring constant, a mechanical instability occurs. Mathematically speaking, we demand that for a stable measurement:

$$\frac{\partial F_{ts}}{\partial z}|_z > k \text{ for all points } z \tag{6.5}$$

The phenomenon of mechanical instability is often referred to as the *jump-to-contact*.

Looking at Fig. 6.2, we realize that large parts of the force curve cannot be measured if the jump-to-contact phenomenon occurs. We will not be able to measure the point at which the attractive forces reach their maximum, representing the temporary chemical bonding of the tip and the surface atoms. Secondly, the sudden instability, the jump-to-contact, will often cause the tip to change the very last tip or surface atoms. A smooth, careful approach needed to measure the full force curve does not seem feasible. Our goal of measuring the chemical interaction forces of two single molecules may become impossible.

There are several solutions to the jump-to-contact problem. On one hand, we can simply choose a sufficiently stiff spring, so that (6.5) is fulfilled at all points of the force curve. On the other hand, we can resort to a trick to enhance the counteracting force of the cantilever: we can oscillate the cantilever with large amplitude, thereby making it virtually stiffer at the point of strong force interaction.

Consider the first solution, which seems simpler at first glance. Chemical bonding forces extend over a distance range of about 0.1 nm. Typical binding energies of a couple of eV will lead to adhesion forces on the order of some nN. Force gradients will, therefore, reach values of some 10 N/m. A spring for stable force measurements will have to be as stiff as 100 N/m to ensure that no instability occurs (a safety factor of ten seems to be a minimum requirement, since usually one cannot be sure a priori that only one atom will dominate the interaction). In order to measure the nN interaction force, a static cantilever deflection of 0.01 nm has to be detected. With standard beam deflection AFM setups this becomes a challenging task.

This problem was solved [4,5] using an *in-situ* optical interferometer measuring the beam deflection at liquid-nitrogen temperature in a UHV environment. In order to ensure that the force gradients are smaller than the lever spring constant (50 N/m), the tips were fabricated to terminate in only three atoms, therefore, minimizing the total force interaction. The field ion microscope (FIM) is a tool that allows one to engineer scanning probe microscopy (SPM) tips down to atomic dimensions. This technique not only allows imaging of the tip apex with atomic precision, but can also be used to manipulate the tip atoms by field evaporation [6], as shown in Fig. 6.3. Atomic interaction forces were measured with sub-nanonewton precision, revealing force curves of only a few atoms interacting without mechanical hysteresis. However, the technical effort to achieve this type of measurement is considerable, and most researchers today have resorted to the second solution.

The alternative solution can be visualized in Fig. 6.2. The straight, dashed line now represents the force values of the oscillating cantilever, with amplitude A assuming Hooke's law is valid. This is the tensile force of the cantilever spring pulling the tip away from the sample. The restoring force of the cantilever is at all points stronger than the adhesion force. For example, the total force at point D is still pointing away from the sample, although the spring has the same stiffness as be-

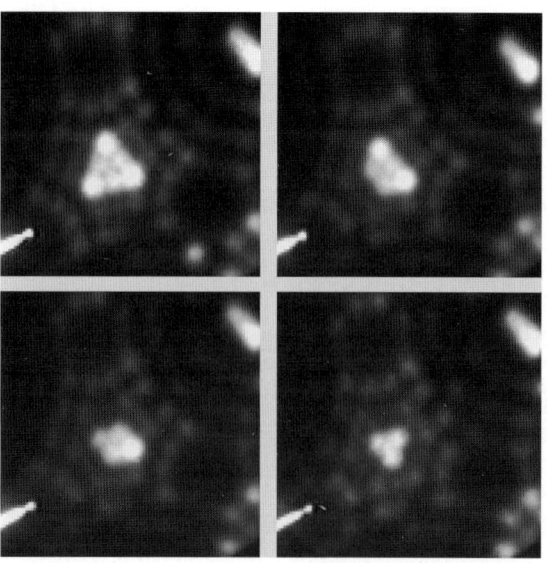

Fig. 6.3. Manipulation of the apex atoms of an AFM tip using field ion microscopy (FIM). Images are acquired at a tip bias of 4.5 kV. The last six atoms of the tip can be inspected in this example. Field evaporation to remove single atoms is performed by increasing the bias voltage for a short time to 5.2 kV. Each of the outer three atoms can be consecutively removed, eventually leaving a trimer tip apex

fore. Mathematically speaking, the measurement is stable as long as the cantilever spring force $F_{cb} = kA$ is larger than the attractive tip–sample force F_{ts} [7]. In the static mode we would already experience an instability at that point. However, in the dynamic mode, the spring is preloaded with a force stronger than the attractive tip–sample force. The equilibrium point of the oscillation is still far away from the point of closest contact of the tip and surface atoms. The total force curve can now be probed by varying the equilibrium point of the oscillation, i.e., by adjusting the z-piezo.

The diagram also shows that the oscillation amplitude has to be quite large if fairly soft cantilevers are to be used. With lever spring constants of $10\,\mathrm{N/m}$, the amplitude must be at least 1 nm to ensure that forces of 1 nN can be reliably measured. In practical applications, amplitudes of 10–100 nm are used to incorporate a safety margin. This means that the oscillation amplitude is much larger than the force interaction range. The above simplification, that the force gradient remains constant within one oscillation cycle, does not hold anymore. Measurement stability is gained at the cost of a simple quantitative analysis of the experiments. In fact, dynamic AFM was first used to obtain atomic resolution images of clean surfaces [8], and it took another six years [1] before quantitative measurements of single bond forces were obtained.

The technical realization of dynamic mode AFMs is based on the same key components as a static AFM set-up. The most common principle is the method of laser deflection sensing (see, e.g., Fig. 6.4). A laser beam is focused on the back side of a microfabricated cantilever. The reflected laser spot is detected with a position-sensitive diode (PSD). This photodiode is sectioned into two parts that are read out separately (usually even a four-quadrant diode is used to detect torsional movements

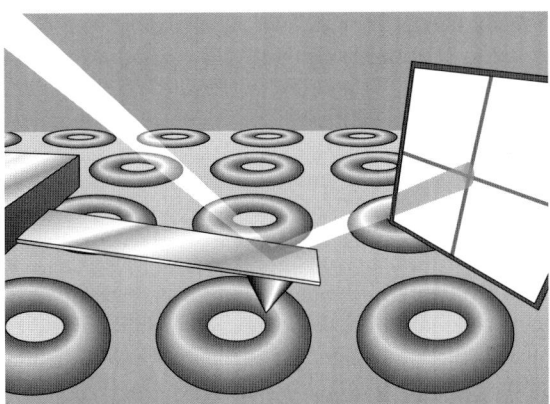

Fig. 6.4. Representation of an AFM set-up with the laser-beam deflection method. Cantilever and tip are microfabricated from silicon wafers. A laser beam is deflected from the backside of the cantilever and again focused onto a photosensitive diode. The diode is segmented into four quadrants, which allows measurement of vertical and torsional bending of the cantilever (artwork from J. Heimel rendered with POV-Ray 3.0)

of the cantilever for lateral friction measurements). With the cantilever in equilibrium, the spot is adjusted such that the two sections show the same intensity. If the cantilever bends up or down, the spot moves, and the difference signal between the upper and lower sections is a measure of the bending.

In order to enhance sensitivity, several groups have adopted an interferometer system to measure the cantilever deflection. A thorough comparison of different measurement methods with analysis of sensitivity and noise level is given in reference [3].

The cantilever is mounted on a device that allows oscillation of the beam. Typically a piezo-element directly underneath the cantilever beam serves this purpose. The reflected laser beam is analyzed for oscillation amplitude, frequency and phase difference. Depending on the mode of operation, a feedback mechanism will adjust the oscillation parameters and/or tip–sample distance during the scanning. The set-up can be operated in air, UHV, and even in fluids. This allows measurement of a wide range of surface properties from atomic-resolution imaging [8] up to studying biological processes in liquids [9, 10].

6.2 Harmonic Oscillator: A Model System for Dynamic AFM

The oscillating cantilever has three degrees of freedom: the amplitude, the frequency, and the phase difference between excitation and oscillation. Let us consider the damped driven harmonic oscillator. The cantilever is mounted on a piezoelectric element that is oscillating with amplitude A_d at frequency ω:

$$z_d(t) = A_d \cos(\omega t). \tag{6.6}$$

We assume that the cantilever spring obeys Hooke's law. Secondly, we introduce a friction force that is proportional to the speed of the cantilever motion, with α denoting the damping coefficient (Amontons's law). With Newton's first law we find for the oscillating system the following equation of motion for the position $z(t)$ of the cantilever tip (see also Fig. 6.1):

$$m\ddot{z}(t) = -\alpha \dot{z}(t) - kz(t) - kz_d(t); \tag{6.7}$$

We define $\omega_0^2 = k/m^*$, which turns out to be the resonant frequency of the free (undamped, i.e., $\alpha = 0$) oscillating beam. We further define the dimensionless quality factor $Q = m^*\omega_0/\alpha$, which is antiproportional to the damping coefficient. The quality factor describes the number of oscillation cycles after which the damped oscillation amplitude decays to $1/e$ of the initial amplitude with no external excitation ($A_d = 0$). After some basic manipulation, this results in the following differential equation:

$$\ddot{z}(t) + \frac{\omega_0}{Q}\dot{z}(t) + \omega_0^2 z(t) = A_d \omega_0^2 \cos(\omega t). \tag{6.8}$$

The solution is a linear combination of two regimes [11]. Starting from rest and switching on the piezo-excitation at $t = 0$, the amplitude will increase from zero to the final magnitude and reach a steady state, where the amplitude, phase, and frequency of the oscillation stay constant over time. The steady-state solution $z_1(t)$ is reached after $2Q$ oscillation cycles and follows the external excitation with amplitude A_0 and phase difference φ:

$$z_1(t) = A_0 \cos(\omega t + \varphi) \tag{6.9}$$

The oscillation amplitude in the transient regime during the first $2Q$ cycles follows:

$$z_2(t) = A_t \cdot e^{-\omega_0 t/2Q} \cdot \sin(\omega_0 t + \varphi_t). \tag{6.10}$$

We emphasize the important fact that the exponential term causes $z_2(t)$ to diminish exponentially with time constant τ:

$$\tau = 2Q/\omega_0 \tag{6.11}$$

In vacuum conditions, only the internal dissipation due to bending of the cantilever is present, and Q reaches values of 10,000 at typical resonant frequencies of 100,000 Hz. This results in a relatively long transient regime of $\tau \cong 30$ ms, which limits the possible operational modes for dynamic AFM (detailed analysis by [11]). Changes in the measured amplitude, which reflect a change of atomic forces, will have a time lag of 30 ms, which is very slow considering one wants to scan a 200 × 200 point image within a few minutes. In air, however, viscous damping due to air friction dominates and Q goes down to less than 1000, resulting in a time constant below the millisecond level. This response time is fast enough to use the amplitude as a measurement parameter.

If we evaluate the steady-state solution $z_1(t)$ of the differential equation, we find the following well-known solution for amplitude and phase of the oscillation as a function of the excitation frequency ω:

$$A_0 = \frac{A_d \cdot Q \cdot \omega_0^2}{\sqrt{\omega^2 \omega_0^2 + Q^2 \left(\omega_0^2 - \omega^2\right)^2}} \tag{6.12}$$

$$\varphi = \arctan\left(\frac{\omega \cdot \omega_0}{Q \cdot \left(\omega_0^2 - \omega^2\right)}\right). \tag{6.13}$$

Amplitude and phase diagrams are depicted in Fig. 6.5. As can be seen from (6.12) the amplitude will reach its maximum at a frequency different from ω_0, if Q has a finite value. The damping term of the harmonic oscillator causes the resonant frequency to shift from ω_0 to ω_0^*:

$$\omega_0^* = \omega_0 \sqrt{1 - \frac{1}{2Q^2}} \tag{6.14}$$

The shift is negligible for Q-factors of 100 and above, which is the case for most applications in vacuum or air. However, for measurements in liquids, Q can be smaller

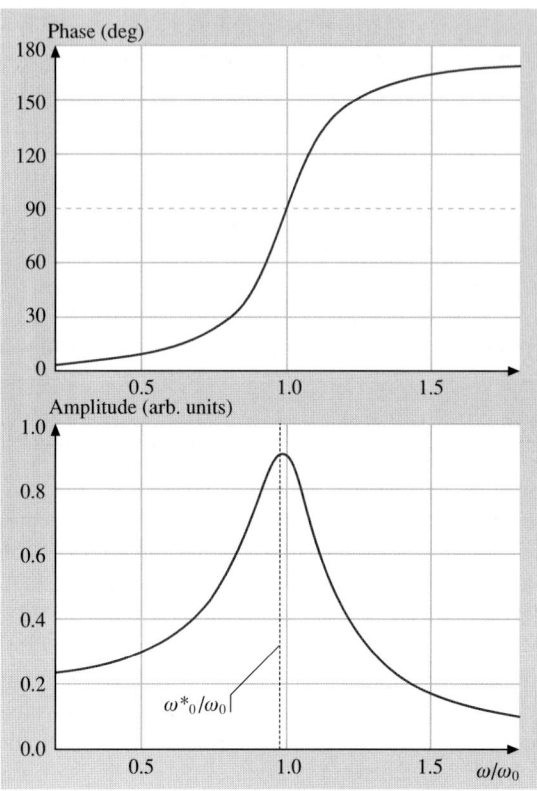

Fig. 6.5. Curves of amplitude and phase versus excitation frequency for the damped harmonic oscillator, with a quality factor of $Q = 4$

than 10 and ω_0 differs significantly from ω_0^*. As we will discuss later, it is also possible to enhance Q by using a special excitation method called *Q-control*.

In the case that the excitation frequency is equal to the resonant frequency of the undamped cantilever $\omega = \omega_0$, we find the useful relation:

$$A_0 = Q \cdot A_d \quad \text{for} \quad \omega = \omega_0. \tag{6.15}$$

Since $\omega_0^* \approx \omega_0$ for most cases, we find that (6.15) holds true when exciting the cantilever at its resonance. From a similar argument, the phase becomes approximately 90 deg for the resonance case. We also see that, in order to reach vibration amplitudes of some 10 nm, the excitation can be as small as 1 pm for typical cantilevers operated in vacuum.

So far, we have not considered an additional force term that describes the interaction between the probing tip and the sample. For typical large-vibration amplitudes of 10–100 nm no general solution for this analytical problem has been found yet. The cantilever tip will experience a whole range of force interactions during one single oscillation cycle, rather than one defined tip–sample force. Only in the special case of a self-excited cantilever oscillation can the problem be solved analytically, as we will see later.

6.3 Dynamic AFM Operational Modes

While the quantitative interpretation of force curves in contact AFM is straightforward using (6.1), we explained in the previous paragraphs that its application to assess short-range attractive interatomic forces is rather limited. The dynamic mode of operation seems to open a viable direction towards achieving this task. However interpretation of the measurements generally appears to be more difficult. Different operational modes are employed in dynamic AFM, and the following paragraphs are intended to distinguish these modes and categorize them in a systematic way.

The oscillation trajectory of a dynamically driven cantilever is determined by three parameters: the amplitude, the phase, and the frequency. Tip–sample interactions can influence all three parameters, which are termed the internal parameters in the following. The oscillation is driven externally, with excitation amplitude A_d and excitation frequency ω. These variables will be referred to as the external parameters. The external parameters are set by the experimentalist, whereas the internal parameters are measured and contain the crucial information about the force interaction. In scanning probe applications, it is common to control the probe–surface distance z_0 in order to keep an internal parameter constant (i.e., the tunneling current in STM or the beam deflection in contact AFM), which represents a certain tip–sample interaction. In z-spectroscopy mode, the distance is varied in a certain range, and the change of the internal parameters is measured as a fingerprint of the tip–sample interactions.

In dynamic AFM the situation is rather complex. Any of the internal parameters can be used for feedback of the tip–sample distance z_0. However, we already realized that, in general, the tip–sample forces can only be fully assessed by measuring all three parameters. Therefore, dynamic AFM images are difficult to interpret. A solution to this problem is to establish additional feedback loops, which keep the internal parameters constant by adjusting the external variables. In the simplest setup, the excitation frequency is set to a predefined value, and the excitation amplitude remains constant by a feedback loop. This is called the amplitude-modulation (AM) or tapping mode. As stated before, in principle, any of the internal parameters can be used for feedback to the tip–sample distance; in AM mode the amplitude signal is used. A certain amplitude (smaller than the free oscillation amplitude) at a frequency close to the resonance of the cantilever is chosen; the tip is approached towards the surface under investigation, and the approach is stopped as soon as the set-point amplitude is reached. The oscillation phase is usually recorded during the scan, however, the shift of the resonant frequency of the cantilever cannot be directly accessed, since this degree of freedom is blocked by the external excitation at a fixed frequency. It turns out that this mode is simple to operate from a technical perspective, but quantitative information about the tip–sample interaction forces has so far not been reliably extracted from AM-mode AFM. Despite this, it is one of the most commonly used modes in dynamic AFM operated in air, and even in liquid. The strength of this mode is the qualitative imaging of a large variety of surfaces.

It is interesting to discuss the AM mode in the situation that the external excitation frequency is much lower than the resonant frequency [12, 13]. This results in a quasistatic measurement, although a dynamic oscillation force is applied, and, therefore, this mode can be viewed as a hybrid between static and dynamic AFM. Unfortunately, it has the drawbacks of the static mode, namely, that stiff spring constants must be used and, therefore, the sensitivity of the deflection measurement must be very good, typically employing a high-resolution interferometer. However, it has the advantage of the static measurement in quantitative interpretation, since in the regime of small amplitudes (< 0.1 nm) a direct interpretation of the experiments is possible. In particular, the force gradient at the tip–sample distance z_0 is given by the change of the amplitude A and the phase angle φ

$$\frac{\partial F_{ts}}{\partial z}\bigg|_{z_0} = k\left(1 - \frac{A_0}{A}\cos\varphi\right). \tag{6.16}$$

In effect, the modulated AFM technique can profit from enhanced sensitivity due to the use of lock-in techniques, which allow the measurement of the amplitude and phase of the oscillation signal with high precision.

As stated before, the internal parameters can be fed back to the external excitation variables. One of the most useful applications in this direction is the self-excitation system. Here the resonant frequency of the cantilever is detected and selected again as the excitation frequency. In a typical set-up, this is done with a phase shift of 90 deg by feeding back the detector signal to the excitation piezo, i.e., the cantilever is always excited in resonance. Tip–sample interaction forces then only influence the resonant frequency, but do not change the two other parameters of the oscillation: amplitude and phase. Therefore, it is sufficient to measure the frequency shift induced by the tip–sample interaction. Since the phase remains at a fixed value, the oscillating system is much better defined than before, and the degrees of freedom for the oscillation are reduced. To reduce the last degree of freedom even further an additional feedback loop can be incorporated to keep the oscillation amplitude A constant by varying the excitation amplitude A_d. Now, all internal parameters have a fixed relation to the external excitation variables, the system is well defined, and all parameters can be assessed during the measurement. As it turns out, this mode is the only dynamic mode in which a quantitative relation between the tip–sample forces and the change of the resonant frequency can be established.

In the following section we want to discuss the two most popular operational modes, the tapping mode and the self-excitation mode, in more detail.

6.3.1 Amplitude-Modulation/Tapping-Mode AFM

In the tapping mode, or AM-AFM, the cantilever is excited externally at a constant frequency close to its resonance. The oscillation amplitude and phase during the approach of the tip and the sample serve as the experimental observation channels. Figure 6.6 shows a diagram of a typical tapping-mode AFM set-up. The oscillation of the cantilever is detected with the photodiode, whose output signal is analyzed with a lock-in amplifier to obtain amplitude and phase information. The amplitude is then compared to the set point, and the resulting difference or error signal is fed into

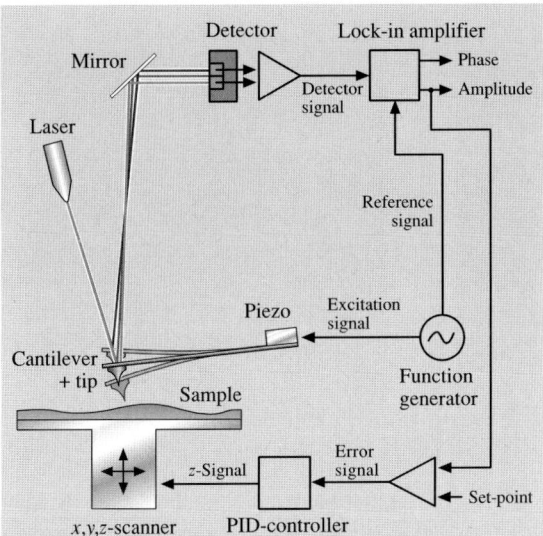

Fig. 6.6. Set-up of a dynamic force microscope operated in the AM or tapping mode. A laser beam is deflected by the backside of the cantilever and the deflection is detected by a split photodiode. The excitation frequency is chosen externally with a modulation unit, which drives the excitation piezo. A lock-in amplifier analyses the phase and amplitude of the cantilever oscillation. The amplitude is used as the feedback signal for the probe–sample distance control

the proportional–integral–derivative (PID) controller, which adjusts the z-piezo, i.e., the probe–sample distance, accordingly. The external modulation unit supplies the signal for the excitation piezo, and, at the same time, the oscillation signal serves as the reference for the lock-in amplifier. As shown by the following applications the tapping mode is typically used to measure surface topography and other material parameters on the nanometer scale. The tapping mode is mostly used in ambient conditions and in liquids.

High-resolution imaging has been extensively performed in the area of material science. Due to its technical relevance the investigation of polymers has been the focus of many studies (see e.g. a recent review about AFM imaging of polymers in [15]. In Fig. 6.7 the topography of a diblock copolymer ($BC_{0.26}$-$3A_{0.53}F8H10$) at different magnifications is shown [14]. On the large scan (a) the large-scale structure of the microphase-separated polystyrene (PS) cylinders [within a polyisoprene (PI) matrix] lying parallel to the substrate can be seen. In the high-resolution image (b) a surface substructure of regular domes can be seen, which were found to be related to the cooling process during the polymer preparation.

Imaging in liquids opens up the avenue for the investigation of biological samples in their natural environment. For example *Möller* et al. [16] have obtained high-resolution images of the topography of hexagonally packed intermediate (HPI) layer of Deinococcus Radiodurans with tapping-mode AFM. Another interesting example is the imaging of deoxyribonucleic acid (DNA) in liquid, as shown in Fig. 6.8. *Jiao* et al. [10] measured the time evolution of a single DNA strand interacting with a molecule, as shown by a sequence of images acquired in liquid over a time period of several minutes.

For a quantitative interpretation of tip–sample forces one has to consider that during one oscillation cycle with amplitudes of 10–100 nm the tip–sample inter-

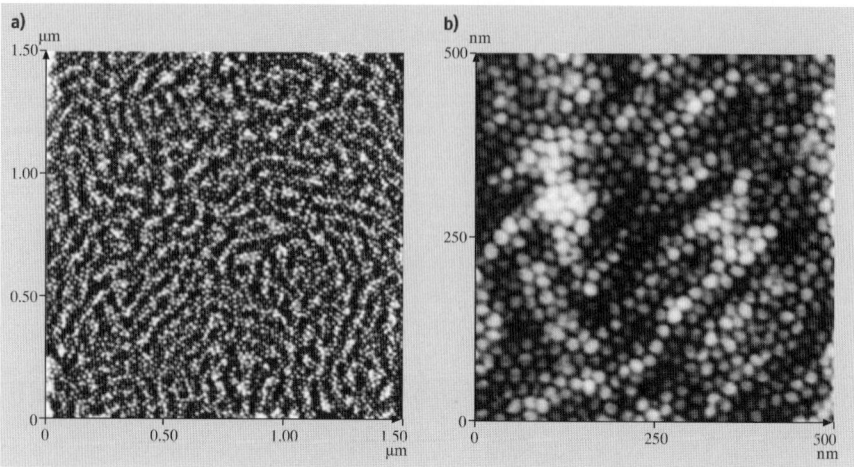

Fig. 6.7. Tapping-mode images of $BC_{0.26}$-$3A_{0.53}F8H10$ at (**a**) low resolution and (**b**) high resolution. The height scale is 10 nm. (Reprinted in part with permission from [14]. Copyright (2001) American Chemical Society)

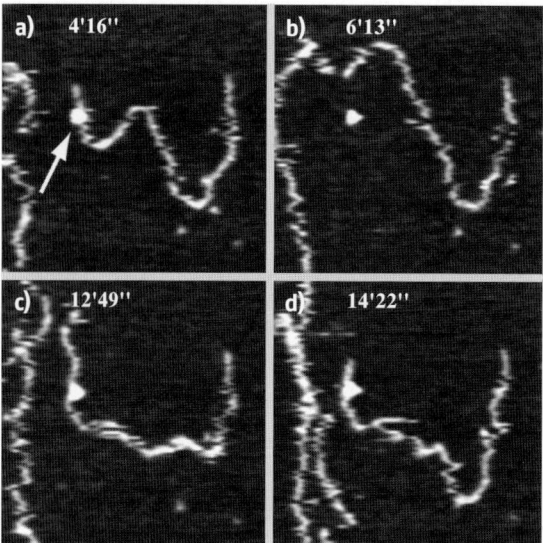

Fig. 6.8a–d. Dynamic p53-DNA interactions observed by time-lapse tapping-mode AFM imaging in solution. Both p53 protein and DNA were weakly adsorbed to a mica surface by balancing the buffer conditions. (**a**) A p53 protein molecule (arrow) was bound to a DNA fragment. The protein (**b**) dissociated from and then (**c**) re-associated with the DNA fragment. (**d**) A downward movement of the DNA with respect the protein occurred, constituting a *sliding* event whereby the protein changes its position on the DNA. Image size: 620 nm. Colorscale (height) range: 4 nm. Time units: minutes, seconds. (Image courtesy of Tilman Schäffer, University of Münster)

action will range over a wide distribution of forces, including attractive as well as repulsive forces. We will, therefore, measure a convolution of the force–distance curve with the oscillation trajectory. This complicates the interpretation of AM-AFM measurements appreciably.

At the same time, the resonant frequency of the cantilever will change due to the appearing force gradients, as could already be seen in the simplified model from (6.4). If the cantilever is excited exactly at its resonant frequency before interaction forces are encountered, it will be excited off-resonance after they are encountered. This, in turn, changes the amplitude and phase (see (6.12) and (6.13)), which serve as the measurement signals. Consequently, a different amplitude will cause a change in the encountered effective force. We can see already from this simple *gedanken*-experiment that the interpretation of force curves will be highly complicated. In fact, no quantitative theory for AM-AFM that allows the experimentalist to convert the experimental data to a force–distance relationship unambiguously is available.

The qualitative behavior for amplitude versus z_0-position curves is depicted in Fig. 6.9. At large distances, where the forces between the tip and the sample are negligible, the cantilever oscillates with its free oscillation amplitude. As the probe approaches the surface the interaction forces cause the amplitude to change, typically resulting in an amplitude that decreases with continuously decreasing tip–sample distance. This is expected, since the force–distance curve will eventually reach the repulsive part and the tip is hindered from indenting further into the sample, resulting in smaller oscillation amplitudes.

However, in order to gain some qualitative insight into the complex relationship between forces and oscillation parameters, we resort to numerical simulations. *Anczykowski* et al. [17, 18] have calculated the oscillation trajectory of the cantilever under the influence of a given force model. Van der Waals interactions were considered the only effective, attractive forces, and the total interaction re-

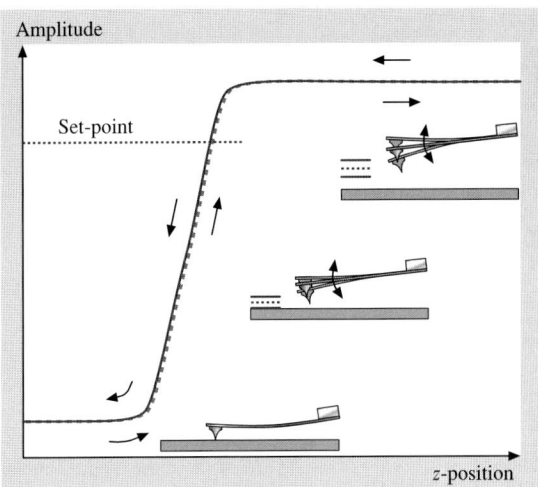

Fig. 6.9. Simplified model showing the oscillation amplitude in tapping-mode AFM for various probe–sample distances

sembled a Lennard–Jones-type potential. Mechanical relaxations of the tip and sample surface were treated in the limits of continuum theory with the numerical Muller-Yushchenko-Derjaguin/Burgess-Hughes-White (MYD/BHW) [19, 20] approach, which allows the simulations to be compared to experiments. Figure 6.10 shows the corresponding force–distance curves used in the simulations for different tip radii.

The cantilever trajectory was analyzed by solving the differential equation (6.7) extended by the force–distance relations from Fig. 6.10 using the numerical Verlet algorithm [21, 22]. The results of the simulation for the amplitude and phase of the tip oscillation as a function of z-position of the probe are presented in Fig. 6.11. One has to keep in mind that the z-position of the probe is not equivalent to the real tip–sample distance at the equilibrium position, since the cantilever might bend statically due to the interaction forces. The behavior of the cantilever can be subdivided into three different regimes. We distinguish the cases in which the beam is oscillated below its resonant frequency ω_0, exactly at ω_0, and above ω_0. In the following, we will refer to ω_0 as the resonant frequency, although the correct resonant frequency is ω_0^* if one takes into account the finite Q-value.

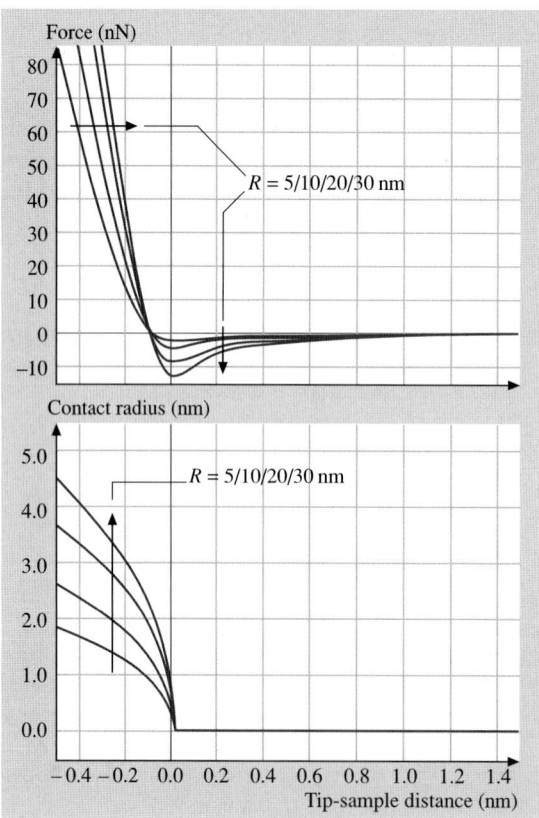

Fig. 6.10. Force curves and corresponding contact radius calculated with the MYD/BHW model as a function of tip radius for a Si–Si contact. These force curves are used for the tapping-mode AFM simulations

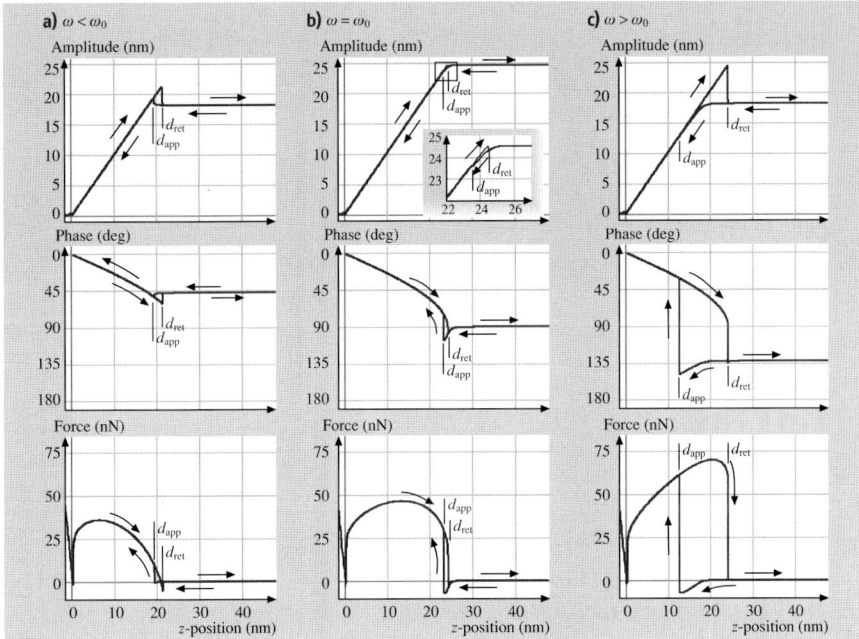

Fig. 6.11. Amplitude and phase diagrams with excitation frequency (**a**) below (**b**) exactly at and (**c**) above the resonant frequency for tapping-mode AFM from the numerical simulations. Additionally, the lower diagrams show the interaction forces at the point of closest tip–sample distance, i.e. the lower turnaround point of the oscillation

Clearly, Fig. 6.12 exhibits more features than were anticipated from the initial, simple arguments. Amplitude and phase seem to change rather abruptly at certain points when the z_0-position is decreased. Besides, the amplitude or phase–distance curves do not resemble the force–distance curves from Fig. 6.11 in a simple, direct manner. Additionally, we find a hysteresis between approach and retraction.

As an example, let us start by discussing the discontinuous features in the AFM spectroscopy curves of the first case, where the excitation frequency is smaller than ω_0. Consider the oscillation amplitude as a function of the excitation frequency in Fig. 6.5 in conjunction with a typical force curve, as depicted in Fig. 6.10. Upon approach of probe and sample, attractive forces will lower the effective resonant frequency of the oscillator. Therefore, the excitation frequency will now be closer to the resonant frequency, causing the vibration amplitude to increase. This, in turn, reduces the tip–sample distance, which again gives rise to a stronger attractive force. The system becomes unstable until the point $z_0 = d_{\text{app}}$ is reached, where repulsive forces stop the self-enhancing instability. This can be clearly observed in Fig. 6.11a. Large parts of the force–distance curve cannot be measured due to this instability.

In the second case, where the excitation equals the free resonant frequency, only a small discontinuity is observed upon reduction of the z-position. Here, a shift of

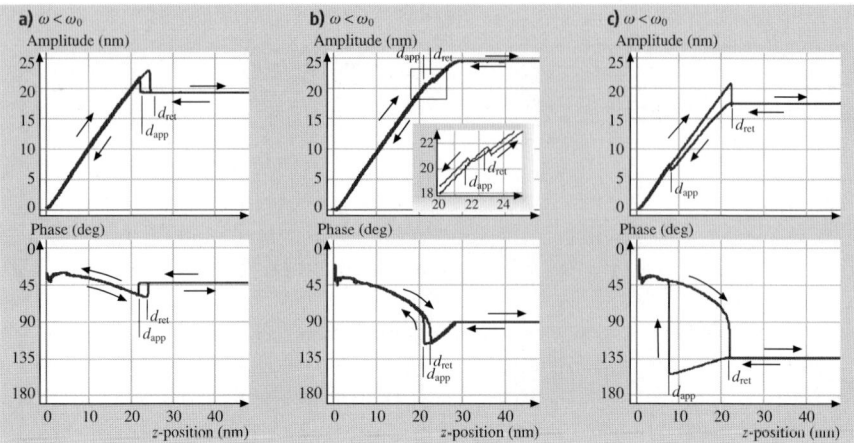

Fig. 6.12. Amplitude and phase diagrams with excitation frequency (**a**) below (**b**) exactly at and (**c**) above the resonant frequency for tapping-mode AFM from experiments with a Si cantilever on a Si wafer in air

the resonant frequency towards smaller values, induced by the attractive force interaction, will reduce the oscillation amplitude. The distance between the tip and sample is, therefore, reduced as well, and the self-amplifying effect with the sudden instability does not occur as long as repulsive forces are not encountered. However, at closer tip–sample distances, repulsive forces will cause the resonant frequency to shift again towards higher values, increasing the amplitude with decreasing tip–sample distance. Therefore, a self-enhancing instability will also occur in this case, but at the crossover from purely attractive forces to the regime where repulsive forces occur. Correspondingly, a small kink in the amplitude curve can be observed in Fig. 6.11b. An even clearer indication of this effect is manifested by the sudden change in the phase signal at d_{app}.

In the last case, with $\omega > \omega_0$, the effect of amplitude reduction due to the resonant frequency shift is even larger. Again, we find no instability in the amplitude signal during approach in the attractive force regime. However, as soon as the repulsive force regime is reached, the instability occurs due to the induced positive frequency shift. Consequently, a large jump in the phase curve from values smaller than 90 deg to values larger than 90 deg is observed. The small change in the amplitude curve is not resolved in the simulated curves in Fig. 6.11c, however, it can be clearly seen in the experimental curves.

Figure 6.12 depicts the corresponding experimental amplitude and phase curves. The measurements were performed in air with a Si cantilever approaching a Si wafer, with a cantilever resonant frequency of 299.95 kHz. Qualitatively, all prominent features of the simulated curves can also be found in the experimental data sets. Hence, this model seems to capture the important factors necessary for an appropriate description of the experimental situation.

But what is the reason for this unexpected behavior? We have to turn to the numerical simulations again, where we have access to all physical parameters, in order to understand the underlying processes. The lower part of Fig. 6.11 also shows the interaction force between the tip and the sample at the point of closest approach, i.e., the sample-side turnaround point of the oscillation. We see that exactly at the points of the discontinuities the total interaction force changes from the net-attractive regime to the attractive–repulsive regime, also termed the intermittent contact regime. The term net-attractive is used to emphasize that the total force is attractive, despite the fact that some minor contributions might still originate from repulsive forces. As soon as a minimum distance is reached, the tip also starts to experience repulsive forces, which completely changes the oscillation behavior. In other words, the dynamic system switches between two oscillatory states.

Directly related to this fact is the second phenomenon: the hysteresis effect. We find separate curves for the approach of the probe towards the surface and the retraction. This seems to be somewhat counterintuitive, since the tip is constantly approaching and retreating from the surface, and the average values of amplitude and phase should be independent of the direction of the average tip–sample distance movement. A hysteresis between approach and retraction within one oscillation due to dissipative processes should directly influence amplitude and phase. However, no dissipation models were included in the simulation. In this case, the hysteresis in Fig. 6.11 is due to the fact that the oscillation jumps into different modes; the system exhibits bistability. This effect is often observed in oscillators under the influence of nonlinear forces [23].

For the interpretation of these effects it is helpful to look at Fig. 6.13, which shows the behavior of the simulated tip trajectory and the force during one oscillation cycle over time. The data is shown for the z-positions where hysteresis is observed, while (a) was taken during the approach and (b) during the retraction. Excitation was in resonance, where the amplitude shows a small hysteresis. Also note that the amplitude is almost exactly the same in (a) and (b). We see that the oscillation at the same z-position exhibits two different modes: while in (a) the experienced force is net-attractive, in (b) the tip is exposed to attractive and repulsive interactions. Experimental and simulated data show that the change between the net-attractive and intermittent contact mode takes place at different z-positions (d_{app} and d_{ret}) for approach and retraction. Between d_{app} and d_{ret} the system is in a bistable mode. Depending on the history of the measurement, e.g., whether the position d_{app} during the approach (or d_{ret} during retraction) has been reached, the system flips to the other oscillation mode. While the amplitude might not be influenced strongly, the phase is a clear indicator of the mode switch. On the other hand, if the point d_{app} is never reached during the approach, the system will stay in the net-attractive regime and no hysteresis is observed, i.e., the system remains stable.

In conclusion, we find that, although a qualitative interpretation of the interaction forces is possible, the AM-AFM is not suitable to gain direct quantitative knowledge of tip–sample force interactions. However, it is a very useful tool for imaging nanometer-sized structures in a wide variety of setups, in air or even in

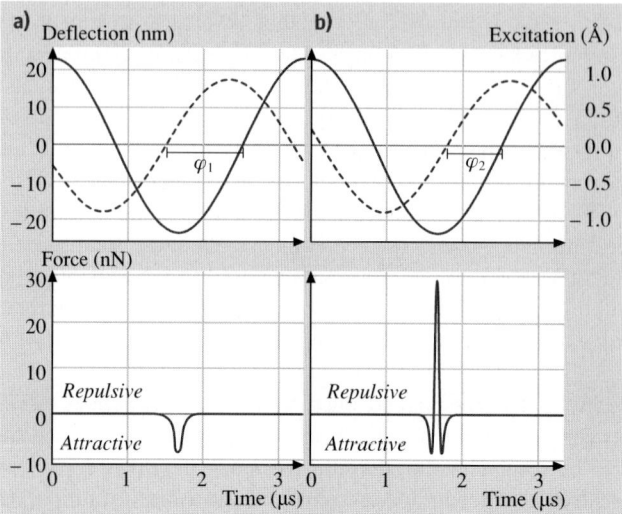

Fig. 6.13. Simulation of the tapping-mode cantilever oscillation in the (**a**) net-attractive and (**b**) the intermittent contact regime. The *dashed line* represents the excitation amplitude and the *solid line* is the oscillation amplitude

liquid. We find that two distinct modes exist for the externally excited oscillation – the net-attractive and the intermittent contact mode – which describe what kind of forces govern the tip–sample interaction. The phase can be used as an indicator of the current mode of the system.

In particular, it can be easily seen that, if the free resonant frequency of the cantilever is higher than the excitation frequency, the system cannot stay in the net-attractive regime due to a self-enhancing instability. Since in many applications involving soft and delicate biological samples strong repulsive forces should be avoided, the tapping-mode AFM should be operated at frequencies equal to or above the free resonant frequency [24]. Even then, statistical changes of tip–sample forces during the scan might induce a sudden jump into the intermittent contact mode, and the previously explained hysteresis will tend to keep the system in this mode. It is therefore of high importance to tune the oscillation parameters in such a way that the AFM stays in the net-attractive regime [25]. A concept that achieves this task is the Q-control system, which will be discussed in some detail in the forthcoming paragraphs.

A last word concerning the overlap of simulation and experimental data: while the qualitative agreement down to the detailed shape of hysteresis and instabilities is rather striking, we still find some quantitative discrepancies between the positions of the instabilities d_{app} and d_{ret}. This is probably due to the simplified force model, which only takes into account van der Waals and repulsive forces. Especially at ambient conditions, an omnipresent water meniscus between the tip and sample will give rise to much stronger attractive and also dissipative forces than considered

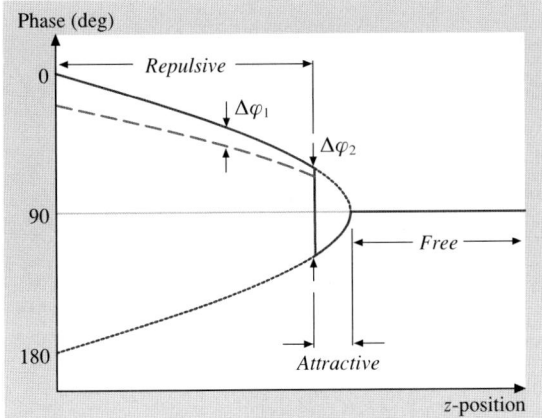

Fig. 6.14. Phase shift in tapping mode as a function of tip–sample distance

in the model. A very interesting feature is that the simulated phase curves in the intermittent contact regime tend to have a steeper slope in the simulation than in the experiments (see also Fig. 6.14). We will later show that this effect is a fingerprint of an effect that was not included in this simulation at all: dissipative processes during the oscillation, giving rise to an additional loss of oscillation energy.

6.3.2 Self-Excitation Modes

Despite the wide range of technical applications of the AM mode of dynamic AFM, it has been found unsuitable for measurements in an environment extremely useful for scientific research: vacuum or ultrahigh vacuum (UHV) with pressures reaching 1×10^{-10} mbar. The STM has already shown how much insight can be gained from some highly defined experiments under those conditions.

Consider (6.11) from the previous section. The time constant τ for the amplitude to adjust to a different tip–sample force scales with $1/Q$. In vacuum applications, Q of the cantilever is on the order of 10,000, which means that τ is in the range of some 10 ms. This is clearly too long for a scan of at least (100×100) data points. On the other hand, the resonant frequency of the system will react instantaneously to tip–sample forces. This has led *Albrecht* et al. [11] to use a modified excitation scheme.

The system is always oscillated at its resonant frequency. This is achieved by feeding back the oscillation signal from the cantilever into the excitation piezo-element. Figure 6.15 illustrates the method in a block diagram. The signal from the PSD is phase-shifted by 90 deg (and, therefore, always exciting in resonance) and used as the excitation signal of the cantilever. An additional feedback loop adjusts the excitation amplitude in such a way that the oscillation amplitude remains constant. This ensures that the tip–sample distance is not influenced by changes in the oscillation amplitude. The only degree of freedom that the oscillation system still has, which can react to the tip–sample forces, is the change of the resonant frequency. This shift of the frequency is detected and used as the set-point signal for

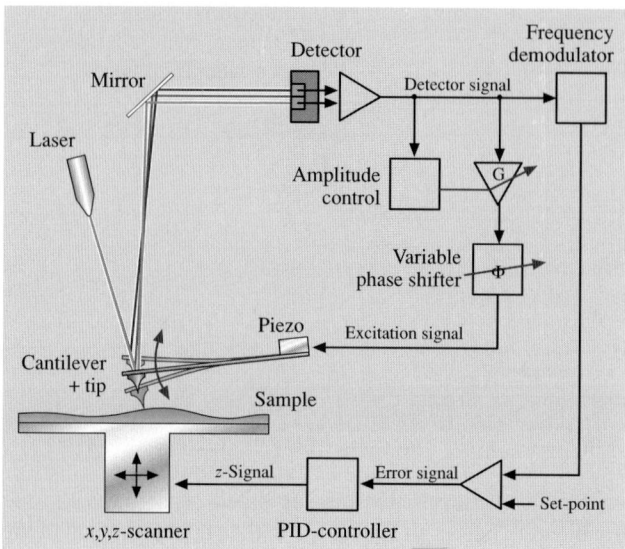

Fig. 6.15. Dynamic AFM operated in the self-excitation mode, where the oscillation signal is directly fed back to the excitation piezo. The detector signal is amplified with the variable gain G and phase shifted by phase φ. The frequency demodulator detects the frequency shift due to tip–sample interactions, which serves as the control signal for the probe–sample distance

surface scans. Therefore, this mode is also called the frequency-modulated (FM) mode.

Let us take a look at the sensitivity of the dynamic AFM. If electronic noise, laser noise, and thermal drift can be neglected, the main noise contribution will come from thermal excitations of the cantilever. A detailed analysis of a dynamic system yields, for the minimum detectable force gradient, the following relation [11]:

$$\frac{\partial F}{\partial z}\bigg|_{\mathrm{MIN}} = \sqrt{\frac{4k \cdot k_\mathrm{B} T \cdot B}{\omega_0 Q \langle z_{\mathrm{osc}}^2 \rangle}} \tag{6.17}$$

Here B is the bandwidth of the measurement, T the temperature, and $\langle z_{\mathrm{osc}}^2 \rangle$ is the mean-square amplitude of the oscillation. Please note that this sensitivity limit was deliberately calculated for the FM mode. A similar analysis of the AM mode, however, yields virtually the same result [26]. We find that the minimum detectable force gradient, i.e., the measurement sensitivity, is inversely proportional to the square root of the Q-factor of the cantilever. This means that it should be possible to achieve very high-resolution imaging under vacuum conditions where the Q-factor is very high.

A breakthrough in high-resolution AFM imaging was the atomic-resolution imaging of the Si(111)–(7×7) surface reconstruction by *Giessibl* [8] under UHV conditions. Moreover, *Sugawara* et al. [27] observed the motion of single atomic defects on InP with true atomic resolution. However, imaging on conducting or semiconducting surfaces is also possible with the scanning tunneling microscope (STM) and these first non-contact AFM (NC-AFM) images provided little new information on surface properties. The true potential of NC-AFM lies in the imaging of nonconducting surface with atomic precision, which was first demonstrated by *Bammerlin* et al. [28] on NaCl. A long-standing question about the surface reconstruction of the technologically relevant material aluminium oxide could be answered by *Barth* et al. [29], who imaged the atomic structure of the high-temperature phase of α-Al$_2$O$_3$ (0001).

The high-resolution capabilities of non-contact atomic force microscopy are nicely demonstrated by the images shown in Fig. 6.16. *Allers* et al. [30] imaged steps and defects on the insulator nickel oxide with atomic resolution. Such a reso-

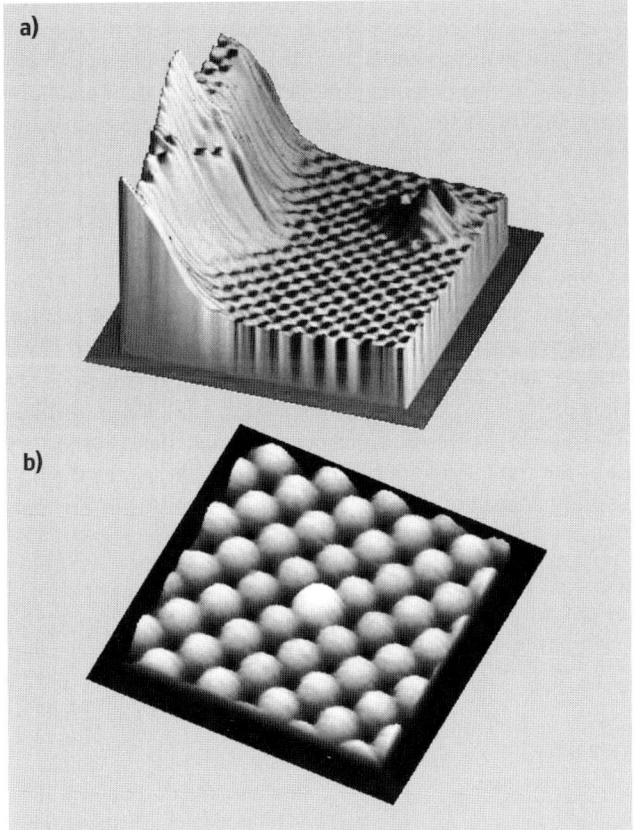

Fig. 6.16. Imaging of a NiO(001) sample surface with non-contact AFM. (**a**) Surface step and an atomic defect. The lateral distance between two atoms is 4.17 Å. (**b**) A dopant atom is imaged as a light protrusion about 0.1 Å higher than the other atoms. (Images courtesy of W. Allers and S. Langkat, University of Hamburg)

lution is routinely obtained today by different research groups (for an overview see, e.g., [31–33]).

However, we are concerned with measuring atomic force potentials of a single pair of molecules. Clearly, FM-mode AFM will allow us to identify single atoms, and with sufficient care we will be able to ensure that only one atom from the tip contributes to the total force interaction. Can we, therefore, fill in the last bit of information and find a quantitative relation between the oscillation parameters and the force?

A good insight into the cantilever dynamics can be drawn from the tip potential displayed in Fig. 6.17 [34]. If the cantilever is far away from the sample surface, the tip moves in a symmetric parabolic potential (dotted line), and the oscillation is harmonic. In such a case, the tip motion is sinusoidal and the resonant frequency is determined by the eigenfrequency f_0 of the cantilever. If the cantilever approaches the sample surface the potential is changed, given by an effective potential V_{eff} (solid line) which is the sum of the parabolic potential and the tip–sample interaction potential V_{ts} (dashed line). This effective potential differs from the original parabolic potential and shows an asymmetric shape. As a result the oscillation becomes anharmonic, and the resonant frequency of the cantilever depends on the oscillation amplitude.

Gotsmann et al. [35] investigated this relation with a numerical simulation. During each oscillation cycle the tip experiences a whole range of forces. For each step during the approach the differential equation for the whole oscillation loop (also including the feedback system) was evaluated and finally the quantitative relation between the force and frequency shift was revealed.

However, there is also an analytical relationship, if some approximations are accepted [7, 36, 37]. Here, we will follow the route indicated by [37], although alternative ways have also been proven successful. Consider the tip oscillation trajectory reaches over a large part of the force gradient curve in Fig. 6.2. We model the tip–sample interaction as a spring constant of stiffness $k_{\text{ts}}(z) = \partial F/\partial z|_{z_0}$ as in Fig. 6.1.

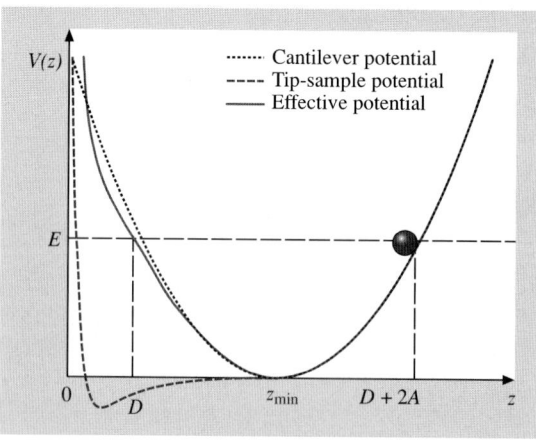

Fig. 6.17. The frequency shift in dynamic force microscopy is caused by the tip–sample interaction potential (*dashed line*), which alters the harmonic cantilever potential (*dotted line*). Therefore, the tip moves in an anharmonic and asymmetric effective potential (*solid line*). Here, z_{min} is the minimum position of the effective potential (from Hölscher et al. [34])

For small oscillation amplitudes we already found that the frequency shift is proportional to the force gradient in (6.4). For large amplitudes, we can calculate an effective force gradient k_{eff} as a convolution of the force and the fraction of time, the tip spends between the positions x and $x + dx$:

$$k_{\text{eff}}(z) = \frac{2}{\pi A^2} \int_z^{z+2A} F(x) \cdot g\left(\frac{x-z}{A} - 1\right) dx \quad \text{with } g(u) = -\frac{u}{\sqrt{1-u^2}}. \tag{6.18}$$

In the approximation that the vibration amplitude is much larger than the range of the tip sample forces the above equation can be simplified to:

$$k_{\text{eff}}(z) = \frac{\sqrt{2}}{\pi} A^{3/2} \cdot \int_z^\infty \frac{F(x)}{\sqrt{x-z}} dx. \tag{6.19}$$

This effective force gradient can now be used in (6.4), the relation between the frequency shift and force gradient. We find:

$$\Delta f = \frac{f_0}{\sqrt{2\pi} \cdot kA^{3/2}} \cdot \int_z^\infty \frac{F(x)}{\sqrt{x-z}} dx. \tag{6.20}$$

If we separate the integral from other parameters, we can define:

$$\Delta f = \frac{f_0}{kA^{3/2} \cdot \gamma(z)} \quad \text{with } \gamma(z) = \frac{1}{\sqrt{2\pi}} \cdot \int_z^\infty \frac{F(x)}{\sqrt{x-z}} dx. \tag{6.21}$$

This means we can define $\gamma(z)$, which is only dependent on the shape of the force curve $F(z)$ but independent of the external parameters of the oscillation. The function $\gamma(z)$ is also referred to as the *normalized frequency shift* [7], a very useful parameter, which allows us to compare measurements independent of resonant frequency, amplitude and spring constant of the cantilever.

The dependence of the frequency shift on the vibration amplitude is an especially useful relation, since this parameter can be easily varied during one experiment. A nice example is depicted in Fig. 6.18, where frequency shift curves for different amplitudes were found to coincide very well in the $\gamma(z)$ diagrams [38].

This relationship has been nicely exploited for the calibration of the vibration amplitude by *Guggisberg* [39], which is a problem often encountered in dynamic AFM operation and worthwhile discussing. One approaches tip and sample and records frequency shift versus distance curves, which show a reproducible shape. Then, with the z-feedback disabled, several curves with different amplitudes are acquired. The amplitudes are typically chosen by adjusting the amplitude set point in volts. One has to take care that drift in the z-direction is negligible. An analysis of the corresponding $\gamma(z)$-curves will show the same curves (as in Fig. 6.18), but the curves will be shifted in the horizontal axis. These shifts correspond to the change in amplitude, allowing one to correlate the voltage values with the z-distances.

For the often-encountered force contributions from electrostatic, van der Waals, and chemical binding forces the frequency shift has been calculated from the force laws. In the approximation that the tip radius R is larger than the tip–sample distance z, an electrostatic potential V will yield a normalized frequency shift of (adapted from *Guggisberg* [40]):

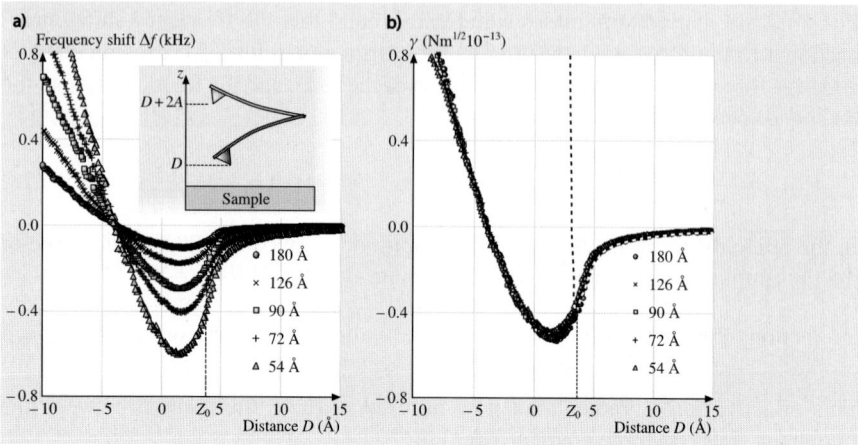

Fig. 6.18. (a) Frequency shift curves for different oscillation amplitudes for a silicon tip on a graphite surface in UHV, (b) γ-curves calculated from the Δf-curves in (a) (reprinted from Hölscher et al. [38] with permission, Copyright (2000) by The American Physical Society)

$$\gamma(z) = \frac{\pi \varepsilon_0 R \cdot V^2}{\sqrt{2}} \cdot z^{-1/2}. \tag{6.22}$$

For van der Waals forces with Hamaker constant H and also with R larger than z we find accordingly:

$$\gamma(z) = \frac{H \cdot R}{12 \sqrt{2}} \cdot z^{-3/2} \tag{6.23}$$

Finally, short-range chemical forces represented by the Morse potential (with the parameters binding energy U_0, decay length λ and equilibrium distance z_{equ} yield:

$$\gamma(z) = \frac{U_0 \sqrt{2}}{\sqrt{\pi \lambda}} \cdot \exp\left(-\frac{(z - z_{\text{equ}})}{\lambda}\right). \tag{6.24}$$

These equations allow the experimentalist to interpret the spectroscopic measurements directly. For example, the contributions of the electrostatic and van der Waals forces can be easily distinguished by their slope in a double logarithmic plot (for an example, see *Guggisberg* et al. [40]).

Alternatively, if the force law is not known beforehand, the experimentalist wants to analyze the experimental frequency-shift data curves and extract the force or energy potential curves. We, therefore, have to invert the integral in (6.21) to find the tip–sample interaction potential V_{ts} from the $\gamma(z)$-curves [37]:

$$V_{\text{ts}}(z) = \sqrt{2} \cdot \int_z^\infty \frac{\gamma(x)}{\sqrt{x - z}} dx. \tag{6.25}$$

Using this method, quantitative force curves were extracted from Δf-spectroscopy measurements on different, atomically resolved sites of the Si(111)-(7×7) reconstruction [1]. Comparison to theoretical molecular dynamics (MD) simulations showed good quantitative agreement with theory and confirmed the assumption that force interactions were governed by a single atom at the tip apex. Our initially formulated goal seems to be achieved: with FM-AFM we have found a powerful method that allows us to measure the chemical bond formation of single molecules. The last uncertainty, the exact shape and identity of the tip apex atom, can possibly be resolved by employing the FIM technique to characterize the tip surface in combination with FM-AFM.

All the above equations are only valid in the approximation that the oscillation amplitudes are much larger than the distance range of the encountered forces. However, for amplitudes of, e.g., 10 nm and long-ranged forces like electrostatic interactions this approximation is no longer valid. Several approaches have been proposed by different authors to solve this issue [41–43]. The *matrix method* [42] uses the fact that in a real experiment the frequency shift curve is not continuous, but rather a set of discrete values acquired at equidistant points. Therefore the integral in (6.18) can be substituted by a sum and the equation can be rewritten as a linear equation system, which in turn can be easily inverted by appropriate matrix operations. This *matrix method* is a very simple and general method for the AFM user to extract force curves from experimental frequency-shift curves without the restrictions of the large-amplitude approximation.

In this context it is worthwhile to point out a slightly different dynamic AFM method. While in the typical FM-AFM set-up the oscillation amplitude is controlled to stay constant by a dedicated feedback circuit, one could simply keep the excitation amplitude constant (this has been termed the constant-excitation (CE) mode as opposed to the constant-amplitude (CA) mode. It is expected that this mode is more gentle to the surface, because any dissipative interaction will reduce the amplitude and therefore prevent a further reduction of the effective tip–sample distance. This mode has been employed to image soft biological molecules like DNA or thiols in UHV [44].

At first glance, quantitative interpretation of the obtained frequency spectra seems more complicated, since the amplitude as well as the tip–sample distance is altered during the measurement. However, it was found by *Hölscher* et al. [45] that for the CE mode in the large-amplitude approximation the distance and the amplitude channel can be decoupled by calculating the effective tip–sample distance from the piezo-controlled tip–sample distance z and the change in the amplitude with distance $A(z)$: $z_{\text{eff}}(z) = z - A(z)$. As a result, (6.21) can then be directly used to calculate the normalized frequency shift $\gamma(z_{\text{eff}})$ and consequently the force curve can be obtained from (6.25). This concept has been verified in experiments by *Schirmeisen* et al. [46] through a direct comparison of spectroscopy curves acquired in the CE and CA modes.

Until now, we have always associated the self-excitation scheme with vacuum applications. Although it is difficult to operate FM-AFM in constant-amplitude mode in air, since large dissipative effects make it difficult to ensure a constant amplitude, it is indeed possible to use constant-excitation FM-AFM in air or even in liquids. However, only a few applications of FM-AFM under ambient or liquid conditions have been reported so far. Interestingly, a low-budget construction set (employing a tuning-fork force sensor) for a CE-mode dynamic AFM set-up has been published on the internet (http://www.sxm4.uni-muenster.de).

If it is possible to measure atomic scale forces with the NC-AFM, it should vice versa also be possible to exert forces with a similar precision. In fact, the new and exciting field of nanomanipulation could be driven to a whole new dimension, if defined forces can be reliably applied to single atoms or molecules. In this respect, *Loppacher* et al. [47] were able to push on different parts of an isolated Cu-tetrabromobisphenol (TBBP) molecule, which is known to possess four rotatable legs. They measured the force–distance curves while pushing one of the legs with the AFM tip. From the force curves they we able to determine the energy which was dissipated during the *switching* process of the molecule. The manipulation of single silicon atoms with NC-AFM was demonstrated by *Oyabu* et al. [48], who removed single atoms from a Si(111)-7×7 surface with an AFM tip and could subsequently deposit atoms from the tip onto the surface again. The possibility to exert and measure forces simultaneously during single-atom or molecule manipulation is an exciting new application of high-resolution NC-AFM experiments.

6.4 Q-Control

We have already discussed the virtues of a high Q value for high-sensitivity measurements: the minimum detectable force gradient was inversely proportional to the square root of Q. In vacuum, Q mainly represents the internal dissipation of the cantilever during oscillation, an internal damping factor. Low damping is obtained by using high-quality cantilevers, which are cut (or etched) from defect-free single-crystal silicon wafers. Under ambient or liquid conditions, the quality factor is dominated by dissipative interactions between the cantilever and the surrounding medium, and Q values can be as low as 100 for air or even 5 in liquid. Still, we ask if it is somehow possible to compensate for the damping effect by exciting the cantilever in a sophisticated way?

It turns out that the shape of the resonance curves in Fig. 6.5 can be influenced towards higher (or lower) Q values by an amplitude feedback loop. In principle, there are several mechanisms to couple the amplitude signal back to the cantilever, e.g., by the photothermal effect [49] or capacitive forces [50]. Figure 6.19 shows a method in which the amplitude feedback is mediated directly by the excitation piezo [51]. This has the advantage that no additional mechanical setups are necessary.

The working principle of the feedback loop can be understood by analyzing the equation of motion of the modified dynamic system:

$$m^*\ddot{z}(t) + \alpha\dot{z}(t) + kz(t) - F_{ts}(z_0 + z(t)) = F_{ext}\cos(\omega t) + Ge^{i\phi}z(t). \tag{6.26}$$

This ansatz takes into account the feedback of the detector signal through a phase shifter, amplifier and adder as an additional force, which is linked to the cantilever

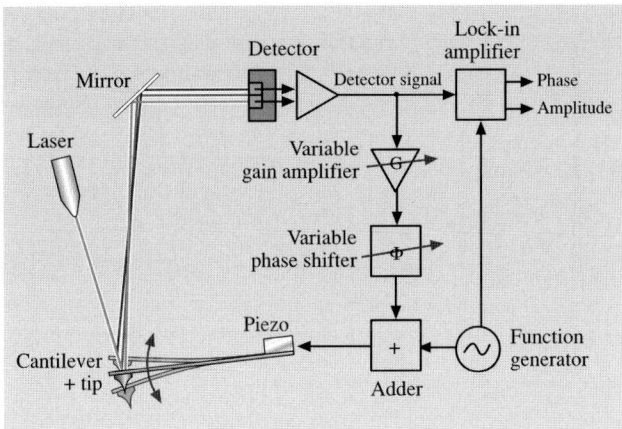

Fig. 6.19. Schematic diagram of operating a *Q-control* feedback circuit with an externally driven dynamic AFM. The tapping-mode set-up is in effect extended by an additional feedback loop

deflection $z(t)$ through the gain G and the phase shift $e^{i\phi}$. We assume that the oscillation can be described by a harmonic oscillation trajectory. With a phase shift of $\phi = \pm \pi/2$ we find:

$$e^{\pm i\pi/2} z(t) = \pm \frac{1}{\omega} \dot{z}(t). \tag{6.27}$$

This means, that the additional feedback force signal $Ge^{i\phi} z(t)$ is proportional to the velocity of the cantilever, just like the damping term in the equation of motion. We can define an effective damping constant α_{eff}, which combines the two terms:

$$m^* \ddot{z}(t) + \alpha_{\text{eff}} \dot{z}(t) + kz(t) - F_{\text{ts}}(z_0 + z(t)) = F_{\text{ext}} \cos(\omega t), \tag{6.28}$$

with $\alpha_{\text{eff}} = \alpha \pm \frac{1}{\omega} G$ for $\phi = \pm \frac{\pi}{2}$.

Equation (6.28) shows that the damping of the oscillator can be enhanced or weakened by choosing $\phi = +\pi/2$ or $\phi = -\pi/2$, respectively. The feedback loop therefore allows us to vary the effective quality factor $Q_{\text{eff}} = m\omega_0/\alpha_{\text{eff}}$ of the complete dynamic system. Hence, this system was termed *c-Control*. Figure 6.20 shows experimental data regarding the effect of *Q-control* on the amplitude and phase as a function of the external excitation frequency [51]. In this example, *Q-control* was able to increase the *Q*-value by a factor of over 40.

The effect of improved image contrast is demonstrated in Fig. 6.21. Here, a computer hard disk was analyzed with a magnetic tip in tapping mode, where the magnetic contrast is observed in the phase image. The upper part shows the recorded magnetic data structures in standard mode, whereas in the lower part of the image *Q-control* feedback was activated, giving rise to an improved signal, i.e., magnetic contrast. A more detailed analysis of measurements on a magnetic tape shows that

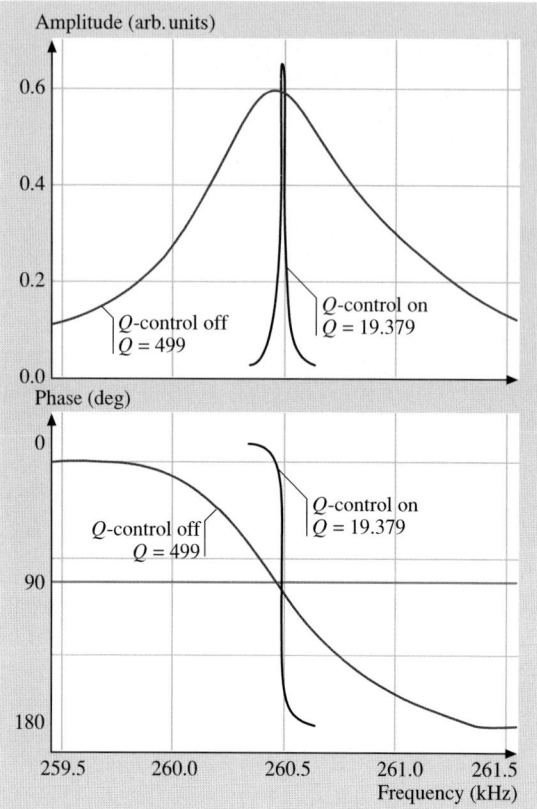

Fig. 6.20. Amplitude and phase diagrams measured in air with a Si cantilever far away from the sample. The quality factor can be increased from 450 to 20,000 by using the *Q-control* feedback method

the signal amplitude (upper diagrams in Fig. 6.22) was increased by a factor of 12.4 by the *Q-control* feedback. The lower image shows a noise analysis of the signal, indicating an improvement of the signal-to-noise ratio by a factor of 2.3.

Note that the diagrams represent measurements in air with an AFM operated in AM mode. Only then can we make a distinction between excitation and vibration frequency, since in the FM mode these two frequencies are equal by definition. Although the relation between sensitivity and Q-factor in (6.17) is the same for AM and FM mode, it must be critically investigated whether the enhanced quality factor by *Q-control* can be inserted in the equation for FM-mode AFM. In vacuum applications, Q is already very high, which makes it unnecessary to operate an additional *Q-control* module.

As stated before, we can also use *Q-control* to enhance the damping in the oscillating system. This would decrease the sensitivity of the system. But on the other hand, the response time of the amplitude change is decreased as well. For tapping-mode applications, where high-speed scanning is the goal, *Q-control* was able to reduce the scan speed that limits relaxation time [52].

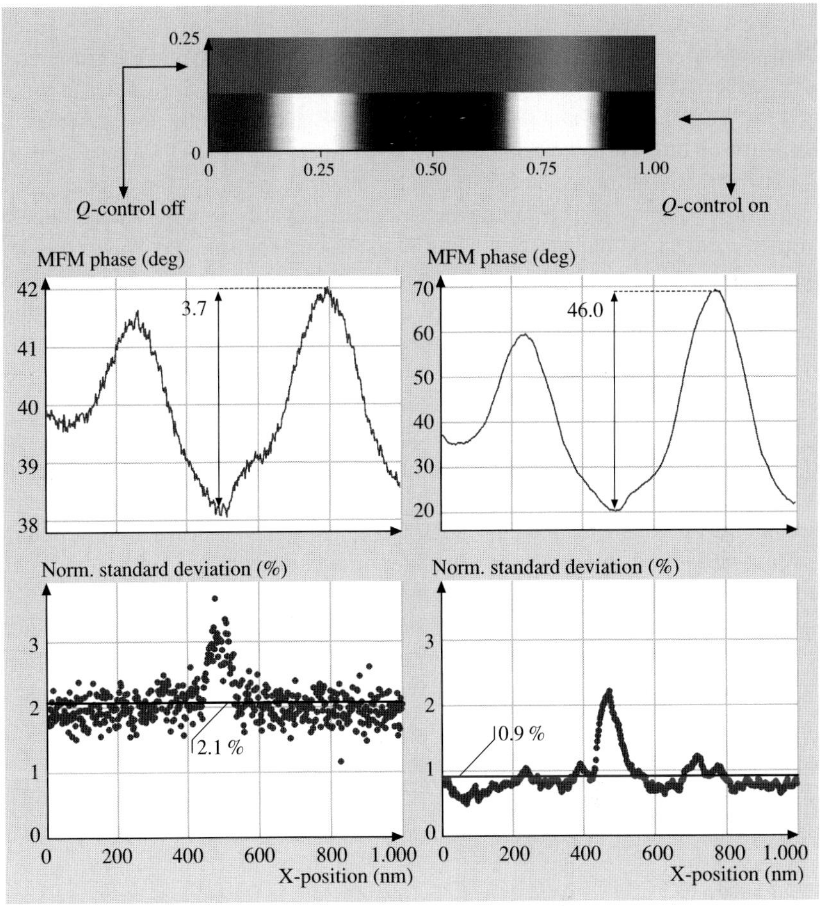

Fig. 6.21. Signal-to-noise analysis with a magnetic tip in tapping mode AFM on a magnetic tape sample with *Q-control*

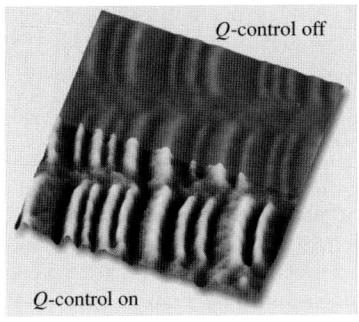

Fig. 6.22. Enhancement of the contrast in the phase channel due to *Q-control* on a magnetic hard disk measured with a magnetic tip in tapping-mode AFM in air. Scan size 5×5 μm, phase range 10 deg (www.nanoanalytics.com)

A large quality factor Q does not only have the virtue of increasing the force sensitivity of the instrument. It also has the advantage of increasing the parameter space of stable AFM operation in AM-mode AFM. Consider the resonance curve of Fig. 6.5. When approaching the tip towards the surface there are two competing mechanisms: on one hand, we bring the tip closer to the sample, which results in an increase in attractive forces (see Fig. 6.2). On the other hand, for the case $\omega > \omega_0$, the resonant frequency of the cantilever is shifted towards smaller values due to the attractive forces, which causes the amplitude to become smaller. This is the desirable regime, where stable operation of the AFM is possible in the net-attractive regime. But as explained before, below a certain tip–sample separation d_{app}, the system switches suddenly into the intermittent contact mode, where surface modifications are more likely due to the onset of strong repulsive forces. The steeper the amplitude curve the larger the regime of stable, net-attractive AFM operation. Looking at Fig. 6.20 we find that the slope of the amplitude curve is governed by the quality factor Q. A high Q, therefore, facilitates stable operation of the AM-AFM in the net-attractive regime.

An example can be found in Fig. 6.23. Here, a surface scan of an ultrathin organic film is acquired in tapping mode under ambient conditions. First, the inner square is scanned without the Q enhancement, and then a wider surface area was scanned with *Q-control* applied. The high quality factor provides a larger parameter space for operating the AFM in the net-attractive regime, allowing good resolution of the delicate organic surface structure. Without the *Q-control* the surface struc-

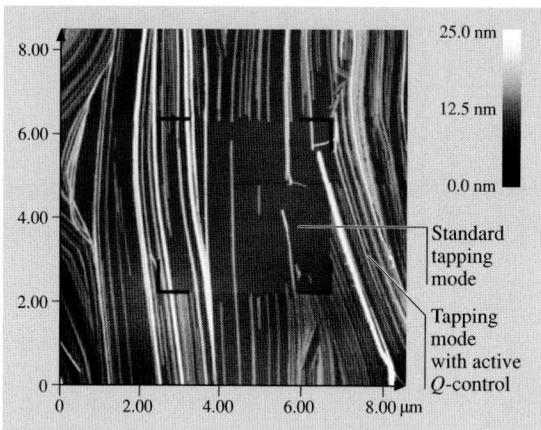

Fig. 6.23. Imaging of a delicate organic surface with Q-Control. Sample was a Langmuir–Blodgett film (ethyl-2,3-dihydroxyoctadecanoate) on a mica substrate. The topographical image clearly shows that the highly sensitive sample surface can only be imaged non-destructively with active *Q-control*, whereas the periodic repulsive contact with the probe in standard operation without *Q-control* leads to a significant modification or destruction of the surface structure (data courtesy of Lifeng Chi and coworkers, University of Münster, Germany)

tures are deformed and even destroyed due to the strong repulsive tip–sample interactions [53–55]. The *Q-control* feedback also allowed imaging of DNA structures without predominantly depressing the soft material during imaging. It was then possible to observe a DNA diameter close to the theoretical value [56].

The same technique has been successfully employed to minimize the interaction forces during scanning in liquids. This is of special relevance for imaging delicate biological samples in environments such as water or buffer solution. When the AFM probe is submerged in a liquid medium, the oscillation of the AFM cantilever is strongly affected by hydrodynamic damping. This typically leads to quality factors below 10 and, accordingly, to a loss in force sensitivity. However, the *Q-control* technique allows one to increase the effective quality factor by about three orders of magnitude in liquids. Figure 6.24 shows results of scanning DNA structures on a mica substrate under a buffer solution. Comparison of the topographic data obtained in standard tapping mode and under *Q-Control*, in particular the difference in the observed DNA height, indicates that the imaging forces were successfully reduced by employing *Q-control*.

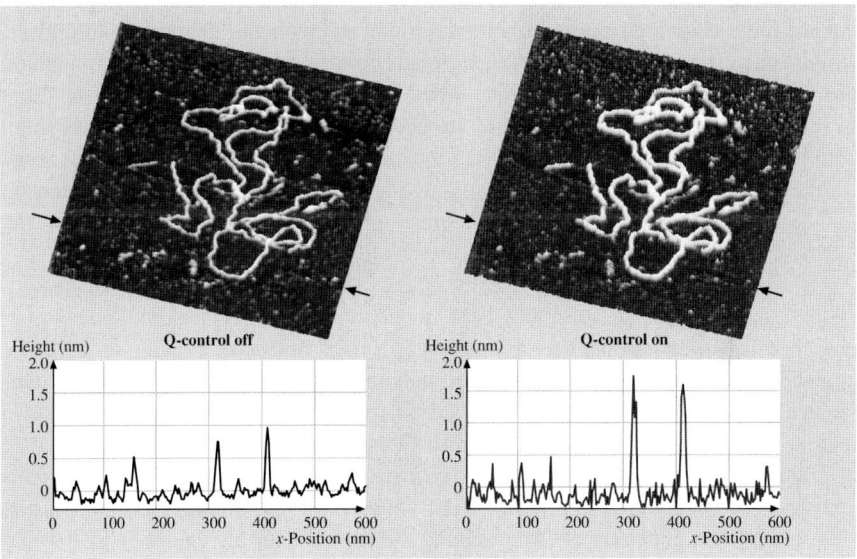

Fig. 6.24. AFM images of DNA on mica scanned in buffer solution (600 nm × 600 nm). Each scan line was scanned twice: in standard tapping mode during the first scan of the line (*left data*) and with *Q-control* activated by a trigger signal during the subsequent scan of the same line (*right data*). This interleave technique allows a direct comparison of the results of the two modes obtained on the same surface area while minimizing drift effects. Cross sections of the topographic data reveal that the observed DNA height is significantly higher in the case of imaging under *Q-control* (data courtesy of Daniel Ebeling and Hendrik Hölscher, University of Münster, Germany)

In conclusion, we have shown that, by applying an additional feedback circuit to the dynamic AFM system, it is possible to influence the quality factor Q of the oscillator system. High-resolution, high-speed, or low-force scanning is then possible.

6.5 Dissipation Processes Measured with Dynamic AFM

Dynamic AFM methods have proven their great potential for imaging surface structures at the nanoscale, and we have also discussed methods that allow the assessment of forces between distinct single molecules. However, there is another physical mechanism that can be analyzed with the dynamic mode and has been mentioned in some previous paragraphs: energy dissipation.

In Fig. 6.12, we have already shown an example, where the phase signal in tapping mode cannot be explained by conservative forces alone; dissipative processes must also play a role. In constant-amplitude FM mode, where the quantitative interpretation of experiments has proven to be less difficult, an intuitive distinction between conservative and dissipative tip–sample interaction is possible. We have shown the correlation between forces and frequency shifts of the oscillating system, but we have neglected one experimental input channel; the excitation amplitude, which is necessary to keep the oscillation amplitude constant, is a direct indication of the energy dissipated during one oscillation cycle. *Dürig* [57] has shown that in self-excitation mode (with an excitation–oscillation phase difference of 90 deg), conservative and dissipative interactions can be strictly separated. Part of this energy is dissipated in the cantilever itself, another part is due to external viscous forces in the surrounding medium. But more interestingly, some energy is dissipated at the tip–sample junction. This is the focus of the following paragraphs.

In contrast to conservative forces acting at the tip–sample junction, which at least in vacuum can be understood in terms of van der Waals, electrostatic, and chemical interactions, the dissipative processes are poorly understood. *Stowe* et al. [58] have shown that, if a voltage potential is applied between tip and sample, charges are induced in the sample surface, which will follow the tip motion (in their set-up the oscillation was parallel to the surface). Due to the finite resistance of the sample material, energy will be dissipated during the charge movement. This effect has been exploited to image the doping level of semiconductors. Energy dissipation has also been observed in imaging magnetic materials. *Liu* et al. [59] found that energy dissipation due to magnetic interactions was enhanced at the boundaries of magnetic domains, which was attributed to domain-wall oscillations, and even a simple system such as two clean metal surfaces which are moved in close proximity can give rise to frictional forces. *Stipe* et al. [60] have measured the energy dissipation due to fluctuating electromagnetic fields between two closely spaced gold surfaces, which was later interpreted by *Volokitin* and *Persson* [61] in terms of van der Waals friction.

Energy dissipation was also observed, in the absence of external electromagnetic fields, when the tip and sample were in close proximity, within 1 nm. Clearly, mechanical surface relaxations must give rise to energy losses. One could model the

AFM tip as a small hammer, hitting the surface at high frequency, possibly resulting in phonon excitations. From a continuum mechanics point of view, we assume that the mechanical relaxation of the surface is not only governed by elastic responses. Viscoelastic effects of soft surfaces will also render a significant contribution to energy dissipation. The whole area of phase imaging in tapping mode is concerned with those effects [62–64].

In the atomistic view, the last tip atom can be envisaged as changing position while experiencing the tip–sample force field. A strictly reversible change of position would not result in a loss of energy. Still, it has been pointed out by *Sasaki* et al. [65] that a change in atom position would result in a change in the force interaction itself. Therefore, it is possible that the tip atom changes position at different tip–surface distances during approach and retraction, effectively causing an atomic-scale hysteresis to develop. *Hoffmann* et al. [13] and *Hembacher* et al. [66] have measured the short-range energy dissipation for different combinations of tip and surface materials in UHV. For atomic-resolution experiments at low temperatures on graphite [66] it was found that the energy dissipation is a step-like function. A similar shape of dissipation curves was found in a theoretical analysis by *Kantorovich* and *Trevethan* [67], where the energy dissipation was directly associated with atomic instabilities at the sample surface.

The dissipation channel has also been used to image surfaces with atomic resolution [68]. Instead of feeding back the distance on the frequency shift, the excitation amplitude in FM mode has been used as the control signal. The Si(111)-(7 × 7) reconstruction was successfully imaged in this mode. The step edges of mono-atomic NaCl islands on single-crystalline copper have also rendered atomic-resolution contrast in the dissipation channel [69]. The dissipation processes discussed so far are mostly in the configuration in which the tip is oscillated perpendicular to the surface. Friction is usually referred to as the energy loss due to lateral movement of solid bodies in contact. It is interesting to note in this context that *Israelachivili* [70] has pointed out a quantitative relationship between lateral and vertical (with respect to the surface) dissipation. He states that the hysteresis in vertical force–distance curves should equal the energy loss in lateral friction. An experimental confirmation of this conjecture at the molecular level is still missing.

Physical interpretation of energy dissipation processes at the atomic scale seems to be a daunting task at this point. Nevertheless, we can find a quantitative relation between the energy loss per oscillation cycle and the experimental parameters in dynamic AFM, as will be shown in the following section.

In static AFM it was found that permanent changes of the sample surface by indentations can cause a hysteresis between approach and retraction. The area between the approach and retraction curves in a force–distance diagram represents the lost or dissipated energy caused by the irreversible change of the surface structure. In dynamic-mode AFM, the oscillation parameters such as amplitude, frequency, and phase must contain the information about the dissipated energy per cycle. So far, we have resorted to a treatment of the equation of motion of the cantilever vibration in order to find a quantitative correlation between forces and the experimental parameters. For the dissipation it is useful to treat the system from the point of view of energy conservation.

Assuming that a dynamic system is in equilibrium, the average energy input must equal the average energy output or dissipation. Applying this rule to an AFM running in a dynamic mode means that the average power fed into the cantilever oscillation by an external driver, denoted by $\overline{P}_{in}$, must equal the average power dissipated by the motion of the cantilever beam $\overline{P}_0$ and by the tip–sample interaction $\overline{P}_{tip}$:

$$\overline{P}_{in} = \overline{P}_0 + \overline{P}_{tip}. \tag{6.29}$$

The term $\overline{P}_{tip}$ is what we are interested in, since it gives us a direct physical quantity to characterize the tip–sample interaction. Therefore, we first have to calculate and then measure the two other terms in (6.29) in order to determine the power dissipated when the tip periodically probes the sample surface. This requires an appropriate rheological model to describe the dynamic system. Although there are investigations in which the complete flexural motion of the cantilever beam has been considered [71], a simplified model, comprising a spring and two dashpots (Fig. 6.25), represents a good approximation in this case [72].

The spring, characterized by the constant k according to Hooke's law, represents the only channel through which power P_{in} can be delivered to the oscillating tip $z(t)$ by the external driver $z_d(t)$. Therefore, the instantaneous power fed into the dynamic system is equal to the force exerted by the driver times the velocity of the driver (the force that is necessary to move the base side of the dashpot can be neglected, since this power is directly dissipated and, therefore, does not contribute to the power delivered to the oscillating tip):

$$P_{in}(t) = F_d(t)\dot{z}_d(t) = k[z(t) - z_d(t)]\dot{z}_d(t). \tag{6.30}$$

Assuming a sinusoidal steady-state response and that the base of the cantilever is driven sinusoidally (see (6.6)) with amplitude A_d and frequency ω, the deflection

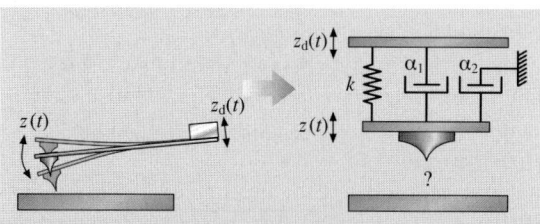

Fig. 6.25. Rheological models applied to describe the dynamic AFM system, comprising the oscillating cantilever and tip interacting with the sample surface. The movement of the cantilever base and the tip is denoted by $z_d(t)$ and $z(t)$, respectively. The cantilever is characterized by the spring constant k and the damping constant α. In a first approach damping is broken into two pieces α_1 and α_2: firstly, intrinsic damping caused by the movement of the cantilever's tip relative to its base, and secondly, damping related to the movement of the cantilever body in a surrounding medium, e. g. air damping

from equilibrium of the end of the cantilever follows (6.9), where A and $0 \le \varphi \le \pi$ are the oscillation amplitude and phase shift, respectively. This allows us to calculate the average power input per oscillation cycle by integrating (6.30) over one period $T = 2\pi/\omega$:

$$\overline{P}_{\text{in}} = \frac{1}{T}\int_0^T P_{\text{in}}(t)\,dt = \frac{1}{2}k\omega A_{\text{d}} A \sin\varphi. \tag{6.31}$$

This contains the familiar result that the maximum power is delivered to an oscillator when the response is 90 deg out of phase with the drive.

The simplified rheological model, as depicted in Fig. 6.25, exhibits two major contributions to the damping term $\overline{P}_0$. Both are related to the motion of the cantilever body and assumed to be well modeled by viscous damping with coefficients α_1 and α_2. The dominant damping mechanism in UHV conditions is intrinsic damping, caused by the deflection of the cantilever beam, i.e., the motion of the tip relative to the cantilever base. Therefore the instantaneous power dissipated by such a mechanism is given by

$$P_{01}(t) = |F_{01}(t)\dot{z}(t)| = |\alpha_1[\dot{z}(t) - \dot{z}_{\text{d}}(t)]\dot{z}(t)|. \tag{6.32}$$

Note that the absolute value has to be calculated, since all dissipated power is *lost* and, therefore, cannot be returned to the dynamic system.

However, when running an AFM in ambient conditions an additional damping mechanism has to be considered; damping due to the motion of the cantilever body in the surrounding medium, e.g., air damping, is in most cases the dominant effect. The corresponding instantaneous power dissipation is given by

$$P_{02}(t) = |F_{02}(t)\dot{z}(t)| = \alpha_2 \dot{z}^2(t). \tag{6.33}$$

In order to calculate the average power dissipation, (6.32) and (6.33) have to be integrated over one complete oscillation cycle. This yields

$$\overline{P}_{01} = \frac{1}{T}\int_0^T P_{01}(t)\,dt$$

$$= \frac{1}{\pi}\alpha_1\omega^2 A\left[(A - A_{\text{d}}\cos\varphi)\arcsin\left(\frac{A - A_{\text{d}}\cos\varphi}{\sqrt{A^2 + A_{\text{d}}^2 - 2AA_{\text{d}}\cos\varphi}}\right) + A_{\text{d}}\sin\varphi\right] \tag{6.34}$$

and

$$\overline{P}_{02} = \frac{1}{T}\int_0^T P_{02}(t)\,dt = \frac{1}{2}\alpha_2\omega^2 A^2. \tag{6.35}$$

Considering the fact that commonly used cantilevers exhibit a quality factor of at least several hundreds (in UHV even several ten thousands), we can assume that the oscillation amplitude is significantly larger than the drive amplitude when the dynamic system is driven at or near its resonance frequency: $A \gg A_{\text{d}}$. Therefore (6.34) can be simplified in a first-order approximation to an expression similar to

(6.35). Combining the two equations yields the total average power dissipated by the oscillating cantilever

$$\overline{P}_0 = \frac{1}{2}\alpha\omega^2 A^2 \quad \text{with} \quad \alpha = \alpha_1 + \alpha_2, \tag{6.36}$$

where α denotes the overall effective damping constant.

We can now solve (6.29) for the power dissipation localized to the small interaction volume of the probing tip with the sample surface, represented by the question mark in Fig. 6.25. Furthermore by expressing the damping constant α in terms of experimentally accessible quantities such as the spring constant k, the quality factor Q and the natural resonant frequency ω_0 of the free oscillating cantilever, $\alpha = \frac{k}{Q \cdot \omega_0}$, we obtain:

$$\overline{P}_{\text{tip}} = \overline{P}_{\text{in}} - \overline{P}_0 = \frac{1}{2}\frac{k\omega}{Q}\left[Q_{\text{cant}} A_{\text{d}} A \sin\varphi - A^2 \frac{\omega}{\omega_0}\right]. \tag{6.37}$$

Note that so far no assumptions have been made on how the AFM is operated, except that the motion of the oscillating cantilever has to remain sinusoidal to a good approximation. Therefore (6.37) is applicable to a variety of different dynamic AFM modes.

For example, in FM-mode AFM the oscillation frequency ω changes due to tip–sample interaction while at the same time the oscillation amplitude A is kept constant by adjusting the drive amplitude A_{d}. By measuring these quantities, one can apply (6.37) to determine the average power dissipation related to the tip–sample interaction. In spectroscopy applications $A_{\text{d}}(z)$ is usually not measured directly, but a signal $G(z)$ proportional to $A_{\text{d}}(z)$ is acquired that represents the gain factor applied to the excitation piezo. With the help of (6.15) we can write:

$$A_{\text{d}}(z) = \frac{A_0 \cdot G(z)}{Q \cdot G_0}, \tag{6.38}$$

while A_0 and G_0 are the amplitude and gain at large tip–sample distances where the tip–sample interactions are negligible.

Now let us consider the tapping-mode AFM. In this case the cantilever is driven at a fixed frequency and with constant drive amplitude, while the oscillation amplitude and phase shift may change when the probing tip interacts with the sample surface. Assuming that the oscillation frequency is chosen to be ω_0, (6.37) can be further simplified by employing (6.15) for the free oscillation amplitude A_0. This yields

$$\overline{P}_{\text{tip}} = \frac{1}{2}\frac{k\omega_0}{Q_{\text{cant}}}\left[A_0 A \sin\varphi - A^2\right]. \tag{6.39}$$

Equation (6.39) implies that, if the oscillation amplitude A is kept constant by a feedback loop, as is commonly done in tapping-mode, simultaneously acquired phase data can be interpreted in terms of energy dissipation [63, 64, 73, 74]. When analyz-

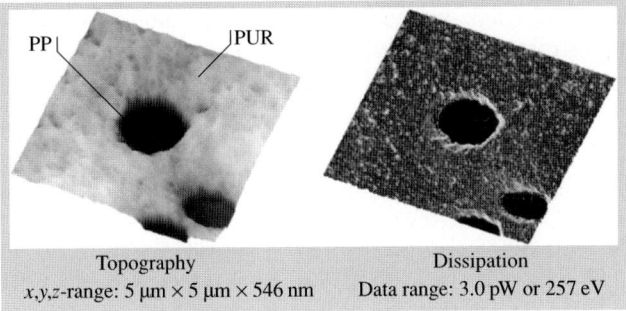

Fig. 6.26. Topography and phase image in tapping-mode AFM of a polymer blend composed of polypropylene (PP) particles embedded in a polyurethane (PUR) matrix. The dissipation image shows a strong contrast between the harder PP (*little dissipation, dark*) to the softer PUR (*large dissipation, bright*) surface

ing such phase images [75–77] one also has to consider the fact that the phase may also change due to the transition from the net-attractive ($\varphi > 90$ deg) to intermittent contact ($\varphi < 90$ deg) interaction between the tip and the sample [18, 51, 78, 79]. For example, consider the phase shift in tapping mode as a function of z-position, Fig. 6.12. If phase measurements are performed close to the point where the oscillation switches from the net-attractive to the intermittent contact regime, a large contrast in the phase channel is observed. However, this contrast is not due to dissipative processes. Only a variation of the phase signal within the intermittent contact regime will give information about the tip–sample dissipative processes.

An example of a dissipation measurement is depicted in Fig. 6.26. The surface of a polymer blend was imaged in air, simultaneously acquiring the topography and dissipation. The dissipation on the softer polyurethane matrix is significantly larger than on the embedded, mechanically stiffer polypropylene particles.

6.6 Conclusion

Dynamic force microscopy is a powerful tool, which is capable of imaging surfaces with atomic precision. It also allows us to look at surface dynamics and it can operate in vacuum, air or even in liquid. However, the oscillating cantilever system introduces a level of complexity, which prevents a straightforward interpretation of acquired images. An exception is the self-excitation mode, where tip–sample forces can be successfully extracted from spectroscopic experiments. However, not only conservative forces can be investigated with dynamic AFM, energy dissipation also influences the cantilever oscillation and can therefore serve as a new information channel.

Open questions remain, concerning the exact geometric and chemical identity of the probing tip, which significantly influences the imaging and spectroscopic results. Using predefined tips like single-walled nanotubes or using atomic-resolution techniques like field ion microscopy to image the tip itself are possible approaches to address this issue.

References

1. M. A. Lantz, H. J. Hug, R. Hoffmann, P. J. A. van Schendel, P. Kappenberger, S. Martin, A. Baratoff, H.-J. Güntherodt: Quantitative measurement of short-range chemical bonding forces, Science **291**, 2580–2583 (2001)
2. G. Binnig, C. F. Quate, Ch. Gerber: Atomic force microscope, Phys. Rev. Lett. **56**, 930–933 (1986)
3. O. Marti: AFM instrumentation and tips. In: *Handbook of Micro/Nanotribology*, 2nd edn., ed. by B. Bushan (CRC, Boca Raton 1999) pp. 81–144
4. G. Cross, A. Schirmeisen, A. Stalder, P. Grütter, M. Tschudy, U. Dürig: Adhesion interaction between atoically defined tip and sample, Phys. Rev. Lett. **80**, 4685–4688 (1998)
5. A. Schirmeisen, G. Cross, A. Stalder, P. Grütter, U. Dürig: Metallic adhesion and tunneling at the atomic scale, New J. Phys. **2**, 1–29 (2000)
6. A. Schirmeisen: PhD thesis, Metallic Adhesion and Tunneling at the Atomic Scale, McGill University, Montréal, Canada, 29-38
7. F. J. Giessibl: Forces and frequency shifts in atomic-resolution dynamic-force microscopy, Phys. Rev. B **56**, 16010–16015 (1997)
8. F. J. Giessibl: Atomic resolution of the silicon (111)-(7×7) surface by atomic force microscopy, Science **267**, 68–71 (1995)
9. M. Bezanilaa, B. Drake, E. Nudler, M. Kashlev, P. K. Hansma, H. G. Hansma: Motion and enzymatic degradation of DNA in the atomic force microscope, Biophys. J. **67**, 2454–2459 (1994)
10. Y. Jiao, D. I. Cherny, G. Heim, T. M. Jovin, T. E. Schäffer: Dynamic interactions of p53 with DNA in solution by time-lapse atomic force microscopy, J. Mol. Biol. **314**, 233–243 (2001)
11. T. R. Albrecht, P. Grütter, D. Horne, D. Rugar: Frequency modulation detection using high-Q cantilevers for enhanced force microscopy sensitivity, J. Appl. Phys. **69**, 668–673 (1991)
12. S. P. Jarvis, M. A. Lantz, U. Dürig, H. Tokumoto: Off resonance AC mode force spectroscopy and imaging with an atomic force microscope, Appl. Surf. Sci. **140**, 309–313 (1999)
13. P. M. Hoffmann, S. Jeffery, J. B. Pethica, H. Ö. Özer, A. Oral: Energy dissipation in atomic force microscopy and atomic loss processes, Phys. Rev. Lett. **87**, 265502–265505 (2001)
14. E. Sivaniah, J. Genzer, G. H. Fredrickson, E. J. Kramer, M. Xiang, X. Li, C. Ober, S. Magonov: Periodic surface topology of three-arm semifluorinated alkane monodendron diblock copolymers, Langmuir **17**, 4342–4346 (2001)
15. S. N. Magonov: Visualization of polymer structures with atomic force microscopy. In: *Applied Scanning Probe Methods*, ed. by H. Fuchs, M. Hosaka, B. Bhushan (Springer, Berlin 2004) pp. 207–250
16. C. Möller, M. Allen, V. Elings, A. Engel, D. J. Müller: Tapping-mode atomic force microscopy produces faithful high-resolution images of protein surfaces, Biophys. J. **77**, 1150–1158 (1999)

17. B. Anczykowski, D. Krüger, H. Fuchs: Cantilever dynamics in quasinoncontact force microscopy: spectroscopic aspects, Phys. Rev. B **53**, 15485–15488 (1996)
18. B. Anczykowski, D. Krüger, K.L. Babcock, H. Fuchs: Basic properties of dynamic force spectroscopy with the scanning force microscope in experiment and simulation, Ultramicroscopy **66**, 251–259 (1996)
19. V.M. Muller, V.S. Yushchenko, B.V. Derjaguin: On the influence of molecular forces on the deformation of an elastic sphere and its sticking to a rigid plane, J. Coll. Interf. Sci. **77**, 91–101 (1980)
20. B.D. Hughes, L.R. White: 'Soft' contact problems in linear elasticity, Quart. J. Mech. Appl. Math. **32**, 445–471 (1979)
21. L. Verlet: Computer experiments on classical fluids. I. Thermodynamical properties of Lennard–Jones Molecules, Phys. Rev. **159**, 98–103 (1967)
22. L. Verlet: Computer experiments on classical fluids. II. Equilibrium correlation functions, Phys. Rev. **165**, 201–214 (1968)
23. P. Gleyzes, P.K. Kuo, A.C. Boccara: Bistable behavior of a vibrating tip near a solid surface, Appl. Phys. Lett. **58**, 2989–2991 (1991)
24. A. San Paulo, R. Garcia: High-resolution imaging of antibodies by tapping-mode atomic force microscopy: Attractive and repulsive tip–sample interaction regimes, Biophys. J. **78**, 1599–1605 (2000)
25. D. Krüger, B. Anczykowski, H. Fuchs: Physical properties of dynamic force microscopies in contact and noncontact operation, Ann. Phys. **6**, 341–363 (1997)
26. Y. Martin, C.C. Williams, H.K. Wickramasinghe: Atomic force microscope–force mapping and profiling on a sub 100-A scale, J. Appl. Phys. **61**, 4723–4729 (1987)
27. Y. Sugawara, M. Otha, H. Ueyama, S. Morita: Defect motion on an InP(110) surface observed with noncontact atomic force microscopy, Science **270**, 1646–1648 (1995)
28. M. Bammerlin, R. Lüthi, E. Meyer, A. Baratoff, J. Lue, M. Guggisberg, Ch. Gerber, L. Howald, H.-J. Güntherodt: True atomic resolution on the surface of an insulator via ultrahigh vacuum dynamic force microscopy, Probe Microsc. **1**, 3–9 (1996)
29. C. Barth, M. Reichling: Imaging the atomic arrangement on the high-temperature reconstructed α-Al$_2$O$_3$(0001) Surface, Nature **414**, 54–57 (2001)
30. W. Allers, S. Langkat, R. Wiesendanger: Dynamic low-temperature scanning force microscopy on nickel oxyde(001), Appl. Phys. A [Suppl.] **72**, S27–S30 (2001)
31. S. Morita, R. Wiesendanger, E. Meyer: *Noncontact Atomic Force Microscopy* (Springer, Berlin, Heidelberg 2002)
32. R. Garcia, R. Pérez: Dynamic atomic force microscopy methods, Surf. Sci. Rep. **47**, 197–301 (2002)
33. F.J. Giessibl: Advances in atomic force microscopy, Rev. Mod. Phys. **75**, 949–983 (2003)
34. H. Hölscher, U.D. Schwarz, R. Wiesendanger: Calculation of the frequency shift in dynamic force microscopy, Appl. Surf. Sci. **140**, 344–351 (1999)
35. B. Gotsmann, H. Fuchs: Dynamic force spectroscopy of conservative and dissipative forces in an Al-Au(111) tip–sample system, Phys. Rev. Lett. **86**, 2597–2600 (2001)
36. H. Hölscher, W. Allers, U.D. Schwarz, A. Schwarz, R. Wiesendanger: Determination of tip–sample interaction potentials by dynamic force spectroscopy, Phys. Rev. Lett. **83**, 4780–4783 (1999)
37. U. Dürig: Relations between interaction force and frequency shift in large-amplitude dynamic force microscopy, Appl. Phys. Lett. **75**, 433–435 (1999)
38. H. Hölscher, A. Schwarz, W. Allers, U.D. Schwarz, R. Wiesendanger: Quantitative analysis of dynamic-force-spectroscopy data on graphite (0001) in the contact and noncontact regime, Phys. Rev. B **61**, 12678–12681 (2000)

39. M. Guggisberg (2000): PhD thesis, Lokale Messung von atomaren Kräften, University of Basel, Switzerland, 9-11
40. M. Guggisberg, M. Bammerlin, E. Meyer, H.-J. Güntherodt: Separation of interactions by noncontact force microscopy, Phys. Rev. B **61**, 11151–11155 (2000)
41. U. Dürig: Extracting interaction forces and complementary observables in dynamic probe microscopy, Appl. Phys. Lett. **76**, 1203–1205 (2000)
42. F.J. Giessibl: A direct method to calculate tip–sample forces from frequency shifts in frequency-modulation atomic force microscopy, Appl. Phys. Lett. **78**, 123–125 (2001)
43. J.E. Sader, S.P. Jarvis: Accurate formulas for interaction force and energy in frequency modulation force spectroscopy, Appl. Phys. Lett. **84**, 1801–1803 (2004)
44. T. Uchihasi, T. Ishida, M. Komiyama, M. Ashino, Y. Sugawara, W. Mizutani, K. Yokoyama, S. Morita, H. Tokumoto, M. Ishikawa: High-resolution imaging of organic monolayers using noncontact AFM, Appl. Surf. Sci. **157**, 244–250 (2000)
45. H. Hölscher, B. Gotsmann, A. Schirmeisen: Dynamic force spectroscopy using the frequency modulation technique with constant excitation, Phys. Rev. B **68**, 153401/1–4 (2003)
46. A. Schirmeisen, H. Hölscher, B. Anczykowski, D. Weiner, M.M. Schäfer, H. Fuchs: Dynamic force spectroscopy using the constant-excitation and constant-amplitude modes, Nanotechnology **16**, 13–17 (2005)
47. Ch. Loppacher, M. Guggisberg, O. Pfeiffer, E. Meyer, M. Bammerlin, R. Lüthi, R. Schlittler, J.K. Gimzewski, H. Tang, C. Joachim: Direct determination of the energy required to operate a single molecule switch, Phys. Rev. Lett. **90**, 066107/1–4 (2003)
48. N. Oyabu, O. Custance, I. Yi, Y. Sugawara, S. Morita: Mechanical vertical manipulation of selected single atoms by soft nanoindentation using near contact atomic force microscopy, Phys. Rev. Lett. **90**, 176102 (2003)
49. J. Mertz, O. Marti, J. Mlynek: Regulation of a microcantilever response by force feedback, Appl. Phys. Lett. **62**, 2344–2346 (1993)
50. D. Rugar, P. Grütter: Mechanical parametric amplification and thermomechanical noise squeezing, Phys. Rev. Lett. **67**, 699–702 (1991)
51. B. Anczykowski, J.P. Cleveland, D. Krüger, V.B. Elings, H. Fuchs: Analysis of the interaction mechanisms in dynamic mode SFM by means of experimental data and computer simulation, Appl. Phys. A **66**, 885 (1998)
52. T. Sulchek, G.G. Yaralioglu, C.F. Quate, S.C. Minne: Characterization and optimisation of scan speed for tapping-mode atomic force microscopy, Rev. Sci. Instr. **73**, 2928–2936 (2002)
53. L.F. Chi, S. Jacobi, B. Anczykowski, M. Overs, H.-J. Schäfer, H. Fuchs: Supermolecular periodic structures in monolayers, Adv. Mater. **12**, 25–30 (2000)
54. S. Gao, L.F. Chi, S. Lenhert, B. Anczykowski, C. Niemeyer, M. Adler, H. Fuchs: High-quality mapping of DNA–protein complexes by dynamic scanning force microscopy, ChemPhysChem **6**, 384–388 (2001)
55. B. Zou, M. Wang, D. Qiu, X. Zhang, L.F. Chi, H. Fuchs: Confined supramolecular nanostructures of mesogen-bearing amphiphiles, Chem. Commun. **9**, 1008–1009 (2002)
56. B. Pignataro, L.F. Chi, S. Gao, B. Anczykowski, C. Niemeyer, M. Adler, H. Fuchs: Dynamic scanning force microscopy study of self-assembled DNA–protein nanostructures, Appl. Phys. A **74**, 447–452 (2002)
57. U. Dürig: Interaction sensing in dynamic force microscopy, New J. Phys. **2**, 1–5 (2000)
58. T.D. Stowe, T.W. Kenny, D.J. Thomson, D. Rugar: Silicon dopant imaging by dissipation force microscopy, Appl. Phys. Lett. **75**, 2785–2787 (1999)
59. Y. Liu, P. Grütter: Magnetic dissipation force microscopy studies of magnetic materials, J. Appl. Phys. **83**, 7333–7338 (1998)

60. B.C. Stipe, H.J. Mamin, T.D. Stowe, T.W. Kenny, D. Rugar: Noncontact friction and force fluctuations between closely spaced bodies, Phys. Rev. Lett. **87**, 96801/1–4 (2001)
61. A.I. Volokitin, B.N.J. Persson: Resonant photon tunneling enhancement of the van der Waals friction, Phys. Rev. Lett. **91**, 106101/1–4 (2003)
62. J. Tamayo, R. Garcia: Effects of elastic and inelastic interactions on phase contrast images in tapping-mode scanning force microscopy, Appl. Phys. Lett. **71**, 2394–2396 (1997)
63. J.P. Cleveland, B. Anczykowski, A.E. Schmid, V.B. Elings: Energy dissipation in tapping-mode atomic force microscopy, Appl. Phys. Lett. **72**, 2613–2615 (1998)
64. B. Anczykowski, B. Gotsmann, H. Fuchs, J.P. Cleveland, V.B. Elings: How to measure energy dissipation in dynamic mode atomic force microscopy, Appl. Surf. Sci. **140**, 376–382 (1999)
65. N. Sasaki, M. Tsukada: Effect of microscopic nonconservative process on noncontact atomic force microscopy, Jpn. J. Appl. Phys. **39**, 1334 (2000)
66. S. Hembacher, F.J. Giessibl, J. Mannhart, C.F. Quate: Local spectroscopy and atomic imaging of tunneling current, forces, and dissipation on graphite, Phys. Rev. Lett. **94**, 056101/1–4 (2005)
67. L.N. Kantorovich, T. Trevethan: General theory of microscopic dynamical response in surface probe microscopy: From imaging to dissipation, Phys. Rev. Lett. **93**, 236102/1–4 (2004)
68. R. Lüthi, E. Meyer, M. Bammerlin, A. Baratoff, L. Howald, C. Gerber, H.-J. Güntherodt: Ultrahigh vacuum atomic force microscopy: True atomic resolution, Surf. Rev. Lett. **4**, 1025–1029 (1997)
69. R. Bennewitz, A.S. Foster, L.N. Kantorovich, M. Bammerlin, Ch. Loppacher, S. Schär, M. Guggisberg, E. Meyer, A.L. Shluger: Atomically resolved edges and kinks of NaCl islands on Cu(111): Experiment and theory, Phys. Rev. B **62**, 2074–2084 (2000)
70. J. Israelachvili: *Intermolecular and Surface Forces* (Academic, London 1992)
71. U. Rabe, J. Turner, W. Arnold: Analysis of the high-frequency response of atomic force microscope cantilevers, Appl. Phys. A **66**, 277 (1998)
72. T.R. Rodriguez, R. Garcia: Tip motion in amplitude modulation (tapping-mode) atomic-force microscopy: Comparison between continuous and point-mass models, Appl. Phys. Lett **80**, 1646–1648 (2002)
73. J. Tamayo, R. Garcia: Relationship between phase shift and energy dissipation in tapping-mode scanning force microscopy, Appl. Phys. Lett. **73**, 2926–2928 (1998)
74. R. Garcia, J. Tamayo, A. San Paulo: Tapping mode scanning force microsopy, Surf. Interface Anal. **27**, 312–316 (1999)
75. S.N. Magonov, V.B. Elings, M.H. Whangbo: Phase imaging and stiffness in tapping-mode atomic force microscopy, Surf. Sci. **375**, 385–391 (1997)
76. J.P. Pickering, G.J. Vancso: Apparent contrast reversal in tapping mode atomic force microscope images on films of polystyrene-b-polyisoprene-b-polystyrene, Polymer Bulletin **40**, 549–554 (1998)
77. X. Chen, S.L. McGurk, M.C. Davies, C.J. Roberts, K.M. Shakesheff, S.J.B. Tendler, P.M. Williams, J. Davies, A.C. Dwakes, A. Domb: Chemical and morphological analysis of surface enrichment in a biodegradable polymer blend by phase-detection imaging atomic force microscopy, Macromolecules **31**, 2278–2283 (1998)
78. A. Kühle, A.H. Sørensen, J. Bohr: Role of attractive forces in tapping tip force microscopy, J. Appl. Phys. **81**, 6562–6569 (1997)
79. A. Kühle, A.H. Sørensen, J.B. Zandbergen, J. Bohr: Contrast artifacts in tapping tip atomic force microscopy, Appl. Phys. A **66**, 329–332 (1998)

7

Molecular Recognition Force Microscopy: From Simple Bonds to Complex Energy Landscapes

Peter Hinterdorfer and Ziv Reich

Summary. Atomic force microscopy (AFM), developed in the late eighties to explore atomic details on hard material surfaces, has evolved to an imaging method capable of achieving fine structural details on biological samples. Its particular advantage in biology is that the measurements can be carried out in aqueous and physiological environment, which opens the possibility to study the dynamics of biological processes in vivo. The additional potential of the AFM to measure ultra-low forces at high lateral resolution has paved the way for measuring inter- and intra-molecular forces of bio-molecules on the single molecule level. Molecular recognition studies using AFM open the possibility to detect specific ligand–receptor interaction forces and to observe molecular recognition of a single ligand–receptor pair. Applications include biotin–avidin, antibody–antigen, NTA nitrilotriacetate–hexahistidine 6, and cellular proteins, either isolated or in cell membranes.

The general strategy is to bind ligands to AFM tips and receptors to probe surfaces (or vice versa), respectively. In a force–distance cycle, the tip is first approached towards the surface whereupon a single receptor–ligand complex is formed, due to the specific ligand receptor recognition. During subsequent tip–surface retraction a temporarily increasing force is exerted to the ligand–receptor connection thus reducing its lifetime until the interaction bond breaks at a critical force (unbinding force). Such experiments allow for estimation of affinity, rate constants, and structural data of the binding pocket. Comparing them with values obtained from ensemble-average techniques and binding energies is of particular interest. The dependences of unbinding force on the rate of load increase exerted to the receptor–ligand bond reveal details of the molecular dynamics of the recognition process and energy landscapes. Similar experimental strategies were also used for studying intra-molecular force properties of polymers and unfolding–refolding kinetics of filamentous proteins. Recognition imaging, developed by combing dynamic force microscopy with force spectroscopy, allows for localization of receptor sites on surfaces with nanometer positional accuracy.

7.1 Introduction

Molecular recognition plays a pivotal role in nature. Signaling cascades, enzymatic activity, genome replication and transcription, cohesion of cellular structures, interaction of antigens and antibodies and metabolic pathways all rely critically on specific recognition. In fact, every process in which molecules interact with each other in a specific manner depends on this trait.

Molecular recognition studies emphasize specific interactions between receptors and their cognitive ligands. Despite a growing body of literature on the structure and function of receptor ligand complexes, it is still not possible to predict reaction kinetics or energetics for any given complex formation, even when the structures are known. Additional insights, in particular about the molecular dynamics (MD) within the complex during the association and dissociation processes, are needed. The high-end strategy is to probe the energy landscape that underlies the interactions between molecules, whose structure is known at atomic resolution.

Receptor ligand complexes are usually formed by a few, non-covalent weak interactions between contacting chemical groups in complementary determining regions, supported by framework residues providing structurally conserved scaffolding. Both the complementary determining regions and the framework have a considerable amount of plasticity and flexibility, allowing for conformational movements during association and dissociation. In addition to knowledge about structure, energies, and kinetic constants, information about these movements is required for the understanding of the recognition process. Deeper insight into the nature of these movements as well as the spatiotemporal action of the many weak interactions, in particular the cooperativity of bond formation, is the key for the understanding of receptor ligand recognition.

For this, experiments at the single-molecule level, on time scales typical for receptor ligand complex formation and dissociation appear to be required. The potential of the atomic force microscope (AFM) [1] to measure ultra-low forces at high lateral resolution has paved the way for single-molecule recognition force microscopy studies. The particular advantage of AFM in biology is that the measurements can be carried out in an aqueous and physiological environment, which opens the possibility for studying biological processes in vivo. The methodology described in this paper for investigating the molecular dynamics of receptor ligand interactions, molecular recognition force microscopy (MRFM [2–4], is based on scanning probe microscopy (SPM) technology [1]. A force is exerted on a receptor ligand complex and the dissociation process is followed over time. Dynamic aspects of recognition are addressed in force spectroscopy (FS) experiments, where distinct force–time profiles are applied to give insight into changes of conformations and states during receptor ligand dissociation. It will be shown that MRFMs a versatile tool to explore kinetic and structural details of receptor ligand recognition.

7.2 Ligand Tip Chemistry

In MRFMxperiments, the binding of ligands immobilized on AFM tips to surface-bound receptors (or vice versa) is studied by applying a force to the receptor ligand complex that reduces its lifetime until the bond breaks at a measurable unbinding force. This requires a careful AFM tip sensor design, including tight attachment of the ligands to the tip surface. In the first pioneering demonstrations of

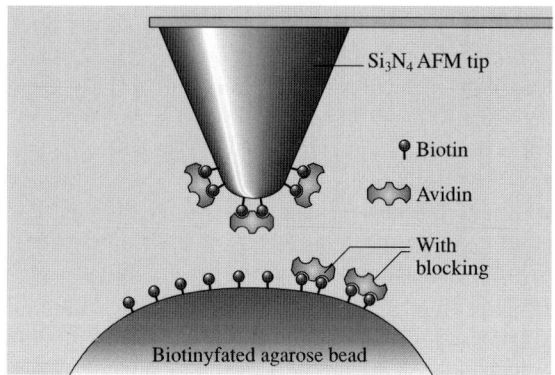

Fig. 7.1. Avidin-functionalized AFM tip. A dense layer of biotinylated BSA was adsorbed to the tip and subsequently saturated with avidin. The biotinylated agarose bead opposing the tip also contained a high surface density of reactive sites. These were partly blocked with avidin to achieve single-molecule binding events. After [3]

single-molecule recognition force measurements [2, 3], strong physical adsorption of bovine serum albumin (BSA) was used to directly coat the tip [3] or a glass bead glued to it [2]. This physisorbed protein layer may then serve as a matrix for biochemical modifications with chemically active ligands (Fig. 7.1). In spite of the large number of probe molecules on the tip (10^3–10^4 nm^{-2}) the low fraction of properly oriented molecules, or internal blocks of most reactive sites (see Fig. 7.1), allowed the measurement of single receptor ligand unbinding forces. Nevertheless, parallel breakage of multiple bonds was predominately observed with this configuration.

To measure interactions between isolated receptor ligand pairs, strictly defined conditions need to be fulfilled. Covalently coupling ligands to gold-coated tip surfaces via freely accessible SH groups guarantees a sufficiently stable attachment because these bonds are about ten times stronger than typical ligand–receptor interactions [5]. This chemistry has been used to detect the forces between complementary DNA strands [6] as well as between isolated nucleotides [7]. Self-assembled monolayers of dithio-bis(succinimidylundecanoate) were formed to enable covalent coupling of biomolecules via amines [8] and were used to study the binding strength between cell adhesion proteoglycans [9] and between biotin-directed IgG antibodies and biotin [10]. A vectorial orientation of Fab molecules on gold tips was achieved by site-directed chemical binding via their SH groups [11], without the need for additional linkers. To this end, antibodies were digested with papain and subsequently purified to generate Fab fragments with freely accessible SH groups in the hinge region.

Gold surfaces provide a unique and selective affinity for thiols, although the adhesion strength of the resulting bonds is comparatively weak [5]. Since all commercially available AFM tips are etched from silicon nitride or silicon oxide material, the deposition of a gold layer onto the tip surface is required prior to using this chemistry. Therefore, designing a sensor with covalent attachments of biomolecules to the silicon surface may be more straightforward. Amine functionalization procedures, a strategy widely used in surface biochemistry, were applied using ethanolamine [4,12] and various silanization methods [13–15] as a first step in thoroughly developed surface-anchoring protocols suitable for single-molecule experiments. Since the amine surface density determines, to a large extent, the number of ligands on the tip that can specifically bind to the receptors on the surface, it has to be sufficiently low to guarantee single-molecular recognition events [4, 12]. Typically, these densities are kept between 200 and 500 molecules per square micron, which, for AFM tips with radii of $\approx 5-20$ nm, amounts to about one molecule per effective tip area. A striking example of a minimally ligated tip was given by *Wong* et al. [16] who derivatized a few carboxyl groups present at the open end of carbon nanotubes attached to the tips of gold-coated Si cantilevers.

In a number of laboratories, a distensible and flexible linker was used to distance the ligand molecule from the tip surface (e. g. [4,14]) (Fig. 7.2). At a given low number of spacer molecules per tip, the ligand can then freely orient and diffuse within a certain volume, provided by the length of the tether, to achieve unconstrained binding to its receptor. The unbinding process occurs with little torque and the ligand molecule escapes the danger of being squeezed between the tip and the surface. It also opens the possibility of site-directed coupling for a defined orientation of the ligand relative to the receptor at receptor ligand unbinding. As a crosslinking element, polyethylene glycol (PEG), a water-soluble, nontoxic polymer with a wide range of applications in surface technology and clinical research, was often used [17]. PEG is known to prevent surface adsorption of proteins and lipid structures and appears therefore ideally suited for this purpose. Glutaraldehyde [13] and DNA [6] were also successfully applied in recognition force studies as molecular spacers. Crosslinker lengths, ideally arriving at a good compromise between high tip-molecule mobility and narrow lateral resolution of the target recognition site, varied from 2 to 100 nm.

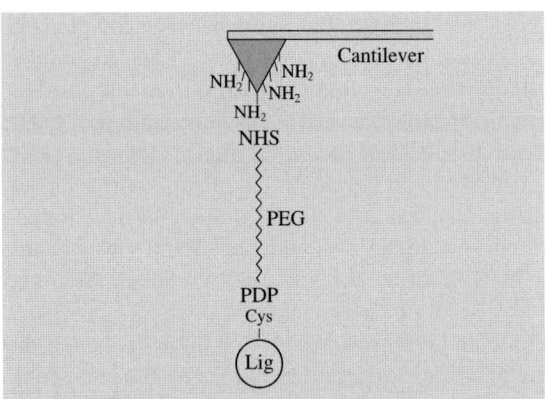

Fig. 7.2. Linkage of ligands to AFM tips. Ligands were covalently coupled to AFM tips via a heterobifunctional polyethylene glycol (PEG) derivative of 8 nm length. Silicon tips were first functionalized with ethanolamine ($NH_2-C_2H_4OHHCl$). Then, the N-hydroxysuccinimide (NHS)-end of the PEG linker was covalently bound to amines on the tip surface before ligands were attached to the pyridoyldithiopropionate (PDP)-end via a free thiol or cysteine

For coupling to the tip surface and to the ligand, the crosslinker typically carries two different functional ends, e. g. an amine reactive N-hydroxysuccinimidyl (NHS) group on one end, and 2-pyridyldithiopropionyl (PDP) [18] or vinyl sulfone [19] groups, which can be covalently bound to thiols, on the other (Fig. 7.2). This sulfur chemistry is highly advantageous since it is very reactive and readily enables site-directed coupling. However, free thiols are hardly available on native ligands and must, therefore, be added.

Different strategies have been used to achieve this goal. Lysine residues were derivatized with the short heterobifunctional linker N-succinimidyl-3-(S-acetylthio)-propionate (SATP) [18]. Subsequent deprotection with NH_2OH led to reactive SH groups. Alternatively, lysins can be directly coupled via aldehyde groups (manuscript in preparation). A problem with the latter two methods is that they do not allow for site-specific coupling of the crosslinker, since lysine residues are quite abundant. Several protocols are commercially available (Pierce, Rockford, IL) to generate active antibody fragments with free cysteines. Half-antibodies are produced by cleaving the two disulfide bonds in the central region of the heavy chain using 2-mercaptoethylamine HCl [20] and Fab fragments are generated from papain digestion [11]. The most elegant methods are to introduce a cysteine into the primary sequence of proteins or to append a thiol group to the end of a DNA strand [6], allowing for a well-defined sequence-specific coupling of the ligand to the crosslinker.

An attractive alternative for covalent coupling is provided by the widely used nitrilotriacetate (NTA)-His_6 system. The strength of binding in this system, which is routinely used in chromatographic and biosensor matrices, is significantly larger than that between most ligand–receptor pairs [21–23]. Since a His_6 tag can be readily appended to proteins, a crosslinker containing an NTA residue is ideally suited for coupling proteins to the AFM tip. This generic, site-specific coupling strategy also allows rigorous and ready control of binding specificity by using Ni^{++} as a molecular switch of the NTA-His_6 bond.

7.3 Immobilization of Receptors onto Probe Surfaces

To enable force detection, the receptors recognized by the ligand-functionalized tip need to be firmly attached to the probed surface. Loose association will unavoidably lead to a pull-off of the receptors from the surface by the tip-immobilized ligands, precluding detection of the interaction force.

Freshly cleaved muscovite mica is a perfectly pure and atomically flat surface and, therefore, ideally suited for MRFMtudies. The strong negative charge of mica also accomplishes very tight electrostatic binding of various biomolecules. For example, lysozyme [20] and avidin [24] strongly adhere to mica at pH < 8. In such cases, simple adsorption of the receptors from the solution is sufficient, since attachment is strong enough to withstand pulling. Nucleic acids can also be firmly bound to mica through mediatory divalent cations such as Zn^{2+}, Ni^{2+} or Mg^{2+} [25]. The strongly acidic sarcoplasmic domain of the skeletal-muscle calcium-release channel (RYR1) was likewise absorbed to mica via Ca^{2+} bridges [26]. Carefully optimizing buffer conditions, similar strategies were used to deposit protein crystals and bacterial layers onto mica in defined orientations [27, 28].

The use of nonspecific electrostatic-mediated binding is however quite limited and generally offers no means to orient the molecules over the surface in a desirable direction. Immobilization through covalent attachment must therefore be frequently explored. When glass, silicon or mica are used as probe surfaces, immobilization is essentially the same as described above for tip functionalization. The number of reactive SiOH groups of the chemically relatively inert mica can optionally be increased by water plasma treatment [29]. As with tips, crosslinkers are also often used to provide receptors with motional freedom and to prevent surface-induced protein denaturation [4]. Immobilization can be controlled, to some extent, by using photoactivatable crosslinkers, such as N-5-azido-2-nitrobenzoyloxysuccinimide [30].

A major limitation of silicon chemistry is that it does not allow for high surface densities, i.e., $> 1000/\mu m^2$. By comparison, the surface density of a monolayer of streptavidin is about 60,000 molecules per μm^2 and that of a phospholipid monolayer may exceed 10^6 molecules per μm^2. The latter high density is also achievable by chemisorption of alkanethiols to gold. Tightly bound functionalized alkanethiol monolayers formed on ultraflat gold surfaces provide excellent probes for AFM [10] and readily allow for covalent and non-covalent attachment of biomolecules [10, 31] (Fig. 7.3).

Recently, *Kada* et al. [32] reported on a new strategy to immobilize proteins on gold surfaces using phosphatidyl choline or phosphatidyl ethanolamine analogues containing dithio-phospholipids at their hydrophobic tail. Phosphatidyl

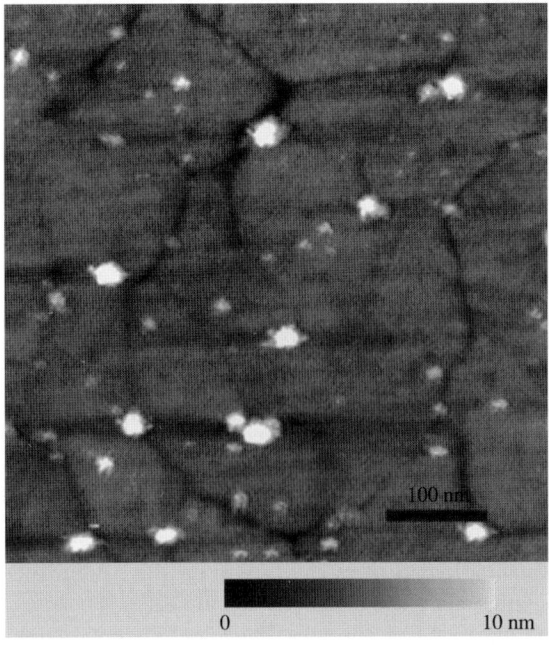

Fig. 7.3. AFM image of his-RNAP molecules specifically bound to nickel-NTA domains on a functionalized gold surface. Alkanethiols terminated with ethylene glycol groups to resist unspecific protein adsorption served as a host matrix and were doped with 10% nickel-NTA alkenthiols. The sample was prepared to achieve full monolayer coverage. Ten individual hisRNAP molecules can be clearly visualized bound to the surface. The more abundant, smaller, lower features are NTA islands with no bound molecules. The underlying morphology of the gold can also be distinguished (after [31])

ethanolamine, which is chemically reactive, was derivatized with a long-chain biotin for the molecular recognition of streptavidin molecules in an initial study [32]. These self-assembled phospholipid monolayers closely mimic the cell surface and minimize nonspecific adsorption. Additionally, they can be spread as insoluble monolayers at an air/water interface. Thereby, the ratio of functionalized thio-lipids to host lipids accurately defines the surface density of bioreactive sites in the monolayer. Subsequent transfer onto gold substrates leads to covalent and, hence, tight attachment of the monolayer.

MRFMas also been used to study the interactions between ligands and cell surface receptors in situ, on fixed or unfixed cells. In these studies it was found that the immobilization of cells strongly depends on cell type. Adherent cells are readily usable for MRFMhereas cells that grow in suspension need to be adsorbed onto the probe surface. Various protocols for tight immobilization of cells over a surface are available. For adherent cells, the easiest way is to grow the cells directly on glass or other surfaces suitable for MRFM [33]. Firm immobilization of non- and weakly adhering cells can be achieved by various adhesive coatings such as Cell-Tak [34], gelatin, or polylysine. Hydrophic surfaces like gold or carbon are also very useful to immobilize non-adherent cells or membranes [35]. Covalent attachment of cells to surfaces can be accomplished by crosslinkers that carry reactive groups, such as those used for immobilization of molecules [34]. Alternatively, one can use crosslinkers carrying a fatty-acid moiety that can penetrate into the lipid bilayer of the cell membrane. Such linkers provide sufficiently strong fixation without interference with membrane proteins [34].

7.4 Single-Molecule Recognition Force Detection

Measurements of interaction forces traditionally rely on ensemble techniques such as shear flow detachment (SFD) [36] and the surface force apparatus (SFA) [37]. In SFD, receptors are fixed to a surface to which ligands carried by beads or presented on the cell surface bind specifically. The surface-bound particles are then subjected to a fluid shear stress that disrupts the ligand–receptor bonds. However, the force acting between single molecular pairs can only be estimated because the net force applied to the particles can only be approximated and the number of bonds per particle is unknown.

SFA measures the forces between two surfaces to which different interacting molecules are attached using a cantilever spring as a force probe and interferometry for detection. The technique, which has a distance resolution of ≈ 1 Å, allows one to measure adhesive and compressive forces and to follow rapid transient effects in real time. However, the force sensitivity of the technique (≈ 10 nN) does not allow for single-molecule measurements of non-covalent interaction forces.

The biomembrane force probe (BFP) technique uses pressurized membrane capsules rather than mechanical springs as the force transducer (Fig. 7.4; for a recent paper see [38]). To form the transducer, a red blood cell or a lipid bilayer vesicle

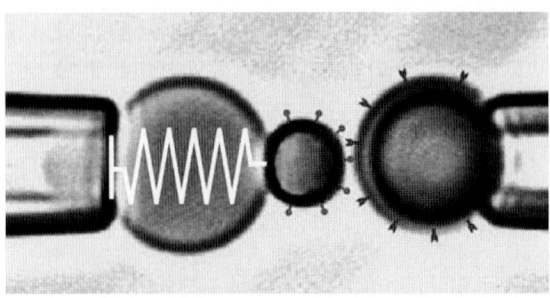

Fig. 7.4. Experimental set-up of the biomembrane force probe (BFP). The spring in the BFP is a pressurized membrane capsule. Its spring constant is set by membrane tension, which is controlled by micropipette suction. The BFP tip is formed by a glass microbead with a diameter of 1–2 µm chemically glued to the membrane. The BFP (*on the left*) was kept stationary and the test surface, formed by another microbead (*on the right*), was translated to or from contact with the BFP tip by precision piezo control (after [38])

is pressurized into the tip of a glass micropipette. The spring constant of the capsule can then be varied over several orders of magnitude by suction. This simple, but highly effective configuration, enables the measurement of forces ranging from 0.1 pN to 1000 pN with a force resolution of about 1 pN, allowing probing of single molecular bonds.

Optical tweezers (OT) belong to a family of techniques that rely on external fields to manipulate single particles. Thus, unlike mechanical transducers, force is exerted from a distance. In optical tweezers (OT), small particles (beads) are manipulated by optical traps [39]. Three-dimensional light-intensity gradients of a focused laser beam are used to pull or push particles with positional accuracy of a few nanometers. Using this technique, forces in the range of 10^{-13}–10^{-10} N can accurately be measured. Optical tweezers have been used extensively to measure the force-generating properties of various molecular motors at the single-molecule level [40] and to obtain force-extension profiles of single DNA [41] or protein [42] molecules. Defined, force-controlled twisting of DNA using rotating magnetically manipulated particles gave even further insights into DNA viscoelastic properties [43].

The atomic force microscope (AFM; [1]) is the force-measuring method with the smallest force sensor and therefore provides the highest lateral resolution. Radii of commercially available AFM tips vary between 2 and 50 nm. In contrast, the particles used for force sensing in SFD, BFP, and OT are in the 1–10 µm range, and the surfaces used in SFA exceed millimeter extensions. The small apex of the AFM tip allows the visualization of single biomolecules with molecular to submolecular resolution [25, 27, 28].

In addition to imaging, AFM has successfully been used to measure interaction forces between various single molecular pairs [2–4]. In these measurements, one of the binding partners is immobilized onto a tip mounted at the end of a flexible cantilever that functions as a force transducer and the other is immobilized over

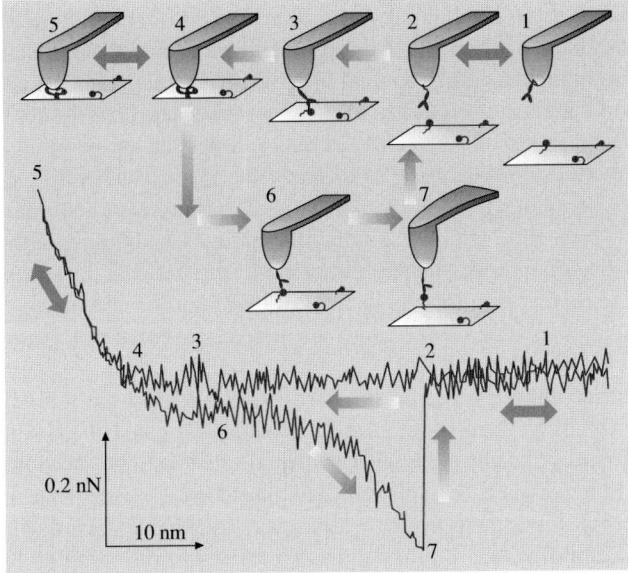

Fig. 7.5. Single-molecule recognition event detected with AFM. A force–distance cycle, measured with and amplitude of 100 nm at a sweep frequency of 1 Hz, for an antibody antigen pair in PBS. Binding of the antibody immobilized on the tip to the antigen on the surface, which occurs during the approach (*trace points 1 to 5*), results in a parabolic retract force curve (*points 6 to 7*) reflecting the extension of the distensible crosslinker antibody antigen connection. The force increases until unbinding occurs at a force of 268 pN (*points 7 to 2*) (after [4])

a hard surface such as mica or glass. The tip is initially brought to, and subsequently retracted from the surface, and the interaction (unbinding) force is measured by following the cantilever deflection, which is monitored by measuring the reflection of a laser beam focused on the back of the cantilever using a split photodiode. Approach and retract traces obtained from the unbinding of a single molecular pair are shown in Fig. 7.5 [4]. In this experiment, the binding partners were immobilized onto their respective surfaces through a distensible PEG tether.

Cantilever deflection, Δx, relates directly to the force F, acting on it directly through Hook's law $F = k\Delta x$, where k is the spring constant of the cantilever. During most of the approach phase (trace, and lines 1 to 5), when the tip and the surface are sufficiently far away from each other (1 to 4), cantilever deflection remains zero because the molecules are as yet unbound to each other. Upon contact (4) the cantilever bends upwards (4 to 5) due to a repulsive force that increases linearly as the tip is pushed further into the surface. If the cycle was futile, and no binding had occurred, retraction of the tip from the surface (retrace, 5 to 7) will lead to a gradual relaxation of the cantilever to its rest position (5 to 4). In such cases, the retract curve will look very much like the approach curve. On the other hand, if binding had occurred, the cantilever will bend downwards as the cantilever is retracted from the surface (retrace, 4 to 7). Since the receptor and ligand were tethered to the surfaces through flexible crosslinkers, the shape of the attractive force–distance profile is nonlinear, in contrast to the profile obtained during contact (4 to 7). The exact shape of the retract curve depends on the elastic properties of the crosslinker used for immobilization [17, 44] and exhibits parabolic-like characteristics, reflecting an increase of the spring constant of the crosslinker during extension. The downward bending of the retracting cantilever continues until the ramping force reaches a critical value that dissociates the ligand–receptor complex (unbinding force, 7). Unbinding of the complex is indicated by a sharp spike in the retract curve that reflects an abrupt recoil of the cantilever to its rest position. Specificity of binding is usually demonstrated by block experiments in which free ligands are added to mask receptor sites over the surface.

The force resolution of the AFM, $\Delta F = (k_B T k)^{1/2}$, is limited by the thermal noise of the cantilever which, in turn, is determined by its spring constant. A way to reduce thermal fluctuations of cantilevers without changing their stiffness or lowering the temperature is to increase the apparent damping constant. Applying an actively controlled external dissipative force to cantilevers to achieve such an increase, *Liang* et al. [45] reported a 3.4-fold decrease in thermal noise amplitude. The smallest forces that can be detected with commercially available cantilevers are in the few piconewton range. Decreasing cantilever dimensions enables one to push the range of detectable forces to smaller forces since small cantilevers have lower coefficients of viscous damping [46]. Such miniaturized cantilevers also have much higher resonance frequencies than conventional cantilevers and, therefore, allow for faster measurements.

Besides the detection of intermolecular forces, the AFM also shows great potential in measuring forces acting within molecules. In these experiments, the molecule is clamped between the tip and the surface and its viscoelastic properties are studied by force–distance cycles.

7.5 Principles of Molecular Recognition Force Spectroscopy

Molecular recognition is mediated by a multitude of non-covalent interactions the energy of which is only slightly higher than thermal energy. Due to the power law dependence of these interactions on the distance, the attractive forces between non-covalently interacting molecules are extremely short-ranged. A close geometrical and chemical fit within the binding interface is therefore a prerequisite for productive association. The weak bonds that govern molecular cohesion are believed to be formed in a spatially and temporarily correlated fashion. Protein binding often involves structural rearrangements that can be either localized or global. These rearrangements often bear functional significance by modulating the activity of the interactants. Signaling pathways, enzyme activity, and the activation and inactivation of genes all depend on conformational changes induced in proteins by ligand binding.

The strength of binding is usually given by the binding energy E_B, which amounts to the free-energy difference between the bound and the free state, and which can readily be determined by ensemble measurements. E_B determines the ratio of bound complexes [RL] to the product of free reactants [R][L] at equilibrium and is related to the equilibrium dissociation constant K_D through $E_B = -RT \ln(K_D)$, where R is the gas constant. K_D itself is related to the empirical association (k_{on}) and dissociation (k_{off}) rate constants through $K_D = k_{off}/k_{on}$. In order to get an estimate for the interaction forces, f, from the binding energies E_B, the depth of the binding pocket may be used as a characteristic length scale l. Using typical values of $E_B = 20 k_B T$ and $l = 0.5$ nm, an order-of-magnitude estimate of $f(= E_B/l) \approx 170$ pN is obtained for the binding strength of a single molecular pair. Classical mechanics describes bond strength as the gradient in energy along the direction of separation. Unbinding therefore occurs when the applied force exceeds the steepest gradient in energy. This purely mechanical description of molecular bonds, however, does not provide insights into the microscopic determinants of bond formation and rupture.

Non-covalent bonds have limited lifetimes and will therefore break even in the absence of external force on characteristic time scales needed for spontaneous dissociation ($\tau(0) = k_{off}^{-1}$). Pulled faster than $\tau(0)$, however, bonds will resist detachment. Notably, the unbinding force may approach and even exceed the adiabatic limit given by the steepest energy gradient of the interaction potential, if rupture occurs in less time than needed for diffusive relaxation (10^{-10}–10^{-9} s for biomolecules in viscous aqueous medium) and friction effects become dominant [48]. Therefore, unbinding forces do not resemble unitary values and the dynamics of the experiment critically affects the measured bond strengths. At the time scale of AFM experiments (milliseconds to seconds), thermal impulses govern the unbinding process. In the thermal activation model, the lifetime of a molecular complex in solution is described by a Boltzmann ansatz, $\tau(0) = \tau_{osc} \exp(E_b/k_B T)$ [49], where τ_{osc} is the inverse of the natural oscillation frequency and E_b is the height of the energy barrier for dissociation. This gives a simple Arrhenius dependency of dissociation rate on barrier height.

A force acting on a complex deforms the interaction free-energy landscape and lowers barriers for dissociation (Fig. 7.6). As a result of the latter, bond lifetime is shortened. The lifetime $\tau(f)$ of a bond loaded with a constant force f is given by: $\tau(f) = \tau_{osc} \exp[(E_b - x_\beta f)/k_B T]$ [49], where x_β marks the thermally averaged projection of the energy barrier along the direction of the force. A detailed analysis of the relation between bond strength and lifetime was performed by *Evans* et al. [50], using Kramers' theory for overdamped kinetics. For a sharp barrier, the lifetime $\tau(f)$ of a bond subjected to a constant force f relates to its characteristic lifetime, $\tau(0)$, according to: $\tau(f) = \tau(0) \exp(-x_\beta f/k_B T)$ [4]. However, in most pulling experiments the applied force is not constant. Rather, it increases in a complex, nonlinear manner, which depends on the pulling velocity, the spring constant of the cantilever, and the force–distance profile of the molecular complex. Nevertheless, contributions arising from thermal activation manifest themselves mostly near the point of detachment. Therefore, the change of force with time or the loading rate $r(= df/dt)$ can be de-

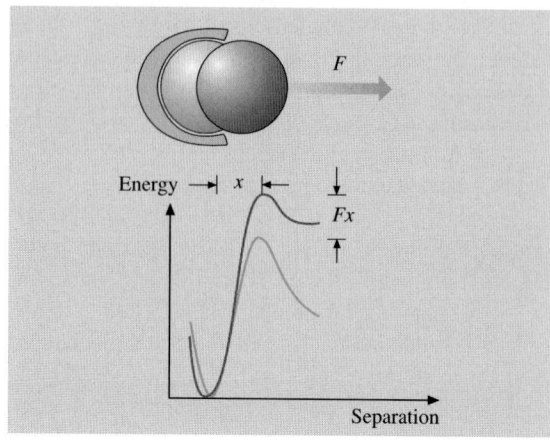

Fig. 7.6. Dissociation over a single sharp energy barrier. Under a constant force, the barrier is lowered by the applied force F. This gives rise to a characteristic length scale x_β that is interpreted as the distance of the energy barrier from the energy minimum along the projection of the force (after [47])

rived from the product of the pulling velocity and the effective spring constant at the end of the force curve, just before unbinding occurs.

The dependence of the rupture force on the loading rate (force spectrum), in the thermally activated regime was first derived by *Evans* and *Ritchie* [50] and described further by *Strunz* et al. [47]. Forced dissociation of receptor ligand complexes using AFM or BFP can often be regarded as an irreversible process because the molecules are kept moving away from each other after unbinding had occurred (rebinding can be safely neglected when measurements are made with soft springs). Rupture itself is a stochastic process and the likelihood of bond survival is expressed in the master equation as a time-dependent probability $N(t)$ to be in the bound state under a steady ramp of force, namely $dN(t)/dt = -k_{\text{off}}(rt)N(t)$ [47]. This results in a distribution of unbinding forces $P(F)$ parameterized by the loading rate [47, 50, 51]. The most probable force for unbinding f^*, given by the maximum of the distribution, relates to the loading rate through $f^* = f_\beta \ln(rk_{\text{off}}^{-1}/f_\beta)$, where the force scale f_β is set by the ratio of thermal energy to x_β [47, 50]. Thus, the unbinding force scales linearly with the logarithm of the loading rate. For a single barrier, this would give rise to a simple, linear force spectrum f^* versus $\log(r)$. In cases where the escape path is traversed by several barriers, the curve will follow a sequence of linear regimes, each marking a particular barrier [38, 50, 51]. Transition from one regime to the other is associated with an abrupt change of slope determined by the characteristic barrier length scale and signifying that a crossover between barriers has occurred.

Dynamic force spectroscopy (DFS) exploits the dependence of bond strength on the loading rate to obtain detailed insights into intra- and intermolecular interactions. By measuring bond strength over a broad range of loading rates, length scales and relative heights of energy barriers traversing the free-energy surface can be readily obtained. The lifetime of a bond at any given force is likewise contained in the complete force distribution [4]. Finally, one may attempt to extract dissociation rate constants by extrapolation to zero force [52]. However, the application of force acts to select the dissociation path. Since the kinetics of reactions is pathway-dependent, such a selection implies that kinetic parameters extracted from force-probe experiments may differ from those obtained from assays conducted in the absence of external force. Under extremely fast complexation/decomplexation kinetics the forces can be independent of the loading rate, indicating that the experiments were carried out under thermodynamic equilibrium [53].

7.6 Recognition Force Spectroscopy: From Isolated Molecules to Biological Membranes

7.6.1 Forces, Energies, and Kinetic Rates

Conducted at fixed loading rates, pioneering measurements of interaction forces represent single points in continuous spectra of bond strengths [38]. Not unexpectedly, the first interaction studied was that between biotin and its extremely high-affinity receptors avidin [3] and streptavidin [2]. The unbinding forces measured for these interactions were 250–300 pN and 160 pN for streptavidin and avidin, respectively. During this initial phase it was also revealed that different unbinding forces can be obtained for the same pulling velocity if the spring constant of the cantilever is varied [2], which is consistent with the aforementioned dependency of bond strength on the loading rate. The interaction force between several biotin analogues and avidin or streptavidin [54] and between biotin and a set of streptavidin mutants [55] was investigated and found generally to correlate with the equilibrium binding enthalpy and the enthalpic activation barrier. No correlation with the equilibrium free energy of binding or the activation free-energy barrier to dissociation was observed, suggesting that internal energies rather than entropic contributions were probed by the force measurements [55].

In another pioneering study, *Lee* et al. [6] measured the forces between complementary 20-base DNA strands covalently attached to a spherical probe and surface. The interaction forces fell into three different distributions amounting to the rupture of duplexes consisting of 12, 16, and 20 base pairs. The average rupture force per base pair was ≈ 70 pN. When a long, single-stranded DNA was analyzed, both intra- and interchain forces were observed, the former probing the elastic properties of the molecule. Hydrogen bonds between nucleotides have been probed for all 16 combinations of the four DNA bases [7]. Directional hydrogen-bonding interactions were measured only when complementary bases were present on the tip and probe surfaces, indicating that AFM can be used to follow specific pairing of DNA strands.

Strunz et al. [15] measured the forces required to separate individual double-stranded DNA molecules of 10, 20, and 30 base pairs (Fig. 7.7). The parameters describing the energy landscape, i.e., the distance of the energy barrier from the minimum energy along the separation path and the logarithm of the thermal dissociation rate, were found to be proportional to the number of base pairs of the DNA duplex. Such scaling suggests that unbinding proceeds in a highly cooperative manner characterized by one length scale and one time scale. Studying the dependence of rupture forces on the temperature, it was proposed by *Schumakovitch* et al. [56]

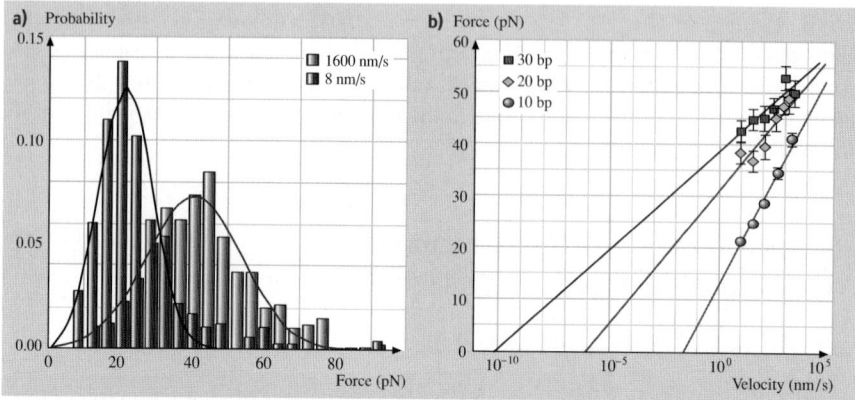

Fig. 7.7. Dependence of the unbinding force between DNA single-strand duplexes on the retract velocity. In addition to the expected logarithmic behavior on the loading rate, the unbinding force scales with the length of the strands, increasing from the 10- to 20- to 30-base-pair duplexes (after [15])

that entropic contributions play an important role in the unbinding of complementary DNA strands [56].

Prevalent as it is, molecular recognition has mostly been discussed in the context of the interactions between antibodies and antigens. To maximize motional freedom and to overcome problems associated with misorientation and steric hindrance, antibodies and antigens were immobilized onto AFM tip and probe surfaces via flexible molecular spacers [4, 10, 13, 14]. Optimizing antibody density over the AFM tip [4, 12], the interaction between individual antibody antigen pairs could be examined. Binding of antigen to the two Fab fragments of the antibody was shown to occur independently and with equal probability. Single antibody antigen recognition events were also recorded with tip-bound antigens interacting with intact antibodies [10, 13] or with single-chain wild-type (Fv) fragments [14]. The latter study also showed that an Fv mutant whose affinity to the antigen was attenuated by about 10 folds dissociated from the antigen under applied forces that were 20% lower than those required to unbind the wild-type (Fv) antibody.

Besides measurements of interaction forces, single-molecule force spectroscopy also allows estimation of association and dissociation rate constants with the concern stated above withstanding [4, 12, 22, 52, 57, 58], and measurement of structural parameters of the binding pocket [4, 12, 15, 57, 58]. Quantification of the association rate constant k_{on} requires determination of the interaction time needed for half-maximal probability of binding ($t_{1/2}$). This can be obtained from experiments where the encounter time between receptor and ligand is varied over a broad range [57]. Given that the concentration of ligand molecules on the tip available for interaction with the surface-bound receptors c_{eff} is known, the association rate constant can be derived from $k_{on} = t_{0.5}^{-1} c_{eff}^{-1}$. Determination of the effective ligand concentration requires knowledge of the effective volume V_{eff} explored by the tip-tethered ligand which, in turn, depends on the tether length. Therefore, only order-of-magnitude estimates of k_{on} can be gained from such measurements [57].

Additional information about the unbinding process is contained in the distributions of the unbinding forces. Concomitant with the shift of maxima to higher unbinding forces, increasing the loading rate also leads to an increase in the width σ of the distributions [22, 38], indicating that, at lower loading rates, the system adjusts closer to equilibrium. The lifetime $\tau(f)$ of a bond under an applied force was estimated by the time the cantilever spends in a regime the force window spanned by the standard deviation of the most probable force for unbinding [4]. In the case of Ni^{2+}-His_6, the lifetime of the complex decreased from 17 to 2.5 ms when the force was increased from 150 to 194 pN [22]. The data fit well to Bell's model, confirming the predicted exponential dependence of bond lifetime on the applied force, and yielded an estimated lifetime at zero force of about 15 seconds. A more direct measurement of τ is afforded by force-clamp experiments in which the applied force is kept constant by a feedback loop. This configuration was first adapted for use with AFM by *Oberhauser* et al. [59], who employed it to study the force dependence of the unfolding probability of the I27 and I28 modules of cardiac titin as well as of the full-length protein [59].

However, as discussed above, in most experiments the applied force is not constant but varies with time, and the measured bond strength depends on the loading rate [48, 50, 60]. In accordance with this, experimentally measured unbinding forces do not assume unitary values but rather vary with both pulling velocity [52, 57] and cantilever spring constant [2]. The predicted logarithmic dependence of the unbinding force on the loading rate in the thermally activated regime was likewise confirmed by a large number of unbinding and unfolding experiments [15, 22, 38, 52, 57, 58, 61]. The slopes of the force–loading rate curves contain information about the length scale x_β of prominent energy barriers along the force-driven dissociation pathway, which may be related to the depth of the binding pocket of the interaction [57].

The force spectra may also be used to derive the dissociation rate constant k_{off} by extrapolation to zero force [52, 57, 58]. As mentioned above, values derived in this manner may differ from those obtained from bulk measurements because only a subset of dissociation pathways defined by the force is sampled. Nevertheless, a simple correlation between unbinding forces and thermal dissociation rates was obtained for a set consisting of nine different Fv fragments constructed from point mutations of three unrelated anti-fluorescein antibodies [58]. This correlation, which implies a close similarity between the force- and thermally driven pathways explored during dissociation, was probably due to the highly rigid nature of the interaction, which proceeds in a lock-and-key fashion. The force spectra obtained for the different constructs exhibited a single linear regime, indicating that in all cases unbinding was governed by a single prominent energy barrier (Fig. 7.8). Interestingly, the position of the energy barrier along the forced-dissociation pathway was found to be proportional to the height of the barrier and, thus, most likely includes contributions arising from elastic stretching of the antibodies during the unbinding process.

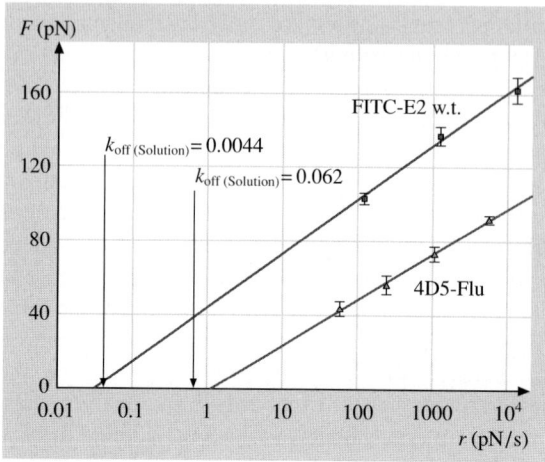

Fig. 7.8. The dependence of the unbinding force on the loading rate for two anti-fluorescein antibodies. For both FITC-E2 w.t. and 4D5-Flu a strictly single-exponential dependence was found in the range accessed, indicating that only a single energy barrier was probed. The same energy barrier dominates dissociation without forces applied because extrapolation to zero force matches kinetic off-rates determined in solution (indicated by the arrow) (after [58])

7.6.2 Complex Bonds and Energy Landscapes

The energy landscapes that describe proteins are generally not smooth. Rather, they are traversed by multiple energy barriers of various heights that render them highly corrugated or rugged. All these barriers affect the kinetics and conformational dynamics of proteins and any one of them may govern interaction lifetime and strength on certain time scales. Dynamic force spectroscopy provides an excellent tool to detect energy barriers which are difficult or impossible to detect by conventional, near-equilibrium assays and to probe the free-energy surface of proteins and protein complexes. It also provides a natural means to study interactions which are normally subjected to varying mechanical loads [52, 57, 62–64].

A beautiful demonstration of the ability of dynamic force spectroscopy to reveal hidden barriers was provided by *Merkel* et al. [38] who used BFP to probe bond formation between biotin and streptavidin or avidin over a broad range of loading rates. In contrast to early studies, which reported fixed values of bond strength [54, 55], a continuous spectrum of unbinding forces ranging from 5 to 170 pN was obtained (Fig. 7.9). Concomitantly, interaction lifetime decreased from about 1 min to 0.001 s, revealing the reciprocal relation between bond strength and lifetime expected for thermally activated kinetics under a rising force. Most notably, depending on the loading rate, unbinding kinetics was dominated by different activation energy barrier positioned along the force-driven unbinding pathway. Barriers emerged sequentially, with the outermost barrier appearing first, each giving rise to a distinct linear regime in the force spectrum. Going from one linear regime to the next was associated with an abrupt change in slope, indicating that a crossover between an outer

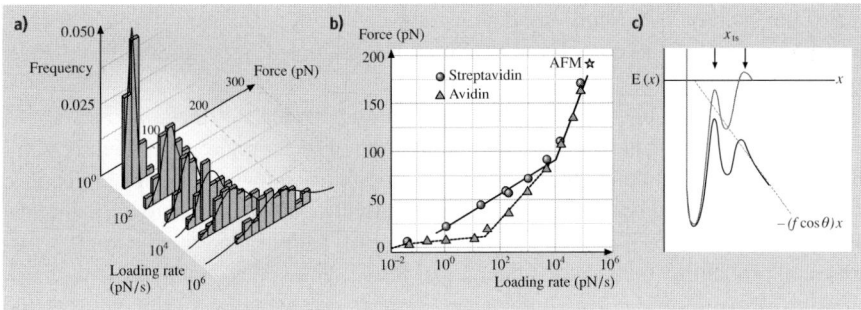

Fig. 7.9. Unbinding force distributions and energy landscape of a complex molecular bond. (**a**) Force histograms of single biotin streptavidin bonds recorded at different loading rates. The shift in peak location and the increase in width with increasing loading rate is clearly demonstrated. (**b**) Dynamic force spectra for biotin streptavidin (*open circles*) and biotin avidin (*closed triangles*). The slopes of the linear regimes mark distinct activation barriers along the direction of force. (**c**) Conceptual energy landscape traversed along a reaction coordinate under force. The external force f adds a mechanical potential that tilts the energy landscape and lowers the barriers. The inner barrier starts to dominate when the outer has fallen below it due to the applied force (after [38])

to (more) inner barrier had occurred. The position of two of the three barriers identified in the force spectra was consistent with the location of prominent transition states revealed by molecular dynamics simulations [48, 60]. However, as mentioned earlier, unbinding is not necessarily confined to a single, well-defined path, and may take different routes even when directed by an external force. Molecular dynamics simulations of force-driven unbinding of an antibody antigen complex characterized by a highly flexible binding pocket revealed a large heterogeneity of enforced dissociation pathways [65].

The rolling of leukocytes on activated endothelium is a first step in the emergence of leukocytes out of the blood stream into sites of inflammation. This rolling, which occurs under hydrodynamic shear forces, is mediated by selectins, a family of extended, calcium-dependent lectin receptors present on the surface of endothelial cells. To fulfill their function, selectins and their ligands exhibit a unique combination of mechanical properties: they associate rapidly and avidly and can tether cells over very long distances by their long, extensible structure. In addition, complexes made between selectins and their ligands can withstand high tensile forces and dissociate in a controllable manner, which allows them to maintain rolling without being pulled out of the cell membrane.

Fritz et al. [52] used dynamic force spectroscopy to study the interaction between P-selectin and its leukocyte-expressed surface ligand P-selectin glycoprotein ligand-1 (PSGL-1). Modeling both intermolecular and intramolecular forces, as well as adhesion probability, they were able to obtain detailed information about rupture forces, elasticity, and the kinetics of the interaction. Complexes were able to withstand forces up to 165 pN and exhibited a chain-like elasticity with a molecular spring constant of $5.3\,\mathrm{pN\,nm^{-1}}$ and a persistence length of $0.35\,\mathrm{nm}$. Rupture forces and the lifetime of the complexes exhibited the predicted logarithmic dependence on the loading rate.

An important characteristics of the interaction between P-selectin and PSGL-1, which is highly relevant to the biological function of the complex, was found by investigating the dependence of the adhesion probability between the two molecules on the velocity of the AFM probe. Counterintuitively and in contrast to experiments with avidin–biotin [54], antibody antigen [4], or cell adhesion proteoglycans [9], the adhesion probability between P-selectin and PSGL-1 was found to *increase* with increasing velocities [52]. This unexpected dependency explains the increase in the leukocyte tethering probability with increased shear flow observed in rolling experiments. Since the adhesion probability approached 1 at fast pulling velocities, it was concluded that binding occurs instantaneously as the tip reaches the surface and, thus, proceeds with a very fast on-rate. The complex also exhibited a fast forced off-rate. Such fast-on fast-off kinetics is probably important for the ability of leukocytes to rapidly bind and detach from the endothelial cell surface. Likewise, the long contour length of the complex together with its high elasticity reduces the mechanical loading on the complex upon binding and allows leukocyte rolling even at high shear rates.

Evans et al. [62] used BFP to study the interaction between PSGL-1 and another member of the selectin family, L-selectin. The force spectra, obtained over a range of loading rates extending from 10 to $100{,}000\,\mathrm{pN\,s^{-1}}$, revealed two prominent energy barriers along the unbinding pathway: an outer barrier, probably constituted by an array of hydrogen bonds, that impeded dissociation under slow detachment and an inner, Ca^{2+}-dependent barrier that dominated dissociation under rapid detachment. The observed hierarchy of inner and outer activation barriers was proposed to be important for multibond recruitment during selectin-mediated function.

Using force clamp AFM [59], bond lifetimes were directly measured as a function of a constantly applied force. For this, lifetime force relations of P-selectin complexed to two forms of P-selectin glycoprotein ligand-1 (PSGL-1) and to G1, a blocking monoclonal antibody against P-selectin, respectively, were determined [64]. Both monomeric (sPSGL-1) and dimeric PSGL-1 exhibited a biphasic relationship between lifetime and force in their interaction to P-selectin (Fig. 7.10a,b). The bond lifetimes initially increased, indicating the presence of catch bonds. After reaching a maximum, the lifetimes decreased with force, indicating a catch bond. In contrast, the P-selectin/G1 bond lifetimes decreased exponentially with force (Fig. 7.10c), displaying typical slip-bond characteristics that are well described by the single-energy-barrier Bell model. The curves of lifetime against force for the two forms of PSGL1-1 had similar biphasic shapes (Fig. 7.10a,b), but the PSGL-1 curve (Fig. 7.10b) was shifted relative to the sPSGL-1 curve (Fig. 7.10a), approximately doubling the force and the lifetime. These data suggest that that sPSGL-1 forms monomeric bonds with P-selectin, whereas PSGL-1 forms dimeric bonds with P-selectin. In agreement with the studies described above, it was concluded that the use of force-induced switching from catch to slip bonds

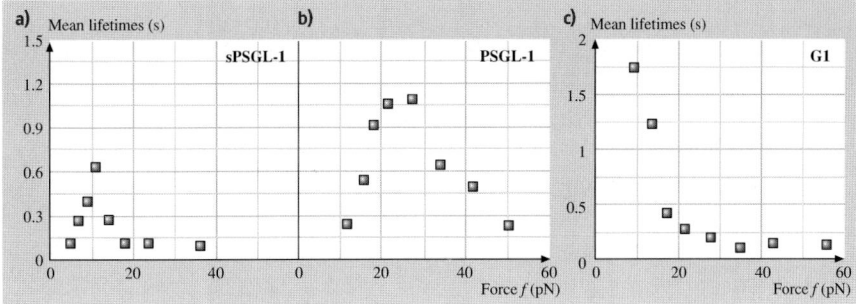

Fig. 7.10. Lifetimes of bonds of single-molecular complexes, depending on a constantly applied force. (**a**) sPSGL-1/P-selectin: catch bond and slip bond. (**b**) PSGL-1/P-selectin: catch bond and slip bond. (**c**) G1/P-selectin: slip bond only (after [64])

might be physiologically relevant for the tethering and rolling process of leukocytes on selectins [64].

Baumgartner et al. [57]) used AFM to probe specific trans-interaction forces and conformational changes of recombinant vascular/enothelial (VE)-cadherin strand dimers. Vascular endothelial-cadherins (VE-cadherin) are cell surface proteins that mediate the adhesion of cells in the vascular endothelium through Ca^{2+}-dependent homophilic interactions of their N-terminal extracellular domains. Acting as such they play an important role in the regulation of intercellular adhesion and communication in the inner surface of blood vessels. Unlike selectin-mediated adhesion, association between trans-interacting VE dimers was slow and independent of probe velocity, and complexes were ruptured at relatively low forces. These differences were attributed to the fact that, as opposed to selectins, cadherins mediate adhesion between resting cells. Mechanical stress on the junctions is thus less intense and high-affinity binding is not required to establish and maintain intercellular adhesion. Determination of the Ca^{2+}-dependency of recognition events between tip- and surface-bound VE-cadherins revealed a surprisingly high K_D (1.15 mM), which is very close to the free extracellular Ca^{2+} concentration in the body. Binding also revealed a strong dependence on calcium concentrations, giving rise to an unusually high Hill coefficient of ≈ 5. This steep dependency suggests that local changes of free extracellular Ca^{2+} in the narrow intercellular space may facilitate rapid remodeling of intercellular adhesion and permeability.

Nevo et al. [66, 67] used single-molecule force spectroscopy to discriminate between alternative mechanisms of protein activation (Fig. 7.11). The activation of proteins by other proteins, protein domains, or small ligands is a central process in biology for signalling pathways, and enzyme activity. Moreover, activation and inactivation of genes all depend on the switching of proteins between alternative functional states. Two general mechanisms have been proposed. The induced-fit model assigns changes in protein activity to conformational changes triggered by effector binding. The population-shift model, on the other hand, ascribes these changes to a redistribution of *pre-existing* conformational isomers. According to this model,

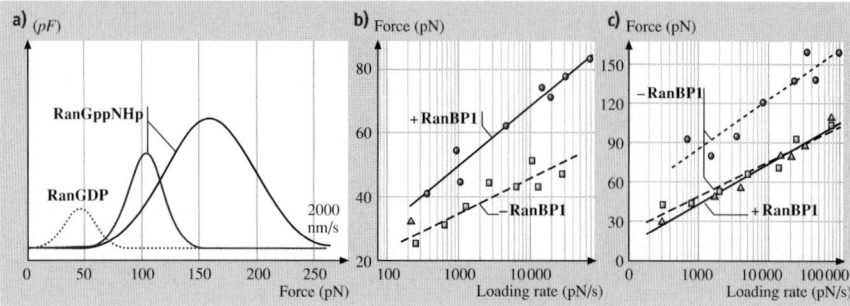

Fig. 7.11. Protein activation revealed by force spectroscopy. Ran and importinβ (impβ) were immobilized onto the AFM cantilevered tip and mica, respectively, and the interaction force was measured at different loading rates in the absence or presence of RanBP1, which was added as a mobile substrate to the solution in the AFM liquid cell. Unbinding force distributions obtained for impβ-Ran complexes at pulling velocity of 2000 nm/s. Association of impβ with Ran loaded with GDP (**a**) or with non-hydrolyzable GTP analogue (GppNHp) (**b**) gives rise to uni- or bimodal force distributions, respectively, reflecting the presence of one and two bound states. (**b–c**) Force spectra obtained for complexes of impβ with RanGDP or with RanGppNHp, in the absence (*dashed lines*) or presence (*solid lines*) of RanBP1. The results indicate that activation of impβ-RanGDP and imp-RanGTP complexes by RanBP1 proceeds through induced-fit and dynamic population-shift mechanisms, respectively (*see text for details*) (after [66, 67])

known also as the pre-equilibrium or conformational selection model, protein structure is regarded as an ensemble of conformations existing in equilibrium. The ligand binds to one of these conformations, i.e., the one to which it is most complementary, thus shifting the equilibrium in favor of this conformation. Discrimination between the two models of activation requires that the distribution of conformational isomers in the ensemble is known. Such information, however, is very hard to obtain from conventional bulk methods because of ensemble averaging.

Using AFM, Nevo and coworkers measured the unbinding forces of two related protein complexes in the absence or presence of a common effector. The complexes consisted of the nuclear transport receptor importin β (impβ) and the small GTPase Ran. The difference between them was the nucleotide bound state of Ran, which was either guanosine diphosphate (GDP) or guanosine triphosphate (GTP). The effector molecule was the Ran-binding protein RanBP1. Loaded with GDP, Ran associated weakly with impβ to form a single bound state characterized by unimodal distributions of small unbinding forces (Fig. 7.11a, dotted line). Addition of RanBP1 resulted in a marked shift of the distribution to higher unbinding forces (Fig. 7.11b, dotted to solid line). These results were interpreted to be consistent with an induced-fit mechanism where binding of RanBP1 induces a conformational change in the complex, which, in turn, strengthens the interaction between impβ and Ran(GDP). In contrast, association of RanGTP with impβ was found to lead to alternative bound states of relatively low and high adhesion strength represented by partially overlapping force distributions (Fig. 7.11a, solid line). When RanBP1 was added to the solution, the higher-strength population, which predominated the ensemble in the absence of the effector (Fig. 7.11c, dotted lines), was diminished, and the lower-strength conformation became correspondingly more populated (Fig. 7.11c, solid line). The means of the distributions, however, remain unchanged, indicating that the strength of the interaction in the two states of the complex has not been altered by the effector. These data fit a dynamic population-shift mechanism in which RanBP1 binds selectively to the lower-strength conformation of RanGTP-impβ, changing the properties and function of the complex by shifting the equilibrium between its two states.

The complex between impβ and RanGTP was also used in studies aimed to measure the energy landscape roughness of proteins. The roughness of the energy landscapes that describe proteins has numerous effects on their folding and binding as well as on their behavior at equilibrium, since undulations in the free-energy surface can attenuate diffusion dramatically. Thus, to understand how proteins fold, bind and function, one needs to know not only the energy of their initial and final states, but also the roughness of the energy surface that connects them. However, for a long time, knowledge of protein energy-landscape roughness came solely from theory and simulations of small model proteins.

Adopting *Zwanzig*'s theory of diffusion in rough potentials [68], *Hyeon* and *Thirumalai* [69] proposed that the energy-landscape roughness of proteins can be measured from single-molecule mechanical unfolding experiments conducted at different temperatures. In particular, their simulations showed that, at constant loading rate, the most probable force for unfolding increases because of roughness that acts to attenuate diffusion. Because this effect is temperature dependent, an overall energy scale of roughness, ε, can be derived from plots of force versus loading rate acquired at two arbitrary temperatures. Extending this theory to the case of unbinding, and performing single-molecule force spectroscopy measurements, *Nevo* et al. [70] extracted the overall energy scale of roughness ε for RanGTP-impβ. The results yielded $\varepsilon > 5k_BT$, indicating a bumpy energy surface, which is consistent with the unusually high structural flexibility of impβ and its ability to interact with different, structurally distinct ligands in a highly specific manner. This mechanistic principle may also be applicable to other proteins whose function demands highly specific and regulated interactions with multiple ligands.

7.6.3 Live Cells and Membranes

Thus far, there have been only a few attempts to apply recognition force spectroscopy to cells. In one of the early studies, *Lehenkari* et al. [71] measured the unbinding forces between integrin receptors present on the surface of intact cells and several RGD-containing (Arg-Gly-Asp) ligands. The unbinding forces measured were found to be cell- and amino acid sequence-specific and sensitive to the pH and divalent cation composition of the cellular culture medium. In contrast to short linear RGD hexapeptides, larger peptides and proteins containing the RGD sequence showed different binding affinities, demonstrating that the context of the RGD motif within a protein has considerable influence upon its interaction with the receptor. In another study, *Chen* et al. [72] used AFM to measure the adhesive strength between concanavalin A (Con A) coupled to an AFM tip and Con A receptors on the surface of NIH3T3 fibroblasts. Crosslinking of receptors with either glutaraldehyde or 3,3'-dithio-bis(sulfosuccinimidylproprionate) (DTSSP) led to an increase in adhesion that was attributed to enhanced cooperativity among adhesion complexes. The results support the notion that receptor crosslinking can increase adhesion strength by creating a shift towards cooperative binding of receptors. *Pfister* et al. [73] investigated the surface localization of HSP60 on stressed and unstressed human umbilical venous endothelial cells (HUVECs). By detecting specific single-molecule binding events between the monoclonal antibody AbII-13 tethered to AFM tips and HSP60 molecules on cells, clear evidence was found for the occurrence of HSP60 on the surface of stressed HUVECs, but not on unstressed HUVECs.

The sidedness and accessibility of protein epitopes of the Na^{2+} D-glucose cotransporter 1 (SGLT1) was probed in intact brush border membranes by a tip-bound antibody directed against an amino acid sequence close to the glucose binding site [35]. Binding of glucose and transmembrane transport altered both the binding probability and the most probable unbinding force, suggesting changes in the orientation and conformation of the transporter. These studies were extended to live SGLT1-transfected CHO cells *Puntheeranurak* et al. [74] Using AFM tips carrying the substrate 1-β-thio D-glucose, direct evidence could be obtained that, in the presence of sodium, a sugar binding site appears on the SGLT1 surface. It was shown that this binding site accepts the sugar residue of the glucoside phlorizin, free d-Glucose and D-galactose, but not free L-glucose. The data indicate the importance of stereo-selectivity for sugar binding and transport.

Zhang et al. [63] studied the interaction between leukocyte function-associated antigen-1 (LFA-1) and its cognate ligand, intercellular adhesion molecule-1 (ICAM-1), which play a crucial role in leukocyte adhesion. The experimental system consisted of an LFA-1-expressing T cell hybridoma attached to the end of the AFM cantilever and an apposing surface expressing ICAM-1. The force spectra revealed fast and slow loading regimes, amounting to a sharp, inner energy barrier and a shallow, outer barrier, respectively. Addition of Mg^{2+} led to an increase of the unbinding force in the slow loading regime whereas ethylene-diaminetetraacidic acid (EDTA) suppressed the inner barrier. These results suggest that the dissociation of LFA-1/ICAM-1 is governed by the outer activation barrier of the complex, while the ability of the complex to resist a pulling force is determined by the divalent cation-dependent inner barrier.

7.7 Recognition Imaging

Besides measuring interaction strengths, locating binding sites over biological surfaces such as cells or membranes is of great interest. To achieve this goal, force detection must be combined with high-resolution imaging.

Ludwig et al. [75] used chemical force microscopy to image a streptavidin pattern with a biotinylated tip. An approach–retract cycle was performed at each point of a raster and topography, adhesion, and sample elasticity were extracted from the local force ramps. This strategy was also used to map binding sites on cells [76, 77] and to differentiate between red blood cells of different blood groups (A and 0) using AFM tips functionalized with a group-A-specific lectin [78].

Identification and localization of single antigenic sites was achieved by recording force signals during the scanning of an AFM tip coated with antibodies along a single line across a surface immobilized with a low density of antigens [4, 12]. Using this method, antigens could be localized over the surface with positional accuracy of 1.5 nm. A similar configuration used by *Willemsen* et al. [79] enabled the simultaneous acquisition of height and adhesion-force images with near-molecular resolution.

The aforementioned strategies of force mapping either lack high lateral resolution [75] and/or are much slower [4, 12, 79] than conventional topographic imaging since the frequency of the force-sensing retract–approach cycles is limited by hydrodynamic damping. In addition, the ligand needs to be detached from the receptor in each retract approach cycle, necessitating large working amplitudes (50 nm). Therefore, the surface-bound receptor is inaccessible to the tip-immobilized ligand on the tip during most of the time of the experiment. This problem however, should be overcome with the use of small cantilevers [46] which should increase the speed for force mapping because the hydrodynamic forces are significantly reduced and the resonance frequency is higher than that of commercially available cantilevers. Short cantilevers were recently applied to follow the association and dissociation of individual chaperonin proteins, GroES to GroEL, in real time using dynamic force microscopy topography imaging [80].

An imaging method for mapping antigenic sites on surfaces was developed [20] by combining molecular recognition force spectroscopy [4] with dynamic force microscopy (DFM) [25, 81]. In DFM, the AFM tip is oscillated across a surface and the amplitude reduction arising from tip–surface interactions is held constant by a feedback loop that lifts or lowers the tip according to the detected amplitude signal. Since the tip contacts the surface only intermittently, this technique provides very gentle tip–surface interactions and the specific interaction of the antibody on the tip with the antigen on the surface can be used to localize antigenic sites for recording recognition images. The AFM tip is magnetically coated and oscillated by an alternating magnetic field at very small amplitudes while being scanned along the surface. Since the oscillation frequency is more than a hundred times faster than typical frequencies in conventional force mapping, the data acquisition rate is much higher. This method was recently extended to yield fast, simultaneous acquisition of two independent maps, i.e. a topography image and a lateral map of recognition sites, recorded with nm resolution at experimental times equivalent to normal AFM imaging [82–84].

Topography and recognition images were simultaneously obtained (TREC imaging) using a special electronic circuit (PicoTrec, Molecular Imaging, Tempe, AZ) (Fig. 7.12a). Maxima (U_{up}) and minima (U_{down}) of each sinusoidal cantilever deflection period were depicted in a peak detector, filtered, and amplified. Direct-current

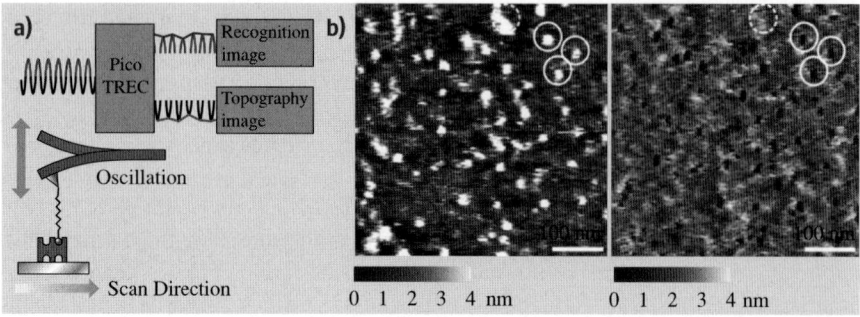

Fig. 7.12. Simultaneous topography and recognition (TREC) imaging (**a**) Principle. The cantilever oscillation is split into lower and upper parts, resulting in simultaneously acquired topography and recognition images. (**b**) Avidin was electrostatically adsorbed to mica and imaged with a biotin-tethered tip. A good correlation between topography (*left image, bright spots*) and recognition (*right image, dark spots*) was found (*solid circles*). Topographical spots without recognition denote structures lacking specific interaction (*dashed circle*). Scan size was 500 nm (after [84])

(DC) offset signals were used to compensate for the thermal drifts of the cantilever. U_{up} and U_{down} were fed into the AFM controller, with U_{down} driving the feedback loop to record the height (i.e. topography) image and U_{up} providing the data for constructing the recognition image (Fig. 7.12a). Since we used cantilevers with low Q-factor ($\approx$ 1 in liquid) driven at frequencies below resonance the two types of information were independent. In this way, topography and recognition image were recorded simultaneously and independently.

The circuit was applied to mica containing singly distributed avidin molecules using a biotinylated AFM tip [84]. The sample was imaged with an antibody-containing tip, yielding the topography (Fig. 7.12b, left image) and the recognition image (Fig. 7.12b, right image) at the same time. The tip oscillation amplitude (5 nm) was chosen to be slightly smaller than the extended crosslinker length (8 nm), so that both the antibody remained bound while passing a binding site and the reduction of the upwards deflection was significant compared to the thermal noise. Since the spring constant of the polymeric crosslinker increases nonlinearly with the tip–surface distance (Fig. 7.5), the binding force is only sensed close to full extension of the crosslinker (given at the maxima of the oscillation period). Therefore, the recognition signals were well separated from the topographic signals arising from the surface, both in space ($\Delta z \approx 5$ nm) and time (half oscillation period ≈ 0.1 ms).

The bright dots, with a height of 2–3 nm and diameter of 15–20 nm, that are visible in the topography image (Fig. 7.12b, left image) represent single avidin molecules stably adsorbed onto the flat mica surface. The recognition image shows black dots at positions of avidin molecules (Fig. 7.12b, right image) because the oscillation maxima are lowered due to the physical avidin–biotin connection established during recognition. Spatial correlation between the lateral positions of the avidin molecules obtained in the topography image and the recognition signals of the recognition image is indicated by solid circles in the images (Fig. 7.12). Recognition between the antibody on the tip and the avidin on the surface took place for almost all avidin molecules, most likely because avidin contains four biotin binding sites, two on either side. Thus, one would assume that there will always be binding epitopes oriented away from the mica surface and accessible to the biotinylated tip, resulting in a high binding efficiency. Structures observed in the topography image and not detected in the recognition image were very rare (dotted circle in Fig. 7.12b).

It is important to note that topography and recognition images were recorded at speeds typical for standard AFM imaging and were therefore considerably faster than conventional force mapping. With this methodology, topography and recognition images can be obtained at the same time and distinct receptor sites in the recognition image can be assigned to structures from the topography image. It is applicable with any ligand and, therefore, it should prove possible to recognize many types of proteins or protein layers and carry out epitope mapping on the nm scale on membranes and cells, and complex biological structures. In a striking recent example, histone proteins H3 were identified and localized in a complex chromatin preparation [83].

7.8 Concluding Remarks

Atomic force microscopy has evolved into an imaging method that yields the greatest structural details on live, biological samples in their native, aqueous environment at ambient conditions. Due to its high lateral resolution and sensitive force detection capability, it is now possible to measure molecular forces of biomolecules on the single-molecule level. Well beyond the proof-of-principle stage of the pioneering experiments, AFM has now developed into a high-end analysis method for exploring the kinetic and structural details of the interactions underlying protein folding and molecular recognition. The information obtained from force spectroscopy, being on a single-molecule level, includes physical parameters not accessible by other methods. In particular, it opens up new perspectives to explore the dynamics of biological processes and interactions.

References

1. G. Binnig, C. F. Quate, Ch. Gerber: Atomic force microscope, Phys. Rev. Lett. **56**, 930–933 (1986)
2. G. U. Lee, D. A. Kidwell, R. J. Colton: Sensing discrete streptavidin–biotin interactions with atomic force microscopy, Langmuir **10**, 354–357 (1994)
3. E. L. Florin, V. T. Moy, H. E. Gaub: Adhesion forces between individual ligand receptor pairs, Science **264**, 415–417 (1994)
4. P. Hinterdorfer, W. Baumgartner, H. J. Gruber, K. Schilcher, H. Schindler: Detection and localization of individual antibody–antigen recognition events by atomic force microscopy, Proc. Natl. Acad. Sci. U.S.A. **93**, 3477–3481 (1996)

5. M. Grandbois, W. Dettmann, M. Benoit, H.E. Gaub: How strong is a covalent bond, Science **283**, 1727–1730 (1999)
6. G.U. Lee, A.C. Chrisey, J.C. Colton: Direct measurement of the forces between complementary strands of DNA, Science **266**, 771–773 (1994)
7. T. Boland, B.D. Ratner: Direct measurement of hydrogen bonding in DNA nucleotide bases by atomic force microscopy, Proc. Natl. Acad. Sci. USA **92**, 5297–5301 (1995)
8. P. Wagner, M. Hegner, P. Kernen, F. Zaugg, G. Semenza: Covalent immobilization of native biomolecules onto Au(111) via N-hydroxysuccinimide ester functionalized self assembled monolayers for mscanning probe microscopy, Biophys. J **70**, 2052–2066 (1996)
9. U. Dammer, O. Popescu, P. Wagner, D. Anselmetti, H.-J. Güntherodt, G.M. Misevic: Binding strength between cell adhesion proteoglycans measured by atomic force microscopy, Science **267**, 1173–1175 (1995)
10. U. Dammer, M. Hegner, D. Anselmetti, P. Wagner, M. Dreier, H.J. Güntherodt, W. Huber: Specific antigen/antibody interactions measured by force microscopy, Biophys. J. **70**, 2437–2441 (1996)
11. Y. Harada, M. Kuroda, A. Ishida: Specific and quantized antibody–antigen interaction by atomic force microscopy, Langmuir **16**, 708–715 (2000)
12. P. Hinterdorfer, K. Schilcher, W. Baumgartner, H.J. Gruber, H. Schindler: A mechanistic study of the dissociation of individual antibody–antigen pairs by atomic force microscopy, Nanobiology **4**, 39–50 (1998)
13. S. Allen, X. Chen, J. Davies, M.C. Davies, A.C. Dawkes, J. C. Edwards, C.J. Roberts, J. Sefton, S.J.B. Tendler, P.M. Williams: Spatial mapping of specific molecular recognition sites by atomic force microscopy, Biochem. **36**, 7457–7463 (1997)
14. R. Ros, F. Schwesinger, D. Anselmetti, M. Kubon, R. Schäfer, A. Plückthun, L. Tiefenauer: Antigen binding forces of individually addressed single-chain Fv antibody molecules, Proc. Natl. Acad. Sci. USA **95**, 7402–7405 (1998)
15. T. Strunz, K. Oroszlan, R. Schäfer, H.-G. Güntherodt: Dynamic force spectroscopy of single DNA molecules, Proc. Natl. Acad. Sci. USA **96**, 11277–11282 (1999)
16. S.S. Wong, E. Joselevich, A.T. Woolley, C.L. Cheung, C. M. Lieber: Covalently functionalyzed nanotubes as nanometre-sized probes in chemistry and biology, Nature **394**, 52–55 (1998)
17. P. Hinterdorfer, F. Kienberger, A. Raab, H.J. Gruber, W. Baumgartner, G. Kada, C. Riener, S. Wielert-Badt, C. Borken, H. Schindler: Poly(ethylene glycol): An ideal spacer for molecular recognition force microscopy/spectroscopy, Single Mol. **1**, 99–103 (2000)
18. Th. Haselgrübler, A. Amerstorfer, H. Schindler, H.J. Gruber: Synthesis and applications of a new poly(ethylene glycol) derivative for the crosslinking of amines with thiols, Bioconjugate Chem. **6**, 242–248 (1995)
19. C.K. Riener, G. Kada, C. Borken, F. Kienberger, P. Hinterdorfer, H. Schindler, G.J. Schütz, T. Schmidt, C.D. Hahn, H.J. Gruber: Bioconjugation for biospecific detection of single molecules in atomic force microscopy (AFM) and in single dye tracing (SDT), Recent Res. Devel. Bioconj. Chem. **1**, 133–149 (2002)
20. A. Raab, W. Han, D. Badt, S.J. Smith-Gill, S.M. Lindsay, H. Schindler, P. Hinterdorfer: Antibody recognition imaging by force microscopy, Nature Biotech. **17**, 902–905 (1999)
21. M. Conti, G. Falini, B. Samori: How strong is the coordination bond between a histidine tag and Ni-Nitriloacetate? An experiment of mechanochemistry on single molecules, Angew. Chem. **112**, 221–224 (2000)
22. F. Kienberger, G. Kada, H.J. Gruber, V.Ph. Pastushenko, C. Riener, M. Trieb, H.-G. Knaus, H. Schindler, P. Hinterdorfer: Recognition force spectroscopy studies of the NTA-His6 bond, Single Mol. **1**, 59–65 (2000)

23. L. Schmitt, M. Ludwig, H. E. Gaub, R. Tampé: A metal-chelating microscopy tip as a new toolbox for single-molecule experiments by atomic force microscopy, Biophys. J. **78**, 3275–3285 (2000)
24. C. Yuan, A. Chen, P. Kolb, V. T. Moy: Energy landscape of avidin–biotin complexes measured by atomic force microscopy, Biochemistry **39**, 10219–10223 (2000)
25. W. Han, S. M. Lindsay, M. Dlakic, R. E. Harrington: Kinked DNA, Nature **386**, 563 (1997)
26. G. Kada, L. Blaney, L. H. Jeyakumar, F. Kienberger, V. Ph. Pastushenko, S. Fleischer, H. Schindler, F. A. Lai, P. Hinterdorfer: Recognition force microscopy/spectroscopy of ion channels: Applications to the skeletal muscle Ca^{2+} release channel (RYR1), Ultramicroscopy **86**, 129–137 (2001)
27. D. J. Müller, W. Baumeister, A. Engel: Controlled unzipping of a bacterial surface layer atomic force microscopy, Proc. Natl. Acad. Sci. USA **96**, 13170–13174 (1999)
28. F. Oesterhelt, D. Oesterhelt, M. Pfeiffer, A. Engel, H. E. Gaub, D. J. Müller: Unfolding pathways of individual bacteriorhodopsins, Science **288**, 143–146 (2000)
29. E. Kiss, C.-G. Gölander: Chemical derivatization of muscovite mica surfaces, Coll. Surf. **49**, 335–342 (1990)
30. S. Karrasch, M. Dolder, F. Schabert, J. Ramsden, A. Engel: Covalent binding of biological samples to solid supports for scanning probe microscopy in buffer solution, Biophys. J. **65**, 2437–2446 (1993)
31. N. H. Thomson, B. L. Smith, N. Almquist, L. Schmitt, M. Kashlev, E. T. Kool, P. K. Hansma: Oriented, active *escherichia coli* RNA polymerase: An atomic force microscopy study, Biophys. J. **76**, 1024–1033 (1999)
32. G. Kada, C. K. Riener, P. Hinterdorfer, F. Kienberger, C. M. Stroh, H. J. Gruber: Dithiophospholipids for biospecific immobilization of proteins on gold surfaces, Single Mol. **3**, 119–125 (2002)
33. C. LeGrimellec, E. Lesniewska, M. C. Giocondi, E. Finot, V. Vie, J. P. Goudonnet: Imaging of the surface of living cells by low-force contact-mode atomic force microscopy, Biophys. J. **75**(2), 695–703 (1998)
34. K. Schilcher, P. Hinterdorfer, H. J. Gruber, H. Schindler: A non-invasive method for the tight anchoring of cells for scanning force microscopy, Cell. Biol. Int. **21**, 769–778 (1997)
35. S. Wielert-Badt, P. Hinterdorfer, H. J. Gruber, J.- T. Lin, D. Badt, H. Schindler, R. K.-H. Kinne: Single molecule recognition of protein binding epitopes in brush border membranes by force microscopy, Biophys. J. **82**, 2767–2774 (2002)
36. P. Bongrand, C. Capo, J.-L. Mege, A.-M. Benoliel: *Use of Hydrodynamic Flows to Study Cell Adhesion*, ed. by P. Bongrand (CRC, Boca Raton, Florida 1988) pp. 125–156
37. J. N. Israelachvili: *Intermolecular and Surface Forces*, 2nd edn. (Academic Press, London & New York 1991) p. 2
38. R. Merkel, P. Nassoy, A. Leung, K. Ritchie, E. Evans: Energy landscapes of receptor-ligand bonds explored by dynamic force spectroscopy, Nature **397**, 50–53 (1999)
39. A. Askin: Optical trapping and manipulation of neutral particles using lasers, Proc. Natl. Acad. Sci. USA **94**, 4853–4860 (1997)
40. K. Svoboda, C. F. Schmidt, B. J. Schnapp, S. M. Block: Direct observation of kinesin stepping by optical trapping interferometry, Nature **365**, 721–727 (1993)
41. S. Smith, Y. Cui, C. Bustamante: Overstretching B-DNA: The elastic response of individual double-stranded and single-stranded DNA molecules, Science **271**, 795–799 (1996)

42. M.S.Z. Kellermayer, S.B. Smith, H.L. Granzier, C. Bustamante: Folding–unfolding transitions in single titin molecules characterized with laser tweezwers, Sience **276**, 1112–1216 (1997)
43. T.R. Strick, J.F. Allemend, D. Bensimon, A. Bensimon, V. Croquette: The elasticity of a single supercoiled DNA molecule, Biophys. J. **271**, 1835–1837 (1996)
44. F. Kienberger, V.Ph. Pastushenko, G. Kada, H.J. Gruber, C. Riener, H. Schindler, P. Hinterdorfer: Static and dynamical properties of single poly(ethylene glycol) molecules investigated by force spectroscopy, Single Mol. **1**, 123–128 (2000)
45. S. Liang, D. Medich, D.M. Czajkowsky, S. Sheng, J.-Y. Yuan, Z. Shao: Thermal noise reduction of mechanical oscillators by actively controlled external dissipative forces, Ultramicroscopy **84**, 119–125 (2000)
46. M.B. Viani, T.E. Schäffer, A. Chand, M. Rief, H.E. Gaub, P.K. Hansma: Small cantilevers for force spectroscopy of single molecules, J. Appl. Phys. **86**, 2258–2262 (1999)
47. T. Strunz, K. Oroszlan, I. Schumakovitch, H.-G. Güntherodt, M. Hegner: Model energy landscapes and the force-induced dissociation of ligand–receptor bonds, Biophys. J. **79**, 1206–1212 (2000)
48. H. Grubmüller, B. Heymann, P. Tavan: Ligand binding: Molecular mechanics calculation of the streptavidin-biotin rupture force, Science **271**, 997–999 (1996)
49. G.I. Bell: Models for the specific adhesion of cells to cells, Science **200**, 618–627 (1978)
50. E. Evans, K. Ritchie: Dynamic strength of molecular adhesion bonds, Biophys. J. **72**, 1541–1555 (1997)
51. E. Evans, K. Ritchie: Strength of a weak bondconnecting flexible polymer chains, Biophys. J. **76**, 2439–2447 (1999)
52. J. Fritz, A.G. Katopidis, F. Kolbinger, D. Anselmetti: Force-mediated kinetics of single P-selectin/ligand complexes observed by atomic force microscopy, Proc. Natl. Acad. Sci. USA **95**, 12283–12288 (1998)
53. T. Auletta, M.R. de Jong, A. Mulder, F.C.J.M. van Veggel, J. Huskens, D.N. Reinhoudt, S. Zou, S. Zapotocny, H. Schönherr, G.J. Vancso, L. Kuipers: β-cyclodextrin host-guest complexes probed under thermodynamic equilibrium: Thermodynamics and force spectroscopy, J. Am. Chem. Soc. **126**, 1577–1584 (2004)
54. V.T. Moy, E.-L. Florin, H.E. Gaub: Adhesive forces between ligand and receptor measured by AFM, Science **266**, 257–259 (1994)
55. A. Chilkoti, T. Boland, B. Ratner, P.S. Stayton: The relationship between ligand-binding thermodynamics and protein-ligand interaction forces measured by atomic force microscopy, Biophys. J. **69**, 2125–2130 (1995)
56. I. Schumakovitch, W. Grange, T. Strunz, P. Bertoncini, H.-J. Güntherodt, M. Hegner: Temperature dependence of unbinding forces between complementary DNA strands, Biophys. J. **82**, 517–521 (2002)
57. W. Baumgartner, P. Hinterdorfer, W. Ness, A. Raab, D. Vestweber, H. Schindler, D. Drenckhahn: Cadherin interaction probed by atomic force microscopy, Proc. Natl. Acad. Sci. USA **8**, 4005–4010 (2000)
58. F. Schwesinger, R. Ros, T. Strunz, D. Anselmetti, H.-J. Güntherodt, A. Honegger, L. Jermutus, L. Tiefenauer, A. Plückthun: Unbinding forces of single antibody–antigen complexes correlate with their thermal dissociation rates, Proc. Natl. Acad. Sci. USA **29**, 9972–9977 (2000)
59. A.F. Oberhauser, P.K. Hansma, M. Carrion-Vazquez, J.M. Fernandez: Stepwise unfolding of titin under force-clamp atomic force microscopy, Proc. Natl. Acad. Sci. USA **16**, 468–472 (2000)
60. S. Izraelev, S. Stepaniants, M. Balsera, Y. Oono, K. Schulten: Molecular dynamics study of unbinding of the avidin–biotin complex, Biophys. J. **72**, 1568–1581 (1997)

61. M. Rief, F. Oesterhelt, B. Heyman, H. E. Gaub: Single molecule force spectroscopy on polysaccharides by atomic force microscopy, Science **275**, 1295–1297 (1997)
62. E. Evans, E. Leung, D. Hammer, S. Simon: Chemically distinct transition states govern rapid dissociation of single L-selectin bonds under force, Proc. Natl. Acad. Sci. USA **98**, 3784–3789 (2001)
63. X. Zhang, E. Woijcikiewicz, V. T. Moy: Force spectroscopy of the leukocyte function-associated antigen-1/intercellular adhesion molecule-1 interaction, Biophys. J. **83**, 2270–2279 (2002)
64. B. T. Marshall, M. Long, J. W. Piper, T. Yago, R. P. McEver, Z. Zhu: Direct observation of catch bonds involving cell adhesion molecules, Nature **423**, 190–193 (2003)
65. B. Heymann, H. Grubmüller: Molecular dynamics force probe simulations of antibody/antigen unbinding: Entropic control and non additivity of unbinding forces, Biophys. J. **81**, 1295–1313 (2001)
66. R. Nevo, C. Stroh, F. Kienberger, D. Kaftan, V. Brumfeld, M. Elbaum, Z. Reich, P. Hinterdorfer: A molecular switch between two bound states in the RanGTP-importinβ1 interaction, Nat. Struct. & Mol. Biol. **10**, 553–557 (2003)
67. R. Nevo, V. Brumfeld, M. Elbaum, P. Hinterdorfer, Z. Reich: Direct discrimination between models of protein activation by single-molecule force measurements, Biophys. J. **87**, 2630–2634 (2004)
68. R. Zwanzig: Diffusion in a rough potential, Proc. Natl. Acad. Sci. USA **85**:, 2029–2030 (1988)
69. C. B. Hyeon, D. Thirumalai: Can energy landscape roughness of proteins and RNA be measured by using mechanical unfolding experiments?, Proc. Natl. Acad. Sci. USA **100**, 10249–10253 (2003)
70. R. Nevo, V. Brumfeld, P. Hinterdorfer, Z. Reich: Direct measurement of protein energy landscape roughness, EMBO Rep. **6**, 482–486 (2005)
71. P. P. Lehenkari, M. A. Horton: Single integrin molecule adhesion forces in intact cells measured by atomic force microscopy, Biochem. Biophys. Res. Com. **259**, 645–650 (1999)
72. A. Chen, V. T. Moy: Cross-linking of cell surface receptors enhances cooperativity of molecular adhesion, Biophys. J. **78**, 2814–2820 (2000)
73. G. Pfister, C. M. Stroh, H. Perschinka, M. Kind, M. Knoflach, P. Hinterdorfer, G. Wick: Detection of HSP60 on the membrane surface of stressed human endothelial cells by atomic force and confocal microscopy, J. Cell. Sci. **118**, 1587–1594 (2005)
74. T. Puntheeranurak, L. Wildling, H. J. Gruber, R. K. H. Kinne, P. Hinterdorfer: Ligands on the string: Single molecule studies on the interaction of antibodies and substrates with the surface of the Na$^+$-glucose cotransporter SGLT1 *in vivo*, J. Cell Sci., in press
75. M. Ludwig, W. Dettmann, H. E. Gaub: Atomic force microscopy imaging contrast based on molecuar recognition, Biophys. J. **72**, 445–448 (1997)
76. P. P. Lehenkari, G. T. Charras, G. T. Nykänen, M. A. Horton: Adapting force microscopy for cell biology, Ultramicroscopy **82**, 289–295 (2000)
77. N. Almqvist, R. Bhatia, G. Primbs, N. Desai, S. Banerjee, R. Lal: Elasticity and adhesion force mapping reveals real-time clustering of growth factor receptors and associated changes in local cellular rheological properties, Biophys. J. **86**, 1753–1762 (2004)
78. M. Grandbois, M. Beyer, M. Rief, H. Clausen-Schaumann, H. E. Gaub: Affinity imaging of red blood cells using an atomic force microscope, J. Histochem. Cytochem. **48**, 719–724 (2000)
79. O. H. Willemsen, M. M. E. Snel, K. O. van der Werf, B. G. de Grooth, J. Greve, P. Hinterdorfer, H. J. Gruber, H. Schindler, Y. van Kyook, C. G. Figdor: Simultaneous height and

adhesion imaging of antibody antigen interactions by atomic force microscopy, Biophys. J. **57**, 2220–2228 (1998)
80. B. V. Viani, L. I. Pietrasanta, J. B. Thompson, A. Chand, I. C. Gebeshuber, J. H. Kindt, M. Richter, H. G. Hansma, P. K. Hansma: Probing protein–protein interactions in real time, Nature Struct. Biol. **7**, 644–647 (2000)
81. W. Han, S. M. Lindsay, T. Jing: A magnetically driven oscillating probe microscope for operation in liquid, Appl. Phys. Lett. **69**, 1–3 (1996)
82. C. M. Stroh, A. Ebner, M. Geretschläger, G. Freudenthaler, F. Kienberger, A. S. M. Kamruzzahan, S. J. Smith-Gill, H. J. Gruber, P. Hinterdorfer: Simultaneous topography and recognition imaging using force microscopy, Biophys. J. **87**, 1981–1990 (2004)
83. C. Stroh, H. Wang, R. Bash, B. Ashcroft, J. Nelson, H. J. Gruber, D. Lohr, S. M. Lindsay, P. Hinterdorfer: Single-molecule recognition imaging microscope, Proc. Natl. Acad. Sci. **101**, 12503–12507 (2004)
84. A. Ebner, F. Kienberger, G. Kada, C. M. Stroh, M. Geretschläger, A. S. M. Kamruzzahan, L. Wildling, W. T. Johnson, B. Ashcroft, J. Nelson, S. M. Lindsay, H. J. Gruber, P. Hinterdorfer: Localization of single avidin biotin interactions using simultaneous topography and molecular recognition imaging, Chem. Phys. Chem. **6**, 897–900 (2005)

Part II

Nanotribology and Nanomechanics: Fundamental Studies

8

Nanotribology, Nanomechanics and Materials Characterization

Bharat Bhushan

Summary. Nanotribology and nanomechanics studies are needed to develop fundamental understanding of interfacial phenomena on a small scale and to study interfacial phenomena in micro/nanoelectromechanical systems (MEMS/NEMS), magnetic storage devices, and other applications. Friction and wear of lightly loaded micro/nanocomponents are highly dependent on the surface interactions (few atomic layers). These structures are generally coated with molecularly thin films. Nanotribology and nanomechanics studies are also valuable in fundamental understanding of interfacial phenomena in macrostructures, and provide a bridge between science and engineering. An atomic force microscope (AFM) tip is used to simulate a single asperity contact with a solid or lubricated surface. AFMs are used to study the various tribological phenomena, which include surface roughness, adhesion, friction, scratching, wear, detection of material transfer, and boundary lubrication. In situ surface characterization of local deformation of materials and thin coatings can be carried out using a tensile stage inside an AFM. Mechanical properties such as hardness, Young's modulus of elasticity and creep/relaxation behavior can be determined on the micro- to picoscales using a depth-sensing indentation system in an AFM. Localized surface elasticity and viscoelastic mapping can be obtained of near-surface regions with nanoscale lateral resolution. Finally, an AFM can be used for nanofabrication/nanomachining.

8.1 Introduction

The mechanisms and dynamics of the interactions of two contacting solids during relative motion, ranging from atomic- to microscale, need to be understood in order to develop fundamental understanding of adhesion, friction, wear, indentation, and lubrication processes. At most solid–solid interfaces of technological relevance, contact occurs at many asperities. Consequently the importance of investigating single asperity contacts in studies of the fundamental micro/nanomechanical and micro/nanotribological properties of surfaces and interfaces has long been recognized. The recent emergence and proliferation of proximal probes, in particular scanning probe microscopies (the scanning tunneling microscope and the atomic force microscope), the surface force apparatus, and of computational techniques for simulating tip–surface interactions and interfacial properties, has allowed systematic investigations of interfacial problems with high resolution as well as means for modifying and manipulating nanoscale structures. These advances have led to the appearance of the new field of nanotribology, which pertains to experimental and theoretical investigations of interfacial processes on scales ranging from the atomic- and molecular- to the microscale, occurring during adhesion, friction, scratching, wear, indentation, and thin-film lubrication at sliding surfaces [1–14]. Proximal probes have also been used for mechanical characterization, in situ characterization of local deformation, and other nanomechanics studies.

Nanotribological and nanomechanics studies are needed to develop fundamental understanding of interfacial phenomena on a small scale and to study interfacial phenomena in micro/nanostructures used in magnetic storage devices, micro/nanoelectromechanical systems (MEMS/NEMS), and other applications [3–12, 15–17]. Friction and wear of lightly loaded micro/nanocomponents are highly dependent on the surface interactions (a few atomic layers). These structures are generally coated with molecularly thin films. Nanotribological and nanomechanics studies are also valuable in fundamental understanding of interfacial phenomena in macrostructures, and provide a bridge between science and engineering.

The surface force apparatus (SFA), the scanning tunneling microscope (STM), atomic force and friction force microscopes (AFM and FFM) are widely used in nanotribological and nanomechanics studies. Typical operating parameters are compared in Table 8.1. The SFA was developed in 1968 and is commonly employed to study both static and dynamic properties of molecularly thin films sandwiched between two molecularly smooth surfaces. The STM, developed in 1981, allows imaging of electrically conducting surfaces with atomic resolution, and has been used for imaging of clean surfaces as well as of lubricant molecules. The introduction of the AFM in 1985 provided a method for measuring ultra-small forces between a probe tip and an engineering (electrically conducting or insulating) surface, and has been used for morphological and surface roughness measurements of surfaces on the nanoscale, as well as for adhesion measurements. Subsequent modifications of the AFM led to the development of the FFM, designed for atomic-scale and microscale studies of friction. This instrument measures forces in the scanning direction. The AFM is also being used for various investigations including scratching, wear, indentation, detection of transfer of material, boundary lubrication, and fabrication and machining [13, 18, 19]. Meanwhile, significant progress in understanding the fundamental nature of bonding and interactions in materials, combined with advances in computer-based modeling and simulation methods, has allowed theoretical studies of complex interfacial phenomena with high resolution in space and time. Such simulations provide insights into atomic-scale energetics, structure, dynamics, thermodynamics, transport and rheological aspects of tribological processes.

The nature of interactions between two surfaces brought close together, and those between two surfaces in contact as they are separated, have been studied experimentally with the surface force apparatus. This has led to a basic understanding of the normal forces between surfaces, and the way in which these are modified by

Table 8.1. Comparison of typical operating parameters in SFA, STM and AFM/FFM used for micro/nanotribological studies

Operating parameter	SFA	STM[a]	AFM/FFM
Radius of mating surface/tip	≈ 10 mm	5–100 nm	5–100 nm
Radius of contact area	10–40 μm	N/A	0.05–0.5 nm
Normal load	10–100 mN	N/A	< 0.1 nN – 500 nN
Sliding velocity	0.001–100 μm/s	0.02–200 μm/s (scan size ≈ 1 nm × 1 nm to 125 μm × 125 μm; scan rate < 1–122 Hz)	0.02–200 μm/s (scan size ≈ 1 nm × 1 nm to 125 μm × 125 μm; scan rate < 1–122 Hz)
Sample limitations	Typically atomically smooth, optically transparent mica; opaque ceramic, smooth surfaces can also be used	Electrically conducting samples	None

[a] Can only be used for atomic-scale imaging

the presence of a thin liquid or a polymer film. The frictional properties of such systems have been studied by moving the surfaces laterally, and such experiments have provided insights into the molecular-scale operation of lubricants such as thin liquid or polymer films. Complementary to these studies are those in which the AFM tip is used to simulate a single asperity contact with a solid or lubricated surface, Fig. 8.1. These experiments have demonstrated that the relationship between friction and surface roughness is not always simple or obvious. AFM studies have also revealed much about the nanoscale nature of intimate contact during wear, indentation, and lubrication.

In this chapter, we present a review of significant aspects of nanotribological, nanomechanical and materials characterization studies conducted using AFM/FFM.

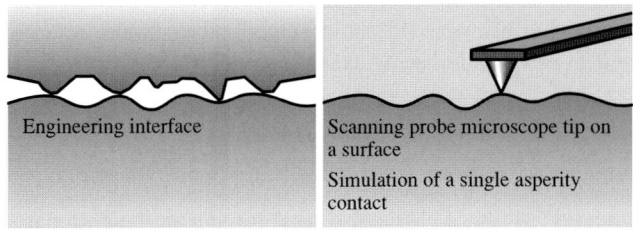

Fig. 8.1. Schematics of an engineering interface and scanning probe microscope tip in contact with an engineering interface

8.2 Description of AFM/FFM and Various Measurement Techniques

An AFM was developed by *Binnig* and his colleagues in 1985. It is capable of investigating surfaces of scientific and engineering interest on an atomic scale [20, 21]. The AFM relies on a scanning technique to produce very-high-resolution, three-dimensional images of sample surfaces. It measures ultra-small forces (less than 1 nN) present between the AFM tip surface mounted on a flexible cantilever beam, and a sample surface. These small forces are obtained by measuring the motion of a very flexible cantilever beam having an ultra-small mass, by a variety of measurement techniques including optical deflection, optical interference, capacitance, and tunneling current. The deflection can be measured to within 0.02 nm, so for a typical cantilever spring constant of 10 N/m, a force as low as 0.2 nN can be detected. To put these numbers in perspective, individual atoms and human hair are typically a fraction of a nanometer and about 75 μm in diameter, respectively, and a drop of water and an eyelash have a mass of about 10 μN and 100 nN, respectively. In the operation of high-resolution AFM, the sample is generally scanned rather than the tip because any cantilever movement would add vibrations. AFMs are available for measurement of large samples, where the tip is scanned and the sample is stationary. To obtain atomic resolution with the AFM, the spring constant of the cantilever should be weaker than the equivalent spring between atoms. A cantilever beam with a spring constant of about 1 N/m or lower is desirable. For high lateral resolution, tips should be as sharp as possible. Tips with a radius in the range 10–100 nm are commonly available. Interfacial forces, adhesion, and surface roughness, including atomic-scale imaging, are routinely measured using the AFM.

A modification to AFM, providing a sensor to measure the lateral force, led to the development of the friction force microscope (FFM) or the lateral force microscope (LFM), designed for atomic-scale and microscale studies of friction [3–5, 7, 8, 13, 22–37] and lubrication [38–43]. This instrument measures lateral or friction forces (in the plane of the sample surface and in the scanning direction). By using a standard or a sharp diamond tip mounted on a stiff cantilever beam, AFM is used in investigations of scratching and wear [6, 9, 13, 27, 44–47], indentation [9, 13, 17, 27, 48–51], and fabrication/machining [4, 13, 27]. An oscillating cantilever is used for localized surface elasticity and viscoelastic mapping, referred to as dynamic AFM [35, 52–60]. In situ surface characterization of local deformation of materials and thin coatings has been carried out by imaging the sample surfaces using an AFM, during tensile deformation using a tensile stage [61–63].

8.2.1 Surface Roughness and Friction Force Measurements

Surface height imaging down to atomic resolution of electrically conducting surfaces is carried out using an STM. An AFM is also used for surface height imaging and roughness characterization down to the nanoscale. Commercial AFM/FFM are routinely used for simultaneous measurements of surface roughness and friction force [4, 11]. These instruments are available for measurement of small and large samples. In a small-sample AFM, shown in Fig. 8.2a, the sample, generally no larger than 10 mm × 10 mm, is mounted on a piezoelectric crystal in the form of a cylindrical tube [referred to as a lead zirconate titanate (PZT) tube scanner] which consists of separate electrodes to scan the sample precisely in the x–y plane in a raster pattern and to move the sample in the vertical (z) direction. A sharp

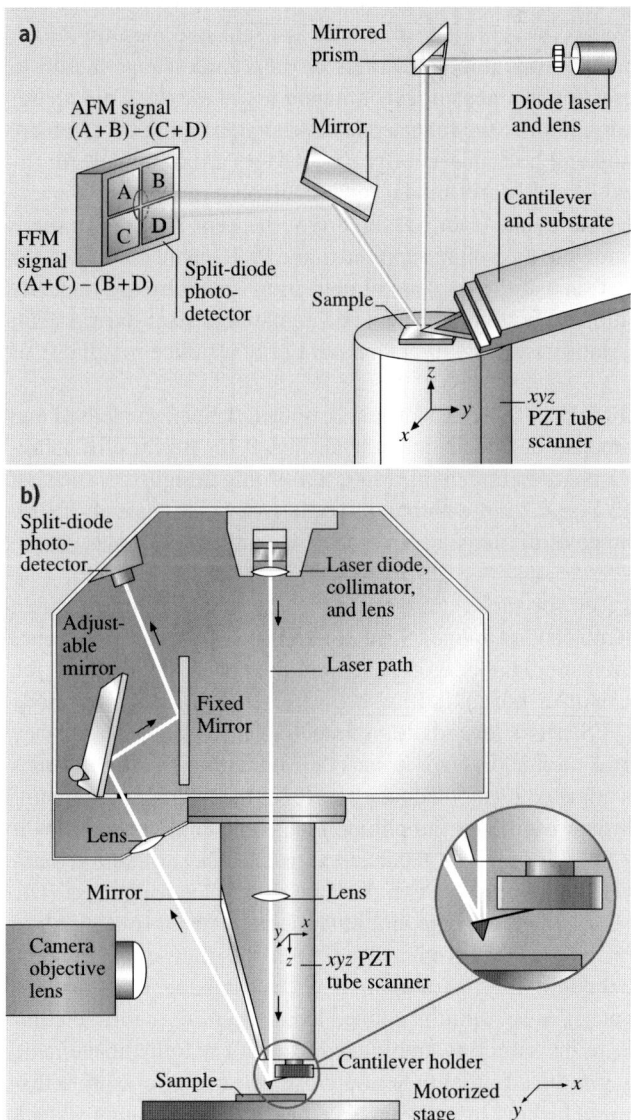

Fig. 8.2. Schematics (**a**) of a commercial small-sample atomic force microscope/friction force microscope (AFM/FFM), and (**b**) of a large-sample AFM/FFM

tip at the free end of a flexible cantilever is brought in contact with the sample. Normal and frictional forces being applied at the tip–sample interface are measured using a laser beam deflection technique. A laser beam from a diode laser is directed by a prism onto the back of a cantilever near its free end, tilted downward at about $10°$ with respect to the horizontal plane. The reflected beam from the vertex of the cantilever is directed through a mirror onto a quad photodetector (split photodetector with four quadrants). The differential signal from the top and bottom photodiodes provides the AFM signal, which is a sensitive measure of the cantilever vertical deflection. Topographic features of the sample cause the tip to deflect in the vertical direction as the sample is scanned under the tip. This tip deflection will change the direction of the reflected laser beam, changing the intensity difference between the top and bottom sets of photodetectors (AFM signal). In the AFM operating mode called the height mode, for topographic imaging or for any other operation in which the applied normal force is to be kept constant, a feedback circuit is used to modulate the voltage applied to the PZT scanner to adjust the height of the PZT, so that the cantilever vertical deflection (given by the intensity difference between the top and bottom detector) will remain constant during scanning. The PZT height variation is thus a direct measure of the surface roughness of the sample.

In a large-sample AFM, both force sensors using optical deflection method and scanning unit are mounted on the microscope head, Fig. 8.2b. Because of vibrations added by cantilever movement, lateral resolution of this design is somewhat poorer than the design in Fig. 8.2a in which the sample is scanned instead of cantilever beam. The advantage of the large-sample AFM is that large samples can be measured readily.

Most AFMs can be used for surface roughness measurements in the so-called tapping mode (intermittent contact mode), also referred to as dynamic (atomic) force microscopy. In the tapping mode, during scanning over the surface, the cantilever/tip assembly, with a normal stiffness of 20–100 N/m (tapping mode etched Si probe or DI TESP) is sinusoidally vibrated at its resonance frequency (350–400 kHz) by a piezo mounted above it, and the oscillating tip slightly taps the surface. The piezo is adjusted using feedback control in the z-direction to maintain a constant (20–100 nm) oscillating amplitude (set point) and constant average normal force (Fig. 8.3) [4, 11]. The feedback signal to the z-direction sample piezo (to keep the set point constant) is a measure of surface roughness. The cantilever/tip assembly is vibrated at some amplitude, here referred to as the free amplitude, before the tip engages the sample. The tip engages the sample at some set point, which may be thought of as the amplitude of the cantilever as influenced by contact with the sample. The set point is defined as a ratio of the vibration amplitude after engagement to the vibration amplitude in free air before engagement. A lower set point gives a reduced amplitude and closer mean tip-to-sample distance. The amplitude should be kept large enough so that the tip does not get stuck to the sample because of adhesive attractions. Also the oscillating amplitude applies a lower average (normal) load compared to the contact mode and reduces sample

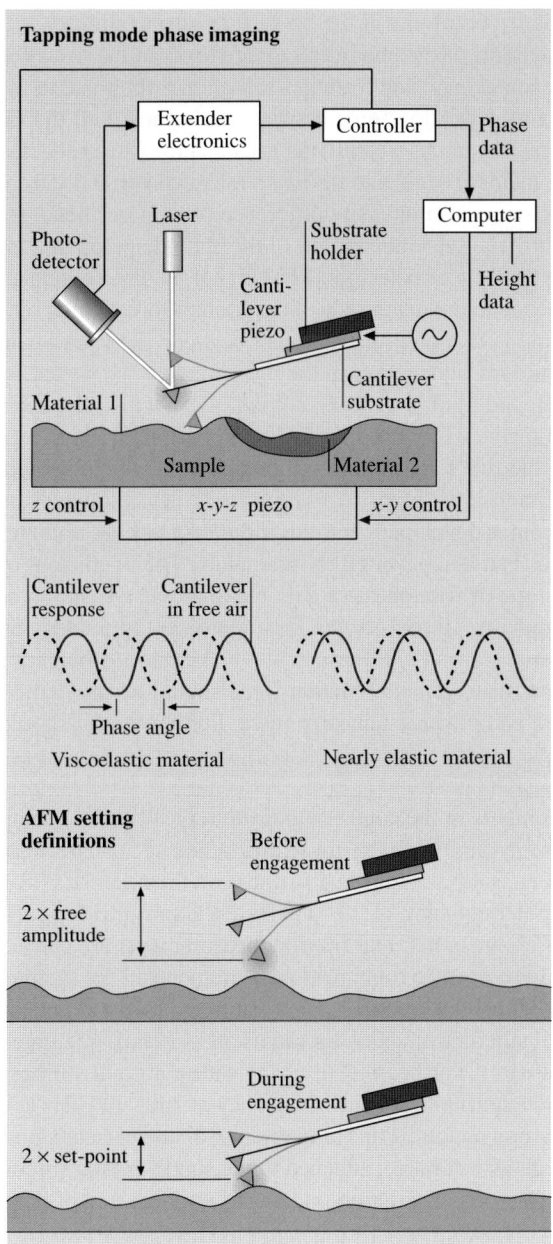

Fig. 8.3. Schematic of tapping mode used to obtain height and phase data and definitions of free amplitude and set point. During scanning, the cantilever is vibrated at its resonance frequency and the sample x–y–z piezo is adjusted by feedback control in the z-direction to maintain a constant set point. The computer records height (which is a measure of surface roughness) and phase angle (which is a function of the viscoelastic properties of the sample) data

damage. The tapping mode is used in topography measurements to minimize the effects of friction and other lateral forces and to measure the topography of soft surfaces.

For the measurement of the friction force at the tip surface during sliding, left-hand and right-hand sets of quadrants of the photodetector are used. In the so-called friction mode, the sample is scanned back and forth in a direction orthogonal to the long axis of the cantilever beam. A friction force between the sample and the tip will produce a twisting of the cantilever. As a result, the laser beam will be reflected out of the plane defined by the incident beam and the beam reflected vertically from an untwisted cantilever. This produces an intensity difference of the laser beam received in the left-hand and right-hand sets of quadrants of the photodetector. The intensity difference between the two sets of detectors (FFM signal) is directly related to the degree of twisting and hence to the magnitude of the friction force. One problem associated with this method is that any misalignment between the laser beam and the photodetector axis would introduce error in the measurement. However, by following the procedures developed by *Ruan* and *Bhushan* [24], in which the average FFM signal for the sample scanned in two opposite directions is subtracted from the friction profiles of each of the two scans, the misalignment effect is eliminated. This method provides three-dimensional maps of friction force. By following the friction force calibration procedures developed by *Ruan* and *Bhushan* [24], voltages corresponding to friction forces can be converted to force units. The coefficient of friction is obtained from the slope of friction force data measured as a function of normal loads typically ranging from 10 to 150 nN. This approach eliminates any contributions due to the adhesive forces [27]. For calculation of the coefficient of friction based on a single-point measurement, friction force should be divided by the sum of applied normal load and intrinsic adhesive force. Furthermore it should be pointed out that, for a single asperity contact, the coefficient of friction is not independent of load (see discussion later).

Surface roughness measurements in the contact mode are typically made using a sharp, microfabricated square-pyramidal Si_3N_4 tip with a radius of 30–50 nm on a triangular cantilever beam (Fig. 8.4a) with normal stiffness on the order of 0.06–0.58 N/m with a normal natural frequency of 13–40 kHz (silicium nitride probe or DI NP) at a normal load of about 10 nN, and friction measurements are carried out in the load range of 1–100 nN. Surface roughness measurements in the tapping mode utilize a stiff cantilever with high resonance frequency; typically a square-pyramidal etched single-crystal silicon tip, with a tip radius of 5–10 nm, integrated with a stiff rectangular silicon cantilever beam (Fig. 8.4a) with a normal stiffness on the order of 17–60 N/m and a normal resonance frequency of 250–400 kHz (DI TESP), is used. Multiwalled carbon nanotube tips having small diameter (a few nm) and a length of about 1 μm (high aspect ratio), attached to the single-crystal silicon square-pyramidal tips are used for high-resolution imaging of surfaces and of deep trenches in the tapping mode (noncontact mode) (Fig. 8.4b) [64]. To study the effect of the radius of a single asperity (tip) on adhesion and friction, microspheres of silica with radii ranging from about 4 to 15 μm are attached at the end of cantilever beams. Optical micrographs of two of the microspheres at the ends of triangular cantilever beams are shown in Fig. 8.4c.

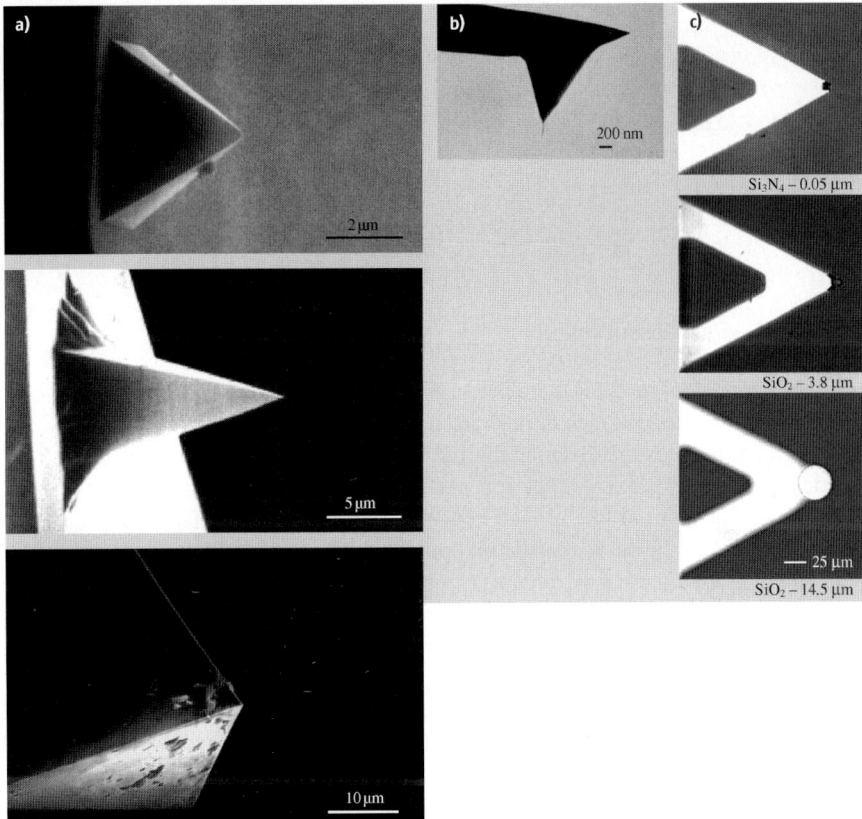

Fig. 8.4. (a) SEM micrographs of a square-pyramidal PECVD Si$_3$N$_4$ tip with a triangular cantilever beam, a square-pyramidal etched single-crystal silicon tip with a rectangular silicon cantilever beam, and a three-sided pyramidal natural diamond tip with a square stainless-steel cantilever beam, (b) SEM micrograph of a multiwalled carbon nanotube (MWNT) physically attached to the single-crystal silicon, square-pyramidal tip, and (c) optical micrographs of commercial Si$_3$N$_4$ tip and two modified tips showing SiO$_2$ spheres mounted over the sharp tip, at the end of the triangular Si$_3$N$_4$ cantilever beams (radii of the tips are given in the figure)

The tip is scanned in such a way that its trajectory on the sample forms a triangular pattern, Fig. 8.5. Scanning speeds in the fast and slow scan directions depend on the scan area and scan frequency. Scan sizes ranging from less than 1 nm × 1 nm to 125 μm × 125 μm and scan rates from less than 0.5 to 122 Hz can typically be used. Higher scan rates are used for smaller scan lengths. For example, scan rates in the fast and slow scan directions for an area of 10 μm × 10 μm scanned at 0.5 Hz are 10 μm/s and 20 nm/s, respectively.

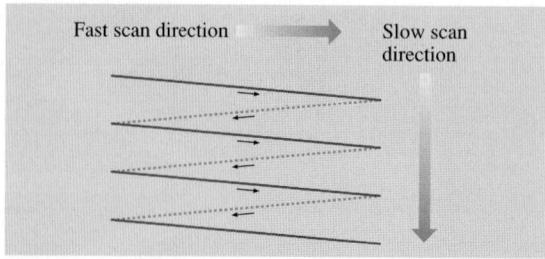

Fig. 8.5. Schematic of triangular pattern trajectory of the tip as the sample (or the tip) is scanned in two dimensions. During scanning, data are recorded only during scans along the *solid scan lines*

8.2.2 Adhesion Measurements

Adhesive force measurements are performed in the so-called force calibration mode. In this mode, force–distance curves are obtained; for an example see Fig. 8.6. The horizontal axis gives the distance the piezo (and hence the sample) travels and the vertical axis gives the tip deflection. As the piezo extends, it approaches the tip, which is at this point in free air and hence shows no deflection. This is indicated by the flat portion of the curve. As the tip approaches the sample within a few nanometers (point A), an attractive force exists between the atoms of the tip surface and the atoms of the sample surface. The tip is pulled towards the sample and contact occurs at point B on the graph. From this point on, the tip is in contact with the surface and, as the piezo further extends, the tip is further deflected. This is represented by the sloped portion of the curve. As the piezo retracts, the tip goes beyond the zero-deflection (flat) line because of attractive forces (van der Waals forces and long-range meniscus forces), into the adhesive regime. At point C on the graph, the tip snaps free of the adhesive forces, and is again in free air. The horizontal distance between points B and C along the retrace line gives the distance moved by the tip in the adhesive regime. This distance multiplied by the stiffness of the cantilever gives the adhesive force. Incidentally, the horizontal shift between the loading and unloading curves results from the hysteresis in the PZT tube [4, 11].

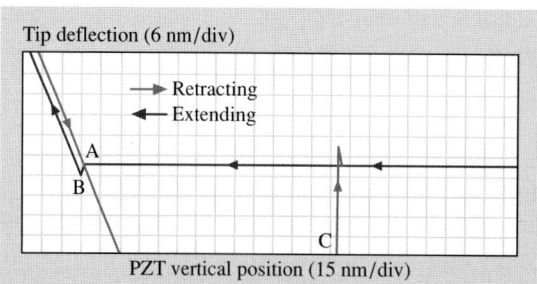

Fig. 8.6. Typical force–distance curve for a contact between Si_3N_4 tip and single-crystal silicon surface in measurements made in the ambient environment. Contact between the tip and silicon occurs at point B; the tip breaks free of adhesive forces at point C as the sample moves away from the tip

8.2.3 Scratching, Wear and Fabrication/Machining

For microscale scratching, microscale wear, nanofabrication/nanomachining and nanoindentation hardness measurements, an extremely hard tip is required. A three-sided pyramidal single-crystal natural-diamond tip with an apex angle of 80° and a radius of about 100 nm mounted on a stainless-steel cantilever beam with normal stiffness of about 25 N/m is used at relatively higher loads (1–150 μN), Fig. 8.4a. For scratching and wear studies, the sample is generally scanned in a direction orthogonal to the long axis of the cantilever beam (typically at a rate of 0.5 Hz) so that friction can be measured during scratching and wear. The tip is mounted on the cantilever such that one of its edges is orthogonal to the long axis of the beam; therefore, wear during scanning along the beam axis is higher (about 2 to 3 times) than that during scanning orthogonal to the beam axis. For wear studies, an area on the order of 2 μm × 2 μm is scanned at various normal loads (ranging from 1 to 100 μN) for a selected number of cycles [4, 11, 45].

Scratching can also be performed at ramped loads and the coefficient of friction can be measured during scratching [47]. A linear increase in the normal load approximated by a large number of normal load increments of small magnitude is applied using a software interface (lithography module in Nanoscope III) that allows the user to generate controlled movement of the tip with respect to the sample. The friction signal is tapped out of the AFM and is recorded on a computer. A scratch length on the order of 25 μm and a velocity on the order of 0.5 μm/s are used and the number of loading steps is usually taken to be 50.

Nanofabrication/nanomachining is conducted by scratching the sample surface with a diamond tip at specified locations and scratching angles. The normal load used for scratching (writing) is on the order of 1–100 μN with a writing speed on the order of 0.1–200 μm/s [4, 6, 11–13, 27, 65].

8.2.4 Surface Potential Measurements

To detect wear precursors and to study the early stages of localized wear, the multimode AFM can be used to measure the potential difference between the tip and the sample by applying a DC bias potential and an oscillating (AC) potential to a conducting tip over a grounded substrate in a Kelvin probe microscopy or so-called *nano-Kelvin probe* technique [66–68].

Mapping of the surface potential is made in the so-called *lift mode* (Fig. 8.7). These measurements are made simultaneously with the topography scan in the tapping mode, using an electrically conducting (nickel-coated single-crystal silicon) tip. After each line of the topography scan is completed, the feedback loop controlling the vertical piezo is turned off, and the tip is lifted from the surface and traced over the same topography at a constant distance of 100 nm. During the lift mode, a DC bias potential and an oscillating potential (3–7 Volts) is applied to the tip. The frequency of oscillation is chosen to be equal to the resonance frequency of the cantilever ($\approx$ 80 kHz). When a DC bias potential equal to the negative value of surface potential of the sample (on the order of ±2 Volts) is applied to the tip, it does not

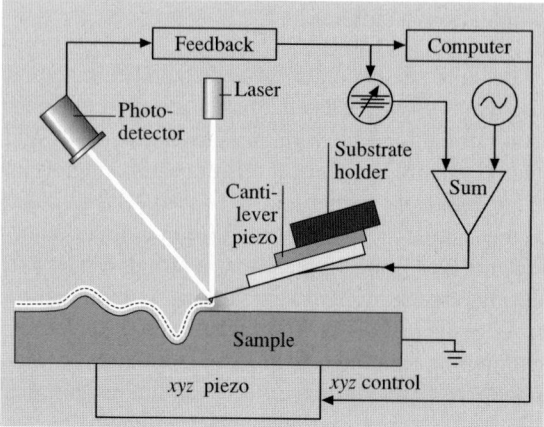

Fig. 8.7. Schematic of lift mode used to make surface potential measurement. The topography is collected in tapping mode in the primary scan. The cantilever piezo is deactivated. Using topography information of the primary scan, the cantilever is scanned across the surface at a constant height above the sample. An oscillating voltage at the resonant frequency is applied to the tip and a feedback loop adjusts the DC bias of the tip to maintain the cantilever amplitude at zero. The output of the feedback loop is recorded by the computer and becomes the surface potential map

vibrate. During scanning, a difference between the DC bias potential applied to the tip and the potential of the surface will create DC electric fields that interact with the oscillating charges (as a result of the AC potential), causing the cantilever to oscillate at its resonance frequency, as in the tapping mode. However, a feedback loop is used to adjust the DC bias on the tip to exactly nullify the electric field, and thus the vibrations of the cantilever. The required bias voltage follows the localized potential of the surface. The surface potential was obtained by reversing the sign of the bias potential provided by the electronics [67, 68]. Surface and subsurface changes of structure and/or chemistry can cause changes in the measured potential of a surface. Thus, mapping of the surface potential after sliding can be used to detect wear precursors and study the early stages of localized wear.

8.2.5 In Situ Characterization of Local Deformation Studies

In situ characterization of local deformation of materials can be carried out by performing tensile, bending, or compression experiments inside an AFM and by observing nanoscale changes during the deformation experiment [17]. In these experiments, small deformation stages are used to deform the samples inside an AFM. In tensile testing of the polymeric films carried out by *Bobji* and *Bhushan* [61, 62] and *Tambe* and *Bhushan* [63] a tensile stage was used (Fig. 8.8). The stage with a left–right combination lead screw (which helps to move the slider in the opposite direction) was used to stretch the sample to minimize the movement of the scanning area, which was kept close to the center of the tensile specimen. One end of

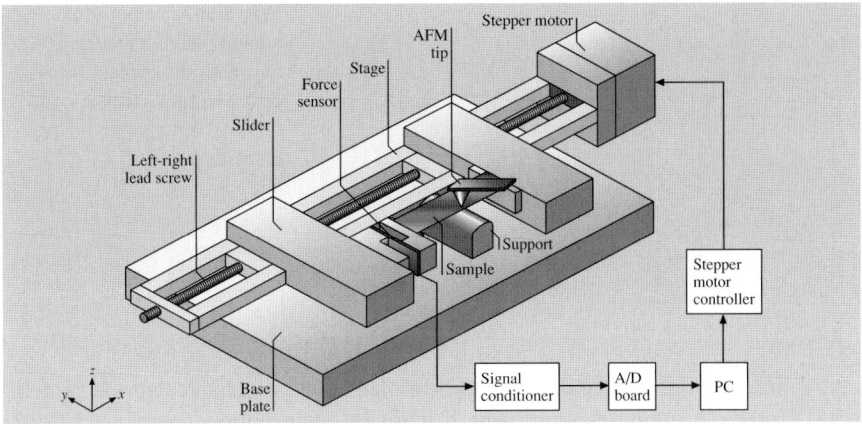

Fig. 8.8. Schematic of the tensile stage to conduct in situ tensile testing of the polymeric films in AFM

the sample was mounted on the slider via a force sensor to monitor the tensile load. The samples were stretched for various strains using a stepper motor and the same control area at different strains was imaged. In order to better locate the control area for imaging, a set of four markers was created at the corners of a 30 μm × 30 μm square at the center of the sample by scratching the sample with a sharp silicon tip. The scratching depth was controlled such that it did not affect the cracking behavior of the coating. A minimum displacement of 1.6 μm could be obtained. This corresponded to a strain increment of $8 \times 10^{-3}\%$ for a sample length of 38 mm. The maximum travel was about 100 mm. The resolution of the force sensor was 10 mN with a capacity of 45 N. During stretching, a stress–strain curve was obtained during the experiment to study any correlation between the degree of plastic strain and propensity of cracks.

8.2.6 Nanoindentation Measurements

For nanoindentation hardness measurements the scan size is set to zero and then a normal load is applied to make the indents using the diamond tip (see Sect. 8.2.5). During this procedure, the tip is continuously pressed against the sample surface for about two seconds at various indentation loads. The sample surface is scanned before and after the scratching, wear or indentation to obtain the initial and the final surface topography, at a low normal load of about 0.3 μN using the same diamond tip. An area larger than the indentation region is scanned to observe the indentation marks. Nanohardness is calculated by dividing the indentation load by the projected residual area of the indents [50].

Direct imaging of the indent allows one to quantify piling up of ductile material around the indenter. However, it becomes difficult to identify the boundary of the indentation mark with great accuracy. This makes the direct measurement of contact area somewhat inaccurate. A technique with the dual capability of depth-sensing as

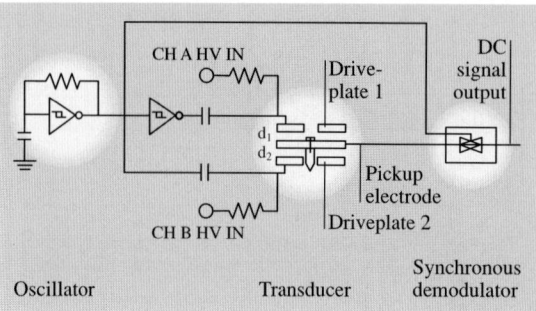

Fig. 8.9. Schematic of a nano/picoindentation system with three-plate transducer with electrostatic actuation hardware and capacitance sensor [49]

well as in situ imaging, which is most appropriate in nanomechanical property studies, is used for accurate measurement of hardness with shallow depths [4, 11, 49]. This nano/picoindentation system is used to make load–displacement measurements and subsequently carry out in situ imaging of the indent, if required. The indentation system, shown in Fig. 8.9, consists of a three-plate transducer with electrostatic actuation hardware used for direct application of normal load and a capacitive sensor used for measurement of vertical displacement. The AFM head is replaced with this transducer assembly while the specimen is mounted on the PZT scanner, which remains stationary during indentation experiments. The transducer consists of a three (Be–Cu) plate capacitive structure and the tip is mounted on the center plate. The upper and lower plates serve as drive electrodes and the load is applied by applying appropriate voltage to the drive electrodes. Vertical displacement of the tip (indentation depth) is measured by measuring the displacement of the center plate relative to the two outer electrodes using capacitance technique. The indent area and consequently hardness value can be obtained from the load–displacement data. The Young's modulus of elasticity is obtained from the slope of the unloading curve.

8.2.7 Localized Surface Elasticity and Viscoelasticity Mapping

Localized Surface Elasticity

Indentation experiments provide a single-point measurement of the Young's modulus of elasticity calculated from the slope of the indentation curve during unloading. Localized surface elasticity maps can be obtained using dynamic force microscopy in which an oscillating tip is scanned over the sample surface in contact under steady and oscillating load. The lower-frequency operation modes in the kHz range, such as the force modulation mode [52, 54] or the pulsed force mode [69], are well suited for soft samples such as polymers. However, if the tip–sample contact stiffness becomes significantly higher than the cantilever stiffness, the sensitivity of these techniques strongly decreases. In this case, the sensitivity of the measurement of stiff materials can be improved by using high-frequency operation modes in the MHz range with a lateral motion, such as acoustic (ultrasonic) force microscopy, referred to as atomic force acoustic microscopy (AFAM) or contact resonance spectroscopy [55, 56, 70].

Inclusion of vibration frequencies other than only the first cantilever flexural or torsional resonance frequency, also allows additional information to be obtained.

In the negative lift mode force modulation technique, height data is recorded during primary scanning in the tapping mode, as described earlier. During interleave scanning, the entire cantilever/tip assembly is moved up and down at the force modulation holder's bimorph resonance frequency (about 24 kHz) at some amplitude, here referred to as the force modulation amplitude, and the z-direction feedback control for the sample x–y–z piezo is deactivated, Fig. 8.10a [52, 54, 57]. During this scanning, height information from the primary scan is used to maintain a constant lift scan height. This eliminates the influence of height on the measured signals during the interleave scan. Lift scan height is the mean tip-to-sample distance between the tip and sample during the interleave scan. The lift scan height is set such that the tip is in constant contact with the sample, i.e. a constant static load is applied. (A higher lift scan height gives a closer mean tip-to-sample distance.) In addition, the tip motion caused by the bimorph vibration results in a modulating periodic force. The sample surface resists the oscillations of the tip to a greater or lesser extent depending upon the sample's stiffness. The computer records amplitude (which is a function of the elastic stiffness of the material). Contact analyses can be used to obtain a quantitative measure of localized elasticity of soft surfaces [54]. Etched single-crystal silicon cantilevers with integrated tips (force modulation etched Si probe or DI FESP) with a radius of 25–50 nm, a stiffness of 1–5 N/m, and a natural frequency of 60–100 kHz, are commonly used for the measurements. Scanning is normally set to a rate of 0.5 Hz along the fast axis.

In the AFAM technique [55, 56, 70], the cantilever/tip assembly is moved either in the normal or lateral mode and the contact stiffness is evaluated by comparing the resonance frequency of the cantilever in contact with the sample surface to those of the free vibrations of the cantilever. Several free resonance frequencies are measured. Based on the shift of the measured frequencies, the contact stiffness is determined by solving the characteristic equation for the tip vibrating in contact with the sample surface. The elastic modulus is calculated from contact stiffness using Hertz analysis for a spherical tip indenting a plane. Contact stiffness is equal to 8× contact radius × reduced shear modulus in the shear mode.

In the lateral mode using the AFAM technique, the sample is glued onto cylindrical pieces of aluminum which serve as ultrasonic delay lines coupled to an ultrasonic shear wave transducer, Fig. 8.10b [33, 55, 56]. The transducer is driven with frequency sweeps to generate in-plane lateral sample surface vibrations. These couple to the cantilever via the tip–sample contact. To measure torsional vibrations of the cantilever at frequencies up to 3 MHz, the original electronic circuit of the lateral channel of the AFM (using a low-pass filter with limited bandwidth to a few hundred kHz) was replaced by a high-speed scheme which bypasses the low-pass filter. The high-frequency signal was fed to a lock-in amplifier, digitized using a fast analogue-to-digital (A/D) card and fed into a broadband amplifier, followed by an rms-to-dc converter and read by a computer. Etched single-crystal silicon cantilevers (normal stiffness of 3.8–40 N/m) integrated tips are used.

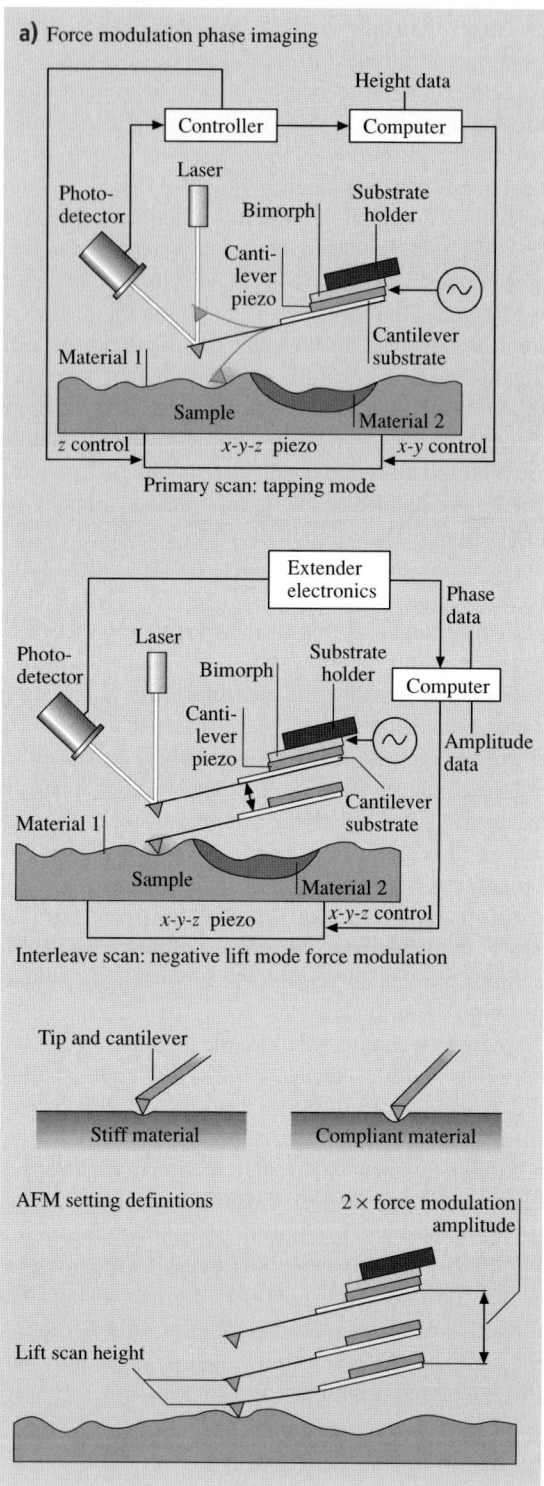

Fig. 8.10. (a) Schematic of force modulation mode used to obtain amplitude (stiffness) and definitions of force modulation amplitude and lift scan height. During primary scanning, height data is recorded in tapping mode. During interleave scanning, the entire cantilever/tip assembly is vibrated at the bimorph's resonance frequency and the z-direction feedback control for the sample x–y–z piezo is deactivated. During this scanning, height information from the primary scan is used to maintain a constant lift scan height. The computer records amplitude (which is a function of material stiffness) during the interleave scan

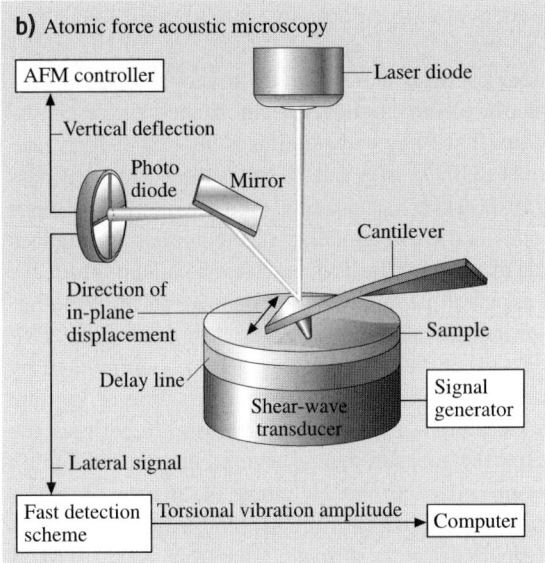

Fig. 8.10. (b) Schematic of an AFM incorporating shear wave transducer which generates in-plane lateral sample surface vibrations. Because of the forces between the tip and the surface, torsional vibrations of the cantilever are excited [33]. The shift in contact resonance frequency is a measure of contact stiffness

Viscoelastic Mapping

Another form of dynamic force microscopy, phase contrast microscopy, is used to detect the contrast in viscoelastic (viscous energy dissipation) properties of the different materials across the surface [53, 57–60, 71, 72]. In these techniques, both deflection amplitude and phase angle contrasts are measured, which are measure of the relative stiffness and viscoelastic properties, respectively. Two phase measurement techniques – tapping mode and torsional resonance (TR) mode – have been developed, which we now describe.

In the tapping-mode (TM) technique, as described earlier, the cantilever/tip assembly is sinusoidally vibrated at its resonant frequency and the sample x–y–z piezo is adjusted using feedback control in the z direction to maintain a constant set point, Fig. 8.3 [57,58]. The feedback signal to the z-direction sample piezo (to keep the set point constant) is a measure of surface roughness. The extender electronics is used to measure the phase-angle lag between the cantilever piezo drive signal and the cantilever response during sample engagement. As illustrated in Fig. 8.3, the phase-angle lag (at least partially) is a function of the viscoelastic properties of the sample material. A range of tapping amplitudes and set points can be used for measurements. Commercially etched single-crystal silicon tip (DI TESP) used for tapping mode, with a radius of 5–10 nm, a stiffness of 20–100 N/m, and a natural frequency of 350 to 400 kHz, is normally used. Scanning is normally set to a rate of 1 Hz along the fast axis.

In the torsional mode (TR mode), a tip is vibrated in the torsional mode at high frequency at the resonance frequency of the cantilever beam. Etched single-crystal silicon cantilever with integrated tip (DI FESP) with a radius of about 5–10 nm, normal stiffness of 1–5 N/m, torsional stiffness of about 30 times the normal stiffness and torsional natural frequency of 800 kHz is normally used. A major difference between the TM and the TR modes is the directionality of the applied oscillation: a normal (compressive) amplitude exerted for the TM, and a torsional amplitude for the TR mode. The TR mode is expected to provide good contrast in the tribological and mechanical properties of the near surface region as compared to the TM. Two of the reasons are as follows. (1) In the TM, the interaction is dominated by the vertical properties of the sample, so the tip spends a small fraction of its time in the near-field interaction with the sample. Furthermore, the distance between the tip and the sample changes during the measurements, which changes interaction time and forces, and affects measured data. In the TR mode, the distance remains nearly constant. (2) The lateral stiffness of a cantilever is typically about two orders of magnitude larger than the normal (flexural) stiffness. Therefore, in the TM, if the sample is relatively rigid, and much of the deformation occurs in the cantilever beam, whereas in the TR mode, much of the deformation occurs in the sample. A few comments on the special applications of the TR mode are made next. Since most of the deformation occurs in the sample, the TR mode can be used to measure stiff and hard samples. Furthermore, the properties of thin films can be measured more readily with the TR mode. For both the TM and TR modes, if the cantilever is driven to vibrate at frequencies above resonance, it would have less motion (high apparent stiffness), leading to higher sample deformation and better contrast. It should be further noted that the TM exerts compressive force, whereas the TR mode exerts torsional force, therefore normal and shear properties are measured in the TM and TR modes, respectively.

In the TR mode, the torsional vibration of the cantilever beam is achieved using a specially designed cantilever holder. It is equipped with a piezo system mounted in a cantilever holder, in which two piezos vibrate out of phase with respect to each other. A tuning process prior to scanning is used to select the torsional vibration frequency. The piezo system excites torsional vibration at the cantilever's resonance frequency. The torsional vibration amplitude of the tip (TR amplitude) is detected by the lateral segments of the split-diode photodetector, Fig. 8.11 [59]. The TR mode measures surface roughness and phase angle as follows. During the measurement, the cantilever/tip assembly is first vibrated at its resonance at some amplitude dependent upon the excitation voltage, before the tip engages the sample. Next, the tip engages the sample at some set point. A feedback system coupled to a piezo stage is used to keep a constant TR amplitude during scanning. This is done by controlling the vertical position of the sample using a piezo moving in the z direction, which changes the degree of tip interaction. The displacement of the sample z piezo gives a roughness image of the sample. A phase-angle image can be obtained by meas-

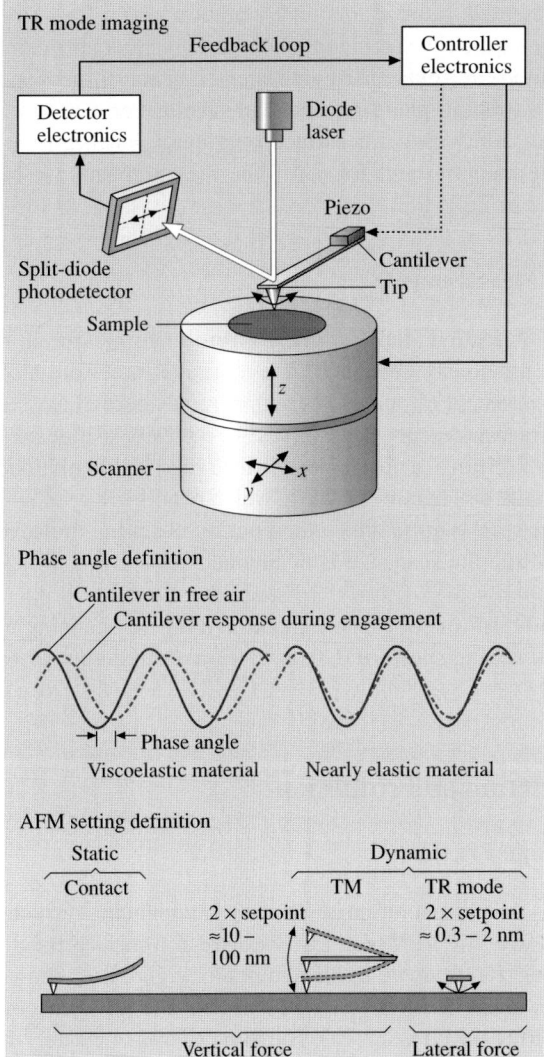

Fig. 8.11. Schematic of torsional resonance mode shown at the top. Two examples of the phase-angle response are shown in the *middle*. One is for materials exhibiting viscoelastic (*left*) and the other nearly elastic properties (*right*). Three AFM settings are compared at the *bottom*: contact, tapping mode (TM), and TR modes. The TR mode is a dynamic approach with a laterally vibrating cantilever tip that can interact with the surface more intensively than other modes. Therefore, more detailed near-surface information is available

uring the phase lag of the cantilever vibration response in the torsional mode during engagement with respect to the cantilever vibration response in free air before engagement. The control feedback of the TR mode is similar to that of tapping, except that the torsional resonance amplitude replaces flexural resonance amplitude [59].

Chen and *Bhushan* [60] have used a variation to the approach just described (referred to as mode I here). They performed measurements at constant normal cantilever deflection (constant load) (mode II) instead of using constant set point in the *Kasai* et al. [59] approach. Their approach overcomes the meniscus adhesion problem present in mode I and reveals true surface properties.

Song and *Bhushan* [73] presented a forced torsional vibration model for a tip–cantilever assembly under viscoelastic tip–sample interaction. This model provides a relationship of torsional amplitude and phase shift with lateral contact stiffness and viscosity which can be used to extract in-plane interfacial mechanical properties.

Various operating modes of AFM used for surface roughness, localized surface elasticity, and viscoelastic mapping and friction force measurements (to be discussed later) are summarized in Table 8.2.

8.2.8 Boundary Lubrication Measurements

To study nanoscale boundary lubrication studies, adhesive forces are measured in the force calibration mode, as previously described. The adhesive forces are also calculated from the horizontal intercept of curves of friction versus normal load at zero friction force. For friction measurements, the samples are typically scanned using an Si_3N_4 tip over an area of 2×2 μm at a normal load in the range 5–130 nN. The samples are generally scanned with a scan rate of 0.5 Hz, which results in a scanning speed of 2 μm/s. Velocity effects on friction are studied out by changing the scan frequency from 0.1 to 60 Hz, while the scan size is maintained at 2×2 μm, which allows the velocity to vary from 0.4 to 240 μm/s. To study the durability properties, the friction force and coefficient of friction are monitored during scanning at a normal load of 70 nN and a scanning speed of 0.8 μm/s, for a desired number of cycles [39, 40, 42].

8.3 Surface Imaging, Friction and Adhesion

8.3.1 Atomic-Scale Imaging and Friction

Surface height imaging down to atomic resolution of electrically conducing surfaces can be carried out using an STM. An AFM can also be used for surface height imaging and roughness characterization down to the nanoscale. Figure 8.12 shows a sequence of STM images at various scan sizes of solvent-deposited C_{60} film on a 200 nm thick gold coated freshly cleaved mica [74]. The film consists of clusters of C_{60} molecules with a diameter of 8 nm. The C_{60} molecules within a cluster appear to pack in a hexagonal array with a spacing of about 1 nm, however, they do not follow any long-range order. The measured cage diameter of the C_{60} molecule is about 0.7 nm, which is very close to the projected diameter of 0.71 nm.

In an AFM measurement during surface imaging, the tip comes into intimate contact with the sample surface and leads to surface deformation with finite tip–sample contact area (typically a few atoms). The finite size of the contact area prevents the imaging of individual point defects, and only the periodicity of the atomic lattice can be imaged. Figure 8.13a shows the topography image of the freshly cleaved surface of highly oriented pyrolytic graphite (HOPG) [30]. The periodicity of the graphite is clearly observed.

Table 8.2. Summary of various operating modes of AFM for surface roughness, stiffness, phase angle, and friction

Operating mode	Direction of cantilever vibration	Vibration frequency of cantilever (kHz)	Vibration amplitude (nm)	Feedback control	Data obtained
Contact	n/a			Constant normal load	Surface height, friction
Tapping	Vertical	350–400	10–100	Set point (constant tip amplitude)	Surface height, phase angle (normal viscoelasticity)
Force modulation	Vertical	10–20 (bimorph)	10–100	Constant normal load	Surface height, amplitude (normal stiffness)
Lateral	Lateral (AAFM)	100–3000 (sample)	≈ 5 (sample)	Constant normal load	Shift in contact resonance (normal stiffness, friction)
TR mode I	Torsional	≈ 800	0.3–2	Set point (constant tip amplitude)	Surface height, phase angle (lateral viscoelasticity)
TR mode II	Torsional	≈ 800	0.3–2	Constant normal load	Surface height, amplitude and phase angle (lateral stiffness and lateral viscoelasticity)
TR mode III	Torsional	Higher than 800 in contact	0.3–2	Constant normal load	Shift in contact resonance (friction)

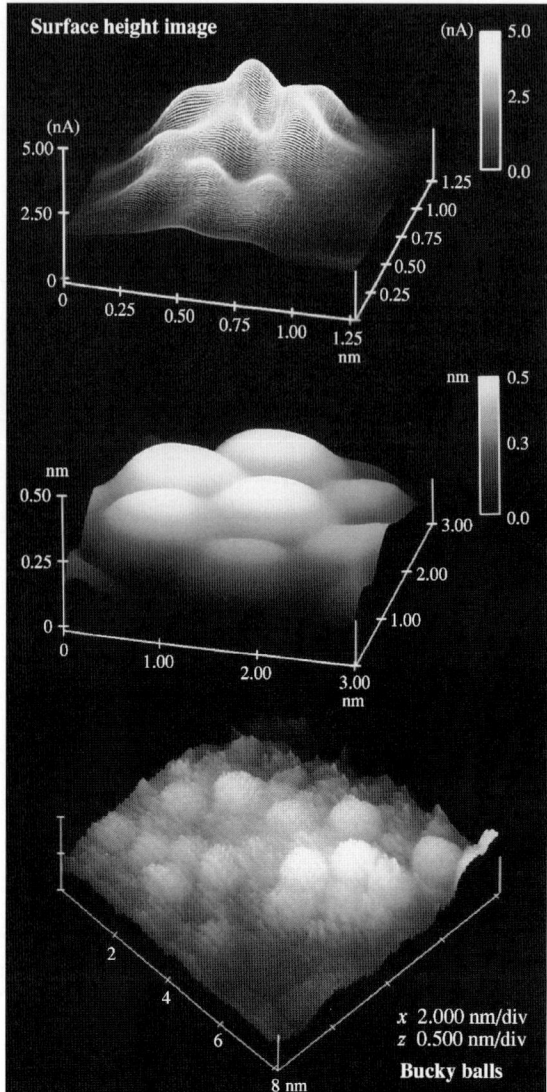

Fig. 8.12. STM images of solvent-deposited C_{60} film on a gold-coated freshly cleaved mica at various scan sizes

To study friction mechanisms on an atomic scale, a freshly cleaved HOPG has been studied by *Mate* et al. [22] and *Ruan and Bhushan* [25]. Figure 8.14a shows the atomic-scale friction force map (raw data) and Fig. 8.13a shows the atomic-scale topography and friction force maps (after 2D spectrum filtering with high-frequency noise truncation) [25]. Figure 8.14a also shows a line plot of friction force profiles along some crystallographic direction. The actual shape of the friction profile depends upon the spatial location of axis of tip motion. Note that a portion of the atomic-scale lateral force is conservative. *Mate* et al. [22] and *Ruan and*

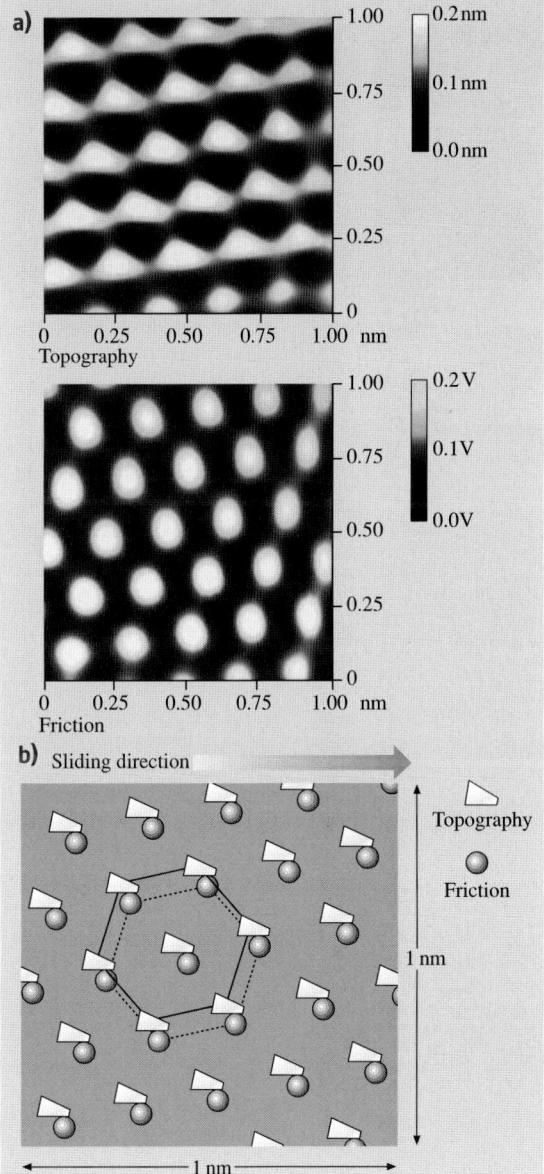

Fig. 8.13. (a) Greyscale plots of surface topography and friction force maps (2D spectrum filtered), measured simultaneously, of a 1 nm × 1 nm area of freshly cleaved HOPG, showing the atomic-scale variation of topography and friction, and (b) schematic of superimposed topography and friction maps from (a); the *symbols* correspond to maxima. Note the spatial shift between the two plots [25]

Bhushan [25] reported that the average friction force increased linearly with normal load and was reversible with load. Friction profiles were similar during sliding of the tip in either direction.

During scanning, the tip moves discontinuously over the sample surface and jumps with discrete steps from one potential minimum (well) to the next. This leads

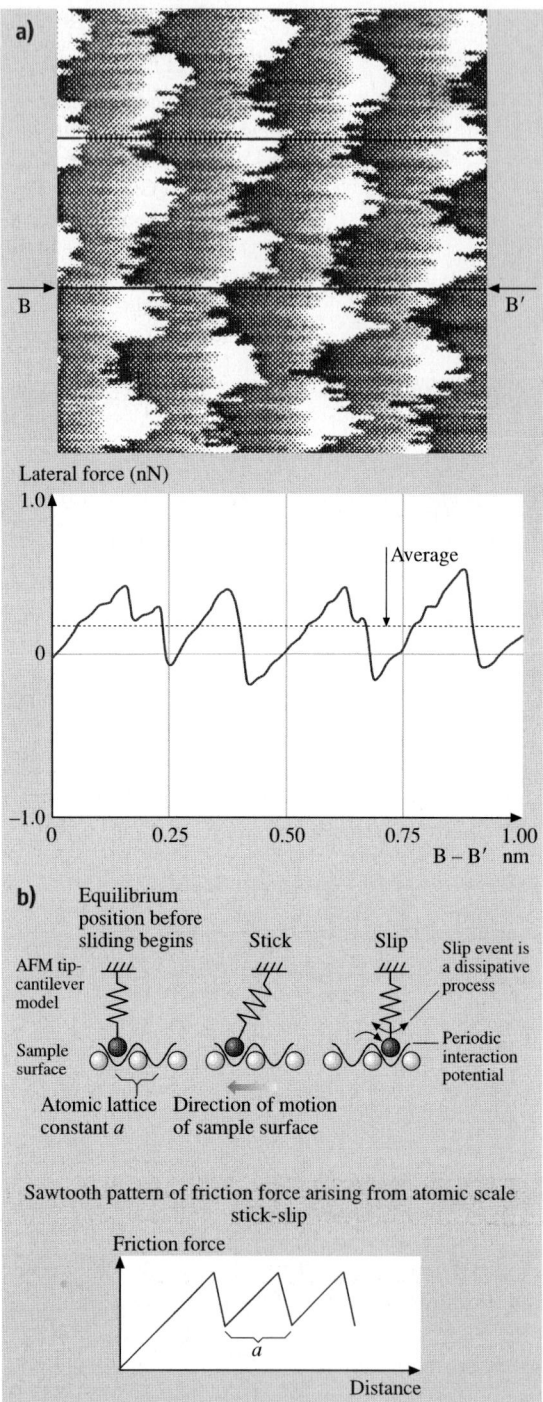

Fig. 8.14. (a) Greyscale plot of friction force map (raw data) of a 1×1 nm^2 area of freshly cleaved HOPG, showing atomic-scale variation of friction force. High points are shows by lighter color. Also shown is a line plot of friction force profile along the line indicated by *arrows*. The normal load was 25 nN and the cantilever normal stiffness was 0.4 N/m [25]. (b) Schematic of a model for a tip atom sliding on an atomically flat periodic surface. The schematic shows the tip jumping from one potential minimum to another, resulting in stick–slip behavior

to a sawtooth-like pattern for the lateral motion (force) with a periodicity of the lattice constant. This motion is called stick–slip movement of the tip [5, 10, 22, 25]. The observed friction force includes two components: conservative and periodic, and nonconservative and constant. If the relative motion of the sample and tip were simply that of two rigid collections of atoms, the effective force would be a conservative force oscillating about zero. Slow reversible elastic deformation would also contribute to the conservative force. The origin of the nonconservative direction-dependent force component would be phonon generation, viscous dissipation, or plastic deformation.

As just mentioned, stick–slip on the atomic scale is the result of the energy barrier that must be overcome to jump over the atomic corrugations on the sample surface. This corresponds to the energy required for the jump of the tip from a stable equilibrium position on the surface into a neighboring position. The perfect atomic regularity of the surface guarantees the periodicity of the lateral force signal, independent of actual atomic structure of tip apex. Few atoms (based on the magnitude of the friction force, less than 10) on a tip sliding over an array of atoms on the sample are expected to go through the stick–slip. For simplicity, Fig. 8.14b shows a simplified model for one atom on a tip with a one-dimensional spring–mass system. As the sample surface slides against the AFM tip, the tip remains *stuck* initially until it can overcome the energy (potential) barrier, which is illustrated by a sinusoidal interaction potential as experienced by the tip. After some motion, there is enough energy stored in the spring which leads to *slip* into the neighboring stable equilibrium position. During the slip and before attaining stable equilibrium, stored energy is converted into vibrational energy of the surface atoms in the range of 10^{13} Hz (phonon generation) and decays within the range of 10^{-11} s into heat. (A wave of atoms vibrating in concert are termed a phonon.) The stick–slip phenomenon, resulting from irreversible atomic jumps, can be theoretically modeled with classical mechanical models [75, 76]. The Tomanek–Zhong–Thomas model [76] is the starting point for determining the friction force during atomic-scale stick–slip. The AFM model describes the total potential as the sum of the potential acting on the tip due to interaction with the sample and the elastic energy stored in the cantilever. Thermally activated stick–slip behavior can explain the velocity effects on friction, to be presented later.

Finally, based on Fig. 8.13a, the atomic-scale friction force of HOPG exhibited the same periodicity as that of the corresponding topography, but the peaks in friction and those in topography are displaced relative to each other (Fig. 8.13b). A Fourier expansion of the interatomic potential was used by *Ruan* and *Bhushan* [25] to calculate the conservative interatomic forces between atoms of the FFM tip and those of the graphite surface. Maxima in the interatomic forces in the normal and lateral directions do not occur at the same location, which explains the observed shift between the peaks in the lateral force and those in the corresponding topography.

8.3.2 Microscale Friction

Local variations in the microscale friction of cleaved graphite are observed, Fig. 8.15. These arise from structural changes that occur during the cleaving process [26]. The cleaved HOPG surface is largely atomically smooth but exhibits line-shaped regions in which the coefficient of friction is more than an order of magnitude larger. Transmission electron microscopy indicates that the line-shaped regions consist of graphite planes of different orientation, as well as of amorphous carbon. Differences in friction have also been observed for multiphase ceramic materials [45]. Figure 8.16 shows the surface roughness and friction force maps of $Al_2O_3 - TiC$ (70–30 wt%). TiC grains have a Knoop hardness of about $2800\,kg/mm^2$ and Al_2O_3 has $2100\,kg/mm^2$, therefore, TiC grains do not polish as much and result in a slightly higher elevation (about 2–3 nm higher than that of Al_2O_3 grains). TiC grains exhibit higher friction force than Al_2O_3 grains. The coefficients of friction of TiC and Al_2O_3 grains are 0.034 and 0.026, respectively, and the coefficient of friction of $Al_2O_3 - TiC$ composite is 0.03. Local variation in friction force also arises from the scratches present on the $Al_2O_3 - TiC$ surface. *Meyer* et al. [77] also used FFM to measure structural variations of organic mono- and multilayer films. All of these measurements suggest that the FFM can be used for structural mapping of the surfaces. FFM measurements can also be used to map

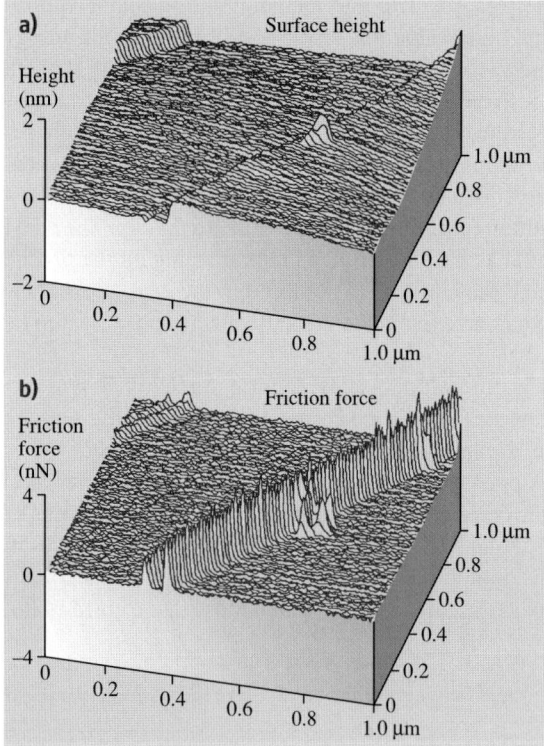

Fig. 8.15. (a) Surface roughness and (b) friction force maps at a normal load of 42 nN of freshly cleaved HOPG surface against an Si_3N_4 FFM tip. Friction in the line-shaped region is over an order of magnitude larger than in the smooth areas [25]

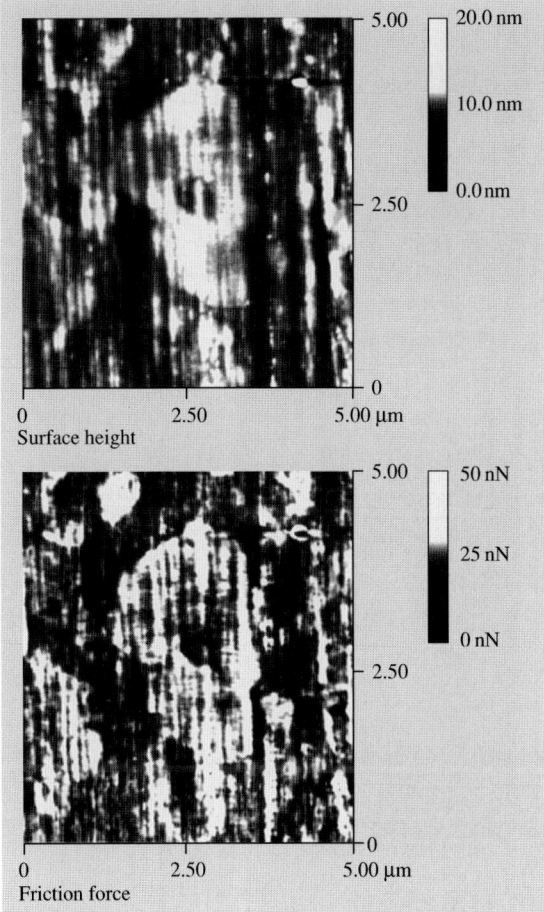

Fig. 8.16. Greyscale surface roughness ($\sigma = 0.80$ nm) and friction force maps (mean = 7.0 nN, $\sigma = 0.90$ nN) for Al_2O_3–TiC (70–30 wt%) at a normal load of 138 nN [45]

chemical variations, as indicated by the use of the FFM with a modified probe tip to map the spatial arrangement of chemical functional groups in mixed organic monolayer films [78]. Here, sample regions that had stronger interactions with the functionalized probe tip exhibited larger friction.

Local variations in the microscale friction of nominally rough, homogeneous-material surfaces can be significant, and are seen to depend on the local surface slope rather than the surface height distribution, Fig. 8.17. This dependence was first reported by *Bhushan* and *Ruan* [23], *Bhushan* et al. [27], and *Bhushan* [65] and later discussed in more detail by *Koinkar* and *Bhushan* [79] and *Sundararajan* and *Bhushan* [80]. In order to show elegantly any correlation between local values of friction and surface roughness, surface roughness and friction force maps of a gold-coated ruler with somewhat rectangular grids and a silicon grid with square pits were obtained, Fig. 8.18 [80]. Figures 8.17 and 8.18 show the surface roughness map, the slopes of the roughness map taken along the sliding direction (surface slope map)

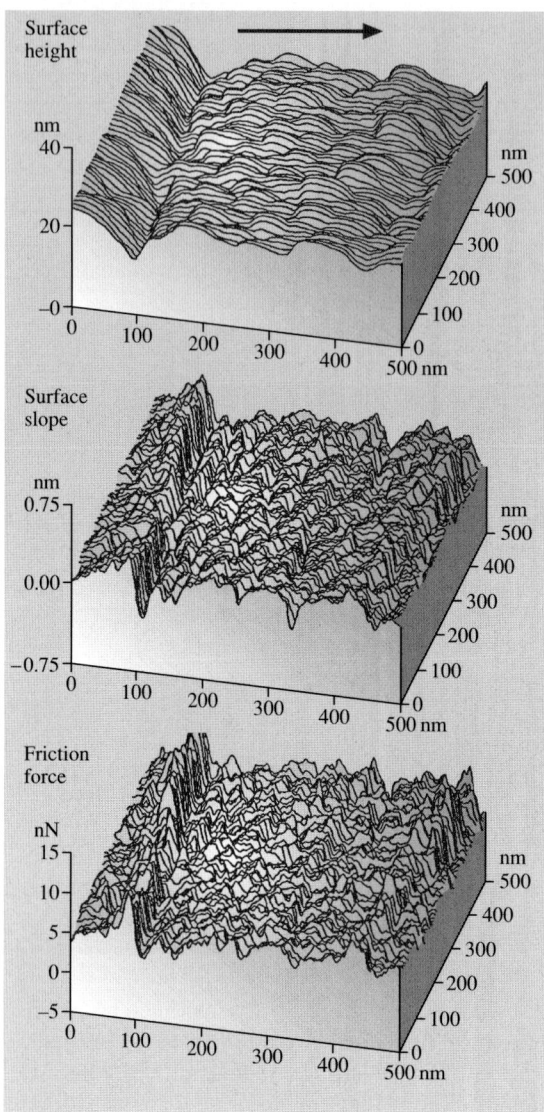

Fig. 8.17. Surface roughness map ($\sigma = 4.4$ nm), surface slope map taken in the sample sliding direction (the horizontal axis; mean = 0.023, $\sigma = 0.197$), and friction force map (mean = 6.2 nN, $\sigma = 2.1$ nN) for a lubricated thin-film magnetic rigid disk for a normal load of 160 nN [27]

and the friction force map for various samples. There is a strong correlation between the surface slopes and friction forces. For example, in Fig. 8.18, the friction force is high locally at the edge of the grids and pits with a positive slope and is low at the edges with a negative slope.

We now examine the mechanism of microscale friction, which may explain the resemblance between the slope of surface roughness maps and the corresponding friction force maps [4,5,11,23,25–27,79,80]. There are three dominant mechanisms of friction; adhesive, ratchet, and plowing [10, 17]. At first order, we may assume

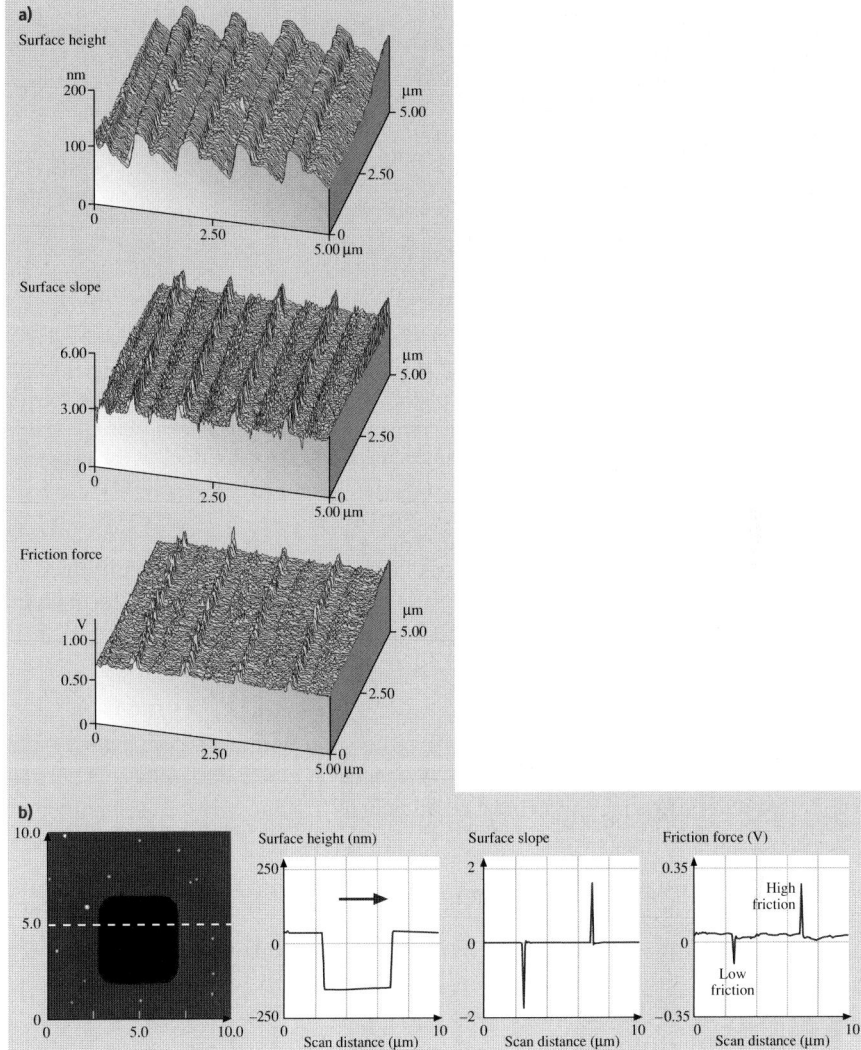

Fig. 8.18. Surface roughness map, surface slope map taken in the sample sliding direction (the horizontal axis), and friction force map for (**a**) a gold-coated ruler (with somewhat rectangular grids with a pitch of 1 μm and a ruling step height of about 70 nm) at a normal load of 25 nN and (**b**) a silicon grid (with 5-μm square pits of depth 180 nm and a pitch of 10 μm) [80]

these to be additive. The adhesive mechanism cannot explain the local variation in friction. Next we consider the ratchet mechanism. We consider a small tip sliding over an asperity making an angle θ with the horizontal plane, Fig. 8.19. The normal force W (normal to the general surface) applied by the tip to the sample surface is constant. The friction force F on the sample would be a constant for a smooth

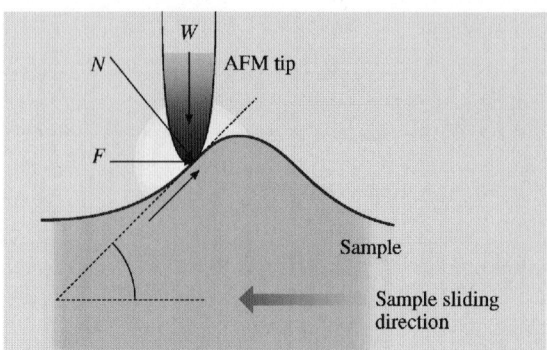

Fig. 8.19. Schematic illustration showing the effect of an asperity (making an angle θ with the horizontal plane) on the surface in contact with the tip on local friction in the presence of adhesive friction mechanism. W and F are the normal and friction forces, respectively, and S and N are the force components along and perpendicular to the local surface of the sample at the contact point, respectively

surface if the friction mechanism does not change. For a rough surface, shown in Fig. 8.19, if the adhesive mechanism does not change during sliding, the local value of the coefficient of friction remains constant,

$$\mu_0 = S/N, \tag{8.1}$$

where S is the local friction force and N is the local normal force. However, the friction and normal forces are measured with respect to global horizontal and normal axes, respectively. The measured local coefficient of friction μ_1 in the ascending part is

$$\mu_1 = F/W = (\mu_0 + \tan\theta)/(1 - \mu_0 \tan\theta) \approx \mu_0 + \tan\theta, \quad \text{for small } \mu_0 \tan\theta \tag{8.2}$$

indicating that, in the ascending part of the asperity, one may simply add the friction force and the asperity slope to one another. Similarly, on the right-hand side (descending part) of the asperity,

$$\mu_2 = (\mu_0 - \tan\theta)/(1 + \mu_0 \tan\theta) \approx \mu_0 - \tan\theta, \quad \text{for small } \mu_0 \tan\theta. \tag{8.3}$$

For a symmetrical asperity, the average coefficient of friction experienced by the FFM tip traveling across the whole asperity is

$$\mu_{ave} = (\mu_1 + \mu_2)/2 = \mu_0 \left(1 + \tan^2\theta\right)/\left(1 - \mu_0^2 \tan^2\theta\right)$$
$$\approx \mu_0 \left(1 + \tan^2\theta\right), \quad \text{for small } \mu_0 \tan\theta. \tag{8.4}$$

Finally we consider the plowing component of friction with the tip sliding in either direction, which is [10, 17]

$$\mu_p \sim \tan\theta. \tag{8.5}$$

Because in FFM measurements we notice little damage of the sample surface, the contribution by plowing is expected to be small and the ratchet mechanism is believed to be the dominant mechanism for the local variations in the friction force map. With the tip sliding over the leading (ascending) edge of an asperity, the surface slope is positive; it is negative during sliding over the trailing (descending) edge of an asperity. Thus, measured friction is high at the leading edge of asperities and low at the trailing edge. In addition to the slope effect, the collision of the tip when encountering an asperity with a positive slope produces additional torsion of the cantilever beam leading to higher measured friction force. When encountering an asperity with the same negative slope, however, there is no collision effect and hence no effect on torsion. This effect also contributes to the difference in friction forces when the tip scans up and down on the same topography feature. The ratchet mechanism and the collision effects thus semiquantitatively explain the correlation between the slopes of the roughness maps and friction force maps observed in Figs. 8.17 and 8.18. We note that, in the ratchet mechanism, the FFM tip is assumed to be small compared to the size of asperities. This is valid since the typical radius of curvature of the tips is about 10–50 nm. The radii of curvature of the asperities of the samples measured here (the asperities that produce most of the friction variation) are found to be typically about 100–200 nm, which is larger than that of the FFM tip [81]. It is important to note that the measured local values of friction and normal forces are measured with respect to global (and not local) horizontal and vertical axes, which are believed to be relevant in applications.

8.3.3 Directionality Effect on Microfriction

During friction measurements, the friction force data from both the forward (trace) and backward (retrace) scans are useful in understanding the origins of the observed friction forces. Magnitudes of material-induced effects are independent of the scanning direction whereas topography-induced effects are different between forward and backward scanning directions. Since the sign of the friction force changes as the scanning direction is reversed (because of the reversal of torque applied to the end of the tip), addition of the friction force data of the forward and backward scan eliminates the material-induced effects while topography-induced effects still remain. Subtraction of the data between forward and backward scans does not eliminate either effect, Fig. 8.20 [80].

Owing to the reversal of the sign of the retrace (R) friction force with respect to the trace (T) data; the friction force variations due to topography are in the same direction (peaks in trace correspond to peaks in retrace). However, the magnitudes of the peaks in trace and retrace at a given location are different. An increase in the friction force experienced by the tip when scanning up a sharp change in topography is more than the decrease in the friction force experienced when scanning down the same topography change, partly because of the collision effects discussed earlier. Asperities on engineering surfaces are asymmetrical, which also affects the magnitude of friction force in the two directions. Asymmetry in tip shape may also have an

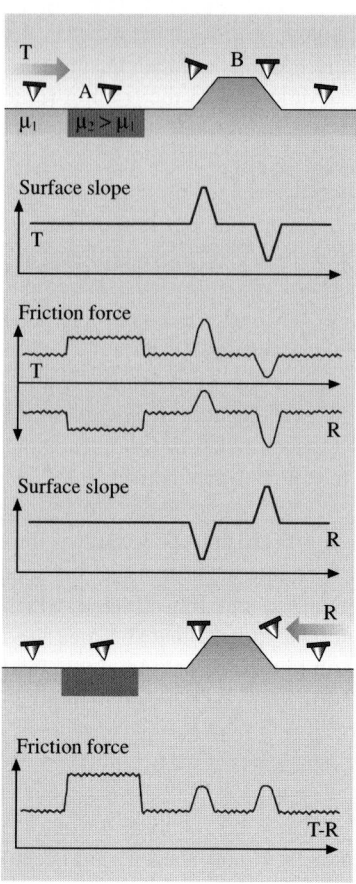

Fig. 8.20. Schematic of friction forces expected when a tip traverses a sample that is composed of different materials and sharp changes in topography. A schematic of surface slope is also shown

effect on the directionality effect of friction. We will note later that the magnitude of the surface slopes are virtually identical, therefore, the tip shape asymmetry should not have much effect.

Figure 8.21 shows surface height and friction force data for a gold ruler and a silicon grid in the trace and retrace directions. Subtraction of the two friction data yields a residual peak because of the differences in the magnitudes of the friction forces in the two directions. This effect is observed at all locations where there is significant change in topography.

In order to facilitate comparison of the effect of directionality on friction, it is important to take into account the change of sign of the surface slope and friction force in the trace and retrace directions. Figure 8.22 shows surface height, surface slope, and friction force data for the two samples in the trace and retrace directions. The correlations between surface slope and friction forces are clear. The third column in the figures shows the retrace slope and friction data with an inverted sign (−retrace). Now we can compare the trace data with the −retrace data. It is

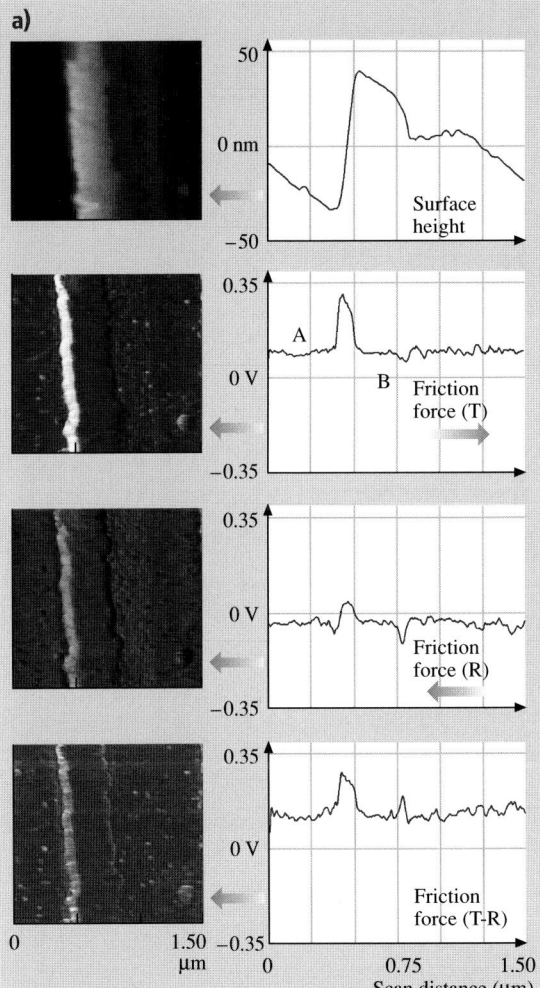

Fig. 8.21. (a) Greyscale images and two-dimensional profiles of surface height and friction forces across a single ruling of the gold-coated ruling

clear that the friction experienced by the tip is dependent upon the scanning direction because of the surface topography. In addition to the effect of topographical changes discussed earlier, during surface-finishing processes material can be transferred preferentially onto one side of the asperities, which also causes asymmetry and direction dependence. Reduction in local variations and in the directionality of friction properties requires careful optimization of surface-roughness distributions and of surface-finishing processes.

The directionality as a result of the surface asperities effect will also manifest itself in macroscopic friction data, i.e., the coefficient of friction may be different in one sliding direction than that in the other direction. Asymmetrical shape of asperities accentuates this effect. Frictional directionality can also exist in materials

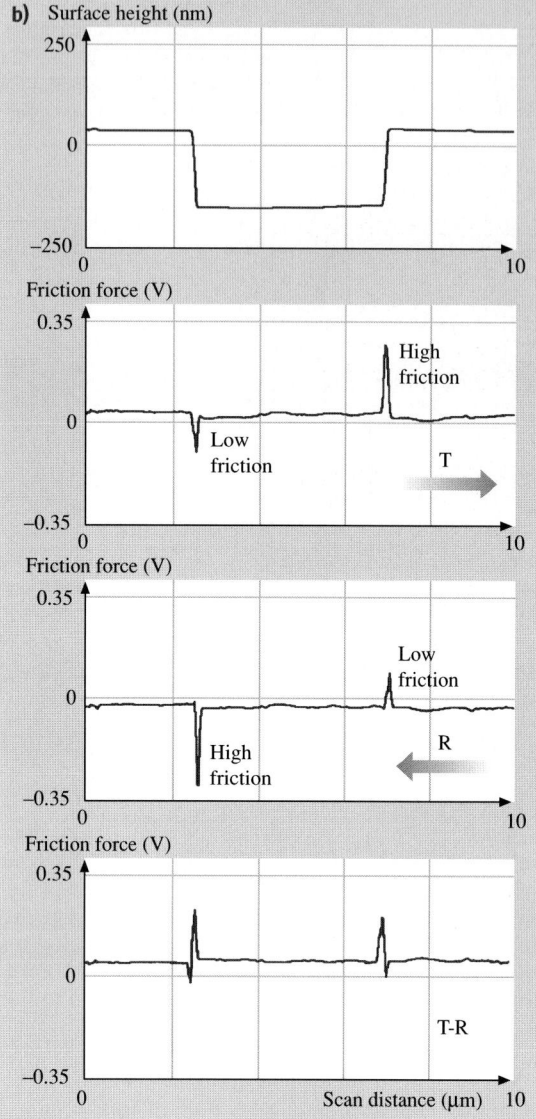

Fig. 8.21. (b) two dimensional profiles of surface height and friction forces across a silicon grid pit. Friction force data in trace and retrace directions, and subtracted force data are presented

with particles having a preferred orientation. The directionality effect in friction on a macroscale is observed in some magnetic tapes. In a macroscale test, a 12.7-mm-wide polymeric magnetic tape was wrapped over an aluminum drum and slid in a reciprocating motion with a normal load of 0.5 N and a sliding speed of about 60 mm/s [3]. The coefficient of friction as a function of sliding distance in either direction is shown in Fig. 8.23. We note that the coefficient of friction on a macroscale for this tape is different in different directions. Directionality in friction is some-

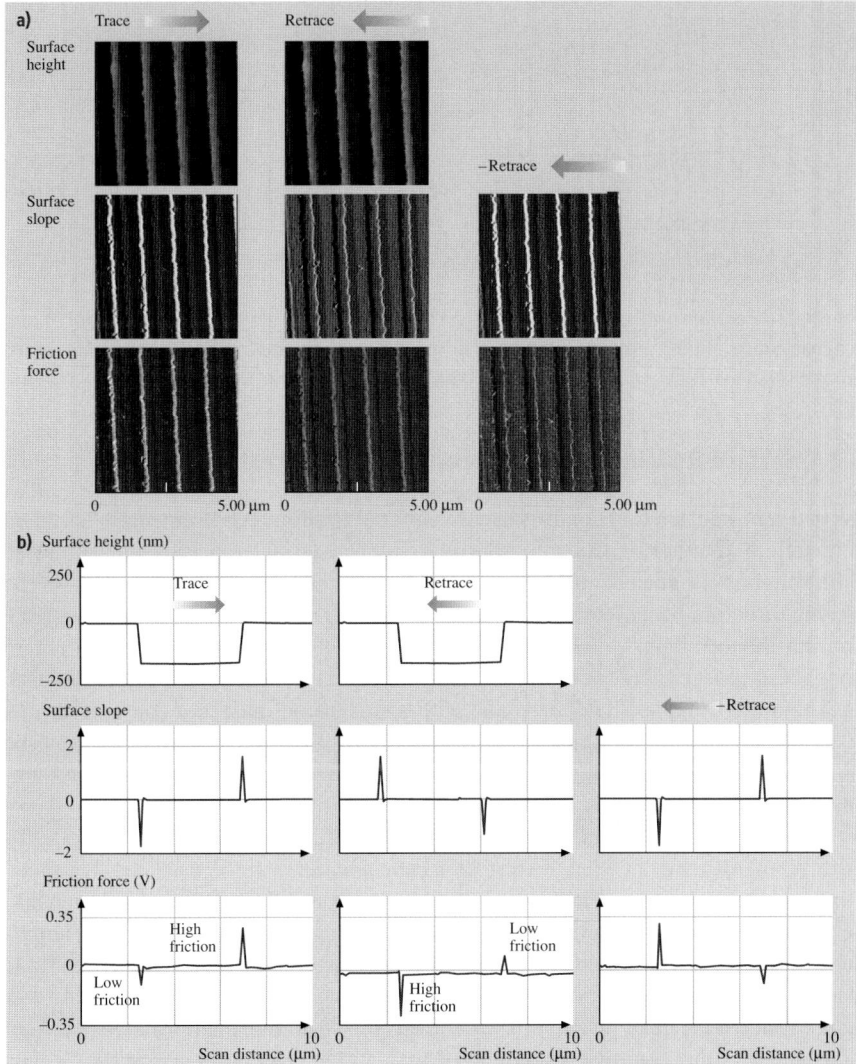

Fig. 8.22. (a) Greyscale images of surface heights, surface slopes and friction forces for scans across a gold-coated ruling, and (b) two-dimensional profiles of surface heights, surface slopes and friction forces for scans across the silicon grid pit. *Arrows* indicate the tip sliding direction [80]

times observed on the macroscale; on the microscale this is the norm [4, 15]. On the macroscale, the effect of surface asperities is normally averaged out over a large number of contacting asperities.

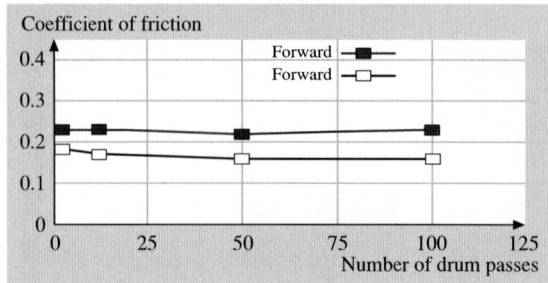

Fig. 8.23. Coefficient of macroscale friction as a function of drum passes for a polymeric magnetic tape sliding over an aluminum drum in a reciprocating mode in both directions. Normal load = 0.5 N over 12.7-mm-wide tape, sliding speed = 60 mm/s [65]

8.3.4 Surface-Roughness-Independent Microscale Friction

As just reported, friction contrast in conventional friction measurements is based on interactions that are dependent upon interfacial material properties superimposed with roughness-induced lateral forces, and the cantilever twist is dependent on the sliding direction because of the local surface slope. Hence it is difficult to separate friction-induced from roughness-induced cantilever twist in the image. To obtain roughness-independent friction, lateral or torsional modulation techniques are used in which the tip is oscillated in-plane with a small amplitude at a constant normal load, and changes in the shape and magnitude of the cantilever resonance are used as a measure of the friction force [31–36, 82]. These techniques also allow measurements over a very small region (a few nm to a few µm).

Scherer et al. [32] and *Reinstaedtler* et al. [33, 34] used the lateral mode for friction measurements (Fig. 8.10b). *Bhushan* and *Kasai* [36] used the TR mode for these measurements (Fig. 8.11). Before engagement, the cantilever is driven into torsional motion of the cantilever/tip assembly with a given normal vibration amplitude (the vibration amplitude in free air). After engagement, the vibration amplitude decreases due to the interaction between the tip and the sample, the vibration frequency increases and a phase shift occurs. During scanning, the normal load is kept constant, and the vibration amplitude of the cantilever is measured at the contact frequency.

As mentioned earlier, the shift in the contact resonance frequency in both the lateral and the TR modes is a measure of the contact stiffness, as shown schematically in Fig. 8.24. At excitation voltage above a certain value, as a result of micro-slip at the interface, a flattening of the resonance frequency spectra occurs (Fig. 8.22). At the lowest excitation voltage, the AFM tip sticks to the sample surface and follows the motion like an elastic contact with viscous damping, and the resonance curve is Lorentzian with a well-defined maximum. The excitation voltage should be high enough to initiate micro-slip. The maximum torsional amplitude at a given resonance frequency is a function of friction force and sample stiffness, so the technique

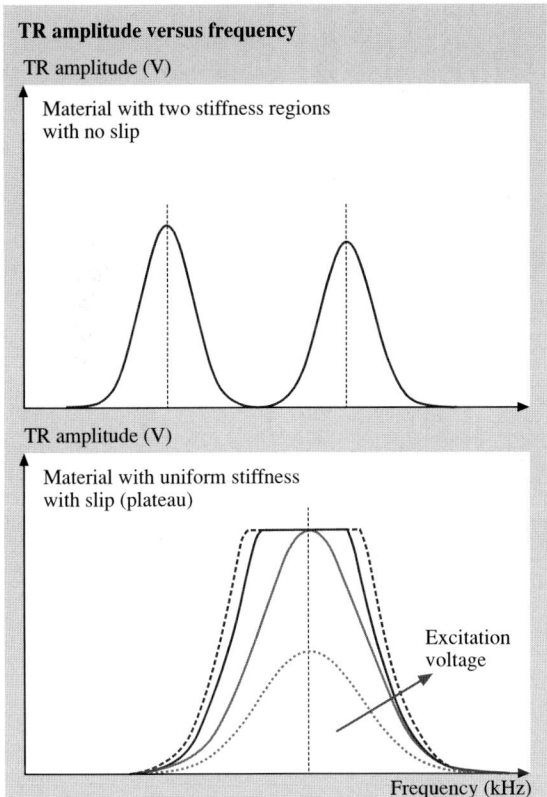

Fig. 8.24. Schematic showing frequency profiles of the TR amplitude for materials with two phases and a single phase. The maximum TR amplitude at the contact resonance frequency of the resonance curve with a flattened top, resulting from slip, can be used for friction force measurement

is not valid for inhomogeneous samples. If the torsional stiffness of the cantilever is very high compared to the sample stiffness, the technique should work.

Reinstaedtler et al. [33] performed lateral-mode experiments on bare Si and Si lubricated with 5-nm-thick chemically bonded perfluoropolyether (Z-DOL) lubricant film. Figure 8.25a shows the amplitude of the cantilever torsional vibration as a function of frequency on a bare silicon sample. The frequency sweep was adjusted such that a contact resonance frequency was covered. The different curves correspond to different excitation voltages applied to the shear-wave transducer. At low amplitudes, the shape of the resonance curve is Lorentzian. Above a critical excitation amplitude of the transducer (excitation voltage = 4 V corresponding to ≈ 0.2 nm lateral surface amplitude, as measured by interferometry), the resonance curve flattens out and the frequency span of the flattened part increases further with the excitation amplitude. Here, the static force applied was 47 nN and the adhesion force was 15 nN. The resonance behavior of the tip–cantilever system in contact with the lubricated silicon sample (Fig. 8.25b) was similar to that of the bare silicon sample. By increasing the static load, the critical amplitude for the appearance of the flattening increases. The deviations from the Lorentzian resonance curve be-

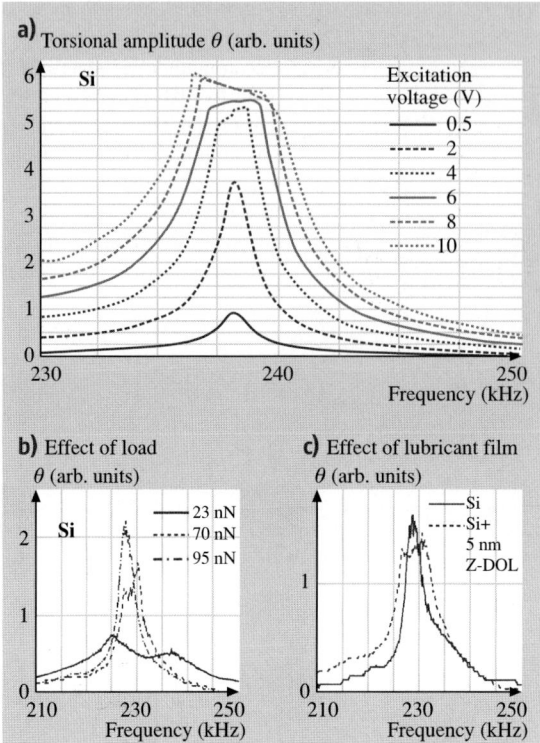

Fig. 8.25. Torsional vibration amplitude of the cantilever as a function of excitation frequency. (**a**) Measurement on bare silicon. The different curves correspond to increasing excitation voltages applied to the transducer and, hence, increasing surface amplitudes. (**b**) Measurement on silicon lubricated with a 5-nm-thick Z-DOL layer. Curves for three different static loads are shown. The transducer was excited with an amplitude of 5 V. (**c**) Measurement with a static load of 70 nN and an excitation amplitude of 7 V. The two curves correspond to bare silicon and lubricated silicon, respectively [33]

came visible at static loads lower than 95 nN. As shown in Fig. 8.25c, the resonance curve obtained at the same normal load of 70 nN and at the same excitation voltage (7 V) is more flattened on the lubricated sample than on the bare silicon, which led us to conclude that the critical amplitude is lower on the lubricated sample than on the bare sample. These experiments clearly demonstrate that torsional vibration of an AFM cantilever at ultrasonic frequencies leads to stick–slip phenomena and sliding friction. Above a critical vibration amplitude, sliding friction sets in.

Bhushan and *Kasai* [36] performed friction measurements on a silicon ruler and demonstrated that friction data in the TR mode is essentially independent of surface roughness and sliding direction. Figure 8.26a shows surface height and friction force maps on a silicon ruler obtained using the TR mode and contact-mode techniques. A comparison is made between the TR mode and contact-mode friction force maps. For easy comparison, the line scan profiles near the central area are shown on top of the greyscale maps. The vertical scales of the friction force profiles in the two graphs are selected to cover the same range of friction force so that direct comparison can be made, i.e., 0.25 V in full scale for the TR mode corresponds to 0.5 V for the contact mode in these measurements. As expected, for the trace scan, small downward peaks in the TR mode map and large upward and downward peaks in the contact mode map are observed. The positions of these peaks coincide with those of the surface slope;

8 Nanotribology, Nanomechanics and Materials Characterization 349

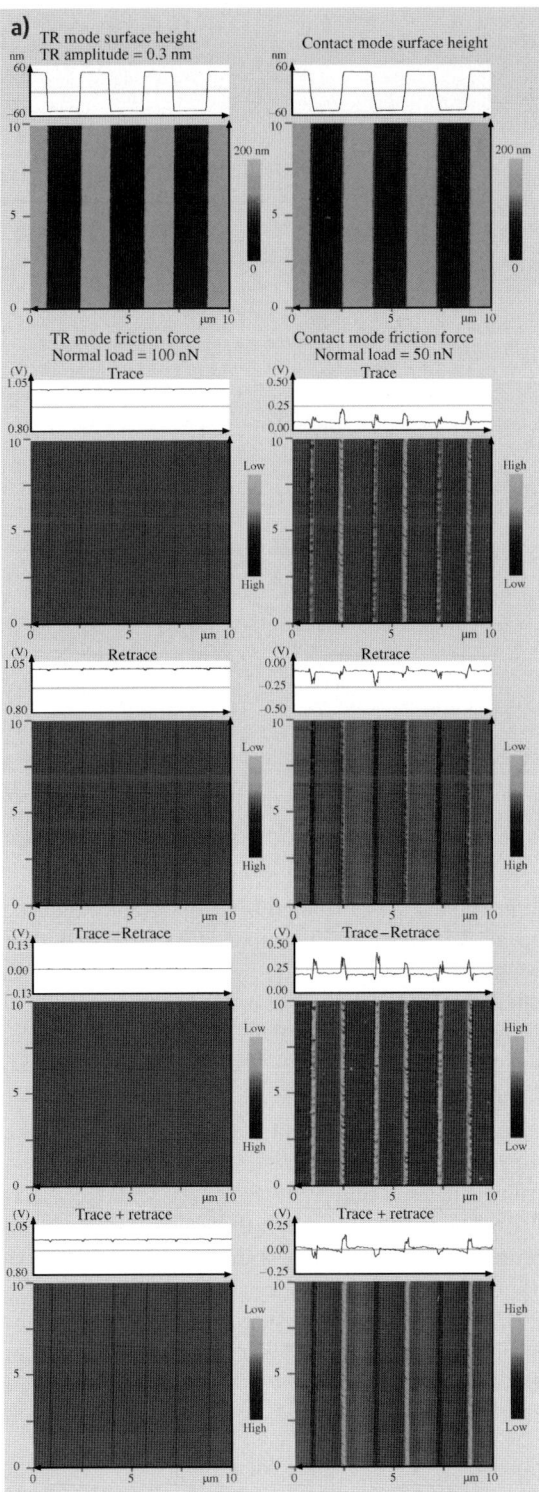

Fig. 8.26. (a) A comparison between the TR-mode friction and contact-mode friction maps together with line scans, on the silicon ruler. The TR-mode surface height and contact-mode surface height images are also shown

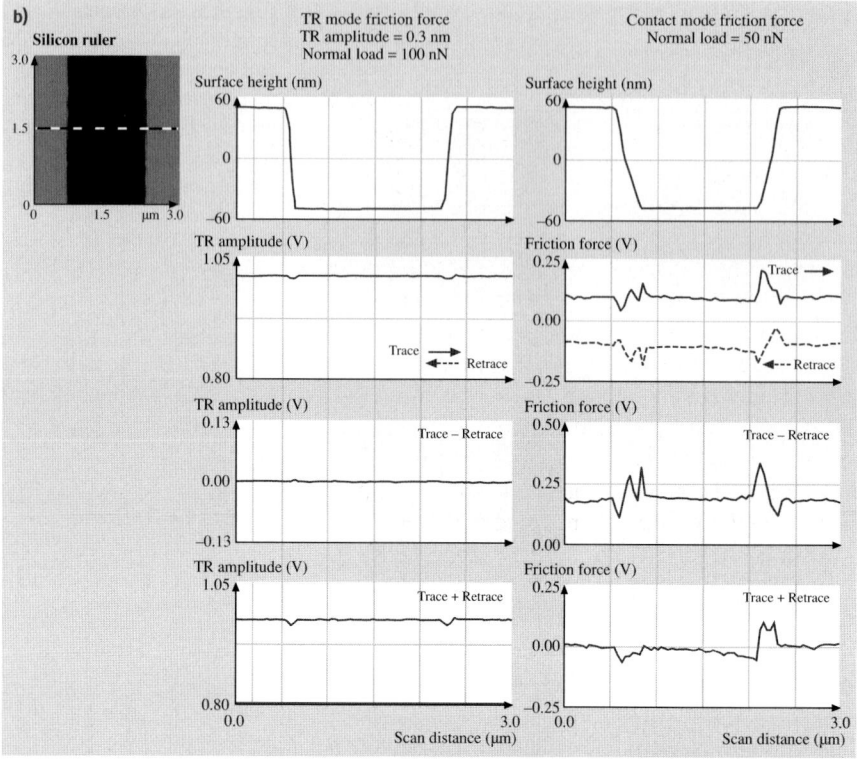

Fig. 8.26. (**b**) A comparison of the line scans of the TR-mode friction and contact-mode friction on a selected pitch of the silicon ruler [36]

therefore, the peaks in the friction signals are attributed to a topography-induced effect. For the retrace scan, the peak pattern for the TR mode stays similar, but for the contact mode, the pattern becomes reversed.

The subtraction image for the TR mode shows almost flat contrast, since the trace and retrace friction data profiles are almost identical. For the contact mode, the subtraction image shows that the topography-induced contribution still exists. As stated earlier, the addition image of the TR mode and the addition image of the contact mode enhance the topography-induced effect, which is observed in the figure.

A closer look at the silicon ruler images at one pitch was taken, and the associated images are shown in Fig. 8.26b. The surface height profiles in the TR mode and contact mode are somewhat different. The TR mode shows sharper edges than those in the contact mode. The ratios of the change in amplitude at the steps to the change in the mean amplitude in the TR mode and in the contact mode are a measure of topography effects. The ratio in the contact mode ($\approx 85\%$) is about seven times larger than that of the TR mode ($\approx 12\%$).

8.3.5 Velocity Dependence on Micro/Nanoscale Friction

AFM/FFM experiments are generally conducted at relative velocities as high as about 200 μm/s. To simulate applications, it is of interest to conduct friction experiments at higher velocities. Furthermore, high-velocity experiments (up to 1 m/s) would be useful to study the dependence of friction and wear on velocity. One approach has been to mount samples on a shear-wave transducer (ultrasonic transducer) and then drive it at very high frequencies (in the MHz range) as reported earlier, see Fig. 8.10 [31–35, 82, 83]. The coefficient of friction is estimated based on the contact resonance frequency and requires the solution of the characteristic equations for the tip vibrating in contact with the sample surface. The approach is complex and depends on various assumptions.

An alternative approach is to utilize piezo stages with a large amplitude (≈ 100 μm) and a relatively low resonance frequency (a few kHz) and measure the friction force directly using the FFM signal without any analysis, with the assumptions used in the previous approaches using shear-wave transducers. The commercial AFM set-up modified with this approach yields sliding velocities up to 10 mm/s [37]. In the modified set-up, the single-axis piezo stage is oriented such that the scanning axis is perpendicular to the long axis of the AFM cantilever (which corresponds to the 90° scan-angle mode of the commercial AFM; see Fig. 8.27). The displacement is monitored using an integrated capacitive feedback sensor, located diametrically opposite to the piezo crystal, as shown in Fig. 8.27. The capacitive change, corresponding to the stage displacement, gives a measure of the amount of displacement and can be used as feedback to the piezo controller for better guiding and tracking accuracy during scanning. The closed-loop position control of the piezoelectric-driven stages using capacitive feedback sensors provides linearity of motion better than 0.01% with nanometer resolution and a stable drift-free motion [37].

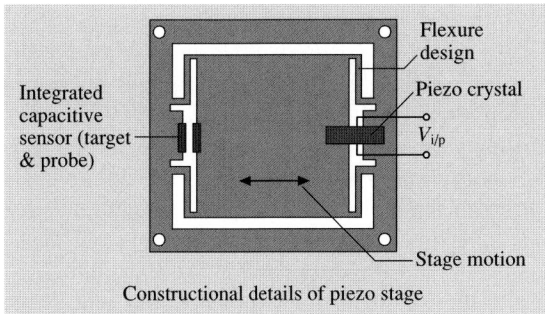

Fig. 8.27. Schematic of cross-sectional view showing construction details of the piezo stage. The integrated capacitive sensors are used as feedback sensors to drive the piezo. The piezo stage is mounted on the standard motorized AFM base and operated using independent amplifier and controller units driven by a frequency generator (not shown in the schematic) [37]

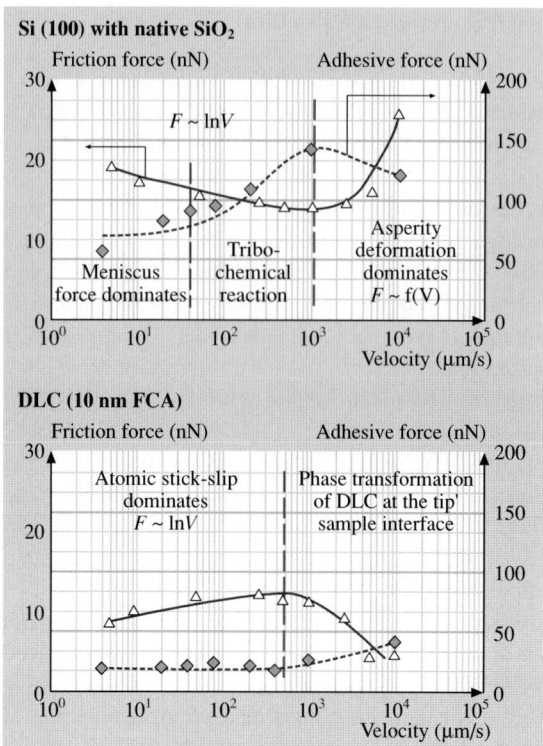

Fig. 8.28. Effect of velocity on friction force at a normal load of 70 nN and adhesive force over a wide range of velocities between 1 μm/s and 10 mm/s measured over a 25 μm scan size on Si(100) and DLC samples. The dominant friction mechanisms acting at different velocities are marked on the figure [84]

Figure 8.28 shows the friction force and adhesive force dependence on sliding velocity for single-crystal silicon, Si(100), and diamond-like carbon (DLC) with 10-nm thickness deposited using the filtered cathode arc (FCA) deposition technique [84–87]. The friction force and adhesive force are seen to vary with a change in velocity and both materials exhibit a reversal in friction behavior at certain critical sliding velocities. These reversals correspond to definitive transitions between different dominant friction mechanisms. For Si(100), which is hydrophilic, meniscus forces are dominant. The initial decrease in friction force with velocity corresponds to diminishing meniscus contributions. Beyond a certain critical velocity the residence time of the tip is not sufficient to form meniscus bridges at the sliding interface and the meniscus force contribution to the friction force drops out. At moderate velocities, tribochemical reactions at the tip–sample interface, in which a low-shear-strength Si(OH)$_4$ layer is formed [42, 88], also reduce friction. At high sliding velocities, deformation of asperities resulting from the high-velocity impacts becomes dominant and governs the friction behavior (see Fig. 8.29a). For DLC, being partially hydrophobic, meniscus forces are not dominant. The increase in friction force with velocity arises from atomic-scale stick–slip. Based on a thermally-activated Eyring model incorporated by *Bouhacina* et al. [89], the potential barrier divided by absolute temperature required for making jumps during stick–slip follows the

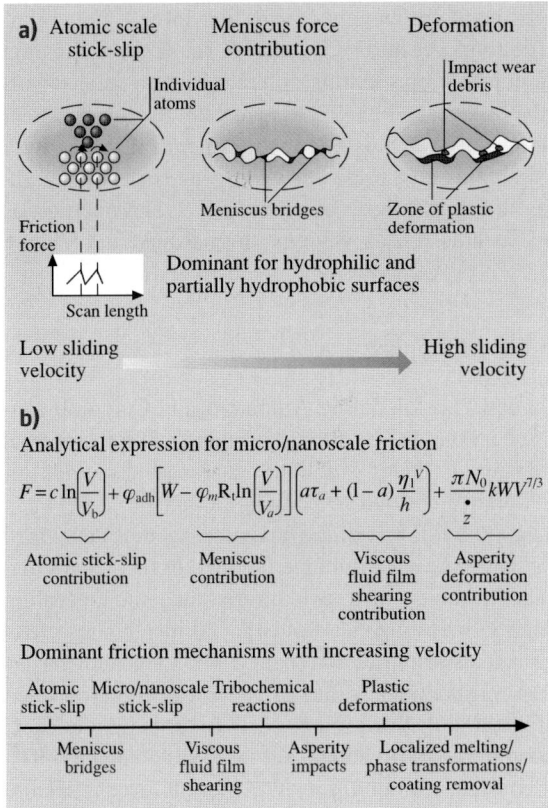

Fig. 8.29. (a) Schematic illustrating various dominant regimes of friction force at different relative sliding velocities from atomic scale stick–slip at low velocities to meniscus force contribution at mid velocities and deformation-related energy dissipation at high velocities, and (b) comprehensive analytical expression for velocity dependence of nanoscale friction with the dominant friction mechanisms [85]

logarithm of the velocity, and is responsible for the logarithmic dependence on velocity shown by friction force. Further work by *Gnecco* et al. [90] showed that this increase continues up to a certain sliding velocity and then levels off. At high velocities, the large frictional-energy dissipation results in a phase transformation of DLC to a graphite-like phase which is responsible for low friction.

For Si(100), the adhesive force is constant initially but then starts to increase beyond a certain velocity. The increase in the adhesive force for Si(100) is believed to be the result of a tribochemical reaction. This layer gets replenished continuously during sliding and results in a higher adhesive force between the tip and sample surface. For DLC, the adhesive force starts to increase beyond a critical velocity and is believed to be the result of phase transformation of DLC to a graphitic phase at the tip–sample interface [87, 92, 93].

For the development of a comprehensive friction model that encompasses various friction mechanisms and accounts for the velocity dependence of each mechanism, it is necessary to take into consideration the atomic-scale stick–slip, the adhesive contributions arising from meniscus forces and the deformation of asperities at the sliding interface resulting from high-velocity impacts [85]. Figure 8.29b gives the analytical expression with different components of friction force. The nonlinear nature of micro/nanoscale friction force and its velocity dependence is apparent from this expression. The different terms representing different friction mechanisms are marked on the analytical expression in Fig. 8.29b and their relative order of precedence with respect to sliding velocity is illustrated schematically. Each of these mechanisms will be dominant depending on the specific sliding interface, the roughness distribution on the surfaces sliding past each other and material properties, as well as the operating conditions such as sliding velocity, normal load and relative humidity.

8.3.6 Nanoscale Friction and Wear Mapping

Contrary to the classical friction laws postulated by Amontons and Coulomb centuries ago, nanoscale friction force is found to be strongly dependent on the normal load and sliding velocity. Many materials, coatings and lubricants that have wide applications show reversals in friction behavior corresponding to transitions between different friction mechanisms [37, 84–86]. Most of the analytical models developed for explaining nanoscale friction behavior have remained limited in their focus and have left investigators short-handed when trying to explain friction behavior scaling multiple regimes. Nanoscale friction maps provide fundamental insights into friction behavior. They help identify and classify the dominant friction mechanisms, as well as determine the critical operating parameters that influence transitions between different mechanisms [85, 86]. Figure 8.30 shows a nanoscale friction map for DLC with friction mapped as a function of the normal load and the sliding velocity [91]. The contours represent lines of constant friction force. The friction force is seen to increase with normal load as well as velocity. The increase in friction force with velocity is the result of atomic-scale stick–slip and is a result of thermal activation

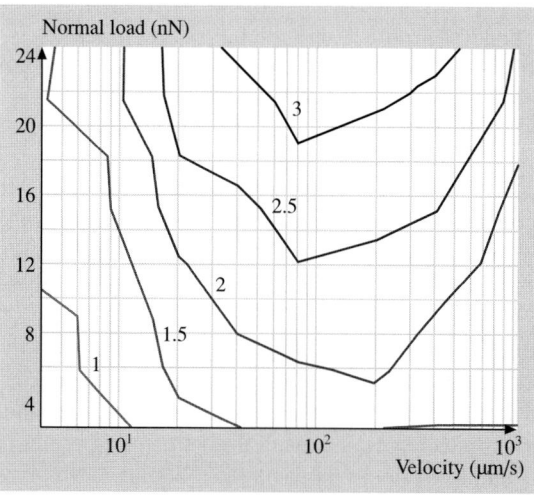

Fig. 8.30. Contour map showing friction force dependence on normal load and sliding velocity for DLC. Contour lines are constant friction force lines [91]

of the irreversible jumps of the AFM tip that arise from overcoming of the energy barrier between adjacent atomic positions, as described earlier. The concentric contour lines, corresponding to constant friction force predicts a peak point, a point where the friction force reaches a maxima and beyond which any further increase in the normal load or the sliding velocity results in a decrease in the friction force. This characteristic behavior for DLC is the result of a phase transformation of DLC into a graphite-like phase by sp^3-to-sp^2 phase transition, as described earlier. During the AFM experiments, the Si_3N_4 tip gives rise to contact pressures in the range of 1.8–4.4 GPa for DLC for normal loads of 10–150 nN [87]. A combination of the high contact pressures that are encountered on the nanoscale and the high frictional-energy dissipation arising from the asperity impacts at the tip–sample interface due to the high sliding velocities accelerates a phase-transition process whereby a low-shear-strength graphite-like layer is formed at the sliding interface.

Similar to friction mapping, one way of exploring the broader wear patterns is to construct wear mechanism maps that summarize data and models for wear, thereby showing mechanisms for any given set of conditions to be identified [94–96]. Wear of sliding surfaces can occur by one or more wear mechanisms, including adhesive,

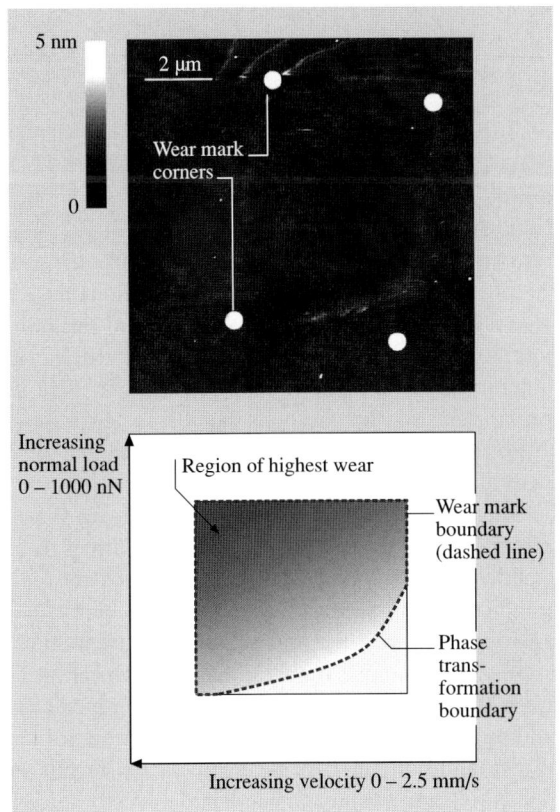

Fig. 8.31. Nanowear map (AFM image and schematic) illustrating the effect of sliding velocity and normal load on the wear of DLC resulting from phase transformation. The *curved area* shows debris lining and is indicative of the minimum frictional energy needed for phase transformation. For clarity, the wear mark corners are indicated by *white dots* in the upper AFM image and the various zones of interest over the entire wear mark are schematically illustrated in the lower image [94]

abrasive, fatigue, impact, corrosive, and fretting [5, 10]. *Tambe* and *Bhushan* [87, 94] performed AFM experiments to develop nanoscale wear maps. Figure 8.31 shows a nanowear map generated for a DLC sample by simultaneously varying the normal load and the sliding velocity over the entire scan area. The wear map was generated for a normal load range of 0–1000 nN and sliding velocity range of 0–2.5 mm/s. Wear debris, believed to result from phase transformation of DLC by an sp^3-to-sp^2 phase transition, was seen to form only for high values of sliding velocities times normal loads, i.e., only beyond certain threshold friction-energy dissipation [87, 94]. Hence the wear region exhibits a transition line indicating that for low velocities and low normal loads there is no phase transformation. For clarity, the wear mark corners are indicated by white dots in the AFM image (top) and the two zones of interest over the entire wear mark are illustrated schematically in Fig. 8.31 (top).

Nanoscale friction and wear mapping are novel techniques for investigating friction force and wear behavior on the nanoscale over a range of operating parameters. By simultaneously varying the sliding velocity and normal load over a large range of values, nanoscale friction and wear behavior can be mapped and the transitions between different wear mechanisms can be investigated. These maps help identify and demarcate critical operating parameters for different wear mechanisms and are very important tools in the process of design and selection of materials/coatings.

8.3.7 Adhesion and Friction in a Wet Environment

Experimental Observations

The tip radius and relative humidity affect adhesion and friction. The relative humidity affects adhesion and friction for dry and lubricated surfaces [30, 97, 98]. Figure 8.32 shows the variation of single-point adhesive force measurements as a function of tip radius on a Si(100) sample for several humidities. The adhesive force data are also plotted as a function of relative humidity for several tip radii. The general trend at humidities up to the ambient is that a 50-nm-radius Si_3N_4 tip exhibits a lower adhesive force compared to the other microtips of larger radii; in the latter case, the values are similar. Thus for the microtips there is no appreciable variation in adhesive force with tip radius at a given humidity up to ambient. The adhesive force increases as relative humidity increases for all tips.

Sources of adhesive force between a tip and a sample surface are van der Waals attraction and meniscus formation [5, 10, 98]. Relative magnitudes of the forces from the two sources are dependent upon various factors, including the distance between the tip and the sample surface, their surface roughness, their hydrophobicity, and relative humidity [99]. For most rough surfaces, the meniscus contribution dominates at moderate to high humidities, which arise from capillary condensation of water vapor from the environment. If enough liquid is present to form a meniscus bridge, the meniscus force should increase with increasing tip radius (proportional to tip radius for a spherical tip). In addition, an increase in tip radius results in increased contact area, which leads to higher values of the van der Waals forces.

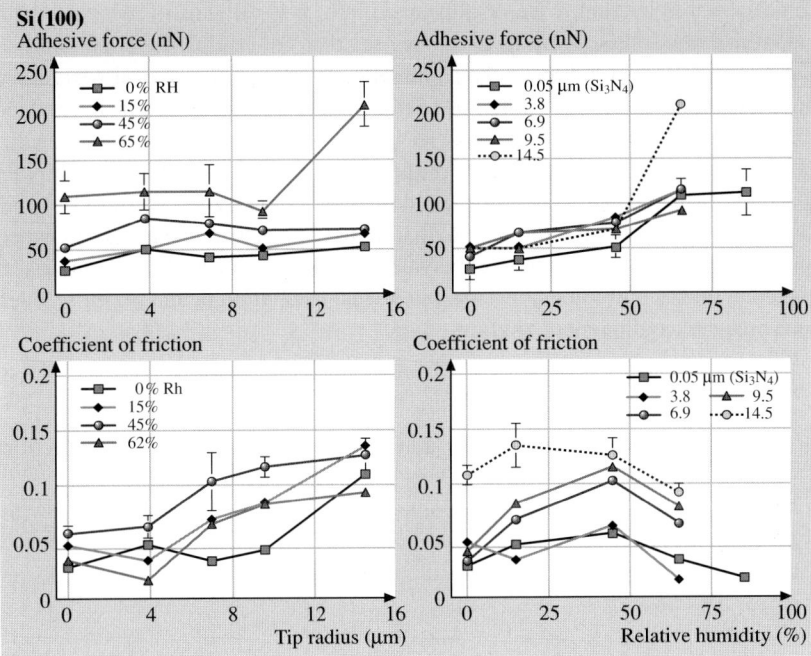

Fig. 8.32. Adhesive force and coefficient of friction as a function of tip radius at several humidities and as a function of relative humidity at several tip radii on Si(100) [30]

However, if nano-asperities on the tip and the sample are considered then the number of contacting and near-contacting asperities forming meniscus bridges increases with increasing humidity, leading to an increase in meniscus forces. These explain the trends observed in Fig. 8.32. From the data, the tip radius has little effect on the adhesive forces at low humidities but increases with tip radius at high humidity. Adhesive force also increases with increasing humidity for all tips. This observation suggests that the thickness of the liquid film at low humidities is insufficient to form continuous meniscus bridges that would affect adhesive forces in the case of all tips.

Figure 8.32 also shows the variation in the coefficient of friction as a function of tip radius at given humidity, and as a function of relative humidity for a given tip radius for Si(100). It can be observed that, for 0% RH, the coefficient of friction is about the same for the tip radii except for the largest tip, which shows a higher value. At all other humidities, the trend consistently shows that the coefficient of friction increases with tip radius. An increase in friction with tip radius at low to moderate humidities arises from increased contact area (higher van der Waals forces) and higher values of shear forces required for larger contact area. At high humidities, similarly to the adhesive force data, an increase with tip radius occurs because of both contact-area and meniscus effects. Although AFM/FFM measurements are able to measure the combined effect of the contribution of van der Waals and meniscus forces towards friction force or adhesive force, it is difficult to measure their individual contributions separately. It can be seen that, for all tips, the coefficient of friction increases with humidity to about ambient, beyond which it starts to decrease. The initial increase in the coefficient of friction with humidity arises from the fact that the thickness of the water film increases with increasing humidity, which results in a larger number of nano-asperities forming meniscus bridges and leads to higher friction (larger shear force). The same trend is expected with the microtips beyond 65% RH. This is attributed to the fact that, at higher humidities, the adsorbed water film on the surface acts as a lubricant between the two surfaces. Thus the interface is changed at higher humidities, resulting in lower shear strength and hence a lower friction force and coefficient of friction.

Adhesion and Friction Force Expressions for a Single-Asperity Contact

We now obtain the expressions for the adhesive force and coefficient of friction for a single-asperity contact with a meniscus formed at the interface, Fig. 8.33. For a spherical asperity of radius R in contact with a flat and smooth surface with a composite modulus of elasticity E^* and in the presence of liquid with a concave meniscus, the attractive meniscus force (adhesive force), designated as F_m or W_{ad} is given by [6, 10]:

$$W_{ad} = 2\pi R \gamma (\cos\theta_1 + \cos\theta_2) , \qquad (8.6)$$

where γ is the surface tension of the liquid and θ_1 and θ_2 are the contact angles of the liquid with surfaces 1 and 2, respectively. For an elastic contact for both extrinsic (W) and intrinsic (W_{ad}) normal load, the friction force is given as,

$$F_e = \pi\tau \left[\frac{3(W + W_{ad})R}{4E^*} \right]^{2/3} , \qquad (8.7)$$

where W is the external load, and τ is the average shear strength of the contacts. (Surface energy effects are not considered here.) Note that the adhesive force increases linearly with increasing the tip radius, and the friction force increases with increasing tip radius as $R^{2/3}$ and with normal load as $(W + W_{ad})^{2/3}$. The experimental data in support of the $W^{2/3}$ dependence on the friction force can be found in various

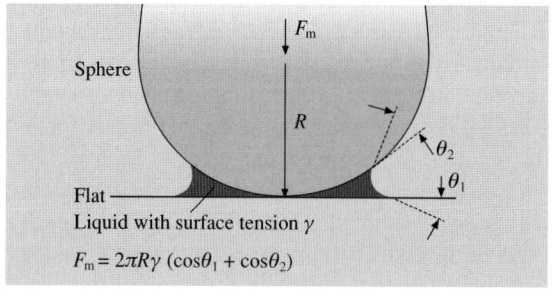

Fig. 8.33. Meniscus formation from a liquid condensate at the interface for a sphere in contact with a plane surface

references (see e.g., *Schwarz* et al. [100]). The coefficient of friction μ_e is obtained from (8.7) as

$$\mu_e = \frac{F_e}{(W+W_{ad})} = \pi\tau \left[\frac{3R}{4E^*}\right]^{2/3} \frac{1}{(W+W_{ad})^{1/3}}. \tag{8.8}$$

In the plastic contact regime [6], the coefficient of friction μ_p is obtained as

$$\mu_p = \frac{F_p}{(W+W_{ad})} = \frac{\tau}{H_s}, \tag{8.9}$$

where H_s is the hardness of the softer material. Note that, in the plastic contact regime, the coefficient of friction is independent of the external load, adhesive contributions and surface geometry.

For comparisons, for multiple-asperity contacts in the elastic contact regime the total adhesive force W_{ad} is the summation of adhesive forces at n individual contacts,

$$W_{ad} = \sum_{i=1}^{n}(W_{ad})_i \quad \text{and} \quad \mu_e \approx \frac{3.2\tau}{E^*\left(\sigma_p/R_p\right)^{1/2} + (W_{ad}/W)}, \tag{8.10}$$

where σ_p and R_p are the standard deviation of the summit heights and average summit radius, respectively. Note that the coefficient of friction depends upon the surface roughness. In the plastic contact regime, the expression for μ_p in (8.9) does not change.

The source of the adhesive force, in a wet contact in the AFM experiments being performed in an ambient environment, includes mainly attractive meniscus force due to capillary condensation of water vapor from the environment. The meniscus force for a single contact increases with increasing tip radius. A sharp AFM tip in contact with a smooth surface at low loads (on the order of a few nN) for most materials can be simulated as a single-asperity contact. At higher loads, for rough and soft surfaces, multiple contacts would occur. Furthermore, at low loads (nN range) for most materials, the local deformation would be primarily elastic. Assuming that the shear strength of contacts does not change, the adhesive force for smooth and hard surfaces at low normal load (on the order of a few nN) (for a single-asperity contact in the elastic contact regime) would increase with increasing tip radius, and the coefficient of friction would decrease with increasing total normal load as $(W+W_{ad})^{-1/3}$ and would increase with increasing tip radius as $R^{2/3}$. In this case, the Amontons law of friction, which states that the coefficient of friction is independent of normal load and is independent of apparent area of contact, does not hold. For a single-asperity plastic contact and multiple-asperity plastic contacts, neither the normal load nor the tip radius come into play in the calculation of the coefficient of friction. In the case of multiple-asperity contacts, the number of contacts increase with increasing normal load; therefore the adhesive force increases with increasing load.

In the data presented earlier in this section, the effect of tip radius and humidity on the adhesive forces and coefficient of friction is investigated for experiments with

Si(100) surface at loads in the range of 10–100 nN. The multiple-asperity elastic-contact regime is relevant for this study involving large tip radii. An increase in humidity generally results in an increase of the number of meniscus bridges, which would increase the adhesive force. As was suggested earlier, this increase in humidity may also decrease the shear strength of contacts. A combination of an increase in adhesive force and a decrease in shear strength would affect the coefficient of friction. An increase in tip radius would increase the meniscus force (adhesive force). A substantial increase in the tip radius may also increase interatomic forces. These effects influence the coefficient of friction with increasing tip radius.

Roughness Optimization for Superhydrophobic Surfaces

One of the crucial surface properties for surfaces in wet environments is non-wetting or hydrophobicity. It is usually desirable to reduce wetting in fluid flow applications and some conventional applications, such as glass windows and automotive windshields, in order for liquid to flow away along their surfaces. Reduction of wetting is also important in reducing meniscus formation, consequently reducing stiction, friction, and wear. Wetting is characterized by the contact angle, which is the angle between the solid and liquid surfaces. If the liquid wets the surface (referred to as a wetting liquid or hydrophilic surface), the value of the contact angle is $0° \le \theta \le 90°$, whereas if the liquid does not wet the surface (referred to as a non-wetting liquid or hydrophobic surface), the value of the contact angle is $90° < \theta \le 180°$. Superhydrophobic surfaces should also have very low water contact angle hysteresis. A surface is considered superhydrophobic if θ is close to $180°$. One of the ways to increase the hydrophobic or hydrophilic properties of the surface is to increase surface roughness. It has been demonstrated experimentally that roughness changes contact angle. Some natural surfaces, including leaves of water-repellent plants such as lotus, are known to be superhydrophobic due to high roughness and the presence of a wax coating. This phenomenon is called in the literature the *lotus effect* [102].

If a droplet of liquid is placed on a smooth surface, the liquid and solid surfaces come together under equilibrium at a characteristic angle called the static contact angle θ_0, Fig. 8.34a. The contact angle can be determined from the condition of the total energy of the system being minimized. It can be shown that

$$\cos\theta_0 = dA_{LA}/dA_{SL}, \qquad (8.11)$$

where θ_0 is the contact angle for smooth surface, A_{SL} and A_{LA} are the solid–liquid and liquid–air contact areas. Next, let us consider a rough solid surface with a typical size of roughness details smaller than the size of the droplet, Fig. 8.34. For a rough surface, the roughness affects the contact angle due to an increased contact area A_{SL}. For a droplet in contact with a rough surface without air pockets, referred to as a homogeneous interface, based on the minimization of the total surface energy of the system, the contact angle is given as [103]

$$\cos\theta = dA_{LA}/dA_F = \left(\frac{A_{SL}}{A_F}\right)(dA_{LA}/dA_{SL}) = R_f \cos\theta_0, \qquad (8.12)$$

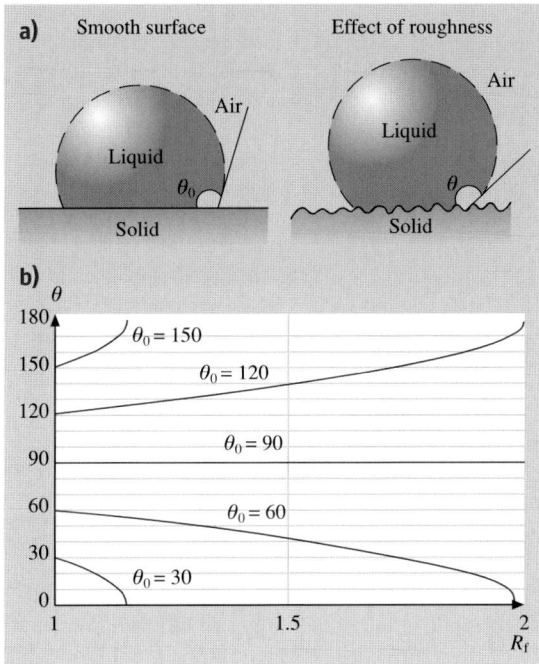

Fig. 8.34. (a) Droplet of liquid in contact with a smooth solid surface (contact angle θ_0) and rough solid surface (contact angle θ), and (b) contact angle for rough surface (θ) as a function of the roughness factor (R_f) for various contact angles for smooth surface (θ_0) [101]

where A_F is the flat solid–liquid contact area (or a projection of the solid–liquid area A_{SL} onto the horizontal plane). R_f is a roughness factor defined as

$$R_f = \frac{A_{SL}}{A_F}. \tag{8.13}$$

Equation (8.13) shows that, if the liquid wets a surface ($\cos\theta_0 > 0$), it will also wet the rough surface with a contact angle $\theta < \theta_0$, and for non-wetting liquids ($\cos\theta_0 < 0$), the contact angle with a rough surface will be greater than that with the flat surface, $\theta > \theta_0$. The dependence of the contact angle on the roughness factor is presented in Fig. 8.34b for different values of θ_0, based on (8.12). It should be noted that (8.12) is valid only for moderate roughness, when $R_f \cos\theta_0 < 1$ [102].

For high roughness, air pockets (composite solid–liquid–air interface) will be formed in the cavities of the surface [104]. In the case of partial contact, the contact angle is given by

$$\cos\theta = R_f f_{SL} \cos\theta_0 - f_{LA}, \tag{8.14}$$

where f_{SL} and f_{LA} are fractional solid–liquid and liquid–air contact areas. The homogeneous and composite interfaces are two metastable states of the system. In reality, some cavities will be filled with liquid, and others with air, and the value of the contact angle is between the values predicted by (8.12) and (8.14). If the distance is large between the asperities or if the slope changes slowly, the liquid–air interface can easily be destabilized due to imperfectness of the profile shape or due to the

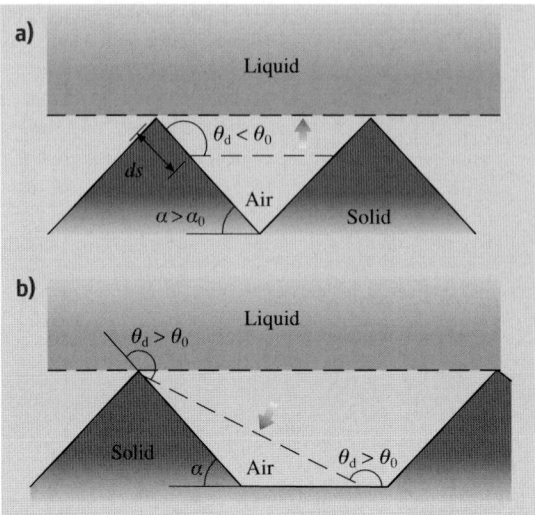

Fig. 8.35. (a) Formation of a composite solid–liquid–air interface for sawtooth and smooth profiles, and (b) destabilization of the composite interface for the sawtooth and smooth profiles due to dynamic effects. Dynamic contact angle $\theta_d > \theta_0$ corresponds to advancing liquid–air interface, whereas $\theta_d < \theta_0$ corresponds to the receding interface [101]

dynamic effects, such as surface waves (Fig. 8.35). *Nosonovsky* and *Bhushan* [101] proposed a stochastic model, which relates the contact angle to the roughness and takes into account the possibility of destabilization of the composite interface due to imperfectness of the shape of the liquid–air interface, caused by effects such as capillary waves.

In addition to the surface roughness, sharp edges of asperities may affect wetting, because they can pin the line of contact of the solid, liquid, and air (also known as the *triple line*) and resist liquid flow. *Nosonovsky* and *Bhushan* [102] considered the effect of the surface roughness and sharp edges and found the optimum roughness distribution for non-wetting. They formulated five requirements for roughness-induced superhydrophobic surfaces. First, asperities must have a high aspect ratio to provide a high surface area. Second, sharp edges should be avoided, to prevent pinning of the triple line. Third, asperities should be tightly packed to minimize the distance between them and avoid destabilization of the composite interface. Fourth, asperities should be small compared to the typical droplet size (on the order of a few hundred microns or larger). And fifth, in the case of hydrophilic surface, a hydrophobic film must be applied in order to have initially $\theta > 90°$. These recommendations can be utilized for producing superhydrophobic surfaces. Remarkably, all these conditions are satisfied by biological water-repellent surfaces, such as some leaves: they have tightly packed hemispherically topped papillae with high (on the order of unity) aspect ratios and a wax coating [102]. Figure 8.36 shows two recommended geometries which use either hemispherically topped asperities with a hexagonal packing pattern or pyramidal asperities with a rounded top. These geometries can be used for producing superhydrophobic surfaces.

Burton and *Bhushan* [105] have provided indirect evidence of an increase in contact angle and a decrease in adhesive force with the presence of discrete asper-

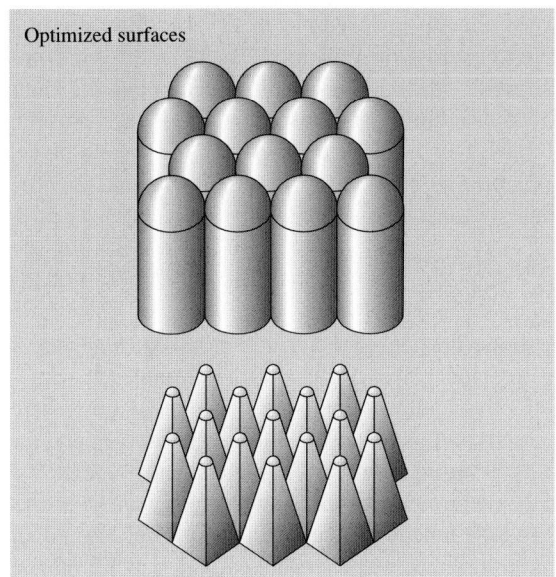

Fig. 8.36. Optimized roughness distribution – hemispherically topped cylindrical asperities (*upper*) and pyramidal asperities (*lower*) with square foundation and rounded tops. Square bases provide higher packing density but introduces undesirable sharp edges [102]

ities with high aspect ratios based on measurements on smooth hydrophobic films ($\cos\theta_0 < 0$) and those with discrete asperities in a humid environment.

8.3.8 Separation Distance Dependence of Meniscus and van der Waals Forces

When two surfaces are in close proximity, sources of adhesive forces are weak van der Waals attraction and meniscus formation. The relative magnitudes of the forces from the two sources are dependent upon various factors including the interplanar separation, their surface roughness, their hydrophobicity, and relative humidity (liquid volume) [99]. The meniscus contribution dominates at moderate to high humidities and van der Waals forces dominate at asperities a few nm apart. In some micro/nanocomponents, it is important to know the relative contribution of the two sources as a function of a given interplanar separation to design an interface for low adhesion. For example, if two ultrasmooth surfaces come in close proximity with interplanar separation on the order of a nm, van der Waals forces may dominate and their magnitude may be reduced by creating bumps on one of the interfaces. This analysis is also of interest in AFM studies to understand the distance dependence of adhesive forces as the tip goes in and out of contact.

Stifter et al. [99] modeled the contact of a parabolic-shaped tip and a flat, smooth sample surface. The tip may represent a surface asperity on an interface or an AFM tip in an AFM experiment. They calculated van der Waals and meniscus forces as a function of different parameters, namely, tip geometry, tip–sample starting distance, relative humidity, surface tension, and contact angles. They compared the meniscus forces with van der Waals forces to understand their relative importance in various operating conditions.

The interacting force between the tip and sample in dry conditions is the Lennard–Jones force derived from Lennard–Jones potential. The Lennard–Jones potential is composed of two interactions – the van der Waals attraction and the Pauli repulsion. Van der Waals forces are significant because they are always present. For a parabolic tip above a half plane with a distance D between the tip and plane, the Lennard–Jones potential is obtained by integrating the atomic potential over the volume of the tip and sample. It is given as [99]

$$V(D) = \frac{c}{12}\left(-\frac{A}{D} + \frac{B}{210D^7}\right), \tag{8.15}$$

where c is the width of the parabolic tip (the diameter in the case of a spherical tip), A and B are two potential parameters where A is Hamakar constant. This equation provides expressions for attractive and repulsive parts. The calculations were made for Lennard–Jones force (total) and van der Waals force (attractive part) for two Hamaker constants 0.04×10^{-19} J (representative of polymers) and 3.0×10^{-19} J (representative of ceramics) and the meniscus force for a water film ($\gamma_\ell = 72.5$ N/m). Figure 8.37 shows various forces as a function of separation distance. The effect of two relative humidities and three tip radii was also studied which affect meniscus forces. The two dashed curves indicate the spread of possible van der Waals forces for two Hamaker constants. The figure shows that meniscus forces exhibit weaker distance dependence. The meniscus forces can be stronger or weaker than van der Waals forces for distances smaller than about 0.5 nm. For longer distances, the meniscus forces are stronger than the van der Waals forces. van der Waals forces must be considered for a tip–sample distance up to a few nm ($D < 5$ nm). The meniscus forces operate until the meniscus breaks, in the range 5–20 nm [99].

8.3.9 Scale Dependence in Friction

Table 8.3 presents adhesive force and coefficient of friction data obtained on the nanoscale and microscale [24, 106, 107]. Adhesive force and coefficient of friction values on the nanoscale are about half to one order of magnitude lower than that

Table 8.3. Micro- and nanoscale values of adhesive force and coefficient of friction in micro- and nanoscale measurements [106]

Sample	Adhesive force		Coefficient of friction	
	Microscale[a] (μN)	Nanoscale[b] (nN)	Microscale[a]	Nanoscale[b]
Si(100)	685	52	0.47	0.06
DLC	325	44	0.19	0.03
Z-DOL	315	35	0.23	0.04
HDT	180	14	0.15	0.006

[a] Versus 500-μm-radius Si(100) ball,
[b] versus 50-nm-radius Si$_3$N$_4$ tip

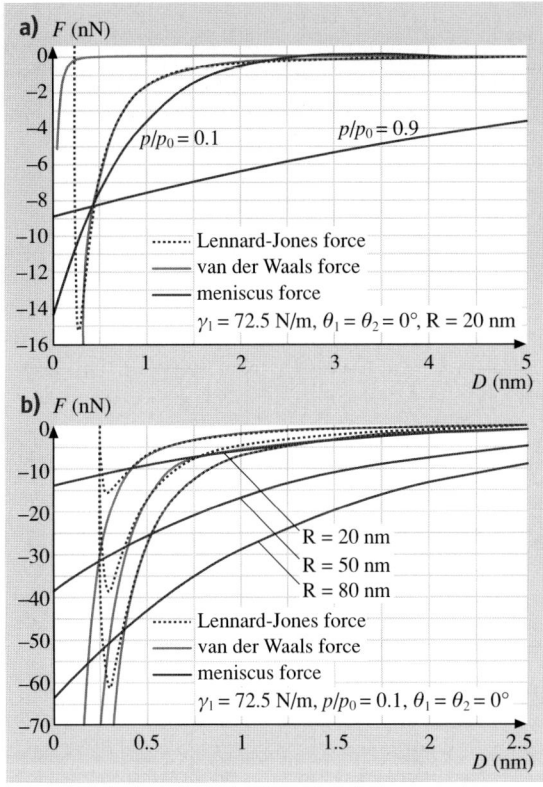

Fig. 8.37. Relative contribution of meniscus, van der Waals and Lennard–Jones forces (F) as a function of separation distance (D) and at (**a**) two values of relative humidity (p/p_0) for tip radius of 20 nm and Hamakar constants of 0.04×10^{-19} J and 3.0×10^{-19} J, and (**b**) three tip radii (R) and Hamakar constant of 3.0×10^{-19} J [99]

on the microscale. Scale dependence is clearly observed in this data. As a further evidence of scale dependence, Table 8.4 shows the coefficient of friction measured for Si(100), HOPG, natural diamond, and DLC on the nanoscale and microscales. It is clearly observed that friction values are scale dependent.

To estimate the scale length, apparent contact radius at test loads are calculated and presented in the table. Mean apparent pressures are also calculated and presented. For nanoscale AFM experiments, it is assumed that an AFM tip coming into contact with a flat surface represents a single asperity and elastic contact, and Hertz analysis was used for the calculations. In the microscale experiments, a ball coming into contact with a flat surface represents multiple-asperity contacts due to the roughness, and the contact pressure of the asperity contacts is higher than the apparent pressure. For calculation of a characteristic scale length for multiple-asperity contacts, which is equal to the apparent length of contact, Hertz analysis was also used. This analysis provide an upper limit on the apparent radius and a lower limit on the mean contact pressure.

There are several factors responsible for the differences in the coefficients of friction at the micro- and nanoscale. Among them are the contributions from wear and contaminant particles, transition from elasticity to plasticity, and meniscus ef-

Table 8.4. Micro- and nanoscale values of the coefficient of friction, typical physical properties of specimen, and calculated apparent contact radii and apparent contact pressures at loads used in micro- and nanoscale measurements. For calculation purposes it is assumed that contacts on micro- and nanoscale are single-asperity elastic contacts [114]

Sample	Coefficient of friction		Elastic modulus (GPa)	Poisson's ratio	Hardness (GPa)	Apparent contact radius at test load for		Mean apparent pressure at test load for	
	Microscale	Nanoscale				Microscale (µm) (upper limit)	Nanoscale (nm)	Microscale (GPa) (lower limit)	Nanoscale (GPa)
Si(100) wafer	0.47[a]	0.06[c]	130[e,f]	0.28[f]	9–10[e,f]	0.8–2.2[a]	1.6–3.4[c]	0.05–0.13[a]	1.3–2.8[c]
Graphite (HOPG)	0.1[b]	0.006[c]	9–15[g] (9)	– (0.25)	0.01[j]	62[b]	3.4–7.4[c]	0.082[b]	0.27–0.58[c]
Natural diamond	0.2[b]	0.05[c]	1140[h]	0.07[h]	80–104[g,h]	21[b]	1.1–2.5[c]	0.74[b]	2.5–5.3[c]
DLC film	0.19[a]	0.03[d]	280[i]	0.25[i]	20–30[i]	0.7–2.0[a]	1.3–2.9[d]	0.06–0.16[a]	1.8–3.8[d]

[a] 500-µm-radius Si(100) ball at 100–2000 µN and 720 µm/s in dry air [106]
[b] 3-mm-radius Si$_3$N$_4$ ball (elastic modulus 310 GPa, Poisson's ratio 0.22 [108]) at 1 N and 800 µm/s [24]
[c] 50-nm-radius Si$_3$N$_4$ tip at load range from 10–100 nN and 0.5 nm/s, in dry air [24]
[d] 50-nm-radius Si$_3$N$_4$ tip at load range from 10–100 nN in dry air [106]
[e] [109] [f] [110] [g] [108] [h] [111] [i] [112] [j] [113]

fect. The contribution of wear and contaminant particles is more significant at the macro/microscale because of the larger number of trapped particles, referred to as the third-body contribution. It can be argued that, for the nanoscale AFM experiments, the asperity contacts are predominantly elastic (with average real pressure less than the hardness of the softer material) and adhesion is the main contribution to the friction, whereas for the microscale experiments the asperity contact are predominantly plastic and deformation is an important factor. It will be shown later that hardness has a scale effect; it increases with decreasing scale and is responsible for less deformation on a smaller scale. The meniscus effect results in an increase of friction with increasing tip radius (Fig. 8.32). Therefore, the third-body contribution, scale-dependent hardness and other properties, the transition from elastic contacts in nanoscale contacts to plastic deformation in microscale contacts, and the meniscus contribution, play an important role [114–116].

Friction is a complex phenomenon, which involves asperity interactions involving adhesion and deformation (plowing). Adhesion and plastic deformation imply energy dissipation, which is responsible for friction, Fig. 8.38 [5, 10]. A contact between two bodies takes place on high asperities, and the real area of contact (A_r) is a small fraction of the apparent area of contact. During contact of two asperities, a lateral force may be required for asperities of a given slope to climb against each other. This mechanism is known as the ratchet mechanism, and it also contributes to friction. Wear and contaminant particles present at the interface, referred to as the *third body*, also contribute to the friction, Fig. 8.38. In addition, during contact even at low humidity, a meniscus is formed (Fig. 8.33). Generally any liquid that wets or has a small contact angle on surfaces will condense from vapor into cracks and pores on the surfaces as bulk liquid and in the form of annular-shaped capillary condensate in the contact zone. A quantitative theory of scale effects in friction should consider the effect of scale on physical properties relevant to these various contributions.

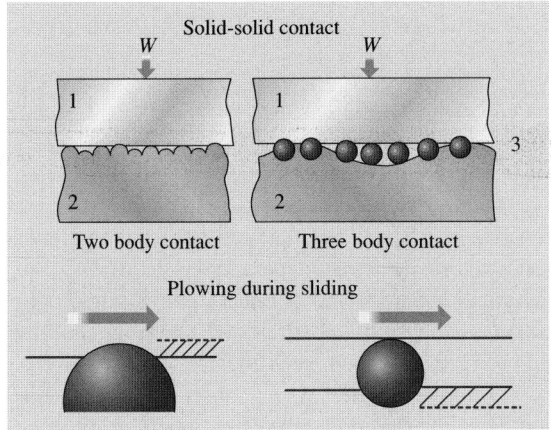

Fig. 8.38. Schematic of two-bodies and three-bodies during dry contact of rough surfaces

According to the adhesion and deformation model of friction, the coefficient of dry friction μ is a sum of an adhesion component μ_a and a deformation (plowing) component μ_d. The later, in the presence of particles, is the sum of the asperity-summit deformation component μ_{ds} and the particles deformation component μ_{dp} so that the total coefficient of friction is [115]

$$\mu = \mu_a + \mu_{ds} + \mu_{dp} = \frac{F_a + F_{ds} + F_{dp}}{W} = \frac{A_{ra}\tau_a + A_{ds}\tau_{ds} + A_{dp}\tau_{dp}}{W}, \quad (8.16)$$

where W is the normal load, F is the friction force, A_{ra}, A_{ds}, A_{dp} are the real areas of contact during adhesion, two-body deformation and with particles, respectively, and τ is the shear strength. The subscripts a, ds, and dp correspond to adhesion, summit deformation, and particle deformation, respectively.

The adhesional component of friction depends on the real area of contact and adhesion shear strength. The real area of contact is scale dependent due to the scale dependence of surface roughness (for elastic and plastic contact) and due to the scale dependence of hardness (for plastic contact) [115]. We limit the analysis here to multiple-asperity contact. For this case, the scale L is defined as the apparent size of the contact between two bodies. (For completeness, for single-asperity contact, the scale is defined as the contact diameter.) It is suggested by *Bhushan* and *Nosonovsky* [117] that, for many materials, dislocation-assisted sliding (microslip) is the main mechanism responsible for the shear strength. They considered dislocation-assisted sliding based on the assumption that contributing dislocations are located in a subsurface volume. The thickness of this volume is limited by the distance which dislocations can climb ℓ_s (material parameter) and by the radius of contact a. They showed that τ_a is scale dependent. Based on this, the adhesional components of the coefficient of friction in the case of elastic contact μ_{ae} and in the case of plastic contact μ_{ap} are given as [117]

$$\mu_{ae} = \frac{\mu_{ae0}}{\sqrt{\ell + (\ell_s/\overline{a_0})}} \left(\frac{L}{L_{\ell_c}}\right)^{m-n} \sqrt{1 + (L_s/L)^m}, \quad L < L_{\ell_c}, \quad (8.17)$$

$$\mu_{ap} = \mu_{ap0} \sqrt{\frac{1 + (\ell_d/\overline{a_0})}{1 + (\ell_s/\overline{a_0})}} \sqrt{\frac{1 + (L_s/L)^m}{1 + (L_d/L)^m}}, \quad L < L_{\ell_c}, \quad (8.18)$$

where μ_{ae0} and μ_{ap0} are values of the coefficient of friction at the macroscale ($L \geq L_{\ell_c}$), m and n are indices that characterize the scale dependence of surface parameters, $\overline{a_0}$ is the macroscale value of the mean contact radius, L_{ℓ_c} is the long-wavelength limit for scale dependence of the contact parameters, ℓ_s and ℓ_d are material-specific characteristic-length parameters, and L_s and L_d are length parameters related to ℓ_s and ℓ_d. The scale dependence of the adhesional component of the coefficient of friction is presented in Fig. 8.39, based on (8.17) and (8.18).

Based on the assumption that multiple asperities of two rough surfaces in contact have a conical shape, the two-body deformation component of friction can be

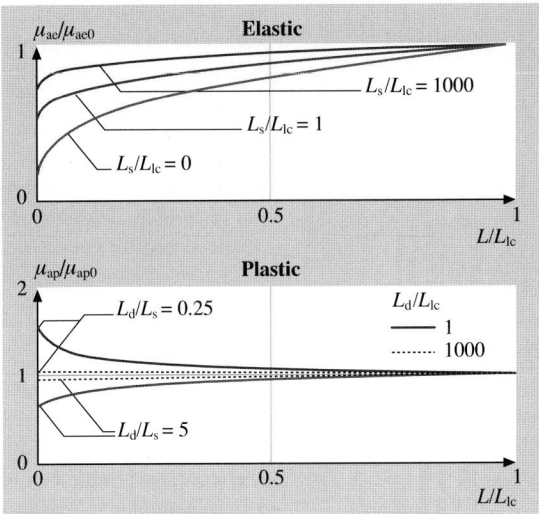

Fig. 8.39. Normalized results for the adhesional component of coefficient of friction, as a function of L/L_{ℓ_c} for multiple-asperity elastic contact. Data are presented for $m = 0.5$, $n = 0.2$. For multiple-asperity plastic contact, data are presented for two values of L_d/L_{ℓ_c} [115]

determined as [5, 10]

$$\mu_{ds} = \frac{2\tan\theta_r}{\pi}, \qquad (8.19)$$

where θ_r is the roughness angle (or attack angle) of a conical asperity. Mechanical properties affect the real area of contact and shear strength and these cancel out in (8.16) [115]. Based on statistical analysis of a random Gaussian surface [115]

$$\mu_{ds} = \frac{2\sigma_0}{\pi\beta_0^*}\left(\frac{L}{L_{\ell_c}}\right)^{n-m} = \mu_{ds0}\left(\frac{L}{L_{\ell_c}}\right)^{n-m}, \qquad L < L_{\ell c}, \qquad (8.20)$$

where μ_{ds0} is the value of the coefficient of the summit-deformation component of the coefficient of friction at the macroscale ($L \geq L_{\ell_c}$), and σ_0 and β_0^* are the macroscale values of the standard deviation of surface heights and correlation length, respectively, for a Gaussian surface. The scale dependence for the two-body deformation component of the coefficient of friction is presented in Fig. 8.40a for $m = 0.5$, $n = 0.2$, based on (8.20). The coefficient of friction increases with decreasing scale, according to (8.20). This effect is a consequence of increasing average slope or roughness angle.

For three-body deformation, it is assumed that wear and contaminant particles at the borders of the contact region are likely to leave the contact region, while the particles in the center are likely to stay (Fig. 8.41). The plowing three-body deformation is plastic and, assuming that particles are harder than the bodies, the shear strength τ_{dp} is equal to the shear yield strength of the softer body τ_Y, the

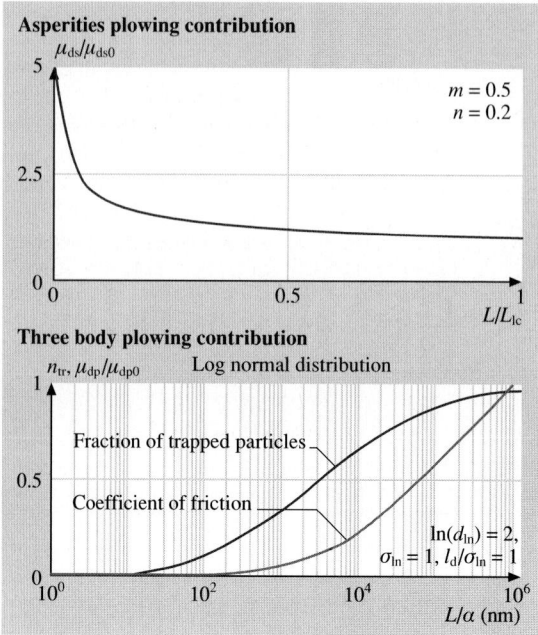

Fig. 8.40. *Top* Normalized results for the two-body deformation component of the coefficient of friction, and *bottom* the number of trapped particles divided by the total number of particles and three-body deformation component of the coefficient of friction, normalized by the macroscale value for log-normal distribution of debris size, where α is the probability of a particle in the border zone to leave the contact region. Various constants given in the figure correspond to the log-normal distribution [115]

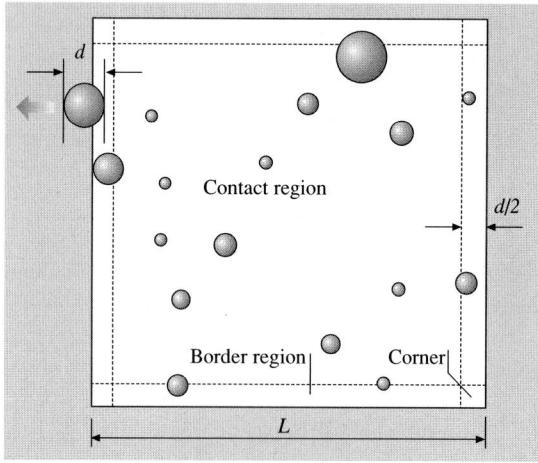

Fig. 8.41. Schematics of debris at the contact zone and at its border region. A particle of diameter d in the border region of $d/2$ is likely to leave the contact zone [115]

three-body deformation component of the coefficient of friction is given by [116]

$$\mu_{dp} = \mu_{dp0} n_{tr} \frac{\overline{d^2}}{\overline{d_0^2}} \frac{\sqrt{1+2\ell_d/\overline{d}}}{\sqrt{1+2\ell_d/\overline{d_0}}}, \qquad (8.21)$$

where $\bar{d}$ is the mean particle diameter, $\bar{d_0}$ is the macroscale value of the mean particle diameter, n_{tr} is the number of trapped particles divided by the total number of particles, and μ_{dp0} is the macroscale ($L \to \infty, n_{tr} \to 1$) value of the third-body deformation component of the coefficient of friction. The scale dependence of μ_{dp} is shown in Fig. 8.40 based on (8.21). Based on the scale-effect predictions presented in Figs. 8.39 and 8.40, trends in the experimental results in Table 8.3 can be explained.

The scale dependence of meniscus effects in friction, wear and interface temperature can be analyzed in a similar way [116].

To demonstrate the load dependence of friction at the nano/microscale, the coefficient of friction as a function of normal load is presented in Fig. 8.42. The coefficient of friction was measured by *Bhushan* and *Kulkarni* [28, 29] for a Si_3N_4 tip versus Si, SiO_2, and natural diamond using an AFM. They reported that, for low loads, the coefficient of friction is independent of load and increases with increasing load after a certain load. It is noted that the critical value of loads for Si and SiO_2

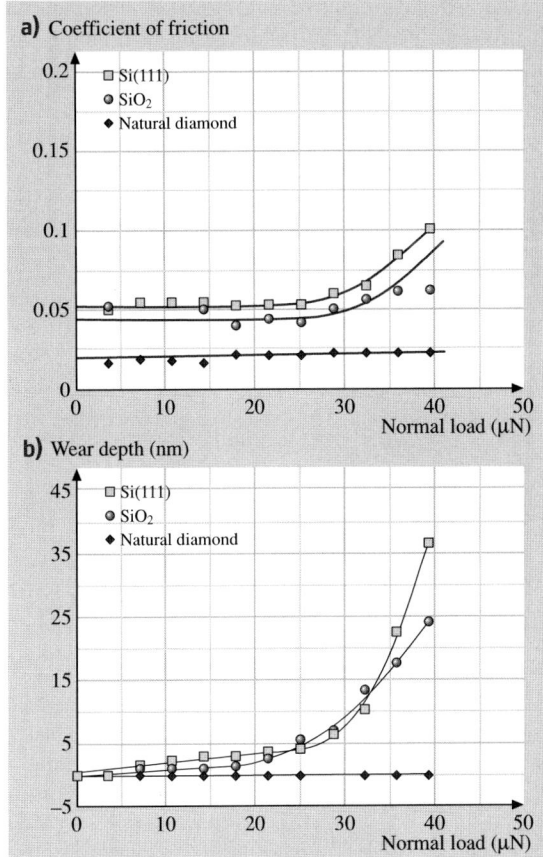

Fig. 8.42. Coefficient of friction as a function of normal load and for Si(111), SiO_2 coating and natural diamond. Inflections in the curves for silicon and SiO_2 correspond to the contact stresses equal to the hardness of these materials [28]

corresponds to stresses equal to their hardness values, which suggests that the transition to plasticity plays a role in this effect. The friction values at higher loads for Si and SiO_2 approach that of macroscale values.

8.4 Wear, Scratching, Local Deformation, and Fabrication/Machining

8.4.1 Nanoscale Wear

Bhushan and *Ruan* [23] conducted nanoscale wear tests on polymeric magnetic tapes using conventional silicon nitride tips, at two different loads of 10 and 100 nN, Fig. 8.43. For a low, normal load of 10 nN, measurements were made twice. There was no discernible difference between consecutive measurements for this load. However, as the load was increased from 10 nN to 100 nN, topographical changes were observed during subsequent scanning at a normal load of 10 nN; material was pushed in the sliding direction of the AFM tip relative to the sample. The material movement is believed to occur as a result of plastic deformation of the tape surface. Thus, deformation and movement of the soft materials on a nanoscale can be observed.

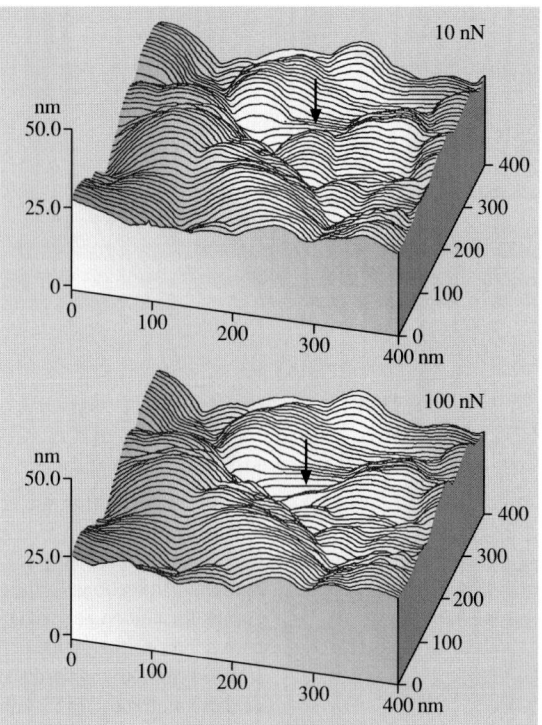

Fig. 8.43. Surface roughness maps of a polymeric magnetic tape at the applied normal load of 10 nN and 100 nN. Location of the change in surface topography as a result of nanowear is indicated by the *arrows* [23]

8.4.2 Microscale Scratching

The AFM can be used to investigate how surface materials can be moved or removed on the micro- to nanoscales, for example, in scratching and wear [4, 11] (where these things are undesirable), and nanofabrication/nanomachining (where they are desirable). Figure 8.44a shows microscratches made on Si(111) at various loads and a scanning velocity of 2 μm/s after 10 cycles [27]. As expected, the scratch depth increases linearly with load. Such microscratching measurements can be used to study failure mechanisms on the microscale and to evaluate the mechanical integrity (scratch resistance) of ultra-thin films at low loads.

To study the effect of scanning velocity, unidirectional scratches, 5 μm in length, were generated at scanning velocities ranging from 1 to 100 μm/s at various normal loads ranging from 40 to 140 μN. There is no effect of scanning velocity obtained at a given normal load. For representative scratch profiles at 80 μN, see Fig. 8.44b. This may be because of a small effect of frictional heating with the change in scanning velocity used here. Furthermore, for a small change in interface temperature, there is a large underlying volume to dissipate the heat generated during scratching.

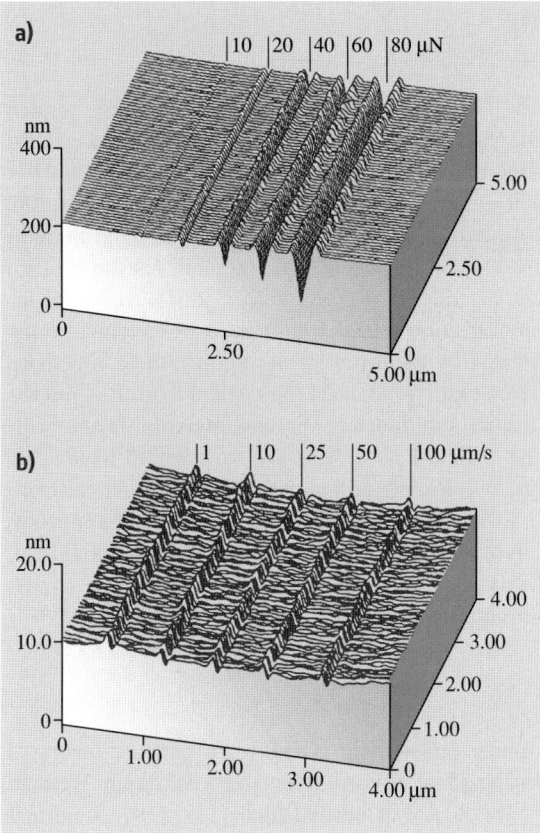

Fig. 8.44. Surface plots of (**a**) Si(111) scratched for ten cycles at various loads and a scanning velocity of 2 μm/s. Note that x and y axes are in μm and the z axis is in nm, and (**b**) Si(100) scratched in one unidirectional scan cycle at a normal force of 80 μN and different scanning velocities

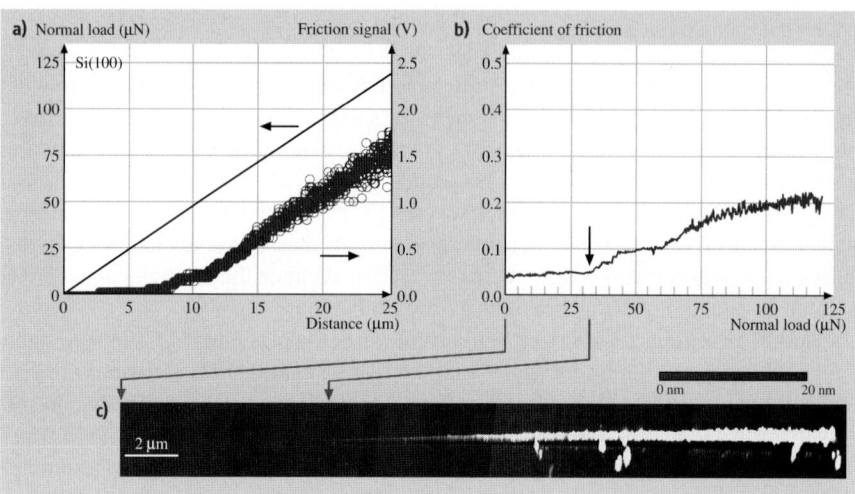

Fig. 8.45. (a) Applied normal load and friction signal measured during the microscratch experiment on Si(100) as a function of scratch distance, (b) friction data plotted in the form of coefficient of friction as a function of normal load, and (c) AFM surface height image of scratch obtained in tapping mode [47]

Scratching can be performed under ramped loading to determine the scratch resistance of materials and coatings. The coefficient of friction is measured during scratching and the load at which the coefficient of friction increases rapidly is known as the *critical load*, which is a measure of scratch resistance. In addition, post-scratch imaging can be performed in situ with the AFM in tapping mode to study failure mechanisms. Figure 8.45 shows data from a scratch test on Si(100) with a scratch length of 25 µm and a scratching velocity of 0.5 µm/s. At the beginning of the scratch, the coefficient of friction is 0.04, which indicates a typical value for silicon. At about 35 µN (indicated by the arrow in the figure), there is a sharp increase in the coefficient of friction, which indicates the critical load. Beyond the critical load, the coefficient of friction continues to increase steadily. In the post-scratch image, we note that at the critical load a clear groove starts to form. This implies that Si(100) was damaged by plowing at the critical load, associated with the plastic flow of the material. At and after the critical load, small and uniform debris is observed and the amount of debris increases with increasing normal load. *Sundararajan* and *Bhushan* [47] have also used this technique to measure the scratch resistance of diamond-like-carbon coatings with thicknesses of 3.5–20 nm.

8.4.3 Microscale Wear

By scanning the sample in two dimensions with the AFM, wear scars are generated on the surface. Figure 8.46 shows the effect of normal load on wear depth. We note that wear depth is very small below 20 µN of normal load [118, 119]. A normal

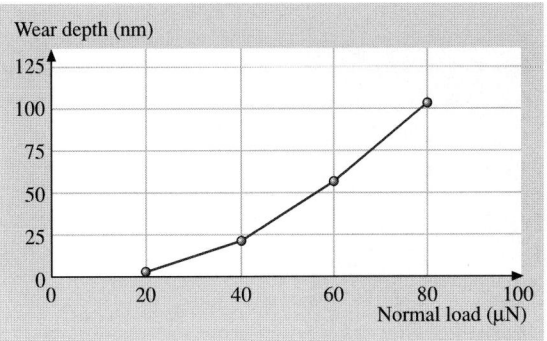

Fig. 8.46. Wear depth as a function of normal load for Si(100) after one cycle [119]

load of 20 μN corresponds to contact stresses comparable to the hardness of the silicon. Primarily, elastic deformation at loads below 20 μN is responsible for low wear [28, 29].

A typical wear mark, of the size 2 μm × 2 μm, generated at a normal load of 40 μN for one scan cycle and imaged using AFM with a scan size of 4 μm × 4 μm at 300 nN load, is shown in Fig. 8.47a [118]. The inverted map of wear marks shown in Fig. 8.47b indicates the uniform material removal at the bottom of the wear mark.

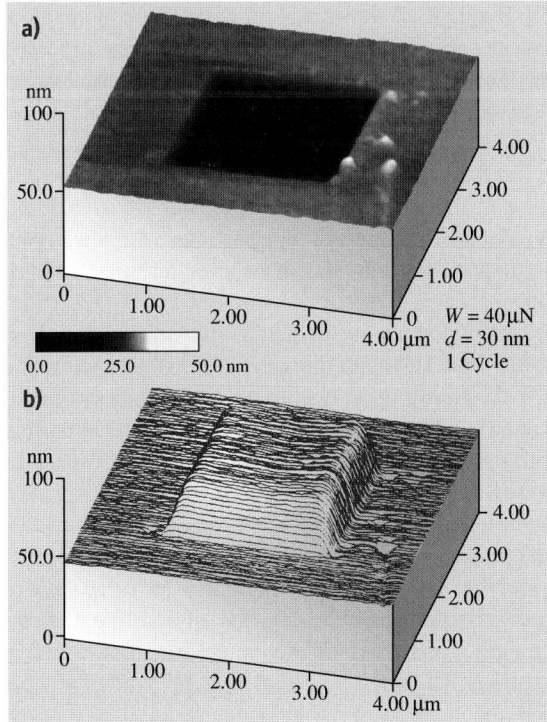

Fig. 8.47. (a) Typical greyscale and (b) inverted AFM images of wear mark created using a diamond tip at a normal load of 40 μN and one scan cycle on a Si(100) surface

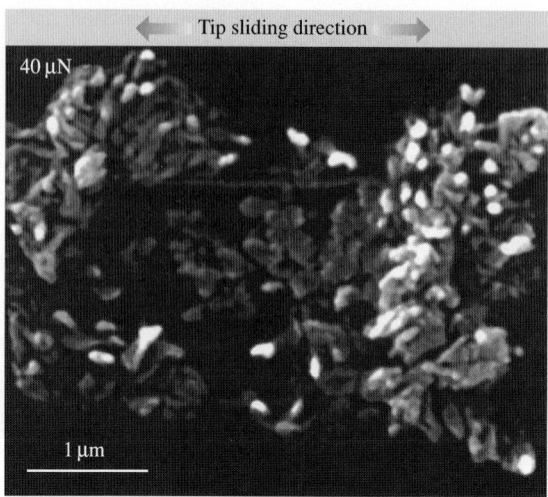

Fig. 8.48. Secondary electron image of wear mark and debris for Si(100) produced at a normal load of 40 μN and one scan cycle

An AFM image of the wear mark shows small debris at the edges, swiped during AFM scanning. Thus the debris is loose (not sticky) and can be removed during the AFM scanning.

Next we examine the mechanism of material removal on a microscale in AFM wear experiments [30, 118, 119]. Figure 8.48 shows a secondary-electron image of the wear mark and associated wear particles. The specimen used for the scanning electron microscope (SEM) was not scanned with the AFM after initial wear, in order to retain wear debris in the wear region. Wear debris is clearly observed. In the SEM micrographs, the wear debris appears to be agglomerated because of the high surface energy of the fine particles. Particles appear to be a mixture of rounded and so-called cutting type (feather-like or ribbon-like material). *Zhao* and *Bhushan* [119] reported an increase in the number and size of cutting-type particles with the normal load. The presence of cutting-type particles indicates that the material is removed primarily by plastic deformation.

To understand the material removal mechanisms better, transmission electron microscopy (TEM) has been used. The TEM micrograph of the worn region and associated diffraction pattern are shown in Fig. 8.49a,b. The bend contours are observed to pass through the wear mark in the micrograph. The bend contours around and inside the wear mark are indicative of a strain field, which in the absence of applied stresses can be interpreted as plastic deformation and/or elastic residual stresses. Often, localized plastic deformation during loading would lead to residual stresses during unloading; therefore, bend contours reflect a mix of elastic and plastic strains. The wear debris is observed outside the wear mark. The enlarged view of the wear debris in Fig. 8.49c shows that much of the debris is ribbon-like, indicating that material is removed by a cutting process via plastic deformation, which is consistent with the SEM observations. The diffraction pattern from inside the wear mark is similar to that of virgin silicon, showing no evidence of any phase trans-

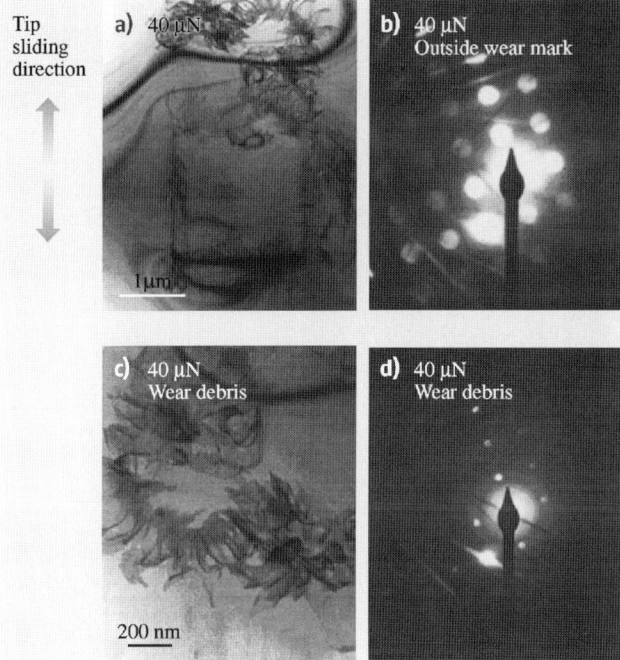

Fig. 8.49. Bright-field TEM micrographs (*left*) and diffraction patterns (*right*) of wear mark (**a**), (**b**) and wear debris (**c**), (**d**) in Si(100) produced at a normal load of 40 μN and one scan cycle. Bend contours around and inside wear mark are observed

formation (amorphization) during wear. A selected area diffraction pattern of the wear debris shows some diffuse rings, which indicates the existence of amorphous material in the wear debris, confirmed as silicon oxide products from chemical analysis. It is known that plastic deformation occurs by generation and propagation of dislocations. No dislocation activity or cracking was observed at 40 μN. However, dislocation arrays could be observed at 80 μN. Figure 8.50 shows the TEM micrographs of the worn region at 80 μN; for better observation of the worn surface, wear debris was moved out of the wear mark by using AFM with a large-area scan at 300 nN after the wear test. The existence of dislocation arrays confirms that material removal occurs by plastic deformation. This corroborates the observations made in scratch tests with a ramped load in the previous section. It is concluded that the material on the microscale at high loads is removed by plastic deformation with a small contribution from elastic fracture [119].

To understand wear mechanisms, the evolution of wear can be studied using AFM. Figure 8.51 shows evolution of wear marks of a DLC-coated disk sample. The data illustrate how the microwear profile for a load of 20 μN develops as a function of the number of scanning cycles [27]. Wear is not uniform, but is initiated at the nanoscratches. Surface defects (with high surface energy) present at nanoscratches act as initiation sites for wear. Coating deposition may also not be uniform on and

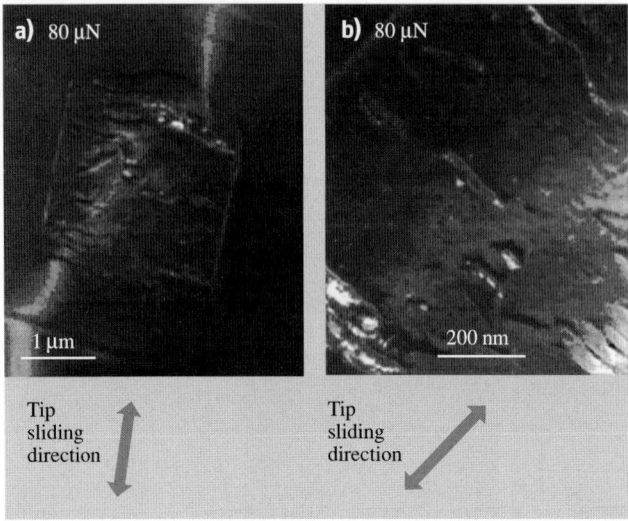

Fig. 8.50. (a) Bright-field and (b) weak-beam TEM micrographs of wear mark in Si(100) produced at a normal load of 80 µN and one scan cycle showing bend contours and dislocations [119]

near nanoscratches, which may lead to coating delamination. Thus, scratch-free surfaces will be relatively resistant to wear.

Wear precursors (precursors to measurable wear) can be studied by making surface potential measurements [66–68]. The contact potential difference or simply the surface potential between two surfaces depends on a variety of parameters such as the electronic work function, adsorption, and oxide layers. The surface potential map of an interface gives a measure of changes in the work function which, is sensitive to both physical and chemical conditions of the surfaces including structural and chemical changes. Before material is actually removed in a wear process, the surface experiences stresses that result in surface and subsurface changes of structure and/or chemistry. These can cause changes in the measured potential of a surface. An AFM tip allows mapping of surface potential with nanoscale resolution. Surface height and change in surface potential maps of a polished single-crystal aluminum (100) sample abraded using a diamond tip at loads of 1 µN and 9 µN, are shown in Fig. 8.52a [Note that the sign of the change in surface potential is reversed here from that in *DeVecchio* and *Bhushan* [66]]. It is evident that both abraded regions show a large potential contrast (≈ 0.17 V), with respect to the non-abraded area. The black region in the lower right-hand part of the topography scan shows a step that was created during the polishing phase. There is no potential contrast between the high region and the low region of the sample, indicating that the technique is independent of surface height. Figure 8.52b shows a close-up scan of the upper (low-load) wear region in Fig. 8.52a. Notice that, while there is no detectable change in the surface topography, there is nonetheless a large change in the potential of the surface in the

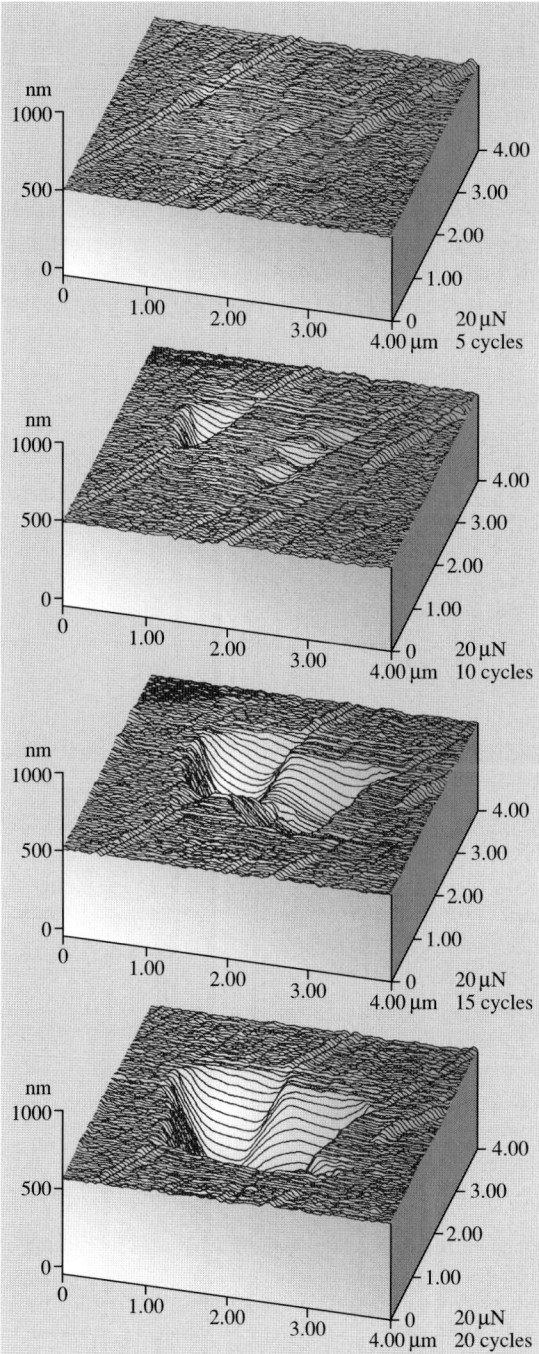

Fig. 8.51. Surface plots of diamond-like carbon-coated thin-film disk showing the worn region; the normal load and number of test cycles are indicated [27]

worn region. Indeed, the wear mark of Fig. 8.52b might not be visible at all in the topography map were it not for the noted absence of wear debris generated nearby and then swept off during the low-load scan. Thus, even in the case of zero wear (no measurable deformation of the surface using AFM), there can be a significant change in the surface potential inside the wear mark, which is useful for the study of wear precursors. It is believed that the removal of the thin contaminant layer including the natural oxide layer gives rise to the initial change in surface potential. The structural changes, which precede generation of wear debris and/or measurable wear scars, occur under ultra-low loads in the top few nanometers of the sample, and are primarily responsible for the subsequent changes in surface potential.

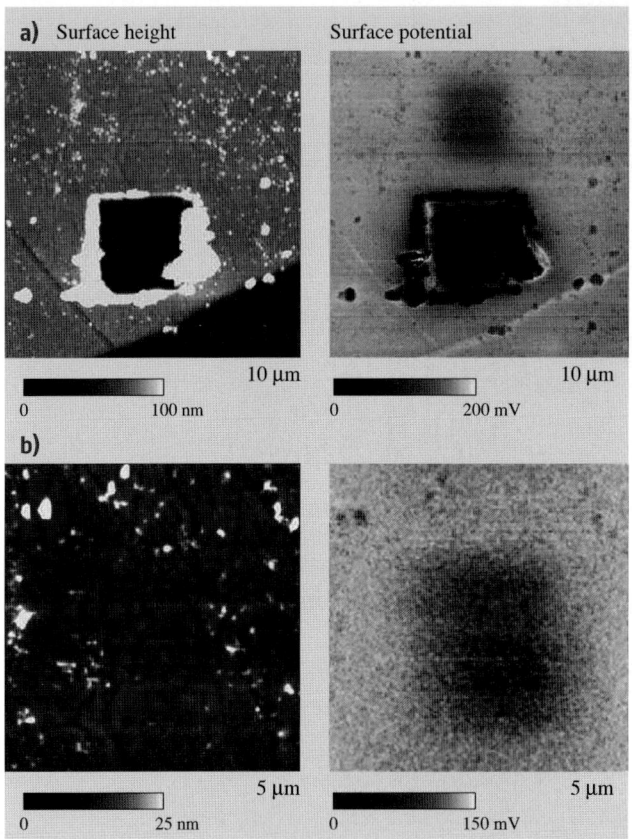

Fig. 8.52. (a) Surface height and change in surface potential maps of wear regions generated at 1 μN (*top*) and 9 μN (*bottom*) on a single-crystal aluminum sample showing bright contrast in the surface potential map on the worn regions. (b) Close-up of upper (low-load) wear region [66]

8.4.4 In Situ Characterization of Local Deformation

In situ surface characterization of local deformation of materials and thin films is carried out using a tensile stage inside an AFM. Failure mechanisms of polymeric thin films under tensile load were studied by *Bobji* and *Bhushan* [61,62]. The specimens were strained at a rate of $4 \times 10^{-3}\%$ per second and AFM images were captured at different strains up to about 10% to monitor generation and propagation of cracks and deformation bands.

Bobji and *Bhushan* [61,62] studied three magnetic tapes with thickness ranging from 7 to 8.5 μm. One of these was with acicular-shaped metal particle (MP) coating and the other two with metal-evaporated (ME) coating and with and without a thin diamond-like carbon (DLC) overcoat both on a polymeric substrate and all with particulate back-coating [15]. They also studied a polyethylene terephthalate (PET) substrate with a thickness of 6 μm. They reported that cracking of the coatings started at about 1% strain for all tapes, well before the substrate starts to yield

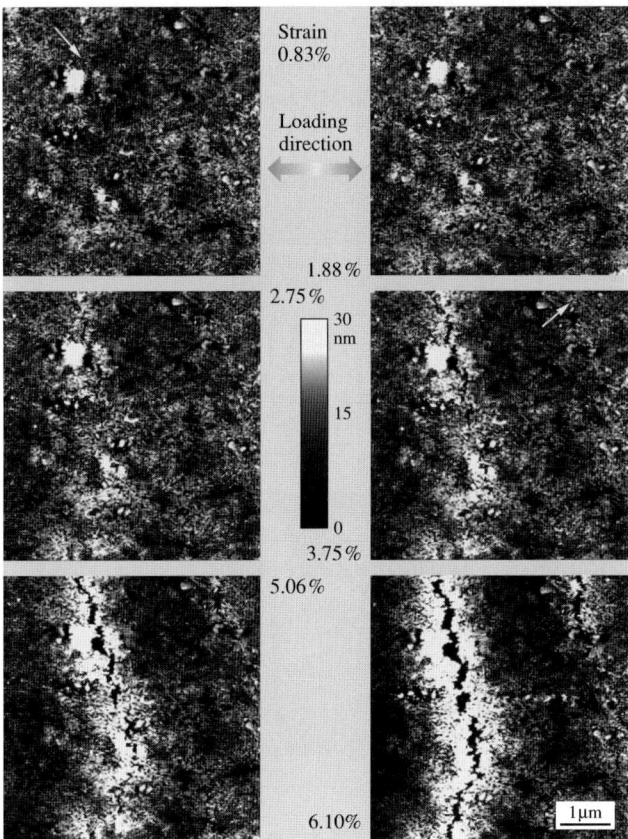

Fig. 8.53. Topographical images of the MP magnetic tape at different strains [61]

at about 2% strain. Figure 8.53 shows the topographical images of the MP tape at different strains. At 0.83% strain, a crack can be seen, originating at the marked point. As the tape is further stretched along the direction, as shown in Fig. 8.53, the crack propagates along the shorter boundary of the ellipsoidal particle. However, the general direction of the crack propagation remains perpendicular to the direction of the stretching. The length, width, and depth of the cracks increase with strain, and at the same time newer cracks keep on nucleating and propagating with reduced crack spacing. At 3.75% strain, another crack can be seen nucleating. This crack continues to grow parallel to the first one. When the tape is unloaded after stretching up to a strain of about 2%, i.e. within the elastic limit of the substrate, the cracks rejoin perfectly and it is impossible to determine the difference from the unstrained tape.

Figure 8.54 shows topographical images of the three magnetic tapes and the PET substrate after being strained to 3.75%, which is well beyond the elastic limit of the substrate. MP tape develops numerous short cracks perpendicular to the direction of loading. In tapes with metallic coating, the cracks extend throughout the tape width. In ME tape with DLC coating, there is a bulge in the coating around the primary cracks that are initiated when the substrate is still elastic, like crack A in the figure. The white band on the right-hand side of the figure is the bulge of another crack. The secondary cracks, such as B and C, are generated at higher strains and are straighter compared to the primary cracks. In ME tape which has a Co – O film

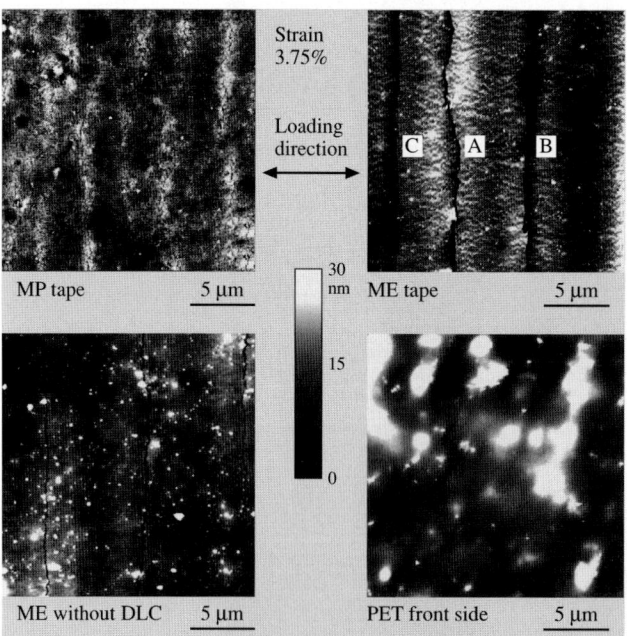

Fig. 8.54. Comparison of crack morphologies at 3.75% strain in three magnetic tapes and PET substrate. Cracks B and C, nucleated at higher strains, are more linear than crack A [62]

on a PET substrate, with a thickness ratio of 0.03, both with and without DLC coating, no difference is observed in the rate of growth between primary and secondary cracks. The failure is cohesive with no bulging of the coating. This seems to suggest that the DLC coating has residual stresses that relax when the coating cracks, causing delamination. Since the stresses are already relaxed, the secondary crack does not result in delamination. The presence of the residual stress is confirmed by the fact that a free-standing ME tape curls up (in a cylindrical form with its axis perpendicular to the tape length) with a radius of curvature of about 6 mm and the ME tape without the DLC does not curl. The magnetic coating side of the PET substrate is much smoother at shorter scan lengths. However, in 20 μm scans it has a lot of bulging out, which appears as white spots in the figure. These spots change shape even while scanning the samples in tapping mode at very low contact forces.

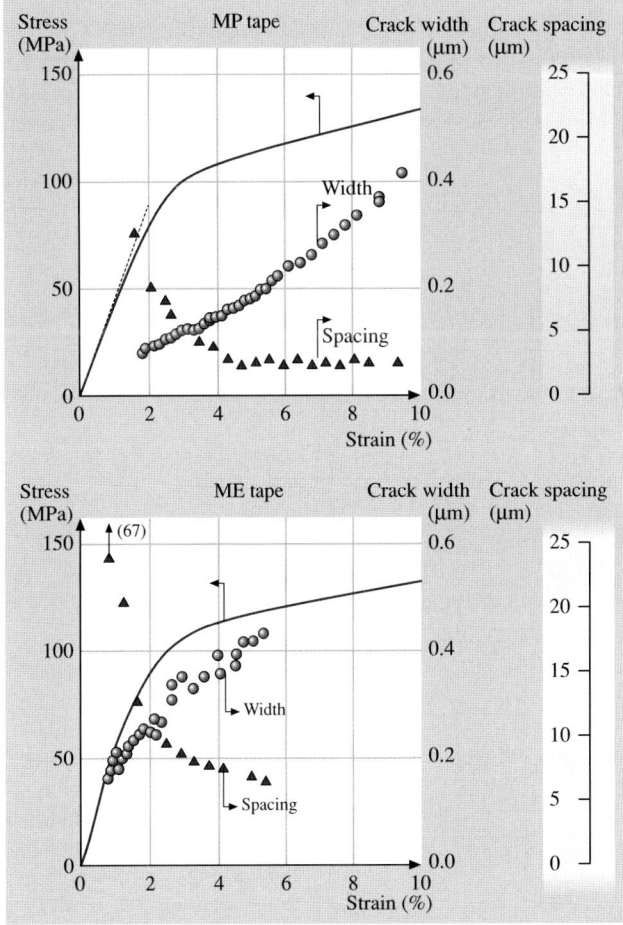

Fig. 8.55. Variation of stress, crack width, and crack spacing with strain in two magnetic tapes [61]

The variation of average crack width and average crack spacing with strain is plotted in Fig. 8.55. The crack width is measured at a spot along a given crack over a distance of 1 μm in the 5 μm scan image at different strains. The crack spacing is obtained by averaging the inter-crack distance measured in five separate 50 μm scans at each strain. It can be seen that the cracks nucleate at a strain of about 0.7–1.0%, well within the elastic limit of the substrate. There is a definite change in the slope of the load–displacement curve at the strain where cracks nucleate and the slope after that is closer to the slope of the elastic portion of the substrate. This would mean that most of the load is supported by the substrate once the coating fails by cracking.

Fatigue experiments can be performed by applying a cyclic stress amplitude with a certain mean stress [63]. Fatigue life was determined by the first occurrence of cracks. Experiments were performed at various constant mean stresses and with a range of cyclic stress amplitudes for each mean stress value for various magnetic tapes. The number of cycles to failure were plotted as a function of stress state to obtain a so-called S–N (stress–life) diagram. As the stress is decreased, there is a stress value for which no failure occurs. This stress is termed the endurance limit or simply the fatigue limit. Figure 8.56 shows the S–N curve for an ME tape and

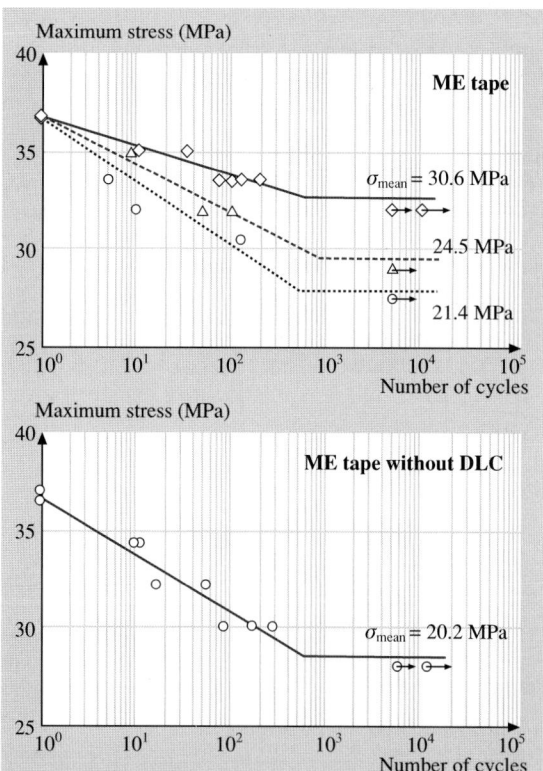

Fig. 8.56. S–N curve for two magnetic tapes with maximum stress plotted on the ordinate and number of cycles to failure on the abscissa. The data points marked with *arrows* indicate tests for which no failure (cracking) was observed in the scan area, even after a large number of cycles (10,000)

an ME tape without DLC. For the ME tape, the endurance limit is seen to go down with decreasing mean stress. This is consistent with the literature and is because for lower mean stress the corresponding stress amplitude is relatively high and this causes failure. The endurance limit is found to be almost the same for all three mean stresses. In the case of ME tape without DLC, the critical number of cycles is also found to be in the same range.

In situ surface characterization of unstretched and stretched films has been used to measure the Poisson's ratio of polymeric thin films by *Bhushan* [120]. Uniaxial tension is applied by the tensile stage. Surface height profiles obtained from the AFM images of unstretched and stretched samples are used to monitor the changes in displacements of the polymer films in the longitudinal and lateral directions simultaneously.

8.4.5 Nanofabrication/Nanomachining

An AFM can be used for nanofabrication/nanomachining by extending the microscale scratching operation [4, 13, 27, 65]. Figure 8.57 shows two examples of nanofabrication. The patterns were created on a single-crystal silicon (100) wafer by scratching the sample surface with a diamond tip at specified locations and scratching angles. Each line is scribed manually at a normal load of 15 µN and a writing speed of 0.5 µm/s. The separation between lines is about 50 nm and the variation in line width is due to the tip asymmetry. Nanofabrication parameters – normal load, scanning speed, and tip geometry – can be controlled precisely to control depth and length of the devices.

Nanofabrication using mechanical scratching has several advantages over other techniques. Better control over the applied normal load, scan size, and scanning speed can be used for nanofabrication of devices. Using the technique, nanofabrication can be performed on any engineering surface. Use of chemical etching or reactions is not required and this dry nanofabrication process can be used where the use of chemicals and electric fields is prohibited. One disadvantage of this technique is the formation of debris during scratching. At light loads, debris formation is not a problem compared to high-load scratching. However, debris can be removed easily from the scan area at light loads during scanning.

8.5 Indentation

Mechanical properties on relevant scales are needed for the analysis of friction and wear mechanisms. Mechanical properties, such as hardness and Young's modulus of elasticity can be determined on the micro- to picoscales using the AFM [23, 27, 44, 50] and a depth-sensing indentation system used in conjunction with an AFM [49, 121–123].

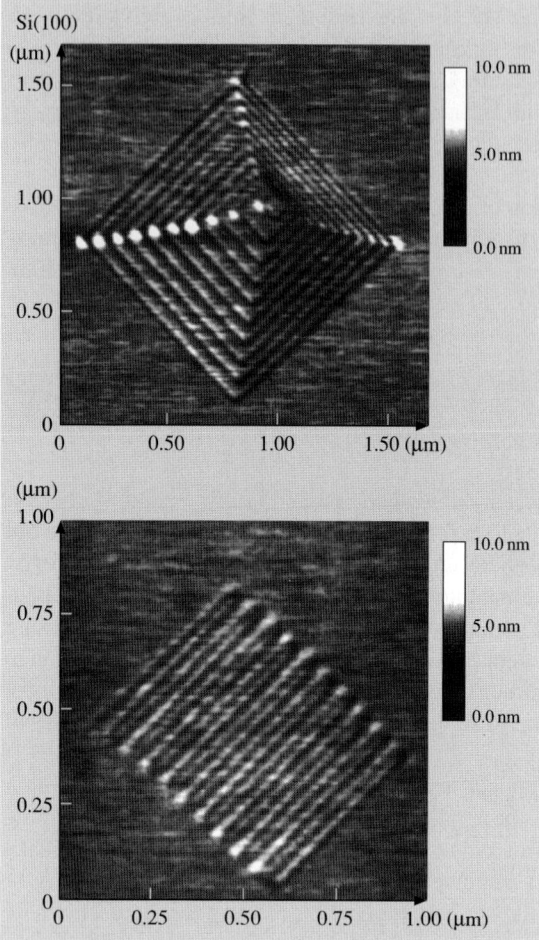

Fig. 8.57. *Top* Trim and *bottom* spiral patterns generated by scratching a Si(100) surface using a diamond tip at a normal load of 15 µN and writing speed of 0.5 µm/s

8.5.1 Picoindentation

Indentability on the sub-nanometer scale of soft samples can be studied in the force calibration mode (Fig. 8.6) by monitoring the slope of cantilever deflection as a function of sample traveling distance after the tip is engaged and the sample is pushed against the tip. For a rigid sample, cantilever deflection equals the sample traveling distance, but the former quantity is smaller if the tip indents the sample. In an example for a polymeric magnetic tape shown in Fig. 8.58, the line in the left portion of the figure is curved with a slope of less than 1 shortly after the sample touches the tip, which suggests that the tip has indented the sample [23]. Later, the slope is equal to 1, suggesting that the tip no longer indents the sample. This observation indicates that the tape surface is soft locally (polymer-rich) but hard (as a result of magnetic particles) underneath. Since the curves in extending and retract-

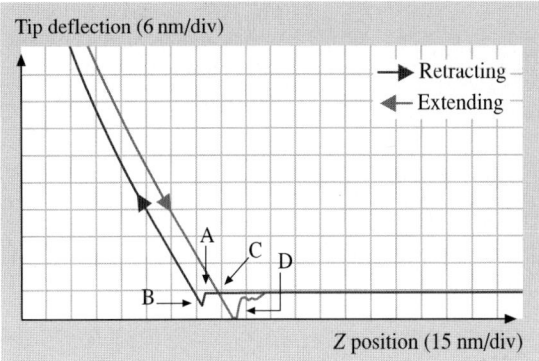

Fig. 8.58. Tip deflection (normal load) as a function of the z (separation distance) curve for a polymeric magnetic tape [23]

ing modes are identical, the indentation is elastic up to the maximum load of about 22 nN used in the measurements.

Detection of transfer of material on a nanoscale is possible with the AFM. Indentation of C_{60}-rich fullerene films with an AFM tip has been shown [48] to result in the transfer of fullerene molecules to the AFM tip, as indicated by discontinuities in the cantilever deflection as a function of sample traveling distance in subsequent indentation studies.

8.5.2 Nanoscale Indentation

The indentation hardness of surface films with an indentation depth as small as about 1 nm can be measured using an AFM [13, 49, 50]. Figure 8.59 shows the greyscale plots of indentation marks made on Si(111) at normal loads of 60, 65, 70 and 100 µN. Triangular indents can be clearly observed with very shallow depths. At a normal load of 60 µN, indents are observed and the depth of penetration is about 1 nm. As the normal load is increased, the indents become clearer and the indentation depth increases. For the case of hardness measurements at shallow depths on the same order as variations in surface roughness, it is desirable to subtract the original (unindented) map from the indent map for accurate measurement of the indentation size and depth [27].

To make accurate measurements of hardness at shallow depths, a depth-sensing nano/picoindentation system (Fig. 8.9) is used [49]. Figure 8.60 shows the load–displacement curves at different peak loads for Si(100). Loading/unloading curves often exhibit sharp discontinuities, particularly at high loads. Discontinuities, also referred to as pop-ins, during the initial part of the loading part of the curve mark a sharp transition from pure elastic loading to a plastic deformation of the specimen surface, correspond to an initial yield point. The sharp discontinuities in the unloading part of the curves are believed to be due to the formation of lateral cracks which form at the base of the median crack, which results in the surface of the specimen being thrust upward. Load–displacement data at residual depths as low as about 1 nm can be obtained. The indentation hardness of surface films has been

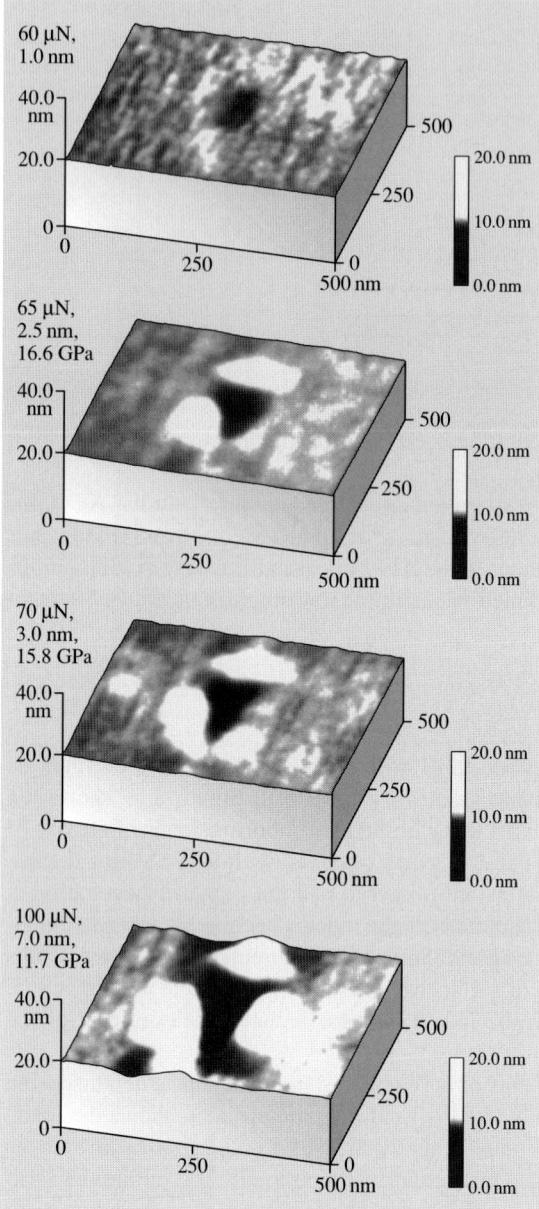

Fig. 8.59. Greyscale plots of indentation marks on the Si(111) sample at various indentation loads. Loads, indentation depths and hardness values are listed in the figure [50]

measured for various materials including Si(100) up to a peak load of 500 μN and Al(100) up to a peak load of 2000 μN by *Bhushan* et al. [49] and *Kulkarni* and *Bhushan* [121–123]. The hardnesses of single-crystal of silicon and single-crystal aluminum on a nanoscale are found to be higher than on a microscale, Fig. 8.61.

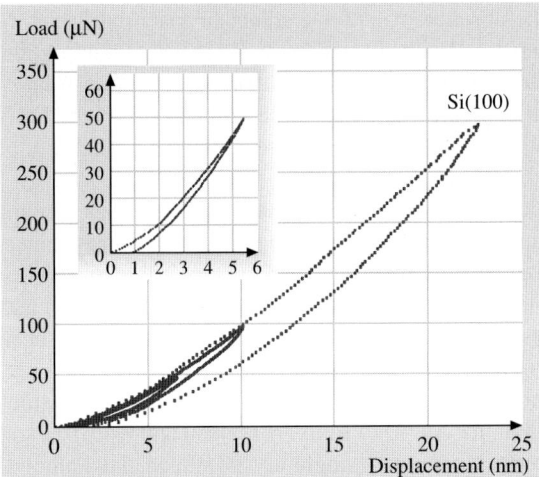

Fig. 8.60. Load–displacement curves at various peak loads for Si(100) [49]

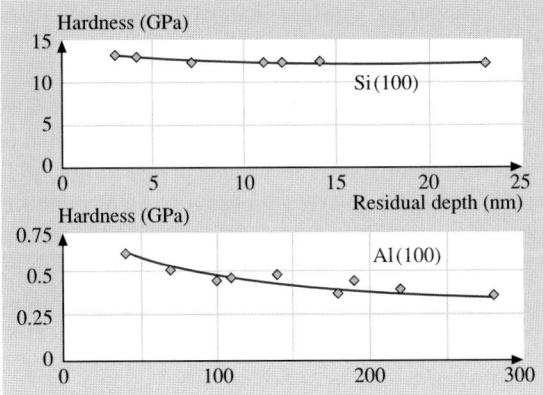

Fig. 8.61. Indentation hardness as a function of residual indentation depth for Si(100) [49], and Al(100) [121]

Microhardness has also been reported to be higher than that on the millimeter scale by several investigators. The data reported to date show that hardness exhibits the scale (size) effect.

During loading, generation and propagation of dislocations is responsible for plastic deformation. A strain gradient plasticity theory has been developed for micro/nanoscale deformations, and is based on randomly created statistically stored and geometrically necessary dislocations [124, 125]. Large strain gradients inherent to small indentations lead to the accumulation of geometrically necessary dislocations, located in a certain subsurface volume, for strain compatibility reasons that cause enhanced hardening. The large strain gradients in small indentations require these dislocations to account for the large slope at the indented surface. These are a function of strain gradient, whereas statistically, stored dislocations are a function of strain. Based on this theory, scale-dependent hardness is given as

$$H = H_0 \sqrt{1 + \ell_d a}, \tag{8.22}$$

where H_0 is the hardness in the absence of strain gradient or macrohardness, ℓ_d is the material-specific characteristic-length parameter, and a is the contact radius. In addition to the role of the strain gradient plasticity theory, an increase in hardness with a decrease in indentation depth can possibly be rationalized on the basis that, as the volume of deformed material decreases, there is a lower probability of encountering material defects.

Bhushan and *Koinkar* [44] have used AFM measurements to show that ion implantation of silicon surfaces increases their hardness and thus their wear resistance. Formation of surface alloy films with improved mechanical properties by ion implantation is of growing technological importance as a means of improving the mechanical properties of materials. Hardness of 20-nm-thick DLC films have been measured by *Kulkarni* and *Bhushan* [123].

The creep and strain-rate effects (viscoelastic effects) of ceramics can be studied using a depth-sensing indentation system. *Bhushan* et al. [49] and *Kulkarni* and *Bhushan* [121–123] have reported that ceramics (single-crystal silicon and diamond-like carbon) exhibit significant plasticity and creep on a nanoscale. Figure 8.62a shows the load–displacement curves for single-crystal silicon at various peak loads held for 180 s. To demonstrate the creep effects, the load–displacement curves for a 500 μN peak load held for 0 and 30 s are also shown as an inset. Note that significant creep occurs at room temperature. Nanoindenter experiments conducted by *Li* et al. [126] exhibited significant creep only at high temperatures (greater than or equal to 0.25 times the melting point of silicon). The mechanism of dislocation glide plasticity is believed to dominate the indentation creep process on the macroscale. To study the strain-rate sensitivity of silicon, data at two different (constant) rates of loading are presented in Fig. 8.62b. Note that a change in the loading rate by a factor of about five results in a significant change in the load–displacement data. The viscoelastic effects observed here for silicon at ambient temperature could arise from the size effects mentioned earlier. Most likely, creep and strain-rate experi-

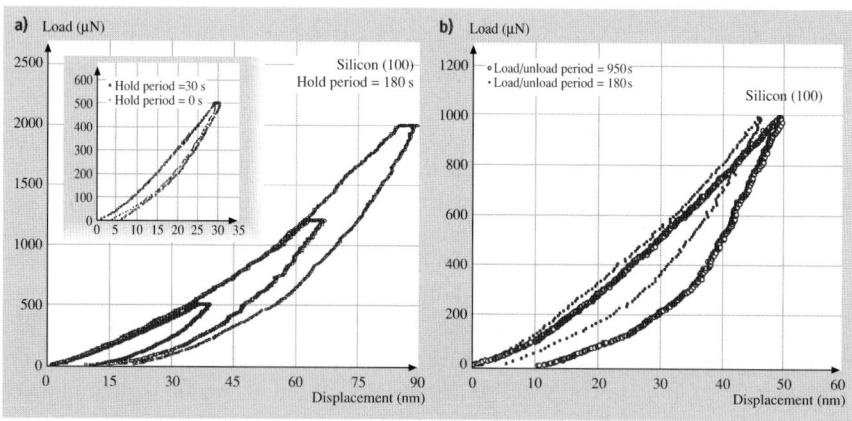

Fig. 8.62. (a) Creep behavior and (b) strain-rate sensitivity of Si(100) [49]

ments are being conducted on the hydrated films present on the silicon surface in ambient environment, and these films are expected to be viscoelastic.

8.5.3 Localized Surface Elasticity and Viscoelasticity Mapping

The Young's modulus of elasticity can be calculated from the slope of the indentation curve during unloading. However, these measurements provide a single-point measurement. By using the force modulation technique, it is possible to get localized elasticity maps of soft and compliant materials of near-surface regions with nanoscale lateral resolution. This technique has been successfully used for polymeric magnetic tapes, which consist of magnetic and nonmagnetic ceramic particles in a polymeric matrix. Elasticity maps of a tape can be used to identify relative distribution of hard magnetic and nonmagnetic ceramic particles on the tape surface, which has an effect on friction and stiction at the head–tape interface [15]. Figure 8.63 shows surface height and elasticity maps on a polymeric magnetic tape [54]. The elasticity image reveals sharp variations in surface elasticity due to the composite nature of the film. As can be clearly seen, regions of high elasticity do not always correspond to high or low topography. Based on a Hertzian elastic-contact analysis, the static indentation depth of these samples during the force modulation scan is estimated to be about 1 nm. We conclude that the contrast seen is influenced most strongly by material properties in the top few nanometers, independent of the composite structure beneath the surface layer.

By using phase contrast microscopy, it is possible to get phase contrast maps or the contrast in viscoelastic properties of near-surface regions with nanoscale lateral resolution. This technique has been successfully used for polymeric films and magnetic tapes which consist of ceramic particles in a polymeric matrix [57–60].

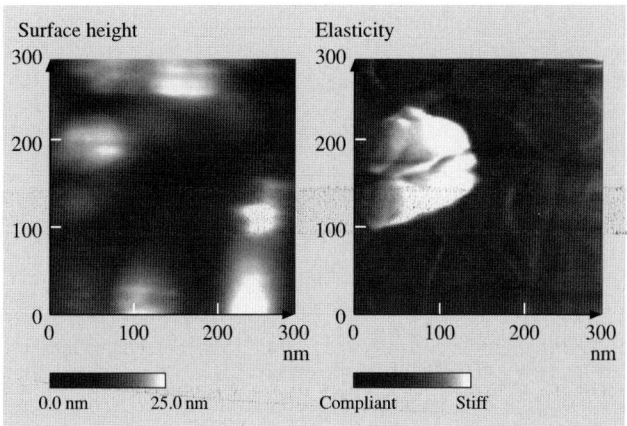

Fig. 8.63. Surface height and elasticity maps on a polymeric magnetic tape ($\sigma = 6.7$ nm and P–V = 32 nm; σ and P–V refer to the standard deviation of surface heights and peak-to-valley distance, respectively). The greyscale on the elasticity map is arbitrary [54]

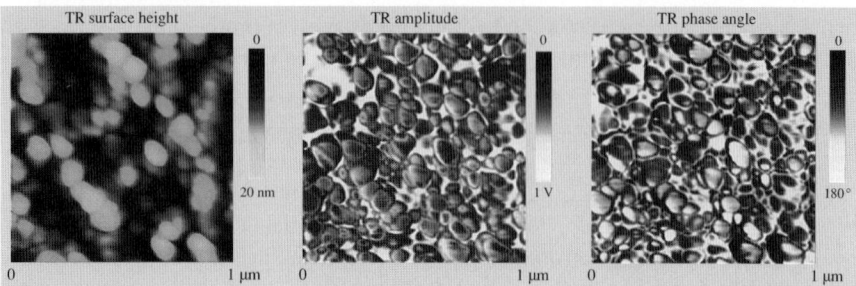

Fig. 8.64. Images of an MP tape obtained with TR mode II (constant deflection). TR mode II amplitude and phase angle images have the largest contrast among tapping, TR mode I and TR mode II techniques [60]

Figure 8.64 shows typical surface height, TR amplitude and TR phase-angle images for a MP tape using the TR mode II, described earlier. TR amplitude image provides contrast in lateral stiffness and TR phase-angle image provides contrast in viscoelastic properties. In TR amplitude and phase-angle images, the distribution of magnetic particles can be clearly seen which have better contrast than that in TR surface height image. MP tape samples show granular structure with elliptical shape magnetic particle aggregates (50–100 nm in diameter). Studies by *Scott* and *Bhushan* [57], *Bhushan* and *Qi* [58], and *Kasai* et al. [59] have indicated that the phase shift can be related to the energy dissipation through the viscoelastic deformation process between the tip and the sample. Recent theoretical analysis has established a quantitative correlation between the lateral surface properties (stiffness and viscoelasticity) of materials and amplitude/phase-angle shift in TR measurements [73]. The contrast in the TR amplitude and phase-angle images is due to the in-plane (lateral) heterogeneity of the surface. Based on the TR amplitude and phase-angle images, the lateral surface properties (lateral stiffness and viscoelasticity) mapping of materials can be obtained.

8.6 Boundary Lubrication

8.6.1 Perfluoropolyether Lubricants

The classical approach to lubrication uses freely supported multimolecular layers of liquid lubricants [5, 10, 15, 127]. The liquid lubricants are sometimes chemically bonded to improve their wear resistance [5, 10, 15]. Partially chemically bonded, molecularly thick perfluoropolyether (PFPE) films are used for lubrication of magnetic storage media because of their thermal stability and extremely low vapor pressure [15]. These are considered as potential candidate lubricants for MEMS/NEMS. Molecularly thick PFPEs are well suited because of the following properties: low surface tension and low contact angle, which allow easy spreading on surfaces and provide hydrophobic properties; chemical and thermal stability which minimize degradation under use; low vapor pressure, which provides low out-gassing; high adhesion to substrate via organic functional bonds; and good lubricity, which reduces contact surface wear.

For boundary lubrication studies, friction, adhesion and durability experiments have been performed on virgin Si(100) surfaces and silicon surfaces lubricated with various PFPE lubricants [39, 40, 42, 128]. Results of two of the PFPE lubricants will be presented here: Z-15 (with $-CF_3$ nonpolar end groups), $CF_3-O-(CF_2-CF_2-O)_m-(CF_2-O)_n-CF_3$ ($m/n \approx 2/3$) and Z-DOL (with $-OH$ polar end groups), $HO-CH_2-CF_2-O-(CF_2-CF_2-O)_m-(CF_2-O)_n-CF_2-CH_2-OH$ ($m/n \approx 2/3$). Z-DOL film was thermally bonded at 150 °C for 30 minutes and the unbonded fraction was removed by a solvent (bonded washed or BW) [15]. The thicknesses of Z-15 and Z-DOL films were 2.8 nm and 2.3 nm, respectively. Lubricant chain diameters of these molecules are about 0.6 nm and molecularly thick films generally lie flat on surfaces with high coverage.

The adhesive forces of Si(100), Z-15 and Z-DOL (BW) measured by a force calibration plot and plots of friction force versus normal load are summarized in

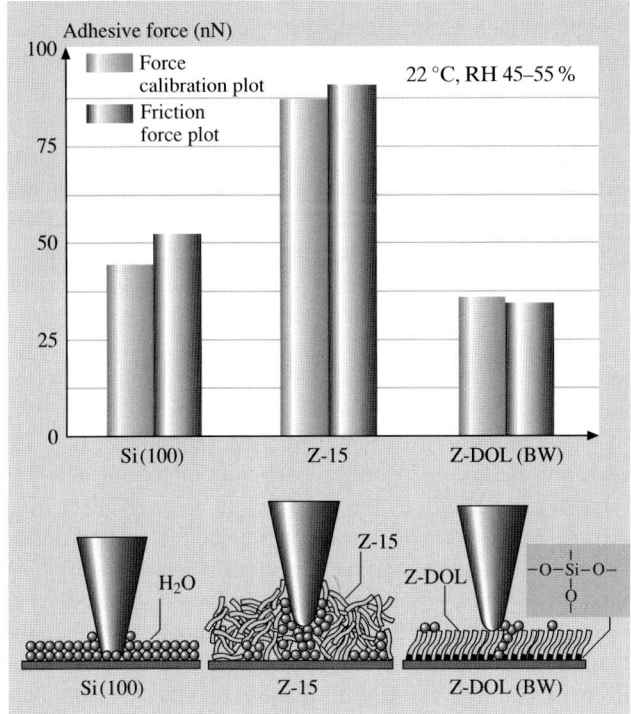

Fig. 8.65. Summary of the adhesive forces of Si(100) and Z-15 and Z-DOL (BW) films measured by force calibration plots and friction force versus normal-load plots in ambient air. The schematic (*bottom*) showing the effect of meniscus, formed between AFM tip and the surface sample, on the adhesive and friction forces [42]

Fig. 8.65 [42]. The results measured by these two methods are in good agreement. Figure 8.65 shows that the presence of the mobile Z-15 lubricant film increases the adhesive force compared to that of Si(100) by meniscus formation. Whereas, the presence of solid-phase Z-DOL (BW) film reduces the adhesive force compared to that of Si(100) because of the absence of mobile liquid. The schematic (bottom) in Fig. 8.65 shows the relative size and sources of meniscus. It is well known that the native oxide layer (SiO_2) on top of a Si(100) wafer exhibits hydrophilic properties, and some water molecules can be adsorbed onto this surface. The condensed water will form meniscus as the tip approaches the sample surface. The larger adhesive force in Z-15 is not only caused by the Z-15 meniscus, the nonpolarized Z-15 liquid does not have good wettability and strong bonding with Si(100). Consequently, in the ambient environment, the condensed water molecules from the environment will permeate through the liquid Z-15 lubricant film and compete with the lubricant molecules present on the substrate. The interaction of the liquid lubricant with the substrate is weakened, and a boundary layer of the liquid lubricant forms puddles [39, 40]. This dewetting allows water molecules to be adsorbed onto the Si(100) surface as aggregates along with Z-15 molecules. And both of them can form a meniscus while the tip approaches to the surface. Thus the dewetting of liquid Z-15 film results in higher adhesive force and poorer lubrication performance. In addition, as the Z-15 film is quite soft compared to the solid Si(100) surface, and penetration of the tip in the film occurs while pushing the tip down. This leads to the large area of the tip involved to form the meniscus at the tip–liquid (mixture of Z-15 and water) interface. It should also be noted that Z-15 has a higher viscosity than water, therefore Z-15 film provides higher resistance to motion and coefficient of friction. In the case of Z-DOL (BW) film, both of the active groups of Z-DOL molecules are mostly bonded on Si(100) substrate, thus the Z-DOL (BW) film has low free surface energy and cannot be displaced readily by water molecules or readily adsorb water molecules. Thus, the use of Z-DOL (BW) can reduce the adhesive force.

To study the velocity effect on friction and adhesion, the variation of friction force, adhesive force, and coefficient of friction of Si(100), Z-15 and Z-DOL(BW) as a function of velocity are summarized in Fig. 8.66 [42]. It indicates that, for silicon wafers, the friction force decreases logarithmically with increasing velocity. For Z-15, the friction force decreases with increasing velocity up to 10 μm/s, after which it remains almost constant. The velocity has a very small effect on the friction force of Z-DOL (BW); it reduced slightly only at very high velocity. Figure 8.66 also indicates that the adhesive force of Si(100) is increased when the velocity is higher than 10 μm/s. The adhesive force of Z-15 is reduced dramatically when the velocity increases to 20 μm/s, after which it is reduced slightly. The adhesive force of Z-DOL (BW) also decreases at high velocity. In the tested velocity range, only the coefficient of friction of Si(100) decreases with velocity, while the coefficients of friction of Z-15 and Z-DOL (BW) almost remain constant. This implies that the friction mechanisms of Z-15 and Z-DOL (BW) do not change with the variation of velocity.

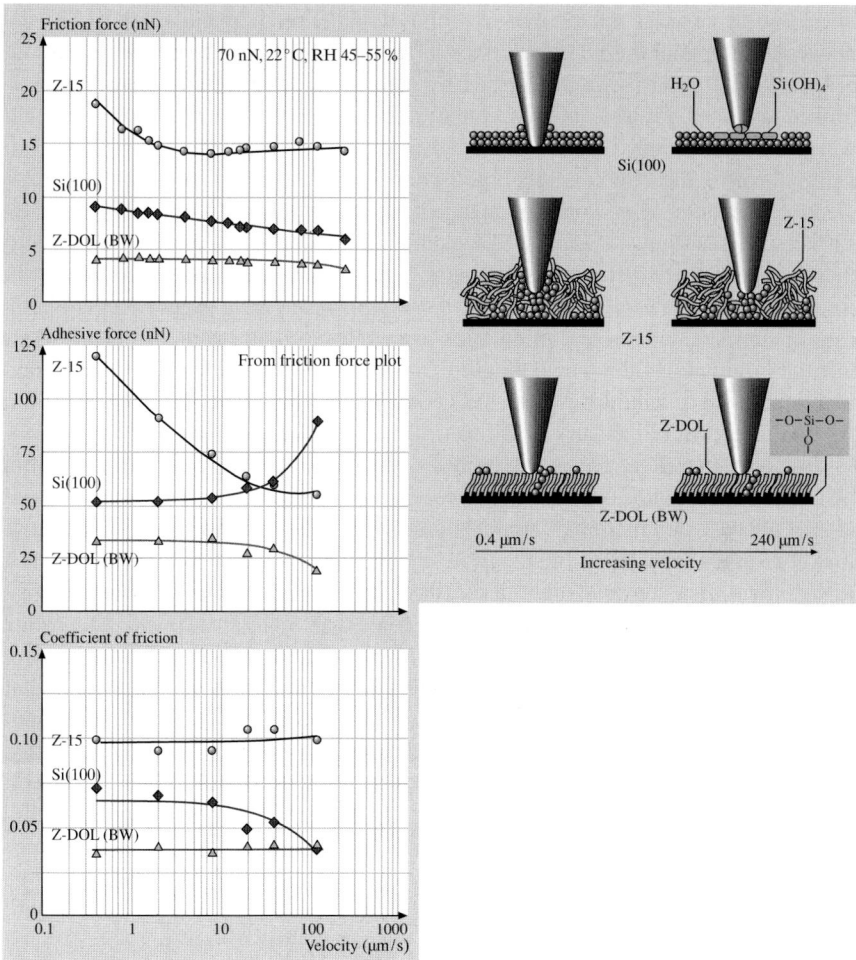

Fig. 8.66. The influence of velocity on the friction force, adhesive force and coefficient of friction of Si(100) and Z-15 and Z-DOL (BW) films at 70 nN, in ambient air. The schematic (*right*) shows the change of surface composition (by tribochemical reaction) and formation of meniscus while increasing the velocity [42]

The mechanisms of the effect of velocity on the adhesion and friction are explained based on the schematics shown in Fig. 8.66 (right) [42]. For Si(100), tribochemical reactions play a major role. Although, at high velocity, the meniscus is broken and does not have enough time to rebuild, the contact stresses and high velocity lead to tribochemical reactions of the Si(100) wafer (which has native oxide (SiO_2)), and Si_3N_4 tip with water molecules, forming $Si(OH)_4$. The $Si(OH)_4$ is removed and continuously replenished during sliding. The $Si(OH)_4$ layer between the tip and Si(100) surface is known to be of low shear strength and causes a decrease in friction force and coefficient of friction [10, 17]. The chemical bonds of Si – OH between the tip and the Si(100) surface induce large adhesive force. For Z-15 film, at high velocity the meniscus formed by condensed water and Z-15 molecules is broken and does not have enough time to rebuild; therefore, the adhesive force and consequently friction force is reduced. The friction mechanisms for the Z-15 film still is shearing the same viscous liquid even at high velocity range, thus the coefficient of friction of Z-15 does not change with velocity. For Z-DOL (BW) film, the surface can adsorb few water molecules in ambient condition, and at high velocity these molecules are displaced, which is responsible for a slight decrease in friction force and adhesive force. *Koinkar* and *Bhushan* [40, 40] have suggested that, in the case of samples with mobile films, such as condensed water and Z-15 films, alignment of liquid molecules (shear thinning) is responsible for the drop in friction force with increasing scanning velocity. This could be another reason for the decrease in friction force for Si(100) and Z-15 film with velocity in this study.

To study the effect of relative humidity on friction and adhesion, the variation of friction force, adhesive force, and coefficient of friction of Si(100), Z-15, and Z-DOL (BW) as a function of relative humidity is shown in Fig. 8.67 [42]. It shows that, for Si(100) and Z-15 film, the friction force increases with relative humidity up to 45%, and then shows a slight decrease with further increases in the relative humidity. Z-DOL (BW) has a smaller friction force than Si(100) and Z-15 in the whole testing range and its friction force shows an apparent relative ncrease when the relative humidity is higher than 45%. For Si(100), Z-15 and Z-DOL (BW), their adhesive forces increase with relative humidity, and their coefficients of friction increase with relative humidity up to 45%, after which they decrease with further increases of relative humidity. It is also observed that the effect of humidity on Si(100) depends on the history of the Si(100) sample. As the surface of Si(100) wafer readily adsorb water in air, without any pretreatment the Si(100) used in our study almost reaches to its saturate stage of adsorbed water, and is responsible for a smaller effect during increasing relative humidity. However, once the Si(100) wafer was thermally treated by baking at 150 °C for 1 hour, a bigger effect was observed.

The schematic (left) in Fig. 8.67 shows that Si(100), because of its high free surface energy, can adsorb more water molecules while increasing relative humidity [42]. As discussed earlier, for the Z-15 film in the humid environment, the condensed water from the humid environment competes with the lubricant film present on the sample surface, and the interaction of the liquid lubricant film with the silicon substrate is weakened and a boundary layer of the liquid lubricant forms puddles. This dewetting allows water molecules to be adsorbed onto the Si(100) substrate mixed with Z-15 molecules [39, 40]. Obviously, more water molecules can be adsorbed onto the Z-15 surface with increasing relative humidity. The higher number of adsorbed water molecules in the case of Si(100), along with the lubricant molecules in the Z-15 film case, form a larger water meniscus, which leads to an increase of friction force, adhesive force, and coefficient of friction of Si(100) and Z-15 with humidity. However, at the very high humidity of 70%, large quantities of adsorbed water can form a continues water layer that separates the tip

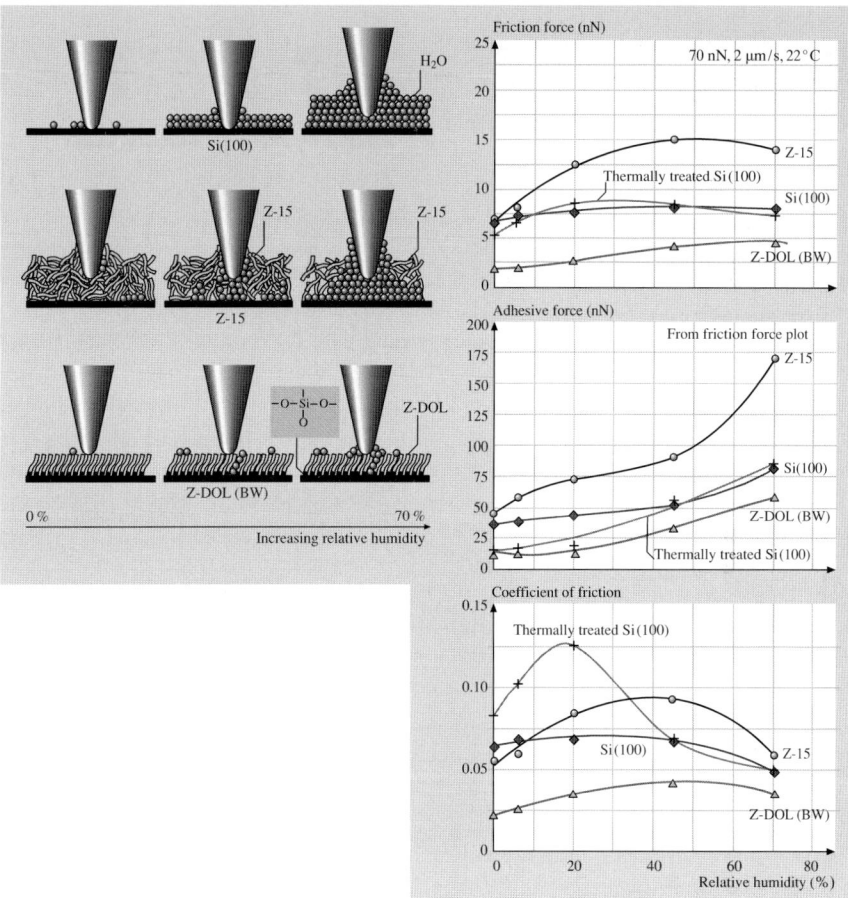

Fig. 8.67. The influence of relative humidity on the friction force, adhesive force, and coefficient of friction of Si(100) and Z-15 and Z-DOL (BW) films at 70 nN, 2 μm/s, and in 22 °C air. The schematic (*left*) shows the change of meniscus while increasing the relative humidity. In this figure, the thermally treated Si(100) represents the Si(100) wafer that was baked at 150 °C for 1 hour in an oven (in order to remove the adsorbed water) just before it was placed in the 0% RH chamber [42]

and sample surface and acts as a kind of lubricant, which causes a decrease in the friction force and coefficient of friction. For the Z-DOL (BW) film, because of its hydrophobic surface properties, water molecules can be adsorbed at humidity higher than 45%, and this causes an increase in the adhesive force and friction force.

To study the effect of temperature on friction and adhesion, the variation of friction force, adhesive force, and coefficient of friction of Si(100), Z-15 and Z-DOL (BW) as a function of temperature are summarized in Fig. 8.68 [42]. It shows that increasing temperature causes a decrease of the friction force, adhesive force and

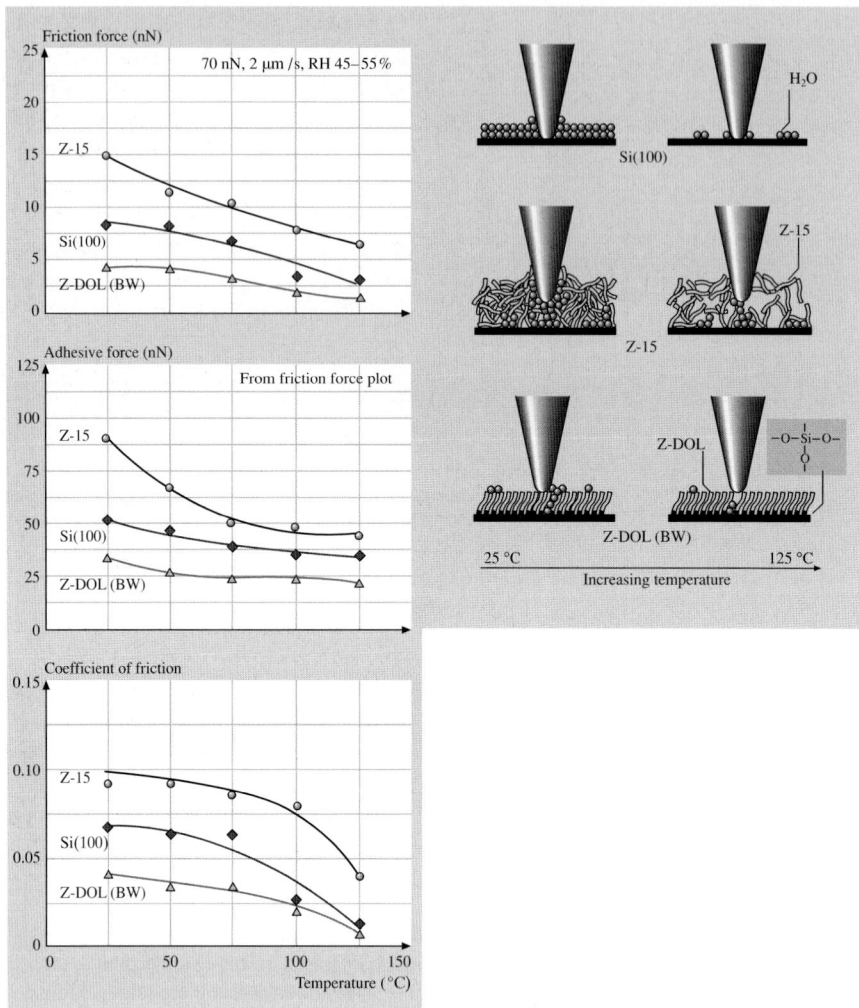

Fig. 8.68. The influence of temperature on the friction force, adhesive force, and coefficient of friction of Si(100) and Z-15 and Z-DOL (BW) films at 70 nN, at 2 μm/s, and in RH 40–50% air. The schematic (*right*) shows that, at high temperature, desorption of water decreases the adhesive forces. And the reduced viscosity of Z-15 leads to the decrease of coefficient of friction. High temperature facilitates orientation of molecules in Z-DOL (BW) film, which results in a lower coefficient of friction [42]

coefficient of friction of Si(100), Z-15 and Z-DOL (BW). The schematic (right) in Fig. 8.68 indicates that, at high temperature, desorption of water leads to a decrease of the friction force, adhesive forces and coefficient of friction for all of the samples. For Z-15 film, the reduction of viscosity at high temperature also makes a contribution to the decrease of friction force and coefficient of friction. In the case of the

Z-DOL (BW) film, molecules are more easily oriented at high temperatures, which may be partly responsible for the low friction force and coefficient of friction.

As a brief summary, the influence of velocity, relative humidity, and temperature on the friction force of mobile Z-15 film is presented in Fig. 8.69 [42]. The changing trends are also addressed in this figure.

To study the durability of lubricant films at nanoscale, the friction of Si(100), Z-15, and Z-DOL (BW) as a function of the number of scanning cycles are shown

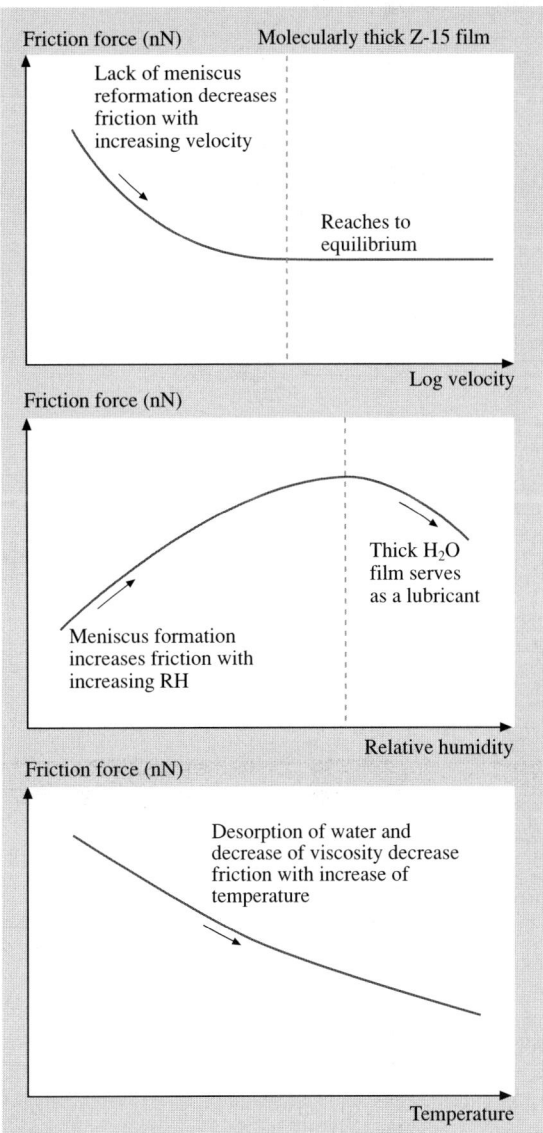

Fig. 8.69. Schematic shows the change of friction force of molecularly thick Z-15 films with log velocity, relative humidity, and temperature. The changing trends are also addressed in this figure [42]

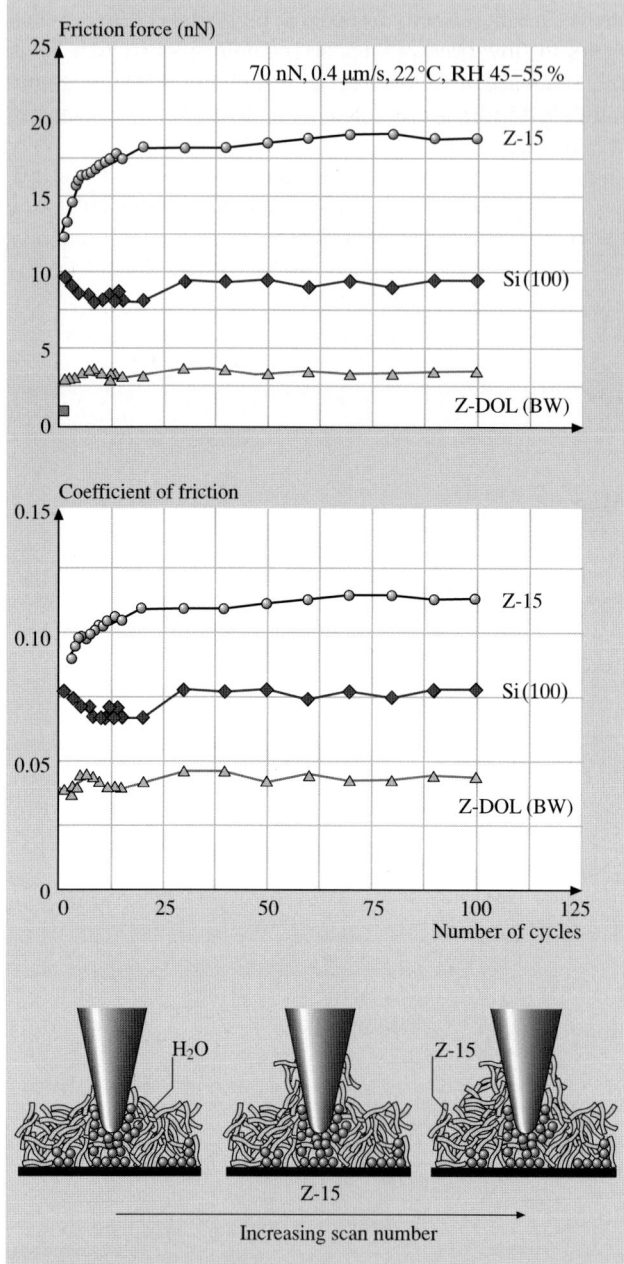

Fig. 8.70. Friction force and coefficient of friction versus number of sliding cycles for Si(100) and Z-15 and Z-DOL (BW) films at 70 nN, 0.8 μm/s, and in ambient air. The schematic (*bottom*) shows that some liquid Z-15 molecules can be attached onto the tip. The molecular interaction between the attached molecules onto the tip with the Z-15 molecules in the film results in an increase of the friction force with multiple scanning [42]

in Fig. 8.70 [42]. As observed earlier, the friction force and coefficient of friction of Z-15 are higher than that of Si(100), with the lowest values for Z-DOL(BW). During cycling, the friction force and coefficient of friction of Si(100) show a slight decrease during the initial few cycles and then remain constant. This is related to the removal of the top adsorbed layer. In the case of the Z-15 film, the friction force and coefficient of friction show an increase during the initial few cycles and then approach higher, stable values. This is believed to be caused by the attachment of Z-15 molecules to the tip. The molecular interaction between these attached molecules on the tip and molecules on the film surface is responsible for an increase in the friction. However, after several scans, this molecular interaction reaches an equilibrium and the friction force and coefficient of friction then remain constant. In the case of the Z-DOL (BW) film, the friction force and coefficient of friction start low and remain low during the entire test for 100 cycles. This suggests that Z-DOL (BW) molecules do not become attached or displaced as readily as Z-15.

8.6.2 Self-Assembled Monolayers

For lubrication of MEMS/NEMS, another effective approach involves the deposition of organized and dense molecular layers of long-chain molecules. Two common methods to produce monolayers and thin films are the Langmuir–Blodgett (LB) deposition and self-assembled monolayers (SAMs) by chemical grafting of molecules. LB films are physically bonded to the substrate by weak van der Waals attraction, while SAMs are chemically bonded via covalent bonds to the substrate. Because of the choice of chain length and terminal linking group that SAMs offer, they hold great promise for boundary lubrication of MEMS/NEMS. A number of studies have been conducted to study tribological properties of various SAMs [38, 41, 43, 129–135].

Bhushan and *Liu* [41] studied the effect of film compliance on adhesion and friction. They used hexadecane thiol (HDT), 1,1,biphenyl-4-thiol (BPT), and crosslinked BPT (BPTC) solvent deposited on Au(111) substrate, Fig. 8.71a. The average values and standard duration of the adhesive force and coefficient of friction are presented in Fig. 8.71b. Based on the data, the adhesive force and coefficient of frictions of SAMs are less than those of corresponding substrates. Among various films, HDT exhibits the lowest values. Based on stiffness measurements of various SAMs, HDT was the most compliant, followed by BPT and BPTC. Based on friction and stiffness measurements, SAMs with high-compliance long carbon chains exhibit low friction; chain compliance is desirable for low friction. Friction mechanism of SAMs is explained by a so-called *molecular spring* model (Fig. 8.72). According to this model, the chemically adsorbed self-assembled molecules on a substrate are just like assembled molecular springs anchored to the substrate. An asperity sliding on the surface of SAMs is like a tip sliding on the top of *molecular springs* or a *brush*. The molecular spring assembly has compliant features and can experience orientation and compression under load. The orientation of the molecular springs or brush under normal load reduces the shearing force at the interface, which in turn reduces the friction force. The orientation is determined by the spring constant of a single

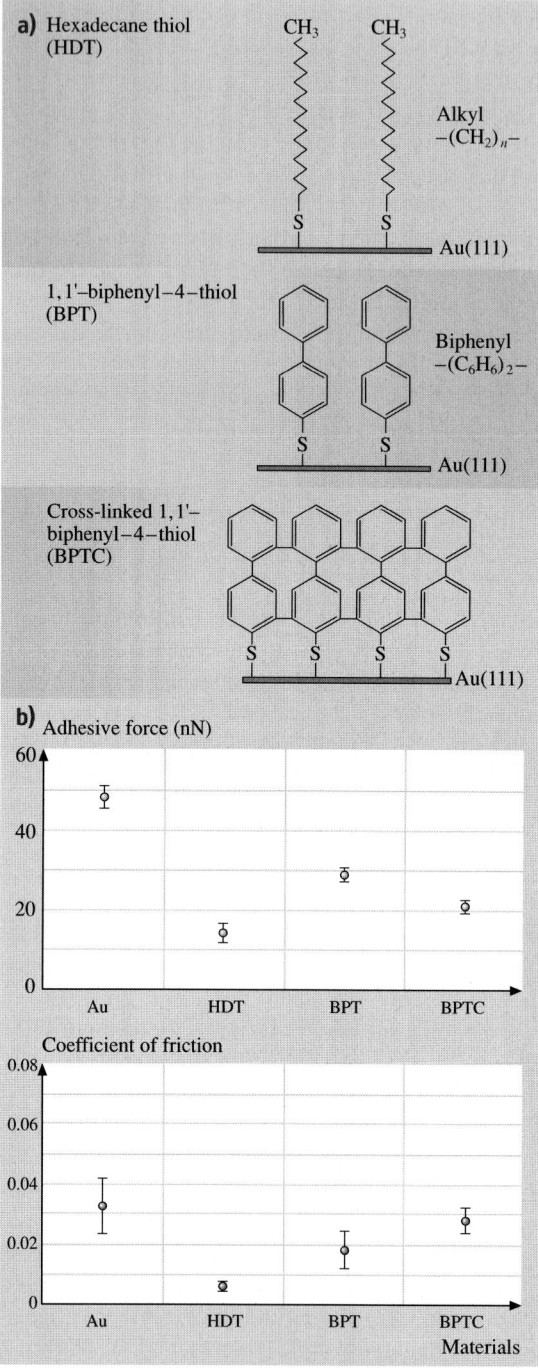

Fig. 8.71. (a) Schematics of the structures of hexadecane thiol and biphenyl thiol SAMs on Au(111) substrates, and (b) adhesive force and coefficient of friction of Au(111) substrate and various SAMs

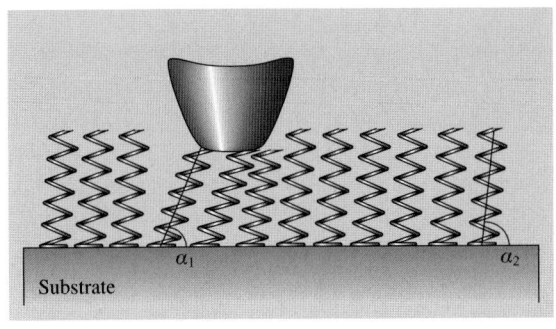

Fig. 8.72. Molecular spring model of SAMs. In this figure, $\alpha_1 < \alpha_2$, which is caused by the further orientation under the normal load applied by an asperity tip [41]

molecule as well as the interaction between the neighboring molecules, which can be reflected by packing density or packing energy. It should be noted that the orientation can lead to conformational defects along the molecular chains, which lead to energy dissipation.

The SAMs with high-compliance long carbon chains also exhibit the best wear resistance [41, 130]. In wear experiments, the wear depth as a function of normal load curves show a critical normal load. A representative curve is shown in Fig. 8.73. Below the critical normal load, SAMs undergo orientation; at the critical load SAMs wear away from the substrate due to weak interface bond strengths, while above the critical normal load severe wear takes place on the substrate.

Bhushan et al. [43], *Kasai* et al. [131], and *Tambe* and *Bhushan* [134] studied perfluorodecyltricholorosilane (PFTS), *n*-octyldimethyl (dimethylamino) silane (ODMS) ($n = 7$), and *n*-octadecylmethyl (dimethylamino) silane (ODDMS) ($n = 17$) vapor deposited on Si substrate, and octylphosphonate (OP) and octadecylphosphonate (ODP) on Al substrate, Fig. 8.74a. Figure 8.74b presents the contact angle, adhesive force, friction force, and coefficient of friction of two substrates and with

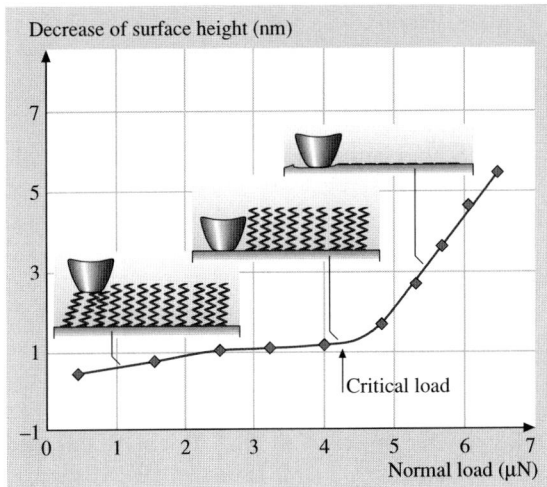

Fig. 8.73. Illustration of the wear mechanism of SAMs with increasing normal load [130]

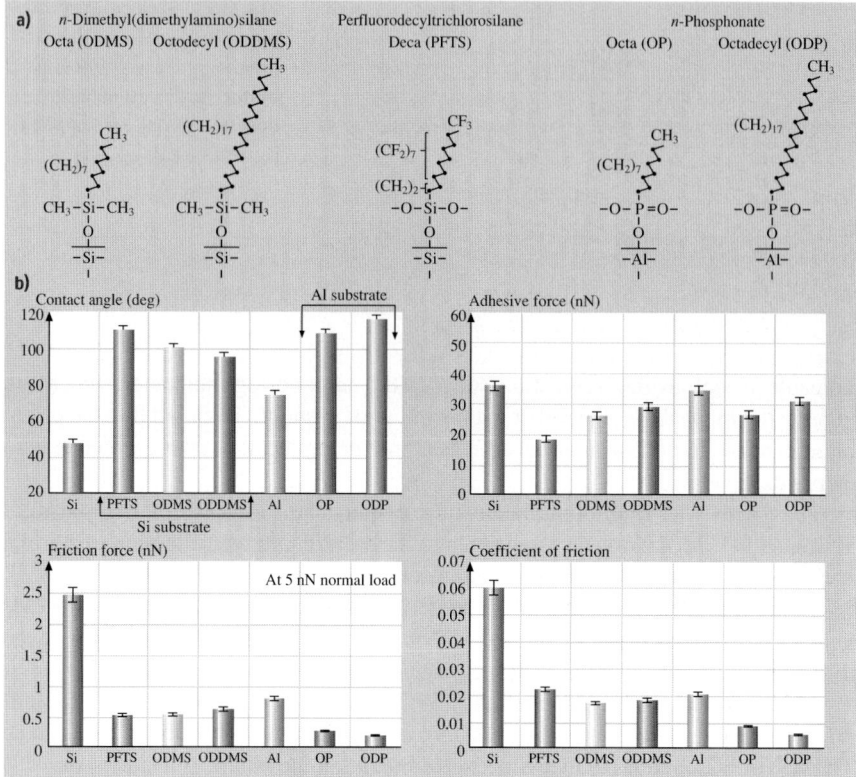

Fig. 8.74. (a) Schematics of structures of perfluoroalkylsilane and alkylsilane SAMs on Si with native oxide substrates, and alkylphosphonate SAMs on Al with native oxide, and (b) contact angle, adhesive force, friction force, and coefficient of friction of Si with native oxide and Al with native oxide substrates and with various SAMs

various SAMs. Based on the data, PFTS/Si exhibits higher contact angle and lower adhesive force as compared to that of ODMS/Si and ODDMS/Si. Data of ODMS and ODDMS on Si substrate are comparable to that of OP and ODP on Al substrate. Therefore, the substrate had little effect. The coefficient of friction of various SAMs were comparable.

For wear performance studies, experiments were conducted on various films. Figure 8.75a shows the relationship between the decrease of surface height and in the normal load for various SAMs and corresponding substrates [37, 131]. As shown in the figure, the SAMs exhibit a critical normal load beyond which the surface height drastically decreases. Unlike SAMs, the substrates show a monotonic decrease in surface height with increasing normal load with wear initiating from the very beginning, i.e., even for low normal loads. The critical loads corresponding to the sudden failure are shown in Fig. 8.75b. Amongst all the SAMs, ODDMS and ODP show the best performance in wear tests. ODDMS/Si and ODP/Al showed

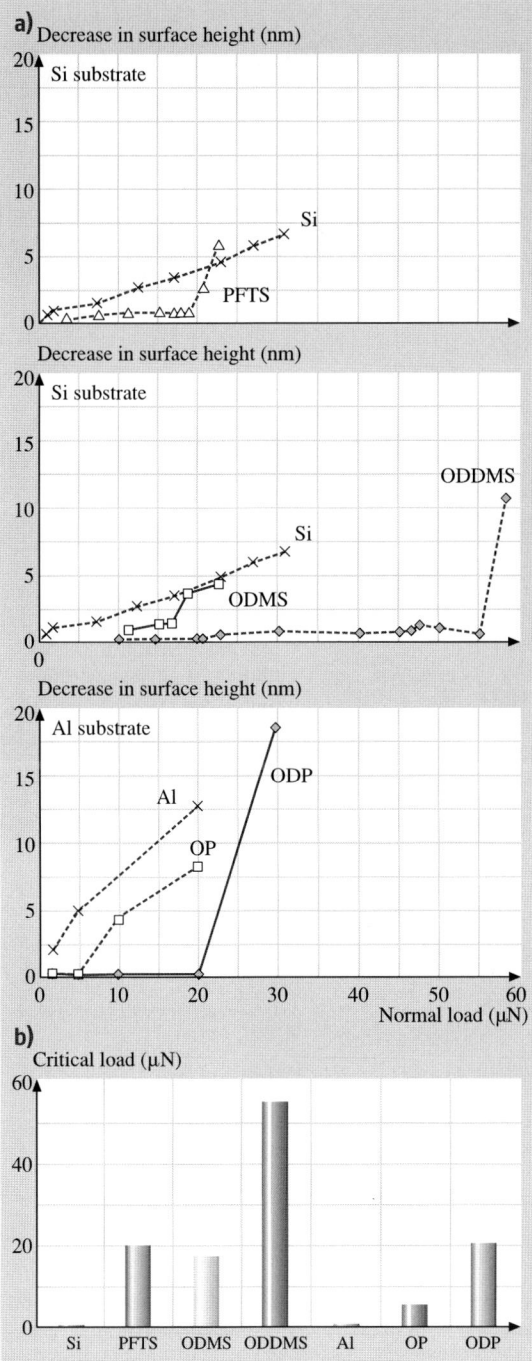

Fig. 8.75. (a) Decrease of surface height as a function of normal load after one scan cycle for various SAMs on Si and Al substrates, and (b) comparison of critical loads for failure during wear tests for various SAMs

a better wear resistance than ODMS/Si and OD/Al due to the chain-length effect. Wear behavior of the SAMs is reported to be mostly determined by the molecule–substrate bond strengths.

8.6.3 Liquid Film Thickness Measurements

Liquid film thickness mapping of ultra-thin films (on the order of couple of 2 nm) can be obtained using friction force microscopy [39] and adhesive force mapping [97]. Figure 8.76 shows greyscale plots of the surface topography and fric-

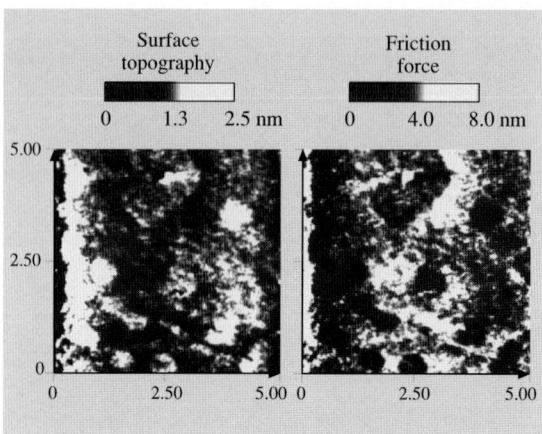

Fig. 8.76. Greyscale plots of the surface topography and friction force obtained simultaneously for unbonded Demnum-type perfluoropolyether lubricant film on silicon [39]

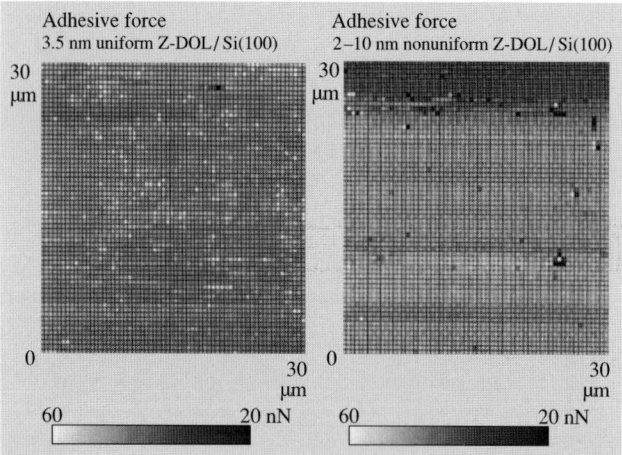

Fig. 8.77. Greyscale plots of the adhesive force distribution of a uniformly-coated, 3.5-nm-thick unbonded Z-DOL film on silicon and 3- to 10-nm-thick unbonded Z-DOL film on silicon that was deliberately coated nonuniformly by vibrating the sample during the coating process [97]

tion force obtained simultaneously for unbonded Demnum S-100 type PFPE lubricant film on silicon. The friction force plot shows well-distinguished low- and high-friction regions roughly corresponding to high and low regions in the surface topography (thick and thin lubricant regions). A uniformly lubricated sample does not show such a variation in the friction. Friction force imaging can thus be used to measure the lubricant uniformity on the sample surface, which cannot be identified by surface topography alone. Figure 8.77 shows the greyscale plots of the adhesive force distribution for silicon samples coated uniformly and nonuniformly with Z-DOL-type PFPE lubricant. It can be clearly seen that there exists a region that has adhesive force that are distinctly different from the other region for the nonuniformly coated sample. This implies that the liquid film thickness is nonuniform, which gives rise to a difference in the meniscus forces.

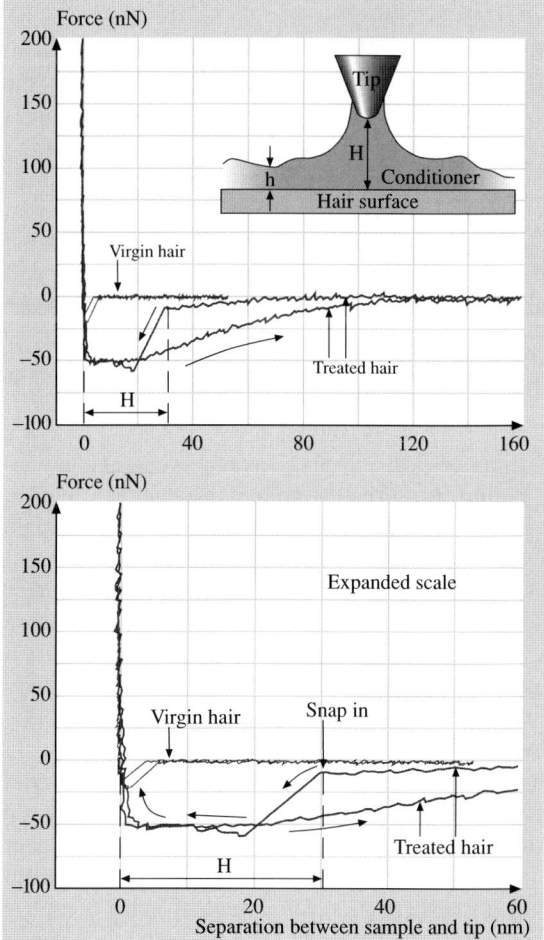

Fig. 8.78. Forces between tip and hair surface as a function of tip–sample separation for virgin hair and conditioner-treated hair. A schematic of measurement for the localized conditioner thickness is shown in the inset at the top. The expanded scale view of the force curve at small separation is shown (*bottom*) [60]

Quantitative measurements of liquid film thickness of thin lubricant films (on the order of a few nm) with nanometer lateral resolution can be made with the AFM [4, 11, 60, 81]. The liquid film thickness is obtained by measuring the force on the tip as it approaches, contacts and pushes through the liquid film and ultimately contacts the substrate. The distance between the sharp snap-in (owing to the formation of a liquid meniscus and van der Waals forces between the film and the tip) at the liquid surface and the hard repulsion at the substrate surface is a measure of the liquid film thickness. Figure 8.78 shows a plot of the forces between the tip and virgin hair, and hair treated with hair conditioner. The hair sample was first brought into contact with the tip and then pulled away at a velocity of 400 nm/s. The zero tip–sample separation is defined to be the position where the force on the tip is zero and the tip is not in contact with the sample. As the tip approaches the sample, a negative force exists, which indicates an attractive force. The treated hair surface shows much a longer range of interaction with the tip compared to the very short range of interaction between the virgin hair surfaces and the tip. Typically, the tip suddenly snaps into contact with the conditioner layer at a finite separation H (about 30 nm), which is proportional to the conditioner thickness h. As the tip contacts the substrate, the tip travels with the sample. When the sample is withdrawn, the forces on the tip slowly decrease to zero once the meniscus of liquid is drawn out from the hair surface. It should be noted that the distance H between the sharp snap-in at the liquid surface and the hard wall contact with the substrate is not the real conditioner thickness h. Due to the interaction of the liquid with the tip at some spacing distance, H tends to be thicker than the actual film thickness, but can still provide an estimate of the actual film thickness and an upper limit on the thickness.

8.7 Closure

At most solid–solid interfaces of technological relevance, contact occurs at many asperities. A sharp AFM/FFM tip sliding on a surface simulates just one such contact. However, asperities come in all shapes and sizes. The effect of the radius of a single asperity (tip) on the friction/adhesion performance can be studied using tips of different radii. AFM/FFM are used to study various tribological phenomena, which include surface roughness, adhesion, friction, scratching, wear, indentation, detection of material transfer, and boundary lubrication. Measurement of atomic-scale friction of a freshly cleaved highly oriented pyrolytic graphite exhibits the same periodicity as that of the corresponding topography. However, the peaks in friction and those in the corresponding topography are displaced relative to each other. Variations in atomic-scale friction and the observed displacement can be explained by the variation in interatomic forces in the normal and lateral directions; the relevant friction mechanism is atomic-scale stick–slip. Local variations in microscale friction occur and are found to correspond to the local slopes, suggesting that a ratchet mechanism and collision effects are responsible for this variation. Directionality in the friction is observed on both micro- and macroscales, which results from the surface roughness and surface preparation. Anisotropy in surface roughness accentuates this effect.

8 Nanotribology, Nanomechanics and Materials Characterization

The friction contrast in conventional frictional measurements is based on interactions dependent upon interfacial material properties superimposed by roughness-induced lateral forces. To obtain roughness-independent friction, lateral or torsional modulation techniques can be used. These techniques also allow measurements over a small region. AFM/FFM experiments are generally conducted at relative velocities up to about 200 µm/s. High-velocity experiments can be performed using either by mounting a sample on a shear-wave transducer driven at very high frequencies or by mounting it on a high-velocity piezo stage. By using these techniques, friction and wear experiments can be performed at a range of sliding velocities as well as normal loads, and the data have been used to develop nanoscale friction and wear maps. Relevant friction mechanisms are different for different ranges of sliding velocities and normal loads.

Adhesion and friction in wet environments depends on the tip radius, surface roughness, and relative humidity. Superhydrophobic surfaces can be designed by roughness optimization.

Nanoscale friction is generally found to be smaller than the microscale friction. There are several factors that are responsible for the differences, including wear and contaminant particles, the transition from elasticity to plasticity, scale-dependent roughness and mechanical properties, and meniscus effects. Nanoscale friction values increase with an increase in the normal load above a certain critical load (stress), approaching to the macroscale friction. The critical contact stress corresponds to the hardness of the softer material.

The mechanism of material removal on the microscale is studied. Wear rate for single-crystal silicon is negligible below 20 µN and is much higher and remains approximately constant at higher loads. Elastic deformation at low loads is responsible for negligible wear. Most of the wear debris is loose. SEM and TEM studies of the wear region suggest that the material on the microscale is removed by plastic deformation with a small contribution from elastic fracture; this observation corroborates the scratch data. Evolution of wear has also been studied using AFM. Wear is found to be initiated at nanoscratches. For a sliding interface requiring near-zero friction and wear, contact stresses should be below the hardness of the softer material to minimize plastic deformation and surfaces should be free of nanoscratches. Further, wear precursors can be detected at early stages of wear by using surface potential measurements. It is found that, even in the case of zero wear (no measurable deformation of the surface using AFM), there can be a significant change in the surface potential inside the wear mark, which is useful for study of wear precursors. Detection of material transfer on the nanoscale is possible with AFM.

In situ surface characterization of local deformation of materials and thin coatings can be carried out using a tensile stage inside an AFM. An AFM can also be used for nanofabrication/nanomachining.

Modified AFM can be used to obtain load–displacement curves and for the measurement of nanoindentation hardness and Young's modulus of elasticity, with an indentation depth as low as 1 nm. The hardness of ceramics on the nanoscale is found to be higher than that on the microscale. Ceramics exhibit significant plasticity and creep on a nanoscale. By using the force modulation technique, localized surface elasticity maps of composite materials with penetration depths as low as 1 nm can be obtained. By using phase contrast microscopy in tapping or torsional mode, it is possible to get phase contrast maps or the contrast in viscoelastic properties of near surface regions. Scratching and indentation on the nanoscale are powerful ways to screen for adhesion and resistance to deformation of ultrathin films.

Boundary lubrication studies and measurement of lubricant-film thickness with a lateral resolution on the nanoscale can be conducted using AFM. Chemically bonded lubricant films and self-assembled monolayers are superior in friction and wear resistance. For chemically bonded lubricant films, the adsorption of water, the formation of meniscus and its change during sliding, and surface properties play an important role on the adhesion, friction, and durability of these films. Sliding velocity, relative humidity and temperature affect adhesion and friction. For SAMs, their friction mechanism is explained by a so-called *molecular spring* model. Films with high-compliance long carbon chains exhibit low friction and wear. Also perfluoroalkylsilane SAMs on Si appear to be more hydrophobic with lower adhesion than alkylsilane SAMs on Si.

Investigations of adhesion, friction, wear, scratching and indentation on the nanoscale using the AFM can provide insights into failure mechanisms of materials. Coefficients of friction, wear rates and mechanical properties such as hardness have been found to be different on the nanoscale than on the macroscale; generally, coefficients of friction and wear rates on the micro- and nanoscales are smaller, whereas hardness is greater. Therefore, micro/nanotribological studies may help define the regimes for ultra-low friction and near-zero wear. These studies also provide insight into the atomic origins of adhesion, friction, wear, and lubrication mechanisms.

References

1. I.L. Singer, H.M. Pollock: *Fundamentals of Friction: Macroscopic and Microscopic Processes* (Kluwer Academic, Dordrecht 1992) p. 220
2. B.N.J. Persson, E. Tosatti: *Physics of Sliding Friction* (Kluwer Academic, Dordrecht 1996) p. E311
3. B. Bhushan: *Micro/Nanotribology and its Applications*, Vol. E (Kluwer Academic, Dordrecht 1997) p. 330
4. B. Bhushan: *Handbook of Micro/Nanotribology*, Vol. 2nd (CRC, Boca Raton 1999)
5. B. Bhushan: *Principles and Applications of Tribology* (Wiley, New York 1999)
6. B. Bhushan: Nanoscale tribophysics and tribomechanics, Wear **225-229**, 465–492 (1999)
7. B. Bhushan: *Modern Tribology Handbook, Vol. 1: Principles of Tribology* (CRC, Boca Raton 2001)
8. B. Bhushan: *Fundamentals of Tribology and Bridging the Gap Between the Macro- and Micro/Nanoscales*, NATO Science Series II, Vol. 10 (Kluwer Academic, Dordrecht 2001)
9. B. Bhushan: Nano- to microscale wear and mechanical characterization studies using scanning probe microscopy, Wear **251**, 1105–1123 (2001)

10. B. Bhushan: *Introduction to Tribology* (Wiley, New York 2002)
11. B. Bhushan: *Nanotribology and Nanomechanics – An Introduction* (Springer, Berlin, Heidelberg 2005)
12. B. Bhushan: Nanotribology and nanomechanics, Wear **259**, 1507–1531 (2005)
13. B. Bhushan, J. N. Israelachvili, U. Landman: Nanotribology: Friction, wear and lubrication at the atomic scale, Nature **374**, 607–616 (1995)
14. H. J. Guntherodt, D. Anselmetti: *Forces in Scanning Probe Methods* (Kluwer Academic, Dordrecht 1995) p. E286
15. B. Bhushan: *Tribology and Mechanics of Magnetic Storage Devices*, 2nd edn. (Springer, New York 1996)
16. B. Bhushan: *Tribology Issues and Opportunities in MEMS* (Kluwer Academic, Dordrecht 1998)
17. B. Bhushan: Wear and mechanical characterisation on micro- to picoscales using AFM, Int. Mater. Rev. **44**, 105–117 (1999)
18. B. Bhushan, H. Fuchs, S. Hosaka: *Applied Scanning Probe Methods* (Springer, Berlin, Heidelberg 2004)
19. B. Bhushan, H. Fuchs: *Applied Scanning Probe Methods II* (Springer, Berlin, Heidelberg 2006)
20. G. Binnig, C. F. Quate, Ch. Gerber: Atomic force microscopy, Phys. Rev. Lett. **56**, 930–933 (1986)
21. G. Binnig, Ch. Gerber, E. Stoll, T. R. Albrecht, C. F. Quate: Atomic resolution with atomic force microscope, Europhys. Lett. **3**, 1281–1286 (1987)
22. C. M. Mate, G. M. McClelland, R. Erlandsson, S. Chiang: Atomic-scale friction of a tungsten tip on a graphite surface, Phys. Rev. Lett. **59**, 1942–1945 (1987)
23. B. Bhushan, J. Ruan: Atomic-scale friction measurements using friction force microscopy: Part II—application to magnetic media, ASME J. Trib. **116**, 389–396 (1994)
24. J. Ruan, B. Bhushan: Atomic-scale friction measurements using friction force microscopy: Part I—general principles and new measurement techniques, ASME J. Tribol. **116**, 378–388 (1994)
25. J. Ruan, B. Bhushan: Atomic-scale and microscale friction of graphite and diamond using friction force microscopy, J. Appl. Phys. **76**, 5022–5035 (1994)
26. J. Ruan, B. Bhushan: Frictional behavior of highly oriented pyrolytic graphite, J. Appl. Phys. **76**, 8117–8120 (1994)
27. B. Bhushan, V. N. Koinkar, J. Ruan: Microtribology of magnetic media, Proc. Inst. Mech. Eng., Part J: J. Eng. Tribol. **208**, 17–29 (1994)
28. B. Bhushan, A. V. Kulkarni: Effect of normal load on microscale friction measurements, Thin Solid Films **278**, 49–56 (1996)
29. B. Bhushan, A. V. Kulkarni: Effect of normal load on microscale friction measurements, Thin Solid Films **293**, 333 (1996)
30. B. Bhushan, S. Sundararajan: Micro/nanoscale friction and wear mechanisms of thin films using atomic force and friction force microscopy, Acta Mater. **46**, 3793–3804 (1998)
31. V. Scherer, W. Arnold, B. Bhushan: Active Friction Control Using Ultrasonic Vibration. In: *Tribology Issues and Opportunities in MEMS*, ed. by B. Bhushan (Kluwer Academic, Dordrecht 1998) pp. 463–469
32. V. Scherer, W. Arnold, B. Bhushan: Lateral force microscopy using acoustic friction force microscopy, Surf. Interface Anal. **27**, 578–587 (1999)
33. M. Reinstaedtler, U. Rabe, V. Scherer, U. Hartmann, A. Goldade, B. Bhushan, W. Arnold: On the nanoscale measurement of friction using atomic-force microscope cantilever torsional resonances, Appl. Phys. Lett. **82**, 2604–2606 (2003)

34. M. Reinstaedtler, U. Rabe, A. Goldade, B. Bhushan, W. Arnold: Investigating ultra-thin lubricant layers using resonant friction force microscopy, Tribol. Int. **38**, 533–541 (2005)
35. M. Reinstaedtler, T. Kasai, U. Rabe, B. Bhushan, W. Arnold: Imaging and measurement of elasticity and friction using the TR mode, J. Phys. D: Appl. Phys. **38**, R269–R282 (2005)
36. B. Bhushan, T. Kasai: A surface topography-independent friction measurement technique using torsional resonance mode in an AFM, Nanotechnology **15**, 923–935 (2004)
37. N. S. Tambe, B. Bhushan: A new atomic force microscopy based technique for studying nanoscale friction at high sliding velocities, J. Phys. D: Appl. Phys. **38**, 764–773 (2005)
38. B. Bhushan, A. V. Kulkarni, V. N. Koinkar, M. Boehm, L. Odoni, C. Martelet, M. Belin: Microtribological characterization of self-assembled and Langmuir–Blodgett monolayers by atomic and friction force microscopy, Langmuir **11**, 3189–3198 (1995)
39. V. N. Koinkar, B. Bhushan: Micro/nanoscale studies of boundary layers of liquid lubricants for magnetic disks, J. Appl. Phys. **79**, 8071–8075 (1996)
40. V. N. Koinkar, B. Bhushan: Microtribological studies of unlubricated and lubricated surfaces using atomic force/friction force microscopy, J. Vac. Sci. Technol. **14**, 2378–2391 (1996)
41. B. Bhushan, H. Liu: Nanotribological properties and mechanisms of alkylthiol and biphenyl thiol self-assembled monolayers studied by AFM, Phys. Rev. B **63**, 245412-1–245412-11 (2001)
42. H. Liu, B. Bhushan: Nanotribological characterization of molecularly-thick lubricant films for applications to MEMS/NEMS by AFM, Ultramicroscopy **97**, 321–340 (2003)
43. B. Bhushan, T. Kasai, G. Kulik, L. Barbieri, P. Hoffmann: AFM study of perfluorosilane and alkylsilane self-assembled monolayers for anti-stiction in MEMS/NEMS, Ultramicroscopy **105**, 176–188 (2005)
44. B. Bhushan, V. N. Koinkar: Tribological studies of silicon for magnetic recording applications, J. Appl. Phys. **75**, 5741–5746 (1994)
45. V. N. Koinkar, B. Bhushan: Microtribological studies of Al_2O_3 – TiC, polycrystalline and single-crystal Mn – Zn ferrite and SiC head slider materials, Wear **202**, 110–122 (1996)
46. V. N. Koinkar, B. Bhushan: Microtribological properties of hard amorphous carbon protective coatings for thin film magnetic disks and heads, Proc. Inst. Mech. Eng. Part J: J. Eng. Tribol. **211**, 365–372 (1997)
47. S. Sundararajan, B. Bhushan: Development of a continuous microscratch technique in an atomic force microscope and its application to study scratch resistance of ultra-thin hard amorphous carbon coatings, J. Mater. Res. **16**, 75–84 (2001)
48. J. Ruan, B. Bhushan: Nanoindentation studies of fullerene films using atomic force microscopy, J. Mater. Res. **8**, 3019–3022 (1993)
49. B. Bhushan, A. V. Kulkarni, W. Bonin, J. T. Wyrobek: Nano/picoindentation measurement using a capacitance transducer system in atomic force microscopy, Philos. Mag. **74**, 1117–1128 (1996)
50. B. Bhushan, V. N. Koinkar: Nanoindentation hardness measurements using atomic force microscopy, Appl. Phys. Lett. **64**, 1653–1655 (1994)
51. B. Bhushan, X. Li: Nanomechanical characterisation of solid surfaces and thin films (invited), Intern. Mater. Rev. **48**, 125–164 (2003)
52. P. Maivald, H. J. Butt, S. A. C. Gould, C. B. Prater, B. Drake, J. A. Gurley, V. B. Elings, P. K. Hansma: Using force modulation to image surface elasticities with the atomic force microscope, Nanotechnology **2**, 103–106 (1991)

53. B. Anczykowski, D. Kruger, K. L. Babcock, H. Fuchs: Basic properties of dynamic force microscopy with the scanning force microscope in experiment and simulation, Ultramicroscopy **66**, 251–259 (1996)
54. D. DeVecchio, B. Bhushan: Localized surface elasticity measurements using an atomic force microscope, Rev. Sci. Instrum. **68**, 4498–4505 (1997)
55. V. Scherer, B. Bhushan, U. Rabe, W. Arnold: Local elasticity and lubrication measurements using atomic force and friction force microscopy at ultrasonic frequencies, IEEE Trans. Magn. **33**, 4077–4079 (1997)
56. S. Amelio, A. V. Goldade, U. Rabe, V. Scherer, B. Bhushan, W. Arnold: Measurements of elastic properties of ultra-thin diamond-like carbon coatings using atomic force acoustic microscopy, Thin Solid Films **392**, 75–84 (2001)
57. W. W. Scott, B. Bhushan: Use of phase imaging in atomic force microscopy for measurement of viscoelastic contrast in polymer nanocomposites and molecularly-thick lubricant films, Ultramicroscopy **97**, 151–169 (2003)
58. B. Bhushan, J. Qi: Phase contrast imaging of nanocomposites and molecularly-thick lubricant films in magnetic media, Nanotechnology **14**, 886–895 (2003)
59. T. Kasai, B. Bhushan, L. Huang, C. Su: Topography and phase imaging using the torsional resonance mode, Nanotechnology **15**, 731–742 (2004)
60. N. Chen, B. Bhushan: Morphological, nanomechanical and cellular structural characterization of human hair and conditioner distribution using torsional resonance mode in an AFM, J. Micros. **220**, 96–112 (2005)
61. M. S. Bobji, B. Bhushan: Atomic force microscopic study of the micro-cracking of magnetic thin films under tension, Scripta Mater. **44**, 37–42 (2001)
62. M. S. Bobji, B. Bhushan: In-situ microscopic surface characterization studies of polymeric thin films during tensile deformation using atomic force microscopy, J. Mater. Res. **16**, 844–855 (2001)
63. N. Tambe, B. Bhushan: In situ study of nano-cracking of multilayered magnetic tapes under monotonic and fatigue loading using an AFM, Ultramicroscopy **100**, 359–373 (2004)
64. B. Bhushan, T. Kasai, C. V. Nguyen, M. Meyyappan: Multiwalled carbon nanotube AFM probes for surface characterization of micro/nanostructures, Microsys. Technol. **10**, 633–639 (2004)
65. B. Bhushan: Micro/nanotribology and its applications to magnetic storage devices and MEMS, Tribol. Int. **28**, 85–95 (1995)
66. D. DeVecchio, B. Bhushan: Use of a nanoscale Kelvin probe for detecting wear precursors, Rev. Sci. Instrum. **69**, 3618–3624 (1998)
67. B. Bhushan, A. V. Goldade: Measurements and analysis of surface potential change during wear of single crystal silicon (100) at ultralow loads using Kelvin probe microscopy, Appl. Surf. Sci **157**, 373–381 (2000)
68. B. Bhushan, A. V. Goldade: Kelvin probe microscopy measurements of surface potential change under wear at low loads, Wear **244**, 104–117 (2000)
69. H. U. Krotil, T. Stifter, H. Waschipky, K. Weishaupt, S. Hild, O. Marti: Pulse force mode: A new method for the investigation of surface properties, Surf. Interface Anal. **27**, 336–340 (1999)
70. U. Rabe, K. Janser, W. Arnold: Vibrations of free and surface-coupled atomic force microscope cantilevers: Theory and experiment, Rev. Sci. Instrum. **67**, 3281–3293 (1996)
71. J. Tamayo, R. Garcia: Deformation, contact time, and phase contrast in tapping mode scanning force microscopy, Langmuir **12**, 4430–4435 (1996)
72. R. Garcia, J. Tamayo, M. Calleja, F. Garcia: Phase contrast in tapping-mode scanning force microscopy, Appl. Phys. A **66**, 309–312 (1998)

73. Y. Song, B. Bhushan: Quantitative extraction of in-plane surface properties using torsional resonance mode in atomic force microscopy, J. Appl. Phys. **87**, 83533 (2005)
74. B. Bhushan, J. Ruan, B. K. Gupta: A scanning tunnelling microscopy study of Fullerene films, J. Phys. D: Appl. Phys. **26**, 1319–1322 (1993)
75. G. A. Tomlinson: A molecular theory of friction, Phil. Mag. Ser. **7**, 905–939 (1929)
76. D. Tomanek, W. Zhong, H. Thomas: Calculation of an atomically modulated friction force in atomic force microscopy, Europhys. Lett. **15**, 887–892 (1991)
77. E. Meyer, R. Overney, R. Luthi, D. Brodbeck, L. Howald, J. Frommer, H. J. Guntherodt, O. Wolter, M. Fujihira, T. Takano, Y. Gotoh: Friction force microscopy of mixed Langmuir–Blodgett films, Thin Solid Films **220**, 132–137 (1992)
78. C. D. Frisbie, L. F. Rozsnyai, A. Noy, M. S. Wrighton, C. M. Lieber: Functional group imaging by chemical force microscopy, Science **265**, 2071–2074 (1994)
79. V. N. Koinkar, B. Bhushan: Effect of scan size and surface roughness on microscale friction measurements, J. Appl. Phys. **81**, 2472–2479 (1997)
80. S. Sundararajan, B. Bhushan: Topography-induced contributions to friction forces measured using an atomic force/friction force microscope, J. Appl. Phys. **88**, 4825–4831 (2000)
81. B. Bhushan, G. S. Blackman: Atomic force microscopy of magnetic rigid disks and sliders and its applications to tribology, ASME J. Tribol. **113**, 452–458 (1991)
82. K. Yamanaka, E. Tomita: Lateral force modulation atomic force microscope for selective imaging of friction forces, Jpn. J. Appl. Phys. **34**, 2879–2882 (1995)
83. O. Marti, H.-U. Krotil: Dynamic Friction Measurement With the Scanning Force Microscope. In: *Fundamentals of Tribology and Bridging the Gap Between the Macro- and Micro/Nanoscales*, ed. by B. Bhushan (Kluwer, Dordrecht 2001) pp. 121–135
84. N. S. Tambe, B. Bhushan: Scale dependence of micro/nano-friction and adhesion of MEMS/NEMS materials, coatings and lubricants, Nanotechnology **15**, 1561–1570 (2004)
85. N. S. Tambe, B. Bhushan: Friction model for the velocity dependence of nanoscale friction, Nanotechnology **16**, 2309–2324 (2005)
86. N. S. Tambe, B. Bhushan: Durability studies of micro/nanoelectromechanical system materials, coatings, and lubricants at high sliding velocities (up to 10 mm/s) using a modified atomic force microscope, J. Vac. Sci. Technol. A **23**, 830–835 (2005)
87. N. S. Tambe, B. Bhushan: Nanoscale friction-induced phase transformation of diamond-like carbon, Scripta Materiala **52**, 751–755 (2005)
88. K. Mizuhara, S. M. Hsu: Tribochemical Reaction of Oxygen and Water on Silicon Surfaces. In: *Wear Particles*, ed. by D. Dowson (Elsevier Science, Amsterdam 1992) pp. 323–328
89. T. Bouhacina, J. P. Aime, S. Gauthier, D. Michel: Tribological behavior of a polymer grafted on silanized silica probed with a nanotip, Phys. Rev. B. **56**, 7694–7703 (1997)
90. E. Gnecco, R. Bennewitz, T. Gyalog, Ch. Loppacher, M. Bammerlin, E. Meyer, H.-J. Guntherodt: Velocity dependence of atomic friction, Phys. Rev. Lett. **84**, 1172–1175 (2000)
91. N. S. Tambe, B. Bhushan: Nanoscale friction mapping, Appl. Phys. Lett. **86**, 193102-1–193102-3 (2005)
92. A. Grill: Tribology of diamondlike carbon and related materials: An updated review, Surf. Coat. Technol. **94-95**, 507–513 (1997)
93. N. S. Tambe, B. Bhushan: Identifying materials with low friction and adhesion for nanotechnology applications, Appl. Phys. Lett **86**, 061906-1–061906-3 (2005)

94. N. S. Tambe, B. Bhushan: Nanowear mapping: A novel atomic force microscopy based approach for studying nanoscale wear at high sliding velocities, Tribol. Lett. **20**, 83–90 (2005)
95. S. C. Lim, M. F. Ashby: Wear mechanism maps, Acta Metall. **35**, 1–24 (1987)
96. S. C. Lim, M. F. Ashby, J. H. Brunton: Wear-rate transitions and their relationship to wear mechanisms, Acta Metall. **35**, 1343–1348 (1987)
97. B. Bhushan, C. Dandavate: Thin-film friction and adhesion studies using atomic force microscop, J. Appl. Phys. **87**, 1201–1210 (2000)
98. B. Bhushan: Adhesion and stiction: Mechanisms, measurement techniques, and methods for reduction, (invited), J. Vac. Sci. Technol. B **21**, 2262–2296 (2003)
99. T. Stifter, O. Marti, B. Bhushan: Theoretical investigation of the distance dependence of capillary and van der Waals forces in scanning probe microscopy, Phys. Rev. B **62**, 13667–13673 (2000)
100. U. D. Schwarz, O. Zwoerner, P. Koester, R. Wiesendanger: Friction Force Spectroscopy in the Low-load Regime with Well-defined Tips. In: *Micro/Nanotribology and Its Applications*, ed. by B. Bhushan (Kluwer Academic, Dordrecht 1997) pp. 233–238
101. M. Nosonovsky, B. Bhushan: Stochastic model for metastable wetting of roughness-induced superhydrophobic surfaces, Microsyst. Technol. **12**, 231–237 (2005)
102. M. Nosonovsky, B. Bhushan: Roughness optimization for biomimetic superhydrophobic surfaces, Microsyst. Technol. **11**, 535–549 (2005)
103. R. N. Wenzel: Resistance of solid surfaces to wetting by water, Indus. Eng. Chem. **28**, 988–994 (1936)
104. A. Cassie, S. Baxter: Wetting of porous surfaces, Trans. Faraday Soc. **40**, 546–551 (1944)
105. Z. Burton, B. Bhushan: Hydrophobicity, adhesion and friction properties with nanopatterned roughness and scale dependence, Nano Lett. **5**, 1607–1613 (2005)
106. B. Bhushan, H. Liu, S. M. Hsu: Adhesion and friction studies of silicon and hydrophobic and low friction films and investigation of scale effects, ASME J. Tribol. **126**, 583–590 (2004)
107. H. Liu, B. Bhushan: Adhesion and friction studies of microelectromechanical systems/nanoelectromechanical systems materials using a novel microtriboapparatus, J. Vac. Sci. Technol. A **21**, 1528–1538 (2003)
108. B. Bhushan, B. K. Gupta: *Handbook of Tribology: Materials, Coatings and Surface Treatments* (McGraw-Hill, New York 1991) reprinted Krieger, Malabar Florida, 1997
109. B. Bhushan, S. Venkatesan: Mechanical and tribological properties of silicon for micromechanical applications: A Review, Adv. Info. Storage Sys. **5**, 211–239 (1993)
110. Anonymous: *Properties of Silicon*, EMIS Data Reviews Series No. 4. INSPEC, Institution of Electrical Engineers, London. See also Anonymous, MEMS Materials Database, http://www.memsnet.org/material/ (2002)
111. J. E. Field: *The properties of natural and synthetic diamond* (Academic, London 1992)
112. B. Bhushan: Chemical, mechanical and tribological characterization of ultra-thin and hard amorphous carbon coatings as thin as 3.5 nm: Recent developments, Diamond and Related Materials **8**, 1985–2015 (1999)
113. Anonymous: *The Industrial Graphite Engineering Handbook* (National Carbon Company, New York 1959)
114. M. Nosonovsky, B. Bhushan: Scale effects in dry friction during multiple-asperity contact, ASME J. Tribol. **127**, 37–46 (2005)
115. B. Bhushan, M. Nosonovsky: Comprehensive model for scale effects in friction due to adhesion and two- and three-body deformation (plowing), Acta Mater. **52**, 2461–2474 (2004)

116. B. Bhushan, M. Nosonovsky: Scale effects in dry and wet friction, wear, and interface temperature, Nanotechnology **15**, 749–761 (2004)
117. B. Bhushan, M. Nosonovsky: Scale effects in friction using strain gradient plasticity and dislocation-assisted sliding (microslip), Acta Mater. **51**, 4331–4345 (2003)
118. V. N. Koinkar, B. Bhushan: Scanning and transmission electron microscopies of single-crystal silicon microworn/machined using atomic force microscopy, J. Mater. Res. **12**, 3219–3224 (1997)
119. X. Zhao, B. Bhushan: Material removal mechanism of single-crystal silicon on nanoscale and at ultralow loads, Wear **223**, 66–78 (1998)
120. B. Bhushan, P. S. Mokashi, T. Ma: A new technique to measure Poisson's ratio of ultrathin polymeric films using atomic force microscopy, Rev. Sci. Instrum. **74**, 1043–1047 (2003)
121. A. V. Kulkarni, B. Bhushan: Nanoscale mechanical property measurements using modified atomic force microscopy, Thin Solid Films **290-291**, 206–210 (1996)
122. A. V. Kulkarni, B. Bhushan: Nano/picoindentation measurements on single-crystal aluminum using modified atomic force microscopy, Mater. Lett. **29**, 221–227 (1996)
123. A. V. Kulkarni, B. Bhushan: Nanoindentation measurement of amorphous carbon coatings, J. Mater. Res. **12**, 2707–2714 (1997)
124. N. A. Fleck, G. M. Muller, M. F. Ashby, J. W. Hutchinson: Strain gradient plasticity: Theory and experiment, Acta Metall. Mater. **42**, 475–487 (1994)
125. W. D. Nix, H. Gao: Indentation size effects in crystalline materials: A law for strain gradient plasticity, J. Mech. Phys. Solids **46**, 411–425 (1998)
126. W. B. Li, J. L. Henshall, R. M. Hooper, K. E. Easterling: The mechanism of indentation creep, Acta Metall. Mater. **39**, 3099–3110 (1991)
127. F. P. Bowden, D. Tabor: *The Friction and Lubrication of Solids* (Clarendon, Oxford 1950)
128. Z. Tao, B. Bhushan: Bonding, degradation, and environmental effects on novel perfluoropolyether lubricants, Wear **259**, 1352–1361 (2005)
129. H. Liu, B. Bhushan, W. Eck, V. Staedtler: Investigation of the adhesion, friction, and wear properties of biphenyl thiol self-assembled monolayers by atomic force microscopy, J. Vac. Sci. Technol. A **19**, 1234–1240 (2001)
130. H. Liu, B. Bhushan: Investigation of nanotribological properties of self-assembled monolayers with alkyl and biphenyl spacer chains, Ultramicroscopy **91**, 185–202 (2002)
131. T. Kasai, B. Bhushan, G. Kulik, L. Barbieri, P. Hoffman: Nanotribological study of perfluorosilane SAMs for anti-stiction and low wear, J. Vac. Sci. Technol. B **23**, 995–1003 (2005)
132. K. K. Lee, B. Bhushan, D. Hansford: Nanotribological characterization of perfluoropolymer thin films for biomedical micro/nanoelectrone chemical systems applications, J. Vac. Sci. Technol. A **23**, 804–810 (2005)
133. B. Bhushan, D. Hansford, K. K. Lee: Surface modification of silicon and polymethylsiloxane surfaces with vapor-phase-deposited ultrathin fluorosilane films for biomedical nanodevices, J. Vac. Sci. Technol. A **24**, 1197–1202 (2006)
134. N. S. Tambe, B. Bhushan: Nanotribological characterization of self assembled monolayers deposited on silicon and aluminum substrates, Nanotechnology **16**, 1549–1558 (2005)
135. Z. Tao, B. Bhushan: Degradation mechanisms and environmental effects on perfluoropolyether, self assembled monolayers, and diamondlike carbon films, Langmuir **21**, 2391–2399 (2005)

9

Surface Forces and Nanorheology of Molecularly Thin Films

Marina Ruths and Jacob N. Israelachvili

Summary. In this chapter, we describe the static and dynamic normal forces that occur between surfaces in vacuum or liquids and the different modes of friction that can be observed between: (i) bare surfaces in contact (dry or interfacial friction), (ii) surfaces separated by a thin liquid film (lubricated friction), and (iii) surfaces coated with organic monolayers (boundary friction).

Experimental methods suitable for measuring normal surface forces, adhesion and friction (lateral or shear) forces of different magnitude at the molecular level are described. We explain the molecular origin of van der Waals, electrostatic, solvation and polymer-mediated interactions, and basic models for the contact mechanics of adhesive and nonadhesive elastically deforming bodies. The effects of interaction forces, molecular shape, surface structure and roughness on adhesion and friction are discussed.

Simple models for the contributions of the adhesion force and external load to interfacial friction are illustrated with experimental data on both unlubricated and lubricated systems, as measured with the surface forces apparatus. We discuss rate-dependent adhesion (adhesion hysteresis) and how this is related to friction. Some examples of the transition from wearless friction to friction with wear are shown.

Lubrication in different lubricant thickness regimes is described together with explanations of nanorheological concepts. The occurrence of and transitions between smooth and stick–slip sliding in various types of dry (unlubricated and solid boundary lubricated) and liquid lubricated systems are discussed based on recent experimental results and models for stick–slip involving memory distance and dilatancy.

9.1 Introduction: Types of Surface Forces

In this chapter, we discuss the most important types of surface forces and the relevant equations for the force and friction laws. Several different attractive and repulsive forces operate between surfaces and particles. Some forces occur in vacuum, for example, attractive van der Waals and repulsive hard-core interactions. Other types of forces can arise only when the interacting surfaces are separated by another condensed phase, which is usually a liquid. The most common types of surface forces and their main characteristics are listed in Table 9.1.

Table 9.1. Types of surface forces in vacuum versus in liquid (colloidal forces)

Type of force	Subclasses or alternative names	Main characteristics
	Attractive forces	
van der Waals	Debye induced dipole force (v & s) London dispersion force (v & s) Casimir force (v & s)	Ubiquitous, occurs both in vacuum and in liquids
Electrostatic	Ionic bond (v) Coulombic force (v & s) Hydrogen bond (v) Charge-exchange interaction (v & s) Acid–base interaction (s) "Harpooning" interaction (v)	Strong, long-range, arises in polar solvents; requires surface charging or charge-separation mechanism
Ion correlation	van der Waals force of polarizable ions (s)	Requires mobile charges on surfaces in a polar solvent
Quantum mechanical	Covalent bond (v) Metallic bond (v) Exchange interaction (v)	Strong, short-range, responsible for contact binding of crystalline surfaces
Solvation	Oscillatory force (s) Depletion force (s)	Mainly entropic in origin, the oscillatory force alternates between attraction and repulsion
Hydrophobic	Attractive hydration force (s)	Strong, apparently long-range; origin not yet understood
Specific binding	"Lock-and-key" or complementary binding (v & s) Receptor–ligand interaction (s) Antibody–antigen interaction (s)	Subtle combination of different non-covalent forces giving rise to highly specific binding; main recognition mechanism of biological systems
	Repulsive forces	
van der Waals	van der Waals disjoining pressure (s)	Arises only between dissimilar bodies interacting in a medium
Electrostatic	Coulombic force (v & s)	Arises only for certain constrained surface charge distributions
Quantum mechanical	Hard-core or steric repulsion (v) Born repulsion (v)	Short-range, stabilizing attractive covalent and ionic binding forces, effectively determine molecular size and shape
Solvation	Oscillatory solvation force (s) Structural force (s) Hydration force (s)	Monotonically repulsive forces, believed to arise when solvent molecules bind strongly to surfaces
Entropic	Osmotic repulsion (s) Double-layer force (s) Thermal fluctuation force (s) Steric polymer repulsion (s) Undulation force (s) Protrusion force (s)	Due to confinement of molecular or ionic species; requires mechanism that keeps trapped species between the surfaces

Table 9.1. (continued)

Type of force	Subclasses or alternative names	Main characteristics
Non-equilibrium	**Dynamic interactions** Hydrodynamic forces (s) Viscous forces (s) Friction forces (v & s) Lubrication forces (s)	Energy-dissipating forces occurring during relative motion of surfaces or bodies

Note: (v) applies only to interactions in vacuum, (s) applies only to interactions in solution (or to surfaces separated by a liquid), and (v & s) applies to interactions occurring both in vacuum and in solution

In *vacuum*, the two main long-range interactions are the attractive van der Waals and electrostatic (Coulomb) forces. At smaller surface separations (corresponding to molecular contact at surface separations of $D \approx 0.2$ nm), additional attractive interactions can be found such as covalent or metallic bonding forces. These attractive forces are stabilized by the hard-core repulsion. Together they determine the surface and interfacial energies of planar surfaces, as well as the strengths of materials and adhesive junctions. Adhesion forces are often strong enough to elastically or plastically deform bodies or particles when they come into contact.

In *vapors* (e.g., atmospheric air containing water and organic molecules), solid surfaces in, or close to, contact will generally have a surface layer of chemisorbed or physisorbed molecules, or a capillary condensed liquid bridge between them. A surface layer usually causes the adhesion to decrease, but in the case of capillary condensation, the additional Laplace pressure or attractive "capillary" force may make the adhesion between the surfaces stronger than in an inert gas or vacuum.

When totally immersed in a *liquid*, the force between particles or surfaces is completely modified from that in vacuum or air (vapor). The van der Waals attraction is generally reduced, but other forces can now arise that can qualitatively change both the range and even the sign of the interaction. The attractive force in such a system can be either stronger or weaker than in the absence of the intervening liquid. For example, the overall attraction can be stronger in the case of two hydrophobic surfaces separated by water, but weaker for two hydrophilic surfaces. Depending on the different forces that may be operating simultaneously in solution, the overall force law is not generally monotonically attractive even at long range; it can be repulsive, or the force can change sign at some finite surface separation. In such cases, the potential-energy minimum, which determines the adhesion force or energy, does not occur at the true molecular contact between the surfaces, but at some small distance further out.

The forces between surfaces in a liquid medium can be particularly complex at *short range*, i.e., at surface separations below a few nanometers or 4–10 molecular diameters. This is partly because, with increasing confinement, a liquid ceases to behave as a structureless continuum with bulk properties; instead, the size and shape of its molecules begin to determine the overall interaction. In addition, the surfaces themselves can no longer be treated as inert and structureless walls (i.e., mathematically flat) and their physical and chemical properties at the atomic scale must now be taken into account. The force laws will then depend on whether the surfaces are amorphous or crystalline (and whether the lattices of crystalline surfaces are matched or not), rough or smooth, rigid or soft (fluid-like), and hydrophobic or hydrophilic.

It is also important to distinguish between *static* (i.e., equilibrium) interactions and *dynamic* (i.e., nonequilibrium) forces such as viscous and friction forces. For example, certain liquid films confined between two contacting surfaces may take a surprisingly long time to equilibrate, as may the surfaces themselves, so that the short-range and adhesion forces appear to be time-dependent, resulting in "aging" effects.

9.2 Methods Used to Study Surface Forces

9.2.1 Force Laws

The full force law $F(D)$ between two surfaces, i.e., the force F as a function of surface separation D, can be measured in a number of ways [2–6]. The simplest is to move the base of a spring by a known amount, ΔD_0. Figure 9.1 illustrates this method when applied to the interaction of two magnets. However, the method is also applicable at the microscopic or molecular level, and it forms the basis of all direct force-measuring apparatuses such as the surface forces apparatus (SFA; [3, 7]) and the atomic force microscope (AFM; [8–10]). If there is a detectable force between the surfaces, this will cause the force-measuring spring to deflect by ΔD_s, while the surface separation changes by ΔD. These three displacements are related by

$$\Delta D_s = \Delta D_0 - \Delta D . \tag{9.1}$$

The difference in force, ΔF, between the initial and final separations is given by

$$\Delta F = k_s \Delta D_s , \tag{9.2}$$

where k_s is the spring constant. The equations above provide the basis for measurements of the force difference between any two surface separations. For example, if a force-measuring apparatus with a known k_s can measure D (and thus ΔD), ΔD_0, and ΔD_s, the force difference ΔF can be measured between a large initial or reference separation D, where the force is zero ($F = 0$), and another separation $D - \Delta D$. By working one's way in increasing increments of $\Delta D = \Delta D_0 - \Delta D_s$, the full force law $F(D)$ can be constructed over any desired distance regime.

In order to measure an equilibrium force law, it is essential to establish that the two surfaces have stopped moving before the displacements are measured. When displacements are measured while two surfaces are still in relative motion, one also measures a viscous or frictional contribution to the total force. Such dynamic force

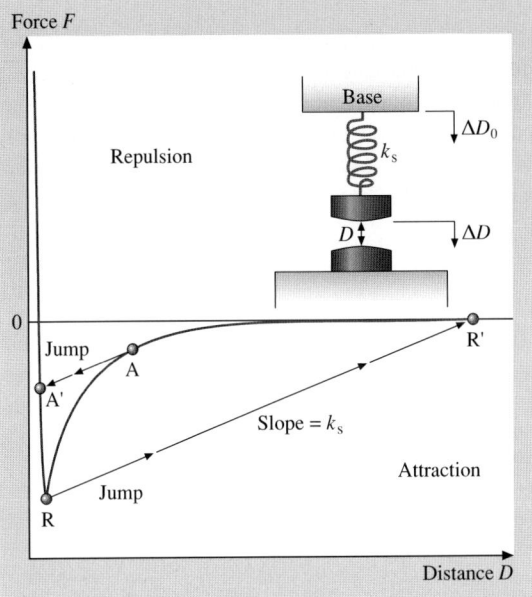

Fig. 9.1. Schematic attractive force law between two macroscopic objects such as two magnets, or between two microscopic objects such as the van der Waals force between a metal tip and a surface. On lowering the base supporting the spring, the latter will expand or contract such that, at any equilibrium separation D, the attractive force balances the elastic spring restoring force. If the gradient of the attractive force dF/dD exceeds the gradient of the spring's restoring force (defined by the spring constant k_s), the upper surface will jump from A into contact at A' (A for "advancing"). On separating the surfaces by raising the base, the two surfaces will jump apart from R to R' (R for "receding"). The distance R–R' multiplied by k_s gives the adhesion force, i.e., the value of F at the point R (after [1] with permission)

measurements have enabled the viscosities of liquids near surfaces and in thin films to be accurately determined [11–13].

In practice, it is difficult to measure the forces between two perfectly flat surfaces, because of the stringent requirement of perfect alignment for making reliable measurements at distances of a few tenths of a nanometer. It is far easier to measure the forces between curved surfaces, e.g., two spheres, a sphere and a flat surface, or two crossed cylinders. Furthermore, the force $F(D)$ measured between two curved surfaces can be directly related to the energy per unit area $E(D)$ between two flat surfaces at the same separation, D, by the so-called Derjaguin approximation [14]:

$$E(D) = \frac{F(D)}{2\pi R}, \tag{9.3}$$

where R is the radius of the sphere (for a sphere and a flat surface) or the radii of the cylinders (for two crossed cylinders, cf. Table 9.2).

9.2.2 Adhesion Forces

The most direct way to measure the adhesion of two solid surfaces (such as two spheres or a sphere on a flat) is to suspend one of them on a spring and measure the adhesion or "pull-off" force needed to separate the two bodies, using the deflection of the spring. If k_s is the stiffness of the force-measuring spring and ΔD is the distance the two surfaces jump apart when they separate, then the adhesion force F_{ad} is given by

$$F_{ad} = F_{max} = k_s \Delta D , \qquad (9.4)$$

where we note that, in liquids, the maximum or minimum in the force may occur at some nonzero surface separation (see Fig. 9.7). From F_{ad} and a known surface geometry, and assuming that the surfaces were everywhere in molecular contact, one may also calculate the surface or interfacial energy γ. For an elastically deformable sphere of radius R on a flat surface, or for two crossed cylinders of radius R, we have [3, 15]

$$\gamma = \frac{F_{ad}}{3\pi R} , \qquad (9.5)$$

while for two spheres of radii R_1 and R_2

$$\gamma = \frac{F_{ad}}{3\pi} \left(\frac{1}{R_1} + \frac{1}{R_2} \right) , \qquad (9.6)$$

where γ is in units of $J m^{-2}$ (see Sect. 9.5.2).

9.2.3 The SFA and AFM

In a typical force-measuring experiment, at least two of the above displacement parameters – ΔD_0, ΔD, and ΔD_s – are directly or indirectly measured, and from these the third displacement and the resulting force law $F(D)$ are deduced using (9.1) and (9.2) together with a measured value of k_s. For example, in SFA experiments, ΔD_0 is changed by expanding or contracting a piezoelectric crystal by a known amount or by moving the base of the spring with sensitive motor-driven mechanical stages. The resulting change in surface separation ΔD is measured optically, and the spring deflection ΔD_s can then be obtained according to (9.1). In AFM experiments, ΔD_0 and ΔD_s are measured using a combination of piezoelectric, optical, capacitance or magnetic techniques, from which the change in surface separation ΔD is deduced. Once a force law is established, the geometry of the two surfaces (e. g., their radii) must also be known before the results can be compared with theory or with other experiments.

The SFA (Fig. 9.2) is used for measurements of adhesion and force laws between two curved molecularly smooth surfaces immersed in liquids or controlled vapors [3, 7, 16]. The surface separation is measured by multiple-beam interferometry with an accuracy of ± 0.1 nm. From the shape of the interference fringes one

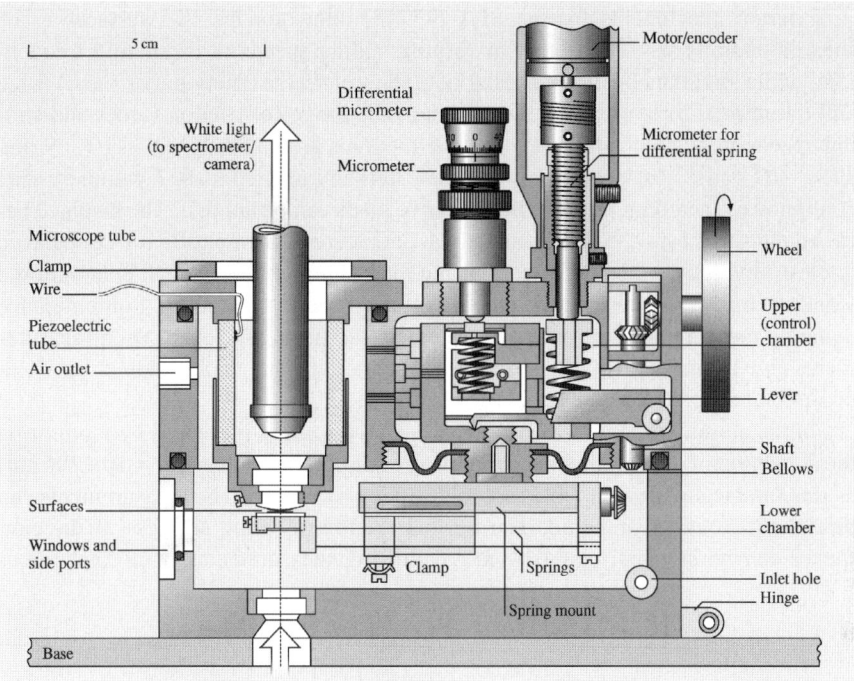

Fig. 9.2. A surface forces apparatus (SFA) where the intermolecular forces between two macroscopic, cylindrical surfaces of local radius R can be directly measured as a function of surface separation over a large distance regime from tenths of a nanometer to micrometers. Local or transient surface deformations can be detected optically. Various attachments for moving one surface laterally with respect to the other have been developed for friction measurements in different regimes of sliding velocity and sliding distance (after [16] with permission)

also obtains the radius of the surfaces, R, and any surface deformation that arises during an interaction [17–19]. The resolution in the lateral direction is about 1 µm. The surface separation can be independently controlled to within 0.1 nm, and the force sensitivity is about 10^{-8} N. For a typical surface radius of $R \approx 1$ cm, γ values can be measured to an accuracy of about 10^{-3} mJ m^{-2}.

Several different materials have been used to form the surfaces in the SFA, including mica [20, 21], silica [22], sapphire [23], and polymer sheets [24]. These materials can also be used as supporting substrates in experiments on the forces between adsorbed or chemically bound polymer layers [13, 25–30], surfactant and lipid monolayers and bilayers [31–34], and metal and metal oxide layers [35–42]. The range of liquids and vapors that can be used is almost endless, and have thus far included aqueous solutions, organic liquids and solvents, polymer melts, various petroleum oils and lubricant liquids, dyes, and liquid crystals.

Friction attachments for the SFA [43–48] allow for the two surfaces to be sheared laterally past each other at varying sliding speeds or oscillating frequencies, while simultaneously measuring both the transverse (frictional or shear) force and the normal force (load) between them. The ranges of friction forces and sliding speeds that can be studied with such methods are currently 10^{-7}–10^{-1} N and 10^{-13}–10^{-2} m s^{-1}, respectively [49]. The externally applied load, L, can be varied continuously, and both positive and negative loads can be applied. The distance between the surfaces, D, their true molecular contact area, their elastic (or viscoelastic or elasto-hydrodynamic) deformation, and their lateral motion can all be monitored simultaneously by recording the moving interference-fringe pattern. Equipment for dynamic measurements of normal forces has also been developed. Such measurements give information on the viscosity of the medium and the location of the shear or slip planes relative to the surfaces [11–13, 50, 51].

In the atomic force microscope (Fig. 9.3), the force is measured by monitoring the deflection of a soft cantilever supporting a sub-microscopic tip ($R \approx 10$–200 nm) as this interacts with a flat, macroscopic surface [8, 52, 53]. The measurements can be done in a vapor or liquid. The normal (bending) spring stiffness of the cantilever can be as small as 0.01 N m^{-1}, allowing measurements of normal forces

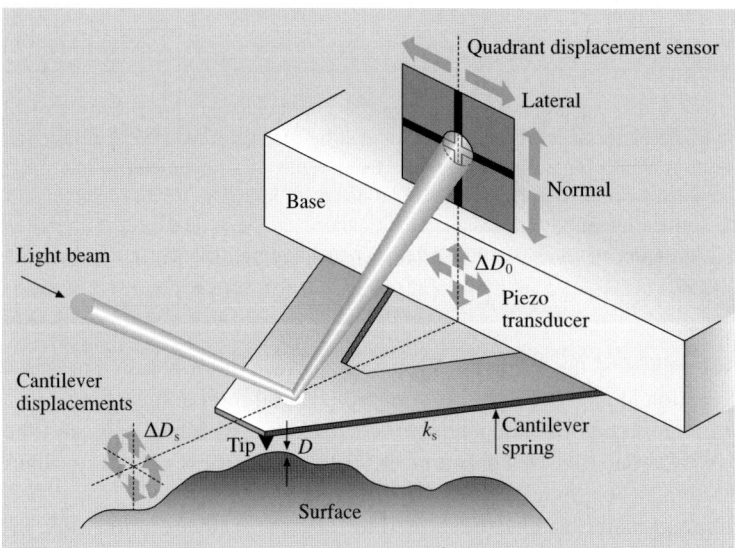

Fig. 9.3. Schematic drawing of an atomic force microscope (AFM) tip supported on a triangular cantilever interacting with an arbitrary solid surface. The normal force and topology are measured by monitoring the calibrated deflection of the cantilever as the tip is moved across the surface by means of a piezoelectric transducer. Various designs have been developed that move either the sample or the cantilever during the scan. Friction forces can be measured from the torsion of the cantilever when the scanning is in the direction perpendicular to its long axis (after [60] with permission)

as small as 1 pN (10^{-12} N), which corresponds to the bond strength of single molecules [54–57]. Distances can be inferred with an accuracy of about 1 nm, and changes in distance can be measured to about 0.1 nm. Since the contact area can be small when using sharp tips, different interaction regimes can be resolved on samples with a heterogeneous composition on lateral scales of a few nanometers. Height differences and the roughness of the sample can be measured directly from the cantilever deflection or, alternatively, by using a feedback system to raise or lower the sample so that the deflection (the normal force) is kept constant during a scan over the area of interest. Weak interaction forces and larger (microscopic) interaction areas can be investigated by replacing the tip with a micrometer-sized sphere to form a "colloidal probe" [9].

The atomic force microscope can also be used for friction measurements (lateral force microscopy, LFM, or friction force microscopy, FFM) by monitoring the torsion of the cantilever as the sample is scanned in the direction perpendicular to the long axis of the cantilever [10, 53, 58, 59]. Typically, the stiffness of the cantilever to lateral bending is much larger than to bending in the normal direction and to torsion, so that these signals are decoupled and height and friction can be detected simultaneously. The torsional spring constant can be as low as 0.1 N m^{-1}, giving a lateral (friction) force sensitivity of 10^{-11} N.

Rapid technical developments have facilitated the calibrations of the normal [61, 62] and lateral spring constants [59, 63–65], as well as in situ measurements of the macroscopic tip radius [66, 67]. Cantilevers of different shapes with a large range of spring constants, tip radii, and surface treatments (inorganic or organic coatings) are commercially available. The flat surface, and also the particle in the colloidal probe technique, can be any material of interest. However, remaining difficulties with this technique are that the distance between the tip and the substrate, D, and the deformations of the tip and sample, are not directly measurable. Another important difference between the AFM/LFM and SFA techniques is the different size of the contact area, and the related observation that, even when a cantilever with a very low spring constant is used in the AFM, the pressure in the contact zone is typically much higher than in the SFA. Hydrodynamic effects in liquids also affect the measurements of normal forces differently on certain time scales [68–71].

9.2.4 Some Other Force-Measuring Techniques

A large number of other techniques are available for the measurements of the normal forces between solid or fluid surfaces (see [5, 60]). The techniques discussed in this section are not used for lateral (friction) force measurements, but are commonly used to study normal forces, particularly in biological systems.

Micropipette aspiration is used to measure the forces between cells or vesicles, or between a cell or vesicle and another surface [72–74]. The cell or vesicle is held by suction at the tip of a glass micropipette and deforms elastically in response to the net interactions with another surface and to the applied suction. The shape of the deformed surface (cell membrane) is measured and used to deduce the force between the surfaces and the membrane tension [73]. The membrane tension, and thus the stiffness of the cell or vesicle, is regulated by applying different hydrostatic pressures. Forces can be measured in the range of 0.1 pN to 1 nN, and the distance resolution is a few nanometers. The interactions between a colloidal particle and another surface can be studied by attaching the particle to the cell membrane [75].

In the osmotic stress technique, pressures are measured between colloidal particles in aqueous solution, membranes or bilayers, or other ordered colloidal structures (viruses, DNA). The separation between the particle surfaces and the magnitude of membrane undulations are measured by X-ray or neutron scattering techniques. This is combined with a measurement of the osmotic pressure of the solution [76–79]. The technique has been used to measure repulsive forces, such as Derjaguin–Landau–Verwey–Overbeek (DLVO) interactions, steric forces, and hydration forces [80]. The sensitivity in pressure is $0.1\,\mathrm{mN\,m^{-2}}$, and distances can be resolved to 0.1 nm.

The optical tweezers technique is based on the trapping of dielectric particles at the center of a focused laser beam by restoring forces arising from radiation pressure and light-intensity gradients [81, 82]. The forces experienced by particles as they are moved toward or away from one another can be measured with a sensitivity in the pN range. Small biological molecules are typically attached to a larger bead of a material with suitable refractive properties. Recent development allows determinations of position with nanometer resolution [83], which makes this technique useful for studying the forces during the extension of single molecules.

In total internal reflection microscopy (TIRM), the potential energy between a micrometer-sized colloidal particle and a flat surface in aqueous solution is deduced from the average equilibrium height of the particle above the surface, measured from the intensity of scattered light. The average height ($D \approx 10-100\,\mathrm{nm}$) results from a balance of gravitational force, radiation pressure from a laser beam focused at the particle from below, and intermolecular forces [84]. The technique is particularly suitable for measuring weak forces (sensitivity ca. 10^{-14} N), but is more difficult to use for systems with strong interactions. A related technique is reflection interference contrast microscopy (RICM), where optical interference is used to also monitor changes in the shape of the approaching colloidal particle or vesicle [85].

An estimate of bond strengths can be obtained from the hydrodynamic shear force exerted by a fluid on particles or cells attached to a substrate [86, 87]. At a critical force, the bonds are broken and the particle or cell will be detached and move with the velocity of the fluid. This method requires knowledge of the contact area and the flow-velocity profile of the fluid. Furthermore, a uniform stress distribution in the contact area is generally assumed. At low bond density, this technique can be used to determine the strength of single bonds (1 pN).

9.3 Normal Forces Between Dry (Unlubricated) Surfaces

9.3.1 Van der Waals Forces in Vacuum and Inert Vapors

Forces between macroscopic bodies (such as colloidal particles) across vacuum arise from interactions between the constituent atoms or molecules of each body across the gap separating them. These intermolecular interactions are electromagnetic forces between permanent or induced dipoles (van der Waals forces), and between ions (electrostatic forces). In this section, we describe the van der Waals forces, which occur between all atoms and molecules and between all macroscopic bodies (see [3]).

The interaction between two permanent dipoles with a fixed relative orientation can be attractive or repulsive. For the specific case of two freely rotating permanent dipoles in a liquid or vapor (orientational or Keesom interaction), and for a permanent dipole and an induced dipole in an atom or polar or nonpolar molecule (induction or Debye interaction), the interaction is on average always attractive. The third type of van der Waals interaction, the fluctuation or London dispersion interaction, arises from instantaneous polarization of one nonpolar or polar molecule due to fluctuations in the charge distribution of a neighboring nonpolar or polar molecule (Fig. 9.4a). Correlation between these fluctuating induced dipole moments gives an attraction that is present between any two molecules or surfaces across vacuum. At very small separations, the interaction will ultimately be repulsive as the electron clouds of atoms and molecules begin to overlap. The total interaction is thus a combination of a short-range repulsion and a relatively long-range attraction.

Except for in highly polar materials such as water, London dispersion interactions give the largest contribution (70–100 %) to the van der Waals attraction. The interaction energy of the van der Waals force between atoms or molecules depends on the separation r as

$$E(D) = \frac{-C_{\text{vdW}}}{r^6}, \tag{9.7}$$

where the constant C_{vdW} depends on the dipole moments and polarizabilities of the molecules. At large separations (> 10 nm), the London interaction is weakened by

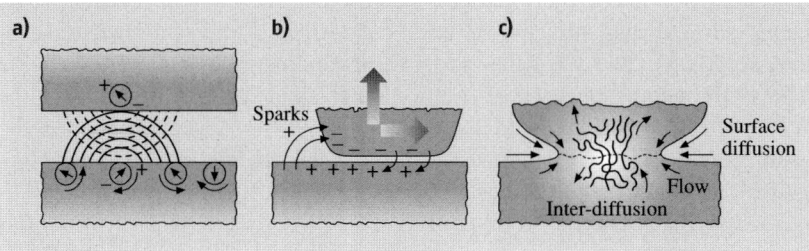

Fig. 9.4a–c. Schematic representation of (**a**) van der Waals interaction (dipole–induced dipole interaction), (**b**) charge exchange, which acts to increase adhesion and friction forces, and (**c**) sintering between two surfaces

a randomizing effect caused by the rapid fluctuations. That is, the induced temporary dipole moment of one molecule may have changed during the time needed for the transmission of the electromagnetic wave (photon) generated by its fluctuating charge density to another molecule and the return of the photon generated by the induced fluctuation in this second molecule. This phenomenon is called retardation and causes the interaction energy to decay as r^{-7} at large separations [88].

Dispersion interactions are to a first approximation additive, and their contribution to the interaction energy between two macroscopic bodies (such as colloidal particles) across vacuum can be found by summing the pairwise interactions [89]. The interaction is generally described in terms of the Hamaker constant, A_H. Another approach is to treat the interacting bodies and an intervening medium as continuous phases and determine the strength of the interaction from bulk dielectric properties of the materials [90, 91]. Unlike the pairwise summation, this method takes into account the screening of the interactions between molecules inside the bodies by the molecules closer to the surfaces and the effects of the intervening medium. For the interaction between material 1 and material 3 across material 2, the non-retarded Hamaker constant given by the Lifshitz theory is approximately [3]:

$$A_{H,123} = A_{H,\nu=0} + A_{H,\nu>0}$$

$$\approx \frac{3}{4} k_B T \left(\frac{\varepsilon_1 - \varepsilon_2}{\varepsilon_1 + \varepsilon_2} \right) \left(\frac{\varepsilon_3 - \varepsilon_2}{\varepsilon_3 + \varepsilon_2} \right)$$

$$+ \frac{3 h \nu_e}{8\sqrt{2}} \frac{(n_1^2 - n_2^2)(n_3^2 - n_2^2)}{\sqrt{(n_1^2 + n_2^2)} \sqrt{(n_3^2 + n_2^2)} \left(\sqrt{(n_1^2 + n_2^2)} + \sqrt{(n_3^2 + n_2^2)} \right)},$$

(9.8)

where the first term ($\nu = 0$) represents the permanent dipole and dipole–induced dipole interactions and the second ($\nu > 0$) the London (dispersion) interaction. ε_i and n_i are the static dielectric constants and refractive indexes of the materials, respectively. ν_e is the frequency of the lowest electron transition (around 3×10^{15} s^{-1}). Either one of the materials 1, 2, or 3 in (9.8) can be vacuum or air ($\varepsilon = n = 1$). A_H is typically 10^{-20}–10^{-19} J (the higher values are found for metals) for interactions between solids and liquids across vacuum or air.

The interaction energy between two macroscopic bodies is dependent on the geometry and is always attractive between two bodies of the same material [A_H positive, see (9.8)]. The van der Waals interaction energy and force laws ($F = -dE(D)/dD$) for some common geometries are given in Table 9.2. Because of the retardation effect, the equations in Table 9.2 will lead to an overestimation of the dispersion force at large separations. It is, however, apparent that the interaction energy between macroscopic bodies decays more slowly with separation (i.e., has a longer range) than between two molecules.

For inert nonpolar surfaces, e.g., consisting of hydrocarbons or van der Waals solids and liquids, the Lifshitz theory has been found to apply even at molecular contact, where it can be used to predict the surface energies (surface tensions) of such solids and liquids. For example, for hydrocarbon surfaces, $A_H = 5 \times 10^{-20}$ J.

Table 9.2. Van der Waals interaction energy and force between macroscopic bodies of different geometries

Geometry of bodies with surfaces D apart ($D \ll R$)	van der Waals interaction Energy, E	Force, F
Two atoms or small molecules $r \gg \sigma$	$\dfrac{-C_{vdW}}{r^6}$	$\dfrac{-6C_{vdW}}{r^7}$
Two flat surfaces (per unit area) $r \gg D$	$\dfrac{-A_H}{12\pi D^2}$	$\dfrac{-A_H}{6\pi D^3}$
Two spheres or macromolecules of radii R_1 and R_2 $R_1, R_2 \gg D$	$\dfrac{-A_H}{6D}\left(\dfrac{R_1 R_2}{R_1 + R_2}\right)$	$\dfrac{-A_H}{6D^2}\left(\dfrac{R_1 R_2}{R_1 + R_2}\right)$
Sphere or macromolecule of radius R near a flat surface $R \gg D$	$\dfrac{-A_H R}{6D}$	$\dfrac{-A_H R}{6D^2}$
Two parallel cylinders or rods of radii R_1 and R_2 (per unit length) $R_1, R_2 \gg D$	$\dfrac{-A_H}{12\sqrt{2}\, D^{3/2}}\left(\dfrac{R_1 R_2}{R_1 + R_2}\right)^{1/2}$	$\dfrac{-A_H}{8\sqrt{2}\, D^{5/2}}\left(\dfrac{R_1 R_2}{R_1 + R_2}\right)^{1/2}$
Cylinder of radius R near a flat surface (per unit length) $R \gg D$	$\dfrac{-A_H \sqrt{R}}{12\sqrt{2}\, D^{3/2}}$	$\dfrac{-A_H \sqrt{R}}{8\sqrt{2}\, D^{5/2}}$
Two cylinders or filaments of radii R_1 and R_2 crossed at $90°$ $R_1, R_2 \gg D$	$\dfrac{-A_H \sqrt{R_1 R_2}}{6D}$	$\dfrac{-A_H \sqrt{R_1 R_2}}{6D^2}$

A negative force (A_H positive) implies attraction, a positive force means repulsion (A_H negative) (after [60], with permission)

Inserting this value into the equation for two flat surfaces (Table 9.2) and using a "cut-off" distance of $D_0 \approx 0.165$ nm as an effective separation when the surfaces are in contact [3], we obtain for the surface energy γ (which is defined as half the interaction energy)

$$\gamma = \frac{E}{2} = \frac{A_H}{24\pi D_0^2} \approx 24 \, \text{mJ m}^{-2}, \qquad (9.9)$$

a value that is typical for hydrocarbon solids and liquids [92].

If the adhesion force is measured between two crossed-cylindrical surfaces of $R = 1$ cm (a geometry equivalent to a sphere with $R = 1$ cm interacting with a flat surface, cf. Table 9.2) using an SFA, we expect the adhesion force to be (see Table 9.2) $F = A_H R/(6D_0^2) = 4\pi R\gamma \approx 3.0$ mN. Using a spring constant of $k_s = 100$ N m^{-1}, such an adhesive force will cause the two surfaces to jump apart by $\Delta D = F/k_s = 30$ μm, which can be accurately measured. (For elastic bodies that deform in adhesive contact, R changes during the interaction and the measured adhesion force is 25% lower, see Sect. 9.5.2). Surface energies of solids can thus be directly measured with the SFA and, in principle, with the AFM if the contact geometry can be quantified. The measured values are in good agreement with calculated values based on the known surface energies γ of the materials, and for nonpolar low-energy solids they are well accounted for by the Lifshitz theory [3].

9.3.2 Charge-Exchange Interactions

Electrostatic interactions are present between ions (Coulomb interactions), between ions and permanent dipoles, and between ions and nonpolar molecules in which a charge induces a dipole moment. The interaction energy between ions or between a charge and a fixed permanent dipole can be attractive or repulsive. For an induced dipole or a freely rotating permanent dipole in vacuum or air, the interaction energy with a charge is always attractive.

Spontaneous charge transfer may occur between two dissimilar materials in contact [93–97]. The phenomenon, called contact electrification, is especially prominent in contact between a metal and a material with low conductivity (including organic liquids) [95, 97], but is also observed, for example, between two different polymer layers. It is believed that when two different materials are in static contact, charge transfer might occur due to quantum tunneling of electrons or, in some cases, transfer of surface ions. The charging is generally seen to be stronger with increasing difference in work function (or electron affinity) between the two materials [95, 97]. During separation, rolling or sliding of one body over the other, the surfaces experience both charge transition from one surface to the other and charge transfer (conductance) along each surface (Fig. 9.4b). The latter process is typically slower, and, as a result, charges remain on the surfaces as they are separated in vacuum or dry nitrogen gas. The charging gives rise to a strong adhesion with adhesion energies of over 1000 mJ m^{-2}, similar to fracture or cohesion energies of the solid bodies themselves [93, 94, 98]. Upon separating the surfaces further apart, a strong, long-range electrostatic attraction is observed. The charging can be decreased through discharges across the gap between the surfaces (which requires a high charging) or through conduction in the solids. The discharge may give rise to triboluminescence [99], but can also cause sparks that may ignite combustible materials [100]. It has been suggested that charge-exchange interactions are particularly important in rolling friction between dry surfaces (which can simplistically be thought of as an adhesion–separation process), where the distance dependence of forces acting normally to the surfaces plays a larger role than in sliding friction. In the case of sliding friction, charge transfer is also observed between identical materials [98, 101]. Mechanisms such as bond formation and breakage (polymer scission), slip at the wall between a flowing liquid and a solid [102], or material transfer and the creation of wear particles have been suggested. However, friction electrification or triboelectrification also occurs during wearless sliding, i.e., when the surfaces are not damaged. Other explanations such as the creation or translation of defects on or near the surface have been put forward [98]. Attempts have been made to correlate the amount of charging with the normal force or load and with the polarizability of the sliding materials [96, 103]. Recent experiments on the sliding friction between metal–insulator surfaces indicate that stick–slip would be accompanied by charge-transfer events [104, 105].

Photoinduced charge transfer, or harpooning, involves the transfer of an electron between an atom in a molecular beam or at a solid surface (typically an alkali or transition metal) to an atom or molecule in a gas (typically a halide) to form a negatively charged molecular ion in a highly excited vibrational state. This transfer process can occur at atomic distances of 0.5–0.7 nm, which is far from molecular contact. The formed molecular ion is attracted to the surface and chemisorbs onto it. Photoinduced charge-transfer processes also occur in the photosynthesis in green plants and in photo-electrochemical cells (solar cells) at the junction between two semiconductors or between a semiconductor and an electrolyte solution [106].

9.3.3 Sintering and Cold Welding

When macroscopic particles in a powder or in a suspension come into molecular contact, they can bond together to form a network or solid body with very different density and shear strength compared to the powder (a typical example is porcelain). The rate of bonding is dependent on the surface energy (causing a stress at the edge of the contact) and the atomic mobility (diffusion rate) of the contacting materials. To increase the diffusion rate, objects formed from powders are heated to about one half of the melting temperature of the components in a process called sintering, which can be done in various atmospheres or in a liquid.

In the sintering process, the surface energy of the system is lowered due to the reduction of total surface area (Fig. 9.4c). In metal and ceramic systems, the most important mechanism is solid-state diffusion, initially surface diffusion. As the surface area decreases and the grain boundaries increase at the contacts, grain boundary diffusion and diffusion through the crystal lattice become more important [107]. The grain boundaries will eventually migrate, so that larger particles are formed (coarsening). Mass can also be transferred through evaporation and condensation, and through viscous and plastic flow. In liquid-phase sintering, the materials can melt, which increases the mass transport. Amorphous materials like polymers and glasses do not have real grain boundaries and sinter by viscous flow [108].

Some of these mechanisms (surface diffusion and evaporation–condensation) reduce the surface area and increase the grain size (coarsening) without densification, in contrast to bulk transport mechanisms such as grain-boundary diffusion and plastic and viscous flow. As the material becomes denser, elongated pores collapse to form smaller, spherical pores with a lower surface energy. Models for sintering typically consider the size and growth rate of the grain boundary (the "neck") formed between two spherical particles. At a high stage of densification, the sintering stress σ at the curved neck between two particles is given by [108]

$$\sigma = \frac{2\gamma_{SS}}{G} + \frac{2\gamma_{SV}}{r_p} , \qquad (9.10)$$

where γ_{SS} is the solid–solid grain boundary energy, γ_{SV} is the solid–vapor surface energy, G is the grain size, and r_p is the radius of the pore.

A related phenomenon is cold welding, which is the spontaneous formation of strong junctions between clean (unoxidized) metal surfaces with a mutual solubility when they are brought into contact, with or without an applied pressure. The plastic deformations accompanying the formation and breaking of such contacts on a molecular scale during motion of one surface normally (see Fig. 9.10c,d) or laterally (shearing) with respect to the other have been studied both experimentally [109, 110] and theoretically [111–116]. The breaking of a cold-welded contact is generally associated with damage or deformation of the surface structure.

9.4 Normal Forces Between Surfaces in Liquids

9.4.1 Van der Waals Forces in Liquids

The dispersion interaction in a medium will be significantly lower than in vacuum, since the attractive interaction between two solute molecules in a medium (solvent) involves displacement and reorientation of the nearest-neighbor solvent molecules. Even though the surrounding medium may change the dipole moment and polarizability from that in vacuum, the interaction between two identical molecules remains attractive in a binary mixture. The extension of the interactions to the case of two macroscopic bodies is the same as described in Sect. 9.3.1. Typically, the Hamaker constants for interactions in a medium are an order of magnitude lower than in vacuum. Between macroscopic surfaces in liquids, van der Waals forces become important at distances below 10–15 nm and may at these distances start to dominate interactions of different origin that have been observed at larger separations.

Figure 9.5 shows the measured van der Waals forces between two crossed-cylindrical mica surfaces in water and various salt solutions. Good agreement is

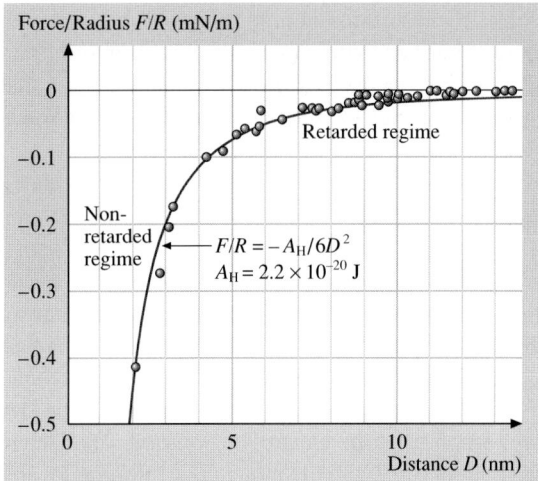

Fig. 9.5. Attractive van der Waals force F between two curved mica surfaces of radius $R \approx$ 1 cm measured in water and various aqueous electrolyte solutions. The electrostatic interaction has been subtracted from the total measured force. The measured non-retarded Hamaker constant is $A_H = 2.2 \times 10^{-20}$ J. Retardation effects are apparent at distances larger than 5 nm, as expected theoretically (after [3]; copyright 1991, with permission from Elsevier Science)

obtained between experiment and theory. At larger surface separations, above about 5 nm, the measured forces fall off more rapidly than D^{-2}. This retardation effect (see Sect. 9.3.1) is also predicted by the Lifshitz theory and is due to the time needed for propagation of the induced dipole moments over large distances.

From Fig. 9.5, we may conclude that, at separations above about 2 nm, or 8 molecular diameters of water, the *continuum* Lifshitz theory is valid. This would mean that water films as thin as 2 nm may be expected to have bulk-like properties, at least as far as their interaction forces are concerned. Similar results have been obtained with other liquids, where in general continuum properties are manifested, both as regards their interactions and other properties such as viscosity, at a film thickness larger than five or ten molecular diameters. In the absence of a solvent (in vacuum), the agreement of measured van der Waals forces with the continuum Lifshitz theory is generally good at all separations down to molecular contact ($D = D_0$).

Van der Waals interactions in a system of three or more different materials (see (9.8)) can be attractive or repulsive, depending on their dielectric properties. Numerous experimental studies show the attractive van der Waals forces in various systems [3], and repulsive van der Waals forces have also been measured directly [117, 118]. A practical consequence of the repulsive interaction obtained across a medium with intermediate dielectric properties is that the van der Waals forces will give rise to preferential, nonspecific adsorption of molecules with an intermediate dielectric constant. This is commonly seen as adsorption of vapors or solutes to a solid surface. It is also possible to diminish the attractive interaction between dispersed colloidal particles by adsorption of a thin layer of material with dielectric properties close to those of the surrounding medium (matching of refractive index), or by adsorption of a polymer that gives a steric repulsive force that keeps the particles separated at a distance where the magnitude of the van der Waals attraction is negligible. Thermal motion will then keep the particles dispersed.

9.4.2 Electrostatic and Ion Correlation Forces

Most surfaces in contact with a highly polar liquid (such as water) acquire a surface charge, either by dissociation of ions from the surface into the solution or by preferential adsorption of certain ions from the solution. The surface charge is balanced by a layer of oppositely charged ions (counterions) in the solution at some small distance from the surface (see [3]). In dilute solution, this distance is the Debye length, κ^{-1}, which is purely a property of the electrolyte *solution*. The Debye length falls with increasing ionic strength (i.e., with the molar concentration M_i and valency z_i) of the ions in solution:

$$\kappa^{-1} = \left(\frac{\varepsilon \varepsilon_0 k_B T}{e^2 N_A \sum_i z_i^2 M_i} \right)^{1/2}, \qquad (9.11)$$

where e is the electronic charge. For example, for 1:1 electrolytes at 25 °C, $\kappa^{-1} = 0.304\,\text{nm}/\sqrt{M_{1:1}}$, where M_i is given in M (mol dm^{-3}). κ^{-1} is thus about 10 nm in a 1 mM NaCl solution and 0.3 nm in a 1 M solution. In totally pure water at pH 7, where $M_i = 10^{-7}$ M, κ^{-1} is 960 nm, or about 1 µm. The Debye length also relates the surface charge density σ of a surface to the electrostatic surface potential ψ_0 via the Grahame equation, which for 1:1 electrolytes can be expressed as:

$$\sigma = \sqrt{8\varepsilon\varepsilon_0 k_B T}\sinh(e\psi_0/2k_B T) \times \sqrt{M_{1:1}}. \qquad (9.12)$$

Since the Debye length is a measure of the thickness of the diffuse atmosphere of counterions near a charged surface, it also determines the range of the electrostatic "double-layer" interaction between two charged surfaces. The electrostatic double-layer interaction is an entropic effect that arises upon decreasing the thickness of the liquid film containing the dissolved ions. Because of the attractive force between the dissolved ions and opposite charges on the surfaces, the ions stay between the surfaces, but an osmotic repulsion arises as their concentration increases. The long-range electrostatic interaction energy at large separations (weak overlap) between two similarly charged molecules or surfaces is typically repulsive and is roughly an exponentially decaying function of D:

$$E(D) \approx +C_{\text{ES}}\,e^{-\kappa D}, \qquad (9.13)$$

where C_{ES} is a constant that depends on the geometry of the interacting surfaces, on their surface charge density, and the solution conditions (Table 9.3). We see that the

Table 9.3. Electrical double-layer interaction energy $E(D)$ and force ($F = -dE/dD$) between macroscopic bodies

Geometry of bodies with surfaces D apart ($D \ll R$)	Electric double-layer interaction Energy E	Force F
Two ions or small molecules $r > \sigma$	$\dfrac{+z_1 z_2 e^2}{4\pi\varepsilon\varepsilon_0 r} \dfrac{e^{-\kappa(r-\sigma)}}{(1+\kappa\sigma)}$	$\dfrac{+z_1 z_2 e^2}{4\pi\varepsilon\varepsilon_0 r^2} \dfrac{(1+\kappa r)}{(1+\kappa\sigma)} e^{-\kappa(r-\sigma)}$
Two flat surfaces (per unit area) $r \gg D$	$(\kappa/2\pi) Z e^{-\kappa D}$	$(\kappa^2/2\pi) Z e^{-\kappa D}$
Two spheres or macromolecules of radii R_1 and R_2 $R_1, R_2 \gg D$	$\left(\dfrac{R_1 R_2}{R_1 + R_2}\right) Z e^{-\kappa D}$	$\kappa\left(\dfrac{R_1 R_2}{R_1 + R_2}\right) Z e^{-\kappa D}$
Sphere or macromolecule of radius R near a flat surface $R \gg D$	$R Z e^{-\kappa D}$	$\kappa R Z e^{-\kappa D}$
Two parallel cylinders or rods of radii R_1 and R_2 (per unit length) $R_1, R_2 \gg D$	$\dfrac{\kappa^{1/2}}{\sqrt{2\pi}} \left(\dfrac{R_1 R_2}{R_1 + R_2}\right)^{1/2} Z e^{-\kappa D}$	$\dfrac{\kappa^{3/2}}{\sqrt{2\pi}} \left(\dfrac{R_1 R_2}{R_1 + R_2}\right)^{1/2} Z e^{-\kappa D}$
Cylinder of radius R near a flat surface (per unit length) $R \gg D$	$\kappa^{1/2} \sqrt{\dfrac{R}{2\pi}} Z e^{-\kappa D}$	$\kappa^{3/2} \sqrt{\dfrac{R}{2\pi}} Z e^{-\kappa D}$
Two cylinders or filaments of radii R_1 and R_2 crossed at 90° $R_1, R_2 \gg D$	$\sqrt{R_1 R_2} Z e^{-\kappa D}$	$\kappa \sqrt{R_1 R_2} Z e^{-\kappa D}$

The interaction energy and force for bodies of different geometries is based on the Poisson–Boltzmann equation (a continuum, mean-field theory). Equation (9.14) gives the interaction constant Z (in terms of the surface potential ψ_0) for the interaction between similarly charged (ionized) surfaces in aqueous solutions of monovalent electrolyte. It can also be expressed in terms of the surface charge density σ by applying the Grahame equation (9.12) (after [60], with permission)

Debye length is the decay length of the interaction energy between two surfaces (and of the mean potential away from one surface). C_{ES} can be determined by solving the so-called Poisson–Boltzmann equation or by using other theories [119–123]. The equations in Table 9.3 are expressed in terms of a constant, Z, defined as

$$Z = 64\pi\varepsilon\varepsilon_0(k_BT/e)^2 \tanh^2[ze\psi_0/(4k_BT)], \qquad (9.14)$$

which depends only on the properties of the *surfaces*.

The above approximate expressions are accurate only for surface separations larger than about one Debye length. At smaller separations one must use numerical solutions of the Poisson–Boltzmann equation to obtain the exact interaction potential, for which there are no simple expressions. In the limit of small D, it can be shown that the interaction energy depends on whether the surfaces remain at constant potential ψ_0 (as assumed in the above equations) or at constant charge σ (when the repulsion exceeds that predicted by the above equations), or somewhere between these limits. In the "constant charge limit" the total *number* of counterions in the compressed film does not change as D is decreased, whereas at constant potential, the *concentration* of counterions is constant. The limiting pressure (or force per unit area) at constant charge is the osmotic pressure of the confined ions:

$$F = k_BT \times \text{ion number density} = 2\sigma k_BT/(zeD), \quad \text{for } D \ll \kappa^{-1}. \qquad (9.15)$$

That is, as $D \to 0$ the double-layer pressure at constant surface charge becomes infinitely repulsive and independent of the salt concentration (at constant potential the force instead becomes a constant at small D). However, at small separations, the van der Waals attraction (which goes as D^{-2} between two spheres or as D^{-3} between two planar surfaces, see Table 9.2) wins out over the double-layer repulsion, unless some other short-range interaction becomes dominant (see Sect. 9.4.4). This is the theoretical prediction that forms the basis of the so-called Derjaguin–Landau–Verwey–Overbeek (DLVO) theory [119, 124], illustrated in Fig. 9.6.

Because of the different distance dependence of the van der Waals and electrostatic interactions, the total force law, as described by the DLVO theory, can show several minima and maxima. Typically, the depth of the outer (secondary) minimum is a few k_BT, which is enough to cause reversible flocculation of particles from an aqueous dispersion. If the force barrier between the secondary and primary minimum is lowered, for example, by increasing the electrolyte concentration, particles can be irreversibly coagulated in the primary minimum. In practice, other forces (described in the following sections) often appear at very small separations, so that the full force law between two surfaces or colloidal particles in solution can be more complex than might be expected from the DLVO theory.

There are situations when the double-layer interaction can be attractive at short range even between surfaces of similar charge, especially in systems with charge regulation due to dissociation of chargeable groups on the surfaces [123, 125]; ion condensation [126], which may lower the effective surface charge density in systems containing di-and trivalent counterions; or ion correlation, which is an additional van der Waals-like attraction due to mobile and therefore highly polarizable counterions

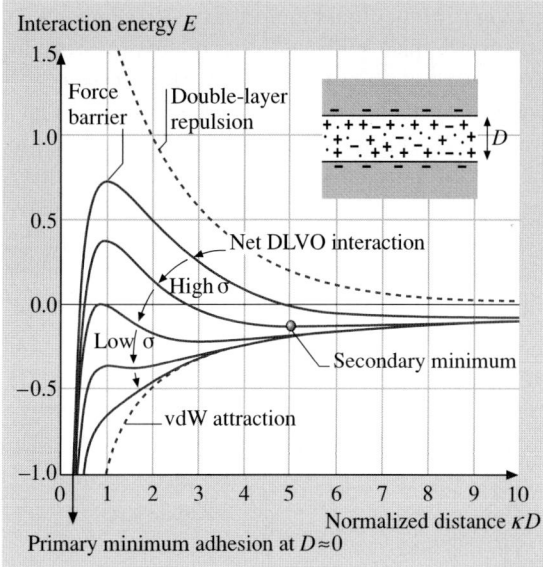

Fig. 9.6. Schematic plots of the DLVO interaction potential energy E between two flat, charged surfaces [or, according to the Derjaguin approximation, (9.3), the force F between two curved surfaces] as a function of the surface separation normalized by the Debye length, κ^{-1}. The van der Waals attraction (inverse power-law dependence on D) together with the repulsive electrostatic "double-layer" force (roughly exponential) at different surface charge σ [or potential, see (9.12)] determine the net interaction potential in aqueous electrolyte solution (after [60] with permission)

located at the surface [127]. The ion correlation (or charge fluctuation) force becomes significant at separations below 4 nm and increases with the surface charge density σ and the valency z of the counterions. Computer simulations have shown that, at high charge density and for monovalent counterions, the ion correlation force can reduce the effective double-layer repulsion by 10–15 %. With divalent counterions, the ion correlation force was found to exceed the double-layer repulsion and the total force then became attractive at a separation below 2 nm even in dilute electrolyte solution [128]. Experimentally, such short-range attractive forces have been found between charged bilayers [129, 130] and also in other systems [131].

9.4.3 Solvation and Structural Forces

When a liquid is confined within a restricted space, for example, a very thin film between two surfaces, it ceases to behave as a structureless continuum. At small surface separations (below about ten molecular diameters), the van der Waals force between two surfaces or even two solute molecules in a liquid (solvent) is no longer a smoothly varying attraction. Instead, there arises an additional "solvation" force that generally oscillates between attraction and repulsion with distance, with a periodicity equal to some mean dimension σ of the liquid molecules [132]. Figure 9.7a shows the force law between two smooth mica surfaces across the hydrocarbon liquid tetradecane, whose inert, chain-like molecules have a width of $\sigma \approx 0.4$ nm.

The short-range oscillatory force law is related to the "density distribution function" and "potential of mean force" characteristic of intermolecular interactions in liquids. These forces arise from the confining effects that the two surfaces have on liquid molecules, forcing them to order into quasi-discrete layers. Such layers are energetically or entropically favored and correspond to the minima in the free energy, whereas fractional layers are disfavored (energy maxima). This effect is quite general and arises in all simple liquids when they are confined between

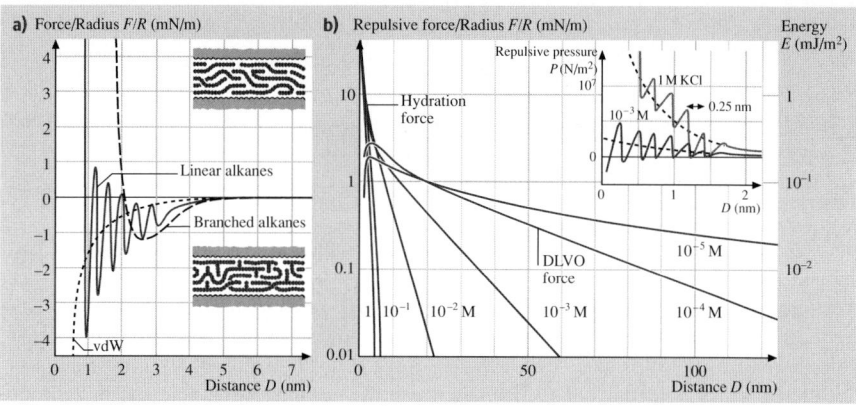

Fig. 9.7. (**a**) *Solid curve*: Forces measured between two mica surfaces across saturated linear chain alkanes such as *n*-tetradecane and *n*-hexadecane [133, 134]. The 0.4-nm periodicity of the oscillations indicates that the molecules are preferentially oriented parallel to the surfaces, as shown schematically in the *upper insert*. The theoretical continuum van der Waals attraction is shown as a *dotted curve*. *Dashed curve*: Smooth, non-oscillatory force law exhibited by irregularly shaped alkanes (such as 2-methyloctadecane) that cannot order into well-defined layers (*lower insert*) [134, 135]. Similar non-oscillatory forces are also observed between "rough" surfaces, even when these interact across a saturated linear chain liquid. This is because the irregularly shaped surfaces (rather than the liquid) now prevent the liquid molecules from ordering in the gap. (**b**) Forces measured between charged mica surfaces in KCl solutions of varying concentrations [20]. In dilute solutions (10^{-5} and 10^{-4} M), the measured forces are excellently described by the DLVO theory, based on exact solutions to the nonlinear Poisson–Boltzmann equation for the electrostatic forces and the Lifshitz theory for the van der Waals forces (using a Hamaker constant of $A_H = 2.2 \times 10^{-20}$ J). At higher concentrations, as more hydrated K^+ cations adsorb onto the negatively charged surfaces, an additional hydration force appears superimposed on the DLVO interaction at distances below 3–4 nm. This force has both an oscillatory and a monotonic component. *Insert*: Short-range hydration forces between mica surfaces shown as pressure versus distance. The lower and upper curves show surfaces 40% and 95% saturated with K^+ ions. At larger separations, the forces are in good agreement with the DLVO theory (after [3]; copyright 1991, with permission from Elsevier Science)

two smooth, rigid surfaces, both flat and curved. Oscillatory forces do not require any attractive liquid–liquid or liquid–wall interaction, only two hard walls confining molecules whose shape is not too irregular and that are free to exchange with molecules in a bulk liquid reservoir. In the absence of any attractive pressure between the molecules, the bulk liquid density could be maintained by an external hydrostatic pressure – in real liquids attractive van der Waals forces play the role of such an external pressure.

Oscillatory forces are now well understood theoretically, at least for simple liquids, and a number of theoretical studies and computer simulations of various confined liquids (including water) that interact via some form of Lennard–Jones potential have invariably led to an oscillatory solvation force at surface separations below a few molecular diameters [136–144]. In a first approximation, the oscillatory force law may be described by an exponentially decaying cosine function of the form

$$E \approx E_0 \cos(2\pi D/\sigma) e^{-D/\sigma}, \tag{9.16}$$

where both theory and experiments show that the oscillatory period and the characteristic decay length of the envelope are close to σ.

Once the solvation zones of the two surfaces overlap, the mean liquid density in the gap is no longer the same as in the bulk liquid. Since the van der Waals interaction depends on the optical properties of the liquid, which in turn depends on the density, the van der Waals and the oscillatory solvation forces are not strictly additive. It is more correct to think of the solvation force as *the* van der Waals force at small separations with the molecular properties and density variations of the medium taken into account. It is also important to appreciate that solvation forces do not arise simply because liquid molecules tend to structure into semi-ordered layers at surfaces. They arise because of the disruption or *change* of this ordering during the approach of a second surface. The two effects are related; the greater the tendency toward structuring at an isolated surface the greater the solvation force between two such surfaces, but there is a real distinction between the two phenomena that should be borne in mind.

Oscillatory forces lead to different adhesion values depending on the energy minimum from which two surfaces are being separated. For an interaction energy described by (9.16), "quantized" adhesion energies will be E_0 at $D = 0$ (primary minimum), E_0/e at $D = \sigma$, E_0/e^2 at $D = 2\sigma$, etc. E_0 can be thought of as a depletion force (see Sect. 9.4.5) that is approximately given by the osmotic limit $E_0 \approx -k_B T/\sigma^2$, which can exceed the contribution to the adhesion energy in contact from the van der Waals forces (at $D_0 \approx 0.15-0.20$ nm, as discussed in Sect. 9.3.1, keeping in mind that the Lifshitz theory fails to describe the force law at *intermediate* distances). Such multivalued adhesion forces have been observed in a number of systems, including the interactions of fibers.

Measurements of oscillatory forces between different surfaces across both aqueous and nonaqueous liquids have revealed their richness of properties [145–149], for example, their great sensitivity to the shape and rigidity of the solvent molecules, to the presence of other components, and to the structure of the confining surfaces (see Sects. 9.5.3 and 9.9). In particular, the oscillations can be smeared out if the molecules are irregularly shaped (e. g., branched) and therefore unable to pack into ordered layers, or when the interacting surfaces are rough or fluid-like (see Sect. 9.4.6).

It is easy to understand how oscillatory forces arise between two flat, plane parallel surfaces. Between two curved surfaces, e. g., two spheres, one might imagine the molecular ordering and oscillatory forces to be smeared out in the same way that they are smeared out between two randomly rough surfaces (see Sect. 9.5.3); however, this is not the case. Ordering can occur as long as the curvature or roughness is itself regular or uniform, i.e., not random. This is due to the Derjaguin approximation (9.3). If the energy between two flat surfaces is given by a decaying oscillatory function (for example, a cosine function as in (9.16)), then the force (and energy) between two curved surfaces will also be an oscillatory function of distance with some phase shift. Likewise, two surfaces with regularly curved regions will also retain their oscillatory force profile, albeit modified, as long as the corrugations are truly regular, i.e., periodic. On the other hand, surface roughness, even on the nanometer scale, can smear out oscillations if the roughness is random and the confined molecules are smaller than the size of the surface asperities [150, 151]. If an organic liquid contains small amounts of water, the expected oscillatory force can be replaced by a strongly attractive capillary force (see Sect. 9.5.1).

9.4.4 Hydration and Hydrophobic Forces

The forces occurring in water and electrolyte solutions are more complex than those occurring in nonpolar liquids. According to continuum theories, the attractive van der Waals force is always expected to win over the repulsive electrostatic "double-layer" force at small surface separations (Fig. 9.6). However, certain surfaces (usually oxide or hydroxide surfaces such as clays or silica) swell spontaneously or repel each other in aqueous solution, even at high salt concentrations. Yet in all these systems one would expect the surfaces or particles to remain in strong adhesive contact or coagulate in a primary minimum if the only forces operating were DLVO forces.

There are many other aqueous systems in which the DLVO theory fails and where there is an additional short-range force that is not oscillatory but monotonic. Between hydrophilic surfaces this force is exponentially repulsive and is commonly referred to as the *hydration*, or *structural*, force. The origin and nature of this force has long been controversial, especially in the colloidal and biological literature. Repulsive hydration forces are believed to arise from strongly hydrogen-bonding surface groups, such as hydrated ions or hydroxyl (−OH) groups, which modify the hydrogen-bonding network of liquid water adjacent to them. Because this network is quite extensive in range [152], the resulting interaction force is also of relatively long range.

Repulsive hydration forces were first extensively studied between clay surfaces [153]. More recently, they have been measured in detail between mica and silica surfaces [20–22, 154], where they have been found to decay exponentially with decay lengths of about 1 nm. Their effective range is 3–5 nm, which is about twice the range of the oscillatory solvation force in water. Empirically, the hydration repulsion between two hydrophilic surfaces appears to follow the simple equation

$$E = E_0 e^{-D/\lambda_0}, \tag{9.17}$$

where $\lambda_0 \approx 0.6-1.1$ nm for 1:1 electrolytes and $E_0 = 3-30$ mJ m^{-2} depending on the hydration (hydrophilicity) of the surfaces, higher E_0 values generally being associated with lower λ_0 values.

The interactions between molecularly smooth mica surfaces in dilute electrolyte solutions obey the DLVO theory (Fig. 9.7b). However, at higher salt concentrations, specific to each electrolyte, hydrated cations bind to the negatively charged surfaces and give rise to a repulsive hydration force [20, 21]. This is believed to be due to the energy needed to dehydrate the bound cations, which presumably retain some of their water of hydration on binding. This conclusion was arrived at after noting that the strength and range of the hydration forces increase with the known hydration numbers of the electrolyte cations in the order: $Mg^{2+} > Ca^{2+} > Li^+ \sim Na^+ > K^+ > Cs^+$. Similar trends are observed with other negatively charged colloidal surfaces.

While the hydration force between two mica surfaces is overall repulsive below a distance of 4 nm, it is not always monotonic below about 1.5 nm but exhibits oscillations of mean periodicity of 0.25 ± 0.03 nm, roughly equal to the diameter of the water molecule. This is shown in the insert in Fig. 9.7b, where we may note that the first three minima at $D = 0, 0.28$, and 0.56 nm occur at negative energies, a result that rationalizes observations on certain colloidal systems. For example, clay platelets such as montmorillonite often repel each other increasingly strongly as they come closer together, but they are also known to stack into stable aggregates with water interlayers of typical thickness 0.25 and 0.55 nm between them [155, 156], suggestive of a turnabout in the force law from a monotonic repulsion to discretized attraction. In chemistry we would refer to such structures as stable hydrates of fixed stoichiometry, whereas in physics we may think of them as experiencing an oscillatory force.

Both surface force and clay swelling experiments have shown that hydration forces can be modified or "regulated" by exchanging ions of different hydration on surfaces, an effect that has important practical applications in controlling the stability of colloidal dispersions. It has long been known that colloidal particles can be precipitated (coagulated or flocculated) by increasing the electrolyte concentration, an effect that was traditionally attributed to the reduced screening of the electrostatic double-layer repulsion between the particles due to the reduced Debye length. However, there are many examples where colloids are stabilized at high salt concentrations, not at low concentrations. This effect is now recognized as being due to the increased hydration repulsion experienced by certain surfaces when they bind highly hydrated ions at higher salt concentrations. Hydration regulation of adhesion and interparticle forces is an important practical method for controlling various processes such as clay swelling [155, 156], ceramic processing and rheology [157, 158], material fracture [157], and colloidal particle and bubble coalescence [159].

Water appears to be unique in having a solvation (hydration) force that exhibits both a monotonic and an oscillatory component. Between hydrophilic surfaces the monotonic component is repulsive (Fig. 9.7b), but between hydrophobic surfaces it is attractive and the final adhesion is much greater than expected from the Lifshitz theory.

A hydrophobic surface is one that is inert to water in the sense that it cannot bind to water molecules via ionic or hydrogen bonds. Hydrocarbons and fluorocarbons are hydrophobic, as is air, and the strongly attractive hydrophobic force has many important manifestations and consequences such as the low solubility or miscibility of water and oil molecules, micellization, protein folding, strong adhesion and rapid coagulation of hydrophobic surfaces, non-wetting of water on hydrophobic surfaces, and hydrophobic particle attachment to rising air bubbles (the basic principle of froth flotation).

In recent years, there has been a steady accumulation of experimental data on the force laws between various hydrophobic surfaces in aqueous solution [160–178]. These studies have found that the force law between two macroscopic hydrophobic surfaces is of surprisingly long range, decaying exponentially with a characteristic decay length of 1–2 nm in the separation range of 0–10 nm, and then more gradually further out. The hydrophobic force can be far stronger than the van der Waals attraction, especially between hydrocarbon surfaces in water, for which the Hamaker constant is quite small. The magnitude of the hydrophobic attraction has been found to decrease with the decreasing hydrophobicity (increasing hydrophilicity) of lecithin lipid bilayer surfaces [31] and silanated surfaces [168], whereas examples of the opposite trend have been shown for some Langmuir–Blodgett-deposited monolayers [179]. An apparent correlation has been found between high stability of the hydrophobic surface (as measured by its contact angle hysteresis) and the absence of a long-range part of the attractive force [180].

For two surfaces in water the purely hydrophobic interaction energy (ignoring DLVO and oscillatory forces) in the range 0–10 nm is given by

$$E = -2\gamma e^{-D/\lambda_0} , \tag{9.18}$$

where typically $\lambda_0 = 1-2$ nm, and $\gamma = 10-50$ mJ m^{-2}. The higher value corresponds to the interfacial energy of a pure hydrocarbon–water interface.

At a separation below 10 nm, the hydrophobic force appears to be insensitive or only weakly sensitive to changes in the type and concentration of electrolyte ions in the solution. The absence of a "screening" effect by ions attests to the nonelectrostatic origin of this interaction. In contrast, some experiments have shown that, at separations greater than 10 nm, the attraction does depend on the intervening electrolyte, and that in dilute solutions, or solutions containing divalent ions, it can continue to exceed the van der Waals attraction out to separations of 80 nm [165, 181]. Recent research suggests that the interactions at very long range might not be a "hydrophobic" force since they are influenced by the presence of dissolved gas in the solution [176, 177], the stability of the hydrophobic surface [178, 180], and, on some types of surfaces, bridging submicroscopic bubbles [172–174].

The long-range nature of the hydrophobic interaction has a number of important consequences. It accounts for the rapid coagulation of hydrophobic particles in water and may also account for the rapid folding of proteins. It also explains the ease with which water films rupture on hydrophobic surfaces. In this case, the van der Waals force across the water film is repulsive and therefore favors wetting, but this is more than offset by the attractive hydrophobic interaction acting between the two hydrophobic phases across water. Hydrophobic forces are increasingly being implicated in the adhesion and fusion of biological membranes and cells. It is known that both osmotic and electric-field stresses enhance membrane fusion, an effect that may be due to the concomitant increase in the hydrophobic area exposed between two adjacent surfaces.

From the previous discussion we can infer that hydration and hydrophobic forces are not of a simple nature. These interactions are probably the most important, yet the least understood of all the forces in aqueous solutions. The unusual properties of water and the nature of the surfaces (including their homogeneity and stability) appear to be equally important. Some particle surfaces can have their hydration forces regulated, for example, by ion exchange. Others appear to be intrinsically hydrophilic (e.g., silica) and cannot be coagulated by changing the ionic condition, but can be rendered hydrophobic by chemically modifying their surface groups. For example, on heating silica to above 600 °C, two adjacent surface silanol (−OH) groups release a water molecule and form a hydrophobic siloxane (−O−) group, whence the repulsive hydration force changes into an attractive hydrophobic force.

How do these exponentially decaying repulsive or attractive forces arise? Theoretical work and computer simulations [138, 140, 182, 183] suggest that the solvation forces in water should be purely oscillatory, whereas other theoretical studies [184–191] suggest a monotonically exponential repulsion or attraction, possibly superimposed on an oscillatory force. The latter is consistent with experimental findings, as shown in the inset to Fig. 9.7b, where it appears that the oscillatory force is simply additive with the monotonic hydration and DLVO forces, suggesting that these arise from essentially different mechanisms. It has been suggested that for a sufficiently solvophilic surface, there could be "hydration"-like forces also in nonaqueous systems [190].

It is probable that the short-range hydration force between all smooth, rigid, or crystalline surfaces (e.g., mineral surfaces such as mica) has an oscillatory component. This may or may not be superimposed on a monotonic force due to image interactions [186], dipole–dipole interactions [191], and/or structural or hydrogen-bonding interactions [184, 185].

Like the repulsive hydration force, the origin of the hydrophobic force is still unknown. *Luzar* et al. [188] carried out a Monte Carlo simulation of the interaction between two hydrophobic surfaces across water at separations below 1.5 nm. They obtained a decaying oscillatory force superimposed on a monotonically attractive curve. In more recent computational and experimental work [192–195], it has been suggested that hydrophobic surfaces generate a depleted region of water around them, and that a long-range attractive force due to depletion arises between two such surfaces. Such a difference in density might also cause boundary slip of water at hydrophic surfaces [51, 196, 197].

It is questionable whether the hydration or hydrophobic force should be viewed as an ordinary type of solvation or structural force that reflects the packing of water molecules. The energy (or entropy) associated with the hydrogen-bonding network, which extends over a much larger region of space than the molecular correlations, is probably at the root of the long-range interactions of water. The situation in water appears to be governed by much more than the molecular packing effects that dominate the interactions in simpler liquids.

9.4.5 Polymer-Mediated Forces

Polymers or macromolecules are chain-like molecules consisting of many identical segments (monomers or repeating units) held together by covalent bonds. The size of a polymer coil in solution or in the melt is determined by a balance between van der Waals attraction (and hydrogen bonding, if present) between polymer segments, and the entropy of mixing, which causes the polymer coil to expand. In polymer melts above the glass transition temperature, and at certain conditions in solution, the attraction between polymer segments is exactly balanced by the entropy effect. The polymer solution will then behave virtually ideally, and the density distribution of segments in the coil is Gaussian. This is called the theta (θ) condition, and it occurs at the theta or Flory temperature for a particular combination of polymer and solvent or solvent mixture. At lower temperatures (in a poor or bad solvent), the polymer–polymer interactions dominate over the entropic, and the coil will shrink or precipitate. At higher temperatures (good solvent conditions), the polymer coil will be expanded.

High-molecular-weight polymers form large coils, which significantly affect the properties of a solution even when the total mass of polymer is very low. The radius of the polymer coil is proportional to the segment length, a, and the number of segments, n. At theta conditions, the hydrodynamic radius of the polymer coil (the root-mean-square separation of the ends of one polymer chain) is theoretically given by $R_h = a n^{1/2}$, and the unperturbed radius of gyration (the average root-mean-square distance of a segment from the center of mass of the molecule) is $R_g = a (n/6)^{1/2}$. In a good solvent the perturbed size of the polymer coil, the Flory radius R_F, is proportional to $n^{3/5}$.

Polymers interact with surfaces mainly through van der Waals and electrostatic interactions. The physisorption of polymers containing only one type of segment is reversible and highly dynamic, but the rate of exchange of adsorbed chains with free chains in the solution is low, since the polymer remains bound to the surface as long as one segment along the chain is adsorbed. The adsorption energy per segment is on the order of $k_B T$. In a good solvent, the conformation of a polymer on a surface is very different from the coil conformation in bulk solution. Polymers adsorb in "trains", separated by "loops" extending into solution and dangling "tails" (the ends of the chain). Compared to adsorption at lower temperatures, good solvent conditions favor more of the polymer chain being in the solvent, where it can attain its optimum conformation. As a result, the extension of the polymer is longer, even though the total amount of adsorbed polymer is lower. In a good solvent, the polymer chains can also be effectively repelled from a surface, if the loss in conformational entropy close to the surface is not compensated for by a gain in enthalpy from adsorption of segments. In this case, there will be a layer of solution (thickness $\approx R_g$) close to the surfaces that is depleted of polymer.

The interaction forces between two surfaces across a polymer solution will depend on whether the polymer adsorbs onto the surfaces or is repelled from them, and also on whether the interaction occurs at "true" or "restricted" thermodynamic equilibrium. At true or full equilibrium, the polymer between the surfaces can equilibrate (exchange) with polymer in the bulk solution at all surface separations. Some theories [198, 199] predict that, at full equilibrium, the polymer chains would move from the confined gap into the bulk solution where they could attain entropically more favorable conformations, and that a monotonic attraction at all distances would result from bridging and depletion interactions (which will be discussed below). Other theories suggest that the interaction at small separations would be ultimately repulsive, since some polymer chains would remain in the gap due to their attractive interactions with many sites on the surface (enthalpic) – more sites would be available to the remaining polymer chains if some others desorbed and diffused out from the gap [73, 200–202].

At restricted equilibrium, the polymer is kinetically trapped, and the adsorbed amount is thus constant as the surfaces are brought toward each other, but the chains can still rearrange on the surfaces and in the gap. Experimentally, the true equilibrium situation is very difficult to attain, and most experiments are done at restricted equilibrium conditions. Even the equilibration of conformations assumed in theoretical models for restricted equilibrium conditions can be so slow that this condition is difficult to reach experimentally.

In systems of adsorbing polymer, bridging of chains from one surface to the other can give rise to a long-range attraction, since the bridging chains would gain conformational entropy if the surfaces were closer together. In poor solvents, both bridging and intersegment interactions contribute to an attraction [26]. However, regardless of solvent and equilibrium conditions, a strong repulsion due to the osmotic interactions is seen at small surface separations in systems of adsorbing polymers at restricted equilibrium.

In systems containing high concentrations of non-adsorbing polymer, the difference in solute concentration in the bulk and between the surfaces at separations smaller than the approximate polymer coil diameter ($2R_g$, i.e., when the polymer has been squeezed out from the gap between the surfaces) may give rise to an attractive osmotic force (the "depletion attraction") [207–212]. In addition, if the polymer coils become initially compressed as the surfaces approach each other, this can give rise to a repulsion ("depletion stabilization") at large separations [210]. For a system of two cylindrical surfaces or radius R, the maximum depletion force, F_{dep}, is expected to occur when the surfaces are in contact and is given by multiplying the depletion (osmotic) pressure, $P_{\text{dep}} = \rho k_B T$, by the contact area πr^2, where r is given by the chord theorem: $r^2 = (2R - R_g)R_g \approx 2RR_g$ [3]:

$$F_{\text{dep}}/R = -2\pi R_g \rho k_B T \,, \tag{9.19}$$

where ρ is the number density of the polymer in the bulk solution.

If a part of the polymer (typically an end group) is different from the rest of the chain, this part may preferentially adsorb to the surface. End-adsorbed polymer is attached to the surface at only one point, and the extension of the chain is dependent on the grafting density, i.e., the average distance, s, between adsorbed end groups on the surface (Fig. 9.8). One distinguishes between different regions of increasing overlap of the chains (stretching) called pancake, mushroom, and brush

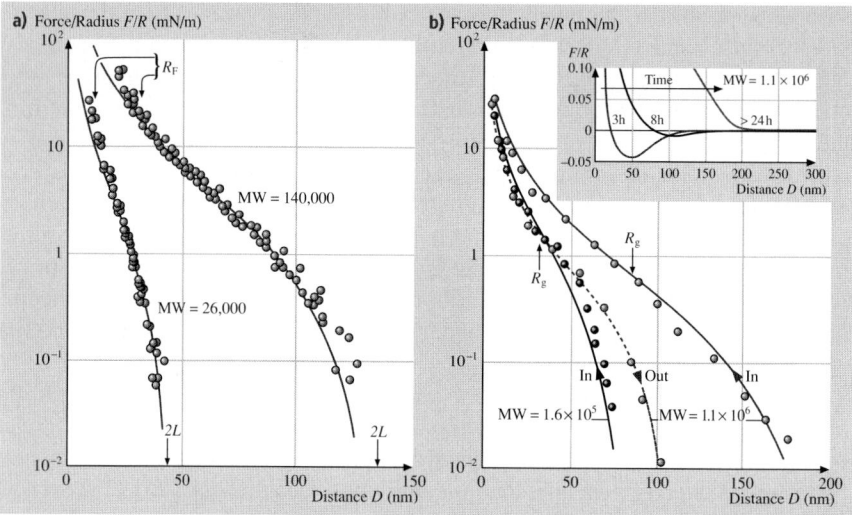

Fig. 9.8a,b. Experimentally determined forces in systems of two interacting polymer layers: (a) Polystyrene brush layers grafted via an adsorbing chain-end group onto mica surfaces in toluene (a good solvent for polystyrene). *Left curve*: MW = 26,000 g/mol, $R_F = 12$ nm. *Right curve*: MW = 140,000 g/mol, $R_F = 32$ nm. Both force curves were reversible on approach and separation. The *solid curves* are theoretical fits using the Alexander–de Gennes theory with the following measured parameters: spacing between attachments sites: $s = 8.5$ nm, brush thickness: $L = 22.5$ nm and 65 nm, respectively. (adapted from [203]). (**b**) Polyethylene oxide layers physisorbed onto mica from 150 μg/ml solution in aqueous 0.1 M KNO$_3$ (a good solvent for polyethylene oxide). *Main figure*: Equilibrium forces at full coverage after ~ 16 h adsorption time. *Left curve*: MW = 160,000 g/mol, $R_g = 32$ nm. *Right curve*: MW = 1,100,000 g/mol, $R_g = 86$ nm. Note the hysteresis (irreversibility) on approach and separation for this *physisorbed* layer, in contrast to the absence of hysteresis with *grafted* chains in case (**a**). The *solid curves* are based on a modified form of the Alexander–de Gennes theory. *Inset* in (**b**): evolution of the forces with the time allowed for the higher MW polymer to adsorb from solution. Note the gradual reduction in the attractive bridging component. (adapted from [204–206]; after [3]; copyright 1991, with permission from Elsevier Science)

regimes [213]. In the mushroom regime, where the coverage is sufficiently low that there is no overlap between neighboring chains, the thickness of the adsorbed layer is proportional to $n^{1/2}$ (i.e., to R_g) at theta conditions and to $n^{3/5}$ in a good solvent.

Several models [213–218] have been developed for the extension and interactions between two brushes (strongly stretched grafted chains). They are based on a balance between osmotic pressure within the brush layers (uncompressed and compressed) and the elastic energy of the chains and differ mainly in the assumptions of the segment density profile, which can be a step function or parabolic. At high coverage (in the brush regime), where the chains will avoid overlapping each other, the thickness of the layer is proportional to n.

Experimental work on both monodisperse [27,28,203,219] and polydisperse [30, 220] systems at different solvent conditions has confirmed the expected range and magnitude of the repulsive interactions resulting from compression of densely packed grafted layers.

9.4.6 Thermal Fluctuation Forces

If a surface is not rigid but very soft or even fluid-like, this can act to smear out any oscillatory solvation force. This is because the thermal fluctuations of such interfaces make them dynamically "rough" at any instant, even though they may be perfectly smooth on a time average. The types of surfaces that fall into this category are fluid-like amphiphilic surfaces of micelles, bilayers, emulsions, soap films, etc., but also solid colloidal particle surfaces that are coated with surfactant monolayers, as occurs in lubricating oils, paints, toners, etc.

These thermal fluctuation forces (also called entropic or steric forces) are usually short range and repulsive and are very effective at stabilizing the attractive van der Waals forces at some small but finite separation. This can reduce the adhesion energy or force by up to three orders of magnitude. It is mainly for this reason that fluid-like micelles and bilayers, biological membranes, emulsion droplets, or gas bubbles adhere to each other only very weakly.

Because of their short range it was, and still is, commonly believed that these forces arise from water ordering or "structuring" effects at surfaces, and that they reflect some unique or characteristic property of water. However, it is now known that these repulsive forces also exist in other liquids [221, 222]. Moreover, they appear to become stronger with increasing temperature, which is unlikely if the force originated from molecular ordering effects at surfaces. Recent experiments, theory, and computer simulations [223–226] have shown that these repulsive forces have an entropic origin arising from the osmotic repulsion between exposed thermally mobile surface groups once these overlap in a liquid. These phenomena include undulating and peristaltic forces between membranes or bilayers, and, on the molecular scale, protrusion and head-group overlap forces where the interactions are also influenced by hydration forces.

9.5 Adhesion and Capillary Forces

9.5.1 Capillary Forces

When considering the adhesion of two solid surfaces or particles in air or in a liquid, it is easy to overlook or underestimate the important role of capillary forces, i.e., forces arising from the Laplace pressure of curved menisci formed by condensation of a liquid between and around two adhering surfaces (Fig. 9.9).

The adhesion force between a non-deformable spherical particle of radius R and a flat surface in an inert atmosphere (Fig. 9.9a) is

$$F_{ad} = 4\pi R \gamma_{SV} \,. \tag{9.20}$$

But in an atmosphere containing a condensable vapor, the expression above is replaced by

$$F_{ad} = 4\pi R(\gamma_{LV} \cos\theta + \gamma_{SL}) \,, \tag{9.21}$$

where the first term is due to the Laplace pressure of the meniscus and the second is due to the direct adhesion of the two contacting solids within the liquid. Note that the above equation does not contain the radius of curvature, r, of the liquid meniscus (see Fig. 9.9b). This is because for smaller r the Laplace pressure γ_{LV}/r increases, but the area over which it acts decreases by the same amount, so the two effects cancel out. Experiments with inert liquids, such as hydrocarbons, condensing between two mica surfaces indicate that (9.21) is valid for values of r as small as 1–2 nm,

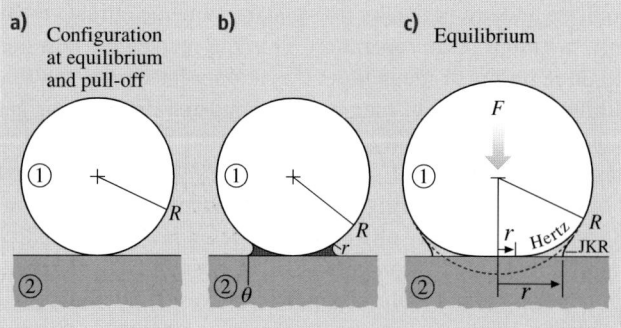

Fig. 9.9a–c. Adhesion and capillary forces: (**a**) a non-deforming sphere on a rigid, flat surface in an inert atmosphere and (**b**) in a vapor that can "capillary condense" around the contact zone. At equilibrium, the concave radius, r, of the liquid meniscus is given by the Kelvin equation. For a concave meniscus to form, the contact angle θ has to be less than 90°. In the case of hydrophobic surfaces surrounded by water, a vapor cavity can form between the surfaces. As long as the surfaces are perfectly smooth, the contribution of the meniscus to the adhesion force is independent of r (after [1] with permission). (**c**) Elastically deformable sphere on a rigid flat surface in the absence (Hertz) and presence (JKR) of adhesion ((**a**) and (**c**) after [3]; copyright 1991, with permission from Elsevier Science)

corresponding to vapor pressures as low as 40% of saturation [148,227,228]. Capillary condensation also occurs in binary liquid systems, e. g., when water dissolved in hydrocarbon liquids condenses around two contacting hydrophilic surfaces or when a vapor cavity forms in water around two hydrophobic surfaces. In the case of water condensing from vapor or from oil, it also appears that the bulk value of γ_{LV} is applicable for meniscus radii as small as 2 nm.

The capillary condensation of liquids, especially water, from vapor can have additional effects on the physical state of the contact zone. For example, if the surfaces contain ions, these will diffuse and build up within the liquid bridge, thereby changing the chemical composition of the contact zone, as well as influencing the adhesion. In the case of surfaces covered with surfactant or polymer molecules (amphiphilic surfaces), the molecules can turn over on exposure to humid air, so that the surface nonpolar groups become replaced by polar groups, which renders the surfaces hydrophilic. When two such surfaces come into contact, water will condense around the contact zone and the adhesion force will also be affected – generally increasing well above the value expected for inert hydrophobic surfaces. It is apparent that adhesion in vapor or a solvent is often largely determined by capillary forces arising from the condensation of liquid that may be present only in very small quantities, e. g., 10–20 % of saturation in the vapor, or 20 ppm in the solvent.

9.5.2 Adhesion Mechanics

Two bodies in contact deform as a result of surface forces and/or applied normal forces. For the simplest case of two interacting elastic spheres (a model that is easily extended to an elastic sphere interacting with an undeformable surface, or vice versa) and in the absence of attractive surface forces, the vertical central displacement (compression) was derived by *Hertz* [229] (Fig. 9.9c). In this model, the displacement and the contact area are equal to zero when no external force (load) is applied, i. e., at the points of contact and of separation. The contact area A increases with normal force or load as $L^{2/3}$.

In systems where attractive surface forces are present between the surfaces, the deformations are more complicated. Modern theories of the adhesion mechanics of two contacting solid surfaces are based on the Johnson–Kendall–Roberts (JKR) theory [15, 230], or on the Derjaguin–Muller–Toporov (DMT) theory [231–233]. The JKR theory is applicable to easily deformable, large bodies with high surface energy, whereas the DMT theory better describes very small and hard bodies with low surface energy [234]. The intermediate regime has also been described [235].

In the JKR theory, two spheres of radii R_1 and R_2, bulk elastic modulus K, and surface energy γ will flatten due to attractive surface forces when in contact at no external load. The contact area will increase under an external load L or normal force F, such that at mechanical equilibrium the radius of the contact area, r, is given by

$$r^3 = \frac{R}{K}\left[F + 6\pi R\gamma + \sqrt{12\pi R\gamma F + (6\pi R\gamma)^2}\right], \tag{9.22}$$

where $R = R_1 R_2/(R_1 + R_2)$. In the absence of surface energy, γ, equation (9.22) is reduced to the expression for the radius of the contact area in the Hertz model. Another important result of the JKR theory gives the adhesion force or "pull-off"

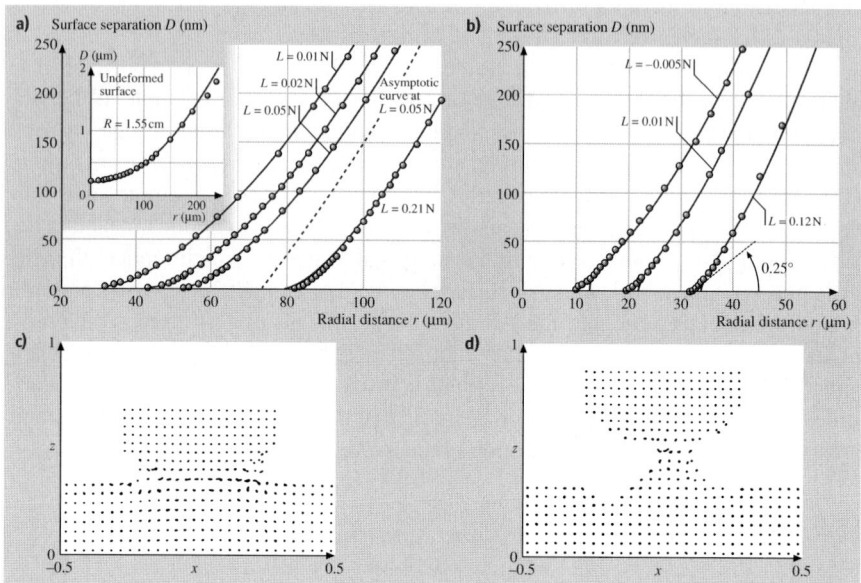

Fig. 9.10a–d. Experimental and computer simulation data on contact mechanics for ideal Hertz and JKR contacts. (**a**) Measured profiles of surfaces in nonadhesive contact (*circles*) compared with Hertz profiles (*continuous curves*). The system was mica surfaces in a concentrated KCl solution in which they do not adhere. When not in contact, the surface shape is accurately described by a sphere of radius $R = 1.55$ cm (*inset*). The applied loads were 0.01, 0.02, 0.05, and 0.21 N. The last profile was measured in a different region of the surfaces where the local radius of curvature was 1.45 cm. The Hertz profiles correspond to central displacements of $\delta = 66.5, 124, 173,$ and 441 nm. The *dashed line* shows the shape of the undeformed sphere corresponding to the curve at a load of 0.05 N; it fits the experimental points at larger distances (not shown). (**b**) Surface profiles measured with adhesive contact (mica surfaces adhering in dry nitrogen gas) at applied loads of $-0.005, 0.01,$ and 0.12 N. The continuous lines are JKR profiles obtained by adjusting the central displacement in each case to get the best fit to points at larger distances. The values are $\delta = -4.2, 75.6,$ and 256 nm. Note that the scales of this figure exaggerate the apparent angle at the junction of the surfaces. This angle, which is insensitive to load, is only about $0.25°$. (**c**) and (**d**) Molecular dynamics simulation illustrating the formation of a connective neck between an Ni tip (topmost eight layers) and an Au substrate. The figures show the atomic configuration in a slice through the system at indentation (**c**) and during separation (**d**). Note the crystalline structure of the neck. Distances are given in units of x and z, where $x = 1$ and $z = 1$ correspond to 6.12 nm. ((**a**) and (**b**) after [236]; copyright 1987, with permission from Elsevier Science; (**c**) and (**d**) after [112], with kind permission from Kluwer Academic Publishers)

force:

$$F_{ad} = -3\pi R\gamma_S , \qquad (9.23)$$

where the surface energy, γ_S, is defined through $W = 2\gamma_S$, where W is the reversible work of adhesion. Note that, according to the JKR theory, a finite elastic modulus, K, while having an effect on the load–area curve, has no effect on the adhesion force, an interesting and unexpected result that has nevertheless been verified experimentally [15, 236–238].

Equation (9.22) and (9.23) provide the framework for analyzing results of adhesion measurements (Fig. 9.10) of contacting solids, known as contact mechanics [230, 239], and for studying the effects of surface conditions and time on adhesion energy hysteresis (see Sect. 9.5.4).

The JKR theory has been extended [240, 241] to consider rigid or elastic substrates separated by thin compliant layers of very different elastic moduli, a situation commonly encountered in SFA and AFM experiments. The deformation of the system is then strongly dependant on the ratio of r to the thickness of the confined layer. At small r (low L), the deformation occurs mostly in the thin confined layer, whereas at large r (large L), it occurs mainly in the substrates. Because of the changing distribution of traction across the contact, the adhesion force in a layered system is also modified from that of isotropic systems (9.23) so that it is no longer independent of the elastic moduli.

9.5.3 Effects of Surface Structure, Roughness, and Lattice Mismatch

In a contact between two rough surfaces, the real area of contact varies with the applied load in a different manner than between smooth surfaces [243, 244]. For non-adhering surfaces exhibiting an exponential distribution of *elastically* deforming asperities (spherical caps of equal radius), it has been shown that the contact area for rough surfaces increases approximately linearly with the applied normal force (load), L, instead of as $L^{2/3}$ for smooth surfaces [243]. It has also been shown that for *plastically* deforming metal microcontacts the real contact area increases with load as $A \propto L$ [245, 246]. In systems with attractive surface forces, there is a competition between this attraction and repulsive forces arising from compression of high asperities. As a result, the adhesion in such systems can be very low, especially if the surfaces are not easily deformed [247–249]. The opposite is possible for soft (viscoelastic) surfaces where the real (molecular) contact area might be larger than for two perfectly smooth surfaces [250]. The size of the real contact area at a given normal force is also an important issue in studies of nanoscale friction, both of single-asperity contacts (cf. Sect. 9.7) and of contacts between rough surfaces (cf. Sect. 9.9.2).

Adhesion forces may also vary depending on the commensurability of the crystallographic lattices of the interacting surfaces. *McGuiggan* and *Israelachvili* [251] measured the adhesion between two mica surfaces as a function of the orientation (twist angle) of their surface lattices. The forces were measured in air, water, and

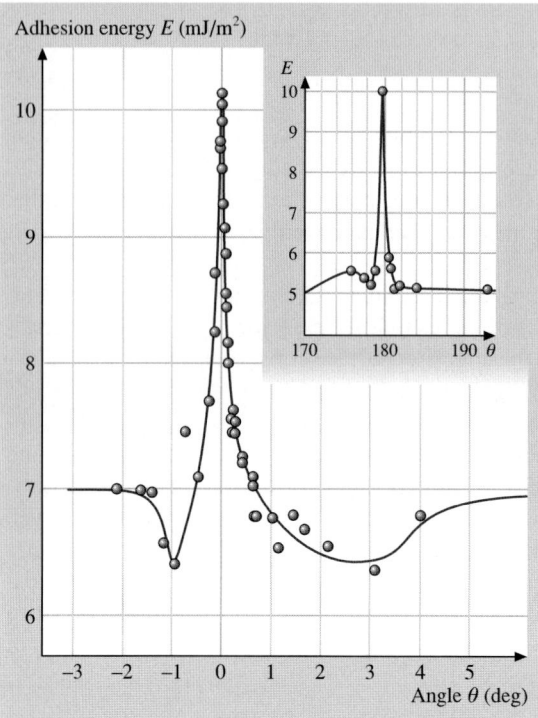

Fig. 9.11. Adhesion energy for two mica surfaces in contact in water (in the primary minimum of an oscillatory force curve) as a function of the mismatch angle θ about $\theta = 0°$ and $180°$ between the mica surface lattices (after [242] with permission)

an aqueous salt solution where oscillatory structural forces were present. In humid air, the adhesion was found to be relatively independent of the twist angle θ due to the adsorption of a 0.4 nm thick amorphous layer of organics and water at the interface. In contrast, in water, sharp adhesion peaks (energy minima) occurred at $\theta = 0°$, ±60°, ±120° and 180°, corresponding to the "coincidence" angles of the surface lattices (Fig. 9.11). As little as ±1° away from these peaks, the energy decreased by 50%. In aqueous KCl solution, due to potassium ion adsorption the water between the surfaces becomes ordered, resulting in an oscillatory force profile where the adhesive minima occur at discrete separations of about 0.25 nm, corresponding to integral numbers of water layers. The whole interaction potential was now found to depend on the orientation of the surface lattices, and the effect extended at least four molecular layers.

It has also been appreciated that the structure of the confining surfaces is just as important as the nature of the liquid for determining the solvation forces [111, 150, 151, 252–256]. Between two surfaces that are completely flat but "unstructured", the liquid molecules will order into layers, but there will be no lateral ordering within the layers. In other words, there will be positional ordering normal but not parallel to the surfaces. If the surfaces have a crystalline (periodic) lattice, this may induce ordering parallel to the surfaces, as well, and the oscillatory force then also depends on the structure of the surface lattices. Further, if the two lattices have different dimensions ("mismatched" or "incommensurate" lattices), or if the lattices are similar but are not in register relative to each other, the oscillatory force law is further modified [251, 257] and the tribological properties of the film are also influenced, as discussed in Sect. 9.9 [257, 258].

As shown by the experiments, these effects can alter the magnitude of the adhesive minima found at a given separation within the last one or two nanometers of a thin film by a factor of two. The force barriers (maxima) may also depend on orientation. This could be even more important than the effects on the minima. A high barrier could prevent two surfaces from coming closer together into a much deeper adhesive well. Thus the maxima can effectively contribute to determining not only the final separation of two surfaces, but also their final adhesion. Such considerations should be particularly important for determining the thickness and strength of intergranular spaces in ceramics, the adhesion forces between colloidal particles in concentrated electrolyte solution, and the forces between two surfaces in a crack containing capillary condensed water.

For surfaces that are *randomly* rough, oscillatory forces become smoothed out and disappear altogether, to be replaced by a purely monotonic solvation force [134, 150, 151]. This occurs even if the liquid molecules themselves are perfectly capable of ordering into layers. The situation of *symmetric* liquid molecules confined between *rough* surfaces is therefore not unlike that of *asymmetric* molecules between *smooth* surfaces (see Sect. 9.4.3 and Fig. 9.7a). To summarize, for there to be an oscillatory solvation force, the liquid molecules must be able to be correlated over a reasonably long range. This requires that both the liquid molecules and the surfaces have a high degree of order or symmetry. If either is missing, so will the oscillations. Depending on the size of the molecules to be confined, a roughness of only a few tenths of a nanometer is often sufficient to eliminate any oscillatory component of the force law [42, 150].

9.5.4 Nonequilibrium and Rate-Dependent Interactions: Adhesion Hysteresis

Under ideal conditions the adhesion energy is a well-defined thermodynamic quantity. It is normally denoted by E or W (the work of adhesion) or γ (the surface tension, where $W = 2\gamma$) and gives the reversible work done on bringing two surfaces together or the work needed to separate two surfaces from contact. Under ideal, equilibrium conditions these two quantities are the same, but under most realistic conditions they are not; the work needed to separate two surfaces is always greater than that originally gained by bringing them together. An understanding of the molecular mechanisms underlying this phenomenon is essential for understanding many adhesion phenomena, energy dissipation during loading–unloading cycles, contact angle hysteresis, and the molecular mechanisms associated with many frictional processes.

It is wrong to think that hysteresis arises because of some imperfection in the system such as rough or chemically heterogeneous surfaces, or because the supporting material is viscoelastic. Adhesion hysteresis can arise even between perfectly smooth and chemically homogenous surfaces supported by perfectly elastic materials. It can be responsible for such phenomena as rolling friction and elastoplastic adhesive contacts [239, 263–266] during loading–unloading and adhesion–decohesion cycles.

Adhesion hysteresis may be thought of as being due to mechanical effects such as instabilities, or chemical effects such as interdiffusion, interdigitation, molecular reorientations and exchange processes occurring at an interface after contact, as illustrated in Fig. 9.12. Such processes induce roughness and chemical heterogeneity even though initially (and after separation and re-equilibration) both surfaces are perfectly smooth and chemically homogeneous. In general, if the energy change, or work done, on separating two surfaces from adhesive contact is not fully recoverable on bringing the two surfaces back into contact again, the adhesion hysteresis

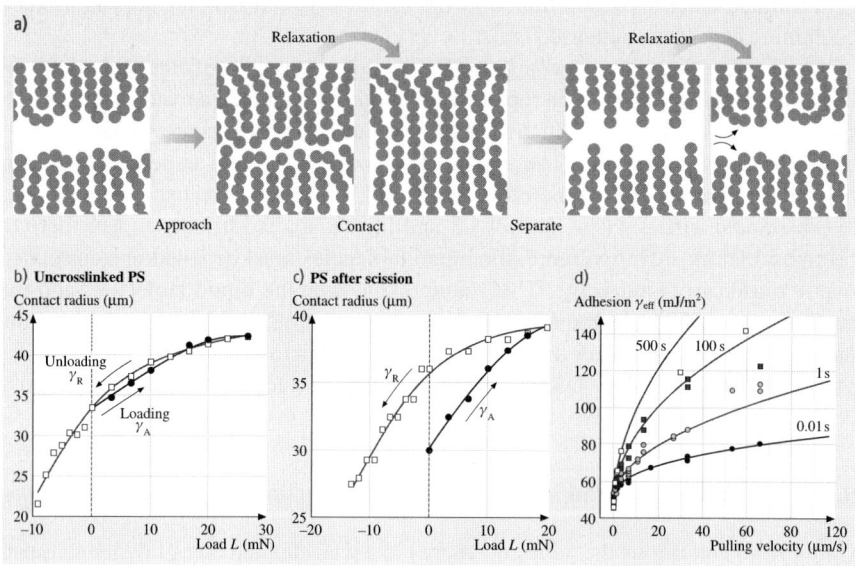

Fig. 9.12. (a) Schematic representation of interpenetrating chains. (b) and (c): JKR plots (contact radius r as a function of applied load L) showing small adhesion hysteresis for uncrosslinked polystyrene and larger adhesion hysteresis after chain scission at the surfaces after 18 h irradiation with ultraviolet light in an oxygen atmosphere. The adhesion hysteresis continues to increase with the irradiation time. (d) Rate-dependent adhesion of hexadecyl trimethyl ammonium bromide (CTAB) surfactant monolayers. The *solid curves* [259] are fits to experimental data on CTAB adhesion after different contact times [260] using an approximate analytical solution for a JKR model, including crack tip dissipation. Due to the limited range of validity of the approximation, the fits rely on the part of the experimental data with low effective adhesion energy only. From the fits one can determine the thermodynamic adhesion energy, the characteristic dissipation velocity, and the intrinsic dissipation exponent of the model. ((a) after [261]. Copyright 1993 American Chemical Society; (b) and (c) after [262]. Copyright 2002 American Association for the Advancement of Science; (d) after [259]. Copyright 2000 American Chemical Society)

may be expressed as

$$W_R > W_A$$
Receding Advancing

or

$$\Delta W = (W_R - W_A) > 0, \tag{9.24}$$

where W_R and W_A are the adhesion or surface energies for receding (separating) and advancing (approaching) two solid surfaces, respectively.

Hysteresis effects are also commonly observed in wetting/dewetting phenomena [267]. For example, when a liquid spreads and then retracts from a surface the advancing contact angle θ_A is generally larger than the receding angle θ_R. Since the contact angle, θ, is related to the liquid–vapor surface tension, γ_L, and the solid–liquid adhesion energy, W, by the Dupré equation,

$$(1 + \cos\theta)\gamma_L = W, \tag{9.25}$$

we see that *wetting hysteresis* or *contact angle hysteresis* ($\theta_A > \theta_R$) actually implies adhesion hysteresis, $W_R > W_A$, as given by (9.24).

Energy-dissipating processes such as adhesion and contact angle hysteresis arise because of practical constraints of the *finite time* of measurements and the *finite elasticity* of materials. This prevents many loading–unloading or approach–separation cycles from being thermodynamically reversible, even though they would be if carried out infinitely slowly. By thermodynamically irreversible one simply means that one cannot go through the approach–separation cycle via a continuous series of equilibrium states, because some of these are connected via spontaneous – and therefore thermodynamically irreversible – instabilities or transitions where energy is liberated and therefore "lost" via heat or phonon release [268]. This is an area of much current interest and activity, especially regarding the fundamental molecular origins of adhesion and friction in polymer and surfactant systems, and the relationships between them [239, 259, 260, 262, 264, 269–272].

9.6 Introduction: Different Modes of Friction and the Limits of Continuum Models

Most frictional processes occur with the sliding surfaces becoming damaged in one form or another [263]. This may be referred to as "normal" friction. In the case of brittle materials, the damaged surfaces slide past each other while separated by relatively large, micron-sized wear particles. With more ductile surfaces, the damage remains localized to nanometer-sized, plastically deformed asperities. Some features of the friction between damaged surfaces will be described in Sect. 9.7.4.

There are also situations in which sliding can occur between two perfectly smooth, undamaged surfaces. This may be referred to as "interfacial" sliding or "boundary" friction and is the focus of the following sections. The term "boundary lubrication" is more commonly used to denote the friction of surfaces that contain a thin protective lubricating layer such as a surfactant monolayer, but here we shall use the term more broadly to include any molecularly thin solid, liquid, surfactant, or polymer film.

Experiments have shown that, as a liquid film becomes progressively thinner, its physical properties change, at first quantitatively and then qualitatively [44, 47, 273, 274, 276, 277]. The quantitative changes are manifested by an increased viscosity, non-Newtonian flow behavior, and the replacement of normal melting by a glass transition, but the film remains recognizable as a liquid (Fig. 9.13). In tribology, this regime is commonly known as the "mixed lubrication" regime, where the rheological properties of a film are intermediate between the bulk and boundary properties. One may also refer to it as the "intermediate" regime (Table 9.4).

For even thinner films, the changes in behavior are more dramatic, resulting in a qualitative change in properties. Thus first-order phase transitions can now occur to solid or liquid-crystalline phases [46, 255, 261, 275, 278–281], whose properties can no longer be characterized even qualitatively in terms of bulk or continuum

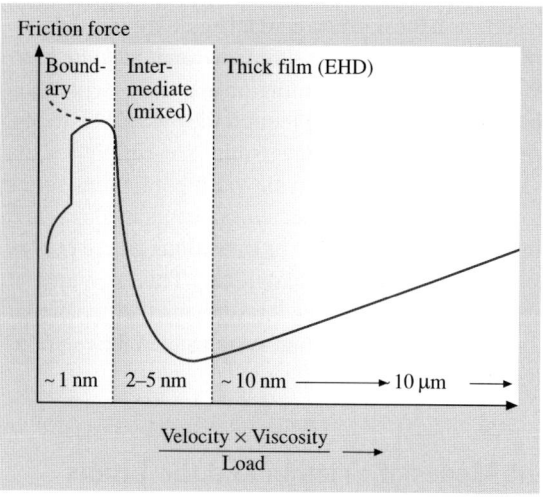

Fig. 9.13. Stribeck curve: an empirical curve giving the trend generally observed in the friction forces or friction coefficients as a function of sliding velocity, the bulk viscosity of the lubricating fluid, and the applied load (normal force). The three friction/lubrication regimes are known as the boundary lubrication regime (see Sect. 9.7), the intermediate or mixed lubrication regime (Sect. 9.8.2), and thick film or elasto-hydrodynamic (EHD) lubrication regime (Sect. 9.8.1). The film thicknesses believed to correspond to each of these regimes are also shown. For thick films, the friction force is purely viscous, e.g., Couette flow at low shear rates, but may become complicated at higher shear rates where EHD deformations of surfaces can occur during sliding (after [1] with permission)

Table 9.4. The three main tribological regimes characterizing the changing properties of liquids subjected to increasing confinement between two solid surfaces[a]

Regime	Conditions for getting into this regime	Static/equilibrium properties[b]	Dynamic properties[c]
Bulk	• Thick films (> 10 molecular diameters, $\gg R_g$ for polymers) • Low or zero loads • High shear rates	Bulk (continuum) properties: • Bulk liquid density • No long-range order	Bulk (continuum) properties: • Newtonian viscosity • Fast relaxation times • No glass temperature • No yield point • Elastohydrodynamic lubrication
Intermediate mixed	• Intermediately thick films (4–10 molecular diameters, $\sim R_g$ for polymers) • Low loads or pressure	Modified fluid properties include: • Modified positional and orientational order[a] • Medium- to long-range molecular correlations • Highly entangled states	Modified rheological properties include: • Non-Newtonian flow • Glassy states • Long relaxation times • Mixed lubrication
Boundary	• Molecularly thin films (< 4 molecular diameters) • High loads or pressure • Low shear rates • Smooth surfaces or asperities	Onset of non-fluidlike properties: • Liquid-like to solid-like phase transitions • Appearance of new liquid-crystalline states • Epitaxially induced long-range ordering	Onset of tribological properties: • No flow until yield point or critical shear stress reached • Solid-like film behavior characterized by defect diffusion, dislocation motion, shear melting • Boundary lubrication

Based on work by *Granick* [273], *Hu* and *Granick* [274], and others [38, 261, 275] on the dynamic properties of short chain molecules such as alkanes and polymer melts confined between surfaces

[a] Confinement can lead to an increased or decreased order in a film, depending both on the surface lattice structure and the geometry of the confining cavity

[b] In each regime both the static and dynamic properties change. The static properties include the film density, the density distribution function, the potential of mean force, and various positional and orientational order parameters

[c] Dynamic properties include viscosity, viscoelastic constants, and tribological yield points such as the friction coefficient and critical shear stress

liquid properties such as viscosity. These films now exhibit yield points (characteristic of fracture in solids) and their molecular diffusion and relaxation times can be ten orders of magnitude longer than in the bulk liquid or even in films that are just slightly thicker. The three friction regimes are summarized in Table 9.4.

9.7 Relationship Between Adhesion and Friction Between Dry (Unlubricated and Solid Boundary Lubricated) Surfaces

9.7.1 Amontons' Law and Deviations from It Due to Adhesion: The Cobblestone Model

Early theories and mechanisms for the dependence of friction on the applied normal force or load, L, were developed by *da Vinci, Amontons, Coulomb* and *Euler* [282]. For the macroscopic objects investigated, the friction was found to be directly proportional to the load, with no dependence on the contact area. This is described by the so-called Amontons' law:

$$F = \mu L, \tag{9.26}$$

where F is the shear or friction force and μ is a constant defined as the coefficient of friction. This friction law has a broad range of applicability and is still the principal means of quantitatively describing the friction between surfaces. However, particularly in the case of adhering surfaces, Amontons' law does not adequately describe the friction behavior with load, because of the finite friction force measured at zero and even negative applied loads.

When a lateral force, or shear stress, is applied to two surfaces in adhesive contact, the surfaces initially remain "pinned" to each other until some critical shear force is reached. At this point, the surfaces begin to slide past each other either smoothly or in jerks. The frictional force needed to initiate sliding from rest is known as the *static* friction force, denoted by F_s, while the force needed to maintain smooth sliding is referred to as the *kinetic* or *dynamic* friction force, denoted by F_k. In general, $F_s > F_k$. Two sliding surfaces may also move in regular jerks, known as stick–slip sliding, which is discussed in more detail in Sect. 9.8.3. Such friction forces cannot be described by models used for thick films that are viscous (see Sect. 9.8.1) and, therefore, shear as soon as the smallest shear force is applied.

In Sects. 9.7 and 9.8 we will be concerned mainly with single-asperity contacts. Experimentally, it has been found that during both smooth and stick–slip sliding at small film thicknesses the local geometry of the contact zone remains largely unchanged from the static geometry [45]. In an adhesive contact, the contact area as a function of load is thus generally well described by the JKR equation, (9.22). The friction force between two molecularly smooth surfaces sliding in *adhesive* contact is not simply proportional to the applied load, L, as might be expected from Amontons' law. There is an additional adhesion contribution that is proportional to the area of contact, A. Thus, in general, the interfacial friction force of dry, unlubricated surfaces sliding smoothly past each other in adhesive contact is given by

$$F = F_k = S_c A + \mu L, \tag{9.27}$$

where S_c is the "critical shear stress" (assumed to be constant), $A = \pi r^2$ is the contact area of radius r given by (9.22), and μ is the coefficient of friction. For low loads we have:

$$F = S_c A = S_c \pi r^2 = S_c \pi \left[\frac{R}{K} \left(L + 6\pi R\gamma + \sqrt{12\pi R\gamma L + (6\pi R\gamma)^2} \right) \right]^{2/3} ; \tag{9.28}$$

whereas for high loads (or high μ), or when γ is very low [283–287], (9.27) reduces to Amontons' law: $F = \mu L$. Depending on whether the friction force in (9.27) is dominated by the first or second term, one may refer to the friction as *adhesion-controlled* or *load-controlled*, respectively.

The following friction model, first proposed by *Tabor* [288] and developed further by *Sutcliffe* et al. [289], *McClelland* [290], and *Homola* et al. [45], has been quite successful at explaining the interfacial and boundary friction of two solid crystalline surfaces sliding past each other in the absence of wear. The surfaces may be unlubricated, or they may be separated by a monolayer or more of some boundary lubricant or liquid molecules. In this model, the values of the critical shear stress S_c, and the coefficient of friction μ, in (9.27) are calculated in terms of the energy needed to overcome the attractive intermolecular forces and compressive externally applied load as one surface is raised and then slid across the molecular-sized asperities of the other.

This model (variously referred to as the *interlocking asperity model*, *Coulomb friction*, or the *cobblestone model*) is similar to pushing a cart over a road of cobblestones where the cartwheels (which represent the molecules of the upper surface or film) must be made to roll over the cobblestones (representing the molecules of the lower surface) before the cart can move. In the case of the cart, the downward force of gravity replaces the attractive intermolecular forces between two material surfaces. When at rest, the cartwheels find grooves between the cobblestones where they sit in potential-energy minima, and so the cart is at some stable mechanical equilibrium. A certain lateral force (the "push") is required to raise the cartwheels against the force of gravity in order to initiate motion. Motion will continue as long as the cart is pushed, and rapidly stops once it is no longer pushed. Energy is dissipated by the liberation of heat (phonons, acoustic waves, etc.) every time a wheel hits the next cobblestone. The cobblestone model is not unlike the *Coulomb* and *interlocking asperity* models of friction [282] except that it is being applied at the molecular level and for a situation where the external load is augmented by attractive intermolecular forces.

There are thus two contributions to the force pulling two surfaces together: the externally applied load or pressure, and the (internal) attractive intermolecular forces that determine the adhesion between the two surfaces. Each of these contributions affects the friction force in a different way, which we will discuss in more detail below.

9.7.2 Adhesion Force and Load Contribution to Interfacial Friction

Adhesion Force Contribution

Consider the case of two surfaces sliding past each other, as shown in Fig. 9.14a. When the two surfaces are initially in adhesive contact, the surface molecules will adjust themselves to fit snugly together [291], in an analogous manner to the self-positioning of the cartwheels on the cobblestone road. A small tangential force applied to one surface will therefore not result in the sliding of that surface relative to the other. The attractive van der Waals forces between the surfaces must first be overcome by having the surfaces separate by a small amount. To initiate motion, let the separation between the two surfaces increase by a small amount ΔD, while the lateral distance moved is $\Delta\sigma$. These two values will be related via the geometry of the two surface lattices. The energy put into the system by the force F acting over a lateral distance $\Delta\sigma$ is

Input energy: $F \times \Delta\sigma$. (9.29)

Fig. 9.14. (a) Schematic illustration of how one molecularly smooth surface moves over another when a lateral force F is applied (the "cobblestone model"). As the upper surface moves laterally by some fraction of the lattice dimension $\Delta\sigma$, it must also move up by some fraction of an atomic or molecular dimension ΔD before it can slide across the lower surface. On impact, some fraction ε of the kinetic energy is "transmitted" to the lower surface, the rest being "reflected" back to the colliding molecule (upper surface) (after [292], with permission). (b) Difference in the local distribution of the total applied external load or normal adhesive force between load-controlled non-adhering surfaces and adhesion-controlled surfaces. In the former case, the total friction force F is given either by $F = \mu L$ for one contact point (*left side*) or by $F = \frac{1}{3}\mu L + \frac{1}{3}\mu L + \frac{1}{3}\mu L = \mu L$ for three contact points (*right side*). Thus the load-controlled friction is always proportional to the applied load, independently of the number of contacts and of their geometry. In the case of adhering surfaces, the effective "internal" load is given by kA, where A is the real local contact area, which is proportional to the number of intermolecular bonds being made and broken across each single contact point. The total friction force is now given by $F = \mu k A$ for one contact point (*left side*), and $F = \mu(kA_1 + kA_2 + kA_3) = \mu k A_{\text{tot}}$ for three contact points (*right side*). Thus, for adhesion-controlled friction, the friction is proportional to the real contact area, at least when no additional external load is applied to the system (after [287], with permission; copyright 2004 American Chemical Society). (c),(d) Friction force between benzyltrichlorosilane monolayers chemically bound to glass or Si, measured in ethanol ($\gamma < 1$ mJ/m^2). (c) SFA measurements where both glass surfaces were covered with a monolayer. *Circles* and *squares* show two different experiments: one with $R = 2.6$ cm, $v = 0.15$ μm/s, giving $\mu = 0.33 \pm 0.01$; the other with $R = 1.6$ cm, $v = 0.5$ μm/s, giving $\mu = 0.30 \pm 0.01$. (d) Friction force microscopy (FFM) measurements of a monolayer-functionalized Si tip ($R = 11$ nm) sliding on a monolayer-covered glass surface at $v = 0.15$ μm/s, giving $\mu = 0.30 \pm 0.01$. Note the different scales in (c) and (d) (after [286], with permission; copyright 2003 American Chemical Society)

This energy may be equated with the change in interfacial or surface energy associated with separating the surfaces by ΔD, i.e., from the equilibrium separation $D = D_0$ to $D = (D_0 + \Delta D)$. Since $\gamma \propto D^{-2}$ for two flat surfaces (cf. Sect. 9.3.1 and Table 9.2), the energy cost may be approximated by:

Surface energy change × area: $2\gamma A \left[1 - D_0^2/(D_0+\Delta D)^2\right] \approx 4\gamma A(\Delta D/D_0)$, (9.30)

where γ is the surface energy, A the contact area, and D_0 the surface separation at equilibrium. During steady-state sliding (kinetic friction), not all of this energy will be "lost" or absorbed by the lattice every time the surface molecules move by one lattice spacing: some fraction will be reflected during each impact of the "cartwheel" molecules [290]. Assuming that a fraction ε of the above surface energy is "lost" every time the surfaces move across the characteristic length $\Delta\sigma$ (Fig. 9.14a), we obtain after equating (9.29) and (9.30)

$$S_c = \frac{F}{A} = \frac{4\gamma\varepsilon\Delta D}{D_0 \Delta\sigma}.$$ (9.31)

For a typical hydrocarbon or a van der Waals surface, $\gamma \approx 25$ mJ m^{-2}. Other typical values would be: $\Delta D \approx 0.05$ nm, $D_0 \approx 0.2$ nm, $\Delta\sigma \approx 0.1$ nm, and $\varepsilon \approx 0.1$–0.5. Using

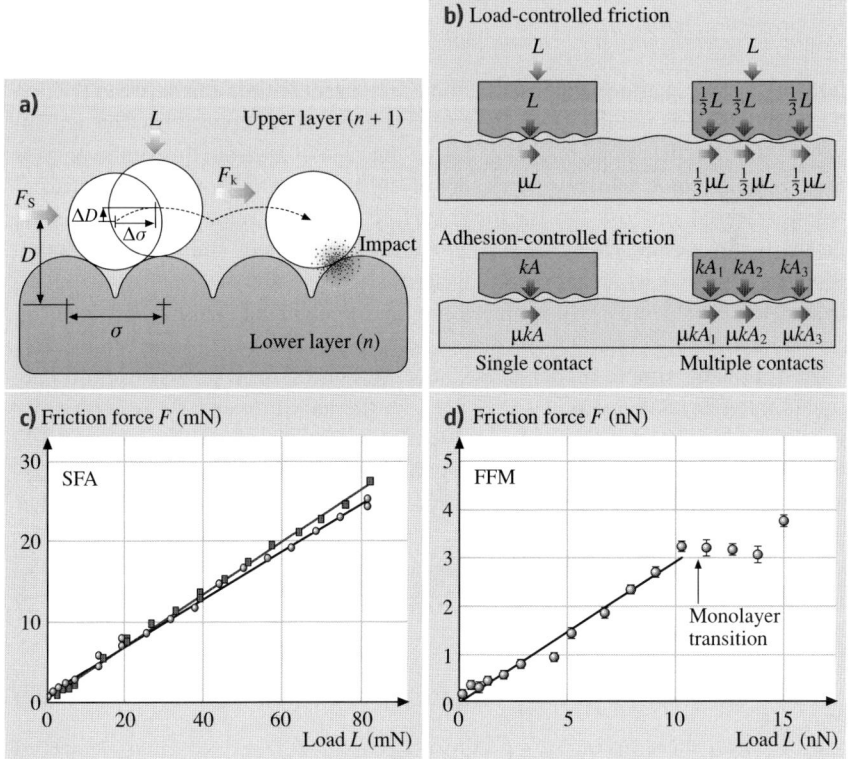

the above parameters, (9.31) predicts $S_c \approx (2.5-12.5) \times 10^7$ N m^{-2} for van der Waals surfaces. This range of values compares very well with typical experimental values of 2×10^7 N m^{-2} for hydrocarbon or mica surfaces sliding in air (see Fig. 9.16) or separated by one molecular layer of cyclohexane [45].

The above model suggests that all interfaces, whether dry or lubricated, dilate just before they shear or slip. This is a small but important effect: the dilation provides the crucial extra space needed for the molecules to slide across each other or flow. This dilation is known to occur in macroscopic systems [293, 294] and for nanoscopic systems it has been computed by *Thompson* and *Robbins* [255] and *Zaloj* et al. [295] and measured by *Dhinojwala* et al. [296].

This model may be extended, at least semiquantitatively, to lubricated sliding, where a thin liquid film is present between the surfaces. With an increase in the number of liquid layers between the surfaces, D_0 increases while ΔD decreases, hence the friction force decreases. This is precisely what is observed, but with more than one liquid layer between two surfaces the situation becomes too complex to analyze analytically (actually, even with one or no interfacial layers, the calculation of the fraction of energy dissipated per molecular collision ε is not a simple matter). Furthermore, even in systems as simple as linear alkanes, interdigitation and interdiffusion have been found to contribute strongly to the properties of the system [143, 297]. Sophisticated modeling based on computer simulations is now required, as discussed in the following section.

Relation Between Boundary Friction and Adhesion Energy Hysteresis

While the above equations suggest that there is a direct correlation between friction and adhesion, this is not the case. The correlation is really between friction and adhesion hysteresis, described in Sect. 9.5.4. In the case of friction, this subtle point is hidden in the factor ε, which is a measure of the amount of energy absorbed (dissipated, transferred, or "lost") by the lower surface when it is impacted by a molecule from the upper surface. If $\varepsilon = 0$, all the energy is reflected, and there will be no kinetic friction force or any adhesion hysteresis, but the absolute magnitude of the adhesion force or energy will remain finite and unchanged. This is illustrated in Figs. 9.17 and 9.19.

The following simple model shows how adhesion hysteresis and friction may be quantitatively related. Let $\Delta \gamma = \gamma_R - \gamma_A$ be the adhesion energy hysteresis per unit area, as measured during a typical loading–unloading cycle (see Figs. 9.17a and 9.19c,d). Now consider the same two surfaces sliding past each other and assume that frictional energy dissipation occurs through the same mechanism as adhesion energy dissipation, and that both occur over the same characteristic molecular length scale σ. Thus, when the two surfaces (of contact area $A = \pi r^2$) move a distance σ, equating the frictional energy $(F \times \sigma)$ to the dissipated adhesion energy $(A \times \Delta \gamma)$, we obtain

$$\text{Friction force: } F = \frac{A \times \Delta \gamma}{\sigma} = \frac{\pi r^2}{\sigma} (\gamma_R - \gamma_A) \,, \tag{9.32}$$

$$\text{or Critical shear stress: } S_c = F/A = \Delta \gamma / \sigma \,, \tag{9.33}$$

which is the desired expression and has been found to give order-of-magnitude agreement between measured friction forces and adhesion energy hysteresis [261]. If we equate (9.33) with (9.31), since $4\Delta D/(D_0 \Delta \sigma) \approx 1/\sigma$, we obtain the intuitive relation

$$\varepsilon = \frac{\Delta \gamma}{\gamma}. \tag{9.34}$$

External Load Contribution to Interfacial Friction

When there is no interfacial adhesion, S_c is zero. Thus, in the absence of any adhesive forces between two surfaces, the only "attractive" force that needs to be overcome for sliding to occur is the externally applied load or pressure, as shown in Fig. 9.14b.

For a preliminary discussion of this question, it is instructive to compare the magnitudes of the *externally* applied pressure to the *internal* van der Waals pressure between two smooth surfaces. The internal van der Waals pressure between two flat surfaces is given (see Table 9.2) by $P = A_H/6\pi D_0^3 \approx 1$ GPa (10^4 atm), using a typical Hamaker constant of $A_H = 10^{-19}$ J, and assuming $D_0 \approx 2$ nm for the equilibrium interatomic spacing. This implies that we should not expect the externally applied load to affect the interfacial friction force F, as defined by (9.27), until the externally applied pressure L/A begins to exceed ~ 100 MPa (10^3 atm). This is in agreement with experimental data [298] where the effect of load became dominant at pressures in excess of 10^3 atm.

For a more general semiquantitative analysis, again consider the cobblestone model used to derive (9.31), but now include an additional contribution to the surface-energy change of (9.30) due to the work done against the external load or pressure, $L\Delta D = P_{\text{ext}} A \Delta D$ (this is equivalent to the work done against gravity in the case of a cart being pushed over cobblestones). Thus:

$$S_c = \frac{F}{A} = \frac{4\gamma\varepsilon\Delta D}{D_0 \Delta \sigma} + \frac{P_{\text{ext}}\varepsilon\Delta D}{\Delta \sigma}, \tag{9.35}$$

which gives the more general relation

$$S_c = F/A = C_1 + C_2 P_{\text{ext}}, \tag{9.36}$$

where $P_{\text{ext}} = L/A$ and C_1 and C_2 are characteristic of the surfaces and sliding conditions. The constant $C_1 = 4\gamma\varepsilon\Delta D/(D_0\Delta\sigma)$ depends on the mutual adhesion of the two surfaces, while both C_1 and $C_2 = \varepsilon\Delta D/\Delta\sigma$ depend on the topography or atomic bumpiness of the surface groups (Fig. 9.14a). The smoother the surface groups the smaller the ratio $\Delta D/\Delta\sigma$ and hence the lower the value of C_2. In addition, both C_1 and C_2 depend on ε (the fraction of energy dissipated per collision), which depends on the relative masses of the shearing molecules, the sliding velocity, the temperature, and the characteristic molecular relaxation processes of the surfaces. This is by far the most difficult parameter to compute, and yet it is the most important since it represents the energy-transfer mechanism in any friction process, and since ε can vary between 1 and 0, it determines whether a particular friction force will be large or close to zero. Molecular simulations offer the best way to understand and predict the magnitude of ε, but the complex multibody nature of the problem makes simple conclusions difficult to draw [299–302]. Some of the basic physics of the energy transfer and dissipation of the molecular collisions can be drawn from simplified models such as a 1D three-body system [268].

Finally, the above equation may also be expressed in terms of the friction force F:

$$F = S_c A = C_1 A + C_2 L. \tag{9.37}$$

Equations similar to (9.36) and (9.37) were previously derived by *Derjaguin* [303, 304] and by *Briscoe* and *Evans* [305], where the constants C_1 and C_2 were interpreted somewhat differently than in this model.

In the absence of any attractive interfacial force, we have $C_1 \approx 0$, and the second term in (9.36) and (9.37) should dominate (Fig. 9.15). Such situations typically arise when surfaces repel each other across the lubricating liquid film. In such cases, the total frictional force should be low and should increase *linearly* with the external load according to

$$F = C_2 L. \tag{9.38}$$

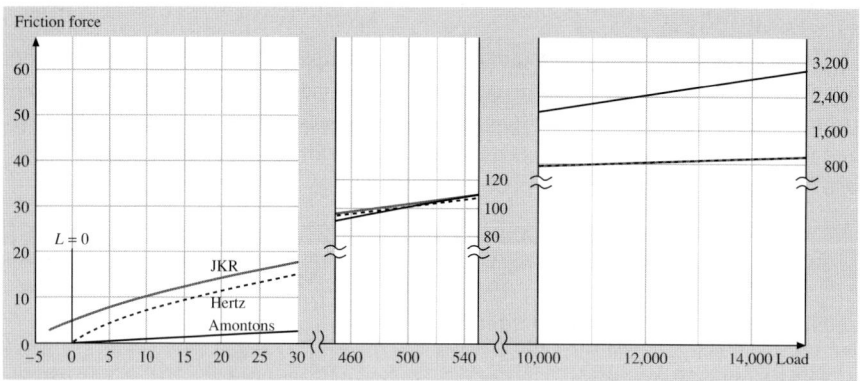

Fig. 9.15. Friction as a function of load for smooth surfaces. At low loads, the friction is dominated by the $C_1 A$ term of (9.37). The adhesion contribution (JKR curve) is most prominent near zero load where the Hertzian and Amontons' contributions to the friction are minimal. As the load increases, the adhesion contribution becomes smaller as the JKR and Hertz curves converge. In this range of loads, the linear $C_2 L$ contribution surpasses the area contribution to the friction. At much higher loads the explicit load dependence of the friction dominates the interactions, and the observed behavior approaches Amontons' law. It is interesting to note that for smooth surfaces the pressure over the contact area does not increase as rapidly as the load. This is because as the load is increased, the surfaces deform to increase the surface area and thus moderate the contact pressure (after [307], with permission of Kluwer Academic Publishers)

An example of such lubricated sliding occurs when two mica surfaces slide in water or in salt solution (see Fig. 9.20a), where the short-range "hydration" forces between the surfaces are repulsive. Thus, for sliding in 0.5 M KCl it was found that $C_2 = 0.015$ [283]. Another case where repulsive surfaces eliminate the adhesive contribution to friction is for polymer chains attached to surfaces at one end and swollen by a good solvent [219]. For this class of systems, $C_2 < 0.001$ for a finite range of polymer layer compressions (normal loads, L). The low friction between the surfaces in this regime is attributed to the entropic repulsion between the opposing brush layers with a minimum of entanglement between the two layers. However, with higher normal loads, the brush layers become compressed and begin to entangle, which results in higher friction (see [306]).

It is important to note that (9.38) has exactly the same form as Amontons' Law

$$F = \mu L, \tag{9.39}$$

where μ is the coefficient of friction.

Figure 9.14c,d shows the kinetic friction force measured with both SFA and FFM (friction force microscopy, using AFM) on a system where both surfaces were covered with a chemically bound benzyltrichlorosilane monolayer [286]. When immersed in ethanol, the adhesion in this system is low, and very different contact areas and loads give a linear dependence of F on L with the same friction coefficients, and $F \to 0$ as $L \to 0$. In the FFM measurements (Fig. 9.14d), the plateau in the data at higher loads suggest a transition in the monolayers, similar to previous observations on other monolayer systems. The pressure in the contact region in the SFA is much lower than in the FFM, and no transitions in the friction forces or in the thickness of the confined monolayers were observed in the SFA experiments (and no damage to the monolayers or the underlying substrates was observed during the experiments, indicating that the friction was "wearless"). Despite the difference of more than six orders of magnitude in the contact radii, pressure, loads, and friction forces, the measured friction coefficients are practically the same.

At the molecular level a thermodynamic analog of the Coulomb or cobblestone models (see Sect. 9.7.1) based on the "contact value theorem" [3, 283, 307] can explain why $F \propto L$ also holds at the microscopic or molecular level. In this analysis we consider the surface molecular groups as being momentarily compressed and decompressed as the surfaces move along. Under irreversible conditions, which always occur when a cycle is completed in a finite amount of time, the energy "lost" in the compression–decompression cycle is dissipated as heat. For two non-adhering surfaces, the stabilizing pressure P_i acting locally between any two elemental contact points i of the surfaces may be expressed by the contact value theorem [3]:

$$P_i = \rho_i k_B T = k_B T / V_i, \tag{9.40}$$

where $\rho_i = V_i^{-1}$ is the local number density (per unit volume) or activity of the interacting entities, be they molecules, atoms, ions or the electron clouds of atoms. This equation is essentially the osmotic or entropic pressure of a gas of confined molecules. As one surface moves across the other, local regions become compressed and decompressed by a volume ΔV_i. The work done per cycle can be written as $\varepsilon P_i \Delta V_i$, where ε ($\varepsilon \leq 1$) is the fraction of energy per cycle "lost" as heat, as defined earlier. The energy balance shows that, for each compression–decompression cycle, the dissipated energy is related to the friction force by

$$F_i x_i = \varepsilon P_i \Delta V_i, \tag{9.41}$$

where x_i is the lateral distance moved per cycle, which can be the distance between asperities or the distance between surface lattice sites. The pressure at each contact junction can be expressed in terms of the local normal load L_i and local area of contact A_i as $P_i = L_i/A_i$. The volume change over a cycle can thus be expressed as $\Delta V_i = A_i z_i$, where z_i is the vertical distance of confinement. Inserting these into (9.41), we get

$$F_i = \varepsilon L_i (z_i/x_i), \tag{9.42}$$

which is independent of the local contact area A_i. The total friction force is thus

$$F = \sum F_i = \sum \varepsilon L_i(z_i/x_i) = \varepsilon \langle z_i/x_i \rangle \sum L_i = \mu L, \tag{9.43}$$

where it is assumed that on average the local values of L_i and P_i are independent of the local *slope* z_i/x_i. Therefore, the friction coefficient μ is a function only of the average surface topography and the sliding velocity, but is independent of the local (real) or macroscopic (apparent) contact areas.

While this analysis explains non-adhering surfaces, there is still an additional explicit contact area contribution for the case of adhering surfaces, as in (9.37). The distinction between the two cases arises because the initial assumption of the contact value theorem, (9.40), is incomplete for adhering systems. A more appropriate starting equation would reflect the full intermolecular interaction potential, including the attractive interactions, in addition to the purely repulsive contributions of (9.40), much as the van der Waals equation of state modifies the ideal gas law.

9.7.3 Examples of Experimentally Observed Friction of Dry Surfaces

Numerous model systems have been studied with a surface forces apparatus (SFA) modified for friction experiments (see Sect. 9.2.3). The apparatus allows for control of load (normal force) and sliding speed, and simultaneous measurement of surface separation, surface shape, true (molecular) area of contact between smooth surfaces, and friction forces. A variety of both unlubricated and solid- and liquid-lubricated surfaces have been studied both as smooth single-asperity contacts and after they have been roughened by shear-induced damage.

Figure 9.16 shows the contact area, A, and friction force, F, both plotted against the applied load, L, in an experiment in which two molecularly smooth surfaces of mica in adhesive contact were slid past each other in an atmosphere of dry nitrogen

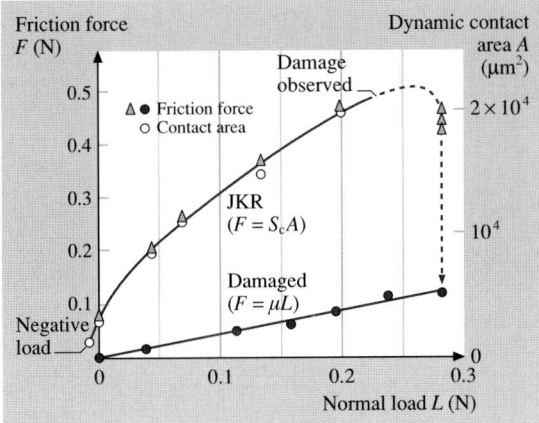

Fig. 9.16. Friction force F and contact area A versus load L for two mica surfaces sliding in adhesive contact in dry air. The contact area is well described by the JKR theory, (9.22), even during sliding, and the friction force is found to be directly proportional to this area, (9.28). The *vertical dashed line* and *arrow* show the transition from interfacial to normal friction with the onset of wear (*lower curve*). The sliding velocity is $0.2\,\mu\mathrm{m\,s^{-1}}$ (after [45], with permission, copyright 1989 American Society of Mechanical Engineers)

gas. This is an example of the low-load adhesion-controlled limit, which is excellently described by (9.28). In a number of different experiments, S_c was measured to be $2.5 \times 10^7\,\mathrm{N\,m^{-2}}$ and to be independent of the sliding velocity [45, 308]. Note that there is a friction force even at negative loads, where the surfaces are still sliding in adhesive contact.

Figure 9.17 shows the correlation between adhesion hysteresis and friction for two surfaces consisting of silica films deposited on mica substrates [41]. The friction between undamaged hydrophobic silica surfaces showed stick–slip both at dry conditions and at 100% relative humidity. Similar to the mica surfaces in Figs. 9.16, 9.18, and 9.20a, the friction of damaged silica surfaces obeyed Amontons' law with a friction coefficient of 0.25–0.3 both at dry conditions and at 55% relative humidity.

The high friction force of unlubricated sliding can often be reduced by treating the solid surface with a boundary layer of some other solid material that exhibits lower friction, such as a surfactant monolayer, or by ensuring that during sliding a thin liquid film remains between the surfaces (as will be discussed in Sect. 9.8). The effectiveness of a solid boundary lubricant layer on reducing the forces of friction is illustrated in Fig. 9.18. Comparing this with the friction of the unlubricated/untreated surfaces (Fig. 9.16) shows that the critical shear stress has been reduced by a factor of about ten: from 2.5×10^7 to $3.5 \times 10^6\,\mathrm{N\,m^{-2}}$. At much higher applied loads or pressures, the friction force is proportional to the load, rather than the area of contact [298], as expected from (9.27).

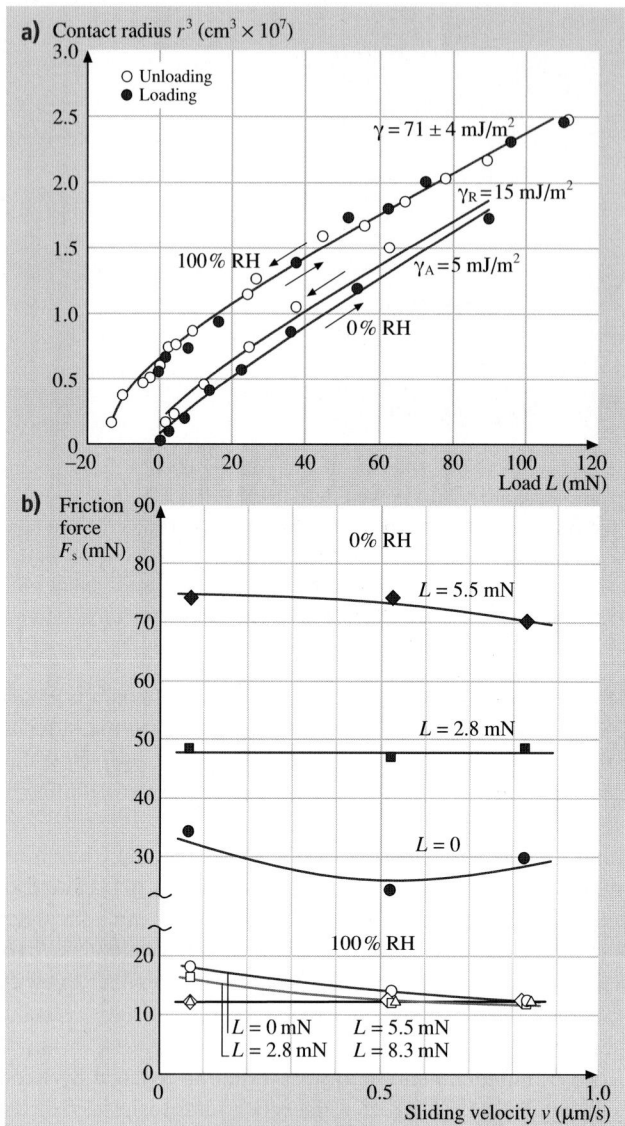

Fig. 9.17. (a) Contact radius r versus externally applied load L for loading and unloading of two hydrophilic silica surfaces exposed to dry and humid atmospheres. Note that, while the adhesion is higher in humid air, the *hysteresis* in the adhesion is higher in dry air. (b) Effect of velocity on the static friction force F_s for hydrophobic (heat-treated electron-beam-evaporated) silica in dry and humid air. The effects of humidity, load, and sliding velocity on the friction forces, as well as the stick–slip friction of the hydrophobic surfaces, are qualitatively consistent with a "friction" phase diagram representation as in Fig. 9.28 (after [41]; copyright 1994, with permission from Elsevier Science)

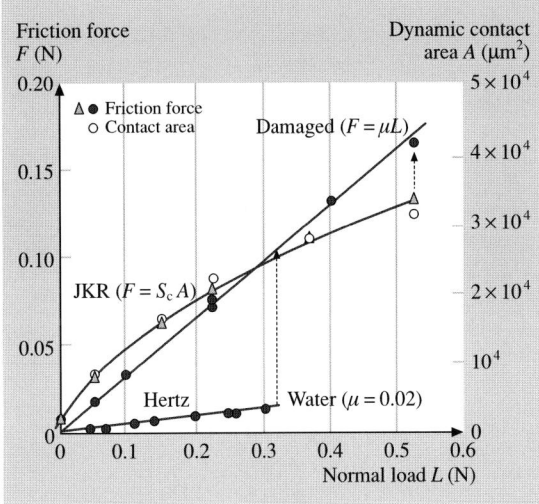

Fig. 9.18. Sliding of mica surfaces, each coated with a 2.5 nm thick monolayer of calcium stearate surfactant, in the absence of damage (obeying JKR-type boundary friction) and in the presence of damage (obeying Amontons-type normal friction). Note that both for this system and for the bare mica in Figs. 9.16 and 9.20a, the friction force obeys Amontons' law with a friction coefficient of $\mu \approx 0.3$ after damage occurs. At much higher applied loads, the undamaged surfaces also follow Amontons-type sliding, but for a different reason: the dependence on adhesion becomes smaller. *Lower line*: interfacial sliding with a monolayer of water between the mica surfaces (load-controlled friction, cf. Fig. 9.20a), shown for comparison (after [308]; copyright 1990, with permission from Elsevier Science)

Yamada and *Israelachvili* [309] studied the adhesion and friction of fluorocarbon surfaces (surfactant-coated boundary lubricant layers), which were compared to those of hydrocarbon surfaces. They concluded that well-ordered fluorocarbon surfaces have high friction, in spite of their lower adhesion energy (in agreement with previous findings). The low friction coefficient of Teflon (polytetrafluoroethylene, PTFE) must, therefore, be due to some effect other than low adhesion. For example, the softness of PTFE, which allows material to flow at the interface, which thus behaves like a fluid lubricant. On a related issue, *Luengo* et al. [310] found that C_{60} surfaces also exhibited low adhesion but high friction. In both cases the high friction appears to arise from the bulky surface groups – fluorocarbon compared to hydrocarbon groups in the former, large fullerene spheres in the latter. Apparently, the fact that C_{60} molecules rotate *in their lattice* does not make them a good lubricant: the molecules of the opposing surface must still climb over them in order to slide, and this requires energy that is independent of whether the surface molecules are fixed or freely rotating. Larger particles such as ~ 25 nm sized nanoparticles (also known as "inorganic fullerenes") do appear to produce low friction by behaving like molecular ball bearings, but the potential of this promising new class of solid lubricant has still to be explored [311].

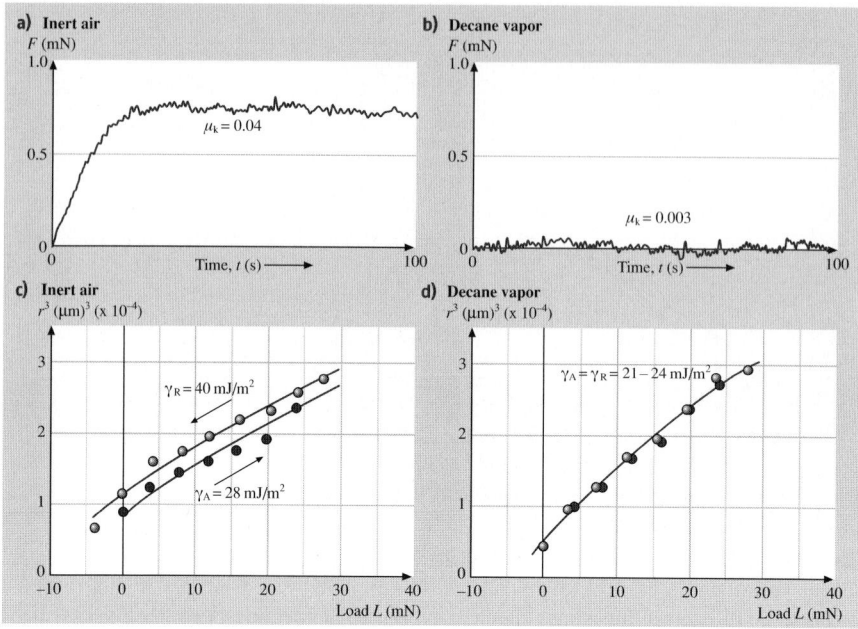

Fig. 9.19. *Top*: friction traces for two fluid-like calcium alkylbenzene sulfonate monolayer-coated surfaces at 25 °C showing that the friction force is much higher between dry monolayers (**a**) than between monolayers whose fluidity has been enhanced by hydrocarbon penetration from vapor (**b**). *Bottom*: Contact radius vs. load (r^3 vs. L) data measured for the same two surfaces as above and fitted to the JKR equation (9.22), shown by the *solid curves*. For dry monolayers (**c**) the adhesion energy on unloading ($\gamma_R = 40 \, \mathrm{mJ\,m^{-2}}$) is greater than that on loading ($\gamma_R = 28 \, \mathrm{mJ\,m^{-2}}$), which is indicative of an adhesion energy hysteresis of $\Delta\gamma = \gamma_R - \gamma_A = 12 \, \mathrm{mJ\,m^{-2}}$. For monolayers exposed to saturated decane vapor (**d**) their adhesion hysteresis is zero ($\gamma_A = \gamma_R$), and both the loading and unloading data are well fitted by the thermodynamic value of the surface energy of fluid hydrocarbon chains, $\gamma = 24 \, \mathrm{mJ\,m^{-2}}$ (after [261], with permission; copyright 1993 American Chemical Society)

Figure 9.19 illustrates the relationship between adhesion hysteresis and friction for surfactant-coated surfaces under different conditions. This effect, however, is much more general and has been shown to hold for other surfaces as well [41, 262, 292, 312].

Direct comparisons between absolute adhesion energies and friction forces show little correlation. In some cases, higher adhesion energies for the same system under different conditions correspond to lower friction forces. For example, for hydrophilic silica surfaces (Fig. 9.17) it was found that with increasing relative humidity the adhesion energy *increases*, but the adhesion energy hysteresis measured in a loading–unloading cycle *decreases*, as does the friction force [41]. For hydrophobic silica surfaces under dry conditions, the friction at load $L = 5.5 \, \mathrm{mN}$ was $F = 75 \, \mathrm{mN}$. For the same sample, the adhesion energy hysteresis was $\Delta\gamma = 10 \, \mathrm{mJ\,m^{-2}}$, with a contact area of $A \approx 10^{-8} \, \mathrm{m^2}$ at the same load. Assuming a value for the characteristic distance σ on the order of one lattice spacing, $\sigma \approx 1$ nm, and inserting these values into (9.32), the friction force is predicted to be $F \approx 100$ mN for the kinetic friction force, which is close to the measured value of 75 mN. Alternatively, we may conclude that the dissipation factor is $\varepsilon = 0.75$, i.e., that almost all the energy is dissipated as heat at each molecular collision.

A liquid lubricant film (Sect. 9.8.3) is usually much more effective at lowering the friction of two surfaces than a solid boundary lubricant layer. However, to use a liquid lubricant successfully, it must "wet" the surfaces, that is, it should have a high affinity for the surfaces, so that not all the liquid molecules become squeezed out when the surfaces come close together, even under a large compressive load. Another important requirement is that the liquid film remains a liquid under tribological conditions, i.e., that it does not epitaxially solidify between the surfaces.

Effective lubrication usually requires that the lubricant be injected between the surfaces, but in some cases the liquid can be made to condense from the vapor. This is illustrated in Fig. 9.20a for two untreated mica surfaces sliding with a thin layer of water between them. A monomolecular film of water (of thickness 0.25 nm per surface) has reduced S_c from its value for dry surfaces (Fig. 9.16) by a factor of more than 30, which may be compared with the factor of ten attained with the boundary lubricant layer (of thickness 2.5 nm per surface) in Fig. 9.18. Water appears to have unusual lubricating properties and usually gives wearless friction with no stick–slip [313].

The effectiveness of a water film only 0.25 nm thick to lower the friction force by more than an order of magnitude is attributed to the "hydrophilicity" of the mica surface (mica is "wetted" by water) and to the existence of a strongly repulsive short-range hydration force between such surfaces in aqueous solutions, which effectively removes the adhesion-controlled contribution to the friction force [283]. It is also interesting that a 0.25 nm thick water film between two mica surfaces is sufficient to bring the coefficient of friction down to 0.01–0.02, a value that corresponds to the unusually low friction of ice. Clearly, a single monolayer of water can be a very good lubricant – much better than most other monomolecular liquid films – for reasons that will be discussed in Sect. 9.9. A linear dependence of F on L has also been observed for mica surfaces separated by certain hydrocarbon liquids [275,285]. Figure 9.20b shows the kinetic friction forces measured at a high velocity across thin films of squalane, a branched hydrocarbon liquid ($C_{30}H_{62}$), which is a model for lubricating oils. Very low adhesive forces are measured between mica surfaces across this liquid [285] and the film thickness decreased monotonically with load. The friction force at a given load was found to be velocity-dependent, whereas the contact area was not [285].

Dry polymer layers (Fig. 9.21) typically show a high initial static friction ("stiction") as sliding commences from rest in adhesive contact. The development of the friction force after a change in sliding direction, a gradual transition from stick–slip to smooth sliding, is shown in Fig. 9.21. A correlation between adhesion hysteresis and friction similar to that observed for silica surfaces in Fig. 9.17 can be seen for dry polymer layers below their glass-transition temperature. As shown in

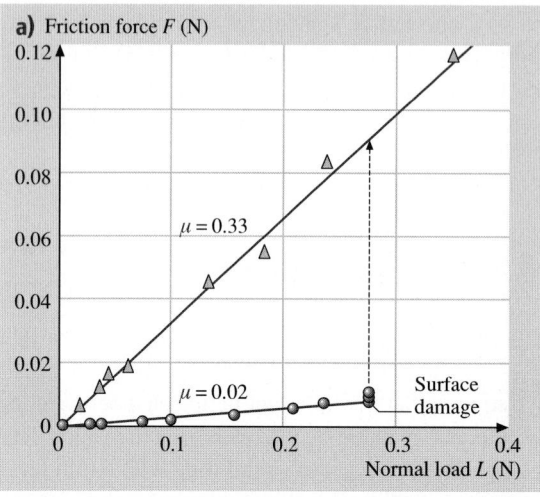
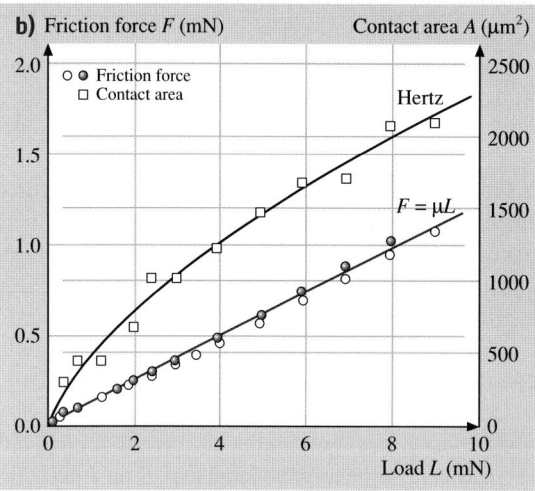

Fig. 9.20. Load-controlled friction. (**a**) Two mica surfaces sliding past each other while immersed in a 0.01 M KCl salt solution (nonadhesive conditions). The water film is molecularly thin, 0.25 to 0.5 nm thick, and the interfacial friction force is very low: $S_c \approx 5 \times 10^5$ N m^{-2}, $\mu \approx 0.02$ (before damage occurs). After the surfaces have become damaged, the friction coefficient is about 0.3 (after [308]; copyright 1990, with permission from Elsevier Science) (**b**) Steady-state friction force and contact area measured on a confined squalane film between two undamaged mica surfaces at $v = 0.6$ μm/s in the smooth sliding regime (no stick–slip). *Open circles* show F obtained on loading (increasing L), *solid circles* show unloading. Both data sets are *straight lines* passing through the origin, as shown by the *brown line* ($\mu = 0.12$). The *black curve* is a fit of the Hertz equation (cf. Sect. 9.5.2 and [3]) to the A vs. L data (*open squares*) using $K = 10^{10}$ N/m^2, $R = 2$ cm. The thickness D varies monotonically from $D = 2.5$ to $D = 1.7$ nm as the load increases from $L = 0$ to $L = 10$ mN (adapted from [285]; copyright 2003 American Physical Society)

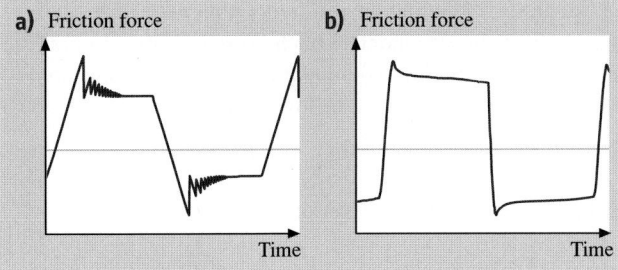

Fig. 9.21. Typical friction traces showing how the friction force varies with the sliding time for two symmetric, glassy polymer films under dry conditions. Qualitative features that are common to both polystyrene and polyvinyl benzyl chloride: (**a**) Decaying stick–slip motion is observed until smooth sliding is attained if the motion continues for a sufficiently long distance. (**b**) Smooth sliding observed at sufficiently high speeds. Similar observations have been made by *Berthoud* et al. [314] in measurements on polymethyl methacrylate (after [262], with permission; copyright 2002 American Association for the Advancement of Science)

Fig. 9.12b,c, the adhesion hysteresis for polystyrene surfaces can be increased by irradiation to induce scission of chains, and it has been found that the steady-state friction force (kinetic friction) shows a similar increase with irradiation time [262].

Figure 9.22 shows an example of a computer simulation of the sliding of two unlubricated silicon surfaces (modeled as a tip sliding over a planar surface) [112]. The sliding proceeds through a series of stick–slip events, and information on the friction force and the local order of the initially crystalline surfaces can be obtained. Similar studies for cold-welding systems [112] have demonstrated the occurrence of shear or friction damage within the sliding surface (tip) as the lowest layer of it adheres to the bottom surface. Recent computer simulations have addressed many of

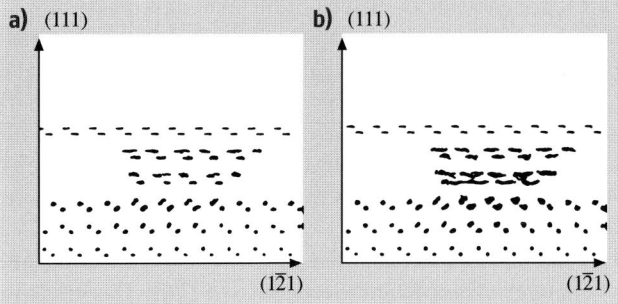

Fig. 9.22. Computer simulation of the sliding of two contacting Si surfaces (a tip and a flat surface). Shown are particle trajectories in a constant-force simulation, $F_{z,\text{external}} = -2.15 \times 10^{-8}$ N, viewed along the $(10\bar{1})$ direction just before (**a**) and after (**b**) a stick–slip event for a large, initally ordered, dynamic tip (after [112] with permission of Kluwer Academic Publishers)

the phenomena seen experimentally, including the differences between adhesive and non-adhesive systems, the issue of the dependence of the observed friction on the real contact area (a parameter that is difficult to define or measure at the nanoscale), and the molecular origin of friction responses that follow Amontons' law [287, 302, 315–317].

9.7.4 Transition from Interfacial to Normal Friction with Wear

Frictional damage can have many causes, such as adhesive tearing at high loads or overheating at high sliding speeds. Once damage occurs, there is a transition from "interfacial" to "normal" or load-controlled friction as the surfaces become forced apart by the torn-out asperities (wear particles). For low loads, the friction changes from obeying $F = S_c A$ to obeying Amontons' law, $F = \mu L$, as shown in Figs. 9.16 and 9.18, and sliding now proceeds smoothly with the surfaces separated by a 10–100 nm forest of wear debris (in this case, mica flakes). The wear particles keep the surfaces apart over an area that is much greater than their size, so that even one submicroscopic particle or asperity can cause a significant reduction in the area of contact and, therefore, in the friction [308]. For this type of frictional sliding, one can no longer talk of the molecular contact area of the two surfaces, although the macroscopic or "apparent" area is still a useful parameter.

One remarkable feature of the transition from interfacial to normal friction of brittle surfaces is that, while the strength of interfacial friction, as reflected in the values of S_c, is very dependent on the type of surface and on the liquid film between the surfaces, this is not the case once the transition to normal friction has occurred. At the onset of damage, the material properties of the underlying substrates control the friction. In Figs. 9.16, 9.18, and 9.20a the friction for the damaged surfaces is that of any damaged mica–mica system, $\mu \approx 0.3$, *independent of the initial surface coatings or liquid films between the surfaces*. A similar friction coefficient was found for damaged silica surfaces [41].

In order to modify the frictional behavior of such brittle materials practically, it is important to use coatings that will both alter the interfacial tribological character and remain intact and protect the surfaces from damage during sliding. *Berman* et al. [318] found that the friction of a strongly bound octadecyl phosphonic acid monolayer on alumina surfaces was higher than for untreated, undamaged α-alumina surfaces, but the bare surfaces easily became damaged upon sliding, resulting in an ultimately higher friction system with greater wear rates than the more robust monolayer-coated surfaces.

Clearly, the mechanism and factors that determine *normal* friction are quite different from those that govern *interfacial* friction (Sects. 9.7.1–9.7.2). This effect is not general and may only apply to brittle materials. For example, the friction of ductile surfaces is totally different and involves the continuous plastic deformation of contacting surface asperities during sliding, rather than the rolling of two surfaces on hard wear particles [263]. Furthermore, in the case of ductile surfaces, water and other surface-active components do have an effect on the friction coefficients under "normal" sliding conditions.

9.8 Liquid Lubricated Surfaces

9.8.1 Viscous Forces and Friction of Thick Films: Continuum Regime

Experimentally, it is usually difficult to unambiguously establish which type of sliding mode is occurring, but an empirical criterion, based on the Stribeck curve (Fig. 9.13), is often used as an indicator. This curve shows how the friction force or the coefficient of friction is expected to vary with sliding speed, depending on which type of friction regime is operating. For thick liquid lubricant films whose behavior can be described by bulk (continuum) properties, the friction forces are essentially the hydrodynamic or viscous drag forces. For example, for two plane parallel surfaces of area A separated by a distance D and moving laterally relative to each other with velocity v, if the intervening liquid is *Newtonian*, i.e., if its viscosity η is independent of the shear rate, the frictional force experienced by the surfaces is given by

$$F = \frac{\eta A v}{D}, \tag{9.44}$$

where the shear rate $\dot{\gamma}$ is defined by

$$\dot{\gamma} = \frac{v}{D}. \tag{9.45}$$

At higher shear rates, two additional effects often come into play. First, certain properties of liquids may change at high $\dot{\gamma}$ values. In particular, the effective viscosity may become non-Newtonian, one form given by

$$\eta \propto \dot{\gamma}^n, \tag{9.46}$$

where $n = 0$ (i.e., η_{eff} = constant) for Newtonian fluids, $n > 0$ for shear-thickening (dilatant) fluids, and $n < 0$ for shear-thinning (pseudoplastic) fluids (the latter become less viscous, i.e., flow more easily, with increasing shear rate). An additional effect on η can arise from the higher local stresses (pressures) experienced by the liquid film as $\dot{\gamma}$ increases. Since the viscosity is generally also sensitive to the pressure (usually increasing with P), this effect also acts to increase η_{eff} and thus the friction force.

A second effect that occurs at high shear rates is surface deformation, arising from the large hydrodynamic forces acting on the sliding surfaces. For example, Fig. 9.23 shows how two surfaces deform elastically when the sliding speed increases to a high value. These deformations alter the hydrodynamic friction forces, and this type of friction is often referred to as *elasto-hydrodynamic lubrication* (EHD or EHL), as mentioned in Table 9.4.

How thin can a liquid film be before its dynamic, e.g., viscous flow, behavior ceases to be described by bulk properties and continuum models? Concerning the static properties, we have already seen in Sect. 9.4.3 that films composed of simple liquids display continuum behavior down to thicknesses of 4–10 molecular diameters. Similar effects have been found to apply to the dynamic properties, such as

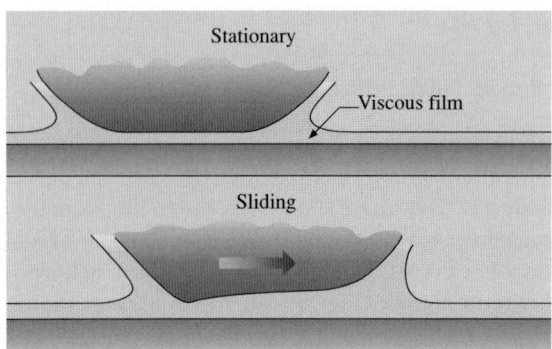

Fig. 9.23. *Top:* Stationary surfaces (one more deformable and one rigid) separated by a thick liquid film. *Bottom:* Elasto-hydrodynamic deformation of the upper surface during sliding (after [1], with permission)

the viscosity, of simple liquids in thin films. Concerning viscosity measurements, a number of dynamic techniques were recently developed [11–13, 43, 51, 319, 320] for directly measuring the viscosity as a function of film thickness and shear rate across very thin liquid films between two surfaces. By comparing the results with theoretical predictions of fluid flow in thin films, one can determine the effective positions of the shear planes and the onset of non-Newtonian behavior in very thin films.

The results show that, for simple liquids including linear chain molecules such as alkanes, the viscosity in thin films is the same, within 10%, as the bulk even for films as thin as 10 molecular diameters (or segment widths) [11–13, 319, 320]. This implies that the shear plane is effectively located within one molecular diameter of the solid–liquid interface, and these conclusions were found to remain valid even at the highest shear rates studied (of $\sim 2 \times 10^5$ s^{-1}). With water between two mica or silica surfaces [22, 313, 319–321] this has been found to be the case (to within $\pm 10\%$) down to surface separations as small as 2 nm, implying that the shear planes must also be within a few tenths of a nanometer of the solid–liquid interfaces. These results appear to be independent of the existence of electrostatic "double-layer" or "hydration" forces. For the case of the simple liquid toluene confined between surfaces with adsorbed layers of C_{60} molecules, this type of viscosity measurement has shown that the traditional no-slip assumption for flow at a solid interface does not always hold [322]. The C_{60} layer at the mica–toluene interface results in a "full-slip" boundary, which dramatically lowers the viscous drag or effective viscosity for regular Couette or Poiseuille flow.

With polymeric liquids (polymer melts) such as polydimethylsiloxanes (PDMS) and polybutadienes (PBD), or with polystyrene (PS) adsorbed onto surfaces from solution, the far-field viscosity is again equal to the bulk value, but with the non-slip plane (hydrodynamic layer thickness) being located at $D = 1 - 2R_g$ away from each surface [11, 47], or at $D = L$ or less for polymer brush layers of thickness L per surface [13, 323]. In contrast, the same technique was used to show that, for non-adsorbing polymers in solution, there is actually a depletion layer of nearly pure solvent that exists at the surfaces that affects the confined solution flow properties [321]. These effects are observed from near contact to surface separations in excess of 200 nm.

Further experiments with surfaces closer than a few molecular diameters ($D < 2-4$ nm for simple liquids, or $D < 2-4R_g$ for polymer fluids) indicate that large deviations occur for thinner films, described below. One important conclusion from these studies is, therefore, that the dynamic properties of simple liquids, including water, near an *isolated* surface are similar to those of the bulk liquid *already within the first layer of molecules adjacent to the surface*, only changing when another surface approaches the first. In other words, the viscosity and position of the shear plane near a surface are not simply a property of that surface, but of how far that surface is from another surface. The reason for this is that, when two surfaces are close together, the constraining effects on the liquid molecules between them are much more severe than when there is only one surface. Another obvious consequence of the above is that one should not make measurements on a single, isolated solid–liquid interface and then draw conclusions about the state of the liquid or its interactions in a thin film *between* two surfaces.

9.8.2 Friction of Intermediate Thickness Films

For liquid films in the thickness range between 4 and 10 molecular diameters, the properties can be significantly different from those of bulk films. Still, the fluids remain recognizable as fluids; in other words, they do not undergo a phase transition into a solid or liquid-crystalline phase. This regime has recently been studied by *Granick* et al. [44, 273, 274, 276, 277], who used a friction attachment [43, 44] to the SFA where a sinusoidal input signal to a piezoelectric device makes the two surfaces slide back and forth laterally past each other at small amplitudes. This method provides information on the real and imaginary parts (elastic and dissipative components, respectively) of the shear modulus of thin films at different shear rates and film thickness. *Granick* [273] and *Hu* et al. [277] found that films of simple liquids become non-Newtonian in the 2.5–5 nm regime (about 10 molecular diameters, see Fig. 9.24). Polymer melts become non-Newtonian at much larger film thicknesses, depending on their molecular weight [47].

Klein and *Kumacheva* [46, 280, 324] studied the interaction forces and friction of small quasi-spherical liquid molecules such as cyclohexane between molecularly smooth mica surfaces. They concluded that surface epitaxial effects can cause the liquid film to solidify already at six molecular diameters, resulting in a sudden (discontinuous) jump to high friction at low shear rates. Such dynamic first-order transitions, however, may depend on the shear rate.

A generalized friction map (Fig. 9.24c,d) has been proposed by *Luengo* et al. [325] that illustrates the changes in η_eff from bulk Newtonian behavior ($n = 0$, $\eta_\text{eff} = \eta_\text{bulk}$) through the transition regime where n reaches a minimum of -1 with decreasing shear rate to the solid-like creep regime at very low $\dot{\gamma}$ where n returns to 0. A number of results from experimental, theoretical, and computer simulation work have shown values of n from $-1/2$ to -1 for this transition regime for a variety of systems and assumptions [273, 274, 299, 326–332].

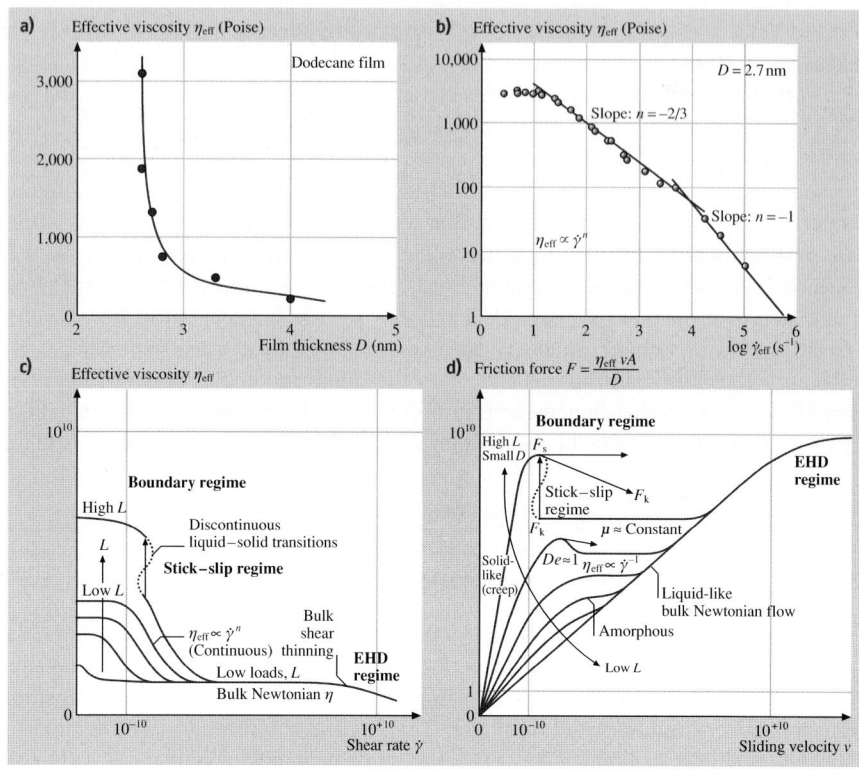

The intermediate regime appears to extend over a narrow range of film thickness, from about 4 to 10 molecular diameters or polymer radii of gyration. Thinner films begin to adopt boundary or interfacial friction properties (described below, see also Table 9.5). Note that the intermediate regime is actually a very narrow one when defined in terms of film thickness, for example, varying from about $D = 2$ to 4 nm for hexadecane films [273].

A fluid's effective viscosity η_{eff} in the intermediate regime is usually higher than in the bulk, but η_{eff} usually *decreases* with increasing sliding velocity, v (known as *shear thinning*). When two surfaces slide in the intermediate regime, the motion tends to thicken the film (dilatancy). This sends the system into the bulk EHD regime where, as indicated by (9.44), the friction force now *increases* with velocity. This initial decrease, followed by an increase, in the frictional forces of many lubricant systems is the basis for the empirical Stribeck curve of Fig. 9.13. In the transition from bulk to boundary behavior there is first a quantitative change in the material properties (viscosity and elasticity), which can be continuous, to discontinuous qualitative changes that result in yield stresses and non-liquidlike behavior.

Fig. 9.24a–d. Typical rheological behavior of liquid films in the mixed lubrication regime. (**a**) Increase in effective viscosity of dodecane film between two mica surfaces with decreasing film thickness. At distances larger than 4–5 nm, the effective viscosity η_eff approaches the bulk value η_bulk and does not depend on the shear rate $\dot{\gamma}$ (after [273]; copyright 1991 American Association for the Advancement of Science). (**b**) Non-Newtonian variation of η_eff with shear rate of a 2.7-nm-thick dodecane film at a net normal pressure of 0.12 MPa and at 28 °C. The effective viscosity decays as a power law, as in (9.46). In this example, $n = 0$ at the lowest $\dot{\gamma}$ and changes to $n = -2/3$ and -1 at higher $\dot{\gamma}$. For films of bulk thickness, dodecane is a low-viscosity Newtonian fluid ($n = 0$). (**c**) Proposed general friction map of effective viscosity η_eff (arbitrary units) as a function of effective shear rate $\dot{\gamma}$ (arbitrary units) on logarithmic scales. Three main classes of behavior emerge: (i) Thick films: elasto-hydrodynamic sliding. At $L = 0$, approximating bulk conditions, η_eff is independent of shear rate except when shear thinning might occur at sufficiently large $\dot{\gamma}$. (ii) Boundary layer films, intermediate regime. A Newtonian regime is again observed [η_eff = constant, $n = 0$ in (9.46)] at low loads and low shear rates, but η_eff is much higher than the bulk value. As the shear rate $\dot{\gamma}$ increases beyond $\dot{\gamma}_\mathrm{min}$, the effective viscosity starts to drop with a power-law dependence on the shear rate [see panel (**b**)], with n in the range $-1/2$ to -1 most commonly observed. As the shear rate $\dot{\gamma}$ increases still more, beyond $\dot{\gamma}_\mathrm{max}$, a second Newtonian plateau is encountered. (iii) Boundary layer films, high load. The η_eff continues to grow with load and to be Newtonian provided that the shear rate is sufficiently low. Transition to sliding at high velocity is discontinuous ($n < -1$) and usually of the stick–slip variety. (**c**) Proposed friction map of friction force as a function of sliding velocity in various tribological regimes. With increasing load, Newtonian flow in the elasto-hydrodynamic (EHD) regimes crosses into the boundary regime of lubrication. Note that even EHD lubrication changes, at the highest velocities, to limiting shear stress response. At the highest loads (L) and smallest film thickness (D), the friction force goes through a maximum (the static friction, F_s), followed by a regime where the friction coefficient (μ) is roughly constant with increasing velocity (meaning that the kinetic friction, F_k, is roughly constant). Non-Newtonian shear thinning is observed at somewhat smaller load and larger film thickness; the friction force passes through a maximum at the point where $De = 1$. De, the Deborah number, is the point at which the applied shear rate exceeds the natural relaxation time of the boundary layer film. The velocity axis from 10^{-10} to 10^{10} (arbitrary units) indicates a large span (Panels (**b**)–(**d**) after [325]; copyright 1996, with permission from Elsevier Science)

The rest of this section is devoted to friction in the boundary regime. Boundary friction may be thought of as applying to the case where a lubricant film is present, but where this film is of molecular dimensions – a few molecular layers or less.

9.8.3 Boundary Lubrication of Molecularly Thin Films: Nanorheology

When a liquid is confined between two surfaces or within any narrow space whose dimensions are less than 4 to 10 molecular diameters, both the static (equilibrium) and dynamic properties of the liquid, such as its compressibility and viscosity, can no longer be described even qualitatively in terms of the bulk properties. The molecules confined within such molecularly thin films become ordered into layers ("out-of-plane" ordering), and within each layer they can also have lateral order ("in-plane" ordering). Such films may be thought of as behaving more like a liquid crystal or a solid than a liquid.

As described in Sect. 9.4.3, the measured *normal* forces between two solid surfaces across molecularly thin films exhibit exponentially decaying oscillations, varying between attraction and repulsion with a periodicity equal to some molecular dimension of the solvent molecules. Thus most liquid films can sustain a finite normal stress, and the adhesion force between two surfaces across such films is "quantized", depending on the thickness (or number of liquid layers) between the surfaces. The structuring of molecules in thin films and the oscillatory forces it gives rise to are now reasonably well understood, both experimentally and theoretically, at least for simple liquids.

Work has also recently been done on the dynamic, e.g., viscous or shear, forces associated with molecularly thin films. Both experiments [38, 46, 257, 275, 280, 281, 335, 336] and theory [254, 255, 326, 337] indicate that, even when two surfaces are in steady-state sliding, they still prefer to remain in one of their stable potential-energy minima, i.e., a sheared film of liquid can retain its basic layered structure. Thus even during motion the film does not become totally liquid-like. Indeed, if there is some "in-plane" ordering within a film, it will exhibit a yield point before it begins to flow. Such films can therefore sustain a finite shear stress, in addition to a finite normal stress. The value of the yield stress depends on the number of layers comprising the film and represents another "quantized" property of molecularly thin films [254].

The dynamic properties of a liquid film undergoing shear are very complex. Depending on whether the film is more liquid-like or solid-like, the motion will be smooth or of the stick–slip type illustrated schematically in Fig. 9.25. During sliding, transitions can occur between n layers and $(n-1)$ or $(n+1)$ layers (see Fig. 9.27). The details of the motion depend critically on the externally applied load, the temperature, the sliding velocity, the twist angle between the two surface lattices, and the sliding direction relative to the lattices.

Smooth and Stick–Slip Sliding

Recent advances in friction-measuring techniques have enabled the interfacial friction of molecularly thin films to be measured with great accuracy. Some of these advances have involved the surface forces apparatus technique [38, 44–47, 274, 275, 280, 281, 285, 286, 296, 297, 308, 313, 335, 336, 338] while others have involved the atomic force microscope [10, 58, 59, 284, 290, 339, 340]. In addition, computer simulations [111, 151, 254, 255, 287, 295, 299–302, 315–317, 333, 337, 341] have become sufficiently sophisticated to enable fairly complex tribological systems to be studied. All these advances are necessary if one is to probe such subtle effects as smooth or stick–slip friction, transient and memory effects, and ultralow friction mechanisms at the molecular level.

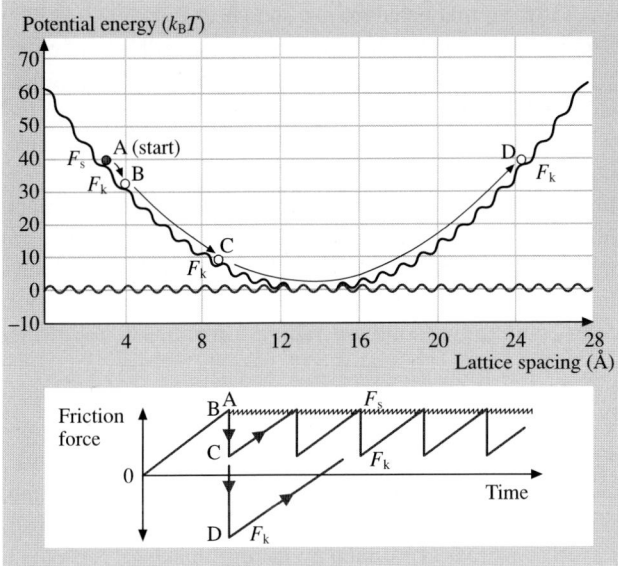

Fig. 9.25. Simple schematic illustration of the most common molecular mechanism leading from smooth to stick–slip sliding in terms of the efficiency of the energy transfer from mechanical to kinetic to phonons. The potential energy of the corrugated surface lattice is shown by the horizontal sine wave. Let the depth of each minimum be ε which is typically $> k_B T$. At equilibrium, a molecule will 'sit' at one of these minima. When the molecule is connected to a horizontal spring, a smooth parabolic curve must be added to the horizontal curve. If this spring is now pushed or pulled laterally at a constant velocity v, the sine curve will move like a wave along the parabola carrying the molecule up with it (towards point A). When the point of inflection at A is reached the molecule will drop and acquire a kinetic energy greater than ε even before it reaches the next lattice site. This energy can be "released" at the next lattice site (i. e., on the first collision), in which case the processes will now be repeated – each time the molecule reaches point A it will fall to point B. This type of motion will give rise to periodic changes in temperature at the interface, as predicted by computer simulations [333]. The stick–slip here will have a magnitude of the lattice dimension and, except for AFM measurements that can detect such small atomic-scale jumps [59, 334], the measured macro- and microscopic friction forces will be smooth and independent of v. If the energy dissipation (or "transfer") mechanism is less than 100% efficient on each collision, the molecule will move further before it stops. In this case the stick–slip amplitude can be large (point C), and the kinetic friction F_k can even be negative in the case of an overshoot (point D) (after [287], with permission; copyright 2004 American Chemical Society)

The theoretical models presented in this section will be concerned with a situation commonly observed experimentally: stick–slip occurs between a static state with high friction and a low-friction kinetic state, and a transition from this sliding regime to smooth sliding can be induced by an increase in velocity. Experimental data on various systems showing this behavior are shown in Figs. 9.27, 9.30b, and 9.31a. Recent studies on adhesive systems have revealed the possibility of other dynamic responses such as inverted stick–slip between two kinetic states of higher and lower friction and with a transition from smooth sliding to stick–slip with increasing velocity, as shown in Fig. 9.30c [342]. Similar friction responses have also been seen in computer simulations [343].

With the added insights provided by computer simulations, a number of distinct molecular processes have been identified during smooth and stick–slip sliding in model systems for the more familiar static-to-kinetic stick–slip and transition from stick–slip to smooth sliding. These are shown schematically in Fig. 9.26 for the case of spherical liquid molecules between two solid crystalline surfaces. The following regimes may be identified:

Surfaces at rest (Fig. 9.26a): even with no externally applied load, solvent–surface epitaxial interactions can cause the liquid molecules in the film to attain a solid-like state. Thus at rest the surfaces are stuck to each other through the film.

Sticking regime (frozen, solid-like film) (Fig. 9.26b): a progressively increasing lateral shear stress is applied. The solid-like film responds elastically with a small lateral displacement and a small increase or dilatancy in film thickness (less than

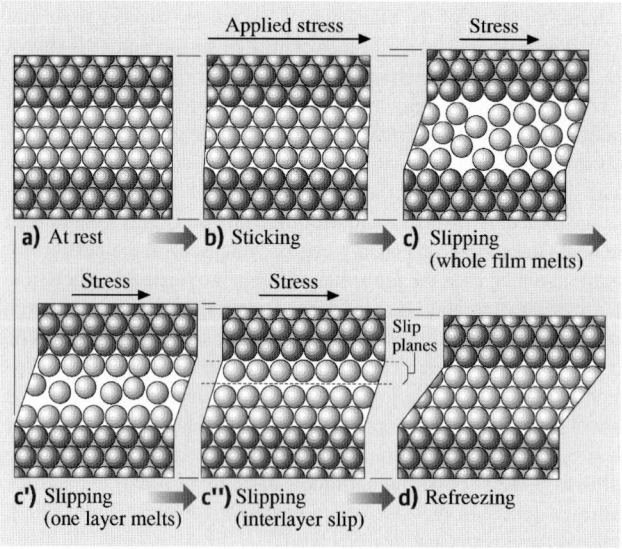

Fig. 9.26. Idealized schematic illustration of molecular rearrangements occurring in a molecularly thin film of spherical or simple chain molecules between two solid surfaces during shear. Depending on the system, a number of different molecular configurations within the film are possible during slipping and sliding, shown here as stages (**c**): total disorder as the whole film melts; (**c′**): partial disorder; and (**c″**): order persists even during sliding with slip occurring at a single slip plane either within the film or at the walls. A dilation is predicted in the direction normal to the surfaces (after [278], with permission)

a lattice spacing or molecular dimension, σ). In this regime the film retains its frozen, solid-like state: all the strains are elastic and reversible, and the surfaces remain effectively stuck to each other. However, slow creep may occur over long time periods.

Slipping and sliding regimes (molten, liquid-like film) (Fig. 9.26c,c',c''): when the applied shear stress or force has reached a certain critical value, the *static* friction force, F_s, the film suddenly melts (known as "shear melting") or rearranges to allow for wall slip or slip within the film to occur at which point the two surfaces begin to slip rapidly past each other. If the applied stress is kept at a high value, the upper surface will continue to slide indefinitely.

Refreezing regime (resolidification of film) (Fig. 9.26d): In many practical cases, the rapid slip of the upper surface relieves some of the applied force, which eventually falls below another critical value, the *kinetic* friction force F_k, at which point the film resolidifies and the whole stick–slip cycle is repeated. On the other hand, if the slip rate is smaller than the rate at which the external stress is applied, the surfaces will continue to slide smoothly in the kinetic state and there will be no more stick–slip. The critical velocity at which stick–slip disappears is discussed in more detail in Sect. 9.8.3.

Experiments with linear chain (alkane) molecules show that the film thickness remains quantized during sliding, so that the structure of such films is probably more like that of a nematic liquid crystal where the liquid molecules have become shear-aligned in some direction, enabling shear motion to occur while retaining some order within the film [344]. Experiments on the friction of two molecularly smooth mica surfaces separated by three molecular layers of the liquid octamethylcyclotetrasiloxane (OMCTS, see Fig. 9.27) show how the friction increases to higher values in a quantized way when the number of layers falls from $n = 3$ to $n = 2$ and then to $n = 1$.

Computer simulations for simple spherical molecules [255] further indicate that during slip the film thickness is roughly 15% higher than at rest (i.e., the film density falls), and that the order parameter within the film drops from 0.85 to about 0.25. Such dilatancy has been investigated both experimentally [296] and in further computer simulations [295]. The changes in thickness and in the order parameter are consistent with a disorganized state for the whole film during the slip [337], as illustrated schematically in Fig. 9.26c. At this stage, we can only speculate on other possible configurations of molecules in the slipping and sliding regimes. This probably depends on the shapes of the molecules (e.g., whether they are spherical or linear or branched), on the atomic structure of the surfaces, on the sliding velocity, etc. [345]. Figure 9.26c,c',c'' shows three possible sliding modes wherein the shearing film either totally melts, or where the molecules retain their layered structure and where slip occurs between two or more layers. Other sliding modes, for example, involving the movement of dislocations or disclinations are also possible, and it is unlikely that one single mechanism applies in all cases.

Both friction and adhesion hysteresis vary nonlinearly with temperature, often peaking at some particular temperature, T_0. The temperature dependence of these

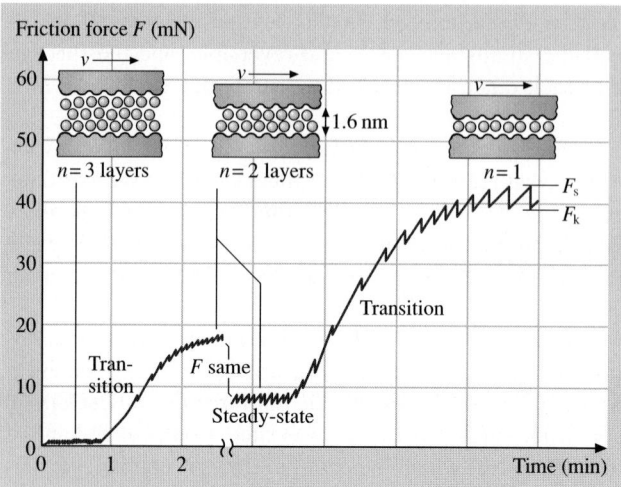

Fig. 9.27. Measured change in friction during interlayer transitions of the silicone liquid octamethylcyclotetrasiloxane (OMCTS), an inert liquid whose quasi-spherical molecules have a diameter of 0.8 nm. In this system, the shear stress $S_c = F/A$ was found to be constant as long as the number of layers, n, remained constant. Qualitatively similar results have been obtained with other quasi-spherical molecules such as cyclohexane [335]. The shear stresses are only weakly dependent on the sliding velocity v. However, for sliding velocities above some critical value, v_c, the stick–slip disappears and sliding proceeds smoothly at the kinetic value (after [275], with permission)

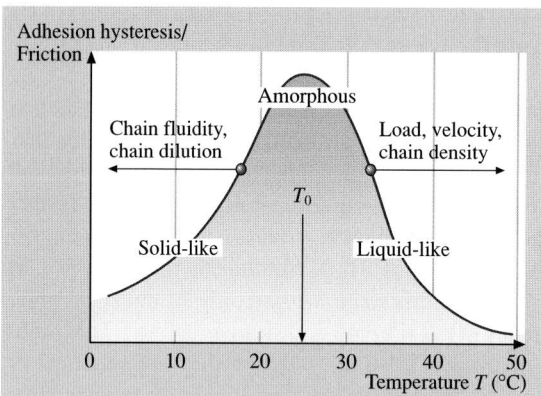

Fig. 9.28. Schematic friction phase diagram representing the trends observed in the boundary friction of a variety of different surfactant monolayers. The characteristic bell-shaped curve also correlates with the adhesion energy hysteresis of the monolayers. The *arrows* indicate the direction in which the whole curve is dragged when the load, velocity, etc., is increased (after [292], with permission)

forces can, therefore, be represented on a friction phase diagram such as the one shown in Fig. 9.28. Experiments have shown that T_0, and the whole bell-shaped curve, are shifted along the temperature axis (as well as in the vertical direction) in a systematic way when the load, sliding velocity, etc., are varied. These shifts also appear to be highly correlated with one another, for example, an increase in temperature produces effects that are similar to *decreasing* the sliding speed or load.

Such effects are also commonly observed in other energy-dissipating phenomena such as polymer viscoelasticity [346], and it is likely that a similar physical mechanism is at the heart of all such phenomena. A possible molecular process underlying the energy dissipation of chain molecules during boundary-layer sliding is illustrated in Fig. 9.29, which shows the three main dynamic phase states of boundary monolayers.

In contrast to the characteristic relaxation time associated with fluid lubricants, it has been established that for unlubricated (dry, solid, rough) surfaces, there is a characteristic memory distance that must be exceeded before the system loses all memory of its initial state (original surface topography). The underlying mechanism for a characteristic distance was first used to successfully explain rock mechanics and earthquake faults [347] and, more recently, the tribological behavior of unlubricated surfaces of ceramics, paper and elastomeric polymers [314, 348]. Recent experiments [285, 344, 345, 349] suggest that fluid lubricants composed of complex branched-chained or polymer molecules may also have characteristic distances (in addition to characteristic relaxation times) associated with their tribological behavior – the characteristic distance being the total sliding distance that must be exceeded before the system reaches its steady-state tribological conditions (see Sect. 9.8.3).

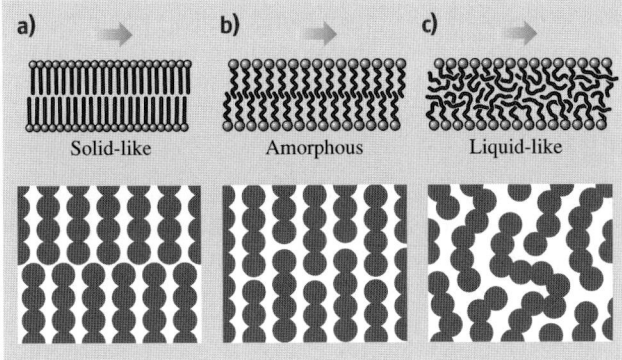

Fig. 9.29. Different dynamic phase states of boundary monolayers during adhesive contact and/or frictional sliding. Solid-like (**a**) and liquid-like monolayers (**c**) exhibit low adhesion hysteresis and friction. Increasing the temperature generally shifts a system from the left to the right. Changing the load, sliding velocity, or other experimental conditions can also change the dynamic phase state of surface layers, as shown in Fig. 9.28 (after [292], with permission)

Abrupt versus Continuous Transitions Between Smooth and Stick–Slip Sliding

An understanding of stick–slip is of great practical importance in tribology [350], since these spikes are the major cause of damage and wear of moving parts. Stick–slip motion is a very common phenomenon and is also the cause of sound generation (the sound of a violin string, a squeaking door, or the chatter of machinery), sensory perception (taste texture and feel), earthquakes, granular flow, nonuniform fluid flow such as the spurting flow of polymeric liquids, etc. In the previous section, the stick–slip motion arising from freezing–melting transitions in thin interfacial films was described. There are other mechanisms that can give rise to stick–slip friction, which will now be considered. However, before proceeding with this, it is important to clarify exactly what one is measuring during a friction experiment.

Most tribological systems and experiments can be described in terms of an equivalent mechanical circuit with certain characteristics. The friction force F_0, which is generated at the surfaces, is generally measured as F at some other place in the set-up. The mechanical coupling between the two may be described in terms of a simple elastic stiffness or compliance, K, or as more complex nonelastic coefficients, depending on the system. The distinction between F and F_0 is important because, in almost all practical cases, the applied, measured, or detected force, F, is *not* the same as the "real" or "intrinsic" friction force, F_0, generated at the surfaces. F and F_0 are coupled in a way that depends on the mechanical construction of the system, for example, the axle of a car wheel that connects it to the engine. This coupling can be modeled as an elastic spring of stiffness K and mass m. This is the simplest type of mechanical coupling and is also the same as in SFA- and AFM-type experiments. More complicated real systems can be reduced to a system of springs and dashpots, as described by *Peachey* et al. [351] and *Luengo* et al. [47].

We now consider four different models of stick–slip friction, where the mechanical couplings are assumed to be of the simple elastic spring type. The first three mechanisms may be considered traditional or classical mechanisms or models [350], the fourth is essentially the same as the freezing–melting phase-transition model described in Sect. 9.8.3.

Rough Surfaces or Surface Topology Model Rapid slips can occur whenever an asperity on one surface goes over the top of an asperity on the other surface. The extent of the slip will depend on asperity heights and slopes, on the speed of sliding, and on the elastic compliance of the surfaces and the moving stage. As in all cases of stick–slip motion, the driving velocity v may be constant, but the resulting motion at the surfaces v_0 will display large slips. This type of stick–slip has been described by *Rabinowicz* [350]. It will not be of much concern here since it is essentially a noise-type fluctuation, resulting from surface imperfections rather than from the intrinsic interaction between two surfaces. Actually, at the atomic level, the regular atomic-scale corrugations of surfaces can lead to periodic stick–slip motion of the type shown here. This is what is sometimes measured by AFM tips [10, 58, 59, 290, 339, 340].

Distance-Dependent or Creep Model Another theory of stick–slip, observed in solid-on-solid sliding, is one that involves a characteristic *distance*, but also a characteristic time, τ_s, this being the characteristic time required for two asperities to increase their adhesion strength after coming into contact. Originally proposed by *Rabinowicz* [350, 352], this model suggests that two rough macroscopic surfaces adhere through their microscopic asperities of a characteristic length. During shearing, each surface must first creep this distance (the size of the contacting junctions) after which the surfaces continue to slide, but with a lower (kinetic) friction force than the original (static) value. The reason for the decrease in the friction force is that even though, on average, new asperity junctions should form as rapidly as the old ones break, the time-dependent adhesion and friction of the new ones will be lower than the old ones.

The friction force, therefore, remains high during the creep stage of the slip. However, once the surfaces have moved the characteristic distance, the friction rapidly drops to the kinetic value. For any system where the kinetic friction is less than the static force (or one that has a negative slope over some part of its curve of F_0 versus v_0) will exhibit regular stick–slip sliding motion for certain values of K, m, and driving velocity, v.

This type of friction has been observed in a variety of dry (unlubricated) systems such as paper-on-paper [353, 354] and steel-on-steel [352, 355, 356]. This model is also used extensively in geologic systems to analyze rock-on-rock sliding [357, 358]. While originally described for adhering macroscopic asperity junctions, the distance-dependent model may also apply to molecularly smooth surfaces. For example, for polymer lubricant films, the characteristic length would now be the chain–chain entanglement length, which could be much larger in a confined geometry than in the bulk.

Velocity-Dependent Friction Model In contrast to the two friction models mentioned above, which apply mainly to unlubricated, solid-on-solid contacts, the stick–slip of surfaces with thin liquid films between them is better described by other mechanisms. The velocity-dependent friction model is the most studied mechanism of stick–slip and, until recently, was considered to be the only cause of intrinsic stick–slip. If a friction force decreases with increasing sliding velocity, as occurs with boundary films exhibiting shear thinning, the force (F_s) needed to initiate motion will be higher than the force (F_k) needed to maintain motion.

A decreasing intrinsic friction force F_0 with sliding velocity v_0 results in the sliding surface or stage moving in a periodic fashion, where during each cycle rapid acceleration is followed by rapid deceleration. As long as the drive continues to move at a fixed velocity v, the surfaces will continue to move in a periodic fashion punctuated by abrupt stops and starts whose frequency and amplitude depend not only on the function $F_0(v_0)$, but also on the stiffness K and mass m of the moving stage, and on the starting conditions at $t = 0$.

More precisely, the motion of the sliding surface or stage can be determined by solving the following differential equation:

$$m\ddot{x} = (F_0 - F) = F_0 - (x_0 - x)K \quad \text{or} \quad m\ddot{x} + (x_0 - x)K - F_0 = 0 \, , \tag{9.47}$$

where $F_0 = F_0(x_0, v_0, t)$ is the intrinsic or "real" friction force at the shearing surfaces, F is the force on the spring (the externally applied or measured force), and $(F_0 - F)$ is the force on the stage. To solve (9.47) fully, one must also know the initial (starting) conditions at $t = 0$, and the driving or steady-state conditions at finite t. Commonly, the driving condition is: $x = 0$ for $t < 0$ and $x = vt$ for $t > 0$, where $v = $ constant. In other systems, the appropriate driving condition may be $F = $ constant.

Various, mainly phenomenological, forms for $F_0 = F_0(x_0, v_0, t)$ have been proposed to explain various kinds of stick–slip phenomena. These models generally assume a particular functional form for the friction as a function of velocity only, $F_0 = F_0(v_0)$, and they may also contain a number of mechanically coupled elements comprising the stage [359, 360]. One version is a two-state model characterized by two friction forces, F_s and F_k, which is a simplified version of the phase transitions model described below. More complicated versions can have a rich F–v spectrum, as proposed by *Persson* [361]. Unless the experimental data is very detailed and extensive, these models cannot generally distinguish between different types of mechanisms. Neither do they address the basic question of the *origin* of the friction force, since this is assumed to begin with.

Experimental data has been used to calculate the friction force as a function of velocity *within* an individual stick–slip cycle [363]. For a macroscopic granular material confined between solid surfaces, the data shows a velocity-weakening friction force during the first half of the slip. However, the data also shows a hysteresis loop in the friction–velocity plot, with a different behavior in the deceleration half of the slip phase. Similar results were observed for a 1–2 nm liquid lubricant film between mica surfaces [345]. These results indicate that a purely velocity-dependent friction law is insufficient to describe such systems, and an additional element such as the *state* of the confined material must be considered.

Phase Transitions Model In recent molecular dynamics computer simulations it has been found that thin interfacial films undergo first-order phase transitions between solid-like and liquid-like states during sliding [255, 364], as illustrated in Fig. 9.30. It has been suggested that this is responsible for the observed stick–slip behavior of simple isotropic liquids between two solid crystalline surfaces. With this interpretation, stick–slip is seen to arise because of the abrupt change in the flow properties of a film at a transition [278, 279, 326], rather than the gradual or continuous change, as occurs in the previous example. Other computer simulations indicate that it is the stick–slip that induces a disorder ("shear melting") in the film, not the other way around [365].

An interpretation of the well-known phenomenon of decreasing coefficient of friction with increasing sliding velocity has been proposed by *Thompson* and *Robbins* [255] based on their computer simulation. This postulates that it is not the friction that changes with sliding speed v, but rather the time various parts of the system spend in the sticking and sliding modes. In other words, at any instant during sliding, the friction at any local region is always F_s or F_k, corresponding to the "static" or "kinetic" values. The measured frictional force, however, is the sum of all these discrete values averaged over the whole contact area. Since as v increases each local region spends more time in the sliding regime (F_k) and less in the sticking regime (F_s), the overall friction coefficient falls. One may note that this interpretation reverses the traditional way that stick–slip has been explained. Rather than invoking a decreasing friction with velocity to explain stick–slip, it is now the more fundamental stick–slip phenomenon that is producing the apparent decrease in the friction force with increasing sliding velocity. This approach has been studied analytically by *Carlson* and *Batista* [366], with a comprehensive rate- and state-dependent friction force law. This model includes an analytic description of the freezing–melting transitions of a film, resulting in a friction force that is a function of sliding velocity in a natural way. This model predicts a full range of stick–slip behavior observed experimentally.

An example of the rate- and state-dependent model is observed when shearing thin films of OMCTS between mica surfaces [367, 368]. In this case, the static friction between the surfaces is dependent on the time that the surfaces are at rest with respect to each other, while the intrinsic kinetic friction $F_{k,0}$ is relatively constant over the range of velocities. At slow driving velocities, the system responds with stick–slip sliding with the surfaces reaching maximum static friction before each slip event, and the amplitude of the stick–slip, $F_s - F_k$, is relatively constant. As the driving velocity increases, the static friction decreases as the time at relative rest becomes shorter with respect to the characteristic time of the lubricant film. As the static friction decreases with increasing drive velocity, it eventually equals the intrinsic kinetic friction $F_{k,0}$, which defines the critical velocity v_c, above which the surfaces slide smoothly without the jerky stick–slip motion.

The above classifications of stick–slip are not exclusive, and molecular mechanisms of real systems may exhibit aspects of different models simultaneously. They do, however, provide a convenient classification of existing models and indicate which experimental parameters should be varied to test the different models.

Critical Velocity for Stick–Slip For any given set of conditions, stick–slip disappears above some critical sliding velocity v_c, above which the motion continues smoothly in the liquid-like or kinetic state [261, 285, 342, 345, 362]. The critical velocity is well described by two simple equations. Both are based on the phase transition model, and both include some parameter associated with the inertia of the measuring instrument. The first equation is based on both experiments and simple theoretical modeling [362]:

$$v_c \approx \frac{(F_s - F_k)}{5K\tau_0}, \tag{9.48}$$

where τ_0 is the *characteristic nucleation time* or freezing time of the film. For example, inserting the following typically measured values for a ~ 1 nm thick hexadecane film between mica: $(F_s - F_k) \approx 5$ mN, spring constant $K \approx 500\,\text{N m}^{-1}$, and nucleation time [362] $\tau_0 \approx 5$ s, we obtain $v_c \approx 0.4\,\mathrm{\mu m\,s^{-1}}$, which is close to typically measured values (Fig. 9.30b).

Fig. 9.30. (a) "Phase transitions" model of stick–slip where a thin liquid film alternately freezes and melts as it shears, shown here for 22 spherical molecules confined between two solid crystalline surfaces. In contrast to the velocity-dependent friction model, the intrinsic friction force is assumed to change abruptly (at the transitions), rather than smooth or continuously. The resulting stick–slip is also different, for example, the peaks are sharper and the stick–slip disappears above some critical velocity v_c. Note that, while the slip displacement is here shown to be only two lattice spacings, in most practical situations it is much larger, and that freezing and melting transitions at surfaces or in thin films may not be the same as freezing or melting transitions between the bulk solid and liquid phases. (b) Exact reproduction of a chart-recorder trace of the friction force for hexadecane between two untreated mica surfaces at increasing sliding velocity v, plotted as a function of time. In general, with increasing sliding speed, the stick–slip spikes increase in frequency and decrease in magnitude. As the critical sliding velocity v_c is approached, the spikes become erratic, eventually disappearing altogether at $v = v_c$. At higher sliding velocities the sliding continues smoothly in the kinetic state. Such friction traces are fairly typical for simple liquid lubricants and dry boundary lubricant systems (see Fig. 9.31a) and may be referred to as the "conventional" type of static–kinetic friction (in contrast to panel (c)). Experimental conditions: contact area $A = 4 \times 10^{-9}$ m^2, load $L = 10$ mN, film thickness $D = 0.4$–0.8 nm, $v = 0.08$–$0.4\,\mu\mathrm{m\,s}^{-1}$, $v_c \approx 0.3\,\mu\mathrm{m\,s}^{-1}$, atmosphere: dry N$_2$ gas, $T = 18$ °C. ((a) and (b) after [362] with permission; copyright 1993 American Chemical Society.) (c) Transition from smooth sliding to "inverted" stick–slip and to a second smooth-sliding regime with increasing driving velocity during shear of two adsorbed surfactant monolayers in aqueous solution at a load of $L = 4.5$ mN and $T = 20$ °C. The smooth sliding (*open circles*) to inverted stick–slip (*squares*) transition occurs at $v_c \sim 0.3$ μm/s. Prior to the transition, the kinetic stress levels off at after a logarithmic dependence on velocity. The quasi-smooth regime persists up to the transition at v_c. At high driving velocities (*filled circles*), a new transition to a smooth sliding regime is observed between 14 and 17 μm/s (after [342] with permission). (d) Friction response of a thin squalane (a branched hydrocarbon) film at different loads and a constant sliding velocity $v = 0.08\,\mu\mathrm{m\,s}^{-1}$, slightly above the critical velocity for this system at low loads. Initially, with increasing load, the stick–slip amplitude and the mean friction force decrease with sliding time or sliding distance. However, at high loads or pressures, the mean friction force increases with time, and the stick–slip takes on a more symmetrical, sinusoidal shape. At all loads investigated, the stick–slip component gradually decayed as the friction proceeded towards smooth sliding (after [285] with permission; copyright 2003 American Physical Society)

The second equation is based on computer simulations [364]:

$$v_c \approx 0.1 \sqrt{\frac{F_s \sigma}{m}}, \qquad (9.49)$$

where σ is a molecular dimension and m is the mass of the stage. Again, inserting typical experimental values into this equation, viz., $m \approx 20$ g, $\sigma \approx 0.5$ nm, and $(F_s - F_k) \approx 5$ mN as before, we obtain $v_c \approx 0.3\,\mu\mathrm{m\,s}^{-1}$, which is also close to measured values.

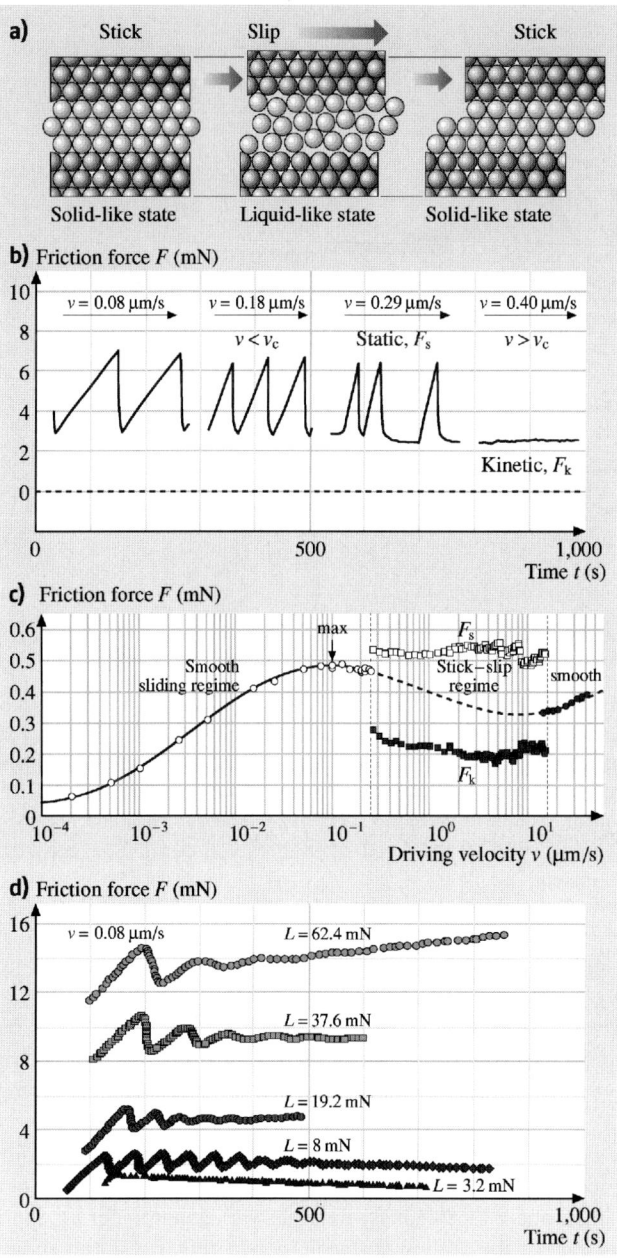

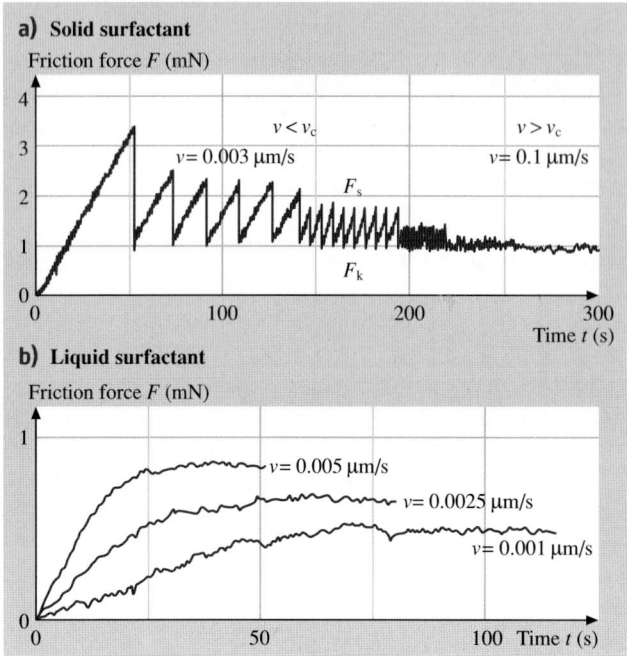

Fig. 9.31. (a) Exact reproduction of chart-recorder trace for the friction of closely packed surfactant monolayers (L-α-dimirystoyl-phosphatidyl-ethanolamine, DMPE) on mica (dry boundary friction) showing qualitatively similar behavior to that obtained with a liquid hexadecane film (Fig. 9.30b). In this case, $L = 0$, $v_c \approx 0.1\,\mu\mathrm{m\,s^{-1}}$, atmosphere: dry N_2 gas, $T = 25\,°\mathrm{C}$. (b) Sliding typical of liquid-like monolayers, here shown for calcium alkylbenzene sulfonate in dry N_2 gas at $T = 25\,°\mathrm{C}$ and $L = 0$ (after [261], with permission; copyright 1993 American Chemical Society)

Stick–slip also disappears above some critical temperature T_c, which is not the same as the melting temperature of the bulk fluid [285]. Certain correlations have been found between v_c and T_c and between various other tribological parameters that appear to be consistent with the principle of time–temperature superposition (see Sect. 9.8.3), similar to that occurring in viscoelastic polymer fluids [346, 369, 370].

Recent work on the coupling between the mechanical resonances of the sliding system and molecular-scale relaxations [295, 338, 341, 371] has resulted in a better understanding of a phenomenon previously noted in various engineering applications: the vibration of one of the sliding surfaces perpendicularly to the sliding direction can lead to a significant reduction of the friction. At certain oscillation amplitudes and a frequency higher than the molecular-scale relaxation frequency, stick–slip friction can be eliminated and replaced by an ultralow kinetic-friction state.

9.9 Effects of Nanoscale Texture on Friction

The above scenario is already quite complicated, and yet this is the situation for the simplest type of experimental system. The factors that appear to determine the critical velocity v_c depend on the type of liquid between the surfaces, as well as on the surface lattice structure.

9.9.1 Role of the Shape of Confined Molecules

Small spherical molecules such as cyclohexane and OMCTS have been found to have very high v_c, which indicates that these molecules can rearrange relatively quickly in thin films. Chain molecules and especially branched-chain molecules have been found to have much lower v_c, which is to be expected, and such liquids tend to slide smoothly or with erratic stick–slip [345], rather than in a stick–slip fashion (see Table 9.5). With highly asymmetric molecules, such as multiply branched isoparaffins and polymer melts, no regular spikes or stick–slip behavior occurs at any speed, since these molecules can never order themselves sufficiently to solidify.

Table 9.5. Effect of molecular shape and short-range forces on tribological properties[a]

Liquid	Short-range force	Type of friction	Friction coefficient	Bulk liquid viscosity (cP)
Organic (water-free)				
Cyclohexane	Oscillatory	Quantized stick–slip	$\gg 1$	0.6
OMCTS[b]	Oscillatory	Quantized stick–slip	$\gg 1$	2.3
Octane	Oscillatory	Quantized stick–slip	1.5	0.5
Tetradecane	Oscillatory $\leftrightarrow$ smooth	stick–slip $\leftrightarrow$ smooth	1.0	2.3
Octadecane (branched)	Oscillatory $\leftrightarrow$ smooth	stick–slip $\leftrightarrow$ smooth	0.3	5.5
PDMS[b] ($M = 3700$ g mol^{-1}, melt)	Oscillatory $\leftrightarrow$ smooth	Smooth	0.4	50
PBD[b] ($M = 3500$ g mol^{-1}, branched)	Smooth	Smooth	0.03	800
Water				
Water (KCl solution)	Smooth	Smooth	0.01–0.03	1.0

[a] For molecularly thin liquid films between two shearing mica surfaces at 20 °C
[b] OMCTS: Octamethylcyclotetrasiloxane, PDMS: Polydimethylsiloxane, PBD: Polybutadiene

Examples of such liquids are some perfluoropolyethers and polydimethylsiloxanes (PDMS).

Recent computer simulations [144, 151, 287, 315, 372] of the structure, interaction forces, and tribological behavior of chain molecules between two shearing surfaces indicate that both linear *and* singly or doubly branched-chain molecules order between two flat surfaces by aligning into discrete layers parallel to the surfaces. However, in the case of the weakly branched molecules, the expected oscillatory forces do not appear because of a complex cancelation of entropic and enthalpic contributions to the interaction free energy, which results in a monotonically smooth interaction, exhibiting a weak energy minimum rather than the oscillatory force profile that is characteristic of linear molecules. During sliding, however, these molecules can be induced to further align, which can result in a transition from smooth to stick–slip sliding.

Table 9.5 shows the trends observed with some organic and polymeric liquids between smooth mica surfaces. Also listed are the bulk viscosities of the liquids. From the data of Table 9.5 it appears that there is a direct correlation between the shapes of molecules and their coefficient of friction or effectiveness as lubricants (at least at low shear rates). Small spherical or chain molecules have high friction with stick–slip, because they can pack into ordered solid-like layers. In contrast, longer chained and irregularly shaped molecules remain in an entangled, disordered, fluid-like state even in very thin films, and these give low friction and smoother sliding. It is probably for this reason that irregularly shaped branched chain molecules are usually better lubricants. It is interesting to note that the friction coefficient generally decreases as the bulk viscosity of the liquids *increases*. This unexpected trend occurs because the factors that are conducive to low friction are generally conducive to high viscosity. Thus molecules with side groups such as branched alkanes and polymer melts usually have higher bulk viscosities than their linear homologues for obvious reasons. However, in thin films the linear molecules have higher shear stresses, because of their ability to become ordered. The only exception to the above correlations is water, which has been found to exhibit both low viscosity *and* low friction (see Fig. 9.20a, and Sect. 9.7.3). In addition, the presence of water can drastically lower the friction and eliminate the stick–slip of hydrocarbon liquids when the sliding surfaces are hydrophilic.

If an "effective" viscosity, η_{eff}, were to be calculated for the liquids of Table 9.5, the values would be many orders of magnitude higher than those of the bulk liquids. This can be demonstrated by the following simple calculation based on the usual equation for Couette flow (see (9.44)):

$$\eta_{\text{eff}} = F_k D / A v \,, \tag{9.50}$$

where F_k is the kinetic friction force, D is the film thickness, A the contact area, and v the sliding velocity. Using typical values for experiments with hexadecane [362]: $F_k = 5$ mN, $D = 1$ nm, $A = 3 \times 10^{-9}$ m^2, and $v = 1$ μm s^{-1}, one gets $\eta_{\text{eff}} \approx 2000$ Ns m^{-2}, or 20,000 Poise, which is $\sim 10^6$ times higher than the bulk viscosity, η_{bulk}, of the liquid. It is instructive to consider that this very high effective viscosity nevertheless still produces a low friction force or friction coefficient μ of about 0.25. It is interesting to speculate that, if a 1 nm film were to exhibit bulk viscous behavior, the friction coefficient under the same sliding conditions would be as low as 0.000001. While such a low value has never been reported for any tribological system, one may consider it a theoretical lower limit that could conceivably be attained under certain experimental conditions.

9.9.2 Effects of Surface Structure

Various studies [44, 273, 274, 276, 284–286] have shown that confinement and load generally increase the effective viscosity and/or relaxation times of molecules, suggestive of an increased glassiness or solid-like behavior (Figs. 9.32 and 9.33). This is in marked contrast to studies of liquids in small confining capillaries where the opposite effects have been observed [373, 374]. The reason for this is probably because the two modes of confinement are different. In the former case (confinement of molecules between two structured solid surfaces), there is generally little opposition to any lateral or vertical displacement of the two surface lattices relative to each other. This means that the two lattices can shift in the x–y–z planes (Fig. 9.32a)

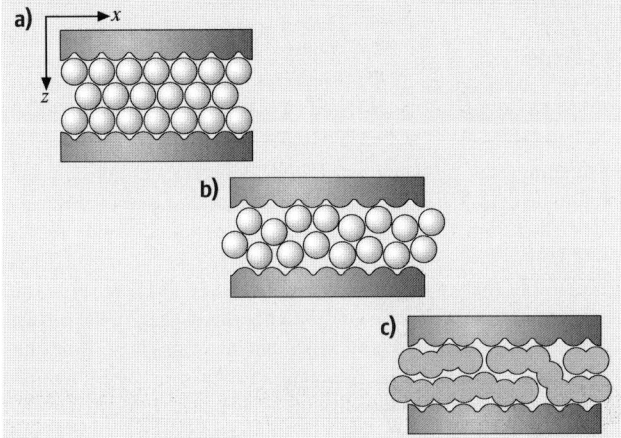

Fig. 9.32. Schematic view of interfacial film composed of spherical molecules under a compressive pressure between two solid crystalline surfaces. (**a**) If the two surface lattices are free to move in the x–y–z directions, so as to attain the lowest energy state, they could equilibrate at values of x, y, and z, which induce the trapped molecules to become "epitaxially" ordered into a "solid-like" film. (**b**) Similar view of trapped molecules between two solid surfaces that are not free to adjust their positions, for example, as occurs in capillary pores or in brittle cracks. (**c**) Similar to (**a**), but with chain molecules replacing the spherical molecules in the gap. These may not be able to order as easily as do spherical molecules even if x, y, and z can adjust, resulting in a situation that is more akin to (**b**)(after [362] with permission; copyright 1993 American Chemical Society)

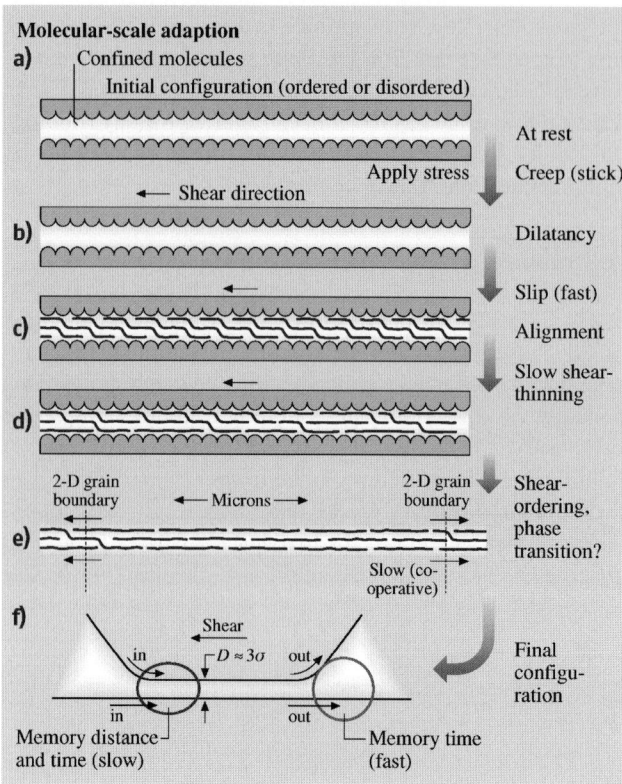

Fig. 9.33. Schematic representation of the film under shear. (**a**) The lubricant molecules are just confined, but not oriented in any particular direction. Because of the need to shear, the film dilates (**b**). The molecules disentangle (**c**) and get oriented in a certain direction related to the shear direction (**d**). (**e**) Slowly evolving domains grow inside the contact region. These macroscopic domains are responsible for the long relaxation times. (**f**) At the steady-state, a continuous gradient of confinement time and molecular order is established in the contact region, which is different for molecules adsorbed on the upper and lower surfaces. Molecules entering into the contact are not oriented or ordered. The required sliding distance to modify their state defines a characteristic distance. Molecules leaving the contact region need some (short) characteristic time to regain their bulk, unconfined configuration (after [344], with permission; copyright 2000 American Chemical Society)

to accommodate the trapped molecules in the most crystallographically commensurate or epitaxial way, which would favor an ordered, solid-like state. In contrast, the walls of capillaries are rigid and cannot easily move or adjust to accommodate the confined molecules (Fig. 9.32b), which will therefore be forced into a more disordered, liquid-like state (unless the capillary wall geometry and lattice are *exactly* commensurate with the liquid molecules, as occurs in certain zeolites [374]).

Experiments have demonstrated the effects of surface lattice mismatch on the friction between surfaces [257, 258, 375]. Similar to the effects of lattice mismatch on adhesion (Fig. 9.11), the static friction of a confined liquid film is maximum when the lattices of the confining surfaces are aligned. For OMCTS confined between mica surfaces [258] the static friction was found to vary by more than a factor of 4, while for bare mica surfaces the variation was by a factor of 3.5 [375]. In contrast to the sharp variations in adhesion energy over small twist angles, the variations in friction as a function of twist angle were much broader both in magnitude and angular spread. Similar variations in friction as a function of twist or misfit angles have also been observed in computer simulations [376].

Robbins and coworkers [315] computed the friction forces of two clean crystalline surfaces as a function of the angle between their surface lattices. They found that, for all non-zero angles (finite "twist" angles), the friction forces fell to zero due to incommensurability effects. They further found that sub-monolayer amounts of organic or other impurities trapped between two incommensurate surfaces can generate a finite friction force. They therefore concluded that any finite friction force measured between incommensurate surfaces is probably due to such "third-body" effects.

The reason why surface texture (lattice structure, roughness, granularity, topography, etc.) has a larger effect on the lateral (shear or friction) forces between two surfaces than on their normal (adhesion) forces is because friction is proportional to the adhesion hysteresis (Sect. 9.7.2), which can be low even when the adhesion force is high. It is also important to recognize that a system might be defined by more than one length scale. Some systems have well-defined dimensions or size (e.g., a perfect lattice, monodisperse nanoparticles), while others have different lateral and vertical dimensions and macroscopic curvature [249]. Furthermore, the morphology or texture of many systems, such as asperities that are randomly distributed over a surface, affects adhesion and tribological properties [244, 249, 287, 317, 377–379].

With rough surfaces, i.e., those that have *random* protrusions rather than being periodically structured, we expect a smearing out of the correlated intermolecular interactions that are involved in film freezing and melting (and in phase transitions in general). This should effectively eliminate the highly regular stick–slip and may also affect the location of the slipping planes [151, 287, 314, 348]. The stick–slip friction of "real" surfaces, which are generally rough, may, therefore, be quite different from those of perfectly smooth surfaces composed of the same material. We should note, however, that even between rough surfaces, most of the contacts occur between the tips of microscopic asperities, which may be smooth over their microscopic contact area [380].

References

1. J.N. Israelachvili: Surface Forces and Microrheology of Molecularly Thin Liquid Films. In: *Handbook of Micro/Nanotribology*, ed. by B. Bhushan (CRC, Boca Raton 1995) pp. 267–319
2. K.B. Lodge: Techniques for the measurement of forces between solids, Adv. Colloid Interface Sci. **19**, 27–73 (1983)

3. J.N. Israelachvili: *Intermolecular and Surface Forces*, 2nd edn. (Academic, London 1991)
4. P.F. Luckham, B.A. de L. Costello: Recent developments in the measurement of interparticle forces, Adv. Colloid Interface Sci. **44**, 183–240 (1993)
5. P.M. Claesson, T. Ederth, V. Bergeron, M.W. Rutland: Techniques for measuring surface forces, Adv. Colloid Interface Sci. **67**, 119–183 (1996)
6. V.S.J. Craig: An historical review of surface force measurement techniques, Colloids Surf. A **129–130**, 75–94 (1997)
7. J.N. Israelachvili, G.E. Adams: Measurements of forces between two mica surfaces in aqueous electrolyte solutions in the range 0–100 nm, J. Chem. Soc. Faraday Trans. I **74**, 975–1001 (1978)
8. G. Binnig, C.F. Quate, C. Gerber: Atomic force microscope, Phys. Rev. Lett. **56**, 930–933 (1986)
9. W.A. Ducker, T.J. Senden, R.M. Pashley: Direct measurement of colloidal forces using an atomic force microscope, Nature **353**, 239–241 (1991)
10. E. Meyer, R.M. Overney, K. Dransfeld, T. Gyalog: *Nanoscience: Friction and Rheology on the Nanometer Scale* (World Scientific, Singapore 1998)
11. J.N. Israelachvili: Measurements of the viscosity of thin fluid films between two surfaces with and without adsorbed polymers, Colloid Polym. Sci. **264**, 1060–1065 (1986)
12. J.P. Montfort, G. Hadziioannou: Equilibrium and dynamic behavior of thin films of a perfluorinated polyether, J. Chem. Phys. **88**, 7187–7196 (1988)
13. A. Dhinojwala, S. Granick: Surface forces in the tapping mode: Solvent permeability and hydrodynamic thickness of adsorbed polymer brushes, Macromolecules **30**, 1079–1085 (1997)
14. B.V. Derjaguin: Untersuchungen über die Reibung und Adhäsion, IV. Theorie des Anhaftens kleiner Teilchen, Kolloid Z. **69**, 155–164 (1934)
15. K.L. Johnson, K. Kendall, A.D. Roberts: Surface energy and the contact of elastic solids, Proc. R. Soc. London A **324**, 301–313 (1971)
16. J.N. Israelachvili, P.M. McGuiggan: Adhesion and short-range forces between surfaces. Part 1: New apparatus for surface force measurements, J. Mater. Res. **5**, 2223–2231 (1990)
17. J.N. Israelachvili: Thin film studies using multiple-beam interferometry, J. Colloid Interface Sci. **44**, 259–272 (1973)
18. Y.L. Chen, T. Kuhl, J. Israelachvili: Mechanism of cavitation damage in thin liquid films: Collapse damage vs. inception damage, Wear **153**, 31–51 (1992)
19. M. Heuberger, G. Luengo, J. Israelachvili: Topographic information from multiple beam interferometry in the surface forces apparatus, Langmuir **13**, 3839–3848 (1997)
20. R.M. Pashley: DLVO and hydration forces between mica surfaces in Li^+, Na^+, K^+, and Cs^+ electrolyte solutions: A correlation of double-layer and hydration forces with surface cation exchange properties, J. Colloid Interface Sci. **83**, 531–546 (1981)
21. R.M. Pashley: Hydration forces between mica surfaces in electrolyte solution, Adv. Colloid Interface Sci. **16**, 57–62 (1982)
22. R.G. Horn, D.T. Smith, W. Haller: Surface forces and viscosity of water measured between silica sheets, Chem. Phys. Lett. **162**, 404–408 (1989)
23. R.G. Horn, D.R. Clarke, M.T. Clarkson: Direct measurements of surface forces between sapphire crystals in aqueous solutions, J. Mater. Res. **3**, 413–416 (1988)
24. W.W. Merrill, A.V. Pocius, B.V. Thakker, M. Tirrell: Direct measurement of molecular level adhesion forces between biaxially oriented solid polymer films, Langmuir **7**, 1975–1980 (1991)

25. J. Klein: Forces between mica surfaces bearing adsorbed macromolecules in liquid media, J. Chem. Soc. Faraday Trans. I **79**, 99–118 (1983)
26. S. S. Patel, M. Tirrell: Measurement of forces between surfaces in polymer fluids, Annu. Rev. Phys. Chem. **40**, 597–635 (1989)
27. H. Watanabe, M. Tirrell: Measurements of forces in symmetric and asymmetric interactions between diblock copolymer layers adsorbed on mica, Macromolecules **26**, 6455–6466 (1993)
28. T. L. Kuhl, D. E. Leckband, D. D. Lasic, J. N. Israelachvili: Modulation and modeling of interaction forces between lipid bilayers exposing terminally grafted polymer chains. In: *Stealth Liposomes*, ed. by D. Lasic, F. Martin (CRC, Boca Raton 1995) pp. 73–91
29. J. Klein: Shear, friction, and lubrication forces between polymer-bearing surfaces, Annu. Rev. Mater. Sci. **26**, 581–612 (1996)
30. M. Ruths, D. Johannsmann, J. Rühe, W. Knoll: Repulsive forces and relaxation on compression of entangled, polydisperse polystyrene brushes, Macromolecules **33**, 3860–3870 (2000)
31. C. A. Helm, J. N. Israelachvili, P. M. McGuiggan: Molecular mechanisms and forces involved in the adhesion and fusion of amphiphilic bilayers, Science **246**, 919–922 (1989)
32. Y. L. Chen, C. A. Helm, J. N. Israelachvili: Molecular mechanisms associated with adhesion and contact angle hysteresis of monolayer surfaces, J. Phys. Chem. **95**, 10736–10747 (1991)
33. D. E. Leckband, J. N. Israelachvili, F.-J. Schmitt, W. Knoll: Long-range attraction and molecular rearrangements in receptor–ligand interactions, Science **255**, 1419–1421 (1992)
34. J. Peanasky, H. M. Schneider, S. Granick, C. R. Kessel: Self-assembled monolayers on mica for experiments utilizing the surface forces apparatus, Langmuir **11**, 953–962 (1995)
35. C. J. Coakley, D. Tabor: Direct measurement of van der Waals forces between solids in air, J. Phys. D **11**, L77–L82 (1978)
36. J. L. Parker, H. K. Christenson: Measurements of the forces between a metal surface and mica across liquids, J. Chem. Phys. **88**, 8013–8014 (1988)
37. C. P. Smith, M. Maeda, L. Atanasoska, H. S. White, D. J. McClure: Ultrathin platinum films on mica and the measurement of forces at the platinum/water interface, J. Phys. Chem. **92**, 199–205 (1988)
38. S. J. Hirz, A. M. Homola, G. Hadziioannou, C. W. Frank: Effect of substrate on shearing properties of ultrathin polymer films, Langmuir **8**, 328–333 (1992)
39. J. M. Levins, T. K. Vanderlick: Reduction of the roughness of silver films by the controlled application of surface forces, J. Phys. Chem. **96**, 10405–10411 (1992)
40. S. Steinberg, W. Ducker, G. Vigil, C. Hyukjin, C. Frank, M. Z. Tseng, D. R. Clarke, J. N. Israelachvili: Van der Waals epitaxial growth of α-alumina nanocrystals on mica, Science **260**, 656–659 (1993)
41. G. Vigil, Z. Xu, S. Steinberg, J. Israelachvili: Interactions of silica surfaces, J. Colloid Interface Sci. **165**, 367–385 (1994)
42. M. Ruths, M. Heuberger, V. Scheumann, J. Hu, W. Knoll: Confinement-induced film thickness transitions in liquid crystals between two alkanethiol monolayers on gold, Langmuir **17**, 6213–6219 (2001)
43. J. Van Alsten, S. Granick: Molecular tribometry of ultrathin liquid films, Phys. Rev. Lett. **61**, 2570–2573 (1988)
44. J. Van Alsten, S. Granick: Shear rheology in a confined geometry: Polysiloxane melts, Macromolecules **23**, 4856–4862 (1990)

45. A.M. Homola, J.N. Israelachvili, M.L. Gee, P.M. McGuiggan: Measurements of and relation between the adhesion and friction of two surfaces separated by molecularly thin liquid films, J. Tribol. **111**, 675–682 (1989)
46. J. Klein, E. Kumacheva: Simple liquids confined to molecularly thin layers. I. Confinement-induced liquid-to-solid phase transitions, J. Chem. Phys. **108**, 6996–7009 (1998)
47. G. Luengo, F.-J. Schmitt, R. Hill, J. Israelachvili: Thin film rheology and tribology of confined polymer melts: Contrasts with bulk properties, Macromolecules **30**, 2482–2494 (1997)
48. L. Qian, G. Luengo, D. Douillet, M. Charlot, X. Dollat, E. Perez: New two-dimensional friction force apparatus design for measuring shear forces at the nanometer scale, Rev. Sci. Instrum. **72**, 4171–4177 (2001)
49. E. Kumacheva: Interfacial friction measurements in surface force apparatus, Prog. Surf. Sci. **58**, 75–120 (1998)
50. F. Restagno, J. Crassous, E. Charlaix, C. Cottin-Bizonne, M. Monchanin: A new surface forces apparatus for nanorheology, Rev. Sci. Instrum. **73**, 2292–2297 (2002)
51. C. Cottin-Bizonne, B. Cross, A. Steinberger, E. Charlaix: Boundary slip on smooth hydrophobic surfaces: Intrinsic effects and possible artifacts, Phys. Rev. Lett. **94**, 056102/1–4 (2005)
52. A.L. Weisenhorn, P.K. Hansma, T.R. Albrecht, C.F. Quate: Forces in atomic force microscopy in air and water, Appl. Phys. Lett. **54**, 2651–2653 (1989)
53. G. Meyer, N.M. Amer: Simultaneous measurement of lateral and normal forces with an optical-beam-deflection atomic force microscope, Appl. Phys. Lett. **57**, 2089–2091 (1990)
54. E.L. Florin, V.T. Moy, H.E. Gaub: Adhesion forces between individual ligand–receptor pairs, Science **264**, 415–417 (1994)
55. G.U. Lee, D.A. Kidwell, R.J. Colton: Sensing discrete streptavidin–biotin interactions with atomic force microscopy, Langmuir **10**, 354–357 (1994)
56. H. Skulason, C.D. Frisbie: Detection of discrete interactions upon rupture of Au microcontacts to self-assembled monolayers terminated with $-S(CO)CH_3$ or $-SH$, J. Am. Chem. Soc. **122**, 9750–9760 (2000)
57. M. Carrion-Vazquez, A.F. Oberhauser, S.B. Fowler, P.E. Marszalek, S.E. Broedel, J. Clarke, J.M. Fernandez: Mechanical and chemical unfolding of a single protein: A comparison, Proc. Nat. Acad. Sci. USA **96**, 3694–3699 (1999)
58. C.M. Mate, G.M. McClelland, R. Erlandsson, S. Chiang: Atomic-scale friction of a tungsten tip on a graphite surface, Phys. Rev. Lett. **59**, 1942–1945 (1987)
59. R.W. Carpick, M. Salmeron: Scratching the surface: Fundamental investigations of tribology with atomic force microscopy, Chem. Rev. **97**, 1163–1194 (1997)
60. D. Leckband, J. Israelachvili: Intermolecular forces in biology, Quart. Rev. Biophys. **34**, 105–267 (2001)
61. J.P. Cleveland, S. Manne, D. Bocek, P.K. Hansma: A nondestructive method for determining the spring constant of cantilevers for scanning force microscopy, Rev. Sci. Instrum. **64**, 403–405 (1993)
62. J.E. Sader, J.W.M. Chon, P. Mulvaney: Calibration of rectangular atomic force microscope cantilevers, Rev. Sci. Instrum. **70**, 3967–3969 (1999)
63. Y. Liu, T. Wu, D.F. Evans: Lateral force microscopy study on the shear properties of self-assembled monolayers of dialkylammonium surfactant on mica, Langmuir **10**, 2241–2245 (1994)

64. A. Feiler, P. Attard, I. Larson: Calibration of the torsional spring constant and the lateral photodiode response of frictional force microscopes, Rev. Sci. Instrum. **71**, 2746–2750 (2000)
65. C. P. Green, H. Lioe, J. P. Cleveland, R. Proksch, P. Mulvaney, J. E. Sader: Normal and torsional spring constants of atomic force microscope cantilevers, Rev. Sci. Instrum. **75**, 1988–1996 (2004)
66. C. Neto, V. S. J. Craig: Colloid probe characterization: Radius and roughness determination, Langmuir **17**, 2097–2099 (2001)
67. G. M. Sacha, A. Verdaguer, J. Martinez, J. J. Saenz, D. F. Ogletree, M. Salmeron: Effective tip radius in electrostatic force microscopy, Appl. Phys. Lett. **86**, 123101/1–3 (2005)
68. R. G. Horn, J. N. Israelachvili: Molecular organization and viscosity of a thin film of molten polymer between two surfaces as probed by force measurements, Macromolecules **21**, 2836–2841 (1988)
69. R. G. Horn, S. J. Hirz, G. Hadziioannou, C. W. Frank, J. M. Catala: A reevaluation of forces measured across thin polymer films: Nonequilibrium and pinning effects, J. Chem. Phys. **90**, 6767–6774 (1989)
70. O. I. Vinogradova, H.-J. Butt, G. E. Yakubov, F. Feuillebois: Dynamic effects on force measurements. I. Viscous drag on the atomic force microscope cantilever, Rev. Sci. Instrum. **72**, 2330–2339 (2001)
71. V. S. J. Craig, C. Neto: In situ calibration of colloid probe cantilevers in force microscopy: Hydrodynamic drag on a sphere approaching a wall, Langmuir **17**, 6018–6022 (2001)
72. E. Evans, D. Needham: Physical properties of surfactant bilayer membranes: Thermal transitions, elasticity, rigidity, cohesion, and colloidal interactions, J. Phys. Chem. **91**, 4219–4228 (1987)
73. E. Evans, D. Needham: Attraction between lipid bilayer membranes in concentrated solutions of nonadsorbing polymers: Comparison of mean-field theory with measurements of adhesion energy, Macromolecules **21**, 1822–1831 (1988)
74. S. E. Chesla, P. Selvaraj, C. Zhu: Measuring two-dimensional receptor-ligand binding kinetics by micropipette, Biophys. J. **75**, 1553–1572 (1998)
75. E. Evans, K. Ritchie, R. Merkel: Sensitive force technique to probe molecular adhesion and structural linkages at biological interfaces, Biophys. J. **68**, 2580–2587 (1995)
76. D. M. LeNeveu, R. P. Rand, V. A. Parsegian: Measurements of forces between lecithin bilayers, Nature **259**, 601–603 (1976)
77. A. Homola, A. A. Robertson: A compression method for measuring forces between colloidal particles, J. Colloid Interface Sci. **54**, 286–297 (1976)
78. V. A. Parsegian, N. Fuller, R. P. Rand: Measured work of deformation and repulsion of lecithin bilayers, Proc. Nat. Acad. Sci. USA **76**, 2750–2754 (1979)
79. R. P. Rand, V. A. Parsegian: Hydration forces between phospholipid bilayers, Biochim. Biophys. Acta **988**, 351–376 (1989)
80. S. Leikin, V. A. Parsegian, D. C. Rau, R. P. Rand: Hydration forces, Annu. Rev. Phys. Chem. **44**, 369–395 (1993)
81. S. Chu, J. E. Bjorkholm, A. Ashkin, A. Cable: Experimental observation of optically trapped atoms, Phys. Rev. Lett. **57**, 314–317 (1986)
82. A. Ashkin: Optical trapping and manipulation of neutral particles using lasers, Proc. Nat. Acad. Sci. USA **94**, 4853–4860 (1997)
83. K. Visscher, S. P. Gross, S. M. Block: Construction of multiple-beam optical traps with nanometer-resolution positioning sensing, IEEE J. Sel. Top. Quantum Electron. **2**, 1066–1076 (1996)

84. D.C. Prieve, N.A. Frej: Total internal reflection microscopy: A quantitative tool for the measurement of colloidal forces, Langmuir **6**, 396–403 (1990)
85. J. Rädler, E. Sackmann: On the measurement of weak repulsive and frictional colloidal forces by reflection interference contrast microscopy, Langmuir **8**, 848–853 (1992)
86. P. Bongrand, C. Capo, J.-L. Mege, A.-M. Benoliel: Use of hydrodynamic flows to study cell adhesion. In: *Physical Basis of Cell–Cell Adhesion*, ed. by P. Bongrand (CRC, Boca Raton 1988) pp. 125–156
87. G. Kaplanski, C. Farnarier, O. Tissot, A. Pierres, A.-M. Benoliel, A.-C. Alessi, S. Kaplanski, P. Bongrand: Granulocyte–endothelium initial adhesion: Analysis of transient binding events mediated by E-selectin in a laminar shear flow, Biophys. J. **64**, 1922–1933 (1993)
88. H.B.G. Casimir, D. Polder: The influence of retardation on the London–van der Waals forces, Phys. Rev. **73**, 360–372 (1948)
89. H.C. Hamaker: The London–van der Waals attraction between spherical particles, Physica **4**, 1058–1072 (1937)
90. E.M. Lifshitz: The theory of molecular attraction forces between solid bodies, Sov. Phys. JETP (English translation) **2**, 73–83 (1956)
91. I.E. Dzyaloshinskii, E.M. Lifshitz, L.P. Pitaevskii: The general theory of van der Waals forces, Adv. Phys. **10**, 165–209 (1961)
92. H.W. Fox, W.A. Zisman: The spreading of liquids on low-energy surfaces. III. Hydrocarbon surfaces, J. Colloid Sci. **7**, 428–442 (1952)
93. B.V. Derjaguin, V.P. Smilga: Electrostatic component of the rolling friction force moment, Wear **7**, 270–281 (1964)
94. R.G. Horn, D.T. Smith: Contact electrification and adhesion between dissimilar materials, Science **256**, 362–364 (1992)
95. W.R. Harper: *Contact and frictional electrification* (Laplacian, Morgan Hill 1998)
96. J.A. Wiles, B.A. Grzybowski, A. Winkleman, G.M. Whitesides: A tool for studying contact electrification in systems comprising metals and insulating polymers, Anal. Chem. **75**, 4859–4867 (2003)
97. J. Lowell, A.C. Rose-Innes: Contact electrification, Adv. Phys. **29**, 947–1023 (1980)
98. C. Guerret-Piecourt, S. Bec, F. Seguault, D. Juve, D. Treheux, A. Tonck: Adhesion forces due to nano-triboelectrification between similar materials, Eur. Phys. J.: Appl. Phys. **28**, 65–72 (2004)
99. T. Miura, N. Hirokawa, K. Enokido, I. Arakawa: Spatially resolved spectroscopy of gas discharge during sliding friction between diamond and quartz in N_2 gas, Appl. Surf. Sci. **235**, 114–118 (2004)
100. T.E. Fischer: Tribochemistry, Annu. Rev. Mater. Sci. **18**, 303–323 (1988)
101. J. Lowell, W.S. Truscott: Triboelectrification of identical insulators. I. An experimental investigation, J. Phys. D: Appl. Phys. **19**, 1273–80 (1986)
102. L. Perez-Trejo, J. Perez-Gonzalez, L. de Vargas, E. Moreno: Triboelectrification of molten linear low-density polyethylene under continuous extrusion, Wear **257**, 329–337 (2004)
103. K. Ohara, T. Tonouchi, S. Uchiyama: Frictional electrification of thin films deposited by the Langmuir-Blodgett method, J. Phys. D: Appl. Phys. **23**, 1092–1096 (1990)
104. R. Budakian, S.J. Putterman: Correlation between charge transfer and stick–slip friction at a metal–insulator interface, Phys. Rev. Lett. **85**, 1000–1003 (2000)
105. J.V. Wasem, P. Upadhyaya, S.C. Langford, J.T. Dickinson: Transient current generation during wear of high-density polyethylene by a stainless steel stylus, J. Appl. Phys. **93**, 719–730 (2003)

106. M. Grätzel: Photoelectrochemical cells, Nature **414**, 338–344 (2001)
107. X. Xu, Y. Liu, R. M. German: Reconciliation of sintering theory with sintering practice, Adv. Powder Metallurgy Particulate Mater. **5**, 67–78 (2000)
108. R. M. German: *Sintering Theory and Practice* (Wiley, New York 1996)
109. R. Budakian, S. J. Putterman: Time scales for cold welding and the origins of stick–slip friction, Phys. Rev. B **65**, 235429/1–5 (2002)
110. D. H. Buckley: Influence of various physical properties of metals on their friction and wear behavior in vacuum, Metals Eng. Quart. **7**, 44–53 (1967)
111. U. Landman, W. D. Luedtke, N. A. Burnham, R. J. Colton: Atomistic mechanisms and dynamics of adhesion, nanoindentation, and fracture, Science **248**, 454–461 (1990)
112. U. Landman, W. D. Luedtke, E. M. Ringer: Molecular dynamics simulations of adhesive contact formation and friction, NATO Sci. Ser. E **220**, 463–510 (1992)
113. W. D. Luedtke, U. Landman: Solid and liquid junctions, Comput. Mater. Sci. **1**, 1–24 (1992)
114. U. Landman, W. D. Luedtke: Interfacial junctions and cavitation, MRS Bull. **18**, 36–44 (1993)
115. B. Bhushan, J. N. Israelachvili, U. Landman: Nanotribology: Friction, wear and lubrication at the atomic scale, Nature **374**, 607–616 (1995)
116. M. R. Sørensen, K. W. Jacobsen, P. Stoltze: Simulations of atomic-scale sliding friction, Phys. Rev. B **53**, 2101–2113 (1996)
117. A. Meurk, P. F. Luckham, L. Bergström: Direct measurement of repulsive and attractive van der Waals forces between inorganic materials, Langmuir **13**, 3896–3899 (1997)
118. S.-w. Lee, W. M. Sigmund: AFM study of repulsive van der Waals forces between Teflon AF thin film and silica or alumina, Colloids Surf. A **204**, 43–50 (2002)
119. E. J. W. Verwey, J. T. G. Overbeek: *Theory of the Stability of Lyophobic Colloids*, 1st edn. (Elsevier, Amsterdam 1948)
120. D. Y. C. Chan, R. M. Pashley, L. R. White: A simple algorithm for the calculation of the electrostatic repulsion between identical charged surfaces in electrolyte, J. Colloid Interface Sci. **77**, 283–285 (1980)
121. J. E. Sader, S. L. Carnie, D. Y. C. Chan: Accurate analytic formulas for the double-layer interaction between spheres, J. Colloid Interface Sci. **171**, 46–54 (1995)
122. D. Harries: Solving the Poisson–Boltzmann equation for two parallel cylinders, Langmuir **14**, 3149–3152 (1998)
123. P. Attard: Recent advances in the electric double layer in colloid science, Curr. Opin. Colloid Interface Sci. **6**, 366–371 (2001)
124. B. Derjaguin, L. Landau: Theory of the stability of strongly charged lyophobic sols and of the adhesion of strongly charged particles in solutions of electrolytes, Acta Physicochim. URSS **14**, 633–662 (1941)
125. D. Chan, T. W. Healy, L. R. White: Electrical double layer interactions under regulation by surface ionization equilibriums – dissimilar amphoteric surfaces, J. Chem. Soc. Faraday Trans. 1 **72**, 2844–2865 (1976)
126. G. S. Manning: Limiting laws and counterion condensation in polyelectrolyte solutions. I. Colligative properties, J. Chem. Phys. **51**, 924–933 (1969)
127. L. Guldbrand, V. Jönsson, H. Wennerström, P. Linse: Electrical double-layer forces: A Monte Carlo study, J. Chem. Phys. **80**, 2221–2228 (1984)
128. H. Wennerström, B. Jönsson, P. Linse: The cell model for polyelectrolyte systems. Exact statistical mechanical relations, Monte Carlo simulations, and the Poisson–Boltzmann approximation, J. Chem. Phys. **76**, 4665–4670 (1982)
129. J. Marra: Effects of counterion specificity on the interactions between quaternary ammonium surfactants in monolayers and bilayers, J. Phys. Chem. **90**, 2145–2150 (1986)

130. J. Marra: Direct measurement of the interaction between phosphatidylglycerol bilayers in aqueous-electrolyte solutions, Biophys. J. **50**, 815–825 (1986)
131. B. Jönsson, H. Wennerström: Ion–ion correlations in liquid dispersions, J. Adh. **80**, 339–364 (2004)
132. R.G. Horn, J.N. Israelachvili: Direct measurement of structural forces between two surfaces in a nonpolar liquid, J. Chem. Phys. **75**, 1400–1411 (1981)
133. H.K. Christenson, D.W.R. Gruen, R.G. Horn, J.N. Israelachvili: Structuring in liquid alkanes between solid surfaces: Force measurements and mean-field theory, J. Chem. Phys. **87**, 1834–1841 (1987)
134. M.L. Gee, J.N. Israelachvili: Interactions of surfactant monolayers across hydrocarbon liquids, J. Chem. Soc. Faraday Trans. **86**, 4049–4058 (1990)
135. J.N. Israelachvili, S.J. Kott, M.L. Gee, T.A. Witten: Forces between mica surfaces across hydrocarbon liquids: Effects of branching and polydispersity, Macromolecules **22**, 4247–4253 (1989)
136. I.K. Snook, W. van Megen: Solvation forces in simple dense fluids I, J. Chem. Phys. **72**, 2907–2913 (1980)
137. W.J. van Megen, I.K. Snook: Solvation forces in simple dense fluids II. Effect of chemical potential, J. Chem. Phys. **74**, 1409–1411 (1981)
138. R. Kjellander, S. Marcelja: Perturbation of hydrogen bonding in water near polar surfaces, Chem. Phys. Lett. **120**, 393–396 (1985)
139. P. Tarazona, L. Vicente: A model for the density oscillations in liquids between solid walls, Mol. Phys. **56**, 557–572 (1985)
140. D. Henderson, M. Lozada-Cassou: A simple theory for the force between spheres immersed in a fluid, J. Colloid Interface Sci. **114**, 180–183 (1986)
141. J.E. Curry, J.H. Cushman: Structure in confined fluids: Phase separation of binary simple liquid mixtures, Tribol. Lett. **4**, 129–136 (1998)
142. M. Schoen, T. Gruhn, D.J. Diestler: Solvation forces in thin films confined between macroscopically curved substrates, J. Chem. Phys. **109**, 301–311 (1998)
143. F. Porcheron, B. Rousseau, M. Schoen, A.H. Fuchs: Structure and solvation forces in confined alkane films, Phys. Chem. Chem. Phys. **3**, 1155–1159 (2001)
144. J. Gao, W.D. Luedtke, U. Landman: Layering transitions and dynamics of confined liquid films, Phys. Rev. Lett. **79**, 705–708 (1997)
145. H.K. Christenson: Forces between solid surfaces in a binary mixture of non-polar liquids, Chem. Phys. Lett. **118**, 455–458 (1985)
146. H.K. Christenson, R.G. Horn: Solvation forces measured in non-aqueous liquids, Chem. Scr. **25**, 37–41 (1985)
147. J. Israelachvili: Solvation forces and liquid structure, as probed by direct force measurements, Acc. Chem. Res. **20**, 415–421 (1987)
148. H.K. Christenson: Non-DLVO forces between surfaces – solvation, hydration and capillary effects, J. Disp. Sci. Technol. **9**, 171–206 (1988)
149. V. Franz, H.-J. Butt: Confined liquids: Solvation forces in liquid alcohols between solid surfaces, J. Phys. Chem. B **106**, 1703–1708 (2002)
150. L.J.D. Frink, F. van Swol: Solvation forces between rough surfaces, J. Chem. Phys. **108**, 5588–5598 (1998)
151. J. Gao, W.D. Luedtke, U. Landman: Structures, solvation forces and shear of molecular films in a rough nano-confinement, Tribol. Lett. **9**, 3–13 (2000)
152. H.E. Stanley, J. Teixeira: Interpretation of the unusual behavior of H_2O and D_2O at low temperatures: Tests of a percolation model, J. Chem. Phys. **73**, 3404–3422 (1980)
153. H. van Olphen: *An Introduction to Clay Colloid Chemistry*, 2nd edn. (Wiley, New York 1977) Chap. 10

154. N. Alcantar, J. Israelachvili, J. Boles: Forces and ionic transport between mica surfaces: Implications for pressure solution, Geochimica et Cosmochimica Acta **67**, 1289–1304 (2003)
155. U. Del Pennino, E. Mazzega, S. Valeri, A. Alietti, M. F. Brigatti, L. Poppi: Interlayer water and swelling properties of monoionic montmorillonites, J. Colloid Interface Sci. **84**, 301–309 (1981)
156. B. E. Viani, P. F. Low, C. B. Roth: Direct measurement of the relation between interlayer force and interlayer distance in the swelling of montmorillonite, J. Colloid Interface Sci. **96**, 229–244 (1983)
157. R. G. Horn: Surface forces and their action in ceramic materials, J. Am. Ceram. Soc. **73**, 1117–1135 (1990)
158. B. V. Velamakanni, J. C. Chang, F. F. Lange, D. S. Pearson: New method for efficient colloidal particle packing via modulation of repulsive lubricating hydration forces, Langmuir **6**, 1323–1325 (1990)
159. R. R. Lessard, S. A. Zieminski: Bubble coalescence and gas transfer in aqueous electrolytic solutions, Ind. Eng. Chem. Fundam. **10**, 260–269 (1971)
160. J. Israelachvili, R. Pashley: The hydrophobic interaction is long range, decaying exponentially with distance, Nature **300**, 341–342 (1982)
161. R. M. Pashley, P. M. McGuiggan, B. W. Ninham, D. F. Evans: Attractive forces between uncharged hydrophobic surfaces: Direct measurements in aqueous solutions, Science **229**, 1088–1089 (1985)
162. P. M. Claesson, C. E. Blom, P. C. Herder, B. W. Ninham: Interactions between water-stable hydrophobic Langmuir–Blodgett monolayers on mica, J. Colloid Interface Sci. **114**, 234–242 (1986)
163. Ya. I. Rabinovich, B. V. Derjaguin: Interaction of hydrophobized filaments in aqueous electrolyte solutions, Colloids Surf. **30**, 243–251 (1988)
164. J. L. Parker, D. L. Cho, P. M. Claesson: Plasma modification of mica: Forces between fluorocarbon surfaces in water and a nonpolar liquid, J. Phys. Chem. **93**, 6121–6125 (1989)
165. H. K. Christenson, J. Fang, B. W. Ninham, J. L. Parker: Effect of divalent electrolyte on the hydrophobic attraction, J. Phys. Chem. **94**, 8004–8006 (1990)
166. K. Kurihara, S. Kato, T. Kunitake: Very strong long range attractive forces between stable hydrophobic monolayers of a polymerized ammonium surfactant, Chem. Lett. **19**, 1555–1558 (1990)
167. Y. H. Tsao, D. F. Evans, H. Wennerstrom: Long-range attractive force between hydrophobic surfaces observed by atomic force microscopy, Science **262**, 547–550 (1993)
168. Ya. I. Rabinovich, R.-H. Yoon: Use of atomic force microscope for the measurements of hydrophobic forces between silanated silica plate and glass sphere, Langmuir **10**, 1903–1909 (1994)
169. V. S. J. Craig, B. W. Ninham, R. M. Pashley: Study of the long-range hydrophobic attraction in concentrated salt solutions and its implications for electrostatic models, Langmuir **14**, 3326–3332 (1998)
170. P. Kékicheff, O. Spalla: Long-range electrostatic attraction between similar, charge-neutral walls, Phys. Rev. Lett. **75**, 1851–1854 (1995)
171. H. K. Christenson, P. M. Claesson: Direct measurements of the force between hydrophobic surfaces in water, Adv. Colloid Interface Sci. **91**, 391–436 (2001)
172. P. Attard: Nanobubbles and the hydrophobic attraction, Adv. Colloid Interface Sci. **104**, 75–91 (2003)
173. J. L. Parker, P. M. Claesson, P. Attard: Bubbles, cavities, and the long-ranged attraction between hydrophobic surfaces, J. Phys. Chem. **98**, 8468–8480 (1994)

174. T. Ederth, B. Liedberg: Influence of wetting properties on the long-range "hydrophobic" interaction between self-assembled alkylthiolate monolayers, Langmuir **16**, 2177–2184 (2000)
175. S. Ohnishi, V. V. Yaminsky, H. K. Christenson: Measurements of the force between fluorocarbon monolayer surfaces in air and water, Langmuir **16**, 8360–8367 (2000)
176. Q. Lin, E. E. Meyer, M. Tadmor, J. N. Israelachvili, T. L. Kuhl: Measurement of the long- and short-range hydrophobic attraction between surfactant-coated surfaces, Langmuir **21**, 251–255 (2005)
177. E. E. Meyer, Q. Lin, J. N. Israelachvili: Effects of dissolved gas on the hydrophobic attraction between surfactant-coated surfaces, Langmuir **21**, 256–259 (2005)
178. E. E. Meyer, Q. Lin, T. Hassenkam, E. Oroudjev, J. Israelachvili: Origin of the long-range attraction between surfactant-coated surfaces, Proc. Nat. Acad. Sci. USA **102**, 6839–6842 (2005)
179. M. Hato: Attractive forces between surfaces of controlled "hydrophobicity" across water: A possible range of "hydrophobic interactions" between macroscopic hydrophobic surfaces across water, J. Phys. Chem. **100**, 18530–18538 (1996)
180. H. K. Christenson, V. V. Yaminsky: Is long-range hydrophobic attraction related to the mobility of hydrophobic surface groups?, Colloids Surf. A **129-130**, 67–74 (1997)
181. H. K. Christenson, P. M. Claesson, J. Berg, P. C. Herder: Forces between fluorocarbon surfactant monolayers: Salt effects on the hydrophobic interaction, J. Phys. Chem. **93**, 1472–1478 (1989)
182. N. I. Christou, J. S. Whitehouse, D. Nicholson, N. G. Parsonage: A Monte Carlo study of fluid water in contact with structureless walls, Faraday Symp. Chem. Soc. **16**, 139–149 (1981)
183. B. Jönsson: Monte Carlo simulations of liquid water between two rigid walls, Chem. Phys. Lett. **82**, 520–525 (1981)
184. S. Marcelja, D. J. Mitchell, B. W. Ninham, M. J. Sculley: Role of solvent structure in solution theory, J. Chem. Soc. Faraday Trans. II **73**, 630–648 (1977)
185. D. W. R. Gruen, S. Marcelja: Spatially varying polarization in water: A model for the electric double layer and the hydration force, J. Chem. Soc. Faraday Trans. 2 **79**, 225–242 (1983)
186. B. Jönsson, H. Wennerström: Image-charge forces in phospholipid bilayer systems, J. Chem. Soc. Faraday Trans. 2 **79**, 19–35 (1983)
187. D. Schiby, E. Ruckenstein: The role of the polarization layers in hydration forces, Chem. Phys. Lett. **95**, 435–438 (1983)
188. A. Luzar, D. Bratko, L. Blum: Monte Carlo simulation of hydrophobic interaction, J. Chem. Phys. **86**, 2955–2959 (1987)
189. P. Attard, M. T. Batchelor: A mechanism for the hydration force demonstrated in a model system, Chem. Phys. Lett. **149**, 206–211 (1988)
190. J. Forsman, C. E. Woodward, B. Jönsson: Repulsive hydration forces and attractive hydrophobic forces in a unified picture, J. Colloid Interface Sci. **195**, 264–266 (1997)
191. E. Ruckenstein, M. Manciu: The coupling between the hydration and double layer interactions, Langmuir **18**, 7584–7593 (2002)
192. K. Leung, A. Luzar: Dynamics of capillary evaporation. II. Free energy barriers, J. Chem. Phys. **113**, 5845–5852 (2000)
193. D. Bratko, R. A. Curtis, H. W. Blanch, J. M. Prausnitz: Interaction between hydrophobic surfaces with metastable intervening liquid, J. Chem. Phys. **115**, 3873–3877 (2001)
194. J. R. Grigera, S. G. Kalko, J. Fischbarg: Wall–water interface. A molecular dynamics study, Langmuir **12**, 154–158 (1996)

195. M. Mao, J. Zhang, R.-H. Yoon, W. A. Ducker: Is there a thin film of air at the interface between water and smooth hydrophobic solids?, Langmuir **20**, 1843–1849 (2004). Erratum: Langmuir **20** 4310 (2004)
196. Y. Zhu, S. Granick: Rate-dependent slip of Newtonian liquid at smooth surfaces, Phys. Rev. Lett. **87**, 096105/1–4 (2001)
197. D. Andrienko, B. Dunweg, O. I. Vinogradova: Boundary slip as a result of a prewetting transition, J. Chem. Phys. **119**, 13106–13112 (2003)
198. P. G. de Gennes: Polymers at an interface. 2. Interaction between two plates carrying adsorbed polymer layers, Macromolecules **15**, 492–500 (1982)
199. J. M. H. M. Scheutjens, G. J. Fleer: Interaction between two adsorbed polymer layers, Macromolecules **18**, 1882–1900 (1985)
200. E. A. Evans: Force between surfaces that confine a polymer solution: Derivation from self-consistent field theories, Macromolecules **22**, 2277–2286 (1989)
201. H. J. Ploehn: Compression of polymer interphases, Macromolecules **27**, 1627–1636 (1994)
202. J. Ennis, B. Jönsson: Interactions between surfaces in the presence of ideal adsorbing block copolymers, J. Phys. Chem. B **103**, 2248–2255 (1999)
203. H. J. Taunton, C. Toprakcioglu, L. J. Fetters, J. Klein: Interactions between surfaces bearing end-adsorbed chains in a good solvent, Macromolecules **23**, 571–580 (1990)
204. J. Klein, P. Luckham: Forces between two adsorbed poly(ethylene oxide) layers immersed in a good aqueous solvent, Nature **300**, 429–431 (1982)
205. J. Klein, P. F. Luckham: Long-range attractive forces between two mica surfaces in an aqueous polymer solution, Nature **308**, 836–837 (1984)
206. P. F. Luckham, J. Klein: Forces between mica surfaces bearing adsorbed homopolymers in good solvents, J. Chem. Soc. Faraday Trans. **86**, 1363–1368 (1990)
207. S. Asakura, F. Oosawa: Interaction between particles suspended in solutions of macromolecules, J. Polym. Sci. **33**, 183–192 (1958)
208. J. F. Joanny, L. Leibler, P. G. de Gennes: Effects of polymer solutions on colloid stability, J. Polym. Sci. Polym. Phys. **17**, 1073–1084 (1979)
209. B. Vincent, P. F. Luckham, F. A. Waite: The effect of free polymer on the stability of sterically stabilized dispersions, J. Colloid Interface Sci. **73**, 508–521 (1980)
210. R. I. Feigin, D. H. Napper: Stabilization of colloids by free polymer, J. Colloid Interface Sci. **74**, 567–571 (1980)
211. P. G. de Gennes: Polymer solutions near an interface. 1. Adsorption and depletion layers, Macromolecules **14**, 1637–1644 (1981)
212. G. J. Fleer, R. Tuinier: Concentration and solvency effects on polymer depletion and the resulting pair interaction of colloidal particles in a solution of non-adsorbing polymer, Polymer Prepr. **46**, 366 (2005)
213. P. G. de Gennes: Polymers at an interface; a simplified view, Adv. Colloid Interface Sci. **27**, 189–209 (1987)
214. S. Alexander: Adsorption of chain molecules with a polar head. A scaling description, J. Phys. (France) **38**, 983–987 (1977)
215. P. G. de Gennes: Conformations of polymers attached to an interface, Macromolecules **13**, 1069–1075 (1980)
216. S. T. Milner, T. A. Witten, M. E. Cates: Theory of the grafted polymer brush, Macromolecules **21**, 2610–2619 (1988)
217. S. T. Milner, T. A. Witten, M. E. Cates: Effects of polydispersity in the end-grafted polymer brush, Macromolecules **22**, 853–861 (1989)

218. E. B. Zhulina, O. V. Borisov, V. A. Priamitsyn: Theory of steric stabilization of colloid dispersions by grafted polymers, J. Colloid Interface Sci. **137**, 495–511 (1990)
219. J. Klein, E. Kumacheva, D. Mahalu, D. Perahia, L. J. Fetters: Reduction of frictional forces between solid surfaces bearing polymer brushes, Nature **370**, 634–636 (1994)
220. D. Goodman, J. N. Kizhakkedathu, D. E. Brooks: Evaluation of an atomic force microscopy pull-off method for measuring molecular weight and polydispersity of polymer brushes: Effect of grafting density, Langmuir **20**, 6238–6245 (2004)
221. T. J. McIntosh, A. D. Magid, S. A. Simon: Range of the solvation pressure between lipid membranes: Dependence on the packing density of solvent molecules, Biochemistry **28**, 7904–7912 (1989)
222. P. K. T. Persson, B. A. Bergenståhl: Repulsive forces in lecithin glycol lamellar phases, Biophys. J. **47**, 743–746 (1985)
223. J. N. Israelachvili, H. Wennerström: Hydration or steric forces between amphiphilic surfaces?, Langmuir **6**, 873–876 (1990)
224. J. N. Israelachvili, H. Wennerström: Entropic forces between amphiphilic surfaces in liquids, J. Phys. Chem. **96**, 520–531 (1992)
225. M. K. Granfeldt, S. J. Miklavic: A simulation study of flexible zwitterionic monolayers. Interlayer interaction and headgroup conformation, J. Phys. Chem. **95**, 6351–6360 (1991)
226. G. Pabst, J. Katsaras, V. A. Raghunathan: Enhancement of steric repulsion with temperature in oriented lipid multilayers, Phys. Rev. Lett. **88**, 128101/1–4 (2002)
227. L. R. Fisher, J. N. Israelachvili: Direct measurements of the effect of meniscus forces on adhesion: A study of the applicability of macroscopic thermodynamics to microscopic liquid interfaces, Colloids Surf. **3**, 303–319 (1981)
228. H. K. Christenson: Adhesion between surfaces in unsaturated vapors – a reexamination of the influence of meniscus curvature and surface forces, J. Colloid Interface Sci. **121**, 170–178 (1988)
229. H. Hertz: Über die Berührung fester elastischer Körper, J. Reine Angew. Math. **92**, 156–171 (1881)
230. H. M. Pollock, D. Maugis, M. Barquins: The force of adhesion between solid surfaces in contact, Appl. Phys. Lett. **33**, 798–799 (1978)
231. B. V. Derjaguin, V. M. Muller, Yu. P. Toporov: Effect of contact deformations on the adhesion of particles, J. Colloid Interface Sci. **53**, 314–326 (1975)
232. V. M. Muller, V. S. Yushchenko, B. V. Derjaguin: On the influence of molecular forces on the deformation of an elastic sphere and its sticking to a rigid plane, J. Colloid Interface Sci. **77**, 91–101 (1980)
233. V. M. Muller, B. V. Derjaguin, Y. P. Toporov: On 2 methods of calculation of the force of sticking of an elastic sphere to a rigid plane, Colloids Surf. **7**, 251–259 (1983)
234. D. Tabor: Surface forces and surface interactions, J. Colloid Interface Sci. **58**, 2–13 (1977)
235. D. Maugis: Adhesion of spheres: The JKR–DMT transition using a Dugdale model, J. Colloid Interface Sci. **150**, 243–269 (1992)
236. R. G. Horn, J. N. Israelachvili, F. Pribac: Measurement of the deformation and adhesion of solids in contact, J. Colloid Interface Sci. **115**, 480–492 (1987)
237. V. Mangipudi, M. Tirrell, A. V. Pocius: Direct measurement of molecular level adhesion between poly(ethylene terephthalate) and polyethylene films: Determination of surface and interfacial energies, J. Adh. Sci. Technol. **8**, 1251–1270 (1994)
238. H. K. Christenson: Surface deformations in direct force measurements, Langmuir **12**, 1404–1405 (1996)

239. M. Barquins, D. Maugis: Fracture mechanics and the adherence of viscoelastic bodies, J. Phys. D: Appl. Phys. **11**, 1989–2023 (1978)
240. I. Sridhar, K.L. Johnson, N.A. Fleck: Adhesion mechanics of the surface force apparatus, J. Phys. D: Appl. Phys. **30**, 1710–1719 (1997)
241. I. Sridhar, Z.W. Zheng, K.L. Johnson: A detailed analysis of adhesion mechanics between a compliant elastic coating and a spherical probe, J. Phys. D: Appl. Phys. **37**, 2886–2895 (2004)
242. P.M. McGuiggan, J. Israelachvili: Measurements of the effects of angular lattice mismatch on the adhesion energy between two mica surfaces in water, Mat. Res. Soc. Symp. Proc. **138**, 349–360 (1989)
243. J.A. Greenwood, J.B.P. Williamson: Contact of nominally flat surfaces, Proc. R. Soc. London A **295**, 300–319 (1966)
244. B.N.J. Persson: Elastoplastic contact between randomly rough surfaces, Phys. Rev. Lett. **87**, 116101/1–4 (2001)
245. F.P. Bowden, D. Tabor: *An Introduction to Tribology* (Anchor/Doubleday, Garden City 1973)
246. D. Maugis, H.M. Pollock: Surface forces, deformation and adherence at metal microcontacts, Acta Metallurgica **32**, 1323–1334 (1984)
247. K.N.G. Fuller, D. Tabor: The effect of surface roughness on the adhesion of elastic solids, Proc. R. Soc. London A **345**, 327–342 (1975)
248. D. Maugis: On the contact and adhesion of rough surfaces, J. Adh. Sci. Technol. **10**, 161–175 (1996)
249. B.N.J. Persson, E. Tosatti: The effect of surface roughness on the adhesion of elastic solids, J. Chem. Phys. **115**, 5597–5610 (2001)
250. H.-C. Kim, T.P. Russell: Contact of elastic solids with rough surfaces, J. Polym. Sci. Polym. Phys. **39**, 1848–1854 (2001)
251. P.M. McGuiggan, J.N. Israelachvili: Adhesion and short-range forces between surfaces. Part II: Effects of surface lattice mismatch, J. Mater. Res. **5**, 2232–2243 (1990)
252. C.L. Rhykerd, Jr., M. Schoen, D.J. Diestler, J.H. Cushman: Epitaxy in simple classical fluids in micropores and near-solid surfaces, Nature **330**, 461–463 (1987)
253. M. Schoen, D.J. Diestler, J.H. Cushman: Fluids in micropores. I. Structure of a simple classical fluid in a slit-pore, J. Chem. Phys. **87**, 5464–5476 (1987)
254. M. Schoen, C.L. Rhykerd, Jr., D.J. Diestler, J.H. Cushman: Shear forces in molecularly thin films, Science **245**, 1223–1225 (1989)
255. P.A. Thompson, M.O. Robbins: Origin of stick–slip motion in boundary lubrication, Science **250**, 792–794 (1990)
256. K.K. Han, J.H. Cushman, D.J. Diestler: Grand canonical Monte Carlo simulations of a Stockmayer fluid in a slit micropore, Mol. Phys. **79**, 537–545 (1993)
257. M. Ruths, S. Granick: Influence of alignment of crystalline confining surfaces on static forces and shear in a liquid crystal, 4′-n-pentyl-4-cyanobiphenyl (5CB), Langmuir **16**, 8368–8376 (2000)
258. A.D. Berman: Dynamics of molecules at surfaces. Ph.D. Thesis (Univ. of California, Santa Barbara 1996)
259. E. Barthel, S. Roux: Velocity dependent adherence: An analytical approach for the JKR and DMT models, Langmuir **16**, 8134–8138 (2000)
260. M. Ruths, S. Granick: Rate-dependent adhesion between polymer and surfactant monolayers on elastic substrates, Langmuir **14**, 1804–1814 (1998)
261. H. Yoshizawa, Y.L. Chen, J. Israelachvili: Fundamental mechanisms of interfacial friction. 1: Relation between adhesion and friction, J. Phys. Chem. **97**, 4128–4140 (1993)

262. N. Maeda, N. Chen, M. Tirrell, J.N. Israelachvili: Adhesion and friction mechanisms of polymer-on-polymer surfaces, Science **297**, 379–382 (2002)
263. F.P. Bowden, D. Tabor: *The Friction and Lubrication of Solids* (Clarendon, London 1971)
264. J.A. Greenwood, K.L. Johnson: The mechanics of adhesion of viscoelastic solids, Phil. Mag. A **43**, 697–711 (1981)
265. D. Maugis: Subcritical crack growth, surface energy, fracture toughness, stick–slip and embrittlement, J. Mater. Sci. **20**, 3041–3073 (1985)
266. F. Michel, M.E.R. Shanahan: Kinetics of the JKR experiment, C. R. Acad. Sci. II (Paris) **310**, 17–20 (1990)
267. C.A. Miller, P. Neogi: *Interfacial Phenomena: Equilibrium and Dynamic Effects* (Dekker, New York 1985)
268. J. Israelachvili, A. Berman: Irreversibility, energy dissipation, and time effects in intermolecular and surface interactions, Israel J. Chem. **35**, 85–91 (1995)
269. A.N. Gent, A.J. Kinloch: Adhesion of viscoelastic materials to rigid substrates. III. Energy criterion for failure, J. Polym. Sci. A-2 **9**, 659–668 (1971)
270. A.N. Gent: Adhesion and strength of viscoelastic solids. Is there a relationship between adhesion and bulk properties?, Langmuir **12**, 4492–4496 (1996)
271. H.R. Brown: The adhesion between polymers, Annu. Rev. Mater. Sci. **21**, 463–489 (1991)
272. M. Deruelle, M. Tirrell, Y. Marciano, H. Hervet, L. Léger: Adhesion energy between polymer networks and solid surfaces modified by polymer attachment, Faraday Discuss. **98**, 55–65 (1995)
273. S. Granick: Motions and relaxations of confined liquids, Science **253**, 1374–1379 (1991)
274. H.W. Hu, S. Granick: Viscoelastic dynamics of confined polymer melts, Science **258**, 1339–1342 (1992)
275. M.L. Gee, P.M. McGuiggan, J.N. Israelachvili, A.M. Homola: Liquid to solidlike transitions of molecularly thin films under shear, J. Chem. Phys. **93**, 1895–1906 (1990)
276. J. Van Alsten, S. Granick: The origin of static friction in ultrathin liquid films, Langmuir **6**, 876–880 (1990)
277. H.-W. Hu, G.A. Carson, S. Granick: Relaxation time of confined liquids under shear, Phys. Rev. Lett. **66**, 2758–2761 (1991)
278. J. Israelachvili, M. Gee, P. McGuiggan, P. Thompson, M. Robbins: Melting–freezing transitions in molecularly thin liquid films during shear, Fall Meeting of the Mater. Res. Soc., Boston, MA. 1990, ed. by J.M. Drake, J. Klafter, R. Kopelman (MRS Publications, Pittsburgh, PA 1990) 3–6
279. J. Israelachvili, P. McGuiggan, M. Gee, A. Homola, M. Robbins, P. Thompson: Liquid dynamics in molecularly thin films, J. Phys.: Condens. Matter **2**, SA89–SA98 (1990)
280. J. Klein, E. Kumacheva: Confinement-induced phase transitions in simple liquids, Science **269**, 816–819 (1995)
281. A.L. Demirel, S. Granick: Glasslike transition of a confined simple fluid, Phys. Rev. Lett. **77**, 2261–2264 (1996)
282. D. Dowson: *History of Tribology*, 2nd edn. (Professional Engineering Publishing, London 1998)
283. A. Berman, C. Drummond, J. Israelachvili: Amontons' law at the molecular level, Tribol. Lett. **4**, 95–101 (1998)
284. M. Ruths: Boundary friction of aromatic self-assembled monolayers: Comparison of systems with one or both sliding surfaces covered with a thiol monolayer, Langmuir **19**, 6788–6795 (2003)

285. D. Gourdon, J.N. Israelachvili: Transitions between smooth and complex stick-slip sliding of surfaces, Phys. Rev. E **68**, 021602/1–10 (2003)
286. M. Ruths, N.A. Alcantar, J.N. Israelachvili: Boundary friction of aromatic silane self-assembled monolayers measured with the surface forces apparatus and friction force microscopy, J. Phys. Chem. B **107**, 11149–11157 (2003)
287. J. Gao, W.D. Luedtke, D. Gourdon, M. Ruths, J.N. Israelachvili, U. Landman: Frictional forces and Amontons' law: From the molecular to the macroscopic scale, J. Phys. Chem. B **108**, 3410–3425 (2004)
288. D. Tabor: The role of surface and intermolecular forces in thin film lubrication, Tribol. Ser. **7**, 651–682 (1982)
289. M.J. Sutcliffe, S.R. Taylor, A. Cameron: Molecular asperity theory of boundary friction, Wear **51**, 181–192 (1978)
290. G.M. McClelland: Friction at weakly interacting interfaces. In: *Adhesion and Friction*, ed. by M. Grunze, H.J. Kreuzer (Springer, Berlin, Heidelberg 1989) pp. 1–16
291. D.H. Buckley: The metal-to-metal interface and its effect on adhesion and friction, J. Colloid Interface Sci. **58**, 36–53 (1977)
292. J.N. Israelachvili, Y.-L. Chen, H. Yoshizawa: Relationship between adhesion and friction forces, J. Adh. Sci. Technol. **8**, 1231–1249 (1994)
293. J.-C. Géminard, W. Losert, J.P. Gollub: Frictional mechanics of wet granular material, Phys. Rev. E **59**, 5881–5890 (1999)
294. R.G. Cain, N.W. Page, S. Biggs: Microscopic and macroscopic aspects of stick–slip motion in granular shear, Phys. Rev. E **64**, 016413/1–8 (2001)
295. V. Zaloj, M. Urbakh, J. Klafter: Modifying friction by manipulating normal response to lateral motion, Phys. Rev. Lett. **82**, 4823–4826 (1999)
296. A. Dhinojwala, S.C. Bae, S. Granick: Shear-induced dilation of confined liquid films, Tribol. Lett. **9**, 55–62 (2000)
297. L.M. Qian, G. Luengo, E. Perez: Thermally activated lubrication with alkanes: The effect of chain length, Europhys. Lett. **61**, 268–274 (2003)
298. B.J. Briscoe, D.C.B. Evans, D. Tabor: The influence of contact pressure and saponification on the sliding behavior of steric acid monolayers, J. Colloid Interface Sci. **61**, 9–13 (1977)
299. M. Urbakh, L. Daikhin, J. Klafter: Dynamics of confined liquids under shear, Phys. Rev. E **51**, 2137–2141 (1995)
300. M.G. Rozman, M. Urbakh, J. Klafter: Origin of stick–slip motion in a driven two-wave potential, Phys. Rev. E **54**, 6485–6494 (1996)
301. M.G. Rozman, M. Urbakh, J. Klafter: Stick–slip dynamics as a probe of frictional forces, Europhys. Lett. **39**, 183–188 (1997)
302. M. Urbakh, J. Klafter, D. Gourdon, J. Israelachvili: The nonlinear nature of friction, Nature **430**, 525–528 (2004)
303. B.V. Derjaguin: Molekulartheorie der äußeren Reibung, Z. Physik **88**, 661–675 (1934)
304. B.V. Derjaguin: Mechanical properties of the boundary lubrication layer, Wear **128**, 19–27 (1988)
305. B.J. Briscoe, D.C.B. Evans: The shear properties of Langmuir–Blodgett layers, Proc. R. Soc. London A **380**, 389–407 (1982)
306. P.F. Luckham, S. Manimaaran: Investigating adsorbed polymer layer behaviour using dynamic surface force apparatuses – a review, Adv. Colloid Interface Sci. **73**, 1–46 (1997)
307. A. Berman, J. Israelachvili: Control and minimization of friction via surface modification, NATO ASI Ser. E: Appl. Sci. **330**, 317–329 (1997)

308. A.M. Homola, J.N. Israelachvili, P.M. McGuiggan, M.L. Gee: Fundamental experimental studies in tribology: The transition from "interfacial" friction of undamaged molecularly smooth surfaces to "normal" friction with wear, Wear **136**, 65–83 (1990)
309. S. Yamada, J. Israelachvili: Friction and adhesion hysteresis of fluorocarbon surfactant monolayer-coated surfaces measured with the surface forces apparatus, J. Phys. Chem. B **102**, 234–244 (1998)
310. G. Luengo, S.E. Campbell, V.I. Srdanov, F. Wudl, J.N. Israelachvili: Direct measurement of the adhesion and friction of smooth C_{60} surfaces, Chem. Mater. **9**, 1166–1171 (1997)
311. L. Rapoport, Y. Bilik, Y. Feldman, M. Homyonfer, S.R. Cohen, R. Tenne: Hollow nanoparticles of WS_2 as potential solid-state lubricants, Nature **387**, 791–793 (1997)
312. J. Israelachvili, Y.-L. Chen, H. Yoshizawa: Relationship between adhesion and friction forces. In: *Fundamentals of Adhesion and Interfaces*, ed. by D.S. Rimai, L.P. DeMejo, K.L. Mittal (VSP, Utrecht, The Netherlands 1995) pp. 261–279
313. U. Raviv, S. Perkin, P. Laurat, J. Klein: Fluidity of water confined down to subnanometer films, Langmuir **20**, 5322–5332 (2004)
314. P. Berthoud, T. Baumberger, C. G'Sell, J.M. Hiver: Physical analysis of the state- and rate-dependent friction law: Static friction, Phys. Rev. B **59**, 14313–14327 (1999)
315. G. He, M. Müser, M. Robbins: Adsorbed layers and the origin of static friction, Science **284**, 1650–1652 (1999)
316. M.H. Müser, L. Wenning, M.O. Robbins: Simple microscopic theory of Amontons's laws for static friction, Phys. Rev. Lett. **86**, 1295–1298 (2001)
317. B. Luan, M.O. Robbins: The breakdown of continuum models for mechanical contacts, Nature **435**, 929–932 (2005)
318. A. Berman, S. Steinberg, S. Campbell, A. Ulman, J. Israelachvili: Controlled microtribology of a metal oxide surface, Tribol. Lett. **4**, 43–48 (1998)
319. D.Y.C. Chan, R.G. Horn: The drainage of thin liquid films between solid surfaces, J. Chem. Phys. **83**, 5311–5324 (1985)
320. J.N. Israelachvili, S.J. Kott: Shear properties and structure of simple liquids in molecularly thin films: The transition from bulk (continuum) to molecular behavior with decreasing film thickness, J. Colloid Interface Sci. **129**, 461–467 (1989)
321. T.L. Kuhl, A.D. Berman, S.W. Hui, J.N. Israelachvili: Part 1: Direct measurement of depletion attraction and thin film viscosity between lipid bilayers in aqueous polyethylene glycol solutions, Macromolecules **31**, 8250–8257 (1998)
322. S.E. Campbell, G. Luengo, V.I. Srdanov, F. Wudl, J.N. Israelachvili: Very low viscosity at the solid–liquid interface induced by adsorbed C_{60} monolayers, Nature **382**, 520–522 (1996)
323. J. Klein, Y. Kamiyama, H. Yoshizawa, J.N. Israelachvili, G.H. Fredrickson, P. Pincus, L.J. Fetters: Lubrication forces between surfaces bearing polymer brushes, Macromolecules **26**, 5552–5560 (1993)
324. E. Kumacheva, J. Klein: Simple liquids confined to molecularly thin layers. II. Shear and frictional behavior of solidified films, J. Chem. Phys. **108**, 7010–7022 (1998)
325. G. Luengo, J. Israelachvili, A. Dhinojwala, S. Granick: Generalized effects in confined fluids: New friction map for boundary lubrication, Wear **200**, 328–335 (1996) Erratum. Wear **205** (1997) 246
326. P.A. Thompson, G.S. Grest, M.O. Robbins: Phase transitions and universal dynamics in confined films, Phys. Rev. Lett. **68**, 3448–3451 (1992)
327. Y. Rabin, I. Hersht: Thin liquid layers in shear: Non-Newtonian effects, Physica A **200**, 708–712 (1993)

328. P.A. Thompson, M.O. Robbins, G.S. Grest: Structure and shear response in nanometer-thick films, Israel J. Chem. **35**, 93–106 (1995)
329. M. Urbakh, L. Daikhin, J. Klafter: Sheared liquids in the nanoscale range, J. Chem. Phys. **103**, 10707–10713 (1995)
330. A. Subbotin, A. Semenov, E. Manias, G. Hadziioannou, G. ten Brinke: Rheology of confined polymer melts under shear flow: Strong adsorption limit, Macromolecules **28**, 1511–1515 (1995)
331. A. Subbotin, A. Semenov, E. Manias, G. Hadziioannou, G. ten Brinke: Rheology of confined polymer melts under shear flow: Weak adsorption limit, Macromolecules **28**, 3901–3903 (1995)
332. J. Huh, A. Balazs: Behavior of confined telechelic chains under shear, J. Chem. Phys. **113**, 2025–2031 (2000)
333. H. Xie, K. Song, D.J. Mann, W.L. Hase: Temperature gradients and frictional energy dissipation in the sliding of hydroxylated α-alumina surfaces, Phys. Chem. Chem. Phys. **4**, 5377–5385 (2002)
334. U. Landman, W.D. Luedtke, A. Nitzan: Dynamics of tip-substrate interactions in atomic force microscopy, Surf. Sci. **210**, L177–L184 (1989)
335. J.N. Israelachvili, P.M. McGuiggan, A.M. Homola: Dynamic properties of molecularly thin liquid films, Science **240**, 189–191 (1988)
336. A.M. Homola, H.V. Nguyen, G. Hadziioannou: Influence of monomer architecture on the shear properties of molecularly thin polymer melts, J. Chem. Phys. **94**, 2346–2351 (1991)
337. M. Schoen, S. Hess, D.J. Diestler: Rheological properties of confined thin films, Phys. Rev. E **52**, 2587–2602 (1995)
338. M. Heuberger, C. Drummond, J. Israelachvili: Coupling of normal and transverse motions during frictional sliding, J. Phys. Chem. B **102**, 5038–5041 (1998)
339. G.M. McClelland, S.R. Cohen: *Chemistry and Physics of Solid Surfaces VIII* (Springer, Berlin, Heidelberg 1990) pp. 419–445
340. E. Gnecco, R. Bennewitz, T. Gyalog, E. Meyer: Friction experiments on the nanometre scale, J. Phys. Cond. Matter **13**, R619–R642 (2001)
341. J. Gao, W.D. Luedtke, U. Landman: Friction control in thin-film lubrication, J. Phys. Chem. B **102**, 5033–5037 (1998)
342. C. Drummond, J. Elezgaray, P. Richetti: Behavior of adhesive boundary lubricated surfaces under shear: A new dynamic transition, Europhys. Lett. **58**, 503–509 (2002)
343. A.E. Filippov, J. Klafter, M. Urbakh: Inverted stick–slip friction: What is the mechanism?, J. Chem. Phys. **116**, 6871–6874 (2002)
344. C. Drummond, J. Israelachvili: Dynamic behavior of confined branched hydrocarbon lubricant fluids under shear, Macromolecules **33**, 4910–4920 (2000)
345. C. Drummond, J. Israelachvili: Dynamic phase transitions in confined lubricant fluids under shear, Phys. Rev. E **63**, 041506/1–11 (2001)
346. J.D. Ferry: *Viscoelastic Properties of Polymers*, 3rd edn. (Wiley, New York 1980)
347. A. Ruina: Slip instability and state variable friction laws, J. Geophys. Res. **88**, 10359–10370 (1983)
348. T. Baumberger, P. Berthoud, C. Caroli: Physical analysis of the state- and rate-dependent friction law. II. Dynamic friction, Phys. Rev. B **60**, 3928–3939 (1999)
349. J. Israelachvili, S. Giasson, T. Kuhl, C. Drummond, A. Berman, G. Luengo, J.-M. Pan, M. Heuberger, W. Ducker, N. Alcantar: Some fundamental differences in the adhesion and friction of rough versus smooth surfaces, Tribol. Ser. **38**, 3–12 (2000)
350. E. Rabinowicz: *Friction and Wear of Materials*, 2nd edn. (Wiley, New York 1995) Chap. 4

351. J. Peachey, J. Van Alsten, S. Granick: Design of an apparatus to measure the shear response of ultrathin liquid films, Rev. Sci. Instrum. **62**, 463–473 (1991)
352. E. Rabinowicz: The intrinsic variables affecting the stick–slip process, Proc. Phys. Soc. **71**, 668–675 (1958)
353. T. Baumberger, F. Heslot, B. Perrin: Crossover from creep to inertial motion in friction dynamics, Nature **367**, 544–546 (1994)
354. F. Heslot, T. Baumberger, B. Perrin, B. Caroli, C. Caroli: Creep, stick–slip, and dry-friction dynamics: Experiments and a heuristic model, Phys. Rev. E **49**, 4973–4988 (1994)
355. J. Sampson, F. Morgan, D. Reed, M. Muskat: Friction behavior during the slip portion of the stick–slip process, J. Appl. Phys. **14**, 689–700 (1943)
356. F. Heymann, E. Rabinowicz, B. Rightmire: Friction apparatus for very low-speed sliding studies, Rev. Sci. Instrum. **26**, 56–58 (1954)
357. J.H. Dieterich: Time-dependent friction and the mechanics of stick–slip, Pure Appl. Geophys. **116**, 790–806 (1978)
358. J.H. Dieterich: Modeling of rock friction. 1. Experimental results and constitutive equations, J. Geophys. Res. **84**, 2162–2168 (1979)
359. G.A. Tomlinson: A molecular theory of friction, Phil. Mag. **7**, 905–939 (1929)
360. J.M. Carlson, J.S. Langer: Mechanical model of an earthquake fault, Phys. Rev. A **40**, 6470–6484 (1989)
361. B.N.J. Persson: Theory of friction: The role of elasticity in boundary lubrication, Phys. Rev. B **50**, 4771–4786 (1994)
362. H. Yoshizawa, J. Israelachvili: Fundamental mechanisms of interfacial friction. 2: Stick–slip friction of spherical and chain molecules, J. Phys. Chem. **97**, 11300–11313 (1993)
363. S. Nasuno, A. Kudrolli, J.P. Gollub: Friction in granular layers: Hysteresis and precursors, Phys. Rev. Lett. **79**, 949–952 (1997)
364. M.O. Robbins, P.A. Thompson: Critical velocity of stick–slip motion, Science **253**, 916 (1991)
365. P. Bordarier, M. Schoen, A. Fuchs: Stick–slip phase transitions in confined solidlike films from an equilibrium perspective, Phys. Rev. E **57**, 1621–1635 (1998)
366. J.M. Carlson, A.A. Batista: Constitutive relation for the friction between lubricated surfaces, Phys. Rev. E **53**, 4153–4165 (1996)
367. A.D. Berman, W.A. Ducker, J.N. Israelachvili: Origin and characterization of different stick–slip friction mechanisms, Langmuir **12**, 4559–4563 (1996)
368. A.D. Berman, W.A. Ducker, J.N. Israelachvili: Experimental and theoretical investigations of stick–slip friction mechanisms, NATO ASI Ser. E: Appl. Sci. **311**, 51–67 (1996)
369. K.G. McLaren, D. Tabor: Viscoelastic properties and the friction of solids. Friction of polymers and influence of speed and temperature, Nature **197**, 856–858 (1963)
370. K.A. Grosch: Viscoelastic properties and friction of solids. Relation between friction and viscoelastic properties of rubber, Nature **197**, 858–859 (1963)
371. L. Bureau, T. Baumberger, C. Caroli: Shear response of a frictional interface to a normal load modulation, Phys. Rev. E **62**, 6810–6820 (2000)
372. J.P. Gao, W.D. Luedtke, U. Landman: Structure and solvation forces in confined films: Linear and branched alkanes, J. Chem. Phys. **106**, 4309–4318 (1997)
373. J. Warnock, D.D. Awschalom, M.W. Shafer: Orientational behavior of molecular liquids in restricted geometries, Phys. Rev. B **34**, 475–478 (1986)
374. D.D. Awschalom, J. Warnock: Supercooled liquids and solids in porous glass, Phys. Rev. B **35**, 6779–6785 (1987)

375. M. Hirano, K. Shinjo, R. Kaneko, Y. Murata: Anisotropy of frictional forces in muscovite mica, Phys. Rev. Lett. **67**, 2642–2645 (1991)
376. T. Gyalog, H. Thomas: Friction between atomically flat surfaces, Europhys. Lett. **37**, 195–200 (1997)
377. B.N.J. Persson, F. Bucher, B. Chiaia: Elastic contact between randomly rough surfaces: comparison of theory with numerical results, Phys. Rev. B **65**, 184106/1–7 (2002)
378. S. Hyun, L. Pei, J.-F. Molinari, M.O. Robbins: Finite-element analysis of contact between elastic self-affine surfaces, Phys. Rev. E **70**, 026117/1–12 (2004)
379. J. Israelachvili, N. Maeda, K.J. Rosenberg, M. Akbulut: Effects of sub-ångstrom (picoscale) structure of surfaces on adhesion, friction and bulk mechanical properties, J. Mater. Res. **20**, 1952–1972 (2005)
380. T.R. Thomas: *Rough Surfaces*, 2nd edn. (Imperial College Press, London 1999)

10

Interfacial Forces and Spectroscopic Study of Confined Fluids

Y. Elaine Zhu, Ashis Mukhopadhyay, and Steve Granick

Summary. In this chapter we discuss three specific issues which are relevant for liquids in intimate contact with solid surfaces. (1) Studies of the hydrodynamic flow of simple and complex fluids within ultra-narrow channels show the effects of flow rate, surface roughness and fluid–surface interaction on the determination of the boundary condition. We draw attention to the importance of the microscopic particulars to the discovery of what boundary condition is appropriate for solving continuum equations and the potential to capitalize on *slip at the wall* for purposes of materials engineering. (2) We address the long-standing question of the structure of aqueous films near a hydrophobic surface. When water was confined between adjoining hydrophobic and hydrophilic surfaces (a Janus interface), giant fluctuations in shear responses were observed, which implies some kind of flickering, fluctuating complex at the water–hydrophobic interface. (3) Finally we discuss recent experiments that augment friction studies by measurement of diffusion, using fluorescence correlation spectroscopy (FCS). Here spatially resolved measurements showed that translation diffusion slows exponentially with increasing mechanical pressure from the edges of a Hertzian contact toward the center, accompanied by increasingly heterogeneous dynamical responses. This dynamical probe of how liquids order in molecularly thin films fails to support the hypothesis that shear produces a melting transition.

10.1 Introduction

Confinement of fluids, ranging from water, hydrocarbon oil, polymer melts and solutions, to DNA, protein and other bio-macromolecules, can strongly affect the structures and dynamics of the fluid molecules, particularly when the thickness of fluid films becomes comparable to the length scale of the molecular dimension. Phase transitions, such as solidification or glass transition, can be induced, and confinement can not only result in structural changes in terms of molecular orientation and packing, but changes of dynamical processes, such as molecular translational as well as rotational diffusion. These questions about confined fluid structure and dynamics involve deep scientific puzzles. On the applied side, they also pertain directly to the understanding of many fundamental issues in nature and technology, including various applications in adhesion, colloidal stabilization, lubrication, micro/nanofluidics, and micro/nano-electromechanical systems (MEMS/NEMS) devices, among others.

The techniques that have been used to study confined fluids are mainly surface forces apparatus (SFA) [1–4] and recently atomic force microscopy (AFM) and its derivatives [5, 6]. SFA, which was originally designed to directly measure the van der Waals and Derjaguin–Landau–Verwey–Overbeek (DLVO) forces in simple liquids and in colloidal systems, was later modified by several groups to study the frictional forces of molecularly thin films confined between rigid surfaces [2–4]. AFM is an alternative method to study surface force, friction, adhesion and lubrication of thin film. With these two techniques many new phenomena about confined fluids, such as layering structure, solidification, stick–slip motion, etc. have been experimentally observed and interpreted.

The previous Chap. 9 reviewed the advances in the studies of surface forces, adhesion and shear interaction of confined liquids. In this chapter, we review recent studies on some specific interfacial forces of thin liquid films by using modified SFA devices and the results from an integrated experimental platform, which enables direct fluorescence-based observation of the dynamics of individual molecules under confinement.

This chapter is organized as follows. This introductory section is followed by a discussion of the validity of the no-slip boundary for fluid molecules moving past a solid surface, determined by measuring the hydrodynamic force in flow geometry. A specific interfacial force – the long-range hydrophobic interaction – and its related hydrophobic effect on the behavior of water molecules near a hydrophobic interface are discussed in Sect. 10.3. In Sects. 10.4–10.6, recent developments in combining ultrafast spectroscopy with the SFA to investigate the dynamic behavior of individual confined molecules, instead of their ensemble-averaged behavior, are reviewed.

10.2 Hydrodynamic Force of Fluids Flowing in Micro- to Nanofluidics: A Question About No-Slip Boundary Condition

Viscous flow is familiar and useful, yet the underlying physics is surprisingly subtle and complex. A tenet of textbook continuum fluid dynamics is the *no-slip* boundary condition, which means that fluid molecules immediately adjacent to the surface of a solid move with exactly the same velocity as that solid. It stands at the bedrock of our understanding of the flow of simple low-viscosity fluids and comprises a springboard for much sophisticated calculation. While it is true that at some level of detail this continuum description must fail and demand description at the molecular level, it is tremendously successful as the basis for continuum-based calculations. This section is adapted from discussions in some previous primary accounts [7–9].

As expressed in a prominent fluid dynamics textbook [10]:

"In other words there is no relative motion between the fluid and the solid. This fact may seem surprising but it is undoubtedly true. No matter how smooth the solid

surface or how small the viscosity of the fluid, the particles immediately adjacent to the surface do not move relative to it. It is perhaps not without interest that Newton's term for viscosity was 'defectus lubricitatis' – 'lack of slipperiness'. Even for a fluid that does not 'wet' the surface this rule is not violated."

Feynman noted in his lectures that the no-slip boundary condition explains why large particles are easy to remove by blowing past a surface, but small particles are not.

However, for many years it was observed that the *no-slip* boundary condition is not intuitively obviously and has been controversial for centuries. A pantheon of great scientists – among them, Bernoulli, Coulomb, Navier, Poisson, Stokes, Taylor, Debye, de Gennes – has worried about it. The compelling rational arguments – for and against – were divided by the pragmatic observation that predictions agree with experiments. The possibility of slip was discussed until recently, in mainstream literature, only in the context of the flow of polymer melts [11, 12], though over the years persistent doubts were expressed [13]. *No-slip* contrasts with *slip* characteristic of highly viscous polymers [11, 12], monolayers of gas condensed on vibrated solids [14], superfluid helium [15], moving contact lines of liquid droplets on solids [16], and kinetic friction of liquid films less than 5–10 molecular dimensions thick [17,18]. However, experimental capability and technical needs have changed – especially so with the emerging interest in microfluidics and microelectromechanical system (MEMS)-based devices.

The situation has changed but the enormous and enduring success of the no-slip assumption for modeling must be emphasized. It works beautifully provided that certain assumptions are met: a single-component fluid, a wetted surface, and low levels of shear stress. Then careful experiments imply that the fluid comes to rest within 1–2 molecular diameters of the surface [19–21]. But the necessary assumptions are more restrictive, and their applicability is more susceptible to intentional control than is widely appreciated. Generally, one may argue from the fact that fluid molecules are ("obviously") stuck to walls by intermolecular forces. The traditional explanation is that, since most surfaces are rough, the viscous dissipation as fluid flows past surface irregularities brings it to rest, regardless of how weakly or strongly molecules are attracted to the surface [7, 8, 22–24]. This has been challenged by accumulating evidence that, if molecularly smooth surfaces are wet only partially, hydrodynamic models work better when one uses instead *partial slip* boundary conditions [13, 21, 25–30]. Then the main issue would be whether the fluid molecules attract the surface or the fluid more strongly [9, 13, 27–34].

10.2.1 How to Quantify the Amount of Slip

To know what boundary condition is appropriate in solving continuum equations requires inquiry into microscopic particulars. Attention is drawn to unresolved topics of investigation and to the potential to purposively capitalize on *slip at the wall* for purposes of materials engineering.

The formal idea of a *slip length* is common. *Slip* signifies that, in the continuum model of flow, the fluid velocity at the surface is finite (the slip velocity, v_s) and

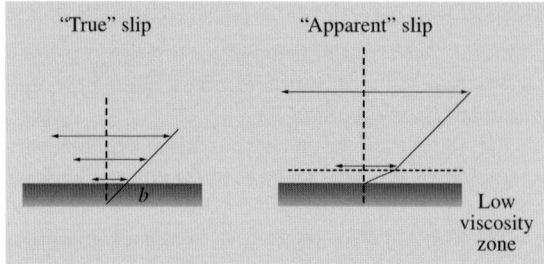

Fig. 10.1. Schematic illustration of the distinction between true slip (*left*) and apparent slip (*right*) in oscillatory flow, although for both cases the velocity of the moving fluid extrapolates to zero at a notional distance inside the wall and is finite where it crosses the wall. For true slip, this is literally so. It may also happen that a low-viscosity component in the fluid facilitates flow because it segregates near the surface. The velocity gradient is then larger nearest the surface because the viscosity is smaller. When specific real systems are investigated, structural and chemical materials analysis at the microscopic scale are needed to distinguish between these possibilities. After [7] with permission

increases linearly with distance normal to it. The slip velocity is assumed to be proportional to the shear stress at the surface, σ_s in the relation:

$$\eta v_s = b\sigma_s, \tag{10.1}$$

where η is the viscosity and b, the slip length, is the notional distance inside the surface at which the velocity equals zero, if the velocity profile is extrapolated inside the surface until it reaches zero. In much of the literature the slip length has been assumed to be a constant that characterizes the material response of a given fluid–surface pair, but evidence of additional dependence on velocity is discussed below.

It is unreasonable to expect this continuum description to yield microscopic information. One example of this was already given – the appearance of no-slip conditions when the microscopic reason is that flow irregularities pin the fluid to the wall [7, 8, 22–24]. Another example is apparent slip when a low-viscosity component in the fluid facilitates flow because it segregates near the surface [35, 36]. When conventional continuum mechanics contends with situations that are more complex than the model allows, one should resist the temptation to interpret literally the parameters in which the continuum mechanics model is couched. These examples emphasize instances where the notions of *stick* and *slip* are numerical conveniences not to be interpreted literally in terms of molecular mechanism. The appearance of slip owing to surface segregation of a low-viscosity component is illustrated in Fig. 10.1.

10.2.2 The Mechanisms that Control Slip in Low-Viscosity Fluids

Partial slip of so-called Newtonian fluids such as alkanes and water is predicted by an increasing number of computer simulations [25–27, 37–39] and, in the laboratory when forces are measured, systematic deviations from the predictions based on *stick* are found [7–9, 28–34, 40, 41]. Some sense of urgency comes from potential practical applications. Typical magnitudes of the slip length reported in the literature are submicron, so small that the practical consequences of slip would be minimal for flow in channels whose size is macroscopic. But if the channel size is very small, there are potential ramifications for microfluidics.

The simulations must be believed because they are buttressed by direct measurements. In the past, all laboratory reports of slip were based on comparing mechanical measurements of force to fluid mechanics models and hence were indirect inferences. Recently, optical methods were introduced to measure fluid velocity directly. For example, *Léger* and coworkers photobleached tracer fluorescent dyes and from the time rate of fluorescence recovery, measured in attenuated total reflection in order to focus on the region within an optical wavelength of the surface, the velocity of flow near the surface was inferred [28]. They reported slip of hexadecane near an oleophobic surface provided that it was smooth, but not when it was rough. *Tretheway* and *Meinhart* used laser particle image velocimetry of tracer latex particles to infer the velocity of water flow in microchannels [34]. They reported slip when the surface was hydrophobic but stick when it was hydrophilic. *Callahan* and coworkers demonstrated the feasibility of using nuclear magnetic resonance (NMR) velocity imaging, though this method has been used to date only for multicomponent fluids [42].

An important hint about the mechanism comes from the repeated observation that the amount of slip depends on the flow rate, in measurements using not only the surface forces apparatus (SFA) [8, 9, 31, 32] but also atomic force microscopy (AFM) [29, 41]. The main idea of all of these experiments is that two solids of mean radius of curvature R, at spacing D, experience a hydrodynamic force, F_H as they approach one another (or retreat from one another) in a liquid medium, thereby squeezing fluid out of (or into) the intervening gap. This force F_H is proportional to the rate at which the spacing changes, dD/dt (where t denotes time), is proportional to the viscosity, η (assumed to be constant), and is inversely proportional to D. The no-slip boundary condition combined with the Navier–Stokes equations gives to first order the following expression, known as the Reynolds equation:

$$F_H = f \frac{6\pi R^2 \eta}{D} \frac{dD}{dt}. \tag{10.2}$$

Higher-order solutions of the Navier–Stokes equations confirm that the lowest-order term is enough. The deviation of the dimensionless number f^* from unity quantifies the deviation from the classical no-slip prediction. The classical prediction is analogous when the surface spacing is vibrated. In that case a sinusoidal oscillatory drive generates an oscillatory hydrodynamic force whose peak we denote as $F_{H,peak}$. The peak velocity of vibration is $v_{peak} = d \cdot \omega$, where d is the vibration amplitude and ω the angular frequency of vibration. Studies show that, when the frequency and amplitude of oscillatory flow are varied, results depend on their product, v_{peak} and the deviations from (10.2) depend on v_{peak}/D. This ratio, the flow rate, is the ratio suggested by the form of (10.2).

Without necessarily assigning physical meaning to this quantity, it can be used as an alternative expression of the same data. Mathematical manipulation [43] shows that f^* and the slip length, b are related as:

$$f^* = 2\frac{D}{6b}\left[\left(1+\frac{D}{6b}\right)\ln\left(1+\frac{6b}{D}\right)-1\right]. \tag{10.3}$$

10.2.3 Experimental

The slippery question was studied experimentally by testing the limits of ideas about interfacial force and surface roughness. Figure 10.2 sketches the experimental strategy and shows AFM (atomic force microscopy) images of some of the surfaces studied. Three strategies were used to vary surface roughness systematically. The first was based on using collapsed polymers. Narrow-distribution samples of diblock copolymers of polystyrene (M_w = 55,400) and polyvinylpyridine (M_w = 28,200), PS/PVP, were allowed to adsorb for a limited time from dilute (5×10^3 µg/ml) toluene solution onto freshly cleaved muscovite mica. They appeared to aggregate during the drying process. The remaining bare regions of mica were then coated with an organic monolayer of condensed octadecyltriethoxysilane (OTE) [44]. The contact angles against water and tetradecane were stable in time, which shows that the liquids did not penetrate them. AFM images in Fig. 10.2 quantify the roughness. To produce larger roughness values, mica was coated with condensed octadecyltrichlorosilane (OTS) after first saturating the cyclohexane deposition solution with H_2O to encourage partial polymerization in solution before surface attachment. In the third method, self-assembled monolayers of alkane thiols were formed on silver whose roughness was controlled by a direct-current (DC) bias applied during sput-

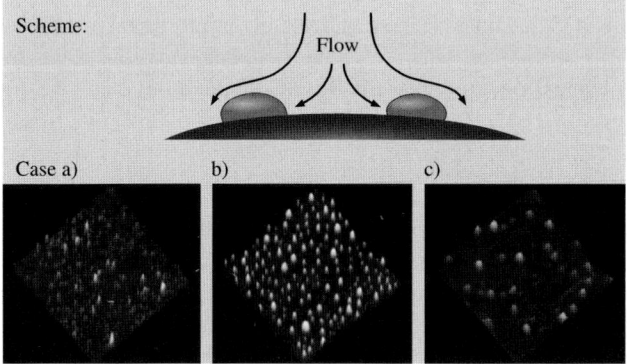

Fig. 10.2a–c. The scheme of flow over a rough surface is shown schematically in the top portion. In the bottom panel, AFM images are shown of the following case: (**a**) self-assembled OTS layers; (**b**) PS/PVP-OTE layers (surface coverage ≈ 80%); (**c**) PS/PVP-OTE layers (surface coverage ≈ 20%). Each AFM image concerns an area 3 µm × 3 µm. After [8] with permission

ter deposition. The silver was coated with octadecanethiol deposited from ethanol solution and placed in opposition with a molecularly smooth OTE surface.

According to (10.2), the *stick* prediction ($f^* = 1$) corresponds to a horizontal line and one observes in Fig. 10.3 that deviations from this prediction decreased systematically as the surface roughness increased. In addition, deviations from the predictions of the no-slip boundary condition are alternatively often represented as the *slip length* discussed above, the fictitious distance inside the solid at which the no-slip flow boundary condition would hold; the equivalent representation of this data in

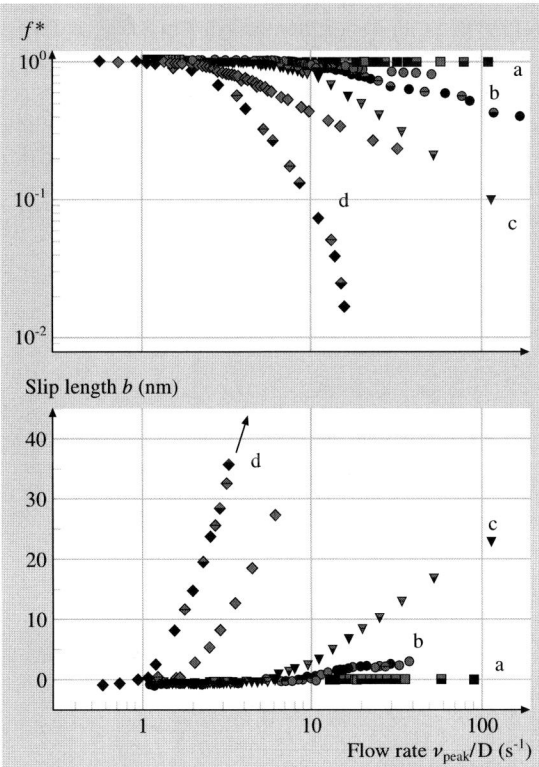

Fig. 10.3. f^* as a function of logarithmic flow rate v_{peak}/D (*top panel*) where f^* is defined in (10.2), and the equivalent slip length (*bottom panel*), for deionized water (*filled symbols*) and tetradecane (*open symbols*) between surfaces of different levels of rms surface roughness, specifically: 6 nm (case a; *squares*), 3.5 nm (case b; *circles*), 2 nm (case c; *down triangles*), 1.2 nm (case f; *hexagons*), 0.6 nm (case e, *upper triangles*) and molecularly-smooth (case d; *diamonds*). The data, taken at different amplitudes in the range of 0.3–1.5 nm and frequencies in the range 6.3–250 rad/s, are mostly not distinguished in order to avoid clutter. To illustrate the similarly successful collapse at these rough surfaces, data taken at the two frequencies 6.3 rad/s (*cross filled symbols*) and 31 rad/s (*semi-filled symbols*) for water are included explicitly. After [8] with permission

terms of the slip length is included in Fig. 10.3. While it was known previously that a very large amount of roughness is sufficient to generate *no-slip* [22–24], an experimental study in which roughness was varied systematically [7, 8] succeeded in quantifying how much actual roughness was needed in an actual system. The critical level of 6 nm considerably exceeded the size of the fluid molecules. As methods are known to achieve greater smoothness in MEMS devices, and potentially in microfluidic devices, this offers the promise of practical applications.

Observation of rate-dependent slip suggests that fluid is pinned up to some critical shear stress, beyond which it slips. However, some laboratories report slip *regardless* of flow rate [27, 29, 32–39]. Perhaps, the essential difference is that the magnitude of shear stress is larger in the latter experiments [45]. But in cases where slip is rate dependent, this affords a potential strategy by which to effect purposeful mixing in a microfluidic device. The idea would be simply to make some patches on the surface hydrophobic and other patches hydrophilic, so that when flow was slow enough it would be smooth, but when it was fast enough, mixing would result from jerkiness at the hydrophobic patches.

While it is true that slip at smooth surfaces is predicted based on computer simulations [25–27, 37–39], the shear rate of molecular dynamics (MD) simulations so

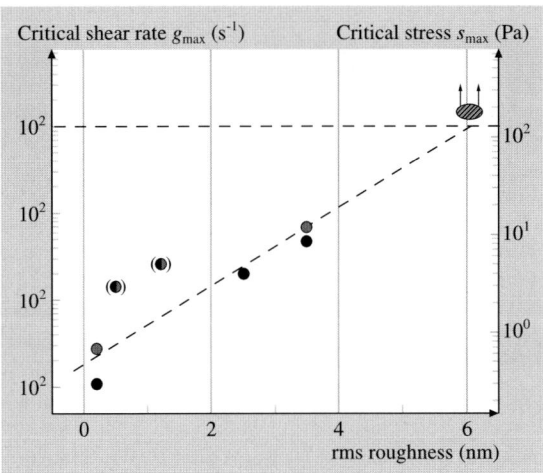

Fig. 10.4. Illustration that deviations from the traditional no-slip boundary condition depend systematically on surface roughness. Here the critical shear rate for onset of slip (*left ordinate*) and critical shear stress (*right ordinate*) are plotted semilogarithmically against rms surface roughness for flow of deionized water (*solid symbols*) and tetradecane (*open symbols* and *semi-filled symbols* as identified in [32]. The data in parentheses indicate the asymmetric situation – one surface was rough and the opposed surface was atomically smooth, as discussed in [32]. Maximum shear rate and shear stress on the crossed cylinders were calculated using known relations based on continuum hydrodynamics from [13, 42]. Specifically, $\dot{\gamma}_{max} = A\sqrt{\frac{R}{D}}\frac{v_{peak}}{D}$ and $\sigma_{peak} = 1.378\eta R^{1/2} \frac{v_{peak}}{D^{3/2}}$. After [7] with permission

much exceeds shear rate in those laboratory experiments that the direct connection to experiment is not evident. To quantify the influence of surface roughness, Fig. 10.4 considers the limit up to which predictions based on the classical no-slip boundary condition still described the data in Fig. 10.3. Since the no-slip boundary condition still held, it was valid [7–9] to calculate the shear rate and shear stress by known equations.

The data show that deviations from the no-slip prediction began at very low levels of hydrodynamic stress – on the order of only 1–10 Pa. Beyond this point, in some sense the moving fluid was depinned from the surface.

Slip need not necessarily be predicated on having surfaces coated with self-assembled monolayers to render them partially wetted, though this was the case in most of the studies cited so far. The no-slip boundary condition switches to partial slip when the fluid contains a small amount of adsorbing surfactant [30, 32].

10.2.4 Slip Can Be Modulated by Dissolved Gas

When experimental observations deviate from expectations based on the *stick* boundary condition, there are at least two alternative scenarios with microscopic interpretations. One might argue that the fluid viscosity depends on distance from the wall, but for Newtonian fluids this would not be realistic. Why then do experimental data appear to undergo shear thinning with increasing values of the parameter v_{peak}/D, if it is unreasonable to suppose that the viscosity really diminished? Inspection shows that the data for smooth surfaces at high flow rates are consistent with a two-layer-fluid model in which a layer < 1 nm thick, but with viscosity 10–20 times *less* than the bulk fluid, adjoins each solid surface [9]. A possible mechanism to explain its genesis was proposed by *de Gennes* [46], who conjectured that shear may induce nucleation of vapor bubbles; once the nucleation barrier is exceeded the bubbles grow to cover the surface, and flow of liquid is over this thin gas film rather the solid surface itself. Indeed, it is likely that incomplete air removal from the solid surfaces can profoundly influence findings in these situations where surface roughness is so low. This has been identified by recent research as a likely source of the misnamed *long-range hydrophobic attraction* [47, 48]; gases also appear to influence the sedimentation rate of small particles in liquid [49].

Accordingly, similar experiments were performed, in which the surface forces apparatus was used to measure hydrodynamic forces of Newtonian fluids that had been purged with various gases. Dissolved gas strongly influences hydrodynamic forces, in spite of the fact that gas solubility is low.

Figure 10.5 (top panel) illustrates experiments in which a simple nonpolar fluid (tetradecane) was placed between a wetted mica surface on one side, and a partially wetted methyl-terminated surface on the other, using methods described in detail elsewhere [8, 9]. The surface–surface spacing of 10–100 nm substantially exceeded the size of the fluid molecules. The spacings were vibrated with small amplitude at these spacings where the fluid responded as a continuum, and the magnitude of hydrodynamic force was measured as a function of the ratio v_{peak}/D suggested by (10.2). The experiments showed that, whereas textbook behavior [10] was nearly

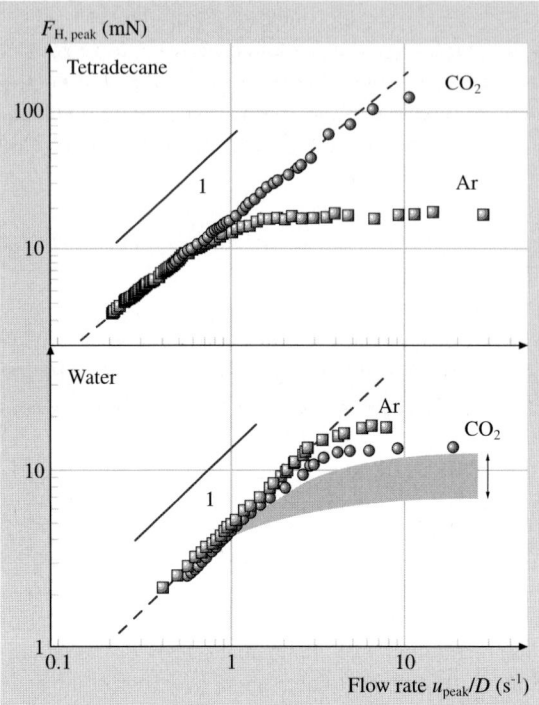

Fig. 10.5. Illustration that the onset of slip depends on dissolved gas, when simple Newtonian fluids flow past atomically smooth surfaces, either wetted or partially wetted. On log–log scales, the hydrodynamic force $F_{\text{H,peak}}$ is plotted against reduced flow rate, v_{peak}/D, such that a straight line of slope unity would indicate the *no-slip* condition assumed by (10.1). The vibration frequency is 9 Hz. *Top panel*: tetradecane flowing between the asymmetric case of a wetted mica surface on one side, a partially wetted surface of methyl-terminated self-assembled monolayer on the other side, prepared as described elsewhere [33]. *Filled symbols*, tetradecane saturated with carbon dioxide; *open symbols*, tetradecane saturated with argon. *Bottom panel*: deionized water flowing between mica surfaces that are wetted by this fluid. The hatched region of the graph shows the range of irreproducible results obtained when the gas dissolved in the water was not controlled. *Filled symbols*, water saturated with carbon dioxide; *open symbols*, water saturated with argon. After [7] with permission

followed when the tetradecane had been saturated with carbon dioxide gas, massive deviations from this prediction were found when the tetradecane was saturated with argon. This makes it seem likely that argon segregated to the solid walls, creating a low-viscosity boundary layer, in this way greasing the flow of fluid past that surface. Presumably, the amount of segregation is a material property of the fluid, the chemical makeup of the surface, and the chemical identity of the dissolved gas. In this example, the fact that argon possesses low solubility in tetradecane may have made it more prone to segregate to the surfaces.

Indeed, when a solid wall is hydrophobic and immersed in water, the ideas of *Chandler* and coworkers [50] suggest that thermodynamics may assist the formation of a vapor phase near the wall. Recent force measurements support this idea [51,52].

10.2.5 Slip Past Wetted Surfaces

The influence of dissolved gas (just discussed) casts doubt on a traditional assumption of work in this field, which is that slip arises because fluid–fluid intermolecular interactions are stronger than those between fluid and surface, i.e. that the surface must be wetted only partially. Yet for several years, there have been prominent counterexamples [28, 41]. Recent experiments show that dissolved gases can mediate apparent slip even for solid surfaces that are fully wetted by the flowing fluid.

Figure 10.5 (bottom panel) summarizes experiments in which deionized water was placed between wetted surfaces of mica and the surface–surface spacing of 10–100 nm was vibrated with small amplitude in the manner described previously [7–9, 31–34]. Hydrodynamic force is plotted against the ratio, v_{peak}/D. It is obvious that the prediction based on (10.2), a straight line on the log–log plot with a slope of unity, was violated systematically when the hydrodynamic force reached a critical level. An intriguing point is that initial findings were found to be irreproducible (they varied within the range marked by the hatched lines in the graph) but became reproducible when the water was first deliberately saturated with gas. One observes that water saturated with argon appeared to *slip* at a slightly higher level of shear stress than water saturated with carbon dioxide, and that in both cases the limiting hydrodynamic force was larger than when the nature of the dissolved gas was not controlled.

This rich and complex sensitivity to the detailed materials chemistry of the system disappears, unfortunately, when surfaces are so rough that the *stick* boundary condition is produced trivially by the influence of surface roughness [7, 8, 22–24]. Therefore for scientific and practical reasons alike, these issues of flow past nearly smooth surfaces comprise fertile ground for future work.

10.2.6 The Purposeful Generation of Slip

Inspired by these ideas to design new engineering structures, one might strive to "grease" the flow of liquids past solid surfaces by altering the boundary condition. One strategy is to make the surfaces ultra-smooth [7–9]. Another (also mentioned already) is to add processing aids that segregate to the surface [30, 32, 36]. A third way is to purposefully use multicomponent fluids to generate concentration gradients and differential wetting to generate slip, as can occur even if there is no velocity gradient in the fluid [38]. These methods could potentially be used in nanomotors or nanopumps.

Alternatively, one may seek to maximize contact with air, which is exceedingly solvophobic. Readers will have noticed that water ubiquitously beads up on the leaves of plants. Some plants can display a contact angle that approaches 180°, even though water at a smooth surface of the same chemical makeup displays a much lower contact angle. A beautiful recent series of experiments from the Kao Corporation in Japan provided insight into why [53] – the surfaces are rough on many length scales [54, 55] and trap air beneath them. Readers will have noticed that, if one tilts a leaf, a drop of water on it rolls smoothly, because it rides mainly on a cushion of air, whose effect will be further discussed in the next section. It is the principle of an ingenious method introduced recently to lower the viscous drag when fluids [56] are caused to flow through pipes whose diameter is macroscopic. Of course, given a long enough period of time it is likely that the trapped gas would dissolve into the flowing fluid, but perhaps this effect could be enhanced by placing air nozzles along the walls of the tube and replenishing the trapped gas with a stream of inlet air.

A final method by which flow of a Newtonian fluid past surfaces may be facilitated is to *ciliate* the surfaces by coating with chain molecules – polymers, proteins, or sugars. Recent experiments using a surface forces apparatus (SFA) suggest a similar (but less dramatic) rate-dependent slip in this case also [31]. This is possibly related to fluid flow in biological organs whose surfaces are also extensively ciliated, such as blood vessels and the kidney [57].

10.2.7 Outlook

The textbook presentation of engineering fluid mechanics is often of a subject thoroughly understood, but recent experiments and simulations using smooth surfaces show behavior that is richer and more complex than had been supposed. The correct boundary condition appears to depend on physical chemical properties of the solid–liquid interface that are susceptible to rational control.

10.3 Hydrophobic Interaction and Water at a Hydrophobicity Interface

The role of water in physical situations from biology to geology is almost universally thought to be important but the details are disputed [1, 48, 50–52, 58–72]. For example, as concerns proteins, the side-chains of roughly half the amino acids are polar while the other half are hydrophobic; the non-mixing of the two is a major mechanism steering the folding of proteins and other self-assembly processes. As a second example, it is an everyday occurrence to observe beading-up of raindrops, on raincoats or leaves of plants. Moreover, it is observed theoretically and experimentally that, when the gap between two hydrophobic surfaces becomes critically small, water is ejected spontaneously [50, 51, 70–72] whereas water films confined between symmetric hydrophilic surfaces are stable [1]. Despite its importance, water exhibits many anomalous behaviors when compared to other fluids. Particularly, it presents some even more puzzling behaviors near hydrophobic surfaces. This section is adapted from discussions in several primary accounts published previously [51, 52].

In its liquid form, water consists of an ever-changing three-dimensional network of hydrogen bonds. Hydrophobic surfaces cannot form hydrogen bonds, and the hydrogen-bonding network must be disrupted. So what happens when water is compelled to be close to a hydrophobic surface? Energetically, it is expected that the system forms as many hydrogen bonds as possible, resulting in a preferential ordering of the water. Entropically, it is expected that the system orients randomly and thus samples the maximum number of states. Which of these two competing interactions dominates? What effect does the competition have on the dynamic and equilibrium properties of the system? The answers to these questions are still hotly debated. To help resolve this debate, static and dynamic interaction of water confined to a hydrophobic surface is studied by SFA.

10.3.1 Experimental

The atomically smooth clay surfaces used in this study, muscovite mica (hydrophilic) and muscovite mica blanketed with a methyl-terminated organic self-assembled monolayer (hydrophobic), allowed the surface separation to be measured, by multiple beam interferometry, with a resolution of $\pm 2-5$ Å. Surfaces were brought to the spacings described below using a surface forces apparatus modified for dynamic oscillatory shear [44, 73]. A droplet of water was placed between the two surfaces oriented in a crossed cylinder geometry. Piezoelectric bimorphs were used to produce and detect controlled shear motions. The deionized water was previously passed through a purification system, Barnstead Nanopure II (control experiments with water containing dissolved salt were similar). In experiments using degassed water, the water was either first boiled, then cooled in a sealed container, or vacuumed for 5–10 h in an oven at room temperature. The temperature of measurements was 25 °C.

In order to determine firmly that findings did not depend on details of surface preparation, three methods were used to render one surface hydrophobic. In order of increasing complexity, these were: (a) atomically smooth mica coated with a self-assembled monolayer of condensed octadecyltriethoxysilane (OTE), using methods described previously [44]; (b) mica coated using Langmuir–Blodgett methods with a monolayer of condensed octadecyltriethoxysilane; and (c) a thin film of silver sputtered onto atomically smooth mica, then coated with a self-assembled thiol monolayer. In method (a), the monolayer quality was improved by distilling the OTE before self-assembly. In method (b), OTE was spread onto aqueous HCl (pH = 2.5), 0.5 h was allowed for hydrolysis, the film was slowly compressed to the surface pressure $\pi = 20$ mN/m (3–4 h), and the close-packed film was transferred onto mica by the Langmuir–Blodgett technique at a creep-up speed of 2 mm/min. Finally the transferred films were vacuum baked at 120 °C for 2 h. In method (c), 650 Å of silver were sputtered at 120 V (1 Å/s) onto mica that was held at room temperature, and then octadecanethiol was deposited from 0.5 mM ethanol solution. In this case, AFM (atomic force microscopy; Nanoscope II) showed the rms (root mean square) roughness to be 0.5 nm. All three methods led to the same conclusions, summarized below.

10.3.2 Hydrophobic Interaction

A puzzling aspect of the hydrophobic attraction is that its intensity and range appear to be qualitatively different as concerns extended surfaces of large area, and small molecules of modest size [50, 60, 67, 74]. One difference is fundamental: the hydrogen-bond network of water is believed for theoretical reasons to be less disrupted near a single alkane molecule than near an extended surface [50, 60, 67, 74]. A second difference is phenomenological: direct measurement shows attractive forces between extended surfaces starting at separations too large to be reasonably explained by disruption of the hydrogen-bond network. This conclusion comes from 20 years of research using the surface forces apparatus (SFA) and, more recently, atomic force microscopy (AFM). The onset of attraction, ≈ 10 nm in the first experiments [69, 75–78], soon increased by nearly an order of magnitude [79–81] and has been reported, in the most recent work, to begin at separations as large as 500 nm [71]. This has engendered much speculation because it is unreasonably large compared to the size of the water molecule (≈ 0.25 nm). The range of interaction decreases if the system (water and the hydrophobic surfaces) are carefully degassed [48, 82–86]. Water in usual laboratory experiments is not degassed, however, so it is relevant to understand the origin of long-range attraction in that environment. A recent review summarizes the experimental and theoretical situation [81].

In the course of experiments intended to probe the predicted slip of water over hydrophobic surfaces [9, 45] (see the previous section), weakening of the long-range hydrophobic force to the point of vanishing was observed when the solid surfaces experienced low-level vibrations around a mean static separation.

The attraction recorded during the approach of OTE surfaces with a droplet of deionized water in between is plotted in Fig. 10.6 as a function of surface–surface separation (D). $D = 0$ here refers to a monolayer–monolayer contact in air. In water, the surfaces jumped into adhesive contact at 0 ± 2 Å. This *jump in* was very slow to develop, however. The pull-off force to separate the surfaces from contact at rest (113 mN/m in Fig. 10.6) implies, from the Johnson–Kendall–Roberts (JKR) theory [76], the surface energy of about 12 mJ/m^2 (and up to about 30% less than this when oscillations were applied). The onset of attraction at 650 nm for the hydrophobic surfaces at rest is somewhat larger than in any past study of which we are aware. However, we emphasize that the level of pull-off force was consistent with the prior findings of other groups using other systems [1, 60–72].

These observations clearly imply some kind of rate-dependent process. As shown in Fig. 10.7, the force F diminished with increasing velocity and its magnitude at a given D appeared in every instance to extrapolate smoothly to zero. The possible role of hydrodynamic forces was considered but discarded as a possible explanation. Similar results (not shown) were also obtained when the surfaces were vibrated parallel to one another rather than in the normal direction. Some precedence is found in a recent AFM study that reported weakened hydrophobic adhesion force with increasing approach rate [85].

These observations remove some of the discrepancy between the range of hydrophobic forces between extended surfaces of macroscopic size [1, 60–72] and the

10 Interfacial Forces and Spectroscopic Study of Confined Fluids 531

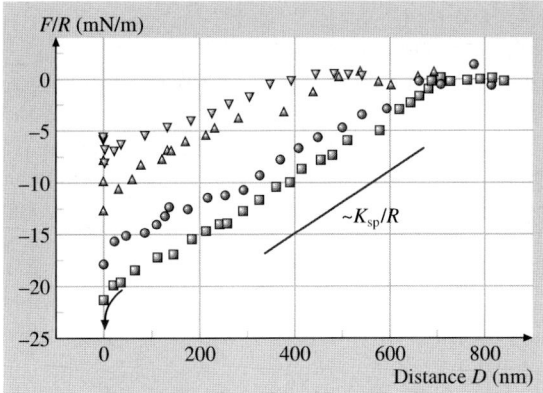

Fig. 10.6. Force–distance profiles of deionized water between hydrophobic surfaces (OTE monolayers on mica). Force F, normalized by the mean radius of curvature ($R \approx 2$ cm) of the crossed cylinders, is plotted against surface separation. Forces were measured during approach from static deflection of the force-measuring spring, while simultaneously applying small-amplitude harmonic oscillations in the normal direction with peak velocity $v_{peak} = d \times 2\pi f$ where d denotes displacement amplitude and f denotes frequency. This velocity was zero (*solid squares*), 7.6 nm/s ($d = 1.6$ nm, $f = 0.76$ Hz; *circles*), 26 nm/s ($d = 3.2$ nm, $f = 1.3$ Hz; *up triangles*), 52 nm/s ($d = 3.2$ nm, $f = 2.6$ Hz; *down triangles*). The pull-off adhesion forces ("jump out"), measured at rest and with oscillation, are indicated by the respective *semi-filled symbols*. The approach data follow the straight line with slope K_{sp}/R (*drawn separately as a guide to the eye*), indicating that they represented a spring instability ("jump in") such that the gradient of the attractive force exceeded the spring constant (K_{sp}), 930 N/m. After [51] with permission

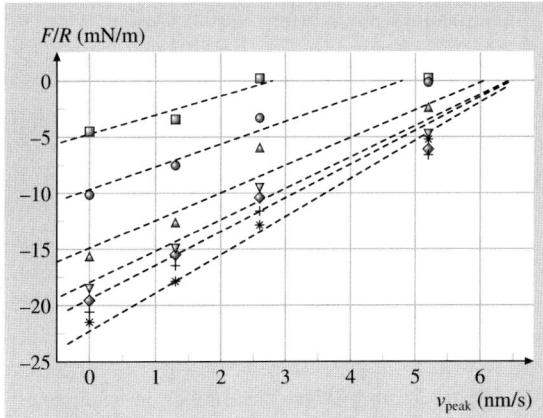

Fig. 10.7. The attractive force (F/R) at seven different surface separations (D) is plotted against peak velocity. The film thickness was $D = 720$ nm (*squares*), 540 nm (*circles*), 228 nm (*up triangles*), 116 nm (*down triangles*), 63 nm (*diamonds*), 17 nm (*crosses*), 5 nm (*stars*). After [51] with permission

range that is expected theoretically [50, 60, 67, 74]. A tentative explanation is based on the frequent suggestion that the long-range hydrophobic attraction between extended surfaces stems from the action of microscopic or submicron-sized bubbles that arise either from spontaneous cavitation or the presence of adventitious air droplets that form bridges between the opposed surfaces [48, 81, 83–86]. The experiments reported here show that this effect required time to develop. Hydrophobic attraction at long range was softened to the point of vanishing when the solid surfaces were not stationary.

10.3.3 Hydrophobicity at a Janus Interface

Due to the strong propensity to repel water completely out of two hydrophobic surfaces, it is then interesting to consider the antisymmetric situation, a hydrophilic surface on one side (A) and a hydrophobic surface (B) on the other. The surface A prevents water from being expelled, to successfully retain a stabilized aqueous thin film at intimately close contact. The surface B introduces the hydrophobic effect. This Janus situation is shown in the cartoon of Fig. 10.8.

Similar to the force between two OTE surfaces, long-range attraction was also observed at the Janus interface, as shown in the inset of Fig. 10.8. The opposed surfaces ultimately sprang into contact from $D \approx 5$ nm and upon pulling the surfaces apart, an attractive minimum was observed at $D = 5.4$ nm. The surfaces could be squeezed to a lower thickness, $D \approx 2.0$ nm. Knowing that the linear dimension of a water molecule is ≈ 0.25 nm [1], the thickness of the resulting aqueous films amounted to the order of 5–20 water molecules, although (see below) it is not clear that molecules were distributed evenly across this space. Below we discuss shear results.

In the shear measurements, the sinusoidal shear deformations were gentle – the significance of the resulting linear response being that the act of measurement did not perturb their equilibrium structure (linear responses were verified from the absence of harmonics in the time dependence of shear motion). Using techniques that are well known in rheology, from the phase lag and amplitude during oscillatory excitation the responses to shear excitation were decomposed into one in-phase component (the elastic force, f') and one out-of-phase component (the viscous force, f'') [87]. Figure 10.8 (main portion) illustrates responses at a single frequency and variable thickness. The shear forces stiffened by more than one order of magnitude as the films were squeezed. It is noteworthy that, when molecularly thin aqueous films are confined between clay surfaces that are symmetrically hydrophilic, deviations from the response of bulk water appear only at smaller separations [88]; evidently the physical origin is different here. Moreover, at each separation the elastic and viscous forces were nearly identical. The equality of elastic and viscous forces proved to be general, not an accident of the shear frequency chosen. Again this contrasts with recent studies of molecularly thin water films between surfaces that are symmetrically hydrophilic [88]. The magnitudes of the shear moduli in Fig. 10.8 are "soft" – something like those of agar or jelly. They were considerably softer than for molecularly thin aqueous films between symmetrically hydrophilic

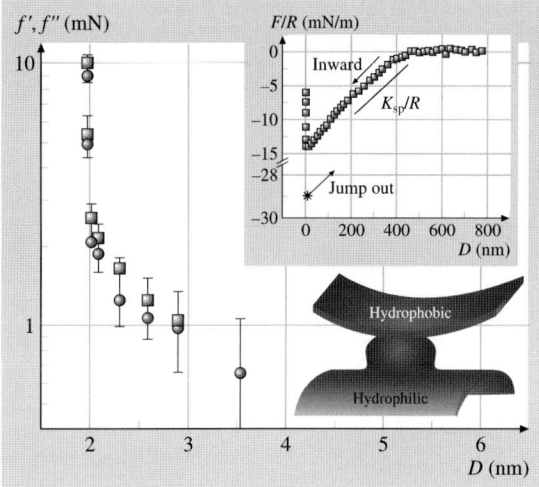

Fig. 10.8. Deionized water confined between a hydrophilic surface on one side and a hydrophobic surface on the other (*cartoon*). The cartoon is not to scale because the gap thickness is nanometers at closest approach and the droplet size (≈ 2 mm on a side) vastly exceeds the contact zone (≈ 10 μm on a side). The main figure shows the time-averaged viscous (*circles*) and elastic (*squares*) shear forces measured at 1.3 Hz and 0.3-nm deflection, plotted semilogarithmically against surface separation for deionized water confined between OTE deposited onto mica using the Langmuir–Blodgett (LB) technique (shear impulses were applied to this hydrophobic surface). The *inset* shows the static force–distance relations. Force, normalized by the mean radius of curvature ($R ≈ 2$ cm) of the crossed cylinders, is plotted against the thickness of the water film ($D = 0$ refers to contact in air). The pull-off adhesion at $D ≈ 5.4$ nm is indicated by a *star*. The *straight line* with slope K_{sp}/R indicates the onset of a spring instability where the gradient of attractive force exceeds the spring constant (K_{sp}), 930 N/m. Following this *jump* into contact, films of stable thickness resulted, whose thickness could be varied in the range $D = 1$–4 nm with application of compressive force. After [52] with permission

surfaces. This again emphasizes the different physical origin of shear forces in the present Janus situation.

Figure 10.9 illustrates the unusual result that the shear forces scaled in magnitude with the *same* power law, the square root of excitation frequency. This behavior, which is intermediate between solid and liquid, is often associated in other systems with dynamical heterogeneity [89, 90]. By known arguments it indicates a broad distribution of relaxation times rather than any single dominant one [87]. The slope of 1/2 is required mathematically by the Kramers–Kronig relations if $G'(\omega) = G''(\omega)$ [87]; its observation lends credibility to the measurements. Figure 10.9 (main panel) illustrates this scaling for an experiment in which data were averaged over a long time. The inset shows that the same was observed using other methods to prepare a hydrophobic surface. In all of these instances, $\omega^{1/2}$ scaling

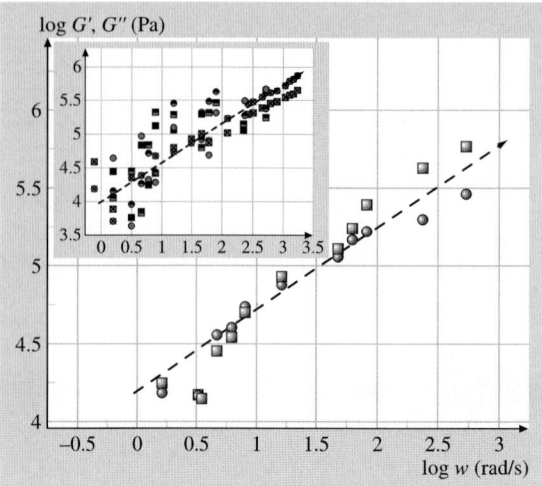

Fig. 10.9. The frequency dependence of the momentum transfer between the moving surface (hydrophobic) and the aqueous film with an adjoining hydrophilic surface is plotted on log–log scales. Time-averaged quantities are plotted. On the *right-hand ordinate* scale are the viscous, g'' (*circles*), and elastic, g' (*squares*) spring constants. The equivalent loss moduli, G''_{eff} (*circles*), and elastic moduli, G'_{eff} (*squares*), are on the *left-hand ordinate*. All measurements were made just after the jump into contact shown in Fig. 10.6, i.e. at nearly the same compressive pressure ≈ 3 MPa. The main panel, representing LB-deposited OTE, shows $\omega^{1/2}$ scaling after long-time averaging, 0.5–1 h per datum. The *inset* shows comparisons with using other methods to produce a hydrophobic surface. In those experiments the thickness was generally $D = 1.5$–1.6 nm but occasionally as large as 2.5 nm when dealing with octadecanethiol monolayers. Symbols in the *inset* show data averaged over only 5–10 min with hydrophobic surfaces prepared with (**a**) a self-assembled monolayer of OTE (*half-filled symbols*), (**b**) Langmuir–Blodgett deposition of OTE (*crossed symbols*), or (**c**) deposition of octadecanethiol on Ag (*open symbols*). As in the main figure, *circles* denote viscous forces, *squares* denote elastic forces. Scatter in this data reflects shorter averaging times than in the main part of this figure (Fig. 10.8). After [52] with permission

was observed regardless of the method of rendering the surface hydrophobic. But to observe clean $\omega^{1/2}$ scaling required extensive time averaging – see later.

In repeated measurements at the same frequency, we observed giant fluctuations (±30–40%) around a definite mean, as illustrated in Fig. 10.10, although water confined between symmetric hydrophilic–hydrophilic surfaces (bottom panel) did not display this. It is extraordinary that fluctuations did not average out over the large contact area (≈ 10 μm on a side) that far exceeded any molecular size. The structural implication is that the confined water film comprised some kind of fluctuating complex – seeking momentarily to dewet the hydrophobic side by a thermal fluctuation in one direction, but unable to because of the nearby hydrophilic side; seeking next to dewet the hydrophobic side by a thwarted fluctuation in another direction; and so on. Apparently, nearby hydrophobic and hydrophilic surfaces may produce

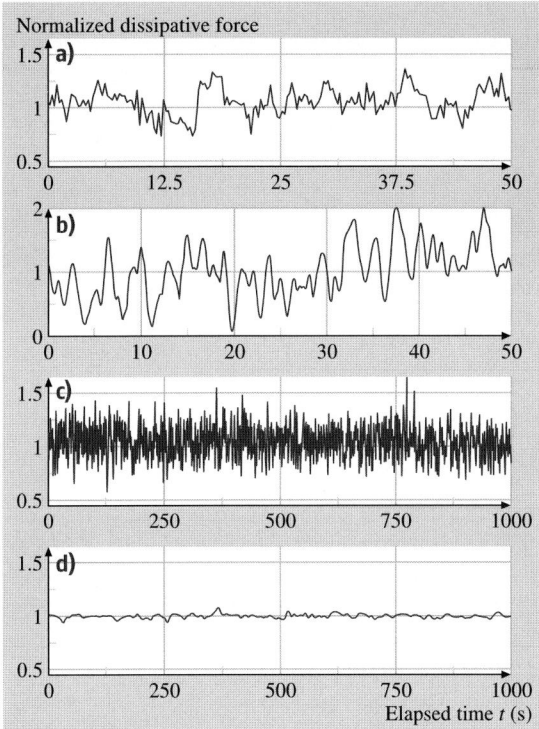

Fig. 10.10a–d. The prominence of fluctuations for water confined in a Janus interface is illustrated. In panels (**a**)–(**c**), the viscous forces, normalized to the mean (at 1.3 Hz), are plotted against time elapsed. In panel (**a**), the surfaces were first wetted with ethanol to remove adsorbed gas, then flushed with degassed, deionized water. In panel (**b**), the ethanol rinse was omitted. Panel (**c**) shows a long-time trace for data taken under the same conditions as for panel (**b**). Panel (**d**) illustrates that water confined between symmetric hydrophilic–hydrophilic surfaces (panel (**d**)) did not display noisy responses. Confined ethanol films likewise failed to display noisy responses (not shown). After [52] with permission

a quintessential instance of competing terms in the free energy, to satisfy which there may be many metastable states that are equally bad (or almost equally bad) compromises [91,92]. This suggests the physical picture of flickering capillary-type waves, sketched hypothetically in Fig. 10.11. These proposed long-wavelength capillary fluctuations would differ profoundly from those at the free liquid–gas interface because they would be constricted by the nearby solid surface.

The power spectrum is the decomposition of the traces into their Fourier components whose squared amplitudes are plotted, on log–log scales, against Fourier frequency in Fig. 10.11. In the power spectrum one observes, provided the Fourier frequency is sufficiently low, a high level of "white" frequency-independent noise. But the amplitude began to decrease beyond a threshold Fourier frequency ($f \approx 0.001$ Hz), about 10^3 times less than the drive frequency. Other experiments

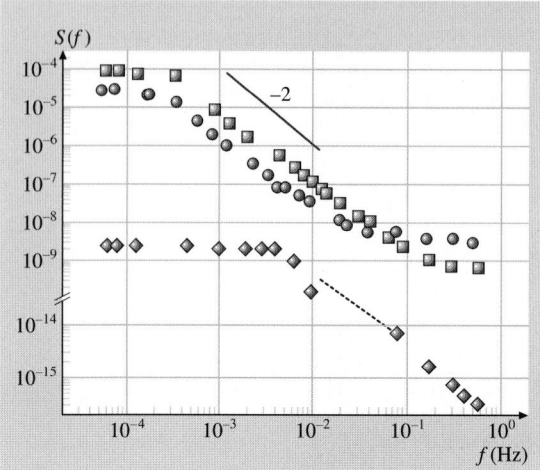

Fig. 10.11. The power spectrum for deionized water and the hydrophobized surface comprised of OTE deposited by the LB technique (*squares*); degassed deionized water and the hydrophobized surface comprised of an octadecanethiol monolayer (*circles*); and water containing 25 mM NaCl confined between symmetric hydrophilic–hydrophilic surfaces (*diamonds*). To calculate the power spectra the time elapsed was at least 10^5 s (the complete time series is not shown). The power-law exponent is close to -2. After [52] with permission

(not shown) showed the same threshold Fourier frequency when the drive frequency was raised from 1.3 to 80 Hz, and therefore it appears to be a characteristic feature of the system. It defines a characteristic time for rearrangement of some kind of structure, $\approx 10^3/2\pi$s; we tentatively identify this with the lifetime of bubbles or vapor (see below). The subsequent decay of the power spectrum as roughly $1/f^2$ suggests that these fluctuations reflect discrete entities, as smooth variations would decay more rapidly. Noise again appeared to become "white" but with an amplitude 10^4 times smaller at a Fourier frequency of $f > 0.1$ Hz but the physical origin of this is not evident at this time.

The possible role of dissolved gas is clear in the context of our proposed physical explanation. Indeed, submicron-sized bubbles resulting from dissolved air have been proposed to explain the anomalously long range of the hydrophobic attraction observed between extended surfaces [48, 51, 70–72]. To test this idea, we performed control experiments using degassed water. The power spectrum, shown in Fig. 10.12, was nearly the same. Although we cannot exclude that a small amount of residual dissolved gas was responsible, this method of degassing is reported to remove long-range hydrophobic attraction [71], whereas the comparison in Fig. 10.12 shows the consequence in the present situation to be minor. We conclude that the effects reported in this chapter do not appear to stem from the presence of dissolved gas.

From recent theoretical analysis of the hydrophobic interaction the expectation of dewetting emerges – it is predicted that an ultrathin gas gap, with a thickness on

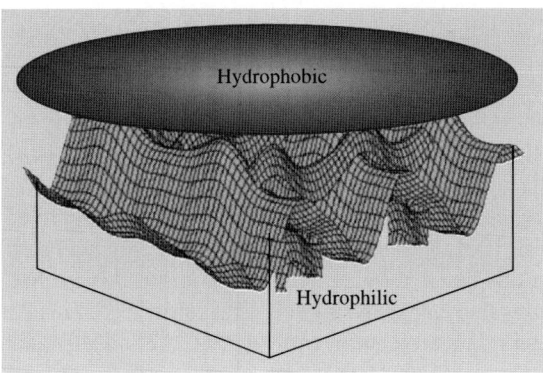

Fig. 10.12. Schematic illustration of the capillary waves of water meeting the hydrophobic surface with a flickering vapor phase in between

the order of 1 nm, forms spontaneously when an extended hydrophobic surface is immersed in water [50, 61, 67]. The resulting capillary-wave spectrum does not appear to have yet been considered theoretically, but for the related case of the free water–vapor interface, measurements confirm that capillary waves with a broad spectrum of wavelengths up to micrometers in size contribute to its width [93]. On physical grounds, the thin gas gap suggested by our measurements should also be expected to possess soft modes with fluctuations whose wavelength ranges from small to large. From this perspective, we then expect that the experimental geometry of a Janus-type water film, selected for experimental convenience, was incidental to the main physical effect.

This has evident connections to understanding the long-standing question of the structure of aqueous films near a hydrophobic surface and may have a bearing on understanding the structure of water films near the patchy hydrophilic–hydrophobic surfaces that are so ubiquitous in proteins.

10.4 Ultrafast Spectroscopic Study of Confined Fluids: Combining Ultra-Fast Spectroscopy with Force Apparatus

The surface forces apparatus (SFA) modified to measure interfacial rheology has been used widely in last few years to study the viscoelasticity of molecularly thin fluid films [1–4, 94–98]. A recent application of this technique is described in Sect. 10.3. Though the force-based techniques are powerful and sensitive, they are indirect. The observation of structure-related transitions, e. g., oscillatory forces [1], confinement-induced solidification [4, 94, 95], and stick–slip motion in SFA experiments [17, 96, 97] have not been directly verified experimentally using an independent technique.

The power of scattering, microscopy and spectroscopic techniques in the studies of confined fluids has been of speculative interest for a long time. However, there are only a few reported successes, primarily because of technical difficulties of combining these forms of techniques with SFA. Neutron and X-ray diffraction methods are very powerful for direct determination of structures at the nanoscale. Recently developed X-ray surface forces apparatus permits simultaneous X-ray diffraction and direct normal and lateral forces measurements of complex fluids under shear and confinement [98]. *Safinya* et al. have investigated the structure of thin liquid crystal films under confinement using this apparatus [99]. The deep penetrating power of neutrons and the ability to substitute hydrogen with deuterium in many liquid systems can be exploited to measure the molecular density and orientation of confined fluids by using neutron diffraction [100]. The structure of end-grafted polymer brush layers have been investigated in this manner. Successful utilization of this method requires one to devise an apparatus that can keep single-crystal substrates of quartz or sapphire with areas up to tens of square centimeters parallel at controlled and well-defined separations [101]. So far, both neutron and X-ray confinement cells are limited to confining gaps ranging from several hundred angstroms to millimeters and are not capable of studying ultrathin ($\approx$ nm) liquid films. This intermediate length scale is more suited to study complex fluids, e. g., long polymer chains, colloidal particles and biological cells, where self-organized structures of larger scale come into play. For simple fluids, the difficulty arises because, as the film thickness decreases, the total number of scatterers decreases and the signal-to-noise ratio presents severe limitations. It is difficult to distinguish with sufficient confidence the structure of a molecularly thin fluid film from that of the thicker solids that envelop it. Synchrotron X-ray sources, such as the advanced photon source at Argonne National Laboratory have sufficient flux for experiments of extremely confined liquid films possible. Recent X-ray reflectivity experiments have confirmed the expected layering in the direction perpendicular to the confining surfaces [102], but questions about in-plane order and responses to external fields remain conjectural.

The interaction of light with matter – for example Raman and infrared – has impressive potential, but the problem is to distinguish the signal (the fluid monolayer) from all the noise (the sliding surfaces and the solids beneath them). The microrheometer developed by *Dhinojwala* et al. can readily be combined with spectroscopy (Fourier-transform infrared spectroscopy and dielectric spectroscopy) or scattering (X-ray and neutron) techniques [103]. It uses two parallel optically flat windows plates whose separation can be controlled from a few tens of nm to tens of μm, but is more suited to thicker (0.1–10 μm) films. It has been used for in situ study of shear-induced molecular orientation of nematic liquid crystals by using Fourier-transform infrared time-resolved spectroscopy (FTIR-TRS) synchronized with the shear motion [104]. By replacing one of the plates with a prism, recently it has been shown that this rheometer can be combined with the surface-sensitive technique of infrared-visible sum-frequency generation (SFG) in the total internal reflection (TIR) geometry [105]. This combination can be used to probe the orientation, alignment and relaxation modes of organic molecules at the buried interface in a condition of flow or shear. Some years back it was shown that SFG can be combined with the surface forces apparatus to study nanometer-thick films of self-assembled monolayer confined between atomically smooth mica surfaces [106], but implementation of this approach to other experimental situations, such as confined fluids still presents significant challenges.

Another problematic issue arises in interpreting experiments that measure mechanical properties, such as the SFA or AFM. The measurement generates a single number, the force, but although the surface separations are molecular, the areas of interaction are macroscopic. So this force is the result of the average response of a large collection of molecules. Recent advances in optical spectroscopy and microscopy have made it possible not only to detect and image single molecules, but to conduct spectroscopic measurements and monitor dynamic processes as well at the level of single (or a handful of) molecules [107, 108]. These studies illustrate their utility to dissect the distributions around the average for processes such as diffusion or chemical reactions. In many of these experiments, a fluorescent molecule is doped into the sample, which acts as a probe of its local environment [109]. Monitoring motions of the probe over time and in the presence of external fields can offer insights into changes in this local environment within which the dye molecule is embedded. However, to integrate force measurements using SFA with concurrent measurements using fluorescence spectroscopy required significant modification of the usual methods [110]. In the following we discuss the challenges of combining SFA with single-molecule-sensitive spectroscopy techniques. This section is adapted from discussions in several primary accounts published previously [109, 110].

10.4.1 Challenges

One of the major difficulties is to detect and collect fluorescence efficiently and to separate it from background noise. Background originates from many sources: Raman and Rayleigh scattering, fluorescence owing to impurities in the solvent, and from the substrates, which includes the lens, glue and mica (the glue attaches a cleaved mica sheet onto the supporting cylindrical lens in SFA experiments). Typical background counts can far exceed those from a dilute concentration of fluorophore molecules doped inside the sample of interest.

Another type of challenge comes from the limited photochemical lifetime of a fluorophore. A common fluorophore photobleaches irreversibly after emitting a finite number of photons (10^5–10^6). This problem becomes severe in ultrathin films, where the dynamics can become slower and a dye molecule resides for a long time within the laser focus.

A third difficulty is the necessity to perform spectroscopy at the same time as multiple-beam interferometry (MBI) to determine the film thickness. Traditionally a silver coating of thickness ≈ 63 nm is used at the back side of mica for the purpose of MBI, but the high reflectivity of silver from the infrared to UV regime excludes this possibility here.

The final challenge is to incorporate the SFA and the needed optics. As the signal must be as large as possible, the maximum possible amount of fluorescence from the fluorophore of interest should be detected for a successful experiment. A high numerical aperture (N.A.) objective is desirable but such objectives have a very small working distance ($\approx 1-2$ mm). This requires significant modification of the traditional SFA in order to make it possible to focus the laser beam on the sample.

We recently succeeded in implementing the technique of fluorescence correlation spectroscopy (FCS), which can measure the translational diffusion with surface forces measurement and friction studies within the SFA [111]. The scientific objective of building this integrated platform was to answer questions such as: how is the rate of molecular probe diffusion, within a confined fluid, related to the stress relaxation time? Is there significant collective molecular motion or dynamical heterogeneity? What happens to the molecules during the stick–slip motion? The principle of the FCS technique and the experimental set-up of the combined platform are described below.

10.4.2 Principles of FCS Measurement

Fluorescence correlation spectroscopy (FCS) is an experimental method to extract information on dynamical processes from the fluctuation of fluorescence intensity [112]. The technique has enjoyed widespread application recently in the field of chemical biology because of its ability to access to a multitude of parameters with biological relevance [113,114]. The fluctuation of fluorescence, when dye molecules are dilute, can in principle result from diffusion, aggregation, or chemical reaction. Compared to other techniques for studying diffusion problems, such as quasi-elastic light scattering (QELS), fluorescence recovery after photobleaching (FRAP), and laser-induced transient grating spectroscopy, FCS presents the unique capability of measuring extremely dilute systems with high spatial resolution (down to the optical diffraction limit). On the average there can be as few as 1–5 dye molecules within the ≈ 1 fl volume element of the focused laser beam. However, these dye molecules move in and out due to Brownian motion, causing intensity fluctuations which can be observed as low-frequency noise on the mean fluorescence signal (Fig. 10.13). By inspecting the autocorrelation function of this fluctuation,

$$G(\tau) \equiv \langle \delta I(t) \delta I(t+\tau) \rangle / \langle \delta I(t) \rangle^2 , \qquad (10.4)$$

(here I denotes fluorescence intensity and t is the time variable), and by choosing a suitable model to analyze it, the rate of dynamic process is obtained [112]. If the primary reason for fluctuation is translational diffusion, and assuming that the fluorescence characteristics of the diffusing molecules do not change while traversing the laser volume, one can use Fick's second law to calculate the translational diffusion coefficient (D) from the autocorrelation function by using the relation [115],

$$G(\tau) = G(0)/\left(1 + 8D\tau/\omega_0^2\right) . \qquad (10.5)$$

This result follows from the convolution of the concentration correlation with the spatial profile of the laser focus, which has been assumed to be a two-dimensional Gaussian of width ω_0. The magnitude of the autocorrelation function at time zero, $G(0)$, is related to the average number of fluorophores (N) in the observation volume by the relation [116]

$$G(0) = 1/(2\sqrt{2}N).\qquad(10.6)$$

Molecular mobilities can be measured over a wide range of characteristic time constants from $\approx 10^{-3}$ to 10^3 ms by using this technique.

Fluctuation analysis is best performed if the system under observation is restricted to very small ensembles and if the background is efficiently suppressed. These can be accomplished by a combination of very low sample concentrations ($\approx$ nanomolar) with extremely small measurements volumes ($\approx$femtoliter). The excitation of the fluorophores can be performed either with two photons using a pulsed laser or with one photon using continuous-wave lasers [115]. In one-photon FCS, spatial resolution is obtained with a confocal set-up, in which a small pinhole inserted into the image plane can reject the out-of-focus fluorescence. For two-photon excitation on the other hand, simultaneous (within $\approx 10^{-15}$ s) absorption of two lower-energy photons of approximately twice the wavelength is required for a transition to the excited state. Mode-locked lasers providing short pulses ($\approx 10^{-13}$ s) with a high repetition rate (10^8 Hz) can provide sufficient photon flux densities for two-photon processes. As the excitation probability is proportional to the mean square of the intensity, it results in inherent depth discrimination. Additionally, two-photon excitation improves the signal-to-background ratio considerably. As the most prominent scattering came from the incident light, which is well separated in wavelength from the induced fluorescence, this makes it easy to separate the fluorescence emission from the excitation light and the scattered light. However, the photobleaching rates with two-photon excitation are significantly enhanced with respect to one-photon excitation at comparable photon-emission yields [117].

10.4.3 Experimental Set-up

A schematic diagram of the method of combing fluorescence correlation spectroscopy with the surface forces apparatus is shown in Fig. 10.12. The FCS portion of the set-up consists of three major parts: light source, microscope and data acquisition [110]. A femtosecond Ti:sapphire laser, which typically generates laser pulses with full width at half maximum (FWHM) of 100 fs at a repetition rate of 80 MHz can be used for the two-photon excitation of the fluorophores. An inverted microscope serves as the operational platform for the whole experiment. The excitation light is focused onto the sample with an objective lens of high N.A. and the emitted light is collected through the same objective and is detected by a photomultiplier tube (PMT) or avalanche photodiode (APD). The photon counting output is recorded by an integrated FCS data-acquisition board and data analysis can be performed with commercial or home-written software. By introducing the laser beam through the objective lens, a small excitation volume (≈ 1 fl) is generated within the sample. The lateral dimension of the excitation spot is about $\approx 0.5\,\mu$m, which can be determined by a calibration experiment using widely used dyes, such as fluorescein, whose diffusion coefficient in water is known to be $\approx 300\,\mu\text{m}^2/\text{s}$. The excitation power at the sample needs to be less than 1 mW to avoid photobleaching and heating effects of the sample.

The modified surface forces apparatus sat directly on the microscope stage. The traditional crossed-cylindrical geometry produced a circular contact of parallel plates when the crossed cylinders were squeezed together such that they flattened at the apex. Using an inchworm motor, separation of the surfaces can be controlled from nanometers to millimeters. To determine the separation between the surfaces, the traditional silver sheets for interferometric measurements of surface spacing in the SFA need to be replaced by multilayer dielectric coatings [118]. These multilayers can be produced by successive evaporation of layers (typically 13 or 15) of TiO_x and Al_2O_3 by electron-beam evaporation. The optical thickness of each layer was approximately $\lambda/4$ ($\lambda \approx 650$ nm), as determined by the optical monitor within the coating chamber. The thickness of each coating determines the windows of reflectivity and translucency. This approach can produce high reflectivity in the region 600–700 nm, as well as translucent windows in the region ≈ 800 nm (to allow fluorescence excitation) and 400–550 nm (to detect fluorescence). The reflectivity as a function of wavelength is shown in Fig. 10.13 for the bare mica surface and for surfaces with different numbers of multilayers.

The same set-up (Fig. 10.14) with some modification can be used to probe molecular rotational diffusion. In the ground state fluorophores are all randomly oriented. When excited by polarized light, only those fluorophores that have their dipole moments oriented along the electric vector of the incident light are prefer-

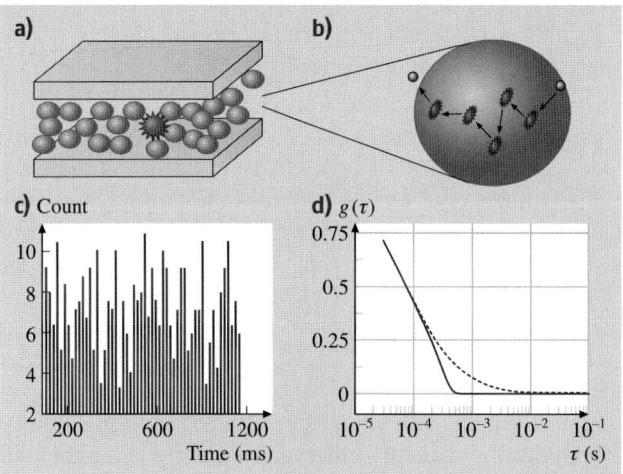

Fig. 10.13a–d. Schematic illustration of the utility of fluorescence correlation spectroscopy in a confined geometry: (**a**) A fluorescent molecule is doped within an ultrathin film of fluids (e. g., simple alkanes, polymers, colloidal particles) confined between two solid surfaces. Photon emission counts can fluctuate with time (**c**) resulting from the diffusion of fluorophores through the laser focus (**b**). From the autocorrelation function of this fluctuation $G(\tau)$, the rate of dynamic process can be obtained (**d**). Calculated $G(\tau)$ for pure Brownian diffusion (*dashed curve*) and flow superimposed with diffusion (*solid curve*) are shown

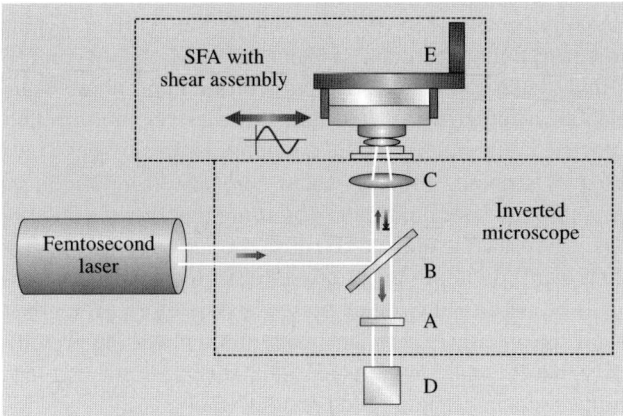

Fig. 10.14. Schematic diagram of the assembly used to perform fluorescence correlation spectroscopy within a surface forces apparatus equipped for shear experiments. A miniaturized surface forces apparatus sits on a microscope stage. A femtosecond pulsed laser excites fluorescent dye molecules within a molecularly thin liquid film contained between two opposed surfaces of muscovite mica. Colored filter (A) to remove the residual excitation light ($\lambda = 800$ nm) from the fluorescence light (400–550 nm) which is collected by the single-photon counting module (D). Dichroic mirror (B) and the objective lens (C) used to focus and collect the light. An inchworm motor (E) controls the separation of the surfaces from nanometers up to several micrometers

entially excited. So the excited-state population is not randomly oriented, instead there is a larger number of excited molecules having their transition moments oriented along the polarization direction of the incident light. This anisotropy of orientation can be determined by measuring the intensity of light polarized parallel to the incident light and perpendicular to the incident light. This preferential anisotropy at time zero decays due to the rotational diffusion of the dye molecule, following an exponential law with a *characteristic rotation time* (τ). As the typical rotational time ranges from picoseconds to nanoseconds, the dynamics on much smaller length scales, that of only one or two nm, can be investigated by this method. Translational diffusion experiments by FCS, on the other hand, (as discussed in the following sections) involves a large distance – the probe molecules travel hundreds of nm into and out of the laser spot – to produce a signal. Therefore, it is more sensitive to the *global* environment of the molecules.

10.5 Contrasting Friction with Diffusion in Molecularly Thin Films

In FCS experiments, the magnitude of the fluctuation autocorrelation function scales inversely with the number of fluorescent molecules in the observation volume (10.6). Large fluorophore concentrations more than micromolar are not efficient in

FCS, because $G(0)$ eventually becomes too small for fluctuation analysis. Typical dye concentration for confined fluid experiments is kept at 50 nM. A key point for these experiments is to find systems in which adsorption of the fluorophore would not complicate the situation. In other words, the fluids themselves, not the fluorophores, should be attracted preferentially to the confining solid surfaces [111]. This point can be verified by scanning the laser focus vertically from within the mica, through the surface, into the bulk fluids, and observing that there is no jump in fluorescence counts as the surface is crossed. Finally, one needs to be sensitive to the concern that, when using fluorophores to probe local micro-environments, micro-environments might be perturbed by their presence. Therefore, it is essential to perform normal and shear force experiments with and without the presence of dye molecules to verify that these are not affected. This section is adapted from discussions in several primary accounts published previously [111].

In SFA experiments, a drop of the fluid for study is placed between the two mica sheets, oriented as crossed cylinders so that in projection the geometry os a sphere against a flat (Fig. 10.16). In the study of surface forces, it is well known that, as surfaces separated by fluid are brought together, fluid drains smoothly until a thickness of 5–10 molecular dimensions, at which point the fluid supports stress owing to packing of molecules at the surface [1]. When rounded surfaces of this kind are pressed together, separated by fluid, the curved surfaces flatten at the apex to form parallel plates. The resulting inhomogeneous pressure distribution over the contact region is well known in the field of tribology. It is approximately *Hertzian*; zero at the edges of the contact zone and, at the center, 3/2 the mean value [119]. The Hertzian model is generally a good approximation in the absence of strongly attractive forces.

Figure 10.15 shows results for two different fluid systems: (a) propane diol containing $\approx$ 50 nM rhodamine 123, and (b) octamethylcyclotetrasiloxane (OMCTS), containing $\approx$ 50 nM coumarin. Propane diol is a low-viscosity fluid ($\approx$ 0.4 Poise) with a glass-transition temperature far below room temperature ($T_g = -105\,°C$). The OMCTS molecule is ring-shaped; it is the cyclic tetramer of dimethylsiloxane. It has a viscosity much like water ($\approx$ 0.002 Poise) and possesses the intriguing feature that it crystallizes at 1 atm near room temperature ($T_m = 17\,°C$), thus enhancing the possibility that a confinement-induced elevation of the melting temperature might be detected. There is a long tradition of considering it to constitute a model system when studying friction and surface-induced structure of nonpolar confined fluids because numerous computer simulations designed to model lubrication have concerned particles of spherical shape [2, 95, 120]. As the typical size of the contact area is $\approx$ 50 μm and the size of the laser spot is $\approx$ 0.5 μm, it is possible to scan the laser focus laterally, as sketched in Fig. 10.14. From time series of fluctuations of the fluorescence intensity, the intensity–intensity autocorrelation function can be calculated and is plotted against logarithmic time lag. From Fig. 10.15 it is obvious that the characteristic diffusion time increased with increasing proximity to the center of the contact. Their physical meaning is to describe the time to diffuse through the spot of calibrated diameter, $\approx$ 0.5 μm, at which the interrogatory laser beam was

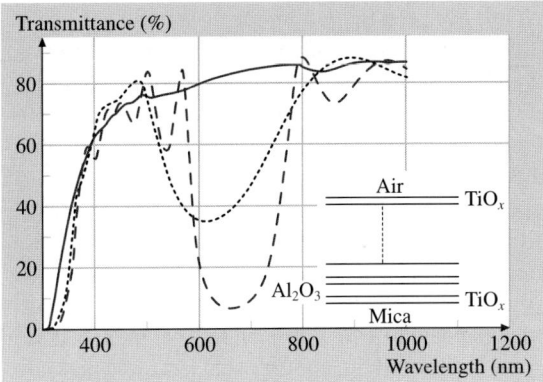

Fig. 10.15. Optical transmittance of mica coated with a multilayer dielectric, showing the feasibility of performing optical spectroscopies in the region 400–600 nm while also measuring surface–surface separation by multiple beam interferometry in the region 650–750 nm. Bare mica (*solid line*), mica coated with 7 layers of coating (*dotted line*) and with 13 layers of coating (*dash-dot line*). The window of optical transmission is adjustable through variations of multilayer dielectric. (*Inset*) The schematic diagram of the dielectric coating, composed of alternate layers of TiO_x and Al_2O_3.

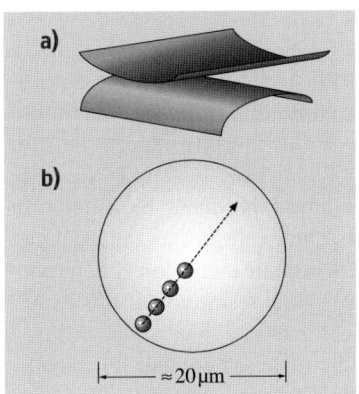

Fig. 10.16a,b. Experimental scheme. (a) Crossed-cylinder configuration. Droplets of fluids were placed between crossed cylinders of mica as a drop and force is applied in the normal direction, causing the formation of a circular area of flattened contact. (b) Fluorescence correlation spectroscopy (FCS) is performed after compressing the films with a mean normal pressure of 2–3 MPa. The sequence of smaller circles illustrate that the focus of the laser beam (diameter $\approx 0.5\,\mu m$) was scanned across the contact (radius $\approx 10\,\mu m$), enabling spatially resolved measurements. After [111] with permission

focused. The time scale of these processes, which can be estimated as the time at which the autocorrelation function decayed to one half of its initial value, slowed for example from 2 ms (rhodamine 123 in bulk propane diol) to 800 ms (the slowest curve shown in Fig. 10.15a).

For quantitative analysis, one can fit these curves to the standard model for two-dimensional Fickian diffusion and assume a single diffusion process; the imperfect fits towards the center of the contact zone are discussed below. Figure 10.18 shows D/D_{bulk}, the relative diffusion coefficient, plotted against position within the contact on semi-logarithmic scales; the logarithmic scale is needed because the dependence was so strong. Equivalently, assuming a Hertzian contact-pressure distribution, D/D_{bulk} is plotted against relative pressure within the contact in Fig. 10.18 (inset). A separate control experiment in which the bulk pressure was varied showed the diffusion in propane diol slowed by a much smaller amount, a factor of only 1.5, when the pressure was raised to 7 MPa. Evidently the findings described in this report are significantly larger than those produced by compressibility of the bulk fluid under isotropic pressure, and a different sort of explanation is needed.

Postulating that D_{eff} was proportional to a Boltzmann factor in energy ($\Delta E/kT$, where E is the energy, k the Boltzmann constant, and T the absolute temperature, which was constant), ΔE can be regarded as the net differential normal pressure, Δp, times an activation volume, ΔV_{act}. Figure 10.18 (inset) shows that the data are consistent with the implied exponential decrease of D_{eff} with p, in spite of the fact that p is mechanical pressure squeezing the confined fluid, not the usual isotropic pressure. From the slope in Fig. 10.18, one deduces that $\Delta V_{\text{act}} = 15-20$ nm^3. It is intriguingly close to the activation volume obtained some years ago from independent friction measurements [121]. In the bulk, by contrast, the activation volume for diffusion is only ≈ 0.2 nm^3, the size of a fluid molecule. This analysis highlights one of the key conclusions that diffusion appeared to involve cooperative rearrangements of many molecules. But this concept assumes a single reaction coordinate and a fully equilibrated homogeneous system. Therefore, it may not seem physically meaningful to identify the deduced activation volume with the lateral size of cooperatively rearranging regions within the confined films.

The inflections in the intensity–intensity autocorrelation functions of Fig. 10.17 are quantitatively reproducible on different days with fresh samples. They refer to the same data-acquisition time, ≈ 45 min. Near the edges of the contact this time is enough to produce an autocorrelation curve that conformed well to expectations for a diffusion process with a single diffusion rate. But in the same system, at the same film thickness, at the same temperature, and in the same experiment, the autocorrelation curves deviated from this expectation more and more, the higher the local mechanical pressure. The results suggest that the system became increasingly heterogeneous with increasing pressure, so much so that the spatial resolution of this experiment sampled zones of slightly different dynamical response. Molecules diffused at one rate through a certain micro-environment, then entered another. Physically, this signifies some kind of long-lived but quantitatively reproducible heterogeneity, which becomes increasingly intense when moving from the edges of the contact towards the center. To make this more precise, it is worth remembering that molecules in confined fluid films are known to organize into layers parallel to the confining surfaces, in which local density differs. In this scenario, the multiple processes suggested in the autocorrelation curves signify that translational diffusion rates differed in these layers. For example, for OMCTS the slowest curve in Fig. 10.17b can be decomposed into sub-processes with $D_{\text{eff}} = 0.1, 0.7, 2.9,$ and $40\,\mu\text{m}^2/\text{s}$, the average being $5.5\,\mu\text{m}^2/\text{s}$ because the slowest two processes had the smallest amplitude and contributed least. This interpretation is consistent with the observation that the heterogeneity was more pronounced for OMCTS than for 1,2-propane diol, which is not expected to organize so definitively into distinct layers. From another perspec-

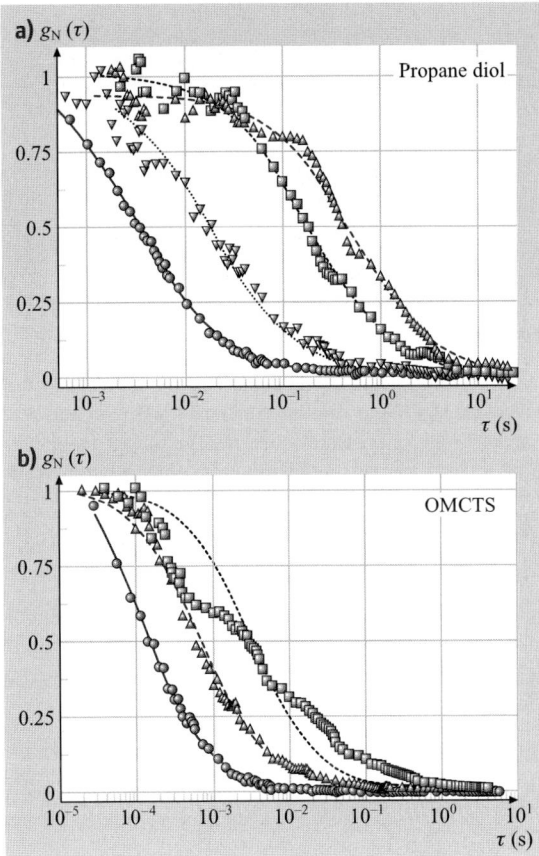

Fig. 10.17a,b. Fluorescence autocorrelation functions, normalized to unity, are plotted against logarithmic time for rhodamine 123 in 1,2-propane diol (**a**) and coumarin 153 in OMCTS (**b**). The focus was at distance a from the center of the contact and the ratio $x \equiv a/r$ was considered, where r is the contact radius. In panel (**a**), $x \approx 0.95$ (*circles*), 0.82 (*down triangles*), 0.7 (*squares*), and 0.6 (*up triangles*). In panel (**b**), $x \approx 0.95$ (*circles*), 0.8 (*up triangles*), and 0.7 (*squares*). Lines through the autocorrelation curves are the least-squares fit to a single Fickian diffusion process. After [111] with permission

tive, these data are qualitatively reminiscent of the 'cage' slowing down observed in autocorrelation curves from dynamic light scattering (DLS) studies of colloidal glasses – some kind of incipient but incomplete solidification [122]. From the available data it is difficult so far to assess the relative importance of these two scenarios, but one conclusion is firm: the scale of heterogeneity must have been impressively large, when one considers that these heterogeneities did not average out in spite of the long averaging time and the fact that the laser beam spot (≈ 500 – nm diameter) exceeded the size of the diffusing molecules by so much.

Concerning the question of how the diffusion of individual molecules is related to friction, these results suggest that the slit-averaged retardation of diffusion is much less than the confinement-induced enhancement of effective viscosity, which is known to be at least 6–7 orders of magnitude [94, 95]. However, the generality of the observation has been cast into doubt recently by mechanical experiments, which showed that the solidity of molecularly thin films using mica as substrates depends on a particular method of surface preparation [124]. Mica recleaved immediately before an experiment to minimize potential exposure to airborne contam-

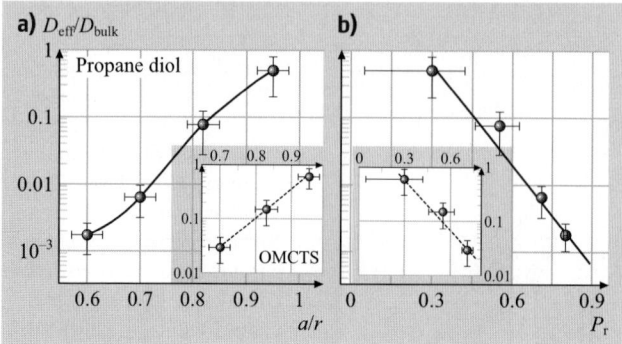

Fig. 10.18a,b. Quantification of the effective diffusion coefficients (D_{eff}) inferred from the raw data in the previous figure. (**a**) The ratio, $D_{\text{eff}}/D_{\text{bulk}}$, is plotted against focus position, a/r. (**b**) The ratio, $D_{\text{eff}}/D_{\text{bulk}}$, is plotted against relative pressure squeezing the surfaces together, $P_{\text{r}} \equiv P/P_{\text{max}}$ ($P_{\text{max}} \approx 4\,\text{MPa}$) assuming a Hertzian pressure distribution. *Open circles* (main figures) denote rhodamine 123 in 1,2-propane diol ($D_{\text{bulk}} \approx 8\,\mu\text{m}^2/\text{s}$). *Filled circles* (*insets*) denote coumarin 153 in OMCTS ($D_{\text{bulk}} \approx 190\,\mu\text{m}^2/\text{s}$). After [123] with permission

inants, is reported to produce very low friction [125]. These results are consistent with the spectroscopy experiments, with molecular dynamics simulation and dielectric measurements, none of which had observed divergence of relaxation times, or confinement induced solidification.

10.6 Diffusion of Confined Molecules During Shear

In most previous friction experiments using the traditional mica cleaving procedure, static friction (a mechanically solid state) was found to give way, under sliding, to kinetic friction (a mechanically molten state) [126]. The transition from mechanical *solidity* to *fluidity* has been interpreted to suggest a shear-induced phase transition. The premise was that, if the hypothesis of shear-melting held, shear should induce considerable quickening of the intensity–intensity autocorrelation function time scale of the dye molecule. Shear forces in SFA experiments are generated by sliding the top cylindrical lens, which is suspended from the upper portion of the apparatus by two piezoelectric bimorphs, mounted symmetrically to the two ends of the mount (Fig. 10.14). Shear forces were generated by one of the piezoelectric bimorphs (the "sender"), and the response of the device induced a voltage across the right-hand bimorph (the "receiver"). A typical frequency range is 0.1–256 Hz, with shear displacement amplitude of 0.1–10 µm. These forces were usually sinusoidal in time but sometimes were a triangular ramp of constant slope, for better comparison with nanotribology measurements and simulations that employed a constant sliding velocity. The triangular ramp produced nearly constant velocity except for a slight acceleration to prevent stick–slip. This section is adapted from discussions in several primary accounts published previously [123].

Figure 10.19 shows representative autocorrelation curves of the fluorescence fluctuations. In Fig. 10.20, the ratio of D_{eff} during sliding to D_{eff} at rest is plotted against the shear rate; the main figure shows the entire range of shear rate and the inset magnifies the regime of small shear rate. One sees that shear speeded up this measure of the time scale of the autocorrelation function by less than a factor of 5 in every case [123]. The significance is that this affords a direct (negative) test of the common hypothesis that the transition from rest to sliding reflects shear melting that is analogous to the melting transition of a solid. The reason that the autocorrelation functions were unaffected by motion is believed to be that the fluid undergoes partial slip when the moving surface is smooth; the same conclusion follows from friction measurements in similar systems.

Taken together, these ultrafast spectroscopy experiments show that these complex molecularly thin systems retain a high degree of fluidity at the molecular level. While it is true that the structure factor of in-plane correlation is predicted from computer simulations to be enhanced relative to the bulk and that the activation volume for diffusion exceeds that in the bulk liquid by about three orders of magnitude, indicating a higher degree of order than in the bulk fluid, shear does not appear to substantially modify the degree of fluidity inferred by direct measurements of the diffusion of individual molecules. An agenda for future investigation will be to un-

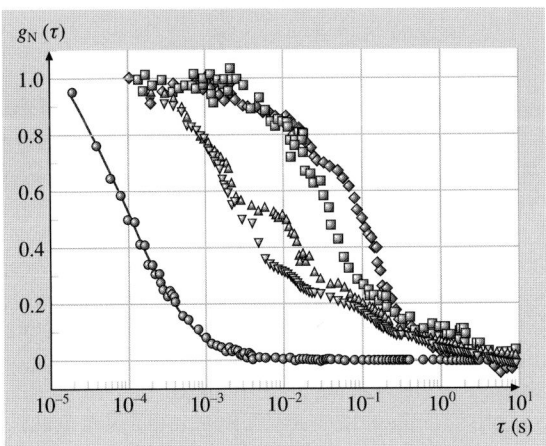

Fig. 10.19. Illustrative fluorescence intensity–intensity autocorrelation functions, $g_N(\tau)$, normalized to unity at short time and plotted against logarithmic time lag τ. The thickness of the liquid film was 3.0 ± 0.4 nm relative to air, corresponding to $\approx 3-4$ molecular layers. The curves compare diffusion at rest in the unconfined bulk (*circles*; $D = 180$ μm^2/s) and at two radial positions ("a") within the contact zone of radius $r \approx 25$ μm; $a/r = 0.7$ (*triangles*) and $a/r = 0.5$ (*squares*). These data are taken at rest (*open symbols*) and while sliding (*filled symbols*) at shear rates of 10^4 and 10^2 s^{-1}, respectively. Sliding was performed at 1–256 Hz such that it was unidirectional for half the period then reversed direction, and so on repetitively; for the data shown here, it was 256 and 32 Hz, respectively. Lines through the data for the bulk are fits to a single diffusion process. After [111] with permission

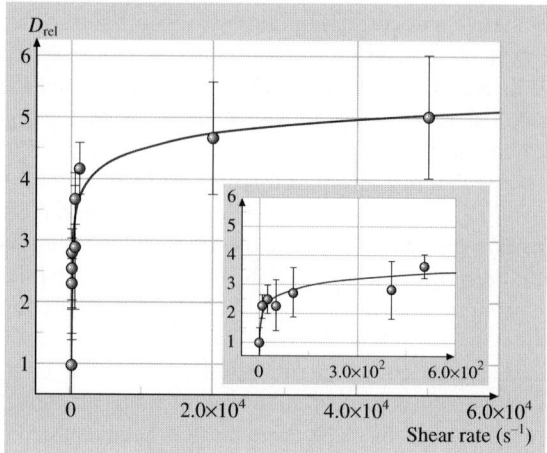

Fig. 10.20. The ratio (D_{rel}) of the effective diffusion coefficient during sliding (defined in text) to that at rest, plotted against peak shear rate. The data represent the average of more than 30 experiments, with error bars indicated. The *inset* shows the low-shear part of this data. After [123] with permission

derstand better the relationship between the mechanical friction response, which is an ensemble average, and measurements such as those presented here, which refer to the motion of individual molecules.

10.7 Summary

We have reviewed some recent advances in the experimental study of confined fluids. Some important questions involving thin liquid films with broad applications from biology to tribology have been addressed with complementary approaches. Surface and interfacial forces of liquids in intimate contact with surfaces can be directly measured by SFA, AFM and other force-based techniques. Meanwhile, scattering, microscopic and spectroscopic techniques combined with the surface forces apparatus have been developed to provide unique experimental platforms in order to understand the structure, the phase behavior, and the dynamical responses of fluids confined between surfaces whose spacing becomes comparable to the correlation length of short-range packing, the size of supramolecular structures, and/or the size of the molecules. Taking all the efforts together, the study of fluid molecules in close proximity to surfaces and under confinement gets us to the nitty-gritty of the beauty and distinction of the nano-world and nanotechnology.

Acknowledgement. Y. Elaine Zhu gratefully acknowledges the financial support from the US Department of Energy, Division of Materials Science (Grant No. DE-FG02-07ER46390) and the National Science Foundation (Grant No. CBET-0651408 and CBET-0730813). Ashis Mukhopadhyay acknowledges the supports of the American Chemical Society Petroleum Research fund (PRF No. 44953-G5) and National Science Foundation (Grant No. DMR-0605900). Steve Granick appreciates financial support from the NSF (Surface Engineering program) and also from the NSF (Polymers Program, Grant No. DMR-0605947).

References

1. J.N. Israelachvili: *Intermolecular and Surface Forces*, 2nd edn. (Academic, New York 1991)
2. B. Bhushan, J.N. Israelachvili, U. Landman: Nanotribology—Friction, wear and lubrication at the atomic-scale, Nature **374**, 607–616 (1995)
3. J.M. Drake, J. Klafter, P.E. Levitz, R.M. Overney, M. Urbakh: *Dynamics in Small Confining Systems V* (Materials Research Society, Warrendale 2000)
4. S. Granick: Soft matter in a tight spot, Phys. Today **52**, 26–31 (1999)
5. C.M. Mate, G.M. McClelland, R. Erlandsson, S. Chiang: Atomic-scale friction of a tungsten tip on a graphite surface, Phys. Rev. Lett. **59**, 1942–1945 (1987)
6. G. Meyer, N.M. Amer: Simultaneous measurement of lateral and normal forces with an optical-beam-deflection atomic force microscope, Appl. Phys. Lett. **57**, 2089–2091 (1990)
7. S. Granick, Y. Zhu, H. Lee: Slippery questions about complex fluids flowing past solids, Nature Mater. **2**, 221–227 (2003)
8. Y. Zhu, S. Granick: Limits of the hydrodynamic no-slip boundary condition, Phys. Rev. Lett. **88**, 106102–(1–4) (2002)
9. Y. Zhu, S. Granick: Rate-dependent slip of Newtonian liquid at smooth surfaces, Phys. Rev. Lett. **87**, 096105–(1–4) (2001)
10. B.S. Massey: *Mechanics of Fluids*, 6th edn. (Chapman Hall, London 1989)
11. P.-G. de Gennes: Viscometric flows of tangled polymers, C. R. Acad. Sci. B. Phys. **288**, 219 (1979)
12. L. Léger, E. Raphael, H. Hervet: Surface-anchored polymer chains: Their role in adhesion and friction, Adv. Polym. Sci. **138**, 185–225 (1999)
13. O.I. Vinogradova: Slippage of water over hydrophobic surfaces, Int. J. Miner. Process **56**, 31–60 (1999)
14. C. Mak, J. Krim: Quartz-crystal microbalance studies of the velocity dependence of interfacial friction, Phys. Rev. B **58**, 5157–5179 (1998)
15. S.M. Tholen, J.M. Parpia: Slip and the effect of He-4 at the He-3–silicon interface, Phys. Rev. Lett. **67**, 334–337 (1991)
16. C. Huh, L.E. Scriven: Hydrodynamic model of steady movement of a solid/liquid/fluid contact line, J. Colloid Interface Sci. **35**, 85–101 (1971)
17. G. Reiter, A.L. Demirel, S. Granick: From static to kinetic friction in confined liquid-films, Science **263**, 1741–1744 (1994)
18. G. Reiter, A.L. Demirel, J.S. Peanasky, L. Cai, S. Granick: Stick to slip transition and adhesion of lubricated surfaces in moving contact, J. Chem. Phys. **101**, 2606–2615 (1994)
19. N.V. Churaev, V.D. Sobolev, A.N. Somov: Slippage of liquids over lyophobic solid surfaces, J. Colloid Interface Sci. **97**, 574–581 (1984)
20. D.Y.C. Chan, R.G. Horn: The drainage of thin liquid films between solid surfaces, J. Chem. Phys. **83**, 5311–5324 (1985)
21. J.N. Israelachvili: Measurement of the viscosity of liquids in very thin films, J. Colloid Interface Sci. **110**, 263–271 (1986)
22. J.F. Nye: A calculation on the sliding of ice over a wavy surface using a Newtonian viscous approximation, Proc. Roy. Soc. A **311**, 445–467 (1969)
23. S. Richardson: On the no-slip boundary condition, J. Fluid Mech. **59**, 707–719 (1973)
24. K.M. Jansons: Determination of the macroscopic (partial) slip boundary condition for a viscous flow over a randomly rough surface with a perfect slip microscopic boundary condition, Phys. Fluids **31**, 15–17 (1988)

25. P. A. Thompson, M. O. Robbins: Shear flow near solids: epitaxial order and flow boundary condition, Phys. Rev. A **41**, 6830–6839 (1990)
26. P. A. Thompson, S. Troian: A general boundary condition for liquid flow at solid surfaces, Nature **389**, 360–362 (1997)
27. J.-L. Barrat, L. Bocquet: Large slip effect at a nonwetting fluid–solid interface, Phys. Rev. Lett. **82**, 4671–4674 (1999)
28. R. Pit, H. Hervet, L. Léger: Direct experimental evidence of slip in hexadecane–solid interfaces, Phys. Rev. Lett. **85**, 980–983 (2000)
29. V. S. J. Craig, C. Neto, D. R. M. Williams: Shear-dependent boundary slip in aqueous Newtonian liquid, Phys. Rev. Lett. **87**, 54504–(1–4) (2001)
30. O. A. Kiseleva, V. D. Sobolev, N. V. Churaev: Slippage of the aqueous solutions of cetyltriimethylammonium bromide during flow in thin quartz capillaries, Colloid J. **61**, 263–264 (1999)
31. Y. Zhu, S. Granick: Apparent slip of Newtonian fluids past adsorbed polymer layers, Macromolecules **36**, 4658–4663 (2002)
32. Y. Zhu, S. Granick: The no slip boundary condition switches to partial slip when the fluid contains surfactant, Langmuir **18**, 10058–10063 (2002)
33. J. Baudry, E. Charlaix, A. Tonck, D. Mazuyer: Experimental evidence of a large slip effect at a nonwetting fluid–solid interface, Langmuir **17**, 5232–5236 (2002)
34. D. C. Tretheway, C. D. Meinhart: Apparent fluid slip at hydrophobic microchannel walls, Phys. Fluids **14**, L9–L12 (2002)
35. H. A. Barnes: A review of the slip (wall depletion) of polymer solutions, emulsions and particle suspensions in viscometers: Its cause, character, and cure, J. Non-Newtonian Fluid Mech. **56**, 221–251 (1995)
36. E. C. Achilleos, G. Georgiou, S. G. Hatzikiriakos: Role of processing aids in the extrusion of molten polymers, J. Vinyl Additive Technol. **8**, 7–24 (2002)
37. H. Brenner, V. Ganesan: Molecular wall effects: Are conditions at a boundary 'boundary conditions'?, Phys. Rev. E. **61**, 6879–6897 (2000)
38. J. Gao, W. D. Luedtke, U. Landman: Structures, solvation forces and shear of molecular films in a rough nano-confinement, Tribology Lett. **9**, 3–134 (2000)
39. C. Denniston, M. O. Robbins: Molecular and continuum boundary conditions for a miscible binary fluid, Phys. Rev. Lett. **87**, 178302(1–4) (2001)
40. S. E. Campbell, G. Luengo, V. I. Srdanov, F. Wudl, J. N. Israelachvili: Very low viscosity at the solid–liquid interface induced by adsorbed C-60 monolayers, Nature **382**, 520–522 (1996)
41. E. Bonaccurso, M. Kappl, H.-J. Butt: Hydrodynamic force measurements: Boundary slip of water on hydrophilic surfaces and electrokinetic effects, Phys. Rev. Lett. **88**, 076103(1–4) (2002)
42. M. M. Britton, P. T. Callaghan: Two-phase shear band structures at uniform stress, Phys. Rev. Lett. **78**, 4930–4933 (1997)
43. O. I. Vinogradova: Drainage of a thin liquid-film confined between hydrophobic surfaces, Langmuir **11**, 2213–2220 (1995)
44. J. S. Peanasky, H. M. Schneider, S. Granick, C. R. Kessel: Self-assembled monolayers on mica for experiments utilizing the surface forces apparatus, Langmuir **11**, 953–962 (1995)
45. H. A. Spikes: The half-wetted bearing. Part 2: Potential application to low load contacts, Proc. Inst. Mech. Eng. Part J **217**, 15–26 (2003)
46. P.-G. de Gennes: On fluid/wall slippage, Langmuir **18**, 3413–3414 (2002)

47. J.W.G. Tyrrell, P. Attard: Atomic force microscope images of nanobubbles on a hydrophobic surface and corresponding force–separation data, Langmuir **18**, 160–167 (2002)
48. N. Ishida, T. Inoue, N. Miyahara, K. Higashitani: Nano bubbles on a hydrophobic surface in water observed by tapping-mode atomic force microscopy, Langmuir **16**, 6377–6380 (2000)
49. U.C. Boehnke, T. Remmler, H. Motschmann, S. Wurlitzer, J. Hauwede, T.M. Fischer: Partial air wetting on solvophobic surfaces in polar liquids, J. Coll. Int. Sci. **211**, 243–251 (1999)
50. K. Lum, D. Chandler, J.D. Weeks: Hydrophobicity at small and large length scales, J. Phys. Chem. B **103**, 4570–4577 (1999)
51. X. Zhang, Y. Zhu, S. Granick: Softened hydrophobic attraction between macroscopic surfaces in relative motion, J. Am. Chem. Soc. **123**, 6736–6737 (2001)
52. X. Zhang, Y. Zhu, S. Granick: Hydrophobicity at a Janus Interface, Science **295**, 663–666 (2002)
53. T. Onda, S. Shibuichi, N. Satoh, K. Tsuji: Super-water-repellent fractal surfaces, Langmuir **12**, 2125–2127 (1996)
54. J. Bico, C. Marzolin, D. Quéré: Pearl drops, Europhys. Lett. **47**, 220–226 (1999)
55. S. Herminghaus: Roughness-induced non-wetting, Europhys. Lett. **52**, 165–170 (2000)
56. K. Watanabe, Y. Udagawa, H. Udagawa: Drag reduction of Newtonian fluid in a circular pipe with a highly water-repellent wall, J. Fluid Mech. **381**, 225–238 (1999)
57. D.W. Bechert, M. Bruse, W. Hage, R. Meyer: Fluid mechanics of biological surfaces and their technological application, Naturwiss. **87**, 157–171 (2000)
58. W. Kauzmann: Some forces in the interpretation of protein denaturation, Adv. Prot. Chem. **14**, 1 (1959)
59. C. Tanford: *The Hydrophobic Effect – Formation of Micelles and Biological Membranes* (Wiley-Interscience, New York 1973)
60. F.H. Stillinger: Structure in aqueous solutions of nonpolar solutes from the standpont of scaled-particle theory, J. Solution Chem. **2**, 141 (1973)
61. E. Ruckinstein, P. Rajora: On the no-slip boundary-condition of hydrodynamics, J. Colloid Interface Sci. **96**, 488–491 (1983)
62. L.R. Pratt, D. Chandler: Theory of hydrophobic effect, J. Chem. Phys. **67**, 3683–3704 (1977)
63. A. Ben-Naim: *Hydrophobic Interaction* (Kluwer, New York 1980)
64. A. Wallqvist, B.J. Berne: Computer-simulation of hydrophobic hydration forces stacked plates at short-range, J. Phys. Chem. **99**, 2893–2899 (1995)
65. G. Hummer, S. Garde, A.E. Garcia, A. Pohorille, L.R. Pratt: An information theory model of hydrophobic interactions, Proc. Nat. Acad. Sci. USA **93**, 8951–8955 (1996)
66. Y.K. Cheng, P.J. Rossky: The effect of vicinal polar and charged groups on hydrophobic hydration, Biopolymers **50**, 742–750 (1999)
67. D.M. Huang, D. Chandler: Temperature and length scale dependence of hydrophobic effects and their possible implications for protein folding, Proc. Nat. Acad. Sci. USA **97**, 8324–8327 (2000)
68. G. Hummer, S. Garde, A.E. Garcia, L.R. Pratt: New perspectives on hydrophobic effects, Chem. Phys. **258**, 349–370 (2000)
69. D. Bratko, R.A. Curtis, H.W. Blanch, J.M. Prausnitz: Interaction between hydrophobic surfaces with metastable intervening liquid, J. Chem. Phys. **115**, 3873–3877 (2001)
70. Y.-H. Tsao, D.F. Evans, H. Wennerstöm: Long-range attractive force between hydrophobic surfaces observed by atomic force microscopy, Science **262**, 547–550 (1993) and references therein

71. R.F. Considine, C.J. Drummond: Long-range force of attraction between solvophobic surfaces in water and organic liquids containing dissolved air, Langmuir **16**, 631–635 (2000)
72. J.W.G. Tyrrell, P. Attard: Images of nanobubbles on hydrophobic surfaces and their interactions, Phys. Rev. Lett. **87**, 176104 (2001)
73. J. Peachey, J. Van Alsten, S. Granick: Design of an apparatus to measure the shear response of ultrathin liquid-films, Rev. Sci. Instrum. **62**, 463–473 (1991)
74. C.Y. Lee, J.A. McCammon, P.J. Rossky: The structure of liquid water at an extended hydrophobic surface, J. Chem. Phys. **80**, 4448–4455 (1984)
75. J.N. Israelachvili, R.M. Pashley: The hydrophobic interaction is long-range, decaying exponentially with distance, Nature **300**, 341–342 (1982)
76. J.N. Israelachvili, R.M. Pashley: Measure of the hydrophobic interaction between 2 hydrophobic surfaces in aqueous-electrolyte solutions, J. Colloid Interface Sci. **98**, 500–514 (1984)
77. R.M. Pashley, P.M. McGuiggan, B.W. Ninham, D.F. Evans: Attractive forces between uncharged hydrophobic surfaces-direct measurement in aqueous-solution, Science **229**, 1088–1089 (1985)
78. P.M. Claesson, C.E. Blom, P.C. Herder, B.W. Ninham: Interactions between water-stable hydrophobic Langmuir–Blodgett monolayers on mica, J. Colloid Interface Sci. **114**, 234–242 (1986)
79. P.M. Claesson, H.K. Christenson: Very long-range attractive forces between uncharged hydrocarbon and fluorocarbon surfaces in water, J. Phys. Chem. **92**, 1650–1655 (1988)
80. H.K. Christenson, P.M. Claesson, J. Berg, P.C. Herder: Forces between fluorocarbon surfactant monolayers – salt effects on the hydrophobic interact, J. Phys. Chem. **93**, 1472–1478 (1989)
81. O. Spalla: Long-range attraction between surfaces: Existence and amplitude?, Curr. Opin. Colloid Interface Sci. **5**, 5–12 (2000) and references therein
82. J. Wood, R. Sharma: How long is the long-range hydrophobic attraction?, Langmuir **11**, 4797–4802 (1995)
83. J.L. Parker, P.M. Claesson, P. Attard: Bubbles, cavities, and the long-range attraction between hydrophobic surfaces, J. Phys. Chem. **98**, 8468–8480 (1994)
84. A. Carambassis, L.C. Jonker, P. Attard, M.W. Rutland: Forces measured between hydrophobic surfaces due to a submicroscopic bridging bubble, Phys. Rev. Lett. **80**, 5357–5360 (1998)
85. V.S.J. Craig, B.W. Ninham, R.M. Pashley: Direct measurement of hydrophobic forces: A study of dissolved gas, approach rate, and neutron irradiation, Langmuir **15**, 1562–1569 (1999)
86. R.F. Considine, R.A. Hayes, R.G. Horn: Forces measured between latex spheres in aqueous electrolyte: Non-DLVO behavior and sensitivity to dissolved gas, Langmuir **15**, 1657–1659 (1999)
87. J.D. Ferry: *Viscoelastic Properties of Polymers*, 3rd edn. (Wiley, New York 1982)
88. Y. Zhu, S. Granick: Viscosity of interfacial water, Phys. Rev. Lett. **87**, 096104 (2001)
89. H.H. Winter, F. Chambon: Analysis of linear viscoelasticity of a cross-linking polymer at the gel point, J. Rheol. **30**, 367–382 (1986)
90. R. Yamamoto, A. Onuki: Dynamics of highly supercooled liquids: Heterogeneity, rheology, and diffusion, Phys. Rev. E **58**, 3515–3529 (1998) and references therein
91. A.O. Parry, R. Evans: Influence of wetting on phase-equilibra – a novel mechanism for critical-point shifts in films, Phys. Rev. Lett. **64**, 439–442 (1990)
92. K. Binder, D.P. Landau, A.M. Ferrenberg: Thin ising films with completing walls – a Monte Carlo study, Phys. Rev. E **51**, 2823–2838 (1995)

93. D.K. Schwartz, M.L. Schlossman, E.H. Kawamoto, G.J. Kellog, P.S. Perhan, B.M. Ocko: Thermal diffuse X-ray-scattering studies of the water–vapor interface, Phys. Rev. A **41**, 5687–5690 (2000)
94. S. Granick: Motions and relaxations of confined liquids, Science **253**, 1374–1379 (1991)
95. J. Klein, E. Kumacheva: Simple liquids confined to molecularly thin layers. I. Confinement-induced liquid-to-solid phase transitions, J. Chem. Phys. **108**, 6996–7009 (1998)
96. E. Kumacheva, J. Klein: Simple liquids confined to molecularly thin layers. II. Shear and frictional behavior of solidified films, J. Chem. Phys. **108**, 7010–7022 (1998)
97. C. Drummond, J. Israelachvili: Dynamic phase transitions in confined lubricant fluids under shear, Phys. Rev. E **63**, 041506-1–11 (2001)
98. Y. Golan, M. Seitz, C. Luo, A. Martin-Herranz, M. Yasa, Y.L. Li, C.R. Safinya, J. Israelachvili: The X-ray surface forces apparatus for simultaneous X-ray diffraction and direct normal and lateral force measurements, Rev. Sci. Instrum. **73**, 2486–248 (2002)
99. Y. Golan, A. Martin-Herranz, Y. Li, C.R. Safinya, J. Israelachvili: Direct observation of shear-induced orientational phase coexistence in a lyotropic system using a modified X-ray surface forces apparatus, Phys. Rev. Lett. **86**, 1263–1266 (2001)
100. S.M. Baker, G. Smith, R. Pynn, P. Butler, J. Hayter, W. Hamilton, L. Magid: Shear cell for the study of liquid–solid interfaces by neutron scattering, Rev. Sci. Instrum. **65**, 412–416 (1994)
101. T.L. Kuhl, G.S. Smith, J.N. Israelachvili, J. Majewski, W. Hamilton: Neutron confinement cell for investigating complex fluids, Rev. Sci. Instrum. **72**, 1715–1720 (2001)
102. O.H. Seeck, H. Kim, D.R. Lee, D. Shu, I.D. Kaendler, J.K. Basu, S.K. Sinha: Observation of thickness quantization in liquid films confined to molecular dimension, Europhys. Lett. **60**, 376–382 (2002)
103. A. Dhinojwala, S. Granick: Micron-gap rheo-optics with parallel plates, J. Chem. Phys. **107**, 8664–8667 (1998)
104. I. Soga, A. Dhinojwala, S. Granick: Optorheological studies of sheared confined fluids with mesoscopic thickness, Langmuir **4**, 1156–1161 (1998)
105. S. Mamedov, A.D. Schwab, A. Dhinojwala: A device for surface study of confined micron thin films in a total internal reflection geometry, Rev. Sci. Instrum. **73**, 2321–2324 (2002)
106. P. Frantz, F. Wolf, X.D. Xiao, Y. Chen, S. Bosch, M. Salmeron: Design of surface forces apparatus for tribolgy studies combined with nonlinear optical spectroscopy, Rev. Sci. Instrum. **68**, 2499–2504 (1997)
107. X.S. Xie, J.K. Trautman: Optical studies of single molecules at room temperature, Annu. Rev. Phys. Chem. **49**, 441–480 (1998)
108. W.E. Moerner, M. Orritt: Illuminating single molecules, Science **283**, 670–1676 (1999)
109. L.A. Deschenes, D.A. Vanden Bout: Single molecule studies of heterogeneous dynamics in polymer melts near the glass transition, Science **292**, 255–258 (2001)
110. A. Mukhopadhyay, S. Granick: An integrated platform for surface force measurements and fluorescence correlation spectroscopy, Rev. Sci. Instrum. **74**, 3067–3072 (2003)
111. A. Mukhopadhyay, J. Zhao, S.C. Bae, S. Granick: Contrasting friction and diffusion in molecularly-thin films, Phys. Rev. Lett. **89**, 136103 (2002)
112. K.M. Berland, P.T.C. So, E. Gratton: 2-Photon fluorescence correlation spectroscopy – method and application to the intracellular environment, Biophys. J. **68**, 694–701 (1995)

113. U. Kettling, A. Koltermann, P. Schwille, M. Eigen: Real-time enzyme kinetics monitored by dual-color fluorescence cross-correlation spectroscopy, Proc. Natl. Acad. Sci. USA **95**, 1416–1420 (1998)
114. A.M. Lieto, R.C. Cush, N.L. Thompson: Ligand-receptor kinetics measured by total internal reflection with fluorescence correlation spectroscopy, Biophys. J. **85**, 3294–3302 (2003)
115. P. Schwille, U. Haupts, S. Maiti, W.W. Webb: Molecular dynamics in living cells observed by fluorescence correlation spectroscopy with one- and two-photon excitation, Biophys. J **77**, 2251–2265 (1999)
116. W.W. Webb: Fluorescence correlation spectroscopy: inception, biophysical experimentations, and prospectus, Appl. Opt. **40**, 3969–3983 (2001)
117. P.S. Dittrich, P. Schwill: Photobleaching and stabilization of fluorophores used for single-molecule analysis with one- and two-photon excitation, Appl. Phys. B **73**, 829–837 (2001)
118. M. Born, E. Wolf: *Principles of Optics* (Cambridge Univ. Press, Cambridge 1999) p. 7
119. I. Sridhar, K.L. Johnson, N.A. Fleck: Adhesion mechanics of the surface force apparatus, J. Appl. Phys. D **30**, 1710–1719 (1997)
120. Y. Zhu, S. Granick: Reassessment of solidification in fluids confined between mica sheets, Langmuir **19**, 8148–8151 (2003)
121. H.-W. Hu, G. Carson, S. Granick: Relaxation-time of confined liquids under shear, Phys. Rev. Lett. **66**, 2758–2761 (1991)
122. K.N. Pham, A.M. Puertas, J. Bergenholtz, S.U. Egelhaaf, A. Moussaid, P.N. Pusey, B. Schofield, M.E. Cates, M. Fuchs, W.C.K. Poon: Multiple glassy states in a simple model system, Science **296**, 104–106 (2002)
123. A. Mukhopadhyay, S.C. Bae, J. Zhao, S. Granick: How confined lubricants diffuse during shear, Phys. Rev. Lett. **93**, 236105 (2004)
124. Z.Q. Lin, S. Granick: Platinum nanoparticles at mica surfaces, Langmuir **19**, 7061–7070 (2003)
125. Y. Zhu, S. Granick: Superlubricity: A paradox about confined fluids resolved, Phys. Rev. Lett. **93**, 096101 (2004)
126. M. Urbakh, J. Klafteer, D. Gourdon, J. Israelachvili: The nonlinear nature of friction, Nature **430**, 525–528 (2004)

11

Friction and Wear on the Atomic Scale

Enrico Gnecco, Roland Bennewitz, Oliver Pfeiffer, Anisoara Socoliuc, and Ernst Meyer

Summary. Friction has long been the subject of research: the empirical da Vinci–Amontons friction laws have been common knowledge for centuries. Macroscopic experiments performed by the school of *Bowden* and *Tabor* revealed that macroscopic friction can be related to the collective action of small asperities. Over the last 20 years, experiments performed with the atomic force microscope have provided new insights into the physics of single asperities sliding over surfaces. This development, together with the results from complementary experiments using surface force apparatus and the quartz microbalance, have led to the new field of *nanotribology*. At the same time, increasing computing power has permitted the simulation of processes that occur during sliding contact involving several hundreds of atoms. It has become clear that atomic processes cannot be neglected when interpreting nanotribology experiments. Even on well-defined surfaces, experiments have revealed that atomic structure is directly linked to friction force. This chapter will describe friction force microscopy experiments that reveal, more or less directly, atomic processes during sliding contact.

We will begin by introducing friction force microscopy, including the calibration of cantilever force sensors and special aspects of the ultrahigh vacuum environment. The empirical Tomlinson model often used to describe atomic stick-slip results is therefore presented in detail. We review experimental results regarding atomic friction, including thermal activation, velocity dependence and temperature dependence. The geometry of the contact is crucial to the interpretation of experimental results, such as the calculation of the lateral contact stiffness, as we shall see. The onset of wear on the atomic scale has recently been sudied experimentally and it is described here. In order to compare results, we present molecular dynamics simulations that are directly related to atomic friction experiments. The chapter ends with a discussion of dissipation measurements performed in noncontact force microscopy, which may become an important complementary tool for the study of mechanical dissipation in nanoscopic devices.

11.1 Friction Force Microscopy in Ultrahigh Vacuum

The *friction force microscope* (FFM, also called the lateral force microscope, LFM) exploits the interaction of a sharp tip sliding on a surface in order to quantify dissipative processes down to the atomic scale (Fig. 11.1).

11.1.1 Friction Force Microscopy

The relative motion of tip and surface is realized by a *scanner* created from piezo-electric elements, which moves the surface perpendicularly to the tip with a certain periodicity. The scanner can be also extended or retracted in order to vary the normal force (F_N) that is applied to the surface. This force is responsible for the deflection of the *cantilever* that supports the tip. If the normal force F_N increases while scanning due to the local slope of the surface, the scanner is retracted by a feedback loop. On the other hand, if F_N decreases, the surface is brought closer to the tip by extending the scanner. In this way, the surface topography can be determined line-by-line from the vertical displacement of the scanner. Accurate control of such vertical movement is made possible by a light beam reflected from the rear of the lever into a photodetector. When the cantilever bends, the light spot on the detector moves up or down and causes the photocurrent to vary, which in turn triggers a corresponding change in the normal force F_N applied.

The relative sliding of tip and surface is usually also accompanied by *friction*. A lateral force (F_L), which acts in the opposite direction to the scan velocity, v, hinders the motion of the tip. This force causes torsion in the cantilever, which can be observed along with the topography if the photodetector can detect not only the normal deflection but also the lateral movement of the lever while scanning. In practice this is achieved using a four-quadrant photodetector, as shown in Fig. 11.1. We should note that friction forces also cause lateral bending of the cantilever, but this effect is negligible if the thickness of the lever is much less than the width.

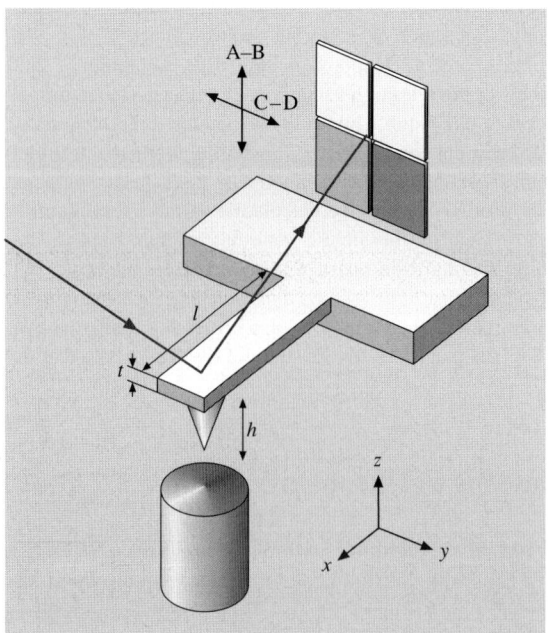

Fig. 11.1. Schematic diagram of a beam-deflection friction force microscope

The FFM was first used by *Mate* et al. in 1987 to study the friction associated with atomic features [1] (just one year after *Binnig* et al. introduced the atomic force microscope [2]). In their experiment, Mate used a tungsten wire and a slightly different technique to that described above to detect lateral forces (nonfiber interferometry). Optical beam deflection was introduced later by *Marti* et al. and *Meyer* et al. [3, 4]. Other methods of measuring the forces between tip and surface include capacitance detection [5], dual fiber interferometry [6] and piezolevers [7]. In the first method, two plates close to the cantilever reveal the capacitance while scanning. The second technique uses two optical fibers to detect the cantilever deflection along two orthogonal directions aligned 45° with respect to the surface normal. Finally, in the third method, cantilevers with two Wheatstone bridges at their bases reveal normal and lateral forces, which are respectively proportional to the sum and the difference of both bridge signals.

11.1.2 Force Calibration

Force calibration is relatively simple if rectangular cantilevers are used. Due to possible discrepancies with the geometric values provided by manufacturers, one should use optical and electron microscopes to determine the width, thickness and length of the cantilever (w, t, l), the tip height h and the position of the tip with respect to the cantilever. The thickness of the cantilever can also be determined from the resonance frequency of the lever, f_0, using the relation [8]:

$$t = \frac{2\sqrt{12}\pi}{1.875^2}\sqrt{\frac{\rho}{E}} f_0 l^2, \tag{11.1}$$

Here ρ is the density of the cantilever and E is its Young's modulus. The normal spring constant (c_N) and the lateral spring constant (c_L) of the lever are given by

$$c_N = \frac{Ewt^3}{4l^3}, \quad c_L = \frac{Gwt^3}{3h^2 l}, \tag{11.2}$$

where G is the shear modulus. Figure 11.2 shows some SEM images of rectangular silicon cantilevers used for FFM. In the case of silicon, $\rho = 2.33 \times 10^3$ kg/m³, $E = 1.69 \times 10^{11}$ N/m² and $G = 0.5 \times 10^{11}$ N/m². Thus, for the cantilever shown in Fig. 11.2, $c_N = 1.9$ N/m and $c_L = 675$ N/m.

The next force calibration step consists of measuring the sensitivity of the photodetector, S_z (nm/V). For beam-deflection FFMs, the sensitivity (S_z) can be determined by force versus distance curves measured on hard surfaces (such as Al₂O₃), where elastic deformations are negligible and the vertical movement of the scanner equals the deflection of the cantilever. A typical relation between the difference between the vertical signals on the four-quadrant detector (V_N) and the distance from the surface (z) is sketched in Fig. 11.3. When the tip is approached, no signal is revealed until the tip jumps into contact at $z = z_1$. Further extension or retraction of the scanner results in elastic behavior until the tip jumps out of contact again

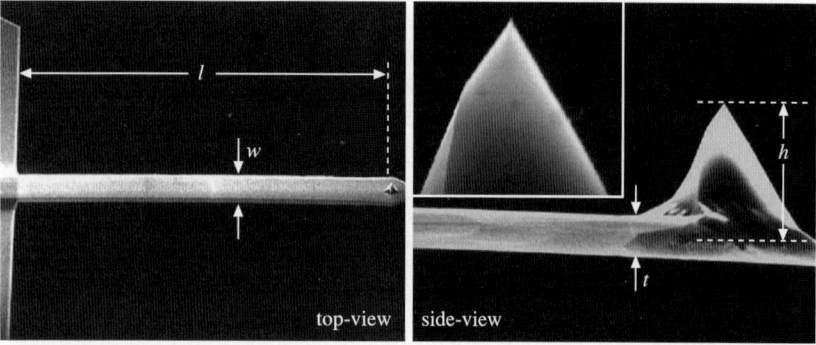

Fig. 11.2. SEM images of a rectangular cantilever. The relevant dimensions are $l = 445$ µm, $w = 43$ µm, $t = 4.5$ µm, $h = 14.75$ µm. Note that h is given by the sum of the tip height and half of the cantilever thickness. (After [9])

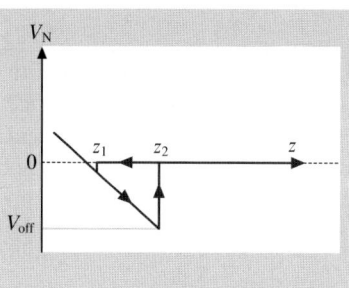

Fig. 11.3. Sketch of a typical force versus distance curve

at a distance $z_2 > z_1$. The slope of the elastic part of the curve gives the required sensitivity, S_z.

The normal and lateral forces are related to the voltage V_N, and the difference between the horizontal signals, V_L, as follows:

$$F_N = c_N S_z V_N, \quad F_L = \frac{3}{2} c_L \frac{h}{l} S_z V_L, \tag{11.3}$$

It is assumed here that the light beam is positioned above the probing tip.

The normal spring constant c_N can also be calibrated using other methods. *Cleveland* et al. [10] attached tungsten spheres to the tip, which changes the resonance frequency f_0 according to the formula

$$f_0 = \frac{1}{2\pi} \sqrt{\frac{c_N}{M + m^*}}. \tag{11.4}$$

M is the mass of the added object, and m^* is an effective mass of the cantilever, which depends on its geometry [10]. The spring constant can be extrapolated from the frequency shifts corresponding to the different masses attached.

As an alternative, *Hutter* et al. observed that the spring constant c_N can be related to the area of the power spectrum of the thermal fluctuations of the cantilever, P [11].

The correct relation is $c_N = 4k_B T/3P$, where $k_B = 1.38 \times 10^{-3}$ J/K is Boltzmann's constant and T is the temperature [12].

Cantilevers with different shapes require finite element analysis, although analytical formulas can be derived in a few cases. For V-shaped cantilevers, *Neumeister* et al. derived the following approximation for the lateral spring constant c_L [13]:

$$c_L = \frac{Et^3}{3(1+\nu)h^2} \times \left(\frac{1}{\tan\alpha} \ln\frac{w}{d\sin\alpha} + \frac{L\cos\alpha}{w} - \frac{3\sin 2\alpha}{8} \right)^{-1}, \qquad (11.5)$$

The geometrical quantities L, w, α, d, t and h are defined in Fig. 11.4. The expression for the normal constant is more complex and can be found in the cited reference.

Surfaces with well-defined profiles permit an alternative in situ calibration of lateral forces [14]. We present a slightly modified version of the method [15]. Figure 11.5 shows a commercial grating formed by alternate faces with opposite inclinations with respect to the scan direction. When the tip slides on the inclined planes, the normal force (F_N) and the lateral force (F_L) with respect to the surface are different from the two components $F_\perp$ and $F_\parallel$, which are separated by the photodiode (see Fig. 11.6a).

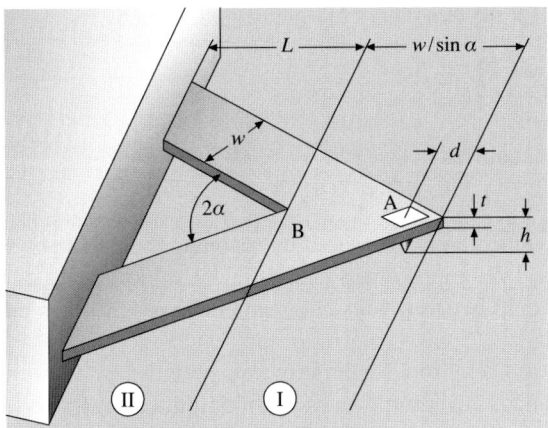

Fig. 11.4. Geometry of a V-shaped cantilever. (After [13])

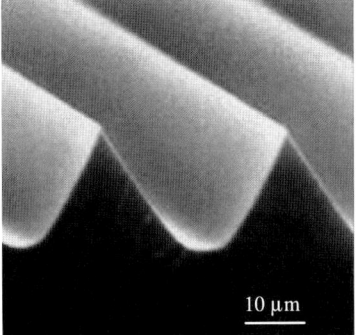

Fig. 11.5. Silicon grating formed by alternated faces angled at ±55° from the surface (courtesy Silicon-MDT Ltd., Moscow, Russia)

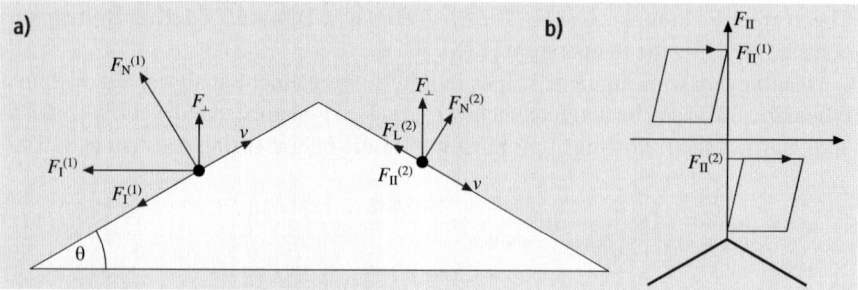

Fig. 11.6. (a) Forces acting on a FFM tip sliding on the grating shown in Fig. 11.5; (b) Friction loops acquired on the two faces

If the linear relation $F_L = \mu F_N$ holds (see Sect. 11.5), the component $F_\parallel$ can be expressed in terms of $F_\perp$:

$$F_\parallel = \frac{\mu + \tan\theta}{1 - \mu\tan\theta} F_\perp . \tag{11.6}$$

The component $F_\perp$ is kept constant by the feedback loop. The sum of and the difference between the $F_\parallel$ values for the two planes (1) and (2) are given by

$$F_+ \equiv F_\parallel^{(1)} + F_\parallel^{(2)} = \frac{2\mu(1 + \tan^2\theta)}{1 - \mu^2 \tan^2\theta} F_\perp$$

$$F_- \equiv F_\parallel^{(1)} - F_\parallel^{(2)} = \frac{2(1 + \mu^2)\tan\theta}{1 - \mu^2 \tan^2\theta} F_\perp . \tag{11.7}$$

The values of F_+ and F_- (in volts) can be measured by scanning the profile back and forth (Fig. 11.6b). If F_+ and F_- are recorded with different values of $F_\perp$, one can determine the conversion ratio between volts and nanonewtons as well as the coefficient of friction, μ.

An accurate error analysis of lateral force calibration was provided by *Schwarz* et al., who revealed the importance of the cantilever oscillations induced by the feedback loop and the so-called "pull-off force" (Sect. 11.5) in friction measurements, aside from the geometrical positioning of the cantilevers and laser beams [16].

An adequate estimation of the radius of curvature of the tip, R, is also important for some applications (Sect. 11.5.2). This quantity can be evaluated with a scanning electron microscope. This allows well-defined structures such as step sites [17,18] or whiskers [19] to be imaged. Images of these high aspect ratio structures are convolutions with the tip structure. A deconvolution algorithm that allows for the extraction of the probe tip's radius of curvature was suggested by *Villarrubia* [20].

11.1.3 The Ultrahigh Vacuum Environment

Atomic friction studies require well-defined surfaces and – whenever possible – tips. For the surfaces, established methods of surface science performed in ultra-high vacuum (UHV) can be employed. Ionic crystals such as NaCl have become standard materials for friction force microscopy on the atomic scale. Atomically clean and flat surfaces can be prepared by cleavage in UHV. The crystal has to be heated to about 150 °C for 1 h in order to remove charge after the cleavage process. Metal surfaces can be cleaned and flattened by cycles of sputtering with argon ions and annealing. Even surfaces prepared in air or liquids, such as self-assembled molecular monolayers, can be transferred into the vacuum and studied after careful heating procedures that remove water layers.

Tip preparation in UHV is more difficult. Most force sensors for friction studies have silicon nitride or pure silicon tips. Tips can be cleaned and oxide layers removed by sputtering with argon ions. However, the sharpness of the tip is normally reduced by sputtering. As an alternative, tips can be etched in fluoric acid directly before transfer to the UHV. The significance of tip preparation is limited by the fact that the chemical and geometrical structure of the tip can undergo significant changes when sliding over the surface.

Using the friction force microscope in UHV conditions requires some additional effort. First of all, only materials with low vapor pressures can be used, which excludes most plastics and lubricants. Beam-deflection force microscopes employ either a light source in the vacuum chamber or an optical fiber guiding the light into the chamber. The positioning of the light beam on the cantilever and the reflected beam on the position-sensitive detector is achieved by motorized mirrors [21] or by moving the light source or detector [22]. Furthermore, a motorized sample approach must be realized.

The quality of the force sensor's electrical signal can seriously deteriorate when it is transferred out of the vacuum chamber. Low noise and high bandwidth can be preserved using a preamplifier in the vacuum. Again, the choice of materials for printing and devices is limited by the need for low vapor pressure. Stronger heating of the electrical circuitry in vacuum, therefore, may be needed.

11.1.4 A Typical Microscope Operated in UHV

A typical AFM used in UHV is shown in Fig. 11.7. The housing (1) contains the light source and a set of lenses that focus the light onto the cantilever. Alternatively, the light can be guided via an optical fiber into the vacuum. By using light emitting diodes with low coherency it is possible to avoid interference effects often found in instruments that use a laser as the light source. A plane mirror fixed on the spherical rotor of a first stepping motor (2) can be rotated around vertical and horizontal axes in order to guide the light beam onto the rear of the cantilever, which is mounted on a removable carrier plate (3). The light is reflected off the cantilever toward a second motorized mirror (4) that guides the beam to the center of the quadrant photodiode (5), where the light is then converted into four photocurrents. Four preamplifiers in close vicinity to the photodiode allow low-noise measurements with a bandwidth of 3 MHz.

The two motors with spherical rotors, used to realign the light path after the cantilever has been exchanged, work as *inertial stepping motors*: the sphere rests

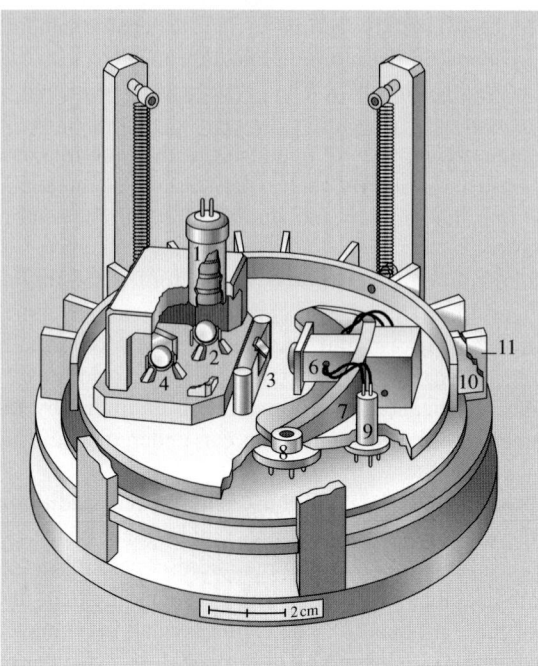

Fig. 11.7. Schematic view of the UHV-AVM realized at the University of Basel. (After [21])

on three piezoelectric legs that can be moved in small amounts tangentially to the sphere. Each step of the motor consists of the slow forward motion of two legs followed by an abrupt jump backwards. During the slow forward motion, the sphere follows the legs due to friction, whereas it cannot follow the sudden jump due to its inertia. A series of these tiny steps rotates the sphere macroscopically.

The sample, which is also placed on an exchangeable carrier plate, is mounted at the end of a tube scanner (6), which can move the sample in three dimensions over several micrometers. The whole scanning head (7) is the slider of a third inertial stepping motor for coarse positioning of the sample. It rests with its flat and polished bottom on three supports. Two of them are symmetrically placed piezoelectric legs (8), whereas the third central support is passive. The slider (7) can be moved in two dimensions and rotated about a vertical axis by several millimeters (rotation is achieved by antiparallel operation of the two legs). The slider is held down by two magnets, close to the active supports, and its travel is limited by two fixed posts (9) that also serve as cable attachments. The whole platform is suspended by four springs. A ring of radial copper lamellae (10), floating between a ring of permanent magnets (11) on the base flange, acts to efficiently damp eddy currents.

11.2 The Tomlinson Model

In Sect. 11.3, we show that the FFM can reveal friction forces down to the atomic scale, which are characterized by a typical sawtooth pattern. This phenomenon can be seen as a consequence of a *stick-slip* mechanism, first discussed by *Tomlinson* in 1929 [23].

11.2.1 One-Dimensional Tomlinson Model

In the Tomlinson model, the motion of the tip is influenced by both the interaction with the atomic lattice of the surface and the elastic deformations of the cantilever. The shape of the tip–surface potential, $V(r)$, depends on several factors, such as the chemical composition of the materials in contact and the atomic arrangement at the tip end. For the sake of simplicity, we will start the analysis in the one-dimensional case considering a sinusoidal profile with an atomic lattice periodicity a and a peak-to-peak amplitude E_0. In Sect. 11.5, we will show how the elasticity of the cantilever and the contact area can be described in a unique framework by introducing an effective lateral spring constant, k_{eff}. If the cantilever moves with a constant velocity v along the x-direction, the total energy of the system is

$$E_{\mathrm{tot}}(x,t) = -\frac{E_0}{2}\cos\frac{2\pi x}{a} + \frac{1}{2}k_{\mathrm{eff}}(vt - x)^2 . \tag{11.8}$$

Figure 11.8 shows the energy profile $E_{\mathrm{tot}}(x,t)$ at two different instants. When $t = 0$, the tip is localized in the absolute minimum of E_{tot}. This minimum increases with time due to the cantilever motion, until the tip position becomes unstable when $t = t^*$.

At a given time t, the position of the tip can be determined by equating the first derivative of $E_{\mathrm{tot}}(x,t)$ with respect to x to zero:

$$\frac{\partial E_{\mathrm{tot}}}{\partial x} = \frac{\pi E_0}{a}\sin\frac{2\pi x}{a} - k_{\mathrm{eff}}(vt - x) = 0 . \tag{11.9}$$

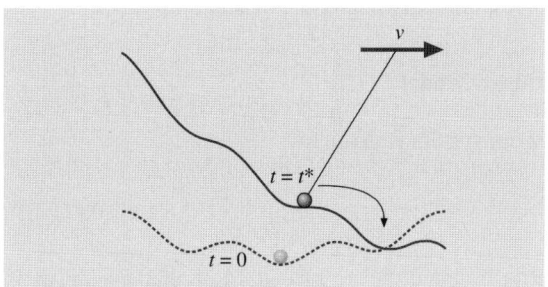

Fig. 11.8. Energy profile experienced by the FFM tip (*black circle*) at $t = 0$ (*dotted line*) and $t = t^*$ (*continuous line*)

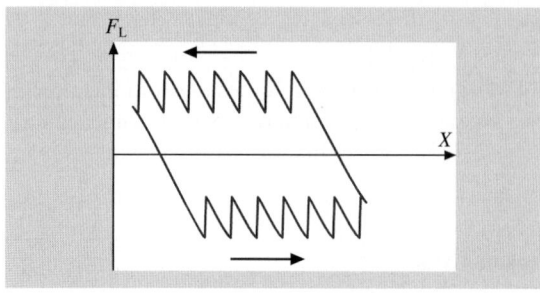

Fig. 11.9. Friction loop obtained by scanning back and forth in the 1-D Tomlinson model. The effective spring constant k_{eff} is the slope of the sticking part of the loop (if $\gamma \gg 1$)

The critical position x^* corresponding to $t = t^*$ is determined by equating the second derivative $\partial^2 E_{\text{tot}}(x,t)/\partial x^2$ to zero, which gives

$$x^* = \frac{a}{4}\arccos\left(-\frac{1}{\gamma}\right), \quad \gamma = \frac{2\pi^2 E_0}{k_{\text{eff}} a^2}. \tag{11.10}$$

The coefficient γ compares the strength of the interaction between the tip and the surface with the stiffness of the system. When $t = t^*$ the tip suddenly *jumps* into the next minimum of the potential profile. The lateral force $F^* = k_{\text{eff}}(vt - x^*)$ that induces the jump can be evaluated from (11.9) and (11.10):

$$F^* = \frac{k_{\text{eff}} a}{2\pi}\sqrt{\gamma^2 - 1}. \tag{11.11}$$

Thus the stick-slip is observed only if $\gamma > 1$: when the system is not too stiff or when the tip–surface interaction is strong enough. Figure 11.9 shows the lateral force F_L as a function of the cantilever position, X. When the cantilever is moved to the right, the lower part of the curve in Fig. 11.9 is obtained. If, at a certain point, the cantilever's direction of motion is suddenly inverted, the force has the profile shown in the upper part of the curve. The area of the *friction loop* obtained by scanning back and forth gives the total energy dissipated.

On the other hand, when $\gamma < 1$, the stick-slip is suppressed. The tip slides in a continuous way on the surface and the lateral force oscillates between negative and positive values. Instabilities vanish in this regime, which leads to the disappearance of lateral force hysteresis and correspondingly negligible dissipation losses.

11.2.2 Two-Dimensional Tomlinson Model

In two dimensions, the energy of our system is given by

$$E_{\text{tot}}(\mathbf{r},t) = U(\mathbf{r}) + \frac{k_{\text{eff}}}{2}(\mathbf{v}t - \mathbf{r})^2, \tag{11.12}$$

where $\mathbf{r} \equiv (x,y)$ and $\mathbf{v}$ is arbitrarily oriented on the surface (note that $\mathbf{v} \neq d\mathbf{r}/dt$!). Figure 11.10 shows the total energy corresponding to a periodic potential of the

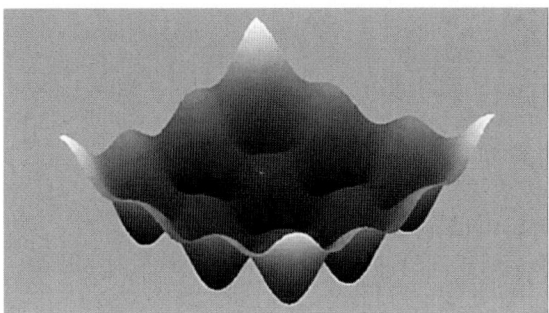

Fig. 11.10. Energy landscape experienced by the FFM tip in 2-D

form

$$U(x,y,t) = -\frac{E_0}{2}\left(\cos\frac{2\pi x}{a} + \cos\frac{2\pi y}{a}\right) + E_1 \cos\frac{2\pi x}{a} \cos\frac{2\pi y}{a}. \quad (11.13)$$

The equilibrium condition becomes

$$\nabla E_{\text{tot}}(\mathbf{r},t) = \nabla U(\mathbf{r}) + k_{\text{eff}}(\mathbf{r} - \mathbf{v}t) = 0. \quad (11.14)$$

The stability of the equilibrium can be described by introducing the Hessian matrix

$$H = \begin{pmatrix} \dfrac{\partial^2 U}{\partial x^2} + k_{\text{eff}} & \dfrac{\partial^2 U}{\partial x \partial y} \\ \dfrac{\partial^2 U}{\partial y \partial x} & \dfrac{\partial^2 U}{\partial y^2} + k_{\text{eff}} \end{pmatrix}. \quad (11.15)$$

When both eigenvalues $\lambda_{1,2}$ of the Hessian are positive, the position of the tip is stable. Figure 11.11 shows these regions for a potential of the form (11.13). The tip

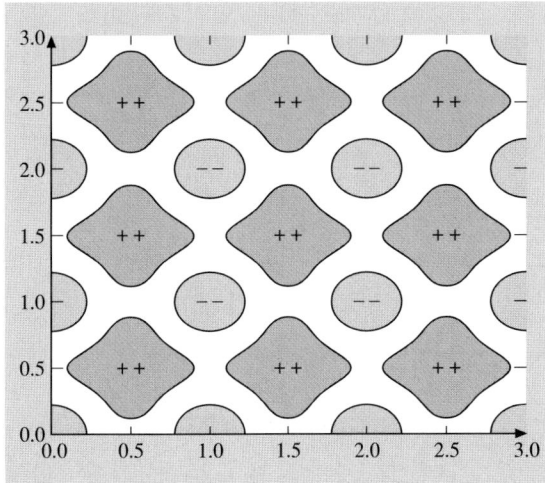

Fig. 11.11. Regions on the tip plane labeled according to the signs of the eigenvalues of the Hessian matrix. (After [24])

follows the cantilever adiabatically as long as it remains in the (++)-region. When the tip is dragged to the border of the region, it suddenly jumps into the next (++)-region. A comparison between a theoretical friction map deduced from the 2-D Tomlinson model and an experimental map acquired by UHV-FFM is given in the next section.

11.2.3 Friction Between Atomically Flat Surfaces

So far we have implicitly assumed that the tip is terminated by only one atom. It is also instructive to consider the case of a periodic surface sliding on another periodic surface. In the Frenkel–Kontorova–Tomlinson (FKT) model, the atoms of one surface are harmonically coupled with their nearest neighbors. We will restrict ourselves to the case of quadratic symmetries, with lattice constants a_1 and a_2 for the upper and lower surfaces, respectively (Fig. 11.12). In this context, the role of *commensurability* is essential. It is well known that any real number z can be represented as a continued fraction:

$$z = N_0 + \cfrac{1}{N_1 + \cfrac{1}{N_2 + \ldots}} . \tag{11.16}$$

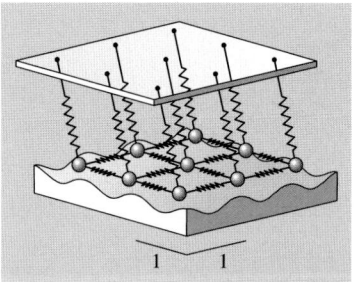

Fig. 11.12. The FKT model in 2-D. (After [25])

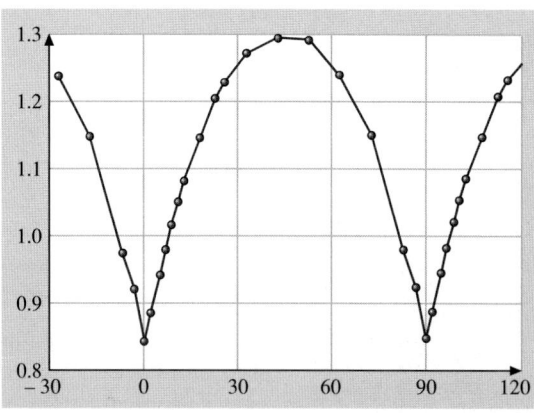

Fig. 11.13. Friction as a function of the sliding angle φ in the 2-D FKT model. (After [25])

The sequence that converges most slowly is obtained when all $N_i = 1$, which corresponds to the *golden mean* $\bar{z} = (\sqrt{5}-1)/2$. In 1-D, *Weiss* and *Elmer* predicted that friction should decrease with decreasing commensurability, the minimum friction being reached when $a_1/a_2 = \bar{z}$ [26].

In 2-D, *Gyalog* and *Thomas* studied the case $a_1 = a_2$, with a misalignment between the two lattices given by an angle θ [25]. When the sliding direction changes, friction also varies from a minimum value (corresponding to the sliding angle $\varphi = \theta/2$) to a maximum value (which is reached when $\varphi = \theta/2 + \pi/4$; see Fig. 11.13). The misfit angle θ is related to the commensurability. Since the misfit angles that give rise to commensurate structure form a dense subset, the dependence of friction on θ should be discontinuous. The numerical simulations performed by Gyalog are in agreement with this conclusion.

11.3 Friction Experiments on the Atomic Scale

Figure 11.14 shows the first atomic-scale friction map, as observed by Mate. The periodicity of the lateral force is the same as that of the atomic lattice of graphite. The series of friction loops in Fig. 11.15 reveals the stick-slip effect discussed in the previous section. The applied loads are in the range of tens of μN. According to the continuum models discussed in Sect. 11.5, these values correspond to contact diameters of 100 nm. A possible explanation for the atomic features observed at such high loads is that graphite flakes may have detached from the surface and adhered to the tip [27]. Another explanation is that the contact between tip and surface consisted of few nm-scale asperities and that the corrugation was not entirely averaged out while sliding. The load dependence of friction as found by Mate is rather linear, with a small friction coefficient $\mu = 0.01$ (Fig. 11.16).

The UHV environment reduces the influence of contaminants on the surface and leads to more precise and reproducible results. *Meyer* et al. [28] obtained a series of interesting results on ionic crystals using the UHV-FFM apparatus de-

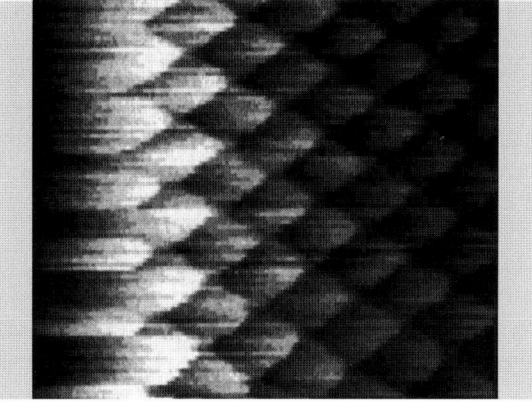

Fig. 11.14. First atomic friction map acquired on graphite with a normal force $F_N = 56\,\mu\text{N}$. Frame size: 2 nm. (After [1])

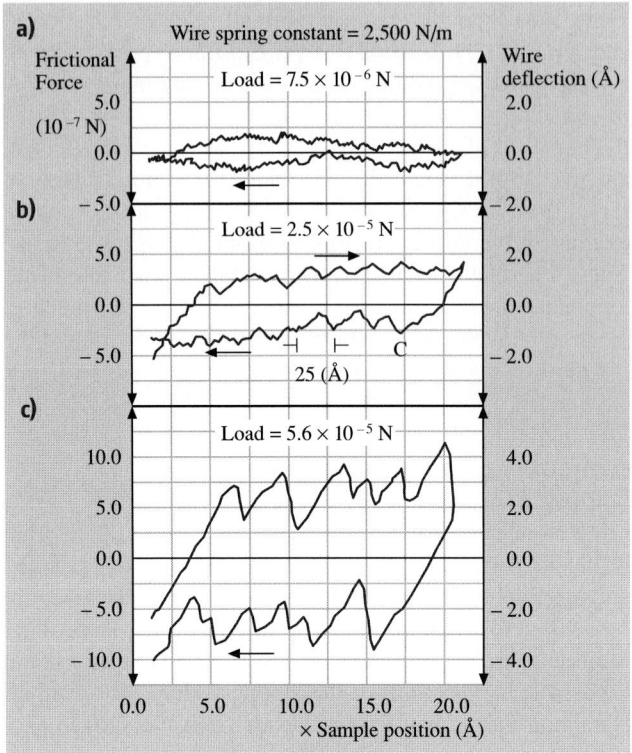

Fig. 11.15. Friction loops on graphite acquired with (**a**) $F_N = 7.5$ μN, (**b**) 24 μN and (**c**) 75 μN. (After [1])

scribed in Sect. 11.1.4. By applying subnanonewton loads to a NaCl surface, *Socoliuc* et al. observed the transition from stick-slip to continuous sliding discussed in Sect. 11.2.1 [29]. In Fig. 11.17, a friction map recorded on KBr(100) is compared with a theoretical map obtained with the 2-D Tomlinson model. The periodicity $a = 0.47$ nm corresponds to the spacing between equally charged ions. No individual defects were observed. One possible reason is that the contact realized by the FFM tip is always formed by many atoms, which superimpose and average their effects. Molecular dynamics (MD) calculations (Sect. 11.7) show that even single-atom contact may cause rather large stresses in the sample, which lead to the motion of defects far away from the contact area. In a rather picturesque analogy, we can say that "defects behave like dolphins that swim away in front of an ocean cruiser" [28].

Lüthi et al. [30] even detected atomic-scale friction on a reconstructed Si(111) 7×7 surface. However, uncoated Si tips and tips coated with Pt, Au, Ag, Cr and Pt/C damaged the sample irreversibly, and the observation of atomic features was achieved only after coating the tips with polytetrafluoroethylene (PTFE), which has

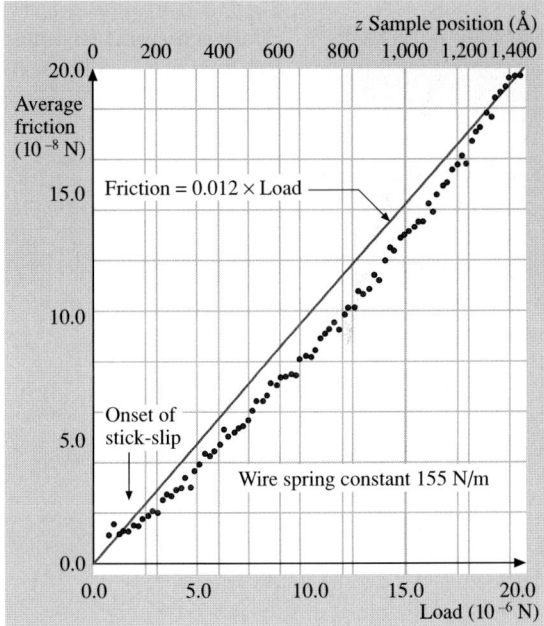

Fig. 11.16. Load dependence of friction on graphite. (After [1])

lubricant properties and does not react with the dangling bonds of Si(111)7 × 7 (Fig. 11.18).

Recently friction was even resolved on the atomic scale on metallic surfaces in UHV [31]. In Fig. 11.19a a reproducible stick-slip process on Cu(111) is shown. Sliding on the (100) surface of copper produced irregular patterns, although atomic features were recognized even in this case (Fig. 11.19b). Molecular dynamics suggests that wear should occur more easily on the Cu(100) surface than on the close-packed Cu(111) (Sect. 11.7). This conclusion was achieved by adopting copper tips in computer simulations. The assumption that the FFM tip used in the experiments was covered by copper is supported by current measurements performed at the same time.

Atomic stick-slip on diamond was observed by *Germann* et al. with an apposite diamond tip prepared by chemical vapor deposition [33] and, a few years later, by *van der Oetelaar* et al. [34] with standard silicon tips. The values of friction vary dramatically depending on the presence or absence of hydrogen on the surface.

Fujisawa et al. [36] measured friction on mica and on MoS_2 with a 2-D FFM apparatus that could also reveal forces perpendicular to the scan direction. The features in Fig. 11.20 correspond to a zigzag tip walk, which is predicted by the 2-D Tomlinson model [37]. Two-dimensional stick-slip on NaF was detected with normal forces below 14 nN, whereas loads of up to 10 μN could be applied to layered materials. The contact between tip and NaF was thus formed by one or a few atoms. A zigzag walk on mica was also observed by *Kawakatsu* et al. using an original 2-D FFM with two laser beams and two quadrant photodetectors [38].

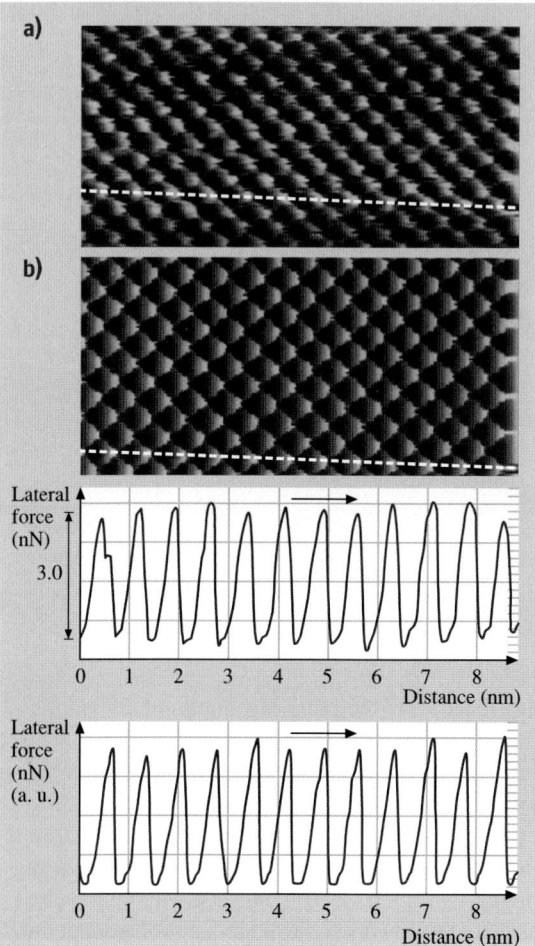

Fig. 11.17. (a) Measured and (b) theoretical friction map on KBr(100). (After [32])

11.3.1 Anisotropy of Friction

The importance of the misfit angle in the reciprocal sliding of two flat surfaces was first observed experimentally by *Hirano* et al. in the contact of two mica sheets [39]. The friction increased when the two surfaces formed commensurate structures, in agreement with the discussion in Sect. 11.2.3. In more recent measurements with a monocrystalline tungsten tip on Si(001), *Hirano* et al. observed *superlubricity* in the incommensurate case [40].

Overney et al. [44] studied the effects of friction anisotropy on a bilayer lipid film. In this case, different molecular alignments resulted in significant variations in the friction. Other measurements of friction anisotropy on single crystals of stearic acid were reported by *Takano* and *Fujihira* [45]. An impressive confirmation of this effect recently came from a dedicated force microscope developed by *Frenken* and

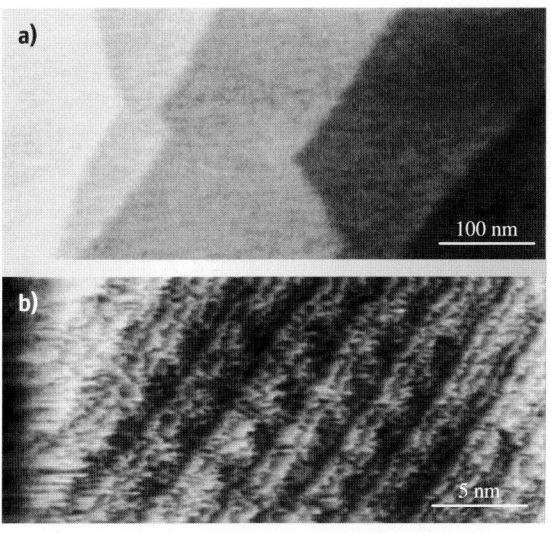

Fig. 11.18. (a) Topography and (b) friction image of Si(111)7×7 measured with a PTFE-coated Si tip. (After [30])

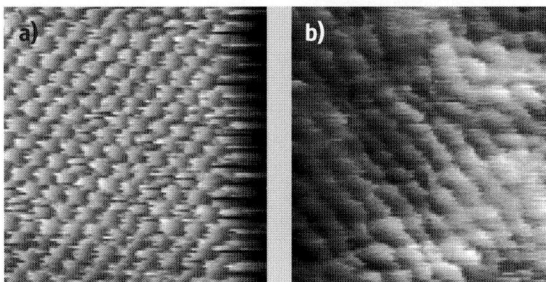

Fig. 11.19. Friction images of (a) Cu(111) and (b) Cu(100). Frame size: 3 nm. (After [35])

coworkers, the Tribolever, which allows quantitative tracking of the scanning force in three dimensions [46]. With this instrument, a flake from a graphite surface was picked up and the lateral forces between the flake and the surface were measured at different angles of rotation. Stick-slip and energy dissipation were only clearly revealed at rotation angles of 0° and 60°, when the two lattices are in registry.

Liley et al. [41] observed flower-shaped islands of a lipid monolayer on mica, which consisted of domains with different molecular orientations (Fig. 11.21). The angular dependence of friction reflects the tilt direction of the alkyl chains of the monolayer, as revealed by other techniques.

Lüthi et al. [42] used the FFM tip to move C_{60} islands, which slide on sodium chloride in UHV without disruption (Fig. 11.22). In this experiment the friction was found to be independent of the sliding direction. This was not the case in other experiments performed by *Sheehan* and *Lieber*, who observed that the misfit angle is relevant when MoO_3 islands are dragged on the MoS_2 surface [47]. In these experiments, sliding was possible only along low index directions. The weak orientation dependence found by *Lüthi* et al. [42] is probably due to the large mismatch of C_{60} on NaCl.

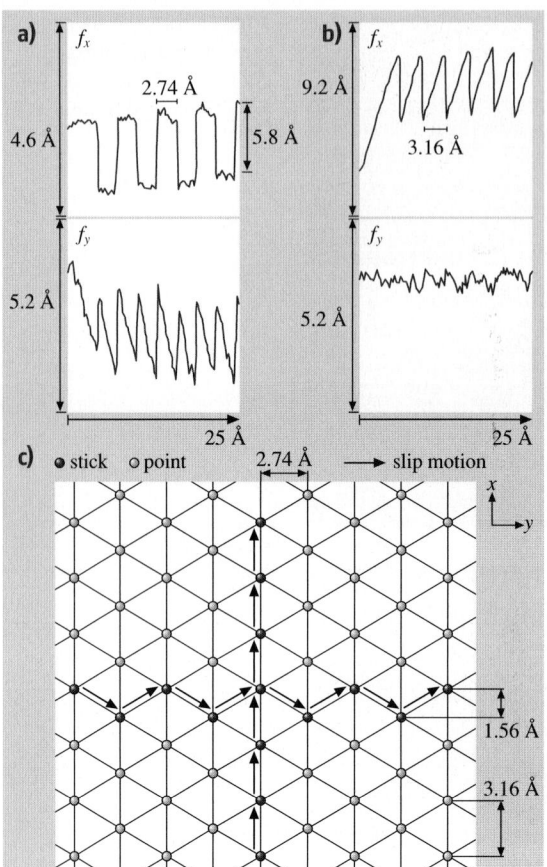

Fig. 11.20. (a) Friction force on MoS$_2$ acquired by scanning along the cantilever and (b) across the cantilever. (c) Motion of the tip on the sample. (After [36])

A recent example of friction anisotropy is related to carbon nanotubes. *Falvo* et al. [43] manipulated nanotubes on graphite using a FFM tip (Fig. 11.23). A dramatic increase in the lateral force was found in directions corresponding to commensurate contact. At the same time, the nanotube motion changed from sliding/rotating to stick-roll.

11.4 Thermal Effects on Atomic Friction

Although the Tomlinson model gives a good interpretation of the basic mechanism of the atomic stick-slip discussed in Sect. 11.2, it cannot explain some minor features observed in the atomic friction. For example, Fig. 11.24 shows a friction loop acquired on NaCl(100). The peaks in the sawtooth profile have different heights, which is in contrast to the result in Fig. 11.9. Another effect is observed if the scan velocity v is varied: the mean friction force increases with the logarithm of v

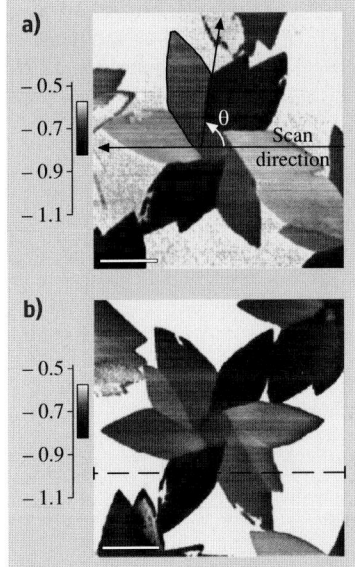

Fig. 11.21. Friction images of a thiolipid monolayer on a mica surface. (After [41])

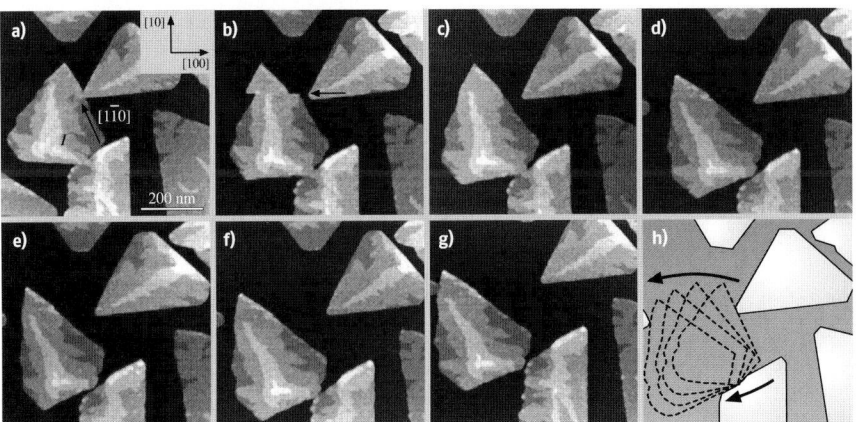

Fig. 11.22. Sequence of topography images of C_{60} islands on NaCl(100). (After [42])

(Fig. 11.25). This effect cannot be interpreted within the mechanical approach in Sect. 11.2 without further assumptions.

11.4.1 The Tomlinson Model at Finite Temperature

Let us focus again on the energy profile discussed in Sect. 11.2.1. For the sake of simplicity, we will assume that $\gamma \gg 1$. At a given time $t < t^*$, the tip jump is prevented by the energy barrier $\Delta E = E(x_{max}, t) - E(x_{min}, t)$, where x_{max} corresponds to the first maximum observed in the energy profile and x_{min} is the actual position of

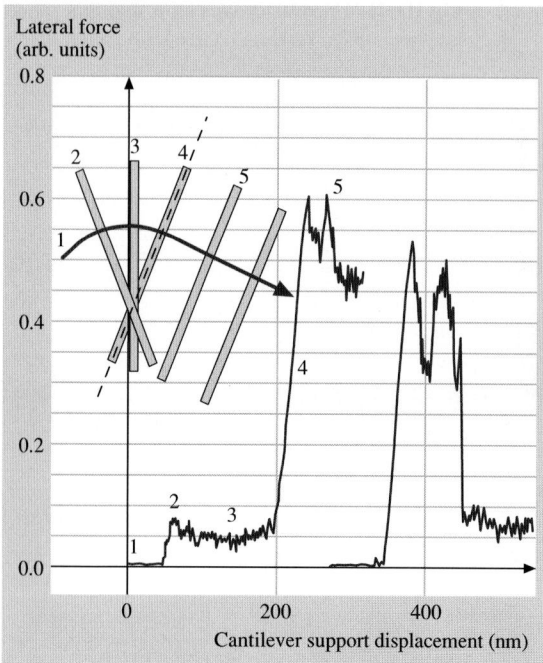

Fig. 11.23. Friction force experienced as a carbon nanotube is rotated into (*left trace*) and out of (*right trace*) commensurate contact. (After [43])

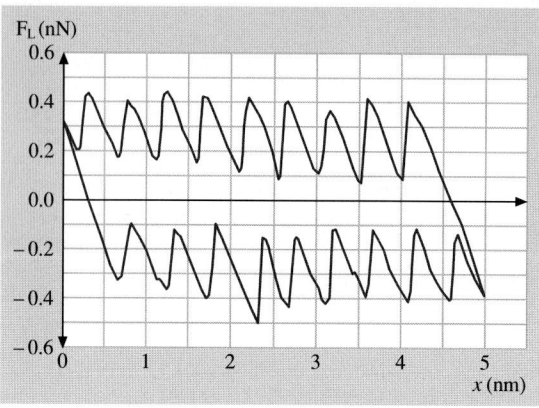

Fig. 11.24. Friction loop on NaCl(100). (After [48])

the tip (Fig. 11.26). The quantity ΔE decreases with time or, equivalently, with the frictional force F_L until it vanishes when $F_L = F^*$ (Fig. 11.27). Close to the critical point, the energy barrier can be written approximately as

$$\Delta E = \lambda(\tilde{F} - F_L), \tag{11.17}$$

where $\tilde{F}$ is close to the critical value $F^* = \pi E_0/a$.

At finite temperature T, the lateral force required to induce a jump is lower than F^*. To estimate the most probable value of F_L at this point, we first consider

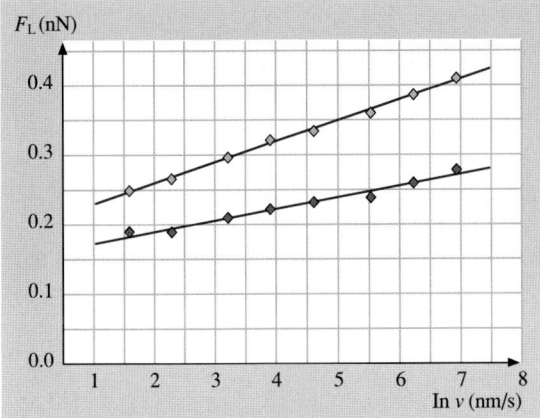

Fig. 11.25. Mean friction force versus scanning velocity on NaCl(100) at $F_N = 0.44\,\text{nN}$ (+) and $F_N = 0.65\,\text{nN}$ (×). (After [48])

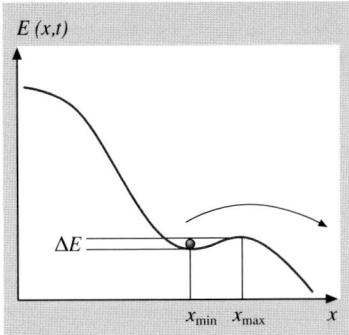

Fig. 11.26. Energy barrier that hinders the tip jump in the Tomlinson model

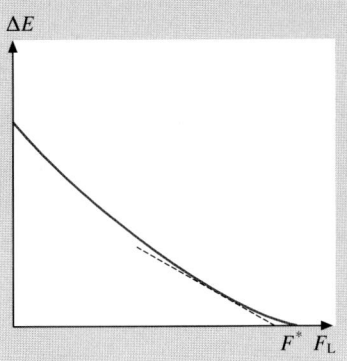

Fig. 11.27. Energy barrier ΔE as a function of the lateral force F_L. The *dashed line* close to the critical value corresponds to the linear approximation (11.17)

the probability p that the tip does *not* jump. The probability p changes with time t according to the master equation

$$\frac{dp(t)}{dt} = -f_0 \exp\left(-\frac{\Delta E(t)}{k_B T}\right) p(t), \tag{11.18}$$

where f_0 is a characteristic frequency of the system. The physical meaning of this frequency is discussed in Sect. 11.4.2. We should note that the probability of a reverse jump is neglected, since in this case the energy barrier that must be overcome is much higher than ΔE. If time is replaced by the corresponding lateral force, the master equation becomes

$$\frac{dp(F_L)}{dF_L} = -f_0 \exp\left(-\frac{\Delta E(F_L)}{k_B T}\right)\left(\frac{dF_L}{dt}\right)^{-1} p(F_L). \tag{11.19}$$

At this point, we substitute

$$\frac{dF_L}{dt} = \frac{dF_L}{dX}\frac{dX}{dt} = k_{\text{eff}} v \tag{11.20}$$

and use the approximation (11.17). The maximum probability transition condition $d^2 p(F)/dF^2 = 0$ then yields

$$F_L(v) = F^* - \frac{k_B T}{\lambda} \ln \frac{v_c}{v} \tag{11.21}$$

with

$$v_c = \frac{f_0 k_B T}{k_{\text{eff}} \lambda}. \tag{11.22}$$

Thus, the lateral force depends logarithmically on the sliding velocity, as observed experimentally. However, approximation (11.17) does not hold when the tip jump occurs very close to the critical point $x = x^*$, which is the case at high velocities. In this instance, the factor $(dF_L dt)^{-1}$ in (11.19) is small and, consequently, the probability $p(t)$ does not change significantly until it suddenly approaches 1 when $t \rightarrow t^*$. Thus friction is constant at high velocities, in agreement with the classical Coulomb's law of friction [28].

Sang et al. [50] observed that the energy barrier close to the critical point is better approximated by a relation like

$$\Delta E = \mu(F^* - F_L)^{3/2}. \tag{11.23}$$

The same analysis performed using approximation (11.23) instead of (11.17) leads to the expression [51]

$$\frac{\mu(F^* - F_L)^{3/2}}{k_B T} = \ln \frac{v_c}{v} - \ln \sqrt{1 - \frac{F^*}{F_L}}, \tag{11.24}$$

where the critical velocity v_c is now

$$v_c = \frac{\pi\sqrt{2}}{2} \frac{f_0 k_B T}{k_{\text{eff}} a}. \tag{11.25}$$

The velocity v_c discriminates between two different regimes. If $v \ll v_c$, the second logarithm in (11.24) can be neglected, which leads to the logarithmic dependence

$$F_L(v) = F^* - \left(\frac{k_B T}{\mu}\right)^{2/3} \left(\ln \frac{v_c}{v}\right)^{2/3}. \tag{11.26}$$

In the opposite case, $v \gg v_c$, the term on the left in (11.23) is negligible and

$$F_L(v) = F^* \left(1 - \frac{v_c}{v}\right)^2. \tag{11.27}$$

In such a case, the lateral force F_L tends to F^*, as expected.

In a recent work, *Reimann* et al. distinguished between the dissipation that occurs in the tip apex and that in the substrate volume in contact with the tip [52]. After the initial logarithmic increase, the velocity dependence of friction changes in different ways, depending on the relative contribution of the tip apex to the total dissipation. A 'friction plateau' is only predicted when $\theta \approx 0.5$ over a limited velocity range. At lower or higher values of θ, friction is expected to increase, or, respectively, decrease beyond the critical velocity v_c.

The thermally activated Tomlinson model has been recently extended to two dimensions by *Fasolino* and coworkers [53].

11.4.2 Velocity Dependence of Friction

The velocity dependence of friction was only recently studied by FFM. *Zwörner* et al. observed that friction between silicon tips and diamond, graphite or amorphous carbon is constant with scan velocities of a few μm/s [54]. The friction decreased when v was reduced below 1 μm/s. In their experiment on lipid films on mica (Sect. 11.3.1), *Gourdon* et al. [49] explored a range of velocities from 0.01 to 50 μm/s and found a critical velocity $v_c = 3.5$ μm/s that discriminates between an increasing friction and a constant friction regime (Fig. 11.28). Although these results were not explained by thermal activation, we argue that the previous theoretical discussion gives the correct interpretative key. A clear observation of a logarithmic dependence of friction on the micrometer scale was reported by *Bouhacina* et al., who studied friction on triethoxysilane molecules and polymers grafted on silica with sliding velocities of up to $v = 300$ μm/s [55]. The result was explained with a thermally activated Eyring model, which does not differ significantly from the model discussed in the previous subsection [56, 57].

The first measurements on the atomic scale were performed by *Bennewitz* et al. on copper and sodium chloride [31, 48]; in both cases a logarithmic dependence

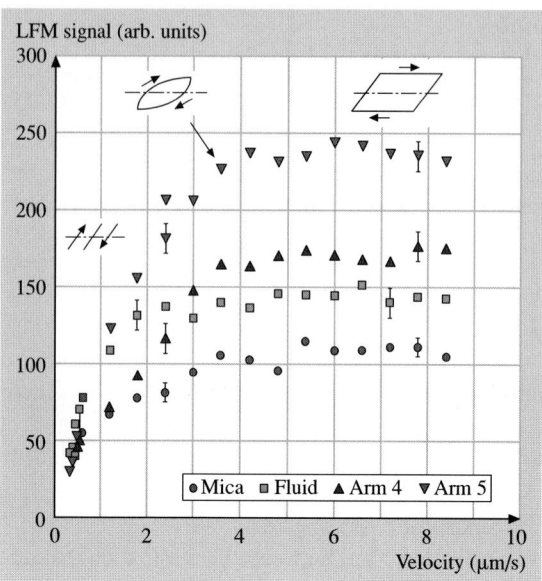

Fig. 11.28. Velocity dependence of friction on mica and on lipid films with different orientations (*arms 4 and 5*) and in a fluid phase. (After [49])

of friction was revealed up to $v < 1$ μm/s (Fig. 11.25), in agreement with (11.21). Higher values of velocities were not explored, due to the limited range of the scan frequencies possible with FFM on the atomic scale. The same limitation does not allow a clear distinction between (11.21) and (11.26) when interpreting the experimental results.

At this point we would like to discuss the physical meaning of the characteristic frequency f_0. With a lattice constant a of a few angstroms and an effective spring constant k_{eff} of about 1 N/m, which are typical of FFM experiments, (11.25) gives a value of a few hundred kHz for f_0. This is the characteristic range in which the torsional eigenfrequencies of the cantilevers are located in both contact and noncontact modes (Fig. 11.29). Future work may clarify whether or not f_0 must be identified with these frequencies.

To conclude this section, we should emphasize that the increase in friction with increasing velocity is ultimately related to the materials and the environment in which the measurements are realized. In a humid environment, *Riedo* et al. observed that the surface wettability plays an important role [59]. Friction *decreases*

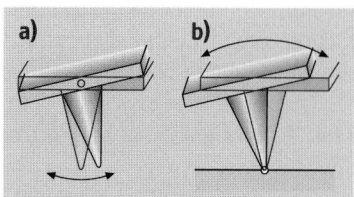

Fig. 11.29. Torsional modes of cantilever oscillation (**a**) when the tip is free and (**b**) when the tip is in contact with a surface. (After [58])

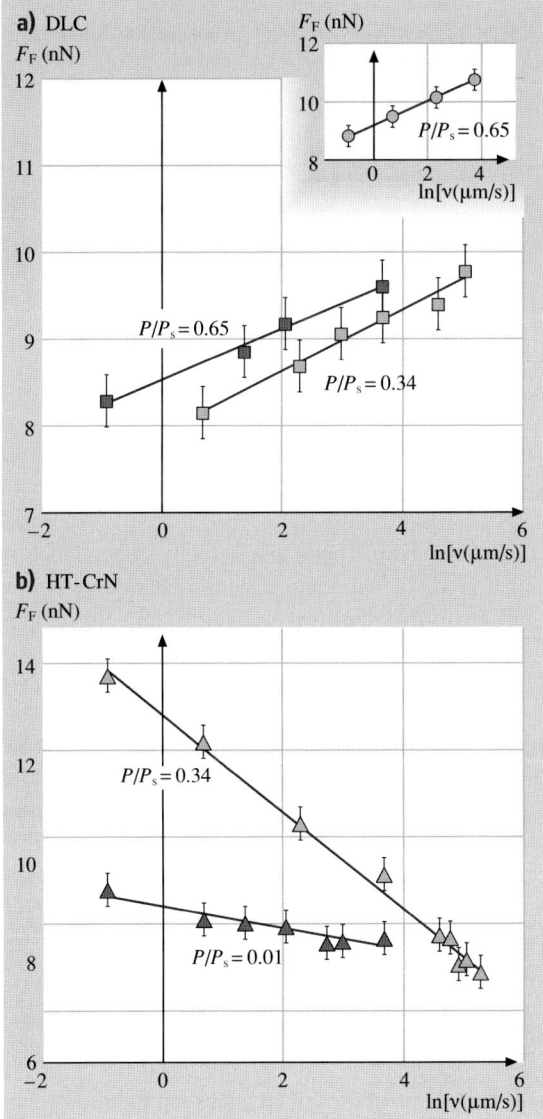

Fig. 11.30. Friction versus sliding velocity (**a**) on hydrophobic surfaces and (**b**) on hydrophilic surfaces. (After [59])

with increasing velocity on hydrophilic surfaces, and the rate of this decrease depends drastically on humidity. A logarithmic increase is again found on partially hydrophobic surfaces (Fig. 11.30). These results were interpreted considering the thermally activated nucleation of water bridges between tip and sample asperities, as discussed in the cited reference.

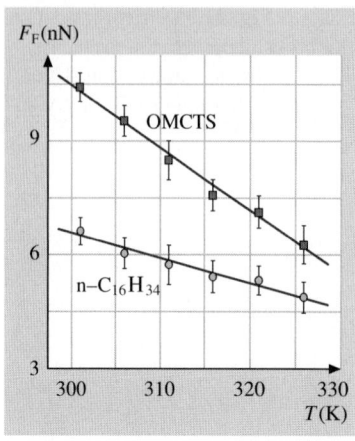

Fig. 11.31. Temperature dependence of friction on *n*-hexadecane and octamethylcyclotetrasiloxane. (After [60])

11.4.3 Temperature Dependence of Friction

Thus far we have used thermal activation to explain the velocity dependence of friction. The same mechanism also predicts that friction should change with temperature. The master equation (11.18) shows that the probability of a tip jump is reduced at low temperatures T until it vanishes when $T = 0$. Within this limit case, thermal activation is excluded, and the lateral force F_L is equal to F^*, independent of the scanning velocity v.

To our knowledge, stick-slip processes at low temperatures have not been reported. A significant increase in the mean friction with decreasing temperature was recentlymeasured by *He* et al. [60] by FFM (Fig. 11.31). Neglecting logarithmic contributions, (11.21) and (11.26) predict that $(F^* - F_L) \approx T$ and $(F^* - F_L) \approx T^{2/3}$ respectively for the temperature dependence of friction. Although He et al. applied a linear fit to their data, the 30 K range that they considered is again not large enough to prove that a $T^{2/3}$ fit would be preferable.

11.5 Geometry Effects in Nanocontacts

Friction is ultimately related to the real shape of the contact between the sliding surfaces. On the macroscopic scale, the contact between two bodies is studied within the context of continuum mechanics, which is based on the elasticity theory developed by Hertz in the nineteenth century. Various FFM experiments have shown that continuum mechanics is still valid down to contact areas just a few nanometers in size. Only when contact is reduced to few atoms does the continuum frame become unsuitable, and other approaches like molecular dynamics become necessary. This section will deal with continuum mechanics theory; molecular dynamics will be discussed in Sect. 11.7.

11.5.1 Continuum Mechanics of Single Asperities

The lateral force F_L between two surfaces in reciprocal motion depends on the size of the real area of contact, A, which can be a few orders of magnitude smaller than the apparent area of contact. The simplest assumption is that friction is proportional to A; the proportionality factor is called the *shear strength* σ [61]:

$$F_L = \sigma A. \tag{11.28}$$

For plastic deformation, the asperities are compressed until the pressure (p) equals a certain yield value, p^*. The resulting contact area is thus $A = F_N/p^*$, and the well-known Amontons' law is obtained: $F_L = \mu F_N$, where $\mu = \sigma/p^*$ is the *coefficient of friction*. The same idea can be extended to contacts formed by many asperities, and it leads again to Amontons' law. The simplicity of this analysis explains why most friction processes were related to plastic deformation for a long time. Such a mechanism, however, should provoke quick disruption of surfaces, which is not observed in practice.

Elastic deformation can be easily studied in the case of a sphere of radius R pressed against a flat surface. In this case, the contact area is

$$A(F_N) = \pi \left(\frac{R}{K}\right)^{2/3} F_N^{2/3}, \tag{11.29}$$

where $K = 3E^*/4$ and E^* is an effective Young's modulus, related to the Young's moduli (E_1 and E_2) and the Poisson numbers (ν_1 and ν_2) of sphere and plane, by the following relation [62]:

$$\frac{1}{E^*} = \frac{1-\nu_1^2}{E_1} + \frac{1-\nu_2^2}{E_2}. \tag{11.30}$$

The result $A \propto F_N^{2/3}$ contrasts with Amontons' law. However, a linear relation between F_L and F_N can be obtained for contacts formed from several asperities in particular cases. For example, the area of contact between a flat surface and a set of asperities with an exponential height distribution and the same radius of curvature R depends linearly on the normal force F_N [63]. The same conclusion holds approximately even for a Gaussian height distribution. However, the hypothesis that the radii of curvature are the same for all asperities is not realistic. A general model was recently proposed by *Persson*, who analytically derived the proportionality between contact area and load for a large variety of elastoplastic contacts formed by surfaces with arbitrary roughnesses [64]. However, this discussion is not straightforward and goes beyond the purposes of this section.

Further effects are observed if adhesive forces between the asperities are taken into account. If the range of action of these forces is smaller than the elastic deformation, (11.29) is extended to the Johnson–Kendall–Roberts (JKR) relation

$$A(F_N) = \pi \left(\frac{R}{K}\right)^{2/3} \times \left(F_N + 3\pi\gamma R + \sqrt{6\pi\gamma R F_N + (3\pi\gamma R)^2}\right)^{2/3}, \tag{11.31}$$

where γ is the surface tension of the sphere and plane [66]. The real contact area at zero load is finite and the sphere can be detached only by pulling it away with a certain force. This is also true in the opposite case, in which the range of action of adhesive forces is larger than the elastic deformation. In this case, the relation between contact area and load takes the simple form

$$A(F_N) = \pi \left(\frac{R}{K}\right)^{2/3} (F_N - F_{\text{off}})^{2/3} , \qquad (11.32)$$

where F_{off} is the negative load required to break the contact. The Hertz-plus-offset relation (11.32) can be derived from the Derjaguin–Muller–Toporov (DMT) model [67]. To discriminate between the JKR or DMT models, *Tabor* introduced a nondimensional parameter

$$\Phi = \left(\frac{9R\gamma^2}{4K^2 z_0^3}\right)^{1/3} , \qquad (11.33)$$

where z_0 is the equilibrium distance during contact. The JKR model can be applied if $\Phi > 5$; the DMT model holds when $\Phi < 0.1$ [68]. For intermediate values of Φ, the Maugis–Dugdale model [69] could reasonably explain experimental results (Sect. 11.5.3).

11.5.2 Dependence of Friction on Load

The FFM tip represents a single asperity sliding on a surface. The previous discussion suggests a nonlinear dependence of friction on the applied load, provided that continuum mechanics is applicable. Schwarz et al. observed the Hertz-plus-offset relation (11.32) on graphite, diamond, amorphous carbon and C_{60} in an argon atmosphere (Fig. 11.32). In their measurements, they used well-defined spherical tips with radii of curvature of tens of nanometers, obtained by contaminating silicon tips with amorphous carbon in a transmission electron microscope. In order to compare the tribological behavior of different materials, *Schwarz* et al. suggested the introduction of an effective coefficient of friction, $\tilde{C}$, which is independent of the tip curvature [65].

Meyer et al., *Carpick* et al., and *Polaczyc* et al. performed friction measurements in UHV in agreement with JKR theory [17, 70, 71]. Different materials were considered (ionic crystals, mica and metals) in these experiments. In order to correlate the lateral and normal forces with improved statistics, Meyer et al. applied an original 2-D histogram technique (Fig. 11.33). Carpick et al. extended the JKR relation (11.32) to include nonspherical tips. In the case of an axisymmetric tip profile $z \propto r^{2n}$ ($n > 1$), it can be proven analytically that the increase in the friction becomes less pronounced with increasing n (Fig. 11.34).

11.5.3 Estimation of the Contact Area

In contrast to other tribological instruments, such as the surface force apparatus [73], the contact area cannot be measured directly by FFM. Indirect methods are provided

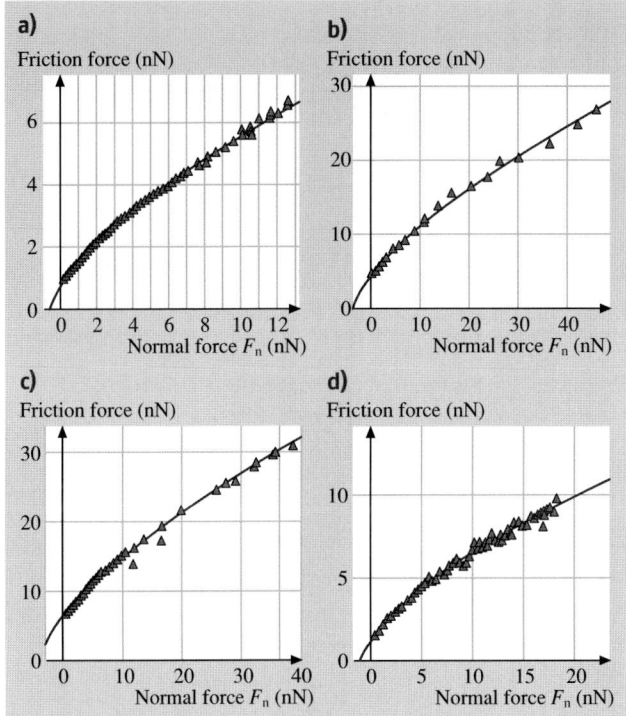

Fig. 11.32. Friction versus load curve on amorphous carbon in argon atmosphere. Curves (**a**)–(**d**) refer to tips with different radii of curvature. (After [65])

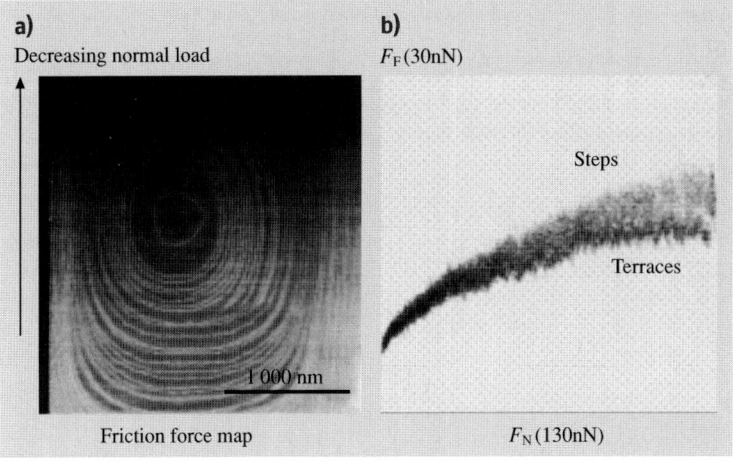

Fig. 11.33. (**a**) Friction force map on NaCl(100). The load is decreased from 140 to 0 nN (jump-off point) during imaging. (**b**) 2-D-histogram of (**a**). (After [17])

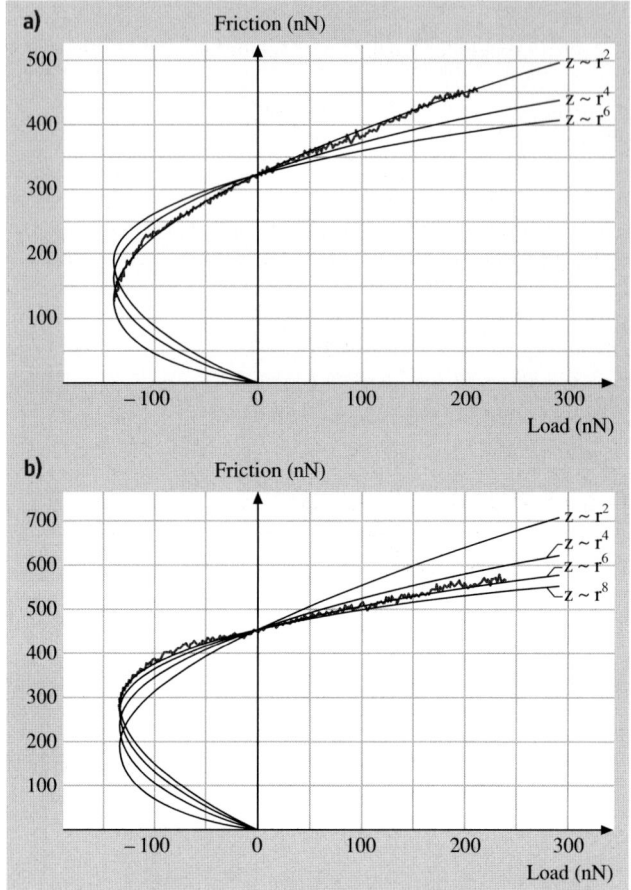

Fig. 11.34. Friction versus load curves (**a**) for a spherical tip and (**b**) for a blunted tip. The *solid curves* are determined using the JKR theory. (After [70])

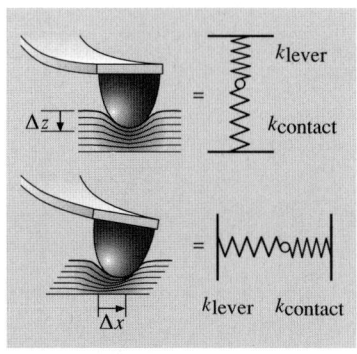

Fig. 11.35. Sketch of normal and lateral stiffness of the contact between tip and surface. (After [72])

by contact stiffness measurements. The contact between the FFM tip and the sample can be modeled by a series of two springs (Fig. 11.35). The effective constant k_{eff}^z of the series is given by

$$\frac{1}{k_{\text{eff}}^z} = \frac{1}{k_{\text{contact}}^z} + \frac{1}{c_N}, \tag{11.34}$$

where c_N is the normal spring constant of the cantilever and k_{contact}^z is the normal stiffness of the contact. This quantity is related to the radius of the contact area (a) by the simple relation

$$k_{\text{contact}}^z = 2aE^*, \tag{11.35}$$

where E^* is the effective Young's modulus introduced previously [74]. Typical values of k_{contact}^z are an order of magnitude larger than c_N, however, and practical application of (11.34) is not possible.

Carpick et al. independently suggested an alternative method [72,75]. According to various models, the *lateral* contact stiffness of the contact between a sphere and a flat surface is [76]

$$k_{\text{contact}}^x = 8aG^*, \tag{11.36}$$

where the effective shear stress G^* is defined by

$$\frac{1}{G^*} = \frac{2 - v_1^2}{G_1} + \frac{2 - v_2^2}{G_2} \tag{11.37}$$

G_1, G_2 are the shear moduli of the sphere and the plane, respectively. The contact between the FFM tip and the sample can again be modeled by a series of springs (Fig. 11.35). The effective constant k_{eff}^x of the series is given by

$$\frac{1}{k_{\text{eff}}^x} = \frac{1}{k_{\text{contact}}^x} + \frac{1}{k_{\text{tip}}^x} + \frac{1}{c_L}, \tag{11.38}$$

where c_L is the lateral spring constant of the cantilever and k_{contact}^x is the lateral stiffness of the contact. As suggested by Lantz, (11.38) also includes the lateral stiffness of the tip, k_{tip}^x, which can be comparable to the lateral spring constant. The effective spring constant k_{eff}^x is simply given by the slope dF_L/dx of the friction loop (Sect. 11.2.1). Once k_{contact}^x is determined, the contact radius a is easily estimated by (11.36).

The lateral stiffness method was applied to contacts between silicon nitride and muscovite mica in air and between $NbSe_2$ and graphite in UHV. The dependences of both the spring constant k_{eff}^x and the lateral force F_L on the load F_N were explained within the same models (JKR and Maugis–Dugdale, respectively), which confirms that friction is proportional to the contact area for the range of loads applied (up to $F_N = 40$ nN in both experiments).

Enachescu et al. estimated the contact area by measuring the contact conductance on diamond as a function of the applied load [77, 78]. Their experimental data were fitted with the DMT model, which was also used to explain the dependence of friction on load. Since the contact conductance is proportional to the contact area, the validity of the hypothesis (11.28) was confirmed again.

11.6 Wear on the Atomic Scale

If the normal force F_N applied by the FFM exceeds a critical value, which depends on the tip shape and on the material under investigation, the surface topography is permanently modified. In some cases wear is exploited to create patterns with well-defined shapes. Here we will focus on the mechanisms that act at the nanometer scale, where recent experiments have demonstrated the unique ability of the FFM to both scratch and image surfaces down to the atomic scale.

11.6.1 Abrasive Wear on the Atomic Scale

Lüthi et al. observed the appearance of wear at very low loads, i.e. $F_N = 3$ nN, for ionic crystals [32]. Atomically resolved images of the damage produced by scratching the FFM tip area on potassium bromide were obtained very recently by *Gnecco* et al. [79]. In Fig. 11.36, a small mound that has piled up at the end of a groove on KBr(100) is shown at different magnifications. The groove was created a few minutes before imaging by repeatedly scanning with the normal force $F_N = 21$ nN. The image shows a lateral force map acquired with a load of about 1 nN; no atomic features were observed in the corresponding topographic signal. Figure 11.36a,b shows that the debris extracted from the groove recrystallized with the same atomic arrangement of the undamaged surface, which suggests that the wear process occurred in a similar way to epitaxial growth, assisted by the microscope tip.

Although it is not that easy to understand how wear is initiated and how the tip transports the debris, important indications are given by the profile of the lateral force F_L recorded while scratching. Figure 11.37 shows some friction loops acquired when the tip was scanned laterally on areas of size 5×5 nm^2. The mean lateral force multiplied by the scanned length gives the total energy dissipated in the

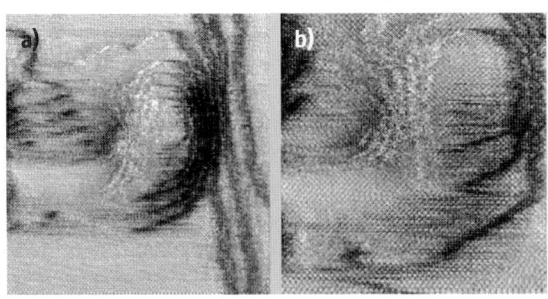

Fig. 11.36. Lateral force images acquired at the end of a groove scratched 256 times with a normal force $F_N = 21$ nN. Frame sizes: (**a**) 39 nm, (**b**) 25 nm

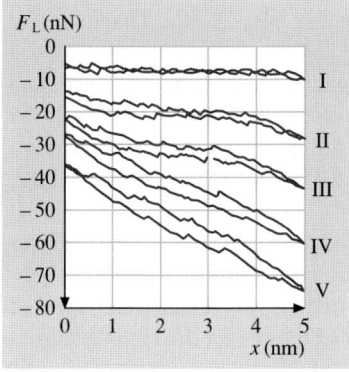

Fig. 11.37. Friction loops acquired while scratching the KBr surface on 5 nm long lines with different loads $F_N = 5.7$ to 22.8 nN. (After [79])

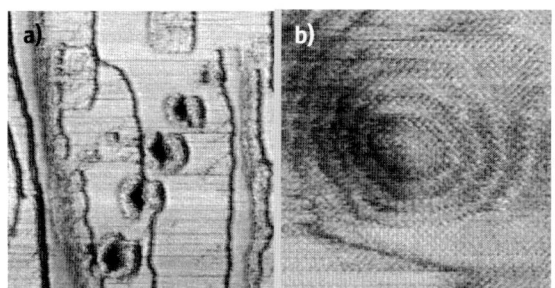

Fig. 11.38. (a) Lateral force images of the pits produced with $F_N = 5.7$ to 22.8 nN. Frame size: 150 nm; (b) Detailed image of the fourth pit from the top with pseudo-atomic resolution. Frame size: 20 nm

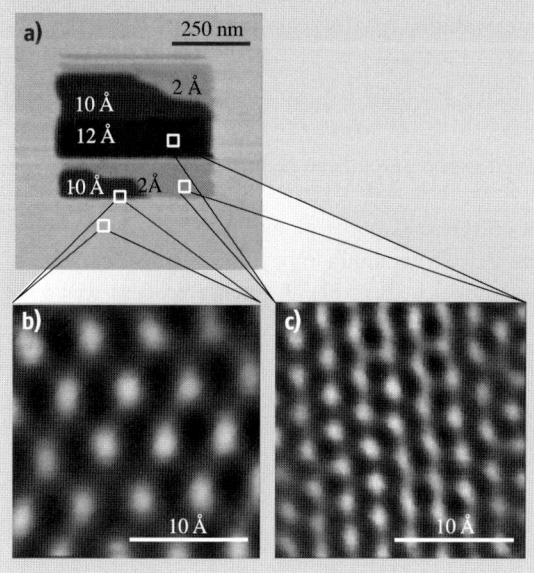

Fig. 11.39. (a) Topography image of an area scratched on muscovite mica with $F_N = 230$ nN; (b),(c) Fourier-filtered images of different regions. (After [80])

process. The tip movement produces the pits shown in Fig. 11.38a. Thanks to the pseudo-atomic resolution obtained by FFM (Fig. 11.38b), the number of removed atoms can be determined from lateral force images, which allow us to estimate that 70% of the dissipated energy went into wearless friction [79]. Figures 11.37 and 11.38 clearly show that the damage increases with increasing load. On the other hand, changing the scan velocity v between 25 and 100 nm/s did not produce any significant variation in the wear process.

A different kind of wear was observed on layered materials. *Kopta* et al. [80] removed layers from a muscovite mica surface by scratching with normal force $F_N = 230$ nN (Fig. 11.39a). Fourier-filtered images acquired on very small areas revealed the different periodicities of the underlying layers, which reflect the complex structure of the muscovite mica (Fig. 11.39b,c).

11.6.2 Contribution of Wear to Friction

The mean lateral force detected while scratching a KBr(100) surface with a fixed load $F_N = 11$ nN is shown in Fig. 11.40. A rather continuous increase in "friction" with the number of scratches N is observed, which can be approximated with the following exponential law:

$$F_L = F_0 \, e^{-N/N_0} + F_\infty \left(1 - e^{-N/N_0}\right). \tag{11.39}$$

Equation (11.39) is easily interpreted by assuming that friction is proportional to contact area $A(N)$, and that time evolution of $A(N)$ can be described by

$$\frac{dA}{dN} = \frac{A_\infty - A(N)}{N_0}. \tag{11.40}$$

Here A_∞ is the limit area in which the applied load can be balanced without scratching.

To interpret their experiment on mica, Kopta et al. assumed that wear is initiated by atomic defects. When the defects accumulate beyond a critical concentration, they grow to form the scars shown in Fig. 11.39. Such a process was once

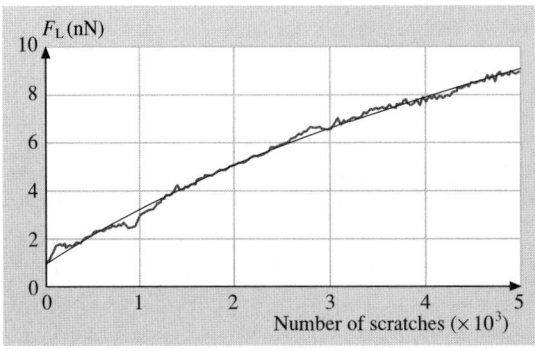

Fig. 11.40. Mean value of the lateral force during repeated scratching with $F_N = 11$ nN on a 500 nm line. (After [79])

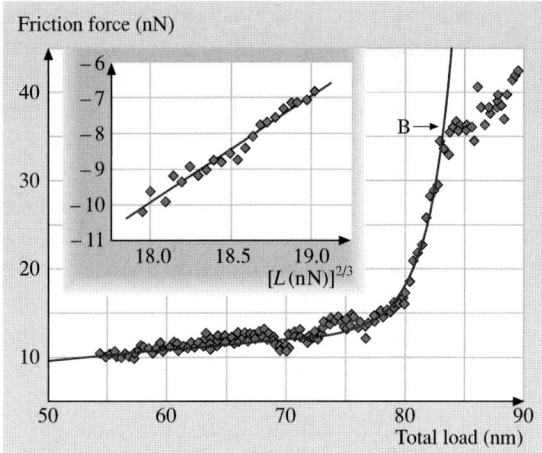

Fig. 11.41. Friction versus load curve during the creation of a hole in the muscovite mica. (After [80])

again related to thermal activation. The number of defects created in the contact area $A(F_N)$ is

$$N_{\text{def}}(F_N) = t_{\text{res}} n_0 A(F_N) f_0 \exp\left(-\frac{\Delta E}{k_B T}\right), \tag{11.41}$$

where t_{res} is the residence time of the tip, n_0 is the surface density of atoms, and f_0 is the frequency of attempts to overcome the energy barrier ΔE to break a Si–O bond, which depends on the applied load. When the defect density reaches a critical value, a hole is nucleated. The friction force during the creation of a hole was also estimated via thermal activation by Kopta et al., who derived the formula

$$F_L = c(F_N - F_{\text{off}})^{2/3} + \gamma F_N^{2/3} \exp\left(B_0 F_N^{2/3}\right). \tag{11.42}$$

The first term on the right gives the wearless dependence of friction in the Hertz-plus-offset model (Sect. 11.5.1); the second term is the contribution of the defect production. The agreement between (11.42) and experiment can be observed in Fig. 11.41.

11.7 Molecular Dynamics Simulations of Atomic Friction and Wear

Section 11.5 mentioned that small sliding contacts can be modeled by continuum mechanics. This modeling has several limitations. The first and most obvious is that continuum mechanics cannot account for atomic-scale processes like atomic stick-slip. While this limit can be overcome by semiclassical descriptions like the Tomlinson model, one definite limit is the determination of contact stiffness for contacts with a radius of a few nanometers. Interpreting experimental results with the methods introduced in Sect. 11.5.3 regularly yields contact radii of atomic or even smaller size, in clear contradiction to the minimal contact size given by adhesion forces. Macroscopic quantities such as shear modulus or pressure fail to describe the mechanical behavior of these contacts. Microscopic modeling that includes the atomic structure of the contact is therefore required. This is usually achieved through a *molecular dynamics* (MD) simulation of the contact. In such simulations, the sliding contact is set up by boundaries of fixed atoms in relative motion and the atoms of the contact, which are allowed to relax their positions according to interactions between each pair of atoms. Methods of computer simulation used in tribology are discussed elsewhere in this book. In this section we will discuss simulations that can be directly compared to experimental results showing atomic friction processes. The major outcome of the simulations beyond the inclusion of the atomic structure is the importance of including displacement of atoms in order to correctly predict forces. Then we present simulation studies that include wear of the tip or the surface.

11.7.1 Molecular Dynamics Simulations of Friction Processes

The first experiments that exhibited the features of atomic friction were performed on layered materials, often graphite. A theoretical study of forces between an atomically sharp diamond tip and the graphite surface has been reported by *Tang* et al. [81]. The authors found that the forces were significantly dependent on distance. The strongest contrasts appeared at different distances for normal and lateral forces due to the strong displacement of surface atoms. The order of magnitude found in this study was one nanonewton, much less than in most experimental reports, which indicated that contact areas of far larger dimensions were realized in such experiments. Tang et al. determined that the distance dependence of the forces could even change the symmetrical appearance of the lateral forces observed. The experimental situation has also been studied in numerical simulations using a simplified one-atom potential for the tip–surface interaction but including the spring potential of the probing force sensor [37]. The motivation for these studies was the observation of a hexagonal pattern in the friction force, while the surface atoms of graphite are ordered in a honeycomb structure. The simulations revealed how the jump path of the tip under lateral force is dependent on the force constant of the probing force sensor.

Surfaces of ionic crystals have become model systems for studies in atomic friction. Atomic stick-slip behavior has been observed by several groups with a lateral force modulation of the order of 1 nN. Pioneering work in atomistic simulation of sliding contacts has been done by *Landman* et al. The first ionic system studied was a CaF_2 tip sliding over a $CaF_2(111)$ surface [82]. In MD calculations with controlled temperature, the tip was first moved toward the surface up to the point at which an attractive normal force of −3 nN acted on the tip. Then the tip was moved laterally, and the lateral force determined. An oscillation with a periodicity corresponding to the atomic periodicity of the surface and with an amplitude decreasing from 8 nN was found. Inspection of the atomic positions revealed a wear process from shear cleavage of the tip. This transfer of atoms between tip and surface plays a crucial role in atomic friction studies, as was shown by *Shluger* et al. [83]. These authors simulated a MgO tip scanning laterally over a LiF(100) surface. Initially an irregular oscillation of the system's energy is found together with transfer of atoms between surface and tip. After a while, the tip apex structure is changed by adsorption of Li and F ions in such a way that nondestructive sliding with perfectly regular energy oscillations correlating with the periodicity of the surface was observed. The authors called this effect self-lubrication and speculate that, in general, dynamic self-organization of the surface material on the tip might promote the observation of periodic forces. In a less costly molecular mechanics study, in which the forces were calculated for each fixed tip–sample configuration, *Tang* et al. produced lateral and normal force maps for a diamond tip over a NaCl(100) surface, including such defects as vacancies and a step [84]. As with the studies mentioned before, they found that significant atomic force contrast can be expected for tip–sample distances of less than 0.35 nm, while distances below 0.15 nm result in destructive forces. For the idealized conditions of scanning at constant height in this regime, the authors predict that even atomic-sized defects could be imaged. Experimentally, these conditions cannot be stabilized in the static modes used so far in lateral force measurements. However, dynamic modes of force microscopy have given atomic resolution of defects within the distance regime of 0.2 and 0.4 nm [85]. Recent experimental progress in atomic friction studies of surfaces of ionic crystals include the velocity dependence of lateral forces and atomic-scale wear processes. Such phenomena are not yet accessible by MD studies: the experimental scanning timescale is too far from the atomic relaxation timescales that govern MD simulations. Furthermore, the number of freely transferable atoms that can be included in a simulation is simply limited by meaningful calculation time.

Landman et al. also simulated a system of high reactivity, namely a silicon tip sliding over a silicon surface [86]. A clear stick-slip variation in the lateral force was observed for this situation. Strong atom displacements created an interstitial atom

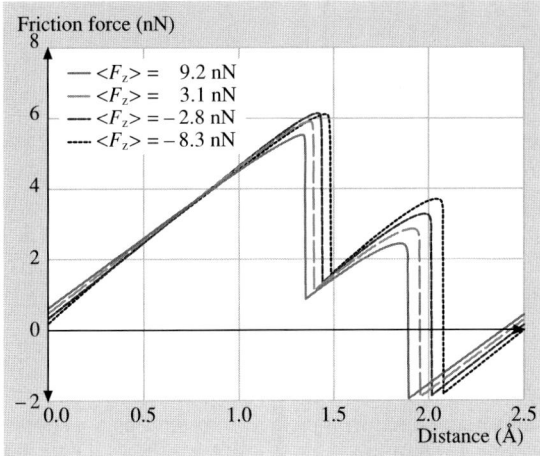

Fig. 11.42. Lateral force acting on a Cu(111) tip in matching contact with a Cu(111) substrate as a function of the sliding distance at different loads. (After [87])

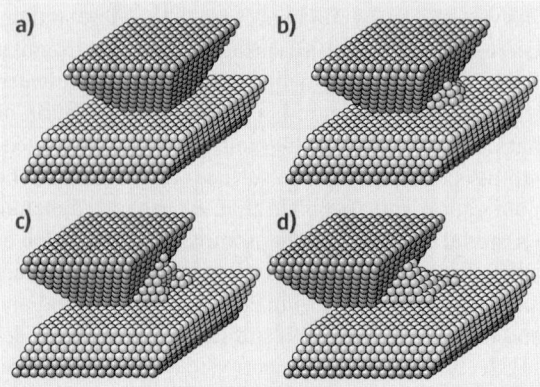

Fig. 11.43. Snapshot of a Cu(100) tip on a Cu(100) substrate during sliding. (**a**) Starting configuration; (**b**)–(**d**) snapshots after two, four, and six slips. (After [87])

under the influence of the tip, which was annealed as the tip moved on. Permanent damage was predicted, however, when the tip enters the repulsive force regime. Although the simulated Si(111) surface is not experimentally accessible, it should be mentioned that the tip had to be passivated by a Teflon layer on the Si(111)7×7 reconstructed surface before nondestructive contact mode measurements became possible (Sect. 11.3). It is worth noting that the simulations for the Cu(111) surface revealed a linear relation between contact area and mean lateral force, similar to classical macroscopic laws.

Wear processes are predicted by several MD studies of metallic sliding over metallic surfaces, which will be discussed in the following section. For a (111)-terminated copper tip sliding over a Cu(111) surface, however, Sørensen et al. found that nondestructive sliding is possible while the lateral force exhibits the sawtooth-like shape characteristic of atomic stick-slip (Fig. 11.42). In contrast, a Cu(100) surface would be disordered by a sliding contact (Fig. 11.43). This difference between the (100) and (111) plane (as well as the absolute lateral forces) has been confirmed experimentally (Sect. 11.3).

11.7.2 Molecular Dynamics Simulations of Abrasive Wear

The long timescales characteristic of wear processes and the large amount of material involved make any attempt to simulate these mechanisms on a computer a tremendous challenge. Despite this, MD can provide useful insights on the mechanisms of removal and deposition of single atoms by the FFM tip, which is not the kind of information directly observable experimentally. Complex processes like abrasive wear and nanolithography can be investigated only within approximate classical mechanics.

The observation made by Livshits and Shluger, that the FFM tip undergoes a process of self-lubrication when scanning ionic surfaces (Sect. 11.7.1), proves that friction and wear are strictly related phenomena. In their MD simulations on copper, *Sørensen* et al. considered not only ordered (111)- and (100)-terminated tips, but also amorphous structures obtained by "heating" the tip to high temperatures [87]. The lateral motion of the neck thus formed revealed stick-slip behavior due to combined sliding and stretching, as well as ruptures, accompanied by deposition of debris on the surface (Fig. 11.44).

To our knowledge, only a few examples of abrasive wear simulations on the atomic scale have been reported. *Buldum* and *Ciraci*, for instance, studied nanoindentation and sliding of a sharp Ni(111) tip on Cu(110) and a blunt Ni(001) tip on Cu(100) [89]. In the case of the sharp tip, quasiperiodic variations of the lateral force were observed, due to stick-slip involving phase transition. One layer of the asperity was deformed to match the substrate during the first slip and then two asperity layers merged into one in a structural transition during the second slip. In the case of the blunt tip, the stick-slip was less regular.

Different results have been reported in which the tip is harder than the underlying sample. *Komanduri* et al. considered an infinitely hard Ni indenter scratching single crystal aluminium at extremely low depths (Fig. 11.45) [88]. A linear relation between friction and load was found, with a high coefficient of friction $\mu = 0.6$, independent of the scratch depth. Nanolithography simulations were recently performed by *Fang* et al. [90], who investigated the role of the displacement of the FFM tip along the direction of slow motion between a scan line and the next one. They found a certain correlation with FFM experiments on silicon films coated with aluminium.

11.8 Energy Dissipation in Noncontact Atomic Force Microscopy

Historically, the measurement of energy dissipation induced by tip–sample interaction has been the domain of friction force microscopy, where the sharp AFM tip slides over a sample that it is in gentle contact with. The origins of dissipation in friction are related to phonon excitation, electronic excitation and irreversible changes of the surface. In a typical stick-slip experiment, the energy dissipated in a single atomic slip event is of the order of 1 eV.

However, the lateral resolution of force microscopy in the contact mode is limited by a minimum contact area of several atoms due to adhesion between tip and sample.

This problem has been overcome in noncontact dynamic force microscopy. In the dynamic mode, the tip oscillates with a constant amplitude A of typically 1–20 nm at the eigenfrequency f of the cantilever, which shifts by Δf due to interaction forces between tip and sample. This technique is described in detail in Part B of this book.

Dissipation also occurs in the noncontact mode of force microscopy, where the atomic structure of tip and sample are reliably preserved. In this dynamic mode, the damping of the cantilever oscillation can be deduced from the excitation amplitude A_{exc} required to maintain the constant tip oscillation amplitude on resonance.

Compared to friction force microscopy, the interpretation of noncontact AFM (nc-AFM) experiments is complicated due to the perpendicular oscillation of the

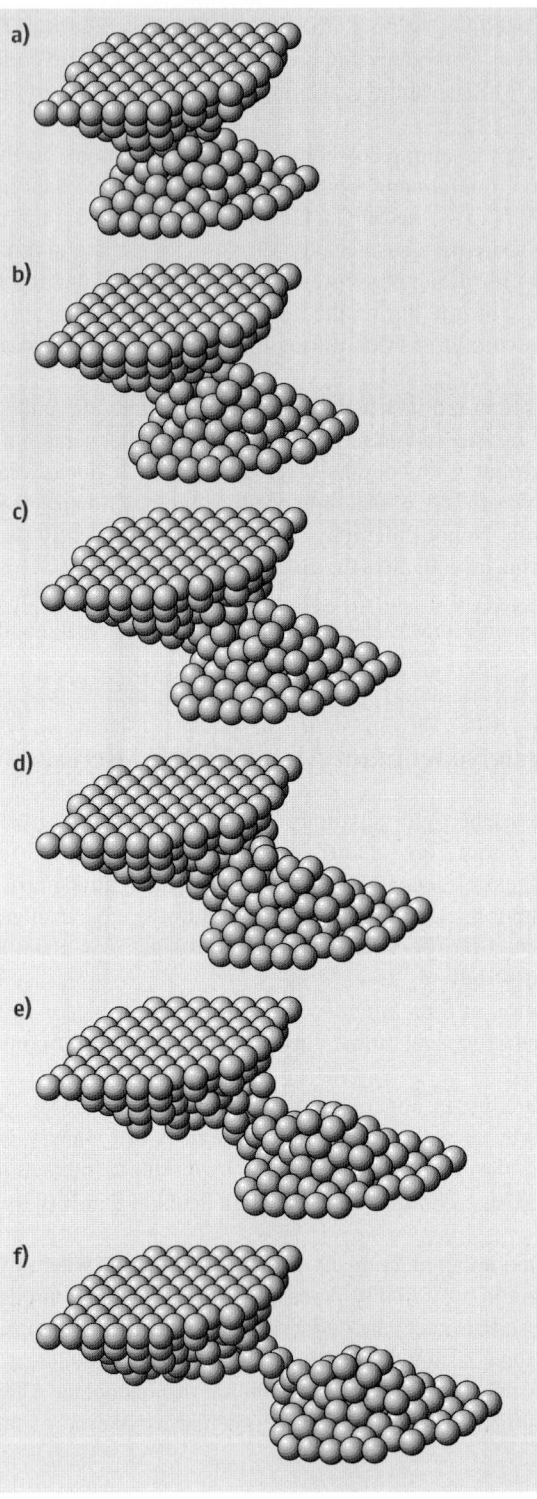

Fig. 11.44. Snapshot of a Cu(100) neck during shearing starting from configuration (**a**). The upper substrate was displaced 4.2 Å between subsequent pictures. (After [87])

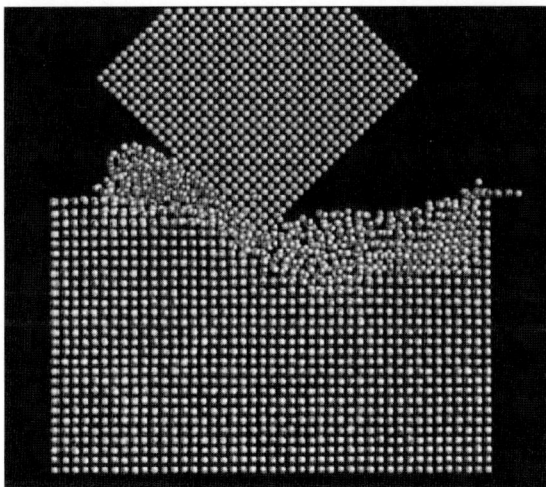

Fig. 11.45. MD simulation of a scratch realized with an infinitely hard tool. (After [88])

tip, typically with an amplitude that is large compared to the minimum tip–sample separation. Another problem is to relate the measured damping of the cantilever to the different origins of dissipation.

In all dynamic force microscopy measurements, a power dissipation P_0 caused by internal friction of the freely oscillating cantilever is observed, which is proportional to the eigenfrequency ω_0 and to the square of the amplitude A and is inversely proportional to the known Q value of the cantilever. When the tip–sample distance is reduced, the tip interacts with the sample and therefore additional damping of the oscillation is encountered. This extra dissipation P_{ts} caused by the tip–sample interaction can be calculated from the excitation signal A_{exc} [91].

The observed energy losses per oscillation cycle (100 meV) [92] are roughly similar to the 1 eV energy loss in the contact slip process. When estimating the contact area in the contact mode for a few atoms, the energy dissipation per atom that can be associated with a bond being broken and reformed is also around 100 meV.

The idea of relating the additional damping of the tip oscillation to dissipative tip–sample interactions has recently attracted much attention [93]. The origins of this additional dissipation are manifold: one may distinguish between apparent energy dissipation (for example from inharmonic cantilever motion, artefacts from the phase controller, or slow fluctuations round the steady state solution [93, 94]), velocity-dependent dissipation (for example electric and magneticfield-mediated Joule dissipation [95, 96]) and hysteresis-related dissipation (due to atomic instabilities [97, 98] or hysteresis due to adhesion [99]).

Forces and dissipation can be measured by recording Δf and A_{exc} simultaneously during a typical AFM experiment. Many experiments show true atomic contrast in topography (controlled by Δf) and in the dissipation signal A_{exc} [100]; however, the origin of the atomic energy dissipation process is still not completely resolved.

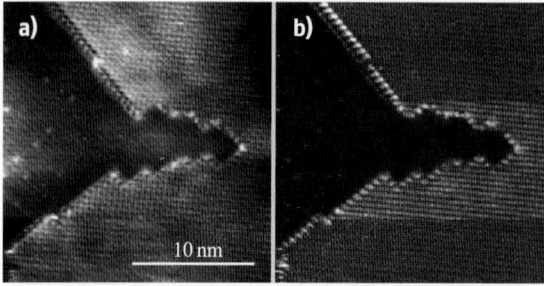

Fig. 11.46. (a) Topography and (b) A_{exc} images of a NaCl island on Cu(111). The tip changes after 1/4 of the scan, thereby changing the contrast in the topography and enhancing the contrast in A_{exc}. After 2/3 of the scan, the contrast from the *lower part* of the image is reproduced, indicating that the tip change was reversible. (After [85])

To prove that the observed atomic-scale variation in the damping is indeed due to atomic-scale energy dissipation and not an artefact of the distance feedback, *Loppacher* et al. [92] carried out a nc-AFM experiment on Si(111)-7×7 at constant height (with distance feedback stopped). Frequency-shift and dissipation exhibit atomic-scale contrast, demonstrating true atomic-scale variations in force and dissipation.

Strong atomic-scale dissipation contrast at step edges has been demonstrated in a few experiments (NaCl on Cu [85] or measurements on KBr [101]). In Fig. 11.46, ultrathin NaCl islands grown on Cu(111) are shown. As shown in Fig. 11.46a, the island edges have a higher contrast than the NaCl terrace and they show atomically resolved corrugation. The strongly enhanced contrast of the step edges and kink sites could be attributed to a slower decay of the electric field and to easier relaxation of the positions of the ions at these locations. The dissipation image shown in Fig. 11.46b was recorded at the same time. To establish a direct spatial correspondence between the excitation and the topography signal, the match between topography and A_{exc} has been studied on many images. Sometimes the topography and A_{exc} are in phase, sometimes they are shifted a little bit, and sometimes A_{exc} is at a minimum when topography is at a maximum. The local contrast formation thus depends strongly on the atomic tip structure. In fact, the strong dependence of the dissipation contrast on the atomic state of the tip apex is impressively confirmed by the tip change observed in the experimental images shown in Fig. 11.46b. The dissipation contrast is seriously enhanced, while the topography contrast remains almost unchanged. The dissipation clearly depends strongly on the state of the tip and exhibits more short-range character than the frequency shift.

More directly related to friction measurements, where the tip is sliding in contact with the sample, are nc-AFM experiments, where the tip is oscillating parallel to the surface. *Stowe* et al. [102] oriented cantilever beams with in-plane tips perpendicular to the surface, so that the tip motion was approximately parallel to the surface. The noncontact damping of the lever was used to measure localized electrical Joule dissipation. They were able to image the dopant density for n- and p-type silicon samples with 150 nm spatial resolution. A dependence of U_{ts}^2 on the tip–sample voltage was found for the dissipation, as proposed by *Denk* and *Pohl* [95] for electric field Joule dissipation.

Stipe et al. [103] measured the noncontact friction between a Au(111) surface and a gold-coated tip with the same set-up. They observed the same U_{ts}^2 dependence of the bias voltage and a distance dependence that follows the power law $1/d^n$, where n is between 1.3 and 3 [103, 104]. A substantial electric-field is present even when the external bias voltage is zero. The presence of inhomogeneous tip–sample electric fields is difficult to avoid, even under the best experimental conditions. Although this dissipation is electrical in origin, the detailed mechanism is not totally clear. The most straightforward mechanism is to assume that inhomogeneous fields emanating from the tip and the sample induce surface charges in the nearby metallic sample. When the tip moves, currents are induced, causing ohmic dissipation [95, 102]. But in metals with good electrical conductivity, ohmic dissipation is insufficient to account for the observed effect [105]. Thus the tip–sample electric field must have an additional effect, such as driving the motions of adsorbates and surface defects.

When exciting the torsional oscillation of commercial, rectangular AFM cantilevers, the tip is oscillating approximately parallel to the surface. In this mode, it was possible to measure lateral forces acting on the tip at step edges and near impurities quantitatively [58]. Enhanced energy dissipation was also observed at the impurities. When the tip is moved further toward the sample, contact formation transforms the nearly free torsional oscillation of the cantilever into a different mode, with the tip–sample contact acting as a hinge. When this contact is formed, a rapid increase in the power required to maintain a constant tip oscillation amplitude and a positive frequency shift are found. The onsets of the simultaneously recorded damping and positive frequency shift are sharp and essentially coincide. It is assumed that these changes indicate the formation of a tip–sample contact. Two recent studies [106, 107] report on the use of the torsional eigenmode to measure the elastic properties of the tip–sample contact, where the tip is in contact with the sample and the shear stiffness depends on the normal load.

Kawagishi et al. [108] scanned with lateral amplitudes of the order of 10 pm to 3 nm; their imaging technique showed up contrast between graphite terraces, silicon and silicon dioxide, graphite and mica. Torsional self-excitation showed nanometric features of self-assembled monolayer islands due to different lateral dissipations.

Giessibl et al. [109] recently established true atomic resolution of both conservative and dissipative forces by lateral force microscopy. The interaction between a single tip atom oscillated parallel to an Si(111)-7×7 surface was measured. A dissipation energy of up to 4 eV per oscillation cycle was found, which is explained by the plucking action of one atom onto the other, as described by *Tomlinson* in 1929 [23].

A detailed review of dissipation phenomena in noncontact force microscopy has been given by *Hug* [110].

11.9 Conclusion

Over the last 20 years, two instrumental developments have stimulated scientific activities in the field of nanotribology. On the one hand, the invention and development of friction force microscopy has allowed us to quantitatively study single-asperity friction. As we have discussed in this chapter, atomic processes are observed using forces of around 1 nN (forces related to single chemical bonds). On the other hand, the enormous increase in achievable computing power has provided the basis for molecular dynamics simulations of systems containing several hundreds of atoms. These methods allow the development of the atomic structure in a sliding contact to be analyzed and the forces to be predicted.

The most prominent observation of atomic friction is stick-slip behavior with the periodicity of the surface atomic lattice. Semiclassical models can explain experimental findings, including the velocity dependence, which is a consequence of the thermal activation of slip events. Classical continuum mechanics can also describe the load dependence of friction in contacts with an extension of several tens of nanometers. However, when we try to apply continuum mechanics to contacts formed at just ten atoms, obviously wrong numbers result (for the contact radius for instance). Only comparison with atomistic simulations can provide a full, meaningful picture of the physical parameters of such sliding contacts. These simulations predict a close connection between wear and friction, in particular the transfer of atoms between surface and tip, which in some cases can even lower the friction in a process of self-lubrication.

First experiments have succeeded in studying the onset of wear with atomic resolution. Research into microscopic wear processes will certainly grow in importance as nanostructures are produced and their mechanical properties exploited. Simulations of such processes involving the transfer of thousands of atoms will become feasible with further increases in computing power. Another aspect of nanotribology is the expansion of atomic friction experiments toward surfaces with well-defined roughnesses. In general, the problem of bridging the gap between single-asperity experiments on well-defined surfaces and macroscopic friction should be approached, both experimentally and via modeling.

References

1. C. M. Mate, G. M. McClelland, R. Erlandsson, S. Chiang: Atomic-scale friction of a tungsten tip on a graphite surface, Phys. Rev. Lett. **59**, 1942–1945 (1987)
2. G. Binnig, C. F. Quate, Ch. Gerber: Atomic force microscope, Phys. Rev. Lett. **56**, 930–933 (1986)
3. O. Marti, J. Colchero, J. Mlynek: Combined scanning force and friction microscopy of mica, Nanotechnology **1**, 141–144 (1990)
4. G. Meyer, N. Amer: Simultaneous measurement of lateral and normal forces with an optical-beam-deflection atomic force microscope, Appl. Phys. Lett. **57**, 2089–2091 (1990)

5. G. Neubauer, S.R. Cohen, G.M. McClelland, D.E. Horn, C.M. Mate: Force microscopy with a bidirectional capacitance sensor, Rev. Sci. Instrum. **61**, 2296–2308 (1990)
6. G.M. McClelland, J.N. Glosli: Friction at the atomic scale. In: *NATO ASI Series E*, Vol. 220, ed. by L. Singer, H.M. Pollock (Kluwer, Dordrecht 1992) pp. 405–425
7. R. Linnemann, T. Gotszalk, I.W. Rangelow, P. Dumania, E. Oesterschulze: Atomic force microscopy and lateral force microscopy using piezoresistive cantilevers, J. Vacuum Sci. Technol. B **14**, 856–860 (1996)
8. M. Nonnenmacher, J. Greschner, O. Wolter, R. Kassing: Scanning force microscopy with micromachined silicon sensors, J. Vacuum Sci. Technol. B **9**, 1358–1362 (1991)
9. R. Lüthi: Untersuchungen zur Nanotribologie und zur Auflösungsgrenze im Ultrahochvakuum mittels Rasterkraftmikroskopie. Ph.D. Thesis (Univ. of Basel, Basel 1996)
10. J. Cleveland, S. Manne, D. Bocek, P.K. Hansma: A nondestructive method for determining the spring constant of cantilevers for scanning force microscopy, Rev. Sci. Instrum. **64**, 403–405 (1993)
11. J.L. Hutter, J. Bechhoefer: Calibration of atomic-force microscope tips, Rev. Sci. Instrum. **64**, 1868–1873 (1993)
12. H.J. Butt, M. Jaschke: Calculation of thermal noise in atomic-force microscopy, Nanotechnology **6**, 1–7 (1995)
13. J.M. Neumeister, W.A. Ducker: Lateral, normal, and longitudinal spring constants of atomic-force microscopy cantilevers, Rev. Sci. Instrum. **65**, 2527–2531 (1994)
14. D.F. Ogletree, R.W. Carpick, M. Salmeron: Calibration of frictional forces in atomic force microscopy, Rev. Sci. Instrum. **67**, 3298–3306 (1996)
15. E. Gnecco: AFM study of friction phenomena on the nanometer scale. Ph.D. Thesis (Univ. of Genova, Genova 2001)
16. U.D. Schwarz, P. Köster, R. Wiesendanger: Quantitative analysis of lateral force microscopy experiments, Rev. Sci. Instrum. **67**, 2560–2567 (1996)
17. E. Meyer, R. Lüthi, L. Howald, M. Bammerlin, M. Guggisberg, H.-J. Güntherodt: Site-specific friction force spectroscopy, J. Vacuum Sci. Technol. B **14**, 1285–1288 (1996)
18. S.S. Sheiko, M. Möller, E.M.C.M. Reuvekamp, H.W. Zandberger: Calibration and evaluation of scanning-force microscopy probes, Phys. Rev. B **48**, 5675 (1993)
19. F. Atamny, A. Baiker: Direct imaging of the tip shape by AFM, Surf. Sci. **323**, L314 (1995)
20. J.S. Villarrubia: Algorithms for scanned probe microscope image simulation, surface reconstruction, and tip estimation, J. Res. Natl. Inst. Stand. Technol. **102**, 425–454 (1997)
21. L. Howald, E. Meyer, R. Lüthi, H. Haefke, R. Overney, H. Rudin, H.-J. Güntherodt: Multifunctional probe microscope for facile operation in ultrahigh vacuum, Appl. Phys. Lett. **63**, 117–119 (1993)
22. Q. Dai, R. Vollmer, R.W. Carpick, D.F. Ogletree, M. Salmeron: A variable temperature ultrahigh vacuum atomic force microscope, Rev. Sci. Instrum. **66**, 5266–5271 (1995)
23. G.A. Tomlinson: A molecular theory of friction, Philos. Mag. Ser. **7**, 905 (1929)
24. T. Gyalog, M. Bammerlin, R. Lüthi, E. Meyer, H. Thomas: Mechanism of atomic friction, Europhys. Lett. **31**, 269–274 (1995)
25. T. Gyalog, H. Thomas: Friction between atomically flat surfaces, Europhys. Lett. **37**, 195–200 (1997)
26. M. Weiss, F.J. Elmer: Dry friction in the Frenkel–Kontorova–Tomlinson model: Static properties, Phys. Rev. B **53**, 7539–7549 (1996)
27. J.B. Pethica: Comment on "Interatomic forces in scanning tunneling microscopy: Giant corrugations of the graphite surface", Phys. Rev. Lett. **57**, 3235 (1986)

28. E. Meyer, R. M. Overney, K. Dransfeld, T. Gyalog: *Nanoscience, Friction and Rheology on the Nanometer Scale* (World Scientific, Singapore 1998)
29. A. Socoliuc, R. Bennewitz, E. Gnecco, E. Meyer: Transition from stick-slip to continuous sliding in atomic friction: Entering a new regime of ultralow friction, Phys. Rev. Lett. **92**, 134301 (2004)
30. L. Howald, R. Lüthi, E. Meyer, H.-J. Güntherodt: Atomic-force microscopy on the Si(111)7×7 surface, Phys. Rev. B **51**, 5484–5487 (1995)
31. R. Bennewitz, T. Gyalog, M. Guggisberg, M. Bammerlin, E. Meyer, H.-J. Güntherodt: Atomic-scale stick-slip processes on Cu(111), Phys. Rev. B **60**, R11301–R11304 (1999)
32. R. Lüthi, E. Meyer, M. Bammerlin, L. Howald, H. Haefke, T. Lehmann, C. Loppacher, H.-J. Güntherodt, T. Gyalog, H. Thomas: Friction on the atomic scale: An ultrahigh vacuum atomic force microscopy study on ionic crystals, J. Vacuum Sci. Technol. B **14**, 1280–1284 (1996)
33. G. J. Germann, S. R. Cohen, G. Neubauer, G. M. McClelland, H. Seki: Atomic-scale friction of a diamond tip on diamond (100) and (111) surfaces, J. Appl. Phys. **73**, 163–167 (1993)
34. R. J. A. van den Oetelaar, C. F. J. Flipse: Atomic-scale friction on diamond(111) studied by ultrahigh vacuum atomic force microscopy, Surf. Sci. **384**, L828–L835 (1997)
35. R. Bennewitz, E. Gnecco, T. Gyalog, E. Meyer: Atomic friction studies on well-defined surfaces, Tribol. Lett. **10**, 51–56 (2001)
36. S. Fujisawa, E. Kishi, Y. Sugawara, S. Morita: Atomic-scale friction observed with a two-dimensional frictional-force microscope, Phys. Rev. B **51**, 7849–7857 (1995)
37. N. Sasaki, M. Kobayashi, M. Tsukada: Atomic-scale friction image of graphite in atomic-force microscopy, Phys. Rev. B **54**, 2138–2149 (1996)
38. H. Kawakatsu, T. Saito: Scanning force microscopy with two optical levers for detection of deformations of the cantilever, J. Vacuum Sci. Technol. B **14**, 872–876 (1996)
39. M. Hirano, K. Shinjo, R. Kaneko, Y. Murata: Anisotropy of frictional forces in muscovite mica, Phys. Rev. Lett. **67**, 2642–2645 (1991)
40. M. Hirano, K. Shinjo, R. Kaneko, Y. Murata: Observation of superlubricity by scanning tunneling microscopy, Phys. Rev. Lett. **78**, 1448–1451 (1997)
41. M. Liley, D. Gourdon, D. Stamou, U. Meseth, T. M. Fischer, C. Lautz, H. Stahlberg, H. Vogel, N. A. Burnham, C. Duschl: Friction anisotropy and asymmetry of a compliant monolayer induced by a small molecular tilt, Science **280**, 273–275 (1998)
42. R. Lüthi, E. Meyer, H. Haefke, L. Howald, W. Gutmannsbauer, H.-J. Güntherodt: Sled-type motion on the nanometer scale: Determination of dissipation and cohesive energies of C_{60}, Science **266**, 1979–1981 (1994)
43. M. R. Falvo, J. Steele, R. M. Taylor, R. Superfine: Evidence of commensurate contact and rolling motion: AFM manipulation studies of carbon nanotubes on HOPG, Tribol. Lett. **9**, 73–76 (2000)
44. R. M. Overney, H. Takano, M. Fujihira, W. Paulus, H. Ringsdorf: Anisotropy in friction and molecular stick-slip motion, Phys. Rev. Lett. **72**, 3546–3549 (1994)
45. H. Takano, M. Fujihira: Study of molecular scale friction on stearic acid crystals by friction force microscopy, J. Vacuum Sci. Technol. B **14**, 1272–1275 (1996)
46. M. Dienwiebel, G. Verhoeven, N. Pradeep, J. Frenken, J. Heimberg, H. Zandbergen: Superlubricity of graphite, Phys. Rev. Lett. **92**, 126101 (2004)
47. P. E. Sheehan, C. M. Lieber: Nanotribology and nanofabrication of MoO_3 structures by atomic force microscopy, Science **272**, 1158–1161 (1996)

48. E. Gnecco, R. Bennewitz, T. Gyalog, Ch. Loppacher, M. Bammerlin, E. Meyer, H.-J. Güntherodt: Velocity dependence of atomic friction, Phys. Rev. Lett. **84**, 1172–1175 (2000)
49. D. Gourdon, N. A. Burnham, A. Kulik, E. Dupas, F. Oulevey, G. Gremaud, D. Stamou, M. Liley, Z. Dienes, H. Vogel, C. Duschl: The dependence of friction anisotropies on the molecular organization of LB films as observed by AFM, Tribol. Lett. **3**, 317–324 (1997)
50. Y. Sang, M. Dubé, M. Grant: Thermal effects on atomic friction, Phys. Rev. Lett. **87**, 174301 (2001)
51. E. Riedo, E. Gnecco, R. Bennewitz, E. Meyer, H. Brune: Interaction potential and hopping dynamics governing sliding friction, Phys. Rev. Lett. **91**, 084502 (2003)
52. P. Reimann, M. Evstigneev: Nonmonotonic velocity dependence of atomic friction, Phys. Rev. Lett. **93**, 230802 (2004)
53. C. Fusco, A. Fasolino: Velocity dependence of atomic-scale friction: A comparative study of the one- and two-dimensional Tomlinson model, Phys. Rev. B **71**, 45413 (2005)
54. O. Zwörner, H. Hölscher, U. D. Schwarz, R. Wiesendanger: The velocity dependence of frictional forces in point-contact friction, Appl. Phys. A **66**, 263–267 (1998)
55. T. Bouhacina, J. P. Aimé, S. Gauthier, D. Michel, V. Heroguez: Tribological behavior of a polymer grafted on silanized silica probed with a nanotip, Phys. Rev. B **56**, 7694–7703 (1997)
56. H. J. Eyring: The activated complex in chemical reactions, J. Chem. Phys. **3**, 107 (1937)
57. J. N. Glosli, G. M. McClelland: Molecular dynamics study of sliding friction of ordered organic monolayers, Phys. Rev. Lett. **70**, 1960–1963 (1993)
58. O. Pfeiffer, R. Bennewitz, A. Baratoff, E. Meyer, P. Grütter: Lateral-force measurements in dynamic force microscopy, Phys. Rev. B **65**, 161403 (2002)
59. E. Riedo, F. Lévy, H. Brune: Kinetics of capillary condensation in nanoscopic sliding friction, Phys. Rev. Lett. **88**, 185505 (2002)
60. M. He, A. S. Blum, G. Overney, R. M. Overney: Effect of interfacial liquid structuring on the coherence length in nanolubrucation, Phys. Rev. Lett. **88**, 154302 (2002)
61. F. P. Bowden, F. P. Tabor: *The Friction and Lubrication of Solids* (Oxford Univ. Press, Oxford 1950)
62. L. D. Landau, E. M. Lifshitz: *Introduction to Theoretical Physics* (Nauka, Moscow 1998) Vol. 7
63. J. A. Greenwood, J. B. P. Williamson: Contact of nominally flat surfaces, Proc. R. Soc. Lond. A **295**, 300 (1966)
64. B. N. J. Persson: Elastoplastic contact between randomly rough surfaces, Phys. Rev. Lett. **87**, 116101 (2001)
65. U. D. Schwarz, O. Zwörner, P. Köster, R. Wiesendanger: Quantitative analysis of the frictional properties of solid materials at low loads, Phys. Rev. B **56**, 6987–6996 (1997)
66. K. L. Johnson, K. Kendall, A. D. Roberts: Surface energy and contact of elastic solids, Proc. R. Soc. Lond. A **324**, 301 (1971)
67. B. V. Derjaguin, V. M. Muller, Y. P. Toporov: Effect of contact deformations on adhesion of particles, J. Colloid Interf. Sci. **53**, 314–326 (1975)
68. D. Tabor: Surface forces and surface interactions, J. Colloid Interf. Sci. **58**, 2–13 (1977)
69. D. Maugis: Adhesion of spheres: the JKR-DMT transition using a Dugdale model, J. Colloid Interf. Sci. **150**, 243–269 (1992)
70. R. W. Carpick, N. Agraït, D. F. Ogletree, M. Salmeron: Measurement of interfacial shear (friction) with an ultrahigh vacuum atomic force microscope, J. Vacuum Sci. Technol. B **14**, 1289–1295 (1996)

71. C. Polaczyk, T. Schneider, J. Schöfer, E. Santner: Microtribological behavior of Au(001) studied by AFM/FFM, Surf. Sci. **402**, 454–458 (1998)
72. R.W. Carpick, D.F. Ogletree, M. Salmeron: Lateral stiffness: A new nanomechanical measurement for the determination of shear strengths with friction force microscopy, Appl. Phys. Lett. **70**, 1548–1550 (1997)
73. J.N. Israelachvili, D. Tabor: Measurement of van der Waals dispersion forces in range 1.5 to 130 nm, Proc. R. Soc. Lond. A **331**, 19 (1972)
74. S.P. Jarvis, A. Oral, T.P. Weihs, J.B. Pethica: A novel force microscope and point-contact probe, Rev. Sci. Instrum. **64**, 3515–3520 (1993)
75. M.A. Lantz, S.J. O'Shea, M.E. Welland, K.L. Johnson: Atomic-force-microscope study of contact area and friction on $NbSe_2$, Phys. Rev. B **55**, 10776–10785 (1997)
76. K.L. Johnson: *Contact Mechanics* (Cambridge Univ. Press, Cambridge 1985)
77. M. Enachescu, R.J.A. van den Oetelaar, R.W. Carpick, D.F. Ogletree, C.F.J. Flipse, M. Salmeron: Atomic force microscopy study of an ideally hard contact: the diamond(111)/tungsten carbide interface, Phys. Rev. Lett. **81**, 1877–1880 (1998)
78. M. Enachescu, R.J.A. van den Oetelaar, R.W. Carpick, D.F. Ogletree, C.F.J. Flipse, M. Salmeron: Observation of proportionality between friction and contact area at the nanometer scale, Tribol. Lett. **7**, 73–78 (1999)
79. E. Gnecco, R. Bennewitz, E. Meyer: Abrasive wear on the atomic scale, Phys. Rev. Lett. **88**, 215501 (2002)
80. S. Kopta, M. Salmeron: The atomic scale origin of wear on mica and its contribution to friction, J. Chem. Phys. **113**, 8249–8252 (2000)
81. H. Tang, C. Joachim, J. Devillers: Interpretation of AFM images – the graphite surface with a diamond tip, Surf. Sci. **291**, 439–450 (1993)
82. U. Landman, W.D. Luedtke, E.M. Ringer: Atomistic mechanisms of adhesive contact formation and interfacial processes, Wear **153**, 3–30 (1992)
83. A.I. Livshits, A.L. Shluger: Self-lubrication in scanning force microscope image formation on ionic surfaces, Phys. Rev. B **56**, 12482–12489 (1997)
84. H. Tang, X. Bouju, C. Joachim, C. Girard, J. Devillers: Theoretical study of the atomic-force microscopy imaging process on the NaCl(100) surface, J. Chem. Phys. **108**, 359–367 (1998)
85. R. Bennewitz, A.S. Foster, L.N. Kantorovich, M. Bammerlin, Ch. Loppacher, S. Schär, M. Guggisberg, E. Meyer, A.L. Shluger: Atomically resolved edges and kinks of NaCl islands on Cu(111): Experiment and theory, Phys. Rev. B **62**, 2074–2084 (2000)
86. U. Landman, W.D. Luetke, M.W. Ribarsky: Structural and dynamical consequences of interactions in interfacial systems, J. Vacuum Sci. Technol. A **7**, 2829–2839 (1989)
87. M.R. Sørensen, K.W. Jacobsen, P. Stoltze: Simulations of atomic-scale sliding friction, Phys. Rev. B **53**, 2101–2113 (1996)
88. R. Komanduri, N. Chandrasekaran, L.M. Raff: Molecular dynamics simulation of atomic-scale friction, Phys. Rev. B **61**, 14007–14019 (2000)
89. A. Buldum, C. Ciraci: Contact, nanoindentation and sliding friction, Phys. Rev. B **57**, 2468–2476 (1998)
90. T.H. Fang, C.I. Weng, J.G. Chang: Molecular dynamics simulation of a nanolithography process using atomic force microscopy, Surf. Sci. **501**, 138–147 (2002)
91. B. Gotsmann, C. Seidel, B. Anczykowski, H. Fuchs: Conservative and dissipative tip–sample interaction forces probed with dynamic AFM, Phys. Rev. B **60**, 11051–11061 (1999)
92. C. Loppacher, R. Bennewitz, O. Pfeiffer, M. Guggisberg, M. Bammerlin, S. Schär, V. Barwich, A. Baratoff, E. Meyer: Experimental aspects of dissipation force microscopy, Phys. Rev. B **62**, 13674–13679 (2000)

93. M. Gauthier, M. Tsukada: Theory of noncontact dissipation force microscopy, Phys. Rev. B **60**, 11716–11722 (1999)
94. J.P. Aimé, R. Boisgard, L. Nony, G. Couturier: Nonlinear dynamic behavior of an oscillating tip-microlever system and contrast at the atomic scale, Phys. Rev. Lett. **82**, 3388–3391 (1999)
95. W. Denk, D.W. Pohl: Local electrical dissipation imaged by scanning force microscopy, Appl. Phys. Lett. **59**, 2171–2173 (1991)
96. S. Hirsekorn, U. Rabe, A. Boub, W. Arnold: On the contrast in eddy current microscopy using atomic force microscopes, Surf. Interf. Anal. **27**, 474–481 (1999)
97. U. Dürig: Atomic-scale metal adhesion. In: *Forces in Scanning Probe Methods*, NATO ASI, Ser. E, Vol. 286, ed. by H.J. Güntherodt, D. Anselmetti, E. Meyer (Kluwer, Dordrecht 1995) pp. 191–234
98. N. Sasaki, M. Tsukada: Effect of microscopic nonconservative process on noncontact atomic force microscopy, Jpn. J. Appl. Phys. **39**, L1334–L1337 (2000)
99. B. Gotsmann, H. Fuchs: The measurement of hysteretic forces by dynamic AFM, Appl. Phys. A **72**, 55–58 (2001)
100. M. Guggisberg, M. Bammerlin, A. Baratoff, R. Lüthi, C. Loppacher, F.M. Battiston, J. Lü, R. Bennewitz, E. Meyer, H.J. Güntherodt: Dynamic force microscopy across steps on the Si(111)-(7×7) surface, Surf. Sci. **461**, 255–265 (2000)
101. R. Bennewitz, S. Schär, V. Barwich, O. Pfeiffer, E. Meyer, F. Krok, B. Such, J. Kolodzej, M. Szymonski: Atomic-resolution images of radiation damage in KBr, Surf. Sci. **474**, 197–202 (2001)
102. T.D. Stowe, T.W. Kenny, D. Thomson, D. Rugar: Silicon dopant imaging by dissipation force microscopy, Appl. Phys. Lett. **75**, 2785–2787 (1999)
103. B.C. Stipe, H.J. Mamin, T.D. Stowe, T.W. Kenny, D. Rugar: Noncontact friction and force fluctuations between closely spaced bodies, Phys. Rev. Lett. **87**, 96801 (2001)
104. B. Gotsmann, H. Fuchs: Dynamic force spectroscopy of conservative and dissipative forces in an Al-Au(111) tip–sample system, Phys. Rev. Lett. **86**, 2597–2600 (2001)
105. B.N.J. Persson, A.I. Volokitin: Comment on "Brownian motion of microscopic solids under the action of fluctuating electromagnetic fields", Phys. Rev. Lett. **84**, 3504 (2000)
106. K. Yamanaka, A. Noguchi, T. Tsuji, T. Koike, T. Goto: Quantitative material characterization by ultrasonic AFM, Surf. Interf. Anal. **27**, 600–606 (1999)
107. T. Drobek, R.W. Stark, W.M. Heckl: Determination of shear stiffness based on thermal noise analysis in atomic force microscopy: Passive overtone microscopy, Phys. Rev. B **64**, 045401 (2001)
108. T. Kawagishi, A. Kato, Y. Hoshi, H. Kawakatsu: Mapping of lateral vibration of the tip in atomic force microscopy at the torsional resonance of the cantilever, Ultramicroscopy **91**, 37–48 (2002)
109. F.J. Giessibl, M. Herz, J. Mannhart: Friction traced to the single atom, Proc. Natl. Acad. Sci. USA **99**, 12006–12010 (2002)
110. H.-J. Hug, A. Baratoff: Measurement of dissipation induced by tip–sample interactions. In: *Noncontact Atomic Force Microscopy*, ed. by S. Morita, R. Wiesendanger, E. Meyer (Springer, Berlin, Heidelberg 2002) p. 395

12

Nanomechanical Properties of Solid Surfaces and Thin Films

Adrian B. Mann

Summary. Instrumentation for the testing of mechanical properties on the submicron scale has developed enormously in recent years. This has enabled the mechanical behavior of surfaces, thin films, and coatings to be studied with unprecedented accuracy. In this chapter, the various techniques available for studying nanomechanical properties are reviewed with particular emphasis on nanoindentation. The standard methods for analyzing the raw data obtained using these techniques are described, along with the main sources of error. These include residual stresses, environmental effects, elastic anisotropy, and substrate effects. The methods that have been developed for extracting thin-film mechanical properties from the often convoluted mix of film and substrate properties measured by nanoindentation are discussed. Interpreting the data is frequently difficult, as residual stresses can modify the contact geometry and, hence, invalidate the standard analysis routines. Work hardening in the deformed region can also result in variations in mechanical behavior with indentation depth. A further unavoidable complication stems from the ratio of film to substrate mechanical properties and the depth of indentation in comparison to film thickness. Even very shallow indentations may be influenced by substrate properties if the film is hard and very elastic but the substrate is compliant. Under these circumstances nonstandard methods of analysis must be used. For multilayered systems many different mechanisms affect the nanomechanical behavior, including Orowan strengthening, Hall–Petch behavior, image force effects, coherency and thermal stresses, and composition modulation.

The application of nanoindentation to the study of phase transformations in semiconductors, fracture in brittle materials, and mechanical properties in biological materials are described. Recent developments such as the testing of viscoelasticity using nanoindentation methods are likely to be particularly important in future studies of polymers and biological materials. The importance of using a range of complementary methods such as electron microscopy, in situ AFMimaging, acoustic monitoring, and electrical contact measurements is emphasized. These are especially important on the nanoscale because so many different physical and chemical processes can affect the measured mechanical properties.

12.1 Introduction

When two bodies come into contact their surfaces experience the first and usually largest mechanical loads. Hence, characterizing and understanding the mechanical properties of surfaces is of paramount importance in a wide range of engineering applications. Obvious examples of where surface mechanical properties are important are in wear-resistant coatings on reciprocating surfaces and hard coatings for machine tool bits. This chapter details the current methods for measuring the mechanical properties of surfaces and highlights some of the key experimental results that have been obtained.

The experimental technique that is highlighted in this chapter is nanoindentation. This is for the simple reason that it is now recognized as the preferred method for testing thin film and surface mechanical properties. Despite this recognition, there are still many pitfalls for the unwary researcher when performing nanoindentation tests. The commercial instruments that are currently available all have attractive, user-friendly software, which makes the performance and analysis of nanoindentation tests easy. Hidden within the software, however, are a myriad of assumptions regarding the tests that are being performed and the material that is being examined. Unless the researcher is aware of these, there is a real danger that the results obtained will say more about the analysis routines than they do about the material being tested.

12.2 Instrumentation

The instruments used to examine nanomechanical properties of surfaces and thin films can be split into those based on point probes and those complimentary methods that can be used separately or in conjunction with point probes. The complimentary methods include a wide variety of techniques ranging from optical tests such as micro-Raman spectroscopy to high-energy diffraction studies using X-rays, neutrons, or electrons to mechanical tests such as bulge or blister testing.

Point-probe methods have developed from two historically different methodologies, namely, scanning probe microscopy [1] and microindentation [2]. The two converge at a length scale between 10–1000 nm. Point-probe mechanical tests in this range are often referred to as nanoindentation.

12.2.1 AFM and Scanning Probe Microscopy

Atomic force microscopy (AFM) and other scanning probe microscopies are covered in detail elsewhere in this volume, but it is worth briefly highlighting the main features in order to demonstrate the similarities to nanoindentation. There are now a myriad of different variants on the basic scanning probe microscope. All use piezoelectric stacks to move either a probe tip or the sample with subnanometer precision in the lateral and vertical planes. The probe itself can be as simple as a tungsten wire electrochemically polished to give a single atom at the tip, or as complex as an AFM tip that is bio-active with, for instance, antigens attached. A range of scanning probes have been developed with the intention of measuring specific physical properties such as magnetism and heat capacity.

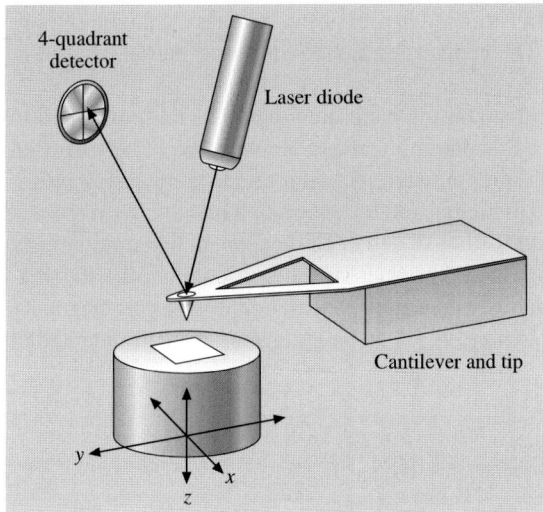

Fig. 12.1. Diagram of a commercial AFM. The AFM tip is mounted on a compliant cantilever, and a laser light is reflected off the back of the cantilever onto a position-sensitive detector (4-quadrant detector). Any movements of the cantilever beam cause a deflection of the laser light that the detector senses. The sample is moved using piezoelectric stack, and forces are calculated from the cantilever's stiffness and the measured deflection

To measure mechanical properties with an AFM, the standard configuration is a hard probe tip (such as silicon nitride or diamond) mounted on a cantilever (see Fig. 12.1). The elastic deflection of the cantilever is monitored either directly or via a feedback mechanism to measure the forces acting on the probe. In general, the forces experienced by the probe tip split into attractive or repulsive forces. As the tip approaches the surface, it experiences intermolecular forces that are attractive, although they can be repulsive under certain circumstances [3]. Once in contact with the surface the tip usually experiences a combination of attractive intermolecular forces and repulsive elastic forces. Two schools of thought exist regarding the attractive forces when the tip is in contact with the surface. The first is often referred to as the DMT or Bradley model. It holds that attractive forces only act outside the region of contact [4–7]. The second theory, usually called the JKR model, assumes that all the forces experienced by the tip, whether attractive or repulsive, act in the region of contact [8]. Most real nanoscale contacts lie somewhere between these two theoretical extremes.

12.2.2 Nanoindentation

The fundamental difference between AFM and nanoindentation is that during a nanoindentation experiment an external load is applied to the indenter tip. This load enables the tip to be pushed into the sample, creating a nanoscale impression on the surface, otherwise referred to as a nanoindentation or nanoindent.

Conventional indentation or microindentation tests involve pushing a hard tip of known geometry into the sample surface using a fixed peak load. The area of indentation that is created is then measured, and the mechanical properties of the sample, in particular its hardness, is calculated from the peak load and the indentation area.

Various types of indentation testing are used in measuring hardness, including Rockwell, Vickers, and Knoop tests. The geometries and definitions of hardness used in these tests are shown by Fig. 12.2.

When indentations are performed on the nanoscale there is a basic problem in measuring the size of the indents. Standard optical techniques cannot easily be used to image anything smaller than a micron, while electron microscopy is simply impractical due to the time involved in finding and imaging small indents. To overcome these difficulties, nanoindentation methods have been developed that continuously record the load, displacement, time, and contact stiffness throughout the indentation process. This type of continuously recording indentation testing was originally developed in the former Soviet Union [9–13] as an extension of microindentation

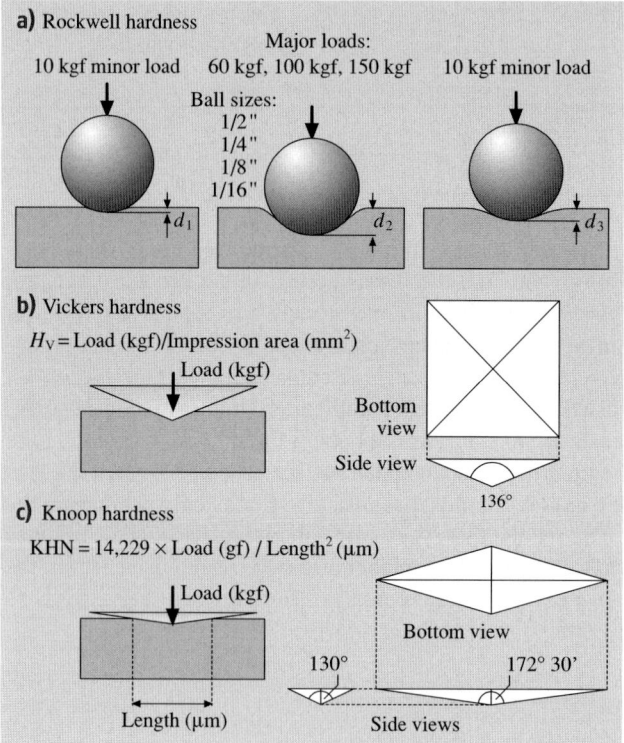

Fig. 12.2. (a) The standard Rockwell hardness test involves pushing a ball into the sample with a minor load, recording the depth, d_1, then applying a major load and recording the depth, d_2, then returning to the lower load and recording the depth, d_3. Using the depths, the hardness is calculated. (b) Vickers hardness testing uses a four-sided pyramid pushed into the sample with a known load. The area of the resulting indentation is measured optically and the hardness calculated as the load is divided by area. (c) Knoop indentation uses the same definition of hardness as the Vickers test, load divided by area, but the indenter geometry has one long diagonal and one short diagonal

tests. It was applied to nanoscale indentation testing in the early 1980s [14, 15], hence, giving rise to the field of nanoindentation testing.

In general, nanoindentation instruments include a loading system that may be electrostatic, electromagnetic, or mechanical, along with a displacement measuring system that may be capacitive or optical. Schematics of several commercial nanoindentation instruments are shown in Fig. 12.3a–c.

Among the many advantages of nanoindentation over conventional microindentation testing is the ability to measure the elastic, as well as the plastic properties of the test sample. The elastic modulus is obtained from the contact stiffness (S) using the following equation that appears to be valid for all elastic contacts [16, 17]:

$$S = \frac{2}{\sqrt{\pi}} E_r \sqrt{A} . \tag{12.1}$$

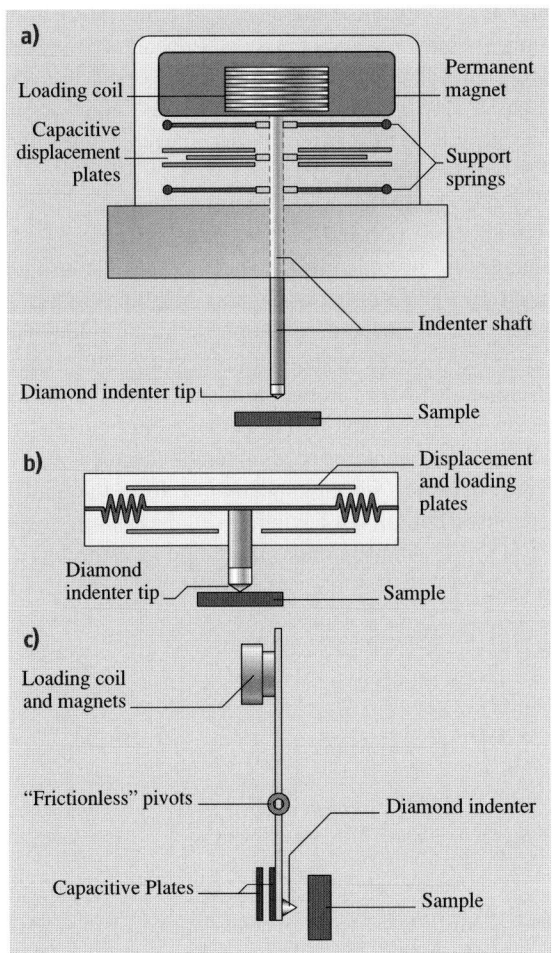

Fig. 12.3. Schematics of three commercial nanoindentation devices made by (**a**) MTS Nanoinstruments, Oak Ridge, Tennessee, (**b**) Hysitron Inc., Minneapolis, Minnesota, (**c**) Micro Materials Limited, Wrexham, UK. Instruments and use electromagnetic loading, while uses electrostatic loads

A is the contact area and E_r is the reduced modulus of the tip and sample as given by:

$$\frac{1}{E_r} = \frac{(1-v_t^2)}{E_t} + \frac{(1-v_s^2)}{E_s}, \qquad (12.2)$$

where E_t, v_t and E_s, v_s are the elastic modulus and Poissons ratio of the tip and sample, respectively.

12.2.3 Adaptations of Nanoindentation

Several adaptations to the basic nanoindentation setup have been used to obtain additional information about the processes that occur during nanoindentation testing, for example, in situ measurements of acoustic emissions and contact resistance. Environmental control has also been used to examine the effects of temperature and surface chemistry on the mechanical behavior of nanocontacts. In general, it is fair to say that the more information that can be obtained and the greater the control over the experimental parameters the easier it will be to understand the nanoindentation results. Load-displacement curves provide a lot of information, but they are only part of the story.

During nanoindentation testing discontinuities are frequently seen in the load–displacement curve. These are often called "pop-ins" or "pop-outs", depending on their direction. These sudden changes in the indenter displacement, at a constant load (see Fig. 12.4), can be caused by a wide range of events, including fracture, delamination, dislocation multiplication, or nucleation and phase transformations. To help distinguish between the various sources of discontinuities, acoustic transducers have been placed either in contact with the sample or immediately behind the indenter tip. For example, the results of nanoindentation tests that monitor acoustic emissions have shown that the phase transformations seen in silicon during nanoindentation are not the sudden events that they would appear to be from the load–displacement curve. There is no acoustic emission associated with the pop-out seen in the unloading curve of silicon [18]. An acoustic emission would be expected if

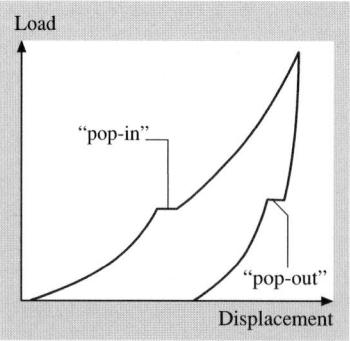

Fig. 12.4. Sketch of a load/displacement curve showing a pop-in and a pop-out

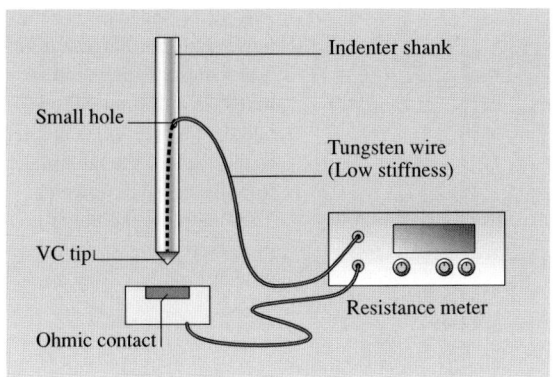

Fig. 12.5. Schematic of the basic setup for making contact resistance measurements during nanoindentation testing

there were a very rapid phase transformation causing a sudden change in volume. Fracture and delamination of films, however, give very strong acoustic signals [19], but the exact form of the signal appears to be more closely related to the sample geometry than to the event [20].

Additional information about the nature of the deformed region under the nanoindentation can be obtained by performing in situ measurements of contact resistance. The basic setup for this type of testing is shown in Fig. 12.5. An electrically conductive tip is needed to study contact resistance. Consequently, a conventional diamond tip is of limited use. Elastic, hard, and metallically conductive materials such as vanadium carbide can be used as substitutes for diamond [21, 22], or a thin conductive film (e. g., Ag) can be deposited on the diamond's surface (such a film is easily transferred to the indented surface so great care must be taken if multiple indents are performed). Measurements of contact resistance have been most useful for examining phase transformations in semiconductors [21, 22] and the dielectric breakdown of oxide films under mechanical loading [23].

One factor that is all too frequently neglected during nanoindentation testing is the effect of the experimental environment. Two obvious ways in which the environment can affect the results of nanoindentation tests are increases in temperature, which give elevated creep rates, and condensation of water vapor, which modifies the tip-sample interactions. Both of these environmental effects have been shown to significantly affect the measured mechanical properties and the modes of deformation that occur during nanoindentation [24–27]. Other environmental effects, for instance, those due to photoplasticity or hydrogen ion absorption, are also possible, but they are generally less troublesome than temperature fluctuations and variations in atmospheric humidity.

12.2.4 Complimentary Techniques

Nanoindentation testing is probably the most important technique for characterizing the mechanical properties of thin films and surfaces, but there are many alternative or additional techniques that can be used. One of the most important alternative

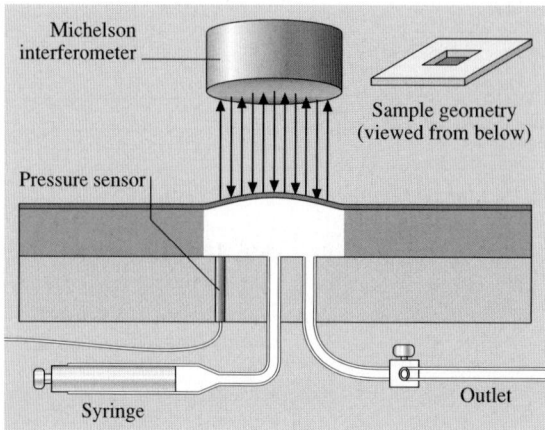

Fig. 12.6. Schematic of the basic setup for bulge testing. The sample is prepared so that it is a thin membrane, and then a pressure is applied to the back of the membrane to make it bulge upwards. The height of the bulge is measured using an interferometer

methods for measuring the mechanical properties of thin films uses bulge or blister testing [28]. Bulge tests are performed on thin films mounted on supporting substrates. A small area of the substrate is removed to give a window of unsupported film. A pressure is then applied to one side of the window causing it to bulge. By measuring the height of the bulge, the stress-strain curve and the residual stress are obtained. The basic configuration for bulge testing is shown in Fig. 12.6.

12.2.5 Bulge Tests

The original bulge tests used circular windows because they are easier to analyze mathematically, but now square and rectangular windows have become common [29]. These geometries tend to be easier to fabricate. Unfortunately, there are several sources of errors in bulge testing that can potentially lead to large errors in the measured mechanical properties. These errors at one time led to the belief that multilayer films can show a "super modulus" effect, where the elastic modulus of the multilayer is several times that of its constituent layers [30]. It is now accepted that any enhancement to the elastic modulus in multilayer films is small, on the order of 15% [31]. The main sources of error stem from compressive stresses in the film (tensile stress is not a problem), small variations in the dimensions of the window, and uncertainty in the exact height of the bulge. Despite these difficulties, one advantage of bulge testing over nanoindentation testing is that the stress state is biaxial, so that only properties in the plane of the film are measured. In contrast, nanoindentation testing measures a combination of in-plane and out-of-plane properties.

12.2.6 Acoustic Methods

Acoustic and ultrasonic techniques have been used for many years to study the elastic properties of materials. Essentially, these techniques take advantage of the fact that the velocity of sound in a material is dependent on the inter-atomic or intermolecular forces in the material. These, of course, are directly related to the material's elastic constants. In fact, any nonlinearity of inter-atomic forces enables slight variations in acoustic signals to be used as a measure of residual stress.

An acoustic method ideally suited to studying surfaces is scanning acoustic microscopy (SAM) [32]. There are also several other techniques that have been used to study surface films and multilayers, but we will first consider SAM in detail. In a SAM, a lens made of sapphire is used to bring acoustic waves to focus via a coupling fluid on the surface. A small piezoelectric transducer at the top of the lens generates the acoustic signal. The same transducer can be used to detect the signal when the SAM is used in reflection mode. The use of a transducer as both generator and detector, a common imaging mode, necessitates the use of a pulsed rather than a continuous acoustic signal. Continuous waves can be used if phase changes are used to build up the image. The transducer lens generates two types of acoustic waves in the material: longitudinal and shear. The ability of a solid to sustain both types of wave (liquids can only sustain longitudinal waves) gives rise to a third type of acoustic wave called a Rayleigh, or surface, wave. These waves are generated as a result of superposition of the shear and longitudinal waves with a common phase velocity. The stresses and displacements associated with a Rayleigh wave are only of significance to a depth of ≈ 0.6 Rayleigh wavelengths below the solid surface. Hence, using SAM to examine Rayleigh waves in a material is a true surface characterization technique.

Using a SAM in reflection mode gives an image where the contrast is directly related to the Rayleigh wave velocity, which is in turn a function of the material's elastic constants. The resolution of the image depends on the frequency of the transducer used, i.e., for a 2 GHz signal a resolution better than 1 μm is achievable. The contrast in the image results from the interference of two different waves in the coupling fluid. Rayleigh waves that are excited in the surface "leak" into the coupling fluid and interfere with the acoustic signal that is directly reflected back from the surface. It is usually assumed that the properties of the coupling fluid are well characterized. The interference of the two waves gives a characteristic $V(z)$ curve, as illustrated by Fig. 12.7, where z is the separation between the lens and the surface. Analyzing the periodicity of the $V(z)$ curves provides information on the Rayleigh wave velocity. As with other acoustic waves, the Rayleigh velocity is related to the elastic constants of the material. When using the SAM for a material's characterization, the lens is usually held in a fixed position on the surface. By using a lens designed specifically to give a line-focus beam, rather than the standard spherical lens, it is possible to use SAM to look at anisotropy in the wave velocity [33] and hence in elastic properties by producing waves with a specific direction.

One advantage of using SAM in conjunction with nanoindentation to characterize a surface is that the measurements obtained with the two methods have a slightly different dependence on the test material's elastic properties, E_s and v_s (the elastic modulus and Poisson's ratio). As a result, it is possible to use SAM and nanoinden-

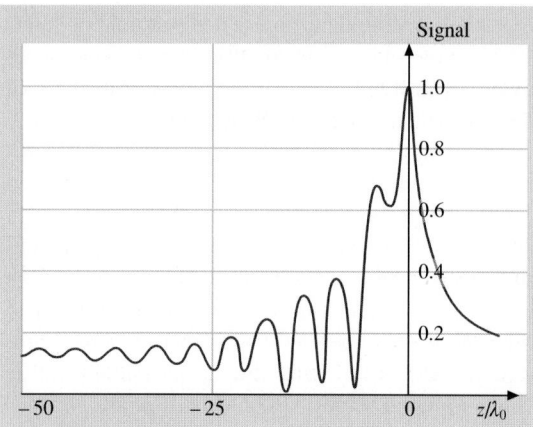

Fig. 12.7. A typical $V(z)$ curve obtained with a SAM when testing fused silica

tation combined to find both E_s and ν_s, as illustrated by Fig. 12.8 [34]. This is not possible when using only one of the techniques alone.

In addition to measuring surface properties, SAM has been used to study thin films on a surface. However, the Rayleigh wave velocity can be dependent on a complex mix of the film and substrate properties. Other acoustic methods have been utilized to study freestanding films. A freestanding film can be regarded as a plate, and, therefore, it is possible to excite Lamb waves in the film. Using a pulsed laser to generate the waves and a heterodyne interferometer to detect the arrival of the Lamb wave, it is possible to measure the flexural modulus of the film [35]. This has

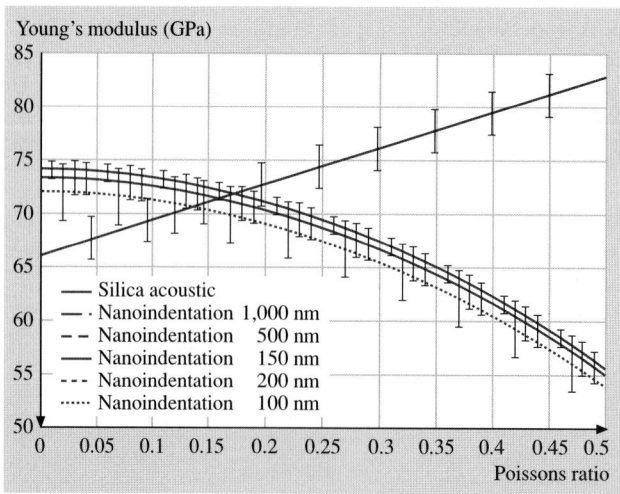

Fig. 12.8. Because SAM and nanoindentation have different dependencies on Young's modulus, E, and Poissons ratio, ν, it is possible to use the two techniques in combination to find E and ν [34]. On the graph, the intersection of the curves gives E and ν

been successfully demonstrated for multilayer films with a total thickness < 10 μm. In the plate configuration, due to the nonlinearity of elastic properties, it is also possible to measure stress. This has been demonstrated for horizontally polarized shear waves in plates [36], but thin plates require very high frequency transducers or laser sources.

12.2.7 Imaging Methods

When measuring the mechanical properties of a surface or thin film using nanoindentation, it is not always easy to visualize what is happening. In many instances there is a risk that the mechanical data can be completely misinterpreted if the geometry of the test is not as expected. To expedite the correct interpretation of the mechanical data, it is generally worthwhile to use optical, electron, or atomic force microscopy to image the nanoindentations. Obviously, optical techniques are only of use for larger indentations, but they will often reveal the presence of median or lateral cracks [37]. Electron microscopy and AFM, however, can be used to examine even the smallest nanoindentations. The principle problem with these microscopy techniques is the difficulty in finding the nanoindentations. It is usually necessary to make large, "marker" indentations in the vicinity of the nanoindentations to be examined in order to find them [38].

It is possible to see features such as extrusions with a scanning electron microscope (SEM) [39], as well as pile-up and sink-in around the nanoindents, though AFM is generally better for this. Transmission electron microscopy (TEM) is useful for examining what has happened subsurface, for instance, the indentation induced dislocations in a metal [40] or the phases present under a nanoindent in silicon [21]. However, with TEM there is the added difficulty of sample preparation and the associated risk of observing artifacts. Recently, there has been considerable interest in the use of focused ion beams to cut cross sections through nanoindents [41]. When used in conjunction with SEM or TEM this provides an excellent means to see what has happened in the subsurface region.

One other technique that has proved to be useful in studying nanoindents is micro-Raman spectroscopy. This involves using a microscope to focus a laser on the sample surface. The same microscope is also used to collect the scattered laser light, which is then fed into a spectroscope. The Raman peaks in the spectrum provide information on the bonding present in a material, while small shifts in the wave number of the peaks can be used as a measure of strain. Micro-Raman has proven to be particularly useful for examining the phases present around nanoindentations in silicon [42].

12.3 Data Analysis

The analysis of nanoindentation data is far from simple. This is mostly due to the lack of effective models that are able to combine elastic and plastic deformation under a contact. However, provided certain precautions are taken, the models for perfectly elastic deformation and ideal plastic materials can be used in the analysis of nanoindentation data. For this reason, it is worth briefly reviewing the models for perfect contacts.

12.3.1 Elastic Contacts

The theoretical modeling of elastic contacts can be traced back many years, at least to the late nineteenth century and the work of *Hertz* (1882) [43] and *Boussinesq* (1885) [44]. These models, which are still widely used today, consider two axisymmetric curved surfaces in contact over an elliptical region (see Fig. 12.9). The contact region is taken to be small in comparison to the radius of curvature of the contacting surfaces, which are treated as elastic half-spaces. For an elastic sphere, radius R, in contact with a flat, elastic half-space, the contact region will be circular and the Hertz model gives the following relationships:

$$a = \sqrt[3]{\frac{3PR}{4E_r}}, \tag{12.3}$$

$$\delta = \sqrt[3]{\frac{9P^2}{16RE_r^2}}, \tag{12.4}$$

$$P_0 = \sqrt[3]{\frac{6PE_r^2}{\pi^3 R^2}}, \tag{12.5}$$

where a is the radius of the contact region, E_r is given by (12.2), δ is the displacement of the sphere into the surface, P is the applied load and P_0 is the maximum pressure under the contact (in this case at the center of the contact).

The work of Hertz and Boussinesq was extended by *Love* [45, 46] and later by *Sneddon* [47], who simplified the analysis using Hankel transforms. Love showed

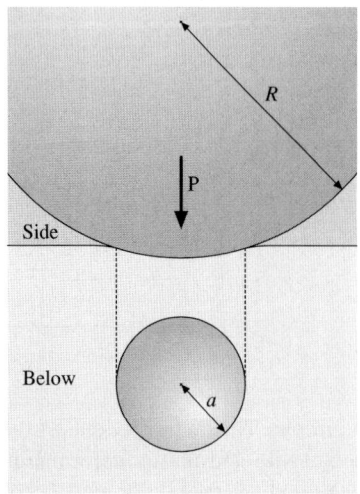

Fig. 12.9. Hertzian contact of a sphere, radius R, on a semi-infinite, flat surface. The contact in this case is a circular region of radius a

how Boussinesq's model could be used for a flat-ended cylinder and a conical indenter, while Sneddon produced a generalized relationship for any rigid axisymmetric punch pushed into an elastic half-space. Sneddon applied his new analysis to punches of various shapes and derived the following relationships between the applied load, P, and displacement, δ, into the elastic half-space for, respectively, a flat-ended cylinder, a cone of semi-vertical angle ϕ, and a parabola of revolution where $a^2 = 2k\delta$:

$$P = \frac{4\mu a \delta}{1-\nu}, \tag{12.6}$$

$$P = \frac{4\mu \cot\phi}{\pi(1-\nu)}\delta^2, \tag{12.7}$$

$$P = \frac{8\mu}{3(1-\nu)}\left(2k\delta^3\right)^{1/2}, \tag{12.8}$$

where μ and ν are the shear modulus and Poisson's ratio of the elastic half-space, respectively.

The key point to note about (12.6), (12.7), and (12.8) is that they all have the same basic form, namely:

$$P = \alpha \delta^m, \tag{12.9}$$

where α and m are constants for each geometry.

Equation (12.9) and the relationships developed by Hertz and his successors, (12.3–12.8), form the foundation for much of the current nanoindentation data analysis routines.

12.3.2 Indentation of Ideal Plastic Materials

Plastic deformation during indentation testing is not easy to model. However, the indentation response of ideal plastic metals was considered by *Tabor* in his classic text, "The Hardness of Metals" [48]. An ideal plastic material (or more accurately an ideal elastic-plastic material) has a linear stress-strain curve until it reaches its elastic limit and then yields plastically at a yield stress, Y_0, that remains constant even after deformation has commenced. In a 2-D problem, the yielding occurs because the Huber-Mises [49] criterion has been reached. In other words, the maximum shear stress acting on the material is around $1.15Y_0/2$.

First, we consider a 2-D flat punch pushed into an ideal plastic material. By using the method of slip lines it is found that the mean pressure, P_m, across the end of the punch is related to the yield stress by:

$$P_m = 3Y_0. \tag{12.10}$$

If the *Tresca* criterion [50] is used, then P_m is closer to $2.6Y_0$. In general, for both 2-D and three-dimensional punches pushed into ideal plastic materials, full plasticity across the entire contact region can be expected when $P_m = 2.6$ to $3.0Y_0$.

However, significant deviations from this range can be seen if, for instance, the material undergoes work-hardening during indentation, or the material is a ceramic, or there is friction between the indenter and the surface.

The apparently straightforward relationship between P_m and Y_0 makes the mean pressure a very useful quantity to measure. In fact, P_m is very similar to the Vickers hardness, H_V, of a material:

$$H_V = 0.927 P_m . \tag{12.11}$$

During nanoindentation testing it is the convention to take the mean pressure as the nanohardness. Thus, the "nanohardness", H, is defined as the peak load, P, applied during a nanoindentation divided by the projected area, A, of the nanoindentation in the plane of the surface, hence:

$$H = \frac{P}{A} . \tag{12.12}$$

12.3.3 Adhesive Contacts

During microindentation testing and even most nanoindentation testing the effects of intermolecular and surface forces can be neglected. Very small nanoindentations, however, can be influenced by the effects of intermolecular forces between the sample and the tip. These adhesive effects are most readily seen when testing soft polymers, but there is some evidence that forces between the tip and sample may be important in even relatively strong materials [51, 52].

Contact adhesion is usually described by either the JKR or DMT model, as discussed earlier in this chapter. Both the models consider totally elastic spherical contacts under the influence of attractive surface forces. The JKR model considers the surface forces in terms of the associated surface energy, whereas the DMT model considers the effects of adding van der Waals forces to the Hertzian contact model. The differences between the two models are illustrated by Fig. 12.10.

For nanoindentation tests conducted in air the condensation of water vapor at the tip-sample interface usually determines the size of the adhesive force acting during unloading. The effects of water vapor on a single nanoasperity contact have been studied using force-controlled AFM techniques [53] and, more recently, nanoindentation methods [26]. Unsurprisingly, it has also been found that water vapor can affect the deformation of surfaces during nanoindentation testing [27].

In addition to water vapor, other surface adsorbates can cause dramatic changes in the nanoscale mechanical behavior. For instance, oxygen on a clean metal surface can cause an increase in the apparent strength of the metal [54]. These effects are likely to be related to, firstly, changes in the surface and intermolecular forces acting between the tip and the sample and, secondly, changes in the mechanical stability of surface nanoasperities and ledges. Adsorbates can help stabilize atomic-scale variations in surface morphology, thereby making defect generation at the surface more difficult.

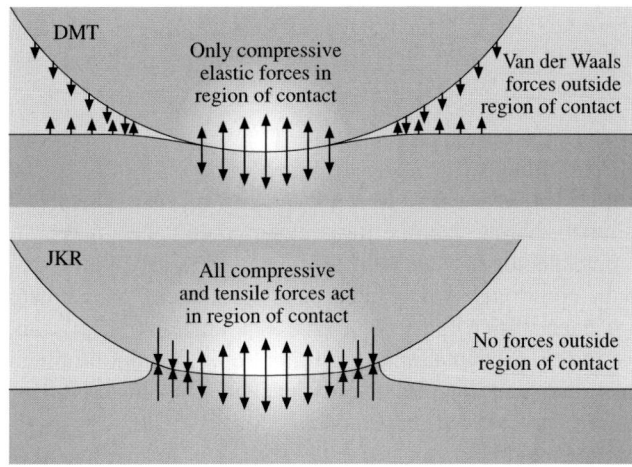

Fig. 12.10. The contact geometry for the DMT and JKR models for adhesive contact. Both models are based on the Hertzian model. In the DMT, model van der Waals forces outside the region of contact introduce an additional load in the Hertz model. But for the JKR model, it is assumed that tensile, as well as compressive stresses can be sustained within the region of contact

12.3.4 Indenter Geometry

All of the indenter geometries considered up to this point have been axisymmetric, largely because they are easier to deal with theoretically. Unfortunately, fabricating axisymmetric nanoindentation tips is extremely difficult, because shaping a hard tip on the scale of a few nanometers is virtually impossible. Despite these problems, there has been considerable effort put into the use of spherical nanoindentation tips [55]. This clearly demonstrates that the spherical geometry can be useful at larger indentation depths.

Because of the problems associated with creating axisymmetric nanoindentation tips, pyramidal indenter geometries have now become standard during nanoindentation testing. The most common geometries are the three-sided Berkovich pyramid and cube-corner (see Fig. 12.11). The Berkovich pyramid is based on the four-sided

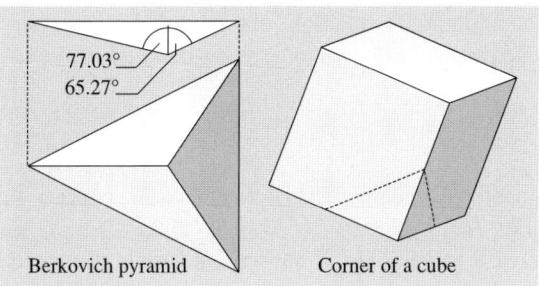

Fig. 12.11. The ideal geometry for the three-sided Berkovich pyramid and cube corner tips

Vickers pyramid, the opposite sides of which make an 136° angle. For both the Vickers and Berkovich pyramids the cross-sectional area of the pyramid's base, A, is related to the pyramid's height, D, by:

$$A = 24.5D^2. \qquad (12.13)$$

The cube-corner geometry is now widely used for making very small nanoindentations, because it is much sharper than the Berkovich pyramid. This makes it easier to initiate plastic deformation at very light loads, but great care should be taken when using the cube-corner geometry. Sharp cube-corners can wear down quickly and become blunt, hence the cross-sectional area as a function of depth can change over the course of several indentations. There is also a potential problem with the standard analysis routines [56], which were developed for much blunter geometries and are based on the elastic contact models outlined earlier. The elastic contact models all assume the displacement into the surface is small compared to the tip radius. For the cube-corner geometry this is probably only the case for nanoindentations that are no more than a few nanometers deep.

12.3.5 Analyzing Load/Displacement Curves

The load/displacement curves obtained during nanoindentation testing are deceptively simple. Most newcomers to the area will see the curves as being somewhat akin to the stress/strain curves obtained during tensile testing. There is also a real temptation just to use the values of hardness, H, and elastic modulus, E, obtained from standard analysis software packages as the "true" values. This may be the case in many instances, but for very shallow nanoindents and tests on thin films the geometry of the contact can differ significantly from the geometry assumed in the analysis routines. Consequently, experimentalists should think very carefully about the test itself before concluding that the values of H and E are correct.

The basic shape of a load/displacement curve can reveal a great deal about the type of material being tested. Figure 12.12 shows some examples of ideal curves for materials with different elastic moduli and yield stresses. Discontinuities in the load/displacement curve can also provide information on such processes as fracture, dislocation nucleation, and phase transformations. Initially, though, we will consider ideal situations such as those illustrated by Fig. 12.12.

The loading section of the load/displacement curve approximates a parabola [57] whose width depends on a combination of the material's elastic and plastic properties. The unloading curve, however, has been shown to follow a more general relationship [56] of the form:

$$P = \alpha(\delta - \delta_i)^m, \qquad (12.14)$$

where δ is the total displacement and δ_i is the intercept of the unloading curve with the displacement axis shown in Fig. 12.13.

Equation (12.14) is essentially the same as (12.9) but with the origin displaced. Since (12.9) is obtained by considering purely elastic deformation, it follows that

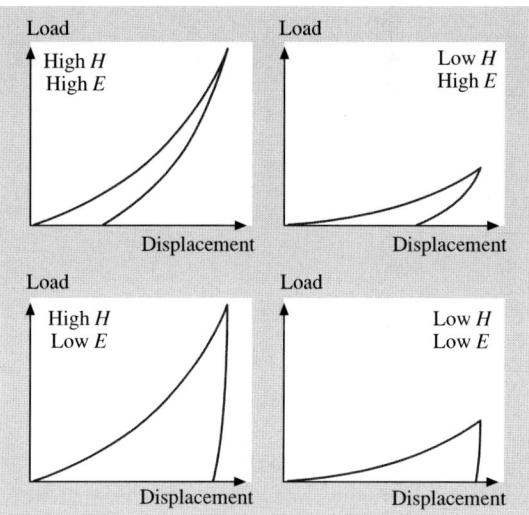

Fig. 12.12. Examples of load/displacement curves for idealized materials with a range of hardness and elastic properties

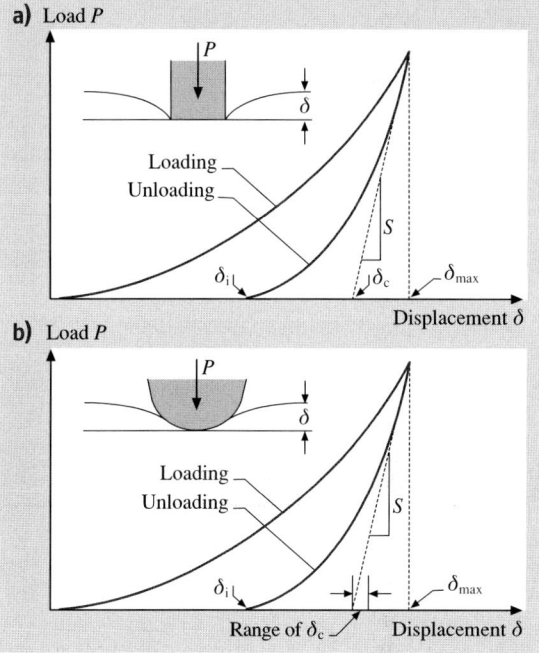

Fig. 12.13. Analysis of the load/displacement curve gives the contact stiffness, S, and the contact depth, δ_c. These can then be used to find the hardness, H, and elastic, or Young's modulus, E. (**a**) The first method of analysis [58–60] assumed the unloading curve could be approximated by a flat punch on an elastic half-space. (**b**) A more refined analysis [56] uses a paraboloid on an elastic half-space

the unloading curve is exhibiting purely elastic behavior. Since the shape of the unloading curve is determined by the elastic recovery of the indented region, it is not entirely surprising that its shape resembles that found for purely elastic deformation. What is fortuitous is that the elastic analysis used for an elastic half-space seems to be valid for a surface where there is a plastically formed indentation crater present under the contact. However, the validity of this analysis may only hold when the crater is relatively shallow and the geometry of the surface does not differ significantly from that of a flat, elastic half-space. For nanoindentations with a Berkovich pyramid, this is generally the case.

Before *Oliver* and *Pharr* [56] proposed their now standard method for analyzing nanoindentation data, the analysis had been based on the observation that the initial part of the unloading curve is almost linear. A linear unloading curve, equivalent to $m = 1$ in (12.14), is expected when a flat punch is used on an elastic half-space. The flat punch approximation for the unloading curve was used in [58–60] to analyze nanoindentation data. When Oliver and Pharr looked at a range of materials they found m was typically larger than 1, and that $m = 1.5$, or a paraboloid, was a better approximation than a flat punch. Oliver and Pharr used (12.1) and (12.12) to obtain the values for a material's elastic modulus and hardness. Equation (12.1) relates the contact stiffness during the initial part of the unloading curve (see Fig. 12.13) to the reduced elastic modulus and the contact area at the peak load. Equation (12.12) gives the hardness as the peak load divided by the contact area. It is immediately obvious that the key to measuring the mechanical properties of a material is knowing the contact area at the peak load. This is the single most important factor in analyzing nanoindentation data. Most mistakes in the analysis come from incorrect assumptions about the contact area.

To find the contact area, a function relating the contact area, A_c, to the contact depth, δ_c, is needed. For a perfect Berkovich pyramid this would be the same as (12.13). But since making a perfect nanoindenter tip is impossible, an expanded equation is used:

$$A_c(\delta_c) = 24.5\delta_c^2 + \sum_{j=1}^{7} C_j \sqrt[2^j]{\delta_c}, \tag{12.15}$$

where C_j are calibration constants of the tip.

There is a crucial step in the analysis before A_c can be calculated, namely, finding δ_c. The contact depth is not the same as the indentation depth, because the surface around the indentation will be elastically deflected during loading, as illustrated by Fig. 12.14. *Sneddon*'s analysis [47] provides a way to calculate the deflection of the surface at the edge of an axisymmetric contact. Subtracting the deflection from the total indentation depth at peak load gives the contact depth. For a paraboloid, as used by *Oliver* and *Pharr* [56] in their analysis, the elastic deflection at the edge of

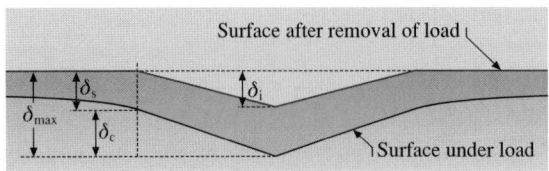

Fig. 12.14. Profile of surface under load and unloaded showing how δ_c compares to δ_i and δ_{max}

the contact is given by:

$$\delta_s = \varepsilon \frac{P}{S} = 0.75 \frac{P}{S},\qquad(12.16)$$

where S is the contact stiffness and P the peak load. The constant ε is 0.75 for a paraboloid, but ranges between 0.72 (conic indenter) and 1 (flat punch). Figure 12.15 shows how the contact depth depends on the value of ε. The contact depth at the peak load is, therefore:

$$\delta_c = \delta - \delta_s.\qquad(12.17)$$

Using the load/displacement data from the unloading curve and (12.1), (12.2), (12.12), (12.14–12.17), the hardness and reduced elastic modulus for the test sample can be calculated. To find the elastic modulus of the sample, E_s, it is also necessary to know Poisson's ratio, ν_s, for the sample, as well as the elastic modulus, E_t, and Poisson's ratio, ν_t, of the indenter tip. For diamond these are 1141 GPa and 0.07, respectively.

There also remains the issue of calibrating the tip shape, or finding the values for C_j in (12.15). Knowing the exact expansion of $A_c(\delta_c)$ is vital if the values for E_s and H are to be accurate. Several methods for calibrating the tip shape have been used, including imaging the tip with an electron microscope, measuring the size of nanoindentations using SEM or TEM of negative replicas, and using scanning probes to examine either the tip itself or the nanoindentations made with the tip. There are strengths and weaknesses to each of these methods. In general, however, the accuracy and usefulness of the methods depends largely on how patient and rigorous the experimentalist is in performing the calibration.

Because of the experimental difficulties and time involved in calibrating the tip shape by these methods, *Oliver* and *Pharr* [56] developed a method for calibration based on standard specimens. With a standard specimen that is mechanically isotropic and has a known E and H that does not vary with indentation depth, it should be possible to perform nanoindentations to a range of depths, and then use the analysis routines in reverse to deduce the tip area function, $A_c(\delta_c)$. In other words, if you perform a nanoindentation test, you can find the contact stiffness, S, at the peak load, P, and the contact depth, δ_c, from the unloading curve. Then if you know E

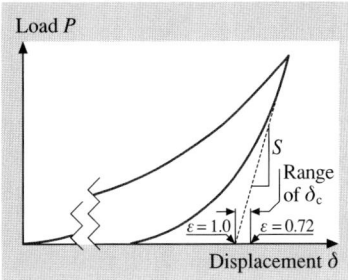

Fig. 12.15. Load-displacement curve showing how δ_c varies with ε

a priori, (12.1) can be used to calculate the contact area, A, and, hence, you have a value for A_c at a depth δ_c. Repeating this procedure for a range of depths will give a numerical version of the function $A_c(\delta_c)$. Then, it is simply a case of fitting (12.15) to the numerical data. If the hardness, H, is known and not a function of depth, and the calibration specimen was fully plastic during testing, then essentially the same approach could be used but based on (12.12). Situations where a constant H is used to calibrate the tip are extremely rare.

In addition to the tip shape function, the machine compliance must be calibrated. Basic Newtonian mechanics tells us that for the tip to be pushed into a surface the tip must be pushing off of another body. During nanoindentation testing the other body is the machine frame. As a result, during a nanoindentation test it is not just the sample, but the machine frame that is being loaded. Consequently, a very small elastic deformation of the machine frame contributes to the total stiffness obtained from the unloading curve. The machine frame is usually very stiff, $> 10^6$ N/m, so the effect is only important at relatively large loads.

To calibrate the machine frame stiffness or compliance, large nanoindentations are made in a soft material such as aluminum with a known, isotropic elastic modulus. For very deep nanoindentations made with a Berkovich pyramid, the contact area, $A_c(\delta_c)$, can be reasonably approximated to $24.5\delta_c^2$, thus (12.1) can be used to find the expected contact stiffness for the material. Any difference between the expected value of S and the value measured from the unloading curve will be due to the compliance of the machine frame. Performing a number of deep nanoindentations enables an accurate value for the machine frame compliance to be obtained.

Currently, because of its ready availability and predictable mechanical properties, the most popular calibration material is fused silica ($E = 72$ GPa, $v = 0.17$), though aluminum is still used occasionally.

12.3.6 Modifications to the Analysis

Since the development of the analysis routines in the early 1990s, it has become apparent that the standard analysis of nanoindentation data is not applicable in all situations, usually because errors occur in the calculated contact depth or contact area. *Pharr* et al. [61–64] have used finite element modeling (FEM) to help understand and overcome the limitations of the standard analysis. Two important sources of errors have been identified in this way. The first is residual stress at the sample surface. The second is the change in the shape of nanoindents after elastic recovery.

The effect of residual stresses at a surface on the indentation properties has been the subject of debate for many years [65–67]. The perceived effect was that compressive stresses increased hardness, while tensile stresses decreased hardness. Using FEM it is possible to model a pointed nanoindenter being pushed into a model material that is in residual tension or compression. An FEM model of nanoindentation into aluminum alloy 8009 [61] has confirmed earlier experimental observations [68] indicating that the contact area calculated from the unloading curve is incorrect if there are residual stresses. In the FEM model of an aluminum alloy the mechanical behavior of the material is modeled using a stress-strain curve, which resembles that

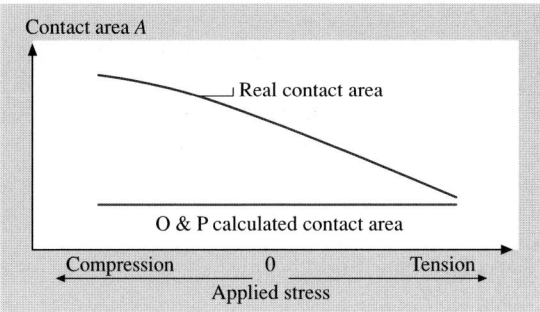

Fig. 12.16. When a surface is in a state of stress there is a significant difference between the contact area calculated using the Oliver and Pharr method and the actual contact area [61]. For an aluminum alloy this can lead to significant errors in the calculated hardness and elastic modulus

of an elastic-perfectly-plastic metal with a flow stress of 425.6 MPa. Yielding starts at 353.1 MPa and includes a small amount of work hardening. The FEM model was used to find the contact area directly and using the simulated unloading curve in conjunction with Oliver and Pharr's method. The results as a function of residual stress are illustrated in Fig. 12.16. Note that the differences between the two measured contact areas lead to miscalculations of E and H.

Errors in the calculated contact area stem from incorrect assumptions about the pile-up and sink-in at the edge of the contact, as illustrated by Fig. 12.17. The Oliver and Pharr analysis assumes the geometry of the sample surface is the same as that given by *Sneddon* [47] in his analytical model for the indentation of elastic surfaces. Clearly, for materials where there is significant plastic deformation, it is possible that there will be large deviations from the surface geometry found using Sneddon's elastic model. In reality, the error in the contact area depends on how much the geometry of the test sample surface differs from that of the calibration material (typically fused silica). It is possible that a test sample, even without a residual stress, will have a different surface geometry and, hence, contact area at a given depth, when compared to the calibration material. This is often seen for thin films

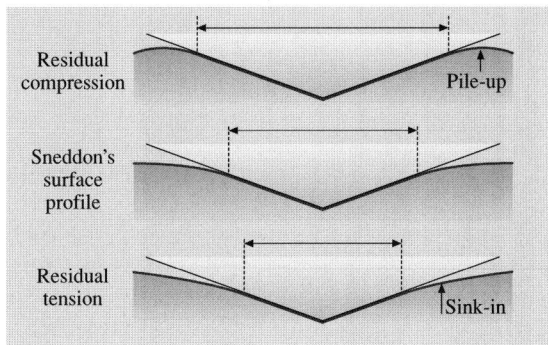

Fig. 12.17. Pile-up and sink-in are affected by residual stresses, and, hence, errors are introduced into standard Oliver and Pharr analysis

on a substrate (e.g., *Tsui* et al. [69, 70]). Residual stresses increase the likelihood that the contact area calculated using Oliver and Pharr's method will be incorrect.

The issue of sink-in and pile-up is always a factor in nanoindentation testing. However, there is still no effective way to deal with these phenomena other than reverting to imaging of the indentations to identify the true contact area. Even this is difficult, as the edge of an indentation is not easy to identify using AFM or electron microscopy. One approach that has been used [71] with some success is measuring the ratio E_r^2/H, rather than E_r and H separately. Because E_r is proportional to $1/\sqrt{A}$ and H is proportional to $1/A$, E_r^2/H should be independent of A and, hence, unaffected by pile-up or sink-in. While this does not provide quantitative values for mechanical properties, it does provide a way to identify any variations in mechanical properties with indentation depth or between similar samples with different residual stresses.

Another source of error in the Oliver and Pharr analysis is due to incorrect assumptions about the nanoindentation geometry after unloading [63]. Once again, this is due to differences between the test sample and the calibration material. The exact shape of an unloaded nanoindentation on a material exhibiting elastic recovery is not simply an impression of the tip shape; rather, there is some elastic recovery of the nanoindentation sides giving them a slightly convex shape (see Fig. 12.18). The shape actually depends on Poisson's ratio, so the standard Oliver and Pharr analysis will only be valid for a material where $v = 0.17$, the value for fused silica, assuming it is used for the calibration.

To deal with the variations in the recovered nanoindentation shape, it has been suggested [63] that a modified nanoindenter geometry with a slightly concave side be used in the analysis (see Fig. 12.18). This requires a modification to (12.1):

$$S = \gamma 2 E_r \sqrt{\frac{A}{\pi}}, \qquad (12.18)$$

where γ is a correction term dependent on the tip geometry. For a Berkovich pyramid the best value is:

$$\gamma = \frac{\frac{\pi}{4} + 0.15483073 \cot \Phi \left(\frac{(1-2v_s)}{4(1-v_s)} \right)}{\left[\frac{\pi}{2} - 0.83119312 \cot \Phi \left(\frac{(1-2v_s)}{4(1-v_s)} \right) \right]^2}, \qquad (12.19)$$

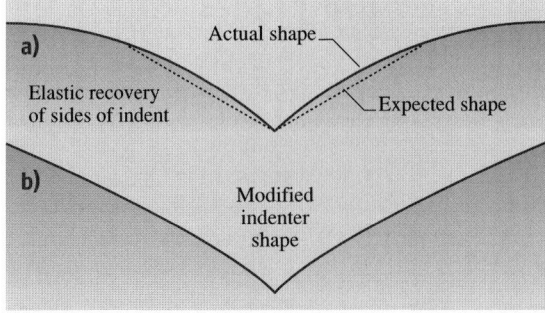

Fig. 12.18. (a) *Hay* et al. [63] found from experiments and FEM simulations that the actual shape of an indentation after unloading is not as expected. (b) They introduced a γ term to correct for this effect. This assumes the indenter has slightly concave sides

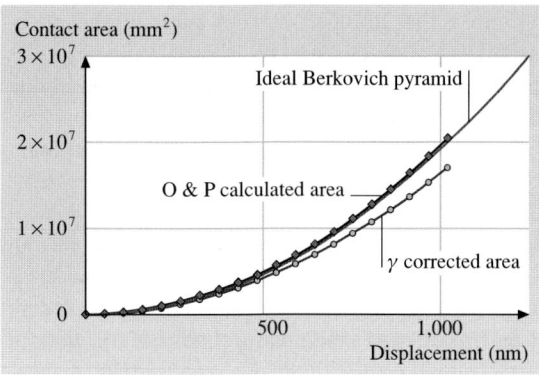

Fig. 12.19. For a real Berkovich tip the γ corrected area [63] is less at a given depth than the area calculated using the Oliver and Pharr method

where $\Phi = 70.32°$. For a cube corner the correction can be even larger and γ is given by:

$$\gamma = 1 + \left(\frac{(1-2\nu_s)}{4(1-\nu_s)\tan\Phi}\right), \tag{12.20}$$

where $\Phi = 42.28°$. Figure 12.19 shows how the modified contact area varies with depth for a real diamond Berkovich pyramid.

The validity of the γ-modified geometry is questionable from the perspective of contact mechanics since it relies on assuming an incorrect geometry for the nanoindenter tip to correct for an error in the geometry of the nanoindentation impression. The values for E and H obtained using the γ-modification are, however, good and can be significantly different to the values obtained with the standard Oliver and Pharr analysis.

12.3.7 Alternative Methods of Analysis

All of the preceding discussion on the analysis of nanoindentation curves has focused on the unloading curve, virtually ignoring the loading curve data. This is for the simple reason that the unloading curve can in many cases be regarded as purely elastic, whereas the shape of the loading curve is determined by a complex mix of elastic and plastic properties.

It is clear that there is substantially more data in the loading curve if it can be extracted. *Page* et al. [57, 72] have explored the possibility of curve fitting to the loading data using a combination of elastic and plastic properties. By a combination of analysis and empirical fitting to experimental data, it was suggested that the loading curve is of the following form:

$$P = E\left(\psi\sqrt{\frac{H}{E}} + \phi\sqrt{\frac{E}{H}}\right)^{-2}\delta^2, \tag{12.21}$$

where ψ and ϕ are determined experimentally to be 0.930 and 0.194, respectively. For homogenous samples this equation gives a linear relationship between P and δ^2.

Coatings, thin film systems, and samples that strain-harden can give significant deviations from linearity. Analysis of the loading curve has yet to gain popularity as a standard method for examining nanoindentation data, but it should certainly be regarded as a prime area for further investigation.

Another alternative method of analysis is based on the work involved in making an indentation. In essence, the nanoindentation curve is a plot of force against distance indicating integration under the loading curve will give the total work of indentation, or the sum of the elastic strain energy and the plastic work of indentation. Integrating under the unloading curve should give only the elastic strain energy. Thus, the work involved in both elastic and plastic deformation during nanoindentation can be found. *Cheng* and *Cheng* [73] combined measurements of the work of indentation with a dimensional analysis that deals with the effects of scaling in a material that work-hardens to estimate H/E_r. They subsequently evaluated H and E using the Oliver and Pharr approach to find the contact area.

12.3.8 Measuring Contact Stiffness

As discussed earlier, it is possible to add a small AC load on top of the DC load used during nanoindentation testing, providing a way to measure the contact stiffness throughout the entire loading and unloading cycle [74, 75]. The AC load is typically at a frequency of $\approx 60\,\text{Hz}$ and creates a dynamic system, with the sample acting as a spring with stiffness S (the contact stiffness), and the nanoindentation system acting as a series of springs and dampers. Figure 12.20 illustrates how the small AC load is added to the DC load. Figure 12.21 shows how the resulting dynamic system can be modeled. An analysis of the dynamic system gives the following relationships for S based on the amplitude of the AC displacement oscillation and the phase difference between the AC load and displacement signals:

$$\left|\frac{P_{os}}{\delta(\omega)}\right| = \sqrt{\left[(S^{-1}+C_f)^{-1}+K_s-m\omega^2\right]^2+\omega^2 D^2}, \quad (12.22)$$

$$\tan(\chi) = \frac{\omega D}{(S^{-1}+C_f)^{-1}+K_s-m\omega^2}, \quad (12.23)$$

where C_f is the load frame compliance (the reciprocal of the load frame stiffness), K_s is the stiffness of the support springs (typically in the region of 50–100 N/m),

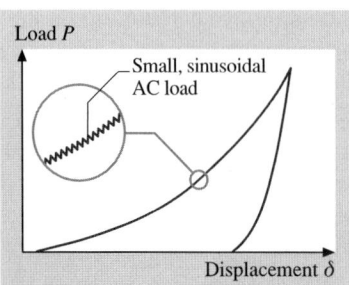

Fig. 12.20. A small AC load can be added to the DC load. This enables the contact stiffness, S, to be calculated throughout the indentation cycle

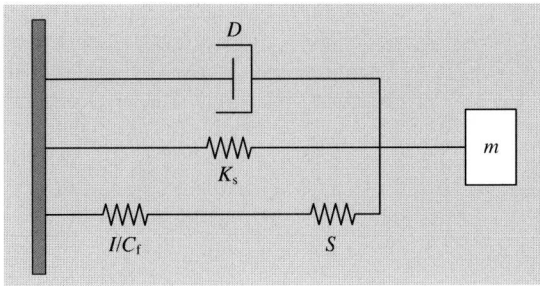

Fig. 12.21. The dynamic model used in the analysis of the AC response of a nanoindentation device

D is the damping coefficient, P_{os} is the magnitude of the load oscillation, $\delta(\omega)$ is the magnitude of the displacement oscillation, ω is the oscillation frequency, m is the mass of the indenter, and χ is the phase angle between the force and the displacement.

In order to find S using either (12.22) or (12.23), it is necessary to calibrate the dynamic response of the system when the tip is not in contact with a sample ($S^{-1} = 0$). This calibration combined with the standard DC calibrations will provide the values for all of the constants in the two equations. All that needs to be measured in order to obtain S is either $\delta(\omega)$ or χ, both of which are measured by the lock-in amplifier used to generate the AC signal. Since the S obtained is the same as the S in (12.1), it follows that the Oliver and Pharr analysis can be applied to obtain E_r and H throughout the entire nanoindentation cycle.

The dynamic analysis detailed here was developed for the MTS Nanoindenter™ (Oakridge, Tennessee), but a similar analysis has been applied to other commercial instruments such as the Hysitron Triboscope™ (Minneapolis, Minnesota) [76]. For all instruments, an AC oscillation is used in addition to the DC voltage, and a dynamic model is used to analyze the response.

12.3.9 Measuring Viscoelasticity

Using an AC oscillation in addition to the DC load introduces the possibility of measuring viscoelastic properties during nanoindentation testing. This has recently been the subject of considerable interest with researchers looking at the loss modulus, storage modulus, and loss tangent of various polymeric materials [25, 77]. Recording the displacement response to the AC force oscillation enables the complex modulus (including the loss and storage modulus) to be found. If the modulus is complex, it is clear from (12.1) that the stiffness also becomes complex. In fact, the stiffness will have two components: S', the component in phase with the AC force and S'', the component out of phase with the AC force.

The dynamic model illustrated in Fig. 12.21 is no longer appropriate for this situation, as the contact on the test sample also includes a damping term, shown in Fig. 12.22. Equations (12.22) and (12.23) must also be revised. Neglecting the load frame compliance, C_f, which in most real situations is negligible, (12.22) and

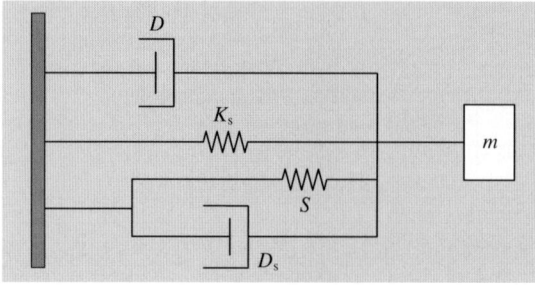

Fig. 12.22. The simplified dynamic model used when the sample is viscoelastic. It is assumed that the load frame compliance is negligible

(12.23) when the sample damping, D_s, is included become:

$$\left|\frac{P_{os}}{\delta(\omega)}\right| = \sqrt{\{S+K_s-m\omega^2\}^2+\omega^2(D+D_s)^2}, \tag{12.24}$$

$$\tan(\chi) = \frac{\omega(D+D_s)}{S+K_s-m\omega^2}. \tag{12.25}$$

In order to find the loss modulus and storage modulus, (12.1) is used to relate S' (storage component) and S'' (loss component) to the complex modulus.

This method for measuring viscoelastic properties using nanoindentation has now been proven in principal, but has still only been applied to a very small range of polymers and remains an area of future growth.

12.4 Modes of Deformation

As described earlier, the analysis of nanoindentation data is based firmly on the results of elastic continuum mechanics. In reality, this idealized, purely elastic situation rarely occurs. For very shallow contacts on metals with thin surface films such as oxides, carbon layers, or organic layers [78, 79], the contact can initially be very similar to that modeled by Hertz and, later, Sneddon. It is very important to realize that this in itself does not constitute proof that the contact is purely elastic, because in many cases a small number of defects are present. These may be preexisting defects that move in the strain-field generated beneath the contact. Alternatively, defects can be generated either when the contact is first made or during the initial loading [52, 80]. When defects such as short lengths of dislocation are present the curves may still appear to be elastic even though inelastic processes like dislocation glide and cross-slip are taking place.

12.4.1 Defect Nucleation

Nucleation of defects during nanoindentation testing has been the subject of many experimental [81, 82] and theoretical studies [83, 84]. This is probably because nanoindentation is seen as a way to deform a small, defect-free volume of material to its elastic limit and beyond in a highly controlled geometry. There are, unfortunately, problems in comparing experimental results with theoretical predictions, largely because the kinetic processes involved in defect nucleation are difficult to model. Simulations conducted at 0 K do not permit kinetic processes, and molecular dynamics simulations are too fast (nanoseconds or picoseconds). Real nanoindentation experiments take place at ≈ 293 K and last for seconds or even minutes.

Kinetic effects appear in many forms, for instance, during the initial contact between the indenter tip and the surface when defects can be generated by the combined action of the impact velocity and surface forces [51]. A second example of a kinetic effect occurs during hold cycles at large loads when what appears to be an elastic contact can suddenly exhibit a large discontinuity in the displacement data [80]. Figure 12.23 shows how these kinetic effects can affect the nanoindentation data and the apparent yield point load.

During the initial formation of a contact, the deformation of surface asperities [51] and ledges [85] can create either point defects or short lengths of dislocation line. During the subsequent loading, the defects can help in the nucleation and multiplication of dislocations. The large strains present in the region surrounding the contact, coupled with the existence of defects generated on contact, can result in the extremely rapid multiplication of dislocations and, hence, pronounced discontinuities in the load-displacement curve. It is important to realize that the discontinuities are due to the rapid multiplication of dislocations, which may or may not occur at the same time that the first dislocation is nucleated. Dislocations may have been present for some time with the discontinuity only occurring when the existing defects are configured appropriately, as a Frank–Read source, for instance. Even under large strains, the time taken for a dislocation source to form from preexisting defects may be long. It is, therefore, not surprising that large discontinuities can be seen during hold cycles or unloading.

The generation of defects at the surface and the initiation of yielding is a complex process that is extremely dependent on surface asperities and surface forces.

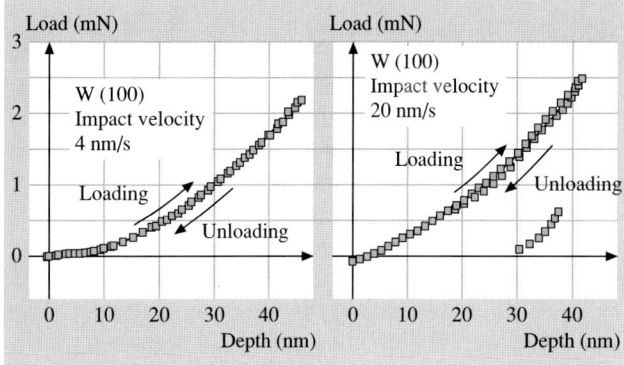

Fig. 12.23. Load-displacement curves for W(100) showing how changes in the impact velocity can cause a transition from perfectly elastic behavior to yielding during unloading

These, in turn, are closely related to the surface chemistry. It is not only the magnitude of surface forces, but also their range in comparison to the height of surface asperities that determines whether defects are generated on contact. Small changes in the surface chemistry or the velocity of the indenter tip when it first contacts the surface, can cause a transition from a situation in which defects are generated on contact to one where the contact is purely elastic [52].

When the generation of defects during the initial contact is avoided and the deformed region under the contact is truly defect free, then the yielding of the sample should occur at the yield stress of a perfect crystal lattice. The load at which plastic deformation commences under these circumstances becomes very reproducible [86]. Unfortunately, nanoindenter tips on the near-atomic scale are not perfectly smooth or axisymmetric. As a result, accurately measuring the yield stress is very difficult. In fact, a slight rotation in the plane of the surface of either the sample or the tip can give a substantial change in the observed yield point load. Coating the surface in a cushioning self-assembled monolayer [87] can alleviate some of these variations, but it also introduces a large uncertainty in the contact area. Surface oxide layers, which may be several nanometers thick, have also been found to enhance the elastic behavior seen for very shallow nanoindentations on metallic surfaces [78]. Removal of the oxide has been shown to alleviate the initial elastic response.

While nanoindentation testing is ideal for examining the mechanical properties of defect-free volumes and looking at the generation of defects in perfect crystal lattices, it should be clear from the preceding discussion that great care must be taken in examining how the surface properties and the loading rate affect the results, particularly when comparisons are being made to theoretical models for defect generation.

12.4.2 Variations with Depth

Ideal elastic-plastic behavior, as described by *Tabor* [48], can be seen during indentation testing, provided the sample has been work-hardened so that the flow stress is a constant. However, it is often the case that the mechanical properties appear to change as the load (or depth) is increased. This apparent change can be a result of several processes, including work-hardening during the test. This is a particularly important effect for soft metals like copper. These metals usually have a high hardness at shallow depths, but it decreases asymptotically with increasing indentation depth to a hardness value that may be less than half that observed at shallow depths. This type of behavior is due to the increasing density of geometrically necessary dislocations at shallow depths [88]. Hence the effects of work-hardening are most pronounced at shallow depths. For hard materials the effect is less obvious.

Work-hardening is one of the factors that contribute to the so-called indentation size effect (ISE), whereby at shallow indentation depths the material appears to be harder. The ISE has been widely observed during microindentation testing, with at least part of the effect appearing to result from the increased difficulty in optically measuring the area of an indentation when it is small. During nanoindentation testing the ISE can also be observed, but it is often due to the tip area function, $A_c(\delta_c)$, being incorrectly calibrated. However, there are physical reasons other than work-hardening for expecting an increase in mechanical strength in small volumes. As described in the previous section, small volumes of crystalline materials can have either no defects or only a small number of defects present, making plastic yielding more difficult. Also, because of dislocation line tension, the shear stress required to make a dislocation bow out increases as the radius of the bow decreases. Thus, the shear stress needed to make a dislocation bow out in a small volume is greater than it is in a large volume. These physical reasons for small volumes appearing stronger than large volumes are particularly important in thin film systems, as will be discussed later. Note, however, that these physical reasons for increased hardness do not apply for an amorphous material such as fused silica, which partially explains its value as a calibration material.

12.4.3 Anisotropic Materials

The analysis methods detailed earlier were concerned primarily with the interpretation of data from nanoindentations in isotropic materials where the elastic modulus is assumed to be either independent of direction or a polycrystalline average of a material's elastic constants. Many crystalline materials exhibit considerable anisotropy in their elastic constants, hence, these analysis techniques may not always be appropriate. The theoretical problem of a rigid indenter pressed into an elastic, anisotropic half-space has been considered by *Vlassak* and *Nix* [89]. Their aim was to identify the feasibility of interpreting data from a depth-sensing indentation apparatus for samples with elastic constants that are anisotropic. Nanoindentation experiments [90] have shown the validity of the elastic analysis for crystalline zinc, copper, and beta-brass. The observed indentation modulus for zinc, as predicted, varied by as much as a factor of 2 between different orientations. The variations in the observed hardness values for the same materials were smaller, with a maximum variation with orientation of 20% detected in zinc. While these variations are clearly detectable with nanoindentation techniques, the variations are small in comparison to the actual anisotropy of the test material's elastic properties. This is because the indentation modulus is a weighted average of the stiffness in all directions.

At this time the effects of anisotropy on the hardness measured using nanoindentation have not been fully explored. For materials with many active slip planes it is likely that the small anisotropy observed by Vlassak and Nix is correct once plastic flow has been initiated. It is possible, however, that for defect-free crystalline specimens with a limited number of active slip planes that very shallow nanoindentations may show a much larger anisotropy in the observed hardness and initial yield point load.

12.4.4 Fracture and Delamination

Indentation testing has been widely used to study fracture in brittle materials [91], but the lower loads and smaller deformation regions of nanoindentation tests make it harder to initiate cracks and, hence, less useful as a way to evaluate fracture toughness. To overcome these problems the cube corner geometry, which generates larger shear stresses than the Berkovich pyramid, has been used with nanoindentation testing to study fracture [92]. These studies have had mixed success, because the cube corner geometry blunts very quickly when used on hard materials. In many cases, brittle materials are very hard.

Depth sensing indentation is better suited to studying delamination of thin films. Recent work extends the research conducted by *Marshall* et al. [93, 94], who examined the deformation of residually stressed films by indentation. A schematic of their analysis is given by Fig. 12.24. Their indentations were several microns deep, but the basic analysis is valid for nanoindentations. The analysis has been extended to multilayers [95], which is important since it enables a quantitative assessment of adhesion energy when an additional stressed film has been deposited on top of the film and substrate of interest. The additional film limits the plastic deformation of the film of interest and also applies extra stress that aids in the delamination. After indentation, the area of the delaminated film is measured optically or with an AFM to assess the extent of the delamination. This measurement, coupled with the load-displacement data, enables quantitative assessment of the adhesion energy to be made for metals [96] and polymers [97].

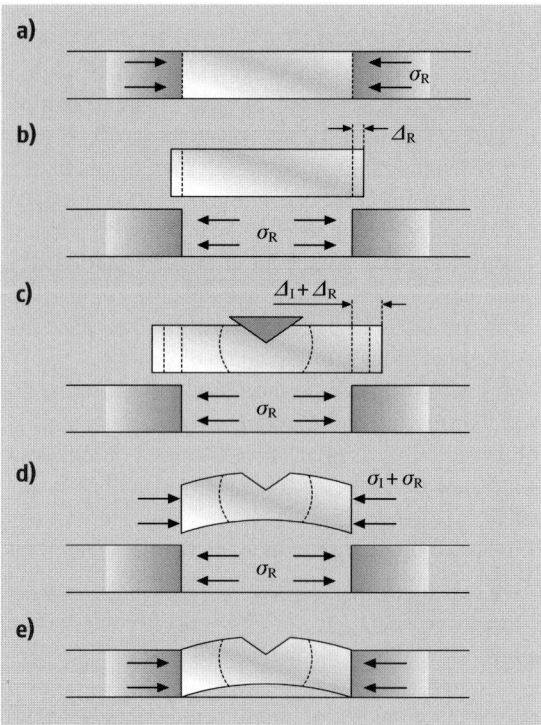

Fig. 12.24. To model delamination *Kriese* et al. [95] adapted the model developed by *Marshall* and *Evans* [93]. The model considers a segment of removed stressed film that is allowed to expand and then indented, thereby expanding it further. Replacing the segment in its original position requires an additional stress, and the segment bulges upwards

12.4.5 Phase Transformations

The pressure applied to the surface of a material during indentation testing can be very high. Equation (12.10) indicates that the pressure during plastic yielding is about three times the yield stress. For many materials, high hydrostatic pressures can cause phase transformations, and provided the transformation pressure is less than the pressure required to cause plastic yielding, it is possible during indentation testing to induce a phase transformation. This was first reported for silicon [98], but it has also been speculated [99] that many other materials may show the same effects. Most studies still focus on silicon because of its enormous technological importance, although there is some evidence that germanium also undergoes a phase transformation during the nanoindentation testing [100].

Recent results [21,22,41,101,102] indicate that there are actually multiple phase transformations during the nanoindentation of silicon. TEM of nanoindentations in diamond cubic silicon have shown the presence of amorphous-Si and the body-centered cubic BC-8 phase (see Fig. 12.25). Micro-Raman spectroscopy has indicated the presence of a further phase, the rhombhedral R-8 (see Fig. 12.26). For

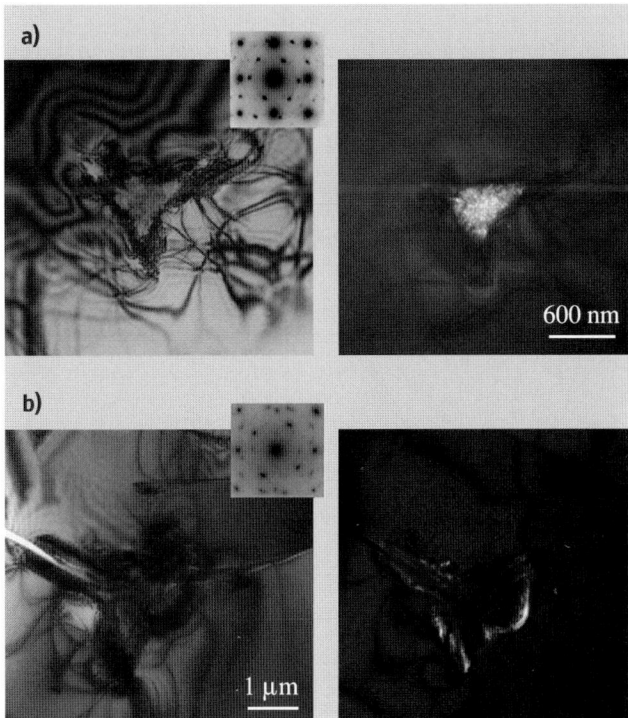

Fig. 12.25. Bright-field and dark-field TEM of (**a**) small and (**b**) large nanoindentations in Si. In small nanoindents the metastable phase BC-8 is seen in the center, but for large nanoindents BC-8 is confined to the edge of the indent, while the center is amorphous

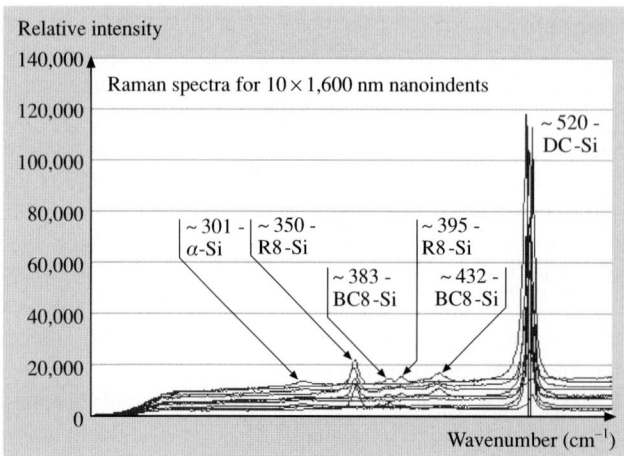

Fig. 12.26. Micro-Raman generally shows the BC-8 and R-8 Si phases that are at the edge of the nanoindents, but the amorphous phase in the center is not easy to detect, as it is often subsurface and the Raman peak is broad

many nanoindentations on silicon there is a characteristic discontinuity in the unloading curve (see Fig. 12.27), which seems to correlate with a phase transformation. The exact sequence in which the phases form is still highly controversial with

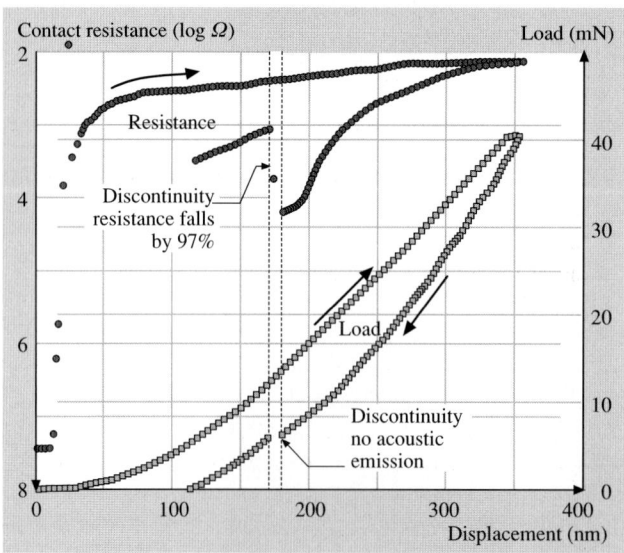

Fig. 12.27. Nanoindentation curves for deep indents on Si show a discontinuity during unloading and simultaneously a large drop in contact resistance

some [42], suggesting that the sequence during loading and unloading is:

Increasing load →
Diamond cubic Si → β-Sn Si

← **Decreasing load**
BC-8 Si and R-8 Si ← β-Sn Si

Other groups [21, 22] suggest that the above sequence is only valid for shallow nanoindentations that do not exhibit an unloading discontinuity. For large nanoindentations that show an unloading discontinuity, they suggest the sequence will be:

Increasing load →
Diamond cubic Si → β-Sn Si

← **Decreasing load**
α-Si ← BC-8 Si and R-8 Si ← β-Sn Si

The disagreement is over the origin of the unloading discontinuity. *Mann* et al. [21] suggest it is due to the formation of amorphous silicon, while *Gogotsi* et al. [42] believe it is the β-Sn Si to BC-8 or R-8 transformation. Mann et al. argue that the high contact resistance before the discontinuity and the low contact resistance afterwards rule out the discontinuity being the metallic β-Sn Si transforming to the more resistive BC-8 or R-8. The counterargument is that amorphous Si is only seen with micro-Raman spectroscopy when the unloading is very rapid or there is a large nanoindentation with no unloading discontinuity. The importance of unloading rate and cracking in determining the phases present are further complications. The controversy will remain until in situ characterization of the phases present is undertaken.

12.5 Thin Films and Multilayers

In almost all real applications, surfaces are coated with thin films. These may be intentionally added such as hard carbide coatings on a tool bit, or they may simply be native films such as an oxide layer. It is also likely that there will be adsorbed films of water and organic contaminants that can range from a single molecule in thickness up to several nanometers. All of these films, whether native or intentionally placed on the surface, will affect the surface's mechanical behavior on the nanoscale. Adsorbates can have a significant impact on the surface forces [3] and, hence, the geometry and stability of asperity contacts. Oxide films can have dramatically different mechanical properties to the bulk and will also modify the surface forces. Some of the effects of native films have been detailed in the earlier sections on dislocation nucleation and adhesive contacts.

The importance of thin films in enhancing the mechanical behavior of surfaces is illustrated by the abundance of publications on thin film mechanical properties (see for instance *Nix* [88] or *Cammarata* [31] or *Was* and *Foecke* [103]). In the following sections, the mechanical properties of films intentionally deposited on the surface will be discussed.

12.5.1 Thin Films

Measuring the mechanical properties of a single thin surface film has always been difficult. Any measurement performed on the whole sample will inevitably be dominated by the bulk substrate. Nanoindentation, since it looks at the mechanical properties of a very small region close to the surface, offers a possible solution to the problem of measuring thin film mechanical properties. However, there are certain inherent problems in using nanoindentation testing to examine the properties of thin films. The problems stem in part from the presence of an interface between the film and substrate. The quality of the interface can be affected by many variables, resulting in a range of effects on the apparent elastic and plastic properties of the film. In particular, when the deformation region around the indent approaches the interface, the indentation curve may exhibit features due to the thin film, the bulk, the interface, or a combination of all three. As a direct consequence of these complications, models for thin-film behavior must attempt to take into account not only the properties of the film and substrate, but also the interface between them.

If, initially, the effect of the interface is neglected, it is possible to divide thin-coated systems into a number of categories that depend on the values of E (elastic modulus) and Y (the yield stress) of the film and substrate. These categories are typically [104, 105]:

1. coatings with high E and high Y, substrates with high E and high Y;
2. coatings with high E and high Y, substrates with high or low E and low Y;
3. coatings with high or low E and low Y, substrates with high E and high Y;
4. coatings with high or low E and low Y, substrates with high or low E and low Y.

The reasons for splitting thin film systems into these different categories have been amply demonstrated experimentally by *Whitehead* and *Page* [104, 105] and theoretically by *Fabes* et al. [106]. Essentially, hard, elastic materials (high E and Y) will possess smaller plastic zones than soft, inelastic (low E and Y) materials. Thus, when different combinations of materials are used as film and substrate, the overall plastic zone will differ significantly. In some cases, the plasticity is confined to the film, and in other cases, it is in both the film and substrate, as shown by Fig. 12.28. If the standard nanoindentation analysis routines are to be used, it is essential that the plastic zone and the elastic strain field are both confined to the film and do not reach the substrate. Clearly, this is difficult to achieve unless extremely shallow nanoindentations are used. There is an often quoted 10% rule, that says nanoindents in a film must have a depth of less than 10% of the film's thickness if only the film properties are to be measured. This has no real validity [107]. There are film/substrate combinations for which 10% is very conservative, while for other combinations even 5% may be too deep. The effect of the substrate for different combinations of film and substrate properties has been studied using FEM [108], which has shown that the maximum nanoindentation depth to measure film only properties decreases in moving from soft on hard to hard on soft combinations. For a very soft film on a hard substrate, nanoindentations of 50% of the film thickness are alright, but this drops to < 10% for a hard film on a soft substrate. For a very

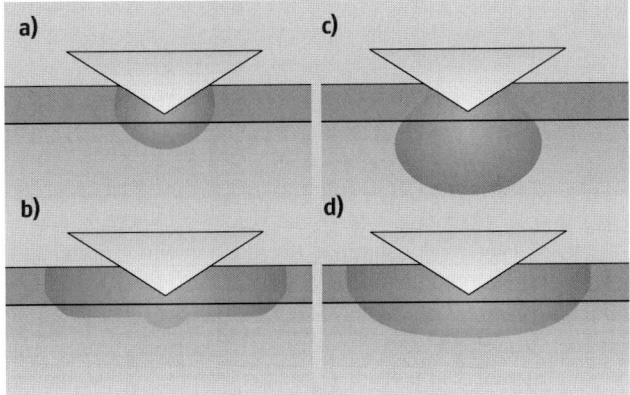

Fig. 12.28. Variations in the plastic zone for indents on films and substrates of different properties. (**a**) Film and substrate have high E and Y, (**b**) film has a high E and Y, substrate has a high or low E and low Y, (**c**) film has a high or low E and a low Y, and substrate has a high E and Y, (**d**) film has a high or low E and a low Y, and substrate has a high or low E and a low Y

strong film on a soft substrate, the surface film behaves like an elastic membrane or a bending plate.

Theoretical analysis of thin-film mechanical behavior is difficult. One theoretical approach that has been adopted uses the volumes of plastically deformed material in the film and substrate to predict the overall hardness of the system. However, it should be noted that this method is only really appropriate for soft coatings and indentation depths below the thickness of the coating (see cases c and d of Fig. 12.28), otherwise the behavior will be closer to that detailed later and shown by Fig. 12.29.

The technique of combining the mechanical properties of the film and substrate to evaluate the overall hardness of the system is generally referred to as the rule of mixtures. It stems from work by *Burnett* et al. [109–111] and *Sargent* [112], who derived a weighted average to relate the "composite" hardness (H) to the volumes of plastically deformed material in the film (V_f) and substrate (V_s) and their respective

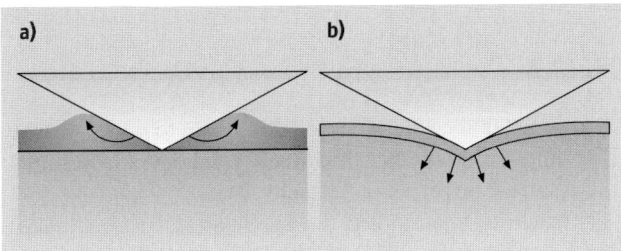

Fig. 12.29. Two different modes of deformation during nanoindentation of films. In (**a**) materials move upwards and outwards, while in (**b**) the film acts like a membrane and the substrate deforms

values of hardness, H_f and H_s. Thus,

$$H = \frac{H_f V_f + H_s V_s}{V_{total}}, \qquad (12.26)$$

where V_{total} is $V_f + V_s$.

Equation (12.26) was further developed by *Burnett* and *Page* [109] to take into account the indentation size effect. They replaced H_s with $K\delta_c^{n-2}$, where K and n are experimentally determined constants dependent on the indenter and sample, and δ_c is the contact depth. This expression is derived directly from Meyer's law for spherical indentations, which gives the relationship $P = Kd^n$ between load, P, and the indentation dimension, d. Burnett and Page also employed a further refinement to enable the theory to fit experimental results from a specific sample, ion-implanted silicon. This particular modification essentially took into account the different sizes of the plastic zones in the two materials by multiplying H_s by a dimensionless factor (V_s/V_{total}). While this seems to be a sensible approach, it is mostly empirical, and the physical justification for using this particular factor is not entirely clear. Later, *Burnett* and *Rickerby* [110, 111] took this idea further and tried to generalize the equations to take into account all of the possible scenarios. Thus, the following equations were suggested:

$$H = \frac{H_f(\Omega^3) V_f + H_s V_s}{V_{total}}, \qquad (12.27)$$

$$H = \frac{H_f V_f + H_s(\Omega^3) V_s}{V_{total}}. \qquad (12.28)$$

The first of these, (12.27), deals with the case of a soft film on a hard substrate, and the second, (12.28), with a hard film on a soft substrate. The Ω term expresses the variation of the total plastic zone from the ideal hemispherical shape. This was taken still further by *Bull* and *Rickerby* [113], who derived an approximation for Ω based on the film and substrate zone radii being related to their respective hardness and elastic modulii [114, 115]. Hence:

$$\Omega = (E_f H_s / E_s H_f)^l, \qquad (12.29)$$

where l is determined empirically. E_f and H_f and E_s and H_s are the elastic modulus and hardness of the film and substrate, respectively.

Experimental data [116] indicate that the effect of the substrate on the elastic modulus of the film can be quite different than the effect on hardness, due to the zones of the elastic and plastic strain fields being different sizes.

Chechechin et al. [117] have recently studied the behavior of Al_2O_3 films of various thicknesses on different substrates. Their results indicate that many of the models correctly predict the transition between the properties of the film and those of the substrate, but do not always fit the observed hardness against depth curves. This group have also studied the pop-in behavior of Al_2O_3 films [118] and have attempted to model the range of loads and depths at which they occur via a Weibull-type distribution, as utilized in fracture analysis.

A point raised by Burnett and Rickerby should be emphasized. They state that there are two very distinct modes of deformation during nanoindentation testing. The first, referred to as *Tabor*'s [48] model for low Y/E materials, involves the buildup of material at the side of the indenter through movement of material at slip lines. The second, for materials with large Y/E does not result in surface pile-up. The displaced material is then accommodated by radial displacements [115]. The point is that a thin, strong, and well-bonded surface film can cause a substrate that would normally deform by Tabor's method to behave more like a material with high Y/E (see Fig. 12.29). It should be noted that this only applies as long as the film does not fail.

In recent theoretical and experimental studies the importance of material pile-up and sink-in has been investigated extensively. As discussed in an earlier section, pile-up can be increased by residual compressive stresses, but even in the absence of residual stresses pile-up can introduce a significant error in the calculated contact area. This is most pronounced in materials that do not work-harden [62]. For these materials using the Oliver and Pharr method fails to account for the pile-up and results in a large error in the values for E and H. For thin films *Tsui* et al. have used a focused ion beam to section through Knoop indentations in both soft films on hard substrates [69] and hard films on soft substrates [70]. The soft films, as expected, exhibit pile-up, while the hard film acts more like a membrane and the indentation exhibits sink-in with most of the plasticity in the substrate. Thus, there are three clearly identifiable factors affecting the pile-up and sink-in around nanoindents during testing of thin films:

1. Residual stresses
2. Degree of work-hardening
3. Ratio of film and substrate mechanical properties.

The bonding or adhesion between the film and substrate could also be added to this list. And it should not be forgotten that the depth of the nanoindentation relative to the film thickness also affects pile-up. For a very deep nanoindentation into a thin, soft film on a hard substrate pile-up is reduced, due to the combined constraints on the film of the tip and substrate [119]. Due to all of these complications, using nanoindentation to study thin film mechanical properties is fraught with danger. Many unprepared researchers have misguidedly taken the values of E and H obtained during nanoindentation testing to be absolute values only to find out later that the values contain significant errors.

Many of the problems associated with nanoindentation testing are related to incorrectly calculating the contact area, A. The *Joslin* and *Oliver* method [71] is one way that A can be removed from the calculations. This approach has been used with some success to look at strained epitaxial II/VI semiconductor films [120], but there is evidence that the lattice mismatch in these films can cause dramatic changes in the mechanical properties of the films [121]. This may be due to image forces and the film/substrate interface acting as a barrier to dislocation motion. Recently, it has been shown that using films and substrates with known matching elastic moduli, it is possible to use the assumption of constant elastic modulus with depth to evaluate H [122]. In effect, this is using (12.1) to evaluate A from the contact stiffness data, and then substituting the value for A into (12.12). The value of E is measured independently, for instance, using acoustic techniques.

12.5.2 Multilayers

Multilayered materials with individual layers that are a micron or less in thickness, sometimes referred to as superlattices, can exhibit substantial enhancements in hardness or strength. This should be distinguished from the super modulus effect discussed earlier, which has been shown to be largely an artifact. The enhancements in hardness can be as much as 100% when compared to the value expected from the rule of mixtures, which is essentially a weighted average of the hardness for the constituents of the two layers [123]. Table 12.1 shows how the properties of isostructural multilayers can show a substantial increase in hardness over that for fully interdiffused layers. The table also shows how there can be a substantial enhancement in hardness for non-isostructural multilayers compared to the values for the same materials when they are homogeneous.

There are many factors that contribute to enhanced hardness in multilayers. These can be summarized as [103]:

1. Hall-Petch behavior
2. Orowan strengthening
3. Image effects
4. Coherency and thermal stresses
5. Composition modulation

Hall-Petch behavior is related to dislocations piling-up at grain boundaries. (Note that pile-up is used to describe two distinct effects: One is material building up at the side of an indentation, the other is an accumulation of dislocations on

Table 12.1. Results for some experimental studies of multilayer hardness

Study	Multilayer	Maximum hardness and multilayer repeat length	Reference hardness value	Range of hardness values for multilayers
Isostructural Knoop hardness [124]	Cu/Ni	524 at 11.6 nm	284 (interdiffused)	295–524
Non-isostructural Nanoindentation [125]	Mo/NbN	33 GPa at 2 nm	NbN – 17 GPa Mo – 2.7 GPa Wo – 7 GPa	12–33 GPa
	W/NbN	29 GPa at 3 nm	(individual layer materials)	23–29 GPa

a slip-plane.) The dislocation pile-up at grain boundaries impedes the motion of dislocations. For materials with a fine grain structure there are many grain boundaries, and, hence, dislocations find it hard to move. In polycrystalline multilayers, it is often the case that the size of the grains within a layer scales with the layer thickness so that reducing the layer thickness reduces the grain size. Thus, the Hall-Petch relationship (below) should be applicable to polycrystalline multilayer films with the grain size, d_g, replaced by the layer thickness.

$$Y = Y_0 + k_{HP} d_g^{-0.5}, \tag{12.30}$$

where Y is the enhanced yield stress, Y_0 is the yield stress for a single crystal, and k_{HP} is a constant.

There is an ongoing argument about whether Hall-Petch behavior really takes place in nanostructured multilayers. The basic model assumes many dislocations are present in the pile-up, but such large dislocation pile-ups are not seen in small grains [126] and are unlikely to be present in multilayers. As a direct consequence, studies have found a range of values, between 0 and -1, for the exponent in (12.30), rather than the -0.5 predicted for Hall-Petch behavior.

Orowan strengthening is due to dislocations in layered materials being effectively pinned at the interfaces. As a result, the dislocations are forced to bow out along the layers. In narrow films, dislocations are pinned at both the top and bottom interfaces of a layer and bow out parallel to the plane of the interface [127, 128]. Forcing a dislocation to bow out in a layered material requires an increase in the applied shear stress beyond that required to bow out a dislocation in a homogeneous sample. This additional shear stress would be expected to increase as the film thickness is reduced.

Image effects were suggested by *Koehler* [129] as a possible source of enhanced yield stress in multilayered materials. If two metals, A and B, are used to make a laminate and one of them, A, has a high dislocation line energy, but the other, B, has a low dislocation line energy, then there will be an increased resistance to dislocation motion due to image forces. However, if the individual layers are thick enough that there may be a dislocation source present within the layer, then dislocations could pile-up at the interface. This will create a local stress concentration point and the enhancement to the strength will be very limited. If the layers are thin enough that there will be no dislocation source present, the enhanced mechanical strength may be substantial. In Koehler's model only nearest neighbor layers were taken to contribute to the image forces. However, this was extended to include more layers [130] without substantial changes in the results. The consequence on image effects of reducing the thickness of the individual layers in a multilayer is that it prevents dislocation sources from being active within the layer.

For many multilayer systems there is an increase in strength as the bilayer repeat length is reduced, but there is often a critical repeat length (e.g., 3 nm for the W/NbN multilayer of Table 12.1) below which the strength falls. One explanation for the fall in strength involves the effects of coherency and thermal stresses on dislocation energy. Unlike image effects where the energy of dislocations are a maximum or minimum in the center of layers, the energy maxima and minima are at the interfaces for coherency stresses. Combining the effects of varying moduli and coherency stresses shows that the dependence of strength on layer thickness has a peak near the repeat period where coherency strains begin decreasing [131].

Another source of deviations in behavior at very small repeat periods is the imperfect nature of interfaces. With the exception of atomically perfect epitaxial films, interfaces are generally not atomically flat and there is some interdiffusion. For the Cu/Ni film of Table 12.1, the effects of interdiffusion on hardness were examined [124] by annealing the multilayers. The results were in agreement with a model by *Krzanowski* [132] that predicted the variations in hardness would be proportional to the amplitude of the composition modulation.

It is interesting to note that the explanations for enhanced mechanical properties in multilayered materials are all based on dislocation mechanisms. So it would seem natural to assume that multilayered materials that do not contain dislocations will show no enhanced hardness over their rule of mixtures values. This has been verified by studies on amorphous metal multilayers [133], which shows that the hardness of the multilayers, firstly, lies between that of the two individual materials and, secondly, has almost no variation with repeat period.

12.6 Developing Areas

Over the past 20 to 30 years, the driving force for studying nanomechanical behavior of surfaces and thin films has been largely, though not exclusively, the microelectronics industry. The importance of electronics to the modern world is only likely to grow in the foreseeable future, but other technological areas may overtake microelectronics as the driving force for research, including the broad fields of biomaterials and nanotechnology. In several places in this chapter, a number of developing areas have been mentioned. These include nanoscale measurements of viscoelasticity and the study of environmental effects (temperature and surface chemistry) on nanomechanical properties. Both of these topics will be vital in the study of biological systems and, as a result, will be increasingly important from a research point of view.

We still have a relatively rudimentary understanding of the nanomechanics of complex biological systems such as bone cells (osteoblasts and osteoclasts) and skin cells (fibroblasts), or even, for that matter, simpler biological structures such as dental enamel. For example, Fig. 12.30 shows how the mechanical properties of dental enamel can vary within a single tooth [134]. But this is still a relatively large-scale measure of mechanical behavior. The prismatic structure of enamel means that there are variations in mechanical properties on a range of scales from millimeters down to nanometers.

In terms of data analysis, there remains much to be done. If an analysis method can be developed that deals with the problems of pile-up and sink-in, the utility of nanoindentation testing will be greatly enhanced.

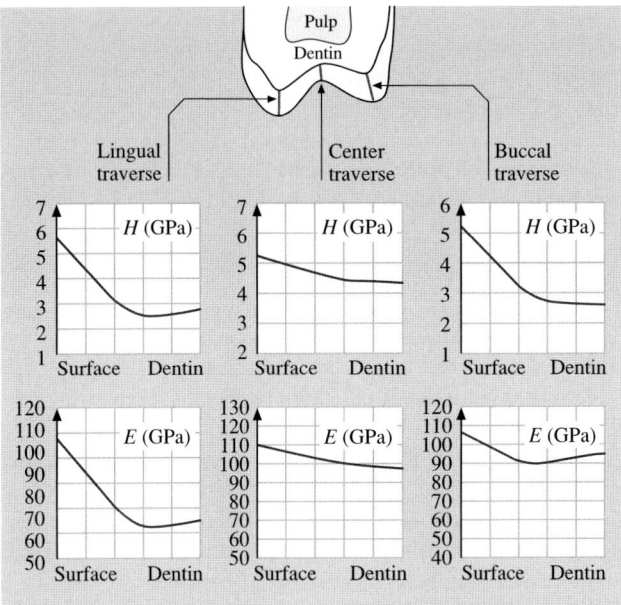

Fig. 12.30. Variations in E and H across human dental enamel. The sample is an upper 2nd molar cut in cross section from the lingual to the buccal side. Nanoindentations are performed across the surface to examine how the mechanical properties vary

References

1. G. Binnig, C.F. Quate, C. Gerber: Atomic force microscope, Phys. Rev. Lett. **56**, 930–933 (1986)
2. *Microindentation Techniques in Materials Science and Engineering*, ed. by P.J. Blau, B.R. Lawn (ASTM, Pennsylvannia 1986)
3. J.N. Israelachvili: *Intermolecular and Surface Forces* (Academic, London 1992)
4. R.S. Bradley: The cohesive force between solid surfaces and the surface energy of solids, Philos. Mag. **13**, 853–862 (1932)
5. B.V. Derjaguin, V.M. Muller, Yu.P. Toporov: Effect of contact deformations on the adhesion of particles, J. Coll. Interface Sci. **53**, 314–326 (1975)
6. V.M. Muller, V.S. Yuschenko, B.V. Derjaguin: On the influence of molecular forces on the deformation of an elastic sphere and its sticking to a rigid plane, J. Coll. Interface Sci. **77**, 91–101 (1980)
7. V.M. Muller, B.V. Derjaguin, Yu.P. Toporov: On two methods of calculation of the force of sticking of an elastic sphere to a rigid plane, Coll. Surf. **7**, 251–259 (1983)
8. K.L. Johnson, K. Kendal, A.D. Roberts: Surface energy and the contact of elastic solids, Proc. R. Soc. A **324**, 301–320 (1971)
9. A.P. Ternovskii, V.P. Alekhin, M.Kh. Shorshorov, M.M. Khrushchov, V.N. Skvortsov: Zavod Lab. **39**, 1242 (1973)
10. S.I. Bulychev, V.P. Alekhin, M.Kh. Shorshorov, A.P. Ternovskii, G.D. Shnyrev: Determining Young's modulus from the indentor penetration diagram, Zavod Lab. **41**, 1137 (1975)

11. S.I. Bulychev, V.P. Alekhin, M.Kh. Shorshorov, A.P. Ternovskii: Mechanical properties of materials studied from kinetic diagrams of load versus depth of impression during microimpression, Prob. Prochn. **9**, 79 (1976)
12. S.I. Bulychev, V.P. Alekhin: Zavod Lab. **53**, 76 (1987)
13. M.Kh. Shorshorov, S.I. Bulychev, V.P. Alekhin: Sov. Phys. Doklady **26**, 769 (1982)
14. J.B. Pethica: Microhardness tests with penetration depths less than ion implanted layer thickness. In: *Ion Implantation into Metals*, ed. by V. Ashworth, W. Grant, R. Procter (Pergamon, Oxford 1982) p. 147
15. D. Newey, M.A. Wilkens, H.M. Pollock: An ultra-low-load penetration hardness tester, J. Phys. E: Sci. Instrum. **15**, 119 (1982)
16. D. Kendall, D. Tabor: An ultrasonic study of the area of contact between stationary and sliding surfaces, Proc. R. Soc. A **323**, 321–340 (1971)
17. G.M. Pharr, W.C. Oliver, F.R. Brotzen: On the generality of the relationship among contact stiffness, contact area and elastic-modulus during indentation, J. Mater. Res. **7**, 613 (1992)
18. T.P. Weihs, C.W. Lawrence, B. Derby, C.B. Scruby, J.B. Pethica: Acoustic emissions during indentation tests, MRS Symp. Proc. **239**, 361–366 (1992)
19. D.F. Bahr, J.W. Hoehn, N.R. Moody, W.W. Gerberich: Adhesion and acoustic emission analysis of failures in nitride films with 14 metal interlayer, Acta Mater. **45**, 5163 (1997)
20. D.F. Bahr, W.W. Gerberich: Relationships between acoustic emission signals and physical phemomena during indentation, J. Mat. Res. **13**, 1065 (1998)
21. A.B. Mann, D. van Heerden, J.B. Pethica, T.P. Weihs: Size-dependent phase transformations during point-loading of silicon, J. Mater. Res. **15**, 1754 (2000)
22. A.B. Mann, D. van Heerden, J.B. Pethica, P. Bowes, T.P. Weihs: Contact resistance and phase transformations during nanoindentation of silicon, Philos. Mag. A **82**, 1921 (2002)
23. S. Jeffery, C.J. Sofield, J.B. Pethica: The influence of mechanical stress on the dielectric breakdown field strength of SiO_2 films, Appl. Phys. Lett. **73**, 172 (1998)
24. B.N. Lucas, W.C. Oliver: Indentation power-law creep of high-purity indium, Metall. Trans. A **30**, 601 (1999)
25. S.A. Syed Asif: Time dependent micro deformation of materials. Ph.D. Thesis (Oxford Univ., Oxford 1997)
26. S.A. Syed Asif, R.J. Colton, K.J. Wahl: Nanoscale Surface Mechanical Property Measurements: Force Modulation Techniques Applied to Nanoindentation. In: *Interfacial Properties on the Submicron Scale*, ed. by J. Frommer, R. Overney (ACS Books, Whashington 2000)
27. A.B. Mann, J.B. Pethica: Nanoindentation studies in a liquid environment, Langmuir **12**, 4583 (1996)
28. J.W. Beams: Mechanical properties of thin films of gold and silver. In: *Structure and Properties of Thin Films*, ed. by C.A. Neugebauer, J.B. Newkirk, D.A. Vermilyea (Wiley, New York 1959) pp. 183–192
29. J.J. Vlassak, W.D. Nix: A new bulge test technique for the determination of Youngs modulus and Poissons ratio of thin-films, J. Mater. Res. **7**, 3242 (1992)
30. W.M.C. Yang, T. Tsakalakos, J.E. Hilliard: Enhanced elastic modulus in composition modulated gold-nickel and copper-palladium foils, J. Appl. Phys. **48**, 876 (1977)
31. R.C. Cammarata: Mechanical properties of nanocomposite thin-films, Thin Solid Films **240**, 82 (1994)
32. G.A.D. Briggs: *Acoustic Microscopy* (Clarendon, Oxford 1992)

33. J. Kushibiki, N. Chubachi: Material characterization by line-focus-beam acoustic microscope, IEEE Trans. Sonics Ultrasonics **32**, 189–212 (1985)
34. M.J. Bamber, K.E. Cooke, A.B. Mann, B. Derby: Accurate determination of Young's modulus and Poisson's ratio of thin films by a combination of acoustic microscopy and nanoindentation, Thin Solid Films **398**, 299–305 (2001)
35. S.E. Bobbin, R.C. Cammarata, J.W. Wagner: Determination of the flexural modulus of thin-films from measurement of the 1st arrival of the symmetrical Lamb wave, Appl. Phys. Lett. **59**, 1544–1546 (1991)
36. R.B. King, C.M. Fortunko: Determination of inplane residual-stress state in plates using horizontally polarized shear waves, J. Appl. Phys. **54**, 3027–3035 (1983)
37. R.F. Cook, G.M. Pharr: Direct observation and analysis of indentation cracking in glasses and ceramics, J. Am. Ceram. Soc. **73**, 787–817 (1990)
38. T.F. Page, W.C. Oliver, C.J. McHargue: The deformation-behavior of ceramic crystals subjected to very low load (nano)indentations, J. Mater. Res. **7**, 450–473 (1992)
39. G.M. Pharr, W.C. Oliver, D.S. Harding: New evidence for a pressure-induced phase-transformation during the indentation of silicon, J. Mater. Res. **6**, 1129–1130 (1991)
40. C.F. Robertson, M.C. Fivel: The study of submicron indent-induced plastic deformation, J. Mater. Res. **14**, 2251–2258 (1999)
41. J.E. Bradby, J.S. Williams, J. Wong-Leung, M.V. Swain, P. Munroe: Transmission electron microscopy observation of deformation microstructure under spherical indentation in silicon, Appl. Phys. Lett. **77**, 3749–3751 (2000)
42. Y.G. Gogotsi, V. Domnich, S.N. Dub, A. Kailer, K.G. Nickel: Cyclic nanoindentation and Raman microspectroscopy study of phase transformations in semiconductors, J. Mater. Res. **15**, 871–879 (2000)
43. H. Hertz: Über die Berührung fester elastischer Körper, J. reine angew. Math. **92**, 156–171 (1882)
44. J. Boussinesq: *Application des potentiels à l'étude de l'équilibre et du mouvement des solides élastiques* (Blanchard, Paris 1885) Reprint (1996)
45. A.E.H. Love: The stress produced in a semi-infinite solid by pressure on part of the boundary, Philos. Trans. R. Soc. **228**, 377–420 (1929)
46. A.E.H. Love: Boussinesq's problem for a rigid cone, Quarter. J. Math. **10**, 161 (1939)
47. I.N. Sneddon: The relationship between load and penetration in the axisymmetric Boussinesq problem for a punch of arbitrary profile, Int. J. Eng. Sci. **3**, 47–57 (1965)
48. D. Tabor: *Hardness of Metals* (Oxford Univ. Press, Oxford 1951)
49. R. von Mises: Mechanik der festen Körper in plastisch deformablen Zustand, Goettinger Nachr. Math.-Phys. **K1**, 582–592 (1913)
50. H. Tresca: Sur l'ecoulement des corps solids soumis s fortes pression, Compt. Rend. **59**, 754 (1864)
51. A.B. Mann, J.B. Pethica: The role of atomic-size asperities in the mechanical deformation of nanocontacts, Appl. Phys. Lett. **69**, 907–909 (1996)
52. A.B. Mann, J.B. Pethica: The effect of tip momentum on the contact stiffness and yielding during nanoindentation testing, Philos. Mag. A **79**, 577–592 (1999)
53. S.P. Jarvis: Atomic force microscopy and tip-surface interactions. Ph.D. Thesis (Oxford Univ., Oxford 1993)
54. J.B. Pethica, D. Tabor: Contact of characterised metal surfaces at very low loads: Deformation and adhesion, Surf. Sci. **89**, 182 (1979)
55. J.S. Field, M.V. Swain: Determining the mechanical-properties of small volumes of materials from submicrometer spherical indentations, J. Mater. Res. **10**, 101–112 (1995)

56. W.C. Oliver, G.M. Pharr: An improved technique for determining hardness and elastic-modulus using load and displacement sensing indentation experiments, J. Mater. Res. **7**, 1564–1583 (1992)
57. S.V. Hainsworth, H.W. Chandler, T.F. Page: Analysis of nanoindentation load-displacement loading curves, J. Mater. Res. **11**, 1987–1995 (1996)
58. J.L. Loubet, J.M. Georges, O. Marchesini, G. Meille: Vickers indentation curves of magnesium oxide (MgO), Mech. Eng. **105**, 91–92 (1983)
59. J.L. Loubet, J.M. Georges, O. Marchesini, G. Meille: Vickers indentation curves of magnesium oxide (MgO), J. Tribol. Trans. ASME **106**, 43–48 (1984)
60. M.F. Doerner, W.D. Nix: A method for interpreting the data from depth sensing indentation experiments, J. Mater. Res. **1**, 601–609 (1986)
61. A. Bolshakov, W.C. Oliver, G.M. Pharr: Influences of stress on the measurement of mechanical properties using nanoindentation. 2. Finite element simulations, J. Mater. Res. **11**, 760–768 (1996)
62. A. Bolshakov, G.M. Pharr: Influences of pileup on the measurement of mechanical properties by load and depth sensing instruments, J. Mater. Res. **13**, 1049–1058 (1998)
63. J.C. Hay, A. Bolshakov, G.M. Pharr: A critical examination of the fundamental relations used in the analysis of nanoindentation data, J. Mater. Res. **14**, 2296–2305 (1999)
64. G.M. Pharr, T.Y. Tsui, A. Bolshakov, W.C. Oliver: Effects of residual-stress on the measurement of hardness and elastic-modulus using nanoindentation, MRS Symp. Proc. **338**, 127–134 (1994)
65. T.R. Simes, S.G. Mellor, D.A. Hills: A note on the influence of residual-stress on measured hardness, J. Strain Anal. Eng. Des. **19**, 135–137 (1984)
66. W.R. Lafontaine, B. Yost, C.Y. Li: Effect of residual-stress and adhesion on the hardness of copper-films deposited on silicon, J. Mater. Res. **5**, 776–783 (1990)
67. W.R. Lafontaine, C.A. Paszkiet, M.A. Korhonen, C.Y. Li: Residual stress measurements of thin aluminum metallizations by continuous indentation and X-ray stress measurement techniques, J. Mater. Res. **6**, 2084–2090 (1991)
68. T.Y. Tsui, W.C. Oliver, G.M. Pharr: Influences of stress on the measurement of mechanical properties using nanoindentation. 1. Experimental studies in an aluminum alloy, J. Mater. Res. **11**, 752–759 (1996)
69. T.Y. Tsui, J. Vlassak, W.D. Nix: Indentation plastic displacement field: Part I. The case of soft films on hard substrates, J. Mater. Res. **14**, 2196–2203 (1999)
70. T.Y. Tsui, J. Vlassak, W.D. Nix: Indentation plastic displacement field: Part II. The case of hard films on soft substrates, J. Mater. Res. **14**, 2204–2209 (1999)
71. D.L. Joslin, W.C. Oliver: A new method for analyzing data from continuous depth-sensing microindentation tests, J. Mater. Res. **5**, 123–126 (1990)
72. M.R. McGurk, T.F. Page: Using the P-delta(2) analysis to deconvolute the nanoindentation response of hard-coated systems, J. Mater. Res. **14**, 2283–2295 (1999)
73. Y.T. Cheng, C.M. Cheng: Relationships between hardness, elastic modulus, and the work of indentation, Appl. Phys. Lett. **73**, 614–616 (1998)
74. J.B. Pethica, W.C. Oliver: Mechanical properties of nanometer volumes of material: Use of the elastic response of small area indentations, MRS Symp. Proc. **130**, 13–23 (1989)
75. W.C. Oliver, J.B. Pethica: Method for continuous determination of the elastic stiffness of contact between two bodies, United States Patent Number 4,848,141, (1989)
76. S.A.S. Asif, K.J. Wahl, R.J. Colton: Nanoindentation and contact stiffness measurement using force modulation with a capacitive load-displacement transducer, Rev. Sci. Instrum. **70**, 2408–2413 (1999)

77. J.L. Loubet, W.C. Oliver, B.N. Lucas: Measurement of the loss tangent of low-density polyethylene with a nanoindentation technique, J. Mater. Res. **15**, 1195–1198 (2000)
78. W.W. Gerberich, J.C. Nelson, E.T. Lilleodden, P. Anderson, J.T. Wyrobek: Indentation induced dislocation nucleation: The initial yield point, Acta Mater. **44**, 3585–3598 (1996)
79. J.D. Kiely, J.E. Houston: Nanomechanical properties of Au(111), (001), and (110) surfaces, Phys. Rev. B **57**, 12588–12594 (1998)
80. D.F. Bahr, D.E. Wilson, D.A. Crowson: Energy considerations regarding yield points during indentation, J. Mater. Res. **14**, 2269–2275 (1999)
81. D.E. Kramer, K.B. Yoder, W.W. Gerberich: Surface constrained plasticity: Oxide rupture and the yield point process, Philos. Mag. A **81**, 2033–2058 (2001)
82. S.G. Corcoran, R.J. Colton, E.T. Lilleodden, W.W. Gerberich: Anomalous plastic deformation at surfaces: Nanoindentation of gold single crystals, Phys. Rev. B **55**, 16057–16060 (1997)
83. E.B. Tadmor, R. Miller, R. Phillips, M. Ortiz: Nanoindentation and incipient plasticity, J. Mater. Res. **14**, 2233–2250 (1999)
84. J.A. Zimmerman, C.L. Kelchner, P.A. Klein, J.C. Hamilton, S.M. Foiles: Surface step effects on nanoindentation, Phys. Rev. Lett. **87**, article 165507 (1–4) (2001)
85. J.D. Kiely, R.Q. Hwang, J.E. Houston: Effect of surface steps on the plastic threshold in nanoindentation, Phys. Rev. Lett. **81**, 4424–4427 (1998)
86. A.B. Mann, P.C. Searson, J.B. Pethica, T.P. Weihs: The relationship between near-surface mechanical properties, loading rate and surface chemistry, Mater. Res. Soc. Symp. Proc. **505**, 307–318 (1998)
87. R.C. Thomas, J.E. Houston, T.A. Michalske, R.M. Crooks: The mechanical response of gold substrates passivated by self-assembling monolayer films, Science **259**, 1883–1885 (1993)
88. W.D. Nix: Elastic and plastic properties of thin films on substrates: Nanoindentation techniques, Mater. Sci. Eng. A **234**, 37–44 (1997)
89. J.J. Vlassak, W.D. Nix: Indentation modulus of elastically anisotropic half-spaces, Philos. Mag. A **67**, 1045–1056 (1993)
90. J.J. Vlassak, W.D. Nix: Measuring the elastic properties of anisotropic materials by means of indentation experiments, J. Mech. Phys. Solids **42**, 1223–1245 (1994)
91. B.R. Lawn: *Fracture of Brittle Solids* (Cambridge Univ. Press, Cambridge 1993)
92. G.M. Pharr: Measurement of mechanical properties by ultra-low load indentation, Mater. Sci. Eng. A **253**, 151–159 (1998)
93. D.B. Marshall, A.G. Evans: Measurement of adherence of residually stressed thin-films by indentation. 1. Mechanics of interface delamination, J. Appl. Phys. **56**, 2632–2638 (1984)
94. C. Rossington, A.G. Evans, D.B. Marshall, B.T. Khuriyakub: Measurement of adherence of residually stressed thin-films by indentation. 2. Experiments with ZnO/Si, J. Appl. Phys. **56**, 2639–2644 (1984)
95. M.D. Kriese, W.W. Gerberich, N.R. Moody: Quantitative adhesion measures of multilayer films: Part I. Indentation mechanics, J. Mater. Res. **14**, 3007–3018 (1999)
96. M.D. Kriese, W.W. Gerberich, N.R. Moody: Quantitative adhesion measures of multilayer films: Part II. Indentation of W/Cu, W/W, Cr/W, J. Mater. Res. **14**, 3019–3026 (1999)
97. M. Li, C.B. Carter, M.A. Hillmyer, W.W. Gerberich: Adhesion of polymer-inorganic interfaces by nanoindentation, J. Mater. Res. **16**, 3378–3388 (2001)

98. D.R. Clarke, M.C. Kroll, P.D. Kirchner, R.F. Cook, B.J. Hockey: Amorphization and conductivity of silicon and germanium induced by indentation, Phys. Rev. Lett. **60**, 2156–2159 (1988)
99. J.J. Gilman: Insulator-metal transitions at microindentation, J. Mater. Res. **7**, 535–538 (1992)
100. G.M. Pharr, W.C. Oliver, R.F. Cook, P.D. Kirchner, M.C. Kroll, T.R. Dinger, D.R. Clarke: Electrical-resistance of metallic contacts on silicon and germanium during indentation, J. Mater. Res. **7**, 961–972 (1992)
101. A. Kailer, Y.G. Gogotsi, K.G. Nickel: Phase transformations of silicon caused by contact loading, J. Appl. Phys. **81**, 3057–3063 (1997)
102. J.E. Bradby, J.S. Williams, J. Wong-Leung, M.V. Swain, P. Munroe: Mechanical deformation in silicon by micro-indentation, J. Mater. Res. **16**, 1500–1507 (2000)
103. G.S. Was, T. Foecke: Deformation and fracture in microlaminates, Thin Solid Films **286**, 1–31 (1996)
104. A.J. Whitehead, T.F. Page: Nanoindentation studies of thin-film coated systems, Thin Solid Films **220**, 277–283 (1992)
105. A.J. Whitehead, T.F. Page: Nanoindentation studies of thin-coated systems, NATO ASI Ser. E **233**, 481–488 (1993)
106. B.D. Fabes, W.C. Oliver, R.A. McKee, F.J. Walker: The determination of film hardness from the composite response of film and substrate to nanometer scale indentations, J. Mater. Res. **7**, 3056–3064 (1992)
107. T.F. Page, S.V. Hainsworth: Using nanoindentation techniques for the characterization of coated systems – a critique, Surface Coat. Technol. **61**, 201–208 (1993)
108. X. Chen, J.J. Vlassak: Numerical study on the measurement of thin film mechanical properties by means of nanoindentation, J. Mater. Res. **16**, 2974–2982 (2001)
109. P.J. Burnett, T.F. Page: Surface softening in silicon by ion-implantation, J. Mater. Sci. **19**, 845–860 (1984)
110. P.J. Burnett, D.S. Rickerby: The mechanical-properties of wear resistant coatings. 1. Modeling of hardness behavior, Thin Solid Films **148**, 41–50 (1987)
111. P.J. Burnett, D.S. Rickerby: The mechanical-properties of wear resistant coatings. 2. Experimental studies and interpretation of hardness, Thin Solid Films **148**, 51–65 (1987)
112. P.M. Sargent: A better way to present results from a least-squares fit to experimental-data – an example from microhardness testing, J. Test. Eval. **14**, 122–127 (1986)
113. S.J. Bull, D.S. Rickerby: Evaluation of coatings, Brit. Ceram. Trans. J. **88**, 177–183 (1989)
114. B.R. Lawn, A.G. Evans, D.B. Marshall: Elastic/plastic indentation damage in ceramics: The median/radial crack system, J. Am. Ceram. Soc. **63**, 574–581 (1980)
115. R. Hill: *The Mathematical Theory of Plasticity* (Clarendon, Oxford 1950)
116. W.C. Oliver, C.J. McHargue, S.J. Zinkle: Thin-film characterization using a mechanical-properties microprobe, Thin Solid Films **153**, 185–196 (1987)
117. N.G. Chechechin, J. Bottiger, J.P. Krog: Nanoindentation of amorphous aluminum oxide films. 1. Influence of the substrate on the plastic properties, Thin Solid Films **261**, 219–227 (1995)
118. N.G. Chechechin, J. Bottiger, J.P. Krog: Nanoindentation of amorphous aluminum oxide films. 2. Critical parameters for the breakthrough and a membrane effect in thin hard films on soft substrates, Thin Solid Films **261**, 228–235 (1995)
119. D.E. Kramer, A.A. Volinsky, N.R. Moody, W.W. Gerberich: Substrate effects on indentation plastic zone development in thin soft films, J. Mater. Res. **16**, 3150–3157 (2001)

120. A. B. Mann: Nanomechanical measurements: Surface and environmental effects. Ph.D. Thesis (Oxford Univ., Oxford 1995)
121. A. B. Mann, J. B. Pethica, W. D. Nix, S. Tomiya: Nanoindentation of epitaxial films: A study of pop-in events, Mater. Res. Soc. Symp. Proc. **356**, 271–276 (1995)
122. R. Saha, W. D. Nix: Effects of the substrate on the determination of thin film mechanical properties by nanoindentation, Acta Mater. **50**, 23–38 (2002)
123. S. A. Barnett: Deposition and mechanical properties of superlattice thin films. In: *Physics of Thin Films*, ed. by M. H. Francombe, J. L. Vossen (Academic, New York 1993)
124. R. R. Oberle, R. C. Cammarata: Dependence of hardness on modulation amplitude in electrodeposited Cu-Ni compositionally modulated thin-films, Scripta Metall. **32**, 583–588 (1995)
125. A. Madan, Y. Y. Wang, S. A. Barnett, C. Engstrom, H. Ljungcrantz, L. Hultman, M. Grimsditch: Enhanced mechanical hardness in epitaxial nonisostructural Mo/NbN and W/NbN superlattices, J. Appl. Phys. **84**, 776–785 (1998)
126. R. Venkatraman, J. C. Bravman: Separation of film thickness and grain-boundary strengthening effects in Al thin-films on Si, J. Mater. Res. **7**, 2040–2048 (1992)
127. J. D. Embury, J. P. Hirth: On dislocation storage and the mechanical response of fine-scale microstructures, Acta Mater. **42**, 2051–2056 (1994)
128. D. J. Srolovitz, S. M. Yalisove, J. C. Bilello: Design of multiscalar metallic multilayer composites for high-strength, high toughness, and low CTE mismatch, Metall. Trans. A **26**, 1805–1813 (1995)
129. J. S. Koehler: Attempt to design a strong solid, Phys. Rev. B **2**, 547–551 (1970)
130. S. V. Kamat, J. P. Hirth, B. Carnahan: Image forces on screw dislocations in multilayer structures, Scripta Metall. **21**, 1587–1592 (1987)
131. M. Shinn, L. Hultman, S. A. Barnett: Growth, structure, and microhardness of epitaxial TiN/NbN superlattices, J. Mater. Res. **7**, 901–911 (1992)
132. J. E. Krzanowski: The effect of composition profile on the strength of metallic multi-layer structures, Scripta Metall. **25**, 1465–1470 (1991)
133. J. B. Vella, R. C. Cammarata, T. P. Weihs, C. L. Chien, A. B. Mann, H. Kung: Nanoindentation study of amorphous metal multilayered thin films, MRS Symp. Proc. **594**, 25–29 (2000)
134. J. L. Cuy, A. B. Mann, K. J. Livi, M. F. Teaford, T. P. Weihs: Nanoindentation mapping of the mechanical properties of human molar tooth enamel, Arch. Oral Biol. **47**, 281–291 (2002)

13

Computer Simulations of Nanometer-Scale Indentation and Friction

Susan B. Sinnott, Seong-Jun Heo, Donald W. Brenner, and Judith A. Harrison

Summary. Engines and other machines with moving parts are often limited in their design and operational lifetime by friction and wear. This limitation has motivated the study of fundamental tribological processes with the ultimate aim of controlling and minimizing their impact. The recent development of miniature apparatus, such as microelectromechanical systems (MEMS) and nanometer-scale devices, has increased interest in atomic-scale friction, which has been found to, in some cases, be due to mechanisms that are distinct from the mechanisms that dominate in macroscale friction.

Presented in this chapter is a review of computational studies of tribological processes at the atomic and nanometer scale. In particular, a review of the findings of computational studies of nanometer-scale indentation, friction and lubrication is presented, along with a review of the salient computational methods that are used in these studies, and the conditions under which they are best applied.

13.1 Introduction

Engines and other machines with moving parts are often limited in their design and operational lifetime by friction and wear. This limitation has motivated the study of tribological processes with the aim of controlling and minimizing the impact of these processes. There are numerous historical examples that illustrate the importance of friction to the development of civilizations, including the ancient Egyptians who invented technologies to move the stones used to build the pyramids [1]; Coulomb, who was motivated to study friction by the need to move ships easily and without wear from land to the water [1]; and *Johnson* et al. [2], who developed an improved understanding of contact mechanics and surface energies through the study of automobile windshield wipers. At present, substantial research and development is aimed at microscale and nanoscale machines with moving parts that at times challenge our fundamental understanding of friction and wear. This has motivated the study of atomic-scale friction and has, consequently, led to new discoveries such as self-lubricating surfaces and wear-resistant materials. While there are similarities between friction at the macroscale and the atomic scale, in many instances the mechanisms that lead to friction at these two scales are quite different. Thus, as devices such as magnetic storage disks and microelectromechanical systems (MEMS) [3] continue to shrink in size, it is expected that new phenomena associated with atomic-scale friction, adhesion and wear will dominate the functioning of these devices.

The last two decades have seen considerable scientific effort expended on the study of atomic-scale friction [4–17]. This effort has been facilitated by the development of new advanced experimental tools to measure friction over nanometer-scale distances at low loads, rapid improvements in computer power, and the maturation of computational methodologies for the modeling of materials at the atomic scale. For example, friction-force and atomic-force microscopes (FFM and AFM) allow the frictional properties of solids to be characterized with atomic-scale resolution under single-asperity indentation and sliding conditions [18–21]. In addition, the surface force apparatus (SFA) provides data about the tribological and lubrication responses of many liquid and solid systems with atomic resolution [22], and the quartz crystal microbalance (QCM) provides information about the atomic-scale origins of friction [4,23]. These and related experimental methods allow researchers to study sliding surfaces at the atomic scale and relate the observed phenomena to macroscopically observed friction, lubrication and wear.

Analytic models and computational simulations have played an important role in characterizing and understanding friction. They can, for example, assist in the interpretation of experimental data or provide predictions that subsequent experiments can confirm or refute. Analytic models have long been used to study friction, including early studies by *Tomlinson* [24] and *Frenkel* and *Kontorova* [25] and more recent studies by *McClelland* et al. [26], *Sokoloff* [13, 27–33], *Persson* [34–37] and others [38–44]. Most of these idealized models divide the complex motions that create friction into more fundamental components defined by quantities such as spring constants, the curvature and magnitude of potential wells, and bulk phonon frequencies. While these simplifications provide these approaches with some predictive capabilities, many assumptions must be made in order to be able to apply these models to study friction, which may lead to incorrect or incomplete results.

In atomic-scale molecular dynamics (MD) simulations, atom trajectories are calculated by numerically integrating coupled classical equations of motion. Interatomic forces that enter these equations are typically calculated either from total energy methods that include electronic degrees of freedom, or from simplified mathematical expressions that give the potential energy as a function of interatomic displacements. MD simulations can be considered numerical experiments that provide a link between analytic models and experiments. The main strength of MD simulations is that they can reveal unanticipated phenomena or unexpected mechanisms for well-known observations. Weaknesses include a lack of quantum effects in classical atomistic dynamics, and perhaps more importantly, the fact that meaningless results can be obtained if the simulation conditions are chosen incorrectly. The next section contains a review of MD simulations, including the approximations that are inherent in their application to the study of friction, and the conditions under which they should and should not be applied.

13.2 Computational Details

Molecular dynamics simulations are straightforward to describe: given a set of initial conditions and a way of mathematically modeling interatomic forces, Newton's (or equivalent) classical equation of motion is numerically integrated [45]

$$F = ma, \quad (13.1a)$$
$$-\nabla E = m(\partial^2 r/\partial t^2), \quad (13.1b)$$

where F is the force on each atom, m is the atomic mass, a is the atomic acceleration, E is the potential energy felt by each atom, r is the atomic position, and t is time. The forces acting on any given atom are calculated, and then the atoms move a short increment ∂t (called a time step) forward in time in response to these applied forces. This is accompanied by a change in atomic positions, velocities and accelerations. The process is then repeated for some specified number of time steps.

The output of these simulations includes new atomic positions, velocities, and forces that allow additional quantities such as temperature and pressure to be determined. As the size of the system increases, it is useful to render the atomic positions in animated movies that reveal the responses of the system in a qualitative manner. Quantitative data can be obtained by analyzing the numerical output directly.

The following sections review the way in which energies and forces are calculated in MD simulations and the important approximations that are used to realistically model the friction that occurs in experiments with smaller systems of only a few tens of thousands of atoms in simulations. The reader is referred to additional sources [46–52] for a more comprehensive overview of MD simulations (including computer algorithms) and the potentials that are used to calculate energies and forces in MD simulations.

13.2.1 Energies and Forces

There are several different approaches by which interatomic energies and forces are determined in MD simulations. The most theoretically rigorous methods are those that are classified as *ab initio* or first principles. These techniques, which include density functional theory [53,54] and quantum chemical *ab initio* [55] methods, are derived from quantum mechanical principles and are generally both the most accurate and the most computationally intensive. They are therefore limited to a small number of atoms (< 500), which has limited their use in the study of friction. Alternatively, empirical methods are functions containing parameters that are determined by fitting to experimental data or the results of *ab initio* calculations [50]. These techniques can usually be relied on to correctly describe qualitative trends and are often the only choice available for modeling systems containing tens of thousands, millions, or billions of atoms. Empirical methods have therefore been widely used in studies of friction. Semi-empirical methods, including tight-binding methods, include some elements of both empirical methods and *ab initio* methods. For instance, they require quantum mechanical information in the form of, for example, on-site and hopping matrix elements, and include fits to experimental data [56].

Empirical methods simplify the modeling of materials by treating the atoms as spheres that interact with each other via repulsive and attractive terms that can be either pairwise additive or many-body in nature. In this approach, electrons are not treated explicitly, although it is understood that the interatomic interactions are ultimately dependent on them. As discussed in this section, some empirical methods explicitly include charge through classical electrostatic interactions, although most methods assume charge-neutral systems. The repulsive and attractive functional forms generally depend on interatomic distances and/or angles and contain adjustable parameters that are fit to *ab initio* results and/or experimental data.

The main strength of empirical potentials is their computational speed. Recent simulations with these approaches have modeled billions of atoms [57], something that is not possible with *ab initio* or semi-empirical approaches at this time. The main weakness of empirical potentials is their lack of quantitative accuracy, especially if they are poorly formulated or applied to systems that are too far removed from the fitting database used in their construction. Furthermore, because of the differences in the nature of chemical bonding in various materials, such as covalent bonding in carbon versus metallic bonding in gold, empirical methods have been historically derived for particular classes of materials. They are therefore generally nontransferable, although some methods have been shown to be theoretically equivalent [51, 58], and in recent years there has been progress towards the development of empirical methods that can model heterogeneous material systems [59–64].

Several of the most important and common general classes of empirical methods used for calculating interatomic energies and forces in materials, the so-called potentials, are reviewed here. The first to be considered are the potentials that are used to model covalently bound materials, including the bond-order potential and the Stillinger–Weber potential.

The bond-order potential was first formulated by *Abell* [65] and subsequently developed and parameterized by *Tersoff* for silicon and germanium [66, 67], *Brenner* and coworkers for hydrocarbons [52, 68, 69], *Dyson* and *Smith* for carbon–silicon–hydrogen systems [70], *Sinnott* and coworkers for carbon–oxygen–hydrogen systems [71], and *Graves* and coworkers [72] and *Sinnott* and coworkers [73] for fluorocarbons.

The bond-order potential has the general functional form

$$E = \sum_i \sum_{j(>i)} [V_R(r_{ij}) - b_{ij} V_A(r_{ij})] \qquad (13.2)$$

where $V_R(r)$ and $V_A(r)$ are pair-additive interactions that model the interatomic repulsion and electron–nuclear attraction, respectively. The quantity r_{ij} is the distance between pairs of nearest-neighbor atoms i and j, and b_{ij} is a bond-order term that takes into account the many-body interactions between atoms i and j, including those due to nearest neighbors and angle effects. The potential is short-ranged and only considers nearest neighbor bonds. To model long-range nonbonded interactions, the bond-order potential is combined with pair-wise potentials either directly through splines [74] or indirectly with more sophisticated functions [75].

The *Stillinger–Weber* potential [76] potential was formulated to model silicon, with a particular emphasis on the liquid phases of silicon. It includes many-body interactions in the form of a sum of two- and three-body interactions

$$E = \sum_{ij} V^2_{ij}(r_{ij}) + \sum_{jik} V^3_{jik}(r_{ij}, r_{ik}), \quad (13.3)$$

where V^2 is a pair-additive interaction and V^3 is a three-body term. The three-body term includes an angular interaction that minimizes the potential energy for tetrahedral angles. This term favors the formation of open structures, such as the diamond cubic crystal structure.

The second potential is the embedded atom method (EAM) approach [77, 78] and related methods [79], which were initially developed for modeling metals and alloys. The functional form in the EAM is

$$E = \sum_i F(\rho_i) + \sum_{i>j} \Phi(r_{ij}), \quad (13.4)$$

where F is called the embedding energy. This term models the energy due to embedding an atom into a uniform electron gas with a uniform compensating positive background (jellium) of density ρ_i that is equal to the actual electron density of the system. The term $\Phi(r_{ij})$ is a pairwise functional form that corrects for the jellium approximation. Several parameterizations of the EAM exist (see, for example, [77,78,80–82]) and it has recently been extended to model nonmetallic systems. For example, the modified EAM (MEAM) approach [64, 83, 84] was developed so that EAM could be applied to metal oxides [60] and covalently bound materials [84].

The third method is the general class of Coulomb or multipole interaction potentials used to model charged ionic materials or molecules [85]. In this formalism, an energy term is given as

$$E = \sum_i \sum_{j(>i)} [(q(r_i)q(r_j)/r_{ij})], \quad (13.5)$$

where $q(r_i)$ is the charge on atom i and r_{ij} is the distance between atoms i and j. More complex formalisms that take into account, say, the Madelung constant in the case of ionic crystals, are used in practice. In general, the charges are held fixed, but methods that allow charge to vary in a realistic manner have been developed [61, 86].

Lastly, long-range van der Waals or related forces are typically modeled with pairwise additive potentials. A widely used approximation is the Lennard–Jones (LJ) potential [87], which has the following functional form:

$$E = 4\varepsilon \sum_i \sum_{j(>i)} \left[(\sigma/r_{ij})^{12} - (\sigma/r_{ij})^6 \right]. \quad (13.6)$$

In this approach ε and σ are parameters and r_{ij} is the distance between atoms i and j.

All of these potentials are widely used in MD simulations of materials, including studies of friction, lubrication, and wear.

13.2.2 Important Approximations

Several approximations are typically used in MD simulations of friction. The first is the use of periodic boundary conditions (PBCs) and the minimum image convention for interatomic interactions [48]. In both cases the simulation supercell is surrounded by replicas of itself so that atoms (or phonons, etc.) that exit one side of the supercell remerge into the simulation through the opposite side of the supercell. In the minimum image convention an atom interacts either with another atom in the supercell or its equivalent atom in a surrounding cell depending on which distance to the atom is shortest. This process is illustrated in Fig. 13.1. In this convention supercells must be large enough that atoms do not interact with themselves over the periodic boundaries. In computational studies of friction and wear, PBCs are usually applied in the two dimensions within the plane(s) of the sliding surface(s). The strength of this approach is that it allows a finite number of atoms to model an infinite system. However, the influence of boundaries on system dynamics is not completely eliminated; for example, phonon scattering due to the periodic boundaries can influence heat transport and therefore frictional properties of sliding interfaces.

Another important tool that is often used in MD simulations of friction is thermostats to regulate system temperature. In macroscopic systems, heat that is generated from friction is dissipated rapidly from the surface to the bulk phonon modes. Because atomistic computer simulations are limited systems that are many orders of magnitude smaller than systems that are generally studied experimentally, thermostats are needed to prevent the system temperature from rising in a nonphysical manner. Typically in simulations of indentation or friction, the thermostat is applied

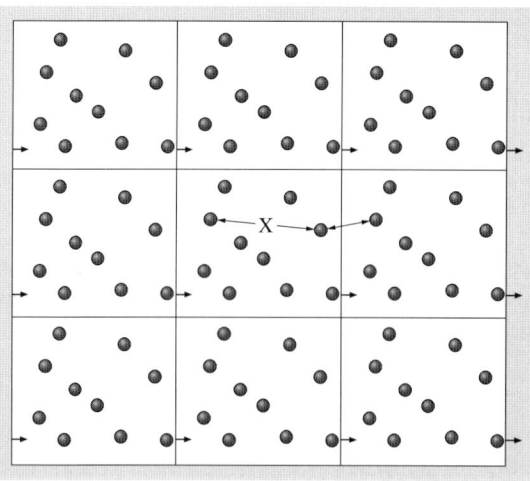

Fig. 13.1. Illustration of periodic boundary conditions consisting of a central simulation cell surrounded by replica systems. The *solid arrows* indicate an atom leaving the central box and re-entering on the opposite side. The *dotted arrows* illustrate the minimum image convention

to a region of the simulation cell that is well removed from the interface where friction and indentation is taking place. In this way, local heating of the interface that occurs as work is done on the system, but excess heat is efficiently dissipated from the system as a whole. In this manner the adjustment of atomic temperatures occurs away from the processes of interest, and simplified approximations for the friction term can be used without unduly influencing the dynamics produced by the interatomic forces.

There are several different formalisms for atomistic thermostats. The simplest of these controls the temperature by intermittently rescaling the atomic velocities to values corresponding to the desired temperature [88] such that

$$\left(\frac{v_{\text{new}}}{v_{\text{old}}}\right)^2 = \frac{T}{T_{\text{ins}}}, \tag{13.7}$$

where v_{new} is the rescaled velocity, and v_{old} is the velocity before the rescaling. This approach, which is called the velocity rescaling method, is both simple to implement and effective at maintaining a given temperature over the course of an MD simulation. It was consequently widely used in early MD simulations. The velocity rescaling approach does have some significant disadvantages, however. First, there is little theoretical basis for the adjustment of atomic velocities, and the system dynamics are not time-reversible, which is inconsistent with classical mechanics. Second, the rate and mode of heat dissipation are disconnected from system properties, which may affect system dynamics. Lastly, for typical MD simulation system sizes, the averaged quantities that are obtained, such as pressure for instance, do not correspond to values in any thermodynamic ensemble.

For these reasons, more sophisticated methods for maintaining system temperatures in MD simulations have been developed. The Langevin dynamics approach [48], which was originally developed from the theory of Brownian motion, falls into this category. In this approach, terms are added to the interatomic forces that correspond to a random force and a frictional term [46, 89, 90]. Therefore, Newton's equation of motion for atoms subjected to Langevin thermostats is given by the following equation rather than (13.1a, 13.1b):

$$m\boldsymbol{a} = \boldsymbol{F} - m\xi \boldsymbol{v} + R(t), \tag{13.8}$$

where $\boldsymbol{F}$ are the forces due to the interatomic potential, ξ is a friction coefficient, m and v are the particle's mass and velocity, respectively, and $R(t)$ is a random force that acts as "white noise". The friction term can be formulated in terms of a memory kernal, typically for harmonic solids [91–93], or a friction coefficient can be approximated using the Debye frequency. The random force can be given by a Gaussian distribution where the width, which is chosen to satisfy the fluctuation-dissipation theorem, is determined from the equation

$$\langle R(0) R(t) \rangle = 2 m k_\text{B} T \xi \delta(t). \tag{13.9}$$

Here, the function R is the random force in (13.8), m is the particle mass, T is the desired temperature, k_B is Boltzmann's constant, t is time, and ξ is the friction coefficient. It should be noted that the random forces are uncoupled from those at previous steps, which is denoted by the delta function. Additionally, the width of the Gaussian distribution from which the random force is obtained varies with temperature. Thus, the Langevin approach does not require any feedback from the current temperature of the system as the random forces are determined solely from (13.9).

In the early 1980s, *Nosé* developed a new thermostat that corresponds directly to a canonical ensemble (system with constant temperature, volume and number of atoms) [94,95], which is a significant advance from the methods described so far. In this approach, Nosé introduces a degree of freedom s that corresponds to the heat bath and acts as a time scaling factor, and adds a parameter Q that may be regarded as the heat bath "mass". A simplified form of Nosé's method was subsequently implemented by *Hoover* [46] that eliminated the time scaling factor whilst introducing a thermodynamic friction coefficient ζ. Hoover's formulation of Nosé's method is therefore easy to use and is commonly referred to as the Nosé–Hoover thermostat.

When this thermostat is applied to a system containing N atoms, the equations of motion are written as (dots denote time derivatives):

$$\dot{r}_i = \frac{p_i}{m_i}, \quad \dot{p}_i = F_i - \zeta p_i, \quad \dot{\zeta} = \frac{1}{Q}\left(\sum_{i=1}^{N} \frac{p_i^2}{m_i} - N_f k_B T\right), \quad (13.10)$$

where r_i is the position of atom i, p_i is the momentum and F_i is the force applied to each atom. The last equation in (13.10) contains the temperature control mechanism in the Nosé–Hoover thermostat. In particular, the term between the parentheses on the right-hand side of this equation is the difference between the system's instantaneous kinetic energy and the kinetic energy at the desired temperature. If the instantaneous value is higher than the desired one, the friction force will increase to lower it and vice versa.

It should be pointed out that the choice of the heat bath "mass" Q is arbitrary but crucial to the successful performance of the thermostat. For example, a small value of Q leads to rapid temperature fluctuation while large Q values result in inefficient sampling of phase space. Nosé recommended that Q should be proportional to $N_f k_B T$ and should allow the added degree of freedom s to oscillate around its averaged value at a frequency of the same order as the characteristic frequency of the physical system [94,95]. If ergodic dynamic behavior is assumed, the Nosé–Hoover thermostat will maintain a well-defined canonical distribution in both momentum and coordinate space. However, for small systems where the dynamic is not ergodic, the Nosé–Hoover thermostat fails to generate a canonical distribution. Therefore, more sophisticated algorithms based on the Nosé–Hoover thermostat have been proposed to fix its ergodicity problem; for example, the "Nosé–Hoover chain" method of *Martyna* et al. [96]. However, these complex thermostats are not as easy to apply as the Nosé–Hoover thermostat due to the difficult evaluation of the coupling parameters for each different case and the significant computational cost [97]. From a practical point of view, if the molecular system is large enough that the movements of the atoms are sufficiently chaotic, ergodicity is guaranteed and the performance of the Nosé–Hoover thermostat is satisfactory [25].

In an alternative approach, *Schall* et al. recently introduced a hybrid continuum-atomistic thermostat [98]. In this method, an MD system is divided into grid regions, and the average kinetic energy in the atomistic simulation is used to define a temperature for each region. A continuum heat transfer equation is then solved stepwise on the grid using a finite difference approximation, and the velocities of the atoms in each grid region are scaled to match the solution of the continuum equation. To help account for a time lag in the transfer of kinetic to potential energy, Hoover constraining forces are added to those from the interatomic potential. This process is continued, leading to an ad hoc feedback between the continuum and atomistic simulations. The main advantage of this approach is that the experimental thermal diffusivity can be used in the continuum expression, leading to heat transfer behavior that matches experimental data. For example, in metals the majority of the thermal properties at room temperature arise from electronic degrees of freedom that are neglected with strictly classical potentials. This thermostat is relatively straightforward to implement, and requires only the interatomic potential and the bulk thermal diffusivity as input. It is also appropriate for nonequilibrium heat transfer, such as occurs as heat is dissipated from sliding surfaces moving at high relative velocities.

Cushman et al. [99, 100] developed a unique alternative to the grand canonical ensemble by performing a series of grand canonical Monte Carlo simulations [48, 101] at various points along a hypothetical sliding trajectory. The results from these simulations are then used to calculate the correct particle numbers at a fixed chemical potential, which are then used as inputs to nonsliding, constant-*NVE* MD simulations at each of the chosen trajectory points. The sliding speed can be assumed to be infinitely slow because the system is fully equilibrated at each step along the sliding trajectory. This approach offers a useful alternative to continuous MD simulations that are restricted to sliding speeds that are orders of magnitude larger than most experimental studies (about 1 m/s or greater).

To summarize, this section provides a brief review and description of components that are used in atomistic, molecular dynamics simulation of many of the processes related to friction, such as indentation, sliding, and wear. The components discussed here include the potential energy expression used to calculate energies and forces in the simulations, periodic boundary conditions and thermostats. Each of these components has their own strengths and weaknesses that should be well-understood both prior to their use and in the interpretation of results. For example, general principles related to liquid lubrication in confined areas may be most easily understood and generalized from simulations that use pair potentials and may not require a thermostat. On the other hand, if one wants to study the wear or indentation of a surface of a particular metal, then EAM or other semiempirical potentials, together with a thermostat, would be expected to yield more reliable results. If one requires information on electronic effects, *ab initio* or semi-empirical approaches that include the evaluation of electronic degrees of freedom must be used. Thus, the best combination of components for a particular study depends on the chemical na-

13.3 Indentation

It is critical to understand the nanometer-scale properties of materials that are being considered for use as new coatings with specific friction and wear behavior. Experimental determination of these properties is most frequently done with the AFM, which provides a variety of data related to the interaction of the microscope tips with the sample surface [102–104]. In AFM experiments, the tip has a radius of about 1–100 nm and is pressed against the surface under ambient conditions (in air), ultrahigh vacuum (UHV) conditions, or in a liquid. The microscope tip can either move in the direction normal to the surface, which is the case in nanoindentation studies, or raster across the surface, which is the case in surface imaging or friction studies. Sliding rates of 1 nm/s–1 µ/s are typically used, which are many orders of magnitude slower than the rates used in MD simulations of sliding or indentation of around 1–100 m/s. As discussed in the previous section, the higher rates used in computational simulations are a consequence of modeling full atomic motion, which occurs on a femto- to picosecond timescale, and the stepwise solution of the classical equations of motion, which makes the large number of simulation steps needed to reach experimental timescales computationally impossible with current processor speeds.

As the tip moves either normal to or across the surface, the forces acting upon it as a result of its interactions with the surface are measured. When the tip is moved in the surface normal direction, it can penetrate the surface on the nanometer scale and provide information on the nanometer-scale mechanical properties of the surface [105, 106]. The indentation process also causes the force on the tip to increase, and the rate of increase is related to both the depth of indentation and the properties of the surface. The region of the force curve that reflects this high force is known as the repulsive wall region [102], or, when considered without any lateral motion of the tip, an indentation curve. When the tip is retracted after indentation, enhanced adhesion between the tip and surface relative to the initial contact can result. This phenomenon is indicated by hysteresis in the force curve.

Tip–surface adhesion can result from the formation of chemical bonds between the tip and the sample, or from the formation of liquid capillaries between the microscope tip and the surface caused by the interaction of the tip with a layer of liquid contamination on the surface. The latter case is especially prevalent in AFM studies conducted in ambient environments. In the case of clean metallic systems, the sample can wet the tip or the tip can wet the sample in the form of a connective "neck" of metal atoms between the surface and the tip that can lead to adhesion. In the case of polymeric or molecular systems, entanglement of molecules that are anchored on the tip with molecules anchored on the sample can be responsible for force curve hysteresis.

In the case of horizontal rastering of AFM tips across surfaces, the force curve data provide a map of the surface that is indicative of the surface topography [107]. If the deflection of the tip in the lateral direction is recorded while the tip is being rastered, a friction map of the surface [20] is produced.

The rest of this section discusses some of the important insights and findings that have been obtained from MD simulations of nanoindentation. These studies have not only provided insight into the physical phenomena responsible for the qualitative shapes of AFM force curves, they have also revealed a wealth of atomic-scale phenomena that occur during nanoindentation that was not previously known.

13.3.1 Surfaces

The nature of adhesive interactions between clean, deformable metal tips indenting metal surfaces have been identified and clarified over the course of the last decade through the use of MD simulations [104, 108–112]. In particular, the high surface energies associated with clean metal surfaces can lead to strongly attractive interactions between surfaces in contact. The strength of this attraction can be so large that when the tip gets close enough to the surface to interact with it, surface atoms "jump" upwards to wet the tip in a phenomena known as jump-to-contact (JC). This wetting mechanism was first discovered in MD simulations [111] and has been confirmed experimentally [105, 113–115] using the AFM, as shown in Fig. 13.2.

The MD simulations of *Landman* et al. [111, 116–118] using EAM potentials revealed that the JC phenomenon in metallic systems is driven by the need of the atoms at the tip-surface interface to optimize their interaction energies while maintaining their individual material cohesive binding. When the tip advances past the JC point it indents the surface, which causes the force to increase. This behavior is indicated in Fig. 13.3, points D to M. This region of the computer-generated force curve has a maximum not present in the force curve generated from experimental data (Fig. 13.3, point L). This is due to tip-induced flow of the metal atoms in the surface that causes "pile-up" of the surface atoms around the edges of the indenter. Hysteresis on the withdrawal of the tip, shown in Fig. 13.3, points M to X, is present due to adhesion between the tip and the substrate. In particular, as the tip retracts from the sample, a connective "neck" or nanowire of atoms forms between the tip and the substrate that is primarily composed of metal atoms from the surface with some atoms from the metal indenter that have diffused into the structure. A snapshot from the MD simulations that illustrates this behavior is shown in Fig. 13.4.

As the tip is withdrawn farther, the magnitude of the force increases (becomes more negative) until, at a critical force, the atoms in adjacent layers of the connective nanowire rearrange so that an additional row of atoms is created. This process causes elongation of the connective nanowire and is responsible for the fine structure (apparent as a series of maxima) present in the retraction portion of the force curve. These elongation and rearrangement steps are repeated until the connection between the tip and the surface is broken. Similar elongation events have been observed experimentally. For example, scanning tunneling microscopy (STM) exper-

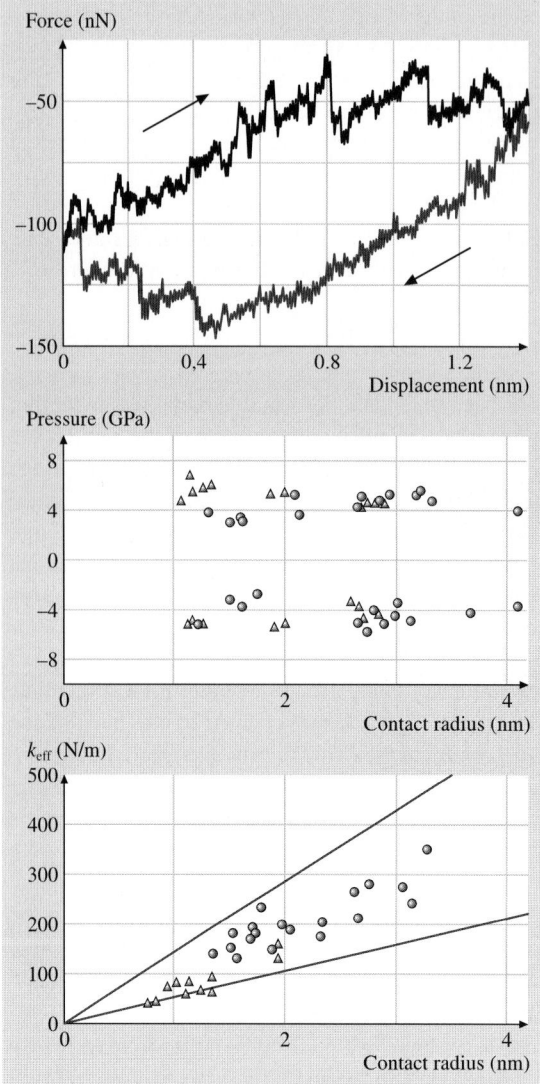

Fig. 13.2. *Top*: The experimental values for the force between a tip and a surface that have a connective neck between them. The neck contracts and extends without breaking on the scales shown. *Bottom*: The effective spring constant k_{eff} determined experimentally for the connective necks and corresponding maximum pressures, versus contact radius of the tip. The *triangles* indicate measurements taken at room temperature; the *circles* are the measurements taken at liquid He temperatures. After [115], with permission of the ACS (1996)

iments demonstrate that the metal nanowires between metal tips and surfaces can elongate approximately 2500 Å without breaking [119].

The JC process has been shown to affect the temperature at the tip–surface interface. For instance, the constant-energy MD simulations of *Tomagnini* et al. [121] predicted that the energy released due to the wetting of the tip by surface atoms increases the temperature of the tip by about 15 K at room temperature and is accompanied by significant structural rearrangement. At temperatures high enough to cause the first few metal surface layers to be liquid, the distance at which the JC

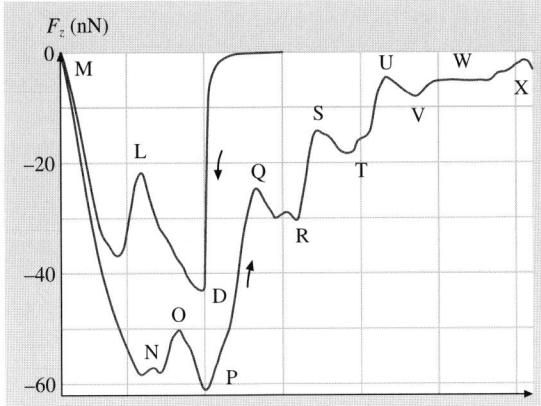

Fig. 13.3. Computationally derived force F_z versus tip-to-sample distance d_{hs} curves for approach, contact, indentation, then separation using the same tip–sample system shown in Fig. 13.4. These data were calculated from an MD simulation. After [111] with the permission of the AAAS (1990)

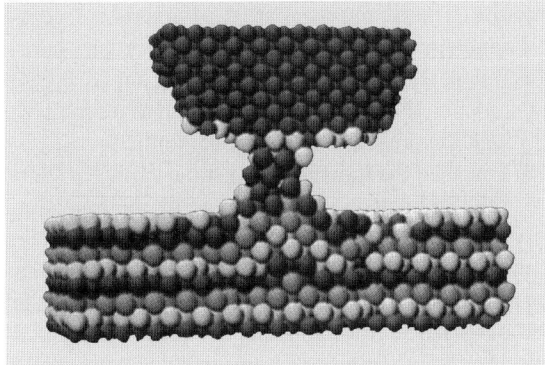

Fig. 13.4. Illustration of atoms in the MD simulation of a Ni tip being pulled back from an Au substrate. This causes the formation of a connective neck of atoms between the tip and the surface. After [111] with the permission of the AAAS (1990)

occurs increases, as does the contact area between the tip and the surface and the amount of nanowire elongation prior to breakage.

Simulations by *Komvopoulos* and *Yan* [122] using LJ potentials showed how metallic surfaces respond to single and repeated indentation by metallic, or covalently bound, rigid tips. The simulations predicted that a single indentation event produces hysteresis in the force curve as a result of surface plastic deformation and heating. The repulsive force decreases abruptly during surface penetration by the tip and surface plastic deformation. Repeated indentation results in the continuous decrease of the elastic stiffness, surface heating, and mean contact pressures at maximum penetration depths to produce behaviors that are similar to cyclic work hardening and softening by annealing observed in metals at the macroscale.

When the tip is much stiffer than the surface, pile-up of surface atoms around the tip occurs to relieve the stresses induced by nanoindentation. In contrast, when the surface is much stiffer than the tip, the tip can be damaged or destroyed. Simulations by *Belak* et al. [123] using perfectly rigid tips showed the mechanism by which the surface yields plastically after its elastic threshold is exceeded. The simulations showed how nanoindentation causes surface atoms to move on to the surface but under the tip and thus cause atomic pile-up. In this study, variations in the indentation rate reveal that point defects created as a result of nanoindentation relax by moving through the surface if the rate of indentation is slow enough. If the indentation rate is too high, there is no time for the point defects to relax and move away from the indentation area and so strain builds up more rapidly. The rigid indenters considered in these MD simulations are analogous to experiments that use surface passivation to prevent JC between the tip and the surface [120, 124], the results of which agree with the predicted results of pile-up and crater formation, as shown in Fig. 13.5 [120].

In short, MD simulations are able to explain the atomic-scale mechanisms behind measured experimental force curves produced when metal tips indent homogeneous metal surfaces to nanometer-scale depths. This preliminary work has spawned much of the current interest in using the JC to produce metal nanowires [126–128].

MD simulations have also been used to examine the relationship between nanoindentation and surface structure. This is most apparent in a series of computational studies that consider the indentation of a surface with a "virtual" hard-sphere indenter in a manner that is independent of the rate of indentation, as shown in Fig. 13.6. The virtual indenter is modeled through the application of a repulsive force to the surface rather than through the presence of an actual atomic tip. *Kelchner* et al. [129], rather than use MD, pushed the indenter against the surface a short distance and then allowed the system to relax using standard energy minimization methods in combination with EAM potentials. The system is fully relaxed when the energy of the surface system is minimized. After relaxation, the tip is pushed further into the surface and the process is repeated. As the tip generates more stresses in the surface, dislocations are generated and plastic deformation occurs. If the tip is pulled back after indenting less than a specific critical value, the atoms that were plastically deformed are healed during the retraction and the surface recovers its original structure. In contrast, if the tip is indented past the critical depth, additional dislocations

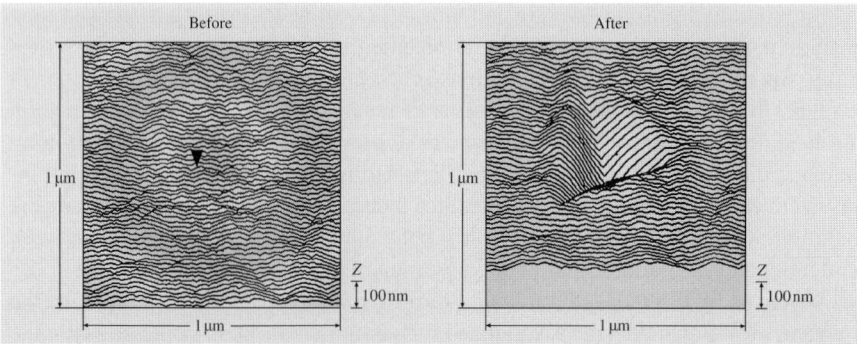

Fig. 13.5. Images of a gold surface before and after being indented with a pyramidal shaped diamond tip in air. The indentation created a surface crater. Note the pile-up around the crater edges. After [120] with the permission of Elsevier (1993)

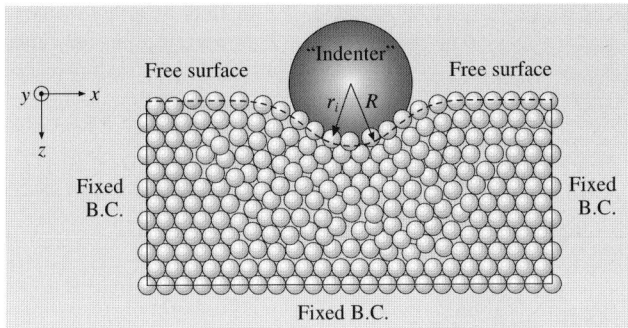

Fig. 13.6. A schematic of a spherical, virtual tip indenting a metal surface. After [125] with the permission of Elsevier (1993)

are created that interfere with the surface healing process on tip withdrawal. In this case, a surface crater is left on the surface following nanoindentation.

A similar study by *Lilleodden* et al. [125] considered the generation of dislocations in perfect crystals and near grain boundaries in gold. Analysis of the relationship between the load and the tip displacement in the perfect crystal shows discrete load drops followed by elastic behavior. These load drops are shown to correspond to the homogeneous nucleation of dislocations, as illustrated in Fig. 13.7, which is a snapshot taken just after the load drop. When nanoindentation occurs close to a grain boundary, similar relationships between the load and tip displacement are predicted to occur as were seen for the perfect crystal. However, the dislocations responsible for the load drop are preferentially emitted from the grain boundaries, as illustrated in Fig. 13.8.

Simulations can also show how atomic structure and stresses are affected by nanoindentation. For instance, MD simulations with a virtual indenter by *Hasnaoui*

Fig. 13.7. Snapshot of two partial dislocations separated by a stacking fault. The *dark spheres in the center* of the structure indicate atoms in perfect crystal positions after both partial dislocations have passed. After [125] with the permission of Elsevier (1993)

Fig. 13.8. Snapshot of the high-energy atoms only after a load drop caused by dislocation generation during the nanoindentation of gold near a grain boundary. After [125] with the permission of Elsevier (1993)

et al. [130] using semi-empirical tight-binding methods showed the interaction between the grain boundaries under the indenter and the dislocations generated by the indentation, as illustrated in Fig. 13.9. This study shows that if the size of the indenter is smaller than the grain size, the grain boundaries can emit, absorb, and reflect the dislocations in a manner that depends on atomic structure and the distribution of stresses.

Zimmerman et al. considered the indentation of a single-crystal gold substrate both near and far from a surface step [131]. The results of these simulations, which used EAM potentials, showed that the onset of plastic deformation depends to a significant degree on the distance of indentation from the step, and whether the indentation is on the plane above or below the step. In a related set of simulations, *Shenderova* et al. [132] examined whether ultrashallow elastic nanoindentation can nondestructively probe surface stress distributions associated with surface structures such as a trench and a dislocation intersecting a surface. The simulations carried out the nanoindentation to a constant depth. They predicted maximum loads that reflect the in-plane stresses at the point of contact between the indenter and the substrate, as illustrated in Fig. 13.10.

Since the 1930s, studies have been performed using hardness measurement techniques [133–136] and indentation methods [137] that suggest that the hardness of a material depends on applied in-plane uni- and bi-axial strain. In general, tensile strain appeared to decrease hardness while increases in hardness under compressive in-plane strain were reported. This behavior had traditionally been attributed to the contribution of stresses from the local strain from the indentation to the resolved shear stresses and the in-plane strain [134, 136]. However, in 1996, *Pharr* and coworkers determined that changes in elastic modulus determined from unloading curves of strained substrates using contact areas estimated via an elastic model are too large to have physical significance, a result that brought into question the interpretation of prior hardness data [137, 138]. They hypothesized that

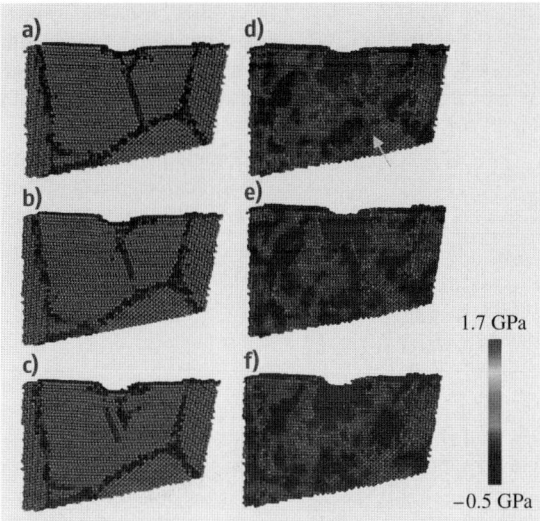

Fig. 13.9a–f. Snapshots showing the atomic stress distribution and atomic structures in a gold surface. Figures (**a**)–(**c**) show the atomic structure at indentation depths of 7.9, 8.6, and 9.6 Å, respectively, with a virtual spherical indenter. A dislocation is represented by the two parallel {111} planes (*two dark lines*) that show the stacking fault left behind after the leading partial dislocation has passed. Figures (**d**)–(**f**) show the atomic stress distribution of the same system at the same indentation depths. Here the *dark color* indicates compressive hydrostatic pressures of 1.7 GPa and higher while the *gray* color indicates tensile pressures of −0.5 GPa and lower. The *arrow* in (**d**) shows the region of the system where a dislocation interacts with a grain boundary. After [130] with the permission of Elsevier (2004)

the apparent change in modulus (and hardness) with in-plane strain is mainly due to changes in contact area that are not typically taken into consideration in elastic half-space models. This hypothesis was based on experimental nanoindentation studies of a strained polycrystalline aluminium alloy and finite element calculations on an isotropic solid [137, 138]. They further suggested that in-plane compression increases pile-up around the indenter that, when not taken into account in the analysis of unloading curves, implies a nonphysical increase in modulus. Likewise, they suggested that in-plane tensile strain reduces the amount of material that is piled up around an indenter, which leads to a corresponding reduced (nonphysical) modulus when interpreting unloading curves using elastic models.

To explore in more detail the issue of pile-up and its influence on the interpretation of loading curves, *Schall* and *Brenner* used MD simulations and EAM potentials to model the plastic nanoindentation of a single-crystal gold surface under an applied in-plane strain [139]. These simulations predicted that the mean pressure, calculated from true contact areas that take into account plastic pile-up around the indenter, varies only slightly with applied pre-stress. They also predicted that the higher values occur in compression rather than in tension, and that the modu-

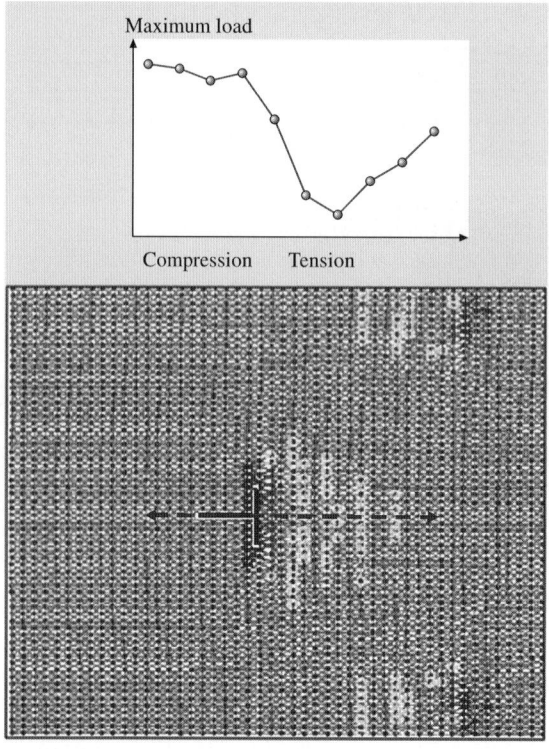

Fig. 13.10. Data and system illustration from a simulation of a gold surface containing a dislocation. *Top*: Maximum load for simulated shallow indentation at several points along the *dotted line* in the *bottom* illustration. *Bottom*: Top view of the simulated surface. The dislocation is denoted by the *solid black lines*

lus calculated from the true contact area is essentially independent of the pre-stress level in the substrate. In contrast, if the contact area is estimated from approximate elastic formulae, the contact area is underestimated, which leads to a strong, incorrect dependence of apparent modulus on the pre-stress level. The simulations also showed larger pile-up in compression than in tension, in agreement with the Pharr model, and both regimes produced contact areas larger than those typically assumed in elastic analyses. These findings are illustrated in Fig. 13.11.

Nanometer-scale indentation of ceramic systems has also been investigated with MD simulations. Ceramics are stiffer and more brittle than metals at the macroscale and examining the nanoindentation of ceramic surfaces provides information about the nanometer-scale properties. They also reveal the manner by which defects form in covalent and ionic materials. For example, *Landman* et al. [110, 140] considered the interaction of a CaF_2 tip with a CaF_2 substrate in MD simulations using empirical potentials. As the tip approaches the surface, the attractive force between them steadily increases. This attractive force increases dramatically at the critical distance of 2.3 Å as the interlayer spacing of the tip increases (the tip is elongated) in a process that is similar to the JC phenomenon observed in metals. An important difference, however, is the amount of elongation, which is 0.35 Å in the case of the ionic ceramics and several angstroms in the case of metals. As the distance between

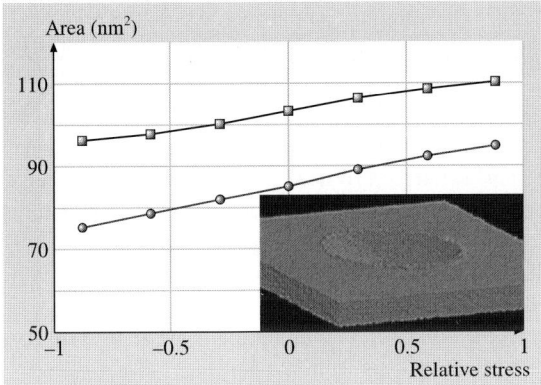

Fig. 13.11. Contact area projected in the plane at a maximum load for simulated indention of a gold surface as a function of in-plane biaxial stress. The stress is normalized to the theoretical yield stress. The *top curve* is from an atomistic simulation; the *bottom curve* is from an elastic model. *Inset*: Illustration of the region near the indention from the simulation. The tip is not shown for clarity. Initial formation of pile-up around the edge of the indentation is apparent

the tip and the surface decreases further, the attractive nature of their interaction increases until a maximum value is reached. Indentation beyond this point results in a repulsive tip–substrate interaction, compression of the tip, and ionic bonding between the tip and substrate. These bonds are responsible for the hysteresis predicted to occur in the force curve on retraction, which ultimately leads to plastic deformation of the tip followed by fracture.

The responses of covalently bound ceramics such as diamond and silicon to nanoindentation have been heavily studied with MD simulations. One of the first of these computational studies was carried out by *Kallman* et al. who used the Stillinger–Weber potential to examine the indentation of amorphous and crystalline silicon [141]. The motivation for this study came from experimental data that indicated a large change in electrical resistivity during indentation of silicon, which led to the suggestion of a load-induced phase transition below the indenter. *Clarke* et al., for example, reported forming an Ohmic contact under load, and using transmission electron microscopy they observed an amorphous phase at the point of contact after indentation [142]. Using micro-Raman microscopy, *Kailer* et al. identified a metallic β-Sn phase in silicon near the interface of a diamond indenter during hardness loading [143]. Furthermore, upon rapid unloading they detected amorphous silicon as in the *Clarke* et al. [142] experiments, while slow unloading resulted in a mixture of high-pressure polymorphs near the indent point. At the highest indentation rate and the lowest temperature, the simulations by *Kallman* et al. [141] showed that amorphous and crystalline silicon have similar yield strengths of 138 and 179 kbar, respectively. In contrast, at temperatures near the melting temperature and at the slowest indentation rate, both amorphous and crystalline silicon are predicted to have lower yield strengths of 30 kbar. The simulations thus show how the predicted yield strength of silicon at the nanometer scale depends on structure, rate of deformation, and surface temperature.

Interestingly, *Kallman* et al. [141] found that amorphous silicon does not crystallize upon indentation, but indentation of crystalline silicon at temperatures near the melting point transforms the surface structure near the indenter to the amorphous phase. The simulations do not predict transformation to the β-Sn structure under any of the conditions considered. These results agree with the outcomes of scratching experiments [144] that showed that amorphous silicon emerges from room-temperature scratching of crystalline silicon.

Kaxiras and coworkers revisited the silicon nanoindentation issue using a quasi-continuum model that couples interatomic forces from the Stillinger–Weber potential to a finite element grid [145]. They report good agreement between simulated loading curves and experiment provided that the curves are scaled by the indenter size. Rather than the β-Sn structure, however, atomic displacements suggest formation of a metallic structure with fivefold coordination below the indenter upon loading, and a residual simple cubic phase near the indentation site after the load is released rather than the mix of high-pressure phases characterized experimentally. *Smith* et al. attribute this discrepancy to shortcomings of the Stillinger–Weber potential in adequately describing the high-pressure phases of silicon. They also used a simple model for changes in electrical resistivity with loading involving contributions from both a Schottky barrier and spreading resistance. Simulated resistance-versus-loading curves agree well with experiment despite possible discrepancies between the high-pressure phases under the indenter, suggesting that the salient features of the experiment are not dependent on the details of the high-pressure phases produced.

Additional MD simulations of the indentation of silicon were carried out by *Cheong* and *Zhang* [146]. Their simulations provide further details about the phase transformations that occur in silicon as a result of nanoindentation. In particular, they find that the diamond cubic silicon is transformed into a body-centered tetragonal structure (β-Si) upon loading of the indenter, as illustrated in Fig. 13.12. Figure 13.13 shows that the coordination numbers of silicon atoms also coincide with that of the theoretical β-Si structure. The body-centered tetragonal structure is transformed into amorphous silicon during the unloading stage. A second indentation simulation again predicted that that this is a reversible process. Atomistic simulations by *Sanz-Navarro* et al. [147] shows the relation between the indentation of silicon and the hydrostatic pressure on surface cells due to the nanoindentation, as illustrated in Fig. 13.14. These simulations further predict that the transformation of diamond silicon into the β-Si structure can occur if the hydrostatic pressure is somewhat over 12 GPa.

Multimillion atom simulations of the indentation of silicon nitride were recently carried out by *Walsh* et al. [148]. The elastic modulus and hardness of the surface was calculated using load–displacement relationships. Snapshots from the simulations, illustrated in Fig. 13.15, show that pile-up occurs on the surface along the

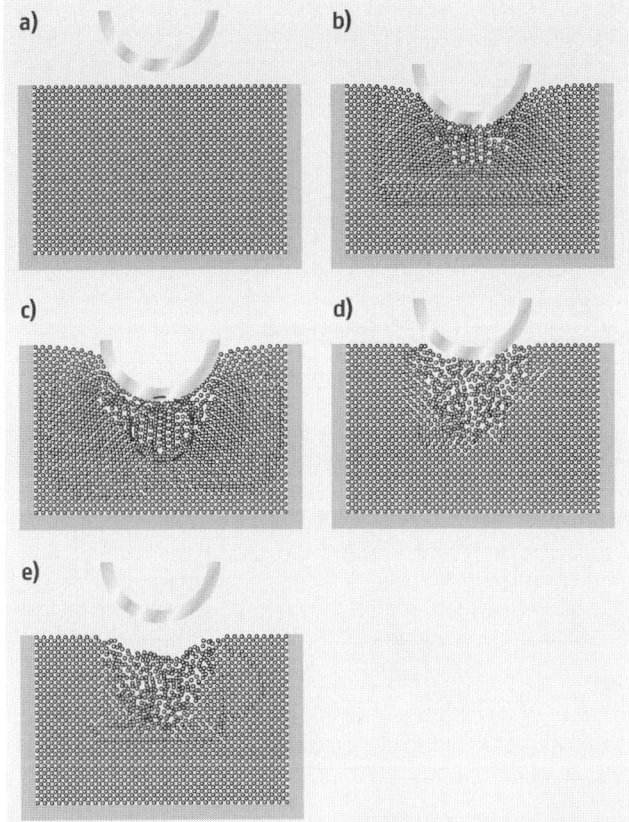

Fig. 13.12a–e. Snapshots of a silicon sample during indentation. The *smaller dots* are diamond atoms. (**a**) Crystalline silicon prior to indentation. (**b**) Atoms beneath the indenter are displaced as a result of indentation. (**c**) The system at maximum indentation. Some of the atoms are in a crystalline arrangement (*circled region*) that is different from the diamond structure. (**d**) The surface structure is largely amorphous as the tip is withdrawn. (**e**) The surface after indentation. Note the amorphous region at the site of the indentation process. After [130] with the permission of the IOP (2000)

edges of the tip. Plastic deformation of the surface is predicted to extend a significant distance beyond the actual contact area of the indenter, as illustrated in Fig. 13.15.

The indentation of bare and hydrogen-terminated diamond (111) surfaces beyond the elastic limit was investigated by *Harrison* et al. [149] using a hydrogen-terminated sp^3-bonded tip in MD simulations that utilized bond-order potentials. The simulations identified the depth and applied force at which the diamond (111) substrate incurred plastic deformation due to indentation. At low indentation forces, the tip–surface interaction is purely elastic, as illustrated in Fig. 13.16. This finding agrees with the findings of *Cho* et al. [150], who examined the atomic-scale

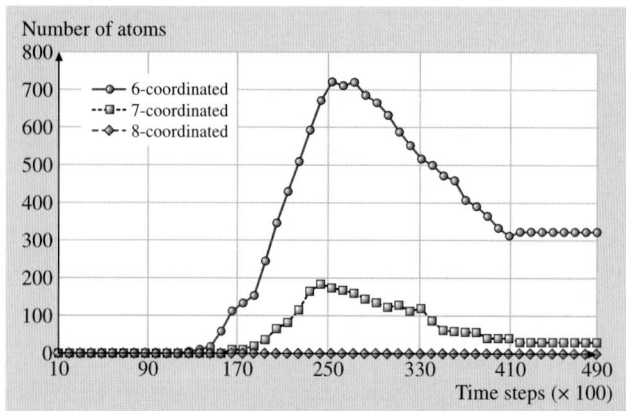

Fig. 13.13. The coordination of the silicon atoms shown in Fig. 13.12 as a function of time during nanoindentation. After [146] with the permission of the IOP (2000)

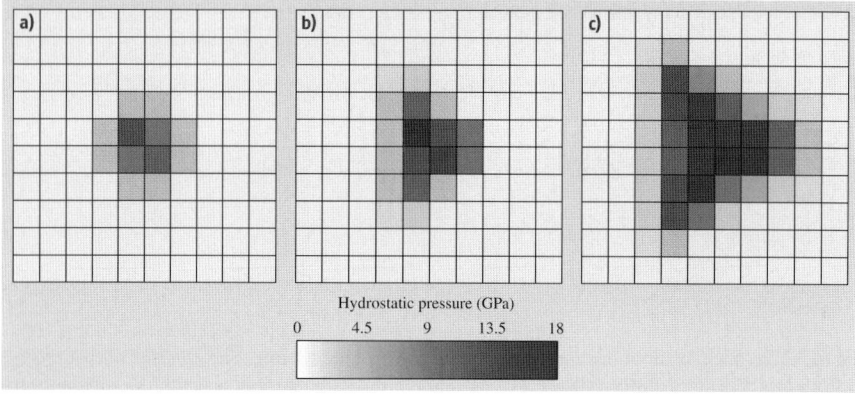

Fig. 13.14. Calculated hydrostatic pressure of surface cells at indentation depths of (**a**) 8.9 Å, (**b**) 15.7 Å, and (**c**) 25.3 Å. After [147] with the permission of the IOP (2000)

mechanical hysteresis experienced by an AFM tip indenting Si(100) with density functional theory. The calculations predicted that at low rates it is possible to cycle repeatedly between two buckled configurations of the surface without adhesion.

When the nanoindentation process of diamond (111) is plastic, connective strings of atoms are formed between the tip and the surface, as illustrated in Fig. 13.17. These strings break as the distance between the tip and crystal increases and each break is accompanied by a sudden drop in the potential energy at large positive values of tip–substrate separation. The simulations further predict that the tip end twists to minimize interatomic repulsive interactions between the hydrogen atoms on the surface and the tip. This behavior is predicted to lead to new covalent bond formation between the tip and the carbon atoms below the first layer of the surface and connective strings of atoms between the tip and the surface when the tip is

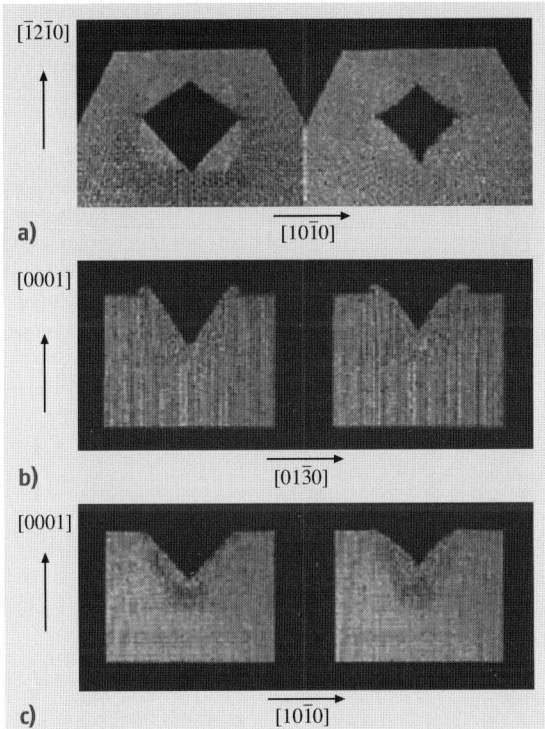

Fig. 13.15. Snapshots of the silicon nitride (**a**) surface, (**b**) slide parallel to the edges of the indenter, and (**c**) slide across the indenter diagonal. The *left-hand side* shows the surface when it is fully loaded, while the *right-hand side* shows the surface after the tip has been withdrawn. After [148] with the permission of the AIOP (2003)

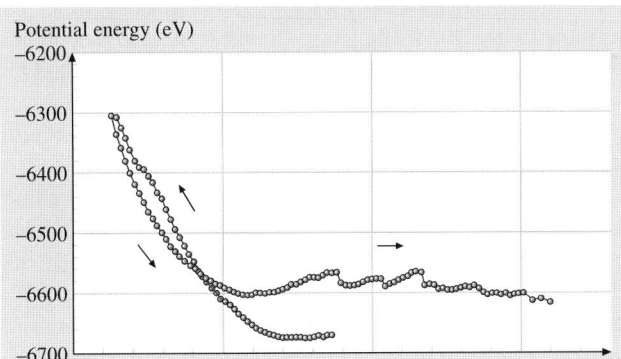

Fig. 13.16. Potential energy as a function of rigid-layer separation generated from an MD simulation of an elastic (nonadhesive) indentation of a hydrogen-terminated diamond (111) surface using a hydrogen-terminated, sp^3-hybridized tip. After [149] with the permission of Elsevier (1992)

retracted. Not surprisingly, when the surface is bare and not terminated with hydrogen atoms, the repulsive interactions between the tip and the surface are minimized and the tip indents the substrate without twisting [149]. Because carbon–carbon

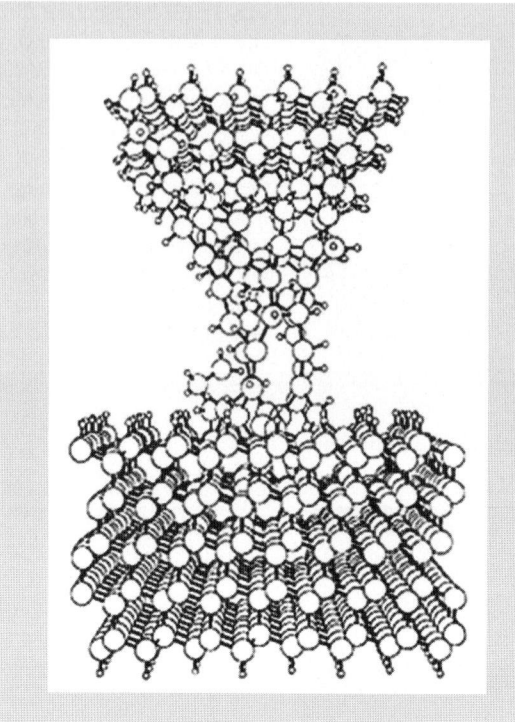

Fig. 13.17. Illustration of atoms in the MD simulation of the indentation of a hydrogen-terminated diamond (111) substrate with a hydrogen-terminated, sp^3-hybridized tip at selected time intervals. The figure illustrates the tip–substrate system as the tip was being withdrawn from the sample. *Large* and *small* spheres represent carbon and hydrogen atoms, respectively. After [149] with the permission of Elsevier (1992)

bonds are formed between the tip and the first layer of the substrate, the indentation is ordered (the surface is not disrupted as much by interacting with the tip) and the eventual fracture of the tip during retraction results in minimal damage to the substrate. The concerted fracture of all bonds in the tip gives rise to a single maximum in the potential versus distance curve at large distances.

Harrison et al. [152, 153] and *Garg* et al. [151, 154] considered the indentation of hydrogen-terminated diamond and graphene surfaces with AFM tips of carbon nanotubes and nanotube bundles using MD simulations and bond-order potentials. Tips consisting of both single-wall nanotubes and multiwall nanotubes were considered. The simulations predicted that nanotubes do not plastically deform during tip crashes on these surfaces. Rather, they elastically deform, buckle, and slip as shown in Fig. 13.18. However, as is the case for diamond tips indenting reactive diamond surfaces discussed above, strong adhesion can occur between the nanotube and the surface that destroys the nanotube in the case of highly reactive surfaces, as illustrated in Fig. 13.19.

To summarize, MD simulations reveal the properties of ceramic tips and surfaces with covalent or ceramic bonding that are most important for nanometer-scale indentation. They predict that brittle fracture of the tip can occur that is sometimes

13 Computer Simulations of Nanometer-Scale Indentation and Friction 679

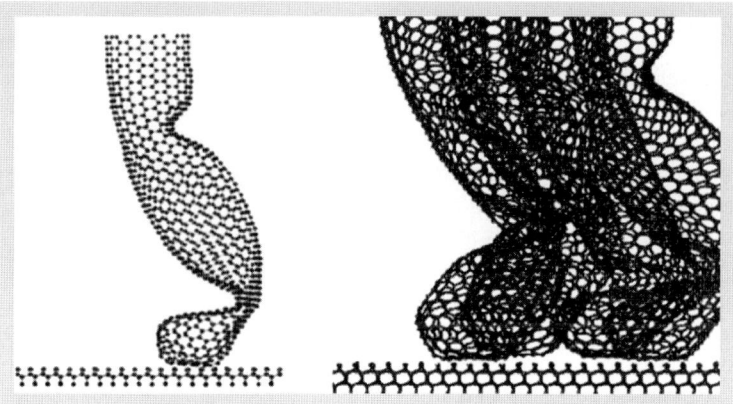

Fig. 13.18. Snapshots of the indentation of a single-wall nanotube (*left-hand image*) and a bundle of nanotubes (*right-hand image*) on hydrogen-terminated diamond (111)

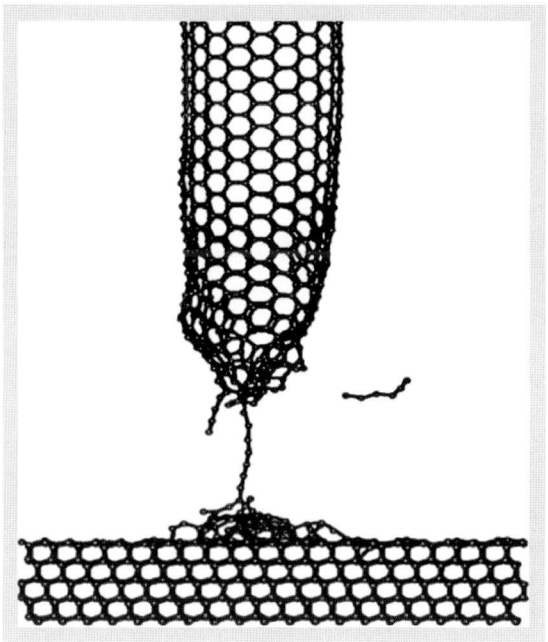

Fig. 13.19. Snapshot of a single-wall carbon nanotube as it is withdrawn following indentation on a bare diamond (111) surface. After [151] with the permission of APS (1999)

accompanied by strong adhesion with the surface. They also reveal the conditions under which neither the tip nor the surface is affected by the nanoindentation process. The insight gained from these simulations helps in the interpretation of experimental data, and it also reveals the nanometer-scale mechanisms by which, for example, tip buckling and permanent modification of the surface occur.

680   Susan B. Sinnott et al.

13.3.2 Thin Films

In many instances, surfaces are covered with thin films that can range in thickness from a few atomic layers to several μm. These films are more likely to have properties that differ from the properties of bulk materials of similar composition, and the likelihood of this increases as the film thickness decreases. Nanoindentation is one of the best approaches to determining the properties of these films. Consequently, numerous computational simulations of this process have been carried out.

For example, MD simulations have been used to study the indentation of metal surfaces covered with liquid *n*-hexadecane films, as illustrated in Fig. 13.20. As the metal tip touches the film, some of the molecules from the surface transfer to the tip and this causes the film to "swell". As the tip continues to push against the surface, the hydrocarbon film wets the side of the tip. The simulations show how the hydrocarbon film passivates the surface and prevents the strong attractive interactions discussed above for clean metal surfaces and tips from occurring.

In a series of MD simulations, *Tupper* and *Brenner* modeled the compression of a thiol self-assembled monolayer (SAM) on a rigid gold surface using both

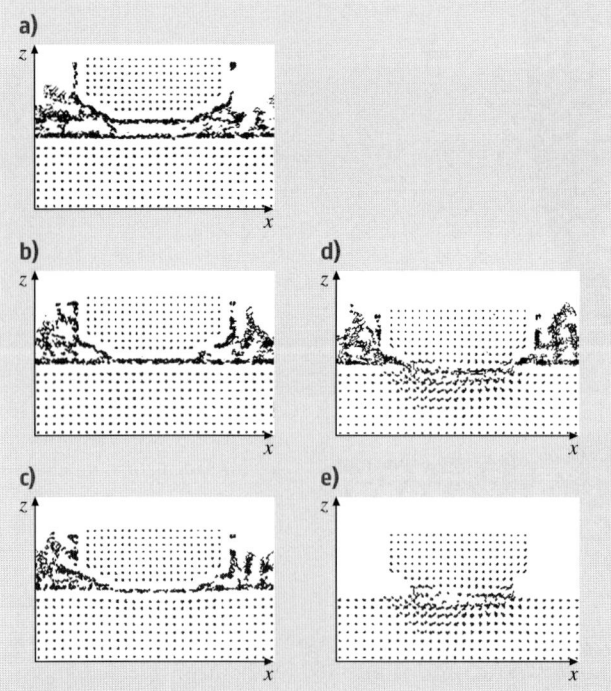

Fig. 13.20a–e. Cutaways of the side view from molecular dynamics simulations of a Ni tip indenting a Au (001) surface covered with a hexadecane film. In (**e**) only the metal atoms are shown. Note how the hexadecane is forced out from between the metal surfaces. After [110] with the permission of Elsevier (1995)

a smooth compressing surface [155] and a compressing surface with an asperity [156]. These simulations showed that compression with the smooth surface produced a compression-induced structural change that led to a change in slope of the simulated force versus compression curve. This transition is reversible and involves a change in the ordered arrangement of the sulfur head groups on the gold surface. A similar change in slope seemed to be present in the experimental indentation curves of *Houston* and coworkers [157], but was not discussed by the authors. The simulations with the asperity showed that the asperity is able to penetrate the tail groups of the SAM, as illustrated in Fig. 13.21, before an appreciable load is apparent on the compressing surface. This result indicates that it is possible to image the head groups of a thiol self-assembled monolayer that are adsorbed onto the surface of a gold substrate using STM, and consequently ordered images of these systems may not be indicative of the arrangement of the tail groups.

Zhang et al. [158] used a hybrid MD simulation approach, where a dynamic element model for the AFM cantilever was merged with a MD relaxation approach for the rest of the system, to study the frictional properties of alkanethiol SAMs on gold. They investigated the effect of several variables like chain length, terminal group, scan direction, and scan velocity. Their results show that friction forces decrease as the chain length of the SAMs increase. In the case of shorter chains such

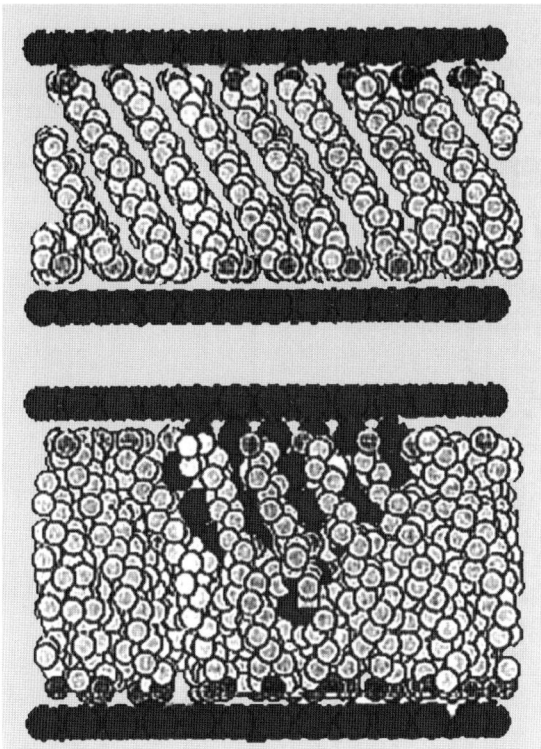

Fig. 13.21. Snapshots illustrating the compression of a self-assembled thiol film on gold for a smooth surface (*top*) and a surface containing an asperity (*bottom*). The asperity can penetrate and disorder the film tail groups before appreciable load occurs

as C_7CH_3, the SAMs near the tip can be deformed by indentation, as illustrated in Fig. 13.22. This behavior is predicted to be the cause of higher friction that occurs for the short-length chains.

Harrison and coworkers have used classical MD simulations [153, 159] to examine the indentation of monolayers composed of linear hydrocarbon chains that are chemically bound (or anchored) to a diamond substrate. Both flexible and rigid single-wall, capped nanotubes were used as tips. The simulations showed that indentation causes the ordering of the monolayer to be disrupted regardless of the type of tip used. Indentation results in the formation of gauche defects within the monolayer and, for deep indents, results in the pinning of selected hydrocarbon chains beneath the tube. Flexible nanotubes tilt slightly as they begin to indent the softer monolayers. This small distortion is due to the fact that nanotubes are stiff along their axial direction and more flexible in the transverse direction. In contrast, when the nanotubes encounter the hard diamond substrate, after "pushing" through the monolayer, they buckle. This process is illustrated in Fig. 13.23 and the force curves are shown in Fig. 13.24. The buckling of the nanotube was previously observed when single-wall, capped nanotubes were brought into contact with hydrogen-terminated diamond (111) surfaces [151, 152]. In the absence of the monolayer, the nanotube tips encounter the hard substrate in an almost vertical position. This interaction with the diamond substrate causes the cap of the nanotubes to be "pushed" inside the nanotube (they invert). Increasing the load on the nanotubes causes the walls of the tube to buckle. Both the cap inversion and the buckling are reversible processes. That is, when the load on the tube is removed, it recovers its original shape.

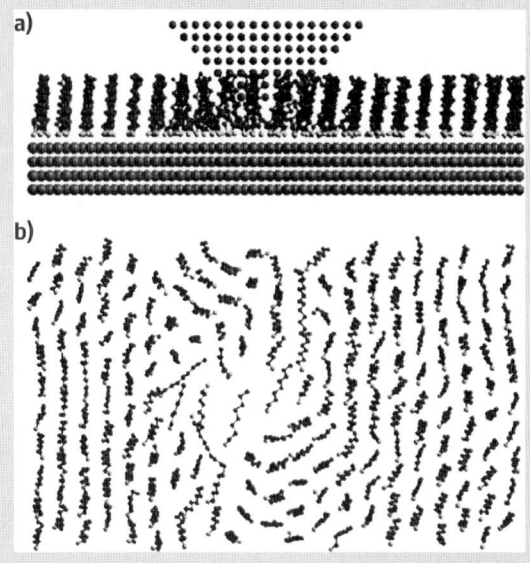

Fig. 13.22. (a) Side and (b) top views of the final configuration of a C_7CH_3 self-assembled monolayer on Au (111) under a high normal load of 1.2 nN at 300 K. The tip is not shown in (b) for clarity. After [158] with the permission of the ACS (2003)

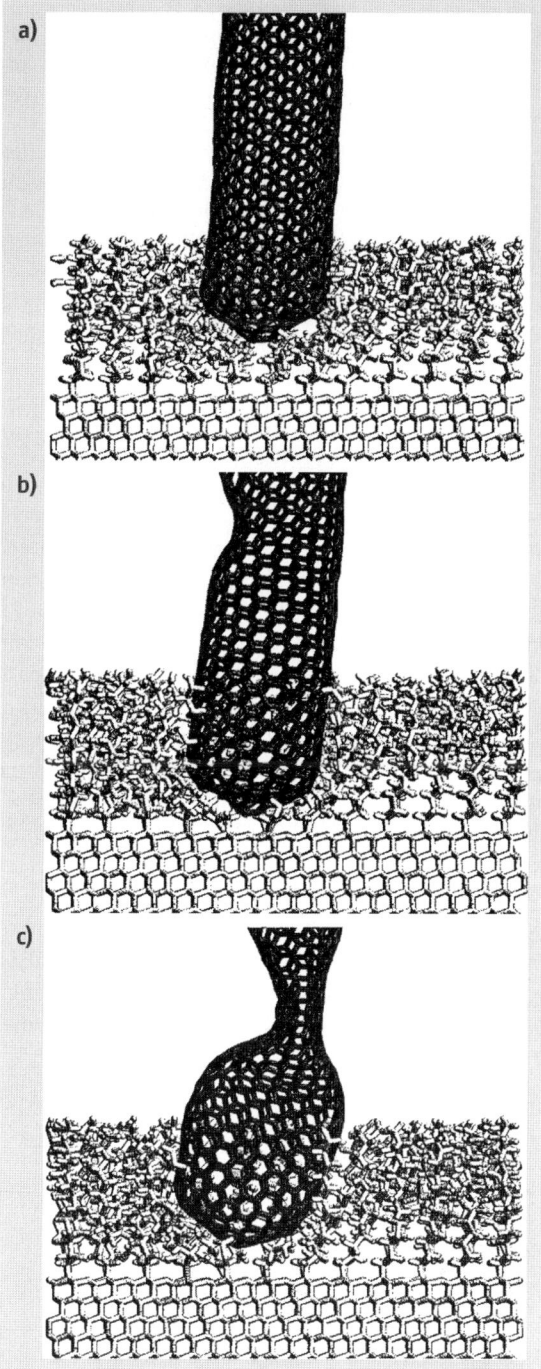

Fig. 13.23a–c. Snapshots from the simulation of the interaction of a flexible single-walled carbon nanotube with a monolayer of C_{13} chains on diamond. The loads are (**a**) 19.8 nN, (**b**) 41.2 nN, and (**c**) 36.0 nN. After [159] with the permission of the ACS (2003)

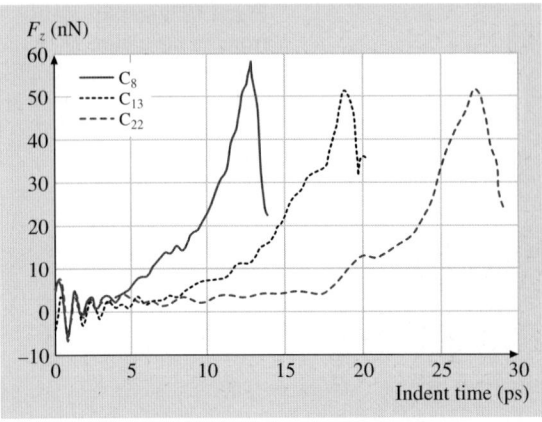

Fig. 13.24. The load on the upper two layers of the flexible carbon nanotube indenter shown in Fig. 13.23 as a function of indentation time for the nanoindentation of the indicated hydrocarbon monolayes on diamond. After [159] with the permission of the ACS (1999)

Deep indents of the hydrocarbon monolayers using rigid nanotubes result in rupture of chemical bonds. The simulations also show that the number of gauche defects generated by the indentation is a linear function of penetration depth and equal for C_{13} and C_{22} monolayers. Thus, it is the tip that governs the number of gauche defects generated.

Leng and *Jiang* [160] investigated the effect of using tips coated with SAMs containing hydrophobic methyl (CH_3) or hydrophilic hydroxyl (OH) terminal groups to nanoindent gold surfaces that also are covered with SAMs with the identical terminal groups as the tip. Figure 13.25 contains snapshots for the indentation process predicted to occur for terminal OH/OH interactions during compression and the pull-off. The adhesion force of OH/OH pairs is calculated to be about four times larger than that of CH_3/CH_3 pairs, as shown in Fig. 13.26. This is due to the formation of hydrogen bonding between OH/OH pairs. This interaction is also expected to increase the frictional force between monolayers with OH terminations.

Related MD simulations by *Mate* predict that the end groups on polymer lubricants have a significant influence on the lubrication properties of polymers [161]. For instance, fluorinated end groups are predicted to be less reactive than regular alcohol end groups. When fluorinated films are indented, the normal force becomes

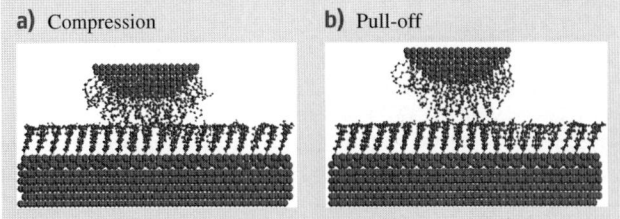

Fig. 13.25. Snapshots from the OH/OH pair interaction during (**a**) compression and (**b**) pull-off. After [160] with permission of the ACS (2002)

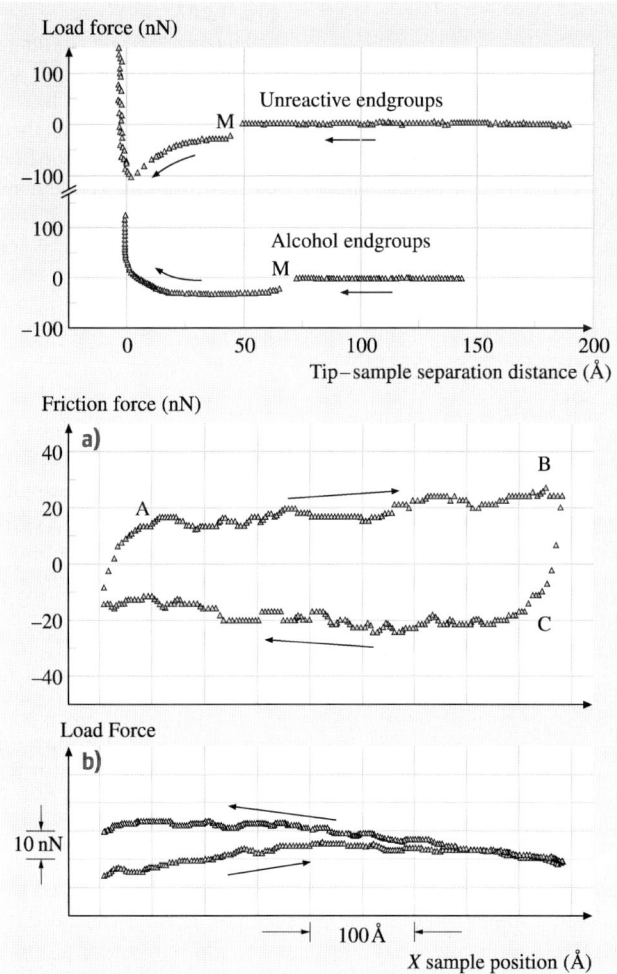

Fig. 13.26. *Top*: The force versus distance curve (indentation part only) for unbonded perfluoropolyether on Si (100). The unreactive end groups were from a 10 Å-thick film; the reactive alcohol end groups were from a 30 Å-thick film. The negative forces represent attractive interactions between the tip and the surface. *Bottom*: Measured plots of friction and load forces of the tip as it slides over the sample with the alcohol end groups. After [161] with the permission of the APS (1992)

more attractive as the distance between the tip and film decreases until the hard wall limit is reached and the interactions become repulsive. In contrast, when AFM tips indent hydrogenated films, the forces become increasingly repulsive as the distance between them decreases, as shown in Fig. 13.27 and Fig. 13.28. This predicted behavior is due to the compression of the end group beneath the tip. For the lubricant molecules to be squeezed out from between the tip and the surface, the hydrogen

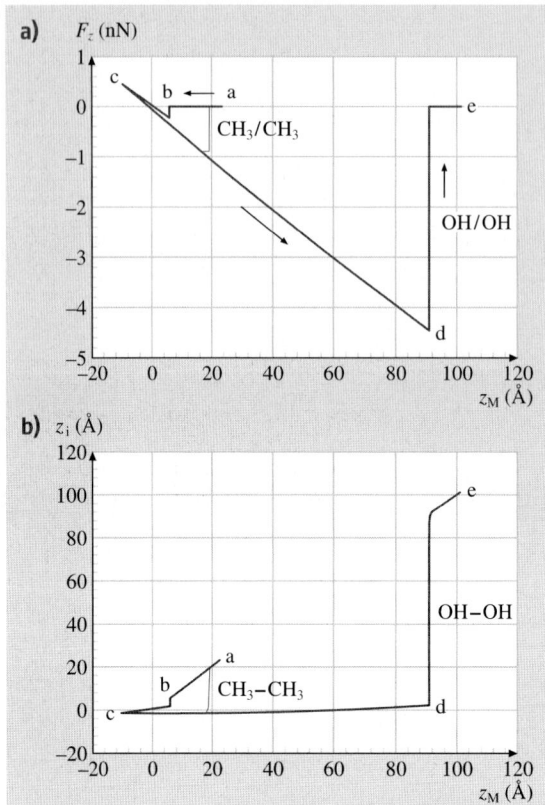

Fig. 13.27. (a) Force–distance curves and (b) tip position (z_i) versus support position (z_M) for the OH/OH contact pair and the CH_3/CH_3 contact pair. After [160] with the permission of the ACS (2002)

bonding between the two must first be broken and this increases the force needed to indent the system. As a result, a major effect of the presence of alcohol end groups is to dramatically increase the load that a liquid lubricant can support before failure (solid–solid contact) occurs.

When atomically sharp tips are used to indent solid-state thin films where there is a large mismatch in the mechanical properties of the film and the substrate, it is difficult to determine the true contact area between the tip and the surface during nanoindentation. In the case of soft films on hard substrates, pile-up can occur around the tip that effectively increases the contact area. In contrast, with hard films on soft substrates, "sink-in" is experienced around the tip that decreases the true contact area.

A class of coatings that has received much attention is diamond-like amorphous carbon (DLC) coatings. DLC coatings are almost as hard as crystalline diamond and may have very low friction coefficients (< 0.01) depending upon the growth conditions [162–165]. They have therefore generated much interest in the tribological community and there have been several MD simulation studies to determine the

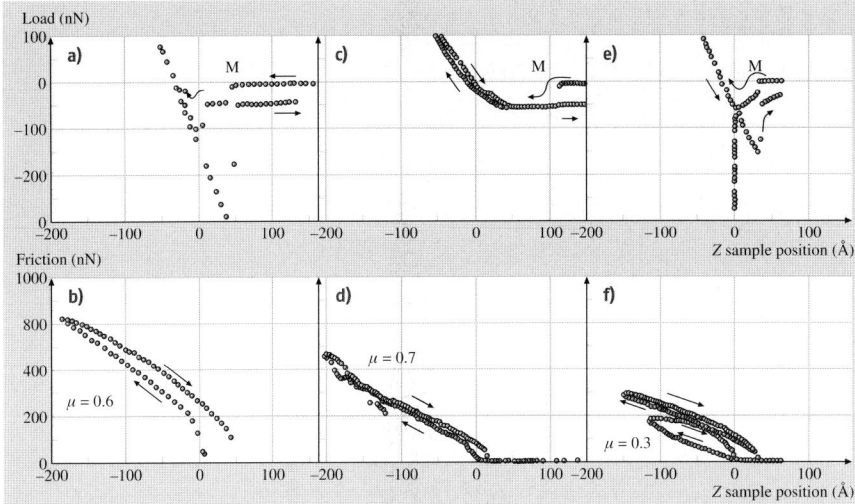

Fig. 13.28a–f. Measured values for friction and load as an atomic-force microscope tip is scanned across a 30 Å-thick sample of perfluoropolyether on Si(100). (**a**) and (**b**) The unbonded polymer with unreactive end groups. (**c**) and (**d**) The unbonded polymer with alcohol end groups. (**e**) and (**f**) A bonded polymer. After [161] with the permission of the APS (1992)

mechanical and atomic-scale frictional properties of DLC coatings. MD simulations with bond-order potentials by *Sinnott* et al. [166] examined the differences in indentation behavior of a hydrogen-terminated diamond tip on hydrogen-terminated single-crystal diamond surfaces and diamond surfaces covered with DLC. In the former case, the tip goes through shear and twist deformations at low loads that change to plastic deformation and adhesion with the surface at high loads. When the surface is covered with the DLC film, the tip easily penetrates the film, as illustrated in Fig. 13.29, which "heals" easily when the tip is retracted so that no crater or other evidence of the indentation is left behind.

MD simulations by *Glosli* et al. [167] of the indentation of DLC films that are about 20 nm-thick give similar results. In this case a larger, rigid diamond tip was used in the indentations and was also slid across the surface. During sliding, the tip plows the surface, which causes some changes to the film not seen during indentation. However, because the tip is perfectly rigid, adhesion between the film and surface is not allowed which influences the results.

This section shows that repulsive interactions between surfaces covered with molecular films and proximal probe tips are minimized relative to interactions between bare surfaces and indentation tips. The lubrication properties of polymers and SAMs can vary with chain length, the rigidity of the tip, and the chemical properties of the end groups. In some cases, indentations can disrupt the initial ordering of polymers and SAMs, which affects their responses to nanoindentation and friction.

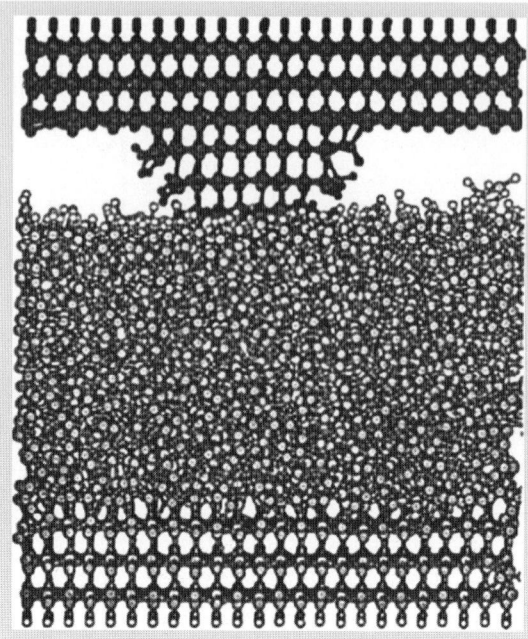

Fig. 13.29. Snapshot from a molecular dynamics simulation where a pyramidal diamond tip indented an amorphous carbon thin film that is 20 layers thick. The simulation took place at room temperature and the carbon atoms in the film were 21% sp^3-hybridized and 58% sp^2-hybridized (the remaining atoms were on the surface and were not counted). After [166] with the permission of the AIP (1997)

13.4 Friction and Lubrication

Work is required to slide two surfaces against one another. When the work of sliding is converted to a less ordered form, as required by the first law of thermodynamics, friction will occur. For instance, if the two surfaces are strongly adhering to one another, the work of sliding can be converted to damage that extends beyond the surfaces and into the bulk. If the adhesive force between the two surfaces is weaker, the conversion of work results in damage that is limited to the area at or near the surface and produces transfer films or wear debris [169, 170]. While the thermodynamic principles of the conversion of work to heat are well known, the mechanisms by which this takes place at sliding surfaces are much less well established despite their obvious importance for a wide variety of technological applications.

Atomic-scale simulations of friction are therefore important tools for achieving this understanding. They have consequently been applied to numerous materials in a wide variety of structures and configurations, including atomically flat and atomically rough diamond surfaces [171–173], rigid substrates covered with monolayers of alkane chains [174], perfluorocarboxylic acid and hydrocarboxylic Langmuir–Blodgett (LB) monolayers [175], between contacting copper surfaces [168, 176], between a silicon tip and a silicon substrate [116, 140], and between contacting diamond surfaces that have organic molecules absorbed on them [177]. These and several other studies are discussed below.

13.4.1 Bare Surfaces

Sliding friction that takes place between two surfaces in the absence of lubricant is termed "dry" friction even if the process occurs in an ambient environment. Simple models have been developed to model dry sliding friction that, for example, consider the motion of a single atom over a monoatomic chain [178]. Results from these models reveal how elastic deformation of the substrate from the sliding atom affects energy dissipation and how the average frictional force varies with changes in the force constant of the substrate in the direction normal to the scan direction. Much of the correct behavior involved in dry sliding friction is captured by these types of simple models. However, more detailed models and simulations, such as MD simulations, are required to provide information about more complex phenomena.

MD simulations have been used to study the sliding of metal tips across clean metal surfaces by numerous groups [168, 179–183]. An illustrative case is shown in Fig. 13.30 for a copper tip sliding across a copper surface [168]. Adhesion and wear occur when the attractive force between the atoms on the tip and the atoms at the surface becomes greater than the attractive forces within the tip itself. Atomic-scale stick and slip can occur through nucleation and subsequent motion of dislocations, and wear can occur if part of the tip gets left behind on the surface (Fig. 13.30). The simulations can further provide data on how the characteristic 'stick-slip' friction motion can depend on the area of contact, the rate of sliding, and the sliding direction (Fig. 13.31).

An additional study of stick-slip in the sliding of much larger, square-shaped metal tips across metal surfaces was carried out by *Li* et al. [184] using EAM potentials. The initial structure of a NiAl tip and surface system is shown in Fig. 13.32. This study predicted that collective elastic deformation of the surface layers in response to sliding is the main cause of the stick-slip behavior shown in Fig. 13.33. The simulations also predicted that stick-slip produces phonons that propagate through the surface slab.

Large-scale simulations using pairwise Morse potentials that are similar in form to (13.6) were used to study the wear of metal surfaces caused by metal tips that

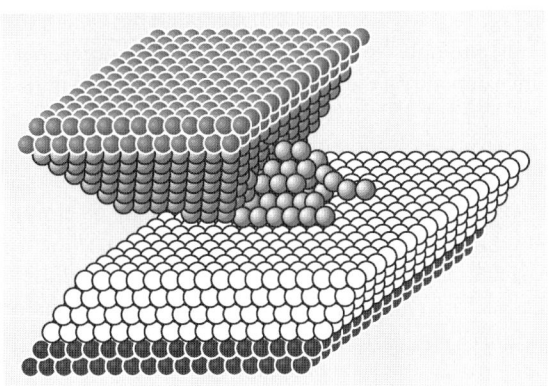

Fig. 13.30. Snapshot from a molecular dynamics simulation of a copper tip sliding across a Cu(100) surface. A connective neck between the two is sheared during the sliding, leading to wear of the tip. The simulation was performed at a temperature of 0 K. After [168] with the permission of the APS (1996)

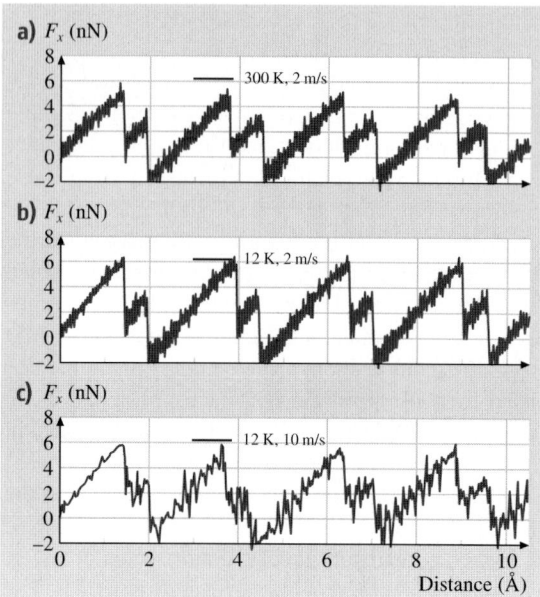

Fig. 13.31a–c. Plots of the lateral force versus distance from a simulation similar to that shown in Fig. 13.30. The plots illustrate the dependence of the force on temperature and sliding velocity. After [168] with the permission of the APS (1996)

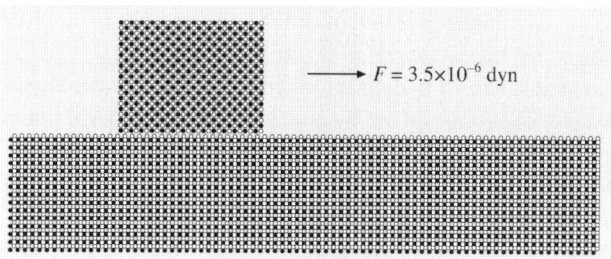

Fig. 13.32. Starting configuration of sliding NiAl tip on a NiAl surface. After [184] with the permission of the AIP (2001)

plow the surface, as illustrated in Fig. 13.34. They provide insight into the wear track dependence of the sliding rate [186] and how variations in the scratching force, friction coefficient, and other quantities depend on the scratch depth [185], as illustrated in Fig. 13.35.

On the whole, the results of experimental studies show good agreement with the results of the computational studies described above. This is true despite the fact that all of these MD simulations use empirical potentials that do not include electronic effects and thus effectively assume that the electronic contributions to friction on metal surfaces are negligible. However, experiments have measured a non-negligible contribution of conduction electrons to friction [188]. Thus, future simulations of metal-tip–metal-substrate interactions using more sophisticated tight-binding or first principles methods that include electronic effects are encouraged.

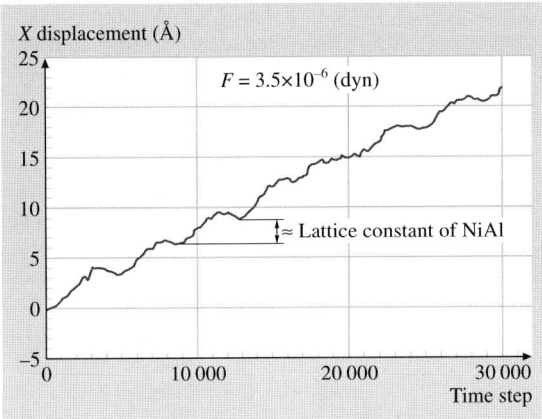

Fig. 13.33. A structured curve of frictional dynamics of an atom in the *upper right corner* that is indicative of stick-slip. After [184] with the permission of the AIP (2001)

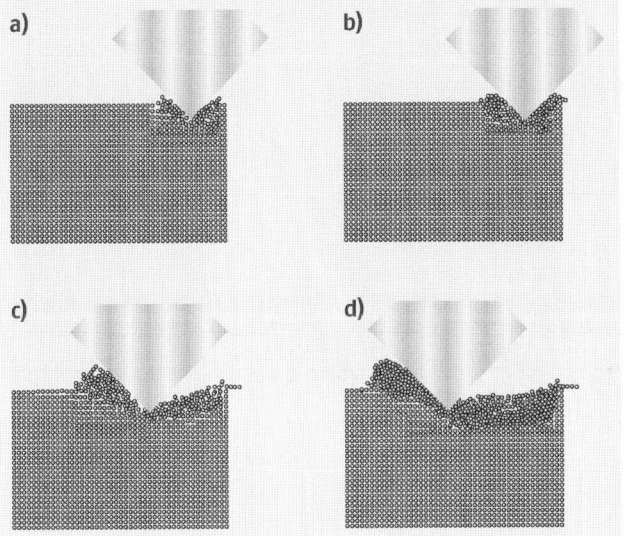

Fig. 13.34a–d. Snapshots of the scratching of an aluminium surface with a rigid tip at a depth of 0.8 nm. After [185] with the permission of the APS (2000)

Layered ceramics, such as mica, graphite and MoS_2, that have structures that include strongly bound layers that interact with one another through weak van der Waals bonds, have long been known to have good lubricating properties because of the ease with which the layers slide over one another. They have, therefore, been the focus of some of the earliest experimental studies of nanometer-scale friction [19, 189]. The results of these early studies lead researchers to hypothesize that at high loads measured friction forces were related to "incipient sliding" [190, 191] caused by a small flake from the surface becoming attached to the end of the tip. If true, this would mean that all measured interactions were between the surface and the flake,

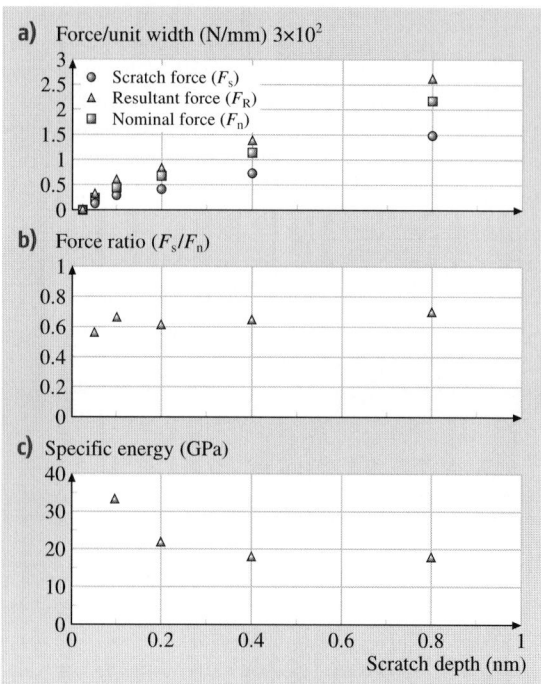

Fig. 13.35. Variation in (**a**) the scratching force, the normal force, and the resultant force, and (**b**) the friction coefficient, and (**c**) the specific energy during scratch processes similar to those shown in Fig. 13.34 at scratch depths ranging from 0.8 nm to almost 0 nm. After [185] with the permission of the APS (2000)

which has a larger contact area than the clean tip. However, subsequent simulations of constant force AFM images of graphite by *Tang* et al. [192] showed that there is no need for the assumption of a graphite flake under the tip to reproduce the experimental images of a graphite surface.

Surprisingly strong localized fluctuations in atomic-scale friction are displayed by layered ceramics [187, 194–196]. For instance, square-well signals with sub-angstrom lateral width are obtained in FFM scans on $MoS_2(001)$ in the direction across the scan direction, while sawtooth signals are detected along the scan direction, as shown in Fig. 13.36. This finding can be explained by a stick-slip model by *Mate* [19] and *Erlandsson* [189] that assumes that the tip does a zigzag walk along the scan. Measured variations in the frictional force with the periodicity of cleavage planes [189] are consistent with the results of this simple model. However, additional experiments indicate a more complex tip–surface interaction, such as changes in the intrinsic lateral force between the substrate and the AFM tip [197] or sliding-induced chemistry between the tip and the surface [198, 199].

Crystalline ceramics differ from layered ceramics in that they are held together by relatively strong covalent or ionic bonds. In the case of ionic systems, *Shluger* et al. [193] used a mixture of atomistic and macroscopic modeling methods to study the interaction of a MgO tip and a LiF surface. In particular, the tip–surface interaction was treated atomistically and the cantilever deflection was treated with a macroscopic approach. The results, shown in Fig. 13.37, show that if the tip is

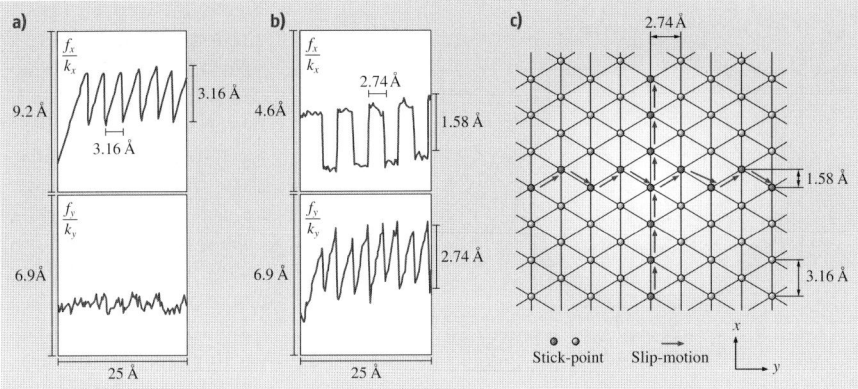

Fig. 13.36a–c. Displacement data from a scan across a MoS$_2$(001) surface. The data in (**a**) and (**b**) are form scans along the x-and y-directions, respectively, on the surface shown in (**c**). After [187] with the permission of the APS (1996)

charged and in hard contact with the surface, tip and surface distortions are possible that can lead to motion of the surface ions within the surface plane and the transfer of some of the ions onto the tip.

In the case of covalently bound ceramics, there is extensive literature related to friction of diamond [200, 201] because, while it is the hardest material known, it also exhibits relatively low friction. The "ratchet mechanism" has been proposed for energy dissipation during friction on the macroscale in diamond, where energy is released by the transfer of normal force from one surface asperity to another. The elastic mechanism is another mechanism that has been proposed, where the released energy comes from elastic strain in an asperity. Atomic-scale friction has been measured experimentally [20] for diamond tips with near atomic-scale radii sliding over hydrogen-terminated diamond surfaces. These experiments are sensitive enough to detect the 2×1 reconstruction on the diamond (100) surface. Furthermore, the average friction coefficient determined with an AFM on H-terminated diamond (111) surfaces is about two orders of magnitude smaller than the value measured on bare, 2×1 diamond (111) surfaces, indicating greater adhesion in the latter case [202]. More recently, the friction between a tungsten carbide tip and hydrogen-terminated diamond (111) was examined with AFM in UHV by *Enachescu* et al. [203]. The friction between these two hard surfaces was shown to obey Derjaguin–Muller–Toporov or DMT [204] contact mechanics and the shear strength of the interface was determined to be 246 MPa.

Extensive MD simulations have been carried out by *Harrison* and coworkers that examine the friction between hydrogen-terminated diamond (111) surfaces [171, 208] and diamond (100) surfaces [205] in sliding contact. The simulations of sliding between the diamond (111) surfaces reveal that the potential energy, load, and friction are all periodic functions of the sliding distance (Fig. 13.38). Maxima in these quantities occur when the hydrogen atoms on opposing surfaces in-

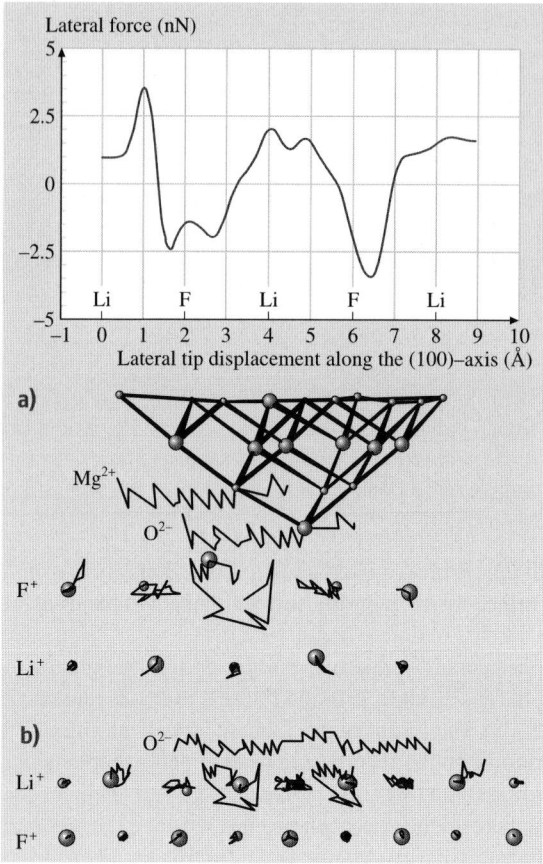

Fig. 13.37. *Top*: The lateral force calculated for a MgO tip scanning in the ⟨001⟩-direction on LiF(001). *Bottom*: A view of the side of the surface plane along the scan direction. The surface Li^+ and F^- atoms are seen to relax to relieve the frictional energy and this relaxation motion is indicated in the figure by the *category lines*. (**a**) How a F^- ion on the surface can be moved into an interstitial site by the tip and then it returns to its original position. (**b**) How the relaxation of the surface atoms is reversible. After [193] with the permission of Elsevier (1995)

teract strongly. Recent *ab initio* studies by *Neitola* and *Park* of the friction between hydrogen-terminated diamond (111) surfaces also show that the potential energy is periodic with sliding distance (Fig. 13.39) [206]. Because the results of the *ab initio* studies and the MD simulations are in good agreement, Neitola and Park conclude that the potential model used in the MD studies is accurate.

As mentioned previously, the maxima in the load and the friction values during sliding are caused by the interactions of hydrogen atoms on opposing surfaces. When sliding in the [11$\bar{2}$] direction, the H atoms "revolve" around one another, thus decreasing the repulsive interaction between the sliding surfaces because the hydrogen atoms are not forced to pass directly over one another [171]. Increasing the load causes increased stress at the interface. The opposing hydrogen atoms become "stuck". Once the stress at the interface becomes large enough to overcome the hydrogen–hydrogen interaction between opposing surfaces, the hydrogen atoms "slip" past one another with the same "revolving" motion observed at low loads. This phenomenon is known as atomic-scale stick-slip and has the periodicity of the diamond lattice. It should be noted that due to the alignment of the opposing sur-

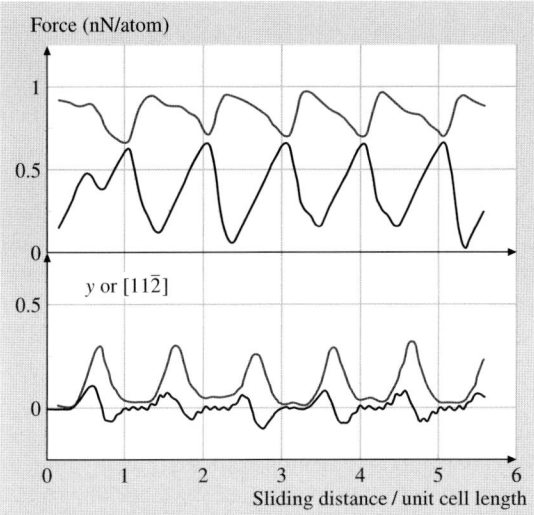

Fig. 13.38. Calculated frictional force (*lower lines*) and normal force (*upper lines*) felt by a hydrogen-terminated (111) surface as it slides against another hydrogen-terminated diamond (111) surface in a MD simulation. The sliding direction is given in the legend. The sliding speed is 1 Å/ps and the simulation temp is 300 K. The two plots show how the simulated stick-slip motion changes as a function of the applied load. The load is high and low in the *upper* and *lower* panels, respectively. After [205] with the permission of the ACS (1995)

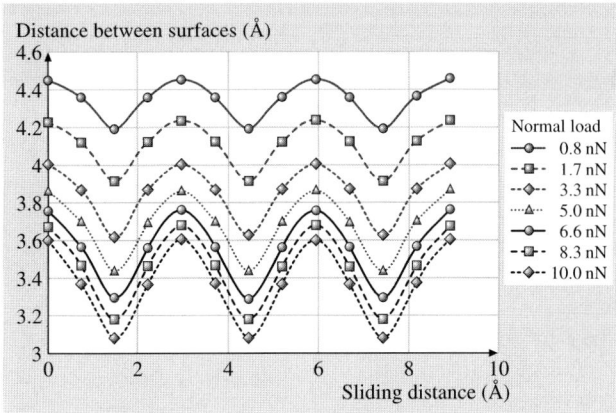

Fig. 13.39. Distance between hydrogen-terminated (111) crystals as a function of sliding distance. After [206] with the permission of the ACS (2001)

faces, the hydrogen atoms are directly in line with each other when sliding in the [11$\bar{2}$] direction. However, the hydrogen atoms are not "aligned" with each other when sliding in the [1$\bar{1}$0], so the friction in this direction is lower than in the [11$\bar{2}$]

direction. It should be noted, however, that experimentally all initial alignments are likely to be probed.

Harrison and coworkers have further shown that the peaks in the frictional force are correlated with peaks in the temperature of the atoms at the interface when two hydrogen-terminated diamond (111) surfaces are in sliding contact [208]. Figure 13.40 shows the vibrational energy (or temperature) between diamond layers as a function of sliding distance. These data clearly show that layers close to the sliding interface can be vibrationally excited during sliding. When the hydrogen atoms are "stuck" or interacting with each other strongly, the stress and friction force at the interface build up. When the hydrogen atoms "slip" past one another, the stress at the interface is relieved and the energy is transferred to the diamond in the form of vibration or heat. Thus, the peaks in the temperature occur slightly after the peaks in the frictional force.

It should be noted that atomic-scale stick slip is observed in other systems. *Harrison* and coworkers used MD simulations to demonstrate that two hydrogen-terminated diamond (100) (2×1) surfaces in sliding contact also exhibit stick-slip [205]. In addition, it was shown that the shape of the friction versus sliding distance curves is influenced slightly by the speed of the sliding, with features in the curves becoming more pronounced at slower speeds. Stick-slip behavior was also observed in AFM studies of diamond (100) (2×1) surfaces [202]. However, in this

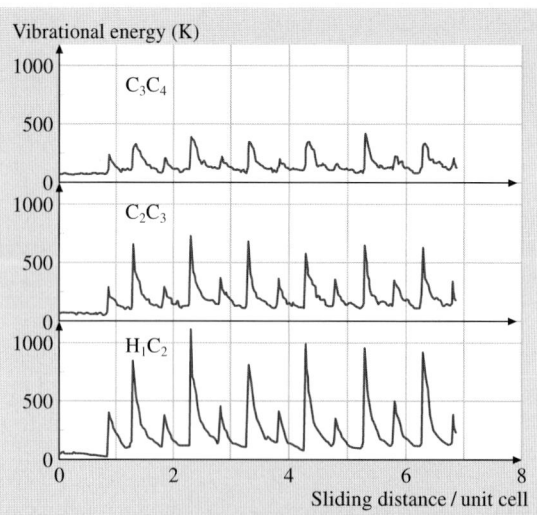

Fig. 13.40. Average vibrational energy of oscillators between diamond layers as a function of sliding distance. These energies are derived from a molecular dynamics simulation of the sliding of a hydrogen-terminated diamond (111) surface over another hydrogen-terminated diamond (111) surface. The vibrational energy between the first and second layers of the lower diamond surface is shown in the *lower panel*, between the second and third layers in the *middle panel*, and between the third and fourth layers in the *upper panel*. After [207] with the permission of Elsevier (1995)

case the stick-slip was over a much longer length scale and may be due to the fact that the surfaces were not hydrogen-terminated.

Mulliah et al. [210] used MD simulations with bond-order potentials [211] to model interactions between indenter atoms, EAM potentials [212] to model interactions between substrate atoms, and the Ziegler–Biersack–Littmack potential [213] to model interactions between indenter and substrate atoms to study the atomic-scale stick-slip phenomenon of a pyramidal diamond tip interacting with a silver surface at several sliding rates and vertical support displacements. These simulations showed that dislocations are related to the stick events emitting a dislocation in the substrate near the tip. The scratch in the substrate is discrete due to the tip jumping over the surface in the case of small vertical displacements. In contrast, large displacements of 15 Å or more result in a continuous scratch. These simulations also showed how the dynamic friction coefficient and the static friction coefficient increase with increasing tip depth. The tip moves continuously through a stick and slip motion at large depths, whereas it comes to a halt in the case of shallow indents. Although the sliding rate can change the exact points of stick and slip, the range of sliding rates over the range of values considered in this study (1.0 to 5.0 m/s) has no influence on the damage to the substrate, the atomistic stick-slip mechanisms, or the calculated friction coefficients.

The effect of the way in which the tip is rastered across the surface in MD simulations was considered by *Cai* and *Wang* [209,214] using bond-order potentials. In particular, they dragged silicon tips across several silicon surfaces, as illustrated in Fig. 13.41, in two different ways. In the first, they moved the tip every MD step while in the other they advanced the tip every 1000 steps. In both cases, the overall sliding rate is the same and equals 1.67 m/s. In both cases, wear of the tip such as is illustrated in Fig. 13.41 occurs. However, the mechanisms by which the wear occurs are found to depend on the approach used, and the latter approach is found to be in better agreement with experimental data.

In many studies, diamond tips or diamond-decorated tips are used in friction measurements. Diamond is an attractive material for an FFM tip because of its high mechanical strength and the belief that such tips are wear-resistant. However, diamond tips that were used to scratch diamond and silicon surfaces and then imaged showed significant wear that increased with the increasing hardness of the tested material [215,216]. This wear altered the shape of the tip and hence influenced the contact area that is used to determine friction coefficients.

In summary, MD simulations provide insight into dry sliding friction and the sliding of metal tips across clean metal, crystalline ceramics, and layered ceramics surfaces. Stick-slip friction or wear can occur depending on the sliding conditions. The good lubricating properties of layered ceramics are observed in the simulations along with localized fluctuations in atomic-scale friction. Crystalline ceramics, such as diamond, exhibit relatively low friction and the simulations show how stick-slip atomic-scale motion changes with the conditions of sliding and the way in which the simulation is performed.

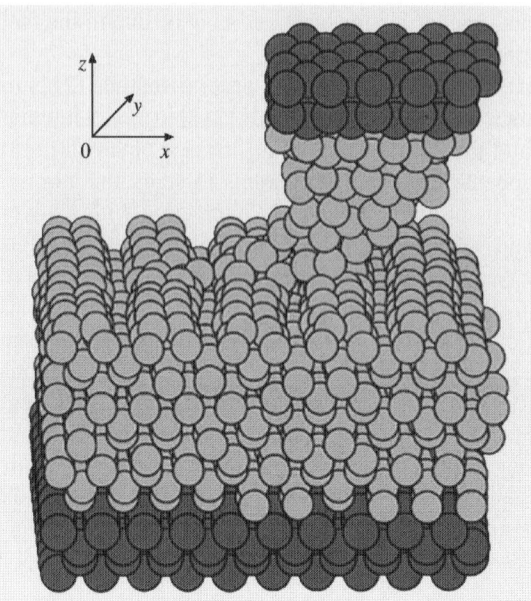

Fig. 13.41. Snapshots of a Si(111) tip interacting with a Si(001) 2×1 surface. The tip is rastering along the surface in the x-direction and starts off at a distance of 9 Å from the surface. After [209] with the permission of the CCLRC (2002)

13.4.2 Decorated Surfaces

While dry sliding friction in vacuum assumes that ambient gas particles have no direct effect on the results, MD simulations show that free particles between two surfaces in sliding contact influence friction to a surprisingly large degree. These so-called third-body molecules have been studied extensively by *Perry* et al. [177, 217, 218] using MD simulations with bond-order and LJ potentials. These simulations focus on the effect of trapped small hydrocarbon molecules on the atomic-scale friction of two (111) crystal faces of diamond with hydrogen termination. These molecules might represent hydrocarbon contamination trapped between contacting surfaces prior to a sliding experiment in dry friction, or hydrocarbon debris formed during sliding.

In particular, the effects on friction of methane (CH_4), ethane (C_2H_6), and isobutane ($(CH_3)_3CH$) trapped between diamond (111) surfaces in sliding contact were examined in separate studies (Fig. 13.42). The frictional force for all these systems generally increases as the load increases, as illustrated in Fig. 13.43. The simulations predict that the third-body molecules markedly reduce the average frictional force compared to the results for pristine hydrogen-terminated surfaces. This is particularly true at high loads, where the third-body molecules act as a boundary layer between the two diamond surfaces. That is, the third-body molecules reduce the interaction of hydrogen atoms on opposing surfaces [218]. This is demonstrated by examining the vibrational energy excited in the diamond lattice during the sliding (Fig. 13.44). Significant vibrational excitation of the diamond outer layer (C–H) occurs in the absence of the methane molecules. Thus, the friction is approximately 3.5

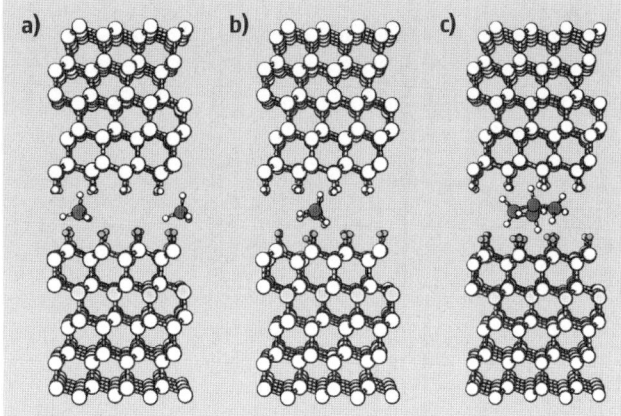

Fig. 13.42a–c. Initial configuration at low load for the diamond plus third-body molecule systems. These systems are composed of two diamond surfaces, viewed along the [$\bar{1}$10] direction, and two methane molecules in (**a**), one ethane molecule in (**b**), and one isobutane molecule in (**c**). Large *white* and *dark gray spheres* represent carbon atoms of the diamond surfaces and the third-body molecules, respectively. *Small gray spheres* represent hydrogen atoms of the lower diamond surface. Hydrogen atoms of the upper diamond surface and the third-body molecules are both represented by *small white spheres*. Sliding is achieved by moving the rigid layers of the upper surface from *left to right* in the figure. After [177] with the permission of the ACS (1997)

times larger when methane is not present. The application of load to the diamond surfaces causes the normal mode vibrations of the trapped methane molecules to change. Power spectra calculated from MD simulations [217, 218] show that even under low loads, the peaks in the power spectra are significantly broadened. Peaks in the low-energy region of the spectrum almost disappear with the additional application of load.

The size of the methane molecules allows them to be "pushed" in-between hydrogen atoms on the diamond surfaces while sliding [218]. However, steric considerations cause the larger ethane and isobutane molecules to change orientation during sliding. Conformations that lead to increased interactions with the diamond surfaces increase the average frictional force. Thus, despite the fact that the two diamond surfaces are farther apart when ethane and isobutane are present compared to when methane is present, the friction is larger because these molecules do not "fit" nicely into potential energy valleys between hydrogen atoms when sliding.

When similar hydrocarbon molecules (methyl, ethyl, and *n*-propyl groups) are chemisorbed to one of the sliding diamond surfaces, instead of trapped between the surfaces, different behavior is observed by *Harrison* et al. [172, 173, 208, 219]. Simulations show that methyl-termination does not decrease friction significantly but results in frictional forces that are nearly the same as they are for hydrogen-terminated diamond surfaces [207]. While the methane third-body molecules de-

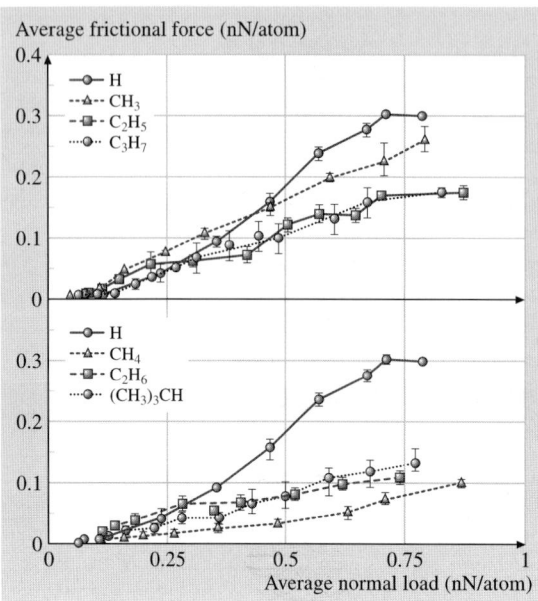

Fig. 13.43. Average frictional force per rigid-layer atom as a function of average normal load per rigid-layer atom for sliding the upper diamond surface in the [11$\bar{2}$] crystallographic direction. Data for the methane (CH_4) system (*open triangles*), the ethane (C_2H_6) system (*open squares*), the isobutane $(CH_3)_3CH$ system (*filled circles*), and diamond surfaces in the absence of third-body molecules (*open circles*) are shown in the lower panel. Data for the methyl-terminated $-CH_3$ system (*open triangles*), the ethyl-terminated ($-C_2H_5$) system (*open squares*), the n-propyl-terminated ($-C_3H_7$) system (*filled circles*), and diamond surfaces in the absence of third-body molecules (*open circles*) are shown in the *upper panel*. *Lines* have been drawn to aid the eye. After [177] with the permission of the ACS (1997)

crease the frictional force to a greater extent than the chemisorbed methyl groups, friction as a function of load is comparable for the ethyl-terminated and ethane systems, with the former giving slightly higher frictional forces. Attaching the hydrocarbon groups to the diamond surfaces causes them to have less freedom to move between hydrogen atoms on opposing diamond surfaces during sliding. This generally increases their repulsive interaction with the diamond counterface.

MD simulations can also provide insight into the rich, nonequilibrium tribochemistry that occurs between surfaces in sliding contact. *Harrison* and *Brenner* examined the tribochemistry that occurs when ethane molecules are trapped between diamond surfaces in sliding contact, as illustrated in Fig. 13.45 [220]. This simulation was the first to show the atomic-scale mechanisms for the degradation of lubricant molecules due to friction. The type of debris formed during the sliding simulation is similar to the types of debris molecules that were observed in macroscopic experiments that examined the friction between diamond surfaces [221].

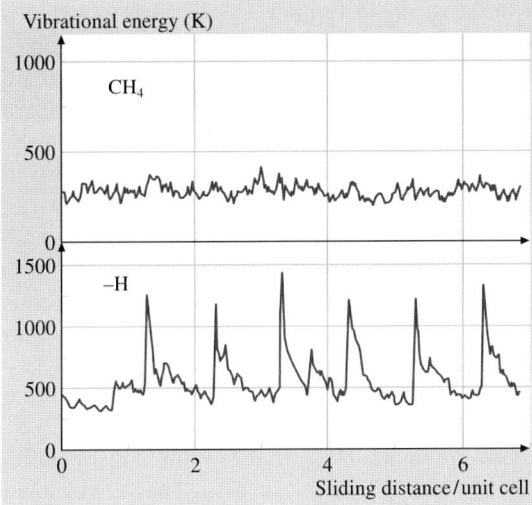

Fig. 13.44. Average vibrational energy between the (C–H) bonds of the upper diamond (111) surface versus sliding distance for hydrogen-terminated diamond (111) surfaces, with (CH$_4$) and without methane (H), trapped between them. The average normal load is approximately the same in both simulations and is in the range 0.8–0.85 nN/atom. The average frictional force on the upper surface is about 3.5 times smaller in the presence of the methane third-body molecules. After [218] with the permission of Elsevier (1996)

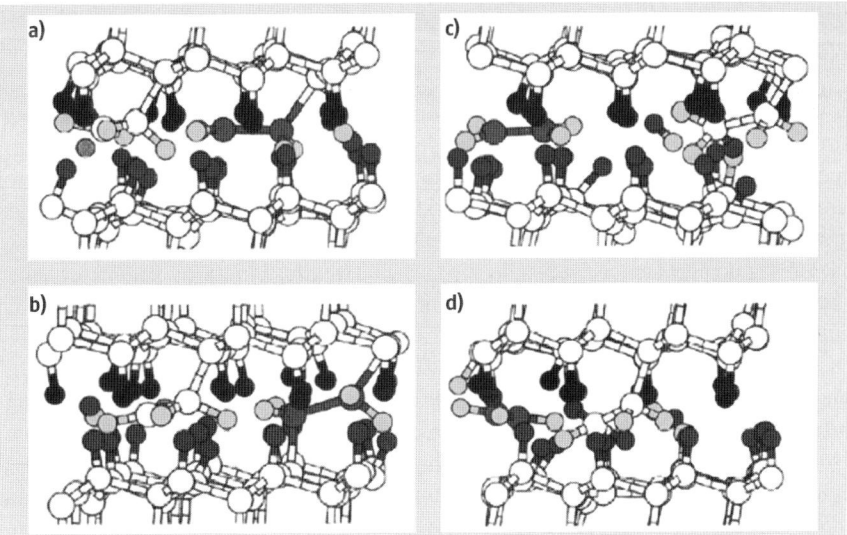

Fig. 13.45a–d. Snapshots from a molecular dynamics simulation of the sliding of two hydrogen-terminated diamond (111) surfaces against one another in the [11$\bar{2}$] direction. The upper surface has two ethyl fragments chemisorbed to it. The simulation shows how sliding can induce chemistry at the interface. After [220] with the permission of the ACS (1994)

In the case of sliding metal surfaces, impurity molecules or atoms (both adsorbed and absorbed) on thin metal films can be expected to affect the film's properties. For example, calculations have shown that resistivity changes in the metal are strongly dependent on the nature of the adsorption bond [222]. When this result is used to interpret atomic-scale friction results obtained with the QCM, the sliding of adsorbate structures on metal surfaces are shown to be a combination of electron excitation and lattice vibrations. Additionally, other interesting quantum effects can come into play when the adsorbate is very different chemically from the surface on which it is sliding. For instance, the electronic frictional forces acting on small, inert atoms and molecules, such as C_2H_6 and Xe, sliding on metal surfaces have been calculated by *Persson* [223], where the metal surface was approximated by a electron gas (jellium) model. The calculations showed that the Pauli repulsive and attractive van der Waals forces are of similar magnitudes. In addition, the calculated electronic friction contributions agree well with the values derived from surface resistivity by *Grabhorn* et al. [224] and QCM measurements. These studies showed that parallel friction is mainly due to electronic effects while perpendicular friction is phononic in nature in this system.

In summary, MD simulations show that the average frictional force decreases significantly in systems with third-body molecules, especially at high loads. Simulations also provide information about the details of tribochemical interactions that can occur between lubricants and sliding surfaces. Additionally, the effect of the presence of small molecules on thin metal films can influence film properties, such as resistivity.

13.4.3 Thin Films

As discussed at the beginning of this section, the conversion of the work of sliding into some other less ordered form is responsible for friction at sliding solid interfaces. In the case of adhering systems, the work of sliding may be converted into damage within the bulk (plastic deformation), while in the case of weakly adhesive forces, friction can occur through the conversion of work to heat at the interface that causes no permanent damage to the surface (wearless friction). The latter case, when it is achieved through the presence of lubricating thin films, is the topic of this section.

There are several types of lubricating thin films, the simplest of which consist of small molecules that are analogous to wear debris that can "roll" between the sliding surfaces or that represent very short-chain bonded lubricants. These thin films were discussed in the previous subsection. The rest of this subsection will, therefore, focus on the effects of liquids, larger nanoparticles, self-assembled monolayers and solid thin films on lubrication and friction.

Liquids

Liquids are common lubricants and so they have been studied in great depth at the macroscale. At the nanoscale, the tribological response of spherical liquid molecules

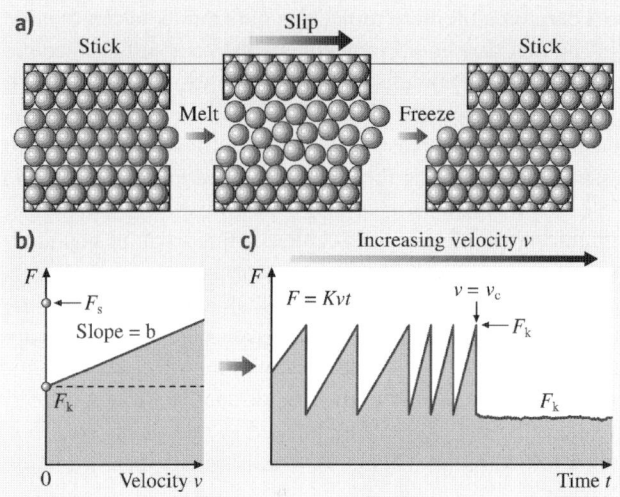

Fig. 13.46a–c. The stick/slip transition that occurs for thin films of liquid between two sliding solid surfaces. F is the intrinsic friction and F_s is the friction where the liquid is in the rigid state; F_k is the friction where the liquid is in the liquid-like state. After [225] with the permission of the ACS (1996)

has been well-characterized experimentally using the SFA and computationally with MD simulations by *Berman* et al. [225]. The SFA experiments considered one to three liquid layers and the stick-slip motion at the interface is found to increase in a quantized fashion as the number of lubricant layers decrease. When no external forces are applied to the system, the sliding stops and the solid–lubricant interactions are strong enough to force the liquid molecules to form a close-packed structure that is ordered. The transformation of the liquid into this solid-like structure causes the two surfaces to effectively bond to each other through the lubricant. When the surfaces start to slide again, lateral shear forces are introduced that steadily increase, which causes the molecules in the liquid to undergo small lateral displacements that change the film thickness. If these shear forces become greater than a critical value, the film disorders in a manner that is analogous to melting. This allows the surfaces to slide easily past each other in a manner that is still quantized. This sequence of events is nicely illustrated in Fig. 13.46 and can be reproduced multiple times for the same system.

Persson et al. [226] used MD simulations with pairwise potentials similar to those in (13.6) to examine the mechanism by which this sharp transition occurs. They find that in the case of sliding on insulating crystal surfaces, the solid-state lubricant may be in a "superlubric" state where the friction is negligible. It is clear from the simulations, however, that any surface defects, even in low concentrations, will disrupt this state and transform the lubricant back into a fluid. In addition, when sliding occurs on metallic surfaces above cryogenic temperatures, the electronic contributions to friction are no longer zero and no superlubric state is possible.

High applied pressures can force the fluid molecules out from between the two confining surfaces [227]. The fact that liquid molecules close to a stiff surface are strongly layered in the direction perpendicular to the surface explains the experimental observation of a ($n \rightarrow n-1$) layer transition, where n is number of monolayers, that is observed as the normal load increases [228]. Nucleation theory is used to calculate the critical pressure and determine the spreading dynamics of the ($n-1$) island.

The reactivity of the liquid molecules are also critically important to boundary layer friction. MD studies by *Persson* et al. [229] show that inert molecules interact weakly with sliding surfaces. Consequently, as the rate of sliding increases, the molecular conversion from the solid state to the liquid state occurs in an abrupt manner. However, when the molecules interact strongly with the surfaces, they undergo a more gradual transition from the solid to the liquid state. *Persson* et al. [34] also considered systems where the molecules are attached to one of the surfaces, which causes the transitions to be abrupt. This is especially true if there are large separations between the chains.

While the studies discussed so far have focused on spherical liquids, most widely used liquid lubricants consist of long-chain hydrocarbons. Nonspherical liquid molecules have more difficulty aligning and solidifying. This is borne out in MD simulations by *Thompson* et al. [231] that show that spherical molecules have higher critical velocities than branched molecules. In particular, the simulations show that when the molecules are branched, the amount of time various parts of the system spend in the sticking and sliding modes changes with sliding rate. The critical velocity can also depend on the number of liquid layers in the film, the structure and relative orientation of the two sliding surfaces, the applied load and the stiffness of the surfaces.

Additional studies by *Landman* et al. [230] used MD simulations with bond-order and EAM potentials coupled with pair-wise potentials similar to (13.6) to study the sliding of two gold surfaces with pyramidal asperities that have straight chain $C_{16}H_{34}$ lubricant molecules trapped between them, as illustrated in Fig. 13.47. An important aspect of this study is that the sliding rate in the simulations is about 10 m/s, which is the same order of magnitude as the scanning speed in a computer disk. As the asperities approach each other, the hydrocarbon molecules begin to form layers. This is reflected in the oscillations in the frictional force shown in Fig. 13.48. When the asperities overlap in height and approach each other laterally, the pressure of the lubricant molecules increases to about 4 GPa which leads to the deformation of the gold asperities.

Glosli and *McClelland* [174] modeled the sliding of two ordered monolayers of alkane chains that are attached to two rigid substrates. This system is shown schematically in Fig. 13.49. The simulations predicted that energy dissipation occurs by a discontinuous plucking mechanism (sudden release of shear strain) or a viscous mechanism (continuous collisions of atoms of opposite films). The specific mechanism that occurs depends on the interfacial interaction strength. In particular, the "pluck" occurs when mechanical energy stored as strain is converted into

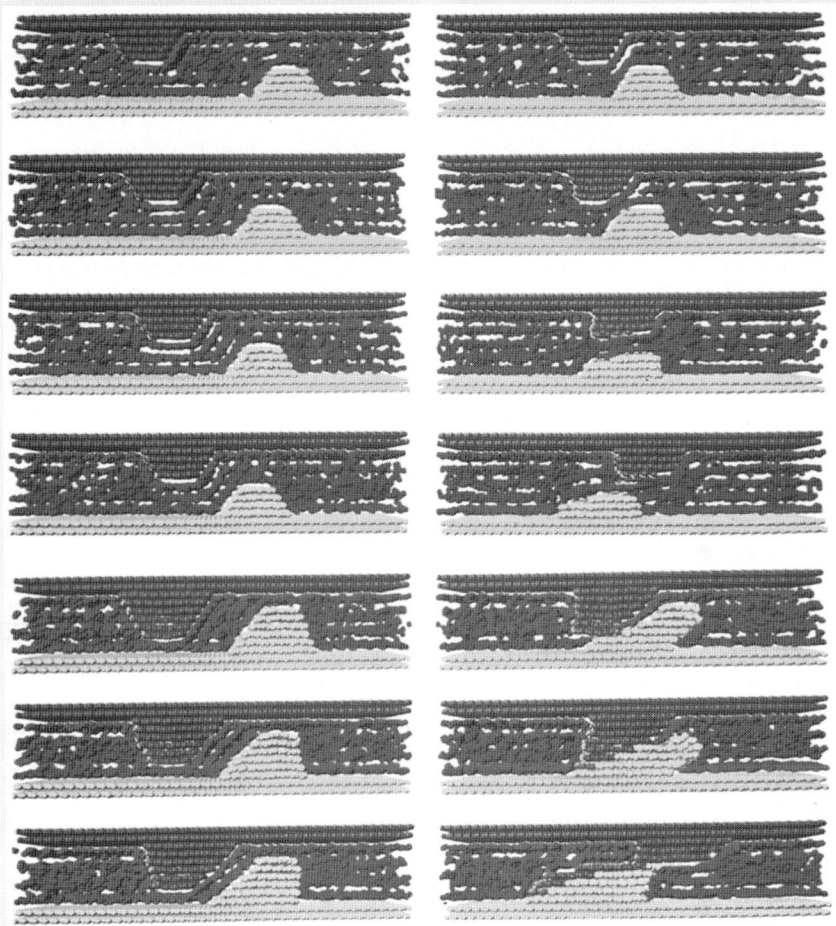

Fig. 13.47. Stills from a molecular dynamics simulation where Au (111) surfaces with surface roughness slide over one another while separated by hexadecane molecules. The scanning velocity is 10 m/s. Layering of the lubricant and asperity deformation occurs as the sliding continues. The *top three rows* show the results when the asperity heights are separated by 4.6 Å. The *bottom three rows* show the results when the asperity heights are separated by −6.7 Å. After [230] with the permission of the ACS (1996)

thermal energy that leads to low friction forces at low temperatures. On the other hand, at higher temperatures some of the energy of sliding is dissipated through phonon excitations, which results in higher frictional forces. Interestingly, this trend reverses again at the highest temperatures considered when the molecules move so much that they slide easily over the surfaces, which decreases the frictional force. These results are summarized in Figs. 13.50 and 13.51.

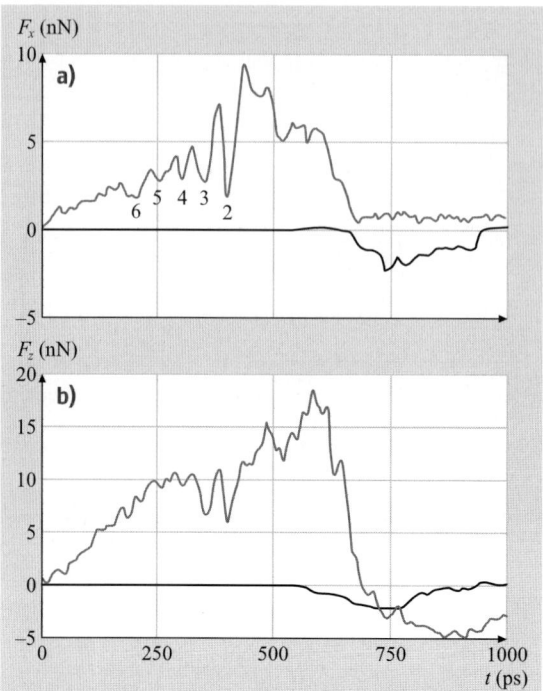

Fig. 13.48a,b. The lateral force (F_x) and normal force (F_z) from the molecular dynamics simulations shown in Fig. 13.35 as a function of time. The forces between the two metal surfaces are shown by the *dashed line*. The force oscillations correspond to the structural changes of the lubricant in Fig. 13.35. After [230] with the permission of the ACS (1996)

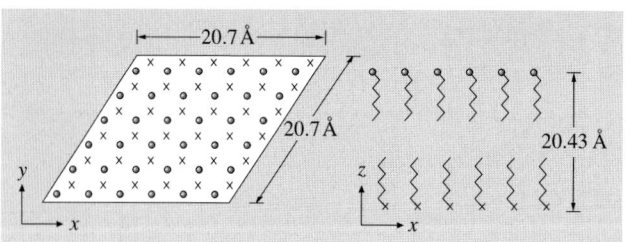

Fig. 13.49. Top and side views of the alkane chains attached to surfaces that are sliding against each other. After [174] with the permission of the APS (1993)

Other studies of sliding surfaces with attached organic chains include MD simulations with LJ potentials by *Müser* and coworkers, [41, 232, 233] which considered friction between polymer "brushes" in sliding contact with one another. In particular, they considered the effect of sliding rate on the tilting of polymers and the effect of steady-state sliding versus non-steady-state ("transient") sliding. The simulations find that shear forces are lower for chains that tilt in a direction that is parallel to the shear direction. This tilting effect is significant for grafted polymers, as illustrated in Fig. 13.52, and less significant for absorbed polymers. This is due to the decrease in the differential frictional coefficient for the grafted polymers as well as the increase in the friction coefficient for absorbed polymers under shear. The tilting is also af-

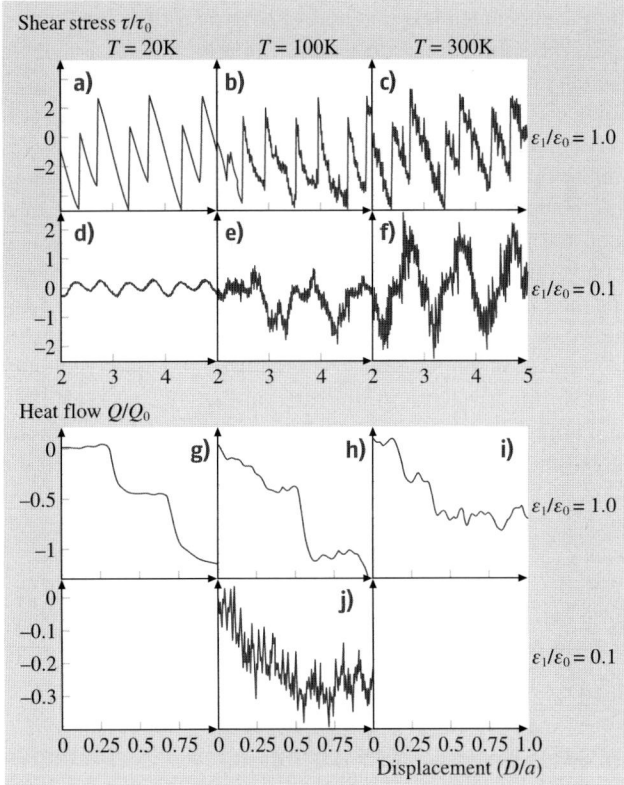

Fig. 13.50a–j. Data from molecular dynamics simulations of the sliding of the surfaces shown in Fig. 13.32. (**a**) – (**f**) The shear stress and (**g**) – (**j**) the heat flow as a function of sliding for normal and reduced interfacial strengths. The *plots* show how the calculated values change with system temperature. After [174] with the permission of the APS (1993)

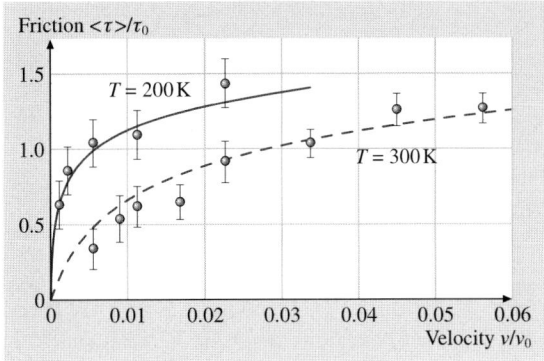

Fig. 13.51. A plot of calculated values of the average interfacial shear stress as a function of the velocity of sliding of the two surfaces shown in Fig. 13.33. After [174] with the permission of the APS (1993)

fected by the rate of sliding and is much larger at high sliding rates than small rates, as indicated in Fig. 13.52. The simulations further show that the inclination angle of the chains decreases much more slowly than the shear stress, and the shear stress maximum is more pronounced if there is hysteresis in the chain orientations.

Typical friction loops for tips that are functionalized and sliding against surfaces that are functionalized in the same manner as illustrated in Fig. 13.25 are shown in Fig. 13.53. The friction force between the OH/OH pairs is significantly larger than the friction force between the CH_3/CH_3 pairs. This is due to the formation and breaking of hydrogen bonds during the shearing for the OH/OH pairs. The mean forces vs. load forces for the OH/OH and CH_3/CH_3 pairs given in Fig. 13.54 are reduced by the tip radius.

MD simulations by *Manias* et al. considered the shearing of entangled oligomer chains that are attached to sliding surfaces, as illustrated in Fig. 13.55 [234]. They find that slip takes place within the film and that this occurs through changes in the chain conformations. Increased viscosity is predicted at the film–surface interface

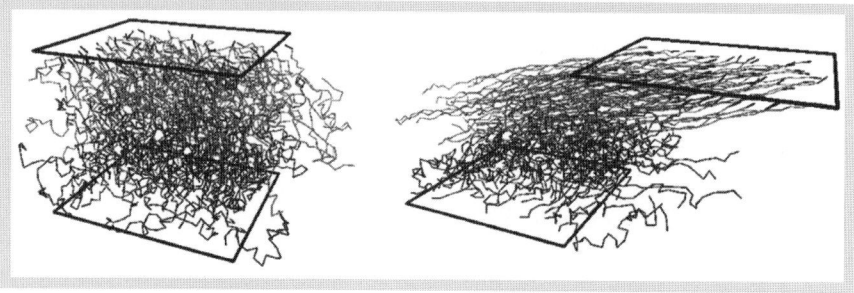

Fig. 13.52. Snapshots of sliding walls with attached polymers in a solvent. *Right-hand figure* illustrates the sliding process at low sliding rates while the *left-hand figure* illustrates the sliding process at high sliding rates. After [41] with the permission of the CCLR (2002)

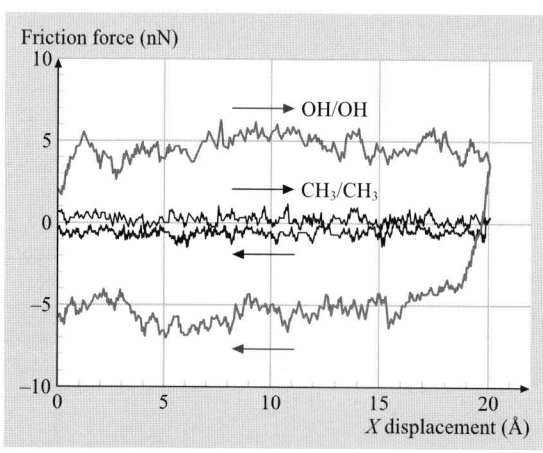

Fig. 13.53. Typical friction loops for the systems shown in Fig. 13.25 for CH_3/CH_3 and OH/OH pairs under a contact load of 0.2 nN. After [160] with the permission of the ACS (2002)

compared to the middle of the film, which results in a range of viscosities across the film as one moves away from the points of sliding contact.

To summarize this section, experiments and MD simulations show similar stick/slip transitions that occur for thin films of liquid between two sliding solid surfaces. Frictional properties are found to depend to a significant degree on molecular shape, whether the molecules are grafted on the surfaces or merely absorbed on them, and the degree of tilting in the case of molecular chains. In the case of long-chain molecules, temperature is found to affect the frictional force because the

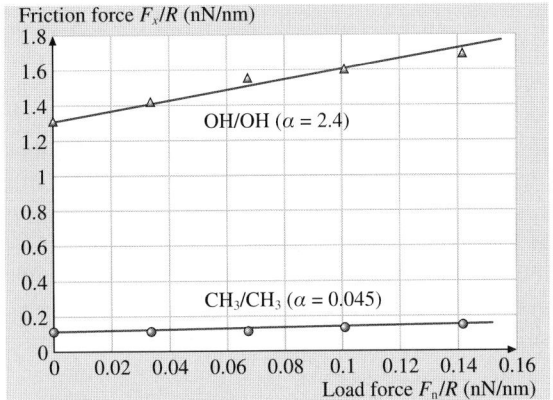

Fig. 13.54. Friction force versus contact load from the systems shown in Fig. 13.25 for CH_3/CH_3 and OH/OH. After [160] with the permission of the ACS (2002)

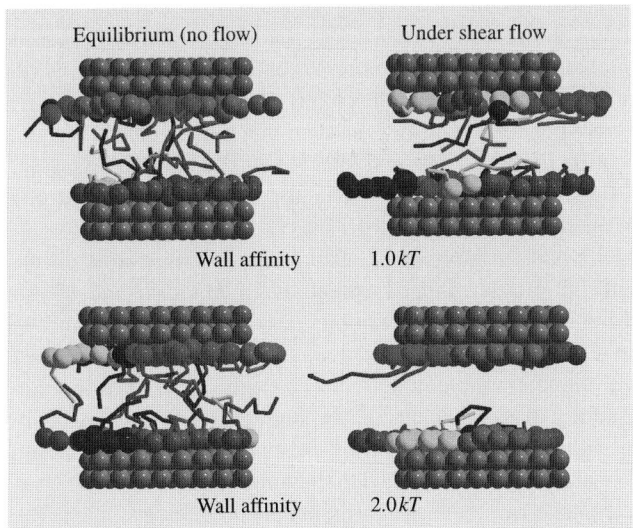

Fig. 13.55. Changes in the conformation of adsorbed hydrocarbon chains on weakly (*top*) and strongly (*bottom*) physisorbing surfaces at equilibrium and under shear. After [234] with the permission of the ACS (1996)

mechanical energy stored in long-chains can be converted into thermal energy by friction.

Self-Assembled Structures

There have been numerous experimental studies of friction on SAMs on solid surfaces with AFM and FFM. The experimental results reveal relationships among elastic compliance, topography and friction on thin LB films [235]. For example, they have detected differences in the adhesive interactions between the microscope tips and CH_3 and CF_3 end groups [106]. Fluorocarbon domains generally exhibit higher friction than the hydrocarbon films, which the authors attribute to the lower elasticity modulus of the fluorocarbon films that results in a larger contact area between the tip and the sample [235–237]. *Perry* and coworkers examined the friction of alkanethiols terminated with $-CF_3$ and $-CH_3$ [238]. The lattice constants for both films are similar and the films are well-ordered. The friction of the SAMs with chains that are terminated with fluorine end groups is larger than the friction of the SAMs with chains that are terminated with hydrogen end groups. However, the pull-off force is similar in both systems, which implies that these end groups have similar contact areas. The authors speculate that the larger $-CF_3$ groups interact more strongly with adjacent chains than the $-CH_3$-terminated chains. Therefore, the fluorinated chains have more modes of energy dissipation within the plane of the monolayer and, thus, have larger friction.

Molecular disorder of the alkyl chains at the surface can also affect the frictional properties of self-assembled films if the layers are not packed too closely together [240]. Indentation can induce disorder in the chains that then compress as the tip continues to press against them. If the tip presses hard enough, the film hardens as a result of the repulsive forces between the chains. However, if the chains are tilted, they bend or deform when the tip pushes on them in a mostly elastic fashion that produces long lubrication lifetimes. At low contact loads of about 10^{-8} N, wear usually occurs at defect sites, such as steps. Wear can also occur if there are strong adhesive forces between the film and the surface [241].

The friction of model SAMs composed of alkane chains was examined using MD simulations with bond-order and LJ potentials by *Mikulski* and *Harrison* [239, 242]. These simulations show that periodicities observed in a number of system quantities are the result of the synchronized motion of the chains when they are in sliding contact with the diamond counterface. The tight packing of the monolayer and commensurability of the counterface are both needed to achieve synchronized motion when sliding in the direction of chain tilt. The tightly packed monolayer is composed of alkane chains attached to diamond (111) in the (2×2) arrangement and the loosely packed system has approximately 30% fewer chains. The average friction at low loads is similar in both the tightly and loosely packed systems at low loads. Increasing the load, however, causes the tightly packed monolayer to have significantly lower friction than the loosely packed monolayer (Fig. 13.56). While the movement of chains is somewhat restricted in both systems, the tightly packed monolayer under high loads is more constrained with respect to the move-

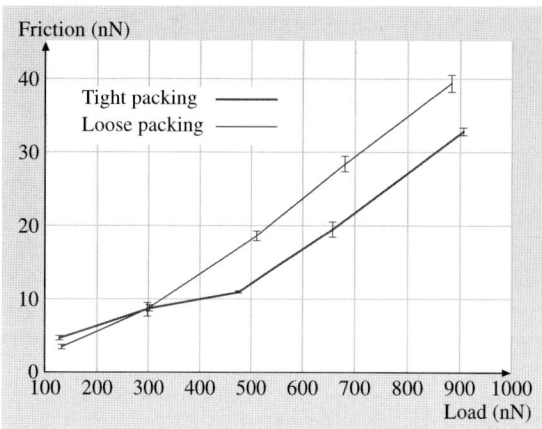

Fig. 13.56. Friction as a function of load when a hydrogen-terminated counterface is in sliding contact with C_{18} alkane monolayers. After [239] with the permission of the ACS (2001)

ment of individual chains than the loosely packed monolayer, as illustrated in Fig. 13.57. Therefore, sliding initiates larger bond-length fluctuations in the loosely packed system, which ultimately lead to more energy dissipation via vibration and, thus, higher friction. Thus, the efficient packing of the chains is responsible for the lower friction observed for tightly packed monolayers under high loads.

Several AFM experiments have examined the friction of SAMs composed of chains of mixed lengths [243]. For example, the friction of SAMs composed of

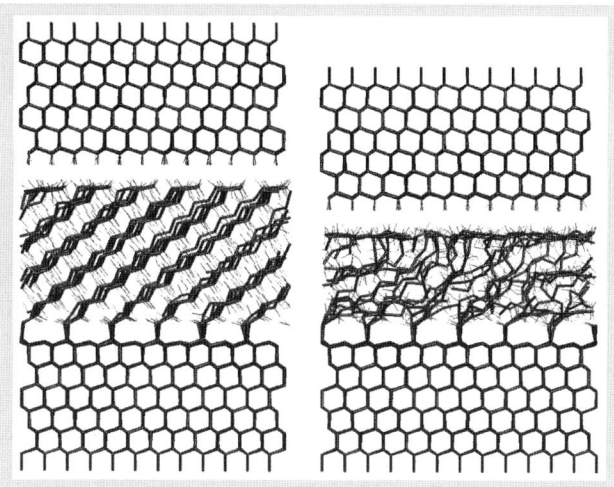

Fig. 13.57. Snapshots of tightly packed C_{18} alkane monolayers on the *left*, and loosely packed monolayers on the *right* under a load of about 500 nN. The chains in both systems are arranged in a (2×2) arrangement on diamond (111). The loosely packed system has 30% of the chains randomly removed. The sliding direction is from *left to right*. After [239] with the permission of the ACS (2001)

spiroalkanedithiols was examined by *Perry* and coworkers [244]. The effects of crystalline order at the sliding interface were examined by systematically shortening some of the chains. The resulting increase in disorder at the sliding interface causes an increase in friction.

The link between friction and disorder in monolayers composed of *n*-alkane chains was recently examined using MD simulations by *Harrison* and coworkers [245]. The tribological behavior of monolayers of 14 carbon atom-containing alkane chains, or pure monolayers, was compared to monolayers that randomly combine equal amounts of 12 and 16 carbon-atom chains, or mixed monolayers. Pure monolayers consistently show lower friction than mixed monolayers when sliding under repulsive (positive) loads in the direction of chain tilt. These MD simulations reproduce trends observed in AFM experiments of mixed-length alkanethiols [243] and spiroalkanedithiols on Au [246].

Because the force on individual atoms is known as a function of time in MD simulations, it is possible to calculate the contact forces between individual monolayer chain groups and the tip, where contact force is defined as the force between the tip and a $-CH_3$ or a $-CH_2-$ group in the alkane chains. The distribution of contact forces between individual monolayer chain groups and the tip are shown in Fig. 13.58. It is clear from these contact force data that the magnitude, or scale of the forces, is similar in both the pure and the mixed monolayers. In addition, it is also apparent that the pure and mixed monolayers resist tip motion in the same way. That is, the shape of the histograms in the positive force intervals is similar. In contrast, the contact forces "pushing" the tip along differ in the two monolayers. The pure monolayers exhibit a high level of symmetry between resisting and pushing forces. Because the net friction is the sum of the resisting and pushing forces, the symmetry in these distributions of the pure monolayers results in a lower net friction than the mixed monolayers. Thus, the ordered, densely packed nature of the pure monolayers allows the energy stored when the monolayer is resisting tip motion (positive forces) to be regained efficiently when the monolayer "pushes" on the tip (negative forces). The distribution of negative contact forces in the mixed monolayers is different from the distribution of the positive forces. For this reason, mechanical energy is not efficiently channeled back into the mixed monolayer as the tip passes over the chains and, as a result, the friction is higher. The range of motion of the chains is monitored by computing the deviation in a chain group's position compared to its starting position, as illustrated in Fig. 13.59. It is clear from analyzing these data that the increased range of motion is linked to large contact forces. The increased range of motion of the protruding tails in the mixed system prevents the efficient recovery of energy during sliding (negative contact force distribution) and allows for the dissipation of energy.

The pure monolayers exhibit marked friction anisotropy. The contact force distribution changes dramatically as a result of the change in sliding direction, resulting in an increase in friction (Fig. 13.58). Sliding in the direction perpendicular to chain tilt can cause both types of monolayers to transition to a state where the chains are

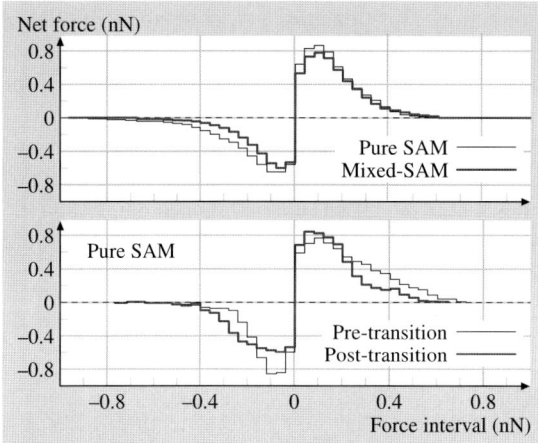

Fig. 13.58. The distribution of contact forces along the sliding direction (friction force). In the *upper panel*, the forces for the mixed and pure system sliding in the direction of chain tilt are shown. The forces for the pure system sliding in the transverse direction to the chain tilt are shown in the *lower panel*. Positive force intervals correspond to chain groups that resist tip motion while negative intervals correspond to chain groups that "push" the tip in the sliding direction. Forces from four runs with independent starting configurations are binned for all sets of data

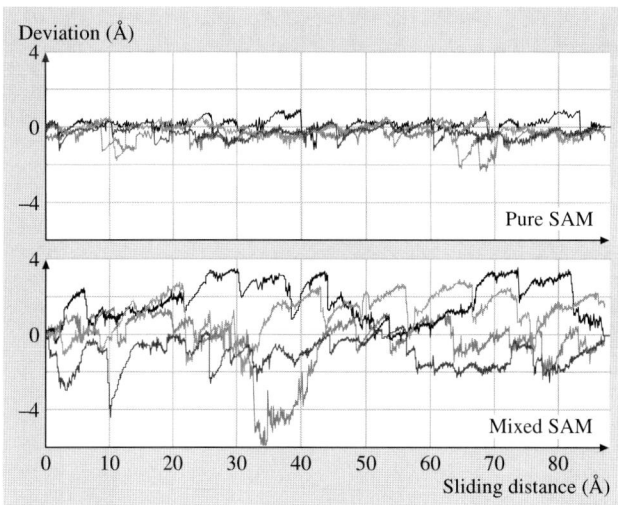

Fig. 13.59. Trajectories of individual chain groups that generate the largest contact forces when sliding in the direction of chain tilt for both the pure and mixed monolayer systems. The deviation is defined as the change in position along the sliding direction relative to the chain group's starting position. (The positions are averaged over 2000 simulation steps)

primarily tilted along the sliding direction. This transition is accompanied by a large change in the distribution of contact forces and a reduction in friction.

Recently, the response of monolayers composed of alkyne chains, which contain diacetylene moieties, to compression and shear [247] was examined using MD simulations. These are the only simulations to date that show that compression and shear can result in cross-linking, or polymerization, between chains. The vertical positioning of the diacetylene moieties within the alkyne chains (spacer length) and the sliding direction both have an influence on the pattern of cross-linking and friction. Compression and shear cause irregular polymerization patterns to be formed among

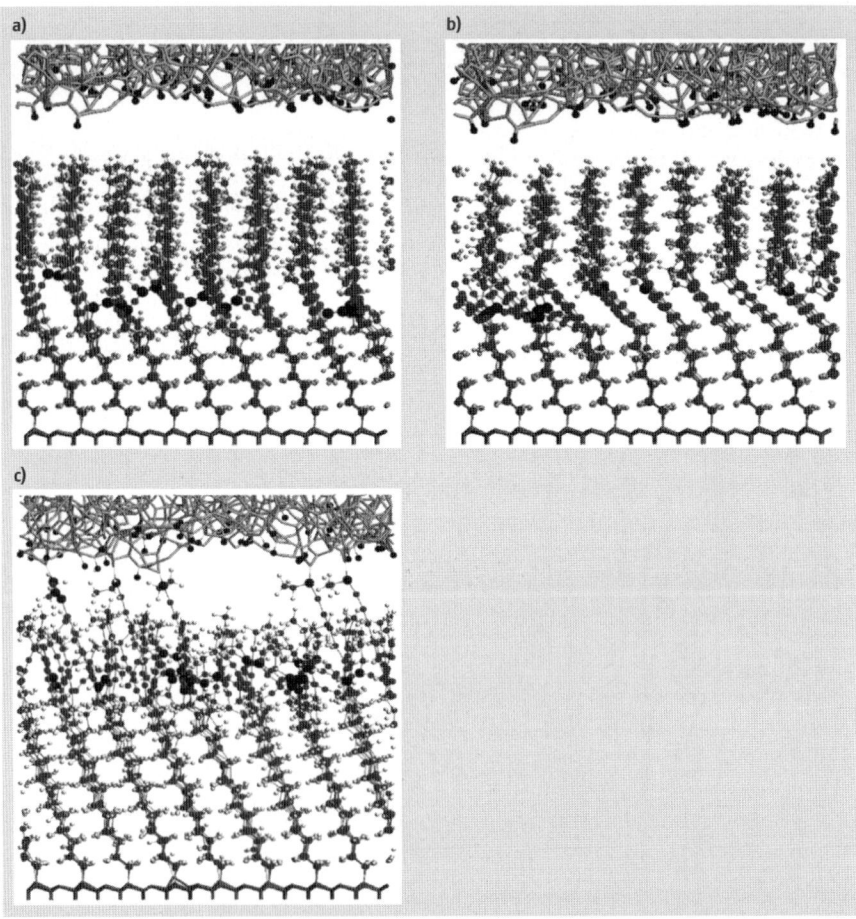

Fig. 13.60. (a) Perpendicular-, (b) tilted-, and (c) end-chain monolayer systems after compression to 200 nN and pull-back of the hydrogen-terminated tip. *Large, dark spheres* in the hydrocarbon monolayers represent cross-linked atoms with sp^2 hybridization. *Dark, small spheres* represent hydrogen atoms that are initially on the hydrogen-terminated amorphous carbon tip. After [247] with the permission of the ACS (2004)

the carbon backbones, as illustrated in Fig. 13.60. When diacetylene moieties are located at the ends of the chains closest to the tip, chemical reactions between the chains of the monolayer and the amorphous carbon tip occur causing the friction to increase 100 times, as indicated in Fig. 13.61. The friction between the amorphous carbon tip and all of the diacetylene-containing chains is larger than the friction between a hydrogen-terminated diamond counterface and tightly packed monolayers composed of *n*-alkane chains [239]. This is attributed to the disorder at the interface caused by the irregular counterface.

Zhang et al. [248] used MD simulations to study the effect of confined water between alkyl monolayers terminated with $-CH_3$ (hydrophobic) and $-OH$ (hydrophilic) groups on Si(111), as illustrated in Fig. 13.62. For the hydrophobic molecules, the friction coefficient is almost constant independent of the number of water molecules. For the hydrophilic molecules, the friction coefficient decreases rapidly with an increase in the number of water molecules, as shown in Fig. 13.63. These results are in good agreement with surface force microscopy (SFM) experimental results. *Zhang* et al. [249] also studied the friction of alkanethiol SAMs on gold using hybrid molecular simulations at the same time scales as are used in AFM and FFM experiments. Various quantities were varied in the simulations, including chain length, terminal group, scan direction and scan velocity. The simulations showed that the frictional force decreases as the chain length increases and is smallest when scanned along the tilt direction. They also predicted a maximum friction coefficient for hydrophobic $-CH_3$-terminated SAMs and low friction coefficients for hydrophilic $-OH$-terminated SAMs as the scan velocity increases. The simulat-

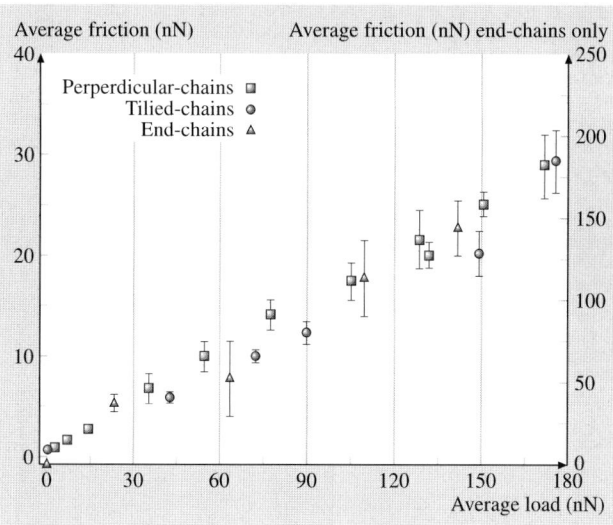

Fig. 13.61. Average friction on the tip as a function of load for the monolayer systems shown in Fig. 13.60. The *scale* for the average friction in the end-chain system is shown on the *right-hand side* of the figure. After [247] with the permission of the ACS (2004)

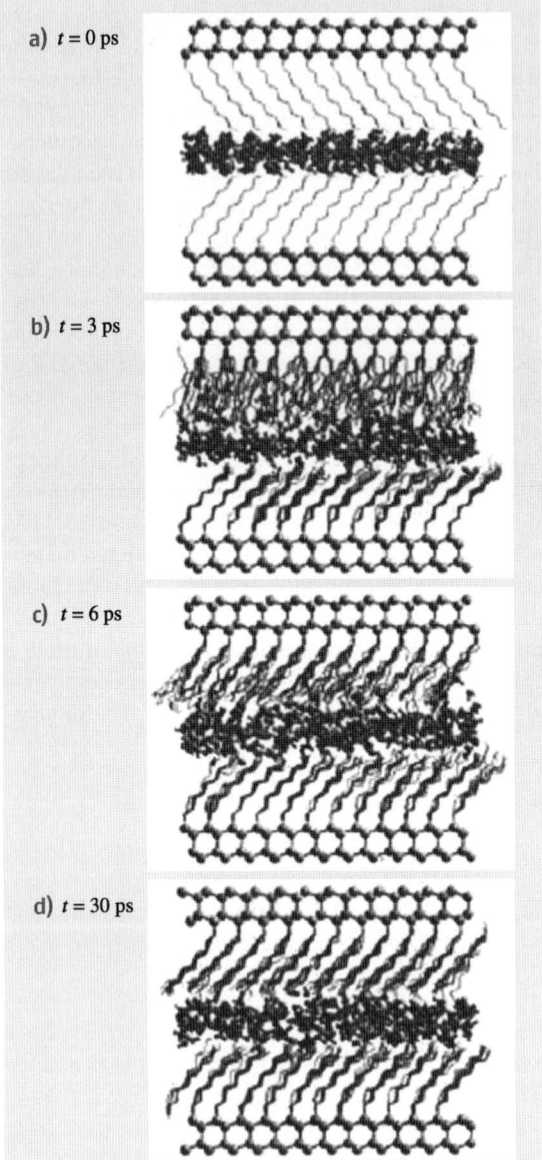

Fig. 13.62a–d. Snapshots of hydrophilic monolayers and confined water molecules from MD simulations at 300 K. The tilt direction of monolayers on the top plate changed after $t = 10.0$ ps. After [248] with the permission of the AIP (2005)

ions further predicted a saturated constant value at high scan rates for both surfaces. These results are summarized in Figs. 13.64 and 13.65.

The work of *Chandross* et al. [250, 251] illustrates the effects of chain length on friction and stick-slip behavior between two ordered SAMs consisting of alkylsilane chains over a range of shear rates at various separation distances or pressure, as illustrated in Fig. 13.66. The adhesion forces between the two SAMs at the same

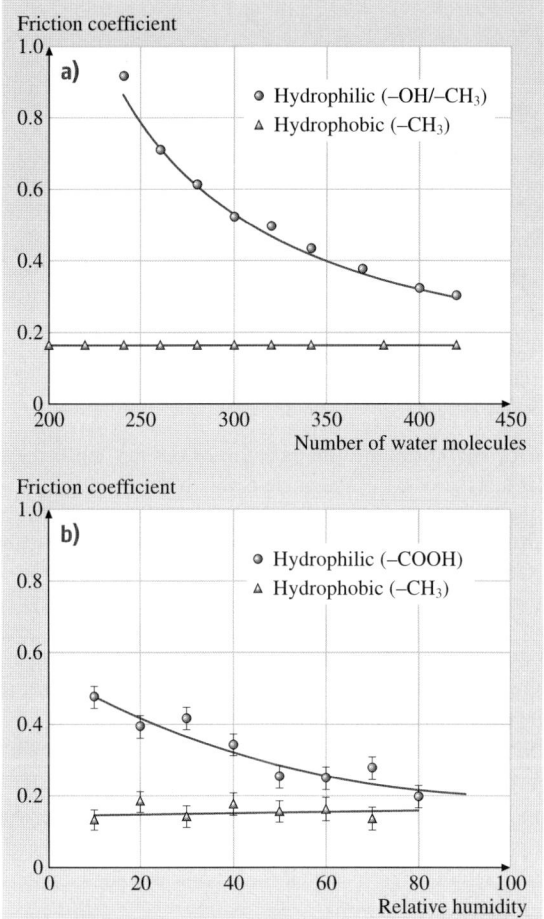

Fig. 13.63. (a) Friction coefficients for hydrophobic ($-CH_3$) and hydrophilic (50% mixed $-CH_3/-OH$) monolayers as a function of water molecules from MD simulations at 300 K ($H = 6.0$ Å), and (b) scanning force microscopy measurements of frictional forces of difference surfaces under various relative humidities. After [248] with the permission of the AIP (2005)

separation distance decrease as the chain length increases from 6 to 18 carbon atoms. However, the friction forces are independent of the chain length and the shear velocity. The system size is shown to have an effect on the sharpness of the slip transitions but not on the dynamical events, as shown in Fig. 13.67.

In short, atomic-scale simulations show the relationship between elastic properties, degree of molecular disorder and friction of self-assembled thin films that illuminates the origin of the properties that are measured experimentally.

Nanoparticles

Nanoparticles are being considered for a wide variety of applications, including as fillers for nanocomposite materials, novel catalysts or catalytic supports, and components for nanometer-scale electronic devices [252]. They have also generated considerable interest as possible new lubricant materials that have the potential to func-

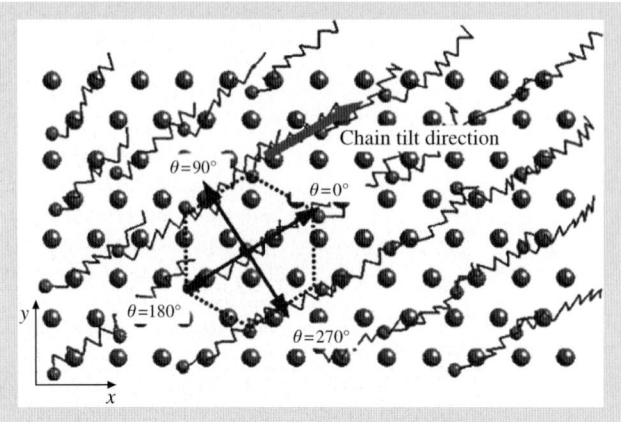

Fig. 13.64. Schematic illustration of the chain tilt and scan directions on alkanethiol SAMs/Au(111) in hybrid molecular simulations; θ is the angle between the tip moving direction and the chain tilt direction. The *larger spheres* represent substrate Au atoms, smaller spheres sulfur atoms in molecular chains, and *zigzag lines* molecular chains. After [249] with the permission of the ACS (2003)

tion as "nanoballbearings" with exceptionally low friction coefficients. The nanoparticles of most interest for tribological applications include C_{60} [253–265], carbon nanotubes [266–273], and MoO_3 nanoparticles [274, 275], among others [276].

The experiments report wide variations in frictional coefficients (for instance, values of 0.06 to 0.9 have been measured for C_{60}) that may be caused by differences in the experimental methods used, the thickness of the nanoparticle layer or island, the atmosphere (argon versus air, levels of humidity) used, and the transfer of nanoparticles to the FFM tips. As a result, there is much that remains to be clarified about the tribological behavior of nanoparticle films.

In the case of C_{60}, the mechanistic response to applied shear forces has not been definitively determined. For example, some experimental studies show evidence of C_{60} molecules rolling against the substrate, each other, or the sliding surfaces [253, 258, 260, 263, 265] while others hypothesize that the low friction of C_{60} films is due in part to blunting of the tip by transfer of fullerene molecules to the tip apex. Fullerene films are found experimentally to have dissipation energies and shear strengths that are a full order of magnitude lower than the values that are typical for boundary lubricants [277]. Experimental testing of the frictional properties of fullerenes reveal low mechanical stability accompanied by progressive wear and transfer of fullerene materials when they are only physisorbed on a solid surface [278]. Furthermore, measurements with a FFM show that under certain conditions, adsorbed fullerene films deteriorate at pressures as low as about 0.1 GPa [279]. The challenge is therefore to obtain mechanically stable, ordered molecular films of fullerenes firmly attached to a solid substrate.

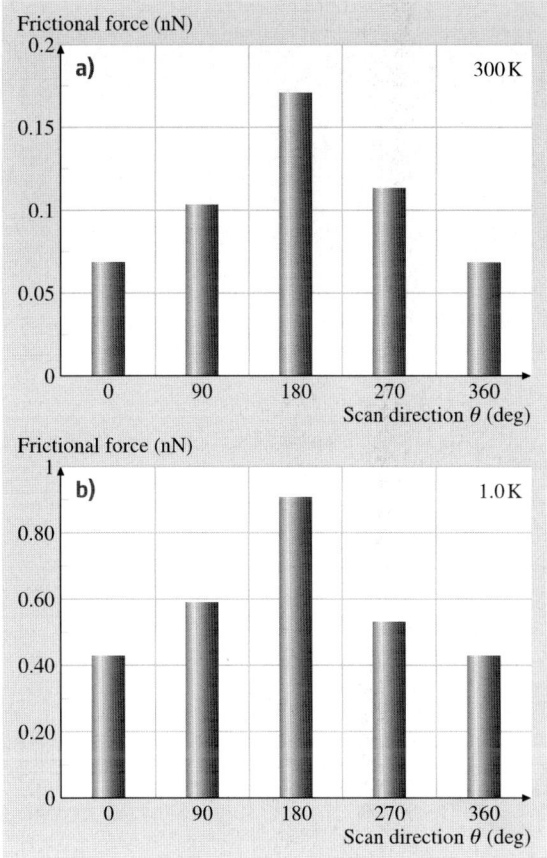

Fig. 13.65. Frictional force as a function of scan direction from hybrid simulations for $C_{11}CH_3$ SAMs on Au(111) at (**a**) 300 K and (**b**) 1.0 K. Frictional force is the smallest when scanned along the tilt direction, the largest when scanned against the tilt direction, and between when scanned perpendicular to the tilt direction at both temperatures. After [249] with the permission of the ACS (2003)

There have been several MD simulation studies to investigate the tribological properties of fullerenes. A representative study by *Legoas* et al. [280] investigated the experimentally observed low-friction system of C_{60} molecules positioned on highly oriented pyrolytic graphite. The results show that decreasing the van der Waals interaction between a C_{60} monolayer and graphite sheets, and the characteristic movements of graphite flakes over C_{60} monolayers, explains the measured ultralow friction of C_{60} molecules and graphite sheets.

Several MD simulation studies have also been carried out on the tribological properties of carbon nanotubes. For example, simulations by *Buldum* et al. and *Schall* et al. [266, 269] indicate that single-wall carbon nanotubes roll when their honeycomb lattice is "in registry" with the honeycomb lattice of the graphite. If this registry is not present, the carbon nanotubes respond to applied forces from an AFM by sliding. This behavior is nicely summarized in Fig. 13.68. These MD simulation findings were simultaneously confirmed in experimental studies by *Falvo* et al. [268]. Experimental studies of multiwall carbon nanotubes on graphite [272] show similar evidence of nanotube rolling when the outer tube is pushed.

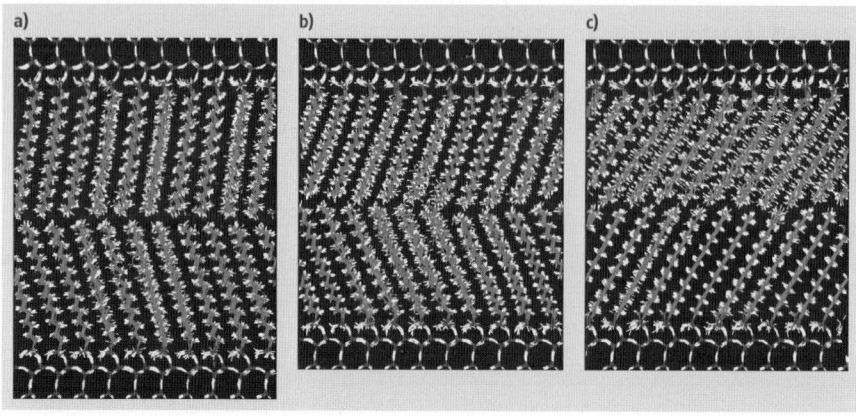

Fig. 13.66. Wireframe images of $n = 18$ SAMs at fixed separations of (**a**) $d = -5.2$ Å (low pressure, under compression only) (**b**) $d = -10.2$ Å (high pressure, under compression only) and (**c**) $d = -10.2$ Å (high pressure, under shear). After [251] with the permission of the ACS (2005)

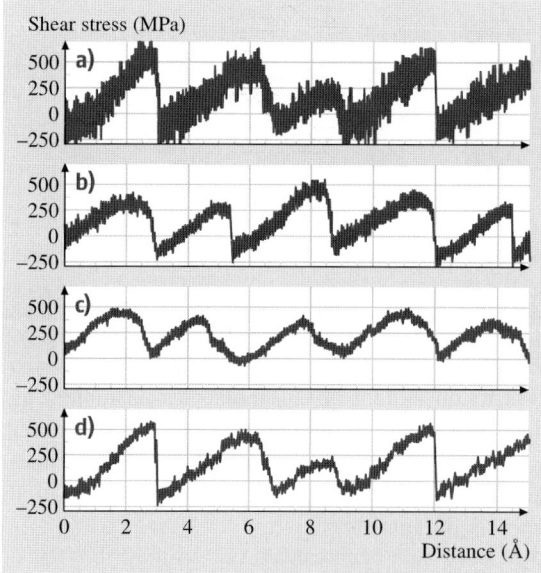

Fig. 13.67a–d. Shear stress σ_s as a function of system size for $n = 6$ SAMs corresponding to a pressure of 200 MPa at $v = 1.0$ m/s: (**a**) 100 chains per surface, (**b**) 400 chains per surface, (**c**) 1600 chains per surface, and (**d**) 16 point box average of system with 100 chains per surface. After [251] with the permission of the ACS (2002)

The tribological properties of nanotube bundles are important, as it is well-known that carbon nanotubes agglomerate together very readily to form bundles and are often grown in bundle form [252]. An experimental study by *Miura* et al. [273] of carbon nanotube bundles being pushed around on a KCl surface with an AFM

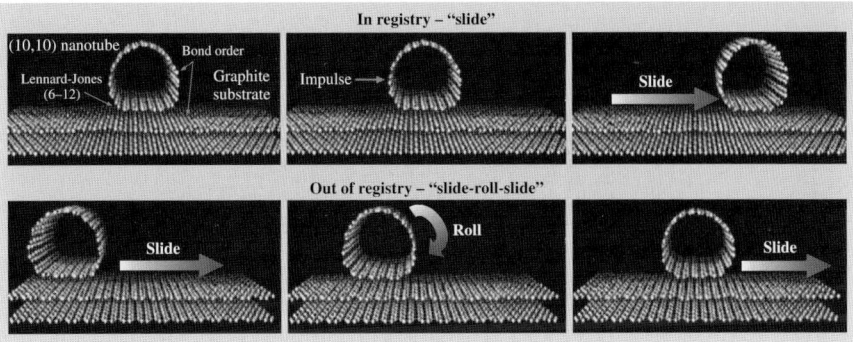

Fig. 13.68. Dynamics of a nanotube on a graphite surface. When the nanotube and graphite plane are out of registry, the nanotube slides as it slows down from an initial impulse (*upper right panel*). When the nanotube is oriented such that it is in registry with the graphite, it slows by a combination of rolling and sliding

tip indicates that bundles of single-wall carbon nanotubes can be induced to roll in a manner that is similar to the rolling observed for multiwall nanotubes.

MD simulations by *Ni* et al. [270, 271] considered the responses of horizontally and vertically aligned single-wall carbon nanotubes between two hydrogen-terminated diamond surfaces, where the top surface is slid relative to the bottom surface. The movement of the carbon nanotubes in response to the shear forces was predicted to be simple sliding for both orientations. Interestingly, the simulations do not predict rolling of the horizontally arranged carbon nanotubes even when they are aligned with each other in two-layer and three-layer structures. Instead, at low compressive forces, illustrated in Fig. 13.69, the nanotube bundles slide as a single unit, and at high compressive forces, also illustrated in Fig. 13.69, the deformed carbon nanotubes closest to the topmost moving diamond surface start to slide in a motion reminiscent of the movement of a tank or bulldozer wheel belt. However, when these moving carbon nanotube atoms would have turned the first corner at the top of the ellipse, they encounter the neighboring nanotube and cannot slide past it. This causes them to deform even further, form cross-links with one another, and, in some cases, move in the reverse direction to the sliding motion of the diamond surface. This causes the large oscillations in the normal and lateral forces plotted in Fig. 13.69.

The responses of the horizontally arranged carbon nanotubes are substantially different from the responses of the vertically arranged nanotubes at high compression, as can be seen by comparing Figs. 13.69 and 13.70. The vertical, capped carbon nanotubes are quite flexible and bend and buckle in response to applied forces. As the buckle is forming, the normal force decreases then stabilizes in the buckled structure, as illustrated in Fig. 13.70. As the topmost diamond surface slides, the buckled nanotubes swing around the buckle "neck" which helps dissipate the applied stresses. For this reason, the magnitudes of the lateral forces are not signifi-

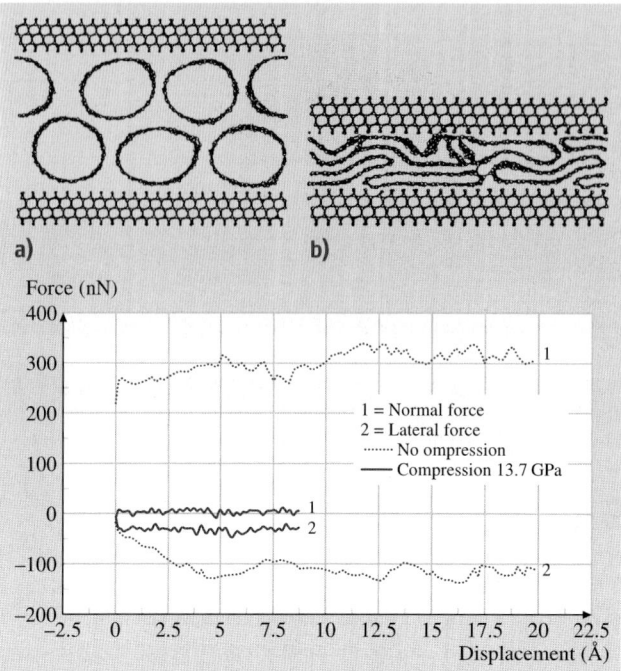

Fig. 13.69. *Top*: Snapshots from simulations that examine the sliding of the topmost diamond surface on horizontally arranged nanotubes at different compressions; (**a**) is at a pressure of ≈ 0 GPa; (**b**) is with a pressure of 13.7 GPa. *Bottom*: Plots of the normal and lateral components of force during sliding of the top diamond's surface on horizontally arranged nanotubes as a function of the displacement of the top diamond surface with respect to the diamond surface on the bottom

cantly different for the vertical nanotubes at low and high compression, as indicated in Fig. 13.70.

When the ratio of the frictional (lateral) force to the normal force is taken to calculate friction coefficients for these systems, high, nonintuitive values were obtained. As outlined by *Ni* et al. [270], this is because the actual contact area of the nanotubes is not proportional to the sliding force. In the case of the horizontal nanotube bundles, the tubes are able to deform and significantly change their contact area with the sliding surface with minimal change in the normal force, as shown in Fig. 13.69. In the case of the vertical nanotubes, the contact area remains approximately the same regardless of the initial loading force because of the flexibility of the nanotubes. This causes the lateral forces to change only slightly with significant changes in the normal force, as shown in Fig. 13.70. This analysis indicates that care must be taken in calculating friction coefficients for nanotube systems. Recent experiments by *Dickrell* et al. [281] show good agreement with these predictions, as shown in Fig. 13.71.

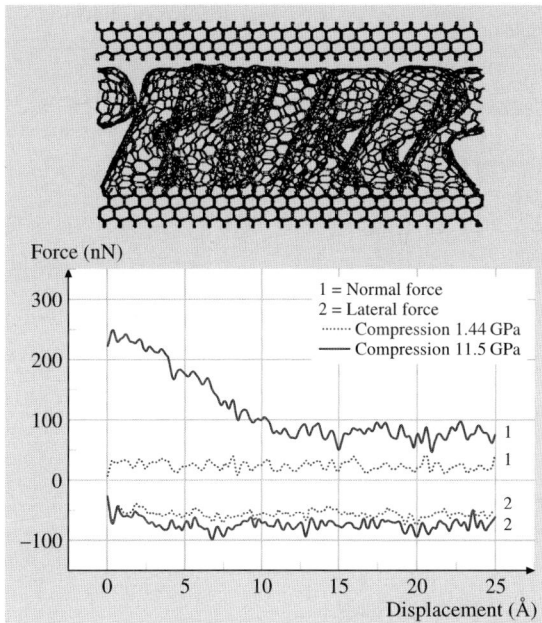

Fig. 13.70. *Top*: Snapshots from simulations that examine the sliding of the topmost diamond surface on vertically arranged nanotubes with one set of capped ends compressed at a pressure of 11.5 GPa. *Bottom*: Plots of the normal and lateral components of force during sliding of the top diamond's surface on vertically arranged nanotubes as a function of the displacement of the top diamond surface with respect to the diamond surface on the bottom

To summarize, this section shows that nanoparticles show some promise as lubricating materials due to their exceptionally low friction coefficients in experiments and simulations. Some nanopaticles show lattice-directed sliding on substrates due to their unique atomic structures. However, there is much that remains to be done before the nanometer-scale friction of these materials is well understood.

Solid State

Surfaces are able to slide over each other at high loads with a minimum of resistance from friction in the presence of liquid lubricants. Some solid thin films can also fulfill these functions and, when they do, are termed solid lubricants. Solid lubricants are generally defined as having friction coefficients of 0.3 or less and low wear.

Bowden and *Tabor* showed how thin solid films can reduce friction as follows [283]. The total friction force F_f is given as

$$F_f = AF_S + F_p, \tag{13.11}$$

where F_p is the plowing term, A is the area of contact and F_S is the shear strength of the interface. If the surfaces are soft, F_S will be reduced while the other parameters will increase. However, if the surfaces under the solid film are very stiff, A and F_p will decrease thereby decreasing friction. The properties specific to the film will also have an effect on friction. For instance, if the films are less than 1 μm thick, the surface asperities will be able to break through the film to eventually cause wear between the surfaces under normal circumstances. On the other hand, if the lubricant

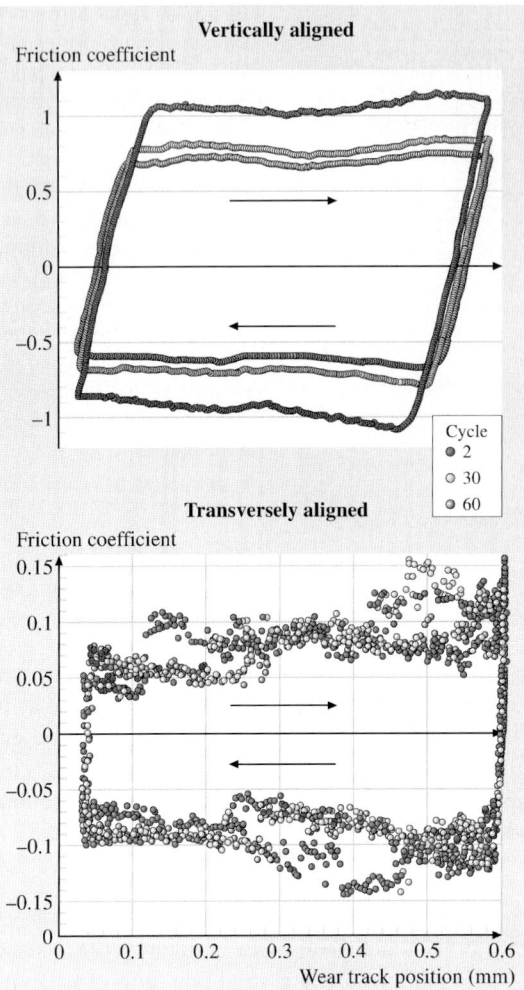

Fig. 13.71. Coefficient of friction data versus track position collected for one full cycle of reciprocating sliding for nanotubes that are vertically and transversely aligned. After [281]

film is too thick, there will be increased plowing and wear that causes the frictional forces to increase. It is important that the lubricant not delaminate in response to the frictional forces, so strong bonds between the lubricant and the surface are required for a solid state lubricant to be effective.

The most common materials used as solid lubricants have layered structures like graphite or MoS_2, that, as discussed above, experience low friction. It is not necessary for the lubricant film to have a layered structure to give low friction. For example, diamond-like carbon has some of the lowest coefficients of friction measured and yet does not have a layered structure. Similarly, not all layered structures are lubricants. For instance, mica gives a relatively high coefficient of friction (> 1).

The atomic-scale tribological behavior that occurs when a hydrogen-terminated diamond (111) counterface is in sliding contact with amorphous, hydrogen-free,

DLC films was examined using MD simulations by *Gao* et al. [282]. Two films, with approximately the same ratio of sp^3–sp^2 carbon but different thicknesses, were examined. Similar average friction was obtained from both films in the load range examined. A series of tribochemical reactions occur above a critical load that result in a significant restructuring of the film, which is analogous to the "run-in" observed in macroscopic friction experiments, and reduces the friction. The contribution of adhesion between the counterface and the sample to friction is examined by varying the saturation of the counterface. The friction increases when the degree of saturation of the diamond counterface is reduced by randomly removing hydrogen atoms. Lastly, two potential energy functions that differ only in their long-range forces are used to examine the contribution of long-range interactions to friction in the same system (as illustrated in Figs. 13.72 and 13.73).

MD simulations were also recently used by *Gao* et al. [284] to examine the effects of the sp^2–sp^3 carbon ratio and surface hydrogen on the mechanical and tribological properties of amorphous carbon films. This work showed that, in addition to the sp^2–sp^3 ratio of carbon, the three-dimensional structures of the films are

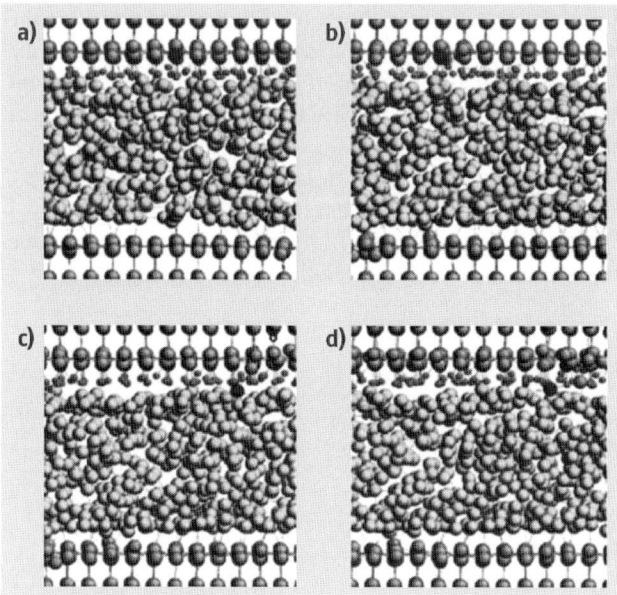

Fig. 13.72a–d. A series of chemical reactions induced by sliding of the counterface over the thin film under an average load of 300 nN. (**a**) The sliding causes the rupture of a carbon–hydrogen bond in the counterface. (**b**) The hydrogen atom is incorporated into the film and forms a bond to a carbon atom in the film. (**c**) A bond is formed between the unsaturated carbon atoms in the film and the carbon that suffered the bond rupture in the counterface, and continued sliding causes this carbon to be transferred into the film. (**d**) The transferred carbon forms a bond with another carbon in the counterface. The counterface has slid 0.0 (**a**), 15.9 (**b**), 26.1 (**c**), and 30.5 Å (**d**). After [282] with the permission of the ACS (2002)

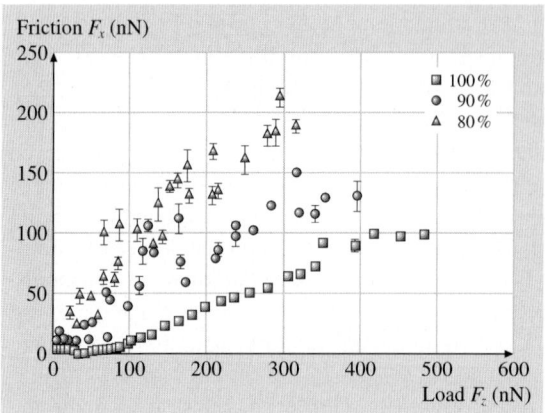

Fig. 13.73. Friction curves for the thin film system with a counterface that is 100% hydrogen-terminated (*open squares*), 90% hydrogen-terminated (*filled squares*), and 80% hydrogen-terminated (*open circles*). After [282] with the permission of the ACS (2002)

important when determining the mechanical properties of the films. For example, it is possible to have high sp^2-carbon content, which is normally associated with softer films, and large elastic constants. This occurs when sp^2-ringlike structures are oriented perpendicular to the compression direction. The layered nature of the amorphous films examined leads to novel mechanical behavior that influences the shape of the friction versus load data, as illustrated in Fig. 13.74. When load is applied to the films, the film layer closest to the interface is compressed. This results in the very low friction of films I and II up to approximately 300 nN and the response of films IV and V up to 100 nN. Once the outer film layers have been compressed, additional application of load causes an almost linear increase in friction for films I and II as well as IV and V. Film III has an erratic friction versus load response due to the early onset of tribochemical reactions between the tip and the film.

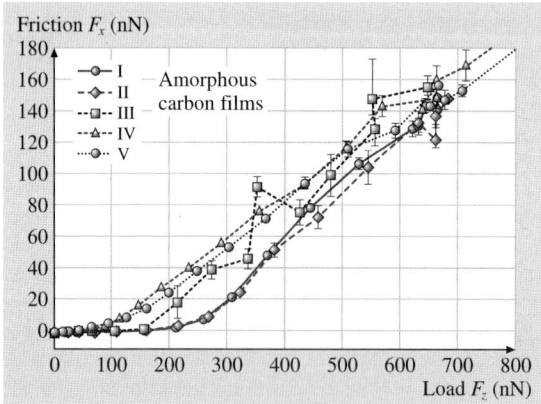

Fig. 13.74. Average friction versus load for five amorphous carbon films. Films I–III are hydrogen-free and contain various ratios of sp^2- to-sp^3 carbon. Films IV and V are both over 90% sp^3 carbon and have surface hydrogenation

13.5 Conclusions

This chapter provides a wide-ranging discussion of the background of MD and related simulation methods, their role in the study of nanometer-scale indentation and friction, and their contributions to these fields. Specific, illustrative examples are presented that show how these approaches are providing new and exciting insights into mechanisms responsible for nanoindentation, atomic-scale friction, wear, and related atomic-scale and molecular scale processes. The examples also illustrate how the results from MD and related simulations are complementary to experimental studies, serve to guide experimental work, and assist in the interpretation of experimental data. The ability of these simulations and experimental techniques such as the surface force apparatus and proximal probe microscopes to study nanometer-scale indentation and friction at approximately the same scale is revolutionizing our understanding of the origin of friction at its most fundamental atomic level.

References

1. D. Dowson: *History of Tribology* (Longman, London 1979)
2. K.L. Johnson, K. Kendell, A.D. Roberts: Surface energy and the contact of elastic solids, Proc. R. Soc. Lond. A **324**, 301–313 (1971)
3. M. Gad-el-Hak (Ed.): *The MEMS Handbook*, Mech. Eng. Handbook Ser. (CRC, Boca Raton 2002)
4. J. Krim: Friction at the atomic scale, Sci. Am. **275**, 74–80 (1996)
5. J. Krim: Atomic-scale origins of friction, Langmuir **12**, 4564–4566 (1996)
6. J. Krim: Progress in nanotribology: experimental probes of atomic-scale friction, Comment Cond. Mat. Phys. **17**, 263–280 (1995)
7. A.P. Sutton: Deformation mechanisms, electronic conductance and friction of metallic nanocontacts, Curr. Opin. Sol. St. Mater. Sci. **1**, 827–833 (1996)
8. C.M. Mate: Force microscopy studies of the molecular origins of friction and lubrication, IBM J. Res. Dev. **39**, 617–627 (1995)
9. A.M. Stoneham, M.M.D. Ramos, A.P. Sutton: How do they stick together – the statics and dynamics of interfaces, Philos. Mag. A **67**, 797–811 (1993)
10. I.L. Singer: Friction and energy dissipation at the atomic scale: A review, J. Vacuum Sci. Technol. A **12**, 2605–2616 (1994)
11. B. Bhushan, J.N. Israelachvili, U. Landman: Nanotribology – friction, wear and lubrication at the atomic scale, Nature **374**, 607–616 (1995)
12. J.A. Harrison, D.W. Brenner: *Handbook of Micro/Nanotechnology*, ed. by B. Bhushan (CRC, Boca Raton 1995)
13. J.B. Sokoloff: Theory of atomic level sliding friction between ideal crystal interfaces, J. Appl. Phys. **72**, 1262–1270 (1992)
14. W. Zhong, G. Overney, D. Tomanek: Theory of atomic force microscopy on elastic surfaces. In: *The Structure of Surfaces III: Proc. 3rd Int. Conf. on the Structure of Surfaces*, Vol. 24, ed. by S.Y. Tong, M.A.V. Hove, X. Xide, K. Takayanagi (Springer, Berlin, Heidelberg 1991) p. 243
15. J.N. Israelachvili: Adhesion, friction and lubrication of molecularly smooth surfaces. In: *Fundamentals of Friction: Macroscopic and Microscopic processes*, ed. by I.L. Singer, H.M. Pollock (Kluwer, Dordrecht 1992) pp. 351–385

16. S.B. Sinnott: Theory of atomic-scale friction. In: *Handbook of Nanostructured Materials and Nanotechnology*, Vol. 2, ed. by H. Nalwa (Academic, San Diego 2000) pp. 571–618
17. S.-J. Heo, S.B. Sinnott, D.W. Brenner, J.A. Harrison: Computational modeling of nanometer-scale tribology. In: *Nanotribology and Nanomechanics*, ed. by B. Bhushan (Springer, Berlin, Heidelberg 2005)
18. G. Binnig, C.F. Quate, C. Gerber: Atomic force microscope, Phys. Rev. Lett. **56**, 930–933 (1986)
19. C.M. Mate, G.M. Mcclelland, R. Erlandsson, S. Chiang: Atomic-scale friction of a tungsten tip on a graphite surface, Phys. Rev. Lett. **59**, 1942–1945 (1987)
20. G.J. Germann, S.R. Cohen, G. Neubauer, G.M. Mcclelland, H. Seki, D. Coulman: Atomic-scale friction of a diamond tip on diamond (100) surface and (111) surface, J. Appl. Phys. **73**, 163–167 (1993)
21. R.W. Carpick, M. Salmeron: Scratching the surface: Fundamental investigations of tribology with atomic force microscopy, Chem. Rev. **97**, 1163–1194 (1997)
22. J.N. Israelachvili: *Intermolecular and surface forces: With applications to colloidal and biological systems* (Academic, London 1992)
23. J. Krim, D.H. Solina, R. Chiarello: Nanotribology of a Kr monolayer – a quartz crystal microbalance study of atomic-scale friction, Phys. Rev. Lett. **66**, 181–184 (1991)
24. G.A. Tomlinson: A molecular theory of friction, Philos. Mag. Ser. 7 **7**, 905–939 (1929)
25. F.C. Frenkel, T. Kontorova: On the theory of plastic deformation and twinning, Zh. Eksp. Teor. Fiz. **8**, 1340 (1938)
26. G.M. McClelland, J.N. Glosli: Friction at the atomic scale. In: *Fundamentals of friction: Macroscopic and microscopic processes*, ed. by I.L. Singer, H.M. Pollock (Kluwer, Dordrecht 1992) pp. 405–422
27. J.B. Sokoloff: Theory of dynamical friction between idealized sliding surfaces, Surf. Sci. **144**, 267–272 (1984)
28. J.B. Sokoloff: Theory of energy dissipation in sliding crystal surfaces, Phys. Rev. B **42**, 760–765 (1990)
29. J.B. Sokoloff: Possible nearly frictionless sliding for mesoscopic solids, Phys. Rev. Lett. **71**, 3450–3453 (1993)
30. J.B. Sokoloff: Microscopic mechanisms for kinetic friction: Nearly frictionless sliding for small solids, Phys. Rev. B **52**, 7205–7214 (1995)
31. J.B. Sokoloff: Theory of electron and phonon contributions to sliding friction. In: *Physics of Sliding Friction*, ed. by B.N.J. Persson, E. Tosatti (Kluwer, Dordrecht 1996) pp. 217–229
32. J.B. Sokoloff: Static friction between elastic solids due to random asperities, Phys. Rev. Lett. **86**, 3312–3315 (2001)
33. J.B. Sokoloff: Possible microscopic explanation of the virtually universal occurrence of static friction, Phys. Rev. B **65**, 115415 (2002)
34. B.N.J. Persson, D. Schumacher, A. Otto: Surface resistivity and vibrational damping in adsorbed layers, Chem. Phys. Lett. **178**, 204–212 (1991)
35. A.I. Volokitin, B.N.J. Persson: Resonant photon tunneling enhancement of the van der Waals friction, Phys. Rev. Lett. **91**, 106101 (2003)
36. A.I. Volokitin, B.N.J. Persson: Noncontact friction between nanostructures, Phys. Rev. B **68**, 155420 (2003)
37. A.I. Volokitin, B.N.J. Persson: Adsorbate-induced enhancement of electrostatic noncontact friction, Phys. Rev. Lett. **94**, 086104 (2005)
38. J.S. Helman, W. Baltensperger, J.A. Holyst: Simple model for dry friction, Phys. Rev. B **49**, 3831–3838 (1994)

39. T. Kawaguchi, H. Matsukawa: Dynamical frictional phenomena in an incommensurate two-chain model, Phys. Rev. B **56**, 13932–13942 (1997)
40. M. H. Müser: Nature of mechanical instabilities and their effect on kinetic friction, Phys. Rev. Lett. **89**, 224301 (2002)
41. M. H. Müser: Towards an atomistic understanding of solid friction by computer simulations, Comput. Phys. Commun. **146**, 54–62 (2002)
42. P. Reimann, M. Evstigneev: Nonmonotonic velocity dependence of atomic friction, Phys. Rev. Lett. **93**, 230802 (2004)
43. C. Ritter, M. Heyde, B. Stegemann, K. Rademann, U. D. Schwarz: Contact area dependence of frictional forces: Moving adsorbed antimony nanoparticles, Phys. Rev. B **71**, 085405 (2005)
44. C. Fusco, A. Fasolino: Velocity dependence of atomic-scale friction: A comparative study of the one- and two-dimensional Tomlinson model, Phys. Rev. B **71**, 045413 (2005)
45. C. W. Gear: *Numerical Initial Value Problems in Ordinary Differential Equations* (Prentice-Hall, Englewood Cliffs 1971)
46. W. G. Hoover: *Molecular Dynamics* (Springer, Berlin, Heidelberg 1986)
47. D. W. Heermann: *Computer Simulation Methods in Theoretical Physics* (Springer, Berlin, Heidelberg 1986)
48. M. P. Allen, D. J. Tildesley: *Computer Simulation of Liquids* (Clarendon, Oxford 1987)
49. J. M. Haile: *Molecular Dynamics Simulation: Elementary Methods* (Wiley, New York 1992)
50. M. Finnis: *Interatomic Forces in Condensed Matter* (Oxford University Press, Oxford 2003)
51. D. W. Brenner: Relationship between the embedded-atom method and Tersoff potentials, Phys. Rev. Lett. **63**, 1022–1022 (1989)
52. D. W. Brenner: The art and science of an analytic potential, Phys. Stat. Sol. B **217**, 23–40 (2000)
53. R. G. Parr, W. Yang: *Density Functional Theory of Atoms and Molecules* (Oxford Univ. Press, New York 1989)
54. R. Car, M. Parrinello: Unified approach for molecular dynamics and density functional theory, Phys. Rev. Lett. **55**, 2471–2474 (1985)
55. C. Cramer: *Essentials of Computational Chemistry, Theories and Models* (Wiley, Chichester 2004)
56. A. P. Sutton: *Electronic Structure of Materials* (Clarendon, Oxford 1993)
57. K. Kadau, T. C. Germann, P. S. Lomdahl: Large-scale molecular dynamics simulation of 19 billion particles, Int. J. Mod. Phys. C **15**, 193–201 (2004)
58. B. J. Thijsse: Relationship between the modified embedded-atom method and Stillinger–Weber potentials in calculating the structure of silicon, Phys. Rev. B **65**, 195207 (2002)
59. M. I. Baskes, J. S. Nelson, A. F. Wright: Semiempirical modified embedded atom potentials for silicon and germanium, Phys. Rev. B **40**, 6085–6100 (1989)
60. T. Ohira, Y. Inoue, K. Murata, J. Murayama: Magnetite scale cluster adhesion on metal oxide surfaces: Atomistic simulation study, Appl. Surf. Sci. **171**, 175–188 (2001)
61. F. H. Streitz, J. W. Mintmire: Electrostatic potentials for metal oxide surfaces and interfaces, Phys. Rev. B **50**, 11996–12003 (1994)
62. A. Yasukawa: Using an extended Tersoff interatomic potential to analyze the static fatigue strength of SiO_2 under atmospheric influence, JSME Int. J. A **39**, 313–320 (1996)
63. T. Iwasaki, H. Miura: Molecular dynamics analysis of adhesion strength of interfaces between thin films, J. Mater. Res. **16**, 1789–1794 (2001)

64. B.-J. Lee, M. I. Baskes: Second nearest-neighbor modified embedded-atom method potential, Phys. Rev. B **62**, 8564–8567 (2000)
65. G. C. Abell: Empirical chemical pseudopotential theory of molecular and metallic bonding, Phys. Rev. B **31**, 6184–6196 (1985)
66. J. Tersoff: New empirical approach for the structure and energy of covalent systems, Phys. Rev. B **37**, 6991–7000 (1988)
67. J. Tersoff: Modeling solid-state chemistry: Interatomic potentials for multicomponent systems, Phys. Rev. B **39**, 5566–5569 (1989)
68. D. W. Brenner: Empirical potential for hydrocarbons for use in simulating the chemical vapor deposition of diamond films, Phys. Rev. B **42**, 9458–9471 (1990)
69. D. W. Brenner, O. A. Shenderova, J. A. Harrison, S. J. Stuart, B. Ni, S. B. Sinnott: Second generation reactive empirical bond order (REBO) potential energy expression for hydrocarbons, J. Phys. C **14**, 783–802 (2002)
70. A. J. Dyson, P. V. Smith: Extension of the Brenner empirical interactomic potential to C-Si-H, Surf. Sci. **355**, 140–150 (1996)
71. B. Ni, K.-H. Lee, S. B. Sinnott: Development of a reactive empirical bond order potential for hydrocarbon-oxygen interactions, J. Phys. C **16**, 7261–7275 (2004)
72. J. Tanaka, C. F. Abrams, D. B. Graves: New C-F interatomic potential for molecular dynamics simulation of fluorocarbon film formation, Nucl. Instrum. Meth. B **18**, 938–945 (2000)
73. I. Jang, S. B. Sinnott: Molecular dynamics simulations of the chemical modification of polystyrene through $C_xF_y^+$ beam deposition, J. Phys. Chem. B **108**, 9656–9664 (2004)
74. S. B. Sinnott, O. A. Shenderova, C. T. White, D. W. Brenner: Mechanical properties of nanotubule fibers and composites determined from theoretical calculations and simulations, Carbon **36**, 1–9 (1998)
75. S. J. Stuart, A. B. Tutein, J. A. Harrison: A reactive potential for hydrocarbons with intermolecular interactions, J. Chem. Phys. **112**, 6472–6486 (2000)
76. F. H. Stillinger, T. A. Weber: Computer simulation of local order in condensed phases of silicon, Phys. Rev. B **31**, 5262–5271 (1985)
77. S. M. Foiles: Application of the embedded-atom method to liquid transition metals, Phys. Rev. B **32**, 3409–3415 (1985)
78. M. S. Daw, M. I. Baskes: Semiempirical, quantum mechanical calculation of hydrogen embrittlement in metals, Phys. Rev. Lett. **50**, 1285–1288 (1983)
79. T. J. Raeker, A. E. Depristo: Theory of chemical bonding based on the atom-homogeneous electron gas system, Int. Rev. Phys. Chem. **10**, 1–54 (1991)
80. R. W. Smith, G. S. Was: Application of molecular dynamics to the study of hydrogen embrittlement in Ni–Cr–Fe alloys, Phys. Rev. B **40**, 10322–10336 (1989)
81. R. Pasianot, D. Farkas, E. J. Savino: Empirical many-body interatomic potential for bcc transition metals, Phys. Rev. B **43**, 6952–6961 (1991)
82. R. Pasianot, E. J. Savino: Embedded-atom method interatomic potentials for hcp metals, Phys. Rev. B **45**, 12704–12710 (1992)
83. M. I. Baskes, J. S. Nelson, A. F. Wright: Semiempirical modified embedded-atom potentials for silicon and germanium, Phys. Rev. B **40**, 6085–6100 (1989)
84. M. I. Baskes: Modified embedded-atom potentials for cubic materials and impurities, Phys. Rev. B **46**, 2727–2742 (1992)
85. K. Ohno, K. Esfarjani, Y. Kawazoe: *Computational Materials Science from Ab Initio to Monte Carlo Methods* (Springer, Berlin, Heidelberg 1999)
86. A. K. Rappe, W. A. Goddard III: Charge equilibration for molecular dynamics simulations, J. Phys. Chem. **95**, 3358–3363 (1991)

87. D. Frenkel, B. Smit: *Understanding Molecular Simulation: From Algorithms to Applications* (Academic, San Diego 1996)
88. L. V. Woodcock: Isothermal molecular dynamics calculations for liquid salts, Chem. Phys. Lett. **10**, 257–261 (1971)
89. T. Schneider, E. Stoll: Molecular dynamics study of a three-dimensional one-component model for distortive phase transitions, Phys. Rev. B **17**, 1302–1322 (1978)
90. K. Kremer, G. S. Grest: Dynamics of entangled linear polymer melts – a molecular dynamics simulation, J. Chem. Phys. **92**, 5057–5086 (1990)
91. S. A. Adelman, J. D. Doll: Generalized Langevin equation approach for atom-solid-surface scattering – general formulation for classical scattering off harmonic solids, J. Chem. Phys. **64**, 2375–2388 (1976)
92. S. A. Adelman: Generalized Langevin equations and many-body problems in chemical dynamics, Adv. Chem. Phys. **44**, 143–253 (1980)
93. J. C. Tully: Dynamics of gas-surface interactions – 3D generalized Langevin model applied to fcc and bcc surfaces, J. Chem. Phys. **73**, 1975–1985 (1980)
94. S. Nosé: A unified formulation of the constant temperature molecular dynamics methods, J. Chem. Phys. **81**, 511–519 (1984)
95. S. Nosé: A molecular dynamics method for simulations in the canonical ensemble, Mol. Phys. **52**, 255–268 (1984)
96. G. J. Martyna, M. L. Klein, M. Tuckerman: Nose-Hoover chains – the canonical ensemble via continuous dynamics, J. Chem. Phys. **97**, 2635–2643 (1992)
97. M. D'Alessandro, M. D'Abramo, G. Brancato, A. Di Nola, A. Amadei: Statistical mechanics and thermodynamics of simulated ionic solutions, J. Phys. Chem. B **106**, 11843–11848 (2002)
98. J. D. Schall, C. W. Padgett, D. W. Brenner: Ad hoc continuum-atomistic thermostat for modeling heat flow in molecular dynamics simulations, Mol. Simul. **31**, 283–288 (2005)
99. M. Schoen, C. L. Rhykerd, D. J. Diestler, J. H. Cushman: Shear forces in molecularly thin films, Science **245**, 1223–1225 (1989)
100. J. E. Curry, F. S. Zhang, J. H. Cushman, M. Schoen, D. J. Diestler: Transient coexisting nanophases in ultrathin films confined between corrugated walls, J. Chem. Phys. **101**, 10824–10832 (1994)
101. D. J. Adams: Grand canonical ensemble Monte Carlo for a Lennard–Jones fluid, Mol. Phys. **29**, 307–311 (1975)
102. N. A. a. C., R. J. Burnham: Force microscopy. In: *Scanning Tunneling Microscopy and Spectroscopy: Theory, Techniques, and Applications*, ed. by D. A. Bonnell (VCH, New York 1993) pp. 191–249
103. E. Meyer: *Nanoscience: Friction and Rheology on the Nanometer Scale* (World Scientific, Hackensack 1998)
104. G. E. Totten, H. Liang: *Mechanical Tribology: Materials Characterization and Applications* (Marcel Dekker, New York 2004)
105. N. A. Burnham, R. J. Colton: Measuring the nanomechanical properties and surface forces of materials using an atomic force microscope, J. Vacuum Sci. Technol. A **7**, 2906–2913 (1989)
106. N. A. Burnham, D. D. Dominguez, R. L. Mowery, R. J. Colton: Probing the surface forces of monolayer films with an atomic force microscope, Phys. Rev. Lett. **64**, 1931–1934 (1990)
107. E. Meyer, R. Overney, D. Brodbeck, L. Howald, R. Luthi, J. Frommer, H. J. Guntherodt: Friction and wear of Langmuir-Blodgett films observed by friction force microscopy, Phys. Rev. Lett. **69**, 1777–1780 (1992)

108. A.P. Sutton, J.B. Pethica, H. Rafii-Tabar, J.A. Nieminen: Mechanical properties of metals at the nanometer scale. In: *Electron Theory in Alloy Design*, ed. by D.G. Pettifor, A.H. Cottrell (Institute of Materials, London 1992) pp. 191–233
109. H. Raffi-Tabar, A.P. Sutton: Long-range Finnis–Sinclair potentials for fcc metallic alloys, Philos. Mag. Lett. **63**, 217–224 (1991)
110. U. Landman, W.D. Luedtke, E.M. Ringer: Atomistic mechanisms of adhesive contact formation and interfacial processes, Wear **153**, 3–30 (1992)
111. U. Landman, W.D. Luedtke, N.A. Burnham, R.J. Colton: Atomistic mechanisms and dynamics of adhesion, nanoindentation and fracture, Science **248**, 454–461 (1990)
112. O. Tomagnini, F. Ercolessi, E. Tosatti: Microscopic interaction between a gold tip and a Pb(110) surface, Surf. Sci. **287/288**, 1041–1045 (1991)
113. N. Ohmae: Field ion microscopy of microdeformation induced by metallic contacts, Philos. Mag. A **74**, 1319–1327 (1996)
114. N.A. Burnham, R.J. Colton, H.M. Pollock: Interpretation of force curves in force microscopy, Nanotechnology **4**, 64–80 (1993)
115. N. Agrait, G. Rubio, S. Vieira: Plastic deformation in nanometer-scale contacts, Langmuir **12**, 4505–4509 (1996)
116. U. Landman, W.D. Luedtke, A. Nitzan: Dynamics of tip-substrate interactions in atomic force microscopy, Surf. Sci. **210**, L177–L182 (1989)
117. U. Landman, W.D. Luedtke: Nanomechanics and dynamics of tip substrate interactions, J. Vacuum Sci. Technol. B **9**, 414–423 (1991)
118. U. Landman, W.D. Luedtke, J. Ouyang, T.K. Xia: Nanotribology and the stability of nanostructures, Jpn. J. Appl. Phys. Pt. 1 **32**, 1444–1462 (1993)
119. J.W.M. Frenken, H.M. Vanpinxteren, L. Kuipers: New views on surface melting obtained with STM and ion scattering, Surf. Sci. **283**, 283–289 (1993)
120. T. Yokohata, K. Kato: Mechanism of nanoscale indentation, Wear **168**, 109–114 (1993)
121. O. Tomagnini, F. Ercolessi, E. Tosatti: Microscopic interaction between a gold tip and a Pb(110) surface, Surf. Sci. **287**, 1041–1045 (1993)
122. K. Komvopoulos, W. Yan: Molecular dynamics simulation of single and repeated indentation, J. Appl. Phys. **82**, 4823–4830 (1997)
123. J. Belak, I.F. Stowers: *A Molecular Dynamics Model of the Orthogonal Cutting Process* (Proc. Am. Soc. Precision Eng. Annu. Conf., 1990) pp. 76–79
124. M. Fournel, E. Lacaze, M. Schott: Tip-surface interactions in STM experiments on Au(111): Atomic-scale metal friction, Europhys. Lett. **34**, 489–494 (1996)
125. E.T. Lilleodden, J.A. Zimmerman, S.M. Foiles, W.D. Nix: Atomistic simulations of elastic deformation and dislocation nucleation during nanoindentation, J. Mech. Phys. Sol. **51**, 901–920 (2003)
126. J.L. Costakramer, N. Garcia, P. Garciamochales, P.A. Serena: Nanowire formation in macroscopic metallic contacts – quantum-mechanical conductance tapping a table top, Surf. Sci. **342**, L1144–L1149 (1995)
127. A.I. Yanson, J.M. van Ruitenbeek, I.K. Yanson: Shell effects in alkali metal nanowires, Low Temp. Phys. **27**, 807–820 (2001)
128. A.I. Yanson, I.K. Yanson, J.M. van Ruitenbeek: Crossover from electronic to atomic shell structure in alkali metal nanowires, Phys. Rev. Lett. **8721**, 216805 (2001)
129. C.L. Kelchner, S.J. Plimpton, J.C. Hamilton: Dislocation nucleation and defect structure during surface indentation, Phys. Rev. B **58**, 11085–11088 (1998)
130. A. Hasnaoui, P.M. Derlet, H.V. Swygenhoven: Interaction between dislocations, grain boundaries under an indenter – a molecular dynamics simulation, Acta Mater. **52**, 2251–2258 (2004)

131. O.R. de la Fuente, J.A. Zimmerman, M.A. Gonzalez, J. de la Figuera, J.C. Hamilton, W.W. Pai, J.M. Rojo: Dislocation emission around nanoindentations on a (001) fcc metal surface studied by scanning tunneling microscopy and atomistic simulations, Phys. Rev. Lett. **88**, 036101 (2002)
132. O.A. Shenderova, J.P. Mewkill, D.W. Brenner: Nanoindentation as a probe of nanoscale residual stresses, Mol. Simul. **25**, 81–92 (2000)
133. S. Kokubo: On the change in hardness of a plate caused by bending, Sci. Rep. Tohoku Imperial University **21**, 256–267 (1932)
134. G. Sines, R. Calson: Hardness measurements for determination of residual stresses, ASTM Bull. **180**, 35–37 (1952)
135. G.U. Oppel: Biaxial elasto-plastic analysis of load and residual stresses, Exp. Mech. **21**, 135–140 (1964)
136. T.R. Simes, S.G. Mellor, D.A. Hills: A note on the influence of residual stress on measured hardness, J. Strain Anal. Eng. Des. **19**, 135–137 (1984)
137. T.Y. Tsui, G.M. Pharr, W.C. Oliver, C.S. Bhatia, C.T. White, S. Anders, A. Anders, I.G. Brown: Nanoindentation and nanoscratching of hard carbon coatings for magnetic disks, Mat. Res. Soc. Symp. Proc. **383**, 447–452 (1995)
138. A. Bolshakov, W.C. Oliver, G.M. Pharr: Influences of stress on the measurement of mechanical properties using nanoindentation. 2. Finite element simulations, J. Mater. Res. **11**, 760–768 (1996)
139. J.D. Schall, D.W. Brenner: Atomistic simulation of the influence of pre-existing stress on the interpretation of nanoindentation data, J. Mater. Res. **19**, 3172–3180 (2004)
140. U. Landman, W.D. Luedtke, M.W. Ribarsky: Structural and dynamical consequences of interactions in interfacial systems, J. Vacuum Sci. Technol. A **7**, 2829–2839 (1989)
141. J.S. Kallman, W.G. Hoover, C.G. Hoover, A.J. Degroot, S.M. Lee, F. Wooten: Molecular-dynamics of silicon indentation, Phys. Rev. B **47**, 7705–7709 (1993)
142. D.R. Clarke, M.C. Kroll, P.D. Kirchner, R.F. Cook, B.J. Hockey: Amorphization and conductivity of silicon and germanium induced by indentation, Phys. Rev. Lett. **60**, 2156–2159 (1988)
143. A. Kailer, K.G. Nickel, Y.G. Gogotsi: Raman microspectroscopy of nanocrystalline and amorphous phases in hardness indentations, J. Raman Spec. **30**, 939–961 (1999)
144. K. Minowa, K. Sumino: Stress-induced amorphization of a silicon crystal by mechanical scratching, Phys. Rev. Lett. **69**, 320–322 (1992)
145. G.S. Smith, E.B. Tadmor, E. Kaxiras: Multiscale simulation of loading and electrical resistance in silicon nanoindentation, Phys. Rev. Lett. **84**, 1260–1263 (2000)
146. W.C.D. Cheong, L.C. Zhang: Molecular dynamics simulation of phase transformations in silicon monocrystals due to nano-indentation, Nanotechnology **11**, 173–180 (2000)
147. C.F. Sanz-Navarro, S.D. Kenny, R. Smith: Atomistic simulations of structural transformations, Nanotechnology **15**, 692–697 (2004)
148. P. Walsh, A. Omeltchenko, R.K. Kalia, A. Nakano, P. Vashishta, S. Saini: Nanoindentation of silicon nitride: A multimillion-atom molecular dynamics study, Appl. Phys. Lett. **82**, 118–120 (2003)
149. J.A. Harrison, C.T. White, R.J. Colton, D.W. Brenner: Nanoscale investigation of indentation, adhesion and fracture of diamond (111) surfaces, Surf. Sci. **271**, 57–67 (1992)
150. K. Cho, J.D. Joannopoulos: Mechanical hysteresis on an atomic-scale, Surf. Sci. **328**, 320–324 (1995)
151. A. Garg, J. Han, S.B. Sinnott: Interactions of carbon-nanotubule proximal probe tips with diamond and graphene, Phys. Rev. Lett. **81**, 2260–2263 (1998)

152. J. A. Harrison, S. J. Stuart, D. H. Robertson, C. T. White: Properties of capped nanotubes when used as SPM tips, J. Phys. Chem. B **101**, 9682–9685 (1997)
153. J. A. Harrison, S. J. Stuart, A. B. Tutein: A new, reactive potential energy function to study indentation and friction of C_{13} n-alkane monolayers. In: *Interfacial Properties on the Submicron Scale*, ed. by J. E. Frommer, R. Overney (ACS, Washington 2001) pp. 216–229
154. A. Garg, S. B. Sinnott: Molecular dynamics of carbon nanotubule proximal probe tip-surface contacts, Phys. Rev. B **60**, 13786–13791 (1999)
155. K. J. Tupper, D. W. Brenner: Compression-induced structural transition in a self-assembled monolayer, Langmuir **10**, 2335–2338 (1994)
156. K. J. Tupper, R. J. Colton, D. W. Brenner: Simulations of self-assembled monolayers under compression – effect of surface asperities, Langmuir **10**, 2041–2043 (1994)
157. S. A. Joyce, R. C. Thomas, J. E. Houston, T. A. Michalske, R. M. Crooks: Mechanical relaxation of organic monolayer films measured by force microscopy, Phys. Rev. Lett. **68**, 2790–2793 (1992)
158. L. Zhang, Y. Leng, S. Jiang: Tip-based hybrid simulation study of frictional properties of self-assembled monolayers: Effects of chain length, terminal group, and scan direction, scan velocity, Langmuir **19**, 9742–9747 (2003)
159. A. B. Tutein, S. J. Stuart, J. A. Harrison: Indentation analysis of linear-chain hydrocarbon monolayers anchored to diamond, J. Phys. Chem. B **103**, 11357–11365 (1999)
160. Y. Leng, S. Jiang: Dynamic simulations of adhesion and friction in chemical force microscopy, J. Am. Chem. Soc. **124**, 11764–11770 (2002)
161. C. M. Mate: Atomic force microscope study of polymer lubricants on silicon surfaces, Phys. Rev. Lett. **68**, 3323–3326 (1992)
162. K. Enke, H. Dimigen, H. Hubsch: Frictional properties of diamond-like carbon layers, Appl. Phys. Lett. **36**, 291–292 (1980)
163. K. Enke: Some new results on the fabrication of and the mechanical, electrical, optical properties of I-carbon layers, Thin Solid Films **80**, 227–234 (1981)
164. S. Miyake, S. Takahashi, I. Watanabe, H. Yoshihara: Friction and wear behavior of hard carbon films, ASLE Trans. **30**, 121–127 (1987)
165. A. Erdemir, C. Donnet: Tribology of diamond, diamond-like carbon, and related films. In: *Modern Tribology Handbook*, Vol. II, ed. by B. Bhushan (CRC, Boca Raton 2000) pp. 871–908
166. S. B. Sinnott, R. J. Colton, C. T. White, O. A. Shenderova, D. W. Brenner, J. A. Harrison: Atomistic simulations of the nanometer-scale indentation of amorphous carbon thin films, J. Vacuum Sci. Technol. A **15**, 936–940 (1997)
167. J. N. Glosli, M. R. Philpott, G. M. McClelland: Molecular dynamics simulation of mechanical deformation of ultra-thin amorphous carbon films, Mater. Res. Soc. Symp. Proc. **383**, 431–435 (1995)
168. M. R. Sorensen, K. W. Jacobsen, P. Stoltze: Simulations of atomic-scale sliding friction, Phys. Rev. B **53**, 2101–2113 (1996)
169. I. L. Singer: A thermochemical model for analyzing low wear-rate materials, Surf. Coat. Technol. **49**, 474–481 (1991)
170. I. L. Singer, S. Fayeulle, P. D. Ehni: Friction and wear behavior of tin in air – the chemistry of transfer films and debris formation, Wear **149**, 375–394 (1991)
171. J. A. Harrison, C. T. White, R. J. Colton, D. W. Brenner: Molecular dynamics simulations of atomic-scale friction of diamond surfaces, Phys. Rev. B **46**, 9700–9708 (1992)
172. J. A. Harrison, R. J. Colton, C. T. White, D. W. Brenner: Effect of atomic-scale surface roughness on friction – a molecular dynamics study of diamond surfaces, Wear **168**, 127–133 (1993)

173. J. A. Harrison, C. T. White, R. J. Colton, D. W. Brenner: Atomistic simulations of friction at sliding diamond interfaces, MRS Bull. **18**, 50–53 (1993)
174. J. N. Glosli, G. M. Mcclelland: Molecular dynamics study of sliding friction of ordered organic monolayers, Phys. Rev. Lett. **70**, 1960–1963 (1993)
175. A. Koike, M. Yoneya: Molecular dynamics simulations of sliding friction of Langmuir-Blodgett monolayers, J. Chem. Phys. **105**, 6060–6067 (1996)
176. J. E. Hammerberg, B. L. Holian, S. J. Zhou: Studies of sliding friction in compressed copper, Conference of the American Physical Society Topical Group on Shock Compression of Condensed Matter, Seattle, WA 1995, ed. by S. C. Schmidt, W. C. Tao (AIP, New York 1995) 370
177. M. D. Perry, J. A. Harrison: Friction between diamond surfaces in the presence of small third-body molecules, J. Phys. Chem. B **101**, 1364–1373 (1997)
178. A. Buldum, S. Ciraci: Atomic-scale study of dry sliding friction, Phys. Rev. B **55**, 2606–2611 (1997)
179. A. P. Sutton, J. B. Pithica: Inelastic flow processes in nanometre volumes of solids, J. Phys. Cond. Matter **2**, 5317–5326 (1990)
180. S. Akamine, R. C. Barrett, C. F. Quate: Improved atomic force microscope images using microcantilevers with sharp tips, Appl. Phys. Lett. **57**, 316–318 (1990)
181. J. A. Nieminen, A. P. Sutton, J. B. Pethica: Static junction growth during frictional sliding of metals, Acta Metall. Mater. **40**, 2503–2509 (1992)
182. J. A. Niemienen, A. P. Sutton, J. B. Pethica, K. Kaski: Mechanism of lubrication by a thin solid film on a metal surface, Model. Simul. Mater. Sci. Eng **1**, 83–90 (1992)
183. V. V. Pokropivny, V. V. Skorokhod, A. V. Pokropivny: Atomistic mechanism of adhesive wear during friction of atomic sharp tungsten asperity over (114) bcc-iron surface, Mater. Lett. **31**, 49–54 (1997)
184. B. Li, P. C. Clapp, J. A. Rifkin, X. M. Zhang: Molecular dynamics simulation of stick-slip, J. Appl. Phys. **90**, 3090–3094 (2001)
185. R. Komanduri, N. Chandrasekaran: Molecular dynamics simulation of atomic-scale friction, Phys. Rev. B **61**, 14007–14019 (2000)
186. T.-H. Fang, C.-I. Weng, J.-G. Chang: Molecular dynamics simulation of nanolithography process using atomic force microscopy, Surf. Sci. **501**, 138–147 (2002)
187. S. Morita, S. Fujisawa, Y. Sugawara: Spatially quantized friction with a lattice periodicity, Surf. Sci. Rep. **23**, 1–41 (1996)
188. A. Dayo, W. Alnasrallah, J. Krim: Superconductivity-dependent sliding friction, Phys. Rev. Lett. **80**, 1690–1693 (1998)
189. R. Erlandsson, G. Hadziioannou, C. M. Mate, G. M. Mcclelland, S. Chiang: Atomic scale friction between the muscovite mica cleavage plane and a tungsten tip, J. Chem. Phys. **89**, 5190–5193 (1988)
190. K. L. Johnson: *Contact Mechanics* (Cambridge Univ. Press, Cambridge 1985)
191. J. B. Pethica: Interatomic forces in scanning tunneling microscopy – giant corrugations of the graphite surface – comment, Phys. Rev. Lett. **57**, 3235–3235 (1986)
192. H. Tang, C. Joachim, J. Devillers: Interpretation of AFM images – the graphite surface with a diamond tip, Surf. Sci. **291**, 439–450 (1993)
193. A. L. Shluger, R. T. Williams, A. L. Rohl: Lateral and friction forces originating during force microscope scanning of ionic surfaces, Surf. Sci. **343**, 273–287 (1995)
194. S. Fujisawa, Y. Sugawara, S. Morita: Localized fluctuation of a two-dimensional atomic-scale friction, Jpn. J. Appl. Phys. Pt. 1 **35**, 5909–5913 (1996)
195. S. Fujisawa, Y. Sugawara, S. Ito, S. Mishima, T. Okada, S. Morita: The two-dimensional stick-slip phenomenon with atomic resolution, Nanotechnology **4**, 138–142 (1993)

196. S. Fujisawa, Y. Sugawara, S. Morita, S. Ito, S. Mishima, T. Okada: Study on the stick-slip phenomenon on a cleaved surface of the muscovite mica using an atomic-force lateral force microscope, J. Vacuum Sci. Technol. B **12**, 1635–1637 (1994)
197. J. A. Ruan, B. Bhushan: Atomic-scale and microscale friction studies of graphite and diamond using friction force microscopy, J. Appl. Phys. **76**, 5022–5035 (1994)
198. R. W. Carpick, N. Agrait, D. F. Ogletree, M. Salmeron: Variation of the interfacial shear strength and adhesion of a nanometer-sized contact, Langmuir **12**, 3334–3340 (1996)
199. R. W. Carpick, N. Agrait, D. F. Ogletree, M. Salmeron: Measurement of interfacial shear (friction) with an ultrahigh vacuum atomic force microscope, J. Vacuum Sci. Technol. B **14**, 1289,2772 (1996)
200. B. Samuels, J. Wilks: The friction of diamond sliding on diamond, J. Mater. Sci. **23**, 2846–2864 (1988)
201. T. Cagin, J. W. Che, M. N. Gardos, A. Fijany, W. A. Goddard: Simulation and experiments on friction, wear of diamond: A material for MEMS and NEMS application, Nanotechnology **10**, 278–284 (1999)
202. R. J. A. van den Oetelaar, C. F. J. Flipse: Atomic-scale friction on diamond(111) studied by ultra-high vacuum atomic force microscopy, Surf. Sci. **384**, L828–L835 (1997)
203. M. Enachescu, R. J. A. van den Oetelaar, R. W. Carpick, D. F. Ogletree, C. F. J. Flipse, M. Salmeron: Atomic force microscopy study of an ideally hard contact: The diamond(111) tungsten carbide interface, Phys. Rev. Lett. **81**, 1877–1880 (1998)
204. B. V. Derjaguin, V. M. Muller, Y. Toporov: Effect of contact deformations on adhesion of particles, J. Colloid Interf. Sci. **53**, 314–326 (1975)
205. M. D. Perry, J. A. Harrison: Universal aspects of the atomic-scale friction of diamond surfaces, J. Phys. Chem. B **99**, 9960–9965 (1995)
206. R. Neitola, T. A. Pakannen: Ab initio studies on the atomic-scale origin of friction between diamond (111) surfaces, J. Phys. Chem. B **105**, 1338–1343 (2001)
207. J. A. Harrison, R. J. Colton, C. T. White, D. W. Brenner: Atomistic simulation of the nanoindentation of diamond and graphite surfaces, Mater. Res. Soc. Sym. Proc. **239**, 573–578 (1992)
208. J. A. Harrison, C. T. White, R. J. Colton, D. W. Brenner: Investigation of the atomic-scale friction and energy dissipation in diamond using molecular dynamics, Thin Solid Films **260**, 205–211 (1995)
209. J. Cai, J.-S. Wang: Friction between Si tip and (001)–2×1 surface: A molecular dynamics simulation, Comput. Phys. Commun. **147**, 145–148 (2002)
210. D. Mulliah, S. D. Kenny, R. Smith: Modeling of stick-slip phenomena using molecular dynamics, Phys. Rev. B **69**, 205407 (2004)
211. D. W. Brenner: Empirical potential for hydrocarbons for use in simulating the chemical vapor deposition of diamond films, Phys. Rev. B **42**, 9458–9471 (1990)
212. G. J. Ackland, G. Tichy, V. Vitek, M. W. Finnis: Simple n-body potentials for the noble metals and nickel, Philos. Mag. A **56**, 735–756 (1987)
213. J. P. Biersack, J. Ziegler, U. Littmack: *The Stopping and Range of Ions in Solids* (Pergamon, Oxford 1985)
214. J. Cai, J. S. Wang: Friction between a Ge tip and the (001)–2×1 surface: A molecular dynamics simulation, Phys. Rev. B **64**, 113313 (2001)
215. A. G. Khurshudov, K. Kato, H. Koide: Nano-wear of the diamond AFM probing tip under scratching of silicon, studied by AFM, Tribol. Lett. **2**, 345–354 (1996)
216. A. Khurshudov, K. Kato: Volume increase phenomena in reciprocal scratching of polycarbonate studied by atomic-force microscopy, J. Vacuum Sci. Technol. B **13**, 1938–1944 (1995)

217. M.D. Perry, J.A. Harrison: Molecular dynamics studies of the frictional properties of hydrocarbon materials, Langmuir **12**, 4552–4556 (1996)
218. M.D. Perry, J.A. Harrison: Molecular dynamics investigations of the effects of debris molecules on the friction and wear of diamond, Thin Solid Films **291**, 211–215 (1996)
219. J.A. Harrison, C.T. White, R.J. Colton, D.W. Brenner: Effects of chemically-bound, flexible hydrocarbon species on the frictional properties of diamond surfaces, J. Phys. Chem. **97**, 6573–6576 (1993)
220. J.A. Harrison, D.W. Brenner: Simulated tribochemistry – an atomic-scale view of the wear of diamond, J. Am. Chem. Soc. **116**, 10399–10402 (1994)
221. Z. Feng, J.E. Field: Friction of diamond on diamond and chemical vapor deposition diamond coatings, Surf. Coat. Technol. **47**, 631–645 (1991)
222. B.N.J. Persson: Applications of surface resistivity to atomic scale friction, to the migration of hot adatoms, and to electrochemistry, J. Chem. Phys. **98**, 1659–1672 (1993)
223. B.N.J. Persson, A.I. Volokitin: Electronic friction of physisorbed molecules, J. Chem. Phys. **103**, 8679–8683 (1995)
224. H. Grabhorn, A. Otto, D. Schumacher, B.N.J. Persson: Variation of the dc-resistance of smooth and atomically rough silver films during exposure to C_2H_6 and C_2H_4, Surf. Sci. **264**, 327–340 (1992)
225. A.D. Berman, W.A. Ducker, J.N. Israelachvili: Origin and characterization of different stick-slip friction mechanisms, Langmuir **12**, 4559–4563 (1996)
226. B.N.J. Persson: Theory of friction – dynamical phase transitions in adsorbed layers, J. Chem. Phys. **103**, 3849–3860 (1995)
227. B.N.J. Persson, E. Tosatti: Layering transition in confined molecular thin films – nucleation and growth, Phys. Rev. B **50**, 5590–5599 (1994)
228. H. Yoshizawa, J. Israelachvili: Fundamental mechanisms of interfacial friction. 2. Stick-slip friction of spherical and chain molecules, J. Phys. Chem. **97**, 11300–11313 (1993)
229. B.N.J. Persson: Theory of friction: Friction dynamics for boundary lubricated surfaces, Phys. Rev. B **55**, 8004–8012 (1997)
230. U. Landman, W.D. Luedtke, J.P. Gao: Atomic-scale issues in tribology: Interfacial junctions and nano-elastohydrodynamics, Langmuir **12**, 4514–4528 (1996)
231. P.A. Thompson, M.O. Robbins: Origin of stick-slip motion in boundary lubrication, Science **250**, 792–794 (1990)
232. T. Kreer, M.H. Müser, K. Binder, J. Klein: Frictional drag mechanisms between polymer-bearing surfaces, Langmuir **17**, 7804–7813 (2001)
233. T. Kreer, K. Binder, M.H. Müser: Friction between polymer brushes in good solvent conditions: Steady-state sliding versus transient behavior, Langmuir **19**, 7551–7559 (2003)
234. E. Manias, G. Hadziioannou, G. ten Brinke: Inhomogeneities in sheared ultrathin lubricating films, Langmuir **12**, 4587–4593 (1996)
235. R.M. Overney, T. Bonner, E. Meyer, M. Reutschi, R. Luthi, L. Howald, J. Frommer, H.J. Guntherodt, M. Fujihara, H. Takano: Elasticity, wear, and friction properties of thin organic films observed with atomic-force microscopy, J. Vacuum Sci. Technol. B **12**, 1973–1976 (1994)
236. R.M. Overney, E. Meyer, J. Frommer, D. Brodbeck, R. Luthi, L. Howald, H.J. Guntherodt, M. Fujihira, H. Takano, Y. Gotoh: Friction measurements on phase-separated thin-films with a modified atomic force microscope, Nature **359**, 133–135 (1992)
237. R.M. Overney, E. Meyer, J. Frommer, H.J. Guntherodt, M. Fujihira, H. Takano, Y. Gotoh: Force microscopy study of friction and elastic compliance of phase-separated organic thin-films, Langmuir **10**, 1281–1286 (1994)

238. H.I. Kim, T. Koini, T.R. Lee, S.S. Perry: Systematic studies of the frictional properties of fluorinated monolayers with atomic force microscopy: Comparison of CF_3- and CH_3-terminated films, Langmuir **13**, 7192–7196 (1997)
239. P.T. Mikulski, J.A. Harrison: Packing density effects on the friction of n-alkane monolayers, J. Am. Chem. Soc. **123**, 6873–6881 (2001)
240. M. GarciaParajo, C. Longo, J. Servat, P. Gorostiza, F. Sanz: Nanotribological properties of octadecyltrichlorosilane self-assembled ultrathin films studied by atomic force microscopy: Contact and tapping modes, Langmuir **13**, 2333–2339 (1997)
241. R.M. Overney, H. Takano, M. Fujihira, E. Meyer, H.J. Guntherodt: Wear, friction and sliding speed correlations on Langmuir-Blodgett films observed by atomic force microscopy, Thin Solid Films **240**, 105–109 (1994)
242. P.T. Mikulski, J.A. Harrison: Periodicities in the properties associated with the friction of model self-assembled monolayers, Tribol. Lett. **10**, 29–35 (2001)
243. E. Barrena, C. Ocal, M. Salmeron: A comparative AFM study of the structural and frictional properties of mixed and single component films of alkanethiols on Au(111), Surf. Sci. **482**, 1216–1221 (2001)
244. Y.-S. Shon, S. Lee, R. Colorado, S.S. Perry, T.R. Lee: Spiroalkanedithiol-based SAMS reveal unique insight into the wettabilities and frictional properties of organic thin films, J. Am. Chem. Soc. **122**, 7556–7563 (2000)
245. P.T. Mikulski, G. Gao, G.M. Chateauneuf, J.A. Harrison: Contact forces at the sliding interface: Mixed versus pure model alkane monolayers, J. Chem. Phys. **122**, 024701 (2005)
246. S. Lee, Y.S. Shon, R. Colorado, R.L. Guenard, T.R. Lee, S.S. Perry: The influence of packing densities, surface order on the frictional properties of alkanethiol self-assembled monolayers (SAMs) on gold: A comparison of SAMs derived from normal and spiroalkanedithiols, Langmuir **16**, 2220–2224 (2000)
247. G.M. Chateauneuf, P.T. Mikulski, G.T. Gao, J.A. Harrison: Compression- and shear-induced polymerization in model diacetylene-containing monolayers, J. Phys. Chem. B **108**, 16626–16635 (2004)
248. L. Zhang, S. Jiang: Molecular simulation study of nanoscale friction for alkyl monolayers on Si(111), J. Chem. Phys. **117**, 1804–1811 (2002)
249. L.Z. Zhang, Y.S. Leng, S.Y. Jiang: Tip-based hybrid simulation study of frictional properties of self-assembled monolayers: Effects of chain length, terminal group, scan direction, and scan velocity, Langmuir **19**, 9742–9747 (2003)
250. M. Chandross, E.B.W. III, M.J. Stevens, G.S. Grest: Systematic study of the effect of disorder on nanotribology of self-assembled monolayers, Phys. Rev. Lett. **93**, 166103 (2004)
251. M. Chandross, G.S. Grest, M.J. Stevens: Friction between alkylsilane monolayers: Molecular simulation of ordered monolayers, Langmuir **18**, 8392–8399 (2002)
252. S.B. Sinnott, R. Andrews: Carbon nanotubes: Synthesis, properties and applications, Crit. Rev. Sol. St. Mater. Sci. **26**, 145–249 (2001)
253. B. Bhushan, B.K. Gupta, G.W. Van Cleef, C. Capp, J.V. Coe: Sublimed C_{60} films for tribology, Appl. Phys. Lett. **62**, 3253–3255 (1993)
254. T. Thundat, R.J. Warmack, D. Ding, R.N. Compton: Atomic force microscope investigation of C_{60} adsorbed on silicon and mica, Appl. Phys. Lett. **63**, 891–893 (1993)
255. C.M. Mate: Nanotribology studies of carbon surfaces by force microscopy, Wear **168**, 17–20 (1993)
256. R. Lüthi, E. Meyer, H. Haefke: Sled-type motion on the nanometer scale: Determination of dissipation and cohesive energies of C_{60}, Science **266**, 1979–1981 (1993)

257. R. Lüthi, H. Haefke, E. Meyer, L. Howald, H.-P. Lang, G. Gerth, H.J. Güntherodt: Frictional and atomic-scale study of C_{60} thin films by scanning force microscopy, Z. Phys. B **95**, 1–3 (1994)
258. Q.-J. Xue, X.-S. Zhang, F.-Y. Yan: Study of the structural transformations of C_{60}/C_{70} crystals during friction, Chin. Sci. Bull. **39**, 819–822 (1994)
259. W. Allers, U.D. Schwarz, G. Gensterblum, R. Wiesendanger: Low-load friction behavior of epitaxial C_{60} monolayers, Z. Phys. B **99**, 1–2 (1995)
260. U.D. Schwarz, W. Allers, G. Gensterblum, R. Wiesendanger: Low-load friction behavior of epitaxial C_{60} monolayers under Hertzian contact, Phys. Rev. B **52**, 14976–14984 (1995)
261. J. Ruan, B. Bhushan: Nanoindentation studies of sublimed fullerene films using atomic force microscopy, J. Mater. Res. **8**, 3019–3022 (1996)
262. U.D. Schwarz, O. Zworner, P. Koster, R. Wiesendanger: Quantitative analysis of the frictional properties of solid materials at low loads. I. Carbon compounds, Phys. Rev. B **56**, 6987–6996 (1997)
263. S. Okita, M. Ishikawa, K. Miura: Nanotribological behavior of C_{60} films at an extremely low load, Surf. Sci. **442**, L959–L963 (1999)
264. S. Okita, K. Miura: Molecular arrangement in C_{60} and C_{70} films on graphite and their nanotribological behavior, Nano Lett. **1**, 101–103 (2001)
265. K. Miura, S. Kamiya, N. Sasaki: C_{60} molecular bearings, Phys. Rev. Lett. **90**, 055509 (2003)
266. A. Buldum, J.P. Lu: Atomic scale sliding and rolling of carbon nanotubes, Phys. Rev. Lett. **83**, 5050–5053 (1999)
267. M.R. Falvo, R.M. Taylor, A. Helser, V. Chi, F.P. Brooks, S. Washburn, R. Superfine: Nanometer-scale rolling and sliding of carbon nanotubes, Nature **397**, 236–238 (1999)
268. M.R. Falvo, J. Steele, R.M.T. II, R. Superfine: Gearlike rolling motion mediated by commensurate contact: Carbon nanotubes on HOPG, Phys. Rev. B **62**, R10664–R10667 (2000)
269. J.D. Schall, D.W. Brenner: Molecular dynamics simulations of carbon nanotube rolling and sliding on graphite, Mol. Simul. **25**, 73–80 (2000)
270. B. Ni, S.B. Sinnott: Tribological properties of carbon nanotube bundles, Surf. Sci. **487**, 87–96 (2001)
271. B. Ni, S.B. Sinnott: Mechanical and tribological properties of carbon nanotubes investigated with atomistic simulations, *Nanotubes and related materials*. In: *Nanotubes and Related Materials*, MRS Proceedings, Vol. 633 (Materials Research Society, Pittsburgh, PA 2001) pp. A17.13.11–A17.13.15
272. K. Miura, T. Takagi, S. Kamiya, T. Sahashi, M. Yamauchi: Natural rolling of zigzag multiwalled carbon nanotubes on graphite, Nano Lett. **1**, 161–163 (2001)
273. K. Miura, M. Ishikawa, R. Kitanishi, M. Yoshimura, K. Ueda, Y. Tatsumi, N. Minami: Bundle structure and sliding of single-walled carbon nanotubes observed by friction-force microscopy, Appl. Phys. Lett. **78**, 832–834 (2001)
274. P.E. Sheehan, C.M. Lieber: Nanotribology and nanofabrication of MoO_3 structures by atomic force microscopy, Science **272**, 1158–1161 (1996)
275. J. Wang, K.C. Rose, C.M. Lieber: Load-independent friction: MoO_3 nanocrystal lubricants, J. Phys. Chem. B **103**, 8405–8408 (1999)
276. Q. Ouyang, K. Okada: Nanoballbearing effect of ultra-fine particles of cluster diamond, Appl. Surf. Sci. **78**, 309–313 (1994)
277. R. Luthi, E. Meyer, H. Haefke, L. Howald, W. Gutmannsbauer, H.J. Guntherodt: Sled-type motion on the nanometer-scale – determination of dissipation and cohesive energies of C_{60}, Science **266**, 1979–1981 (1994)

278. B. Bhushan, B.K. Gupta, G.W. Vancleef, C. Capp, J.V. Coe: Fullerene (C_{60}) films for solid lubrication, Tribol. Trans. **36**, 573–580 (1993)
279. U.D. Schwarz, W. Allers, G. Gensterblum, R. Wiesendanger: Low-load friction behavior of epitaxial C_{60} monolayers under Hertzian contact, Phys. Rev. B **52**, 14976–14984 (1995)
280. S.B. Legoas, R. Giro, D.S. Galvao: Molecular dynamics simulations of C_{60} nanobearings, Chem. Phys. Lett. **386**, 425–429 (2004)
281. P.L. Dickrell, S.B. Sinnott, D.W. Hahn, N.R. Raravikar, L.S. Schadler, P.M. Ajayan, W.G. Sawyer: Frictional anisotropy of oriented carbon nanotube surfaces, Tribol. Lett. **18**, 59–62 (2005)
282. G.T. Gao, P.T. Mikulski, J.A. Harrison: Molecular-scale tribology of amorphous carbon coatings: Effects of film thickness, adhesion, and long-range interactions, J. Am. Chem. Soc. **124**, 7202–7209 (2002)
283. F.P. Bowden, D. Tabor: *The Friction and Lubrication of Solids, Part 2* (Clarenden, Oxford 1964)
284. G.T. Gao, P.T. Mikulski, G.M. Chateauneuf, J.A. Harrison: The effects of film structure and surface hydrogen on the properties of amorphous carbon films, J. Phys. Chem. B **107**, 11082–11090 (2003)

14

Mechanical Properties of Nanostructures

Bharat Bhushan

Summary. Structural integrity is of paramount importance in all devices. Load applied during the use of devices can result in component failure. Cracks can develop and propagate under tensile stresses, leading to failure. Knowledge of the mechanical properties of nanostructures is necessary for designing realistic MEMS/NEMS and BioMEMS/BioNEMS devices. Elastic and inelastic properties are needed to predict deformation from an applied load in the elastic and inelastic regimes, respectively. The strength property is needed to predict the allowable operating limit. Some of the properties of interest are hardness, elastic modulus, bending strength, fracture toughness and fatigue strength. Many of the mechanical properties are scale dependent therefore these should be measured at relevant scales. Atomic force microscopy and nanoindenters can be used satisfactorily to evaluate the mechanical properties of micro/nanoscale structures.

Commonly used materials in MEMS/NEMS are single-crystal silicon and silicon-based materials, e.g., SiO_2 and polysilicon films deposited by low-pressure chemical vapor deposition. An early study showed silicon to be a mechanically resilient material in addition to its favorable electronic properties. Single-crystal SiC deposited on large-area silicon substrates is used for high-temperature micro/nanosensors and actuators. Amorphous alloys can be formed on both metal and silicon substrates by sputtering and plating techniques, providing more flexibility in surface-integration. Electroless deposited Ni–P amorphous thin films have been used to construct microdevices, especially using the so-called LIGA techniques. Micro/nanodevices need conductors to provide power, as well as electrical/magnetic signals to make them functional. Electroplated gold films have found wide applications in electronic devices because of their ability to make thin films and process simply.

Polymers, such as poly (methyl methacrylate) (PMMA) and poly (dimethylsiloxane) (PDMS) are commonly used in BioMEMS/BioNEMS, such as micro/nanofluidic devices, because of ease of manufacturing and reduced cost. The use of polymers also offers a wide range of material properties to allow tailoring of biological interactions for improved biocompatibility.

This chapter presents a review of mechanical property measurements on the micro/nanoscale of various materials of interest and stress and deformation analyses of nanostructures.

14.1 Introduction

Microelectromechanical systems (MEMS) refer to microscopic devices that have a characteristic length of less than 1 mm but more than 100 nm (1 µm), and nanoelectromechanical systems (NEMS) refer to nanoscopic devices that have a characteristic length of less than 100 nm (or 1 µm). These are referred to as an intelligent miniaturized system comprising sensing, processing, and/or actuating functions and combine electrical and mechanical components. The acronym MEMS originated in the U.S.A. The term commonly used in Europe is microsystem technology (MST) and in Japan is micromachines. Another term generally used is micro/nanodevices. MEMS/NEMS terms are also now used in a broad sense and include electrical, mechanical, optical, biological, and/or fluidic functions. To put the dimensions in perspective, individual atoms are typically fraction of a nanometer in diameter, DNA molecules are about 2.5 nm wide, biological cells are in the range of thousands of nm in diameter, and human hair is about 75 µm in diameter. The mass of a micromachined silicon structure can be as low as 1 nN and NEMS can be built with mass as low as 10^{-20} N with cross sections of about 10 nm. In comparison, the mass of a drop of water is about 10 µN and the mass of an eyelash is about 100 nN.

A wide variety of MEMS, including Si-based devices, chemical and biological sensors and actuators, and miniature non-silicon structures (e.g., devices made from plastics or ceramics) have been fabricated with dimensions in the range of a couple to a few thousand microns (see e.g., [1–13]). A variety of NEMS have also been produced (see e.g., [14–19]). MEMS/NEMS technology and fabrication processes have found a variety of applications in biology and biomedicine, leading to the establishment of an entirely new field known as BioMEMS/BioNEMS [20–28]. The ability to use micro/nanofabrication processes to develop precision devices that can interface with biological environments at the cellular and molecular level has led to advances in the fields of biosensor technology [27, 29–32], drug delivery [33–35], and tissue engineering [36–38]. The miniaturization of fluidic systems using micro/nanofabrication techniques has led to new and more efficient devices for medical diagnostics and biochemical analysis [39]. Largest "killer" industrial applications of MEMS include accelerometers (over a billion dollars a year in 2004), pressure sensors for manifold absolute pressure sensing for engines (more than 30 million units in 2004), and tire pressure measurements, inkjet printer heads (more than 500 million units in 2004), and digital micromirror devices (about $700 million revenues in 2004). BIOMEMS and BIONEMS are increasingly used in commercial applications. Largest applications of BIOMEMS include silicon-based disposable blood pressure sensor chips for blood pressure monitoring (more than 25 million units in 2004), and a variety of biosensors.

Structural integrity is of paramount importance in all devices. Load applied during the use of devices can result into component failure. Cracks can develop and propagate under tensile stresses leading to failure [31, 40]. Friction/stiction and wear limit the lifetimes and compromise the performance and reliability of the devices involving relative motion [4, 5, 41]. Most MEMS/NEMS applications demand extreme reliability. Stress and deformation analyses are carried out for an optimal design. MEMS/NEMS designers require mechanical properties on the nanoscale. Mechanical properties include elastic, inelastic (plastic, fracture or viscoelastic), and strength. Elastic and inelastic properties are needed to predict deformation from an applied load in the elastic and inelastic regimes, respectively. The strength property is needed to predict the allowable operating limit. Some of the properties of interest are hardness, elastic modulus, creep, bending strength (fracture stress), fracture toughness, and fatigue strength. Micro/nanostructures have some surface topography and local scratches dependent upon the manufacturing process. Surface roughness and local scratches may compromise the reliability of the devices and their effect needs to be studied.

Most mechanical properties are scale dependent [5, 40, 42]. Several researchers have measured mechanical properties of silicon and silicon-based milli- to microscale structures including tensile tests and bending tests [43–52], resonant structure tests for measurement of elastic properties [53], fracture toughness tests [44, 46, 54–58], and fatigue tests [56, 59, 60]. Most recently, few researchers have measured mechanical properties of nanoscale structures using atomic force microscopy (AFM) [61, 62] and nanoindentation [63–65]. For stress and deformation analyses of simple geometries and boundary condition, analytical models can be used. For analysis of complex geometries, numerical models are needed. Conventional finite element method (FEM) can be used down to few tens of nanometer scale although its applicability is questionable at nanoscale. FEM has been used for simulation and prediction of residual stresses and strains induced in MEMS devices during fabrication [66], to perform fault analysis in order to study MEMS faulty behavior [67], to compute mechanical strain resulting from doping of silicon [68], analyze micromechanical experimental data [46, 69, 70], and nanomechanical experimental data [62]. FEM analysis of nanostructures has been carried out to analyze the effect of types of surface roughness and scratches on stresses in nanostructures [71, 72].

Commonly used materials for MEMS/NEMS are single-crystal silicon and silicon-based materials (e.g., SiO_2 and polysilicon films deposited by low pressure chemical vapor deposition (LPCVD) process) [5]. An early study showed silicon to be a mechanically resilient material in addition to its favorable electronic properties [73]. Single-crystal 3C-SiC (cubic or β-SiC) films, deposited by atmospheric pressure chemical vapor deposition (APCVD) process on large-area silicon substrates are produced for high-temperature micro/nanosensor and actuator applications [74–76]. Amorphous alloys can be formed on both metal and silicon substrates by sputtering and plating techniques, providing more flexibilities in surface-integration. Electroless deposited Ni−P amorphous thin films have been used to construct microdevices, especially using the so-called LIGA techniques [5, 64]. Micro/nanodevices need conductors to provide power as well as electrical/magnetic signals to make them functional. Electroplated gold films have found wide applications in electronic devices because of their ability to make thin films and process simplicity [64].

As the field of MEMS/NEMS has progressed, alternative materials, especially polymers have established an important role in the advancement of the technology. This trend has been driven by the reduced cost associated with polymer materials. Polymer microfabrication processes, including the micromolding and hot embossing techniques [77], can be orders of magnitude less expensive than traditional silicon photolithography processes. The use of polymers in the BioMEMS/BioNEMS field has additional functional advantages, as polymers offer a wide range of material properties to allow tailoring of biological interactions for improved biocompatibility. Polymer BioMEMS structures involving microbeams have been designed to measure cellular forces [65]. Polymer materials most commonly used for biomedical applications include poly(methyl methacrylate) (PMMA), and poly(dimethylsiloxane) (PDMS) [77, 78]. Another material of interest due to ease of fabrication is poly(propyl methacrylate) (PPMA), which has lower glass transition temperature (T_g) (35–43 °C) [79] than PMMA (104–106 °C) [80, 81], which permits low temperature thermal processing.

This chapter presents a review of mechanical property measurements on nanoscale of various materials of interest and stress and deformation analyses of nanostructures.

14.2 Experimental Techniques for Measurement of Mechanical Properties of Nanostructures

14.2.1 Indentation and Scratch Tests Using Micro/Nanoindenters

A nanoindenter is commonly used to measure hardness, elastic modulus, and fracture toughness, and to perform micro/nanoscratch studies to get a measure of scratch/wear resistance of materials [5, 82].

Hardness and Elastic Modulus

The nanoindenter monitors and records the dynamic load and displacement of the three-sided pyramidal diamond (Berkovich) indenter during indentation with a force resolution of about 75 nN and displacement resolution of about 0.1 nm. Hardness and elastic modulus are calculated from the load-displacement data [5, 82]. The peak indentation load depends on mechanical properties of the specimen; a harder material requires higher load for a reasonable indentation depth.

Fracture Toughness

The indentation technique for fracture toughness measurement of brittle samples, on the microscale, is based on the measurement of the lengths of median-radial cracks produced by indentation. A Vickers indenter (a four-sided diamond pyramid) is used in a microhardness tester. A load on the order of 0.5 N is typically used in making the Vickers indentations. The indentation impressions are examined using an optical microscope with Nomarski interference contrast to measure the length of median-radial cracks, c. The fracture toughness (K_{IC}) is calculated by the following relation [83],

$$K_{IC} = \alpha \left(\frac{E}{H}\right)^{\frac{1}{2}} \left(\frac{P}{c^{\frac{3}{2}}}\right), \qquad (14.1)$$

where α is an empirical constant depending on the geometry of the indenter, E and H are hardness and elastic moduli, and P is the peak indentation load. For Vickers indenters, α has been empirically found based on experimental data and is equal to 0.016 [5]. Both E and H values are obtained from the nanoindentation data. The crack length is measured from the center of the indent to the end of crack using an optical microscope. For one indent, all crack lengths are measured. The crack length c is obtained from the average values of several indents.

Indentation Creep

For indentation creep test of polymer samples, the test is performed using a continuous stiffness measurements (CSM) technique [82]. In a study by Wei et al. [65], the indentation load was typically 30 µN and the loading rate was 3 µN/s. The tip was held typically for 600 s after the indentation load reached 30 µN. To measure the mean stress and contact stiffness, during the hold segment the indenter was oscillated at a peak-to-peak load amplitude of 1.2 µN and a frequency of 45 Hz.

Scratch Resistance

In micro/nanoscratch studies, in a nanoindenter, a conical diamond indenter having a tip radius of about 1 µm and an included angle of 60°, is drawn over the sample surface, and the load is ramped up until substantial damage occurs [5, 82]. The coefficient of friction is monitored during scratching. In order to obtain scratch depths during scratching, the surface profile of the sample surface is first obtained by translating the sample at a low load of about 0.2 mN, which is insufficient to damage a hard sample surface. The 500 µm long scratches are made by translating the sample while ramping the loads on the conical tip over different loads dependent upon the material hardness. The actual depth during scratching is obtained by subtracting the initial profile from the scratch depth measured during scratching. In order to measure the scratch depth after the scratch, the scratched surface is profiled at a low load of 0.2 mN and is subtracted from the actual surface profile before scratching.

14.2.2 Bending Tests of Nanostructures Using an AFM

Quasi-static bending tests of fixed nanobeam arrays are carried out using an AFM [62, 84]. A three-sided pyramidal diamond tip (with a radius of about 200 nm) mounted on a rectangular stainless steel cantilever is used for the bending tests. The beam stiffness is selected based on the desired load range. The stiffness of the cantilever beams for application of normal load up to 100 µN is about 150–200 N/m.

The wafer with nanobeam array is fixed onto a flat sample chuck using double-stick tape [62]. For the bending test, the tip is brought over the nanobeam array with

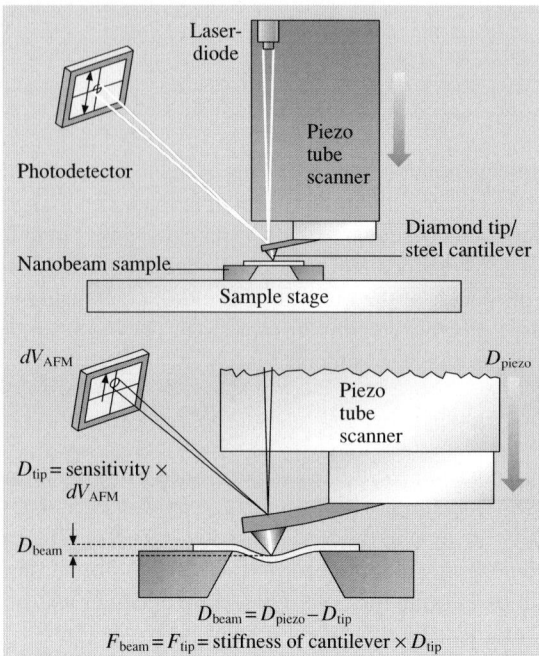

Fig. 14.1. Schematic showing the details of a nanoscale bending test using an AFM. The AFM tip is brought to the center of the nanobeam and the piezo is extended over a known distance. By measuring the tip displacement, a load displacement curve of the nanobeam can be obtained [62]

the help of the sample stage of the AFM and a built-in high magnification optical microscope Fig. 14.1. For the fine positioning of the tip over a chosen beam, the array is scanned in contact mode at a contact load of about 2 to 4 µN, which results in a negligible damage to the sample. After scanning, the tip is located at one end of a chosen beam. To position the tip at the center of the beam span, the tip is moved to the other end of the beam by giving the X-piezo an offset voltage. The value of this offset is determined after several such attempts have been made in order to minimize effects of piezo drift. Half of this offset is then applied to the X-piezo after the tip is positioned at one end of the beam, which usually results in the tip being moved to the center of the span. Once the tip is positioned over the center of the beam span, the tip is held stationary without scanning and the Z-piezo is extended by a known distance, typically about 2.5 µm, at a rate of 10 nm/s, as shown in Fig. 14.1. During this time, the vertical deflection signal (dV_{AFM}), which is proportional to the deflection of the cantilever (D_{tip}), is monitored. The displacement of the piezo is equal to the sum of the displacements of the cantilever and the nanobeam. Hence the displacement of the nanobeam (D_{beam}) under the point of load can be determined as

$$D_{beam} = D_{piezo} - D_{tip} \,. \tag{14.2}$$

The load (F_{beam}) on the nanobeam is the same as the load on the tip/cantilever (F_{tip}) and is given by

$$F_{beam} = F_{tip} = D_{tip} \times k \,, \tag{14.3}$$

where k is the stiffness of the tip/cantilever. In this manner, a load displacement curve for each nanobeam can be obtained.

The photodetector sensitivity of the cantilever needs to be calibrated to obtain D_{tip} in nm. For this calibration, the tip is pushed against a smooth diamond sample by moving the Z-piezo over a known distance. For the hard diamond material, the actual deflection of the tip can be assumed to be the same as the Z-piezo travel (D_{piezo}), and the photodetector sensitivity (S) for the cantilever setup is determined as

$$S = D_{piezo}/dV_{AFM} \text{nm/V}. \tag{14.4}$$

In the measurements, D_{tip} is given as $dV_{AFM} \times S$.

Since a sharp tip would result in an undesirable large local indentation, *Sundararajan* and *Bhushan* [62] used a diamond tip which was a worn (blunt) diamond tip. Indentation experiments using this tip on a silicon substrate yielded a residual depth of less than 8 nm at a maximum load of 120 μN, which is negligible compared to displacements of the beams (several hundred nm). Hence we can assume that negligible local indentation or damage is created during the bending process of the beams and that the displacement calculated from (14.2) is from the beam structure.

Elastic Modulus and Bending Strength

Elastic modulus and bending strength (fracture stress) of the beams can be estimated by equations based on the assumption that the beams follow linear elastic theory of an isotropic material. This is probably valid since the beams have high length-to-width ℓ/w and length-to-thickness ℓ/t ratios and also since the length direction is along the principal stress direction during the test. For a fixed elastic beam loaded at the center of the span, the elastic modulus is expressed as

$$E = \frac{\ell^3}{192 I} m, \tag{14.5}$$

where ℓ is the beam length, I is the area moment of inertia for the beam cross-section and m is the slope of the load-displacement curve during bending [85]. The area moment of inertia for a beam with trapezoidal cross section, is calculated from the following equation

$$I = \frac{w_1^2 4 w_1 w_2 + w_2^2}{36(w_1 + w_2)} t^3, \tag{14.6}$$

where w_1 and w_2 are the upper and lower widths, respectively, and t is the thickness of the beam. According to linear elastic theory, for a centrally loaded beam, the moment diagram is shown in Fig. 14.2. The maximum moments are generated at the ends (negative moment) and under the loading point (positive moment) as shown in Fig. 14.2. The bending stresses generated in the beam are proportional to the moments and are compressive or tensile about the neutral axis (line of zero stress).

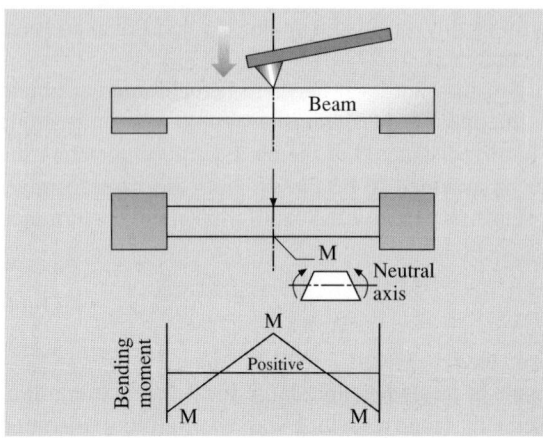

Fig. 14.2. A schematic of the bending moments generated in the beam during a quasi-static bending experiment, with the load at the center of the span. The maximum moments occur under the load and at the fixed ends. Due to the trapezoidal cross section, the maximum tensile bending stresses occur at the top surfaces at the fixed ends

The maximum tensile stress (σ_b, which is the bending strength or fracture stress) is produced on the top surface at both the ends and is given by [85]

$$\sigma_b = \frac{F_{\max} \ell e_1}{8I}, \tag{14.7}$$

where $F_{\max}$ is the applied load at failure and e_1 is the distance of the top surface from the neutral plane of the beam cross-section and is given by [85]

$$e_1 = \frac{t(w_1 + 2w_2)}{3(w_1 + w_2)}. \tag{14.8}$$

Although the moment value at the center of the beam is the same as at the ends, the tensile stresses at the center (generated on the bottom surface) are less than those generated at the ends (per (14.7)) because the distance from the neutral axis to the bottom surface is less than e_1. This is because of the trapezoidal cross section of the beam, which results in the neutral axis being closer to the bottom surface than the top (Fig. 14.2).

In the preceding analysis, the beams were assumed to have fixed ends. However, in the nanobeams used by *Sundararajan* and *Bhushan* [62], the underside of the beams was pinned over some distance on either side of the span. Hence a finite element model of the beams was created to see if the difference in the boundary conditions affected the stresses and displacements of the beams. It was found that the difference in the stresses was less than 1%. This indicates that the boundary conditions near the ends of the actual beams are not that different from that of fixed ends. Therefore the bending strength values can be calculated from (14.7).

Fracture Toughness

Fracture toughness is another important parameter for brittle materials such as silicon. In the case of the nanobeam arrays, these are not best suited for fracture toughness measurements because they do not possess regions of uniform stress during

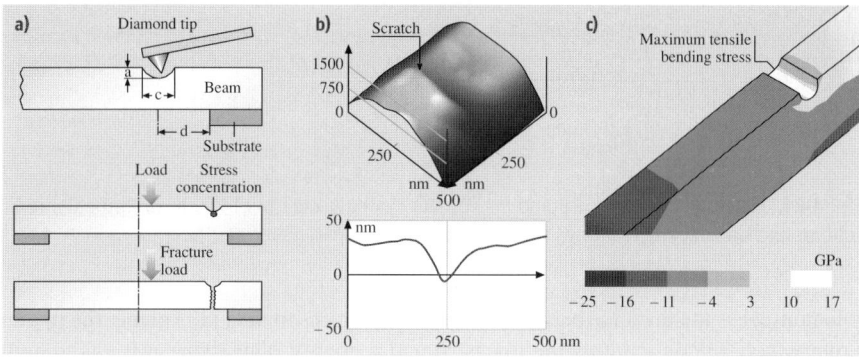

Fig. 14.3. (a) Schematic of technique to generate a defect (crack) of known dimensions in order to estimate fracture toughness. A diamond tip is used to generate a scratch across the width of the beam. When the beam is loaded as shown, a stress concentration is formed at the bottom of the scratch. The fracture load is then used to evaluate the stresses using FEM. (b) AFM 3-D image and 2-D profile of a typical scratch. (c) Finite element model results verifying that the maximum bending stress occurs at the bottom of the scratch [62]

bending. *Sundararajan* and *Bhushan* [62] developed a methodology and its steps are outlined schematically in Fig. 14.3a. First, a crack of known geometry is introduced in the region of maximum tensile bending stress, i.e. on the top surface near the ends of the beam. This is achieved by generating a scratch at high normal load across the width (w_1) of the beam using a sharp diamond tip (radius < 100 nm). A typical scratch thus generated is shown in Fig. 14.3b. By bending the beam as shown, a stress concentration will be formed under the scratch. This will lead to failure of the beam under the scratch once a critical load (fracture load) is attained. The fracture load and relevant dimensions of the scratch are input into the FEM model, which is used to generate the fracture stress plots. Figure 14.3c shows an FEM simulation of one such experiment, which reveals that the maximum stress does occur under the scratch.

If we assume that the scratch tip acts as a crack tip, a bending stress will tend to open the crack in Mode I. In this case, the stress field around the crack tip can be described by the stress intensity parameter K_I (for Mode *I*) for linear elastic materials [86]. In particular the stresses corresponding to the bending stresses are described by

$$\sigma = \frac{K_I}{\sqrt{2\pi r}} \cos\left(\frac{\theta}{2}\right)\left[1 + \sin\left(\frac{\theta}{2}\right)\sin\left(\frac{3\theta}{2}\right)\right] \tag{14.9}$$

for every point $p(r,\theta)$ around the crack tip as shown in Fig. 14.4. If we substitute the fracture stress (σ_f) into the left hand side of (14.9), then the K_I value can be substituted by its critical value, which is the fracture toughness K_{IC}. Now, the fracture stress can be determined for the point ($r = 0, \theta = 0$), i.e. right under the crack tip as explained above. However, we cannot substitute $r = 0$ in (14.9). The alternative is to substitute a value for r, which is as close to zero as possible. For silicon, a reason-

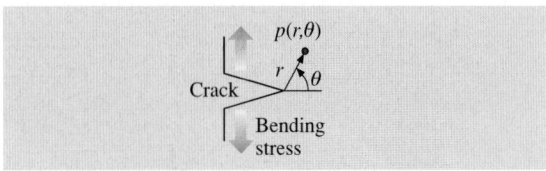

Fig. 14.4. Schematic of crack tip and coordinate systems used in (14.9) to describe a stress field around the crack tip in terms of the stress intensity parameter, K_I [62]

able number is the distance between neighboring atoms in the (111) plane, the plane along which silicon exhibits the lowest fracture energy. This value was calculated from silicon unit cell dimensions of 0.5431 nm [87] to be 0.4 nm (half of the face diagonal). This assumes that Si displays no plastic zone around the crack tip, which is reasonable since in tension, silicon is not known to display much plastic deformation at room temperature. *Sundararajan* and *Bhushan* [62] used values $r = 0.4$ to 1.6 nm (i.e. distances up to 4 times the distance between the nearest neighboring atoms) to estimate the fracture toughness for both Si and SiO_2 according to the following equation

$$K_{IC} = \sigma_f \sqrt{2\pi r} \quad r = 0.4 \text{ to } 1.6 \text{ nm}. \tag{14.10}$$

Fatigue Strength

In addition to the properties mentioned so far that can be evaluated from quasi-static bending tests, the fatigue properties of nanostructures are also of interest. This is especially true for MEMS/NEMS involving vibrating structures such as oscillators and comb drives [88] and hinges in digital micromirror devices [89]. To study the fatigue properties of the nanobeams, *Sundararajan* and *Bhushan* [62] applied monotonic cyclic stresses using an AFM, Fig. 14.5a. Similar to the bending test, the diamond tip is first positioned at the center of the beam span. In order to ensure that the tip is always in contact with the beam (as opposed to impacting it), the piezo is first extended by a distance D_1, which ensures a minimum stress on the beam. After this extension, a cyclic displacement of amplitude D_2 is applied continuously until failure of the beam occurs. This results in the application of a cyclic load to the beam. The maximum frequency of the cyclic load that could be attained using the AFM by *Sundararajan* and *Bhushan* [62] was 4.2 Hz. The vertical deflection signal of the tip is monitored throughout the experiment. The signal follows the pattern of the piezo input up to failure, which is indicated by a sudden drop in the signal. During initial runs, piezo drift was observed that caused the piezo to gradually move away from the beam (i.e. to retract), resulting in a continuous decrease in the applied normal load. In order to compensate for this, the piezo is given a finite extension of 75 nm every 300 seconds as shown in Fig. 14.5a. This results in keeping the applied loads fairly constant. The normal load variation (calculated from the vertical deflection signal) from a fatigue test is shown in Fig. 14.5b. The cyclic stress amplitudes (cor-

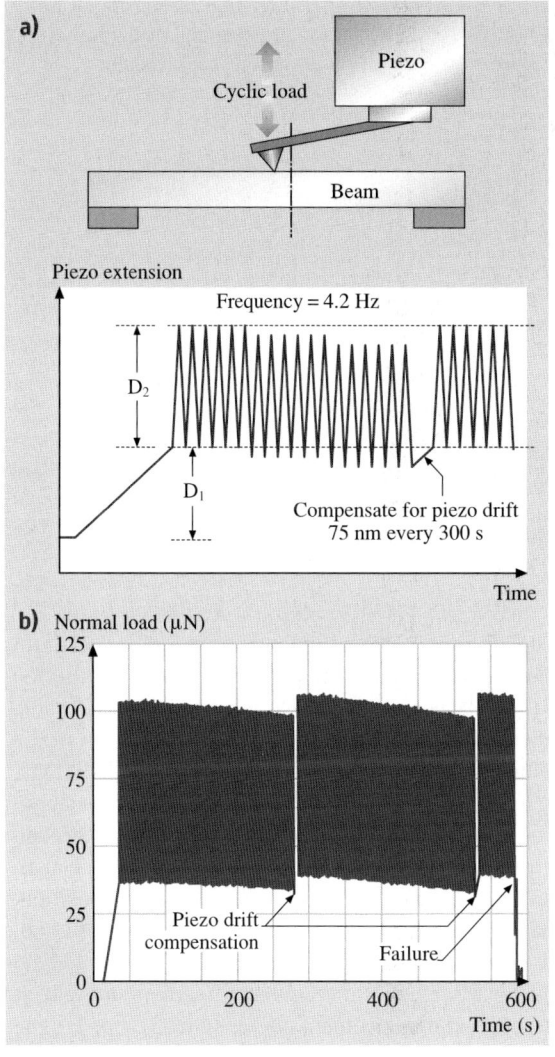

Fig. 14.5. (a) Schematic showing the details of the technique to study fatigue behavior of the nanobeams. The diamond tip is located at the middle of the span and a cyclic load at 4.2 Hz is applied to the beam by forcing the piezo to move in the pattern shown. An extension is made every 300 s to compensate for the piezo drift to ensure that the load on the beam is kept fairly constant. (b) Data from a fatigue experiment on a nanobeam until failure. The normal load is computed from the raw vertical deflection signal. The compensations for piezo drift keep the load fairly constant [62]

responding to D_2) and fatigue lives are recorded for every sample tested. Values for D_1 are set such that minimum stress levels are about 20% of the bending strengths.

14.2.3 Bending Tests of Micro/Nanostructures Using a Nanoindenter

Quasi-static bending tests of micro/nanostructures are also carried out using a nanoindenter (Fig. 14.6) [63–65]. The advantage of nanoindenter is that loads up to about 400 mN, higher than that in AFM (up to about 100 μN), can be used for structures requiring high loads for experiments. To avoid the indenter tip pushing into the specimen, a blunt tip is used in the bending and fatigue tests. For ceramic and metallic

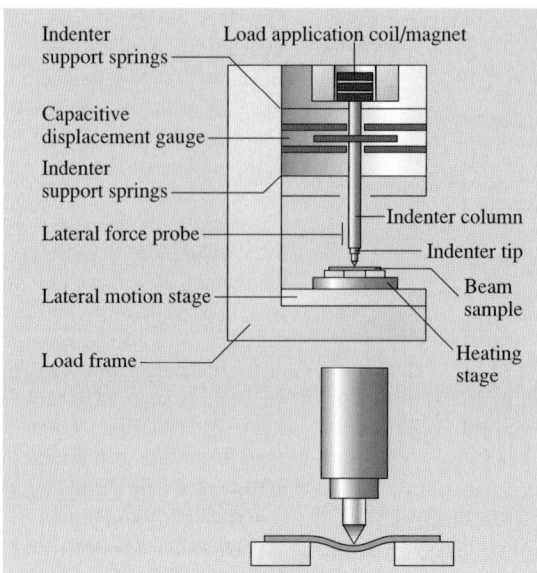

Fig. 14.6. Schematic of micro/nanoscale bending test using a nanoindenter

beam samples, *Li* et al. [64] used a diamond conical indenter with a radius of 1 μm and an included angle of 60°. For polymer beam samples, *Wei* et al. [65] reported that the diamond tip penetrated the polymer beams easily and caused considerable plastic deformation during the bending test, which led to significant errors in the measurements. To avoid this issue, the diamond tip was dip-coated with PMMA (about 1–2 μm thick) by dipping the tip in the 2% PMMA (wt/wt) solution for about 5 seconds. Load position used was at the center of the span for the bridge beams and at 10 μm off from the free end of cantilever beams. An optical microscope with a magnification of 1500 × or an in-situ AFM is used to locate the loading position. Then the specimen is moved under the indenter location with a resolution of about 200 nm in longitudinal direction and less than 100 nm in lateral direction.

Using the analysis presented earlier, elastic modulus and bending strength of the beams can be obtained from the load displacement curves [64, 65]. For fatigue tests, an oscillating load is applied and contact stiffness is measured during the tests. A significant drop in the contact stiffness during the test is a measure of number of cycles to failure [63].

14.3 Experimental Results and Discussion

14.3.1 Indentation and Scratch Tests of Various Ceramic and Metals Using a Micro/Nanoindenter

Studies have been conducted on five different materials: undoped single-crystal Si(100), undoped polysilicon film, SiO_2 film, SiC film, electroless deposited Ni–

11.5 wt % P amorphous film, and electroplated Au film [64,75,76]. A 3 μm thick polysilicon film was deposited by a low pressure chemical vapor deposition (LPCVD) process on an Si(100) substrate. The 1 μm thick SiO_2 film was deposited by a plasma enhanced chemical vapor deposition (PECVD) process on an Si(111) substrate. A 3 μm thick 3C-SiC film was epitaxially grown using an atmospheric pressure chemical vapor deposition (APCVD) process on Si(100) substrate. A 12 μm thick Ni–P film was electroless plated on a 0.8 mm thick Al – 4.5 wt % Mg alloy substrate. A 3 μm thick Au film was electroplated on an Si(100) substrate.

Hardness and Elastic Modulus

Hardness and elastic modulus measurements are made using a nanoindenter [64]. The hardness and elastic modulus values of various materials at a peak indentation depth of 50 nm are summarized in Fig. 14.7 and Table 14.1. The SiC film exhibits the highest hardness of about 25 GPa an elastic modulus of about 395 GPa among the samples examined, followed by the undoped Si(100), undoped polysilicon film, SiO_2 film, Ni–P film, and Au film. The hardness and elastic modulus data of the undoped Si(100) and undoped polysilicon film are comparable. For the metal alloy films, the Ni–P film exhibits higher hardness and elastic modulus than the Au film.

Fracture Toughness

The optical images of Vickers indentations made using a microindenter, at a normal load of 0.5 N held for 15 s on the undoped Si(100), undoped polysilicon film, and SiC film are shown in Fig. 14.8 [76]. The SiC film exhibits the smallest indentation mark, followed by the undoped polysilicon film and undoped Si(100). These Vickers indentation depths are smaller than one-third of the film thickness. Thus, the influence of substrate on the fracture toughness of the films can be ignored. In addition to the indentation marks, radial cracks are observed, emanating from the indentation corners. The SiC film shows the longest radial crack length, followed by the undoped Si(100) and undoped polysilicon film. The radial cracks for the undoped Si(100) are straight whereas those for the SiC, undoped polysilicon film are

Table 14.1. Hardness, elastic modulus, fracture toughness and critical load results of the bulk single-crystal Si(100) and thin films of undoped polysilicon, SiO_2, SiC, Ni–P and Au

Samples	Hardness (GPa)	Elastic modulus (GPa)	Fracture toughness (MPa m$^{1/2}$)	Critical load (mN)
Undoped Si(100)	12	165	0.75	11
Undoped polysilicon film	12	167	1.11	11
SiO_2 film	9.5	144	0.58 (Bulk)	9.5
SiC film	24.5	395	0.78	14
Ni–P film	6.5	130		0.4 (Plowing)
Au film	4	72		0.4 (Plowing)

not straight but go in a zigzag manner. The fracture toughness (K_{IC}) is calculated using (14.1).

The fracture toughness values of all samples are summarized in Fig. 14.7 and Table 14.1. The SiO_2 film used in this study is about 1 μm thick, which is not thick enough for fracture toughness measurement. The fracture toughness value of bulk silica are listed instead for a reference. The Ni–P and Au films exhibit very high fracture toughness values that cannot be measured by indentation methods. For other samples, the undoped polysilicon film has the highest value, followed by the

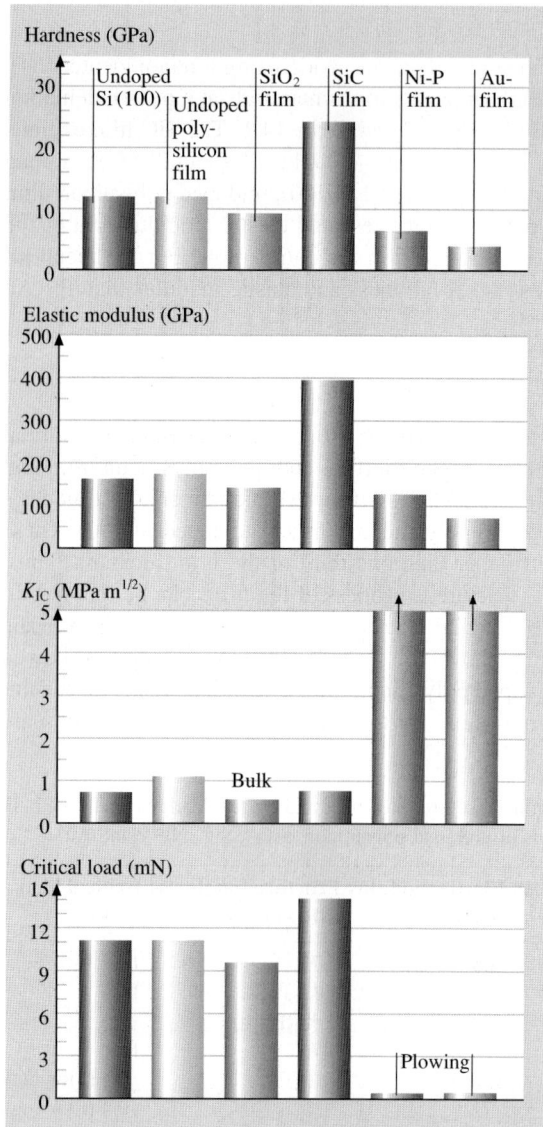

Fig. 14.7. Bar chart summarizing the hardness, elastic modulus, fracture toughness, and critical load (from scratch tests) results of the bulk undoped single-crystal Si(100) and thin films of undoped polysilicon, SiO_2, SiC, Ni–P, and Au [64]

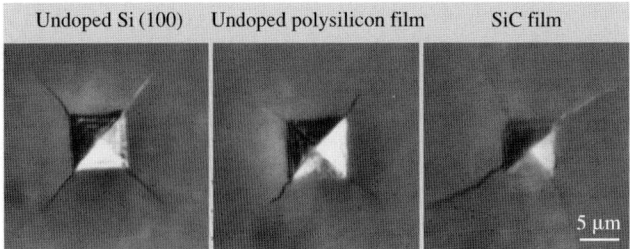

Fig. 14.8. Optical images of Vickers indentations made at a normal load of 0.5 N held for 15 s on the undoped Si(100), undoped polysilicon film, and SiC film [76]

undoped Si(100), SiC film, and SiO_2 film. For the undoped polysilicon film, the grain boundaries can stop the radial cracks and change the propagation directions of the radial cracks, making the propagation of these cracks more difficult. Values of fracture toughness for the undoped Si(100) and SiC film are comparable. Since the undoped Si(100) and SiC film are single crystal, no grain boundaries are present to stop the radial cracks and change the propagation directions of the radial cracks. This is why the SiC film shows a lower fracture toughness value than the bulk polycrystal SiC materials of 3.6 MPam$^{1/2}$ [90].

Scratch Resistance

Scratch resistance of various materials have been studied using a nanoindenter by Li et al. [64]. Figure 14.9 compares the coefficient of friction and scratch depth profiles as a function of increasing normal load and optical images of three regions over scratches: at the beginning of the scratch (indicated by A on the friction profile), at the point of initiation of damage at which the coefficient of friction increases to a high value or increase abruptly (indicated by B on the friction profile), and towards the end of the scratch (indicated by C on the friction profile) for all samples. Note that the ramp loads for the Ni–P and Au range are from 0.2 to 5 mN whereas the ramp loads for other samples are from 0.2 to 20 mN. All samples exhibit a continuous increase in the coefficient of friction with increasing normal load from the beginning of the scratch. The continuous increase in the coefficient of friction during scratching is attributed to the increasing plowing of the sample by the tip with increasing normal load, as shown in the SEM images in Fig. 14.9. The abrupt increase in the coefficient of friction is associated with catastrophic failure as well as significant plowing of the tip into the sample. Before the critical load, the coefficient of friction of the undoped polysilicon, SiC and SiO_2 films increased at a slower rate, and was smoother than that of the other samples. The undoped Si(100) exhibits some bursts in the friction profiles before the critical load. At the critical load, the SiC and undoped polysilicon films exhibit a small increase in the coefficient of friction whereas the undoped Si(100) and undoped polysilicon film exhibit a sudden increase in the coefficient of friction. The Ni–P and Au films show a continuous increase in the coefficient of friction, indicating the behavior of a ductile metal. The

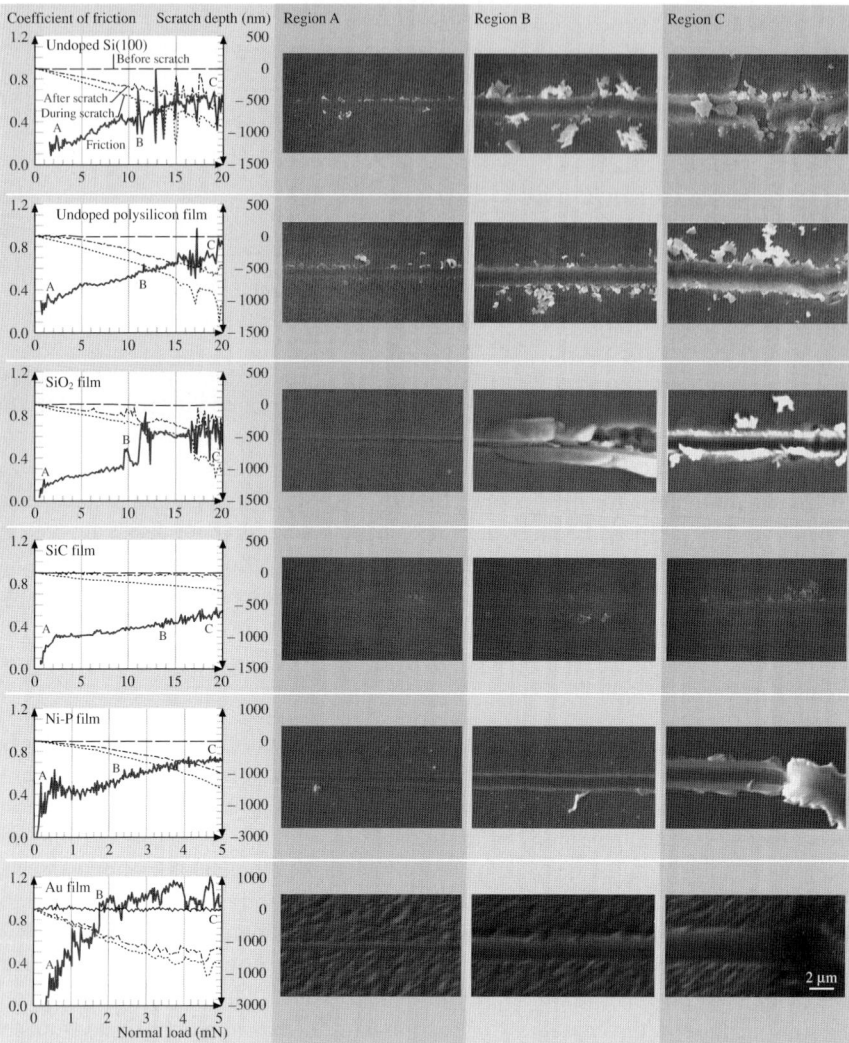

Fig. 14.9. Coefficient of friction and scratch depth profiles as a function of increasing normal load and optical images of three regions over scratches: at the beginning of the scratch (indicated by A on the friction profile), at the point of initiation of damage at which the coefficient of friction increases to a high value or increase abruptly (indicated by B on the friction profile), and towards the end of the scratch (indicated by C on the friction profile) for all samples [64]

bursts in the friction profile might result from the plastic deformation and material pile-up in front of the scratch tip. The Au film exhibits a higher coefficient of friction than the Ni–P film. This is because the Au film has lower hardness and elastic modulus values than the Ni–P film.

The SEM images show that below the critical loads the undoped Si(100) and undoped polysilicon film were damaged by plowing, associated with the plastic flow of the material and formation of debris on the sides of the scratch. For the SiC and SiO$_2$ films, in region A, a plowing scratch track was found without any debris on the side of the scratch, which is probably responsible for the smoother curve and slower increase in the coefficient of friction before the critical load. After the critical load, for the SiO$_2$ film, delamination of the film from the substrate occurred, followed by cracking along the scratch track. For the SiC film, only several small debris particles were found without any cracks on the side of the scratch, which is responsible for the small increase in the coefficient of friction at the critical load. For the undoped Si(100), cracks were found on the side of the scratch right from the critical load and up, which is probably responsible for the big bursts in the friction profile. For the undoped polysilicon film, no cracks were found on the side of the scratch at the critical load. This might result from grain boundaries which can stop the propagation of cracks. At the end of the scratch, some of the surface material was torn away and cracks were found on the side of the scratch in the undoped Si(100). A couple of small cracks were found in the undoped polysilicon and SiO$_2$ films. No crack was found in the SiC film. Even at the end of the scratch, less debris was found in the SiC film. A curly chip was found at the end of the scratch in both Ni−P and Au films. This is a typical characteristic of ductile metal alloys. The Ni−P and Au films were damaged by plowing right from the beginning of the scratch with material pile-up at the side of the scratch.

The scratch depth profiles obtained during and after the scratch on all samples with respect to initial profile, after the cylindrical curvature is removed, are plotted in Fig. 14.9. Reduction in scratch depth is observed after scratching as compared to that of during scratching. This reduction in scratch depth is attributed to an elastic recovery after removal of the normal load. The scratch depth after scratching indicates the final depth which reflects the extent of permanent damage and plowing of the tip into the sample surface, and is probably more relevant for visualizing the damage that can occur in real applications. For the undoped Si(100), undoped polysilicon film, and SiO$_2$ film, there is a large scatter in the scratch depth data after the critical loads, which is associated with the generation of cracks, material removal and debris. The scratch depth profile is smooth for the SiC film. It is noted that the SiC film exhibits the lowest scratch depth among the samples examined. The scratch depths of the undoped Si(100), undoped polysilicon film and SiO$_2$ film are comparable. The Ni−P and Au films exhibit much lager scratch depth than other samples. The scratch depth of the Ni−P film is smaller than that of the Au film.

The critical loads estimated from friction profiles for all samples are compared in Fig. 14.7 and Table 14.1. The SiC film exhibits the highest critical load of about 14 mN, as compared to other samples. The undoped Si(100) and undoped polysilicon film show comparable critical load of about 11 mN whereas the SiO$_2$ film shows a low critical load of about 9.5 mN. The Ni−P and Au films were damaged by plowing right from the beginning of the scratch.

14.3.2 Bending Tests of Ceramic Nanobeams Using an AFM

Bending tests have been performed on Si and SiO_2 nanobeam arrays [62, 84]. The single-crystal silicon bridge nanobeams were fabricated by bulk micromachining incorporating enhanced-field anodization using an AFM [61]. The Si nanobeams are oriented along the [110] direction in the (001) plane. Subsequent thermal oxidation of the beams results in formation of SiO_2 beams. The cross section of the nanobeams is trapezoidal owing to the anisotropic wet etching process. SEM micrographs of Si and SiO_2 nanobeam arrays and a schematic of the shape of a typical nanobeam are shown in Fig. 14.10. The actual widths and thicknesses of nanobeams were measured using an AFM in tapping mode prior to tests using a standard Si tapping mode tip (tip radius < 10 nm). Surface roughness measurements of the nanobeam surfaces in tapping mode yielded a σ of 0.7 ± 0.2 nm and peak-to-valley (P–V) distance of 4 ± 1.2 nm for Si and a σ of 0.8 ± 0.3 nm and a P–V distance of 3.1 ± 0.8 nm for SiO_2. Prior to testing, the Si nanobeams were cleaned by immersing them in a "piranha etch" solution (3:1 solution by volume of 98% sulphuric acid and 30% hydrogen peroxide) for 10 minutes to remove any organic contaminants.

Bending Strength

Figure 14.11 shows typical load displacement curves for Si and SiO_2 beams that were bent to failure [62, 84]. The upper width (w_1) of the beams is indicated in the figure. Also indicated in the figure are the elastic modulus values obtained from the slope of the load displacement curve (14.5). All the beams tested showed linear elastic behavior followed by abrupt failure, which is suggestive of brittle fracture. Figure 14.12 shows the scatter in the values of elastic modulus obtained for both Si and SiO_2 along with the average values (± standard deviation). The scatter in the values may be due to differences in orientation of the beams with respect to the trench and the loading point being a little off-center with respect to the beam span. The average values are a little higher than the bulk values (169 GPa for Si[110] and 73 GPa for SiO_2 in Table 14.2). However the values of E obtained from (14.5) have an error of about 20% due to the uncertainties in beam dimensions and spring constant of the tip/cantilever (which affects the measured load). Hence the elastic modulus values on the nanoscale can be considered to be comparable to bulk values.

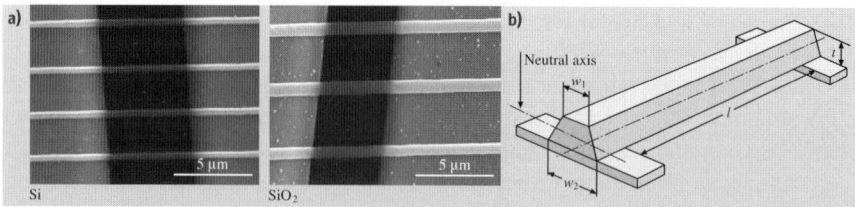

Fig. 14.10. (a) SEM micrographs of nanobeam arrays, and (b) a schematic of the shape of a typical nanobeam. The trapezoidal cross-section is due to the anisotropic wet etching during the fabrication [84]

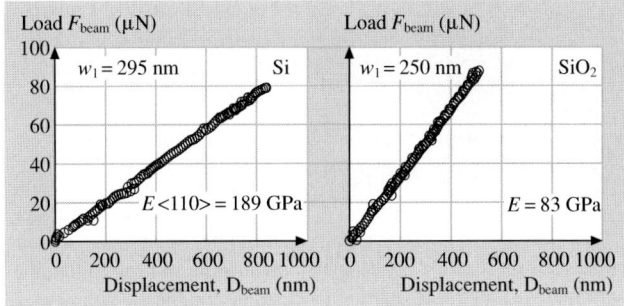

Fig. 14.11. Typical load displacement curves of silicon and SiO$_2$ nanobeams. The curves are linear until sudden failure, indicative of brittle fracture of the beams. The elastic modulus (E) values calculated from the curves are shown. The dimensions of the Si beam were $w_1 =$ 295 nm, $w_2 = 484$ nm and $t = 255$ nm, while those of the SiO$_2$ beam were $w_1 = 250$ nm, $w_2 = 560$ nm and $t = 425$ nm [84]

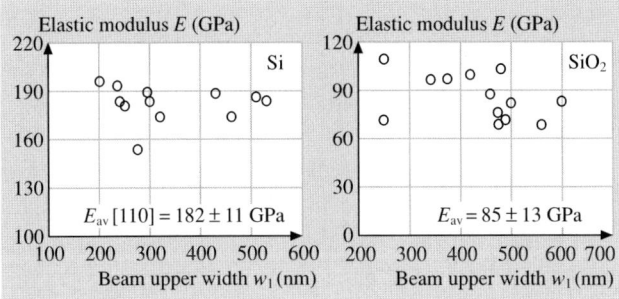

Fig. 14.12. Elastic modulus values measured for Si and SiO$_2$. The average values are shown. These are comparable to bulk values, which shows that elastic modulus shows no specimen size dependence [62]

Table 14.2. Summary of measured parameters from quasi-static bending tests

Sample	Elastic modulus E (GPa)		Bending strength σ_b (GPa)		Fracture toughness K_{IC} (MPa $\sqrt{m}$)		
	Measured	Bulk value	Measured	Reported (microscale)	Estimated	Reported (microscale)	Bulk value
Si	182±11	169[a]	18±3	< 10[c]	1.67±0.4	0.6–1.65[e]	0.9[f]
SiO$_2$	85±13	73[b]	7.6±2	< 2[d]	0.60±0.2	0.5–0.9[4]	–

[a] Si[110] [91]   [b] [92]   [c] [43,44,46–49,70,93,94]   [d] [58]   [e] [54–57]   [f] [87]

Most of the beams when loaded quasi-statically at the center of the span broke at the ends as shown in Fig. 14.13a, which is consistent with the fact that maximum tensile stresses occur on the top surfaces near the ends. (See FEM stress distribution results in Fig. 14.13b.) Figure 14.14 shows the values of bending strength

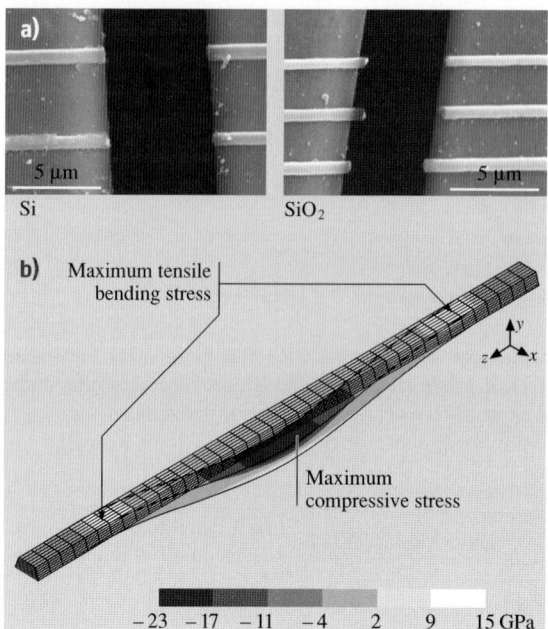

Fig. 14.13. (a) SEM micrographs of nanobeams that failed during quasi-static bending experiments. The beams failed at or near the ends, which is the location of maximum tensile bending stress [84], and (b) bending stress distribution for silicon nanobeam indicating that the maximum tensile stresses occur on the top surfaces near the fixed ends

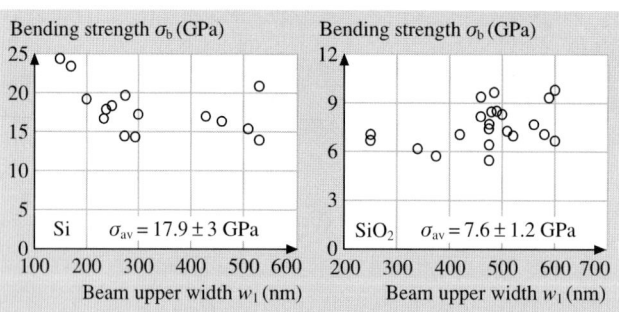

Fig. 14.14. Bending strength values obtained from bending experiments. Average values are indicated. These values are much higher than values reported for microscale specimens, indicating that bending strength shows a specimen size effect [62]

obtained for different beams. There appears to be no trend in bending strength with the upper width (w_1) of the beams. The large scatter is expected for the strength of brittle materials, since they are dependent on pre-existing flaw population in the material and hence are statistical in nature. The Weibull distribution, a statistical analysis, can be used to describe the scatter in the bending strength values. The means of the Weibull distributions were found to be 17.9 GPa and 7.6 GPa for Si and SiO_2, respectively. Previously reported numbers of strengths range from 1–6 GPa for silicon [44, 46–49, 51, 54, 70, 93, 94] and about 1 GPa for SiO_2 [58] microscale specimens. This clearly indicates that bending strength shows a specimen

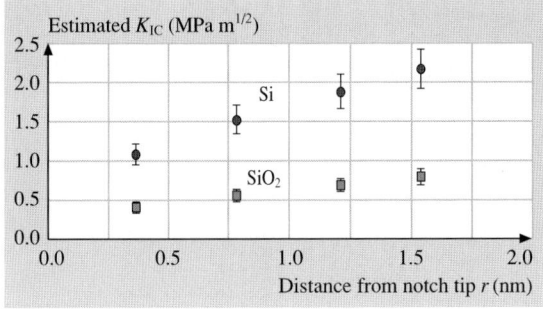

Fig. 14.15. Fracture toughness (K_{IC}) values of for increasing values of r corresponding to distance between neighboring atoms in {111} planes of silicon (0.4 nm). Hence r values between 0.4 and 1.6 nm are chosen. The K_{IC} values thus estimated are comparable to values reported by others for both Si and SiO_2 [62]

size dependence. Strength of brittle materials is dependent on pre-existing flaws in the material. Since for nanoscale specimens, the volume is smaller than for micro and macroscale specimens, the flaw population will be smaller as well, resulting in higher values of strength.

Fracture Toughness

Estimates of fracture toughness calculated using (14.10) for Si and SiO_2 are shown in Fig. 14.15 [62]. The results show that the K_{IC} estimate for Si is about 1–2 MPa $\sqrt{m}$ whereas for SiO_2 the estimate is about 0.5–0.9 MPa $\sqrt{m}$. These values are comparable to values reported by others on larger specimens for Si [54–57] and SiO_2 [58]. The high values obtained for Si could be due to the fact that the scratches, despite being quite sharp, still have a finite radius of about 100 nm. The bulk value for silicon is about 0.9 MPa $\sqrt{m}$. Fracture toughness is considered to be a material property and is believed to be independent of specimen size. The values obtained in this study, given its limitations, appear to show that fracture toughness is comparable, if not a little higher on the nanoscale.

Fatigue Strength

Fatigue strength measurements of Si nanobeams have been carried out by *Sundararajan* and *Bhushan* [62] using an AFM and *Li* and *Bhushan* [63] using a nanoindenter. Various stress levels were applied to nanobeams by *Sundararajan* and *Bhushan* [62]. The minimum stress was 3.5 GPa for Si beams and 2.2 GPa for SiO_2 beams. The frequency of applied load was 4.2 Hz. In general, the fatigue life decreased with increasing mean stress as well as increasing stress amplitude. When the stress amplitude was less than 15% of the bending strength, the fatigue life was greater than 30,000 cycles for both Si and SiO_2. However, the mean stress had to be less than 30% of the bending strength for a life of greater than 30,000 for Si whereas even at a mean stress of 43% of the bending strength, SiO_2 beams showed

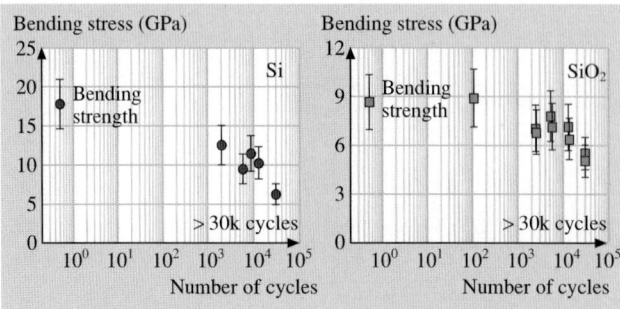

Fig. 14.16. Fatigue test data showing applied bending stress as a function of number of cycles. A single load–unload sequence is considered as 1 cycle. The bending strength data points are therefore associated with 1/2 cycle, since failure occurs upon loading [62]

a life greater than 30,000. During fatigue, the beams broke under the loading point or at the ends, when loaded at the center of the span. This was different from the quasi-static bending tests, where the beams broke at the ends almost every time. This could be due to the fact that the stress levels under the load and at the ends are not that different and fatigue crack propagation could occur at either location. Figure 14.16 shows a nanoscale S–N curve, with bending stress (S) as a function of fatigue in cycles (N) with an apparent endurance life at lower stress. This study clearly demonstrates that fatigue properties of nanoscale specimens can be studied.

SEM Observations of Fracture Surfaces

Figure 14.17 shows SEM images of the fracture surfaces of nanobeams broken during quasi-static bending as well as fatigue [62]. In the quasi-static cases, the maximum tensile stresses occur on the top surface, so it is reasonable to assume that fracture initiated at or near the top surface and propagated downward. The fracture surfaces of the beams suggest a cleavage type of fracture. Silicon beam surfaces show various ledges or facets, which is typical for crystalline brittle materials. Silicon usually fractures along the (111) plane due to this plane having the lowest surface energy to overcome by a propagating crack. However, failure has also been known to occur along the (110) planes in microscale specimens, despite the higher energy required as compared to the (111) planes [70]. The plane normal to the beam direction in these samples is the (110) plane while (111) planes will be oriented at 35° from the (110) plane. The presence of facets and irregularities on the silicon surface in Fig. 14.17a suggest that it is a combination of these two types of fractures that has occurred. Since the stress levels are very high for these specimens, it is reasonable to assume that crack propagating forces will be high enough to result in (110) type failures.

In contrast, the silicon fracture surfaces under fatigue, shown in Fig. 14.17b, appear very smooth without facets or irregularities. This is suggestive of low energy fracture, i.e. of (111) type fracture. We do not see evidence of fatigue crack propagation in the form of steps or striations on the fracture surface. We believe that for

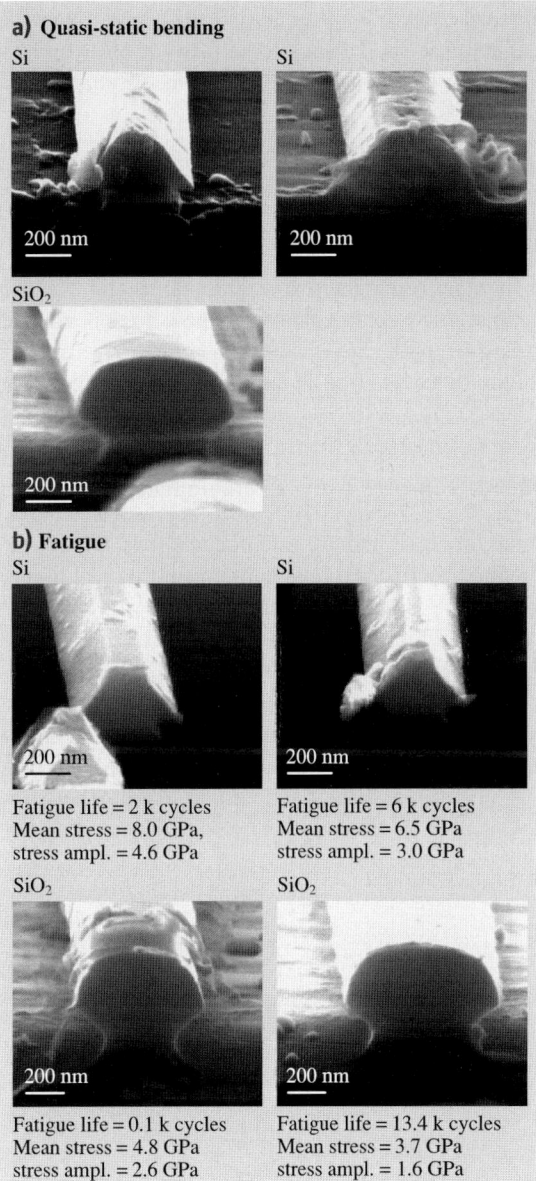

Fig. 14.17. SEM micrographs of fracture surfaces of silicon and SiO_2 beams subjected to (**a**) quasi-static bending and (**b**) fatigue [62]

the stress levels applied in these fatigue experiments, failure in silicon occurred via cleavage associated with 'static fatigue' type of failures.

SiO_2 shows very smooth fracture surfaces for both quasi-static bending and fatigue. This is in contrast to the hackled surface one might expect for the brittle failure of an amorphous material on the macroscale. However, in larger scale fracture surfaces for such materials, the region near the crack initiation usually appears smooth or mirror-like. Since the fracture surface here is so small and very near the crack initiation site, it is not unreasonable to see such a smooth surface for SiO_2 on this scale. There appears to be no difference between the fracture surfaces obtained by quasi-static bending and fatigue for SiO_2.

Summary of Mechanical Properties Measured Using Quasi-Static Bending Tests

Table 14.2 summarizes the various properties measured via quasi-static bending in this study [62]. Also shown are bulk values of the parameters along with values reported on larger scale specimens by other researchers. Elastic modulus and fracture toughness values appear to be comparable to bulk values and show no dependence on specimen size. However bending strength shows a clear specimen size dependence with nanoscale numbers being twice as large as numbers reported for larger scale specimens.

14.3.3 Bending Tests of Metallic Microbeams Using a Nanoindenter

Bending tests have been performed on Ni–P and Au microbeams [64]. The Ni–P cantilever microbeams were fabricated by focused ion beam machining technique. The dimensions were $10, \times 12, \times 50\,\mu m^3$. Notches with a depth of 3 µm and a tip radius of 0.25 µm were introduced in the microbeams to facilitate failure at a lower load in the bending tests. The Au bridge microbeams were fabricated by electroplating technique.

Figure 14.18 shows the SEM images, load displacement curve and FEM stress contour for the notched Ni–P cantilever microbeam that was bent to failure [64]. The distance between the loading position and the fixed end is 40 µm. The 3 µm

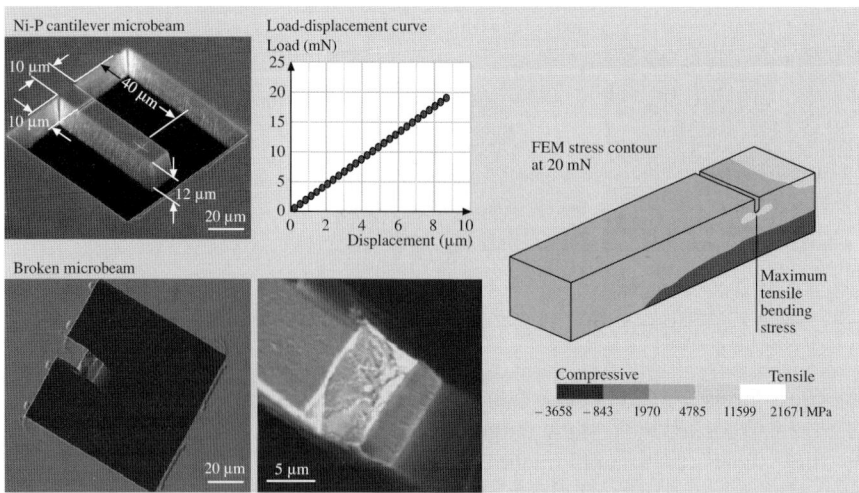

Fig. 14.18. SEM micrographs of the new and broken beams, load displacement curve and FEM stress contour for the notched Ni–P cantilever microbeam [64]

deep notch is 10 μm from the fixed end. The notched beam showed linear behavior followed by abrupt failure. The FEM stress contour shows that there is higher stress concentration at the notch tip. The maximum tensile stress σ_m at the notch tip can be analyzed by using Griffith fracture theory as follows [83],

$$\sigma_m \approx 2\sigma_0 \left(\frac{c}{\rho}\right)^{\frac{1}{2}}, \qquad (14.11)$$

where σ_0 is the average applied tensile stress on the beam, c is the crack length, and ρ is the crack tip radius. Therefore, elastic–plastic deformation will first occur locally at the end of the notch tip, followed by abrupt fracture failure after the σ_m reaches the ultimate tensile strength of Ni–P, even though the rest of the beam is still in the elastic regime. The SEM image of the fracture surface shows that the fracture started right from the notch tip with plastic deformation characteristic. This indicates that although local plastic deformation occurred at the notch tip area, the whole beam failed catastrophically. The present study shows that FEM simulation can predict well the stress concentration, and helps in understanding the failure mechanism of the notched beams.

Figure 14.19 shows the SEM images, load displacement curve and FEM stress contour for the Au bridge microbeam that was deformed by the indenter [64]. The recession gap between the beam and substrate is about 7 μm, which is not large enough to break the beam at the load applied. From the load–displacement curve, we note that the beam experienced elastic–plastic deformation. The FEM stress contour shows that the maximum tensile stress is located at the fixed ends whereas the minimum compressive stress is located around the center of the beam. The SEM

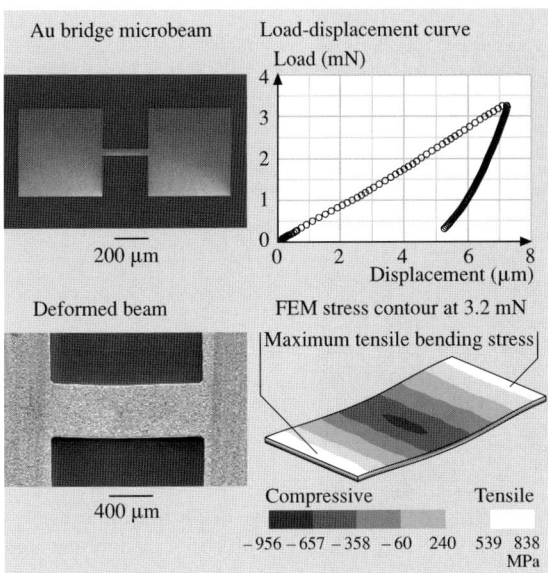

Fig. 14.19. SEM micrographs of the new and deformed beams, load displacement curve and FEM stress contour for the Au bridge microbeam [64]

image shows that the beam has been permanently deformed. No crack was found on the beam surface. The present study shows a possibility for mechanically forming the Au film into the shape as needed. This may help in designing/fabricating functionally complex smart micro/nanodevices which need conductors for power supply and input/output signals.

14.3.4 Indentation and Scratch Tests of Polymeric Microbeams Using a Nanoindenter

Studies have been conducted on two polymer microbeams made of PMMA and PPMA with thickness of about 2 to 5 µm [65]. PPMA was chosen due to its relatively low glass transition temperature, allowing easy thermal processing of the material during the microfabrication process. PMMA was chosen due to its wide use in commercial biomedical applications. Table 14.3 summarizes published data on these materials.

Hardness and Elastic Modulus

The hardness (H), elastic modulus (E) and creep of PPMA and PMMA beams were measured at the supported region of the beams [65]. In Fig. 14.20, the indentation location, where the H, E and creep were measured, is indicated by an arrow. Figure 14.21a shows the H and E of 2.9 µm PPMA and 3.4 µm PMMA beams as a function of contact depth [65]. The H and E were calculated by averaging the H and E

Fig. 14.20. SEM images of PPMA beams at low and high magnifications

Table 14.3. Summary of physical properties of PPMA and PMMA materials

Structure	T_g (°C)	Water absorption after 24h (%)	Thermal expansion coefficient (K^{-1})	Thermal Conduc. (0–50°C) (W/mK)	Elastic modulus (E) (GPa)				Hardness (H) (measured) (MPa)
					Bulk (literature)	Nano-indentation (literature)	Nano-indentation (measured)	Beam bending (measured)	
PMMA [−CH$_2$−C(CH$_3$)(COOCH$_3$)−]	104–106[1]	0.3–0.4[3]	2–3×10^{-4} (<T_g)[5] 6×10^{-4} (>T_g)[4]	0.193[6]	3.1–3.3[1]	4.43[7]	5.0	2.0	410
PPMA [−CH$_2$C(CH$_3$)(COCH$_2$CH$_3$)−]	35–43[2]	0.1–0.3[4]	–	–	–	–	3.7	0.7	190

[1] [80, 81] [2] [79] [3] [95] [4] [96] [5] [97] [6] [98, 99] [7] [100]

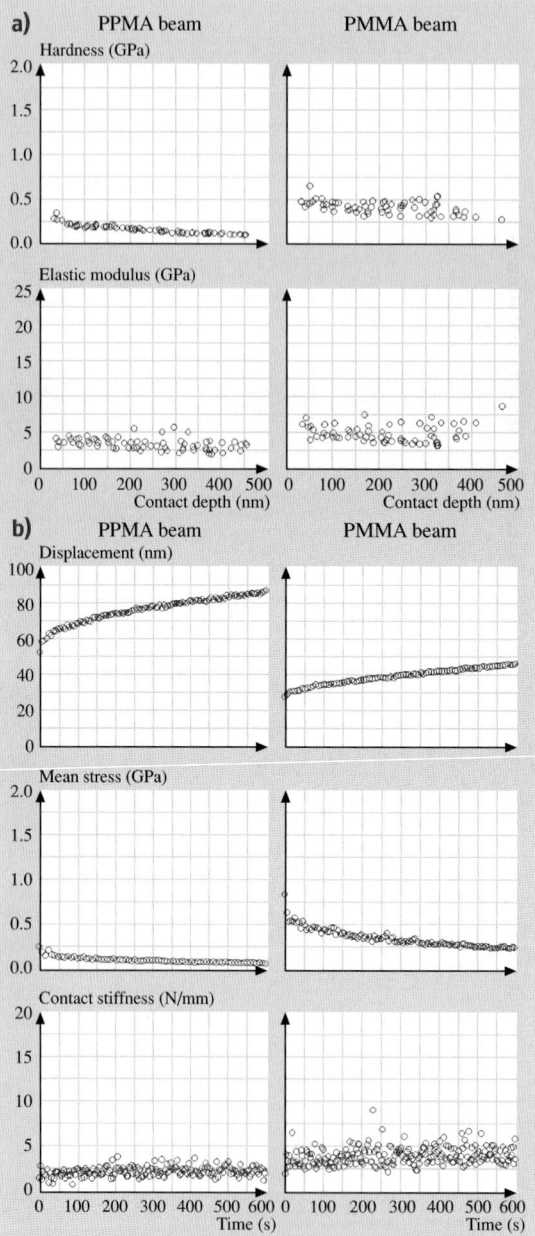

Fig. 14.21. (a) Hardness and elastic modulus of 2.9 μm thick PPMA and 3.4 μm thick PMMA beams as a function of contact depth, (b) indentation displacement, mean stress and contact stiffness as a function of time for 2.9 μm thick PPMA and 3.4 μm thick PMMA beams

values obtained at contact depth of 100 nm from five indents. The H (0.41 GPa) and E (5.0 GPa) of the PMMA beams are both higher than H (0.19 GPa) and E (3.7 GPa) of the PPMA beams. Table 14.3 compares the present data with the published data. We note that the E of PMMA is comparable to the published data. It should be

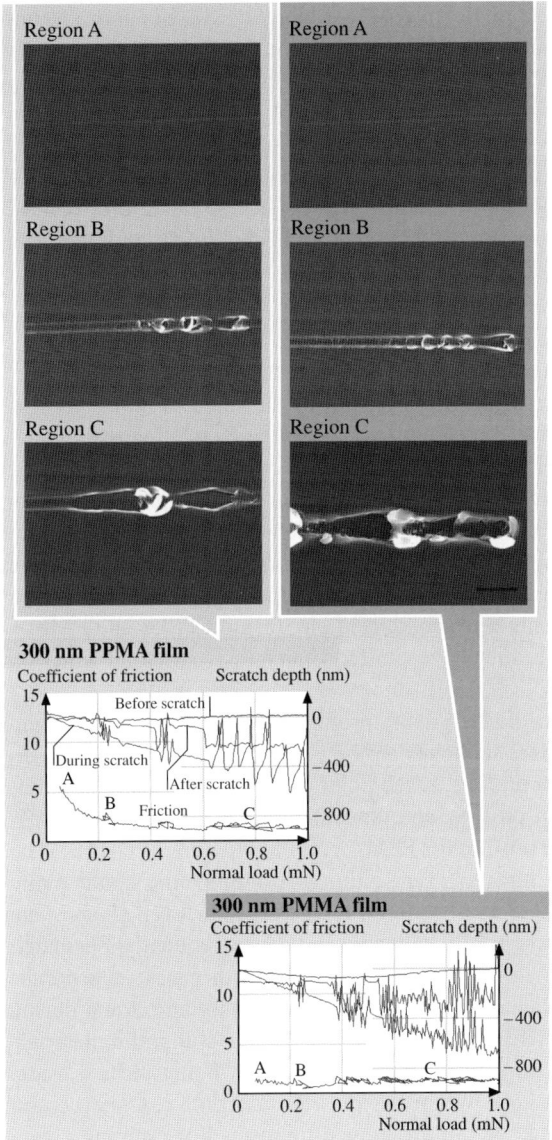

Fig. 14.21. (continued) (c) scratch depth profiles and coefficient of friction as a function of increasing normal load and SEM images of three regions over scratches: at the beginning of the scratch (indicated by A on the scratch depth profile), and at the point of initiation of damage (indicated by B on the scratch depth profile) and at the end of scratch (indicated by C on the scratch depth profile) for PPMA and PMMA thin films (after [65])

noted that the H and E were also measured on 3 μm PPMA and 3 μm PMMA thin films/SU8-25, and the results were very close to the values obtained from the beams.

Indentation Creep

In the creep tests, the changes in indentation displacement, mean stress (hardness) and contact stiffness were monitored [65]. Figure 14.21b shows the indentation displacement, mean stress and contact stiffness as a function of time for PPMA and

PMMA beams [65]. The indentation depth corresponding to 30 μN is about 60 nm for PPMA and about 30 nm for PMMA. The indentation depths of both polymer beams increase with time. The PPMA beam exhibits a faster increase in indentation depth (from about 60 nm to about 90 nm) than the PMMA beam (from about 30 nm to about 50 nm). This indicates that a higher hardness is associated with a higher creep resistance. In contrast with indentation displacement, the mean stresses of both polymer beams decrease with time, indicating that stress relaxation occurred during the hold segment. The creep tests were also conducted on the polymer thin films and the results were similar to the polymer beams.

Scratch Resistance

The polymer films were annealed at 95 °C and 175 °C to PPMA and PMMA respectively to simulate the thermal processing used in the micromolding process for polymer beam fabrication. The annealing treatment seemed to effectively simulate the beam fabrication process, because the H, E and creep measurements show that the PPMA and PMMA thin films have similar mechanical properties with PPMA and PMMA beams. In order to evaluate the scratch resistance and adhesion of PPMA and PPMA beams/SU8-25 substrate, nanoscratch measurements were needed. Since polymer beams were too narrow (width 5 μm) to perform nanoscratch tests, the nanoscratch experiments were conducted on 300 nm PPMA and PMMA thin films/SU8-25, assuming that they have similar scratch resistance and adhesion with PPMA and PMMA beams.

Figure 14.21c shows that the scratch depth profiles, coefficient of friction, and SEM images of three regions over scratches: at the beginning of the scratch (indicated by A on the scratch depth profile), at the point of initiation of damage (indicated by B on the scratch depth profile), and towards the end of the scratch (indicated by C on the scratch depth profile) for PPMA and PMMA thin films [65]. The scratch results of polymer thin films are very different from metallic and ceramic thin films [82]. Firstly, the coefficient of friction value is very high, even greater than one, and the coefficient of friction profile jumps up and down frequently. Secondly, after the polymer thin films were damaged, the scratch depth profile also jumps up and down considerably, indicating the scratch tip moved up and down during scratching. The SEM images of PPMA show that during scratching, the materials were plowed and accumulated in front of the tip, instead of being pushed aside, because the polymers were so soft. When the plowed materials reached certain amount, the tip was almost stuck, so the tip jumped up instead of ramping down. This may explain the oscillation in the scratch depth profile. At region B., the coefficient of friction increased and the in situ scratch depth also changed abruptly, indicating that the film was delaminated. The critical load of PPMA and PMMA are about the same, around 0.22 mN.

14.3.5 Bending Tests of Polymeric Microbeams Using a Nanoindenter

As mentioned earlier, to avoid tip penetration into polymer beams, the tip was dip-coated with PMMA. Figure 14.22 shows the load–displacement curves of PPMA

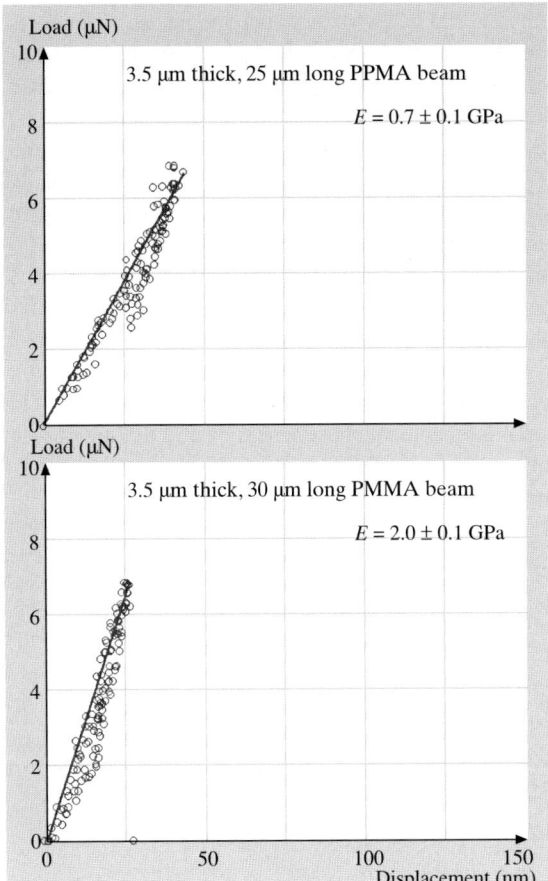

Fig. 14.22. Bending results on PPMA and PMMA beams with PMMA coated diamond conical tip (after [65])

and PMMA beams measured with the PMMA coated tip [65] loading curve was linear and the unloading curve does not exhibit much plastic deformation, which means that the bending tests were performed mostly in the elastic region of the samples and the elastic modulus can be calculated from the loading curve. Any slight degree of deformation of the tip coating during the bending tests was assumed negligible compared to beam bending. Based on Fig. 14.22, the calculated elastic modulus for PPMA and PMMA beams are 0.7 ± 0.1 GPa and 2.0 ± 0.1 GPa, respectively. The standard deviation was calculated based on five measurements. Calculated elastic moduli of PPMA and PMMA beams from bending tests are lower than the values obtained from indentation.

Effect of Soaking and Temperature

The primary operating environment for polymer BioMEMS is an aqueous solution (e.g., cell culture medium). Absorption of water by the polymer matrix has the potential to change the properties of the material. In order to determine if immersion in water has a significant effect on material behavior, tests need to be run on samples that have been subjected to an aqueous environment for a considerable time. In addition, in the case of employing BioMEMS for in vivo or elevated temperature in vitro applications, the effect of human body temperature (37.5 °C) on polymer properties may be important. The T_g of PPMA (35–43 °C) is around the body temperature and the T_g of PMMA (104–109 °C) is above body temperature. It is necessary to study how the mechanical properties of PPMA and PMMA are affected when operating at body temperature as compared to room temperature.

To simulate the aqueous working environment of the polymer BioMEMS, the bending tests were also performed on soaked samples in addition to the dry samples and the results were compared. To make the soaked samples, the PMMA and PPMA beams were soaked in DI water for 36 hours before the bending tests. In addition, the nanoindentation tests were conducted in a range of temperature (22–37.5 °C) to study the effect of temperature on nanomechanical properties of polymer beams. In order to heat the sample, the polymer beam sample was placed on a heating stage, which can increase the temperature up to 200 °C.

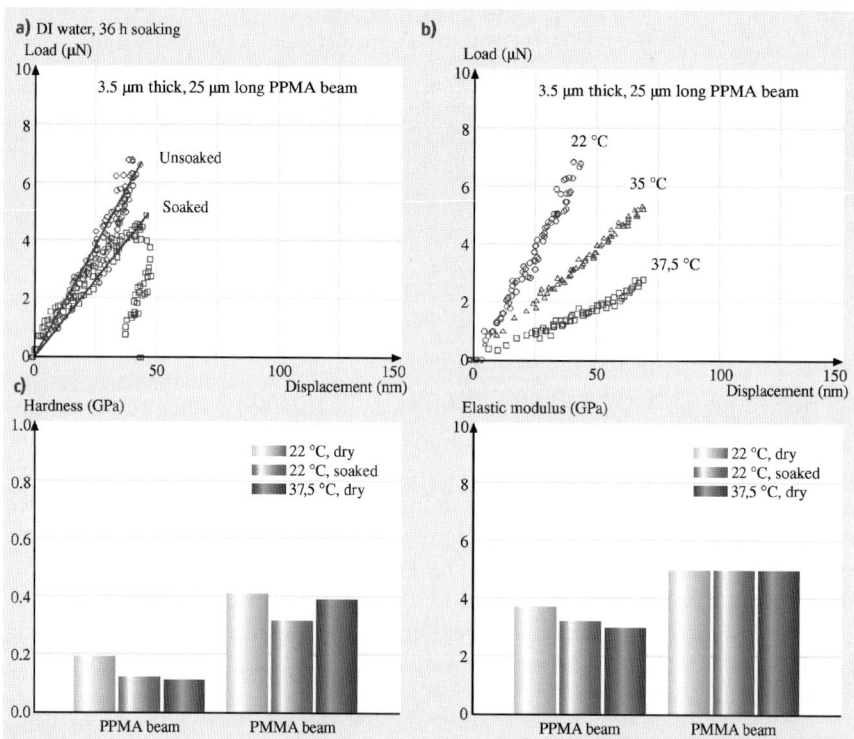

Fig. 14.23. (a) Soaking effect on bending of PMMA beam, (b) temperature effect on bending of PPMA beam, and (c) soaking and temperature effects on H and E of PPMA and PMMA beams [65]

Figure 14.23a shows the effect of soaking on elastic modulus of PPMA beams. After soaking, the calculated elastic modulus (0.5 ± 0.1 GPa) of PPMA was lower than the elastic modulus (0.7 ± 0.1 GPa) obtained at dry conditions [65]. But for PMMA, after soaking, the calculated elastic modulus does not change (figure not shown). The more open structure of PPMA versus PMMA might allow more water to penetrate into the matrix, thus having a greater effect on the modulus than with PMMA (closed, hydrophobic structure). The open structure of PPMA is due to the larger side chain compared to the side chain of PMMA.

Figure 14.23b shows the effect of temperature on load–displacement curves of PPMA beams. As the temperature increased from room temperature (22 °C) to human body temperature (37.5 °C), the slope of the loading curve decreased [65]. The T_g of PMMA is 35–43 °C, so it is understandable that the elastic modulus, thus the slope of the loading curve, would decrease when the temperature increased beyond 35 °C. However, since the T_g of PMMA is about 100 °C, at 37.5 °C, the load–displacement curve of PMMA beam was similar to 22 °C (figure not shown).

The soaking and temperature effects on hardness and elastic modulus of PPMA and PMMA beams were also studied and the results are shown in Figure 14.23c [65]. It can be seen that for PPMA beam, the hardness and elastic modulus decreased after 36 h soaking in DI water, and also decreased at 37.5 °C, which is consistent with the bending test results. For PMMA beam, the hardness decreased after soaking, and also decreased slightly at 37.5 °C. However, the elastic modulus of PMMA beam did not change after soaking or heating up to 37.5 °C, which is in agreement with the bending results. The lower elastic modulus of PPMA compared to PMMA makes it an attractive option for certain applications, such as cantilevers. However, the changes in the properties of PPMA as a function of aqueous environment and temperature would require careful calibration at operating conditions when implementing PPMA microstructures in a biological setting.

14.4 Finite Element Analysis of Nanostructures with Roughness and Scratches

Micro/nanostructures have some surface topography and local scratches dependent upon the manufacturing process. Surface roughness and local scratches may compromise the reliability of the devices and their effect needs to be studied. Finite element modeling is used to perform parametric analysis to study the effect of surface roughness and scratches in different well defined forms on tensile stresses which are responsible for crack propagation [71, 72]. The analysis has been carried out on trapezoidal beams supported at the bottom whose data (on Si and SiO_2 nanobeams) have been presented earlier.

The finite element analysis has been carried out by using the static analysis of ANSYS 5.7 which calculates the deflections and stresses produced by the applied loading. The type of element selected for the present study was SOLID95 type which allows the use of different shapes without much loss of accuracy. This

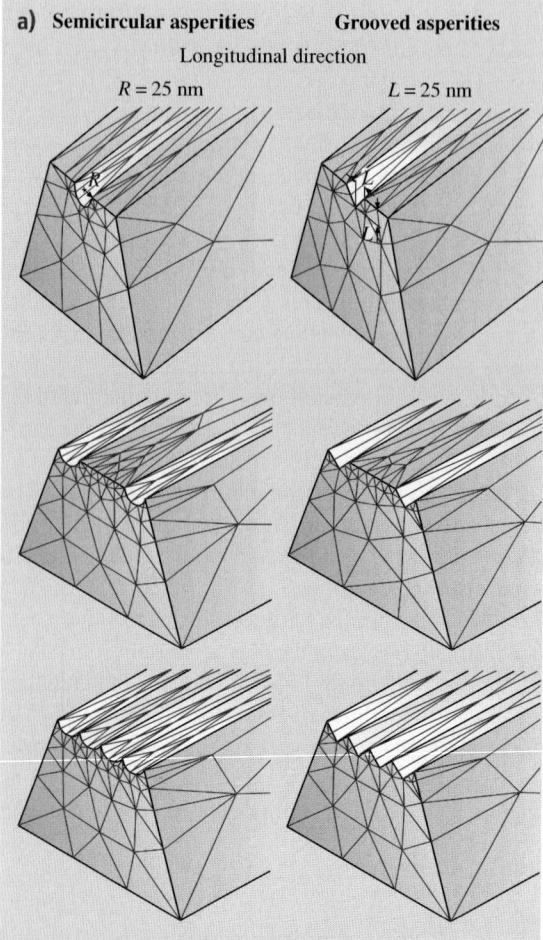

Fig. 14.24. (a) Plots showing the geometries of modeled roughness–semicircular and grooved asperities along the nanobeam length with defined geometrical parameters

element is 3-D with 20 nodes, each node having three degrees of freedom which implies translation in the x, y and z directions. The nanobeam cross section is divided into six elements each along the width and the thickness and forty elements along the length. SOLID95 has plasticity, creep, stress stiffening, large deflection and large strain capabilities. The large displacement analysis is used for large loads. The mesh is kept finer near the asperities and the scratches in order to take into account variation in the bending stresses. The beam materials studied are made of single-crystal silicon (110) and SiO_2 films whose data have been presented earlier. Based on bending experiments presented earlier, the beam materials can be assumed to be linearly elastic isotropic materials. Young's modulus of elasticity (E) and Poisson's ratio (v) for Si and SiO_2 are 169 GPa [91] and 0.28 [87], and 73 GPa [92] and 0.17 [101], respectively. A sample nanobeam of silicon was chosen for performing most of the analysis as silicon is the most widely used MEMS/NEMS material. The cross sec-

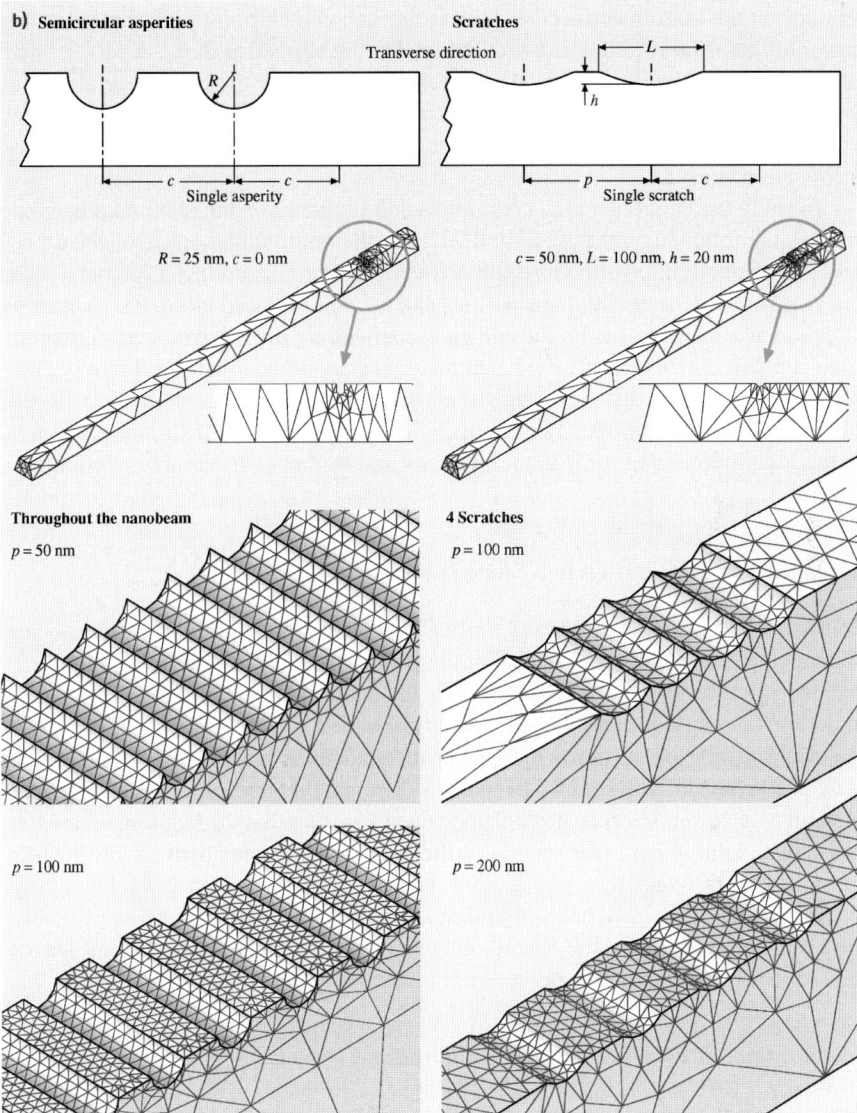

Fig. 14.24. (continued) (**b**) Schematic showing semicircular asperities and scratches in the transverse direction followed by the illustration of the mesh created on the beam with fine mesh near the asperities and the scratches. Also shown are the semicircular asperities and scratches at different pitch values

tion of the fabricated beams used in the experiment is trapezoidal and supported at the bottom so nanobeams with trapezoidal cross section is modeled, Fig. 14.10. The following dimensions are used $w_1 = 200$ nm, $w_2 = 370$ nm, $t = 255$ nm, and $P = 6\,\mu$m. In the boundary conditions, the displacements are constrained in all directions on the bottom surface for 1 μm from each end. A point load applied at the center of the beam is simulated with the load being applied at three closely located central nodes on the beam used. It has been observed from the experimental results that the Si nanobeam breaks at around 80 μN. Therefore, in this analysis, a nominal load of 70 μN is selected. At this load, deformations are large, and large displacement option is used.

To study the effect of surface roughness and scratches on the maximum bending stresses the following cases were studied. First the semicircular and grooved asperities in the longitudinal direction with defined geometrical parameters are analyzed, Fig. 14.24a. Next semicircular asperities and scratches placed along the transverse direction at a distance c from the end and separated by pitch p from each other are analyzed, Fig. 14.24b. Lastly the beam material is assumed to be either purely elastic, elastic–plastic, or elastic–perfectly plastic. In the following, we begin with the stress distribution in smooth nanobeams followed by the effect of surface roughness in the longitudinal and transverse directions and scratches in the transverse direction.

14.4.1 Stress Distribution in a Smooth Nanobeam

Figure 14.25 shows the stress and vertical displacement contours for a nanobeam supported at the bottom and loaded at the center, [71, 72]. As expected, the maximum tensile stress occurs at the ends while the maximum compressive stress occurs under the load at the center. Stress contours obtained at a section of the beam from the front and side are also shown. In the beam cross section, the stresses remain constant at a given vertical distance from one side to another and change with a change in vertical location. This can be explained due to the fact that the bending moment is constant at a particular cross section so the stress is only dependent on the distance from the neutral axis. However, in cross section A-A the high tensile and compressive stresses are localized near the end of the beam at top and bottom, respectively, whereas the lower values are spread out away from the ends. High value of tensile stresses occurs near the ends because of high bending moment.

14.4.2 Effect of Roughness in the Longitudinal Direction

The roughness in the form of semicircular and grooved asperities in the longitudinal direction on the maximum bending stresses are analyzed [72]. The radius R and depth L are kept fixed at 25 nm while the number of asperities is varied and their effect is observed on the maximum bending stresses. Figure 14.26 shows the variation of maximum bending stresses as a function of asperity shape and the number of asperities. The maximum bending stresses increase as the asperity number increases for both semicircular and grooved asperities. This can be attributed to the fact that as asperity number increases, the moment of inertia decreases for that cross section. Also the distance from the neutral axis increases because neutral axis shifts downwards. Both these factors lead to the increase in the maximum bending

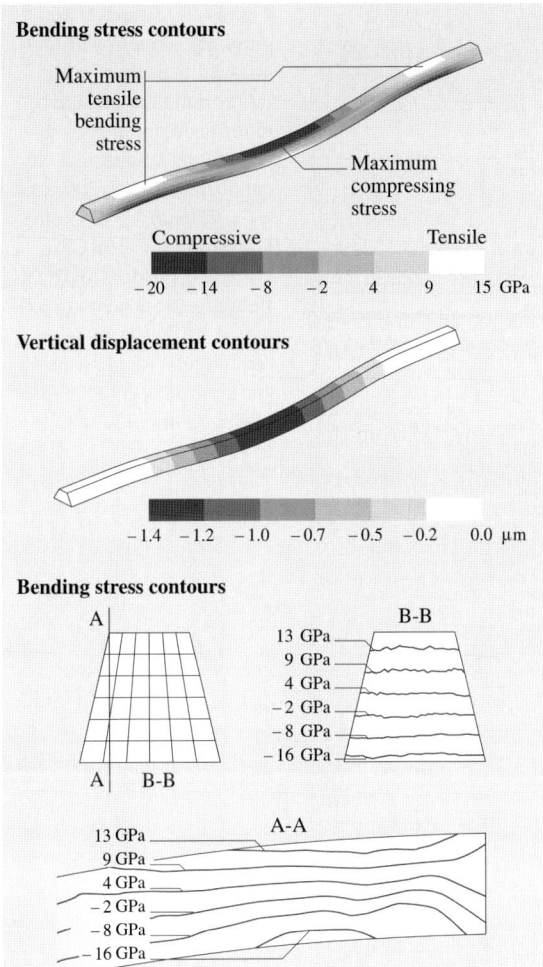

Fig. 14.25. Bending stress contours, vertical displacement contours, and bending stress contours after loading trapezoidal Si nanobeam ($w_1 = 200$ nm, $w_2 = 370$ nm, $t = 255$ nm, $\ell = 6\,\mu$m, $E = 169$ GPa, $v = 0.28$) at 70 μN load [72]

stresses and this effect is more pronounced in the case of semicircular asperity as it exhibits a higher value of maximum bending stress than that in grooved asperity. Figure 14.26 shows the stress contours obtained at a section of the beam from front and from the side for both cases when we have a single semicircular asperity and when four adjacent semicircular asperities are present. Trends are similar to that observed earlier for a smooth nanobeam (Fig. 14.25).

14.4.3 Effect of Roughness in the Transverse Direction and Scratches

We analyze semicircular asperities when placed along the transverse direction followed by the effect of scratches on the maximum bending stresses in varying numbers and different pitch [72]. In the analysis of semicircular transverse asperities

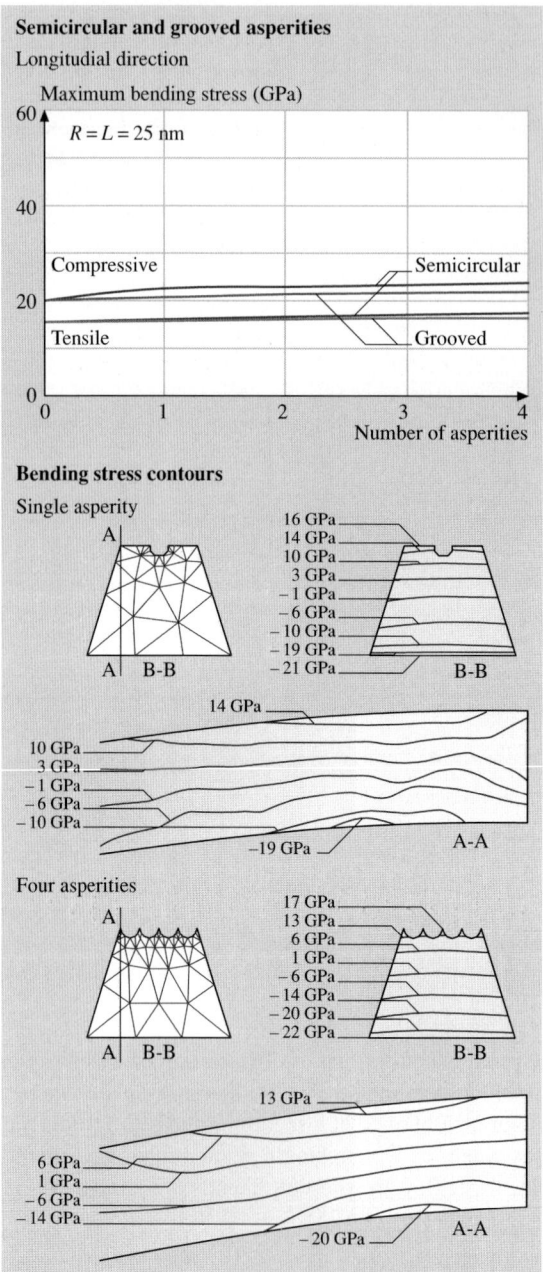

Fig. 14.26. Effect of longitudinal semicircular and grooved asperities in different numbers on maximum bending stresses after loading trapezoidal Si nanobeams ($w_1 = 200$ nm, $w_2 = 370$ nm, $t = 255$ nm, $\ell = 6\,\mu$m, $E = 169$ GPa, $v = 0.28$, Load = 70 μN). Bending stress contours obtained in the beam with semicircular single asperity and four adjacent asperities of $R = 25$ nm [72]

three cases were considered which included a single asperity, asperities throughout the nanobeam surface separated by pitch equal to 50 nm and pitch equal to 100 nm. In all of these cases, c value was kept equal to 0 nm. Figure 14.27 shows that the

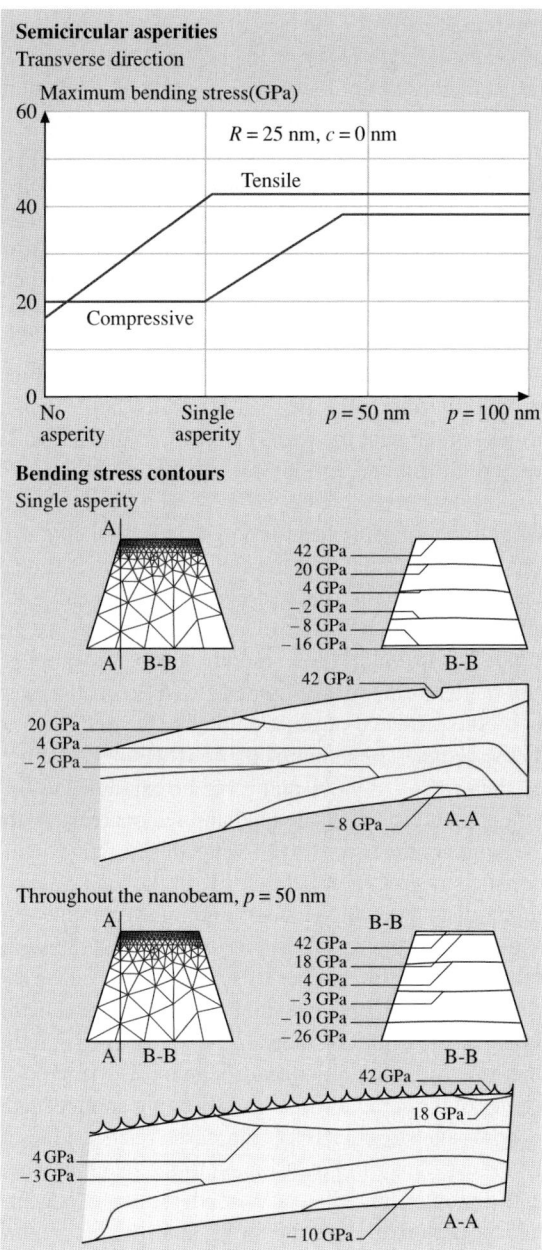

Fig. 14.27. Effect of transverse semicircular asperities located at different pitch values on the maximum bending stresses after loading trapezoidal Si nanobeams ($w_1 = 200$ nm, $w_2 = 370$ nm, $t = 255$ nm, $\ell = 6\,\mu\text{m}$, $E = 169$ GPa, $v = 0.28$, Load = 70 μN). Bending stress contours obtained in the beam with semicircular single asperity and semicircular asperities throughout the nanobeam surface at $p = 50$ nm [72]

value of maximum tensile stress is 42 GPa which is much larger than the maximum tensile stress value with no asperity of 16 GPa or when the semicircular asperity is present in the longitudinal direction. It is also observed that the maximum tensile stress does not vary with the number of asperities or the pitch while the maximum compressive stress does increase dramatically for the asperities present throughout the beam surface from its value when a single asperity is present. Maximum tensile stress occurs at the ends and an increase in p does not add any asperities at the ends whereas asperities are added in the central region where compressive stresses are maximum. The semicircular asperities present at the center cause the local perturbation in the stress distribution at the center of the asperity where load is being applied leading to a high value of maximum compressive stress [101]. Figure 14.27 also shows the stress contours obtained at a section of the beam from front and from the side for both cases when there is a single semicircular asperity and when asperities are present throughout the beam surface at a pitch equal to 50 nm. Trends are similar to that observed earlier for a smooth nanobeam (Fig. 14.25).

In the study pertaining to scratches, the number of scratches are varied along with the variation in the pitch as well. Furthermore the load is applied at the center of the beam and at the center of the scratch near the end as well for all the cases. In all of these c value was kept equal to 50 nm and L value was equal to 100 nm with h value being 20 nm. Figure 14.28 shows that the value of maximum tensile stress remains almost the same with the number of scratches for both types of loading that is when load is applied at the center of the beam and at the center of the scratch near the end. This is because that the maximum tensile stress occurs at the beam ends no matter where the load gets applied. But the presence of scratch does increase the maximum tensile stress as compared to its value for a smooth nanobeam although the number of scratches no longer matter as the maximum tensile stress occurring at the nanobeam end is unaffected by the presence of more scratches beyond the first scratch in the direction towards the center. The value of the tensile stress is much lower when the load is applied at the center of scratch and it can be explained as follows. The negative bending moment at the end near the load applied decreases with load offset after two-thirds of the length of the beam [102]. Since this negative bending moment is responsible for tensile stresses, their behavior with the load offset is the same as the negative bending moment. Also the value of maximum compressive stress when load is applied at the center of the nanobeam remains almost the same as the center geometry is unchanged due to the number of scratches and hence the maximum compressive stress occurring below the load at the center is same. On the other hand when the load is applied at the center of the scratch we observe that the maximum compressive stress increases dramatically because the local perturbation in the stress distribution at the center of the scratch where load is being applied leads to a high value of maximum compressive stress [101]. It increases further with the number of scratches and then levels off. This can be attributed to the fact that when there is another scratch present close to the scratch near the end, the stress concentration is more as the effect of local perturbation in the stress distribution is more significant. However, this effect is insignificant when more than two scratches are present.

Now we address the effect of pitch on the maximum compressive stress when the load is applied at the center of the scratch near the end. When the pitch is up to a value of 200 nm the maximum compressive stress increases with the number of

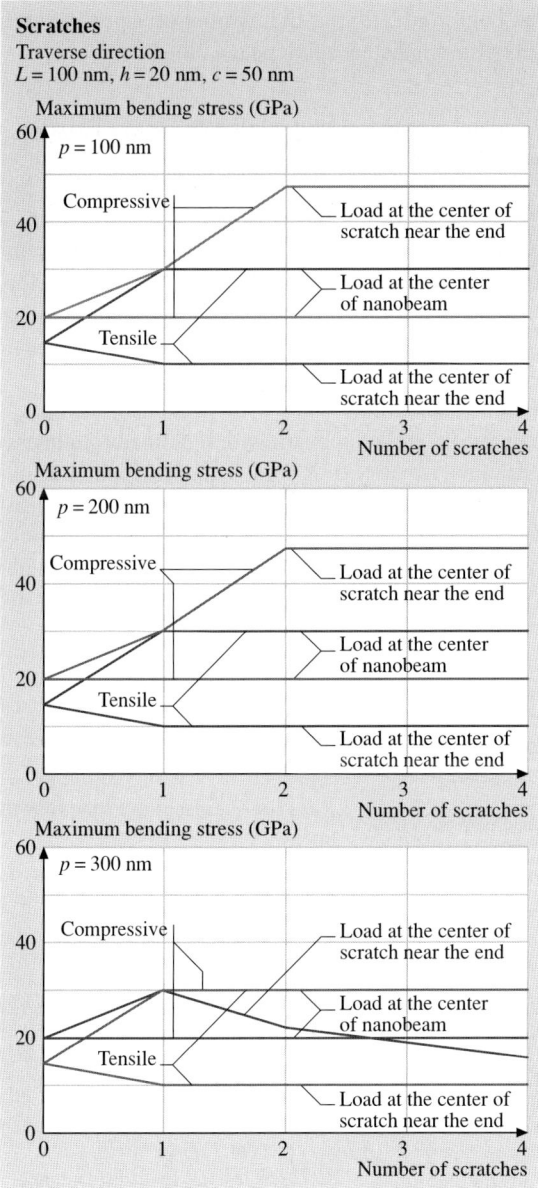

Fig. 14.28. Effect of number of scratches along with the variation in the pitch on the maximum bending stresses after loading trapezoidal Si nanobeams (w_1 = 200 nm, w_2 = 370 $v\mu$, τ = 255 $v\mu$, ℓ = 6 µm, E = 169 GPa, v = 0.28, Load = 70 µN.) Also shown is the effect of load when applied at the center of the beam and at the center of the scratch near the end [72]

scratches as discussed earlier. On the other hand, when the pitch value goes beyond 225 nm this effect is reversed. This is because the presence of another scratch no longer affects the local perturbation in the stress distribution at the scratch near the end. Instead more scratches at a fair distance distribute the maximum compressive stress at the scratch near the end and the stress starts going down. Such observations of maximum bending stresses can help in identifying the number of asperities and scratches allowed separated by an optimum distance from each other.

14.4.4 Effect on Stresses and Displacements for Materials Which are Elastic, Elastic-Plastic or Elastic-Perfectly Plastic

This section deals with the beam modeled as elastic, elastic–plastic and elastic–perfectly plastic to observe the variation in the stresses and displacements from an elastic model used so far [72]. Figure 14.29 shows the typical stress-strain curves for the three types of deformation regimes and their corresponding load–displacement curves obtained from the model of an Si nanobeam which are found to exhibit same trends.

Table 14.4 shows the comparison of maximum von Mises stress and maximum displacements for both smooth nanobeam and nanobeam with a defined roughness which is single semicircular longitudinal asperity of R value equal to 25 nm for the three different models. It is observed that the maximum value of stress is obtained at a given load for elastic material whereas the displacement is maximum for elastic–perfectly plastic material. Also the pattern that the maximum bending stress value increases for a rough nanobeam still holds true in the other models as well.

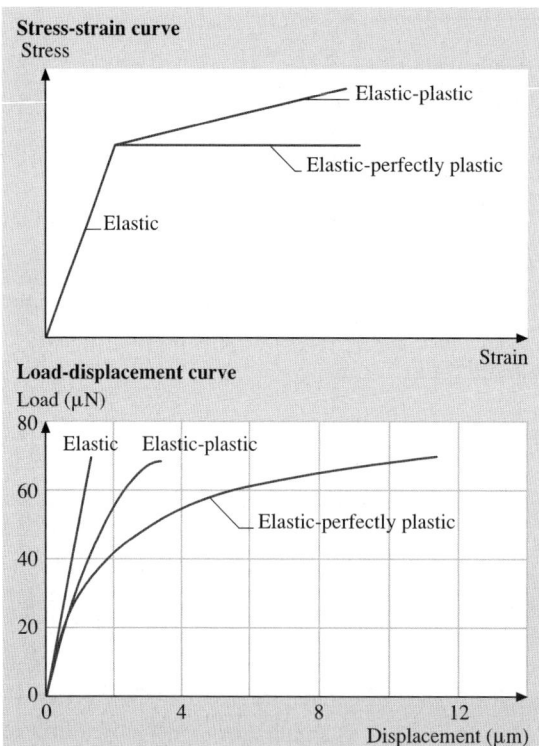

Fig. 14.29. Schematic representation of stress–strain curves and load-displacement curves for material when it is elastic, elastic-plastic, or elastic-perfectly plastic for a Si nanobeam ($w_1 = 200$ nm, $w_2 = 370$ nm, $t = 255$ nm, $\ell = 6\,\mu\text{m}$, $E = 169$ GPa, tangent modulus in plastic range $= 0.5E$, $\nu = 0.28$) [72]

Table 14.4. Stresses and displacements for materials that are elastic, elastic–plastic or elastic–perfectly plastic (load = 70 μN, w_1 = 200 nm, w_2 = 370 nm, t = 255 nm, ℓ = 6 μm, R = 25 nm, E = 169 GPa, tangential modulus in plastic range = $0.5E$, v = 0.28)

	Elastic		Elastic–plastic	
	Smooth nanobeam	Single semicircular longitudinal asperity	Smooth nanobeam	Single semicircular longitudinal asperity
Max. von Mises stress (GPa)	18.2	19.3	13.5	15.2
Max. displacement (μm)	1.34	1.40	3.35	3.65

	Elastic–perfectly plastic	
	Smooth nanobeam	Single semicircular longitudinal asperity
Max. von Mises stress (GPa)	7.8	9.1
Max. displacement (μm)	11.5	12.3

14.5 Closure

Mechanical properties of nanostructures are necessary in designing realistic MEMS/NEMS and BioMEMS/BioNEMS devices. Most mechanical properties are scale dependent. Micro/Nanomechanical properties, hardness, elastic modulus, and scratch resistance of bulk materials of undoped single-crystal silicon (Si) and thin films of undoped polysilicon, SiO_2, SiC, Ni–P, and Au are presented. It is found that the SiC film exhibits higher hardness, elastic modulus and scratch resistance as compared to other materials.

Bending tests have been performed on the Si and SiO_2 nanobeams and Ni–P, Au, PPMA and PMMA microbeams using an AFM and a depth-sensing nanoindenter, respectively. The bending tests were used to evaluate elastic modulus, bending strength (fracture stress), fracture toughness (K_{IC}), and fatigue strength of the beam materials. The Si and SiO_2 nanobeams exhibited elastic linear response with sudden brittle fracture. The notched Ni–P beam showed linear deformation behavior followed by abrupt failure. The Au beam showed elastic–plastic deformation behavior. Elastic modulus values of 182 ± 11 GPa for Si(110) and 85 ± 3 GPa for SiO_2 were obtained, which are comparable to bulk values. Bending strength values of 18 ± 3 GPa for Si and 7.6 ± 2 GPa for SiO_2 were obtained, which are twice as large as values reported on larger scale specimens. This indicates that bending strength shows a specimen size dependence. Fracture toughness value estimates obtained were 1.67 ± 0.4 MPa $\sqrt{m}$ for Si and $0.6. \pm 0.2$ MPa $\sqrt{m}$ for SiO_2, which are also comparable to values obtained on larger specimens. At stress amplitudes less than 15% of their bending strength and at mean stresses of less than 30% of the bending strength, Si and SiO_2 displayed an apparent endurance life of greater than 30,000 cycles. SEM observations of the fracture surfaces revealed a cleavage type of fracture for both materials when subjected to bending as well as fatigue.

The hardness, elastic modulus and creep behavior of PPMA and PMMA polymeric beams are also presented. The hardness (0.41 GPa), elastic modulus (5.0 GPa), and creep resistance of PMMA beams are higher than the hardness (0.19 GPa, elastic modulus (3.7 GPa), and creep resistance of PPMA beams. The scratch behavior of PPMA and PMMA films is different from metallic and ceramic films in that the in situ displacement oscillated and coefficient of friction are greater than one during the scratch test. This might be because the polymer is so soft and easily plowed. After 36 hours soaking in DI water, the hardness and elastic modulus of PPMA beam were decreased. The soaking seems to have had no effect on elastic modulus of PMMA beam. At 37.5 °C, the hardness and elastic modulus of PPMA beam were decreased as compared to room temperature, while this temperature appears to have no effect on elastic modulus of PMMA beam.

The AFM and nanoindenters used in this study can be satisfactorily used to evaluate the mechanical properties of micro/nanoscale structures for use in MEMS/NEMS.

FEM simulations are used to predict the stress and deformation in nanostructures. The FEM has been used to analyze the effect of type of surface roughness and scratches on stresses and deformation of nanostructures. We find that roughness affects the maximum bending stresses. The maximum bending stresses increase as the asperity number increases for both semicircular and grooved asperities in longitudinal direction. When the semicircular asperity is present in the transverse direction the maximum tensile stress is much larger than the maximum tensile stress value with no asperity or when the semicircular asperity is present in the longitudinal direction. This observation suggests that the asperity in the transverse direction is more detrimental. The presence of scratches increases the maximum tensile stress. The maximum tensile stress remains almost the same with the number of scratches for two types of loading, that is, when load is applied at the center of the beam or at the center of the scratch near the end, although the value of the tensile stress is much lower when the load is applied at the center of the scratch. This means that the load applied at the ends is less damaging. This analysis shows that FEM simulations can be useful to designers to develop the most suitable geometry for nanostructures.

References

1. R. S. Muller, R. T. Howe, S. D. Senturia, R. L. Smith, R. M. White: *Microsensors* (IEEE, New York 1990)
2. I. Fujimasa: *Micromachines: A New Era in Mechanical Engineering* (Oxford Univ. Press, Oxford 1996)
3. W. S. Trimmer (ed.): *Micromachines and MEMS, Classic and Seminal Papers to 1990* (IEEE, New York 1997)
4. B. Bhushan: *Tribology Issues and Opportunities in MEMS* (Kluwer, Dordrecht 1998)
5. B. Bhushan: *Handbook of Micro/Nanotribology*, 2nd edn. (CRC, Boca Raton 1999)
6. B. Bhushan: *Nanotribology and Nanomechanics – An Introduction* (Springer, Heidelberg 2005)
7. G. T. A. Kovacs: *Micromachined Transducers Sourcebook* (WCB McGraw-Hill, Boston 1998)
8. S. D. Senturia: *Microsystem Design* (Kluwer, Boston 2001)
9. M. Elwenspoek, R. Wiegerink: *Mechanical Microsensors* (Springer, Berlin 2001)
10. M. Gad-el-Hak: *The MEMS Handbook* (CRC, Boca Raton 2002)
11. T. R. Hsu: *MEMS and Microsystems: Design and Manufacture* (McGraw-Hill, Boston 2002)
12. M. Madou: *Fundamentals of Microfabrication: The Science of Miniaturization*, 2nd edn. (CRC, Boca Raton 2002)
13. A. Hierlemann: *Integrated Chemical Microsensor Systems in CMOS Technology* (Springer, Berlin 2005)
14. K. E. Drexler: *Nanosystems: Molecular Machinery, Manufacturing and Computation* (Wiley, New York 1992)
15. G. Timp (ed.): *Nanotechnology* (Springer, Berlin, Heidelberg 1999)
16. M. S. Dresselhaus, G. Dresselhaus, Ph. Avouris: *Carbon Nanotubes – Synthesis, Structure, Properties and Applications* (Springer, Berlin 2001)
17. E. A. Rietman: *Molecular Engineering of Nanosystems* (Springer, Berlin, Heidelberg 2001)
18. H. S. Nalwa (ed.): *Nanostructured Materials and Nanotechnology* (Academic, San Diego 2002)
19. W. A. Goddard, D. W. Brenner, S. E. Lyshevski, G. J. Iafrate: *Handbook of Nanoscience, Engineering, and Technology* (CRC, Boca Raton 2003)
20. A. Manz, H. Becker (eds.): *Microsystem Technology in Chemistry and Life Sciences, Topics in Current Chemistry 194* (Springer, Heidelberg 1998)
21. J. Cheng, L. J. Kricka (eds.): *Biochip Technology* (Harwood Academic Publishers, Philadelphia 2001)
22. M. J. Heller, A. Guttman (eds.): *Integrated Microfabricated Biodevices* (Dekker, New York, N.Y. 2001)
23. C. Lai Poh San, E. P. H. Yap (eds.): *Frontiers in Human Genetics* (World Scientific, Singapore 2001)
24. C. H. Mastrangelo, H. Becker (eds.): *Microfluidics and BioMEMS, Proc. of SPIE*, Vol. 4560 (SPIE, Bellingham, Washington 2001)
25. H: Becker, L. E. Lacascio: Polymer Microfluidic Devices, Talanta **56**, 267–287 (2002)
26. D. J. Beebe, G. A. Mensing, G. M. Walker: Physics and applications of microfluidics in biology, Annu. Rev. Biomed. Eng. **4**, 261–286 (2002)
27. A. van der Berg (ed.): *Lab-on-a-Chip: Chemistry in Miniaturized Synthesis and Analysis Systems* (Elsevier, Amsterdam, Netherlands 2003)

28. C. P. Poole, F. J. Owens: *Introduction to Nanotechnology* (Wiley, Hoboken, New Jersey 2003)
29. J. V. Zoval, M. J. Madou: Centrifuge-based fluidic platforms, Proc. IEEE **92**, 140–153 (2000)
30. R. Raiteri, M. Grattarola, H. Butt, P. Skladal: Micromechanical cantilever-based biosensors, Sensors and Actuators B: Chemical **79**, 115–126 (2001)
31. W. C. Tang, A. P. Lee: Defense applications of MEMS, MRS Bulletin **26**, 318–319 (2001)
32. M. R. Taylor, P. Nguyen, J. Ching, K. E. Peterson: Simulation of microfluidic pumping in a genomic DNA blood-processing cassette, J. Micromech. Microeng. **13**, 201–208 (2003)
33. K. Park (ed.): *Controlled Drug Delivery: Challenges and Strategies* (American Chemical Society, Washington, DC 1997)
34. R. S. Shawgo, A. C. R. Grayson, Y. Li, M. J. Cima: BioMEMS for drug delivery, Current Opinion in Solid State & Mat. Sci. **6**, 329–334 (2002)
35. P. A. Oeberg, T. Togawa, F. A. Spelman: *Sensors in Medicine and Health Care* (Wiley, New York 2004)
36. S. N. Bhatia, C. S. Chen: Tissue engineering at the micro-scale, Biomedical Microdevices **2**, 131–144 (1999)
37. R. P. Lanza, R. Langer, J. Vacanti (eds.): *Principles of Tissue Engineering* (Academic Press, San Diego 2000)
38. E. Leclerc, K. S. Furukawa, F. Miyata, T. Sakai, T. Ushida, T. Fujii: Fabrication of microstructures in photosensitive biodegradable polymers for tissue engineering applications, Biomaterials **25**, 4683–4690 (2004)
39. T. H. Schulte, R. L. Bardell, B. H. Weigl: Microfluidic technologies in clinical diagnostics, Clinica Chimica Acta **321**, 1–10 (2002)
40. B. Bhushan: *Introduction to Tribology* (Wiley, New York 2002)
41. B. Bhushan: Macro- and microtribology of MEMS materials. In: *Modern Tribology Handbook*, ed. by B. Bhushan (CRC, Boca Raton 2001) pp. 1515–1548
42. B. Bhushan: *Principles and Applications of Tribology* (Wiley, New York 1999)
43. S. Johansson, J. A. Schweitz, L. Tenerz, J. Tiren: Fracture testing of silicon microelements in-situ in a scanning electron microscope, J. Appl. Phys. **63**, 4799–4803 (1988)
44. F. Ericson, J. A. Schweitz: Micromechanical fracture strength of silicon, J. Appl. Phys. **68**, 5840–5844 (1990)
45. E. Obermeier: Mechanical and thermophysical properties of thin film materials for MEMS: Techniques and devices, Micromechan. Struct. Mater. Res. Symp. Proc. **444**, 39–57 (1996)
46. C. J. Wilson, A. Ormeggi, M. Narbutovskih: Fracture testing of silicon microcantilever beams, J. Appl. Phys. **79**, 2386–2393 (1995)
47. W. N. Sharpe, Jr., B. Yuan, R. L. Edwards: A new technique for measuring the mechanical properties of thin films, J. Microelectromech. Syst. **6**, 193–199 (1997)
48. K. Sato, T. Yoshioka, T. Anso, M. Shikida, T. Kawabata: Tensile testing of silicon film having different crystallographic orientations carried out on a silicon chip, Sens. Actuators A **70**, 148–152 (1998)
49. S. Greek, F. Ericson, S. Johansson, M. Furtsch, A. Rump: Mechanical characterization of thick polysilicon films: Young's modulus and fracture strength evaluated with microstructures, J. Micromech. Microeng. **9**, 245–251 (1999)
50. D. A. LaVan, T. E. Buchheit: Strength of polysilicon for MEMS devices, Proc. SPIE **3880**, 40–44 (1999)

51. E. Mazza, J. Dual: Mechanical behavior of a μm-sized single crystal silicon structure with sharp notches, J. Mech. Phys. Solids **47**, 1795–1821 (1999)
52. T. Yi, C.J. Kim: Measurement of mechanical properties for MEMS materials, Meas. Sci. Technol. **10**, 706–716 (1999)
53. H. Kahn, M.A. Huff, A.H. Heuer: Heating effects on the Young's modulus of films sputtered onto micromachined resonators, Microelectromech. Struct. Mater. Res. Symp. Proc. **518**, 33–38 (1998)
54. S. Johansson, F. Ericson, J.A. Schweitz: Influence of surface-coatings on elasticity, residual-stresses, and fracture properties of silicon microelements, J. Appl. Phys. **65**, 122–128 (1989)
55. R. Ballarini, R.L. Mullen, Y. Yin, H. Kahn, S. Stemmer, A.H. Heuer: The fracture toughness of polysilicon microdevices: A first report, J. Mater. Res. **12**, 915–922 (1997)
56. H. Kahn, R. Ballarini, R.L. Mullen, A.H. Heuer: Electrostatically actuated failure of microfabricated polysilicon fracture mechanics specimens, Proc. R. Soc. London A **455**, 3807–3823 (1999)
57. A.M. Fitzgerald, R.H. Dauskardt, T.W. Kenny: Fracture toughness and crack growth phenomena of plasma-etched single crystal silicon, Sensor. Actuat. A **83**, 194–199 (2000)
58. T. Tsuchiya, A. Inoue, J. Sakata: Tensile testing of insulating thin films: Humidity effect on tensile strength of SiO_2 films, Sens. Actuators A **82**, 286–290 (2000)
59. J.A. Connally, S.B. Brown: Micromechanical fatigue testing, Exp. Mech. **33**, 81–90 (1993)
60. K. Komai, K. Minoshima, S. Inoue: Fracture and fatigue behavior of single-crystal silicon microelements and nanoscopic AFM damage evaluation, Microsyst. Technol. **5**, 30–37 (1998)
61. T. Namazu, Y. Isono, T. Tanaka: Evaluation of size effect on mechanical properties of single-crystal silicon by nanoscale bending test using AFM, J. Microelectromech. Syst. **9**, 450–459 (2000)
62. S. Sundararajan, B. Bhushan: Development of AFM-based techniques to measure mechanical properties of nanoscale structures, Sens. Actuators A **101**, 338–351 (2002)
63. X. Li, B. Bhushan: Fatigue studies of nanoscale structures for MEMS/NEMS applications using nanoindentation techniques, Surf. Coat. Technol. **163-164**, 521–526 (2003)
64. X. Li, B. Bhushan, K. Takashima, C.W. Baek, Y.K. Kim: Mechanical characterization of micro/nanoscale structures for MEMS/NEMS applications using nanoindentation techniques, Ultramicroscopy **97**, 481–494 (2003)
65. G. Wei, B. Bhushan, N. Ferrell, D. Hansford: Microfabrication and nanomechanical characterization of polymer MEMS for biological applications, J. Vac. Sci. Technol. A **23**, 811–819 (2005)
66. T. Hsu, N. Sun: Residual stresses/strains analysis of MEMS. In: *Proc. Int. Conf. on Modeling and Simulation of Microsystems, Semiconductors, Sensors and Actuators*, ed. by M. Laudon, B. Romanowicz (Computational Publications, Cambridge 1998) pp. 82–87
67. A. Kolpekwar, C. Kellen, R.D. (Shawn) Blanton: Fault model generation for MEMS. In: *Proc. Int. Conf. on Modeling and Simulation of Microsystems, Semiconductors, Sensors and Actuators*, ed. by M. Laudon, B. Romanowicz (Computational Publications, Cambridge 1998) pp. 111–116
68. H.A. Rueda, M.E. Law: Modeling of strain in boron-doped silicon cantilevers. In: *Proc. Int. Conf. on Modeling and Simulation of Microsystems, Semiconductors, Sensors and Actuators*, ed. by M. Laudon, B. Romanowicz (Computational Publications, Cambridge 1998) pp. 94–99

69. M. Heinzelmann, M. Petzold: FEM analysis of microbeam bending experiments using ultra-micro indentation, Comput. Mater. Sci. **3**, 169–176 (1994)
70. C.J. Wilson, P.A. Beck: Fracture testing of bulk silicon microcantilever beams subjected to a side load, J. Microelectromech. Syst. **5**, 142–150 (1996)
71. B. Bhushan, G.B. Agrawal: Stress analysis of nanostructures using a finite element method, Nanotechnology **13**, 515–523 (2002)
72. B. Bhushan, G.B. Agrawal: Finite element analysis of nanostructures with roughness and scratches, Ultramicroscopy **97**, 495–501 (2003)
73. K.E. Petersen: Silicon as a mechanical material, Proc. IEEE **70**, 420–457 (1982)
74. B. Bhushan, S. Sundararajan, X. Li, C.A. Zorman, M. Mehregany: Micro/nanotribological studies of single-crystal silicon and polysilicon and SiC films for use in MEMS devices. In: *Tribology Issues and Opportunities in MEMS*, ed. by B. Bhushan (Kluwer, Dordrecht 1998) pp. 407–430
75. S. Sundararajan, B. Bhushan: Micro/nanotribological studies of polysilicon and SiC films for MEMS applications, Wear **217**, 251–261 (1998)
76. X. Li, B. Bhushan: Micro/nanomechanical characterization of ceramic films for microdevices, Thin Solid Films **340**, 210–217 (1999)
77. H: Becker, C. Gaertner: Polymer Microfabrication Methods for Microfluidic Analytical Applications, Electrophoresis **21**, 12–26 (2000)
78. J.C. McDonald, D.C. Duffy, J.R. Anderson, D.T. Chiu, H. Wu, O.J.A. Schueller, G.M Whitesides: Fabrication of microfluidic systems in poly(dimethylsiloxane), Electrophoresis **21**, 27–40 (2000)
79. B. Ellis: *Polymers: A Property Database, Available on compact disk* (CRC Press, Boca Raton 2000) also see www.polymersdatabase.com
80. J. Brandrup, E.H. Immergut: *Polymer Handbook* (Wiley, New York 1989)
81. J.E. Mark: *Polymers Data Handbook* (Oxford Univ. Press, Oxford 1999)
82. B. Bhushan, X. Li: Nanomechanical characterization of solid surfaces and thin films, Int. Mater. Rev. **48**, 125–164 (2003)
83. B.R. Lawn, A.G. Evans, D.B. Marshall: Elastic/plastic indentation damage in ceramics: The median/radial system, J. Am. Ceram. Soc. **63**, 574 (1980)
84. S. Sundararajan, B. Bhushan, T. Namazu, Y. Isono: Mechanical property measurements of nanoscale structures using an atomic force microscope, Ultramicroscopy **91**, 111–118 (2002)
85. R.J. Roark: *Formulas for Stress and Strain*, 6th edn. (McGraw-Hill, New York 1989)
86. R.W. Hertzberg: *Deformation and Fracture Mechanics of Engineering Materials*, 3rd edn. (Wiley, New York 1989) pp. 277–278
87. T.K. Ning (ed): *Properties of Silicon*, EMIS Datareviews Series No. 4 (INSPEC, Institution of Electrical Engineers, London 1988)
88. C.T.C. Nguyen, R.T. Howe: An integrated CMOS micromechanical resonator high-Q oscillator, IEEE J. Solid-State Circ. **34**, 440–455 (1999)
89. L.J. Hornbeck: A digital light processing ^{TM} update – status and future applications, Proc. Soc. Photo-Opt. Eng. **3634**, 158–170 (1999)
90. M. Tanaka: Fracture toughness and crack morphology in indentation fracture of brittle materials, J. Mater. Sci. **31**, 749 (1996)
91. B. Bhushan, S. Venkatesan: Mechanical and tribological properties of silicon for micromechanical applications: A review, Adv. Info. Stor. Syst. **5**, 211–239 (1993)
92. B. Bhushan, B.K. Gupta: *Handbook of Tribology: Materials, Coatings, and Surface Treatments* (McGraw-Hill, New York 1997)

93. T. Tsuchiya, O. Tabata, J. Sakata, Y. Taga: Specimen size effect on tensile strength of surface-micromachined polycrystalline silicon thin films, J. Microelectromech. Syst. **7**, 106–113 (1998)
94. T. Yi, L. Li, C. J. Kim: Microscale material testing of single crystalline silicon: Process effects on surface morphology and tensile strength, Sens. Actuators A **83**, 172–178 (2000)
95. F. W. J. Billimeyer: *Textbook of Polymer Science* (Wiley, New York 1984)
96. Anonymous: Rohm and Haas General Information on PMMA
97. W. Wunderlich (ed): *Physical Constants of Poly(methyl methacrylate)*, 2nd edn. (Wiley, New York 1975)
98. E. Calvet, J. P. Bros, H. Pinelle: Compt. Rend. **260**, 1164 (1965)
99. K. Eiermann: Kolloid-Z. **198**, 5 (1965)
100. G. Hochstetter, A. Jimenez, J. P. Cano, E. Felder: An attempt to determine the true stress-strain curves of amorphous polymer by nanoindentation, Tribol. Inter. **36**, 973–985 (2003)
101. S. P. Timoshenko, J. N. Goodier: *Theory of Elasticity*, 3rd edn. (McGraw-Hill, New York 1970)
102. J. E. Shigley, L. D. Mitchell: *Mechanical Engineering Design*, 4th edn. (McGraw-Hill, New York 1993)

15

Scale Effect in Mechanical Properties and Tribology

Bharat Bhushan and Michael Nosonovsky

Summary. A model, which explains scale effects in mechanical properties and tribology is presented. Mechanical properties are scale dependent based on the strain gradient plasticity and the effect of dislocation-assisted sliding. Both single asperity and multiple asperity contacts are considered. The relevant scaling length is the nominal contact length – contact diameter for a single-asperity contact, and scan length for multiple-asperity contacts. For multiple asperity contacts, based on an empirical power-rule for scale dependence of roughness, contact parameters are calculated. The effect of load on the contact parameters and the coefficient of friction is also considered. During sliding, adhesion and two- and three-body deformation, as well as ratchet mechanism, contribute to the dry friction force. These components of the friction force depend on the relevant real areas of contact (dependent on roughness and mechanical properties), average asperity slope, number of trapped particles, and shear strength during sliding. Scale dependence of the components of the coefficient of friction is studied. A scale dependent transition index, which is responsible for transition from predominantly elastic adhesion to plastic deformation has been proposed. Scale dependence of the wet friction, wear, and interface temperature has been also analyzed. The proposed model is used to explain the trends in the experimental data for various materials at nanoscale and microscale, which indicate that nanoscale values of coefficient of friction are lower than the microscale values due to an increase of the three-body deformation and transition from elastic adhesive contact to plastic deformation.

15.1 Nomenclature

$a, \bar{a}, \bar{a}_0, a_{\max}, \bar{a}_{\max}$: Contact radius, mean contact radius, macroscale value of mean contact radius, maximum contact radius, mean value of maximum contact radius.
$A_a, A_r, A_{ra}, A_{re}, A_{re0}, A_{rp}, A_{rp0}, A_{ds}, A_{dp}$: Apparent area of contact, real area of contact, real area of contact during adhesion, real area of elastic contact, macroscale value of real area of elastic contact, real area of plastic contact, macroscale value of real area of plastic contact, real area of contact during asperity summit deformation, area of contact with particles
b: Burgers vector
c: Constant, specified by crystal structure

C_0: Constant required for normalization of $p(d)$

$d, d_e, d_n, d_{ln}, \bar{d}, \bar{d}_0$: Particle diameter, minimum for exponential distribution, mean for normal distribution, exponential of mean of $\ln(d)$ for log-normal distribution, mean trapped particles diameter, macroscale value of mean trapped particles diameter

D: Interface zone thickness

E_1, E_2, E^*: Elastic moduli of contacting bodies, effective elastic modulus

$F, F_a, F_d, F_{ae}, F_{ap}, F_a, F_{ds}, F_{dp}, F_m, F_{m0}$: Friction force, friction force due to adhesion, friction force due to deformation, friction force during elastic adhesional contact, plastic adhesional contact, summit deformation, particles deformation respectively, meniscus force for wet contact, macroscale value of meniscus force

G: Elastic shear modulus

h: Indentation depth

h_f: Liquid film thickness

H, H_0: Hardness, hardness in absence of strain gradient

k, k_0: Wear coefficient, macroscale value of wear coefficient

l_s, l_d: Material-specific characteristic length parameters

$L, L_{lwl}, L_{lc}, L_s, L_d$: Length of the nominal contact zone, long wavelength limit for roughness parameters, long wavelength limit for contact parameters, length parameters related to l_s and l_d

L_p: Peclet number

m, n: Indices of exponents for scale-dependence of σ and β^*

n_{tr}: Number of trapped particles divided by the total number of particles

p_a, p_{ac}: Apparent pressure, critical apparent pressure

$p(d), p_{tr}(d)$: Probability density function for particle size distribution, probability density function for trapped particle size distribution

$P(d)$: Cumulative probability distribution for particle size

$R, R_p, \overline{R_p}, \overline{R_{p0}}$: Effective radius of summit tips, radius of summit tip, mean radius of summit tips, macroscale value of the mean radius of summit tips

$R(\tau)$: Autocorrelation function

s: Spacing between slip steps on the indentation surface

s_d: Separation distance between reference planes of two surfaces in contact

N, N_0: Total number of contacts, macroscale value of total number of contacts

T, T_0: Maximum flash temperature rise, macroscale value of temperature rise

x: Sliding distance

v: Volume of worn material

V: Sliding velocity

W: Normal load

z, z_{min}, z_{max}: Random variable, minimum and maximum value of z

α: Probability for a particle in the border zone to leave the contact region

β^*, β_0^*: Correlation length, macroscale value of correlation length

γ: Surface tension

Γ: Gamma function

ε: Strain

η: Density of particles per apparent area of contact

η_{int}, η_{cr}: Density of dislocation lines per interface area, critical density of dislocation lines per interface area

κ: Curvature

κ_t: Thermal diffusivity

θ: Contact angle between the liquid and surface

θ_i: Indentation angle

θ_r: Roughness angle

$\mu, \mu_a, \mu_{ae}, \mu_{ae0}, \mu_{ap}, \mu_{ap0}, \mu_d, \mu_{ds}, \mu_{ds0}, \mu_{dp}, \mu_{dp0}, \mu_r, \mu_{r0}, \mu_{re}, \mu_{re0}, \mu_{rp}, \mu_{rp0}, \mu_{wet}$: Coefficient of friction, coefficient of adhesional friction, coefficient of adhesional elastic friction, macroscale value of coefficient of adhesional elastic friction, coefficient of adhesional plastic friction, macroscale value of coefficient of adhesional plastic friction, coefficient of deformation friction, coefficient of summits deformation friction, macroscale value of coefficient of summits deformation friction, coefficient of particles deformation friction, macroscale value of coefficient of particles deformation friction, ratchet component of the coefficient of friction, macroscale value of ratchet component of the coefficient of friction, ratchet component of the coefficient of elastic friction, macroscale value of ratchet component of the coefficient of elastic friction, ratchet component of the coefficient of plastic friction, macroscale value of ratchet component of the coefficient of plastic friction, and coefficient of wet friction

ν_1, ν_2: Poisson's ratios of contacting bodies

ρc_p: Volumetric specific heat

$\sigma, \sigma_0, \sigma_e, \sigma_n, \sigma_{ln}$: Standard deviation of rough surface profile height, macroscale value of standard deviation of rough surface profile height, standard deviation for the exponential distribution, standard deviation for the normal distributions, standard deviation for $\ln(d)$ of the log normal distribution

ρ, ρ_G, ρ_S: Total density of dislocation lines per volume, density of GND per volume, density of SSD per volume

ϕ, ϕ_0: Transition index, macroscale value of transition index

τ, τ_0: Spatial parameter, value at which the autocorrelation function decays

$\tau_a, \tau_{a0}, \tau_Y, \tau_{Y0}, \tau_{ds}, \tau_{ds0}, \tau_{dp}, \tau_{dp0}, \tau_p$: Adhesional shear strength during sliding, macroscale value of adhesional shear strength, shear yield strength, shear yield strength in absence of strain gradient, shear strength during summits deformation, macroscale value of shear strength during summits deformation, shear strength during particles deformation, macroscale value of shear strength during particles deformation, Peierls stress

15.2 Introduction

Microscale and nanoscale measurements of tribological properties, which became possible due to the development of the Surface Force Apparatus (SFA), Atomic Force Microscope (AFM), and Friction Force Microscope (FFM) demonstrate scale dependence of adhesion, friction, and wear as well as mechanical properties including hardness [1–4]. Advances of micro/nanoelectromechanical systems (MEMS/NEMS) technology in the past decade make understanding of scale effects in adhesion, friction, and wear especially important, since surface to volume ratio grows with miniaturization and surface phenomena dominate. Dimensions of MEMS/NEMS devices range from about 1 mm to few nm.

Experimental studies of scale dependence of tribological phenomena have been conducted recently. AFM experiments provide data on nanoscale [5–10] whereas microtriboapparatus [11, 12] and SFA [13] provide data on microscale. Experimental data indicate that wear mechanisms and wear rates are different at macro- and micro/nanoscales [14, 15]. During sliding, the effect of operating conditions such as load and velocity on friction and wear are frequently manifestations of the effect of temperature rise on the variable under study. The overall interface temperature rise is a cumulative result of numerous flash temperature rises at individual asperity contacts. The temperature rise at each contact is expected to be scale dependent, since it depends on contact size, which is scale dependent.

Friction is a complex phenomenon, which involves asperity interactions involving adhesion and deformation (plowing), Fig. 15.1. Adhesion and plastic deformation imply energy dissipation, which is responsible for friction. A contact between two bodies takes place on high asperities, and the real area of contact (A_r) is a small fraction of the apparent area of contact [16]. During contact of two asperities, a lateral force may be required for asperities of a given slope to climb against each other. This mechanism is known as ratchet mechanism, and it also contributes to the friction. Wear and contaminant particles present at the interface, referred as the "third body", also contribute to friction, Fig. 15.2a. In addition, during contact, even at low humidity, a meniscus is formed. Generally, any liquid that wets or has a small contact angle on surfaces will condense from vapor into cracks and pores on surfaces as bulk liquid and in the form of annular-shaped capillary condensate in the contact zone. Figure 15.2b shows a random rough surface in contact with a smooth surface with a continuous liquid film on the smooth surface. The presence of the liquid film of the condensate or pre-existing film of the liquid can significantly increase the adhesion between the solid bodies [16]. The effect of meniscus is scale-dependent.

A quantitative theory of scale effects in friction should consider scale effect on physical properties relevant to these contributions. However, conventional theories of contact and friction lack characteristic length parameters, which would be responsible for scale effects. The linear elasticity and conventional plasticity theories are scale-invariant and do not include any material length scales. A strain gradient plasticity theory has been developed, for microscale deformations, by *Fleck* et al. [17], *Nix* and *Gao* [18] and *Hutchinson* [19]. Their theory predicts a dependence of mechanical properties on the strain gradient, which is scale dependent: the smaller is the size of the deformed region, the greater is the gradient of plastic strain, and, the greater is the yield strength and hardness.

A comprehensive model of scale effect in friction including adhesion, two- and three-body deformations and the ratchet mechanism, has recently been proposed by

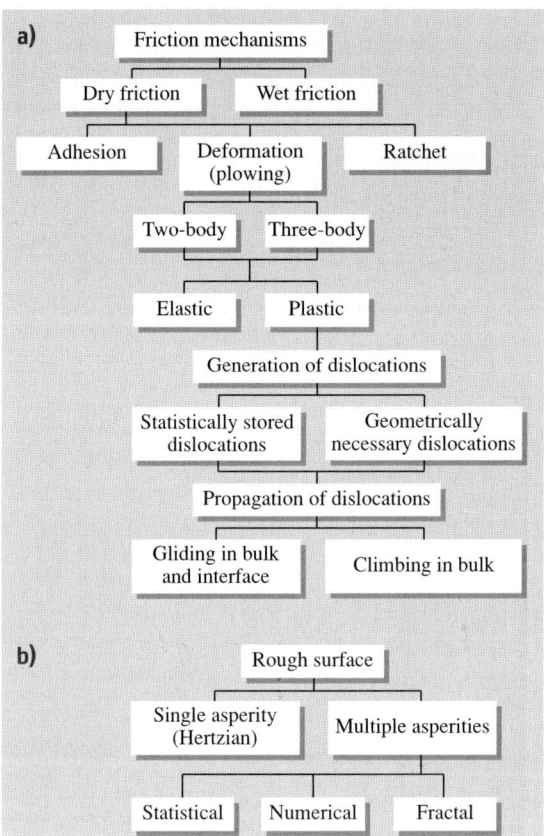

Fig. 15.1. (a) A block diagram showing friction mechanisms and generation and propagation of dislocations during sliding, (b) a block diagram of rough contact models

Bhushan and *Nosonovsky* [20–22] and *Nosonovsky* and *Bhushan* [23]. The model for adhesional friction during single and multiple asperity contact was developed by *Bhushan* and *Nosonovsky* [20] and is based on the strain gradient plasticity and dislocation assisted sliding (gliding dislocations at the interface or microslip). The model for the two-body and three-body deformation was proposed by *Bhushan* and *Nosonovsky* [21] and for the ratchet mechanism by *Nosonovsky* and *Bhushan* [23]. The model has been extended for wet contacts, wear and interface temperature by *Bhushan* and *Nosonovsky* [22]. The detailed model is presented in this chapter.

The chapter is organized as follows. In the next section of this chapter, the scale effect in mechanical properties is considered, including yield strength and hardness based on the strain gradient plasticity and shear strength at the interface based on the dislocation assisted sliding (microslip). In the fourth section, scale effect in surface roughness and contact parameters is considered, including the real area of contact, number of contacts, and mean size of contact. Load dependence of contact parameters is also studied in this section. In the fifth section, scale effect in friction is considered, including adhesion, two- and three-body deformation, ratchet mech-

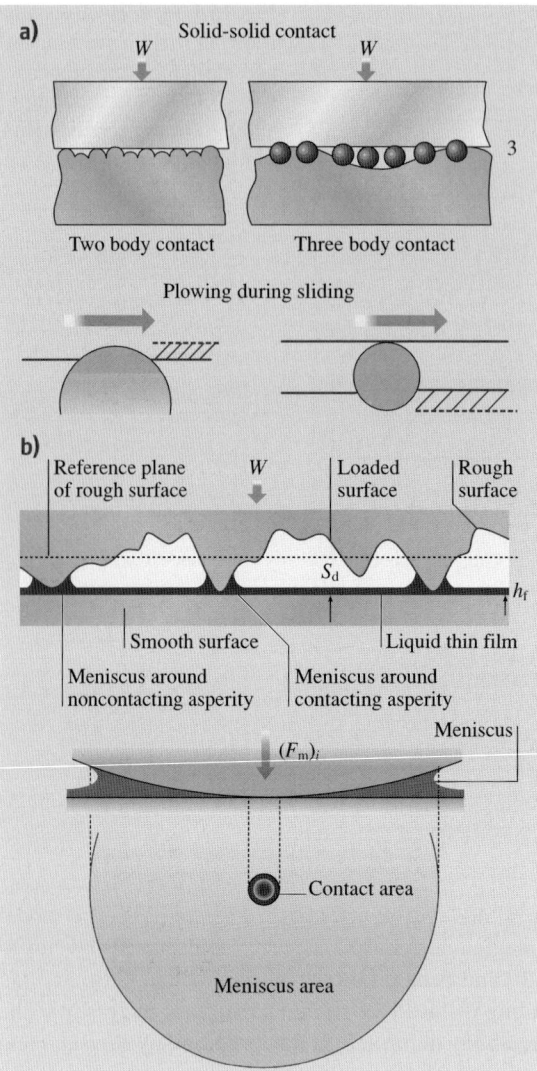

Fig. 15.2. Schematics of (**a**) two-bodies and three-bodies during dry contact of rough surfaces, (**b**) formation of menisci during wet contact

anism, meniscus analysis, total value of the coefficient of friction and comparison with the experimental data. In the sixth and seventh sections, scale effects in wear and interface temperature are analyzed, respectively.

15.3 Scale Effect in Mechanical Properties

In this section, scale dependence of hardness and shear strength at the interface is considered. A strain gradient plasticity theory has been developed, for microscale deformations, by *Fleck* et al. [17], *Nix* and *Gao* [18], *Hutchinson* [19], and others, which is based on statistically stored and geometrically necessary dislocations (to be described later). Their theory predicts a dependence of mechanical properties on the strain gradient, which is scale dependent: the smaller is the size of the deformed region, the greater is the gradient of plastic strain, and, the greater is the yield strength and hardness. *Gao* et al. [24] and *Huang* et al. [25] proposed a mechanism-based strain gradient (MSG) plasticity theory, which is based on a multiscale framework, linking the microscale (10–100 nm) notion of statistically stored and geometrically necessary dislocations to the mesoscale (1–10 μm) notion of plastic strain and strain gradient. *Bazant* [26] analyzed scale effect based on the MSG plasticity theory in the limit of small scale, and found that corresponding nominal stresses in geometrically similar structures of different sizes depend on the size according to a power exponent law.

It was recently suggested also, that relative motion of two contacting bodies during sliding takes place due to dislocation-assisted sliding (microslip), which results in scale-dependent shear strength at the interface [20]. Scale effects in mechanical properties (yield strength, hardness, and shear strength at the interface) based on the strain gradient plasticity and dislocation-assisted sliding models are considered in this section.

15.3.1 Yield Strength and Hardness

Plastic deformation occurs during asperity contacts because a small real area of contact results in high contact stresses, which are often beyond the limits of the elasticity. As stated earlier, during loading, generation and propagation of dislocations is responsible for plastic deformation. Because dislocation motion is irreversible, plastic deformation provides a mechanism for energy dissipation during friction. The strain gradient plasticity theories [17–19] consider two types of dislocations: randomly created Statistically Stored Dislocations (SSD) and Geometrically Necessary Dislocations (GND). The GND are required for strain compatibility reasons. Randomly created SSD during shear and GND during bending are presented in Fig. 15.3a. The density of the GND (total length of dislocation lines per volume) during bending is proportional to the curvature κ and to the strain gradient

$$\rho_G = \frac{\kappa}{b} = \frac{1}{b}\frac{\partial \varepsilon}{\partial z} \propto \nabla \varepsilon, \tag{15.1}$$

where ε is strain, b is the Burgers vector, and $\nabla \varepsilon$ is the strain gradient.

The GND during indentation, Fig. 15.3b, are located in a certain sub-surface volume. The large strain gradients in small indentations require GND to account for the large slope at the indented surface. SSD, not shown here, also would be created and would contribute to deformation resistance, and are function of strain rather than strain gradient. According to *Nix* and *Gao* [18], we assume that indentation is accommodated by circular loops of GND with Burgers vector normal to the plane

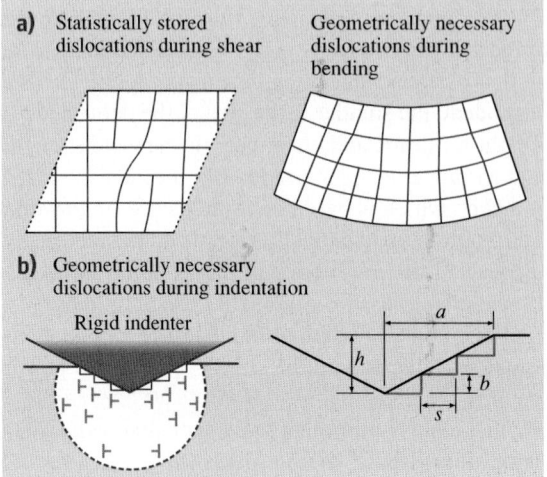

Fig. 15.3. (a) Illustration of statistically stored dislocations during shear and geometrically necessary dislocations during bending, (b) geometrically necessary dislocations during indentation

of the surface. If we think of the individual dislocation loops being spaced equally along the surface of the indentation, then the surface slope

$$\tan\theta_i = \frac{h}{a} = \frac{b}{s}, \tag{15.2}$$

where θ_i is the angle between the surface of the conical indenter and the plane of the surface, a is the contact radius, h is the indentation depth, b is the Burgers vector, and s is the spacing between individual slip steps on the indentation surface (Fig. 15.3b). They reported that for geometrical (strain compatibility) considerations, the density of the GND is

$$\rho_G = \frac{3}{2bh}\tan^2\theta_i = \frac{3}{2b}\left(\frac{\tan\theta_i}{a}\right) = \frac{3}{2b}\nabla\varepsilon. \tag{15.3}$$

Thus ρ_G is proportional to strain gradient (scale dependent) whereas the density of SSD, ρ_S is dependent upon the average strain in the indentation, which is related to the slope of the indenter ($\tan\theta_i$). Based on experimental observations, ρ_S is approximately proportional to strain [17].

According to the Taylor model of plasticity [30], dislocations are emitted from Frank–Read sources. Due to interaction with each other, the dislocations may become stuck in what is called the Taylor network, but when externally applied stress reaches the order of Peierls stress for the dislocations, they start to move and the plastic yield is initiated. The magnitude of the Peierls stress τ_p is proportional to the dislocation's Burgers vector b divided by a distance between dislocation lines s [30, 31]

$$\tau_p = Gb/(2\pi s), \tag{15.4}$$

where G is the elastic shear modulus. An approximate relation of the shear yield strength τ_Y to the dislocations density at a moment when yield is initiated is given by [30]

$$\tau_{Y0} = cGb/s = cGb\sqrt{\rho}, \tag{15.5}$$

where c is a constant on the order of unity, specified by the crystal structure and ρ is the total length of dislocation lines per volume, which is a complicated function of strain ε and strain gradient ($\nabla \varepsilon$)

$$\rho = \rho_S(\varepsilon) + \rho_G(\nabla \varepsilon). \tag{15.6}$$

The shear yield strength τ_Y can be written now as a function of SSD and GND densities [30]

$$\tau_Y = cGb\sqrt{\rho_S + \rho_G} = \tau_{Y0}\sqrt{1 + (\rho_G/\rho_S)}, \tag{15.7}$$

where

$$\tau_{Y0} = cGb\sqrt{\rho_S} \tag{15.8}$$

is the shear yield strength value in the limit of small ρ_G/ρ_S ratio (large scale) that would arise from the SSD, in the absence of GND. Note that the ratio of the two densities is defined by the problem geometry and is scale dependent. Based on the relationships for ρ_G (15.3) and ρ_S, the ratio ρ_G/ρ_S is inversely proportional to a and (15.7) reduces to

$$\tau_Y = \tau_{Y0}\sqrt{1 + (l_d/a)}, \tag{15.9}$$

where l_d is a plastic deformation length that characterizes depth dependence on shear yield strength. According to *Hutchinson* [19], this length is physically related to an average distance a dislocation travels, which was experimentally determined to be between 0.2 µm and 5 µm for copper and nickel. Note that l_d is a function of the material and the asperity geometry and is dependent on SSD.

Using von Mises yield criterion, hardness $H = 3\sqrt{3}\tau_Y$. From (15.9) the hardness is also scale-dependent [18]

$$H = H_0\sqrt{1 + (l_d/a)}, \tag{15.10}$$

where H_0 is hardness in absence of strain gradient. Equation (15.9) provides dependence of the resistance force to deformation upon the scale in a general case of plastic deformation [20].

Scale dependence of yield strength and hardness has been well established experimentally. *Bhushan* and *Koinkar* [32] and *Bhushan* et al. [27] measured hardness of single-crystal silicon (100) up to a peak load of 500 µN. *Kulkarni* and *Bhushan* [28] measured hardness of single crystal aluminium (100) up to 2000 µN and *Nix* and *Gao* [18] presented data for single crystal copper; using a three-sided pyramidal (Berkovich) diamond tip. The hardness on nanoscale is found to be higher than on microscale (Fig. 15.4). Similar results have been reported in other tests, including indentation tests for other materials [29, 33, 34], torsion and tension experiments on copper wires [17, 19], and bending experiments on silicon and silica beams [35].

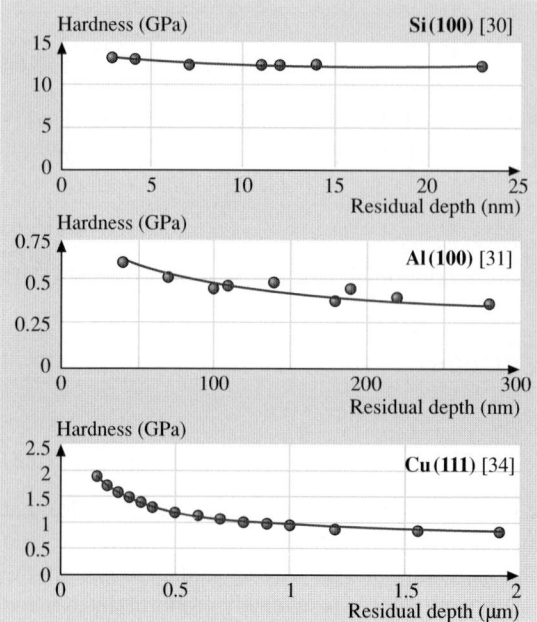

Fig. 15.4. Indentation hardness as a function of residual indentation depth for Si(100) [27], Al(100) [28], Cu(111) [29]

15.3.2 Shear Strength at the Interface

Mechanism of slip involves motion of large number of dislocations, which is responsible for plastic deformation during sliding. Dislocations are generated and stored in the body and propagate under load. There are two modes of possible line (or edge) dislocation motion: gliding, when dislocation moves in the direction of its Burgers vector b by a unit step of its magnitude, and climbing, when dislocation moves in a direction, perpendicular to its Burgers vector (Fig. 15.5a). Motion of dislocations can take place in the bulk of the body or at the interface. Due to periodicity of the lattice, a gliding dislocation experiences a periodic force, known as the Peierls force [31]. The Peierls force is responsible for keeping the dislocation at a central position between symmetric lattice lines and it opposes dislocation's gliding (Fig. 15.5b). Therefore, an external force should be applied to overcome Peierls force resistance against dislocation's motion. *Weertman* [36] showed that a dislocation or a group of dislocations can glide uniformly along an interface between two bodies of different elastic properties. In continuum elasticity formulation, this motion is equivalent to a propagating interface slip pulse, however the physical nature of this deformation is plastic, because dislocation motion is irreversible. The local plastic deformation can occur at the interface due to concentration of dislocations even in the predominantly elastic contacts. Gliding of a dislocation along the interface results in a relative displacement of the bodies for a distance equal to the Burgers vector of the dislocation, whereas a propagating set of dislocations effectively results in dislocation-assisted sliding, or microslip (Fig. 15.6).

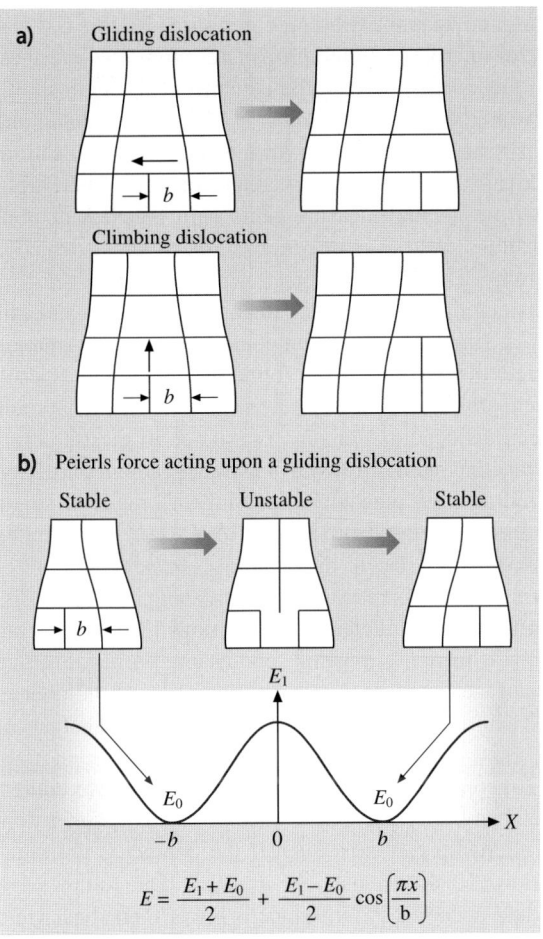

Fig. 15.5. (a) Schematics of gliding and climbing dislocations motion by a unit step of Burgers vector b. (b) Origin of the periodic force acting upon a gliding dislocation (Peierls force). Gliding dislocation passes locations of high and low potential energy

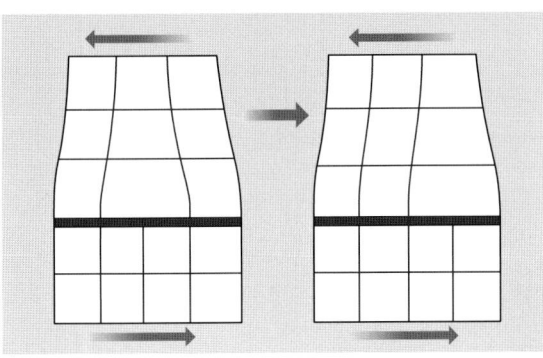

Fig. 15.6. Schematic showing microslip due to gliding dislocations at the interface

Several types of microslip are known in the tribology literature [16], the dislocation-assisted sliding is one type of microslip, which propagates along the interface. Conventional mechanism of sliding is considered to be concurrent slip with simultaneous breaking of all adhesive bonds. Based on *Johnson* [37] and *Bhushan* and *Nosonovsky* [20], for contact sizes on the order of few nm to few μm, dislocation-assisted sliding is more energetically profitable than a concurrent slip. Their argument is based on the fact that experimental measurements with the SFA demonstrated that, for mica, frictional stress is of the same order as Peierls stress, which is required for gliding of dislocations.

Polonsky and *Keer* [38] considered the pre-existing dislocation sources and carried out a numerical microcontact simulation based on contact plastic deformation representation in terms of discrete dislocations. They found that when the asperity size decreases and becomes comparable with the characteristic length of materials microstructure (distance between dislocation sources), resistance to plastic deformation increases, which supports conclusions drawn from strain gradient plasticity. *Deshpande* et al. [39] conducted discrete plasticity modeling of cracks in single crystals and considered dislocation nucleation from Frank–Read sources distributed randomly in the material. Pre-existing sources of dislocations, considered by all of these authors, are believed to be a more realistic reason for increasing number of dislocations during loading, rather than newly nucleated dislocations [30]. In general, dislocations are emitted under loads from pre-existing sources and propagate along slip lines (Fig. 15.7). As shown in the figure, in regions of higher loads, number of emitted dislocations is higher. Their approach was limited to numerical analysis of special cases.

Bhushan and *Nosonovsky* [20] considered a sliding contact between two bodies. Slip along the contact interface is an important special case of plastic deformation. The local dislocation-assisted microslip can exist even if the contact is predominantly elastic due to concentration of dislocations at the interface. Due to these dislocations, the stress at which yield occurs at the interface is lower than shear yield strength in the bulk. This means that average shear strength at the interface is lower than in the bulk.

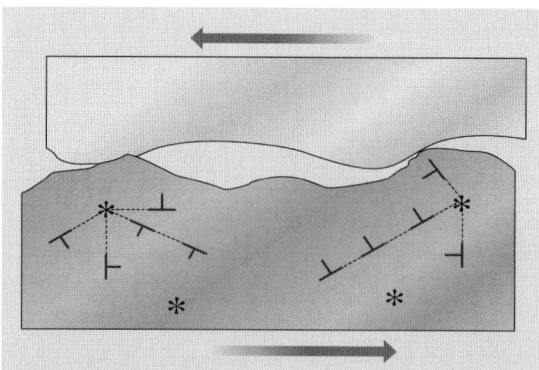

Fig. 15.7. Generation of dislocations from sources (∗) during plowing due to plastic deformation

An assumption that all dislocations produced by externally applied forces are distributed randomly throughout the volume would result in vanishing small probability for a dislocation to be exactly at the interface. However, many traveling (gliding and climbing) dislocations will be stuck at the interface as soon as they reach it. As a result of this, a certain number of dislocations will be located at the interface. In order to account for a finite dislocation density at the interface, *Bhushan* and *Nosonovsky* [20] assumed, that the interface zone has a finite thickness D. Dislocations within the interface zone may reach the contact surface due to climbing and contribute into the microslip. In the case of a small contact radius a, compared to interface zone thickness D, which is scale dependent, and is approximately equal to a. However, in the case of a large contact radius, the interface zone thickness is approximately equal to the average distance dislocations can climb l_s. An illustration of this is provided in Fig. 15.8. The depth of the subsurface volume, from which dislocations have a high chance to reach the interface is limited by l_s and by a, respectively, for the two cases considered here. Based on these geometrical considerations, an approximate relation can be written as

$$D = \frac{al_s}{l_s + a}.\tag{15.11}$$

The interface density of dislocations (total length of dislocation lines per interface area) is related to the volume density as

$$\eta_{int} = \rho D = \rho \left(\frac{al_s}{l_s + a} \right).\tag{15.12}$$

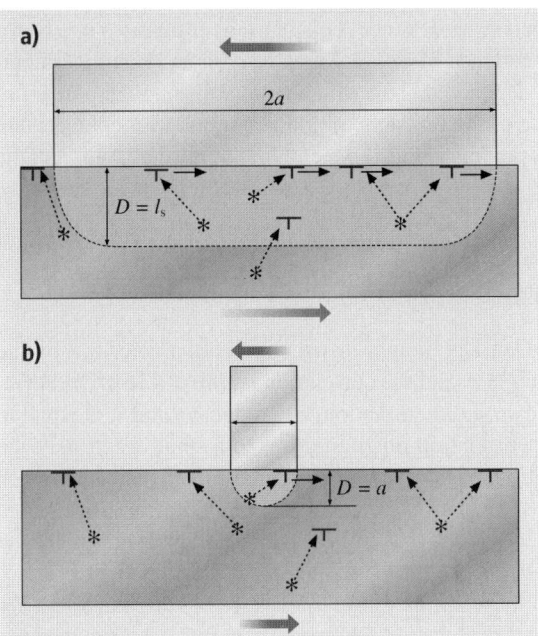

Fig. 15.8. Gliding dislocations at the interface generated from sources (∗). Only dislocations generated within the interface zone can reach the interface. (**a**) For a large contact radius a, thickness of this zone D is approximately equal to an average distance dislocations climb l_s. (**b**) For small contact radius a, the thickness of the interface zone is approximately equal to a

During sliding, dislocations must be generated at the interface with a certain critical density $\eta_{int} = \eta_{cr}$. The corresponding shear strength during sliding can be written following (15.9) as

$$\tau_a = \tau_{a0} \sqrt{1 + (l_s/a)}, \tag{15.13}$$

where

$$\tau_{a0} = cGb\sqrt{\frac{\eta_{cr}}{l_s}} \tag{15.14}$$

is the shear strength during sliding in the limit of $a \gg l_s$.

Equation (15.13) gives scale-dependence of the shear strength at the interface and is based on the following assumptions. First, it is assumed that only dislocations in the interface zone of thickness D, given by (15.11), contribute into sliding. Second, it is assumed, that a critical density of dislocations at the interface η_{cr} is required for sliding. Third, the shear strength is equal to the Peierls stress, which is related to the volume density of the dislocations $\rho = \eta/D$ according to (15.4), with the typical distance between dislocations $s = 1/\sqrt{\rho}$. It is noted, that proposed scaling rule for the dislocation assisted sliding mechanism (15.13) has a similar form to that for the yield strength (15.9), since both results are consequences of scale dependent generation and propagation of dislocations under load [20].

15.4 Scale Effect in Surface Roughness and Contact Parameters

During multiple-asperity contact, scale dependence of surface roughness is a factor which contributes to scale dependence of the real area of contact. Roughness parameters are known to be scale dependent [16], which results, during the contact of two bodies, in scale dependence of the real area of contact, number of contacts and mean contact size. The contact parameters also depend on the normal load, and the load dependence is similar to the scale dependence [23]. Both effects are analyzed in this section.

15.4.1 Scale Dependence of Roughness and Contact Parameters

A random rough surface with Gaussian height distribution is characterized by the standard deviation of surface height σ and the correlation length β^* [16]. The correlation length is a measure of how quickly a random event decays and it is equal to the length, over which the autocorrelation function drops to a small fraction of the value at the origin. The correlation length can be considered as a distance, at which two points on a surface have just reached the condition where they can be regarded as being statistically independent. Thus, σ is a measure of height distribution and β^* is a measure of spatial distribution.

A surface is composed of a large number of length scales of roughness that are superimposed on each other. According to AFM measurements on glass-ceramic

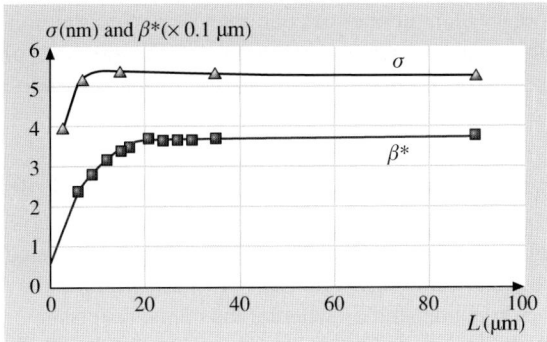

Fig. 15.9. Roughness parameters as a function of scan size for a glass-ceramic disk measured using AFM [16]

disk surface, both σ and β^* initially increase with the scan size and then approach a constant value, at certain scan size (Fig. 15.9). This result suggests that disk roughness has a long wavelength limit, L_{lwl}, which is equal to the scan size at which the roughness values approach a constant value [16]. It can be assumed that σ and β^* depend on the scan size according to an empirical power rule

$$\sigma = \sigma_0 \left(\frac{L}{L_{\text{lwl}}}\right)^n, \quad L < L_{\text{lwl}},$$

$$\beta^* = \beta_0^* \left(\frac{L}{L_{\text{lwl}}}\right)^m, \quad L < L_{\text{lwl}}, \tag{15.15}$$

where n and m are indices of corresponding exponents and σ_0 and β_0^* are macroscale values [20]. Based on the data, presented in Fig. 15.9, it is noted that for glass-ceramic disk, long-wavelength limit for σ and β^* is about 17 μm and 23 μm, respectively. The difference is expected to be due to measurement errors. An average value $L_{\text{lwl}} = 20$ μm is taken here for calculations. The values of the indices are found as $m = 0.5$, $n = 0.2$, and the macroscale values are $\sigma_0 = 5.3$ nm, $\beta_0^* = 0.37$ μm [23].

For two random surfaces in contact, the length of the nominal contact size L defines the characteristic length scale of the problem. The contact problem can be simplified by considering a rough surface with composite roughness parameters in contact with a flat surface. The composite roughness parameters σ and β^* can be obtained based on individual values for the two surfaces [16]. For Gaussian surfaces, the contact parameters of interest, to be discussed later, are the real area of contact A_r, number of contacts N, and mean contact radius $\bar{a}$. The long wavelength limit for scale dependence of the contact parameters L_{lc}, which is not necessarily equal to that of the roughness, L_{lwl}, will be used for normalization of length parameters. The scale dependence of the contact parameters exists if $L < L_{\text{lc}}$ [23].

The mean of surface height distribution corresponds to so-called reference plane of the surface. Separation s_d is a distance between reference planes of two surfaces in contact, normalized by σ. For a given s_d and statistical distribution of surface heights, the total real area of contact (A_r), number of contacts (N), and elastic normal load W_e can be found, using statistical analysis of contacts. The real area of contact, number of contacts and elastic normal load are related to the separation distance s_d [40]

$$A_r \propto F_A(s_d), \quad N \propto \frac{1}{(\beta^*)^2} F_N(s_d), \quad W_e \propto \frac{E^*\sigma}{\beta^*} F_W(s_d), \tag{15.16}$$

where $F_A(s_d)$, $F_N(s_d)$, and $F_W(s_d)$, are integral functions defined by *Onions* and *Archard* [40]. It should be noted, that A_r and N as functions of s_d are prescribed by the contact geometry (σ, β^*) and do not depend on whether the contact is elastic or plastic. Based on Onions and Archard data, it is observed that the ratio F_W/F_A is almost constant for moderate $s_d < 1.4$ and increases slightly for $s_d > 1.4$. The ratio F_A/F_N decreases rapidly with s_d and becomes almost constant for $s_d > 2.0$. For moderate loads, the contact is expected to occur on the upper parts of the asperities ($s_d > 2.0$), and a linear proportionality of $F_A(s_d)$, $F_N(s_d)$, and $F_W(s_d)$ can be assumed [20].

Based on (15.16) and the observation that F_W/F_A is almost constant, for moderate loads, A_{re} (the real area of elastic contact), N, and $\bar{a}$ are related to the roughness, based on the parameter L_{lc}, as

$$A_{re} \propto \frac{\beta^*}{\sigma E^*} W = A_{re0} \left(\frac{L}{L_{lc}}\right)^{m-n}, \quad L < L_{lc}, \tag{15.17}$$

$$N \propto \frac{W}{\beta^*\sigma E^*} = N_0 \left(\frac{L}{L_{lc}}\right)^{-m-n}, \quad L < L_{lc}, \tag{15.18}$$

$$\bar{a} \propto \beta^* = \sqrt{\frac{A_r}{N}} = \bar{a}_0 \left(\frac{L}{L_{lc}}\right)^m, \quad L < L_{lc}. \tag{15.19}$$

The mean radius of summit tips $\overline{R_p}$ is given, according to *Whitehouse* and *Archard* [41]

$$\overline{R_p} \propto \frac{(\beta^*)^2}{\sigma} = \overline{R}_{p0} \left(\frac{L}{L_{lwl}}\right)^{2m-n}, \quad L < L_{lwl}, \tag{15.20}$$

where $\bar{a}_0$, N_0 and $\overline{R}_{p0}$ are macroscale values, E^* is the effective elastic modulus of contacting bodies [22], which is related to the elastic moduli E_1, E_2 and Poisson's ratios ν_1, ν_2 of the two bodies as $1/E^* = (1-\nu_1^2)/E_1 + (1-\nu_2^2)/E_2$ and which is known to be scale independent, and variables with the subscript "0" are corresponding macroscale values (for $L \geq L_{lc}$).

Dependence of the real area of plastic contact A_{rp} on the load is given by

$$A_{rp} = \frac{W}{H}, \tag{15.21}$$

where H is hardness. According to the strain gradient plasticity model [17,18], the yield strength τ_Y is given by (15.9) and hardness H is given by (15.10). In the

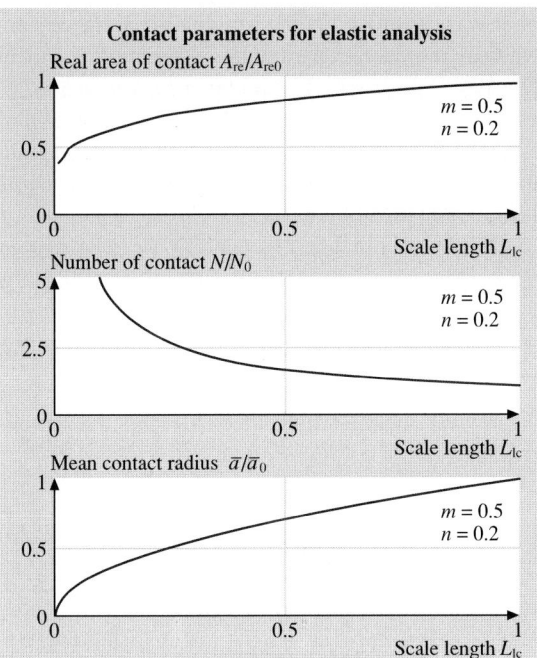

Fig. 15.10. Scale length dependence of normalized contact parameters ($m = 0.5$, $n = 0.2$) (**a**) real area of contact, (**b**) number of contacts, and (**c**) mean contact radius

case of plastic contact, the mean contact radius can be determined from (15.19), which is based on the contact geometry and independent of load [20]. Assuming the contact radius as its mean value from (15.19) based on elastic analysis, and combining (15.10), (15.19) and (15.21), the real area of plastic contact is given as

$$A_{rp} = \frac{W}{H_0 \sqrt{1+(l_d/\bar{a})}} = \frac{W}{H_0 \sqrt{1+(L_d/L)^m}}, \quad L < L_{lc}, \tag{15.22}$$

where L_d is a characteristic length parameter related to l_d, $\bar{a}$, and L_{lc} [20]

$$L_d = L_{lc} \left(\frac{l_d}{\bar{a}_0} \right)^{1/m}. \tag{15.23}$$

The scale dependence of A_{re}, N, and $\bar{a}$ is presented in Fig. 15.10.

15.4.2 Dependence of Contact Parameters on Load

The effect of short and long wavelength details of rough surfaces on contact parameters also depends on the normal load. For low loads, the ratio of real to apparent areas of contact A_r/A_a, is small, contact spots are small, and long wavelength details are irrelevant. For higher A_r/A_a, long wavelength details become important, whereas small wavelength details of the surface geometry become irrelevant. The effect of increased load is similar to the effect of increased scale length [23].

In the preceding subsections, it was assumed that the roughness parameters are scale-dependent for $L < L_{\text{lwl}}$, whereas the contact parameters are scale-dependent for $L < L_{\text{lc}}$. The upper limit of scale dependence for the contact parameters, L_{lc}, depends on the normal load, and it is reasonable to assume that L_{lc} is a function of A_r/A_a, and the contact parameters are scale-dependent when A_r/A_a is below a certain critical value. It is convenient to consider the apparent pressure p_a, which is equal to the normal load divided by the apparent area of contact [23].

For elastic contact, based on (15.15) and (15.17), this condition can be written as

$$\frac{A_{\text{re}}}{A_a} \propto \frac{\beta^* p_a}{\sigma} = p_a \frac{\beta_0^*}{\sigma_0} \left(\frac{L}{L_{\text{lwl}}} \right)^{m-n} < p_{\text{ac}}, \tag{15.24}$$

where p_{ac} is a critical apparent pressure, below which the scale dependence occurs [23]. From (15.24) one can find

$$L < L_{\text{lwl}} \left(\frac{\beta_0^*}{\sigma_0} \frac{p_a}{p_{\text{ac}}} \right)^{1/(n-m)}. \tag{15.25}$$

The right-hand expression in (15.24) is defined as L_{lc}

$$L_{\text{lc}} = L_{\text{lwl}} \left(\frac{\beta_0^*}{\sigma_0} \frac{p_a}{p_{\text{ac}}} \right)^{1/(n-m)}. \tag{15.26}$$

For plastic contact, based on (15.22)

$$\frac{A_{\text{rp}}}{A_a} \propto \frac{p_a}{\sqrt{1 + (L_d/L)^m}} < p_{\text{ac}}. \tag{15.27}$$

In a similar manner to the elastic case, (15.27) yields [23]

$$L_{\text{lc}} = L_d \left[\left(\frac{p_a}{p_{\text{ac}}} \right)^2 - 1 \right]^{-1/m}. \tag{15.28}$$

Load dependence of the long wavelength limit for contact parameters, L_{lc} is presented in Fig. 15.11 for an elastic contact based on (15.28), and for a plastic contact based on (15.28), for $m = 0.5$, $n = 0.2$ [23]. The load (apparent pressure) is normalized by $\beta_0^*/(p_{\text{ac}}\sigma_0)$ for the elastic contact and by p_{ac} for the plastic contact. In the case of elastic contact, it is observed, that the long wavelength limit decreases with increasing load. For a problem, characterized by a given scale length L, increase of load will result in decrease of L_{lc} and, eventually, the condition $L < L_{\text{lc}}$ will be violated; thus the contact parameters, including the coefficient of friction, will reach the macroscale values. Decrease of L_{lc} with increasing load is also observed in the case of plastic contact, the data presented for $p_a/p_{\text{ac}} > 1$.

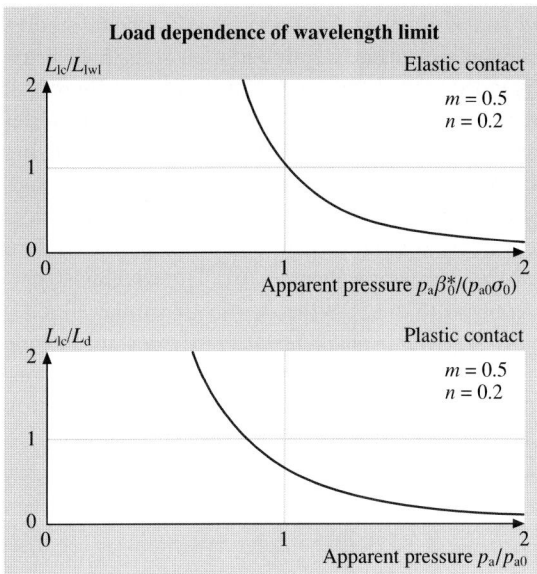

Fig. 15.11. Dependence of the normalized long wavelength limit for contact parameters on load (normalized apparent pressure) for elastic and plastic contacts ($m = 0.5$, $n = 0.2$)

15.5 Scale Effect in Friction

According to the adhesion and deformation model of friction [16], the coefficient of dry friction μ can be presented as a sum of adhesion component μ_a and deformation (plowing) component μ_d. The later, in the presence of particles, is a sum of asperity summits deformation component μ_{ds} and particles deformation component μ_{dp}, so that the total coefficient of friction is [21]

$$\mu = \mu_a + \mu_{ds} + \mu_{dp} = \frac{F_a + F_{ds} + F_{dp}}{W} = \frac{A_{ra}\tau_a + A_{ds}\tau_{ds} + A_{dp}\tau_{dp}}{W}, \quad (15.29)$$

where W is the normal load, F is the friction force, A_{ra}, A_{ds}, A_{dp} are the real areas of contact during adhesion, two body deformation and with particles, respectively, and τ is the shear strength. The subscripts a, ds, and dp correspond to adhesion, summit deformation and particle deformation.

In the presence of meniscus, the friction force is given by

$$F = \mu(W + F_m), \quad (15.30)$$

where F_m is the meniscus force [16]. The coefficient of friction in the presence of the meniscus force, μ_{wet}, is calculated using only the applied normal load, as normally measured in the experiments [22]

$$\mu_{wet} = \mu\left(1 + \frac{F_m}{W}\right) = \frac{A_{ra}\tau_a + A_{ds}\tau_{ds} + A_{dp}\tau_{dp}}{W}\left(1 + \frac{F_m}{W}\right). \quad (15.31)$$

The (15.31) shows that μ_{wet} is greater than μ, because F_m is not taken into account for calculation of the normal load in the wet contact.

It was shown by *Greenwood* and *Williamson* [42] and by subsequent modifications of their model, that for contacting surfaces with common statistical distributions of asperity heights, the real area of contact is almost linearly proportional to the normal load. This linear dependence, along with (15.29), result in linear dependence of the friction force on the normal load, or coefficient of friction being independent of the normal load. For a review of the numerical analysis of rough surface contacts, see *Bhushan* [43, 44] and *Bhushan* and *Peng* [45]. The statistical and numerical theories of contact involve roughness parameters – e.g. the standard deviation of asperity heights and the correlation length [16]. The roughness parameters are scale dependent. In contrast to this, the theory of self-similar (fractal) surfaces solid contact developed by *Majumdar* and *Bhushan* [46] does not include length parameters and are scale-invariant in principle. The shear strength of the contacts in (15.29) is also scale dependent. In addition to the adhesional contribution to friction, elastic and plastic deformation on nano- to macroscale contributes to friction [16]. The deformations are also scale dependent.

15.5.1 Adhesional Friction

The adhesional component of friction depends on the real area of contact and adhesion shear strength. Here we derive expressions for scale dependence of adhesional friction during single-asperity and multiple-asperity contacts.

Single-Asperity Contact

The scale length during single-asperity contact is the nominal contact length, which is equal to the contact diameter $2a$. In the case of predominantly elastic contacts, the real area of contact A_{re} depends on the load according to the Hertz analysis [47]

$$A_{re} = \pi a^2 , \tag{15.32}$$

and

$$a = \left(\frac{3WR}{4E^*}\right)^{1/3} , \tag{15.33}$$

where R is effective radius of curvature of summit tips, and E^* is the effective elastic modulus of the two bodies. In the case of predominantly plastic contact, the real area of contact A_{rp} is given by (15.21), whereas the hardness is given by (15.10).

Combining (15.10), (15.13), (15.29), and (15.32), the adhesional component of the coefficient of friction can be determined for the predominantly elastic contact as

$$\mu_{ae} = \mu_{ae0} \sqrt{1 + (l_s/a)} \tag{15.34}$$

and for the predominantly plastic contact as

$$\mu_{ap} = \mu_{ap0} \sqrt{\frac{1 + (l_s/a)}{1 + (l_d/a)}} , \tag{15.35}$$

where μ_{ae0} and μ_{ap0} are corresponding values at the macroscale [20].

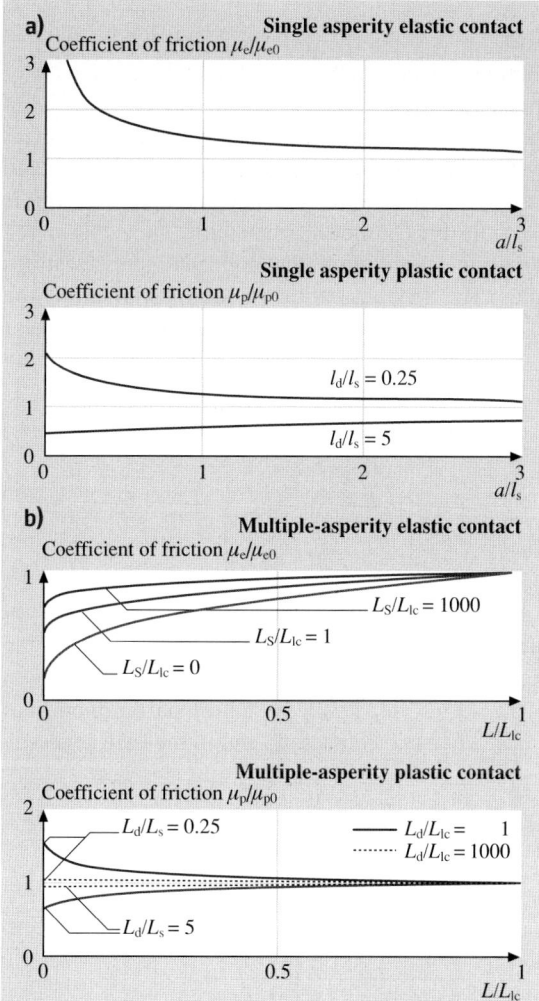

Fig. 15.12a,b. Normalized results for the adhesional component of the coefficient of friction, as a function of scale (a/l_s for single asperity contact and L/L_{lc} for multi-asperity contact). (**a**) In the case of single asperity plastic contact, data are presented for two values of l_d/l_s. (**b**) In the case of multi-asperity contact, data are presented for $m = 0.5$, $n = 0.2$. For multi-asperity elastic contact, data are presented for three values of L_S/L_{lc}. For multi-asperity plastic contact, data are presented for two values of L_d/L_s

The scale dependence of adhesional friction in single-asperity contact is presented in Fig. 15.12a. In the case of single asperity elastic contact, the coefficient of friction increases with decreasing scale (contact diameter), because of an increase in the adhesion strength, according to (15.34). In the case of single asperity plastic contact, the coefficient of friction can increase or decrease with decreasing scale, because of an increased hardness or increase in adhesional strength. The competition of these two factors is governed by l_d/l_s, according to (15.35). There is no direct way to measure l_d and l_s. We will see later, from experimental data, that the coefficient of friction tends to decrease with decreasing scale, therefore, it must be assumed that $l_d/l_s > 1$ for the data reported in the paper [20].

Multiple-Asperity Contact

The adhesional component of friction depends on the real area of contact and adhesion shear strength. Scale dependence of the real area of contact was considered in the preceding section. Here we derive expressions for scale-dependence of the shear strength at the interface during adhesional friction. It is suggested by *Bhushan and Nosonovsky* [20] that, for many materials, dislocation-assisted sliding (microslip) is the main mechanism, which is responsible for the shear strength. They considered dislocation assisted sliding based on the assumption, that contributing dislocations are located in a subsurface volume. The thickness of this volume is limited by the distance which dislocations can climb l_s (material parameter) and by the radius of contact a. They showed that τ_a is scale dependent according to (15.13). Assuming the contact radii equal to the mean value given by (15.19)

$$\tau_a = \tau_{a0} \sqrt{1 + (L_s/L)^m}, \quad L < L_{lc}, \tag{15.36}$$

where

$$L_s = L_{lc} \left(\frac{l_s}{\bar{a}_0}\right)^{1/m}. \tag{15.37}$$

In the case of absence of the microslip (e.g., for an amorphous material), it should be assumed in (15.34–15.36), $L_s = l_s = 0$.

Based on (15.9, 15.17, 15.24, 15.29, 15.36, 15.37), the adhesional component of the coefficient of friction in the case of elastic contact μ_{ae} and in the case of plastic contact μ_{ap}, is given as [20]

$$\mu_{ae} = \frac{\tau_a A_{re}}{W} = \frac{\tau_{a0} A_{re0}}{W} \left(\frac{L}{L_{lc}}\right)^{m-n} \sqrt{1 + (L_s/L)^m}$$

$$= \frac{\mu_{ae0}}{\sqrt{1 + (l_s/\bar{a}_0)}} \left(\frac{L}{L_{lc}}\right)^{m-n} \sqrt{1 + (L_s/L)^m}, \quad L < L_{lc}; \tag{15.38}$$

$$\mu_{ap} = \frac{\tau_{a0}}{H_0} \sqrt{\frac{1 + (L_s/L)^m}{1 + (L_d/L)^m}} = \mu_{ap0} \sqrt{\frac{1 + (l_d/\bar{a}_0)}{1 + (l_s/\bar{a}_0)}} \sqrt{\frac{1 + (L_s/L)^m}{1 + (L_d/L)^m}}, \quad L < L_{lc}, \tag{15.39}$$

where μ_{ae0} and μ_{ap0} are values of the coefficient of friction at macroscale ($L \geq L_{lc}$).

The scale dependence of adhesional friction in multiple-asperity elastic contact is presented in Fig. 15.12b, which is based on (15.38), for various values of L_s/L_{lc}. The change of scale length L affects the coefficient of friction in two different ways: through the change of A_{re} (15.17) and τ_a (15.36) below L_{lc}. Further, τ_a is controlled by the ratio L_s/L. Based on (15.36), for small ratio of L_s/L_{lc}, scale effects on τ_a is insignificant for $L/L_{lc} > 0$. As it is seen from Fig. 15.12b by comparison of the curve with $L_s/L_{lc} = 0$ (insignificant scale effect on τ_a), $L_s/L_{lc} = 1$, and $L_s/L_{lc} = 1000$ (significant scale effect on τ_a), the results for the normalized coefficient of friction are close, thus, the main contribution to the scaling effect is due to change of A_{re}.

Table 15.1. Scaling factors for the coefficient of adhesional friction [20]

Single asperity elastic contact	Single asperity plastic contact	Multiple-asperity elastic contact	Multiple-asperity plastic contact
$\mu_e =$ $\mu_{e0}\sqrt{1+(l_s/a)}$	$\mu_e =$ $\mu_{e0}\sqrt{\frac{1+(l_s/a)}{1+(l_d/a)}}$	$\mu_e =$ $\mu_{e0} C_E L^{m-n}$ $\times \sqrt{1+(L_s/L)^m}$	$\mu_p =$ $\mu_{p0} C_P \sqrt{\frac{1+(L_s/L)^m}{1+(L_d/L)^m}}$

In the case of multiple-asperity plastic contact, the results, based on (15.39), are presented in Fig. 15.12b for $L_d/L_s = 0.25$, $L_d/L_s = 5$ and $L_d/L_{lc} = 1$ and $L_d/L_{lc} = 1000$. The change of scale affects the coefficient of friction through the change of A_{rp} (15.34), which is controlled by L_d, and τ_a (15.36), which is controlled by L_s. It can be observed from Fig. 15.12b, that for $L_d > L_s$, the change of A_{rp} prevails over the change of τ_a, with decreasing scale, and the coefficient of friction decreases. For $L_d < L_s$, the change of τ_a prevails, with decreasing scale, and the coefficient of friction increases [20]. Expressions for the coefficient of adhesional friction are presented in Table 15.1.

15.5.2 Two-Body Deformation

Based on the assumption that multiple asperities of two rough surfaces in contact have conical shape, the two-body deformation component of friction can be determined as

$$\mu_{ds} = \frac{2\tan\theta_r}{\pi}, \qquad (15.40)$$

where θ_r is the roughness angle (or attack angle) of a conical asperity [16,48]. Mechanical properties affect the real area of contact and shear strength and these cancel out in (15.29).

The roughness angle is scale-dependent and is related to the roughness parameters [41]. Based on statistical analysis of a random Gaussian surface,

$$\tan\theta_r \propto \frac{\sigma}{\beta^*}. \qquad (15.41)$$

From (15.40) it can be interpreted that stretching the rough surface in the vertical direction (increasing vertical scale parameter σ) increases $\tan\theta_r$, and stretching in the horizontal direction (increasing vertical scale parameter β^*) decreases $\tan\theta_r$.

Using (15.40) and (15.41), the scale dependence of the two-body deformation component of the coefficient of friction is given as [21]

$$\mu_{ds} = \frac{2\sigma_0}{\pi\beta^*}\left(\frac{L}{L_{lc}}\right)^{n-m} = \mu_{ds0}\left(\frac{L}{L_{lc}}\right)^{n-m}, \quad L < L_{lc}, \qquad (15.42)$$

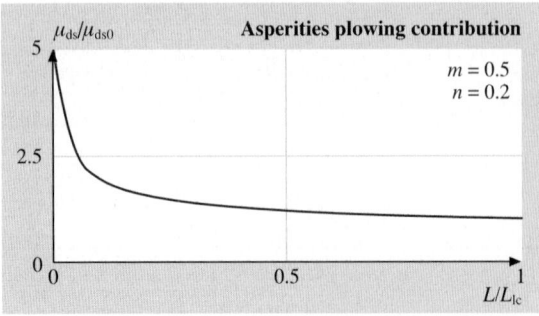

Fig. 15.13. Normalized results for the two-body deformation component of the coefficient of friction

where μ_{ds0} is the value of the coefficient of summits deformation component of the coefficient of friction at macroscale ($L \geq L_{lc}$).

The scale dependence for the two-body deformation component of the coefficient of friction is presented in Fig. 15.13 for $m = 0.5$, $n = 0.2$. The coefficient of friction increases with decreasing scale, according to (15.42). This effect is a consequence of increasing average slope or roughness angle.

15.5.3 Three-Body Deformation Friction

In this sections of the paper, size distribution of particles will be idealized according to the exponential, normal, and log normal density functions, since these distributions are the most common in nature and industrial applications (see Sect. 15.A). The probability for a particle of a given size to be trapped at the interface depends on the size of the contact region. Particles at the edge of the region of contact are likely to leave the contact area, whereas those in the middle are likely to be trapped. The ratio of the edge region area to the total apparent area of contact increases

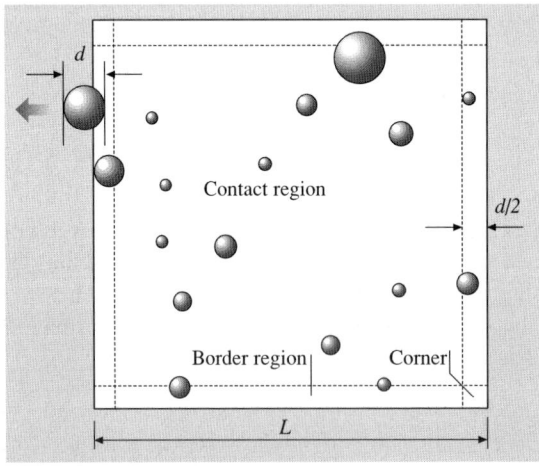

Fig. 15.14. Schematics of debris at the contact zone and at its border region. A particle of diameter d in the border region of $d/2$ is likely to leave the contact zone

with decreasing scale size. Therefore, the probability for a particle to be trapped decreases, as well as the three-body component of the coefficient of friction [21].

Let us consider a square region of contact of two rough surfaces with a length L (relevant scale length), with the density of debris of η particles per unit area (Fig. 15.14). We assume that the particles have the spherical form and that $p(d)$ is the probability density function of particles size. It is also assumed that, for a given diameter, particles at the border region of the contact zone of the width $d/2$ are likely to leave the contact zone, with a certain probability α, whereas particles at the center of the contact region are likely to be trapped. It should be noted, that particles in the corners of the contact region can leave in two different directions, therefore, for them the probability to leave is 2α. The total nominal contact area is equal to L^2, the area of the border region, without the corners, is equal to $4(L-d)d/2$, and the area of the corners is equal to d^2.

The probability density of size distribution for the trapped particles $p_{tr}(d)$ can be calculated by multiplying $p(d)$ by one minus the probability of a particle with diameter d to leave; the later is equal to the ratio of the border region area, multiplied by a corresponding probability of the particle to leave, divided by the total contact area [21]

$$p_{tr}(d) = p(d)\left(1 - \frac{2\alpha(L-d)d + 2\alpha d^2}{L^2}\right) = p(d)\left(1 - \frac{2\alpha d}{L}\right), \quad d < \frac{L}{2\alpha}. \quad (15.43)$$

The ratio of the number of trapped particles to the total number of particles, average radius of a trapped particle $\overline{d}$, and average square of trapped particles $\overline{d^2}$, as functions of L, can be calculated as

$$n_{tr} = \frac{\int_0^{L/2} p_{tr}(d)\,dd}{\int_0^{\infty} p(d)\,dd} = \frac{\int_0^{L/2} p(d)\left(1 - \frac{2\alpha d}{L}\right)dd}{\int_0^{\infty} p(d)\,dd},$$

$$\overline{d} = \frac{\int_0^{L/2} d\, p_{tr}(d)\,dd}{\int_0^{L/2} p_{tr}(d)\,dd}, \quad \overline{d^2} = \frac{\int_0^{L/2} d^2 p_{tr}(d)\,dd}{\int_0^{L/2} p_{tr}(d)\,dd}. \quad (15.44)$$

Let us assume an exponential distribution of particles' size (15.A7) with $d_e = 0$. Substituting (15.A7) into (15.44) and integrating yields for the ratio of trapped particles [21]

$$n_{tr} = \frac{\int_0^{L/(\alpha 2)} \frac{1}{\sigma_e} \exp\left(-\frac{d}{\sigma_e}\right)\left(1 - \frac{2\alpha d}{L}\right)dd}{\int_0^{\infty} \frac{1}{\sigma_e} \exp\left(-\frac{d}{\sigma_e}\right)dd} = \exp\left(-\frac{d}{\sigma_e}\right)\frac{\sigma_e - L/(2\alpha) + d}{L/(2\alpha)}\Big|_0^{L/(2\alpha)}$$

$$= \frac{2\alpha\sigma_e}{L}\left[\exp\left(-\frac{L}{2\alpha\sigma_e}\right) - 1\right] + 1 \quad (15.45)$$

whereas the mean diameter of the trapped particles is

$$\bar{d} = \frac{\int_0^{L/(2\alpha)} d \exp\left(-\frac{d}{\sigma_e}\right)\left(1 - \frac{2\alpha d}{L}\right) dd}{\int_0^{L/(2\alpha)} \exp\left(-\frac{d}{\sigma_e}\right)\left(1 - \frac{2\alpha d}{L}\right) dd} = \sigma_e \frac{\exp\left(-\frac{L}{2\alpha\sigma_e}\right)\left(1 + \frac{4\alpha\sigma_e}{L}\right) + 1 - \frac{4\alpha\sigma_e}{L}}{\frac{2\alpha\sigma_e}{L}\left[\exp\left(-\frac{L}{2\alpha\sigma_e}\right) - 1\right] + 1}$$
(15.46)

and the mean square radius of the trapped particles is

$$\overline{d^2} = \frac{\int_0^{L/(2\alpha)} d^2 \exp\left(-\frac{d}{\sigma_e}\right)\left(1 - \frac{2\alpha d}{L}\right) dd}{\int_0^{L/(2\alpha)} \exp\left(-\frac{d}{\sigma_e}\right)\left(1 - \frac{2\alpha d}{L}\right) dd}$$

$$= \sigma_e^2 \frac{\exp\left(-\frac{L}{2\alpha\sigma_e}\right)\left(\frac{L}{2\alpha\sigma_e} + 4 + \frac{12\alpha\sigma_e}{L}\right) + 2 - \frac{12\alpha\sigma_e}{L}}{\frac{2\alpha\sigma_e}{L}\left(\exp\left(-\frac{L}{2\alpha\sigma_e}\right) - 1\right) + 1}.$$
(15.47)

For the normal and log normal distributions, similar calculations can be conducted numerically.

The area, supported by particles can be found as the number of trapped particles $\eta L^2 n_{tr}$ multiplied by average particle contact area

$$A_{dp} = \eta L^2 n_{tr} \frac{\pi \overline{d^2}}{4},$$
(15.48)

where $\overline{d^2}$ is mean square of particle diameter, η is particle density per apparent area of contact (L^2) and n_{tr} is a number of trapped particles divided by the total number of particles [21].

The plowing deformation is plastic and, assuming that particles are harder than the bodies, the shear strength τ_{dp} is equal to the shear yield strength of the softer body τ_Y which is given by the (15.9) with $a = \bar{d}/2$. Combining (15.29) with (15.9) and (15.48)

$$\mu_{dp} = \frac{A_{dp}\tau_{dp}}{W} = \eta \frac{L^2}{W} \frac{\pi \overline{d^2}}{4} n_{tr} \tau_{Y0} \sqrt{1 + 2l_d/\bar{d}} = \mu_{dp0} n_{tr} \frac{\overline{d^2}}{d_0^2} \frac{\sqrt{1 + 2l_d/\bar{d}}}{\sqrt{1 + 2l_d/d_0}},$$
(15.49)

where $\bar{d}$ is mean particle diameter, $\bar{d}_0$ is the macroscale value of mean particle diameter, and μ_{dp0} is macroscale ($L \to \infty$, $n_{tr} \to 1$) value of the third-body deformation component of the coefficient of friction given as

$$\mu_{dp0} = \eta \frac{L^2}{W} \frac{\pi \overline{d_0^2}}{4} \tau_{Y0} \sqrt{1 + 2l_d/\bar{d}_0}.$$
(15.50)

Scale dependence of the three-body deformation component of the coefficient of friction is presented in Fig. 15.15, based on (15.49). The number of trapped particles divided by the total number of particles, as well as the three-body deformation

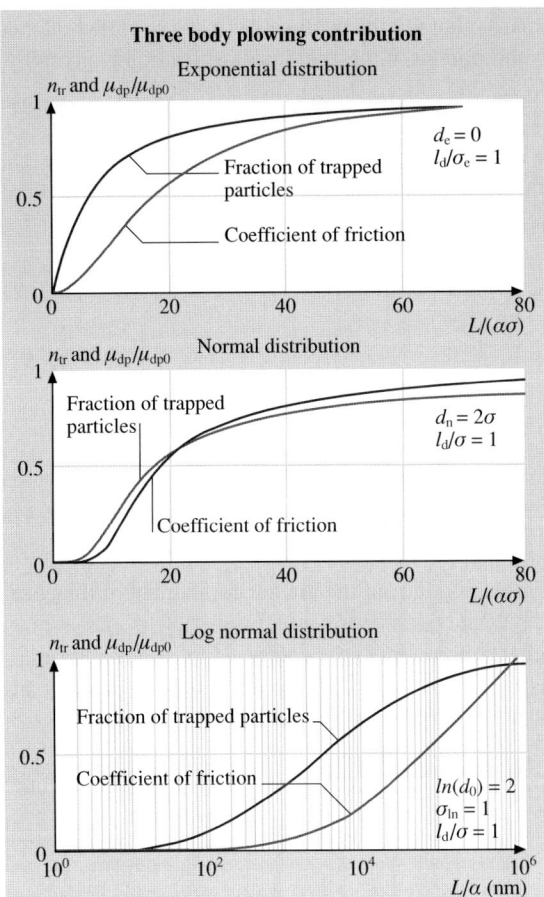

Fig. 15.15. The number of trapped particles divided by the total number of particles and three-body deformation component of the coefficient of friction, normalized by the macroscale value, for three different distributions of debris size: **(a)** exponential, **(b)** normal, and **(c)** log-normal distributions

component of the coefficient of friction, are presented as a function of scale size divided by α for the exponential, normal, and log normal distributions. The dependence of μ_d/μ_{d0} is shown as a function of $L/(\alpha\sigma_e)$ for the exponential distribution and normal distribution, for $d_n = d_e = 2\sigma_e$ and $l_d/\sigma_e = 1$, whereas for the log normal distribution the results are presented as a function of L/α, for $(\ln d_{\ln}) = 2$, $\sigma_{\ln} = 1$, and $l_d/\sigma_{\ln} = 1$. This component of the coefficient of friction decreases for all of the three distributions. The results are shown for $l_d/\sigma_{\ln} = 1$, however, variation of $l_d/\sigma_{\ln}$ in the range between 0.1 and 10 does not change significantly the shape of the curve. The decrease of the three-body deformation friction force with decreasing scale results with this component being small at the nanoscale.

15.5.4 Ratchet Mechanism

Surface roughness can have an appreciable influence on friction during adhesion. If one of the contacting surfaces has asperities of much smaller lateral size, such that a small tip slides over an asperity, having the average angle θ_r (so called ratchet mechanism), the corresponding component of the coefficient of friction is given by

$$\mu_r = \mu_a \tan^2 \theta_r, \tag{15.51}$$

where μ_r is the ratchet mechanism component of friction [16]. Combining (15.15, 15.41, 15.38, 15.39) yields for the scale dependence of the ratchet component of the coefficient of friction in the case of elastic, μ_{re}, and plastic contact, μ_{rp}

$$\mu_{re} = \mu_{ae} \left[\frac{2\sigma_0}{\pi \beta_0^*} \left(\frac{L}{L_{lc}}\right)^{n-m} \right]^2 = \frac{\mu_{re0}}{\sqrt{1+(l_s/\bar{a}_0)}} \left(\frac{L}{L_{lc}}\right)^{n-m} \sqrt{1+(L_s/L)^m}, \quad L < L_{lc}, \tag{15.52}$$

$$\mu_{rp} = \mu_{ap} \left[\frac{2\sigma_0}{\pi \beta_0^*} \left(\frac{L}{L_{lc}}\right)^{n-m} \right]^2$$

$$= \mu_{rp0} \left(\frac{L}{L_{lc}}\right)^{2(n-m)} \sqrt{\frac{1+(l_d/\bar{a}_0)}{1+(l_s/\bar{a}_0)}} \sqrt{\frac{1+(L_s/L)^m}{1+(L_d/L)^m}}, \quad L < L_{lc}, \tag{15.53}$$

where μ_{re0} and μ_{rp0} are the macroscale values of the ratchet component of the coefficient of friction for elastic and plastic contact correspondingly [23].

Scale dependence of the ratchet component of the coefficient of friction, normalized by the macroscale value, is presented in Fig. 15.16, for scale independent adhesional shear strength, $\tau_a = \text{const}$, ($L_s = 0$) and for scale dependent τ_a ($L_s = 10L_d$), based on (15.51) and (15.53). The ratchet component during adhesional elastic friction, μ_{re}, is presented in Fig. 15.16a. It is observed, that, with decreasing scale, μ_{re} increases. The ratchet component during adhesional plastic friction, μ_{rp}, is pre-

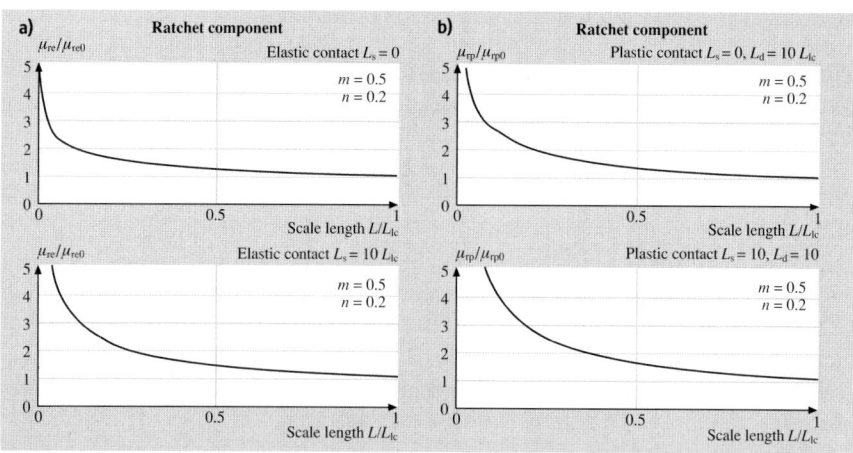

Fig. 15.16a,b. Normalized results for the ratchet component of the coefficient of friction, as a function of scale, for scale independent ($L_s = 0$) and scale dependent ($L_s = 10L_{lc}$) shear strength ($m = 0.5$, $n = 0.2$). (**a**) contact, (**b**) Plastic contact ($L_d = 10L_{lc}$)

sented in Fig. 15.16b. It is observed, that, for $L_s = 0$, with decreasing scale, μ_{rp} increases [23].

15.5.5 Meniscus Analysis

During contact, if a liquid is introduced at the point of asperity contact, the surface tension results in a pressure difference across a meniscus surface, referred to as capillary pressure or Laplace pressure. The attractive force for a sphere in contact with a plane surface is proportional to the sphere radius R_p, for a sphere close to a surface with separation s or for a sphere close to a surface with continuous liquid film [16]

$$F_m \propto R_p . \tag{15.54}$$

The case of multiple-asperity contact is shown in Fig. 15.1b. Note, that both contacting and near-contacting asperities wetted by the liquid film contribute to the total meniscus force. A statistical approach can be used to model the contact. In general, given the interplanar separation s_d, the mean peak radius $\overline{R_p}$, the thickness of liquid film h_f, the surface tension γ, liquid contact angle between the liquid and surface θ, and the total number of summits in the nominal contact area N,

$$F_m = 2\pi \overline{R_p} \gamma (1 + \cos\theta) N . \tag{15.55}$$

In (15.54), γ and θ are material properties, which are not expected to depend on scale, whereas $\overline{R_p}$ and N depend on surface topography, and are scale-dependent, according to (15.18) and (15.20).

$$F_m \propto \overline{R_p} N = F_{m0} \left(\frac{L}{L_{lwl}} \right)^{m-2n} , \quad L < L_{lwl} , \tag{15.56}$$

where F_{m0} is the macroscale value of the meniscus force ($L \geq L_{lwl}$).

Scale dependence of the meniscus force is presented in Fig. 15.17, based on (15.56) for $m = 0.5$, $n = 0.2$. It may be observed that, depending on the value of D, the meniscus force may increase or decrease with decreasing scale size.

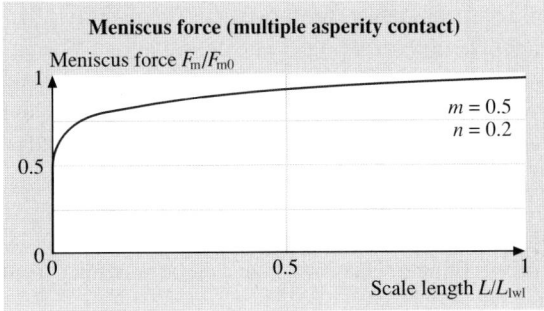

Fig. 15.17. Meniscus force for $m = 0.5$, $n = 0.2$

15.5.6 Total Value of Coefficient of Friction and Transition from Elastic to Plastic Regime

During transition from elastic to plastic regime, contribution of each of the three components of the coefficient of friction in (15.29) changes. In the elastic regime, the dominant contribution is expected to be adhesion involving elastic deformation, and in the plastic regime the dominant contribution is expected to be deformation. Therefore, in order to study transition from elastic to plastic regime, the ratios of deformation to adhesion component should be considered. The expression for the total value of the coefficient of friction, which includes meniscus force contribution, based on (15.29) and (15.31) can be rewritten as [21]

$$\mu_{\text{wet}} = \mu_a \left(1 + \frac{\mu_{ds}}{\mu_a} + \frac{\mu_{dp}}{\mu_a}\right)\left(1 + \frac{F_m}{W}\right). \tag{15.57}$$

The ratchet mechanism component is ignored here since it is present only in special cases. Results in the preceding subsection provide us with data about the adhesion and two-body and three-body deformation components of the coefficient of friction, normalized by their values at the macroscale. However, that analysis does not provide any information about their relation to each other or about transition from the elastic to plastic regime. In order to analyze the transition from pure adhesion involving elastic deformation to plastic deformation, a transition index ϕ can be considered [21]. The transition index is equal to the ratio of average pressure in the elastic regime (normal load per real area of elastic contact) to hardness or simply the ratio of the real area of plastic contact divided by the real area of elastic contact

$$\phi = \frac{W}{A_{\text{re}} H} = \frac{A_{\text{rp}}}{A_{\text{re}}}. \tag{15.58}$$

Using (15.17) and (15.22), the scale-dependence of ϕ is

$$\phi = \frac{W}{A_{\text{re0}}(L/L_{\text{lc}})^{m-n} H_0 \sqrt{1+(L_s/L)^m}} = \phi_0 \frac{\sqrt{1+(l_s/\overline{a})}(L/L_{\text{lc}})^{n-m}}{\sqrt{1+(L_s/L)^m}}, \quad L < L_{\text{lc}}, \tag{15.59}$$

where ϕ_0 is the macroscale value of the transition index [21].

With a low value of ϕ close to zero, the contacts are mostly elastic and only adhesion contributes to the coefficient of friction involving elastic deformation. Whereas with increasing ϕ approaching unity, the contacts become predominantly plastic and deformation becomes a dominant contributor. It can be argued that A_{ds}/A_{re} and A_{dp}/A_{re} will also be a direct function of ϕ, and in the paper these will be assumed to have linear relationship.

Next, the ratio of adhesion and deformation components of the coefficient of friction in terms of ϕ is obtained. In this relationship, τ_{ds} and τ_{dp} are equal to the shear yield strength, which is proportional to hardness and can be obtained from

(15.9), using (15.19) and (15.36)

$$\frac{\mu_{ds}}{\mu_{ae}} = \frac{A_{ds}\tau_{ds}}{A_{re}\tau_a} \propto \phi \frac{\tau_{ds}}{\tau_a} = \phi \frac{\tau_{ds0}\sqrt{1+(L_d/L)^m}}{\tau_{a0}\sqrt{1+(L_s/L)^m}}, \qquad L < L_{lc}, \qquad (15.60)$$

$$\frac{\mu_{dp}}{\mu_{ae}} = \frac{A_{dp}\tau_{dp}}{A_{re}\tau_a} \propto \phi \frac{\tau_{dp}}{\tau_a} = \phi \frac{\tau_{Y0}\sqrt{1+2l_d/d}}{\tau_{a0}\sqrt{1+(L_s/L)^m}}, \qquad L < L_{lc}. \qquad (15.61)$$

The sum of adhesion and deformation components [21]

$$\mu_{wet} = \mu_{ae}\left[1+\phi\left(\frac{\tau_{ds0}\sqrt{1+(L_d/L)^m}}{\tau_{a0}\sqrt{1+(L_s/L)^m}} + \frac{\tau_{Y0}\sqrt{1+2l_d/d}}{\tau_{a0}\sqrt{1+(L_s/L)^m}}\right)\right]$$

$$\times \left[1+\frac{F_{m0}}{W}\left(\frac{L}{L_{lwl}}\right)^{m-2n}\right], \qquad L < L_{lc}. \qquad (15.62)$$

Note that ϕ itself is a complicated function of L, according to (15.59).

Scale dependence of the transition index, normalized by the macroscale value, is presented in Fig. 15.18, based on (15.59). It is observed that, for $L_s = 0$, the transition index decreases with increasing scale. For $L_s = 10L_{lc}$, the same trend is observed for $m > 2n$, but, in the case $m < 2n$, ϕ decreases. An increase of the transition index means that the ratio of plastic to elastic real areas of contact increases. With decreasing scale, the mean radius of contact decreases, causing hardness enhancement and decrease of the plastic area of contact. Based on this, the model may predict an increase or decrease of the transition index, depending on whether elastic or plastic area decreases faster.

The dependence of the coefficient of friction on ϕ is illustrated in Fig. 15.19, based on (15.62). It is assumed in the figure that the slope for the dependence of μ_{dp}

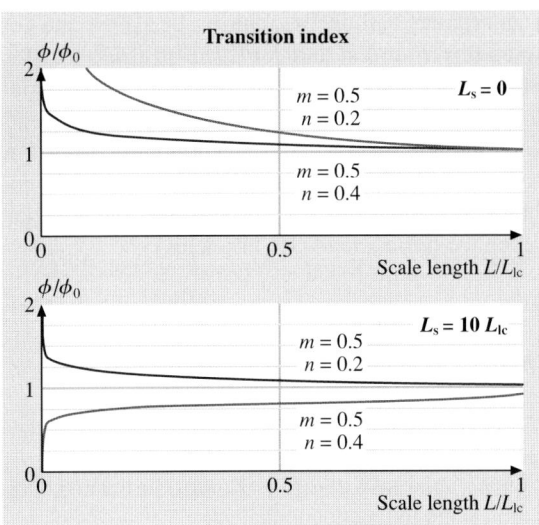

Fig. 15.18. The transition index as a function of scale. Presented for $m = 0.5, n = 0.2$ and $m = 0.5, n = 0.4$

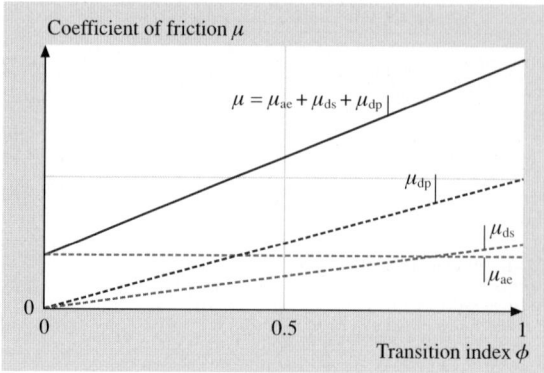

Fig. 15.19. The coefficient of friction (dry contact) as a function of the transition index for given scale length L. With increasing ϕ and onset of plastic deformation, both μ_{ds} and μ_{dp} grow, as a result of this, the total coefficient of friction μ grows as well

on ϕ is greater than the slope for the dependence of μ_{ds} on ϕ. For ϕ close to zero, the contact is predominantly elastic, whereas for ϕ approaching unity the contact is predominantly plastic.

15.5.7 Comparison with the Experimental Data

Experimental data on friction at micro- and nanoscale are presented in this subsection and compared with the model. First, a single-asperity predominantly elastic contact is considered [20], then transition to plastic deformation involving multiple asperity contacts is analyzed [23].

Single-Asperity Predominantly-Elastic Contact

Nanoscale dependence of friction force upon the normal load was studied for Pt-coated AFM tip versus mica in ultra-high vacuum (UHV) by *Carpick* et al. [7], for Si tip versus diamond and amorphous carbon by *Schwarz* et al. [8] and for Si_3N_4 tip on Si, SiO_2, and diamond by *Bhushan* and *Kulkarni* [6] (Fig. 15.20a). *Homola* et al. [13] conducted SFA experiments with mica rolls with a single contact zone (before onset of wear), Fig. 15.21b. Contacts relevant in these experiments can be considered as single-asperity, predominantly elastic in all of these cases. For a single-asperity elastic contact of radius a, expression for μ is given by (15.17). For the limit of a small contact radius $a \ll l_s$, the (15.13) combined with the Hertzian dependence of the contact area upon the normal load (15.33) yields

$$F_e \approx \pi a^2 \tau_0 \sqrt{l_s/a} \propto a^{3/2} \propto W^{1/2} \ . \tag{15.63}$$

If an adhesive pull-off force W_0 is large, (15.63) can be modified as

$$F_e = C_0 \sqrt{W + W_0} \ , \tag{15.64}$$

where C_0 is a constant. Friction force increases with square root of the normal load, opposed to the two third exponent in scale independent analysis.

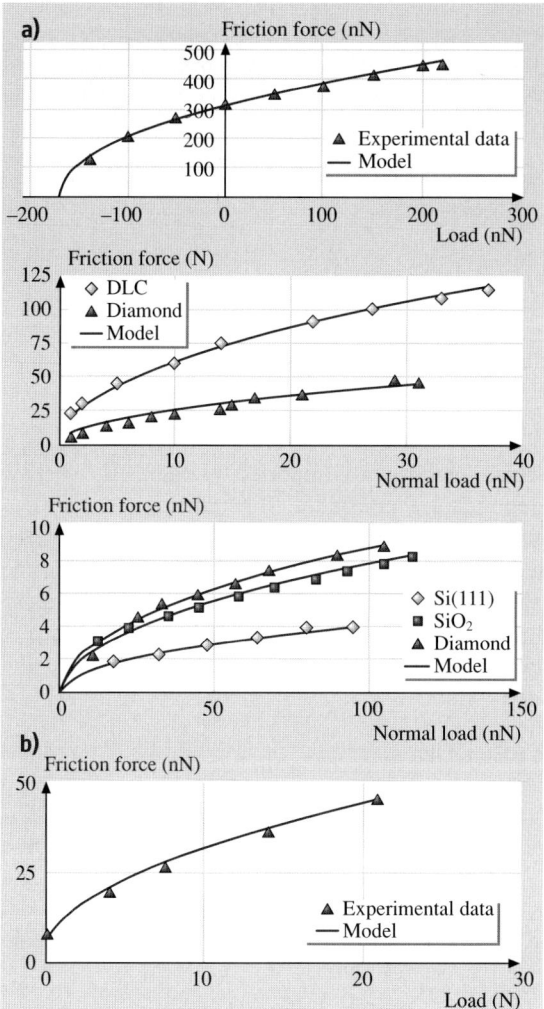

Fig. 15.20. Summary of (**a**) AFM data (*upper*: Pt-coated tip on mica in UHV [7], *middle*: Si tip on DLC and diamond in UHV [8], *lower*: Si_3N_4 tip on various materials [6]) and (**b**) SFA data (on mica vs. mica in dry air [13]) for friction force as a function of normal load

The results in Fig. 15.20 demonstrate a reasonable agreement of the experimental data with the model. The platinum-coated tip versus mica [7] has a relatively high pull-off force and the data fit with $C_0 = 23.7$ $(nN)^{1/2}$ and $W_0 = 170$ nN. For the silicon tip vs. amorphous carbon and natural diamond, the fit is given by $C_0 = 8.0$, 19.3 $(nN)^{1/2}$ and small W_0. For the virgin Si(111), SiO_2, and natural diamond sliding versus Si_3N_4 tip [8], the fit is given by $C_0 = 0.40, 0.76, 0.86$ $(nN)^{1/2}$ for Si(111), SiO_2, and diamond, respectively and small W_0. For two mica rolls [13], the fit is given by $C_0 = 10$ $N^{1/2}$ and $W_0 = 0.5$ N [20].

AFM experiments provide data on nanoscale, whereas SFA experiments provide data on microscale. Next we study scale dependence on the shear strength based on these data. In the AFM measurements by *Carpick* et al. [7], the average shear

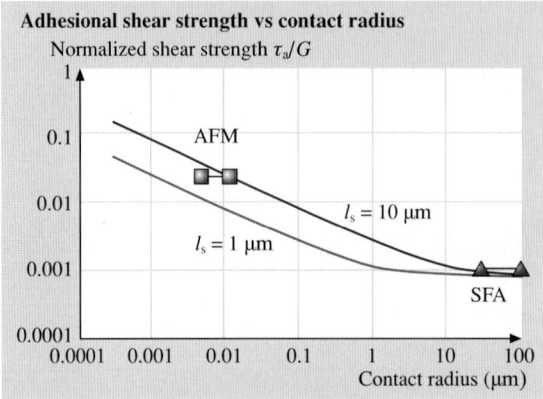

Fig. 15.21. Shear stress as a function of contact radius. Microscale and nanoscale data compared with the model for $l_s = 1\,\mu\text{m}$ and $l_s = 10\,\mu\text{m}$

strength during sliding for Pt–mica interface was reported as 0.86 GPa, whereas the pull-off contact radius was reported as 7 nm. In the SFA measurements by Homola et al. [13], the average shear strength during sliding for mica–mica interface was reported as 25 MPa, whereas the contact area during high loads was on the order of $10^{-8}\,\text{m}^2$, which corresponds to a contact radius on the order 100 μm. To normalize shear strength, we need shear modulus. The shear modulus for mica is $G_{\text{mica}} = 25.7\,\text{GPa}$ [49] and for Pt is $G_{\text{Pt}} = 63.7\,\text{GPa}$ [50]. For mica–Pt interface, the effective shear modulus is

$$G = 2G_{\text{mica}}G_{\text{Pt}}/(G_{\text{mica}} + G_{\text{Pt}}) = 36.6\,\text{GPa}\,. \tag{15.65}$$

This yields the value of the shear stress normalized by the shear modulus $\tau_a/G = 2.35 \times 10^{-2}$ for *Carpick* et al. [7] AFM data and 9.73×10^{-4} for the SFA data. These values are presented in the Fig. 15.21 together with the values predicted by the model for assumed values of $l_s = 1\,\mu\text{m}$ and 10 μm. It can be seen that the model (15.13) provides an explanation of adhesional shear strength increase with a scale decrease [20].

Transition to Predominantly Plastic Deformation Involving Multiple Asperity Contacts

Next, we analyze the effect of transition from predominantly elastic adhesion to predominantly plastic deformation involving multiple asperity contacts [23]. The data on nano- and microscale friction for various materials, are presented in Table 15.2, based on *Ruan* and *Bhushan* [5], *Liu* and *Bhushan* [11], and *Bhushan* et al. [12], for Si(100), graphite (HOPG), natural diamond, and diamond-like carbon (DLC). There are several factors responsible for the differences in the coefficients of friction at micro- and nanoscale. Among them are the contributions from ratchet mechanism, meniscus effect, wear and contamination particles, and transition from elasticity to plasticity. The ratchet mechanism and meniscus effect result in an increase of friction with decreasing scale and cannot explain the decrease of friction found in the

Table 15.2. Micro- and nanoscale values of the coefficient of friction, typical physical properties of specimens, and calculated apparent contact radii and apparent contact pressures at loads used in micro- and nanoscale measurements. For calculation purposes it is assumed that contacts on micro- and nanoscale are single-asperity elastic contacts [23]

Specimen	Coefficient of friction		Elastic modulus (GPa)	Poisson's ratio	Hardness (GPa)	Apparent contact radius at test load for		Mean apparent pressure at test load for	
	Microscale	Nanoscale				Microscale (μm)	Nanoscale (nm)	Microscale (GPa)	Nanoscale (GPa)
Si(100) wafer	0.47[a]	0.06[c]	130[e,f]	0.28[f]	9–10[e,f]	0.8–2.2	1.6–3.4	0.05–0.13[a]	1.3–2.8[c]
Graphite (HOPG)	0.1[b]	0.006[c]	9–15[g] (9) – (0.25)	0.01[j]	62	3.4–7.4	0.082[b]	0.27–0.58[c]
Natural diamond	0.2[b]	0.05[c]	1140[h]	0.07[h]	80–104[g,h]	21	1.1–2.5	0.74[b]	2.5–5.3[c]
DLC film	0.19[a]	0.03[d]	280[i]	0.25[i]	20–30[i]	0.7–2.0	1.3–2.9	0.06–0.16[a]	1.8–3.8[d]

[a] 500 μm radius Si(100) ball at 100–2000 μN and 720 μm/s in dry air [12]
[b] 3 mm radius Si₃N₄ ball (Elastic modulus 310 GPa, Poisson's ratio 0.22 [50]), at 1 N and 800 μm/s [5]
[c] 50 nm radius Si₃N₄ tip at load range from 10–100 nN and 0.5 nm/s, in dry air [5]
[d] 50 nm radius Si₃N₄ tip at load range from 10–100 nN in dry air [12]
[e] [54], [f] [52], [g] [50], [h] [53], [i] [55], [j] [51]

experiments. The contribution of wear and contamination particles is more significant at macro/microscale because of larger number of trapped particles (Fig. 15.15). It can be argued, that for the nanoscale AFM experiments the contacts are predominantly elastic and adhesion is the main contribution to the friction, whereas for the microscale experiments the contacts are predominantly plastic and deformation is an important factor. Therefore, transition from elastic contacts in nanoscale contacts to plastic deformation in microscale contacts is an important effect [23].

According to (15.29), the friction force depends on the shear strength and a relevant real area of contact. For calculation of contact radii and contact pressures, the elastic modulus, Poisson's ratio, and hardness for various samples, are required and presented in Table 15.2 [50–55]. In the nanoscale AFM experiments a sharp tip was slid against a flat sample. The apparent contact size and mean contact pressures are calculated based on the assumption, that the contacts are single asperity, elastic contacts (contact pressures are small compared to hardness). Based on the Hertz equation [47], for spherical asperity of radius R in contact with a flat surface, with an effective elastic modulus E^*, under normal load W, the contact radius a and mean apparent contact pressure p_a are given by

$$a = \left(\frac{3WR}{4E^*}\right)^{1/3}, \tag{15.66}$$

$$p_a = \frac{W}{\pi a^2}. \tag{15.67}$$

The surface energy effect [16] was neglected in (15.66) and (15.67), because the experimental value of the normal adhesion force was small, compared to W [5]. The calculated values of a and p_a for the relevant normal load are presented in Table 15.2 [23].

In the microscale experiments, a ball was slid against a nominally flat surface. The contact in this case is multiple-asperity contact due to the roughness, and the contact pressure of the asperity contacts is higher than the apparent pressure. For calculation of a characteristic scale length for the multiple asperity contacts, which is equal to the apparent length of contact, (15.66) was also used. Apparent radius and mean apparent contact pressure for microscale contacts at relevant load ranges are also presented in Table 15.2 [23].

A quantitative estimate of the effect of the shear strength and the real area of contact on friction is presented in Table 15.3. The friction force at mean load (average of maximum and minimum loads) is shown, based on the experimental data presented in Table 15.2. For microscale data, the real area of contact was estimated based on the assumption that the contacts are plastic and based on (15.33) for mean loads given in Table 15.2. For nanoscale data, the apparent area of contact was on the order of several square nanometers, and it was assumed that the real area of contact is comparable with the mean apparent area of contact, which can be calculated for the mean apparent contact radius, given in Table 15.2. The estimate provides with the upper limit of the real area of contact. The lower limit of the shear strength is calculated as friction force, divided by the upper limit of the real area of con-

Table 15.3. Mean friction force, the real area of contact and lower limit of shear strength [23]

Specimen	Friction force at mean load [a]		Upper limit of real area of contact at mean load		Lower limit of mean shear strength (GPa)	
	Microscale (mN)	Nanoscale (nN)	Microscale[b] (μm^2)	Nanoscale[c] (nm^2)	Microscale[d]	Nanoscale[d]
Si(100) wafer	0.49	3.3	0.11	19	4.5	0.17
Graphite (HOPG)	100	0.33	10^5	92	0.001	0.004
Natural diamond	200	2.7	10.9	10	18.4	0.27
DLC film	0.2	1.7	0.042	14	4.8	0.12

[a] Based on the data from Table 15.2. Mean load at microscale is 1050 μN for Si(100) and DLC film and 1 N for HOPG and natural diamond, and 55 nN for all samples at nanoscale

[b] For plastic contact, based on hardness values from Table 15.2. Scale-dependent hardness value will be higher at relevant scale, presented values of the real area of contact are an upper estimate

[c] Upper limit for the real area is given by the apparent area of contact calculated based on the radius of contact data from Table 15.2

[d] Lower limit for the mean shear strength is obtained by dividing the friction force by the upper limit of the real area of contact

tact, and presented in Table 15.3 [23]. Based on the data in Table 15.3, for Si(100), natural diamond and DLC film, the microscale value of shear strength is about two orders of magnitude higher, than the nanoscale value, which indicates, that transition from adhesion to deformation mechanism of friction and the third-body effect are responsible for an increase of friction at microscale. For graphite, this effect is less pronounced due to molecularly smooth structure of the graphite surface [23].

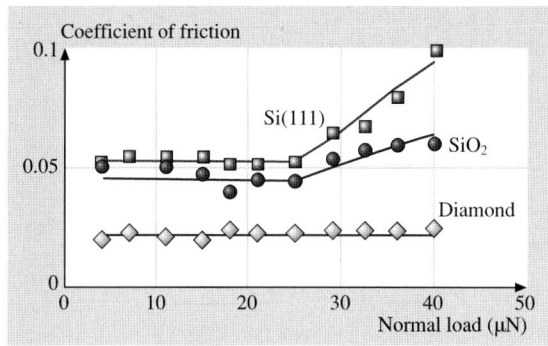

Fig. 15.22. Coefficient of friction as a function of normal load [6]

Based on data available in the literature [6], load dependence of friction at nano/microscale as a function of normal load is presented in Fig. 15.22. Coefficient of friction was measured for Si_3N_4 tip versus Si, SiO_2, and natural diamond using an AFM. They reported that for low loads the coefficient of friction is independent of load and increases with increasing load after a certain load. It is noted that the critical value of loads for Si and SiO_2 corresponds to stresses equal to their hardness values, which suggests that transition to plasticity plays a role in this effect. The friction values at higher loads for Si and SiO_2 approach to that of macroscale values. This result is consistent with predictions of the model for plastic contact (Fig. 15.11), which states that, with increasing normal load, the long wavelength limit for the contact parameters decreases. This decrease results in violation of the condition $L < L_{lc}$, and the contact parameters and the coefficient of friction reach the macroscale values, as was discussed earlier. It must be noted, that the values of $m = 0.5$ and $n = 0.2$ are taken based on available data for the glass-ceramic disk (Fig. 15.9), these parameters depend on material and on surface preparation and may be different for Si, SiO_2, and natural diamond, however, no experimental data on scale dependence of roughness parameters for the materials of interest are available.

15.6 Scale Effect in Wear

The amount of wear during adhesive or abrasive wear involving plastic deformation is proportional to the load and sliding distance x, divided by hardness [16]

$$v = k_0 \frac{Wx}{H}, \tag{15.68}$$

where v is volume of worn material and k_0 is a nondimensional wear coefficient. Using (15.10) and (15.19), the relationships can be obtained for scale dependence of the coefficient of wear in the case of the fractal surface and power-law dependence

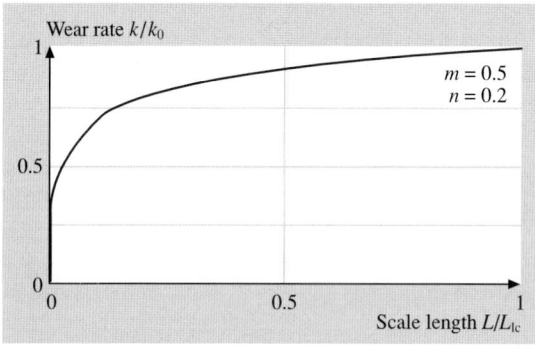

Fig. 15.23. The wear coefficient as a function of scale, presented for $m = 0.5, n = 0.2$

of roughness parameters

$$v = k\frac{Wx}{H_0} \tag{15.69}$$

and

$$k = \frac{k_0}{\sqrt{1+(l_d/\bar{a})}} = \frac{k_0}{\sqrt{1+(L_d/L)^m}}, \quad L < L_{lwl}, \tag{15.70}$$

where k is scale-dependent wear coefficient, and k_0 corresponds to the macroscale limit of the value of k [22].

Scale dependence of the wear coefficient is presented in Fig. 15.23 for $m = 0.5$ and $n = 0.2$, based on (15.70). It is observed, that the wear coefficient decreases with decreasing scale; this is due to the fact that the hardness increases with decreasing mean contact size.

15.7 Scale Effect in Interface Temperature

Frictional sliding is a dissipative process, and frictional energy is dissipated as heat over asperity contacts. Therefore, a high amount of heat per unit area is generated during sliding. A contact is formed and destroyed as one asperity passes the other at a given velocity. When an asperity comes into contact with another asperity, the real area of contact starts to grow, when the asperities are directly above each other, the area is at maximum, as they move away from each other, the area starts to get smaller. There are number of contacts at a given time during sliding. For each individual asperity contact, a flash temperature rise can be calculated. High temperature rise affects mechanical and physical properties of contacting bodies.

For thermal analysis, a dimensionless Peclet number is used

$$L_p = \frac{6Va_{max}}{16\kappa_t}, \tag{15.71}$$

where V is sliding velocity, a_{max} is maximum radius of contact for a given contact spot, and κ_t is thermal diffusivity. This parameter indicates whether the sliding is high-speed or low-speed. If $L_p > 10$, the contact falls into the category of high speed; if $L_p < 0.5$, it falls into the category of low speed; if $0.5 \le L_p \le 10$, a transition regime should be considered [16]. For high L_p, there is not enough time for the heat to flow to the sides during the lifetime of the contact and the heat flows only in the direction, perpendicular to the sliding surface. Based on the numerical calculations for flash temperature rise of as asperity contact for adhesional contact [16], the following relation holds for the maximum temperature rise T, normalized by the rate at which heat is generated q, divided by the volumetric specific heat ρc_p

$$\frac{T\rho c_p V}{q} = 0.95\left(\frac{2Va_{max}}{\kappa_t}\right)^{1/2}, \quad L_p > 10$$

$$= 0.33\left(\frac{2Va_{max}}{\kappa_t}\right), \quad L_p < 0.5. \tag{15.72}$$

The rate at which heat generated per time per unit area depends on the coefficient of friction μ, sliding velocity V, apparent normal pressure p_a, and ratio of the apparent to real areas of contact (A_a/A_r)

$$q = \mu p_a V \frac{A_a}{A_r}. \tag{15.73}$$

Based on (15.72) and (15.73),

$$\frac{T\rho c_p}{p_a} = 0.95 \frac{A_r}{A_a} \mu \left(\frac{2Va_{max}}{\kappa_t}\right)^{1/2}, \quad L_p > 10$$

$$= 0.33 \frac{A_r}{A_a} \mu \left(\frac{2Va_{max}}{\kappa_t}\right), \quad L_p < 0.5. \tag{15.74}$$

For a multiple asperity contact, mean temperature in terms of average of maximum contact size can be written as

$$\frac{\overline{T}\rho c_p}{p_a} = 0.95 \frac{A_r}{A_a} \mu \left(\frac{2V\overline{a}_{max}}{\kappa_t}\right)^{1/2}, \quad L_p > 10$$

$$= 0.33 \frac{A_r}{A_a} \mu \left(\frac{2V\overline{a}_{max}}{\kappa_t}\right), \quad L_p < 0.5. \tag{15.75}$$

In (15.75) $\overline{a}_{max}$, μ and A_a/A_r are scale dependent parameters. During adhesional contact, the maximum radius $\overline{a}_{max}$ is proportional to the contact radius $\overline{a}$, and the scale dependence for $\overline{a}_{max}$ is given by (15.19), for μ by (15.38–15.39), and for A_{re} and A_{rp} by (15.17) and (15.21). The scale dependence of q, involving μ and A_r, and $\overline{a}_{max}$ in (15.72) can be considered separately and then combined. For the sake of simplicity, we only consider the scale dependence of $\overline{a}_{max}$. For the empirical rule dependence of surface roughness parameters and the fractal model, in the case of high and low velocity, (15.75) yields [22]

$$\frac{\overline{T}\rho c_p V}{q} = 0.95 \left(\frac{2VC_A L^m}{\kappa}\right)^{1/2}, \quad L < L_{lwl}, \; L_p > 10 \tag{15.76}$$

$$= 0.33 \left(\frac{2VC_A L^m}{\kappa}\right), \quad L < L_{lwl}, \; L_p < 0.5.$$

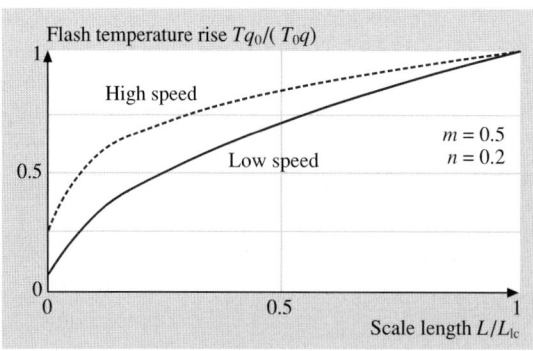

Fig. 15.24. Ratio of the flash temperature rise to the amount of heat generated per unit time per unit area, for a given sliding velocity, as a function of scale. Presented for $m = 0.5$, $n = 0.2$

Scale dependence for the ratio of the flash temperature rise to the amount of heat generated per unit time per unit area, for a given sliding velocity, as a function of scale, is presented in Fig. 15.24, based on (15.76), for the high-speed and low-speed cases. For the empirical rule dependence of roughness parameters, the results are shown for $m = 0.5$, $n = 0.2$.

15.8 Closure

A model, which explains scale effects in mechanical properties (yield strength, hardness, and shear strength at the interface) and tribology (surface roughness, contact parameters, friction, wear, and interface temperature), has been presented in this chapter.

Both mechanical properties and roughness parameters are scale-dependent. According to the strain gradient plasticity, the scale dependence of the so-called geometrically necessary dislocations causes enhanced yield strength and hardness with decreasing scale. The shear strength at the interface is scale dependent due to the effect of dislocation-assisted sliding. An empirical rule for scale dependence of the roughness parameters has been proposed, namely, it was assumed, that the standard deviation of surface height and autocorrelation length depend on scale according to a power law when scale is less than the long wavelength limit value.

Both single asperity and multiple asperity contacts were considered. For multiple asperity contacts, based on the empirical power-rule for scale dependence of roughness, contact parameters were calculated. The effect of load on the contact parameters was also studied. The effect of increasing load is similar to that of increasing scale because it results in increased relevance of longer wavelength details of roughness of surfaces in contact.

During sliding, adhesion and two- and three-body deformation, as well as ratchet mechanism, contribute to the friction force. These components of the friction force depend on the relevant real areas of contact (dependent on roughness, mechanical properties, and load), average asperity slope, number of trapped particles, and relevant shear strength during sliding. The relevant scaling length is the nominal contact length – contact diameter ($2a$) for a single-asperity contact, only considered in adhesion, and scan length (L) for multiple-asperity contacts, considered in adhesion and deformation.

For the adhesional component of the coefficient of friction, the shear yield strength and hardness increase with decreasing scale. In the case of elastic contact, the real area of contact is scale independent for single-asperity contact, and may increase or decrease depending on roughness parameters, for multiple-asperity contact. In the case of plastic contact, enhanced hardness results in a decrease in the real area of contact. The adhesional shear strength at the interface may remain constant or increase with decreasing scale, due to dislocation-assisted sliding (or microslip). The model predicts that the adhesional component of the coefficient of friction may increase or decrease with scale, depending on the material parameters and roughness. The coefficient of friction during two-body deformation and the ratchet component depend on the average slope of the rough surface. The average slope increases with scale due to scale dependence of the roughness parameters. As a result, the two-body deformation component of the coefficient of friction increases with decreasing scale. The three-body component of the coefficient of friction depends on the concentrations of particles, trapped at the interface, which decreases with decreasing scale.

The transition index, which is responsible for transition from predominantly elastic adhesional friction to plastic deformation was proposed and was found to change with scale, due to scale dependence of roughness parameters. For the transition index close to zero, the contact is predominantly elastic and the dominant contribution to friction is adhesion involving elastic deformation. The increase of the transition index leads to an increase in plastic deformation with increasing contribution of the deformation component of friction, which results in larger value of the total coefficient of friction.

In presence of the meniscus force, the measured value of the coefficient of friction is greater than the value of the coefficient of dry friction. The difference is especially important for small loads, when the normal load is comparable with the meniscus force. The meniscus force depends on peak radii and may either increase or decrease with scale, depending on the surface parameters.

The wear coefficient and the ratio of the maximum flash temperature rise to the amount of heat generated per unit time per unit area, for a given sliding velocity, as a function of scale, decrease with decreasing scale due to decrease in the mean contact size.

The proposed model is used to explain the trends in the experimental data for various materials at nanoscale and microscale, which indicate that nanoscale values of coefficient of friction are lower than the microscale values (Tables 15.2 and 15.3). The two factors responsible for this trend are the increase of the three-body deformation and transition from elastic adhesive contact to plastic deformation. Experimental data show that the coefficient of friction increases with increasing load after a certain load and reaches the macroscale value. This is due to the onset of plastic deformation with increasing load and the effect of load on contact parameters, which affect the coefficient of friction.

15.A Statistics of Particle Size Distribution

15.A.1 Statistical Models of Particle Size Distribution

Particle size analysis is an important field for different areas of engineering, environmental, and biomedical studies. In general, size distribution of particles depends on how the particles were formed and sorted. Several statistical distributions, which govern distribution of random variables including particle size, have been suggested (Fig. 15.25), [56–60]. Statistical distributions commonly used are either the probability density (or frequency) function (PDF), $p(z)$, or cumulative distribution function (CDF), $P(h)$. $P(h)$ associated with random variable $z(x)$, which can take any

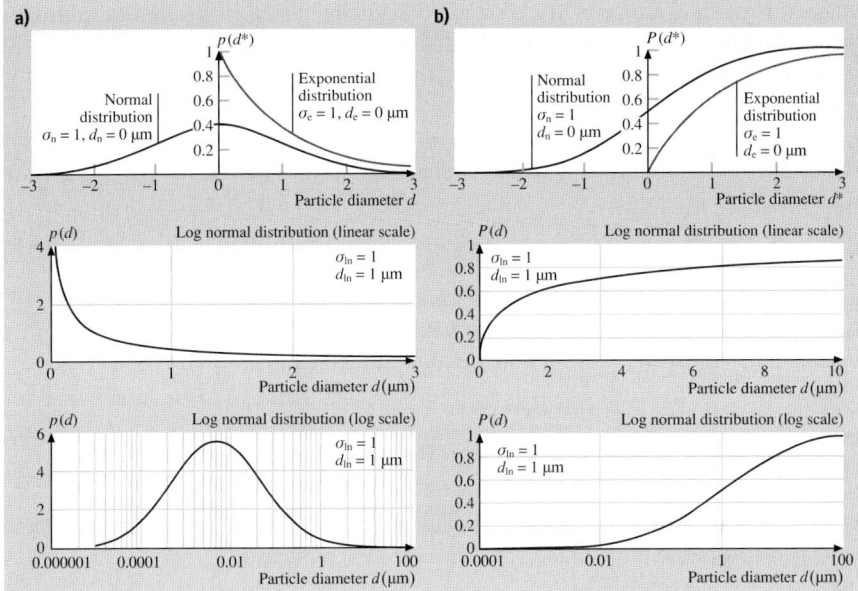

Fig. 15.25a,b. Common statistical distributions of particle size. (**a**) Probability density distributions. (**b**) Cumulative distributions

value between $-\infty$ and $+\infty$ or z_{min} and z_{max}, is defined as the probability of the event $z(x) \leq z'$ and is written as [61]

$$P(z) = Prob(z \leq z') \tag{15.A1}$$

with $P(-\infty) = 0$ and $P(\infty) = 1$.

The pdf is the slope of the CDF given by its derivative

$$P(z) = \frac{dP(z)}{dz} \tag{15.A2}$$

or

$$P(z \leq z') = P(z') = \int_{-\infty}^{z'} p(z) \, dz . \tag{15.A3}$$

Furthermore, the total area under the probability density function must be unity; that is, it is certain that the value of z at any x must fall somewhere between plus and minus infinity or z_{min} and z_{max}. The definition of $p(z)$ is phrased as that the random variable $z(x)$ is distributed as $p(z)$.

The probability density (or frequency) function, $p(d)$, in the exponential form is the simplest distribution mathematically

$$p(d) = \frac{1}{\sigma_e} \exp\left(-\frac{d - d_e}{\sigma_e}\right), \quad d \geq d_0 , \tag{15.A4}$$

where d is particle diameter, σ_e is standard deviation, and d_e is minimum value (for this distribution). For convenience, the density function can be normalized by σ_e in terms of a normalized variable, d^*, equal to $(d - d_e)/\sigma_e$

$$p(d^*) = \exp(-d^*), \quad d^* \geq 0 \tag{15.A5}$$

which has zero minimum and unity standard deviation. The cumulative distribution function $P(d')$ is given as

$$P(d') = P(d^* \leq d') = 1 - \exp(-d'). \tag{15.A6}$$

The Gaussian or normal distribution is used to represent data for a wide collection of random physical phenomena in practice such as surface roughness. The probability density and cumulative distribution functions are given as

$$p(d) = \frac{1}{\sqrt{2\pi}\sigma_n} \exp\left(-\frac{(d-d_n)^2}{2\sigma_n^2}\right), \quad -\infty < d < \infty, \; -\infty < d_n < \infty, \; \sigma_e > 0, \tag{15.A7}$$

where d_n is the mean value. The integral of $p(d)$ in the interval $-\infty < d < \infty$ is equal to 1. In terms of the normalized variables, (15.A6) reduces to

$$p(d^*) = \frac{1}{\sqrt{2\pi}} \exp\left(-\frac{d^{*2}}{2}\right) \tag{15.A8}$$

and

$$P(d') = P(d^* \leq d') = \frac{1}{\sqrt{2\pi}} \int_{-\infty}^{d'} \exp\left[-(d^*)^2/2\right] dd^* = \text{erf}(d'), \tag{15.A9}$$

where $\text{erf}(d')$ is called the "error function" and its values are listed in most statistical handbooks. The pdf is bell-shaped and the CDF is S-shaped.

For particle size distribution, of interest here, the diameter cannot be less than zero. For this condition, (15.A7) must be modified by using a constant on the right side

$$p(d) = \frac{C_0}{\sqrt{2\pi}\sigma_e} \exp\left(-\frac{(d-d_n)^2}{2\sigma_e^2}\right), \quad 0 \leq d < \infty, \tag{15.A10}$$

where

$$C_0 = \left[\frac{1}{\sqrt{2\pi}} \int_{-d_0/\sigma}^{\infty} \exp\left(-\frac{t^2}{2}\right) dt\right]^{-1}.$$

The constant is calculated by integrating $p(d)$ in the interval $0 \leq d \leq \infty$ and equating to one

$$\int_0^{\infty} p(d) \, dd = 1. \tag{15.A11}$$

The log normal distribution is commonly used to describe particle size distribution. A variable d is log normally distributed if $\ln d$ is normally distributed. Log normal probability density function for variable d, for which $\ln(d)$ has a Gaussian distribution with a mean $\ln(d)_{\ln}$ and standard deviation $\sigma_{\ln}$, is given as

$$p(d) = \frac{1}{\sqrt{2\pi}\sigma_{\ln}} \left(\frac{1}{d}\right) \exp\left(-\frac{[\ln(d/d_{\ln})]^2}{2\sigma_{\ln}^2}\right), \quad 0 < d < \infty. \quad (15.A12)$$

The mean of the log normal distribution is $\exp(\ln d_{\ln} + \sigma_{\ln}^2/2)$, the standard deviation is $\exp(2\ln d_{\ln} + \sigma_{\ln}^2)[\exp(\sigma_{\ln}^2) - 1]$, the skewness is $[\exp(\sigma_{\ln})^2 + 2][\exp(\sigma_{\ln}^2) - 1]^{1/2}$, and kurtosis is $\exp[4(\sigma_{\ln})^2] + 2\exp[3(\sigma_{\ln})^2] + 3\exp[2(\sigma_{\ln})^2] - 3$ [58]. The case where $d_{\ln} = 0$ is called the standard log normal distribution. The density function can be normalized by $\sigma_{\ln}$ in terms of a normalized variable, $d^* = (\ln d - d_{\ln})/\sigma_{\ln}$

$$p(d^*) = \frac{1}{\sqrt{2\pi}} \left(\frac{1}{d^*}\right) \exp\left(-\frac{(d^*)^2}{2}\right) \quad (15.A13)$$

and

$$P(d') = P(d^* \le d') = \frac{1}{2}\left[1 + \mathrm{erf}\left(\frac{d'}{\sqrt{2}}\right)\right]. \quad (15.A14)$$

The log normal distribution of particle size occurs when the dispersion is attained by comminution (milling, grinding, crushing). The size distribution of pulverized silica, granite, calcite, limestone, quartz, soda, ash, alumina, clay, as well as of wear particles is often governed by the log-normal distribution [62]. A size distribution is usually presented either as probability density or frequency $p(d)$, or as cumulative percent (percent of particles greater than given size) $P(d)$, or as cumulative mass vs. particle size. All these presentations are interrelated [62].

15.A.2 Typical Particle Size Distribution Data

Typical experimental data for size distributions of atmospheric (dust), sand, and abrasive diamond particles are presented in Fig. 15.26a. It can be seen, that the atmospheric particles [63] follow the normal distribution function. The dune sand is low in heavy mineral content, so the curve is concaved downward. Micaceous dune sand is sorted by gravity slide on a sharp mountain slope and appears to follow log normal distribution, as many distributions of sediments, which are sorted by gravity. Whereas beach sand distribution curve is concaved upward, due to richness in smaller size component [62]. The abrasive diamond particles follow the log normal distribution [64].

Size distribution of wear particles has been studied actively since 1970s, when the ferrography was introduced [65, 66]. The data for wear particles is presented in Fig. 15.26b. *Xuan* et al. [67] studied the size distribution of submicrometer particles during sliding of steel-steel using a Falex 3, pin-on-disk machine, using a laser particle counter for various sliding distances. They found a distribution, which is close

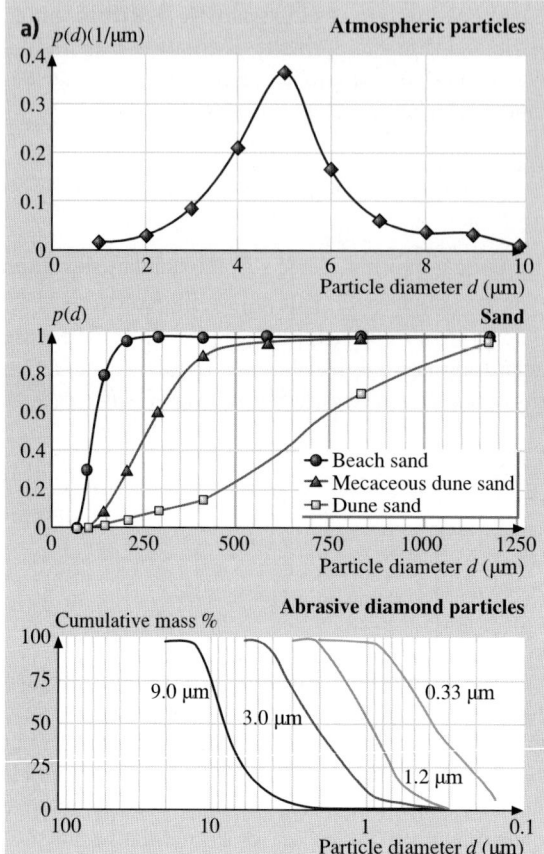

Fig. 15.26. (a) Experimental data for atmospheric [63], sand [62], and abrasive diamond [64] particle size distribution. (b) Experimental data for wear particle size distribution (steel–steel [67], steel–diamond [68], steel–polyethylene [69]). (c) Change with time of wear debris production rate during lubricated sliding as a function of particle size [71]

to the log normal function. *Mizumoto* and *Kato* [68] studied size distribution of particles generated during pin-on-disk test, for diamond, sapphire, silicon carbide, and tungsten carbide pins vs. steel disk, using a laser particle counter. They found that the probability density function is exponential for particles greater than 1 μm diameter, however for smaller particles a linear law was assumed. *Shanbhag* et al. [69] studied wear particles for ultrahigh molecular weight polyethylene (UHMWPE) versus titanium in biomedical applications (total knee replacement) using a scanning electron microscope. They found that the distribution is close to that of the normal distribution. Numerous data for wear particles are presented by *Anderson* [70]. *Hunt* [71] discusses various techniques of debris measurement and analysis in lubricants. A typical change in wear debris generation rate, which occurs with time, is presented in Fig. 15.26c. Change in the size distribution of wear particles in lubricant indicates an onset of mechanical failure.

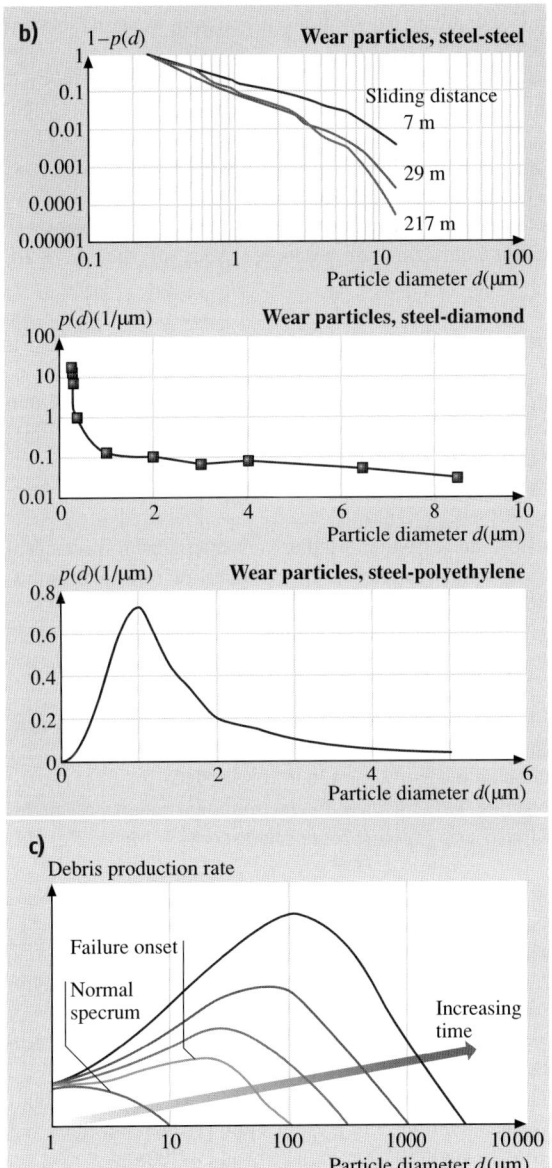

Fig. 15.26. (continued)

References

1. B. Bhushan: *Handbook of Micro/Nanotribology*, 2nd edn. (CRC, Boca Raton 1999)
2. B. Bhushan: Nanoscale tribophysics and tribomechanics, Wear **225–229**, 465–492 (1999)
3. B. Bhushan: *Springer Handbook of Nanotechnology* (Springer, Berlin, Heidelberg 2004)

4. B. Bhushan, J. N. Israelachvili, U. Landman: Nanotribology: Friction, wear and lubrication at the atomic scale, Nature **374**, 607–616 (1995)
5. J. Ruan, B. Bhushan: Atomic-scale friction measurements using friction force microscopy: Part I – General principles and new measurement technique, ASME J. Tribol. **116**, 378–388 (1994)
6. B. Bhushan, A. V. Kulkarni: Effect of normal load on microscale friction measurements, Thin Solid Films **278**, 49–56 (1996); Erratum: **293** 333
7. R. W. Carpick, N. Agrait, D. F. Ogletree, M. Salmeron: Measurement of interfacial shear (friction) with an ultrahigh vacuum atomic force microscope, J. Vac. Sci. Technol. B **14**, 1289–1295 (1996)
8. U. D. Schwarz, O. Zwörner, P. Köster, R. Wiesendanger: Quantitative analysis of the frictional properties of solid materials at low loads. 1. Carbon compounds, Phys. Rev. B **56**, 6987–6996 (1997)
9. B. Bhushan, S. Sundararajan: Micro/nanoscale friction and wear mechanisms of thin films using atomic force and friction force microscopy, Acta Mater. **46**, 3793–3804 (1998)
10. B. Bhushan, C. Dandavate: Thin-film friction and adhesion studies using atomic force microscopy, J. Appl. Phys. **87**, 1201–1210 (2000)
11. H. Liu, B. Bhushan: Adhesion and friction studies of microelectromechanical systems/nanoelectromechanical systems materials using a novel microtriboapparatus, J. Vac. Sci. Technol. A **21**, 1538–1538 (2003)
12. B. Bhushan, H. Liu, S. M. Hsu: Adhesion and friction studies of silicon and hydrophobic and low friction films and investigation of scale effects, ASME J. Tribol. **126**, 583–590 (2004)
13. A. W. Homola, J. N. Israelachvili, P. M. McGuiggan, M. L. Gee: Fundamental experimental studies in tribology: The transition from interfacial friction of undamaged molecularly smooth surfaces to normal friction with wear, Wear **136**, 65–83 (1990)
14. V. N. Koinkar, B. Bhushan: Scanning and transmission electron microscopies of single-crystal silicon microworn/machined using atomic force microscopy, J. Mater. Res. **12**, 3219–3224 (1997)
15. X. Zhao, B. Bhushan: Material removal mechanisms of single-crystal silicon on nanoscale and at ultralow loads, Wear **223**, 66–78 (1998)
16. B. Bhushan: *Introduction to Tribology* (Wiley, New York 2002)
17. N. A. Fleck, G. M. Muller, M. F. Ashby, J. W. Hutchinson: Strain gradient plasticity: Theory and experiment, Acta Metall. Mater. **42**, 475–487 (1994)
18. W. D. Nix, H. Gao: Indentation size effects in crystalline materials: A law for strain gradient plasticity, J. Mech. Phys. Solids **46**, 411–425 (1998)
19. J. W. Hutchinson: Plasticity at the micron scale, Int. J. Solids Struct. **37**, 225–238 (2000)
20. B. Bhushan, M. Nosonovsky: Scale effects in friction using strain gradient plasticity and dislocation-assisted sliding (microslip), Acta Mater. **51**, 4331–4345 (2003)
21. B. Bhushan, M. Nosonovsky: Comprehensive model for scale effects in friction due to adhesion and two- and three-body deformation (plowing), Acta Mater. **52**, 2461–2474 (2004)
22. B. Bhushan, M. Nosonovsky: Scale effects in dry and wet friction, wear, and interface temperature, Nanotechnol. **15**, 749–761 (2004)
23. M. Nosonovsky, B. Bhushan: Scale effect in dry friction during multiple asperity contact, ASME J. Tribol. **127**, 37–46 (2005)
24. H. Gao, Y. Huang, W. D. Nix, J. W. Hutchinson: Mechanism-based strain-gradient plasticity – I. theory, J. Mech. Phys. Solids **47**, 1239–1263 (1999)

25. Y. Huang, H. Gao, W.D. Nix, J.W. Hutchinson: Mechanism-based strain-gradient plasticity – II. analysis, J. Mech. Phys. Solids **48**, 99–128 (2000)
26. Z.P. Bazant: Scaling of dislocation-based strain-gradient plasticity, J. Mech. Phys. Solids **50**, 435–448 (2002)
27. B. Bhushan, A.V. Kulkarni, W. Bonin, J.T. Wyrobek: Nano/picoindentation measurement using a capacitive transducer system in atomic force microscopy, Philos. Mag. **74**, 1117–1128 (1996)
28. A.V. Kulkarni, B. Bhushan: Nanoscale mechanical property measurements using modified atomic force microscopy, Thin Solid Films **290-291**, 206–210 (1996)
29. K.W. McElhaney, J.J. Vlassak, W.D. Nix: Determination of indenter tip geometry and indentation contact area of depth-sensing indentation experiments, J. Mater. Res. **13**, 1300–1306 (1998)
30. J. Friedel: *Dislocations* (Pergamon, New York 1964)
31. J. Weertman, J.R. Weertman: *Elementary Dislocations Theory* (MacMillan, New York 1966)
32. B. Bhushan, A.V. Koinkar: Nanoindentation hardness measurements using atomic force microscopy, Appl. Phys. Lett. **64**, 1653–1655 (1994)
33. N. Gane, J.M. Cox: The micro-hardness of metals at very low loads, Philos. Mag. **22**, 881–891 (1970)
34. M.A. Stelmashenko, M.G. Walls, L.M. Brown, Y.V. Miman: Microindentation on W and Mo oriented single crystal an SEM study, Acta Met. Mater. **41**, 2855–2865 (1993)
35. S. Sundararajan, B. Bhushan: Development of AFM-based techniques to measure mechanical properties of nanoscale structures, Sensors Actuator A **101**, 338–351 (2002)
36. J.J. Weertman: Dislocations moving uniformly on the interface between isotropic media of different elastic properties, J. Mech. Phys. Solids **11**, 197–204 (1963)
37. K.L. Johnson: Adhesion and friction between a smooth elastic spherical asperity and a plane surface, Proc. R. Soc. London A **453**, 163–179 (1997)
38. I.A. Polonsky, L.M. Keer: Scale effects of elastic-plastic behavior of microscopic asperity contact, ASME J. Tribol. **118**, 335–340 (1996)
39. V.S. Deshpande, A. Needleman, E. Van der Giessen: Discrete dislocation plasticity modeling of short cracks in single crystals, Acta Mater. **51**, 1–15 (2003)
40. R.A. Onions, J.F. Archard: The contact of surfaces having a random structure, J. Phys. D **6**, 289–304 (1973)
41. D.J. Whitehouse, J.F. Archard: The properties of random surfaces of significance in their contact, Proc. R. Soc. London A **316**, 97–121 (1970)
42. J.A. Greenwood, J.B.P. Williamson: Contact of nominally flat surfaces, Proc. R. Soc. London A **295**, 300–319 (1966)
43. B. Bhushan: Contact mechanics of rough surfaces in tribology: Single asperity contact, Appl. Mech. Rev. **49**, 275–298 (1996)
44. B. Bhushan: Contact mechanics of rough surfaces in tribology: Multiple asperities contact, Tribol. Lett. **4**, 1–35 (1998)
45. B. Bhushan, W. Peng: Contact modeling of multilayered rough surfaces, Appl. Mech. Rev. **55**, 435–480 (2002)
46. A. Majumdar, B. Bhushan: Fractal model of elastic-plastic contact between rough surfaces, ASME J. Tribol. **113**, 1–11 (1991)
47. K.L. Johnson: *Contact Mechanics* (Clarendon, Oxford 1985)
48. E. Rabinowicz: *Friction and Wear of Materials*, 2nd edn. (Wiley, New York 1995)
49. H.R. Clauser (Ed.): *The Encyclopedia of Engineering Materials and Processes* (Reinhold, London 1963)

50. B. Bhushan, B.K. Gupta: *Handbook of Tribology: Materials, Coatings, and Surface Treatments* (McGraw-Hill, New York 1991; Krieger, Malabar, New York 1997)
51. National Carbon Comp.: *The Industrial Graphite Engineering Handbook* (National Carbon Company, New York 1959)
52. INSPEC: *Properties of Silicon*, EMIS Data Rev. Ser. No. 4 (INSPEC, Institution of Electrical Engineers, London 2002) See also, MEMS Materials Database, http://www.memsnet.org/material/
53. J.E.. Field (Ed.): *The Properties of Natural and Synthetic Diamond* (Academic, London 1992)
54. B.. Bhushan, S. Venkatesan: Mechanical and tribological properties of silicon for micromechanical applications: A review, Adv. Info. Storage Syst. **5**, 211–239 (1993)
55. B. Bhushan: Chemical, mechanical and tribological characterization of ultra-thin and hard amorphous carbon coatings as thin as 3.5 nm: Recent developments, Diam. Relat. Mater. **8**, 1985–2015 (1999)
56. C. Bernhardt: *Particle Size Analysis* (Chapman Hall, London 1994)
57. J.L. Devoro: *Probability and Statistics for Engineering and the Sciences* (Duxbury, New York 1995)
58. B.S. Everitt: *The Cambridge Dictionary of Statistics* (Cambridge Univ. Press, Cambridge 1998)
59. D. Zwillinger, S. Kokoska: *CRC Standard Probability and Statistics Tables and Formulas* (CRC, Boca Raton 2000)
60. S. Wolfram: *The Mathematica Book*, 5th edn. (Wolfram Media, Champaign 2003)
61. J.S. Bendet, A.G. Piersol: *Engineering Applications of Correlation and Spectral Analysis*, 2nd edn. (Wiley, New York 1986)
62. G. Herdan: *Small Particle Statistics* (Butterworth, London 1960)
63. R.D. Cadle: *Particle Size – Theory and Industrial Applications* (Reinhold, New York 1965)
64. Y. Xie, B. Bhushan: Effect of particle size, polishing pad and contact pressure in free abrasive polishing, Wear **200**, 281–295 (1996)
65. W.W. Seifert, V.C. Westcott: A method for the study of wear particles in lubricating oil, Wear **21**, 27–42 (1972)
66. D. Scott, V.C. Westcott: Predictive maintenance by ferrography, Wear **44**, 173–182 (1977)
67. J.L. Xuan, H.S. Cheng, R.J. Miller: Generation of submicrometer particles in dry sliding, ASME J. Tribol. **112**, 664–691 (1990)
68. M. Mizumoto, K. Kato: Size distribution and number of wear particles generated by the abrasive sliding of a model asperity in the SEM-tribosystem. In: *Wear Particles: From the Cradle to the Grave*, ed. by D. Dowson et al. (Elsevier, Amsterdam 1992) pp. 523–530
69. A.S. Shanbhag, H.O. Bailey, D.S. Hwang, C.W. Cha, N.G. Eror, H.E. Rubash: Quantitative analysis of ultrahigh molecular weight polyethylene (UHMWPE) wear debris associated with total knee replacements, J. Biomed. Mater. Res. **53**, 100–110 (2000)
70. D.P. Anderson: *Wear Particle Atlas*, 2nd edn. (Spectro Inc. Industrial Tribology Systems, Littleton 1991)
71. T.M. Hunt: *Handbook of Wear Debris Analysis and Particle Detection in Liquids* (Elsevier Applied Science, London 1993)

Part III

Molecularly-Thick Films for Lubrication

Part III

16

Nanotribology of Ultrathin and Hard Amorphous Carbon Films

Bharat Bhushan

Summary. One of the best materials to use in applications that require very low wear and reduced friction is diamond, especially in the form of a diamond coating. Unfortunately, true diamond coatings can only be deposited at high temperatures and on selected substrates, and they require surface finishing. However, hard amorphous carbon – commonly known as diamond-like carbon or a DLC coating – has similar mechanical, thermal and optical properties to those of diamond. It can also be deposited at a wide range of thicknesses using a variety of deposition processes on various substrates at or near room temperature. The coatings reproduce the topography of the substrate, removing the need for finishing. The friction and wear properties of some DLC coatings make them very attractive for some tribological applications. The most significant current industrial application of DLC coatings is in magnetic storage devices.

In this chapter, the state-of-the-art in the chemical, mechanical and tribological characterization of ultrathin amorphous carbon coatings is presented.

EELS and Raman spectroscopies can be used to characterize amorphous carbon coatings chemically. The prevailing atomic arrangement in the DLC coatings is amorphous or quasi-amorphous, with small diamond (sp^3), graphite (sp^2) and other unidentifiable micro- or nanocrystallites. Most DLC coatings, except for those produced using a filtered cathodic arc, contain from a few to about 50 at. % hydrogen. Sometimes hydrogen is deliberately incorporated into the sputtered and ion-plated coatings in order to tailor their properties.

Amorphous carbon coatings deposited by different techniques exhibit different mechanical and tribological properties. Thin coatings deposited by filtered cathodic arc, ion beam and ECR-CVD hold much promise for tribological applications. Coatings of 5 nm or even less provide wear protection. A nanoindenter can be used to measure DLC coating hardness, elastic modulus, fracture toughness and fatigue life. Microscratch and microwear tests can be performed on the coatings using either a nanoindenter or an AFM, and along with accelerated wear testing, can be used to screen potential industrial coatings. For the examples shown in this chapter, the trends observed in such tests were similar to those found in functional tests.

16.1 Introduction

Carbon exists in both crystalline and amorphous forms and exhibits both metallic and nonmetallic characteristics [1–3]. Forms of crystalline carbon include graphite, diamond, and a family of fullerenes, Fig. 16.1. The graphite and diamond are infinite periodic network solids with a planar structure, whereas the fullerenes are a molecular form of pure carbon with a finite network and a nonplanar structure. Graphite has a hexagonal, layered structure with weak interlayer bonding forces and it exhibits excellent lubrication properties. The graphite crystal may be visualized as infinite parallel layers of hexagons stacked 0.34 nm apart with an interatomic distance of 0.1415 nm between the carbon atoms in the basal plane. Each atom lying in the basal planes is trigonally coordinated and closely packed with strong σ (covalent) bonds to its three carbon neighbors via hybrid sp^2 orbitals. The fourth electron lies in a p_z orbital lying normal to the σ bonding plane and forms a weak π bond by overlapping side-to-side with a p_z orbital of an adjacent atom to which the carbon is attached by a σ bond. The layers (basal planes) themselves are relatively far apart and the forces that bond them are weak van der Waals forces. These layers can align themselves parallel to the direction of the relative motion and slide over one another with relative ease, meaning low friction. Strong interatomic bonding and packing in each layer is thought to help reduce wear. The operating environment has a significant influence on the lubrication – low friction and low wear – properties of graphite.

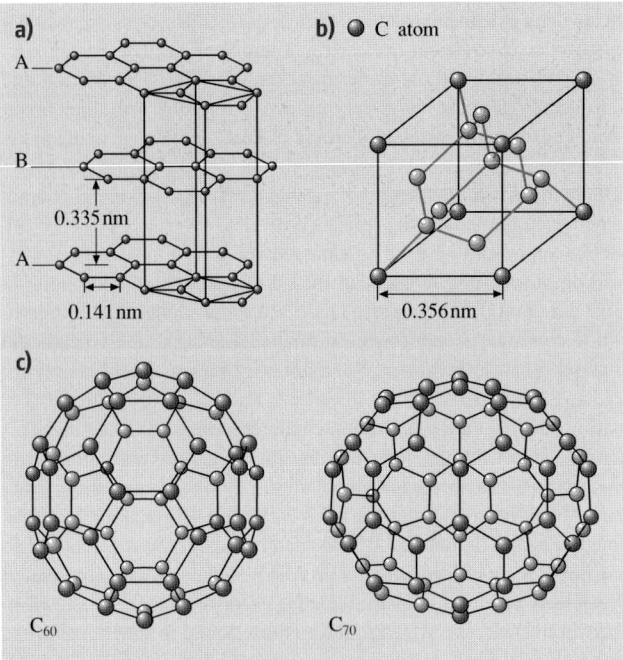

Fig. 16.1a–c. The structures of the three known forms of crystalline carbon: (**a**) hexagonal structure of graphite, (**b**) modified face-centered cubic (fcc) structure (two interpenetrating fcc lattices displaced by a quarter of the cube diagonal) of diamond (each atom is bonded to four others that form the corners of a tetrahedron), and (**c**) the structures of the two most common fullerenes: a soccer ball C_{60} and a rugby ball C_{70} molecules

It lubricates better in a humid environment than a dry one, due to the adsorption of water vapor and other gases from the environment, which further weakens the interlayer bonding forces and results in easy shear and transfer of the crystallite platelets to the mating surface. Thus, transfer plays an important role in controlling friction and wear. Graphite oxidizes at high operating temperatures and can be used up to about 430 °C.

One of the most well-known fullerene molecules is C_{60}, commonly known as buckyballs. Since these C_{60} molecules are very stable and do not require additional atoms to satisfy chemical bonding requirements, they are expected to have low adhesion to the mating surface and low surface energy. Since the C_{60} molecule, which has a perfect spherical symmetry, bonds only weakly to other molecules, C_{60} clusters readily become detached , similar to other layered lattice structures, and either get transferred to the mating surface by mechanical compaction or are present as loose wear particles that may roll like tiny ball bearings in a sliding contact, resulting in low friction and wear. The wear particles are expected to be harder than as-deposited C_{60} molecules, because of their phase transformation at the high-asperity contact pressures present in a sliding interface. The low surface energy, the spherical shapes of C_{60} molecules, the weak intermolecular bonding, and the high load bearing capacity offer vast potential for various mechanical and tribological applications. Sublimed C_{60} coatings and fullerene particles used as an additive to mineral oils and greases have been reported to be good solid lubricants comparable to graphite and MoS_2 [4–6].

Diamond crystallizes in a modified face-centered cubic (fcc) structure with an interatomic distance of 0.154 nm. The diamond cubic lattice consists of two interpenetrating fcc lattices displaced by a quarter of the cube diagonal. Each carbon atom is tetrahedrally coordinated, making strong σ (covalent) bonds to its four carbon neighbors using the hybrid sp^3 atomic orbitals, which accounts for it having the highest hardness (80–104 GPa) and thermal conductivity (900–2100 W/mK, on the order of five times that of copper) of any known solid, as well as high electrical resistivity and optical transmission and a large optical band gap. It is relatively chemically inert, and it exhibits poor adhesion with other solids, enhancing its low friction and wear properties. Its high thermal conductivity permits the dissipation of frictional heat during sliding and it protects the interface, and the dangling carbon bonds on the surface react with the environment to form hydrocarbons that act as good lubrication films. These are some of the reasons for the low friction and wear of the diamond. Diamond and its coatings find many industrial applications: tribological applications (low friction and wear), optical applications (exceptional optical transmission, high abrasion resistance), and thermal management or heat sink applications (high thermal conductivity). Diamond can be used at high temperatures; it starts to graphitize at about 1000 °C in ambient air and at about 1400 °C in vacuum. Diamond is an attractive material for cutting tools, abrasives for grinding wheels and lapping compounds, and other extreme wear applications.

Natural diamond – particularly in large quantities – is very expensive, and so diamond coatings – a low-cost alternative – are attractive. True diamond coatings are deposited by chemical vapor deposition (CVD) processes at high substrate temperatures (on the order of 800 °C). They adhere best on silicon substrate and require an interlayer for other substrates. One major hindrance to the widespread use of true diamond films in tribological, optical and thermal management applications is their surface roughness. Growth of the diamond phase on a non-diamond substrate is initiated by nucleation at either randomly seeded sites or at thermally favored sites, due to statistical thermal fluctuations at the substrate surface. Depending on the growth temperature and pressure conditions, certain favored crystal orientations dominate the competitive growth process. As a result, the films grown are polycrystalline in nature with a relatively large grain size (> 1 μm) and they terminate in very rough surfaces, with RMS roughnesses ranging from a few tenths of a micron to tens of microns. Techniques for polishing these films have been developed. It has been reported that laser polished films exhibit friction and wear properties almost comparable to those of bulk polished diamond [7, 8].

Amorphous carbon has no long-range order, and the short-range order of the carbon atoms in it can have one or more of three bonding configurations: sp^3 (diamond), sp^2 (graphite), or sp^1 (with two electrons forming strong σ bonds, and the remaining two electrons left in orthogonal p_y and p_z orbitals, that form weak π bonds). Short-range order controls the properties of amorphous materials and coatings. Hard amorphous carbon (a-C) coatings, commonly known as diamond-like carbon or DLC (implying high hardness) coatings, are a class of coatings that are mostly metastable amorphous materials, but that include a micro- or nanocrystalline phase. The coatings are random networks of covalently bonded carbon in hybridized tetragonal (sp^3) and trigonal (sp^2) local coordination with some of the bonds terminated by hydrogen. These coatings have been successfully deposited by a variety of vacuum deposition techniques on a variety of substrates at or near room temperature. These coatings generally reproduce substrate topography and do not require any post-finishing. However, these coatings mostly adhere best on silicon substrates. The best adhesion is obtained on substrates that form carbides, such as Si, Fe and Ti. Based on depth profile analyses (using Auger and XPS) of DLC coatings deposited on silicon substrates, it has been reported that a substantial amount of silicon carbide (on the order of 5–10 nm in thickness) is present at the carbon–silicon interface, giving good adhesion and hardness (see [9]). For good adhesion of DLC coatings to other substrates, in most cases, an interlayer of silicon is required in most cases, except for coatings deposited by a cathodic arc.

There is significant interest in DLC coatings due to their unique combination of desirable properties. These properties include high hardness and wear resistance, chemical inertness to both acids and alkalis, lack of magnetic response, and an optical band gap ranging from zero to a few eV, depending upon the deposition conditions. These are used in a wide range of applications, including tribological, optical, electronic and biomedical applications [1, 10, 11]. The high hardness, good friction and wear properties, versatility in deposition and substrates, and no requirement for post-deposition finishing make them very attractive for tribological applications. Two primary examples include overcoats for magnetic media (thin film disks and ME tapes) and MR-type magnetic heads for magnetic storage devices, Fig. 16.2 [12–20], and the emerging field of microelectromechanical systems, Fig. 16.3 [21–24]. The largest industrial application of the family of amorphous carbon coatings, typically deposited by DC/RF magnetron sputtering, plasma-enhanced chemical vapor deposition or ion beam deposition techniques, is in magnetic storage devices. These are employed to protect magnetic coatings on thin film rigid disks and metal evaporated tapes and the thin film head structure of a read/write disk head against wear and corrosion (Fig. 16.2). Thicknesses ranging from 3 to 10 nm are employed to maintain low physical spacing between the magnetic element of a read/write head and the magnetic layer of the storage media. Mechanical properties affect friction wear and therefore need to be optimized. In 1998, Gillette introduced Mach 3 razor blades with ultrathin DLC coatings, which could potentially become a very large industrial application. DLC coatings are also used in other commercial applications such as the glass windows of supermarket laser barcode scanners and sunglasses. These coatings are actively pursued in microelectromechanical systems (MEMS) components [23].

In this chapter, a state-of-the-art review of recent developments in the field of chemical, mechanical, and tribological characterization of ultrathin amorphous carbon coatings is presented. An overview of the most commonly used deposition techniques is provided, followed by typical chemical and mechanical characterization data and typical tribological data from both coupon-level testing and functional testing.

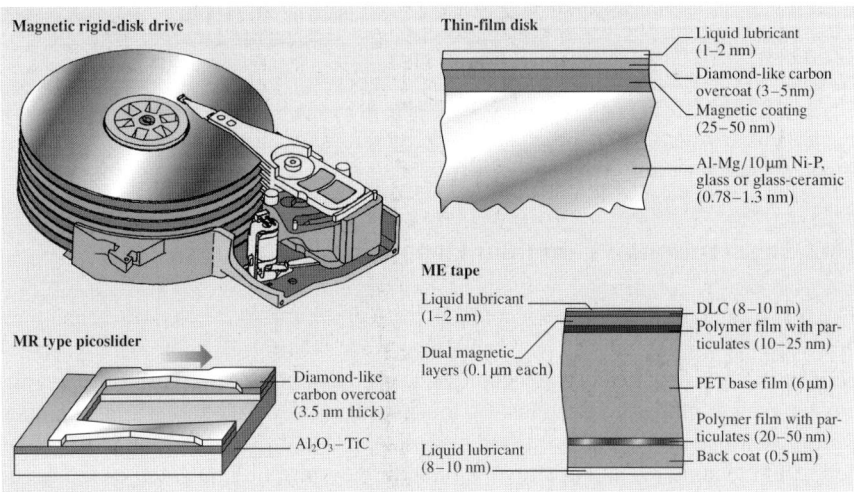

Fig. 16.2. Schematic of a magnetic rigid-disk drive and MR type picoslider, and cross-sectional schematics of a magnetic thin film rigid disk and a metal evaporated (ME) tape

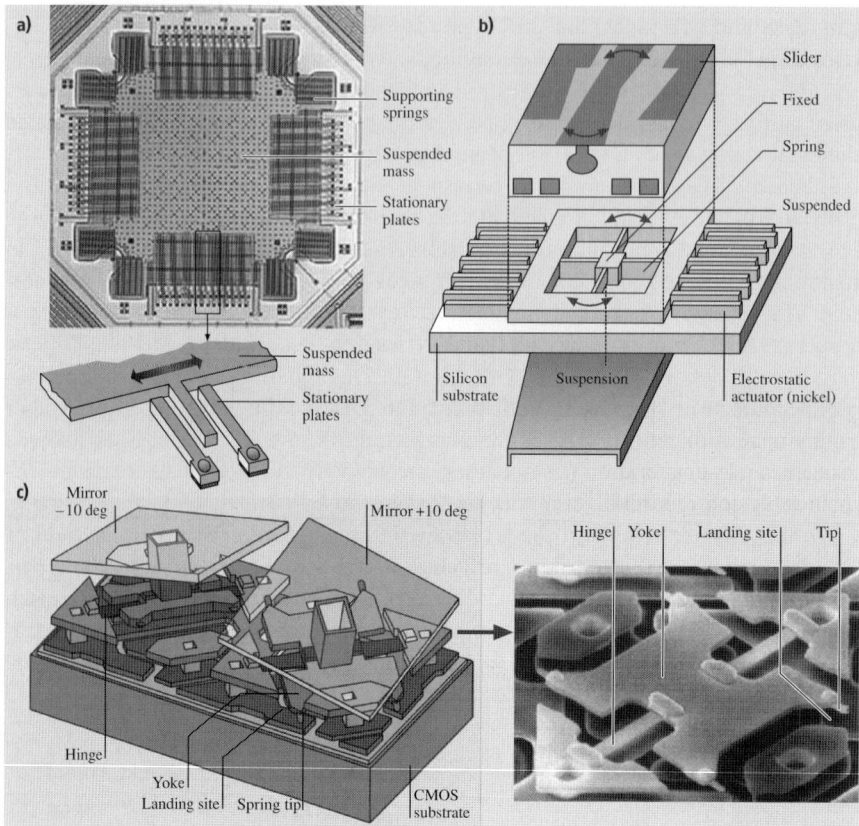

Fig. 16.3. Schematics of (**a**) a capacitive-type silicon accelerometer for automotive sensory applications, (**b**) digital micrometer devices for high-projection displays, and (**c**) a polysilicon rotary microactuator for a magnetic disk drives

16.2 Description of Common Deposition Techniques

The first hard amorphous carbon coatings were deposited by a beam of carbon ions produced in an argon plasma on room temperature substrates, as reported by *Aisenberg* and *Chabot* [25]. Subsequent confirmation by *Spencer* et al. [26] led to the explosive growth of this field. Following this first work, several alternative techniques were developed. Amorphous carbon coatings have been prepared by a variety of deposition techniques and precursors, including evaporation, DC, RF or ion beam sputtering, RF or DC plasma-enhanced chemical vapor deposition (PECVD), electron cyclotron resonance chemical vapor deposition (ECR-CVD), direct ion beam deposition, pulsed laser vaporization and vacuum arc, from a variety of carbon-bearing solids or gaseous source materials [1, 27]. Coatings with both graphitic and diamond-like properties have been produced. Evaporation and ion plating techniques have been used to produce coatings with graphitic properties (low hardness, high electrical conductivity, very low friction, and so on, and all of the techniques have been used to produce coatings with diamond-like properties.

The structure and properties of a coating are dependent upon the deposition technique and parameters. High-energy surface bombardment has been used to produce harder and denser coatings. It is reported that the sp^3/sp^2 fraction decreases in the order: cathodic arc deposition, pulsed laser vaporization, direct ion beam deposition, plasma-enhanced chemical vapor deposition, ion beam sputtering, DC/RF sputtering [12, 28, 29]. A common feature of these techniques is that the deposition is energetic; in other words the carbon species arrive with an energy significantly greater than that represented by the substrate temperature. The resultant coatings are amorphous in structure, with hydrogen contents of up to 50%, and display a high degree of sp^3 character. From the results of previous investigations, it has been proposed that deposition of sp^3-bonded carbon requires that the depositing species have kinetic energies on the order of 100 eV or higher, well above those obtained in thermal processes like evaporation (0–0.1 eV). The species must then be quenched into the metastable configuration via rapid energy removal. Excess energy, such as that provided by substrate heating, is detrimental to the achievement of a high sp^3 fraction. In general, the higher the fraction of sp^3-bonded carbon atoms in the amorphous network, the greater the hardness [29–36]. The mechanical and tribological properties of a carbon coating depend on the sp^3/sp^2-bonded carbon ratio, the amount of hydrogen in the coating, and the adhesion of the coating to the substrate, which are influenced by the precursor material, the kinetic energy of the carbon species prior to deposition, the deposition rate, the substrate temperature, the substrate biasing, and the substrate itself [29, 33, 35, 37–46]. The kinetic energies and deposition rates involved in selected deposition processes used in the deposition of DLC coatings are compared in Table 16.1 [1, 28].

In the studies by *Gupta* and *Bhushan* [12, 47], *Li* and *Bhushan* [48, 49], and *Sundararajan* and *Bhushan* [50], DLC coatings typically ranging in thickness from 3.5 nm to 20 nm were deposited on single-crystal silicon, magnetic Ni-Zn ferrite, and Al_2O_3-TiC substrates (surface roughness ≈ 1–3 nm RMS) by filtered cathodic arc (FCA) deposition, (direct) ion beam deposition (IBD), electron cyclotron resonance chemical vapor deposition (ECR-CVD), plasma-enhanced chemical vapor deposition (PECVD), and DC/RF planar magnetron sputtering (SP) deposition techniques [51]. In this chapter, we will limit the presentation of data to coatings deposited by FCA, IBD, ECR-CVD and SP deposition techniques.

16.2.1 Filtered Cathodic Arc Deposition

When the filtered cathodic arc deposition technique is used to create carbon coatings [29, 52–59], a vacuum arc plasma source is used to form the carbon film. In the FCA technique used by *Bhushan* et al. (see [12]), energetic carbon ions are produced by a vacuum arc discharge between a planar graphite cathode and a grounded anode, Fig. 16.4a. The cathode is a 6 mm-diameter high-density graphite disk mounted on a water-cooled copper block. The arc is driven at an arc current of 200 A, with an arc duration of 5 ms and an arc repetition rate of 1 Hz. The plasma beam is guided by

Table 16.1. Summary of common deposition techniques, the kinetic energies of the depositing species and deposition rates

Deposition technique	Process	Kinetic energy (eV)	Deposition rate (nm/s)
Filtered cathodic arc (FCA)	Energetic carbon ions produced by a vacuum arc discharge between a graphite cathode and a grounded anode	100–2500	0.1–1
Direct ion beam (IB)	Carbon ions produced from methane gas in an ion source and accelerated toward a substrate	50–500	0.1–1
Plasma-enhanced chemical vapor deposition (PECVD)	Hydrocarbon species produced by plasma decomposition of a hydrocarbon gas (such as acetylene) are accelerated toward a DC-biased substrate	1–30	1–10
Electron cyclotron resonance plasma chemical vapor deposition (ECR-CVD)	Hydrocarbon ions produced by the plasma decomposition of ethylene gas in the presence of a plasma at the electron cyclotron resonance condition are accelerated toward a RF-biased substrate	1–50	1–10
DC/RF sputtering	Sputtering of graphite target by argon ion plasma	1–10	1–10

a magnetic field that transports current between the electrodes to form tiny, rapidly moving spots on the cathode surface. The source is coupled to a 90° bent magnetic filter to remove the macroparticles produced concurrently with the plasma in the cathode spots. The ion current density at the substrate is in the range of 10–50 mA/cm^2. The base pressure is less than 10^{-4} Pa. A much higher plasma density is achieved using a powerful arc discharge than using electron beam evaporation with auxiliary discharge. In the discharge process, the cathodic material suffers a complicated transition from the solid phase to an expanding, nonequilibrium plasma via liquid and dense equilibrium nonideal plasma phases [58]. The carbon ions in the vacuum arc plasma have a direct kinetic energy of 20–30 eV. The high voltage pulses are applied to the substrate which is mounted on a water-cooled sample holder, and ions are accelerated through the sheath and arrive at the substrate with an additional energy given by the potential difference between the plasma and the substrate. The substrate holder is pulsed-biased to a voltage of up to −2 kV with a pulse duration of 1 µs. The negative biasing of −2 kV corresponds to a kinetic energy of 2 keV for the carbon ions. The use of a pulsed bias instead of a DC bias enables much higher voltages to be applied and it permits a surface potential to be created on nonconducting films. The ion energy is varied during the deposition. For the first 10% of the deposition, the substrates are pulsed-biased to −2 keV with a pulse duty cycle of 25%, so for 25% of the time the energy is 2 keV and for the remaining 75% it is 20 eV, which is the "natural" energy of carbon ions in a vacuum discharge. For the last 90% of the deposition, the pulsed-biased voltage is reduced to −200 eV with a pulsed bias duty cycle of 25%, so the energy is 200 eV for 25% and 20 eV for 75% of the deposition. The high energy at the beginning leads to good intermixing and adhesion of the films, whereas the lower energy at the later stage leads to hard films. Under the conditions described, the deposition rate at the substrate is about 0.1 nm/s, which is slow. Compared with most gaseous plasma, the cathodic arc plasma is almost fully ionized, and the ionized carbon atoms have high kinetic energies which promotes the formation of a high fraction of sp^3-bonded carbon ions, which in turn results in high hardness and higher interfacial adhesion. *Cuomo et al.* [42] have reported, based on electron energy loss spectroscopy (EELS) analysis, that the sp^3-bonded carbon fraction of a cathodic arc coating is 83% compared to 38% for ion beam sputtered carbon. These coatings are reported to be *nonhydrogenated*.

This technique does not require an adhesion underlayer for non-silicon substrates. However, adhesion of the DLC coatings on the electrically insulating substrate is poor, as negative pulsed biasing forms an electrical sheath that accelerates depositing ions to the substrate and enhances the adhesion of the coating to the substrate with associated ion implantation. It is difficult to build potential on an insulating substrate, and lack of biasing results in poor adhesion.

16.2.2 Ion Beam Deposition

In the direct ion beam deposition of a carbon coating [60–64], as used by *Bhushan et al.* (see [12]), the carbon coating is deposited from an accelerated carbon ion beam. The sample is precleaned by ion etching. In the case of non-silicon substrates, a 2–3 nm-thick amorphous silicon adhesion layer is deposited by ion beam sputtering using an ion beam containing a mixture of methane and argon at 200 V. For the carbon deposition, the chamber is pumped to about 10^{-4} Pa, and methane gas is fed through the cylindrical ion source and ionized by energetic electrons produced by a hot-wire filament, Fig. 16.4b. Ionized species then pass through a grid with a bias voltage of about 50 eV, where they gain a high acceleration energy and reach a hot-wire filament, emitting thermionic electrons that neutralize the incoming ions. The discharging of ions is important when insulating ceramics are used as substrates. The species are then deposited on a water-cooled substrate. Operating conditions are adjusted to give an ion beam with an acceleration energy of about 200 eV and a current density of about 1 mA/cm^2. At these operating conditions, the deposition rate is about 0.1 nm/s, which is slow. Incidentally, tough and soft coatings are deposited at a high acceleration energy of about 400 eV and at a deposition rate of about 1 nm/s. The ion beam-deposited carbon coatings are reported to be hydrogenated (30–40 at. % hydrogen).

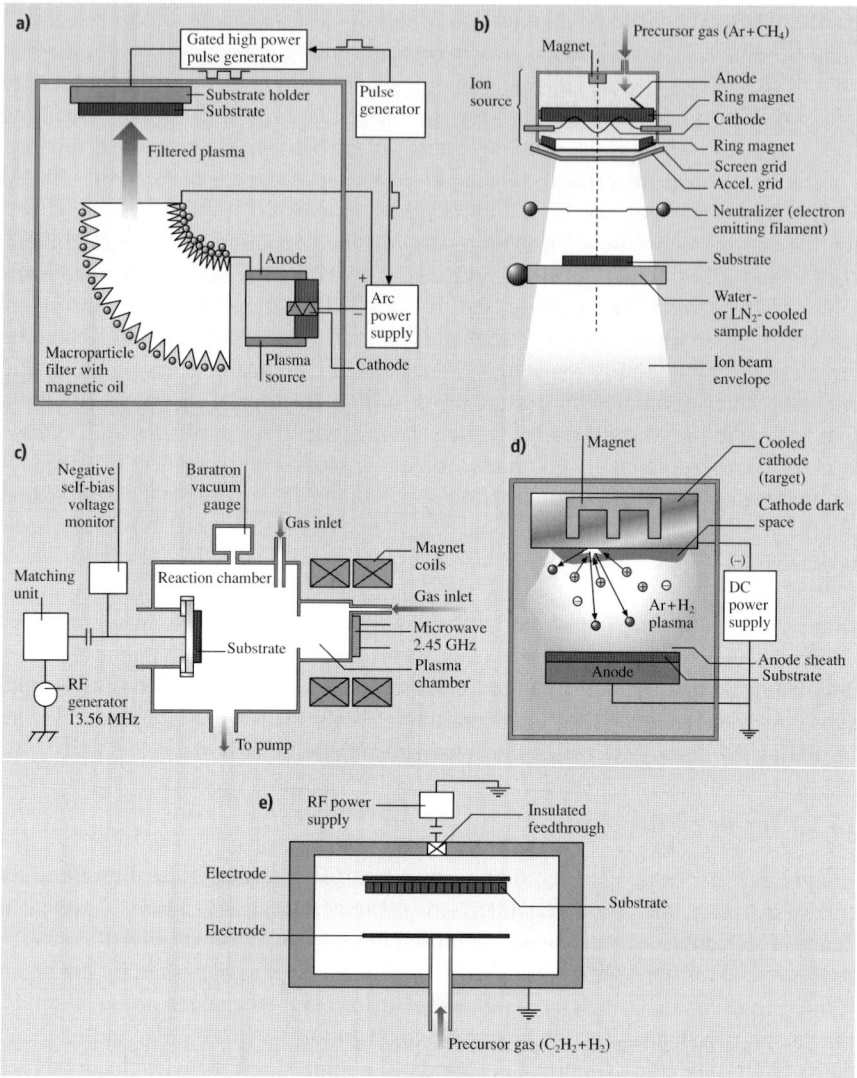

Fig. 16.4. Schematic diagrams of deposition by (**a**) filtered cathodic arc deposition, (**b**) ion beam deposition, (**c**) electron cyclotron resonance chemical vapor deposition (ECR-CVD), (**d**) DC planar magnetron sputtering, and (**e**) plasma-enhanced chemical vapor deposition (PECVD)

16.2.3 Electron Cyclotron Resonance Chemical Vapor Deposition

The lack of electrodes in the ECR-CVD technique and its ability to create high densities of charged and excited species at low pressures ($\leq 10^{-4}$ Torr) make it attractive for coating deposition [65]. In the ECR-CVD carbon deposition process described by *Suzuki* and *Okada* [66] and used by *Li* and *Bhushan* [48, 49] and *Sundararajan* and *Bhushan* [50], microwave power is generated by a magnetron operating in continuous mode at a frequency of 2.45 GHz, Fig. 16.4c. The plasma chamber functions as a microwave cavity resonator. The magnetic coils arranged around the plasma chamber generate a magnetic field of 875 G, necessary for electron cyclotron resonance. The substrate is placed on a stage that is connected capacitively to a 13.56 MHz RF generator. The process gas is introduced into the plasma chamber and the hydrocarbon ions generated are accelerated by a negative self-bias voltage, which is generated by applying RF power to the substrate. Both the substrate stage and the plasma chamber are water-cooled. The process gas used is 100% ethylene and its flow rate is held constant at 100 sccm. The microwave power is 100–900 W. The RF power is 30–120 W. The pressure during deposition is kept close to the optimum value of 5.5×10^{-3} Torr. Before the deposition, the substrates are cleaned using Ar ions generated in the ECR plasma chamber.

16.2.4 Sputtering Deposition

In DC planar magnetron carbon sputtering [13, 33, 37, 40, 67–71], the carbon coating is deposited by the sputtering of a graphite target with Ar ion plasma. In the glow discharge, positive ions from the plasma strike the target with sufficient energy to dislodge the atoms by momentum transfer, which are intercepted by the substrate. An ~5 nm-thick amorphous silicon adhesion layer is initially deposited by sputtering if the deposition is to be carried out on a non-silicon surface. In the process used by *Bhushan* et al. (see [12]), the coating is deposited by the sputtering of a 200 mm-diameter graphite target with Ar ion plasma at 300 W power and a pressure of about 0.5 Pa (6 mTorr), Fig. 16.4d. Plasma is generated by applying a DC potential between the substrate and a target. *Bhushan* et al. [35] reported that the sputtered carbon coating contains about 35 at. % hydrogen. The hydrogen comes from the hydrocarbon contaminants present in the deposition chamber. In order to produce a hydrogenated carbon coating with a larger concentration of hydrogen, the deposition is carried out in Ar and hydrogen plasma.

16.2.5 Plasma-Enhanced Chemical Vapor Deposition

In the RF-PECVD deposition of carbon, as used by *Bhushan* et al. (see [12]), the carbon coating is deposited by adsorbing hydrocarbon free radicals onto the substrate and then via chemical bonding to other atoms on the surface. The hydrocarbon species are produced by the RF plasma decomposition of hydrocarbon precursors such as acetylene (C_2H_2), Fig. 16.4e [27, 69, 72–75]. Instead of requiring thermal energy, as in thermal CVD, the energetic electrons in the plasma (at a pressure of $1–5 \times 10^2$ Pa, and typically less than 10 Pa) can activate almost any reaction among the gases in the glow discharge at relatively a low substrate temperature of 100 to 600 °C (typically less than 300 °C). To deposit the coating on non-silicon substrates, an amorphous silicon adhesion layer about 4 nm-thick is first deposited under similar conditions from a gas mixture of 1% silane in argon in order to improve adhesion [76]. In the process used by Bhushan and coworkers [12], the plasma is sustained in a parallel-plate geometry by a capacitive discharge at 13.56 MHz, at a surface power density of around 100 mW/cm^2. The deposition is performed at a flow rate on the order of 6 sccm and a pressure on the order of 4 Pa (30 mTorr) on a cathode-mounted substrate maintained at a substrate temperature of 180 °C. The cathode bias is held fixed at about −120 V with an external DC power supply attached to the substrate (powered electrode). The carbon coatings deposited by PECVD usually contain hydrogen at levels of up to 50% [35, 77].

16.3 Chemical and Physical Coating Characterization

The chemical structures and properties of amorphous carbon coatings depend on the deposition conditions employed when they are formed. It is important to understand the relationship between the chemical structure of a coating and its properties since it allows useful deposition parameters to be defined. Amorphous carbon films are metastable phases formed when carbon particles are condensed on a substrate. The prevailing atomic arrangement in the DLC coatings is amorphous or quasi-amorphous, with small diamond (sp^3), graphite (sp^2) and other unidentifiable micro- or nanocrystallites. The coating is dependent upon the deposition process and the deposition conditions used because these influence the sp^3/sp^2 ratio and the proportion of hydrogen in the coating. The sp^3/sp^2 ratios of DLC coatings typically range from 50% to close to 100%, and hardness increases with the sp^3/sp^2 ratio. Most DLC coatings, except those produced by a filtered cathodic arc, contain from a few to about 50 at. % hydrogen. Sometimes hydrogen and nitrogen are deliberately added to produce hydrogenated (a-C:H) and nitrogenated amorphous carbon (a-C:N) coatings, respectively. Hydrogen helps to stabilize sp^3 sites (most of the carbon atoms attached to hydrogen have a tetrahedral structure), so the sp^3/sp^2 ratio for hydrogenated carbon is higher [30]. The optimum sp^3/sp^2 ratio for a random covalent network composed of sp^3 and sp^2 carbon sites (N_{sp^2} and N_{sp^3}) and hydrogen is [30]:

$$\frac{N_{sp^3}}{N_{sp^2}} = \frac{6X_H - 1}{8 - 13X_H}, \tag{16.1}$$

where X_H is the atomic fraction of hydrogen. The hydrogenated carbon has a larger optical band gap, higher electrical resistivity (semiconductor), and a lower optical absorption or high optical transmission. Hydrogenated coatings have lower densities, probably because of the reduced cross-linking due to hydrogen incorporation. However, the hardness decreases with increasing hydrogen, even though the proportion of sp^3 sites increases (that is, as the local bonding environment becomes more diamond-like) [78, 79]. It is speculated that the high hydrogen content introduces frequent terminations in the otherwise strong 3-D network, and hydrogen increases the soft polymeric component of the structure more than it enhances the cross-linking sp^3 fraction.

A number of investigations have been performed to identify the microstructure of amorphous carbon films using a variety of techniques, such as Raman spectroscopy, EELS, nuclear magnetic resonance, optical measurements, transmission electron microscopy, and X-ray photoelectron spectroscopy [33]. The structure of diamond-like amorphous carbon is amorphous or quasi-amorphous, with small graphitic (sp^2), tetrahedrally coordinated (sp^3) and other types of nanocrystallites (typically on the order of a couple of nm in size, randomly oriented) [33, 80, 81]. These studies indicate that the chemical structure and physical properties of the coatings are quite variable, depending on the deposition techniques and film growth conditions. It is clear that both sp^2- and sp^3-bonded atomic sites are incorporated in diamond-like amorphous carbon coatings and that the physical and chemical properties of the coatings depend strongly on their chemical bonding and microstructures. Systematic studies have been conducted to carry out chemical characterization and to investigate how the physical and chemical properties of amorphous carbon coatings vary as a function of deposition conditions (see [33, 35, 40]). EELS and Raman spectroscopy are commonly used to characterize the chemical bonding and microstructure. The hydrogen concentration in the coating is obtained via forward recoil spectrometry (FRS). A variety of physical properties relevant to tribological performance are measured.

In order to give the reader a feel for typical data obtained when characterizing amorphous carbon coatings and their relationships to physical properties, we present data on several sputtered coatings, RF-PECVD amorphous carbon and microwave-PECVD (MPECVD) diamond coatings [33, 35, 40]. The sputtered coatings were DC magnetron sputtered at a chamber pressure of 10 mTorr under sputtering power densities of 0.1 and 2.1 W/cm^2 (labeled as coatings W1 and W2, respectively) in a pure Ar plasma. These coatings were prepared at a power density of 2.1 W/cm^2 with various hydrogen fractions (0.5, 1, 3, 5, 7 and 10%) of Ar/H; the gas mixtures were labeled as H1, H2, H3, H4, H5, and H6, respectively.

16.3.1 EELS and Raman Spectroscopy

EELS and Raman spectra of four sputtered (W1, W2, H1, and H3) carbon samples and one PECVD carbon sample were obtained. Figure 16.5 shows the EELS spectra of these carbon coatings. EELS spectra (up to 50 eV) for bulk diamond and polycrystalline graphite are also shown in Fig. 16.5. One prominent peak is seen at 35 eV in diamond, while two peaks are seen at 27 eV and 6.5 eV in graphite, which are called the ($\pi + \sigma$) and (π) peaks, respectively. These peaks are produced by the loss of transmitted electron energy to plasmon oscillations of the valence electrons. The $\pi + \sigma$ peak in each coating is positioned at a lower energy region than that of graphite. The π peaks in the W series and PECVD samples also occur at a lower energy region than that of the graphite. However, the π peaks in the H-series are comparable to or higher than those of graphite (see Table 16.2). The plasmon oscillation frequency is proportional to the square root of the corresponding electron density to a first approximation. Therefore, the samples in the H-series most likely have a higher density of π electrons than the other samples.

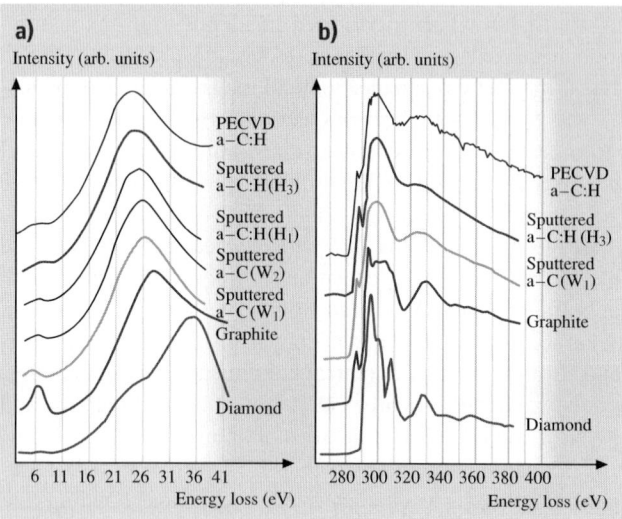

Fig. 16.5. (a) Low-energy and (b) high-energy EELS of DLC coatings produced by the DC magnetron sputtering and RF-PECVD techniques. Data for bulk diamond and polycrystalline graphite are included for comparison [35]

Amorphous carbon coatings contain (mainly) a mixture of sp^2- and sp^3-bonds, even though there is some evidence for the presence of sp-bonds as well [82]. The PECVD coatings and the H-series coatings in this study have almost the same mass density (as seen in Table 16.4, discussed in more detail later), but the former have a lower concentration of hydrogen (18.1%) than the H-series (35–39%) (as seen in Table 16.3, also discussed in more detail later). The relatively low-energy positions of the π peaks of the PECVD coatings compared to those of the H-series indicates that the PECVD coatings contain a higher fraction of sp^3-bonds than the sputtered hydrogenated carbon coatings (H-series).

Figure 16.5b shows EELS spectra associated with the inner-shell (K-shell) ionization. Again, the spectra for diamond and polycrystalline graphite are included for comparison. Sharp peaks are observed at 285.5 eV and 292.5 eV in graphite, while no peak is seen at 285.5 eV in diamond. The general features of the K-shell EELS spectra for the sputtered and PECVD carbon samples resemble those of graphite, but with the higher energy features smeared. The peak at 285.5 eV in the sputtered and PECVD coatings also indicates the presence of sp^2-bonded atomic sites in the coatings. All of these spectra peak at 292.5 eV, similar to the spectra of graphite, but the peak in graphite is sharper.

Raman spectra from samples W1, W2, H1 and PECVD are shown in Fig. 16.6. Raman spectra could not be observed in specimens H2 and H3 due to high fluorescence signals. The Raman spectra of single-crystal diamond and polycrystalline graphite are also shown for comparison in Fig. 16.6. The results from the spectral fits are summarized in Table 16.2. We will focus on the position of the G-band,

Table 16.2. Experimental results from EELS and Raman spectroscopy [35]

Sample	EELS peak position π (eV)	$\pi+\sigma$ (eV)	Raman peak position G-band[b] (cm^{-1})	D-band[c] (cm^{-1})	Raman FWHM[a] G-band (cm^{-1})	D-band (cm^{-1})	I_D/I_G^d
Sputtered a-C coating (W1)	5.0	24.6	1541	1368	105	254	2.0
Sputtered a-C coating (W2)	6.1	24.7	1560	1379	147	394	5.3
Sputtered a-C:H coating (H1)	6.3	23.3	1542	1334	95	187	1.6
Sputtered a-C:H coating (H3)	6.7	22.4	e	e	e	e	e
PECVD a-C:H coating	5.8	24.0	1533	1341	157	427	1.5
Diamond coating	...	...	1525[f]	1333[g]	...	8[g]	...
Graphite (for reference)	6.4	27.0	1580	1358	37	47	0.7
Diamond (for reference)	...	37.0	...	1332[g]	...	2[g]	...

[a] Full width at half maximum
[b] Peak associated with sp^2 "graphite" carbon
[c] Peak associated with sp^2 "disordered" carbon (not sp^3-bonded carbon)
[d] Intensity ratio of the D-band to the G-band
[e] Fluorescence
[f] Includes D and G band, signal too weak to analyze
[g] Peak position and width for diamond phonon

which has been shown to be related to the fraction of sp^3-bonded sites. Increasing the power density in the amorphous carbon coatings (W1 and W2) results in a higher G-band frequency, implying a smaller fraction of sp^3-bonding in W2 than in W1. This is consistent with the higher density of W1. H1 and PEVCD have even lower G-band positions than W1, implying an even higher fraction of sp^3-bonding, which is presumably caused by the incorporation of H atoms into the lattice. The high hardness of H3 might be attributed to efficient sp^3 cross-linking of small sp^2-ordered domains.

The Raman spectrum of a MPECVD diamond coating is shown in Fig. 16.6. The Raman peak of diamond is at 1333 cm^{-1}, with a line width of 7.9 cm^{-1}. There is a small broad peak at around 1525 cm^{-1}, which is attributed to a small amount of a-C:H. This impurity peak is not intense enough to be able to separate the G- and D-bands. The diamond peak frequency is very close to that of natural diamond (1332.5 cm^{-1}, see Fig. 16.6), indicating that the coating is not under stress [83].

Table 16.3. Experimental results of FRS analysis [35]

Sample	Ar/H ratio	C (at.% ±0.5)	H (at.% ±0.5)	Ar (at.% ±0.5)	O (at.% ±0.5)
Sputtered a-C coating (W2)	100/ 0	90.5	9.3	0.2	...
Sputtered a-C:H coating (H2)	99/ 1	63.9	35.5	0.6	...
Sputtered a-C:H coating (H3)	97/ 3	56.1	36.5	...	7.4
Sputtered a-C:H coating (H4)	95/ 5	53.4	39.4	...	7.2
Sputtered a-C:H coating (H5)	93/ 7	58.2	35.4	0.2	6.2
Sputtered a-C:H coating (H6)	90/10	57.3	35.5	...	7.2
PECVD a-C:H coating	99.5% CH_4	81.9	18.1	...	...
Diamond coating	H_2-1 mole % CH_4	94.0	6.0	...	...

Table 16.4. Experimental results of physical properties [35]

Sample	Mass density (g/cm^3)	Nano-hardness (GPa)	Elastic modulus (GPa)	Electrical resistivity (Ohm-cm)	Compressive residual stress (GPa)
Sputtered a-C coating (W1)	2.1	15	141	1300	0.55
Sputtered a-C:H coating (W2)	1.8	14	136	0.61	0.57
Sputtered a-C:H coating (H1)	...	14	96	...	> 2
Sputtered a-C:H coating (H3)	1.7	7	35	> 10^6	0.3
PECVD a-C:H coating	1.6–1.8	33–35	~ 200	> 10^6	1.5–3.0
Diamond coating	...	40–75	370–430	...	...
Graphite (for reference)	2.267	Soft	9–15	5×10^{-5}[a], 4×10^{-3}[b]	0
Diamond (for reference)	3.515	70–102	900–1050	10^7–10^{20}	0

[a] Parallel to layer planes
[b] Perpendicular to layer planes

The large line width compared to that of natural diamond (2 cm^{-1}) indicates that the microcrystallites probably have a high concentration of defects [84].

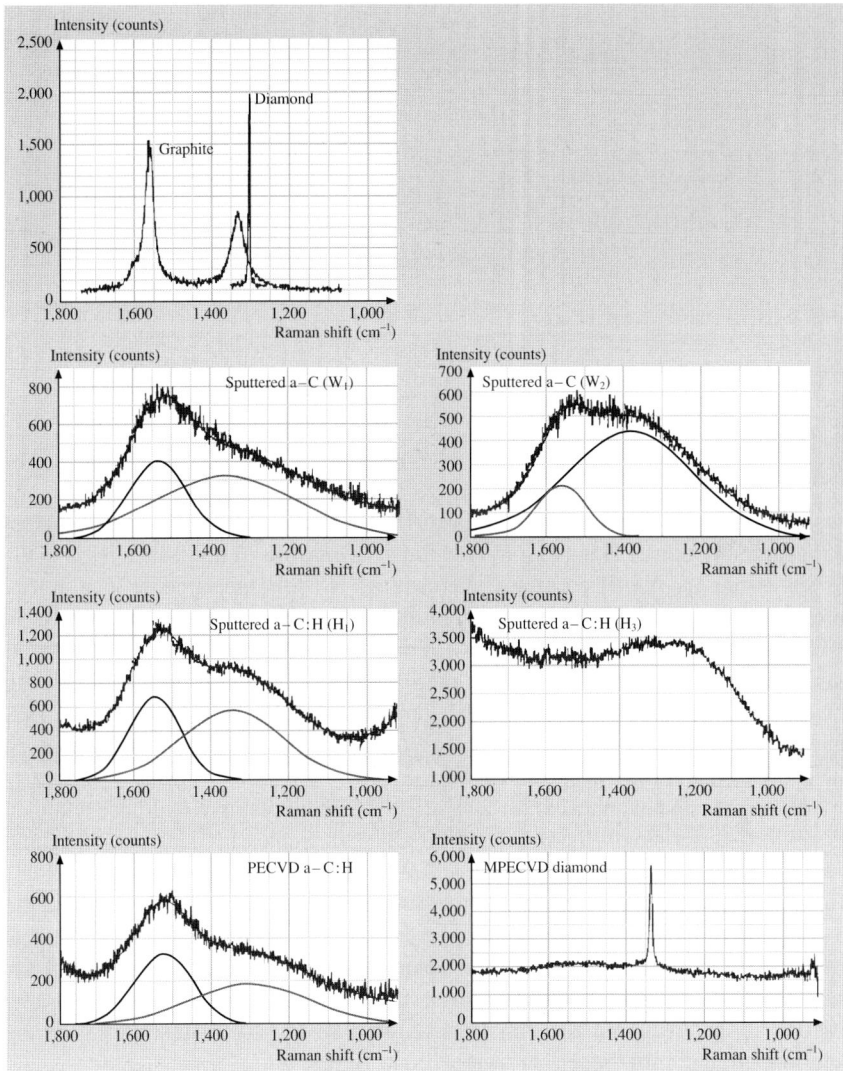

Fig. 16.6. Raman spectra of the DLC coatings produced by DC magnetron sputtering and RF-PECVD techniques and a diamond film produced by the MPE-CVD technique. Data for bulk diamond and microcrystalline graphite are included for comparison [35]

16.3.2 Hydrogen Concentrations

A FRS analysis of six sputtered (W2, H2, H3, H4, H5, and H6) coatings, one PECVD coating, and one diamond coating was performed. Figure 16.7 shows an overlay of the spectra from the six sputtered samples. Similar spectra were obtained from the PECVD and the diamond films. Table 16.3 shows the H and C fractions as

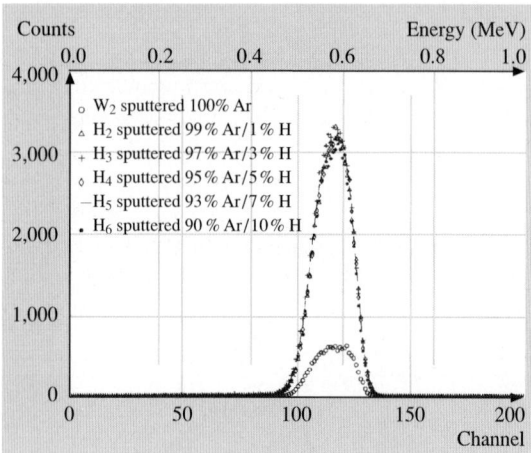

Fig. 16.7. FRS spectra for six DLC coatings produced by DC magnetron sputtering [35]

well as the amount of impurities (Ar and O) in the films in atomic %. Most apparent is the large fraction of H in the sputtered films. Regardless of how much H_2 is in the Ar sputtering gas, the H content of the coatings is about the same, ≈ 35 at. %. Interestingly, there is still $\approx 10\%$ H present in the coating sputtered in pure Ar (W2). It is interesting to note that Ar is present only in coatings grown using Ar carrier gas with a low ($< 1\%$) H content. The presence of O in the coatings combined with the fact that the coatings were prepared approximately nine months before the FRS analysis caused suspicion that they had absorbed water vapor, and that this may be the cause of the H peak in specimen W2.

All samples were annealed for 24 h at 250 °C in a flowing He furnace and then reanalyzed. Surprisingly, the H contents of all coatings measured increased slightly, even though the O content decreased, and W2 still had a substantial amount of H_2. This slight increase in H concentration is not understood. However, the fact that the H concentration did not decrease with the oxygen as a result of annealing suggests that high H concentration is not due to adsorbed water vapor. The PECVD film has more H ($\approx 18\%$) than the sputtered films initially, but after annealing it has the same fraction as specimen W2, the film sputtered in pure Ar. The diamond film has the smallest amount of hydrogen, as seen in Table 16.3.

16.3.3 Physical Properties

The physical properties of the four sputtered (W1, W2, H1, and H3) coatings, one PECVD coating, one diamond coating, and bulk diamond and graphite are presented in Table 16.4. The hydrogenated carbon and the diamond coatings have very high resistivity compared to unhydrogenated carbon coatings. It appears that unhydrogenated carbon coatings have higher densities than hydrogenated carbon coatings, although both groups are less dense than graphite. The density depends upon the deposition technique and the deposition parameters. It appears that unhydrogenated sputtered coatings deposited at low power exhibit the highest density. The nanohardness of hydrogenated carbon is somewhat lower than that of unhydrogenated carbon. PECVD coatings are significantly harder than sputtered coatings. The nanohardness and modulus of elasticity of a diamond coating is very high compared to that of a DLC coating, even though the hydrogen contents are similar. The compressive residual stresses of the PECVD coatings are substantially higher than those of sputtered coatings, which is consistent with the results for the hardness.

Figure 16.8a shows the effect of hydrogen in the plasma on the residual stresses and the nanohardness for the sputtered coatings W2 and H1 to H6. The coatings made with a hydrogen flow of between 0.5 and 1.0% delaminate very quickly, even when they are only a few tens of nm thick. In pure Ar and at H_2 flows that are greater than 1%, the coatings appear to be more adhesive. The tendency of some coatings to delaminate can be caused by intrinsic stress in the coating, which is measured by substrate bending. All of the coatings in the figure are in compressive stress. The maximum stress occurs between 0 and 1% H_2 flow, but the stress cannot be quantified in this range because the coatings instantly delaminate upon exposure to air. At higher hydrogen concentrations the stress gradually diminishes. A generally decreasing trend is observed for the hardness of the coatings as the hydrogen content is increased. The hardness decreases slightly, going from 0% H_2 to 0.5% H_2, and then decreases sharply. These results are probably lower than the true values because of local delamination around the indentation point. This is especially likely for the

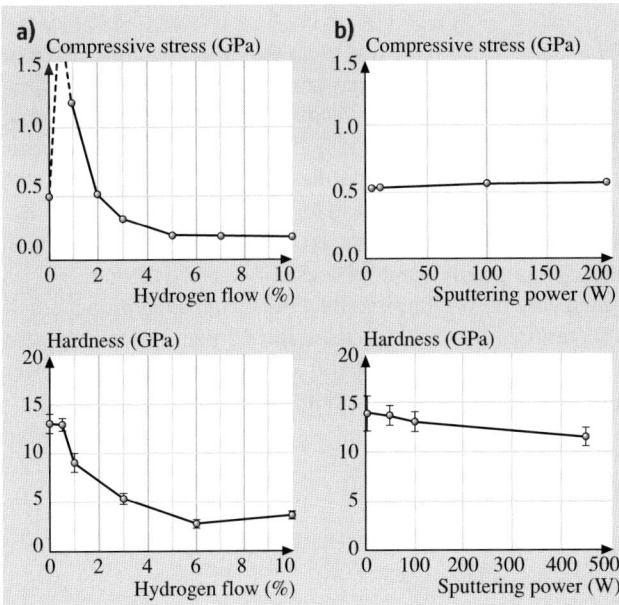

Fig. 16.8. Residual compressive stresses and nanohardness (**a**) as a function of hydrogen flow rate, where the sputtering power is 100 W and the target diameter is 75 mm (power density = 2.1 W/cm^2), and (**b**) as a function of sputtering power over a 75 mm diameter target with no hydrogen added to the plasma [40]

0.5% and 1.0% coatings, where delamination is visually apparent, but may also be true to a lesser extent for the other coatings. Such an adjustment would bring the hardness profile into closer correlation with the stress profile. *Weissmantel* et al. [68] and *Scharff* et al. [85] observed a downturn in hardness for high bias and a low hydrocarbon gas pressure for ion-plated carbon coating, and, therefore, presumably low hydrogen content in support of the above contention.

Figure 16.8b shows the effect of sputtering power (with no hydrogen added to the plasma) on the residual stresses and nanohardness for various sputtered coatings. As the power decreases, the compressive stress does not seem to change while the nanohardness slowly increases. The rate of change becomes more rapid at very low power levels.

The addition of H_2 during sputtering of the carbon coatings increases the H concentration in the coating. Hydrogen causes the character of the C–C bonds to shift from sp^2 to sp^3, and a rise in the number of C–H bonds, which ultimately relieves stress and produces a softer "polymer-like" material. Low power deposition, like the presence of hydrogen, appears to stabilize the formation of sp^3 C–C bonds, increasing hardness. These coatings relieve stress and lead to better adhesion. Increasing the temperature during deposition at high power density results in graphitization of the coating material, producing a decrease in hardness with an increase in power density. Unfortunately, low power also means impractically low deposition rates.

16.3.4 Summary

Based on analyses of EELS and Raman data, it is clear that all DLC coatings have both sp^2 and sp^3 bonding characteristics. The sp^2/sp^3 bonding ratio depends upon the deposition technique and parameters used. DLC coatings deposited by sputtering and PECVD contain significant concentrations of hydrogen, while diamond coatings contain only small amounts of hydrogen impurities. Sputtered coatings with no deliberate addition of hydrogen in the plasma contain a significant amount of hydrogen. Regardless of how much hydrogen is in the Ar sputtering gas, the hydrogen content of the coatings increases initially but then does not increase further.

Hydrogen flow and sputtering power density affect the mechanical properties of these coatings. Maximum compressive residual stress and hardness occur between 0 and 1% hydrogen flow, resulting in rapid delamination. Low sputtering power moderately increases hardness and also relieves residual stress.

16.4 Micromechanical and Tribological Coating Characterization

16.4.1 Micromechanical Characterization

Common mechanical characterizations include measurements of hardness and elastic modulus, fracture toughness, fatigue life, and scratch and wear testing. Nanoindentation and atomic force microscopy (AFM) are used for the mechanical characterization of ultrathin films.

Hardness and elastic modulus are calculated from the load displacement data obtained by nanoindentation at loads of typically 0.2 to 10 mN using a commercially available nanoindenter [23, 86]. This instrument monitors and records the dynamic load and displacement of the three-sided pyramidal diamond (Berkovich) indenter during indentation. For fracture toughness measurements of ultrathin films 100 nm to a few μm thick, a nanoindentation-based technique is used in which through-thickness cracking in the coating is detected from a discontinuity observed in the load–displacement curve, and the energy released during the cracking is obtained from the curve [87–89]. Based on the energy released, fracture mechanics analysis is then used to calculate the fracture toughness. An indenter with a cube-corner tip geometry is preferred because the through-thickness cracking of hard films can be accomplished at lower loads. In fatigue measurement, a conical diamond indenter with a tip radius of about one micron is used and load cycles with sinusoidal shapes are applied [90, 91]. The fatigue behavior of a coating is studied by monitoring the change in contact stiffness, which is sensitive to damage formation.

Hardness and Elastic Modulus

For materials that undergo plastic deformation, high hardness and elastic modulus are generally needed for low friction and wear, whereas for brittle materials, high fracture toughness is needed [2, 3, 21]. The DLC coatings used for many applications are hard and brittle, and values of hardness and fracture toughness need to be optimized.

Representative load–displacement plots of indentations made at 0.2 mN peak indentation load on 100 nm-thick DLC coatings deposited by the four deposition techniques on a single-crystal silicon substrate are compared in Fig. 16.9. The indentation depths at the peak load range from about 18 to 26 nm, smaller than that of the coating thickness. Many of the coatings exhibit a discontinuity or pop-in marks in the loading curve, which indicate a sudden penetration of the tip into the sample. A nonuniform penetration of the tip into a thin coating possibly results from formation of cracks in the coating, formation of cracks at the coating–substrate interface, or debonding or delamination of the coating from the substrate.

The hardness and elastic modulus values for a peak load of 0.2 mN on the various coatings and single-crystal silicon substrate are summarized in Table 16.5 and Fig. 16.10 [47,49,89,90]. Typical values for the peak and residual indentation depths range from 18 to 26 nm and 6 to 12 nm, respectively. The FCA coating exhibits the greatest hardness of 24 GPa and the highest elastic modulus of 280 GPa of the various coatings, followed by the ECR-CVD, IB and SP coatings. Hardness and elastic modulus have been known to vary over a wide range with the sp^3-to-sp^2 bonding ratio, which depends on the kinetic energy of the carbon species and the amount of hydrogen [6, 30, 47, 92, 93]. The high hardness and elastic modulus of the FCA coatings are attributed to the high kinetic energy of the carbon species involved in the FCA deposition [12,47]. *Anders* et al. [57] also reported a high hardness, measured by nanoindentation, of about 45 GPa for cathodic arc carbon coatings. They observed a change in hardness from 25 to 45 GPa with a pulsed bias voltage and

Table 16.5. Hardness, elastic modulus, fracture toughness, fatigue life, critical load during scratch, coefficient of friction during accelerated wear testing and residual stress for various DLC coatings on single-crystal silicon substrate

Coating	Hardness[a] [48] (GPa)	Elastic modulus[a] [48] (GPa)	Fracture toughness[a] [89] (MPa m$^{1/2}$)	Fatigue life[b], N_f^d [90] ×10^4	Critical load during scratch[b] [48] (mN)	Coefficient of friction during accelerated wear testing[b] [48]	Compressive residual stress[c] [47] (GPa)
Cathodic arc carbon coating (a-C)	24	280	11.8	2.0	3.8	0.18	12.5
Ion beam carbon coating (a-C:H)	19	140	4.3	0.8	2.3	0.18	1.5
ECR-CVD carbon coating (a-C:H)	22	180	6.4	1.2	5.3	0.22	0.6
DC sputtered carbon coating (a-C:H)	15	140	2.8	0.2	1.1	0.32	2.0
Bulk graphite (for comparison)	Very soft	9–15	–	–	–	–	–
Diamond (for comparison)	80–104	900–1050	–	–	–	–	–
Si(100) substrate	11	220	0.75	–	0.6	0.55	0.02

[a] Measured on 100 nm-thick coatings
[b] Measured on 20 nm-thick coatings
[c] Measured on 400 nm-thick coatings
[d] N_f was obtained for a mean load of 10 µN and a load amplitude of 8 µN

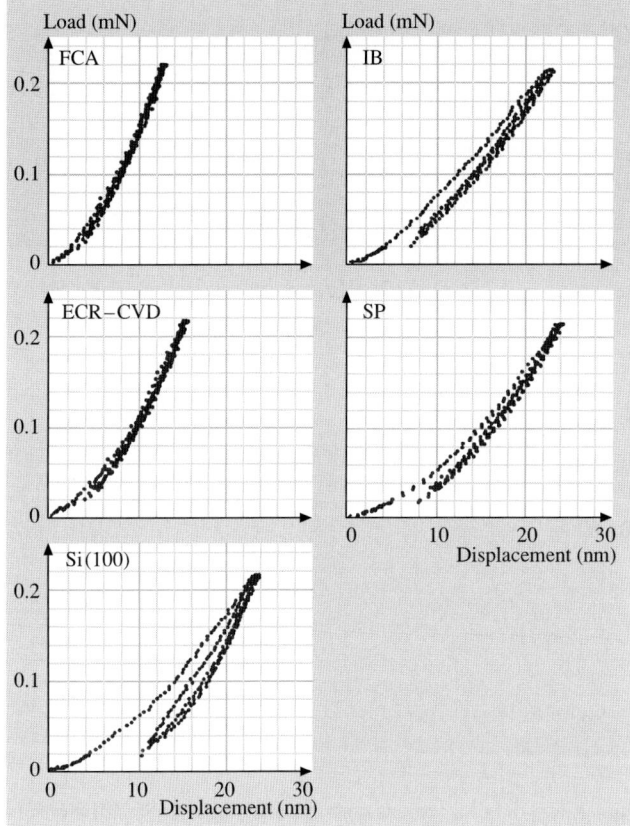

Fig. 16.9. Load versus displacement plots for various 100 nm-thick amorphous carbon coatings on single-crystal silicon substrate and bare substrate

bias duty cycle. The high hardness of cathodic arc carbon was attributed to the high percentage (more than 50%) of sp^3 bonding. *Savvides* and *Bell* [94] reported an increase in hardness from 12 to 30 GPa and an increase in elastic modulus from 62 to 213 GPa with an increase in the sp^3-to-sp^2 bonding ratio, from 3 to 6, for a C:H coating deposited by low-energy ion-assisted unbalanced magnetron sputtering of a graphite target in an Ar-H_2 mixture.

Bhushan et al. [35] reported hardnesses of about 15 and 35 GPa and elastic moduli of about 140 and 200 GPa, measured by nanoindentation, for a-C:H coatings deposited by DC magnetron sputtering and RF-plasma-enhanced chemical vapor deposition techniques, respectively. The high hardness of RF-PECVD a-C:H coatings is attributed to a higher concentration of sp^3 bonding than in a sputtered hydrogenated a-C:H coating. Hydrogen is believed to play a crucial role in the bonding configuration of carbon atoms by helping to stabilize the tetrahedral coordination (sp^3 bonding) of carbon species. *Jansen* et al. [78] suggested that the incorpora-

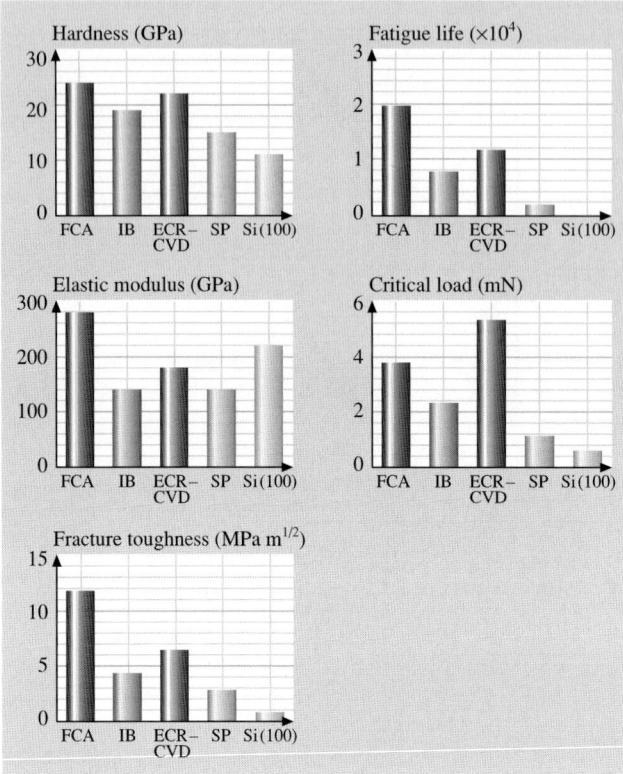

Fig. 16.10. Bar charts summarizing data for various coatings and single-crystal silicon substrate. Hardnesses, elastic moduli, and fracture toughnesses were measured on 100 nm-thick coatings, and fatigue lifetimes and critical loads during scratch were measured on 20 nm-thick coatings

tion of hydrogen efficiently passivates the dangling bonds and saturates the graphitic bonding to some extent. However, a large concentration of hydrogen in the plasma in sputter deposition is undesirable. *Cho* et al. [33] and *Rubin* et al. [40] observed that the hardness decreased from 15 to 3 GPa with increased hydrogen content. *Bhushan* and *Doerner* [95] reported a hardness of about 10–20 GPa and an elastic modulus of about 170 GPa, measured by nanoindentation, for 100 nm-thick DC magnetron sputtered a-C:H on the silicon substrate.

Residual stresses measured using a well-known curvature measurement technique are also presented in Table 16.5. The DLC coatings are under significant compressive internal stresses. Very high compressive stresses in FCA coatings are believed to be partly responsible for their high hardness. However, high stresses result in coating delamination and buckling. For this reason, the coatings that are thicker than about 1 µm have a tendency to delaminate from the substrates.

Fracture Toughness

Representative load–displacement curves of indentations on 400 nm-thick cathodic arc carbon coating on silicon for various peak loads are shown in Fig. 16.11. Steps are found in all of the curves, as shown by arrows in Fig. 16.11a. In the 30 mN SEM micrograph, in addition to several radial cracks, ring-like through-thickness cracking is observed with small lips of material overhanging the edge of indentation. The steps at about 23 mN in the loading curves of indentations made with 30 and 100 mN peak indentation loads result from the ring-like through-thickness cracking. The step at 175 mN in the loading curve of the indentation made with 200 mN peak indentation load is caused by spalling and second ring-like through-thickness cracking.

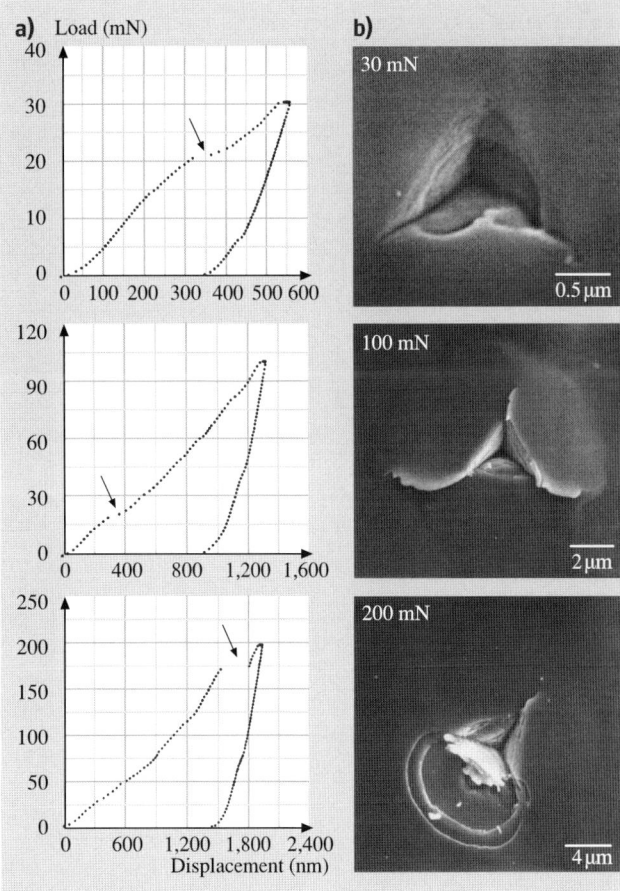

Fig. 16.11. (**a**) Load–displacement curves of indentations made with 30, 100, and 200 mN peak indentation loads using the cube corner indenter, and (**b**) SEM micrographs of indentations on a 400 nm-thick cathodic arc carbon coating on silicon. *Arrows* indicate steps during the loading portion of the load–displacement curve [87]

According to Li et al. [87], the fracture process progresses in three stages: (1) ring-like through-thickness cracks form around the indenter due to high stresses in the contact area, (2) delamination and buckling occur around the contact area at the coating/substrate interface due to high lateral pressure, and (3) second ring-like through-thickness cracks and spalling are generated by high bending stresses at the edges of the buckled coating, see Fig. 16.12a. In the first stage, if the coating

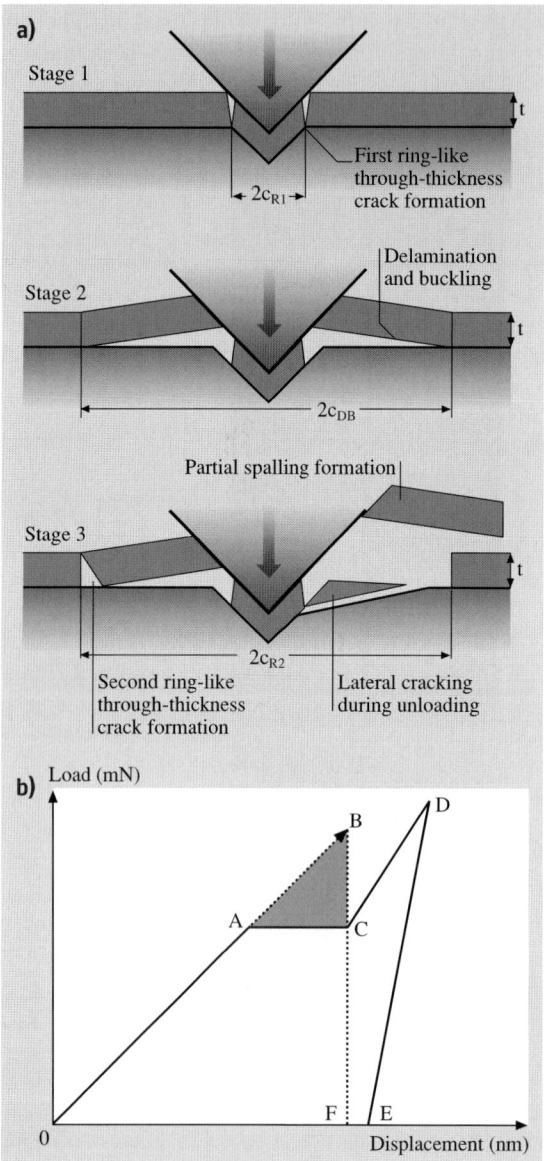

Fig. 16.12. (a) Schematic of various stages in nanoindentation fracture for the coating/substrate system, and (b) schematic of a load–displacement curve showing a step during the loading cycle and the associated energy release

under the indenter is separated from the bulk coating via the first ring-like through-thickness cracking, a corresponding step will be present in the loading curve. If discontinuous cracks form and the coating under the indenter is not separated from the remaining coating, no step appears in the loading curve, because the coating still supports the indenter and the indenter cannot suddenly advance into the material. In the second stage, for the coating used in the present study, the advances of the indenter during the radial cracking delamination, and buckling are not big enough to form steps in the loading curve, because the coating around the indenter still supports the indenter, but they generate discontinuities that change the slope of the loading curve with increasing indentation load. In the third stage, the stress concentration at the end of the interfacial crack cannot be relaxed by the propagation of the interfacial crack. With an increase in indentation depth, the height of the bulged coating increases. When the height reaches a critical value, the bending stresses caused by the bulged coating around the indenter will result in second ring-like through-thickness crack formation and spalling at the edge of the buckled coating, as shown in Fig. 16.12a, which leads to a step in the loading curve. This is a single event and it results in the separation of the part of the coating around the indenter from the bulk coating via cracking through coatings. The step in the loading curve results (completely) from the coating cracking and not from the interfacial cracking or the substrate cracking.

The area under the load–displacement curve is the work performed by the indenter during the elastic–plastic deformation of the coating/substrate system. The strain energy release in the first/second ring-like cracking and spalling can be calculated from the corresponding steps in the loading curve. Figure 16.12b shows a modeled load–displacement curve. OACD is the loading curve and DE is the unloading curve. The first ring-like through-thickness crack should be considered. It should be emphasized that the edge of the buckled coating is far from the indenter and, therefore, it does not matter if the indentation depth exceeds the coating thickness, or if deformation of the substrate occurs around the indenter when we measure the fracture toughness of the coating from the energy released during the second ring-like through-thickness cracking (spalling). Suppose that the second ring-like through-thickness cracking occurs at AC. Now, let us consider the loading curve OAC. If the second ring-like through-thickness crack does not occur, OA will extend to OB to reach the same displacement as OC. This means that crack formation changes the loading curve OAB into OAC. For point B, the elastic–plastic energy stored in the coating/substrate system should be OBF. For point C, the elastic–plastic energy stored in the coating/substrate system should be OACF. Therefore, the energy difference before and after the crack generation is the area of ABC, so this energy stored in ABC will be released as strain energy, creating the ring-like through-thickness crack. According to the theoretical analysis by Li et al. [87], the fracture toughness of a thin film can be written as

$$K_{Ic} = \left[\left(\frac{E}{(1-v^2)2\pi C_R}\right)\left(\frac{U}{t}\right)\right]^{1/2}, \qquad (16.2)$$

where E is the elastic modulus, v is the Poisson ratio, $2\pi C_R$ is the crack length in the coating plane, t is the coating thickness, and U is the strain energy difference before and after cracking.

The fracture toughness of the coatings can be calculated using (16.2). The loading curve is extrapolated along the tangential direction of the loading curve from the starting point of the step up to reach the same displacement as the step. The area between the extrapolated line and the step is the estimated strain energy difference before and after cracking. C_R is measured from SEM micrographs or AFM images of indentations. The second ring-like crack is where the spalling occurs. For example, for the 400 nm-thick cathodic arc carbon coating data presented in Fig. 16.11, the U value of 7.1 nNm is assessed from the steps in Fig. 16.11a at peak indentation loads of 200 mN. For a C_R value of 7.0 μm, from Fig. 16.11b, with $E = 300$ GPa (measured using a nanoindenter and an assumed value of 0.25 for v), fracture toughness values are calculated as 10.9 MPa $\sqrt{m}$ [87, 88]. The fracture toughness and related data for various 100 nm-thick DLC coatings are presented in Fig. 16.10 and Table 16.5.

Nanofatigue

Delayed fracture resulting from extended service is called fatigue [96]. Fatigue fracturing progresses through a material via changes within the material at the tip of a crack, where there is a high stress intensity. There are several situations: cyclic fatigue, stress corrosion and static fatigue. Cyclic fatigue results from cyclic loading of machine components. In a low-flying slider in a magnetic head-disk interface, isolated asperity contacts occur during use and the fatigue failure occurs in the multilayered thin film structure of the magnetic disk [13]. Impact occurs in many MEMS components and the failure mode is cyclic fatigue. Asperity contacts can be simulated using a sharp diamond tip in oscillating contact with the component.

Figure 16.13 shows the schematic of a fatigue test on a coating/substrate system using a continuous stiffness measurement (CSM) technique. Load cycles are applied to the coating, resulting in cyclic stress. P is the cyclic load, P_{mean} is the mean load, P_o is the oscillation load amplitude, and ω is the oscillation frequency.

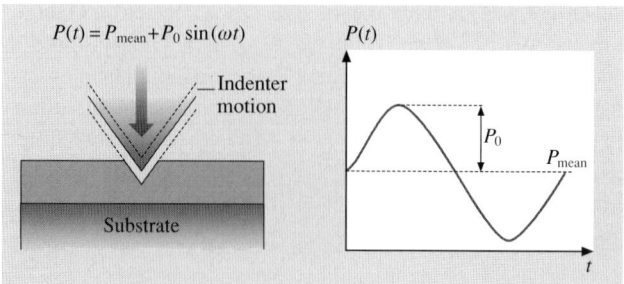

Fig. 16.13. Schematic of a fatigue test on a coating/substrate system using the continuous stiffness measurement technique

The following results can be obtained: (1) endurance limit (the maximum load below which there is no coating failure for a preset number of cycles); (2) number of cycles at which the coating failure occurs; and (3) changes in contact stiffness (measured using the unloading slope of each cycle), which can be used to monitor the propagation of interfacial cracks during a cyclic fatigue process.

Figure 16.14a shows the contact stiffness as a function of the number of cycles for 20 nm-thick FCA coatings cyclically deformed by various oscillation load amplitudes with a mean load of 10 μN at a frequency of 45 Hz. At 4 μN load amplitude, no change in contact stiffness was found for all coatings. This indicates that 4 μN load amplitude is not high enough to damage the coatings. At 6 μN load amplitude, an abrupt decrease in contact stiffness was found after a certain number of cycles for each coating, indicating that fatigue damage had occurred. With increasing load amplitude, the number of cycles to failure, N_f, decreases for all coatings. Load amplitude versus N_f, a so-called S–N curve, is plotted in Fig. 16.14b. The critical load amplitude below which no fatigue damage occurs (an endurance limit), was identified for each coating. This critical load amplitude, together with the mean load, are of critical importance to the design of head-disk interfaces or MEMS/NEMS device interfaces.

To compare the fatigue lives of the different coatings studied, the contact stiffness is shown as a function of the number of cycles for 20 nm-thick FCA, IB, ECR-CVD and SP coatings cyclically deformed by an oscillation load amplitude of 8 μN with a mean load of 10 μN at a frequency of 45 Hz in Fig. 16.14c. The FCA coating has the largest N_f, followed by the ECR-CVD, IB and SP coatings. In addition, after N_f, the contact stiffness of the FCA coating shows a slower decrease than the other coatings. This indicates that the FCA coating was less damaged than the others after N_f. The fatigue behaviors of FCA and ECR-CVD coatings of different thicknesses are compared in Fig. 16.14d. For both coatings, N_f decreases with decreasing coating thickness. At 10 nm, FCA and ECR-CVD have almost the same fatigue life. At 5 nm, the ECR-CVD coating shows a slightly longer fatigue life than the FCA coating. This indicates that the microstructure and residual stresses are not uniform across the thickness direction, even for nanometer-thick DLC coatings. Thinner coatings are more influenced by interfacial stresses than thicker coatings.

Figure 16.15a shows high-magnification SEM images of 20 nm-thick FCA coatings before, at, and after N_f. In the SEM images, the net-like structure is the gold film coated on the DLC coating, which should be ignored when analyzing the indentation fatigue damage. Before N_f, no delamination or buckling was found except for the residual indentation mark at magnifications of up to 1,200,000× using SEM. This suggests that only plastic deformation occurred before N_f. At N_f, the coating around the indenter bulged upwards, indicating delamination and buckling. Therefore, it is believed that the decrease in contact stiffness at N_f results from delamination and buckling of the coating from the substrate. After N_f, the buckled coating was broken down around the edge of the buckled area, forming a ring-like crack. The remaining coating overhung at the edge of the buckled area. It is noted that the indentation size

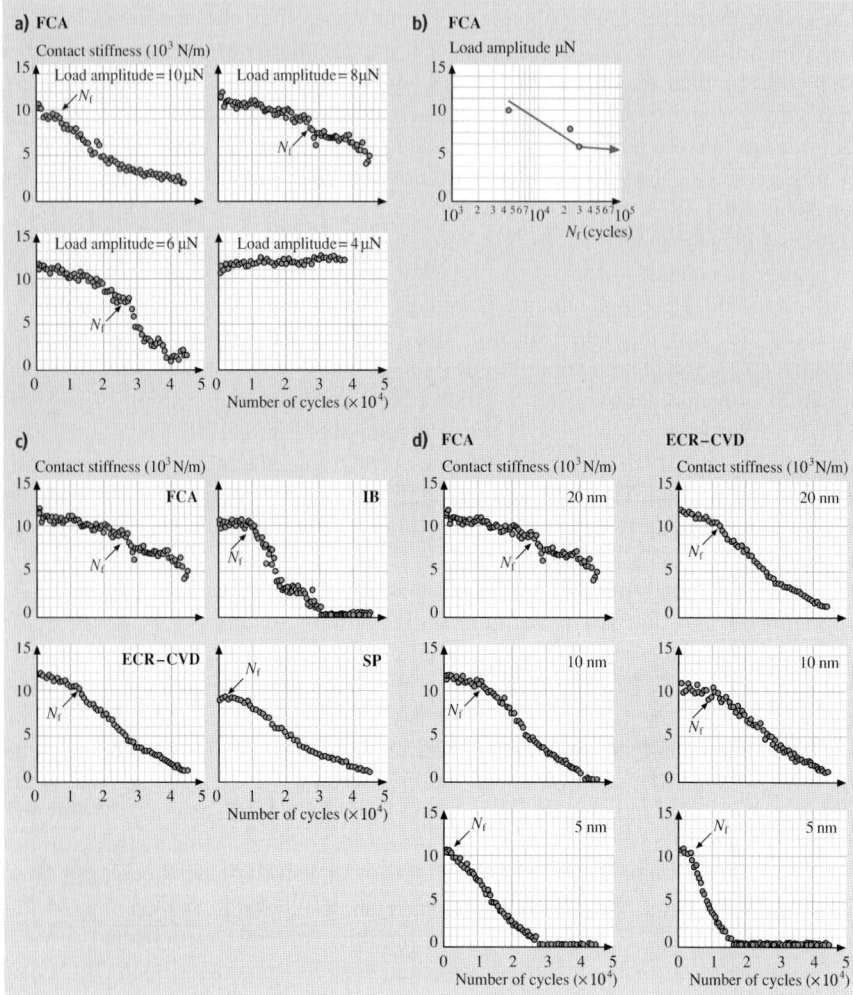

Fig. 16.14. (a) Contact stiffness as a function of the number of cycles for 20 nm-thick FCA coatings cyclically deformed by various oscillation load amplitudes with a mean load of 10 μN at a frequency of 45 Hz; (b) plot of load amplitude versus N_f; (c) contact stiffness as a function of the number of cycles for four different 20 nm-thick coatings with a mean load of 10 μN and a load amplitude of 8 μN at a frequency of 45 Hz; and (d) contact stiffness as a function of the number of cycles for two coatings of different thicknesses at a mean load of 10 μN and a load amplitude of 8 μN at a frequency of 45 Hz

increases with the number of cycles. This indicates that deformation, delamination and buckling as well as ring-like crack formation occurred over a period of time.

The schematic in Fig. 16.15b shows various stages of indentation fatigue damage for a coating/substrate system. Based on this study, three stages of indentation

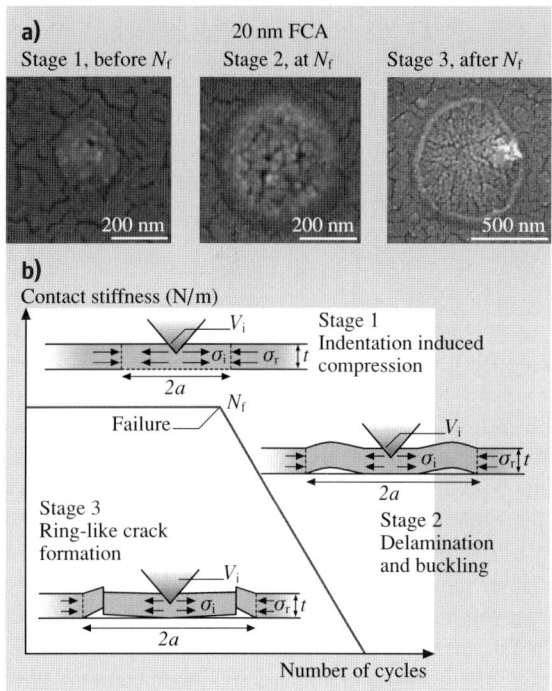

Fig. 16.15. (a) High-magnification SEM images of a coating before, at, and after N_f, and (b) schematic of various stages of indentation fatigue damage for a coating/substrate system [90]

fatigue damage appear to exist: (1) indentation-induced compression; (2) delamination and buckling; (3) ring-like crack formation at the edge of the buckled coating. The deposition process often induces residual stresses in coatings. The model shown in Fig. 16.15b considers a coating with uniform biaxial residual compression σ_r. In the first stage, indentation induces elastic/plastic deformation, exerting a pressure (acting outward) on the coating around the indenter. Interfacial defects like voids and impurities act as original cracks. These cracks propagate and link up as the indentation compressive stress increases. At this stage, the coating, which is under the indenter and above the interfacial crack (with a crack length of $2a$), still maintains a solid contact with the substrate; the substrate still fully supports the coating. Therefore, this interfacial crack does not lead to an abrupt decrease in contact stiffness, but gives rise to a slight decrease in contact stiffness, as shown in Fig. 16.14. The coating above the interfacial crack is treated as a rigidly clamped disk. We assume that the crack radius, a, is large compared with the coating thickness t. Since the coating thickness ranges from 5 to 20 nm, this assumption is easily satisfied in this study (the radius of the delaminated and buckled area, shown in Fig. 16.15a, is on the order of 100 nm). The compressive stress caused by indentation is given as [97]:

$$\sigma_i = \frac{E}{(1-v)}\varepsilon_i = \frac{EV_i}{2\pi t a^2 (1-v)}, \tag{16.3}$$

where ν and E are the Poisson ratio and elastic modulus of the coating, V_i is the indentation volume, t is the coating thickness, and a is the crack radius. As the number of cycles increases, so does the indentation volume V_i. Therefore, the indentation compressive stress σ_i increases accordingly. In the second stage, buckling occurs during the unloading segment of the fatigue testing cycle when the sum of the indentation compressive stress σ_i and the residual stress σ_r exceed the critical buckling stress σ_b for the delaminated circular section, as given by [98]

$$\sigma_b = \frac{\mu^2 E}{12(1-\nu^2)} \left(\frac{t}{a}\right)^2, \tag{16.4}$$

where the constant μ equals 42.67 for a circular clamped plate with a constrained center point and 14.68 when the center is unconstrained. The buckled coating acts as a cantilever. In this case, the indenter indents a cantilever rather than a coating/substrate system. This ultrathin coating cantilever has much less contact stiffness than the coating/substrate system. Therefore, the contact stiffness shows an abrupt decrease at N_f. In the third stage, with more cycles, the delaminated and buckled size increases, resulting in a further decrease in contact stiffness since the cantilever beam length increases. On the other hand, a high bending stress acts at the edge of the buckled coating. The larger the buckled size, the higher the bending stress. The cyclic bending stress causes fatigue damage at the end of the buckled coating, forming a ring-like crack. The coating under the indenter is separated from the bulk coating (caused by the ring-like crack at the edge of the buckled coating) and the substrate (caused by the delamination and buckling in the second stage). Therefore, the coating under the indenter is not constrained; it is free to move with the indenter during fatigue testing. At this point, the sharp nature of the indenter is lost, because the coating under the indenter gets stuck on the indenter. The indentation fatigue experiment results in the contact of a (relatively) huge blunt tip with the substrate. This results in a low contact stiffness value.

Compressive residual stresses result in delamination and buckling. A coating with a higher adhesion strength and a lower compressive residual stress is required for a higher fatigue life. Interfacial defects should be avoided in the coating deposition process. We know that ring-like crack formation occurs in the coating. Formation of fatigue cracks in the coating depends upon the hardness and the fracture toughness. Cracks are more difficult to form and propagate in the coating with higher strength and fracture toughness.

It is now accepted that long fatigue life in a coating/substrate almost always involves "living with a crack", that the threshold or limit condition is associated with the nonpropagation of existing cracks or defects, even though these cracks may be undetectable [96]. For all of the coatings studied at 4 μN, the contact stiffness does not change much. This indicates that delamination and buckling did not occur within the number of cycles tested in this study. This is probably because the indentation-induced compressive stress was not high enough to allow the cracks to propagate and link up under the indenter, or the sum of the indentation compressive stress σ_i and the residual stress σ_r did not exceed the critical buckling stress σ_b.

Figure 16.10 and Table 16.5 summarize the hardnesses, elastic moduli, fracture toughnesses, and fatigue lifetimes of all of the coatings studied. A good correlation exists between the fatigue life and other mechanical properties. Higher mechanical properties result in a longer fatigue life. The mechanical properties of DLC coatings are controlled by the sp^3-to-sp^2 ratio. An sp^3-bonded carbon exhibits the outstanding properties of diamond [51]. Higher kinetic energy during deposition will result in a larger fraction of sp^3-bonded carbon in an amorphous network. Thus, the higher kinetic energy for the FCA could be responsible for its enhanced carbon structure and mechanical properties [48–50, 99]. Higher adhesion strength between the FCA coating and substrate makes the FCA coating more difficult to delaminate from the substrate.

16.4.2 Microscratch and Microwear Studies

For microscratch studies, a conical diamond indenter (that has a tip radius of about one micron and a cone angle of 60° for example) is drawn over the sample surface, and the load is ramped up (typically from 2 mN to 25 mN) until substantial damage occurs. The coefficient of friction is monitored during scratching. Scratch-induced coating damage, specifically fracture or delamination, can be monitored by in situ friction force measurements and using optical and SEM imaging of the scratches after tests. A gradual increase in friction is associated with plowing, and an abrupt increase in friction is associated with fracture or catastrophic failure [100]. The load corresponding to an abrupt increase in friction or an increase in friction above a certain value (typically $2\times$ the initial value) provides a measure of the scratch resistance or the adhesive strength of a coating, and is called the "critical load". The depths of scratches are measured with increasing scratch length or normal load using an AFM, typically with an area of $10 \times 10\,\mu\text{m}$ [48, 49, 101].

Microscratch and microwear studies are also conducted using an AFM [23, 50, 99, 102, 103]. A square pyramidal diamond tip (tip radius ≈ 100 nm) or a three-sided pyramidal diamond (Berkovich) tip with an apex angle of 60° and a tip radius of about 100 nm, mounted on a platinum-coated, rectangular stainless steel cantilever of stiffness of about 40 N/m, is scanned orthogonal to the long axis of the cantilever to generate scratch and wear marks. During the scratch test, the normal load is either kept constant or is increased (typically from 0 to 100 µN) until damage occurs. Topographical images of the scratch are obtained in situ with the AFM at a low load. By scanning the sample during scratching, wear experiments can be conducted. Wear is monitored as a function of the number of cycles at a constant load. Normal loads (10–80 µN) are typically used.

Microscratch

Scratch tests conducted with a sharp diamond tip simulate a sharp asperity contact. In a scratch test, the cracking or delamination of a hard coating is signaled by a sudden increase in the coefficient of friction [23]. The load associated with this event is called the "critical load".

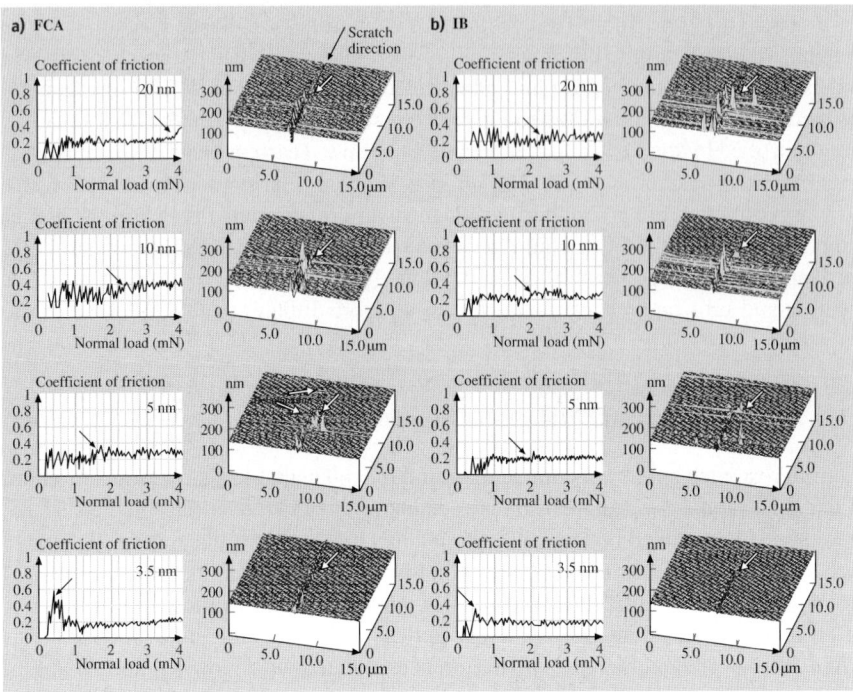

Fig. 16.16a,b. Coefficient of friction profiles as a function of normal load, as well as corresponding AFM surface height maps of regions over scratches at the respective critical loads (indicated by the *arrows* in the friction profiles and AFM images), for coatings of different thicknesses deposited by various deposition techniques: (**a**) FCA, (**b**) IB

Wu [104], *Bhushan* et al. [70], *Gupta* and *Bhushan* [12, 47], and *Li* and *Bhushan* [48, 49, 101] have used a nanoindenter to perform microscratch (mechanical durability) studies of various carbon coatings. The coefficient of friction profiles as a function of increasing normal load as well as AFM surface height maps of regions over scratches at the respective critical loads (indicated by the arrows in the friction profiles and AFM images) observed for coatings with different thicknesses and single-crystal silicon substrate using a conical tip are compared in Figs. 16.16 and 16.17. *Bhushan* and *Koinkar* [102], *Koinkar* and *Bhushan* [103], *Bhushan* [23], and *Sundararajan* and *Bhushan* [50, 99] used an AFM to perform microscratch studies. Data obtained for coatings with different thicknesses and silicon substrate using a Berkovich tip are compared in Figs. 16.18 and 16.19. Critical loads for various coatings tested using a nanoindenter and AFM are summarized in Fig. 16.20. Selected data for 20 nm-thick coatings obtained using nanoindenter are also presented in Fig. 16.10 and Table 16.5.

It is clear that a well-defined critical load exists for each coating. The AFM images clearly show that below the critical loads the coatings were plowed by the scratch tip, associated with the plastic flow of materials. At and after the criti-

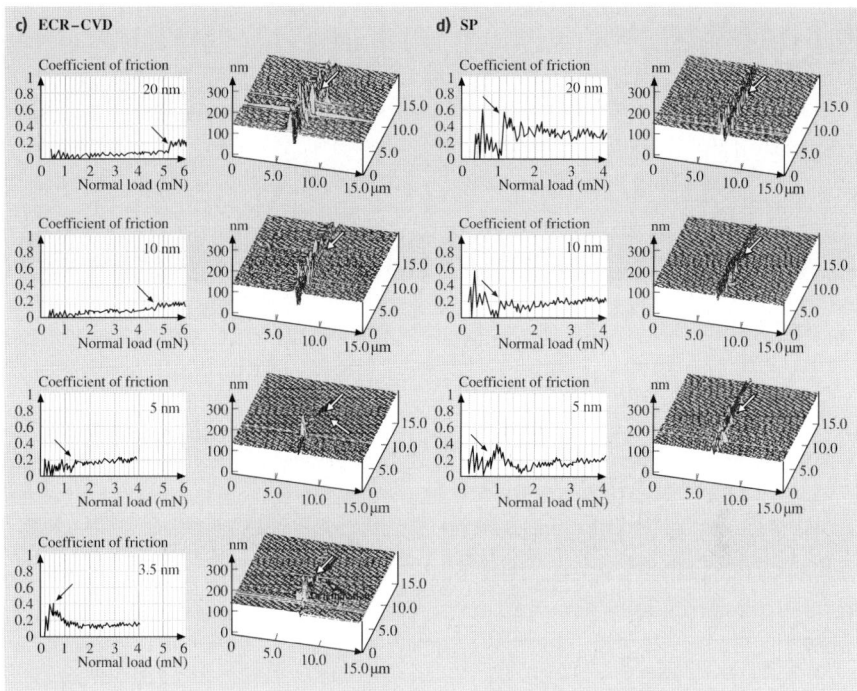

Fig. 16.16c,d. Coefficient of friction profiles as a function of normal load, as well as corresponding AFM surface height maps of regions over scratches at the respective critical loads (indicated by the *arrows* in the friction profiles and AFM images), for coatings of different thicknesses deposited by various deposition techniques: (**c**) ECR-CVD, (**d**) SP

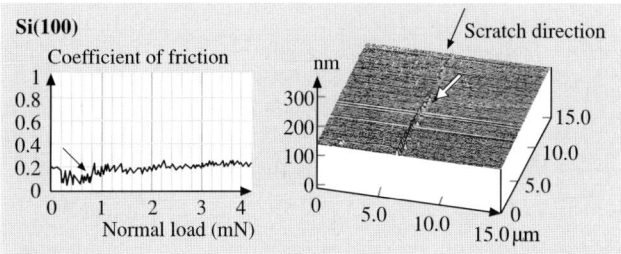

Fig. 16.17. Coefficient of friction profiles as a function of normal load as well as corresponding AFM surface height maps of regions over scratches at the respective critical loads (indicated by the *arrows* in the friction profiles and AFM images) for Si(100)

cal loads, debris (chips) or buckling was observed on the sides of the scratches. Delamination or buckling can be observed around or after the critical loads, which suggests that the damage starts from delamination and buckling. For the 3.5- and 5 nm-thick FCA coatings, before the critical loads small debris is observed on the sides of the scratches. This suggests that the thinner FCA coatings are not so durable.

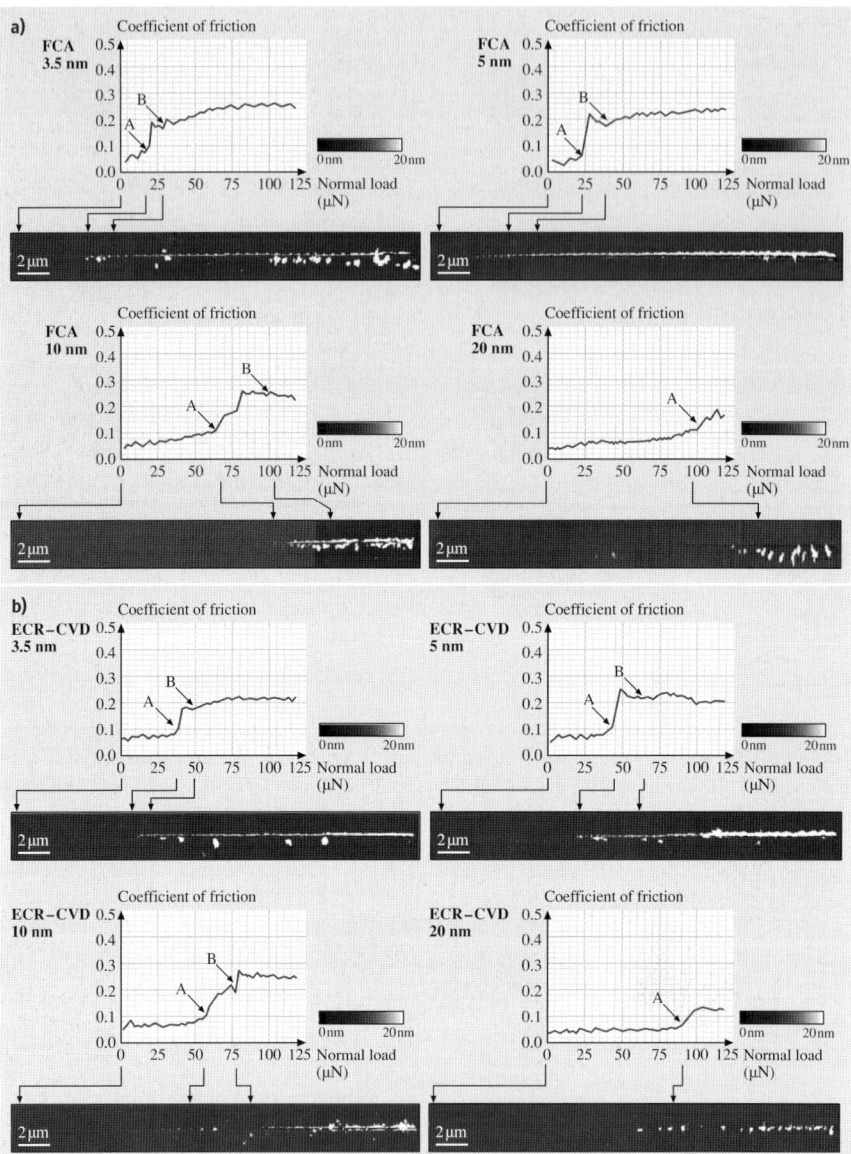

Fig. 16.18. Coefficient of friction profiles during scratch as a function of normal load and corresponding AFM surface height maps for (**a**) FCA, (**b**) ECR–CVD

It is obvious that, for a given deposition method, the critical loads increase with increasing coating thickness. This indicates that the critical load is determined not only by the strength of adhesion to the substrate, but also by the coating thickness. We note that more debris generated on the thicker coatings than thinner coatings.

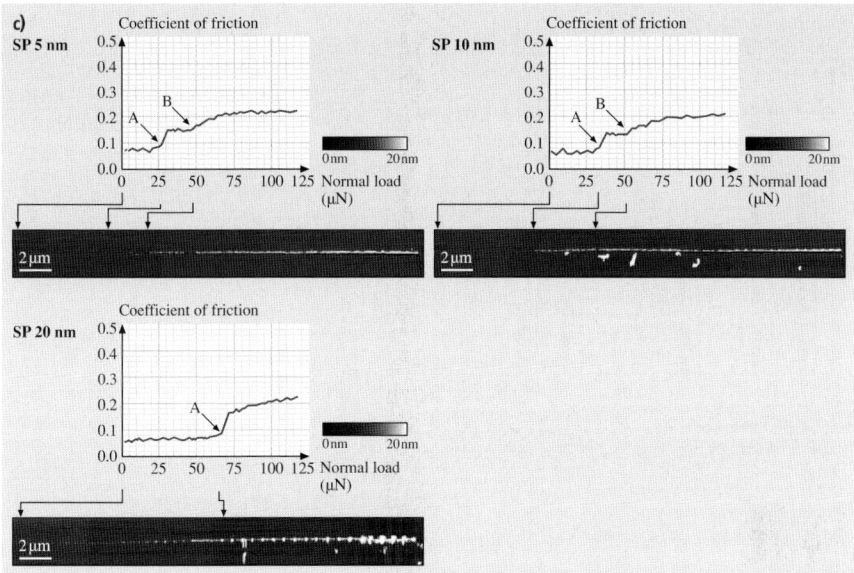

Fig. 16.18. (continued) Coefficient of friction profiles during scratch as a function of normal load and corresponding AFM surface height maps for (**c**) SP coatings [99]

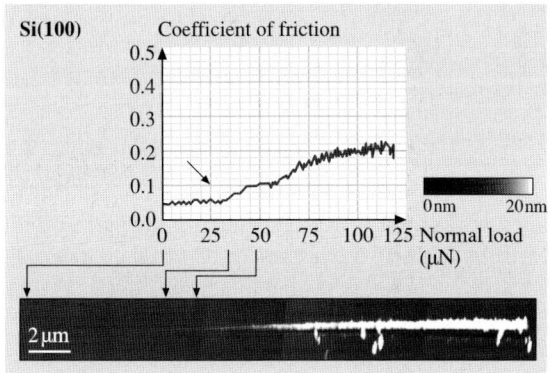

Fig. 16.19. Coefficient of friction profiles during scratch as a function of normal load and corresponding AFM surface height maps for Si(100) [99]

A thicker coating is more difficult to break; the broken coating chips (debris) seen for a thicker coating are larger than those for the thinner coatings. The different residual stresses of coatings of different thicknesses may also affect the size of the debris. The AFM image shows that the silicon substrate was damaged by plowing, associated with the plastic flow of material. At and after the critical load, a small amount of uniform debris is observed and the amount of debris increases with increasing normal load.

Since the damage mechanism at the critical load appears to be the onset of plowing, harder coatings with more fracture toughness will therefore require a higher load for deformation and hence a higher critical load. Figure 16.21 gives critical

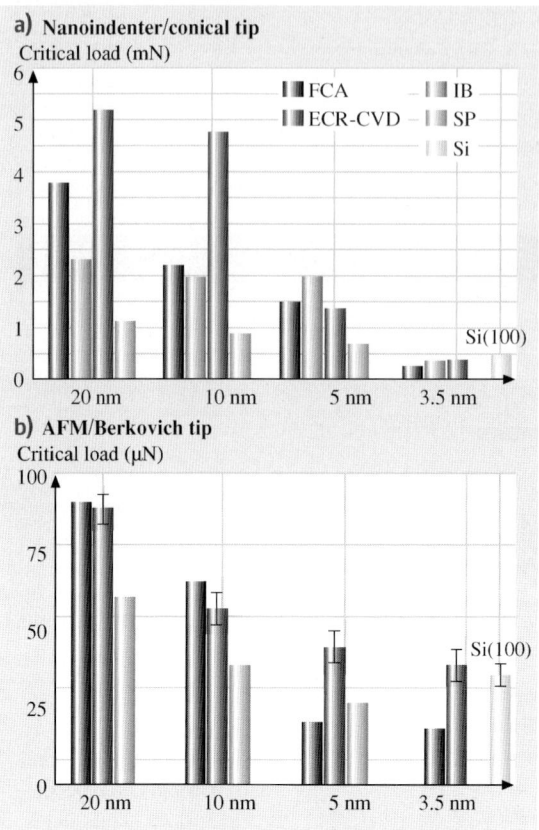

Fig. 16.20. Critical loads estimated from the coefficient of friction profiles from (**a**) nanoindenter and (**b**) AFM tests for various coatings of different thicknesses and Si(100) substrate

loads of various coatings, obtained with AFM tests, as a function of the coating hardness and fracture toughness (from Table 16.5). It can be seen that, in general, increasing coating hardness and fracture toughness results in a higher critical load. The only exceptions are the FCA coatings at 5 and 3.5 nm thickness, which show the lowest critical loads despite their high hardness and fracture toughness. The brittleness of the thinner FCA coatings may be one reason for their low critical loads. The mechanical properties of coatings that are less than 10 nm thick are not known. The FCA process may result in coatings with low hardness at low thickness due to differences in coating stoichiometry and structure compared to coatings of higher thickness. Also, at these thicknesses stresses at the coating–substrate interface may affect coating adhesion and load-carrying capacity.

Based on the experimental results, a schematic of the scratch damage mechanisms encountered for the DLC coatings used in this study is shown in Fig. 16.22. Below the critical load, if a coating has a good combination of strength and fracture toughness, plowing associated with the plastic flow of materials is responsible for the coating damage (Fig. 16.22a). However, if the coating has a low frac-

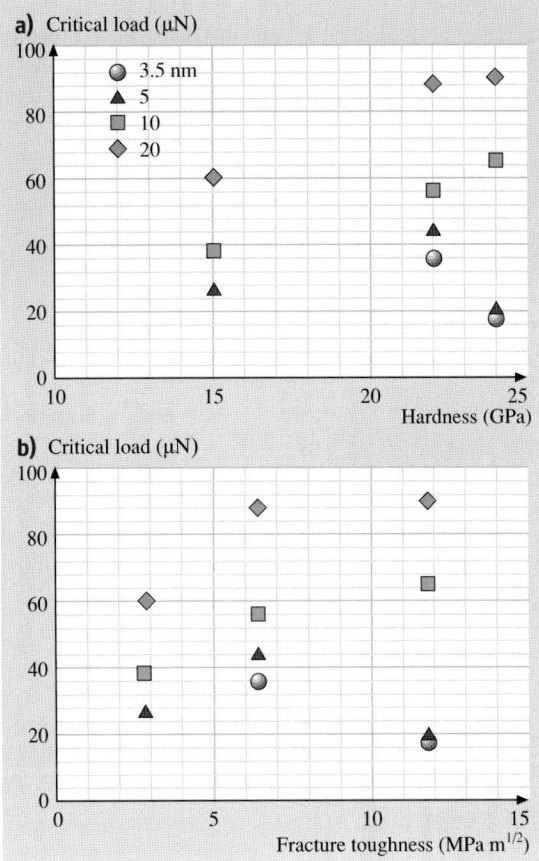

Fig. 16.21. Measured critical loads estimated from the coefficient of friction profiles from AFM tests as a function of (**a**) coating hardness and (**b**) fracture toughness. Coating hardness and fracture toughness values were obtained using a nanoindenter on 100 nm-thick coatings (Table 16.5)

ture toughness, cracking could occur during plowing, resulting in the formation of small amounts of debris (Fig. 16.22b). When the normal load is increased to the critical load, delamination or buckling will occur at the coating/substrate interface (Fig. 16.22c). A further increase in normal load will result in coating breakdown via through-coating thickness cracking, as shown in Fig. 16.22d. Therefore, adhesion strength plays a crucial role in the determination of critical load. If a coating adheres strongly to the substrate, the coating is more difficult to delaminate, which will result in a higher critical load. The interfacial and residual stresses of a coating may also greatly affect the delamination and buckling [1]. A coating with higher interfacial and residual stresses is more easily delaminated and buckled, which will result in a low critical load. It was reported earlier that FCA coatings have higher residual stresses than other coatings [47]. Interfacial stresses play an increasingly important role as the coating gets thinner. A large mismatch in elastic modulus between the FCA coating and the silicon substrate may cause large interfacial stresses. This may be why thinner FCA coatings show lower critical loads than thicker FCA coatings,

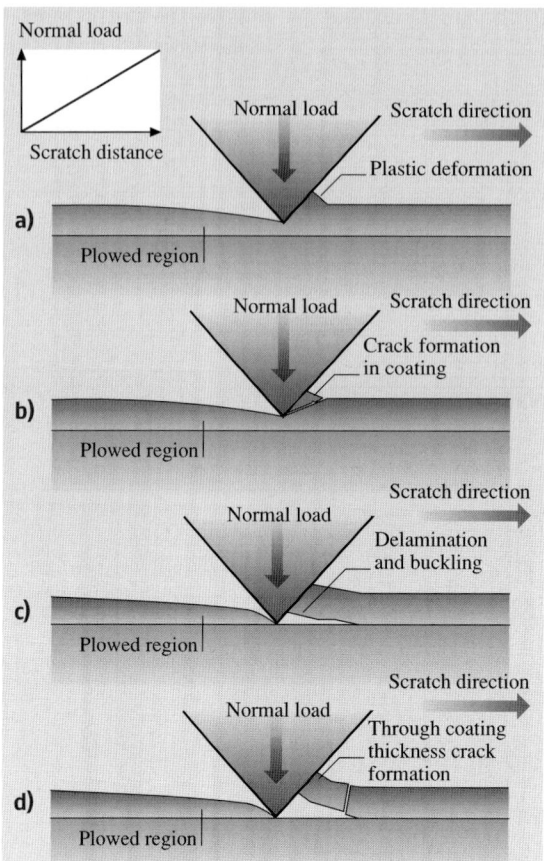

Fig. 16.22. Schematic of scratch damage mechanisms for DLC coatings: (**a**) plowing associated with the plastic flow of materials, (**b**) plowing associated with the formation of small debris, (**c**) delamination and buckling at the critical load, and (**d**) breakdown via through-coating thickness cracking at and after the critical load [48]

even though the FCA coatings have higher hardness and elastic moduli. The brittleness of thinner FCA coatings may be another reason for the lower critical loads. The strength and fracture toughness of a coating also affect the critical load. Greater strength and fracture toughness will make the coating more difficult to break after delamination and buckling. The high scratch resistance/adhesion of FCA coatings is attributed to the atomic intermixing that occurs at the coating–substrate interface due to the high kinetic energy (2 keV) of the plasma formed during the cathodic arc deposition process [57]. This atomic intermixing provides a graded compositional transition between the coating and the substrate material. In all other coatings used in this study, the kinetic energy of the plasma is insufficient for atomic intermixing.

Gupta and *Bhushan* [12, 47] and *Li* and *Bhushan* [48, 49] measured the scratch resistances of DLC coatings deposited on Al_2O_3-TiC, Ni-Zn ferrite and single-crystal silicon substrates. An interlayer of silicon is required to adhere the DLC coating to other substrates, except in the case of cathodic arc-deposited coatings.

The best adhesion with cathodic arc carbon coating is obtained on electrically conducting substrates such as Al_2O_3-TiC and silicon rather than Ni-Zn ferrite.

Microwear

Microwear studies can be conducted using an AFM [23]. For microwear studies, a three-sided pyramidal single-crystal natural diamond tip with an apex angle of about 80° and a tip radius of about 100 nm is used at relatively high loads of 1–150 µN. The diamond tip is mounted on a stainless steel cantilever beam with a normal stiffness of about 30 N/m. The sample is generally scanned in a direction orthogonal to the long axis of the cantilever beam (typically at a rate of 0.5 Hz). The tip is mounted on the beam such that one of its edges is orthogonal to the beam axis. In wear studies, an area of 2 µm × 2 µm is typically scanned for a selected number of cycles.

Microwear studies of various types of DLC coatings have been conducted [50, 102, 103]. Figure 16.23a shows a wear mark on uncoated Si(100). Wear occurs uniformly and material is removed layer-by-layer via plowing from the first cycle, resulting in constant friction force during the wear (Fig. 16.24a). Figure 16.23b shows AFM images of the wear marks on 10 nm-thick coatings. It is clear that the coatings wear nonuniformly. Coating failure is sudden and accompanied by a sudden rise

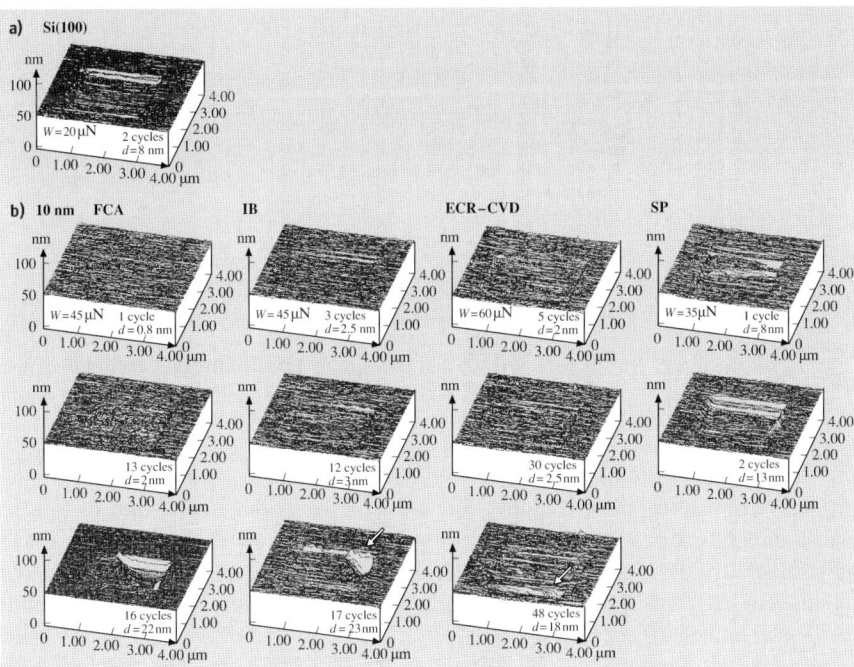

Fig. 16.23. AFM images of wear marks on (**a**) bare Si(100), and (**b**) various 10 nm-thick DLC coatings [50]

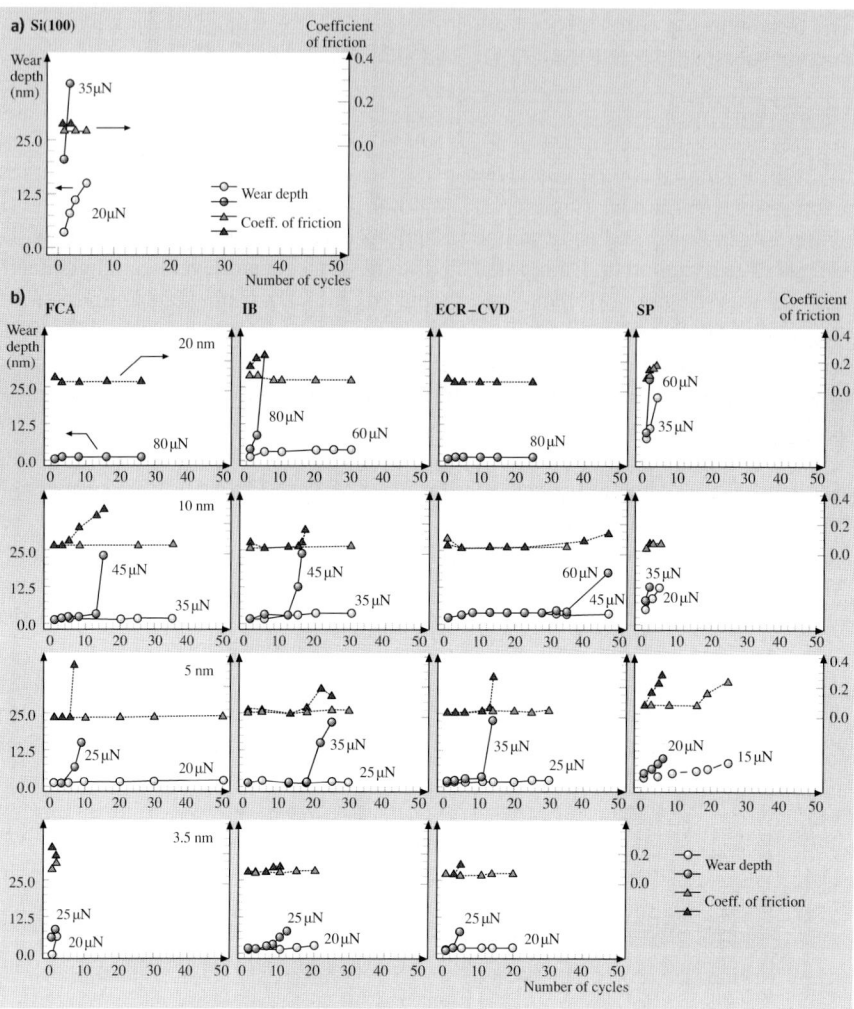

Fig. 16.24. Wear data for (**a**) bare Si(100) and (**b**) various DLC coatings. Coating thickness is constant along each row in (**b**). Both the wear depth and the coefficient of friction during wear are plotted for a given cycle [50]

in the friction force (Fig. 16.24b). Figure 16.24 shows the wear depth of Si(100) substrate and various DLC coatings at two different loads. FCA- and ECR-CVD-deposited 20 nm-thick coatings show excellent wear resistance up to 80 μN, the load that is required for the IB 20 nm coating to fail. In these tests, "failure" of a coating occurs when the wear depth exceeds the quoted coating thickness. The SP 20 nm coating fails at a much lower load of 35 μN. At 60 μN, the coating hardly provides any protection. Moving on to the 10 nm coatings, the ECR-CVD coating requires about 45 cycles at 60 μN to fail, whereas the IB and FCA, coatings fail at 45 μN.

The FCA coating exhibits slight roughening in the wear track after the first few cycles, which leads to an increase in the friction force. The SP coating continues to exhibit poor resistance, failing at 20 μN. For the 5 nm coatings, the load required to make the coatings fail continues to decrease, but IB and ECR-CVD still provide adequate protection compared to bare Si(100) in that order, with the silicon failing at 35 μN, the FCA coating at 25 μN and the SP coating at 20 μN. Almost all of the 20, 10, and 5 nm coatings provide better wear resistance than bare silicon. At 3.5 nm, FCA coating provides no wear resistance, failing almost instantly at 20 μN. The IB and ECR-CVD coatings show good wear resistance at 20 μN compared to bare Si(100). However, IB only lasts about ten cycles and ECR-CVD about three cycles at 25 μN.

The wear tests highlight the differences between the coatings more vividly than the scratch tests. At higher thicknesses (10 and 20 nm), the ECR-CVD and FCA coatings appear to show the best wear resistance. This is probably due to the greater hardness of the coatings (see Table 16.5). At 5 nm, the IB coating appears to be the best. FCA coatings show poorer wear resistance with decreasing coating thickness. This suggests that the trends in hardness seen in Table 16.5 no longer hold at low thicknesses. SP coatings show consistently poor wear resistance at all thicknesses. The 3.5 nm IB coating does provide reasonable wear protection at low loads.

16.4.3 Macroscale Tribological Characterization

So far, we have presented data on mechanical characterization and microscratch and microwear studies using a nanoindenter and an AFM. Mechanical properties affect the tribological performance of a coating, and microwear studies simulate a single asperity contact, which helps us to understand the wear process. These studies are useful when screening various candidate coatings, and also aid our understanding of the relationships between deposition conditions and properties of the samples. In the next step, macroscale friction and wear tests need to be conducted to measure the tribological performance of the coating.

Macroscale accelerated friction and wear tests have been conducted to screen a large number of candidates, as have functional tests on selected candidates. An accelerated test is designed to accelerate the wear process such that it does not change the failure mechanism. The accelerated friction and wear tests are generally conducted using a ball-on-flat tribometer under reciprocating motion [70]. Typically, a diamond tip with a tip radius of 20 μm or a sapphire ball with a diameter of 3 mm and a surface finish of about 2 nm RMS is slid against the coated substrates at selected loads. The coefficient of friction is monitored during the tests.

Functional tests are conducted using an actual machine running at close to the actual operating conditions for which the coatings have been developed. The tests are generally accelerated somewhat to fail the interface in a short time.

Accelerated Friction and Wear Tests

Li and *Bhushan* [48] conducted accelerated friction and wear tests on DLC coatings deposited by various deposition techniques using a ball-on-flat tribometer. The

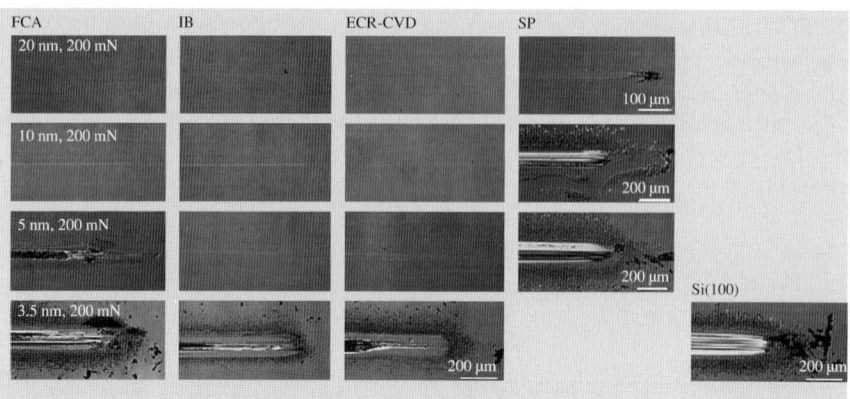

Fig. 16.25. Optical micrographs of wear tracks and debris formed on various coatings of different thicknesses and silicon substrate when slid against a sapphire ball after a sliding distance of 5 nm. The end of the wear track is on the right-hand side of the image

average coefficient of friction values observed are presented in Table 16.5. Optical micrographs of wear tracks and debris formed on all samples when slid against a sapphire ball after a sliding distance of 5 m are presented in Fig. 16.25. The normal load used for the 20 and 10 nm-thick coatings was 200 mN, and the normal load used for the 5 and 3.5 nm-thick coatings and the silicon substrate was 150 mN.

Among the 20 nm-thick coatings, the SP coating exhibits a higher coefficient of friction (about 0.3) than for the other coatings coefficient of friction (all of which were about 0.2). The optical micrographs show that the SP coating has a larger wear track and more debris than the IB coating. No wear track or debris were found on the 20 nm-thick FCA and ECR-CVD coatings. The optical micrographs of 10 nm-thick coatings show that the SP coating was severely damaged, showing a large wear track with scratches and lots of debris. The FCA and ECR-CVD coatings show smaller wear tracks and less debris than the IB coatings.

For the 5 nm-thick coatings, the wear tracks and debris of the IB and ECR-CVD coatings are comparable. The bad wear resistance of the 5 nm-thick FCA coating is in good agreement with the low critical scratch load, which may be due to the higher interfacial and residual stresses as well as the brittleness of the coating.

At 3.5 nm, all of the coatings exhibit wear. The FCA coating provides no wear resistance, failing instantly like the silicon substrate. Large block-like debris is observed on the sides of the wear track of the FCA coating. This indicates that large region delamination and buckling occurs during sliding, resulting in large block-like debris. This block-like debris, in turn, scratches the coating, making the damage to the coating even more severe. The IB and ECR-CVD coatings are able to provide some protection against wear at 3.5 nm.

In order to better evaluate the wear resistance of various coatings, based on an optical examination of the wear tracks and debris after tests, a bar chart of the wear damage index for various coatings of different thicknesses and an uncoated

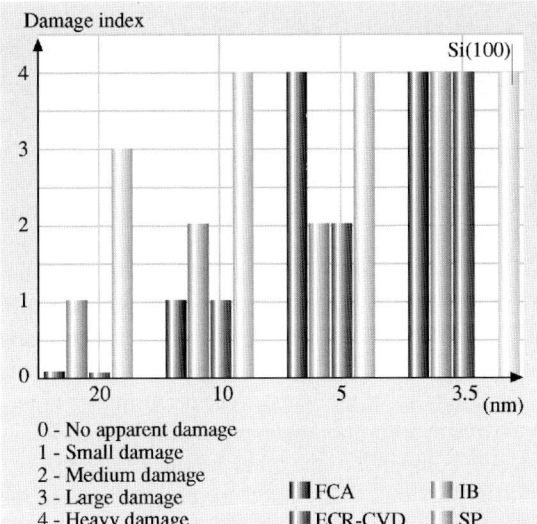

Fig. 16.26. Bar chart of the wear damage indices for various coatings of different thicknesses and Si(100) substrate based on optical examination of the wear tracks and debris

silicon substrate is presented in Fig. 16.26. Among the 20 and 10 nm-thick coatings, the SP coatings show the worst damage, followed by FCA/ECR-CVD. At 5 nm, the FCA and SP coatings show the worst damage, followed by the IB and ECR-CVD coatings. All of the 3.5 nm-thick coatings show the same heavy damage as the uncoated silicon substrate.

The wear damage mechanisms of the thick and thin DLC coatings studied are believed to be as illustrated in Fig. 16.27. In the early stages of sliding, deformation zone, Hertzian and wear fatigue cracks that have formed beneath the surface extend within the coating upon subsequent sliding [1]. Formation of fatigue cracks depends on the hardness and subsequent cycles. These are controlled by the sp^3-to-sp^2 ratio. For thicker coatings, the cracks generally do not penetrate the coating. For a thinner coating, the cracks easily propagate down to the interface aided by the interfacial

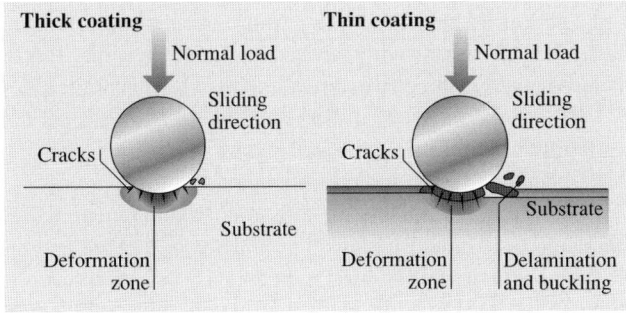

Fig. 16.27. Schematic of wear damage mechanisms for thick and thin DLC coatings [48]

stresses and get diverted along the interface just enough to cause local delamination of the coating. When this happens, the coating experiences excessive plowing. At this point, the coating fails catastrophically, resulting in a sudden rise in the coefficient of friction. All 3.5 nm-thick coatings failed much quicker than the thicker coatings. It appears that these thin coatings have very low load-carrying capacities and so the substrate undergoes deformation almost immediately. This generates stresses at the interface that weaken the coating adhesion and lead to delamination of the coating. Another reason may be that the thickness is insufficient to produce a coating that has the DLC structure. Instead, the bulk may be made up of a matrix characteristic of the interface region where atomic mixing occurs with the substrate and/or any interlayer used. This would also result in poor wear resistance and silicon-like behavior of the coating, especially for FCA coatings, which show the worst performance at 3.5 nm. SP coatings show the worst wear performance at any thickness (Fig. 16.25). This may be due to their poor mechanical properties, such as lower hardness and scratch resistance, compared to the other coatings.

Comparison of Figs. 16.20 and 16.26 shows a very good correlation between the wear damage and critical scratch loads. Less wear damage corresponds to a higher critical scratch load. Based on the data, thicker coatings do show better scratch and wear resistance than thinner coatings. This is probably due to the better load-carrying capacities of the thick coatings compared to the thinner ones. For a given coating thickness, increased hardness and fracture toughness and better adhesion strength are believed to be responsible for the superior wear performance.

Effect of Environment

The friction and wear performance of an amorphous carbon coating is known to be strongly dependent on the water vapor content and partial gas pressure in the test environment. The friction data for an amorphous carbon film on a silicon substrate sliding against steel are presented as a function of the partial pressure of water vapor in Fig. 16.28 [1, 13, 69, 105, 106]. Friction increases dramatically above a relative humidity of about 40%. At high relative humidity, condensed water vapor forms meniscus bridges at the contacting asperities, and the menisci result in an intrinsic attractive force that is responsible for an increase in the friction. For completeness, data on the coefficient of friction of bulk graphitic carbon are also presented in Fig. 16.28. Note that the friction decreases with increased relative humidity [107]. Graphitic carbon has a layered crystal lattice structure. Graphite absorbs polar gases (such as H_2O, O_2, CO_2, NH_3) at the edges of the crystallites, which weakens the interlayer bonding forces facilitating interlayer slip and results in lower friction [1].

A number of tests have been conducted in controlled environments in order to better study the effects of environmental factors on carbon-coated magnetic disks. *Marchon* et al. [108] conducted tests in alternating environments of oxygen and nitrogen gases, Fig. 16.29. The coefficient of friction increases as soon as oxygen is added to the test environment, whereas in a nitrogen environment the coefficient of friction reduces slightly. Tribochemical oxidation of the DLC coating in the oxidizing environment is responsible for an increase in the coefficient of friction, implying

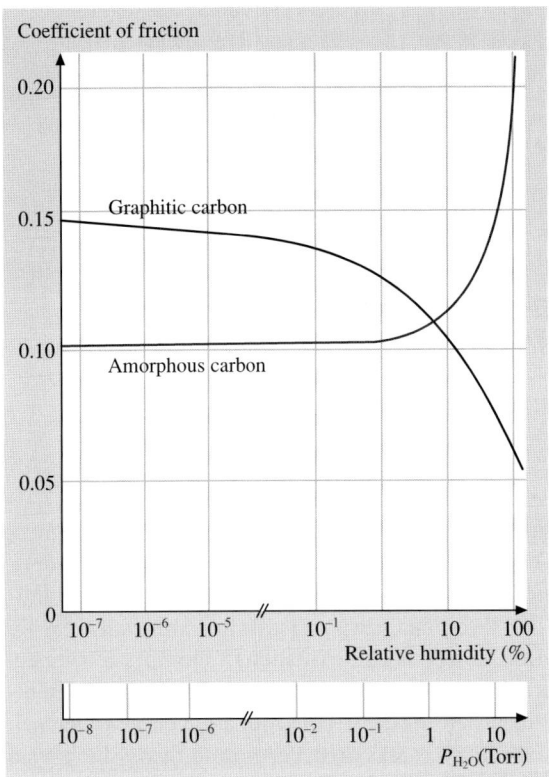

Fig. 16.28. Coefficient of friction as a function of relative humidity and water vapor partial pressure for a RF-plasma deposited amorphous carbon coating and a bulk graphitic carbon coating sliding against a steel ball

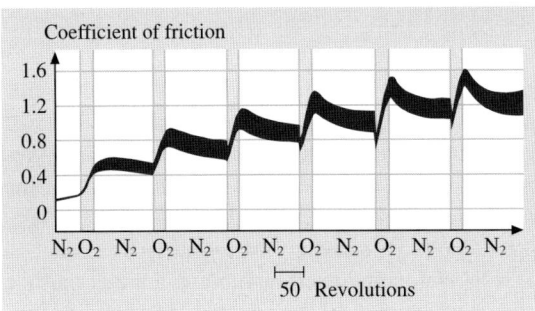

Fig. 16.29. Coefficient of friction as a function of sliding distance for a ceramic slider against a magnetic disk coated with a 20 nm-thick DC magnetron sputtered DLC coating, measured at a speed of 0.06 m/s for a load of 10 g. The environment was alternated between oxygen and nitrogen gases [108]

wear. *Dugger* et al. [109], *Strom* et al. [110], *Bhushan* and *Ruan* [111] and *Bhushan* et al. [71] conducted tests with DLC-coated magnetic disks (with about 2 nm-thick perfluoropolyether lubricant film) in contact with Al_2O_3-TiC sliders in different gaseous environments, including a high vacuum of 2×10^{-7} Torr, Fig. 16.30. The wear lives are the shortest in high vacuum and the longest in atmospheres of mostly nitrogen and argon with the following order (from best to worst): argon or nitrogen, Ar + H_2O, ambient, Ar + O_2, Ar + H_2O, vacuum. From this sequence of wear

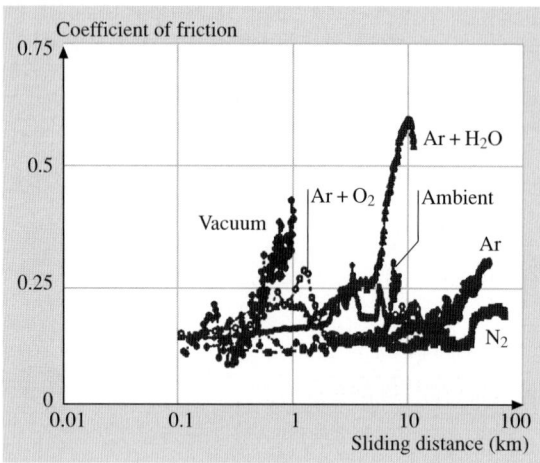

Fig. 16.30. Durability, measured by sliding a Al$_2$O$_3$-TiC magnetic slider against a magnetic disk coated with a 20 nm-thick DC sputtered amorphous carbon coating and 2 nm-thick perfluoropolyether film, measured at a speed of 0.75 m/s and for a load of 10 g. Vacuum refers to 2×10^{-7} Torr [71]

performance, we can see that having oxygen and water in an operating environment worsens the wear performance of the coating, but having a vacuum is even worse. Indeed, failure mechanisms differ in different environments. In high vacuum, intimate contact between the disk and the slider surface results in significant wear. In ambient air, Ar + O$_2$ and Ar + H$_2$O, tribochemical oxidation of the carbon overcoat is responsible for interface failure. For experiments performed in pure argon and nitrogen, mechanical shearing of the asperities causes the formation of debris, which is responsible for the formation of scratch marks on the carbon surface, which were observed with an optical microscope [71].

Functional Tests

Magnetic thin film heads made with Al$_2$O$_3$-TiC substrate are used in magnetic storage applications [13]. A multilayered thin film pole-tip structure present on the head surface wears more rapidly than the much harder Al$_2$O$_3$-TiC substrate. Pole-tip recession (PTR) is a serious concern in magnetic storage [15–19, 112]. Two of the diamond-like carbon coatings with superior mechanical properties – ion beam and cathodic arc carbon coatings – were deposited on the air-bearing surfaces of Al$_2$O$_3$-TiC head sliders [15]. Functional tests were conducted by running a metal particle (MP) tape in a computer tape drive. The average PTR as a function of sliding distance is presented in Fig. 16.31. We note that the PTR increases for the uncoated head, whereas there is a slight increase in PTR for the coated heads during early sliding followed by little change. Thus, the coatings provide protection.

The micromechanical as well as the accelerated and functional tribological data presented here clearly suggest that there is a good correlation between the scratch resistance and wear resistance measured using accelerated tests and functional tests. Thus, scratch tests can be successfully used to screen coatings for wear applications.

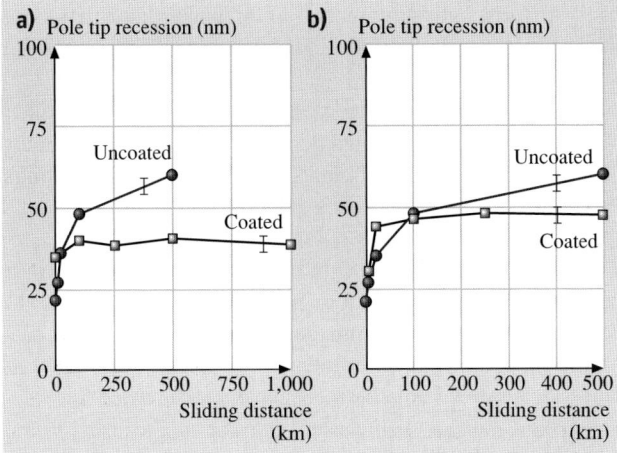

Fig. 16.31. Pole-tip recession as a function of sliding distance, measured with an AFM, for (a) uncoated and 20 nm-thick ion beam carbon coated, and (b) uncoated and 20 nm-thick cathodic arc carbon coated Al_2O_3-TiC heads run against MP tapes [15]

16.4.4 Coating Continuity Analysis

Ultrathin (less than 10 nm) coatings may not uniformly coat the sample surface. In other words, the coating may be discontinuous and deposited in the form of islands on the microscale. Therefore, one possible reason for poor wear protection and the nonuniform failure of thin coatings may be poor coverage of the substrate. Coating continuity can be studied using surface analytical techniques such as

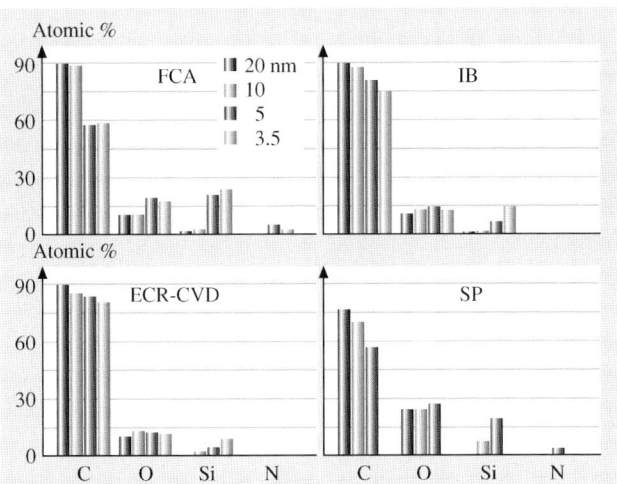

Fig. 16.32. Quantified XPS data for various DLC coatings on Si(100) substrate [50]. Atomic concentrations are shown

Auger and/or XPS analyses. Any discontinuity in coating thickness that is less than the sampling depth of the instrument result in the local detection of the substrate species [49, 50, 102].

The results from an XPS analysis of 1.3 mm^2 regions (single point measurement with spot diameter of 1300 μm) on various coatings deposited on Si(100) substrates are shown in Fig. 16.32. The sampling depth is about 2–3 nm. The poor SP coatings and the poor 5 nm and 3.5 nm FCA coatings () show much lower carbon contents (atomic concentrations of < 75% and < 60% respectively) than the IB and ECR-CVD coatings. Silicon is detected in all of the 5 nm coatings. From the data it is hard to infer whether the Si is from the substrate or from exposed regions due to discontinuous coating. Based on the sampling depth, any Si detected in 3.5 nm coatings would likely be from the substrate. The other interesting observation is that all poor coatings (all SP and FCA 5 and 3.5 nm coatings) have almost twice the oxygen content of the other coatings. Any oxygen present may be due to leaks in the deposition chamber, and it is present as silicon oxides.

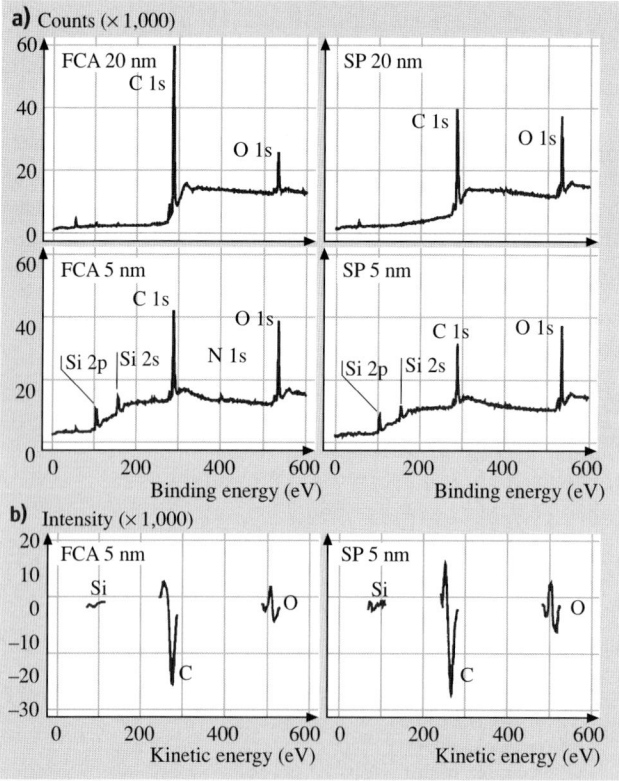

Fig. 16.33. (a) XPS spectra for 5 nm- and 20 nm-thick FCA and SP coatings on Si(100) substrate, and (b) AES spectra for FCA and SP coatings of 5 nm thickness on Si(100) substrate [50]

AES measurements averaged over a scan area of 900 µm² were conducted on FCA and SP 5 nm coatings at six different regions on each sample. Very little silicon was detected on this scale, and the detected peaks were characteristic of oxides. The oxygen levels were comparable to those seen for good coatings via XPS. These results contrast with the XPS measurements performed at a larger scale, suggesting that the coatings only possess discontinuities at isolated areas and that the 5 nm coatings are generally continuous on the microscale. Figure 16.33 shows representative XPS and AES spectra of selected samples.

16.5 Closure

Diamond material and its smooth coatings are used for very low wear and relatively low friction. Major limitations of the true diamond coatings are that they need to be deposited at high temperatures, can only be deposited on selected substrates, and require surface finishing. Hard amorphous carbon (a-C) or commonly known as DLC coatings exhibit mechanical, thermal and optical properties close to that of diamond. These can be deposited with a large range of thicknesses by using a variety of deposition processes, on variety of substrates at or near room temperature. The coatings reproduce substrate topography avoiding the need of post finishing. Friction and wear properties of some DLC coatings can be very attractive for tribological applications. The largest industrial application of these coatings is in magnetic storage devices. They are expected to be used in MEMS/NEMS.

EELS and Raman spectroscopies can be successfully used for chemical characterization of amorphous carbon coatings. The prevailing atomic arrangement in the DLC coatings is amorphous or quasi-amorphous with small diamond (sp^3), graphite (sp^2) and other unidentifiable micro- or nanocrystallites. Most DLC coatings except those produced by filtered cathodic arc contain from a few to about 50 at% hydrogen. Sometimes hydrogen is deliberately incorporated in the sputtered and ion plated coatings to tailor their properties.

Amorphous carbon coatings deposited by various techniques exhibit different mechanical and tribological properties. The nanoindenter can be successfully used for measurement of hardness, elastic modulus, fracture toughness, and fatigue life. Microscratch and microwear experiments can be performed using either a nanoindenter or an AFM. Thin coatings deposited by filtered cathodic arc, ion beam and ECR-CVD hold a promise for tribological applications. Coatings as thin as 5 nm or even thinner in thickness provide wear protection. Microscratch, microwear, and accelerated wear testing, if simulated properly can be successfully used to screen coating candidates for industrial applications. In the examples shown in this chapter, trends observed in the microscratch, microwear, and accelerated macrofriction wear tests are similar to that found in functional tests.

References

1. B. Bhushan, B. K. Gupta: *Handbook of Tribology: Materials, Coatings, and Surface Treatments*, reprint edn. (Krieger, Malabar 1997)
2. B. Bhushan: *Principles and Applications of Tribology* (Wiley, New York 1999)
3. B. Bhushan: *Introduction to Tribology* (Wiley, New York 2002)
4. B. Bhushan, B. K. Gupta, G. W. VanCleef, C. Capp, J. V. Coe: Fullerene (C_{60}) films for solid lubrication, Tribol. Trans. **36**, 573–580 (1993)
5. B. K. Gupta, B. Bhushan, C. Capp, J. V. Coe: Material characterization and effect of purity and ion implantation on the friction and wear of sublimed fullerene films, J. Mater. Res. **9**, 2823–2838 (1994)
6. B. K. Gupta, B. Bhushan: Fullerene particles as an additive to liquid lubricants and greases for low friction and wear, Lubr. Eng. **50**, 524–528 (1994)
7. B. Bhushan, V. V. Subramaniam, A. Malshe, B. K. Gupta, J. Ruan: Tribological properties of polished diamond films, J. Appl. Phys. **74**, 4174–4180 (1993)
8. B. Bhushan, B. K. Gupta, V. V. Subramaniam: Polishing of diamond films, Diam. Films Technol. **4**, 71–97 (1994)
9. P. Sander, U. Kaiser, M. Altebockwinkel, L. Wiedmann, A. Benninghoven, R. E. Sah, P. Koidl: Depth profile analysis of hydrogenated carbon layers on silicon by X-ray photoelectron spectroscopy, auger electron spectroscopy, electron energy-loss spectroscopy, and secondary ion mass spectrometry, J. Vacuum Sci. Technol. A **5**, 1470–1473 (1987)
10. A. Matthews, S. S. Eskildsen: Engineering applications for diamond-like carbon, Diam. Relat. Mater. **3**, 902–911 (1994)
11. A. H. Lettington: Applications of diamond-like carbon thin films, Carbon **36**, 555–560 (1998)
12. B. K. Gupta, B. Bhushan: Mechanical and tribological properties of hard carbon coatings for magnetic recording heads, Wear **190**, 110–122 (1995)
13. B. Bhushan: *Tribology and Mechanics of Magnetic Storage Devices*, 2nd edn. (Springer, Berlin, Heidelberg 1996)
14. B. Bhushan: *Mechanics and Reliability of Flexible Magnetic Media*, 2nd edn. (Springer, Berlin, Heidelberg 2000)
15. B. Bhushan, S. T. Patton, R. Sundaram, S. Dey: Pole tip recession studies of hard carbon-coated thin-film tape heads, J. Appl. Phys. **79**, 5916–5918 (1996)
16. J. Xu, B. Bhushan: Pole tip recession studies of thin-film rigid disk head sliders II: Effects of air bearing surface and pole tip region designs and carbon coating, Wear **219**, 30–41 (1998)
17. W. W. Scott, B. Bhushan: Corrosion and wear studies of uncoated and ultra-thin DLC coated magnetic tape-write heads and magnetic tapes, Wear **243**, 31–42 (2000)
18. W. W. Scott, B. Bhushan: Loose debris and head stain generation and pole tip recession in modern tape drives, J. Info. Stor. Proc. Syst. **2**, 221–254 (2000)
19. W. W. Scott, B. Bhushan, A. V. Lakshmikumaran: Ultrathin diamond-like carbon coatings used for reduction of pole tip recession in magnetic tape heads, J. Appl Phys **87**, 6182–6184 (2000)
20. B. Bhushan: Macro- and microtribology of magnetic storage devices. In: *Modern Tribology Handbook*, ed. by B. Bhushan (CRC, Boca Raton 2001) pp. 1413–1513
21. B. Bhushan: *Nanotribology and nanomechanics of MEMS devices*, Proc. Ninth Annual Workshop on Micro Electro Mechanical Systems (IEEE, New York 1996) pp. 91–98

22. B. Bhushan (ed): *Tribology Issues and Opportunities in MEMS* (Kluwer, Dordrecht 1998)
23. B. Bhushan: *Handbook of Micro/Nanotribology*, 2nd edn. (CRC, Boca Raton 1999)
24. B. Bhushan: Macro- and microtribology of MEMS materials. In: *Modern Tribology Handbook*, ed. by B. Bhushan (CRC, Boca Raton 2001) pp. 1515–1548
25. S. Aisenberg, R. Chabot: Ion beam deposition of thin films of diamond-like carbon, J. Appl. Phys. **49**, 2953–2958 (1971)
26. E. G. Spencer, P. H. Schmidt, D. C. Joy, F. J. Sansalone: Ion beam deposited polycrystalline diamond-like films, Appl. Phys. Lett. **29**, 118–120 (1976)
27. A. Grill, B. S. Meyerson: Development and status of diamondlike carbon. In: *Synthetic Diamond: Emerging CVD Science and Technology*, ed. by K. E. Spear, J. P. Dismukes (Wiley, New York 1994) pp. 91–141
28. Y. Catherine: Preparation techniques for diamond-like carbon. In: *Diamond and Diamond-Like Films and Coatings*, ed. by R. E. Clausing, L. L. Horton, J. C. Angus, P. Koidl (Plenum, New York 1991) pp. 193–227
29. J. J. Cuomo, D. L. Pappas, J. Bruley, J. P. Doyle, K. L. Seagner: Vapor deposition processes for amorphous carbon films with sp^3 fractions approaching diamond, J. Appl. Phys. **70**, 1706–1711 (1991)
30. J. C. Angus, C. C. Hayman: Low pressure metastable growth of diamond and diamond-like phase, Science **241**, 913–921 (1988)
31. J. C. Angus, F. Jensen: Dense diamondlike hydrocarbons as random covalent networks, J. Vacuum Sci. Technol. A **6**, 1778–1782 (1988)
32. D. C. Green, D. R. McKenzie, P. B. Lukins: The microstructure of carbon thin films, Mater. Sci. Forum **52-53**, 103–124 (1989)
33. N. H. Cho, K. M. Krishnan, D. K. Veirs, M. D. Rubin, C. B. Hopper, B. Bhushan, D. B. Bogy: Chemical structure and physical properties of diamond-like amorphous carbon films prepared by magnetron sputtering, J. Mater. Res. **5**, 2543–2554 (1990)
34. J. C. Angus: Diamond and diamondlike films, Thin Solid Films **216**, 126–133 (1992)
35. B. Bhushan, A. J. Kellock, N. H. Cho, J. W. Ager III: Characterization of chemical bonding and physical characteristics of diamond-like amorphous carbon and diamond films, J. Mater. Res. **7**, 404–410 (1992)
36. J. Robertson: Properties of diamond-like carbon, Surf. Coat. Technol. **50**, 185–203 (1992)
37. N. Savvides, B. Window: Diamondlike amorphous carbon films prepared by magnetron sputtering of graphite, J. Vacuum Sci. Technol. A **3**, 2386–2390 (1985)
38. J. C. Angus, P. Koidl, S. Domitz: Carbon thin films. In: *Plasma Deposited Thin Films*, ed. by J. Mort, F. Jensen (CRC, Boca Raton 1986) pp. 89–127
39. J. Robertson: Amorphous carbon, Adv. Phys. **35**, 317–374 (1986)
40. M. Rubin, C. B. Hooper, N. H. Cho, B. Bhushan: Optical and mechanical properties of DC sputtered carbon films, J. Mater. Res. **5**, 2538–2542 (1990)
41. G. J. Vandentop, M. Kawasaki, R. M. Nix, I. G. Brown, M. Salmeron, G. A. Somorjai: Formation of hydrogenated amorphous carbon films of controlled hardness from a methane plasma, Phys. Rev. B **41**, 3200–3210 (1990)
42. J. J. Cuomo, D. L. Pappas, R. Lossy, J. P. Doyle, J. Bruley, G. W. Di Bello, W. Krakow: Energetic carbon deposition at oblique angles, J. Vacuum Sci. Technol. A **10**, 3414–3418 (1992)
43. D. L. Pappas, K. L. Saegner, J. Bruley, W. Krakow, J. J. Cuomo: Pulsed laser deposition of diamondlike carbon films, J. Appl. Phys. **71**, 5675–5684 (1992)
44. H. J. Scheibe, B. Schultrich: DLC film deposition by laser-arc and study of properties, Thin Solid Films **246**, 92–102 (1994)

45. C. Donnet, A. Grill: Friction control of diamond-like carbon coatings, Surf. Coat. Technol. **94-95**, 456 (1997)
46. A. Grill: Tribological properties of diamondlike carbon and related materials, Surf. Coat. Technol. **94-95**, 507 (1997)
47. B. K. Gupta, B. Bhushan: Micromechanical properties of amorphous carbon coatings deposited by different deposition techniques, Thin Solid Films **270**, 391–398 (1995)
48. X. Li, B. Bhushan: Micro/nanomechanical and tribological characterization of ultra-thin amorphous carbon coatings, J. Mater. Res. **14**, 2328–2337 (1999)
49. X. Li, B. Bhushan: Mechanical and tribological studies of ultra-thin hard carbon overcoats for magnetic recording heads, Z. Metallkd. **90**, 820–830 (1999)
50. S. Sundararajan, B. Bhushan: Micro/nanotribology of ultra-thin hard amorphous carbon coatings using atomic force/friction force microscopy, Wear **225-229**, 678–689 (1999)
51. B. Bhushan: Chemical, mechanical, and tribological characterization of ultra-thin and hard amorphous carbon coatings as thin as 3.5 nm: Recent developments, Diam. Relat. Mater. **8**, 1985–2015 (1999)
52. I. I. Aksenov, V. E. Strel'Nitskii: Wear resistance of diamond-like carbon coatings, Surf. Coat. Technol. **47**, 252–256 (1991)
53. D. R. McKenzie, D. Muller, B. A. Pailthorpe, Z. H. Wang, E. Kravtchinskaia, D. Segal, P. B. Lukins, P. J. Martin, G. Amaratunga, P. H. Gaskell, A. Saeed: Properties of tetrahedral amorphous carbon prepared by vacuum arc deposition, Diam. Relat. Mater. **1**, 51–59 (1991)
54. R. Lossy, D. L. Pappas, R. A. Roy, J. J. Cuomo: Filtered arc deposition of amorphous diamond, Appl. Phys. Lett. **61**, 171–173 (1992)
55. I. G. Brown, A. Anders, S. Anders, M. R. Dickinson, I. C. Ivanov, R. A. MacGill, X. Y. Yao, K. M. Yu: Plasma synthesis of metallic and composite thin films with atomically mixed substrate bonding, Nucl. Instrum. Meth. B **80-81**, 1281–1287 (1993)
56. P. J. Fallon, V. S. Veerasamy, C. A. Davis, J. Robertson, G. A. J. Amaratunga, W. I. Milne, J. Koskinen: Properties of filtered-ion-beam-deposited diamond-like carbon as a function of ion energy, Phys. Rev. B **48**, 4777–4782 (1993)
57. S. Anders, A. Anders, I. G. Brown, B. Wei, K. Komvopoulos, J. W. Ager III, K. M. Yu: Effect of vacuum arc deposition parameters on the properties of amorphous carbon thin films, Surf. Coat. Technol. **68-69**, 388–393 (1994)
58. S. Anders, A. Anders, I. G. Brown, M. R. Dickinson, R. A. MacGill: Metal plasma immersion ion implantation and deposition using arc plasma sources, J. Vacuum Sci. Technol. B **12**, 815–820 (1994)
59. S. Anders, A. Anders, I. G. Brown: Transport of vacuum arc plasma through magnetic macroparticle filters, Plasma Sources Sci. **4**, 1–12 (1995)
60. D. M. Swec, M. J. Mirtich, B. A. Banks: *Ion Beam and Plasma Methods of Producing Diamondlike Carbon Films*, Report No. NASA TM102301 (NASA, Cleveland 1989)
61. A. Erdemir, M. Switala, R. Wei, P. Wilbur: A tribological investigation of the graphite-to-diamond-like behavior of amorphous carbon films ion beam deposited on ceramic substrates, Surf. Coat. Technol. **50**, 17–23 (1991)
62. A. Erdemir, F. A. Nicols, X. Z. Pan, R. Wei, P. J. Wilbur: Friction and wear performance of ion-beam deposited diamond-like carbon films on steel substrates, Diam. Relat. Mater. **3**, 119–125 (1993)
63. R. Wei, P. J. Wilbur, M. J. Liston: Effects of diamond-like hydrocarbon films on rolling contact fatigue of bearing steels, Diam. Relat. Mater. **2**, 898–903 (1993)
64. A. Erdemir, C. Donnet: Tribology of diamond, diamond-like carbon, and related films. In: *Modern Tribology Handbook*, ed. by B. Bhushan (CRC, Boca Raton 2001) pp. 871–908

65. J. Asmussen: Electron cyclotron resonance microwave discharges for etching and thin-film deposition, J. Vacuum Sci. Technol. A **7**, 883–893 (1989)
66. J. Suzuki, S. Okada: Deposition of diamondlike carbon films using electron cyclotron resonance plasma chemical vapor deposition from ethylene gas, Jpn. J. Appl. Phys. **34**, L1218–L1220 (1995)
67. B. A. Banks, S. K. Rutledge: Ion beam sputter deposited diamond like films, J. Vacuum Sci. Technol. **21**, 807–814 (1982)
68. C. Weissmantel, K. Bewilogua, K. Breuer, D. Dietrich, U. Ebersbach, H. J. Erler, B. Rau, G. Reisse: Preparation and properties of hard i-C and i-BN coatings, Thin Solid Films **96**, 31–44 (1982)
69. H. Dimigen, H. Hubsch: Applying low-friction wear-resistant thin solid films by physical vapor deposition, Philips Tech. Rev. **41**, 186–197 (1983/84)
70. B. Bhushan, B. K. Gupta, M. H. Azarian: Nanoindentation, microscratch, friction and wear studies for contact recording applications, Wear **181-183**, 743–758 (1995)
71. B. Bhushan, L. Yang, C. Gao, S. Suri, R. A. Miller, B. Marchon: Friction and wear studies of magnetic thin-film rigid disks with glass-ceramic, glass and aluminum-magnesium substrates, Wear **190**, 44–59 (1995)
72. L. Holland, S. M. Ojha: Deposition of hard and insulating carbonaceous films of an RF target in butane plasma, Thin Solid Films **38**, L17–L19 (1976)
73. L. P. Andersson: A review of recent work on hard i-C films, Thin Solid Films **86**, 193–200 (1981)
74. A. Bubenzer, B. Dischler, B. Brandt, P. Koidl: R.F. plasma deposited amorphous hydrogenated hard carbon thin films, preparation, properties and applications, J. Appl. Phys. **54**, 4590–4594 (1983)
75. A. Grill, B. S. Meyerson, V. V. Patel: Diamond-like carbon films by RF plasma-assisted chemical vapor deposition from acetylene, IBM J. Res. Dev. **34**, 849–857 (1990)
76. A. Grill, B. S. Meyerson, V. V. Patel: Interface modification for improving the adhesion of a-C:H to metals, J. Mater. Res. **3**, 214 (1988)
77. A. Grill, V. V. Patel, B. S. Meyerson: Optical and tribological properties of heat-treated diamond-like carbon, J. Mater. Res. **5**, 2531–2537 (1990)
78. F. Jansen, M. Machonkin, S. Kaplan, S. Hark: The effect of hydrogenation on the properties of ion beam sputter deposited amorphous carbon, J. Vacuum Sci. Technol. A **3**, 605–609 (1985)
79. S. Kaplan, F. Jansen, M. Machonkin: Characterization of amorphous carbon-hydrogen films by solid-state nuclear magnetic resonance, Appl. Phys. Lett. **47**, 750–753 (1985)
80. H. C. Tsai, D. B. Bogy, M. K. Kundmann, D. K. Veirs, M. R. Hilton, S. T. Mayer: Structure and properties of sputtered carbon overcoats on rigid magnetic media disks, J. Vacuum Sci. Technol. A **6**, 2307–2315 (1988)
81. B. Marchon, M. Salmeron, W. Siekhaus: Observation of graphitic and amorphous structures on the surface of hard carbon films by scanning tunneling microscopy, Phys. Rev. B **39**, 12907–12910 (1989)
82. B. Dischler, A. Bubenzer, P. Koidl: Hard carbon coatings with low optical absorption, Appl. Phys. Lett. **42**, 636–638 (1983)
83. D. S. Knight, W. B. White: Characterization of diamond films by Raman spectroscopy, J. Mater. Res. **4**, 385–393 (1989)
84. J. W. Ager, D. K. Veirs, C. M. Rosenblatt: Spatially resolved Raman studies of diamond films grown by chemical vapor deposition, Phys. Rev. B **43**, 6491–6499 (1991)
85. W. Scharff, K. Hammer, O. Stenzel, J. Ullman, M. Vogel, T. Frauenheim, B. Eibisch, S. Roth, S. Schulze, I. Muhling: Preparation of amorphous i-C films by ion-assisted methods, Thin Solid Films **171**, 157–169 (1989)

86. B. Bhushan, X. Li: Nanomechanical characterization of solid surfaces and thin films, Int. Mater. Rev. **48**, 125–164 (2003)
87. X. Li, D. Diao, B. Bhushan: Fracture mechanisms of thin amorphous carbon films in nanoindentation, Acta Mater. **45**, 4453–4461 (1997)
88. X. Li, B. Bhushan: Measurement of fracture toughness of ultra-thin amorphous carbon films, Thin Solid Films **315**, 214–221 (1998)
89. X. Li, B. Bhushan: Evaluation of fracture toughness of ultra-thin amorphous carbon coatings deposited by different deposition techniques, Thin Solid Films **355-356**, 330–336 (1999)
90. X. Li, B. Bhushan: Development of a nanoscale fatigue measurement technique and its application to ultrathin amorphous carbon coatings, Scripta Mater. **47**, 473–479 (2002)
91. X. Li, B. Bhushan: Nanofatigue studies of ultrathin hard carbon overcoats used in magnetic storage devices, J. Appl. Phys. **91**, 8334–8336 (2002)
92. J. Robertson: Deposition of diamond-like carbon, Philos. Trans. R. Soc. Lond. A **342**, 277–286 (1993)
93. S. J. Bull: Tribology of carbon coatings: DLC, diamond and beyond, Diam. Relat. Mater. **4**, 827–836 (1995)
94. N. Savvides, T. J. Bell: Microhardness and Young's modulus of diamond and diamond-like carbon films, J. Appl. Phys. **72**, 2791–2796 (1992)
95. B. Bhushan, M. F. Doerner: Role of mechanical properties and surface texture in the real area of contact of magnetic rigid disks, ASME J. Tribol. **111**, 452–458 (1989)
96. S. Suresh: *Fatigue of Materials* (Cambridge Univ. Press, Cambridge 1991)
97. D. B. Marshall, A. G. Evans: Measurement of adherence of residual stresses in thin films by indentation. I. Mechanics of interface delamination, J. Appl. Phys. **15**, 2632–2638 (1984)
98. A. G. Evans, J. W. Hutchinson: On the mechanics of delamination and spalling in compressed films, Int. J. Solids Struct. **20**, 455–466 (1984)
99. S. Sundararajan, B. Bhushan: Development of a continuous microscratch technique in an atomic force microscope and its application to study scratch resistance of ultrathin hard amorphous carbon coatings, J. Mater. Res. **16**, 437–445 (2001)
100. B. Bhushan, B. K. Gupta: Micromechanical characterization of Ni-P coated aluminum-magnesium, glass and glass-ceramic substrates and finished magnetic thin-film rigid disks, Adv. Info. Stor. Syst. **6**, 193–208 (1995)
101. X. Li, B. Bhushan: Micromechanical and tribological characterization of hard amorphous carbon coatings as thin as 5 nm for magnetic recording heads, Wear **220**, 51–58 (1998)
102. B. Bhushan, V. N. Koinkar: Microscale mechanical and tribological characterization of hard amorphous coatings as thin as 5 nm for magnetic disks, Surf. Coat. Technol. **76-77**, 655–669 (1995)
103. V. N. Koinkar, B. Bhushan: Microtribological properties of hard amorphous carbon protective coatings for thin-film magnetic disks and heads, Proc. Inst. Mech. Eng. Part J **211**, 365–372 (1997)
104. T. W. Wu: Microscratch and load relaxation tests for ultra-thin films, J. Mater. Res. **6**, 407–426 (1991)
105. R. Memming, H. J. Tolle, P. E. Wierenga: Properties of polymeric layers of hydrogenated amorphous carbon produced by plasma-activated chemical vapor deposition: tribological and mechanical properties, Thin Solid Films **143**, 31–41 (1986)
106. C. Donnet, T. Le Mogne, L. Ponsonnet, M. Belin, A. Grill, V. Patel: The respective role of oxygen and water vapor on the tribology of hydrogenated diamond-like carbon coatings, Tribol. Lett. **4**, 259 (1998)

107. F.P. Bowden, J.E. Young: Friction of diamond, graphite and carbon and the influence of surface films, Proc. R. Soc. Lond. **208**, 444–455 (1951)
108. B. Marchon, N. Heiman, M.R. Khan: Evidence for tribochemical wear on amorphous carbon thin films, IEEE Trans. Magn. **26**, 168–170 (1990)
109. M.T. Dugger, Y.W. Chung, B. Bhushan, W. Rothschild: Friction, wear, and interfacial chemistry in thin film magnetic rigid disk files, ASME J. Tribol. **112**, 238–245 (1990)
110. B.D. Strom, D.B. Bogy, C.S. Bhatia, B. Bhushan: Tribochemical effects of various gases and water vapor on thin film magnetic disks with carbon overcoats, ASME J. Tribol. **113**, 689–693 (1991)
111. B. Bhushan, J. Ruan: Tribological performance of thin film amorphous carbon overcoats for magnetic recording rigid disks in various environments, Surf. Coat. Technol. **68/69**, 644–650 (1994)
112. B. Bhushan, G.S.A.M. Theunissen, X. Li: Tribological studies of chromium oxide films for magnetic recording applications, Thin Solid Films **311**, 67–80 (1997)

17

Self-Assembled Monolayers (SAMs) for Controlling Adhesion, Friction, and Wear

Bharat Bhushan

Summary. Making micro- and nanodevices as well as magnetic storage devices reliable necessitates the use of protective hydrophobic lubricating films that can minimize the adhesion, friction, and wear of sliding surfaces. Because of the small clearances associated with these devices, such films need to be very thin (on the order of a few molecules thick). Chemically-bonded low surface tension liquid films are suitable for this purpose, as are a select number of hydrophobic solid films. Highly hydrophobic ordered molecular assemblies can also be used; these are engineered by chemically grafting various polymer molecules with suitable functional head groups, spacer chains and nonpolar surface terminal groups to the surface involved.

In this chapter, we focus on the use of self-assembled monolayers (SAMs) for high hydrophobicity and/or low adhesion, friction and wear applications. SAMs are produced by various organic precursors, so the chapter starts with a primer for the organic chemistry associated with this field. This is followed by an overview of selected SAMs with various spacer chains and terminal groups in their molecular chains on a variety of substrates, and a summary of the tribological properties of SAMs. The adhesion, friction and wear properties of SAMs with various spacer chains and surface terminal and head groups (hexadecane thiol, biphenyl thiol, alkylsilane, perfluoroalkylsilane and alkylphosphonate) on various substrates (Au, Si, and Al) are then surveyed. Degradation mechanisms and environmental effects are studied. Nanotribological studies of various SAM films by atomic force microscopy (AFM), show that perfluoroalkylsilane SAMs in particular exhibit attractive hydrophobic and tribological properties.

17.1 Introduction

Maximizing reliability is an important issue in the field of micro/nanodevices, commonly referred to as micro/nanoelectromechanical systems (MEMS/NEMS) and BioMEMS/BioNEMS, as well as for magnetic storage devices (which include magnetic rigid disk drives, flexible disk drives and tape drives). It often necessitates the application of molecular films to sliding surfaces in order to lubricate and protect them [1–8]. A solid or liquid film is generally required in order to obtain acceptable tribological properties for sliding interfaces. However, the presence of a small quantity of liquid between smooth surfaces can substantially increase the adhesion, friction and wear due to the formation of menisci or adhesive bridges [9, 10]. This becomes a major concern in micro/nanodevices operating at ultralow loads, as the magnitude of the liquid-mediated adhesive force may be similar to that of the external load.

The liquid film may be pre-existing and/or it can be a capillary condensate of water vapor from the environment. If the liquid wets the surface ($0 \leq \theta < 90°$, where θ is the contact angle between the liquid–vapor interface and the liquid–solid interface for a liquid droplet sitting on a solid surface, Fig. 17.1a), the surface of the liquid is constrained to lie parallel with the solid surface [11–13], and so the surface of the liquid must be concave, Fig. 17.1b. The direct measurement of the contact angle θ is often achieved using sessile drops. The angle is generally measured by finding the tangent to the drop profile at the point of contact with the solid surface using a telescope equipped with a goniometer eyepiece.

Surface tension results in a pressure difference across any meniscus surface, referred to as the capillary pressure or the Laplace pressure, and this is negative for a concave meniscus [9, 10]. The negative Laplace pressure results in an intrinsic attractive (adhesive) force that depends on the roughness (the local geometries of interacting asperities and the number of asperities) of the interface, the surface tension and the contact angle. During sliding, frictional effects must be overcome, not only because of external load but also because of intrinsic adhesive force. A high static friction force deriving largely from liquid-mediated adhesion (contribution of the meniscus) is generally referred to as "stiction". There are three basic ways to minimize the effect of liquid-mediated adhesion: to increase the surface roughness, to use hydrophobic (water-fearing) rather than hydrophilic (water-loving) surfaces, and/or to use a liquid with low surface tension [9, 10, 14–16].

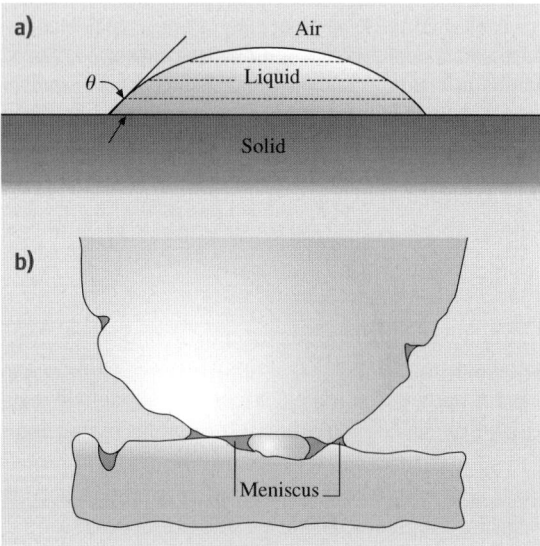

Fig. 17.1. (a) Schematic of a sessile drop on a solid surface and the definition of contact angle, and (b) the formation of meniscus bridges due to the presence of liquid at an interface

As an example, bulk silicon and polysilicon films used in the construction of micro/nanodevices can be dipped in hydrofluoric (HF) acid to make them hydrophobic [17]. In the etching of silicon with HF, hydrogen passivates the silicon surface by saturating the dangling bonds, resulting in a hydrogen-terminated silicon surface that adsorbs less water. However, after the treated surface has been exposed to the environment it reoxidizes, which means that it can adsorb water and so the surface again becomes hydrophilic.

The surfaces can also be treated or coated with a liquid with relatively low surface tension or with certain types of solid film to make them hydrophobic and/or to control adhesion, friction and wear. The liquid lubricant film should be thin (about half the composite roughness of the interface) to minimize the liquid-mediated adhesion [9, 10, 14, 15, 18, 19]. Thus, for ultra-smooth surfaces with RMS roughnesses of a few nm, molecularly thick liquid films are required for liquid lubrication. The classical approach to lubrication uses freely supported multimolecular layers of liquid lubricants [2, 4, 9, 10, 19–23]. Boundary lubricant films are formed by either physisorption, chemisorption or chemical reaction. The thickness of the physisorbed film can be either monomolecular or polymolecular. The chemisorbed films are monomolecular, but stoichiometric films formed by chemical reaction can be multilayers. In general, the stabilities and durabilities of surface films decrease in the following order: chemically reacted films, chemisorbed films and physisorbed films. A good boundary lubricant should have a high degree of interaction between its molecules and the sliding surface. As a general rule, liquids will perform better when they are polar and thus able to grip onto solid surfaces (or when they can be adsorbed). Polar lubricants contain reactive functional end groups. Boundary lubrication properties are also dependent upon the molecular conformation and lubricant spreading. It should be noted that a liquid film with a thickness of a few nanometers may be discontinuous and may deposit in "islands" where the layer has nonuniform thickness with a lateral resolution on the nanometer scale.

Solid films are also commonly used to control hydrophobicity and/or adhesion, friction and wear. Hydrophobic films have nonpolar surface terminal groups (described later) that repel water. These films have low surface energies (15 to 30 dyn/cm) and high contact angles ($\theta \geq 90°$) which minimize wetting [21, 24, 25]. Multimolecularly thick (a few tenths of a nanometer in thickness) films of conventional solid lubricants have been studied. *Hansma* et al. [26] reported the deposition of multimolecularly thick, highly oriented PTFE films from the melt, vapor phase or from solution by a mechanical deposition technique achieved by dragging the polymer at controlled temperature, pressure and speed against a smooth glass substrate. *Scandella* et al. [27] reported that the coefficient of nanoscale friction for MoS_2 platelets on mica, obtained by exfoliation of lithium-intercalated MoS_2 in water, was a factor of 1.4 less than that of mica itself. However, MoS_2 reacts with water and its friction and wear properties degrade with increasing humidity [9, 10]. Amorphous diamondlike carbon (DLC) coatings can be produced with extremely high hardnesses and are used commercially as wear-resistant coatings [28, 29]. They are widely used in magnetic storage devices [2]. Doping the DLC matrix with elements like hydrogen, nitrogen, oxygen, silicon and fluorine influences its hydrophobicity and tribological properties [28, 30, 31]. Nitrogen and oxygen reduce the contact angle (or increase the surface energy) due to the strong polarity induced when these elements are bonded to carbon. On the other hand, silicon and fluorine increase the contact angle to 70–100° (or reduce the surface energy to 20–40 dyn/cm) making them hydrophobic [32, 33]. Nanocomposite coatings with diamondlike carbon (a-C:H) networks and glasslike a-Si:O networks are generally deposited using a PECVD (plasma-enhanced chemical vapor deposition) technique in which plasma is formed from a siloxane precursor using a hot filament. For a fluorinated DLC, CF_4 is added to acetylene plasma to provide the fluorocarbon source. In addition, fluorination of DLC can be achieved by the post-deposition treatment of DLC coatings in a CF_4 plasma. Silicon- and fluorine-containing DLC coatings usually have reduced polarity due to the loss of sp^2-bonded carbon (resulting in reduced polarization potential from π electrons) and the dangling bonds of the DLC network. As silicon and fluorine are unable to form double bonds, they force carbon into an sp^3 bonding state [33]. The friction and wear properties of both silicon-containing and fluorinated DLC coatings have been reported to be superior to those of conventional DLC coatings [34, 35]. However, DLC coatings require a line-of-sight deposition process, which prevents deposition on complex geometries. Furthermore, it has been reported that some self-assembled monlayers (SAMs) are superior to DLC coatings in terms of hydrophobicity and tribological performance [36, 37].

Organized and dense molecular scale layers of, preferably, long-chain organic molecules are known to be better lubricants on macro-, micro- and nanoscales than freely supported multimolecular layers [4, 5, 38–48]. Common techniques used to produce molecular scale organized layers include Langmuir–Blodgett (LB) deposition and chemical grafting of organic molecules to the surface to realize SAMs [24, 25]. In the LB technique, organic molecules from suitable amphiphilic molecules are first organized at the air–water interface and then physisorbed onto a solid surface to form mono- or multimolecular layers [49]. On the other hand, SAMs are produced when functional groups of molecules chemisorb on a solid surface, which results in the spontaneous formation of robust, highly ordered, oriented and dense monolayers [25]. In both cases, the organic molecules used have well distinguished amphiphilic properties (a hydrophilic functional head and a hydrophobic aliphatic tail) so that adsorption of the molecules on an active inorganic substrate leads to their firm attachment to the surface. Direct organization of SAMs on the solid surfaces allows tight areas such as the bearing and journal surfaces in an assembled bearing to be coated. The weak adhesion of classical LB films to the substrate surface restricts their lifetimes during sliding, whereas certain SAMs can be very durable. As a result, SAMs are of great interest for tribological applications.

Much research into the application of SAMs has been associated with the so-called soft lithographic technique [50, 51]. This is a nonphotolithographic technique. Photolithography is based on a projection printing system used to project an image from a mask to a thin film photoresist, and its resolution is limited by optical diffraction limits. In soft lithography, an elastomeric stamp or mold is used to generate micropatterns of SAMs using either contact printing (known as microcontact printing or μCP [52]), embossing (nanoimprint lithography) [53], or replica molding [54], which all circumvent the diffraction limits of photolithography. The stamps are generally cast from photolithographically generated patterned masters, and the stamp material is generally polydimethylsiloxane (PDMS). In μCP, the ink is a SAM precursor, and nanometer-thick resists with lines thinner than 100 nm are produced. Soft lithography requires little capitol investment. μCP and embossing techniques can be used to produce microdevices that are substantially cheaper and more flexible in terms of construction materials than conventional photolithography (examples include SAMs and non-SAM entities for μCP, and elastomers for embossing).

The largest industrial application of SAMs is in digital micromirror devices (DMD) used in optical projection displays [55, 56]. The chip set of a DMD consists of half a million to more than two million independently controlled reflective aluminium alloy micromirrors each about 12 μm square. These micromirrors switch forward and backward at a frequency of around 5–7 kHz with a rotation of ±12° with respect to the horizontal plane; the movement is limited by a mechanical stop. Mechanical contact leads to stiction and wear of the contacting surfaces. A SAM of vapor-deposited perfluorinated n-alkanoic acid ($C_nF_{2n-1}O_2H$) (such as perfluorodecanoic acid, or PFDA, $CF_3(CF_2)_8COOH$), is used to coat the contacting surfaces to make them hydrophobic in order to minimize meniscus formation. Furthermore, the entire DMD chip set is hermetically sealed in order to prevent particulate contamination and excessive condensation of water at the contacting surfaces. A so-called "getter" strip of PFDA is included inside the hermetically sealed enclosure containing the chip, which acts as a reservoir that maintains PFDA vapor within the package. Degradation mechanisms of SAMs leading to stiction have been studied by *Liu* and *Bhushan* [57, 58].

Other industrial applications of SAMs are in the areas of bio/chemical and optical sensors, devices for use as drug delivery vehicles, and in the construction of electronic components [59–62]. Bio/chemical sensors require highly sensitive organic layers with tailored biological properties that can be incorporated into electronic, optical or electrochemical devices. Self-assembled microscopic vesicles are being developed to ferry potentially life-saving drugs to cancer patients. By assembling organic, metal and phosphonate molecules (complexes of phosphorous and oxygen atoms) into conductive materials, these can be produced as sandwiches for use as electronic components. Several applications have been proposed based on silicon, glass or polymer nanochannels, including cell immunoisolation chambers, protective DNA separation devices, and biocapsules for drug delivery. Using hydrophobic surfaces in gas-based separations based on nanochannels provides several advantages, including low fouling and high gas transport rates.

An overview of molecularly thick layers of liquid lubricants and conventional solid lubricants can be found in various references, such as *Bhushan* [2, 4, 9, 10, 18, 28], *Bhushan* and *Zhao* [19], and *Liu* and *Bhushan* [23]. In this chapter, we focus on the use of SAMs for high hydrophobicity, low adhesion, friction and wear applications. SAMs are produced by various organic precursors, and so we first present a primer to the organic chemistry associated with this topic. This is followed by an overview of suitable substrates, spacer chains and molecular chain end groups, a summary of the tribological properties of various SAMs, and finally some concluding remarks.

17.2 A Brief Organic Chemistry Primer

All organic compounds contain the carbon (C) atom. The fact that carbon can be combined with hydrogen, oxygen, nitrogen, sulfur and phosphorus results in a vast number of potential organic compounds. The atomic number of carbon is six, and its electron structure is $1s^2\ 2s^2\ 2p^2$. Two stable isotopes of carbon, ^{12}C and ^{13}C, exist. With four electrons in its outer shell, carbon forms four covalent bonds, with each bond resulting from two atoms sharing a pair of electrons. The number of electron pairs that two atoms share determines whether or not the bond is single or multiple. In a single bond, only one pair of electrons is shared by the atoms. Carbon can also form multiple bonds by sharing two or three pairs of electrons between the atoms. For example, the double bond formed by sharing two electron pairs is stronger than a single bond and it is shorter than a single bond. An organic compound is classified as saturated if it contains only a single bond and unsaturated if the molecules possess one or more multiple carbon–carbon bonds.

17.2.1 Electronegativity/Polarity

When two different kinds of atoms share a pair of electrons a bond is formed in which electrons are shared unequally; one atom assumes a partial positive charge and the other a negative charge with respect to each other. This difference in charge occurs because the two atoms exert different levels of attraction on the pair of shared electrons. The attractive force that an atom of a particular element has for shared electrons in a molecule or polyatomic ion is known as its electronegativity. Elements differ in their electronegativities. A scale of relative electronegativities, in which the most electronegative element, fluorine, is assigned a value of 4.0, was developed by Pauling. Relative electronegativities of the elements in the periodic table can be found in most undergraduate chemistry textbooks (see [63] for example). The relative electronegativities of nonmetals are high compared to those of metals. The relative electronegativities of selected elements of interest (those with high electronegativities) are presented in Table 17.1.

The polarity of a bond is determined by the difference in electronegativity between the atoms forming the bond. If the electronegativities are the same, the bond is completely nonpolar and the electrons are shared equally. In this type of bond there is no separation of positive and negative charge between atoms. If the atoms have vastly different electronegativities the bond is very polar. A dipole is a molecule that is electrically asymmetrical, causing it to be oppositely charged at two points. As an example, in hydrogen chloride (HCl), both hydrogen and chlorine need one electron to form stable electron configurations. They share a pair of electrons. Chlorine is

Table 17.1. Relative electronegativities of selected elements

Element	Relative electronegativity
F	4.0
O	3.5
N	3.0
Cl	3.0
C	2.5
S	2.5
P	2.1
H	2.1

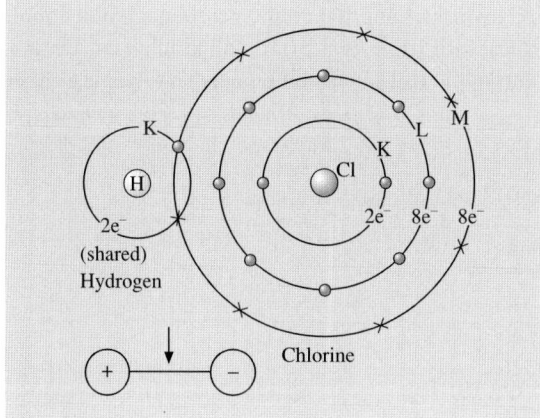

Fig. 17.2. Schematic representation of the formation of a polar HCl molecule

more electronegative and therefore has a greater attraction for the shared electrons than hydrogen does. As a result, the pair of electrons is displaced towards the chlorine atom, giving it a partial negative charge and leaving the hydrogen atom with a partial positive charge, (Fig. 17.2). However, the entire molecule, HCl, is electrically neutral. The hydrogen atom with its partial positive charge (exposed proton on one end) can be easily attracted to the negative charge of another molecule, and this is responsible for the polarity of the molecule. A partial charge is usually indicated by δ, and the electronic structure of HCl is given as:

$$\overset{\delta+}{H}:\overset{\delta-}{\ddot{C}l}: \, .$$

Similar to the HCl molecule, HF is polar and behaves as a small dipole. On the other hand, methane (CH_4), carbon tetrachloride (CCl_4) and carbon dioxide (CO_2) are nonpolar. In CH_4 and CCl_4, the four C–H and C–Cl polar bonds are identical, and because these bonds emanate from the center to the corners of a tetrahedron in the molecule, their polarities cancel when considering the whole molecule. CO_2(O=C=O) is nonpolar because the carbon–oxygen dipoles cancel each other by acting in opposite directions. Water (H–O–H) is a polar molecule. If the atoms in water were linear as in CO_2, the two O–H dipoles would cancel each other, and the molecule would be nonpolar. However, water has a bent structure, with an angle of 105° between the two bonds, which causes water to be a polar molecule.

17.2.2 Classification and Structures of Organic Compounds

Table 17.2 presents selected organic compounds grouped into classes.

Hydrocarbons

Hydrocarbons are compounds that are composed entirely of carbon and hydrogen atoms bonded to each other by covalent bonds. Saturated hydrocarbons (alkanes) contain single bonds. Unsaturated hydrocarbons that contain carbon–carbon double bonds are called alkenes, and ones with triple bonds are called alkynes. Unsaturated hydrocarbons that contain aromatic rings, including benzene rings, are called aromatic hydrocarbons.

Table 17.2. Names and formulae of selected hydrocarbons

Name	Formula
Saturated hydrocarbons	
Straight-chain alkanes	C_nH_{2n+2}
e.g., methane	CH_4
ethane	C_2H_6 or CH_3CH_3
Alkyl groups	C_nH_{2n+1}
e.g., methyl	$-CH_3$
ethyl	$-CH_2CH_3$
Unsaturated hydrocarbons	
Alkenes	$(CH_2)_n$
e.g., ethene	C_2H_4 or $CH_2=CH_2$
propene	C_3H_6 or $CH_3CH=CH_2$
Alkynes	
e.g., acetylene	$HC\equiv CH$
Aromatic hydrocarbons	
e.g., benzene	C_6H_5OH or (structure shown) 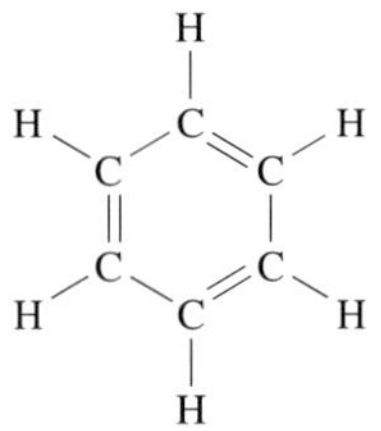

Saturated Hydrocarbons: Alkanes The alkanes, also known as paraffins, are saturated hydrocarbons: straight- or branched-chain hydrocarbons with only single covalent (saturated) bonds between the carbon atoms. The general molecular formula for alkanes is C_nH_{2n+2}, where n is the number of carbon atoms in the molecule. Each carbon atom is connected to four other atoms by four single covalent bonds. These bonds are separated by angles of 109.5° (the angle subtended by lines drawn from the center of a regular tetrahedron to its corners). Alkane molecules contain only carbon–carbon and carbon–hydrogen bonds, which are symmetrically directed towards the corners of a tetrahedron. Alkane molecules are therefore essentially nonpolar.

Common alkyl groups have the general formula C_nH_{2n+1} (one hydrogen atom less than the corresponding alkane). The missing H atom can be detached from any carbon in the alkane. The name of the group is formed from the name of the corresponding alkane by replacing -ane with -yl. Some examples are shown in Table 17.2.

Unsaturated Hydrocarbons Unsaturated hydrocarbons consist of three families of compounds that contain fewer hydrogen atoms than the alkane with the corresponding number of carbon atoms, and contain multiple bonds (unsaturation) between carbon atoms. These include alkenes (with carbon–carbon double bonds), alkynes (with carbon–carbon triple bonds), and aromatic compounds (with benzene rings, which are arranged in a six-membered ring with one hydrogen atom bonded to each carbon atom). Some examples are shown in Table 17.2.

Alcohols, Ethers, Phenols and Thiols

Organic molecules with certain functional groups are synthesized because they have desirable properties. Alcohols, ethers and phenols are derived from the structure of water by replacing the hydrogen atoms of water with hydroxy groups (OH), alkyl groups (R) or aromatic rings (Ar), respectively. For example, phenol is a class of compounds that has a hydroxy group attached to an aromatic ring (benzene ring). Organic compounds that contain the −SH group are analogs of alcohols, and are known as thiols. Some examples are shown in Table 17.3.

Aldehydes and Ketones

Both aldehydes and ketones contain the carbonyl group (>C=O), a carbon–oxygen double bond (C=O). Aldehydes have at least one hydrogen atom bonded to the carbon of the carbonyl group, whereas ketones have only alkyl or aromatic groups bonded to the carbon of the carbonyl group. The general formula for the saturated homologous series of aldehydes and ketones is $C_nH_{2n}O$. Some examples are shown in Table 17.4.

Carboxylic Acids and Esters

The functional group of the carboxylic acids is known as the carboxyl group. This group is essentially a carbonyl group where a hydroxy group is bonded to the carbon (−COOH). Carboxylic acids can be either aliphatic (RCOOH) or aromatic

Table 17.3. Names and formulae of selected alcohols, ethers, phenols and thiols

Name	Formula
Alcohols	R–OH
e.g., methanol	CH_3OH
ethanol	CH_3CH_2OH
Ethers	R–O–R'
e.g., dimethyl ether	CH_3-O-CH_3
diethyl ether	$CH_3CH_2-O-CH_2CH_3$
Phenols	C_6H_5OH or [phenol ring structure with OH]
Thiols	–SH
e.g., methanethiol	CH_3SH

The letters R– and R'– represent an alkyl group. The R– groups in ethers can be the same or different and can be alkyl or aromatic (Ar) groups

Table 17.4. Names and formulae of selected aldehydes and ketones

Name	Formula
Aldehydes	RCHO or $R-\overset{\overset{\displaystyle O}{\|\|}}{C}-H$
	ArCHO or $Ar-\overset{\overset{\displaystyle O}{\|\|}}{C}-H$
e.g., methanal or formaldehyde	HCHO
ethanal or acetaldehyde	CH_3CHO
Ketones	RCOR' or $R-\overset{\overset{\displaystyle O}{\|\|}}{C}-R'$
	RCOAr or $R-\overset{\overset{\displaystyle O}{\|\|}}{C}-Ar$
	ArCOAr or $Ar-\overset{\overset{\displaystyle O}{\|\|}}{C}-Ar$
e.g., butanone or methyl ethyl ketone	$CH_3COCH_2CH_3$

The letters R and R' represent an alkyl group and Ar represents an aromatic group

Table 17.5. Names and formulae of selected carboxylic acids and esters

Name	Formula
Carboxylic acid[a]	RCOOH or R—C(=O)—OH
	ArCOOH or Ar—C(=O)—OH
e.g., methanoic acid (formic acid)	HCOOH
ethanoic acid (acetic acid)	CH_3COOH
octadecanoic acid (stearic acid)	$CH_3(CH_2)_{16}COOH$
Esters[b]	RCOOR' or R—C(=O)—O—R' (acid)(alcohol)
e.g., methyl propanoate	$CH_3CH_2COOCH_3$

[a] The letter R represents an alkyl group and Ar represents an aromatic group

[b] The letter R represents a hydrogen, an alkyl group or an aromatic group and R' represents an alkyl group or an aromatic group

Table 17.6. Names and formulae of selected organic nitrogen compounds (amides and amines)

Name	Formula
Amides	$RCONH_2$ or R—C(=O)—NH_2
e.g., methanamide (formamide)	$HCONH_2$
ethanamide (acetamide)	CH_3CONH_2
Amines	RNH_2 or R—N(H)(H)
	R_2NH
	R_3N
e.g., methylamine	CH_3NH_2
ethylamine	$CH_3CH_2NH_2$

The letter R represents an alkyl group or aromatic group

(ArCOOH). Carboxylic acids with even numbers of carbon atoms, n (ranging from 4 to about 20), are called fatty acids (for example, $n = 10, 12, 14, 16$ and 18 are called capric acid, lauric acid, myristic acid, palmitic acid, and stearic acid, respectively).

Esters are alcohol derivatives of carboxylic acids, where the hydrogen of the hydroxyl group is replaced with an alkyl or aryl group. Therefore, their general formula is RCOOR′, where R can be a hydrogen, alkyl group or aromatic group, and R′ may be an alkyl group or aromatic group but not a hydrogen. Esters are found in fats and oils. Some examples are shown in Table 17.5.

Table 17.7. Some examples of polar (hydrophilic) and nonpolar (hydrophobic) groups

Name	Formula
Polar	
Alcohol (hydroxyl)	−OH
Carboxylic acid	−COOH
Aldehyde	−COH
Ketone	R−C(=O)−R
Ester	−COO−
Carbonyl	>C=O
Ether	R−O−R
Amine	−NH$_2$
Amide	−C(=O)−NH$_2$
Phenol	C$_6$H$_5$−OH
Thiol	−SH
Trichlorosilane	SiCl$_3$
Nonpolar	
Methyl	−CH$_3$
Trifluoromethyl	−CF$_3$
Aryl (benzene ring)	C$_6$H$_5$−

The letter R represents an alkyl group

Table 17.8. Organic groups listed in increasing order of polarity

---
Alkanes
Alkenes
Aromatic hydrocarbons
Ethers
Trichlorosilanes
Aldehydes, ketones, esters, carbonyls
Thiols
Amines
Alcohols, phenols
Amides
Carboxylic acids

Amides and Amines

Amides and amines are organic compounds containing nitrogen. Amides are nitrogen derivatives of carboxylic acids where the carbon atom of the carbonyl group is bonded directly to the nitrogen atom of an $-NH_2$, $-NHR$, or $-NR_2$ group instead of the oxygen of a hydroxyl group. The characteristic structure of an amide is $RCONH_2$.

An amine is a substituted ammonia molecule where at least one of the hydrogens attached to nitrogen is replaced with an alkyl or aryl group. It therefore has a general structure of RNH_2, R_2NH or R_3N, where R is an alkyl or aromatic group. Some examples are shown in Table 17.6.

17.2.3 Polar and Nonpolar Groups

Table 17.7 summarizes the polar and nonpolar groups commonly used to construct hydrophobic and hydrophilic molecules. Table 17.8 lists the relative polarities of selected polar groups [64]. Thiol, silane, carboxylic acid, and alcohol (hydroxyl) groups are the polar anchor groups most commonly used to attach to surfaces. Silane anchor groups are commonly used for Si or SiO_2 surfaces because $-Si-O-$ bonds are strong. Methyl and trifluoromethyl are commonly used as end groups for hydrophobic film surfaces.

17.3 Self-Assembled Monolayers: Substrates, Spacer Chains; and End Groups in the Molecular Chains

SAMs are formed as a result of the spontaneous self-organization of functionalized organic molecules into stable, well defined structures on substrate surfaces, Fig. 17.3. The final structure is close to or at thermodynamic equilibrium and so it tends to form spontaneously and it rejects defects. SAMs consist of three building

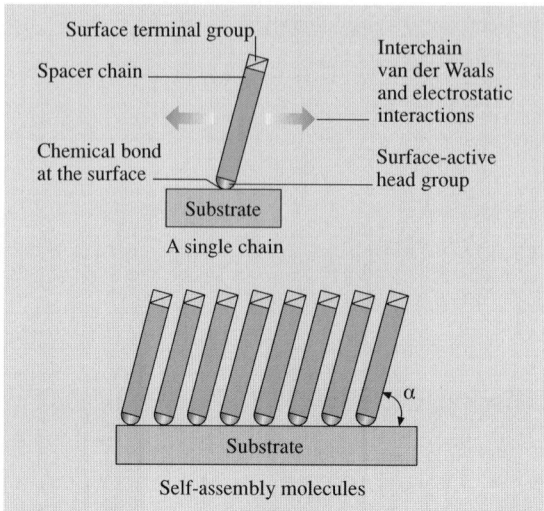

Fig. 17.3. Schematic of a self-assembled monolayer on a surface and associated forces

groups: a head group that binds strongly to a substrate, a surface terminal (tail or end) group that constitutes the outer surface of the film, and a spacer chain (backbone chain) that connects the head and surface terminal groups. The SAMs are named according to the surface terminal group (as opposed to the spacer chain and the head group), or the type of compound formed at the surface. In order to control hydrophobicity, adhesion, friction, and wear, it should strongly adhere to the substrate and the surface terminal group of the organic molecular chain should be nonpolar. To obtain strong attachment of the organic molecules to the substrate, the head group of the molecular chain should contain a polar terminal group that associates with the surface in an exothermic process (energies on the order of tens of kcal/mol); in other words it results in an apparent pinning of the head group to a specific site on the surface through a chemical bond. Furthermore, their molecular structures and the presence of crosslinking both have a significant effect on their friction and wear performance. The substrate surface should have a high surface energy (hydrophilic), so that there is a strong tendency for molecules to adsorb on the surface. The surface should be highly functional, with polar groups and dangling bonds (generally unpaired electrons), so that it can react with organic molecules and provide a strong bond. Because of the exothermic head group–substrate interactions, molecules try to occupy every available binding site on the surface, and during this process they generally push together molecules that have already been adsorbed. This process results into the formation of ordered molecular assemblies. The interactions between molecular chains are van der Waals or electrostatic in nature, with energies on the order of a few (< 10) kcal/mol (exothermic). The molecular chains in SAMs are not perpendicular to the surface; the tilt angle depends on the anchor group as well as on the substrate and the spacer group. For example, the tilt angle for alkanethiolate on Au is typically about $30-35°$ with respect to the substrate normal.

Table 17.9. Selected substrates and precursors commonly used to create SAMs

Substrate	Precursor	Binding with Substrate
Au	R SH (thiol)	RS–Au
Au	Ar SH (thiol)	ArS–Au
Au	RSSR′ (disulfide)	RS–Au
Au	RSR′ (sulfide)	
Si/SiO$_2$, glass	RSiCl$_3$ (trichlorosilane)	Si–O–Si (siloxane)
Si/Si–H	RCOOH (carboxyl)	R–Si
Metal oxides (e.g., Al$_2$O$_3$, SnO$_2$, TiO$_2$)	RCOOH (carboxyl)	RCOO–...MO$_n$

R represents alkane (C$_n$H$_{2n+2}$) and Ar represents aromatic hydrocarbon. These consist of various surface-active headgroups, usually with a methyl terminal group

Table 17.9 lists selected systems that have been used to form SAMs [51]. The spacer chain of the SAM is often an alkyl chain ((−CH$_2$)$_n$) or a derivatized alkyl group. By attaching different terminal groups to the surface, the film surface can be made to attract or repel water. Commonly used surface terminal groups in hydrophobic films with low surface energy, in the case of a single alkyl chain, include nonpolar methyl (−CH$_3$) or trifluoromethyl (−CF$_3$) groups. For hydrophilic films, commonly used surface terminal groups include alcohol (−OH) or carboxylic acid (−COOH) groups. Commonly used surface-active head groups include thiol (−SH), silane (such as trichlorosilane or −SiCl$_3$) and carboxyl (−COOH) groups. The substrates most commonly used are gold, silver, platinum, copper, hydroxylated (activated) surfaces of SiO$_2$ on Si, Al$_2$O$_3$ on Al, and glass.

The substrate surface should be clean before deposition. For silicon substrates, a concentrated HF solution (typically 49% HF) is commonly used to remove the oxide layer, and then the surface is rinsed with deionized water [45, 46]. Hydrogen passivates the surface by saturating the dangling bonds resulting in a hydrogen-terminated silicon surface with hydrophobic properties. In the deposition of multimolecularly thick polymer films with nonpolar ends, hydrophobic substrates can lead to a coated surface with a high contact angle, and are therefore preferred. The substrate should be hydrophilic for SAM deposition in order to ensure strong interfacial bonds with head groups. Hydroxylation of oxide surfaces is carried out to make them hydrophilic. Silicon and other metals get oxidized and hydroxylated to some degree when exposed to the environment. Bulk silicon, polysilicon film or SiO$_2$ film surfaces are commonly treated by immersing them in Piranha solution (a mixture of typically 3 : 1 v/v 98% H$_2$SO$_4$: 30% H$_2$O$_2$) at temperatures of about 90 °C for about 30 min, followed by rinsing in deionized water, to produce a hydroxylated silica surface [23, 45, 46]. Piranha solution also removes any organic and metallic contaminants, whereas HF would not necessarily remove organics. Oxygen plasma is another technique used to hydroxylate SiO$_2$ as well as polymer surfaces [43, 44]. For complex silicon geometries, oxygen plasma may be preferable. Figure 17.4 shows schematics of surfaces after various surface treatments. Surfaces

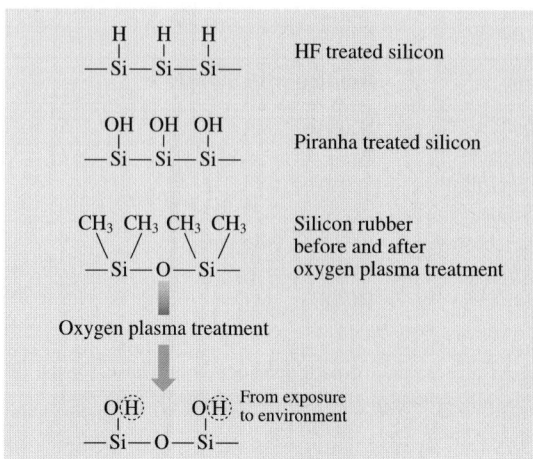

Fig. 17.4. Schematic showing HF-treated silica and the hydroxylation process that occurs on silica and elastomeric surfaces treated with Piranha solution and oxygen plasma, respectively

after piranha or oxygen plasma treatment remain hydrophilic for a few hours to about a day and become hydrophobic when they come into contact with carbon. They can retain hydrophilicity longer in dry nitrogen. To retain their hydrophobicity, the polymers are generally stored in DI water. Surfaces treated with HF remain hydrophobic for about 2–3 h and retain their hydrophobicity longer in dry nitrogen.

Epitaxial Au film on glass, mica or single-crystal silicon (produced by e-beam evaporation) substrates are commonly used because they can be deposited on smooth surfaces as an atomically flat and defect-free film. To get the organic molecules to pack together and provide a better ordering, the substrate should be chosen such that the cross-sectional diameter of the spacer chain is equal to or smaller than the distance between the anchor groups attached to the substrate. In the case of alkanethiol film, the advantage of Au substrate over SiO_2 substrate is that it results in better ordering because the cross-sectional diameter of the alkane molecule is slightly smaller than the distance between the sulfur atoms attached to the Au substrate (≈ 0.53 nm). The thickness of the film can be controlled by varying the length of the hydrocarbon chain, and the surface properties of the film can be modified by the terminal group.

SAMs are usually produced by immersing a substrate in the solution containing a precursor (ligand) that reacts with the substrate surface or by exposing the substrate to the vapor of the reactive chemical precursor [24]. A schematic of a vapor deposition system is shown in Fig. 17.5. Samples are placed in a quartz reaction tube. A silane bubbler is used to introduce gaseous silane into the quartz reaction tube, which is placed in an oven at a controlled temperature. An inert gas (N_2) is used as the carrier gas. A by-product condenser is used to trap the by-products and/or unreacted silane.

Some SAMs have been widely reported. SAMs of long-chain fatty acids $C_nH_{2n+1}COOH$ or $(CH_3)(CH_2)_nCOOH$ ($n = 10, 12, 14$ or 16) on glass or alumina substrates have been widely studied since the 1950s [12, 20, 21, 24, 25]. Probably the

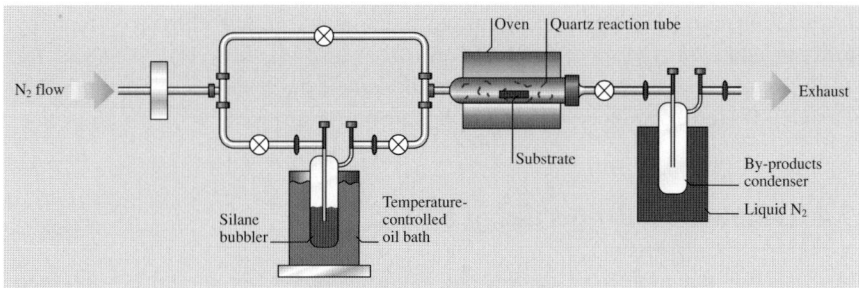

Fig. 17.5. Schematic showing vapor phase deposition system for silane SAMs [43]

most studied SAMs to date are n-alkanethiolate (n-alkyl and n-alkane are used interchangeably) monolayers $CH_3(CH_2)_nS-$ prepared by the adsorption of alkanethiol $-(CH_2)_nSH$ solution onto Au film [39–41, 51], and n-alkylsiloxane monolayers produced by the adsorption of n-alkyltrichlorosilane $(-CH_2)_nSiCl_3$ onto hydroxylated Si/SiO_2 substrate with siloxane (Si–O–Si) binding; see Fig. 17.6 [43, 44, 65]. (Note that "siloxane" (Si–O–Si) refers to the bond, while "silane" (Si_nX_{2n+2}), which describes covalently bonded compounds containing the elements Si and other atoms or groups such as H and Cl to form SiH_4 and $SiCl_4$, refers to the head group of the precursor. These terms are used interchangeably.) *Jung* et al. [66] have produced organosulfur monolayers – decanethiol $(CH_3)(CH_2)_9SH$ and didecyl sulfide $CH_3(CH_2)_9-S-(CH_2)_9CH_3$ – on Au films. *Geyer* et al. [67], *Bhushan* and *Liu* [39], *Liu* et al. [40], and *Liu* and *Bhushan* [41] have produced monolayers of 1,1′-biphenyl-4-thiol (BPT) on Au surfaces, where the spacer chain of the film consists of two phenyl rings with hydrogen end groups. *Bhushan* and *Liu* [39] and

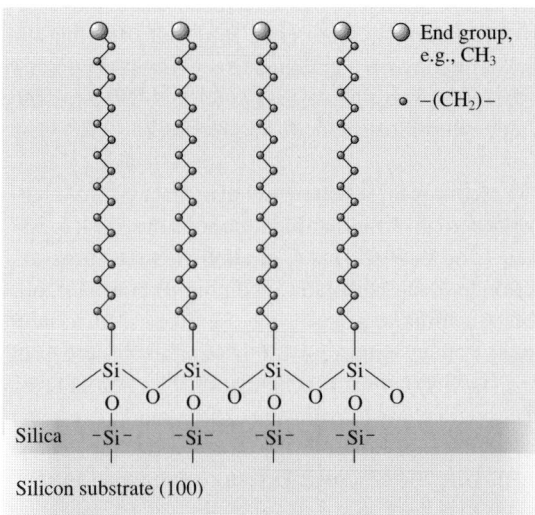

Fig. 17.6. Schematics of a methyl-terminated, n-alkylsiloxane monolayer on Si/SiO_2

Liu and *Bhushan* [41] have also reported monolayers of 4,4′-dihydroxybiphenyl on Si surfaces. *Bhushan* et al. [43, 46], *Kasai* et al. [44], *Lee* et al. [45] and *Tao* and *Bhushan* [48] have produced perfluoroalkylsilane on Si surfaces and *Tambe* and *Bhushan* [47] have produced alkylphosphonate on Al surfaces.

17.4 Tribological Properties of SAMs

The basis for the molecular design and tailoring of SAMs should be complete knowledge of the interrelationships between the molecular structures and tribological properties of SAMs, as well as a deep understanding of the adhesion, friction and wear mechanisms of SAMs at the molecular level. Friction and wear properties of SAMs have been studied on the macro- and nanoscale. Macroscale tests are conducted using a so-called pin-on-disk tribotester apparatus, in which a ball specimen slides against a lubricated flat specimen [9, 10]. Nanoscale tests are conducted using an atomic force/friction force microscope (AFM/FFM) [4, 5, 9, 10]. In the AFM/FFM experiments, a sharp tip of radius 5–50 nm slides against a SAM specimen. A Si_3N_4 tip is commonly used for friction studies and a Si or natural diamond tip is commonly used for scratch, wear and indentation studies.

In early studies, the effect of the chain length of the fatty acid monolayers on the coefficient of friction and wear on the macroscale was studied by *Bowden* and *Tabor* [20] and *Zisman* [21]. *Zisman* [21] reported that there is a steady decrease in friction with increasing chain length for monolayers deposited on a glass surface sliding against a stainless steel surface. At a significantly long chain length, the coefficient of friction reaches a lower limit, Fig. 17.7a. He also reported that monolayers with a chain length of below 12 carbon atoms behave as liquids (poor durability), those with chain lengths of 12–15 carbon atoms behave like plastic solids (medium durability), whereas those with chain lengths of above 15 carbon atoms behave like crystalline solids (high durability). Investigations by *Ruhe* et al. [68] indicated that the lifetime of the alkylsilane monolayer coating on a silicon surface increases greatly with increasing chain length of the alkyl substituent. *DePalma* and *Tillman* [69] showed that a monolayer of *n*-octadecyltrichlorosilane (n-$C_{18}H_{37}SiCl_3$, OTS) is an effective lubricant on silicon.

With the development of AFM techniques, researchers have successfully characterized the nanotribological properties of self-assembled monolayers [1, 4, 5, 38–44, 47]. Studies by *Bhushan* et al. [38] showed that C_{18} alkylsiloxane films exhibit the lowest coefficient of friction and can withstand much higher normal loads during sliding than LB films, soft Au films and hard SiO_2 coatings. *McDermott* et al. [70] studied the effect of alkyl chain length on the frictional properties of methyl-terminated *n*-alkylthiolate $CH_3(CH_2)_nS-$ films chemisorbed on Au(111) using an AFM. They reported that the longer chain monolayers exhibit markedly lower friction and reduced propensity to wear than shorter chain monolayers, Fig. 17.7b. These results are in good agreement with the macroscale results published by *Zisman* [21]. They also conducted infrared reflection spectroscopy in order to measure

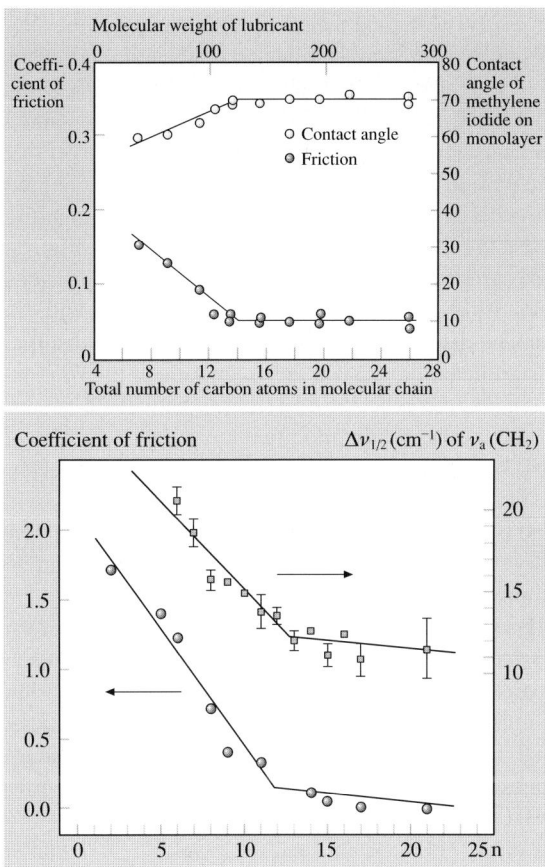

Fig. 17.7. (a) Effect of chain length (or molecular weight) on coefficient of macroscale friction of stainless steel sliding on glass lubricated with a monolayer of fatty acid, and contact angle of methyl iodide on condensed monolayers of fatty acid on glass [21]. (b) Effect of chain length of methyl-terminated, n-alkanethiolate over Au film AuS(CH$_2$)$_n$CH$_3$ on the coefficient of microscale friction and peak bandwidth at half-maximum ($\Delta\nu_{1/2}$) for the bandwidth of the methylene stretching mode [ν_a (CH$_2$)] [70]

the bandwidth of the methylene stretching mode [ν_a(CH$_2$)], which exhibits a qualitative correlation with the packing density of the chains. It was found that the chain structures of monolayers prepared with longer chain lengths are more ordered and more densely packed than those of monolayers prepared with shorter chain lengths. They further reported that the ability of the longer chain monolayers to retain molecular scale order during shear leads to a decrease in friction. Monolayers with a chain length of more than 12 carbon atoms – preferably 18 or more – are desirable for tribological applications. (Incidentally, monolayers with 18-carbon- atom octadecanethiol films are often studied.)

Xiao et al. [71] and *Lio* et al. [72] also studied the effect of the length of the alkyl chain on the frictional properties of n-alkanethiolate films on gold and n-alkylsilane films on mica. Friction was found to be particularly high for short chains of less than eight carbon atoms. Thiols and silanes exhibit similar friction force for the same n when $n > 11$; for $n < 11$, silanes exhibit higher friction, about three times larger than that for thiols for $n = 6$. The increase in friction was attributed to the large number of dissipative modes in the less ordered chains that occur when going from a thiol to a silane anchor or when decreasing n. Longer chains ($n > 11$), stabilized by van der Waals attractions, form more compact and rigid layers and are better lubricants. *Schonherr* and *Vancso* [73] also correlated the magnitude of the friction with the order among the alkane chains. The disorder in short-chain hydrocarbon disulfide SAMs was found to result in a significant increase in the magnitude of the friction.

Tsukruk and *Bliznyuk* [74] studied the adhesion and friction between a Si sample and a Si_3N_4 tip, in which both surfaces were modified by $-CH_3-$, $-NH_2-$ and $-SO_3H$-terminated silane-based SAMs. Various polymer molecules were used for the backbone. They reported a very broad maximum adhesive force at pH 4–8, with minimum adhesion at pH > 9 and pH < 3, for all of the studied mating surfaces. This observation can be understood by considering a balance of electrostatic and van der Waals interactions between composite surfaces with multiple isoelectric points. The friction coefficients of NH_2/NH_2- and SO_3H/SO_3H-mating SAMs are very high in aqueous solution. Capping NH_2-modified surfaces (3-aminopropyltriethoxysilane) with rigid and soft polymer layers resulted in a significant reduction in adhesion, to a level lower than that of the untreated surface [75]. *Fujihira* et al. [76] studied the influence of surface terminal groups of SAMs and functional tips on adhesive force. It was found that the adhesive force measured in air increases in the order CH_3/CH_3, $CH_3/COOH$, $COOH/COOH$.

Bhushan and *Liu* [39], *Liu* et al. [40], and *Liu* and *Bhushan* [41, 42] studied adhesion, friction and wear properties of alkylthiol and biphenylthiol SAMs. They explained the friction mechanisms using a molecular spring model in which local stiffness and intermolecular forces govern friction properties. They studied the influence of relative humidity, temperature and velocity on adhesion and friction. They also investigated the wear mechanisms of SAMs using a continuous microscratch AFM technique.

Fluorinated carbon (fluorocarbon) molecules are known to have low surface energy and are commonly used for lubrication [2, 9, 10]. *Bhushan* et al. [43, 46], *Kasai* et al. [44] and *Lee* et al. [45] studied the friction and wear properties of methyl- and perfluoro-terminated alkylsilanes on silicon. *Bhushan* et al. [5] and *Kasai* et al. [44] reported that perfluoroalkylsilane SAMs exhibit lower surface energies, higher contact angles, lower adhesive forces, and lower wear than those of alkylsilanes. *Kasai* et al. [44] also reported the influence of relative humidity, temperature and velocity on adhesion and friction. *Tambe* and *Bhushan* [47] studied the nanotribological properties of methyl-terminated alkylphosphonate on aluminium, which is of industrial interest. They found that these SAMs perform as well on aluminium as on silicon. *Tao* and *Bhushan* [48] studied degradation mechanisms of SAMs. They reported that oxygen from the air causes thermal oxidation of SAMs.

To date, the nanotribological properties of alkanethiol, biphenylthiol, alkylsilane and perfluoroalkylsilane SAMs have been widely studied. In this chapter, we review, in some detail, the nanotribological properties of various SAMs that have alkyl and biphenyl spacer chains with different surface terminal groups ($-CH_3$, $-CF_3$) and head groups ($-S-H$, $-Si-O-$, $-OH$, and $P-O-$), which were investigated by AFM at various operating conditions, Fig. 17.8a,b [5, 39–42, 44, 47, 48]. Hexadecane thiol (HDT), 1,1′-biphenyl-4-thiol (BPT) and crosslinked BPT (BPTC) were deposited on Au(111) substrates by immersing the substrate in a solution containing a precursor (ligand) that reacts with the substrate surface. Crosslinked BPTC was produced by irradiating BPT monolayers with low-energy electrons. Perfluorodecyltricholorosilane (PFTS), $CF_3-(CF_2)_7-(CH_2)_2-SiCl_3$, n-octyldimethyl (dimethylamino) silane (ODMS), $CH_3-(CH_2)_n-Si(CH_3)_2-N(CH_3)_2$ ($n=7$), and n-octadecylmethyl (dimethylamino) silane ($n=17$) (ODDMS) were deposited on Si by exposing the substrate to the vapor of the reactive chemical precursor. Octylphosphonate (OP)

$$CH_3 - (CH_2)_n - \overset{\overset{O}{|}}{\underset{\underset{O}{\|}}{P}} - OH (n=7)$$

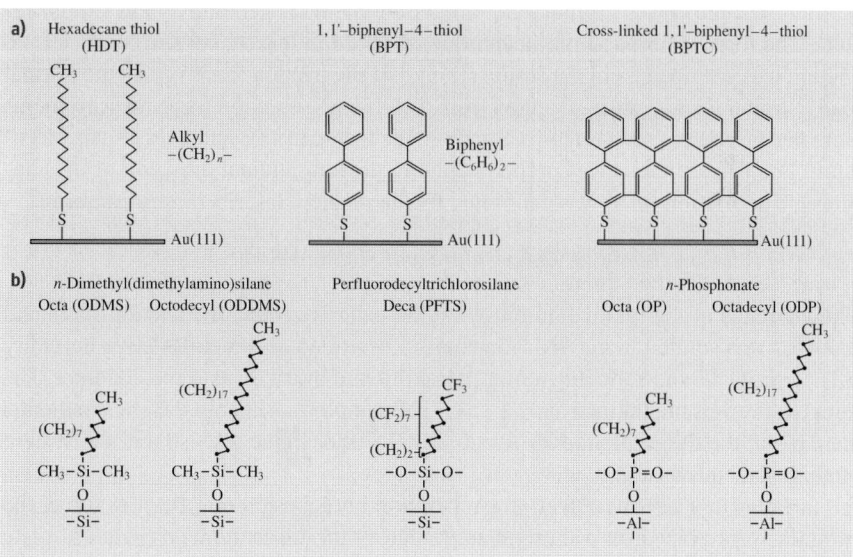

Fig. 17.8. Schematics of the structures of (**a**) hexadecane and biphenyl thiol SAMs on Au(111) substrates, and (**b**) perfluoroalkylsilane and alkylsilane SAMs on Si with native oxide substrates, as well as alkylphosphonate SAMs on Al with native oxide

and octadecylphosponate ($n = 17$) (ODP) were deposited on Al by liquid deposition. Thermally evaporated Au(111) films on Si(111) substrate were selected because the epitaxial film is smooth and defect-free, which is desirable for SAM applications. Bulk Si(100) and Al with natural oxide layers were selected because they are used in the construction of MEMS/NEMS.

17.4.1 Measurement Techniques

Static Contact Angle Measurement Using DI Water

The static contact angle, a measure of how water repellent a material is, was measured using a Rame–Hart model 100 contact angle goniometer (Mountain Lakes, NJ, USA) [77, 78]. Ten microliter droplets of DI water were typcially used for the contact angle measurements. At least two measurements of the contact angle were taken. The contact angles were reproducible within $\pm 2°$.

AFM Adhesion and Friction Measurements

Adhesion and friction tests were conducted using a commercial AFM system (Dimension 3000, Nanoscope IIIa controller, DI, Santa Barbara, USA). Square-pyramidal Si_3N_4 tips with a 30–50 nm tip radius were used on a gold back-coated triangular Si_3N_4 cantilever with a typical spring constant of 0.58 N/m. The adhesion can be calculated using either force calibration plots or from the negative intercepts on friction force versus normal load plots. Both methods generally yield similar results. The force calibration plot technique was used in this study. The coefficient of friction was obtained from the slope of a plot of the friction force versus the normal load. Normal loads typically ranged from 5 to 100 nN. Friction force measurements were performed at a scan rate of 1 Hz along the fast scan axis and over a scan size of 2 µm. The fast scan axis was perpendicular to the longitudinal direction of the cantilever. The friction force was calibrated by the method described in *Bhushan* [4].

Effects of Relative Humidity, Temperature and Sliding Velocity

The influence of the relative humidity on the adhesive force, the friction force and the wear was studied in an environmentally controlled chamber. Relative humidity was controlled by introducing a mixture of dry and moist air into the chamber. The temperature was maintained at $22 \pm 1\,°C$. The sample was kept in the environmental chamber at the desired humidity for at least 2 h prior to the tests so that the system could reach equilibrium.

In order to study the effect of temperature on adhesion and friction force, the samples were placed on a thermal stage during the measurements. A glass plate was placed under the thermal stage to prevent the heat from being transported away. The temperature range studied was from 20 to $110\,°C$. The relative humidity was maintained at $50 \pm 5\%$ during the measurements.

The effect of sliding velocity on friction force was monitored in ambient conditions using a high-velocity piezo stage designed to achieve high relative sliding velocities on a commercial AFM set-up [79]. The traveling distance of the sample (the scan size) was set at 25 μm, while the scan frequency was varied between 0.1 Hz (5 μm/s) and 100 Hz (5000 μm/s).

AFM Wear Measurements

Wear tests were conducted using a diamond tip with a nominal radius of 50 nm and a cantilever with a nominal stiffness of 10 N/m. Wear tests were performed on a scan area of 1 μm × 1 μm at the desired normal load and at a scan rate of 1 Hz. After each wear test, an area of 3 μm × 3 μm was imaged and the average wear depth was calculated.

Degradation and Environmental Studies

The lubricant degradation experiments were carried out in a high-vacuum tribotest apparatus [82, 83]. The system was equipped with a mass spectrometer so that gaseous emissions from the interface could be monitored in situ during the sliding in high vacuum. The normal loads and friction forces at the contacting interface were measured using resistive-type strain-gauge transducers. Sliding tests were conducted by rubbing the sample against a flat sample of Si(100) at a vacuum pressure of 2×10^{-7} Torr at a sliding speed of 0.3 m/s. The environmental effects were investigated in high vacuum (2×10^{-7} Torr), argon, dry air (less than 2% RH), ambient air (30% RH) and in high humidity air (70% RH).

17.4.2 Hexadecane Thiol and Biphenyl Thiol SAMs on Au(111)

Hexadecane thiol on Au(111) was selected as it is a widely studied film. Biphenyl thiol was selected to study the effect of rigidity on nanotribological performance. The biphenyl thiol film was crosslinked to further increase its stiffness.

Surface Roughness, Adhesion and Friction

Surfaceheight and friction force images of SAMs were recorded simultaneously on an area of 1 μm × 1 μm by an AFM, and adhesive forces were measured using an AFM in force calibration mode [39].

Detailed analysis is presented later in this chapter, but the measured roughnesses, thickness, tilt angles and spacer chain lengths of Si(111), Au(111) and various SAMs are listed in Table 17.10 [39]. The roughness of BPT is very close to that of Au(111), but the roughness of BPTC is lower than that of Au(111) and BPT; this is caused by electron irradiation. Table 17.10 indicates that the roughness of HDT is much higher than the substrate roughness of Au(111). This is caused by local aggregation of organic compounds on the substrates during SAM deposition. Table 17.5 also indicates that the thicknesses of biphenyl thiol SAMs are generally smaller than those of the alkylthiol, because of the shorter spacer chain in biphenyl thiol.

Table 17.10. R_a roughnesses, thicknesses, tilt angles and spacer chain lengths of SAMs

Samples	R_a roughness[a] (nm)	Thickness[b] (nm)	Tilt angle[b] (degrees)	Spacer length[c] (nm)
Si(111)	0.07			
Au(111)	0.37			
HDT	0.92	1.89	30	1.91
BPT	0.36	1.25	15	0.89
BPTC	0.14	1.14	25	0.89

[a] Measured by an AFM with a 1 μm × 1 μm scan size, using a Si_3N_4 tip under 3.3 nN normal load
[b] The thickness and tilt angles of BPT and BPTC are reported by *Geyer* et al. [67]. The thickness and tilt angles of HDT are reported by *Ulman* [25]
[c] The spacer chain lengths of alkylthiols were calculated by the method reported by *Miura* et al. [80]. The spacer chain lengths of biphenyl thiols were calculated from the data reported by *Ratajczak-Sitarz* et al. [81]

Average values and standard deviations for the adhesive force and coefficient of friction are presented in Fig. 17.9 [39]. Based on the data, the adhesive forces and friction coefficients of SAMs are less than those of their corresponding substrates. Among the various films, HDT exhibits the lowest values. The adhesive force F_a is ranked as follows: $F_{a-Au} > F_{a-BPT} > F_{a-BPTC} > F_{a-HDT}$. The rankings for the friction coefficients μ are: $\mu_{Au} > \mu_{BPTC} > \mu_{BPT} > \mu_{HDT}$. Note that many SAMs have similar rankings for both adhesive force and coefficient of friction. This suggests that alkylthiol and biphenyl SAMs would both make effective molecular lubricants for micro/nanodevices.

Liquid capillary condensation is one source of adhesion and friction in micro/nanoscale contact. For a sphere in contact with a flat surface, the attractive Laplace force caused by a water capillary is

$$F_L = 2\pi R \gamma_{la}(\cos\theta_1 + \cos\theta_2), \tag{17.1}$$

where R is the radius of the sphere, γ_{la} is the surface tension of the liquid against air, and θ_1 and θ_2 are the contact angles between the liquid and flat and spherical surfaces, respectively [9, 10]. In an AFM adhesive study, the tip–flat sample contact is just like a sphere in contact with a flat surface, and the liquid is water. Since a single tip is used in the adhesion measurements, $\cos\theta_2$ can be treated as a constant. Therefore,

$$\begin{aligned}F_L &= 2\pi R \gamma_{la}(1+\cos\theta_1) - 2\pi R \gamma_{la}(1-\cos\theta_2) \\ &= 2\pi R \gamma_{la}(1+\cos\theta_1) - C,\end{aligned} \tag{17.2}$$

where C is a constant.

Based on the following Young–Dupre equation, the work of adhesion W_a (the work required to pull apart a unit area of the solid–liquid interface) can be written

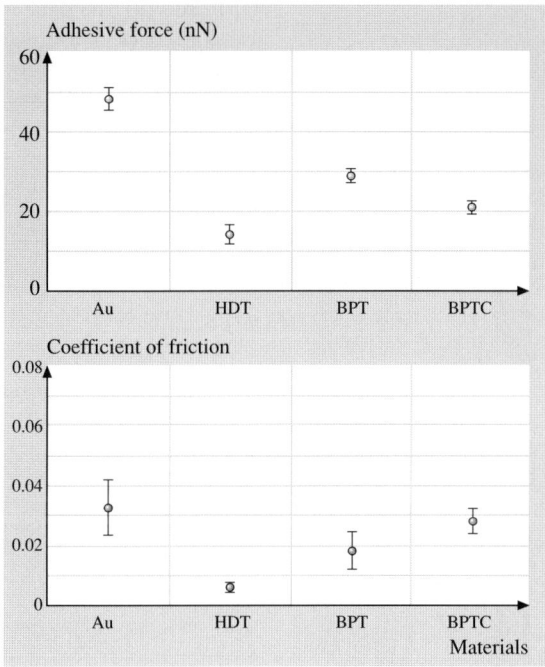

Fig. 17.9. Adhesive forces and coefficients of friction for Au(111) and various SAMs

as [84]

$$W_a = \gamma_{la}(1 + \cos\theta_1). \tag{17.3}$$

This indicates that W_a is determined by the SAM contact angle; in other words it is influenced by the surface chemistry properties (polarization and hydrophobicity) of the SAM. By substituting (17.3) into (17.2), F_L can be expressed as

$$F_L = 2\pi R W_a - C \tag{17.4}$$

When the influence of other factors, such as van der Waals force, on the adhesive force is very small, then the adhesive force $F_a \approx F_L$. Thus the adhesive force F_a should be proportional to the work of adhesion W_a.

The contact angle is a measure of the wettability of a solid by a liquid, and it determines the W_a value [77, 78]. The contact angles for distilled water on Au(111) and SAMs have been measured, and are summarized in Fig. 17.10a [39]. For water, $\gamma_{la} = 72.6$ mJ/m^2 at 22 °C. Therefore, using this value and (17.3), it is possible to obtain W_a data, and these are presented in Fig. 17.10b. The W_a values can be ranked in the following order: $W_{a-Au}(97.1) > W_{a-BPT}(86.8) > W_{a-BPTC}(82.1) > W_{a-HDT}(61.4)$. Except for W_{a-Au}, this order mimics the order of adhesion force in Fig. 17.9. The relationship between F_a and W_a is summarized in Fig. 17.11 [39]. It indicates that the adhesive force F_a (nN) increases with the work of adhesion W_a (mJ/m^2) as follows:

$$F_a = 0.57 W_a - 22. \tag{17.5}$$

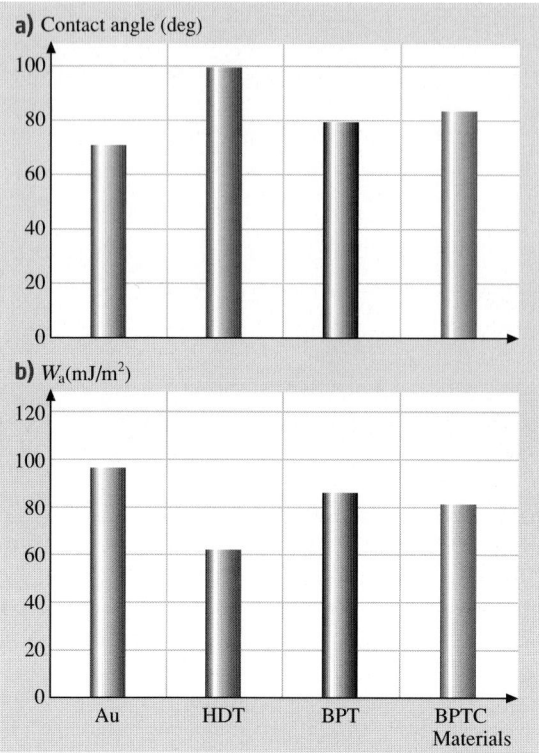

Fig. 17.10. (a) The static advancing contact angle, and (b) the work of adhesion for Au(111) and various SAMs. Each point in this figure represents the mean value of six measurements. The uncertainty associated with the average contact angle is ±2°

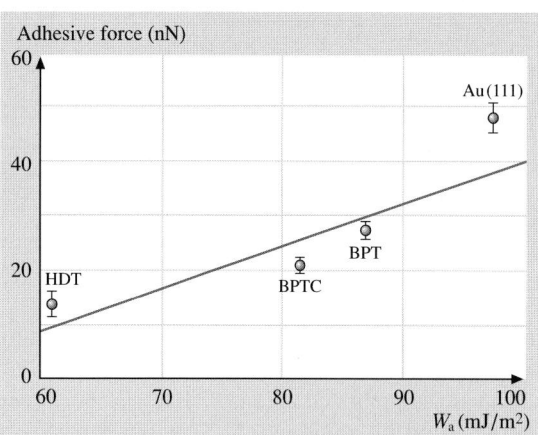

Fig. 17.11. Relationship between the adhesive force and the work of adhesion for different specimens

These experimental results agree well with the modeling prediction presented earlier in (17.4). It proves that, on the nanoscale, and under ambient conditions, the adhesive forces of SAMs are mainly influenced by the water capillary force. Though neither HDT nor BPT has polar surface groups, the surface terminal group of HDT has a symmetrical structure, which causes a smaller electrostatic attractive force and yields a smaller adhesive force than BPT. It is believed that the easy attachment of Au to the tip is one of the reasons for the unexpectedly large adhesive force, which means that it does not fit the linear relationship described by (17.5).

Stiffness, Molecular Spring Model and Micropatterned SAMs

The friction mechanisms of SAMs were also examined. Monte Carlo simulation of the mechanical relaxation of a $CH_3(CH_2)_{15}SH$ self-assembled monolayer, as performed by *Siepman* and *McDonald* [85], indicated that SAMs compress and respond nearly elastically to microindentation by an AFM tip when the load is below a critical normal load. Compression can lead to major changes in the mean molecular tilt (the orientation), but the original structure is recovered as the normal load is removed.

Stiffness properties were measured by an AFM in force modulation mode [4, 5, 41]. They reported that BPT was stiffer than HDT. Since BPT has rigid benzene structure, it is more difficult to compress than HDT. Figure 17.12 shows the variation in the displacement with normal load during indentation mode. It clearly indicates that SAMs can be compressed. At a given normal load, SAMs with long carbon chain structures such as HDT are easy to compress compared to SAMs with rigid benzene ring structures, such as BPT. *Garcia-Parajo* et al. [86] have also reported on the compression and relaxation of octadecyltrichlorosilane (OTS) film obtained from loading and unloading tests.

A molecular spring model is presented in Fig. 17.13 to explain the difference in friction between the SAMs observed in the friction and stiffness measurements by AFM and the Monte Carlo simulation. It is believed that the self-assembled molecules on a substrate act just like assembled molecular springs anchored to the substrate [39]. A Si_3N_4 tip sliding on the surface of a SAM is like a tip sliding on the top of molecular springs or a brush. The molecular spring assembly has compliant features and can experience compression and orientation under normal load.

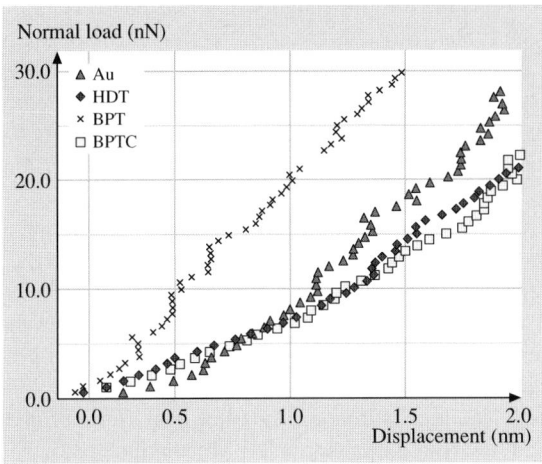

Fig. 17.12. Normal load versus displacement curves for Au(111) and various SAMs

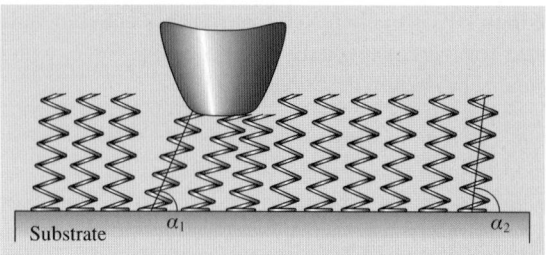

Fig. 17.13. Molecular spring model for SAMs. In this figure, $\alpha_1 < \alpha_2$, which is caused by reorientation under the normal load applied by the AFM tip. The reorientation of the molecular springs reduces the shearing force at the interface, which in turn reduces the friction force. The molecular spring constant and the intermolecular forces determine the magnitude of the coefficient of friction. In this figure, the size of the tip and the molecular springs are not drawn to the same scale [39]

The orientation of the molecular springs or brush reduces the shearing force at the interface, which in turn reduces the friction force. The possibility of orientation is determined by the spring constant of a single molecule (local stiffness), as well as the interactions between neighboring molecules, which is reflected in the packing density or packing energy. It should be noted that the orientation can lead to conformational defects along the molecular chains, which lead to energy dissipation. In the study of BPT by AFM, it was found that the friction force is significantly reduced after the first several scans, but the surface height does not show any apparent change. This suggests that molecular orientation can be facilitated by initial sliding and is reversible [42].

Based on the stiffness measurement results presented in Fig. 17.12 and the view of molecular structures given in Fig. 17.13, biphenyl is a more rigid structure due to the contribution of the two rigid benzene rings. Therefore, the spring constant of BPT is larger than that of HDT. The hydrogen (H^+) in the biphenyl chain has an electrostatic attraction to the π electrons in the neighboring benzene ring. Thus the intermolecular forces between biphenyl chains are stronger than those between alkyl chains. The larger spring constant of BPT and stronger intermolecular forces mean that it requires a larger external force to allow it to orient, thus causing a higher coefficient of friction. The crosslinking of BPT leads to a larger packing energy for BPTC. Therefore BPTC orientation requires a larger external force: the coefficient of BPTC is higher than BPT.

An elegant way to demonstrate the influence of molecular stiffness on friction is to investigate SAMs with different structures on the same wafer. A micropatterned SAM was prepared for this purpose. First biphenyldimethylchlorosilane (BDCS) was deposited on the silicon using a typical self-assembly method [41]. Then the film was partially crosslinked using a mask technique by low-energy electron irradiation. Finally, micropatterned BDCS films that had the different coating regions on the same wafer were realized. The local stiffness properties of these micropatterned samples were investigated by a force modulation AFM technique [87]. The

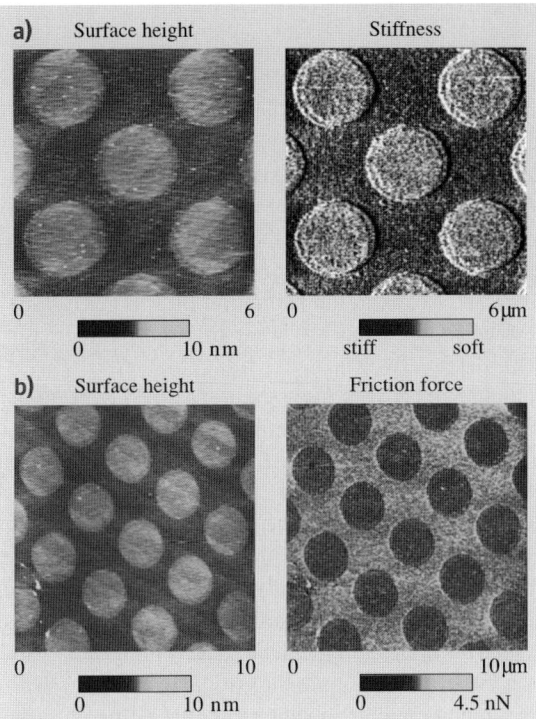

Fig. 17.14. (a) AFM grayscale surface height and stiffness images, and (b) AFM grayscale surface height and friction force images of micropatterned BDCS [41]

variation in the deflection amplitude provides a measure of the relative local stiffness of the surface. Surface height, stiffness and friction images of the micropatterened biphenyldimethylchlorosilane (BDCS) specimen were obtained and are presented in Fig. 17.14 [41]. The circular areas correspond to the as-deposited film, and the remaining area to the crosslinked film. Figure 17.14a indicates that crosslinking caused by the low-energy electron irradiation leads to a decrease of about 0.5 nm in the surface height of the BDCS film. The corresponding stiffness images indicate that the crosslinked area has a higher stiffness than the as-deposited area. Figure 17.14b indicates that the as-deposited area has lower friction force. Obviously, these data for the micropatterned sample prove that the local stiffness of the SAM influences its friction performance. Higher stiffness leads to larger friction force. These results correlate well with the suggested molecular spring model.

In summary, it was found that SAMs exhibit compliance and can experience compression and orientation under normal load. SAM orientation reduces the shear stress at the interface, so SAMs provide good lubricants. The molecular spring constant (local stiffness) and intermolecular forces can both influence the friction coefficient of a SAM.

Wear and Scratch Resistance

Wear resistance was studied on an area of 1 μm × 1 μm. The variation of wear depth with normal load is presented in Fig. 17.15 [39]. HDT exhibits the best wear resistance. A critical normal load (marked by arrows in Fig. 17.15) appears in the curves for all of the SAMs tested. When the normal load is smaller than the critical normal load, the monolayer only changes in height very slightly in the scan area. When the normal load is higher than the critical value, the change in height of the SAM increases dramatically. Relocation and accumulation of BPT molecules is observed during the first few scans, which leads to the formation of a larger terrace. Wear studies of a single BPT terrace indicate that the wear life of BPT increases exponentially with terrace size [40, 41].

The scratch resistances of Au(111) and SAMs were studied using a continuous AFM microscratch technique. Figure 17.16a shows coefficient of friction profiles as a function of increasing normal load, and corresponding tapping mode AFM surface height images of the scratches captured on Au(111) and SAMs [41]. Figure 17.16a indicates that there is an abrupt increase in the coefficient of friction for all of the tested samples. The normal load associated with this event is termed the critical load (indicated by the arrows labeled "A"). Initially during scratching, all of the samples exhibit a low coefficient of friction, indicating that the friction force is dominated by the shear component. This is in agreement with AFM image analysis, which shows negligible damage on the surfaces prior to the critical load. At the critical load, a clear groove is formed, which is accompanied by material piling up at the sides of the scratch. This suggests that the initial damage that occurs at the critical load is due to plowing associated with plastic deformation, and this causes the sharp rise in the coefficient of friction. Beyond the critical load, debris can be seen as well as material pile-up at the sides of the scratch. Figure 17.16b summarizes the critical loads for the various samples obtained in this study. It clearly indicates that all of the SAMs increase the critical load of the corresponding substrate.

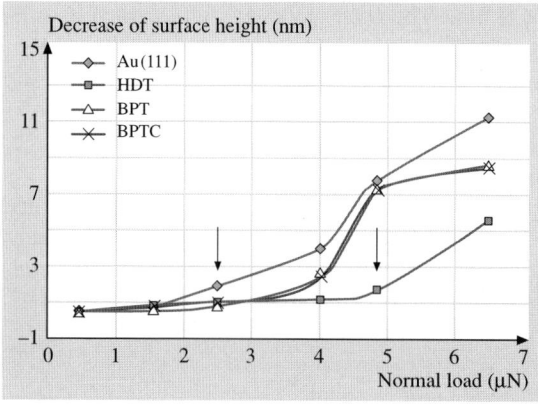

Fig. 17.15. Wear depth as a function of normal load after one scan cycle

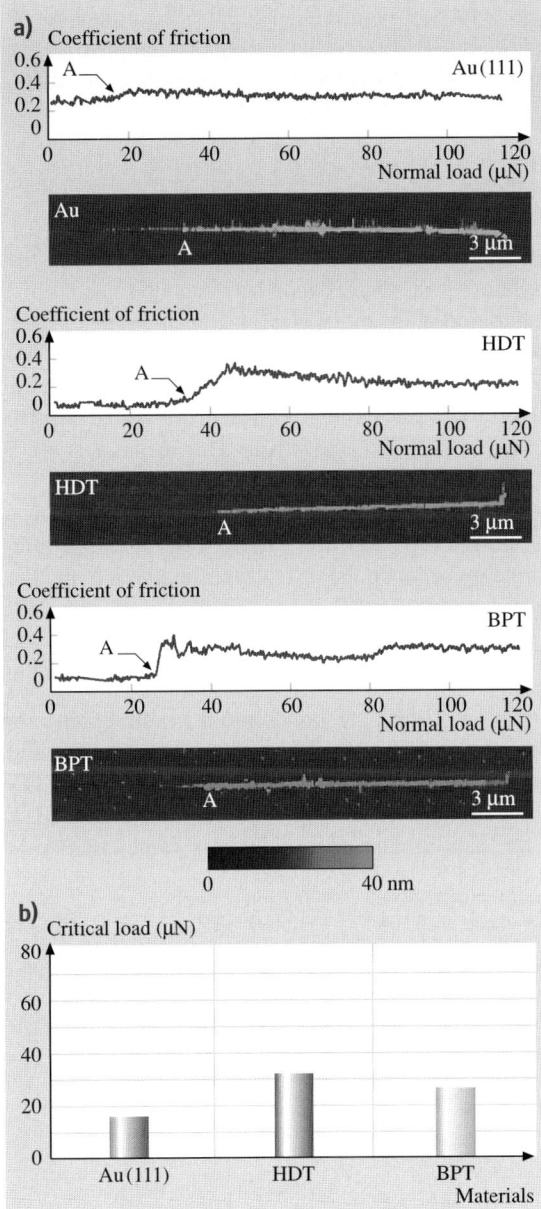

Fig. 17.16. (a) Coefficient of friction profiles during scratch as a function of normal load, and corresponding AFM surface height images as well as (b) critical loads estimated from coefficient of friction profiles and AFM images for Au(111), HDT/Au(111) and BPT/Au(111) [41]

The mechanisms responsible for the sudden drop in surface height with increasing load during the wear and scratch tests need to be understood. *Barrena* et al. [88] observed that the heights of self-assembled alkylsilane layers decrease in discrete amounts with normal load. This step-like behavior is due to discrete molecular tilts, which are dictated by the geometrical requirements of the close packing of molecules. Only certain angles are allowed due to the zigzag arrangement of the carbon atoms. The relative height of the monolayer under pressure can be calculated by the following equation

$$\left(\frac{h}{L}\right) = \left[1 + \left(\frac{na}{d}\right)^2\right]^{-1/2}, \tag{17.6}$$

where L is the total length of the molecule, h is the height of the SAMs in the tilt configuration (monolayer thickness), a is the distance between alternate carbon atoms in the molecule, d is the separation of the molecules, and n is the step number. The values of a (0.25 nm) and d (0.47 nm) are used in the calculation of HDT. The calculated and measured relative heights of HDT are listed in Table 17.11. When the normal loads are smaller than the critical values in Fig. 17.15, the measured relative height values of HDT are very close to the calculated values. This means that HDT underwent step tilting below the critical normal load.

The residual SAM thickness after wear under critical normal load was measured by profiling the worn film using AFM. The results are listed in Table 17.12. For an alkanethiol monolayer, the relationship between the monolayer thickness h and the intercept length L_0 can be expressed as (Fig. 17.17)

$$h = b\cos(\alpha)n + L_0, \tag{17.7}$$

where b is the length of the projection of the C–C bond onto the main chain axis ($b = 0.127$ nm for alkanethiol), n is the chain length defined by $CH_3(CH_2)_n SH$, and α is the tilt angle [80]. The L_0 values have also been calculated for BPT and BPTC, based on the same principle, and using the bond lengths reported in reference [81], Table 17.7. This indicates that the measured residual thickness values of SAMs

Table 17.11. Calculated $[1 + (\frac{na}{d})^2]^{-\frac{1}{2}}$ and measured $\left(\frac{h}{L}\right)$ relative heights of an HDT self-assembled monolayer [39]

Steps (n)	Calculated [a] $[1 + (\frac{na}{d})^2]^{-\frac{1}{2}}$	Measured $\left(\frac{h}{L}\right)$
1	0.883	
2	0.685	0.674[b]
3	0.531	0.532[c]
4	0.425	0.416[d]
5	0.352	0.354[e]
6	0.299	

[a] Calculations are based on the assumption that the molecules tilt in discrete steps (n) upon compression with a diamond AFM tip [88]

[b,c,d,e] These measured values correspond to normal loads of 0.50 μN, 1.57 μN, 2.53 μN and 4.03 μN, respectively

Table 17.12. Calculated L_0 values and measured residual film thicknesses of SAMs under critical load

	L_0 [a] (nm)	Residual thickness[b] (nm)
HDT	0.24	0.25
BPT	0.39	0.42
BPTC	0.33	0.38

[a] Calculated using the equation $h = b\cos(\alpha)n + L_0$ [80]
[b] Measured by AFM using a diamond tip under critical normal load. All of the data are the mean values of three tests

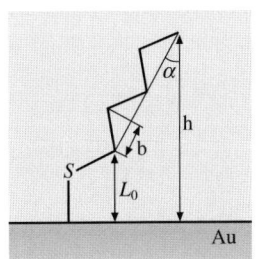

Fig. 17.17. Illustration of the relationship between the components of the equation $h = b\cos(\alpha)n + L_0$ [39]

under critical load are very close to the calculated intercept length (L_0) values. This means that the Si_3N_4 tip approaches the interface and SAMs are severely worn away from the substrate under the critical normal load. This is because chemical adsorption bond strength at the interface (S–Au) is generally smaller than other chemical bond strengths in SAM spacer chains (Table 17.8).

The change in the wear mechanism of a SAM with increasing normal load is therefore illustrated in Fig. 17.18. Below the critical normal load SAMs undergo

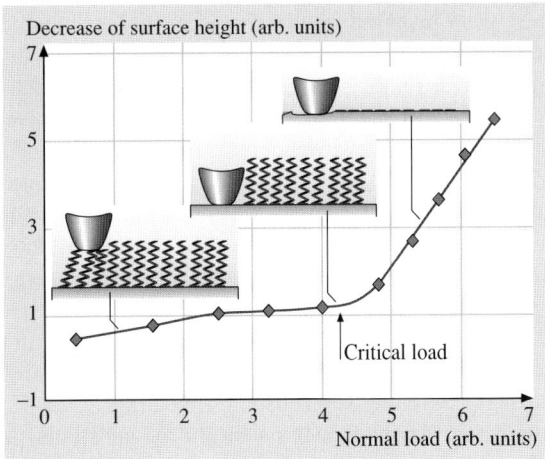

Fig. 17.18. Illustration of the wear mechanisms of SAMs with increasing normal load [41]

step orientation; at the critical load SAMs wear away from the substrate due to the weak interface bond strengths; while above the critical normal load severe wear takes place on the substrate. To improve the wear resistance, the interface bond must be enhanced; a rigid spacer chain and a hard substrate are also preferable in this regard.

17.4.3 Alkylsilane and Perfluoroalkylsilane SAMs on Si(100) and Alkylphosphonate SAMS on Al

The nanotribological parameters of perfluorodecyltricholorosilane (PFTS), $CF_3-(CF_2)_7-(CH_2)_2-SiCl_3$, n-octyldimethyl (dimethylamino)silane (ODMS), $CH_3-(CH_2)_n-Si(CH_3)_2-N(CH_3)_2$ ($n = 7$), and n-octadecylmethyl(dimethylamino)-silane ($n = 17$) (ODDMS) vapor deposited on Si substrate were investigated, as were those of octylphosphonate (OP),

$$CH_3-(CH_2)_n-\overset{\overset{\displaystyle O}{|}}{\underset{\underset{\displaystyle O}{\|}}{P}}-OH$$

($n = 7$) and octadecylphosponate (ODP) ($n = 17$) on Al substrate. The perfluoroalkylsilane SAM was selected because fluorinated films are known to have low surface energy. Alkylsilanes with two different chain lengths (with 8 and 18 carbon atoms) were selected in order to compare their nanotribological performance with that of PFTS as well as to study the influence of chain length. The alkylphosphonate SAMs (with 8 and 18 carbon atoms) on Al were selected due to their industrial use in applications such as digital projection displays.

Surface Free Energy and Contact Angle Measurement

As stated earlier, adhesive force arises from the presence of a thin liquid film such as a mobile lubricant or an adsorbed water layer that causes meniscus bridges to build up around asperities due to surface energy effects. The intrinsic attractive force arising from meniscus contributions depends on the surface tension of the film and the contact angle and may result in high friction and wear. Surface energies and contact angles were measured in order to evaluate the hydrophobicity. Figure 17.19 shows a Zisman plot for the SAMs deposited on Si and their surface energies obtained using various alkane liquids [43]. Zisman analysis data was not available for the Si substrate because the alkane liquids used for the measurement spreads instantly across such surfaces. A significantly lower critical surface tension or surface energy was observed for PFTS (12.9 mN/m for PFTS/Si) than for ODMS (24.7 mN/m for ODMS/Si) or ODDMS (23.9 mN/m for ODDMS/Si). The surface energies for ODMS and ODDMS were comparable. This suggests that the surface was covered by these SAMs to a comparable degree without bare substrate appearing.

Contact angles were measured for various SAMs on Si and Al substrates so that they could be compared. The measured values for the samples are shown in

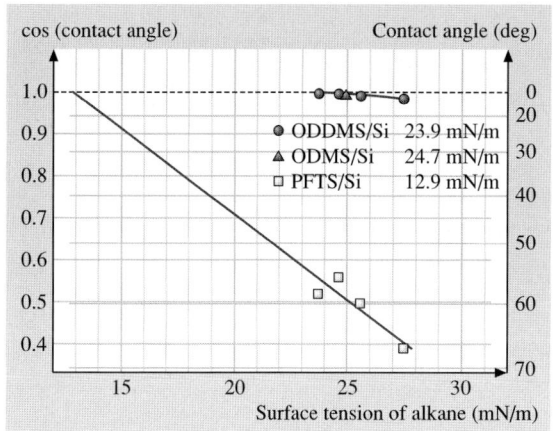

Fig. 17.19. Zisman plots for PFTS/Si, ODMS/Si and ODDMS/Si used to calculate the critical surface tension, a measure of surface energy, which is given by the x-intercept [cos(contact angle) = 1] of the line fitted to the data

Table 17.13. A summary of RMS roughness, contact angle and film thicknesses of various SAMs

SAM/substrate	Acronym	RMS roughness (nm)	Contact angle (deg.)	Film thickness (nm)
Silicon(111)	Si	0.07	48	–
Perfluorodecyltricholoro-silane/Si	PFTS/Si	0.09	112	1.8[a]
Octyldimethyl(dimethyl-amino)silane/Si	ODMS/Si	0.10	99	1.9[a]
Octadecyldimethyl-(dimethylamino)silane/Si	ODDMS/Si	0.10	92	2.1[a]
Aluminium	Al	1.73	74	–
Octylphosphonate/Al	OP/Al	–	108	≈ 1.9[b]
Octadecylphosphonate/Al	ODP/Al	–	115	≈ 2.1[b]

[a] Kasai et al. [44]
[b] Kulik (personal communication)

Fig. 17.20a [43, 47]. A summary of RMS roughness measured using an AFM, contact angles and film thickness measured using an ellipsmeter are summarized in Table 17.13. Significantly better water repellency was observed for PFTS compared to bare Si with natural oxide. The contact angle for PFTS/Si was ≈ 110° whereas it was ≈ 50° for Si. The contact angle generally increases with decreasing surface energy [89], which is consistent with the data obtained. The water contact angles of ODMS and ODDMS were also large (≈ 100°), implying high surface hydrophobicity. These contact angles can be influenced by the packing density as well as the sample roughness [90], which probably accounts for the slightly higher contact angles for the SAMs deposited on Al substrates compared to those on Si substrate.

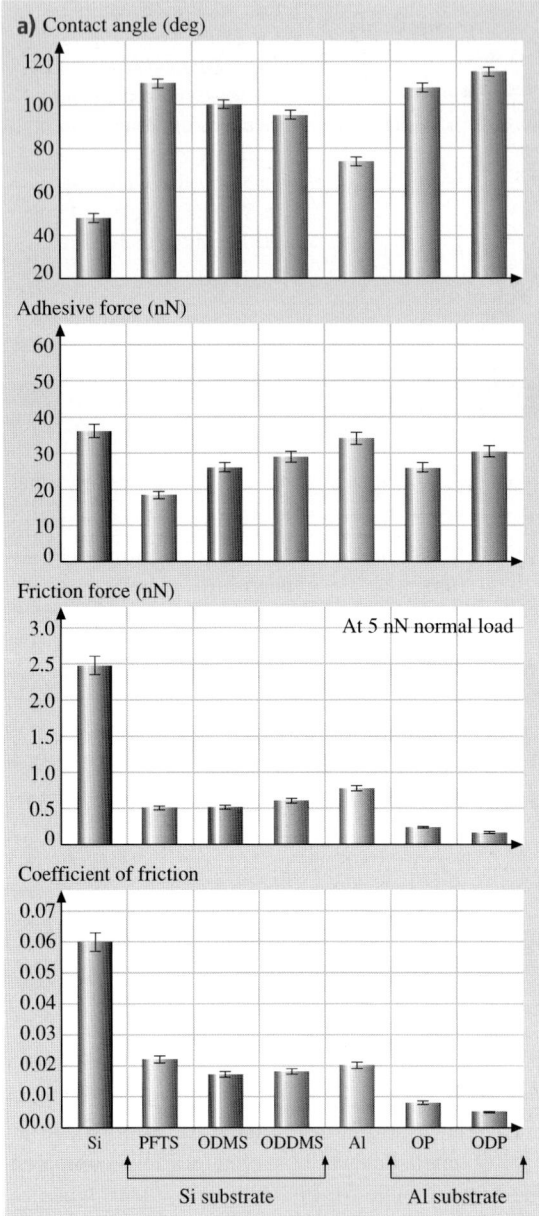

Fig. 17.20. (a) The static contact angle, adhesive force, friction force and coefficient of friction measured using an AFM for various SAMs on Si and Al substrates

The −CH₃ groups in ODMS and ODDMS are nonpolar and are known to contribute to the water repellency; however, OP and ODP (which have different head groups to ODMS and ODDMS) exhibited greater surface hydrophobicity. Among the SAMs, PFTS and ODP exhibited the highest contact angle.

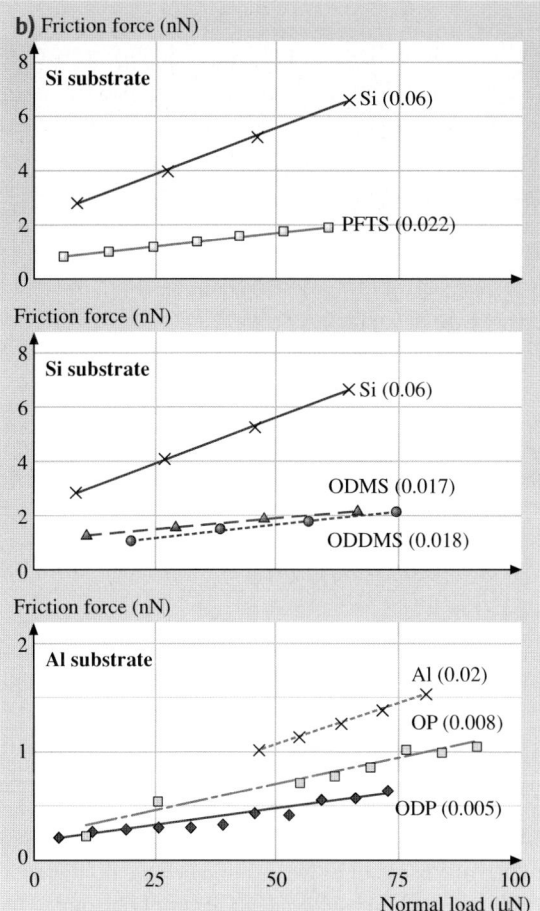

Fig. 17.20. (continued) (**b**) friction force vs. normal load plots for various SAMs on Si and Al substrates

AFM Adhesion and Friction Measurements under Ambient Conditions

Figure 17.20a gives the adhesive force, friction force and the coefficient of friction measured under ambient conditions using AFM for various SAMs deposited onto Si and Al substrates, while Fig. 17.20 shows friction force–normal load plots for these systems [43, 47]. Figure 17.21 shows surface height and friction force maps for Si and PFTS, ODMS and ODDMS on Si [43].

The bare substrates gave much higher adhesive force than the SAM coatings. ODMS and ODDMS show adhesive forces comparable to OP and ODP despite their lower water contact angles. These SAMs have the same tail groups, and during AFM measurements the AFM tip only interacts with the tail groups, whereas the contact angles can also be influenced by the head groups in these SAMs. This is probably why the adhesive forces for these SAMs are comparable. PFTS, which has one of the highest contact angles, showed the lowest adhesion.

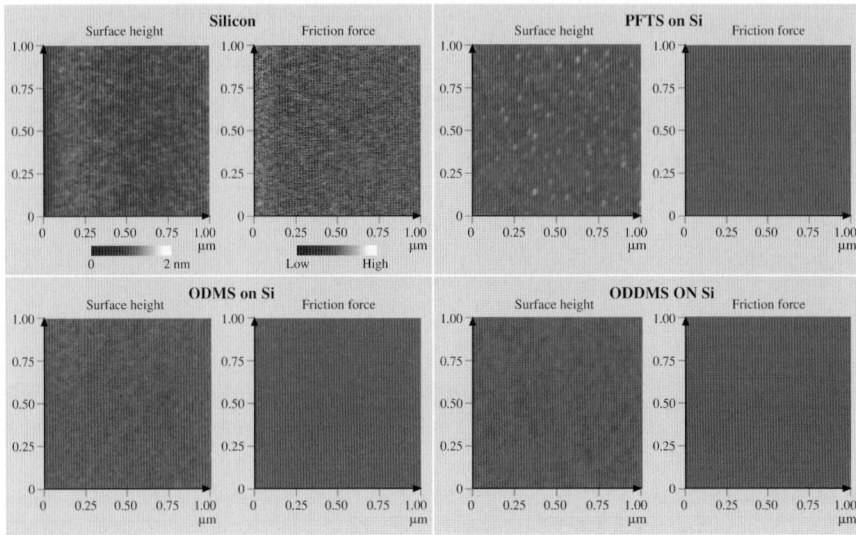

Fig. 17.21. Surface height (*left*) and friction force (*right*) maps for (a)Si, PFTS/Si, ODMS/Si and ODDMS/Si [43]

Friction force images of SAMs are more uniform than those of Si. The coefficient of friction was higher for the bare substrates than for the corresponding SAMs deposited on them. The SAMs deposited on the Si substrate showed higher coefficients of friction than those deposited on the Al substrates. The primary reason for this is believed to be the greater roughness of the Al substrates and possibly higher packing densities of SAMs on Al as mentioned earlier. For the SAMs deposited on Si substrates, the ones with fluorocarbon backbones were found to have higher coefficients of friction than those with hydrocarbon backbones. This might be due to the higher stiffness of the fluorocarbon backbone [44]. It is harder to rotate a fluorocarbon backbone because the F atom is larger than the H atom [91]. The C–C bonds of hydrocarbon chains, on the other hand, have more freedom to rotate. We presented a molecular spring or brush model earlier that explained why less compliant SAMs exhibit more friction. SAMs with higher spring constants or stiffer backbones may need more energy to be elastically deformed during sliding, so these SAMS show more friction. In terms of the influence of the chain length, it has been reported that the friction coefficients for SAM surfaces decrease with the carbon backbone chain length (n) up to 12 carbon atoms ($n \approx 12$) [70]. However, this effect of chain length on the coefficient of friction was not apparent in these data.

Effect of Relative Humidity, Temperature and Sliding Velocity on AFM Adhesion and Friction

The influence of relative humidity was studied for various SAMs. Its effects on adhesive force, friction force at a normal load of 5 nN, the coefficient of friction and

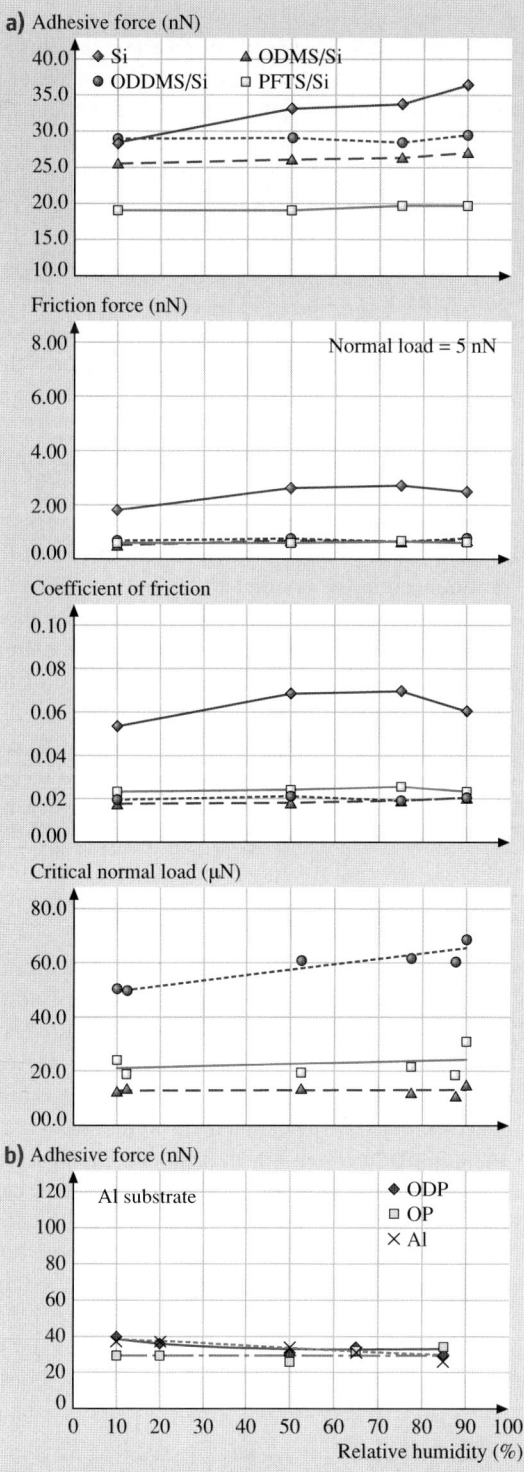

Fig. 17.22. Relative effect of humidity on the adhesive force, the friction force, the coefficient of friction and the microwear for various SAMs on (**a**) Si substrates [44], and (**b**) Al substrates

on microwear are shown in Fig. 17.22 [44,47]. The adhesive force of silicon showed an increase with relative humidity, Fig. 17.22a. This is expected since the surface of silicon is hydrophilic, as shown in Fig. 17.20a. Greater condensation of water at the tip–sample interface increases the adhesive force due to the capillary effect. On the other hand, the adhesive forces of the SAMs showed very weak dependencies on humidity. This may be because the surfaces of the SAMs are hydrophobic. The adhesive forces of ODMS/Si and ODDMS/Si showed slight increases from 75 to 90% RH. This increase was absent for PFTS/Si. This may result from the more hydrophobic surface properties of PFTS/Si. The Al substrate is partially hydrophobic and hence does not show much of a dependence on humidity, Fig. 17.22b. The OP and ODP SAMs deposited on Al substrates showed almost no change in adhesive force with humidity. The highly hydrophobic nature of these monolayers means that the contribution of the water menisci to the overall adhesive force is negligible at all humidities.

The friction force of silicon showed an increase with relative humidity up to about 75% RH and then a slight decrease beyond this point, see Fig. 17.22a. The initial increase could result from the increase in adhesive force. The decrease in friction force at higher humidities could be attributed to the lubricant effects of the water layer. This effect is more pronounced in the coefficient of friction. Since the adhesive force increased and the coefficient of friction decreased in this range, those effects cancelled each other out and the resulting friction force showed only slight changes. On the other hand, the friction forces and coefficients of friction of the SAMs showed very small changes with relative humidity; similar behavior to that observed for adhesive force. This suggests that the adsorbed water layer on the surface maintained a similar thickness throughout the relative humidity range tested. The differences among the SAM types were small within measurement errors, but a closer look at the coefficient of friction for ODMS/Si showed a slight increase from 75 to 90% RH compared to PFTS/Si, possibly due to the same reasons as for the increase in adhesive force. The inherent hydrophobicity of the SAMs means that they do not show much relative humidity dependence.

Figure 17.23 shows the effect of temperature on the adhesive force, the friction force at a normal load of 5 nN and the coefficient of friction [44, 47]. The adhesive force for silicon increased with the temperature from room temperature (RT) to 55 °C, and then decreased from 55 to 75 °C before eventually leveling off from 75 to 110 °C, see Fig. 17.23a. The adhesive forces of the SAMs showed a similar tendency except that the initial increase was not pronounced. The initial increase in adhesive force for silicon at low temperatures is not understood. The decrease observed for silicon could be attributed to the desorption of water molecules on the surface. After the water layer has been almost completely depleted, the adhesive force may stay constant. The SAMs with hydrocarbon backbone chains (OP, ODP, ODMS and ODDMS) showed similar behavior to the Al substrate but the initial increase in the adhesive force with temperature was much smaller. The SAMs with fluorocarbon backbone chains showed almost no temperature dependence. The adhesive force shows some temperature dependence for the SAMs with hydrocar-

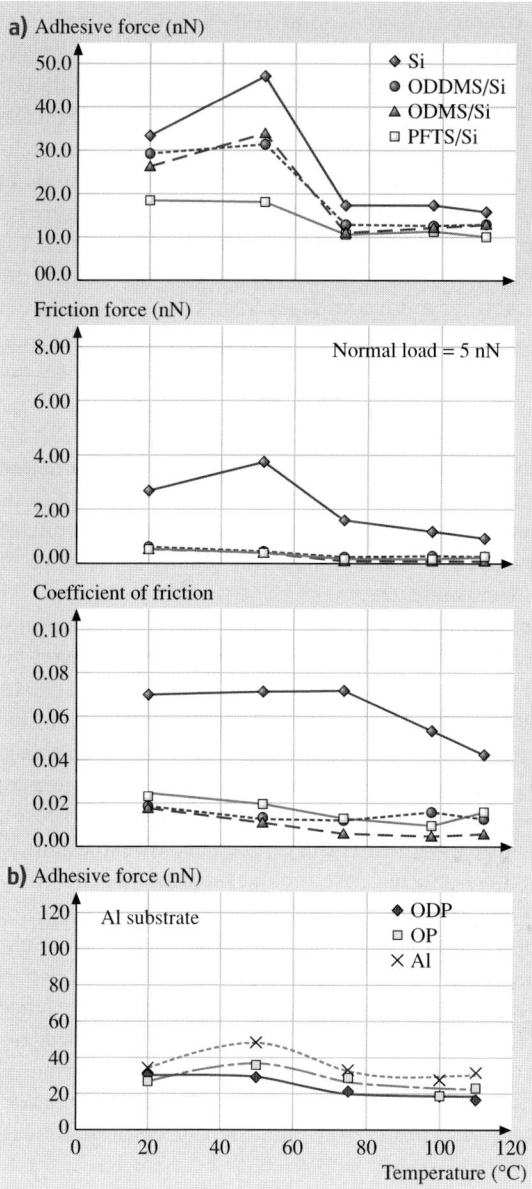

Fig. 17.23. Effect of temperature on the adhesive force, the friction force and the coefficient of friction for various SAMs on (a) Si substrates [44], and (b) Al substrates

bon backbone chains. This increase in adhesive force is believed to be caused by the melting of the SAM film. The typical melting point for a linear carbon chain molecule such as $CH_3(CH_2)_{14}CH_2OH$ is 50 °C [92]. As the temperature increases, the SAM film softens, thereby increasing the real area of contact and consequently the adhesive force. Once the temperature is higher than the melting point, the lubrication regime is changed from boundary lubrication in the solid SAM to liquid lubrication in the melted SAM [41].

The friction force of silicon increased with temperature and then steadily decreased. The friction force is highly affected by changes in adhesion. The decrease in friction could therefore result from the depletion of the water layer. The coefficient of friction for silicon remained constant and then decreased, starting at about 80 °C. For the SAMs, the coefficient of friction exhibited a monotonic decrease with temperature. The decrease in the friction and the coefficient of friction for the SAMs possibly results from the decrease in stiffness. As mentioned before, the spring model suggests lower friction for more compliant SAMs [41]. The different types of SAM types behaved reasonably similarly. PFTS kept its stiffness more than ODMS and ODDMS when the temperature was increased [93], but this was not a particularly pronounced effect in the results.

Figure 17.23b shows the effect of temperature on the adhesive force for SAMs on Al. The adhesive force increased for Al substrate up to 50 °C and then decreased to a stable value for higher temperatures. This initial increase in the adhesive force with temperature for Al is not understood. The inherently high hydrophobicity of the SAMs over the corresponding substrates meant that they did not show much of a relative humidity or temperature dependence.

Figure 17.24 shows the effect of sliding velocity on the adhesive force, the friction force and the coefficient of friction [44, 47]. The adhesive force for silicon remained relatively constant at low sliding velocities, and then increased rapidly: see Fig. 17.24a. A similar trend was observed for the SAMs. The increase in adhesive force for silicon is believed to be due to a tribochemical reaction at the interface between the tip and sample [23] and the increased contact area due to mechanical plowing. For the SAMs, the increased adhesive force at high velocities may result from viscous drag of SAM molecules [94]. SAMs can be detached from the surface and attached to an AFM tip. In addition, the increased contact area may be caused by greater penetration of the AFM tip into the SAM. The rate of increase was larger for ODMS than for PFTS, presumably because of the higher stiffness and more dense structure of PFTS.

The coefficient of friction increased with sliding velocity and then reached a plateau for Si, ODMS/Si and ODDMS/Si. As the sliding speed is increased, extra work is needed for SAM reorientation, which may lead to increased friction. For PFTS, the coefficient of friction decreased at large sliding velocities, resulting in a peak. This peak structure may result from the viscoelastic properties of SAMs [95].

Figure 17.24b shows the effect of sliding velocity on the friction force. Friction force was found to remain constant over a range of sliding velocities for Al substrate as well as the OP and ODP SAMs deposited on Al substrates. The increase in friction force at high velocities (> 1 mm/s) is the result of asperity impacts and correspondingly high frictional energy dissipation at the sliding interface for Al [96]. For the OP and ODP SAMs, the increase in friction force is believed to result from the SAM molecules reorienting under the tip load and during the tip motion. The

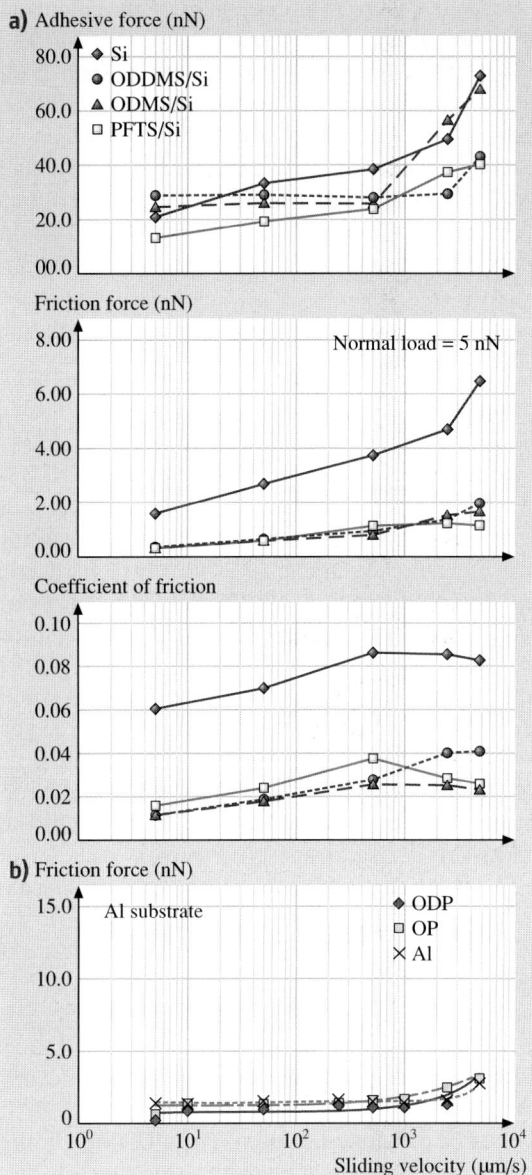

Fig. 17.24. Effect of sliding velocity on the adhesive force, the friction force and the coefficient of friction for various SAMs on (**a**) Si substrates [44], and (**b**) Al substrates (22 °C, 50% RH, 70 nN normal load)

SAM reorientation can act as an additional hindrance to tip motion when the AFM tip reverses during scanning, resulting in higher friction. The molecules can get entangled and/or get detached from the substrate and attach to the AFM tip. *Tambe* and *Bhushan* [94] extended the molecular spring model presented by *Bhushan* and

Liu [39] to explain this velocity-dependent increase in friction force for compliant SAM molecules.

Overall, the SAMs deposited on the Al substrate showed much lower friction at all sliding velocities than those deposited on the Si substrates. As discussed, the primary reason for this is believed to be the ease with which the molecules on Al substrate can rotate due to the absence of either the $-CH_3$ groups or any crosslinking at the head groups. The higher stiffness of the fluorocarbon backbone chains than the hydrocarbon backbone chains [44] is also believed to result in the higher friction for PFTS than ODMS and ODDMS.

AFM Wear Measurements

Figure 17.25a shows the relationships between surface height and normal load found for various SAMs during wear tests [44, 47]. As shown in the figure, the SAMs exhibit a critical normal load beyond which the surface height decreases drastically. Figure 17.25a also shows the wear behavior of the Al and Si substrates. Unlike the SAMs, the substrates show a monotonic decrease in surface height with increasing normal load, with wear initiating from the very beginning, even for low normal loads. Si (Young's modulus of elasticity, $E = 130$ GPa [97], hardness, $H = 11$ GPa [28]) is relatively hard in comparison to Al ($E = 77$ GPa, $H = 0.41$ GPa) and hence the decrease in surface height for Al is much larger than that for Si for similar normal loads.

The critical loads corresponding to the sudden failure of the SAM are shown in Fig. 17.25b. Amongst all the SAMs, ODDMS shows the best performance in the wear tests, and this is believed to be due to the chain length effect (it has a longer chain). OP and ODP show very similar wear behavior to ODMS and ODDMS. ODP exhibits a higher critical load than OP because of its longer chain length. The mechanism of failure for compliant SAMs during wear tests was presented earlier, in Fig. 17.18. It is believed that the SAMs usually fail due to shearing of the molecule at the head group; that is, the molecules are sheared off the substrate. Table 17.14 gives the bond strengths for various intermolecular bonds. The weakest bonds are at the interface, and hence failure is expected to occur at the interface first.

To study the effect of relative humidity on wear, wear tests were performed at various humidities. The bottom of Fig. 17.22a shows the critical normal load as a function of relative humidity. The critical normal load shows a weak dependency on the relative humidity for ODMS/Si and PFTS/Si, and was larger for ODMS/Si than for PFTS/Si throughout the range of humidities used. For ODDMS/Si, the critical normal load showed an increase with relative humidity. This suggests that water molecules can penetrate into ODDMS, which then might work as lubricant [41, 100]. This effect was absent for PFTS/Si and ODMS/Si.

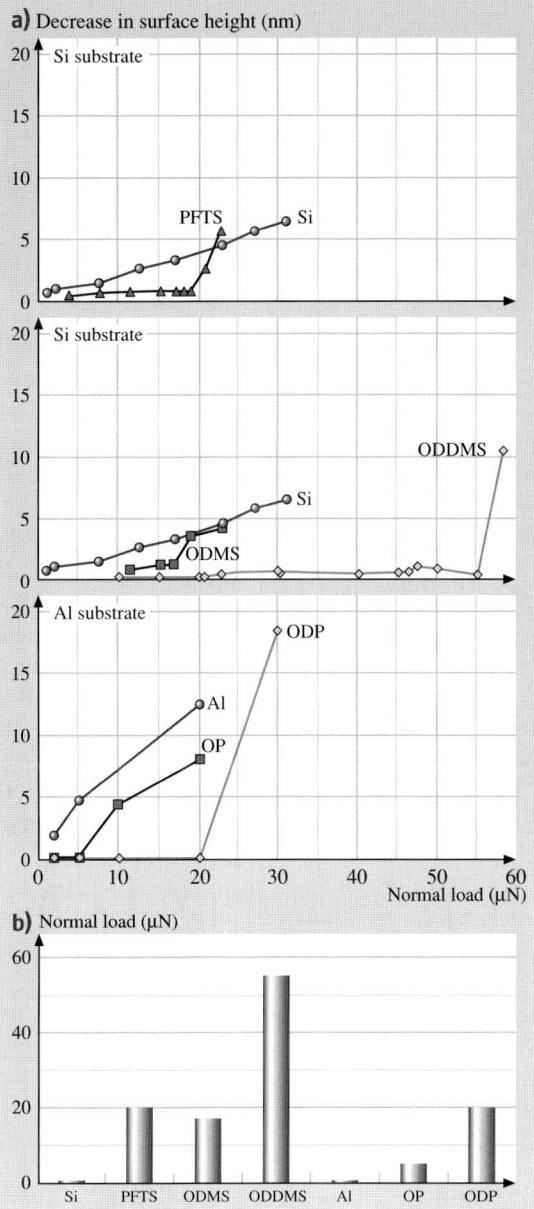

Fig. 17.25. (a) Decrease in surface height as a function of normal load after one scan cycle for various SAMs on Si and Al substrates, and (b) comparison of critical loads for failure during wear tests for various SAMs

Table 17.14. Typical bond strengths[a] in SAMs

Bond	Hexadecanethiol (HDT) (kJ/mol)	Biphenylthiol (BPT) (kJ/mol)	Bond	Perfluoroalkylsilane (PFTS) (kJ/mol)	Alkylsilane (ODMS or ODDMS)	Akylphosphonate (OP and ODP)
Interfacial bonds						
S–Au	184[b]	184[b]	Si–O	242[c]	242[c]	242[c]
S–C	286[a]		Si–C	800[d]	800[d]	800[d]
C_6H_5–S		362[a]	Al–O	414[a]	414[a]	511[a]
			P–C			513[a]
			P–O			599[a]
Bonds in backbone						
C–C			C–C			
CH_2–CH_2	326[e]		CH_2–CH_2	326[e]	326[e]	
CH_3–CH_2	≈ 305[a]		CF_2–CF_2	≈ 326[f]		
C_6H_5		strong	CF_2–CH_2	≈ 326[f]		
			CF_3–CF_2	≈ 326[f]		
			CH_3–CH_2		≈ 305[a]	

[a] *Lide* [92]
[b] Chemical adsorption bond from *Lio* et al. [72]
[c] Chemical adsorption bond from *Hoshino* [98]
[d] In diatomic molecules
[e] *Cottrell* [99]
[f] Because of the C–C bond it is expected to be close to that of CH_2–CH_2

17.4.4 Degradation and Environmental Studies

Degradation Studies

The coefficient of friction and the gaseous products detected for HDT/Au are shown in Fig. 17.26a [48]. A normal pressure of 50 kPa was applied to the HDT films. The coefficient of friction increased after a sliding distance of about 10 m. During sliding, $(CH_2)_{15}S$, C_2H_3, CH_3, CH_2 and H_2 were detected by a mass spectrometer. The partial pressure of the HS fragments is of interest since it corresponds to the interface bonds, and so it is reported here. The increase in $(CH_2)_{15}S$ was much

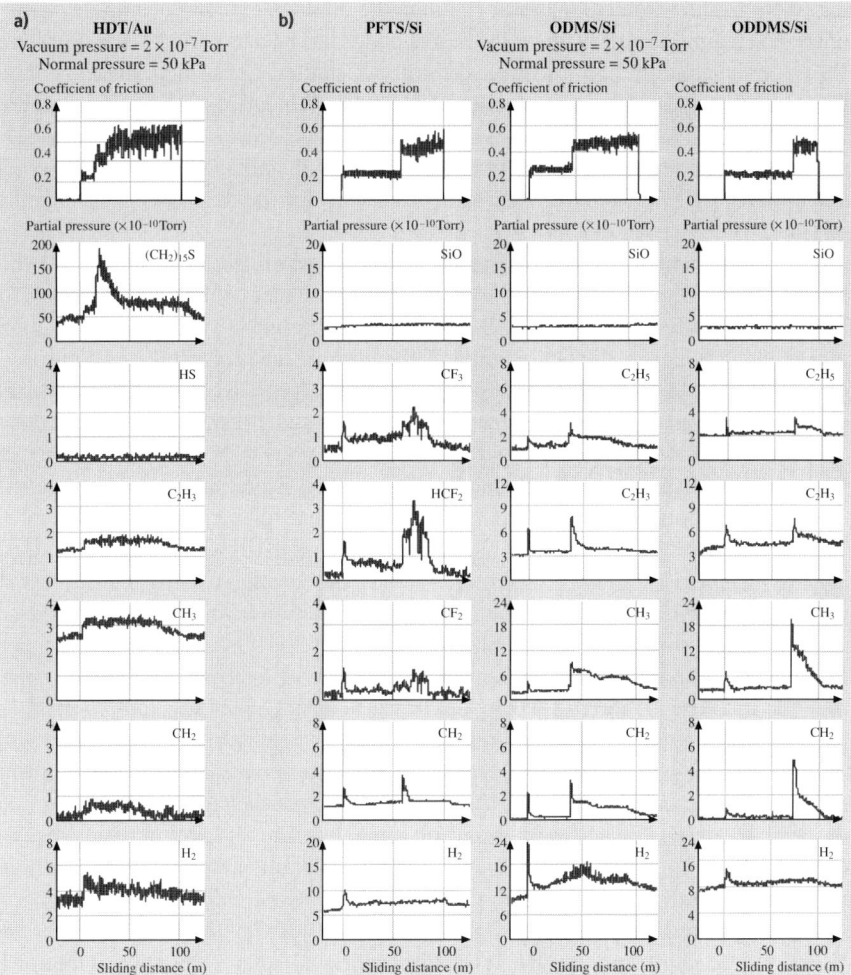

Fig. 17.26. Coefficients of friction and mass spectral data on (**a**) HDT/Au (1.9 nm), (**b**) PFTS/Si (1.8 nm), ODMS/Si ($\approx$ 1.9 nm) and ODDMS/Si ($\approx$ 2.1 nm) in high vacuum [48]

more than that of other species, due to the breaking of the S–Au bond. The partial pressures of C_2H_3, CH_3, CH_2, and H_2 were also found to increase during the sliding. There was no noticeable change in the partial pressure of HS.

The HDT film was deposited on an Au(111) layer. The bond strength of S–Au is 184 kJ/mol (Table 17.14), which is lower than those of the C–C bonds (425 kJ/mol), C–H bonds (422 kJ/mol), and C–S bonds (286 kJ/mol) in the alkyl chains. Since the S–Au bond is the weakest bond in the alkanethiol chain, the whole chain should be sheared away from the substrate. Because the upper atomic mass unit (amu) limit of the mass spectrometer used is 250, we monitored $(CH_2)_{15}S$ (amu = 242), which is the chain with CH_3 sheared away. The generation rate of $(CH_2)_{15}S$ is much larger than that of other species. This suggests that the mechanical shear of the whole alkanethiol chain be the dominant factor causing the failure of the HDT film. The cleavage of the S–Au bonds has been reported in the literature. Based on the bond strengths, as well as the above studies, mechanical shearing of the C–C bonds and C–H bonds probably does not happen during sliding. The reaction induced by low-energy electrons, generated by triboelectrical emission during the sliding, could be responsible for the degradation of the alkanethiol chain. Thermal desorption of HDT from Au is another possibility for the degradation mechanism of HDT.

The coefficient of friction and gaseous products generated for PFTS/Si, ODMS/Si and ODDMS/SI are shown in Fig. 17.26b [48]. The coefficients of friction for PFTS/Si, ODMS/Si, and ODDMS/Si increase sharply after a certain sliding distance, which indicates the degradation of the film. At the same time, gaseous products of CF_3, HCF_2, CF_2, CH_2 and H_2 were detected for PFTS/Si, and C_2H_5, C_2H_3, CH_3, CH_2 and H_2 were detected for ODMS/Si and ODDMS/Si.

PFTS/Si showed lower friction than ODMS/Si in the tests. ODDMS/Si showed lower friction than both PFTS/Si and ODMS/Si. This is because of the chain length effect; as mentioned earlier, it has been reported that for SAMs the coefficient of friction decreases with the carbon backbone chain length (n) when the carbon atoms are less than 12. For chains with more than 12 carbons, increasing the number of carbon atoms will not influence the coefficient of friction to any noticeable extent.

PFTS/Si showed greater durability than ODMS/Si. It is harder to rotate a perfluorinated carbon backbone (due to the larger size of F versus H) which implies that this structure is more rigid than a hydrocarbon backbone [91]. *Chambers* [101] has reported that the C–C bond strength increases when hydrogen is replaced with fluorine. This suggests that the rigid perfluorinated carbon backbone may be responsible for the increased durability. The length of the alkyl chain also influences the desorption energies of alkanes. Based on studies of the adsorption of alkanes on Cu(100), Au(111), Pt(110) and others, the physisorption energy increases with the alkyl chain length [102–104]. Therefore, ODDMS are more durable than ODMS.

During sliding on PFTS films, gaseous products of CF_3, HCF_2, CF_2, CH_2 and H_2 were detected. From the structure of perfluoroalkylsilane, the only source of H on the molecular chain which would cause a partial pressure increase of H_2 is the $(CH_2)_2$, which is located at the bottom of the chain. Since the partial pressure of

H_2 increases immediately after sliding and remains high until the end of sliding, it is probably generated by low-energy electrons arising from triboelectrical emission. The partial pressure of CH_2 exhibits a sharp peak at the beginning of sliding and at the moment when friction changes. Meanwhile, the partial pressures of CH_3, HSF_2 and CF_2 increased significantly when the friction increased. For ODMS and ODDMS, C_2H_5, C_2H_3, CH_3, CH_2 and H_2 were detected during sliding. The partial pressures of the carbon-related products increase considerably when the friction is increased. SiO, which is associated with interface bonds, shows no noticeable change during sliding.

Perfluoroalkylsilanes and alkylsilanes are attached to the naturally oxidized silicon by Si–O bonds. The Si–O bond strength varies widely (Table 17.14) depending on the formation conditions. In the alkylsilane chain, the C–Si bond strength (414 kJ/mol) is slightly lower than the C–C bond strength. Based on Table 17.14, the interfacial bonds (Si–O) are weaker than the C–C bonds in the backbone. Therefore, it is believed that film cleavage occurs at the interface. We have previously reported evidence of the cleavage of interfacial bonds using an AFM. To explain the hydrogen, C_1 and C_2 hydrocarbon (in the tests for PFTS/Si, ODMS/Si and ODDMS/Si) or fluorocarbon (in the tests for PFTS/Si) products, *Kluth* et al. [105] suggested that the alkylsilane (or perfluoroalkylsilane) chains break and create rad-

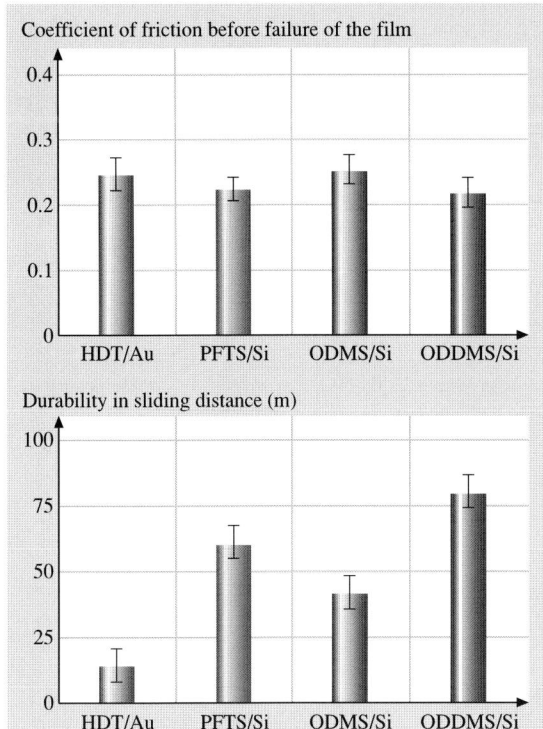

Fig. 17.27. Coefficient of friction (*upper*) and durability (*lower*) comparisons for HDT/Au, PFTS/Si, ODMS/Si and ODDMS/Si in high vacuum. *Error bars* represent $\pm 3\sigma$ based on five measurements (normal pressure 50 kPa) (after [48])

icals. The radical could remain on the surface and decompose to generate a short radical and an alkene. The radical can repeatedly decompose to ever shorter radicals and alkenes so long as it remains on the surface.

A summary of the coefficients of friction and the durability of all films in vacuum is presented in Fig. 17.27.

Environmental Studies

To study the effect of the environment, friction tests were conducted in high vacuum, argon, dry air (less than 2% RH), air with 30% RH, and air with 70% RH; Fig. 17.28. The applied normal pressure was 50 kPa for HDT/Au, PFTS/Si, ODMS/Si and ODDMS/Si, which is the same as used in the degradation tests. By comparing the coefficients of friction in different environments, it was found that the friction is the lowest in argon for the SAMs tested. The intimate contact induced by high vacuum leads to high friction. Friction is higher in dry air than in argon. This shows that oxygen appears to effect SAM performance. *Kim* et al. [106] studied the thermal stability of alkylsiloxane SAMs in air. They found that the alkylsiloxane decomposes at about 200 °C, which is much lower than the decomposition temperature of 470 °C

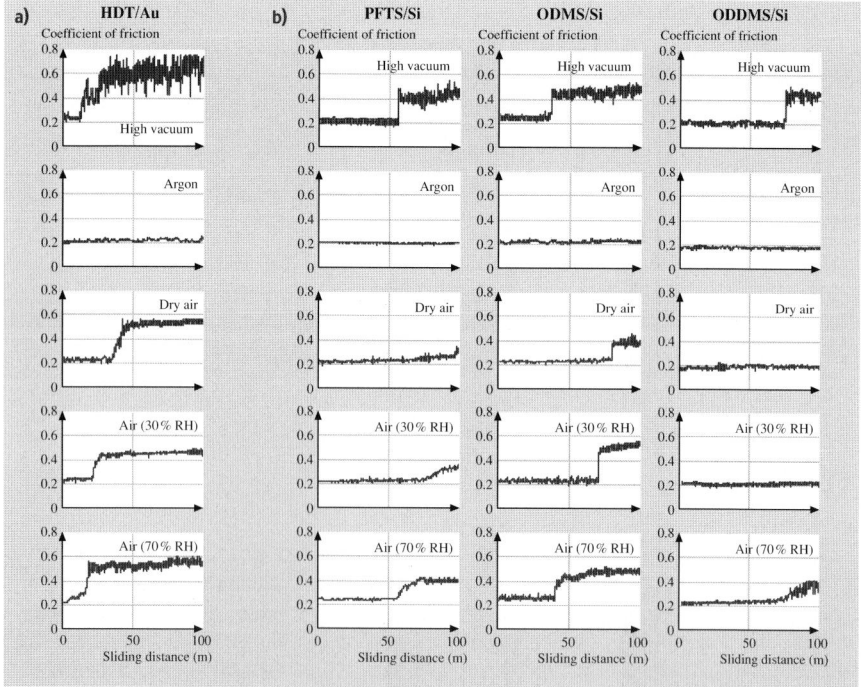

Fig. 17.28. Coefficient of friction data taken in high vacuum, argon and air with different humidity levels for (**a**) HDT/Au (1.9 nm), (**b**) PFTS/Si (1.8 nm), ODMS/Si ($\approx$ 1.9 nm) and ODDMS/Si ($\approx$ 2.1 nm) (RH: relative humidity, normal pressure 50 kPa) (after [48])

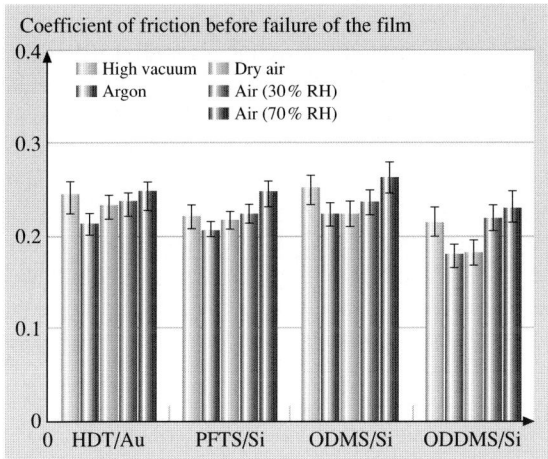

Fig. 17.29. Coefficient of friction comparison for HDT/Au, PFTS/Si, ODMS/Si and ODDMS/Si in high vacuum, argon and air with different humidity levels. *Error bars* represent ±3σ based on five measurements (normal pressure 50 kPa) (after [48])

in vacuum reported by *Kluth* et al. [105]. This difference could be attributed to the oxygen in air. Water in the air is found to have significant influence on the friction of SAMs. A study of humidity effects for alkylsilane on mica substrate performed by *Tian* et al. [100] indicated that the water molecules can penetrate the alkylsilane film, altering their molecular chain ordering and also detaching the alkylsilane molecules from the substrate.

A summary of friction coefficients observed before lubricant film failure in various environments is presented in Fig. 17.29. The data in Fig. 17.29 are average values based on five measurements.

17.5 Closure

Exposure of devices to a humid environment results in condensates of water vapor. Condensed water or a pre-existing film of liquid results in the formation of concave meniscus bridges between the hydrophilic mating surfaces. The negative Laplace pressure present in the meniscus results in an adhesive force which depends on the roughness of the interface, the surface tension and the contact angle. The adhesive force can be significant in an interface with ultrasmooth surfaces, and it can be on the same order as the external load if the latter is small, such as in micro/nanodevices. Surfaces with high hydrophobicity can be produced by surface treatment. In many applications, hydrophobic films are expected to provide low adhesion, friction and wear. Because of the small clearances inherent to micro/nanodevices, these films should be molecularly thick. Liquid films with low surface tension or hydrophobic solid films can be used. Ordered molecular assemblies with high hydrophobicity can be engineered by chemically grafting various polymer molecules with suitable functional head groups and nonpolar surface terminal groups.

The adhesion, friction and wear properties of SAMs with alkyl, biphenyl and perfluoroalkyl spacer chains and different surface terminal ($-CH_3$ and $-CF_3$) and head groups (−SH, −Si−O−, −OH, and P−O−), studied using an AFM, are reported in this chapter. It was found that the adhesive force varies linearly with the W_a value of the SAM, which indicates that capillary condensation of water plays an important role in the adhesion of SAMs on the nanoscale at ambient conditions. SAMs with long high-compliance carbon spacer chains exhibit the smallest adhesive and friction forces. The friction data are explained using a molecular spring model, in which the local stiffness and intermolecular force govern frictional performance. Results from stiffness and friction characterizations of a micropatterned sample with different structures on it support this model. Perfluoroalkylsilane SAMs exhibit lower surface energies, higher contact angles and lower adhesive forces than alkylsilane SAMs. The substrate had little effect. The coefficients of friction for various SAMs were comparable.

The influence of the relative humidity on the adhesion and the friction of SAMs is dominated by the thickness of the adsorbed water layer. At higher humidity, water increases friction due to the increased adhesion caused by the meniscus effect in the contact zone. As the temperature is increased, in the case of Si(111), the desorption of the adsorbed water layer and the reduced surface tension of water reduces the adhesive and friction forces. Decreases in adhesion and friction with temperature were found for PFTS/Si, ODMS/Si and ODDMS/Si. PFTS showed the lowest adhesion at low temperature ($\approx 70\,°C$). Differences among the SAMs were small at high temperature ($\approx 100\,°C$). Increases in adhesive force and friction with sliding velocity were observed for PFTS/Si, ODMS/Si and ODDMS/Si. A peak in the coefficient of friction appeared for PFTS.

PFTS/Si showed better wear resistance than ODMS/Si. ODDMS/Si showed better wear resistance than ODMS/Si due to the chain length effect. SAM wear behavior is mostly determined by the molecule–substrate bond strength. The long chain molecules of ODP/Al and ODDMS/Si showed higher critical loads for failure.

The results from nanotribological characterization studies of SAMs deposited on Si and Al substrates are summarized in Table 17.15 [47]. SAMs deposited on Si

Table 17.15. Summary of nanotribological characterization studies of SAMs on Si and Al substrates

SAM property		Friction force	Adhesive force	Wear
Substrate	Hard	High	Low	Low
	Soft	Low	Low	High
Chemical structure	Linear chain molecule	High	Low	High
	Ring molecule	High	High	Low
Backbone	Fluorocarbon backbone	Low	Low	Low
	Hydrocarbon backbone	Low	High	High
Chain length	Long backbone chain		High	High
	Short backbone chain		Low	Low

and Al substrates show low friction and low adhesion, both of which are desirable for MEMS/NEMS applications.

Based on studies in high vacuum (2×10^{-7} Torr), the friction coefficients for SAMs have been found to follow the order (from low to high): ODDMS/Si, PFTS/Si, HDT/Au, ODMS/Si. HDT on Au is less durable than perfluoroalkylsilane and alkylsilane on Si due to its weak interfacial bonding. PFTS/Si is more durable than ODMS/Si. This indicates that fluorinating alkylsilane can improve durability. ODDMS/Si is more durable than ODMS/Si and PFTS/Si because of the chain length effect. SAM friction is higher in high vacuum than in argon because of the intimate contact involved. Based on studies in argon and air with various relative humidities, oxygen can increase the friction and durability of SAMs. The water molecule can detach SAM molecules from the substrate, resulting in high friction and low durability.

In summary, nanotribological studies of SAM films using AFM demonstrate that they exhibit attractive hydrophobic and tribological properties. Fluorinated SAMs appear to be the most performant. SAM films should find application in many fields, including micro/nanodevices.

References

1. B. Bhushan, J. N. Israelachvili, U. Landman: Nanotribology: friction, wear and lubrication at the atomic scale, Nature **374**, 607–616 (1995)
2. B. Bhushan: *Tribology and Mechanics of Magnetic Storage Devices*, 2nd edn. (Springer, Berlin, Heidelberg 1996)
3. B. Bhushan (Ed.): *Tribology Issues and Opportunities in MEMS* (Kluwer Academic, Dordrecht 1998)
4. B. Bhushan (Ed.): *Handbook of Micro/Nanotribology*, 2nd edn. (CRC, Boca Raton 1999)
5. B. Bhushan: *Nanotribology and Nanomechanics – An Introduction* (Springer, Berlin, Heidelberg 2005)
6. K. F. Man, B. H. Stark, R. Ramesham: *A Resource Handbook for MEMS Reliability* (JPL Press, Jet Propulsion Laboratory, California Institute of Technology, Pasadena 1998)
7. K. F. Man: MEMS reliability for space applications by elimination of potential failure modes through testing and analysis, http://www.rel.jpl.nasa.gov/Org/5053/atpo/products/Prod-map.html, (2002)
8. D. M. Tanner, N. F. Smith, L. W. Irwin et al.: *MEMS Reliability: Infrastructure, Test Structure, Experiments, and Failure Modes*, SAND2000-0091 (Sandia National Laboratories, Albuquerque 2000)
9. B. Bhushan: *Principles and Applications of Tribology* (Wiley, New York 1999)
10. B. Bhushan: *Introduction to Tribology* (Wiley, New York 2002)
11. M. E. Schrader, G. I. Loeb (Eds.): *Modern Approaches to Wettability* (Plenum, New York 1992)
12. A. Ulman (Eds.): *Characterization of Organic Thin Films* (Butterworth–Heinemann, Boston 1995)
13. A. W. Neumann, J. K. Spelt (Eds.): *Applied Surface Thermodynamics* (Marcel Dekker, New York 1996)

14. B. Bhushan: Contact mechanics of rough surfaces in tribology: multiple asperity contact, Tribol. Lett. **4**, 1–35 (1998)
15. B. Bhushan, W. Peng: Contact mechanics of multilayered rough surfaces, Appl. Mech. Rev. **55**, 435–480 (2002)
16. M. Nosonovsky, B. Bhushan: Roughness optimization for biomimetic superhydrophobic surfaces, Microsyst. Technol. **11**, 535–549 (2005)
17. R. Maboudian: Surface processes in MEMS technology, Surf. Sci. Rep. **30**, 209–269 (1998)
18. B. Bhushan (Ed.): *Modern Tribology Handbook, Vol. 1 – Principles of Tribology; Vol. 2 – Materials, Coatings, and Industrial Applications* (CRC, Boca Raton 2001)
19. B. Bhushan, Z. Zhao: Macro- and microscale tribological studies of molecularly-thick boundary layers of perfluoropolyether lubricants for magnetic thin-film rigid disks, J. Info. Stor. Proc. Syst. **1**, 1–21 (1999)
20. F. P. Bowden, D. Tabor: *The Friction and Lubrication of Solids, Part I* (Clarendon, Oxford 1950)
21. W. A. Zisman: Friction, durability and wettability properties of monomolecular films on solids. In: *Friction and Wear*, ed. by R. Davies (Elsevier, Amsterdam 1959) pp. 110–148
22. V. N. Koinkar, B. Bhushan: Microtribological studies of unlubricated and lubricated surfaces using atomic force/friction force microscopy, J. Vac. Sci. Technol. A **14**, 2378–2391 (1996)
23. H. Liu, B. Bhushan: Nanotribological characterization of molecularly-thick lubricant films for applications to MEMS/NEMS by AFM, Ultramicroscopy **97**, 321–340 (2003)
24. A. Ulman: *An Introduction to Ultrathin Organic Films: From Langmuir-Blodgett to Self-Assembly* (Academic, San Diego 1991)
25. A. Ulman: Formation and structure of self-assembled monolayers, Chem. Rev **96**, 1533–1554 (1996)
26. H. Hansma, F. Motamedi, P. Smith, P. Hansma, J. C. Wittman: Molecular resolution of thin, highly oriented poly(tetrafluoroethylene) films with the atomic force microscope, Polym. Commun. **33**, 647–649 (1992)
27. L. Scandella, A. Schumacher, N. Kruse, R. Prins, E. Meyer, R. Luethi, L. Howald, M. Scherge, J. A. Schaefer: Surface modification and mechanical properties of bulk silicon. In: *Tribology Issues and Opportunities in MEMS*, ed. by B. Bhushan (Kluwer, Dordrecht 1998) pp. 529–537
28. B. Bhushan: Chemical, mechanical and tribological characterization of ultra-thin and hard amorphous carbon coatings as thin as 3.5 nm: recent developments, Diamond Relat. Mater. **8**, 1985–2015 (1999)
29. A. Erdemir, C. Donnet: Tribology of diamond, diamond-like carbon, and related films. In: *Modern Tribology Handbook*, Vol. 2: Materials, Coatings, and Industrial Applications, ed. by B. Bhushan (CRC, Boca Raton 2001) pp. 871–908
30. V. F. Dorfman: Diamond-like nanocomposites (DLN), Thin Solid Films **212**, 267–273 (1992)
31. M. Grischke, K. Bewilogua, K. Trojan, H. Dimigan: Application-oriented modification of deposition process for diamond-like carbon based coatings, Surf. Coat. Technol. **74-75**, 739–745 (1995)
32. R. S. Butter, D. R. Waterman, A. H. Lettington, R. T. Ramos, E. J. Fordham: Production and wetting properties of fluorinated diamond-like carbon coatings, Thin Solid Films **311**, 107–113 (1997)

33. M. Grischke, A. Hieke, F. Morgenweck, H. Dimigan: Variation of the wettability of DLC coatings by network modification using silicon and oxygen, Diamond Relat. Mater. **7**, 454–458 (1998)
34. C. Donnet, J. Fontaine, A. Grill, V. Patel, C. Jahnes, M. Belin: Wear-resistant fluorinated diamondlike carbon films, Surf. Coat. Technol. **94-95**, 531–536 (1997)
35. D. J. Kester, C. L. Brodbeck, I. L. Singer, A. Kyriakopoulos: Sliding wear behavior of diamond-like nanocomposite coatings, Surf. Coat. Technol. **113**, 268–273 (1999)
36. H. Liu, B. Bhushan: Adhesion and friction studies of microelectromechanical systems/nanoelectromechanical systems materials using a novel microtriboapparatus, J. Vac. Sci. Technol. A **21**, 1528–1538 (2003)
37. B. Bhushan, H. Liu, S. M. Hsu: Adhesion and friction studies of silicon and hydrophobic and low friction films and investigation of scale effects, ASME J. Tribol. **126**, 583–590 (2004)
38. B. Bhushan, A. V. Kulkarni, V. N. Koinkar, M. Boehm, L. Odoni, C. Martelet, M. Belin: Microtribological characterization of self-assembled and Langmuir–Blodgett monolayers by atomic and friction force microscopy, Langmuir **11**, 3189–3198 (1995)
39. B. Bhushan, H. Liu: Nanotribological properties and mechanisms of alkylthiol and biphenyl thiol self-assembled monolayers studied by atomic force microscopy, Phys. Rev. B **63**, 245412–1–11 (2001)
40. H. Liu, B. Bhushan, W. Eck, V. Stadler: Investigation of the adhesion, friction, and wear properties of biphenyl thiol self-assembled monolayers by atomic force microscopy, J. Vac. Sci. Technol. A **19**, 1234–1240 (2001)
41. H. Liu, B. Bhushan: Investigation of nanotribological properties of alkylthiol and biphenyl thiol self-assembled monolayers, Ultramicroscopy **91**, 185–202 (2002)
42. H. Liu, B. Bhushan: Orientation and relocation of biphenyl thiol self-assembled monolayers, Ultramicroscopy **91**, 177–183 (2002)
43. B. Bhushan, T. Kasai, G. Kulik, L. Barbieri, P. Hoffmann: AFM study of perfluorosilane and alkylsilane self-assembled monolayers for anti-stiction in MEMS/NEMS, Ultramicroscopy **105**, 176–188 (2005)
44. T. Kasai, B. Bhushan, G. Kulik, L. Barbieri, P. Hoffmann: Nanotribological study of perfluorosilane SAMs for anti-stiction and low wear, J. Vac. Sci. Technol. B **23**, 995–1003 (2005)
45. K. K. Lee, B. Bhushan, D. Hansford: Nanotribological characterization of perfluoropolymer thin films for BioMEMS applications, J. Vac. Sci. Technol. A **23**, 804–810 (2005)
46. B. Bhushan, D. Hansford, K. K. Lee: Surface modification of silicon surfaces with vapor phase deposited ultrathin fluorosilane films for biomedical devices, J. Vac. Sci. Technol. A **24**, 1197–1202 (2006)
47. N. S. Tambe, B. Bhushan: Nanotribological characterization of self assembled monolayers deposited on silicon and aluminum substrates, Nanotechnology **16**, 1549–1558 (2005)
48. Z. Tao, B. Bhushan: Degradation mechanisms and environmental effects on perfluoropolyether self assembled monolayers and diamondlike carbon films, Langmuir **21**, 2391–2399 (2005)
49. J. A. Zasadzinski, R. Viswanathan, L. Madsen, J. Garnaes, D. K. Schwartz: Langmuir–Blodgett films, Science **263**, 1726–1733 (1994)
50. J. Tian, Y. Xia, G. M. Whitesides: Microcontact printing of SAMs. In: *Thin Films – Self-Assembled Monolayers of Thiols*, Vol. 24, ed. by A. Ulman (Academic, San Diego 1998) pp. 227–254

51. Y. Xia, G. M. Whitesides: Soft lithography, Angew. Chem. Int. Ed. **37**, 550–575 (1998)
52. A. Kumar, G. M. Whitesides: Features of gold having micrometer to centimeter dimensions can be formed through a combination of stamping with an elastomeric stamp and an alkanethiol ink followed by chemical etching, Appl. Phys. Lett. **63**, 2002–2004 (1993)
53. S. Y. Chou, P. R. Krauss, P. J. Renstrom: Imprint lithography with 25-nanometer resolution, Science **272**, 85–87 (1996)
54. Y. Xia, E. Kim, X. M. Zhao, J. A. Rogers, M. Prentiss, G. M. Whitesides: Complex optical surfaces formed by replica molding against elastomeric masters, Science **273**, 347–349 (1996)
55. L. J. Hornbeck: The DMD^{TM} projection display chip: a MEMS-based technology, MRS Bull. **26**, 325–328 (2001)
56. M. R. Douglass: Lifetime estimates and unique failure mechanisms of the digital micromirror device (DMD). In: *1998 International Reliability Physics Proceedings*, IEEE Catalog No. 98 CH 36173 (, 1998) pp. 9–16 Presented at the 36th Annual International Reliability Physics Symposium, Reno
57. H. Liu, B. Bhushan: Nanotribological characterization of digital micromirror devices using an atomic force microscope, Ultramicroscopy **100**, 391–412 (2004)
58. H. Liu, B. Bhushan: Investigation of nanotribological and nanomechanical properties of the digital micromirror device by atomic force microscope, J. Vac. Sci. Technol. A **22**, 1388–1396 (2004)
59. A. Manz, H. Becker (Eds.): *Microsystem Technology in Chemistry and Life Sciences* (Springer, Berlin, Heidelberg 1998)
60. J. Cheng, L. J. Krica (Eds.): *Biochip Technology* (Harwood, New York 2001)
61. M. J. Heller, A. Guttman (Eds.): *Integrated Microfabricated Biodevices* (Marcel Dekker, New York 2001)
62. A. van der Berg (Ed.): *Lab-on-a-Chip: Chemistry in Miniaturized Synthesis and Analysis Systems* (Elsevier, Amsterdam 2003)
63. M. Hein, L. R. Best, S. Pattison, S. Arena: *Introduction to General, Organic, and Biochemistry*, 6th edn. (Brooks/Cole, Pacific Grove 1997)
64. J. R. Mohrig, C. N. Hammond, T. C. Morrill, D. C. Neckers: *Experimental Organic Chemistry* (W. H. Freeman, New York 1998)
65. S. R. Wasserman, Y. T. Tao, G. M. Whitesides: Structure and reactivity of alkylsiloxane monolayers formed by reaction of alkylchlorosilanes on silicon substrates, Langmuir **5**, 1074–1089 (1989)
66. C. Jung, O. Dannenberger, Y. Xu, M. Buck, M. Grunze: Self-assembled monolayers from organosulfur compounds: a comparison between sulfides, disulfides, and thiols, Langmuir **14**, 1103–1107 (1998)
67. W. Geyer, V. Stadler, W. Eck, M. Zharnikov, A. Golzhauser, M. Grunze: Electron-induced crosslinking of aromatic self-assembled monolayers: negative resists for nanolithography, Appl. Phys. Lett. **75**, 2401–2403 (1999)
68. J. Ruhe, V. J. Novotny, K. K. Kanazawa, T. Clarke, G. B. Street: Structure and tribological properties of ultrathin alkylsilane films chemisorbed to solid surfaces, Langmuir **9**, 2383–2388 (1993)
69. V. DePalma, N. Tillman: Friction and wear of self-assembled tricholosilane monolayer films on silicon, Langmuir **5**, 868–872 (1989)
70. M. T. McDermott, J. B. D. Green, M. D. Porter: Scanning force microscopic exploration of the lubrication capabilities of n-alkanethiolate monolayers chemisorbed at gold: structural basis of microscopic friction and wear, Langmuir **13**, 2504–2510 (1997)

71. X. Xiao, J. Hu, D.H. Charych, M. Salmeron: Chain length dependence of the frictional properties of alkylsilane molecules self-assembled on mica studied by atomic force microscopy, Langmuir **12**, 235–237 (1996)
72. A. Lio, D.H. Charych, M. Salmeron: Comparative atomic force microscopy study of the chain length dependence of frictional properties of alkanethiol on gold and alkylsilanes on mica, J. Phys. Chem. B **101**, 3800–3805 (1997)
73. H. Schonherr, G.J. Vancso: Tribological properties of self-assembled monolayers of fluorocarbon and hydrocarbon thiols and disulfides on Au(111) studied by scanning force microscopy, Mater. Sci. Eng. C **8-9**, 243–249 (1999)
74. V.V. Tsukruk, V.N. Bliznyuk: Adhesive and friction forces between chemically modified silicon and silicon nitride surfaces, Langmuir **14**, 446–455 (1998)
75. V.V. Tsukruk, T. Nguyen, M. Lemieux, J. Hazel, W.H. Weber, V.V. Shevchenko, N. Klimenko, E. Sheludko: Tribological properties of modified MEMS surfaces. In: *Tribology Issues and Opportunities in MEMS*, ed. by B. Bhushan (Kluwer, Dordrecht 1998) pp. 607–614
76. M. Fujihira, Y. Tani, M. Furugori, U. Akiba, Y. Okabe: Chemical force microscopy of self-assembled monolayers on sputtered gold films patterned by phase separation, Ultramicroscopy **86**, 63–73 (2001)
77. R.J. Good, C.J.V. Oss: *Modern Approaches to Wettability – Theory and Applications* (Plenum, New York 1992)
78. M.H.V.C. Adao, B.J.V. Saramago, A.C. Fernandes: Estimation of the surface properties of styrene-acrylonitrile random copolymers from contact angle measurements, J. Colloid Interf. Sci. **217**, 94–106 (1999)
79. N.S. Tambe, B. Bhushan: A new atomic force microscopy based technique for studying nanoscale friction at high sliding velocities, J. Phys. D **38**, 764–773 (2005)
80. Y.F. Miura, M. Takenga, T. Koini, M. Graupe, N. Garg, R.L. Graham, T.R. Lee: Wettability of self-assembled monolayers generated from CF_3-terminated alkanethiols on gold, Langmuir **14**, 5821–5825 (1998)
81. M. Ratajczak-Sitarz, A. Katrusiak, Z. Kaluski, J. Garbarczyk: 4,4′-biphenyldithiol, Acta Crystallogr. **C.43**, 2389–2391 (1987)
82. B. Bhushan, J. Ruan: Tribological performance of thin film amorphous carbon overcoats for magnetic recording disks in various environments, Surf. Coat. Technol. **68/69**, 644–650 (1994)
83. B. Bhushan, L. Yang, C. Gao, S. Suri, R.A. Miller, B. Marchon: Friction and wear studies of magnetic thin film rigid disks with glass-ceramic, glass and aluminum-magnesium substrates, Wear **190**, 44–59 (1995)
84. J.N. Israelachvili: *Intermolecular and Surface Forces*, 2nd edn. (Academic, London 1992)
85. J.I. Siepman, I.R. McDonald: Monte Carlo simulation of the mechanical relaxation of a self-assembled monolayer, Phys. Rev. Lett. **70**, 453–456 (1993)
86. M. Garcia-Parajo, C. Longo, J. Servat, P. Gorostiza, F. Sanz: Nanotribological properties of octadecyltrichlorosilane self-assembled ultrathin films studied by atomic force microscopy: contact and tapping modes, Langmuir **13**, 2333–2339 (1997)
87. D. DeVecchio, B. Bhushan: Localized surface elasticity measurements using an atomic force microscope, Rev. Sci. Instrum. **68**, 4498–4505 (1997)
88. E. Barrena, S. Kopta, D.F. Ogletree, D.H. Charych, M. Salmeron: Relationship between friction and molecular structure: alkysilane lubricant films under pressure, Phys. Rev. Lett. **82**, 2880–2883 (1999)
89. N. Eustathopoulos, M. Nicholas, B. Drevet: *Wettability at High Temperature* (Pergamon, Amsterdam 1999)

90. S. Ren, S. Yang, Y. Zhao, T. Yu, X. Xiao: Preparation and characterization of ultrahydrophobic surface based on a stearic acid self-assembled monolayer over polyethyleneimine thin films, Surf. Sci. **546**, 64–74 (2003)
91. E.S. Clark: The molecular conformations of polytetrafluoroethylene: forms II and IV, Polymer **40**, 4659–4665 (1999)
92. D.R. Lide: *CRC Handbook of Chemistry and Physics*, 85th edn. (CRC, Boca Raton 2004)
93. W.D. Callister: *Mater. Sci. Eng.*, 4th edn. (Wiley, New York 1997) p.4
94. N.S. Tambe, B. Bhushan: Friction model for velocity dependence of nanoscale friction, Nanotechnology **16**, 2309–2324 (2005)
95. S.C. Clear, P.F. Nealey: The effect of chain density on the frictional behavior of surfaces modified with alkylsilanes and immersed in n-Alcohols, J. Chem. Phys. **114**, 2802–2811 (2001)
96. N.S. Tambe, B. Bhushan: Durability studies of micro/nanoelectromechanical systems materials, coatings, and lubricants at high sliding velocities (up to 10 mm/s) using a modified atomic force microscope, J. Vac. Sci. Technol. A **23**, 830–835 (2005)
97. INSPEC: *Properties of Silicon*, EMIS Data Reviews Series No. 4 (INSPEC, Institution of Electrical Engineers, London 1988)
98. T. Hoshino: Adsorption of atomic and molecular oxygen and desorption of silicon monoxide on Si(111) surfaces, Phys. Rev. B **59**, 2332–2340 (1999)
99. T.L. Cottrell: *The Strength of Chemical Bonds*, 2nd edn. (Butterworths, London 1958)
100. F. Tian, X. Xiao, M.M.T. Loy, C. Wang, C. Bai: Humidity and temperature effect on frictional properties of mica and alkylsilane monolayer self-assembled on mica, Langmuir **15**, 244–249 (1999)
101. R.D. Chambers: *Fluorine in Organic Chemistry* (Wiley, New York 1973)
102. B.A. Sexton, A.E. Hughes: A comparison of weak molecular adsorption of organic-molecules on clean copper and platinum surfaces, Surf. Sci. **140**, 227–248 (1984)
103. L.H. Dubois, B.R. Zegarski, R.G. Nuzzo: Fundamental studies of microscopic wetting on organics surfaces 2. Interaction of secondary adsorbates with chemically textured organic monolayers, J. Am. Chem. Soc. **112**, 570–579 (1990)
104. M.C. McMaster, S.L.M. Schroeder, R.J. Madix: Molecular propane adsorption dynamics on Pt(110)-(1×2), Surf. Sci. **297**, 253–271 (1993)
105. G.J. Kluth, M. Sander, M.M. Sung, R. Maboudian: Study of the desorption mechanism of alkylsiloxane self-assembled monolayers through isotopic labeling and high resolution electron energy-loss spectroscopy experiments, J. Vac. Sci. Technol. A **16**, 932–936 (1998)
106. H.K. Kim, J.P. Lee, C.R. Park, H.T. Kwak, M.M. Sung: Thermal decomposition of alkylsiloxane self-assembled monolayers in air, J. Phys. Chem. B **107**, 4348–4351 (2003)

18

Nanoscale Boundary Lubrication Studies

Bharat Bhushan and Huiwen Liu

Summary. Boundary films are formed by physisorption, chemisorption, and chemical reaction. With physisorption, no exchange of electrons takes place between the molecules of the adsorbate and those of the adsorbant. The physisorption process typically involves van der Waals forces, which are relatively weak. In chemisorption, there is an actual sharing of electrons or electron interchange between the chemisorbed species and the solid surface. The solid surfaces bond very strongly to the adsorption species through covalent bonds. Chemically reacted films are formed by the chemical reaction of a solid surface with the environment. The physisorbed film can be either monomolecularly or polymolecularly thick. The chemisorbed films are monomolecular, but stoichiometric films formed by chemical reaction can have a large film thickness. In general, the stability and durability of surface films decrease in the following order: chemically reacted films, chemisorbed films, and physisorbed films. A good boundary lubricant should have a high degree of interaction between its molecules and the sliding surface. As a general rule, liquids are good lubricants when they are polar and, thus, able to grip solid surfaces (or be adsorbed). In this chapter, we focus on perfluoropolyethers (PFPEs). We first introduce details of the commonly used PFPE lubricants; then present a summary of nanodeformation, molecular conformation, and lubricant spreading studies; followed by an overview of nanotribological properties of polar and nonpolar PFPEs studied by atomic force microscopy (AFM) and some concluding remarks.

18.1 Introduction

Boundary films are formed by physisorption, chemisorption, and chemical reaction. With physisorption, no exchange of electrons takes place between the molecules of the adsorbate and those of the adsorbant. The physisorption process typically involves van der Waals forces, which are relatively weak. In chemisorption, there is an actual sharing of electrons or electron interchange between the chemisorbed species and the solid surface. The solid surfaces bond very strongly to the adsorption species through covalent bonds. Chemically reacted films are formed by the chemical reaction of a solid surface with the environment. The physisorbed film can be either monomolecularly or polymolecularly thick. The chemisorbed films are monomolecular, but stoichiometric films formed by chemical reaction can have a large film thickness. In general, the stability and durability of surface films decrease in the following order: chemically reacted films, chemisorbed films, and physisorbed films. A good boundary lubricant should have a high degree of interaction between its molecules and the sliding surface. As a general rule, liquids are good lubricants when they are polar and, thus, able to grip solid surfaces (or be adsorbed). Polar lubricants contain reactive functional groups with low ionization potential, or groups having high polarizability [1–3]. Boundary lubrication properties of lubricants are also dependent upon the molecular conformation and lubricant spreading [4–7].

Self-assembled monolayers (SAMs), Langmuir–Blodgett (LB) films, and perfluoropolyether (PFPE) films can be used as boundary lubricants [2, 3, 8–10]. PFPE films are commonly used for lubrication of magnetic rigid disks and metal evaporated magnetic tapes to reduce friction and wear of a head–medium interface [10]. PFPEs are well suited for this application because of the following properties: low surface tension and low contact angle, which allow easy spreading on surfaces and provide a hydrophobic property; chemical and thermal stability, which minimizes degradation under use; low vapor pressure, which provides low out-gassing; high adhesion to substrate via organofunctional bonds; and good lubricity, which reduces the interfacial friction and wear [10–12]. While the structure of the lubricants employed at the head–medium interface has not changed substantially over the past decade, the thickness of the PFPE film used to lubricate the disk has steadily decreased from multilayer thicknesses to the sub-monolayer thickness regime [11, 13]. Molecularly thick PFPE films are also being considered for lubrication purposes of the evolving microelectromechanical systems (MEMS) industry [14]. It is well known that the properties of molecularly thick liquid films confined to solid surfaces can be dramatically different from those of the corresponding bulk liquid. In order to efficiently develop lubrication systems that meet the requirements of the advanced rigid disk drive and MEMS industries, the impact of thinning the PFPE lubricants on the resulting nanotribology should be fully understood [15, 16]. It is also important to understand lubricant–substrate interfacial interactions and the influence of the operating environment on the nanotribological performance of molecularly thick PFPEs.

An overview of nanotribological properties of SAMs and LB films can be found in many references, such as [17]. In this chapter, we focus on PFPEs. We first introduce details of the commonly used PFPE lubricants; then present a summary of nanodeformation, molecular conformation, and lubricant spreading studies; followed by an overview of nanotribological properties of polar and nonpolar PFPEs studied by atomic force microscopy (AFM) and some concluding remarks.

18.2 Lubricants Details

Properties of two commonly used PFPE lubricants (Z-15 and Z-DOL) are reviewed here. Their molecular structures are shown schematically in Fig. 18.1. Z-15 has nonpolar $-CF_3$ end groups, whereas Z-DOL is a polar lubricant with hydroxyl (−OH)

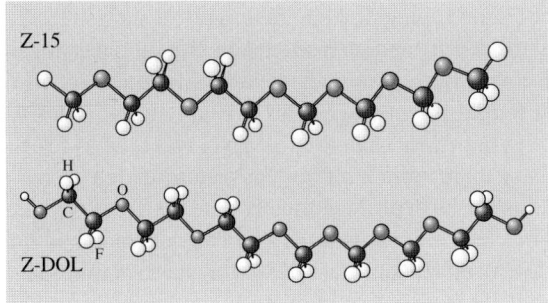

Fig. 18.1. Schematics of the molecular structures of Z-15 and Z-DOL. In this figure the m/n value, shown in Table 18.1, equals 2/3

Table 18.1. Typical properties of Z-15 and Z-DOL (data from Montefluous S.P.A., Milan, Italy)

	Z-15	Z-DOL (2000)
Formula	$CF_3-O-(CF_2-CF_2-O)_m-(CF_2-O)_n-CF_3{}^*$	$HO-CH_2-CF_2-O-(CF_2-CF_2-O)_m-(CF_2-O)_n-CF_2-CH_2-OH^*$
Molecular weight (Daltons)	9100	2000
Density *(ASTM D891)* 20 °C (g/cm^3)	1.84	1.81
Kinematic viscosity *(ASTM D445)* (cSt) 20 °C 38 °C 99 °C	148 90 25	85 34 –
Viscosity index *(ASTM D2270)*	320	–
Surface tension *(ASTM D1331)* (dyn/cm) 20 °C	24	24
Vapor pressure (torr) 20 °C 100 °C	1.6×10^{-6} 1.7×10^{-5}	2×10^{-5} 6×10^{-4}
Pour point *(ASTM D972)* °C	−80	–
Evaporation weight loss *(ASTM D972)* 149 °C, 22 h (%)	0.7	–
Oxidative stability (°C)	–	320
Specific heat (cal/g °C) 38 °C	0.21	–

* $m/n \sim 2/3$

end groups. Their typical properties are summarized in Table 18.1; it shows that Z-15 and Z-DOL have almost the same density and surface tension. But Z-15 has larger molecular weight and higher viscosity. Both of them have low surface tension, low vapor pressure, low evaporation weight loss, and good oxidative stability [10, 12]. Generally, a single-crystal Si(100) wafer with a native oxide layer was used as a substrate for deposition of molecularly thick lubricant films for nanotribological characterization. Z-15 and Z-DOL films can be deposited directly onto the Si(100) wafer by the dip-coating technique. The clean silicon wafer is vertically submerged into a dilute solution of lubricant in hydrocarbon solvent (HT-70) for a certain time. The silicon wafers are pulled up vertically from the solution with a motorized stage at a constant speed for deposition of the desired thicknesses of Z-15 and Z-DOL lubricants. The lubricant film thickness obtained in dip coating is a function of the concentration and pull-up speed, among other factors. The Z-DOL film is bonded to the silicon substrate by heating the as-deposited Z-DOL samples in an oven at 150 °C for about 30 minutes. The native oxide layer of Si(100) wafer reacts with the −OH groups of the lubricants during thermal treatment [18–21]. Subsequently, fluorocarbon solvent (FC-72) washing of the thermally treated specimen removes loosely absorbed species, leaving the chemically bonded phase on the substrate. The chemical bonding between Z-DOL molecules and silicon substrate is illustrated in Fig. 18.2. The bonded and washed Z-DOL film is referred to as Z-DOL(BW) in this chapter. The as-deposited Z-15 and Z-DOL films are mobile-phase lubricants (i.e., liquid-like lubricants), whereas the Z-DOL(BW) films are fully bonded soft solid phase (i.e., solid-like) lubricants. This will be further discussed in the next section.

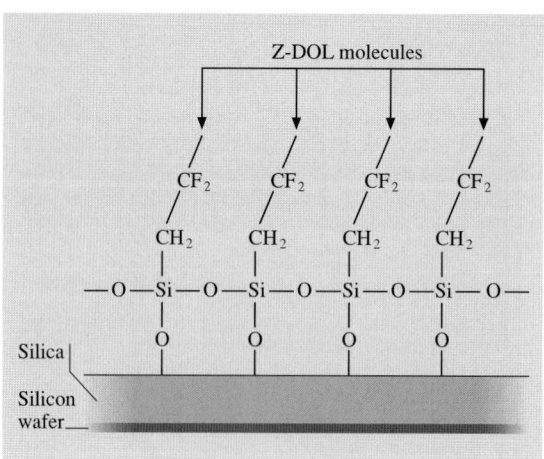

Fig. 18.2. Schematic of Z-DOL molecules that are chemically bonded on Si(100) substrate surface (which has native oxide) after thermal treatment at 150 °C for 30 min

18.3 Nanodeformation, Molecular Conformation, and Lubricant Spreading

Nanodeformation behavior of Z-DOL lubricants was studied using an AFM by *Blackman* et al. [22, 23]. Before bringing a tungsten tip into contact with a molecular overlayer, it was brought into contact with a bare clean-silicon surface, Fig. 18.3. As the sample approaches the tip, the force initially is zero, but at point A the force suddenly becomes attractive (top curve), which increases until point B, where the sample and tip come into intimate contact and the force becomes repulsive. As the sample is retracted, a pull-off force of 5×10^{-8} N (point D) is required to overcome adhesion between the tungsten tip and the silicon surface. When an AFM tip is brought into contact with an unbonded Z-DOL film, a sudden jump into adhesive contact is also observed. A much larger pull-off force is required to overcome the adhesion. The adhesion is initiated by the formation of a lubricant meniscus surrounding the tip. This suggests that the unbonded Z-DOL lubricant shows liquid-like behavior. However, when the tip was brought into contact with a lubricant film, which was firmly bonded to the surface, the liquid-like behavior disappears. The initial attractive force (point A) is no longer sudden, as with the liquid film, but, rather, gradually increases as the tip penetrates the film.

According to *Blackman* et al. [22,23], if the substrate and tip were infinitely hard with no compliance and/or deformation in the tip and sample supports, the line for B to C would be vertical with an infinite slope. The tangent to the force–distance curve at a given point is referred to as the stiffness at that point and was determined

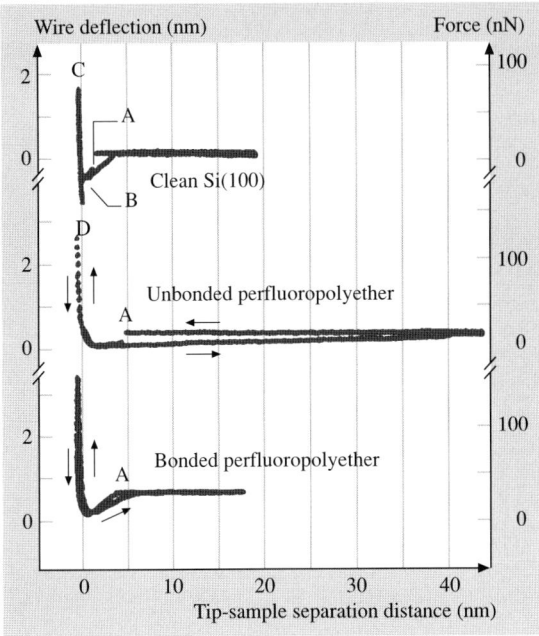

Fig. 18.3. Deflection (normal load) as a function of tip–sample separation-distance curves comparing the behavior of clean Si(100) surface to a surface lubricated with free and unbonded PFPE lubricant, and a surface where the PFPE lubricant film was thermally bonded to the surface [22]

by fitting a least-squares line through the nearby data points. For silicon, the deformation is reversible (elastic), since the retracting (outgoing) portion of the curve (C to D) follows the extending (ingoing) portion (B to C). For bonded lubricant film, at the point where the slope of the force changes gradually from attractive to repulsive, the stiffness changes gradually, indicating compression of the molecular film. As the load is increased, the slope of the repulsive force eventually approaches that of the bare surface. The bonded film was found to respond elastically up to the highest loads of 5 µN that could be applied. Thus, bonded lubricant behaves as a soft polymer solid.

Figure 18.4 illustrates two extremes for the conformation on a surface of a linear liquid polymer without any reactive end groups and at submonolayer coverages [4, 6]. At one extreme, the molecules lie flat on the surface, reaching no more than their chain diameter δ above the surface. This would be the case if a strong attractive interaction exists between the molecules and the solid. On the other extreme, when a weak attraction exists between polymer segments and the solid, the molecules adopt a conformation close to that of the molecules in the bulk, with the microscopic thickness equal to about the radius of gyration R_g. Mate and Novotny [6] used AFM to study conformation of 0.5–1.3 nm-thick Z-15 molecules

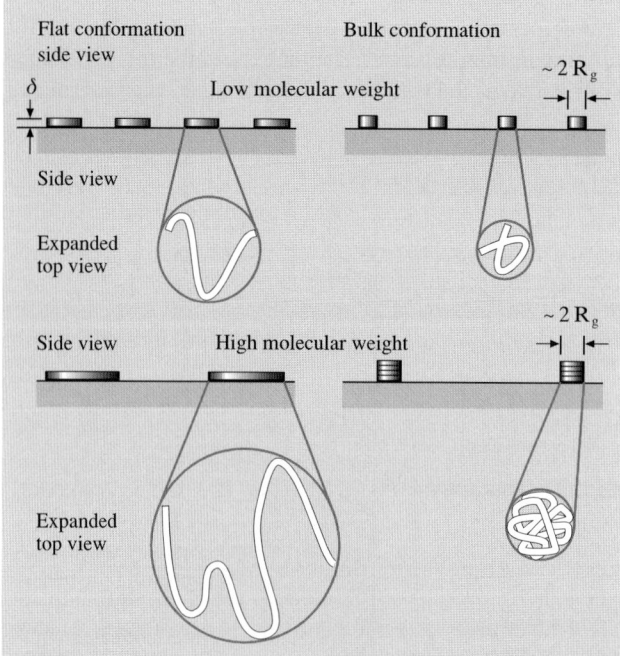

Fig. 18.4. Schematic representation of two extreme liquid conformations at the surface of the solid for low and high molecular weights at low surface coverage. δ is the cross-sectional diameter of the liquid chain, and R_g is the radius of gyration of the molecules in the bulk [6]

on clean Si(100) surfaces. They found that the thickness measured by AFM is thicker than that measured by ellipsometry, with the offset ranging from 3–5 nm. They found that the offset was the same for very thin submonolayer coverages. If the coverage is submonolayer and inadequate to make a liquid film, the relevant thickness is then the height (h_e) of the molecules extended above the solid surface. The offset should be equal to $2h_e$, assuming that the molecules extend the same height above both the tip and silicon surfaces. They therefore concluded that the molecules do not extend more than 1.5–2.5 nm above a solid or liquid surface, much smaller than the radius of gyration of the lubricants, which ranges between 3.2 and 7.3 nm, and to the approximate cross-sectional diameter of 0.6–0.7 nm for the linear polymer chain. Consequently, the height that the molecules extend above the surface is considerably less than the diameter of gyration of the molecules and only a few molecular diameters in height, implying that the physisorbed molecules on a solid surface have an extended, flat conformation. They also determined the disjoining pressure of these liquid films from AFM measurements of the distance needed to break the liquid meniscus that forms between the solid surface and the AFM tip. (Also see [7].) For a monolayer thickness of about 0.7 nm, the disjoining pressure is about 5 MPa, indicating strong attractive interaction between the liquid molecules and the solid surface. The disjoining pressure decreases with increasing film thickness in a manner consistent with a strong attractive van der Waals interaction between the liquid molecules and the solid surface.

Rheological characterization shows that the flow activation energy of PFPE lubricants is weakly dependent on chain length and is strongly dependent on the functional end groups [25]. PFPE lubricant films that contain polar end groups have lower mobility than those with nonpolar end groups of similar chain length [26]. The mobility of PFPE also depends on the surface chemical properties of the substrate. The spreading of Z-DOL on amorphous carbon surface has been studied as a function of hydrogen or nitrogen content in the carbon film, using scanning microellipsometry [24]. The diffusion coefficient data presented in Fig. 18.5 is thickness-dependent. It shows that the surface mobility of Z-DOL increased as the hydrogen content increased, but decreased as nitrogen content increased. The enhancement of Z-DOL surface mobility by hydrogenation may be understood from the fact that the interactions between Z-DOL molecules and the carbon surface can be significantly weakened, due to a reduction of the number of high-energy binding sites on the carbon surface. The stronger interactions between the Z-DOL molecules and carbon surface, as the nitrogen content in the carbon coating increases, leads to the lowering of Z-DOL surface mobility.

Molecularly thick films may be sheared at very high shear rates, on the order of 10^8-10^9 s^{-1} during sliding, such as during magnetic disk drive operation. During such shear, lubricant stability is critical to the protection of the interface. For proper lubricant selection, viscosity at high shear rates and associated shear thinning need to be understood. Viscosity measurements of eight different types of PFPE films show that all eight lubricants display Newtonian behavior and their viscosity remains constant at shear rates up to 10^7 s^{-1} [27, 28].

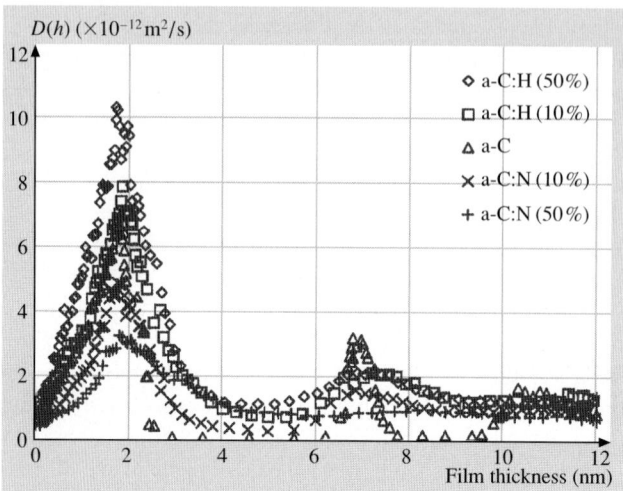

Fig. 18.5. Diffusion coefficient $D(h)$ as a function of lubricant film thickness for Z-DOL on different carbon films [24]

18.4 Boundary Lubrication Studies

With the development of AFM techniques, studies have been carried out to investigate the nanotribological performance of PFPEs. *Mate* [29, 30], *O'Shea* et al. [31, 32], *Bhushan* et al. [15, 33], *Koinkar* and *Bhushan* [20, 34], *Bhushan* and *Sundararajan* [35], *Bhushan* and *Dandavate* [36], and *Liu* and *Bhushan* [21] used an AFM to provide insight into how PFPE lubricants function at the molecular level. *Mate* [29, 30] conducted friction experiments on bonded and unbonded Z-DOL and found that the coefficient of friction of the unbonded Z-DOL is about two times larger than the bonded Z-DOL (also see [31, 32]). *Koinkar* and *Bhushan* [20, 34] and *Liu* and *Bhushan* [21] studied the friction and wear performance of a Si(100) sample lubricated with Z-15, Z-DOL, and Z-DOL(BW) lubricants. They found that using Z-DOL(BW) could significantly improve the adhesion, friction, and wear performance of Si(100). They also discussed the lubrication mechanisms on the molecular level. *Bhushan* and *Sundararajan* [35] and *Bhushan* and *Dandavate* [36] studied the effect of tip radius and relative humidity on the adhesion and friction properties of Si(100) coated with Z-DOL(BW).

In this section, we review, in some detail, the adhesion, friction, and wear properties of Z-15 and Z-DOL at various operating conditions (rest time, velocity, relative humidity, temperature, and tip radius). The experiments were carried out using a commercial AFM system with pyramidal Si_3N_4 and diamond tips. An environmentally controlled chamber and a thermal stage were used to perform relative humidity and temperature effect studies.

18.4.1 Friction and Adhesion

To investigate the friction properties of Z-15 and Z-DOL(BW) films on Si(100), the curves of friction force versus normal load were measured by making friction measurements at increasing normal loads [21]. The representative results of Si(100), Z-15, and Z-DOL(BW) are shown in Fig. 18.6. An approximately linear response of all three samples is observed in the load range of 5–130 nN. The friction force of solid-like Z-DOL(BW) is consistently smaller than that for Si(100), but the friction force of liquid-like Z-15 lubricant is higher than that of Si(100). *Sundararajan* and *Bhushan* [37] have studied the static friction force of silicon micromotors lubricated with Z-DOL by AFM. They also found that liquid-like lubricants of Z-DOL significantly increase the static friction force, whereas solid-like Z-DOL(BW) coatings can dramatically reduce the static friction force. This is in good agreement with the results of *Liu* and *Bhushan* [21]. In Fig. 18.6, the nonzero value of the friction force signal at zero external load is due to the adhesive forces. It is well known that the following relationship exists between the friction force F and external normal load W [2,3]

$$F = \mu(W + W_a), \qquad (18.1)$$

where μ is the coefficient of friction and W_a is the adhesive force. Based on this equation and the data in Fig. 18.6, we can calculate the μ and W_a values. The coefficients of friction of Si(100), Z-15, and Z-DOL are 0.07, 0.09, and 0.04, respectively. Based on (18.1), the adhesive force values are obtained from the horizontal intercepts of the curves of friction force versus normal load at a zero value of friction force. Adhesive force values of Si(100), Z-15, and Z-DOL are 52 nN, 91 nN, and 34 nN, respectively.

The adhesive forces of these samples were also measured using a force calibration plot (FCP) technique. In this technique, the tip is brought into contact with the

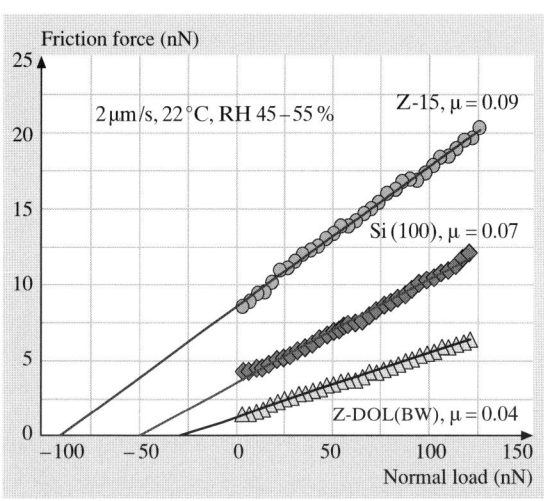

Fig. 18.6. Curves of friction force versus normal load for Si(100), 2.8-nm-thick Z-15 film, and 2.3-nm-thick Z-DOL(BW) film at 2 μm/s, and in ambient air sliding against a Si_3N_4 tip. Based on these curves, the coefficient of friction μ and adhesion force of W_a can be calculated [21]

sample and the maximum force needed to pull the tip and sample apart is measured as the adhesive force. Figure 18.7 shows the typical FCP curves of Si(100), Z-15, and Z-DOL(BW) [21]. As the tip approaches the sample within a few nanometers (point A), an attractive force exists between the tip and the sample surfaces. The tip is pulled toward the sample, and contact occurs at point B on the graph. The adsorption of water molecules and/or presence of liquid lubricant molecules on the sample surface can also accelerate this so-called snap-in, due to the formation of meniscus of the water and/or liquid lubricant around the tip. From this point on, the tip is in contact with surface, and as the piezo extends further, the cantilever is further deflected. This is represented by the slope portion of the curve. As the piezo retracts, at point C the tip goes beyond the zero deflection (flat) line, because of the attractive forces, into the adhesive force regime. At point D, the tip snaps free of the

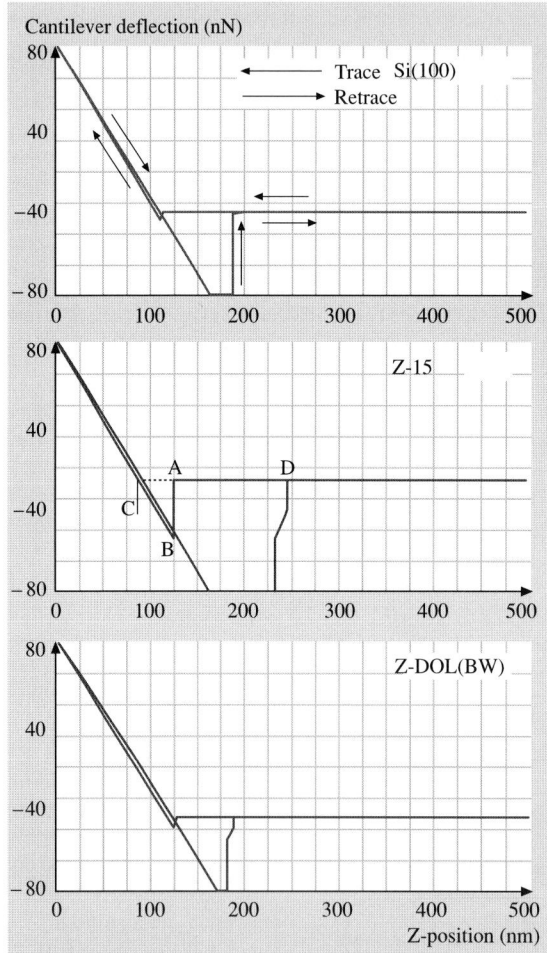

Fig. 18.7. Typical force calibration plots of Si(100), 2.8-nm-thick Z-15 film, and 2.3-nm-thick Z-DOL(BW) film in ambient air. The adhesive forces can be calculated from the horizontal distance between points C and D, and the cantilever spring constant of 0.58 N/m [21]

adhesive forces and is again in free air. The adhesive force (pull-off force) is determined by multiplying the cantilever spring constant (0.58 N/m) by the horizontal distance between points C and D, which corresponds to the maximum cantilever deflection toward the samples before the tip is disengaged. Incidentally, the horizontal shift between the loading and unloading curves results from the hysteresis of the PZT tube.

The adhesive forces of Si(100), Z-15, and Z-DOL(BW) measured by FCP and plots of friction force versus normal load are summarized in Fig. 18.8 [21]. The results measured by these two methods are in good agreement. Figure 18.8 shows that the presence of mobile Z-15 lubricant film increases the adhesive force compared to that of Si(100). In contrast, the presence of solid-phase Z-DOL(BW) film reduces the adhesive force compared to that of Si(100). This result is in good agreement with the results of *Blackman* et al. [22] and *Bhushan* and *Ruan* [38]. Sources of adhesive forces between the tip and the sample surfaces are van der Waals attraction and long-range meniscus forces [2,3,16]. The relative magnitudes of the forces from the two sources are dependent on various factors, including the distance between the tip and the sample surface, their surface roughness, their hydrophobicity, and relative humidity [39]. For most surfaces with some roughness, meniscus contribution dominates at moderate to high humidities.

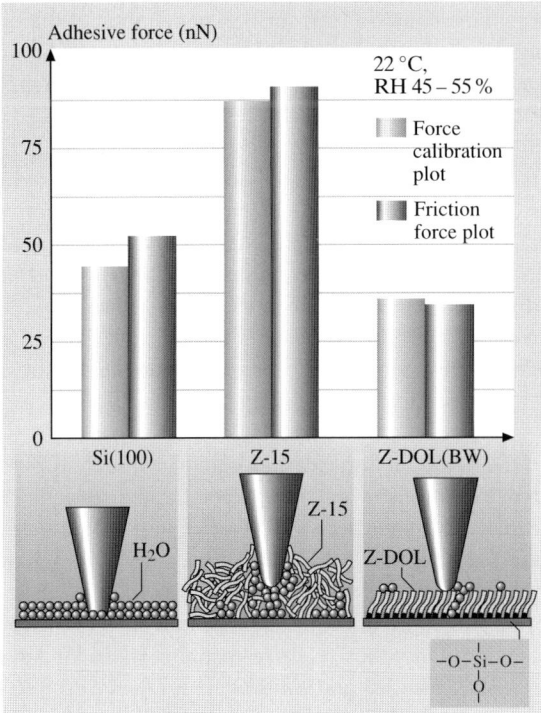

Fig. 18.8. Summary of the adhesive forces of Si(100), 2.8-nm-thick Z-15 film, and 2.3-nm-thick Z-DOL(BW) film measured by force calibration plots and friction force versus normal load in ambient air. The schematic (*bottom*) shows the effect of meniscus formed between the AFM tip and the sample surface on the adhesive and friction forces [21]

The schematic (bottom) in Fig. 18.8 shows relative size and sources of meniscus. The native oxide layer (SiO_2) on the top of Si(100) wafer exhibits hydrophilic properties, and some water molecules can be adsorbed on this surface. The condensed water will form a meniscus as the tip approaches the sample surface. In the case of a sphere (such as a single-asperity AFM tip) in contact with a flat surface, the attractive Laplace force (F_L) caused by capillary is:

$$F_L = 2\pi R \gamma_{la}(\cos\theta_1 + \cos\theta_2), \qquad (18.2)$$

where R is the radius of the sphere, γ_{la} is the surface tension of the liquid against air, θ_1 and θ_2 are the contact angles between liquid and flat and spherical surfaces, respectively [2, 3, 40]. As the surface tension value of Z-15 (24 dyn/cm) is smaller than that of water (72 dyn/cm), the larger adhesive force in Z-15 cannot only be caused by the Z-15 meniscus. The nonpolarized Z-15 liquid does not have complete coverage and strong bonding with Si(100). In the ambient environment, the condensed water molecules will permeate through the liquid Z-15 lubricant film and compete with the lubricant molecules present on the substrate. The interaction of the liquid lubricant with the substrate is weakened, and a boundary layer of the liquid lubricant forms puddles [20, 34]. This dewetting allows water molecules to be adsorbed on the Si(100) surface as aggregates along with Z-15 molecules. And both of them can form meniscus while the tip approaches the surface. In addition, as the Z-15 film is pretty soft compared to the solid Si(100) surface, penetration of the tip in the film occurs while pushing the tip down. This leads to a large area of the tip involved to form the meniscus at the tip–liquid (water aggregates along with Z-15) interface. These two factors of the liquid-like Z-15 film result in higher adhesive force. It should also be noted that Z-15 has a higher viscosity compared to that of water. Therefore, Z-15 film provides higher resistance to sliding motion and results in a larger coefficient of friction. In the case of Z-DOL(BW) film, both of the active groups of Z-DOL molecules are strongly bonded on Si(100) substrate through the thermal and washing treatment. Thus, the Z-DOL(BW) film has relatively low free surface energy and cannot be displaced readily by water molecules or readily adsorb water molecules. Thus, the use of Z-DOL(BW) can reduce the adhesive force. We further believe that the bonded Z-DOL molecules can be orientated under stress (behaving as a soft polymer solid), which facilitates sliding and reduces coefficient of friction.

These studies suggest that, if the lubricant films exist as liquid-like, such as Z-15 films, they easily form meniscus (by themselves and the adsorbed water molecules), and thus have higher adhesive force and higher friction force. Whereas, if the lubricant film exists in solid-like phase, such as Z-DOL(BW) films, they are hydrophobic with low adhesion and friction.

In order to study the uniformity of lubricant film and its influence on friction and adhesion, friction force mapping and adhesive force mapping of PFPE have been carried out by *Koinkar* and *Bhushan* [34] and *Bhushan* and *Dandavate* [36], respectively. Figure 18.9 shows gray scale plots of surface topography and friction force images obtained simultaneously for unbonded Demnum-type PFPE lubricant

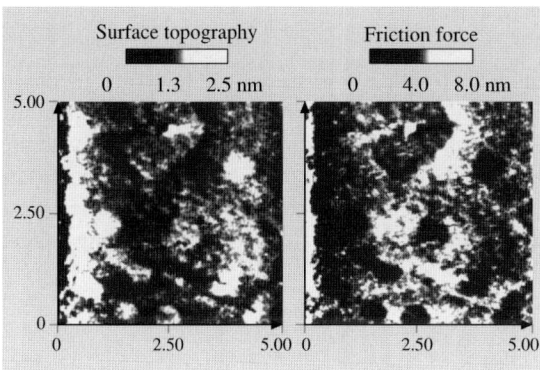

Fig. 18.9. Gray scale plots of the surface topography and friction force obtained simultaneously for unbonded 2.3-nm-thick Demnum-type PFPE lubricant film on silicon [20]

film on silicon [34]. The friction force plot shows well-distinguished low- and high-friction regions corresponding roughly to high- and low-surface-height regions in the topography image (thick- and thin-lubricant regions). A uniformly lubricated sample does not show such a variation in friction. Figure 18.10 shows the gray scale plots of the adhesive force distribution for silicon samples coated uniformly and nonuniformly with Z-DOL lubricant. It can be clearly seen that there exists a region that has an adhesive force distinctly different from the other region for the nonuniformly coated sample. This implies that the liquid film thickness is nonuniform, giving rise to a difference in the meniscus forces.

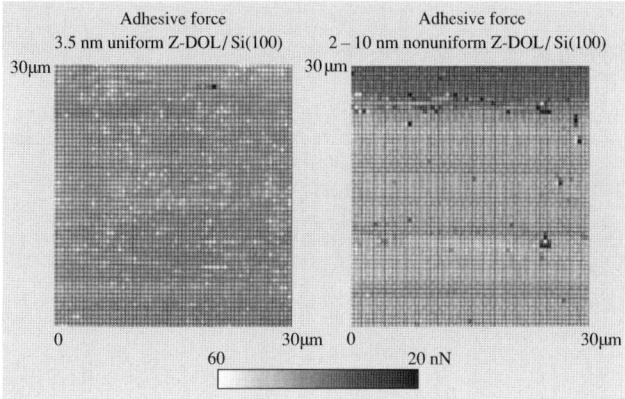

Fig. 18.10. Gray scale plots of the adhesive force distribution of a uniformly coated, 3.5-nm-thick unbonded Z-DOL film on silicon and 3–10 nm-thick unbonded Z-DOL film on silicon that was deliberately coated nonuniformly by vibrating the sample during the coating process [36]

18.4.2 Rest Time Effect

It is well known that, in the computer rigid disk drive, the stiction force increases rapidly with an increase in rest time between the head and magnetic-medium disk [10, 11]. Considering that the stiction and friction are two of the major issues that lead to the failure of computer rigid disk drives and MEMS, it is very important to find out if the rest time effect also exists on the nanoscale. First, the rest time effect on the friction force, adhesive force, and coefficient of Si(100) sliding against

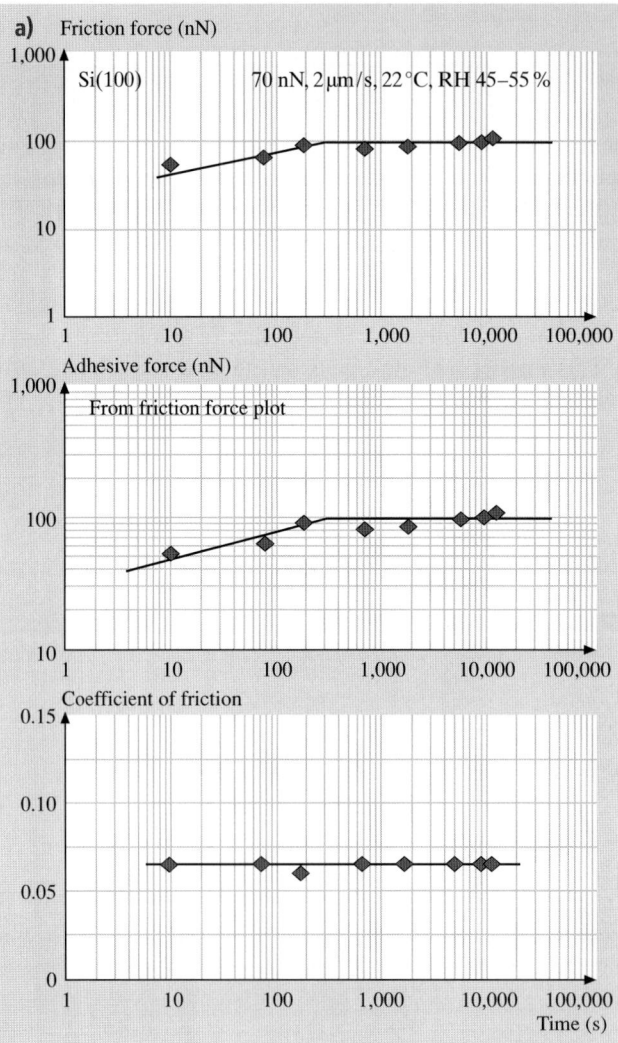

Fig. 18.11. (a) Rest time effect on friction force, adhesive force, and coefficient of friction of Si(100).

Si$_3$N$_4$ tip was studied, Fig. 18.11a [21]. It was found that the friction and adhesive forces logarithmically increase up to a certain equilibrium time after which they remain constant. Figure 18.11a also shows that the rest time does not affect the coefficient of friction. These results suggest that the rest time can result in the growth of the meniscus, which causes a higher adhesive force, and in turn, a higher friction force. But in the whole testing range the friction mechanisms do not change with

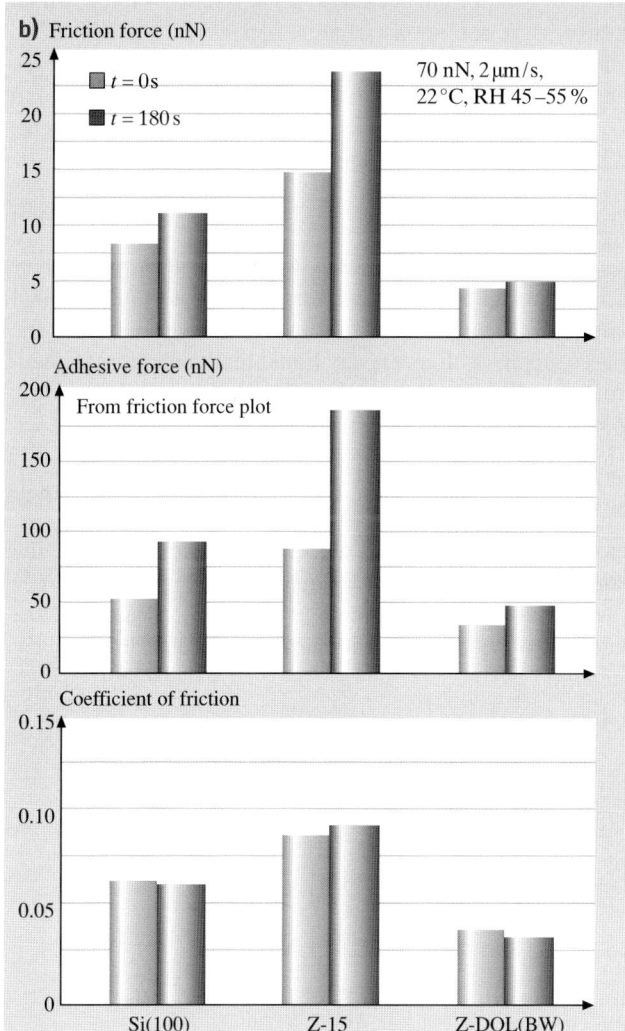

Fig. 18.11. (b) Summary of the rest time effect on friction force, adhesive force, and coefficient of friction of Si(100), 2.8-nm-thick Z-15 film, and 2.3-nm-thick Z-DOL(BW) film. All of the measurements were carried out at 70 nN, 2 µm/s, and in ambient air [21]

the rest time. Similar studies were also performed on Z-15 and Z-DOL(BW) films. The results are summarized in Fig. 18.11b [21]. It is seen that a similar time effect has been observed on Z-15 film, but not on Z-DOL(BW) film.

An AFM tip in contact with a flat sample surface can be treated as a single-asperity contact. Therefore, a Si_3N_4 tip in contact with Si(100) or Z-15/Si(100) can be modeled as a sphere in contact with a flat surface covered by a layer of liquid (adsorbed water and/or liquid lubricant), Fig. 18.12a. Meniscus forms around the contacting asperity and grows with time until equilibrium occurs [41]. The meniscus force, which is the product of meniscus pressure and meniscus area, depends on the flow of liquid phase toward the contact zone. The flow of the liquid toward the contact zone is governed by the capillary pressure P_c, which draws liquid into the meniscus, and the disjoining pressure Π, which tends to draw the liquid away from the meniscus. Based on the Young and Laplace equation, the capillary pressure, P_c, is:

$$P_c = 2K\gamma, \tag{18.3}$$

where $2K$ is the mean meniscus curvature ($= K_1 + K_2$, where K_1 and K_2 are the curvatures of the meniscus in the contact plane and perpendicular to the contact plane) and γ is the surface tension of the liquid. *Mate* and *Novotny* [6] have shown that the disjoining pressure decreases rapidly with increasing liquid film thickness in a manner consistent with a strong van der Waals attraction. The disjoining pressure, Π, for these liquid films can be expressed as:

$$\Pi = \frac{A}{6\pi h^3}, \tag{18.4}$$

where A is the Hamaker constant and h is the liquid film thickness. The driving forces that cause the lubricant flow that results in an increase in the meniscus force are the disjoining pressure gradient, due to a gradient in film thickness, and the capillary pressure gradient, due to the curved liquid–air interface. The driving pressure, P, can then be written as:

$$P = -2K\gamma - \Pi. \tag{18.5}$$

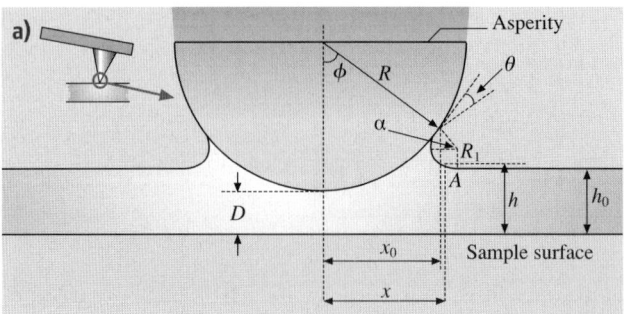

Fig. 18.12. (a) Schematic of a single asperity in contact with a smooth flat surface in the presence of a continuous liquid film when ϕ is large

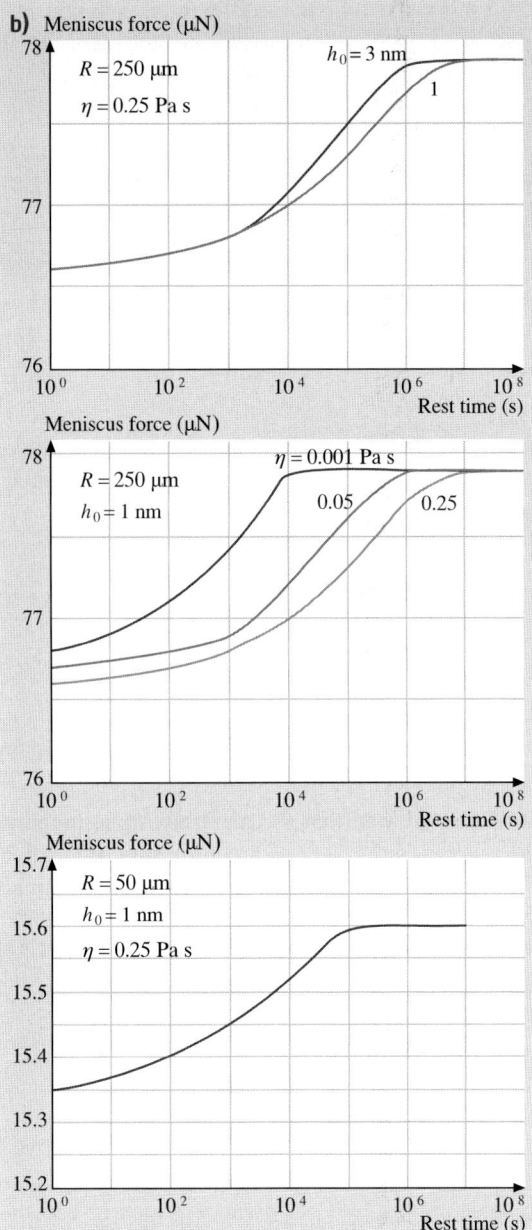

Fig. 18.12. (b) Results of the single-asperity model. Effect of viscosity of the liquid, radius of the asperity, and film thickness is studied with respect to the time-dependent meniscus force [41]

Based on these three basic relationships, the following differential equation has been derived by *Chilamakuri* and *Bhushan* [41], which can describe the meniscus at time t:

$$2\pi x_0 \left(D + \frac{x_0^2}{2R} - h_0\right) \frac{dx_0}{dt}$$
$$= \frac{2\pi h_0^3 \gamma (1 + \cos\theta)}{3\eta} \frac{1}{D + a - h_0} - \frac{A x_0}{3\eta h} \cot\alpha, \quad (18.6)$$

where η is the viscosity of the liquid and a is given as

$$a = R(1 - \cos\phi) \sim \frac{R\phi^2}{2} \sim \frac{x_0^2}{2R}. \quad (18.7)$$

The differential equation (18.4) was solved numerically using Newton's iteration method. The meniscus force at any time t less than the equilibrium time is proportional to the meniscus area and meniscus pressure ($2K\gamma$), and it is given by

$$f_m(t) = 2\pi R\gamma(1 + \cos\theta) \left(\frac{x_0}{(x_0)_{eq}}\right)^2 \left(\frac{K}{K_{eq}}\right), \quad (18.8)$$

where $(x_0)_{eq}$ is the value of x_0 at the equilibrium time

$$[(x_0)_{eq}]^2 = 2R\left[\frac{-6\pi h_0^3 \gamma(1 + \cos\theta)}{A} + (h_0 - D)\right]. \quad (18.9)$$

This modeling work (at the microscale) showed that the meniscus force initially increases logarithmically with the rest time up to a certain equilibrium time, after which it remains constant. Equilibrium time decreases with an increase in liquid film thickness, a decrease in viscosity, and a decrease in the tip radius, Fig. 18.12b. This early numerical modeling work and the data at the nanoscale in Fig. 18.11a are in good agreement.

18.4.3 Velocity Effect

To investigate the velocity effect on friction and adhesion, the relationships between friction force and normal load for Si(100), Z-15, and Z-DOL(BW) at different velocities were measured, Fig. 18.13 [21]. Based on these data, the adhesive force and coefficient of friction values can be calculated by (18.1). The variation of friction force, adhesive force, and coefficient of friction of Si(100), Z-15, and Z-DOL(BW) as a function of velocity are summarized in Fig. 18.14. This indicates that, for silicon wafer, the friction force decreases logarithmically with increasing velocity. For Z-15, the friction force decreases with increasing velocity up to 10 μm/s, after which it remains almost constant. The velocity has a much smaller effect on the friction force of Z-DOL(BW); it reduced slightly only at very high velocity. Figure 18.14

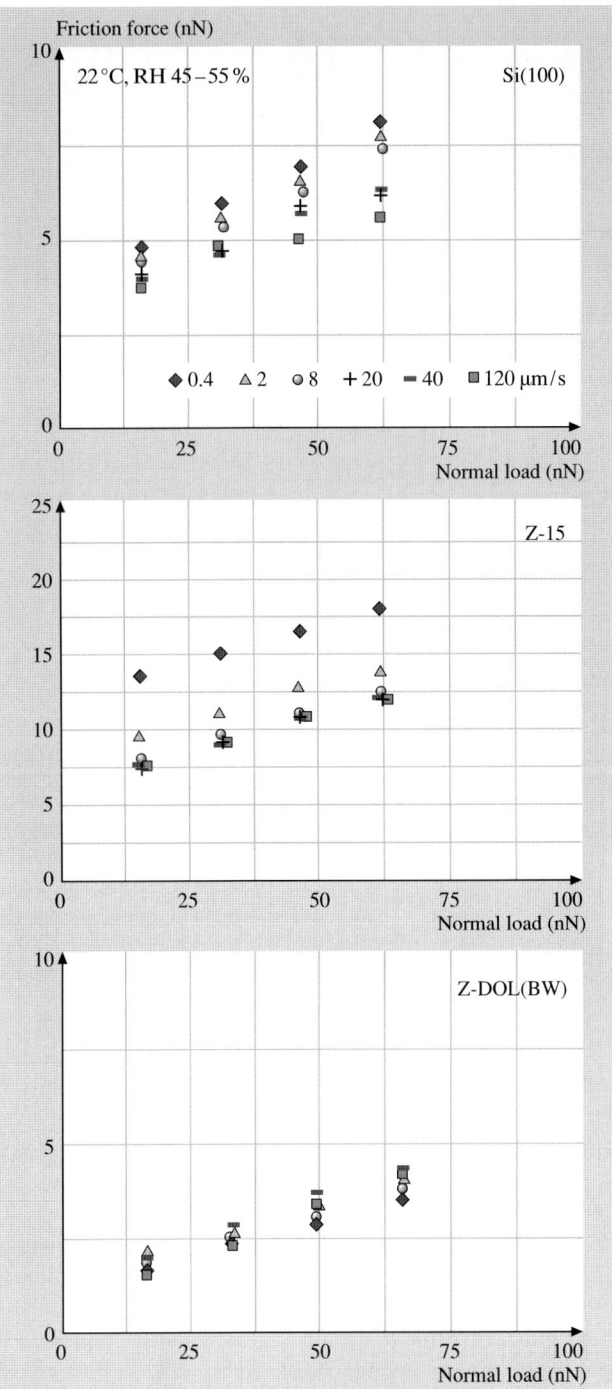

Fig. 18.13. Friction forces versus normal load data of Si(100), 2.8-nm-thick Z-15 film, and 2.3-nm-thick Z-DOL(BW) film at various velocities in ambient air [21]

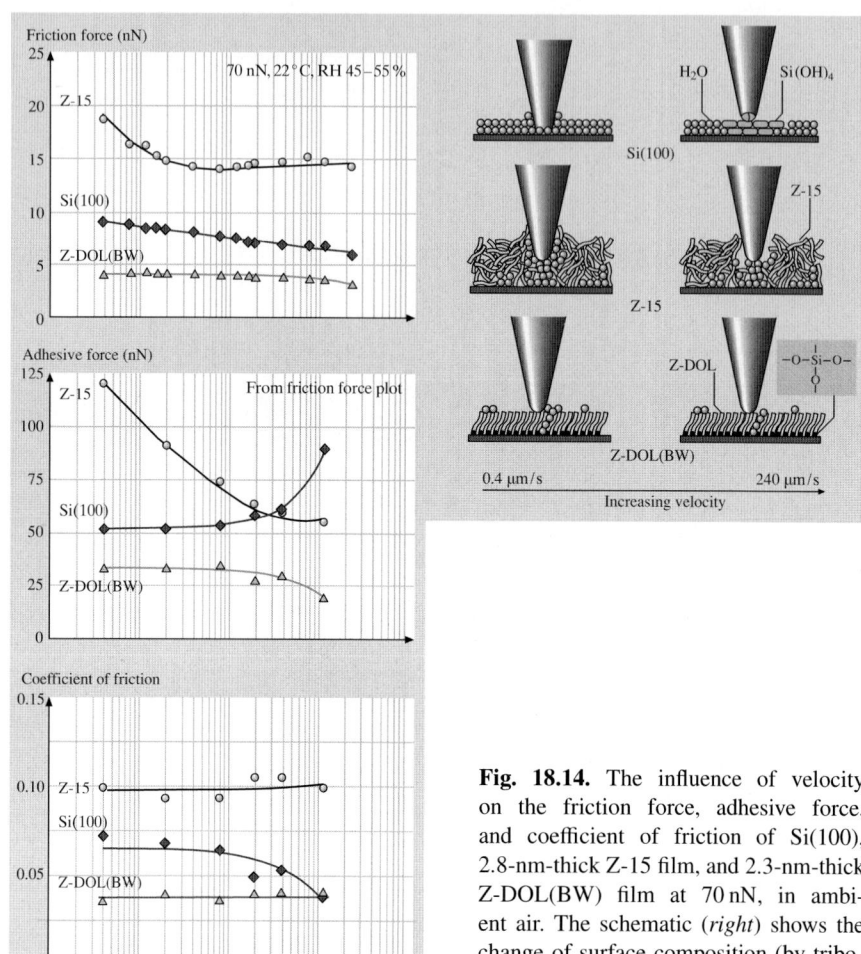

Fig. 18.14. The influence of velocity on the friction force, adhesive force, and coefficient of friction of Si(100), 2.8-nm-thick Z-15 film, and 2.3-nm-thick Z-DOL(BW) film at 70 nN, in ambient air. The schematic (*right*) shows the change of surface composition (by tribochemical reaction) and change of meniscus with increasing velocity [21]

also indicates that the adhesive force of Si(100) is increased when the velocity is higher than 10 μm/s. The adhesive force of Z-15 is reduced dramatically with a velocity increase up to 20 μm/s, after which it is reduced slightly; the adhesive force of Z-DOL(BW) also decreases at high velocity. In the testing range of velocity, only the coefficient of friction of Si(100) decreases with velocity, but the coefficients of friction of Z-15 and Z-DOL(BW) almost remain constant. This implies that the friction mechanisms of Z-15 and Z-DOL(BW) do not change with the variation of velocity.

The mechanisms of the effect of velocity on the adhesion and friction are explained based on the schematics shown in Fig. 18.14 (right). For Si(100), tribochemical reaction plays a major role. Although at high velocity the meniscus is broken and does not have enough time to rebuild, the contact stresses and high velocity lead to tribochemical reactions of the Si(100) wafer and Si_3N_4 tip, which have native oxide (SiO_2) layers with water molecules. The following reactions occur:

$$SiO_2 + 2H_2O \rightarrow Si(OH)_4 \tag{18.10}$$

$$Si_3N_4 + 16H_2O \rightarrow 3Si(OH)_4 + 4NH_4OH \,. \tag{18.11}$$

The $Si(OH)_4$ is removed and continuously replenished during sliding. The $Si(OH)_4$ layer between the tip and Si(100) surface is known to be of low shear strength and causes a decrease in friction force and coefficient of friction in the lateral direction [42–46]. The chemical bonds of Si−OH between the tip and Si(100) surface induce large adhesive force in the normal direction. For Z-15 film, at high velocity the meniscus formed by condensed water and Z-15 molecules is broken and does not have enough time to rebuild. Therefore, the adhesive force and, consequently, friction force is reduced. For Z-DOL(BW) film, the surface can adsorb few water molecules in ambient condition, and at high velocity these molecules are displaced, which is responsible for a slight decrease in friction force and adhesive force. Even in the high-velocity range, the friction mechanisms for Z-15 and Z-DOL(BW) films are still shearing of the viscous liquid and molecular orientation, respectively. Thus the coefficients of friction of Z-15 and Z-DOL(BW) do not change with velocity.

Koinkar and *Bhushan* [20, 34] have suggested that, in the case of samples with mobile films, such as condensed water and Z-15 films, alignment of liquid molecules (shear thinning) is responsible for the drop in friction force with an increase in scanning velocity. This could be another reason for the decrease in friction force with velocity for Si(100) and Z-15 film in this study.

18.4.4 Relative Humidity and Temperature Effect

The influence of relative humidity on friction and adhesion was studied in an environmentally controlled chamber. The friction force was measured by making measurements at increasing relative humidity, the results are presented in Fig. 18.15 [21]. These shows that, for Si(100) and Z-15 film, the friction force increases with a relative humidity increase up to RH 45%, and then it shows a slight decrease with a further increase in relative humidity. Z-DOL(BW) has a smaller friction force than Si(100) and Z-15 in the whole testing range, and its friction force shows a relatively apparent increase when the relative humidity is higher than RH 45%. For Si(100), Z-15, and Z-DOL(BW), adhesive forces increase with relative humidity. And their coefficients of friction increase with relative humidity up to RH 45%, after which they decrease with further increases of the relative humidity. It is also observed that the humidity effect on Si(100) really depends on the history of the Si(100) sample. As the surface of Si(100) wafer readily adsorbs water in air, without any pretreatment the Si(100) used in our study almost reaches its saturated stage of adsorbing water and is responsible for a smaller effect with increasing relative humidity. However, once the Si(100) wafer was thermally treated by baking at 150 °C for 1 h, a bigger effect was observed.

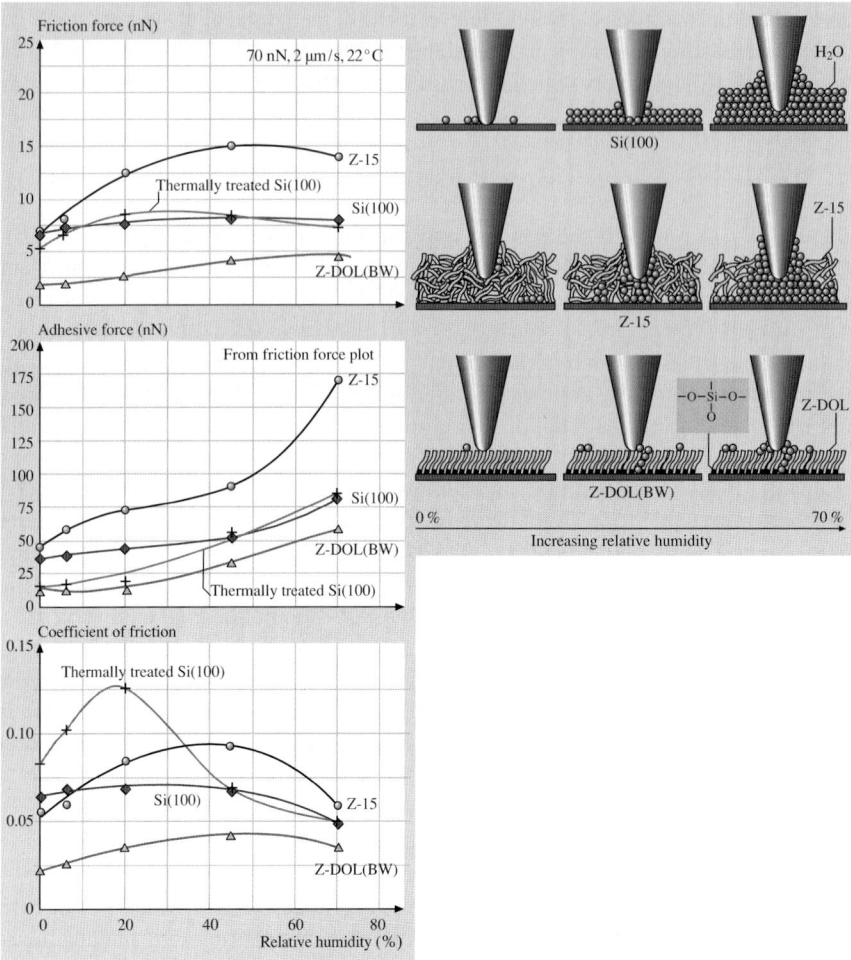

Fig. 18.15. The influence of relative humidity (RH) on the friction force, adhesive force, and coefficient of friction of Si(100), 2.8-nm-thick Z-15 film, and 2.3-nm-thick Z-DOL(BW) film at 70 nN, 2 μm/s, and in 22 °C air. Schematic (*right*) shows the change of meniscus with increasing relative humidity. In this figure, the thermally treated Si(100) represents the Si(100) wafer that was baked at 150 °C for 1 h in an oven (in order to remove the adsorbed water) just before it was placed in the 0% RH chamber [21]

The schematic (right) in Fig. 18.15 shows that, because of its high free surface energy, Si(100) can adsorb more water molecules with increasing relative humidity. As discussed earlier, for Z-15 film in a humid environment, condensed water competes with the lubricant film present on the sample surface. Obviously, more water molecules can also be adsorbed on a Z-15 surface with increasing relative humidity. Thermal adsorbed water molecules in the case of Si(100), along with lubricant molecules in Z-15 film, form a bigger water meniscus, which leads to an increase of the friction force, adhesive force, and coefficient of friction of Si(100) and Z-15 with humidity. But at the very high humidity of RH 70%, large quantities of adsorbed water can form a continuous water layer that separates the tip and sample surface, and acts as a kind of lubricant, which causes a decrease in the friction force and coefficient of friction. For the Z-DOL(BW) film, because of its hydrophobic surface properties, water molecules can only be adsorbed at high humidity (≥ RH 45%), which causes an increase in the adhesive force and friction force.

The effect of temperature on friction and adhesion was studied using a thermal stage attached to the AFM. The friction force was measured at increasing temperature from 22–125 °C. The results are presented in Fig. 18.16 [21]. It shows that the increasing temperature causes a decrease of friction force, adhesive force, and

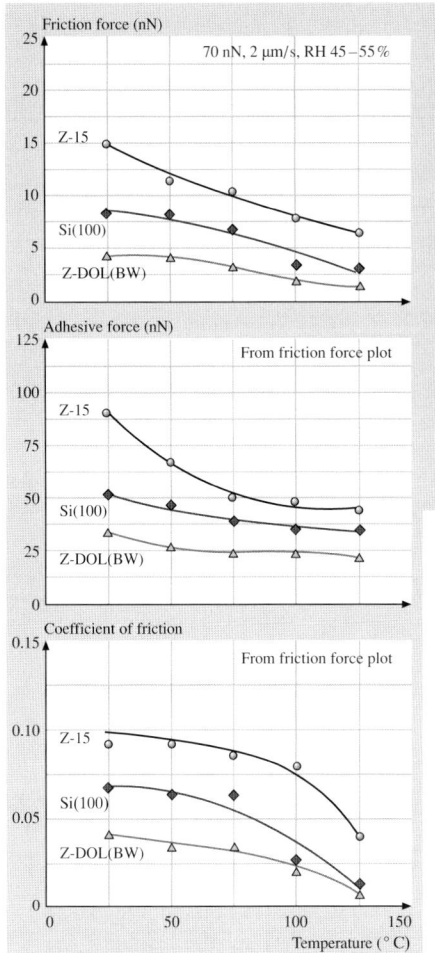

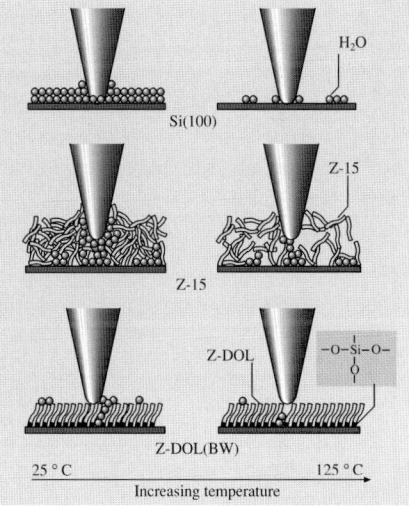

Fig. 18.16. The influence of temperature on the friction force, adhesive force, and coefficient of friction of Si(100), 2.8-nm-thick Z-15 film, and 2.3-nm-thick Z-DOL(BW) film at 70 nN, at 2 μm/s, and in RH 40–50 % air. The schematic (*right*) shows that, at high temperature, desorption of water decreases the adhesive forces. And the reduced viscosity of Z-15 leads to the decrease of coefficient of friction. High temperature facilitates orientation of molecules in Z-DOL(BW) film, which results in a lower coefficient of friction [21]

coefficient of friction of Si(100), Z-15, and Z-DOL(BW). The schematic (right) in Fig. 18.16 indicates that, at high temperature, desorption of water leads to a decrease in the friction force, adhesive force, and coefficient of friction for all of the samples. Besides that, the reduction of surface tension of water also contributes to the decrease of friction and adhesion. For Z-15 film, the reduction of viscosity at high temperature makes an additional contribution to the decrease of friction. In the case of Z-DOL(BW) film, molecules are more easily oriented at high temperature, which may also be responsible for the low friction.

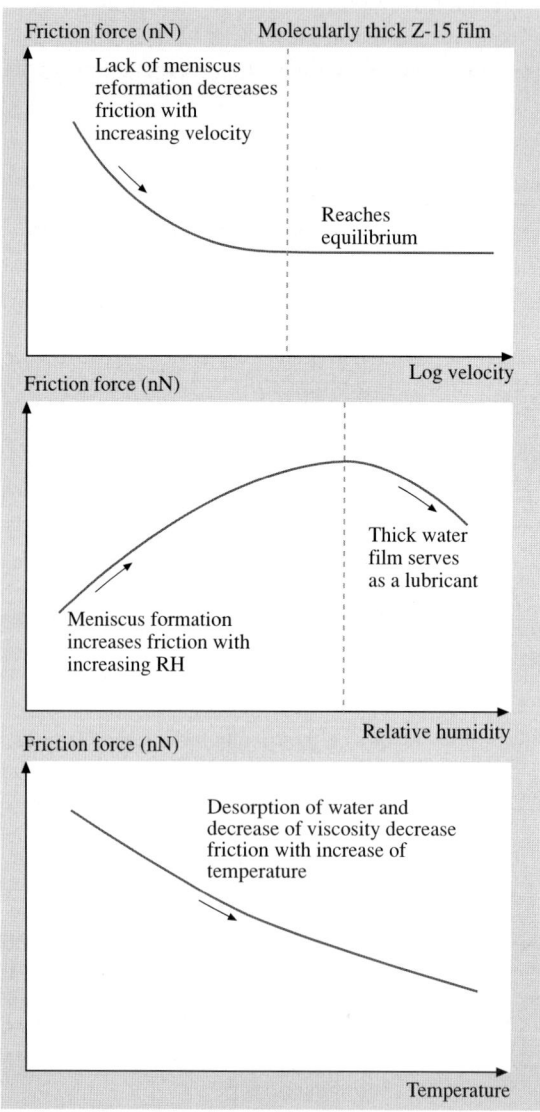

Fig. 18.17. Schematic showing the change of friction force of molecularly thick Z-15 films with log velocity, relative humidity, and temperature [21]

Using a surface force apparatus, *Yoshizawa* and *Israelachvili* [47,48] have shown that a change in the velocity or temperature induces phase transformation (from crystalline solid-like to amorphous, then to liquidlike) in surfactant monolayers, which are responsible for the observed changes in the friction force. Stick–slip is observed in a low-velocity regime of a few μm/s, and adhesion and friction first increase followed by a decrease in the temperature range of 0–50 °C. Stick–slip at low velocity and adhesion and friction curves peaking at some particular temperature (observed in their study), have not been observed in the AFM study. This suggests that the phase transformation may not happen in this study, because PFPEs generally have very good thermal stability [10, 12].

As a brief summary, the influence of velocity, relative humidity, and temperature on the friction force of Z-15 film is presented in Fig. 18.17. The changing trends are also addressed in this figure.

18.4.5 Tip Radius Effect

The tip radius and relative humidity affect adhesion and friction for dry and lubricated surfaces [35, 36]. Figure 18.18a shows the variation of single-point adhesive force measurements as a function of tip radius on a Si(100) sample for several humidities. The adhesive force data are also plotted as a function of relative humidity for various tip radii. Figure 18.18a indicates that the tip radius has little effect on the adhesive forces at low humidities, but the adhesive force increases with tip radius at high humidity. Adhesive force also increases with increasing humidity for all tips. The trend in adhesive forces as a function of tip radii and relative humidity, in Fig. 18.18a, can be explained by the presence of meniscus forces, which arise from capillary condensation of water vapor from the environment. If enough liquid is present to form a meniscus bridge, the meniscus force should increase with an increase in tip radius, based on (18.2). This observation suggests that the thickness of the liquid film at low humidities is insufficient to form continuous meniscus bridges and to affect adhesive forces in the case of all tips.

Figure 18.18a also shows the variation in coefficient of friction as a function of tip radius at a given humidity and as a function of relative humidity for a given tip radius on the Si(100) sample. It can be observed that, for RH 0%, the coefficient of friction is about the same for the tip radii investigated except for the largest tip, which shows a higher value. At all other humidities, the trend consistently shows that the coefficient of friction increases with tip radius. An increase in friction with tip radius at low to moderate humidities arises from increased contact area (i.e., higher van der Waals forces) and the higher values of shear forces required for a larger contact area. At high humidities, similarly to adhesive force data, an increase with tip radius occurs due to both contact area and meniscus effects. It can be seen that, for all tips, the coefficient of friction increases with humidity to about RH 45%, beyond which it starts to decrease. This is attributed to the fact that, at higher humidities, the adsorbed water film on the surface acts as a lubricant between the two surfaces [21]. Thus the interface is changed at higher humidities, resulting in lower shear strength and, hence, a lower friction force and coefficient of friction.

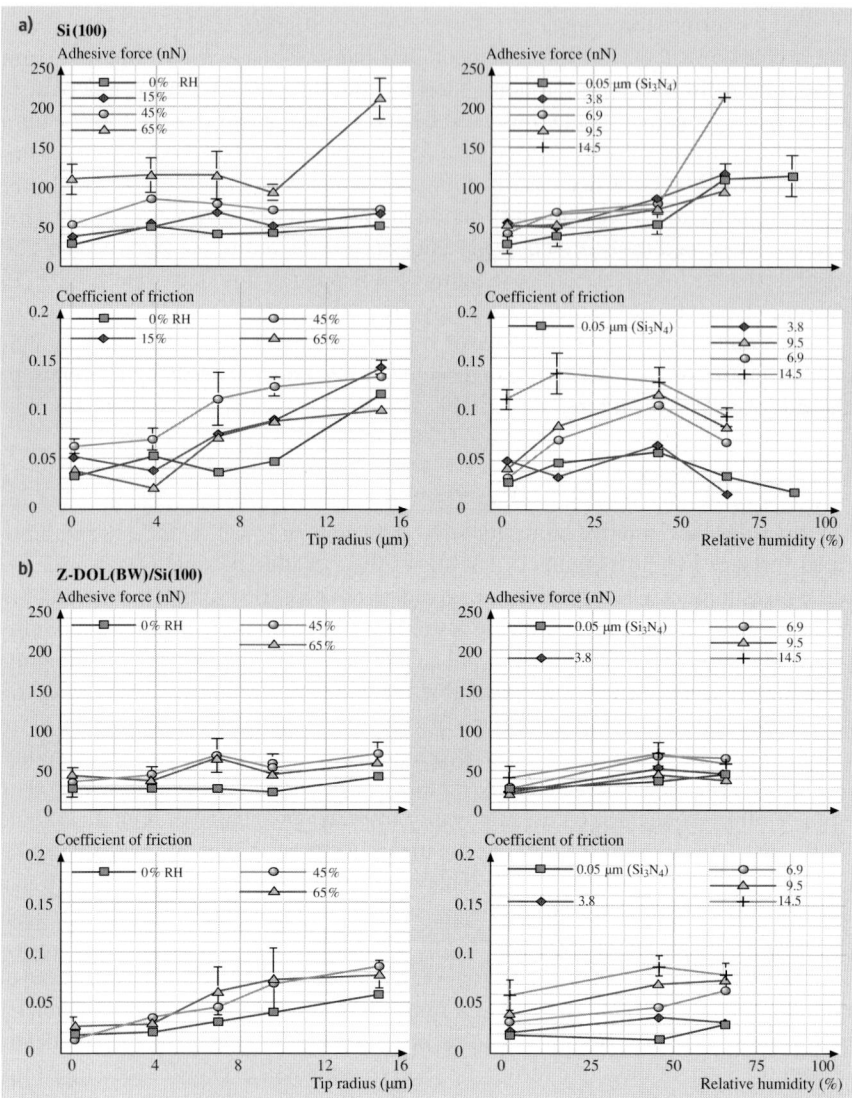

Fig. 18.18. Adhesive force and coefficient of friction as a function of tip radius at several humidities and as a function of relative humidity at several tip radii on (**a**) Si(100) and (**b**) 0.5-nm Z-DOL(BW) films [35]

Figure 18.18b shows the adhesive forces as a function of tip radius and relative humidity on Si(100) coated with 0.5 nm-thick Z-DOL(BW) film. Adhesive forces for all the tips with the Z-DOL(BW)-lubricated sample are much lower than those experienced on unlubricated Si(100), shown in Fig. 18.18a. The data also show that, even with a monolayer thickness of the lubricant, there is very little variation in adhesive forces with tip radius at a given humidity. For a given tip radius, the variation in adhesive forces with relative humidity indicates that these forces slightly increase from RH 0% to RH 45%, but remain more or less the same with further increases in humidity. This is seen even with the largest tip, which indicates that the lubricant is indeed hydrophobic; there is some meniscus formation at humidities higher than RH 0%, but the formation is very minimal and does not increase even up to RH 65%. Figure 18.18b also shows the coefficient of friction for various tips at different humidities for the Z-DOL(BW)-lubricated sample. Again, all the values obtained with the lubricated sample are much lower than the values obtained on unlubricated Si(100), shown in Fig. 18.18a. The coefficient of friction increases with tip radius for all humidities, as was seen on unlubricated Si(100), due to an increase in the contact area. Similarly to the adhesive forces, there is an increase in friction from RH 0% to RH 45%, due to a contribution from an increasing number of menisci bridges. However, thereafter very little additional water film forms, due to the hydrophobicity of the Z-DOL(BW) layer, and, consequentially, the coefficient of friction does not change, even with the largest tip. These findings show that even a monolayer of Z-DOL(BW) offers good hydrophobic performance of the surface.

18.4.6 Wear Study

To study the durability of lubricant films at the nanoscale, the friction of Si(100), Z-15, and Z-DOL(BW) as a function of the number of scanning cycles was measured, Fig. 18.19 [21]. As observed earlier, friction force and coefficient of friction of Z-15 is higher than that of Si(100), and Z-DOL(BW) has the lowest values. During cycling, friction force and coefficient of friction of Si(100) show a slight variation during the initial few cycles then remain constant. This is related to the removal of the top adsorbed layer. In the case of Z-15 film, the friction force and coefficient of friction show an increase during the initial few cycles and then approach higher and stable values. This is believed to be caused by the attachment of the Z-15 molecules to the tip. The molecular interaction between these attached molecules on the tip and molecules on the film surface is responsible for an increase in the friction. However, after several scans, this molecular interaction reaches equilibrium, and after that, the friction force and coefficient of friction remain constant. In the case of Z-DOL(BW) film, the friction force and coefficient of friction start out low and remain low during the entire test for 100 cycles. This suggests that Z-DOL(BW) molecules do not become attached or displaced as readily as Z-15.

Koinkar and *Bhushan* [20, 34] conducted wear studies using a diamond tip at high loads. Figure 18.20 shows the plots of wear depth as a function of normal force, and Fig. 18.21 shows the wear profiles of the worn samples at 40 μN normal load. The 2.3 nm-thick Z-DOL(BW)-lubricated sample exhibits better wear resistance than unlubricated and 2.9 nm-thick Z-15-lubricated silicon samples. Wear resistance of a Z-15-lubricated sample is little better than that of the unlubricated sample. The Z-15-lubricated sample shows debris inside the wear track. Since Z-15 is a liquid lubricant, debris generated is held by the lubricant and they become sticky.

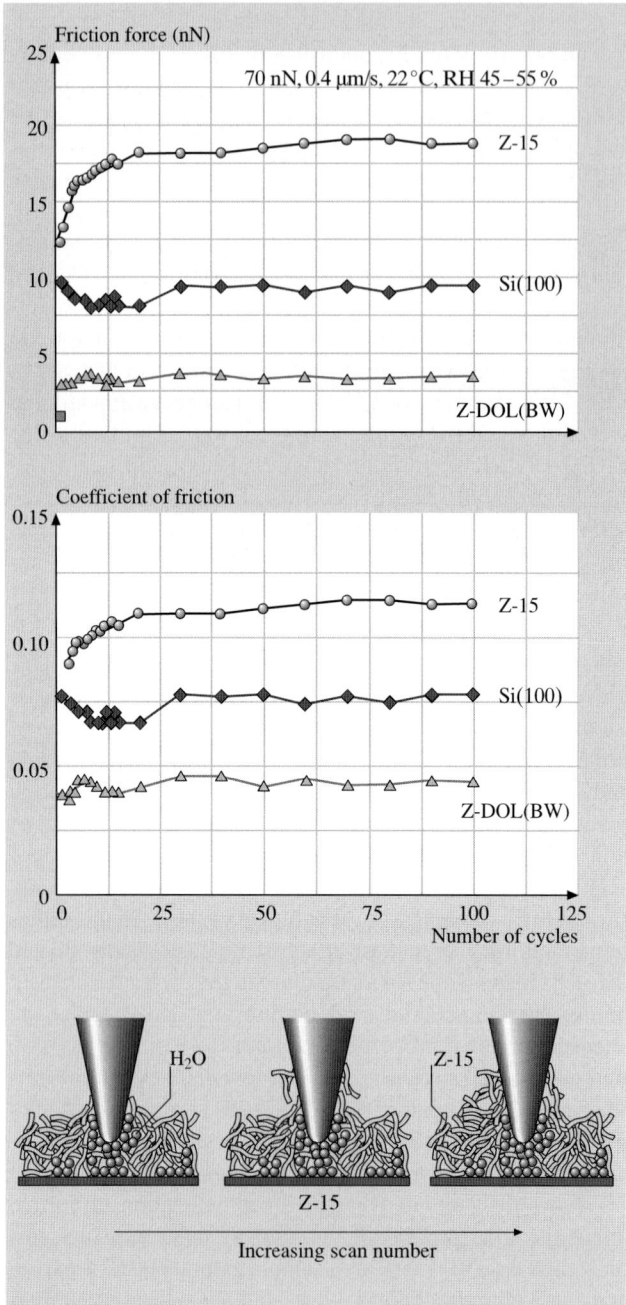

Fig. 18.19. Friction force and coefficient of friction versus number of sliding cycles for Si(100), 2.8-nm-thick Z-15 film, and 2.3-nm-thick Z-DOL(BW) film at 70 nN, 0.8 μm/s, and in ambient air. Schematic (*bottom*) shows that some liquid Z-15 molecules can be attached on the tip. The molecular interaction between the attached molecules on the tip with the Z-15 molecules in the film results in an increase of the friction force with multiple scanning [21]

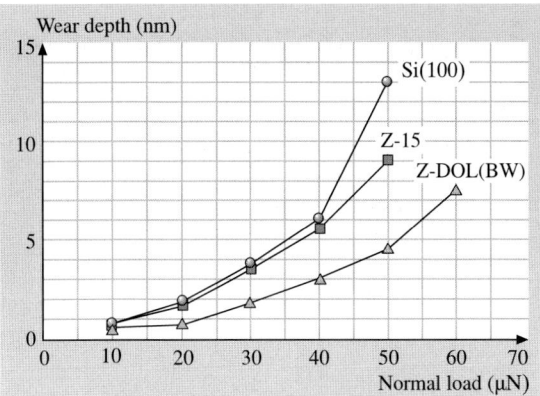

Fig. 18.20. Wear depth as a function of normal load using a diamond tip for Si(100), 2.9-nm-thick Z-15 film, and 2.3-nm-thick Z-DOL(BW) after one cycle [20]

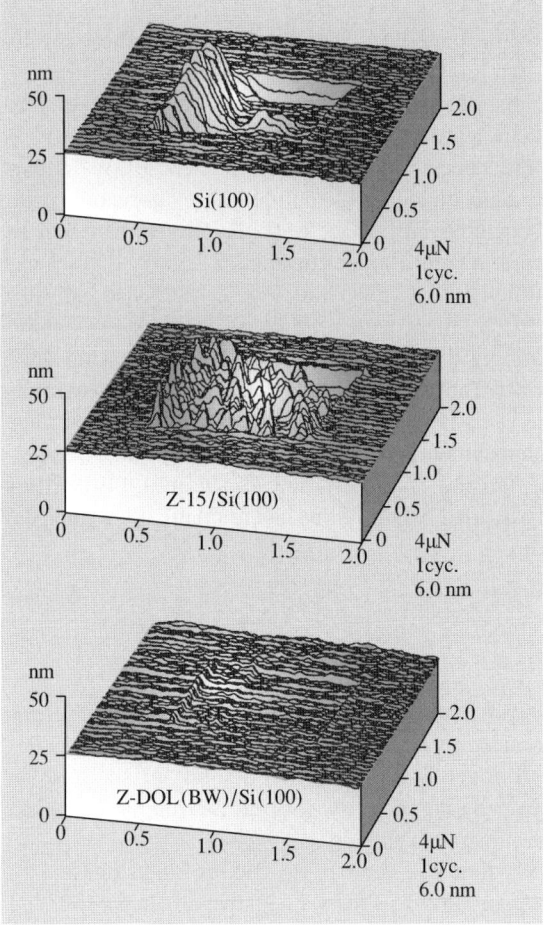

Fig. 18.21. Wear profiles for Si(100), 2.9-nm-thick Z-15 film, and 2.3-nm-thick Z-DOL(BW) film after wear studies using a diamond tip. The normal force used and wear depths are listed in the figure [20]

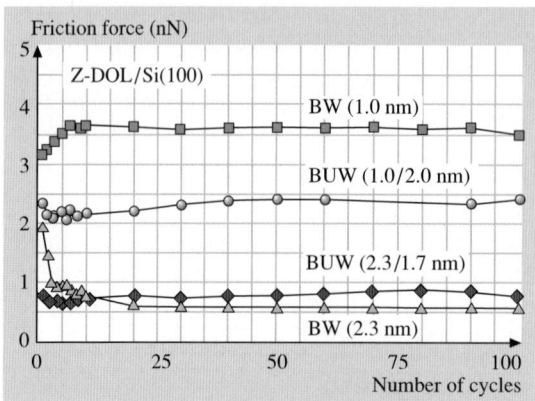

Fig. 18.22. Friction force as a function of number of cycles using a Si_3N_4 tip at a normal load of 300 nN for Z-DOL(BW) and Z-DOL(BUW) films with different film thicknesses [34]

The debris moves inside the wear track and does damage, Fig. 18.20. These results suggest that using Z-DOL(BW) can improve the wear resistance of the substrate.

To study the effect of the degree of chemical bonding, the durability tests were conducted on both fully bonded and partially bonded Z-DOL films. Durability results for Z-DOL(BW) and Z-DOL bonded and unwashed (Z-DOL(BUW), a partially bonded film that contains both bonded and mobile phase lubricants) with different film thicknesses are shown in Fig. 18.22 [34]. Thicker films, such as Z-DOL(BUW), with a thickness of 4.0 nm (bonded/mobile = 2.3 nm/1.7 nm) exhibit behavior similar to 2.3 nm-thick Z-DOL(BW) film. Figure 18.22 also indicates that Z-DOL(BW) and Z-DOL(BUW) films with a thinner film thickness exhibit a higher friction value. Comparing 1.0 nm-thick Z-DOL(BW) with 3.0 nm-thick Z-DOL(BUW) (bonded/mobile = 1.0 nm/2.0 nm), the Z-DOL(BUW) film exhibits a lower and stable friction value. This is because the mobile phase on the surface acts as a source of lubricant replenishment. A similar conclusion has also been reported by *Ruhe* et al. [19], *Bhushan* and *Zhao* [13], and *Eapen* et al. [49]. All of them indicate that using partially bonded Z-DOL films can dramatically reduce the friction and improve the wear life.

18.5 Closure

Nanodeformation studies have shown that fully bonded Z-DOL lubricants behave as polymer coatings, while the unbonded lubricants behave like liquids. AFM studies have shown that the physisorbed nonpolar molecules on a solid surface have an extended, flat conformation. The spreading property of PFPE is strongly dependent on the molecular end groups and substrate chemistry.

Using solid-like Z-DOL(BW) film can reduce the friction and adhesion of Si(100), while using mobile lubricant of Z-15 shows a negative effect. Si(100) and Z-15 film show apparent time effects. The friction and adhesion forces increase as a result of growth of meniscus up to an equilibrium time, after which they remain constant. Using Z-DOL(BW) film can prevent time effects. High velocity leads to the rupture of meniscus and prevents its reformation, which leads to a decrease of friction and adhesive forces of Z-15 and Z-DOL(BW). The influence of relative humidity on the friction and adhesion is dominated by the amount of the adsorbed water molecules. Increasing humidity can either increase friction through increased adhesion by the water meniscus, or reduce friction through an enhanced water-lubricating effect. Increasing temperature leads to desorption of the water layer, decrease of the water surface tension, decrease of viscosity, and easier orientation of the Z-DOL(BW) molecules. These changes cause a decrease of the friction force and adhesion at high temperature. During cycling tests, the molecular interaction between the Z-15 molecules attached to the tip and the Z-15 molecules on the film surface causes the initial rise of friction. Wear tests show that Z-DOL(BW) can improve the wear resistance of silicon. Partially bonded PFPE film appears to be more durable than fully bonded films.

These results suggest that partially/fully bonded films are good lubricants for micro/nanoscale devices operating in different environments and under varying conditions.

References

1. B. Bhushan: Magnetic recording surfaces. In: *Characterization of Tribological Materials*, ed. by W. A. Glaeser (Butterworth–Heinemann, Boston 1993) pp. 116–133
2. B. Bhushan: *Principles and Applications of Tribology* (Wiley, New York 1999)
3. B. Bhushan: *Introduction to Tribology* (Wiley, New York 2002)
4. V. J. Novotny, I. Hussla, J. M. Turlet, M. R. Philpott: Liquid polymer conformation on solid surfaces, J. Chem. Phys. **90**, 5861–5868 (1989)
5. V. J. Novotny: Migration of liquid polymers on solid surfaces, J. Chem. Phys. **92**, 3189–3196 (1990)
6. C. M. Mate, V. J. Novotny: Molecular conformation and disjoining pressures of polymeric liquid films, J. Chem. Phys. **94**, 8420–8427 (1991)
7. C. M. Mate: Application of disjoining and capillary pressure to liquid lubricant films in magnetic recording, J. Appl. Phys. **72**, 3084–3090 (1992)
8. G. G. Roberts: *Langmuir–Blodgett Films* (Plenum, New York 1990)
9. A. Ulman: *An Introduction to Ultrathin Organic Films* (Academic, Boston 1991)
10. B. Bhushan: *Tribology and Mechanics of Magnetic Storage Devices*, 2nd edn. (Springer, New York 1996)
11. B. Bhushan: Macro- and microtribology of magnetic storage devices. In: *Modern Tribology Handbook Vol. 2: Materials, Coatings, and Industrial Applications*, ed. by B. Bhushan (CRC, Boca Raton 2001) pp. 1413–1513
12. Anonymous: Fomblin Z Perfluoropolyethers, Data sheet Montedism Group, Milan (2002)
13. B. Bhushan, Z. Zhao: Macroscale and microscale tribological studies of molecularly thick boundary layers of perfluoropolyether lubricants for magnetic thin-film rigid disks, J. Info. Storage Proc. Syst. **1**, 1–21 (1999)
14. B. Bhushan: *Tribology Issues and Opportunities in MEMS* (Kluwer, Dordrecht 1998)

15. B. Bhushan, J. N. Israelachvili, U. Landman: Nanotribology: Friction, wear and lubrication at the atomic scale, Nature **374**, 607–616 (1995)
16. B. Bhushan: *Handbook of Micro/Nanotribology*, 2nd edn. (CRC, Boca Raton 1999)
17. B. Bhushan: Self-assembled monolayers for controlling hydrophobicity and/or friction and wear. In: *Modern Tribology Handbook Vol. 2: Materials, Coatings, and Industrial Applications*, ed. by B. Bhushan (CRC, Boca Raton 2001) pp. 909–929
18. J. Ruhe, G. Blackman, V. J. Novotny, T. Clarke, G. B. Street, S. Kuan: Thermal attachment of perfluorinated polymers to solid surfaces, J. Appl. Polym. Sci. **53**, 825–836 (1994)
19. J. Ruhe, V. Novotny, T. Clarke, G. B. Street: Ultrathin perfluoropolyether films – influence of anchoring and mobility of polymers on the tribological properties, ASME J. Tribol. **118**, 663–668 (1996)
20. V. N. Koinkar, B. Bhushan: Microtribological studies of unlubricated and lubricated surfaces using atomic force/friction force microscopy, J. Vac. Sci. Technol. A **14**, 2378–2391 (1996)
21. H. Liu, B. Bhushan: Nanotribological characterization of molecularly-thick lubricant films for applications to MEMS/NEMS by AFM, Ultramicroscopy **97**, 321–340 (2003)
22. G. S. Blackman, C. M. Mate, M. R. Philpott: Interaction forces of a sharp tungsten tip with molecular films on silicon surface, Phys. Rev. Lett. **65**, 2270–2273 (1990)
23. G. S. Blackman, C. M. Mate, M. R. Philpott: Atomic force microscope studies of lubricant films on solid surfaces, Vacuum **41**, 1283–1286 (1990)
24. X. Ma, J. Gui, K. J. Grannen, L. A. Smoliar, B. Marchon, M. S. Jhon, C. L. Bauer: Spreading of PFPE lubricants on carbon surfaces: Effect of hydrogen and nitrogen content, Tribol. Lett. **6**, 9–14 (1999)
25. C. A. Kim, H. J. Choi, R. N. Kono, M. S. Jhon: Rheological characterization of perfluoropolyether lubricant, Polym. Prepr. **40**, 647–649 (1999)
26. M. Ruths, S. Granick: Rate-dependent adhesion between opposed perfluoropoly(alkylether) layers: Dependence on chain-end functionality and chain length, J. Phys. Chem. B **102**, 6056–6063 (1998)
27. U. Jonsson, B. Bhushan: Measurement of rheological properties of ultrathin lubricant films at very high shear rates and near-ambient pressure, J. Appl. Phys. **78**, 3107–3109 (1995)
28. C. Hahm, B. Bhushan: High shear rate viscosity measurement of perfluoropolyether lubricants for magnetic thin-film rigid disks, J. Appl. Phys. **81**, 5384–5386 (1997)
29. C. M. Mate: Atomic-force-microscope study of polymer lubricants on silicon surface, Phys. Rev. Lett. **68**, 3323–3326 (1992)
30. C. M. Mate: Nanotribology of lubricated and unlubricated carbon overcoats on magnetic disks studied by friction force microscopy, Surf. Coat. Technol. **62**, 373–379 (1993)
31. S. J. O'Shea, M. E. Welland, T. Rayment: Atomic force microscope study of boundary layer lubrication, Appl. Phys. Lett. **61**, 2240–2242 (1992)
32. S. J. O'Shea, M. E. Welland, J. B. Pethica: Atomic force microscopy of local compliance at solid–liquid interface, Chem. Phys. Lett. **223**, 336–340 (1994)
33. B. Bhushan, T. Miyamoto, V. N. Koinkar: Microscopic friction between a sharp diamond tip and thin-film magnetic rigid disks by friction force microscopy, Adv. Info. Storage Syst. **6**, 151–161 (1995)
34. V. N. Koinkar, B. Bhushan: Micro/nanoscale studies of boundary layers of liquid lubricants for magnetic disks, J. Appl. Phys. **79**, 8071–8075 (1996)
35. B. Bhushan, S. Sundararajan: Micro/nanoscale friction and wear mechanisms of thin films using atomic force and friction force microscopy, Acta Mater. **46**, 3793–3804 (1998)

36. B. Bhushan, C. Dandavate: Thin-film friction and adhesion studies using atomic force microscopy, J. Appl. Phys. **87**, 1201–1210 (2000)
37. S. Sundararajan, B. Bhushan: Static friction and surface roughness studies of surface micromachined electrostatic micromotors using an atomic force/friction force microscope, J. Vac. Sci. Technol. A **19**, 1777–1785 (2001)
38. B. Bhushan, J. Ruan: Atomic-scale friction measurements using friction force microscopy: Part II – application to magnetic media, ASME J. Tribol. **116**, 389–396 (1994)
39. T. Stifter, O. Marti, B. Bhushan: Theoretical investigation of the distance dependence of capillary and van der Waals forces in scanning probe microscopy, Phys. Rev. B **62**, 13667–13673 (2000)
40. J. N. Israelachvili: *Intermolecular and Surface Forces*, 2nd edn. (Academic, London 1992)
41. S. K. Chilamakuri, B. Bhushan: A comprehensive kinetic meniscus model for prediction of long-term static friction, J. Appl. Phys. **15**, 4649–4656 (1999)
42. H. Ishigaki, I. Kawaguchi, M. Iwasa, Y. Toibana: Friction and wear of hot pressed silicon nitride and other ceramics, ASME J. Tribol. **108**, 514–521 (1986)
43. T. E. Fischer: Tribochemistry, Annu. Rev. Mater. Sci. **18**, 303–323 (1988)
44. K. Mizuhara, S. M. Hsu: Tribochemical reaction of oxygen and water on silicon surfaces. In: *Wear Particles*, ed. by D. Dowson (Elsevier, New York 1992) pp. 323–328
45. S. Danyluk, M. McNallan, D. S. Park: Friction and wear of silicon nitride exposed to moisture at high temperatures. In: *Friction and Wear of Ceramics*, ed. by S. Jahanmir (Dekker, New York 1994) pp. 61–79
46. V. A. Muratov, T. E. Fischer: Tribochemical polishing, Annu. Rev. Mater. Sci. **30**, 27–51 (2000)
47. H. Yoshizawa, Y. L. Chen, J. N. Israelachvili: Fundamental mechanisms of interfacial friction I: Relationship between adhesion and friction, J. Phys. Chem. **97**, 4128–4140 (1993)
48. H. Yoshizawa, J. N. Israelachvili: Fundamental mechanisms of interfacial friction II: Stick slip friction of spherical and chain molecules, J. Phys. Chem. **97**, 11300–11313 (1993)
49. K. C. Eapen, S. T. Patton, J. S. Zabinski: Lubrication of microelectromechanical systems (MEMS) using bound and mobile phase of Fomblin Z-DOL, Tibol. Lett. **12**, 35–41 (2002)

Part IV

Biomimetics

19

Lotus Effect: Roughness-Induced Superhydrophobic Surfaces

Bharat Bhushan, Michael Nosonovsky, and Yong Chae Jung

Superhydrophobic surfaces have considerable technological potential for various applications due to their extreme water-repellent properties. These surfaces with high contact angle and low contact angle hysteresis also exhibit a self-cleaning effect and low drag for fluid flow. These surfaces are of interest in various applications, including self-cleaning windows, exterior paints for buildings, navigation ships, textiles, and applications requiring a reduction in fluid flow, e.g., in micro/nanochannels. Superhydrophobic surfaces prevent formation of menisci at a contacting interfaces and can be used to minimize high adhesion and stiction. Certain plant leaves, notably lotus leaves, are known to be superhydrophobic due to their roughness and the presence of a thin wax film on the leaf surface, and the phenomenon is known as the "Lotus effect." Extremely water-repellent superhydrophobic surfaces can be produced by using roughness combined with hydrophobic coatings. In this chapter, the theory of roughness-induced superhydrophobicity is presented followed by the characterization data of natural leaf surfaces and artificial superhydrophobic surfaces. Wetting is studied as a multiscale process involving the macroscale (water droplet size), microscale (surface texture size), and nanoscale (molecular size). This includes fundamental physical mechanisms of wetting, responsible for the transition between various wetting regimes, contact angle and contact angle hysteresis. Practical aspects of design of superhydrophobic surfaces are also discussed.

19.1 Introduction

The primary parameter that characterizes wetting is the static contact angle, which is defined as the measurable angle that a liquid makes with a solid. The contact angle depends on several factors, such as roughness and the manner of surface preparation, and its cleanliness [2,61]. If the liquid wets the surface (referred to as wetting liquid or hydrophilic surface), the value of the static contact angle is $0 \leq \theta \leq 90°$, whereas if the liquid does not wet the surface (referred to as non-wetting liquid or hydrophobic surface), the value of the contact angle is $90° < \theta \leq 180°$. The term hydrophobic/philic, which was originally applied only to water ("hydro-" means "water" in Greek), is often used to describe the contact of a solid surface with any liquid. The term "oleophobic/philic" is used sometimes with regard to the wetting by oil. Surfaces with high energy, formed by polar molecules, tend to be hydrophilic, whereas those with low energy and built of non-polar molecules tend to be hydrophobic.

Surfaces with the contact angle between 150° and 180° are called superhydrophobic. For fluid flow applications, in addition to the high contact angle, superhydrophobic surfaces should also have very low water contact angle hysteresis. In the case of these surfaces, water droplets roll off (with some slip) the surface and take contaminant with them providing self-cleaning ability, known as the Lotus effect. The contact angle hysteresis is the difference between the advancing and receding contact angles, which are two stable values. It occurs due to roughness and surface heterogeneity. Contact angle hysteresis reflects a fundamental asymmetry of wetting and dewetting and the irreversibility of the wetting/dewetting cycle. It is a measure of energy dissipation during the flow of a droplet along a solid surface. For a droplet moving along the solid surface (for example, if the surface is tilted), the contact angle at the front of the droplet (advancing contact angle) is greater than that at the back of the droplet (receding contact angle), due to roughness, resulting in the contact angle hysteresis (Fig. 19.1a). Surfaces with low contact angle hysteresis have a very low water roll-off angle, which denotes the angle to which a surface must be tilted for roll off of water drops (i.e., very low water contact angle hysteresis) [18, 19, 46, 69, 73]. It is understood that during roll-off of water droplets, some slip is associated. Superhydrophobic surfaces have low drag for fluid flow and low tilt angle. The self-cleaning surfaces are of interest in various applications, including self-cleaning windows, windshields, exterior paints for buildings and navigation ships, utensils, roof tiles, textiles, and applications requiring a reduction of drag in fluid flow, e.g., in micro/nanochannels.

Wetting may lead to formation of menisci at the interface between solid bodies during sliding contact, which increases adhesion and friction. As a result of this, the wet friction force is greater than the dry friction force, which is usually undesirable [10, 13–15]. On the other hand, high adhesion is desirable in some applications, such as adhesive tapes and adhesion of cells to biomaterial surfaces; therefore, enhanced wetting roughness would be desirable in these applications. Numerous applications, such as magnetic storage devices and micro/nanoelectromechanical systems (MEMS/NEMS), require surfaces with low adhesion and stiction [10–12, 16, 20]. As the size of these devices decreases, the surface forces tend to dominate over the volume forces, and adhesion and stiction constitute a challenging problem for proper operation of these devices. This makes the development of non-adhesive surfaces crucial for many of these emerging applications.

In the last decade, materials scientists paid attention to natural surfaces, which are extremely hydrophobic. Among them are leaves of water-repellent plants such as Nelumbo nucifera (lotus) and Colocasia esculenta, which have high contact angles with water [8, 17, 32, 92, 131]. First, the surface of the leaves is usually covered with a range of different waxes made from a mixture of hydrocarbon compounds

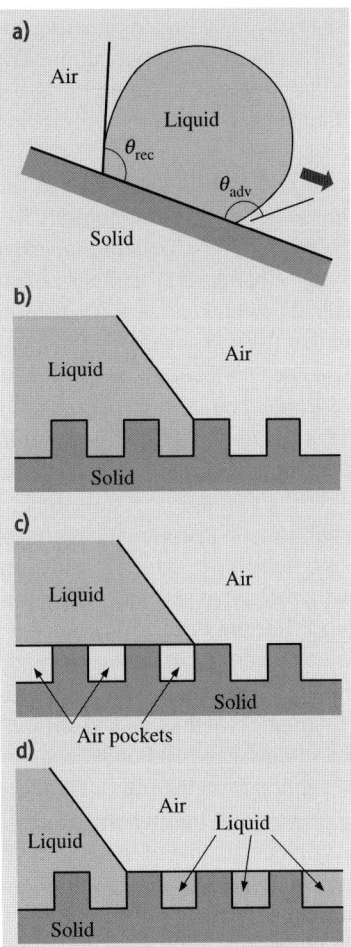

Fig. 19.1. (a) Schematics of a droplet on a tilted substrate showing advancing (θ_{adv}) and receding (θ_{rec}) contact angles. The difference between these angles constitutes the contact angle hysteresis. Configurations described by the (**b**) Wenzel equation for the homogeneous interface (19.6), (**c**) Cassie–Baxter equation for the composite interface with air pockets (19.9), and (**d**) the Cassie equation for the homogeneous interface (19.10)

that have a strong phobia of being wet. Second, the surface is very rough due to so-called papillose epidermal cells, which form asperities or papillae. In addition to the microscale roughness of the leaf due to the papillae, the surface of the papillae is also rough with sub-micron sized asperities composed of the wax [131]. Thus, they have hierarchical micro- and nano-sized structures, which were studied extensively by *Bhushan* and *Jung* [17]. The water droplets on these surfaces readily sit on the apex of nanostructures because air bubbles fill in the valleys of the structure under the droplet. Therefore, these leaves exhibit considerable superhydrophobicity. The water droplets on the leaves remove any contaminant particles from their surfaces when they roll off, leading to self cleaning ability referred to as the Lotus-effect [92]. Other examples of biological surfaces include duck feathers and butterfly wings. Their corrugated surfaces provide air pockets that prevent water from completely touching the surface. Study and simulation of biological objects with desired properties is referred to as "biomimetics". Biomimetics involves taking engineering solutions from nature, mimicking them and implementing them in an application. The word biomimetics is derived from a Greek word "biomimesis" meaning to mimic life. Several other names are used as well, such as bionics and biognosis. Mimicking the lotus effect falls in the field of biomimetics.

One of the ways to increase the hydrophobic or hydrophilic properties of the surface is to increase surface roughness; so roughness-induced hydrophobicity has become a subject of extensive investigation. Extremely water-repellent (superhydrophobic) surfaces have been produced by applying a patterned roughness combined with hydrophobic coatings which may satisfy the need for the non-adhesive surfaces [16, 18, 19, 31, 67–69]. *Wenzel* [133] found that the contact angle of a liquid with a rough surface is different from that with a smooth surface. *Cassie* and *Baxter* [34] showed that gaseous phase including water vapor, commonly referred to as "air" in the literature, may be trapped in the cavities of a rough surface, resulting in a composite solid-liquid-air interface, as opposed to the homogeneous solid-liquid interface. The two models describe two possible wetting regimes or states: the homogeneous (Wenzel) and the composite (Cassie–Baxter or just Cassie) regimes. *Shuttleworth* and *Bailey* [124] studied spreading of a liquid over a rough solid surface and found that the contact angle at the absolute minimum of surface energy corresponds to the values predicted by *Wenzel* [133] or *Cassie* and *Baxter* [34]. *Johnson* and *Dettre* [66] showed that the homogeneous and composite interfaces correspond to the two metastable equilibrium states of a droplet. Many authors have investigated recently the metastability of artificial superhydrophobic surfaces and showed that whether the interface is homogeneous or composite may depend on the history of the system, in particular whether the liquid was applied from the top or from the bottom [24, 56, 79, 86, 87, 108, 109]. *Extrand* [46] pointed out that whether the interface is homogeneous or composite depends on droplet size, due to the gravity. It was suggested also that the so-called two-tiered (or double) roughness, composed by superposition of two roughness patterns at different length-scale [57, 110, 127], and fractal roughness [120] may lead to superhydrophobicity.

Herminghaus [57] showed that certain self-affine profiles may result in superhydrophobic surfaces even for wetting liquids, in the case the local equilibrium condition for the triple line (line of contact between solid, liquid and air) is satisfied. *Nosonovsky* and *Bhushan* [97, 98] pointed out that such configurations, although formally possible, are likely to be unstable. *Nosonovsky* and *Bhushan* [98, 99] proposed a probabilistic model for wetting of rough surfaces with a certain probability associated with every equilibrium state. According to their model, the overall contact angle with a two-dimensional rough profile is calculated by assuming that the overall configuration of a droplet occurs as a result of superposition of numerous metastable states. The probability-based concept is consistent with the experimental data [16, 18, 19, 68, 79], [69], which suggests that transition between the composite and homogeneous interfaces is gradual, rather than instant. *Nosonovsky* and *Bhushan* [100, 101, 103–106] have identified mechanisms, which lead to destabilization of the composite interface, namely, the capillary waves, condensation and accumulation of nanodroplets, and surface inhomogeneity. These mechanisms are scale-dependent with different characteristic scale lengths. To effectively resist these scale-dependent mechanisms, a multiscale (hierarchical) roughness is required. High asperities resist the capillary waves, while nanobumps prevent nanodroplets from filling the valleys between asperities and pin the tripe line in case of a hydrophilic spot.

Various criteria have been formulated to predict the transitions from a metastable composite state to a wetted state [16, 48, 68, 69, 101–104, 106, 109]. *Extrand* [48] formulated the transition criterion referred to as the contact line density criterion which was obtained by balancing the droplet weight and the surface forces along the contact line. *Patankar* [109] proposed a transition criterion based on energy balance. There is an energy barrier in going from a higher energy Cassie–Baxter droplet to a lower energy Wenzel droplet. The most probable mechanism is that the decrease in the gravitational potential energy during the transition helps in overcoming the energy barrier. This energy barrier was estimated by considering an intermediate state in which the water fills the grooves below the contact area of a Cassie–Baxter droplet but the liquid-solid contact is yet to be formed at the bottom of the valleys. These criteria were tested on selected experiments from the literature [23, 56, 107, 137]. *Bhushan* et al. [16] and *Nosonovsky* and *Bhushan* [101, 102] found that the transition occurs at a critical value of the spacing factor, a non-dimensional parameter which is defined as the diameter of the pillars divided by the pitch distance between them for patterned surfaces, and its ratio to the droplet size. *Bhushan* and *Jung* [18, 19] and *Jung* and *Bhushan* [68, 69] proposed the transition criterion based on the pitch distance between the pillars and the curvature of droplet governed by the Laplace equation, which relates pressure inside the droplet to its curvature. In addition, the transition can occur by applying external pressure on the droplet, or by the impact of droplet on the patterned surfaces [9, 33, 79, 115].

It has been demonstrated experimentally that roughness changes contact angle in accordance with the Wenzel model. *Yost* el al. [138] found that roughness enhances wetting of a copper surface with Sn-Pb eutectic solder, which has a contact angle of 15–20° for smooth surface. *Shibuichi* et al. [120] measured the contact angle of various liquids (mixtures of water and 1,4-dioxane) on alkylketen dimmer (AKD) substrate (contact angle not larger than 109° for smooth surface). They found that for wetting liquids, the contact angle decreases with increasing roughness, whereas for non-wetting liquids it increases. *Semal* et al. [117] investigated the effect of surface roughness on contact angle hysteresis by studying a sessile droplet of squalane spreading dynamically on multilayer substrates (behenic acid on glass) and found an increase in microroughness slows the rate of droplet spreading. *Erbil* et al. [44] measured the contact angle of polypropylene (contact angle of 104° for smooth surface) and found that the contact angle increases with increasing roughness. *Burton* and *Bhushan* [31] measured contact angle with roughness of patterned surfaces and found that in the case of hydrophilic surfaces, it decreases with increasing roughness, and for hydrophobic surfaces, it increases with increasing roughness. *Bhushan* and *Jung* [17–19] and *Jung* and *Bhushan* [67–69] studied wetting properties of hydrophobic and hydrophilic leaves and patterned surfaces and found similar trends.

To study superhydrophobicity, patterned surfaces are created and contact angle is measured to understand how roughness structure affects contact angle. Whether the Cassie–Baxter regime is present or not depends on the roughness structure and droplet radius. Evaporation studies are useful in characterizing wetting behavior because droplet with various sizes can be created to evaluate the transition criterion on a patterned surface. Many researchers have considered the evaporation of small droplets of liquid on solid surfaces [29, 43, 68, 69, 102–106, 116]. It has been shown that the wetting state changes from the Cassie–Baxter to Wenzel state as the droplet becomes smaller than a critical value on patterned surfaces during evaporation [68, 69, 89, 102–106].

An environmental scanning electron microscope (ESEM) can be used to condense or evaporate water droplets on surfaces by adjusting the pressure of the water vapor in the specimen chamber and the temperature of the cooling stage. Transfer of the water droplet has been achieved by a specially designed micro-injector device on wool fibers and then imaged at room temperature in ESEM [39]. Images of water droplets show strong topographic contrast in ESEM such that reliable contact angle measurements can be made on the surfaces [125]. Water condensation and evaporation studies on synthetic patterned surfaces were carried out by *Jung* and *Bhushan* [69] and *Nosonovsky* and *Bhushan* [102–106] where the change of contact angle was related with the surface roughness.

In this chapter, numerical models, which provide relationships between roughness and contact angle, and contact angle hysteresis as well as the Cassie–Wenzel regime transition, are discussed. The role of microbumps and nanobumps is examined by analyzing surface characterization on the micro- and nanoscale of hydrophobic and hydrophilic leaves. Along with measuring and characterizing surface roughness, the contact angle and adhesion and friction properties of these leaves are also considered. The knowledge gained by examining these properties of the leaves and by quantitatively analyzing the surface structure, will be helpful in designing superhydrophobic surfaces. Micro- and nanopatterned polymers (hydrophobic and hydrophilic) are fabricated to validate models and to provide design guidelines for superhydrophobic surfaces. These surfaces are examined by measuring their contact angle. To further examine the effect of meniscus force and real area of contact, scale dependence is considered with the use of AFM tips of various radii. Also in this chapter, criteria for the transition from the Cassie–Baxter to Wenzel regime are discussed. To investigate how the droplet size influences the transition, a study of droplet evaporation is conducted on silicon surfaces patterned with pillars of two different diameters and heights and with varying pitch values and deposited with a hydrophobic coating. An environmental scanning electron microscope (ESEM) study on the wetting behavior for a microdroplet with about 20 μm radius on the patterned Si surfaces is presented. The importance of hierarchical roughness structure on destabilization of air pockets is discussed. Finally, the techniques of producing superhydrophobic surfaces are described.

19.2 Modeling of Contact Angle for a Liquid in Contact with a Rough Surface

19.2.1 Contact Angle Definition

In this section, the dependence of the contact angle on the surface tension is considered for a liquid in contact with a solid surface, forming a homogeneous interface. The surface atoms or molecules of liquids or solids have energy above that of similar atoms and molecules in the interior, which results in surface tension or free surface energy being an important surface property. This property is characterized quantitatively by the surface tension or free surface energy γ, which is equal to work, that is required to create a unit area of the surface at constant pressure and temperature. The units of γ are J/m^2 or N/m and it can be interpreted either as energy per unit surface area or as tension force per unit length of a line at the surface. When a solid is in contact with liquid, the molecular attraction will reduce the energy of the system below that for the two separated surfaces. This is expressed by the Dupré equation

$$W_{SL} = \gamma_{SA} + \gamma_{LA} - \gamma_{SL}, \tag{19.1}$$

where W_{SL} is the work of cohesion per unit area between two surfaces, γ_{SA} and γ_{SL} are the surface energies (surface tensions) of the solid against air and liquid, and γ_{LA} is the surface energy (surface tension) of liquid against air [61].

If a droplet of liquid is placed on a solid surface, the liquid and solid surfaces come together under equilibrium at a characteristic angle called the static contact angle θ_0. This contact angle can be determined from the condition of the net surface free energy of the system being minimized [2, 61]. The total energy E_{tot} is given by

$$E_{tot} = \gamma_{LA}(A_{LA} + A_{SL}) - W_{SL}A_{SL}, \tag{19.2}$$

where A_{LA} and A_{SL} are the contact areas of the liquid with the solid and air, respectively. It is assumed that the droplet is small enough so that the gravitational potential energy can be neglected. It is also assumed that the volume and pressure are constant, so that the volumetric energy does not change. At the equilibrium $dE_{tot} = 0$, which yields

$$\gamma_{LA}(dA_{LA} + dA_{SL}) - W_{SL}dA_{SL} = 0. \tag{19.3}$$

For a droplet of constant volume, it is easy to show using geometrical considerations, that

$$dA_{LA}/dA_{SL} = \cos\theta_0. \tag{19.4}$$

Combining (19.1), (19.3) and (19.4), the well-known Young equation for the contact angle is obtained

$$\cos\theta_0 = \frac{\gamma_{SA} - \gamma_{SL}}{\gamma_{LA}}. \tag{19.5}$$

Equation (19.5) provides with the value of the static contact angle for given surface energies. Note that although we use the term "air", the analysis does not change in the case of another gas, such as water vapor.

19.2.2 Heterogeneous Interfaces and the Wenzel and Cassie–Baxter Equations

In this section, we will discuss the so-called heterogeneous interface and introduce the equations that govern the contact angle for the heterogeneous interface.

Contact Angle with a Rough and Heterogeneous Surfaces

The *Wenzel* [133] equation, which was derived using the surface force balance and empirical considerations, relates contact angle of a water droplet upon a rough solid surface, θ, with that upon a smooth surface, θ_0 (Fig. 19.1b), through the non-dimensional surface roughness factor, R_f, equal to the ratio of the surface area to its flat projected area

$$\cos\theta = \frac{dA_{LA}}{dA_F} = \frac{dA_{SL}}{dA_F}\frac{dA_{LA}}{dA_{SL}} = R_f \cos\theta_0 \tag{19.6}$$

$$R_f = \frac{A_{SL}}{A_F}. \tag{19.7}$$

The dependence of the contact angle on the roughness factor is presented in Fig. 19.2 for different values of θ_0. The Wenzel model predicts that a hydrophobic surface ($\theta_0 > 90°$) becomes more hydrophobic with an increase in R_f, while a hydrophilic surface ($\theta_0 < 90°$) becomes more hydrophilic with an increase in R_f [67, 97].

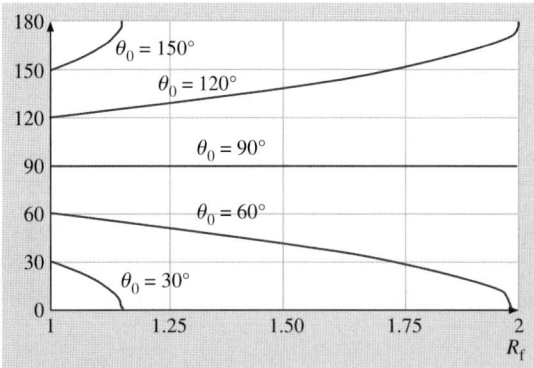

Fig. 19.2. Contact angle for rough surface (θ) as a function of the roughness factor (R_f) for various contact angles for smooth surface (θ_o) [97]

In a similar manner, for a surface composed of two fractions, one with the fractional area f_1 and the contact angle θ_1 and the other with f_2 and θ_2 respectively (so that $f_1 + f_2 = 1$), the contact angle is given by the Cassie equation

$$\cos\theta = f_1 \cos\theta_1 + f_2 \cos\theta_2 . \tag{19.8}$$

For the case of a composite interface (Fig. 19.1c), consisting of the solid-liquid fraction ($f_1 = f_{SL}$, $\theta_1 = \theta_0$) and liquid-air fraction ($f_2 = f_{LA} = 1 - f_{SL}, \cos\theta_2 = -1$), combining (19.7) and (19.8) yields the Cassie–Baxter equation

$$\cos\theta = R_f f_{SL} \cos\theta_0 - 1 + f_{SL}$$

or

$$\cos\theta = R_f \cos\theta_0 - f_{LA}(R_f \cos\theta_0 + 1) . \tag{19.9}$$

The opposite limiting case of $\cos\theta_2 = 1$ ($\theta_2 = 0°$) corresponds to the water-on-water contact) yields

$$\cos\theta = 1 + f_{SL}(\cos\theta_0 - 1) . \tag{19.10}$$

Equation (19.10) is used sometimes [41] for the homogeneous interface instead of (19.6), if the rough surface is covered by holes filled with water (Fig. 19.1d).

The Cassie–Baxter Equation

Two situations in wetting of a rough surface should be distinguished: the homogeneous interface without any air pockets (sometimes called the Wenzel interface, since the contact angle is given by the Wenzel equation or (19.6)) and the composite interface with air pockets trapped between the rough details as shown in Fig. 19.3a (sometimes called the Cassie or Cassie–Baxter interface, since the contact angle is given by (19.9)).

While (19.9) for the composite interface was derived using (19.6) and (19.8), it could also be obtained independently. For this purpose, two sets of interfaces are considered: a liquid-air interface with the ambient and a flat composite interface under the droplet, which itself involves solid-liquid, liquid-air, and solid-air interfaces. For fractional flat geometrical areas of the solid-liquid and liquid-air interfaces under the droplet, f_{SL} and f_{LA}, the flat area of the composite interface is

$$A_C = f_{SL} A_C + f_{LA} A_C = R_f A_{SL} + f_{LA} A_C . \tag{19.11}$$

In order to calculate the contact angle in a manner similar to the derivation of (19.6), the differential area of the liquid-air interface under the droplet, $f_{LA} \, dA_C$, should be subtracted from the differential of the total liquid-air area dA_{LA}, which yields

$$\cos\theta = \frac{dA_{LA} - f_{LA} \, dA_C}{dA_C} = \frac{dA_{SL}}{dA_F} \frac{dA_F}{dA_C} \frac{dA_{LA}}{dA_{SL}} - f_{LA}$$
$$= R_f f_{SL} \cos\theta_0 - f_{LA} = R_f \cos\theta_0 - f_{LA}(R_f \cos\theta_0 + 1) . \tag{19.12}$$

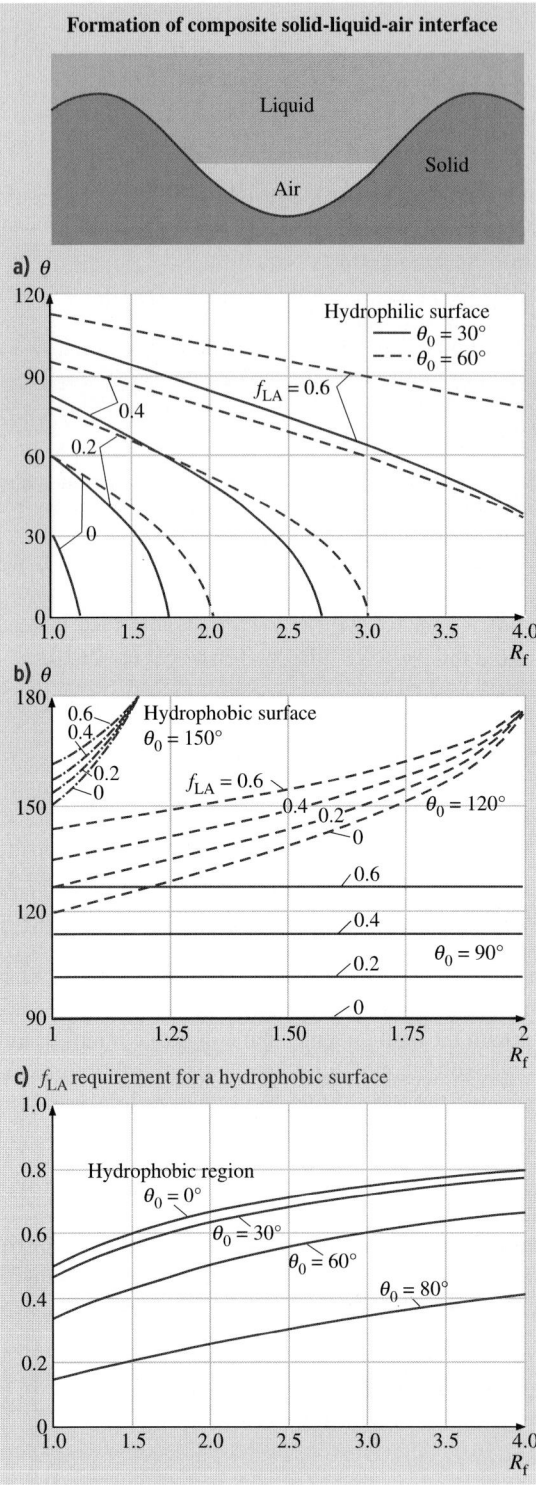

Fig. 19.3. (a) Formation of a composite solid-liquid-air interface for rough surface, (b) contact angle for rough surface (θ) as a function of the roughness factor (R_f) for various f_{LA} values on the hydrophilic surface and the hydrophobic surface, and (c) f_{LA} requirement for a hydrophilic surface to be hydrophobic as a function of the roughness factor (R_f) and θ_0 [67]

The dependence of the contact angle on the roughness factor for hydrophilic and hydrophobic surfaces is presented in Fig. 19.3b.

According to (19.12), even for a hydrophilic surface, the contact angle increases with an increase of f_{LA}. At a high value of f_{LA}, surface can become hydrophobic; however, the value required may be unachievable or formation of air pockets may become unstable. Using the Cassie–Baxter equation, the value of f_{LA} at which a hydrophilic surface could turn into a hydrophobic one, is given by [67]

$$f_{LA} \geq \frac{R_f \cos\theta_0}{R_f \cos\theta_0 + 1} \quad \text{for } \theta_0 < 90°. \tag{19.13}$$

Figure 19.3c shows the value of f_{LA} requirement as a function of R_f for four surfaces with different contact angles, θ_0. Hydrophobic surfaces can be achieved above a certain f_{LA} values as predicted by (19.13). The upper part of each contact angle line is hydrophobic region. For the hydrophobic surface, contact angle increases with an increase in f_{LA} both for smooth and rough surfaces.

Shuttleworth and *Bailey* [124] studied spreading of a liquid over a rough solid surface and found that the contact angle at the absolute minimum of surface energy corresponds to the values, given by (19.6) (for the homogeneous interface) or (19.12) (for composite interface). According to their analysis, spreading of a liquid continues, until simultaneously the (19.5) (the Young equation) is satisfied locally at the triple line and the minimal surface condition is satisfied over the entire liquid-air interface. The minimal surface condition states, that the sum of inversed principal radii of curvature, R_1 and R_2 (mean curvature), is constant at any point, and thus governs the shape of the liquid-air interface.

$$\frac{1}{R_1} + \frac{1}{R_2} = \text{const}. \tag{19.14}$$

The same condition is also the consequence of the Laplace equation, which relates pressure change through an interface with its mean curvature.

Stability of the composite interface is an important issue. Even though it may be geometrically possible for the system to become composite, it may be energetically profitable for the liquid to penetrate into valleys between asperities and to form the homogeneous interface. *Marmur* [86] formulated geometrical conditions for a surface, under which the energy of the system has a local minimum and the composite interface may exist. *Patankar* [109] pointed out that, whether the homogeneous or composite interface exists, depends on the system's history, i.e., on whether the droplet was formed at the surface or deposited. However, the above-mentioned analyses do not provide with an answer, which of the two possible configurations, homogeneous or composite, will actually form.

Limitations of the Wenzel and Cassie Equations

The Cassie equation (19.8) is based on the assumption that the heterogeneous surface is composed of well-separated distinct patches of different material, so that the free surface energy can be averaged. It has been argued also that when the size of the chemical heterogeneities is very small (of atomic or molecular dimensions), the quantity that should be averaged is not the energy, but the dipole moment of a macromolecule [62], and (19.8) should be replaced by

$$(1+\cos\theta)^2 = f_1(1+\cos\theta_1)^2 + f_2(1+\cos\theta_2)^2 . \tag{19.15}$$

Experimental studies of polymers with different functional groups showed a good agreement with (19.15) [130].

Later investigations put the Wenzel and Cassie equations into a thermodynamic framework, however, they showed also that there is no one single value of the contact angle for a rough or heterogeneous surface [66, 83, 86]. The contact angle can be in a range of values between the receding contact angle, θ_{rec}, and the advancing contact angle, θ_{adv}. The system tends to achieve the receding contact angle when liquid is removed (for example, at the rear end of a moving droplet), whereas the advancing contact angle is achieved when the liquid is added (for example, at the front end of a moving droplet). When the liquid is neither added nor removed, the system tends to have static or "most stable" contact angle, which is given approximately by (19.5)–(19.10). The difference between θ_{adv} and θ_{rec} is known as the "contact angle hysteresis" and it reflects a fundamental asymmetry of wetting and dewetting and the irreversibility of the wetting/dewetting cycle. Although for surfaces with the roughness carefully controlled on the molecular scale it is possible to achieve contact angle hysteresis as low as $< 1°$ [54], hysteresis cannot be eliminated completely, since even the atomically smooth surfaces have a certain roughness and heterogeneity. The contact angle hysteresis is a measure of energy dissipation during the flow of a droplet along a solid surface. A water-repellent surface should have a low contact angle hysteresis to allow water to flow easily along the surface.

It is emphasized that the contact angle provided by (19.5)–(19.10) is a macroscale parameter, so it is called sometimes "the apparent contact angle." The actual angle, under which the liquid-air interface comes in contact with the solid surface at the micro- and nanoscale can be different. There are several reasons for that. First, water molecules tend to form a thin layer upon the surfaces of many materials. This is because of a long-distance van der Waals adhesion force that creates the so-called disjoining pressure [42]. This pressure is dependent upon the liquid layer thickness and may lead to formation of stable thin films. In this case, the shape of the droplet near the triple line (line of contact of the solid, liquid and air, shown later in Fig. 19.6) transforms gradually from the spherical surface into a precursor layer, and thus the nanoscale contact angle is much smaller than the apparent contact angle. In addition, adsorbed water monolayers and multilayers are common for many materials. Second, even carefully prepared atomically smooth surfaces exhibit certain roughness and chemical heterogeneity. Water tends to cover at first the hydrophilic spots with high surface energy and low contact angle [35]. The tilt angle due to the roughness can also contribute into the apparent contact angle. Third, the very concept of the static contact angle is not well defined. For practical purposes, the contact angle, which is formed after a droplet is gently placed upon a surface and stops propagating, is considered the static contact angle. However, depositing the droplet involves adding liquid while leaving it may involve evaporation, so it is difficult to avoid dynamic effects. Fourth, for small droplets and curved triple lines, the effect of the contact line tension may be significant. Molecules at the surface of a liquid or solid phase have higher energy because they are bonded to fewer molecules, than those in the bulk. This leads to the surface tension and surface energy. In a similar manner, molecules at the edge have fewer bonds than those at the surface, which leads to the line tension and the curvature dependence of the surface energy. This effect becomes important when the radius of curvature is comparable with the so-called Tolman's length, normally of the molecular size [3]. However, the triple line at the nanoscale can be curved so that the line tension effects become important [111]. The contact angle with account for the contact line effect for a droplet with radius R is given by $\cos\theta = \cos\theta_0 + 2\tau/(R\gamma_{LA})$, where τ is the contact line tension and θ_0 is the value given by the Young equation [28]. Thus while the contact angle is a convenient macroscale parameter, wetting is governed by interactions at the micro- and nanoscale, which determine the contact angle hysteresis and other wetting properties (Table 19.1).

Range of Applicability of the Wenzel and Cassie Equations

Gao and *McCarthy* [53] showed experimentally that the contact angle of a droplet is defined by the triple line and does not depend upon the roughness under the bulk of the droplet. A similar result for chemically heterogeneous surfaces was obtained by *Extrand* [47]. *Gao* and *McCarthy* [53] concluded that the Wenzel and Cassie–Baxter equations "should be used with the knowledge of their fault." The question remained, however, under what circumstances the Wenzel and Cassie–Baxter equations can be safely used and under what circumstances do they become irrelevant.

For a liquid front propagating along a rough two-dimensional profile (Fig. 19.4ab), the derivative of the free surface energy (per liquid front length), W, by the profile length, t, yields the surface tension force $\sigma = dW/dt = \gamma_{SL} - \gamma_{SA}$. The quantity of practical interest is the component of the tension force that corresponds

Table 19.1. Wetting of a superhydrophobic surface as a multiscale process [102, 106]

Scale level	Characteristic length	Parameters	Phenomena	Interface
Macroscale	Droplet radius (mm)	Contact angle, droplet radius	Contact angle hysteresis	2D
Microscale	Roughness detail (µm)	Shape of the droplet, position of the liquid-air interface (h)	Kinetic effects	3D solid surface, 2D liquid surface
Nanoscale	Molecular heterogeneity (nm)	Molecular description	Thermodynamic and dynamic effects	3D

to the advancing of the liquid front in the horizontal direction for dx. This component is given by $dW/dx = (dW/dt)(dt/dx) = (\gamma_{SL} - \gamma_{SA})dt/dx$. It is noted that the derivative $R_f = dt/dx$ is equal to Wenzel's roughness factor in the case when the roughness factor is constant throughout the surface. Therefore, the Young equation, which relates the contact angle with solid, liquid, and air interface tensions, $\gamma_{LA}\cos\theta = \gamma_{SA} - \gamma_{SL}$, is modified as [96]

$$\gamma_{LA}\cos\theta = R_f(\gamma_{SA} - \gamma_{SL}). \tag{19.16}$$

The empirical Wenzel equation (19.6) is a consequence of (19.16) combined with the Young equation.

Nosonovsky [96] showed that for a more complicated case of a non-uniform roughness, given by the profile $z(x)$, the local value of $r(x) = dt/dx = (1+(dz/dx)^2)^{1/2}$ matters. In the cases that were studied experimentally by *Gao and McCarthy* [53] and *Extrand* [47], the roughness was present ($r > 1$) under the bulk of the droplet, but there was no roughness ($r = 0$) at the triple line, and the contact angle was given by (19.6) (Fig. 19.4c). In the general case of a 3D rough surface $z(x,y)$, the roughness factor can be defined as a function of the coordinates $r(x,y) = (1+(dz/dx)^2+(dz/dy)^2)^{1/2}$.

Whereas (19.6) is valid for uniformly rough surfaces, that is, surfaces with $r = \text{const}$, for non-uniformly rough surfaces the generalized Wenzel equation is formulated to determine the local contact angle (a function of x and y) with a rough surfaces at the triple line [96]

$$\cos\theta = r(x,y)\cos\theta_0. \tag{19.17}$$

The difference between the Wenzel equation (19.6) and the Nosonovsky equation (19.17) in that the later is valid for a non-uniform roughness with the roughness factor as the function of the coordinates. Equation (19.17) is consistent with the experimental results of the scholars, who showed that roughness beneath the droplet does not affect the contact angle, since it predicts that only roughness at the triple line matters. It is consistent also with the results of the researchers who confirmed the Wenzel equation (for the case of the uniform roughness) and of those who reported that only the triple line matters (for non-uniform roughness) (Table 19.2).

The Cassie equation for the composite surface can be generalized in a similar manner introducing the spatial dependence of the local densities, f_1 and f_2 of the solid-liquid interface with the contact angle, as a function of x and y, given by

$$\cos\theta_{composite} = f_1(x,y)\cos\theta_1 + f_2(x,y)\cos\theta_2, \tag{19.18}$$

where $f_1 + f_2 = 1, \theta_1$ and θ_2 are contact angles of the two components [96].

The important question remains, what should be the typical size of roughness/heterogeneity details in order for the generalized Wenzel and Cassie equations (19.17)–(19.18) to be valid? Some scholars have suggested that roughness/heterogeneity details should be comparable with the thickness of the liquid-air interface and thus "the roughness would have to be of molecular dimensions to

Table 19.2. Summary of experimental results for uniform and non-uniform rough and chemically heterogeneous surfaces. For non-uniform surfaces, the results shown for droplets larger than the islands of non-uniformity. Detailed quantitative values of the contact angle in various sets of experiments can be found in the referred sources [96]

Experiment	Roughness/ hydropobicity at the triple line and at the rest of the surface	Roughness at the bulk (under the droplet)	Experimental contact angle (compared with that at the rest of the surface)	Theoretical contact angle, Wenzel/ Cassie equations	Theoretical contact angle, generalized Wenzel/ Cassie (19.17)–(19.18)
Gao and *McCarthy* [53]	Hydrophobic Rough Smooth	Hydrophilic Smooth Rough	Not changed Not changed Not changed	Decreased Decreased Increased	Not changed Not changed Not changed
Extrand [47]	Hydrophilic Hydrophobic	Hydrophobic Hydrophilic	Not changed Not changed	Increased Decreased	Not changed Not changed
Bhushan et al. [22]	Rough	Rough	Increased	Increased	Increased
Barbieri et al. [6]	Rough	Rough	Increased	Increased	Increased

alter the equilibrium conditions" [7], whereas others have claimed that roughness/heterogeneity details should be small comparing with the linear size of the droplet [6, 16, 18, 19, 66–69, 83]. The interface in our analysis is an idealized 2D object, which has no thickness. In reality, the triple line zone has two characteristic dimensions: the thickness (of the order of molecular dimensions) and the length (of the order of the droplet size).

The apparent contact angle, given by (19.17)–(19.18), may be viewed as the result of averaging of the local contact angle at the triple line by its length, and thus the size of the roughness/heterogeneity details should be small comparing to the length (and not the thickness) of the triple line. A rigorous definition of the generalized equation requires the consideration of several scale lengths. The length dx needed for averaging of the energy gives the length over which the averaging is performed to obtain $r(x,y)$. This length should be larger than roughness details. However, it is still smaller than the droplet size and the length scale at which the apparent contact angle is observed (at which local variations of the contact angle level out). Since of these lengths (the roughness size, dx, the droplet size) the first and the last are of practical importance, we conclude that the roughness details should be smaller than the droplet size. When the liquid-air interface is studied at the length scale of roughness/heterogeneity details, the local contact angle, θ_0, is given by (19.6)–(19.10). The liquid-air interface at that scale has perturbations, caused by the roughness/heterogeneity, and the scale of the perturbations is the same as the scale of

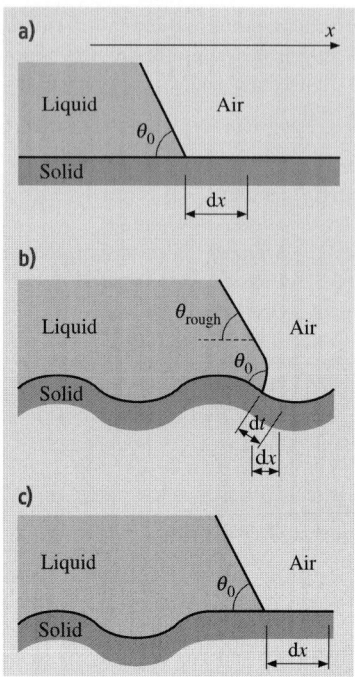

Fig. 19.4. Liquid front in contact with a **(a)** smooth solid surface **(b)** rough solid surface, propagation for a distance dt along the curved surface corresponds to the distance dx along the horizontal surface. **(c)** Surface roughness under the bulk of the droplet does not affect the contact angle

the roughness/heterogeneity details. However, when the same interface is studied at a larger scale, the effect of the perturbation vanishes, and apparent contact angle is given by (19.17)–(19.18) (Fig. 19.4c). This apparent contact angle is defined at the scale length, for which the small perturbations of the liquid-air interface vanish, and the interface can be treated as a smooth surface. The values of $r(x,y)$, $f_1(x,y)$, $f_2(x,y)$ in (19.17)–(19.18) are average values by the area (x,y) with size larger than a typical roughness/heterogeneity detail size. Therefore, the generalized Wenzel and Cassie equations can be used at the scale, at which the effect of the interface perturbations vanish, or, in other words, when the size of the solid surface roughness/heterogeneity details is small comparing with the size of the liquid-air interface, which is of the same order as the size of the droplet.

Nosonovsky and *Bhushan* [106] used the surface energy approach to find the domain of validity of the Wenzel and Cassie equations (uniformly rough surfaces) and generalized it for a more complicated case of non-uniform surfaces. The generalized equations explain a wide range of existing experimental data, which could not be explained by the original Wenzel and Cassie equations.

19.2.3 Contact Angle Hysteresis

The contact angle hysteresis is another important characteristic of a solid-liquid interface (Fig. 19.5a). The contact angle hysteresis occurs due to surface roughness and heterogeneity. Low contact angle hysteresis results in a very low water roll-

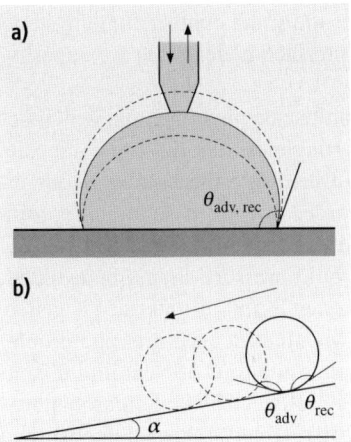

Fig. 19.5. (a) Liquid droplet in contact with rough surface (advancing and receding contact angles are θ_{adv} and θ_{rec}, respectively) and (b) tilted surface profile (the tilt angle is α) with a liquid droplet

off angle, which denotes the angle to which a surface may be tilted for roll-off of water drops (i.e., very low water contact angle hysteresis) [18, 19, 46, 68, 69, 73] (Fig. 19.5b). Low water roll-off angle is important in liquid flow applications such as in micro/nanochannels and surfaces with self cleaning ability.

There is no simple expression for the contact angle hysteresis as a function of roughness; however, certain conclusions about the relation of the contact angle hysteresis to roughness can be made. Using (19.12), the difference of cosines of the advancing and receding angles is related to the difference of those for a nominally smooth surface, θ_{adv0} and θ_{rec0}, as [16, 100]

$$\cos\theta_{adv} - \cos\theta_{rec} = R_f(1 - f_{LA})(\cos\theta_{adv0} - \cos\theta_{rec0}) + H_r, \quad (19.19)$$

where H_r is the effect of surface roughness, which is equal to the total perimeter of the asperity per unit area. It is observed from (19.12) and (19.19), that increasing $f_{LA} \to 1$ results in increasing the contact angle ($\cos\theta \to -1, \theta \to \pi$) and decreasing the contact angle hysteresis ($\cos\theta_{adv} - \cos\theta_{rec} \to 0$). In the limiting case of very small solid-liquid fractional contact area under the droplet, when the contact angle is large ($\cos\theta \approx -1 + (\theta - \pi)^2/2, \sin\theta \approx \pi - \theta$) and the contact angle hysteresis is small ($\theta_{adv} \approx \theta \approx \theta_{rec}$), (19.12) and (19.19) are reduced to

$$\theta - \pi = \sqrt{2(1 - f_{LA})(R_f \cos\theta_0 + 1)} \quad (19.20)$$

$$\theta_{adv} - \theta_{rec} = (1 - f_{LA})R_f \frac{\cos\theta_{a0} - \cos\theta_{r0}}{\sin\theta}$$

$$= \left(\sqrt{1 - f_{LA}}\right) R_f \frac{\cos\theta_{r0} - \cos\theta_{a0}}{\sqrt{2(R_f \cos\theta_0 + 1)}}. \quad (19.21)$$

For the homogeneous interface, $f_{LA} = 0$, whereas for composite interface f_{LA} is not a zero number. It is observed from (19.20)–(19.21) that for homogeneous interface, increasing roughness (high R_f) leads to increasing the contact angle hysteresis (high values of $\theta_{adv} - \theta_{rec}$), while for composite interface, an approach to unity of f_{LA} provides with both high contact angle and small contact angle hysteresis [16, 67, 100, 101]. Therefore, the composite interface is desirable for superhydrophobicity.

Formation of a composite interface is also a multiscale phenomenon, which depends upon relative sizes of the liquid droplet and roughness details. The composite interface is fragile and can be irreversibly transformed into the homogeneous interface, thus damaging superhydrophobicity. In order to form a stable composite interface with air pockets between solid and liquid, the destabilizing factors such as capillary waves, nanodroplet condensation, and liquid pressure should be avoided. For high f_{LA}, nanopattern is desirable because whether liquid-air interface is generated depends upon the ratio of distance between two adjacent asperities and droplet radius. Furthermore, asperities can pin liquid droplets and thus prevent liquid from filling the valleys between asperities. High R_f can be achieved by both micropatterns and nanopatterns. *Nosonovsky* and *Bhushan* [100, 101, 103, 104] have demonstrated that a combination of microroughness and nanoroughness (multiscale roughness) can help to resist the destabilization, with convex surfaces pinning the interface and thus leading to stable equilibrium as well as preventing from filling the gaps between the pillars even in the case of a hydrophilic material. The effect of roughness on wetting is scale dependent and mechanisms that lead to destabilization of a composite interface are also scale-dependent. To effectively resist these scale-dependent mechanisms, it is expected that a multiscale roughness is optimum for superhydrophobicity [100, 101, 103, 104].

A sharp edge can pin the line of contact of the solid, liquid, and air (also known as the "triple line") at a position far from stable equilibrium, i.e. at contact angles different from θ_0 [45]. This effect is illustrated in the bottom sketch of Fig. 19.6, which shows a droplet propagating along a solid surface with grooves. At the edge point, the contact angle is not defined and can have any value between the values corresponding to the contact with the horizontal and inclined surfaces. For a droplet moving from left to right, the triple line will be pinned at the edge point until it will be able to proceed to the inclined plane. As it is observed from Fig. 19.6, the change of the surface slope (α) at the edge is the reason, which causes the pinning. Because of the pinning, the value of the contact angle at the front of the droplet (dynamic maximum advancing contact angle or $\theta_{adv} = \theta_0 + \alpha$) is greater than θ_0, whereas the value of the contact angle at the back of the droplet (dynamic minimum receding contact angle or $\theta_{rec} = \theta_0 - \alpha$) is smaller than θ_0, This phenomenon is known as the contact angle hysteresis [45, 61, 66]. A hysteresis domain of the dynamic contact angle is thus defined by the difference $\theta_{adv} - \theta_{rec}$. The liquid can travel easily along the surface if the contact angle hysteresis is small. It is noted that the static contact angle lies within the hysteresis domain, therefore, increasing the static contact angle up to the values of a superhydrophobic surface (approaching 180° will result also in reduction of the contact angle hysteresis. In a similar manner, the contact angle hysteresis also can exist even if the surface slope changes smoothly, without sharp edges.

Effect of roughness

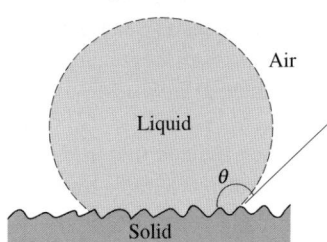

Effect of sharp edges

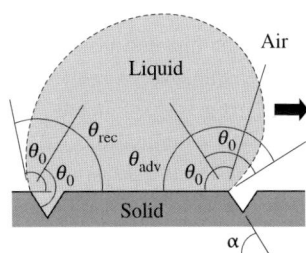

Fig. 19.6. Droplet of liquid in contact with a solid surface–smooth surface, contact angle θ_0; rough surface, contact angle θ; and a surface with sharp edges. For a droplet moving from left to right on a sharp edge (shown by *arrow*), the contact angle at a sharp edge may be any value between the contact angle with the horizontal plane and with the inclined plane. This effect results in difference of advancing ($\theta_{adv} = \theta_0 + \alpha$) and receding ($\theta_{rec} = \theta_0 - \alpha$) contact angles [97]

For a micropatterned surface built of flat-top columns ($R_f = 1$), the contact angle hysteresis involves a term inherent to the nominally smooth surface and the term dependent upon the surface roughness, H_r. Using the same approach as in derivation of 19.12 for the advancing and receding contact angles, one finds

$$\cos\theta_{adv} - \cos\theta_{rec} = f_{SL}(\cos\theta_{adv0} - \cos\theta_{rec0}) + H_r, \qquad (19.22)$$

where θ_{adv0} and θ_{rec0} are the advancing and receding contact angles for the smooth surface [16, 94, 95]. The first term in the right-hand part of the (19.22), which corresponds to the inherent contact angle hysteresis of a smooth surface is proportional to the fraction of the solid-liquid contact area, f_{SL}. The second term, H_r, may be assumed to be proportional to the length density of the pillar edges, or, in other words, to the length density of the triple line [16]. Thus (19.22) involves both the term proportional to the solid-liquid interface area and to the triple line length.

19.2.4 The Cassie–Wenzel Wetting Regime Transition

It is known from experimental observations that the transition from the Cassie–Baxter to Wenzel regime can be an irreversible event. Whereas such a transition can be induced, for example, by applying pressure or force to the droplet [79], electric voltage [4, 78], light for a photocatalytic texture [49], and vibration [27], the opposite transition is not observed, although there is no apparent reason for that.

Several approaches have been proposed for investigation of the transition between the Cassie–Baxter and Wenzel regimes, referred to as "the Cassie-Wenzel transition". *Lafuma* and *Quèrè* [79] suggested that the transition takes place when the net surface energy of the Wenzel regime becomes equal to that of the Cassie–Baxter regime, or, in other words, when the contact angle predicted by the Cassie equation is equal to that predicted by the Wenzel equation. They noticed that in certain case the transition does not occur even when it is energetically profitable and considered such Cassie state metastable. *Extrand* [47] suggested that the weight of the droplet is responsible for the transition and proposed the contact line density model, according to which the transition takes place when the weight exceeds the surface tension force at the triple line. *Patankar* [109] suggested that which of the two states is realized may depend upon how the droplet was formed, that is upon the history of the system. *Quèrè* [114] also suggested that the droplet curvature (which depends upon the pressure difference between inside and outside of the droplet) governs the transition. *Nosonovsky* and *Bhushan* [98] suggested that the transition is a dynamic process of destabilization and identified possible destabilizing factors. It has been also suggested that curvature of multiscale roughness defines the stability of the Cassie–Baxter wetting regime [94, 95, 100, 101, 103–106] and that the transition is a stochastic gradual process [26, 27, 63, 97]. Numerous experimental results support many of these approaches, however, it is not clear which particular mechanism prevails.

There is an asymmetry between the wetting and dewetting processes, since droplet nucleation requires less energy than air bubbles nucleation (cavitation). During wetting, which involves creation of the solid-liquid interface, less energy is released than the amount required for dewetting or destroying the solid-liquid interface due to the adhesion hysteresis. Adhesion hysteresis is one of the reasons that lead to the contact angle hysteresis and it results also in the hysteresis of the Wenzel–Cassie state transition. Figure 19.7 shows the contact angle of a rough surface as a function of surface roughness parameter, given by 19.12. Here it is assumed that $R_f \sim 1$ for Cassie–Baxter regime if the liquid droplet sits flat over the surface. It is noted that at a certain point, the contact angles given by Wenzel and Cassie–Baxter equations are the same, and $R_f = (1 - f_{LA}) - f_{LA}/\cos\theta_0$. At this point, the lines corresponding to the Wenzel and Cassie–Baxter regimes intersect. This point corresponds to an equal net energy of the Cassie–Baxter and Wenzel regimes. For a lower roughness (e.g., larger pitch between the pillars) the Wenzel regime is more energetically profitable, whereas for a higher roughness the Cassie–Baxter regime is more energetically profitable.

It is observed from Fig. 19.7 that an increase of roughness may lead to the transition between the Wenzel and Cassie–Baxter regimes at the intersection point. With decreasing roughness, the system is expected to transit to the Wenzel state. However, experiments show [6, 16, 18, 19, 68, 69] that, despite the energy in the Wenzel regime being lower than that in the Cassie–Baxter regime, the transition does not necessarily occur and the droplet may remain in the metastable Cassie–Baxter regime. This

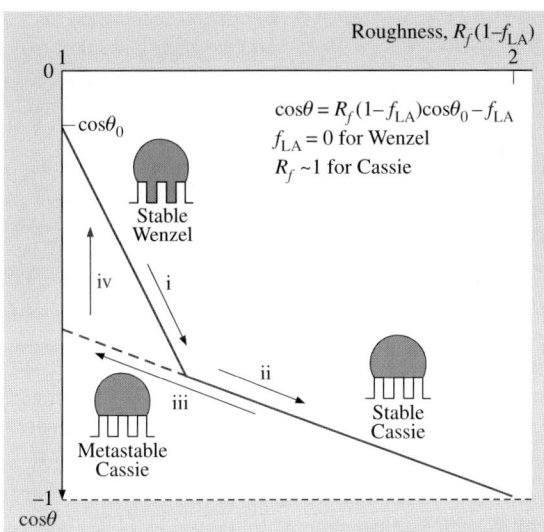

Fig. 19.7. Wetting hysteresis for a superhydrophobic surface. Contact angle as a function of roughness. The stable Wenzel state (i) can transform into the stable Cassie state with increasing roughness (ii). The metastable Cassie state (iii) can abruptly transform (iv) into the stable Wenzel state. The transition i-ii corresponds to equal Wenzel and Cassie states free energies, whereas the transition iv corresponds to a significant energy dissipation and thus it is irreversible [106]

is because there are energy barriers associated with the transition, which occurs due to destabilization by dynamic effects (such as waves and vibration).

In order to understand the contact angle hysteresis and transition between the Cassie–Baxter and Wenzel regimes, the shape of the free surface energy profile can be analyzed. The free surface energy of a droplet upon a smooth surface as a function of the contact angle has a distinct minimum, which corresponds to the most stable contact angle. As shown in Fig. 19.8a, the macroscale profile of the net surface energy allows us to find the contact angle (corresponding to energy minimums), however it fails to predict the contact angle hysteresis and Cassie-Wenzel transition, which are governed by micro- and nanoscale effects. As soon as the microscale substrate roughness is introduced, the droplet shape cannot anymore be considered as an ideal truncated sphere, and energy profiles have multiple energy minimums, corresponding to location of the pillars (Fig. 19.8b). The microscale energy profile (solid line) has numerous energy maxima and minima due to the surface micropattern. While exact calculation of the energy profile for a 3D droplet is complicated, a qualitative shape may be obtained by assuming a periodic sinusoidal dependence [66], superimposed upon the macroscale profile, as shown in Fig. 19.8b. Thus the advancing and receding contact angles can be identified as the maximum and the minimum possible contact angles corresponding to energy minimum points. However, the transition between the Wenzel and Cassie branches still cannot be explained. Note also that Fig. 19.8b explains qualitatively the hysteresis due to the kinetic effect of the pillars, but not the inherited adhesion hysteresis, which is characterized by the molecular scale length and cannot be captured by the microscale model.

The energy profile as a function of the contact angle does not provide any information on how the transition between the Cassie–Baxter and Wenzel regimes

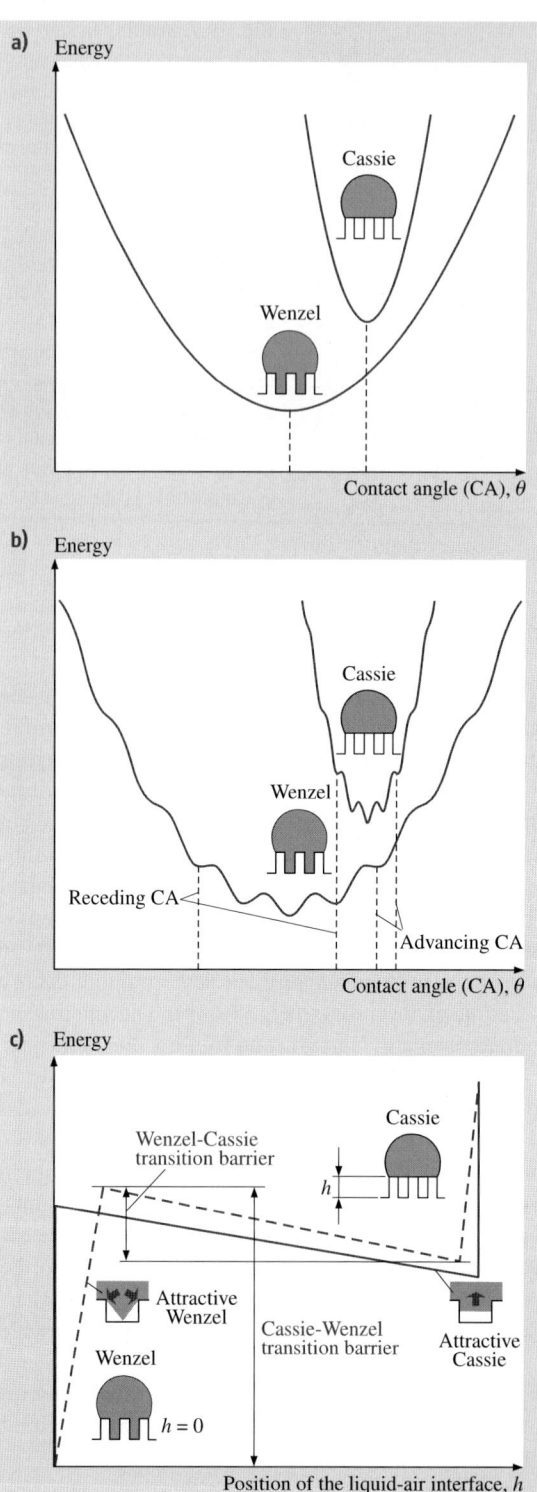

Fig. 19.8. Schematics of net free energy profiles. (**a**) Macroscale description; energy minimums correspond to the Wenzel and Cassie states. (**b**) Microscale description with multiple energy minimums due to surface texture. Largest and smallest values of the energy minimum correspond to the advancing and receding contact angles. (**c**) Origin of the two branches (Wenzel and Cassie) is found when a dependence of energy upon h is considered for the microscale description (*solid line*) and nanoscale imperfectness (*dashed line*) [106]. When the nanoscale imperfectness is introduced, it is observed that the Wenzel state corresponds to an energy minimum and the energy barrier for the Wenzel–Cassie transition is much smaller than for the opposite transition

occurs, because their two states correspond to completely isolated branches of the energy profile in Fig. 19.8ab. However, the energy may depend not only upon the contact angle, but also upon micro/nanoscale parameters, such as for example the vertical position of the liquid-air interface under the droplet, h (assuming that the interface is a horizontal plane) or similar geometrical parameters (assuming a more complicated shape of the interface). In order to investigate the Wenzel–Cassie transition, the dependence of the energy upon these parameters should be studied. We assume that the liquid-air interface under the droplet is a flat horizontal plane. When such air layer thickness or the vertical position of the liquid-air interface h is introduced, the energy can be studied as a function of droplet's shape, the contact angle, and h (Fig. 19.8c). For an ideal situation the energy profile has an abrupt minimum at the point corresponding to the Wenzel state, which corresponds to the sudden net energy change due to destroying solid-air and liquid-air interfaces ($\gamma_{SL} - \gamma_{SA} - \gamma_{LA} = -\gamma_{LA}(\cos\theta + 1)$ times the interface area) (Fig. 19.8c). In a more realistic case, the liquid-air interface cannot be considered horizontal due to nanoscale imperfectness or dynamic effects such as the capillary waves [98]. A typical size of the imperfectness is much smaller than the size of details of the surface texture and thus belongs to the molecular scale level. The height of the interface h can now be treated as an average height. The energy dependence upon h is now not as abrupt as in the idealized case. For example, the "triangular" shape as shown in Fig. 19.8c, the Wenzel state may become the second attractor for the system. It is seen that there are two equilibriums which correspond to the Wenzel and Cassie–Baxter regimes, with the Wenzel state corresponding to a much lower energy level. The energy dependence upon h governs the transition between the two states and is observed that a much larger energy barrier exists for the transition from Wenzel to Cassie–Baxter regime than for the opposite transition. This is why the first transition has never been observed experimentally [102].

To summarize, we showed that the contact angle hysteresis and Cassie–Wenzel transition cannot be determined from the macroscale equations and are governed by micro- and nanoscale phenomena. Our theoretical arguments are supported by our experimental data on micropatterned surfaces.

19.3 Lotus-Effect and Water-Repellent Surfaces in Nature

Many biological surfaces are known to be water-repellent and hydrophobic. The most known example is the leaf of the lotus plant that gave the name to the lotus-effect. In this chapter, we will discuss water-repellent plants, their roughness and wax coatings in relation to their hydrophobic and self-cleaning properties.

19.3.1 Water-Repellent Plants

Hydrophobic and water-repellent abilities of many plant leaves have been known for a long time. Scanning electron microscope (SEM) studies in the past 30 years revealed that the hydrophobicity of the leaf surface is related to its microstructure. All primary parts of plants are covered by a cuticle composed of soluble lipids embedded in a polyester matrix, which makes the cuticle hydrophobic in most cases [8]. The hydrophobicity of the leaves is related to another important effect, the ability of the hydrophobic leaves to remain clean after being immersed into dirty water, known as the self-cleaning. This ability is best known for the Lotus (*Nelumbo nucifera*) leaf that is considered by some oriental cultures as "sacred" due to its purity. Not surprisingly, the ability of lotus-like surfaces for self-cleaning and water repellency was dubbed the "Lotus effect." As far as biological implications of the Lotus-effect, *Barthlott* and *Neinhuis* [8] suggested that self-cleaning plays an important role in the defense against pathogens bounding to the leaf surface. Many spores and conidia of pathogenic organisms–most fungi–require water for germination and can infect leaves in the presence of water.

Neinhuis and *Barthlott* [92] studied systematically surfaces and wetting properties of about 200 water-repellent plants. The outer single-layered group of cells covering a plant, especially the leaf and young tissues is called epidermis. Protective waxy covering produced by the epidermal cells of leaves are called cuticles. The cuticle is composed of an insoluble cuticular membrane covered with epicuticular waxes. They reported several types of epidermal relief features and epicuticular wax crystals. Among the epidermal relief features are the papillose epidermal cells either with every epidermal cell forming a single papilla or cell being divided into papillae. The scale of the epidermal relief ranged from 5 μm in multipapillate cells to 100 μm in large epidermal cells. Some cells also were convex (rather than having real papillae) and/or had hairs (trichomes). They also found various types and shapes of wax crystals at the surface [131]. Interestingly, the hairy surfaces with a very thin film of wax exhibited water-repellency for short periods (minutes), after which water penetrated between the hairs, whereas waxy trichomes led to strong water-repellency. The wax crystal creates a roughness, in addition to the roughness created by the papillae. The chemical structure of the epicuticular waxes has been studied extensively by plant scientists and lipid chemists in recent decades [5, 65]. Apparently, roughness plays the dominant role in the lotus effect since the super-hydrophobicity can be achieved by using some type of wax or other hydrophobic coating.

The SEM study reveals that the lotus leaf surface is covered by "bumps", more exactly called papillae (papillose epidermal cells), which, in turn, are covered by an additional layer of epicuticular waxes [8]. The wax is present in crystalline tubules, composed of a mixture of aliphatic compounds, principally nonacosanol and nonacosanediols [76]. The wax is hydrophobic with the water contact angle of about 95°–110°, whereas the papillae provide with the tool to magnify the contact angle based on the Wenzel model, discussed in the preceding section. The experimental value of the static water contact angle with the lotus leaf was reported about 160° [8]. Indeed, taking the papillae density of 3400 per square millimeter, the average radius of the hemisphereical asperities $r = 10 \mu m$ and the aspect ratio $h/r = 1$, provides, based on (19.6), the value of the roughness factor $R_f \approx 4$ [97]. Taking the value of the contact angle for wax $\theta_0 = 104°$ [70], the naive calculation with the Wenzel equation yields $\theta = 165°$, which is not far from the experimentally observed values [97]. However, the simple Wenzel model may be not sufficient to explain the lotus-effect, since the lotus leaf exhibits also low contact angle hysteresis, apparently, forming the composite interface. Moreover, its structure has hierarchical roughness. So, a number of more sophisticated models has been developed [100–106].

While it is intuitive that the water-repellency and self-cleaning are related to each other, because the ability to repel water is related to ability to repel contaminants, it is difficult to quantify self-cleaning. Therefore, a quantitative relation of the two properties remains to be established. A qualitative explanation of how was suggested by *Barthlott* and *Neinhuis* [8] who suggested, that on a smooth surface contamination particles are mainly redistributed by a water droplet, on a rough surface they adhere to the droplet and removed from the leaves when the droplet roll off. A detailed model of this process has not been developed, but, obviously, whether the particle adheres to the droplet depends upon the interactions at the triple line.

The role of surface hierarchy in the lotus effect is also not completely clear, although a number of explanations why most natural surfaces are hierarchical has been suggested [52, 94, 95, 100, 101, 103–106]. *Nosonovsky* and *Bhushan* [100, 101, 103–106] showed that the mechanisms involved into the superhydrophobicity are scale dependent and thus the roughness must be hierarchical in order to respond to these mechanisms. It may have to do also with the simple fact, that the surface must be able to repel both macroscopic and microscopic droplets. Experiments with artificial fog (microdroplets) and artificial rain show that surfaces with only one scale of roughness repel well rain droplets, however, they cannot repel small fog droplets with are trapped in the valleys between the bumps [50].

19.3.2 Characterization of Hydrophobic and Hydrophilic Leaf Surfaces

In order to understand the mechanisms of hydrophobicity in plant leaves, a comprehensive comparative study of the hydrophobic and hydrophilic leaf surfaces and their properties was carried out by *Bhushan* and *Jung* [17] and *Burton* and *Bhushan* [32]. Below is a discussion of the findings of the study.

Experimental Techniques

The static contact angles were measured using a Rame-Hart model 100 contact angle goniometer with droplets of deionized water [17, 32]. Droplets of about 5 μL in volume (with diameter of a spherical droplet about 2.1 mm) were gently deposited on the substrate using a microsyringe for the static contact angle. All measurements were made by five different points for each sample at $22 \pm 1°C$ and $50 \pm 5\%$ RH. The measurement results were reproducible within $\pm 3°$.

An optical profiler (NT-3300, Wyko Corp., Tuscon, AZ) was used to measure surface roughness for different surface structures [17, 32]. A greater Z-range of the optical profiler of 2 mm is a distinct advantage over the surface roughness measurements with an AFM which has a Z-range of 7 μm, but it has a maximum lateral resolution of approximately 0.6 μm [13, 14]. A commercial AFM (D3100, Nanoscope IIIa controller, Digital Instruments, Santa Barbara, CA) was used for additional surface roughness measurements with a high lateral resolution and for adhesion and friction measurements [17, 31]. The measurements were performed with a square pyramidal Si(100) tip with a native oxide layer which had a nominal radius of 20 nm on a rectangular Si(100) cantilever with a spring constant of $3\,\mathrm{Nm^{-1}}$ in the tapping mode. Adhesion and friction force at various relative humidity (RH) were measured using a 15 μm radius borosilicate ball. A large tip radius was used to measure contributions from several microbumps and a large number of nanobumps. Friction force was measured under a constant load using a 90° scan angle at the velocity of 100 μm/s in 50 μm and at a velocity of 4 μm/s in 2 μm scans. The adhesion force was measured using the single point measurement of a force calibration plot.

Hydrophobic and Hydrophilic Leaves

Figure 19.9 shows the SEM micrographs of two hydrophobic leaves – lotus (*Nelumbo nucifera*) and elephant ear or taro plant (*Colocasia esculenta*) referred to as lotus and Colocasia, respectively – and two hydrophilic leaves – beech (*Fagus sylvatica*) and magnolia (*Magnolia grandiflora*) referred to as Fagus and Magnolia, respectively – [17]. Lotus and Colocasia are characterized by papillose epidermal cells responsible for creation of papillae or bumps on the surfaces, and an additional layer of three-dimensional epicuticular waxes which are a mixture of very long chain fatty acids molecules (compounds with chains > 20 carbon atoms). Fagus and Magnolia are characterized by rather flat tabulor cells with a thin wax film with a 2-D structure [8]. The leaves are not self cleaning, and contaminant particles from ambient are accumulated which make them hydrophilic.

Contact Angle Measurements

Figure 19.10a shows the contact angles for the hydrophobic and hydrophilic leaves before and after applying acetone. The acetone was applied in order to remove any wax present on the surface. As a result, for the hydrophobic leaves the contact angle dramatically reduced. whereas for the hydrophilic leaves, the contact angle was almost unchanged. It is known that there is a 2-D very thin wax layer on the hydrophilic leaves which introduces little roughness. In contrast, hydrophobic leaves are known to have a thin 3-D wax layer on their surface consists of nanoscale roughness over microroughness created by the papillae, which results in a hierarchical roughness. The combination of this wax and the roughness of the leaf creates a superhydrophobic surface.

Bhushan and *Jung* [17] calculated the contact angles for leaves with smooth surfaces using the Wenzel equation and the calculated R_f and the contact angle of

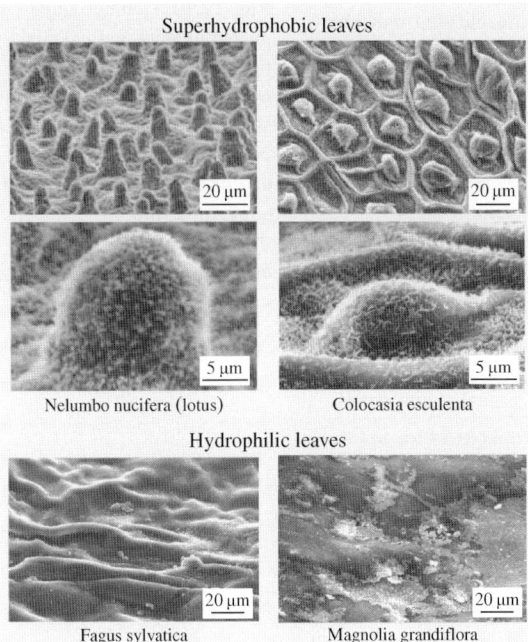

Fig. 19.9. Scanning electron micrographs of the relatively rough, water-repellent leaf surfaces of *Nelumbo nucifera (lotus)* and *Colocasia esculenta* and the relatively smooth, wettable leaf surfaces of *Fagus sylvatica* and *Magnolia grandiflora* [17]

the four leaves. The results are presented in Fig. 19.10a. The approximate values of R_f for lotus and colocasia are 5.6 and 8.4 and for Fagus and Magnolia are 3.4 and 3.8, respectively. Based on the calculations, the contact angles on smooth surface were approximately 99° for lotus and 96° for colocasia. For both Fagus and Magnolia, the contact angles for the smooth surfaces were found as approximately 86° and 88°. A further discussion on the effect of R_f on the contact angle will be presented later.

Figure 19.10b shows the contact angles for both fresh and dried states for the four leaves. There is a decrease in the contact angle for all four leaves when they are dried. For lotus and colocasia, this decrease is present because it is found that a fresh leaf has taller bumps than a dried leaf (data to be presented later), which will give a larger contact angle, according to the Wenzel equation. When the surface area is at a maximum compared to the footprint area, as with a fresh leaf, the roughness factor will be at a maximum and will only reduce when shrinking has occurred after drying. To understand the reason for the decrease of contact angle after drying of hydrophilic leaves, dried magnolia leaves were also measured using an AFM. It is found that the dried leaf (peak-valley (P–V) height = 7 µm, mid-width = 15 µm, and peak radius = 18 µm) has taller bumps than a fresh leaf (P–V height = 3 µm, mid-width = 12 µm, and peak radius = 15 µm), which increases the roughness, and the contact angle decreases, leading to a more hydrophilic surface.

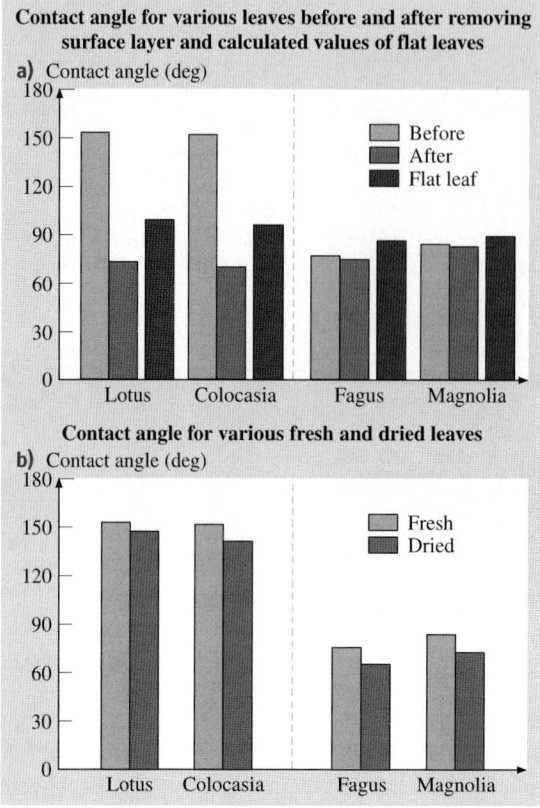

Fig. 19.10. Contact angle measurements and calculations for the leaf surfaces, (**a**) before and after removing surface layer as well as calculated values, and (**b**) fresh and dried leaves. The contact angle on a smooth surface for the four leaves was obtained using the roughness factor calculated [17]

Surface Characterization Using an Optical Profiler

The use of an optical profiler allows measurements to be made on fresh leaves, which have a large P–V distance. Three different surface height maps for hydrophobic and hydrophilic leaves are shown in Fig. 19.11 and 19.12 [17]. In each figure, a 3-D map and a flat map along with a 2-D profile in a given location of the flat 3-D map are shown. A scan size of 60 μm × 50 μm was used to obtain a sufficient amount of bumps to characterize the surface but also to maintain enough resolution to get an accurate measurement.

The structures found with the optical profiler correlate well with the SEM images shown in Fig. 19.9. The bumps on the lotus leaf are distributed on the entire surface, but the colocasia leaf shows a very different structure to that of the lotus. The surface structure for colocasia not only has bumps similar to lotus but surrounding each bump, a ridge is present that keeps the bumps separated. With these ridges, the bumps have a hexagonal (honeycomb) packing geometry that allows for the maximum number of bumps in a given area. The bumps of lotus and both bumps and ridges of colocasia contribute to the hydrophobic nature since they both increase the R_f factor and result in air pockets between the droplet of water and the surface.

In Fagus and Magnolia height maps, short bumps on the surface can be seen. This means that with decreased bump height, the probability of air pocket formation decreases and bumps have a less beneficial effect on the contact angle.

As shown in 2D profiles of hydrophobic and hydrophilic leaves in Figs. 19.11 and 19.12, a curve has been fitted to each profile to show exactly how the bump shape behaves. For each leaf a second order curve fit has been given to the profiles to show how closely the profile is followed. By using the second order curve fitting of the profiles, the radius of curvature can be found [17, 32].

Using these optical surface height maps, different statistical parameters of bumps and ridges can be found to characterize the surface: P–V height, mid-width, and peak radius [13, 14]. The mid-width is defined as the width of the bump at a height equal to half of peak to mean value. Table 19.3 shows these quantities found in the optical height maps for four leaves. Comparing the hydrophobic and hydrophilic leaves it can be seen that the P–V height for bumps of lotus and colocasia is much taller than that for bumps of Fagus and Magnolia. The peak radius for bumps of lotus and colocasia is also smaller than that for bumps of Fagus and Magnolia. However, the values of mid-width for bumps of four leaves are similar.

Table 19.3. Microbump and nanobump map statistics for hydrophobic and hydrophilic leaves, measured both fresh and dried leaves using an optical profiler and AFM [17]

Leaf		Microbump (μm) Scan size (50 × 50 μm)			Nanobump (μm) Scan size (2 × 2 μm)		
		P–V height	Mid-width	Peak radius	P–V height	Mid-width	Peak radius
Lotus							
Fresh		13*	10*	3*	0.78**	0.40**	0.15**
Dried		9**	10**	4**	0.67**	0.25**	0.10**
Colocasia							
Fresh	Bump	9*	15*	5*	0.53**	0.25**	0.07**
	Ridge	8*	7*	4*	0.68**	0.30**	0.12**
Dried	Bump	5**	15**	7**	0.48**	0.20**	0.06**
	Ridge	4**	8**	4**	0.57**	0.25**	0.11**
Fagus							
Fresh		5*	10*	15*	0.18**	0.04**	0.01**
		4**	5**	10**			
Magnolia							
Fresh		4*	13*	17*	0.07**	0.05**	0.04**
		3**	12**	15**			

* Data measured using optical profiler
** Data measured using AFM

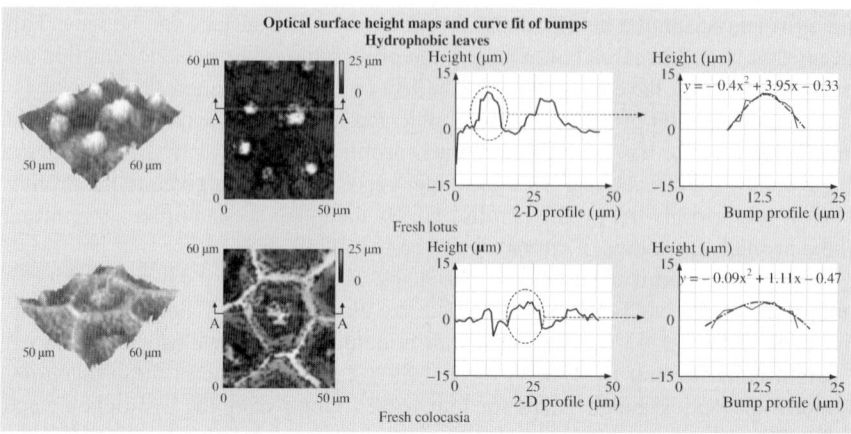

Fig. 19.11. Surface height maps and 2-D profiles of hydrophobic leaves using an optical profiler. For lotus leaf, a microbump is defined as a single, independent microstructure protruding from the surface. For colocasia leaf, a microbump is defined as the single, independent protrusion from the leaf surface, whereas a ridge is defined as the structure that surrounds each bump and is completely interconnected on the leaf. A curve has been fitted to each profile to show exactly how the bump shape behaves. The radius of curvature is calculated from the parabolic curve fit of the bump [17]

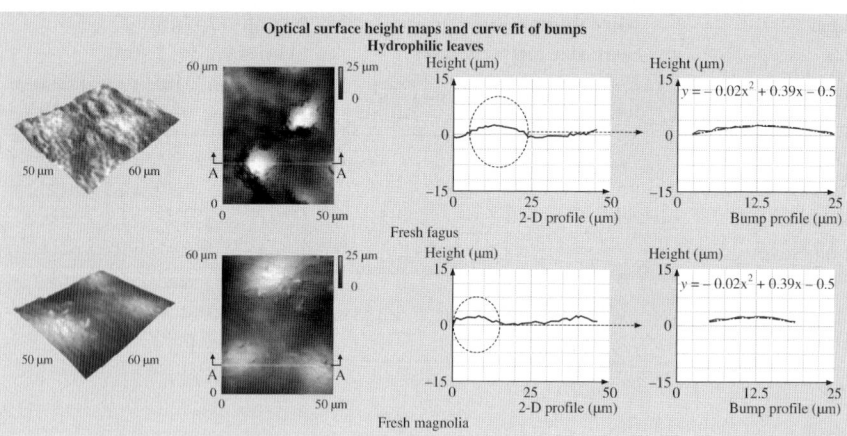

Fig. 19.12. Surface height maps and 2-D profiles of hydrophilic leaves using an optical profiler. For Fagus and Magnolia leaves, a microbump is defined as a single, independent microstructure protruding from the surface. A curve has been fitted to each profile to show exactly how the bump shape behaves. The radius of curvature is calculated from the parabolic curve fit of the bump [17]

Leaf Characterization Using an AFM

Comparison of Two Techniques

To measure topographic images of the leaf surfaces, both the contact and tapping modes were first used [17]. Figure 19.13 shows surface height maps of dried lotus obtained using the two techniques. In the contact mode, local height variation for lotus leaf was observed in 50 μm scan size. However, little height variation was obtained in a 2 μm scan even at loads as low as 2 nN. This could be due to the substantial frictional force generated as the probe scanned over the sample. The frictional force can damage the sample. The tapping mode technique allows high-resolution topographic imaging of sample surfaces that are easily damaged, loosely held to their substrate, or difficult to image by other AFM techniques [13, 14]. As shown in Fig. 19.13, with the tapping mode technique, the soft and fragile leaves can be imaged successfully. Therefore tapping mode technique was used to examine the surface roughness of the hydrophobic and hydrophilic leaves using an AFM.

Surface Characterization

The AFM has a Z-range of about 7 μm, and cannot be used for measurements in conventional way because of high P–V distances of lotus leaf. Burton and Bhushan [32] developed a new method to fully determine the bump profiles. In order to compensate for the large P-V distance, two scans were made for each height: one measurement that scans the tops of the bumps and another measurement that scans the bottom or valleys of the bumps. The total height of the bumps is embedded within the two scans. Figure 19.14 shows the 50 μm surface height maps obtained using this method [17]. The 2-D profiles in the right side column take the profiles from the top scan and the bottom scan for each scan size and splice them together to get the total profile of the leaf. The 2 μm surface height maps for both fresh and dried lotus can also be seen in Fig. 19.14. This scan area was selected on the top of a microbump obtained in the 50 μm surface height map. It can be seen that nanobumps are randomly and densely distributed on the entire surface of lotus.

Bhushan and *Jung* [17] also measured the surface height maps for the hydrophilic leaves in both 50 μm and 2 μm scan sizes as shown in Fig. 19.15. For Fagus and Magnolia, microbumps were found on the surface and the P–V distance of these leaves is lower than that of lotus and colocasia. It can be seen in the 2 μm surface height maps that nanobumps selected on the peak of the microbump have an extremely low P-V distance.

Using the AFM surface height maps, different statistical parameters of bumps and ridges can be obtained: P–V height, mid-width, and peak radius. These quantities for four leaves are listed in Table 19.3. It can be seen that the values correlate well with the values obtained from optical profiler scans except for the bump heights, which decreases by more than half because of leaf shrinkage.

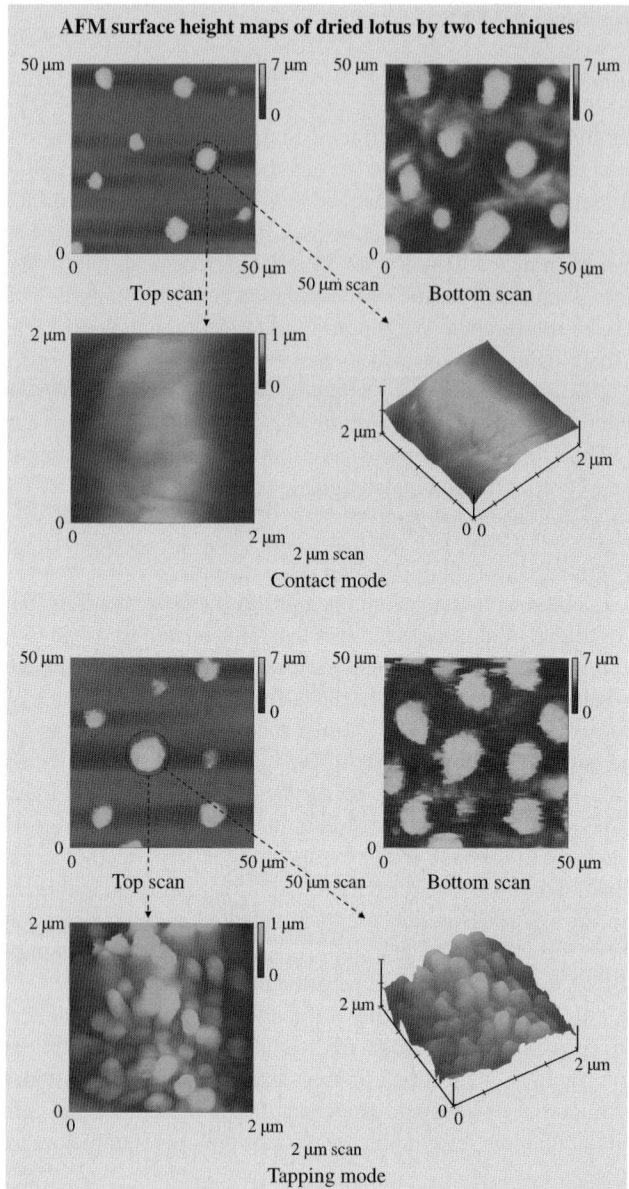

Fig. 19.13. Surface height maps showing the top scan and bottom scan in a 50 μm scan size and the bump peak scan selected in a 2 μm scan size for a lotus leaf in contact mode and tapping mode. Two methods were tested to get high resolution of nanotopography for a lotus leaf [17]

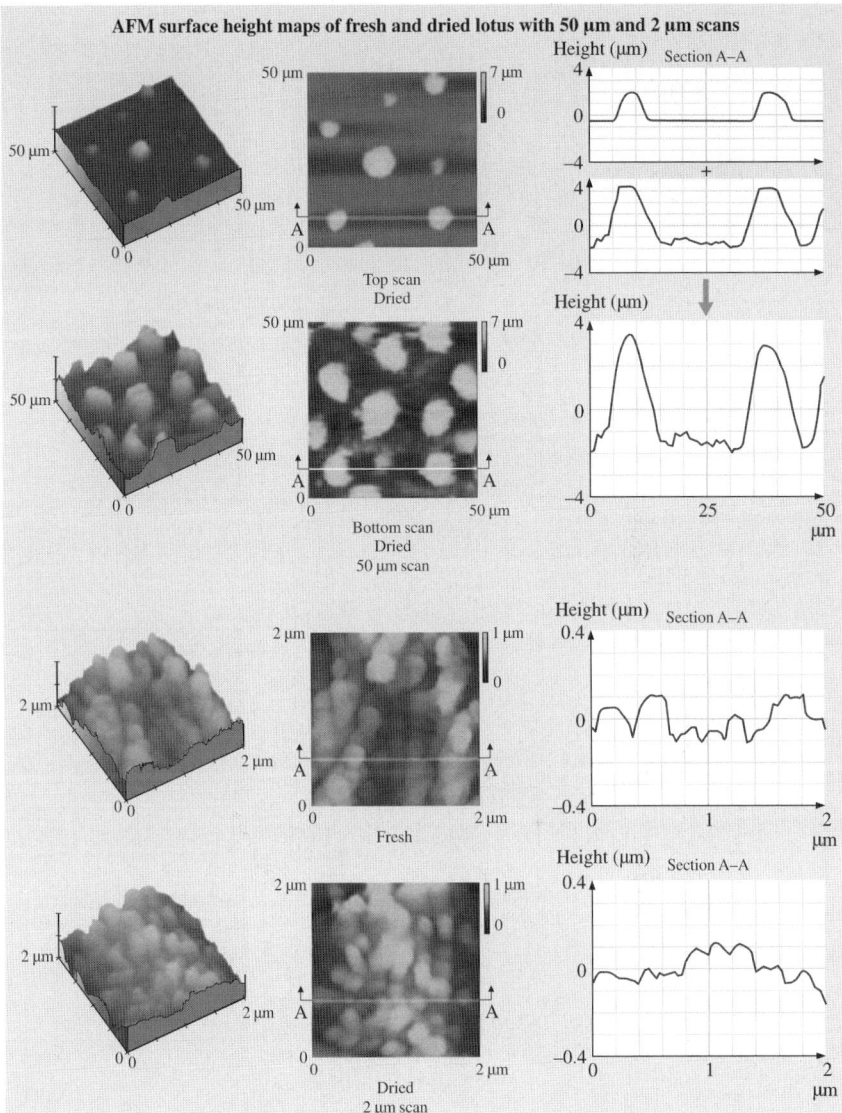

Fig. 19.14. Surface height maps and 2-D profiles showing the top scan and bottom scan of a dried lotus leaf in 50 μm scan (because the P–V distance of a dried lotus leaf is greater than the Z-range of an AFM), and both fresh and dried lotus in a 2 μm scan [17]

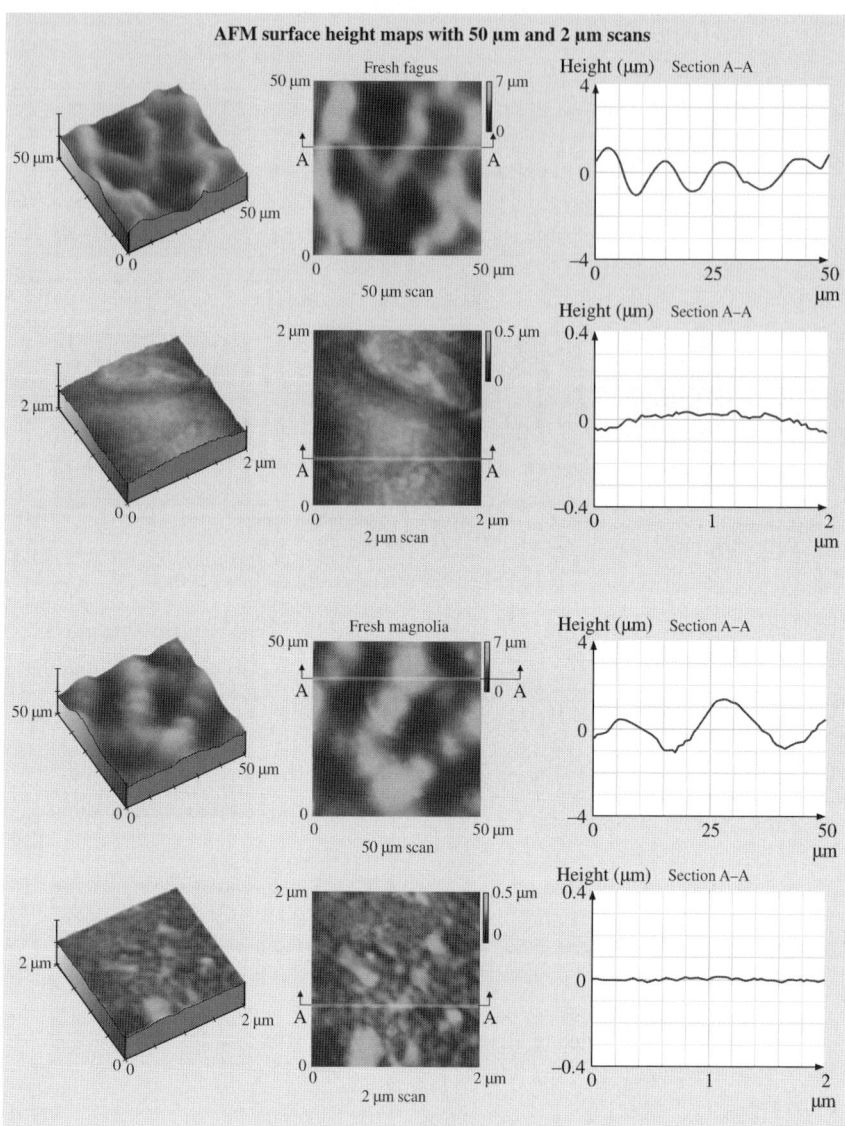

Fig. 19.15. Surface height maps and 2-D profiles of Fagus and Magnolia using an AFM in both 50 μm and 2 μm scans [17]

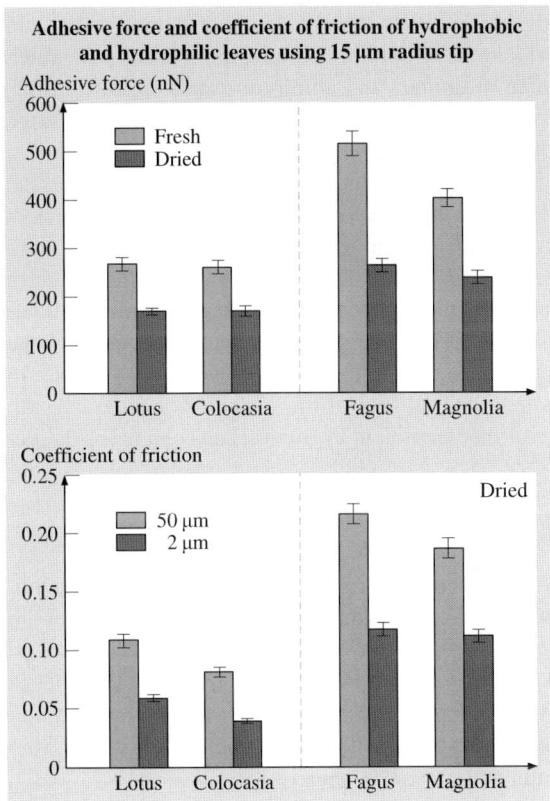

Fig. 19.16. Adhesive force for fresh and dried leaves, and the coefficient of friction for dried leaves for 50 μm and 2 μm scan sizes for hydrophobic and hydrophilic leaves. All measurements were made using a 15 μm radius borosilicate tip. Reproducibility for both adhesive force and coefficient of friction is ±5 % for all measurements [17]

Adhesive Force and Friction

Adhesive force and coefficient of friction of hydrophobic and hydrophilic leaves using AFM are presented in Fig. 19.16 [17]. For each type of leaf, adhesive force measurements were made for both fresh and dried leaves using a 15 μm radius tip. It is found that the dried leaves had a lower adhesive force than the fresh leaves. Adhesive force arises from several sources in changing the presence of a thin liquid film such as adsorbed water layer that causes meniscus bridges to build up around the contacting and near contacting bumps as a result of surface energy effects [13, 14]. When the leaves are fresh there is moisture within the plant material that causes the leaf to be soft and when the tip comes into contact with the leaf sample, the sample will deform and a larger real area of contact between the tip and sample will occur and the adhesive force will increase. After the leaf has dried, the moisture that was in the plant material is gone, and there is not as much deformation of the leaf when the tip comes into contact with the leaf sample. Hence, the adhesive force is decreased because the real area of contact has decreased.

The adhesive force of Fagus and Magnolia is higher than that of Lotus and Colocasia. The reason is that the real area of contact between the tip and leaf sample is expected to be higher in hydrophilic leaves than that in hydrophobic leaves. In addition, the Fagus and Magnolia are hydrophilic and have high affinity to water. The combination of high real area of contact and affinity to water are responsible for higher meniscus forces [13, 14]. The coefficient of friction was only measured on a dried plant surface with the same sliding velocity (10 μm/s) in different scan sizes rather than including the fresh surface because the P-V was too large to scan back and forth with the AFM to obtain friction force. As expected, the coefficient of friction for hydrophobic leaves is lower than that for hydrophilic leaves due to the real area of contact between the tip and leaf sample, similar to the adhesive force results. When the scan size from microscale to nanoscale decreases, the coefficient of friction also decreases in each leaf. The reason for such dependence is the scale dependent nature of the roughness of the leaf surface. Figures 19.14 and 19.15 show AFM topography images and 2-D profiles of the surfaces for different scan sizes. The scan size dependence of the coefficient of friction has been reported previously [77, 112, 128].

Role of the Nanoscale Roughness

The approximation of the roughness factor for the leaves on the micro- and nanoscale was made using AFM scan data [17]. Roughness factors for various leaves are presented in Table 19.4. As mentioned earlier, the open space between asperities on a surface has the potential to collect air, and its probability appears to be higher in nanobumps as the distance between bumps in the nanoscale is smaller than those in microscale. Using roughness factor values, along with the contact angles (θ) from both hydrophobic and hydrophilic surfaces; 153° and 152° in lotus and colocasia, and 76° and 84° in Fagus and Magnolia, respectively, the contact angles (θ_0) for the smooth surfaces can be calculated using the Wenzel equation for microbumps and the Cassie–Baxter equation (19.9) for nanobumps. Contact angle ($\Delta\theta$) calculated using R_f on the smooth surface can be found in Table 19.4. It can be seen that the roughness factors and the differences ($\Delta\theta$) between θ and θ_0 on nanoscale are higher than those in the microscale. This means that nanobumps on the top of a microbump increase contact angle more effectively than microbumps. In the case of hydrophilic leaves, the values of R_f and $\Delta\theta$ change very little on both scales.

Based on the data in Fig. 19.16, the coefficient of friction values in the nanoscale are much lower than those in the microscale. It is clearly observed that friction values are scale dependent. The height of a bump and the distance between bumps in microscale is much larger than those in nanoscale, which may be responsible for larger values of friction force on the microscale.

A difference between microbumps and nanobumps for surface enhancement of water repellency is the effect on contact angle hysteresis, in other words, the ease with which a droplet of water can roll on the surface. It has been stated earlier that contact angle hysteresis decreases and contact angle increases due to the decreased contact with the solid surface caused by the air pockets beneath the droplet. The surface with nanobumps has high roughness factor compared with that of microbumps. With large distances between microbumps, the probability of air pockets formation

Table 19.4. Roughness factor and contact angle ($\Delta\theta = \theta - \theta_0$) calculated using R_f on the smooth surface for hydrophobic and hydrophilic leaves measured using an AFM, both microscale and nanoscale [17]

Leaf (Contact angle)	Scan size	State	R_f	$\Delta\theta$
Lotus (153°)	50 μm	Dried	5.6	54*
	2 μm	Fresh	20	61**
		Dried	16	60**
Colocasia (152°)	50 μm	Dried	8.4	56*
	2 μm bump	Fresh	18	60**
		Dried	14	59**
	2 μm ridge	Fresh	18	60**
		Dried	15	59**
Fagus (76°)	50 μm	Fresh	3.4	−10*
	2 μm	Fresh	5.3	2**
Magnolia (84°)	50 μm	Fresh	3.8	−4*
	2 μm	Fresh	3.6	14**

* Calculations made using Wenzel equation
**Calculations made using Cassie–Baxter equation. We assume that the contact area between the droplet and air is the half of the whole area of the rough surface

decreases, and is responsible for high contact angle hysteresis. Therefore, on the surface with nanobumps, the contact angle is high and contact angle hysteresis is low, and drops rebound easily and can set into a rolling motion with a small tilt angle [17].

19.4 Wetting of Micro- and Nanopatterned Surfaces

In this section, we will discuss experimental observations of wetting properties of micro- and nanopatterned surfaces.

19.4.1 Experimental Techniques

Contact Angle, Surface Roughness, and Adhesion

The static and dynamic (advancing and receding) contact angles were measured using a Rame-Hart model 100 contact angle goniometer and water droplets of deionized water [17, 32, 67]. For the measurement of static contact angle, the droplet size should be small but larger than dimension of the structures present on the surfaces. Droplets of about 5 μL in volume (with diameter of a spherical droplet about 2.1 mm) were gently deposited on the substrate using a microsyringe for the static contact angle. The receding contact angle was measured by the removal of water from a DI water sessile droplet (~ 5 μL) using a microsyringe. The advancing contact angle was measured by adding additional water to the sessile droplet (~ 5 μL) using the microsyringe. The contact angle hysteresis was calculated by the difference between the measured advancing and receding contact angles. The tilt angle was measured by a simple stage tilting experiment with the droplets of 5 μL volume [18, 19]. All measurements were made by five different points for each sample at $22 \pm 1°C$ and $50 \pm 5\%$ RH. The measurements were reproducible to within $\pm 3°$.

For surface roughness, an optical profiler (NT-3300, Wyko Corp., Tuscon, AZ) was used for different surface structures [16–19, 32, 69]. A greater Z-range of the optical profiler of 2 mm is a distinct advantage over the surface roughness measurements using an AFM which has a Z-range of 7 μm, but it has a maximum lateral resolution of approximately 0.6 μm [13, 14]. Experiments were performed using three different radii tips to study the effect of scale dependence. Large radii atomic force microscopy (AFM) tips were primarily used in this study. Borosilicate ball with 15 μm radius and silica ball with 3.8 μm radius were mounted on a gold-coated triangular Si_3N_4 cantilever with a nominal spring constant of $0.58\,Nm^{-1}$. A square pyramidal Si_3N_4 tip with nominal radius 30–50 nm on a triangular Si_3N_4 cantilever with a nominal spring constant of $0.58\,Nm^{-1}$ was used for smaller radius tip. Adhesive force was measured using the single point measurement of a force calibration plot [13–15].

Measurement of Droplet Evaporation

Droplet evaporation was observed and recorded by a digital camcorder (Sony, DCRSR100) with a 10 X optical and 120 X digital zoom for every run of the experiment. Then the decrease in the diameter of the droplets with time was determined [68, 69]. The resolution of the camcorder was 0.03 s per frame. An objective lens placed in front of the camcorder during recording gave a total magnification of between 10 to 20 times. Droplet diameter as small as few hundred microns could be measured with this method. Droplets were gently deposited on the substrate using a microsyringe and the whole process of evaporation was recorded. The evaporation starts right after the deposition of the droplets. Images obtained were analyzed using Imagetool® software (University of Texas Health Science Center) for the contact angle. To find dust trace remaining after droplet evaporation, an optical microscope with a CCD camera (Nikon, Optihot-2) was used. All measurements were made in a controlled environment at $22 \pm 1°C$ and $45 \pm 5\%$ RH [68, 69].

Measurement of Contact Angle Using ESEM

A Philips XL30 ESEM equipped with a Peltier cooling stage was used to study smaller droplets [69]. ESEM uses a gaseous secondary electron detector (GSED) for imaging. The ESEM column is equipped with a multistage differential pressure-pumping unit. The pressure in the upper part is about 10^{-6} to 10^{-7} Torr, but the pressure of about 1 to 15 Torr can be maintained in the observation chamber. When the electron beam (primary electrons) ejects secondary electrons from the surface of the sample, the secondary electrons collide with gas molecules in the ESEM chamber, which in turn acts as a cascade amplifier, delivering the secondary electron signal to the positively biased GSED. The positively charged ions are attracted toward the specimen to neutralize the negative charge produced by the electron beam. Therefore, the ESEM can be used to examine electrically isolated specimens in their natural state. In ESEM, adjusting the pressure of the water vapor in the specimen chamber and the temperature of the cooling stage will allow the water to condense on the sample in the chamber. For the measurement of the static and dynamic contact angles on patterned surfaces, the video images were recorded. The voltage of the electron beam was 15 kV and the distance of the specimen from the final aperture was about 8 mm. If the angle of observation is not parallel to the surface, the electron beam is not parallel to the surface but inclined at an angle, this will produce a distortion in the projection of the droplet profile. A mathematical model to calculate the real contact angle from the ESEM images was used to correct the tilting of the surfaces during imaging [30, 69].

19.4.2 Micro- and Nanopatterned Polymers

Jung and *Bhushan* [67] studied two types of polymers: poly(methyl methacrylate) (PMMA) and polystyrene (PS). PMMA and PS were chosen because they are widely used in MEMS/NEMS devices. Both hydrophilic and hydrophobic surfaces can be produced by using these two polymers, as PMMA has polar (hydrophilic) groups with high surface energy while PS has electrically neutral and nonpolar(hydrophobic) groups with low surface energy. Furthermore, a PMMA structure can be made hydrophobic by treating it appropriately, for example, by coating with a hydrophobic self-assembled monolayer (SAM).

Four types of surface patterns were fabricated from PMMA: a flat film, low aspect ratio asperities (LAR, 1:1 height-to-diameter ratio), high aspect ratio asperities (HAR, 3:1 height-to-diameter ratio), and a replica of the lotus leaf (the lotus pattern). Two types of surface patterns were fabricated from PS: a flat film and the lotus pattern. Figure 19.17 shows SEM images of the two types of nanopatterned structures, LAR and HAR, and the one type of micropatterned structure, lotus pattern, all on a PMMA surface [31,67]. Both micro- and nanopatterned structures were manufactured using soft lithography. For nanopatterned structures, PMMA film was spin-coated on the silicon wafer. A UV cured mold (PUA mold) with nanopatterns of interest was made which enables one to create sub-100 nm patterns with high aspect ratio [36]. The mold was placed on the PMMA film and a slight pressure of ~ 10 g/cm^2 (~ 1 kPa) was applied and annealed at 120°C. Finally, the PUA mold was removed from PMMA film. For micropatterned structures, polydimethylsiloxane (PDMS) mold was first made by casting PDMS against a lotus leaf following by heating. As shown in Fig. 19.17, it can be seen that only microstructures exist on the surface of lotus pattern [67].

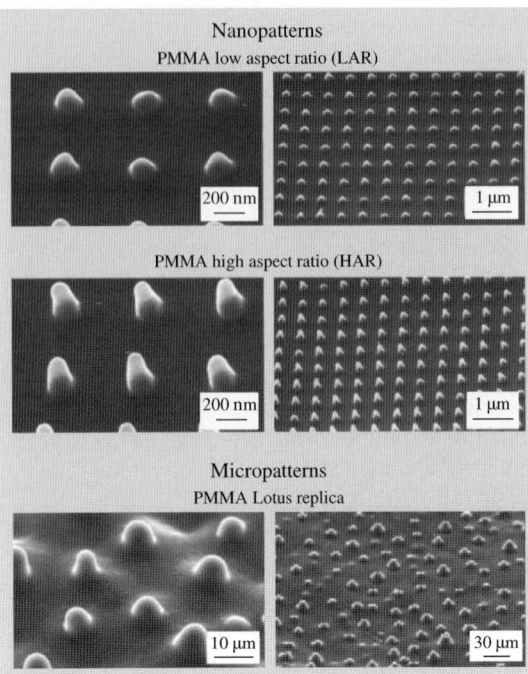

Fig. 19.17. Scanning electron micrographs of the two nanopatterned polymer surfaces (shown using two magnifications to see both the asperity shape and the asperity pattern on the surface) and the micropatterned polymer surface (Lotus pattern, which has only microstructures on the surface) [31, 67]

Since PMMA by itself is hydrophilic, in order to obtain a hydrophobic sample, a self-assembled monolayer (SAM) of perfluorodecyltriethoxysilane (PFDTES) was deposited on the sample surfaces using vapor phase deposition technique. PFDTES was chosen because of its hydrophobic nature. The deposition conditions for PFDTES were 100°C temperature, 400 Torr pressure, 20 min deposition time and 20 min annealing time. The polymer surface was exposed to an oxygen plasma treatment (40 W, O_2 187 Torr, 10 s) prior to coating [21]. The oxygen plasma treatment is necessary to oxidize any organic contaminants on the polymer surface and to also alter the surface chemistry to allow for enhanced bonding between the SAM and the polymer surface.

Contact Angle Measurements

Jung and *Bhushan* [67] measured the static contact angle of water with the patterned PMMA and PS structures; see Fig. 19.18. Since the Wenzel roughness factor is the parameter that often determines wetting behavior, the roughness factor was calculated and it is presented in Table 19.5 for various samples. The data show that contact angle of the hydrophilic materials decreases with an increase in the roughness factor, as predicted by the Wenzel model. When the polymers were coated with PFDTES, the film surface became hydrophobic. Figure 19.18 also shows the contact angle for various PMMA samples coated with PFDTES. For a hydrophobic surface, the standard Wenzel model predicts an increase of contact angle with roughness

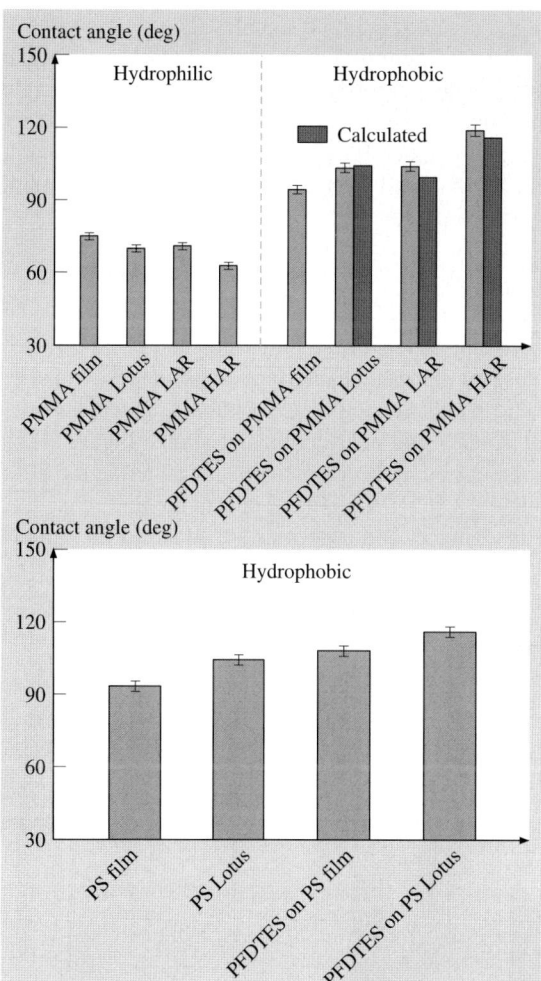

Fig. 19.18. Contact angles for various patterned surfaces on PMMA and PS polymers [67]

Table 19.5. Roughness factor for micro- and nanopatterned polymers [67]

	LAR	HAR	Lotus
R_f	2.1	5.6	3.2

factor, which is what happens in the case of patterned samples. The calculated values of contact angle for various patterned samples based on the contact angle of the smooth film and Wenzel equation are also presented. The measured contact angle values for the lotus pattern were comparable with the calculated values, whereas for the LAR and HAR patterns they are higher. It suggests that nanopatterns benefit from air pocket formation. For the PS material, the contact angle of the lotus pattern also increased with increased roughness factor.

Scale Dependence on Adhesive Force

Jung and *Bhushan* [67] found that scale-dependence of adhesion and friction forces are important for this study because the tip/surface interface area changes with size. The meniscus force will change due to either changing tip radius, the hydrophobicity of the sample, or the number of contact and near-contacting points. Figure 19.19 shows the dependence of the tip radius and hydrophobicity on the adhesive force for PMMA and PFDTES coated on PMMA [67]. When the radius of the tip is changed, the contact angle of the sample is changed, and asperities are added to the sample

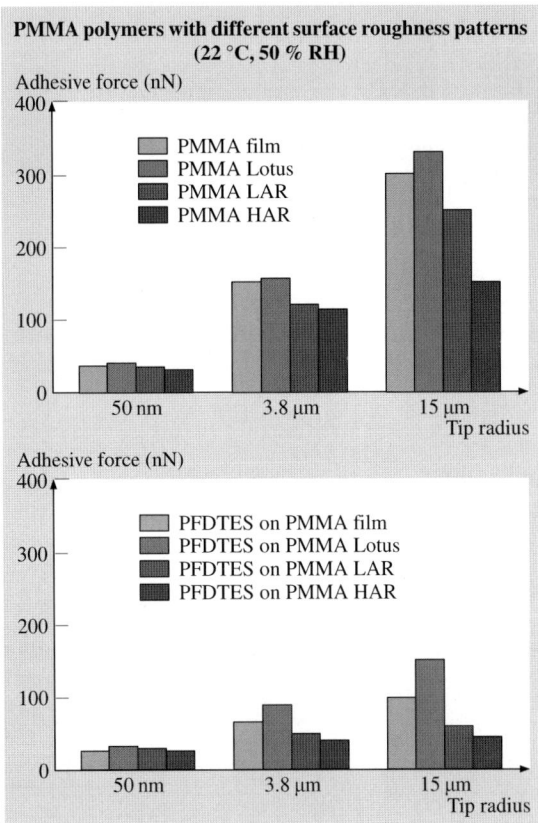

Fig. 19.19. Scale dependent adhesive force for various patterned surfaces measured using AFM tips of various radii [67]

surface, the adhesive force will change due to the change in the meniscus force and the real area of contact.

The two plots in Fig. 19.19 show the adhesive force on a linear scale for the different surfaces with varying tip radius. The left bar chart in Fig. 19.19 is for hydrophilic PMMA film, Lotus pattern, LAR, and HAR, and shows the effect of tip radius and hydrophobicity on adhesive force. For increasing radius, the adhesive force increases for each material. With a larger radius, the real area of contact and the meniscus contribution increase, resulting in the increased adhesion. The right bar chart in Fig. 19.19 shows the results for PFDTES coated on each material. These samples show the same trends as the film samples, but the increase in adhesion is not as dramatic. The hydrophobicity of PFDTES on material reduces meniscus forces, which in turn reduces adhesion from the surface. The dominant mechanism for the hydrophobic material is real area of contact and not meniscus force, whereas with hydrophilic material there is a combination of real area of contact and meniscus forces [67].

19.4.3 Micropatterned Si Surfaces

Micropatterned surfaces produced from a single-crystal silicon (Si) by electrolythography and coated with a self-assembled monolayer (SAM) were used by *Jung* and *Bhushan* [68, 69] in their study. Silicon has traditionally been the most commonly used structural material for micro/nanocomponents. A Si surface can be made hydrophobic by coating with a SAM. One of purposes of this investigation was to study the transition from the Cassie–Baxter to Wenzel regimes by changing the distance between the pillars. To create patterned Si, two series of nine samples each were fabricated using photolithography [6]. Series 1 had 5 µm diameter and 10 µm height flat-top, cylindrical pillars with different pitch values (7, 7.5, 10, 12.5, 25, 37.5, 45, 60, and 75) µm, and Series 2 has 14 µm diameter and 30 µm height flat-top, cylindrical pillars with different pitch values (21, 23, 26, 35, 70, 105, 126, 168, and 210) µm. The pitch is the spacing between the centers of two adjacent pillars. The SAM of 1, 1, −2, 2,-tetrahydroperfluorodecyltrichlorosilane (PF_3) was deposited on the Si sample surfaces using vapor phase deposition technique [6]. PF_3 was chosen because of the hydrophobic nature of the surface. The thickness and rms roughness of the SAM of PF_3 were 1.8 nm and 0.14 nm, respectively [71].

An optical profiler was used to measure the surface topography of the patterned surfaces [18, 19, 69]. One sample each from the two series was chosen to characterize the surfaces. Two different surface height maps can be seen for the patterned Si in Fig. 19.20. In each case, a 3-D map and a flat map along with a 2-D profile in a given location of the flat 3-D map are shown. A scan size of 100 µm × 90 µm was used to obtain a sufficient amount of pillars to characterize the surface but also to maintain enough resolution to get an accurate measurement.

The images found with the optical profiler show the flat-top, cylindrical pillars on the Si surface are distributed on the entire surface in a square grid with different pitch values. Sample in two series had the same values of Wenzel roughness factors

1038 B. Bhushan et al.

($R_f = 1 + \pi DH/P^2$), so that the Cassie–Baxter and Wenzel theoretical models predict exactly the same series of contact angle values for all two series of nine samples.

Contact Angle Relationships for a Geometry of Flat-Top, Cylindrical Pillars

Let us consider a geometry of flat-top, cylindrical pillars of diameter D, height H, and pitch P, distributed in a regular square array as shown in Fig. 19.20. For the special case of the droplet size much larger than P (of interest in this study), a droplet contacts the flat-top of the pillars forming the composite interface, and the cavities are filled with air. For this case, $f_{LA} = 1 - \frac{\pi D^2}{4P^2} = 1 - f_{SL}$. Further assume that the flat tops are smooth with $R_f = 1$. The contact angles for the Wenzel and Cassie–Baxter regimes are given by (19.6) and (19.9) [18].

$$\text{Wenzel:} \quad \cos\theta = \left(1 + \frac{\pi DH}{P^2}\right)\cos\theta_0 \tag{19.23}$$

$$\text{Cassie–Baxter:} \quad \cos\theta = \frac{\pi D^2}{4P^2}(\cos\theta_0 + 1) - 1. \tag{19.24}$$

Geometrical parameters of the flat-top, cylindrical pillars in series 1 and 2 are used for calculating the contact angle for the above-mentioned two cases. Figure 19.21 shows the plot of the predicted values of the contact angle as a function of

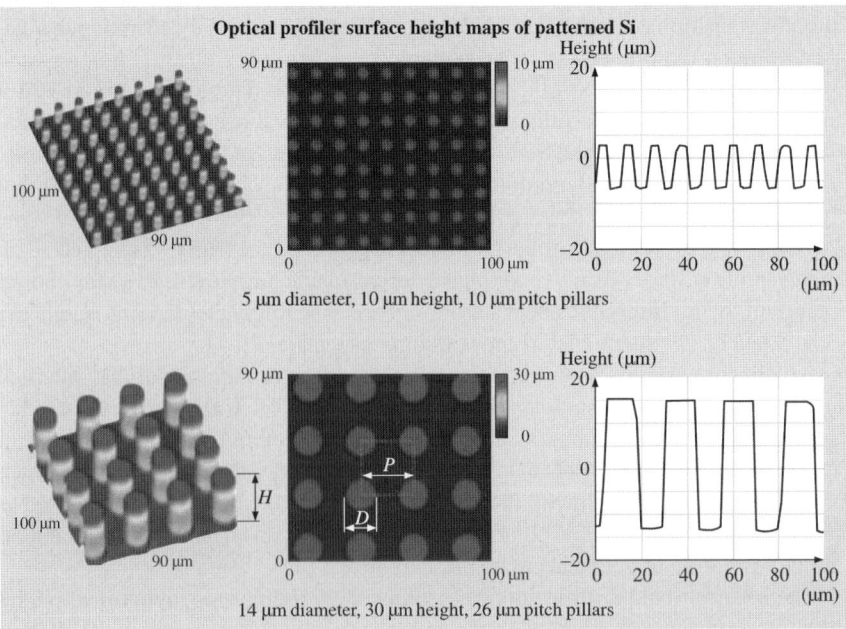

Fig. 19.20. Surface height maps and 2-D profiles of the patterned surfaces using an optical profiler. [18]

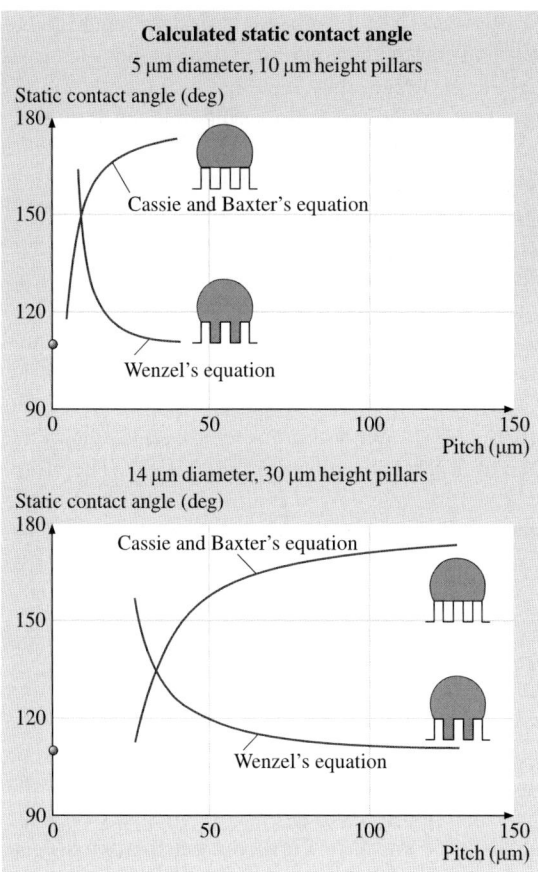

Fig. 19.21. Calculated static contact angle as a function of geometric parameters for a given value of θ_0 using the Wenzel and Cassie–Baxter equations for two series of the patterned surfaces with different pitch values [18]

pitch between the pillars for the two cases. The Wenzel and Cassie–Baxter equations present two possible equilibrium states for a water droplet on the surface. This indicates that there is a critical pitch below which the composite interface dominates and above which the homogeneous interface dominates the wetting behavior. Therefore, one needs to find the critical point that can be used to design the superhydrophobic surfaces. Furthermore, even in cases where the liquid droplet does not contact the bottom of the cavities, the water droplet can be in a metastable state and can become unstable and with the transition from the Cassie–Baxter to Wenzel regime occurring if the pitch is large.

Curvature-based Cassie-Wenzel Transition Criteria

A stable composite interface is essential for the successful design of superhydrophobic surfaces. However, the composite interface is fragile and it may transform into the homogeneous interface. What triggers the transition between the regimes remains a subject of arguments, although a number of explanations have been suggested. *Nosonovsky* and *Bhushan* [100] have studied destabilizing factors for the composite interface and found that convex surface (with bumps) leads to a stable interface and high contact angle. Also, they have been suggested the effects of droplet's weight and curvature among the factors which affect the transition.

Bhushan and *Jung* [18, 19] and *Jung* and *Bhushan* [68, 69] investigated the effect of droplet curvature on the Cassie-Wenzel regime transition. First, they considered a small water droplet suspended on a superhydrophobic surface consisting of a regular array of circular pillars with diameter D, height H, and pitch P as shown in Fig. 19.22. The local deformation for small droplets is governed by surface effects rather than gravity. The curvature of a droplet is governed by the Laplace equation, which relates the pressure inside the droplet to its curvature [2]. Therefore, the curvature is the same at the top and at the bottom of the droplet [79, 101]. For the patterned surface considered here, the maximum droop of the droplet occurs in the center of the square formed by the four pillars as shown in Fig. 19.22a. Therefore, the maximum droop of the droplet (δ) in the recessed region can be found in the middle of two pillars which are diagonally across as shown in Fig. 19.22b, which is $(\sqrt{2}P - D)^2/(8R)$. If the droop is much greater than the depth of the cavity,

$$(\sqrt{2}P - D)^2 / R \geq H \qquad (19.25)$$

then the droplet will just contact the bottom of the cavities between pillars, resulting into the transition from the Cassie–Baxter to Wenzel regime. Furthermore, in the case of large distances between the pillars, the liquid-air interface can easily be destabilized due to dynamic effects, such as surface waves that are formed at the liquid-air interface due to the gravitational or capillary forces. This leads to the formation of the homogeneous solid-liquid interface. However, whether the droplet droop or other mechanisms dominate the transition, remains to be investigated.

Contact Angle Measurements

The initial experiment performed with 1 mm in radius (5 µL volume) on the patterned Si coated with PF_3 was to determine the static contact angle [18, 19, 68, 69]. The contact angles on the prepared surfaces are plotted as a function of pitch between the pillars in Fig. 19.23a. A dotted line represents the transition criteria range obtained using (19.25). The flat Si coated with PF_3 showed the static contact angle of 109°. As the pitch increases up to 45 µm of series 1 and 126 µm of series 2, the static contact angle first increases gradually from 152° to 170°. Then, the contact angle starts decreasing sharply. Initial increase with an increase of pitch has to do with more open air space present which increases the propensity of air pocket formation. In the series 1, the value predicted from the curvature transition criteria (19.25) is a little higher than the experimental observations. However, in the series 2, there is a good agreement between the experimental data and the theoretically predicted by *Jung* and *Bhushan* [68, 69] values for the Cassie–Wenzel transition.

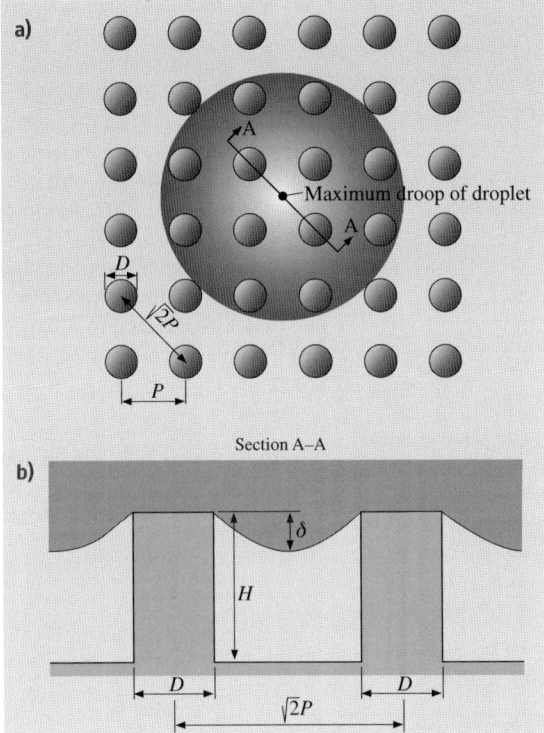

Fig. 19.22. A small water droplet suspended on a superhydrophobic surface consisting of a regular array of circular pillars. (**a**) Plan view. The maximum droop of droplet occurs in the center of square formed by four pillars. (**b**) Side view in section A-A. The maximum droop of droplet (δ) can be found in the middle of two pillars which are diagonally across [68, 69]

Figure 19.23b shows hysteresis and tilt angle as a function of pitch between the pillars [18, 19]. The flat Si coated with PF_3 showed a hysteresis angle of 34° and tilt angle of 37°. The patterned surfaces with low pitch increase the hysteresis and tilt angles compared to the flat surface due to the effect of sharp edges on the pillars, resulting into pinning [97]. Hysteresis for a flat surface can arise from roughness and surface heterogeneity. For a droplet moving down on the inclined patterned surfaces, the line of contact of the solid, liquid and air will be pinned at the edge point until it will be able to move, resulting into increasing hysteresis and tilt angles. Figure 19.24 shows droplets on patterned Si with 5 μm diameter and 10 μm height pillars with different pitch values. The asymmetrical shape of the droplet signifies pinning. The pinning on the patterned surfaces can be observed as compared to the flat surface. The patterned surface with low pitch (7 μm) has more the pinning than the patterned surface with high pitch (37.5 μm), because the patterned surface with low pitch has more sharp edges contacting with a droplet.

For various pitch values, hysteresis and tilt angles show the same trends with varying pitch between the pillars. After an initial increase as discussed above, they gradually decrease with increasing pitch (due to reduced number of sharp edges) and show an abrupt minimum in the value which has the highest contact angle. The lowest hysteresis and tilt angles are 5° and 3°, respectively, which were observed

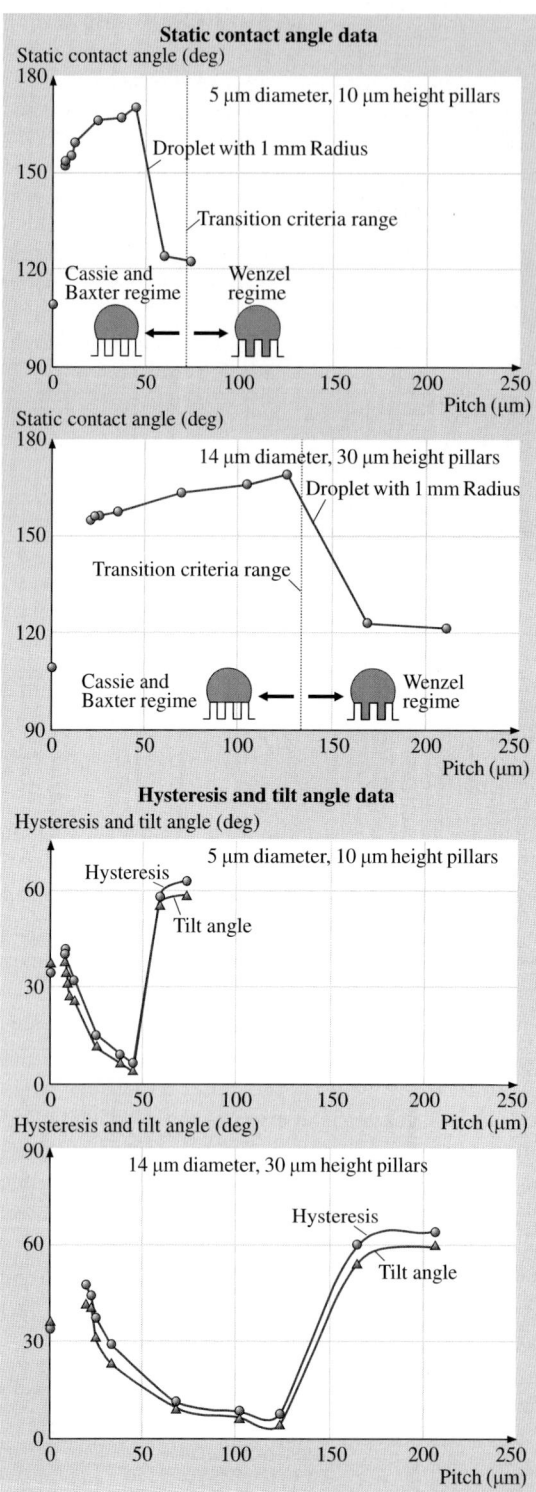

Fig. 19.23. (a) Static contact angle (*a dotted line* represents the transition criteria range obtained using (19.25) and (b) hysteresis and tilt angles as a function of geometric parameters for two series of the patterned surfaces with different pitch values for a droplet with 1 mm in radius (5 µL volume). Data at zero pitch correspond to a flat sample [18, 69]

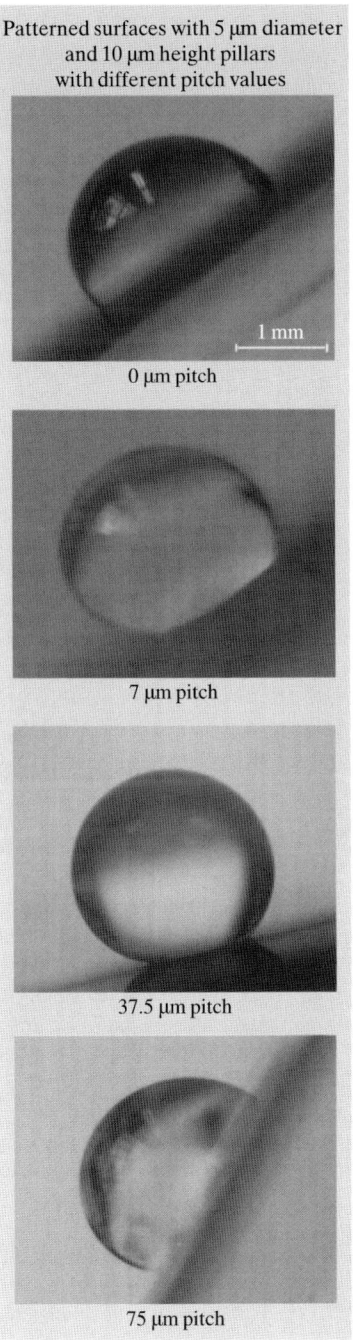

Patterned surfaces with 5 μm diameter and 10 μm height pillars with different pitch values

0 μm pitch

7 μm pitch

37.5 μm pitch

75 μm pitch

Fig. 19.24. Optical micrographs of droplets on the inclined patterned surfaces with different pitch values. The images were taken when the droplet started to move down. Data at zero pitch correspond to a flat sample [18]

on the patterned Si with 45 μm of series 1 and 126 μm of series 2. As discussed earlier, an increase in the pitch value allows the formation of composite interface. At higher pitch values, it is difficult to form the composite interface. The decrease in hysteresis and tilt angles occurs due to formation of composite interface at pitch values raging from 7 μm to 45 μm in series 1 and from 21 μm to 126 μm in series 2. The hysteresis and tilt angles start to increase again due to lack of formation of air pockets at pitch values raging from 60 μm to 75 μm in series 1 and from 168 μm to 210 μm in series 2. These results suggest that the air pocket formation and the

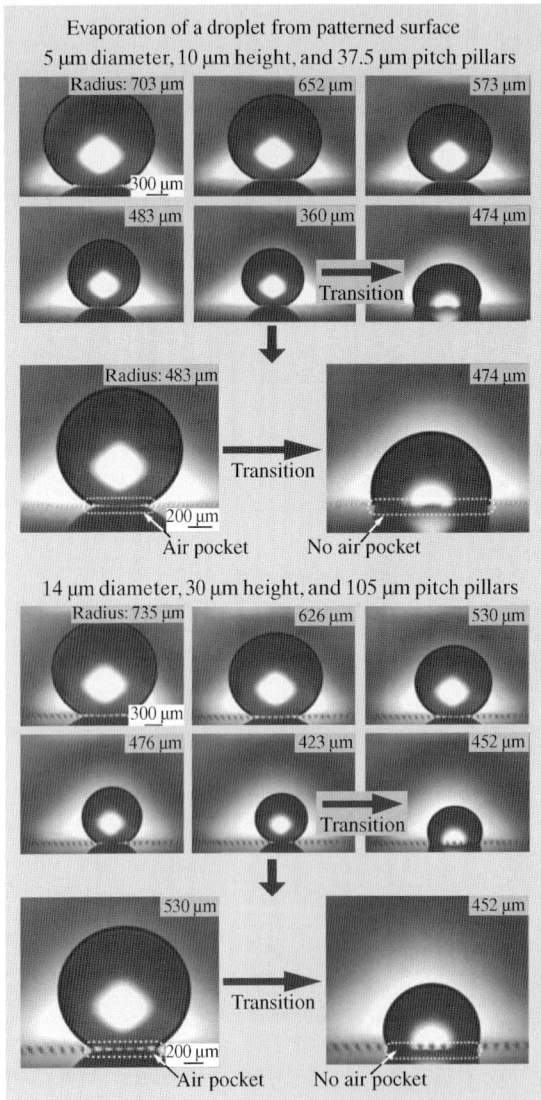

Fig. 19.25. Evaporation of a droplet on two different patterned surfaces. The initial radius of the droplet is about 700 μm, and the time interval between successive photos is 30 s. As the radius of droplet reaches 360 μm on the surface with 5 μm diameter, 10 μm height, and 37.5 μm pitch pillars, and 420 μm on the surface with 14 μm diameter, 30 μm height, and 105 μm pitch pillars, the transition from the Cassie–Baxter regime to Wenzel regime occurs, as indicated by the arrow. Before the transition, air pocket is clearly visible at the bottom area of the droplet, but after the transition, air pocket is not found at the bottom area of the droplet [69]

reduction of pinning in the patterned surface play an important role for a surface with both low hysteresis and tilt angle [18, 19]. Hence, to create superhydrophobic surfaces, it is important that they are able to form a stable composite interface with air pockets between solid and liquid. Capillary waves, nanodroplet condensation, hydrophilic spots due to chemical surface inhomogeneity, and liquid pressure can destroy the composite interface. *Nosonovsky* and *Bhushan* [100, 101, 103–106] suggested that these factors which make the composite interface unstable have different characteristic length scales, so nanostructures or the combination of microstructures and nanostructures is required to resist them.

Observation of Transition During the Droplet Evaporation

Jung and *Bhushan* [68, 69] performed the droplet evaporation experiments to observe the Cassie–Wenzel transition on two different patterned Si surfaces coated with PF_3. The series of six images in Fig. 19.25 show the successive photos of a droplet evaporating on the two patterned surfaces. The initial radius of the droplet was about 700 µm, and the time interval between successive photos was 30 s. In the first five photos, the droplet is shown in a Cassie–Baxter state, and its size gradually decreases with time. However, as the radius of the droplet reached 360 µm on the surface with 5 µm diameter, 10 µm height, and 37.5 µm pitch pillars, and 423 µm on the surface with 14 µm diameter, 30 µm height, and 105 µm pitch pillars, the transition from the Cassie–Baxter to Wenzel regime occurred, as indicated by the arrow. Figure 19.25 also shows a zoom-in of water droplets on two different patterned Si surfaces coated with PF_3 before and after the transition. The light passes below the left droplet, indicating that air pockets exist, so that the droplet is in the Cassie–Baxter state. However, an air pocket is not visible below the bottom right droplet, so it is in the Wenzel state. This could result from an impalement of the droplet in the patterned surface, characterized by a smaller contact angle.

To find the contact angle before and after the transition, the values of the contact angle are plotted against the theoretically predicted value, based on the Wenzel (calculated using (19.6)) and Cassie–Baxter (calculated using (19.9)) models. Figure 19.26 shows the static contact angle as a function of geometric parameters for the experimental contact angles before (circle) and after (triangle) the transition compared with the Wenzel and Cassie–Baxter equations (solid lines) with a given value of θ_0 for two series of the patterned Si with different pitch values coated with PF_3 [69]. The fit is good between the experimental data and the theoretically predicted values for the contact angles before and after transition.

To prove the validity of the transition criteria in terms of droplet size, the critical radius of droplet deposited on the patterned Si with different pitch values coated with PF_3 is measured during the evaporation experiment [68, 69]. Figure 19.27 shows the radius of a droplet as a function of geometric parameters for the experimental results (circle) compared with the transition criterion (19.5) from the Cassie–Baxter regime to Wenzel regime (solid lines) for two series of the patterned Si with different pitch values coated with PF_3. It is found that the critical radius of impalement is in good quantitative agreement with our predictions. The critical radius of the droplet

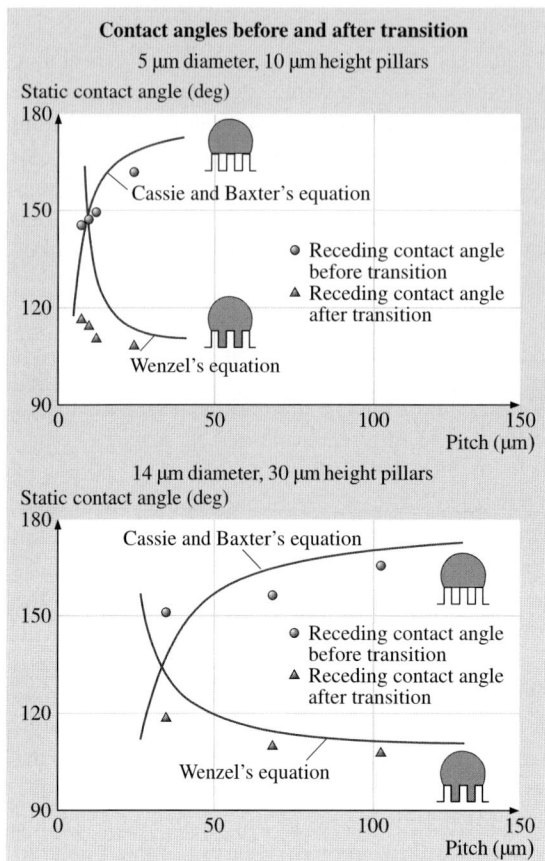

Fig. 19.26. Receding contact angle as a function of geometric parameters before *(circle)* and after *(triangle)* transition compared with predicted static contact angle values obtained using the Wenzel and Cassie–Baxter equations *(solid lines)* with a given value of θ_0 for two series of the patterned surfaces with different pitch values [69]

increases linearly with the geometric parameter (pitch). For the surface with small pitch, the critical radius of droplet can become quite small.

To verify the transition, *Jung* and *Bhushan* [68,69] used another approach using the dust mixed in water. Figure 19.28 presents the dust trace remaining after droplet with 1 mm radius (5 μL volume) evaporation on the patterned Si surface with pillars of 5 μm diameter and 10 μm height with 37.5 μm pitch in which the transition occurred at 360 μm radius of the droplet, and with 7 μm pitch in which the transition occurred at about 20 μm radius of the droplet during the process of evaporation. As shown in the top image, after the transition from the Cassie–Baxter regime to Wenzel regime, the dust particles remained not only at the top of the pillars but also at the bottom with a footprint size of about 450 μm. However, as shown in the bottom image, the dust particles remained on only a few pillars until the end of the evaporation process. The transition occurred at about 20 μm radius of droplet and the dust particles left a footprint of about 25 μm. From Fig. 19.27, it is observed that the transition occurs at about 300 μm radius of droplet on the 5 μm diameter and 10 μm height pillars with 37.5 μm pitch, but the transition does not occur on the patterned

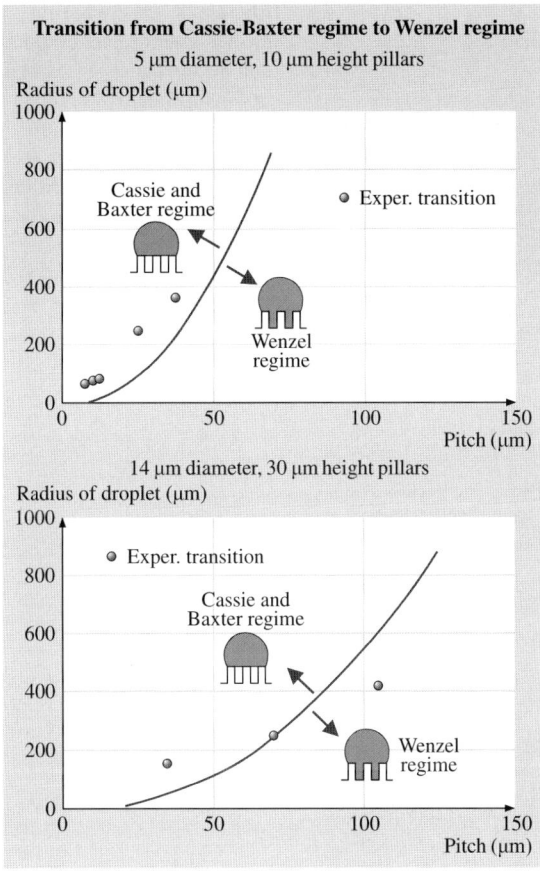

Fig. 19.27. Radius of droplet as a function of geometric parameters for the experimental results (*circle*) compared with the transition criteria from the Cassie–Baxter regime to Wenzel regime (*solid lines*) for two series of the patterned surfaces with different pitch values [69]

Si surface with pitch of less than about 5 μm. These experimental observations are consistent with model predictions. In the literature, it has been shown that on superhydrophobic natural lotus, the droplet remains in the Cassie–Baxter regime during the evaporation process [142]. This indicates that the distance between the pillars should be minimized enough to improve the ability of the droplet to resist sinking.

Scaling of the Cassie-Wenzel Transition for Different Series

Nosonovsky and *Bhushan* [101–104, 106] studied the data for the Cassie–Wenzel transition with the two series of the surfaces using the non-dimensional spacing factor

$$S_f = \frac{D}{P}. \tag{19.26}$$

The values of the droplet radius at which the transition occurs during the evaporation plotted against the spacing factor scale well for the two series of the experimental results, yielding virtually the same straight line. Thus the two series of the

Fig. 19.28. Dust trace remained after droplet evaporation for the patterned surface. In the top image, the transition occurred at 360 μm radius of droplet, and in the bottom image, the transition occurred at about 20 μm radius of droplet during the process of droplet evaporation. The footprint size is about 450 and 25 μm for the top and bottom images, respectively [69]

patterned surfaces scale well with each other and the transition occurs at the same value of the spacing factor multiplied by the droplet radius (Fig. 19.29). The physical mechanism leading to this observation remains to be determined, however, it is noted that this mechanism is different from the one suggested by 19.25. The observation suggests that the transition is a linear "1D" phenomenon and that neither droplet droop (that would involve P^2/H) nor droplet weight (that would involve R^3) are responsible for the transition, but rather linear geometric relations are involved. Note that the experimental values approximately correspond to the values of the ratio $RD/P = 50\,\mu m$, or the total area of the pillar tops under the droplet $(\pi D^2/4)\pi R^2/P^2 = 6200\,\mu m^2$.

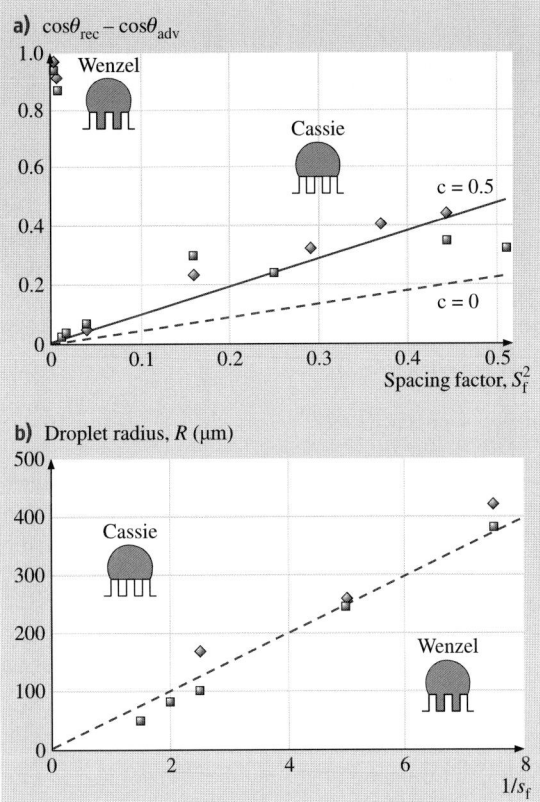

Fig. 19.29. (a) contact angle hysteresis as a function of S_f for the 1st (*squares*) and 2nd (*diamonds*) series of the experiments compared with the theoretically predicted values of $\cos\theta_{adv} - \cos\theta_{rec} = (D/P)^2(\pi/4)(\cos\theta_{adv0} - \cos\theta_{rec0}) + c(D/P)^2$, where c is a proportionality constant. It is observed that when only the adhesion hysteresis/interface energy term is considered ($c = 0$), the theoretical values are underestimated by about a half, whereas $c = 0.5$ provides a good fit. Therefore, the contribution of the adhesion hysteresis is of the same order of magnitude as the contribution kinetic effects. (b) Droplet radius, R, for the Cassie–Wenzel transition as a function of $P/D = 1/S_f$. It is observed that the transition takes place at a constant value of $RD/P \sim 50\,\mu m$ (*dashed line*). This shows that the transition is a linear phenomenon

Contact Angle Hysteresis and Wetting/Dewetting Asymmetry

The contact angle hysteresis can be viewed as a result of two factors that act simultaneously. First, the changing contact area affects the hysteresis, since a certain value of the contact angle hysteresis is inherent for even nominally flat surface. Decreasing the contact area by increasing the pitch between the pillars leads to a proportional decrease of the hysteresis. This effect is clearly proportional to the contact area between the solid surface and the liquid droplet. Second, edges of the pillar tops prevent the motion of the triple line. This roughness effect is proportional to the contact line density and its contribution was, in the experiment, comparable with the contact area effect. Interestingly, the effect of the edges is much more significant for the advancing than for the receding contact angle.

Nosonovsky and *Bhushan* [101–104, 106] studied wetting of two series of patterned Si surfaces with different pitch values coated with PF$_3$ based on the spacing factor (19.26). They found that the contact angle hysteresis involves two terms: the term $S_f^2(\pi/4)(\cos\theta_{adv0} - \cos\theta_{rec0})$ corresponding to the adhesion hysteresis (which is found even at a nominally flat surface and is a result of molecular-scale imperfectness) and the term $H_r \propto D/P^2$ corresponding to microscale roughness and proportional to the edge line density. Thus the contact angle hysteresis is given, based on (19.19), by [16, 100]

$$\cos\theta_{adv} - \cos\theta_{rec} = \frac{\pi}{4}S_f^2(\cos\theta_{adv0} - \cos\theta_{rec0}) + H_r. \tag{19.27}$$

Besides the contact angle hysteresis, the asymmetry of the Wenzel and Cassie states is the result of the wetting/dewetting asymmetry. While fragile metastable Cassie state is often observed, as well as its transition to the Wenzel state, the opposite transition never happens. Using (19.6) and (19.9) the contact angle with the patterned surfaces is given by [16, 100]

$$\cos\theta = \left(1 + 2\pi S_f^2\right)\cos\theta_0 \quad \text{(Wenzel state)} \tag{19.28}$$

$$\cos\theta = \frac{\pi}{4}S_f^2(\cos\theta_0 + 1) - 1 \quad \text{(Cassie state)}. \tag{19.29}$$

For a perfect macroscale system, the transition between the Wenzel and Cassie states should occur only at the intersection of the two regimes (the point at which the contact angle and net energies of the two regimes are equal, corresponding to $S_f = 0.51$). It is observed, however, that the transition from the metastable Cassie to stable Wenzel occurs at much lower values of the spacing factor $0.083 < S_f < 0.111$. As shown in Fig. 19.30a, the stable Wenzel state (i) can transform into the stable Cassie state with increasing S_f (ii). The metastable Cassie state (iii) can abruptly transform (iv) into the stable Wenzel state. The transition (i–ii) corresponds to equal Wenzel and Cassie states free energies, whereas the transition (iv) corresponds to the Wenzel energy much lower than the Cassie energy and thus involves significant energy dissipation and is irreversible. The solid and dashed straight lines correspond to the values of the contact angle, calculated from (19.28)–(19.29) using the contact angle for a nominally flat surface, $\theta_0 = 109°$. The two series of the experimental data are shown with squares and diamonds.

Figure 19.30b shows the values of the advancing contact angle plotted against the spacing factor (19.26). The solid and dashed straight lines correspond to the values of the contact angle for the Wenzel and Cassie states, calculated from (19.28)–(19.29) using the advancing contact angle for a nominally flat surface, $\theta_{adv0} = 116°$. It is observed that the calculated values underestimate the advancing contact angle, especially for big S_f (small distance between the pillars or pitch P). This is understandable, because the calculation takes into account only the effect of the contact

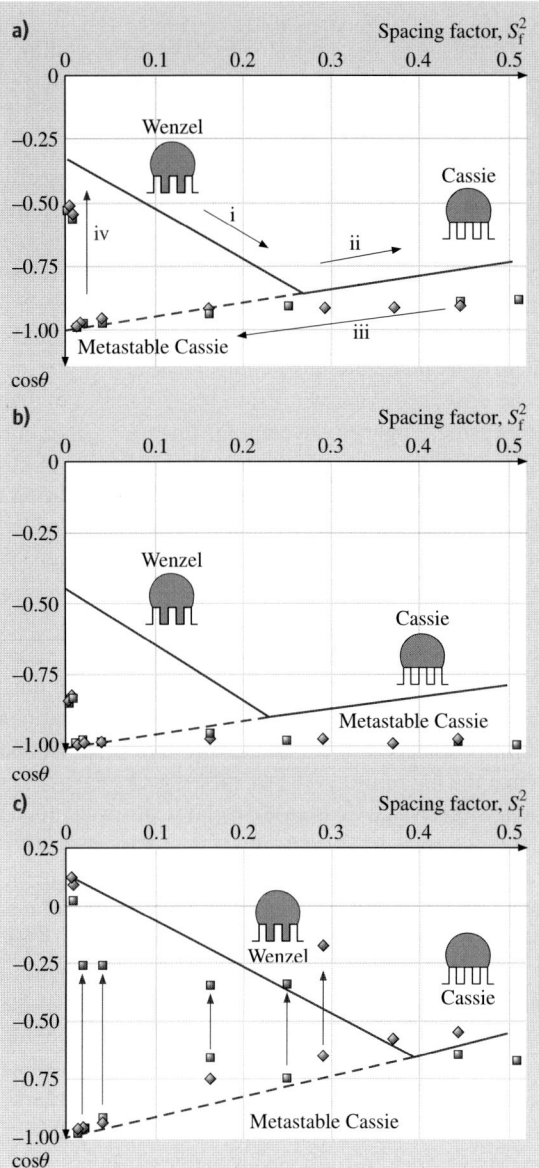

Fig. 19.30. Theoretical (*solid and dashed*) and experimental (*squares* for the 1st series, *diamonds* for the 2nd series) (**a**) contact angle as a function of the spacing factor. (**b**) Advancing contact angle (**c**) Receding contact angle and values of the contact angle observed after the transition during evaporation (*blue*)

area and ignores the effect of roughness and edge line density (it corresponds to $H_r = 0$ in (19.27)), while this effect is more pronounced for high pillar density (big S_f). In a similar manner, the contact angle is underestimated for the Wenzel state, since the pillars constitute a barrier for the advancing droplet.

Figure 19.30c shows the values of the contact angle after the transition took place (dimmed blue squares and diamonds), as it was observed during evaporation.

For the both series, the values almost coincided. For comparison, the values of the receding contact angle measured for millimeter-sized water droplets are also shown (squares and diamonds), since evaporation constitutes removing liquid and thus the contact angle during evaporation should be compared with the receding contact angle. The solid and dashed straight lines correspond to the values of the contact angle, calculated from (19.28)–(19.29) using the receding contact angle for a nominally flat surface, $\theta_{rec0} = 82°$. Figure 19.30c demonstrates a good agreement between the experimental data and (19.28)–(19.29).

In the analysis of the evaporation data of micropatterned surfaces, they found several effects specific for the multiscale character of this process. First, they discussed applicability of the Wenzel and Cassie equations for average surface roughness and heterogeneity. These equations relate the local contact angle with the apparent contact angle of a rough/heterogeneous surface. However, it is not obvious what should be the size of roughness/heterogeneity averaging, since the triple line at which the contact angle is defined has two very different scale lengths: its width is of molecular size scale while its length is of the order of the size of the droplet (that is, microns or millimeters). They presented an argument that in order for the averaging to be valid, the roughness details should be small compared to the size of the droplet (and not the molecular size). They showed that while for the uniform roughness/heterogeneity the Wenzel and Cassie equations can be applied, for a more complicated case of the non-uniform heterogeneity, the generalized equations should be used. The proposed generalized Cassie–Wenzel equations are consistent with a broad range of available experimental data. The generalized equations are valid both in the cases when the classical Wenzel and Cassie equations can be applied as well as in the cases when the later fail.

The macroscale contact angle hysteresis and Cassie–Wenzel transition cannot be determined from the macroscale equations and are governed by micro- and nanoscale effects, so wetting is a multiscale phenomenon [101–104, 106]. The kinetic effects associated with the contact angle hysteresis should be studied at the microscale, whereas the effects of the adhesion hysteresis and the Cassie-Wenzel transition involve processes at the nanoscale. Their theoretical arguments are supported by the experimental data on micropatterned surfaces. The experimental study of the contact angle hysteresis demonstrates that two different processes are involved: the changing solid-liquid area of contact and pinning of the triple line. The later effect is more significant for the advancing than for the receding contact angle. The transition between wetting states was observed for evaporating microdroplets and droplet radius scales well with the geometric parameters of the micropattern.

Observation and Measurement of Contact Angle Using ESEM

Figure 19.31 shows how water droplets grow and merge under ESEM [69]. ESEM was used as a contact angle analysis tool to investigate superhydrophobicity on the patterned surfaces. Microdroplets (with the diameter less than 1 mm) were distributed on the patterned surface coated with PF_3 during increasing condensation by decreasing temperature. Even if the microdroplets were not of the same size, they

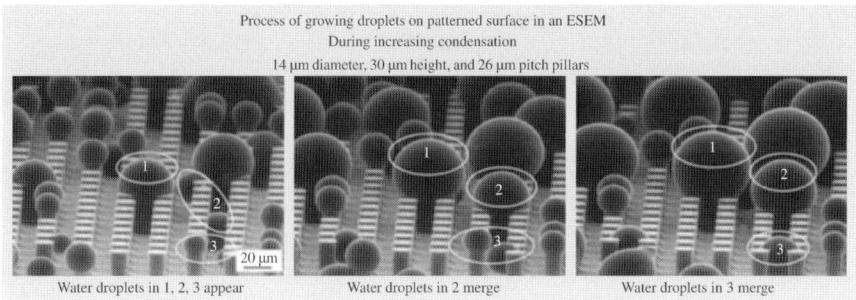

Fig. 19.31. Microdroplet (in dimension of less than 1 mm diameter) growing and merging process under ESEM during increasing condensation by decreasing temperature. *Left image*: Some small water droplets appear at the beginning, i.e. water droplets 1, 2, 3. *Middle image*: Water droplets at locations 1 and 3 increase in size and water droplets at location 2 merge together to form one big droplet. *Right image*: Water droplets at locations 1 and 2 increase in size and water droplets at location 3 merge together to form one big droplet [69]

showed hydrophobic characteristics of the patterned surface. At the beginning, some small water droplets appeared, i.e. water droplets at locations 1, 2 and 3 in the left image. During increasing condensation by decreasing temperature, water droplets at locations 1 and 3 gradually increased in size and water droplets at location 2 merged together to form one big droplet in the middle image. With further condensation, water droplets at locations 1 and 2 increased in size and water droplets at location 3 merge together to one big droplet in right image. In all cases condensation was initiated at the bottom, therefore, as can be observed, the droplets are in the Wenzel regime. This could also be evidence that the droplet on the macroscale used in the conventional contact angle measurement comes from the merging of smaller droplets.

Compared with the conventional contact angle measurement, ESEM is able to provide detailed information about the contact angle of microdroplets on patterned surfaces. The diameter of the water droplets used for the contact angle measurement was more than 10 μm, so that the size limit pointed out by *Stelmashenko* et al. [125] was avoided. For droplet size smaller than 1 μm, substrate backscattering can distort the intensity profile such that the images are inaccurate.

As shown in Fig. 19.32, the static contact angle and hysteresis angle of the microdroplets condensed on flat and two different patterned surfaces were obtained from the images and corrected using methodology mentioned earlier. The difference between the data estimated from the images and corrected θ is about 3%. Once the microdroplet's condensation and evaporation has reached a dynamic equilibrium, static contact angles are determined. The flat Si coated with PF_3 showed a static contact angle of 98°. The patterned surfaces coated with PF_3 increase the static contact angle compared to the flat surface coated with PF_3 due to the effect of roughness. Advancing contact angle was taken after increasing condensation by decreasing the temperature of the cooling stage. Receding contact angle was taken after increasing

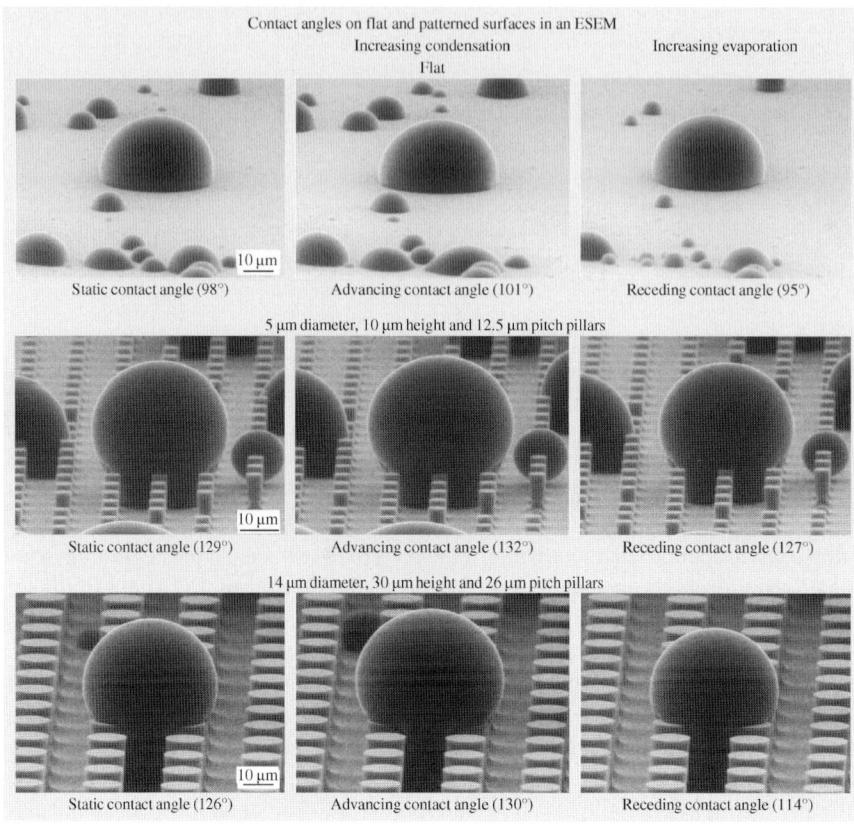

Fig. 19.32. Microdroplets on flat and two patterned surfaces using ESEM. Second set of images were taken during increasing condensation, and the third set of images were taken during increasing evaporation. Static contact angle was measured when the droplet was stable. Advancing contact angle was measured after increasing condensation by decreasing the temperature of the cooling stage. Receding contact angle was measured after decreasing evaporation by increasing the temperature of the cooling stage [69]

evaporation by increasing the temperature of the cooling stage. The hysteresis angle was then calculated [69].

Figure 19.33 shows hysteresis angle as a function of geometric parameters for the microdroplets formed in the ESEM (triangle) for two series of the patterned Si with different pitch values coated with PF_3. Data at zero pitch correspond to a flat Si sample. The droplets with about 20 μm radii, which are larger than the pitch, were selected in order to look at the effect of pillars in contact with the droplet. This data were compared with conventional contact angle measurements obtained with the droplet with 1 mm radius (5 μL volume) (circle and solid lines) [18]. When the distance between pillars increases above a certain value, the contact area be-

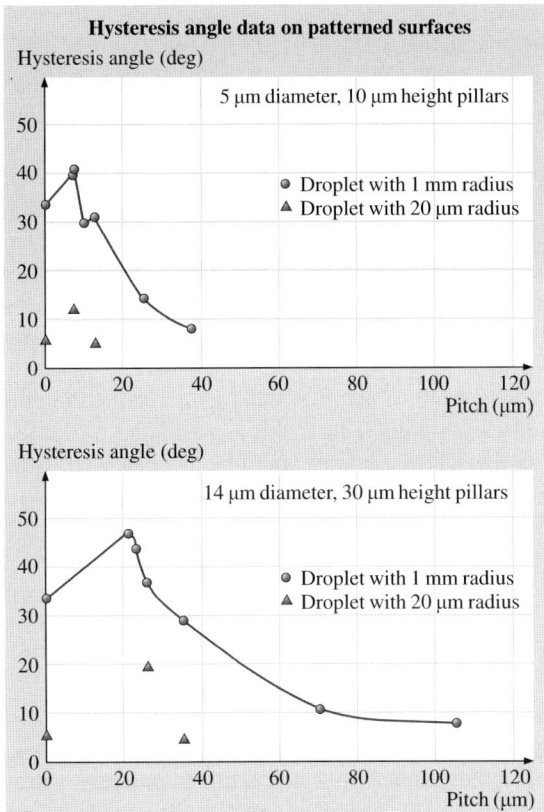

Fig. 19.33. Hysteresis angle as a function of geometric parameters for the microdroplet with about 20 μm radius from ESEM (*triangle*) compared with the droplet with 1 mm radius (5 μL volume) (*circle* and *solid lines*) for two series of the patterned surfaces with different pitch values. Data at zero pitch correspond to a flat sample [69]

tween the patterned surface and the droplet decreases, resulting in the decrease of the hysteresis angle. Both the droplets with 1 mm and 20 μm radii show the same trend. The hysteresis angles for the patterned surfaces with low pitch are higher compared to the flat surface due to the effect of sharp edges on the pillars, resulting in pinning [97]. Hysteresis for a flat surface can arise from roughness and surface heterogeneity. For a droplet advancing forward on the patterned surfaces, the line of contact of the solid, liquid and air will be pinned at the edge point until it is able to move, resulting in increasing hysteresis angle. The hysteresis angle for the microdroplet from ESEM is lower as compared to that for the droplet with 1 mm radius. The difference of hysteresis angle between a microdroplet and a droplet with 1 mm radius could come from the different pinning effects, because the latter has more sharp edges contacting with a droplet compared with the former. The results show how droplet size can affect the wetting properties of patterned Si surfaces [69].

19.4.4 Self-Cleaning

The self-cleaning abilities of patterned surfaces were investigated by *Furstner* et al. [50]. They studied Si wafer specimens with regular patterns of spikes manufactured by X-ray lithography. The specimens were hydrophobized with Au thiol. For comparison, they studied also replicas of plant surfaces, made by a two-component silicon molding mass applied onto the leaf's surface. The negative replica is flexible and rubber-like. Onto this mold a melted hydrophobic polyether was applied. They also studied several metal foil specimens, hydrophobized by means of a fluorinated agent. In order to investigate self-cleaning, a luminescent and hydrophobic powder was used as a contaminant. Following contamination, the specimens were subjected to artificial fog and rain.

Drops of water rolled off easily from Si samples with a microstructure consisting of rather slender and sufficiently high spikes; this is attributed to the fact that the Cassie wetting state existed. These samples could be cleaned after artificial contamination by means of fog treatment almost completely. On surfaces with low spikes and a rather high pitch the behavior of water drops was different; that is, we found a considerable decrease of the contact angles and a distinct rise in the sliding angles; apparently, corresponding to the Wenzel state. Some metal foils and some replicates had two levels of roughness. These specimens did not show a total removal of all contaminating particles when they were subjected to artificial fog, but water drops impinging with sufficient kinetic energy could clean them perfectly. A substrate without structures smaller than 5 µm could not be cleaned by means of fog consisting of water droplets with diameter 8–20 µm because this treatment resulted in a continuous water film on the samples. However, artificial rain removed all the contamination. On the other hand, smooth specimens made of the same material could not be cleaned completely by impinging droplets. This is a clear indication of the different contact phenomena on smooth hydrophobic in contrast to self-cleaning microstructured surfaces. Another interesting observation of this group was that despite the missing structure of the wax crystals, the water contact angle of the Lotus replica was the highest of all the replicates, indicating that the microstructure formed by the papillae alone is already optimized with regard to water repellency [50].

19.5 Role of Hierarchical Roughness for Superhydrophobicity

Natural water-repellent surfaces such as the lotus leaf or water strider leg [51] are hierarchical. It is recognized that the hierarchical structure is beneficial for the superhydrophobicity. However, the functionality of this hierarchical roughness remains a subject of discussions, and several explanations have been suggested, including the simple ideas that the large bumps serve for structural toughness while the small ones decrease the solid-liquid contact area [51] or that the large bumps allow to maintain the composite interface while the small ones enhance the contact angle in accordance to the Wenzel model [52]. *Furstner* et al., [50] pointed out that artificial surfaces with one level of roughness can repel well large "artificial rain" droplets, however, they cannot repel small "artificial fog" droplets trapped in the valleys between the bumps, so the hierarchy may have to do with the ability to repel droplets of various size ranges.

Nosonovsky and *Bhushan* [100, 101, 103–106] showed that the mechanisms involved into the superhydrophobicity are scale-dependent with effects at various scale ranges acting simultaneously, and thus the roughness must be hierarchical in order to respond to these mechanisms. They also suggested that the small roughness can pin the composite interface and thus prevent undesirable Cassie–Wenzel transition [94, 95, 100, 101]. For most superhydrophobic surfaces, it is important that a composite solid-air-liquid interface is formed. The composite interface dramatically reduces the area of solid-liquid contact and, therefore, reduces adhesion of a liquid droplet to the solid surface and contact angle hysteresis. Formation of a composite interface is a multiscale phenomenon, which depends upon relative sizes of the liquid droplets and roughness details. The transition from a composite interface to a homogeneous interface is irreversible; therefore, stability of a composite interface is crucial for superhydrophobicity and should be addressed for successful development of superhydrophobic surfaces. *Nosonovsky* and *Bhushan* [100, 101, 103, 104, 106] have demonstrated that a multiscale (hierarchical) roughness can help to resist the destabilization, with small convex surfaces pinning the interface and thus leading to stable equilibrium as well as preventing from filling the gaps between the pillars even in the case of a hydrophilic material.

Regarding the size of the micro- and nanostructures, the following considerations can be made. The structure of ideal hierarchical surface is shown in Fig. 19.34. The asperities should be high enough so that the droplet does not touch the valleys. For a structure with circular pillars, the following relationship should hold for a composite interface, $(\sqrt{2}P-D)^2/R < H$, (19.25)). As an example, for a droplet with a radius on the order of 1 mm or larger, a value of H on the order of 30 μm, D on the order of 15 μm, a P on the order of 130 μm (Fig. 19.23) is optimum. Nanoasperities can pin the liquid-air interface and thus prevent liquid from filling the valleys between asperities. They are also required to support nanodroplets, which may condense in the valleys between large asperities. Therefore, nanoasperities should have a small pitch to handle nanodroplets, less than 1 mm down to few nm radius. The values of h on the order of 10 nm and d on the order of 100 nm can be easily fabricated.

19.6 How to Make a Superhydrophobic Surface

Fabrication of superhydrophobic surfaces has been an area of active research since mid 1990s. In general, the same techniques that are used for micro- and nanostructure fabrication, such as the lithography, etching, and deposition, have been utilized for producing superhydrophobic surfaces (Fig. 19.35, Table 19.6). Pros and cons of these techniques are summarized in Table 19.7. Among especially interesting development is the creation of switchable surfaces that can be turned from hydrophobic to hydrophilic by applying electric potential, heat or ultraviolet (UV) irradiation [49, 80, 84, 126, 135]. Another important requirement for potential applications

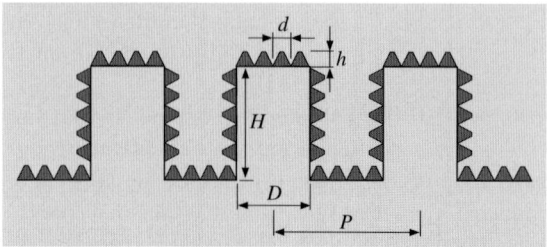

Fig. 19.34. Schematic of structure of an ideal hierarchical surface. Microasperities consist of the circular pillars with diameter D, height H, and pitch P. Nanoasperities consist of pyramidal nanoasperities of height h and diameter d with rounded tops

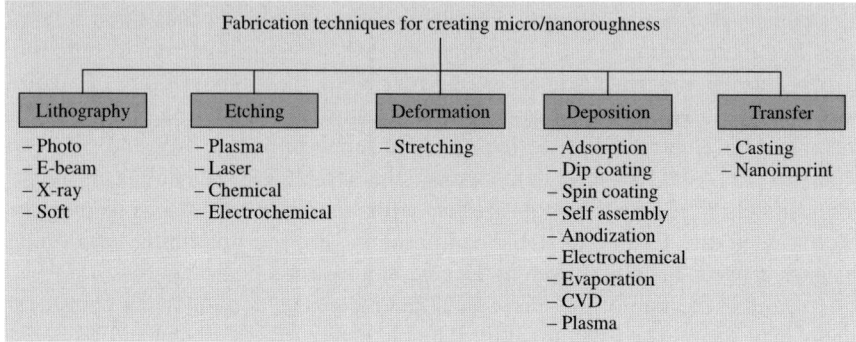

Fig. 19.35. Typical methods to fabricate micro/nanoroughened surface

for optics and self-cleaning glasses is creation of transparent superhydrophobic surfaces. In order for the surface to be transparent, roughness details should be smaller than the wavelength of the visible light (about 400–700 nm) [91].

Two main requirements for a superhydrophobic surface are that the surface should be rough and that it should have a hydrophobic (low surface energy) coating. These two requirements lead to two methods of producing a superhydrophobic surface: first, it is possible to make a rough surface from an initially hydrophobic material and, second, to modify a rough hydrophilic surface by modifying surface chemistry or applying a hydrophobic material upon it. Note that roughness is usually a more critical property than the low surface energy, since both moderately hydrophobic and very hydrophobic materials can exhibit similar wetting behavior when roughened.

19.6.1 Roughening to Create One-level Structure

Lithography is a well-established technique, applied for creating large area of periodic micro/nanopatterns. It includes photo, E-beam, X-ray, and soft lithography. *Bhushan* and *Jung* [18] produced patterned Si using photolithography. To obtain a sample that is hydrophobic, a SAM of −1,1,−2,2,-tetrahydroperfluorodecyl-

Table 19.6. Typical materials and corresponding techniques to produce micro/nanoroughness

Material	Technique	Contact angle	Notes	Source
Teflon	Plasma	168		Zhang et al. [140]; Shiu et al. [123]
Fluorinated block polymer solution	Casting under humid environment	160	Transparent	Yabu and Shimomura [136]
PFOS	Electro- and chemical polymerization	152	Reversible (electric potential)	Xu et al. [135]
PDMS	Laser treatment	166		Khorasani et al. [72]
PS-PDMS Block copolymer	Electrospining	>150		Ma et al. [85]
PS, PC, PMMA	evaporation	>150		Bormashenko et al. [25]
PS nanofiber	Nanoimprint	156		Lee et al. [82]
PET	Oxygen plasma etching	>150		Teshima et al. [129]
Organo-triethoxysilanes	Sol-gel	155	Reversible (temperature)	Shirtcliffe et al. [122]
Al	Chemical etching	>150		Qian and Shen [113]
Copper	Electrodeposition	160	Hierarchical	Shirtcliffe et al. [121]
Si	Photolithography	170		Bhushan and Jung [18]
Si	E-beam lithography	164		Martines et al. [88]
Si	X-ray lithography	>166		Furstner et al. [50]
Si	Casting	158	Plant leaf replica	Sun et al. [127]; Furstner et al. [50]
Si (Black Si)	Plasma etching	>150	For liquid flow	Jansen et al. [64]
Silica	Sol-gel	150		Hikita et al. [58]; Shang et al. [118]

Table 19.6. (continued) Typical materials and corresponding techniques to produce micro/nanoroughness

Material	Technique	Contact angle	Notes	Source
Polyelectrolyte multilayer surface overcoated with silica nanoparticles	Self assembly	168		*Zhai* et al. [139]
Nano-silica spheres	Dip coating	105		*Klein* et al. [75]
Silica colloidal particles in PDMS	Spin coated	165	Hierarchical	*Ming* et al. [90]
Au clusters	Electrochemical deposition	>150		*Zhang* et al. [141]
Carbon nanotubes	CVD	159		*Huang* et al. [60]
ZnO, TiO$_2$ Nanorods	Sol-gel	>150	Reversible (UV irradiation)	*Feng* et al. [49]

Table 19.7. Pros and cons of various fabrication techniques

Techniques	Pros	Cons
Lithography	Accuracy, large area	Slow process, high cost
Etching	Fast	Chemical contamination, less control
Deposition	Flexibility, cheap	Can be high temperature, less control

trichlorosilane (PF_3) was deposited on the sample surfaces using vapor phase deposition technique. They obtained a superhydrophobic surface with the contact angle up to 170°. *Martines* et al. [88] fabricated ordered arrays of nanopits and nanopillars by using electron beam lithography. They obtained a superhydrophobic surface with the contact angle of 164° and hysteresis of 1° for a surface consisting of tall pillars with cusped tops after a hydrophobization with octadecyltrichlorosilane (OTS). *Furstner* et al. [50] created silicon wafers with regular patterns of spikes by X-ray lithography. The wafer was hydrophobized by sputtering a layer of gold and subsequent immersion in a hexadecanethiol solution. *Jung* and *Bhushan* [67] created low aspect ratio asperities (LAR, 1:1 height-to-diameter ratio), high aspect ratio asperities (HAR, 3:1 height-to-diameter ratio), and lotus pattern (replica from the lotus leaf), all on a PMMA surface using soft lithography. A self-assembled monolayer

(SAM) of perfluorodecyltriethoxysilane (PFDTES) was deposited on the patterned surfaces using vapor phase deposition technique.

One well-known and effective way to make rough surfaces is etching using either plasma, laser, chemical or electrochemical techniques [84]. *Jansen* et al. [64] etched a silicon wafer using a fluorine-based plasma by utilizing the black silicon method to obtain isotropic, positively and negatively tapered as well as vertical walls with smooth surfaces. *Coulson* et al. [38] described an approach in plasma chemical roughening of poly(tetrafluoroethylene) (PTFE) substrates followed by the deposition of low surface energy plasma polymer layers, which give rise to high repellency towards polar and nonpolar probe liquids. A different approach was taken by *Shiu* et al. [123], who treated a Teflon film with oxygen plasma and obtained a superhydrophobic surface with contact angle of 168°. Fluorinated materials have a limited solubility, which makes it difficult to roughen them. However, they may be linked or blended with other materials, which are often easier to roughen, in order to make superhydrophobic surfaces. *Teshima* et al. [129] obtained a transparent superhydrophobic surface from a poly(ethylene terephthalate) (PET) substrate via selective oxygen plasma etching followed by plasma-enhanced chemical vapor deposition using tetramethylsilane (TMS) as the precursor. *Khorasani* et al. [72] produced porous PDMS surfaces with the contact angle of 175° using CO_2-pulsed laser etching method as an excitation source for surface. *Qian* and *Shen* [113] described a simple surface roughening method by dislocation selective chemical etching on polycrystalline metals such as aluminum. After treatment with fluoroalkylsilane, the etched metallic surfaces exhibited superhydrophobicity. *Xu* et al. [135] fabricated a reversible superhydrophobic surface with a double-roughened perfluorooctanesulfonate (PFOS) doped conducting polypyrrole (PPy) film by a combination of electropolymerization and chemical polymerization. The reversibility was achieved by switching between superhydrophobic doped or oxidized states and superhydrophilicity dedoped or neutral states with changing the applied electrochemical potential.

Stretching method can be used to produce a superhydrophobic surface. *Zhang* et al. [140] stretched a Teflon film and converted it into fibrous crystals with a large fraction of void space in the surface, leading to high roughness and the superhydrophobicity.

Deposition methods also make a substrate rough from the bulk properties of the material and enlarge potential applications of superhydrophobic surfaces. There are several ways to make a rough surface including adsorption, dip coating, electrospinning, anodization, electrochemical, evaporation, chemical vapor deposition (CVD), and plasma. Solidification of wax can be used to produce a superhydrophobic surface. *Shibuichi* et al. [120] used alkylketene dimer (AKD) wax on a glass plate to spontaneously form a fractal structure in its surfaces. They obtained the surface with a contact angle larger than 170° without any fluorination treatments. *Klein* et al. [75] obtained superhydrophobic surfaces by simply dip-coating a substrate with a slurry containing nano-silica spheres, which adhered to substrate after a low temperature heat treatment. After reaction of the surface with a fluoroalkyltrichlorosilane, the hydrophobicity increased with decreasing area fraction of spheres. Ma et al. [85] produced block copolymer poly(styrene-b-dimethylsiloxane) fibers with submicrometer diameters in the range 150–400 nm by electrospinning from solution in tetrahydrofuran and dimethylformamide. They obtained superhydrophobic nonwoven fibrous mats with a contact angle of 163°. *Shiu* et al. [123] showed that self-organized close-packed superhydrophobic surfaces can be easily achieved by spin-coating the monodispersed polystyrene beads solution on substrate surfaces. The sizes of beads were reduced by controlling the etching conditions. After plasma treatment, the surfaces were coated with a layer of gold and eventually a layer of octadecanethiol SAM to render hydrophobicity. *Abdelsalam* et al. [1] studied the wetting of structured gold surfaces formed by electrodeposition through a template of submicrometer spheres and discussed the role of the pore size and shape in controlling wetting. *Bormashenko* et al. [25] used evaporated polymer solutions of polystyrene (PS), polycarbonate (PC) and polymethylmethacrylate (PMMA) dissolved in chlorinated solvents: dichloromethane (CH_2Cl_2) and chloroform ($CHCl_3$) to obtain self-assembled structure with hydrophobic properties. Chemical/physical vapor deposition (CVD/PVD) has been used for the modification of surface chemistry as well. *Lau* et al. [81] created superhydrophobic carbon nanotube forests by modifying the surface of vertically aligned nanotubes with plasma enhanced chemical vapor deposition (PECVD). Superhydrophobicity was achieved down to the microscopic level where essentially spherical, micrometer-sized water droplets can be suspended on top of the nanotube forest. *Zhu* et al. [145] and *Huang* et al. [60] prepared surfaces with two-scale roughness by controlled growth of carbon nanotube (CNT) arrays by CVD. *Zhao* et al. [144] also synthesized the vertically aligned multiwalled carbon nanotube (MWCNT) arrays by chemical vapor deposition on Si substrates using thin film of iron (Fe) as catalyst layer and aluminum (Al) film.

Attempts to create superhydrophobic surface by casting and nanoimprint methods have been successful. *Yabu* and *Shimomura* [136] prepared a porous superhydrophobic transparent membrane by casting a fluorinated block polymer solution under humid environment. The transparency was achieved because the honeycomb-patterned films had sub-wavelength pores size. *Sun* et al. [127] reported a nanocasting method to make superhydrophobic PDMS surface. They first made a negative PDMS template using lotus leaf as an original template and then used the negative template to make a positive PDMS template–a replica of the original lotus leaf. *Zhao* et al. [143] prepared a superhydrophobic surface by casting a micellar solution of a copolymer poly(styrene-b-dimethylsiloxane) (PS-PDMS) in humid air based on the cooperation of vapor-induced phase separation and surface enrichment of PDMS block. *Lee* et al. [82] produced vertically aligned PS nanofibers by using nanoporous anodic aluminum oxide as a replication template in a heat- and pressure-driven nanoimprint pattern transfer process. As the aspect ratio of the polystyrene (PS) nanofibers increased, the nanofibers could not stand upright but formed twisted bundles resulting in a three-dimensionally rough surface with advancing and receding contact angles of 155.8° and 147.6°, respectively.

19.6.2 Coating to Create One-level Hydrophobic Structures

Modifying the surface chemistry with a hydrophobic coating widens the potential applications of superhydrophobic surfaces. There are several ways to modify the chemistry of a surface including sol-gel, dip coating, self-assembly, electrochemical and chemical/physical vapor deposition. *Shirtcliffe* et al. [122] prepared porous sol-gel foams from organo-triethoxysilanes which exhibited switching between superhydrophobicity and superhydrophilicity when exposed to different temperatures. The critical switching temperature was between 275°C and 550°C for different materials, and when the foam was heated above the critical temperature, complete rejection of water in the cavities to complete filling of the pores. *Hikita* et al. [58] used colloidal silica particles and fluoroalkylsilane as the starting materials and prepared a sol-gel film with superliquid-repellency by hydrolysis and condensation of alkoxysilane compounds. *Feng* et al. [49] produced superhydrophobic surfaces using ZnO nanorods by Sol-gel method. They showed that superhydrophobic surfaces can be switched into hydrophilic surfaces by alternation of ultraviolet (UV) irradiation. *Shang* et al. [118] did not blend low surface energy materials in the sols, but described a procedure to make transparent superhydrophobic surface by modifying silica based gel films with a fluorinated silane. In a similar way, *Wu* et al. [134] made a microstructured ZnO-based surface via a wet chemical process and obtained the superhydrophobicity after coating the surface with long-chain alkanoic acids.

Zhai et al. [139] used an layer-by-layer (LBL) self-assembly technique to create a poly(allylamine hydrochloride)/poly(acrylic acid) (PAH/PAA) multilayer which formed a honeycomb-like structure on the surface after an appropriate combination of acidic treatments. After cross-linking the structure, they deposited silica nanoparticles on the surface via alternating dipping of the substrates into an aqueous suspension of the negatively charged nanoparticles and an aqueous PAH solution, followed by a final dipping into the nanoparticle suspension. Superhydrophobicity was obtained after the surface was modified by a chemical vapor deposition of (tridecafluoro-1, 1, 2, 2-tetrahydrooctyl)-1-trichlorosilane followed by a thermal annealing.

Zhang et al. [141] showed that the surface covered with dendritic gold clusters, which was formed by electrochemical deposition onto indium tin oxide (ITO) electrode modified with a polyelectrolyte multilayer, showed superhydrophobic properties after further deposition of a n-dodecanethiol monolayer. *Han* et al. [55] described the fabrication of lotus leaf-like superhydrophobic metal surfaces by using electrochemical reaction of Cu or Cu–Sn alloy plated on steel sheets with sulfur gas, and subsequent perfluorosilane treatment. The chemical bath deposition (CBD) has also been used to make nanostructured surfaces, thus, *Hosono* et al. [59] fabricated a nanopin film of brucite-type cobalt hydroxide (BCH) and achieved the contact angle of 178° after further modification of lauric acid (LA). *Shi* et al. [119] described the use of galvanic cell reaction as a facile method to chemically deposit Ag nanostructures on the *p*-silicon wafer on a large scale. When the Ag covered silicon wafer was further modified with a self-assembled monolayer of *n*-dodecanethiol, a superhydrophobic surface was obtained with a contact angle of about 154° and a tilt angle lower than 5°.

19.6.3 Methods to Create Two-Level (Hierarchical) Superhydrophobic Structures

A two-level (hierarchical) roughness structures are typical for superhydrophobic surfaces in nature, as it was discussed above. Recently, many efforts have been devoted to fabricate these hierarchical structures in various ways. *Shirtcliffe* et al. [121] prepared a hierarchical (double-roughened) copper surface by electrodeposition from acidic copper sulfate solution onto flat copper and patterning technique of coating with a fluorocarbon hydrophobic layer. Another way to obtain a rough surface for superhydrophobicity is the assembly from colloidal systems. *Ming* et al. [90] prepared a hierarchical (double roughened) surface consisting of silica-based raspberry-like particles which were made by covalently grafting amine-functionalized silica particles of 70 nm to epoxy-functionalized silica particles of 700 nm via the reaction between epoxy and amine groups. The surface became superhydrophobic after being modified with PDMS. *Northen* and *Turner* [93] fabricated arrays of flexible silicon dioxide plat forms supported by single high aspect ratio silicon pillars down to 1 μm in diameter and with heights up to ~ 50 μm. When these platforms were coated with polymeric organorods of approximately 2 μm tall and 50–200 nm in diameter, it showed that the surface is highly hydrophobic with a water contact angle of 145°. *Chong* et al. [37] showed the combination of the porous anodic alumina (PAA) template with microsphere monolayers to fabricate hierarchically ordered nanowire arrays, which have periodic voids at the microscale and hexagonally packed nanowires at the nanoscale. They created by selective electrodeposition using nanoporous anodic alumina as a template and a porous gold film as a working electrode that is patterned by microsphere monolayers. *Wang* et al. [132] also developed a novel precursor hydrothermal redox method with $Ni(OH)_2$ as the precursor to fabricate a hierarchical structure of nickel hollow microspheres with nickel nanoparticles as the in situ formed building units. The created hierarchical hollow structure exhibited enhanced coercivity and remnant magnetization as compared with hollow nickel submicrometer spheres, hollow nickel nanospheres, bulk nickel, and free Ni nanoparticles. *Kim* et al. [74] fabricated a hierarchical structure, which looks like the same structures as lotus leaf. First, the nanoscale porosity was generated by anodic aluminum oxidation and then, the anodized porous alumina surface was replicated by polytetrafluoroethylene. The polymer sticking phenomenon during the replication created the sub-microstructures on the negative polytetrafluoroethylene nanostructure replica. The contact angle of the created hierarchical structure was obtained about 160° and the tilting angle is less than 1°. *Del Campo* and *Greiner* [40] reported that SU-8 hierarchical patterns comprising features with lateral dimensions ranging from 5 mm to 2 mm and heights from 10 to 500 um were obtained by photolithography which comprises a step of layer-by-layer exposure in soft contact printed shadow masks which are embedded into the SU-8 multilayer.

19.7 Closure

In this chapter, the theoretical basis of superhydrophobicity and characterization of natural and artificial superhydrophobic surfaces are presented. While theoretical foundations of wetting of rough surfaces have been developed decades ago, emerging applications and the need to design non-sticky surfaces has led to the intensive modeling and experimental research in the area. In particular, such problems as the transition between the Wenzel and Cassie–Baxter wetting regimes, contact angle hysteresis, the role of hierarchical roughness, possibility of creation of reversible hydrophobicity have been investigated very actively in the past decade. Various experimental and theoretical approaches to understand these phenomena are presented. These findings provide new insights on the fundamental mechanisms of wetting and can lead to creation of successful non-adhesive surfaces.

On the application side, there are several ways to manufacture superhydrophobic surfaces, and new methods continue to emerge. Some methods (such as the lithography) allow scientists to create patterned surfaces with clearly defined and controlled geometrical features. These features have typical size ranging from 1 μm to 100 μm. Other (and often cheaper) methods lead to self-assembled or random rough surfaces. These methods include extending, etching, polymer solution evaporation, sol-gel, etc. There is a technology available to produce transparent superhydrophobic materials, hierarchical surfaces and reversible surfaces which can change from hydrophobic to hydrophilic under an external control. The difference in the superhydrophobic properties of surfaces with a patterned and random structure still has to be investigated.

A proper control of roughness constitutes the main challenge in producing a reliable superhydrophobic surface. If the initial material is hydrophilic, a surface treatment or coating is required, which will decrease the surface energy. While the two factors–roughness and low surface energy–are required for superhydrophobicity, the role of roughness clearly dominates. The function of the hierarchical roughness still remains to be investigated as well as the mechanisms that trigger the Cassie-Wenzel regime transitions.

References

1. Abdelsalam ME, Bartlett PN, Kelf T and Baumberg J, Wetting of regularly structured gold surfaces, *Langmuir*, **21** (2005) 1753–1757
2. Adamson AV, 1990, *Physical Chemistry of Surfaces*, Wiley, NY
3. Anisimov MA, Divergence of Tolman's Length for a Droplet Near the Critical Point, *Phys. Rev. Lett.*, **98** (2007) 035702
4. Bahadur V and Garimella SV, Electrowetting-Based Control of Static Droplet States on Rough Surfaces, *Langmuir*, **23** (2007) 4918–4924
5. Baker EA (1982) Chemistry and Morphology of Plant Epicuticular Waxes in *The Plant Cuticle* (Eds. Cutler DF, Alvin KL and Price CE, Academic Press, 1982), 139–165
6. Barbieri L, Wagner E and Hoffmann P, Water Wetting Transition Parameters of Perfluorinated Substrates with Periodically Distributed Flat-Top Microscale Obstavles, *Langmuir*, **23** (2007) 1723–1734

7. Bartell FE and Shepard JW, Surface Roughness as Related to Hysteresis of Contact Angles *J. Phys. Chem.*, **57** (1953) 455–458
8. Barthlott W and Neinhuis C, Purity of the Sacred Lotus, or Escape from Contamination in Biological Surfaces, *Planta*, **202** (1997) 1–8
9. Bartolo D, Bouamrirene F, Verneuil E, Buguin A, Silberzan P and Moulinet S, Bouncing or sticky droplets: Impalement transitions on superhydrophobic micropatterned surfaces, *Europhys. Lett.*, **74** (2006) 299–305
10. Bhushan B, Adhesion and Stiction: Mechanisms, Measurement Techniques and Methods for Reduction, *J. Vac. Sci. Technol.* B **21** (2003) 2262–2296
11. Bhushan B, 1996, *Tribology and Mechanics of Magnetic Storage Systems*, 2nd ed., Springer-Verlag, New York
12. Bhushan B, 1998, *Tribology Issues and Opportunities in MEMS*, Kluwer Academic Publishers, Dordrecht, Netherlands
13. Bhushan B, 1999, *Principles and Applications of Tribology*, Wiley, NY
14. Bhushan B, 2002, *Introduction to Tribology*, Wiley, NY
15. Bhushan B, 2005, *Nanotribology and Nanomechanics–An Introduction*, Springer-Verlag, Heidelberg, Germany
16. Bhushan B, 2007, *Springer Handbook of Nanotechnology*, 2nd ed., Springer-Verlag, Heidelberg, Germany
17. Bhushan B and Jung YC, Micro and Nanoscale Characterization of Hydrophobic and Hydrophilic Leaf Surface, *Nanotechnology*, **17** (2006) 2758–2772
18. Bhushan B and Jung YC, Wetting study of patterned surfaces for superhydrophobicity, *Ultramicroscopy*, **107** (2007) 1033–1041
19. Bhushan B and Jung YC, Wetting, Adhesion and Friction of Superhydrophobic and Hydrophilic Leaves and Fabricated Micro/nanopatterned Surfaces, *J. Phys.: Condens. Matter* (2008, in press)
20. Bhushan B, Israelachvili JN and Landman U, Nanotribology: Friction, Wear and Lubrication at the Atomic Scale, *Nature*, **374** (1995) 607–616
21. Bhushan B, Hansford D and Lee KK, Surface Modification of Silicon and Polydimethylsiloxane Surfaces with Vapor-Phase-Deposited Ultrathin Fluorosilane Films for Biomedical Nanodevices, *J. Vac. Sci. Technol. A*, **24** (2006) 1197–1202
22. Bhushan B, Nosonovsky M and Jung YC, Towards Optimization of Patterned Superhydrophobic Surfaces *J. Royal Soc. Interface*, **4** (2007) 643–648
23. Bico J, Marzolin C and Quèrè D, Pearl drops, *Europhys. Lett.*, **47** (1999) 220–226
24. Bico J, Thiele U and Quèrè D, Wetting of Textured Surfaces, *Colloids and Surfaces A*, **206** (2002) 41–46
25. Bormashenko E, Stein T, Whyman G, Bormashenko Y and Pogreb E, Wetting Properties of the Multiscaled Nanostructured Polymer and Metallic Superhydrophobic Surfaces, *Langmuir*, **22** (2006) 9982–9985
26. Bormashenko E, Bormashenko Y, Stein T, Whyman G, Pogreb R and Barkay Z, Environmental Scanning Electron Microscope Study of the Fine Structure of the Triple Line and Cassie-Wenzel Wetting Transition for Sessile Drops Deposited on Rough Polymer Substrates, *Langmuir*, **23** (2007) 4378–4382
27. Bormashenko E, Pogreb R, Whyman G and Erlich M, Cassie-Wenzel Wetting Transition in Vibrated Drops Deposited on the Rough Surfaces: Is Dynamic Cassie-Wenzel Transition 2D or 1D Affair? *Langmuir*, **23** (2007) 6501–6503
28. Boruvka L and Neumann AW, Generalization of the Classical Theory of Capillarity, *J. Chem. Phys.*, **66** (1977) 5464–5476
29. Bourges-Monnier C and Shanahan MER, Influence of Evaporation on Contact Angle, *Langmuir*, **11** (1995) 2820–2829

30. Brugnara M, Della Volpe C, Siboni S and Zeni D, Contact Angle Analysis on Polymethylmethacrylate and Commercial Wax by Using an Environmental Scanning Electron Microscope, *Scanning,* **28** (2006) 267–273
31. Burton Z and Bhushan B, Hydrophobicity, Adhesion, and Friction Properties of Nanopatterned Polymers and Scale Dependence for Micro- and Nanoelectromechanical Systems, *Nano Lett.,* **5** (2005) 1607–1613
32. Burton Z and Bhushan B, Surface Characterization and Adhesion and Friction Properties of Hydrophobic Leaf Surfaces, *Ultramicroscopy,* **106** (2006) 709–719
33. Callies M and Quéré D, On water repellency, *Soft Matter,* **1** (2005) 55–61
34. Cassie A and Baxter S, Wettability of Porous Surfaces, *Trans. Faraday Soc.,* **40** (1944) 546–551
35. Checco A, Guenoun P and Daillant J, Nonlinear Dependence of the Contact Angle of Nanodroplets on Contact Line Curvatures, *Phys. Rev. Lett.,* **91** (2003)186101
36. Choi SE, Yoo PJ, Baek SJ, Kim TW, and Lee HH, An ultraviolet-curable mold for sub-100-nm lithography, *J. Am. Chem. Soc.,* **126** (2004) 7744–7745
37. Chong MAS, Zheng YB, Gao H, and Tan LK, Combinational template-assisted fabrication of hierarchically ordered nanowire arrays on substrates for device applications, *Appl. Phys. Lett.,* **89** (2006) 233104
38. Coulson SR, Woodward I, Badyal JPS, Brewer SA, and Willis C, Super-Repellent Composite Fluoropolymer Surfaces, *J. Phys. Chem. B,* **104** (2000) 8836–8840
39. Danilatos GD and Brancik JV, Observation of liquid transport in the ESEM, Proc. 44th Annual Meeting EMSA: (1986) 678–679
40. del Campo A, Greiner C, SU-8: a photoresist for high-aspect-ratio and 3D submicron lithography, *J. Micromech. Microeng.,* **17** (2007) R81–R95
41. de Gennes PG, Brochard-Wyart F, and Quéré D *Capillarity and Wetting Phenomena* (Springer, Berlin, 2003)
42. Derjaguin BV and Churaev NV, Structural Component of Disjoining Pressure, *J. Colloid Interface Sci.,* **49** (1974) 249–255
43. Erbil HY, McHale G, and Newton MI, Drop Evaporation on Solid Surfaces: Constant Contact Angle Mode, *Langmuir,* **18** (2002) 2636–2641
44. Erbil HY, Demirel AL, and Avci Y, Transformation of a Simple Plastic into a Superhydrophobic Surface, *Science,* **299** (2003) 1377–1380
45. Eustathopoulos N, Nicholas MG, Drevet B, 1999, *Wettability at High Temparatures,* Pergamon, Amsterdam
46. Extrand CW, Model for Contact Angle and Hysteresis on Rough and Ultraphobic Surfaces, *Langmuir,* **18** (2002) 7991–7999
47. Extrand CW, Contact Angle Hysteresis on Surfaces with Chemically Heterogeneous Islands, *Langmuir,* **19** (2003) 3793–3796
48. Extrand CW, Criteria for Ultralyophobic Surfaces, *Langmuir,* **20** (2004) 5013–5018
49. Feng XJ, Feng L, Jin MH, Zhai J, Jiang L, Zhu DB, Reversible Super-hydrophobicity to Super-hydrophilicity Transition of Aligned ZnO Nanorod Films, *J. Am. Chem. Soc.,* **126** (2004) 62–63
50. Furstner R, Barthlott W, Neinhuis C, and Walzel P, Wetting and Self-Cleaning Properties of Artificial Superhydrophobic Surfaces, *Langmuir,* **21** (2005) 956–961
51. Gao XF and Jiang L, Biophysics: Water-repellent Legs of Water Striders, *Nature,* **432** (2004) 36
52. Gao L and McCarthy TJ, The Lotus Effect Explained: Two Reasons Why Two Length Scales of Topography are Important, *Langmuir,* **22** (2006) 2966–2967
53. Gao L and McCarthy TJ, How Wenzel and Cassie Were Wrong, *Langmuir,* **23** (2007) 3762–3765

54. Gupta P, Ulman A, Fanfan F, Korniakov A., and Loos K., "Mixed Self-Assembled Monolayer of Alkanethiolates on Ultrasmooth Gold do not Exhibit Contact Angle Hysteresis," *J. Am. Chem. Soc.*, **127** (2005) 4-5
55. Han JT, Jang Y, Lee DY, Park JH, Song SH, Ban DY, and Cho K, Fabrication of a bionic superhydrophobic metal surface by sulfur-induced morphological development, *J. Mater. Chem.*, 15 (2005) 3089–3092
56. He B, Patankar NA, and Lee J, Multiple Equilibrium Droplet Shapes and Design Criterion for Rough Hydrophobic Surfaces, *Langmuir* **19** (2003) 4999–5003
57. Herminghaus S, Roughness-Induced Non-Wetting, *Europhys. Lett.*, 52 (2000) 165–170
58. Hikita M, Tanaka K, Nakamura T, Kajiyama T, and Takahara A, Superliquid-repellent surfaces prepared by colloidal silica nanoparticles covered with fluoroalkyl groups, *Langmuir*, **21** (2005) 7299–7302
59. Hosono E, Fujihara S, Honma I, and Zhou H, Superhydrophobic Perpendicular Nanopin Film by the Bottom-Up Process, *J. Am. Chem. Soc.*, **127** (2005) 13458–13459
60. Huang L, Lau SP, Yang HY, Leong ESP, and Yu SF, Stable Superhydrophobic Surface via Carbon Nanotubes Coated with a ZnO Thin Film, *J. Phys. Chem.*, **109** (2005) 7746–7748
61. Israelachvili JN, 1992, *Intermolecular and Surface Forces*, 2nd edition, Academic Press, London
62. Israelachvili JN and Gee ML, Contact angles on Chemically Heterogeneous Surfaces, *Langmuir*, **5** (1989) 288–289
63. Ishino C and Okumura K, Nucleation scenarios for wetting transition on textured surfaces: The effect of contact angle hysteresis, *Europhys. Lett.*, **76** (2006), 464–470
64. Jansen H, de Boer M, Legtenberg R, and Elwenspoek M, The black silicon method: a universal method for determining the parameter setting of a fluorine-based reactive ion etcher in deep silicon trench etching with profile control, *J. Micromech. Microeng.* **5** (1995) 115–120
65. Jetter R, Kunst L, and Samuels AL, Composition of plant cuticular waxes. In *Biology of the Plant Cuticle* (eds. M. Riederer and C. Müller), Oxford, Blackwell Publishing (2006), 145–181
66. Johnson RE and Dettre RH, 1964, Contact Angle Hysteresis, *Contact Angle, Wettability, and Adhesion, Adv. Chem. Ser.*, Vol. 43, Ed. By F. M. Fowkes American Chemical Society, Washington, D. C., 112–135
67. Jung YC and Bhushan B, Contact Angle, Adhesion, and Friction Properties of Micro- and Nanopatterned Polymers for Superhydrophobicity, *Nanotechnology*, **17** (2006) 4970–4980
68. Jung YC and Bhushan B, Wetting transition of water droplets on superhydrophobic patterned surfaces, *Scripta Mater.*, **57** (2007) 1057–1060
69. Jung YC and Bhushan B, Wetting Behavior During Evaporation and Condensation of Water Microdroplets on Superhydrophobic Patterned Surfaces *J. Micros.*, **229** (2008) 127–140
70. Kamusewitz H, Possart W, and Paul D, The Relation Between Young's Equilibrium Contact Angle and the Hysteresis on Rough Paraffin Wax Surfaces, *Colloids and Surfaces A: Physicochem. Eng. Aspects,* **156** (1999) 271–279
71. Kasai T, Bhushan B, Kulik G, Barbieri L, and Hoffmann P, Micro/nanotribological study of perfluorosilane SAMs for antistiction and low wear, *J. Vac. Sci. Technol. B,* **23** (2005) 995–1003
72. Khorasani MT, Mirzadeh H, and Kermani Z, Wettability of porous polydimethylsiloxane surface: morphology study, *Appl. Surf. Sci.*, **242** (2005) 339–345

73. Kijlstra J, Reihs K, and Klami A, Roughness and Topology of Ultra-Hydrophobic surfaces, *Colloids and Surfaces A: Physicochem. Eng. Aspects,* **206** (2002) 521–529
74. Kim D, Hwang W, Park HC, and Lee KH, Superhydrophobic Micro- and Nanostructures Based on Polymer Sticking, *Key Eng. Mat.*, **334-335** (2007) 897–900
75. Klein RJ, Biesheuvel PM, Yu BC, Meinhart CD, and Lange FF, Producing superhydrophobic surfaces with nano-silica spheres, *Z. Metallkd.*, **94** (2003) 377–380
76. Koch K, Dommisse A, and Barthlott W, Chemistry and crystal growth of plant wax tubules of Lotus (Nelumbo nucifera) and Nasturtium (Tropaeolum majus) leaves on technical substrates, *Crystl. Growth Des.* **6** (2006) 2571–2578
77. Koinkar VN and Bhushan B, Effect of scan size and surface roughness on microscale friction measurements, *J. Appl. Phys.*, **81** (1997) 2472–2479
78. Krupenkin TN, Taylor JA, Schneider TM, and Yang S, From Rolling Ball to Complete Wetting: The Dynamic Tuning of Liquids on Nanostructured Surfaces, *Langmuir*, **20** (2004) 3824–3827
79. Lafuma A and Quéré D, Superhydrophobic states, *Nature Materials*, **2** (2003) 457–460
80. Lahann J, Mitragotri S, Tran T, Kaido H, Sundaram J, Choi IS, Hoffer S, Somorjai GA, and Langer R, A reversibly Switching Surface, *Science*, **299** (2003) 371–374
81. Lau KKS, Bico J, Teo KBK, Chhowalla M, Amaratunga GAJ, Milne WI, McKinley GH, and Gleason KK, Superhydrophobic Carbon Nanotube Forests, *Nano Lett.*, **3** (2003) 1701–1705
82. Lee W, Jin M, Yoo W, and Lee J, Nanostructuring of a polymeric substrate with well-defined nanometer-scale topography and tailored surface wettability, *Langmuir*, **20** (2004) 7665–7669
83. Li W and Amirfazli A, A Thermodynamic Approach for Determining the Contact Angle Hysteresis for Superhydrophobic Surfaces, *J. Colloid. Interface Sci.,* **292** (2006) 195–201
84. Ma M and Hill RM, Superhydrophobic Surfaces, *Current Opinion in Colloid and Interface Science*, **11** (2006) 193–202
85. Ma M, Hill RM, Lowery JL, Fridrikh SV, and Rutledge GC, Electrospun poly(styrene-block-dimethylsiloxane) block copolymer fibers exhibiting superhydrophobicity, *Langmuir*, **21** (2005) 5549–5554
86. Marmur A, Wetting on Hydrophobic Rough Surfaces: to be Heterogeneous or Not to be? *Langmuir*, **19** (2003) 8343–8348
87. Marmur A, The Lotus Effect: Superhydrophobicity and Metastability, *Langmuir*, **20** (2004) 3517–3519
88. Martines E, Seunarine K, Morgan H, Gadegaard N, Wilkinson CDW, and Riehle MO, Superhydrophobicity and superhydrophilicity of regular nanopatterns, *Nano Lett.*, **5** (2005) 2097–2103
89. McHale G, Aqil S, Shirtcliffe NJ, Newton MI, and Erbil HY, Analysis of droplet evaporation on a superhydrophobic surface, *Langmuir,* **21** (2005) 11053–11060
90. Ming W, Wu D, van Benthem R, and de With G, Superhydrophobic films from raspberry-like particles, *Nano Lett.*, **5** (2005) 2298–2301
91. Nakajima A, Fujishima A, Hashimoto K, and Watanabe T, Preparation of Transparent Superhydrophobic Boehmite and Silica Films by Sublimation of Aluminum Acetylacetonate, *Adv. Mater.*, **11** (1999) 1365–1368
92. Neinhuis C, and Barthlott W, Characterization and Distribution of Water-Repellent, Self-Cleaning Plant Surfaces, *Annals of Botany*, **79** (1997) 667–677
93. Northen MT and Turner KL, A batch fabricated biomimetic dry adhesive, *Nanotechnology*, **16** (2005) 1159–1166

94. Nosonovsky M, Multiscale roughness and stability of superhydrophobic biomimetic interfaces, *Langmuir*, **23** (2007) 3157–3161
95. Nosonovsky M, Model for Solid-Liquid and Solid-Solid Friction for Rough Surfaces with Adhesion Hysteresis, *J. Chem. Phys.*, **126** (2007) 224701
96. Nosonovsky M, On the Range of Applicability of the Wenzel and Cassie Equations *Langmuir,* **23** (2007) 9919–9920
97. Nosonovsky M and Bhushan B, Roughness optimization for biomimetic superhydrophobic surfaces, *Microsyst. Technol.*, **11** (2005) 535–549
98. Nosonovsky M and Bhushan B, Stochastic model for metastable wetting of roughness-induced superhydrophobic surfaces, *Microsyst. Technol.*, **12** (2006) 231–237
99. Nosonovsky M and Bhushan B, Wetting of Rough Three-Dimensional Superhydrophobic Surfaces, *Microsyst. Technol.*, **12** (2006) 273–281
100. Nosonovsky M and Bhushan B, Hierarchical Roughness Makes Superhydrophobic Surfaces Stable, *Microelectronic Eng.,* **84** (2007) 382–386
101. Nosonovsky M and Bhushan B, Hierarchical roughness optimization for Biomimetic superhydrophobic surfaces, *Ultramicroscopy,* **107** (2007) 969–979
102. Nosonovsky M and Bhushan B, Biomimetic Superhydrophobic Surfaces: Multiscale Approach, *Nano Lett.*, **7** (2007) 2633–2637
103. Nosonovsky M and Bhushan B, Multiscale Friction Mechanisms and Hierarchical Surfaces in Nano- and Bio-Tribology, *Mater. Sci. Eng.:R,* **58** (2007) 162–193
104. Nosonovsky M and Bhushan B, Roughness-induced superhydrophobicity: a way to design non-adhesive surfaces, *J. Phys.: Condens. Matter*, (2008, in press)
105. Nosonovsky M and Bhushan B, Capillary effects and instabilities in nanocontacts, *Ultramicroscopy*, (2008, in press)
106. Nosonovsky M and Bhushan B, Patterned Non-Adhesive Surfaces: Superhydrophobicity and Wetting Regime Transitions, *Langmuir* **24** (2008) 1525–1533
107. Oner D and McCarthy TJ, Ultrahydrophobic surfaces. Effects of topography length scales on wettability, *Langmuir* **16** (2000) 7777–7778
108. Patankar NA, On the Modeling of Hydrophobic Contact Angles on Rough Surfaces, *Langmuir*, **19** (2003) 1249–1253
109. Patankar NA, Transition Between Superhydrophobic States on Rough Surfaces *Langmuir*, **20** (2004) 7097–7102
110. Patankar NA, Mimicking the Lotus Effect: Influence of Double Roughness Structures and Slender Pillars, *Langmuir*, **20** (2004) 8209–8213
111. Pompe T, Fery A, and Herminghaus S, Measurement of Contact Line Tension by Analysis of the Three-Phase Boundary with Nanometer Resolution, in *Apparent and Microscopic Contact Angles* (Drelich J, Laskowski JS, and Mittal KL, eds., VSP, Utrecht, 2000) 3–12
112. Poon CY and Bhushan B, Comparison of surface Roughness measurements by stylus profiler, AFM and non-contact optical profiler, *Wear*, **190** (1995) 76–88
113. Qian B and Shen Z, Fabrication of Superhydrophobic Surfaces by Dislocation-Selective Chemical Etching on Aluminum, Copper, and Zinc Substrates, *Langmuir*, **21** (2005) 9007–9009
114. Quėrė D, Non-Sticking Drops, *Rep. Prog. Phys.*, **68** (2005) 2495–2535
115. Reyssat M, Pepin A, Marty F, Chen Y, and Quėrė D, Bouncing transitions on microtextured materials, *Europhys. Lett.,* **74** (2006) 306–312
116. Rowan SM, Newton MI, and McHale G, Evaporation of Microdroplets and the Wetting of Solid Surfaces, *J. Phys. Chem.,* **99** (1995) 13268–13271
117. Semal S, Blake TD, Geskin V, de Ruijter ML, Castelein G, and De Coninck J, Influence of Surface Roughness on Wetting Dynamics, *Langmuir*, **15** (1999) 8765–8770

118. Shang HM, Wang Y, Limmer SJ, Chou TP, Takahashi K, and Cao GZ, Optically transparent superhydrophobic silica-based films, *Thin Solid Films*, **472** (2005) 37–43
119. Shi F, Song Y, Niu J, Xia X, Wang Z, and Zhang X, Facile Method To Fabricate a Large-Scale Superhydrophobic Surface by Galvanic Cell Reaction, *Chem. Mater.*, **18** (2006) 1365–1368
120. Shibuichi S, Onda T, Satoh N, and Tsujii K, Super-Water-Repellent Surfaces Resulting from Fractal Structure, *J.Phys. Chem.*, **100** (1996) 19512–19517
121. Shirtcliffe NJ, McHale G, Newton MI, Chabrol G, Perry CC, Dual-scale roughness produces unusually water-repellent surfaces, *Adv. Mater.*, **16** (2004) 1929–1932
122. Shirtcliffe NJ, McHale G, Newton MI, Perry CC, and Roach P, Porous materials show superhydrophobic to superhydrophilic switching, *Chem. Commun.*, (2005) 3135–3137
123. Shiu J, Kuo C, Chen P, and Mou C, Fabrication of Tunable Superhydrophobic Surfaces by Nanosphere Lithography, *Chem. Mater.*, **16** (2004) 561–564
124. Shuttleworth R and Bailey GLJ, The Spreading of a Liquid Over a Rough Solid, *Discussions of the Faraday Society*, **3** (1948) 16–22
125. Stelmashenko NA, Craven JP, Donald AM, Terentjev EM, and Thiel BL, Topographic contrast of partially wetting water droplets in environmental scanning electron microscopy, *J. Micros.*, **204** (2001) 172–183
126. Sun T, Wang G, Feng L, Liu B, Ma Y, Jiang L, and Zhu D, Reversible Switching between Superhydrophilicity and Superhydrophobicity, *Angew, Chem.*, **116** (2004) 361–364
127. Sun M, Luo C, Xu L, Ji H, Ouyang Q, Yu D, and Chen Y, Artificial Lotus Leaf by Nanocasting, *Langmuir*, **21** (2005) 8978–8981
128. Tambe NS and Bhushan B, Scale dependence of micro/nano-friction and adhesion of MEMS/NEMS materials, coatings and lubricants, *Nanotechnology*, **15** (2004) 1561–1570
129. Teshima K, Sugimura H, Inoue Y, Takai O, and Takano A, Transparent ultra water-repellent poly(ethylene terephthalate) substrates fabricated by oxygen plasma treatment and subsequent hydrophobic coating, *Appl. Surf. Sci.*, **244** (2005) 619–622
130. Tretinnikov ON, Wettability and Microstructure of Polymer Surfaces: Stereochemical and Conformational Aspects in *Apparent and Microscopic Contact Angles* (Drelich J, Laskowski JS, and Mittal KL, eds., VSP, Utrecht, 2000) 111–128
131. Wagner P, Furstner R, Barthlott W, and Neinhuis C, Quantitative Assessment to the Structural Basis of Water Repellency in Natural and Technical Surfaces, *J. Exper. Botany*, **54** (2003) 1295–1303
132. Wang Y, Zhu Q, and Zhang H, Fabrication and magnetic properties of hierarchical porous hollow nickel microspheres, *J. Mater. Chem.*, **16** (2006) 1212–1214
133. Wenzel RN, Resistance of Solid Surfaces to Wetting by Water, *Indust. Eng. Chem.*, **28** (1936) 988–994
134. Wu X, Zheng L, and Wu D, Fabrication of superhydrophobic surfaces from microstructured ZnO-based surfaces via a wet-chemical route, *Langmuir*, **21** (2005) 2665–2667
135. Xu L, Chen W, Mulchandani A, and Yan Y, Reversible Conversion of Conducting Polymer Films from Superhydrophobic to Superhydrophilic, *Angew. Chem. Int. Ed.*, **44** (2005) 6009–6012
136. Yabu H and Shimomura M, Single-Step Fabrication of Transparent Superhydrophobic Porous Polymer Films, *Chem. Mater.*, **17** (2005) 5231–5234
137. Yoshimitsu Z, Nakajima A, Watanabe T, and Hashimoto K, Effects of Surface Structure on the Hydrophobicity and Sliding Behavior of Water Droplets, *Langmuir*, **18** (2002) 5818–5822

138. Yost FG, Michael JR, and Eisenmann ET, Extensive Wetting Due to Roughness, *Acta Metall. Mater.*, **45** (1995) 299–305
139. Zhai L, Cebeci FC, Cohen RE, and Rubner MF, Stable superhydrophobic coatings from polyelectrolyte multilayers, *Nano Lett.*, **4** (2004) 1349–1353
140. Zhang JL, Li JA, and Han YC, Superhydrophobic PTFE surfaces by extension, *Macromol. Rapid Commun.*, **25** (2004) 1105–1108
141. Zhang X, Feng S, Yu X, Liu H, Fu Y, Wang Z, Jiang L, and Li X, Polyelectrolyte Multilayer as Matrix for Electrochemical Deposition of Gold Clusters: Toward Super-Hydrophobic Surface, *J. Am. Chem. Soc.*, **126** (2004) 3064–3065
142. Zhang X, Tan S, Zhao N, Guo X, Zhang X, Zhang Y, and Xu J, Evaporation of Sessile Water Droplets on Superhydrophobic Natural Lotus and Biomimetic Polymer Surfaces, *ChemPhysChem*, **7** (2006) 2067–2070
143. Zhao N, Xie QD, Weng LH, Wang SQ, Zhang XY, and Xu J, Superhydrophobic surface from vapor-induced phase separation of copolymer micellar solution, *Macromolecules*, **38** (2005) 8996–8999
144. Zhao Y, Tong T, Delzeit L, Kashani A, Meyyappan M, and Majumdar A, Interfacial energy and strength of multiwalled-carbon-nanotube-based dry adhesive, *J. Vac. Sci. Technol. B*, **24** (2006) 331–335
145. Zhu L, Xiu Y, Xu J, Tamirisa PA, Hess DW, and Wong C, Superhydrophobicity on Two-Tier Rough Surfaces Fabricated by Controlled Growth of Aligned Carbon Nanotube Arrays Coated with Fluorocarbon, *Langmuir*, **21** (2005) 11208–11212

Gecko Feet: Natural Hairy Attachment Systems for Smart Adhesion – Mechanism, Modeling and Development of Bio-Inspired Materials

Bharat Bhushan

Summary. Leg attachment pads of several creatures including many insects, spiders, and lizards, are capable of attaching to a variety of surfaces and are used for locomotion. Geckos, in particular, have the largest mass and have developed the most complex hairy attachment structures capable of smart adhesion – the ability to cling to different smooth and rough surfaces and detach at will. These animals make use of about three million microscale hairs (setae) (about $14,000/mm^2$) that branch off into hundreds of nanoscale spatulae (about a billion spatulae). This hierarchical surface construction gives the gecko the adaptability to create a large real area of contact with surfaces. Modeling of the gecko attachment system as a hierarchical spring model has provided insight into adhesion enhancement generated by this system. Van der Waals forces are the primary mechanism utilized to adhere to surfaces, and capillary forces are a secondary effect that can further increase adhesion force. Preload applied to the setae increases adhesive force. Although a gecko is capable of producing on the order of 20 N of adhesive force, it retains the ability to remove its feet from an attachment surface at will. The adhesive strength of gecko setae is dependent on the orientation; maximum adhesion occurs at 30°. During walking a gecko is able to peel its foot from surfaces by changing the angle at which its setae contact a surface. A man-made fibrillar structure capable of replicating gecko adhesion has the potential for use in dry, superadhesive tapes that would be of use in a wide range of applications. These adhesives could be created using micro/nanofabrication techniques or self-assembly.

20.1 Introduction

Leg attachment pads of several creatures including many insects, spiders and lizards are capable of attaching to a variety of surfaces and are used for locomotion [36]. Biological evolution over a long period of time has led to optimization of their leg attachment systems. The attachment pads have the ability to cling to different smooth and rough surfaces and detach at will. The dynamic attachment ability will be referred to as reversible adhesion or smart adhesion [20]. Many insects (e.g., beetles and flies) and spiders have been the subject of investigation. However, the attachment pads of geckos have been the most widely studied due to the fact that they have the highest body mass and exhibit the most versatile and effective adhesive known in nature. As a result, the vast majority of this chapter will be concerned with gecko feet.

Although there are over 1000 species of geckos [39, 59] that have attachment pads of varying morphology [73], the Tokay gecko (*Gekko gecko*) has been the main focus of scientific research [4, 42, 48]. The Tokay gecko is the second largest gecko species, attaining respective lengths of approximately 0.3–0.4 m and 0.2–0.3 m for males and females. They have a distinctive blue or gray body with orange or red spots and can weigh up to 300 g [88]. These geckos have been the most widely investigated species of gecko due to the availability and size of these creatures.

Almost 2500 years ago, the ability of the gecko to "run up and down a tree in any way, even with the head downwards" was observed by *Aristotle* [2] (Book IX, Part 9). Even though the adhesive ability of geckos has been known since the time of Aristotle, little was understood about this phenomenon until the late nineteenth century when microscopic hairs covering the toes of the gecko were first noted. The development of electron microscopy in the 1950s enabled scientists to view a complex hierarchical morphology that covers the skin on the gecko's toes. Over the past century and a half, scientific studies have been conducted to determine the factors that allow the gecko to adhere to and detach from surfaces at will, including surface structure [3, 5, 48, 73–75, 77, 93]; the mechanisms of adhesion [6, 7, 11, 25, 33, 42, 44, 47, 73, 78, 80, 85, 90]; and adhesion strength [3, 6, 42, 46–48]. Recent work in modeling the gecko attachment system as a system of springs [20, 55–58] has provided valuable insight into adhesion enhancement. Van der Waals forces are widely accepted in literature as the dominant adhesive mechanism utilized by hierarchical attachment systems. Capillary forces created by humidity naturally present in the air can further increase the adhesive force generated by the spatulae. Both experimental and theoretical work support these adhesive mechanisms.

There is great interest among the scientific community to further study the characteristics of gecko feet in the hope that this information could be applied to the production of micro/nanosurfaces capable of recreating the adhesion forces generated by these lizards [18]. Common man-made adhesives such as tape or glue involve the use of wet adhesives that permanently attach two surfaces. However, replication of the characteristics of gecko feet would enable the development of a superadhesive polymer tape capable of clean, dry adhesion (e.g. [17,19,32,37,64,65,81,82,96,97]). These reusable adhesives have the potential for use in everyday objects such as tapes, fasteners, and toys and in high technology such as microelectronic and space applications. Replication of the dynamic climbing and peeling ability of geckos could find use in the treads of wall-climbing robots [8, 63, 83].

20.2 Hairy Attachment Systems

There are two kinds of attachment pads – relatively smooth and hairy types. Relatively smooth pads, so-called arolia and euplantulae are soft deformable and are found in tree frogs, cockroaches, grasshoppers and bugs. The hairy types consist of long deformable setae and are found in many insects (e.g., beetles, flies), spiders and lizards. The microstructures utilized by beetles, flies, spiders and geckos have similar structures and can be seen in Fig. 20.1a. As the size (mass) of the creature increases, the radius of the terminal attachment elements decreases. This allows a greater number of setae to be packed into an area, hence increasing the linear dimension of contact and the adhesion strength. *Arzt* et al. [3] determined that the density of the terminal attachment elements ρ_A per m^{-2} strongly increases with increasing body mass m in kg. In fact, a master curve can be fit for all the different species (Fig. 20.1b)

$$\log \rho_A = 13.8 + 0.669 \log m .\tag{20.1}$$

The correlation coefficient of the master curve is equal to 0.919. Beetles and flies have the largest attachment pads and the lowest density of terminal attachment elements. Spiders have highly refined attachment elements that cover the leg of the spider. Geckos have both the highest body mass and greatest density of terminal elements (spatulae). Spiders and geckos can generate high dry adhesion whereas beetles and flies increase adhesion by secreting liquid stored generally within a spongy layer of cuticle and is delivered at the contacting surface through a system of porous channels [3, 54]. It should be noted that in the smooth attachment system discussed earlier, the secretion is essential for attachment.

Figure 20.2 shows the scanning electron micrographs of the end of the legs of two flies – fruit fly (drosophila melanogaster) and syrphid fly. The fruit fly uses setae with flattened tips (spatulae) on the two hairy rods for attachment to smooth surfaces and two front claws for the attachment to rough surfaces. The front claws are also used for locomotion. The syrphid fly uses setae on the legs for attachment. In both cases, fluid is secreted at the contacting surface to increase adhesion.

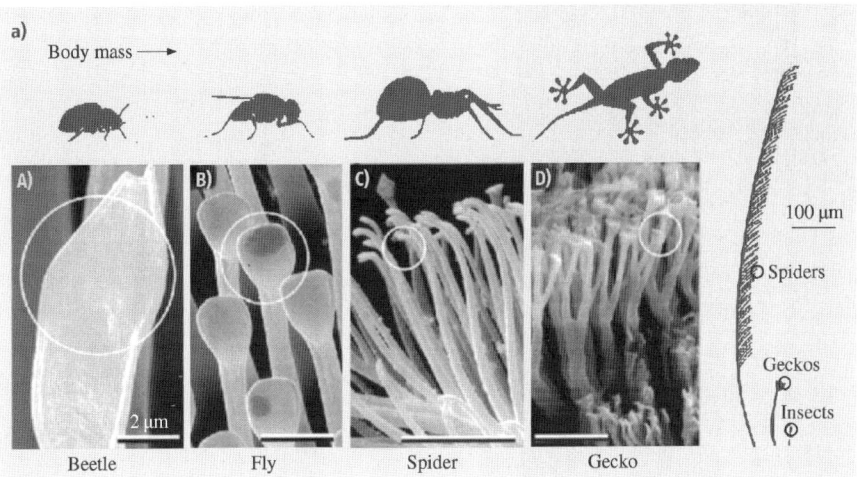

Fig. 20.1. (a) Terminal elements of the hairy attachment pads of a beetle, fly, spider, and gecko shown at two different scales [3]

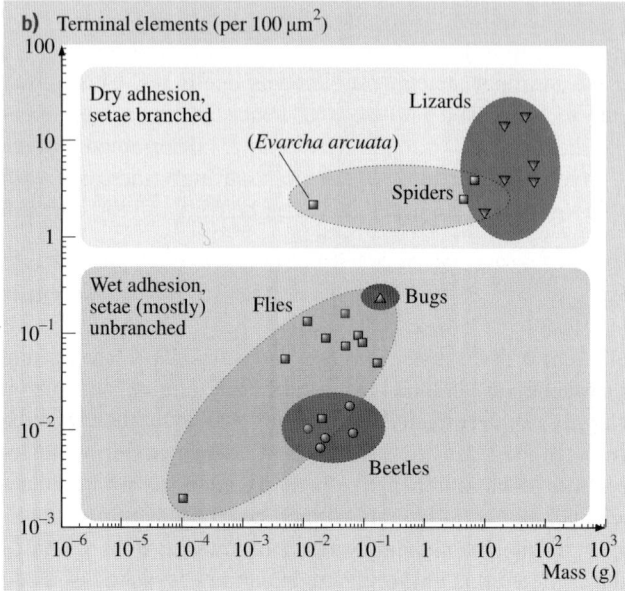

Fig. 20.1. (continued) **(b)** the dependence of terminal element density on body mass [29]. Data are from *Arzt* et al. [3] and *Kesel* et al. [54]

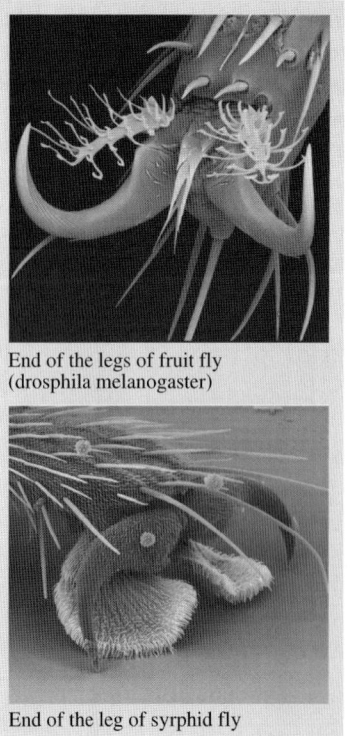

End of the legs of fruit fly (drosphila melanogaster)

End of the leg of syrphid fly

Fig. 20.2. SEM micrographs of the end of the legs of fruit fly (drosophila melanogaster) and syrphid fly [36]

20.3 Tokay Gecko

20.3.1 Construction of Tokay Gecko

The explanation for the adhesive properties of gecko feet can be found in the surface morphology of the skin on the toes of the gecko. The skin is comprised of a complex hierarchical structure of lamellae, setae, branches, and spatulae [73]. Figure 20.3 shows various SEM micrographs of a gecko foot, showing the hierarchical structure down to the nanoscale. Figure 20.4 shows a schematic of the structure and Table 20.1 summarizes the surface characteristics. The gecko attachment system consists of an intricate hierarchy of structures beginning with lamellae, soft ridges, that are 1–2 mm in length [73] that are located on the attachment pads (toes) that compress easily so that contact can be made with rough bumpy surfaces. Tiny curved hairs known as setae extend from the lamellae with a density of approximately 14,000 per square millimeter (Schleich and Kästle, 1986). These setae are typically 30–130 µm in length and 5–10 µm in diameter [42, 73, 74, 93] and are composed primarily of β-keratin [61, 75] with some α-keratin component [72]. At the end of each seta, 100 to 1000 spatulae [42, 73] with a diameter of 0.1–0.2 µm [73] branch out and form the points of contact with the surface. The tips of the spatulae are approximately 0.2–0.3 µm in width [73], 0.5 µm in length and 0.01 µm in thickness [70] and garner their name from their resemblance to a spatula.

The attachment pads on two feet of the Tokay gecko have an area of about 220 mm². About three million setae on their toes can produce a clinging ability of about 20 N (vertical force required to pull a lizard down a nearly vertical (85°) surface) [48] and allow them to climb vertical surfaces at speeds over 1 m/s with capability to attach and detach their toes in milliseconds. In isolated setae, a 2.5 µN

Table 20.1. Surface characteristics of Tokay gecko feet (Young's modulus of surface material, keratin = 1–20 GPa[1,2])

Component	Size	Density	Adhesive force
Seta	30–130[3–6] / 5–10[3–6] length/diameter (µm)	~ 14,000[8,9] setae/mm²	194 µN[10] (in shear) ~ 20 µN[10] (normal)
Branch	20–30[3] / 1–2[3] length/diameter (µm)	–	–
Spatula	2–5[3] / 0.1–0.2[3,7] length/diameter (µm)	100–1000[3,4] spatulae per seta	–
Tip of spatula	~ 0.5[3,7] / 0.2–0.3[3,6] / ~ 0.01[7] length/width/thickness (µm)	–	11 nN[11] (normal)

[1] [75]; [2] [12]; [3] [73]; [4] [42]; [5] [74]; [6] [93]; [7] [70]; [8] [77]; [9] [5]; [10] [6]; [11] [46]

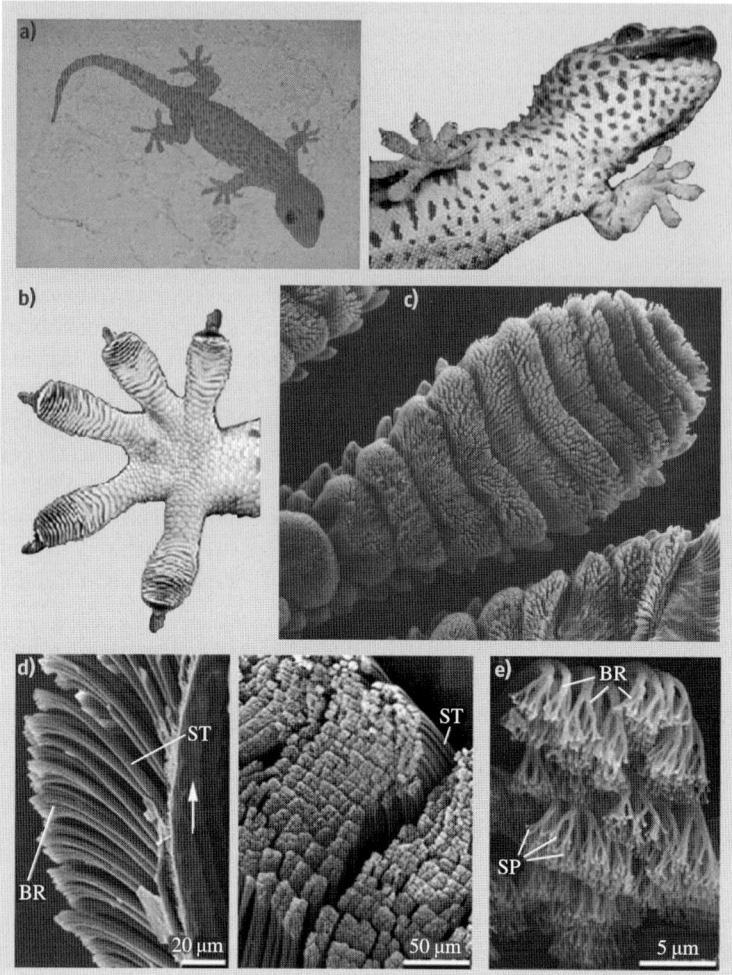

Fig. 20.3. (a) Tokay gecko [6]. The hierarchical structures of a gecko foot; (b) a gecko foot [6] and (c) a gecko toe [4]. Each toe contains hundreds of thousands of setae and each seta contains hundreds of spatulae. Scanning electron microscope (SEM) micrographs of (d) the setae [31] and (e) the spatulae [31]. **ST** seta, **SP** spatula, **BR** branch

preload yielded adhesion of 20 to 40 μN and thus the adhesion coefficient, which represents the strength of adhesion as a function of preload, ranges from 8 to 16 [7].

20.3.2 Adaptation to Rough Surfaces

Typical rough, rigid surfaces are only able to make intimate contact with a mating surface only over a very small portion of the perceived apparent area of contact. In

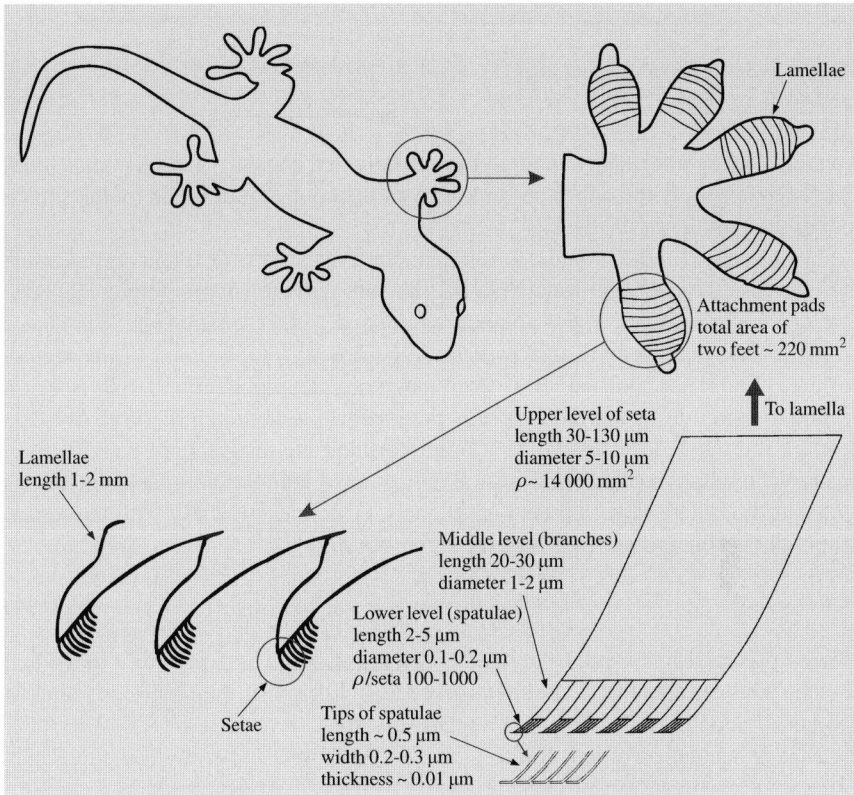

Fig. 20.4. Schematic drawings of a Tokay gecko including the overall body, one foot, a cross-sectional view of the lamellae, and an individual seta. ρ represents number of spatulae

fact, the real area of contact is typically two to six orders of magnitude less than the apparent area of contact [14–16]. Autumn et al. [7] proposed that divided contacts serve as a means for increasing adhesion. Surface energy approach can be used to calculate adhesion force in the dry environment in order to calculate the effect of division of contacts. If the tip of a spatula is considered as a hemisphere with radius R adhesion force of a single contact F_{ad} based on the so called JKR (Johnson–Kendall–Roberts) theory [53], is given as

$$F_{ad} = \frac{3}{2}\pi W_{ad} R \qquad (20.2)$$

where W_{ad} is the work of adhesion (units of energy per unit area). Equation (20.2) shows that adhesion force of a single contact is proportional to a linear dimension of the contact. For a constant area divided into a large number of contacts or setae, n, the radius of a divided contact, R_1, is given by $R_1 = R/\sqrt{n}$ (self-similar scaling) [3]. Therefore, the adhesion force of (20.2) can be modified for multiple contacts such that

$$F'_{ad} = \frac{3}{2}\pi W_{ad}\left(\frac{R}{\sqrt{n}}\right)n = \sqrt{n}F_{ad} \qquad (20.3)$$

where F'_{ad} is the total adhesion force from the divided contacts. Thus the total adhesive force is simply the adhesion force of a single contact multiplied by the square root of the number of contacts.

For a contact in the humid environment, the meniscus (or capillary) forces further increase the adhesion force [14–16]. The attractive meniscus force (F_m) consists of a contribution by both Laplace pressure and surface tension, see Sect. 20.4.2 [15, 66]. Contribution by Laplace pressure is directly proportional to the meniscus area. The other contribution is from the vertical component of surface tension around the circumference. This force is proportional to the circumference as is the case for the work of adhesion [15]. Going through the analysis presented earlier, one can show that the contribution from the component of surface tension increases with splitting into a larger number of contacts. It increases linearly with the square root of the number of contacts n (self-similar scaling) [18, 23]

$$(F'_m)_{\text{surface tension}} = \sqrt{n}(F_m)_{\text{surface tension}}, \qquad (20.4a)$$

where F'_m is the force from the divided contacts and F_m is the force of an individual contact. This component of meniscus force is significant if the meniscus radius is very small and the contact angles are relatively large.

During separation of two surfaces, the viscous force, F_v, of divided contacts is given as [23]

$$F'_v = F_v/n, \qquad (20.4b)$$

where F'_v is the force from the divided contacts and F_v is the force of an individual contact.

The models just presented only consider contact with a flat surface. On natural rough surfaces the compliance and adaptability of setae are the primary sources of high adhesion. Intuitively, the hierarchical structure of gecko setae allows for greater contact with a natural rough surface than non-branched attachment system [82]. Modeling of the contact between gecko setae and rough surfaces is discussed in detail in Sect. 20.6.

Material properties also play an important role in adhesion. A soft material is able to achieve greater contact with a mating surface than a rigid material. Although, gecko skin is primarily comprised of β-keratin, a stiff material with a Young's modulus in the range of 1–20 GPa [12, 75], the effective modulus of the setal arrays on gecko feet is about 100 kPa [4], which is approximately four orders of magnitude lower than the bulk material. Young's modulus of the gecko skin is compared with that of various materials in Table 20.2. The surface of consumer adhesive tape is selected to be very compliant to increase the contact area for high adhesion. Nature has selected a relatively stiff material to avoid clinging to adjacent setae. Division

Table 20.2. Young's modulus of gecko skin and other materials for comparison

Material	Young's modulus
β-keratin, mostly present in gecko skin	1–20 GPa
Steel	210 GPa
Cross-linked rubber	1 MPa
Consumer adhesive tape (uncrosslinked rubber)	1 kPa

of contacts, as discussed earlier, provides high adhesion. By combining optimal surface structure and material properties, mother nature has created an evolutionary superadhesive.

20.3.3 Peeling

Although geckos are capable of producing large adhesion forces, they retain the ability to remove their feet from an attachment surface at will by peeling action. The orientation of the spatulae facilitates peeling. *Autumn* et al. [6] were the first to experimentally show that adhesion force of gecko setae is dependent on the three-dimensional orientation as well as the preload applied during attachment (see Sect. 20.5.1). Due to this fact, geckos have developed a complex foot motion during walking. First the toes are carefully uncurled during attachment. The maximum adhesion occurs at an attachment angle of 30° – the angle between a seta and mating surface. The gecko is then able to peel its foot from surfaces one row of setae at a time by changing the angle at which its setae contact a surface. At an attachment angle greater than 30° the gecko will detach from the surface.

Shah and *Sitti* [79] determined the theoretical preload required for adhesion as well as the adhesion force generated for setal orientations of 30°, 40°, 50°, and 60°. We consider a solid material (elastic modulus, E, Poisson's ratio, v) to make contact with the rough surface described by

$$f(x) = H\sin^2\left(\frac{\pi x}{\chi}\right) \tag{20.5}$$

where H is the amplitude and χ is the wavelength of the roughness profile. For a solid adhesive block to achieve intimate contact with the rough surface neglecting surface forces, it is necessary to apply a compressive stress σ_c [52].

$$\sigma_c = \frac{\pi E H}{2\chi(1-v^2)} \tag{20.6}$$

Equation (20.6) can be modified to account for fibers oriented at an angle, θ. The preload required for contact is summarized in Fig. 20.5a. As the orientation angle decreases, so does the required preload. Similarly, adhesion strength is influenced by fiber orientation. As seen in Fig. 20.5b, the greatest adhesion force occurs at $\theta = 30°$.

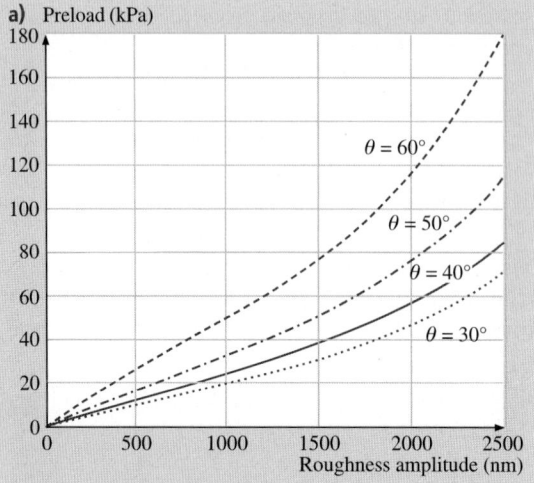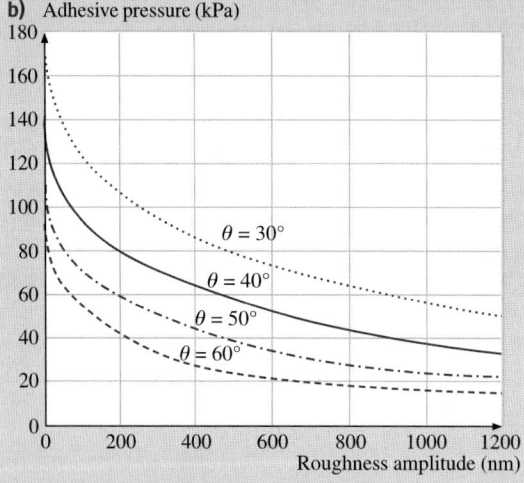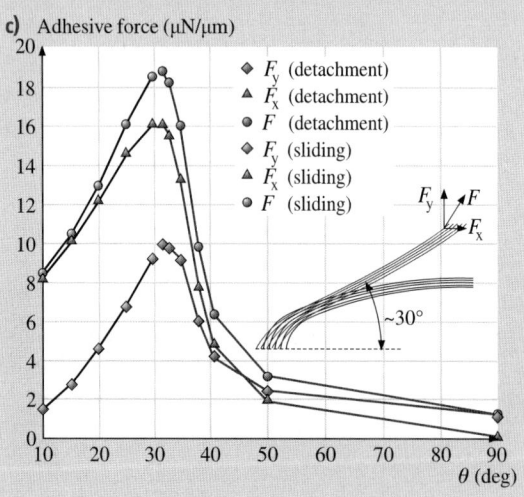

Fig. 20.5. Contact mechanics results for the effect of fiber orientation on **(a)** preload and **(b)** adhesive force for roughness amplitudes ranging from 0–2500 nm [79]. **(c)** Finite element analysis of the adhesive force of a single seta as a function of pull direction [31]

Gao et al. [31] created a finite element model of a single gecko seta in contact with a surface. A tensile force was applied to the seta at various angles, θ, as shown in Fig. 20.5c. For forces applied at an angle less than 30°, the dominant failure mode was sliding. On the contrary, the dominant failure mode for forces applied at angles greater than 30° was detachment. This verifies the results of *Autumn* et al. [6] that detachment occurs at attachment angles greater than 30°.

Tian et al. [87] have suggested that during detachment, angular dependence of both adhesion and friction play a role. The pulling force of a spatula along its shaft with an angle between 0 and 90° to the substrate has a normal adhesion force produced at the spatula-substrate bifurcation zone, and a lateral friction force contribution from the part of the spatula still in contact with the substrate. High net friction and adhesion forces on the whole gecko are obtained by rolling down and gripping the toes inward to realize small pulling angles of the large number of spatulae in contact with the substrate. To detach, the high adhesion/friction is rapidly reduced to a very low value by rolling the toes upward and downward, which, mediated by the lever function of the setal shaft, peels the spatula off perpendicularly from the substrate.

20.3.4 Self Cleaning

Natural contaminants (dirt and dust) as well as man-made pollutants are unavoidable and have the potential to interfere with the clinging ability of geckos. Particles found in the air consist of particulates that are typically less than 10 μm in diameter while those found on the ground can often be larger [43, 51]. Intuitively, it seems that the great adhesion strength of gecko feet would cause dust and other particles to become trapped in the spatulae and that they would have no way of being removed without some sort of manual cleaning action on behalf of the gecko. However, geckos are not known to groom their feet like beetles [86] nor do they secrete sticky fluids to remove adhering particles like ants [30] and tree frogs [40], yet they retain adhesive properties. One potential source of cleaning is during the time when the lizards undergo molting, or the shedding of the superficial layer of epidermal cells. However, this process only occurs approximately once per month [89]. If molting were the sole source of cleaning, the gecko would rapidly lose its adhesive properties as it is exposed to contaminants in nature [41].

Hansen and *Autumn* [41] tested the hypothesis that gecko setae become cleaner with repeated use – a phenomenon known as self-cleaning. The cleaning ability of gecko feet was first tested experimentally by applying 2.5 μm radius silica-alumina ceramic microspheres to clean setal arrays. It was found that a significant fraction of the particles was removed from the setal arrays with each step taken by the gecko.

In order to understand this cleaning process, substrate-particle interactions must be examined. The interaction energy between a dust particle and a wall and spatulae can be modeled as shown in Fig. 20.6. The interaction energy between a spherical

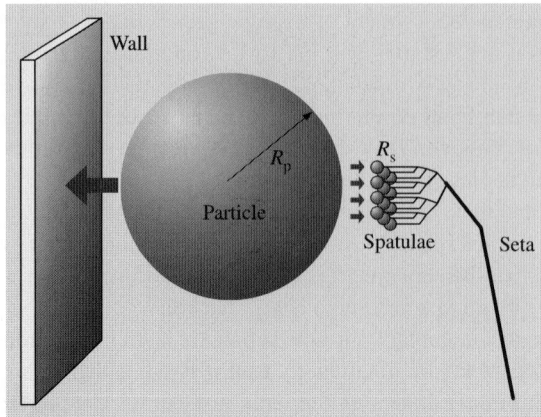

Fig. 20.6. Model of interactions between gecko spatulae of radius R_s, a spherical dirt particle of radius R_p, and a planar wall that enable self cleaning [41]

dust particle and the wall, W_{pw}, can be expressed as [50]

$$W_{pw} = \frac{-H_{pw}R_p}{6D_{pw}}, \tag{20.7}$$

where p and w refer to the particle and wall, respectively. H is the Hamaker constant, R_p is the radius of the particle, and D_{pw} is the separation distance between the particle and the wall. Similarly, the interaction energy between a spherical dust particle and a spatula, s, assuming that the spatula tip is spherical is [50]

$$W_{ps} = \frac{-H_{ps}R_pR_s}{6D_{ps}(R_p+R_s)}. \tag{20.8}$$

The ratio of the two interaction energies, N, can be expressed as

$$N = \frac{W_{pw}}{W_{ps}} = \left(1 + \frac{R_p}{R_s}\right)\frac{H_{pw}D_{ps}}{H_{ps}D_{pw}}. \tag{20.9}$$

When the energy required to separate a particle from the wall is greater than that required to separate it from a spatula, self-cleaning will occur. For small contaminants ($R_p < 0.5\mu m$), there are not enough spatulae available to adhere to the particle. For larger contaminants, the curvature of the particles makes it impossible for enough spatulae to adhere to it. As a result, *Hansen* and *Autumn* [41] concluded that self-cleaning should occur for all spherical spatulae interacting with all spherical particles.

20.4 Attachment Mechanisms

When asperities of two solid surfaces are brought into contact with each other, chemical and/or physical attractions occur. The force developed that holds the two surfaces together is known as adhesion. In a broad sense, adhesion is considered to be either physical or chemical in nature [13–17,21,45,50,99]. Chemical interactions such as electrostatic attraction charges [78] as well as intermolecular forces [42] including van der Waals and capillary forces have all been proposed as potential adhesion mechanisms in gecko feet. Others have hypothesized that geckos adhere to surfaces through the secretion of sticky fluids [80, 90], suction [80], increased frictional force [44], and microinterlocking [25].

Through experimental testing and observations conducted over the last century and a half many potential adhesive mechanisms have been eliminated. Observation has shown that geckos lack glands capable of producing sticky fluids [80, 90], thus ruling out the secretion of sticky fluids as a potential adhesive mechanism. Furthermore, geckos are able to create large adhesive forces normal to a surface. Since friction only acts parallel to a surface, the attachment mechanism of increased frictional force has been ruled out. *Dellit* [25] experimentally ruled out suction and electrostatic attraction as potential adhesive mechanisms. Experiments carried out in vacuum did not show a difference between the adhesive force at low pressures compared to ambient conditions. Since adhesive forces generated during suction are based on pressure differentials, which are insignificant under vacuum, suction was rejected as an adhesive mechanism [25]. Additional testing utilized X-ray bombardment to create ionized air in which electrostatic attraction charges would be eliminated. It was determined that geckos were still able to adhere to surfaces in these conditions and therefore, electrostatic charges could not be the sole cause of attraction [25]. *Autumn* et al. [6] demonstrated the ability of a gecko generate large adhesive forces when in contact with a molecularly smooth SiO_2 MEMS semiconductor. Since surface roughness is necessary for microinterlocking to occur, it has been ruled out as a mechanism of adhesion. Two mechanisms, van der Waals forces and capillary forces, remain as the potential sources of gecko adhesion. These attachment mechanisms are described in detail in the following sections.

20.4.1 Van der Waals Forces

Van der Waals bonds are secondary bonds that are weak in comparison to other physical bonds such as covalent, hydrogen, ionic, and metallic bonds. Unlike other physical bonds, van der Waals forces are always present regardless of separation and are effective from very large separations (~50 nm) down to atomic separation (~0.3 nm). The van der Waals force per unit area between two parallel surfaces, f_{vdW}, is given by [38, 49, 50]

$$f_{vdW} = \frac{H}{6\pi D^3} \quad \text{for } D < 30 nm. \tag{20.10}$$

where H is the Hamaker constant and D is the separation between surfaces.

Hiller [42] showed experimentally that the surface energy of a substrate is responsible for gecko adhesion. One potential adhesive mechanism would then be van der Waals forces [6,85]. Assuming van der Waals forces to be the dominant adhesive mechanism utilized by geckos, the adhesive force of a gecko can be calculated. Typical values of the Hamaker constant range from 4×10^{-20} to 4×10^{-19} J [50]. In calculation, the Hamaker constant is assumed to be 10^{-19} J, the surface area of a spatula is taken to be 2×10^{-14} m^2 [5, 73, 93], and the separation between the spatula and contact surface is estimated to be 0.6 nm. This equation yields the force of a single spatula to be about 0.5 μN. By applying the surface characteristics of Table 20.1, the maximum adhesive force of a gecko is 150–1500 N for varying spatula density of 100–1000 spatulae per seta. If an average value of 550 spatulae/seta is used, the adhesive force of a single seta is approximately 270 μN which is in agreement with the experimental value obtained by *Autumn* et al. [6], which will be discussed in Sect. 20.5.1.

Another approach to calculate adhesive force is to assume that spatulae are cylinders that terminate in hemispherical tips. By using (20.2) and assuming that the radius of each spatula is about 100 nm and that the surface energy is expected to be 50 mJ/m^2 [3], the adhesive force of a single spatula is predicted to be 0.02 μN. This result is an order of magnitude lower than the first approach calculated for the higher value of A. For a lower value of 10^{-20} J for the Hamaker constant, the adhesive force of a single spatula is comparable to that obtained using the surface energy approach.

Several experimental results favor van der Waals forces as the dominant adhesive mechanism including temperature testing [11] and adhesive force measurements of a gecko seta with both hydrophilic and hydrophobic surfaces [6]. This data will be presented in the Sect. 20.5.2 – 20.5.4.

20.4.2 Capillary Forces

It has been hypothesized that capillary forces that arise from liquid mediated contact could be a contributing or even the dominant adhesive mechanism utilized by gecko spatulae [42, 85]. Experimental adhesion measurements (presented in Sect. 20.5.3 and 20.5.4) conducted on surfaces with different hydrophobicities and at various humidities [47] as well as numerical simulations [58] support this hypothesis as a contributing mechanism. During contact, any liquid that wets or has a small contact angle on surfaces will condense from vapor in the form of an annular-shaped capillary condensate. Due to the natural humidity present in the air, water vapor will condense to liquid on the surface of bulk materials. During contact this will cause the formation of adhesive bridges (menisci) due to the proximity of the two surfaces and the affinity of the surfaces for condensing liquid [28, 71, 98].

Capillary force can be divided into two components: the Laplace force F_L and the surface tension force F_s such that the total capillary force F_c is given by the sum of the components

$$F_c = F_L + F_s . \tag{20.11}$$

The Laplace force is caused by the pressure difference across the interface of a curved liquid surface (Fig. 20.7) and depends on pressure difference multiplied by the meniscus area, which can be expressed as [66]

$$F_L = -\pi \kappa \gamma R^2 \sin^2 \phi , \tag{20.12}$$

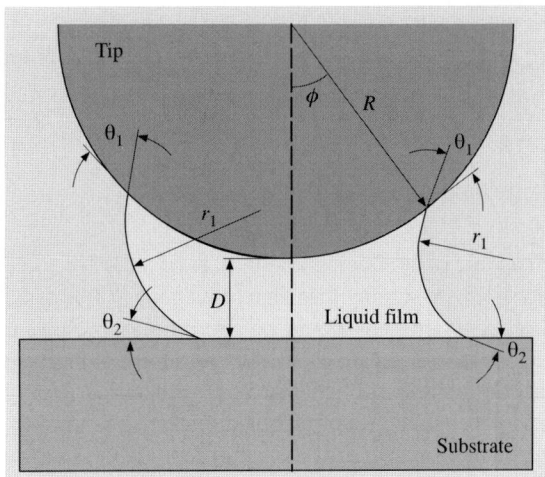

Fig. 20.7. Schematic of a sphere on a plane at distance D with a liquid film in between, forming menisci. In this figure, R is the tip radius, ϕ is the filling angle, θ_1 and θ_2 are contact angles on sphere and plane, respectively, and r_1 and r_2 are the two principal radii of the curved surface [58]

where γ is the surface tension of the liquid, R is the tip radius, ϕ is the filling angle and κ is the mean curvature of meniscus. From the Kelvin equation [50], which is the thermal equilibrium relation, the mean curvature of meniscus can be determined as

$$\kappa = \frac{\Re T}{V\gamma} \ln\left(\frac{p}{p_o}\right), \qquad (20.13)$$

where $\Re$ is the universal gas constant, T is the absolute temperature, V is the molecular volume, p_o is the saturated vapor pressure of the liquid at T, and p is the ambient pressure acting outside the curved surface (p/p_o is the relative humidity). Orr et al. [66] formulated the mean curvature of meniscus between a sphere and a plane in terms of elliptic integrals. The filling angle ϕ can be calculated from the expression just mentioned and (20.13) using iteration method. Then the Laplace force is calculated at a given environment using (20.12).

The surface tension of the liquid results in the formation of a curved liquid-air interface. The surface tension force acting on the sphere is [66]

$$F_s = 2\pi R \gamma \sin\phi \sin(\theta_1 + \phi), \qquad (20.14)$$

where θ_1 is the contact angle on the sphere.

Hence, the total capillary force on the sphere is

$$F_c = \pi R \gamma \{2\sin\phi \sin(\theta_1 + \phi) - \kappa R \sin^2\phi\}. \qquad (20.15)$$

20.5 Experimental Adhesion Test Techniques and Data

Experimental measurements of the adhesion force of a single gecko seta [6] and single gecko spatula [46, 47] have been made. The effect of the environment including temperature [11, 60] and humidity [47] has been studied. Some of the data has been used to understand the adhesion mechanism utilized by the gecko attachment system – van der Waals or capillary forces. The majority of experimental results point towards van der Waals forces as the dominant mechanism of adhesion [6,11]. Recent research suggests that capillary forces can be a contributing adhesive factor [47].

20.5.1 Adhesion under Ambient Conditions

Two feet of a Tokay gecko are capable of producing about 20 N of adhesive force with a pad area of about 220 mm^2 [48]. Assuming that there are about 14,000 setae per mm^2, the adhesion force from a single hair should be approximately 7 µN. It is likely that the magnitude is actually greater than this value because it is unlikely that all setae are in contact with the mating surface [6]. Setal orientation greatly influences adhesive strength. This dependency was first noted by *Autumn* et al. [6]. It was determined that the greatest adhesion occurs at 30°. In order to determine the adhesion mechanism(s) utilized by gecko feet, it is important to know the adhesion force of a single seta. Hence, the adhesion force of gecko foot-hair has been the focus of several investigations [6, 46].

Adhesive Force of a Single Seta

Autumn et al. [6] used both a microelectromechanical (MEMS) force sensor and a wire as a force gauge to determine the adhesion force of a single seta. The MEMS force sensor is a dual-axis atomic force microscope (AFM) cantilever with independent piezoresistive sensors, which allows simultaneous detection of vertical and lateral forces [24]. The wire force gage consisted of an aluminum bonding wire that displaced under a perpendicular pull. *Autumn* et al. [6] discovered that setal force actually depends on the three-dimensional orientation of the seta as well as the preloading force applied during initial contact. Setae that were preloaded vertically to the surface exhibited only one-tenth of the adhesive force (0.6 ± 0.7 µN) compared to setae that were pushed vertically and then pulled horizontally to the surface (13.6 ± 2.6 µN). The dependence of adhesion force of a single gecko spatula on perpendicular preload is illustrated in Fig. 20.8. The adhesion force increases linearly with the preload, as expected [13–15]. The maximum adhesion force of a single gecko foot-hair occurred when the seta was first subjected to a normal preload and then slid 5 µm along the contacting surface. Under these conditions, adhesion force measured 194 ± 25 µN ($\sim$10 atm adhesive pressure).

Adhesive Force of a Single Spatula

Huber et al. [46] used atomic force microscopy to determine the adhesion force of individual gecko spatulae. A seta with four spatulae was glued to an AFM tip. The seta was then brought in contact with a surface and a compressive preload of 90 nN was applied. The force required to pull the seta off of the surface was then measured. As seen in Fig. 20.9, there are two distinct peaks on the graph – one at 10 nN and

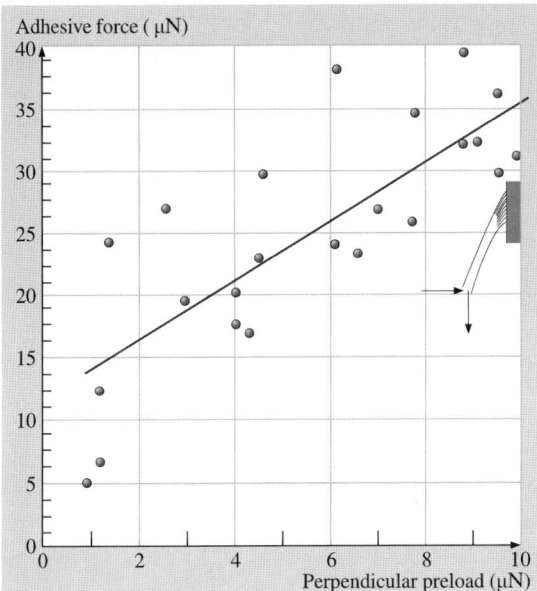

Fig. 20.8. Adhesive force of a single gecko seta as a function of applied preload. The seta was first pushed perpendicularly against the surface and then pulled parallel to the surface [6]

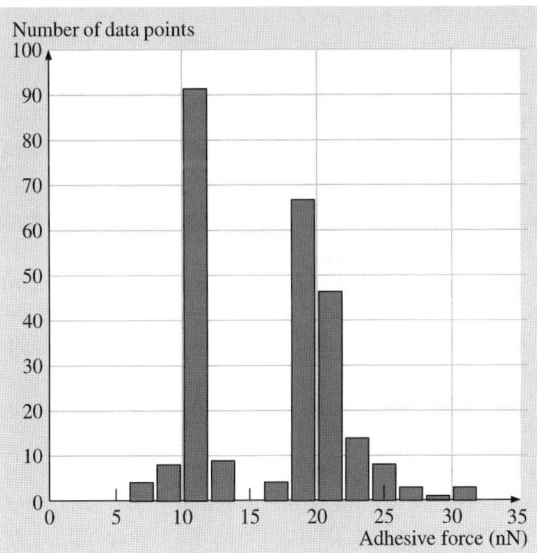

Fig. 20.9. Adhesive force of a single gecko spatula. The peak at 10 nN corresponds to the adhesive force of one spatula and the peak at 20 nN corresponds to the adhesive force of two spatulae [46]

the other at 20 nN. The first peak corresponds to one of the four spatulae adhering to the contact surface while the peak at 20 nN corresponds to two of the four spatulae adhering to the contact surface. The average adhesion force of a single spatula was found to be of 10.8 ± 1 nN. The measured value is in agreement with the measured adhesive strength of an entire gecko (on the order of 10^9 spatulae on a gecko).

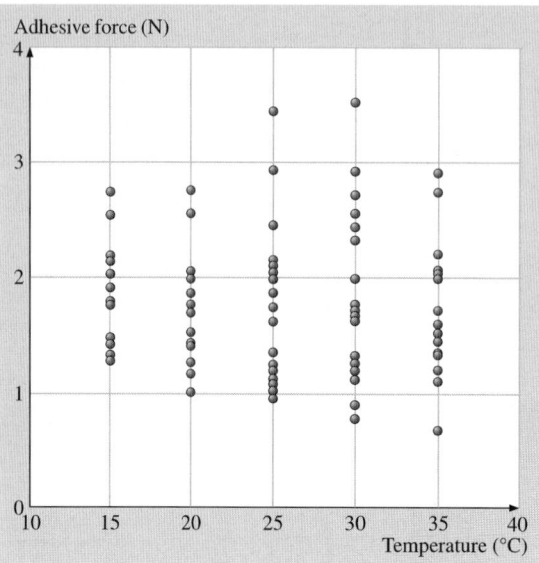

Fig. 20.10. Adhesive force of a gecko as a function of temperature [11]

20.5.2 Effects of Temperature

Environmental factors are known to affect several aspects of vertebrate function, including speed of locomotion, digestion rate and muscle contraction, and as a result several studies have been completed to investigate environmental impact on these functions. Relationships between the environment and other properties such as adhesion are far less studied [11]. Only two known studies exist that examine the affect of temperature on the clinging force of the gecko [11, 60]. *Losos* [60] examined adhesion ability of large live geckos at temperatures up to 17 °C. *Bergmann* and *Irschick* [11] expanded upon this research for body temperatures ranging from 15–35 °C. The geckos were incubated until their body temperature reached a desired level. The clinging ability of these animals was then determined by measuring the maximum exerted force by the geckos as they were pulled off a custom-built force plate. The clinging force of a gecko for the experimental test range is plotted in Fig. 20.10. It was determined that variation in temperature is not statistically significant in the adhesion force of a gecko. From these results, it was concluded that the temperature independence of adhesion supports the hypothesis of clinging as a passive mechanism (i.e. van der Waals forces). Both studies only measured overall clinging ability on the macroscale. There have not been any investigations into effects of temperature on the clinging ability of a single seta on the microscale and therefore testing in this area would be extremely important.

20.5.3 Effects of Humidity

Huber et al. [47] employed similar methods to *Huber* et al. [46] (discussed previously in Sect. 20.5.1) in order to determine the adhesive force of a single spatula

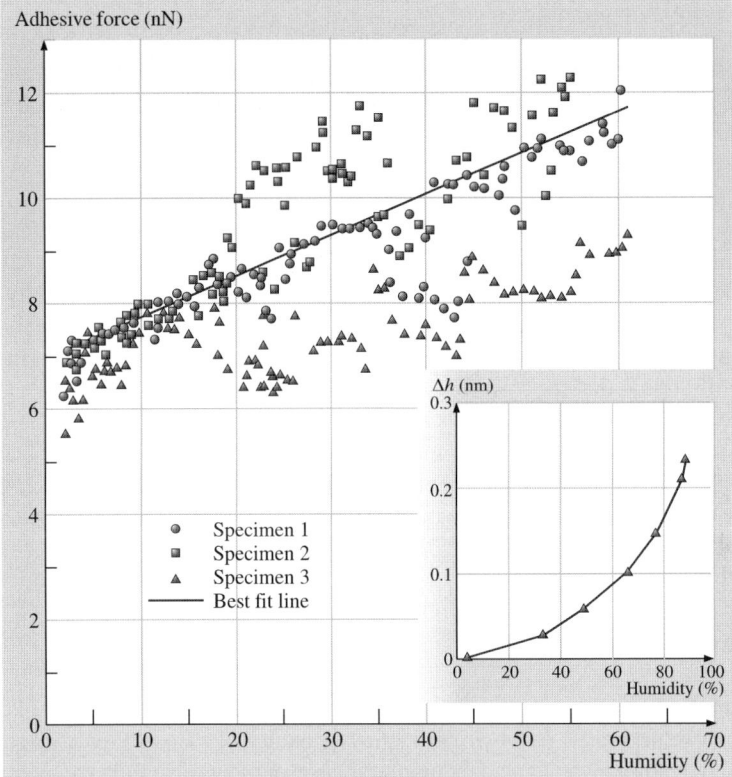

Fig. 20.11. Humidity effects on spatular pull-off force. (Inset) The increase in water film thickness on a Si wafer with increasing humidity [47]

at varying humidity. Measurements were made using an AFM placed in an air-tight chamber. The humidity was adjusted by varying the flow rate of dry nitrogen into the chamber. The air was continuously monitored with a commercially available hygrometer. All tests were conducted at ambient temperature.

As seen in Fig. 20.11, even at low humidity, adhesion force is large. An increase in humidity further increases the overall adhesion force of a gecko spatula. The pull-off force roughly doubled as the humidity was increased from 1.5% to 60%. This humidity effect can be explained in two possible ways: 1. by standard capillarity or 2. by a change of the effective short-range interaction due to absorbed monolayers of water – in other words, the water molecules increase the number of van der Waals bonds that are made. Based on this data, van der Waals forces are the primary adhesion mechanism and capillary forces are a secondary adhesive mechanism.

20.5.4 Effects of Hydrophobicity

To further test the hypothesis capillary forces, plays a role in gecko adhesion, the spatular pull-off force was determined for contact with both hydrophilic and hydrophobic surfaces. As seen in Fig. 20.12a, the capillary adhesion theory predicts that a gecko spatula will generate a greater adhesion force when in contact with a hydrophilic surface as compared to a hydrophobic surface while the van der Waals adhesion theory predicts that the adhesion force between a gecko spatula and a surface will be the same regardless of the hydrophobicity of the surface [7]. Figure 20.12b shows the shear stress of a whole gecko and adhesive force of a single seta on hydrophilic and hydrophobic surfaces. The data shows that the adhesion values are the same on both surfaces. This supports the van der Waals prediction of Fig. 20.12a. *Huber* et al. [47] found that the hydrophobicity of the attachment surface had an effect on the adhesion force of a single gecko spatula as shown in Fig. 20.12c. These results show that adhesion force has a finite value for superhydrophobic surface and increases as the surface becomes hydrophilic. It is concluded that van der Waals forces are the primary mechanism and capillary forces further increase the adhesion force generated.

20.6 Adhesion Modeling

With regard to the natural living conditions of the animals, the mechanics of gecko attachment can be separated into two parts: the mechanics of adhesion of a single contact with a flat surface, and an adaptation of a large number of spatulae to a natural, rough surface. Modeling of the mechanics of adhesion of spatulae to a smooth surface, in the absence of meniscus formation, was developed by *Autumn* et al. [7], *Jagota* and *Bennison* [52] and *Arzt* et al. [3]. As discussed previously in Sect. 20.3.2, the adhesion force of multiple contacts F'_{ad} can be increased by dividing the contact into a large number (n) of small contacts, while the nominal area of the contact remains the same, $F'_{ad} \sim \sqrt{n} F_{ad}$. However, this model only considers contact with a flat surface. On natural, rough surfaces, the compliance and adaptability of setae are the primary sources of high adhesion. As stated earlier, the hierarchical structure of gecko setae allows for a greater contact with a natural, rough surface than a non-branched attachment system [82].

20.6.1 Spring Model

Bhushan et al. [20] and *Kim* and *Bhushan* [55–58] have recently approximated a gecko seta in contact with random rough surfaces using a hierarchical spring model. Each level of springs in their model corresponds to a level of seta hierarchy. The upper level of springs corresponds to the thicker part of gecko seta, the middle spring level corresponds to the branches, and the lower level of springs corresponds to the spatulae. The upper level is the thickest branch of the seta. It is 75 μm in length and 5 μm in diameter. The middle level, referred to as a branch,

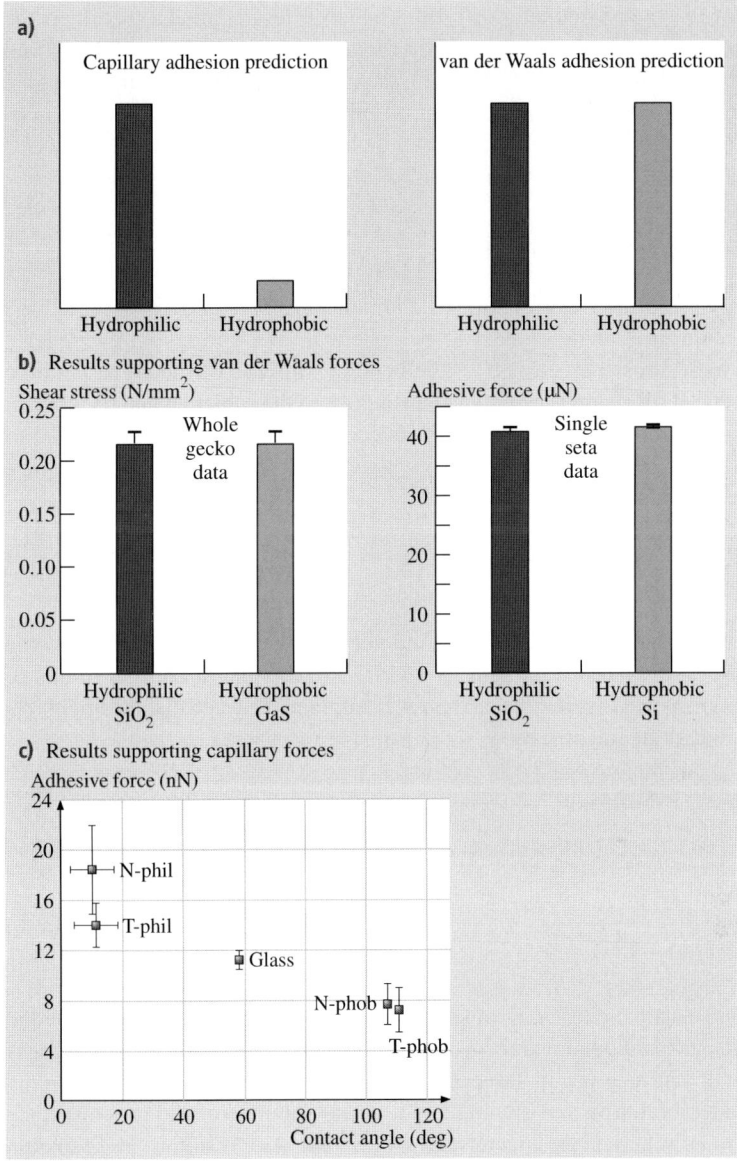

Fig. 20.12. (a) Capillary and van der Waals adhesion predictions for the relative magnitude of the adhesive force of gecko setae with hydrophilic and hydrophobic surfaces [7]. (b) Results of adhesion testing for a whole gecko (*left*) and single seta (*right*) with hydrophilic and hydrophobic surfaces [7] and (c) results of adhesive force testing of a single gecko spatula with surfaces with different contact angles [47]

has a length of 25 μm and diameter of 1 μm. The lower level, called a spatula, is the thinnest branch with a length of 2.5 μm and a diameter of about 0.1 μm (Table 20.2). Autumn et al. [6] showed that the optimal attachment angle between the substrate and a gecko seta is 30° in the single seta pull-off experiment. This finding is supported by the adhesion models of setae as cantilever beams [31,79] (see Sect. 20.3.3 for more details). Therefore, θ was fixed at 30° in the studies [20,55–58].

20.6.2 Single Spring Contact Analysis

In their analysis, *Bhushan* et al. [20] and *Kim* and *Bhushan* [55–58] assumed the tip of the spatula in a single contact to be spherical. The springs on every level of hierarchy have the same stiffness as the bending stiffness of the corresponding branches of seta. If the beam is oriented at an angle θ to the substrate and the contact load F is aligned normal to the substrate, its components along and tangential to the direction of the beam, $F\cos\theta$ and $F\sin\theta$, give rise to bending and compressive deformations, δ_b and δ_c, respectively, as [95]

$$\delta_b = \frac{F \cos\theta \, l_m^3}{3EI}, \quad \delta_c = \frac{F \sin\theta \, l_m}{A_C E} \tag{20.16}$$

where, $I = \pi R_m^4/4$ and $A_C = \pi R_m^2$ are the moments of inertia of the beam and the cross-sectional area, respectively, and l_m and R_m are the length and the radius of seta branches, respectively, and m is the level number. The net displacement, $\delta_\perp$ normal to the substrate is given by

$$\delta_\perp = \delta_c \sin\theta + \delta_b \cos\theta. \tag{20.17}$$

Using (20.16) and (20.17), the stiffness of seta branches, k_m, is calculated as [34].

$$k_m = \frac{\pi R_m^2 E}{l_m \sin^2\theta \left(1 + \frac{4l_m^2 \cot^2\theta}{3R_m^2}\right)} \tag{20.18}$$

For an assumed elastic modulus, E, of seta material of 10 GPa with a load applied at an angle of 60° to spatulae long axis, *Kim* and *Bhushan* [55] calculated the stiffness of every level of seta as given in Table 20.3.

In the model, both tips of a spatula and asperity summits of the rough surface are assumed to be spherical with a constant radius [20]. As a result, a single spatula adhering to a rough surface was modeled as the interaction between two spherical tips. Because β-keratin has a high elastic modulus [12,75], the adhesion force between two round tips was calculated according to the Derjaguin–Muller–Toporov (DMT) theory [26] as

$$F_{ad} = 2\pi R_c W_{ad} \tag{20.19}$$

where R_c is the reduced radius of contact, which is calculated as $R_c = (1/R_1 + 1/R_2)^{-1}$; R_1, R_2 – radii of contacting surfaces; $R_1 = R_2$, $R_c = R/2$. The work of adhesion, W_{ad},

Table 20.3. Geometrical size, calculated stiffness and typical densities of branches of seta for *Tokay gecko* (Kim and Bhushan, 2007a)

Level of seta	Length (μm)	Diameter (μm)	Bending stiffness [1] (N/m)	Typical density (#/mm^2)
III upper	75	5	2.908	14×10^3
II middle	25	1	0.126	–
I lower	2.5	0.1	0.0126	1.4–14 × 10^6

[1] for elastic modulus of 10 GPa with load applied at 60° to spatula long axis

is then calculated using the following equation for two flat surfaces separated by a distance D [50].

$$W_{ad} = -\frac{H}{12\pi D^2} \tag{20.20}$$

where H is the Hamakar constant which depends on the medium between the two surfaces. Typical values of the Hamakar constant for polymers are $H_{air} = 10^{-19} J$ in the air and $H_{water} = 3.7 \times 10^{-20} J$ in the water [50]. For a gecko seta, which is composed of β-keratin, the value of H is assumed to be 10^{-19} J. The works of adhesion of two surfaces in contact separated by an atomic distance $D \approx 0.2$ nm is approximately equal to 66 mJ/m^2 [50]. By assuming that the tip radius, R, is 50 nm, using (20.19), the adhesion force of a single contact is calculated as 10 nN [55]. This value is identical to the adhesion force of a single spatula measured by *Huber* et al. [46]. This adhesion force is used as a critical force in the model for judging whether the contact between the tip and the surface is broken or not during pull-off cycle [20]. If the elastic force of a single spring is less than the adhesion force, the spring is regarded as having been detached.

20.6.3 The Multi-Level Hierarchical Spring Analysis

hierarchical levels in the attachment system on attachment ability, models with one- [20, 55, 56], two- [20, 55, 56] and three- [55, 56] levels of hierarchy were simulated (Fig. 20.13). The one-level model has springs with length $l_I = 2.5$ μm and stiffness $k_I = 0.0126$ N/m. The length and stiffness of the springs in the two-level model are $l_I = 2.5$ μm, $k_I = 0.0126$ N/m and $l_{II} = 25$ μm, $k_{II} = 0.126$ N/m for levels *I* and *II*, respectively. The three-level model has additional upper level springs with $l_{III} = 75$ μm, $k_{III} = 2.908 N/m$ on the springs of the two-level model, which is identical to gecko setae. The base of the springs and the connecting plate between the levels are assumed to be rigid. The distance S_I between neighboring structures of level *I* is 0.35 μm, obtained from the average value of measured spatula density, $8 \times 10^6 mm^{-2}$, calculated by multiplying 14,000 setae/mm^2 by an average of 550 spatula/seta [77] (Table 20.3). A 1 : 10 proportion of the number of springs in the upper level to that in the level below was assumed [20]. This corresoponds to one spring

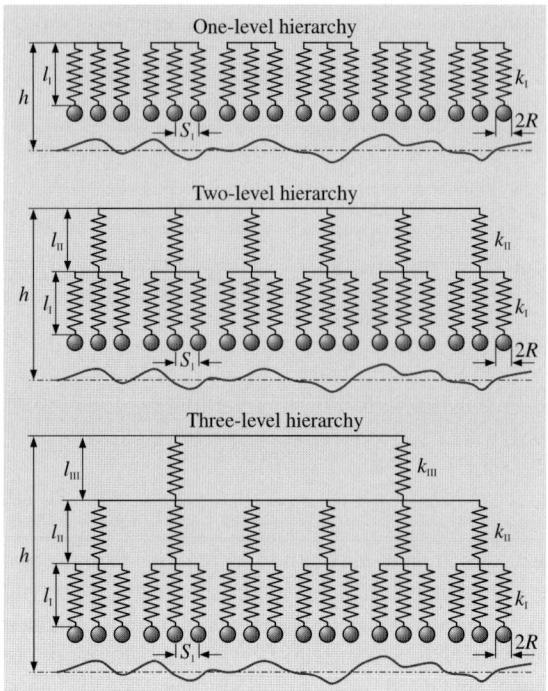

Fig. 20.13. One-, two- and three-level hierarchical spring models for simulating the effect of hierarchical morphology on interaction of a seta with a rough surface. In this figure, $l_{I,II,III}$ are lengths of structures, s_I is space between spatulae, $k_{I,II,III}$ are stiffnesses of structures, I, II and III are level indexes, R is radius of tip, and h is distance between upper spring base of each model and mean line of the rough profile [55]

at level *III* is connected to ten springs on level *II* and each spring on level *II* also has ten springs on level *I*. The number of springs on level *I* considered in the model is calculated by dividing the scan length (2000 μm) with the distance S_I (0.35 μm), which corresponds to 5700.

The spring deflection Δl was calculated as

$$\Delta l = h - l_0 - z, \tag{20.21}$$

where h is the position of the spring base relative to the mean line of surface; l_0 is the total length of a spring structure which is $l_0 = l_I$ for the one-level model, $l_0 = l_I + l_{II}$ for the two-level model, and $l_0 = l_I + l_{II} + l_{III}$ for the three-level model; and z is profile height of the rough surface. The elastic force F_{el} arisen in the springs at a distance h from the surface was calculated for the one-level model as [20]

$$F_{el} = -k_I \sum_{i=1}^{p} \Delta l_i u_i \qquad u_i = \begin{cases} 1 & \text{if contact} \\ 0 & \text{if no contact} \end{cases} \tag{20.22}$$

where p is the number of springs in the level *I* of the model. For the two-level model the elastic force was calculated as [20]

$$F_{el} = -\sum_{j=1}^{q}\sum_{i=1}^{p} k_{ji}(\Delta l_{ji} - \Delta l_j)u_{ji} \qquad u_{ji} = \begin{cases} 1 & \text{if contact} \\ 0 & \text{if no contact} \end{cases} \tag{20.23}$$

where q is the number of springs in the level *II* of the model. For the three-level model the elastic force was calculated as [55]

$$F_{el} = -\sum_{k=1}^{r}\sum_{j=1}^{q}\sum_{i=1}^{p} k_{kji}(\Delta l_{kji} - \Delta l_{kj} - \Delta l_{j})u_{kji} \qquad u_{kji} = \begin{cases} 1 & \text{if contact} \\ 0 & \text{if no contact} \end{cases}$$

(20.24)

where r is the number of springs in the level *III* of the model. The spring force when the springs approach the rough surface is calculated using either (20.22), (20.23), or (20.24) for one-, two- and three-level models, respectively. During pull-off, the same equations are used to calculate the spring force. However, when the applied load is equal to zero, the springs do not detach due to adhesion attraction given by (20.19). The springs are pulled apart until the net force (pull-off force minus attractive adhesion force) at the interface is equal to zero.

The adhesion force is the lowest value of elastic force F_{el} when the seta has detached from the contacting surface. The adhesion energy is calculated as

$$W_{ad} = \int_{\overline{D}}^{\infty} F_{el}(D)\,dD \qquad (20.25)$$

where D is the distance that the spring base moves away from the contacting surface. The lower limit of the distance $\overline{D}$ is the value of D where F_{el} is first zero when the model is pulled away from the contacting surface. Also although the upper limit of the distance is infinity, in practice, the $F_{el}(D)$ curve is integrated to and upper limit where F_{el} increases from a negative value to zero. Figure 20.14 shows the flow chart for the calculation of the adhesion force and the adhesion energy employed by *Kim* and *Bhushan* [55].

The random rough surfaces used in the simulations were generated by a computer program [14, 15]. Two-dimensional profiles of surfaces that a gecko may encounter were obtained using a stylus profiler [20]. These profiles along with the surface selection methods and surface roughness parameters (root mean square (RMS) amplitude σ and correlation length β^*) for scan lengths of 80, 400, and 2000 μm are presented in Appendix 20.A. The roughness parameters are scale dependent, and, therefore, adhesion values also are expected to be scale dependent. As the scan length was increased, the measured values of RMS amplitude and correlation length both increased. The range of values of σ from 0.01 μm to 30 μm and a fixed value of $\beta^* = 200$ μm were used for modeling the contact of a seta with random rough surfaces. The chosen range covers values of roughnesses for relatively smooth, artificial surfaces to natural, rough surfaces. A typical scan length of 2000 μm was also chosen which is comparable to a lamella length of gecko.

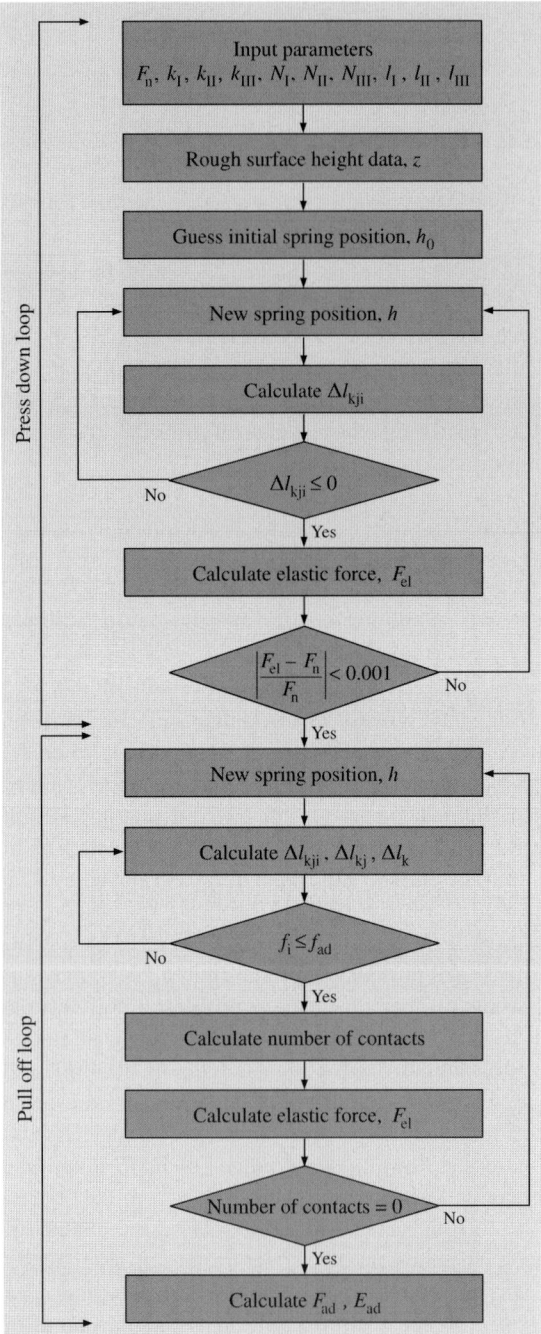

Fig. 20.14. Flow chart for the calculation of the adhesion force (F_{ad}) and the adhesion energy (E_{ad}) for three-level hierarchical spring model. In this figure, F_n is an applied load, $k_{I, II, III}$ and $l_{I, II, III}$ are stiffnesses and lengths of structures, Δl_{kji}, Δl_{ki} and Δl_k are the spring deformations on level I, II and III, respectively, i, j and k are spring indexes on each level, f_i is the elastic force of a single spring and f_{ad} is the adhesion force of a single contact [55]

20.6.4 Adhesion Results of the Multi-level Hierarchical Spring Model

by *Kim* and *Bhushan* [55]. The obtained various useful results and will be presented next. Figure 20.15a shows the calculated spring force–distance curves for the one-, two- and three-level hierarchical models in contact with rough surfaces of different values of root mean square (RMS) amplitude σ anging from σ0.01 µm to σ10 µm at an applied load of 1.6 µN which was derived from the gecko's weight. When the spring model is pressed against the rough surface, contact between the spring and the rough surface occurs at point A; as the spring tip presses into the contacting surface, the force increases up to point B, B' or B". During pull off, the spring relaxes, and the spring force passes an equilibrium state (0 N); tips break free of adhesion forces at point C, C' or C" as the spring moves away from the surface. The perpendicular distance from C, C' or C" to zero is the adhesion force. Adhesion energy stored during contact can be obtained by calculating the area of the triangle during the unloading part of the curves (20.25).

Using the spring force-distance curves, *Kim* and *Bhushan* [55] calculated the adhesion coefficient, the number of contacts per unit length and the adhesion energy per unit length of the one-, two- and three-level models for an applied load of 1.6 µN and a wide range of RMS roughness as seen in the left graphs of Fig. 20.15b. The adhesion coefficient, defined as the ratio of pull-off force to applied preload, represents the strength of adhesion with respect to the preload. For the applied load of 1.6 µN, which corresponds to the weight of a gecko, the maximum adhesion coefficient is about 36 when σ is smaller than 0.01 µm. This means that a gecko can generate enough adhesion force to support 36 times its bodyweight. However, if σ is increased to 1 µm, the adhesion coefficient for the three-level model is reduced to 4.7. It is noteworthy that the adhesion coefficient falls below 1 when the contacting surface has an RMS roughness σ greater than 10 µm. This implies that the attachment system is no longer capable of supporting the gecko's weight. *Autumn* et al. [6,7] showed that in isolated gecko setae contacting with the surface of a single crystalline silicon wafer, a 2.5 µN preload yielded adhesion of 20 to 40 µN and thus a value of adhesion coefficient of 8 to 16, which supports the simulation results of *Kim* and *Bhushan* [55].

Figure 20.15b (top left) shows that the adhesion coefficient for the two-level model is lower than that for the three-level model, but there is only a small difference between the adhesion forces between the two- and three-level models, because the stiffness of level III for the three-level model is calculated to be higher compared to those of levels *I* and II. In order to show the effect of stiffness, the results for the three-level model with springs in level III of which the stiffness is 10 times smaller than that of original level III springs are plotted. It can be seen that the three-level model with a third level stiffness of 0.1 k_{III} has a 20–30% higher adhesion coefficient than the three-level model. The results also show that the trends in the number of contacts are similar to that of the adhesive force. The study also investigated the effect of σn adhesion energy. It was determined that the adhesion energy decreased with an increase of σ. For the smooth surface with σ0.01 µm, the adhesion energies for the two- and three-level hierarchical models are 2 times and

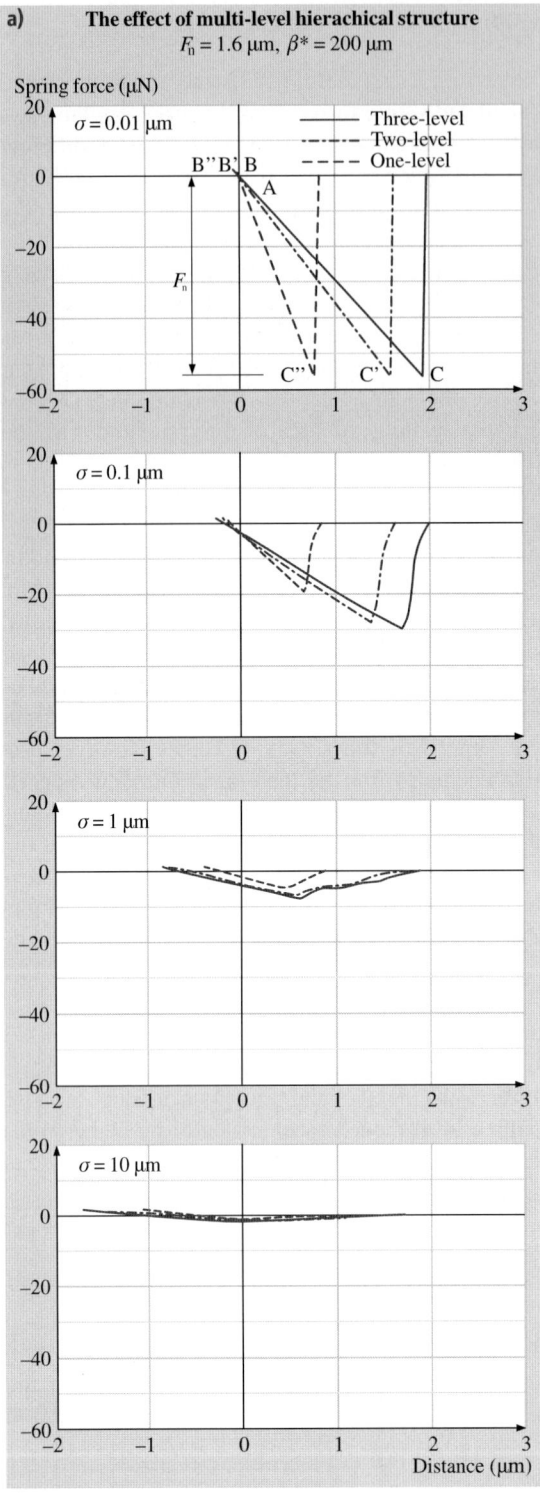

Fig. 20.15. (a) Force-distance curves of one-, two- and three-level models in contact with rough surfaces with different σ values for an applied load of 1.6 μN

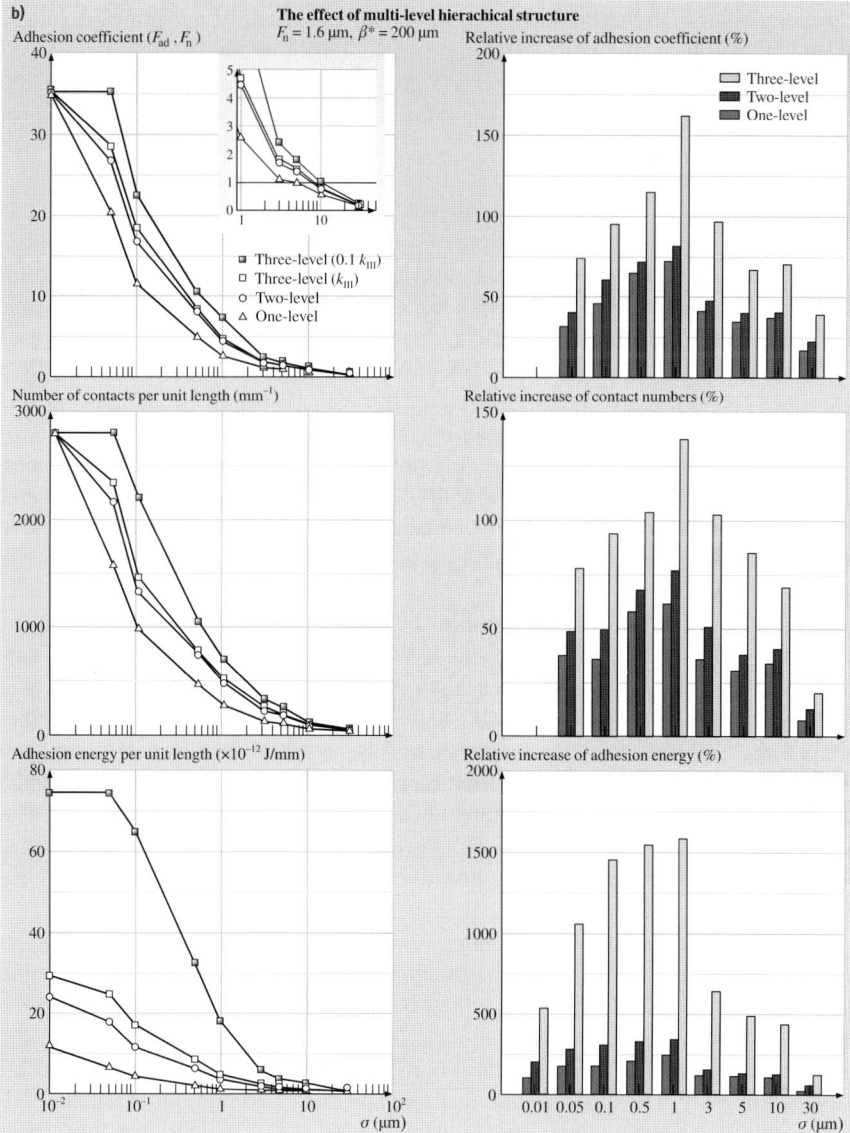

Fig. 20.15. (continued) **(b)** The adhesion coefficient, the number of contacts and the adhesion energy per unit length of profile for one- and multi-level models with an increase of σ value (*left figures*), and relative increases between multi- and one-level models (*right side*) for an applied load of 1.6 µN. The value of k_{III} in the analysis is 2.908 N/m [55]

2.4 times larger than that for the one-level model, respectively, but adhesion energy decreases rapidly at surfaces with σ greater than 0.05 μm; and in every model it finally decreases to zero at surfaces with σ greater than 10 μm. The adhesion energy for the three-level model with 0.1 k_{III} is 2–3 times higher than that for three-level model.

In order to demonstrate the effect of the hierarchical structure on adhesion enhancement, *Kim* and *Bhushan* [55] calculated the increases in the adhesion coefficient, the number of contacts, and the adhesion energy of the two-, three- and three-level (with 0.1 k_{III}) models relative to one-level model. These results are shown in the right side of Fig. 20.15b. It was found for the two- and three-level models, relative increase of the adhesion coefficient increases slowly with an increase of σ nd has the maximum values of about 70% and 80% at σ1 μm, respectively, and then decreases for surfaces with σ greater than 3 μm. On the whole, at the applied load of 1.6 μN, the effect of the variation of σ n the adhesion enhancement for both two- and three-level models is not so large. However, the relative increase of the adhesion coefficient for the three-level model with 0.1 k_{III} has the maximum value of about 170% at σ1 μm, which shows significant adhesion enhancement. Due to the relative increase of adhesion energy, the three-level model with 0.1 k_{III} shows significant adhesion enhancement.

Figure 20.16 shows the variation of adhesion force and adhesion energy as a function of applied load for both one- and three-level models contacting with surface with $\sigma = 1$ μm. It is shown that as the applied load increases, the adhesion force increases up to a certain applied load and then has a constant value, whereas, adhesion energy continues to increase with an increase in the applied load. The one-level model has a maximum value of adhesion force per unit length of about 3 μN/mm at the applied load of 10 μN, and the three-level model has a maximum value of about 7 μN/mm at the applied load of 16 μN. However, the adhesion coefficient continues to decrease at higher applied loads because adhesion force is constant even if the applied load increases.

The simulation results for the three-level model, which is close to gecko setae, presented in Fig. 20.15 show that roughness reduces the adhesion force. At the surface with σ greater than 10 μm, adhesion force by gecko weight cannot support itself. However, in practice, a gecko can cling or crawl on the surface of ceiling with higher roughness. *Kim* and *Bhushan* [55] did not consider the effect of lamellae in their study. The authors state that the lamellae can adapt to the waviness of surface while the setae allow for the adaptation to micro- or nano-roughness and expect that adding the lamellae of gecko skin to the model would lead to higher adhesion over a wider range of roughness. In addition, their hierarchical model considers only normal to surface deformation and motion of seta. It should be noted that measurements of adhesion force of a single gecko seta made by *Autumn* et al. [6] demonstrated that a load applied normal to the surface was insufficient for an effective attachment of seta. The lateral force required to pull parallel to the surface was observed by sliding the seta approximately 5 μm laterally along the surface under a preload.

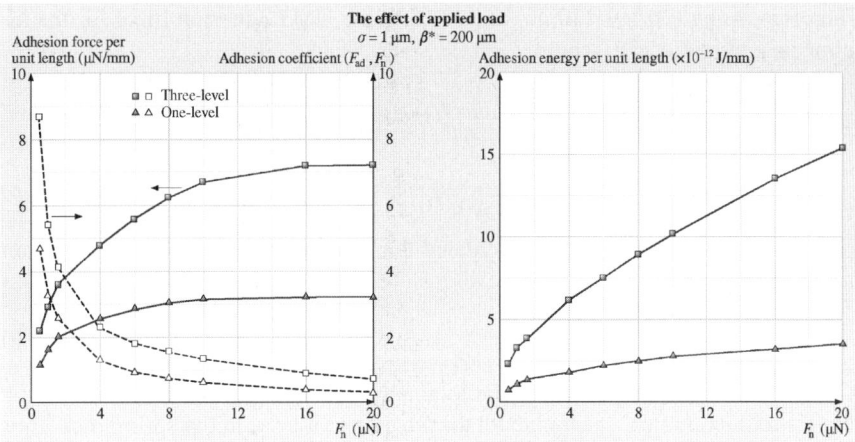

Fig. 20.16. The variation of adhesion force, adhesion coefficient, and adhesion energy as a function of applied loads for both one- and three-level models contacting with surface with $\sigma = 1$ μm. The value of k_{III} in the analysis is 2.908 N/m [55]

20.6.5 Capillary Effects

Kim and *Bhushan* [58] investigated the effects of capillarity on gecko adhesion by considering capillary force as well as the solid-to-solid interaction. The Laplace and surface tension components of the capillary force are treated according to Sect. 20.4.2. The solid to solid adhesive force was calculated by DMT theory according to (20.19) and will be denoted as F_{DMT}.

The work of adhesion was then calculated by (20.20). *Kim* and *Bhushan* [58] assumed typical values of the Hamaker constant to be $H_{\mathrm{air}} = 10^{-19}$ J in the air and $H_{\mathrm{water}} = 6.7 \times 10^{-19}$ J in the water [50]. The work of adhesions of two surfaces in contact separated by an atomic distance $D \approx 0.2$ nm [50] are approximately equal to 66 mJ/m^2 in the air and 44 mJ/m^2 in the water. Assuming tip radius R is 50 nm, the DMT adhesion forces F_{DMT} of a single contact in the air and the water are $F_{\mathrm{DMT}}^{\mathrm{air}} = 11$ nN and $F_{\mathrm{DMT}}^{\mathrm{water}} = 7.3$ nN, respectively. As the humidity increases from zero to 100%, the DMT adhesion force will take a value between $F_{\mathrm{DMT}}^{\mathrm{air}}$ and $F_{\mathrm{DMT}}^{\mathrm{water}}$. To calculate the DMT adhesion force for the intermediate humidity, an approximation method by *Wan* et al. [91] was used. The work of adhesion W_{ad} for the intermediate humidity can be expressed as

$$W_{\mathrm{ad}} = \int_D^\infty \frac{H}{6\pi h^3} dh = \int_D^{h_f} \frac{H_{\mathrm{water}}}{6\pi h^3} dh + \int_{h_f}^\infty \frac{H_{\mathrm{air}}}{6\pi h^3} dh \qquad (20.26)$$

where h is the separation along the plane. h_f is the water film thickness at a filling angle ϕ, which can be calculated as

$$h_f = D + R(1 - \cos\phi) \qquad (20.27)$$

Therefore, using (20.19), (20.26) and (20.27), the DMT adhesion force for the intermediate humidity is given as

$$F_{DMT} = F_{DMT}^{water}\left\{1 - \frac{1}{(1+R(1-\cos\phi)/D)^2}\right\}$$
$$+ F_{DMT}^{air}\left\{\frac{1}{(1+R(1-\cos\phi)/D)^2}\right\} \quad (20.28)$$

Finally, *Kim* and *Bhushan* [58] calculated the total adhesion force F_{ad} as the sum of (20.15) and (20.28)

$$F_{ad} = F_c + F_{DMT} \quad (20.29)$$

Kim and *Bhushan* [58] then used the total adhesion force as a critical force in the three-level hierarchical spring model discussed previously. In the spring model for gecko seta, if the force applied upon spring deformation is greater than the adhesion force, the spring is regarded as having been detached.

20.6.6 Adhesion Results that Account for Capillary Effects

To simulate the capillary contribution to adhesion force for a gecko spatula, *Kim* and *Bhushan* [58] set the contact angle on gecko spatula tip θ_1 equal to 128 [47]. It was assumed that the spatula tip radius $R = 50$ nm, the ambient temperature $T = 25°C$, the surface tension of water $\gamma = 73$ mJ/m^2 and molecular volume of water $V = 0.03$ nm^3 [50].

Figure 20.17a shows total adhesion force as a function of relative humidity for a single spatula in contact with surfaces with different contact angles. Total adhesion force decreases with an increase in the contact angle on the substrate, and the difference of total adhesion force among different contact angles is larger in the intermediate humidity regime. As the relative humidity increases, total adhesion force for the surfaces with contact angle less than 60° has higher value than the DMT adhesion force not considering wet contact, whereas with the value above 60°, total adhesion force has lower values at most relative humidity.

The simulation results of *Kim* and *Bhushan* [58] are compared with the experimental data by *Huber* et al. [47] in Fig. 20.17b. *Huber* et al. [47] measured the pull off force of a single spatula in contact with four different types of Si wafer and glass at the ambient temperature 25°C and the relative humidity 52%. According to their description, wafer families "N" and "T" in Fig. 20.16b differ by the thickness of the top amorphous Si oxide layer. The "Phil" type is the cleaned Si oxide surface which is hydrophilic with a water contact angle $\approx 10°$, whereas the "Phob" type is Si wafer covered hydrophobic monolayer causing water contact angle $> 100°$. The glass has water contact angle of 58°. *Huber* et al. [47] showed that the adhesion force of a gecko spatula rises significantly for substrates with increasing hydrophilicity (adhesive force increases by a factor of two as mating surfaces go from hydrophobic to hydrophilic). As shown in Fig. 20.17b, the simulation results of *Kim* and *Bhushan* [58] closely match the experimental data of *Huber* et al. [47].

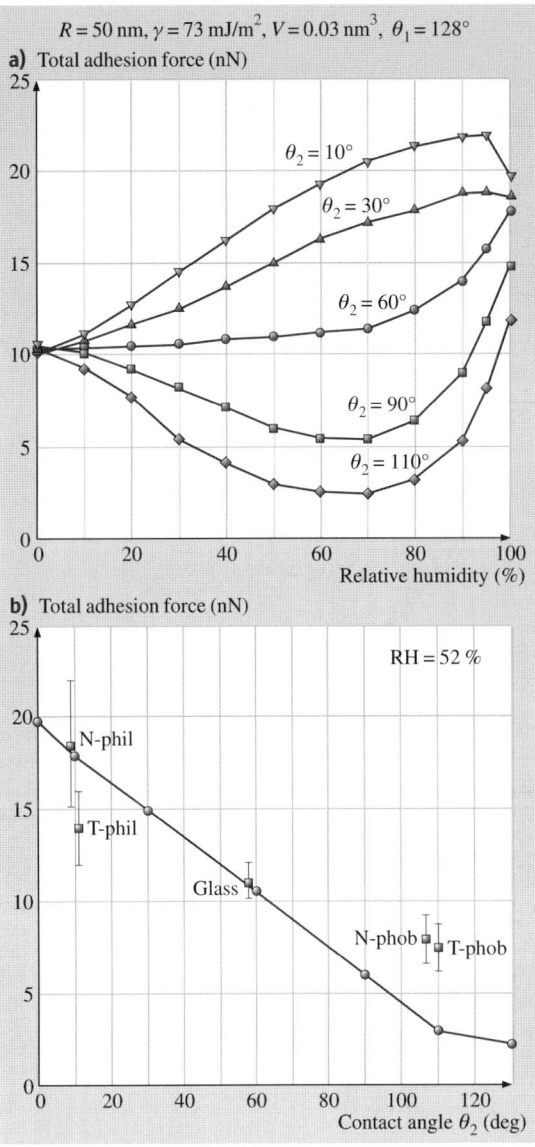

Fig. 20.17. (a) Total adhesion force as a function of relative humidity for a single spatula in contact with surfaces with different contact angles. (b) Comparison of the simulation results of *Kim* and *Bhushan* [58] with the measured data obtained by *Huber* et al. [47] for a single spatula in contact with the hydrophilic and the hydrophobic surfaces [58]

Kim and *Bhushan* [58] carried out adhesion analysis for three-level hierarchical model for gecko seta. Figure 20.18 shows the adhesion coefficient and number of contacts per unit length for the three-level hierarchical model in contact with rough surfaces with different values of the root mean square (RMS) amplitude σ ranging from 0.01 μm to 30 μm for different relative humidities and contact angles of the surface. It can be seen that for the surface with contact angle $\theta_2 = 10°$ the adhesion coefficient is greatly influenced by relative humidity. At 0% relative humidity the

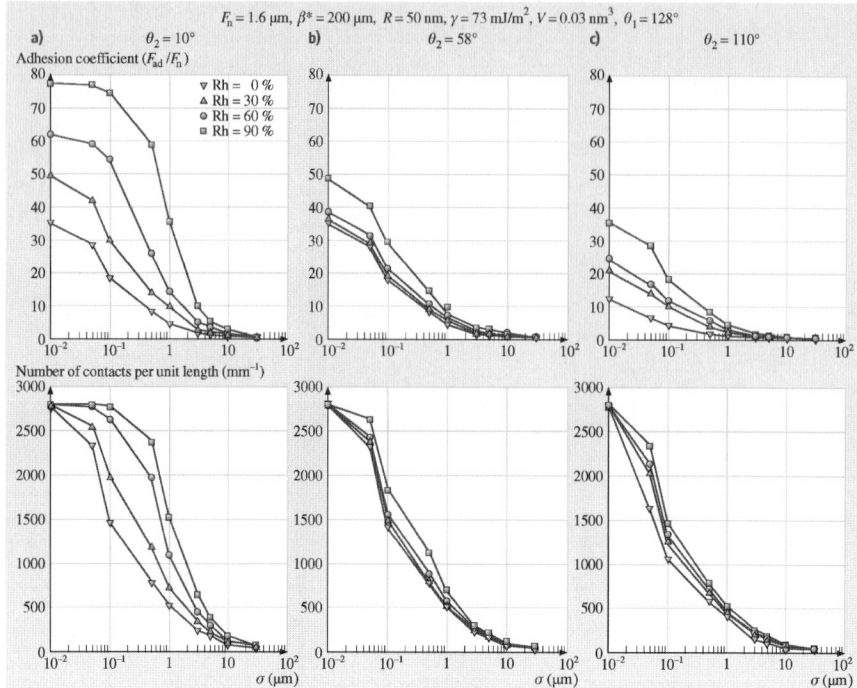

Fig. 20.18. The adhesion coefficient and number of contacts per unit length for three-level hierarchical model in contact with rough surfaces with different values of root mean square amplitudes σ and contact angles for different relative humidities [58]

maximum adhesion coefficient is about 36 at value of σ smaller than 0.01 μm compared to 78 for 90% relative humidity with the same surface roughness. As expected the effect of relative humidity on increasing the adhesion coefficient decreases as the contact angle becomes larger. For hydrophobic surfaces, relative humidity decreases the adhesion coefficient. Similar trends can be noticed in the number of contacts. Thus, the conclusion can be drawn that hydrophilic surfaces are beneficial to gecko adhesion enhancement.

20.7 Modeling of Biomimetic Fibrillar Structures

The mechanics of adhesion between a fibrillar structure and a rough surface as it relates to the design of biomimetic structures has been a topic of investigation by many researchers [31, 34, 35, 52, 56, 57, 69, 82, 94]. In order to better understand the mechanics of adhesion as it relates to design, the approach of *Kim* and *Bhushan* [57] will be described.

Kim and *Bhushan* [57] developed a convenient, general and useful guideline for understanding biological systems and for improving the biomimetic attachment.

This adhesion database was constructed by modeling the fibers as oriented cylindrical cantilever beams with spherical tips. The authors then carried out numerical simulation of the attachment system in contact with random rough surfaces considering three constraint conditions-buckling, fracture and sticking of fiber structure. For a given applied load and roughnesses of contacting surface and fiber material, a procedure to find optimal fiber radius and aspect ratio for the desired adhesion coefficient was developed.

The model of *Kim* and *Bhushan* [57] is used to find the design parameters for fibers of a single-level attachment system capable of achieving desired properties – high adhesion coefficient and durability. The design variables for an attachment system are as follows: fiber geometry (radius and aspect ratio of fibers, tip radius), fiber material, fiber density and fiber orientation. The optimal values for the design variables to achieve the desired properties should be selected for fabrication of a biomimetic attachment system.

20.7.1 Fiber Model

The fiber model of *Kim* and *Bhushan* [57] consists of a simple idealized fibrillar structure consisting of a single-level array of micro/nano beams protruding from a backing as shown in Fig. 20.19. The fibers are modeled as oriented cylindrical cantilever beams with spherical tips. In Fig. 20.18, l is the length of fibers; θ is the fiber orientation; R is the fiber radius; R_t is the tip radius; S is the spacing between fibers; and h is distance between upper spring base of each model and mean line of the rough profile. The end terminal of the fibers is assumed to be a spherical tip with a constant radius and a constant adhesion force.

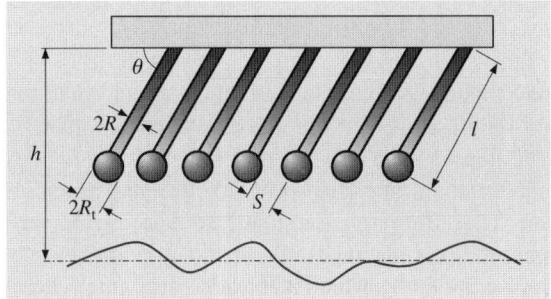

Fig. 20.19. Single-level attachment system with oriented cylindrical cantilever beams with spherical tip. In this figure, l is the length of fibers; θ is the fiber orientation, R is the fiber radius; R_t is the tip radius; S is the spacing between fibers; and h is distance between base of model and mean line of the rough profile [57]

20.7.2 Single Fiber Contact Analysis

Kim and *Bhushan* [57] modeled an individual fiber as a beam oriented at an angle θ to the substrate and the contact load F is aligned normal to the substrate. The net displacement normal to the substrate can be calculated according to (20.16) and (20.17). The fiber stiffness ($k = F/\delta_\perp$) is given by [34]

$$k = \frac{\pi R^2 E}{l \sin^2 \theta \left(1 + \frac{4l^2 \cot^2 \theta}{3R^2}\right)} = \frac{\pi R E}{2\lambda \sin^2 \theta \left(1 + \frac{16\lambda^2 \cot^2 \theta}{3}\right)} \tag{20.30}$$

where $\lambda = l/2R$ is the aspect ratio of the fiber and θ is fixed at 30°.

Two alternative models dominate the world of contact mechanics – the Johnson–Kendall–Roberts (JKR) theory [53] for compliant solids and the Derjaguin-Muller-Toporov (DMT) theory [26] for stiff solids. Although gecko setae are composed of β-keratin with a high elastic modulus [12, 75] which is close to the DMT model, in general the JKR theory prevails for biological or artificial attachment systems. Therefore the JKR theory was applied in the subsequent analysis of *Kim* and *Bhushan* [57] to compare the materials with wide ranges of elastic modulus. The adhesion force between a spherical tip and a rigid flat surface is thus calculated using the JKR theory as [53]

$$F_{ad} = \frac{3}{2}\pi R_t W_{ad} \tag{20.31}$$

where R_t is the radius of spherical tip and W_{ad} is the work of adhesion (calculated according to (20.24)). *Kim* and *Bhushan* [57] used this adhesion force as a critical force. If the elastic force of a single spring is less than the adhesion force, they regarded the spring as having been detached.

20.7.3 Constraints

In the design of fibrillar structures a trade-off exists between the aspect ratio of the fibers and their adaptability to a rough surface. If the aspect ratio of the fibers is too large, they can adhere to each other or even collapse under their own weight as shown in Fig. 20.20a. If the aspect ratio is too small (Fig. 20.20b), the structures will lack the necessary compliance to conform to a rough surface. Spacing between the individual fibers is also important. If the spacing is too small, adjacent fibers can attract each other through intermolecular forces which will lead to bunching. Therefore, *Kim* and *Bhushan* [57] considered three necessary conditions in their analysis; buckling, fracture and sticking of fiber structure, which constrain the allowed geometry.

Non-Buckling Condition

A fibrillar interface can deliver a compliant response while still employing stiff materials because of bending and micro-buckling of fibers. Based on classical Euler

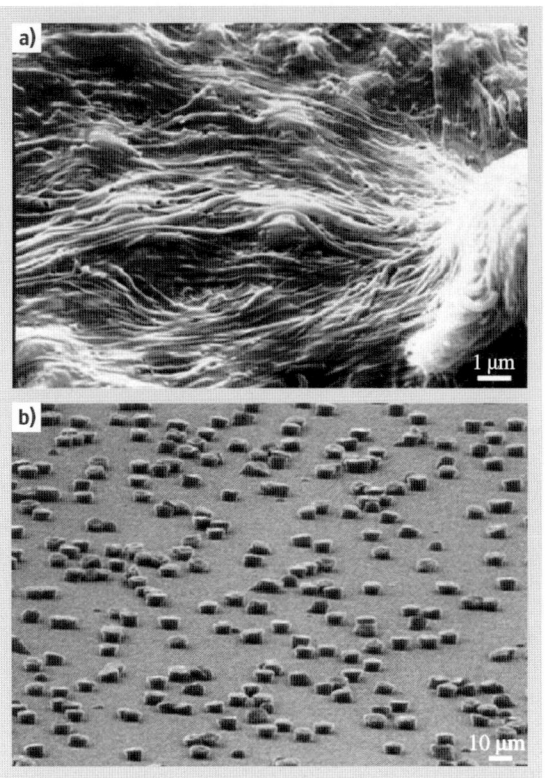

Fig. 20.20. SEM micrographs of (**a**) high aspect ratio polymer fibrils that have collapsed under their own weight and (**b**) low aspect ratio polymer fibrils that are incapable of adapting to rough surfaces [82]

buckling, *Glassmaker* et al. [34] established a stress-strain relationship and a critical compressive strain for buckling, ε_{cr}, for the fiber oriented at an angle, θ, to the substrate,

$$\varepsilon_{cr} = -\frac{b_c \pi^2}{3(Al^2/3I)}\left(1 + \frac{A_c l^2}{3I}\cot^2\theta\right) \quad (20.32)$$

where A_c is the cross-sectional area of the fibril and b_c is a factor that depends on boundary conditions. The factor b_c has a value of 2 for pinned-clamped microbeams. For fibers having a circular cross section, ε_{cr} is calculated as

$$\varepsilon_{cr} = -\frac{b_c \pi^2}{3(4l^2/3R^2)}\left(1 + \frac{4l^2}{3R^2}\cot^2\theta\right) = -b_c\pi^2\left(\frac{1}{16\lambda^2} + \frac{\cot^2\theta}{3}\right) \quad (20.33)$$

In (20.33), ε_{cr} depends on both the aspect ratio, λ, and the orientation, θ, of fibers. If $\varepsilon_{cr} = 1$, which means fiber deforms up to the backing, buckling does not occur. Figure 20.21 plots the critical orientation, θ, as a function of aspect ratio for the case of $\varepsilon_{cr} = 1$. The critical fiber orientation for buckling is 90° at λ less than 1.1. This means that the buckling does not occur regardless of the orientation of fiber at λ less than 1.1. For λ greater than 1.1, the critical fiber orientation for buckling decreases

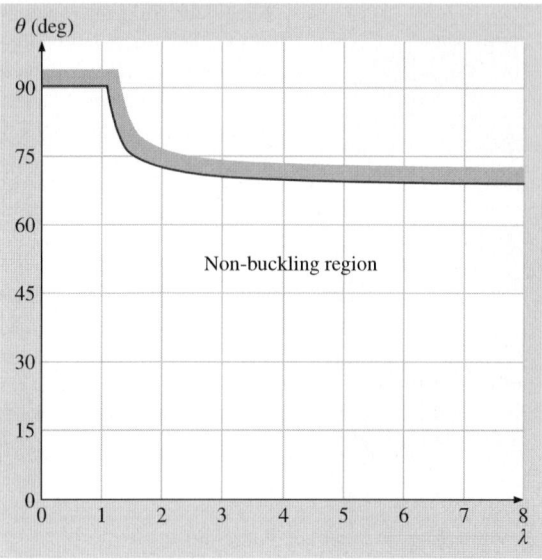

Fig. 20.21. Critical fiber orientation as a function of aspect ratio λ for non-buckling condition for pinned-clamped micro-beams ($b_c = 2$) [57]

with an increase in λ, and has a constant value of 69° at λ greater than 3. *Kim* and *Bhushan* [57] used a fixed value at 30° for θ, because as stated earlier the maximum adhesive force is achieved at this orientation and buckling is not expected to occur.

Non-Fiber Fracture Condition

For small contacts, the strength of the system will eventually be determined by fracture of the fibers. *Spolenak* et al. [84] suggested the limit of fiber fracture as a function of the adhesion force. The axial stress σ_f in a fiber is limited by its theoretical fracture strength σ_{th}^f as

$$\sigma_f = \frac{F_{ad}}{R^2 \pi} \leq \sigma_{th}^f \tag{20.34}$$

Using (20.31), a lower limit for the useful fiber radius R is calculated as

$$R \geq \sqrt{\frac{3R_t W_{ad}}{2\sigma_{th}^f}} \approx \sqrt{\frac{15 R_t W_{ad}}{E}} \tag{20.35}$$

where the theoretical fracture strength is approximated by $E/10$ [27]. The lower limit of fiber radius for fiber fracture by the adhesion force depends on elastic modulus. By assuming $W_{ad} = 66 mJ/m^2$ stated earlier, *Kim* and *Bhushan* [57] calculated the lower limits of fiber radius for $E = 1$ MPa, 0.1 GPa and 10 GPa to be 0.32 µm, 0.032 µm and 0.0032 µm, respectively.

The contact stress cannot exceed the ideal contact strength transmitted through the actual contact area at the instant of tensile instability [84]. *Kim* and *Bhushan* [57] used this condition (20.34) to extract the limit of tip radius, R_t,

$$\sigma_c = \frac{F_{ad}}{a_c^2 \pi} \leq \sigma_{th} \tag{20.36}$$

where σ_c is the contact stress and σ_{th} is the ideal strength of van der Waals bonds which equals approximately W_{ad}/b, b is the characteristic length of surface interaction and a_c is the contact radius. Based on the JKR theory, for the rigid contacting surface, a_c at the instant of pull-off is calculated as

$$a_c = \left(\frac{9\pi W_{ad} R_t^2 (1-v^2)}{8E} \right)^{1/3} \tag{20.37}$$

where v is Poisson ratio. The tip radius can then be calculated by combining (20.36) and (20.37) as

$$R_t \geq \frac{8b^3 E^2}{3\pi^2 (1-v^2)^2 W_{ad}^2} \tag{20.38}$$

The lower limit of tip radius also depends on elastic modulus. Assuming $W_{ad} = 66 \text{ mJ/m}^2$ and $b = 2 \times 10^{-10}$ m [27], the lower limits of tip radius for $E = 1$ MPa, 0.1 GPa and 10 GPa are calculated as 6×10^{-7} nm, 6×10^{-3} nm and 60 nm, respectively. In this study, *Kim* and *Bhushan* [57] fixed the tip radius at 100 nm, which satisfies the tip radius condition throughout a wide range of elastic modulus up to 10 GPa.

Non-Sticking Condition

A high density of fibers is also important for high adhesion. However, if the space S between neighboring fibers is too small, the adhesion forces between them become stronger than the forces required to bend the fibers. Then, fibers might stick to each other and get entangled. Therefore, to prevent fibers from sticking to each other, they must be spaced apart and be stiff enough to prevent sticking or bunching. Several authors (e.g., [82]) have formulated a non-sticking criterion. *Kim* and *Bhushan* [57] adopted the approach of *Sitti* and *Fearing* [82]. Both adhesion and elastic forces will act on bent structures. The adhesion force between neighboring two round tips is calculated as

$$F_{ad} = \frac{3}{2} \pi R_t' W_{ad} \tag{20.39}$$

where R_t' is the reduced radius of contact, which is calculated as $R_t' = (1/R_{t1} + 1/R_{t2})^{-1}$; R_{t1}, R_{t2} – radii of contacting tips; for the case of similar tips, $R_{t1} = R_{t2}$, $R_t' = 2/R_t$.

The elastic force of a bent structure can be calculated by multiplying the bending stiffness ($k_b = 3\pi R^4 E/4l^3$) by a given bending displacement δ as

$$F_{el} = \frac{3}{4} \frac{\pi R^4 E \delta}{l^3} \tag{20.40}$$

The condition for the prevention of sticking is $F_{el} > F_{ad}$. By combining (20.39) and (20.40), a requirement for the minimum distance S between structures which will prevent sticking of the structures is given as [57]

$$S > 2\delta = 2\left(\frac{4}{3}\frac{W_{ad}l^3}{ER^3}\right) = 2\left(\frac{32}{3}\frac{W_{ad}\lambda^3}{E}\right) \tag{20.41}$$

The constant 2 takes into account two nearest structures. Using distance S, the fiber density, ρ, is calculated as

$$\rho = \frac{1}{(S+2R)^2} \tag{20.42}$$

Equation 20.42 was then used to calculate the allowed minimum density of fibers without sticking or bunching. In (20.41), it is shown that the minimum distance, S, depends on both the aspect ratio λ and the elastic modulus E. A smaller aspect ratio and higher elastic modulus allow for greater packing density. However, fibers with a low aspect ratio and high modulus are not desirable for adhering to rough surfaces due to lack of compliance.

20.7.4 Numerical simulation

The simulation of adhesion of an attachment system in contact with random rough surfaces was carried out numerically. In order to conduct 2D simulations it is necessary to calculate applied load F_n as a function of applied pressure P_n as an input condition. Using ρ calculated by non-sticking condition, *Kim* and *Bhushan* [57] calculated F_n as

$$F_n = \frac{P_n p}{\rho} \tag{20.43}$$

where p is the number of springs in scan length L, which equals $L/(S+2R)$.

Fibers of the attachment system are modeled as one-level hierarchy elastic springs (Fig. 20.13) [57]. The deflection of each spring and the elastic force arisen in the springs are calculated according to (20.21) and (20.22), respectively. The adhesion force is the lowest value of elastic force F_{el} when the fiber has detached from the contacting surface. *Kim* and *Bhushan* [57] used an iterative process to obtain optimal fiber geometry – fiber radius and aspect ratio. If the applied load, the roughness of contacting surface and the fiber material are given, the procedure for calculating the adhesion force is repeated iteratively until the desired adhesion force is satisfied. In order to simplify the design problem, fiber material is regarded as a known variable. The next step is constructing the design database. Figure 20.22a shows the flow chart for the construction of adhesion design database and Fig. 20.22b shows the calculation of the adhesion force that is a part of the procedure to construct an adhesion design database.

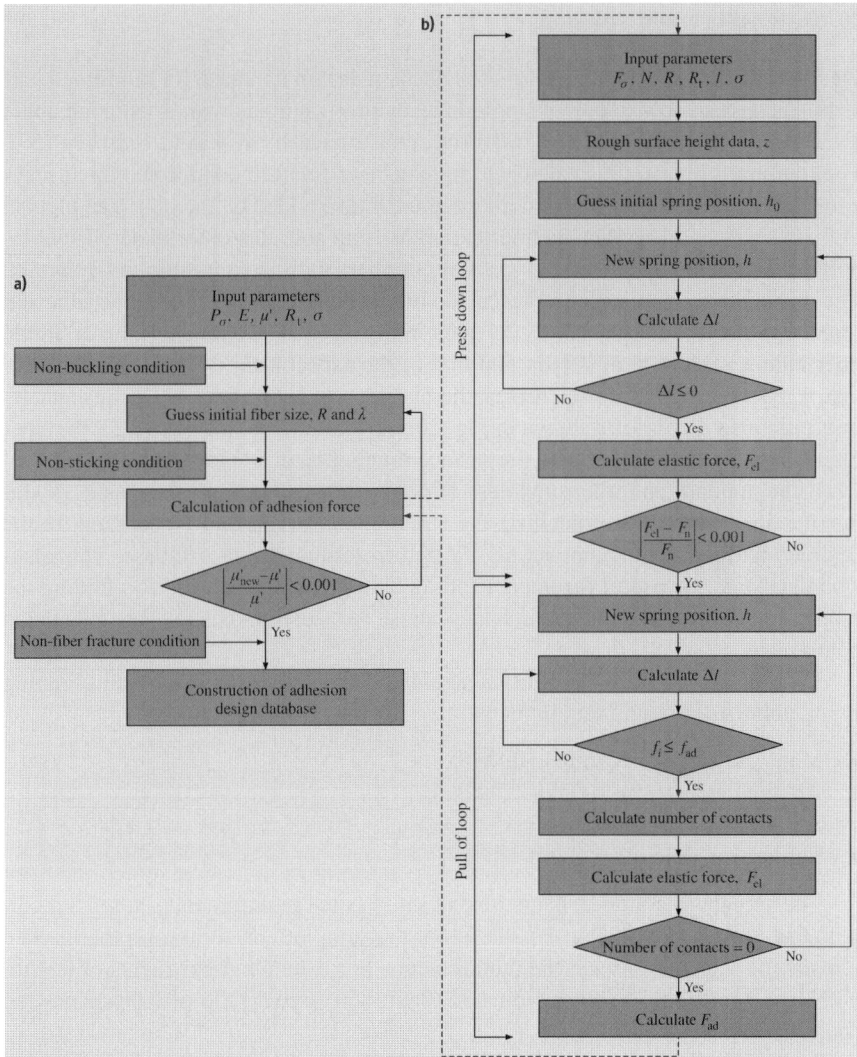

Fig. 20.22. Flow chart for **(a)** the construction of adhesion design database and **(b)** the calculation of the adhesion force. In this figure, P_n is the applied pressure, E is the elastic modulus, μ' is the adhesion coefficient, R_t is the tip radius, σ is root mean square (RMS) amplitude, R is the fiber radius, λ is the aspect ratio of fiber, F_n is the applied load, N is the number of springs, k and l are stiffness and length of structures, Δl is the spring deformation, f_i is the elastic force of a single spring and f_{ad} is the adhesion force of a single contact [57]

20.7.5 Results and discussion

Figure 20.23 shows an example of the adhesion design database for biomimetic attachment systems consisting of single-level cylindrical fibers with orientation angle of 30° and spherical tips of $R_t = 100$ nm constructed by *Kim* and *Bhushan* [57]. The minimum fiber radius calculated by non-fiber fracture condition, which plays a role of lower limit of optimized fiber radius, is also added on the plot. The plots in Fig. 20.22 cover all applicable fiber materials from soft elastomer material such as poly(dimethylsiloxane) (PDMS) to stiffer polymers such as polyimide and β-keratin. The dashed lines in each plot represent the limits of fiber fracture due to the adhesion force. For a soft material with $E = 1$ MPa in Fig. 20.23a, the range of the desirable fiber radius is more than 0.3 μm and that of the aspect ratio is approximately less than 1. As elastic modulus increases, the feasible range of both fiber radius and aspect ratio also increase as shown in Fig. 20.23b and 20.23c. In Fig. 20.23, the fiber radius has a linear relation with the surface roughness on a logarithm scale.

If the applied load, the roughness of contacting surface and the elastic modulus of a fiber material are specified, the optimal fiber radius and aspect ratio for the desired adhesion coefficient can be selected from this design database. The adhesion databases are useful for understanding biological systems and for guiding the fabrication of biomimetic attachment systems. Two case studies [57] are calculated below.

Case study I: Select the optimal size of fibrillar adhesive for a wall climbing robot with the following requirements:

- Material: polymer with $E \approx 100$ MPa
- Applied pressure by weight <10 kPa
- Adhesion coefficient < 5
- Surface roughness $\sigma < 1$ μm.

The subplot of adhesion database that satisfies the requirement is at second column and second row in Fig. 20.23b. From this subplot, any values on the marked line can be selected to meet the requirements. For example, fiber radius of 0.4 μm with aspect ratio of 1 or fiber radius of 10 μm with aspect ratio of 0.8 satisfies the specified requirements.

Case study II: Compare with adhesion test for a single gecko seta [6, 7]:

- Material: β-keratin with $E \approx 10$ GPa
- Applied pressure $= 57$ kPa (2.5 μN on an area of 43.6 μm^2)
- Adhesion coefficient $= 8$ to 16
- Surface roughness $\sigma < 0.01$ μm.

Autumn et al. [6, 7] showed that in isolated gecko setae contacting with surface of a single crystalline silicon wafer, a 2.5 μN preload yielded adhesion of 20 to 40 μN and thus a value of adhesion coefficient of 8 to 16. The region that satisfies the above requirements is marked in Fig. 20.23c. The spatulae of gecko setae have an approximate radius of 0.05 μm with an aspect ratio of 25. However, the radius corresponding to $\lambda = 25$ for the marked line is about 0.015 μm. This discrepancy

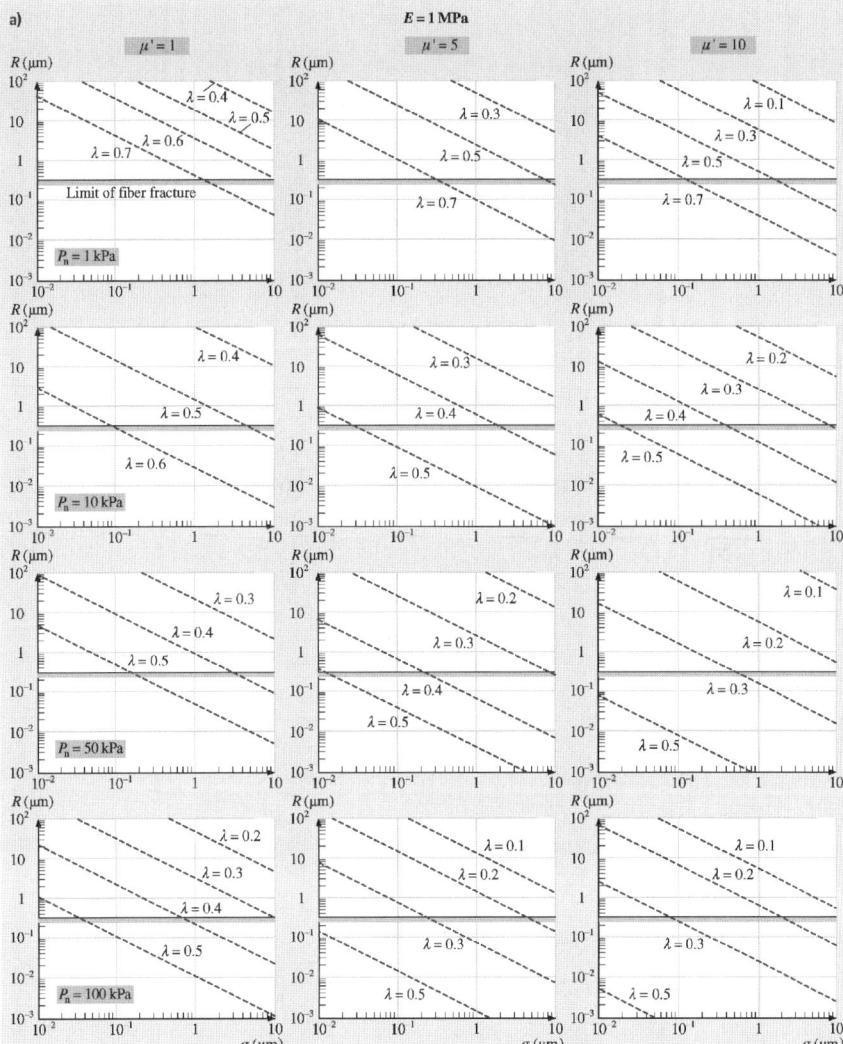

Fig. 20.23. Adhesion design database for biomimetic attachment system consisting of single-level cylindrical fibers with orientation angle of 30° and spherical tips of 100 nm for elastic modulus of (a) 1 MPa, (b) 100 MPa and (c) 10 GPa [57]. The solid lines shown in Figs. (b) and (c) correspond to the cases I and II, respectively, which satisfy the specified requirements

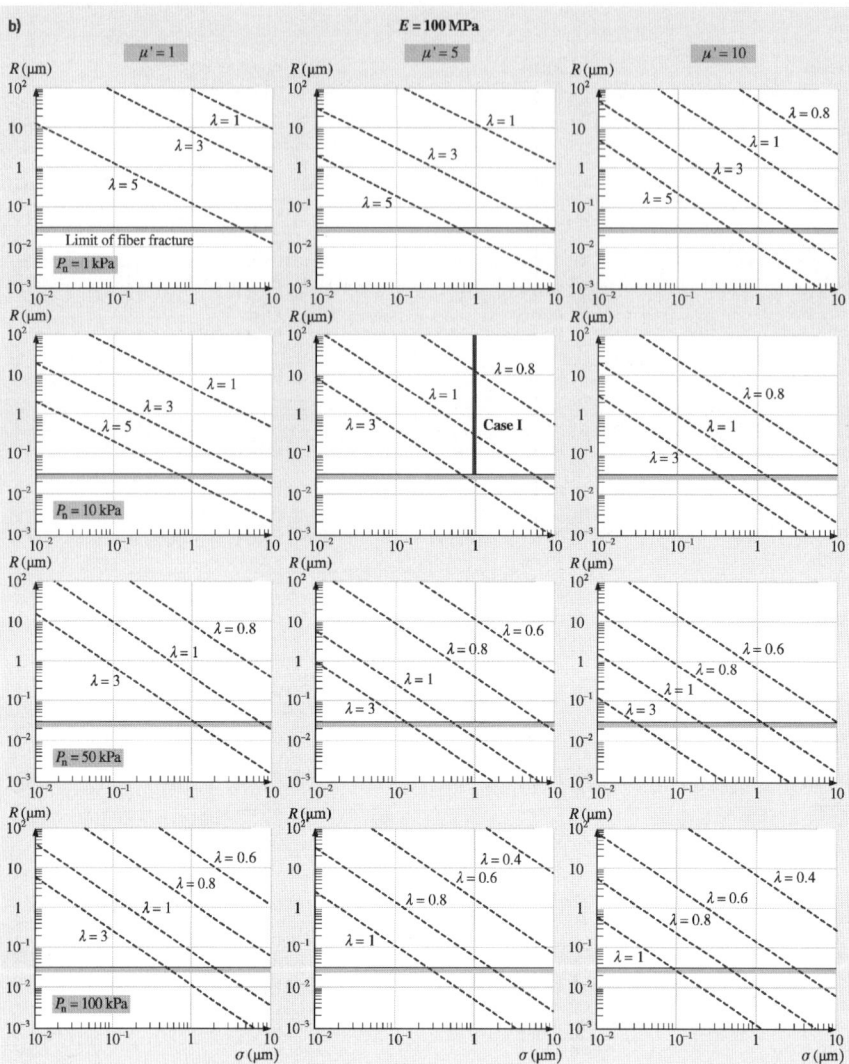

Fig. 20.23. (continued)

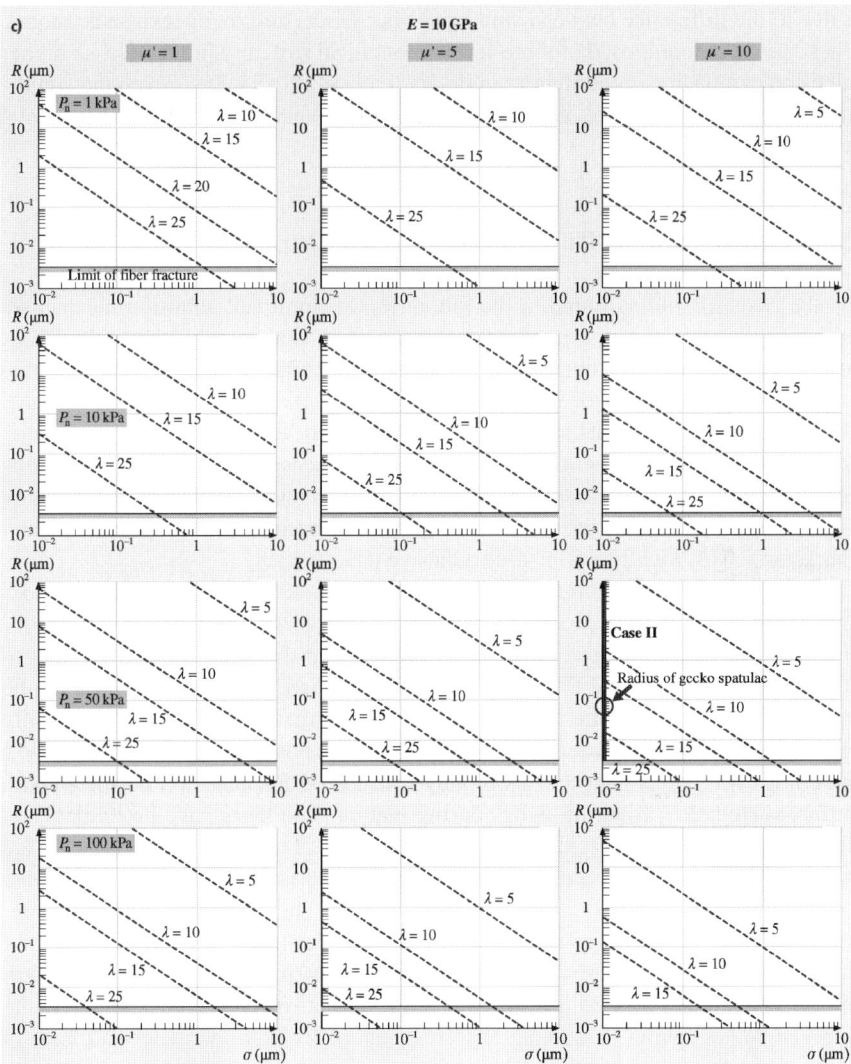

Fig. 20.23. (continued)

is due to the difference between simulated fiber model and real gecko setae model. Gecko setae are composed of three-level hierarchical structure in practice, so higher adhesion can be generated than a single-level model [20, 55, 56]. Given the simplification in the fiber model, this simulation result is very close to the experimental result.

20.8 Fabrication of Biomimetric Gecko Skin

On the basis of studies found in the literature, the dominant adhesion mechanism utilized by geckos and other spider attachment systems appears to be van der Waals forces. The complex divisions of the gecko skin (lamellae-setae-branches-spatulae) enable a large real area of contact between the gecko skin and mating surface. Hence, a hierarchical fibrillar micro/nanostructure is desirable for dry, superadhesive tapes. The development of nanofabricated surfaces capable of replicating this adhesion force developed in nature is limited by current fabrication methods. Many different techniques have been used in an attempt to create [32, 64, 65, 82, 96, 97] and characterize [19, 37, 68] bio-inspired adhesive tapes.

Gorb et al. [37] and *Bhushan* and *Sayer* [19] characterized two polyvinylsiloxane (PVS) samples from Gottlieb Binder Inc., Holzgerlingen, Germany: one consisting of mushroom-shaped pillars (Fig. 20.24a) and the other sample was an unstructured control surface (Fig. 20.24b). The structured sample is inspired by the micropatterns found in the attachment systems of male beetles from the family chrysomelidae, and are easier to fabricate. Both sexes possess adhesive hairs on their tarsi, however, males bear hair extremely specialized for adhesion on the smooth surface of female's covering wings during mating. The hairs have broad flattened tip with grooves under the tip to provide flexibility. The structured samples were produced at room temperature by pouring two-compound polymerizing PVS into the holed template lying on a smooth glass support. The fabricated sample is comprised of pillars that are arranged in a hexagonal order to allow maximum packing density. They are approximately 100 μm in height, 60 μm in base diameter, 35 μm in middle diameter and 25 μm in diameter at the narrowed region just below the terminal contact plates. These plates were of about 40 μm in diameter and 2 μm in thickness at the lip edges. The adhesion force of the two samples in contact with a smooth flat glass substrate was measured by *Gorb* et al. [37] using a home-made microtribometer. Results revealed that the structured specimens featured an adhesion force more than twice that of the unstructured specimens. The adhesion force was also found to be independent of the preload. Moreover, it was found that the adhesive force of the structured sample was more tolerant to contamination compared to the control and it could be easily cleaned with a soap solution.

Bhushan and *Sayer* [19] characterized the surface roughness, friction force, and contact angle of the structured sample and compared the results to an unstructured control. As shown in Fig. 20.25a, the macroscale coefficient of kinetic friction of the structured sample was found to be almost four times greater than the unstructured sample. This increase was determined to be a result of the structured roughness

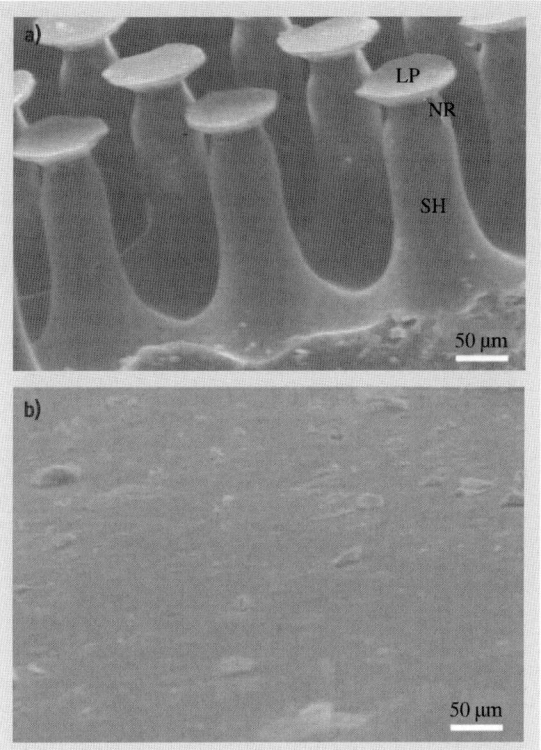

Fig. 20.24. SEM micrographs of the (a) structured and (b) unstructured PVS samples. SH: shaft, NR: neck region, LP: lip [19]

of the sample and not the random nanoroughness. It is also noteworthy that the static and kinetic coefficients of friction are approximately equal for the structured sample. It is believed that the divided contacts allow the broken contacts of the structured sample to constantly recreate contact. As seen in Fig. 20.25b, the pillars also increased the hydrophobicity of the structured sample in comparison to the unstructured sample as expected due to increased surface roughness [22,92]. A large contact angle is important for self cleaning [10], which agrees with the findings of Gorb et al. [37] that the structured sample is more tolerant of contamination than the unstructured sample.

20.8.1 Single Level Hierarchical Structures

One of the simplest approaches to create a single level hierarchical surface employed an AFM tip to create a set of dimples on a wax surface. These dimples served as a mold for creating polymer nanopyramids shown in Fig. 20.26a [82]. The adhesive force to an individual pyramid was measured using another AFM cantilever. The force was found to be about 200 μN. Although each pyramid of the material is capable of producing large adhesive forces, the surface failed to replicate gecko adhesion on a macroscale. This was due to the lack of flexibility in the pyramids. In

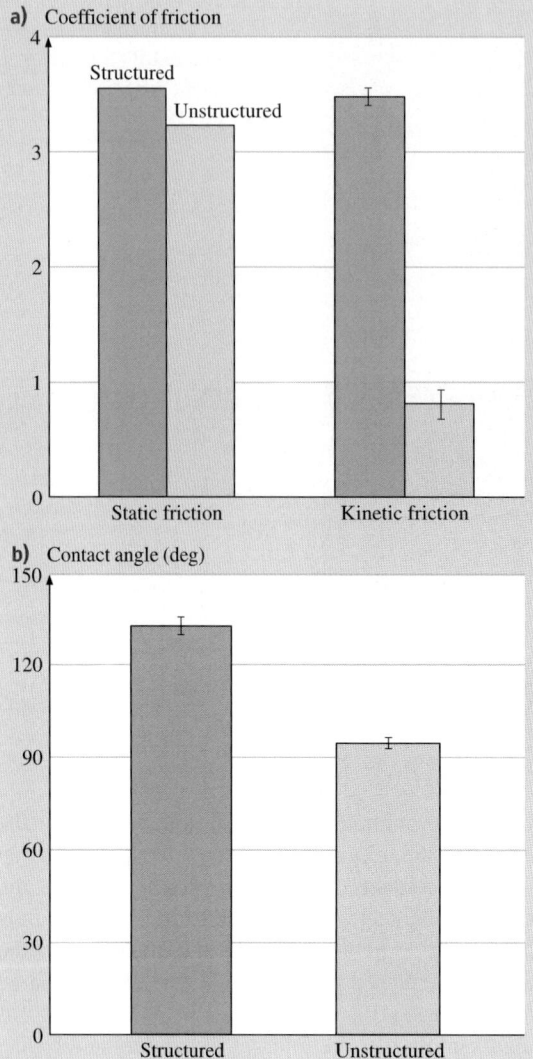

Fig. 20.25. (a) Coefficients of static and kinetic friction for the structured and unstructured samples slid against magnetic tape with a normal load of 130 mN.
(b) Water contact angle for the structured and unstructured samples [19]

order to ensure that the largest possible area of contact occurs between the tape and mating surface, a soft, compliant fibrillar structure would be desired [52]. As shown in previous calculations, the van der Waals adhesive force for two parallel surfaces is inversely proportional to the cube of the distance between two surfaces. Compliant fibrillar structures enable more fibrils to be in close proximity of a mating surface, thus increasing van der Waals forces.

Geim et al. [32] created arrays of nanohairs using electron-beam lithography and dry etching in oxygen plasma (Fig. 20.26b (left)). The original arrays were created on a rigid silicon wafer. This design was only capable of creating 0.01 N

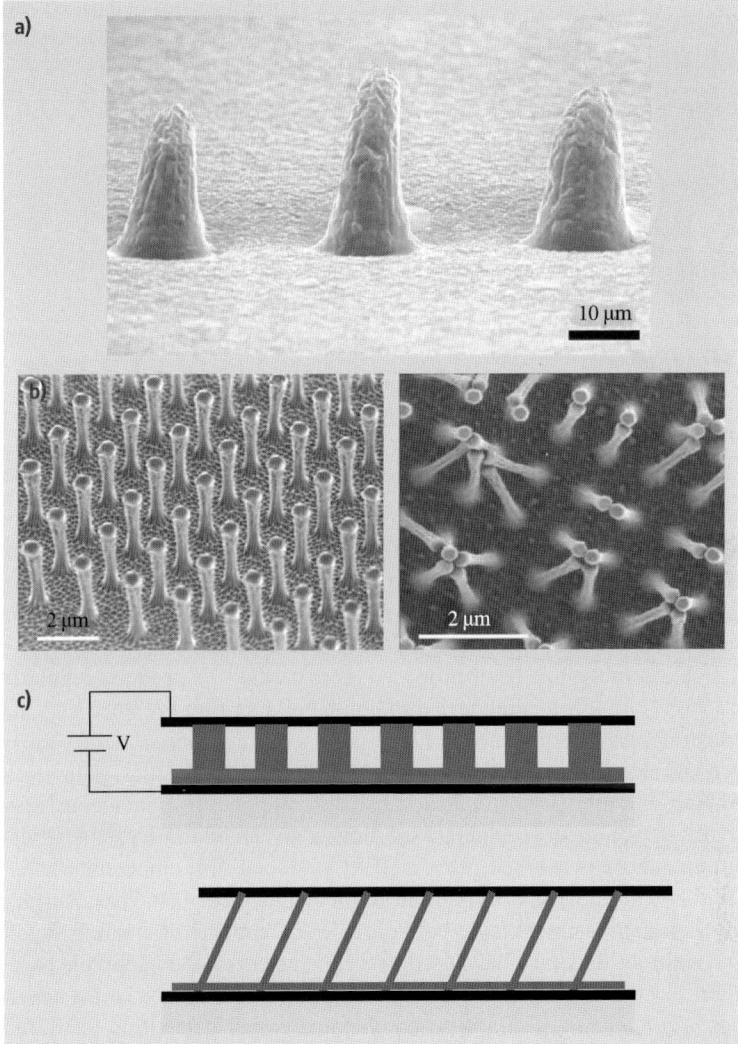

Fig. 20.26. SEM micrographs of (**a**) three pillars created by nano-tip indentation [81], (**b**) (*left*) an array of polyimide nanohairs and (*right*) bunching of the nanohairs, which leads to a reduction in adhesive force [32], and (**c**) directed self-assembly based method of producing high aspect ratio micro/nanohairs [81]

of adhesive force for a 1 cm^2 patch. The nanohairs were then transferred from the silicon wafer to a soft polymer substrate. A 1 cm^2 sample was able to create 3 N of adhesive force under the new arrangement. This is approximately one-third the adhesive strength of a gecko. The fabricated a Spiderman toy (about 0.4 N) with a hand covered with molded polymer nanohairs, Fig. 20.27. The demonstrated that

Fig. 20.27. A Spiderman toy (about 0.4 N) with a hand covered with the molded polymer nanohairs, clinging to a glass plate [32]

it could cling to a glass plate. Bunching of the nanohairs (as described earlier), if they are closely spaced was determined to greatly reduce the both the adhesive strength and durability of the polymer tape. The bunching can be clearly seen in Fig. 20.26b (right). Therefore, an optimal geometry is required.

Directed self-assembly has been proposed as a method to produce regularly spaced fibers [76,81]. In this technique, a thin liquid polymer film is coated on a flat conductive substrate. As demonstrated in Fig. 20.26c, a closely spaced metal plate is used to apply a DC electric field on the polymer film. Due to instabilities, pillars will begin to grow. Self-assembly is desirable because the components spontaneously assemble, typically by bouncing around in a solution or gas phase until a stable structure of minimum energy is reached. This method is crucial in biomolecular nanotechnology, and has the potential to be used in precise devices [1]. These surface coatings have been demonstrated to be both durable and capable of creating superhydrophobic conditions and have been used to form clusters on the nanoscale [67].

Multiwalled carbon nanotube (MWCNT) hairs have been used to create superadhesive tapes. *Yurdumakan* et al. [96] used chemical vapor deposition (CVD) to grow MWCNT that are 50–100 μm in length on quartz or silicon substrates. Patterns are then created using a combination of photolithography and a wet and/or dry etching. SEM images of the nanotube surfaces can be seen in Fig. 20.28. On a small scale (nanometer level), the MWCNT surface was able to achieve adhesive forces two orders of magnitude greater than those of gecko foot-hairs. These structures were only designed to increase adhesion on the nanometer level and were not capable or producing high adhesive forces on the macroscale. *Zhao* et al. [97] created MWCNT arrays that are much more capable of macroscale adhesion. 5–10 μm MWCNT were grown on a Si substrate. This arrangement was able to create and adhesive pressure of 11.7 N/cm^2. Durability of the adhesive tape is an issue as some of the nanotubes detach from the substrate with repeated use. This work does show promise for MWCNT being implemented in bio-inspired tapes.

Fig. 20.28. Multi-walled carbon nanotube structures: (*left*) grown on silicon by vapor deposition, (*right*) transferred into a PMMA matrix and then exposed on the surface after solvent etching [96]

20.8.2 Multi-Level Hierarchical Structures

The aforementioned fabricated surfaces only have one level of hierarchy. Although these surfaces are capable of producing high adhesion on the micro/nanoscale, all have failed in producing large scale adhesion due to a lack of compliance and bunching. In order to overcome these problems, *Northen* and *Turner* [64, 65] created a multi-level compliant system by employing a microelectromechanical based approach. The created a layer of nanorods which they deemed "organorods" (Fig. 20.29a). These organorods are comparable in size to that of gecko spatulae (50–200 nm in diameter and 2 μm tall). They sit atop silicon dioxide chip (approximately 2 μm thick and 100–150 μm across a side), which were created using photolithography (Fig. 20.29b). Each chip is supported on top of a pillar (1 μm in diameter and 50 μm tall) that attaches to a silicon wafer (Fig. 20.29c). The multilevel structures have been created across a 100 mm wafer (Fig. 20.29d).

Adhesion testing was performed using a nanorod surface on a solid substrate and on the multilevel structures. As seen in Fig. 20.30, adhesive pressure of the multilevel structures was several times higher than that of the surfaces with only one level of hierarchy. The durability of the multilevel structure was also much greater than the single level structure. The adhesion of the multilevel structure did not change between iterations one and five. During the same number of iterations, the adhesive pressure of the single level structure decreased to zero.

Sitti [81] proposed a nano-molding technique for creating structures with two levels of structures. In this method two different molds are created – one with pores on the order of magnitude of microns in diameter and a second with pores of nanometer scale diameter. One potential mold material is porous anodic alumina (PAA), which has been demonstrated to produce ordered pores on the nanometer scale of equal size [62]. Pore widening techniques could be used to create micron

Fig. 20.29. Multilevel fabricated adhesive structure composed of (**a**) organorods, (**b**) silicon dioxide chips, and (**c**) support pillars. (**d**) This structure was repeated multiple times over a silicon wafer [64, 65]

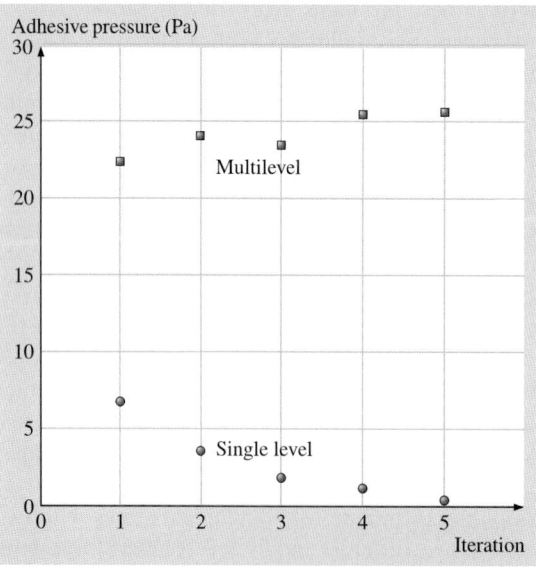

Fig. 20.30. Adhesion test results of a multilevel hierarchical structure (*top*) and a single level hierarchical structure (*bottom*) repeated for five iterations [65]

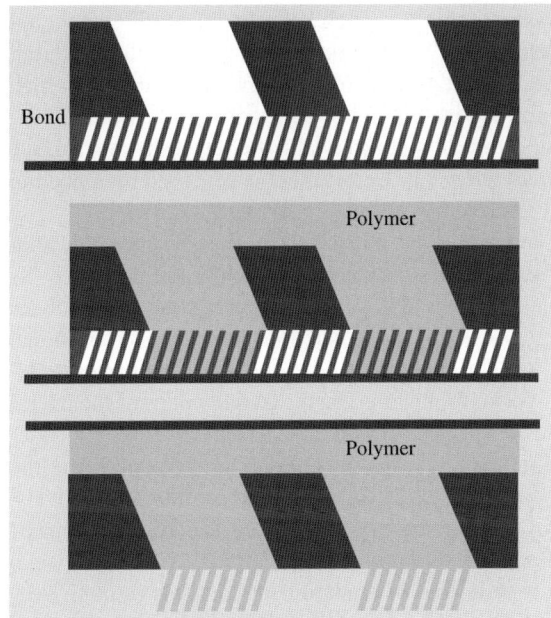

Fig. 20.31. Proposed process of creating multi-level synthetic gecko foot hairs using nano-molding. **(a)** Micron and nanometer sized pore membranes are bonded together and **(b)** filled with liquid polymer. **(c)** The membranes are then etched away leaving the polymer surface [81]

scale pores. As seen in Fig. 20.31, the two molds would be bonded to each other and then filled with a liquid polymer. According to *Sitti* [81], the method would enable the manufacturing of a high volume of synthetic gecko foot hairs at low cost.

Literature clearly indicates that in order to create a dry superadhesive, a fibrillar surface construction is necessary to maximize the van der Waals forces by decreasing the distance between the two surfaces. It is also desirable to have a superhydrophobic surface in order to utilize self cleaning. A material must be soft enough to conform to rough surfaces yet hard enough to avoid bunching, which will decrease the adhesive force.

20.9 Closure

The adhesive properties of geckos and other creatures such as flies, beetles and spiders, are due to the hierarchical structures present on each creature's hairy attachment pads. Geckos have developed the most intricate adhesive structures of any of the aforementioned creatures. The attachment system consists of ridges called lamellae that are covered in microscale setae that branch off into nanoscale spatulae. Each structure plays an important role in adapting to surface roughness bringing the spatulae in close proximity with the mating surface. These structures as well as material properties allow the gecko to obtain a much larger real area of contact between its feet and a mating surface than is possible with a non-fibrillar material. Two feet of a Tokay gecko have about 220 mm^2 of attachment pad area on which the gecko is able to generate approximately 20 N of adhesion force. Although capable of generating high adhesion forces, a gecko is able to detach from a surface at will – an ability known as smart adhesion. Detachment is achieved by a peeling motion of the gecko's feet from a surface.

Experimental results have supported the adhesive theories of intermolecular forces (van der Waals) as a primary adhesion mechanism and capillary forces as a secondary mechanism, and have been used to rule out several other mechanisms of adhesion including the secretion of sticky fluids, suction, and increased frictional forces. Atomic force microscopy has been employed by several investigators to determine the adhesive strength of gecko foot hairs. The measured values of the lateral force required to pull parallel to the surface for of a single seta (194 μN) and adhesive force (normal to the surface) of a single spatula (11 nN) are comparable to the van der Waals prediction of 270 μN and 11 nN for a seta and spatula, respectively. The adhesion force generated by seta increases with preload and reaches a maximum when both perpendicular and parallel preloads are applied. Although gecko feet are strong adhesives, they remain free of contaminant particles through self-cleaning. Spatular size along with material properties enable geckos to easily expel any dust particles that come into contact with their feet.

Recent creation of a three-level hierarchical model for a gecko lamella consisting of setae, branches and spatulae has brought more insight into adhesion of biological attachment systems. One-, two- and three-level hierarchically structured spring models for simulation of a seta contacting with random rough surfaces were considered. The simulation results show that the multi-level hierarchical structure has a higher adhesion force as well as higher adhesion energy than the one-level structure for a given applied load, due to better adaptation and attachment ability. It is concluded that the multi-level hierarchical structure produces adhesion enhancement, and this enhancement increases with an increase in the applied load and a decrease in the stiffness of springs. The condition at which a significant adhesion enhancement occurs appears to be related to the maximum spring deformation. The result shows that significant adhesion enhancement occurs when the maximum spring deformation is greater than two to three times larger than σ value of surface roughness. For the effect of applied load, as the applied load increases, adhesion force increases up to a certain applied load and then has a constant value, whereas, adhesion energy continues to increase with an increase in the applied load. Inclusion of capillary forces in the spring model shows that total adhesion force decreases with an increase in the contact angle of water on the substrate, and the difference of total adhesion force among different contact angles is larger in the intermediate humidity regime. In addition, the simulation results match the measured data for a single spatula in contact with both the hydrophilic and the hydrophobic surfaces which further supports van der Waals forces as the dominant mechanism of adhesion and capillary forces as a secondary mechanism.

There is a great interest among the scientific community to create surfaces that replicate the adhesion strength of gecko feet. These surfaces would be capable of reusable dry adhesion and would have uses in a wide range of applications from everyday objects such as tapes, fasteners, and toys to microelectronic and space applications and even wall-climbing robots. In the design of fibrillar structures, it is necessary to ensure that the fibrils are compliant enough to easily deform to the mating surface's roughness profile, yet rigid enough to not collapse under their own weight. Spacing between the individual fibrils is also important. If the spacing is too small, adjacent fibrils can attract each other through intermolecular forces which will lead to bunching. The adhesion design database developed by *Kim* and *Bhushan* [57] serves as a reference for choosing design parameters.

Nano-indentation, electron-beam lithography, and growing of carbon nanotube arrays are all methods that have been used to create fibrillar structures. The limitations of current machining methods on the micro/nanoscale have resulted in the majority of fabricated surfaces consisting of only one level of hierarchy. Although typically capable of producing high adhesive force with an individual fibril, these surfaces have failed to generate high adhesive forces on the macroscale. Bunching, lack of compliance, and lack of durability are all problems that have arisen with the aforementioned structures. Recently, a multi-layered compliant system was created using a microelectromechanical based approach in combination with nanorods. This method as well as other proposed methods of multilevel nano-molding and directed self assembly show great promise in the creation of adhesive structures with multiple levels of hierarchy, much like those of gecko feet.

20.A Typical Rough Surfaces

Several natural (sycamore tree bark and siltstone) and artificial surfaces (dry wall, wood laminate, steel, aluminum, and glass) were chosen to determine the surface parameters of typical rough surfaces that a gecko might encounter. An Alpha-step® 200 (Tencor Instruments, Mountain View, CA) was used to obtain surface profiles for three different scan lengths, 80 μm, which is approximately the size of a single gecko seta, 2000 μm, which is close to the size of a gecko lamella, and an intermediate scan length of 400 μm. The radius of the stylus tip was 1.5–2.5 μm and the applied normal load was 3 mg. The surface profiles were then analyzed using a specialized computer program to determine the root mean square amplitude, σ, correlation length, β^*, peak to valley distance, P–V, skewness, Sk, and kurtosis, K.

Sample surface profiles and their corresponding parameters at a scan length of 2000 μm can be seen in Fig. 20.32a. The roughness amplitude, σ, varies from as low as 0.01 μm in glass to as high as 30 μm in tree bark. Similarly, correlation length varies from 2 to 300 μm. The scan length dependency of surface parameters is illustrated in Fig. 20.32b. As the scan length of the profile increases, so do the roughness amplitude and correlation length. Table 20.4 summarizes the scan length dependent parameters σ and β^* for all seven sampled surfaces. At a scale length of 80 μm (size of seta), the roughness amplitude does not exceed 5 μm while at a scale length of

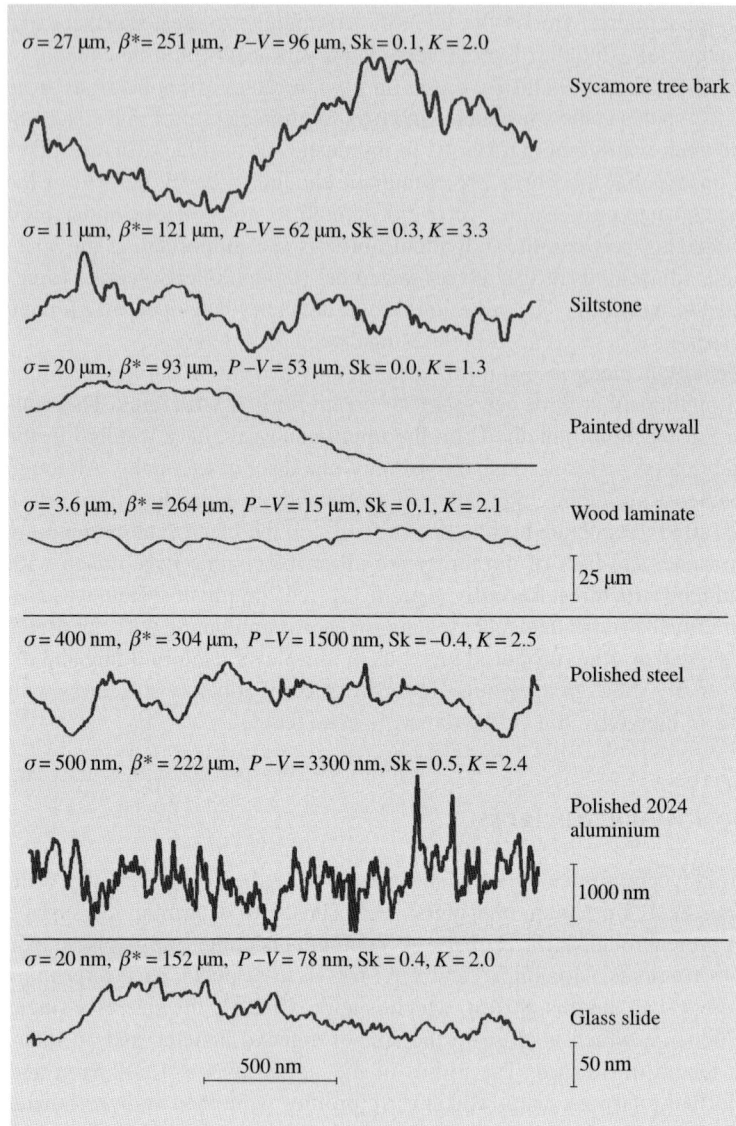

Fig. 20.32. (a) Surface height profiles of various random rough surfaces of interest at a 2000 μm scan length

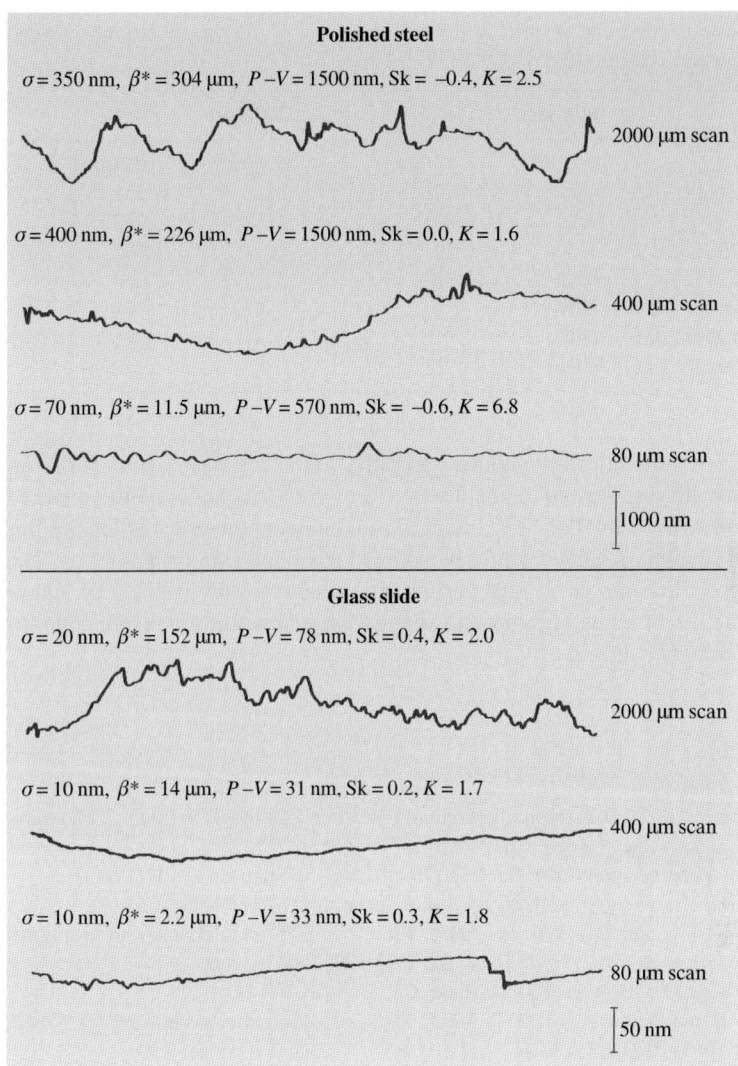

Fig. 20.32. (continued) **(b)** a comparison of the profiles of two surfaces at 80, 400, and 2000 μm scan lengths [20]

Table 20.4. Scale dependence of surface parameters σ and β^* for rough surfaces at scan lengths of 80 µm and 2000 µm [20]

Scan length	80 µm		2000 µm	
Surface	σ (µm)	β^* (µm)	σ (µm)	β^* (µm)
Sycamore tree bark	4.4	17	27	251
Siltstone	1.1	4.8	11	268
Painted drywall	1	11	20	93
Wood laminate	0.11	18	3.6	264
Polished steel	0.07	12	0.40	304
Polished 2024 aluminum	0.40	6.5	0.50	222
Glass	0.01	2.2	0.02	152

2000 µm (size of lamella), the roughness amplitude is as high as 30 µm. This suggests that setae are responsible to adapt to surfaces with roughness on the order of several microns while lamellae must adapt to roughness on the order of tens of microns. Larger roughness values would be adapted to by the skin of the gecko. The spring model of *Bhushan* et al. [20] verifies that setae are only capable of adapting to roughnesses of a few microns and suggests that lamellae are responsible for adaptation to rougher surfaces.

References

1. Anonymous (2002), Self-assembly, *Merriam-Webster's Medical Dictionary*, Merriam-Webster
2. Aristotle, *Historia Animalium*, Book IX, Part 9, trans. Thompson DAW (1918), http://classics.mit.edu/Aristotle/history-anim.html
3. Arzt E, Gorb S and Spolenak R (2003), From micro to nano contacts in biological attachment devices, *Proc. Natl. Acad. Sci. USA* **100**, 10603–10606
4. Autumn K (2006), How gecko toes stick, *Am. Scientist*, **94**, 124–132
5. Autumn K and Peattie AM (2002), Mechanisms of adhesion in geckos, *Integr. Comp. Biol.* **42**, 1081–1090
6. Autumn K, Liang YA, Hsieh ST, Zesch W, Chan WP, Kenny TW, Fearing R and Full RJ (2000), Adhesive force of a single gecko foot-hair *Nature*, **405**, 681–685
7. Autumn K, Sitti M, Liang YA, Peattie AM, Hansen WR, Sponberg S, Kenny TW, Fearing R, Israelachvili JN and Full RJ (2002), Evidence for van der Waals adhesion in gecko setae, *Proc. Natl. Acad. Sci. USA* **99**, 12252–12256
8. Autumn K, Buehler M, Cutkosky M, Fearing R, Full R. J., Goldman D., Groff R., Provancher W., Rizzi A. A., Sranli U., Saunders A. and Koditschek D. E. (2005), Robotics in scansorial environments, *Proc. SPIE*, **5804**, 291–302
9. Autumn K, Majidi C, Groff RE, Dittmore A, and Fearing R (2006), Effective elastic modulus of isolated gecko setal arrays, *J. Exp. Biol.* **209**, 3558–3568
10. Barthlott W and Neinhuis C (1997), Purity of the sacred lotus, or escape from contamination in biological surfaces, *Planta* **202**, 1–8

11. Bergmann PJ and Irschick DJ (2005), Effects of temperature on maximum clinging ability in a diurnal gecko: evidence for a passive clinging mechanism? *J. Exp. Zool.* **303A**, 785–791
12. Bertram JEA and Gosline JM (1987), Functional design of horse hoof keratin: the modulation of mechanical properties through hydration effects, *J. Exp. Biol.* **130**, 121–136
13. Bhushan B (1996), *Tribology and Mechanics of Magnetic Storage Devices*, 2nd Ed., Springer-Verlag, New York
14. Bhushan B (1999), *Principles and Applications of Tribology*, Wiley, New York
15. Bhushan B (2002), *Introduction to Tribology*, Wiley, New York
16. Bhushan B (ed.) (2005), *Nanotribology and Nanomechanics – an Introduction*, Springer-Verlag, Heidelberg, Germany
17. Bhushan B (ed.) (2007), *Springer Handbook of Nanotechnology*, 2nd Ed., Springer-Verlag, Heidelberg, Germany
18. Bhushan B (2007), Adhesion of multi-level hierarchical attachment systems in gecko feet, *J. Adhesion Sci. Technol.* **21**, 1213–1258
19. Bhushan B and Sayer RA (2007), Surface characterization and friction of a bio-inspired reversible adhesive tape, *Microsyst. Technol.* **13**, 71–78
20. Bhushan B, Peressadko AG, and Kim TW (2006) Adhesion analysis of two-level hierarchical morphology in natural attachment systems for "smart adhesion", *J. Adhesion Sci. Technol.* **20**, 1475–1491
21. Bikerman JJ (1961), *The Science of Adhesive Joints*, Academic, New York
22. Burton Z and Bhushan B (2005), Hydrophobicity, adhesion, and friction properties of nanopatterned polymers and scale dependence for micro- and nanoelectromechanical systems, *Nano Letters* **5**, 1607–1613
23. Cai S and Bhushan B (2007), Effects of symmetric and asymmetric contact angles and division of menisci on meniscus and viscous forces during separation, *Philos. Mag.* **87**, 5505–5522
24. Chui BW, Kenny TW, Mamin HJ, Terris BD, and Rugar D (1998), Independent detection of vertical and lateral forces with a sidewall-implanted dual-axis piezoresistive cantilever, *Appl. Phys. Lett.* **72**, 1388–1390
25. Dellit WD (1934), Zur Anatomie und Physiologie der Geckozehe, *Jena. Z. Naturwissen*, **68**, 613–658
26. Derjaguin BV, Muller VM, and Toporov YuP (1975), Effect of contact deformation on the adhesion of particles, *J. Colloid Interface Sci.* **53**, 314–326
27. Dieter GE (1988), *Mechanical Metallurgy*, McGraw Hill, London
28. Fan PL and O'Brien MJ (1975), Adhesion in deformable isolated capillaries, *Adhesion Science and Technology*, ed. Lee L.H., **9A**, 635, Plenum, New York
29. Federle W (2006) Why are so many adhesive pads hairy? *J. Exp. Biol.* **209**, 2611–2621
30. Federle W, Riehle M, Curtis ASG, and Full RJ (2002), An integrative study of insect adhesion: mechanics of wet adhesion of pretarsal pads in ants, *Integr. Comp. Biol.* **42**, 1100–1106
31. Gao H, Wang X, Yao H, Gorb Sm, and Arzt E (2005), Mechanics of hierarchical adhesion structures of geckos, *Mech. Mater.* **37**, 275–285
32. Geim AK, Dubonos SV, Grigorieva IV, Novoselov KS, Zhukov AA, and Shapoval SY (2003), Microfabricated adhesive mimicking gecko foot-hair, *Nat. Mater.* **2**, 461–463
33. Gennaro JGJ (1969), The gecko grip, *Nat. Hist.* **78**, 36–43
34. Glassmaker NJ, Jagota A, Hui CY, and Kim J (2004), Design of biomimetic fibrillar interfaces: 1. Making contact, *J. R. Soc. Interface* **1**, 23–33

35. Glassmaker NJ, Jagota A, and Hui CY (2005), Adhesion enhancement in a biomimetic fibrillar interface, *Acta Biomaterialia* **1**, 367–375
36. Gorb S (2001), *Attachment Devices of Insect Cuticles*, Kluwer, Dordrecht, Netherlands
37. Gorb S, Varenberg M, Peressadko A, and Tuma J (2007), Biomimetic mushroom-shaped fibrillar adhesive microstructures, *J. Royal Soc. Interf.* **4**, 271–275
38. Hamaker HC (1937), London van der Waals attraction between spherical bodies, *Physica*, **4**, 1058
39. Han D, Zhou K, and Bauer AM (2004), Phylogenetic relationships among gekkotan lizards inferred from C-mos nuclear DNA sequences and a new classification of the Gekkota, *Biol. J. Linn. Soc.* **83**, 353–368
40. Hanna G and Barnes WJP (1991), Adhesion and detachment of the toe pads of tree frogs, *J. Exper. Biol.* **155**, 103–125
41. Hansen WR and Autumn K (2005), Evidence for self-cleaning in gecko setae, *Proc. Natl. Acad. Sci. USA* **102**, 385–389
42. Hiller U (1968), Untersuchungen zum Feinbau und zur Funktion der Haftborsten von Reptilien, *Z. Morphol. Tiere*, **62**, 307–362
43. Hinds WC (1982), *Aerosol Technology: Properties, Behavior, and Measurement of Airborne Particles*, Wiley, New York
44. Hora SL (1923), The adhesive apparatus on the toes of certain geckos and tree frogs, *J. Asiat. Soc. Beng.* **9**, 137–145
45. Houwink R and Salomon G (1967), Effect of contamination on the adhesion of metallic couples in ultra high vacuum, *J. Appl. Phys.* **38**, 1896–1904
46. Huber G, Gorb SN, Spolenak R, and Arzt E (2005), Resolving the nanoscale adhesion of individual gecko spatulae by atomic force microscopy. *Biol. Lett.* **1**, 2–4
47. Huber G, Mantz H, Spolenak R, Mecke K, Jacobs K, Gorb SN, and Arzt E (2005), Evidence for capillarity contributions to gecko adhesion from single spatula and nanomechanical measurements, *Proc. Natl. Acad. Sci. USA* **102**, 16293–16296
48. Irschick DJ, Austin CC, Petren K, Fisher RN, Losos JB, and Ellers O (1996), A comparative analysis of clinging ability among pad-bearing lizards, *Biol. J. Linn. Soc.* **59**, 21–35
49. Israelachvili JN and Tabor D (1972), The measurement of Van der Waals dispersion forces in the range of 1.5 to 130 nm, *Proc. R. Soc. Lond. A.* **331**, 19–38
50. Israelachvili JN (1992), *Intermolecular and Surface Forces*, 2nd edition, Academic, San Diego
51. Jaenicke R (1998) Atmospheric aerosol size distribution, *Atmospheric Particles*, RM Harrison and R van Grieken, eds., Wiley, New York, 1–29
52. Jagota A and Bennison SJ (2002), Mechanics of adhesion through a fibrillar microstructure, *Integr. Comp. Biol.* **42**, 1140–1145
53. Johnson KL, Kendall K, and Roberts AD (1971), Surface energy and the contact of elastic solids, *Proc. R. Soc. Lond. A.* **324**, 301–313
54. Kesel AB, Martin A, and Seidl T (2003), Adhesion measurements on the attachment devices of the jumping spider *Evarcha arcuata*, *J. Exp. Biol.* **206**, 2733–2738
55. Kim TW and Bhushan B (2007), The adhesion analysis of multi-level hierarchical attachment system contacting with a rough surface, *J. Adhesion Sci. Technol.* **21**, 1–20
56. Kim TW and Bhushan B (2007), Effect of stiffness of multi-level hierarchical attachment system on adhesion enhancement, *Ultramicroscopy* **107** 902–912
57. Kim TW and Bhushan B (2007), Optimization of biomimetic attachment system contacting with a rough surface, *J. Vac. Sci. Technol. A* **25**, 1003–1012
58. Kim TW and Bhushan B (2008), The adhesion model considering capillarity for gecko attachment system, *J. Royal Soc. Interf.* **5**, 319–327

59. Kluge AG (2001), Gekkotan lizard taxonomy, *Hamadryad* **26**, 1–209
60. Losos JB (1990), Thermal sensitivity of sprinting and clinging performance in the Tokay gecko (*Gekko gecko*), *Asiat. Herp. Res.* **3**, 54–59
61. Maderson PFA (1964), Keratinized epidermal derivatives as an aid to climbing in gekkonid lizards, *Nature* **2003**, 780–781
62. Maschmann MR, Franklin AD, Amama PB, Zakharov DN, Stach EA, Sands TD, and Fisher TS (2006), Vertical single- and double-walled carbon nanotubes grown from modified porous anodic alumina templates, *Nanotechnology*, **17**, 3925–3929
63. Menon C, Murphy M, and Sitti M (2004) Gecko inspired surface climbing robots, *IEEE Int. Conf. on Robotics and Biomimetics*, August 22-26, 431–436
64. Northen MT and Turner KL (2005), A batch fabricated biomimetic dry adhesive, *Nanotechnology,* **16**, 1159–1166
65. Northen MT and Turner KL (2006), Meso-scale adhesion testing of integrated micro- and nano-scale structures, *Sensors and Actuators A*, **130–131**, 583–587
66. Orr FM, Scriven LE, and Rivas AP (1975), Pendular rings between solids: meniscus properties and capillary forces, *J. Fluid. Mech.* **67**, 723–742
67. Pan B, Gao F, Ao L, Tian H, He R, and Cui D (2005), Controlled self-assembly of thiol-terminated poly(amidoamine) dendrimer and gold nanoparticles, *Colloids and Surfaces A*, **259**, 89–94
68. Peressadko A and Gorb SN (2004), When less is more: experimental evidence for tenacity enhancement by division of contact area, *J. Adhesion*, **80**, 247–261
69. Persson BNJ (2003), On the mechanism of adhesion in biological systems, *J. Chem. Phys.* **118**, 7614–7621
70. Persson BNJ and Gorb S (2003), The effect of surface roughness on the adhesion of elastic plates with application to biological systems, *J. Chem. Phys.* **119**, 11437–11444
71. Phipps PB and Rice DW (1979), Role of water in atmospheric corrosion, *ACS Symposium Series* **89**
72. Rizzo N, Gardner K, Walls D, Keiper-Hrynko N, and Hallahan D (2006), Characterization of the structure and composition of gecko adhesive setae, *J. Royal Soc. Interf.* **3**, 441–451
73. Ruibal R and Ernst V (1965), The structure of the digital setae of lizards, *J. Morphol.* **117**, 271–294
74. Russell AP (1975), A contribution to the functional morphology of the foot of the tokay, *Gekko gecko, J. Zool. Lond.* **176**, 437–476
75. Russell AP (1986), The morphological basis of weight-bearing in the scansors of the tokay gecko, *Can. J. Zool.* **64**, 948–955
76. Schäffer E, Thurn-Albrecht T, Russell TP, and Steiner U (2000) Electrically induced structure formation and pattern transfer, *Nature*, **403**, 874–877
77. Schleich HH and Kästle W (1986), Ultrastrukturen an Gecko-Zehen, *Amphibia Reptilia*, **7**, 141–166
78. Schmidt HR (1904), Zur Anatomie und Physiologie der Geckopfote, *Jena. Z. Naturwissen* **39**, 551
79. Shah GJ and Sitti M (2004), Modeling and design of biomimetic adhesives inspired by gecko foot-hairs, *IEEE Int. Conf. on Robotics and Biomimetics*, 873–878
80. Simmermacher G (1884), Untersuchungen über Haftapparate an Tarsalgliedern von Insekten, *Zeitschr. Wissen Zool.* **40**, 481–556
81. Sitti M (2003) High aspect ratio polymer micro/nano-structure manufacturing using nanoembossing, nanomolding and directed self-assembly, *Proc. IEEE/ASME Advanced Mechatronics Conf.* **2**, 886–890

82. Sitti M and Fearing RS (2003) Synthetic gecko foot-hair for micro/nano structures as dry adhesives, *J. Adhesion Sci. Technol.* **17**, 1055–1073
83. Sitti M and Fearing RS (2003) Synthetic gecko foot-hair for micro/nano structures for future wall-climbing robots, *Proc. IEEE Int. Conf. on Robotics and Automation* **1**, 1164–1170
84. Spolenak R, Gorb S, and Arzt E (2005), Adhesion design maps for bio-inspired attachment systems, *Acta Biomaterialia* **1**, 5–13
85. Stork NE (1980), Experimental analysis of adhesion of *Chrysolina polita* on a variety of surfaces, *J. Exp. Biol.* **88**, 91–107
86. Stork NE (1983), *J. Nat. Hist.*, **17**, 829–835
87. Tian Y, Pesika N, Zeng H, Rosenberg K, Zhao B, McGuiggan P, Autumn K, and Israelachvili J (2006), Adhesion and friction in gecko toe attachment and detachment, *Proc. Nat. Acad. Sci. U. S. A.* **103**, 19320–19325
88. Tinkle DW (1992), Gecko, *Encylcopedia Americana*, Grolier, U.K., **12**, 359
89. Van der Kloot WG (1992), Molting, *Encylcopedia Americana*, Grolier, U.K., **19**, 336–337
90. Wagler J (1830), *Naturliches System der Amphibien*, JG Gotta'schen Buchhandlung, Munich
91. Wan KT, Smith DT, and Lawn BR (1992), Fracture and contact adhesion energies of mica-mica, silica-silica, and mica-silica interfaces in dry and moist atmospheres, *J. Am. Ceram. Soc.* **75**, 667–676
92. Wenzel RN (1936), Resistance of solid surfaces to wetting by water, *Ind. Eng. Chem.* **28**, 988–994
93. Williams EE and Peterson JA (1982), Convergent and alternative designs in the digital adhesive pads of scincid lizards, *Science* **215**, 1509–1511
94. Yao H and Gao H (2006), Mechanics of robust and releasable adhesion in biology: bottom-up designed hierarchical structures of gecko, *J. Mech. Phys. Solids*, **54**, 1120–1146
95. Young WC and Budynas R (2001) *Roark's Formulas for Stress and Strain*, 7th ed. McGraw Hill, New York
96. Yurdumakan B, Raravikar NR, Ajayan PM, and Dhinojwala A (2005), Synthetic gecko foot-hairs from multiwalled carbon nanotubes, *Chem. Comm.*, 3799–3801
97. Zhao Y, Tong T, Delzeit L, Kashani A, Meyyappan M, and Majumdar A (2006), Interfacial energy and strength of multiwalled-carbon-nanotube-based dry adhesive, *J. Vac. Sci. Technol. B*, **24**, 331–335
98. Zimon AD (1969), *Adhesion of Dust and Powder*, translated from Russian by M. Corn, Plenum, New York
99. Zisman WA (1963), Influence of constitution on adhesion, *Ind. Eng. Chem.* **55**(10), 18–38

Part V

Applications

21

Micro/Nanotribology and Micro/Nanomechanics of Magnetic Storage Devices

Bharat Bhushan

Summary. A magnetic recording process involves relative motion between a magnetic medium (tape or disk) against a stationary or rotating read/write magnetic head. For ever-increasing, high areal recording density, the linear flux density (number of flux reversals per unit distance) and the track density (number of tracks per unit distance) should be as high as possible. The size of a single bit dimension for current devices is typically less than 1000 nm^2. This dimension places stringent restrictions on the defect size present on the head and medium surfaces.

Reproduced (read-back) magnetic signal amplitude decreases with a decrease in the recording wavelength and/or the track width. The signal loss results from the magnetic coating thickness, read gap length, and head-to-medium spacing (clearance or flying height). It is known that the signal loss as a result of spacing can be reduced exponentially by reducing the separation between the head and the medium. The need for increasingly higher recording densities requires that surfaces be as smooth as possible and the flying height (physical separation or clearance between a head and a medium) be as low as possible. The ultimate objective is to run two surfaces in contact (with practically zero physical separation) if the tribological issues can be resolved. Smooth surfaces in near contact lead to an increase in adhesion, friction, and interface temperatures, and closer flying heights lead to occasional rubbing of high asperities and increased wear. Friction and wear issues are resolved by appropriate selection of interface materials and lubricants, by controlling the dynamics of the head and medium, and the environment. A fundamental understanding of the tribology (friction, wear, and lubrication) of the magnetic head/medium interface, both on macro- and micro/nanoscales, becomes crucial for the continued growth of this more than $ 60 billion a year magnetic storage industry.

In this chapter, initially, the general operation of drives and the construction and materials used in magnetic head and medium components are described. Then the micro/nanotribological and micro/nanomechanics studies including surface roughness, friction, adhesion, scratching, wear, indentation, and lubrication relevant to magnetic storage devices are presented.

21.1 Introduction

21.1.1 Magnetic Storage Devices

Magnetic storage devices used for storage and retrieval are tape, flexible (floppy) disk and rigid disk drives. These devices are used for audio, video and data-storage applications. Magnetic storage industry is some $ 60 billion a year industry with $ 20 billion for audio and video recording (almost all tape drives/media) and $ 40 billion for data storage. In the data-storage industry, magnetic rigid disk drives/media, tape drives/media, flexible disk drives/media, and optical disk drive/media account for about $ 25 B, $ 6 B, $ 3 B, and $ 6 B, respectively. Magnetic recording and playback involves the relative motion between a magnetic medium (tape or disk) against a read-write magnetic head. Heads are designed so that they develop a (load-carrying) hydrodynamic air film under steady operating conditions to minimize head-medium contact. However, physical contact between the medium and head occurs during starts and stops, referred to as contact-start-stops (CSS) technology [1–4]. In the modern magnetic storage devices, the flying heights (head-to-medium separation) are on the order of 5 to 20 nm and roughnesses of head and medium surfaces are on the order of 1–2 nm RMS. The need for ever-increasing recording densities requires that surfaces be as smooth as possible and the flying heights be as low as possible. Smooth surfaces lead to an increase in adhesion, friction, and interface temperatures, and closer flying heights lead to occasional rubbing of high asperities and increased wear. High stiction (static friction) and wear are the limiting technology to future of this industry. Head load/unload (L/UL) technology has recently been used as an alternative to CSS technology in rigid disk drives that eliminates stiction and wear failure mode associated with CSS. In an L/UL drive, a lift tab extending from the suspension load beam engages a ramp or cam structure as the actuator moves beyond the outer radius of the disk. The ramp lifts (or unloads) the head stack from the disk surfaces as the actuator moves to the parking position. Starting and stopping the disk only occur with the head in the unloaded state. Several contact or near contact recording devices are at various stages of development. High stiction and wear are the major impediments to the commercialization of the contact recording [1–7].

Magnetic media fall into two categories: particulate media, where magnetic particles (γ-Fe_2O_3, Co–γFe_2O_3, CrO_2, Fe or metal (MP), or barium ferrite) are dispersed in a polymeric matrix and coated onto a polymeric substrate for flexible media (tape and flexible disks); thin-film media, where continuous films of magnetic materials are deposited by vacuum deposition techniques onto a polymer substrate for flexible media or onto a rigid substrate (typically aluminium and more recently glass or glass ceramic) for rigid disks. The most commonly used thin magnetic films for tapes are evaporated Co–Ni (82–18 at. %) or Co–O dual layer. Typical magnetic films for rigid disks are metal films of cobalt-based alloys (such as sputtered Co–Pt–Ni, Co–Ni, Co–Pt–Cr, Co–Cr and Co–NiCr). For high recording densities, trends have been to use thin-film media. Magnetic heads used to date are either conventional thin-film inductive, magnetoresistive (MR) and giant MR (GMR) heads.

The air-bearing surfaces (ABS) of tape heads are generally cylindrical in shape. For dual-sided flexible-disk heads, two heads are either spherically contoured and slightly offset (to reduce normal pressure) or are flat and loaded against each other. The rigid-disk heads are supported by a leaf spring (flexure) suspension. The ABS of heads are almost made of Mn–Zn ferrite, Ni–Zn ferrite, Al_2O_3–TiC and calcium titanate. The ABS of some conventional heads are made of plasma sprayed coatings of hard materials such as Al_2O_3–TiO_2 and ZrO_2 [2–4].

Figure 21.1 shows the schematic illustrating the tape path with details of tape guides in a data-processing linear tape drive (IBM LTO Gen1) which uses a rectangular tape cartridge. Figure 21.2a shows the sectional views of particulate and thin-film magnetic tapes. Almost exclusively, the base film is made of semicrystalline biaxially-oriented poly (ethylene terephthalate) (or PET) or poly (ethylene 2,6 naphthalate) (or PEN) or Aramid. The particulate coating formulation consists of binder (typically polyester polyurethane), submicron accicular shaped magnetic particles (about 50 nm long with an aspect ratio of about 5), submicron head cleaning agents (typically alumina) and lubricant (typically fatty acid ester). For protection against wear and corrosion and low friction/stiction, the thin-film tape is first coated with a diamondlike carbon (DLC) overcoat deposited by plasma enhanced chemical vapor deposition, topically lubricated with primarily a perfluoropolyether lubricant. Figure 21.2b shows the schematic of an 8-track (along with 2 servo tracks) thin-film read-write head with MR read and inductive write. The head steps up and down to provide 384 total data tracks across the width of the tape. The ABS is made of Al_2O_3–TiC. A tape tension of about 1 N over a 12.7mm wide tape (normal pressure $\approx$ 14 kPa) is used during use. The RMS roughnesses of ABS of the heads and tape surfaces typically are 1–1.5 nm and 5–8 nm, respectively.

Figure 21.3 shows the schematic of a data processing rigid disk drive with 21.6-, 27.4-, 48-, 63.5-, 75-, and 95-mm form factor. Nonremovable stack of multiple disks mounted on a ball bearing or hydrodynamic spindle, are rotated by an electric motor at constant angular speed ranging from about 5000 to in excess of 15,000 RPM, dependent upon the disk size. Head slider-suspension assembly (allowing one slider

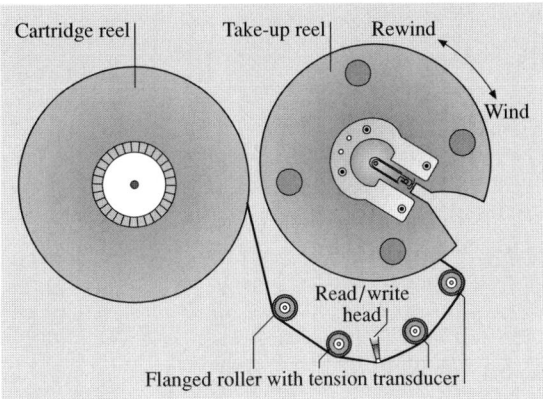

Fig. 21.1. Schematic of tape path in an IBM Linear Tape Open (LTO) tape drive

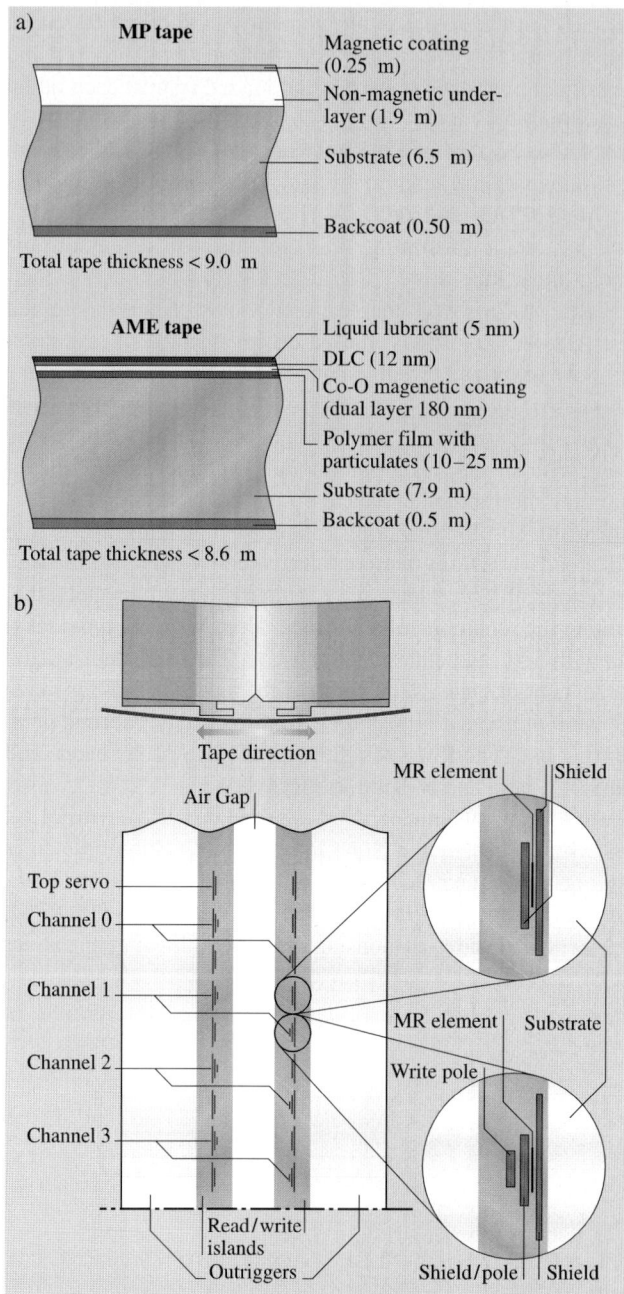

Fig. 21.2. (a) Sectional views of particulate and thin-film magnetic tapes, and (b) schematic of a magnetic thin-film read/write head for an IBM LTO Gen 1 tape drive

Fig. 21.3. Schematic of a data-processing magnetic rigid disk drive

for each disk surface) is actuated by a stepper motor or a voice coil motor using a rotary actuator. Figure 21.4a shows the sectional views of a thin-film rigid disk. The substrate for rigid disks is generally a non heat-treatable aluminium-magnesium alloy 5086, glass or glass ceramic. The protective overcoat commonly used for thin-film disks is sputtered DLC, topically lubricated with perfluoropolyether type of lubricants. Lubricants with polar-end groups are generally used for thin-film disks in order to provide partial chemical bonding to the overcoat surface. The disks used for CSS technology are laser textured in the landing zone. Figure 21.4b shows the schematic of two thin-film head picosliders with a step at the leading edge, and

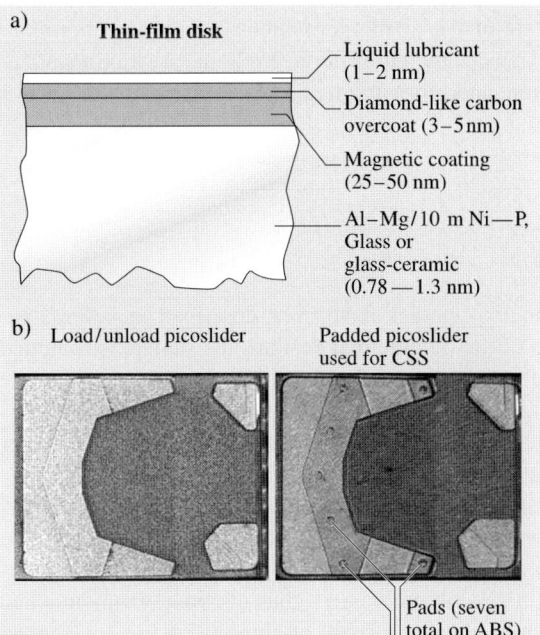

Fig. 21.4. (a) Sectional views of a thin-film magnetic rigid disk, and (b) schematic of two picosliders - load/unload picoslider and padded picoslider used for CSS

GMR read and inductive write. "Pico" refers to the small sizes of 1.25 mm × 1 mm. These sliders use Al_2O_3–TiC (70–30 wt %) as the substrate material with multilayered thin-film head structure coated and with about 3.5nm thick DLC coating to prevent the thin film structure from electrostatic discharge. The seven pads on the padded slider are made of DLC and are about 40 μm in diameter and 50 nm in height. A normal load of about 3 g is applied during use.

21.1.2 Micro/Nanotribology and Micro/Nanomechanics and Their Applications

The micro/nanotribological studies are needed to develop fundamental understanding of interfacial phenomena on a small scale and to study interfacial phenomena in [8–12]. Magnetic storage devices operate under low load and encounter isolated asperity interactions. These use multilayered thin film structure and are generally lubricated with molecularly-thin films. Micro/nanotribological and micro/nanomechanical techniques are ideal to study the friction and wear processes of micro/nanoscale and molecularly thick films. These studies are also valuable in fundamental understanding of interfacial phenomena in macrostructures to provide a bridge between science and engineering. At interfaces of technological applications, contact occurs at multiple asperity contacts. A sharp tip of tip-based microscopes (atomic force/friction force microscopes or AFM/FFM) sliding on a surface simulates a single asperity contact, thus allowing high-resolution measurements of surface interactions at a single asperity contacts. AFMs/FFMs are now commonly used for tribological studies.

In this chapter, we present state-of-the-art of micro/nanotribology and micro/nanomechanics of magnetic storage devices including surface roughness, friction, adhesion, scratching, wear, indentation, and lubrication.

21.2 Experimental

21.2.1 Description of AFM/FFM

AFM/FFMs used in the tribological studies have been described in several papers [10, 12, 13]. Briefly, in one of the commercial designs, the sample is mounted on a PZT tube scanner to scan the sample in the x-y-plane and to move the sample in the vertical (z) direction (Fig. 21.5). A sharp tip at the end of a flexible cantilever is brought in contact with the sample and the sample is scanned in a raster pattern (Fig. 21.6). Normal and frictional forces being applied at the tip–sample interface are simultaneously measured using a laser beam deflection technique. Surface roughness is measured either in the contact mode or the so-called tapping mode (intermittent contact mode). For surface roughness and friction measurements, a microfabricated square pyramidal Si_3N_4 tip with a tip radius of about 30 nm attached to a cantilever beam (with a normal beam stiffness of about 0.5 N/m) for contact

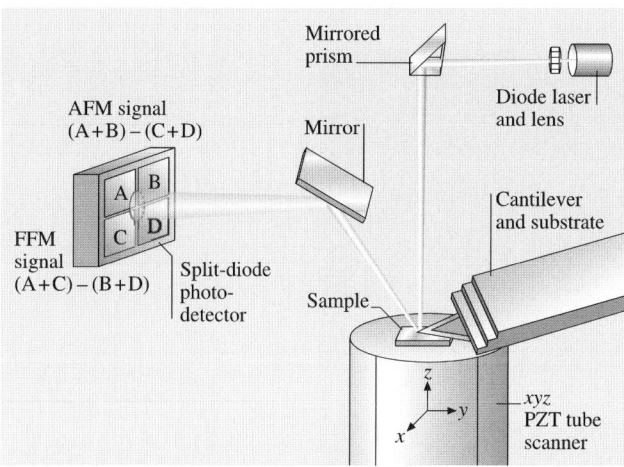

Fig. 21.5. Principles of operation of a commercial small sample AFM/FFM

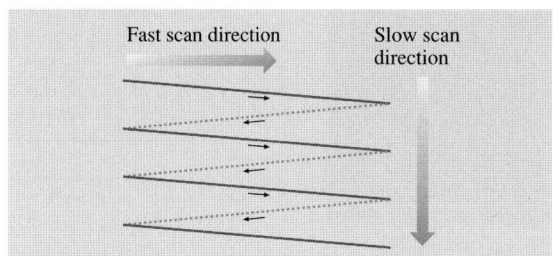

Fig. 21.6. Schematic of triangular pattern trajectory of the AFM tip as the sample is scanned in two dimensions. During imaging, data are recorded only during scans along the solid scan lines

mode or a square-pyramidal etched single-crystal silicon tip with a rectangular silicon cantilever beam (Fig. 21.7) is generally used at normal loads ranging from 10 to 150 nN. A preferred method of measuring friction and calibration procedures for conversion of voltages corresponding to normal and friction forces to force units, are described by *Bhushan* [10,12,13]. The samples are typically scanned over scan areas ranging from 50 nm × 50 nm to 10 μm × 10 μm, in a direction orthogonal to the long axis of the cantilever beam [14]. The scan rate is on the order of 1 Hz. For example, for this rate, the sample scanning speed would be 1 μm/s for a 500 nm × 500 nm scan area. Adhesive force measurements are performed in the so-called friction calibration mode. In this technique, the tip is brought in contact with the sample and then pulled away. The force required to pull the tip off the sample is a measure of adhesive force.

In nanoscale wear studies, the sample is initially scanned twice, typically at 10 nN to obtain the surface profile, then scanned twice at a higher load of typically 100 nN to wear and to image the surface simultaneously, and then rescanned twice at 10 nN to obtain the profile of the worn surface. For magnetic media studied by *Bhushan* and *Ruan* [15], no noticeable change in the roughness profiles was observed between the initial two scans at 10 nN and between profiles scanned at

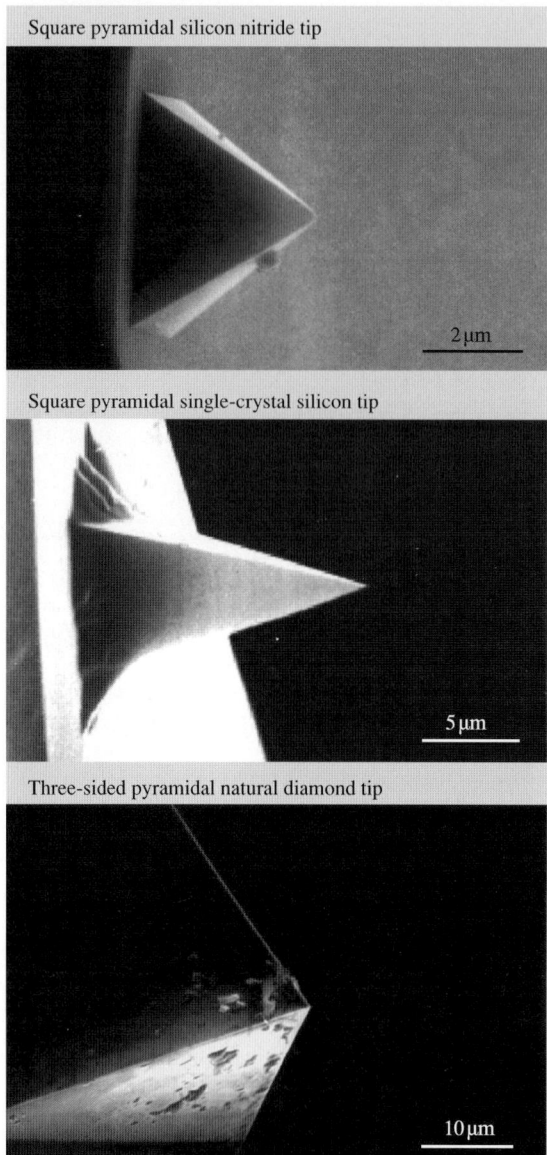

Fig. 21.7. SEM micrographs of a square-pyramidal PECVD Si_3N_4 tip with a triangular cantilever beam (*top*), a square-pyramidal etched single-crystal silicon tip with a rectangular silicon cantilever beam (*middle*), and a three-sided pyramidal natural diamond tip with a square stainless steel cantilever beam (*bottom*)

100 nN and the final scans at 10 nN. Therefore any changes in the topography between the initial scans at 10 nN and the scans at 100 nN (or the final scans at 10 nN) are believed to occur as a result of local deformation of the sample surface. In picoindentation studies, the sample is loaded in contact with the tip. During loading, tip deflection (normal force) is measured as a function of vertical position of the sample. For a rigid sample, the tip deflection and the sample traveling distance (when the tip and sample come into contact) are equal. Any decrease in the tip deflection as compared to vertical position of the sample represents indentation. To ensure that the curvature in the tip deflection-sample traveling distance curve does not arise from PZT hysteresis, measurements on several rigid samples, including single-crystal natural diamond (IIa), were made by *Bhushan* and *Ruan* [15]. No curvature was noticed for the case of rigid samples. This suggests that any curvature for other samples should arise from the indentation of the sample.

For microscale scratching, microscale wear and nanoscale indentation hardness measurements, a three-sided pyramidal single-crystal natural diamond tip with an apex angle of 80° and a tip radius of about 100 nm (determined by scanning electron microscopy imaging) (Fig. 21.7) is used at relatively high loads (1 µN–150 µN). The diamond tip is mounted on a stainless steel cantilever beam with normal stiffness of about 25 N/m [16–19]. For scratching and wear studies, the sample is generally scanned in a direction orthogonal to the long axis of the cantilever beam (typically at a rate of 0.5 Hz). For wear studies, typically an area of 2 µm × 2 µm is scanned at various normal loads (ranging from 1 to 100 µN) for selected number of cycles. Scratching can also be performed at ramped loads [20]. For nanoindentation hardness measurements, the scan size is set to zero and then normal load is applied to make the indents. During this procedure, the diamond tip is continuously pressed against the sample surface for about two seconds at various indentation loads. Sample surface is scanned before and after the scratching, wear, or nanoindentation, to obtain the initial and the final surface topography, at a low normal load of about 0.3 µN using the same diamond tip. An area larger than the scratched, worn or indentation region is scanned to observe the scratch, wear scars, or indentation marks. Nanohardness is calculated by dividing the indentation load by the projected residual area of the indents [17]. Nanohardness and Young's modulus of elasticity (stiffness) at shallow depths as low as 5 nm are measured using a depth-sensing capacitance transducer system in an AFM [19].

Indentation experiments provide a single-point measurement of the Young's modulus of elasticity (stiffness), localized surface elasticity as well as phase contrast maps (to obtain viscoelastic properties map) can be obtained using dynamic force microscopy in which an oscillating tip is scanned over the sample surface in contact under steady and oscillating load [21–24]. Recently, a torsional resonance (TR) mode has been introduced [25, 26] which provides higher resolution. Stiffness and phase contrast maps can provide magnetic particle/polymer distributions in magnetic tapes as well as lubricant film thickness distribution.

Boundary lubrication studies are conducted using either Si_3N_4 or diamond tips [27–30]. The coefficient of friction is monitored as a function of sliding cycles.

All measurements are carried out in the ambient atmosphere (22±1 °C, 45±5% RH, and Class 10000).

21.2.2 Test Specimens

Data on various head slider materials and magnetic media are presented in the chapter. Al_2O_3–TiC (70–30 w/o) and polycrystalline and single-crystal (110) Mn–Zn ferrite are commonly used for construction of disk and tape heads. Al_2O_3, a single-phase material, is also selected for comparisons with the performance of Al_2O_3–TiC, a two-phase material. An α-type SiC is also selected which is a candidate slider material because of its high thermal conductivity and attractive machining and friction and wear properties [31]. Single crystal silicon has also been used in some head sliders but its use is discontinued [32].

Two thin-film rigid disks with polished and textured substrates, with and without a bonded perfluoropolyether are selected. These disks are 95 mm in diameter made of Al–Mg alloy substrate (1.3 mm thick) with a 10 μm thick electroless plated Ni–P coating, 75 nm thick ($Co_{79}Pt_{14}Ni_7$) magnetic coating, 20 nm thick amorphous carbon or diamondlike carbon (DLC) coating (microhardness $\approx 1{,}500\,\text{kg/mm}^2$ as measured using a Berkovich indenter), and with or without a top layer of perfluoropolyether lubricant with polar end groups (Z-DOL) coating. The thickness of the lubricant film is about 2 nm. The metal particle (MP) tape is a 12.7 mm wide and 13.2 μm thick (PET base thickness of 9.8 μm, magnetic coating of 2.9 μm with metal magnetic particles and nonmagnetic particles of Al_2O_3 and Cr_2O_3, and back coating of 0.5 μm). The barium ferrite (BaFe) tape is a 12.7 mm wide and 11 μm thick (PET base thickness of 7.3 μm, magnetic coating of 2.5 μm with barium ferrite magnetic particles and nonmagnetic particles of Al_2O_3, and back coating of 1.2 μm). Metal evaporated (ME) tape is a 12.7 mm wide tape with 10 μm thick base, 0.2 μm thick evaporated Co–Ni magnetic film and about 10 nm thick perfluoropolyether lubricant and a backcoat. PET film is a biaxially-oriented, semicrystalline polymer with particulates. Two sizes of nearly spherical particulates are generally used in the construction of PET: submicron (≈ 0.5 μm) particles of typically carbon and larger particles (2–3 μm) of silica.

21.3 Surface Roughness

Solid surfaces, irrespective of the method of formation, contain surface irregularities or deviations from the prescribed geometrical form. When two nominally flat surfaces are placed in contact, surface roughness causes contact to occur at discrete contact points. Deformation occurs at these points, and may be either elastic or plastic, depending on the nominal stress, surface roughness and material properties. The sum of the areas of all the contact points constitutes the real area that would be in contact, and for most materials at normal loads, this will be only a small fraction of the area of contact if the surfaces were perfectly smooth. In general, real area of contact must be minimized to minimize adhesion, friction and wear [12].

Characterizing surface roughness is therefore important for predicting and understanding the tribological properties of solids in contact. Various measurement techniques are used to measure surface roughness. The AFM is used to measure surface roughness on length scales from nanometers to micrometers. A second technique is noncontact optical profiler (NOP) which is a noncontact technique and does not damage the surface. The third technique is stylus profiler (SP) in which a sharp tip is dragged over the sample surface. These techniques differ in lateral resolution. Roughness plots of a glass-ceramic disk measured using an AFM (lateral resolution ≈ 15 nm), NOP (lateral resolution ≈ 1 μm) and SP (lateral resolution of ≈ 0.2 μm) are shown in Fig. 21.8a. Figure 21.8b compares the profiles of the disk obtained with different instruments at a common scale. The figures show that roughness is found at scales ranging from millimeter to nanometer scales. Measured roughness profile is dependent on the lateral and normal resolutions of the measuring instrument [33–37]. Instruments with different lateral resolutions measure features with different scale lengths. It can be concluded that a surface is composed of a large number of length of scales of roughness that are superimposed on each other.

Surface roughness is most commonly characterized by the standard deviation of surface heights which is the square roots of the arithmetic average of squares of the vertical deviation of a surface profile from its mean plane. Due to the multiscale nature of surfaces, it is found that the variances of surface height and its derivatives and other roughness parameters depend strongly on the resolution of the roughness measuring instrument or any other form of filter, hence not unique for a surface [35–38] (Fig. 21.9). Therefore, a rough surface should be characterized in a way such that the structural information of roughness at all scales is retained. It is necessary to quantify the multiscale nature of surface roughness.

An unique property of rough surfaces is that if a surface is repeatedly magnified, increasing details of roughness are observed right down to nanoscale. In addition, the roughness at all magnifications appear quite similar in structure, as qualitatively shown in Fig. 21.10. The statistical self-affinity is due to similarity in appearance of a profile under different magnifications. Such a behavior can be characterized by fractal analysis [35, 39]. The main conclusion from these studies are that a fractal characterization of surface roughness is *scale independent* and provides information of the roughness structure at all length scales that exhibit the fractal behavior.

Structure function and power spectrum of a self-affine fractal surface follow a power law and can be written as (Ganti and Bhushan model)

$$S(\tau) = C\eta^{(2D-3)}\tau^{(4-2D)}, \tag{21.1}$$

$$P(\omega) = \frac{c_1 \eta^{(2D-3)}}{\omega^{(5-2D)}}, \tag{21.2a}$$

and

$$c_1 = \frac{\Gamma(5-2D)\sin[\pi(2-D)]}{2\pi} C. \tag{21.2b}$$

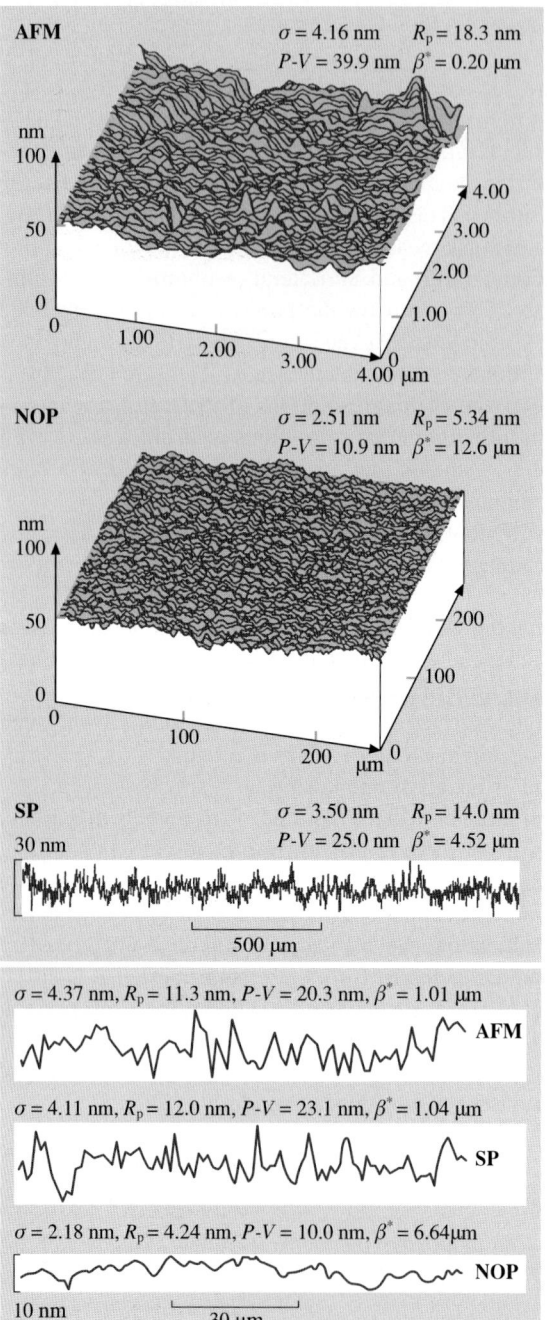

Fig. 21.8. Surface roughness plots of a glass-ceramic disk (**a**) measured using an atomic force microscope (lateral resolution ≈ 15 nm), noncontact optical profiler (NOP) (lateral resolution ≈ 1 μm) and stylus profiler (SP) with a stylus tip of 0.2μm radius (lateral resolution ≈ 0.2 μm), and (**b**) measured using an AFM (≈ 150 nm), SP (≈ 0.2 μm), and NOP (≈ 1 μm) and plotted on a common scale [36]

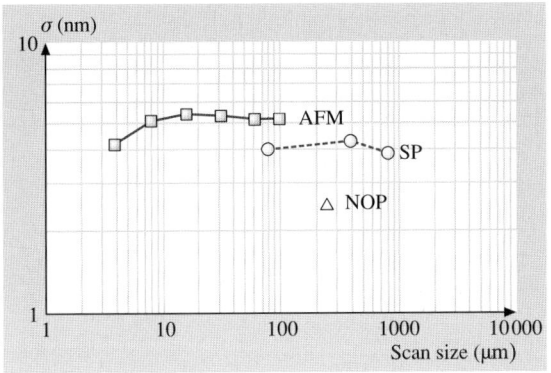

Fig. 21.9. Scale dependence of standard deviation of surface heights for a glass-ceramic disk, measured using atomic force microscope (AFM), stylus profiler (SP), and noncontact optical profiler (NOP)

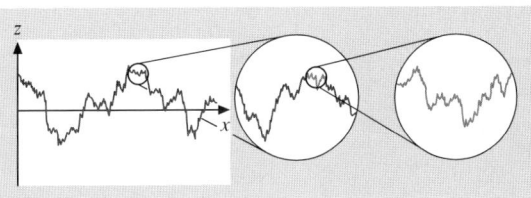

Fig. 21.10. Qualitative description of statistical self-affinity for a surface profile

The fractal analysis allows the characterization of surface roughness by two parameters D and C which are instrument-independent and unique for each surface. D (ranging from 1 to 2 for surface profile) primarily relates to relative power of the frequency contents and C to the amplitude of all frequencies. η is the lateral resolution of the measuring instrument, τ is the size of the increment (distance), and ω is the frequency of the roughness. Note that if $S(\tau)$ or $P(\omega)$ are plotted as a function of τ or ω, respectively, on a log-log plot, then the power law behavior would result into a straight line. The slope of line is related to D and the location of the spectrum along the power axis is related to C.

Figure 21.11 present the structure function of a thin-film rigid disk measured using AFM, non-contact optical profiler (NOP), and stylus profiler (SP). Horizontal shift in the structure functions from one scan to another, arises from the change in the lateral resolution. D and C values for various scan lengths are listed in Table 21.1. We note that fractal dimension of the various scans is fairly constant (1.26 to 1.33); however, C increases/decreases monotonically with σ for the AFM data. The error in estimation of η is believed to be responsible for variation in C. These data show that the disk surface follows a fractal structure for three decades of length scales.

Majumdar and *Bhushan* [40] and *Bhushan* and *Majumdar* [41] developed a fractal theory of contact between two rough surfaces. This model has been used to predict whether contacts experience elastic or plastic deformation and to predict the statistical distribution of contact points. For a review of contact models, see *Bhushan* [42,43] and *Bhushan* and *Peng* [44].

Based on the fractal model of elastic–plastic contact, whether contacts go through elastic or plastic deformation is determined by a critical area which is

Table 21.1. Surface roughness parameters for a polished thin-film rigid disk

Scan size (μm×μm)	σ (nm)	D	C (nm)
1 (AFM)	0.7	1.33	9.8×10^{-4}
10 (AFM)	2.1	1.31	7.6×10^{-3}
50 (AFM)	4.8	1.26	1.7×10^{-2}
100 (AFM)	5.6	1.30	1.4×10^{-2}
250 (NOP)	2.4	1.32	2.7×10^{-4}
4000 (NOP)	3.7	1.29	7.9×10^{-5}

AFM – Atomic force microscope,
NOP – Noncontact optical profiler

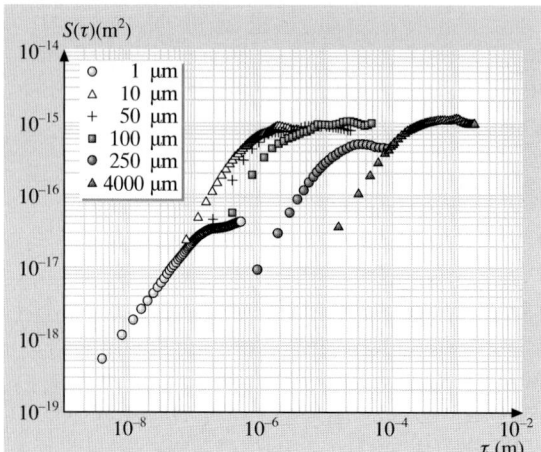

Fig. 21.11. Structure functions for the roughness data measured at various scan sizes using AFM (scan sizes: 1 μm × 1 μm, 10 μm × 10 μm, 50 μm × 50 μm, and 100 μm), NOP (scan size: 250 μm × 250 μm), and SP (scan length: 4000 μm), for a magnetic thin-film rigid disk [35]

a function of D, C, hardness and modulus of elasticity of the mating surfaces. If contact spot is smaller than the critical area, it goes through the plastic deformations and large spots go through elastic deformations. The critical contact area for inception of plastic deformation for a thin-film disk was reported by *Majumdar* and *Bhushan* [40] to be about 10^{-27} m^2, so small that all contact spots can be assumed to be elastic at moderate loads.

The question remains as to how large spots become elastic when they must have initially been plastic spots. The possible explanation is shown in Fig. 21.12. As two surfaces touch, the nanoasperities (detected by AFM type of instruments) first coming into contact have smaller radii of curvature and are therefore plastically deformed instantly, and the contact area increases. When load is increased, nanoasperities in the contact merge, and the load is supported by elastic deformation of the large scale asperities or microasperities (detected by optical profiler type of instruments) [33].

Majumdar and *Bhushan* [40] and *Bhushan* and *Majumdar* [41] have reported relationships for cumulative size distribution of the contact spots, portions of the

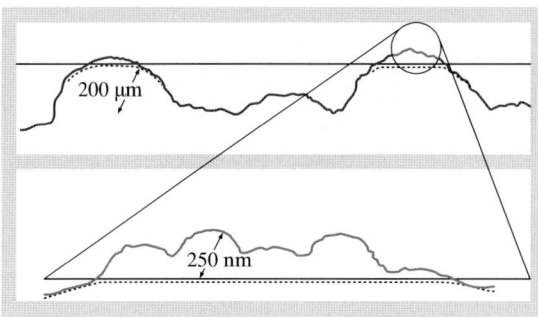

Fig. 21.12. Schematic of local asperity deformation during contact of a rough surface, upper profile measured by an optical profiler and lower profile measured by AFM, typical dimensions are shown for a polished thin-film rigid disk against a flat head slider surface [33]

real area of contact in elastic and plastic deformation modes, and the load-area relationships.

21.4 Friction and Adhesion

Friction and adhesion of magnetic head sliders and magnetic media have been measured by *Bhushan* and *Koinkar* [16, 45–48], *Bhushan* and *Ruan* [15], *Ruan* and *Bhushan* [14], *Bhushan* et al. [27], *Bhushan* [2, 4, 10], *Koinkar* and *Bhushan* [28, 29, 38, 49], *Kulkarni* and *Bhushan* [50], and *Li* and *Bhushan* [51, 52], and *Sundararajan* and *Bhushan* [53].

Koinkar and *Bhushan* [28, 29] and *Poon* and *Bhushan* [36, 37] reported that RMS roughness and friction force increase with an increase in scan size at a given scanning velocity and normal force. Therefore, it is important that while reporting friction force values, scan sizes and scanning velocity should be mentioned. *Bhushan* and *Sundararajan* [54] reported that friction and adhesion forces are a function of tip radius and relative humidity (also see [29]). Therefore, relative humidity should be controlled during the experiments. Care also should be taken to ensure that tip radius does not change during the experiments.

21.4.1 Magnetic Head Materials

Al_2O_3–TiC is a commonly used slider material. In order to study the friction characteristics of this two phase material, friction of Al_2O_3–TiC (70–30 wt %) surface was measured. Figure 21.13 shows the surface roughness and friction force profiles [28]. TiC grains have a Knoop hardness of about $2800\,kg/mm^2$ which is higher than that of Al_2O_3 grains of about $2100\,kg/mm^2$. Therefore, TiC grains do not polish as much and result in a slightly higher elevation (about 2–3 nm higher than that of Al_2O_3 grains). Based on friction force measurements, TiC grains exhibit higher friction force than Al_2O_3 grains. The coefficients of friction of TiC and Al_2O_3 grains are 0.034 and 0.026, respectively and the coefficient of friction of Al_2O_3–TiC composite is 0.03. Local variation in friction force also arises from the scratches present on the Al_2O_3–TiC surface. Thus, local friction values of a two phase materials can

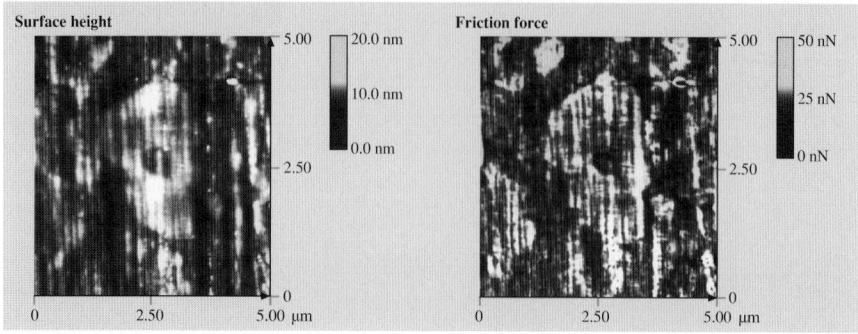

Fig. 21.13. Gray scale surface roughness (σ = 0.97 nm) and friction force map (mean = 7.0 nN, σ = 0.60 nN) for Al_2O_3–TiC (70 to 30 wt %) at a normal load of 166 nN

be measured. *Ruan* and *Bhushan* [55] reported that local variation in the coefficient of friction of cleaved HOP graphite was significant which arises from structural changes occurring during the cleaving process. The cleaved HOPG surface is largely atomically smooth but exhibits line shaped regions in which the coefficient of friction is more than an order of magnitude larger. These measurements suggest that friction measurements can be used for structural mapping of the surfaces.

Table 21.2. Surface roughness (σ and P–V distance), micro- and macro-scale friction, microscratching/wear, and nano- and microhardness data for various samples

Sample	Surface roughness (1 µm × 1 µm)		Coefficient of friction			Scratch depth at 60 µN (nm)	Wear depth at 60 µN (nm)	Hardness	
				Macroscale[b]				Nano at 2 mN (GPa)	Micro at 2 mN (GPa)
	σ (nm)	P–V[a] (nm)	Micro-scale	Initial	Final				
Al_2O_3	0.97	9.9	0.03	0.18	0.2–0.6	3.2	3.7	24.8	15.0
Al_2O_3–TiC	0.80	9.1	0.05	0.24	0.2–0.6	2.8	22.0	23.6	20.2
Polycrystalline Mn–Zn ferrite	2.4	20.0	0.04	0.27	0.24–0.4	9.6	83.6	9.6	5.6
Single-crystal (110) Mn–Zn ferrite	1.9	13.7	0.02	0.16	0.18–0.24	9.0	56.0	9.8	5.6
SiC (α-type)	0.91	7.2	0.02	0.29	0.18–0.24	0.4	7.7	26.7	21.8

[a] Peak-to-valley distance
[b] Obtained using silicon nitride ball with 3 mm diameter in a reciprocating mode at a normal load of 10 mN, reciprocating amplitude of 7 mm and average sliding speed of 1 mm/s. Initial coefficient of friction values were obtained at first cycle (0.007 m sliding distance) and final values at a sliding distance of 5 m

Surface roughness and coefficient of friction of various head slider materials were measured by *Koinkar* and *Bhushan* [29]. For typical values, see Table 21.2. Macroscale friction values for all samples are higher than microscale friction values as there is less plowing contribution in microscale measurements [10, 12, 13].

21.4.2 Magnetic Media

Bhushan and coworkers measured friction properties of magnetic media including polished and textured thin-film rigid disks, metal particle (MP), barium ferrite (BaFe) and metal evaporated (ME) tapes, and poly(ethylene terephthalate)(PET) tape substrate [2, 4, 10]. For typical values of coefficients of friction of thin-film rigid disks and MP, BaFe and ME tapes, PET tape substrate (Table 21.3). In the case of magnetic disks, similar coefficients of friction are observed for both lubricated and unlubricated disks, indicating that most of the lubricant (though partially thermally bonded) is squeezed out from between the rubbing surfaces at high interface pressures, consistent with liquids being poor boundary lubricant [13]. Coefficient of friction values on a microscale are much lower than those on the macroscale. When

Table 21.3. Surface roughness (σ), microscale and macro-scale friction, and nanohardness data of thin-film magnetic rigid disk, magnetic tape and magnetic tape substrate (PET) samples

Sample	σ (nm)			Coefficient of microscale friction		Coefficient of macro-scale friction		Nano-hardness (GPa)/ Normal load (μN)
	NOP 250 μm × 250 μm[a]	AFM 1 μm × 1 μm[a]	10 μm × 10 μm[a]	1 μm × 1 μm[a]	10 μm × 10 μm[a]	Mn–Zn ferrite	Al_2O_3– TiC	
Polished, unlubricated disk	2.2	3.3	4.5	0.05	0.06	–	0.26	21/100
Polished, lubricated disk	2.3	2.3	4.1	0.04	0.05	–	0.19	–
Textured, lubricated disk	4.6	5.4	8.7	0.04	0.05	–	0.16	–
Metal-particle tape	6.0	5.1	12.5	0.08	0.06	0.19	–	0.30/50
Barium-ferrite tape	12.3	7.0	7.9	0.07	0.03	0.18	–	0.25/25
Metal-evaporated tape	9.3	4.7	5.1	0.05	0.03	0.18	–	0.7 to 4.3/75
PET tape substrate	33	5.8	7.0	0.05	0.04	0.55	–	0.3/20 and 1.4/20[b]

[a] Scan area; NOP – Noncontact optical profiler; AFM – Atomic force microscope
[b] Numbers are for polymer and particulate regions, respectively

measured for the small contact areas and very low loads used in microscale studies, indentation hardness and modulus of elasticity are higher than at the macroscale (data to be presented later). This reduces the real area of contact and the degree of wear. In addition, the small apparent areas of contact reduces the number of particles trapped at the interface, and thus minimize the "plowing" contribution to the friction force [8, 14, 18].

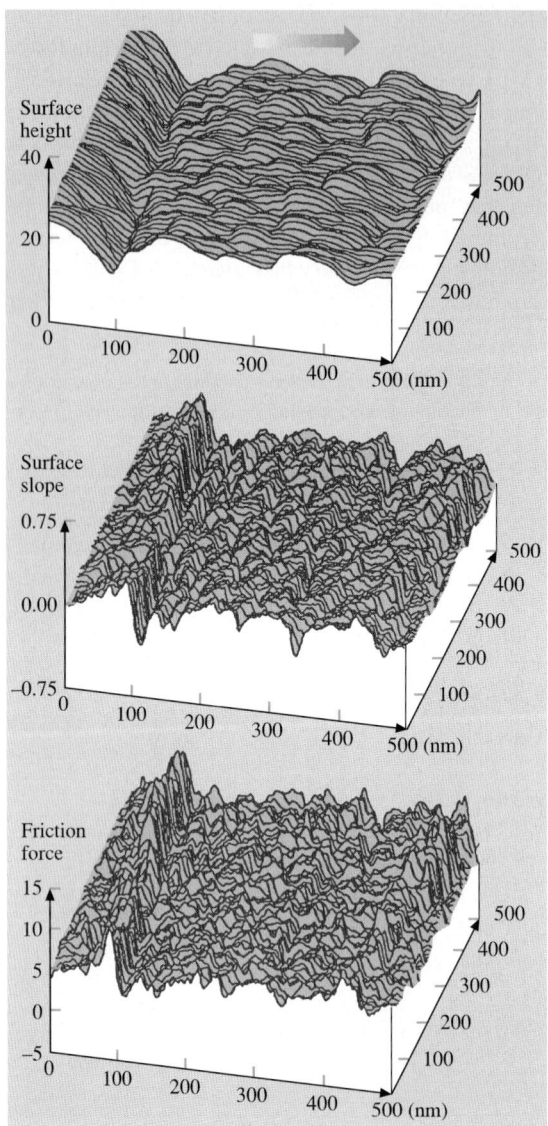

Fig. 21.14. (a) Surface height map ($\sigma = 4.4$ nm), (b) slope of the roughness profiles taken in the sample sliding direction (the *horizontal axis*) (mean = 0.023, $\sigma = 0.18$), and (c) friction force map (mean = 6.2 nN, $\sigma = 2.1$ nN) for a textured thin-film rigid disk at a normal load of 160 nN [18]

Local variations in the microscale friction of rough surfaces can be significant. Figure 21.14 shows the surface height map, the slopes of surface map taken along the sliding direction, and friction force map for a thin-film disk [8, 15, 18, 28, 38, 45–47, 56]. We note that there is no resemblance between the coefficient of friction profiles and the corresponding roughness maps, e.g., high or low points on the friction profile do not correspond to high or low points on the roughness profiles. By comparing the slope and friction profiles, we observe a strong correlation between the two. (For a clearer correlation, see gray-scale plots of slope and friction profiles for FFM tip sliding in either directions, in Fig. 21.15 to be presented in the next paragraph). We have shown that this correlation holds for various magnetic tapes, silicon, diamond, and other materials. This correlation can be explained by a "ratchet" mechanism; based on this mechanism, the local friction is a function of the local slope of the sample surface [10, 12, 13]. The friction is high at the leading edge of asperities and low at the trailing edge. In addition to the slope effect, the collision of tip encountering an asperity with a positive slope produces additional torsion of the cantilever beam leading to higher measured friction force. When encountering an asperity with a negative slope, however, there is no collision effect and hence no effect on friction. The ratchet mechanism and the collision effects thus explain the correlation between the slopes of the roughness maps and friction maps observed in Fig. 21.14.

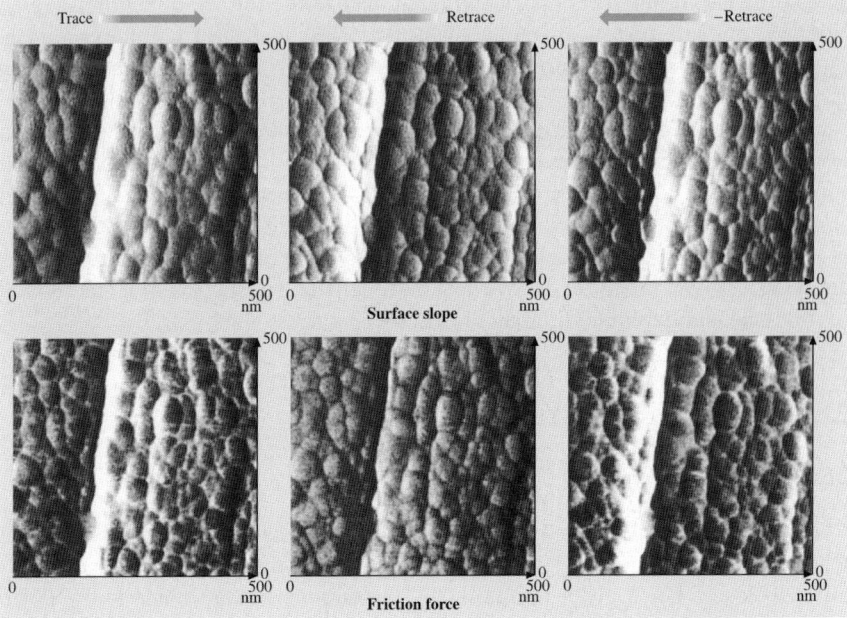

Fig. 21.15. Gray-scale plots of the slope of the surface roughness and the friction force maps for a textured thin-film rigid disk with FFM tip sliding in different directions. *Higher points* are shown by *lighter color*

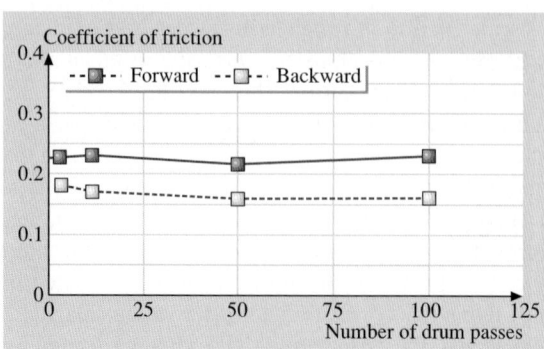

Fig. 21.16. Coefficient of macro-scale of friction as a function of sliding cycles for a metal-particle (MP) tape sliding over an aluminium drum in a reciprocating mode in both directions. Normal load = 0.5 N over 12.7 mm wide tape, sliding speed = 60 mm/s [7]

To study the directionality effect on the friction, gray scale plots of coefficients of local friction of the thin-film disk as the FFM tip is scanned in either direction are shown in Fig. 21.15. Corresponding gray scale plots of slope of roughness maps are also shown in Fig. 21.15. The left hand set of the figures corresponds to the tip sliding from the left towards right (trace direction), and the middle set corresponds to the tip sliding from the right towards left (retrace direction). It is important to take into account the sign change of surface slope and friction force which occur in the trace and retrace directions. In order to facilitate comparison of directionality effect on friction, the last set of the figures in the right hand column show the data with sign of surface slope and friction data in the retrace direction reversed. Now we compare trace and -retrace data. It is clear that the friction experienced by the tip is dependent upon the scanning direction because of surface topography.

The directionality effect in friction on a macroscale is generally averaged out over a large number of contacts. It has been observed in some magnetic tapes. In a macro-scale test, a 12.7 mm wide MP tape was wrapped over an aluminium drum and slid in a reciprocating motion with a normal load of 0.5 N and a sliding speed of about 60 mm/s. Coefficient of friction as a function of sliding distance in either direction is shown in Fig. 21.16. We note that coefficient of friction on a macro-scale for this tape is different in different directions.

21.5 Scratching and Wear

21.5.1 Nanoscale Wear

Bhushan and *Ruan* [15] conducted nanoscale wear tests on MP tapes at a normal load of 100 nN. Figure 21.17 shows the topography of the MP tape obtained at two different loads. For a given normal load, measurements were made twice. There was no discernible difference between consecutive measurements for a given normal load. However, as the load increased from 10 to 100 nN, topographical changes were observed; material (indicated by an arrow) was pushed toward the right side in the sliding direction of the AFM tip relative to the sample. The material movement

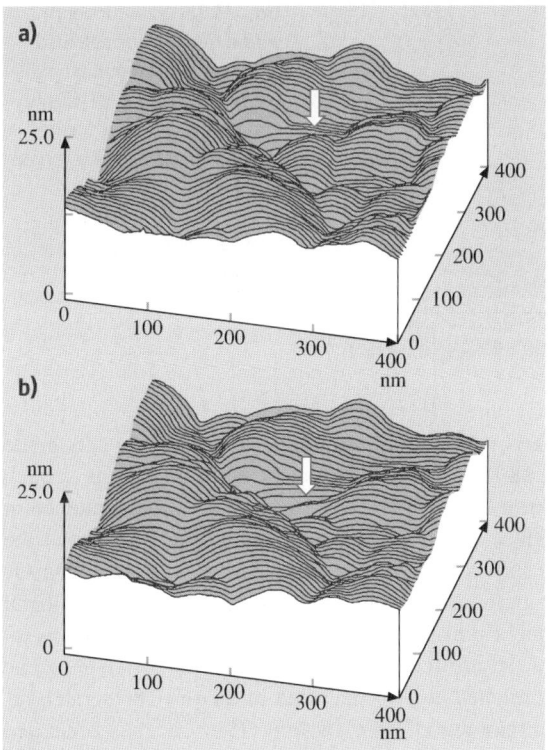

Fig. 21.17. Surface roughness maps of a metal-particle (MP) tape at applied normal load of (**a**) 10 nN and (**b**) 100 nN. Location of the change in surface topography as a result of nanowear is indicated by *arrows* [15]

is believed to occur as a result of plastic deformation of the tape surface. Similar behavior was observed on all tapes studied. Magnetic tape coating is made of magnetic particles and polymeric binder. Any movement of the coating material can eventually lead to loose debris. Debris formation is an undesirable situation as it may contaminate the head which may increase friction and/or wear between the head and tape, in addition to the deterioration of the tape itself. With disks, they did not notice any deformation under a 100 nN normal load.

21.5.2 Microscale Scratching

Microscratches have been made on various potential head slider materials (Al_2O_3, Al_2O_3–TiC, Mn–Zn ferrite and SiC), and various magnetic media (unlubricated polished thin-film disk, MP, BaFe, ME tapes, and PET substrates) and virgin, treated and coated Si(111) wafers at various loads [16, 18, 28, 45–49, 53, 57, 58]. As mentioned earlier, the scratches are made using a diamond tip.

Magnetic Head Materials

Scratch depths as a function of load and representative scratch profiles with corresponding 2-D gray scale plots at various loads after a single pass (unidirectional

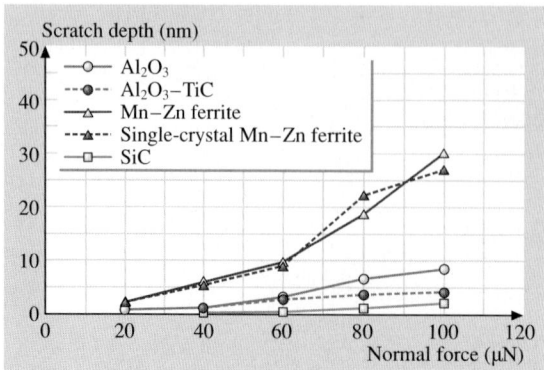

Fig. 21.18. Scratch depth as a function of normal load after one unidirectional cycle for Al$_2$O$_3$, Al$_2$O$_3$–TiC, polycrystalline Mn–Zn ferrite, single-crystal Mn–Zn ferrite and SiC [28]

scratching) for Al$_2$O$_3$, Al$_2$O$_3$–TiC, polycrystalline and single-crystal Mn–Zn ferrite and SiC are shown in Figs. 21.18 and 21.19, respectively. Variation in the scratch depth along the scratch is about ±15%. The Al$_2$O$_3$ surface could be scratched at a normal load of 40 µN. The surface topography of polycrystalline Al$_2$O$_3$ shows the presence of porous holes on the surface. The 2-D gray scale plot of scratched Al$_2$O$_3$ surface shows one porous hole between scratches made at normal loads of 40 µN and 60 µN. Regions with defects or porous holes present, exhibit lower scratch resistance (see region marked by the arrow on 2-D gray scale plot of Al$_2$O$_3$). The Al$_2$O$_3$–TiC surface could be scratched at a normal load of 20 µN. The scratch resistance for TiC grains is higher than that of Al$_2$O$_3$ grains. The scratches generated at normal loads of 80 µN and 100 µN show that scratch depth of Al$_2$O$_3$ grains is higher than that of TiC grains (see corresponding gray scale plot for Al$_2$O$_3$–TiC). Polycrystalline and single-crystal Mn–Zn ferrite could be scratched at a normal load of 20 µN. The scratch width is much larger for the ferrite specimens as compared with other specimens. For SiC, there is no measurable scratch observed at a normal load of 20 µN. At higher normal loads, very shallow scratches are produced. Table 21.2 presents average scratch depth at 60 µN normal load for all specimens. SiC has the highest scratch resistance followed by Al$_2$O$_3$–TiC, Al$_2$O$_3$ and polycrystalline and single-crystal Mn–Zn ferrite. Polycrystalline and single-crystal Mn–Zn ferrite specimens exhibit comparable scratch resistance.

Magnetic Media

Scratch depths as a function of load and scratch profiles at various loads after ten scratch cycles for unlubricated, polished disk and MP tape are shown in Figs. 21.20 and 21.21. We note that scratch depth increases with an increase in the normal load. Tape could be scratched at about 100 nN. With disk, gentle scratch marks under 10 µN load were barely visible. It is possible that material removal did occur at lower load on an atomic scale which was not observable with a scan size of 5 µm square. For disk, scratch depth at 40 µN is less than 10 nm deep. The scratch depth increased slightly at the load of 50 µN. Once the load is increased in excess of 60 µN, the scratch depth increased rapidly. These data suggest that the carbon coating on

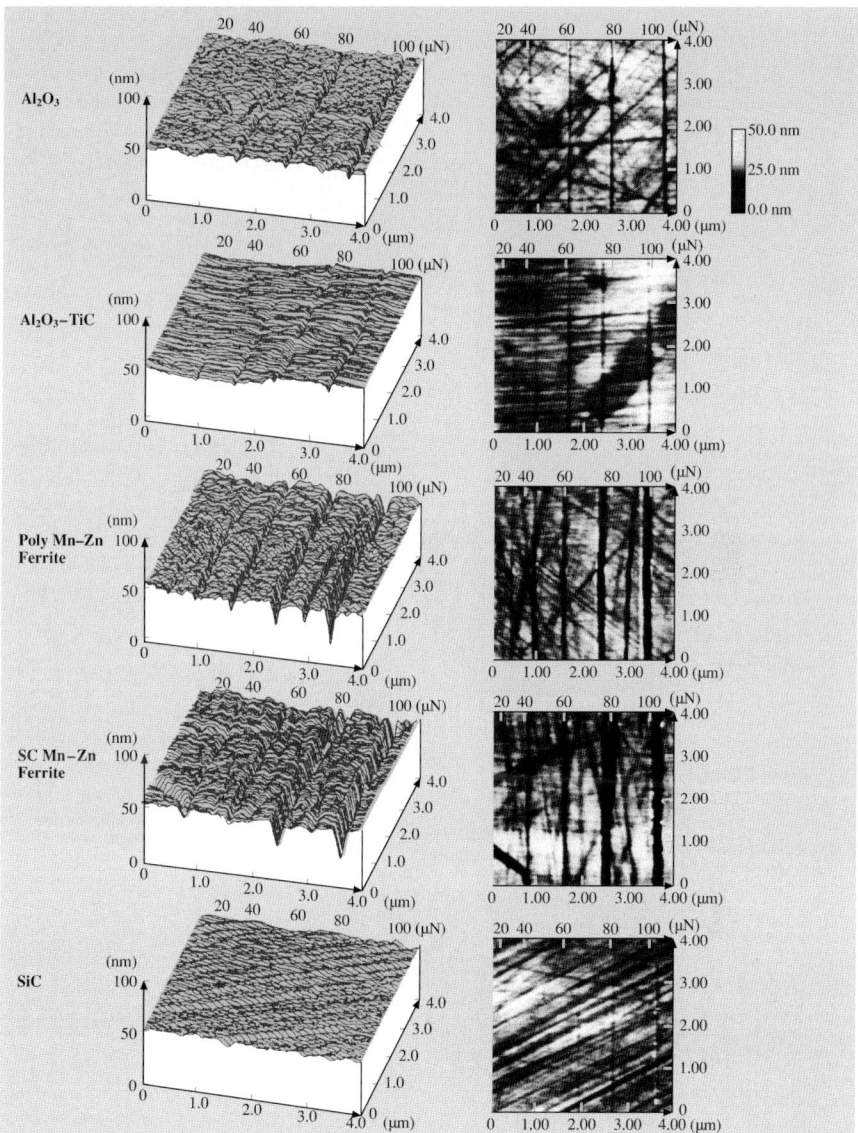

Fig. 21.19. Surface profiles (*left column*) and 2-D gray scale plots (*right column*) of scratched Al_2O_3, Al_2O_3–TiC, polycrystalline Mn–Zn ferrite, single-crystal Mn–Zn ferrite, and SiC surfaces. Normal loads used for scratching for one unidirectional cycle are listed in the figure [28]

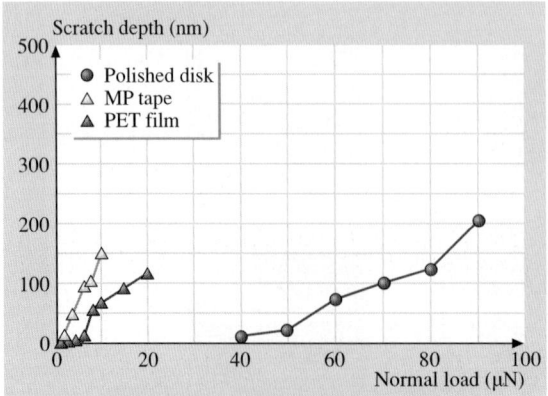

Fig. 21.20. Scratch depth as a function of normal load after ten scratch cycles for unlubricated polished thin-film rigid disk, MP tape and PET film [18, 45, 47]

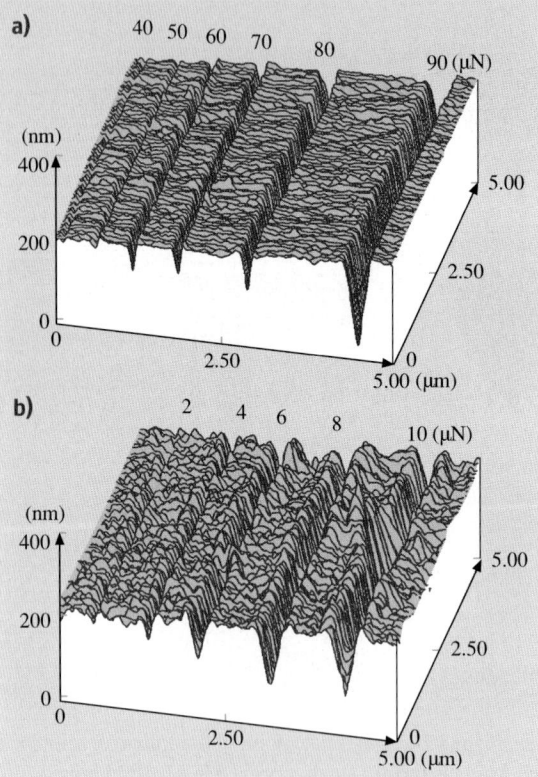

Fig. 21.21. Surface profiles for scratched (**a**) unlubricated polished thin-film rigid disk and (**b**) MP tape. Normal loads used for scratching for ten cycles are listed in the figure [18]

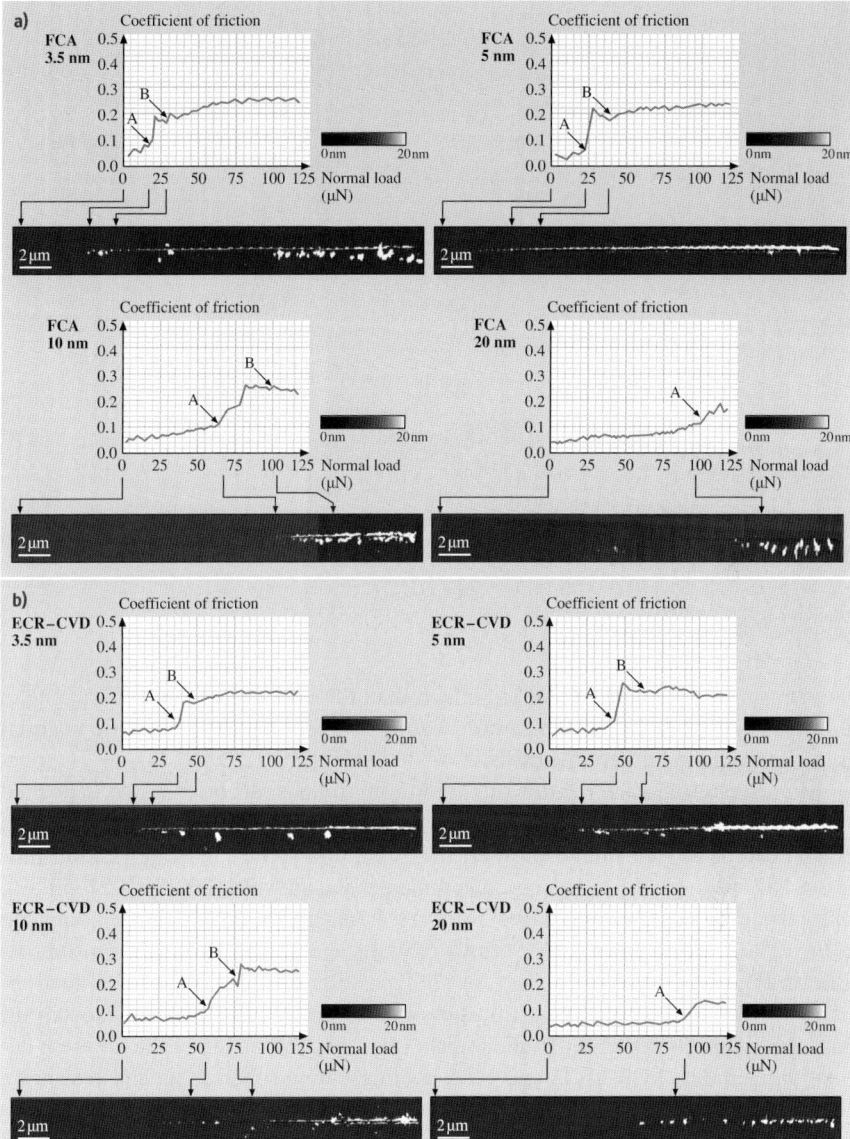

Fig. 21.22. Coefficient of Friction profiles during scratch as a function of normal load and corresponding AFM surface height images for (**a**) FCA, (**b**) ECR-CVD, and (**c**) SP-coatings [20]

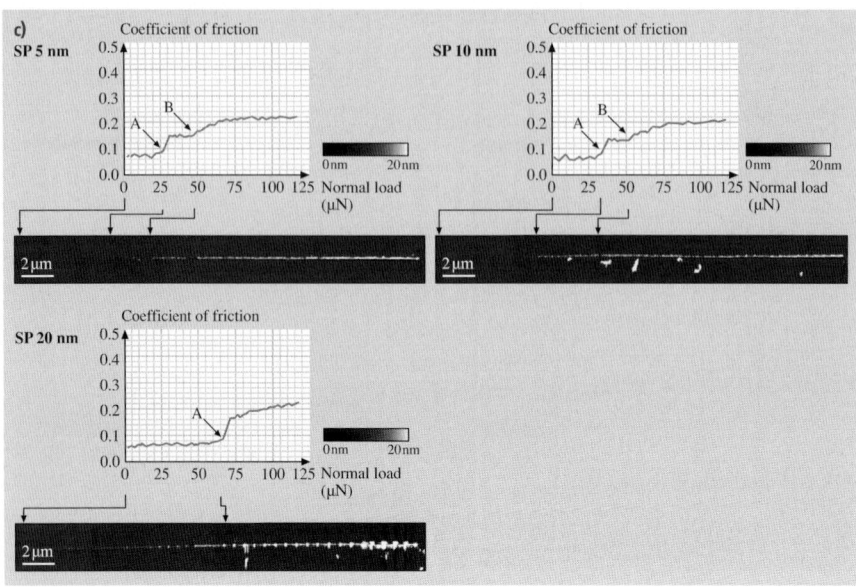

Fig. 21.22. (continued)

the disk surface is much harder to scratch than the underlying thin-film magnetic film. This is expected since the carbon coating is harder than the magnetic material used in the construction of the disks.

Microscratch characterization of ultrathin amorphous carbon coatings, deposited by filtered cathodic arc (FCA) direct ion beam (IB), electron cyclotron resonance plasma chemical vapor deposition (ECR-CVP), and sputter (SP) deposition processes have been conducted using a nanoindenter and an AFM [20,48,49,51–53,59]. Data on various coatings of different thicknesses using a Berkovich tip are compared in Fig. 21.22. Critical loads for various coatings and silicon substrate are summarized in Fig. 21.23. It is clear that, for all deposition methods, the critical load increases with increasing coating thickness due to better load-carrying capacity of thicker coatings as compared to the thinner ones. In comparison of the different deposition methods, ECR-CVD and FCA coatings show superior scratch resistance at 20- and 10nm thicknesses compared to SP coatings. As the coating thickness reduces, ECR-CVD exhibits the best scratch resistance followed by FCA and SP coatings.

Since tapes scratch readily, for comparisons in scratch resistance of various tapes, *Bhushan* and *Koinkar* [47] made scratches on selected three tapes only with one cycle. Figure 21.24 presents the scratch depths as a function of normal load after one cycle for three tapes – MP, BaFe and ME tapes. For the MP and BaFe particulate tapes, *Bhushan* and *Koinkar* [47] noted that the scratch depth along (parallel) and across (perpendicular) the longitudinal direction of the tapes is similar. Between the two tapes, MP tape appears to be more scratch resistant than BaFe tape, which

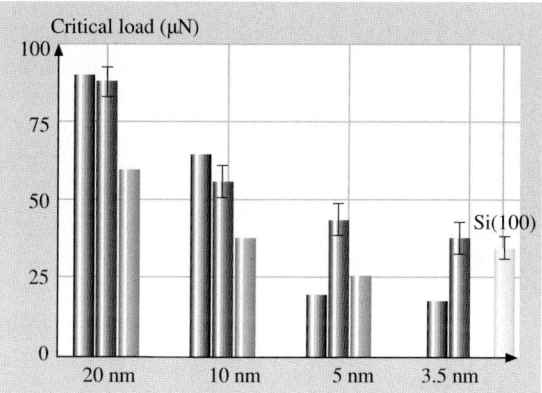

Fig. 21.23. Summary of critical loads estimated from the coefficient of friction profiles and AFM images [20]

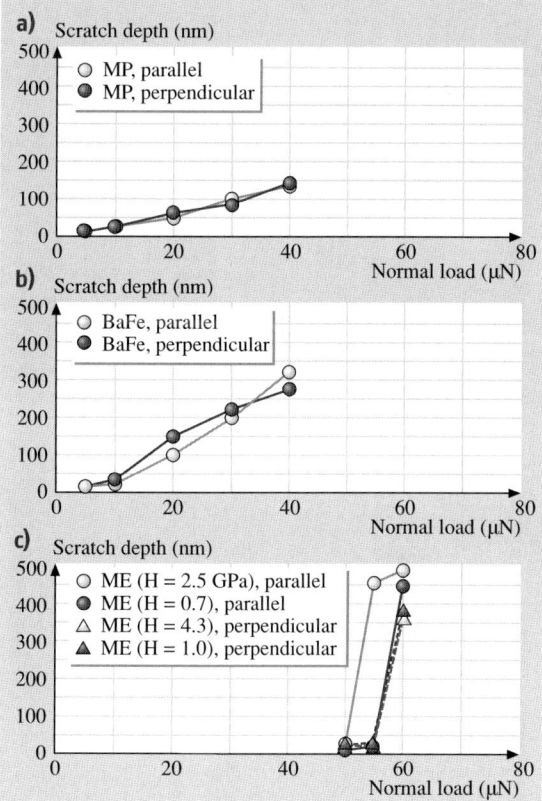

Fig. 21.24. Scratch depth as a function of normal load after one scratch cycle for (**a**) MP, (**b**) BaFe, and (**c**) ME tapes along parallel and perpendicular directions with respect to the longitudinal axis of the tape [47]

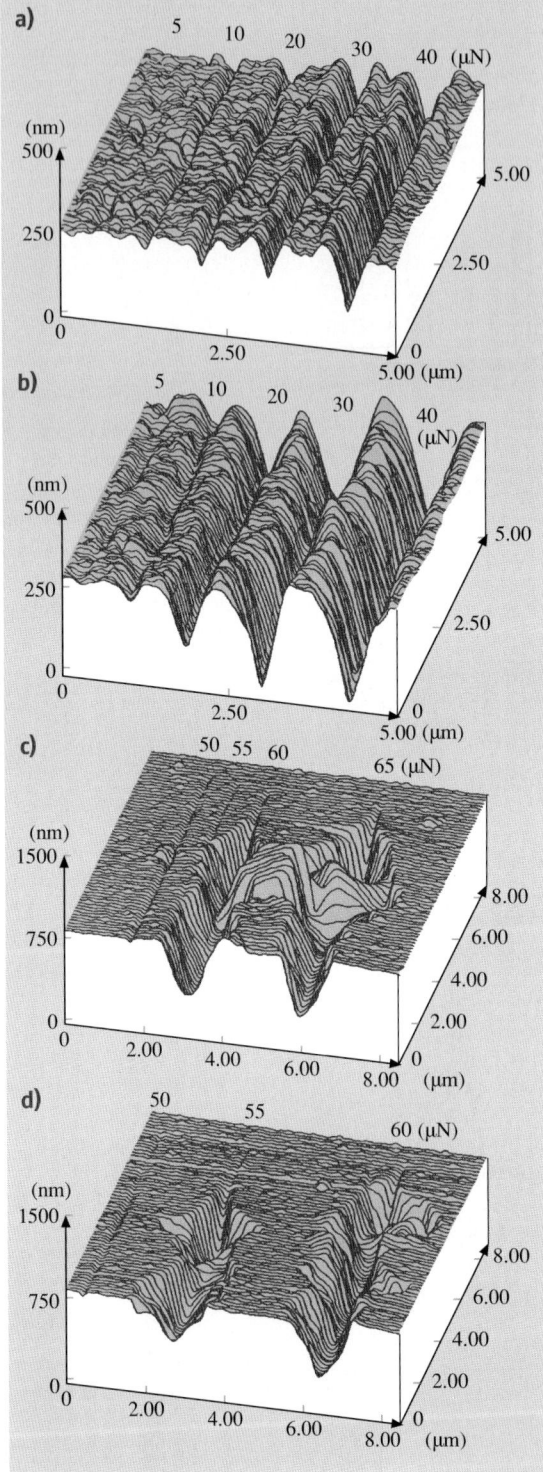

Fig. 21.25. Surface maps for scratched (**a**) MP, (**b**) BaFe, (**c**) ME (H = 0.7 GPa), and (**d**) ME (H= 2.5 GPa) tapes along parallel direction. Normal loads used for scratching for one cycle are listed in the figure [47]

depends on the binder, pigment volume concentration (PVC) and the head cleaning agent (HCA) contents. ME tapes appear to be much more scratch resistant than the particulate tapes. However, the ME tape breaks up catastrophically in a brittle mode at a normal load higher than the 50 μN (Fig. 21.25), as compared to particulate tapes in which the scratch rate is constant. They reported that the hardness of ME tapes is higher than that of particulate tapes, however, a significant difference in the nanoindentation hardness values of the ME film from region to region (Table 21.3) was observed. They systematically measured scratch resistance in the high and low hardness regions along and across the longitudinal directions. Along the parallel direction, load required to crack the coating was lower (implying lower scratch resistance) for a harder region, than that for a softer region. The scratch resistance of high hardness region along the parallel direction is slightly poorer than that for along perpendicular direction. Scratch widths in both low and high hardness regions is about half ($\approx 2\,\mu$m) than that in perpendicular direction ($\approx 1\,\mu$m). In the parallel direction, the material is removed in the form of chips and lateral cracking also emanates from the wear zone. ME films have columnar structure with the columns lined up with an oblique angle of on the order of about 35° with respect to the normal to the coating surface [3,60]. The column orientation may be responsible for directionality effect on the scratch resistance. *Hibst* [60] have reported the directionality effect in the ME tape-head wear studies. They have found that the wear rate is lower when the head moves in the direction corresponding to the column orientation than in the opposite direction.

PET films could be scratched at loads of as low as about 2 μN (Fig. 21.26). Figure 21.26a shows scratch marks made at various loads. Scratch depth along the scratch does not appear to be uniform. This may occur because of variations in the mechanical properties of the film. *Bhushan* and *Koinkar* [45] also conducted scratch studies in the selected particulate regions. Scratch profiles at increasing loads in the particulate region are shown in Fig. 21.26b. We note that the bump (particle) is barely scratched at 5 μN and it can be scratched readily at higher loads. At 20 μN, it essentially disappears.

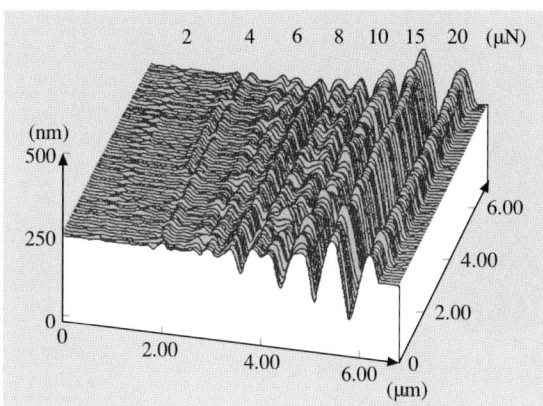

Fig. 21.26. Surface profiles for scratched PET film (**a**) polymer region, (**b**) ceramic particulate region. The loads used for various scratches at ten cycles are indicated in the plots [45]

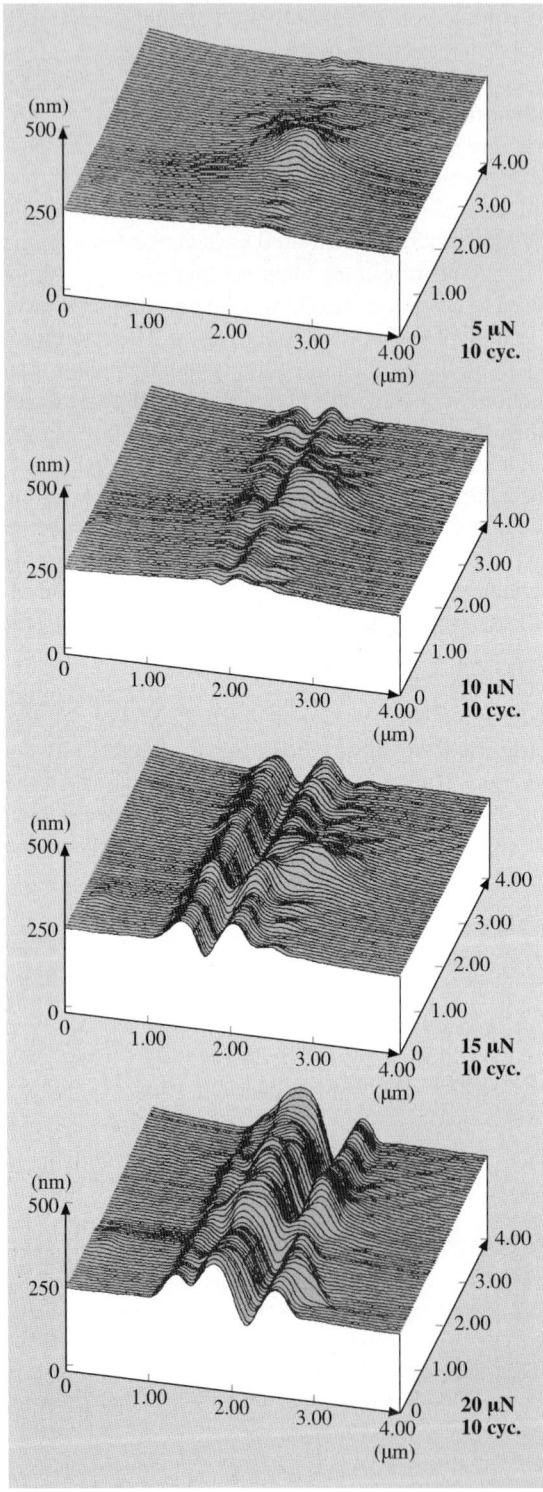

Fig. 21.26. (continued)

21.5.3 Microscale Wear

By scanning the sample (in 2D) while scratching, wear scars are generated on the sample surface [16, 18, 28, 45–49, 53, 54, 57, 58]. The major benefit of a single cycle wear test over a scratch test is that wear data can be obtained over a large area.

Magnetic Head Materials

Figure 21.27 shows the wear depth as a function of load for one cycle for different slider materials. Variation in the wear depth in the wear mark is dependent upon the material. It is generally within ±5%. The mean wear depth increases with the increase in normal load. The representative surface profiles showing the wear marks (central 2 μm × 2 μm region) at a normal load of 60 μN for all specimens are shown in Fig. 21.28. The material is removed uniformly in the wear region for all specimens. Table 21.2 presents average wear depth at 60 μN normal load for all specimens. Microwear resistance of SiC and Al_2O_3 is the highest followed by Al_2O_3–TiC, single-crystal and polycrystalline Mn–Zn ferrite.

Next, wear experiments were conducted for multiple cycles. Figure 21.29 shows the 2-D gray scale plots and corresponding section plot (on tope of each gray scale plot), taken at a location shown by an arrow for Al_2O_3 (left column) and Al_2O_3–TiC (right column) specimen obtained at a normal load of 20 μN and at a different number of scan cycles. The central regions (2 μm × 2 μm) show the wear mark generated after a different number of cycles. Note the difference in the vertical scale of gray scale and section plots. The Al_2O_3 specimen shows that wear initiates at the porous holes or defects present on the surface. Wear progresses at these locations as a function of number of cycles. In the porous hole free region, microwear resistance is higher. In the case of the Al_2O_3–TiC specimen for about five scan cycles, the microwear resistance is higher at the TiC grains and is lower at the Al_2O_3 grains. The TiC grains are removed from the wear mark after five scan cycles. This indicates that microwear resistance of multi-phase materials depends upon the individual grain properties. Evolution of wear is uniform within the wear mark for

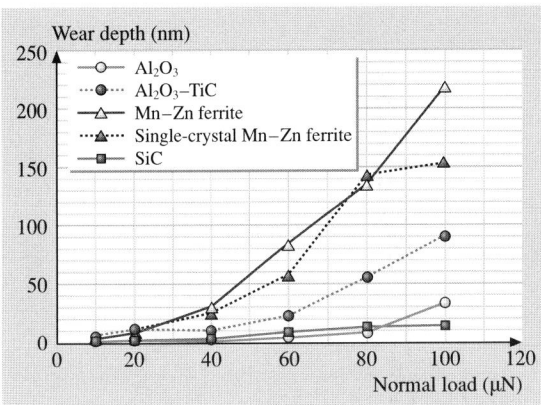

Fig. 21.27. Wear depth as a function of normal load after one scan cycle for Al_2O_3, Al_2O_3–TiC, polycrystalline Mn–Zn ferrite, single-crystal Mn–Zn ferrite and SiC [28]

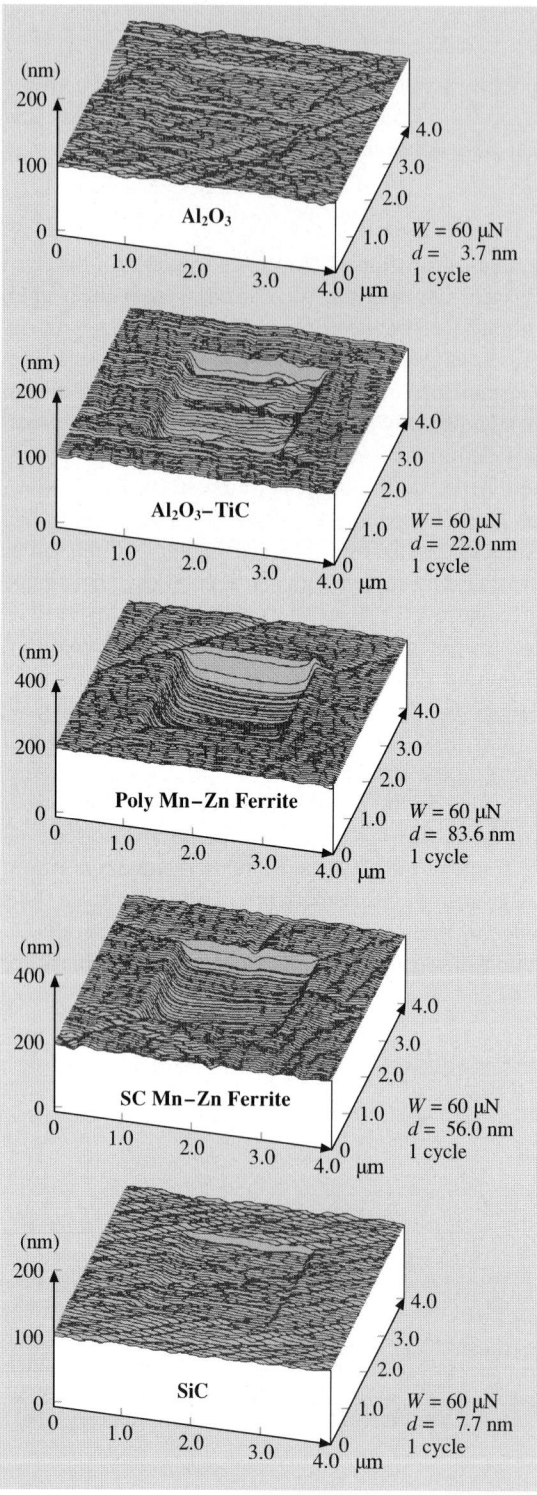

Fig. 21.28. Surface profiles showing the worn region (center $2\,\mu m \times 2\,\mu m$) after one scan cycles at a normal load of $60\,\mu N$ for Al_2O_3, Al_2O_3–TiC, polycrystalline Mn–Zn ferrite, single-crystal Mn–Zn ferrite and SiC [28]

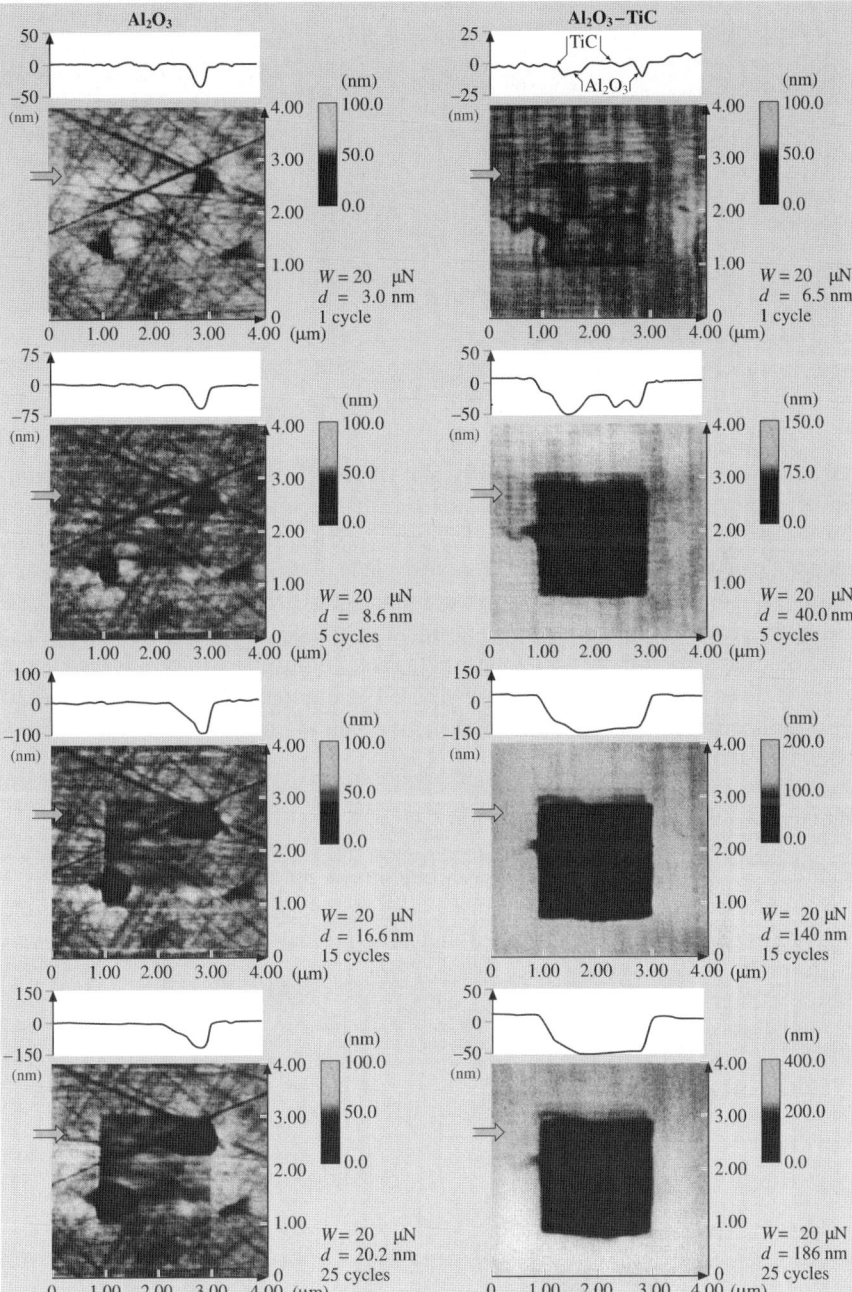

Fig. 21.29. Gray scale 2-D plots showing the worn region (center 2 μm × 2 μm) at a normal load of 20 μN and different number of scan cycles for Al_2O_3 and Al_2O_3-TiC. The 2-D section plots taken at a location shown by an arrow are shown on the top of corresponding gray scale plot. Note the change in vertical scale for gray scale and 2-D section plots [28]

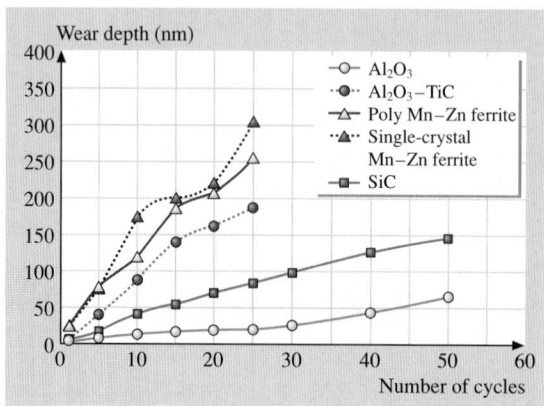

Fig. 21.30. Wear depth as a function of number of cycles at a normal load of 20 μN for Al_2O_3, Al_2O_3–TiC, polycrystalline Mn–Zn ferrite, single-crystal (SC) Mn–Zn ferrite and SiC [28]

ferrite specimens. Figure 21.30 shows plot of wear depth as a function of number of cycles at a normal load of 20 μN for all specimens. The Al_2O_3 specimen reveals highest microwear resistance followed by SiC, Al_2O_3–TiC, polycrystalline and single-crystal Mn–Zn ferrite. Wear resistance of Al_2O_3–TiC is inferior to that of Al_2O_3. Chu et al. [61] studied friction and wear behavior of the single-phase and multi-phase ceramic materials and found that wear resistance of multi-phase materials was poorer than single-phase materials. Multi-phase materials have more material flaws than the single-phase material. The differences in thermal and mechanical properties between the two phases may lead to cracking during processing, machining or use.

Magnetic Media

Figure 21.31 shows the wear depth as a function of load for one cycle for the polished, unlubricated and lubricated disks [18]. Figure 21.32 shows profiles of the wear scars generated on unlubricated disk. The normal force for the imaging was

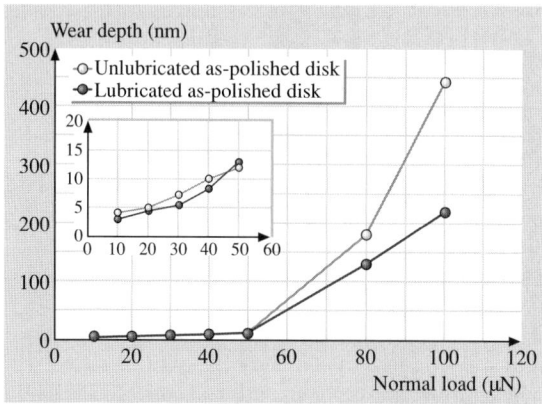

Fig. 21.31. Wear depth as a function of normal load for polished, lubricated and unlubricated thin-film rigid disks after one cycle [18]

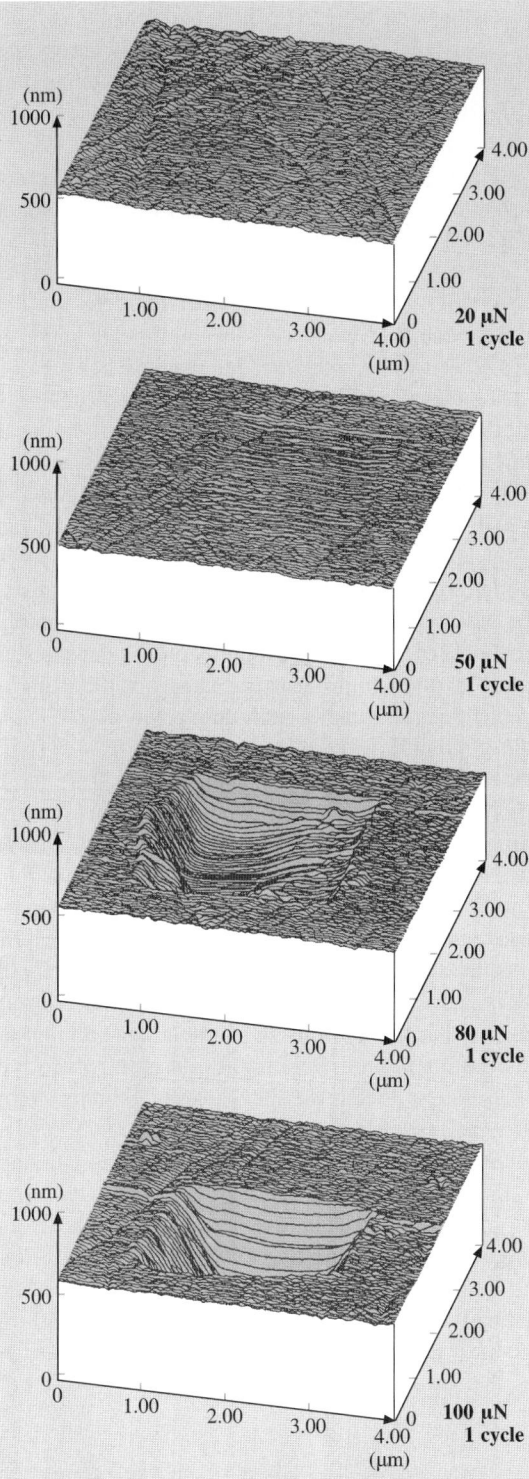

Fig. 21.32. Surface maps of a polished, unlubricated thin-film rigid disk showing the worn region (center 2 μm × 2 μm) after one cycle. The normal loads are indicted in the figure [18]

about 0.5 µN and the loads used for the wear were 20, 50, 80 and 100 µN as indicated in the figure. We note that wear takes place relatively uniformly across the disk surface and essentially independent of the lubrication for the disks studied. For both lubricated and unlubricated disks, the wear depth increases slowly with load at low loads with almost the same wear rate. As the load is increased to about 60 µN, wear increases rapidly with load. The wear depth at 50 µN is about 14 nm, slightly less than the thickness of the carbon film. The rapid increase of wear with load at loads larger than 60 µN is an indication of the breakdown of the carbon coating on the disk surface.

Figure 21.33 shows the wear depth as a function of number of cycles for the polished disks (lubricated and unlubricated). Again, for both unlubricated and lubricated disks, wear initially takes place slowly with a sudden increase between 40 and 50 cycles at 10 µN. The sudden increase occurred after 10 cycles at 20 µN. This rapid increase is associated with the breakdown of the carbon coating. The wear profiles at various cycles are shown in Fig. 21.34 for a polished, unlubricated disk at a normal load of 20 µN. Wear is not uniform and the wear is largely initiated at the texture grooves present on the disk surface. This indicates that surface defects strongly affect the wear rate.

Hard amorphous carbon coatings are used to provide wear and corrosion resistance to magnetic disks and MR/GMR magnetic heads. A thick coating is desirable for long durability; however, to achieve ever increasing high recording densities, it is necessary to use as thin a coating as possible. Microwear data on various amorphous carbon coatings of different thicknesses have been conducted by *Bhushan* and *Koinkar* [48], *Koinkar* and *Bhushan* [49], and *Sundararajan* and *Bhushan* [53]. Figure 21.35 shows a wear mark on an uncoated Si(100) and various 10 nm thick carbon coatings. It is seen that Si(100) wears uniformly, whereas carbon coatings wear nonuniformly. Carbon coating failure is sudden and accompanied by a sudden rise in friction force. Figure 21.36 shows the wear depth of Si(100) substrate and various coatings at two different loads. FCA and ECR-CVD, 20 nm thick coatings show excellent wear resistance up to 80 µN, the load that is required for the IB 20 nm coating to fail. In these tests, failure of a coating results when the wear depth exceeds the quoted coating thickness. The SP 20 nm coating fails at the much lower load of 35 µN. At 60 µN, the coating hardly provides any protection. Moving on to the 10 nm coatings, ECR-CVD coating requires about 40 cycles at 60 µN to fail as compared to IB and FCA, which fail at 45 µN. the FCA coating exhibits slight roughening in the wear track after the first few cycles, which leads to an increase in the friction force. The SP coating continues to exhibit poor resistance, failing at 20 µN. For the 5 nm coatings, the load required to fail the coatings continues to decrease. But IB and ECR-CVD still provide adequate protection as compared to bare Si(100) in that order, failing at 35 µN compared to FCA at 25 µN and SP at 20 µN. Almost all the 20, 10, and 5 nm coatings provide better wear resistance than bare silicon. At 3.5 nm, FCA coating provides no wear resistance, failing almost instantly at 20 µN. The IB and ECR-CVD coating show good wear resistance at 20 µN

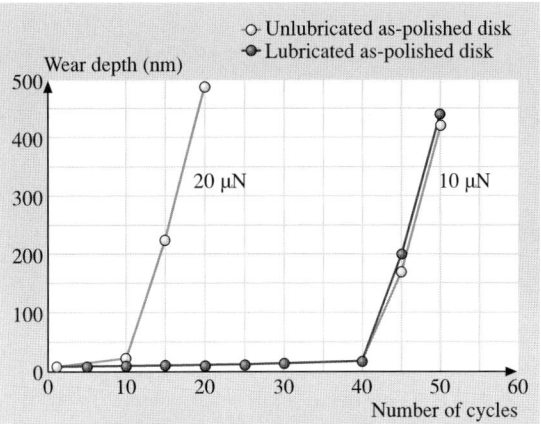

Fig. 21.33. Wear depth as a function of number of cycles for polished, lubricated and unlubricated thin-film rigid disks at 10 μN and for polished, unlubricated disk at 20 μN [18]

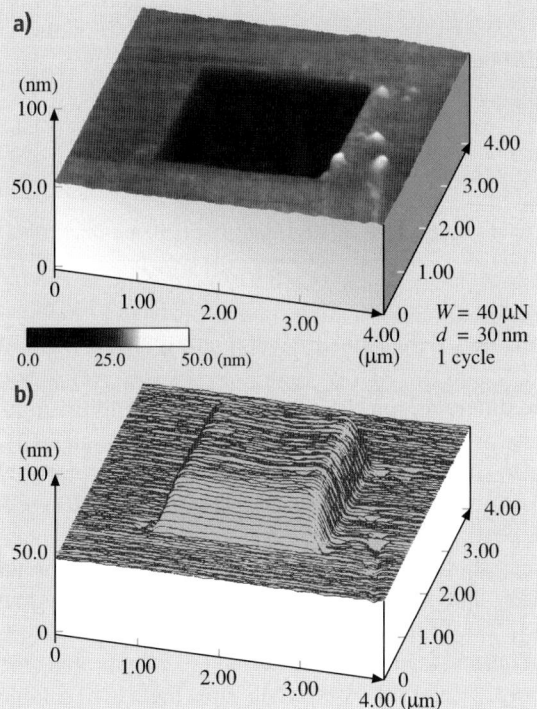

Fig. 21.34. Surface maps of a polished, unlubricated thin-film rigid disk showing the worn region (center 2 μm × 2 μm) at 20 μN. The number of cycles are indicated in the figure [18]

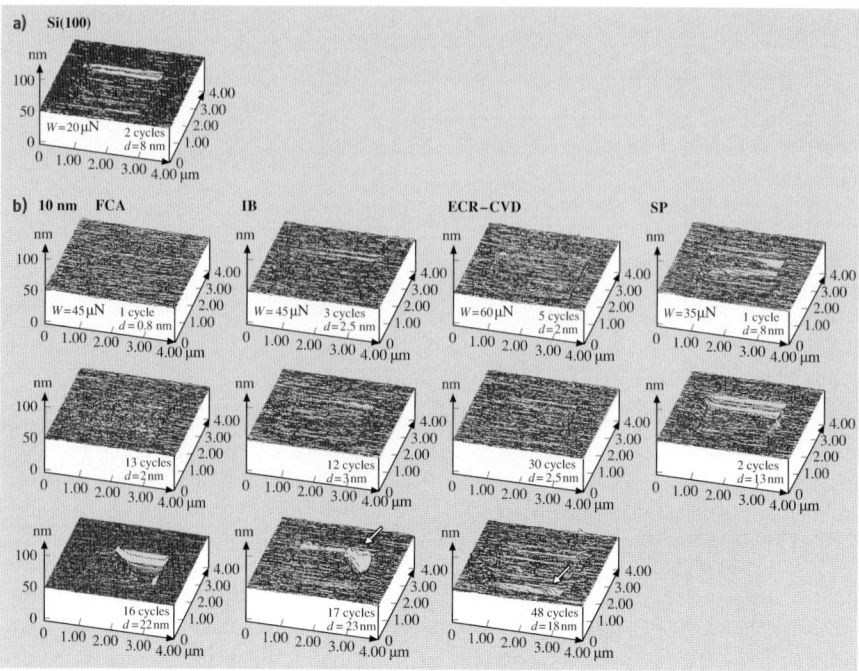

Fig. 21.35. AFM images of wear marks on (**a**) bare Si(100), and (**b**) all 10 nm thick amorphous carbon coatings [53]

compared to bare Si(100). But IB lasts only about 10 cycles and ECR-CVD about 3 cycles at 25 μN.

The wear tests highlight the differences in the coatings more vividly than the scratch tests data presented earlier. At higher thicknesses (10 and 20 nm), the ECR-CVD and FCA coatings appear to show the best wear resistance. This is probably due to higher hardness of the coatings (see data presented later). At 5 nm, IB coating appears to be the best. FCA coatings show poorer wear resisting with decreasing coating thickness. SP coatings showed consistently poor wear resistance at all thicknesses. The IB 2.5 nm coating does provide reasonable wear protection at low loads.

Wear depths as a function of normal load for MP, BaFe and ME tapes along the parallel direction are plotted in Fig. 21.37 [47]. For the ME tape, there is negligible wear until the normal load of about 50 μN, above this load the magnetic coating fails rapidly. This observation is consistent with the scratch data. Wear depths as a function of number of cycles for MP, BaFe, and ME tapes are shown in Fig. 21.38. For the MP and BaFe particulate tapes, wear rates appear to be independent of the particulate density. Again as observed in the scratch testing, wear rate of BaFe tapes is higher than that for MP tapes. ME tapes are much more wear resistant than the particulate tapes. However, the failure of ME tapes are catastrophic as observed in scratch testing. Wear studies were performed along and across the longitudinal

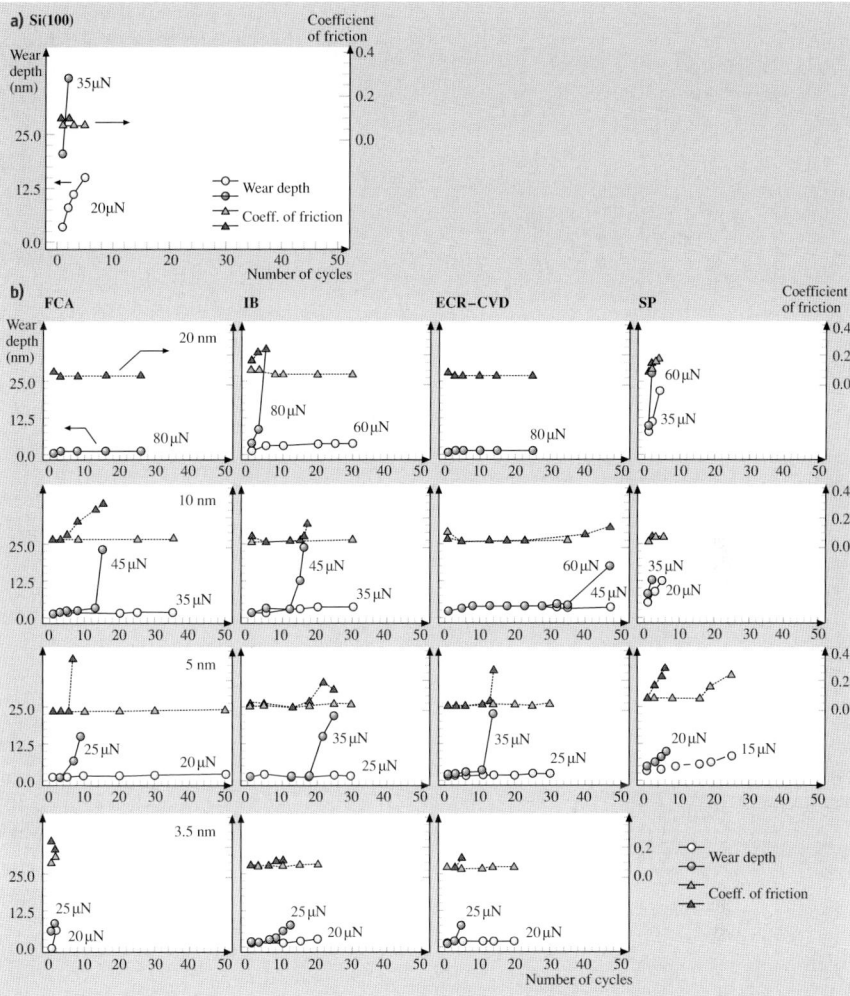

Fig. 21.36. Wear data of (**a**) bare Si(100) and (**b**) all amorphous carbon coatings. Coating thickness is constant along each row in (**b**). Both wear depth and coefficient of friction during wear for a given cycle are plotted [53]

tape direction in high and low hardness regions. At the high hardness regions of the ME tapes, failure occurs at lower loads. Directionality effect again may arise from the columnar structure of the ME films [3, 60]. Wear profiles at various cycles at a normal load of 2 μN for MP and at 20 μN for ME tapes are shown in Fig. 21.39. For the particulate tapes, we note that polymer gets removed before the particulates do (Fig. 21.39a). Based on the wear profiles of the ME tape shown in Fig. 21.39a, we note that most wear occurs between 50 to 60 cycles which shows the catastrophic removal of the coating. It was also observed that wear debris generated during wear

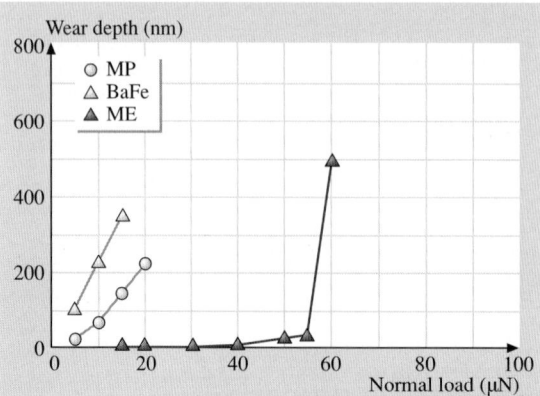

Fig. 21.37. Wear depth as a function of normal load for three tapes in the parallel direction after one cycle [47]

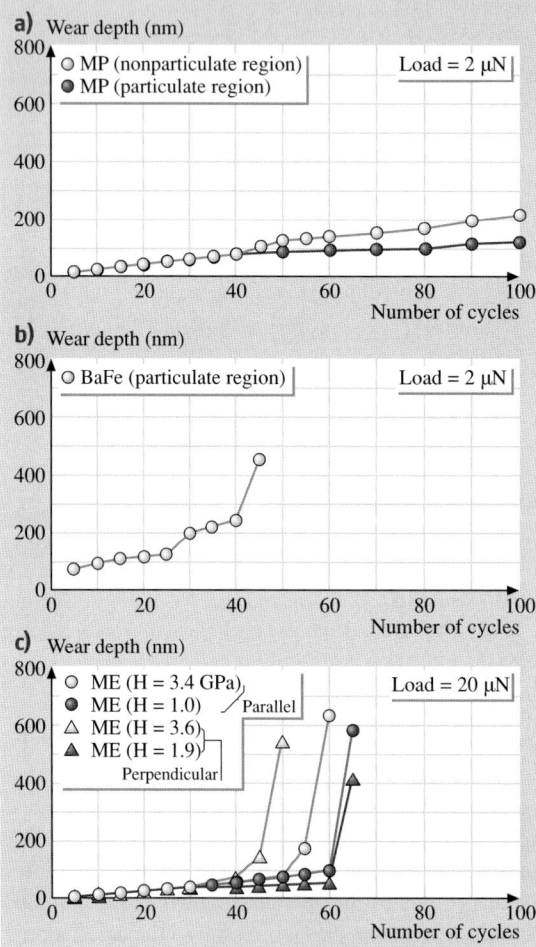

Fig. 21.38. Wear depth as a function of number of cycles for (**a**) MP, (**b**) BaFe, and (**c**) ME tapes in different regions at normal loads indicated in the figure. Note a higher load used for the ME tape in (**c**) [47]

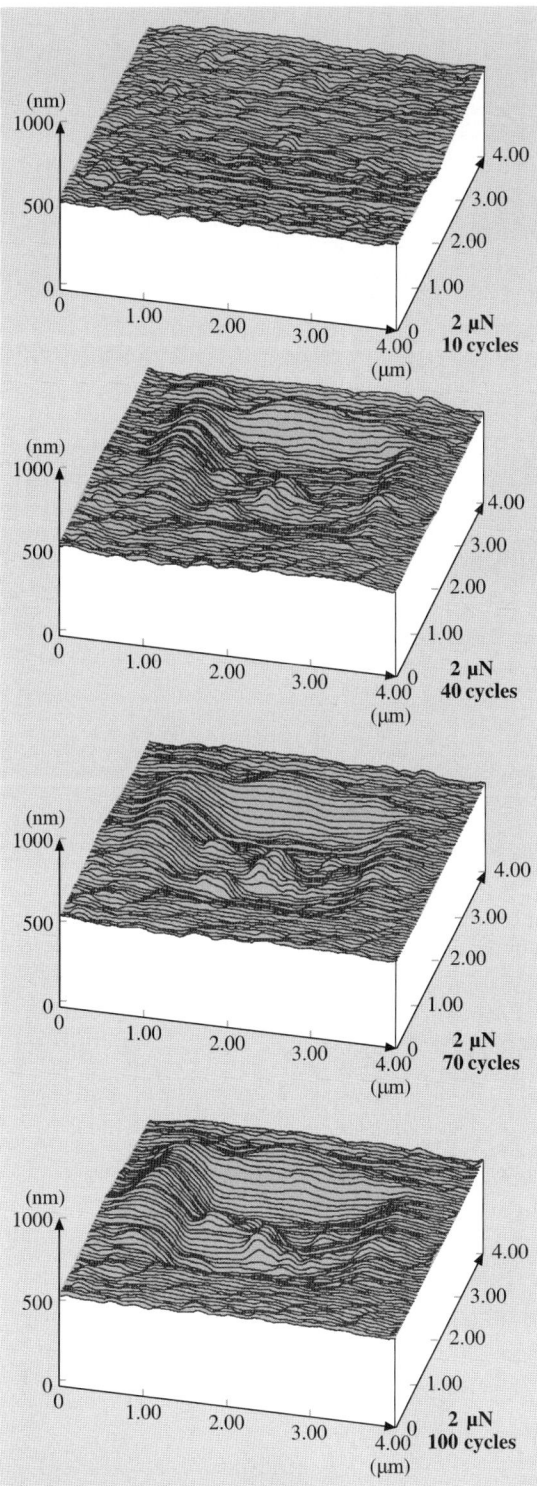

Fig. 21.39. Surface maps showing the worn region (center $2\,\mu m \times 2\,\mu m$) after various cycles of wear at (**a**) $2\,\mu N$ for MP (particulate region) and at (**b**) $20\,\mu N$ for ME (H= 3.4 GPa, parallel direction) tapes. Note a different *vertical scale* for the bottom profile of (**b**) [47]

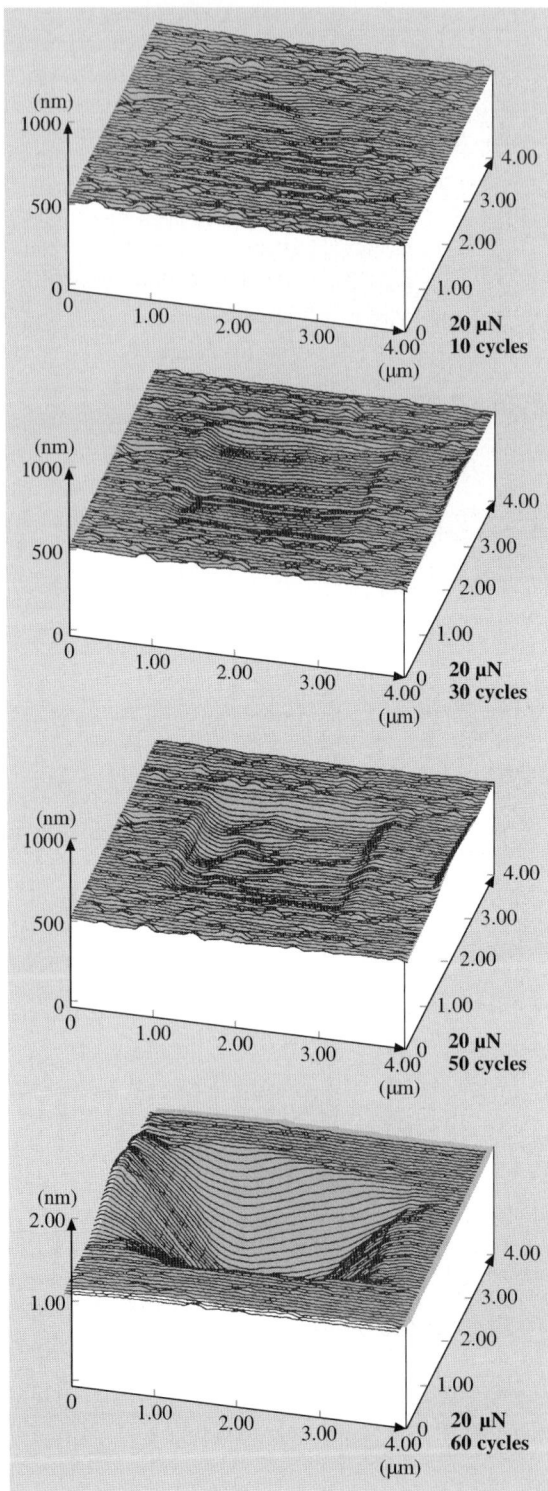

Fig. 21.39. (continued)

test in all cases is loose and can easily be removed away from the scan area at light loads ($\approx 0.3\,\mu N$).

The average wear depth as a function of load for a PET film is shown in Fig. 21.40. Again, the wear depth increases linearly with load. Figure 21.41 shows the average wear depth as a function of number of cycles. The observed wear rate is approximately constant. PET tape substrate consists of particles sticking out on its surface to facilitate winding. Figure 21.42 shows the wear profiles as a function of number of cycles at $1\,\mu N$ load on the PET film in the nonparticulate and particulate regions [45]. We note that polymeric materials tear in microwear tests. The particles do not wear readily at $1\,\mu N$. Polymer around the particles is removed but the particles remain intact. Wear in the particulate region is much smaller than that in the polymer region. We will see later that nanohardness of the particulate region is about 1.4 GPa compared to 0.3 GPa in the nonparticulate region (Table 21.3).

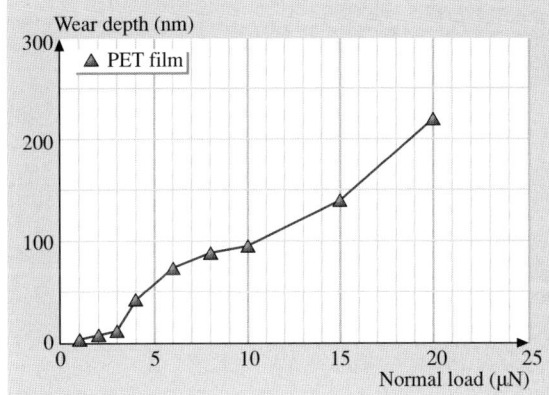

Fig. 21.40. Wear depth as a function of normal load (after one cycle) for a PET film [45]

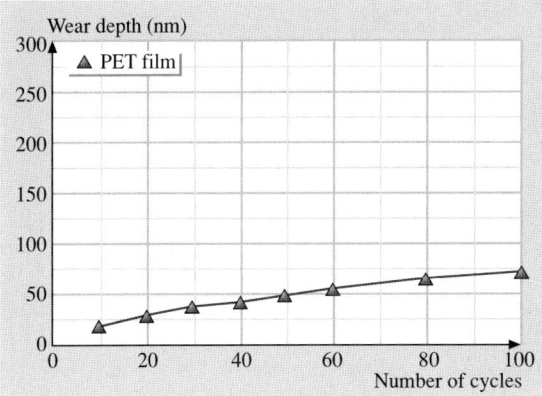

Fig. 21.41. Wear depth as a function of number of cycles at $1\,\mu N$ for a PET film [45]

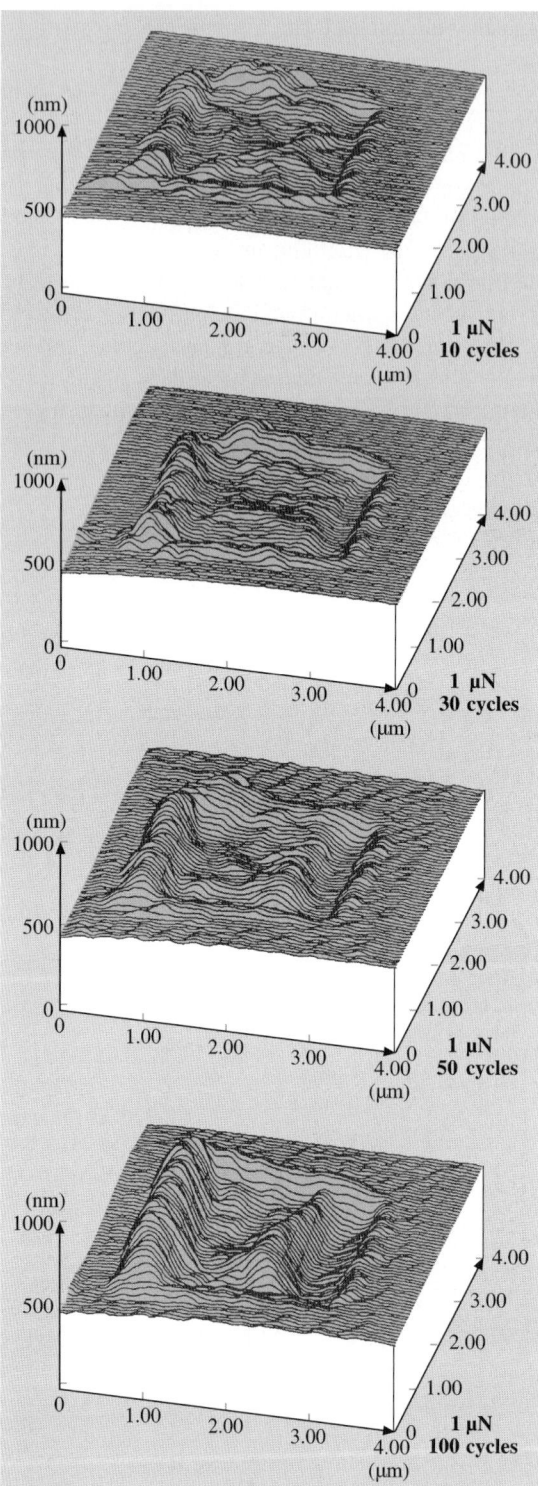

Fig. 21.42. Surface maps showing the worn region (center 2 μm × 2 μm) at 1 μN for a PET film (**a**) in the polymer region, (**b**) in the particulate region. The number of cycles are indicated in the figure [45]

Fig. 21.42. (continued)

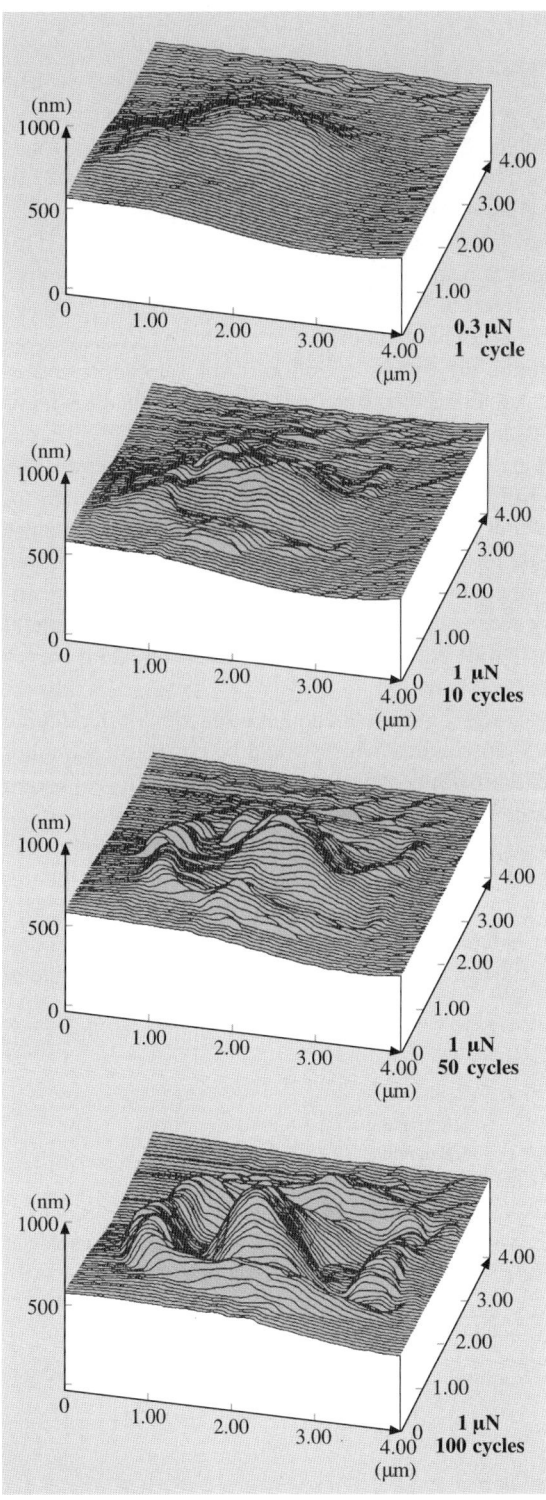

21.6 Indentation

21.6.1 Picoscale Indentation

Bhushan and *Ruan* [15] measured indentability of magnetic tapes at increasing loads on a picoscale, Fig. 21.43. In this figure, the vertical axis represents the cantilever tip deflection and the horizontal axis represents the vertical position (Z) of the sample. The "extending" and "retracting" curves correspond to the sample being moved toward or away from the cantilever tip, respectively. In this experiment, as the sample surface approaches the AFM tip fraction of a nm away from the sample (point A), the cantilever bends toward the sample (part B) because of attractive forces between the tip and sample. As we continue the forward position of the sample, it pushes the cantilever back through its original rest position (point of zero applied load) entering the repulsive region (or loading portion) of the force curve. As the sample is retracted, the cantilever deflection decreases. At point D in the retracting curve, the sample is disengaged from the tip. Before the disengagement, the tip is pulled toward the sample after the zero deflection point of the force curve (point C) because of attractive forces (van der Waals forces and longer range meniscus forces). A thin layer of liquid, such as liquid lubricant and condensations of water vapor from ambient, will give rise to capillary forces that act to draw the tip towards sample at small separations. The horizontal shift between the loading and unloading curves results from the hysteresis in the PZT tube.

The left portion of the curve shows the tip deflection as a function of the sample traveling distance during sample–tip contact, which would be equal to each other for a rigid sample. However, if the tip indents into the sample, the tip deflection would be less than the sample traveling distance, or in other words, the slope of the line would be less than 1. In Fig. 21.43, we note that line in the left portion of the figure is curved with a slope of less than 1 shortly after the sample touches the tip, which suggests that the tip has indented the sample. Later, the slope is equal to 1 suggesting that the tip no longer indents the sample. This observation indicates that the tape surface is soft locally (polymer rich) but it is hard (as a result of magnetic

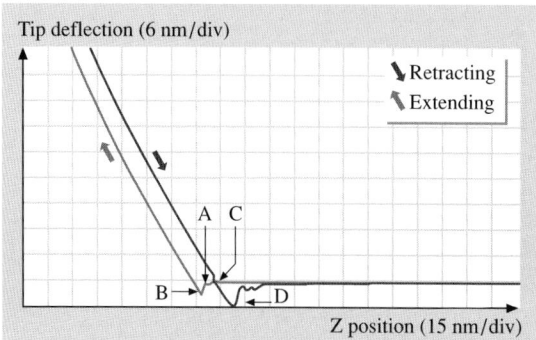

Fig. 21.43. Tip deflection (normal force) as a function of Z (separation distance) curve for a metal-particle (MP) tape. The spring constant of the cantilever used was 0.4 N/m [15]

particles) underneath. Since the curves in extending and retracting modes are identical, the indentation is elastic up to at a maximum load of about 22 nN used in the measurements.

According to *Bhushan* and *Ruan* [15], during indentation of rigid disks, the slope of the deflection curves remained constant as the disks touch and continue to push the AFM tip. The disks were not indented.

21.6.2 Nanoscale Indentation

Indentation hardness with a penetration depth as low as 5 nm can be measured using AFM. *Bhushan* and *Koinkar* [17] measured hardness of thin-film disks at load of 80, 100, and 140 µN. Hardness values were 20 GPa (10 nm), 21 GPa (15 nm) and 9 GPA (40 nm); the depths of indentation are shown in the parenthesis. The hardness value at 100 µN is much higher than at 140 µN. This is expected since the indentation depth is only about 15 nm at 100 µN which is smaller than the thickness of carbon coating (≈ 30 nm). The hardness value at lower loads is primarily the value of the carbon coating. The hardness value at higher loads is primarily the value of the magnetic film, which is softer than the carbon coating [2]. This result is consistent with the scratch and wear data discussed previously.

For the case of hardness measurements made on magnetic thin film rigid disk at low loads, the indentation depth is on the same order at the variation in the surface roughness. For accurate measurements of indentation size and depth, it is desirable to subtract the original (unindented) profile from the indented profile. *Bhushan* et al. [18] developed an algorithm for this purpose. Because of hysteresis, a translational shift in the sample plane occurs during the scanning period, resulting in a shift between images captured before and after indentation. Therefore, the image for perfect overlap needs to be shifted before subtraction can be performed. To accomplish this objective, a small region on the original image was selected and the corresponding region on the indented image was found by maximizing the correlation between the two regions. (Profiles were plane-fitted before subtraction.) Once two regions were identified, overlapped areas between the two images were determined and the original image was shifted with the required translational shift and this subtracted from the indented image. An example of profiles before and after subtraction is shown in Fig. 21.44. It is easier to measure the indent on the subtracted image. At a normal load of 140 mN the hardness value of an unlubricated, as-polish magnetic thin film rigid disk (rms roughness = 3.3 nm) is 9.0 GPa and indentation depth is 40 nm.

For accurate measurement of nanohardness at very shallow indentation depths, depth-sensing capacitance transducer system in an AFM is used [19]. Figure 21.45a shows the hardness as a function of residual depth for three types of 100 nm thick amorphous carbon coatings deposited on silicon by sputtering, ion beam and cathodic arc processes [50]. Data on uncoated silicon are also included for comparisons. The cathodic arc carbon coating exhibits highest hardness of about 24.9 GPa,

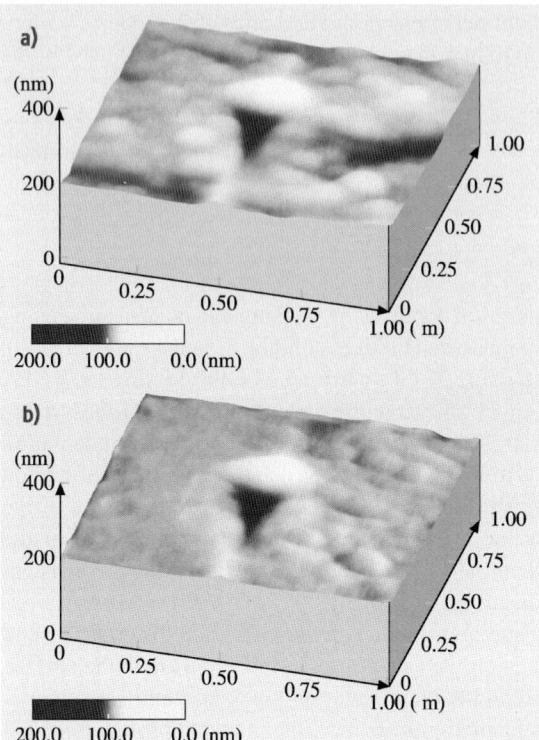

Fig. 21.44. Images with nanoindentation marks generated on a polished, unlubricated thin-film rigid disk at 140 μN (**a**) before subtraction, and (**b**) after subtraction [18]

whereas the sputtered and ion beam carbon coatings exhibit hardness values of 17.2 and 15.2 GPa respectively. The hardness of Si(100) is 13.2 GPa. High hardness of cathodic arc carbon coating explains its high wear resistance, reported earlier. Figure 21.45b shows the elastic modulus as a function of residual depth for various samples. The cathodic arc coating exhibits the highest elastic modulus. Its elastic modulus decreases with an increasing residual depth, while the elastic moduli for the other carbon coatings remain almost constant. In general, hardness and elastic modulus of coatings are strongly influenced by their crystalline structure, stoichiometry and growth characteristics which depend on the deposition parameters. Mechanical properties of carbon coatings have been known to change over a wide range with sp^3–sp^2 bonding ratio and amount of hydrogen. Hydrogen is believed to play an important role in the bonding configuration of carbon atoms by helping to stabilize tetrahedral coordination of carbon atoms. Detailed mechanical characterization of amorphous carbon coatings is presented by *Li* and *Bhushan* [51, 52] and *Bhushan* [59].

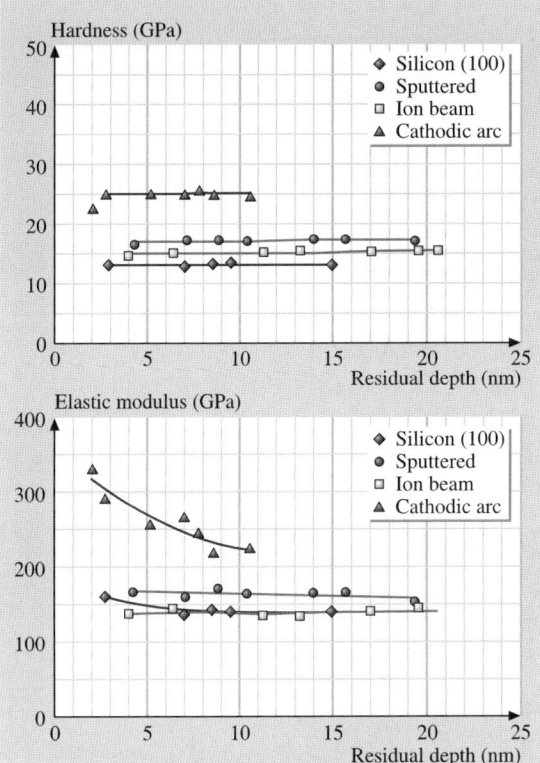

Fig. 21.45. Nanohardness and elastic modulus as a function of residual indentation depth for Si(100) and 100 nm thick coatings deposited by sputtering, ion beam and cathodic arc processes [50]

21.6.3 Localized Surface Elasticity

By using an AFM in a so-called force modulation mode, it is possible to quantitatively measure the elasticity of soft and compliant materials with penetration depths of less than 100 nm [21, 22]. This technique has been successfully used to get localized elasticity maps of particulate magnetic tapes. Elasticity map of a tape can be used to identify relative distribution of hard magnetic/nonmagnetic ceramic particles and the polymeric binder on the tape surface which has an effect on friction and stiction at the head–tape interface. Figure 21.46 shows surface height and elasticity maps on an MP tape. The elasticity image reveals sharp variations in surface elasticity due to the composite nature of the film. As can be clearly seen, regions of high elasticity do not always correspond to high or low topography. Based on a Hertzian elastic-contact analysis, the static indentation depth of these sample during the force modulation scan is estimated to be about 1 nm. The contrast seen is influenced most strongly by material properties in the top few nanometers, independent of the composite structure beneath the surface layer. The trend in number of stiff regions has been correlated to reduced stiction problems in tapes [62].

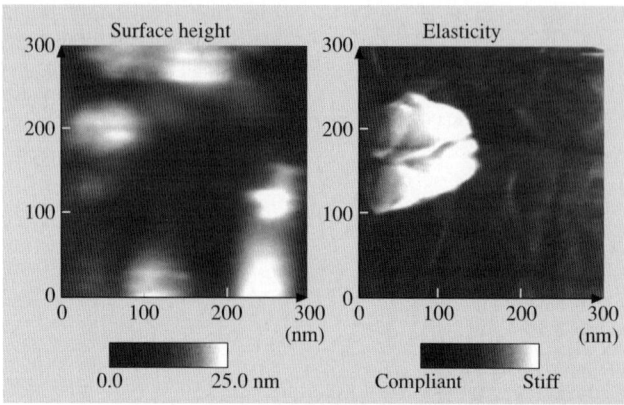

Fig. 21.46. Surface height and elasticity maps for a metal-particle tape A ($\sigma = 6.72$ nm and $P-V = 31.7$ nm). σ and $P-V$ refer to standard deviation of surface heights and peak-to-valley distance, respectively. The *grayscale* on the elasticity map is arbitrary [21]

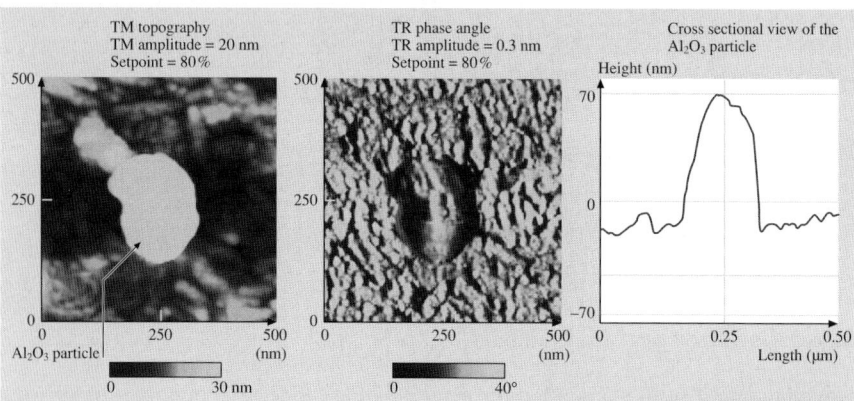

Fig. 21.47. Tapping mode (TM) topography and TR phase angle image of an alumina particle that is used as a head cleaning agent for MP tape. A cross-sectional view of the particle is also shown [25]

Figure 21.47 shows the surface topography and phase image of an alumina particle embedded in the MP tape using a so-called TR mode [25, 26]. The cross-section view of the particle obtained from the topographic image is shown at the bottom as a visual aid. The edges of the particle show up darker in the TR phase angle image, which suggests that it is less viscoelastic compared to the background. The magnetic particles on top of the alumina particle are clearly visible in the TR phase image. These have a brighter contrast, which is the same as that of the background. Phase contrast mapping appears to privide better resolution than stiffness mapping for magnetic tapes.

21.7 Lubrication

The boundary films are formed by physical adsorption, chemical adsorption, and chemical reaction. The physisorbed film can be either monomolecular or polymolecular thick. The chemisorbed films are monomolecular, but stoichiometric films formed by chemical reaction can have a large film thickness. In general, the stability and durability of surface films decrease in the following order: chemical reaction films, chemisorbed films and physisorbed films. A good boundary lubricant should have a high degree of interaction between its molecules and the sliding surface. As a general rule, liquids are good lubricants when they are polar and thus able to grip solid surfaces (or be adsorbed). Polar lubricants contain reactive functional groups with low ionization potential or groups having high polarizability [5]. Boundary lubrication properties of lubricants are also dependent upon the molecular conformation and lubricant spreading [63–66].

Mechanical interactions between the magnetic head and the medium in magnetic storage devices are minimized by the lubrication of the magnetic medium [2,3]. The primary function of the lubricant is to reduce the wear of the medium and to ensure that friction remains low throughout the operation of the drive. The main challenge, though, in selecting the best candidate for a specific surface is to find a material that provides an acceptable wear protection for the entire life of the product, which can be several years in duration. There are many requirements that a lubricant must satisfy in order to guarantee an acceptable life performance. An optimum lubricant thickness is one of these requirements. If the lubricant film is too thick, excessive stiction and mechanical failure of the head-disk is observed. On the other hand, if the film is too thin, protection of the interface is compromised, and high friction and excessive wear will result in catastrophic failure. An acceptable lubricant must exhibit properties such as chemical inertness, low volatility, high thermal, oxidative and hydrolytic stability, shear stability, and good affinity to the magnetic medium surface.

Fatty acid esters are excellent boundary lubricants, and esters such as tridecyl stearate, butyl stearate, butyl palmitate, buryl myristate, stearic acid, and myrstic acid are commonly used as internal lubricants, roughly 1–7% by weight of the magnetic coating in particulate flexible media (tapes and particulate flexible disks) [2,3]. The fatty acids involved include those with acid groups with an even number of carbon atoms between C_{12} and C_{22}, with alcohols ranging from C_3 to C_{13}. These acids are all solids with melting points above the normal surface operating temperature of the magnetic media. This suggests that the decomposition products of the ester via lubrication chemistry during a head-flexible medium contact may be the key to lubrication.

Topical lubrication is used to reduce the wear of rigid disks and thin-film tapes [67]. Perfluoropolyethers (PFPEs) are chemically the most stable lubricants with some boundary lubrication capability, and are most commonly used for topical lubrication of rigid disks. PFPEs commonly used include Fomblin Z lubricants, made by Solvay Solexis Inc., Milan, Italy; and Demnum S, made by Diakin, Japan; and their difunctional derivatives containing various reactive end groups, e.g., hydroxyl or alcohol (Fomblin Z-DOL and Z-TETROL), piperonyl (Fomblin AM 2001), isocyanate (Fomblin Z-DISOC), and ester (Demnum SP). Fomblin Y and Krytox 143AD (made by Dupont USA) have been used in the past for particulate rigid disks. The difunctional derivatives are referred to as reactive (polar) PFPE lubricants. The chemical structures, molecular weights, and viscosities of various types of PFPE lubricants are given in Table 21.4. We note that rheological properties of thin-films of lubricants are expected to be different from their bulk properties. Fomblin Z and Demnum S are linear PFPE, and Fomblin Y and Krytox 143 AD are branched PFPE, where the regularity of the chain is perturbed by $-CF_3$ side groups. The bulk viscosity of Fomblin Y and Krytox 143 AD is almost an order of magnitude higher than the Z type. Fomblin Z is thermally decomposed more rapidly

Table 21.4. Chemical structure, molecular weight, and viscosity of perfluoropolyether lubricants

Lubricant	Formula	Molecular weight (Daltons)	Kinematic viscosity cSt(mm^2/s)
Fomblin Z-25	$CF_3-O-(CF_2-CF_2-O)_m-(CF_2-O)_n-CF_3$	12,800	250
Fomblin Z-15	$CF_3-O-(CF_2-CF_2-O)_m-(CF_2-O)_n-CF_3$ ($m/n \approx 2/3$)	9100	150
Fomblin Z-03	$CF_3-O-(CF_2-CF_2-O)_m-(CF_2-O)_n-CF_3$	3600	30
Fomblin Z-DOL	$HO-CH_2-CF_2-O-(CF_2-CF_2-O)_m-(CF_2-O)_n-CF_2-CH_2-OH$	2000	80
Fomblin AM2001	Piperonyl$-O-CH_2-CF_2-O-(CF_2-CF_2-O)_m-(CF_2-O)_n-CF_2-O-$piperonyl[a]	2300	80
Fomblin Z-DISOC	$O-CN-C_6H_3-(CH_3)-NH-CO-CF_2-O-(CF_2-CF_2-O)_n-(CF_2-O)_m-CF_2-CO-NH-C_6H_3-(CH_3)-N-CO$	1500	160
Fomblin YR	$CF_3-O-(\underset{F}{\overset{CF_3}{C}}-CF_2-O)_m(CF_2-O)_n-CF_3$ ($m/n \approx 40/1$)	6800	1600
Demnum S-100	$CF_3-CF_2-CF_2-O-(CF_2-CF_2-CF_2-O)_m-CF_2-CF_3$	5600	250
Krytox 143AD	$CF_3-CF_2-CF_2-O-(\underset{F}{\overset{CF_3}{C}}-CF_2-O_m)-CF_2-CF_3$	2600	–

[a] 3,4-methylenedioxybenzyl

than Y [5]. The molecular diameter is about 0.8 nm for these lubricant molecules. The monolayer thickness of these molecules depends on the molecular conformations of the polymer chain on the surface [64, 65].

The adsorption of the lubricant molecules on a magnetic disk surface is due to van der Waals forces, which are too weak to offset the spin-off losses, or to arrest displacement of the lubricant by water or other ambient contaminants. Considering that these lubricating films are on the order of a monolayer thick and are required to function satisfactorily for the duration of several years, the task of developing a workable interface is quite formidable. An approach aiming at alleviating these shortcomings is to enhance the attachment of the molecules to the overcoat, which, for most cases, is sputtered carbon. There are basically two approaches which have been shown to be successful in bonding the monolayer to the carbon. The first relies on exposure of the disk lubricated with neutral PFPE to various forms of radiation, such as low-energy X-ray [68], nitrogen plasma [69], or far ultraviolet (e. g., 185 nm) [70]. Another approach is to use chemically active PFPE molecules, where the various functional (reactive) end groups offer the opportunity of strong attachments to specific interface. These functional groups can react with surfaces and bond the lubricant to the disk surface, which reduces its loss due to spin off and evaporation. Bonding of lubricant to the disk surface depends upon the surface cleanliness. After lubrication, the disk is generally heated at 150 °C for 30 min to 1 h to improve the bonding. If only a bonded lubrication is desired, the unbonded fraction can be removed by washing it off for 60 s with a non-freon solvent (FC-72). Their main advantage is their ability to enhance durability without the problem of stiction usually associated with weakly bonded lubricants (*Bhushan*, 1996a).

21.7.1 Boundary Lubrication Studies

Koinkar and *Bhushan* [29] and *Liu* and *Bhushan* [30] studied friction, adhesion, and durability of Z-15 and Z-DOL (bonded and washed, BW) lubricants on Si(100) surface. To investigate the friction properties of Si(100), Z-15, and Z-DOL(BW), the friction force versus normal load curves were obtained by making friction measurements at increasing normal loads, Fig. 21.48. An approximately linear response of all three samples is observed in the load range of 5–130 nN. From the horizontal intercept at zero value of friction force, adhesive force can be obtained. The adhesive forces for three samples were also measured using the force calibration plot technique. The adhesive force data obtained by the two techniques are summarized in Fig. 21.49, and the trends in the data obtained by two techniques are similar. The friction force and adhesive force of solid-like Z-DOL(BW) are consistently smaller than that for Si(100), but these values of liquid-like Z-15 lubricant is higher than that of Si(100). The presence of mobile Z-15 lubricant film increases adhesive force as compared to that of the Si(100) by meniscus formation. Whereas, the presence of Z-DOL(BW) film reduces the adhesive force because of absence of mobile liquid. See schematics at the bottom of Fig. 21.49. It is well known that in computer rigid disk drives, the stiction force increases rapidly with an increase in rest time between head and the disk [2]. The effect of rest time of 180 s on the friction force, adhesive force,

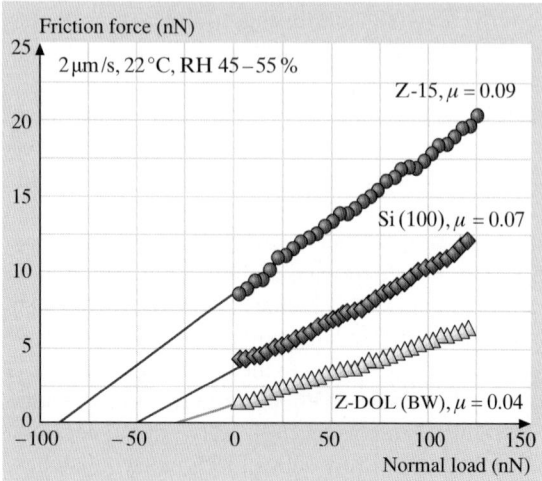

Fig. 21.48. Friction force versus normal load curves for Si(100), 2.8 nm thick Z-15 film, and 2.3 nm thick Z-DOL(BW) film at 2 μm/s, and in ambient air sliding against a Si_3N_4 tip. Based on these curves, coefficient of friction (μ) and adhesive force can be calculated [30]

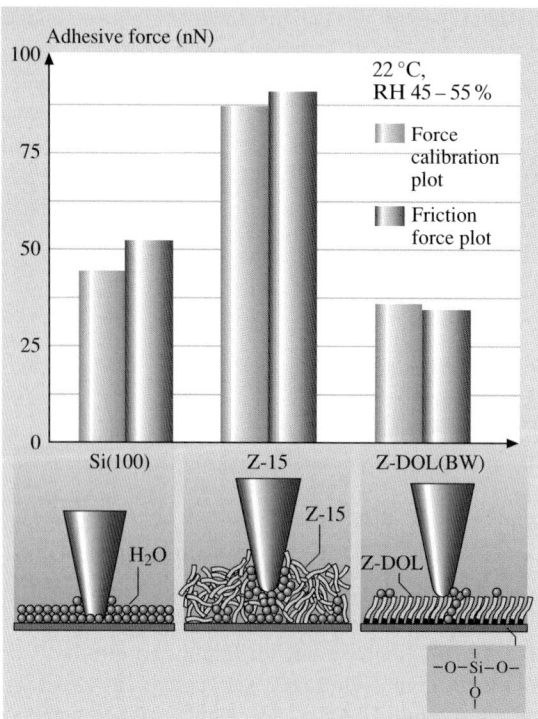

Fig. 21.49. Summary of the adhesive forces of Si(100), 2.8 nm thick Z-15 film, and 2.3 nm thick Z-DOL(BW) film. The schematic (*bottom*) shows the effect of meniscus formation between the AFM tip and the sample surface on the adhesive and friction forces [30]

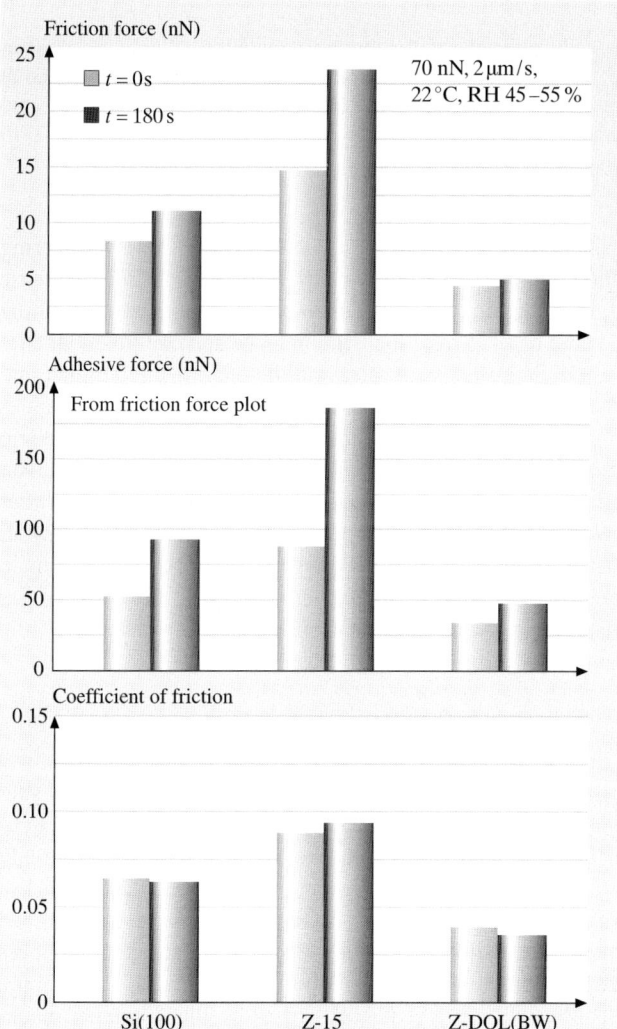

Fig. 21.50. Summary of rest time effect on friction force, adhesive force, and coefficient of friction of Si(100), 2.8 nm thick Z-15 film, and 2.3 nm thick Z-DOL(BW) film [30]

and coefficient of friction for three samples are summarized in Fig. 21.50. It is seen that time effect is present in Si(100) and Z-15 with mobile liquid present. Whereas, time effect is not present for Z-DOL(BW) because of the absence of mobile liquid.

To study lubricant depletion during microscale measurements, nanowear studies were conducted using Si_3N_4 tips. Measured friction as a function of number of cycles for Si(100) and silicon surface lubricated with Z-15 and Z-DOL (BW) lubricants are presented in Fig. 21.51. An area of 2 μm × 2 μm was scanned at a normal force of 70 nN. As observed before, friction force and coefficient of friction of

Fig. 21.51. Friction force and coefficient of friction versus number of sliding cycles for Si(100), 2.8 nm thick Z-15, and 2.3 nm thick Z-DOL (BW) film at 70 nN, 0.8 μm/s, and in ambient air. Schematic bottom shows that some liquid Z-15 molecules can be attached onto the tip. The molecular interaction between the attached molecules onto the tip with the Z-15 molecules in the film results in an increase of the friction force with multiscanning [30]

Z-15 is higher than that of Si(100) with the lowest values for Z-DOL(BW). During cycling, friction force and coefficient of friction of Si(100) show a slight decrease during initial few cycles, then remain constant. This is related to the removal of the top adsorbed layer. In the case of Z-15 film, the friction force and coefficient of friction show an increase during the initial few cycles and then approach to higher and stable values. This is believed to be caused by the attachment of the Z-15 molecules onto the tip. The molecular interaction between these attached molecules to the tip and molecules on the film surface is responsible for an increase in the friction. But after several scans, this molecular interaction reaches to the equilibrium and after that friction force and coefficient of friction remain constant. In the case of Z-DOL (BW) film, the friction force and coefficient of friction start out to be low and remain low during the entire test for 100 cycles. It suggests that Z-DOL (BW) molecules do not get attached or displaced as readily as Z-15.

21.8 Closure

Atomic force microscope/friction force microscope (AFM/FFM) have been successfully used for measurements of surface roughness, friction, adhesion, scratching, wear, indentation, and lubrication on a micro to nanoscales. Commonly measured roughness parameters are scale dependent, requiring the need of scale-independent fractal parameters to characterize surface roughness. A generalized fractal analysis is presented which allows the characterization of surface roughness by two scale-independent parameters. Local variation in microscale friction force is found to correspond to the local surface slope suggesting that a ratchet mechanism is responsible for this variation. Directionality in the friction is observed on both micro- and macro-scales because of surface topography. Microscale friction is found to be significantly smaller than the macro-scale friction as there is less plowing contribution in microscale measurements.

Wear rates for particulate magnetic tapes and polyester tape substrates are approximately constant for various loads and test durations. However, for magnetic disks and magnetic tapes with a multilayered thin-film structure, the wear of the diamondlike amorphous carbon overcoat in the case of disks and magnetic layer in the case of tapes, is catastrophic. Breakdown of thin-films can be detected with AFM. Evolution of the wear has also been studied using AFM. We find that the wear is initiated at nanoscratches. Amorphous carbon films as thin as 3.5 nm are deposited as continuous films and exhibit some wear life. Wear life increases with an increase

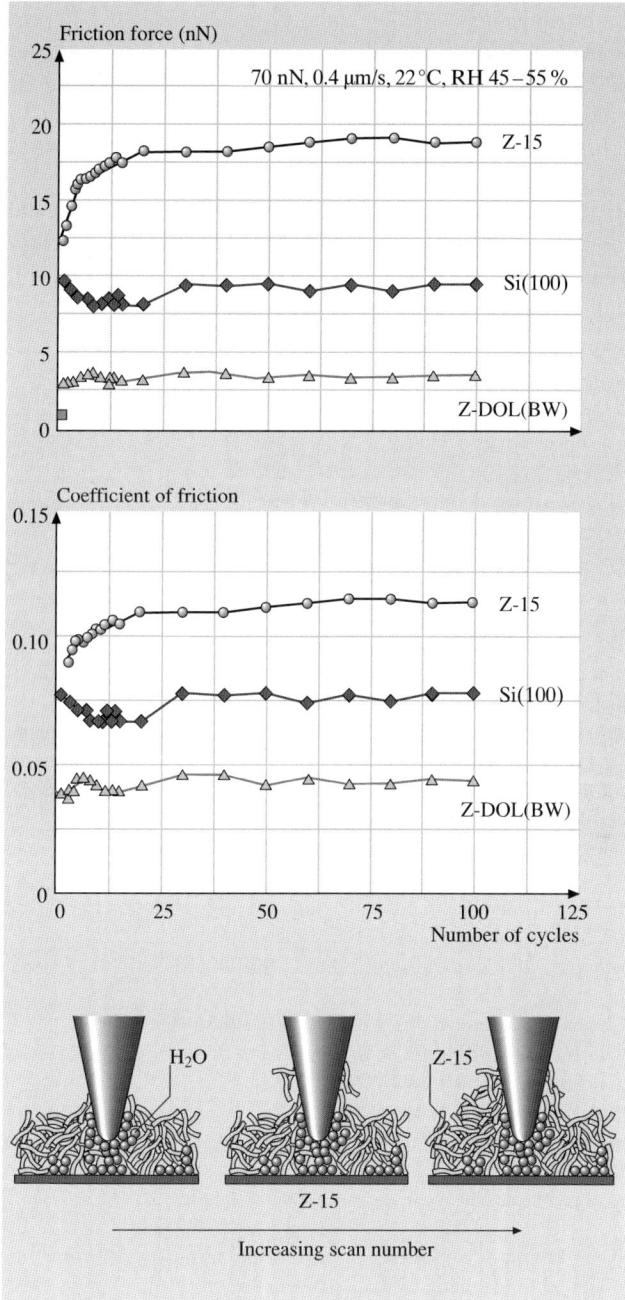

in film thickness. Carbon coatings deposited by cathodic arc and ECR-CVD processes are superior in wear and mechanical properties followed by ion beam and sputtering processes. AFM has been modified for nanoindentation hardness measurements with depth of indentation as low as 5 nm. Scratching and indentation on nanoscales are the powerful ways of evaluation of the mechanical integrity of ultrathin films.

AFM/FFM friction experiments show that lubricants with polar (reactive) end groups dramatically increase the load or contact pressure that a liquid film can support before solid-solid contact and thus exhibit long durability. The lubricants with the absence of mobile liquid exhibit low friction and adhesion and don't exhibit rest time effect.

References

1. B. Bhushan. *Tribology of magnetic storage systems*, volume 3, pages 325–374. CRC, 1994.
2. B. Bhushan. *Tribology and Mechanics of Magnetic Storage Devices*. Springer, 2nd edition, 1996.
3. B. Bhushan. *Mechanics and Reliability of Flexible Magnetic Media*. Springer, 2nd edition, 2000.
4. B. Bhushan. *Macro- and microtribology of magnetic storage devices*, volume 2, Materials, Coatings, and Industrial Applications, pages 1413–1513. CRC, 2001.
5. B. Bhushan. *Magnetic recording surfaces*, pages 116–133. Butterworth-Heinemann, 1993.
6. B. Bhushan. *Nanotribology and its applications to magnetic storage devices and MEMS*, pages 367–395. Kluwer, 1995.
7. B. Bhushan. Micro/nanotribology and its application to magnetic storage devices and mems. *Tribol. Int.*, 28:85–95, 1995.
8. B. Bhushan, J.N. Israelachvili, and U. Landman. Nanotribology: Friction, wear and lubrication at the atomic scale. *Nature*, 374:607–616, 1995.
9. B. Bhushan. *Micro/Nanotribology and its Applications*. NATO ASI Ser. 330. Kluwer, 1997.
10. B. Bhushan. *Handbook of Micro/Nanotribology*. CRC, 2nd edition, 1999.
11. B. Bhushan. *Fundamentals of Tribology and Bridging the Gap Between the Macro- and Micro/Nanoscales*. NATO Sci. Ser. II. Kluwer, 2001.
12. B. Bhushan. *Introduction to Tribology*. Wiley, 2002.
13. B. Bhushan (Ed.). *Springer Handbook of Nanotechnology*. Springer, 2004.
14. J. Ruan and B. Bhushan. Atomic-scale friction measurements using friction force microscopy: Part i – general principles and new measurement techniques. *ASME J. Tribol.*, 116:378–388, 1994.
15. B. Bhushan and J. Ruan. Atomic-scale friction measurements using friction force microscopy: Part ii – application to magnetic media. *ASME J. Tribol.*, 116:389–396, 1994.
16. B. Bhushan and V.N. Koinkar. Tribological studies of silicon for magnetic recording applications. *J. Appl. Phys.*, 75:5741–5746, 1994.
17. B. Bhushan and V.N. Koinkar. Nanoindentation hardness measurements using atomic force microscopy. *Appl. Phys. Lett.*, 64:1653–1655, 1994.

18. B. Bhushan, V.N. Koinkar, and J. Ruan. Microtribology of magnetic media. *Proc. Inst. Mech. Eng., J. Eng. Tribol.*, 208:17–29, 1994.
19. B. Bhushan, A.V. Kulkarni, W. Bonin, and J.T. Wyrobek. Nanoindentation and picoindentation measurements using a capacitance transducer system in atomic force microscopy. *Philos. Mag.*, 74:1117–1128, 1996.
20. S. Sundararajan and B. Bhushan. Development of a continuous microscratch technique in an atomic force microscopy and its applications to study scratch resistance of ultra-thin hard amorphous carbon coatings. *J. Mater. Res.*, 16:437–445, 2001.
21. D. DeVecchio and B. Bhushan. Localized surface elasticity measurements using an atomic force microscope. *Rev. Sci. Instrum.*, 68:4498–4505, 1997.
22. V. Scherer, B. Bhushan, U. Rabe, and W. Arnold. Local elasticity and lubrication measurements using atomic force and friction force microscopy at ultrasonic frequencies. *IEEE Trans. Mag.*, 33:4077–4079, 1997.
23. W.W. Scott and B. Bhushan. Use of phase imaging in atomic force microscopy for measurement of viscoelastic contrast in polymer nanocomposites and molecularly-thick lubricant films. *Ultramicrosc.*, 97:151–169, 2003.
24. B. Bhushan and J. Qi. Phase contrast imaging of nanocomposites and molecularly-thick lubricant films in magnetic media. *Nanotechnol.*, 14:886–895, 2003.
25. T. Kasai, B. Bhushan, L. Huang, and C. Su. Topography and phase imaging using the torsional resonance mode. *Nanotechnol.*, 15:731–742, 2004.
26. B. Bhushan and T. Kasai. A surface topography-independent friction measurement technique using torsional resonance mode in an afm. *Nanotechnol.*, 15:923–935, 2004.
27. B. Bhushan, T. Miyamoto, and V.N. Koinkar. Microscopic friction between a sharp diamond tip and thin-film magnetic rigid disks by friction force microscopy. *Adv. Info. Storage Syst.*, 6:151–161, 1995.
28. V.N. Koinkar and B. Bhushan. Microtribological studies of Al_2O_3, Al_2O_3–TiC, polycrystalline and single-crystal Mn–Zn ferrite and SiC head slider materials. *Wear*, 202:110–122, 1996.
29. V.N. Koinkar and B. Bhushan. Microtribological studies of unlubricated and lubricated surfaces using atomic force/friction force microscopy. *J. Vac. Sci. Technol. A*, 14:2378–2391, 1996.
30. H. Liu and B. Bhushan. Nanotribological characterization of molecularly-thick lubricant films for applications to mems/nems by afm. *Ultramicrosc.*, 97:321–340, 2003.
31. B. Bhushan. Magnetic slider/rigid disk substrate materials and disk texturing techniques – status and future outlook. *Adv. Info. Storage Syst.*, 5:175–209, 1993.
32. B. Bhushan, M. Dominiak, and J.P. Lazzari. Contact- start-stop studies with silicon planar head sliders against thin-film disks. *IEEE Trans. Mag.*, 28:2874–2876, 1992.
33. B. Bhushan and G.S. Blackman. Atomic force microscopy of magnetic rigid disks and sliders and its applications to tribology. *ASME J. Tribol.*, 113:452–458, 1991.
34. P.I. Oden, A. Majumdar, B. Bhushan, A. Padmanabhan, and J.J. Graham. Afm imaging, roughness analysis and contact mechanics of magnetic tape and head surfaces. *ASME J. Tribol.*, 114:666–674, 1992.
35. S. Ganti and B. Bhushan. Generalized fractal analysis and its applications to engineering surfaces. *Wear*, 180:17–34, 1995.
36. C.Y. Poon and B. Bhushan. Comparison of surface roughness measurements by stylus profiler, afm and non-contact optical profiler. *Wear*, 190:76–88, 1995.
37. C.Y. Poon and B. Bhushan. Surface roughness analysis of glass-ceramic substrates and finished magnetic disks, and Ni–P coated Al–Mg and glass substrates. *Wear*, 190:89–109, 1995.

38. V.N. Koinkar and B. Bhushan. Effect of scan size and surface roughness on microscale friction measurements. *J. Appl. Phys.*, 81:2472–2479, 1997.
39. A. Majumdar and B. Bhushan. Role of fractal geometry in roughness characterization and contact mechanics of surfaces. *ASME J. Tribol.*, 112:205–216, 1990.
40. A. Majumdar and B. Bhushan. Fractal model of elastic-plastic contact between rough surfaces. *ASME J. Tribol.*, 113:1–11, 1991.
41. B. Bhushan and A. Majumdar. Elastic-plastic contact model for bifractal surfaces. *Wear*, 153:53–64, 1992.
42. B. Bhushan. Contact mechanics of rough surfaces in tribology: Single asperity contact. *Appl. Mech. Rev.*, 49:275–298, 1996.
43. B. Bhushan. Contact mechanics of rough surfaces in tribology: Multiple asperity contact. *Tribol. Lett.*, 4:1–35, 1998.
44. B. Bhushan and W. Peng. Contact mechanics of multilayered rough surfaces. *Appl. Mech. Rev.*, 55:435–480, 2002.
45. B. Bhushan and V.N. Koinkar. Microtribology of pet polymeric films. *Tribol. Trans.*, 38:119–127, 1995.
46. B. Bhushan and V.N. Koinkar. Macro and microtribological studies of CrO_2 video tapes. *Wear*, 180:9–16, 1995.
47. B. Bhushan and V.N. Koinkar. Microtribology of metal particle, barium ferrite and metal evaporated magnetic tapes. *Wear*, 181–183:360–370, 1995.
48. B. Bhushan and V.N. Koinkar. Microscale mechanical and tribological characterization of hard amorphous carbon coatings as thin as 5 nm for magnetic disks. *Surf. Coat. Tech.*, 76-77:655–669, 1995.
49. V.N. Koinkar and B. Bhushan. Microtribological properties of hard amorphous carbon protective coatings for thin-film magnetic disks and heads. *Proc. Inst. Mech. Eng., J. Eng. Tribol.*, 211:365–372, 1997.
50. A.V. Kulkarni and B. Bhushan. Nanoindentation measurements of amorphous carbon coatings. *J. Mater. Res.*, 12:2707–2714, 1997.
51. X. Li and B. Bhushan. Micro/nanomechanical and tribological characterization of ultra-thin amorphous carbon coatings. *J. Mater. Res.*, 14:2328–2337, 1999.
52. X. Li and B. Bhushan. Mechanical and tribological studies of ultra-thin hard carbon overcoats for magnetic recording heads. *Z. Metallk.*, 90:820–830, 1999.
53. S. Sundararajan and B. Bhushan. Micro/nanotribology of ultra-thin hard amorphous carbon coatings using atomic force/friction force microscopy. *Wear*, 225-229:678–689, 1999.
54. B. Bhushan and S. Sundararajan. Micro/nanoscale friction and wear mechanisms of thin films using atomic force and friction force microscopy. *Acta Mater.*, 46:3793–3804, 1998.
55. J. Ruan and B. Bhushan. Frictional behavior of highly oriented pyrolytic graphite. *J. Appl. Phys.*, 76:8117–8120, 1994.
56. S. Sundararajan and B. Bhushan. Topography-induced contributions to friction forces measured using an atomic force/friction force microscope. *J. Appl. Phys.*, 88:4825–4831, 2000.
57. B. Bhushan and V.N. Koinkar. Microtribological studies of doped single-crystal silicon and polysilicon films for mems devices. *Sensor Actuator A*, 57:91–102, 1997.
58. S. Sundararajan and B. Bhushan. Micro/nanotribological studies of polysilicon and SiC films for mems applications. *Wear*, 217:251–261, 1998.
59. B. Bhushan. Chemical, mechanical and tribological characterization of ultra-thin and hard amorphous carbon coatings as thin as 3.5 nm: Recent developments. *Diam. Relat. Mater.*, 8:1985–2015, 1999.

60. H. Hibst. *Metal evaporated tapes and* Co–Cr *media for high definition video recording*, pages 137–159. Kluwer, 1993.
61. M. Y. Chu, B. Bhushan, and L. DeJonghe. Wear behavior of ceramic sliders in sliding contact with rigid magnetic thin-film disks. *Tribol. Trans.*, 35:603–610, 1992.
62. B. Bhushan, S. Sundararajan, W. W. Scott, and S. Chilamakuri. Stiction analysis of magnetic tapes. *IEEE Trans. Mag.*, 33:3211–3213, 1997.
63. V. J. Novotny, I. Hussla, J. M. Turlet, and M. R. Philpott. Liquid polymer conformation on solid surfaces. *J. Chem. Phys.*, 90:5861–5868, 1989.
64. V. J. Novotny. Migration of liquid polymers on solid surfaces. *J. Chem. Phys.*, 92:3189–3196, 1990.
65. C. M. Mate and V. J. Novotny. Molecular conformation and disjoining pressures of polymeric liquid films. *J. Chem. Phys.*, 94:8420–8427, 1991.
66. C. M. Mate. Application of disjoining and capillary pressure to liquid lubricant films in magnetic recording. *J. Appl. Phys.*, 72:3084–3090, 1992.
67. B. Bhushan and Z. Zhao. Macro- and microscale studies of molecularly-thick boundary layers of perfluoropolyether lubricants for magnetic thin-film rigid disks. *J. Info. Storage Proc. Syst.*, 1:1–21, 1999.
68. R. Heideman and M. Wirth. Transforming the lubricant on a magnetic disk into a solid fluorine compound. *IBM Technol. Disclosure Bull.*, 27:3199–3205, 1984.
69. A. M. Homola, L. J. Lin, and D. D. Saperstein. Process for bonding lubricant to a thin film magnetic recording disk, 1990.
70. D. D. Saperstein and L. J. Lin. Improved surface adhesion and coverage of perfluoropolyether lubricant following far-uv irradiation. *Langmuir*, 6:1522–1524, 1990.

22
Nanotribology and Materials Characterization of MEMS/NEMS and BioMEMS/BioNEMS Materials and Devices

Bharat Bhushan

Summary. Micro/nanoelectromechanical systems (MEMS/NEMS) need to be designed to perform their expected functions with short duration, typically in millisecond to picosecond timescales. Expected life of the devices for high-speed contacts can vary from a few hundred thousand to many billions of cycles, e.g., over a hundred billion cycles for digital micromirror devices (DMDs), which puts serious requirements on materials. For BioMEMS/BioNEMS, adhesion between biological molecular layers and the substrate, and friction and wear of biological layers may be important. There is a need for the development of a fundamental understanding of adhesion, friction/stiction, wear, and the role of surface contamination, and environment. Most mechanical properties are known to be scale-dependent. Therefore, the properties of nanoscale structures need to be measured. MEMS/NEMS materials need to exhibit good mechanical and tribological properties on the micro/nanoscale. There is a need to develop lubricants and identify lubrication methods that are suitable for MEMS/NEMS. Methods need to be developed to enhance adhesion between biomolecules and the device substrate. Component-level studies are required to provide a better understanding of tribological phenomena occurring in MEMS/NEMS. The emergence of micro/nanotribology and techniques based on atomic force microscopy has provided researchers with a viable approach to address these problems. This chapter presents a review of micro/nanoscale adhesion, friction, and wear studies of materials and lubrication studies for MEMS/NEMS and BioMEMS/BioNEMS, and component-level studies of stiction phenomena in MEMS/NEMS devices.

22.1 Introduction

Microelectromechanical systems (MEMS) refers to microscopic devices that have a characteristic length of less than 1 mm but more than 100 nm and combine electrical and mechanical components. Nanoelectromechanical systems (NEMS) refer to nanoscopic devices that have a characteristic length of less than 100 nm and combine electrical and mechanical components. In mesoscale devices, if the functional components are on the micro- or nanoscale, they may be referred to as MEMS or NEMS, respectively. These are referred to as an intelligent miniaturized system consisting of sensing, processing, and/or actuating functions and combining electrical and mechanical components. The acronym MEMS originated in the U.S.A. The term commonly used in Europe is microsystem technology (MST) and in Japan is micromachines. Another term generally used is micro/nanodevices. The terms MEMS/NEMS are also now used in a broad sense and include electrical, mechanical, fluidic, optical, and/or biological functions. MEMS/NEMS for optical applications are referred to as micro/nano-optoelectromechanical systems (MOEMS/NOEMS). MEMS/NEMS for electronic applications are referred to as radio-frequency MEMS/NEMS (RF-MEMS/RF-NEMS). MEMS/NEMS for biological applications are referred to as BioMEMS/BioNEMS.

To put the dimensions of MEMS and NEMS in perspective, see Fig. 22.1 and Table 22.1. Individual atoms are typically a fraction of a nanometer in diameter, deoxyribonucleic acid (DNA) molecules are about 2.5 nm wide, biological cells are in the range of thousands of nm in diameter, and human hair is about 75 μm in diameter. NEMS shown in the figure have a size of 15–300 nm and MEMS have a scale of 12,000 nm. The mass of a micromachined silicon structure can be as low as 1 nN and NEMS can be built with mass as low as 10^{-20} N with cross sections of about 10 nm. In comparison, the mass of a drop of water is about 10 μN and the mass of an eyelash is about 100 nN.

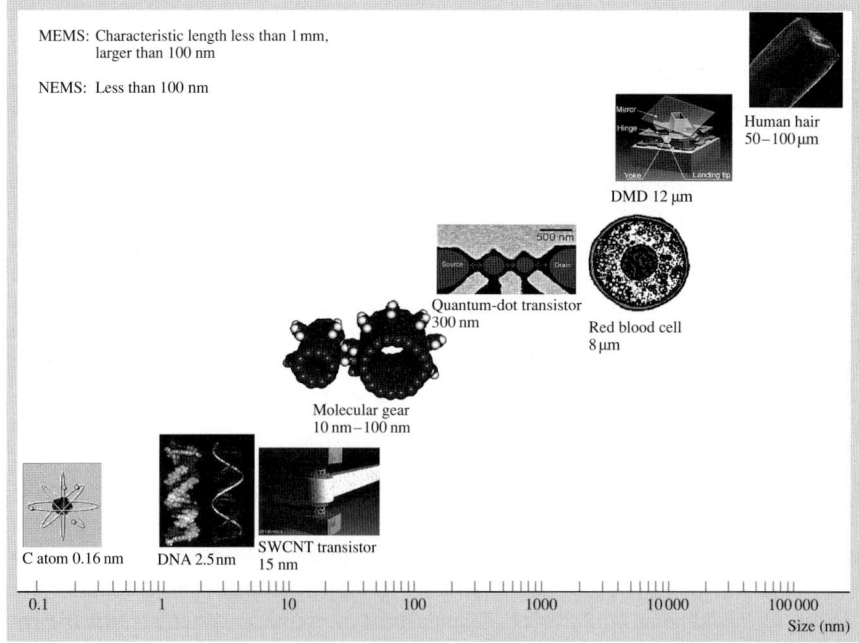

Fig. 22.1. Dimensions of MEMS and NEMS in perspective. MEMS/NEMS examples shown are of a vertical single-walled carbon nanotube (SWCNT) transistor (5 nm wide and 15 nm high) [7], of molecular dynamic simulations of a carbon-nanotube-based gear [8], quantum-dot transistor obtained from *van der Wiel* et al. [9], and DMD obtained from www.dlp.com

Table 22.1. Dimensions and masses in perspective

(a) Dimensions in perspective

NEMS characteristic length: < 100 nm
MEMS characteristic length: < 1 mm and > 100 nm

Molecular gear	≈ 10 nm
Vertical SWCNT transistor	≈ 15 nm
Quantum-dot transistor	300 nm
Digital micromirror	12,000 nm
Individual atoms	typically a fraction of a nm in diameter
DNA molecules	≈ 2.5 nm wide
Biological cells	in the range of thousands of nm in diameter
Human hair	≈ 75,000 nm in diameter

(b) Masses in perspective

NEMS built with cross sections of about 10 nm – as low as 10^{-20} N
Micromachines silicon structure – as low as 1 nN

Water droplet	≈ 10 μN
Eyelash	≈ 100 nN

Micro/nanofabrication techniques include top-down methods, in which one builds down from the large to the small, and bottom-up methods, in which one builds up from the small to the large. Top-down methods include micro/nanomachining methods and methods based on lithography as well as non-lithographic miniaturization, mostly for the fabrication of MEMS and a few NEMS. In bottom-up methods, also referred to as nanochemistry, the devices and systems are assembled from their elemental constituents for NEMS fabrication, much like the way nature uses proteins and other macromolecules to construct complex biological systems. The bottom-up approach has the potential to go far beyond the limits of top-down technology by producing nanoscale features through synthesis and subsequent assembly. Furthermore, the bottom-up approach offers the potential to produce structures with enhanced and/or completely new functions. It allows combination of materials with distinct chemical composition, structure, and morphology. For a brief overview of fabrication techniques, see Sect. 22.A.

MEMS and emerging NEMS are expected to have a major impact on our lives, comparable to that of semiconductor technology, information technology, or cellular and molecular biology [1–4]. MEMS/NEMS are used in electromechanical, electronics, information/communication, chemical, and biological applications. The MEMS industry in 2004 was worth about $4.5 billion, with a projected annual growth rate of 17% [5]. Growth of Si-based MEMS may slow down and non-silicon MEMS may pick up during this decade. The NEMS industry was worth about $10 billion in 2004, mostly in nanomaterials [6]. It is expected to expand in this decade, in nanomaterials, biomedical applications as well as in nanoelectronics or molecular electronics. Due to the enabling nature of these systems and because of the significant impact they can have on both commercial and defense applications, industry as well as federal governments have taken special interest in seeing growth nurtured in this field. MEMS/NEMS are the next logical step in the silicon revolution.

22.1.1 Introduction to MEMS

Advances in silicon photolithographic process technology since the 1960s led to the development of MEMS in the early 1980s. More recently, lithographic processes have also been developed to process non-silicon materials. The lithographic processes are being complemented with non-lithographic processes for fabrication of components or devices made from plastics or ceramics. Using these fabrication processes, researchers have fabricated a wide variety of devices with dimensions in the range of submicron to a few thousand microns (see e. g., [10–20]). MEMS for mechanical applications include acceleration, pressure, flow, and gas sensors, linear and rotary actuators, and other microstructures or microcomponents such as electric motors, gear chains, gas turbine engines, fluid pumps, fluid valves, switches, grippers, and tweezers. MEMS for chemical applications include chemical sensors and various analytical instruments. MOEMS are devices that include optical components, such as micromirror arrays for displays, infrared image sensors, spectrometers, barcode readers, and optical switches. RF-MEMS include inductors, capacitors, antennas, and RF switches. High-aspect-ratio MEMS (HARMEMS) have also been introduced.

A variety of MEMS devices have been produced and some are used commercially [11, 13–16, 18–20]. A variety of sensors are used in industrial, consumer, defense, and biomedical applications. The largest *killer* industrial applications include accelerometers, pressure sensors, thermal and piezoelectric-type inkjet printheads, and digital micromirror devices. Integrated capacitive-type silicon accelerometers have been used in airbag deployment in automobiles since 1991 [21, 22]. Some 90 million units were installed in vehicles in 2004. Accelerometer technology was over a billion-dollar-a-year industry in 2004, dominated by Analog Devices followed by Freescale Semiconductor (formerly Motorola) and Bosch. These accelerometers are being used for many other applications such as vehicle stability, rollover control, and gyro sensors for automotive applications, and various consumer applications, including handheld devices, e. g., laptops (2003), cellular phones (2004), and personal digital assistants (PDAs) Silicon-based piezoresistive pressure sensors were launched in 1990 by GE NovaSensor for manifold absolute pressure (MAP) sensing for engines and for disposable blood-pressure sensors, and their annual sales were more than 30 million units and more than 25 million units, respectively, in 2004. MPA sensors measure the pressure in the intake manifold, which is fed to a computer that determines the optimum air/fuel mixture to maximize fuel economy. Most vehicles have these as part of the electronic engine-control system. Capacitive pressure sensors for tire-pressure measurements were launched by Freescale Semiconductor (formerly Motorola) in early 2000 and are also manufactured by Infineon/SensoNor and GE Novasensor (2003). Piezoresistive-type sensors are also used, manufactured by various companies such as EnTire Solutions (2003). The sensing module is located inside the rim of the wheel and relays the information via radio-frequency link to a central processing unit (CPU) in order to display it to the driver. In 2005, about 9.2 million vehicles were equipped with sensors, which translates to about 37 million units. These sales are expected to grow rapidly as they become required in automobiles. They will be required in the U.S. by 2008, which will affect 17 million vehicles (one in each tire) sold every year. Thermal inkjet printers were developed independently by HP and Cannon and commercialized in 1984 [23–26] and today are made by Canon, Epson, HP, Lexmark, Xerox, and others. They typically cost less initially than dry-toner laser printers and are the solution of choice for low-volume print runs. Annual sales of thermal inkjet printheads with microscale functional components were more than 500 million units in 2004.

Micromirror arrays are used for displays. Commercial digital light processing (DLP) equipments, using digital micromirror devices (DMD), were launched in 1996 by Texas Instruments for digital projection displays in computer projectors, high-definition television (HDTV) sets and movie projectors (DLP cinema) [27–29]. Several million projectors had been sold up to 2004 (about $700 million revenue for TI in 2004). Electrostatically actuated, membrane-type or cantilever-type microswitches have been developed for direct current (DC), RF, and optical applications [30]. There exists two basic forms of RF microswitches: the metal-to-metal contact microswitch (ohmic) and the capacitive microswitch. RF microswitches can be used in a variety of RF applications including cellular phones, phase shifters, smart antennas, multiplexers for data acquisition, and more [31]. Optical microswitches are finding applications in optical networking, telecommunications, and wireless technologies [30, 32].

Other applications of MEMS devices include chemical/biological and gas sensors [20, 33], microresonators, infrared detectors and focal-plane arrays for earth observation, space science and missile defense applications, pico-satellites for space applications, fuel cells, and many hydraulic, pneumatic, and other consumer products. MEMS devices are also being pursued in magnetic storage systems [34], where they are being developed for super-compact ultrahigh-recording-density magnetic disk drives. Several integrated head/suspension microdevices have been fabricated for contact recording applications [35]. High-bandwidth servo-controlled microactuators have been fabricated for ultrahigh-track-density applications which serve as the fine-position control element of a two-stage, coarse/fine servo system, coupled with a conventional actuator [36, 37].

Micro/nano-instruments and micro/nano-manipulators are used to move, position, probe, pattern, and characterize nanoscale objects and nanoscale features [38]. Miniaturized analytical equipments include gas chromatography and mass spectrometry. Other instruments include micro-STM, where STM stands for scanning tunneling microscope.

In some cases, MEMS devices are used primarily for their miniature size, while in others, as in the case of the air bags, because of their low-cost manufacturing techniques. This latter fact has been possible since semiconductor processing costs have reduced drastically over the last decade, allowing the use of MEMS in many fields.

22.1.2 Introduction to NEMS

NEMS are produced by nanomachining in a typical top-down approach (from large to small) and bottom-up approach (from small to large), largely relying on nanochemistry (see e.g., [39–45]). The NEMS field, in addition to fabrication of nanosystems, has provided impetus to the development of experimental and computation tools. Examples of NEMS include microcantilevers with integrated sharp nanotips for STM and atomic force microscopy (AFM) [46,47], quantum corrals formed using STM by placing atoms one by one [48], AFM cantilever arrays (Millipede) for data storage [49], STM and AFM tips for nanolithography, dip-pen nanolithography for printing molecules, nanowires, carbon nanotubes, quantum wires (QWRs), quantum boxes (QBs), quantum transistors [9], nanotube-based sensors [7, 50], biological (DNA) motors, molecular gears made by attaching benzene molecules to the outer walls of carbon nanotubes [8], devices incorporating nm-thick films [e.g., in giant-magnetoresistive (GMR) read/write magnetic heads and magnetic media for magnetic rigid disk and magnetic tape drives], nanopatterned magnetic rigid disks, and nanoparticles (e.g., nanoparticles in magnetic tape substrates and nanomagnetic particles in magnetic tape coatings) [34, 51]. More than 2 billion read/write magnetic heads were shipped for magnetic disk and tape drives in 2004.

Nanoelectronics can be used to build computer memory using individual molecules or nanotubes to store bits of information [52], molecular switches, molecular or nanotube transistors, nanotube flat-panel displays, nanotube integrated circuits, fast logic gates, switches, nanoscopic lasers, and nanotubes as electrodes in fuel cells.

22.1.3 BioMEMS/BioNEMS

BioMEMS/BioNEMS are increasingly used in commercial and defense applications (see e.g., [53–60]). They are used for chemical and biochemical analyses (biosensors) in medical diagnostics [e.g., DNA, ribonucleic acid (RNA), proteins, cells, blood pressure and assays, and toxin identification] [60, 61], tissue engineering [62–64], and implantable pharmaceutical drug delivery [65–67]. Biosensors, also referred to as biochips, deal with liquids and gases. There are two types. A large variety of biosensors are based on micro/nanofluidics [60, 68–70]. Micro/nanofluidic devices offer the ability to work with smaller reagent volumes and shorter reaction times, and perform multiple types of analysis at once. The second type of biosensors include micro/nanoarrays, which perform one type of analysis thousands of times [71–74].

A chip, called a lab-on-a-CD, with micro/nanofluidic technology embedded on the disk can test thousands of biological samples rapidly and automatically [68]). An entire laboratory can be integrated onto a single chip, called a lab-on-a-chip [60, 69, 70]. Silicon-based disposable blood-pressure sensor chips were introduced in the early 1990s by GE NovaSensor for blood-pressure monitoring (about 25 million units in 2004). A blood-sugar monitor, referred to as GlucoWatch, was introduced in 2002. It automatically checks blood sugar every 10 minutes by detecting glucose through the skin, without having to draw blood. If glucose is out of the acceptable range, it sounds an alarm so the diabetic can address the problem quickly. A variety of biosensors, many using plastic substrates, are manufactured by various companies including ACLARA, Agilent Technologies, Calipertech, and I-STAT.

The second type of biochips – micro/nanoarrays – are a tool used in biotechnology research to analyze DNA or proteins to diagnose diseases or discover new drugs. Also called DNA arrays, they can identify thousand of genes simultaneously [56, 71]. They include a microarray of silicon nanowires, roughly a few nm in size, to selectively bind and detect even a single biological molecule such as DNA or a protein by using nanoelectronics to detect the slight electrical charge caused by such binding, or a microarray of carbon nanotubes to detect glucose electrically.

After the tragedy of Sept. 11, 2001, concern over biological and chemical warfare has led to the development of handheld units with bio- and chemical sensors for the detection of biological germs, chemical or nerve agents, and mustard agents, and chemical precursors to protect subways, airports, water supplies, and the population at large [75].

BioMEMS/BioNEMS are also being developed for minimal-invasive surgery including endoscopic surgery, laser angioplasty, and microscopic surgery. Implantable artificial organs can also be produced. Other applications include implantable drug-delivery devices – e. g., micro/nanoparticles with drug molecules encapsulated in functionalized shells for a site-specific targeting applications, and silicon capsules with a nanoporous membrane filled with drugs for long-term delivery [65, 76–78] – nanodevices for sequencing single molecules of DNA in the Human Genome Project [60], cellular growth using carbon nanotubes for spinal-cord repair, nanotubes for nanostructured materials for various applications such as spinal fusion devices, organ growth, and the growth of artificial tissues using nanofibers.

22.1.4 Tribological Issues in MEMS/NEMS and BioMEMS/BioNEMS

Tribological issues are important in MEMS/NEMS and BioMEMS/BioNEMS requiring intended and/or unintended relative motion. In these devices, various forces associated with the device scale down with the size. When the length of the machine decreases from 1 mm to 1 μm, the area decreases by a factor of a million and the volume decreases by a factor of a billion. As a result, surface forces such as adhesion, friction, meniscus forces, viscous drag forces and surface tension that are proportional to area, become a thousand times larger than the forces proportional to the volume, such as inertial and electromagnetic forces. In addition to the consequences of a large surface-to-volume ratio, since these devices are designed for small tolerances, physical contact becomes more likely, which makes them particularly vulnerable to adhesion between adjacent components. Slight particulate or chemical contamination present at the interface can be detrimental. Since the start-up forces and torques involved in operation that are available to overcome retarding forces are small, the increase in resistive forces such as adhesion and friction become a serious tribological concern that limits the durability and reliability of

MEMS/NEMS [13]. A large lateral force required to initiate relative motion between two surfaces, large static friction, is referred to as *stiction*, which has been studied extensively in tribology of magnetic storage systems [34, 46, 47, 79–82]. The source of stiction is generally liquid-mediated adhesion with a source of liquid being process fluid or capillary condensation of water vapor from the environment. Adhesion, friction/stiction (static friction), wear, and surface contamination affect MEMS/NEMS and BioMEMS/BioNEMS performance and, in some cases, can even prevent devices from working. Some examples of devices which experience tribological problems follow.

MEMS

Figure 22.2a shows examples of several microcomponents that can encounter the aforementioned tribological problems. The polysilicon electrostatic micromotor has 12 stators and a four-pole rotor and is produced by surface micromachining. The rotor diameter is 120 μm and the air gap between the rotor and stator is 2 μm [83]. It is capable of continuous rotation up to speeds of 100,000 revolutionsperminute(RPM). The intermittent contacts at the rotor–stator interface and physical contact at the rotor–hub flange interface result in wear issues, and high stiction between the contacting surfaces limits the repeatability of operation or may even prevent operation altogether. Next, a bulk micromachined silicon stator/rotor pair is shown with bladed rotor and nozzle guide vanes on the stator with dimensions of less than a mm [84, 85]. These are being developed for high-temperature micro-gas turbine engine with rotor diameters of 4–6 mm and operating speed of up to 1 million RPM (with a sliding velocity in excess of 500 m/s, comparable to velocities of large turbines operating at high velocities) to achieve high specific power, up to a total of about 10 W. Erosion of blades and vanes, and the design of microbearings required to operate at extremely high speeds used in the turbines are some of the concerns. Ultra-short high-speed micro hydrostatic gas journal bearings with a length-to-diameter (L/D) ratio of less than 0.1 are being developed for operation at surface speeds on the order of 500 m/s, which offer unique design challenges [86].

In Fig. 22.2a is an SEM micrograph of a surface-micromachined polysilicon six-gear chain from Sandia National Laboratory. (For more examples of an early version, see *Mehregany* et al., [87].) As an example of non-silicon components, a milligear system produced using the LIGA process (a German acronym for Lithographie Galvanoformung Abformung) for a DC brushless permanent-magnet millimotor (diameter = 1.9 mm, length = 5.5 mm with an integrated milligear box [88–90]) is also shown. The gears are made of metal (electroplated Ni–Fe) but can also be made from injected polymer materials [e. g., polyoxy-methylene (POM)] using the LIGA process. Even though the torque transmitted at the gear teeth is small, on the order of a fraction of a nN/m, because of the small dimensions of gear teeth, the bending stresses are large where the teeth mesh. Tooth breakage and wear at the contact of gear teeth is a concern.

Figure 22.2b shows a polysilicon, multiple-microgear speed-reduction unit and its components after laboratory wear tests conducted for 600,000 cycles at 1.8%

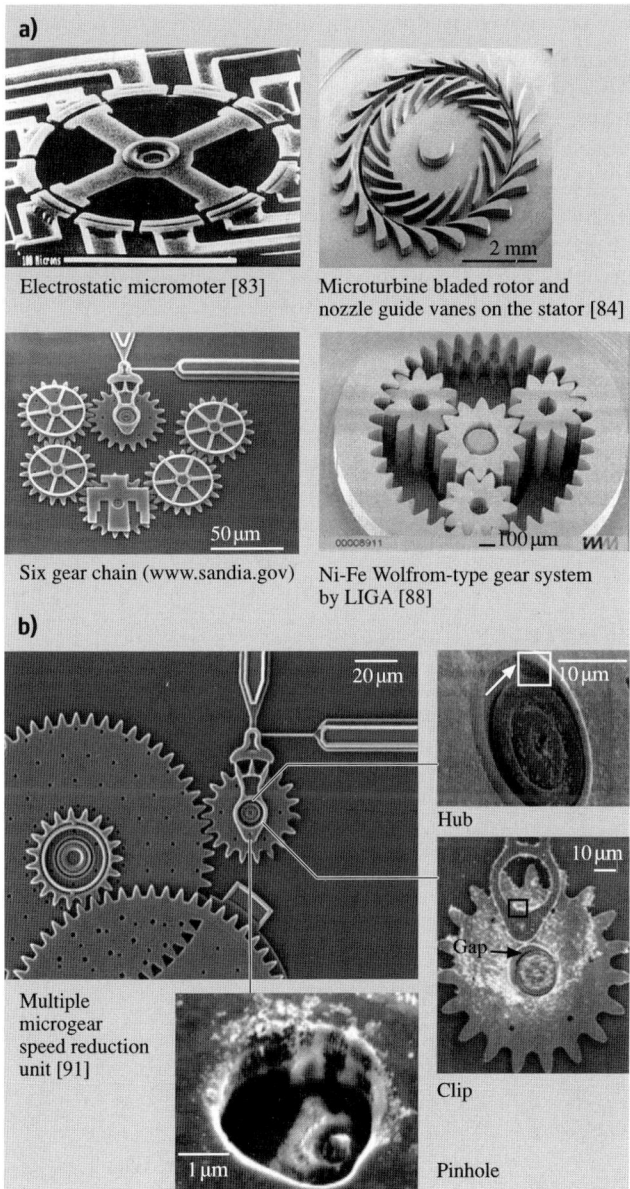

Fig. 22.2a,b. Examples of MEMS devices and components that experience tribological problems: (**a**) several microcomponents, and (**b**) a polysilicon, multiple-microgear speed-reduction unit after laboratory wear testing for 600,000 cycles at 1.8% relative humidity

RH [91]. These units have been developed for electrostatically driven microactuator (microengine) developed at Sandia National Laboratory for operation in the kHz frequency range [92]. Wear of various components is clearly observed in the figure. Humidity was shown to be a strong factor in the wear of rubbing surfaces. In order to improve the wear characteristics of rubbing surfaces, 20-nm-thick tungsten (W) coating deposited at 450 °C using the chemical vapor deposition (CVD) technique was used [93]. Tungsten-coated microengines tested for reliability showed improved wear characteristics with longer lifetimes than polysilicon microengines. However, these coatings have poor yield. Instead, vapor-deposited self-assembled monolayers of fluorinated (dimethylamino) silane are used [94]. They can be deposited with high yield, although durability is not as good.

Figure 22.3 shows a micromachined flow modulator; several micromachined flow channels are integrated in series with electrostatically actuated microvalves [95]. The flow channels lead to a central gas outlet hole drilled in the glass substrate. Gas enters the device through a bulk-micromachined gas inlet hole in the silicon cap.

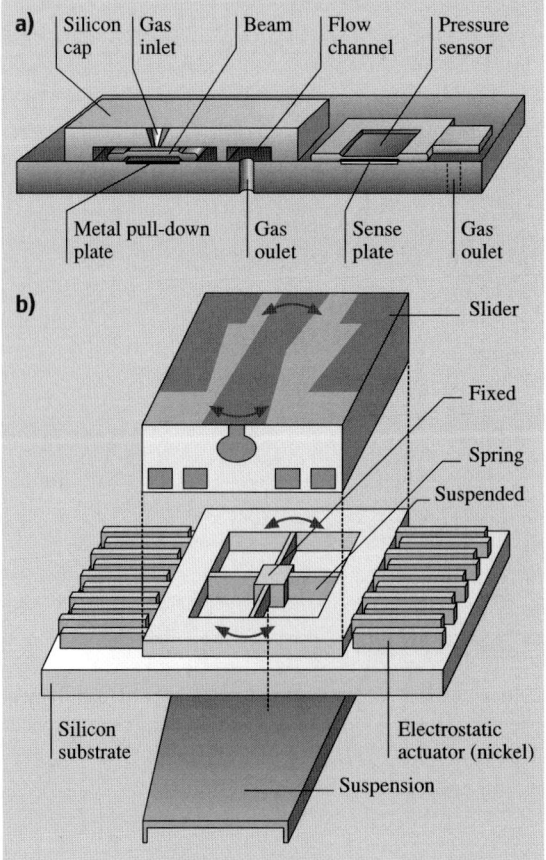

Fig. 22.3a,b. Examples of MEMS devices that experience tribological problems: (**a**) Low pressure flow modulator with electrostatically actuated microvalves [95] (**b**) Electroplated-nickel rotary microactuator for magnetic disc drives [37]

The gas, after passing through an open microvalve, flows parallel to the glass substrate through flow channels and exits the device through an outlet. The normally open valve structure consists of a freestanding double-end-clamped beam, which is positioned beneath the gas inlet orifice. When electrostatically deflected upwards, the beam seals against the inlet orifice and the valve is closed. In these microvalves used for flow control, the mating valve surfaces should be smooth enough to seal while maintaining a minimum roughness to ensure low adhesion [79–81, 96].

A second MEMS device, shown in Fig. 22.3, is an electrostatically driven rotary microactuator for a magnetic disk drive, surface-micromachined by a multilayer electroplating method [37]. This high-bandwidth servo-controlled microactuator, located between a slider and a suspension is being developed for ultrahigh-track-density applications, serves as the fine-position and high-bandwidth control element of a two-stage, coarse/fine servo system when coupled with a conventional actuator [36, 37]. A slider is placed on top of the central block of a microactuator, which gives rotational motion to the slider. The bottom of the silicon substrate is attached to the suspension. The radial flexure beams in the central block give the rotational freedom of motion to the suspended mass (slider), and the electrostatic actuator drives the suspended mass. Actuation is accomplished via interdigitated, cantilevered electrode fingers, which are alternatingly attached to the central body of the moving part and to the stationary substrate to form pairs. A voltage applied across these electrodes results in an electrostatic force, which rotates the central block. The inter-electrode gap width is about 2 µm. Any unintended contacts between the moving and stationary electroplated-nickel electrodes may result in wear and stiction.

Commercially available MEMS devices also exhibit tribological problems. Figure 22.4 shows an integrated capacitive-type silicon accelerometer fabricated using surface micromachining by Analog Devices, with dimensions of a couple of millimeters, which is used for deployment of airbags in automobiles, and more recently for various other consumer electronics markets [21, 97]. The central suspended beam mass (about 0.7 µg) is supported on the four corners by spring structures. The central beam has interdigitated cantilevered electrode fingers (about 125 µm long and 3 µm thick) on all four sides that alternate with those of the stationary electrode fingers as shown, with a gap of about 1.3 µm. Lateral motion of the central beam causes a change in the capacitance between these electrodes, which is used to measure the acceleration. Here stiction between the adjacent electrodes as well as stiction of the beam structure with the underlying substrate are detrimental to the operation of the sensor [21, 97]. Wear during unintended contacts of these polysilicon fingers is also a problem. A molecularly thick diphenyl siloxane lubricant film with high resistance to temperature and oxidation applied by a vapor deposition process is used on the electrodes to reduce stiction and wear [98]. For deposition, a small amount of liquid is dispensed into each package before it is sealed. As the package is heated in the furnace, the liquid evaporates and coats the sensor surface. As sensors are required to sense low-g accelerations, they need to be more compliant and stiction becomes an even bigger concern.

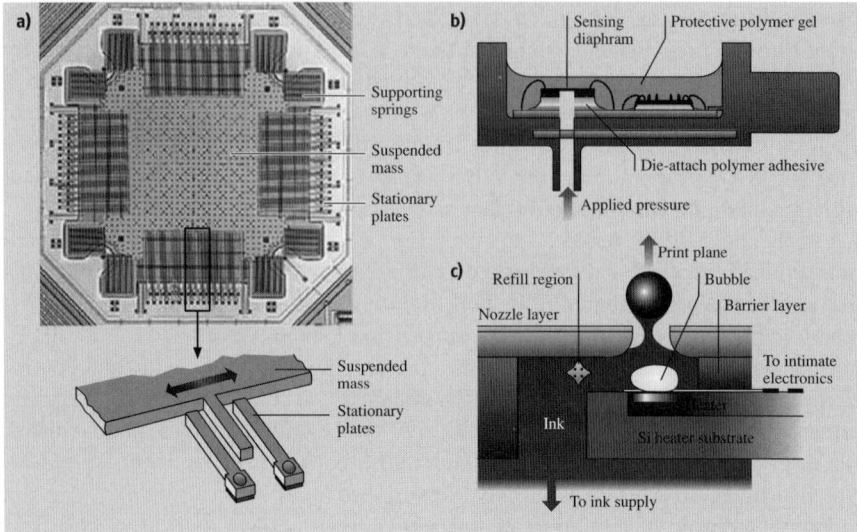

Fig. 22.4a–c. Examples of MEMS devices having commercial use that experience tribological problems. (**a**) Capacitive type silicon accelerometer for automotive sensory applications [97] (**b**) Piezoresistive type pressure sensor [101] (**c**) Thermal inkjet printhead [25]

Figure 22.4 also shows a cross-sectional view of a typical piezoresistive-type pressure sensor, which is used for various applications including manifold absolute pressure (MAP), and tire-pressure measurements, and disposable blood-pressure measurements. The sensing material is a diaphragm formed on a silicon substrate, which bends with applied pressure [99]. The deformation causes a change in the band structure of the piezoresistors that are placed on the diaphragm, leading to a change in the resistivity of the material. The MAP sensors are subjected to drastic conditions - extreme temperatures, vibrations, sensing fluid, and thermal shock. Fluid under extreme conditions could cause corrosive wear. Fluid cavitation could cause erosive wear. The protective gel encapsulent generally used can react with the sensing fluid and result in swelling or dissolution of the gel. Silicon cannot deform plastically, therefore any pressure spikes leading to deformation past its elastic limit will result in fracture and crack propagation. Pressure spikes could also cause the diaphragm to delaminate from the support substrate. Finally, cyclic loading of diaphragm during use can lead to fatigue and wear of silicon diaphragm or delamination.

The bottom schematic in Fig. 22.4 shows a cross-sectional view of a thermal printhead chip (on the order 10–50 cm^3 in volume) used in inkjet printers [25]. They consist of a supply of ink and an array of elements with microscopic heating resistors on a substrate mated to a matching array of ink-injection orifices or nozzles (about 70 µm in diameter) [23, 24, 26]. In each element, a small chamber is heated by the resistor, where a brief electrical impulse vaporizes part of the ink and creates a tiny bubble. The heaters operate at several kHz, and are therefore capable of high-speed printing. As the bubble expands, some of the ink is pushed out of the nozzle onto the paper. When the bubble pops, a vacuum is created and this causes more ink from the cartridge to move into the printhead. Clogged ink ports are the major failure mode. There are also various tribological concerns [23]. The surface of the printhead where the ink is shot out towards the paper can become scratched and damaged as a result of countless trips back and forth across the pages, which are somewhat rough. As a result of repeated heating and cooling, the heated resistors expand and contract. Over time, these elements will experience fatigue and may eventually fail. Bubble formation in the ink reservoir can lead to cavitation erosion of the chamber, which occurs when bubbles formed in the fluid become unstable and implode against the surface of the solid and apply impact energy on that surface. Fluid flow through nozzles may cause erosion and ink particles may also cause abrasive wear. Corrosion of the ink reservoir surfaces can also occur as a result of exposure of the ink to high temperatures as well as due to ink pH. The substrate of the chip consists of silicon with a thermal barrier layer followed by thin film of resistive material and then conducting material. The conductor and resister layers are generally protected by an overcoat layer of a plasma-enhanced chemical vapor deposition (PECVD) α-SiC : H layer, which is 200–500 nm thick [100].

Figure 22.5 shows two digital micromirror device (DMD) pixels used in digital light processing (DLP) technology for digital projection displays in computer projectors, high-definition television (HDTV) sets, and movie projectors [27–29]. The entire array (chip set) consists of a large number of rotatable aluminium alloy micromirrors (digital light switches) which are fabricated on top of a complementary metal–oxide–semiconductor (CMOS) static random-access memory integrated circuit. The surface-micromachined array consists of half of a million to more than two million of these independently controlled reflective, micromirrors (mirror size on the order of 12 μm square and 13 μm pitch) which flip backward and forward at a frequency on the order of 5000–7000 times a second, as a result of electrostatic attraction between the micromirror structure and the underlying electrodes. For the binary operation, micromirror/yoke structure mounted on torsional hinges is rotated $\pm 10°$ (with respect to the horizontal plane), and is limited by a mechanical stop. Contact between the cantilevered spring tips at the end of the yoke (four present on each yoke) with the underlying stationary landing sites is required for true digital (binary) operation. Stiction and wear during contact between the aluminium alloy spring tips and landing sites, hinge memory (metal creep at high operating temperatures), hinge fatigue, shock and vibration failure, and sensitivity to particles in the chip package and operating environment are some of the important issues affecting the reliable operation of a micromirror device [102–104]. Self-assembled monolayers of a fatty acid – perfluorodecanoic acid (PFDA) – applied by a vapor deposition process is used on the surfaces of the tip and landing sites to reduce stiction and wear [105, 106]. However, these films are susceptible to moisture, and to keep moisture out and create a background pressure of PFDA, hermetic chip packages are used. The spring tip is used in order to use the stored spring energy to pop up the tip during pull-off. A lifetime estimate of over one hundred thousand operating hours

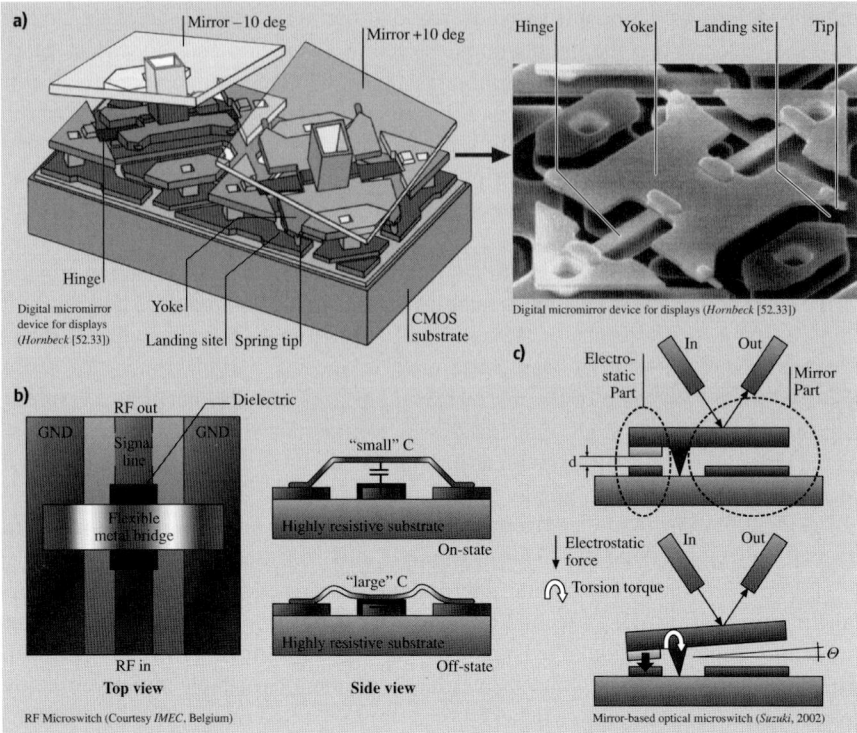

Fig. 22.5. Examples of MOEMS and RF-MEMS devices having commercial use that experience tribological problems

with no degradation in image quality is the norm. At a mirror modulation frequency of 7 kHz, each micromirror element needs to switch about 2.5 trillion cycles.

Figure 22.5 also shows a schematic of an electrostatically actuated capacitive-type RF microswitch for switching of RF signals at microwave and low frequencies [107]. It is a membrane type and consists of a flexible metal (Al) bridge that spans the RF transmission line in the center of a coplanar waveguide. When the bridge is up, the capacitance between the bridge and RF transmission line is small and the RF signal passes without much loss. When a DC voltage is applied between the RF transmission line and the bridge, the latter is pulled down until it touches a dielectric isolation layer. The large capacitance thus created shorts the RF signal to the ground. The failure modes include creep in the metal bridge, fatigue of the bridge, charging and degradation of the dielectric insulator, and stiction of the bridge to the insulator [30, 107]. The stiction occurs due to capillary condensation of water vapor from the environment, van der Waals forces, and/or charging effects. If the restoring force in the bridge of the switch is not large enough to pull the bridge up again after the actuation voltage has been removed, the device fails due to stiction. Humidity-induced stiction can be avoided by hermetically sealing the microswitch. Some roughness of the surfaces reduces the probability of stiction. Selected actuation waveforms can be used to minimize charging effects.

The bottom schematic of Fig. 22.5 shows an electrostatically actuated mirror-based optical microswitch or attenuator [30]. It is a cantilever type and uses a hinged mirror which rotates to reflect light from one (in) fiber to another (out) fiber. It can either change light intensity or can be used as an on/off switch when the deflection angle of the reflective mirror is adjusted. A voltage is applied which creates an electrostatic force that pulls the mirror down until electrical contact is made. The primary source of failure arises from the rubbing present in the hinge, leading to wear issues. Failure in the hinge can occur either by mechanical drift due to wear or complete fracture. Stiction is a major concern after the mirror is kept in contact with the electrode underneath [30]. The dynamic impact of repeated contacts may lead to subsurface fatigue [108]. Bumps are placed on the mirror surface facing the electrode underneath in order to minimize the contact area and stiction. Lubricants are also used to reduce stiction.

NEMS

Figure 22.6 shows an AFM-probe-based nanoscale data-storage system for ultra-high-density recording, which experiences tribological problems [49]. The system uses arrays of 1024 silicon microcantilevers ("Millipede") for thermomechanical recording and playback on a polymer [polymethyl methacrylate (PMMA)] medium about 40 nm thick with a harder Si substrate. The cantilevers consist of integrated tip heaters with tips of nanoscale dimensions. (Sharp tips themselves are also example of NEMS.) Thermomechanical recording is a combination of applying a local force to the polymer layer and softening it by local heating. The tip, heated to about 400 °C, is brought into contact with the polymer for recording. Reading is done using the heater cantilever, originally used for recording, as a thermal read-back sensor by exploiting its temperature-dependent resistance. The principle of thermal sensing is based on the fact that the thermal conductivity between the heater and the storage substrate changes according to the spacing between them. When the spacing between the heater and sample is reduced as the tip moves into a bit, the heater's temperature and hence its resistance will decrease. Thus, changes in temperature of the continuously heated resistor are monitored while the cantilever is scanned over data bits, providing a means of detecting the bits. Erasing for subsequent rewriting is carried out by thermal reflow of the storage field by heating the medium to 150 °C for a few seconds. The smoothness of the reflown medium allows multiple rewriting of the same storage field. Bit sizes in the range 10–50 nm have been achieved by using a 32×32 (1024) array write/read chip (3 mm $\times$ 3 mm). It has been reported that tip wear occurs due to contact between the tip and the Si substrate during writing. Tip wear is considered a major concern for device reliability.

In magnetic data storage, magnetic recording is accomplished by relative motion between the magnetic head slider and a magnetic rigid disk [34]. Magnetic rigid disks and heads used today for magnetic data storage consist of one- to a few-nm-thick nanostructured films. Figure 22.7a shows the sectional view of a conven-

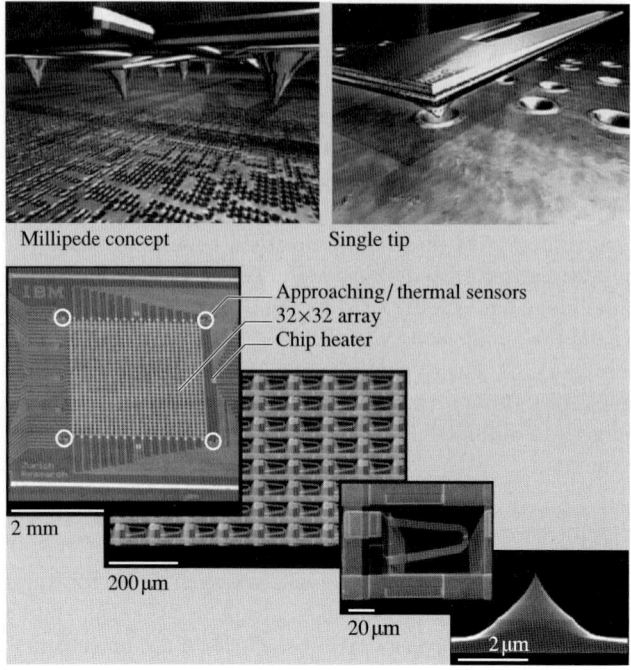

Fig. 22.6. Example of a NEMS device – AFM-probe-based nanoscale data storage system – that experience tribological problems

tional multigrain magnetic rigid disk. The superparamagnetic effect poses a serious challenge for ever increasing areal density of disk drives. One of the promising methods to circumvent the density limitations imposed by this effect is the use of nanopatterned disks Fig. 22.7b. In conventional disks, the thin magnetic layer forms a random mosaic of nanometer-scale grains and each recorded bit consists of many tens of these random grains. In patterned disks, the magnetic layer is created as an ordered array of highly uniform islands, each island capable of storing an individual bit. These islands may be one or a few grains, rather than a collection of random decoupled grains. This increases the density by a couple of orders of magnitude. Figure 22.7c shows a schematic of an inductive-write GMR read head structure. These are constructed from a variety of materials: magnetic alloys, metal conductors, ceramic, and polymer insulators in a complex three-dimensional structure. The multilayered thin-film structures used to construct the sensor and individual films are only a few nm thick. The head slider surface, which flies over the disk surface, is coated with diamond-like carbon coatings that are about 3 nm thick to protect the thin-film structure from electrostatic discharge. Any isolated contacts between the disk and sensor and lubricant pickup pose tribological concerns [34].

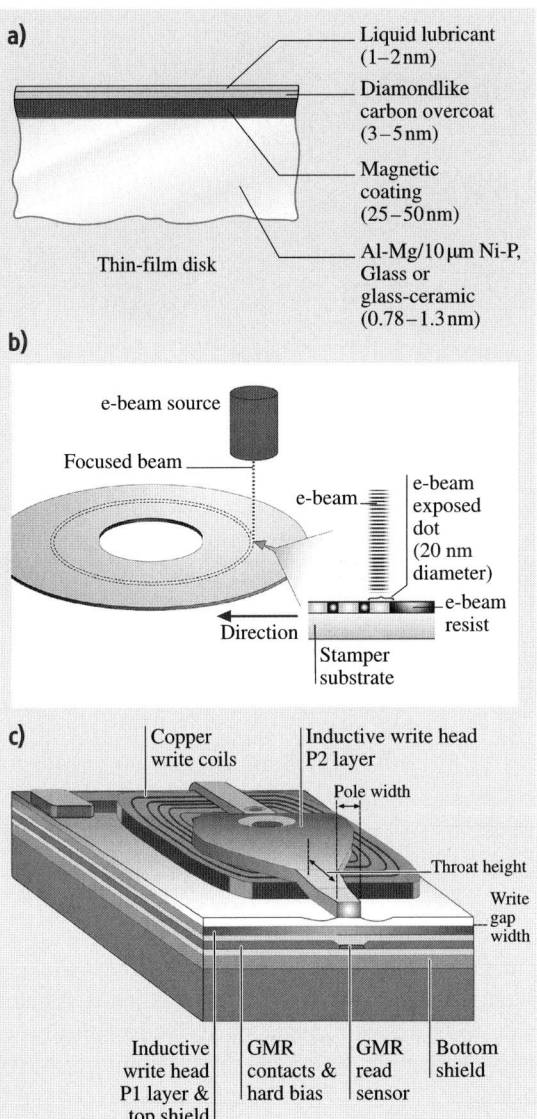

Fig. 22.7. Schematic of (**a**) sectional view of a conventional multigrain magnetic rigid disk, (**b**) nanopatterned magnetic rigid disk, and (**c**) an inductive-write GMR read magnetic head structure for magnetic data storage (Hitachi Global Storage Technologies)

BioMEMS

An example of a wristwatch-type biosensor based on microfluidics, referred to as a lab-on-a-chip system, is shown in Fig. 22.8a [60, 69]. These systems are designed to either detect a single or class of (bio)chemicals, or system-level analytical capabilities for a broad range of (bio)chemical species known as a micro total-analysis system (μTas), and have the advantage of incorporating sample handling, separation, detection, and data analysis onto one platform. The chip relies on microfluidics and

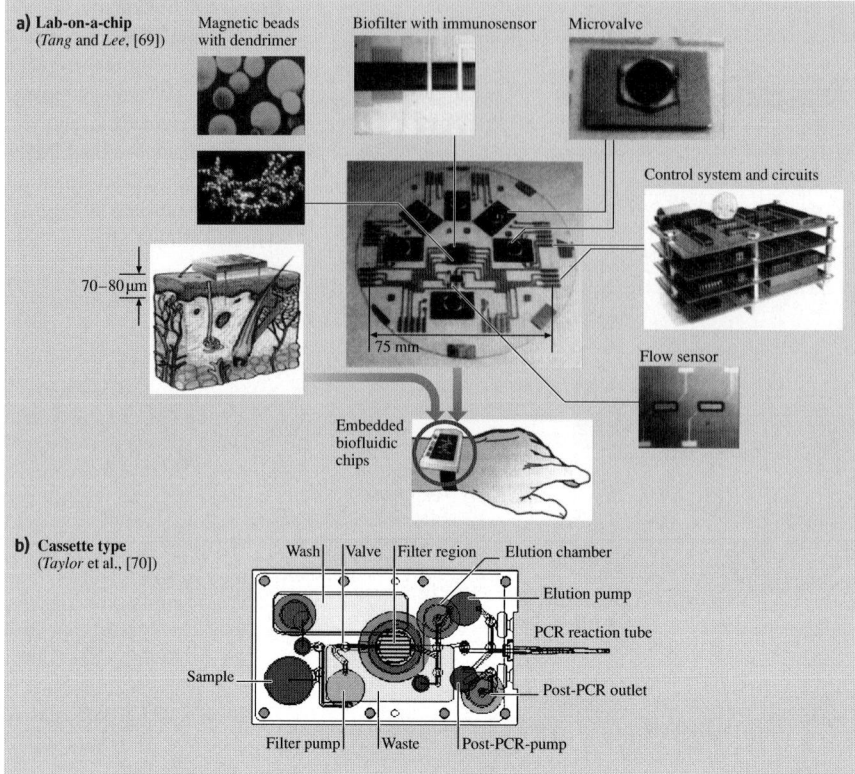

Fig. 22.8. (a) MEMS-based biofluidic chip, commonly known as a lab-on-a-chip, that can be worn like a wristwatch [69]. (b) Cassette-type biosensor used for human genomic DNA analysis [70]

involves manipulation of tiny amounts of fluids in microchannels using microvalves. The test fluid is injected into the chip, generally using an external pump or syringe, for analysis. Some chips have been designed with an integrated electrostatically actuated diaphragm-type micropump. The sample, which can have a volume measured in nanoliters, flows through microfluidic channels via an electric potential and capillary action using microvalves (having various designs including membrane type) for various analyses. The fluid is preprocessed and then analyzed using a biosensor. Another example of a biosensor is the cassette-type biosensor used for human genomic DNA analysis; integrated biological sample preparation is shown in Fig. 22.8b [70]. The implementation of micropumps and microvalves allows for fluid manipulation and multiple sample-processing steps in a single cassette. Blood or other aqueous solutions can be pumped into the system, where various processes are performed.

Microvalves, which are found in most microfluidic components of BioMEMS, can be classified into two categories: active microvalves (with an actuator) for flow regulation in microchannels and passive microvalves integrated with micropumps.

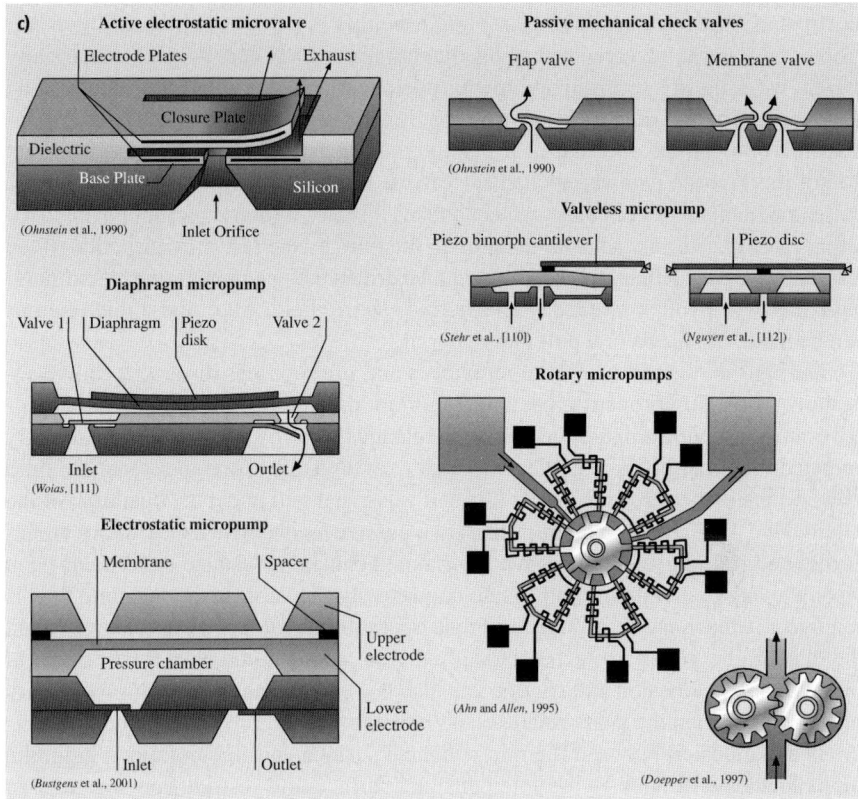

Fig. 22.8. (continued) (**c**) Multiple examples of valves and pumps found in BioMEMS devices. Mechanical check valves, diaphragm micropump, valveless micropump, and rotary micropump

Active microvalves consist of a valve seat and a diaphragm actuated by an external actuator [61, 109]. Different types of actuators are based on piezoelectric, electrostatic, thermopneumatic, electromagnetic, and bimetallic materials, shape-memory alloys and solenoid plungers. An example of an electrostatic cantilever-type active microvalve is shown in Fig. 22.8c. Passive microvalves used in micropumps include mechanical check valves and a diffuser/nozzle [61, 109–112]. Check valves consist of a flap or membrane that is capable of opening and closing with changes in pressure; see Fig. 22.8c for schematics. A diffuser/nozzle uses an entirely different principle and only works with the presence of a reciprocating diaphragm. When one convergent channel works simultaneously with another convergent channel oriented in a specific direction, a change in pressure is possible.

There are four main types of mechanical micropumps, which include diaphragm micropumps that involves mechanical check valves, valveless rectification pumps that use diffuser/nozzle-type valves, valveless pumps without a diffuser/nozzle, electrostatic micropumps, and rotary micropumps [61, 109–112]. Diaphragm micropumps consist of a reciprocating diaphragm, which can be piezoelectrically driven, working in synchronization with two check valves Fig. 22.8c. Electrostatic micropumps have a diaphragm as well, but it is driven using two electrodes. Valveless micropumps also consist of a diaphragm that is piezoelectrically driven, but do not incorporate passive mechanical valves. Instead, these pumps use an elastic buffer mechanism or variable gap mechanisms. Finally, a rotary micropump has a rotating rotor that simply adds momentum to the fluid by the fast-moving action of the blades Fig. 22.8c. Rotary micropumps can be driven using an integrated electromagnetic motor or by the presence of an external electric field. All of these micropumps can be made of silicon or a polymer material.

During the operation of the microvalves and micropumps discussed above, adhesion and friction properties become important in which contacts occur due to relative motion. During operation, active mechanical microvalves have an externally actuated diaphragm which comes into contact with a valve seat to restrict the fluid flow. Adhesion between the diaphragm and valve seat will affect the operation of the microvalve. In diaphragm micropumps, two passive mechanical check valves are incorporated into the design. Passive mechanical check valves also exhibit adhesion when the flap or membrane comes into contact with the valve seat when fluid flow is removed. Adhesion also occurs during the operation of valveless micropumps when the diaphragm, which is piezoelectrically driven, comes into contact with the rigid outlet. Finally, adhesion and friction can also be seen during the operation of rotary micropumps when the gears rotate, come into contact and rub against one another.

If the adhesion between the microchannel surface and the biofluid is high, the biomolecules will stick to the microchannel surface and restrict flow. In order to facilitate flow, microchannel surfaces with low bioadhesion are required. Fluid flow in polymer channels can produce a triboelectric surface potential which may affect the flow. Polymers are known to generate surface potentials and the magnitude of the potential varies from one polymer to another [113–115]. Conductive surface layers on the polymer channels can be deposited to reduce triboelectric effects.

As mentioned, the microfluidic biosensor shown in Fig. 22.8a required the use of micropumps and microvalves. For example, a microdevice with 1000 channels requires 1000 micropumps and 2000 microvalves, which makes it bulky and poses reliability concerns. Two methods can be used to drive the flow of fluids in microchannels: pressure and electrokinetic drive. The electrokinetic flow is based on the movement of molecules in an electric field due to their charges. There are two components to electrokinetic flow: electrophoresis, which results from the accelerating force due to the charge of a molecule in an electric field, and electroosmosis, which uses electrically controlled surface tension to drive uniform liquid flow. Biosensors based on electrokinetic flow have also been developed. In so-called *digital-based microfluidics*, based on the electroosmosis process, electrically controlled surface tension is used to drive liquid droplets, thus eliminating the need for valves and pumps [116, 117]. These microdevices consist of a rectangular grid of gold nanoelectrodes instead of micro/nanochannels. An externally applied electric field enables manipulation of samples of a few nanoliters through the capillary circuitry.

An example of a microarray-type biosensor under development in our laboratory is that based on a field-effect transistor (FET), which is shown in Fig. 22.9 [72,118]. FETs are sensitive to the electrical field produced due to the charge at the surface of the gate insulator. In this sensor, the gate metal of a metal–oxide–semiconductor field-effect transistor (MOSFET) is removed and replaced with a protein (receptor layer) whose cognate is the analyte (e. g., virus or bacteria) that is meant to be sensed. The binding of the receptor layer with the analyte produces a change in the effective charge, which creates a change in the electrical field. This electrical field change may produce a measurable change in the current flow through the device. Adhesion between the protein and silica substrate affect the reliability of the biosensor. In the case of implanted biosensors, the biosensors come in contact with the exterior environment such as tissues and fluids, and any relative motion of the sensor surface with respect to exterior environment such as tissues or fluids may result in surface damage. A schematic of friction and points of wear generation when an implanted biosensor surface comes into contact with living tissue is shown in Fig. 22.9b. The, friction, wear, and adhesion of the biosensor surface may be critical in these applications [72, 119, 120].

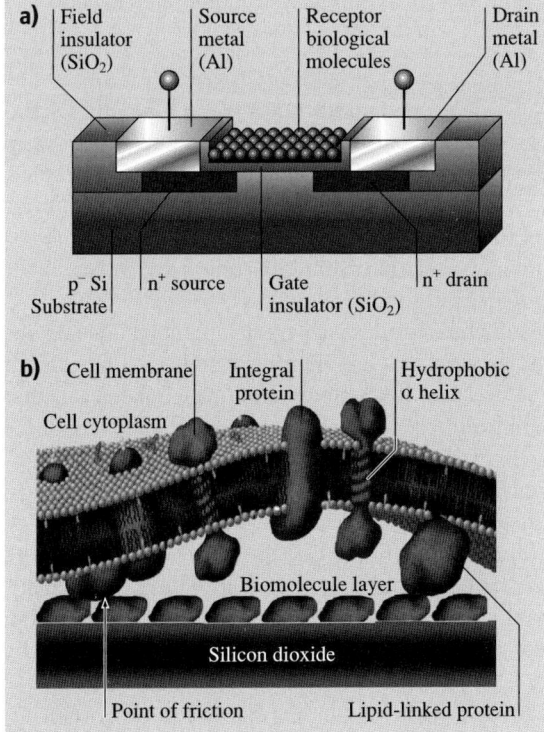

Fig. 22.9. (a) Schematic of MOSFET-based bioFET sensor [72], and (b) schematic showing the generation of friction and wear points due to interaction of implanted biomolecule layer on a biosensor with living tissue

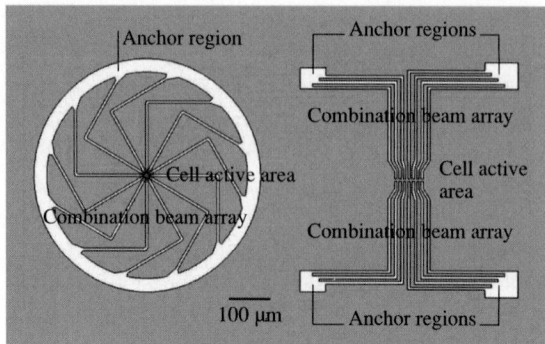

Fig. 22.10. Schematic of two designs for polymer bioMEMS structures to measure cellular forces [121]

Polymer BioMEMS are designed to measure cellular surfaces. For two examples, see Fig. 22.10 [121]. The device on the left shows cantilevers anchored at the periphery of the circular structure, while the device on the right has cantilevers anchored at the two corners on the top and the bottom. The cell adheres to the center of the structure, and the contractile forces generated in the cells cytoskeleton cause the cantilever to deflect. The deflection of the compliant polymer cantilevers is measured optically and related to the magnitude of the forces generated by the cell. Adhesion between cells and polymer beam is desirable. In order to design the sensors, micro- and nanoscale mechanical properties of polymer structures are needed.

BioNEMS

Micro/nanofluidic devices provide a powerful platform for electrophoretic separations for a variety of biochemical and chemical analysis. Electrophoresis is a versatile analytical method which is used for separation of small ions, neutral molecules, and large biomolecules. Figure 22.11 shows an interdigitated micro/nanofluidic silicon array with nanochannels for the separation process. Figure 22.12a shows a schematic of an implantable, immuno-isolation submicroscopic biocapsules, aimed at drug delivery to treat significant medical condition such as type I diabetes [76, 77]. The purpose of the immuno-isolation biocapsules is to create an implantable device capable of supporting foreign living cells that can be transplanted into humans. It is a silicon capsule consisting of two nanofabricated membranes bonded together with the drug (e.g., encapsulated insulin-producing islet cells) contained within the cavities for long-term delivery. Pores or nanochannels in a semipermeable membrane as small as 6 nm are used as flux regulators for the long-term release of drugs. The nanomembrane also protects therapeutic substances from attack by the body's immune system. The pores are large enough to provide the flow of nutrients (e.g., glucose molecules) and drug (e.g., insulin), but small enough to block natural antibodies. Antibodies have the capability to pass through any orifice larger than 18 nm. The 50-nm pores in silicon were etched by using the sacrificial-layer lithography described in Sect. 22.A [77].

The main reliability concerns in the micro/nanofluidic silicon array and implantable biocapsules are biocompatibility and potential biofouling (undesirable ac-

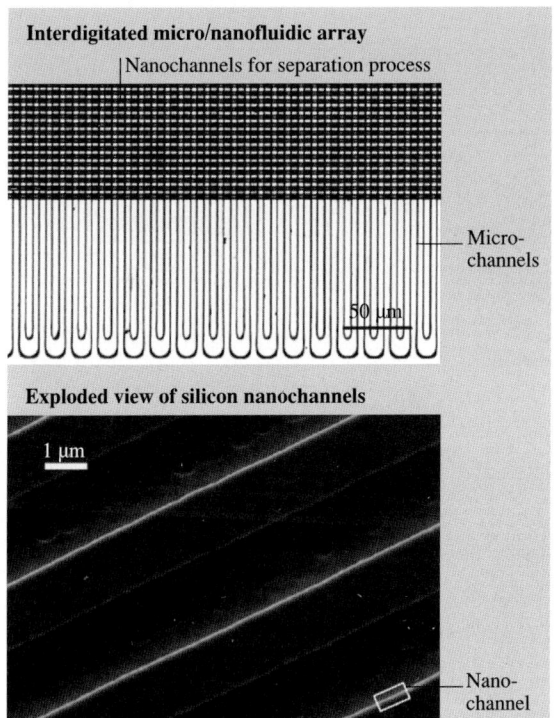

Fig. 22.11. Interdigitated micro/nanofluidic silicon array for the separation process [118]

cumulation of microorganisms) of the channels/membrane by means of protein and cells adsorption from biological fluids. Biofouling can also result in the clogging of the nanochannels/nanopores, which potentially could render the device ineffective. The adhesion of proteins and cells to an implanted device can also cause detrimental results such as inflammation and excessive fibrosis. Deposition of the self-assembled monolayers of selected organic molecules on the channels implants, which makes them hydrophobic, presents an innovative solution to combat the adverse effects of the biological fluids [122–125].

Figure 22.12b shows a conceptual model of an intravascular drug-delivery device – nanoparticles used to search for and destroy disease cells [78]. With lateral dimensions of 1 μm or less, the particles are smaller than any blood cells. These particles can be injected into the blood stream and travel freely through the circulatory system. In order to direct these drug-delivery nanoparticles to cancer sites, their external surfaces are chemically modified to carry molecules that have lock-and-key binding specificity with molecules that support a growing cancer mass. As soon as the particles dock onto the cells, a compound is released that forms a pore on the membrane of the cells, which leads to cell death and ultimately to that of the cancer mass that was being nourished by the blood vessel. Adhesion between nanoparticles and disease cells is required. Furthermore, the particles should travel close to the endothelium lining of vascular arteries. *Decuzzi* et al. [126] recently

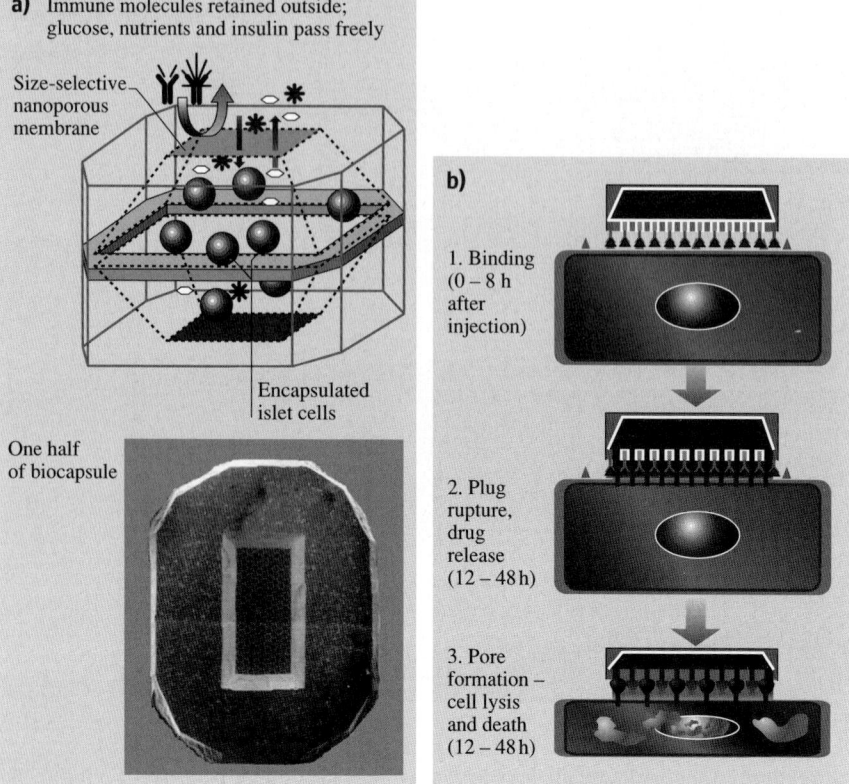

Fig. 22.12. Schematics of (**a**) implantable, immuno-isolation submicroscopic biocapsules (drug-delivery device) [77], and (**b**) intravascular nanoparticles to search for and destroy diseased blood cells [78]

analyzed the margination of a particle circulating in the blood stream and calculated the speed and time for margination (motion of the particles towards the walls of the vessel) as a function of the density and diameter of particle, based on various forces present between the circulating particle and the endothelium lining. Human capillaries can have radii as small as 4–5 µm. They reported that the particles used for drug delivery should have a radius smaller than a critical value (of the order 100 nm).

In summary, adhesion, stiction/friction and wear clearly limit the lifetime and compromise the performance and reliability of MEMS/NEMS and BioMEMS/BioNEMS. Figure 22.13a summarizes tribological problems encountered in some of the MEMS, MOEMS, RF-MEMS, and BioMEMS devices discussed. In addition to in-use stiction, stiction issues are also present in some processes used for fabrication of MEMS/NEMS. For example, the last step in surface micromachining involves the removal of sacrificial layer(s), known as the release process, since the microstructures are released from the surrounding sacrificial layer(s). The re-

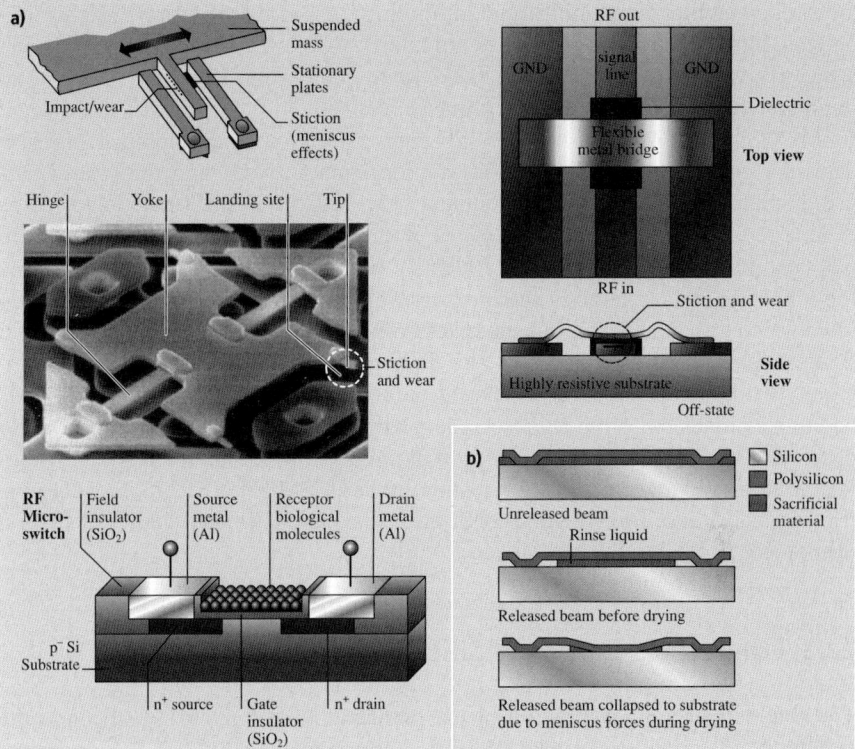

Fig. 22.13. (a) Summary of tribological issues in MEMS, MOEMS and RF-MEMS device operation, and (b) in microfabrication via surface micromachining

lease is accomplished by an aqueous chemical etch, rinsing and drying processes. Due to meniscus effects as a result of wet processes, the suspended structures can sometimes collapse and permanently adhere to the underlying substrate, as shown in Fig. 22.13b [127]. Adhesion is caused by water molecules adsorbed on the adhering surfaces and/or because of the formation of adhesive bonds by silica residues that remain on the surfaces after the water has evaporated. This so-called release stiction is overcome by using dry release methods (e. g., CO_2 critical-point drying or sublimation drying [128]).

Tribological Needs

MEMS/NEMS need to be designed to perform expected functions typically in the millisecond to picosecond range. Expected life of the devices for high-speed contacts can vary from a few hundred thousand to many billions of cycles, e. g., over a hundred billion cycles for DMDs, which puts serious requirements on materials [13, 91, 129–132]. Adhesion between a biological molecular layer and the substrate, referred to as *bioadhesion*, reduction of friction and wear of biological layers, biocompatibility, and biofouling for BioMEMS/BioNEMS are important. Most mechanical properties are known to be scale-dependent [133]. The properties of nanoscale structures need to be measured [134]. Tribology is an important factor affecting the performance and reliability of MEMS/NEMS and BioMEMS/BioNEMS [13,46,47,80]. There is a need for the development of a fundamental understanding of adhesion, friction/stiction, wear, and the role of surface contamination, and the environment [13]. MEMS/NEMS materials need to exhibit good mechanical and tribological properties on the micro/nanoscale. There is a need to develop lubricants and identify lubrication methods that are suitable for MEMS/NEMS. Methods need to be developed to enhance adhesion between biomolecules and the device substrate, referred to as bioadhesion. Component-level studies are required to provide a better understanding of the tribological phenomena occurring in MEMS/NEMS. The emergence of micro/nanotribology and AFM-based techniques has provided researchers with a viable approach to address these problems [46,47,135]. This chapter presents an overview of micro/nanoscale tribological studies of materials and lubrication studies for MEMS/NEMS, bioadhesion, friction and wear of BioMEMS/BioNEMS, and component-level studies of stiction phenomena in MEMS/NEMS devices.

22.2 Tribological Studies of Silicon and Related Materials

The materials of most interest for planar fabrication processes using silicon as the structural material are undoped and boron-doped (p^+-type) single-crystal silicon for bulk micromachining, and phosphorus (n^+-type) doped and undoped low-pressure chemical vapor deposition (LPCVD) polysilicon films for surface micromachining. Silicon-based devices lack high-temperature capabilities with respect to both mechanical and electrical properties. SiC has been developed as a structural material for high-temperature microsensor and microactuator applications [138, 139]. SiC can also be desirable for high-frequency micromechanical resonators, in the GHz range, because of its high ratio of modulus of elasticity to density and consequently high resonance frequency. Table 22.2 compares selected bulk properties of SiC and Si(100). Researchers have found low-cost techniques of producing single-crystalline 3C–SiC (cubic or β-SiC) films via epitaxial growth on large-area silicon substrates for bulk micromachining [140] and polycrystalline 3C–SiC films on polysilicon and silicon dioxide layers for surface micromachining of SiC [141]. Single-crystalline 3C–SiC piezoresistive pressure sensors have been fabricated using bulk micromachining for high-temperature gas turbine applications [142]. Surface-micromachined polycrystalline SiC micromotors have been fabricated and satisfactory operation at high temperatures has been reported [143].

As will be shown, bare silicon exhibits inadequate tribological performance and needs to be coated with a solid and/or liquid overcoat or be surface treated (by, e.g., oxidation and ion implantation, commonly used in semiconductor manufacturing), which exhibits lower friction and wear. SiC films exhibit good tribological

Table 22.2. Selected bulk properties of 3C (β- or cubic) SiC and Si(100)

Sample	Density[a] (kg/m^3)	Hardness[a] (GPa)	Elastic modulus[a] (GPa)	Fracture toughness[a] (MPam$^{1/2}$)
β-SiC	3210	23.5–26.5	440	4.6
Si(100)	2330	9–10	130	0.95

Sample	Thermal conductivity[b] (W/mK)	Coeff. of thermal expansion[b] ($\times 10^{-6}$/°C)	Melting point[a] (°C)	Band gap[a] (eV)
β-SiC	85–260	4.5–6	2830	2.3
Si(100)	155	2–4.5	1410	1.1

[a] Data from *Bhushan* and *Gupta* [136].
[b] Data from *Shackelford* et al. [137].

performance. Both macroscale and microscale tribological properties of virgin and treated/coated silicon, polysilicon films and SiC are presented next.

22.2.1 Virgin and Treated/Coated Silicon Samples

Tribological Properties of Silicon and the Effect of Ion Implantation

Friction and wear of single-crystalline and polycrystalline silicon samples have been studied and the effect of ion implantation with various doses of C$^+$, B$^+$, N$_2^+$ and Ar$^+$ ion species at an energy of 200 keV to improve their friction and wear properties has been studied [144–146]. The coefficient of macroscale friction and wear factor of virgin single-crystal silicon and C$^+$-implanted silicon samples as a function of ion dose are presented in Fig. 22.14 [144]. The macroscale friction and wear tests were conducted using a ball-on-flat tribometer. Each data bar represents the average value of four to six measurements. The coefficient of friction and wear factor for bare silicon are very high and decrease drastically with ion dose. Silicon samples bombarded above an ion dose of 10^{17} C$^+$ cm^{-2} exhibit extremely low values of coefficients of friction (typically 0.03–0.06 in air) and the wear factor (reduced by as much as four orders of magnitude). *Gupta* et al. [144] reported that a decrease in the coefficient of friction and the wear factor of silicon as a result of C$^+$ ion bombardment occurred because of the formation of silicon carbide rather than amorphization of silicon. *Gupta* et al. [145] also reported an improvement in friction and wear with B$^+$ ion implantation.

Microscale friction measurements were performed using an atomic force/friction force microscope (AFM/FFM) [46, 47]. Table 22.3 shows values of surface roughness and coefficients of macroscale and microscale friction for virgin and doped silicon. There is a decrease in the coefficients of microscale and macroscale friction values as a result of ion implantation. When measured for the small contact areas and very low loads used in microscale studies, indentation hardness and

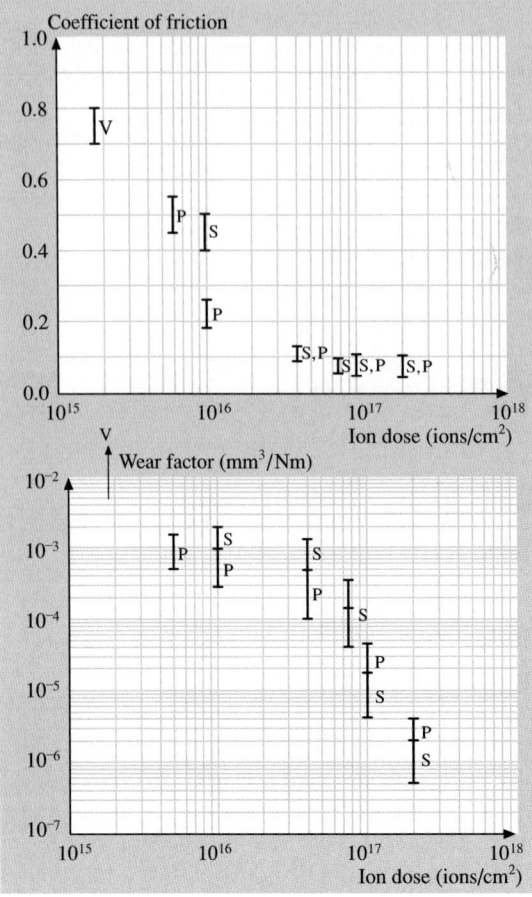

Fig. 22.14. Influence of ion doses on the coefficient of friction and wear factor on C^+-ion-bombarded single-crystal and polycrystalline silicon slid against alumina ball. V corresponds to virgin single-crystal silicon, while S and P denote tests that correspond to doped single-crystal and polycrystalline silicon, respectively [144]

Table 22.3. Surface roughness and micro- and macroscale coefficients of friction of selected samples

Material	RMS roughness (nm)	Coefficient of microscale friction[a]	Coefficient of macroscale friction[b]
Si(111)	0.11	0.03	0.33
C^+-implanted Si(111)	0.33	0.02	0.18

[a] Versus Si_3N_4 tip; tip radius of 50 nm in the load range 10–150 nN (2.5–6.1 GPa) at a scanning speed of 5 µm/s over a scan area of 1 µm × 1 µm in an AFM
[b] Versus Si_3N_4 ball, ball radius of 3 mm at a normal load of 0.1 N (0.3 GPa) at an average sliding speed of 0.8 mm/s using a tribometer

elastic modulus are higher than those at the macroscale. This, added to the effect of the small apparent area of contact reducing the number of trapped particles on the interface, results in less plowing contribution and lower friction in the case of microscale friction measurements. Results of microscale wear resistance studies of ion-implanted silicon samples studied using a diamond tip in an AFM [147] are shown in Figs. 22.15a and 22.15b. For tests conducted at various loads on Si(111)

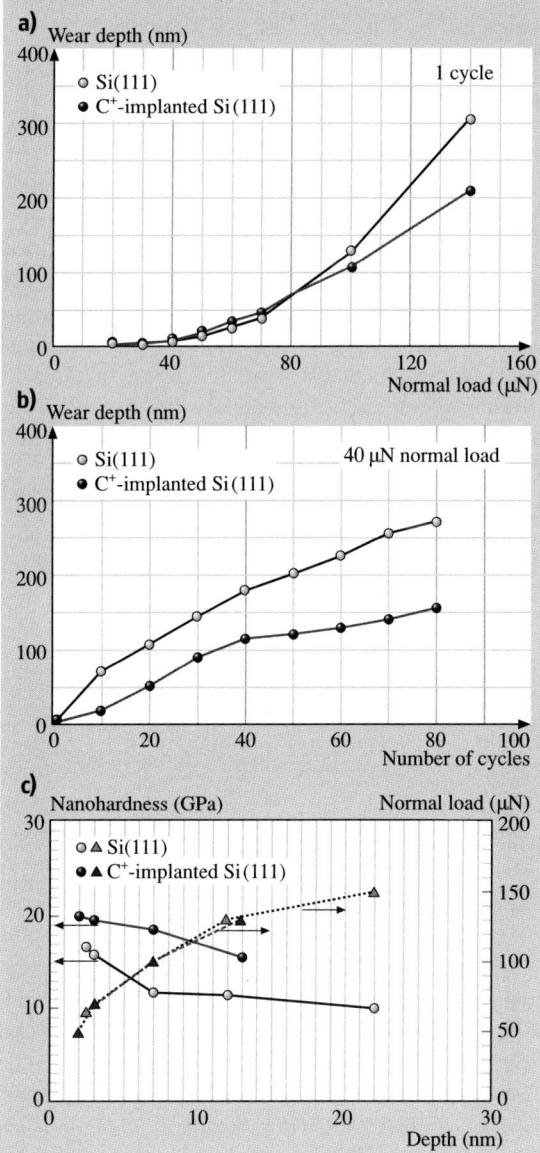

Fig. 22.15. Wear depth as a function of (a) load (after one cycle), and (b) cycles (normal load = 40 mN) for Si(111) and C^+-implanted Si(111). (c) Nanohardness and normal load as function of indentation depth for virgin and C^+-implanted Si(111) [147]

and C^+-implanted Si(111), it is noted that wear resistance of implanted sample is slightly poorer than that of virgin silicon up to about 80 μN. Above 80 μN, the wear resistance of implanted Si improves. As one continues to run tests at 40 μN for a larger number of cycles, the implanted sample, which forms hard and tough silicon carbide, exhibits higher wear resistance than the unimplanted sample. Damage from the implantation in the top layer results in poorer wear resistance, however, the implanted zone at the subsurface is more wear-resistant than the virgin silicon.

Hardness values of virgin and C^+-implanted Si(111) at various indentation depths (normal loads) are presented in Fig. 22.15c [147]. The hardness at a small indentation depth of 2.5 nm is 16.6 GPa and it drops to a value of 11.7 GPa at a depth of 7 nm and a normal load of 100 μN. Higher hardness values obtained in low-load indentation may arise from the observed pressure-induced phase transformation during the nanoindentation [148,149]. Additional increase in the hardness at an the even lower indentation depth of 2.5 nm reported here may arise from the contribution by complex chemical films (not from native oxide films) present on the silicon surface. At small volumes there is a lower probability of encountering material defects (dislocations, etc.). Furthermore, according to the strain-gradient plasticity theory advanced by *Fleck* et al. [150], large strain gradients inherent in small indentations lead to accumulation of geometrically necessary dislocations that cause enhanced hardening. These are some of the plausible explanations for an increase in hardness at smaller volumes. If the silicon material were to be used at very light loads such as in microsystems, the high hardness of surface films would protect the surface until it is worn.

From Fig. 22.15c, hardness values of C^+-implanted Si(111) at a normal load of 50 μN is 20.0 GPa with an indentation depth of about 2 nm, which is comparable to the hardness value of 19.5 GPa at 70 μN, whereas measured hardness value for virgin silicon at an indentation depth of about 7 nm (normal load of 100 μN) is only about 11.7 GPa. Thus, ion implantation with C^+ results in an increase in hardness in silicon. Note that the surface layer of the implanted zone is much harder compared with the subsurface, and may be brittle leading to higher wear on the surface. The subsurface of the implanted zone (SiC) is harder than the virgin silicon, resulting in higher wear resistance, which is also observed in the results of the macroscale tests conducted at high loads.

The Effect of Oxide Films on Tribological Properties of Silicon

Macroscale friction and wear experiments have been performed using a magnetic disk drive with bare, oxidized, and implanted pins sliding against amorphous-carbon-coated magnetic disks lubricated with a thin layer of perfluoropolyether lubricant [151–154]. Representative profiles of the variation of the coefficient of friction with number of sliding cycles for Al_2O_3–TiC slider and bare and dry-oxidized silicon pins are shown in Fig. 22.16. For bare Si(111), after initial increase in the coefficient of friction, it drops to a steady-state value of 0.1 following the increase, as seen in Fig. 22.16. The rise in the coefficient of friction for the Si(111) pin is associated with the transfer of amorphous carbon from the disk to the pin, oxidation-

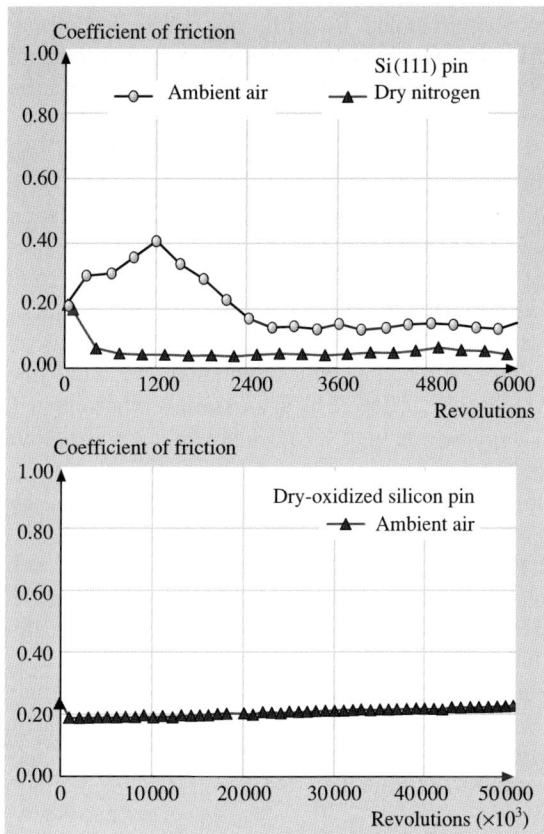

Fig. 22.16. Coefficient of friction as a function of number of sliding revolutions in ambient air for a Si(111) pin in ambient air and dry nitrogen and a dry-oxidized silicon pin in ambient air [151]

enhanced fracture of pin material followed by tribochemical oxidation of the transfer film, while the drop is associated with the formation of a transfer coating on the pin. Dry-oxidized Si(111) exhibits excellent characteristics and no significant increase was observed over 50,000 cycles (Fig. 22.16). This behavior has been attributed to the chemical passivity of the oxide and lack of transfer of diamond-like carbon (DLC) from the disk to the pin. The behavior of PECVD oxide (data not presented here) was comparable to that of dry oxide but for the wet oxide there was some variation in the coefficient of friction (0.26–0.4). The difference between dry and wet oxide was attributed to increased porosity of the wet oxide [151]. Since tribochemical oxidation was determined to be a significant factor, experiments were conducted in dry nitrogen [152, 153]. The variation of the coefficient of friction for a silicon pin sliding against a thin-film disk in dry nitrogen is shown in Fig. 22.16. It is seen that, in a dry nitrogen environment, the coefficient of friction of Si(111) sliding against a disk decreased from an initial value of about 0.05–0.2 with continued sliding. Based on SEM and chemical analysis, this behavior has been attributed to the formation of a smooth amorphous-carbon/lubricant transfer patch and suppression of oxidation in a dry nitrogen environment. Based on macroscale tests using disk drives, it is found that the friction and wear performance of bare silicon is not adequate. With dry-oxidized or PECVD SiO$_2$-coated silicon, no significant friction increase or interfacial degradation was observed in ambient air.

Table 22.4 and Fig. 22.17 show surface roughness, microscale friction and scratch data and nanoindentation hardness for the various silicon samples [147]. Scratch experiments were performed using a diamond tip in an AFM. Results on polysilicon samples are also shown for comparison. Coefficients of microscale friction values for all the samples are about the same. These samples could be scratched at 10 μN load. Scratch depth increased with normal load. Crystalline orientation of silicon has little influence on scratch resistance because natural oxidation of silicon in ambient masks the expected effect of crystallographic orientation. PECVD-oxide samples showed the best scratch resistance, followed by dry-oxidized, wet-oxidized, and ion-implanted samples. Ion implantation with C$^+$ does not appear to improve scratch resistance.

Wear data on the silicon samples are also presented in Table 22.4 [147]. PECVD-oxide samples showed superior wear resistance followed by the dry-oxidized, wet-oxidized, and ion-implanted samples. This agrees with the trends seen in scratch resistance. In PECVD, ion bombardment during the deposition improves the coating properties such as suppression of columnar growth, freedom from pinhole, decrease in crystalline size, and increase in density, hardness and substrate-coating adhesion. These effects may help in improving mechanical integrity of the sample surface. Coatings and treatments improved nanohardness of silicon. Note that dry-oxidized and PECVD films are harder than wet-oxidized films as these films may be porous. The high hardness of oxidized films may be responsible for the high measured scratch/wear resistance.

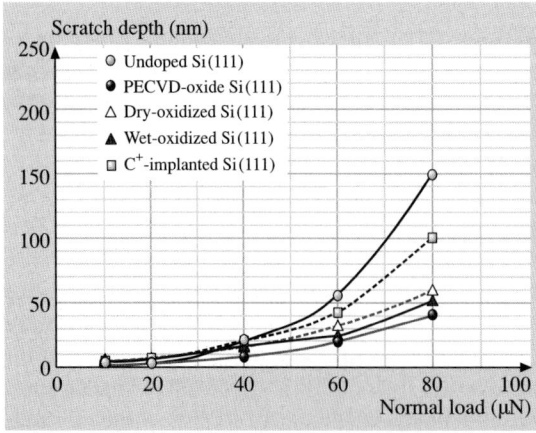

Fig. 22.17. Scratch depth as a function of normal load after 10 cycles for various silicon samples: virgin, treated, and coated [147]

Table 22.4. RMS, microfriction, microscratching/microwear and nanoindentation hardness data for various virgin, coated and treated silicon samples

Material	Rms roughness[a] (nm)	Coefficient of microscale friction[b]	Scratch depth[c] at 40µN (nm)	Wear depth[c] at 40µN (nm)	Nanohardness[c] at 100µN (GPa)
Si(111)	0.11	0.03	20	27	11.7
Si(110)	0.09	0.04	20		
Si(100)	0.12	0.03	25		
Polysilicon	1.07	0.04	18		
Polysilicon (lapped)	0.16	0.05	18	25	12.5
PECVD-oxide coated Si(111)	1.50	0.01	8	5	18.0
Dry-oxidized Si(111)	0.11	0.04	16	14	17.0
Wet-oxidized Si(111)	0.25	0.04	17	18	14.4
C^+-implanted Si(111)	0.33	0.02	20	23	18.6

[a] Scan size of 500 nm × 500 nm using AFM
[b] Versus Si_3N_4 tip in AFM/FFM, radius 50 nm; at 1 µm × 1 µm scan size
[c] Measured using an AFM with a diamond tip of radius 100 nm

22.2.2 Tribological Properties of Polysilicon Films and SiC Film

Studies have also been conducted on undoped polysilicon film, heavily doped (n^+-type) polysilicon film, heavily doped (p^+-type) single-crystal Si(100) and 3C–SiC (cubic or β-SiC) film [155–157]. The polysilicon films studied here are different from those discussed previously.

Table 22.5 presents a summary of the tribological studies conducted on polysilicon and SiC films. Values for single-crystal silicon are also shown for comparison. Polishing of the as-deposited polysilicon and SiC films drastically affect the roughness as the values reduce by two orders of magnitude. Si(100) appears to be the smoothest, followed by polished undoped polysilicon and SiC films, which have comparable roughness. The doped polysilicon film shows higher roughness than the undoped sample, which is attributed to the doping process. Polished SiC film shows the lowest friction followed by polished and undoped polysilicon film, which strongly supports the candidacy of SiC films for use in MEMS/NEMS devices. Macroscale friction measurements indicate that SiC film exhibits one of the lowest friction values as compared to the other samples. Doped polysilicon sample shows

Table 22.5. Summary of micro/nanotribological properties of the sample materials

Sample	Rms roughness[a] (nm)	P-V distance[a] (nm)	Coefficient of friction Micro[b]	Coefficient of friction Macro[c]	Scratch depth[d] (nm)
Undoped Si(100)	0.09	0.9	0.06	0.33	89
Undoped polysilicon film (as deposited)	46	340	0.05		
Undoped polysilicon film (polished)	0.86	6	0.04	0.46	99
n^+-type polysilicon film (as deposited)	12	91	0.07		
n^+-type polysilicon film (polished)	1.0	7	0.02	0.23	61
SiC film (as deposited)	25	150	0.03		
SiC film (polished)	0.89	6	0.02	0.20	6

Sample	Wear depth[e] (nm)	Nano-hardness[f] (GPa)	Young's modulus[f] (GPa)	Fracture toughness[g] K_{IC} (MPa m$^{1/2}$)
Undoped Si(100)	84	12	168	0.75
Undoped polysilicon film (as deposited)				
Undoped polysilicon film (polished)	140	12	175	1.11
n^+-type polysilicon film (as deposited)				
n^+-type polysilicon film (polished)	51	9	95	0.89
SiC film (as deposited)				
SiC film (polished)	16	25	395	0.78

[a] Measured using AFM over a scan size of 10 μm × 10 μm
[b] Measured using AFM/FFM over a scan size of 10 μm × 10 μm
[c] Obtained using a 3-mm-diameter sapphire ball in a reciprocating mode at a normal load of 10 mN and average sliding speed of 1 mm/s after 4 m sliding distance
[d] Measured using AFM at a normal load of 40 μN for 10 cycles; scan length of 5 μm
[e] Measured using AFM at normal load of 40 μN for 1 cycle, wear area of 2 μm × 2 μm
[f] Measured using nanoindenter at a peak indentation depth of 20 nm
[g] Measured using microindenter with Vickers indenter at a normal load of 0.5 N

low friction on the macroscale as compared to the undoped polysilicon sample, possibly due to the doping effect.

Figure 22.18a shows a plot of scratch depth versus normal load for various samples [155, 156]. Scratch depth increases with increasing normal load. Fig. 22.19 shows AFM three-dimensional (3D) maps and averaged two-dimensional (2D) profiles of the scratch marks on the various samples. It is observed that scratch depth increases almost linearly with the normal load. Si(100) and the doped and undoped polysilicon film show similar scratch resistance. From the data, it is clear that the SiC film is much more scratch-resistant than the other samples. Figure 22.18b shows

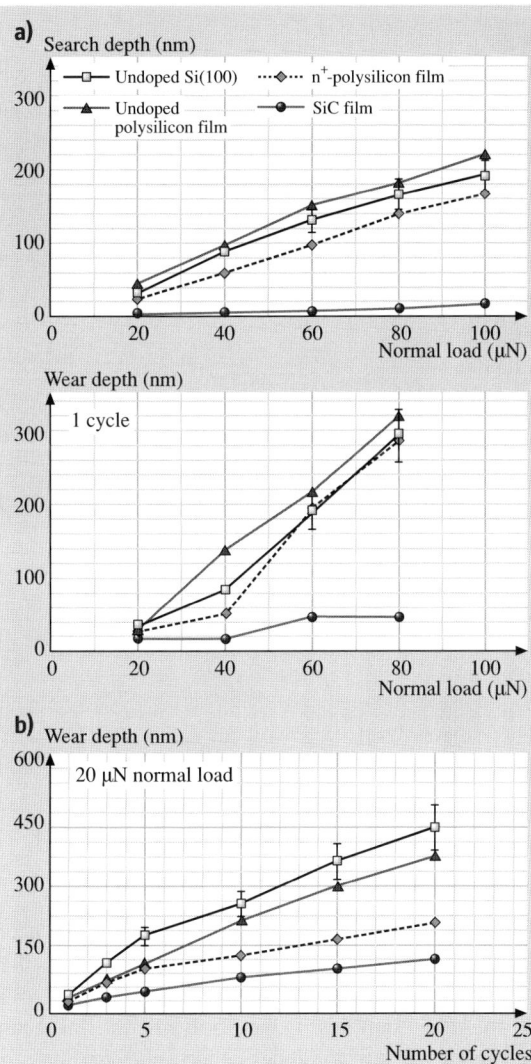

Fig. 22.18. (a) Scratch depths for 10 cycles as a function of normal load and (b) wear depths as a function of normal load and of number of cycles for various samples [155]

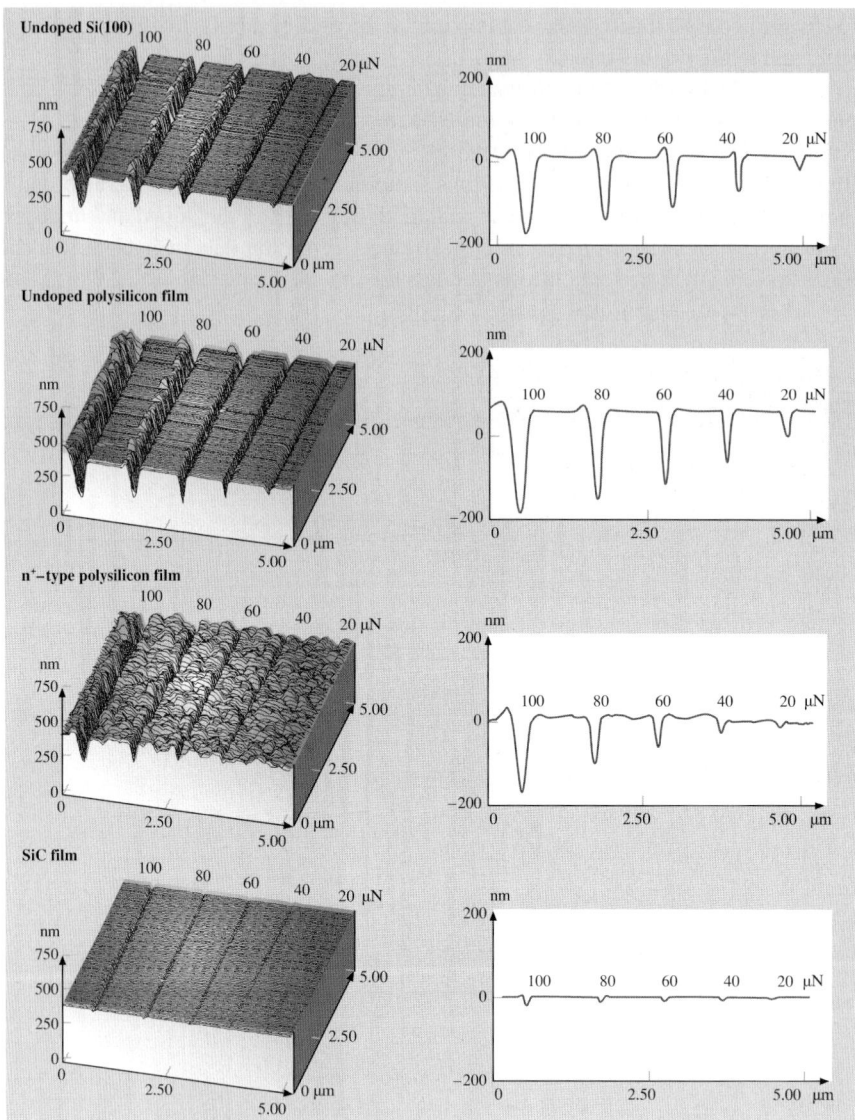

Fig. 22.19. AFM 3D maps and averaged 2D profiles of scratch marks on various samples [155]

results from microscale wear tests on the various films. For all the materials, wear depth increases almost linearly with increasing number of cycles. This suggests that the material is removed layer by layer in all the materials. Here also, SiC film exhibits lower wear depths than the other samples. Doped polysilicon film wears less than the undoped film. The higher fracture toughness and higher hardness of

SiC compared to Si(100) is responsible for its lower wear. Also the higher thermal conductivity of SiC (see Table 22.2 compared to the other materials leads to lower interface temperatures, which generally results in less degradation of the surface [34, 46, 81]). Doping of the polysilicon does not affect the scratch/wear resistance and hardness much. The measurements made on the doped sample are affected by the presence of grain boundaries. These studies indicate that SiC film exhibits desirable tribological properties for use in MEMS devices.

22.3 Lubrication Studies for MEMS/NEMS

Several studies of liquid perfluoropolyether (PFPE) lubricant films, self-assembled monolayers (SAMs), and hard diamond-like carbon (DLC) coatings have been carried out for the purpose of minimizing adhesion, friction, and wear [46, 47, 82, 102, 122–125, 154, 158–166]. Many variations of these films are hydrophobic (low surface tension and high contact angle) and have low shear strength, which provide low adhesion, friction, and wear. Relevant details are presented here.

22.3.1 Perfluoropolyether Lubricants

The classical approach to lubrication uses freely supported multimolecular layers of liquid lubricants [46, 47, 79, 81]. The liquid lubricants are sometimes chemically bonded to improve their wear resistance. Partially chemically bonded, molecularly thick perfluoropolyether (PFPE) lubricants are widely used for lubrication of magnetic storage media because of their thermal stability and extremely low vapor pressure [34], and are found to be suitable for MEMS/NEMS devices.

Adhesion, friction, and durability experiments have been performed on virgin Si (100) surfaces and silicon surfaces lubricated with two commonly used PFPE lubricants – Z-15 (with –CF_3 nonpolar end groups) and Z-DOL (with –OH polar end groups) [46, 47, 158, 160, 161, 165]. Z-DOL film was thermally bonded at 150 °C for 30 minutes and the unbonded fraction was removed by a solvent (bonded washed, BW) [34]. The thicknesses of the Z-15 and Z-DOL (BW) films were 2.8 nm and 2.3 nm, respectively. Nanoscale measurements were made using an AFM. The adhesive forces of Si(100), Z-15 and Z-DOL (BW) measured by plots of force calibration and friction force versus normal load are summarized in Fig. 22.20. The results measured by these two methods are in good agreement. Figure 22.20 shows that the presence of mobile Z-15 lubricant film increases the adhesive force compared to that of Si(100) by meniscus formation [79, 81, 167]. In contrast, the presence of solid phase Z-DOL (BW) film reduces the adhesive force as compared that of Si(100) because of the absence of mobile liquid. The schematic (bottom) in Fig. 22.20 shows the relative size and sources of meniscus. It is well known that the native oxide layer (SiO_2) on the top of Si(100) wafers exhibits hydrophilic properties, and some water molecules can be adsorbed on this surface. The condensed water will form a meniscus as the tip approaches the sample surface. The larger adhesive force in Z-15 is not only caused by the Z-15 meniscus, the nonpolarized Z-15 liquid does

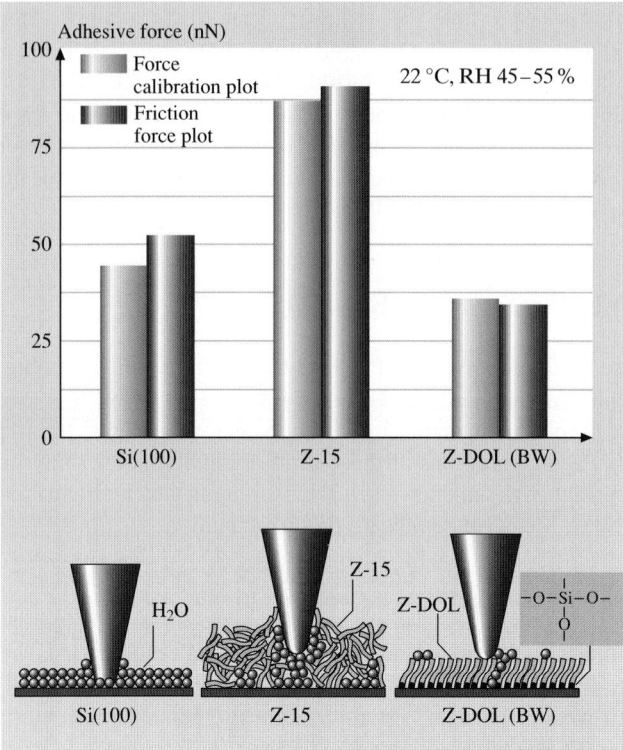

Fig. 22.20. Summary of the adhesive forces of Si(100) and Z-15 and Z-DOL (BW) films measured by plots of force calibration and friction force versus normal load in ambient air. The schematic (*bottom*) showing the effect of meniscus, formed between AFM tip and the surface sample, on the adhesive and friction forces [158]

not have good wettability and strong bonding with Si(100). Consequently, in the ambient environment, the condensed water molecules from the environment permeate through the liquid Z-15 lubricant film and compete with the lubricant molecules present on the substrate. The interaction of the liquid lubricant with the substrate is weakened, and a boundary layer of the liquid lubricant forms puddles [160, 161]. This dewetting allows water molecules to be adsorbed on the Si(100) surface as aggregates along with Z-15 molecules. And both of them can form a meniscus while the tip approaches the surface. Thus, the dewetting of the liquid Z-15 film results in a higher adhesive force and poorer lubrication performance. In addition, as the Z-15 film is fairly soft compared to the solid Si(100) surface, penetration of the tip in the film occurs while pushing the tip down. This leads to the large area of the tip involved to form the meniscus at the tip–liquid (mixture of Z-15 and water) interface. It should also be noted that Z-15 has a higher viscosity than water, therefore Z-15 film provides higher resistance to lateral motion and coefficient of friction. In the case of Z-DOL (BW) film, the active groups of Z-DOL molecules are mostly bonded on Si(100) substrate, thus the Z-DOL (BW) film has low free surface energy and cannot be displaced readily by water molecules or readily adsorb water molecules. Thus, the use of Z-DOL (BW) can reduce the adhesive force.

To study the effect of relative humidity on friction and adhesion, the variation of friction force, adhesive force, and coefficient of friction of Si(100), Z-15, and Z-DOL (BW) as a function of relative humidity are shown in Fig. 22.21. This shows that, for Si(100) and Z-15 film, the friction force increases with a relative humidity increase up to 45%, and then shows a slight decrease with further increases in

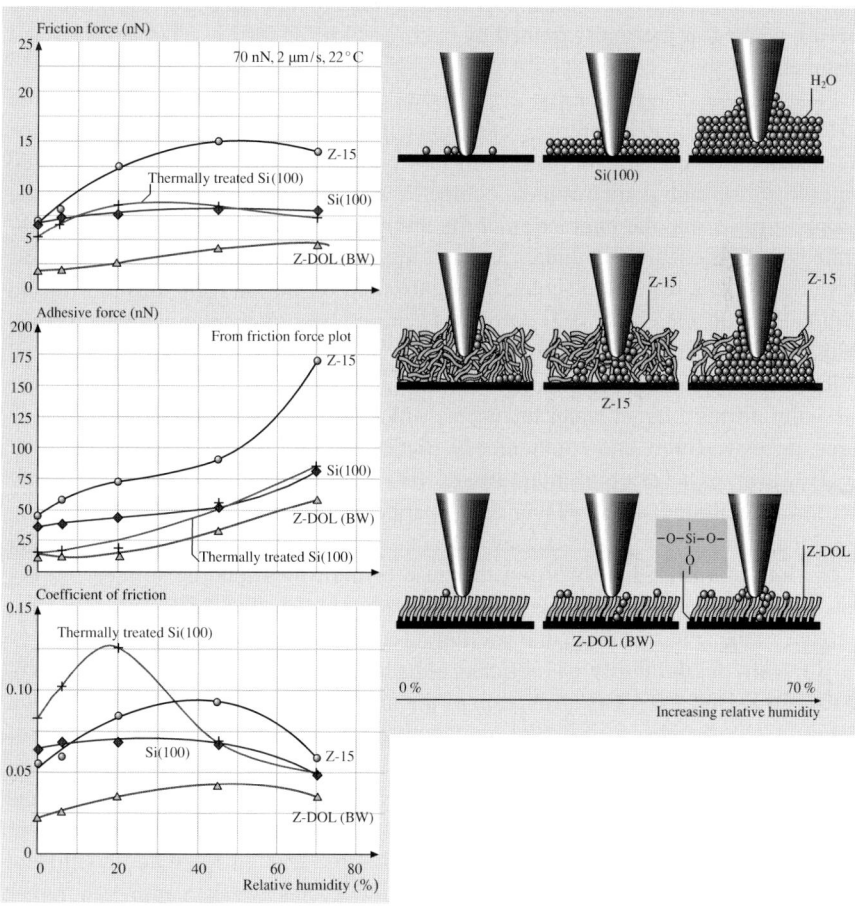

Fig. 22.21. The influence of relative humidity of the friction force, adhesive force, and coefficient of friction of Si(100) and Z-15 and Z-DOL (BW) films at 70 nN, 2 μm/s, and in 22 °C air. Schematic (*right*) shows the change of meniscus while increasing the relative humidity. In this figure, the thermal treated Si(100) represents the Si(100) wafer that was baked at 150 °C for 1 h in an oven (in order to remove the adsorbed water) just before it was placed in the 0% RH chamber [158]

the relative humidity. Z-DOL (BW) has a smaller friction force than Si(100) and Z-15 over the whole testing range, and its friction force shows a relative apparent increase when the relative humidity is higher than 45%. For Si(100), Z-15 and Z-DOL (BW), their adhesive forces increase with relative humidity. And their coefficients of friction increase with a relative humidity up to 45%, after which they decrease with further increasing of the relative humidity. It is also observed that the effect of humidity on Si(100) really depends on the history of the Si(100) sample. As the surface of Si(100) wafer readily adsorb water in air, without any pretreatment the Si(100) used in our study almost reaches its saturated stage of adsorbed water, and shows less effect during increasing relative humidity. However, once the Si(100) wafer was thermally treated by baking at 150 °C for 1 h, a larger effect was observed.

The schematic (right) in Fig. 22.21 shows that Si(100), because of its high free surface energy, can adsorb more water molecules with increasing relative humidity. As discussed earlier, for the Z-15 film in the humid environment, the condensed water from the humid environment competes with the lubricant film present on the sample surface, and the interaction of the liquid lubricant film with the silicon substrate is weakened and a boundary layer of the liquid lubricant forms puddles. This dewetting allows water molecules to be adsorbed on the Si(100) substrate mixed with Z-15 molecules [160, 161]. Obviously, more water molecules can be adsorbed on the Z-15 surface with increasing relative humidity. The larger amount of adsorbed water in the case of Si(100), along with the lubricant molecules in the case of the Z-15 film, forms a larger water meniscus, which leads to an increase of the friction force, adhesive force, and coefficient of friction of Si(100) and Z-15 with humidity. However, at a very high humidity of 70%, large quantities of adsorbed water can form a continuous water layer that separates the tip and sample surface, acting as a kind of lubricant, which causes a decrease in the friction force and coefficient of friction. For Z-DOL (BW) film, because of their hydrophobic surface properties, water molecules can be adsorbed at humidity higher than 45%, and causes an increase in the adhesive force and friction force.

To study the durability of lubricant films at nanoscale, the friction of Si(100), Z-15, and Z-DOL (BW) as a function of the number of scanning cycles are shown in Fig. 22.22. As observed earlier, the friction force and coefficient of friction of Z-15 are higher than that of Si(100) with the lowest values for Z-DOL(BW). During cycling, the friction force and coefficient of friction of Si(100) show a slight decrease during the initial few cycles then remain constant. This is related to the removal of the top adsorbed layer. In the case of Z-15 film, the friction force and coefficient of friction show an increase during the initial few cycles and then approach higher stable values. This is believed to be caused by the attachment of the Z-15 molecules onto the tip. The molecular interaction between these attached molecules to the tip and molecules on the film surface is responsible for an increase in the friction. But after several scans, this molecular interaction reaches equilibrium and after that the friction force and coefficient of friction remain constant. In the case of Z-DOL (BW) film, the friction force and coefficient of friction start out low and remain low

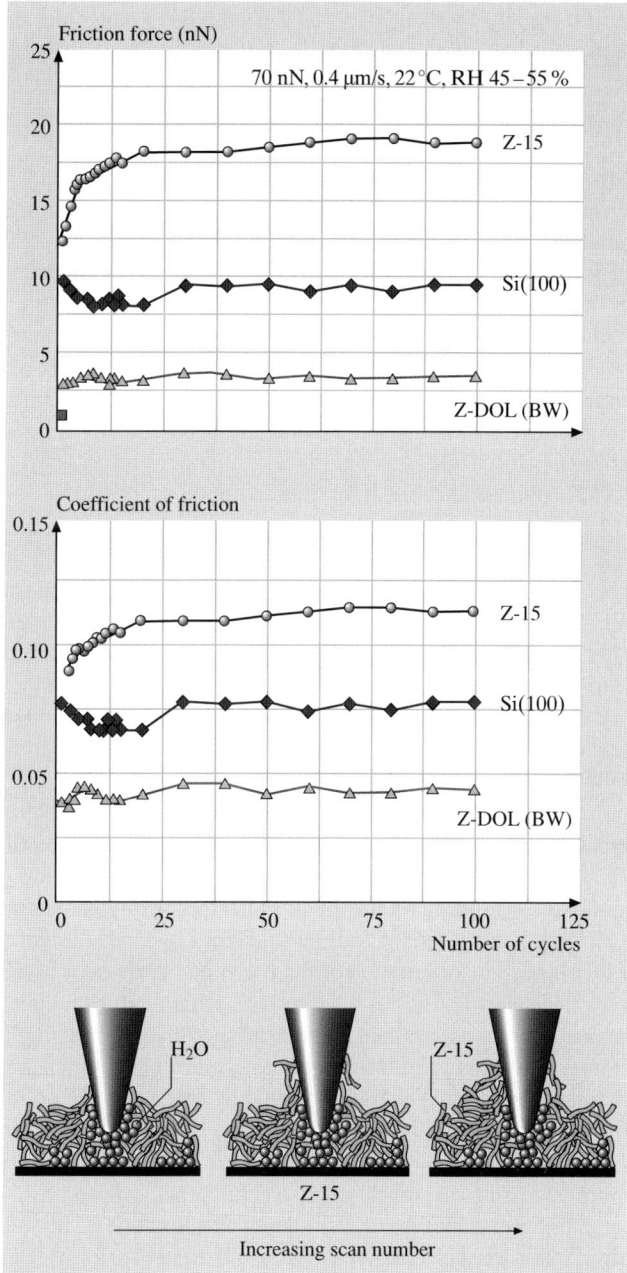

Fig. 22.22. Friction force and coefficient of friction versus number of sliding cycles for Si(100) and Z-15 and Z-DOL (BW) films at 70 nN, 0.4 µm/s, and in ambient air. Schematic (*bottom*) shows that some liquid Z-15 molecules can be attached onto the tip. The molecular interaction between the attached molecules onto the tip with the Z-15 molecules in the film results in an increase of the friction force with multiple scanning [158]

during the entire test for 100 cycles. This suggests that Z-DOL (BW) molecules do not become attached or displaced as readily as Z-15.

22.3.2 Self-Assembled Monolayers (SAMs)

For lubrication of MEMS/NEMS, another effective approach involves the deposition of organized and dense molecular layers of long-chain molecules. Two common methods to produce monolayers and thin films are Langmuir–Blodgett (LB) deposition and self-assembled monolayers (SAMs) by chemical grafting of molecules. LB films are physically bonded to the substrate by weak van der Waals attraction, while SAMs are chemically bonded via covalent bonds to the substrate. Because of the choice of chain length and terminal linking group that SAMs offer, they hold great promise for boundary lubrication of MEMS/NEMS. A number of studies have been conducted to study the tribological properties of various SAMs [122–125, 159, 162–164, 166, 168].

Bhushan and *Liu* [162] studied the effect of film compliance on adhesion and friction. They used hexadecane thiol (HDT), 1,1,biphenyl-4-thiol (BPT), and crosslinked BPT (BPTC) solvent deposited on an Au(111) substrate, Fig. 22.23a. The average values and standard duration of the adhesive force and coefficient of friction are presented in Fig. 22.23b. Based on the data, the adhesive force and coefficient of friction of SAMs are lower than corresponding substrates. Among various films, HDT exhibits the lowest values. Based on stiffness measurements of various SAMs, HDT was the most compliant, followed by BPT and BPTC. Based on friction and stiffness measurements, SAMs with high-compliance long carbon chains exhibit low friction; chain compliance is desirable for low friction. Friction mechanism of SAMs is explained by a so-called *molecular spring* model Fig. 22.24. According to this model, the chemically adsorbed self-assembled molecules on a substrate are just like assembled molecular springs anchored to the substrate. An asperity sliding on the surface of SAMs is like a tip sliding on the top of molecular springs or brush. The molecular spring assembly has compliant features and can experience orientation and compression under load. The orientation of the molecular springs or brush under normal load reduces the shearing force at the interface, which in turn reduces the friction force. The orientation is determined by the spring constant of a single molecule as well as the interaction between the neighboring molecules, which can be reflected by packing density or packing energy. It should be noted that the orientation can lead to conformational defects along the molecular chains, which lead to energy dissipation.

SAMs with high-compliance long carbon chains also exhibit the best wear resistance [162, 163]. In wear experiments, curves of wear depth as a function of normal load show a critical normal load. A representative curve is shown in Fig. 22.25. Below the critical normal load, SAMs undergo orientation, at the critical load SAMs wear away from the substrate due to weak interface bond strengths, while above the critical normal load severe wear takes place on the substrate.

Bhushan et al. [122], *Kasai* et al. [124], and *Tambe* and *Bhushan* [164] studied perfluorodecyltricholorosilane (PFTS), n-octyldimethyl (dimethylamino) silane

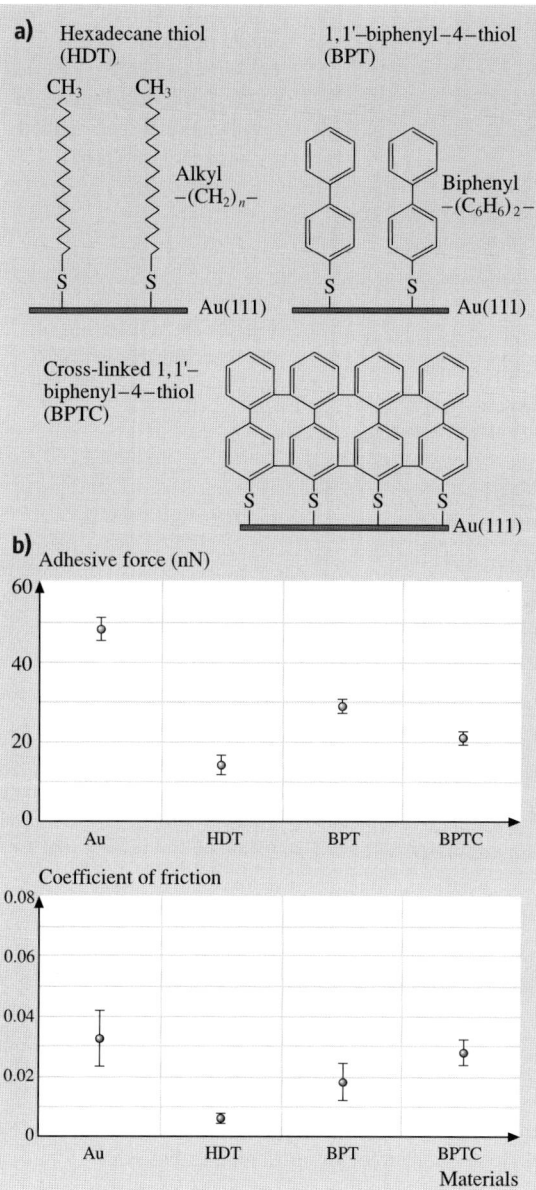

Fig. 22.23. (a) Schematics of structures of hexadecane thiol and biphenyl thiol SAMs on Au(111) substrates, and (b) adhesive force and coefficient of friction of Au(111) substrate and various SAMs

(ODMS) ($n = 7$), and n-octadecylmethyl (dimethylamino) silane (ODDMS) ($n = 17$) vapor-phase-deposited on a Si substrate, and octylphosphonate (OP) and octadecylphosphonate (ODP) on an Al substrate, Fig. 22.26a. Figure 22.26b presents the contact angle, adhesive force, friction force, and coefficient of friction of the two substrates and with various SAMs. Based on the data, PFTS/Si exhibits a higher

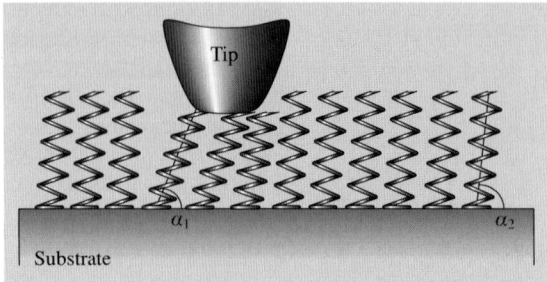

Fig. 22.24. Molecular spring model of SAMs. In this figure, $\alpha_1 < \alpha_2$, which is caused by the further orientation under the normal load applied by an asperity tip [162]

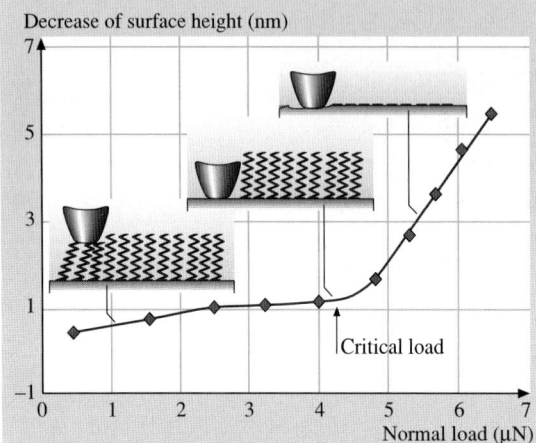

Fig. 22.25. Illustration of the wear mechanism of SAMs with increasing normal load [163]

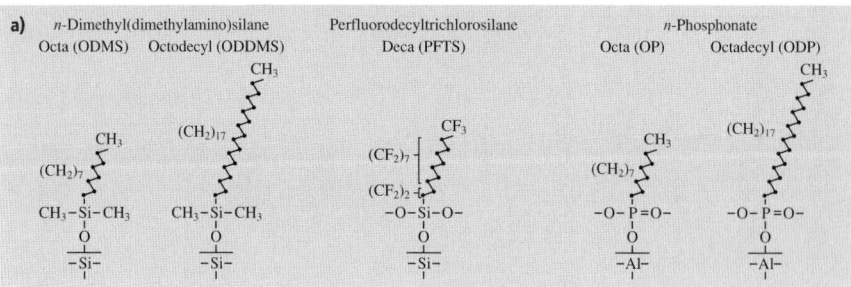

Fig. 22.26. (a) Schematics of structures of perfluoroalkylsilane and alkylsilane SAMs on Si with native oxide substrates, and alkylphosphonate SAMs on Al with native oxide

contact angle and lower adhesive force compared to of ODMS/Si and ODDMS/Si. The data for ODMS and ODDMS on the Si substrate are comparable to those for OP and ODP on the Al substrate. Thus the substrate had little effect. The coefficient of friction of various SAMs were comparable.

For wear performance studies, experiments were conducted on various films. Figure 22.27a shows the relationship between the decrease of surface height and the

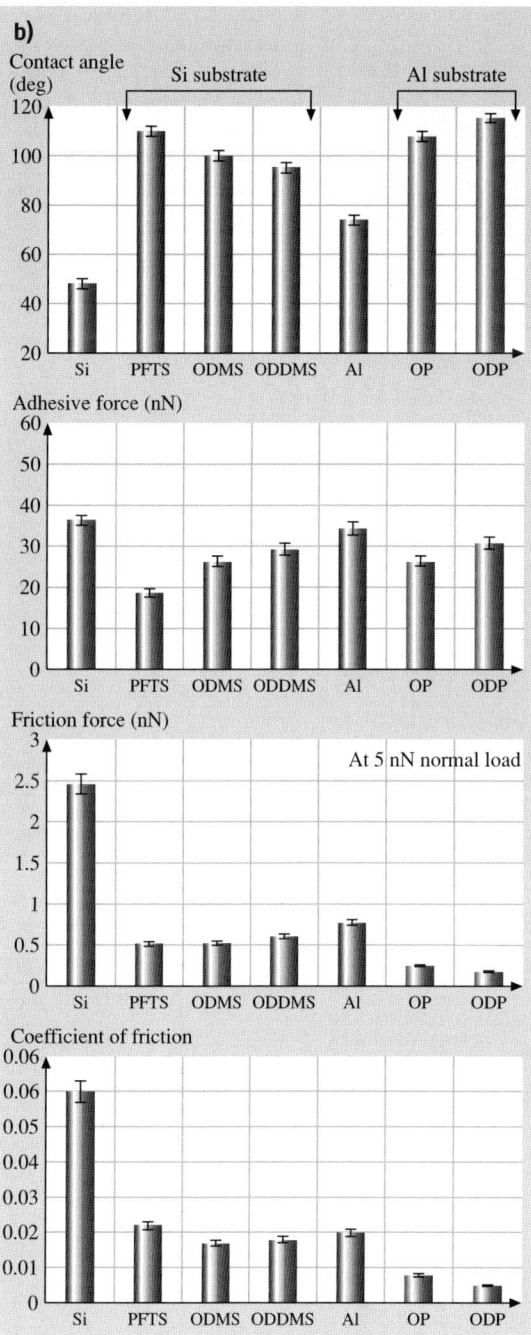

Fig. 22.26. (continued) (**b**) contact angle, adhesive force, friction force, and coefficient of friction of Si with native oxide and Al with native oxide substrates and with various SAMs

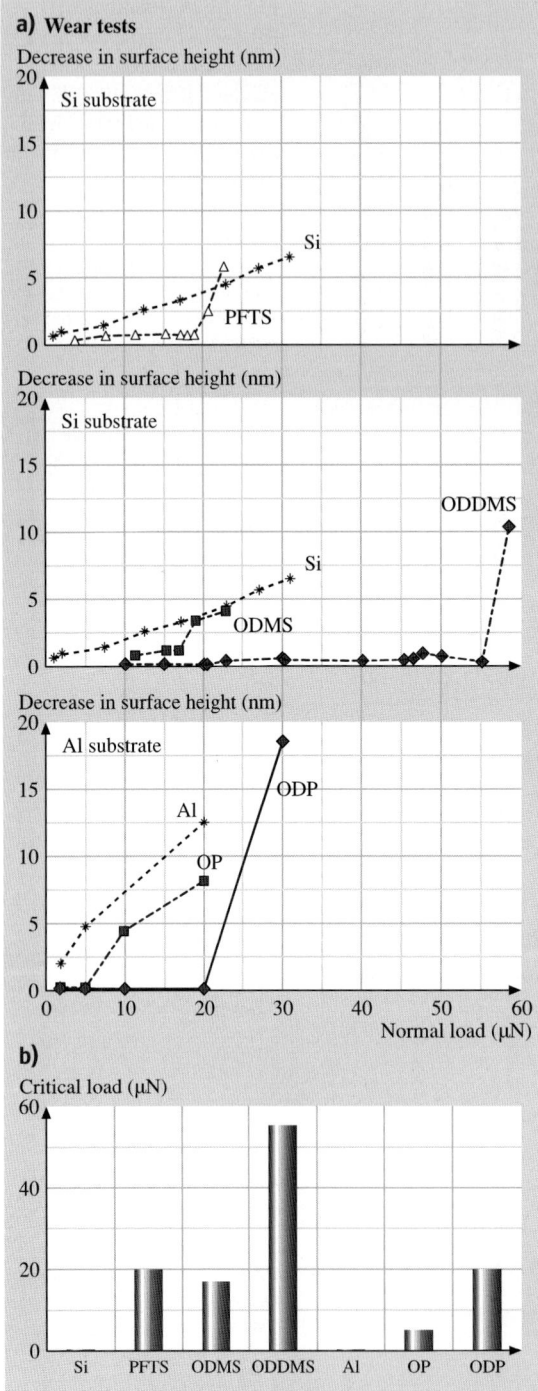

Fig. 22.27. Decrease of surface height as a function of normal load after one scan cycle for various SAMs on Si and Al substrates, and comparison of critical loads for failure during wear tests for various SAMs

normal load for various SAMs and corresponding substrates [124, 164]. As shown in the figure, the SAMs exhibit a critical normal load beyond which the surface height drastically decreases. Unlike SAMs, the substrates show a monotonic decrease in surface height with increasing normal load, with wear initiating from the very beginning, i. e., even for low normal loads. The critical loads corresponding to the sudden failure are shown in Fig. 22.27b. Amongst all the SAMs, ODDMS and ODP show the best performance in the wear tests. Out of the two alkyl SAMs, ODDMS/Si and ODP/Al showed better wear resistance than ODMS/Si and OD/Al due to the effect of chain length. Wear behavior of the SAMs is reported to be mostly determined by the molecule–substrate bond strengths.

Bhushan et al. [122] and *Lee* et al. [125] studied various fluoropolymer multilayers and fluorosilane monolayers on Si and a selected fluorosilane on PDMS surfaces. For nanoscale devices such as in nanochannels, monolayers are preferred. They reported that all fluorosilane films increased the contact angle. The fluorosilane monolayer 1H, 1H, 2H, 2H–perfluorodecyltriethoxysilane (PFDTES) resulted in a contact angle of about 100°.

Based on these studies, a perfluoro SAM with a compliant layer should have optimized tribological performance for MEMS/NEMS and BioMEMS/BioNEMS applications.

22.3.3 Hard Diamond-Like Carbon (DLC) Coatings

Hard amorphous carbon (a-C), commonly known as DLC (implying high hardness) coatings are deposited by a variety of deposition techniques including filtered cathodic arc (FCA), ion beam, electron cyclotron resonance chemical vapor deposition (ECR-CVD), plasma-enhanced chemical vapor deposition (PECVD), and sputtering [136, 154]. These coatings are used in a wide range on applications including tribological, optical, electronic, and biomedical applications. Ultrathin coatings (3–10 nm thick) are employed to protect against wear and corrosion in magnetic storage applications – thin-film rigid disks, metal-evaporated tapes, and thin-film read/write heads –, Gillette Mach 3 razor blades, glass windows, and sunglasses. The coatings exhibit low friction, high hardness and wear resistance, chemical inertness to both acids and alkalis, lack of magnetic response, and optical band gaps ranging from zero to a few eV, depending upon the deposition technique and its conditions. Selected data on DLC coatings relevant for MEMS/NEMS applications is presented in the following section on adhesion measurements.

22.4 Tribological Studies of Biological Molecules on Silicon-Based Surfaces and of Coated Polymer Surfaces

22.4.1 Adhesion, Friction, and Wear of Biomolecules on Si-Based Surfaces

Proteins on silicon-based surfaces are of extreme importance in various applications including silicon microimplants, various bioMEMS such as biosensors, and therapeutics. Silicon is a commonly used substrate in microimplants, but it can have undesired interactions with the human immune system. Therefore, to mimic a biological surface, protein coatings are used on silicon-based surfaces as a passivation layer, so that these implants are compatible with the body and avoid rejection. Whether this surface treatment is applied to a large implant or a bioMEMS, the function of the protein passivation is obtained from the nanoscale 3D structural conformation of the protein. Proteins are also used in bioMEMS because of their function specificity. For biosensor applications, the extensive array of protein activities provides a rich supply of operations that may be performed at the nanoscale. Many antibodies (proteins) have an affinity to specific protein antigens. For example, pathogens (disease causing agents, e. g., virus or bacteria) trigger the production of antigens which can be detected when bound to a specific antibody on the biosensor. The specific binding behavior of proteins that has been applied to laboratory assays may also be redesigned for in vivo use as sensing elements of a bioMEMS. The epitope-specific binding properties of proteins to various antigens are useful in therapeutics. Adhesion between the protein and substrate affects the reliability of an application. Among other things, the morphology of the substrate affects the adhesion. Furthermore, for in vivo environments, the proteins on the biosensor surface should exhibit high wear resistance during direct contact with the tissue and circulatory blood flow without washing off.

Bhushan et al. [72] studied the step-by-step morphological changes and the adhesion of a model protein – streptavidin (STA) – on silicon-based surfaces. Figure 22.28a presents a flow chart showing the sequential modification of a silicon surface. In addition to physical adsorption, they also used nanopatterning and chemical linker methods to improve adhesion. Nanopatterned surfaces contain a large edge surface area, leading to high surface energy, which results in high adhesion. In the chemical linker method, sulfo-NHS-biotin was used as a cross linker because the bonds between the STA and the biotin molecule are some of the strongest noncovalent bonds known (Fig. 22.28b). It was connected to the silica surface through a silane linker, 3-aminopropyltriethoxysilane (3-APTES). In order to make a bond between the silane linker and the silica surface, the silica surface was hydroxylated. Bovine serum albumin (BSA) was used before STA in order to block nonspecific binding sites of the STA protein with silica surface. Figure 22.29 shows the step-by-step morphological changes in the silica surface during the deposition process using the chemical linker method. There is an increase in roughness of the silica surface boiled in de-ionized (DI) water compared to the bare silica surface. After the silanization process, there are many free silane links on the surface which caused higher roughness. Once biotin was coated on the silanized surface, the surface became smoother. Finally, after the deposition of STA, surface shows large and small clumps. Presumably, the large clumps represent BSA and the smaller ones represent STA. To measure adhesion between STA and the corresponding substrates, an STA-coated tip (or functionalized tip) was used and all measurements were made in phosphate buffered saline (PBS) solution, a medium commonly used in protein analysis and to simulate body fluid. Figure 22.30 shows the adhesion values of var-

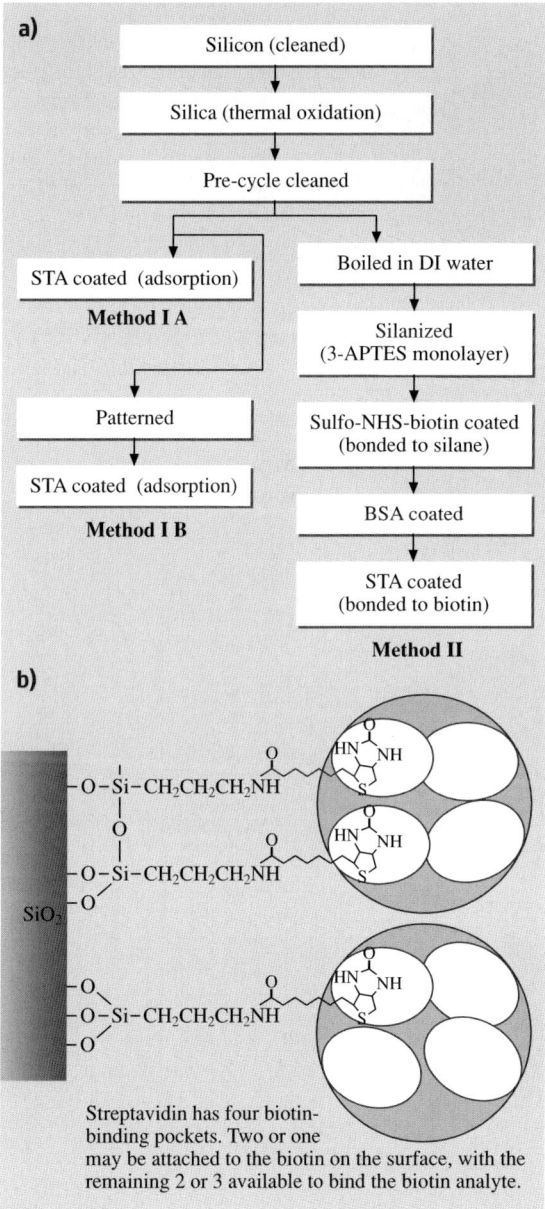

Fig. 22.28. (a) Flow chart showing the samples used and their preparation technique, and (b) a chemical structure showing streptavidin protein binding to the silica substrate by the chemical linker method

ious surfaces. The adhesion value between biotin and STA was higher than that for other samples, which is expected. Edges of patterned silica also exhibited high adhesion values. It appears that both nanopatterned surfaces and chemical linker method increase adhesion with STA.

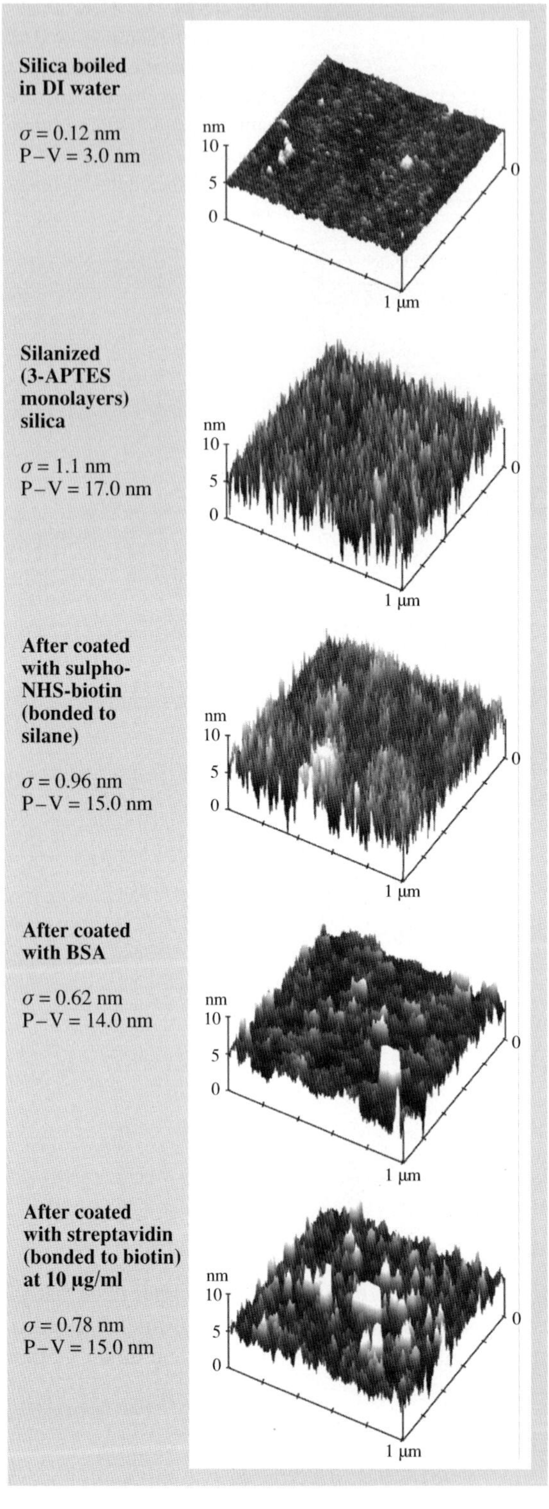
Silica boiled in DI water
σ = 0.12 nm
P–V = 3.0 nm
Silanized (3-APTES monolayers) silica
σ = 1.1 nm
P–V = 17.0 nm
After coated with sulpho-NHS-biotin (bonded to silane)
σ = 0.96 nm
P–V = 15.0 nm
After coated with BSA
σ = 0.62 nm
P–V = 14.0 nm
After coated with streptavidin (bonded to biotin) at 10 μg/ml
σ = 0.78 nm
P–V = 15.0 nm
nm
10
5
0
1 μm

Fig. 22.29. Morphological changes in silica surface during functionalization of silica surface by chemical linker imaged in PBS. Streptavidin is covalently bonded at a concentration of 10 μg/ml [72]

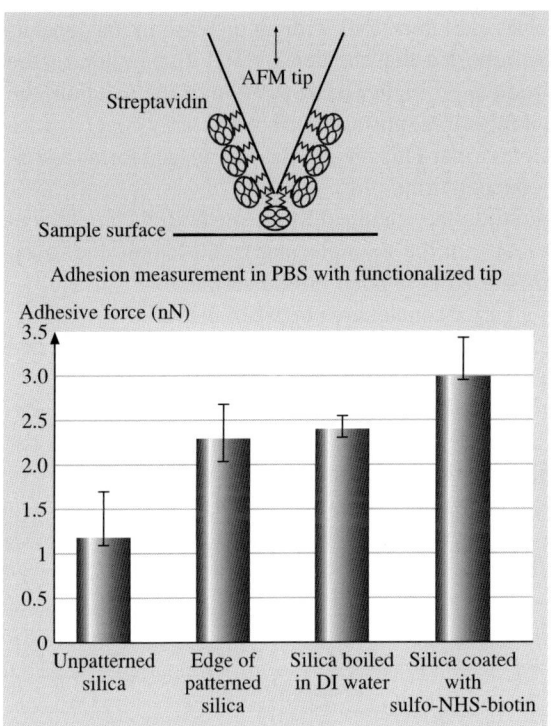

Fig. 22.30. Adhesion measurements of silica, patterned silicon, silica boiled in DI water, and sulfo-NHS-biotin using functionalized (with streptavidin) tips obtained from force–distance curves, captured in PBS

Tokachichu et al. [120] studied friction and wear of STA deposited by physical adsorption and the chemical linker method. Figure 22.31 shows the coefficient of friction between the Si_3N_4 tip and various samples. The coefficient of friction is less for STA-coated silica samples compared to uncoated sample. The streptavidin coating acts as a lubricant film. The coefficient of friction is found to be depen-

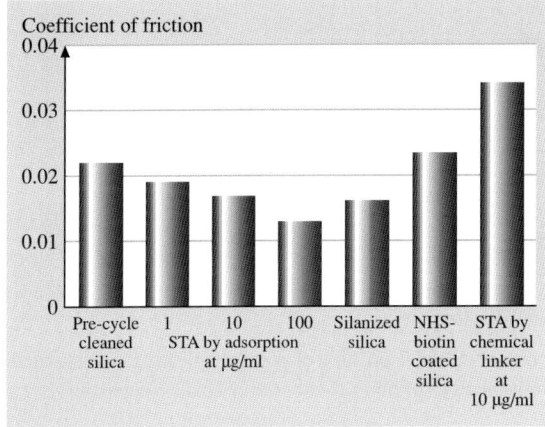

Fig. 22.31. Coefficient of friction for various surfaces with and without biomolecules

dent upon the concentration of STA, and decreases with an increase in the concentration. *Bhushan* et al. [72] have reported that the density and distribution of the biomolecules vary with concentration. At higher concentration of the solution, the coated layer is more uniform and the silica substrate surface is highly covered with biomolecules than at lower concentration. This means that the surface forms a continuous lubricant film at higher concentration.

In the case of samples prepared by the chemical linker method, the coefficient of friction increases with an increase in the biomolecular chain length due to increased compliance. When normal load is applied on the surface, the surface becomes compressed, resulting in a larger contact area between the AFM tip and the biomolecules. Besides that, the size of STA is much larger than that of APTES and biotin. This results in a tightly packed surface with the biomolecules, which results in very little lateral deflection of the linker in the case of STA-coated biotin. Due to this high contact area and low lateral deflection the friction force increases for the same applied normal load compared to directly adsorbed surface. These tests reveal that surfaces coated with biomolecules reduce the friction, but if the biomolecular coating of the surface is too thick or the surface has some cushioning effect, as seen in the chemical linker method, that increases the coefficient of friction.

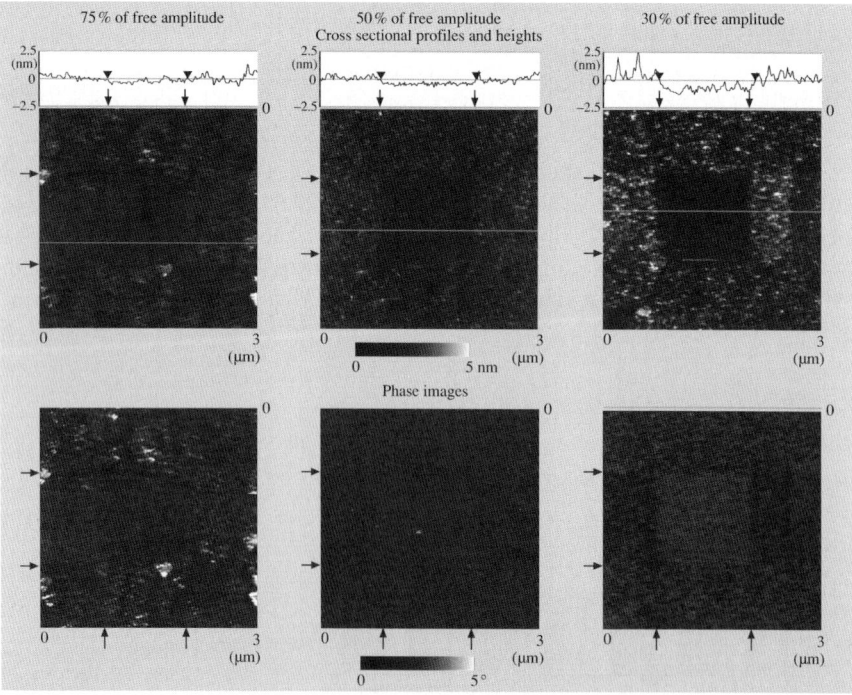

Fig. 22.32. Wear mark images and cross-sectional profiles of precycle cleaned silica coated with streptavidin at 10 µg/ml by physical adsorption at three normal loads (increasing from *left to right*)

Figure 22.32 shows the wear maps of STA deposited by physical adsorption at three normal loads. The wear depth increases with increasing normal load. An increase in normal load causes partial damage to the folding structure of the streptavidin molecules. It is unlikely that the chemical (covalent) bonds within the streptavidin molecule are broken; instead, the folding structure is damaged leading to wear mark. When the load is high, i. e., 30% of the free amplitude ($\approx$ 8 nN), the molecules may have been removed by the AFM tip due to the effect of indentation/ Because of this, there is a significant increase in the wear depth from 50% of the free amplitude ($\approx$ 6 nN) to 30% of the free amplitude ($\approx$ 8 nN). The data show that biomolecules will be damaged during sliding.

22.4.2 Adhesion of Coated Polymer Surfaces

As mentioned in Sect. 22.A, PMMA, PDMS, and other polymers are used in the construction of micro/nanofluidic-based biodevices. Adhesion between the moving parts needs to be minimized. Furthermore, if the adhesion between the microchannel surface and the biofluid is high, the biomolecules will stick to the microchannel surface and restrict flow. In order to facilitate flow, surfaces with low bioadhesion are required.

Tambe and *Bhushan* [169, 170] and *Bhushan* and *Burton* [171] have reported adhesive force data for PMMA and PDMS against an AFM Si_3N_4 tip and a silicon ball. *Tokachichu* and *Bhushan* [172] measured contact angle and adhesion of bare PMMA and PDMS and coated with a perfluoro SAM of perfluorodecyltriethoxysilane (PFDTES). Oxygen plasma treatment was used for hydroxylation of the surface to enhance chemical bonding of the SAM to the polymer surface. They made measurements in ambient and in PBS and fetal bovine serum (FBS); the latter is a blood component. Figs. 22.33 and 22.34 show the contact angle and adhesion data. SAM-coated surfaces have a high contact angle Fig. 22.33, as expected. The adhesion value of PDMS in ambient is high because of electrostatic charge present on the surface. The adhesion values of PDMS are higher than PMMA because PDMS is softer

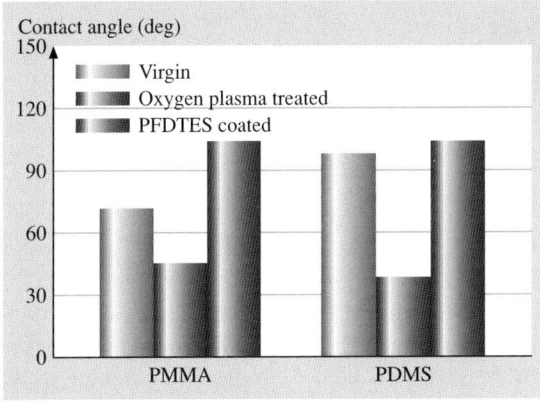

Fig. 22.33. Sessile drop contact-angle measurements of virgin, oxygen-plasma-treated and PFDTES-coated PMMA and PDMS surfaces. The maximum error in the data is ±2° [120]

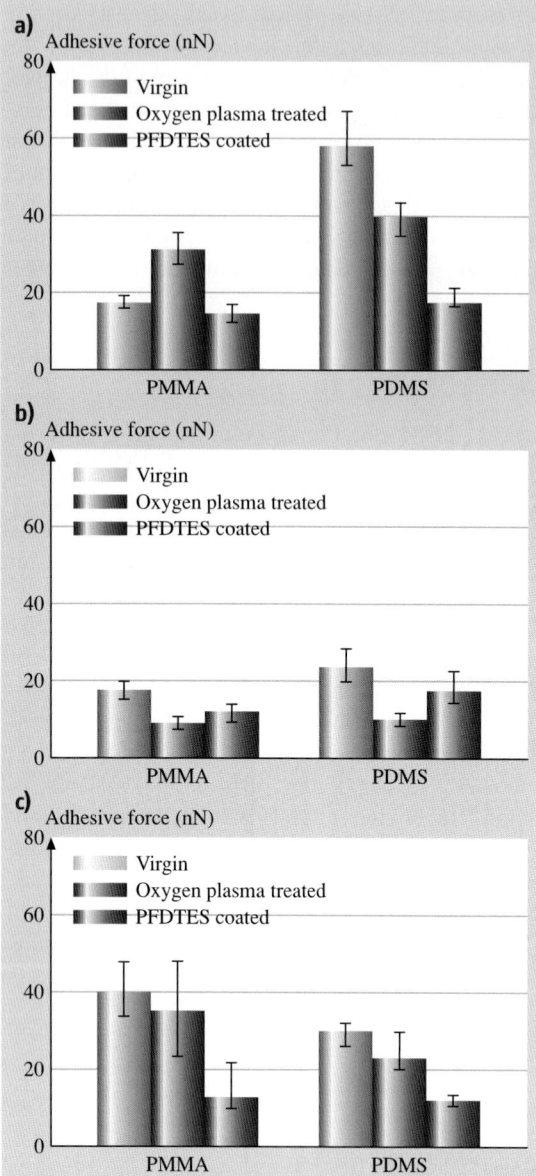

Fig. 22.34. Adhesion measurement of virgin, oxygen-plasma-treated and PFDTES-coated PMMA and PDMS surfaces with bare silicon nitride AFM tip (**a**) in ambient, and (**b**) in PBS environment, and (**c**) dip-coated tip with FBS in a PBS environment [120]

than PMMA (elastic modulus = 5 GPa and hardness = 410 MPa [121]) and results in a higher contact area between the PDMS surface and the AFM tip, and PMMA does not develop electrostatic charge. When SAM is coated on PMMA and PDMS surfaces, the adhesion values are similar, which shows that electrostatic charge on virgin PDMS plays no role when the surface is coated. In the PBS solution, there is a decrease in adhesion values because of the lack of a meniscus contribution. The adhesion values in the FBS-coated tip in PBS are generally lower than for the uncoated tip in PBS. In summary, the adhesion values of SAM-coated surfaces are lower than bare surfaces in various environments.

22.5 Nanopatterned Surfaces

22.5.1 Analytical Model and Roughness Optimization

One of the crucial surface properties for surfaces in wet environments is nonwetting or hydrophobicity. It is usually desirable to reduce wetting in fluid flow applications and some conventional applications, such as glass windows and automotive windshields, in order for liquid to flow away along their surfaces. Reduction of wetting is also important in reducing meniscus formation, consequently reducing stiction, friction, and wear. Wetting is characterized by the contact angle, which is the angle between the solid and liquid surfaces. If the liquid wets the surface (referred to as a wetting liquid or a hydrophilic surface), the value of the contact angle is $0 \le \theta \le 90°$, whereas if the liquid does not wet the surface (referred to as a nonwetting liquid or a hydrophobic surface), the value of the contact angle is $90° < \theta \le 180°$. A surface is considered superhydrophobic if θ is close to 180°. Superhydrophobic surfaces should also have very low water contact angle hysteresis. One of the ways to increase the hydrophobic or hydrophilic properties of the surface is to increase surface roughness. It has been demonstrated experimentally that roughness changes contact angle. Some natural surfaces, including leaves of water-repellent plants such as lotus, are known to be superhydrophobic due to their high roughness and the presence of a wax coating Fig. 22.35. This phenomenon is called in the literature the *lotus effect* [173].

If a droplet of liquid is placed on a smooth surface, the liquid and solid surfaces come together under equilibrium at a characteristic angle called the static contact angle θ_0; see Fig. 22.36. The contact angle can be determined from the condition of the total energy of the system being minimized. It can be shown that

$$\cos\theta_0 = \mathrm{d}A_{\mathrm{LA}}/\mathrm{d}A_{\mathrm{SL}} , \qquad (22.1)$$

where θ_0 is the contact angle for smooth surface, and A_{SL} and A_{LA} are the solid–liquid and liquid–air contact areas. Next, let us consider a rough solid surface with a typical size of roughness details smaller than the size of the droplet (on the order of a few hundred microns or larger), Fig. 22.36. For a rough surface, the roughness affects the contact angle due to the increased contact area A_{SL}. For a droplet in contact with a rough surface without air pockets, referred to as a homogeneous interface, based on the minimization of the total surface energy of the system, the contact angle is given as [174]

$$\cos\theta = \mathrm{d}A_{\mathrm{LA}}/\mathrm{d}A_{\mathrm{F}} = \left(\frac{A_{\mathrm{SL}}}{A_{\mathrm{F}}}\right)(\mathrm{d}A_{\mathrm{LA}}/\mathrm{d}A_{\mathrm{SL}}) = R_{\mathrm{f}}\cos\theta_0 , \qquad (22.2)$$

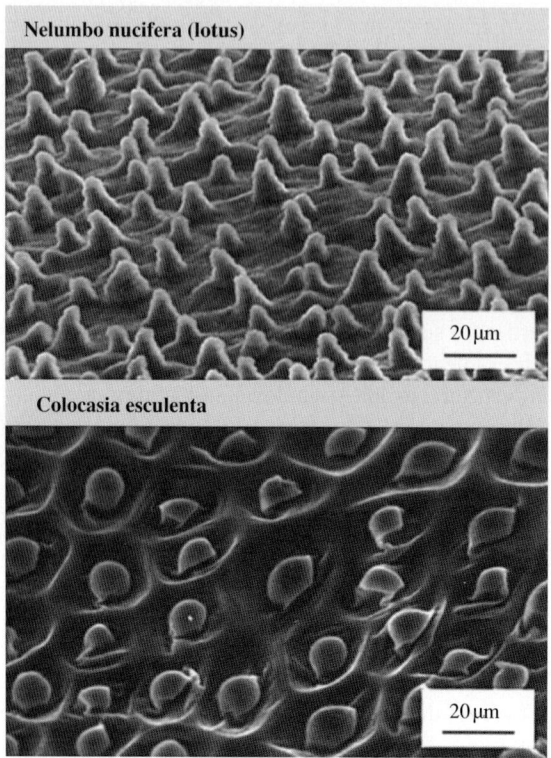

Fig. 22.35. SEM micrographs of two hydrophobic leaves, Nelumbo nucifera (lotus) and Colocasia esculenta

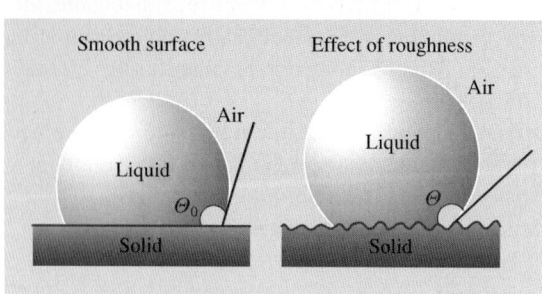

Fig. 22.36. Droplet of liquid in contact with a smooth solid surface (contact angle θ_0) and rough solid surface (contact angle θ) [173]

where A_F is the flat solid–liquid contact area (or a projection of the solid–liquid area A_{SL} onto the horizontal plane). R_f is a roughness factor defined as

$$R_f = \frac{A_{SL}}{A_F}. \tag{22.3}$$

Equation (22.3) shows that, if the liquid wets a surface ($\cos\theta_0 > 0$), it will also wet the rough surface with a contact angle $\theta < \theta_0$, and for nonwetting liquids ($\cos\theta_0 < 0$), the contact angle with a rough surface will be greater than that with the flat surface, $\theta < \theta_0$. The dependence of the contact angle on the roughness factor is resented

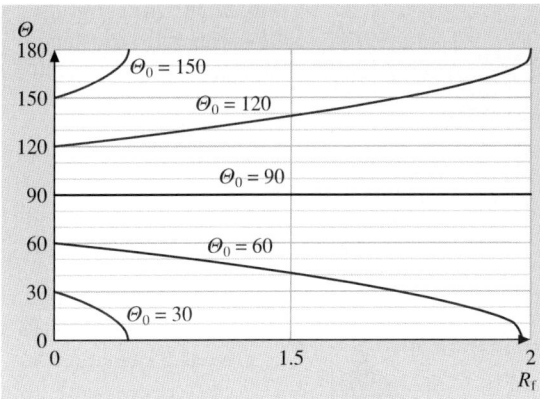

Fig. 22.37. Contact angle for rough surface (θ) as a function of the roughness factor (R_f) for various contact angles for a smooth surface (θ_0) [173]

in Fig. 22.37 for different values of θ_0, based on (22.2). It should be noted that (22.2) is valid only for moderate roughness, when $R_f \cos\theta_0 < 1$ [173].

For higher roughness, air pockets (composite solid–liquid–air interface) will be formed in the cavities of the surface [175]. In the case of partial contact, the contact angle is given by

$$\cos\theta = R_f f_{SL} \cos\theta_0 - f_{LA} ,\qquad(22.4)$$

where f_{SL} and f_{LA} are fractional solid–liquid and liquid–air contact areas. The homogeneous and composite interfaces are two metastable states of the system. In reality, some cavities will be filled with liquid, and others with air, and the value of the contact angle is between the values predicted by Eqs. (22.2) and (22.4). If the distance is large between the asperities or if the slope changes slowly, the liquid–air interface can easily be destabilized due to imperfectness of the profile shape or due to dynamic effects, such as surface waves Fig. 22.38. *Nosonovsky* and *Bhushan* [176] proposed a stochastic model, which relates the contact angle to roughness and takes into account the possibility of destabilization of the composite interface due to imperfectness of the shape of the liquid–air interface, caused by effects such as capillary waves.

In addition to the surface roughness, sharp edges of asperities may affect wetting, because they result in pinning of the solid–liquid–air contact line and resist liquid flow. *Nosonovsky* and *Bhushan* [173] considered the effect of the surface roughness and sharp edges and found the optimum roughness distribution for non-wetting. They formulated five requirements for roughness-induced superhydrophobic surfaces. First, asperities must have a high aspect ratio to provide a high surface area. Second, sharp edges should be avoided, to prevent pinning of the triple line. Third, asperities should be tightly packed to minimize the distance between them and avoid destabilization of the composite interface. Fourth, asperities should be small compared to typical droplet size (on the order of few hundred microns or larger). And fifth, in the case of hydrophilic surfaces, a hydrophobic film must be applied in order to have initial $\theta > 90°$. These recommendations can be utilized for

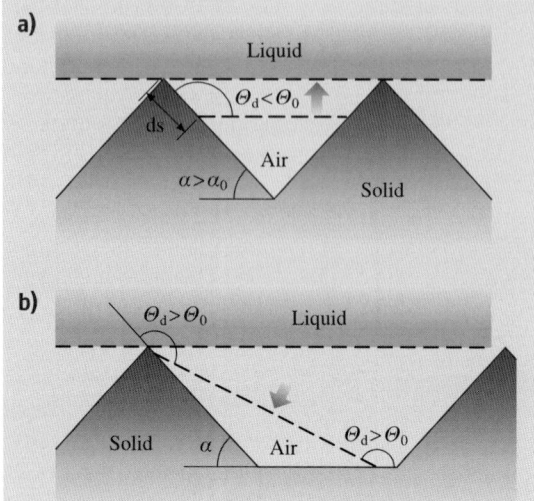

Fig. 22.38. (a) Formation of a composite solid–liquid–air interface for sawtooth and smooth profiles, and (b) destabilization of the composite interface for the sawtooth and smooth profiles due to dynamic effects. The dynamic contact angle $\theta_d > \theta_0$ corresponds to an advancing liquid–air interface, whereas $\theta_d < \theta_0$ corresponds to a receding interface [176]

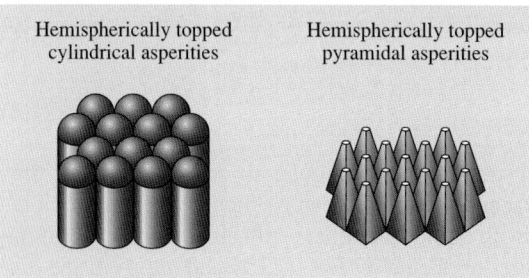

Fig. 22.39. Optimized roughness distribution – hemispherically topped cylindrical asperities and pyramidal asperities with square foundation and rounded tops. The square base gives a higher packing density but introduces undesirable sharp edges [173]

producing superhydrophobic surfaces. Remarkably, all these conditions are satisfied by biological water-repellent surfaces, such as some leaves: they have tightly packed hemispherically topped papillae with high (on the order of unity) aspect ratios and a wax coating [173]. Figure 22.39 shows two recommended geometries which use either hemispherically topped asperities with a hexagonal packing pattern or pyramidal asperities with a rounded top. These geometries can be used for producing superhydrophobic surfaces.

22.5.2 Experimental Validation

To validate the model of contact angle as a function of surface roughness, *Burton* and *Bhushan* [177] measured contact angles and adhesive forces on hydrophilic and hydrophobic polymer films with smooth surfaces and with discrete asperities. PMMA was chosen because it is a polymer often used in BioMEMS/BioNEMS devices. Three types of surface structures were measured: film, low-aspect-ratio asperities (LAR, 1:1 height-to-diameter ratio) and high-aspect-ratio asperities (HAR, 3:1 height-to-diameter ratio). Roughness (σ) and peak-to-valley distance (P–V) for

PMMA film was measured using an AFM with values $\sigma = 0.98$ nm and $P--V = 7.3$ nm. The diameter of the asperities near the top is approximately 100 nm and the pitch of the asperities (distance between each asperity) is approximately 500 nm. Figure 22.40a shows SEM images of the two types of patterned structures, LAR and HAR, on a PMMA surface. According to the model presented earlier, by introducing roughness to a flat surface, the hydrophobicity will either increase or decrease depending on the initial contact angle on a flat surface. The material chosen was initially hydrophilic, so to obtain a sample that is hydrophobic, a self-assembled monolayer (SAM) was deposited on the sample surfaces. The samples chosen for the SAM deposition were the flat film and the HAR for each polymer. The SAM perfluorodecyltriethoxysilane (PFDTES) was deposited on the polymer surface using a vapor-phase deposition technique. PFDTES was chosen because of the hydrophobic nature of the surface. It should be noted that the bumps should be as close as possible to provide a high surface area. In order to benefit from an increase in contact angle and decrease in contact area with an increase in surface roughness, the pitch of the bumps should be smaller than the water droplet and the size of the contacting body.

To study the effect of scale dependence, the adhesive force between the four AFM tips with flat and patterned polymer films was examined. This allowed for complete characterization of the adhesive force of the patterned surfaces with vary-

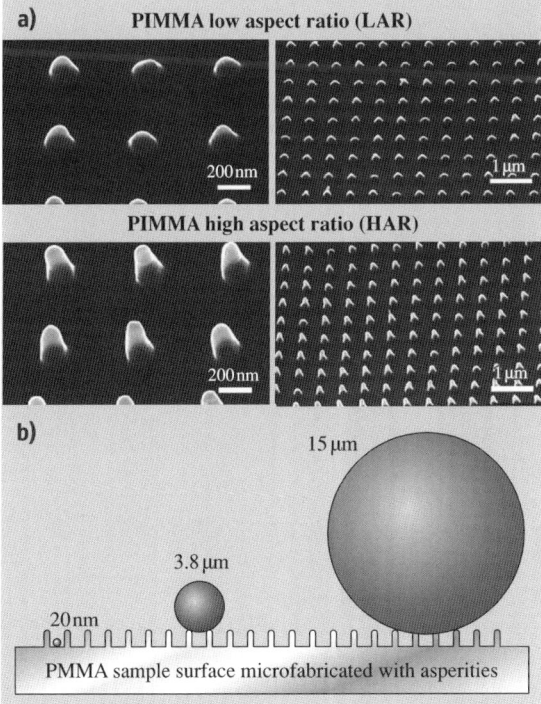

Fig. 22.40. (a) SEM micrographs of the patterned polymer surfaces. Both LAR and HAR are shown at two magnifications to see both the asperity shape and the asperity pattern on the surface. (b) Cartoon showing the effect of different radii on the patterned surface. Small radii can fit between asperities, while large radii rest on top of the asperities

ing tip radii. Figure 22.40b is a cartoon showing the effect of the different radii on the patterned surface. For small radii, such as the 20 nm and 50 nm tips used in this experiment, the tip can easily fit between the asperities and therefore, there is less effect from the asperities. The 3.8-μm- and 15-μm-radii tips will primarily sit on the asperities and will not come into contact with the flat polymer, thus reducing the real area of contact. Experiments in varying relative humidities show the dependence of hydrophobicity for a given surface roughness on adhesion and friction. Dry and wet friction and adhesion can vary dramatically because the dominant mechanism makes a transition from the real area of contact to meniscus forces. Therefore, the effect of relative humidity was also studied by performing measurements at 5, 50, and 80% relative humidity (RH). For these experiments, a tip of radius 15 μm was used to measure both adhesion and coefficient of friction for both the film and patterned polymer surfaces along with the surfaces with the PFDTES coating. Both LAR and HAR were studied to determine the effect of a taller asperity on adhesion.

Contact Angle Measurements

Figure 22.41 shows the static contact angle for various samples. These values correlate well with the model describing roughness with hydrophobicity. The contact angle decreased with increased roughness for the hydrophilic PMMA film. For a hydrophobic surface, the model predicts an increase of contact angle with roughness, which is what happens when PMMA HAR is coated with PFDTES.

Adhesion Studies and Scale Dependence

Scale-dependent effects of adhesion are present because the tip–surface interface changes with size. Meniscus force will change by varying either the tip radius, the hydrophobicity of the sample or the number of contact and near-contacting points. Figure 22.42a shows the dependence of tip radius and hydrophobicity on adhesive

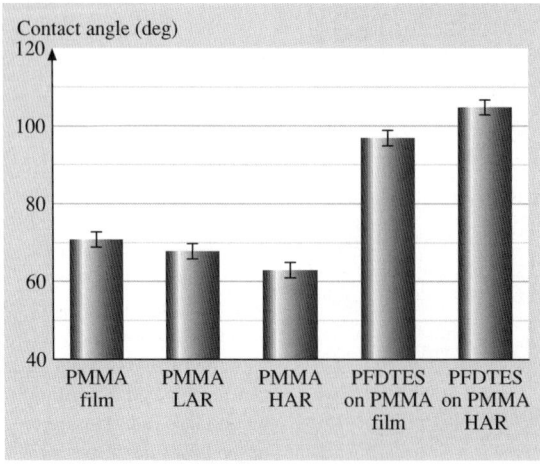

Fig. 22.41. Bar chart showing the contact angles for different materials and for different roughnesses [177]

force for PMMA and PFDTES coated on PMMA. By changing the radius of the tip, the contact angle of the sample, and adding asperities to the sample surface, the adhesive force will change due to the change in the meniscus force and the real area of contact.

Figure 22.42a shows the adhesive force on a linear scale for the different surfaces with varying tip radius. The first bar chart in Fig. 22.42a has PMMA film for the first bar while the second bar is for PFDTES coated on PMMA film and shows the effect of tip radius and hydrophobicity on the adhesive force. For an increase in radius, the adhesive force increases for each material, and decreases for PFDTES on PMMA film compared to PMMA film. With larger radius, the real area of contact and number of meniscus bridges increase and the adhesion is increased. The hydrophobicity of PFDTES on PMMA film reduces meniscus forces, which in turn reduces the adhesion on the PMMA film. The dominant mechanism for the hydrophobic film is the real area of contact and not the meniscus force, whereas with PMMA film there is a combination of real area of contact and meniscus forces.

The second bar chart in Fig. 22.42a shows the results for PMMA HAR and PFDTES coated on PMMA HAR. These samples show the same trends as the film

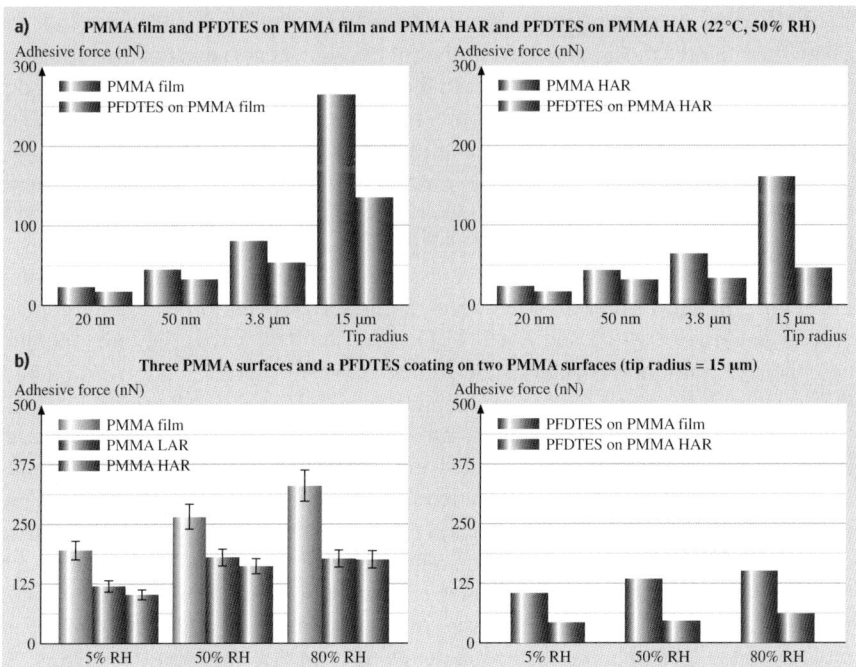

Fig. 22.42. (a) Scale-dependent adhesive force for PMMA film versus PFDTES on PMMA film and PMMA HAR versus PFDTES on PMMA HAR (*top*), and (b) the effect of relative humidity on the adhesive force for PMMA film, LAR and HAR and for PFDTES on PMMA film and HAR [177]

samples, but the increase in adhesion is not as dramatic. This is because of the decrease in real area of contact for each radius from a flat film. Again, meniscus forces do not play a large role in adhesion for PFDTES on PMMA HAR, and the increase in adhesion is due to the real area of contact. With PMMA HAR, a combination of both meniscus forces and real area of contact contribute to the adhesion.

Effect of Relative Humidity on Adhesive Force

The results from varying surface roughness, hydrophobicity and relative humidity are summarized in Fig. 22.42b. Experiments were also run on PMMA film and HAR with a PFDTES coating and are shown in Fig. 22.42b. Film HAR and LAR were used to see the change in adhesion for each type of surface structure. Only film and HAR were used for Fig. 22.42b to show the difference in the two extremes of the surfaces. For these experiments, only the 15 μm tip was used to study the effect from the asperities on the patterned surfaces.

For the adhesive force values, there is a decrease from PMMA film to LAR to HAR for the three humidities. The decrease between LAR and HAR is very small, which means that the contact area is about the same for a single-point measurement. There is, however, a large decrease in adhesive force from film to LAR and HAR. For a flat film, meniscus bridges are the dominant factor in the adhesion, but for a patterned sample the dominant factor is still the contact area and not the formation of menisci. At 5% RH the only factor in the adhesion is the area of contact and not the formation of menisci. The data shows that there is a smaller difference at 5% in adhesion compared to the difference at 80% RH.

Results for PFDTES coating on PMMA film and HAR are also shown in Fig. 22.42b. The adhesion values are much lower than those for PMMA and PMMA HAR and that is primarily due to the lack of meniscus bridge formation because of the hydrophobic contact angle of PFDTES. There is a decrease in adhesion between PFDTES film and PFDTES HAR, which is directly related to the area of contact difference between a film and HAR. Looking at the data across the three humidities, there is not much change in the values. Since the surfaces are hydrophobic, meniscus bridges are not the determining factor in the material adhesion.

In summary, increasing roughness on a hydrophilic surface decreases the contact angle, whereas increasing roughness on a hydrophobic surface increases contact angle. For a flat film, with increasing tip radius, the adhesive force increases due to increased real area of contact between the tip and the flat sample and meniscus force contributions. Introducing a pattern on a flat polymer surface will reduce adhesion because of the reduction of the real area of contact between the tip and the sample surface if the tip is larger than the size of the asperities. In addition, introducing a pattern on a hydrophobic surface increases the contact angle and decreases the number of menisci, which then decreases the adhesive force. Adhesion increases with increasing RH for every sample and decreases from film to LAR to HAR. When PFDTES is coated on the PMMA samples, the adhesion decreases but follows the same trend as the bare polymer. These trends are due to the formation of more menisci at higher relative humidities. In addition, with an increase in relative humidity, the increase in adhesive force for PMMA film is more dramatic than for the patterned samples due to larger menisci formations for a film sample.

22.6 Component-Level Studies

22.6.1 Surface Roughness Studies of Micromotor Components

Most of the friction forces resisting motion in the micromotor are concentrated near the rotor–hub interface where continuous physical contact occurs. The surface roughness of the surfaces usually has a strong influence on the friction characteristics on the micro/nanoscale. A catalog of roughness measurements on various components of a MEMS device does not exist in the literature. Using an AFM, measurements on various component surfaces, for the first time, were made by *Sundararajan* and *Bhushan* [178]. Table 22.6 shows various surface roughness parameters obtained from $5 \times 5\,\mu m$ scans of the various component surfaces of several unlubricated micromotors using the AFM in tapping mode. A surface with a Gaussian height distribution should have a skewness of zero and kurtosis of three. Although the rotor and stator top surfaces exhibit comparable roughness parameters, the underside of the rotors exhibits lower root mean square (RMS) roughness and peak-to-valley distance values. More importantly, the rotor underside shows negative skewness and lower kurtosis than the topsides, both of which are conducive to high real area of contact and hence high friction [79, 81]. The rotor underside also exhibits a higher coefficient of microscale friction than the rotor topside and stator, as shown in Table 22.6. Figure 22.43 shows representative surface-height maps of the various surfaces of a micromotor measured using the AFM in tapping mode. The

Table 22.6. Surface roughness parameters and microscale coefficient of friction for various micromotor component surfaces measured using an AFM. Mean and $\pm 1\sigma$ values are given

	RMS roughness[a] (nm)	Peak-to-valley distance[a] (nm)	Skewness[a] Sk	Kurtosis[a] K	Coefficient of microscale friction[b] (μ)
Rotor topside	21 ± 0.6	225 ± 23	1.4 ± 0.30	6.1 ± 1.7	0.07 ± 0.02
Rotor underside	14 ± 2.4	80 ± 11	-1.0 ± 0.22	3.5 ± 0.50	0.11 ± 0.03
Stator topside	19 ± 1	246 ± 21	1.4 ± 0.50	6.6 ± 1.5	0.08 ± 0.01

[a] Measured from a tapping-mode AFM scan of size $5\,\mu m \times 5\,\mu m$ using a standard Si tip scanning at $5\,\mu m/s$ in a direction orthogonal to the long axis of the cantilever.
[b] Measured using an AFM in contact mode at $5\,\mu m \times 5\,\mu m$ scan size using a standard Si_3N_4 tip scanning at $10\,\mu m/s$ in a direction parallel to the long axis of the cantilever.

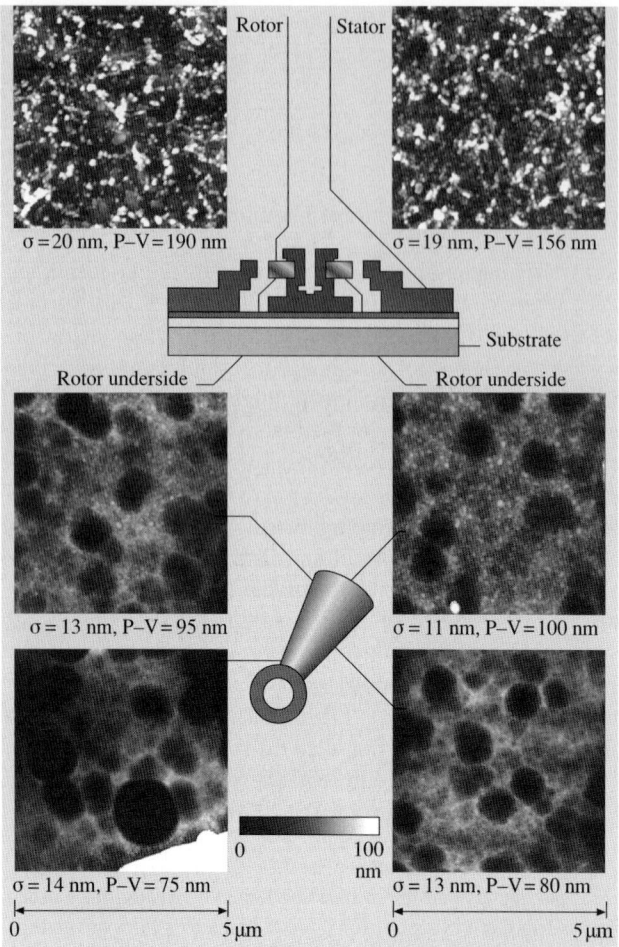

Fig. 22.43. Representative AFM surface-height images obtained in tapping mode (5 μm × 5 μm scan size) of various component surfaces of a micromotor. RMS roughness and peak-to-valley values of the surfaces are given. The underside of the rotor exhibits drastically different topography from the topside [178]

rotor underside exhibits a different topography from the outer edge to the middle and inner edge. At the outer edges, the topography shows smaller circular asperities, similar to the topside. The middle and inner regions show deep pits with fine edges that may have been created by the etchants used for etching of the sacrificial layer. It is known that etching can affect the roughness of surfaces in surface micromachining. The residence time of the etchant near the inner region is high, which is responsible for larger pits. Figure 22.44 shows the roughness of the surface directly beneath the rotors (the base polysilicon layer). There appears to be a difference in the roughness between the portion of this surface that was initially underneath the

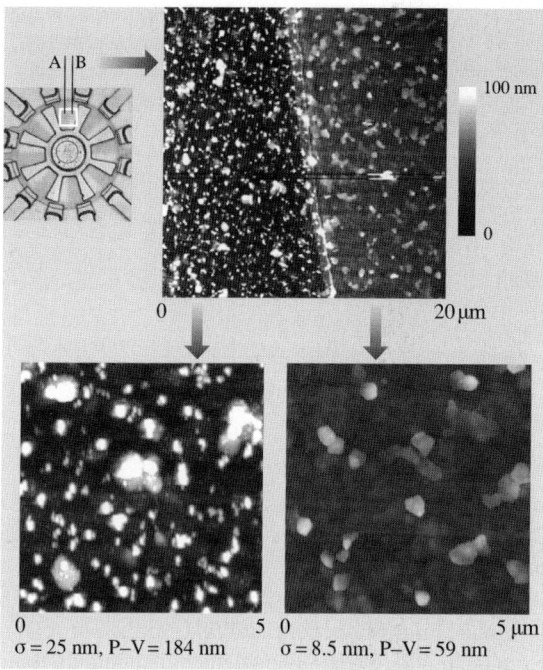

Fig. 22.44. Surface-height images of polysilicon regions directly below the rotor. Region A is away from the rotor while region B was initially covered by the rotor prior to the release etch of the rotor. During this step, slight movement of the rotor caused region B to be exposed [178]

rotor (region B) during fabrication and the portion that was away from the rotor and hence always exposed (region A). The former region shows a lower roughness than the latter region. This suggests that the surfaces at the rotor–hub interface that come into contact at the end of the fabrication process exhibit large real areas of contact, which result in high friction.

22.6.2 Adhesion Measurements of Microstructures

Surface force apparatus (SFA) and AFMs are used to measure adhesion on micro- to nanoscales between two surfaces. In the SFA, adhesion of liquid films sandwiched between two curved and smooth surfaces is measured. In an AFM, as discussed earlier, adhesion between a sharp tip and the surface of interest is measured. The propensity for adhesion between two surfaces can be evaluated by studying the tendency of microstructures with well-defined contact areas, covering a wide of suspension compliances, to stick to the underlying substrate. The test structures which have been used include cantilever beam array (CBA) technique with different lengths [179–182] and stand-off multiple dimples mounted on microstructures with a range of compliances, standing above a substrate [183]. The CBA technique, more commonly used, utilizes an array of micromachined polysilicon beams (for Si MEMS applications), on the mesoscopic length scale, anchored to the substrate at one end and with different lengths parallel to the surface. It relies on peeling and detachment of cantilever beams. Changes in the free energy or reversible work done

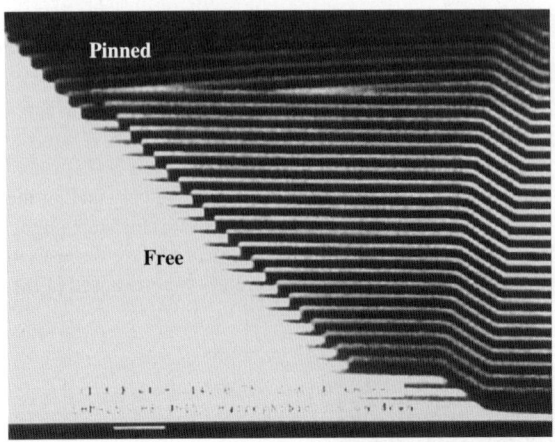

Fig. 22.45. SEM micrograph of micromachined array of polysilicon cantilever beams of increasing length. The micrograph shows the onset of pinning for beams longer than 34 μm [179]

to separate unit areas of the two surfaces from contact is called the work of adhesion. To measure the work of adhesion, electrostatic actuation is used to bring all the beams into contact with the substrate; see Fig. 22.45 [179, 181]. Once the actuation force is removed, the beams begin to peel themselves off the substrate, which can be observed with an optical interference microscope (e. g., a Wyko surface profiler). For beams shorter than a characteristic length, the so-called detachment length, their stiffness is sufficient to free them completely from the substrate underneath. Beams larger than the detachment length remain adhered. The beams at the transition region start to detach and remain attached to the substrate just at the tips. For this case, by equating the elastic energy stored within the beam and the beam–substrate interfacial energy, the work of adhesion, W_{ad}, can be calculated by the following equation [179]

$$W_{ad} = \frac{3Ed^2t^3}{8\ell_d^4} \, , \tag{22.5}$$

where E is the Young's modulus of the beam, d is the spacing between the undeflected beam and the substrate, t is the beam thickness, and P_d is the detachment length. The technique has been used to screen methods for adhesion reduction in polysilicon microstructures.

22.6.3 Microtriboapparatus for Adhesion, Friction, and Wear of Microcomponents

To measure adhesion, friction, and wear between two microcomponents, a microtriboapparatus has been used. Figure 22.46 shows a schematic of a microtriboapparatus, capable of adopting MEMS components [184]. In this apparatus, an upper specimen, mounted on a soft cantilever beam, comes into contact with a lower specimen mounted on a lower specimen holder. The apparatus consists of two piezos (x- and z-piezos), and four fiber-optic sensors (x- and z-displacement sensors, and

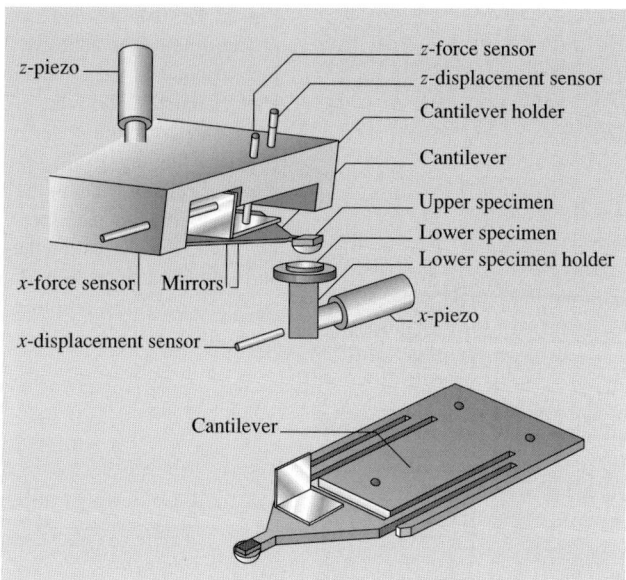

Fig. 22.46. Schematic of the microtriboapparatus including specially designed cantilever (with two perpendicular mirrors attached on the end), lower specimen holder, two piezos (x- and z-piezos), and four fiber-optic sensors (x- and z-displacement sensors and x- and z-force sensors) [184]

x- and z-force sensors). For adhesion and friction studies, z- and x-piezos are used to bring the upper specimen and lower specimen in contact and to apply a relative motion in the lateral direction, respectively. The x- and z-displacement sensors are used to measure the lateral position of the lower specimen and vertical position of the upper specimen, respectively. The x- and z-force sensors are used to measure the friction force and normal load/adhesive force between these two specimens, respectively, by monitoring the deflection of the cantilever.

As most MEMS/NEMS devices are fabricated from silicon, study of silicon-on-silicon contacts is important. This contact was simulated by a flat single-crystal Si(100) wafer (phosphorus-doped) specimen sliding against a single-crystal Si(100) ball (1 mm in diameter, 5×10^{17} atoms/cm^3 boron doped) mounted on a stainless-steel cantilever [184, 185]. Both of them have native oxide layer on their surfaces. The other materials studied were 10-nm-thick DLC deposited by filtered cathodic arc deposition on Si(100), 2.3-nm-thick chemically bonded PFPE (Z-DOL, BW) on Si(100), and hexadecane thiol (HDT) monolayer on evaporated Au(111) film to investigate their anti-adhesion performance.

It is well known that in computer rigid disk drives, the adhesive force increases rapidly with increasing rest time between a magnetic head and a magnetic disk [34]. Considering that adhesion and friction are the major issues that lead to the failure of MEMS/NEMS devices, the effect of rest time on the microscale on Si(100),

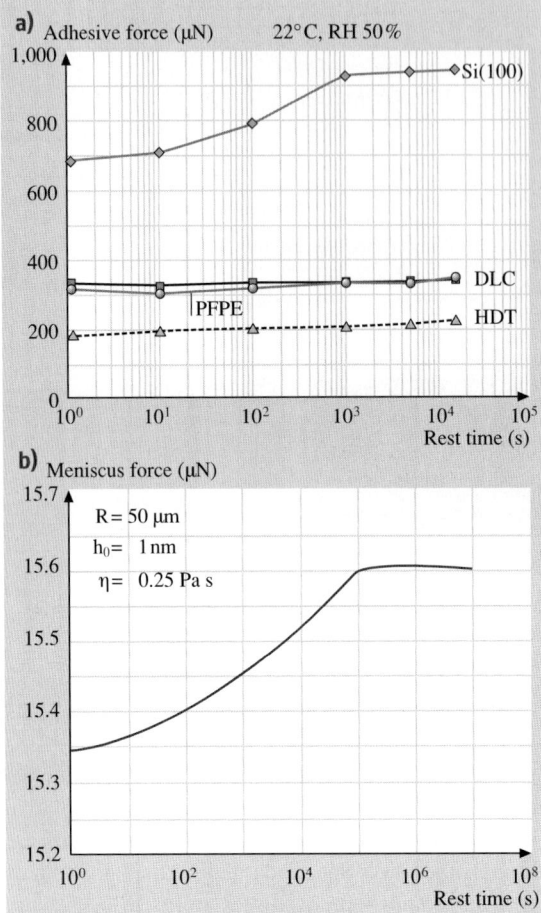

Fig. 22.47. (a) The influence of rest time on the adhesive force of Si(100), DLC, chemically bonded PFPE, and HDT, and (b) Single-asperity contact modeling results of the effect of rest time on the meniscus force for an asperity of R in contact with a flat surface with a water film of thickness of h_0 and absolute viscosity of η_0 [186]

DLC, PFPE, and HDT was studied, and the results are summarized in Fig. 22.47a. It is found that the adhesive force of Si(100) logarithmically increases with the rest time up to a certain equilibrium time ($t = 1000\,\text{s}$), after which it remains constant. Figure 22.47a also shows that the adhesive force of DLC, PFPE, and HDT does not change with rest time. Single-asperity contact modeling of the dependence of meniscus force on the rest time has been carried out by *Chilamakuri* and *Bhushan* [186], and the modeling results Fig. 22.47b verify experimental observations. Due to the presence of thin-film adsorbed water on Si(100), the meniscus forms around the contacting asperities and grows with time until equilibrium occurs, which causes the effect of rest time on its adhesive force. The adhesive forces of DLC, PFPE, and HDT do not change with rest time, which suggests that the water meniscus is not present on their surfaces.

The measured adhesive forces of Si(100), DLC, PFPE, and HDT at a rest time of 1 s are summarized in Fig. 22.48, which shows that the presence of solid films

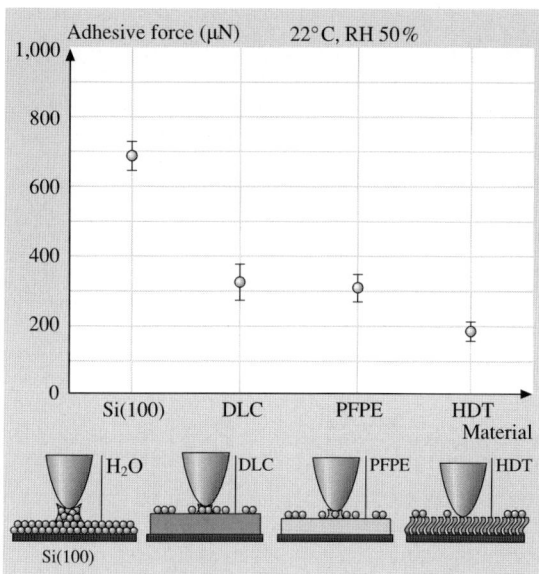

Fig. 22.48. Adhesive forces of Si(100), DLC, chemically bonded PFPE, and HDT at ambient condition and a schematic showing the relative size of the water meniscus on different specimens

of DLC, PFPE, and HDT greatly reduces the adhesive force of Si(100), whereas, the HDT film has the lowest adhesive force. It is well known that the native oxide layer (SiO_2) on the top of Si(100) wafers exhibits hydrophilic properties, and water molecules, produced by capillary condensation of water vapor from the environment, can easily be adsorbed onto this surface. The condensed water will form a meniscus as the upper specimen approaches to the lower specimen surface. The meniscus force is a major contributor to the adhesive force. In the case of DLC, PFPE, and HDT, the films are found to be hydrophobic based on contact angle measurements and the amount of condensed water vapor is low compared to that on Si(100). It should be noted that the measured adhesive force is generally higher than that measured by AFM, because the larger radius of the Si(100) ball compared to that of an AFM tip induces a larger meniscus and van der Waals forces.

To investigate the effect of velocity on friction, the friction force as a function of velocity was measured and is summarized in Fig. 22.49a. It indicates that, for Si(100), the friction force initially decreases with increasing velocity until equilibrium occurs. Figure 22.49a also indicates that the velocity almost has no effect on the friction properties of DLC, PFPE, and HDT. This implies that the friction mechanisms on DLC, PFPE, and HDT do not change with the variation of velocity. For Si(100), at high velocity, the meniscus is broken and does not have enough time to rebuild. In addition, it is also believed that tribochemical reactions plays an important role. The high velocity leads to tribochemical reactions of Si(100) (which has a native oxide SiO_2) with water molecules to form a $Si(OH)_4$ film. This film is removed and continuously replenished during sliding. The $Si(OH)_4$ layer at the sliding surface is known to have low shear strength. The breaking of the water meniscus and the formation of a $Si(OH)_4$ layer results in a decrease in the friction force of

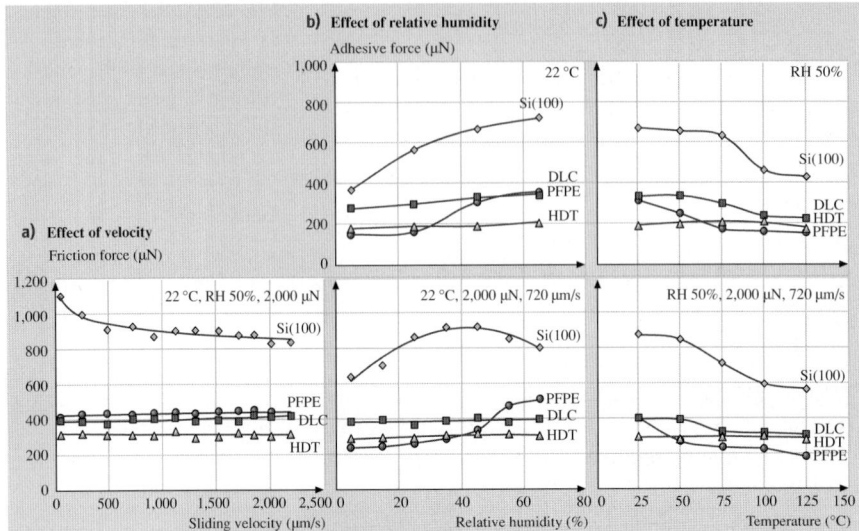

Fig. 22.49. The influence of (**a**) sliding velocity on the friction forces, (**b**) relative humidity on the adhesive and friction forces, and (**c**) temperature on the adhesive and friction forces of Si(100), DLC, chemically bonded PFPE, and HDT

Si(100). For DLC, PFPE, and HDT, their surfaces exhibit hydrophobic properties, and can only adsorb a few water molecules in ambient conditions. The aforementioned meniscus breaking and tribochemical reaction mechanisms do not exist for these films. Therefore, their friction force does not change with velocity.

The influence of relative humidity was studied in an environmentally controlled chamber. The adhesive force and friction force were measured by making measurements at increasing relative humidity, and the results are summarized in Fig. 22.49b, which shows that, for Si(100), the adhesive force increases with relative humidity, but the adhesive forces for DLC and PFPE only show a slight increase when the humidity is higher than 45%, while the adhesive force of HDT does not change with humidity. Figure 22.49b also shows that, for Si(100), the friction force increases with relative humidity increases up to 45%, and then shows a slight decrease with further increases in the relative humidity. For PFPE, there is an increase in the friction force when the humidity is higher than 45%. In the whole testing range, the relative humidity does not have any apparent influence on the friction properties of DLC and HDT . In the case of Si(100), the initial increase in relative humidity up to 45% causes more adsorbed water molecules, and the formation of a larger water meniscus, which leads to an increase of the friction force. However, at very high humidity of 65%, large quantities of adsorbed water can form a continuous water layer that separates the tip and sample surfaces, and acts as a kind of lubricant, which causes a decrease in the friction force. For PFPE, dewetting of the lubricant film at humidities larger than 45% results in an increase in adhesive and friction forces. For

DLC and HDT, their surfaces show hydrophobic properties, and increasing relative humidity does not play a large role in their friction force.

The influence of temperature was studied using a heated stage. The adhesive force and friction force were measured by making measurements at increasing temperatures of 22–125 °C. The results are presented in Fig. 22.49c, which shows that, once the temperature is higher than 50 °C, increasing temperature causes a significant decrease of adhesive and friction forces of Si(100) and a slight decrease in the case of DLC and PFPE. However, the adhesion and friction forces of HDT do not show any apparent change with test temperature. At high temperature, desorption of water, and the reduction of surface tension of water lead to the decrease of adhesive and friction forces of Si(100), DLC, and PFPE. However, in the case of HDT film, as only a few water molecules are adsorbed on the surface, the aforementioned mechanisms do not play a large role. Therefore, the adhesive and friction forces of HDT do not show any apparent change with temperature. Figure 22.49 shows that in the whole velocity, relative humidity, and temperature test range, the adhesive force and friction force of DLC, PFPE, and HDT are always smaller than that of Si(100), and that HDT has the smallest value.

To summarize, several methods can be used to reduce adhesion in microstructures. MEMS/NEMS surfaces can be coated with hydrophobic coatings such as PFPEs, SAMs, and passivated DLC coatings. It should be noted that other methods to reduce adhesion include the formation of dimples on the contact surfaces to reduce contact area [13, 79, 81, 177, 181]. Furthermore, an increase in hydrophobicity of the solid surfaces (high contact angle approaching 180°) can be achieved by using surfaces with suitable roughness, in addition to lowering their surface energy [173, 176]. The hydrophobicity of surfaces is dependent upon a subtle interplay between surface chemistry and mesoscopic topography. The self-cleaning mechanism or so-called lotus effect is closely related to the ultra-hydrophobic properties of the biological surfaces, which usually show microsculptures of specific dimensions.

22.6.4 Static Friction Force (Stiction) Measurements in MEMS

In MEMS devices involving parts in relative motion to each other, such as micromotors, large friction forces become the limiting factor for the successful operation and reliability of the device. It is generally known that most micromotors cannot be rotated as manufactured and require some form of lubrication. It is therefore critical to determine the friction forces present in such MEMS devices. To measure in situ the static friction of a rotor–bearing interface in a micromotor, *Tai* and *Muller* [187] measured the starting torque (voltage) and pausing position for different starting positions under a constant bias voltage. A friction-torque model was used to obtain the coefficient of static friction. To measure the in situ kinetic friction of the turbine and gear structures, *Gabriel* et al. [188] used a laser-based measurement system to monitor the steady-state spins and decelerations. *Lim* et al. [189] designed and fabricated a polysilicon microstructure to measure in situ the static friction of various films. The microstructure consisted of a shuttle suspended above the underlying electrode by a folded beam suspension. A known normal force was applied and the lateral force was measured to obtain the coefficient of static friction. *Beerschwinger* et al. [190] developed a cantilever-deflection rig to measure friction of LIGA-processed micromotors. (Also see [191].) These techniques employ indirect methods to determine the friction forces or involve fabrication of complex structures.

A novel technique to measure the static friction force (stiction) encountered in surface-micromachined polysilicon micromotors using an AFM has been developed by *Sundararajan* and *Bhushan* [178]. Continuous physical contact occurs during rotor movement (rotation) in the micromotors between the rotor and lower hub flange. In addition, contact occurs at other locations between the rotor and the hub surfaces and between the rotor and the stator. Friction forces will be present at these contact regions during motor operation. Although the actual distribution of these forces is not known, they can be expected to be concentrated near the hub where there is continuous contact. If we therefore represent the static friction force of the micromotor as a single force F_s acting at point P_1 (as shown in Fig. 22.50a), then the magnitude of the frictional torque about the center of the motor (O) that must be overcome before rotor movement can be initiated is

$$T_S = F_s \ell_1 , \tag{22.6}$$

where ℓ_1 is the distance OP_1, which is assumed to be the average distance from the center at which the friction force F_s occurs. Now consider an AFM tip moving against a rotor arm in a direction perpendicular to the long axis of the cantilever

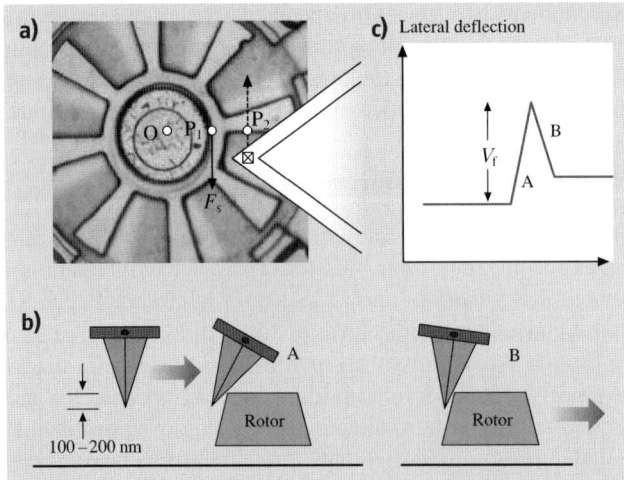

Fig. 22.50. (**a**) Schematic of the technique used to measure the force F_s required to initiate rotor movement using an AFM/FFM. (**b**) As the tip is pushed against the rotor, the lateral deflection experienced by the rotor due to the twisting of the tip prior to rotor movement is a measure of static friction force F_s of the rotors. (**c**) Schematic of lateral deflection expected from the aforementioned experiment. The peak V_f is related to the state of the rotor [178]

beam (the rotor-arm edge closest to the tip is parallel to the long axis of the cantilever beam), as shown in Fig. 22.50a. When the tip encounters the rotor at point P_2, the tip will twist, generating a lateral force between the tip and the rotor (event A in Fig. 22.50b). This reaction force will generate a torque about the center of the motor. Since the tip is trying to move further in the direction shown, the tip will continue to twist to a maximum value at which the lateral force between the tip and the rotor becomes high enough that the resultant torque T_f about the center of the motor equals the static friction torque T_s. At this point, the rotor will begin to rotate and the twist of the cantilever decreases sharply (event B in Fig. 22.50b). The twist of the cantilever is measured in the AFM as a change in the lateral deflection signal (in volts), which is the underlying concept of friction force microscopy (FFM). The change in the lateral deflection signal corresponding to the aforementioned events as the tip approaches the rotor is shown schematically in Fig. 22.50c. The value of the peak V_f is a measure of the force exerted on the rotor by the tip just before the static friction torque is matched and the rotor begins to rotate.

Using this technique, the viability of PFPE lubricants for micromotors has been investigated and the effect of humidity on the friction forces of unlubricated and lubricated devices was studied as well. Figure 22.51 shows static friction forces, normalized over the weight of the rotor, of unlubricated and lubricated micromotors as a function of rest time and relative humidity. Rest time here is defined as the time elapsed between the first experiment conducted on a given motor (solid symbol at time zero) and subsequent experiments (open symbols). Each open-symbol data point is an average of six measurements. It can be seen that, for the unlubricated motor and the motor lubricated with a bonded layer of Z-DOL(BW), the static friction force is highest for the first experiment and then drops to an almost constant level. In the case of the motor with an as-is mobile layer of Z-DOL, the values remain very high up to 10 days after lubrication. In all cases, there is negligible difference in the static friction force at 0% and 45% RH. At 70% RH, the unlubricated motor exhibits a substantial increase in the static friction force, while the motor with bonded Z-DOL shows no increase in static friction force due to the hydrophobicity of the lubricant layer. The motor with an as-is mobile layer of the lubricant shows consistently high values of static friction force that vary little with humidity.

Figure 22.52 summarizes static friction force data for two motors, M1 and M2 along with schematics of the meniscus effects for the unlubricated and lubricated surfaces. Capillary condensation of water vapor from the environment results in the formation of meniscus bridges between contacting and near-contacting asperities of the two surfaces in close proximity to each other, as shown in Fig. 22.52. For unlubricated surfaces, more menisci are formed at higher humidity resulting in a higher friction force between the surfaces. The formation of meniscus bridges is supported by the fact that the static friction force for unlubricated motors increases at high humidity Fig. 22.52. Solid bridging may occur near the rotor–hub interface due to silica residues after the first etching process. In addition the drying process after the final etch can result in liquid bridging, formed by the drying liquid due to

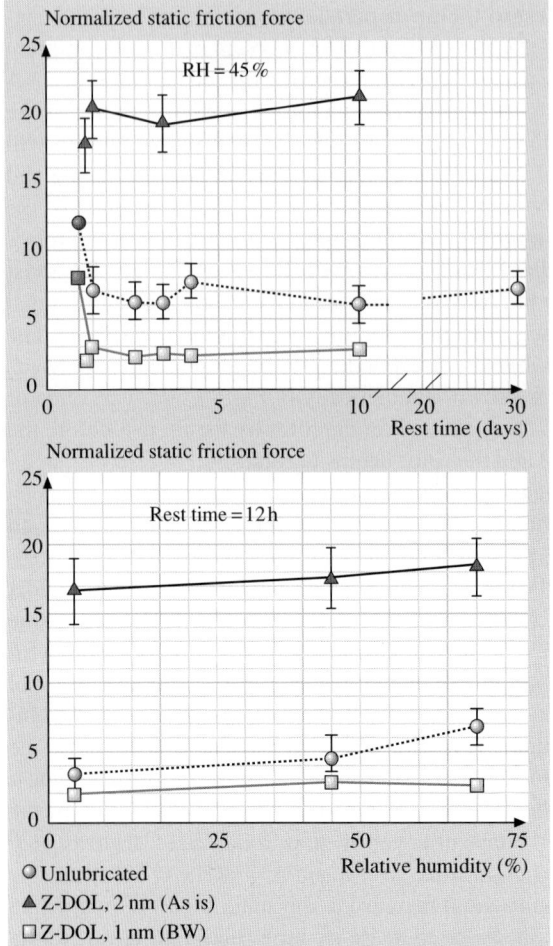

Fig. 22.51. Static friction force values of unlubricated motors and motors lubricated using PFPE lubricants, normalized over the rotor weight, as a function of rest time and relative humidity. Rest time is defined as the time elapsed between a given experiment and the first experiment in which motor movement was recorded (time 0). The motors were allowed to sit at a particular humidity for 12 h prior to measurement [178]

meniscus force at these areas [79, 81, 179, 180]. The initial static friction force therefore will be quite high, as evidenced by the solid data points in Fig. 22.52. Once the first movement of the rotor permanently breaks these solid and liquid bridges, the static friction force of the motors will drop (as seen in Fig. 22.52) to a value dictated predominantly by the adhesive energies of the rotor and hub surfaces, the real area of contact between these surfaces and meniscus forces due to water vapor in the air, at which point, the effect of lubricant films can be observed. Lubrication with a mobile layer, even a thin one, results in very high static friction forces due to meniscus effects of the lubricant liquid itself at and near the contact regions. It should be noted that a motor submerged in a liquid lubricant would result in a fully flooded lubrication regime. In this case there is no meniscus contribution and only the viscous contribution to the friction forces would be relevant. However, submerging the device in a lubricant may not be a practical method. A solid-like

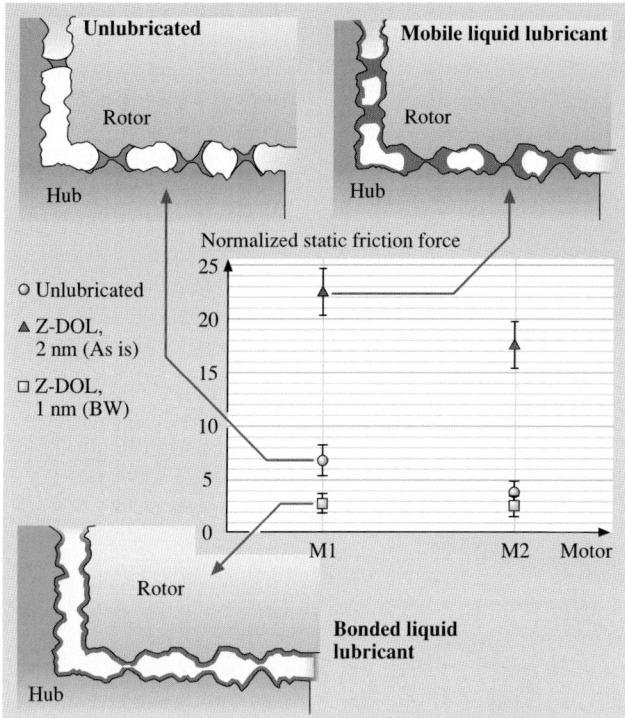

Fig. 22.52. Summary of the effect of liquid and solid lubricants on static friction force of micromotors. Despite the hydrophobicity of the lubricant used (Z-DOL), a mobile liquid lubricant (Z-DOL as is) leads to very high static friction force due to increased meniscus forces whereas a solid-like lubricant (bonded Z-DOL, BW) appears to provide some reduction in static friction force

hydrophobic lubricant layer (such as bonded Z-DOL) results in favorable friction characteristics of the motor. The hydrophobic nature of the lubricant inhibits meniscus formation between the contact surfaces and maintains low friction even at high humidity Fig. 22.52). This suggests that solid-like hydrophobic lubricants are ideal for lubrication of MEMS while mobile lubricants result in increased values of the static friction force.

22.6.5 Mechanisms Associated with Observed Stiction Phenomena in Digital Micromirror Devices (DMD) and Nanomechanical Characterization

DMDs are used in digital projection displays, as described earlier. The DMD has a layered structure, consisting of an aluminium alloy micromirror layer, yoke and hinge layer, and metal layer on a CMOS memory array [27–29]. A blown-up view of the DMD and the corresponding AFM surface-height images are presented in Fig. 22.53 [192]. Single-layered aluminium alloy films are used for the construc-

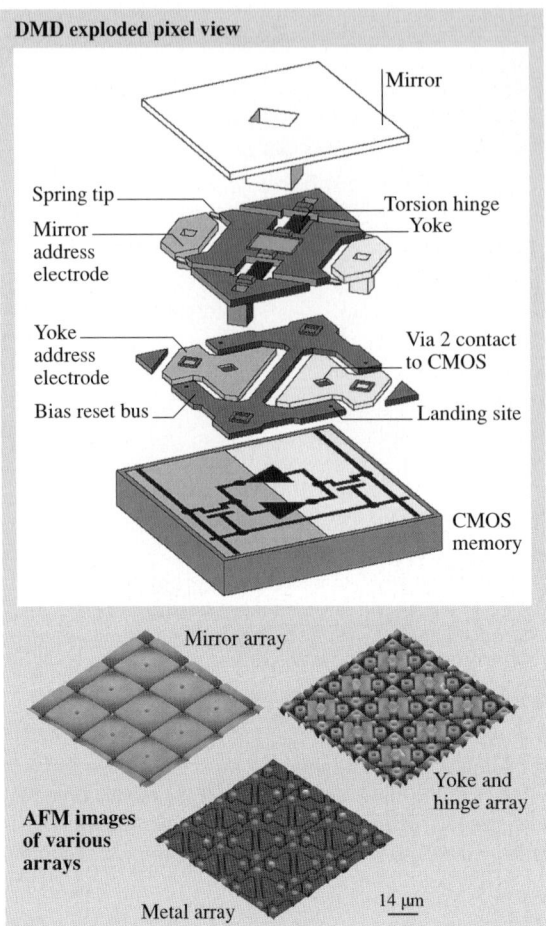

Fig. 22.53. Exploded view of a DMD pixel and AFM surface-height images of various arrays. The DMD layers were removed by an ultrasonic method [192]

tion of micromirrors; these are also sometimes used for the construction of hinges, spring tips, and landing sites. The aluminium alloy films are overwhelmingly comprised of aluminium; trace elements (including Ti and Si) are present to suppress contact spiking and electromigration, which may occur if current densities become high during electrostatic operation. Multilayered sputtered SiO_2TiN/Al alloy films are now generally used for the landing site structure to minimize refraction throughout the visible region of the electromagnetic spectrum in order to increase the contrast ratio in projection display systems [193, 194]. These multilayered films are also generally used for hinges and spring tips. A low-surface-energy SAM is maintained on the surfaces of the DMD, which is packaged in a hermetic environment to minimize stiction during contact between the spring tip and the landing site. An SAM of perfluorinated *n*-alkanoic acid ($C_nF_{2n-1}O_2H$) (e.g., perflurordecanoic acid or PFDA, $CF_3(CF_2)_8COOH$) applied by the vapor-phase deposition process is used. A getter strip of PFDA is included inside the hermetically sealed enclosure containing the chip, which acts as a reservoir in order to maintain a PFDA vapor within the package.

In order to identify a stuck mirror and characterize its nanotribological properties, the chip was scanned using an AFM [192]). It was found that it is hard to tilt the stuck micromirror back to its normal position by adding a normal load at the rotatable corner of the micromirror; thus, this is called a *hard* stuck micromirror. An example of a stuck micromirror is shown in Fig. 22.54a. Once the stuck micromirror was found, the region was repeatedly scanned at a large normal load, up to 300 nN. After several scans, the stuck micromirror was removed. Once the stuck micromirror was removed, the surrounding micromirrors could also be removed by continuous scanning under a large normal load (Fig. 22.54a bottom row). The adhesive force of the landing site underneath the stuck micromirror and the normal micromirror are presented in Fig. 22.54b, which clearly indicates that the landing site underneath the stuck micromirror has much larger adhesion. A 1 μm × 1 μm view of the landing sites under stuck and normal micromirrors are also shown in Fig. 22.54b. The land-

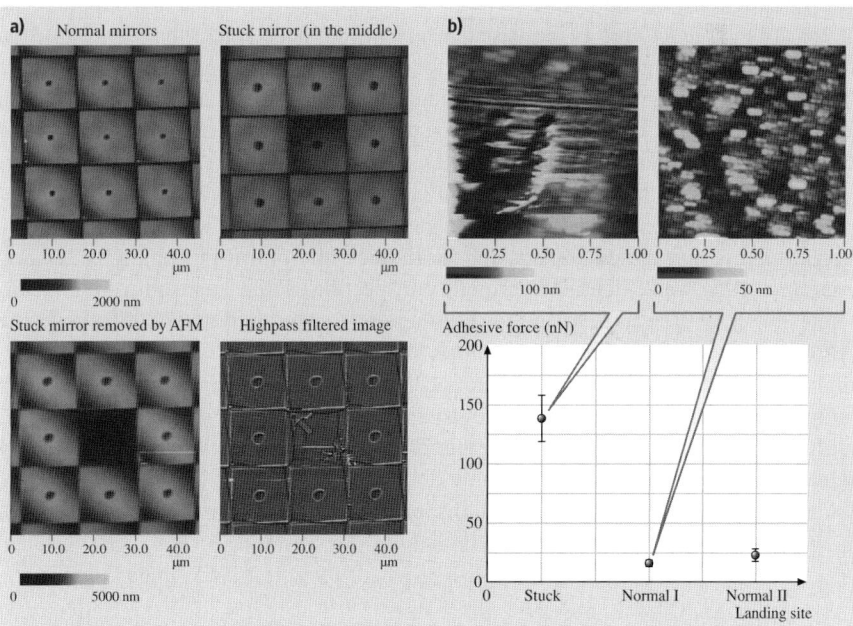

Fig. 22.54. (a) The *top row* shows AFM surface-height images of a stuck micromirror surrounded by eight normal micromirrors. The *left image* in the *bottom row* shows the stuck micromirror, which was removed by an AFM tip after repeated scanning at high normal load. The *right image* in the *bottom row* presents a high-pass-filtered image showing that the residual hinge that sits underneath the removed micromirror is clearly observed. (b) AFM surface-height images and adhesive forces of the landing sites underneath the two normal micromirrors and the stuck micromirror [192]

ing site under the stuck micromirror has an apparent U-shaped wear mark, which is surrounded by a smeared area.

Liu and *Bhushan* [192] calculated contact stresses to examine if the stresses were high enough to cause wear at the spring tip–landing site interface. The calculated contact stress value was about 33 MPa which is substantially lower than the hardness, therefore much plastic deformation and consequently wear was not expected. The wear mark was only found on a very few landing sites on the DMD, which means that the SAM coating can generally endure such high contact stresses. Based on data reported in the literature, the coverage of vapor-deposited SAMs is expected to be about 97%. The bond strength of the molecules close to the boundary of the uncovered sites is expected to be weak. Thus, the uncovered sites and the adjacent molecules are referred to as defects in the SAM coating. Occasionally, if contact occurs at the defect sites, the large cyclic stress may be close to the critical load, and lead to the initial delamination of the SAM coating at the interface. The continuous contact leads to the formation of a high-surface-energy surface by exposure of the fresh substrate and the formation of SAM fragments. This eventually leads to an increase in stiction by the formation of large menisci. Once this happens, the stress at the contact area is increased, which would accelerate the wear. Based on this hypothesis, suggested mechanisms for the wear and stiction of the landing site are summarized in Fig. 22.55. Wear initiates at the defect sites and consequent high stiction can result in high wear. Improving the coverage and wear resistance of SAM coatings could enhance the yield of DMD.

In some cases, the micromirrors are not fully stuck and can be moved by applying a load at the rotatable corner of the micromirror with a discontinuous motion, which is thus called *soft* stiction. Soft-stuck micromirrors studied by *Liu* and *Bhushan* [195] were identified in quality inspection. These micromirrors encountered slow transition from to end to the other end ($+1/-1$). Figure 22.56 shows the AFM surface-height images of a location showing a stuck mirror (S) and surrounding normal micromirrors N_i ($i = 1, 2$, and 3). Surprisingly, the images of the stuck and normal micromirror array are almost the same. On the micromirrors of interest, a tilting test was performed at the corner of the micromirrors, which are marked by white dots in Fig. 22.56. The rotatable direction of the microarray is indicated by an arrow bar in Fig. 22.56. The load–displacement curve for the stuck micromirror

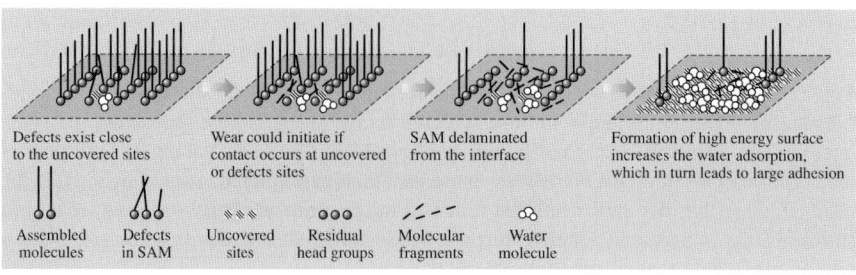

Fig. 22.55. Suggested mechanisms for wear and stiction [192]

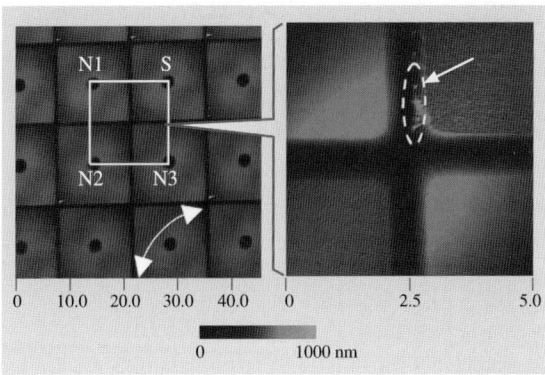

Fig. 22.56. AFM surface-height images of normal micromirrors and a soft-stuck micromirror. The soft-stuck micromirror was labeled S, and the normal micromirrors studied are labeled as N1, N2, and N3 [195]

is presented in Fig. 22.57; it is not smooth and appears serrated. It is clearly indicated that, although the S micromirror can be rotated, it rotates with hesitation. In regimes 1 and 2, as marked in Fig. 22.57, the slopes are much higher. In order to understand the mechanisms for the occurrence of stiction, stiction of landing sites of normal and stuck mirrors were measured. Unlike hard-stuck mirrors, the adhesive forces of soft-stuck and normal mirrors were comparable, which suggests that the SAM coating is intact with stuck mirror. It was found that a high normal load (about 900 nN) and on the order of a couple of hundred scans were required to remove the soft-stuck micromirrors by an AFM. However, only about 300 nN and about ten scans were required to remove a hard-stuck mirror. After careful examination of the AFM images of the micromirror sidewalls in Fig. 22.56 (bottom left), it is noted that there are contaminant particles attached to the sidewalls of the S mirror. It is, therefore, believed that, during the tilting test, the S micromirror (see schematic in Fig. 22.57) a sharper slope regime will occur in the displacement curve. Extra force is required to overcome the resistance that is induced by the sidewall contamination particles. This is believed to be the reason for the slow transition of the micromirror during quality inspection.

Finally, the nanomechanical characterization of various layers used in the construction of landing sites, hinge and micromirror materials have been measured by *Wei* et al. [193, 194]. Bending and fatigue studies of the hinge have been carried out by *Liu* and *Bhushan* [196] and *Bhushan* and *Liu* [197] to measure stiffness and fatigue properties. For these studies, the micromirror was removed. During removal, the micromirror/yoke structure was removed simultaneously, leaving the hinge mounted on one end of the array; see Fig. 22.58. The stiffness of the Al hinge was reported to be comparable to the stiffness of bulk Al. The Al hinge exhibited a higher modulus than the SiO_2 hinge. The fatigue properties depended upon the preparation of the hinge for testing.

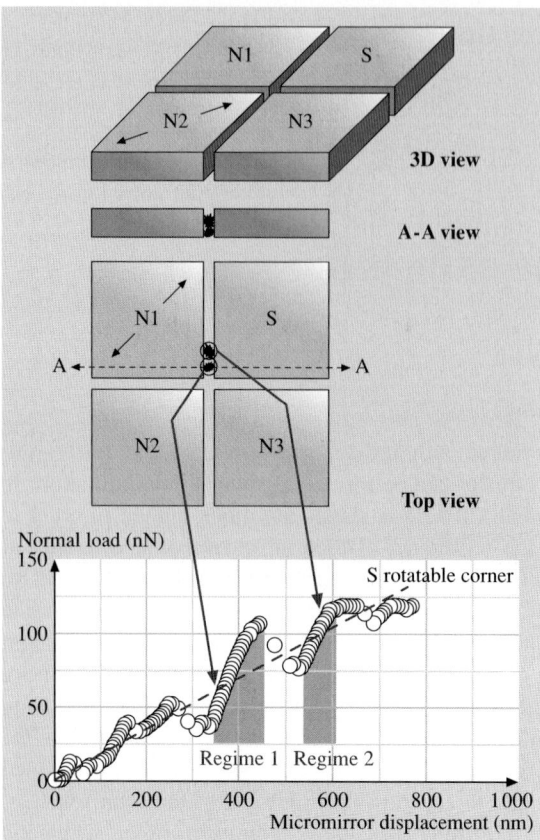

Fig. 22.57. Load–displacement curve obtained on the rotatable corner of the S micromirror and schematic to illustrate the suggested mechanism for the occurrence of soft stiction

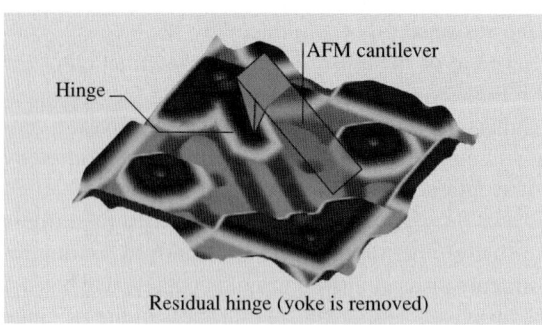

Fig. 22.58. AFM surface-height image of the residual hinge and schematic diagram of the relative position of the hinge and AFM tip during the nanoscale bending and fatigue tests. The tip is located at the free end of the hinge [196]

22.7 Conclusion

The field of MEMS/NEMS has expanded considerably over the last decade. The length scale and large surface-to-volume ratio of the devices result in very high retarding forces such as adhesion and friction, which seriously undermine the performance and reliability of the devices. These tribological phenomena need to be studied and understood at the micro- to nanoscales. In addition, materials for MEMS/NEMS must exhibit good microscale tribological properties. There is a need to develop lubricants and identify lubrication methods that are suitable for MEMS/NEMS. Using AFM-based techniques, researchers have conducted micro/nanotribological studies of materials and lubricants for use in MEMS/NEMS. In addition, component-level testing has also been carried out to aid understanding of the observed tribological phenomena in MEMS/NEMS.

Macroscale and microscale tribological studies of silicon and polysilicon films have been performed. The effect of doping and oxide films and environment on the tribological properties of these popular MEMS/NEMS materials have also been studied. SiC film is found to be a good tribological material for use in high-temperature MEMS/NEMS devices. Perfluoroalkyl self-assembled monolayers and bonded perfluoropolyether lubricants appear to be well suited for lubrication of microdevices under a range of environmental conditions. DLC coatings can also be used for low friction and wear. Adhesion of biomolecules on Si surfaces can be improved by nanopatterning and the chemical linker method. Roughness should be optimized for superhydrophobicity low adhesion and friction. Surface roughness measurements of micromachined polysilicon surfaces have been made using an AFM. The roughness distribution on surfaces is strongly dependent upon the fabrication process. Adhesion and friction of microstructures can be measured using a novel microtriboapparatus. Adhesion and friction measurements of silicon on silicon confirm AFM measurements that hexadecane thiol and bonded perfluoropolyether films exhibit superior adhesion and friction properties. Static friction force measurements of micromotors have been performed using an AFM. The forces are found to vary considerably with humidity. A bonded layer of perfluoropolyether lubricant is found to satisfactorily reduce the friction forces in the micromotor. Tribological failure modes of digital micromirror devices are either *hard* stiction or *soft* stiction. In hard stiction, the tip on the yoke remains stuck to the landing site underneath. The mechanism responsible for the hard stiction is localized damage to the SAM on the landing site. However, in soft stiction, the mirror–yoke assembly rotates with hesitation. The mechanism responsible for soft stiction is contaminant particles present on the mirror sidewalls.

AFM/FFM-based techniques show the capability to study and evaluate micro/nanoscale tribological phenomena related to MEMS/NEMS devices.

22.A Appendix Micro/Nanofabrication Methods

22.A.1 Top-Down Methods

The top-down fabrication methods used in the construction of MEMS include lithographic and non-lithographic techniques to produce micro- and nanostructures. The lithographic techniques fall into three basic categories: bulk micromachining, surface micromachining, and LIGA (a German acronym for Lithographie Galvanoformung Abformung), a German term for lithography, electroplating, and molding.

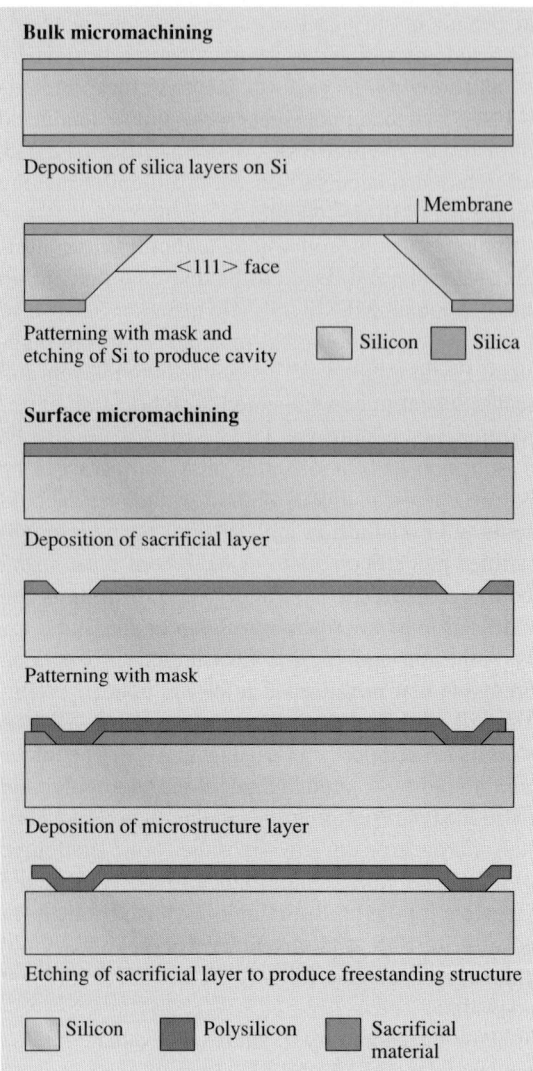

Fig. 22.59. Schematic of the process steps involved in bulk micromachining and surface-micromachining fabrication of MEMS

The first two approaches, bulk and surface micromachining, mostly use planar photolithographic fabrication processes developed for semiconductor devices in producing two-dimensional (2D) structures [13, 19, 198, 199]. The various steps involved in these two fabrication processes are shown schematically in Fig. 22.59. Bulk micromachining employs anisotropic etching to remove sections through the thickness of a single-crystal silicon wafer, typically 250–500 μm thick. Bulk micromachining is a proven high-volume production process and is routinely used to fabricate microstructures such as accelerometers, pressure sensors, and flow sensors. In surface micromachining, structural and sacrificial films are alternatively deposited, patterned, and etched to produce a freestanding structure. These films are typically made of low-pressure chemical vapor deposition (LPCVD) polysilicon film with a thickness of 2–20 μm. Surface micromachining is used to produce sensors, actuators, micromirror arrays, motors, gears, and grippers. The resolution in photolithography is dependent upon the wavelength of light. A commonly used light source is an argon fluoride excimer laser with a wavelength of 193 nm (ultraviolet or UV) used in patterning 90-nm lines and spaces. Deep-UV wavelengths, X-ray lithography, electron beam (e-beam) lithography, focused ion-beam lithography, maskless lithography, liquid-immersion lithography, and STM writing by removing material atom by atom are some of the recent developments for sub-100-nm patterning.

The fabrication of nanostructures such as nanochannels with sub-10-nm resolution can be accomplished through several routes: electron beam lithography and sacrificial-layer lithography (SLL). The process for e-beam lithographic technique is a finely focused electron beam that is exposed over a resist surface, where the exposure duration and location is controlled with the use of a computer [200, 201]. When the resist is exposed to the electron beam, the electrons either break or join the molecules in the resist, so the local characteristics are changed in such a way that further processes can either remove the exposed part (positive resist) or remove the unexposed part (negative resist). The resist material determines if the molecules will either break or join together and thus determines if a positive or negative image is produced. E-beam lithography can either be used to create photolithographic masks for replication or to create the devices directly. The masks that are created can be used for either optical or X-ray lithography. One limitation of e-beam lithography is that throughput is drastically reduced since a single electron beam is used to create the entire exposure pattern on the resist. While this technique is slower than conventional lithographic techniques, it is ideal for prototype fabrication because no masks are required.

In SLL process, the use of a sacrificial layer allows the direct control of nanochannel dimensions as long as there exists a method for removing the sacrificial layer with absolute selectivity to the structural layers. A materials system with such selectivity is the silicon/silicon oxide system used widely in the microfabrication of MEMS devices. The use of sidewall deposition of the sacrificial layer and subsequent etching allows the fabrication of high-density nanochannels for biomedical applications. It is based on surface micromachining [77]. Figure 22.60 shows a schematic of the process steps in sacrificial-layer lithography based on *Hansford* et al.'s [77] work on fabrication of polysilicon membranes with nanochannels. As with all the membrane protocols, the first step in the fabrication is the etching of the support ridge structure into the bulk silicon substrate. A low-stress silicon nitride (LSN or simply nitride), which functions as an etch stop layer, is then deposited using LPCVD. The base structural polysilicon layer (base layer) is deposited on top of the etch stop layer. The plasma etching of holes in the base layer is what defines the shape of the pores. The buried nitride acts as an etch stop for the plasma etching of a polysilicon base layer. After the pore holes are etched through the base layer, the pore sacrificial thermal oxide layer is grown on the base layer. The basic require-

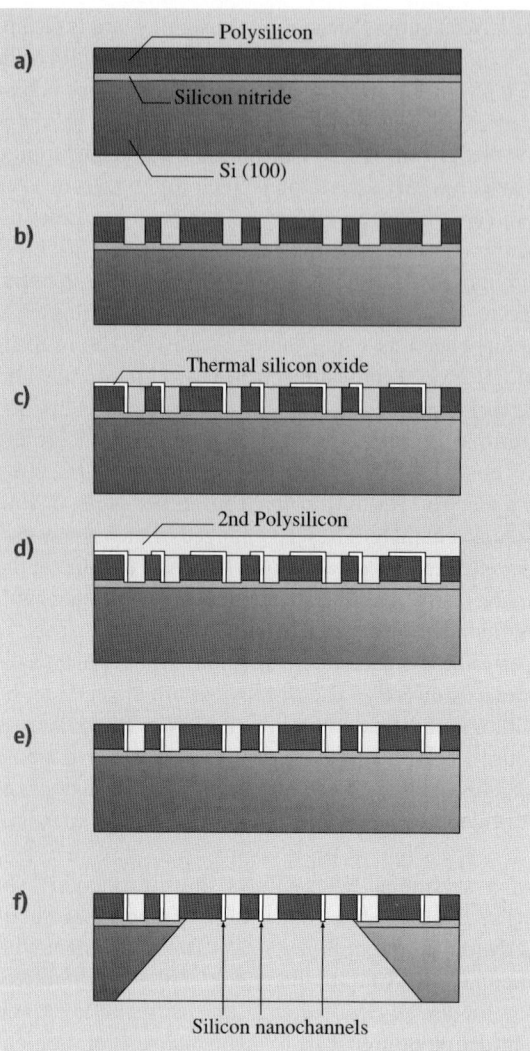

Fig. 22.60. Schematic of process steps involved in sacrificial-layer lithography: (a) Growth of silicon nitride layer (etch stop) and base polysilicon deposition, (b) hole definition in base, (c) growth of thin sacrificial oxide and patterning of anchor points, (d) deposition of plug polysilicon, (e) planarization of plug layer, and (f) deposition and patterning of the protective nitride layer through etch, followed by etching of protective, sacrificial and etch layers before final release of the structure in HF [77]

ment of the sacrificial layer is the ability to control the thickness with high precision across the entire wafer. Anchor points are defined in the sacrificial oxide layer to mechanically connect the base layer with the plug layer (necessary to maintain the pore spacing between layers). This is accomplished by using the same mask shifted from the pore holes. This produces anchors in one or two corners of each pore hole, which provide the desired connection between the structural layers while opening as much pore area as possible. After the anchor points are etched through the sacrificial oxide, the plug polysilicon layer is deposited (using LPCVD) to fill in the holes. To open the pores at the surface, the plug layer is planarized using chemical mechanical polishing (CMP) down to the base layer, leaving the final structure with the plug layer only in the pore hole openings. As the silicon wafer is ready for release, a protective nitride layer is deposited on the wafer (completely covering both sides of the wafer). The back-side etch windows are etched in the protective layer, exposing the silicon wafer in the desired areas, and the wafer is placed in a KOH bath to etch. After the silicon wafer is completely removed up to the membrane (as evidenced by the smooth buried etch-stop layer), the protective, sacrificial, and etch-stop layers are removed by etching in concentrated HF. Etching of sacrificial layer in polysilicon film defines nanochannels.

The LIGA process is based on the combined use of X-ray lithography, electroplating, and molding processes. The steps involved in the LIGA process are shown schematically in Fig. 22.61. LIGA is used to produce high-aspect-ratio MEMS (HARMEMS) devices that are up to 1 mm in height and only a few microns in width or length [202]. The LIGA process yields very sturdy 3D structures due to their increased thickness. One of the limitations of silicon microfabrication processes originally used for fabrication of MEMS devices is the lack of suitable materials that can be processed. With LIGA, a variety of non-silicon materials such as metals, ceramics and polymers can be processed. Non-lithographic micromachining processes, primarily in Europe and Japan, are also being used for the fabrication of millimeter-scale devices using direct material microcutting or micromechanical ma-

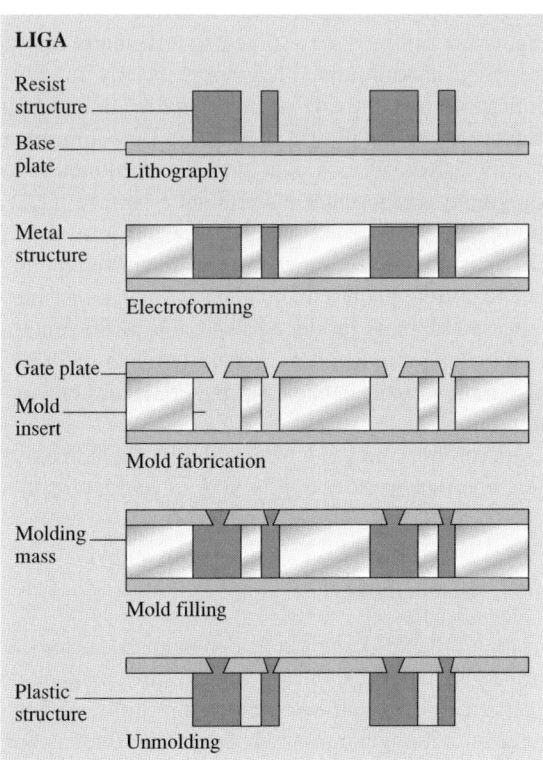

Fig. 22.61. Schematic of the process steps involved in LIGA fabrication of MEMS

chining (such as microturning, micromilling, and microdrilling) or removal by energy beams (such as microspark erosion, focused ion beam, laser ablation, and laser polymerization) [19, 203]. Hybrid technologies including LIGA and high-precision micromachining techniques have been used to produce miniaturized motors, gears, actuators, and connectors [88–90, 204]. These millimeter-scale devices may find more immediate applications.

A micro/nanofabrication technique, so-called *soft lithography*, is a non-lithographic technique [53, 59, 205, 206] in which a master or mold is used to generate patterns, defined by the relief on its surface, on polymers by replica molding [207], embossing (nanoimprint lithography) [208], or by contact printing (known as microcontact printing or μCP) [209]. Soft lithography is faster, less expensive, and more suitable for most biological applications than glass or silicon micromachining. Polymers have established an important role in BioMEMS/BioNEMS because of their reduced cost. The use of polymers also offers a wide range of material properties to allow tailoring of biological interactions for improved biocompatibility. Polymer fabrication is believed to be about an order of magnitude cheaper than silicon fabrication.

Replica molding is the transfer of a topographic pattern by curing or solidifying a liquid precursor against the original patterned mold. The mold or stamp is generally made of a two-part polymer (elastomer and curing agent), called poly(dimethylsiloxane) (PDMS) from photolithographically generated photoresist master. Solvent-based embossing, or imprinting, uses a solvent to restructure a polymer film. Hot embossing, also called nanoimprint lithography, usually refers to the transfer of pattern from a micromachined quartz or metal master to a pliable plastic sheet. Heat and high pressure allow the plastic sheet to become imprinted. These sheets can then be bonded to various plastics such as polymethyl methacrylate (PMMA). Nanoimprint lithography can produce patterns on a surface having 10-nm resolution. Contact printing uses a patterned stamp to transfer ink (mostly self-assembled monolayers) onto a surface in a pattern defined by the raised regions of a stamp. These techniques can be used to pattern line widths as small as 60 nm.

Replica molding is commonly used for mass-produced disposable plastic micro/nanocomponents, for example micro/nanofluidic chips, generally made of PDMS and PMMA [206, 210], and is also more flexible in choice of materials for construction than conventional photolithography.

To assemble microsystems, microrobots are used. Microrobotics include building blocks, such as steering links, microgrippers, conveyor system, and locomotive robots [17].

22.A.2 Bottom-Up Fabrication (Nanochemistry)

The bottom-up approach (from small to large) largely relies on nanochemistry [39, 40, 42–46]. The bottom-up approach includes chemical synthesis, the spontaneous *self-assembly* of molecular clusters (molecular self-assembly) from simple reagents in solution or biological molecules as building blocks to produce three-dimensional nanostructures as done by nature, quantum dots (nanocrystals) of arbitrary diameter (about $1 \times 10^1 - 1 \times 10^5$ atoms), molecular-beam epitaxy (MBE) and organometallic vapor-phase epitaxy (OMVPE) to create specialized crystals one atomic or molecular layer at a time, and manipulation of individual atoms by a scanning tunneling microscope or an atomic force microscope or atom optics. The self-assembly must be encoded, that is, one must be able to precisely assemble one object next to another to form a designed pattern. A variety of nonequilibrium plasma chemistry techniques are also used to produce layered nanocomposites, nanotubes, and nanoparticles. Nanostructures can also be fabricated using mechanosynthesis with proximal probes.

References

1. Anonymous: *Microelectromechanical Systems: Advanced Materials and Fabrication Methods*, NMAB-483 (National Academy, Washington, D.C. 1997)
2. M. Roukes: Nanoelectromechanical systems face the future, Physics World, 25–31 (Feb 2001)
3. Anonymous: *Small Tech 101 – An Introduction to Micro and Nanotechnology* (Small Times, 2003)
4. M. Schulenburg: *Nanotechnology – Innovation for Tomorrow's World* (European Commission, Research DG, Brussels 2004)
5. J.C. Eloy: *Status of the MEMS Industry 2005*, Report of Yole Developpement, France (SPIE Photonic West, San Jose 2005) presented at
6. S. Lawrence: Nanotech grows up, Technol. Rev. **108**(6), 31 (2005)
7. A.P. Graham, G.S. Duesberg, R. Seidel, M. Liebau, E. Unger, F. Kreupl, W. Hoenlein: Towards the integration of carbon nanotubes in microelectronics, Diam. Relat. Mater. **13**, 1296–1300 (2004)
8. D. Srivastava: Computational Nanotechnology of Carbon Nanotubes. In: *Carbon Nanotubes: Science and Applications*, ed. by M. Meyyappan (CRC, Boca Raton 2004) pp. 25–63
9. W.G. van der Wiel, S. De Franceschi, J.M. Elzerman, T. Fujisawa, S. Tarucha, L.P. Kouwenhoven: Electron transport through double quantum dots, Rev. Mod. Phys. **75**, 1–22 (2003)
10. R.S. Muller, R.T. Howe, S.D. Senturia, R.L. Smith, R.M. White: *Microsensors* (IEEE, New York 1990)
11. I. Fujimasa: *Micromachines: A New Era in Mechanical Engineering* (Oxford Univ. Press, Oxford 1996)
12. W.S. Trimmer (ed.): *Micromachines and MEMS, Classic and Seminal Papers to 1990* (IEEE, New York 1997)
13. B. Bhushan: *Tribology Issues and Opportunities in MEMS* (Kluwer Academic, Dordrecht 1998) Netherlands
14. G.T.A. Kovacs: *Micromachined Transducers Sourcebook* (WCB McGraw–Hill, Boston 1998)
15. S.D. Senturia: *Microsystem Design* (Kluwer Academic, Boston 2000)
16. M. Elwenspoek, R. Wiegerink: *Mechanical Microsensors* (Springer, Berlin, Heidelberg 2001)
17. M. Gad-el-Hak: *The MEMS Handbook* (CRC, Boca Raton 2002)

18. T.R. Hsu: *MEMS and Microsystems: Design and Manufacture* (McGraw–Hill, Boston 2002)
19. M. Madou: *Fundamentals of Microfabrication: The Science of Miniaturization*, 2nd edn. (CRC, Boca Raton 2002)
20. A. Hierlemann: *Integrated Chemical Microsensor Systems in CMOS Technology* (Springer, Berlin, Heidelberg 2005)
21. T.A. Core, W.K. Tsang, S.J. Sherman: Fabrication technology for an integrated surface-micromachined sensor, Solid State Technol. **36**, 39–47 (Oct 1993)
22. J. Bryzek, K. Peterson, W. McCulley: Micromachines on the march, IEEE Spectrum, 20–31 (May 1994)
23. J.S. Aden: The third-generation HP thermal inkjet printhead, HP J., 4–45 (Feb. 1994)
24. H. Le: Progress and trends in ink-jet printing technology, J. Imaging Sci. Technol. **42**, 49–62 (1998)
25. R. Baydo, A. Groscup: Getting to the heart of ink jet: Printheads, Beyond Recharger, 10–12 (2001)May 10, also visit
26. E.R. Lee: *Microdrop Generation* (CRC, Boca Raton 2003)
27. L.J. Hornbeck, W.E. Nelson: Bistable Deformable Mirror Device, Technical Digest Series: Spatial Light Modulators and Applications, Vol. 8 (OSA, Washington 1988) pp. 107–110
28. L.J. Hornbeck: A digital light processing™ update – Status and future applications, Proc. Soc. Photo-Opt. Eng. **3634** (Projection Displays V), 158–170 (1999)
29. L.J. Hornbeck: The DMD™ projection display chip: A MEMS-based technology, MRS Bull. **26**, 325–328 (2001)
30. K. Suzuki: Micro electro mechanical systems (MEMS) micro-switches for use in DC, RF, and optical applications, Jpn. J. Appl. Phys. **41**, 4335–4339 (2002)
31. V.M. Lubecke, J.C. Chiao: *MEMS Technologies for Enabling High Frequency Communication Cicuits*, Proc. IEEE 4th Int. Conf. on Telecom. In Modern Satellite, Cable and Broadcasting Services, Nis, Yugoslavia (IEEE, New York 1999) pp. 1–8
32. C.R. Giles, D. Bishop, V. Aksyuk: MEMS for light-wave networks, MRS Bull., 328–329 (April 2001)
33. A. Hierlemann, O. Brand, C. Hagleitner, H. Baltes: Microfabrication Techniques for Chemical/Biosensors. In: *Proc. of the IEEE, Chemical and Biological Microsensors*, Vol. Vol. 91, ed. by S. Casalnuovo, R.B. Brown (IEEE, New York 2003) pp. 839–863
34. B. Bhushan: *Tribology and Mechanics of Magnetic Storage Devices*, 2nd edn. (Springer, New York 1996)
35. H. Hamilton: Contact recording on perpendicular rigid media, J. Mag. Soc. Jpn. **15**((Suppl. S2)), 483–481 (1991)
36. D.A. Horsley, M.B. Cohn, A. Singh, R. Horowitz, A.P. Pisano: Design and fabrication of an angular microactuator for magnetic disk drives, J. Microelectromech. Syst. **7**, 141–148 (1998)
37. T. Hirano, L.S. Fan, D. Kercher, S. Pattanaik, T.S. Pan: HDD Tracking Microactuator and its Integration Issues. In: *Proc. ASME Int. Mech. Eng. Congress and Exp.*, Vol. MEMS- Vol. 2, ed. by A.P. Lee, J. Simon, F.K. Foster, R.S. Keynton (ASME, New York 2000) pp. 449–452
38. T. Fukuda, F. Arai, L. Dong: Assembly of nanodevices with carbon nanotubes through nanorobotic manipulations, Proc. IEEE **91**, 1803–1818 (2003)
39. K.E. Drexler: *Nanosystems: Molecular Machinery, Manufacturing and Computation* (Wiley, New York 1992)
40. G. Timp (ed.): *Nanotechnology* (Springer, New York 1999)

41. M.S. Dresselhaus, G. Dresselhaus, Ph. Avouris: *Carbon Nanotubes – Synthesis, Structure, Properties and Applications* (Springer, Berlin, Heidelberg 2001)
42. E.A. Rietman: *Molecular Engineering of Nanosystems* (Springer, New York 2001)
43. W.A. Goddard, D.W. Brenner, S.E. Lyshevski, G.J. Iafrate (eds): *Handbook of Nanoscience, Engineering, and Technology* (CRC, Boca Raton 2002)
44. H.S. Nalwa (ed.): *Nanostructured Materials and Nanotechnology* (Academic, San Diego 2002)
45. C.P. Poole, F.J. Owens: *Introduction to Nanotechnology* (Wiley, Hoboken 2003)
46. B. Bhushan: *Handbook of Micro/Nanotribology*, 2nd edn. (CRC, Boca Raton 1999)
47. B. Bhushan: *Nanotribology and Nanomechanics – An Introduction* (Springer, Berlin, Heidelberg 2005)
48. J.A. Stroscio, D.M. Eigler: Atomic and molecular manipulation with a scanning tunneling microscope, Science **254**, 1319 (1991)
49. P. Vettiger, J. Brugger, M. Despont, U. Drechsler, U. Duerig, W. Haeberle: Ultrahigh density, high data-rate NEMS based AFM data storage system, Microelec. Eng. **46**, 11–27 (1999)
50. C. Stampfer, A. Jungen, C. Hierold: *Fabrication of Discrete Carbon Nanotube Based Nanoscaled Force Sensor*, Proc. IEEE Sensors 2004, Vienna (IEEE, New York 2004) pp. 1056–1059
51. B. Bhushan: *Mechanics and Reliability of Flexible Magnetic Media*, 2nd edn. (Springer, New York 2000)
52. Anonymous: *International Technology Roadmap for Semiconductors* (2004), http://public.itrs.net/
53. A. Manz, H. Becker (eds): *Microsystem Technology in Chemistry and Life Sciences*, Topics in Current Chemistry 194 (Springer, Berlin, Heidelberg 1998)
54. J. Cheng, L.J. Kricka (eds): *Biochip Technology* (Harwood Academic, Philadelphia 2001)
55. M.J. Heller, A. Guttman (eds): *Integrated Microfabricated Biodevices* (Marcel Dekker, New York 2001)
56. C. Lai Poh San, E.P.H. Yap (eds.): *Frontiers in Human Genetics* (World Scientific, Singapore 2001)
57. C.H. Mastrangelo, H. Becker (eds): *Microfluidics and BioMEMS*, Proc. of SPIE Vol (SPIE, Bellingham 2001) p. 4560
58. H. Becker, L.E. Locascio: Polymer microfluidic devices, Talanta **56**, 267–287 (2002)
59. D.J. Beebe, G.A. Mensing, G.M. Walker: Physcis and applications of microfluidics in biology, Annu. Rev. Biomed. Eng. **4**, 261–286 (2002)
60. A. van der Berg (ed.): *Lab-on-a-Chip: Chemistry in Miniaturized Synthesis and Analysis Systems* (Elsevier, Amsterdam 2003)
61. P. Gravesen, J. Branebjerg, O. Jensen: Microfluidics – A Review, J. Micromech. Microeng. **3**, 168–182 (1993)
62. S.N. Bhatia, C.S. Chen: Tissue engineering at the micro-scale, Biomed. Microdevices **2**, 131–144 (1999)
63. R.P. Lanza, R. Langer, J. Vacanti (eds): *Principles of Tissue Engineering*, 2nd edn. (Academic, San Diego 2000)
64. E. Leclerc, K.S. Furukawa, F. Miyata, T. Sakai, T. Ushida, T. Fujii: Fabrication of microstructures in photosensitive biodegradable polymers for tissue engineering applications, Biomaterials **25**, 4683–4690 (2004)
65. K. Park (ed): *Controlled Drug Delivery: Challenges and Strategies* (American Chemical Society, Washington, D.C. 1997)

66. R. S. Shawgo, A. C. R. Grayson, Y. Li, M. J. Cima: BioMEMS for drug delivery, Curr. Opin. Solid State Mater. Sci. **6**, 329–334 (2002)
67. P. A. Oeberg, T. Togawa, F. A. Spelman: *Sensors in Medicine and Health Care* (Wiley, New York 2004)
68. J. V. Zoval, M. J. Madou: Centrifuge-based fluidic platforms, Proc. IEEE **92**, 140–153 (2000)
69. W. C. Tang, A. P. Lee: Defense applications of MEMS, MRS Bull. **26**, 318–319 (2001)Also see www.darpa.mil/mto/mems
70. M. R. Taylor, P. Nguyen, J. Ching, K. E. Peterson: Simulation of microfluidic pumping in a genomic DNA blood-processing cassette, J. Micromech. Microeng. **13**, 201–208 (2003)
71. R. Raiteri, M. Grattarola, M. Butt, P. Skladal: Micromechanical cantilever-based biosensor, Sens. Actuators B: Chem. **79**, 115–126 (2001)
72. B. Bhushan, D. R. Tokachichu, M. T. Keener, S. C. Lee: Morphology and adhesion of biomolecules on silicon based surfaces, Acta Biomateriala **1**, 327–341 (2005)
73. H. P. Lang, M. Hegner, C. Gerber: Cantilever array sensors, Mater. Today, 30–36 (April 2005)
74. F. Patolsky, C. Lieber: Nanowire nanosensors, Mater. Today, 20–28 (April 2005)
75. M. Scott: *MEMS and MOEMS for National Security ApplicationsReliability, Testing, and Characterization of MEMS/MOEMS II*, Proc. of SPIE, Vol. Vol. 4980 (SPIE, Bellingham 2003) pp. xxxvii–xliv
76. T. A. Desai, D. J. Hansford, L. Kulinsky, A. H. Nashat, G. Rasi, J. Tu, Y. Wang, M. Zhang, M. Ferrari: Nanopore technology for biomedical applications, Biomed. Devices **2**, 11–40 (1999)
77. D. Hansford, T. Desai, M. Ferrari: Nano-Scale Size-Based Biomolecular Separation Technology. In: *Biochip Technology*, ed. by J. Cheng, L. J. Kricka (Harwood Academic, New York 2001) pp. 341–361
78. F. J. Martin, C. Grove: Microfabricated drug delivery systems: concepts to improve clinical benefits, Biomed. Microdev. **3**, 97–108 (2001)
79. B. Bhushan: *Principles and Applications of Tribology* (Wiley, New York 1999)
80. B. Bhushan (ed.): *Modern Tribology Handbook* (CRC, Boca Raton 2001)
81. B. Bhushan: *Introduction to Tribology* (Wiley, New York 2002)
82. B. Bhushan: Adhesion and stiction: Mechanisms, measurement techniques, and methods for reduction, J. Vac. Sci. Technol. B **21**, 2262–2296 (2003)
83. Y. C. Tai, L. S. Fan, R. S. Muller: IC-Processed Micro-Motors: Design, Technology and Testing, Proc. IEEE Micro Electro Mechanical Systems, 1–6 (1989)
84. S. M. Spearing, K. S. Chen: Micro-gas turbine engine materials and structures, Ceramic Eng. Sci. Proc. **18**, 11–18 (2001)
85. L. G. Frechette, S. A. Jacobson, K. S. Breuer: High-speed microfabricated silicon turbomachinery and fluid film bearings, J. MEMS **14**, 141–152 (2005)
86. L. X. Liu, Z. S. Spakovszky: *Effect of Bearing Stiffness Anisotropy on Hydrostatic Micro Gas Journal Bearing Dynamic Behavior*, Proc. ASME Turbo Expo 2005, Paper No. GT-2005–68199 (Reno, Nevada 2005)
87. M. Mehregany, K. J. Gabriel, W. S. N. Trimmer: Integrated fabrication of polysilicon mechanisms, IEEE Trans. Electron. Dev. **35**, 719–723 (1988)
88. H. Lehr, S. Abel, J. Doppler, W. Ehrfeld, B. Hagemann, K. P. Kamper, F. Michel, Ch. Schulz, Ch. Thurigen: Microactuators as Driving Units for Microrobotic Systems. In: *Proc. Microrobotics: Components and Applications*, Vol. Vol. 2906, ed. by A. Sulzmann (SPIE, Bellingham 1996) pp. 202–210

89. H. Lehr, W. Ehrfeld, B. Hagemann, K.P. Kamper, F. Michel, Ch. Schulz, Ch. Thurigen: Development of micro-millimotors, Min. Invas. Ther. Allied Technol. **6**, 191–194 (1997)
90. F. Michel, W. Ehrfeld: Microfabrication Technologies for High Performance Microactuators. In: *Tribology Issues and Opportunities in MEMS*, ed. by B. Bhushan (Kluwer Academic, Dordrecht 1998) pp. 53–72
91. D.M. Tanner, N.F. Smith: *MEMS Reliability: Infrastructure, Test Structures, Experiments, and Failure Modes*, SAND2000-0091 (Sandia National Laboratories, Albuquerque 2000) Download from www.prod.sandia.gov
92. E.J. Garcia, J.J. Sniegowski: Surface micromachined microengine, Sens. Actuators A **48**, 203–214 (1995)
93. S.S. Mani, J.G. Fleming, J.A. Walraven, J.J. Sniegowski: *Effect of W Coating on Microengine Performance*, Proc. 38th Annual Inter. Reliability Phys. Symp. (IEEE, New York 2000) pp. 146–151
94. M.G. Hankins, P.J. Resnick, P.J. Clews, T.M. Mayer, D.R. Wheeler, D.M. Tanner, R.A. Plass: *Vapor Deposition of Amino-Functionalized Self-Assembled Monolayers on MEMS*, Proc. SPIE, Vol. 4980 (SPIE, Bellingham 2003) pp. 238–247
95. J.K. Robertson, K.D. Wise: An electrostatically actuated integrated microflow controller, Sens. Actuators A **71**, 98–106 (1998)
96. B. Bhushan: *Nanotribology and Nanomechanics of MEMS Devices*, Proc. Ninth Annual Workshop on Micro Electro Mechanical Systems (IEEE, New York 1996) pp. 91–98
97. R.E. Sulouff: MEMS Opportunities in Accelerometers and Gyros and the Microtribology Problems Limiting Commercialization. In: *Tribology Issues and Opportunities in MEMS*, ed. by B. Bhushan (Kluwer Academic, Dordrecht 1998) pp. 109–120
98. J.R. Martin, Y. Zhao: Micromachined Device Packaged to Reduce Stiction, US Patent 5,694,740 (1997) Dec. 9
99. G. Smith: The application of microtechnology to sensors for the automotive industry, Microelectron. J. **28**, 371–379 (1997)
100. L.S. Chang, P.L. Gendler, J.H. Jou: Thermal mechanical and chemical effects in the degradation of the plasma-deposited α-SC : H passivation layer in a multlayer thin-film device, J. Mater. Sci. **26**, 1882–1290 (1991)
101. M. Parsons: Design and manufacture of automotive pressure sensors, Sensors **18**(4), 32–46 (2001)
102. S.A. Henck: Lubrication of digital micromirror devices, Tribol. Lett. **3**, 239–247 (1997)
103. M.R. Douglass: *Lifetime Estimates and Unique Failure Mechanisms of the Digital Micromirror Devices (DMD)*, Proc. 36th Annual Inter. Reliability Phys. Symp. (IEEE, New York 1998) pp. 9–16
104. M.R. Douglass: *DMD Reliability: A MEMS Success Story*, Reliability, Testing, and Characterization of MEMS/MOEMS II, Proc. of SPIE Vol. 4980 (SPIE, Bellingham 2003) pp. 1–11
105. L.J. Hornbeck: Low Surface Energy Passivation Layer for Micromechanical Devices, US Patent 5,602,671 (1997) Feb. 11
106. R.A. Robbins, S.J. Jacobs: Lubricant Delivery for Micromechanical Devices, US Patent 6,300,294 B1 (2001) Oct. 9
107. I. DeWolf, W.M. van Spengen: Techniques to study the reliability of metal RF MEMS capacitive switches, Microelectron. Reliab. **42**, 1789–1794 (2002)
108. B. McCarthy, G.G. Adams, N.E. McGruer, D. Potter: A dynamic model, including contact bounce, of an electrostatically actuated microswitch, J. MEMS **11**, 276–283 (2002)

109. S. Shoji, M. Esashi: Microflow devices and systems, J. Micromech. Microeng. **4**, 157–171 (1994)
110. M. Stehr, S. Messner, H. Sandmaier, R. Zenergle: The VAMP – A new device for handing liquids or gases, Sens. Actuators A **57**, 153–157 (1996)
111. P. Woias: Micropumps – Summarizing the First Two Decades. In: *Proc. of SPIE - Microfluidics and BioMEMS*, Vol. Vol. 4560, ed. by C. H. Mastrangelo, H. Becker (SPIE, Bellingham 2001) pp. 39–52
112. N. T. Nguyen, X. Huang, T. K. Chuan: MEMS-micropumps: A review, ASME J. Fluids Eng. **124**, 384–392 (2002)
113. J. Henniker: Triboelectricity in polymers, Nature **196**, 474 (1962)
114. M. Sakaguchi, H. Kashiwabara: A generation mechanism of triboelectricity due to the reaction of mechaniradicals with mechanoions which are produced by mechanical fracture of solid polymer, Colloid Polym. Sci. **270**, 621–626 (1992)
115. G. R. Freeman, N. H. March: Triboelectricity and some associated phenomena, Mater. Sci. Eng. **15**, 1454–1458 (1999)
116. S. K. Cho, H. Moon, C. -J. Kim: Creating, transporting, cutting, and merging liquid droplets by electrowetting-based actuation for digital microfluidic circuits, J. MEMS **12**, 70–80 (2003)
117. A. R. Wheeler, H. Moon, C. A. Bird, R. R. O. Loo, C. -J. Kim, J. A. Loo, R. L. Garrell: Digital microfluidics with in-line sample purification for proteomics analysis with MALDI-MS, Anal. Chem. **77**, 534–540 (2005)
118. S. C. Lee, M. T. Keener, D. R. Tokachichu, B. Bhushan, P. D. Barnes, B. R. Cipriany, M. Gao, L. J. Brillson: Preparation of a thin protein surface on thermally grown silicon dioxide, J. Vac. Sci. Technol. B **23**, 1856–1865 (2005)
119. J. Black: *Biological Performance of Materials: Fundamentals of Biocompatibility* (Marcel Dekker, New York 1999)
120. D. R. Tokachichu, B. Bhushan: Bioadhesion of Polymers for BioMEMS, IEEE Trans. Nanotech. **5**, 228–231 (2006)
121. G. Wei, B. Bhushan, N. Ferrell, D. Hansford: Microfabrication and nanomechanical characterization of polymer microelectromechanical systems for biological applications, J. Vac. Sci. Technol. A **23**, 811–819 (2005)
122. B. Bhushan, T. Kasai, G. Kulik, L. Barbieri, P. Hoffman: AFM study of perfluorosilane and alkylsilane self-assembled monolayers for anti-stiction in MEMS/NEMS, Ultramicroscopy **105**, 176–188 (2005)
123. B. Bhushan, D. Hansford, K. K. Lee: Surface Modification of Silicon and PDMS Surfaces with Vapor Phase Deposited Ultrathin Fluorosilane Films for Biomedical Nanodevices, J. Vac. Sci. Technol. A **24**, 1197–1202 (2005)
124. T. Kasai, B. Bhushan, G. Kulik, L. Barbieri, P. Hoffman: Nanotribological study of perfluorosilane SAMs for anti-stiction and low wear, J. Vac. Sci. Technol. B **23**, 995–1003 (2005)
125. K. K. Lee, B. Bhushan, D. Hansford: Nanotribological characterization of perfluoropolymer thin films for biomems applications, J. Vac. Sci. Technol. A **23**, 804–810 (2005)
126. P. Decuzzi, S. Lee, B. Bhushan, M. Ferrari: A theoretical model for the margination of particles with blood vessels, Annals Biomed. Eng. **33**, 179–190 (2005)
127. H. Guckel, D. W. Burns: Fabrication of micromechanical devices from polysilicon films with smooth surfaces, Sens. Actuators **20**, 117–122 (1989)
128. G. T. Mulhern, D. S. Soane, R. T. Howe: *Supercritical Carbon Dioxide Drying of Microstructures*, Proc. Int. Conf. on Solid-State Sensors and Actuators (IEEE, New York 1993) pp. 296–299

129. K.F. Man, B.H. Stark, R. Ramesham: *A Resource Handbook for MEMS Reliability*, Rev. A. (JPL Press, Jet Propulsion Laboratory, California Institute of Technology, Pasadena 1998)
130. S. Kayali, R. Lawton, B.H. Stark: MEMS reliability assurance activities at JPL, EEE Links **5**, 10–13 (1999)
131. S. Arney: Designing for MEMS reliability, MRS Bull. **26**, 296–299 (2001)
132. K.F. Man: MEMS Reliability for Space Applications by Elimination of Potential Failure Modes Through Testing and Analysis, http://www.rel.jpl.nasa.gov/Org//atop/products/Prod-map.html (2001)
133. B. Bhushan, A.V. Kulkarni, W. Bonin, J.T. Wyrobek: Nano/picoindentation measurement using a capacitance transducer system in atomic force microscopy, Philos. Mag. **74**, 1117–1128 (1996)
134. S. Sundararajan, B. Bhushan: Development of AFM-based techniques to measure mechanical properties of nanoscale structures, Sens. Actuators A **101**, 338–351 (2002)
135. B. Bhushan, J.N. Israelachvili, U. Landman: Nanotribology: Friction, wear and lubrication at the atomic scale, Nature **374**, 607–616 (1995)
136. B. Bhushan, B.K. Gupta: *Handbook of Tribology: Materials, Coatings and Surface Treatments*, Reprint edition edn. (Krieger, Malabar 1997)
137. J.F. Shackelford, W. Alexander, J.S. Park (eds.): *CRC Material Science and Engineering Handbook*, 2nd edn. (CRC, Boca Raton 1994)
138. J.S. Shor, D. Goldstein, A.D. Kurtz: Characterization of n-type β-SiC as a piezoresistor, IEEE Trans. Electron. Dev. **40**, 1093–1099 (1993)
139. M. Mehregany, C.A. Zorman, N. Rajan, C.H. Wu: Silicon Carbide MEMS for Harsh Environments, Proc. IEEE **86**, 1594–1610 (1998)
140. C.A. Zorman, A.J. Fleischmann, A.S. Dewa, M. Mehregany, C. Jacob, S. Nishino, P. Pirouz: Epitaxial growth of 3C–SiC films on 4 in. diam Si(100) silicon wafers by atmospheric pressure chemical vapor deposition, J. Appl. Phys. **78**, 5136–5138 (1995)
141. C.A. Zorman, S. Roy, C.H. Wu, A.J. Fleischman, M. Mehregany: Characterization of polycrystalline silicon carbide films grown by atmospheric pressure chemical vapor deposition on polycrystalline silicon, J. Mater. Res. **13**, 406–412 (1998)
142. C.H. Wu, S. Stefanescu, H.I. Kuo, C.A. Zorman, M. Mehregany: *Fabrication and Testing of Single Crystalline 3C–SiC Piezoresistive Pressure Sensors*, Technical Digest – 11th Int. Conf. Solid State Sensors and Actuators – Eurosensors, Vol. XV (Munich 2001) pp. 514–517
143. A.A. Yasseen, C.H. Wu, C.A. Zorman, M. Mehregany: Fabrication and testing of surface micromachined polycrystalline sic micromotors, IEEE Electron. Dev. Lett. **21**, 164–166 (2000)
144. B.K. Gupta, J. Chevallier, B. Bhushan: Tribology of ion bombarded silicon for micromechanical applications, ASME J. Tribol. **115**, 392–399 (1993)
145. B.K. Gupta, B. Bhushan: Nanoindentation studies of ion implanted silicon, Surf. Coat. Technol. **68-69**, 564–570 (1994)
146. B.K. Gupta, B. Bhushan, J. Chevallier: Modification of tribological properties of silicon by boron ion implantation, Tribol. Trans. **37**, 601–607 (1994)
147. B. Bhushan, V.N. Koinkar: Tribological studies of silicon for magnetic recording applications, J. Appl. Phys. **75**, 5741–5746 (1994)
148. G.M. Pharr: The Anomalous Behavior of Silicon During Nanoindentation. In: *Thin Films: Stresses and Mechanical Properties III*, Vol. 239, ed. by W.D. Nix, J.C. Bravman, E. Arzt, L.B. Freund (Materials Research Soc., Pittsburgh 1991) pp. 301–312
149. D.L. Callahan, J.C. Morris: The extent of phase transformation in silicon hardness indentation, J. Mater. Res. **7**, 1612–1617 (1992)

150. N. A. Fleck, G. M. Muller, M. F. Ashby, J. W. Hutchinson: Strain gradient plasticity: theory and experiment, Acta Metall. Mater. **42**, 475–487 (1994)
151. B. Bhushan, S. Venkatesan: Friction and wear studies of silicon in sliding contact with thin-film magnetic rigid disks, J. Mater. Res. **8**, 1611–1628 (1993)
152. S. Venkatesan, B. Bhushan: The role of environment in the friction and wear of single-crystal silicon in sliding contact with thin-film magnetic rigid disks, Adv. Info Storage Syst. **5**, 241–257 (1993)
153. S. Venkatesan, B. Bhushan: The sliding friction and wear behavior of single-crystal, polycrystalline and oxidized silicon, Wear **171**, 25–32 (1994)
154. B. Bhushan: Chemical, mechanical and tribological characterization of ultra-thin and hard amorphous carbon coatings as thin as 3.5 nm: Recent developments, Diam. Relat. Mater. **8**, 1985–2015 (1999)
155. B. Bhushan, S. Sundararajan, X. Li, C. A. Zorman, M. Mehregany: Micro/Nanotribological Studies of Single-Crystal Silicon and Polysilicon and SiC Films for Use in MEMS Devices. In: *Tribology Issues and Opportunities in MEMS*, ed. by B. Bhushan (Kluwer Academic, Dordrecht 1998) pp. 407–430
156. S. Sundararajan, B. Bhushan: Micro/nanotribological studies of polysilicon and sic films for mems applications, Wear **217**, 251–261 (1998)
157. X. Li, B. Bhushan: Micro/nanomechanical characterization of ceramic films for microdevices, Thin Solid Films **340**, 210–217 (1999)
158. H. Liu, B. Bhushan: Nanotribological characterization of molecularly-thick lubricant films for applications to MEMS/NEMS by AFM, Ultramicroscopy **97**, 321–340 (2003)
159. B. Bhushan, A. V. Kulkarni, V. N. Koinkar, M. Boehm, L. Odoni, C. Martelet, M. Belin: Microtribological characterization of self-assembled and langmuir–blodgett monolayers by atomic force and friction force microscopy, Langmuir **11**, 3189–3198 (1995)
160. V. N. Koinkar, B. Bhushan: Micro/nanoscale studies of boundary layers of liquid lubricants for magnetic disks, J. Appl. Phys **79**, 8071–8075 (1996)
161. V. N. Koinkar, B. Bhushan: Microtribological studies of unlubricated and lubricated surfaces using atomic force/friction force microscopy, J. Vac. Sci. Technol. A **14**, 2378–2391 (1996)
162. B. Bhushan, H. Liu: Nanotribological properties and mechanisms of alkylthiol and biphenyl thiol self-assembled monolayers studied by AFM, Phys. Rev. B **63**, 245412:1–11 (2001)
163. H. Liu, B. Bhushan: Investigation of nanotribological properties of self-assembled monolayers with alkyl and biphenyl spacer chains, Ultramicroscopy **91**, 185–202 (2002)
164. N. S. Tambe, B. Bhushan: Nanotribological characterization of self assembled monolayers deposited on silicon and aluminum substrates, Nanotechnology **16**, 1549–1558 (2005)
165. Z. Tao, B. Bhushan: Bonding, degradation, and environmental effects on novel perfluoropolyether lubrications, Wear **259**, 1352–1361 (2005)
166. Z. Tao, B. Bhushan: Degradation mechanisms and environmental effects on perfluoropolyether, self assembled monolayers, and diamondlike carbon films, Langmuir **21**, 2391–2399 (2005)
167. T. Stifter, O. Marti, B. Bhushan: Theoretical investigation of the distance dependence of capillary and van der waals forces in scanning force microscopy, Phys. Rev. B **62**, 13667–13673 (2000)
168. H. Liu, B. Bhushan, W. Eck, V. Stadler: Investigation of the adhesion, friction, and wear properties of biphenyl thiol self-assembled monolayers by atomic force microscopy, J. Vac. Sci. Technol. A **19**, 1234–1240 (2001)

169. N. S. Tambe, B. Bhushan: Identifying materials with low friction and adhesion for nanotechnology applications, Appl. Phys. Lett. **86**, 061906–1 to –3 (2005)
170. N. S. Tambe, B. Bhushan: Micro/nanotribological characterization of PDMS and PMMA used for bioMEMS/NEMS applications, Ultramicroscopy **105**, 238–247 (2005)
171. B. Bhushan, Z. Burton: Adhesion and friction properties of polymers in microfluidic devices, Nanotechnology **16**, 467–478 (2005)
172. B. Bhushan, D. Tokachichu, M. T. Keener, S. C. Lee: Nanoscale adhesion, friction, and wear studies of biomolecules on silicon based surfaces, Acta Biomaterialia **2**, 39–49 (2005)
173. M. Nosonovsky, B. Bhushan: Roughness optimization for biomimetic superhydrophobic surfaces, Microsyst. Technol. **11**, 535–549 (2005)
174. R. N. Wenzel: Resistance of solid surfaces to wetting by water, Ind. Eng. Chem. **28**, 988–994 (1936)
175. A. Cassie, S. Baxter: Wetting of porous surfaces, Trans. Faraday Soc. **40**, 546–551 (1944)
176. M. Nosonovsky, B. Bhushan: Stochastic model for metastable wetting of roughness-induced superhydrophobic surfaces, Microsyst. Technol. **12**, 231–237 (2006)
177. Z. Burton, B. Bhushan: Hydrophobicity, adhesion and friction properties of nanopatterned polymers and scale dependence for micro- and nanoelectromechanical systems, Nano Lett. **20**, 83–90 (2005)
178. S. Sundararajan, B. Bhushan: Static friction and surface roughness studies of surface micromachined electrostatic micromotors using an atomic force/friction force microscope, J. Vac. Sci. Technol. A **19**, 1777–1785 (2001)
179. C. H. Mastrangelo, C. H. Hsu: Mechanical stability and adhesion of microstructures under capillary forces – Part II: Experiments, J. Microelectromech. Syst. **2**, 44–55 (1993)
180. R. Maboudian, R. T. Howe: Critical review: Adhesion in surface micromechanical structures, J. Vac. Sci. Technol. B **15**, 1–20 (1997)
181. C. H. Mastrangelo: Surface Force Induced Failures in Microelectromechanical Systems. In: *Tribology Issues and Opportunities in MEMS*, ed. by B. Bhushan (Kluwer Academic, Dordrecht 1998) pp. 367–395
182. M. P. De Boer, T. A. Michalske: Accurate method for determining adhesion of cantilever beams, J. Appl. Phys. **86**, 817 (1999)
183. R. L. Alley, G. J. Cuan, R. T. Howe, K. Komvopoulos: In: *Proc. Solid State Sensor and Actuator Workshop*, ed. by C. H. Mastrangelo, C. H. Hsu (IEEE, New York 1992) pp. 202–207
184. H. Liu, B. Bhushan: Adhesion and friction studies of microelectromechanical systems/nanoelectromechanical systems materials using a novel microtriboapparatus, J. Vac. Sci. Technol. A **21**, 1528–1538 (2003)
185. B. Bhushan, H. Liu, S. M. Hsu: Adhesion and friction studies of silicon and hydrophobic and low friction films and investigation of scale effects, ASME J. Tribol. **126**, 583–590 (2004)
186. S. K. Chilamakuri, B. Bhushan: A comprehensive kinetic meniscus model for prediction of long-term static friction, J. Appl. Phys. **15**, 4649–4656 (1999)
187. Y. C. Tai, R. S. Muller: Frictional study of IC processed micromotors, Sens. Actuators A **21-23**, 180–183 (1990)
188. K. J. Gabriel, F. Behi, R. Mahadevan, M. Mehregany: In situ friction and wear measurement in integrated polysilicon mechanisms, Sens. Actuators A. **21-23**, 184–188 (1990)

189. M.G. Lim, J.C. Chang, D.P. Schultz, R.T. Howe, R.M. White: *Polysilicon Microstructures to Characterize Static Friction*, Proc. IEEE Micro Electro Mechanical Systems (IEEE, New York 1990) pp. 82–88
190. U. Beerschwinger, S.J. Yang, R.L. Reuben, M.R. Taghizadeh, U. Wallrabe: Friction measurements on LIGA-processed microstructures, J. Micromech. Microeng. **4**, 14–24 (1994)
191. D. Matheison, U. Beerschwinger, S.J. Young, R.L. Rueben, M. Taghizadeh, S. Eckert, U. Wallrabe: Effect of progressive wear on the friction characteristics of nickel LIGA processed rotors, Wear **192**, 199–207 (1996)
192. H. Liu, B. Bhushan: Nanotribological characterization of digital micromirror devices using an atomic force microscope, Ultramicroscopy **100**, 391–412 (2004)
193. G. Wei, B. Bhushan, S.J. Jacobs: Nanomechanical characterization of digital multilayered thin film structures for digital micromirror devices, Ultramicroscopy **100**, 375–389 (2004)
194. G. Wei, B. Bhushan, S.J. Jacobs: Nanoscale indentation fatigue and fracture toughness measurements of multilayered thin film structures for digital micromirror devices, J. Vac. Sci. Technol. A **22**, 1397–1405 (2004)
195. H. Liu, B. Bhushan: Investigation of nanotribological and nanomechanical properties of the digital micromirror device by atomic force microscope, J. Vac. Sci. Technol. A **22**, 1388–1396 (2004)
196. H. Liu, B. Bhushan: Bending and fatigue study on a nanoscale hinge by an atomic force microscope, Nanotechnology **15**, 1246–1251 (2004)
197. B. Bhushan, H. Liu: Characterization of nanomechanical and nanotribological properties of digital micromirror devices, Nanotechnology **15**, 1785–1791 (2004)
198. R.C. Jaeger: *Introduction to Microelectronic Fabrication*, Vol. Vol. 5 (Addison–Wesley, Reading 1988)
199. J.W. Judy: Microelectromechanical systems (MEMS): Fabrication, design, and applications, Smart Mater. Struct. **10**, 1115–1134 (2001)
200. G. Brewer: *Electron-Bean Technology in Microelectronic Fabrication* (Academic, New York 1980)
201. K. Valiev: *The Physics of Submicron Lithography* (Plenum, New York 1992)
202. E.W. Becker, W. Ehrfeld, P. Hagmann, A. Maner, D. Munchmeyer: Fabrication of microstructures with high aspect ratios and great structural heights by synchrotron radiation lithography, galvanoforming, and plastic moulding (LIGA process), Microelectron. Eng. **4**, 35–56 (1986)
203. C.R. Friedrich, R.O. Warrington: Surface Characterization of Non-Lithographic Micromachining. In: *Tribology Issues and Opportunities in MEMS*, ed. by B. Bhushan (Kluwer Academic, Dordrecht 1998) pp. 73–84
204. M. Tanaka: Development of desktop machining microfactory, Riken Rev. **34**, 46–49 (April 2001)
205. Y. Xia, G.M. Whitesides: Soft lithography, Angew. Chem. Int. Ed. **37**, 550–575 (1998)
206. H. Becker, C. Gaertner: Polymer microfabrication methods for microfluidic analytical applications, Electrophoresis **21**, 12–26 (2000)
207. Y. Xia, E. Kim, X.M. Zhao, J.A. Rogers, M. Prentiss, G.M. Whitesides: Complex optical surfaces formed by replica molding against elastomeric masters, Science **273**, 347–349 (1996)
208. S.Y. Chou, P.R. Krauss, P.J. Renstrom: Imprint lithography with 25-nanometer resolution, Science **272**, 85–87 (1996)

209. A. Kumar, G. M. Whitesides: Features of gold having micrometer to centimeter dimensions can be formed through a combination of stamping with an elastomeric stamp and an alkanethiol ink followed by chemical etching, Appl. Phys. Lett. **63**, 2002–2004 (1993)
210. J. C. McDonald, D. C. Duffy, J. R. Anderson, D. T. Chiu, H. Wu, O. J. A. Schueller, G. M. Whitesides: Fabrication of microfluidic systems in poly(dimethylsiloxane), Electrophoresis **21**, 27–40 (2000)

23

Mechanical Properties of Micromachined Structures

Harold Kahn

Summary. To be able to accurately design structures and make reliability predictions in any field, it is first necessary to know the mechanical properties of the materials that make up the structural components. The devices encountered in the fields of microelectromechanical systems (MEMS) and nanoelectromechanical systems (NEMS), are necessarily very small, and so the processing techniques and the microstructures of the materials used in these devices may differ significantly from bulk structures. Also, the surface-area-to-volume ratios in such structures are much higher than in bulk samples, and so surface properties become much more important. In short, it cannot be assumed that the mechanical properties measured for a bulk specimen of a material will apply when the same material is used in MEMS and NEMS. This chapter will review the techniques that have been used to determine the mechanical properties of micromachined structures, especially residual stress, strength and Young's modulus. The experimental measurements that have been performed will then be summarized, in particular the values obtained for polycrystalline silicon (polysilicon).

23.1 Measuring Mechanical Properties of Films on Substrates

In order to accurately determine the mechanical properties of very small structures, it is necessary to test specimens made from the same materials, processed in the same way, and of the same approximate size. Not surprisingly it is often difficult to handle specimens this small. One solution is to test the properties of films that remain on substrates. Micro- and nanomachined structures are typically fabricated from films that are initially deposited onto a substrate, are subsequently patterned and etched into the appropriate shapes, and are then finally released from the substrate. If the testing is performed on the continuous film, before patterning and release, the substrate can be used as an effective "handle" for the specimen (in this case, the film). Of course, since the films are attached to the substrate, the types of tests possible are severely limited.

23.1.1 Residual Stress Measurements

One common measurement easily performed on films attached to substrates is residual film stress. The curvature of the substrate is measured before and after film deposition. Curvature can be measured in a number of ways. The most common technique is to scan a laser across the surface (or scan the substrate beneath the laser) and detect the angle of the reflected signal. Alternatively, profilometry, optical interferometry or even atomic force microscopy can be used. As expected, tools that map a surface or perform multiple linear scans can give more accurate readings than tools that measure only a single scan.

Assuming that the film is thin compared to the substrate, the average residual stress in the film, σ_f, is given by the Stoney equation,

$$\sigma_f = \frac{1}{6} \frac{E_s}{(1-\nu_s)} \frac{t_s^2}{t_f} \left(\frac{1}{R_1} - \frac{1}{R_2} \right), \tag{23.1}$$

where the subscripts f and s refer to the film and substrate, respectively; t is thickness, E is Young's modulus, ν is Poisson's ratio, and R is the radius of curvature before (R_1) and after (R_2) film deposition [2]. For a typical (100)-oriented silicon substrate, $E/(1-\nu)$ (also known as the biaxial modulus) is equal to 180.5 GPa, independent of in-plane rotation [3]. This investigation can be performed on the as-deposited film or after any subsequent annealing step, provided no changes occur to the substrate.

This measurement will reveal the average residual stress of the film. Typically, however, the residual stresses of deposited films will vary throughout the thickness of the film. One way to detect this, using substrate curvature techniques, is to etch away a fraction of the film and repeat the curvature measurement. This can be iterated any number of times to obtain a residual stress profile for the film [4]. Alternatively, tools have been designed that can measure the substrate curvature dur-

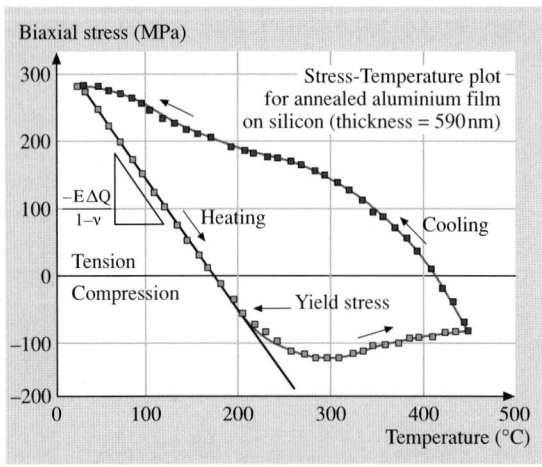

Fig. 23.1. Typical results for residual stress as a function of temperature for an aluminum film on a silicon substrate [1]. The stresses were determined by measuring the curvature of the substrate before and after film deposition, using the reflected signal of a laser scanned across the substrate surface

ing the deposition process itself, in order to obtain information on how the stresses evolve [5].

An additional feature of some of these tools is the ability to heat the substrates while performing the stress measurement. An example of the results obtained in such an experiment is shown in Fig. 23.1 [1], for an aluminum film on a silicon substrate. The slope of the heating curve gives the difference in thermal expansion between the film and the substrate. When the heating curve changes slope and becomes nearly horizontal, the yield strength of the film has been reached.

23.1.2 Mechanical Measurements Using Nanoindentation

Aside from residual stress, it is difficult to measure the mechanical properties of films attached to substrates without the measurement being affected by the presence of the substrate. Recent developments in nanoindentation equipment have allowed this technique to be used in some cases. With specially designed tools, indentation can be performed using very low loads. If the films being investigated are thick and rigid enough, measurements can be made that are not influenced by the presence of the substrate. Of course, this can be verified by depositing the same film onto different substrates. By continuously monitoring the displacement as well as the load during indentation, a variety of properties can be measured, including hardness and Young's modulus [6]. This area is covered in more detail in a separate chapter.

For brittle materials, cracks can be generated by indentation, and strength information can be gathered. But the exact stress fields created during the indentation process are not known exactly, and therefore quantitative values for strength are difficult to determine. Anisotropic etching of single-crystal silicon has been performed to create 30-µm-tall structures that were then indented to examine fracture toughness [7], but this is not possible with most materials.

23.2 Micromachined Structures for Measuring Mechanical Properties

Certainly the most direct way to measure the mechanical properties of small structures is to fabricate structures that would be conducive to such tests. Fabrication techniques are sufficiently advanced that virtually any design can be realized, at least in two dimensions. Two basic types of devices are used for mechanical property testing: "passive" structures and "active" structures.

23.2.1 Passive Structures

As mentioned previously, the main difficulty encountered when testing very small specimens is handling. One way to circumvent this problem is to use passive structures. These structures are designed to act as soon as they are released from the substrate and to provide whatever information they are designed to supply without further manipulation. For all of these passive structures, the forces acting on them come from the internal residual stresses of the structural material. For devices on the micron scale or smaller, gravitational forces can be neglected, and therefore internal stresses are the only source of actuation force.

Stress Measurements

Since internal residual stresses act upon the passive devices when they are released, it is natural to design a device that can be used to measure residual stresses. One such device, a rotating microstrain gage, is shown in Fig. 23.2. There are many different microstrain gauge designs, but all operate via the same principle. In Fig. 23.2a, the large pads, labeled A, will remain anchored to the substrate when the rest of the device is released. Upon release, the device will expand or contract in order to relieve its internal residual stresses. A structure under tension will contract, and a structure under compression will expand. For the structure in Fig. 23.2, compressive stress will cause the legs to lengthen. Since the two opposing legs are not attached to the central beam at the same point – they are offset, they will cause the central beam to rotate when they expand. The device in Fig. 23.2 contains two independent gauges that point to one another. At the ends of the two central beams are two parts of a Vernier scale. By observing this scale, one can measure the rotation of the beams.

If the connections between the legs and the central beams were simple pin connections, the strain, ε, of the legs (the fraction of expansion or contraction) could be

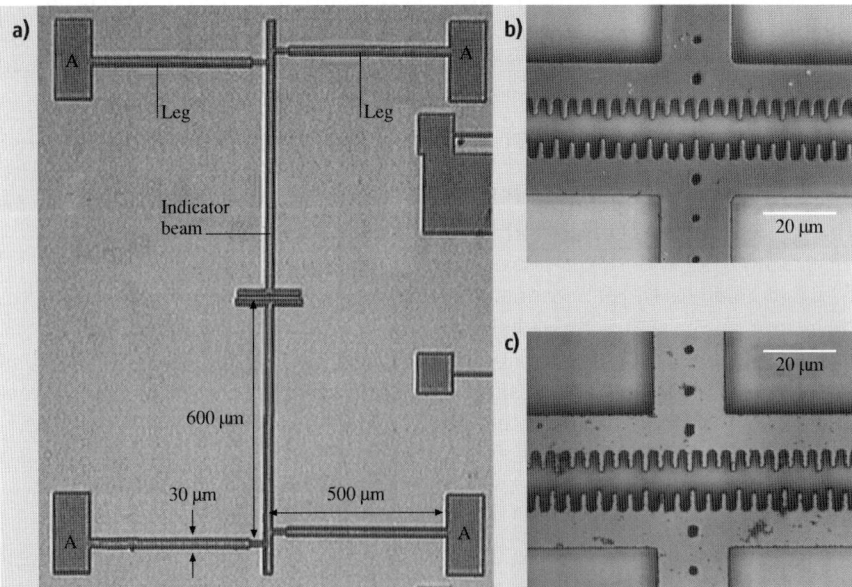

Fig. 23.2. (a) Microstrain gauge fabricated from polysilicon; (b) shows a close-up of the Vernier scale before release, and (c) shows the same area after release

determined simply by the measured rotation and the geometry of the device, namely

$$\varepsilon = \frac{d_{\text{beam}} d_{\text{offset}}}{2 L_{\text{central}} L_{\text{leg}}}, \tag{23.2}$$

where d_{beam} is the lateral deflection of the end of one central beam, d_{offset} is the distance between the connections of the opposing legs, L_{central} is the length of the central beam (measured to the center point between the leg connections), and L_{leg} is the length of the leg. However, since the entire device was fabricated from a single polysilicon film, this cannot be the case; some bending must occur at the connections. As a result, to get an accurate determination of the strain relieved upon release, finite element analysis (FEA) of the structure must be performed. This is a common situation for microdevices. FEA is a powerful tool for determining the displacements and stresses of nonideal geometries. One drawback is that the Young's modulus of the material must be known in order to do the FEA as well as to convert the measured strain into a stress value. But Young's moduli are known for many micromachined materials or they can be measured using other techniques.

Other devices besides rotating strain gauges have been designed that can measure residual stresses. One of the simplest is a doubly clamped beam, a long, narrow beam of constant width and thickness that is anchored to the substrate at both ends. If the beam contains a tensile stress it will remain straight, but if the beam contains a compressive stress it will buckle if its length exceeds a critical value, l_{cr}, according to the Euler buckling criterion [8],

$$\varepsilon_{\text{r}} = -\frac{\pi^2}{3}\left(\frac{h}{l_{\text{cr}}}\right)^2, \tag{23.3}$$

where ε_{r} is the residual strain in the beam and h is the width or thickness of the beam, whichever is less. To determine the residual strain, a series of doubly clamped beams of varying lengths are fabricated. In this way, the critical length for buckling, l_{cr}, can be deduced after release. One problem with this technique is that during the release process, any turbulence in the solution will lead to enhanced buckling of the beams, and a low value for l_{cr} will be obtained.

For films with tensile stresses, a similar analysis can be performed using ring-and-beam structures, also called Guckel rings after their inventor, Henry Guckel. A schematic of this design is shown in Fig. 23.3 [8]. Tensile stress in the outer ring will cause it to contract. This will lead to compressive stress in the central beam, even though the material was originally tensile before release. The amount of compression in the central beam can be determined analytically from the geometry of the device and the residual strain of the material. Again, by changing the length of the central beam it is possible to determine l_{cr}, and then the residual strain can be deduced.

Stress Gradient Measurements

For structures fabricated from thin deposited films, the stress gradient can be just as important as the stress itself. Figure 23.4 shows a portion of a silicon microactuator.

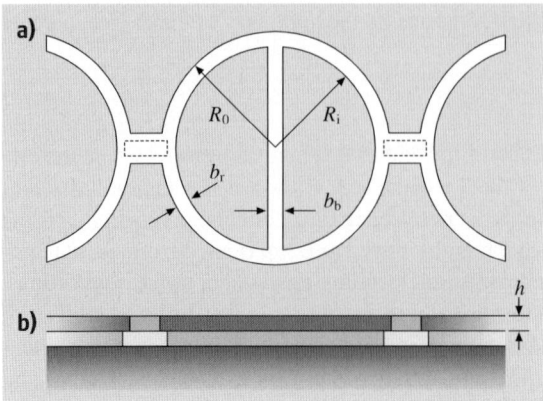

Fig. 23.3. Schematic showing (**a**) top view and (**b**) side view of Guckel ring structures [8]. The *dashed lines* in (**a**) indicate the anchors

The device is designed to be completely planar; however, stress gradients in the film cause the structures to bend. This figure illustrates the importance of characterizing and controlling stress gradients, and it also demonstrates that stress gradients are most easily measured for a simple cantilever beam. By measuring the end deflection δ of a cantilever beam of length l and thickness t, the stress gradient, $d\sigma/dt$ is determined by [9]

$$\frac{d\sigma}{dt} = \frac{2\delta}{l^2} \frac{E}{1-\nu} . \qquad (23.4)$$

The magnitude of the end deflection can be measured by microscopy, optical interferometry, or any other technique.

Another useful structure for measuring stress gradients is a spiral. For this structure, the end of the spiral not anchored to the substrate will move out-of-plane. The diameter of the spiral will also contract, and the free end of the spiral will rotate when released [10].

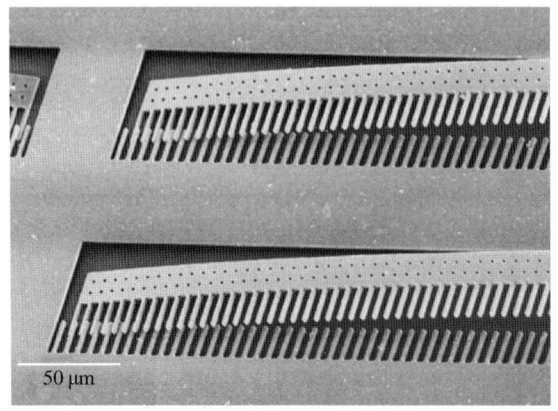

Fig. 23.4. Scanning electron micrograph (SEM) of a portion of a silicon microactuator. Residual stress gradients in the silicon cause the structure to bend

Strength and Fracture Toughness Measurements

As mentioned above, if a doubly clamped beam contains a tensile stress, it will remain taut when released because it cannot relieve any of its stress by contracting. This tensile stress can be thought of as a tensile load being applied at the ends of the beam. If this tensile load exceeds the tensile strength of the material, the beam will break. Since the tensile stress can be measured, as discussed in the Sect. 23.2.1, this technique can be used to gather information on the strength of materials. Figure 23.5 shows two different beam designs that have been used to measure strength. The device shown in Fig. 23.5a was fabricated from a tensile polysilicon film [11]. Different beams were designed with varying lengths of the wider regions (marked l_1 in the figure). In this manner, the load applied to the narrow center beam was varied, even though the entire film contained a uniform residual tensile stress. For l_1 greater than a critical value, the narrow center beam fractured, giving a measurement for the tensile strength of polysilicon.

The design shown in Fig. 23.5b was fabricated from a tensile Si_xN_y film [12]. As seen in the figure, a stress concentration was included in the beam, to ensure the fracture strength would be exceeded. In this case, a notch was etched into one side of the beam. Since the stress concentration is not symmetric with regard to the beam axis, this results in a large bending moment at that position, and the test measures the bend strength of the material. Again, like the beams shown in Fig. 23.5a, the geometry of various beams fabricated from the same film were varied, to vary the maximum stress seen at the notch. By seeing which beams fracture at the stress concentration after release, the strength can be determined.

The fracture toughness of a material can be determined with a similar technique, but an atomically sharp pre-crack is used instead of a stress concentration. Sharp pre-cracks can be introduced into micromachined structures before release by adding a Vickers indent onto the substrate, near the device; the radial crack formed by the indent will propagate into the overlying structure [14]. Accordingly, the beam

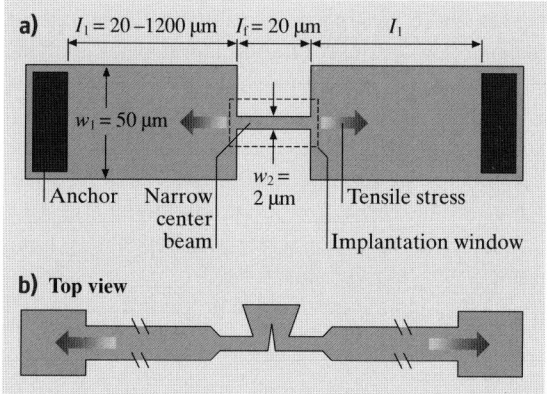

Fig. 23.5a,b. Schematic designs of doubly clamped beams with stress concentrations used for measuring strength. (**a**) was fabricated from polysilicon [11] and (**b**) was fabricated from Si_xN_y [12]

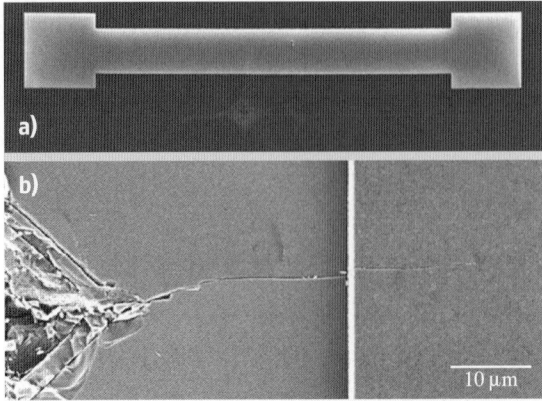

Fig. 23.6. (a) SEM of a 500 μm-long polysilicon beam with a Vickers indent placed near its center; (**b**) higher magnification SEM of the area near the indent showing the pre-crack traveling from the substrate into the beam [13]

with a sharp pre-crack, shown in Fig. 23.6, was fabricated using polysilicon [13]. Due to the stochastic nature of indentation, the initial pre-crack length varies from beam to beam. Because of this, even though the geometry of the beam remains identical, the stress intensity, K, at the pre-crack tip will vary. Upon release, only those pre-cracks whose K exceeds the fracture toughness of the material, K_{Ic}, will propagate, and in this way upper and lower bounds for K_{Ic} can be determined for the material.

For all of the beams discussed in this section, finite element analysis is required to determine the stress concentrations and stress intensities. Even though approximate analytical solutions may exist for these designs, the actual fabricated structure will not have idealized geometries. For example, corners will never be perfectly sharp, and cracks will never be perfectly straight. This reinforces the idea that FEA is a powerful tool when determining mechanical properties of very small structures.

23.2.2 Active Structures

As discussed above, it is very convenient to design structures that act upon release to provide information on the mechanical properties of the structural materials. This is not always possible, however. For example, those passive devices just discussed rely on residual stresses to create the changes (rotation or fracture) that occur upon release, but many materials do not contain high residual stresses as-deposited, or the processing scheme of the device precludes the generation of residual stresses. Also, some mechanical properties, such as fatigue resistance, require motion before they can be studied. Active devices are therefore used. These are acted upon by a force (the source of this force can be integrated into the device itself or can be external to the device) in order to create a change, and the mechanical properties are studied via the response to the force.

Young's Modulus Measurements

Young's modulus, E, is a material property critical to any structural device design. It describes the elastic response of a material and relates stress, σ, and strain, ε, by

$$\sigma = E\varepsilon. \tag{23.5}$$

In bulk samples, E is often measured by loading a specimen under tension and measuring displacement as a function of stress for a given length. While this is far more difficult for small structures, such as those fabricated from thin deposited films, it can be achieved with careful experimental techniques. Figure 23.7 shows a schematic of one such measurement system [15]. The fringe detectors in the figure detect the reflected laser signal from two gold lines deposited onto the polysilicon specimen, which act as gauge markers. This enables the strain in the specimen during loading to be monitored. Besides gold lines, Vickers indents placed in a nickel specimen can also serve as gauge markers [16], or a speckle interferometry technique [17] can be used to determine strain in the specimen. Once the stress-versus-strain behavior is measured, the slope of the curve is equal to E. By using a constant load, such as a dead weight, and resistive heating, high-temperature creep can also be investigated with this method [18].

In addition to the tensile test, Young's modulus can be determined by other measures of stress–strain behavior. As seen in Fig. 23.8, a cantilever beam can be bent by pushing on the free end with a nanoindenter [19]. The nanoindenter can monitor the force applied and the displacement, and simple beam theory can convert the displacement into strain in order to obtain E. A similar technique, shown schematically in Fig. 23.9 [20], involves pulling downward on a cantilever beam by means of an electrostatic force. An electrode is fabricated into the substrate beneath the cantilever beam, and a voltage is applied between the beam and the bottom electrode. The force acting on the beam is equal to the electrostatic force corrected to include the effects of fringing fields acting on the sides of the beam, namely

$$F(x) = \frac{\varepsilon_0}{2}\left(\frac{V}{g+z(x)}\right)^2 \left(1 + \frac{0.65[g+z(x)]}{w}\right), \tag{23.6}$$

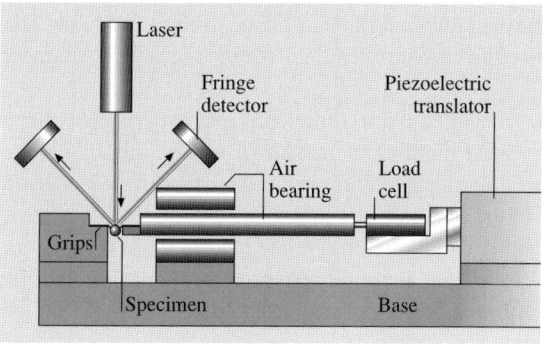

Fig. 23.7. Schematic of a measurement system for tensile loading of micromachined specimens [15]

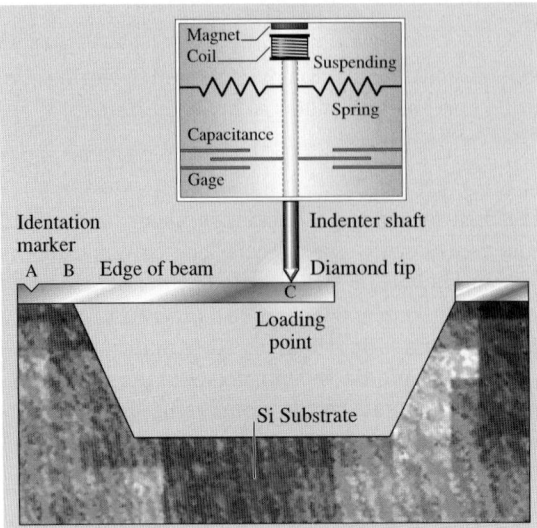

Fig. 23.8. Schematic of a nanoindenter loading mechanism pushing on the end of a cantilever beam [19]

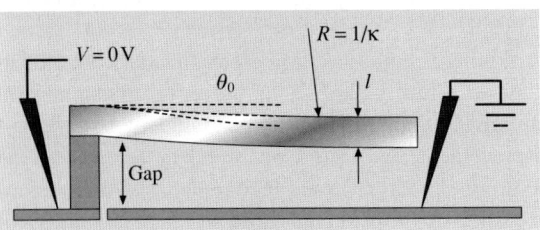

Fig. 23.9. Schematic of a cantilever beam bending test using an electrostatic voltage to pull the beam toward the substrate [20]

where $F(x)$ is the electrostatic force at x, ε_0 is the dielectric constant of air, g is the gap between the beam and the bottom electrode, $z(x)$ is the out-of-plane deflection of the beam, w is the beam width, and V is the applied voltage [20]. In this work, the deflection of the beam as a function of position is measured using optical interferometry. These measurements combine to give stress–strain behavior for the cantilever beam. An extension of this technique uses doubly clamped beams instead of cantilever beams. In this case, the deflection of the beam at a given electrostatic force depends on the residual stress in the material as well as Young's modulus. This method can therefore also be used to measure residual stresses in doubly clamped beams.

Another device that can be fabricated from a thin film and used to investigate stress–strain behavior is a suspended membrane, as shown in Fig. 23.10 [21]. As depicted in the schematic figure, the membrane is exposed to an elevated pressure on one side, causing it to bulge in the opposite direction. The deflection of the membrane is measured by optical or other techniques and related to the strain in the membrane. These membranes can be fabricated in any shape, typically square or circular. Both analytical solutions and finite element analyses have been performed to relate

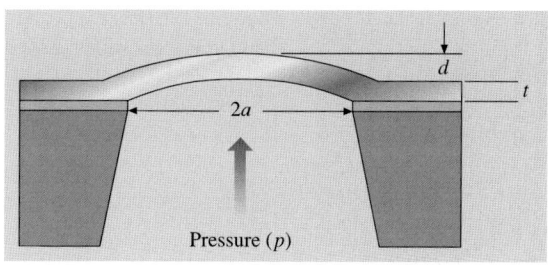

Fig. 23.10. Schematic cross-section of a microfabricated membrane [21]

the deflection to the strain. Like the doubly clamped beams, both Young's modulus and residual stress play a role in the deflected shape. Both of these mechanical properties can therefore be determined by the pressure-versus-deflection performance of the membrane.

Another measurement besides stress–strain behavior that can reveal the Young's modulus of a material is the determination of the natural resonance frequency. For a cantilever, the resonance frequency, f_r, for free undamped vibration is given by

$$f_r = \frac{\lambda_i^2 t}{4\pi l^2}\left(\frac{E}{3\rho}\right)^{1/2}, \tag{23.7}$$

where ρ, l and t are the density, length and thickness of the cantilever; λ_i is the eigenvalue, where i is an integer that describes the resonance mode number; for the first mode $\lambda_1 = 1.875$ [23]. Given the geometry and density, measuring f_r allows E to be determined. The cantilever can be vibrated by a number of techniques, including a laser, loudspeaker or piezoelectric shaker. The frequency that produces the highest amplitude of vibration is the resonance frequency.

A micromachined device that uses an electrostatic comb drive and an AC signal to generate the vibration of the structure is known as a lateral resonator [24]. One example is shown in Fig. 23.11 [22]. When a voltage is applied across either set of the interdigitated comb fingers shown in Fig. 23.11, an electrostatic attraction is generated due to the increase in capacitance as the overlap between the comb fingers increases. The force, F, generated by the comb-drive is given by

$$F = \frac{1}{2}\frac{\partial C}{\partial x}V^2 = n\varepsilon\frac{h}{g}V^2, \tag{23.8}$$

where C is capacitance, x is the distance traveled by one comb-drive toward the other, n is the number of pairs of comb fingers in one drive, ε is the permittivity of the fluid between the fingers, h is the height of the fingers, g is the gap spacing between the fingers, and V is the applied voltage [24]. When an AC voltage at the resonance frequency is applied across either of the two comb drives, the central portion of the device will vibrate. In fact, since force depends on the square of the voltage for electrostatic actuation, for a time t, a dependent drive voltage $v_D(t)$ (given by

$$v_D(t) = V_P + v_d \sin(\omega t), \tag{23.9}$$

Fig. 23.11. SEM of a polysilicon lateral resonator [22]

where V_P is the DC bias and v_d is the AC drive amplitude), the time-dependent portion of the force will scale with

$$2\omega V_P v_d \cos(\omega t) + \omega v_d^2 \sin(2\omega t) \qquad (23.10)$$

[24]. Therefore, if an AC drive signal is used with no DC bias, at resonance, the frequency of the AC drive signal will be one half of the resonance frequency. For this device, the resonance frequency f_r will be

$$f_r = \frac{1}{2\pi}\left(\frac{k_{sys}}{M}\right)^{1/2}, \qquad (23.11)$$

where k_{sys} is the spring constant of the support beams and M is the mass of the portion of the device that vibrates. The spring constant is given by

$$k_{sys} = 24EI/L^3, \qquad (23.12)$$

$$I = \frac{hw^3}{12}, \qquad (23.13)$$

where I is the moment of inertia of the beams, and L, h and w are the length, thickness, and width of each beam. Therefore, by combining these equations and measuring f_r, it is possible to determine E.

One distinct advantage of the lateral resonator technique and the electrostatically pulled cantilever technique for measuring Young's modulus is that they require no external loading sources. Portions of the devices are electrically contacted, and a voltage is applied. For the pure tension tests, as shown in Fig. 23.7, the specimen must be attached to a loading system, which can be extremely difficult for the very small specimens discussed here, and any misalignment or eccentricity in the test could lead to unreliable results. However, the advantage of the externally loaded technique is that there are no limitations on the type of materials that can be tested. Conductivity is not a requirement, nor is any compatibility with electrical actuation.

Strength and Fracture Toughness Measurements

As one might expect, any of the techniques discussed in the previous section that strain specimens in order to measure Young's modulus can also be used to measure fracture strength. The load is simply increased until the specimen breaks. As long as either the load or the strain is measured at fracture, and the geometry of the specimen is known, the maximum stress required for fracture σ_{crit} can be determined, either through analytical analysis or FEA. Depending on the geometry of the test, σ_{crit} will represent the tensile or bend strength of the material.

If the available force is limited, or if a localized fracture site is desired, stress concentrations can be added to the specimens. These are typically notches micromachined into the edges of specimens. Focused ion beams have also been used to carve stress concentrations into fracture specimens.

All of the external loading schemes, such as those shown in Figs. 23.7 and 23.8, have been used to measure fracture strength. Also, the electrostatically loaded doubly clamped beams can be pulled until they fracture. In this case, there is one complication. The electrostatic force is inversely proportional to the distance between the electrodes, and at a certain voltage, called the "pull-in voltage," the attraction between the beam and the substrate will become so great that the beam will immediately be pulled into contact with the bottom electrode. As long as the fracture takes place before the pull-in voltage is reached, the experiment will give valid results.

Other loading techniques have been used to generate fracture of microspecimens. Figure 23.12 [25] shows one device designed to be pushed by the end of a micromanipulated needle. The long beams that extend from the sides of the central shuttle come into contact with anchored posts, and, at a critical degree of bending, the beams will break off. Since the applied force cannot be measured in this technique, the experiment is continuously optically monitored during the test, and the image of the beams just before fracture is analyzed to determine σ_{crit}.

Another loading scheme that has been demonstrated for micromachined specimens utilizes scratch drive actuators to load the specimens [26]. These types of ac-

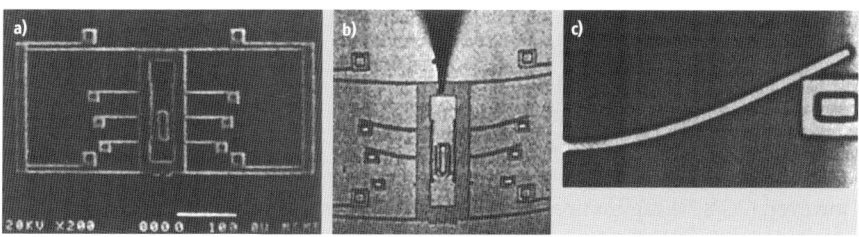

Fig. 23.12. (a) SEM of a device for measuring bend strength of polysilicon beams; (b) image of a test in process; (c) higher magnification view of one beam shortly before breaking [25]

tuators work like inchworms, traveling across a substrate in discrete advances as an electrostatic force is repeatedly applied between the actuator and the substrate. The stepping motion can be made on the nanometer scale, depending on the frequency of the applied voltage, and so it can be an acceptable approximation to continuous loading. One advantage of this scheme is that very large forces can be generated by relatively small devices. The exact forces generated cannot be measured, so (like the technique that used micromanipulated pushing) the test is continuously observed to determine the strain at fracture. Another advantage of this technique is that, like the lateral resonator and the electrically pulled cantilever, the loading takes place on-chip, and therefore the difficulties associated with attaching and aligning an external loading source are eliminated.

Another on-chip actuator used to load microspecimens is shown in Fig. 23.13, along with three different microspecimens [14]. Devices have been fabricated with each of the three microspecimens integrated with the same electrostatic comb-drive actuator. In all three cases, when a DC voltage is applied to the actuator, it moves downward, as oriented in Fig. 23.13. This pulls down on the left end of each of the three microspecimens, which are anchored on the right. The actuator contains 1486 pairs of comb fingers. The maximum voltage that can be applied is limited by the breakdown voltage of the medium in which the test takes place. In air, this limits the voltage to less than 200 V. As a result, given a finger height of 4 μm and a gap of 2 μm, and using (23.8), the maximum force generated by this actuator is limited to about 1 mN. Standard optical photolithography has a minimum feature size of about 2 μm. As a result, the electrostatic actuator cannot generate sufficient force to perform a standard tensile test on MEMS structural materials such as polysilicon. The microspecimens shown in Fig. 23.13 are therefore designed such that the stress is amplified.

The specimen shown in Fig. 23.13b is designed to measure bend strength. It contains a micromachined notch with a root radius of 1 μm. When the actuator pulls downward on the left end of this specimen, the notch serves as a stress concentration, and when the stress at the notch root exceeds σ_{crit}, the specimen fractures. The specimen in Fig. 23.13c is designed to test tensile strength. When the left end of this specimen is pulled downward, a tensile stress is generated in the upper thin horizontal beam near the right end of the specimen. As the actuator continues to move downward, the tensile stress in this beam will exceed the tensile strength, causing fracture. Finally, the specimen in Fig. 23.13d is similar to that in Fig. 23.13b, except that the notch is replaced by a sharp pre-crack that was produced by the Vickers indent placed on the substrate near the specimen. When this specimen is loaded, a stress intensity K is generated at the crack tip. When the stress intensity exceeds a critical value K_{Ic}, the crack propagates. K_{Ic} is also referred to as the fracture toughness.

The force generated by the electrostatic actuator can be calculated using (23.8). However, (23.8) assumes a perfectly planar, two-dimensional device. In fact, when actuated, the electric fields extend out of the plane of the device, and so (23.8) is just an approximation. Instead, like many of the techniques discussed in this section, the

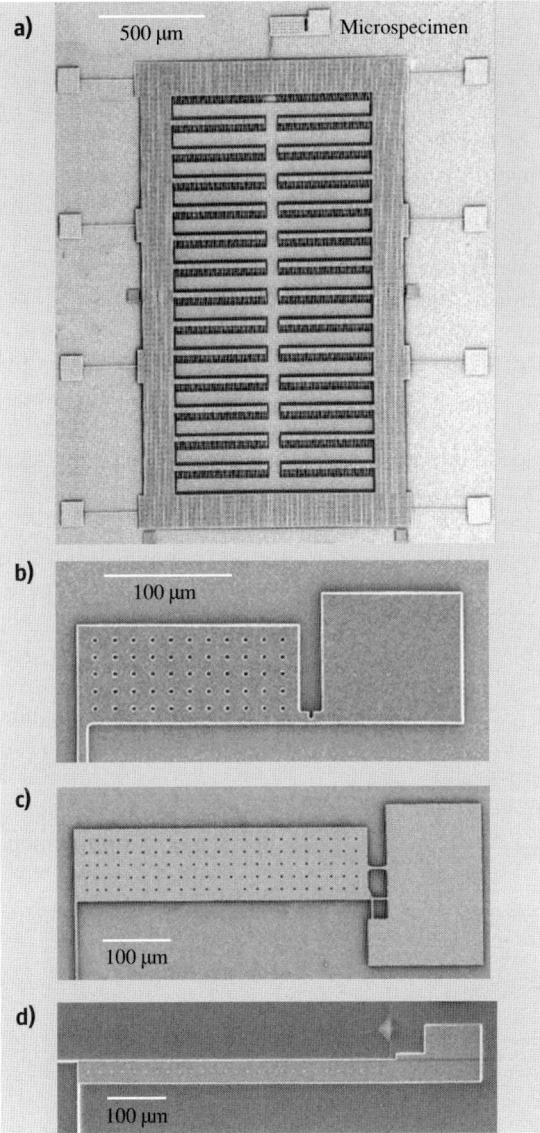

Fig. 23.13. (a) SEM of a micromachined device for conducting strength tests; the device consists of a large comb-drive electrostatic actuator integrated with a microspecimen; (b)–(d) SEMs of various microspecimens for testing bend strength, tensile strength and fracture toughness, respectively [14]

test is continuously monitored, and the actuator displacement at the time of fracture is recorded. Then FEA is used to determine the magnitude of the stress or stress intensity seen by the specimen at the point of fracture.

In order to generate sufficient force to conduct tensile tests, a similar device to that shown in Fig. 23.13 has been designed which uses an array of parallel plate capacitors to provide the force, instead of comb-drives [28]. In this way the available force is increased but the maximum stroke is severely limited.

Fatigue Measurements

A benefit of the electrostatic actuator shown in Fig. 23.13 is that, besides monotonic loading, it can generate cyclic loading. This allows the fatigue resistance of materials to be studied. Simply by using an AC signal instead of a DC voltage, the device can be driven at its resonant frequency. The amplitude of the resonance depends on the magnitude of the AC signal. This amplitude can be increased until the specimen breaks; this will investigate the low-cycle fatigue resistance. Otherwise, the amplitude of resonance can be left constant at a level below that required for fast fracture, and the device will resonate indefinitely until the specimen breaks; this will investigate high-cycle fatigue. It should be noted that the resonance frequency of such a device is about 10 kHz. Therefore, it is possible to stress a specimen for over 10^9 cycles in less than a day. In addition to simple cyclic loading, a mean stress can be superimposed on the cyclic load if a DC bias is added to the AC signal. In this way, nonsymmetric cyclic loading (with a large tensile stress alternating with a small compressive stress, or vice versa) can be studied.

Another device that can be used to investigate fatigue resistance in MEMS materials is shown in Fig. 23.14 [27]. In this case, a large mass is attached to the end of a notched cantilever beam. The mass contains two comb drives on opposite ends. When an AC signal is applied to one comb drive, the device will resonate, cyclically loading the notch. The comb drive on the opposite side is used as a capacitive displacement sensor. This device contains many fewer comb fingers than the device shown in Fig. 23.13. As a result, it can apply cyclic loads by exploiting the resonance frequency of the device, but it cannot supply sufficient force to achieve monotonic loading.

Fatigue loading has also been studied using the same external loading techniques shown in Fig. 23.7. In this case, the frequency of the cyclic load is considerably

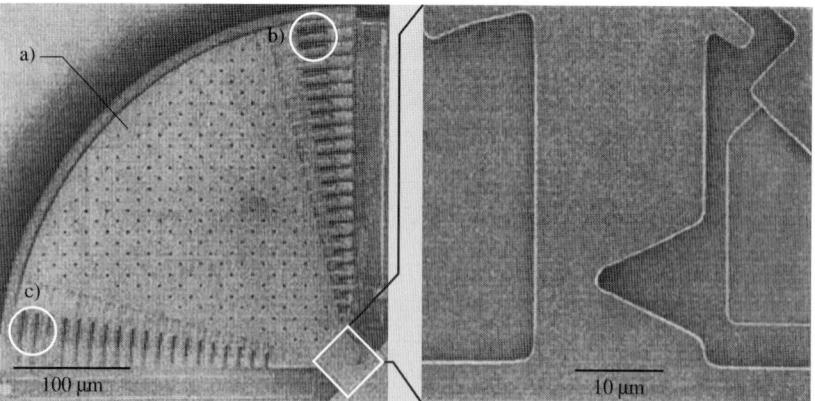

Fig. 23.14. SEM of a device used to investigate fatigue; the image on the right is a higher magnification view of the notch near the base of the moving part of the structure [27]: *a)* mass, *b)* comb drive actuator, *c)* capacitive displacement sensor

lower, since the resonance frequency of the device is not being utilized. This leads to longer high-cycle testing times. Since the force is essentially unlimited, however, this technique allows a variety of frequencies to be studied to determine their effect on the fatigue behavior.

Friction Measurements

Friction is another property that has been studied in micromachined structures. To study friction, of course, two surfaces must be brought into contact with each other. This is usually avoided at all costs for these devices because of the risk of stiction. (Stiction is the term used when two surfaces that come into contact adhere so strongly that they cannot be separated.) Even so, a few devices have been designed that can investigate friction. One of these is shown schematically in Fig. 23.15 [29]. It consists of a movable structure with a comb-drive on one end and a cantilever beam on the other. Beneath the cantilever, on the substrate, is a planar electrode. The device is moved to one side using the comb-drive. Then a voltage is applied between the cantilever beam and the substrate electrode. The voltage on the comb-drive is then released. The device would normally return to its original position, to relax the deflection in the truss suspensions, but the friction between the cantilever and the substrate electrode holds it in place. The voltage to the substrate electrode is slowly decreased until the device starts to slide. Given the electrostatic force generated by the substrate electrode and the stiffness of the truss suspensions, it is possible to determine the static friction. For this device, bumps were fabricated on the bottom of each cantilever beam. This limited the surface area that came into contact with the substrate and so lowered the risk of stiction.

Another device designed to study friction is shown in Fig. 23.16 [30]. This technique uses a hinged cantilever. The portion near the free end acts as the friction test structure, and the portion near the anchored end acts as the driver. The friction

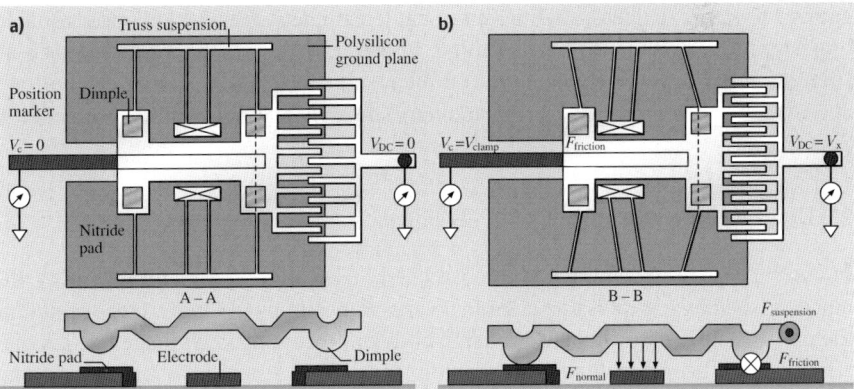

Fig. 23.15a,b. Schematics depicting a device used to study friction. (**a**) shows top and side views of the device in its original position, and (**b**) shows views of the device after it has been displaced using the comb-drive and clamped using the substrate electrode [29]

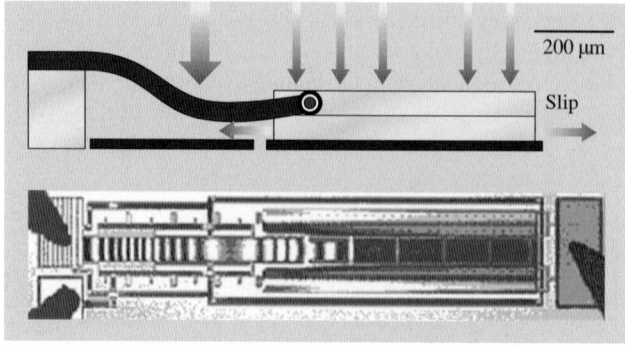

Fig. 23.16. Schematic cross-section and top-view optical micrograph of a hinged-cantilever test structure for measuring friction in micromachined devices [30]

test structure is attracted to the substrate by means of electrostatic actuation, and when a second electrostatic actuator pulls down the driver, the friction test structure slips forward by a length proportional to the forces involved, including the frictional force. This distance, however, has a maximum of 30 nm, so all measurements must be exceedingly accurate in order to investigate a range of forces. This test structure can be used to determine the friction coefficients for surfaces with and without lubricating coatings.

23.3 Measurements of Mechanical Properties

All of the techniques discussed in Sects. 23.1 and 23.2 have been used to measure the mechanical properties of MEMS and NEMS materials. As a general rule, the results from the various techniques have agreed well with each other, and the argument becomes which of the measurement techniques is easiest and most reliable to perform. It is crucial to bear in mind, however, that certain properties (such as strength) are process-dependent, and so the results taken at one laboratory will not necessarily match those taken from another. This will be discussed in more detail in Sect. 23.3.1.

23.3.1 Mechanical Properties of Polysilicon

In current MEMS technology, the most widely used structural material is polysilicon deposited by low-pressure chemical vapor deposition (LPCVD). One reason for the prevalence of polysilicon is the large body of processing knowledge for this material that has been developed by the integrated circuit community. Another reason, of course, is that polysilicon possesses a number of qualities that are beneficial to MEMS devices, in particular high strength and Young's modulus. Therefore, most of the mechanical properties investigations performed on MEMS materials have focused on polysilicon.

Residual Stresses in Polysilicon

The residual stresses of LPCVD polysilicon have been thoroughly characterized using both the wafer curvature technique, discussed in Sect. 23.1.1, and the microstrain gauges, discussed in Sect. 23.2.1. The results from both techniques give consistent values. Figure 23.17 summarizes the residual stress measurements as a function of deposition temperature taken from five different investigations at five different laboratories [31]. All five sets of data show the same trend. The stresses change from compressive at the lowest deposition temperatures to tensile at intermediate temperatures and back to compressive at the highest temperatures. The exact transition temperatures vary somewhat between the different investigations, probably due to differences in the deposition conditions: the silane or dichlorosilane pressure, the gas flow rate, the geometry of the deposition system, and the temperature uniformity. However, in each data set the transitions are easily discernible. The origin of these residual stress changes lies with the microstructure of the LPCVD polysilicon films.

As with all deposited films, the microstructure of the LPCVD polysilicon film is dependent on the deposition conditions. In general, the films are amorphous at the lowest growth temperatures (lower than ∼ 570 °C), display fine (∼ 0.1 µm diameter) grains at intermediate temperatures (∼ 570 °C to ∼ 610 °C), and contain columnar (110)-textured grains with a thin fine-grained nucleation layer at the substrate interface at higher temperatures (∼ 610 °C to ∼ 700 °C) [31]. The fine-grained microstructure results from the homogeneous nucleation and growth of silicon grains within an as-deposited amorphous silicon film. In this regime, the deposition rate is just slightly faster than the crystallization rate. The as-deposited films will be crystalline near the substrate interface and amorphous at the free surface. (The amorphous fraction can be quickly crystallized by annealing above 610 °C.) The columnar microstructure seen at higher growth temperatures results from the formation of

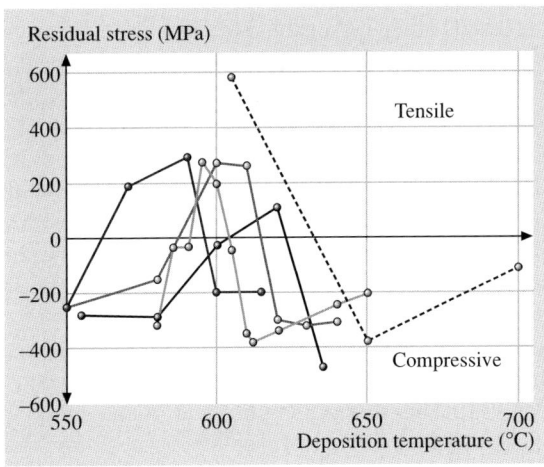

Fig. 23.17. Results for residual stress of LPCVD polysilicon films taken from five different investigations [31]. Data from the same investgation are connected by a line

crystalline silicon films as-deposited, with growth being fastest in the ⟨110⟩ directions.

The origin of the tensile stress in the fine-grained polysilicon arises from the volume decrease that accompanies the crystallization of the as-deposited amorphous material. The origins of the compressive stresses in the amorphous and columnar films are less well understood. One proposed explanation for compressive stress generation during thin film growth postulates that an increase in the surface chemical potential is caused by the deposition of atoms from the vapor; the increase in surface chemical potential induces atoms to flow into newly formed grain boundaries, creating a compressive stress in the film [32].

Stress gradients are also typical of LPCVD polysilicon films. The partially amorphous films contain large stress gradients since they are essentially bilayers of compressive amorphous silicon on top of tensile fine-grained polysilicon. The fully crystalline films also exhibit stress gradients. The columnar compressive films are most highly stressed at the film–substrate interface, with the compressive stresses decreasing as the film thickness increases; the fine-grained films are less tensile at the film–substrate interface, with the tensile stresses increasing as the film thickness increases [31]. Both stress gradients are associated with microstructural variations. For the columnar films, the initial nucleation layer corresponds to a very high compressive stress, which decreases as the columnar morphology develops. For the fine-grained films, the region near the film-substrate interface has a slightly smaller average grain size, due to heterogeneous nucleation at the interface. This region displays a slightly lower tensile stress than the rest of the film, since the increased grain boundary area reduces the local density.

Young's Modulus of Polysilicon

The Young's modulus of polysilicon films has been measured using all of the techniques discussed in Sect. 23.2.2. A good review of the experimental results taken from bulge testing, tensile testing, beam bending and lateral resonators are contained in [33]. All of the reported results are in reasonable agreement, varying from 130 to 175 GPa, though many values are reported with a relatively high experimental scatter. The main origin of the error in these results is the uncertainties involving the geometries of the small specimens used to make the measurements. For example, from (23.13), the Young's modulus determined by the lateral resonators depends on the cube of the tether beam width, typically about 2 μm. In general, the beam width and other dimensions can be measured via scanning electron microscopy to within about 0.1 μm; however, the width of the beam is not perfectly constant along the entire length or even throughout the thickness of the beam. These uncertainties in geometry lead to uncertainties in modulus.

In addition, the various experimental measurements lie close to the Voigt and Reuss bounds for Young's modulus calculated using the elastic stiffnesses and compliances for single-crystal silicon [33]. This strongly implies that Young's moduli of micro- and nanomachined polysilicon structures will be the same as for bulk samples made from polysilicon. This is not unexpected, since Young's modulus is a material property. It is related to interatomic interactions and should have no dependence on the geometry of the sample. It should be noted that polysilicon can display a preferred crystallographic orientation depending on the deposition conditions, and that this could affect the Young's modulus of the material, since the Young's modulus of silicon is not isotropic. However, the anisotropy is fairly small for cubic silicon.

A more recent investigation that utilized electrostatically actuated cantilevers and interferometric deflection detection yielded a Young's modulus of 164 GPa [20]. They found the grains in their polysilicon films to be randomly oriented, and calculated the Voigt and Reuss bounds to be 163.4–164.4 GPa. This appears to be a very reliable value for randomly oriented polysilicon.

Fracture Toughness and Strength of Polysilicon

Using the device shown in Fig. 23.13a and the specimen shown in Fig. 23.13d, the fracture toughness, K_{Ic}, of polysilicon has been shown to be 1.0 ± 0.1 MPa m$^{1/2}$ [34]. Several different polysilicon microstructures were tested, including fine-grained, columnar and multilayered. Amorphous silicon was also investigated. All of the microstructures displayed the same K_{Ic}. This indicates that, like Young's modulus, fracture toughness is a material property, independent of the material microstructure or the geometry of the sample.

A tensile test, such as that shown in Fig. 23.7 but using a sample with sharp indentation-induced pre-cracks, yields a K_{Ic} of 0.86 MPa m$^{1/2}$ [35]. The passive, residual stress loaded beams with sharp pre-cracks shown in Fig. 23.6 gave a K_{Ic} of 0.81 MPa m$^{1/2}$ [13].

Given that K_{Ic} is a material property for polysilicon, the measured fracture strength σ_{crit} is related to K_{Ic} by

$$K_{Ic} = c\sigma_{\text{crit}}(\pi a)^{1/2}, \tag{23.14}$$

where a is the crack-initiating flaw size, and c is a constant of order unity. The value for c will depend on the exact size, shape and orientation of the flaw; for a semicircular flaw, c is equal to 0.71 [36]. Therefore, any differences in the reported fracture strength of polysilicon will be the result of changes in a.

A good review of the experimental results available in the literature for polysilicon strength is contained in [37]. The tensile strength data vary from about 0.5 to 5 GPa. Like many brittle materials, the measured strength of polysilicon is found to obey Weibull statistics. This implies that the polysilicon samples contain a random distribution of flaws of various sizes, and that the failure of any particular specimen will occur at the largest flaw that experiences the highest stress. One consequence of this behavior is that, since larger specimens have a greater probability of containing larger flaws, they will exhibit decreased strengths. More specifically, it was found that the most important geometrical parameter is the surface area of the sidewalls of a polysilicon specimen [37]. The sidewalls, as opposed to the top and bottom surfaces, are those surfaces created by etching the polysilicon film. This is not surprising since LPCVD polysilicon films contain essentially no flaws within the bulk, and the top and bottom surfaces are typically very smooth.

As a result, the etching techniques used to create the structures will have a strong impact on the fracture strength of the material. For single-crystal silicon specimens it was found that the choice of etchant could change the observed tensile strength by a factor of two [38]. In addition, the bend strength of amorphous silicon was measured to be twice that of polysilicon for specimens processed identically [34]. It was found that the reactive ion etching used to fabricate the specimens produced much rougher sidewalls on the polysilicon than on the amorphous silicon.

The tensile strength of single crystal silicon has also been measured using a technique similar to that shown in Fig. 23.8. Sharp notches were introduced into the beams using a focused ion beam, and the apparent fracture toughness was measured for a variety of planes parallel to the notch front, along which the crack propagated. For the {110} notch plane, the fracture toughness was about $1\,\mathrm{MPa\,m^{1/2}}$, and for the {100} notch plane it was about $2\,\mathrm{MPa\,m^{1/2}}$ [39].

Fatigue of Polysilicon

Fatigue failure involves fracture after a number of load cycles, when each individual load is not sufficient by itself to generate catastrophic cracking in the material. For ductile materials, such as metals, fatigue occurs due to accumulated damage at the site of maximum stress and involves local plasticity. As a brittle material, polysilicon would not be expected to be susceptible to cyclic fatigue. However, fatigue has been observed for polysilicon tensile samples [35], polysilicon bend specimens with notches [27, 40], and polysilicon bend specimens with sharp cracks [41]. The exact origins of the fatigue behavior are still subject to debate, but some aspects of the experimental data are that the fatigue lifetime does not depend on the loading frequency [35], the fatigue behavior is affected by the ambient [13, 41], and the fatigue depends on the ratio of compressive to tensile stresses seen in the load cycle [13].

Friction of Polysilicon

The friction of polysilicon structures has been measured using the techniques described in Sect. 23.2.2. The measured coefficient of friction was found to vary from 4.9 [29] to 7.8 [30].

The static and dynamic friction of polysilicon coated with monolayer lubricants has been measured with a device similar to that shown in Fig. 23.15, but using a scratch-drive actuator instead of a comb-drive [42]. The dynamic friction at $0.2\,\mathrm{m/s}$ was approximately 80% of the static friction value; the static friction at zero applied load was due to an adhesive force of $0.95\,\mathrm{nN/\mu m^2}$.

23.3.2 Mechanical Properties of Other Materials

As discussed above, of all the materials used for MEMS and NEMS, polysilicon has generated the most interest as well as the most research in mechanical properties characterization. However, measurements have been taken on other materials, and these are summarized in this section.

As discussed in Sect. 23.2.2, one advantage of the externally loaded tension test, as shown in Fig. 23.7, is that essentially any material can be tested using this technique. As such, tensile strengths have been measured to be 0.6 to 1.9 GPa for SiO_2 [43] and 0.7 to 1.1 GPa for titanium [44]. The yield strength for electrodeposited nickel was found to vary from 370 to 900 MPa, depending on the annealing temperature [16]. In addition, the yield strength was strongly affected by the current density during the electrodeposition process. Both the annealing and current density effects were correlated to changes in the microstructure of the material. Young's moduli were determined to be 100 GPa for titanium [44] and 215 GPa for electrodeposited nickel [16].

The tensile strength, Young's modulus, and Poisson's ratio of silicon nitride were measured to be 5.9 GPa, 0.23 and 257 GPa, respectively [45], and the same properties of amorphous diamond-like carbon were found to be 7.3 GPa, 0.17 and 759 GPa, respectively [46].

The technique of bending cantilever beams, shown in Fig. 23.8, can also be performed on a variety of materials. The yield strength and Young's modulus of gold were found to vary from 260 MPa and 57 GPa, respectively [19], to 300 MPa and 120 GPa, respectively [47], using this method. The same properties in Al were measured to be 150 MPa and 80 GPa, respectively [47], and in silicon nitride to be 6.9 GPa and 260 GPa, respectively [48]. Using a technique similar to that shown in Fig. 23.8, except that a doubly-clamped beam was used instead of a cantilever beam, the fracture toughness of ultrananocrystalline diamond was measured to be 4.5 MPa m$^{1/2}$ [49]. Resonating cantilever beams revealed a Young's modulus for gold of 47 GPa [50], and Young's moduli of 1.8 GPa and 14.4 GPa for silica and alumina aerogel thin films, respectively [51].

Another technique that can be used with a number of materials is the membrane deflection method, shown in Fig. 23.10. Silicon nitride measured with this technique revealed a Young's modulus of 258 [45] to 325 GPa and a burst strength of 7.1 GPa [52]. A polyimide membrane gave a residual stress of 32 MPa, a Young's modulus of 3.0 GPa, and an ultimate strain of about four percent [21]. Membranes were also fabricated from polycrystalline SiC films with two different grain structures [53]. The film with (110)-texture columnar grains had a residual stress of 434 MPa and a Young's modulus of 349 GPa. The film with equiaxed (110)- and (111)-textured grains had a residual stress of 446 MPa and a Young's modulus of 456 GPa.

Other devices that are used to measure mechanical properties require more complicated micromachining, namely patterning, etching and release, in order to operate. These devices are more difficult to fabricate with materials that are not commonly used as MEMS structural materials. However, the following examples demonstrate work in this area. The structure shown in Fig. 23.5b was fabricated from Si_xN_y and revealed a apparent fracture toughness of 1.8 MPa m$^{1/2}$ [12]. The devices shown in Fig. 23.6 revealed a fracture toughness of SiC of 3.1 MPa m$^{1/2}$ [54]. Lateral resonators of the type shown in Fig. 23.11 were processed using polycrystalline SiC, and the Young's modulus was determined to be 426 GPa [55]. The device shown in Fig. 23.12 was fabricated from polycrystalline germanium, and used to measure a bend strength of 1.5 GPa for unannealed Ge and 2.2 GPa for annealed Ge [56]. The same device was also fabricated from SiC and revealed a bend strength of 23 GPa [57]. Devices similar to that shown in Fig. 23.13 revealed the tensile strength of silicon nitride to be 6.4 GPa [52], and the Young's modulus and yield strength of aluminum to be 74.6 GPa and 330 MPa, respectively [58].

References

1. W. Nix: Mechanical properties of thin films, Metall. Trans. A **20**, 2217–2245 (1989)
2. G.G. Stoney: The tension of metallic films deposited by electrolysis, Proc. R. Soc. Lond. A **82**, 172–175 (1909)
3. W. Brantley: Calculated elastic constants for stress problems associated with semiconductor devices, J. Appl. Phys. **44**, 534–535 (1973)
4. A. Ni, D. Sherman, R. Ballarini, H. Kahn, B. Mi, S.M. Phillips, A.H. Heuer: Optimal design of multilayered polysilicon films for prescribed curvature, J. Mater. Sci., **38**, 4169–4173 (2003)
5. J.A. Floro, E. Chason, S.R. Lee, R.D. Twesten, R.Q. Hwang, L.B. Freund: Real-time stress evolution during $Si_{1-x}Ge_x$ heteroepitaxy: Dislocations, islanding, and segregation, J. Electron. Mater. **26**, 969–979 (1997)
6. X. Li, B. Bhusan: Micro/nanomechanical characterization of ceramic films for microdevices, Thin Solid Films **340**, 210–217 (1999)
7. M.P. de Boer, H. Huang, J.C. Nelson, Z.P. Jiang, W.W. Gerberich: Fracture toughness of silicon and thin film micro-structures by wedge indentation, Mater. Res. Soc. Symp. Proc. **308**, 647–652 (1993)
8. H. Guckel, D. Burns, C. Rutigliano, E. Lovell, B. Choi: Diagnostic microstructures for the measurement of intrinsic strain in thin films, J. Micromech. Microeng. **2**, 86–95 (1992)
9. F. Ericson, S. Greek, J. Soderkvist, J.-A. Schweitz: High sensitivity surface micromachined structures for internal stress and stress gradient evaluation, J. Micromech. Microeng. **7**, 30–36 (1997)
10. L.S. Fan, R.S. Muller, W. Yun, R.T. Howe, J. Huang: Spiral microstructures for the measurement of average strain gradients in thin films, Proc. IEEE Micro Electro Mechanical Systems Workshop, Napa Valley 1990 (IEEE, New York 1990) 177–182
11. M. Biebl, H. von Philipsborn: Fracture strength of doped and undoped polysilicon, Proc. Intl. Conf. Solid-State Sensors and Actuators, Stockholm 1995, ed. by S. Middelhoek, K. Cammann (Royal Swedish Academy of Engineering Sciences, Stockholm 1995) 72–75
12. L.S. Fan, R.T. Howe, R.S. Muller: Fracture toughness characterization of brittle films, Sens. Actuators A **21-23**, 872–874 (1990)
13. H. Kahn, R. Ballarini, J.J. Bellante, A.H. Heuer: Fatigue failure in polysilicon is not due to simple stress corrosion cracking, Science **298**, 1215–1218 (2002)
14. H. Kahn, N. Tayebi, R. Ballarini, R.L. Mullen, A.H. Heuer: Wafer-level strength and fracture toughness testing of surface-micromachined MEMS devices, Mater. Res. Soc. Symp. Proc. **605**, 25–30 (2000)
15. W.N. Sharpe Jr., B. Yuan, R.L. Edwards: A new technique for measuring the mechanical properties of thin films, J. Microelectromech. Syst. **6**, 193–199 (1997)

16. H. S. Cho, W. G. Babcock, H. Last, K. J. Hemker: Annealing effects on the microstructure and mechanical properties of LIGA nickel for MEMS, Mater. Res. Soc. Symp. Proc. **657** (2001) EE5.23.1–EE5.23.6
17. W. Suwito, M. L. Dunn, S. J. Cunningham, D. T. Read: Elastic moduli, strength, and fracture initiation at sharp notches in etched single crystal silicon microstructures, J. Appl. Phys. **85**, 3519–3534 (1999)
18. C.-S. Oh, W. N. Sharpe: Techniques for measuring thermal expansion and creep of polysilicon, Sens. Actuators A **112**, 66–73 (2004)
19. T. P. Weihs, S. Hong, J. C. Bravman, W. D. Nix: Mechanical deflection of cantilever microbeams: A new technique for testing the mechanical properties of thin films, J. Mater. Res. **3**, 931–942 (1988)
20. B. D. Jensen, M. P. de Boer, N. D. Masters, F. Bitsie, D. A. La Van: Interferometry of actuated microcantilevers to determine material properties and test structure nonidealities in MEMS, J. Microelectromech. Syst. **10**, 336–346 (2001)
21. M. G. Allen, M. Mehregany, R. T. Howe, S. D. Senturia: Microfabricated structures for the in situ measurement of residual stress, Young's modulus, and ultimate strain of thin films, Appl. Phys. Lett. **51**, 241–243 (1987)
22. H. Kahn, S. Stemmer, K. Nandakumar, A. H. Heuer, R. L. Mullen, R. Ballarini, M. A. Huff: Mechanical properties of thick, surface micromachined polysilicon films, Proc. IEEE Micro Electro Mechanical Systems Workshop, San Diego 1996, ed. by M. G. Allen, M. L. Redd (IEEE, New York 1996) 343–348
23. L. Kiesewetter, J.-M. Zhang, D. Houdeau, A. Steckenborn: Determination of Young's moduli of micromechanical thin films using the resonance method, Sens. Actuators A **35**, 153–159 (1992)
24. W. C. Tang, T.-C. H. Nguyen, R. T. Howe: Laterally driven polysilicon resonant microstructures, Sens. Actuators A **20**, 25–32 (1989)
25. P. T. Jones, G. C. Johnson, R. T. Howe: Fracture strength of polycrystalline silicon, Mater. Res. Soc. Symp. Proc. **518**, 197–202 (1998)
26. P. Minotti, R. Le Moal, E. Joseph, G. Bourbon: Toward standard method for microelectromechanical systems material measurement through on-chip electrostatic probing of micrometer size polysilicon tensile specimens, Jpn. J. Appl. Phys. **40**, L120–L122 (2001)
27. C. L. Muhlstein, E. A. Stach, R. O. Ritchie: A reaction-layer mechanism for the delayed failure of micron-scale polycrystalline silicon structural films subjected to high-cycle fatigue loading, Acta Mater. **50**, 3579–3595 (2002)
28. A. Corigliano, B. De Masi, A. Frangi, C. Comi, A. Villa, M. Marchi: Mechanical characterization of polysilicon through on-chip tensile tests, J. Microelectromech. Syst. **13**, 200–219 (2004)
29. M. G. Lim, J. C. Chang, D. P. Schultz, R. T. Howe, R. M. White: Polysilicon microstructures to characterize static friction, Proc. IEEE Micro Electro Mechanical Systems Workshop, Napa Valley 1990 (IEEE, New York 1990) 82–88
30. B. T. Crozier, M. P. de Boer, J. M. Redmond, D. F. Bahr, T. A. Michalske: Friction measurement in MEMS using a new test structure, Mater. Res. Soc. Symp. Proc. **605**, 129–134 (2000)
31. J. Yang, H. Kahn, A. Q. He, S. M. Phillips, A. H. Heuer: A new technique for producing large-area as-deposited zero-stress LPCVD polysilicon films: The MultiPoly process, J. Microelectromech. Syst. **9**, 485–494 (2000)
32. E. Chason, B. W. Sheldon, L. B. Freund, J. A. Floro, S. J. Hearne: Origin of compressive residual stress in polycrystalline thin films, Phys. Rev. Lett. **88** (2002) 156103-1–156103-4

33. S. Jayaraman, R.L. Edwards, K.J. Hemker: Relating mechanical testing and microstructural features of polysilicon thin films, J. Mater. Res. **14**, 688–697 (1999)
34. R. Ballarini, H. Kahn, N. Tayebi, A.H. Heuer: Effects of microstructure on the strength and fracture toughness of polysilicon: A wafer level testing approach, ASTM STP **1413**, 37–51 (2001)
35. J. Bagdahn, J. Schischka, M. Petzold, W.N. Sharpe Jr.: Fracture toughness and fatigue investigations of polycrystalline silicon, Proc. SPIE **4558**, 159–168 (2001)
36. I.S. Raju, J.C. Newman Jr.: Stress intensity factors for a wide range of semi-elliptical surface cracks in finite-thickness plates, Eng. Fract. Mech. **11**, 817–829 (1979)
37. J. Bagdahn, W.N. Sharpe Jr., O. Jadaan: Fracture strength of polysilicon at stress concentrations, J. Microelectromech. Syst., 302–312 (2003)
38. T. Yi, L. Li, C.-J. Kim: Microscale material testing of single crystalline silicon: Process effects on surface morphology and tensile strength, Sens. Actuators A **83**, 172–178 (2000)
39. X. Li, T. Kasai, S. Nakao, T. Ando, M. Shikida, K. Sato, H. Tanaka: Anisotropy in fracture of single crystal silicon film characterized under uniaxial tensile condition, Sens. Actuators A **117**, 143–150 (2005)
40. H. Kahn, R. Ballarini, R.L. Mullen, A.H. Heuer: Electrostatically actuated failure of microfabricated polysilicon fracture mechanics specimens, Proc. R. Soc. Lond. A **455**, 3807–3923 (1999)
41. W.W. Van Arsdell, S.B. Brown: Subcritical crack growth in silicon MEMS, J. Microelectromech. Syst. **8**, 319–327 (1999)
42. A. Corwin, M.P. de Boer: Effect of adhesion on dynamic and static friction in surface micromachining, Appl. Phys. Lett. **84**, 2451–2453 (2004)
43. T. Tsuchiya, A. Inoue, J. Sakata: Tensile testing of insulating thin films; humidity effect on tensile strength of SiO_2 films, Sens. Actuators A **82**, 286–290 (2000)
44. H. Ogawa, K. Suzuki, S. Kaneko, Y. Nakano, Y. Ishikawa, T. Kitahara: Measurements of mechanical properties of microfabricated thin films, Proc. IEEE Micro Electro Mechanical Systems Workshop (IEEE, New York 1997) 430–435
45. R.L. Edwards, G. Coles, W.N. Sharpe: Comparison of tensile and bulge tests for thin-film silicon nitride, Exp. Mech. **44**, 49–54 (2004)
46. S. Cho, I. Chasiotis, T.A. Friedmann, J.P. Sullivan: Young's modulus, Poisson's ratio and failure properties of tetrahedral amorphous diamond-like carbon for MEMS devices, J. Micromech. Microeng. **15**, 728–735 (2005)
47. D. Son, J.-H. Jeong, D. Kwon: Film-thickness considerations in microcantilever-beam test in measuring mechanical properties of metal thin film, Thin Solid Films **437**, 182–187 (2003)
48. W.-H. Chuang, T. Luger, R.K. Fettig, R. Ghodssi: Mechanical property characterization of LPCVD silicon nitride thin films at cryogenic temperatures, J. Microelectromech. Syst. **13**, 870–879 (2004)
49. H.D. Espinosa, B. Peng: A new methodology to investigate fracture toughness of freestanding MEMS and advanced materials in thin film form, J. Microelectromech. Syst. **14**, 153–159 (2005)
50. C.-W. Baek, Y.-K. Kim, Y. Ahn, Y.-H. Kim: Measurement of the mechanical properties of electroplated gold thin films using micromachined beam structures, Sens. Actuators A **117**, 17–27 (2005)
51. R. Yokokawa, J.-A. Paik, B. Dunn, N. Kitazawa, H. Kotera, C.-J. Kim: Mechanical properties of aerogel-like thin films used for MEMS, J. Micromech. Microeng. **14**, 681–686 (2004)

52. A. Kaushik, H. Kahn, A. H. Heuer: Wafer-level mechanical characterization of silicon nitride MEMS, J. Microelectromech. Syst. **14**, 359–367 (2005)
53. S. Roy, C. A. Zorman, M. Mehregany: The mechanical properties of polycrystalline silicon carbide films determined using bulk micromachined diaphragms, Mater. Res. Soc. Symp. Proc. **657** (2001) EE9.5.1–EE9.5.6
54. J. J. Bellante, H. Kahn, R. Ballarini, C. A. Zorman, M. Mehregany, A. H. Heuer: Fracture toughness of polycrystalline silicon carbide thin films, Appl. Phys. Lett. **86**, 071920–1–071920–3 (2005)
55. A. J. Fleischman, X. Wei, C. A. Zorman, M. Mehregany: Surface micromachining of polycrystalline SiC deposited on SiO_2 by APCVD, Mater. Sci. Forum **264-268**, 885–888 (1998)
56. A. E. Franke, E. Bilic, D. T. Chang, P. T. Jones, T.-J. King, R. T. Howe, G. C. Johnson: Post-CMOS integration of germanium microstructures, Proc. Int. Conf. Solid-State Sensors and Actuators (IEE Japan, Tokyo 1999) 630–637
57. D. Gao, C. Carraro, V. Radmilovic, R. T. Howe, R. Maboudian: Fracture of polycrystalline 3C-SiC films in microelectromechanical systems, J. Microelectromech. Syst. **13**, 972–976 (2004)
58. M. A. Haque, M. T. A. Saif: A review of MEMS-based microscale and nanoscale tensile and bending testing, Exp. Mech. **43**, 248–255 (2003)

24

Structural, Nanomechanical, and Nanotribological Characterization of Human Hair Using Atomic Force Microscopy and Nanoindentation

Bharat Bhushan · Carmen LaTorre

Summary. Human hair is a nanocomposite biological fiber. Maintaining the health, feel, shine, color, softness, and overall aesthetics of the hair is highly desired. Hair care products such as shampoos and conditioners, along with damaging processes such as chemical dyeing and permanent wave treatments, affect the maintenance and grooming process and are important to study because they alter many hair properties. Nanoscale characterization of the cellular structure, mechanical properties, and morphological, frictional, and adhesive properties (tribological properties) of hair are essential to evaluate and develop better cosmetic products, and to advance the understanding of biological and cosmetic science. The atomic/friction force microscope (AFM/FFM) and nanoindenter have become important tools for studying the micro/nanoscale properties of human hair. In this chapter, we present a comprehensive review of structural, mechanical, and tribological properties of various hair and skin as a function of ethnicity, damage, conditioning treatment, and various environments. Various cellular structure of human hair and fine sublamellar structures of the cuticle are identified and studied. Nanomechanical properties such as hardness, elastic modulus, tensile deformation, creep, and scratch resistance are discussed. Nanotribological properties such as roughness, friction, and adhesion are presented, as well as investigations of conditioner distribution, thickness, and binding interactions. To study the electrostatic charge build up on hair, surface potential studies are also presented.

24.1 Introduction

Everybody wants beautiful, healthy hair and skin. For most people, grooming and maintenance of hair and skin is a daily process. The demand for products that improve the look and feel of these surfaces has created a huge industry for hair and skin care. Beauty care technology has advanced the cleaning, protection, and restoration of desirable hair properties by altering the hair surface. For many years, especially in the second half of the twentieth century, scientists have focused on the physical and chemical properties of hair to consistently develop products which alter the health, feel, shine, color, softness, and overall aesthetics of the hair. Hair care products such as shampoos and conditioners aid the maintenance and grooming process. Shampoos clean the hair and skin oils, and conditioners repair hair damage and make the hair easier to comb; prevent flyaway; add feel, shine, and softness. Mechanical processes such as combing, cutting, and blowdrying serve to style the hair. Chemical products and processes such as chemical dyes, colorants, bleaches, and permanent wave treatments enhance the appearance and hue of the hair. Of particular interest is how all these common hair care items deposit onto and change hair properties, since these properties are closely tied to product performance. The fact that companies like Procter & Gamble, L'Oreal, and Unilever have hair care product sales consistently measured in the billions of Dollars [http://www.pg.com; http://www.loreal.com; http://unilever.com] suggests that understanding the science behind human hair has more than just purely academic benefits, as well.

While products and processes such as combing, chemical dyeing, and permanent wave treatment are used to enhance appearance and style of the hair, they also contribute a large amount of chemical and mechanical damage to the fibers, which leads to degradation of structure and mechanical properties. As a result, the fibers become weak and more susceptible to breakage after time, which is undesirable for healthy hair. Shampoos and conditioners, which typically serve cleaning and repairing functions to the hair surface, respectively, have a distinct effect on mechanical properties as well.

The tribology of the hair also changes as a function of the various hair care products and processes. Figure 24.1 illustrates schematically various functions, along with the macro- and micro/nanoscale mechanisms behind these interactions, that make surface roughness, friction, and adhesion very important to hair and skin [45]. Desired features and corresponding tribological attributes of conditioners are listed in Table 24.1 [48]. For a smooth wet and dry feel, friction between hair and skin should be minimized in wet and dry environments, respectively. For a good feel with respect to bouncing and shaking of the hair during walking or running, friction between hair fibers and groups of hair fibers should be low. The friction one feels during combing is a result of interactions between hair and the comb material (generally a plastic), and this too needs to be low to easily maintain, sculpt, and comb the hair. To minimize entanglement, adhesive force (the force required to separate the hair fibers) needs to be low. In other cases, a certain level of adhesion may be acceptable and is often a function of the hair style. For individuals seeking "hair alignment," where hair fibers lay flat and parallel to each other, a small amount of adhesive force between fibers may be desired. For more complex and curly styles, even higher adhesion between fibers may be optimal.

Early research into human hair was done primarily on the chemical and physical properties of the hair fiber itself. Key topics dealt with analysis of chemical composition in the fiber, microstructure, and hair growth, to name a few. Until about 2000, most information about the detailed structure of human hair was obtained from scanning electron microscope (SEM) and transmission electron microscope (TEM) observations [68, 78, 79, 90]. Mechanical properties were also of interest. Most of the mechanical property measurements of human hair were on the macroscale and used conventional methods, such as tension, torsion and bending tests [2, 32, 40, 68, 80, 81]. The mechanical properties obtained from these tests are

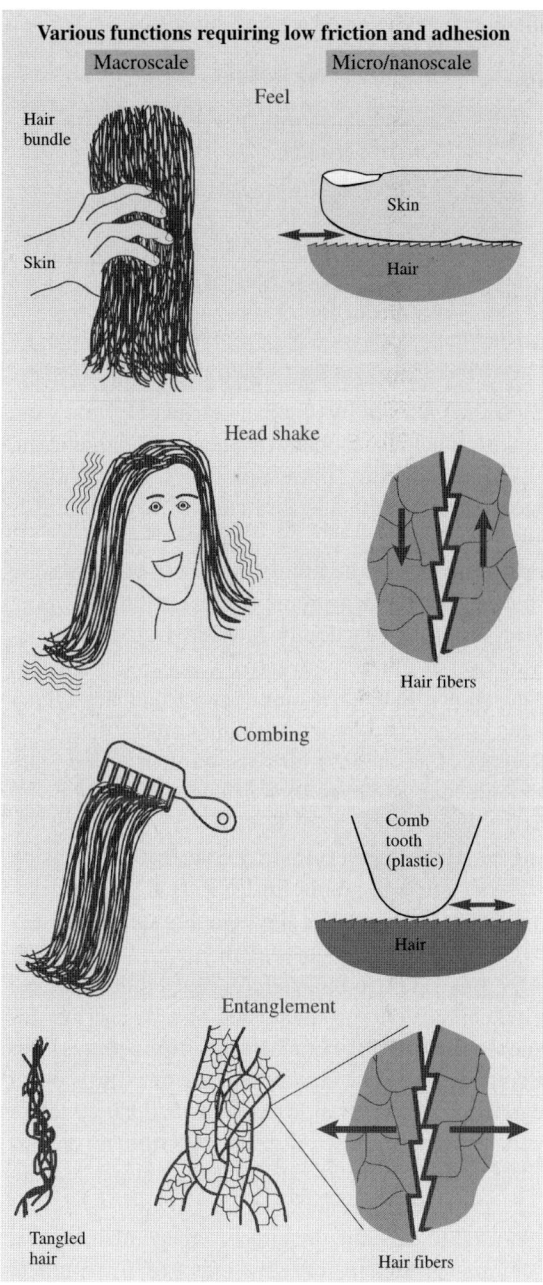

Fig. 24.1. Schematics illustrating various functions with associated macroscale and micro/nanoscale mechanisms of hair and skin friction during feel or touch, shaking and bouncing of the hair, combing, and entanglement [45]

Table 24.1. Desired features and corresponding tribological attributes of conditioners

Desired hair feature	Tribological attributes
Smooth feel in wet and dry environments	Low friction between hair and skin in respective environment
Shaking and bouncing during daily activities	Low friction between hair fibers and groups of hair
Easy combing and styling	Low friction between hair and comb (plastic) and low adhesion. Note: More complex styles may require higher adhesion between fibers.

the overall mechanical properties of the hair, not just the hair surface. Efforts were also made to study the effects of environmental and chemical damage and treatment, such as dyeing, bleaching, and polymer application; these topics have stayed a mainstream area of investigation due to the availability and formation of new chemicals and conditioning ingredients. Tribology has generally been studied via macroscale friction force of hair. As a matter of fact, much of the tribological work performed by the hair care industry today still focuses on the measurement of macroscale friction, particularly between a skin replica and a hair swatch of interest [68]. The intrinsic differences of the hair as a function of ethnicity eventually became a concern as well. For instance, research has shown that African-American hair has higher resistance to combing, higher static charge, and lower moisture content than Caucasian hair [83]. Because of differences like these, a growing number of hair care products specifically tailored for ethnic hair care have been developed and sold with commercial success.

Modern research since the late 1990s has been primarily concerned with using micro/nanoscale experimental methods such as atomic force/friction force microscopies (AFM/ FFM) and nanoindentation to answer the complex questions surrounding the structure and behavior of the hair. Nanoscale characterization of the cellular structure, mechanical properties, and tribological properties of hair are essential to evaluate and develop better cosmetic products, and to advance the understanding of composite biological systems, cosmetic science, and dermatology. AFM/FFM have been used to effectively study the structure of the hair surface and cross-section. AFM provides the potential for being able to see the cellular structure and molecular assembly of hair, for determining various properties of hair, such as elastic modulus and viscoelastic properties, and for investigating the physical behavior of various cellular structure of hair in various environments [13, 26]. As a non-invasive technique, AFM has been used to evaluate the effect of hair treatment and can be operated in ambient conditions in order to study the effect of environment on various physical properties. AFM studies on hair fibers have been carried out for surface topographic imaging [45, 75] and friction, adhesion, and wear properties of hair and skin and the effects of hair care products on hair [27, 45–48, 53]. Roughness parameters have been measured to compare changes due to damaging processes.

Friction force has been measured to understand damage or conditioner distribution and its effect on hair tribology. Adhesive force mapping has shown useful to observe the conditioner distribution as well. Surface charge of human hair has a significant effect on manageability, flyaway behavior, feel and appearance. It is known that interaction of hair with dissimilar materials, such as plastic combs, hands, and latex balloons creates a charge on hair [55, 56]. Physical wear has been shown to cause surface potential change in conductors and semiconductors [15, 16, 31] including hair [55].

The nanoindenter has been used to characterize the nanomechanical behavior of the hair surface and cross-section using nanoindentation and nanoscratch techniques [89, 90]. These properties are important for evaluating cosmetic products by comparing the nanomechanical properties, such as hardness, elastic modulus, and scratch resistance, of the hair surface before and after chemical damage or conditioner treatment. Since hair is a nanocomposite biological fiber with well characterized structures, which will be described in details in the next section, it is a good model to study the role of various structural and chemical components in providing mechanical strength for composite biological fibers [90]. Furthermore, the quantitative determination of the mechanical properties of human hair can also provide the dermatologists with some useful markers for the diagnosis of hair disorders and for the evaluation of their response to therapeutic regimens [64, 77]. Combing results in physical damage such as scratching, and therefore scratch resistance is useful to study, especially when conditioner is applied. Human hair fibers stretch and consequently experience tensile stresses as they are brushed, combed, and go through styling process. Hence the behavior of hair under tension is of interest. On human hair, characterization of fracture after deformation has been carried out using SEM and light/fluorescence microscopy [38, 80]. With in-situ experiments, it is possible to systematically follow the progress of morphological change and deformation in the material, and to accurately pinpoint the initiation of major deformation events. In-situ tensile loading experiments in an AFM have been carried out to study the progress of deformation and morphological changes in the hair fiber, by pausing the straining at regular intervals [74].

In this chapter, we present a comprehensive study of various hair and skin structural, mechanical, and tribological properties as a function of ethnicity, damage, conditioning treatment, and various environments. Various cellular structure (such as the cortex and the cuticle) of human hair and fine sublamellar structures of the cuticle, such as the A-layer, the exocuticle, the endocuticle, and the cell membrane complex are easily identified and studied. Nanomechanical properties including hardness, elastic modulus, tensile deformation, creep, and scratch resistance are discussed. Nanotribological properties including surface roughness, friction, adhesion, and wear are presented, as well as investigations of conditioner localization and thickness. To study the electrostatic charge build up on hair, surface potential studies are presented.

24.2 Human Hair, Skin, and Hair Care Products

24.2.1 Human Hair and Skin

Figure 24.2a shows a schematic of a human hair fiber with its various layers of cellular structure [32, 43, 62, 68, 75, 95]. Hair fibers (about 50 to 100 µm in diameter) consist of the cuticle and cortex, and in some cases medulla in the central region. All are composed of dead cells, which are mainly filled with keratin protein. Table 24.2 displays a summary of the chemical species of hair [26]. Depending on its moisture content, human hair consists of approximately 65-95% keratin proteins, and the remaining constituents are water, lipids (structural and free), pigment, and trace elements. Proteins are made up of long chains of various mixtures of the some 20 or 50 amino acids. Each chain takes up a helical or coiled form. Among numerous amino acids in human hair, cystine is one of the most important amino acids. Every cystine unit contains two cysteine amino acids in different chains which lie near to each other and are linked together by two sulfur atoms, forming a very strong bond known as a disulfide linkage, Fig. 24.2b [37]. In addition to disulfide bonds, hair is also rich in peptide bonds and the abundant CO- and NH- groups present give rise to hydrogen bonds between groups of neighboring chain molecules. The distinct cystine content of various cellular structure of human hair results in a significant effect on their physical properties. A high cystine content corresponds to rich disulfide cross-links, leading to high mechanical properties. The species responsible for color in hair is the pigment – melanin, which is located in the cortex of the hair in granular form.

An average head contains over 100,000 hair follicles, which are the cavities in the skin surface from which hair fibers grow. Each follicle grows about 20 new hair fibers in a lifetime. Each fiber grows for several years until it falls out and is replaced by a new fiber. Hair typically grows at a rate on the order of 10 mm/month.

Table 24.2. Summary of chemical species present in human hair

Keratin protein	65–95%
(Amino acids)	$\overset{\oplus}{N}H_3-CH-R$ | $\overset{\ominus}{C}O_2$ (R: functional group)
Cystine	$\overset{\oplus}{N}H_3-CH-CH_2-S-S-CH_2-CH-\overset{\oplus}{N}H_3$ | $\overset{\ominus}{C}O_2$ | $\overset{\ominus}{C}O_2$
Lipids	Structural and free
18-methyl eicosanoic acid (18-MEA)	$H_3C-\overset{\underset{\|}{CH_3}}{}\!\!\sim\!\!\sim\!\!\sim\!\!\sim\!\!\overset{-(CH_2)_{16}-}{}\!\!\sim\!\!\sim\!\!\sim\!\!\sim\!\!-COOH$
Water	Up to 30%
Pigment and trace elements	Melanin

24 Structural, Nanomechanical, and Nanotribological Characterization 1331

The Cuticle

The cuticle consists of flat overlapping cells (scales). The cuticle cells are attached at the root end and they point forward the tip end of the hair fiber, like tiles on a roof. Each cuticle cell is approximately 0.3 to 0.5 µm thick and the visible length of each cuticle cell is approximately 5 to 10 µm. The cuticle in human hair is generally 5 to 10 scales thick. Each cuticle cell consists of various sublamellar layers (the epicuticle, the A-layer, the exocuticle, the endocuticle and inner layer) and the cell membrane complex (see Fig. 24.2a). Table 24.3 displays the various layers

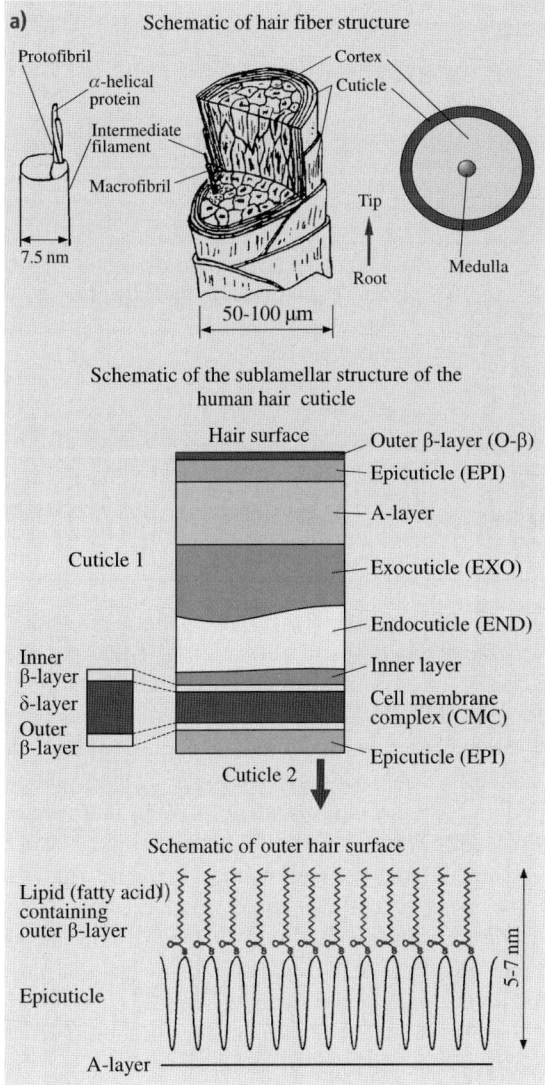

Fig. 24.2. (a) Schematic of hair fiber structure and cuticle sublamellar structure [68, 75]

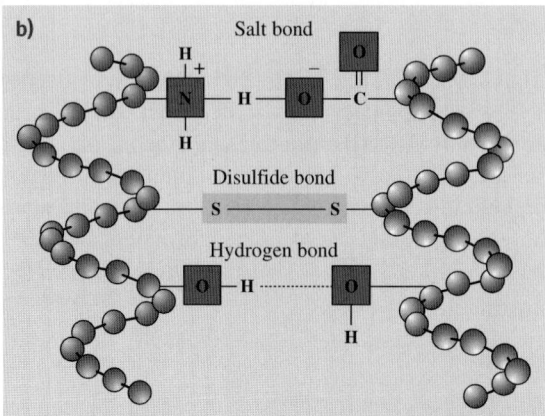

Fig. 24.2. (continued) (**b**) various bonds within hair cellular structure [13, 37]

Table 24.3. Various layers of the cuticle and their details

Cuticle layer	Cystine component	Details
Epicuticle	~ 12%	18-MEA lipid layer attached to outer epicuticle contributes to lubricity of the hair
A-layer	~ 30%	Highly cross-linked
Exocuticle	~ 15%	Mechanically tough Chemically resilient
Endocuticle	~ 3%	
Inner layer	–	
Cell membrane complex (CMC)	~ 2%	Lamellar structure Consists of inner β-layer, δ-layer, and outer β-layer

of the cuticle, their respective cystine levels [68] and other details. The outer epicuticle layer is covered with a thin layer of covalently attached lipid (fatty acid), predominantly 18-Methyl Eicosanoic Acid (18-MEA) (see Table 24.2). This layer constitutes the outer β-layer of the cuticular cell membrane complex, which acts as a boundary lubricant, responsible for low friction and provides a hydrophobic surface. The A-layer is a component of high cystine content ($\sim$ 30%) and located on the outer-facing aspect of each cell. The A-layer is highly crosslinked which gives this layer considerable mechanical toughness and chemical resilience, and the swelling in water is presumed to be minimal. The exocuticle, which is immediately adjacent to the A-layer, is also of high cystine content ($\sim$15%). On the inner facing aspect of each cuticle cell is a thin layer of material which is known as the inner layer. Between the exocuticle and inner layer is the endocuticle which is low in cystine ($\sim$3%). The cell membrane complex (CMC) itself is a lamellar structure, which consists of the inner β-layer, the δ-layer and the outer β-layer. The outer β-layer of the CMC separates the cuticle cells from each other. Low cohesive forces are expected between the lipid-containing outer β-layer and the δ-layer of CMC, which provides a weak bond. It may result into cuticular delamination during mechanical wear, with the potential advantage of revealing a fresh layer of 18-MEA to the newly exposed surface [75].

Figure 24.3 shows the SEM images of virgin Caucasian, Asian and African hair [90]. It can be seen that the Asian hair is the thickest (about 100 μm), followed by African hair (about 80 μm) and Caucasian hair (about 50 μm). The visible cuticle cell is about 5 to 10 μm long for the three hair. A listing of various cross-section dimensional properties are presented in Table 24.4 [90]. While Caucasian and Asian hair typically have similar cross-sectional shape (Asian hair being the most cylindrical), African hair has a highly elliptical shape. African hair is much more curly and wavy along the hair fiber axis than Caucasian or Asian hair.

Chemically, all ethnic hair are found to have similar protein structure and composition [29, 30, 59, 63]. African hair has less moisture content than Caucasian hair. The shape (diameter, ellipticity, and curliness) of various ethnic hair depends on several factors, including the shape of the hair follicle and its opening; these vary from one person to another and also between races [37, 88]. The pronounced ellipsoidal

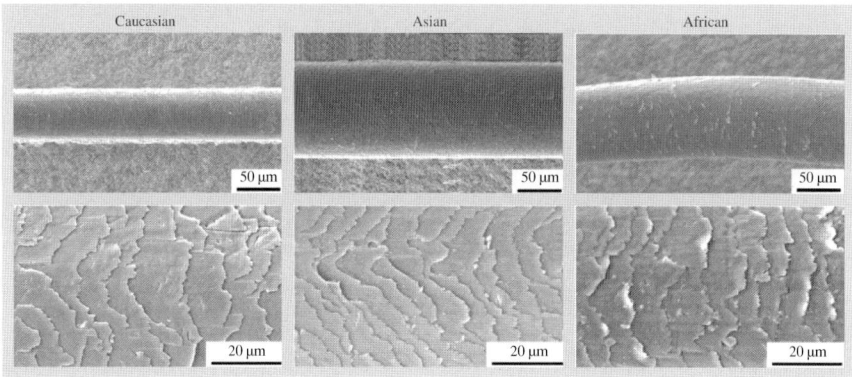

Fig. 24.3. SEM images of various hair [90]

Table 24.4. Variation in cross-sectional dimensions of human hair

	Shape	Maximum diameter (D_1) (μm)	Minimum diameter (D_2) (μm)	Ratio D_1/D_2	Number of cuticle scales	Cuticle scale thickness (μm)
Caucasian	Nearly oval	74	47	1.6	6–7	0.3–0.5
Asian	Nearly round	92	71	1.3	5–6	0.3–0.5
African	Oval-flat	89	44	2.0	6–7	0.3–0.5

Average length of visible cuticle scale: about 5 to 10 μm

cross-section of hair shaft in African hair could be caused by a heterogeneous and asymmetric fiber framework, in addition to internal mechanical stresses [87]. Previously, it was thought that the elliptical cross section of hair was responsible for curl. While straight hair has circular cross sections (Asian and Caucasian), curly hair has a predominantly elliptical cross section (African). However, recent studies suggest that hair follicle shape and not the cross section is responsible for hair curl [88]. This means that if the follicle is straight, even an elliptical cross section could give rise to straight hair. Both in-vitro growth studies and computer aided three dimensional reconstruction [51] support this claim. Curvature of the curly hair is programmed from the basal area of follicle. This bending process is apparently linked to a lack of symmetry in the lower part of the bulb, affecting the hair shaft cuticle.

Figure 24.4 shows the SEM images of virgin Caucasian hair at three locations: near scalp, middle and near tip. Three magnifications were used to show the significant differences. The hair near scalp had complete cuticles, while no cuticles were found on the hair near tip. This may be because that the hair near the tip experienced more mechanical damage during its life than the hair near the scalp. The hair in the middle experienced intermediate damage, i.e., one or more scales of the cuticles were worn away, but many cuticles stayed complete. If some substructures of one cuticle scale, like A-layer or A-layer and exocuticle (see Fig. 24.2a), are gone, or even worse, one or several cuticle scales are worn away, it is impossible to heal the hair biologically, because hair fibers are composed of dead cells. However, it is possible to physically "repair" the damaged hair by using conditioner, one of whose functions is to cover or fill the damaged area of the cuticles. Figure 24.5 shows the

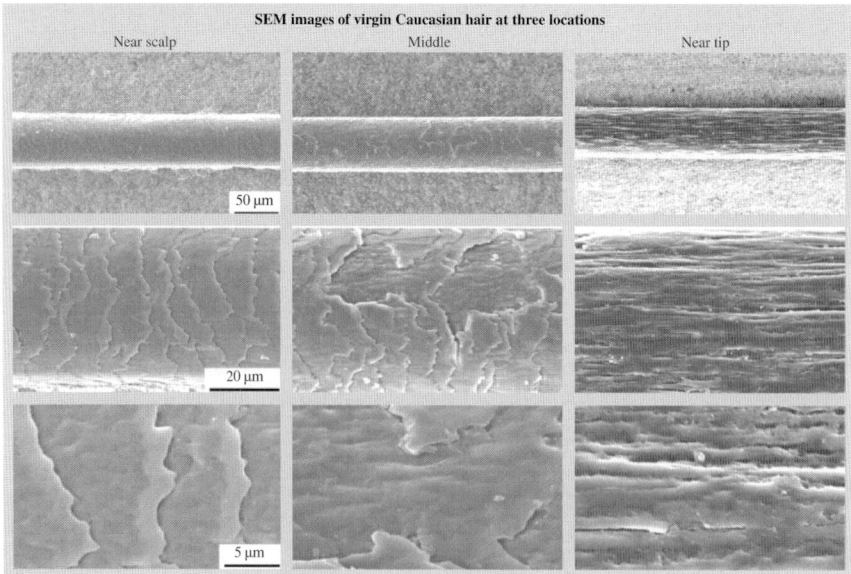

Fig. 24.4. SEM images of virgin Caucasian hair at three locations [90]

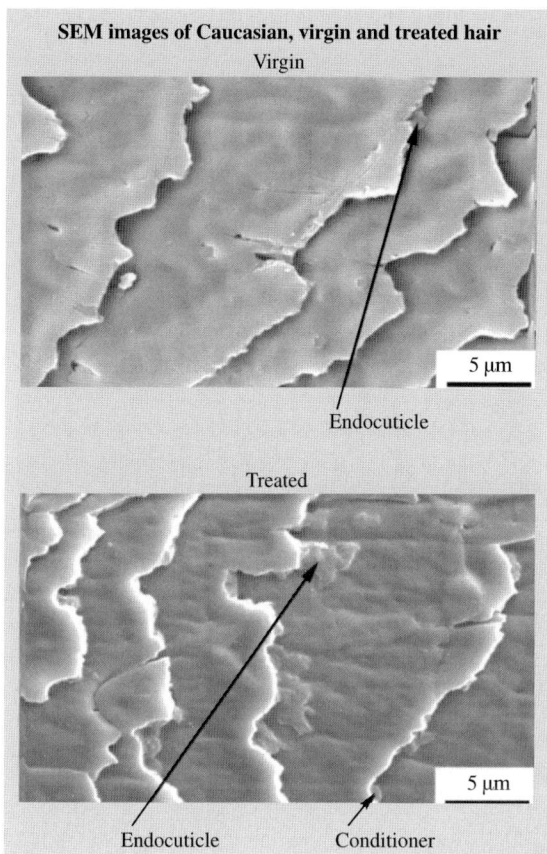

Fig. 24.5. SEM images of Caucasian, virgin and treated hair [90]

high magnification SEM images of virgin and treated Caucasian hair. The endocuticles (pointed by arrows) were found in both hair. In order for the conditioner to physically repair the hair, it is expected for it to cover the endocuticles. In the case of severely damaged hair, for example, an edge of one whole cuticle scale worn away, the conditioner may fill that damaged edge. In the SEM image of the treated hair in Fig. 24.5, the substance which stayed near the cuticle edge is probably the conditioner (pointed by an arrow).

Figure 24.6 shows the AFM images of various virgin hair, along with the section plots [45]. The arrows point to the position where the section plots were taken from. Each cuticle cell is nearly parallel to the underlying cuticle cell, and they all have similar angles to the hair axis, forming a tile-like hair surface structure. The visible cuticle cell is approximately 0.3 to 0.5 μm thick and about 5 to 10 μm long for all three hair.

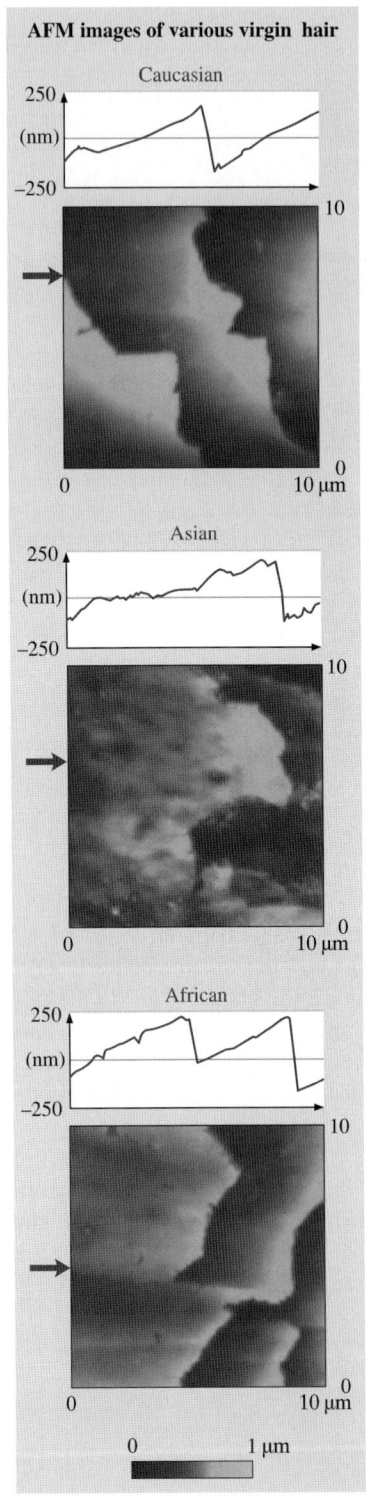

Fig. 24.6. AFM images of various virgin hair [45]

The Cortex and Medulla

The cortex contains cortical cells and the intercellular binding material, or the cell membrane complex. The cortical cells are generally 1 to 6 μm thick and 100 μm long, which run longitudinally along the hair fiber axis and take up the majority of the inner hair fiber composition [67]. The macrofibrils (about 0.1 to 0.4 μm in diameter) comprise a major portion of the cortical cells. Each macrofibril consists of intermediate filaments (about 7.5 nm in diameter), previously called microfibrils, and the matrix. The intermediate filaments are low in cystine (∼ 6%), and the matrix is rich in cystine (∼ 21%). The cell membrane complex consists of cell membranes and adhesive material that binds the cuticle and cortical cells together. The intercellular cement of the cell membrane complex is primarily nonkeratinous protein, and is low in cystine content (∼ 2%). The medulla of human hair, if present, generally makes up only a small percentage of the mass of the whole hair, and is believed to contribute negligibly to the mechanical properties of human hair fibers.

Figure 24.7a shows the SEM images of virgin hair cross-section [90] and Fig. 24.7b shows the TEM images of a cross-section of human hair [79].

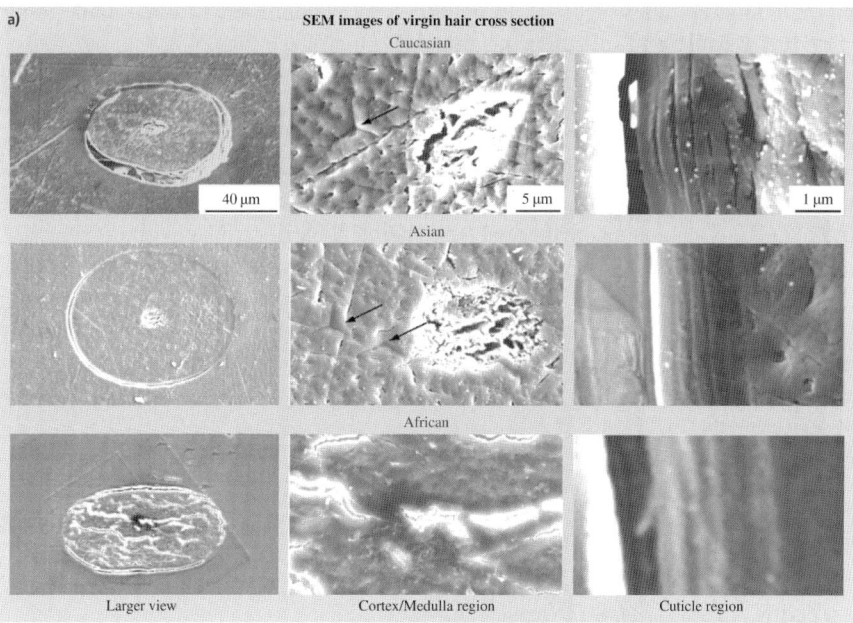

Fig. 24.7. (a) SEM images of virgin hair cross-section [90]

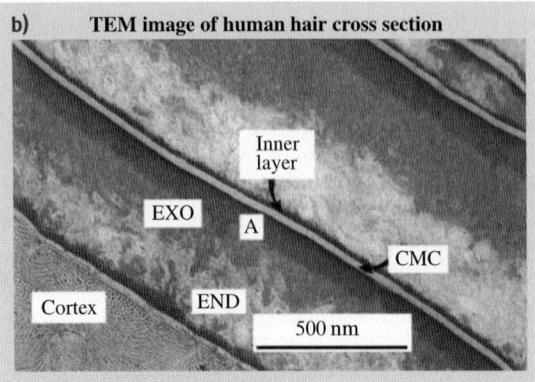

Fig. 24.7. (continued) **(b)** TEM of hair cross-section (in the figure EXO, END, and CMC stand for exocuticle, endocuticle, and cell membrane complex, respectively) [79]

Skin

Skin covers and protects our bodies. The skin at the forehead and scalp areas are of most interest when dealing with human hair, since most of the hair care products are developed specifically for head hair. The skin of the hand and fingers is also of importance because the "feel" of hair is often sensed by physically touching the fibers with these regions. In general, skin is composed of three main parts: epidermis, dermis, and subcutaneous tissue (L'Oreal); see Fig. 24.8.

The epidermis contains four distinct cellular layers: basal, spinous, granular, and horny. In the basal layer, melanocytes deliver the pigment melanin to keratinocytes. Keratinocyte cells that have been cornified are referred to as corneocytes [66]. Hexagonally shapped corneocyte cells compose the horny layer, or stratum corneum. The stratum corneum is the outer layer of the skin; at about 15 μm thick, it acts as a mechanical, thermal, and chemical barrier from environmental factors and contamination. The complex organization of corneocytes and intercellular

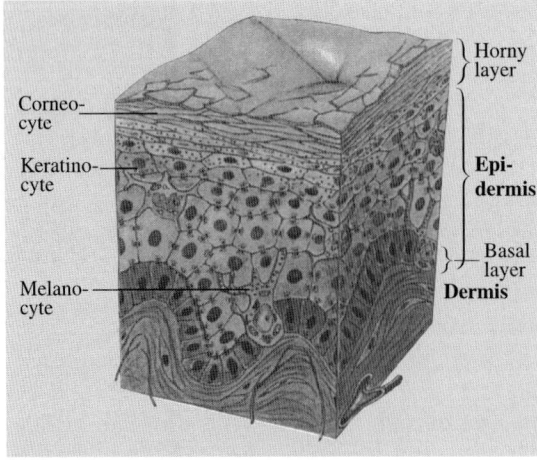

Fig. 24.8. Schematic image of human skin structure with different layers: dermis, epidermis, and horny layer (courtesy L'Oreal Corporation)

matrix contribute to the success of the barrier [91]. In fact, *Wertz* et al. [92] developed a structural model which observes the matrix as a lamellar phase composition of various lipids which provide a glue-like system to provide a barrier effect.

The dermis structure is known for its ability to handle most of the physical stresses imposed on the skin, and takes up roughly 90% of the mass [66]. The dermis is divided into an outermost papillary layer and underlying reticular layer.

24.2.2 Hair Care: Cleaning and Conditioning Treatments, and Damaging Processes

Cleaning and Conditioning Treatments: Shampoo and Conditioner

Shampoos are used primarily to clean the hair and scalp of dirt and other greasy residue that can build up after time. Shampoos also have many secondary functions including controlling dandruff, reducing irritation, and even conditioning. Conditioners, on the other hand, are used primarily to give the hair a soft, smooth feel which results in easier hair combing. Secondary functions include preventing "flyaway" hair due to static electricity, giving the hair a shiny appearance, and protecting the hair from further damage by forming a thin coating over the fibers.

Further developments in marketing and aesthetic factors (brand name, fragrance, feel, and color of the shampoos and conditioners) have created new market segments. In many instances, these factors have become primary reasons for use.

Shampoo: Constitution and Main Functions

The following discussion is based on *Gray* [36, 37]. As stated above, shampoos serve various cleaning functions for the hair and scalp. In the past, typical shampoos were mainly soap based products. However, soaps did not have very good lathering capability, and often left a residual "scum" layer on the hair that was undesirable and could not be rinsed off.

In modern shampoos, advances in chemistry and technology have made it possible to replace the soap bases with complex formulas of cleansing agents, conditioning agents, functional additives, preservatives, aesthetic additives and even medically active ingredients. Table 24.5 shows the most common ingredients of shampoos and their functions.

Cleansing Agents In most modern shampoos, the primary cleansing agents are surfactants. Dirt and greasy residue are removed from the hair and scalp by these surfactants, making them the most important part of the shampoo. Surfactants have great lathering capabilities and rinse off very easily; see Table 24.5 for a full list of features.

Surfactant molecules have two different ends, one which is negatively charged and soluble in water (unable to mix with greasy matter), and another which is soluble in greasy matter (unable to mix with water). In general, surfactants clean the hair by the following process: Surfactant molecules encircle the greasy matter on the hair surface. The molecule end which is soluble in greasy matter buries itself in the

Table 24.5. Components of common shampoos and their functions

Shampoo component	Functions
Cleansing agents	– Produce lather to trap greasy matter, and prevent re-deposition – Remove dirt and grease from hair and scalp – Stabilize the mixture and help keep the ingredient network together – Thicken the shampoo to the desired viscosity
Conditioning agents	Condition the hair
Functional additives	Control the viscosity and pH levels of the shampoo
Preservatives	Prevent decomposition and contamination of shampoo
Aesthetic additives	Enhance color, scent, and luminescence of shampoo
Medically active ingredients	Aid treatment of dandruff or hair loss

grease, which leaves the water soluble molecule end to face outward with a negative charge. Since the hair fibers are negatively charged as well, the two negative charges repel each other. Thus, the greasy matter is easily removed from the hair surface and rinsed off.

Shampoos contain several surfactants, generally up to four, which clean the hair differently depending on the hair type of the individual. Mild cleansing systems, which do not damage or irritate the scalp, hair, and eyes, are now quite common.

Conditioning Agents Many shampoos contain conditioning agents which serve many of the same roles as full conditioners. Conditioning agents are further described in the following subsection.

Functional Additives Functional additives can aid in controlling the thickness and feel of the shampoo itself. Simply stated, the right blend is required so that the shampoo is not too thin and not too thick. Functional additives can also control the acidity of the shampoo by obtaining a goal pH level, typically around a value of 4.

Preservatives Preservatives detract germs and prevent decomposition of the shampoos. They also prevent various other health risks that accompany contamination by germs and bacteria. Typical preservatives in shampoos are sodium benzoate, parabens, DMDM hydantoin and tetrasodium EDTA.

Aesthetic Additives Shampoos contain many aesthetic additives which enhance appearance, color, and smell of the mixture. These additives typically give the shampoo the luminous shine and pleasant fragrance to which many consumers are accustomed.

Medically Active Ingredients For people with dandruff and other more serious hair and scalp disorders, shampoos are available with active ingredients which aim to treat or control these conditions. In the treatment of dandruff, zinc pyrithione is a common shampoo additive. For hair loss issues, panthenol is commonly added to shampoos to aid in hair growth and moisture content.

Conditioner: Constitution and Main Functions

As stated earlier, many shampoos have certain levels of conditioning agents which mimic the functions of a full conditioner product. Conditioner molecules contain cationic surfactants which give a positive electrical charge to the conditioner. The negative charge of the hair is attracted to the positively charged conditioner molecules, which results in conditioner deposition on the hair; see Fig. 24.9. This is especially true for damaged hair, since damaging processes result in hair fibers being even more negatively charged. The attraction of the conditioner to hair results in a reduction of static electricity on the fiber surfaces, and consequently a reduction in the "fly away" behavior. The conditioner layer also flattens the cuticle scales against each other, which improves shine and color. The smooth feel resulting from conditioner use gives easier combing and detangling in both wet and dry conditions (see Table 24.1).

Conditioner consists of a gel network chassis (cationic surfactants, fatty alcohols, and water) for superior wet feel and combination of conditioning actives (cationic surfactants, fatty alcohols, and silicones) for superior dry feel. Figure 24.10 shows the transformation of the cationic surfactants and fatty alcohol mixture into the resulting gel network, which is a frozen lamellar liquid crystal gel phase [13]. The process starts as an emulsion of the surfactants and alcohols in water. The materials then go through a strictly controlled heating and cooling process: the application of heat causes the solid compounds to melt, and the solidification process enables a setting of the lamellar assembly molecules in a fully extended conformation, creating a lamellar gel network. When this network interacts with the hair surface, the high concentration of fatty alcohols make it the most deposited ingredient group, followed by the silicones and cationic surfactants. Typical deposition levels for cationic surfactant, fatty alcohol, and silicone are around 500–800 ppm, 1000–2000 ppm, and 200 ppm, respectively. Typical concentrations are approximately 2–5 weight %, 5–10 weight %, and 1–10 weight %, respectively [48].

The benefits of the conditioner are shown in Table 24.6 [48]. The wet feel benefits are creamy texture, ease of spreading, slippery feel while applying, and soft

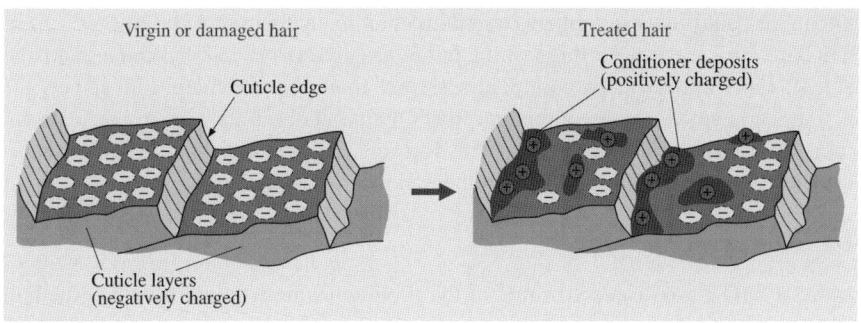

Fig. 24.9. Negatively charged hair and the deposition of positively charged conditioner on the cuticle surface [48]

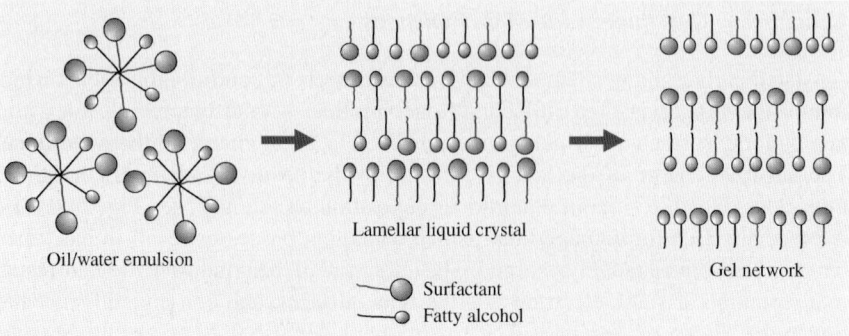

Fig. 24.10. Conditioner formation from emulsion to gel network [13]

Table 24.6. Combinations of conditioner ingredients and their benefits towards wet and dry feel

Gel network chassis for superior wet feel	
Key Ingredients – Cationic surfactant – Fatty alcohols – Water	Benefits – Creamy texture – Ease of spreading – Slippery applying feel – Soft rinsing feel
Combination of "conditioning actives" for superior dry feel	
Key Ingredients – Silicones – Fatty alcohols – Cationic surfactant	Benefits – Moist – Soft – Dry combing ease

rinsing feel. The dry feel benefits are moistness, softness, and easier dry combing. Each of the primary conditioner ingredients also has specific functions and roles that affect performance of the entire product. Table 24.7 displays the functions of the major conditioner ingredients and their chemical structure [48]. Cationic surfactants are critical to the forming of the lamellar gel network in conditioner, and also act as a lubricant and static control agent, since their positive charge aids in counteracting the negative charge of the hair fibers. Fatty alcohols are used to lubricate and moisturize the hair surface, along with forming the gel network. Finally, silicones are the main source of lubrication in the conditioner formulation.

Damaging Processes

In Sect. 24.2.2 we discussed some of the products which aid in "treating" the hair. There are other hair care products and processes which, while creating a desired look or style to the hair, also bring about significant damage to the fibers. Most of these processes occur on some type of periodic schedule, whether it be daily (while

Table 24.7. Chemical structure and purpose/function of conditioner ingredients

Ingredient		Chemical structure	Purpose / Function
Water			
Cationic surfactants	Stearamidopropyl dimethylamine	$H_3C\text{-}N(CH_3)\text{-}(CH_2)_3\text{-}NH\text{-}C(=O)\text{-}(CH_2)_{16}\text{-}CH_3$	– Aids formation of lamellar gel network – Lubricates and static control agent
	Behenyl amidopropyl dimethylamine glutamate (BAPDMA)	$H_3C\text{-}N(CH_3)\text{-}(CH_2)_3\text{-}NH\text{-}C(=O)\text{-}(CH_2)_{20}\text{-}CH_3$	
	Behentrimonium chloride (BTMAC) $CH_3(CH_2)_{21}N(Cl)(CH_3)_3$	$[CH_3(CH_2)_{20}CH_2\text{-}N^+(CH_3)_3]\,Cl^-$	
Fatty Alcohols	Stearyl alcohol ($C_{18}OH$)	$HO\text{-}(CH_2)_{17}\text{-}CH_3$	– Lubricates and moisturizes – Aids formation of lamellar gel network along with cationic surfactant
	Cetyl alcohol ($C_{16}OH$)	$HO\text{-}(CH_2)_{15}\text{-}CH_3$	
Silicones	PDMS blend (Dimethicone)	$H_3C\text{-}Si(CH_3)_2\text{-}O\text{-}[Si(CH_3)_2\text{-}O]_n\text{-}CH_3$	– Primary source of lubrication – Gives hair a soft and smooth feel

combing the hair), or monthly (haircut and coloring at a salon). In general, hair fiber damage occurs most readily by mechanical or chemical means, or by a combination of both (chemo-mechanical).

Mechanical Damage Mechanical damage occurs on a daily basis for many individuals. The damage results from large physical forces or temperatures which degrade and wear the outer cuticle layers. Common causes are

- combing (generally with plastic objects, and often multiple times over the same area lead to scratching and wearing of the cuticle layers)
- scratching (usually with fingernails around the scalp)
- cutting (affects the areas surrounding the fiber tips)
- blowdrying (high temperatures thermally degrade the surface of the hair fibers)

Permanent Wave Treatment Permanent wave treatments saw many advances in the beginning of the 20th century, but have not changed much with the invention of the Cold Wave around the turn of that century. Generally speaking, the Cold Wave uses mercaptans (typically thioglycolic acid) to break down disulphide bridges and style the hair without much user interaction (at least in the period soon after the perm application) [36]. The Cold Wave process does not need increased temperatures (so no thermal damage to the hair), but generally consists of a reduction period (whereby molecular reorientation to the cuticle and cortex occurs via a disulfide-mercaptan interchange pathway [68]) followed by rinsing, setting of the hair to the desired style, and finally neutralization to decrease the mercaptan levels and stabilize the style. The chemical damage brought on by the permanent wave can increase dramatically when not performed with care.

Chemical Relaxation Commonly used as a means of straightening hair (especially in highly curved, tightly curled African hair), this procedure uses an alkaline agent, an oil phase, and a water phase of a high viscosity emulsion to relax and reform bonds in extremely curly hair. A large part of the ability to sculpt the hair to a desired straightness comes from the breakage of disulfide bonds of the fibers.

Coloring and Dyeing Hair coloring and dyeing have become extremely successful hair care procedures, due in part to "over-the-counter" style kits which allow home hair care without professional assistance. The most common dyes are para dyes, which contain paraphenylenediamine (PPD) solutions accompanied by conditioners and antioxidants. Hydrogen peroxide (H_2O_2) is combined with the para dyes to effectively create tinted, insoluble molecules which are contained within the cortex and are not small enough to pass through the cuticle layers, leaving a desired color to the hair. Due to the levels of hydrogen peroxide, severe chemical damage can ensue in the cuticle and cortex.

Bleaching Like dyeing, bleaching consists of using hydrogen peroxide to tint the hair. However, bleaching can only lighten the shade of hair color, as the H_2O_2 releases oxygen to bind hair pigments [36]. Bleaching may also be applied to limited areas of the hair (such as in highlights) to create a desired look. The chemical damage brought on by bleaching leads to high porosity and severe wear of the cuticle layer.

24.3 Experimental

Table 24.8 shows a comparison between the various tools used to study hair on the micro/nanoscale. The SEM has long been the standard means of investigating the surface topography of human hair. The SEM uses an electron beam to give a high resolution image of the sample, but cannot provide quantitative data regarding the surface. SEM requires the hair sample to be covered with a very thin layer of a conductive material and needs to be operated under vacuum during both metallization and measurements. Surface metallization and vacuum exposure could potentially induce modifications to the surface details. TEM examinations provide fine detailed internal structure of human hair. However, thin sections of 50–100 nm thickness and heavy metal compounds staining treatment are required for TEM examinations. The cutting of these thin sections with the aid of an ultra-microtome is not an easy task. Moreover, since both SEM and TEM techniques cannot be used to measure the physical properties (e.g., mechanical and tribological) of various cellular structure of human hair of interest and do not allow ambient imaging conditions, many outstanding issues remain to be answered. For example: How do the various cellular structures of hair behave physically in various environments (temperature, humidity, etc.)? How do they swell in water? For conditioner treated hair, how thick is the conditioner layer and how is the conditioner distribution on hair surface?

AFM is now commonly used for morphological, structural, tribological and mechanical characterization of surfaces [7–10]. As a non-invasive technique, AFM has been used to evaluate the effect of hair treatment and can be used in ambient conditions to study the effect of environment on various physical properties. A schematic of an AFM imaging a hair fiber is shown in Fig. 24.11. AFM/FFM uses a sharp tip with a radius of approximately 10–50 nm. This significant reduction in tip to sample interaction compared to the macroscale allows the simulation of single asperity contact to give detailed surface information. Contact mode allows simultaneous measurement of surface roughness and friction force. Different AFM operating modes, tapping mode and torsional resonance (TR) mode can be used for measurements of material stiffness and viscoelastic properties mapping using amplitude and phase angle imaging. To study the electrostatic charge build up on hair, surface potential studies can be carried out using AFM as a nano Kelvin probe.

When skin comes in contact with hair, actual contact occurs over a large number of asperities. During relative motion, friction and adhesion are governed by the surface interactions which occur at these asperities. Until about 2000, much of the work in the industry has focused on the measurement of macroscale friction, particularly between a skin replica and a hair swatch of interest [68]. Figure 24.12 shows schematics of typical macro- and micro/nanoscale test apparatuses. However, there are many problems associated with performing these types of measurements. Factors such as topographical variations, lumping of the hair fibers, the large size of the synthetic skin, and traditional measurement system errors can all lead to uncertainty in the data.

Depth-sensing nanoindentation measurement techniques are now commonly used to measure nanomechanical properties of surface layers of bulk materials and

Table 24.8. Comparison of methods used to characterize hair on micro/nanoscale

Method	Type of information	Quantitative data	Normal load	Resolution (nm)	Limitations
Scanning electron microscopy (SEM)	Structural	Gross dimensions	None	0.2–1 (spatial)	– Requires thin conductive coating on sample – Requires vacuum environment – Expensive instrumentation[a] – Tedious[a]
Transmission electron microscopy (TEM)	Structural	Gross dimensions	None	0.2–1 (spatial)	– Requires thin sections (<100 nm) and heavy metal compound staining treatment – Requires vacuum environment – Expensive instrumentation[a] – Tedious[a]
Atomic force/friction force microscopy (AFM/FFM)	Structural Mechanical Tribological	– Surface roughness – Elastic modulus – Viscoelasticity – Friction – Adhesion – Conditioner thickness	<0.1 nN–500 nN[a]	0.2–1 (spatial)[a] 0.02 (vertical)[a]	None
Nanoindenter	Mechanical	– Hardness – Elastic modulus – Creep – Scratch resistance	<0.1 mN–350 mN	400 (spatial)[b] 0.1 (vertical)[b]	None

[a] [9]; [b] [18]

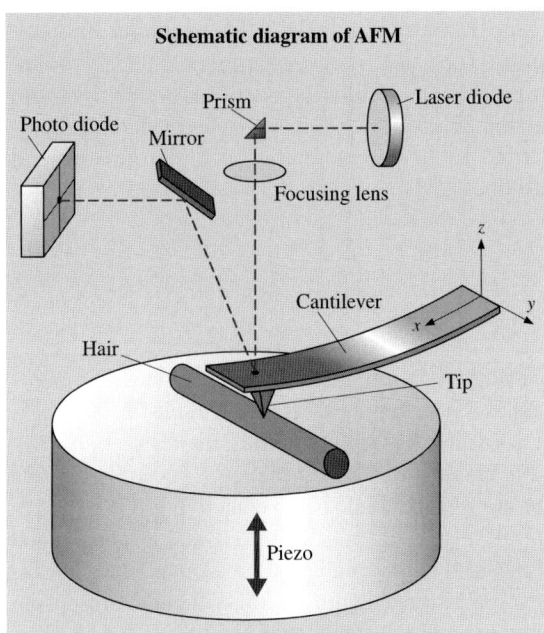

Fig. 24.11. Schematic diagram of AFM operation with human hair sample

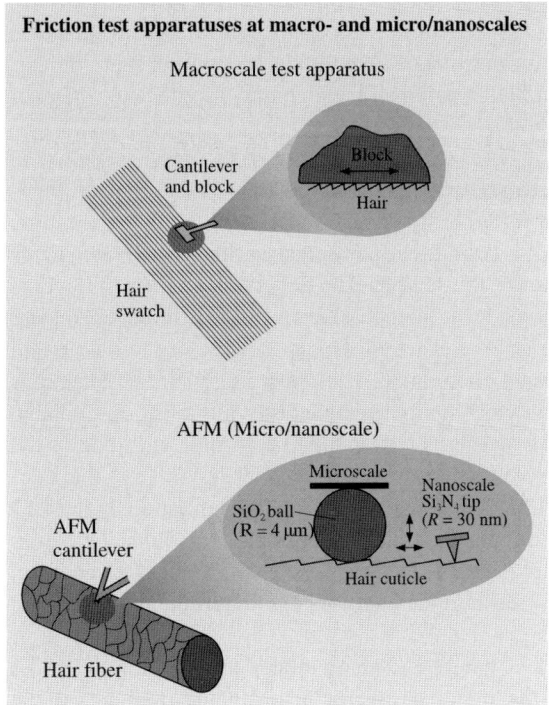

Fig. 24.12. Friction test apparatuses at macro- and micro/nanoscales

of ultrathin coatings [8, 10, 18]. More recently, nanoindentation technique has been used to investigate nanomechanical properties (such as hardness, Young's modulus, creep, and scratch resistance) of various cellular layers of glass fibers, keratin fibers, and hair surface/ cross-section [50, 65, 89, 90]. In-situ tensile loading experiments in the AFM have been carried out to study the progress of deformation and morphological changes in the hair fiber [74].

24.3.1 Experimental Procedure

Structural Characterization Using an AFM

An AFM (Multimode Nanoscope IIIa, DI-Veeco, Santa Barbara, CA) with modifications for the TR mode and a nanoscope extender electronic circuit for the measurement of the phase angle were used in the study reported in the chapter [13, 26]. All measurements are conducted in ambient conditions ($22 \pm 1 °C$, $50 \pm 5\%$ relative humidity). The probes used in this study were single beam etched Si probes (Mikro-Masch) with a fundamental flexural mode frequency of 75 kHz and a fundamental torsional resonance frequency of 835 kHz with a quality factor around 1000. The dimension of the cantilever is typically $230\,\mu m \times 40\,\mu m \times 3\,\mu m$ with a flexural spring constant of 1–5 N/m and a torsional spring constant estimated to be 30–150 N/m. The radius of curvature of the tip is about 10 nm. Surface height images shown in this study were processed using first order planefit command available in the AFM software, which eliminates tilt in the image. Amplitude and phase angle images were processed using zero-order flatten command, which only modifies the offset of the image.

The schematic diagram of a tip-cantilever assembly in an AFM is shown in Fig. 24.13. The cantilever can scan a sample with its tip in constant contact, intermittent contact, or without contact with the sample surface [7–10]. The scanning is implemented by the motion of a cylindrical piezoelectric tube, which can act as the holder of either the cantilever or the sample. The deflection of the cantilever is generally measured using an optical lever method. A laser beam is projected on and reflected from a location on the upper surface of the cantilever close to the tip and led by a mirror into a four-segment photodiode. The normal and lateral deflections of the cantilever at that location can then be obtained by calibrating the voltage output of the photo diode. AFM measurements can be performed with one of the several modes: tapping mode I, torsional resonance mode (TR mode) [44], and contact mode, as shown in Fig. 24.13. The phase angle is also defined in Fig. 24.13 as a phase delay between input/output strain and/or stress profiles. Table 24.9 summarizes the characteristics of tapping mode, torsional resonance modes and contact mode [26]. The contact mode is static mode and uses a nonvibrating tip, therefore a phase analysis is not available.

The TR mode measures surface height and phase angle (and amplitude) images as follows. The tip is vibrated in the torsional mode at the resonance frequency of the cantilever beam in air driven by a specially designed cantilever holder. The torsional vibration amplitude of the tip (TR amplitude) is detected by the lateral segments of

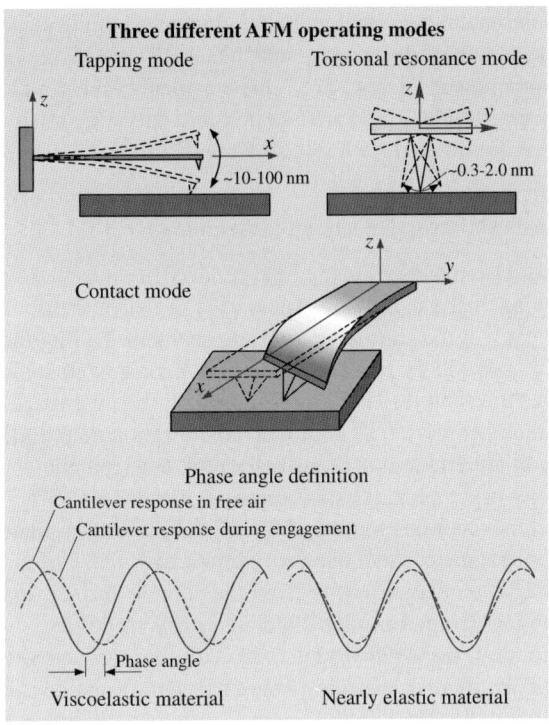

Fig. 24.13. Three different AFM settings are compared at the *top*: tapping mode (TM), torsional resonance (TR) mode and contact mode. The TR mode is a dynamic approach with a laterally vibrating cantilever tip that can interact with the surface more intensively than tapping mode; therefore, more detailed near surface information is available. Phase angle is defined and two examples of the phase angle response are shown at the bottom: one is for materials exhibiting viscoelastic properties (*left*) and the other nearly elastic (*right*)

Table 24.9. Summary of various operating modes of AFM for surface imaging

Operating modes	Direction of cantilever vibration	Parameter controlled	Data obtained
Tapping	Vertical	Setpoint (Constant amplitude)	Surface height, phase angle (normal viscoelasticity)
TR mode I	Torsional (lateral)	Setpoint (Constant amplitude)	Surface height, phase angle (lateral viscoelasticity)
TR mode II	Torsional (lateral)	Constant load	Surface height, amplitude and phase angle (lateral stiffness and viscoelasticity)
Contact	n/a	Constant load	Surface height, friction

the photodiode. A feedback loop system coupled to a piezo-electric scanning stage is used to control the vertical z position of the sample, which changes the degree of in-plane (lateral) tip-sample interaction of interest. The z displacement of the sample gives a surface height image of the sample. There are two possible operating modes depending on which parameter is controlled (see Table 24.9):

1. TR mode I: constant TR amplitude;
2. TR mode II: constant normal cantilever deflection (constant load).

Both modes are operated at the resonance frequency of cantilever in air, which is different from the TR friction mode used in previous study [17] in which the tip is vibrated at the resonance frequency of the cantilever after engagement. During the measurement, the cantilever/tip assembly is first vibrated at its resonance at some amplitude before the tip engages the sample. Then, for TR mode I, the tip engages the sample at some setpoint which is reported as a ratio of the vibration amplitude after engagement to the vibration amplitude in free air before the engagement [19, 44, 73]. For TR mode II, instead of keeping constant setpoint, a constant normal load measured using vertical segments of the photodiode is applied. Under in-plane tip-sample interaction, torsional resonance frequency, amplitude and phase of the cantilever all change from those when it is far away from the sample surface and could be used for contrasting and imaging of in-plane lateral surface properties.

Compared to TM and TR mode I, the AFM tip interacts with the surface more intensively in TR mode II, therefore, more detailed in-plane surface information can be obtained [26]. By using TR mode II, TR amplitude and TR phase angle images show even larger contrast. Previous studies [19,44,73] indicated that the phase shift can be related to the energy dissipation through the viscoelastic deformation process between the tip and the sample. Recent theoretical analysis [76] has established a quantitative correlation between the lateral surface properties (lateral stiffness and viscoelasticity) of materials and amplitude/phase angle shift in TR measurements. The contrast in the TR amplitude and phase angle images is due to the in-plane (lateral) heterogeneity of the surface. Based on the TR amplitude and phase angle images, the lateral surface properties (lateral stiffness and viscoelasticity) mapping of materials can be obtained. In the work presented in the chapter, TR mode II amplitude and phase images were obtained to characterize the cellular structure of human hair. For convenience, TR amplitude is recorded in volt, and 1 V corresponds to about 0.5 nm TR amplitude.

Surface Potential Studies Using AFM Based Kelvin Probe Microscopy

All measurements were taken with a MultiMode atomic force microscope equipped with Extender Electronics modules. The Extender allows for surface potential measurements to be taken. Surface potential measurement is conducted using a two pass method [15, 16, 31, 55, 56]. In the first pass, surface topography is measured using the standard AFM tapping mode, Fig. 24.14a. In the second pass, the tip is scanned over the previously measured topography at a specified distance above the surface (for example, 30 nm), Fig. 24.14b. The piezo normally oscillating the tip in tapping mode is turned off. Instead an oscillating voltage is applied directly to the conducting tip which generates an oscillating electrostatic force. To measure the surface potential, a dc voltage is applied to the tip until the voltage difference between the tip and sample is equal to zero, giving zero oscillating force amplitude. The conductive tip, a silicon tip coated with platinum/iridium, is used.

Hair samples were mounted in two different ways. When mounted in silver paint, there exists a direct path from the sample to ground. When mounted in liquid paper, the sample is electrically isolated from ground.

Nanomechanical Characterization Using Nanoindentation

Nanoindentation

Figure 24.15 shows the schematic of performing nanoindentation and nanoscratch test on human hair surface with a Nano Indenter II® (MTS Systems Corp.). This instrument monitors and records the dynamic load and displacement of the indenter during indentation with a force resolution of about 75 nN and displacement resolution of about 0.1 nm [8, 10, 18].

For hardness, Young's modulus and creep measurement, a three-sided (triangular-based) pyramidal diamond Berkovich indenter tip (radius 100 to 200 nm) was used [89, 90]. For nanoindentation on all the virgin hair surface except African hair, a wide load range (0.1–300 mN) was used, in order to study the mechanical property variation of hair surface depending on the indentation depth. In the case of virgin African hair, the load of 300 mN was not used because the hair was too soft to get reasonable data at 300 mN. For the damaged, treated hair and the hair near scalp,

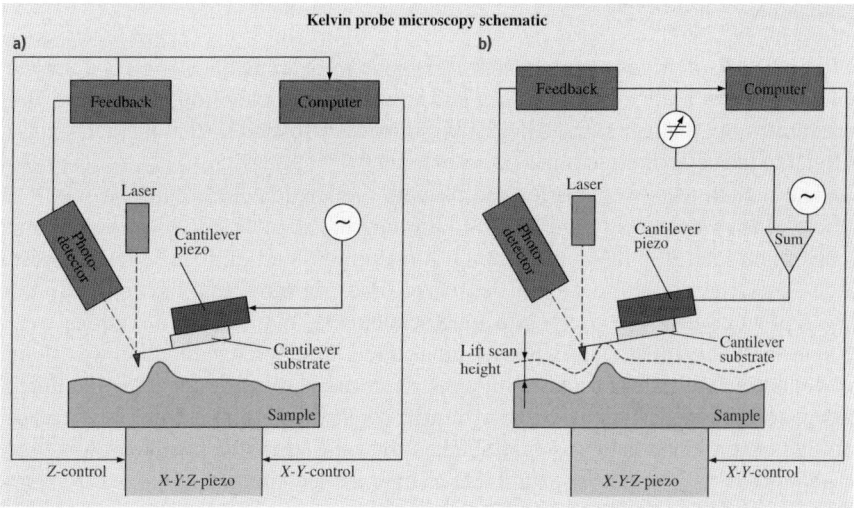

Fig. 24.14. (a) Schematic of first pass of Kelvin probe technique measuring surface height; (b) schematic of second pass of Kelvin probe technique measuring surface potential [15, 16]

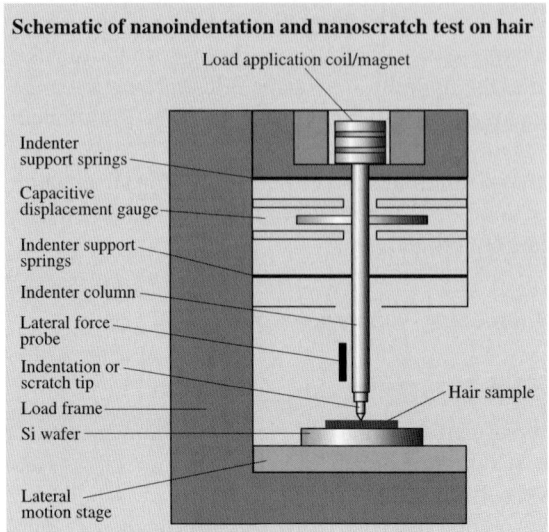

Fig. 24.15. Schematic of nanoindentation and nanoscratch test on hair using nanoindenter

in the middle and near tip, 0.1, 1.0, 10, 100 mN normal loads were used. At each load, five indents were made, and the hardness and elastic modulus values were averaged from them and the standard deviations were calculated. For nanoindentation on virgin hair cross section samples, only one normal load, 1.0 mN, was used to make indents at cuticle, cortex and medulla. To do the creep test, a normal load was applied, and then the tip was held for 600 s. The displacements change during the holding time were recorded. The loads used were 0.1, 1.0 and 10 mN.

Nanoscratch

For coefficient of friction and scratch resistance measurement, a conical diamond tip, having a tip radius of about 1 μm and an included angle of 60°, was used. Before scratching, the hair sample holder was manually rotated so that the hair axis is parallel to the scratch direction.

The scratch tests were performed on both the single cuticle cell and multiple cuticle cells of each hair sample, by controlling the scratch lengths to be 5 μm at a maximum load of 1 mN and 50 μm at a maximum load of 10 mN, respectively. Each scratch test was repeated at least five times on the same hair to verify the data reproducibility. For the 50 μm long scratch test, two scratch directions were used to study the directionality effect of the cuticle. One scratch direction was along cuticle and the other was against cuticle. The coefficient of friction was monitored during scratching. In order to obtain scratch depths during scratching, the surface profile of the human hair was first obtained by translating the sample at low load of about 0.01 mN, which is insufficient to damage the hair surface. The 5 μm and 50 μm long scratches were then made by translating the hair sample at a constant tip velocity of 0.5 μm/s while ramping the loads from 0.01 to 1 mN, and from 0.01 to 10 mN, respectively. The actual depth during scratching was obtained by subtracting the initial profile from the scratch depth measured during scratching. In order to measure the residual depth after scratch, the scratched surface was profiled at a low load of 0.01 mN and was subtracted from the surface profile before scratching. After nanoscratch tests, the scratch wear tracks of the hair samples were measured using a Philips XL-30 ESEM.

In order to study the effect of soaking on hair nanotribological and nanomechanical properties, the hair samples were soaked in de-ionized water for 5 min, which is representative of a typical exposure time when showering/bathing. Then the hair was mounted on Si wafer immediately, and 5 µm and 50 µm long scratch tests were performed on them.

In-situ Tensile Deformation Characterization Using AFM

In-situ tensile testing of human hair fibers in AFM was conducted using a custom-built stage used in place of the regular sample holder [22, 23, 74, 84]. It consists of a linear stepper motor which loads a single hair fiber in tension, see Fig. 24.16. The base plate of this stage attaches to the stepper motor of the AFM base, to enable positioning of the stage in X and Y directions with respect to the AFM tip. During scanning, the sample is held stationery while the cantilever tip mounted on a X-Y-Z piezo, moves back and forth. The hair sample is firmly clamped in between two sliders to prevent slipping on load application. Stage motion is achieved by a left-right combination lead screw that keeps the sample at approximately the same position with respect to the scanning tip. This helps in locating approximately the same control area after each loading increment is applied. With every load increment, there is an increase in the hair length and hence a shift in the location of the control area for the previous position. To locate the same control area in hair, a mark was made on the hair fiber with a nail polish. Because of its reflective nature, it shows up clearly in the AFM optical microscope. A 40 TPI pitch lead screw in combination with 400 steps per revolution stepper motor (model PX245-01AA, using the controller NF-90, both from Velmex Inc) gives a minimum displacement of 1.6 µm. For a sample length of 38 mm (1.5 in), this corresponds to a minimum strain increment of $8.33 \times 10^{-3}\%$. The strain applied was obtained from the total number of steps through which the stepper motor was rotated. The maximum travel was 10 mm. A beam-type strain gauge force sensor (model LCL-010, Omega Engineering, Stamford, CT), with a resolution of 10 mN, was used for measuring stress in the hair samples. The stiffness of the force sensor (18 kN/m) is very high compared to the sample stiffness.

To minimize airborne vibrations during AFM imaging, the hair sample was supported with a smooth aluminum block support having a radius of curvature of 25.4 mm, as shown in Fig. 24.16.

Macroscale Tribological Characterization Using a Friction Test Apparatus

The macroscale tribological (friction and wear) characterization of human hair were conducted using a flat-on-flat tribometer under reciprocating motion [7, 9, 10]

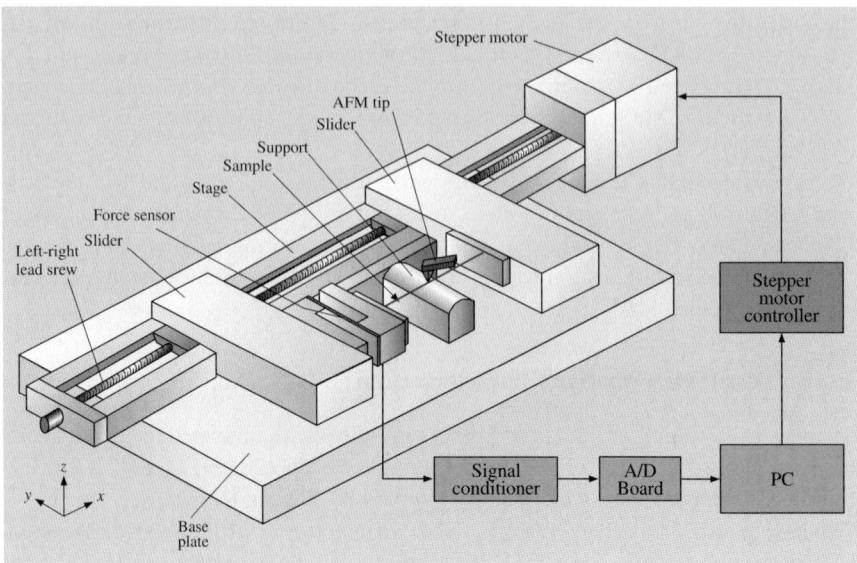

Fig. 24.16. Schematic diagram of the setup used to conduct in-situ tensile testing of human hair in AFM [74]

(Fig. 24.17). A piece of square polyurethane film (film area 25–400 mm²) was fixed at the end of a cantilever beam. The hair strands were mounted on Si wafer in such a way that all strands of hair were separated and parallel to each other. For high temperature studies, the Si wafer was placed on a heating stage, which can increase the temperature of the hair sample up to 100°C [10]. The heat-generating elements of the heating stage were Ohmic resistors encapsulated in a steel holder and kept in good thermal contact by using thermal paste. J-type thermocouples were used to measure the sample temperature. A thermal controller and a solid-state relay were used to control the temperature by adjusting on/off time. A glass plate was attached at the bottom of the heating stage to isolate it thermally from the X-Y axis stage, where the heating stage was mounted. The X-Y axis stage is a motor-driven lead-screw-type stage driven by a stepper motor. The load on the polyurethane film was applied by lowering the beam against the hair sample using a microactuator. Normal and frictional forces were measured with semiconductor strain gage beams mounted on a crossed-I-beam structure as part of the cantilever beam mentioned earlier and data were digitized and collected on a personal computer. The effect of relative humidity on the hair friction was studied in an environmentally controlled chamber, in which humidity was controlled by introducing the combination of dry air and moist air.

In order to select relevant load, velocity, and film area, one needs to be guided by the application. For a feel with a finger with the hair, the normal load applied by the finger was estimated as 50–100 mN, measured by pressing the finger on a microbalance. The estimated apparent contact size and stroke length were 10–100 mm², and

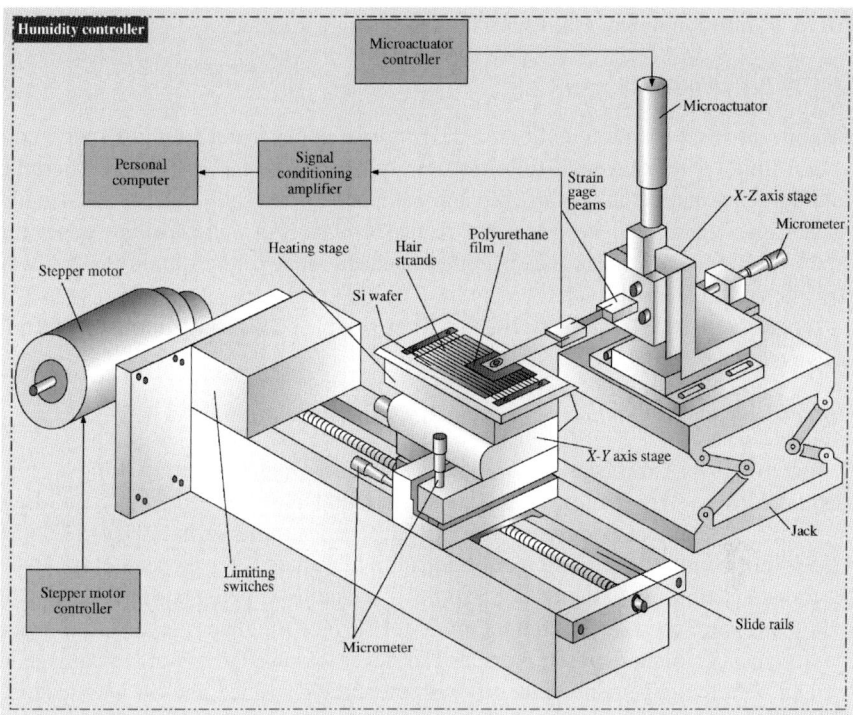

Fig. 24.17. Schematic of the reciprocating tribometer. Normal load is applied by lowering the X-Z stage mounted on a laboratory jack. Normal and friction forces are measured by semiconductor strain gages mounted on a crossed-I-beam structure [10]

5–20 mm, respectively. The sliding velocity was estimated as 5–20 mm/s. To perform a parametric study, tests were performed at a range of operating conditions in the range of interest for the application at a temperature of $22 \pm 1°C$ and $50 \pm 5\%$ RH. The following test conditions were used: stroke length, 3 mm; sliding velocity, 0.4–4.5 mm/s; normal load, 50–100 mN; film size, 25–400 mm^2. For the friction studies of polyurethane film vs. various hair, hair vs. hair, and polyurethane film vs. virgin and treated hair at dry and wet conditions, the following nominal test conditions were chosen: sliding velocity, 1.4 mm/s; normal load, 50 mN; and film area, 100 mm^2. To simulate the wet conditions, plumber's putty was placed around the stage and the hair region was filled with water. For the wear measurements, the polyurethane film rubbed against the virgin and treated Caucasian hair for 24 hours at the above nominal conditions and the coefficient of friction was measured. The hair surface was studied by optical microscope prior to and after wear tests. To study the effect of temperature and humidity on hair friction, the tests were conducted at the nominal test conditions and the followings parameters were used: temperature, 22–80°C; and relative humidity, 35–85%.

Micro/Nanotribological Characterization Using an AFM

Specimen Mounting

Hair specimens were mounted onto AFM sample pucks using Liquid Paper® correction fluid. A thin layer of the fluid was brushed onto the puck, and when the fluid hardened into a tacky state, the hair sample was carefully placed. The Liquid Paper® dries quickly to keep the hair firmly in place. An optical microscope was used to preliminarily image the specimen to ensure none of the Liquid Paper® was deposited on the hair surface.

Synthetic materials were attached to AFM sample pucks using double-sided adhesive tape.

Surface Roughness, Friction Force, and Adhesive Force Measurements

Surface roughness and friction force measurements were performed using a commercial AFM system (MultiMode Nanoscope IIIa, Digital Instruments, Santa Barbara, CA) in ambient conditions (22 ± 1°C, 50 ± 5% relative humidity) [27, 45–48, 53]. For nanoscale measurements, square pyramidal Si_3N_4 tips of nominal 30–50 nm radius attached to the end of a soft Si_3N_4 cantilever beam (spring constant of 0.06 N/m) were used in the contact mode for surface roughness and friction force measurements simultaneously. A softer cantilever was used to minimize damage to the hair. After engagement of the tip with the cuticle surface, the tip was scanned perpendicular to the longitudinal axis of the fiber. The tip was centered over the cross section in order to be at the very top of the fiber, so as to negate effects caused by the AFM tip hitting the sides of the hair and adding error to the measurements. In order to minimize scanning artifacts, a scan rate of 1 Hz was used for all measurements. Topographical images to characterize the shape and structure of the various hair were taken at 5×5, 10×10, and $20 \times 20\,\mu m^2$ scans at a normal load of 5 nN. These scan sizes were suitable for capturing the surface features of multiple scales and scale edges of the cuticle. To characterize roughness, $2 \times 2\,\mu m^2$ scans of the cuticle surface without edges were used. Friction force mapping of the scan area was collected simultaneously with roughness mapping [45–48]. Figure 24.12 (bottom cartoon) and Fig. 24.18 show the AFM tip scanning over the hair surface for untreated and conditioner treated hair, respectively. The effects of the conditioner can be examined by comparing the friction information. A quantitative measure of friction force was calculated by first calibrating the force based on a method by Bhushan [7, 9, 10]. The normal load was varied from 5 to 45 nN in roughly 5 nN increments, and a friction force measurement was taken at each increment. By plotting the friction force as a function of normal force, the average coefficient of friction was determined from the slope of the least squares fit line of the data.

Surface roughness images shown in this study were processed using a first-order "planefit" command available in the AFM software, which eliminates tilt in the image. Roughness data as well as friction force data were taken after the "planefit" command was employed. A first-order "flatten" command was also used on friction force mappings to eliminate scanning artifacts and present a cleaner image.

For completeness, it should be noted that *Lodge* and *Bhushan* [53] have reported some hair damage to occur during imaging in the contact mode. For high resolution surface height imaging, they used the tapping mode AFM. In this method, a Si tip lightly taps the sample surface at a constant amplitude to gather information on the height of the sample surface. They found that the tapping mode is able to capture more high frequency information than the contact mode due to nondestructive nature.

Adhesive force measurements were made with square pyramidal Si_3N_4 tips attached to the end of a Si_3N_4 cantilever beam (spring constant of 0.58 N/m), using the "force calibration plot technique" [27, 45, 46, 48, 53]. In this technique, the AFM tip is brought into contact with the sample by extending the piezo vertically, then retracting the piezo and calculating the force required to separate the tip from the sample. The method is described in detail by *Bhushan* [7–10]. The cantilever deflection is plotted on the vertical axis against the Z-position of the piezo scanner in a force calibration plot, Fig. 24.19. As the piezo extends (from right to left in the graph), it approaches the tip and does not show any deflection while in free air. The tip then approaches within a few nanometers (point A) and becomes attached to the sample via attractive forces at very close range (point B). After initial contact, any extension of the piezo results in a further deflection of the tip. Once the piezo reaches the end of its designated ramp size, it retracts to its starting position (from left to right). The tip goes beyond zero deflection and enters the adhesive regime. At point C, the elastic force of the cantilever becomes equivalent to the adhesive force, causing the cantilever to snap back to point D. The horizontal distance between A and D multiplied by the stiffness of the cantilever results in a quantitative measure of adhesive force [7–10].

The force calibration plot allows for the calculation of an adhesive force at a distinct point on the sample surface. Consequently, by taking a force calibration plot at discrete sampling intervals over an entire scan area, a resulting adhesive force mapping (also known as force-volume mapping) can be created to display the variation in adhesive force over the surface [14]. In this work, plots were taken at 64×64 distinct points over a scan area of $2 \times 2\,\mu m^2$ for all hair types and ethnicities. A custom

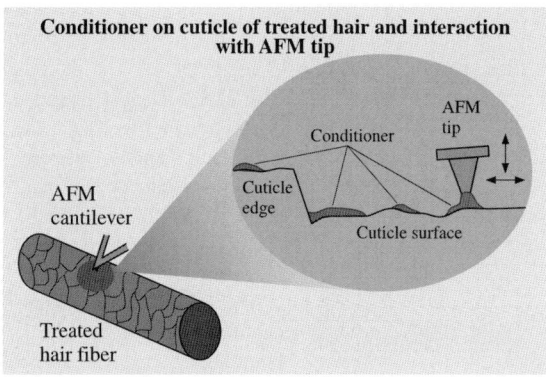

Fig. 24.18. Interactions between the AFM tip and conditioner on the cuticle surface for treated hair [46]

program coded in Matlab was used to display the force-volume maps. The adhesive force for each force calibration plot was obtained by multiplying the spring constant with the horizontal distance (in the retract mode) traveled by the piezo from the point of zero applied load to the point where the tip snaps off.

For microscale adhesion and friction measurements, a 4 μm radius silica ball was mounted on a Si_3N_4 cantilever (spring constant of 0.58 N/m) [47].

Relative Humidity, Temperature, and Soaking, and Durability Measurements

A Plexiglas test chamber enclosing the AFM system was used to contain the environment to be humidity controlled. The humidity inside the chamber was controlled by using desiccant to reduce humidity or by adding humid air to increase the humidity. Measurements were taken at nominal relative humidities of 5, 50, and 80%. Hair fibers were placed at each humidity for approximately 2 hours prior to measurements.

A homemade thermal stage was used to conduct temperature effect measurements at 20, 37, 50, and 80°C. Hair fibers were exposed to each temperature condition for approximately 30 minutes prior to testing.

Soak tests were performed as follows: A dry hair fiber was taken from a switch, and a sample was cut from the fiber (approximately 7 mm long) for coefficient of friction measurements. An adjacent sample was also taken from the fiber and placed in a small beaker filled with de-ionized water. The sample was subjected to the aqueous environment for 5 minutes, which is representative of a typical exposure time when showering/bathing, then immediately analyzed with AFM. It should be noted that hair becomes saturated when wet in about 1 minute and remains saturated

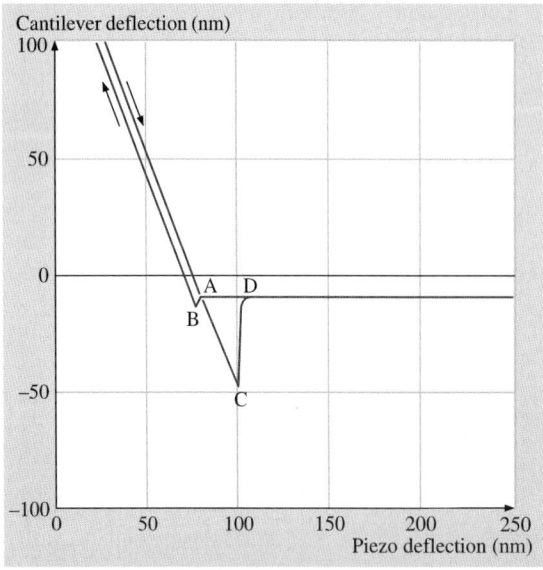

Fig. 24.19. Typical force calibration plot for Caucasian virgin hair. Contact between the tip and hair occurs at point B. At point C, the elastic force of the cantilever becomes equivalent to the adhesive force, causing the cantilever to snap back to point D

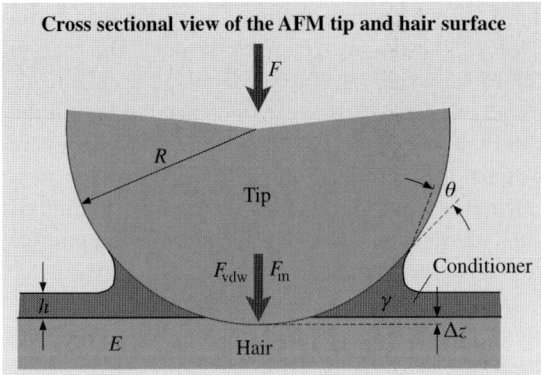

Fig. 24.20. Schematic of an enlarged cross-sectional view of the region around the AFM tip, conditioner and hair surface. The parameters used in the text are defined. Here, R is the tip radius, h is the conditioner film thickness, γ is the surface tension of the conditioner, θ is the contact angle between the tip and conditioner, F is the load applied, F_a is the adhesive force acting on the tip which is the sum of van der Waals force F_{vdw} and the meniscus force F_m, Δz is the indention on the hair surface under the applied load F and the adhesive force F_a, and E is the Young's modulus of the sample

if kept in a humid environment. It was determined from unpublished results that if the wet hair was exposed to the ambient environment for more than 20 minutes while in the AFM, the hair became dry and coefficient of friction became similar to that of dry hair. Thus, coefficient of friction measurements were made within a 20 minute time frame for each sample.

In order to simulate scratching that can occur on the surface of the hair and its effect on the friction force on the cuticle surface, a durability test was conducted using a stiff silicon tip (spring constant of 40 N/m). A load of 10 μN was used on a 2 μm section of the cuticle. A total of 1000 cycles were performed at 2 Hz. Measurements were conducted using an AFM. Friction force signal was recorded with respect to cycling time.

Conditioner Thickness

The measurements were made using the AFM with FESP (force modulation etched Si probe) tips (100 nm radius, spring constant of 5 N/m), and using the force calibration plot technique ("force-volume mode" of the nanoscope software). Typical FESP tip radius is 5–20 nm, but blunt tips were preferred in the study so that when the tip compresses on the surface, the surface tends to deform elastically instead of being indented (plastically deformed). Then the Hertz analysis can be applied to calculate the effective Young's modulus. Figure 24.11 shows a schematic diagram of the AFM, and an enlarged cross-sectional view of the AFM tip, conditioner and hair surface is shown in Fig. 24.20.

The force calibration plot provides the deflection of the cantilever as a function of the piezo position at a distinct point on the sample surface. Figure 24.21 shows a typical force calibration plot curve for commercial conditioner treated hair. From this plot, the conditioner thickness and adhesive force can be extracted. The snap-in distance is proportional to the real film thickness (see further details in Sect. 24.7.1), and the adhesive force F_a (on retract curve), which is the force needed to pull the sample away from the tip and is the sum of van der Waals force F_{vdw} mediated by adsorbed water or conditioner layer and the meniscus force F_m due to Laplace pressure ($F_a = F_{vdw} + F_m$), can be calculated from the force calibration plot by multiplying the spring constant with the horizontal distance between point A and point D in Fig. 24.19. The meniscus force F_m is given by

$$F_m = 2\pi R \gamma (1 + \cos\theta) \,, \tag{24.1}$$

where R is the tip radius, γ is the surface tension of the conditioner, and θ is the contact angle between the tip and conditioner.

One limitation with the Young's modulus measurements using nanoindenter is that it requires loads greater than 1 µN to make accurate measurements. Force calibration plot has been applied to obtain a quantitative measure of the elasticity of compliant samples with an effective Young's modulus as high as a few GPa [13,27]. A knowledge of the adhesive forces is necessary for accurate quantitative measurement of the effective Young's modulus. In these measurements, the applied load is much greater than the measured adhesive force so that uncertainties in the adhesive force do not significantly affect the measurements. The Young's modulus of the sample can be determined using Hertz analysis

$$F + F_a = \frac{4}{3}\sqrt{R}E\Delta z^{3/2} \,, \tag{24.2}$$

where R is the tip radius, $F + F_a$ is the total force acting on the surface on approach curve which is calculated by multiplying the spring constant with the vertical distance between point C and a point on the line during loading in Fig. 24.19, Δz is the indention on the hair surface and E is the Young's modulus of the sample. The total force acting on the surface and the resulting deformation (indentation) of the sample Δz can be extracted from the force calibration plot. Additional details are provided in Sect. 24.7.2.

Consequently, by taking a force calibration plot at discrete sampling intervals over an entire scan area, conditioner thickness, adhesive force and effective Young's modulus mapping can be created to display the distribution and variation over the surface. In this work, the force curves were collected at the same maximum cantilever deflection (relative trigger mode), at each point at 64×64 array (total 4096 measurement points) with each force curve sampled at 64 points over a scan area of $2 \times 2\,\mu m^2$ for all hair samples. A custom program coded in Matlab was used to calculate and display conditioner thickness, adhesive force and Young's modulus maps.

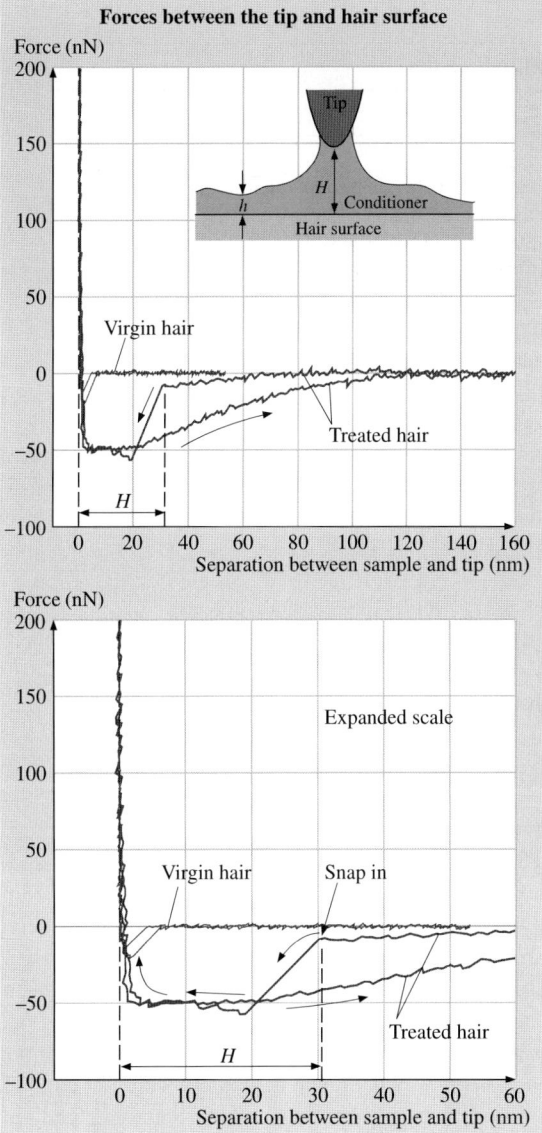

Fig. 24.21. A typical force calibration plot for commercial conditioner treated hair and schematic diagram of the AFM tip, conditioner and hair surface

24.3.2 Hair and Skin Samples

For the research reported in this chapter, all hair samples were prepared per Appendix A. Samples arrived as hair swatches approximately 0.3 m long. Although the exact location from the root is unknown, it is estimated that hair samples used for testing were between 0.1 to 0.2 m from the scalp. Table 24.10 presents a list of all samples used. The main hair categories of interest are: virgin (untreated), vir-

Table 24.10. Hair and skin samples

Sample	Hair and Skin Samples Type
Caucasian hair	– Virgin – Virgin, treated (1 cycle commercial conditioner) – Chemo-mechanically damaged – Chemically damaged – Chemically damaged, treated (1 cycle commercial conditioner) – Chemically damaged, treated (3 cycles commercial conditioner) – Chemically damaged, treated (PDMS blend silicone or amino silicone) – Mechanically damaged
Asian hair	– Virgin – Virgin, treated (1 cycle commercial conditioner) – Chemo-mechanically damaged
African hair	– Virgin – Virgin, treated (1 cycle commercial conditioner) – Chemo-mechanically damaged
Synthetic materials	– Artificial collagen film (hair) – Polyurethane film (skin) – Human skin (putty replica)

gin (treated with 1 cycle of commercial conditioner), chemo-mechanically damaged (untreated), chemically damaged (untreated), mechanically damaged (untreated), and chemo-mechanically or chemically damaged (treated with 1 or 3 cycles of commercial conditioner or experimental conditioners). Virgin samples are considered to be baseline specimens and are defined as having an intact cuticle and absence of chemical damage. Chemo-mechanically damaged hair fibers were exposed to one or more cycles of coloring and permanent wave treatment, washing, and drying, as well as combing (to contribute mechanical damage) which are representative of common hair management and alteration. Chemically damaged hair fibers were not exposed to the combing sequence in their preparation. In the case of African damaged hair samples, chemical damage occurred only by chemical straightening. Mechanically damaged hair fibers were exposed to combing sequence to cause mechanical damage and were observed under an optical microscope at 100× to have mechanical damage. All treated hair samples were treated with either one or three rinse/wash cycles of a conditioner similar to a Procter & Gamble commercial product (with PDMS blend silicone), or were treated with two combinations of silicone types (PDMS, blend of low and high molecular weight silicone, or an amino silicone).

Collagen film is typically used as a synthetic hair material for testing purposes. Polyurethane films represent synthetic human skin. They are cast from human skin and have a similar surface energy, which also makes them suitable test specimens when real skin cannot be used. To characterize surface roughness of human skin, it was also replicated using a two-part silicone elastomer putty (DMR-503 Replication Putty, Dynamold, Inc., Fort Worth, Texas). The thickness of the film was approximately 3 mm.

In order to simulate hair conditioner-skin contact in AFM experiments, it is important to have contact angle and surface energy of an AFM tip close to that of skin. Table 24.11 shows the contact angles and surface energies of materials important to the nanocharacterization of the hair samples: Untreated, damaged and treated human hair; PDMS, which is used in conditioners; skin, which comes into contact with hair; and Si_3N_4 and Si AFM tips used for nanotribological measurements. The dynamic contact angle of various human hair reports in Table 24.11 were measured by *Lodge* and *Bhushan* [53] using the Wilhelmy balance method. This method uses a microbalance to measure the force exerted on a single fiber when it is immersed into the wetting liquid of interest (deionized water). This measured force is related to the wetting force of the liquid on the fiber, and the dynamic contact angle can be calculated. They reported a significant directionality dependence. The values reported here compare well with other values reported in the literature [48, 61]. The data reported for PDMS (bulk) and live skin in-situ were obtained using the standard sessile-drops method using a contact angle goniometer [12, 35, 49, 71]. In the measurement technique developed by *Tao* and *Bhushan* [85, 86] for dynamic contact angle of the AFM, by tips advancing and receding the AFM tip across the water surface, the meniscus force between the tip and the liquid was measured at the tip-water separation. The water contact angle was determined from the meniscus force.

24.4 Structural Characterization Using an AFM

A schematic diagram in Fig. 24.2a provides an overall view of various cellular structure of human hair. Human hair is a complex tissue consisting of several morphological components, and each component consists of several different chemical species. Table 24.2 presented earlier summarizes the main chemical species present in human hair.

Traditionally, most cellular structure characterizations of human hair or wool fiber were done using SEM and TEM. More recently, the cellular structure of human hair has been characterized using AFM and TR mode II (described earlier) [13, 26]. In Sect. 24.4.1, structure of hair cross-section and longitudinal section is presented; and in Sect. 24.4.2, structure of various cuticle layers of human hair is presented.

24.4.1 Structure of Hair Cross-section & Longitudinal Section

Cross-section of Hair

Figure 24.22a shows the AFM images of Caucasian hair cross-section. The hair fiber embedding in epoxy resin can be easily seen. From TR amplitude image and

Table 24.11. Contact angle and surface energy of hair and relevant materials associated with nanotribological characterization of hair

	Contact angle (°)		Surface energy (N/m)
	Dry	**Soaked**	
Virgin Caucasian hair	103[a]	98[a]	0.028[b]
Virgin treated	88[a]	92[a]	
Chemically damaged	70[a]	70[a]	
Chemically damaged, treated	79[a]	84[a]	
Mechanically Damaged	95[a]		
Asian	92[a]		
African	80[a]		
PDMS (bulk)	105[c]	0.020[d]	
Human skin – forehead	55[e]		0.043[e]
– forearm	88[e]		0.038[e]
– finger	84[f]		0.029[f]
	74[f]		0.024[f]
	104[g] (after soap-washing)		0.027[g]
	58[g] (before shop washing)		
Si_3N_4 tip	48[h]		0.047[i]
Si tip	51[j]		

[a] [54]; [b] [48]; [c] [12]; [d] [42]; [e] [49]; [f] [71]; [g] [35]; [h] [85]; [i] [94]; [j] [86]

TR phase image, the cortex region, the cuticle region and epoxy resin region are easily identified. In the cuticle region, 5 layers of cuticle cells are seen, and the total thickness of the cuticle region is about 2 μm for this sample. In the detailed images of the cuticle region, 3 layers of cuticle cells are shown, and various sublamellar structure of the cuticle can be seen. The thickness of the cuticle cell varies from 200 to 500 nm. Cortex region shows very fine circular structure of size about 50 nm, which represents the transverse face of the macrofibril and matrix. At this scale, no intermediate filament structure can be revealed.

Longitudinal Section of Hair

Figure 24.22b shows the AFM images of longitudinal section of virgin Caucasian hair. Different regions (cortex region, the cuticle region and embedding epoxy resin) are easily seen. As shown in the detailed image of the cuticle region, various sublamellar structure of the cuticle, the A-layer, the exocuticle, the endocuticle and the cell membrane complex, which cannot be easily revealed in TR surface height image, are easily resolved in TR amplitude image and TR phase image because of different contrast. Most sublamellar structure features of the cuticle shown in the

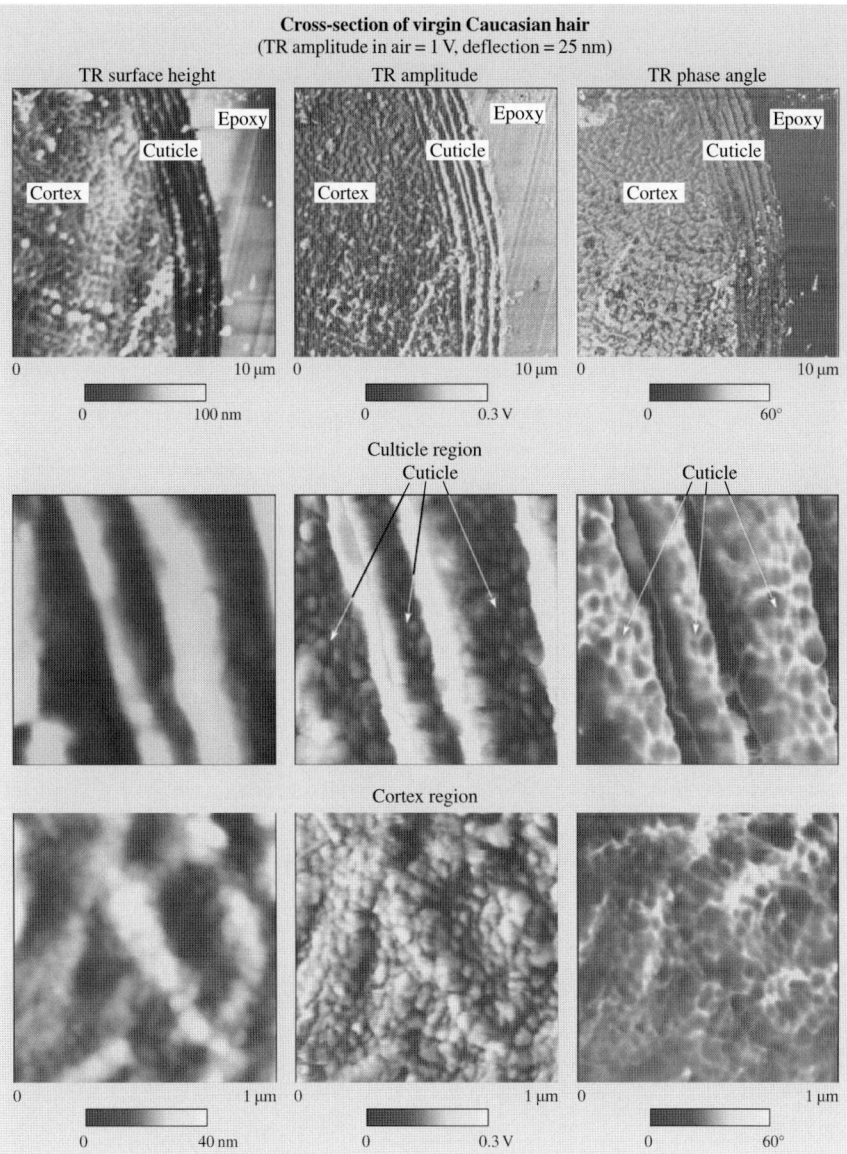

Fig. 24.22. (a) Cross-section images of virgin Caucasian hair and fine detailed images of the cuticle region and cortex region

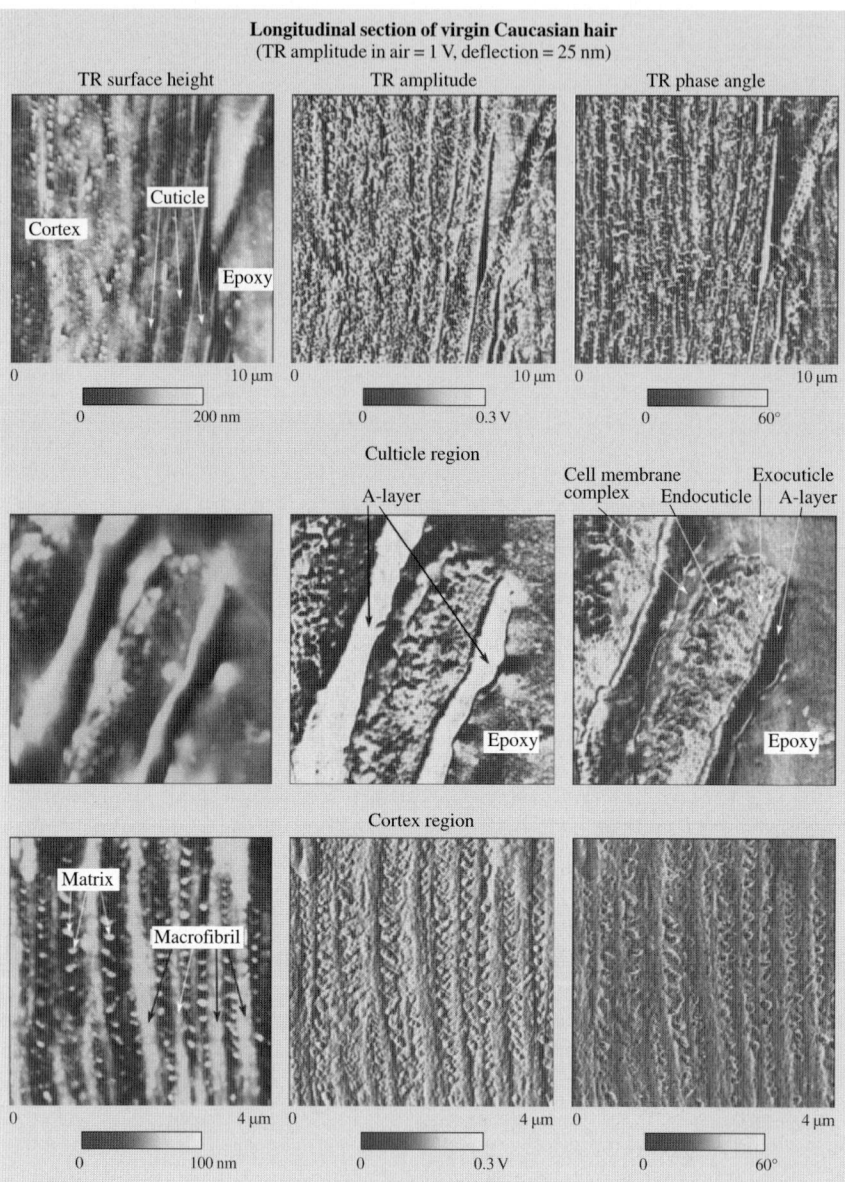

Fig. 24.22. (continued) **(b)** longitudinal section images of virgin Caucasian hair and fine detailed images of the cuticle region and cortex region. Note that the longitudinal section is not perfectly parallel to the long axis of the hair fiber but with a small angle, therefore, the thickness of sublamellar layers of the cuticle is not the real thickness [26]

TEM image of Fig. 24.7b can be identified in the TR amplitude and phase angle images. Previously, these sublamellar structures were only able to be distinguished by TEM [79]. Various cellular sublamellar structure in the cuticle have very different chemical content [68, 79]: the A-layer is rich in disulfide crosslinks due to a very high cystine content of up to 35%; The exocuticle is also rich in disulfide crosslinks (15% cystine); in contrast, the endocuticle is relatively lightly crosslinked containing only about 3% cystine. Consequently, these layers exhibit distinct stiffness and viscoelastic properties, and TR mode II imaging technique (TR amplitude and phase angle images) can easily detect these differences. Note that this longitudinal section is not perfect parallel to the long axis of the hair fiber but with a small angle, therefore, the thickness of various sublamellar layers of the cuticle does not represent the real thickness. In the cortex region, two different morphological regions can be seen: the macrofibril and the matrix. The macrofibril, which is a bundle of the intermediate filament, aligns parallel to each other and looks like a tree trunk; the matrix surrounds macrofibril region. The matrix region has high cystine content of compared to low cystine content of macrofibril region. This chemical content difference between the macrofibril and the matrix make it possible to reveal the fine internal cellular structure of hair using AFM TR mode II technique.

24.4.2 Structure of Various Cuticle Layers

Virgin Hair

Figure 24.23 shows the AFM images of surface of virgin Caucasian hair. Two typical sample positions are shown: position 1 is near root end of hair and position 2 is near tip end of hair. In position 1, one cuticle edge is shown, which is also seen in the TR amplitude and phase images as the black strips because of the topographic effect near the cuticle edge. The topographic effect tends to be significant only when there is a large local geometry change. The cuticle edge shows little natural weathering damage and is still intact with a step height of about 500 nm, and the general cuticle surface which is covered with a lipid layer (the outer β-layer) is relative uniform at large scale. In contrast, the surface near tip of hair (position 2) shows lots of damage which may be simply because of natural weathering and mechanical damage from the effects of normal grooming actions, such as combing, brushing and shampooing. Parts of the cuticle outer sublamellar layers were removed and underneath layers (the A-layer, the endocuticle, the inner layer) are exposed. Because different chemical content of various sublamellar layer of the cuticle results in different stiffness and viscoelastic properties, large contrast can be seen in TR amplitude and TR phase angle images. Note that the surface height within each individual sublamellar layers (the A-layer, the outer β-layer, the inner layer) is relativity uniform, therefore the topographic effect on the TR amplitude and phase is minimum. Detail images of the outer β-layer, the A-layer and the endocuticle are shown at the bottom of Fig. 24.23. All these layers show distinct morphology which can be readily revealed in the TR amplitude and phase angle images: the outer β-layer shows very fine granular structure; the A-layer shows little discriminable features; and the endocuticle has much rougher granular structure [78, 79] than that of the outer β-layer. Previous friction force microscopy (FFM) studies [75] on keratin fibers indicated that for untreated (virgin) fibers, no image contrast was observed on the outer facing surfaces of the scales. These results indicate that the outer lipid layer may form fine domains, which results in the fine granular structure shown in the TR amplitude and phase images (Fig. 24.23). TR mode II technique has higher sensitivity compared to FFM technique, therefore, the fine chemical distribution, which normally cannot be detected by FFM and other techniques, can readily be revealed.

Chemically Damaged Hair

Figure 24.24 shows the AFM images of surfaces of chemically damaged Caucasian hair. Two typical samples are shown. More damage can be seen compared to the surface of virgin hair. More cuticle edges were removed and often larger areas of rough granular endocuticle layer were exposed (see Sample I). Of the components within each cuticle cell (the A-layer, the exocuticle, the endocuticle, etc.), the endocuticle is the least crosslinked [68]. Under wet conditions it will swell preferentially and is the preferred plane for lamellar fracture under mechanical stress. Indeed, many examples have been observed where the cuticle has come off to leave this granular endocuticle layer of approximately half of the thickness of original scale and located at the scale margins. As shown in sample II, the endocuticle layers were further eliminated, entire pieces of cuticles were removed and some fine lines on cuticle surface which delineate the original boundaries of the cuticle edges are clearly seen in TR amplitude images. These lines are referred to cuticle edge "ghost" in literatures [75, 78]. The actual fracture occurs at the interface between the outer β-layer and the δ-layer (see Fig. 24.2). The outer β-layer was originally present, but because of its location, i.e. under the original overlying cuticle but close to the scale edge, it may have undergone oxidative loss through environmental or chemical exposure.

For easy visualization, Fig. 24.25 shows a schematic which illustrates the progress of hair damage. Virgin hair has intact smooth cuticle edge; as damage occurs (natural weathering or mechanical damage), parts of the cuticle outer sublamellar layers wear off and underneath layers (for example, the endocuticle) are exposed. Further damage will cause entire piece of cuticle to be broken off and the ghost which delineate the original boundary of the cuticle edge is seen.

Fig. 24.23. Images of surface of virgin Caucasian hair. Two typical samples are shown: near root end in which intact cuticle edges are seen and near tip end in which damage occur, and part of the cuticle top layers were removed and underneath sublamellar layers are exposed. Detailed images of the outer β-layer, the A-layer and the endocuticle exposed near tip end are shown at the *bottom* [26]

24 Structural, Nanomechanical, and Nanotribological Characterization 1369

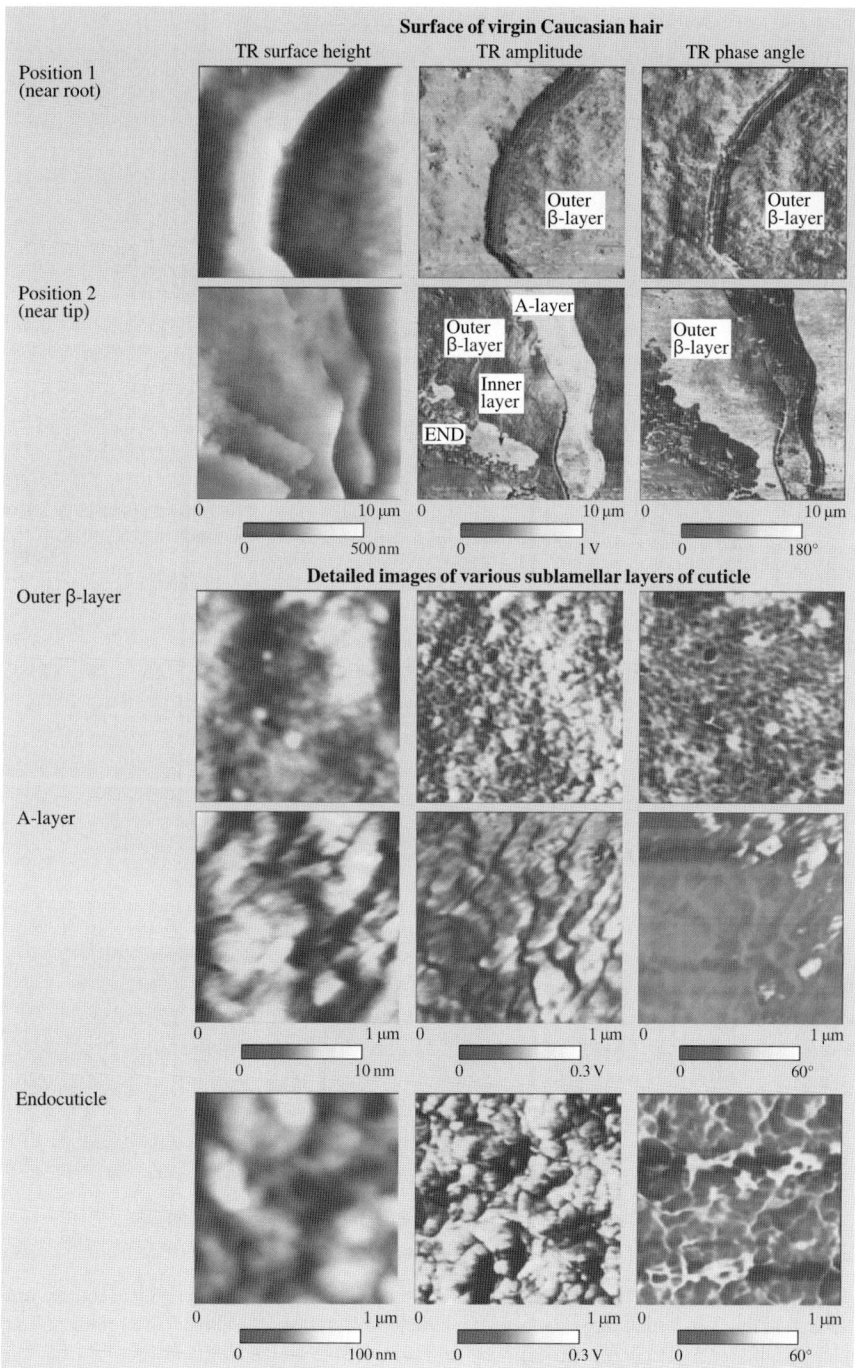

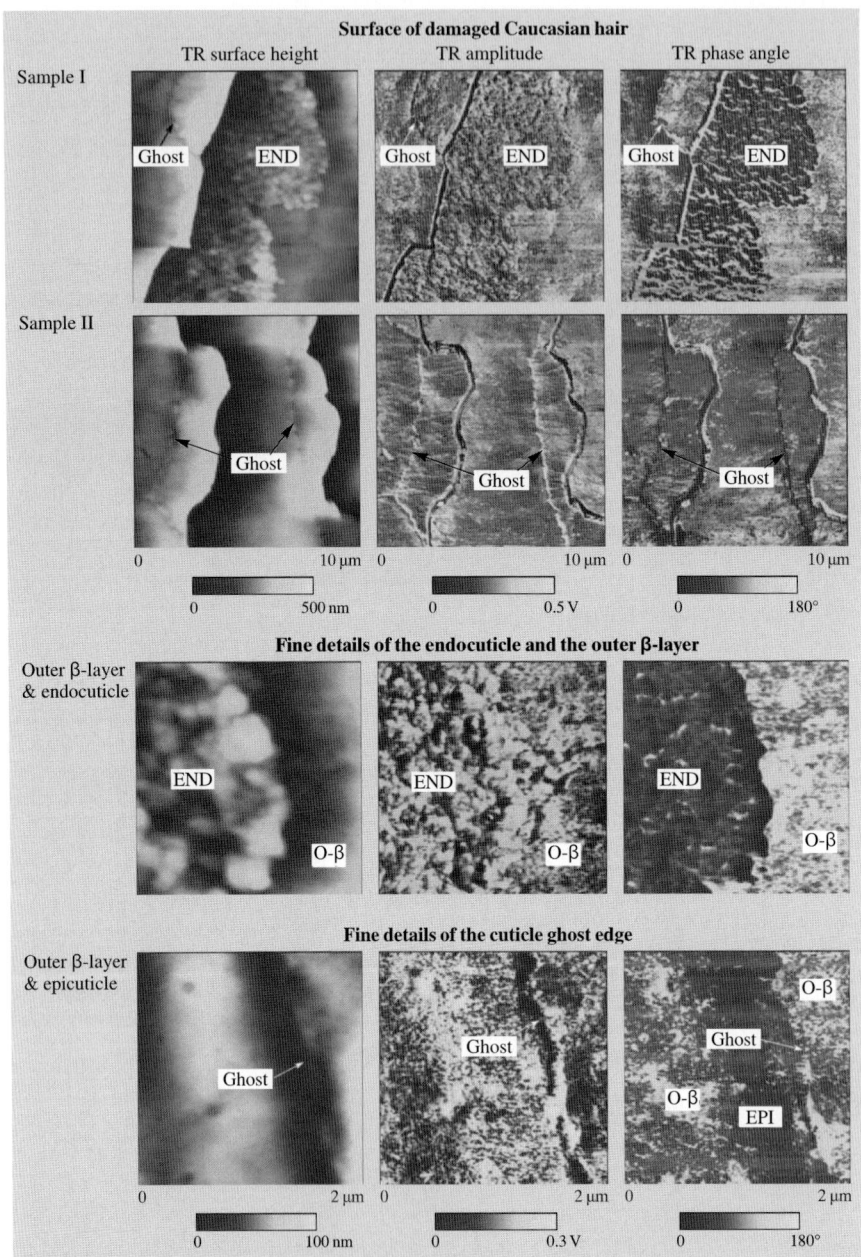

Fig. 24.24. Images of surface of chemically damaged Caucasian hair. Two typical images are shown: sample I in which large areas of the cuticle tops sublamellar layers are removed and much rougher endocuticle layer is exposed; and sample II in which entire pieces of the cuticle are removed and only the cuticle ghost edges are left. Detailed images of the outer β-layer, the endocuticle and the epicuticle are shown at the bottom [26]

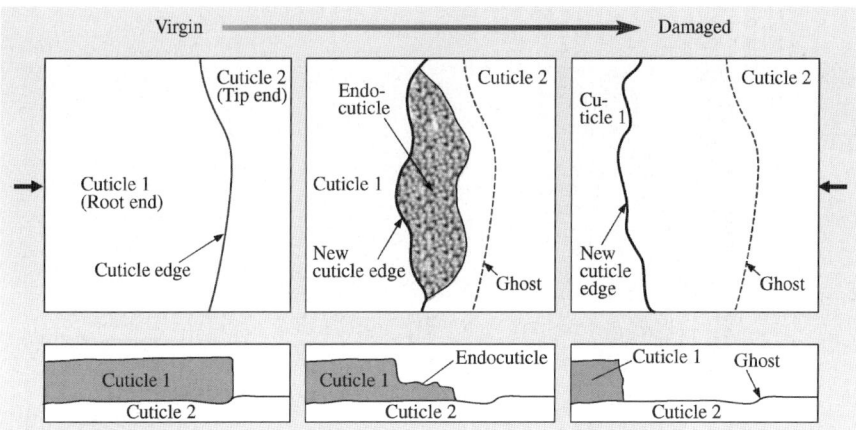

Fig. 24.25. Schematic of the progress of hair damage. The top view images of hair surface at each stage are shown at the top. The cross section views are shown at the bottom taken at the corresponding arrows [26]

Conditioner Treated Hair

Various sublamellar layers of cuticle can be exposed on the surfaces of virgin hair and chemically damaged hair depending on the degree of damage. Because of distinct chemical nature, these layers may have different interaction with conditioner (or other hair care products) which will affect the adsorption of conditioner on hair surface. Figure 24.26 shows the AFM surface images of two samples of chemically damaged treated hair. In sample I, intact cuticle edges can be seen. From TR phase angle image, higher contrast can be seen near cuticle edges. It will be presented later that conditioner is unevenly distributed across the hair surface, and thicker conditioner layer can be found near cuticle edges [45, 46]. This build up of conditioner might be simply caused by the physical entrapment at the steps of the cuticle edges. Uneven thickness of conditioner caused the contrast on TR phase angle images. In sample II, a sharp cuticle edge and a cuticle ghost edge can be readily seen in TR amplitude image. No endocuticle or other sublamellar layers could be found because further treatments removed these layers. As shown in TR phase angle image of sample II, the region between cuticle edge and cuticle ghost edge shows different contrast from other part of hair surface. This newly formed region surface may have exposed the epicuticle layer, while the other parts of hair were still covered with the outer β-layer. The outer β-layer is basically a covalent attached lipid covered layer, which is hydrophobic. The interaction between conditioner and the outer β-layer or other sublamellar layers of the cuticle (such as the epicuticle) could be very different; therefore, the adsorptions of conditioner on these surfaces are different.

Fine details of chemically damaged treated general cuticle surface (the outer β-layer) are shown at the bottom of Fig. 24.26. Compared to the fine granular structure

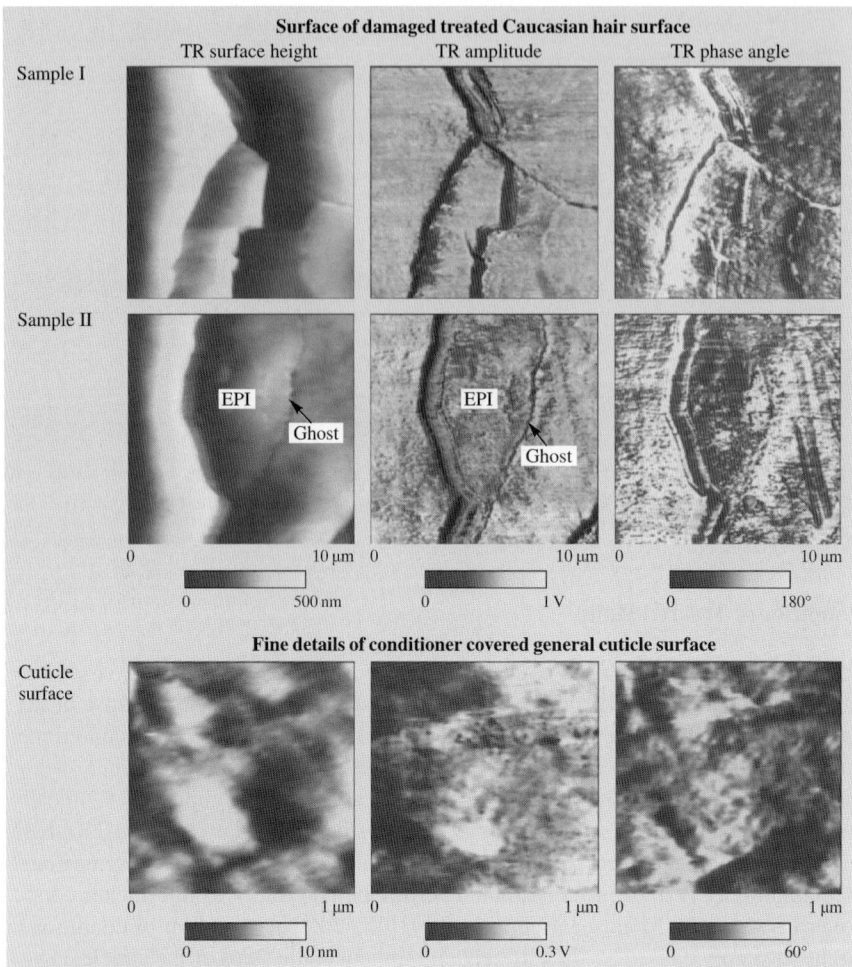

Fig. 24.26. Images of surface of chemically damaged treated Caucasian hair (3 cycles conditioner treatment). Conditioner is unevenly distributed on hair surface. Thicker conditioner layer can be seen near the cuticle edges. Fine details of conditioner covered hair surface are shown at the bottom. No discernible features can be seen [26]

of the outer β-layer shown in the Fig. 24.23, no discriminating features can be seen now since the entire surface is covered with a layer of conditioner.

In summary, although the morphology of the fine cellular structure of human hair has traditionally been investigated using SEM and TEM, these techniques have limited capability to in-situ study environment effects on the physical behavior of hair. AFM TR mode can be used to characterize the fine cellular structure of human hair, and many features previously only seen with SEM and TEM can be identified. The AFM technique provides the possibility to further in-situ studies of the effects of environment (temperature, humidity, etc.) and hair-care product treatment on physical behavior of human hair.

Effect of Humidity on Morphology and Cellular Structure of Hair Surface

Figure 24.27 shows the TR mode phase contrast images of various hair surfaces at different humidities. The phase contrast images of virgin hair surfaces at different humidities show distinct contrast, indicating a very fine granular structure. Interestingly, this fine granular structure is hardly affected by the humidity changes as shown in the TR phase contrast images at different humidities. The hydrophobic nature of the intact lipid layer minimizes the effect of water. On the other hand, damaged hair surface shows some contrast at low (5–10%) and medium (40–50%) humidity, but no longer has the fine granular structure as virgin hair surface. Damaged hair surface underwent chemical treatments, which might have partially destroyed the lipid layer on hair surface, therefore, no fine liquid domain can be observed. Some of the inner cellular structures of hair, which are partially hydrophilic were also exposed. Therefore, damaged hair surface can adsorb more water at high (80–90%) humidity. In this case, TR mode phase contrast measurement can no longer detect the contrast of different surface domain (content), but the relatively uniformly adsorbed water layer.

Treated hair shows different morphology at different humidities. Treated hair surface is covered with a layer of conditioner, which can adsorb large amount of water at medium and high humidity. This thick condition (and water) layer smears the phase contrast of TR mode images as shown in Fig. 24.27, which is similar to the case of damaged hair at high humidity. However, large phase contrast can be observed at low humidity. At low humidity, the conditioner layer will lose water and is no longer able to retain the water content. The large phase contrast indicates that the conditioner gel network has collapsed after staying at low humidity and formed some kind of isolated conditioner gel network domains. Those domains have distinct chemical content and have different lateral stiffness and viscoelastic properties.

The surface height of cross-section of virgin hair at two humidities is shown in Fig. 24.28. The cortex, layers of the cuticle and embedded epoxy can be seen. At high humidity, the cortex and cuticle can adsorb a large amount of water, which will consequently weaken the hydrogen bonds and salt bonds between the protein molecules in hair cellular structure.

24.4.3 Summary

SEM studies of hair cross section and AFM studies of hair surface show that the cuticle is about 5 to 7 scales thick, and each cuticle cell is about 0.3 to 0.5 μm thick. The visible length of each cuticle cell is about 5 to 10 μm long. It appears that the morphology of hair is different from root to tip. That is, the hair near scalp has complete cuticles, and the hair in the middle has worn cuticles, and the hair near tip seldom has cuticles. The size and shape of Caucasian, Asian and African hair

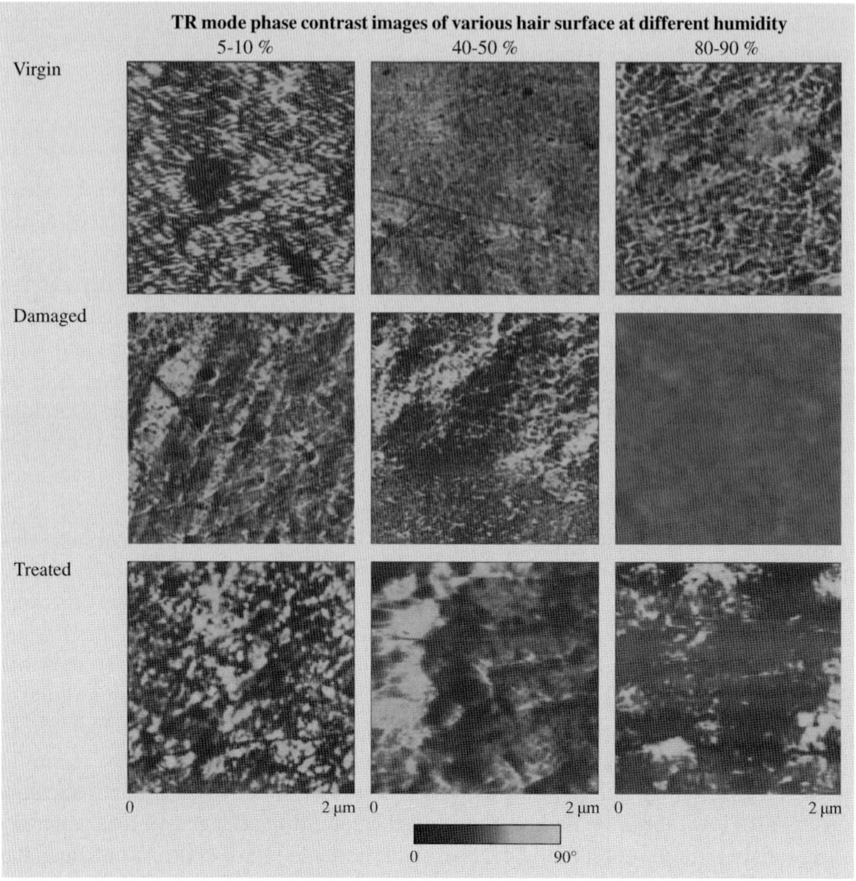

Fig. 24.27. TR mode phase contrast images of Caucasian virgin, chemically damaged, and damaged treated hair surfaces at different humidities [13]

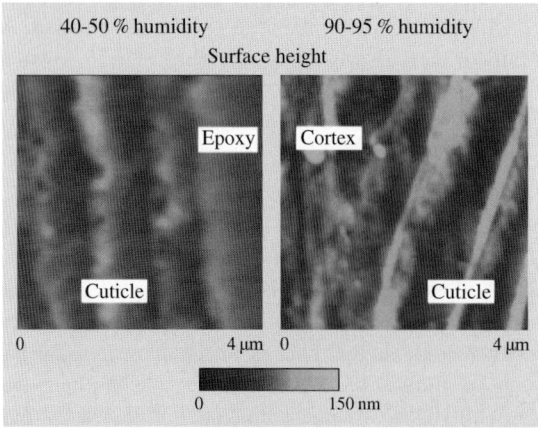

Fig. 24.28. Surface height of virgin Caucasian hair cross-section at different humidities [13]

have been measured from the hair surface and cross section. Asian hair seems to be the thickest (nearly round), followed by African hair (oval-flat) and Caucasian hair (nearly oval).

The cross-section and longitudinal section of virgin Caucasian human hair were investigated using TR mode II. The cortex and the cuticle, the macrofibril and the matrix of human Caucasian hair were readily revealed. Various sublamellar cellular structures of cuticle, such as the A-layer, the exocuticle, the endocuticle, and the cell membrane complex, are easily observed because of their distinct stiffness and viscoelastic properties. The surface features of various Caucasian human hair (virgin, chemically damaged, and chemically damaged treated) were readily revealed. Sublamellar layers show distinct contrast in TR amplitude and phase angle images. The fine granular structure of the outer β-layer, which has not previously been seen by SEM, TEM and other AFM studies, is a result of the fine domain formation of lipid layer. Chemically damaged hair surfaces show much more damage; larger areas of the endocuticle were exposed. The endocuticle has much rougher structure than general cuticle surface (such as the outer β-layer), which could be the part of reason why damaged hair loses shine. Conditioner unevenly distributes on damaged treated hair surface; thicker conditioner films are found near the cuticle edges. At high humidity, the cortex and cuticle can adsorb water, which will consequently weaken the hydrogen bonds and salt bonds between the protein molecules in hair cellular structure. The effect is more significant in chemically damaged and treated hair.

24.5 Nanomechanical Characterization Using Nanoindentation, Nanoscratch, and AFM

Nanomechanical characterization of human hair using nanoindentation and nanoscratch provides valuable information about the hair fiber itself, as well as how damage and treatment affect important mechanical properties of the fiber [89, 90]. In Sect. 24.5.1, the hardness, Young's modulus and creep results for both the hair surface and cross-section are discussed. In Sect. 24.5.2, the coefficient of friction and scratch resistance of the hair surface is presented for unsoaked and soaked hair. In Sect. 24.5.3, stress-strain curves and AFM topographical images of virgin and damaged hair during tensile deformation are presented.

24.5.1 Hardness, Young's Modulus, and Creep

Hair Surface

Figure 24.29 shows the optical micrograph of three indents on virgin Asian hair made at the normal load of 100 mN. The indentation depths and residual depths were about 5 μm and 3 μm, respectively. The sizes of these indents are about 15 to 20 μm in diameter. This image clearly shows that the Nano Indenter II system can successfully make indents on human hair surface.

Figure 24.30a shows the load-displacement curves for virgin, chemo-mechanically damaged and virgin treated Caucasian hair obtained at two loads: 0.1 and 10 mN. (Loads of 1 and 100 mN were also studied; data not shown.). The hardness and elastic modulus values corresponding to each load-displacement curve are listed in the figure boxes. As mentioned in the Experimental section, at each load, five indents were made. Figure 24.30a just presents one representative result for each load. At 0.1 mN, the indentation depths of all these hair were less than 150 nm, which means that the indents were made within one cuticle scale, assuming that the thickness of one cuticle scale is about 0.3 to 0.5 μm for all these hair samples. At 1.0 mN, the indentation depths were about 0.4 to 0.6 μm, indicating that indents were probably made through one to two cuticle scales. At 10 mN, the indenter tip penetrated about three to five cuticle scales. At 100 mN, the indentation depths were about 5 μm, which means that the tip probably reached the cortex of the hair, considering that a hair fiber generally has about 5 to 10 cuticle scales. It is interesting to observe that the loading curves of chemo-mechanically damaged hair obtained at 0.1 mN and 1.0 mN are not so smooth as the virgin and virgin treated hair, especially at the beginning, indicating that the chemo-mechanically damaged hair was very soft at the first 30 to 50 nm or so, probably because of the chemical damage which caused changes to the exposed surface. At 0.1 and 1.0 mN, the hardness and elastic modulus of virgin treated and chemo-mechanically damaged hair are lower than the virgin hair, indicating that chemical damage and conditioner treatment led to the softness of the hair surface. Considering human hair as a polymeric cylinder, the absorption of chemicals used in hair coloring and conditioner ingredients in the first micron or so might plasticize the polymer and hence reduce its mechanical properties. At 10 mN and 100 mN, the hardness and elastic modulus of all three hair look similar. This result indicates that the effective depth of the chemicals/conditioner influence is probably less than 1.5 μm, i.e., the first 3 to 4 scales of the cuticles probably interact with the chemicals and conditioner ingredients more effectively than the rest of the scales.

Figure 24.30b shows the hardness and elastic modulus vs. indentation depth for various hair. Every data point (averaged hardness and elastic modulus value), and every error bar in Fig. 24.30b was calculated from five indentations. According to

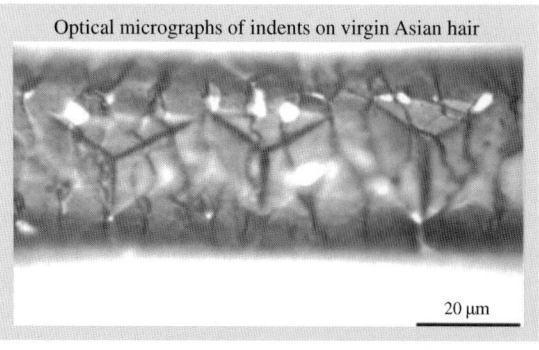

Fig. 24.29. Image of indents on hair surface made using a nanoindenter [90]

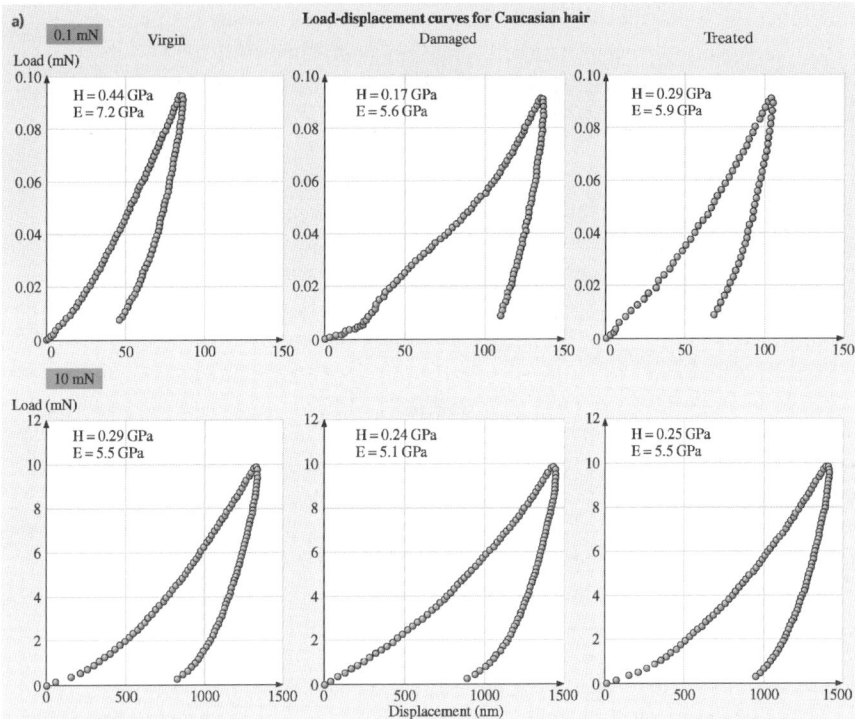

Fig. 24.30. (a) Load-displacement curves for Caucasian virgin, chemo-mechanically damaged, and virgin treated hair at two peak loads;

Fig. 24.30b, the hardness and elastic modulus of hair decreases as the indentation depth increases. In order to explain this, the indentation process is divided into two stages. In the first stage, the indenter tip penetrated the cuticles scales only, in which the indentation depth was probably less than 5 μm. For one cuticle scale, the mechanical properties are expected to decrease from top layer to bottom layer, because the cystine content, thus the disulfide cross-link density, decreases from the A-layer, to the exocuticle, and to the endocuticle (see Fig. 24.2a and Table 24.3). The cuticle scales are bound together by the cell membrane complex, one of weakest parts of the hair fiber in terms of mechanical properties. The intercellular cement of the cell membrane complex is primarily nonkeratinous protein, and is low in cystine content (∼ 2%). When the indenter tip penetrated the cuticle scales one by one continuously, the number of the cell membrane complex layers penetrated by the tip increased. These weak cell membrane complex layers joined together might lead to a deeper displacement upon indentation, contributing to lower mechanical properties. It is also possible that the outer scales of cuticles have higher mechanical properties than the inner scales. In the second stage, the tip began to penetrate the cortex. In general, the cuticles are richer in disulfide cross-links than the cortex [32, 68], so the mechanical properties of the cortex are expected to be lower than the cuticle. Putting the

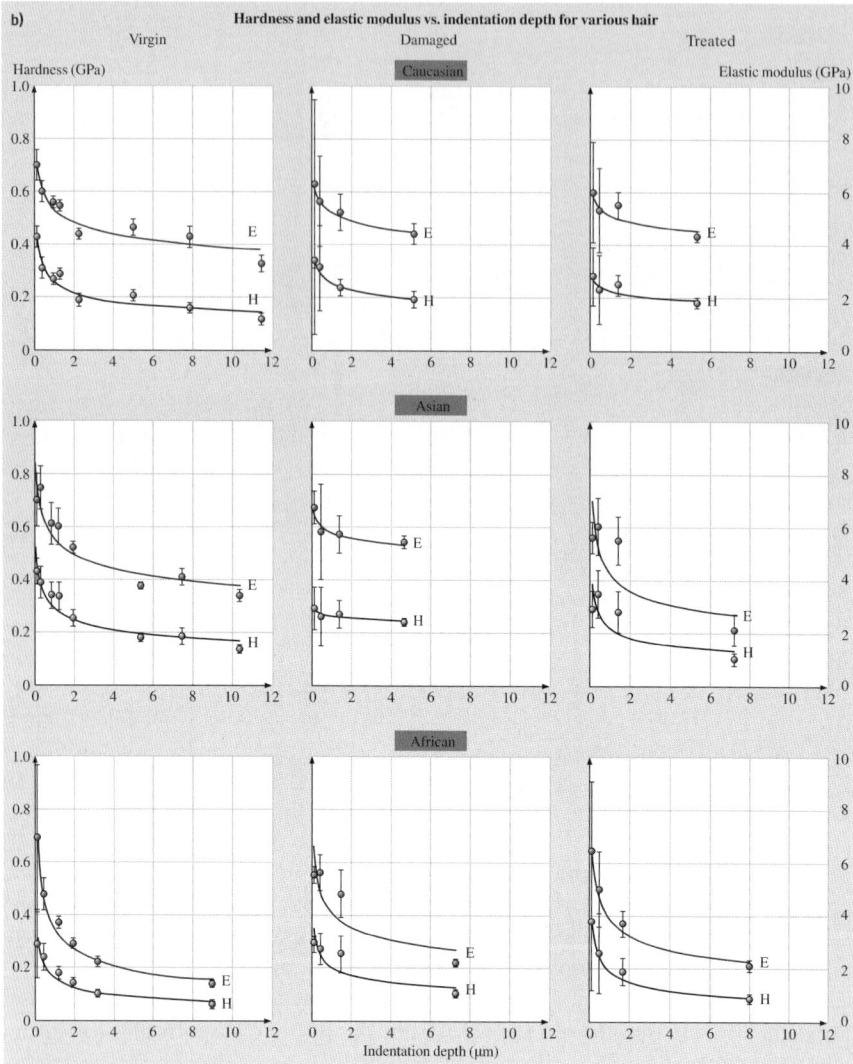

Fig. 24.30. (continued) **(b)** hardness and elastic modulus vs. indentation depth for various virgin, chemo-mechanically damaged and treated hair [90]

two indentation processes together, the hardness and elastic modulus of hair will decrease as a function of the indentation depth.

Figure 24.30b also indicates that at normal loads of 0.1, 1.0 and 10 mN, corresponding to the indentation depth of less than 1.5 μm, the chemo-mechanically damaged and virgin treated hair generally had lower hardness and elastic modulus, but larger error bars, i.e., larger data deviations, than the virgin hair for each ethnicity. This result means that the effective interaction depths were probably less than 1.5 μm for all three ethnicity of hair, and that the effect or distribution of the conditioner on the hair surface were not uniform. It is believed that most of the important interactions between shampoo/conditioner and hair occur at or near the hair surface (the first few micrometers of the fiber periphery). The nanomechanical characterization of hair surface shows that the effective interaction depth (< 1.5 μm) may be shallower than what was thought before. In general, two types of interaction occur between chemical/conditioner ingredients and hair: adsorption and absorption. It has been suggested that for conditioning ingredients in hair conditioners, adsorption is more critical than absorption, because the conditioning ingredients are relatively large species [68]. If this is the case, then the data variation was probably caused by the non-uniform adsorption of the chemical molecules and the conditioning ingredients to the hair surface. Because the interaction affected the hair up to 1.5 μm deep, absorption should also play an important role here. Transcellular and intercellular diffusion are the two theoretical pathways for absorption to occur. The transcellular route involves diffusion across cuticle cells through both high and low cross-linked proteins. The intercellular diffusion involves penetration between cuticles cells through the intercellular cement and the endocuticle that are low in cystine content (low cross-link density regions). The intercellular diffusion is usually the preferred route for entry of most molecules (especially large ones such as surfactants or even species as small as sulfite near neutral pH). However, for small molecules, transcellular diffusion under certain conditions might be the preferred route, especially if the highly crosslinked A-layer and exocuticle are chemomechanically damaged [68]. Depending on the molecular size and the hair condition, the diffusion pathway and diffusion rate might be different from site to site on the hair surface, thus the distribution of conditioner might not be uniform. To sum up, for chemo-mechanically damaged and virgin treated hair, since the adsorption and absorption of chemicals and conditioner ingredients were probably not uniform on the hair surface, the nanomechanical properties of the hair surface (depth < 1.5 μm) were not affected (generally decreased) uniformly, leading to the larger data variation compared to corresponding virgin hair. This implies that the nanoindentation technique can be used to quantitatively evaluate the effective depth of the conditioned hair and distribution of conditioner by measuring the hardness and elastic modulus of the hair surface before and after conditioner treatment as a function of depth and location.

Figure 24.31a summarizes the hardness and elastic modulus of various hair. In general, the chemo-mechanically damaged and virgin treated hair had lower nanomechanical properties and larger error bars than the corresponding virgin hair, as discussed above. The data of African hair was a little strange. For example, the virgin treated African hair seemed to have higher hardness than virgin African hair. It should be noted that the African hair is naturally curly and highly elliptical, and it was very difficult to mount them and make indentations on their surface. The curly and highly elliptical surface of African hair might cause the indentation results vary somewhat from the actual values. If the hardness and elastic modulus measured at 1.0 mN is taken as the hair surface hardness and elastic modulus, then

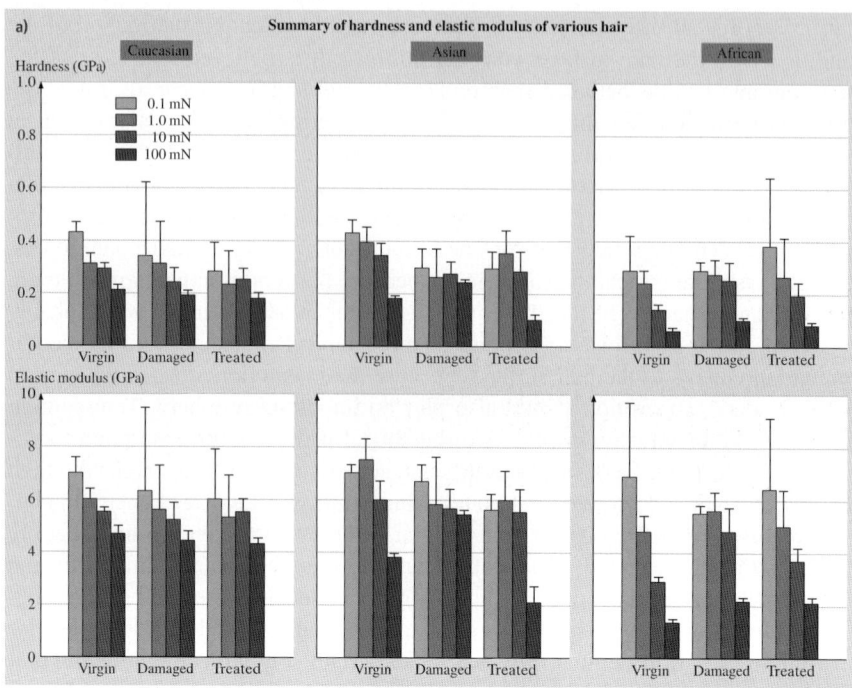

Fig. 24.31. (a) Summary of hardness and elastic modulus of virgin, chemo-mechanically damaged and virgin treated hair at four peak loads

by comparing the virgin Caucasian, Asian and African hair, it is seen that the Asian hair has the highest hardness (0.39 ± 0.06 GPa) and elastic modulus (7.5 ± 0.8 GPa), followed by Caucasian hair with hardness of 0.31 ± 0.04 GPa and elastic modulus of 6.0 ± 0.4 GPa. The African hair seems to have the lowest mechanical properties, whose hardness is 0.24 ± 0.05 GPa and elastic modulus is 4.8 ± 0.6 GPa. Note that all these mechanical properties were measured in the middle part of the hair.

Figure 24.31b summarizes the hardness and elastic modulus for virgin Caucasian hair at three locations: near scalp, middle and near tip. As expected, the hardness and elastic modulus of hair surface decreases from root to tip, because of the cuticle damage. Considering that the hair near scalp has complete cuticles, while the hair near tip only has exposed cortex, it is probably a good way to compare the nanomechanical properties of hair cuticle and cortex in the lateral direction by comparing the nanomechanical properties of hair near scalp and hair near tip. At 1.0 mN, the cuticle (hair near scalp) has higher hardness (0.6 ± 0.29 GPa) and elastic modulus (8.4 ± 1.2 GPa) than cortex (hair near tip), whose hardness is 0.3 ± 0.06 GPa, and elastic modulus is 6.0 ± 0.6 GPa. This result clearly suggests that the cuticles contribute more to the hair lateral mechanical properties than the cortex, which is in good agreement with the theoretical models for wool fibers [93].

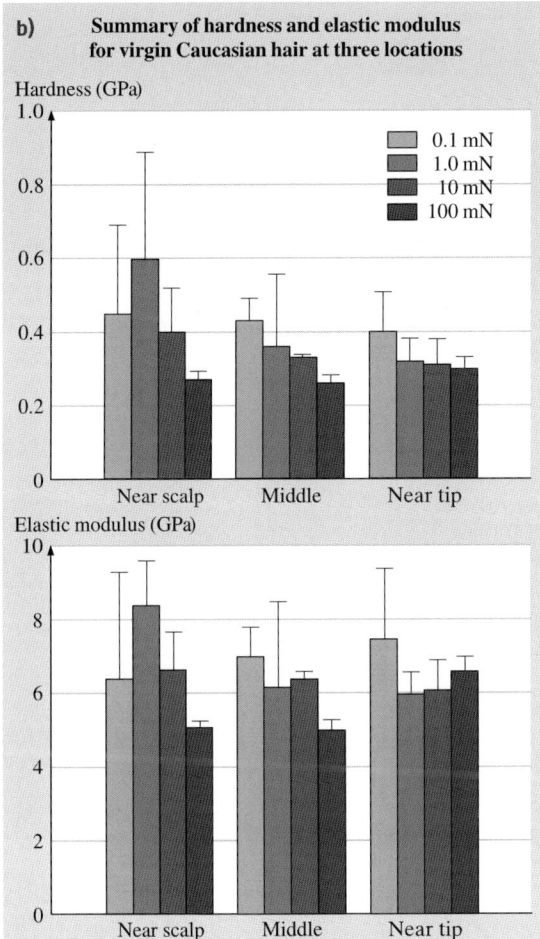

Fig. 24.31. (continued) **(b)** summary of hardness and elastic modulus for virgin Caucasian hair at three locations at four peak loads [90]

Figure 24.32 shows the creep displacement vs. time curves for various hair. The normal load used for creep tests was 10 mN. In all cases, the displacement increased as time passed. The creep behavior of hair may arise from several sources. Hair is rich in peptide bonds and the abundant CO- and NH- groups present give rise to hydrogen bonds between groups of neighboring chain molecules (Fig. 24.2b). Other linkages, such as side-chain interactions of the disulphide type, and the chain folding may also be present in hair. When hair was compressed, the creep behavior was a result of deformation and relaxation of the chemical bonds, the polypeptide chains and the non-crystalline regions [2]. It should be noted that at normal load lower than 10 mN, the creep behavior was not obvious. Assuming that the diameter of Caucasian, Asian and African hair was about 50 μm, 100 μm and 80 μm, respectively, the compression ratio of the indented area of these hair at 10 mN at the beginning of the creep tests were about: Caucasian (∼ 2.6%), Asian (∼ 1.3%) and

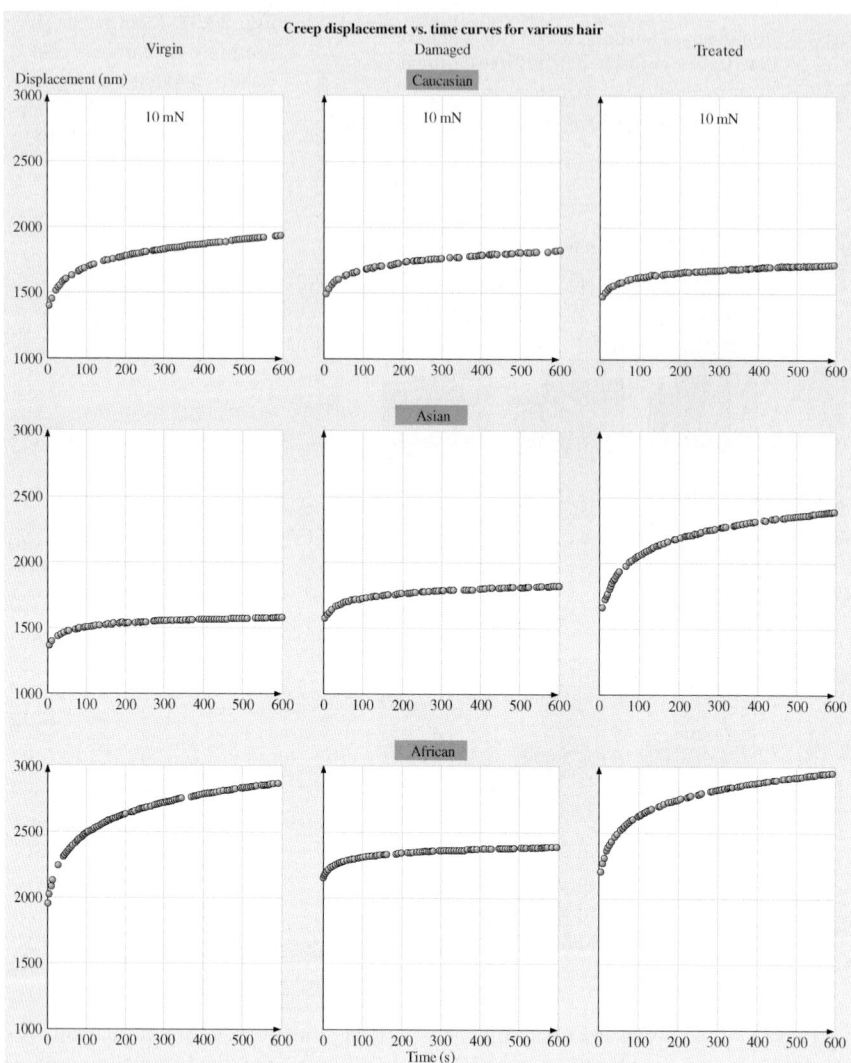

Fig. 24.32. Creep displacement vs. time curves for virgin, chemo-mechanically damaged and virgin treated hair [90]

African (~ 2.5%). This may suggest that if the local compression ratio was less than these values for corresponding hair, the deformation and relaxation of the chemical bonds, the polypeptide chains and the non-crystalline regions might be too small to cause the creep behavior to occur. According to the creep displacement vs. time curves, it is difficult to correlate the creep behavior of each hair with its ethnicity and condition (virgin, chemo-mechanically damaged or virgin treated).

Cross-section

Figure 24.7a presented previously shows the SEM images of virgin hair cross section. These SEM images may represent the typical shape of Caucasian (nearly oval), Asian (nearly round) and African (oval-flat) hair. Regarding the diameter, the Asian hair seems to be the thickest, followed by African and Caucasian hair, which is in good agreement with the SEM studies of the hair surface. The center column of Fig. 24.7a shows the cortex and medulla of each hair. The arrows point to the indents made at the cortex. The medulla of African hair is not so obvious, and it is believed that not all the hair has medulla [68]. The right column shows the images of the cuticles. The top-right image clearly shows that the cuticle of Caucasian hair is about 6 to 7 scales thick, and each cuticle cell is about 0.3 to 0.5 µm thick. Note that the cuticle scales were separated due to polishing, implying that the binding strength of the cell membrane complex between the cuticle scales might not be very strong.

The hardness and elastic modulus of hair cuticle, cortex and medulla were measured from the cross section samples, and Fig. 24.33 shows the hardness and elastic modulus plots across various virgin hair. As expected, the cuticles have the highest mechanical properties, followed by cortex and medulla. Table 24.12 summarizes the hardness and elastic modulus of various hair [90]. The hardness of cuticles was taken from the hair surface measurements (see Fig. 24.31a). By comparing the mechanical properties of Caucasian, Asian and African hair cortex, it can be seen that the Asian cortex appears to have the highest properties, followed by Caucasian and African hair. This trend is in agreement with the trend for the hair surface measurement results (see Fig. 24.31a). Table 24.12 shows that the hardness of cuticles is greater than cortex, but the elastic modulus of cortex is comparable to cuticle [90]. Comparing the hardness (0.3 ± 0.06 GPa) and elastic modulus (6.0 ± 0.6 GPa) of Caucasian cortex in the lateral direction (see Fig. 24.31a) with its hardness (0.27 ± 0.02 GPa) and elastic modulus (6.5 ± 0.5 GPa) in the longitudinal direction, it can be seen that hardness and elastic modulus of the hair cortex in the longitudinal direction are lower and higher respectively, than the lateral direction.

Table 24.12. Summary of hardness and elastic modulus of human hair. Mean and $\pm 1\sigma$ values are presented

	Hardness (GPa)			Elastic modulus (GPa)		
	Cuticle[a]	Cortex[b]	Medulla[b]	Cuticle[a]	Cortex[b]	Medulla[b]
Caucasian	0.32 ± 0.04	0.27 ± 0.02	~ 0.19	6.0 ± 0.4	6.5 ± 0.5	~ 5.5
Asian	0.39 ± 0.06	0.30 ± 0.02	~ 0.18	7.5 ± 0.8	6.7 ± 0.3	~ 5.8
African	0.24 ± 0.05	0.23 ± 0.06	~ 0.16	4.8 ± 0.6	5.8 ± 0.7	~ 5.0

[a]Obtained from the hair surface at normal load of 1.0 mN [b]Obtained from the hair cross section at normal load of 1.0 mN

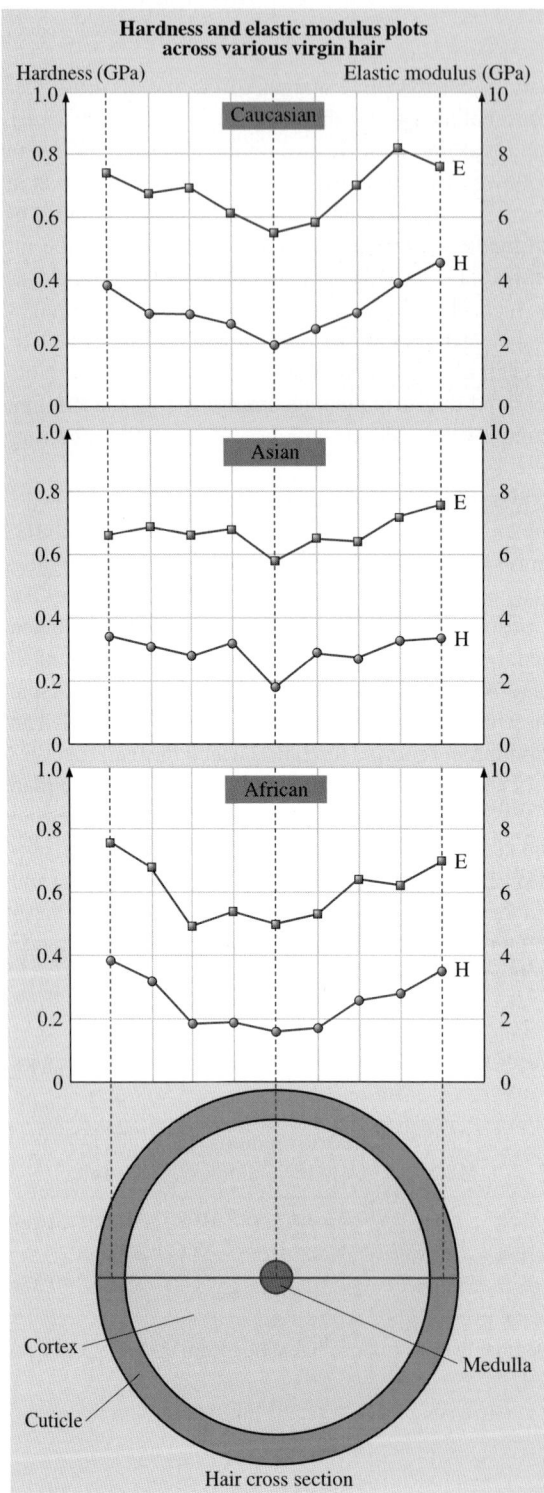

Fig. 24.33. Hardness and elastic modulus plots across various virgin, chemo-mechanically damaged and virgin treated hair [90]

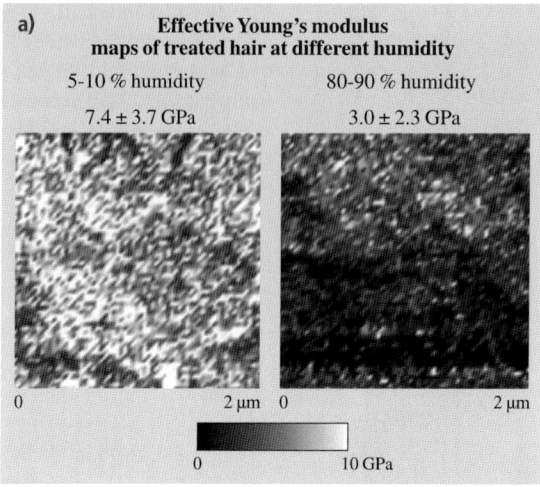

Fig. 24.34. (a) Effective Young's modulus maps of Caucasian chemically damaged treated hair at different humidities [13]

Effect of Humidity and Temperature on Young's Modulus

Figure 24.34a shows effective Young's modulus mappings of chemically damaged treated hair at different humidities obtained from force calibration plots [13]. At different humidities, the maps are distinctly different. As humidity increases, the effective Young's modulus of hair surface decreases significantly. The effective Young's modulus of hair surface at 80–90% humidity is only half the value at 5–10% humidity. At high humidity, the water content in conditioner gel network will increase significantly; therefore, the conditioner layer on hair surface is thicker and softer, consequently a lower value of effective Young's modulus. At low humidity, the conditioner gel network may desorb most of the water content, which results in the collapse of the gel network. Consequently, the conditioner layer no longer acts as a soft protection layer, instead as a hard, thin shell on the hair surface.

Figure 24.34b summarizes the effective Young's modulus of various hair surfaces at different humidities. Young's modulus of damaged hair is lower than virgin hair and it decreases dramatically at high humidity. Since the hydrophobic lipid protection layer of damaged hair has been depleted or damaged, the hydrophilic molecules of the inner cellular structure of hair are exposed to water. Water can adsorb and diffuse easily into hair via the defects on the surface, therefore softening the hair. Treated hair has smaller effective Young's modulus than virgin hair at 50% humidity because of the soft physisorbed conditioner gel network layer on the surface. The layer remains intact at high humidity and protects the hair surface from excess water adsorption and diffusion (penetration). However, at low humidity, the gel network is no longer able to retain the water content. It will lose most of the water content and collapse, and behave as a hard shell on hair surface. Therefore, treated hair surface at low humidity has the same effective Young's modulus as that of virgin hair.

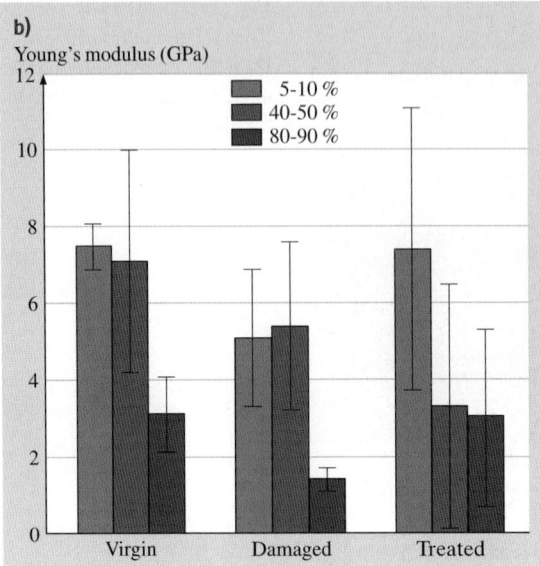

Fig. 24.34. (continued) **(b)** effective Young's modulus of Caucasian virgin, chemically damaged, and damaged treated hair samples at different humidities [13]

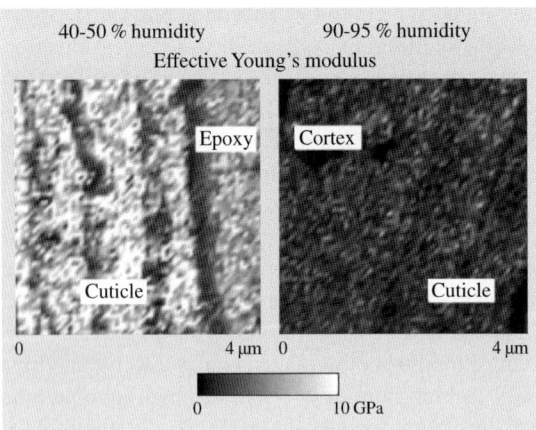

Fig. 24.35. Effective Young's modulus of virgin Caucasian hair cross-section at different humidities [13]

The effective Young's modulus of the cross-section of virgin hair is shown in Fig. 24.35. As humidity increases, the Young's modulus decreases and the differences between various layers (the cortex, the cuticle, and the epoxy) disappear. Human hair consists of various chemical and physical bonds (salt bond, hydrogen bond and disulfide bond). The strength of these bonds will be strongly affected by the level of water content.

The effective Young's modulus of various hair surfaces were also measured at different temperatures. Three different temperatures are studied: room temperature,

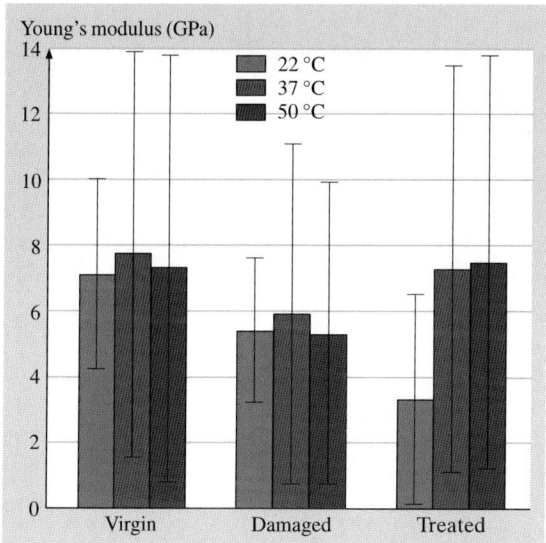

Fig. 24.36. Effective Young's modulus of Caucasian virgin, chemically damaged, and damaged treated hair samples at different temperature [13]

human body temperature and a high temperature which represents the temperature under direct sunshine. Figure 24.36 summarizes the findings for virgin, damaged, and conditioner-treated hair at those temperatures. For virgin or damaged hair surfaces, temperature has little effect on the Young's modulus due to small range of temperature studied. The effective Young's modulus of conditioner-treated hair surface increases as the temperature increases. At high temperature, the conditioner layer is no longer able to retain the water content and collapses to form a hard shell covering the surface. Therefore, conditioner treated hair has high effective Young's modulus at high temperature.

24.5.2 Scratch Resistance

Nanoscratch technique is capable of simulating the scratch phenomena on hair surface on the nanoscale by scratching the hair surface using a conical diamond tip (radius about 1 μm) and recording the coefficient of friction, in-situ scratch depth and residual depth.

Nanoscratch on Single Cuticle Cell

Figure 24.37 shows the coefficient of friction and scratch depth profiles as a function of normal load and tip location on a single cuticle cell of Caucasian and Asian hair (virgin, chemo-mechanically damaged and virgin treated). The scratch direction is from left to right. The scratch length was 5 μm and the normal load was increased from 0.01 to 1 mN during scratching. The coefficient of friction of all the hair samples ranged from 0.3 to 0.6 [89]. The coefficient of friction of virgin treated Caucasian hair (∼ 0.3) is lower than virgin Caucasian hair (∼0.4), and the coefficient of friction of virgin treated Asian hair (~ 0.3) is also lower than virgin Asian hair (~ 0.5). The conditioner acts as a thin layer of lubricant on hair surface and it reduces the coefficient of friction of hair during scratch. The coefficient of friction of chemo-mechanically damaged hair depends on the type and extent of damage. If the chemical damage softens the hair surface, then during scratching, the tip plows into the hair easily, leading to higher coefficient of friction. If the damage hardens the hair surface or does not change the mechanical properties of the hair surface, then the coefficient of friction probably will decrease or stay the same during scratch.

The scratch depth profiles include the profiles obtained before (pre-scratch), during (in-situ scratch) and after (post-scratch) scratching at scratch length of 5 μm and a maximum normal load of 1 mN as indicated in Fig. 24.37. Reduction in scratch depth is observed after scratching as compared to that of during scratching. This reduction in scratch depth is attributed to an elastic recovery after removal of the normal load. The post-scratch depth indicates the final depth which reflects the extent of permanent damage and plowing of the tip into the hair surface. The scratch depth profiles show that at the very beginning of the scratch for all the hair samples,

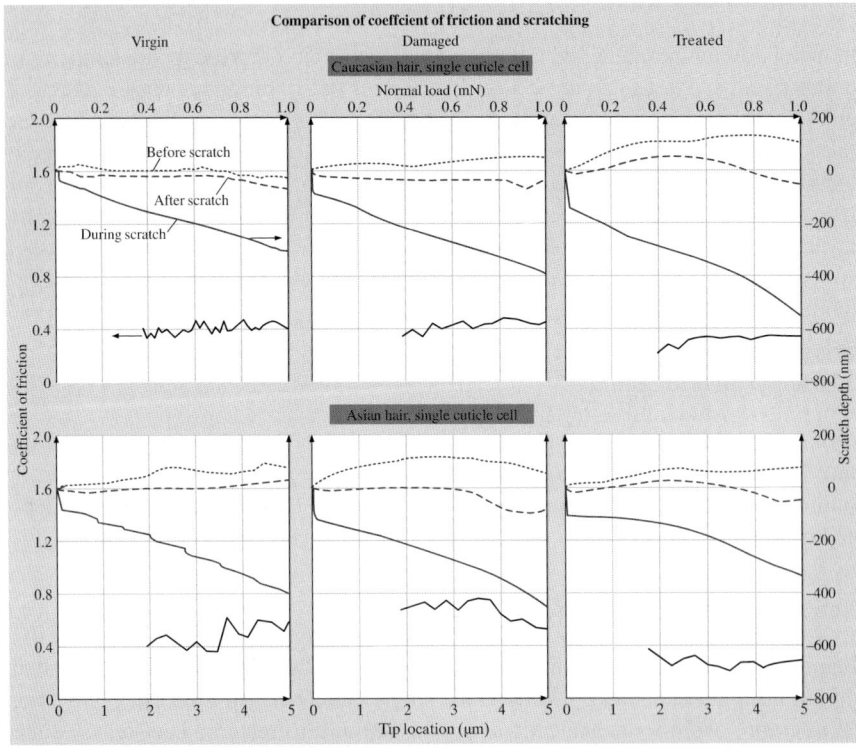

Fig. 24.37. Coefficient of friction and scratch depth profiles as a function of normal load and tip location on single cuticle cell of Caucasian and Asian hair (virgin, chemo-mechanically damaged and virgin treated) [89]

the in-situ displacement (30–200 nm) increased rapidly at very low load. After that, it increased gradually. This observation indicates that the top about 200 nm of the hair surface may be softer than the underlying layer. The scratch depth profiles also show that the reference surface profile before scratch is not very flat, indicating that human hair has a rough surface. AFM studies have shown that the RMS roughness of Caucasian and Asian hair surface ranges from 7 nm to 48 nm [45]. At 1 mN, the in-situ scratch depths of all the hair samples range from 300 to 600 nm, and the residual depths range from 50 to 200 nm. Since the thickness of one cuticle cell is about 300 to 500 nm, during the 1 mN nanoscratch test, the scratch tip might only penetrate one cuticle cell layer.

Nanoscratch on Multiple Cuticle Cells

Most of the time when we comb our hair, the comb is scratching multiple cuticle cells. Figure 24.38a shows the coefficient of friction and scratch depth profiles as a function of normal load and tip location on multiple cuticle cells of chemomechanically damaged Caucasian hair obtained in two scratch tests: scratch along cuticle and scratch against cuticle, and Fig. 24.39b shows the SEM images of the hair surface after scratch. The coefficient of friction obtained when the tip scratched the hair against cuticle is significantly higher than the coefficient of friction obtained when the tip scratched the hair along cuticle, which is known as the "directionality effect". This is understandable because when the tip scratches the hair surface against cuticle, the 300–500 nm high cuticle "wall" resists the tip to move, leading to higher coefficient of friction [89].

By observing the surface profiles (before scratch) of Fig. 24.38a, we can clearly see the shape (height and visible length) of each cuticle cell, i.e., the height is about 300–500 nm, and the visible length is about 5–10 μm, which is in good agreement with SEM and AFM data. The scratch tip acts as a surface profiler before scratching. During scratching, the in-situ displacement increased up to about 3 μm at 10 mN, while the residual depth is about 1.5 μm. Considering that the thickness of cuticle is about 1.5 to 5 μm, it is likely that during the 10 mN scratch test, the tip reached the cortex. The SEM images (Fig. 24.38b) clearly show that in both the along cuticle and against cuticle cases, the cuticle cells were worn away. The topography of the exposed surface is totally different from the cuticle topography and it is believed that the exposed surface is the cortex. It can also be seen from Fig. 24.38b that the scratching against cuticle caused much more damage to the hair than along cuticle.

Given the fact that the "directionality effect" is universal for each human hair from all races and that the scratching along cuticle is more relevant to our daily life, we now focus on the scratch tests along cuticle. Figure 24.39a shows the coefficient of friction and scratch depth profiles as a function of normal load and tip location on multiple cuticle cells of Caucasian and Asian hair (virgin, chemo-mechanically damaged and virgin treated) and the SEM images of the scratch wear tracks. Note that since at least five scratches were made on the same hair, some SEM images may show more than one scratch wear track. For example, the SEM image of chemomechanically damaged Caucasian hair shows two scratches, and the scratch wear

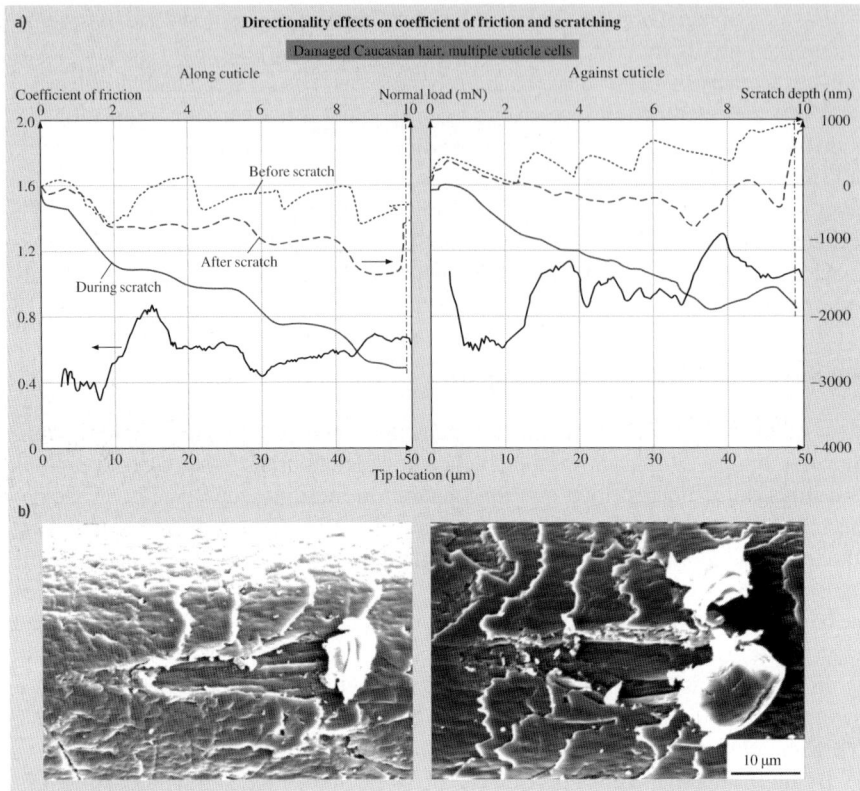

Fig. 24.38. (a) Coefficient of friction and scratch depth profiles as a function of normal load and tip location on multiple cuticle cells of chemo-mechanically damaged Caucasian hair obtained in two scratch tests: scratch along cuticle and scratch against cuticle; (b) SEM images of the hair surface after scratch [89]

track on the right side corresponds to the scratch depth profile. For Caucasian hair, the averaged coefficient of friction of virgin treated hair (~ 0.4) is lower than virgin hair (~ 0.7). For Asian hair, the averaged coefficient of friction of virgin treated hair (~ 0.5) is also lower than virgin hair (~ 0.8). The trend corresponds well with the nanoscratch results on single cuticle. Based on this data, it is clear that the conditioner treatment indeed can reduce the coefficient of friction of hair surface upon scratching. Regarding the chemo-mechanically damaged hair, since the coefficient of friction of chemo-mechanically damaged hair varies depending on the type and extent of damage (as discussed above), it is difficult to make comparison with the virgin or virgin treated hair.

It is worth mentioning that the coefficient of friction of human hair measured by nanoscratch technique is on the micro-scale, and not on the nano-scale, since the tip radius is 1 μm and the normal load range is 1 to 10 mN. It will be shown later that the coefficient of friction of conditioner treated hair measured using an AFM tip (radius

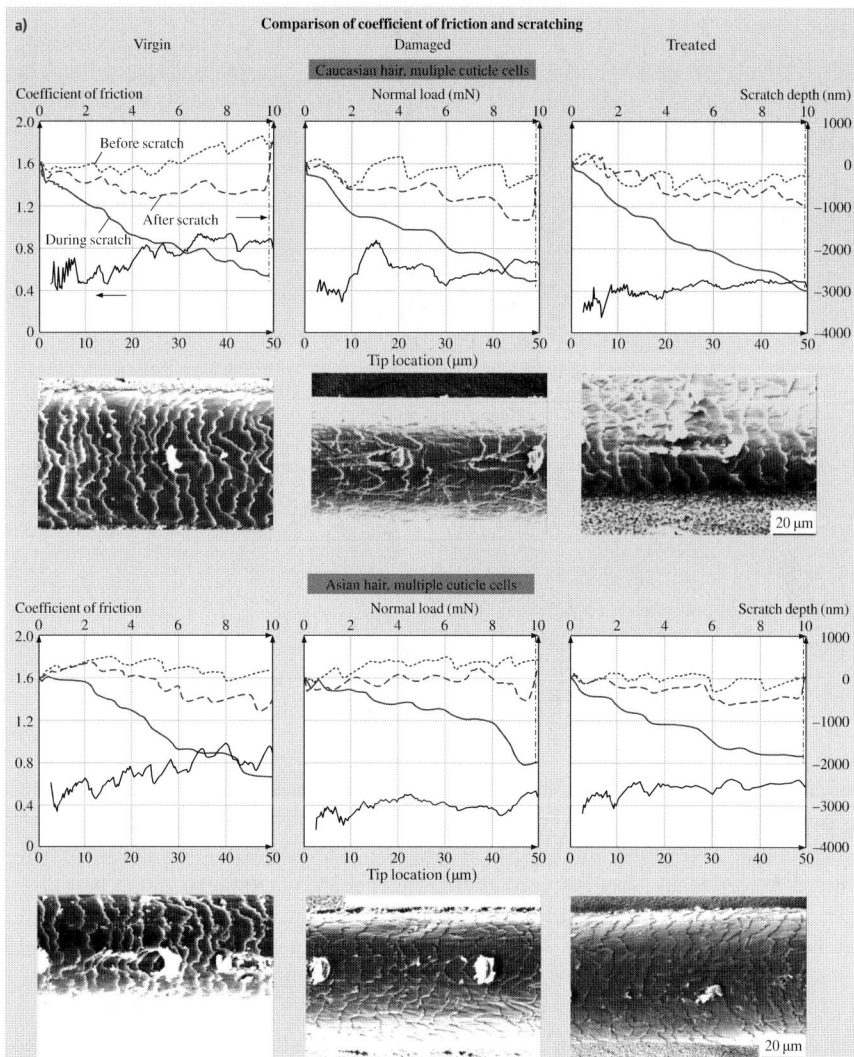

Fig. 24.39. (a) Coefficient of friction, scratch depth profiles and SEM images of Caucasian and Asian hair (virgin, chemo-mechanically damaged and virgin treated)

30–50 nm), is higher than virgin hair [45]. In the nano-scale, the increase in friction force is due in part to an increase in meniscus effects which increase the adhesive force contribution to friction at sites where conditioner is deposited or accumulated on the hair surface. This adhesive force is of the same magnitude as the normal load, which makes the adhesive force contribution to friction rather significant. On the micro-scale, however, the adhesive force is much lower in magnitude than the applied normal load, so the adhesive force contribution to friction is negligible over

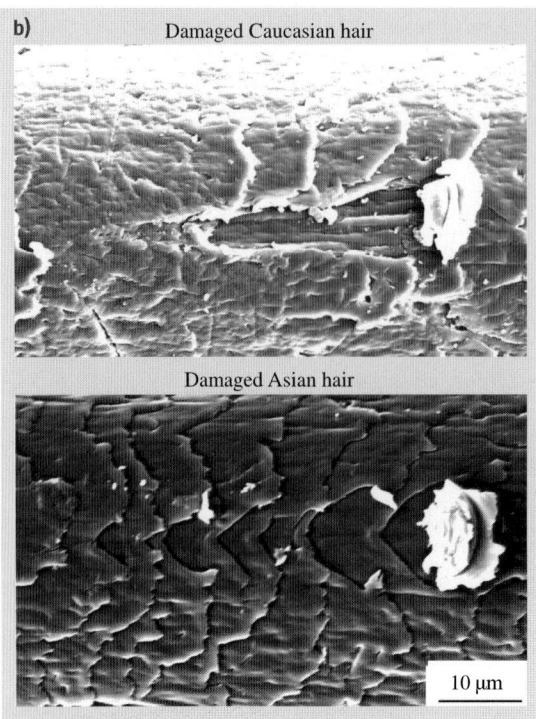

Fig. 24.39. (continued) (b) high magnification SEM images of damaged Caucasian and Asian hair after scratch [89]

the hair surface. On the micro-scale, the conditioner acts as a thin layer of lubricant, decreasing the friction.

Figure 24.39b shows the high magnification SEM images of chemo-mechanically damaged Caucasian hair and chemo-mechanically damaged Asian hair after scratch. It is interesting to find that the failure mechanisms for these two hair are different. For Caucasian hair, it seems that the tip plowed the cuticle cells continuously during scratching and the plowed cuticle cells were accumulated at the end of the scratch. As discussed before (also see Fig. 24.38), the exposed surface of the Caucasian hair is believed to be the cortex. For Asian hair, the tip did not plow the cuticle cells continuously. Instead, the tip only broke the top cuticle cell of each cuticle and carried away the broken cuticle cells until the end of the scratch. In this case, the tip did not reach the cortex during scratching. In order to explain this, we need to look at the nanomechanical properties of Caucasian and Asian hair. According to mechanical property data reported earlier [90], the cuticle of Asian hair has a higher hardness (0.39 ± 0.06 GPa) than Caucasian hair (0.24 ± 0.05 GPa). So the Asian hair may be more "brittle" than Caucasian hair during scratch. That may be the reason why Asian hair is fractured easier than Caucasian hair during scratch. However, it must be noted that our observation is based on limited number samples and experiments. Since human hair varies from one hair to another even in the same race, it is hard to draw a general conclusion on hair failure mechanism in terms of the hair race. What we can say is that the hair fails differently during scratching, depending on the nanomechanical properties of the cuticle of the hair.

Figure 24.40 shows the schematic of the various failure mechanisms during nanoscratching on hair. The top and middle figures show the scratch along cuticle, and the bottom diagram shows the scratch against cuticle. In the case of scratching along cuticle, if the hair cuticle is soft (top figure), then the scratch tip will plow the cuticle and carry away the worn cuticle cells (scales). If the load is high enough, then the tip can reach the cortex during scratching. After scratching, a newly exposed surface will be created and some pileup is formed at the end of the scratch wear track. If the hair cuticle is hard (middle figure), then the scratch tip will fracture the cuticle cells instead of plowing deep into them. After scratching, for each cuticle cell undergoing scratching, part of it is carried away by the tip, resulting in the formation of small pileup at the end of the scratch wear track, and leaving behind a series of incomplete cuticle cells. In the case of scratching against cuticle (bottom figure), the tip will plow the cuticle cells (whether soft or hard) and create a newly exposed surface with large wedge formation and pileup at the end of the scratch wear track.

In the studies reported earlier, nanomechanical measurements have also been performed on African hair [90]. In this study, the curly shape and structure made it very difficult to perform the nanoscratch on its surface.

Soaking Effect

Figure 24.41a,b compare the coefficient of friction and scratch depth profiles of unsoaked and soaked Caucasian obtained on single cuticle cell (at 1 mN) and multiple cuticle cells (10 mN loads), respectively. At 1 mN, (see Fig. 24.41a), the coefficient of friction of virgin and chemo-mechanically damaged Caucasian hair increased from ∼ 0.4 to ∼ 0.7 after soaking, while the coefficient of friction of virgin treated hair (∼ 0.3) does not change much. It is known that the human hair swells in water. In this work, the hair was only soaked in de-ionized water for 5 min. After the sample was soaked, it took a few minutes to mount the sample and run the scratch tests. In this case, it is possible that only a few hundred nanometer of the hair surface contained considerable amount water and were softened. During scratching, it is easier for the tip to plow into the softer hair surface, leading to higher coefficient of friction. This may be the reason that for 1 mN scratch in which the maximum in-situ scratch depths were less than 600 nm, the coefficient of friction of virgin and chemo-mechanically damaged hair increased. For virgin treated hair in the 1 mN scratch test, however, some of the conditioner molecules might occupy the pathways of water molecules so that the swelling of virgin treated hair was not as significant as virgin and chemo-mechanically damaged hair. Therefore, the virgin treated hair shows little change of coefficient of friction after soaking. At 10 mN, (see Fig. 24.41b), the coefficient of friction of all three hair does not change considerably after soaking. This may indicate that the 5 min soaking did not affect the hair surface deeply. Table 24.13 summarizes the coefficient of friction and scratch depths of Caucasian (unsoaked and soaked) and Asian (unsoaked) hair [89].

Table 24.13. Summary of coefficient of friction and scratch depths of Caucasian (unsoaked and soaked) and Asian (unsoaked) hair

Max. normal load/No. of cuticle cells	Unsoaked Caucaisan					
	1 mN/single cuticle cell			10 mN/multiple cuticle cells		
Hair condition	Virgin	Damaged	Treated	Virgin	Damaged	Treated
Average coefficient of friction	0.4	0.4	0.3	0.7	0.6	0.4
Max. in-situ depth (nm)	280	440	650	3250	2500	2750
Max. residual depth (nm)	25	100	150	1100	1000	750
	Soaked Caucasian					
Max. normal load/No. of cuticle cells	1 mN/single cuticle cell			10 mN/multiple cuticle cells		
Hair condition	Virgin	Damaged	Treated	Virgin	Damaged	Treated
Average coefficient of friction	0.7	0.7	0.3	0.8	0.6	0.5
Max. in-situ depth (nm)	570	540	500	3000	2100	2600
Max. residual depth (nm)	130	170	210	1000	500	800
	Unsoaked Asian					
Max. normal load/No. of cuticle cells	1 mN/single cuticle cell			10 mN/multiple cuticle cells		
Hair condition	Virgin	Damaged	Treated	Virgin	Damaged	Treated
Average coefficient of friction	0.5	0.6	0.3	0.8	0.4	0.5
Max. in-situ depth (nm)	480	500	400	2500	2400	1800
Max. residual depth (nm)	60	150	130	700	900	480

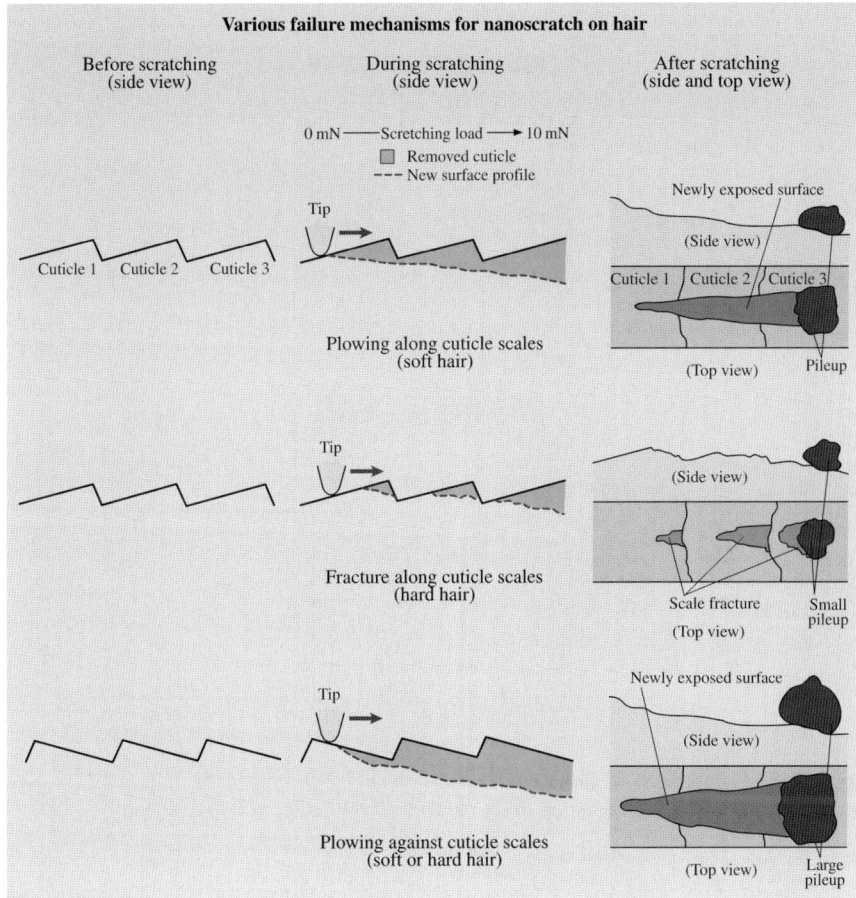

Fig. 24.40. Schematic of the various failure mechanisms during nanoscratching on hair [89]

24.5.3 In-Situ Tensile Deformation Studies on Human Hair Using AFM

Figure 24.42 presents stress strain curves for five types of hair [74]. The stress strain curve of human hair is similar to that of wool and other such keratinous fibers [95]. When a keratin fiber is stretched, the load elongation curve shows three distinct regions as marked in Fig. 24.42 [4, 33]. In the pre-yield region, also referred to as the Hookean region, stress and strain are proportional, and an elastic modulus can be found. In this region, there is the homogenous response of alpha keratin to stretching. The resistance is provided by hydrogen bonds that are present between turns and stabilize the alpha helix of keratin. The yield region represents transition of keratin from the alpha form to the beta form, the chains unfold without any resistance, and hence the stress does not vary with strain. The beta configuration again resists stretching. So, in the post-yield region, the stress again increases with strain until the

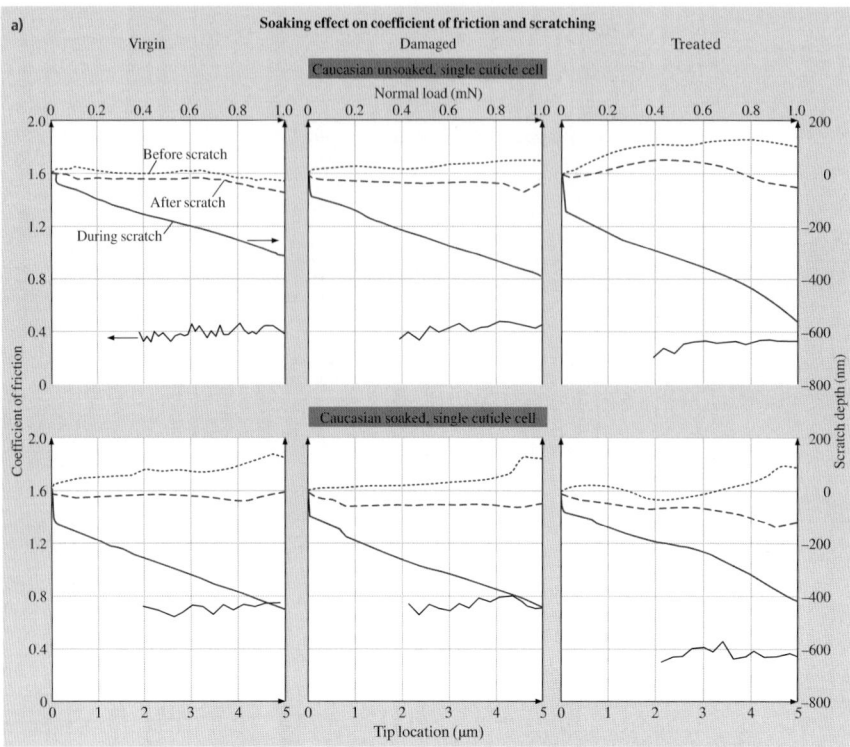

Fig. 24.41. Comparison of nanoscratch results on chemo-mechanically damaged Caucasian unsoaked and soaked hair samples (**a**) Scratch on single cuticle cell (1 mN load)

fiber breaks. This alpha to beta transition of keratin is the reason for the unique shape of the stress strain curve of hair. Typically, yield begins around 5% and post-yield begins at around 15% strain. There is no apparent difference in tensile properties of the different hair types. In the dry state, the tensile properties of hair are significantly contributed by the cortex. Hair fibers oxidized with diperisophthalic acid, which causes almost total removal of cuticle, have shown no significant change in mechanical properties [69]. In the case of conditioner treatment and mechanical damage, the change occurs only in the cuticle. Hence, it is logical that the mechanical properties do not experience measureable change. In the case of chemical damage, apart from cuticle damage, oxidation of the cystine in keratin to cysteic acid residues occurs, and hence disrupts the disulfide cross links. However, the disulfide bonding does not influence tensile properties of keratin fibers in the dry state, though some effect is seen in the wet state [68]. Since all experiments in the present study were carried out in the dry state, it is again logical that no change in the mechanical properties was observed.

Figure 24.43 shows AFM topographical images, and 2-D profiles at indicated planes, of a given control area with increasing strain, of the virgin, chemically dam-

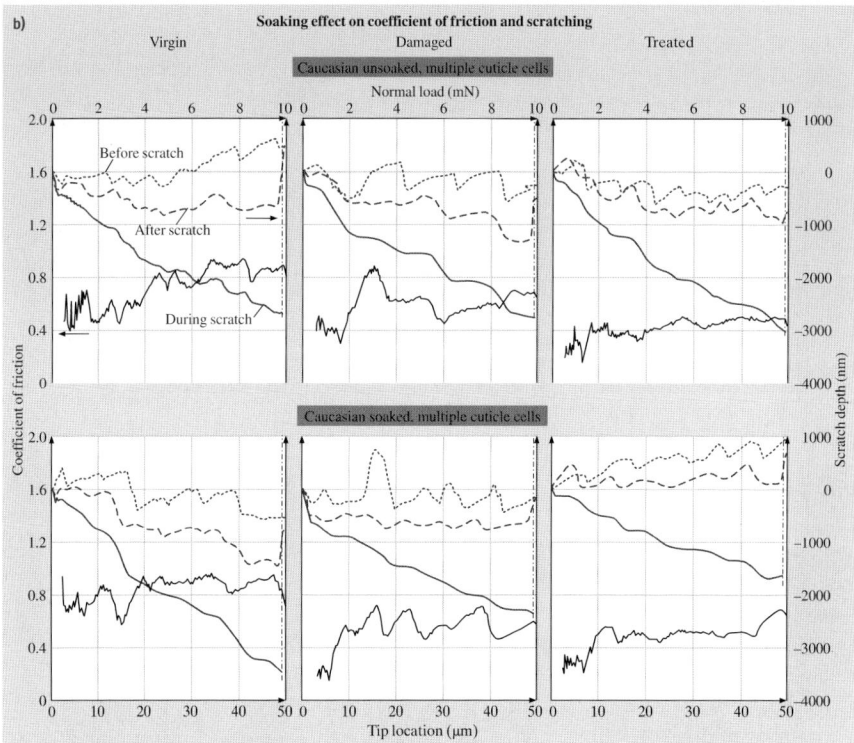

Fig. 24.41. (continued) **(b)** Scratch on multiple cuticle cells (10 mN load) [89]

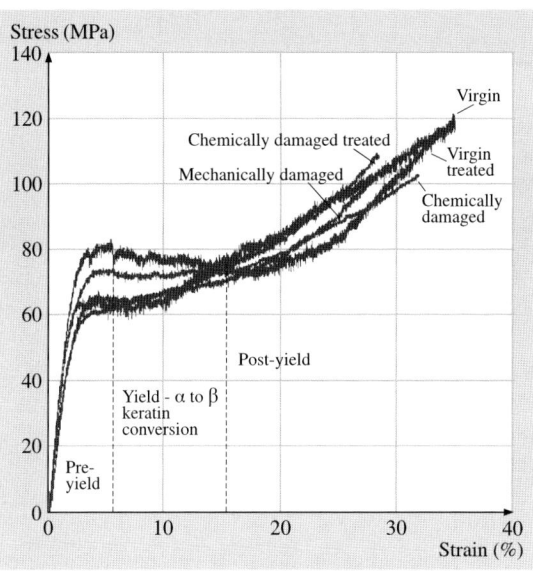

Fig. 24.42. Stress strain curves for five types of hair – Virgin, virgin treated, chemically damaged, chemically damaged treated, mechanically damaged. Transition of alpha keratin to beta keratin in the yield region is the reason for the unique shape of the curves [74]

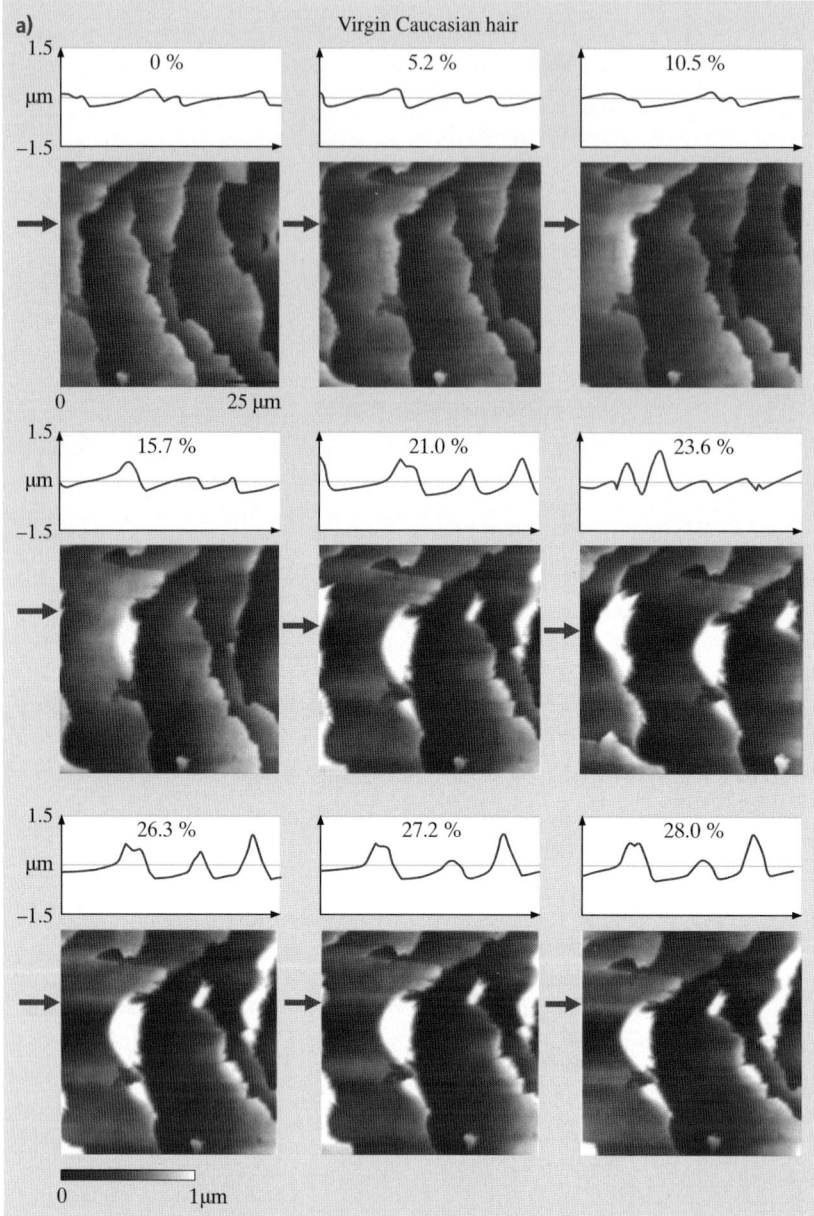

Fig. 24.43. AFM topographical images and 2-D profiles at indicated planes of a control area showing progress of damage with increasing strain in (**a**) virgin hair

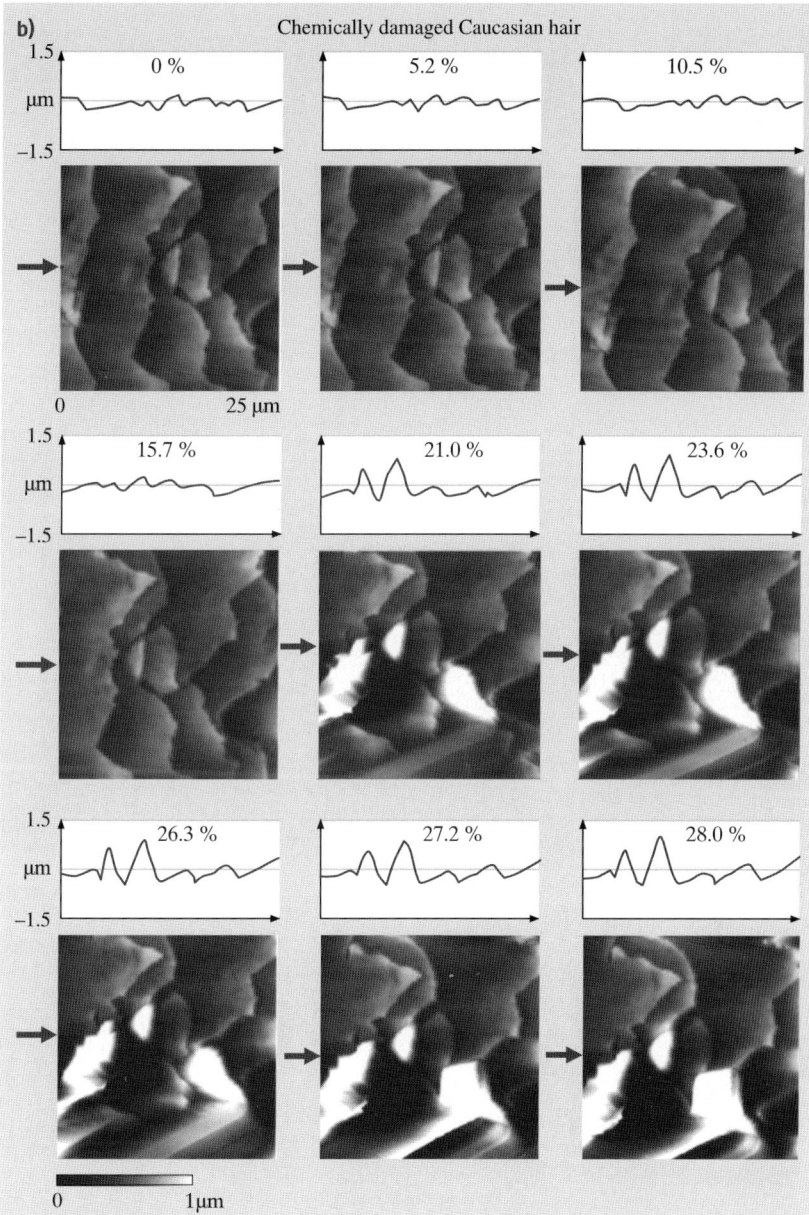

Fig. 24.43. (continued) **(b)** chemically damaged hair

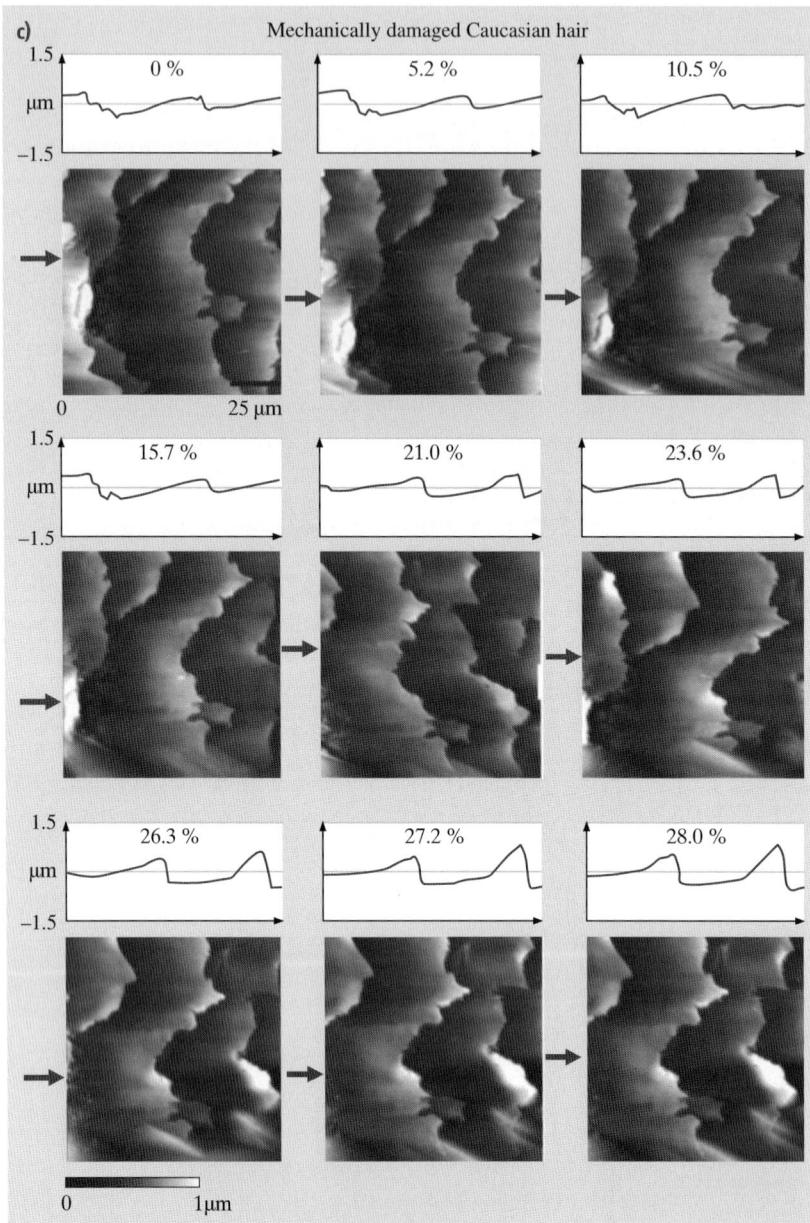

Fig. 24.43. (continued) (**c**) mechanically damaged hair. Cuticles lift off occurs at about 20% strain in all cases. Fracture of cuticle outer layer occur at about 20% strain in chemically damaged hair and about 10% strain in mechanically damaged hair [74]

aged, and mechanically damaged types of hair [74]. The common effect of stretching, observed for all hair types, is the lifting of the outer cuticle layer with increasing strain.

In the case of virgin hair (Fig. 24.43a), this is seen to be the only effect of tension. The cuticle lift off is sudden and occurs consistently at around 20% strain. As mentioned earlier, human hair cuticles have a lamellar structure. The various layers of human hair vary in their cystine content. This causes variation in their mechanical strength. The epicuticle and exocuticle are high in cystine content and are extremely rigid. The endocuticle and cell membrane complex are low in cystine content and are more extensible. Stretching hair, sets up interlayer shear forces due to this difference in extensibility [70]. At 20% strain delamination occurs, and the inner cuticle layers separate from the outer ones. This causes outer cuticle lift off and hence the change in height and slope observed in the AFM images and corresponding cross sectional profiles.

Figures 24.43b and c shows topographical images of chemically damaged hair and mechanically damaged hair, respectively. Apart from the lift off discussed earlier, another noticeable effect in both cases is the disappearance of some cuticle edges and the appearance of sponge like scales instead. Sponge like scales in the cuticle have been observed earlier, these are thought to be remains of cuticle cells, which have split up to the endocuticle [82]. Chemical and mechanical damage cause cuticle damage and weakening. Hence in the case of damaged hair, along with failure and delaminations in the endocuticular region, fracture of the outer cuticle occurs to expose the inner cuticle. This shows up as the sponge like scales observed.

Comparing mechanical damage and chemical damage, it is seen that the fracture occurs earlier in mechanically damaged hair (10%) than in chemically damaged hair (20%). This could be because, in mechanically damaged hair, parts of the cuticle have already broken away, and cuticle damage is more extensive than in chemically damaged hair.

Conditioner in the case of virgin and chemically damaged treated hair has no apparent effect on the lift off or fracture phenomenon (data not shown). Conditioner coats the cuticle and improves its tribological properties. However, it does not chemically or physically alter appreciably the cuticles.

24.5.4 Summary

Nanoindenter has been used to perform nanomechanical studies on human hair. The chemical damage and conditioner treatment caused the hardness and elastic modulus of hair surface to decrease within a depth of less than $1.5\,\mu\text{m}$. That is, the first 3–4 cuticle scales may interact with the chemicals and conditioner ingredients more effectively than the rest of the scales. It is found that the hair cuticle has higher hardness and elastic modulus than cortex in the lateral direction. The hardness and elastic modulus of hair decreased as the indentation depth increased. The cystine content variations in cuticle substructures (A-layer, exocuticle, endocuticle, cell membrane complex) and cortex are proposed to be responsible for the observation. Humidity and temperature had an effect on mechanical properties. The Young's modulus of damaged treated hair decreases dramatically at high humidity. Little temperature effect was observed on the Young's modulus.

Nanoscratch tests were performed on single and multiple cuticles of various hair, in both unsoaked and soaked conditions. The coefficient of friction of virgin treated hair is lower than virgin hair for Caucasian and Asian hair in both cases of single cuticle scratch and multiple cuticle scratch. This thin conditioner layer acts as a layer lubricant, reducing the coefficient of friction during scratching. In-situ displacement (30–200 nm) increased greatly at very low initial load and then increased gradually, indicating that the first approximately 200 nm of the hair surface should be softer than the underlying layer. The nanoscratch tests on multiple cuticles clearly show the directionality effect on the coefficient of friction. It is found that the hair surface fails differently during scratching, depending on the nanomechanical properties of the cuticle of the hair. For a hair with hard cuticle, the cuticle cells tend to be fractured during scratching. For a hair with soft cuticle, the scratch tip usually plows and wears away the cuticle cells continuously until it reaches the cortex. The effect of 5 min soaking in de-ionized water on coefficient of friction and scratch resistance of human hair is limited within a shallow region (about 600 nm deep) of the hair surface. In this case, the coefficient of friction of virgin and chemo-mechanically damaged Caucasian hair increases after soaking because of the swelling of the water, which softens the hair surface.

Human hair shows a stress strain curve typical of keratinous fibers. Transition of alpha keratin to beta keratin in the yield region is the reason for the unique shape of the curves. Chemical damage, mechanical damage and conditioner treatment have no obvious effect on the stress strain curve or tensile properties. This is because such treatments affect the cuticle predominantly and tensile properties of human hair in dry state are governed by the cortex. Tensile stress in general causes lift off of the outer cuticle. The lift off is sudden and occurs consistently at around 20% strain. This lift off occurs in all types of hair studied, and is due to interlayer shear forces and consequent separation between inner and outer cuticle layers at 20% strain. Chemical damage and mechanical damage cause weakening of the outer cuticle. Along with lift off, fracture of the outer cuticle to expose endocuticular layers occurs. Fracture occurs sooner (about 10% strain) in mechanically damaged hair than chemically damaged hair (about 20% strain).

24.6 Multi-Scale Tribological Characterization

24.6.1 Macroscale Tribological Characterization

Tribology is very important to hair care and product development. While current state of the art is to use AFM to measure nanoscale tribological properties of hair in contact with an AFM tip, macroscale tribological measurements provide an excellent simulation of skin-hair and hair-hair contacts [10]. The friction and wear of hair were measured using a flat-on-flat tribometer. Friction and wear studies on various hair are presented, including effect of load, velocity, and skin size. In addition, the effect of humidity and temperature on hair tribological properties is discussed.

Friction and Wear Studies of Various Hair

Figure 24.44a shows the coefficient of friction measured from hair strands sliding against a polyurethane film (simulated skin) [10]. The data shows that the coefficient of friction of virgin Caucasian hair was about 0.14 along cuticle and about 0.23 against cuticle. As with most animal fibers, human hair shows a directionality friction effect; that is, it is easier to move a surface over hair in a root-to-tip direction than in a tip-to-root direction because of anisotropic orientation of hair cuticles [7, 9, 68]. The data shows that the flat-on-flat tribometer can measure the directionality dependence of friction. Note that in Fig. 24.44a, the hair strands were used and all the hair was separated from each other, so during the friction test, there was no interaction between hair and hair. The output signals of normal force and friction force are smooth and the coefficient of friction has small variation.

In industry, many friction tests of hair are performed on bundle of hair, in which some hair is overlapping on each other. So the hair-hair interaction occurs during the friction test and variation in the data is large. Figure 24.44b shows the coefficient of friction obtained from bundle of hair. It can be seen that the output signal of normal force fluctuated a lot, and the output signal of friction force is not smooth, leading to big variation of the coefficient of friction. Both the coefficient of friction along cuticle and against cuticle are greater for bundle of hair compared to hair strands, because of the hair-hair interaction during friction tests. It has been observed that if a bundle of hair is used for friction test, the data has much poorer reproducibility than if hair strands are used. This may be because that when a bundle of hair is used, the hair is placed randomly and the hair to hair position is hard to repeat, but for hair strands, the hair to hair position is easy to control since they are separated and parallel to each other. Therefore, in the paper, hair strands were used to make further measurements.

Figure 24.45a shows the effect of normal load, sliding velocity and film area on coefficient of friction of hair. All the coefficient of friction values were obtained along cuticle. It shows that load and film area has no effect on hair friction, but a higher velocity leads to higher coefficient of friction. According to some observations, coefficient of friction is independent of normal load, apparent area of contact between the contacting bodies and velocity [7, 9]. Macroscale hair friction appears to obey the first two observations. The third rule of friction, which states that friction is independent of velocity, is not valid in the case of hair friction. Changes in the sliding velocity may result in a change in the shear rate, which can influence the mechanical properties of the hair and polyurethane film (shear strength, elastic modulus, yield strength and hardness) [6]. If the mechanical property change can lead to lower strength of hair surface and higher real area of contact, then the coefficient of friction will increase.

Figure 24.45b shows the coefficient of friction of polyurethane film vs. hair and hair vs. hair. In the case of polyurethane film vs. hair, the chemo-mechanically dam-

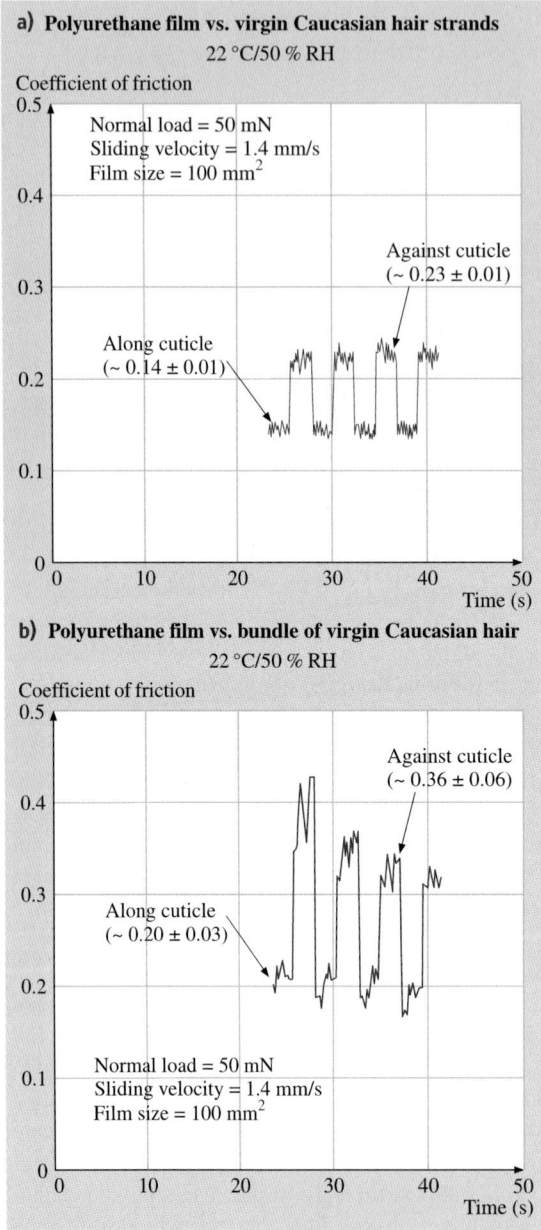

Fig. 24.44. Coefficient of friction measured **(a)** from hair strands, and **(b)** from bundle of hair sliding against polyurethane film [10]

aged hair has the highest coefficient of friction, followed by virgin and virgin treated hair, indicating that the conditioner can reduce the friction of hair. The hair vs. hair results show the same trend as polyurethane film vs. hair, but the coefficient of friction of hair vs. hair is higher than the corresponding polyurethane film vs. hair. Fig-

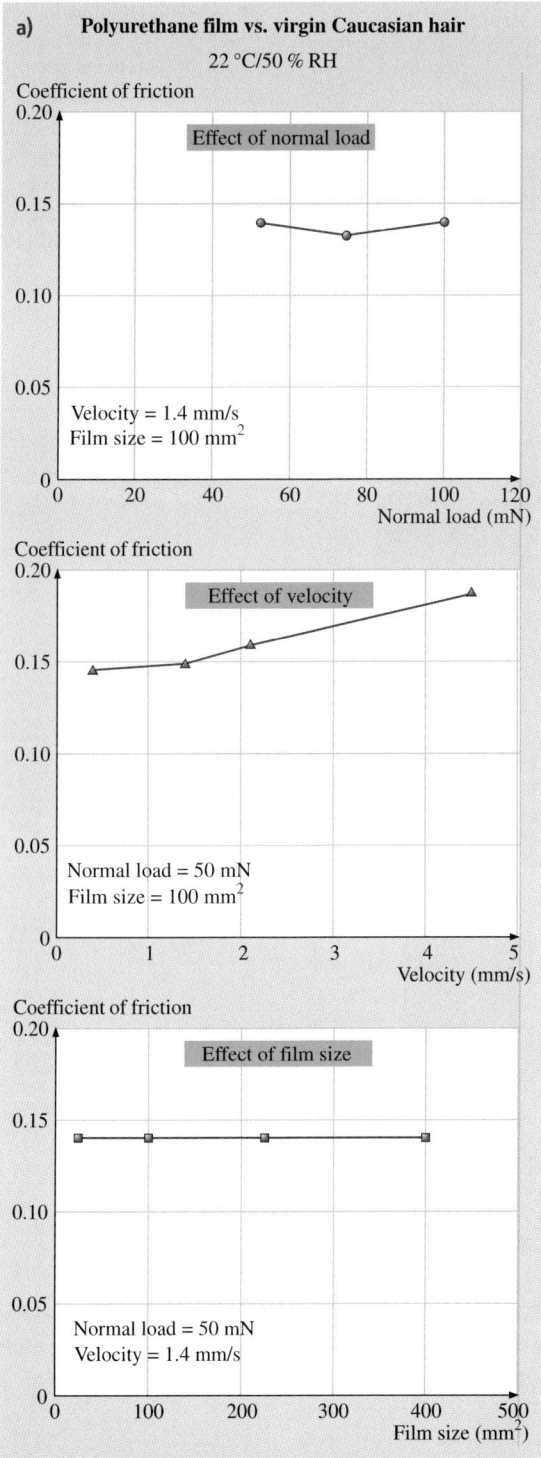

Fig. 24.45. (a) Effect of normal load, velocity and film size on coefficient of friction of virgin Caucasian hair

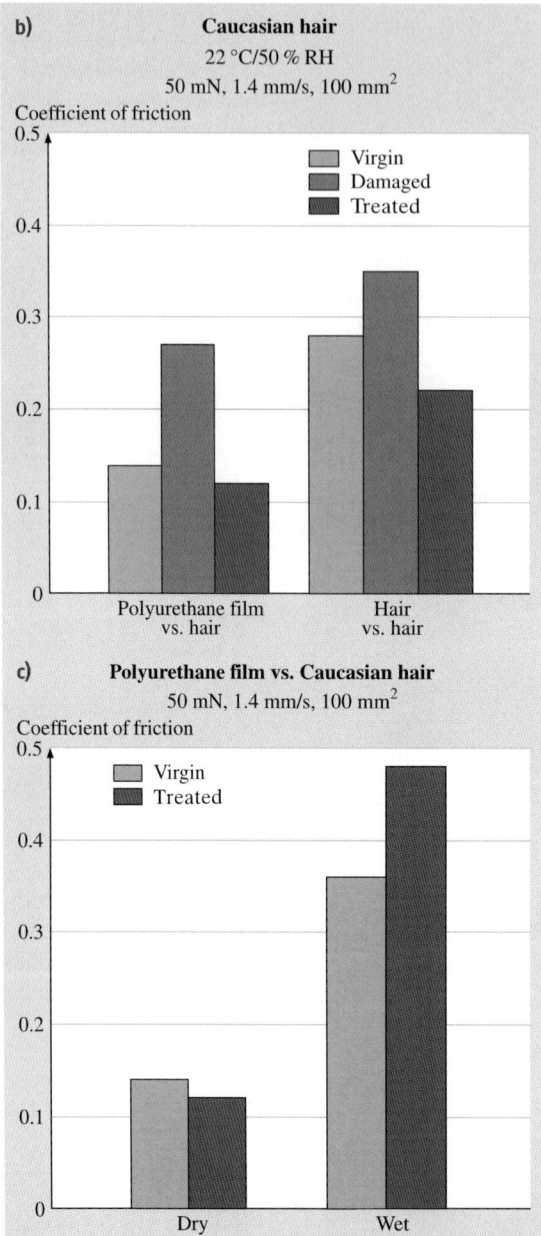

Fig. 24.45. (continued) **(b)** coefficient of friction of polyurethane film vs. Caucasian virgin and treated hair and hair vs. hair; **(c)** coefficient of friction of polyurethane film vs. Caucasian virgin and treated hair at dry and wet conditions [10]

ure 24.45c compares the coefficient of friction of polyurethane film vs. Caucasian hair at dry and wet conditions. Obviously, the coefficient of friction at wet conditions is higher than dry conditions, due to the swelling of hair. When the hair is

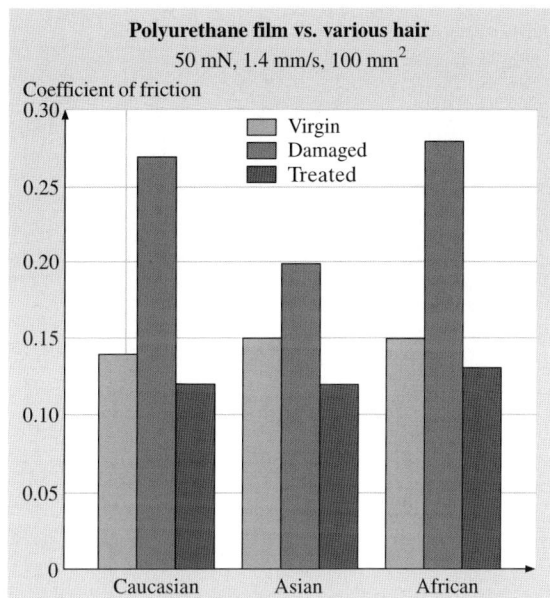

Fig. 24.46. Coefficient of friction of polyurethane film vs. various Caucasian virgin, chemo-mechanically damaged, and virgin treated hair [10]

swollen, the hair cuticle will be lifted up and the real contact area will be increased, leading to higher coefficient of friction.

Figure 24.46 shows the coefficient of friction of polyurethane vs. various hair. For all hair, the chemo-mechanically damaged hair has highest coefficient of friction, followed by virgin hair and virgin treated hair. Note that the coefficient of friction can vary about 10–15% within a given ethnic group. Figure 24.47a shows the coefficient of friction of Caucasian hair during wear tests. The coefficient of friction does not change during 24 hours for both virgin and virgin treated hair. From the optical micrographs in Fig. 24.47b, it can be seen that after wear tests, some cuticles were damaged. Polyurethane film is soft and does not create much damage to hair. Since the observed damaged area was small, it may not affect the overall coefficient of friction.

Effect of Temperature and Humidity on Hair Friction

Figure 24.48 shows the effect of temperature and humidity on hair friction. The coefficient of friction of hair is a strong function of humidity, and it increases as the relative humidity increases. In addition, the differential friction effect also increases with increasing relative humidity, as shown in Fig. 24.48. It is interesting to find that the differential friction effect of virgin treated hair is less dependent on relative humidity than virgin hair. This may be because some of the conditioner molecules occupied the pathways of water molecules so that the swelling of hair was not as significant as virgin hair.

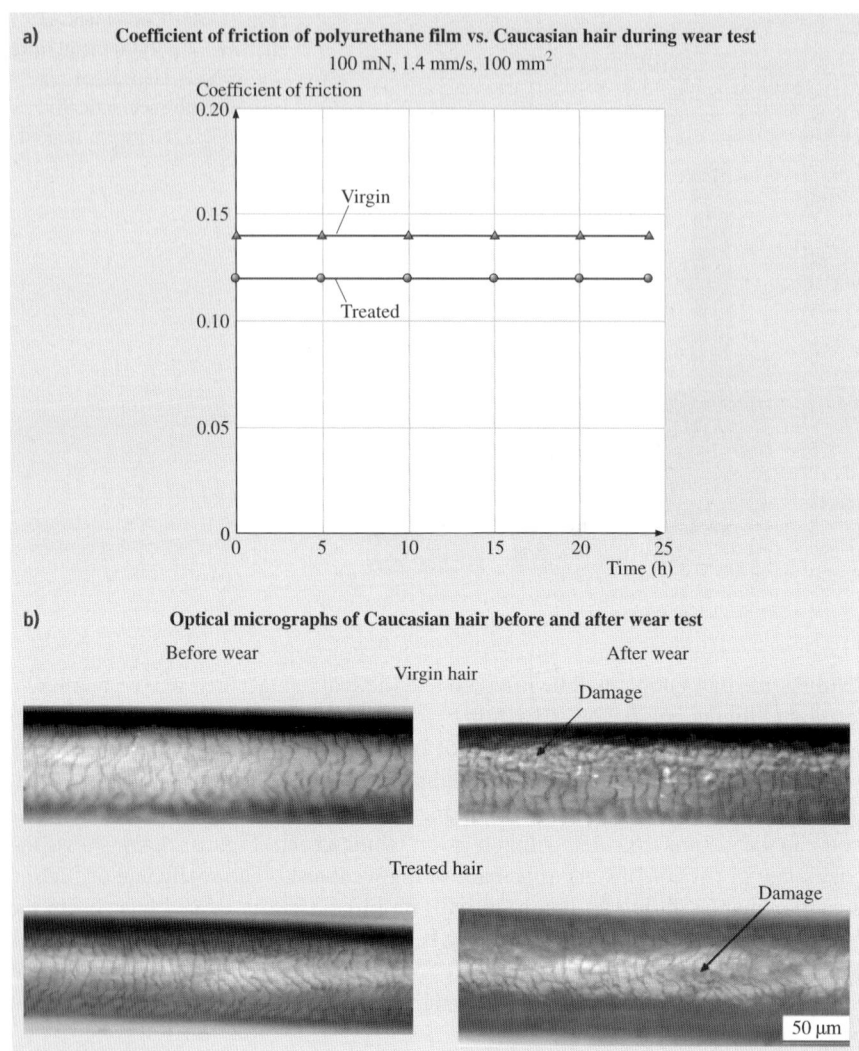

Fig. 24.47. (a) Coefficient of friction of virgin and virgin treated Caucasian hair during wear test, (b) optical micrographs of Caucasian hair before and after wear test [10]

For virgin hair, the temperature has no effect on coefficient of friction. This is in agreement with *Scott* and *Robbins*'s work [72]. For virgin treated hair, it is found that the coefficient of friction increases as the temperature increases. After the hair was treated with conditioner, the hair surface properties might be changed, which could be affected by temperature. For instance, a higher temperature might lead to softer treated hair surface, leading to higher coefficient of friction.

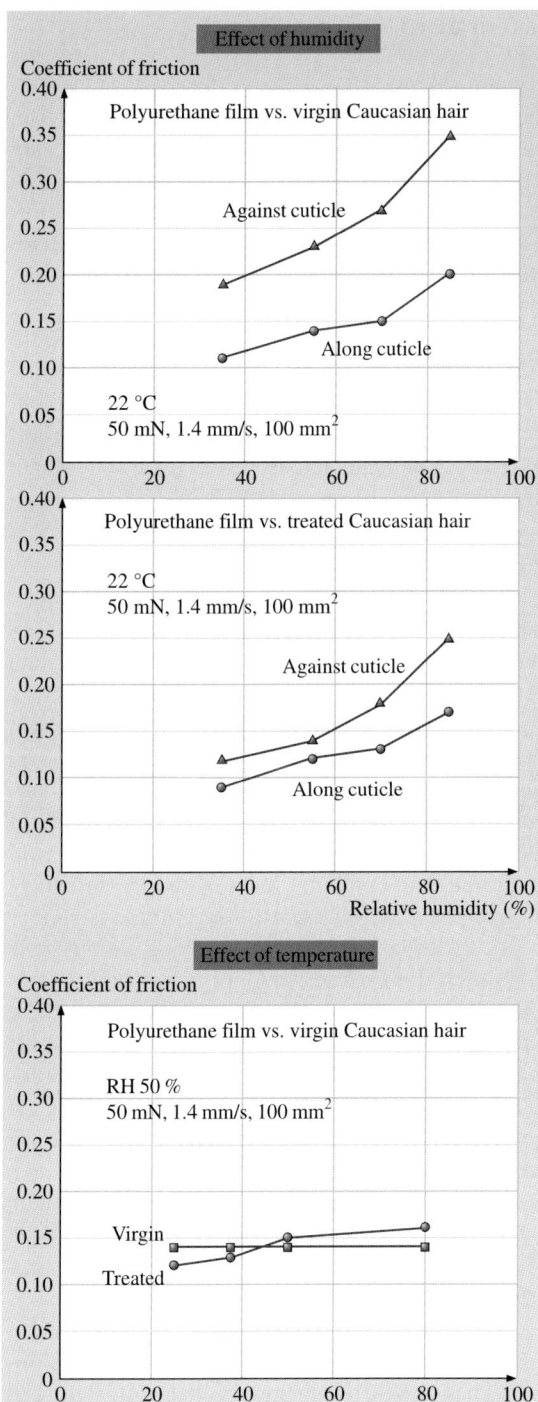

Fig. 24.48. Effect of humidity and temperature on coefficient of friction of virgin Caucasian hair [10]

24.6.2 Nanotribological Characterization Using an AFM

Nanoscale tribological characterization is essential to study the hair and evaluate/develop better cosmetic products. This becomes especially important when studying the effects of damage and conditioner treatment. How common hair care products, such as conditioner, deposit onto and change hair roughness, friction, and adhesion are of interest, since these properties are closely tied to product performance. Other important issues, such as the thickness distribution of conditioner on the hair surface which is important in determining the proper functions of conditioner, have been important to cosmetic scientists for decades.

In this section, the roughness, friction, adhesion, and wear of various hair are studied [45, 46, 48, 53]. In Sect. 24.7, a new method based on the AFM technique for determining thin liquid film thickness is developed, and the conditioner thickness distribution, adhesive force and the effective Young's modulus of various hair samples are presented [26, 27, 53]. The binding interactions between the conditioner molecules and hair surface are discussed as well.

Various Ethnicities

Topographical images of Caucasian, Asian, and African hair were taken with scan sizes of $20 \times 20\,\mu m^2$, as shown in Fig. 24.49. Lighter areas of the images correspond to higher topography, and darker areas correspond to lower topography. Only virgin and chemo-mechanically damaged hair are shown in Fig. 24.49 because virgin treated samples closely resemble virgin hair samples. One can see the variation in cuticle structure even in virgin hair. Cracking and miscellaneous damage at the cuticle edges is evident at both virgin and chemo-mechanically damaged conditions. In virgin hair, the damage is likely to be caused by mechanical damage resulting from daily activities such as washing, drying, and combing. Most of the virgin cuticle scales that were observed, however, were relatively intact. Long striations similar to scratches, and "scale edge ghosts" (outlines of a former overlying cuticle scale edge left on the underlying scale before it was broken away) were found on the surface. In some instances, the areas surrounding the cuticle edges appeared to show residue or debris on the surface, which is most likely remnants of a previous cuticle or the underside of the cuticle edges that are now exposed (such as the endocuticle). Caucasian and Asian virgin hair displayed similar surface structure, while the African hair samples showed more signs of endocuticular remains along the scale edges. One can also see more curvature in the cuticle scales of African hair, which is attributed to its elliptical cross-sectional shape and curliness, which can partially uplift the scales in different places. With respect to chemo-mechanically damaged hair, it is observed that several regions seem to exist in these hair samples, ranging from intact cuticle scales to high levels of wear on the surface. In many cases these regions occur side by side. This wide variation in chemo-mechanically damaged cuticle structure results in a wider range of tribological properties on the micro/nanoscale for these fibers. Caucasian and Asian chemo-mechanically damaged hair showed more worn away cuticle scales than in chemo-mechanically damaged

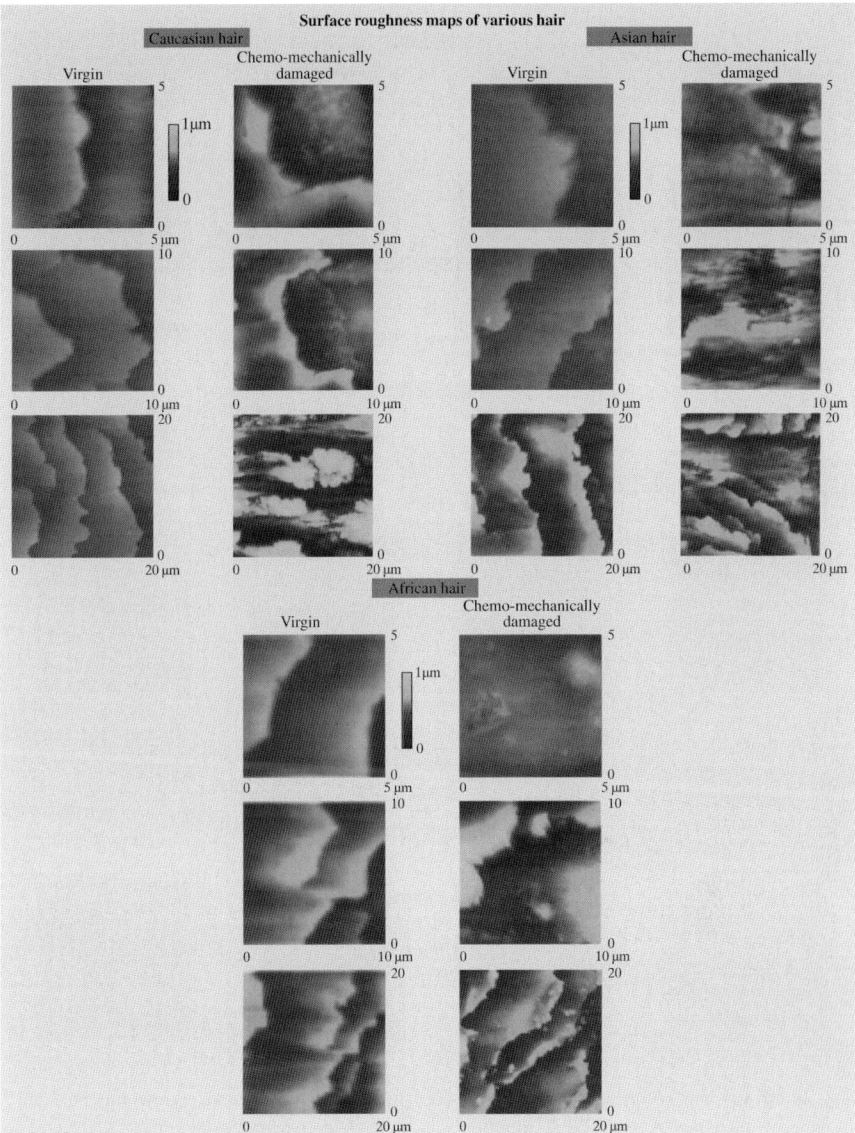

Fig. 24.49. Surface roughness of virgin and chemo-mechanically damaged Caucasian, Asian, and African hair at 5, 10, and 20 μm² scan sizes [45]

African hair, which showed mostly endocuticle remnants. This is most likely due to the different effects that chemical straightening has on the hair versus multiple cycles of perming the hair.

A more focused look into roughness and friction on the cuticle surface can be found by comparing Caucasian, Asian, and African virgin and chemo-mechanically

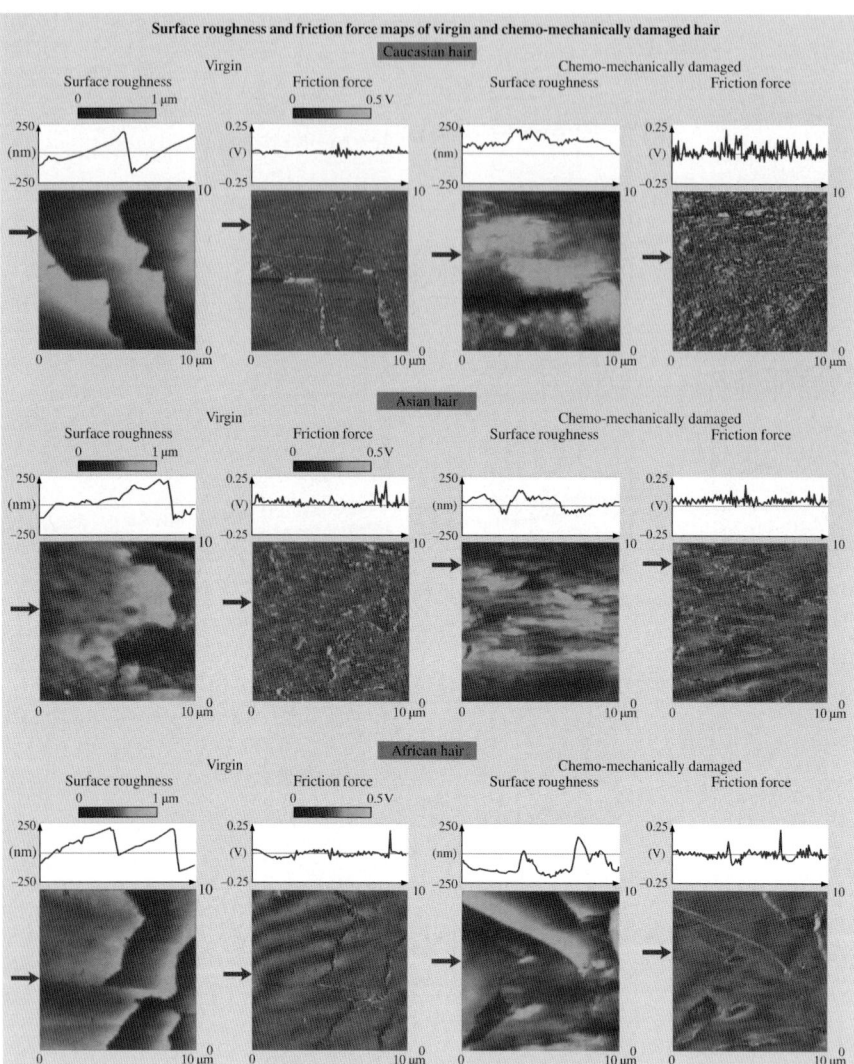

Fig. 24.50. (a) Surface roughness and friction images for virgin and chemo-mechanically damaged Caucasian, Asian, and African hair at 10 µm² scan sizes. Shown above each image is a cross section taken at the corresponding arrows to show roughness and friction force information [45]

damaged hair, Fig. 24.50, and virgin treated hair, Fig. 24.51. Virgin hair was used as the baseline to compare variations in roughness and friction force against modified hair (chemo-mechanically damaged or virgin treated). Scan size of $10 \times 10\,\mu m^2$ is displayed. Above each AFM and FFM image are cross-sectional plots of the surface (taken at the accompanying arrows) corresponding to surface roughness or friction

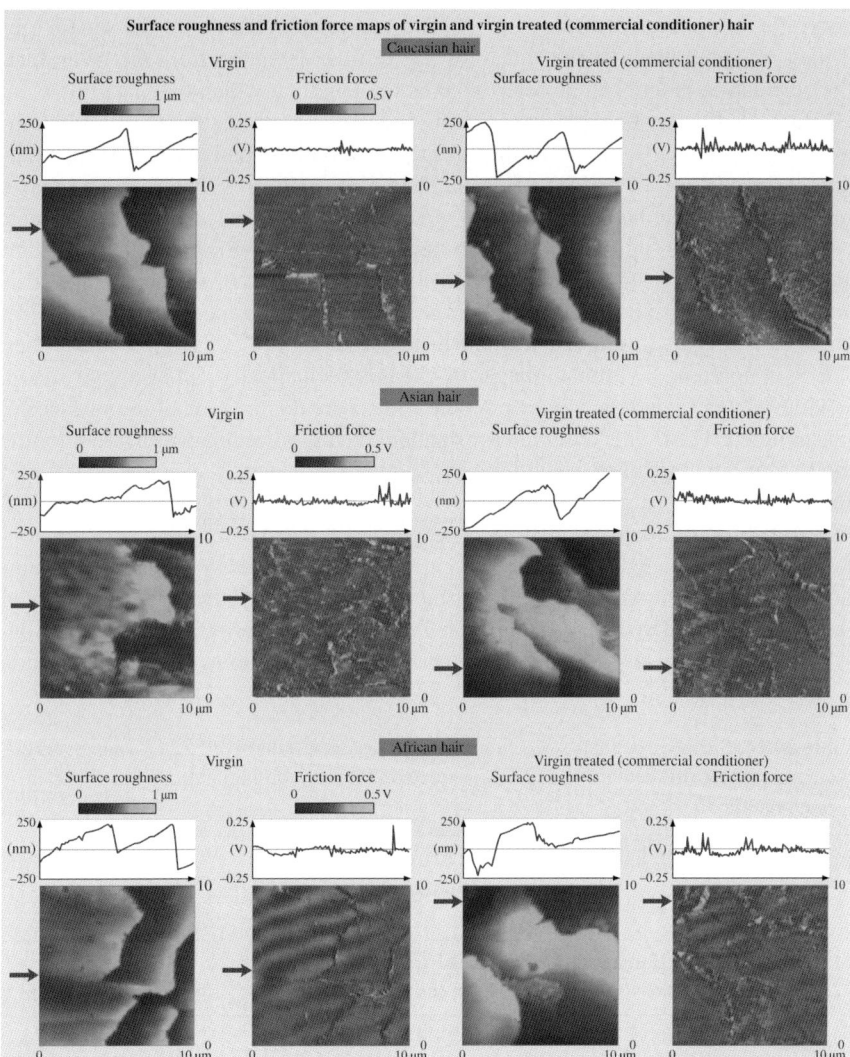

Fig. 24.51. (a) Surface roughness and friction images for virgin and virgin treated Caucasian, Asian, and African hair at 10 μm² scan sizes. Shown above each image is a cross section taken at the corresponding arrows to show roughness and friction force information [45]

force, respectively. From the surface roughness images, the step heights of one or more cuticle edges can be clearly seen. Step heights range from approximately 0.3 to 0.5 μm.

If the surface is assumed to have Gaussian height distribution and an exponential autocorrelation function, then the surface can be statistically characterized by just two parameters: a vertical descriptor, height standard deviation σ, and a spatial descriptor, correlation distance β^* [7,9]. The standard deviation σ is the square root of the arithmetic mean of the square of the vertical deviation from the mean line. The correlation length can be referred to as the length at which two data points on a surface profile can be regarded as being independent, thus serving as a randomness measure [7,9]. Table 24.14 displays these roughness parameters for each ethnicity as a function of hair type (virgin, chemo-mechanically damaged, and virgin treated) [45]. Virgin hair was shown to generally have the lowest roughness values, with virgin treated hair closely resembling virgin hair. Chemo-mechanically damaged hair showed a significantly higher standard deviation of surface height. This variation is expected because of the non-uniformity of the mechanical and chemical damage that occurs throughout a whole head of hair as well as each individual fiber. This is in agreement with the images of chemo-mechanically damaged hair shown previously, where regions of intact cuticle and severe degradation of the surface are intermingled. The trends observed for standard deviation were not as evident for the correlation length β^*. For each ethnicity, chemo-mechanically damaged and virgin treated hair showed similar β^* values.

From Fig. 24.50, friction forces are generally seen to be higher on chemo-mechanically damaged hair than on virgin hair. Although friction forces were similar in magnitude, it was observed that the friction force on the cuticle surface of chemo-mechanically damaged hair showed a much larger variance, which contributed to the higher friction values. Another contribution to the higher friction

Table 24.14. Surface roughness, coefficient of friction, and adhesive force values of virgin, chemo-mechanically damaged, and virgin treated (1 cycle commercial conditioner) hair at each ethnicity

	Virgin hair		Chemo-mechanically damaged hair		Virgin treated hair (commercial conditioner)	
Surface roughness parameters (σ (nm), β^* (µm))						
	σ (nm)	β^* (µm)	σ (nm)	β^* (µm)	σ (nm)	β^* (µm)
Caucasian	12 ± 8	0.61 ± 0.3	17 ± 10	1.0 ± 0.3	12 ± 4	0.90 ± 0.3
Asian	9.7 ± 4	0.73 ± 0.3	33 ± 15	0.94 ± 0.3	7.1 ± 0.1	0.97 ± 0.3
African	12 ± 5	0.92 ± 0.3	21 ± 16	0.78 ± 0.3	11 ± 4	0.89 ± 0.2
Average coefficient of friction μ						
Caucasian	0.02 ± 0.01		0.13 ± 0.05		0.03 ± 0.01	
Asian	0.03 ± 0.01		0.13 ± 0.04		0.06 ± 0.04	
African	0.04 ± 0.02		0.14 ± 0.08		0.05 ± 0.01	
Adhesive force F_m (nN)						
Caucasian	25		16		32	
Asian	31		18		79	
African	35		38		63	

could be that the tiny peaks developed after damage also create a ratchet mechanism on a nanoscale, which affects the friction between the AFM tip and the surface. These peaks could then add to the friction signal. The damage of the hair by chemical and mechanical means have shown high reproducibility in the lab in terms of structure alteration, which explains the similar friction properties no matter the ethnicity for chemo-mechanically damaged hair. With virgin and virgin treated hair, however, it is unknown what prior mechanical damage and sun exposure the fibers have seen, and varies largely on the individuals. Thus, across ethnicity there is variability in friction force for those hair.

Perhaps the most notable difference between virgin and virgin treated hair fibers can be seen in the friction force mappings of Fig. 24.51. Although quite comparable in surface roughness, close examination of the virgin treated hair surface shows an increase in friction force, usually only surrounding the bottom edge of the cuticle. This was unlike virgin hair, where friction generally remained constant along the surface, and unlike chemo-mechanically damaged hair, where there was large variability which was random over the entire surface.

Figure 24.52 presents friction force curves as a function of normal load for Caucasian virgin, chemo-mechanically damaged, and virgin treated hair to further illustrate the previous discussion. One can see a relatively linear relationship between the data points for each type of hair sample. When plotted in such a way, the coefficient of friction is determined by the slope of the least squares fit line through the data. If this line is extended to intercept the horizontal axis, then a value for adhesive force can also be calculated, since friction force F is governed by the relationship

$$F = \mu(W + F_m) \qquad (24.3)$$

where μ is the coefficient of friction, W is the applied normal load, and F_m is the adhesive force [7,9].

One explanation for the increase in friction force of virgin treated hair on the micro/nanoscale is that during tip contact meniscus forces between the tip and the conditioner/cuticle become large as the tip rasters over the surface, causing an increase in the adhesive force. This adhesive force is of the same magnitude as the normal load, which makes the adhesive force contribution to friction rather significant. Thus, at sites where conditioner is accumulated on the surface (namely around the cuticle scale edges), friction force actually increases. On the macroscale, however, the adhesive force is much lower in magnitude than the applied normal load, so the adhesive force contribution to friction is negligible over the hair swatch. As a result, virgin treated hair shows a decrease in friction force on the macroscale, which is opposite the micro/nanoscale trend. The friction and adhesion data on the micro/nanoscale is useful, though, because it relates to the presence of conditioner on the cuticle surface and allows for obtaining an estimate of conditioner distribution.

It is also observed from Fig. 24.52 that while chemo-mechanically damaged hair displays a higher friction force on the application of normal load, and consequently a higher coefficient of friction, chemo-mechanically damaged hair friction

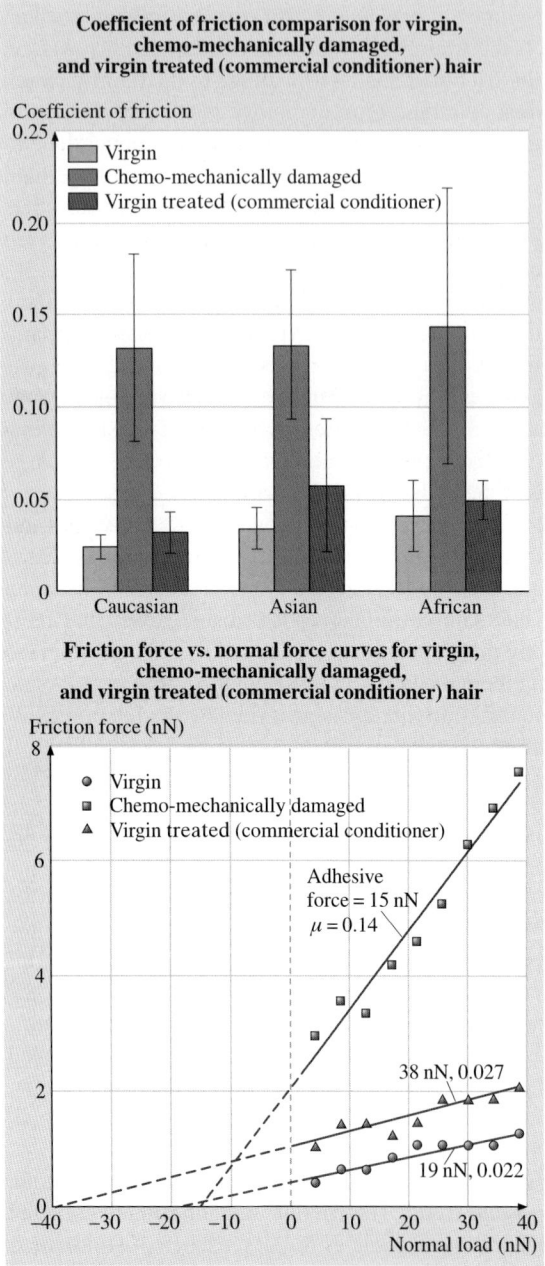

Fig. 24.52. Average coefficient of friction values for virgin, chemo-mechanically damaged, and virgin treated hair of each ethnicity. Error bars represent ±1 σ on the average coefficient value. In the bottom plot, friction force vs. normal load curves showing typical values for virgin, chemo-mechanically damaged, and virgin treated Caucasian hair. While the coefficient to friction is similar in virgin and treated hair, the adhesive force of treated hair is higher than in virgin hair [45]

is not as strongly dependent on adhesive force contribution as the virgin and virgin treated hair. Average values for μ were calculated and are compiled in Figs. 24.52 and 24.53 and Table 24.14 for all hair ethnicities and types [45]. Error bars represent ±1 σ on the average coefficient value. The coefficients of friction for virgin, chemo-mechanically damaged, and virgin treated Caucasian hair are 0.02, 0.13, and 0.03, respectively. For virgin, chemo-mechanically damaged, and virgin treated Asian hair, coefficients of friction are 0.03, 0.13, and 0.06, respectively. Finally, virgin, chemo-mechanically damaged, and virgin treated African hair coefficients of friction are 0.04, 0.14, and 0.05, respectively. Chemo-mechanically damaged hair presents the highest coefficient of friction, but also displays the largest standard deviation, due to the large variations in chemical and mechanical damage that each hair or hair bundle experiences. Coefficient of friction of virgin treated hair is slightly larger than that of virgin hair for all ethnicities. While the coefficient of friction is similar in virgin and virgin treated hair, the adhesive force contribution to friction for Caucasian virgin treated hair is higher than in Caucasian virgin hair, when calculated according to the method described above. However, this was not always the trend for Asian and African virgin treated hair samples. It should be noted that since in friction force measurement the tip moves laterally over the surface, this might cause a smearing out of the conditioner layer which accounts for the inconsistent trend. In the adhesive force mappings described in the next section, where determination of adhesive force does not depend on this lateral movement, all virgin treated hair samples showed higher adhesion than their virgin hair counterparts.

A force calibration plot (FCP) technique and resulting adhesive force maps (commonly called force-volume maps) can be used to understand the adhesive forces between the AFM tip and the sample [8, 10, 14, 52]. Shown in Fig. 24.54 are FV maps and an example of the individual force calibration plots from which the maps were created. Adhesive force distribution for chemo-mechanically damaged hair was shown to be comparable to virgin hair adhesive force values, but slightly lower. A significant increase in adhesive force over the entire mapping was found in all cases of virgin treated hair as compared to virgin hair, especially in Asian and African hair. Conditioner distribution can be seen from these images. This technique shows promise to be very useful in further study of the distribution of materials and hair care products on the surface of the hair.

A typical value for the adhesive force of each FV map was calculated. Values are shown in the plot of Fig. 24.54, along with surface roughness and coefficient of friction data for all hair samples. Adhesive force values are also tabulated in Table 24.14.

Directionality Effects of Friction on the Micro/Nanoscale

The outer surface of human hair is composed of numerous cuticle scales running along the fiber axis, generally stacked on top of each other. As previously discussed, the heights of these step changes are approximately 300 nm. These large changes in topography make the cuticle an ideal surface for investigating the directionality effects of friction using AFM/FFM.

The directionality effect of friction on the macroscale has been well studied in the past. It was shown by *Robbins* [68] and *Bhushan* et al. [21] that rubbing the hair from the tip to the root (against the cuticle steps) results in a higher coeffi-

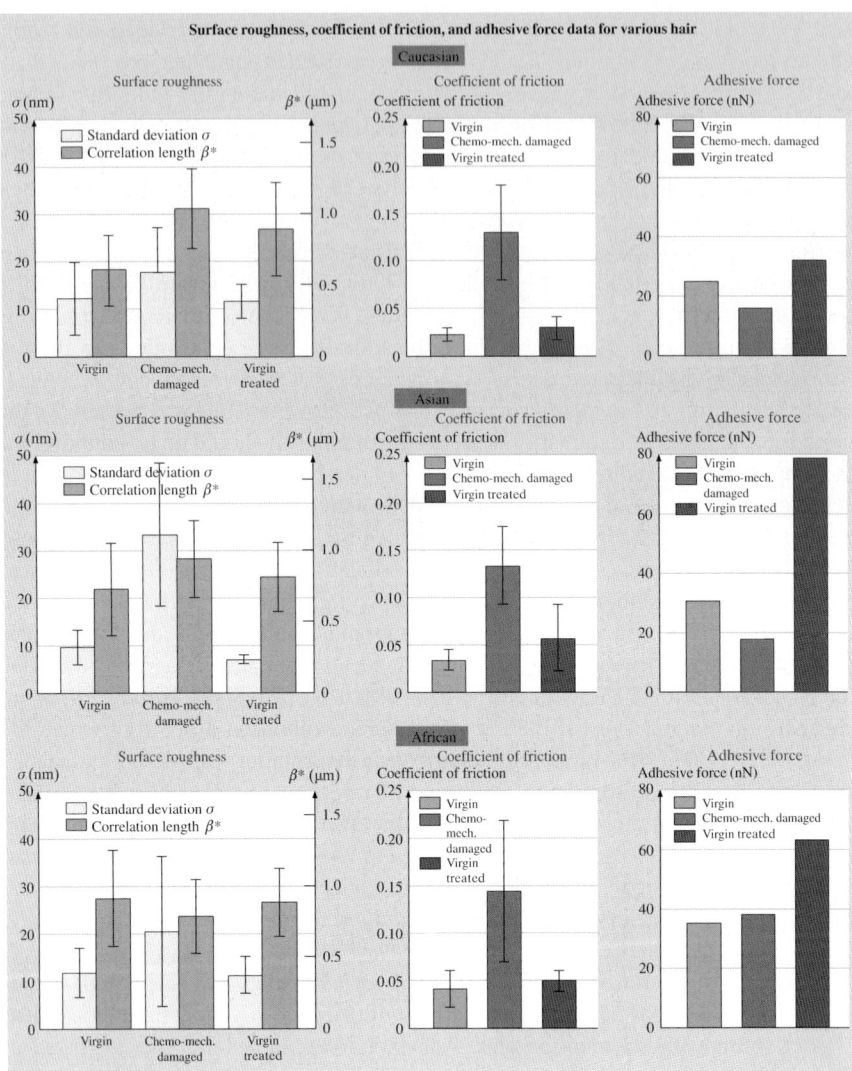

Fig. 24.53. Surface roughness, coefficient of friction, and adhesive force data for virgin, chemo-mechanically damaged, and virgin treated hair at each ethnicity [45]

Fig. 24.54. Force volume maps of virgin, chemo-mechanically damaged, and virgin treated hair at each ethnicity. Examples of the individual force calibration plots, which make up the FV maps, are presented for Caucasian hair of each type [45]

24 Structural, Nanomechanical, and Nanotribological Characterization

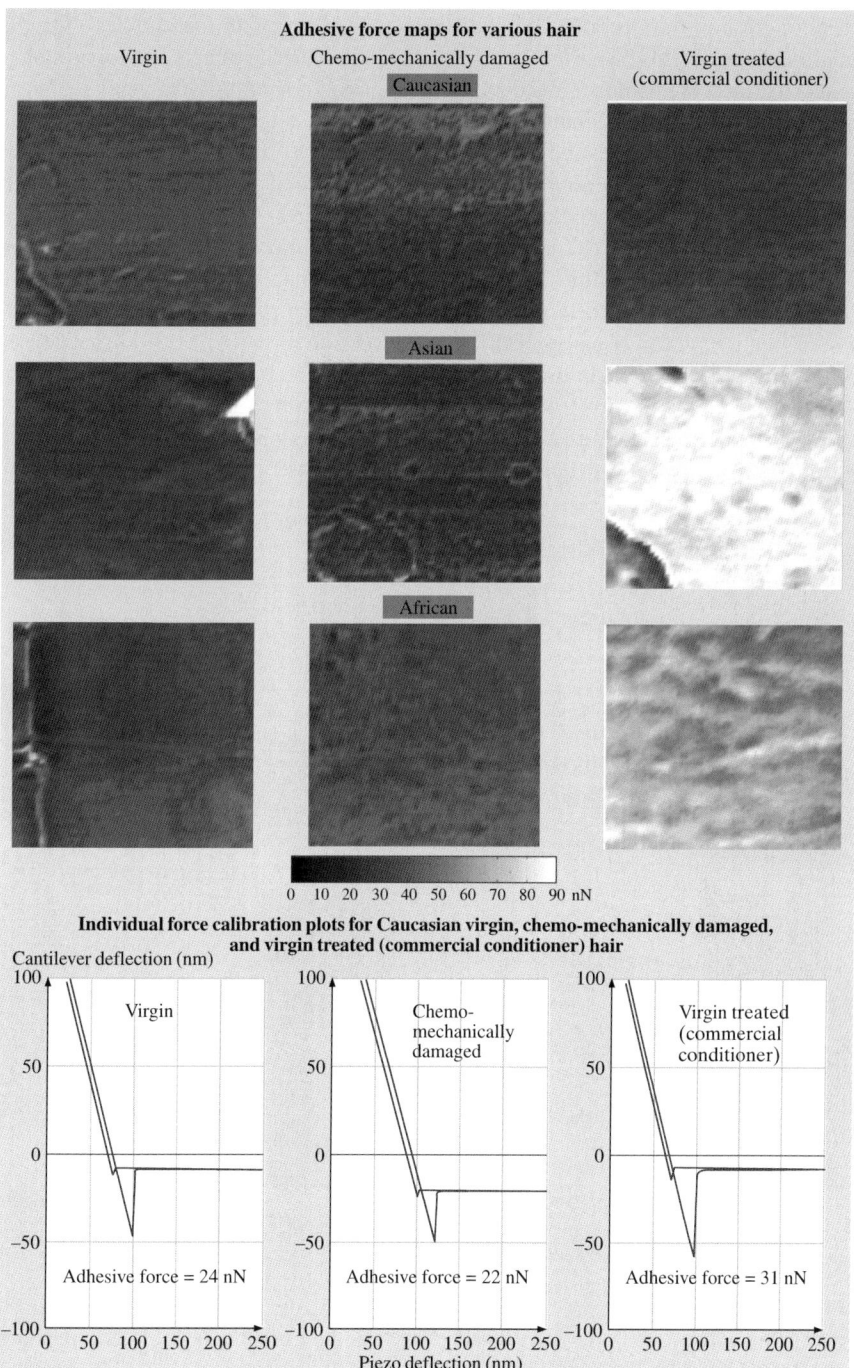

cient of friction than rubbing the hair from root to tip (with the cuticles). On the micro/nanoscale, it is important to distinguish how material effects and topography-induced effects contribute to the directionality effect of friction force when scanning over and back across a small surface region [7–10]. Figure 24.55 shows surface roughness, friction force, and surface slope maps of a Caucasian virgin hair fiber, each coupled to their accompanying 2-D cross-sectional profiles. The scan size of $5 \times 5\,\mu m^2$ provides one cuticle step height to be studied. As the tip rasters over the step in the trace mode (i.e. from left to right), a small decrease in the friction force is observed as the tip follows the step downward. When the tip comes back in the retrace mode (i.e. right to left), climbing up the sharp peak results in a high friction signal. However, because of the sign convention of the AFM/FFM that causes a reversal in the sign when traversing the opposite direction, this signal is now observed to be highly negative. The interesting difference between the two profiles lies in the fact that the magnitude of the decrease in friction when going up the step is much larger than the magnitude of the friction when the tip is going down the

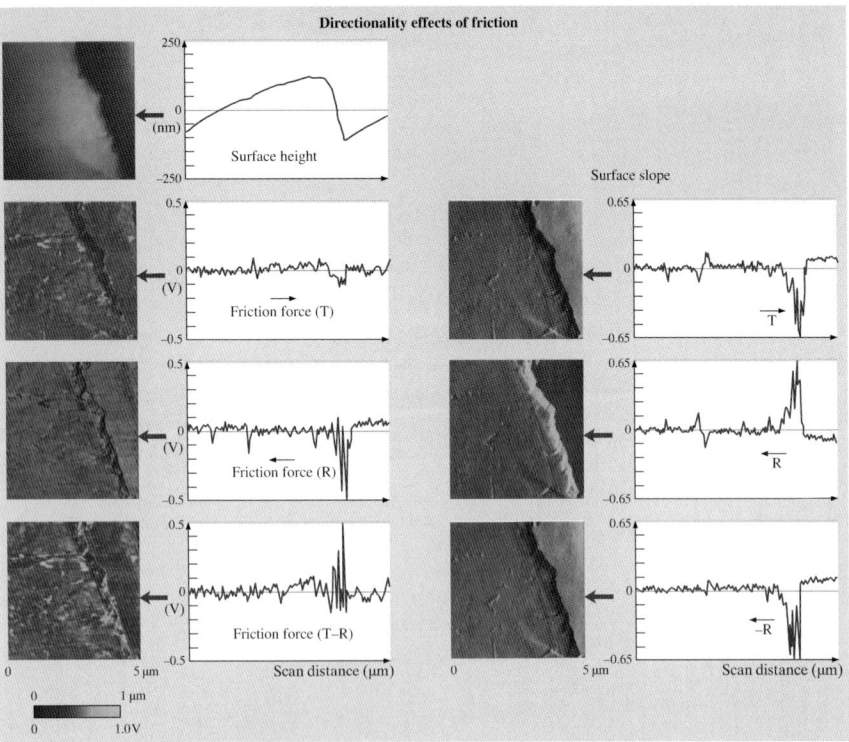

Fig. 24.55. Surface roughness, friction force, and slope across a cuticle scale edge of a Caucasian virgin hair to show the directionality dependence of friction. *Left*: Surface roughness and friction force mappings with accompanying 2-D profiles. *Right*: Surface slope mappings with accompanying 2-D profiles [45]

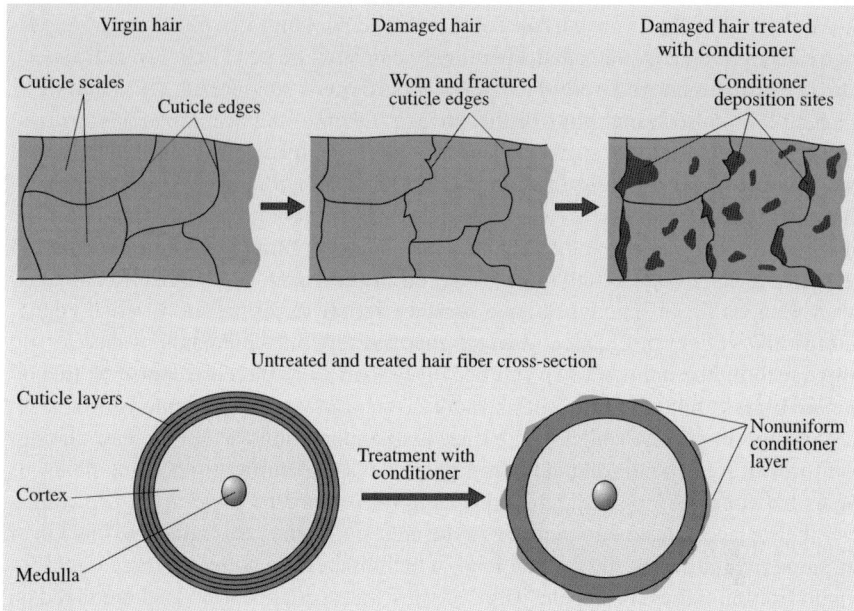

Fig. 24.56. Schematic of the effect of damage to the cuticle scales and the deposition of conditioner on the cuticle surface. The cross-section of the hair with and without conditioner is shown below [48]

step, yet both signals are in the same direction. The important result is that even by subtracting the two signals (T-R), there is still a gross variation in the image due to topography effects. These topography effects yield friction variations in the same direction, whereas material effects show up in opposite directions. It can be shown that the cuticle edge provides a local ratchet and collision mechanism that increases the friction signal at that point. It was concluded that surface slope variation always correlates to friction force variations with respect to topography effects, and the data presented in Fig. 24.55 shows the same trend when comparing trace and negative retrace slope profiles.

Virgin and Chemically Damaged Caucasian Hair (With and Without Commercial Conditioner Treatment)

The hair surface is negatively charged and can be damaged by a variety of chemical (permanent hair waving, chemical relaxation, coloring, bleaching) and mechanical (combing, blowdrying) factors [24, 36, 68]. Figure 24.56 shows the transformation and wear of the cuticles scales before and after damage. Chemical damage causes parts of the scales to fracture and reveal underlying cuticle remnants. Conditioner application provides a protective coating to the hair surface for prevention of future damage.

Shown in Fig. 24.57 are surface roughness and friction force plots for virgin, virgin treated, chemically damaged, chemically damaged treated (1 cycle conditioner), and chemically damaged treated (3 cycles conditioner). Above each AFM and FFM image are cross-sectional plots of the surface (taken at the accompanying arrows) corresponding to surface roughness and friction force, respectively. Although virgin and virgin treated hair are quite comparable in surface roughness maps, examination of the treated hair surface shows an increase in friction force, especially in the area surrounding the scale edge bottom level. These frictional patterns observed in treated hair were not like anything observed in the virgin or chemically damaged cases. Images of all hair types have shown friction variation due to edge contributions and cuticle mechanical damage that has left only remnants of cuticle sub layer (such as the endocuticle). Further investigation of the corresponding treated hair roughness images showed this increase in friction was not due to a significant change in surface roughness, either. One explanation for the increase in friction force of treated hair on the micro/nanoscale is that during tip contact meniscus forces between the tip and the conditioner/cuticle become large as the tip rasters over the surface, causing an increase in the adhesive force. This adhesive force is of the same magnitude as the normal load, which makes the adhesive force contribution to friction rather significant. Thus, at sites where conditioner is accumulated on the surface (namely around the cuticle scale edges), friction force actually increases. On the macroscale, however, the adhesive force is much lower in magnitude than the applied normal load, so the adhesive force contribution to friction is negligible over the hair swatch. As a result, treated hair shows a decrease in friction force on the macroscale, which is opposite the micro/nanoscale trend.

In general, friction forces are higher on chemically damaged hair than on virgin hair. Although friction forces were similar in magnitude, it was observed that the friction force on the cuticle surface of chemically damaged hair showed a much larger variance, which contributed to the higher friction values. Chemically damaged treated hair shows a much stronger affinity to the conditioner. It is widely known that the cuticle surface of hair is negatively charged. This charge becomes even more negative with the application of chemical damage to the hair. As a result, the positively charged particles of conditioner have even stronger attraction to the chemically damaged surface, which explains the increased presence of conditioner (and corresponding higher friction forces) when compared to virgin treated hair. With the application of three conditioner cycles on chemically damaged treated hair, friction force is still higher near the cuticle edge, however it is also increased all over the cuticle surface, showing a more uniform placement of the conditioner.

Figure 24.58 shows adhesive force maps for the various hair, which gives a measurement of adhesive force variation on the surface. Figure 24.59 presents surface roughness, coefficient of friction, and adhesive force plots for the various virgin and chemically damaged hair discussed above. The data is also presented in Table 24.15 [46]. Surface roughness for human hair is characterized by a vertical descriptor, height standard deviation σ, and a spatial descriptor, correlation distance β^* [7, 9]. The standard deviation σ is the square root of the arithmetic mean of the

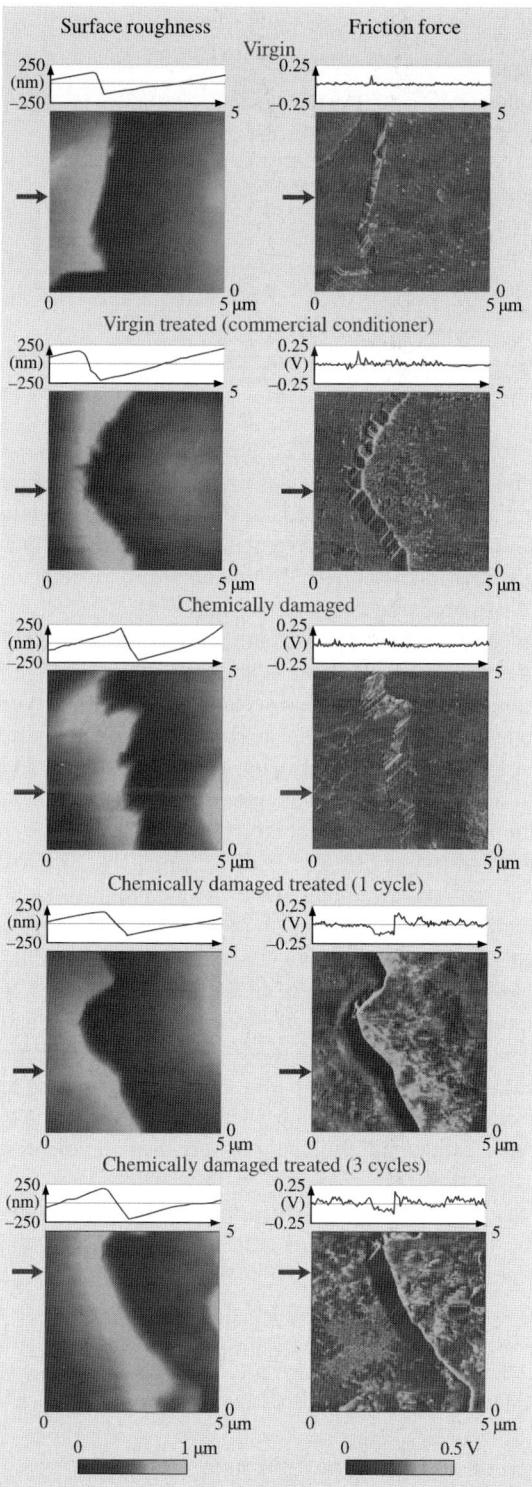

Fig. 24.57. Surface roughness and friction force images for Caucasian virgin, virgin treated, chemically damaged, damaged treated (1 cycle conditioner), and damaged treated (3 cycles conditioner) hair at 5 μm scan sizes. Shown above each image is a cross section taken at the corresponding arrows to show roughness and friction force information [46]

Table 24.15. Surface roughness, coefficient of friction, and adhesive force for virgin and chemically damaged Caucasian hair, with and without conditioner treatment

Hair type	Surface roughness		Coefficient of friction	Adhesive force (nN)
	σ (nm)	β^* (µm)		
Virgin	12 ± 8	0.61 ± 0.2	0.02 ± 0.01	25 ± 5
Virgin, treated	12 ± 4	0.90 ± 0.3	0.03 ± 0.01	32 ± 5
Chemically damaged	8.4 ± 2	0.83 ± 0.2	0.13 ± 0.06	39 ± 0.5
Damaged, treated (1 cycle)	13 ± 4	0.75 ± 0.3	0.05 ± 0.02	66 ± 0.7
Damaged, treated (3 cycles)	11 ± 2	0.80 ± 0.2	0.04 ± 0.02	54 ± 33

square of the vertical deviation from the mean line. The correlation length can be referred to as the length at which two data points on a surface profile can be regarded as being independent, thus serving as a randomness measure. These two parameters are all that is needed if the surface is assumed to have Gaussian height distribution and an exponential autocorrelation function [7, 9]. Virgin and virgin treated hair showed similar σ values, while β^* was higher in virgin treated hair. Chemically damaged hair and both types of chemically damaged treated hair showed similar roughness values, although σ was higher for the treated cases. The chemically damaged hair presented in this work is different from the chemo-mechanically damaged hair studied in [45]. It seems that chemically damaging the surface does not lead to as much wear and surface roughness increase as the combination of both chemical and mechanical damage. Thus, it should be noted (and it is understandable) that there are differences between the chemo-mechanically and chemically damaged hair.

Coefficient of friction of virgin and virgin treated hair is similar, but slightly higher for the treated cases. Chemically damaged hair shows a much higher coefficient of friction, and with more variation in the values since the chemical damage varies throughout each individual fiber. An interesting finding was that, contrary to the virgin and virgin treated hair results, coefficient of friction for chemically damaged hair decreased with application of conditioner treatment (both one and three cycles). One possible explanation is that because the stronger negative charge on chemically damaged hair results in better attraction of conditioner, this leads to higher adhesive force but, more importantly, lower shear strength on the surface. This creates an overall effect of lubrication and ultimately lowers the coefficient of friction.

Average adhesive force values were taken from the adhesive force maps described previously. Virgin treated hair shows a higher adhesive force than virgin hair due to the meniscus effects that come about from AFM tip interaction with the conditioner on the cuticle surface [7, 9]. The same trend is true and even more evident for chemically damaged treated hair compared to chemically damaged hair. A possible reason that one cycle of conditioner on chemically damaged hair showed higher average adhesive force than three cycles could be because the three cycles

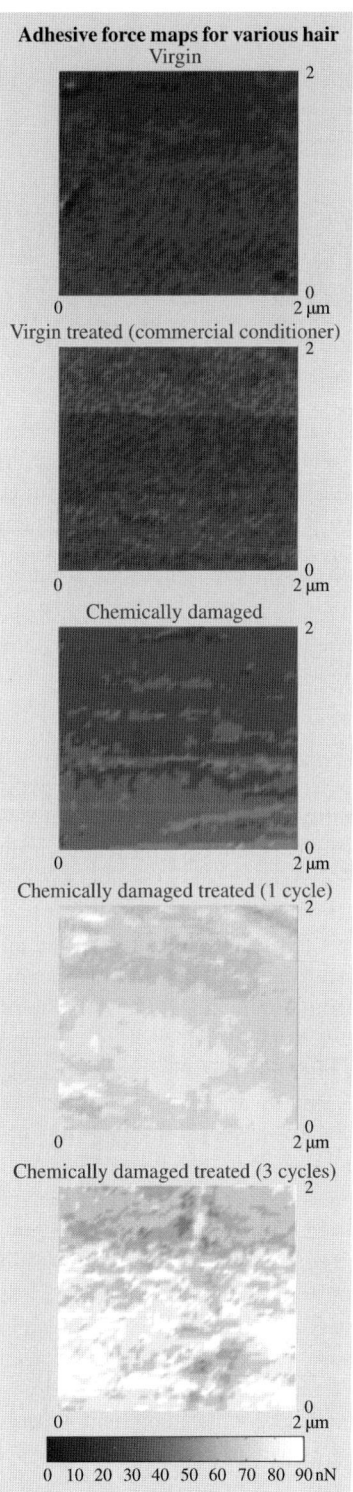

Fig. 24.58. Adhesive force maps displaying variations in adhesive force on the cuticle surface of Caucasian virgin, virgin treated, chemically damaged, damaged treated (1 cycle conditioner), and damaged treated (3 cycles conditioner) hair. Treated hair is shown to have higher adhesive force due to meniscus effects [46]

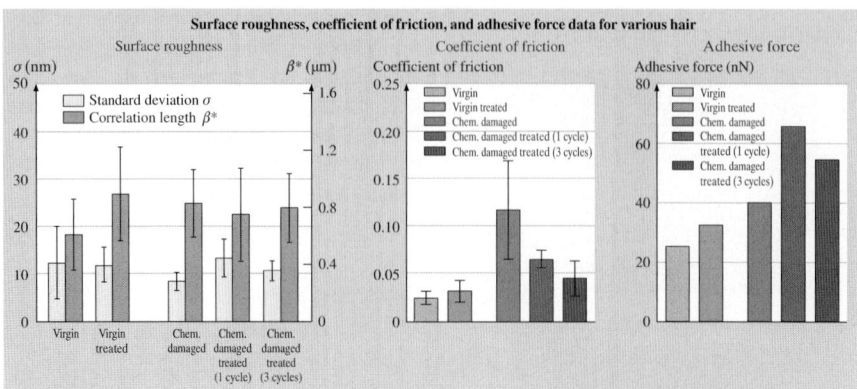

Fig. 24.59. Surface roughness, coefficient of friction, and adhesive force plots for Caucasian virgin, virgin treated, chemically damaged, damaged treated (1 cycle conditioner), and damaged treated (3 cycles conditioner) hair [46]

generally places the conditioner more uniformly on the surface rather than accumulating it most near the bottom surface near the cuticle edge, which is where the adhesive force maps were generally taken. Nevertheless, the increased adhesive force shown in the plots is a clear indication of conditioner present on the hair surface, and its localization can be observed.

Effect of Relative Humidity, Temperature, Soaking, and Durability Measurements

Figure 24.60 displays the effect of relative humidity on friction force and adhesive force. Coefficient of friction remained relatively constant for virgin and virgin treated hair. However, chemically damaged hair experienced a large increase in coefficient of friction at the high humidity, while chemically damaged treated hair experienced the opposite trend. This clearly shows that heavy moisture in the air plays a role on the frictional properties of chemically damaged hair. When combined with conditioner, a lubricating effect once again dominates as the water helps form a liquid layer which is more easily sheared. In terms of adhesive force, most samples showed a decrease in adhesive force with high humidity. It is expected that as water builds up on a surface, meniscus effects diminish and as a result do not readily contribute to adhesive force. Thus, the adhesive force is expected to come down at very high humidity.

Figure 24.61 displays the effect of temperature on friction force and adhesive force. The coefficient of friction generally decreased with increasing temperature. As the hair fiber heats up, conditioner which is present on the surface decreases in viscosity, causing a thinner film and lower friction force. The lower friction force ultimately leads to lower coefficient of friction values. Adhesive force was shown to decrease with increasing temperature as well. This was especially evident for treated hair fibers, whereas large adhesive force at room temperature decreased rapidly to adhesion values similar to non-treated fibers. It is most likely that at higher tempera-

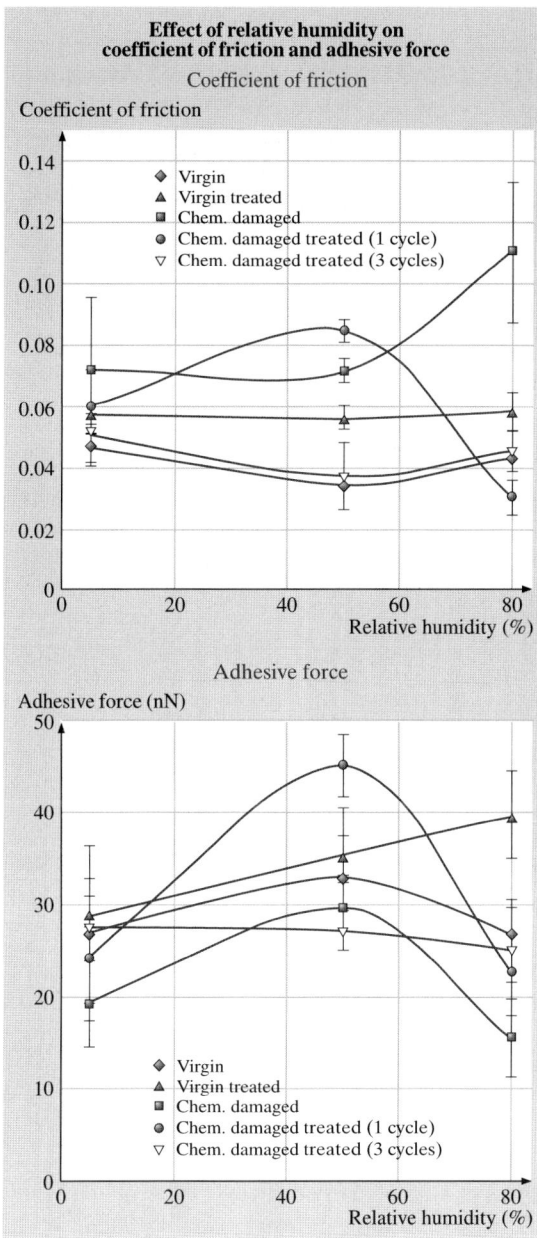

Fig. 24.60. Effect of relative humidity on nanotribological properties of Caucasian virgin, virgin treated, chemically damaged, damaged treated (1 cycle conditioner), and damaged treated (3 cycles conditioner) hair [46]

tures the thinning conditioner layer causes a reduced surface tension, which directly relates to the drop in adhesive force.

Virgin, chemically damaged, and chemically damaged treated hair samples were soaked in de-ionized water for five minutes. Their corresponding coefficient of fric-

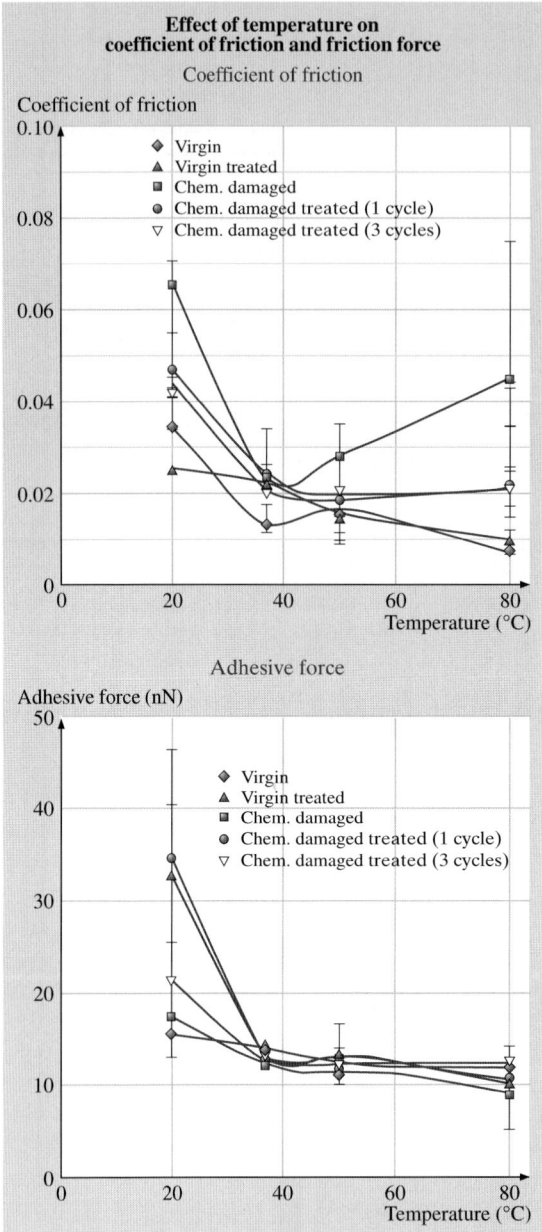

Fig. 24.61. Effect of temperature on nanotribological properties of Caucasian virgin, virgin treated, chemically damaged, damaged treated (1 cycle conditioner), and damaged treated (3 cycles conditioner) hair [46]

tion was measured and compared to coefficient of friction values for dry samples which were adjacent to the wet samples on the respective hair fiber. Figure 24.62 shows the results for two hair samples of each hair type. Virgin hair exhibits a decrease in coefficient of friction after soaking. Virgin hair is more hydrophobic (see

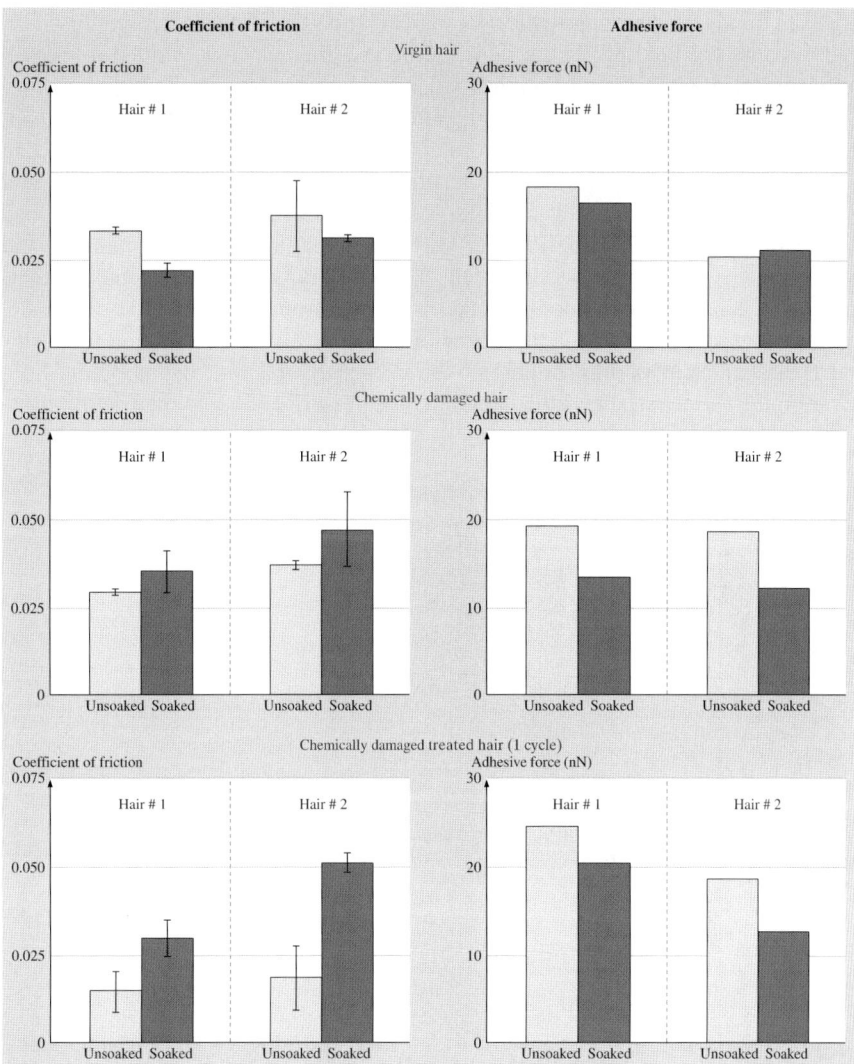

Fig. 24.62. Effect of soaking in de-ionized water on coefficient of friction and adhesive force for Caucasian virgin, chemically damaged, and chemically damaged treated (1 cycle conditioner) hair [48]

Table 24.11), so more of the water is present on the surface and results in a lubrication effect after soaking. Chemically damaged hair tends to be hydrophilic due to the chemical degradation of the cuticle surface, and results in absorption of water after soaking. This softens the hair, which leads to higher friction, even with conditioner treatment. This is yet another indication that virgin and chemically damaged hair have significantly different surface properties which in many cases results in opposite trends for their nanoscale tribological properties. Adhesive force for virgin hair remained approximately the same before and after soaking, while it decreased for chemically damaged and chemically damaged treated hair after soaking.

Figure 24.63 shows the durability effects on friction force for various hair. Above the graph are pictures of unworn and worn virgin hair, with the cuticle edge serving as a reference point. Before testing, the surface is relatively smooth and void of any large debris or wear. After 1000 cycles at approximately 10 µN load with a stiff silicon AFM tip, however, the interaction has caused degradation and wear (scratch) marks on the cuticle scale. This is the type of wear one could potentially see if hair were to come in contact with sand from a day spent at the beach, among other activities. Virgin hair shows an obvious increase in friction force signal as the scratch mark digs further into the surface. By this time the lubricious lipid layer on the surface of the virgin cuticle has been worn away and the friction force

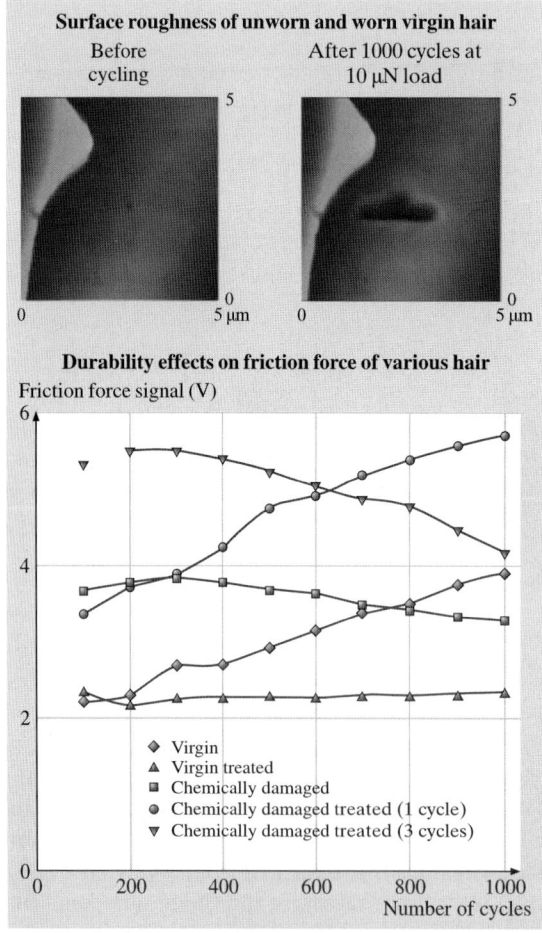

Fig. 24.63. Durability study of friction force change as a function of AFM tip cycling for Caucasian virgin, virgin treated, chemically damaged, damaged treated (1 cycle conditioner), and damaged treated (3 cycles conditioner) hair. The images above the plot signify before and after comparisons of a cuticle surface subjected to cycling at a 10 µN load [46]

24 Structural, Nanomechanical, and Nanotribological Characterization

comes close to the magnitude of chemically damaged hair friction force at the onset of cycling. When conditioner is applied to the virgin hair, however, the wear does not show up as an increase in the friction force. Thus, conditioner serves as a protective covering to the virgin hair and helps protect the tribological properties when wear ensues.

Various Hair Types Treated with Various Conditioner Matrices

LaTorre et al. [48] measured nanotribological properties of various conditioner matrices and compared their performance with a commercial conditioner. As mentioned earlier, a conditioner primarily consists of cationic surfactants, fatty alcohols, and PDMS blend silicone (dimethicone) (Table 24.7). They studied two different silicones – a PDMS (blend of low and high MW) silicone and an amino silicone added to BTMAC surfactant. Motivation for selection of amino silicone was that it may attach to the damaged hair surface and improve nanotribological performance and durability.

Figure 24.64 displays the representative surface roughness and friction force maps for chemically damaged hair, commercially treated hair, PDMS blend silicone treated hair, and amino silicone treated hair [48]. When conditioner is applied to the surface, a pattern of high friction is shown in the area surrounding the bottom

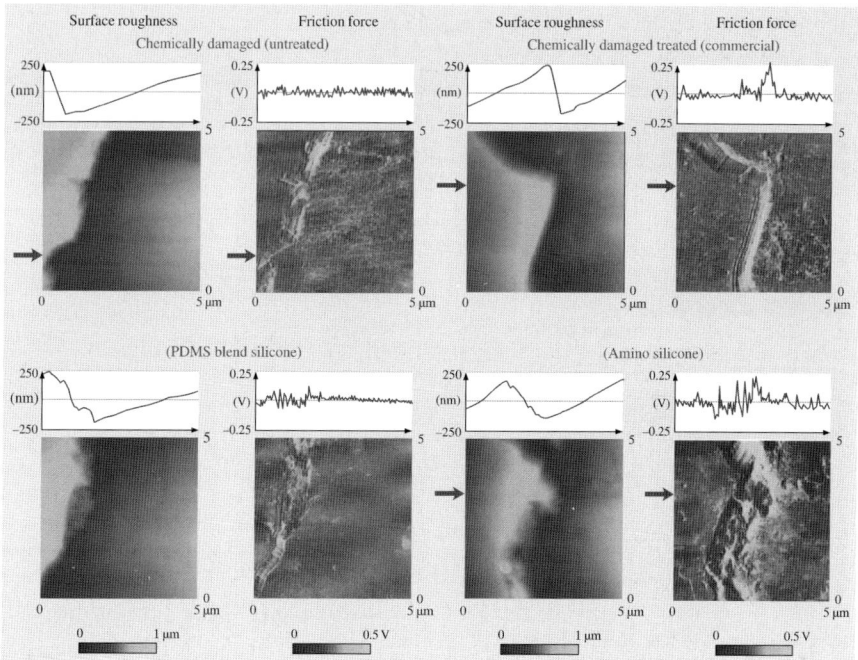

Fig. 24.64. Surface roughness and friction force maps of Caucasian chemically damaged hair with various treatments

edge of the cuticle. This is believed to be an area of conditioner accumulation which causes increased friction due to meniscus effects. Friction maps for PDMS blend silicone do not show this increase as readily, suggesting that this type of silicone is not a contributor of high friction force on the nanoscale. This can be due to the fact that a PDMS type silicone is fairly mobile on the surface and thus does not cause the same meniscus effects as the AFM tip rasters through it. The amino group typically is less mobile and harder to move, which accounts for a different slip plane flow than PDMS silicone.

Figure 24.65a displays the adhesive force maps for chemically damaged hair and the different treatments [48]. As shown in the legend, a lighter area corresponds to a higher tip pull-off force (adhesive force). Chemically damaged untreated hair has relatively low adhesive force, and is more or less consistent over the hair surface. In nearly all cases, addition of a conditioner treatment caused an increase in meniscus forces, which in turn increased the adhesive pull-off force between the AFM tip and the sample. Observing the chemically damaged treated hair, the uneven distribution of the conditioner layer is seen. This uneven distribution is also most evident for the amino silicone images, in which the less mobile silicone brings about a distinguishable change in adhesion over the surface. For the PDMS blend silicone, it is seen that adhesion over the surface is much more consistent than the amino silicone, where various areas of high adhesion occur.

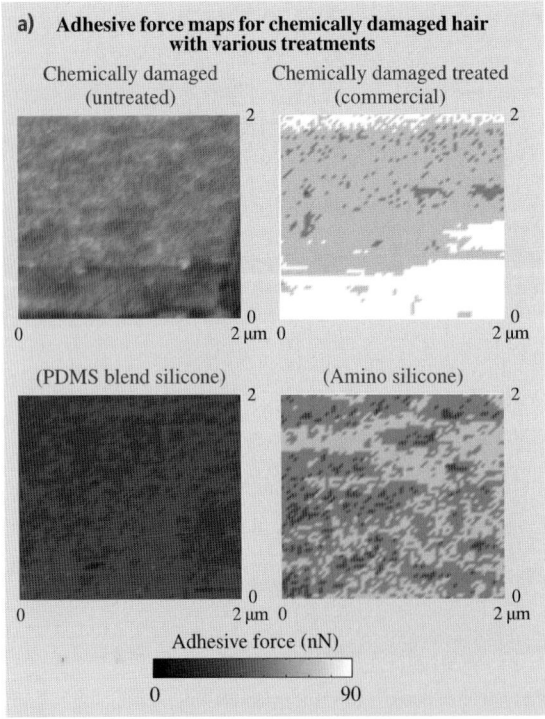

Fig. 24.65. (a) Adhesive force maps for Caucasian chemically damaged hair with various treatments

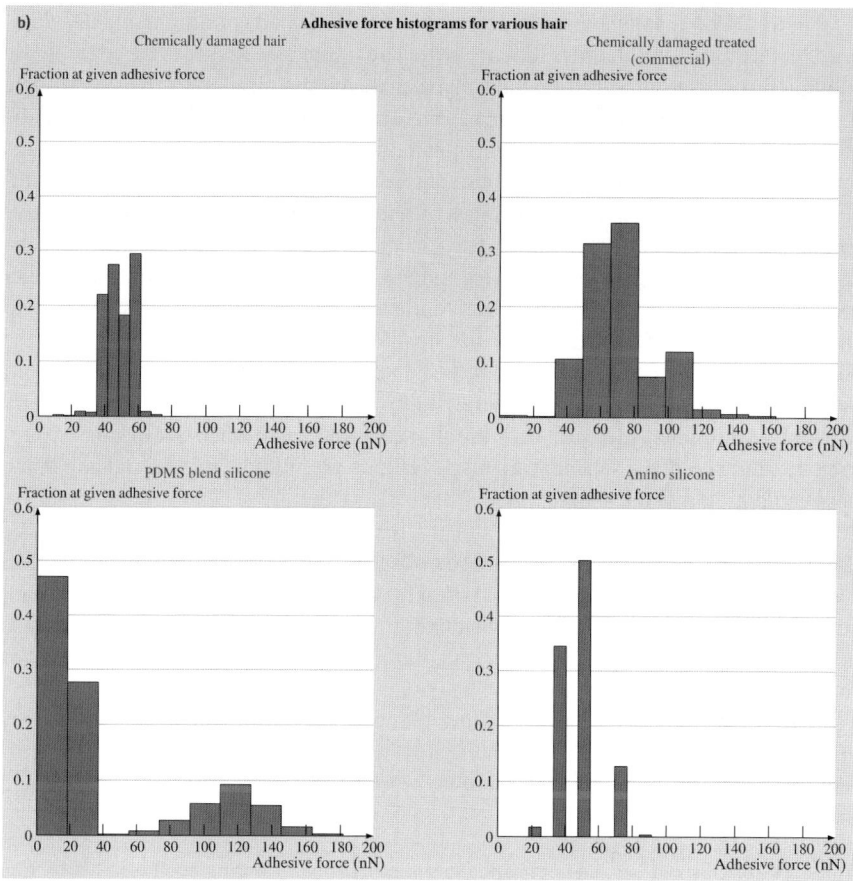

Fig. 24.65. (continued) (**b**) adhesive force histograms [48]

It is important to note that while adhesive force maps presented are representative images for each treatment, adhesive force varies significantly when treatments are applied to the hair surface. Figure 24.65b shows histograms of all adhesive force data for chemically damaged, chemically damaged treated, PDMS blend silicones, and amino silicones [48]. Chemically damaged treated hair shows a much larger range of adhesive force values and a normal distribution, which suggests that the conditioner layer is normally distributed. The histogram for PDMS blend silicone treatment shows a normal distribution at the larger adhesive force values, but also shows another peak at low adhesion values. Amino silicone treated hair follows a normal distribution, but it is interesting to note the distinct groupings of the adhesion values and the spacing between them. This is further evidence that the amino silicone groups most likely attach immediately to the hair surface and are less mobile than PDMS silicone, causing distinct regions of high and low adhesion values over the cuticle surface.

Figure 24.66 displays a summary of the data collected for all chemically damaged hair samples and their treatments [48]. The figure also includes the macroscale coefficient of friction data obtained using a technique similar to the macroscale measurement technique described earlier (Fig. 24.17). Table 24.16 reviews some of the observations and corresponding mechanisms which help to explain the trends found [48]. The application of the commercial conditioner to the chemically damaged hair caused a decreased coefficient of friction and a large increase in adhesive force. The decreased coefficient of friction may be explained by the fact that the chemically damaged hair accumulates much of the positively charged conditioner on the surface due to its highly negative charge, which in turn makes it easier to shear the liquid on the surface, causing lower coefficient of friction. However, the nanoscale pull-off force (adhesive force) is much larger than on the untreated hair because of meniscus effects. In general, adhesive force varied widely, but typically showed a significant increase with the presence of conditioner. As discussed previously, this is a clear sign that meniscus effects are influencing the pull-off force between the tip and the sample.

In most cases, the macroscale and microscale coefficient of friction followed the same trend, in which a decrease was observed with the addition of the PDMS blend or amino silicones to the surfactant. The silicones are typically used as a major source of lubrication and thus give the conditioner more mobility on the hair surface compared to just surfactants and fatty alcohols. The inverse trend was seen only for the amino silicone group.

The amino silicones have a strong electrostatic attraction to the negatively charged hair surface, which in turn creates higher binding forces and less mobility. The dampened mobility of the amino silicone, with respect to hair surface and

Table 24.16. Observations and corresponding mechanisms regarding coefficient of friction and adhesion for various hair treatments

Observation	Mechanism
Damaged vs. damaged treated hair	
Damaged hair shows a decrease in coefficient of friction but an increase in adhesion from the application of commercial conditioner.	The conditioner layer deposited on the surface of the damaged hair results in a lower shear strength which in turn lowers the coefficient of friction, while meniscus effects increase the pull-off (adhesive) force between the tip and hair sample.
PDMS blend vs. amino silicone	
Amino silicones interact strongly with negatively charged hair surface.	A stronger electrostatic attraction exists which results in stronger binding forces (which leads to higher adhesion) for amino silicone.
Amino silicone thickness distribution on hair is less uniform than with PDMS blend.	Less mobility with amino silicones, so molecules attach to hair at contact and do not redistribute easily.

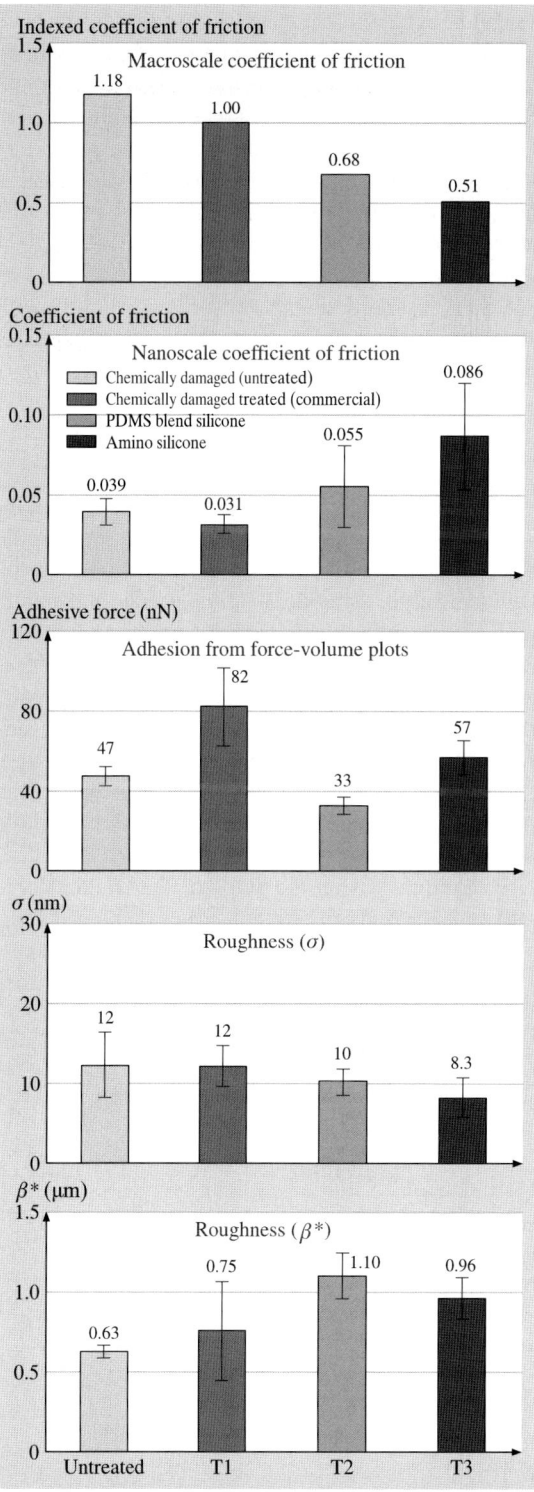

Fig. 24.66. Coefficient of friction, adhesive force, and surface roughness plots for Caucasian chemically damaged hair with various treatments

tip, may account for this wide variation in coefficient of friction and large adhesive force values. In terms of adhesive force, it was previously observed in Fig. 24.65a that the amino silicone treatments showed much more distinct regions of higher and lower adhesion compared to PDMS blend silicones. This nonuniform amino silicone thickness distribution on hair is also believed to be caused by the inhibited mobility, as the molecules immediately attach to the hair at contact and do not redistribute as a uniform coating. The increased polarity of the amino silicones compared to the PDMS blend can also be a major contributor of the higher friction and adhesion at high deposition levels.

In respect to roughness, the vertical standard deviation decreased slightly with most treatments, although standard deviations were similar. The spatial parameter increased slightly with treatments, but the variation also becomes extremely high.

Skin

Synthetic materials were also studied for surface roughness and friction force information, shown in Fig. 24.67a. While macroscale dimples could be seen on the surface of collagen film, it was interesting to find similar pits and dimples on the micro/nanoscale, consequently with a large variation in dimple size and depth. Polyurethane films are shown to have quite different topography and friction forces, while their coefficient of friction is very similar. Human skin shows a rougher texture with higher peaks, Fig. 24.67b. The roughness parameters for the collagen and polyurethane films, and also for human skin, are presented in Table 24.17 [45]. Surface height standard deviation σ was approximately 3 times larger than that of virgin hair for both synthetic materials. However, the correlation length β^* was lower than what was typically observed in hair. The average coefficient of friction for these synthetic materials are shown in Fig. 24.68 plotted next to virgin Caucasian hair as a reference. These values were calculated using the slope of the friction force curves, described previously. Both collagen and polyurethane films displayed similar coefficient of friction values of 0.22 and 0.24, respectively. Virgin hair displays a much lower coefficient of friction than both materials, approximately eight times lower.

Table 24.17. Surface roughness parameters σ, β^* for collagen and polyurethane films and human skin

	σ (nm)	β^* (μm)
Collagen (synthetic hair)	36 ± 11	0.50 ± 0.1
Polyurethane (synthetic skin)	33 ± 6	0.71 ± 0.1
Human skin	80 ± 28	0.59 ± 0.2

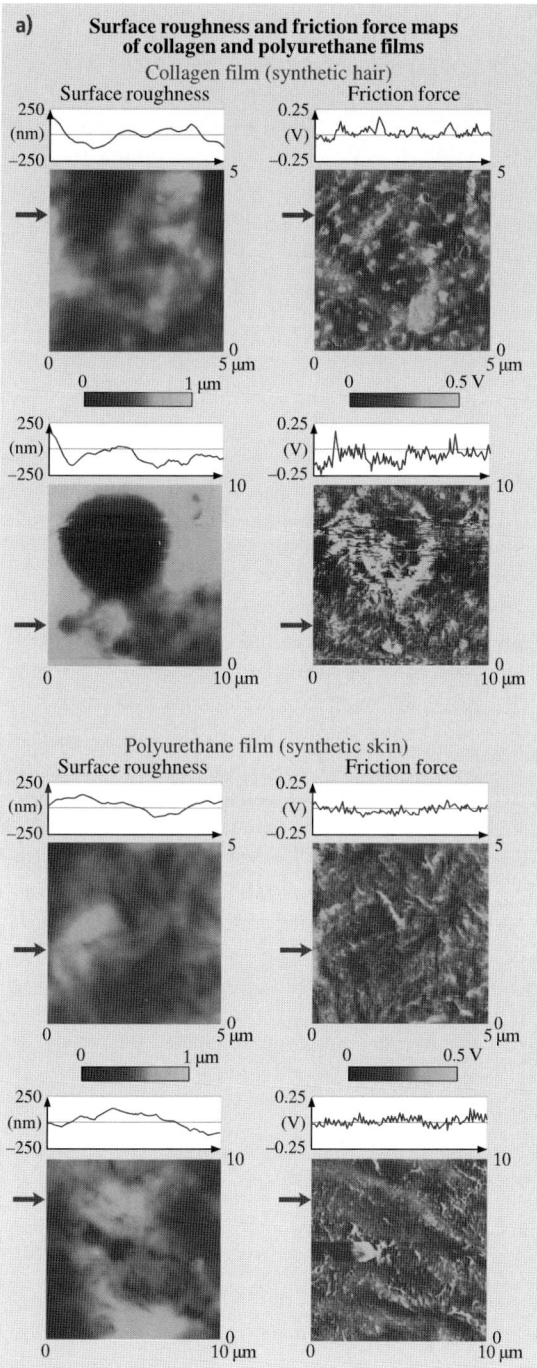

Fig. 24.67. (a) Surface roughness and friction force images for collagen and polyurethane films at 5 and 10 μm² scan sizes. Shown above each image is a cross section taken at the corresponding arrows to show roughness and friction force information

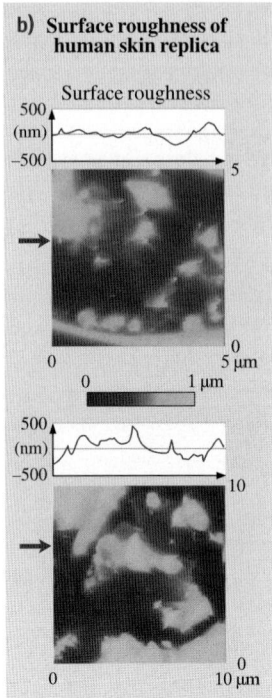

b) Surface roughness of human skin replica

Fig. 24.67. (continued) **(b)** surface roughness for human skin at 5 and 10 μm² scan sizes. Note that vertical scales in 2-D section profiles are doubled [45]

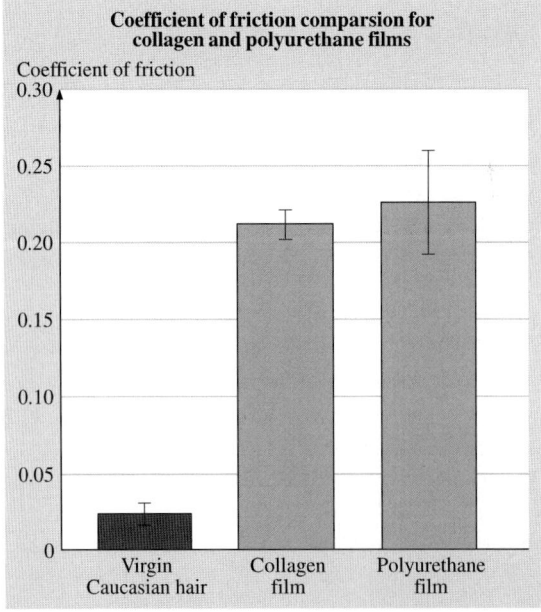

Fig. 24.68. Average coefficient of friction values for collagen and polyurethane films. Error bars represent ±1 σ on the average coefficient value [45]

24.6.3 Scale Effects

Directionality Dependence of Friction

Human hair has been shown in the past to have a directionality friction effect on the macroscale, making it easier to travel over the hair surface from root-to-tip than in the opposite direction due to the tile-like orientation of cuticles [10, 68]. The outer surface of human hair is composed of numerous cuticle scales running along the fiber axis, generally stacked on top of each other. As previously discussed, the heights of these step changes are approximately 300 nm. These large changes in topography make the cuticle an ideal surface for investigating the directionality effects of friction [47]. The first row of Fig. 24.69a displays a low resolution SEM micrograph and a friction profile for macroscale coefficient of friction measurements. Note that this data is taken for measurements where multiple fibers are in contact with the synthetic skin upper specimen at the same time, and does not correspond directly with the SEM micrograph. With an applied normal load of 50 mN and 3 mm travel, it is observed that the friction force produced when scanning from root-to-tip (referred to as "along cuticle") is lower than that when scanning against cuticle. This is a direct consequence of the literally thousands of scale edges which come in contact with the synthetic skin. When traveling against cuticle, these edges act as tiny resistors to motion as they are forced backwards and uplifted from their interface with the underlying cuticle layers. The resistance to motion of so many cuticle edges at the same time becomes "additive" and results in higher values of friction, corresponding to higher coefficient of friction than when traveling along cuticle. For along cuticle travel, these edges are forced downward against the underlying cuticle layers so that the resistance effect of these edges is limited, which results in lower friction values.

The second row of Fig. 24.69a shows an AFM height map corresponding to the microscale friction profile shown to its right. (These measurements were made using an AFM tip mounted with 4 μm radius silica ball.) Due to the size of the hair, it was only possible to capture a rectangular height map as shown. It is evident that the 100 μm travel results in the involvement of several cuticle scales. The applied normal load for the microscale friction profile (about 20 nN) is significantly reduced from that of the macroscale value, which consequently yields much lower friction forces. In the along cuticle direction, we can see small fluctuations in the friction data over the scan distance. These are caused by local variations in surface roughness, system noise, and changes due to traveling over the scale edges. However, when scanning against the cuticle, distinctly large spikes in the data are observed at roughly 5–10 μm intervals. This is clearly the effects of the scale edges coming in contact with the AFM microscale tip and causing local collisions and ratcheting of the tip. Because of the sign convention of the AFM that causes a reversal in the sign when traversing the opposite direction, this signal is now observed to be highly negative. These edge effects are hence the primary item responsible for the higher friction and coefficient of friction observed in the against cuticle direction.

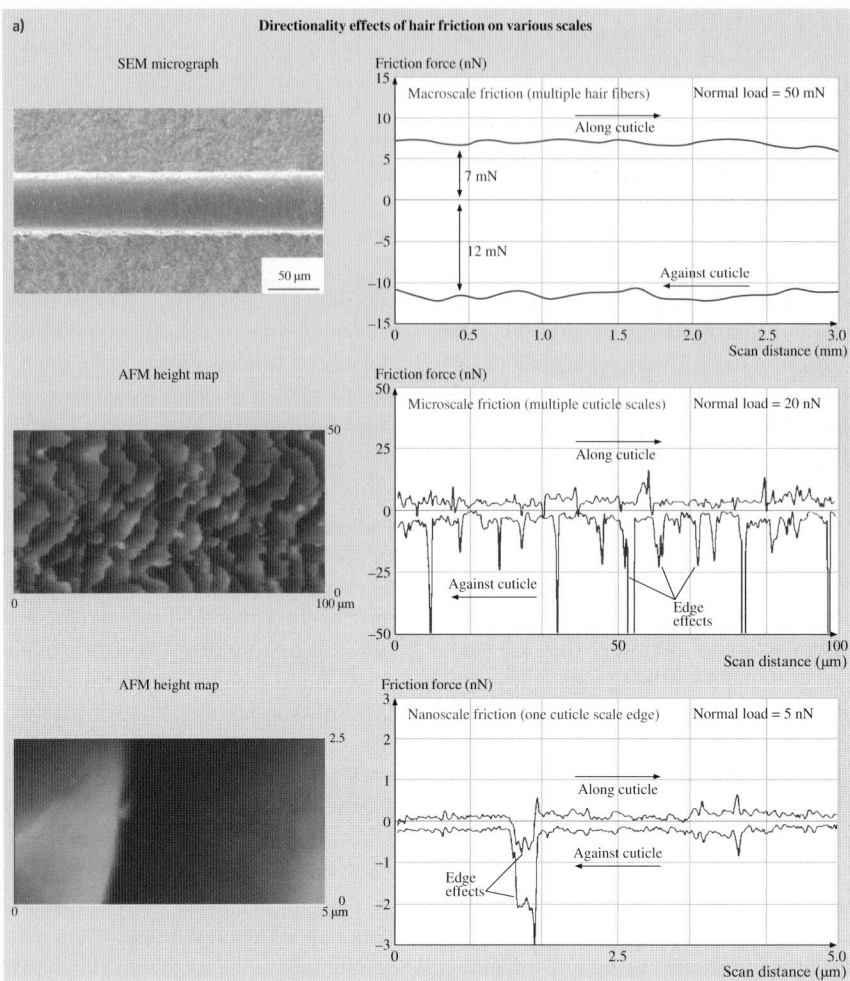

Fig. 24.69. (a) Directionality effects of virgin Caucasian hair friction on various scales

Directionality effects on nanoscale hair friction have previously been reported by *LaTorre* and *Bhushan* [45]. As shown in the bottom row of Fig. 24.69a, the scan size of $5 \times 2.5\,\mu m^2$ provides one cuticle scale edge to be studied (the height map is rectangular only to be consistent with the microscale image in the middle row). As the tip rasters in the along cuticle direction, a small decrease in the friction force is observed as the tip follows the scale edge on a downward slope. When the tip comes back in the against cuticle direction, colliding with the scale edge and climbing up the sharp peak results in a high friction signal. As discussed above, because of the sign convention of the AFM/FFM that causes a reversal in the sign when traversing the opposite direction, this signal is now observed to be highly negative. The interesting difference between the two profiles lies in the fact

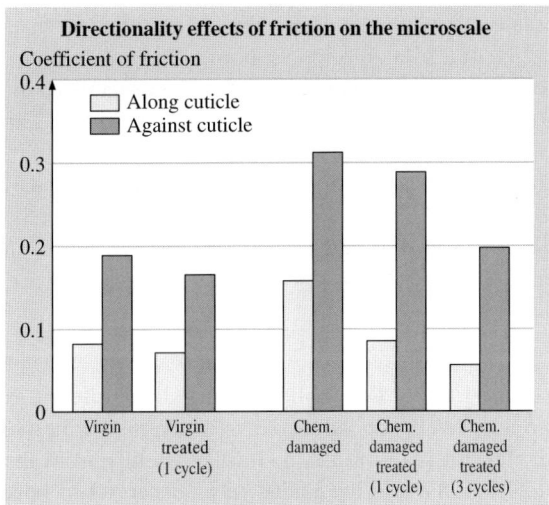

Fig. 24.69. (continued) **(b)** microscale coefficient of friction data for Caucasian virgin, chemically damaged, virgin treated, damaged treated (1 cycle conditioner) and damaged treated (3 cycles conditioner) hair showing directionality effects [47]

that the magnitude of the decrease in friction when going up the step is much larger than the magnitude of the friction when the tip is going down the step, yet both signals are in the same direction. It is thus shown that the cuticle edge provides a local ratchet and collision mechanism that increases the friction signal at that point and clearly shows the directionality dependence caused by edge effects.

Coefficient of friction data showing directionality dependence on the macroscale for various virgin, damaged, and treated hair has been reported previously [10]. Figure 24.69b shows a summary of the microscale coefficient of friction data for the various hair samples of this study. In most cases, the coefficient of friction has more than doubled when scanning in the against cuticle direction. A more in depth discussion on the coefficient of friction trends between the different hair types will follow. For now, however, it is important to realize that there is strong directionality dependence on coefficient of friction data for hair, especially on the microscale. On the nanoscale, while directionality dependence of friction force has been studied, actual coefficient of friction data is generally only measured on a small scan area which does not include the cuticle edge, so that only the true cuticle surface is involved [45]. Hence, nanoscale coefficient of friction directionality data similar to Fig. 24.69b is not shown.

Scale Effects on Coefficient of Friction and Adhesive Force of Various Hair

Given the fact that the "directionality effect" is now observed to be universal for all types of hair and on all scales, and since the along cuticle direction is more relevant to our daily life (i.e. combing), we now focus on coefficient of friction data along cuticle.

As described earlier, microscale and nanoscale coefficient of friction is taken as the slope of the least squares fit line of a friction force vs. normal force data curve. Figure 24.70a shows these types of raw data curves for various representative hair samples using both microscale (top plot) and nanoscale (bottom plot) AFM tips. The nanoscale data was taken from raw data presented by *LaTorre* and *Bhushan* [47]. One can see a relatively linear relationship between the data points for each type of hair sample. If the least squares fit lines of Fig. 24.70a are extended to intercept the horizontal axis (indicated by the dotted lines), then a value for adhesive force can also be calculated, since friction force F is governed by the relationship

$$F = \mu(W + F_a) \tag{24.4}$$

where μ is the coefficient of friction, W is the applied normal load, and F_a is the adhesive force [7, 9]. These adhesive force values serve as average adhesive force values over the course of the full scan profile, and differ slightly from the force calibration plots of Fig. 24.70b. From Fig. 24.70a, it is observed in both plots that treated hair fibers, whether virgin or chemically damaged, have much higher average adhesive force values as compared to their untreated counterparts. Chemically damaged hair is observed to have the highest coefficient of friction (highest slope) on both the micro- and nanoscales. As explained in *LaTorre* and *Bhushan* [45, 46] and *LaTorre* et al. [48], chemical damage to the hair causes the outer lubricious layer of the cuticle to wear off, resulting in an increased coefficient of friction.

Force calibration plots yield adhesive force values at a single point and are considered to be more relevant for measurement of the pull-off force between the tip and hair surface. Since it has been previously shown that treating both virgin and chemically damaged hair with conditioner results in a large increase in adhesive force using both micro- and nanoscale AFM tips, we will focus only on the force calibration plots of the virgin and virgin treated hair to discuss mechanisms for this trend. The first plot in Fig. 24.70b shows a typical virgin hair force calibration plot with the microscale AFM tip. We can see that on the application of 1 cycle conditioner treatment to the virgin hair, the microscale adhesive force jumps to about 230 nN. This is clearly the effect of meniscus forces brought about by the presence of the conditioner layer on the cuticle surface which interacts with the tip [27, 45, 46]. With the nanoscale tip (bottom row of Fig. 24.70b), increase in adhesive force is again seen with conditioner treatment.

It is important to notice that that adhesive force values on the microscale are always larger than those on the nanoscale for a given hair. To explain the scale dependency of adhesive force, we can model the hair-conditioner-tip interaction as a sphere close to a surface with a continuous liquid film [27]. The adhesive force F_a, is the force needed to pull the sample away from the tip (which is the same as the adhesive force calculated with force calibration plots). F_a is the sum of van der Waals force F_{vdw} and the meniscus force F_m due to Laplace pressure ($F_a = F_{vdw} + F_m$).

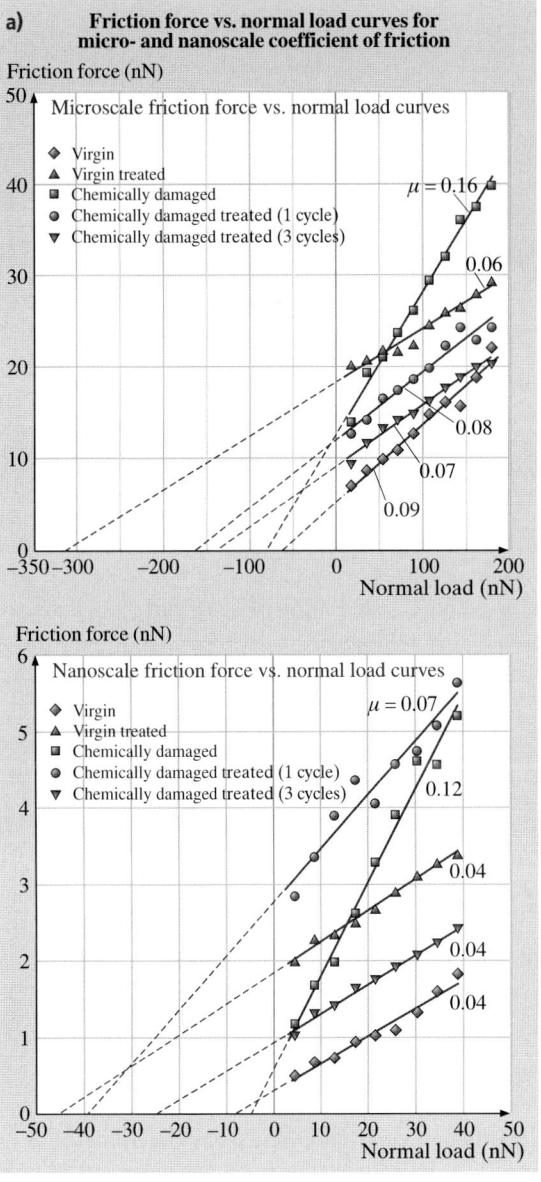

Fig. 24.70. (a) Friction force vs. normal force curves for micro-and nanoscale coefficient of friction of Caucasian virgin, chemically damaged, virgin treated, damaged treated (1 cycle conditioner) and damaged treated (3 cycles conditioner) hair

The meniscus force F_m is given by

$$F_m = 2\pi R \gamma (1 + \cos\theta) \qquad (24.5)$$

where R is the tip radius, γ is the surface tension of the conditioner, and θ is the contact angle between the tip and conditioner [9]. The increase in adhesive force calculated by force calibration plots with the microscale AFM tip compared to the

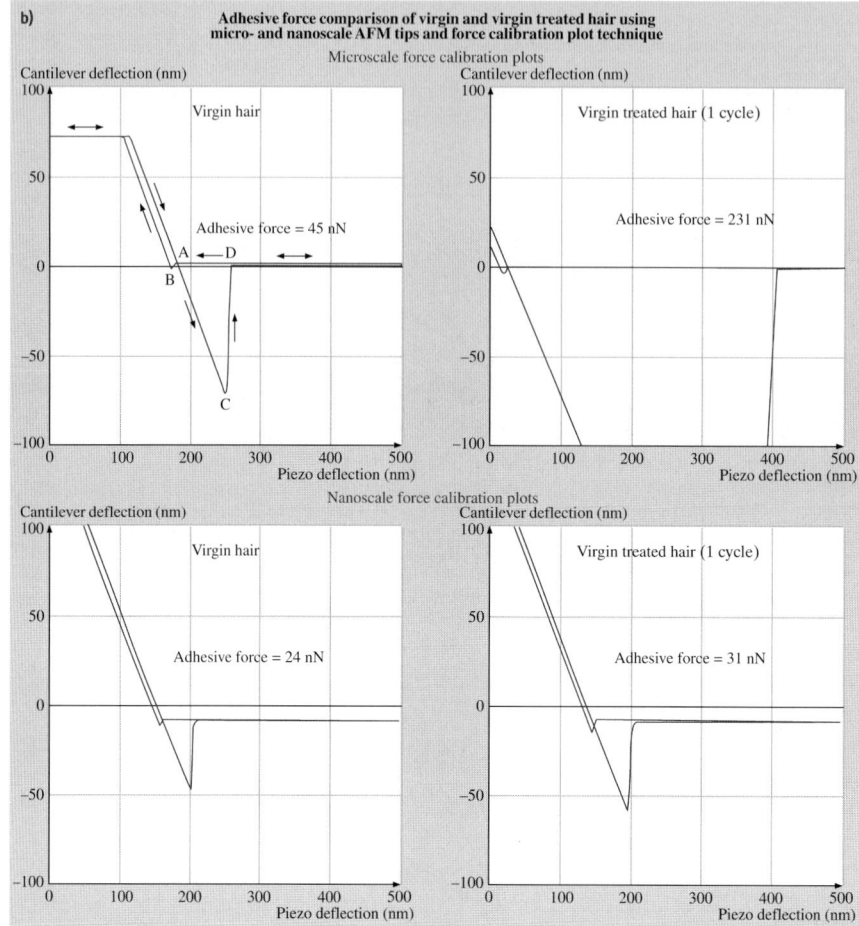

Fig. 24.70. (continued) **(b)** Adhesive force comparison of virgin and virgin treated hair using micro- and AFM tips and force calibration plot technique [47]

nanoscale AFM tip is in large part due to the increased radius R of the microscale ball, which consequently induces larger F_m.

Figure 24.71 displays the coefficient of friction and adhesive force data on macro-, micro-, and nanoscales [47]. Scale dependence is clearly observed. Macroscale data for virgin and virgin treated hair were taken from *Bhushan* et al. [10]. The values for other hair were taken in part from the indexed coefficient of friction values in *LaTorre* and *Bhushan* [46] which were transformed into actual values (as described previously). No adhesive force data is presented for the macroscale data because the adhesive force contribution to friction is considered to be negligible compared to the applied normal load. On the micro- and nanoscales, however, the magnitude of the adhesive force is the same as that of the applied normal

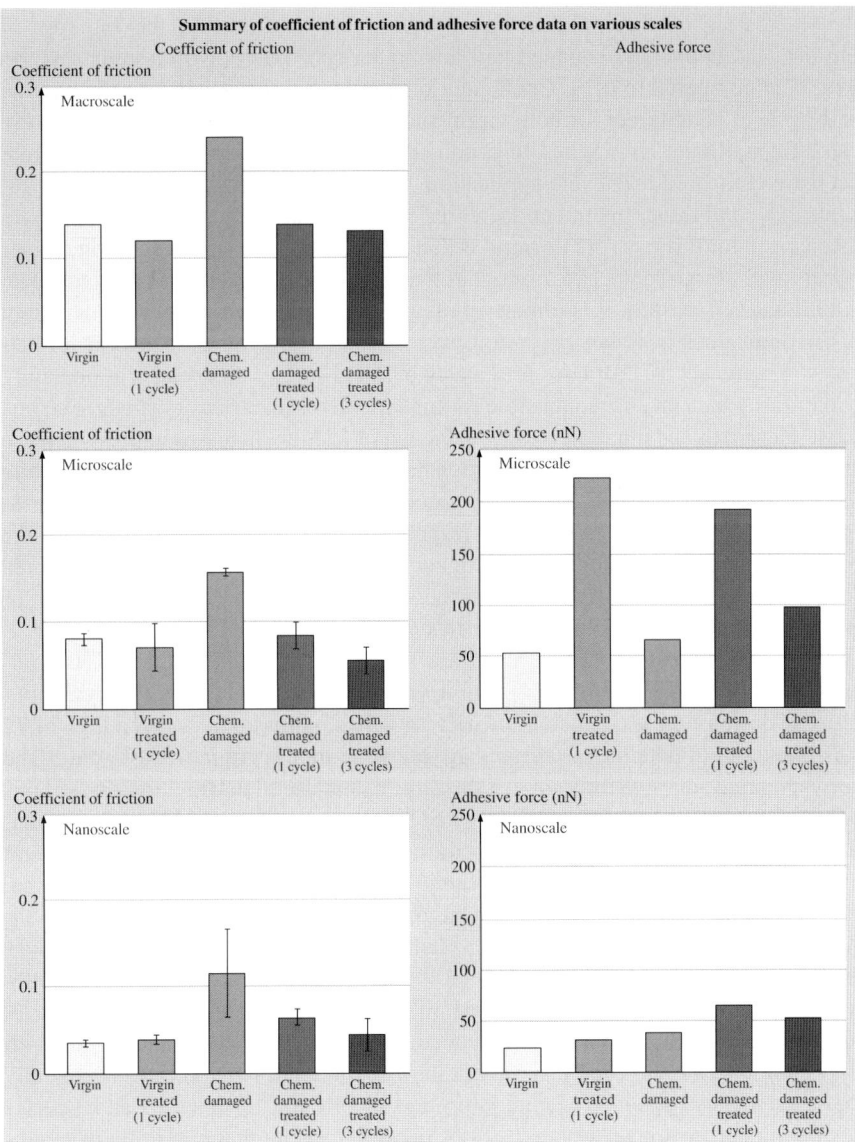

Fig. 24.71. Summary of coefficient of friction and adhesive force data on various scales for Caucasian virgin, chemically damaged, virgin treated, damaged treated (1 cycle conditioner) and damaged treated (3 cycles conditioner) hair [47]

load, so they have significant contributions on the coefficient of friction data, and thus are presented. Microscale values are taken from the raw data represented in Fig. 24.70a,b. Nanoscale data was taken from *LaTorre* and *Bhushan* [46].

Macroscale coefficient of friction (COF) data is shown in the top row of Fig. 24.71. Chemically damaged hair has higher coefficient of friction than virgin hair, 0.24 compared to 0.14. Coefficient of friction decreases with the application of 1 cycle of conditioner for both virgin and chemically damaged hair. When 3 cycles of conditioner are applied to chemically damaged hair, there is only a slight decrease compared to 1 cycle application, 0.14 to 0.13. Thus, the data readily reveals that conditioner treatment decreases coefficient of friction for hair. The main mechanism for this macroscale trend is that lubrication with a thin conditioner layer occurs over a large contact area, and thus the conditioner layer shears easily to create a lubricious effect [46]. It is important to note that the magnitude of the coefficient of friction values is higher on the macroscale than on the other scales for all hair types. *Bhushan* et al. [20] have previously outlined several differences in operating conditions which can be responsible for higher macroscale friction values. The one most relevant to our situation is that coefficient of friction increases with an increase in the AFM tip radius. Nanoscale friction data is taken with a sharp AFM tip, while the macro- and microscale tests have contacts which range from nanoasperities to much larger asperities which may be responsible for larger values of friction force on these scales. The combination of higher normal loads with a larger contact area (due to contact with multiple fibers at the same time) may also be responsible for increased coefficient of friction on the macroscale.

The coefficient of friction trends are similar on the microscale. On the microscale, virgin hair coefficient of friction was 0.08, while application of 1 cycle conditioner decreased the coefficient of friction only slightly to 0.07. From Fig. 24.71 it is important to note the large standard deviation on the virgin treated value. The corresponding adhesive force data in the same row is useful to better understand this behavior. It is shown that the virgin treated hair has a large adhesive force contribution (due to meniscus effects caused by the conditioner layer) which has a significant effect on the variation of friction force, and consequently the coefficient of friction. The chemically damaged hair has the largest coefficient of friction of the set, 0.16. Application of 1 conditioner cycle brought the value down to 0.08, while it was even lower for 3 conditioner cycles, 0.06. Adhesive force increased significantly with conditioner application; for virgin hair, adhesive force increased from about 50 to 220 nN due to the meniscus effect that come about from AFM tip interaction with the conditioner on the cuticle surface. Likewise, the adhesive force for chemically damaged hair was about 65 nN and jumped to 190 nN and 100 nN for 1 and 3 cycles of conditioner treatment, respectively. A possible reason, that one cycle of conditioner on damaged hair showed higher average adhesive force than three cycles, could be that the three cycles generally places the conditioner more uniformly on the surface rather than accumulating it most on the bottom surface near the cuticle edge, which is where the adhesive force maps were generally taken. Nevertheless, the increased adhesive force shown in the plots is a clear indication of conditioner present on the hair surface [46].

On the nanoscale, coefficient of friction of virgin and virgin treated hair is similar, but slightly higher for the treated cases. This is opposite of the trend on the macroscale and slightly different than that observed on the microscale. The meniscus bridges in the treated hair require more force to break through than with untreated hair, which causes the coefficient of friction to be similar to the untreated value, instead of experiencing a significant decrease which is traditionally expected. Damaged hair shows a much higher coefficient of friction, and with more variation in the values since the chemical damage varies throughout each individual fiber. Contrary to the virgin and virgin treated hair nanoscale results, coefficient of friction for damaged hair decreased with application of conditioner treatment (both 1 and 3 cycles), which agrees well with macro- and microscale trends.

There are several reasons that nanoscale coefficient of friction trends for virgin treated and chemically damaged treated hair are different. The effect of chemical damage plays a large role. It is widely known that the cuticle surface of any hair is negatively charged. This charge becomes even more negative with the application of chemical damage to the hair. As a result, the positively charged particles of conditioner have even stronger attraction to the chemically damaged surface, which explains the increased presence of conditioner when compared to virgin treated hair. With the application of three conditioner cycles on damaged hair, there is an even more uniform placement of the conditioner. This leads to high adhesive force due to meniscus effect (similar to that of virgin treated hair) but more importantly, lower shear strength on the surface. This creates an overall effect of lubrication as the tip travels across the cuticle surface and ultimately lowers the coefficient of friction.

Another reason for the difference in nanoscale trends between virgin and chemically damaged hair may have to do with drastic differences in hydrophobicity of the two hair types. Virgin hair has been shown to be hydrophobic, with a contact angle around 100° (Table 24.11). Chemically damaged hair, however, is hydrophilic, with a contact angle around 70°. The conditioner gel network is primarily composed of water, together with fatty alcohols, cationic surfactants, and silicones. Thus, the hydrophobicity of the hair will be relevant to not only how much conditioner is deposited, but also how it diffuses into the hair and bonds to the hair surface. For virgin treated hair, the conditioner deposits in certain locations, especially near the cuticle edge, but due to the hydrophobicity of the cuticle, it does not spread out as readily as with chemically damaged hair. For chemically damaged hair, the conditioner spreads out a bit more uniformly and in more places over the cuticle surface due to both the hydrophilicity and the stronger negative charge which attracts more conditioner deposition. Thus, as the tip scans over the virgin treated surface, the conditioner does not smear as readily, causing the tip to break the tiny meniscus bridges formed with the conditioner. This results in increased adhesive force, which contributes to higher friction force. This ends ups increasing the coefficient of friction to about the same level as the untreated virgin hair, instead of a reduction in coefficient which is typically expected for a lubricated surface. In the case of chemically damaged hair, however, the conditioner layer is already more spread out, especially in the case of 3 cycles of treatment. As the tip scans over the surface, the overall effect is one of reduced shear strength, i.e. the conditioner layer, albeit not fully continuous, smears with tip travel and causes reduced coefficient of friction between the tip and the hair. With 3 cycles, the conditioner thickness increases slightly, and the layer is even more uniformly distributed over the surface, which causes a further reduction in coefficient of friction, much like the results seen on both the micro- and macroscale.

Figure 24.72 shows schematically the mechanisms responsible for the reverse trends seen on the nanoscale for virgin treated hair [47]. In the top cartoon of Fig. 24.72, the thin layer of conditioner acts as a lubricant over the hair fiber, limiting the dry contact with the synthetic skin block and creating easier relative motion, which decreases coefficient of friction compared to the untreated hair. This is true on the macroscale for both virgin and chemically damaged hair. On the microscale, the same trend is experienced; that is, the 4 µm radius of the AFM ball comes in contact with multiple cuticle scales at the same time, causing an overall lubrication effect for both virgin treated and chemically damaged treated hair as the thin conditioner layer is sheared to created easier relative motion. It is important to note, however, that adhesive forces due to meniscus effects are the same magnitude as the applied microscale normal load.

On the nanoscale (bottom cartoon of Fig. 24.72), we see different trends for virgin treated and chemically damaged treated hair. As discussed earlier, the hydrophobicity of the virgin hair causes different deposition of conditioner. The AFM tip breaks the tiny meniscus bridges formed with the conditioner as it scans across

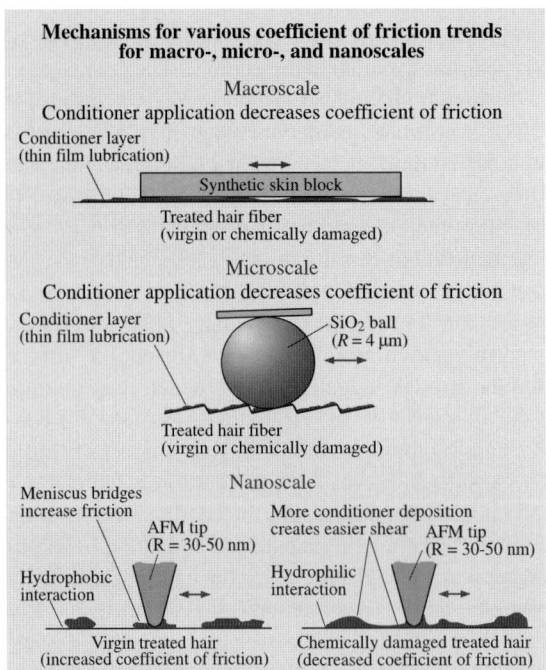

Fig. 24.72. Schematic of mechanisms for various coefficient of friction trends for macro-, micro-, and nanoscales [47]

the hair surface, which increases adhesive force contribution and results in an increased coefficient of friction. For hydrophilic chemically damaged hair, there is more uniform deposition and better smearing of the conditioner layer, which serves to lower coefficient of friction between the tip and the cuticle surface.

24.6.4 Summary

A flat-on-flat tribometer has been used to measure macroscale friction and wear of the polyurethane film (synthetic skin) vs. hair and hair vs. hair. In the case of polyurethane film vs. hair, the chemo-mechanically damaged hair shows highest coefficient of friction, followed by virgin and virgin treated hair. The coefficient of friction obtained in the case of hair vs. hair is greater than that of polyurethane film vs. hair. After 24 hours skin vs. hair wear test, the coefficient of friction did not change, while some of the cuticles were damaged.

AFM contact mode has been used to perform nanotribological studies on various hair and skin. Friction force and the resulting coefficient of friction are seen to be higher on chemo-mechanically damaged hair than on virgin hair, due to the increase in surface roughness and a change in surface properties that results from exposure to chemical damage. Generally speaking, the average coefficient of friction is similar between virgin and virgin treated hair of each ethnicity. However, in virgin treated hair there is an increase in friction forces around the cuticle edges and surrounding area. It is believed that the increase in friction force is due in part to an increase in meniscus effect which increases the adhesive force contribution to friction at sites where conditioner is deposited or accumulated on the surface, namely around the cuticle scale edges. This adhesive force is of the same magnitude as the normal load, which makes the adhesive force contribution to friction rather significant. On the macroscale, however, the adhesive force is much lower in magnitude than the applied normal load, so the adhesive force contribution to friction is negligible over the hair swatch. As a result, treated hair shows a decrease in friction force on the macroscale, which is opposite the micro/nanoscale trend. The friction and adhesion data on the micro/nanoscale is useful, though, because it relates to the presence of conditioner on the cuticle surface and allows for obtaining an estimate of conditioner distribution. Studies using force calibration plot technique showed a decrease in adhesive force with damaged hair, and significantly higher adhesive force for treated hair. This increase on the micro/nanoscale is most likely due to meniscus force contributions from the accumulation and localization of a conditioner layer on the hair surface. Thus, the presence of conditioner can be detected by this increasing adhesive force. The directionality dependence of friction is evident when the cuticle edge is examined using FFM.

Chemically damaged treated hair shows a much stronger affinity to conditioner than virgin hair. The negative charge of hair fibers becomes even more negative with the application of chemical damage to the hair. As a result, the positively charged particles of conditioner have even stronger attraction to the chemically damaged surface, and this results in an increased presence of conditioner (and corresponding higher friction forces) when compared to virgin treated hair. With the application of three conditioner cycles on chemically damaged treated hair, friction force increases all over the cuticle surface, showing a more uniform placement of the conditioner. Contrary to the virgin and virgin treated hair results, coefficient of friction for chemically damaged hair decreased with application of commercial conditioner treatment (both one and three cycles). One possible explanation is that because the stronger negative charge on damaged hair results in better attraction of conditioner, this leads to higher adhesive force but, more importantly, lower shear strength on the surface.

Environmental effects were studied for various hair. The coefficient of friction generally decreased with increasing temperature. After soaking hair in de-ionized water, virgin hair exhibits a decrease in coefficient of friction after soaking. Virgin hair is more hydrophobic (based on contact angle data), so more of the water is present on the surface and results in a lubrication effect after soaking. Chemically damaged hair tends to be hydrophilic due to the chemical degradation of the cuticle surface, and results in absorption of water after soaking. This softens the hair, which leads to higher friction, even with conditioner treatment. Durability tests show that once conditioner is applied to virgin hair, wear does not occur unlike an increase in friction force. Thus, conditioner serves as a protective covering to the virgin hair and helps protect the tribological properties when wear ensues.

In most cases, a decrease in coefficient of friction was observed on chemically damaged hair with the addition of the PDMS blend or amino silicones to the BT-MAC surfactant. The silicones are typically used as a major source of lubrication and thus give the conditioner more mobility on the hair surface compared to just surfactants and fatty alcohols. The inverse trend was seen only for the amino silicone group. The dampened mobility of the amino silicone, with respect to hair surface and tip, may account for this wide variation in coefficient of friction. Adhesive force varied widely, but typically showed a significant increase with the presence of conditioner ingredients. This is a clear sign that meniscus effects are influencing the pull-off force between the tip and the sample. The amino silicones showed much more distinct regions of high and low friction and adhesion, which shows that there is less mobility of these molecules and much less redistribution as they coat the hair. Force calibration plots indicate that commercial conditioner containing only nonpolar silicones and experimental conditioner containing polar amino silicones exhibit distinct affinity for chemically damaged hair surface. Commercial conditioner physically adsorbed on hair surface via van der Waals attractions can be easily squeezed out from the contact region under the load; on the other hand, amino silicones in experimental conditioner are substantive to the hair surface and cannot escape from the contact region under the load because of the strong electrostatic binding between the polar amino groups of silicone molecules and hair surface. Amino silicones provide much better load-bearing capacity and works as a cushion preventing hair surface from further damage.

Coefficient of friction values are largest on the macroscale, followed by microscale and then nanoscale values. In general, coefficient of friction increases with an increase in the AFM tip radius. Nanoscale friction data is taken with a sharp AFM tip, while the macro- and microscale tests have contacts which range from nanoasperities to much larger asperities which may be responsible for larger values of friction force on these scales. The combination of higher normal load with a larger contact area (due to contact with multiple fibers at the same time) may also be responsible for increased coefficient of friction on the macroscale as well.

On all scales, coefficient of friction decreases for chemically damaged treated hair compared to untreated. The same trend occurs for virgin-treated hair on the macro- and microscales, but not on the nanoscale. On the larger scales, the thin layer of conditioner acts as a lubricant over the hair fiber across multiple cuticle contacts. On the nanoscale, the hydrophobicity of the virgin hair causes different deposition of the conditioner which increases the adhesive force contribution and results in an increased coefficient of friction. For chemically damaged hair, there is higher negative charge and a hydrophilic surface, which results in more uniform deposition and better smearing of the conditioner layer, which serves to lower coefficient of friction between the tip and the cuticle surface.

24.7 Conditioner Thickness Distribution and Binding Interactions on Hair Surface

How common hair care products, such as conditioner, deposit onto and change hair properties are of interest in beauty care science, since these properties are closely tied to product performance. Conditioner is one of hair care products which most people use on a daily basis. Conditioner thinly coats hair and can cause drastic changes in the surface properties of hair. Among all the components of conditioner, silicones are the main source of lubrication in the conditioner formulation. Silicone molecules remain as droplets surrounded by water, and their high molecular weight causes them to remain liquid and drain off of hair surface gradually, which creates a long-lasting, soft, and smooth feel for conditioner treated hair. The binding interaction between these molecules and hair surface is one of important factors in determining the conditioner thickness distribution, and consequently the proper functions of conditioner.

Some important issues, such as the thickness and distribution of conditioner on the hair surface which is important in determining the proper functions of conditioner, have been struggled to answer by cosmetic scientist for decades. In order to determine the thickness and distribution of thin liquid film on a substrate, several possible techniques can be applied: Fourier transform infrared spectroscopy (FTIR), ellipsometry, angle-resolved X-ray photon spectroscopy (XPS), and AFM [8, 10]. Ellipsometry and angle-resolved XPS have excellent vertical resolution on the order of 0.1 nm, but their lateral resolutions are on the order of 1 and 0.2 mm, respectively. Therefore, these techniques cannot be used for hair since the diameter of a hair fiber is only 50–100 µm. AFM on the other hand is a versatile tool and has a lateral resolution on the order of the tip radius (few nm), which is difficult to achieve by other techniques. In the study reported here, force calibration plot measurements using an AFM are conducted to obtain the local conditioner thickness distribution on various hair surfaces. The conditioner thickness is extracted by measuring the forces on the

AFM tip as it approaches, contacts, and pushes through the conditioner layer. The conditioner thickness distribution on hair is effectively measured using an AFM. The effective Young's moduli of various hair surfaces are also calculated from the force distance curves using Hertz analysis. The binding interactions of different silicones on the hair surface, as well as the effect on their effective Young's modulus of the hair are also discussed.

It has been reported earlier that, based on surface imaging using TR mode and friction force mapping on couple of cuticles, the conditioner is unevenly distributed across the hair surface, and thicker conditioner layer can be found near cuticle edges.

24.7.1 Conditioner Thickness and Adhesive Force Mapping

Figure 24.21 presented earlier shows a typical force calibration plot for commercial conditioner treated hair, and the detailed interactions of the AFM tip, conditioner and hair surface. Conceptually, the distance that the sample moves after the tip snaps in until it contacts the hair surface should be a measure of the thickness of the conditioner film. Here we define it as snap-in distance H_s. This snap-in distance H_s is not the real conditioner thickness h and it tends to be thicker than the actual film thickness. In previous studies on determination of lubricant film thickness on a particulate disk surface by AFM [11], it has been realized that as the ellipsometry thickness of the lubricant film increases, the thickness as measured by AFM also increases, however, AFM tends to measure a thicker film than measured by ellipsometry. There is a few nm offset between the snap-in distance H_s determined by AFM and film thickness h determined by ellipsometry. The most likely origin of the offset of the AFM thickness was attributed to a thin coating of the lubricant on the surface of the AFM tip that results from the tip being previously in contact with the lubricant film. This might account for part of the offset, but main reason for this offset should be attributed to the deformation of the liquid film due to its interaction with the AFM tip as the liquid film approaches to the tip [26, 27, 53]. The liquid film can only approach the AFM tip to a finite minimal distance, below which the liquid surface is no longer stable due to the van der Waals attractive force between the liquid film and the AFM tip. For smaller distances, surface tension and adhesion to the substrate cannot keep the liquid surface from bulging and jumping into contact with the tip. *Forcada* et al. [34] theoretically analyzed the hydrostatics of the liquid film in the force fields originated by the tip and solid substrate, and indicated that the offset between the snap-in distance and the film thickness measured by ellipsometry arises from the bulging and posterior instability of the liquid film.

A number of theoretical publications [1, 5] have addressed the issue of liquid coalescence in terms of the 'effective stiffness' of a liquid surface or interface, and have concluded that a liquid surface behaves like a Hookian spring with an effective spring constant K_{eff} equal to its surface or interfacial tension γ, viz:

$$K_{\text{eff}} \approx \gamma \qquad (24.6)$$

Therefore, for one of the main components of conditioner, silicone (PDMS) in air, an effective spring constant of only 20 mN/m is expected, as the surface tension of

PDMS is about 20 mN/m [25] and water film has an effective spring constant of 72 mN/m. These are extremely low values compared to the cantilever spring constant of 5 N/m. It suggests that liquid surfaces can deform even by a very weak force. When the hair surface approaches the AFM tip, the weak van der Waals attractive force between the liquid conditioner film and the tip will cause the liquid film to deform and jump-in the tip at a finite distance. The theoretically expected van der Waals force F_{vdw} between a sphere and a flat surface is given by

$$F_{\text{vdw}} = HR/6D^2 \tag{24.7}$$

where H is the Hamaker constant which can be estimated based on the Lifshitz theory [39], R is the radius of the AFM tip, and D is the distance between the AFM tip and the liquid conditioner film. Therefore, the jump-in distance D_J (the minimal distance between a stable liquid film and the AFM tip) can be theoretically calculated based on the criterion for a jump instability:

$$\mathrm{d}F_{\text{vdw}}/\mathrm{d}D = HR/3D_J^3 = K \tag{24.8}$$

where K is the spring constant. For conditioner treated hair surface, K is the effective spring constant of the liquid film K_{eff} since it is much weaker than that of the cantilever. From (24.8), one obtains [27],

$$D_J = (HR/3K_{\text{eff}})^{1/3} \tag{24.9}$$

For a silicon tip covered by SiO_2 (the sphere) interacting with a liquid conditioner film (the flat surface) in air, the Hamaker constant H is estimated to be in the order of 10^{-20} J, R is about 100 nm and K_{eff} is in the range of 20–72 mN/m (the surface tension of PDMS and water are 20 mN/m and 72 mN/m, respectively). Then a jump-in distance D_J about 2 nm can be obtained based on Eq. (24.9). This jump-in distance is basically the offset between the snap-in distance H_s and the actual film thickness h. This 2 nm offset (jump-in distance) is surprisingly close to previous measurements on the localized lubricant-film thickness on a particulate magnetic rigid disk, which indicated that the measured thickness using an AFM is about 2 nm larger than the actual thickness based on the ellipsometry measurements [11]. Note that the offset (jump-in distance D_J) only depends on the radius of the tip and the intrinsic properties of the film (surface tension and Hamaker constant), but is independent of the film thickness. Surface forces apparatus (SFA) experiments performed by *Chen* et al. [28] indicated that for two 25 nm thick liquid PDMS films interacting in air at quasi-equilibrium state (very slow approach rate of about 0.3 nm/s), the jump-in distance is about 200 nm. SFA measurements gave much large jump-in distance because of the much larger value of radius. (In SFA experiments, the radius R is about 20 mm.) Therefore, although the snap-in distance H_s in force calibration plot overestimates the actual film thickness (approximately 2 nm thicker), it still provides a very good measurement for the actual thickness of thin conditioner film.

Figure 24.73 shows typical film thickness (snap-in distance H_s) and adhesive force mappings of virgin, chemically damaged, chemically damaged treated (1 cycle and 3 cycles) hair. The snap-in distance H_s of virgin hair is 2.0 ± 0.6 nm with

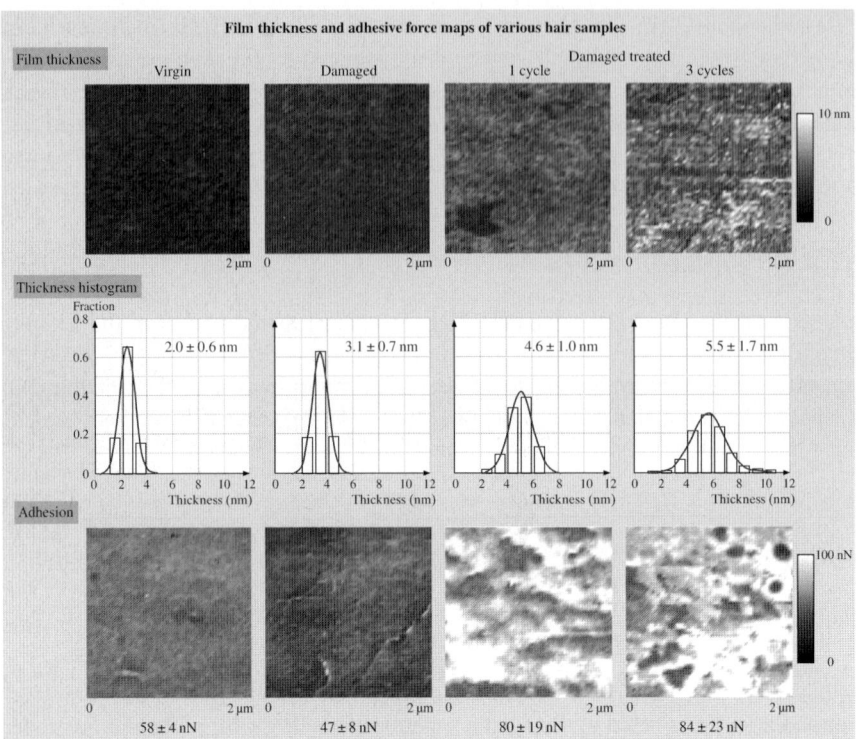

Fig. 24.73. Film thickness maps, histograms, and adhesive force maps of Caucasian virgin, chemically damaged, and damaged treated hair samples. The dotted lines in histograms are the Gaussian fits for film thickness [13]

a very narrow distribution, indicating that virgin hair surface is relative uniform and is not damaged. Virgin hair surface is covered with a saturated fatty acid lipid layer called 18-methyleicosanoic acid (18-MEA) [68], which makes the hair surface hydrophobic (see Table 24.11 for contact angle data). Therefore, virgin hair surface does not consist continuous water film. Instead of the deformation of the liquid film, the AFM tip will jump into contact with the hair surface at a finite tip-sample distance because of the van der Waals attractive force.

The snap-in distance of chemically damaged hair is larger than that of virgin hair, and increased to 3.1 ± 0.7 nm. The outermost surface of hair is cuticle which consists of large amount of cystine. Chemical treatment will partially remove the fatty acid lipid layer (18-MEA) covering the hair surface and break the disulfide bonds in cystine to form new ionic groups (such as cysteic acid residues by oxidation of cystine acids). The hair surface becomes hydrophilic and the contact angle with water decreases [61], consequently, the amount of water adsorbed on the hair surface increases. As the sample approaches the tip, the AFM tip will jump into contact with the hair surface at a finite distance because of van der Waals attractive force, as well as small meniscus bridge formation between the tip and adsorbed water layer. The adhesive force of chemically damaged hair surface is 47 ± 8 nN, less than that of virgin hair (58 ± 4 nN). The existence of thin adsorbed water layer tends to decrease van der Waals force attractive force but increases the meniscus force. Note that the tip and the hair surface are not as smooth as shown in Fig. 24.20 (rough on the molecular scale) and the adsorbed water layer is very thin, therefore, no single large meniscus but many small menisci form between the tip and the hair surface. The total attractive force of chemically damaged hair therefore is smaller than that of virgin hair.

After conditioner treatment, the snap-in distance H_s increases with the number of cycle of treatment. The snap-in distances H_s for 1 cycle treatment and 3 cycle treatments are 4.6 ± 1.0 nm and 5.5 ± 1.7 nm, respectively. Thicker liquid film is unevenly present on conditioner treated hair surface, and the thickness tends to have a broader distribution than that of chemically damaged hair. It is more obvious for 3 cycles damaged treated hair, in which the thickness distribution is much broader and tends to have a long tail to larger thickness. The measured mean thickness varies from 5 nm up to 25 nm (data not shown). Excluding the 2 nm offset due to the jump-in deformation of the liquid conditioner film, the actual film thickness should be around 3 to 23 nm. These values are consistent with the previous estimated conditioner thickness based on the amount of material deposition [48] (see Appendix B). The adhesive forces for 1 cycle and 3 cycles are 80 ± 19 nN and 84 ± 23 nN, respectively. The larger adhesive force is attributed to the formation of large meniscus between the tip and the conditioner layer on the hair surface.

Effect of Humidity and Temperature on Film Thickness and Adhesion

Figure 24.74 shows typical film thickness mapping, histogram of film thickness, and adhesive force mappings of damaged treated hair at different humidities [13]. At different humidities, the film thickness and adhesive force of hair are distinctly different. At low humidity, the treated hair surface has only a very thin layer of film (2.6 ± 0.7 nm). At high humidity, a much thicker film is present on hair surface (4.7 ± 1.8 nm). This film tends to extend a much longer tail into the thicker film regime. At high humidity, hair surface has higher adhesive force compared to that at low humidity. The conditioner-treated hair surface will be covered with a gel network. This gel network can adsorb or desorb water depending on the environmental condition. The conditioner treated hair surface tends to adsorb more water as the humidity increases, which will increase the film thickness on hair surface and consequently increases the adhesive force.

Figure 24.75 summarizes the conditioner thickness and adhesive force of various hair surfaces at different humidities [13]. Film thickness on virgin hair remains relatively constant at different humidities and it only increases slightly at high humidity. The outer layer of the hair is covered with hydrophobic 18-MEA and remains intact in virgin hair. Therefore, water is hardly adsorbed (or desorbed) and penetrated into hair surface and humidity has little effect on film thickness of virgin hair. On the other hand, the hydrophobic lipid layer of damaged hair surface may

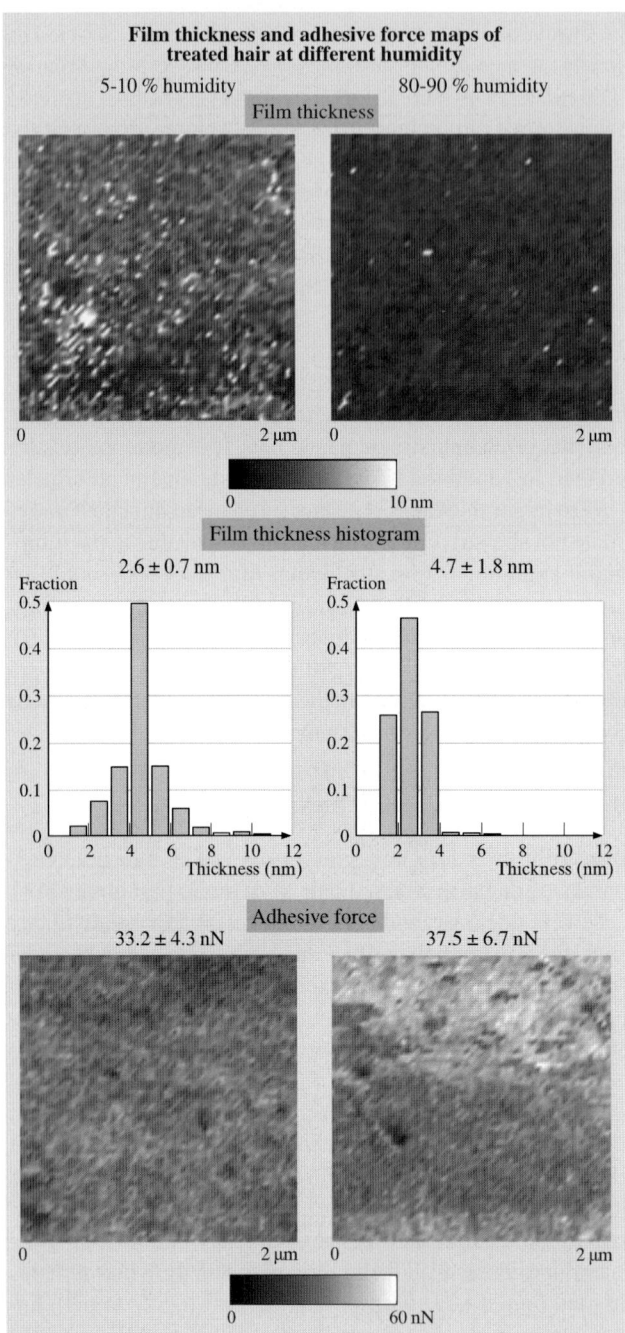

Fig. 24.74. Typical film thickness and adhesive force maps of damaged treated hair at different humidities [13]

be depleted or damaged, therefore, inner cellular structures of hair, which consist of many hydrophilic molecules are now exposed to water. Damaged hair surface is partially hydrophilic and will be more sensitive to the humidity. It tends to adsorb more water at high humidity; therefore, film thickness of damaged hair increases gradually as the humidity increases. Damaged treated hair is covered with a conditioner film; therefore, it has a thicker film thickness. Film thicknesses of treated hair surface typically show large deviations compared to those of virgin and damaged hair, indicating uneven distribution and adsorption of conditioner layer on hair surface. The film thickness increases as the humidity increases because the conditioner

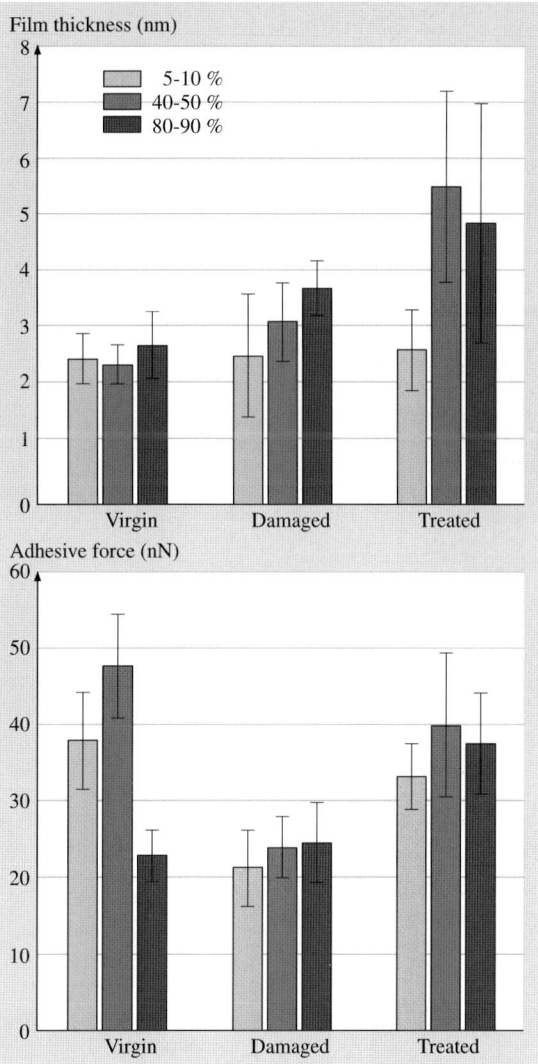

Fig. 24.75. Film thickness and adhesive forces of Caucasian virgin, chemically damaged, and damaged treated hair samples at different humidities. Large deviations compare to average values, especially for conditioner-treated hair surface, indicating uneven distribution and adsorption of conditioner layers [13]

gel network layer can easily adsorb more water at high humidity. However, this conditioner gel network layer fails to retain water content at low humidity. As shown in Fig. 24.75, virgin hair usually has higher adhesive force than damaged hair. Adhesive force includes the van der Waals attraction and meniscus force. Virgin hair has larger van der Waals force than damaged hair because of the hydrophobic nature of the virgin hair surface. Treated hair has higher adhesive force due to conditioner meniscus formation.

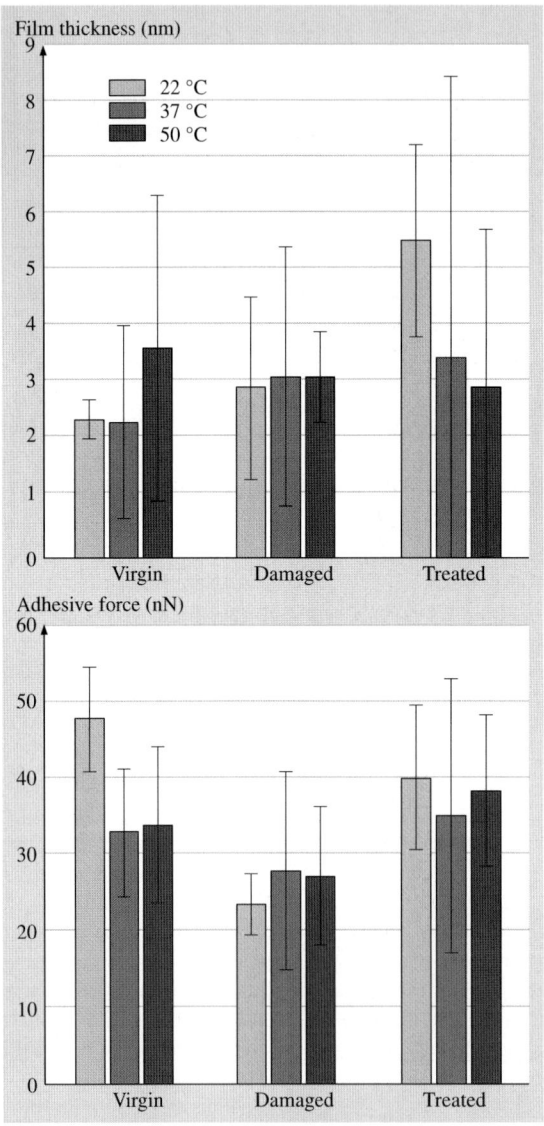

Fig. 24.76. Film thickness and adhesive forces of Caucasian virgin, chemically damaged, and damaged treated hair samples at different temperature [13]

The film thickness and adhesive force were also measured at different temperatures [13]. Figure 24.76 summarizes the data for virgin, damaged, and conditioner treated hair at those temperatures. Temperature has little effect on the film thickness of virgin or damaged hair in the temperature range studied. Conditioner treated hair has thicker film compared to virgin or damaged hair. As temperature increases, the thickness of conditioner layer starts to decrease. At high temperature, conditioner layer will eventually lose all water content. In the studied temperature range, there is little effect on adhesive force of various hair surfaces.

24.7.2 Effective Young's Modulus Mapping

The adhesive force as well as the total force acting on the tip at each measurement point can be accurately measured from the force calibration plot. If the zero tip-sample separation is defined to be the position where the force on the tip is zero when in contact with the sample, the force calibration plot (cantilever deflection vs. piezo position plot as shown in Fig. 24.21) can be converted to a force vs. tip-sample separation curve [27]. Figure 24.77a shows the forces acting on the tip as a function of tip-sample separation for virgin, commercial conditioner (PDMS blend silicone) treated, and experimental conditioner (with amino silicone) treated hair. The lowest point on the approach curve is assumed to be the point that the tip contacts the hair surface. Afterward, the hair surface deforms elastically under the load, from which the deformation (indentation Δz) of the surface can be extracted. Plotting the obtained deformation (indentation Δz) against the total force (on approach curve) acting on surface, gives force vs. indentation (deformation) curves. Figure 24.77b shows the force vs. indentation curves for virgin, commercial conditioner treated and experimental conditioner (amino silicone) treated hair surface which are extracted from the force vs. tip-sample separation curves shown in Fig. 24.77a, and the effective Young's modulus of various hair surfaces can be determined from these curves by fitting them to (24.2). The effective Young's modulus of virgin, commercial conditioner treated and experimental conditioner treated hair surface are 5.3 ± 0.9 GPa, 0.60 ± 0.03 GPa and 0.032 ± 0.002 GPa respectively for these three specific curves. For these calculations, the radius of the tip is measured to be approximately 100 nm. The effective Young's modulus of conditioner treated hair surface can be one to two orders of magnitude less than that of virgin hair.

Repeating these calculations over the whole surface on various hair samples, the maps of the effective Young's modulus of various hair surfaces are obtained as shown in Fig. 24.76. Virgin hair surface has the effective Young's modulus of about 7.1 ± 2.9 GPa, which is the most stiff among all the samples and is consistent with previous nanoindentation measurement results [90]. The rich in the disulfide crosslinks on the outermost layer of the hair surface accounts for this stiffness; the effective Young's modulus of chemically damaged hair surface (6.6 ± 3.3 GPa) is slightly smaller than that of virgin hair surface. The chemical treatment partially breaks the disulfide crosslinks in the outermost layer of the hair surface, and weakens the hair surface. Commercial conditioner treated hair surface tends to have much smaller effective Young's modulus than that of chemically damaged hair surface

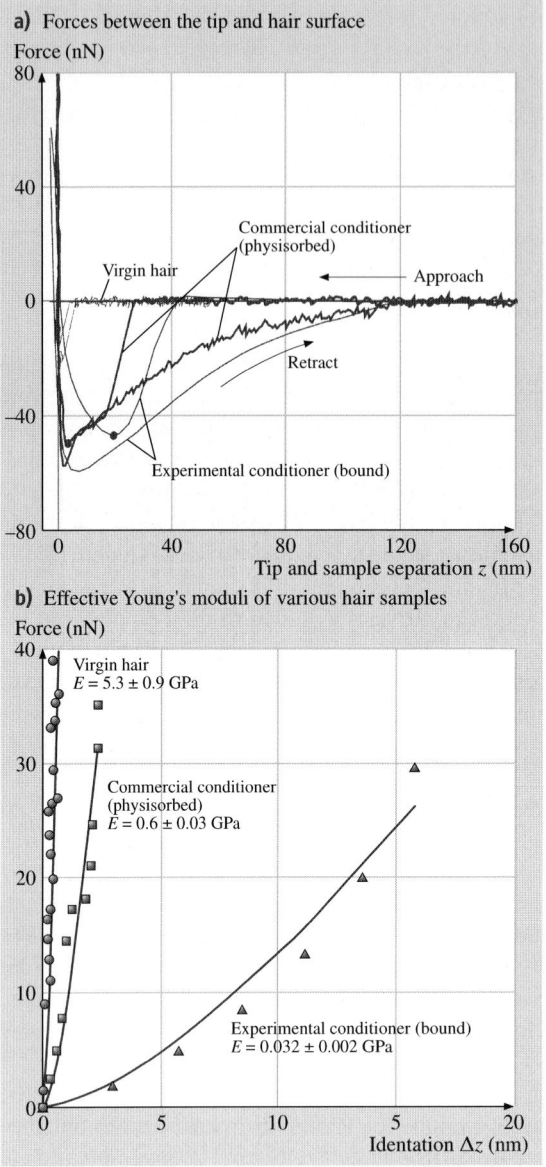

Fig. 24.77. (a) The force acting on the tip as a function of tip sample separation for various hair samples. The sample is moved with a velocity of 200–400 nm/s, and the zero tip sample separation is defined to be the position where the force on the tip is zero when in contact with the sample. The arrows show the direction of motion of the sample relative to the tip. A negative force indicates an attractive force; (b) surface effective Young's modulus extracted from the force distance curves. The solid lines show the best fits for the data points of various hair samples [27]

(5.5 ± 4.6 GPa for 1 cycle treatment, and 3.3 ± 3.2 GPa for 3 cycles treatment). The effective Young's modulus decreases as the number of cycle of treatment increases. Large deviations of the effective Young's modulus of conditioner treated hair surfaces are seen, indicating the uneven distribution and adsorption of conditioner on

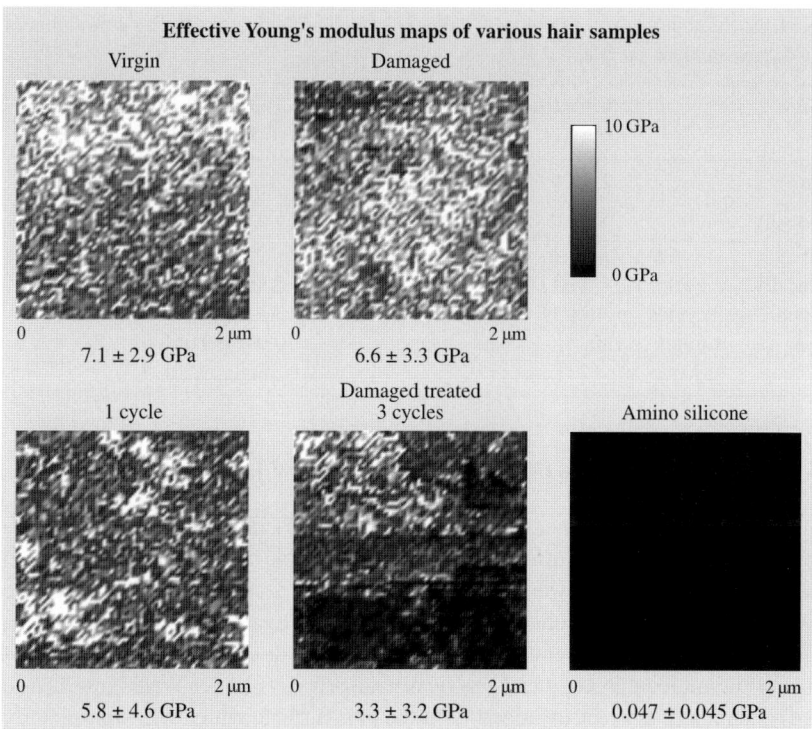

Fig. 24.78. Effective Young's modulus map of Caucasian virgin, chemically damaged, and damaged treated hair samples. The average value and standard deviation for each map are listed. Large deviations compare to average values, especially for conditioner treated hair surface, indicating uneven distribution and adsorption of conditioner layers [27]

hair surface. This large deviation of the effective Young's modulus of conditioner treated hair surface is due to a large area of the hair surface has low effective Young's modulus (dark region). Comparing the conditioner thickness map (see Fig. 24.76) and the effective Young's modulus map (see Fig. 24.78) of 3 cycles treated sample, it is clear that the region with small effective Young's modulus (dark region) is closely correlated to the region with thick conditioner layer (bright region) on the hair surface. The effective Young's modulus decreases as conditioner thickness increases. Therefore, although the average value of the effective Young's modulus of 3 cycles treated hair surface is 3.3 GPa, some part of the surface can have very low effective Young's modulus value (as low as the 0.60 GPa value shown in Fig. 24.77b). Experimental conditioner (amino silicone) treated hair surface has extremely low effective Young's modulus (0.047 ± 0.045 GPa) compared to the value of chemically damaged hair surface. The film thickness, adhesive force and effective Young's modulus of various hair samples are summarized in Table 24.18 [27].

Table 24.18. Summary of film thickness, adhesive force and effective Young's modulus of various Caucasian hair samples

Samples		Film thickness (nm)	Adhesive force (nN)	Young's modulus (GPa)
Virgin		2.0 ± 0.6	58 ± 4	7.1 ± 2.9
Chemically damaged		3.1 ± 0.7	47 ± 8	6.6 ± 3.1
Chemically damaged treated	1 cycle	4.9 ± 1.0	80 ± 19	5.8 ± 4.6
	3 cycles	5.5 ± 1.7	84 ± 23	3.3 ± 3.2
	Amino silicone	n/a	n/a	0.047 ± 0.046

24.7.3 Binding Interactions Between Conditioner and Hair Surface

Small molecules in conditioner, such as water and surfactant, may diffuse into the outermost layers of the hair surface and swell the disulfide crosslink network within these layers, which in a way weakens these layers, consequently, a smaller effective Young's modulus for conditioner treated hair surfaces. Even more significantly, although silicones in conditioner cannot diffuse into the hair surface layers because of the size of the molecules, they can physically adsorb via van der Waals interaction or bind (chemically or electrostatically) on the hair surface, which will significantly affect the physical properties of the hair surface. Although the AFM cannot give an absolute distance as the surface forces apparatus (SFA) [39], the force distance curves obtained using an AFM still can show the interactions between the surface and the tip, and consequently reveal the essence of various surfaces. As shown in Figs. 24.21 and 24.77a, virgin hair and conditioner treated hair behave differently as the tip approaches, compresses and retracts from the surface. Virgin hair surface behaves like a stiff elastic solid surface. As the virgin hair surface approaches to the tip, the tip will jump into contact with the hair surface at a small separation around 4 nm because of the van der Waals attractive interaction between the hair surface and the tip. Then the force quickly goes to repulsive and the tip reaches the hard wall contact (tip hair solid-solid contact) with very small deformation. When the virgin hair surface is withdrawn, the tip will simply jump out and no large liquid deformation occurs because of the lack of liquid layer.

Conditioner treated hair surfaces show much longer range of interactions than virgin hair because of the presence of liquid conditioner layer. The tip will snap in at large separation because of the van der Waals force as well as meniscus formation between tip and conditioner liquid layer. When the sample is pulled away from the tip, the forces on the tip will slowly decrease to zero as a long meniscus bridge of liquid is drawn out from the surface and eventually breaks at a large separation. Another important feature is that before the tip reaches the tip-hair solid-solid contact, commercial conditioner treated hair surface shows larger deformation than virgin hair surface, indicating a thin physically adsorbed layers present on the hair surface. Experimental conditioner treated hair surface shows very large deformation and no tip-hair solid-solid contact is reached under the given experimental conditions, indicating there is a strongly bound conditioner layer. Large hysteresis between approaching and retracting curves indicates that this bound layer deforms viscoelastically under the load. The adhesive force (on the approach curve) of experimental conditioner treated hair surface is smaller than that of commercial conditioner treated hair surface. The meniscus forces in both cases would be similar, and the extra repulsive force on the tip for experimental conditioner treated surface comes from the compression of the molecules underneath the tip. The tip has to overcome the intermolecular polar or hydrogen bonding and strong binding between amino silicone and the hair surface so as to squeeze out silicone molecules from between the tip and the hair surface.

Figure 24.79 illustrates the difference between physisorbed conditioner and bound conditioner. Chemical treatment breaks the disulfide bonds of cystine crosslink network in cuticle and form new ionic groups on the hair surface, therefore, chemically damaged hair surface is negatively charged. Silicones in commercial conditioner are non-polar and do not contain any functional groups, therefore, most of silicone molecules are free and mobile except that the last layer of molecules adjacent to the hair surface may be physically adsorbed via van der Waals interactions. As the tip starts to touch and compress the conditioner film, free conditioner molecules can easily escape from the contact region. The last layer of adsorbed molecules may sustain some load before they are squeezed out eventually at higher load. Then the tip penetrates the conditioner layer and reaches tip hair solid-solid contact and compresses directly on the hair surface. The last layer of physisorbed conditioner molecules is very thin but effectively lowers the Young's modulus of the hair surface by one order of magnitude (see Figs. 24.77b and 24.78).

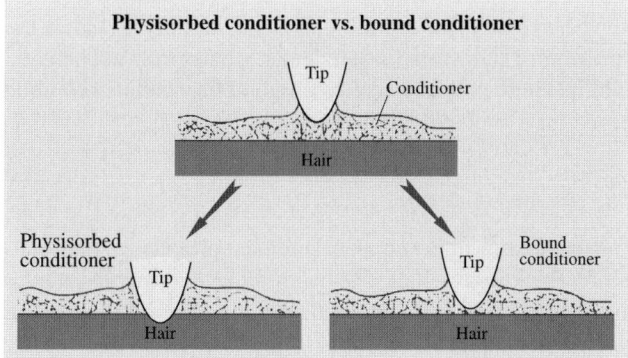

Fig. 24.79. Comparison between physisorbed conditioner and bound conditioner. Physisorbed conditioner can be easily squeezed out from the contact region under the load; on the other hand, bound conditioner cannot escape from the contact region under the load because of the strong electrostatic binding between the silicone molecules and the hair surface (black dots are the binding sites) [27]

On the other hand, experimental conditioner consists of amino silicones, which are positively charged in an aqueous condition and can strongly bind with the anionic groups present on chemically damaged hair surface. The binding between amino silicone molecules and the hair surface, which are much stronger than van der Waals attractions in commercial conditioner treated hair surface, make bound conditioner molecules difficult to escape from the contact region even at high load. Therefore, after the tip touches the conditioner, instead of penetrating the conditioner layer and compressing directly on the hair surface, it compresses on a soft bound conditioner layer. The effective Young's modulus of this layer (silicones) is about 0.047 ± 0.045 GPa. Although it is much softer than hair, it is still much stiffer than bulk PDMS whose Young's modulus is only 680 kPa [58]. Two reasons are attributed to this increase of effective Young's modulus of silicones (PDMS): the bound silicone layer is very thin (a few nm) and it is in a confined geometry during the measurements. It is well known that thin films can exhibit extremely different physical properties compared to bulk materials. Thin liquid film can behave like a solid under confinement [39]. On the other hand, underneath of this thin bound silicones layer is the stiff hair surface whose Young's modulus is much higher (chemically damaged hair surface 6.6 ± 3.3 GPa, see Fig. 24.78). Since the bound layer is very thin, the measured Young's modulus therefore has the contributions from both the soft bound silicone layer and stiff hair surface.

Amino silicones work as a cushion protecting the hair surface. Although this strong affinity to the hair surface can dramatically increase the load or contact pressure that a conditioner film can support before solid-solid contact, an AFM study on hair [48] indicated that treated hair with excess amino silicone has higher coefficient of friction than commercial conditioner (physisorbed) treated hair. It's well known that in general, lubricant film with mobile and immobile fractions (partially bonded lubricant film) is desirable for low friction and high durability [7, 9]. Biolubrication studies using SFA [3] also indicated that a low coefficient of friction is not necessarily a measure of good wear protection. A conditioner layer strongly bound on the hair surface is good for preventing hair from further damage, but it may not be good from the lubrication point of view. A balance between good lubrication and good protection has to be reached for a good conditioner product.

24.7.4 Summary

The snap-in distance H_s in the force calibration measurement provides a good estimate for the conditioner film thickness on hair. The conditioner unevenly distributes on hair surface. Conditioner thickness increases with the number of the cycles of treatment. The mean conditioner thickness is on the order of 5 nm. The adhesive force of damaged hair was smaller than that of virgin hair because of different surface hydrophobicities. The treated hair showed a much higher adhesive force because of meniscus formation at the interface. Humidity had little effect on film thickness and adhesive force for virgin hair, however it had a larger effect on damaged and treated hair. Temperature effect was little on all hair. The effective Young's modulus of virgin hair surface was the highest and damaged hair was slightly smaller and treated hairs had the smallest. Amino silicone conditioner showed strong affinity to hair and strengthened the hair fiber and is expected to work as a cushion for protecting the hair surface from further damage.

24.8 Surface Potential Studies of Human Hair Using Kelvin Probe Microscopy

It is obvious that during combing and running the hands through one's hair that physical damage is likely to occur. These actions also tend to create an electrostatic charge on the hair, and it is of interest whether or not the physical wear and the electrostatic charge are related, or if charge by other mechanism has been created. It is therefore the aim to observe the effects of physical wear on surface potential for both conducting and insulating samples to clarify the mechanism behind the behavior. Further, electrostatic charging of hair is studied by charging the hair with latex, a material that is known to create a static charge on hair. The presence of this static charge in the real world is a major problem concerning hair manageability, flyaway behavior, feel and appearance so understanding the mechanisms behind charge buildup, and how to control it, is thus a focal point in designing effective hair care products such as conditioner. Conditioner is known to affect the surface potential characteristics of hair, and understanding this behavior on the small scale is of great interest. Because surface potential of human hair is of such interest, Kelvin probe may provide great insight into the mechanisms behind the electrostatic behavior of hair. For this reason, Kelvin probe has been used to characterize hairs with varying types and degrees of damage as well as varying types and degrees of conditioner treatments.

24.8.1 Effect of Physical Wear and Rubbing with Latex on Surface Potential

Lodge and *Bhushan* [55] measured surface potential change after performing wear experiments in an AFM on hair fiber using a diamond tip with a tip radius of 60 nm at a load of 5 μN. The wear scars were created at 2×2 μm size for one pass. After samples were worn with the diamond tip, the AFM tip was changed to conductive tip, and the surface potential was mapped in the wear area. All samples were mounted on a sample puck in conductive silver paint. They found that physical wear alone does not cause a measureable electrostatic charge on hair because hair is nonconductive.

The electrostatic charge build up on hair was caused by rubbing with a natural latex finger cot and surface potential change was measured after rubbing [55] First, a surface potential map of the hair sample, which was electrically isolated from ground to prevent quick discharging was obtained for "before rubbing" data. Then samples were wiped lightly with the finger cot 5 times along the hair shaft in the direction from root to tip. Surface potential was then mapped immediately, for "after rubbing" data. Figures 24.80a,b show AFM images for virgin, virgin treated, chemically damaged, chemically damaged treated (1 cycle), and chemically damaged

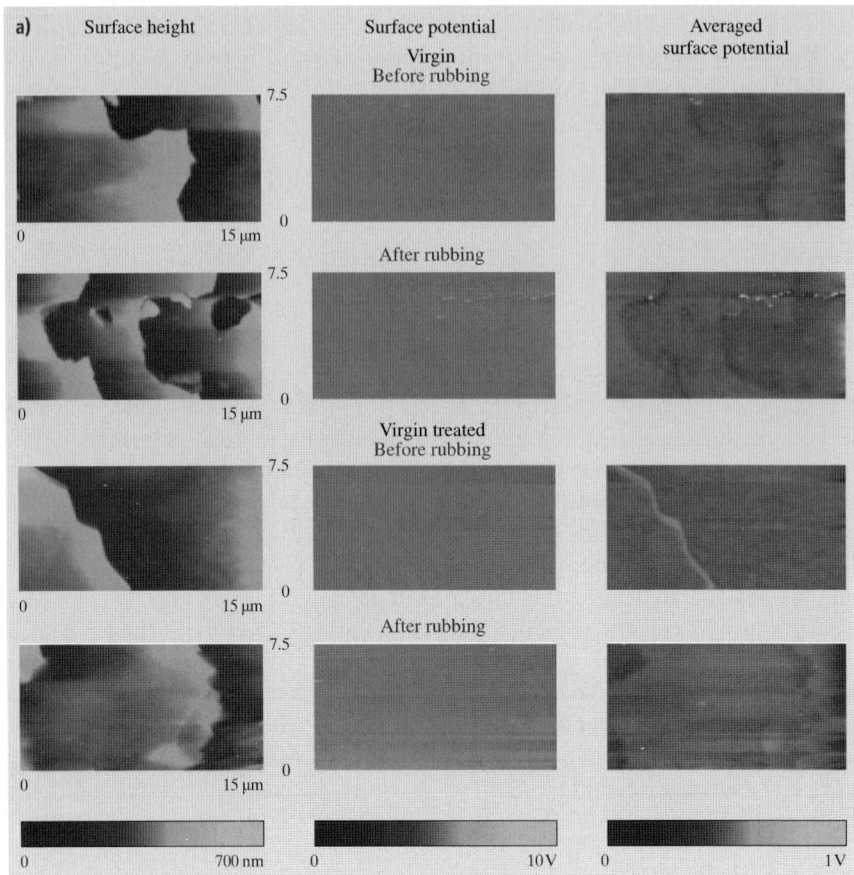

Fig. 24.80. (a) AFM images of virgin and virgin treated hair samples both before and after rubbing with latex

treated (3 cycles) hair samples. These images are presented in three columns. The first is the surface height of the sample and the second is the absolute surface potential. The final column shows the same surface potential data as the second column, but the average surface potential (which is presented in the subsequent bar charts) is subtracted out and the scale is reduced to show more contrast. Before rubbing data was taken as a baseline in order to report a change in potential due to rubbing, and will not be discussed or analyzed further. Absolute quantitative values of surface potential were not reliable for this experiment because the samples were not grounded, thus the necessity to report a potential change. It should be noted that some samples became highly charged as a result of the rubbing, beyond the capability of the microscope (10 V). These samples were not used in this study. However, it is important to note that all of the samples tested are capable of developing high amounts of charge. A method where the force of contact is more accurately controlled may

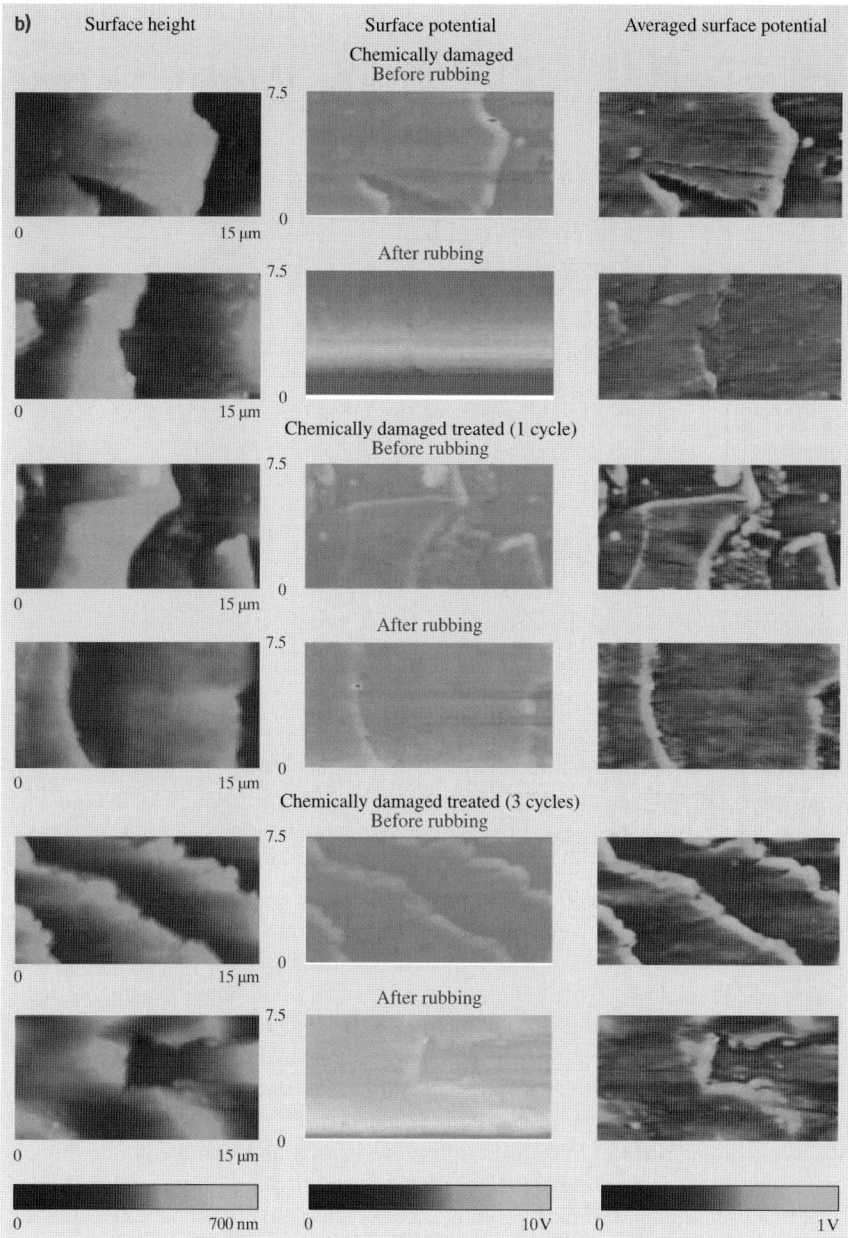

Fig. 24.80. (continued) **(b)** chemically damaged, chemically damaged treated (1 cycle), and chemically damaged treated (3 cycles) hair samples both before and after rubbing with latex

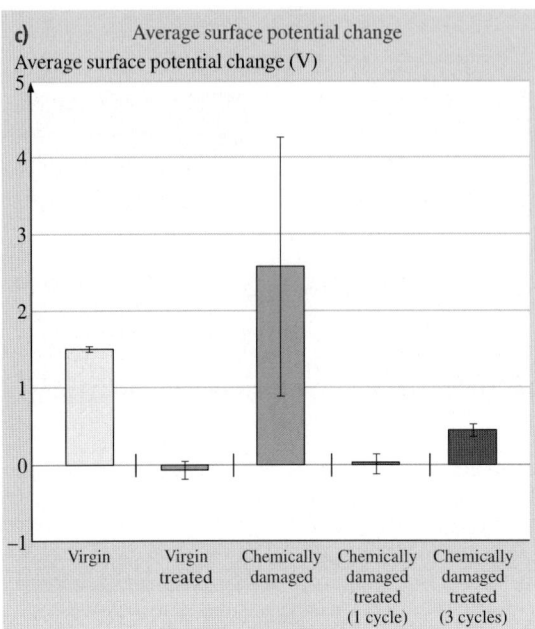

Fig. 24.80. (continued) (c) bar charts of average surface potential change after rubbing with latex [55]

provide better insight into this behavior. Some samples, especially the chemically damaged one (Fig. 24.80a), show a significant amount of change in surface potential during measurement. This is illustrated as a higher potential at the bottom of the image than at the top, as the measurement was started at the bottom. This indicates that the charge dissipates relatively rapidly, and seems to reach a constant value by the time the measurement is finished. The fact that the virgin and treated samples (Fig. 24.80a) seem to have a more constant potential during measurement indicates that these samples dissipate charge faster, and by the time the measurement is taken a constant potential is more or less achieved. The average surface potential change is shown in Fig. 24.80c. This bar chart indicates that conditioner treatment greatly reduces the amount of charge present on the hair surface. It also shows that chemical damage increases the amount of charge built on the hair surface.

Figure 24.81 shows chemically damaged hair that has been treated with an experimental conditioner containing amino silicone, which is believed to chemically attach to the hair surface. Very little difference is seen between the amino silicone conditioner and the commercial PDMS silicone conditioner. Commercial PDMS silicone treated hair data was shown in Fig. 24.80b. This is likely because the latex rubbing occurs over a very large area of the hair, and any difference in conditioner distribution is occurring on a much smaller scale. PDMS silicone conditioner is believed to only physically attach to the hair surface, and it remains mobile. Therefore, rubbing with the finger cot is likely to redistribute the conditioner, effectively spreading it evenly over the hair. However, amino silicone conditioner chemically attached to the surface, and distributes evenly to begin with. Moreover, amino silicone condi-

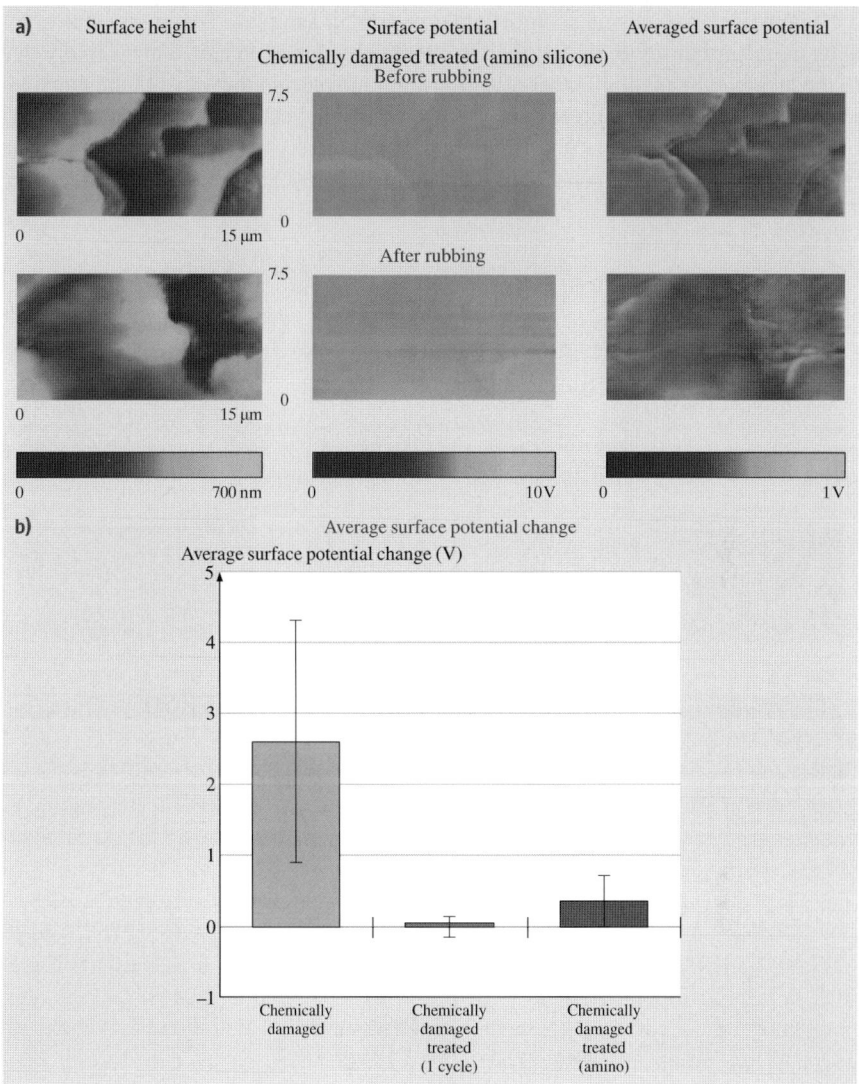

Fig. 24.81. (a) AFM images of chemically damaged treated (amino) hair both before and after rubbing with latex; (b) bar chart showing average surface potential change after rubbing with latex [55]

tioner remains evenly distributed due to the chemical attachment. A method to rub the hair over a smaller area may provide more insight into this behavior, and may elucidate the effect of conditioner distribution on surface potential properties. However, the amino silicone conditioner still shows a significant decrease in charge from the untreated chemically damaged sample.

Figure 24.82 shows mechanically damaged hair samples. This figure shows results more similar to virgin hair than to chemically damaged hair. This is likely due to the fact that the top lipid layer of hair is not removed and that most of the mechanical damage occurs at scale edges. Chemical damage occurs over the entire surface of the hair. For this reason it is likely that mechanically damaged hair will behave more like virgin hair than chemically damaged hair. Because the rubbing procedure

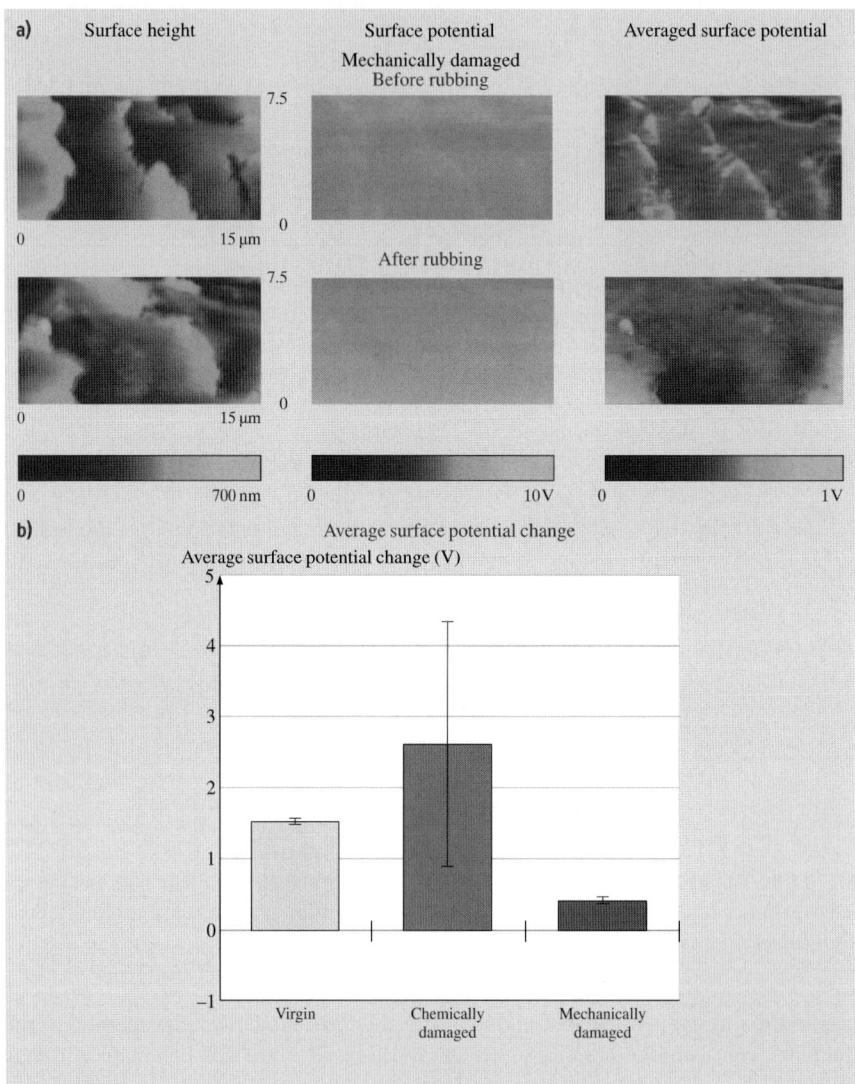

Fig. 24.82. (a) AFM images of mechanically damaged hair both before and after rubbing with latex; (b) bar chart showing average surface potential change after rubbing with latex [55]

is not strictly controlled in terms of location of contact, it is difficult to distinguish a difference between the topmost layer of the hair, and the portions which have been mechanically damaged. However, this indifference between these regions (which generally occur on the surface away from scale edges, and near scale edges, respectively) may also be due to the fact that physical wear does not affect surface potential in insulating samples, as discussed previously.

24.8.2 Effect of External Voltage and Humidity on Surface Potential

Samples were mounted in conductive silver paint and an external power supply was used to develop charge [56]. To determine environmental effects on surface charge, relative humidity was controlled and varied.

To study the effects of external voltage and humidity, virgin, virgin treated, chemically damaged, chemically damaged treated (1 cycle), and chemically damaged treated (3 cycles) were studied at an external voltage of 0, 1, and 2 V at 50% and 10% relative humidities. Representative AFM images for chemically damaged and chemically damaged treated (1 cycle of conditioner) hair samples are shown in Fig. 24.83a. It should be noted that the contrast seen around cuticle scale edges is likely a result of the topography effect [56]. A key point shown by these images is the existence of areas of trapped charge in the low humidity samples. This is not seen in the samples measured in ambient. These areas of trapped charge are seen as bright areas on the sample in the "averaged surface potential" images. In other words, there is more contrast in the averaged surface potential maps under low humidity (neglecting any contrast seen around the scale edge) than is seen for the same samples under ambient humidity. It is important to clarify that the 'bright areas' are only in reference to the increase in contrast in the surface potential map (averaged surface potential) away from the scale edges. This is only considered for the "0 V" samples, where added external charge does not affect the surface potential of the sample. This observation indicates that water vapor in the air contributes significantly to the mobility of surface charges on the hair. This result has been previously reported for macro studies on surface charge of hair [41, 57, 60]. Trapped charges are most pronounced in the untreated samples. This suggests that conditioner treatment has a similar effect as water vapor. Even under very low humidity conditions, conditioner treatment increases the mobility of surface charges, dissipating trapped charges. Figure 24.83b shows bar charts of the average surface potential change. Error bars indicate +/- one standard deviation. It can be observed in these figures that all samples exhibit very similar values in the 50% relative humidity scenario. This again indicates that the water vapor in the air plays a significant role on the surface charge of the hair, and also on the mobility of charge. For the 50% case it is shown that the potential changes nearly 1 V for every 1 V change in applied potential. However, this is not the case for the 10% relative humidity situation. In this case, a higher 0–1 V value indicates more charge mobility and the ability to dissipate charge, whereas a lower value indicates less charge mobility. This being the

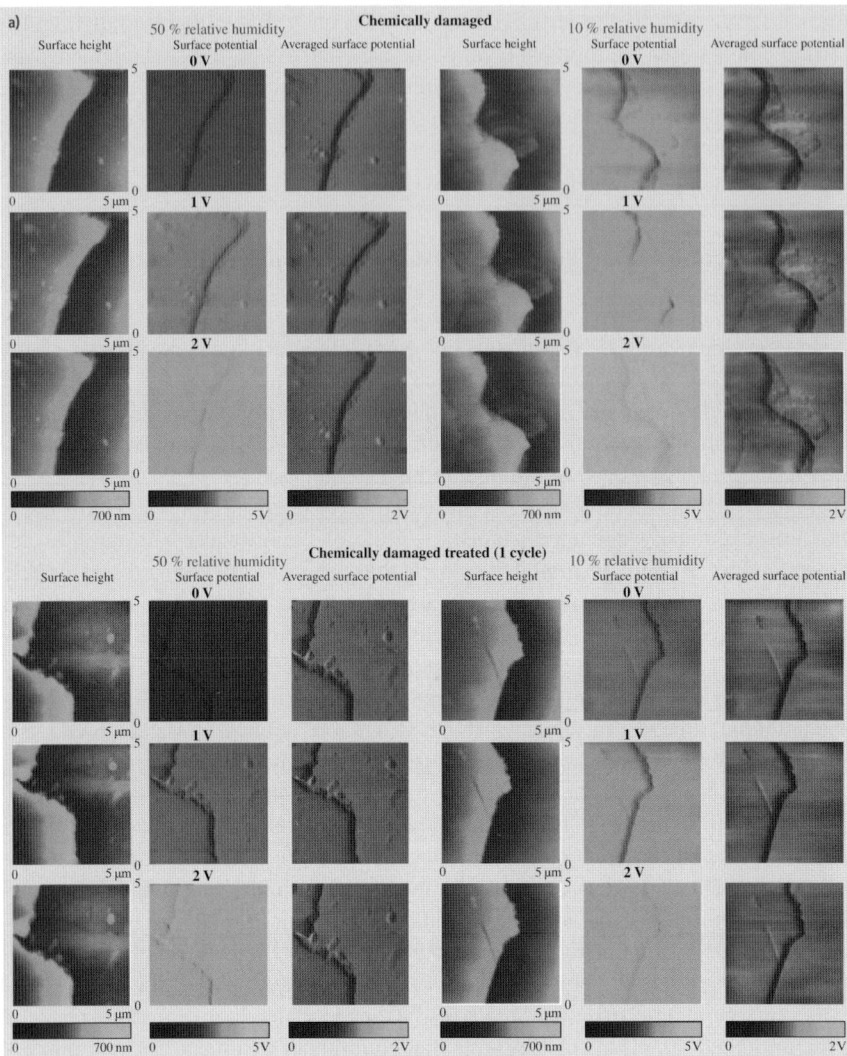

Fig. 24.83. (a) AFM images showing surface height and surface potential of chemically damaged and chemically damaged treated (1 cycle conditioner) hair samples at both 50% and 10% relative humidity; (b) Bar charts showing average surface potential change at both 50% and 10% relative humidities of virgin, virgin treated, chemically damaged, chemically damaged treated (1 cycle conditioner), and chemically damaged treated (3 cycles conditioner) hair samples [56]

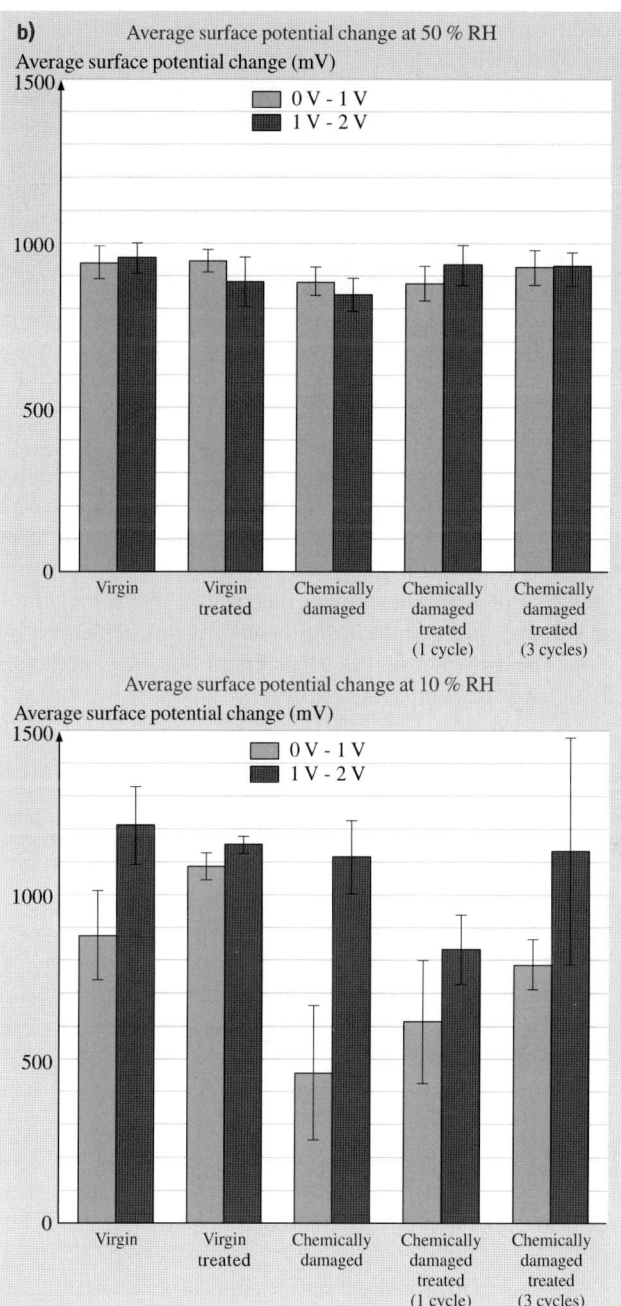

case, it is evident that conditioner treatment greatly increases the mobility of surface charges. It is also shown that virgin hair has a better charge mobility and can therefore dissipate charge more readily than chemically damaged hair. This is likely due to the lipid layer that is intact in virgin hair, but has been removed in damaged hair.

Figure 24.84a shows AFM images of chemically damaged hair treated with amino silicone conditioner. Here it is observed that the surface of the hair is more or less an equipotential surface, similar to the chemically damaged treated (3 cycles) sample. This suggests that the amino silicone conditioner coats the surface fully with one cycle, where commercial conditioner does not coat fully until several cycles. Figure 24.84b shows this behavior as well. Amino silicone conditioner shows potential change in 10% relative humidity that are almost identical to those measured in 50% relative humidity. This indicates that the conditioner is evenly spread over the surface of the hair, which has been previously reported [48, 53]. It also indicates that amino silicone conditioner greatly increases charge mobility. Because commercial conditioner is unevenly distributed on the hair surface, the data for the commercially treated samples exhibit relatively high standard deviations in most cases. Consequently, different areas on the hair surface may contain vastly different amounts of conditioner on the surface. For the amino silicone case, however, different areas of the hair seem to have similar amounts of conditioner as evidenced by the lower standard deviation. The data then suggests that amino silicone conditioner is the superior conditioner in terms of charge dissipation.

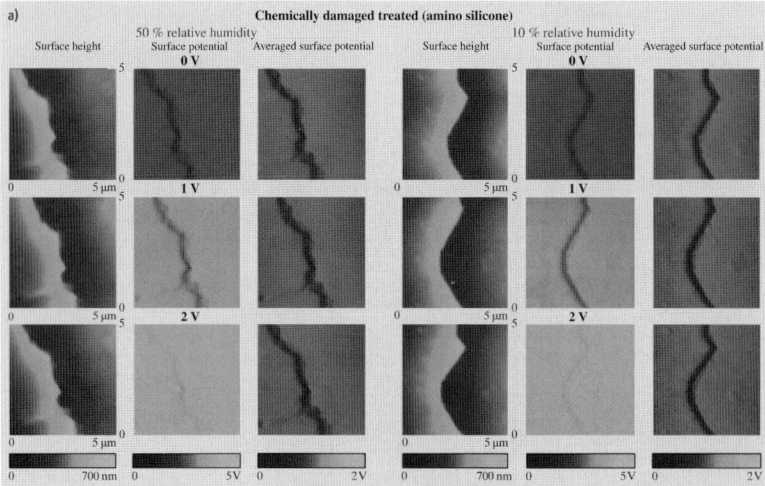

Fig. 24.84. (a) AFM images showing surface height and surface potential of chemically damaged treated (amino silicone) hair samples at both 50% and 10% relative humidities;

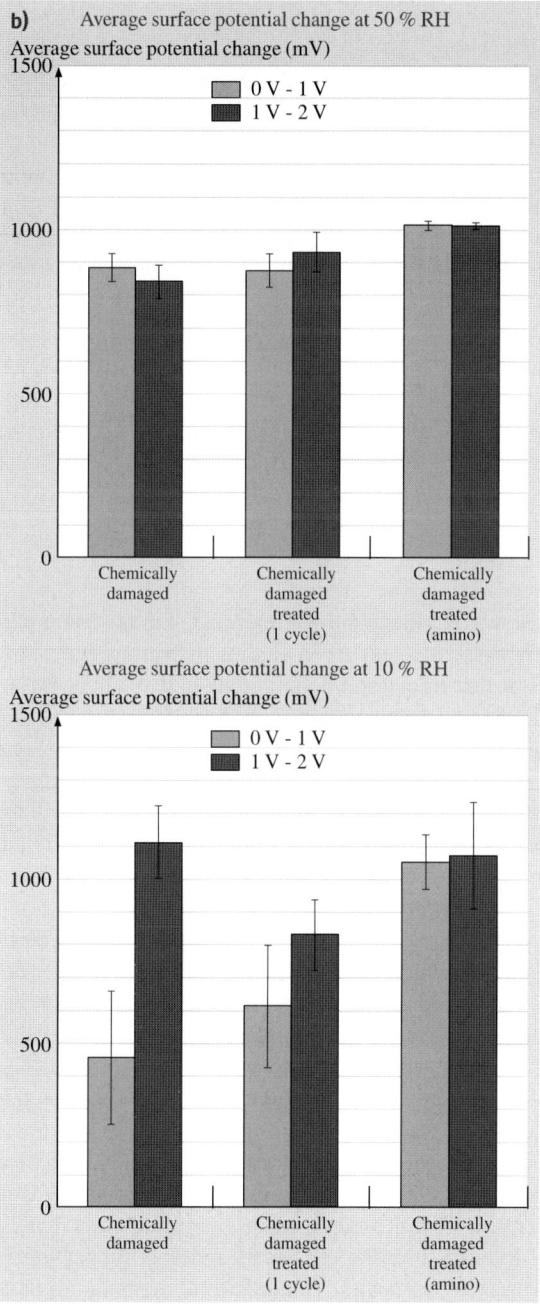

Fig. 24.84. (continued) **(b)** Bar charts showing average surface potential change at both 50% and 10% relative humidities for chemically damaged, chemically damaged treated (1 cycle conditioner) and chemically damaged treated (amino silicone) hair samples [56]

24.8.3 Summary

Kelvin probe microscopy is a powerful tool when used to study electrostatic properties of hair. Physical wear alone does not cause a measureable electrostatic charge on hair because hair is nonconductive. Rubbing with a latex finger cot can build up electrostatic charge on hair. Any electrostatic charge on hair can dissipate rapidly. The natural lipid layer on virgin hair and conditioner treatment may increase the rate of dissipation. Untreated chemically damaged hair, which has no lipid layer, seems to retain charge the longest. Conditioner treatment significantly decreases the charge built on the hair surface. Amino silicone conditioner coats the hair surface uniformly and provides higher charge mobility and less likelihood of trapped charges than the commercial conditioner. Relative humidity plays a significant role in the behavior of surface charges on hair. A low relative humidity decreases charge mobility thereby increasing the likelihood of charges trapped on the surface.

24.9 Closure

A comprehensive nanoscale characterization of human hair and skin has been performed. SEM studies of hair cross section and AFM studies of hair surface show that the cuticle is about 5 to 7 scales thick, and each cuticle cell is about 0.3 to 0.5 µm thick. The visible length of each cuticle cell is about 5 to 10 µm long. Asian hair seems to be the thickest (nearly round), followed by African hair (oval-flat) and Caucasian hair (nearly oval).

The cross-section and longitudinal section of virgin Caucasian human hair have been investigated using TR mode II. The cortex and the cuticle, the macrofibril and the matrix of human Caucasian hair were readily revealed. The fine granular structure of the outer β-layer, which has not previously been seen by SEM, TEM and other AFM studies, is a result of the fine domain formation of lipid layer. Chemically damaged hair surfaces show damage; larger areas of the endocuticle were exposed. Conditioner unevenly distributes on damaged treated hair surface; thicker conditioner films are found near the cuticle edges. At high humidity, the cortex and cuticle can absorb water, which will consequently weaken the hydrogen bonds and salt bonds between the protein molecules in hair cellular structure. The effect is more significant in chemically damaged and treated hair.

Nanoindenter has been used to perform nanomechanical studies on human hair. The chemical damage and conditioner treatment caused the hardness and elastic modulus of hair surface to decrease within a depth of less than 1.5 µm. That is, the first 3-4 cuticle scales may interact with the chemicals and conditioner ingredients more effectively than the rest of the scales. Humidity and temperature have an effect on mechanical properties. The Young's modulus of damaged treated hair decreases dramatically at high humidity. Little temperature effect was observed on the Young's modulus. Nanoscratch tests show that coefficient of friction of virgin treated hair is lower than virgin hair for Caucasian and Asian hair in both cases of single cuticle scratch and multiple cuticle scratch. This thin conditioner layer acts as a layer lubricant, reducing the coefficient of friction during scratching. In-situ displacement (30–200 nm) increased greatly at very low initial load and then increased gradually, indicating that the first approximately 200 nm of the hair surface should be softer than the underlying layer. For a hair with hard cuticle, the cuticle cells tend to be fractured during scratching. For a hair with soft cuticle, the scratch tip usually plows and wears away the cuticle cells continuously until it reaches the cortex. The coefficient of friction of virgin and chemo-mechanically damaged Caucasian hair increases after soaking because of the swelling of the water, which softens the hair surface.

Human hair shows a stress strain curve typical of keratinous fibers. Transition of alpha keratin to beta keratin in yield region is the reason for the unique shape of the curves. Chemical damage, mechanical damage and conditioner treatment have no obvious effect on the stress strain curve or tensile properties. This is because such treatments affect the cuticle predominantly and tensile properties of human hair in dry state are governed by the cortex. Tensile stress in general causes lift off of the outer cuticle. The lift off is sudden and occurs consistently at around 20% strain. Fracture occurs sooner (about 10% strain) in mechanically damaged hair than chemically damaged hair (about 20% strain).

A flat-on-flat tribometer has been used to measure macroscale friction and wear of the polyurethane film (synthetic skin) vs. hair and hair vs. hair. In the case of polyurethane film vs. hair, the chemo-mechanically damaged hair shows highest coefficient of friction, followed by virgin and virgin treated hair. After 24 hours skin vs. hair wear test, the coefficient of friction did not change, while some of the cuticles were damaged.

AFM contact mode has been used to perform nanotribological studies on various hair and skin. Friction force and the resulting coefficient of friction are seen to be higher on chemo-mechanically damaged hair than on virgin hair. In virgin treated hair, there is a slight increase in friction forces around the cuticle edges and surrounding area. It is currently believed that the increase in friction force is due in part to an increase in meniscus effects which increases the adhesive force contribution to friction at sites where conditioner is deposited or accumulated on the surface, namely around the cuticle scale edges. Adhesive force studies showed a decrease in adhesive force with chemo-mechanically damaged hair, and significantly higher adhesive force for virgin treated hair, most likely due to meniscus force contributions from the conditioner layer.

Chemically damaged treated hair shows a much stronger affinity to conditioner than virgin hair. With the application of three conditioner cycles on chemically damaged treated hair, friction force increases all over the cuticle surface, showing a more uniform placement of the conditioner. Contrary to the virgin and virgin treated hair results, coefficient of friction for chemically damaged hair decreased with application of commercial conditioner treatment (both one and three cycles), possibly because the stronger negative charge on damaged hair results in better attraction of conditioner (and lower shear strength), which leads to an overall effect of lubrication and ultimately lower coefficient of friction.

Environmental studies show that coefficient of friction generally decreases with increasing temperature. After soaking hair in de-ionized water, virgin hair exhibits a decrease in coefficient of friction after soaking while chemically damaged hair exhibits an increase. Virgin hair is more hydrophobic so more of the water is present on the surface and results in a lubrication effect after soaking. Chemically damaged hair tends to be hydrophilic due to the chemical degradation of the cuticle surface, and results in absorption of water after soaking which softens the hair, leading to high nanoscale friction even with conditioner treatment. Durability tests show that once conditioner is applied to virgin hair, wear does not occur unlike an increase in friction force. Thus, conditioner serves as a protective covering to the virgin hair and helps protect the tribological properties when wear ensues.

In most cases, a decrease in coefficient of friction was observed on chemically damaged hair with the addition of the PDMS blend or amino silicones. The silicones are typically used as a major source of lubrication and thus give the conditioner more mobility on the hair surface compared to just surfactants and fatty alcohols. The inverse trend was seen for the amino silicone group, due to the dampened mobility of the amino silicone. Commercial conditioner physically adsorbed on hair surface via van der Waals attractions can be easily squeezed out from the contact region under the load; on the other hand, amino silicones in experimental conditioner cannot escape from the contact region under the load because of the strong electrostatic binding between the polar amino groups of silicone molecules and hair surface. Amino silicones provide much better load-bearing capacity and works as a cushion preventing hair surface from further damage.

Coefficient of friction values are largest on the macroscale, followed by microscale and then nanoscale values. In general, coefficient of friction increases with an increase in the AFM tip radius. On all scales, coefficient of friction decreases for chemically damaged treated hair compared to untreated. The same trend occurs for virgin-treated hair on the macro- and microscales, but not on the nanoscale. On the larger scales, the thin layer of conditioner acts as a lubricant over the hair fiber across multiple cuticle contacts. On the nanoscale, the hydrophobicity of the virgin hair causes different deposition of the conditioner which increases the adhesive force contribution and results in an increased coefficient of friction. Chemically damaged hair surface consists of higher negative charge and is a hydrophilic surface, which results in more uniform deposition and better smearing of the conditioner layer, and it is responsible for lower coefficient of friction between the tip and the cuticle surface.

The snap-in distance H_s in the force calibration measurement provides a good estimate for the conditioner film thickness on hair. The mean conditioner thickness is on the order of 5 nm. The adhesive force of damaged hair was smaller than that of virgin hair because of different surface hydrophobicities. The treated hair showed a much higher adhesive force because of meniscus formation at the interface. Humidity has same effect on film thickness and adhesive force, which depends upon hair type. Temperature has little effect. Amino silicones showed strong affinity to hair and are expected to work as a cushion for protecting the hair surface from further damage.

Kelvin probe microscopy is a powerful tool when used to study electrostatic properties of hair. Physical wear alone does not cause a measureable electrostatic charge on human hair because hair is nonconductive. Rubbing with a latex finger cot can build up electrostatic charge on hair. Any electrostatic charge on hair can dissipate rapidly. The natural lipid layer on virgin hair and conditioner treatment may increase the rate of dissipation. Untreated chemically damaged hair, which has no lipid layer, seems to retain charge the longest. Conditioner treatment significantly decreases the charge built on the hair surface. Amino silicone conditioner coats the hair surface uniformly and provides higher charge mobility and less likelihood of trapped charges than the commercial conditioner. Relative humidity plays a significant role in the behavior of surface charges on hair. A low relative humidity decreases charge mobility thereby increasing the likelihood of charges trapped on the surface.

24.A Shampoo and Conditioner Treatment Procedure

This appendix section outlines the steps involved in washing hair swatches with shampoo and/or conditioner.

24.A.1 Shampoo Treatments

Shampoo treatments consisted of applying a commercial shampoo evenly down a hair swatch with a syringe. The hair was lathered for 30 seconds, rinsed with tap water for 30 seconds, and the process was repeated. The amount of shampoo used for each hair swatch was 0.1 cc of shampoo per gram. Swatches were hung to dry in an environmentally controlled laboratory, and then wrapped in aluminum foil.

24.A.2 Conditioner Treatments

A commercial conditioner was applied, 0.1 cc of conditioner per gram of hair. The conditioner was applied in a downward direction (scalp to tip) thoroughly throughout the hair swatch for 30 seconds. The swatch was then rinsed thoroughly for 30 seconds. Swatches were hung to dry in an environmentally controlled laboratory, and then wrapped in aluminum foil.

24.B Conditioner Thickness Approximation

We consider a cylindrical hair fiber of diameter $D = 50\,\mu m$ (radius $R = 25\,\mu m$). For conditioner thickness calculations, the following assumptions are made: (a) hair and the material being added have the same density, (b) the coating of material is uniform on the hair surface, (c) the cross-sectional area of a hair fiber remains constant along the longitudinal axis of the fiber (i.e., from root to tip); the hair fiber is perfectly cylindrical (circular cross section), and (d) the deposited conditioner remains bonded to the cuticle surface (no absorption into the cuticle layer).

The cross-sectional area of an untreated hair fiber is initially calculated. By adding a specified amount of conditioner, this area will increase and cause the radius of the treated hair fiber to increase. This increase in the radius of the treated hair will be equivalent to the thickness of the conditioner layer. The original cross-sectional area A_c of hair fiber is

$$A_c = \pi R^2 = \pi (25\,\mu m)^2 = 1963.4954\,\mu m^2$$

Adding 200 ppm of material to the surface (which is comparable to the amount that commercial conditioners typically deposit) will cause an increase in volume (for a unit fiber length) by 200 ppm, or by 0.0002. Thus, the cross-sectional area $A_{c,\text{conditioner}}$ of the treated hair will increase by the same amount to

$$A_{c,\text{conditioner}} = 1.0002\,A_c = 1963.888\,\mu m^2$$

which results in a new radius $R_{\text{conditioner}}$

$$R_{\text{conditioner}} = \sqrt{\frac{A_{c,\text{conditioner}}}{\pi}} = 25.0025\,\mu m \,.$$

Therefore, subtracting the original radius from the radius after treatment increases the thickness of the hair by 0.0025 microns, or 2.5 nm.

It is important to note that the approximation of the conditioner thickness as 2.5 nm was determined for a particular hair diameter and material deposition amount (with the hair and material having equal densities). Although these are generally realistic approximations, hair diameter often varies by a factor of 2 and the deposition level can vary up to an order of magnitude. The conditioner layer has been shown in previous work to be nonuniform as well. Thus, actual conditioner thickness can deviate significantly from this number.

References

1. P Attard and SJ Miklavcic, Effective spring constant of bubbles and droplets, *Langmuir* **17** (2001) 8217–8223
2. HA Barnes and GP Roberts, The non-linear viscoelastic behaviour of human hair at moderate extensions, *Inter. J. Cosmet. Sci.* **22** (2000) 259–264
3. M Benz, NH Chen, and J Israelachvili, Lubrication and wear properties of grafted polyelectrolytes, hyaluronan and hylan, measured in the surface forces apparatus, *J. Biomed. Mater. Res. A* **71A** (2004) 6–15
4. R Beyak, CF Meyer, and GS Kass, Elasticity and tensile properties of human hair I – Single fiber test method, *J. Soc. Cosmet. Chem* **20** (1969), 615
5. D Bhatt, J Newman, and CJ Radke, Equilibrium force isotherms of a deformable bubble/drop interacting with a solid particle across a thin liquid film, *Langmuir* **17** (2001) 116–130
6. B Bhushan, Influence of test parameters in the measurement of the coefficient of friction of magnetic tapes, *Wear* **93** (1984) 81–89
7. B Bhushan: *Principles and Applications of Tribology* (Wiley, New York 1999)
8. B Bhushan: *Handbook of Micro/Nanotribology*, 2nd Ed. (CRC Press, Boca Raton, FL 1999)
9. B Bhushan: *Introduction to Tribology* (Wiley, New York 2002)
10. B Bhushan: *Nanotribology and Nanomechanics-An Introduction* (Springer-Verlag, Heidelberg, Germany 2005)
11. B Bhushan and GS Blackman, Atomic force microscopy of magnetic rigid disks and sliders and its applications to tribology, *J. Tribol.-Trans. ASME* **113** (1991) 452–457
12. B Bhushan and Z Burton, Adhesion and friction properties of polymers in microfluidic devices, *Nanotechnology* **16** (2005) 467–478
13. B Bhushan and N Chen, AFM studies of environmental effects on nanomechanical properties and cellular structure of human hair, *Ultramicroscopy* **106** (2006) 755–764
14. B Bhushan and C Dandavate, Thin-film friction and adhesion studies using atomic force microscopy, *J. Appl. Phys.* **87** (2000) 1201–1210
15. B Bhushan and AV Goldade, Measurement and analysis of surface potential change during wear of single-crystal silicon (100) at ultralow loads using Kelvin probe microscopy, *Appl. Surf. Sci.* **157** (2000) 373–381
16. B Bhushan and AV Goldade, Kelvin probe microscopy measurements of surface potential change under wear at low loads, *Wear* **244** (2000) 104–117
17. B Bhushan and T Kasai, A surface topography-independent friction measurement technique using torsional resonance mode in an AFM, *Nanotechnology* **15** (2004) 923–935
18. B Bhushan and X Li, Nanomechanical properties of solid surfaces and thin films (invited), *Inter. Mater. Rev.* **48** (2003) 125–164
19. B Bhushan and J Qi, Phase contrast imaging of nanocomposites and molecularly thick lubricant films in magnetic media, *Nanotechnology* **14** (2003) 886–895
20. B Bhushan, H Liu, SM Hsu, Adhesion and friction studies of silicon and hydrophobic and low friction films and investigation of scale effects, *ASME J. Tribol.* **126** (2004) 583–590
21. B Bhushan, G Wei, and P Haddad, Friction and wear studies of human hair and skin, *Wear* **259**, (2005) 1012–1021
22. MS Bobji and B Bhushan, In situ microscopic surface characterization studies of polymeric thin films during tensile deformation using atomic force microscopy, *J. Mater. Res.* **16**, (2001) 844–855

23. MS Bobji and B Bhushan, Atomic force microscopy study of the microcracking of magnetic thin films under tension, *Scripta Mater.* **44**, (2001) 37–42
24. C Bolduc and J Shapiro, Hair care products: waving, straightening, conditioning, and coloring, *Clinics in Dermatology* **19** (2001) 431–436.
25. J Brandup, EH Immergut, and EA Grulke, eds.: *Polymer Handbook* (4th Edition, Wiley, New York 1999)
26. N Chen and B Bhushan, Morphological, nanomechanical and cellular structural characterization of human hair and conditioner distribution using torsional resonance mode in an AFM, *J. Microscopy* **220** (2005) 96–112
27. N Chen and B Bhushan, Atomic force microscopy studies of conditioner thickness distribution and binding interactions on the hair surface, *J. Microscopy* **221** (2006) 203–215
28. NH Chen, T Kuhl, R Tadmor, Q Lin, and JN Israelachvili, Large deformations during the coalescence of fluid interfaces, *Phys. Rev. Lett.* **92** (2004) Art. No.-024501
29. S Dekoi and J Jedoi, Hair low-sulphur protein composition does not differ electrophoretically among different races, *J. Dermatol.* **15** (1988) 393–396
30. S Dekoi and J Jedoi, Amount of fibrous and matrix substances from hair of different races, *J. Dermatol.* **19** (1990) 62–64
31. D DeVecchio and B Bhushan, Use of a nanoscale Kelvin probe for detecting wear precursors, *Rev. Sci. Instrum.* **69** (1998) 3618–3824
32. A Feughelman: *Mechanical Properties and Structure of Alpha-Keratin Fibres: Wool, Human Hair and Related Fibres* (University of South Wales Press, Sydney, 1997)
33. M Feughelman, The physical properties of alpha keratin fibers, *J. Soc. Cosmet. Chem.* **33**, (1982) 385–406
34. ML Forcada, MM Jakas, and A Grasmarti, On liquid-film thickness measurements with the atomic-force microscope, *J. Chem. Phys.* **95** (1991) 706–708
35. ME Ginn, CM Noyes, and E Jungermann, The contact angle of water on viable human skin, *J. Colloid and Interface Science* **26** (1968) 146–151
36. J Gray, Hair care and hair care products, *Clinics in Dermatology* **19** (2001) 227–236
37. J Gray: *The World of Hair* (Online, http://www.pg.com/science/haircare/hair_twh_toc.htm, 2003)
38. GH Henderson, GM Karg, and JJ O'Neill, Fractography of human hair, *J. Soc. Cosmet. Chem.* **29** (1978) 449–467
39. JN Israelachvili: *Intermolecular & surface forces* (2nd Edition, Academic Press, London 1992)
40. J Jachowicz and R McMullen, Mechanical analysis of elasticity and flexibility of virgin and polymer-treated hair fiber assemblies, *J. Cosmet. Sci.* **53** (2002) 345–361
41. J Jachowitz, G Wis-Surel, and ML Garcia, Relationship between triboelectric charging and surface modifications of human hair, *J. Soc. Cosmet. Chem.* **36** (1985) 189–212
42. C Jalbert, JT Koberstein, I Yilgor, P Gallagher, and V Krukonis, Molecular weight dependence and end-group effects on the surface tension of poly(dimethylsiloxane), *Macromolecules* **26** (1993) 3069–3074
43. P Jollès, H Zahn, and H Höcker (Eds.): *Formation and Structure of Human Hair* (Birkhäuser Verlag, Berlin 1997)
44. T Kasai, B Bhushan, L Huang, and CM Su, Topography and phase imaging using the torsional resonance mode, *Nanotechnology* **15** (2004) 731–742
45. C LaTorre and B Bhushan, Nanotribological characterization of human hair and skin using atomic force microscopy, *Ultramicroscopy* **105** (2005) 155–175

46. C LaTorre and B Bhushan, Nanotribological effects of hair care products and environment on human hair using atomic force microscopy, *J. Vac. Sci. Technol. A* **23** (2005) 1034–1045
47. C LaTorre and B Bhushan, Investigation of scale effects and directionality dependence on adhesion and friction of human hair using AFM and macroscale friction test apparatus, *Ultramicroscopy* **106** (2006) 720–734
48. C LaTorre, B Bhushan, JZ Yang, and PM Torgerson, Nanotribological Effects of silicone type, silicone deposition level, and surfactant type on human hair using atomic force microscopy, *J. Cosmetic Sci.* **57** (2006) 37–56
49. G Lerebour, S Cupferman, C Cohen, and MN Bellon-Fontaine, Comparison of surface free energy between reconstructed human epidermis and in-situ human skin, *Skin Research and Technology* **6** (2000) 245–249
50. X Li, B Bhushan, and PB McGinnis, Nanoscale mechanical characterization of glass fibers, *Mater. Lett.* **29** (1996) 215–220
51. B Lindelof, B Forslind, and MS Hedblad, Human hair form. Morphology revieled by light and scanning electron microscopy and computer aided three-dimensional reconstruction, *Arch. Dermatol.* **124** (1988) 1359–1363
52. H Liu and B Bhushan, Nanotribological characterization of molecularly thick lubricant films for applications to MEMS/NEMS by AFM, *Ultramicroscopy* **97** (2003) 321–340
53. RA Lodge and B Bhushan, Surface characterization of human hair using tapping mode atomic force microscopy and measurement of conditioner thickness distribution, *J. Vac. Sci. Technol. A* **24** (2006) 1258–1269
54. RA Lodge and B Bhushan, Wetting properties of human hair by means of dynamic contact angle measurement, *J. Appl. Poly. Sci.* **102** (2006) 5255–5265
55. RA Lodge and B Bhushan, Effect of physical wear and triboelectric interaction on surface charges measured by Kelvin probe microscopy, *J. Colloid Interface Sci.* **310** (2007) 321–330
56. RA Lodge and B Bhushan, Surface potential measurement of human hair using Kelvin probe microscopy, *J. Vac. Sci. Technol. A* **25** (2007) 893–902
57. AC Lunn and RE Evans, The electrostatic properties of human hair, *J. Soc. Cosmet. Chem* **28** (1977) 549–569.
58. JE Mark: *Polymers Data Handbook* (Oxford University Press, Oxford 1999)
59. J Menkart, LJ Wolfram, and I Mao, Caucasian hair, Negro hair, and wool: similarities and differences, *J. Soc. Cosmet. Chem.* **35** (1984) 21–43
60. CM Mills, VC Ester, and H Henkin, Measurement of static charge on hair, *J. Soc. Cosmet. Chem.* **7** (1956) 466–475
61. R Molina, F Comelles, MR Julia, and P Erra, Chemical modifications on human hair studied by means of contact angle determination, *J. Colloid Interface Sci.* **237** (2001) 40–46
62. AP Negri, HJ Cornell, and DE Rivett, A model for the surface of keratin fibers, *Textile Res. J.* **63** (1993) 109–115
63. C Nappe and M Kermici, Electrophoretic analysis of aklylated proteins of human hair from various ethnic groups, *J. Cosmet. Chem.* **40** (1989) 91–99
64. G Nikiforidis, C Balas, and D Tsambaos, Mechanical parameters of human hair: possible applications in the diagnosis and follow-up of hair disorders, *Clin. Phys. Physiol. Meas.* **13** (1992) 281–290
65. AN Parbhu, WG Bryson, and R Lal, Disulfide bonds in the outer layer of keratin fibers confer higher mechanical rigidity: correlative nano-indentation and elasticity measurement with an AFM, *Biochemistry* **38** (1999) 11755–11761

66. PT Pugliese: *Physiology of the Skin* (Allured Publishing Corporation, Carol Stream, Illinois 1996)
67. RJ Randebrook, *J. Soc. Cosmet. Chem.* **15** (1964) 691
68. C Robbins: *Chemical and Physical Behavior of Human Hair* (3rd Edition, Springer-Verlag, New York 1994)
69. CR Robbins and RJ Crawford, Cuticle damage and the tensile properties of human hair, *J. Soc. Cosmet. Chem.* **42** (1991) 59–67
70. SB Ruetsch and HG Weigmann, Mechanism of tensile stress release in the keratin fiber, *J. Soc. Cosmet. Chem.* **47** (1996) 13–26
71. H Schott, Contact Angles and wettability of human skin, *Journal of Pharmaceutical Sciences* **60** (1971) 1893–1895
72. GV Scott and CR Robbins, Effects of surfactant solutions on hair fiber friction, *J. Soc. Cosmet. Chem.* **31** (1980) 179–200
73. WW Scott and B Bhushan, Use of phase imaging in atomic force microscopy for measurement of viscoelastic contrast in polymer nanocomposites and molecularly thick lubricant films, *Ultramicroscopy* **97** (2003) 151–169
74. I Seshadri and B Bhushan, In-situ tensile deformation characterization of human hair with atomic force microscopy, *Acta Mater.* **56** (2008) 774–781
75. JR Smith and JA Swift, Lamellar subcomponents of the cuticular cell membrane complex of mammalian keratin fibres show friction and hardness contrast by AFM, *J. Microscopy* **206** (2002) 182–193
76. Y Song and B Bhushan, Quantitative extraction of in-plane surface properties using torsional resonance mode of atomic force microscopy, *J. Appl. Phys.* **97** (2005) 083533
77. G Swanbeck, J Nyren and L Juhlin, Mechanical properties of hair from patients with different types of hair diseases, *J. Invest. Dermatol.* **54** (1970) 248–251
78. JA Swift, Fine details on the surface of human hair, *Int. J. Cosmetic Sci.* **13** (1991) 143–159
79. JA Swift: Morphology and histochemistry of human hair, *Formation and Structure of Human Hair*, ed. By P Jolles, H Zahn, and H Hocker (Birkhauser Verlag, Berlin 1997) 149–175
80. JA Swift, The mechanics of fracture of human hair, *Inter. J. Cosmet. Sci.* **21** (1999) 227–239
81. JA Swift, The cuticle controls bending stiffness of hair, *J. Cosmet. Sci.* **51** (2000) 37–38
82. JA Swift and B Bews, The chemistry of human hair cuticle-I: A new method for the physical isolation of the cuticle, *J. Soc. Cosmet. Chem.* **25** (1974) 13–22
83. AN Syed, A Kuhajda, H Ayoub, K Ahmad, and EM Frank: African-American hair: Its physical properties and differences relative to Caucasian hair, in *Hair Care* (Cosmetics & Toiletries Applied Research Series) (Allured Publishing Corporation, Carol Stream, IL 1996)
84. NS Tambe and B Bhushan, In situ study of nano-cracking in multilayered magnetic tapes under monotonic and fatigue loading using an AFM, *Ultramicroscopy* **100** (2004) 359–373
85. Z Tao and B Bhushan, Surface modification of AFM Si_3N_4 probes for adhesion/friction reduction and imaging improvement, *ASME J. Tribol.* **128** (2006) 865–875
86. Z Tao and B Bhushan, Wetting properties of AFM probes by means of contact angle measurements, *J. Phys: Appl. Phys.* **39** (2006) 3858–3862
87. S Thibaut and BA Bernard, The biology of hair shape, *Int. J. Dermatol.* **44** (2005) S2–S3
88. S Thibaut, O Gailard, P Bouhanna, DW Cannell, and BA Bernard, Human hair shaped is programmed from the bulb, *Brit. J. Dermatol.* **152** (2005) 632–638

89. G Wei and B Bhushan, Nanotribological and nanomechanical characterization of human hair using a nanoscratch technique, *Ultramicroscopy* **106** (2006) 742–754
90. G Wei, B Bhushan, and PM Torgerson, Nanomechanical characterization of human hair using nanoindentation and SEM, *Ultramicroscopy* **105** (2005) 155–175
91. PW Wertz, DT Downing, Stratum corneum: biological and biochemical considerations, *Transdermal Drug Delivery*, ed. by J Swarbrick and RH Guy (Marcel Dekker, New York, 1989)
92. PW Wertz, KC Madison, DT Downing, Covalently bound lipids of human stratum corneum, *J. Invest. Dermatol.* **92** (1989) 109
93. FJ Wortmann and H Zahn, The stress/strain curve of α-keratin fibers and the structure of the intermediate filament, *Text Res. J.* **64** (1994) 737–743
94. H Yanazawa, Adhesion model and experimental-verification for polymer SIO2 system, *Colloids and Surfaces* **9** (1984) 133–145
95. C Zviak (ed.): *The Science of Hair Care* (Marcel Dekker, New York 1986)

The Editor

Dr. Bharat Bhushan received an M.S. in mechanical engineering from the Massachusetts Institute of Technology in 1971, an M.S. in mechanics and a Ph.D. in mechanical engineering from the University of Colorado at Boulder in 1973 and 1976, respectively, an MBA from Rensselaer Polytechnic Institute at Troy, NY in 1980, Doctor Technicae from the University of Trondheim at Trondheim, Norway in 1990, a Doctor of Technical Sciences from the Warsaw University of Technology at Warsaw, Poland in 1996, and Doctor Honouris Causa from the National Academy of Sciences at Gomel, Belarus in 2000. He is a registered professional engineer. He is presently an Ohio Eminent Scholar and The Howard D. Winbigler Professor in the College of Engineering, and the Director of the Nanoprobe Laboratory for Bio- & Nanotechnology and Biomimetics (NLB^2) at the Ohio State University, Columbus, Ohio. His research interests include fundamental studies with a focus on scanning probe techniques in the interdisciplinary areas of bio/nanotribology, bio/nanomechanics and bio/nanomaterials characterization, and applications to bio/nanotechnology and biomimetics. He is an internationally recognized expert of bio/nanotribology and bio/nanomechanics using scanning probe microscopy, and is one of the most prolific authors. He is considered by some a pioneer of the tribology and mechanics of magnetic storage devices. He has authored 6 technical books, more than 70 handbook chapters, more than 650 technical papers, and more than 60 technical reports, edited more than 45 books, and holds 16 U.S. and foreign patents. He is co-editor of Springer NanoScience and Technology Series and co-editor of Microsystem Technologies. He has given more than 400 invited presentations on five continents and more than 140 keynote/plenary addresses at major international conferences.

Dr. Bhushan is an accomplished organizer. He organized the first symposium on Tribology and Mechanics of Magnetic Storage Systems in 1984 and the first international symposium on Advances in Information Storage Systems in 1990, both of which are now held annually. He is the founder of an ASME Information Storage and Processing Systems Division founded in 1993 and served as the founding chair during 1993–1998. His biography has been listed in over two dozen Who's Who books including Who's Who in the World and has received more than two dozen awards for his contributions to science and technology from professional societies, industry, and U.S. government agencies. He is also the recipient of various international fellowships including the Alexander von Humboldt Research Prize for Senior Scientists, Max Planck Foundation Research Award for Outstanding Foreign Scientists, and the Fulbright Senior Scholar Award. He is a foreign member of the International Academy of Engineering (Russia), Byelorussian Academy of Engineering and Technology and the Academy of Triboengineering of Ukraine, an honorary member of the Society of Tribologists of Belarus, a fellow of ASME, IEEE, STLE, and the New York Academy of Sciences, and a member of ASEE, Sigma Xi and Tau Beta Pi.

Dr. Bhushan has previously worked for the R & D Division of Mechanical Technology Inc., Latham, NY; the Technology Services Division of SKF Industries Inc., King of Prussia, PA; the General Products Division Laboratory of IBM Corporation, Tucson, AZ; and the Almaden Research Center of IBM Corporation, San Jose, CA. He has held visiting professor appointments at University of California at Berkeley, University of Cambridge, UK, Technical University Vienna, Austria, University of Paris, Orsay, ETH Zurich and EPFL Lausanne.

Index

1,1′-biphenyl-4-thiol (BPT)   921
2-D FKT model   568
2-D histogram technique   584
2-DEG (two-dimensional electron gas)   219
2-mercaptoethylamine HCl   283
2-pyridyldithiopropionyl (PDP)   283
$\sqrt{3}$-Ag   151
3-D bulk state   201
3-D force measurements   93
3-aminopropyltriethoxysilane (3-APTES)   1246
α−Al_2O_3(0001)   160
γ-modified geometry   629
μCP (microcontact printing)   905

abrasive wear   588, 594
Abrikosov lattice   222
accelerated friction   885
acceleration energy   851
accelerometer   1280
acetylene (C_2H_2)   853
acoustic emission   612
activation energy barrier   294
actuation, miniaturized   1284
actuator
   conventional   1209
   external   1217
adatoms   147
adhesion   38, 312, 330, 356, 465, 853, 937, 1205, 1218, 1223, 1249
   biodevices   1251
   biomolecules   1279
   biosensor surface   1246
   control   903
   force   421, 422, 430, 448, 450, 453, 1258
   force, quantized   439, 480
   hysteresis   453, 454, 462, 497
   hysteresis, relation to friction   462, 463, 467, 468, 470, 471, 484
   mechanics   448–451
   performance   966
   primary minimum   419, 436, 440
   protein   1221
   rate-dependent   453, 454
   reduction   1264
   scale dependence   1258
   Si(100)   1235
   Si100 wafer   1264
   silicon nitride   1252
   tip-surface   664
adhesion-controlled friction   458–460, 463, 464
adhesive
   coating   285
   friction force   1236
   interaction surface   665
adhesive force   54, 583, 688, 967, 1269
   increase   983
   intrinsic   902
   mapping   970
   measurement   320
   micromirror   1275
   SAM   1240
   tip radius   1259

adiabatic limit 289
adsorbate 620
adsorption 281
　water 981
　water film 983
AES (Auger electron spectroscopy)
　　measurement 893
AFAM (atomic force acoustic microscopy)
　　38
AFM (atomic force microscope) 180
　3D map 1233
　bath cryostat 185
　Binnig design 50
　calculated sensitivity 87
　cantilever 49, 79
　carbon nanotube tip 678
　commercial 50
　contact mode 46
　control electronics 100
　design optimization 74
　designs 119
　diamond tip 1230
　dynamic mode 208, 214
　feedback loop 99
　for UHV application 563
　instrumentation 74
　interferometer 185
　manufacturers 50
　microscratch technique 930
　piezo creep 181
　probe construction 59
　probes 121
　resolution 117
　static mode 208
　surface height map 878, 879
　test 880
　thermal drift 180
　thermal noise 181
　tip 124, 136, 139, 146, 149, 281, 1204, 1213, 1227, 1250–1252, 1257–1260, 1263, 1270, 1275, 1278
　tip radius 365
　tip sensor design 280
　variable temperature 183
　vibration isolation 183
Ag 154
Ag trimer 151
air induced oscillations 74

air/water interface 285
Al_2O_3 559
　grains 336
Al_2O_3-TiC
　composite 336
　head 891
alcohol 909
aldehydes 909
alkali halides 155, 156
alkane 908
alkanethiolate, film 916
alkanethiols 166
all-fiber interferometer 185
aluminum oxide 559
AM-AFM 139
amide 913
Amontons law 359, 453–455, 458, 459, 464, 469, 474, 583
amorphous carbon 846, 1245
　chemical structure 854
　coatings 854, 865
amorphous surfaces 420
amplitude modulation (AM) 139
　mode 47
　SFM (scanning force microscope) 208
anatase $TiO_2(001)$ 160
anchor group 914
angle of twist formula 86
anisotropy 635
　of friction 572
annealing effect 1319
antibody 1246
　antigen 279
　complex 295
　recognition 292
arc discharge 849
Arg-Gly-Asp (RGD) 299
artifacts 182
as-deposited film 1315
assembled nanotube probes 122
association process 280
atomic 655
　friction 655
　manipulation 188
　motion imaging 189
atomic force acoustic microscopy (AFAM)
　　38

atomic force microscope (AFM) 37, 46, 112, 135, 146, 165, 166, 280, 312, 420, 422, 424, 480, 486, 521, 608, 656, 741, 793, 918, 959, 960, 1204
atomic resolution 44, 129, 139–141, 199, 280
  image 150
  imaging 146
atomic-scale
  dissipation 598
  force measurement 73
  friction 655, 692, 693
  image 39, 64, 209
  stick-slip 697
atomically
  rough diamond surfaces 688
  sharp tips 686
attraction, long-range 427, 445
attractive force–distance profile 288
attractive interaction 964
Au(111) 194

backbone chain 914
background noise 539
BaF_2 157
ball-on-flat tribometer 1225
bare surfaces 689
Barnstead Nanopure 529
batch fabrication
  nanotube tip 126
  techniques 114
beam-deflection FFM 559
bending stiffness 77
Berkovich
  pyramid 621
  tip 876
$Bi_2Sr_2CaCu_2O_{8+\delta}$ (BSCCO) 205
bias voltage 42
bimorph 548
binary compounds 155
bioadhesion 1224, 1251
biofet 1219
biological surfaces 300
biomaterials 646
biomembrane force probe (BFP) 285
bioMEMS (biological or biomedical microelectromechanical systems) 1199, 1200, 1204–1206, 1215, 1216, 1220, 1222, 1224, 1245, 1246, 1256, 1284
  mechanical properties 1224
biomolecular coating 1250
biomolecules, adhesion 1279
bioNEMS (biological or biomedical nanoelectromechanical systems) 1199, 1200, 1204–1206, 1222, 1224, 1245, 1256, 1284
  mechanical properties 1224
biosensor 1204, 1215, 1246
  implanted 1219
  microarray 1219
  surface adhesion 1246
biotin 281
biotin-directed IgG 281
biotinylated tip 300
biphenyl 188
birefringent crystal 83
blister test 614
Bloch
  states 199
  wave 199
block-like debris 886
Boltzmann distribution factor 546
bond
  breakage 280
  lifetime 289
  strength 426
bond-order potential 658
bonded lubricant film 964
bonding energy 137
boron ion implantation 46
bottom-up approach 1284
boundary
  film formation 959
  low-viscosity layer 526
  lubricant 959, 960
  lubricant film 903
  lubrication 456, 457, 459, 474, 478, 480, 942, 966
  lubrication measurement 330
  slip 443
bovine serum albumin (BSA) 281, 1246
BPT 1240
BPTC 1240
bridging of polymer chains 445, 446
broken coating chip 879

brush *see* polymer brush
bubble   442
buckling   869, 873, 1301
  force   121
  stress   874
bulge test   614
bulk
  atoms   138
  conduction band   194
  diamond   846, 856, 859
  fluid   525
  graphitic carbon   888
  micromachining   1224, 1279
  phonon modes   660
  state   201
  xenon   214

C_{70} 844
C_{60} 166, 584, 844, 845
  film   44, 718
  island   573, 575
  multilayered film   166
  ultralow friction   719
CaF bilayer on Si(111)   157
CaF_2 157
  (111)   157, 159
  (111) surface   592
  tip   592
calibration   625, 626
cantilever   137, 145, 288
  axis   186
  beam array (CBA)   1263
  deflection   49, 113, 138, 141, 185, 287, 558, 559
  deflection calculation   68
  effective mass   560
  eigenfrequency   181
  elasticity   565
  flexible   48
  foil   114
  material   60, 138
  motion   46
  mount   54
  Q-factor   208, 209
  resonance behavior   78
  resonance frequency   1307
  spring constant   291, 293
  stainless steel   875
  stiffness   58, 324

  thermal noise   208
  thermomechanical noise   208
  thickness   559, 560
  triangular   60, 77, 86
  untwisted   53
cantilever-based probes   208
capacitance detection   89
capacitive
  detection   98
  detector   47
  displacement sensor   1312
capillary
  force   419, 440, 448, 449
  wave spectrum   537
  waves of water   537
capped nanotube   682
  tip   682
carbon   843
  crystalline   843, 844
  diamond-like (DLC)   1229, 1235, 1245
  diamond-like coatings   1214
  film   966
  friction chains   1240
  magnetron sputtered   853
  unhydrogenated coating   860
carbon nanotube (CNT)   197, 198, 718, 1204
  mechanical properties   120
  tip   65, 117, 120, 318
carbon–carbon distance   212
carboxylates ($RCOO^-$)   166
carboxylic acid   909
carrier gas   860
Casimir force   418
catalysis   155
cathodic arc carbon   865
cell adhesion   281
CeO_2(111)   160
ceramic slider   889
CFM (chemical force microscopy)   300
characteristic distance   485
charge
  exchange interaction   418, 427, 430
  fluctuation force *see* ion correlation forces
  transfer   162, 430
  transfer interaction   430

charge density wave (CDW) 202
chemical
    bond   137, 169
    bonding   139, 150, 151, 154, 962
    bonding force   48
    characterization   854
    force   136, 137, 141, 146, 170
    force microscopy (CFM)   300
    grafting   1240
    heterogeneity   454
    mechanical polishing (CMP)   1282
    vapor deposition (CVD)   846, 1208
chemisorption   959
chip, biofluidic   1216
Co clusters   198
CO on Cu(110)   188
Co–O film on PET   383
co-transporter   300
coated tips   684
coating
    biomolecular   1250
    continuity   891
    damage   886
    failure   883
    friction and wear behavior   664
    friction coefficients   686
    hardness   881
    hydrophobic   1279
    mass density   856
    microstructure   855
    substrate interface   882
    thickness   870, 873
coefficient of friction   *see* friction coefficient, 318, 340, 359, 471, 562, 864, 875, 886, 889, 937, 970, 978, 1226
    lubricants   724
    relationship   68
    Si(100)   1238
    Si(111)   1228
    Z-15   1238
    Z-DOL (BW)   1238
coefficient, effective   584
cognitive ligands   280
coherence length   194
cold welding   431, 432, 473
collapsed polymers   522
colloidal
    forces   418
    probe   425
colossal magneto resistive effect   220
comb drive   1307
complex bonds   294
component surface roughness   1261
compressive
    forces nanotube   721
    stress   861, 873, 1316
computational studies, tribological processes   655
computer simulation   474
    forces   432, 437, 439, 443, 447, 450, 480
    friction   432, 462, 464, 473, 479, 480, 482, 483, 488, 494, 497
concentration
    correlation   540
    critical   590
confinement   418, 419, 457, 479, 495, 496
    of fluids   517
conformation   964
conformational defect   928
constant
    amplitude mode   157
    current mode   42
    force mode   100
    height mode   42
constant-*NVE*   663
contact
    angle   904
    angle of SAM   925
    apparent macroscopic   458, 466, 474
    area   425, 590, 592, 628, 643
    conductance   588
    elastic   633
    electrification   430
    mechanics
        see adhesion mechanics
    mode   114
    potential difference   219
    printing   905
    resistance   612
    spiking   1274
    stiffness   611, 625, 630, 871
    stress   979
    true molecular   424, 425, 449, 451, 459, 464, 466, 467, 497
    value theorem   465

contact-start-stops (CSS) 1138
contamination 49
continuous stiffness measurement (CSM) 870
continuum
  model 569
  theory 433, 435, 438, 440, 455, 475
contrast 210
  formation 155
control system 94, 99
controlled
  desorption 188
  geometry (CG) 45
Cooper pairs 203
copper
  surface adhesion 689
  surface wear 689
  tip 594, 689
correlation, in-plane 549
Couette flow 456, 494
Coulomb
  force 154
  interaction potentials 659
  law of friction 578
covalent bonding
  modeling of materials 658
Cr coating 206
Cr(001) 199
crack spacing 382
cracks 867
crater formation, surface 668
creep 182, 613
  effect 390
critical
  concentration 590
  degree of bending 1309
  load 864
  normal load 944
  position 566
  shear stress 457–459, 467, 480, 483
  surface tension 935
  temperature 193, 209
  velocity 484, 489, 490, 493
crosslinked BPT (BPTC) 921
crosstalk 101
cryostat 182
crystal
  structure 49
  surfaces 118
crystalline
  carbon 843, 844
  surfaces 420
CSM (continuous stiffness measurement) 870
Cu(100) 571, 573
Cu(111) 194, 571, 573, 598
  surface states 199
  tip 593
cube corner 621, 636
cuprates 205
current density effect 1319
cut-off distance 430, 463
CVD (chemical vapor deposition) 846
cyclic fatigue 870, 1318

D_2O 196
damage 432, 455, 459, 466, 467, 469, 472–474, 486
damage mechanism 879
damping 157
  pneumatic 185
Deborah number 478
debris 886
Debye
  frequency 661
  length 434, 437
defect
  motion 182
  nucleation 632
  production 591
deflection
  measurement 138
  noise 209
deformation 424, 425
  elastic 449, 475, 476, 584
  plastic 432, 474
delamination 636, 861, 869, 873
Demnum-type PFPE, lubricant film 971
dental enamel 646
depletion
  attraction 418, 445
  interaction 445
  stabilization 418, 445
deposition
  rate 851
  techniques 848, 850
Derjaguin approximation 421, 437, 440

Derjaguin–Landau–Verwey–
  Overbeek (DLVO) theory   436
Derjaguin–Muller–Toporov (DMT)
  model   584
  theory   449
design rule   76
detection systems   49, 79
dewetting   970, 1236, 1268
DFM (dynamic force microscopy)   208,
  301, 316
DFS (dynamic force spectroscopy)   294
diamond   114, 125, 625, 844, 845
  coating   860, 903
  film   846
  friction   693
  like amorphous carbon (DLC) coating
    686, 855
  like carbon (DLC)   381, 824, 1235
  nanoindentation   676
  tip   323, 592, 875, 883, 918, 966, 1227,
    1230
dielectric breakdown   613
diffusion   189
  coefficient   966
  parameters   189
  retardation   547
  rotational   517
  thermally activated   182
diffusion coefficient
  translational   540
diffusive relaxation   289
digital
  feedback   99
  light processing (DLP)   1211
  light switch   1211
  micromirror device (DMD)   1199, 1273
  signal processor (DSP)   42
dilation   462, 482, 483, 496
dimension   137
dimer structure   149
dimer–adatom–stacking (DAS) fault   147
dip-coating technique   962
dipole molecule   906
directly growing nanotubes   124
disjoining pressure   974
dislocation
  line tension   635
  motion   457

nanoindentation   669
nucleation   622
stick-slip   697
dispersion force   170
dissipation measurement   557
dissociation   280
distortion   97
dithio-bis(succinimidylundecanoate)   281
dithio-phospholipids   284
DLC (diamond like carbon)   1245,
  1266–1269, 1279
  coating   382, 846, 849, 864, 882, 891,
    1279
  coating microstructure   871
DLVO interactions   426, 436–438
DMT (Derjaguin–Muller–Toporov)
  model   620
  theory   449
DNA   166, 281
  analysis   1216
domain pattern   220
DOS structure   193
double-layer interaction   418, 434, 435,
  437
doubly clamped beam   1306
drift states   201
driving frequency   80
drug-delivery device   1221, 1222
dry sliding friction   689
dry surfaces   459, 490
  forces   427, 431
  friction   458, 466, 485, 487
Dupré equation   455
durability Si(100)   1235
dynamic
  AFM   139, 141
  force microscopy (DFM)   301, 316
  force spectroscopy (DFS)   294
  friction   see kinetic friction
  interactions   419, 420, 457, 475, 479,
    480
  light scattering (DLS)   547
  mode   47
  operation mode   139
dynamical response   546

EBD (electron beam deposited) tips   119
ECR (electron cyclotron resonance) CVD
  848, 877, 1245

1496   Index

coating   885
edge channels   219
EELS (electron resonance loss spectroscopy)   855
effective
  coefficient of friction   584
  ligand concentration   292
  mass   209
  shear stress   587
  spring constant   566, 580
  viscosity   478, 494
EFM   180, 219
elastic
  contact   633
  deformation   584
  modulus   451, 625, 863, 864
  tip-surface interaction   675
elasticity   209
elasto-hydrodynamic lubrication   456, 457, 475, 476, 478
electric force gradient   58
electrochemical
  AFM   59
  etch   131
  etched tips   131
  etching   114
  STM   44
electromagnetic forces   427
electromigration   188, 1274
electron
  beam deposition (EBD)   119
  cyclotron resonance chemical vapor deposition (ECR-CVD)   848, 852
  energy loss spectroscopy (EELS)   851
  interactions   201
  tunneling   39
electron–electron interaction   194
electron–phonon interaction   201
electronegativity   906
electronic noise   50
electrophoretic separations   1220
electrostatic
  binding   283
  force   137, 217, 418, 419, 427, 430, 434, 438
  force interaction   149
  force microscopy   218
  interaction   156

potential   220
embedded atom method (EAM)   659
embedding energy   659
embossing   905
end deflection   1302
end group   964
endothelial cell surface   296
energies, MD simulations   657
energy
  barrier   289
  dissipation   157, 453–455, 459, 461–463, 465, 466, 471, 485, 597, 928
  resolution   191
entangled states   457
entropic force   447
epitope mapping   303
equilibrium
  interactions   see static interactions
  true (full) or restricted   445
ester   909
ethanolamine   281
ether   909
Euler
  buckling criterion   1301
  equation   76
examples of NEMS   1204
exchange
  carrier plate   564
  force interaction   163
  force microscopy   165
  interaction   163
external
  noise   182
  normal load   967
  vibrations   43
Eyring model   579

Fab molecule   281
fabrication
  methods   1279
  nanostructures   1281
face-centered cubic (fcc)   844
failure mechanism   890
Fano resonance   199
fatigue   870
  crack   887
  damage   871
  failure   1318
  life   864

measurement 1312
  resistance 1304
  test 874
fatty acid monolayer 918
FCA (filtered cathodic arc) 876
  coating 863, 872
Fe(NO$_3$)$_3$ 127
Fe-coated tip 164
feedback
  circuit 50
  loop 102, 113, 293, 301, 302, 562
  network 40
  signal 139
Fermi
  level 191
  points 201
ferromagnetic
  probe 220
  tip 223
fetal bovine serum (FBS) 1251
FFM (friction force microscopy) 559
  dynamic mode 593
  on atomic scale 580
  tip 565, 570
FIB (focused ion beam) tips 118
FIB-milled probe 64
fiber-optical interferometer 82
field
  effect transistor (FET) 1219
  emission tip 122
film
  C$_{60}$ 718
  SiC 1231
  lubricant 1249
  nanoindentation thickness 680
  on substrate 1297
  PMMA 1257, 1259, 1260
  polysilicon 1224, 1231, 1281
  substrate interface 1316
  surface interface viscosity 708
  water 1266
  Z-15 1236
  Z-DOL (BW) 1235
filtered cathodic arc (FCA) 1245
  deposition 849–872
finite element modeling (FEM) 626
first principle calculation 154
first principle simulation 210

first principles, MD simulations 657
FKT (Frenkel–Kontorova–Tomlinson)
    model 568
flat punch 624
flexible cantilever 48
flocculation 436
Flory temperature 444
flow
  activation energy 965
  rate 521
  sensors 1280
fluid
  low viscosity 521
  mechanics 521
  multicomponent 521
  Newtonian 521
fluorescence
  correlation spectroscopy (FCS) 540, 545
  fluctuation 540, 549
  recovery after photobleaching (FRAP) 540
  spectroscopy 539
fluorides 155, 157
fluorinated DLC 904
fluorophore 539
FM (frequency modulation), AFM images 211
FM-AFM 141, 142, 144, 145
focused ion beam (FIB) 45, 64, 118, 617, 643
foil cantilever 114
force
  between macroscopic bodies 429, 435
  between surfaces in liquids 432
  calibration 559
  calibration mode 386
  calibration plot (FCP) 967
  cantilever-based 208
  chemical force 210
  curve 129
  detection 76
  electrostatic force 218
  indentation 675
  long-range 215
  magnetic exchange force 223
  mapping 301
  MD simulations 657

measurement 138
measuring techniques 420, 422, 425
modulation technique 325
repulsive 287
resolution 208, 288
sensing tip 93
sensitivity 285
sensor 48, 142, 563
spectroscopy (FS) 100, 209, 214
undulation 418, 447
van der Waals force 212
force calibration 1236
force field spectroscopy
three-dimensional 217
force spectroscopy (FS) 214, 280, 299
3D-FFS 217
site specific 216
force–distance
curve 57
cycle 288
formate ($HCOO^-$) 166
formate-covered surface 168
formation of SAM 915
four-quadrant photodetector 558
Fourier-transform infrared time-resolved spectroscopy (FTIR-TRS) 538
fracture 635
strength 1318
toughness 863, 864, 867, 869, 881
toughness measurement 1303, 1309
Frank–Read source 633
free surface energy 980
freezing–melting transitions 486, 489
Frenkel–Kontorova–Tomlinson (FKT) 568
frequency
measurement precision 93
modulation (FM) mode 47
modulation SFM (FM-SFM) 181
frequency modulation (FM) 139, 140
friction 38, 312, 330, 356, 540, 557, 655, 688, 845, 937, 1205
anisotropy 572
coefficient 456–459, 466, 469, 471, 474, 478, 489, 493, 495, 690
load dependence 365
control 903
directionality effect 342

effect of humidity on 356
effect of tip on 356
electrification 431
experiments on atomic scale 569
force (FFM) 656
image 573
influence of 1268
kinetic 458, 461, 462, 478, 480, 483, 484, 487–490, 494, 548
loop 566, 569, 588
lubrication 688
map 478, 479, 572
measurement 546, 1240, 1313
measurement methods 66
measuring techniques 424, 425, 466, 478, 480
mechanism 340, 920, 927, 973
metal surface 690
microscale coefficient 1261
molecular dynamics simulation 592
non-contact friction 217
performance 966
relative humidity 1268
SAMs 710
scale dependent 365
Si(100) 1235
Si100 wafer 1264
static 1269
surfaces 688
torque model 1269
velocity effect 1267
friction force 48, 53, 318, 358, 574, 708, 875, 920, 967, 1268, 1269
adhesive 1236
calibration 54, 71
curve 101
decrease of 981
magnitude 69
map 341, 585
mapping 970
microscopy (FFM) 37, 312, 314, 425, 460, 465, 557, 793, 918, 1271
motor 1271
static 1271
friction models
cobblestone model 459, 460, 463
Coulomb model 459
creep model 486

distance-dependent model 486
interlocking asperity model 459
phase transitions model 488–490
rate- and state-dependent 489
rough surfaces model 486
surface topology model 486
velocity-dependent model 487, 490
frictional properties, fullerenes 718
fringe detector 1305
fullerene 844
frictional properties 718
fundamental resonant frequency 78
fused silica 626, 628, 635

g-factor 196
GaAs/AlGaAs
heterostructures 219
gain control circuit 142
gap stability 41
GaP(110) 211
gauge marker 1305
Gd on Nb(110) 203
geometrically necessary dislocations (GND) 797
geometry effects in nanocontacts 582
germanium 637
giant MR (GMR) 1138
glass-transition temperature 457, 471
glassiness 495
GMR (giant magnetic resoanance) 1214
gold 45
coated tip 281
Grahame equation 434
grain
boundary 431, 432, 1316
nanoindentation boundary 669
size 846
graphite 181, 185, 212, 336, 569, 592, 844
(0001) 199
cathode 849
flakes 719
sheets 719
grinding 130

H_2O 196
H-terminated diamond 693
Hall resistance 219
Hall–Petch behavior 644

Hamaker constant 137, 428, 429, 432, 433, 438, 463, 467, 974
Hamilton–Jacobi Method 144
hard amorphous carbon coatings 848
hardness 323, 854, 857, 863, 864, 1228, 1299
harpooning interaction 418, 431
HDT (hexadecanethiol) 921, 1240, 1267, 1268
head group 914
heavy-ion bombardment 222
Hertz model 618
Hertz-plus-offset relation 584
Hertzian contact model 620
heterodyne interferometer 81
hexadecanethiol (HDT) 921, 1241, 1265–1269, 1279
high-aspect-ratio
MEMS (HARMEMS) 1202, 1283
tips 118, 121
high-definition TV (HDTV) 1211
high-pressure phases, silicon 674
high-resolution
FM-AFM 212
imaging 170
spectroscopy 191
tips 124
high-temperature operation STM 183
high-temperature superconductivity (HTCS) 188, 205, 222
higher orbital tip states 193
highest resolution images 129
highly oriented pyrolytic graphite (HOPG) 57, 212, 330, 332, 824
Hill coefficient 297
hinged cantilever 1313
homodyne interferometer 80
honeycomb-chained trimer (HCT) 151
Hooke's law 208, 287
horizontal coupling 43
horizontally arranged nanotube 721
HtBDC (hexa-tert-butyl-decacyclene) on Cu(110) 189
HTCS (high-temperature superconductivity) 188, 205, 222
Huber-Mises 619
humidity 581
hybrid

continuum-atomistic thermostat 663
nanotube tip fabrication 126
hybridization 154
hydration
  forces 418, 426, 438, 440, 447, 465
  regulation 441
hydrocarbon 908
  precursors 853
  saturated 909
  unsaturated 909
hydrodynamic
  force 419, 425, 521
  radius 444
  stress 525
hydrofluoric (HF) 903
hydrogen 149
  bond network 530
  bonding 443, 444
  concentrations 859
  content 861
  end group 917
  flow rate 861
  termination 150
hydrogen-terminated diamond 682
hydrogenated
  carbon 854
  coating 854
hydrophilic
  SAMs 715
  surfaces 419, 441, 442
hydrophobic 1235
  attraction 530, 536
  coating 1279
  force 440, 442
  SAMs 715
  surface 419, 442, 529, 537
hydrophobicity 1253
hydroxylation 915
hysteresis 96, 129, 141, 182, 221
  loop 50, 102

IB (ion beam) coating 885
IBD (ion beam deposition) 849, 851, 876
image
  effects 644, 645
  processing software 102
  topography 48, 168
imaging
  bandwidth 59

electronic wave functions 199
in situ sharpening of the tips 41
in-plane stresses nanoindentation 670
InAs 185
InAs(110) 194, 211
indentation 312, 385
  creep process 390
  depth 387, 863, 869, 1227, 1228
  fatigue damage 873
  hardness 38, 387
  induced compression 873
  low-load 1228
  rate 673
  size 387
  size effect (ISE) 634
inelastic tunneling 187, 188
initial contact 633
inkjet printer 1203, 1210
InP(110) 146
instability 141, 145
integrated tip 61, 115
intensity–intensity autocorrelation 546
interaction
  force 210
  hydrophobic 529
interatomic
  attractive force 68
  force 48
  force constants 138
  interaction 1317
  spring constant 48
intercalated MoS_2 903
intercellular adhesion molecule-1 (ICAM-1) 300
intercept length 932
interdiffusion 454
interdigitation 454
interfacial
  defects 874
  energy (tension) see surface energy (tension)
  force 518
  friction see also boundary lubrication, 458–460, 463, 472, 474, 480, 960
  stress 881
interferometeric detection sensitivity 84
intermediate or mixed lubrication 456, 457, 478, 479

intermittent contact mode   114
intermolecular force   288
internal stress   1300
intraband transitions   194
intrinsic
  adhesive force   902
  stress   861
iodine   188
iodobenzene   188
ion
  correlation forces   418, 434, 437
  implantation   1224
  plating techniques   848
  source   851
ion beam
  deposition (IBD)   849, 851
  sputtered carbon   851
ionic bond   418
isolated nanotubes   127
itinerant nanotube levels   199

jellium approximation   659
JKR (Johnson–Kendall–Roberts)
  model   620
  relation   583
  theory   449
Joule
  dissipation   598
  heating   220
jump-to-contact (JC)   139–141, 215, 665

K-shell EELS spectra   856
KBr   156
KBr(100)   572, 590
$KCl_{0.6}Br_{0.4}(001)$ surface   157
Kelvin
  equation   448
  probe force microscopy (KPFM)   170, 321
ketones   909
kinetic
  friction   458, 461, 462, 478, 480, 483, 484, 487–490, 494, 548
  friction ultralow   492
  processes   633
knife-edge blocking   98
KOH   45
Kondo
  effect   181
  temperature   198

lab-on-a-chip   1204, 1215, 1216
laboratory assays   1246
Lamb waves   616
Landau
  levels   182, 194, 220
  quantization   194
Langevin dynamics approach   661
Langmuir–Blodgett (LB)   533, 688, 960, 1240
  deposition   904
Laplace
  force   970
  pressure   419, 448, 902
laser deflection technique   51
lateral
  contact stiffness   557
  cracks   617
  force   558, 576
  force calculation   86
  force microscopy (LFM)   37, 165, 314, 425, 557
  resolution   41, 48, 282, 316
  resonator   1319
  spring constant   59, 565
  stiffness   100, 120, 587
lattice imaging   48
layered structure, lubricants   724
layers, lubricant   703
Lennard–Jones (LJ) potential   212, 659
leukocyte function-associated antigen-1 (LFA-1)   300
LiF   156
  (100) surface   593
lifetime   289
  broadening   187, 194
  force relation   296
Lifshitz theory   428, 430, 433
lift mode   59, 321
LIGA   1206, 1270, 1283, 1284
  technique   741
ligand concentration, effective   292
ligand–receptor interaction   281
light beam deflection galvanometer   84
linear variable differential transformers (LVDT)   98
linearization, active   98
lipid film   580

liquid   702
　helium   182
　helium operation STM   183
　lubricant   702, 903, 1236, 1272
　lubricated surfaces   459, 466, 471, 490
　mediated adhesion   902
　nitrogen (LN)   183
　solid interface   902
　vapor interface   902
load
　contribution to friction   459, 463, 464
　critical   864
　dependence of friction   584
load–displacement   390
　curve   867, 869
load-carrying capacity   888
load-controlled friction   459, 460, 472, 474
loading curve   629
loading rate   290
local deformation   322, 372
　of material   381
local density of states (LDOS)   191
local stiffness   100
London
　dispersion interaction   418, 427, 428
　penetration depth   203, 222
long-chain hydrocarbon
　lubricant   704
long-range
　attraction   427, 445
　force   141, 156, 215
long-range hydrophobic interaction   518
long-term
　measurements   181
　stability   180
longitudinal piezo-resistive effect   89
loss modulus   632
lotus   1253
low-cycle fatigue resistance   1312
low-noise measurement   563
low-pressure chemical vapor deposition (LPCVD)   1281
low-temperature
　AFM/STM   49
　microscope operation   181
　scanning tunneling spectroscopy (LT-STM)   187
　SFM (LT-SFM)   185

SPM (LT-SPM)   180
low-viscosity   526
　fluid   544
LPCVD (low-pressure chemical vapor deposition)   1314
LT-SFM   185
LT-STM   187
lubricant   1213
　film   1209, 1249
　liquid   1235
　MEMS/NEMS   1279
　meniscus   963
　performance   1236
　PFPE   1235
　spreading   963
lubricating thin films   702
lubrication
　elastohydrodynamic   456, 475, 476, 478
　intermediate or mixed   456, 478, 479
　micromotors   1269
lubrication properties
　ceramics   691
　polymers   684
LVDT (linear variable differential transformers)   98

magnetic
　disk   889
　disk drive   848
　head   1204
　Ni   205
　quantum flux   222
　resonance force microscopy (MRFM)   224
　storage device   847, 901
　tape   344, 382, 960
　thin-film head   890
magnetic force   136, 139, 142
　gradient   58
　microscopy (MFM)   38, 59, 220
magnetically coated tip   301
magnetoresistive (MR)   1138
magnetostatic interaction   220
magnetron sputtered carbon   853
manganites   205
manifold absolute pressure (MAP)   1202
manipulation
　individual atoms   188, 218
　molecules   188

of individual atoms 38
manually assembled MWNT tips 122
mass of cantilever 46
mass, effective 209
material
 lubricants 724
 structural 1304
maximum loads, nanoindentation 670
MBI (multiple-beam interferometry) 422
MD simulations 657
 constant-*NVE* 663
mean-field theory 435
measurement of hardness 324
mechanical
 coupling 486
 dissipation in nanoscopic device 557
 relaxation 927
 resonance 492
 stability
  fullerenes 718
mechanical properties 1297
 characterization 1318
 of DLC coating 875
 of nanostructures 741
mechanically cut tips 130
mechanics of cantilevers 74
mechanism-based strain gradient (MSG) 797
melting point of SAM 942
membrane
 deflection method 1319
 proteins 117
memory distance 485
MEMS 118, 847, 901, 960, 1199–1203, 1206, 1297
 applications 1203
 devices 1207–1210
 Si 1201
 technology 1314
MEMS/NEMS lubricant 1279
meniscus
 bridge 902, 983
 force 974, 1259
 formation 1273
 of liquid 448, 449
metal
 catalyst 124
 deposited Si surface 151

evaporated (ME) tape 847
oxide 155, 160
particle (MP) tape 890
porphyrin (Cu–TBPP) 166
tip, deformable 665
metal–oxide–semiconductor field-effect transistor (MOSFET) 1219
metal-catalyzed chemical vapor deposition (CVD) 124
metallic bonding, modeling of materials 658
methylene stretching mode 919
Meyer's law 642
MFM 180, 220
 AM-MFM 220
 domain imaging 220
 FM-MFM 220
 sensitivity 59
 vortex imaging 221
MgO(001) 160
mica 57, 547, 571
 muscovite 590
 surface 575
micro total-analysis system (μTas) 1215
micro-Raman spectroscopy 617
micro/nanoelectromechanical systems (MEMS/NEMS) 517, 794
microactuator 1208, 1209
microarray 1219
microcantilevers 1204, 1213
microchannel 521
microcomponents 1264
microcontact printing (μCP) 905
microcrystalline graphite 859
microdevice 1301
microelectromechanical system (MEMS) 655, 656
microelectronics 155
microenvironment 544
microfabricated
 cantilever 112
microfabricated silicon cantilevers 137
microfabrication 1223
 techniques 112
microfriction 341, 1231
microimplants 1245
microindentation 609
microindente 1232

micromachined
  notch 1310
  polysilicon 1263
  silicon 1206
micromachining 1223
  bulk 1279
  surface 1279
micromanipulators 1203
micromirror 1202, 1275
  adhesive force 1275
  stuck 1276
micromirror devices (DMD), digital 1203
micromotor
  components 1261
  lubricated 1271
  lubrication 1269
  properties 1261
micropatterned SAM 927
micropipette aspiration 425
micropump
  diaphragm 1217
  diaphragm-type 1216
  rotary 1217
  valveless 1218
microscale
  friction 1225, 1261
  material removal 376
  scratching 321
  wear 321, 374
microscope eigenfrequency 74
microscopic bubbles 532
microscratch 875
microscratching 1231
  measurement 373
microsensor, high-temperature 1224
microstrain gauge 1300, 1315
microswitch 1213
microtriboapparatus 1264
microvalve 1208, 1209, 1216
  active 1216
  cantilever-type 1217
microwear 875, 883, 1231
milligear system 1206
miniaturization 1201
minimal detectable depression 115
misfit angle 569
mismatch of crystalline surfaces 451, 453, 496

mixed lubrication *see* intermediate or mixed lubrication
Mn on Nb(110) 203
Mn on W(110) 206
modeling of materials 658
modulus
  elastic 625, 863, 864
  of elasticity 38, 358
molecular
  beam epitaxy (MBE) 1285
  chain 913
  conformation 963
  dynamics (MD) 280
  dynamics (MD) calculation 570
  dynamics simulation (MDS) 295, 557, 592
  interaction 985
  recognition force spectroscopy principles 288
  resolution 191
  scale layer 904
  shape 438, 440, 483, 493–495
  spring model 920, 927
  springs 1240
  stiffness 928
  translational 517
molecular dynamics (MD) 524, 656
  simulation 548
moment of inertia 1308
monohydride 149
monolayer
  indentation 682
  thickness 932
Monte Carlo simulation 927
MoO_3 nanoparticles 718
$MoO_3(010)$ 160
MoS_2 friction 571, 574
MOSFET 1219
MRFM 223
  magnetic resonance force microscopy 224
  single spin detection 223
multilayer 644
  thin-film 890
multimode AFM 58
multiple-asperity contact 359
multiple-beam interferometry (MBI) 422
multiplication of dislocations 633

multipole interaction potentials   659
multiwall carbon nanotube (MWCNT)   65, 120, 721
   manually assembled tips   122
   purified   122
muscovite mica   590

N-hydroxysuccinimidyl   283
n-octadecylmethyl (dimethylamino) silane (ODDMS)   1241
n-octadecyltrichlorosilane (n-$C_{18}H_{37}SiCl_3$, OTS)   918
n-octyldimethyl (dimethylamino) silane (ODMS)   1240
N-succinimidyl-3-(S-acetylthio)propionate (SATP)   283
Na on Cu (111)   191
NaCl   156
   island on Cu(111)   598
NaCl(001) on Cu(111)   156
NaCl(100)   574–576, 585
NaF   156, 571
nanoarray   1204, 1205
nanoasperity   620
nanochannel   1220, 1281
nanochemistry   1201, 1204, 1284
nanocomposite   1285
   coating   904
   materials   717
nanocrystallites   854
nanodeformation   963, 988
nanoelectronic   1204
nanofabrication   38, 385, 1201
nanofatigue   870
nanofluidic silicon array   1221
nanohardness   620, 861
nanoindentation   608, 609, 624, 664, 862, 1228, 1231, 1299
   MD simulations   665
   measurement   323
nanoindentation relax, surface atom   668
nanoindenter   741, 881, 1232, 1305
nanomachining   38
nanomagnetism   205
nanomanipulators   1203
nanomembrane   1220
nanometer   655
nanometer-scale
   devices   655
   electronic devices   717
   friction   655
   indentation   655
   lubrication   655
nanometer-scale friction
   ceramics   691
nanometer-scale properties
   materials   664
nanomotor   527
nanoparticle   116, 717
   tips   718
nanopatterned disks   1214
nanopatterning   1246
nanopositioning   94
nanopump   527
nanorheology   479
nanoscale supramolecular assemblies   170
nanoscale wear   372
Nanoscope I   42
Nanoscope II   529
nanoscratch   377
nanostructures
   fabrication   1281
   mechanical properties   1297
nanotribological material properties   1232
nanotribology   548, 960
nanotube
   AFM tips   125
   buckling   121
   bundles   720
   length   129
   surface energy   126
   tip fabrication   126
   tips, surface growth   126
nanotube-based sensor   1204
native oxide layer   970
Navier–Stokes equations   521
Nb superconductor   193
$NbSe_2$ 203, 587
NC-AFM (noncontact atomic force microscopy)   208
near-field
   scanning optical microscopy (NSOM)   112
   technique   209
negative contact forces   712
NEMS   901, 1199–1201, 1204, 1297
Newtonian flow   457, 475, 478, 479

Newtonian fluid 528
Ni(001) tip on Cu(100) 595
Ni(111) tip on Cu(110) 595
NIH3T3 fibroblast 300
NiO 185
NiO(001) 155, 160
NiO(001) surface 163
nitrogenated carbon 854
no-slip boundary condition 518
noble metal surfaces 196
noise 137, 146
  electronic 50
  external 182
  performance 146
  source 87
Nomarski interferometer 83
non-Newtonian flow 456, 457, 475, 476, 478, 479
non-wetting 442
nonconducting film 851
nonconductive
  materials 165
  sample 43
noncontact
  AFM 135, 136, 146, 148, 149, 151, 152, 154, 155
  AFM image 151
  atomic force microscopy (NC-AFM) 116, 147, 155, 595
  dynamic force microscopy 595
  friction 217, 599
  imaging 47
  mode 114, 149
noncontant mode 139
nondestructive contact mode measurement 594
nonequilibrium interactions see dynamic interactions
nonliquidlike behavior 479
nonmagnetic Zn 205
nonpolar group 913
nonspherical tip 584
normal friction 455, 469, 474
normal load 967
  external 967
Nosé–Hoover thermostat 662
NTA (nitrilotriacetate)-His$_6$ 283

octadecylphosphonate (ODP) 1241

octadecyltrichlorosilane (OTS) 522
octadecyltriethoxysilane (OTE) 522
octylphosphonate (OP) 1241
on-chip actuator 1310
operation 138
optical
  deflection systems 112
  detector 47
  head 54
  head mount 56
  trap 286
  tweezers (OT) 286, 426
optical lever 84
  angular sensitivity 87
  deflection method 100
  optimal sensitivity 87
optimal beam waist 87
order
  in-plane 452, 480, 496
  long-range 457
  out-of-plane 438–440, 452, 457, 479, 480, 494
  parameter 457, 483
organic compounds 908
organometallic vapor-phase epitaxy (OMVPE) 1285
Orowan strengthening 644, 645
oscillating tip 53
oscillation amplitude 141
oscillatory flow 521
oscillatory force 418, 419, 438–441, 443, 452, 537
oscillatory shear 529
osmotic
  interactions 418, 434, 445, 447
  pressure 436, 446, 465
  stress technique 426
OTS (octadecyltrichlorosilane) 927
oxide layer 634
oxide-sharpened tips 118
oxygen content 892

P-selectin glycoprotein ligand-1 (PSGL-1) 296
packed systems, friction 710
packing density chains 919
packing, short-range 550
paraboloid load displacement 624
paraffins 909

passive
  linearization   97
  structure   1300
patterned
  silicon   1249
  structures   1257
patterning   1281
Pb on Ge(111)   202
PDP (2-pyridyldithiopropionyl)   283
peak indentation load   870
PECVD (plasma enhanced chemical vapor deposition)   904, 1245
  carbon sample   855
Peierls instability   201
perfluorodecyltricholorosilane (PFTS)   1240
perfluorodecyltriethoxysilane (PFDTES)   1245
perfluoropolyether (PFPE)   960, 1235
performance lubricant   1236
periodic
  boundary conditions (PBCs)   660
  potential   566
permanent dipole moment   170
perpendicular scan   69
persistence length   296
perturbation approach   144
perylene   166
PFPE (perfluoropolyether)   963
  lubricant   960, 971
phase transformation   612, 637
  nanoindentation   674
phase transition, shear-induced   548
phenol   909
photoemission spectroscopy (PES)   194
photolithography   1284
physisorbed protein layer   281
physisorption   959
pick-up
  SWNT tips   129
  tip method   127
piezo
  ceramic material   96
  effect   95
  hysteresis   182
  relaxation   182
  stacks   98
  tube   94, 95
piezoelectric
  drive   40
  leg   564
  positioning elements   180
  scanner   113, 182
piezoresistive
  cantilever   88
  coefficients   89
  detection   88, 138
piezoscanner   101, 130
piezotranslator   101
piezotube calibration   43
pile-up
  nanoindent   617, 627, 628, 643, 646
  nanoindentation   671
  surface   674
  surface atoms   667
pin-on-disk tribotester   918
pinning   222
planar fabrication   1224
plasma
  chemistry   1285
  enhanced chemical vapor deposition (PECVD)   61, 848, 853, 1211
plasmon mode   191
plastic
  contact regime   359
  deformation   377, 871, 930
plastic deformation   702
  surface   668
platinum-iridium   45
pneumatic damping   185
point probes   608
Poisson's ratio   625
polar
  group   913
  lubricant   903, 960
polishing   846
polycrystalline SiC   1224
polycrystalline graphite   856
polydimethylsiloxane (PDMS)   905
polyethylene glycol (PEG)   282
polyethylene terephthalate (PET)   381
polymer
  "brushes"   706
  brush   446, 447, 465, 476
  liquids (melts)   444, 476, 479
  magnetic tape   386

mushroom   446
polyoxy-methylene (POM)   1206
polypropylene   166
polysilicon   1206, 1231, 1233
  fatigue   1318
  film   1224, 1231, 1281, 1303
  fracture strength   1317
  fracture toughness   1317
  friction   1318
  mechanical properties   1314
  micromachined beams   1263
  microstructure   1317
  residual stress   1315
  Young's modulus   1316
polytetrafluoroethylene (PTFE)   570
pop-ins   612
pop-outs   612
position accuracy   102
potentials   658
power dissipation   597
power spectra, MD simulations   699
pre-crack length   1304
pressure   974
pressure sensor   1210, 1280
probe
  FIB-milled   64
  fluorophores   544
probe tip   112, 113
  performance   115
process gas   853
processing aids   527
propagation of cracks   381
properties of a coating   849
protection of sliding surface   901
protein   282
  adhesion   1221
prototype fabrication   1281
protrusion force   418, 447
Pt alloy tip   46
Pt-Ir tip   45
PTCDA   166
PTFE (polytetrafluoroethylene)
  coated Si-tip   573
  film   903
pull-off adhesion force   531
pull-off force   *see* adhesion force, 562, 963
pulse-etching   124
purified MWNT   122

pyramidal etch   114
PZT (lead zirconate titanate)
  scanner   316
  tube scanner   42, 51

Q-factor   138, 140, 214
quad photodetector   51
quadrant detector   85
quality factor Q 60, 76, 137, 142
quantum
  corrals   199
  Hall regime   219
  transistor   1204
  well   191
  wire   1204
quartz crystal microbalance (QCM)   656
quasi-continuum model   674
quasi-elastic light scattering (QELS)   540
quasi-optical experiment   191
quasi-static mode   141

radial cracking   869
radius of gyration   444
Raleigh's method   75
Raman
  spectra   856
  spectroscopy   855
randomly oriented polysilicon   1317
ratchet mechanism   693
rate of deformation, simulations   674
rate-dependent slip   524
Rayleigh wave   615
RbBr   156
RC-oscillators   92
reaction kinetics   280
read chip   1213
readout electronics   85
receptor ligand
  interaction   280, 418
  unbinding force   281
recognition
  force microscopy (RFM)   280
  force spectroscopy (RFS)   291
  imaging   300
rectangular cantilever   60, 599
reduced modulus   612
reflection interference contrast microscopy
  (RICM)   426
relative humidity (RH)

effects of   979
influence on   1268
relaxation time   457, 458, 463, 478, 485, 495, 496
relaxation, mechanical   927
release stiction   1223
reliability   901
remote detection system   85
replica molding   905
repulsive
  force   287
  tip–substrate   673
residual
  film stress   1298
  stress   615, 626, 643, 861, 864, 874
  thickness   933
resistance, temperature-dependent   1213
resisting pushing forces   712
resolution   115
  vertical   48
resonance
  curve detection   60
  frequency   114
  mechanical   492
rest time   972
retardation effect   428
Reynolds equation   521
RF magnetron sputtering   847
RF microswitch   1212
rheological characterization   965
rheology, interfacial   537
rheometer   538
rhombhedral R-8   637
rigid disk drive   960, 972
rolling friction   431, 454
root mean square   208
roughness   438, 440, 453, 454, 497
  component surface   1261
  surface   1226
rule of mixtures   641, 644
rutile $TiO_2(100)$ 160

s-like tip states   191
sacrificial-layer lithography (SLL)   1281
SAM (scanning acoustic microscopy)   615, 616
SAM (self-assembled monolayer)
  adhesive force   1240
  coating   1276
  melting point of   942
  wear resistance   1240
sammicroscopy   615, 616
sample holder   183
scan
  area   54
  direction   70
  frequency   54
  head   183
  range   95, 97
  rate   43
  size   56
  speed   97
scanner, piezo   214
scanning
  acoustic microscopy (SAM)   615
  capacitance microscopy (SCM)   38
  chemical potential microscopy (SCPM)   38
  electrochemical microscopy (SEcM)   38
  electron microscopy (SEM)   115, 376
  electrostatic force microscopy (SEFM)   38
  head   42
  ion conductance microscopy (SICM)   38
  Kelvin probe microscopy (SKPM)   38
  magnetic microscopy (SMM)   38
  near field optical microscopy (SNOM)   38
  probe microscopy (SPM)   38, 112, 136
  speed   54
  system   94
  thermal microscopy (SThM)   38
  tunneling microscopy (STM)   37, 112, 129, 135, 136, 154, 155, 312, 665
  velocity   373
scanning force
  acoustic microscopy (SFAM)   38
  microscopy (SFM)   38, 206
  spectroscopy (SFS)   206
scratch
  critical load   888
  damage mechanism   880, 882
  depth   690, 1230
  drive actuator   1309

induced damage 875
profile 373
test 374, 875
scratching force 690
scratching measurement 38
self-assembled
  growth 189
  microscopic vesicles 905
  monolayer (SAM) 281, 529, 634, 680, 913, 960, 1208, 1221
  monolayers (SAMs) 904
  structures 710
self-assembly 189
self-lubrication 593, 594
semiconductor
  quantum dots 199
  surfaces 146
sensitivity 89, 91, 144, 285, 559
SFM (scanning force microscope) 180
  bath cryostat 185
  dynamic mode 208, 214
  interferometer 185
  piezo creep 181
  static mode 208
  thermal drift 180
  thermal noise 181
  variable temperature 183
  vibration isolation 183
SH group 281
sharpened metal wire tip 129
shear
  flow detachment (SFD) 285
  force see kinetic friction, static friction, 533
  melting 457, 483
  modulus 559
  motion 532
  plane 424
  rate 965
  strength 583
  stress 929
  thinning 475, 478, 479, 487
shear stress 524
  critical 458, 459, 467, 480, 483
  effective 587
shearing force 928
short-cut carbon nanotubes 199

short-range
  chemical force 215
  electrostatic attraction 158
  electrostatic interaction 156
  magnetic interaction 155, 162–164
  packing 550
shot noise 87
Si 49, 89, 115, 118, 139, 617, 637
  accelerometer 1209
  adatom 149
  cantilever 115, 138, 559
  chemistry 284
  coated 1231
  grain 1315
  MEMS 1202
  micromotor 967
  nanofluidic array 1221
  NEMS 1202
  nitride 1252
  nitride (Si_3N_4) 49, 281
    cantilever 113
    tip 117, 563
  nitride adhesion 1252
  on-silicon 1265
  oxide 281
  surface 281
  terminated tip 210
  tip 115, 147, 150, 210, 563
  treated 1231
  trimer 151
  virgin 1231
  wafer 114
Si(001)(2×1) 150
Si(001)(2×1)-H surface 150
Si(100) 357, 374, 880, 885, 962, 968, 969, 972, 981, 984, 986
  adhesion 1235
  bulk properties 1224
  coefficient of friction 1238
  hydrophilic properties 1267
  wafer 962
Si(111) 373, 387
  coefficient of friction 1228
  dry-oxidized 1228, 1229
  surface 166
  surface roughness 1230
Si(111)(7×7) 573
Si(111)-($\sqrt{3} \times \sqrt{3}$)-Ag 153

Si–Ag covalent bond   154, 155
Si-SiO$_2$ stress formation   118
Si$_3$N$_4$ 118, 318
  cantilever   114
  layer   114
  tip   61, 115, 918, 966, 967, 974, 988, 1249
SiC
  bulk properties   1224
  film   1231
silanization   1246
siloxane   904
single asperity   583
  AFM tip   970
  contact   359
single crystal
  Si(100)   962
  aluminum (100)   378
  silicon   321, 741, 1299
  silicon cantilever   61
single crystal silicon   1225
single molecule recognition   282
single-particle wave function   191
single-wall carbon nanotube (SWNT)   65, 120, 212
sink-in   617, 627, 628, 643, 646
sintering   427, 431
SiO$_2$ 114, 118
SiOH groups   284
sliding
  contact   592
  direction   480
  distance   490
  induced chemistry   692
  of tip   558
  surface   1267
  velocity   463, 480, 579
  work   702
slip   527
  at the wall   519
  length   522
  partial   549
  plane   424
slipping and sliding regimes   483
small specimen handling   1299
smooth sliding   480
SnO$_2$(110)   160
soft

coatings   641
lithography   904
substrates   686
solid
  boundary lubricated surfaces   458, 459, 466, 467, 471, 490, 492
  thin films   723
  xenon   213
solid–liquid interface   528
solid–liquid–air contact   1255
solid–lubricant interaction   703
solid-like
  behavior   495
  Z-DOL(BW)   967
solidification   471, 478
  incomplete   547
solvation
  forces   418, 437, 439, 443, 452, 453
SP-STM   205
sp^3
  bonded carbon   875
  bonded tip   675
  bonding   865
spacer chain   914, 915, 917
  length   923
spectroscopic resolution in STS   187
spin
  density waves   205
  quantization   194
SPM (scanning probe microscopy)   112, 180
spring
  instability   533
  sheet cantilever   64
  system   41
spring constant   121, 137, 138, 142, 143, 148, 208, 287, 290, 560, 1308
  calculation   72
  changes   181
  effective   566, 580
  lateral   59, 565
  measurement   72
  vertical   59
sputtered coatings, physical properties   860
sputtering
  deposition   853
  power   862

SrF$_2$ 157
SrTiO$_3$(100) 160, 161
static
  AFM 138, 139, 141
  friction 458, 468, 471, 478, 480, 483, 487–490, 497, 548, 1269, 1271, 1313
  friction force 967, 1271
  friction torque 1271
  interactions 420, 479
  mode 138, 208
  mode AFM 85
statistically stored dislocation (SSD) 797
steady-state sliding *see* kinetic friction, 706
step tilting 932
stepping motor 564
steric repulsion 418
stick boundary condition 525
stick–slip 458, 468, 473, 478, 480, 481, 488, 490, 493, 537, 983
  friction 689
  mechanism 565
sticking regime 482
stiction 471, 902, 1213, 1270
  mechanisms 1276
stiffness 883
  continuous measurement (CSM) 870
  measurements 1240
  torsional 77
Stillinger–Weber potential 659, 673
STM (scanning tunneling microscope) 37, 180
  bath cryostat 185
  cantilever 45
  cantilever material 61
  light emission 191
  piezo creep 181
  principle 40
  probe construction 45
  spectroscopy 191
  spin-polarized 205
  thermal drift 180
  thermal noise 181
  tip 38, 183
  variable temperature 183
  vibration isolation 183
Stoney equation 1298
storage modulus 632

strain
  energy difference 870
  gauge, rotating 1301
stray capacitance 91
strength 1297
streptavidin (STA) 284, 1246
  mutants 291
stress 1297
  field 1299
  gradient 1301
  maximum 1303
  measurement 1300
stress–strain behavior 1306
Stribeck curve 456, 479
strip domain 221
structural
  forces 418, 437, 440, 452
  material 1304
structure, active 1304
STS 191
  energy resolution 193
  inelastic tunneling 196
stuck micromirrors 1276
sub-Angstrom deflections 113
submonolayer coverage 965
substrate 913, 915
  curvature technique 1298
sum–frequency generation (SFG) 538
super modulus 614
superconducting
  gap 203
  magnetic levitation 41
  matrix 222
superconductivity 188, 203, 205, 221
  vortices 203, 221
superconductors 203
  type-I 203
  type-II 203
superhydrophobic surfaces 1255
superlubric state 703
superlubricity 572
surface
  amorphous 420
  asperity effect 343
  atom layer 140
  band 194
  charge 437
  charge density 434–436

crystalline 420
elasticity 391
energy 845, 935, 1253
energy (tension) 422, 428, 430, 449, 451, 453, 461, 463, 470
film hardness 1228
forces 417
forces apparatus (SFA) 312
friction 688
hydrophilic 528
hydrophobic 419, 442, 528
micromachining 1206, 1209, 1279
micromechanical properties 311
mobility 965
nanomechanical properties 311
nanometer-scale mechanical properties 664
phonon modes 660
potential 188, 434, 435, 437, 851
potential measurement 321
roughness 451, 1226
roughness parameters 1261
state lifetime 194
stiffness 325
structure 420, 438, 439, 451, 452, 457, 483, 493
tension 358, 902, 924, 960, 962
terminal group 914
topography 558
unlubricated 1271
surface force apparatus (SFA) 285, 420, 422, 423, 430, 466, 480, 486, 530, 656, 793, 1263
calibration 541
surface roughness 130, 923, 1253, 1255, 1260
map 337
measurement 314
Si(111) 1230
surface temperature, simulations 674
surface–surface spacing 527
surfactant monolayers 423, 447, 454–456, 467, 469, 470, 474, 484, 490, 492
suspended membrane 1306
SWNT (single-walled nanotube) 212

tapping mode 48, 114, 316
etched silicon probe (TESP) 63
Teflon layer 594

temperature 481
critical 193, 209
dependence of friction 582
effect 979
tensile
load 1303
stress 1301, 1316
test 1317
test environment 888
tether length 293
thermal
drift 181, 183
effect 574
energy 288
expansion coefficient 139
fluctuation 560
fluctuation forces 447
frequency noise 181
noise 288
printhead chip 1210
vibration amplitude 121
thermomechanical
noise 208
recording 1213
thermostats 660
theta condition 444
thick film lubrication *see* elasto-hydrodynamic lubrication
thin film 630, 640, 680
lubrication 312
thin liquid films 518
thiol 909
third-body molecules 698
frictional force 698
three-dimensional force field spectroscopy 217
through-thickness cracking 867
Ti atoms 166
TiC grains 336
tilt angle of SAMs 932
TiO_2 196
(110) 168
(110) surface 160
(110) surface simultaneously obtained with STM and NC-AFM 162
substrate 167
tip 1213
Fe-coated 164

Si_3N_4 1249
apex 210, 598
artifact 148
atom(s) 182
  atomic structure 131
  containing antibodies 302
  deflection 320
  ferromagnetic 223
  geometry 121
  jumping 697
  material 131
  mount 43
  oscillation 114
  oscillation amplitude 595
  performance 130
  preparation in UHV 563
  preparation method 45
  properties 130
  radius adhesive force 1259
  radius effect 60, 356, 875, 983, 984
  sample force 140
  surface 281
tip–cantilever assembly 119
tip–liquid interface 970
tip–molecule distance 188
tip–sample
  distance 140, 155, 212
  electric field 599
  energy dissipation 142
  force 113, 136, 140, 141, 144, 208, 215
  force gradient 144, 145
  interaction 208, 223, 595
  interface 112
  potential 144
tip–sample interaction 145
  chemical 210
  electrostatic 218
  exchange force 223
  van der Waals 212
tip–surface
  interaction 301, 592, 692
  interface 666, 1258
  potential 565
tip-bound
  antigens 292
tip-broadened image 115
tip-induced atomic relaxation 211
tip-induced quantum dot 196

titin 293
Tomlinson model 557, 565
  finite temperature 575
  one-dimensional 565
  two-dimensional 566
topographic images 48
topographical asymmetry 165
topography measurement 52
torsional resonance (TR) 1145
torus model calculation 96
total internal reflection (TIR) 538
  microscopy (TIRM) 426
transition metal oxides 162
transitions between smooth and stick–slip
  sliding 480
translational diffusion 540
transmission electron microscopy (TEM)
  116, 376, 617
transversel piezo-resistive effect 89
traveling direction of the sample 71
Tresca criterion 619
triangular cantilever 60, 77, 86
tribochemical
  oxidation 1229
  reaction 979, 1267
tribochemistry 700
triboelectric surface potential 1218
triboelectrification 431
tribological
  C_{60} 718
  characterization of coating 862
  material 1279
  performance of coatings 885
  problems 1213
  problems MEMS devices 1209
  properties of SAM 918
triethoxysilane 579
true atomic resolution 142, 146, 210
"true" atomic resolution images of insulators
  155
tungsten 45
  sphere 560
  tip 45
tunneling
  current 39, 135, 136, 140, 142, 160, 196
  detector 47
  tip 205
two-dimensional

electron gas (2-DEG)   219
electron system (2-DES)   201
two-layer-fluid model   525
two-photon excitation   541

UHV environment   569
UHV-AVM   564
ultrahigh vacuum (UHV)   146, 185, 563, 664
ultrasonic lubrication   351
ultrathin DLC coatings   847
unbinding force   280
unconstrained binding   282
unloading curve   670
unlubricated sample   985
unlubricated surfaces   *see* dry surfaces
unperturbed motion   143

V-shaped cantilever   63, 561
van der Waals (vdW)   136, 137, 141, 166, 169, 170
  attraction   356
  force   113, 156, 418, 419, 427–429, 432, 433, 437, 438
  interaction   920, 965
  surfaces   212
vanadium carbide   613
variable
  force mode   100
  temperature STM setup   183
velocity
  critical   484, 489, 490, 493
  effect   976
  of vibration   521
  rescaling   661
velocity dependence of friction   579
vertical
  coupling   43
  nanotube   722
  noise   145
  rms-noise   209
Verwey transition temperature   220
vibration
  amplitude   521
  external   43
Vickers hardness   620
vinylidene fluoride   166
viscoelasticity   631, 646
  mapping   324, 391

viscosity   424
viscous
  drag   528
  force   456, 475, 480
voltage bias   131
vortex in superconductor   221

W tip   46
wafer curvature technique   1315
water
  capillary force   926
  films   532
  vapor   613, 620
  vapor content   888
wear   *see* damage, 312, 655, 845, 864
  Si100 wafer   1264
  contribution to friction   590
  control   903
  damage   888
  damage mechanism   887
  debris   376
  depth   1227, 1233
  fullerenes   718
  map   1251
  measurement   38
  mechanism   377, 1276
  performance   966
  process   885
  region   378
  resistance   885, 886, 934, 985
  resistance SAM   1240
  study   985
  test   885
  tip   697
  track   690
wearless friction   702
Weibull statistics   1317
weight function   144
wettability   925, 1236
wetted surface   527
wetting   455, 1253
  surfaces   665
wire cantilever   60
work
  hardening   634, 643
  of adhesion   925
  of indentation   630
write/read chip   1213
Wyko surface profiler   1264

xenon 185

yield point 457, 458, 480
  load 633
yield strength simulations 674
yield stress 634
Young's modulus 72, 100, 121, 181, 583, 1297, 1298
  measurement 1305
Young–Dupre equation 924

Z-15 960, 961, 980–982, 986, 1235
  coefficient of friction 1238
  film 987, 1236
  lubricant film 969
  properties 961
Z-DOL 960, 961, 971
  partially bonded film 988
  properties 961
Z-DOL(BW) 969, 980, 984, 986, 987, 1235, 1237
  coefficient of friction 1238
  film 969, 1235
  static friction force 1271
zinc 635

Printing: Krips bv, Meppel, The Netherlands
Binding: Stürtz, Würzburg, Germany